CONCISE ENCYCLOPEDIA OF CHEMICAL TECHNOLOGY

Fifth Edition

VOLUME 1

BICENTENNIAL
1807
WILEY
2007
BICENTENNIAL

THE WILEY BICENTENNIAL—KNOWLEDGE FOR GENERATIONS

*E*ach generation has its unique needs and aspirations. When Charles Wiley first opened his small printing shop in lower Manhattan in 1807, it was a generation of boundless potential searching for an identity. And we were there, helping to define a new American literary tradition. Over half a century later, in the midst of the Second Industrial Revolution, it was a generation focused on building the future. Once again, we were there, supplying the critical scientific, technical, and engineering knowledge that helped frame the world. Throughout the 20th Century, and into the new millennium, nations began to reach out beyond their own borders and a new international community was born. Wiley was there, expanding its operations around the world to enable a global exchange of ideas, opinions, and know-how.

For 200 years, Wiley has been an integral part of each generation's journey, enabling the flow of information and understanding necessary to meet their needs and fulfill their aspirations. Today, bold new technologies are changing the way we live and learn. Wiley will be there, providing you the must-have knowledge you need to imagine new worlds, new possibilities, and new opportunities.

Generations come and go, but you can always count on Wiley to provide you the knowledge you need, when and where you need it!

WILLIAM J. PESCE
PRESIDENT AND CHIEF EXECUTIVE OFFICER

PETER BOOTH WILEY
CHAIRMAN OF THE BOARD

KIRK-OTHMER

CONCISE ENCYCLOPEDIA OF CHEMICAL TECHNOLOGY

Fifth Edition

VOLUME 1

WILEY-INTERSCIENCE

A JOHN WILEY & SONS, INC., PUBLICATION

For general information on our other products and services please contact our Customer Care Department within the U.S. at
(800) 762-2974, outside the U.S. at (317) 572-3993 or fax (317) 572-4002.

Wiley also publishes its books in a variety of electronic formats. Some content that appears in print, however, may not be available in
electronic formats. For more information about Wiley pproducts, visit our web site at www.wiley.com.

Wiley Bicentennial Logo: Richard J. Pacifico

Library of Congress Cataloging-in-Publication Data:

Kirk-Othmer concise encyclopedia of chemical technology: 2 volume set. – 5th ed.
      p.   cm.
Includes index.
   ISBN 978-0-470-04748-4 (cloth)
   1. Chemistry, Technical–Encyclopedias.
TP9.K54   2007
660.03–dc22   2007008057

Printed in the United States of America

10   9   8   7   6   5   4   3   2   1

PREFACE

The *Kirk-Othmer Concise Encyclopedia of Chemical Technology, Fifth Edition* contains an abridged version of every article in the renowned, twenty-seven-volume *Kirk-Othmer Encyclopedia of Chemical Technology, Fifth Edition*. It provides coverage of the entire field of chemical technology to professionals and students who prefer a shorter format but still depend on accurate, up-to-date information on both fundamentals and advanced topics. The goal of this concise edition is, therefore, to present key information from the original material in a convenient, accessible, and more affordable format.

The fifth edition of the Encyclopedia is based on the content and design of previous editions, with additions, adjustments, and modernization to reflect changes and developments in the field. It includes information on the properties, manufacturing, and uses of chemicals and materials; scientific and engineering principles; processes and process design and control; and environmental, health, and economic concerns. Special emphasis is on sustainable and environmentally conscious chemical technology, advanced characterization and analytical techniques, synthesis and fabrication methods, and new materials made available with modern technology. New, cutting-edge topics in the Encyclopedia include chemoinformatics, combinatorial chemistry, crystal engineering, fullerenes, green chemistry, ionic liquids, metathesis, microfluidics, nanotechnology, technology transfer, and many more.

Although a large amount of information on chemistry and chemical technology is now available from various other sources, the *Kirk-Othmer Encyclopedia of Chemical Technology* has remained the best and most comprehensive reference of chemical technology for over fifty years. We hope that this concise edition will continue to be a useful resource, whether as a quick guide or as a starting point for more extensive research. *The Kirk-Othmer Encyclopedia of Chemical Technology* is also available online as part of the electronic reference works content of Wiley InterScience.

ACKNOWLEDGMENTS

The *Kirk-Othmer Encyclopedia of Chemical Technology, Fifth Edition* was made possible by the hard work of many hundreds of individuals, including contributors, reviewers, and other advisors, who dedicated their time and expertise to this endeavor. We wish to acknowledge and thank all those—veterans and newcomers alike—who took part in the preparation of this edition.

The Editors

Vice President, STM Books: **Janet Bailey**

Editorial Director, STM Encyclopedias: **Sean Pidgeon**

Editor: **Arza Seidel**

Managing Editor: **Michalina Bickford**

Director, Book Production and Manufacturing: **Camille P. Carter**

Production Manager: **Shirley Thomas**

Senior Production Editor: **Kellsee Chu**

Illustration Manager: **Dean Gonzalez**

Editorial Assistant: **Stephanie Anderson**

CONTRIBUTORS TO
THE ENCYCLOPEDIA

Anthony G. Abatjoglou, *Union Carbide Corporation, South Charleston, WV,* Aldehydes

Marc E. Ackerman, *White & Case LLP, New York, NY,* Trademarks

Rakesh Agrawal, *Air Products & Chemicals, Inc., Allentown, PA,* Cryogenic Technology

Elizabeth Aguinaldo, *FMC Corporation, Princeton, NJ,* Barium Compounds

Jacqueline Akhavan, *Royal Military College of Science, Wiltshire, United Kingdom,* Explosives and Propellants

Danny E. Akin, *USDA Agricultural Research Service, Athens, GA,* Flax Fiber

Fazlul Alam, *U.S. Borax Research Corporation, Anaheim, CA,* Boron Halides

Lyle F. Albright, *Purdue University, West Lafayette, IN,* Nitration

David T. Allen, *University of Texas, Austin, TX,* Air Pollution

Bijan Amini, *E. I. du Pont de Nemours & Co., Inc., Deepwater, NJ,* Aniline and Its Derivatives

Eric Amis, *National Institute of Standards and Technology, Gaithersburg, MD,* Combinatorial Chemistry

Colin Anderson, *International Coatings, Ltd., Gateshead, United Kingdom,* Coatings, Antifoulings

Jerry Andrews, *Noveon Kalama, Inc., Kalama, WA,* Benzaldehyde; Benzoic Acid

George Antaki, *Aiken, SC,* Piping Systems

Anthony Anton, *E. I. du Pont de Nemours & Company, Inc., Wilmington, DE,* Polyamides, Fibers

D. H. Antonsen, *International Nickel, Inc., Wyckoff, NJ,* Nickel Compounds

Joseph M. Antonucci, *National Institute of Standards and Technology, Gaithersburg, MD,* Dental Materials

Kazumi Araki, *University of East Asia, Shimonoseki, Japan,* Amino Acids

Mohammad Aslam, *Hoechst-Celanese Corporation, Corpus Christi, TX,* Esterification

José M. Asua, *The University of the Basque Country, San Sebastian, Spain,* Copolymers

David A. Atwood, *University of Kentucky, Lexington, KY,* Aluminum Halides and Aluminum Nitrate

Kenneth B. Atwood, *E. I. du Pont de Nemours & Company, Inc., Washington, WV,* Polyester Fibers

Amos A. Avidan, *Bechtel Corporation, Houston, TX,* Fluidization

Neil Ayres, *The University of Southern Mississippi, Hattiesburg, MS,* Polymers, Water-Soluble

C. D. Azzara, *The Pennsylvania State University, Hershey Foods Corporation, Hershey, PA,* Chocolate and Cocoa

William F. Baade, *Air Products and Chemicals, Inc., Allentown, PA,* Hydrogen

Peter Bacher, *Loyola University Medical Center, Maywood, IL,* Blood Coagulation and Anticoagulant Drugs

Darlene M. Back, *Dow Chemical Company, Piscataway, NJ,* Ethylene Oxide Polymers

Marvin O. Bagby, *United States Department of Agriculture, Peoria, IL,* Carboxylic Acids

Amalendu Bagchi, *Wisconsin Department of Natural Resources, Madison, WI,* Waste Management, Solid

Coral M. Baglin, *Lawrence Berkeley National Laboratory, Berkeley, CA,* Radioisotopes

Webb I. Bailey, *Air Products and Chemicals, Inc., Allentown, PA,* Halogen Fluorides

Bennett R. Baird, *E. I. du Pont de Nemours & Company, Inc., Wilmington, DE,* Polyamides, Fibers

Malcolm H. I. Baird, *McMaster University, Ontario, Canada,* Extraction, Liquid–Liquid

William X. Bajzer, *Dow Corning Corporation, Midland, MI,* Fluorine Compounds, Organic

Frederick Baker, *Westvaco Corporation, Charleston, SC,* Carbon, Activated

Richard W. Baker, *Membrane Technology & Research, Inc., Menlo Park, CA,* Membrane Technology

Denise H. Bale, *University of Washington, Seattle, WA, Nonlinear Optical Materials*

Vincenzo Balzani, *Università di Bologna, Bologna, Italy,* Dendrimers, Luminescent

Donna Bange, *3M Company, St. Paul, MN,* Abrasives

Jay A. Bardole, *Restoration Chemicals, Inc., Vincennes, IN,* Paint and Finish Removers

Paul T. Barger, *UOP LLC, Des Plaines, IL,* Alkylation

Colin Barker, *The University of Tulsa, Tulsa, OK,* Petroleum

Scott D. Barnicki, *Eastman Chemical Company, Kingsport, TN,* Separations Process Synthesis

Edmund F. Baroch, *International Titanium, Inc., Moses Lake, WA,* Vanadium and Vanadium Alloys

J. F. Barrett, *Merck & Company, Inc., Rahway, NJ,* Quinolone Antibacterials

Calvin H. Bartholomew, *Brigham Young University, Provo, UT,* Catalyst Deactivation and Regeneration

Robert W. Bassemir, *Sun Chemical Corporation, Carlstadt, NJ,* Inks

Joseph J. Batelka, *Consultant, Savannah, GA,* Packaging, Converting

Roger G. Bates, *University of Florida, Gainesville, FL,* Hydrogen-Ion Activity

William Bauer, Jr., *Rohm and Haas Company, Spring House, PA,* Acrylic Acid and its Derivatives

Anthony Bean, *Sun Chemical Corporation, Carlstadt, NJ,* Inks

D. Beaulieu, *Bristol-Myers Squibb, Wallingford, CT,* Quinolone Antibacterials

Thomas C. Bedard, *Lonza, Williamsport, PA,* Ketones

Donald H. Beermann, *University of Nebraska, Lincoln, NE,* Growth Regulators, Animal

Laurence A. Belfiore, *Colorado State University, Fort Collins, CO,* Transport Phenomena for Chemical Reactor Design

L. E. Bell, *UOP LLC, Des Plaines, IL,* Fluid Catalytic Cracking (FCC) Units, Regeneration

James Bellows, *Siemens Power Generation, Inc., Orlando, FL,* Steam

James N. BeMiller, *Purdue University, West Lafayette, IN,* Carbohydrates; Gums

Elizabeth A. Benham, *Chevron-Phillips Chemical Company, Bartlesville, OK,* Polyethylene, High Density (HDPE)

Ronald L. Berglund, *Terracon, Kingwood, TX,* Emission Control, Industrial

Barbara Berrie, *National Gallery of Art, Landover, MD,* Fine Art Examination and Conservation

David A. Berry, *U.S. Department of Energy, Morgantown, WV,* Coal Gasification

Christopher A. Bertelo, *ARKEMA, Inc., King of Prussia, PA,* Hydrofluorocarbons

Noelie R. Bertoniere, *Southern Regional Research Center, USDA, New Orleans, LA,* Cellulose

D. E. Beyer, *The Dow Chemical Company, Midland, MI,* Vinylidene Chloride Polymers

D. Bhattacharyya, *University of Kentucky, Lexington, KY,* Reverse Osmosis

Martin Bide, *University of Rhode Island, Kingston, RI,* Dyeing

H. James Bigalow, *Colin A. Houston & Associates, Inc., Brewster, NY,* Market and Marketing Research

Ernst Billig, *Union Carbide Corporation, South Charleston, WV,* Butyl Alcohols; Butyraldehydes

Wolfgang H. Binder, *Technical University of Vienna, Vienna, Austria,* Melamine Resins

Reinaldo Bittar, *Cia. Brasileira Carbureto de Cálcio, Santos Dumont, Brazil,* Silicon and Silicon Alloys, Chemical and Metallurgical

Mary C. Blackburn, *Chemical Market Associates, Inc., Houston, TX,* Chlorine

Wallace R. Blischke, *University of Southern California (Emeritus), Sherman Oaks, CA,* Reliability

James D. Bliss, *U.S. Geological Survey, Tucson, AZ,* Gallium and Gallium Compounds

John H. Block, *Oregon State University, Corvallis, OR,* Vitamins

George Blomgren, *Eveready Battery Co., Inc. West Lake, Ohio,* Batteries, Primary Cells

Bob Blumenthal, *NOAH Technologies Corporation, San Antonio, TX,* Thallium and Thallium Compounds

John T. Boepple, *Nexant ChemSystems, White Plains, NY,* Petrochemicals, Feedstocks

Christopher Boerner, *Washington University, St. Louis, MO,* Recycling

Claudio Boffito, *Saes Getters SpA, Milan, Italy,* Barium

Claudia T. Bogdanos, *White & Case LLP, New York, NY,* Trademarks

Allen F. Bollmeier, Jr., *ANGUS Chemical Company, Buffalo Grove, IL,* Alkanolamines from Nitro Alcohols

Tilak V. Bommaraju, *Process Technology Optimization, Inc., Grand Island, NY,* Chlorine; Hydrogen Chloride

Werner Bonrath, *Roche Vitamins Ltd, Basel, Switzerland,* Nutraceuticals

Karl Booksh, *Arizona State University, Tempe, AZ,* Chemometrics

John K. Borchardt, *Southhaven Communications, Houston, TX,* Petroleum, Enhanced Oil Recovery; Recycling, Paper; Recycling, Plastics

Torben Vedel Borchert, *Novozymes A/S, Bagsvaerd, Denmark,* Enzyme Applications, Industrial

Walter H. Bortle, *General Chemical Corporation, Claymont, DE,* Sodium Nitrate and Nitrite

John C. Bost, *The Dow Chemical Company, Midland, MI,* Acetic Acid, Halogenated Derivatives

J. Bouzas, *The Pennsylvania State University, Hershey Foods Corporation, Hershey, PA,* Chocolate and Cocoa

C. Douglas Boyette, *U.S. Department of Agriculture, ARS, Stoneville, MS,* Herbicides, Biotechnology

Carl Bozzuto, *ALSTOM Power, Inc., Windsor, CT,* Furnaces, Fuel-Fired

Allan Bradbury, *Kraft Foods Corporation, Roehrmoos, Germany,* Coffee

Judith M. Bradow, *U.S. Department of Agriculture, New Orleans, LA,* Herbicides

Robert F. Brady, Jr., *U.S. Naval Research Laboratory, Washington, DC,* Coatings, Marine

Dario Braga, *Università degli Studi di Bologna, Bologna, Italy,* Crystal Engineering

Michael G. Bramucci, *DuPont Central Research and Development, Wilmington, DE,* Genetic Engineering, Microbes

James Brazdil, *BP Amoco Chemicals, Naperville, IL,* Acrylonitrile

Joseph Breen, *U.S. Environmental Protection Agency, Washington, DC,* Lead and Lead Alloys

Hans Joachim Breunig, *University of Bremen, Bremen, Germany,* Bismuth Compounds

Michael Briggs, *U.S. Borax Research Corporation, Anaheim, CA,* Boron Oxides, Boric Acid, and Borates

Margaret Brimble, *The University of Auckland, Auckland, New Zealand,* Polyether Antibiotics

Dennis Brinkman, *Indiana Wesleyan University, Marion, IN,* Recycling, Oil

Janice K. Britt, *TERRA, Inc., Tallahassee, FL,* Toxicology

Ralph Brodd, *Broddarp of Nevada, Inc., Henderson, NV,* Batteries, Introduction

Aaron L. Brody, *Packaging/Brody, Inc., Duluth, GA,* Packaging, Food

Robert T. Brooker, *Olin Corporation, Charleston, TN,* Perchloric Acid and Perchlorates

Charlie R. Brooks, *The University of Tennessee, Knoxville, TN,* Metal Surface Treatments

William E. Brooks, *U.S. Geological Survey, Reston, VA,* Mercury

Richard K. Brow, *University of Missouri-Rolla, Rolla, MO,* Glass

Edward Brown, *Noveon Kalama, Inc., Kalama, WA,* Benzaldehyde; Benzoic Acid

R. Malcolm Brown, *The University of Texas at Austin, Austin, TX,* Cellulose

J. R. Brummer, *FMC Corporation, Princeton, NJ,* Phosphorus

Daniel J. Brunelle, *General Electric Corporate R&D, Schenectady, NY,* Polycarbonates

Robert G. Bryant, *NASA Langley Research Center, Hampton, VA,* Polyimides

Relva C. Buchanan, *University of Cincinnati, Cincinnati, OH,* Ceramics as Electrical Materials

Daniel B. Bullen, *Exponent Failure Analysis Associates, Wood Dale, IL,* Nuclear Fuel Reserves

Kathryn R. Bullock, *Johnson Controls, Inc., Milwaukee, WI,* Batteries, Secondary Cells

David Burdick, *Roche Vitamins Ltd, Basel, Switzerland,* Nutraceuticals

Thomas H. Burgess, *Consultant, Summerfield, FL,* Flow Measurement

James C. Burnett, *Virginia Commonwealth University, Richmond, VA,* Cardiovascular Agents

Joseph C. Burnett, *Huntsman Performance Products, The Woodlands, TX,* Maleic Anhydride, Maleic Acid, and Fumaric Acid

David R. Bush, *PPG Industries, Inc., New Martinsville, WV,* Sodium Sulfates and Sulfides

Karen Bush, *Johnson & Johnson Pharmaceutical Research & Development, Raritan, NJ,* Antibacterial Agents, Overview

E. P. Butler, *National Institute of Standards and Technology, Gaithersburg, MD,* Ceramic-Matrix Composites

M. A. Butler, *Sandia National Laboratories, Albuquerque, NM,* Sensors

David Butts, *Great Salt Lake Minerals Corporation, Ogden, UT,* Chemicals from Brine; Sodium Sulfates and Sulfides

Matthew Butts, *GE Global Research Center, Niskayuna, NY,* Silicones

Christian E. Butzke, *Purdue University, West Lafayette, IN,* Wine

S. Buzby, *University of Delaware, Newark, DE,* Thin Film Formation Techniques

K. Byrappa, *University of Mysore, Mysore, India,* Hydrothermal Processing

Wensheng Cai, *Auburn University, Auburn, AL,* Hydrogels

Narasimhan Calamur, *Amoco Corporation, Naperville, IL,* Butylenes; Propylene

William Cameron, *William Cameron Consulting, Niagara Falls, ON, Canada,* Calcium Carbide; Cyanamides

Gary J. Capone, *Solutia, Inc., Decatur, AL,* Fibers, Acrylic

C. Robert Cappel, *Eastman Kodak Company, Rochester, NY,* Silver Compounds

S. C. Carapella, *Consultant, Tuckahoe, NY,* Arsenic and Arsenic Alloys; Tellurium and Tellurium Compounds

Clifton M. Carey, *National Institute of Standards and Technology, Gaithersburg, MD,* Dental Materials

Paul Carmichael, *Imperial College London, London, UK,* Aminophenols

Ronald N. Caron, *Olin Corporation, New Haven, CT,* Copper Alloys, Wrought

Dodd S. Carr, *Consultant, Durham, NC,* Lead Compounds

F. Patrick Carr, *OMYA, Inc., Proctor, VT,* Calcium Carbonate

Martin E. Carrera, *Amoco Corporation, Naperville, IL,* Butylenes; Propylene

Richard G. Carter, *The Dow Chemical Company, South Charleston, WV,* Diamines and Higher Amines, Aliphatic

Anthony R. Cartolano, *Air Products and Chemicals, Inc., Allentown, PA,* Toluenediamine

Jeremiah P. Casey, *Air Products and Chemicals, Inc., Allentown, PA,* Amines, Cycloaliphatic

Artur Cavaco-Paulo, *University of Minito, Guimaraes, Portugal,* Bleaching Agents

Giuliano Cecchin, *Basell Polyolefins Italia, Ferrara, Italy,* Ziegler-Natta Catalysts

James Cella, *GE Global Research Center, Niskayuna, NY,* Silicones

Paola Ceroni, *Università di Bologna, Bologna, Italy,* Dendrimers, Luminescent

Mark J. Chagnon, *Atlantic Metals and Alloy, Stratford, CT,* Bismuth and Bismuth Alloys; Indium and Indium Compounds

John G. Chambers, *Unilever R&D, Wirral, United Kingdom,* Soap

Michael D. Champness, *Office of the Secretary of Defense, Homeland Security Task Force, Washington, DC,* Technology Transfer

Henri Chanzy, *CNRS-CERMAV, Grenoble, France,* Cellulose

Shou-Bai Chao, *Wyeth Pharmaceuticals, Collegeville, PA,* Vaccine Technology

A. Ray Chapman, *Rochester Institute of Technology, Rochester, NY,* Packaging, Containers for Industrial Materials

Roger Chapman, *Texon UK Ltd, Leicester, United Kingdom,* Nonwoven Fabrics, Staple Fibers

Robert A. Charvat, *Charvat and Associates, Inc., Cleveland, OH,* Colorants for Plastics

George G. Chase, *The University of Akron, Akron, OH,* Filtration

Mahesh Chaubal, *Baxter Health Corporation, Round Lake, IL,* Drug Delivery Systems

Krishan K. Chawla, *The University of Alabama at Birmingham, Birmingham, AL,* Metal-Matrix Composites

Nikhilesh Chawla, *Arizona State University, Tempe, AZ,* Metal-Matrix Composites

Shiou-Shan Chen, *Raytheon Engineers & Constructors, East Weymouth, MA,* Cumene; Styrene

Wai-Kai Chen, *University of Illinois at Chicago (Emeritus), Fremont, CA,* Dimensional Analysis

W. C. Cheng, *WR Grace & Company, Columbia, MD,* Fluid Catalytic Cracking (FCC) Catalysts and Additives

T. T. Peter Cheung, *Phillips Petroleum Company, Bartlesville, OK,* Cyclopentadiene and Dicyclopentadiene

Kenneth Chilton, *Washington University, St. Louis, MO,* Recycling

David L. Chinkes, *The University of Texas Medical Branch, Galveston, TX,* Radioactive Tracers

Yusuf Chisti, *Massey University, Palmerston North, New Zealand,* Mass Transfer

Hsin-Tien Chiu, *National Chiao Tung University, Hsinchu, Taiwan,* Chemical Vapor Deposition

Kuen-Wai Chiu, *Callery Chemical Company, Pittsburgh, PA,* Potassium

S. M. Cho, *Foster Wheeler Energy Corporation, Clinton, NJ,* Heat Transfer

Young I. Cho, *Drexel University, Philadelphia, PA,* Heat Transfer

Igor Chorvath, *Dow Corning Corporation, Midland, MI,* Poly(fluorosilicones)

Lawrence C. Chow, *National Institute of Standards and Technology, Gaithersburg, MD,* Dental Materials

John R. Christoe, *CSIRO Textile and Fibre Technology, Belmont, Australia,* Wool

Kevin Chronley, *Hammond Lead Products, Inc., Philadelphia, PA,* Lead Compounds

C. C. Chu, *Cornell University, Ithaca, NY,* Sutures

Rasik J. Chudger, *BASF Corporation, Rensselaer, NY,* Dyes, Azo

Deborah D. L. Chung, *University of Buffalo, Buffalo, NY,* Composite Materials

David L. Clark, *Los Alamos National Laboratory, Los Alamos, NM,* Plutonium and Plutonium Compounds; Thorium and Thorium Compounds; Uranium and Uranium Compounds

James H. Clark, *University of York, York, United Kingdom,* Catalysts, Supported

P. D. Clark, *University of Calgary, Alberta, Canada,* Sulfur and Hydrogen Sulfide Recovery

Margaret A. Clarke, *Sugar Processing Research Institute, Inc., New Orleans, LA,* Sugar

Stephen I. Clarke, *Air Products and Chemicals, Inc., Allentown, PA,* Nitric Acid

Michael Cleary, *Imperial Holly Corporation, Colorado Springs, CO,* Sugar

K. J. Coeling, *Nordson Corporation, Westlake, OH,* Coating Processes, Spray

Alan P. Cohen, *UOP, Des Plaines, IL,* Desiccants

Edward D. Cohen, *Consultant, Fountain Hills, AZ,* Coating Processes

Peter J. Collings, *Swarthmore College, Swarthmore, PA,* Liquid Crystalline Materials

William J. Colonna, *American Crystal Sugar Company, Moorhead, MN,* Sugar

B. Conley, *University of Kentucky, Lexington, KY,* Aluminum Halides and Aluminum Nitrate

Robert R. Contrell, *Union Camp Corporation, Wayne, NJ,* Carboxylic Acids

C. David Cooper, *University of Central Florida, Orlando, FL,* Air Pollution Control Methods

James Corbin, *University of Illinois at Urbana-Champaign, Urbana, IL,* Feeds and Feed Additives, Pet Foods

Cajetan F. Cordeiro, *Air Products and Chemicals, Inc., Allentown, PA,* Vinyl Acetate Polymers

Nick Corner-Walker, *Alan Letki, Alfa Laval Inc., Warminster, PA,* Centrifugal Separation

Michael G. Costello, *3M Company, St. Paul, MN,* Fluoroethers and Fluoroamines

J. Kevin Cotchen, *MAN GHH Corporation, Pittsburgh, PA,* Furnaces, Electric

Howard B. Cottam, *University of California, San Diego, CA,* Antiaging Agents

Leroy Covington, Jr., *North Carolina Agricultural and Technical State University, Greensboro, NC,* Nitrides

Joseph A. Cowfer, *The Geon Company, Avon Lake, OH,* Vinyl Chloride

Peter J. Cragg, *University of Brighton, Brighton, United Kingdom,* Supramolecular Chemistry

J. M. Criscione, *UCAR Carbon Company, Inc., Cleveland, OH,* Carbon; Graphite, Artificial

Daniel A. Crowl, *Michigan Technological University, Houghton, MI,* Hazard Analysis and Risk Assessment

Michael J. Cruickshank, *Consultant, Honolulu, HI,* Ocean Raw Materials

Theodore Cruz, *Kraft Foods Corporation, Tarrytown, NY,* Coffee

Mark Csele, *Niagara College, Ontario, Canada,* Lasers

E. L. Cussler, *University of Minnesota, Minneapolis, MN,* Chemical Product Design

Horace G. Cutler, *Mercer University, Atlanta, GA,* Growth Regulators, Plant

Stephen J. Cutler, *Mercer University, Atlanta, GA,* Growth Regulators, Plant

Gregory P. Dado, *Stepan Company, Northfield, IL,* Sulfonation and Sulfation

Dady B. Dadyburjor, *West Virginia University, Morgantown, WV,* Coal Liquefaction

Qizhou Dai, *University of British Columbia, British Columbia, Canada,* Pulp

David R. Dalton, *Temple University, Philadelphia, PA,* Alkaloids

Larry R. Dalton, *University of Washington, Seattle, WA,* Nonlinear Optical Materials

Ture Damhus, *Novozymes A/S, Bagsvaerd, Denmark,* Enzyme Applications, Industrial

Glen Dammann, *Krupp Uhde GmbH, Dortmund, Germany,* Chlorine

R. W. Daniels, *Union Camp Corporation, Wayne, NJ,* Carboxylic Acids

K. V. Darragh, *Rhone-Poulenc, Inc. Cranbury, NJ,* Aluminum Sulfate and Alums

Tapas K. Das, *Washington State Department of Ecology, Olympia, WA,* Disinfection

Purnendu K. Dasgupta, *Texas Tech University, Lubbock, TX,* Capillary Separations

Rathin Datta, *Consultant, Chicago, IL,* Hydroxycarboxylic Acids

Simon Davies, *Davy Process Technology, London, United Kingdom,* Methanol

Edmond de Hoffmann, *University of Louvain, Louvain-la-Neuve, Belgium,* Mass Spectrometry

Philippe Degée, *University of Mons-Hainaut, Mons, Belgium,* Polylactides

Phillip DeLassus, *The University of Texas-Pan American, Edinburg, TX,* Barrier Polymers; Vinylidene Chloride Polymers

Lionel Delaude, *Université de Liège, Liège, Belgium,* Metathesis

Ron J. Denning, *CSIRO Textile and Fibre Technology, Belmont, Australia,* Wool

Maurice Dery, *Akzo Nobel Chemicals, Inc., Dobbs Ferry, NY,* Ammonium Compounds

Siddharth Devarajan, *Rensselaer Polytechnic Institute, Troy, NY,* Semiconductors, Silicon Based

J. P. Dever, *Union Carbide Technical Center, South Charleston, WV,* Ethylene Oxide

Stephen C. DeVito, *U.S. Environmental Protection Agency, Washington, DC,* Lead and Lead Alloys; Mercury; Nitriles

Martin Dexter, *Consultant,* Antioxidants, Polymers

Sandeep S. Dhingra, *The Dow Chemical Company, Midland, MI,* Molecular Sieves

Patrick Dibello, *FMC Corporation, Princeton, NJ,* Barium Compounds

Gary L. Dickerson, *Solutia Inc., Cantonment, FL,* Adipic Acid

Charles Dickert, *Consultant, Yardley, PA,* Ion Exchange

James R. Dickey, *Consultant, Basking Ridge, NJ,* Lubrication and Lubricants

C. M. Dietz, *Consultant,* Acetylene

Ibrahim Dincer, *University of Ontario Institute of Technology, Ontario, Canada,* Refrigeration

Christopher P. Dionigi, *U.S. Department of Agriculture, New Orleans, LA,* Herbicides

Rocco DiSanto, *Cambrex Technical Center, North Brunswick, NJ,* Biocatalysis

Urmila M. Diwekar, *Vishwamitra Research Institute, Westmont, IL,* Sampling Techniques

G. O. Doak, *North Carolina State University, Raleigh, NC,* Antimony Compounds

Michael F. Doherty, *University of California, Santa Barbara, Santa Barbara, CA,* Distillation, Azeotropic, and Extractive

William R. Dolbier, Jr., *University of Florida, Gainesville, FL,* Hydrofluorocarbons

A. J. Domb, *The Hebrew University of Jerusalem, Jerusalem, Israel,* Drug Delivery Systems

Larry Dominey, *OM Group, Inc. (Retired), Chagrin, OH,* Cobalt and Cobalt Alloys

Maurizio Dorini, *Basell Polyolefins, Ferrara, Italy,* Polypropylene

Vishu Dosaj, *Dow Corning Corporation, Midland, MI,* Silicon and Silicon Alloys, Chemical and Metallurgical

Ronald L. Dotson, *Olin Corporation, Charleston, TN,* Perchloric Acid and Perchlorates

T. J. Dougan, *WR Grace & Company, Columbia, MD,* Fluid Catalytic Cracking (FCC) Catalysts and Additives

T. J. Dougherty, *Pfizer, Inc., Groton, CT,* Quinolone Antibacterials

Ross Dowbenko, *PPG Industries, Allison Park, PA,* Allyl Monomers and Polymers

James Downing, *Consultant, Ellicottville, NY,* Manganese and Manganese Alloys

W. H. Dresher, *WHD Consulting, Tucson, AZ,* Copper

Lawrence J. Drew, *U.S. Geological Survey, Reston, VA,* Petroleum

Richard W. Drisko, *U.S. Naval Civil Engineering Laboratory, Port Hueneme, CA,* Coatings, Marine

Philippe Dubois, *University of Mons-Hainaut, Mons, Belgium,* Polylactides

Paul F. Duby, *Columbia University, New York, NY,* Metallurgy, Extractive

Markus Dugal, *Bayer MaterialScience AG, Dormagen, Germany,* Nitrobenzene and Nitrotoluenes

Stephen O. Duke, *U.S. Department of Agriculture, ARS, University, MS,* Herbicides, Biotechnology

Budd L. Duncan, *Olin Corporation, Norwalk, CT,* Chloric Acid and Chlorates

Manfred Dunky, *Dynea Austria GmbH, Krems, Austria,* Melamine Resins

Kenneth L. Dunlap, *Bayer Corporation, New Martinsville, WV,* Phosgene

Irene Durbak, *USDA Forest Service, Madison, WI,* Wood

Katherine A. Durham, *University of Missouri–Rolla Coatings Institute, Rolla, MO,* Paint

Douglas J. Durian, *University of Pennsylvania, Philadelphia, PA,* Foams

Brad Durkin, *MacDermid, Inc., New Hudson, MI,* Electroless Deposition

Richard A. Durst, *Cornell University, Ithaca, NY,* Hydrogen-Ion Activity

Cecil Dybowski, *University of Delaware, Newark, DE,* Chromatography

Paul J. Dyson, *Ecole Polytechnique Fédérale de Lausanne, Lausanne, Switzerland,* Metal Carbonyls

A. J. Dzermejko, *UCAR Carbon Company, Inc., Cleveland, OH,* Graphite, Artificial

Alan C. Eachus, *ANGUS Chemical Company, Buffalo Grove, IL,* Alkanolamines from Nitro Alcohols

Anthony J. East, *Brooklake Polymers, Madison, NJ,* Polyesters, Thermoplastic

Alan D. Eastman, *Phillips Petroleum Research Center, Bartlesville, OK,* Hydrocarbons

Sina Ebnesajjad, *DuPont Fluoro Products, Wilmington, DE,* Poly(vinyl fluoride)

W. Wesley Eckenfelder, Jr., *Eckenfelder Inc., Nashville, TN,* Wastewater Treatment

Martha R. Edens, *The Dow Chemical Company, Freeport, TX,* Alkanolamines from Olefin Oxides and Ammonia

Thomas F. Edgar, *The University of Texas at Austin, Austin, TX,* Process Control

Janice W. Edwards, *Monsanto Company, St. Louis, MO,* Genetic Engineering, Plants

Tim Eggeman, *Neoterics International, Lakewood, CO,* Ammonia; Hydrides; Sodium Carbonate; Sodium Hydroxide

Manfred Eggersdorfer, *Roche Vitamins Ltd, Basel, Switzerland,* Nutraceuticals

Fred C. Eichmiller, *National Institute of Standards and Technology, Gaithersburg, MD,* Dental Materials

Marco Eissen, *Carl von Ossietzky Universität Oldenburg, Oldenburg, Germany,* Sustainable Development and Chemistry

Maher Y. Elsheikh, *ARKEMA, Inc., King of Prussia, PA,* Hydrofluorocarbons

Alan English, *John Brown E&C, Houston, TX,* Methanol

Matthew Ennis, *Stanford University, Stanford, CA,* Nitrogen

Richard A. Eppler, *Consultant, Cheshire, CT,* Colorants for Ceramics

Carolyn A. Ertell, *Rhone-Poulenc, Inc. Cranbury, NJ,* Aluminum Sulfate and Alums

Samuel F. Etris, *Consultant, Wayne, PA,* Gold and Gold Compounds; Silver and Silver Alloys; Silver Compounds

William Etzkorn, *Union Carbide Corporation, South Charleston, WV,* Acrolein and Derivatives

Wilfried Eul, *Degussa AG, Frankfurt, Germany,* Hydrogen Peroxide; Peroxides, Inorganic

David J. Evans, *CSIRO Textile and Fibre Technology, Belmont, Australia,* Wool

Francis Evans, *Honeywell International, Morristown, NJ,* Boron Halides

Tom J. Evans, *Cubic Defense Applications Group, White City, OR,* Chemical Warfare

Kevin G. Ewsuk, *Sandia National Laboratories, Albuquerque, NM,* Ceramics, Processing

Edward F. Ezell, *The BOC Group, Inc., Murray Hill, NJ,* High Purity Gases

James R. Fair, *The University of Texas at Austin, Austin, TX,* Distillation

James Falcone, Jr., *West Chester University, West Chester, PA,* Silicas and Silicates, Anthropogenic

Jawed Fareed, *Loyola University Medical Center, Maywood, IL,* Blood Coagulation and Anticoagulant Drugs

David S. Farinato, *Cytec Technology Corporation, Stamford, CT,* Acrylamide Polymers

Brian E. Farkas, *North Carolina State University, Raleigh, NC,* Food Processing

Daniel F. Farkas, *Oregon State University, Corvallis, OR,* Food Processing

James P. Farr, *The Clorox Company, Pleasanton, CA,* Bleaching Agents

Rudolf Faust, *University of Massachusetts Lowell, Lowell, MA,* Initiators, Cationic

Darrell Fee, *Astaris LLC, St. Louis, MO,* Phosphorus Compounds

Susan R. Feldman, *Salt Institute, Alexandria, VA,* Sodium Chloride

Timothy R. Felthouse, *MECS, Inc., Overland, MO,* Maleic Anhydride, Maleic Acid, and Fumaric Acid

Richard Fengi, *Eastman Chemical Company, Kingsport, TN,* Cellulose Esters, Organic Esters

Audeen W. Fentiman, *Purdue University, West Lafayette, IN,* Waste Management, Radioactive

William Ferguson, *Tanco, Lac du Bonnet, Canada,* Cesium and Cesium Compounds

K. Thomas Finley, *SUNY Brockport, Brockport, NY,* Quinolines and Isoquinolines; Quinones

Frank Fischetti, Jr., *Craftmaster Flavor Technology, Inc., Amityville, NY,* Flavors

F. F. Fisher, *UCAR Carbon Company, Inc., Cleveland, OH,* Graphite, Artificial

Ian A. Fisher, *Shell Oil Company, Houston, TX,* Methyl Isobutyl Ketone

James Fisher, *IBMA, Princeton Junction, NJ,* Titanium Compounds, Inorganic

Richard M. Flynn, *3M Company, St. Paul, MN,* Fluoroethers and Fluoroamines

Peter J. S. Foot, *Kingston University, Kingston, Surrey, United Kingdom,* Conducting Polymers

M. W. Forkner, *Union Carbide Corporation, South Charleston, WV,* Glycols

Ivan Fortelný, *Academy of Sciences of the Czech Republic, Prague, Czech Republic,* Polymer Blends

Peter R. Foster, *SNBTS Protein Fractionation Centre, Edinburgh, United Kingdom,* Fractionation, Plasma

Anna C. Fraker, *National Institute of Standards and Technology, Gaithersburg, MD,* Dental Materials

S. Franklin, *University of Delaware, Newark, DE,* Thin Film Formation Techniques

David K. Frederick, *OMYA, Inc., Proctor, VT,* Calcium Carbonate

Leon D. Freedman, *North Carolina State University, Raleigh, NC,* Antimony Compounds

Harold S. Freeman, *North Carolina State University, Raleigh, NC,* Dyes, Environmental Chemistry

Mira Freiberg, *Dead Sea Bromine Group, Beer Sheva, Israel,* Bromine; Bromine, Inorganic Compounds

Mark B. Freilich, *The University of Memphis, Memphis, TN,* Potassium Compounds

Alfred D. French, *Southern Regional Research Center, USDA, New Orleans, LA,* Cellulose

F. H. Sam Froes, *University of Idaho, Moscow, ID,* Titanium and Titanium Alloys

Katharina M. Fromm, *University of Fribourg, Fribourg, Switzerland,* Coordination Compounds

William Fruscella, *Unocal Corporation El Segundo, CA,* Benzene

Albert J. Fry, *Wesleyan University, Middletown, CT,* Electrochemical Processing, Organic

Gerhard E. Fuchs, *University of Florida, Gainesville, FL,* High Temperature Alloys

Claus Crone Fuglsang, *Novozymes A/S, Bagsvaerd, Denmark,* Enzyme Applications, Industrial

C. F. Fulgenzi, *UCAR Carbon Company, Inc., Cleveland, OH,* Graphite, Artificial

E. R. Fuller, Jr., *National Institute of Standards and Technology, Gaithersburg, MD,* Ceramic-Matrix Composites

Thomas F. Fuller, *University of California, Berkeley, Berkeley, CA,* Electrochemical Processing

Carlo Fumagalli, *LONZA SpA, Scanzorosciate, Italy,* Succinic Acid and Succinic Anhydride

Barbara J. Furches, *The Dow Chemical Company, Midland, MI,* Plastics Testing

S. K. Gaggar, *GE Plastics Technology Center, Pittsfield, MA,* Acrylonitrile-Butadiene-Styrene (ABS) Polymers

John Gallini, *Consultant, Richmond, VA,* Polyamides, Aromatic

Derek W. Gammon, *California Environmental Protection Agency, Sacramento, CA,* Insecticides

Subhash V. Gangal, *E. I. du Pont de Nemours & Company, Inc., Wilmington, DE,* Perfluorinated Polymers

Richard G. Gann, *Fire Research Division, National Institute of Standards and Technology, Gaithersburg, MD,* Flame Retardants

Richard E. Gannon, *Textron Defense Systems, Wilmington, MA,* Acetylene

Fabio Garbassi, *Eni Chem Research Center, Novara, Italy,* Engineering Thermoplastics

David R. Gard, *Astaris LLC, St. Louis, MO,* Phosphoric Acids and Phosphates; Phosphorus Compounds

Jerry D. Gargulak, *LignoTech USA, Inc., Rothschild, WI,* Lignin

Roger Gary, *Milacron Inc., Cincinnati, OH,* Abrasives

Bruce C. Gates, *University of California, Davis, Davis, CA,* Catalysis

Günter Gauglitz, *Universität Tübingen, Tübingen, Germany,* Spectroscopy

Charles C. Gaver, Jr., *Consultant, Mt. Laurel, NJ,* Tin and Tin Alloys

Steven Gedon, *Eastman Chemical Company, Kingsport, TN,* Cellulose Esters, Organic Esters

Gordon H. Geiger, *The University of Arizona, Tucson, AZ,* Iron

Chester H. Gelbert, *DuPont Company, Philadelphia, PA,* Latex Technology

Stanley A. Gembicki, *UOP LLC, Des Plaines, IL,* Adsorption, Liquid Separation

Christine George, *Air Liquide America Corporation, Houston, TX,* Carbon Monoxide

Kathy F. George, *Union Carbide Technical Center, South Charleston, WV,* Ethylene Oxide

Laurie A. George, *National Institute of Standards and Technology, Gaithersburg, MD,* Dental Materials

H. Robert Gerberich, *Hoechst-Celanese Corporation, Corpus Christi, TX,* Formaldehyde

Michael Gernon, *Arkema Group, Inc., King of Prussia, PA,* Sulfur Compounds

Thomas Gessmann, *Rensselaer Polytechnic Institute, Troy, NY,* Light Emitting Diodes

Christen M. Giandomenico, *Johnson Matthey, West Chester, PA,* Platinum-Group Metals, Compounds

D. S. Gibbs, *The Dow Chemical Company, Midland, MI,* Vinylidene Chloride Polymers

Paul Gifford, *Ovonic Battery Company, Troy, MI,* Batteries, Secondary Cells

Gregory Gillette, *GE Global Research Center, Niskayuna, NY,* Silicones

Jeffrey W. Gilman, *Fire Research Division, National Institute of Standards and Technology, Gaithersburg, MD,* Flame Retardants

M. N. Gitlitz, *M&T Chemicals, Inc., Rahway, NJ,* Tin Compounds

S. Jill Glass, *Sandia National Laboratories, Albuquerque, NM,* Ceramics, Mechanical Properties

Wolfgang Glasser, *Virginia Polytechnic Institute and State University, Blacksburg, VA,* **Cellulose**

Furman E. Glenn, *DuPont Dow Elastomers LLC, Louisville, KY,* Polychloroprene

Bartek A. Glowacki, *University of Cambridge, Cambridge, United Kingdom,* Superconductivity and Superconductors

Mary An Godshall, *Sugar Processing Research Institute, Inc., New Orleans, LA,* Sugar

Dasantila Golemi-Kotra, *Wayne State University, Detroit, MI,* Antibiotic Resistance

Pedro Gómez-Romero, *Materials Science Institute, Barcelona, Spain,* Hybrid Nanocomposite Materials

Frank E. Goodwin, *International Lead and Zinc Research Organization, Inc., Research Triangle Park, NC,* Zinc and Zinc Alloys; Zinc Compounds

Regina Goralczyk, *Roche Vitamins Ltd, Basel, Switzerland,* Nutraceuticals

Maximilian B. Gorensek, *The Geon Company, Avon Lake, OH,* Vinyl Chloride

Dena Gorrie, *Tanco, Lac du Bonnet, Canada,* Cesium and Cesium Compounds

Myron Gottlieb, *Gas Research Institute, Chicago, IL,* Gas, Natural

Michael C. Grady, *DuPont Company, Philadelphia, PA,* Latex Technology

G. W. Grames, *Witco Corporation, Oakland, NJ,* Aluminum Halides and Aluminum Nitrate

Claes G. Granqvist, *Uppsala University, Uppsala, Sweden,* Solar Energy Materials

Charles A. Gray, *Cabot Corporation, Billerica, MA,* Carbon Black

Derek Gray, *McGill University, Montreal, Canada,* Cellulose

Floyd Gray, *U.S. Geological Survey, Tucson, AZ,* Gallium and Gallium Compounds

David W. Green, *USDA Forest Service, Madison, WI,* Wood

Peter Gregory, *Zeneca Specialties,* Dyes and Dye Intermediates

Anthony Grenis, *Elite Spice, Inc., Somonauk, IL,* Spices

Fabrizia Grepioni, *Università di Sassari, Sassari, Italy,* Crystal Engineering

John J. Gresens, *Merchant & Gould, Minneapolis, MN,* Patents and Trade Secrets

Baruch Grinbaum, *IMI (TAMI) Institute for Research and Development, Haifa Bay, Israel,* Bromine; Drilling Fluids

Morris P. Grotheer, *Kerr-McGee Corporation, Edmond, OK,* Electrochemical Processing, Inorganic

Karl Grozinger, *Boehringer Ingelheim Pharmaceuticals, Inc., Ridgefield, CT,* Pharmaceuticals, Large-Scale Synthesis

R. Guglielmetti, *Université de la Méditerrannée, Marseille, France,* Chromogenic Materials, Electrochromic; Chromogenic Materials, Photochromic; Chromogenic Materials, Piezochromic; Chromogenic Materials, Thermochromic

Jeroen B. Guinée, *Leiden University, Leiden, The Netherlands,* Life Cycle Assessment

Zhihua Guo, *The Ohio State University, Columbus, OH,* Supercritical Fluids

Ram B. Gupta, *Auburn University, Auburn, AL,* Hydrogels

B. Frank Gupton, *Boehringer Ingelheim Chemicals, Inc., Petersburg, VA,* Pharmaceuticals, Large-Scale Synthesis

Randolph J. Guschl, *DuPont Center for Collaborative Research & Education, Wilmington, DE,* Technology Transfer

Edgar B. Gutoff, *Consultant, Brookline, MA,* Coating Processes

Erich Habegger, *Rohner AG, Pratteln, Switzerland,* Fine Chemicals

C. E. Habermann, *Dow Chemical, Midland, MI,* Acrylamide

Kenneth Hacias, *Parker Amchem, Madison Heights, MI,* Metal Surface Treatments

H. J. Hagameyer, *Texas Eastman Company, Longview, TX,* Acetaldehyde

David S. Hage, *University of Nebraska, Lincoln, NE,* Chromatography, Affinity

Gerald J. Hahn, *General Electric, Schenectady, NY,* Design of Experiments

Juergen Hahn, *Texas A&M University, College Station, TX,* Process Control

Nick Hallale, *AspenTech Ltd., Cheshire, United Kingdom,* Process Integration Technology

James G. Hansel, *Air Products and Chemicals, Inc., Allentown, PA,* Oxygen

Steven M. Hansen, *E. I. du Pont de Nemours & Company, Inc., Washington, WV,* Polyester Fibers

Tomas Tage Hansen, *Novozymes A/S, Bagsvaerd, Denmark,* Enzyme Applications, Industrial

Thomas L. Hardenburger, *Air Liquide America Corporation, Tualatin, OR,* Nitrogen

R. H. Harding, *WR Grace & Company, Columbia, MD,* Fluid Catalytic Cracking (FCC) Catalysts and Additives

Robert Hart, *Parker Amchem, Madison Heights, MI,* Metal Surface Treatments

Gerard L. Hasenhuettl, *Consultant, Port Saint Lucie, FL,* Fats and Fatty Oils

Kazuyuki Hattori, *Kitami Institute of Technology, Kitami, Japan,* Cellulose

Makoto Hattori, *Sumitomo Chemical Company Ltd., Tokyo, Japan,* Dyes, Anthraquinone

Kathryn S. Hayes, *Air Products and Chemicals, Inc., Allentown, PA,* Amines, Lower Aliphatic Amines; Methylamines

Sigfried S. Hecker, *Los Alamos National Laboratory, Los Alamos, NM,* Plutonium and Plutonium Compounds

J. B. Hedge, *UCAR Carbon Company, Inc., Cleveland, OH,* Graphite, Artificial

B. W. Hedrick, *UOP LLC, Des Plaines, IL,* Fluid Catalytic Cracking (FCC) Units, Regeneration

Reinout Heijungs, *Leiden University, Leiden, The Netherlands,* Life Cycle Assessment

Howard I. Heitner, *Cytec Industries, Stamford, CT,* Flocculating Agents

John D. Hem, *U.S. Geological Survey, Menlo Park, CA,* Water

C. L. Hemler, *UOP LLC, Des Plaines, IL,* Fluid Catalytic Cracking (FCC) Units, Regeneration

Ramesh R. Hemrajani, *ExxonMobil Research and Engineering Company, Fairfax, VA,* Mixing and Blending

Larry L. Hench, *University of Florida, Gainesville, FL,* Sol–Gel Technology

Jean-Marie Herrmann, *Université Claude Bernard Lyon 1, Villeurbanne cédex, France,* Photocatalysis

D. Michael Herron, *Air Products & Chemicals, Inc., Allentown, PA,* Cryogenic Technology

Norman Herron, *E. I. Du Pont de Nemours & Co., Inc., Wilmington, DE,* Cadmium Compounds

Martha Hesser, *Noveon Kalama, Inc., Kalama, WA,* Benzaldehyde; Benzoic Acid

John D. Hewes, *Honeywell International, Inc., Washington, DC,* Combinatorial Chemistry

Stephen G. Hibbins, *Timminco Metals, Ontario, Canada,* Strontium and Strontium Compounds

J. C. Hickman, *The Dow Chemical Company, Freeport, TX,* Chloroethylenes and Chloroethanes

Terry L. Highley, *USDA Forest Service, Madison, WI,* Wood

W. D. Hinsberg, *IBM Research Division, San Jose, CA,* Lithographic Resists

Mohamed W. M. Hisham, *Occidental Chemical Corporation, Dallas, TX,* Hydrogen Chloride

Brent Hiskey, *The University of Arizona, Tucson, AZ,* Metallurgy

Drahomíra Hlavatá, *Academy of Sciences of the Czech Republic, Prague, Czech Republic,* Polymer Blends

Albert M. Hochhauser, *ExxonMobil Research and Engineering Company, Paulsboro, NJ,* Gasoline and Other Motor Fuels

F. Galen Hodge, *Haynes International, Inc., Kokomo, IN,* Cobalt and Cobalt Alloys

Darleane Hoffman, *University of California, Berkeley, Oakland, CA,* Actinides and Transactinides

W. C. Hoffman, *Union Carbide Technical Center, South Charleston, WV,* Ethylene Oxide

James E. Hoffmann, *Jan Reimers and Associates USA Inc., Houston, TX,* Selenium and Selenium Compounds; Tellurium and Tellurium Compounds

Siegfried Hofmann, *Max Planck Institute for Metals Research, Stuttgart, Germany,* Surface and Interface Analysis

Michael T. Holbrook, *Dow Chemical USA, Plaquemine, LA,* Chloroform; Methyl Chloride; Methylene Chloride

Geoffrey Holden, *Holden Polymer Consulting, Inc., Prescott, AZ,* Thermoplastic Elastomers

Rawle I. Hollingsworth, *Michigan State University, East Lansing, MI,* Oxazolidinones, Antibacterial

Zdeněk Horák, *Academy of Sciences of the Czech Republic, Prague, Czech Republic,* Polymer Blends

Ben Horrell, *Huntsman Performance Products, The Woodlands, TX,* Maleic Anhydride, Maleic Acid, and Fumaric Acid

Eugene V. Hort, *GAF Corporation, Wayne, NJ,* Acetylene-Derived Chemicals

Ramachandra S. Hosmane, *University of Maryland, Baltimore County, Baltimore, MD,* Antiviral Agents

Kevin L. Houghton, *ICI Chemicals and Polymers Ltd., Aberdeen, United Kingdom,* Chlorinated Paraffins

James L. Howard, *USDA Forest Service, Madison, WI,* Wood

William L. Howard, *Consultant,* Acetone; Chelating Agents

B. A. Howell, *Central Michigan University, Mount Pleasant, MI,* Vinylidene Chloride Polymers

James Hower, *University of Kentucky, Lexington, KY,* Coal

Timothy E. Howson, *Wyman-Gordon, Worcester, MA,* Nickel and Nickel Alloys

Craig A. Hoyme, *Eastman Chemical Company, Kingsport, TN,* Separations Process Synthesis

You-Lo Hsieh, *University of California, Davis, Davis, CA,* Fibers

Chang Samuel Hsu, *Exxon Research and Engineering Company, Annandale, NJ,* Petroleum

Wern-Jir Hsu, *Columbia University, New York, NY,* Microfluidics

Sun-Yi Huang, *Cytec Technology Corporation, Stamford, CT,* Acrylamide Polymers

Colin D. Hubbard, *University of Erlangen-Nürnberg, Erlangen, Germany,* High Pressure Chemistry

Martin A. Hubbe, *North Carolina State University, Raleigh, NC,* Paper

Robert C. Hughes, *Sandia National Laboratories, Albuquerque, NM,* Sensors

Konrad Hungerbühler, *Swiss Federal Institute of Technology, Zürich, Switzerland,* Sustainable Development and Chemistry

James Hunter, *Eveready Battery Co., Inc. West Lake, OH,* Batteries, Primary Cells

William N. Hunter, *Celanes Canada Inc., Edmonton, Alberta, Canada,* Alcohols, Polyhydric

T. R. Hupp, *UCAR Carbon Company, Inc., Cleveland, OH,* Graphite, Artificial

Mickey G. Huson, *CSIRO Textile and Fibre Technology, Belmont, Australia,* Wool

S. Y. Hwang, *Washington Group International, East Weymouth, MA,* Cumene

Shuen-Cheng Hwang, *The BOC Group, Inc., Murray Hill, NJ,* High Purity Gases; Noble Gases

Yng-Long Hwang, *ExxonMobil Research and Engineering, Fairfax, VA,* Ketones

Andrei Ya. Il'chenko, *National Academy of Sciences of Ukraine, Kyiv, Ukraine,* Polymethine Dyes

Edward E. Jaffe, *Consultant, Wilmington, DE,* Pigments, Organic

Shahab Jahromi, *DSM Research, Geleen, The Netherlands,* Melamine Resins

Robert C. James, *TERRA, Inc., Tallahassee, FL,* Toxicology

Sei-Joo Jang, *The Pennsylvania State University, University Park, PA,* Ferroelectrics

Christoph Janiak, *Universität Freiburg, Freiburg, Germany,* Metallocene Catalysts

Linda H. Jansen, *Callery Chemical Co., Pittsburgh, PA,* Boron, Elemental

Gordon D. Jarvinen, *Los Alamos National Laboratory, Los Alamos, NM,* Plutonium and Plutonium Compounds

Walter P. Jeske, *Loyola University Medical Center, Maywood, IL,* Blood Coagulation and Anticoagulant Drugs

Ingegärd Johansson, *Akzo Chemical Corporation, Stenungsund, Sweden,* Amides, Fatty Acid

W. R. Johns, *Chemcept Limited, Reading, United Kingdom,* Computer-Aided Chemical Engineering

James A. Johnson, *UOP LLC, Des Plaines, IL,* Adsorption, Liquid Separation; Alkylation

R. W. Johnson, Jr., *Union Camp Corporation, Wayne, NJ,* Carboxylic Acids

Richard M. Johnson, *U.S. Department of Agriculture, Houma, LA,* Herbicides

Leslie N. Jones, *CSIRO Textile and Fibre Technology, Belmont, Australia,* Wool

Gajanan S. Joshi, *ISBDD/Virginia Biotechnology Research Park, Richmond, VA,* Cardiovascular Agents

Jamie Jerrick Juliette, *Rohm and Haas Company, Spring House, PA,* Methacrylic Acid and Derivatives

Beth Junker, *Merck Research Laboratories, Rahway, NJ,* Fermentation

Donald J. Kaczynski, *Brush Wellman, Inc. Oak Harbor, OH,* Beryllium, Beryllium Alloys and Composites; Beryllium Compounds

John F. Kadla, *University of British Columbia, British Columbia, Canada,* Pulp

J. Kadow, *Bristol-Myers Squibb, Wallingford, CT,* Quinolone Antibacterials

Alan B. Kaiser, *MacDiarmid Institute for Advanced Materials and Nanotechnology, Victoria University of Wellington, Wellington, New Zealand,* Conducting Polymers

Debra Kaiser, *National Institute of Standards and Technology, Gaithersburg, MD,* Combinatorial Chemistry

Mary Kaiser, *E.I. du Pont de Nemours, Wilmington, DE,* Chromatography

Rustu S. Kalyoncu, *U.S. Geological Survey, Reston, VA,* Graphite, Natural

Peter Kamarchik, Jr., *PPG Industries, Allison Park, PA,* Rheology and Rheological Measurements

Sanjay Kamat, *Cambrex Technical Center, North Brunswick, NJ,* Biocatalysis

Conrad W. Kamienski, *Consultant, Gastonia, NC,* Lithium and Lithium Compounds

Arie Kampf, *Dead Sea Bromine Group, Beer Sheva, Israel,* Bromine, Organic Compounds

G. S. Kamali Kannangara, *University of Western Sydney, New South Wales, Australia,* Nanotechnology

Jon Kapecki, *Eastman Kodak Company, Rochester, NY,* Photography, Color

David L. Kaplan, *Tufts University, Medford, MA,* Silk

Lawrence Karas, *ARCO Chemical Company, Newtown Square, PA,* Ethers

Alamgir Karim, *National Institute of Standards and Technology, Gaithersburg, MD,* Combinatorial Chemistry

Subhash Karkare, *Scientia Consulting, Thousand Oaks, CA,* Cell Culture Technology

Joseph J. Katz, *Argonne National Laboratory, Argonne, IL,* Deuterium and Tritium

J. A. Keely, *FMC Corporation, Princeton, NJ,* Phosphorus

M. J. Keenan, *Exxon Chemical Company, Baton Rouge, LA,* **Carboxylic Acids**

Thomas R. Keenan, *Knox Gelatine, Inc., Sioux City, IA,* Gelatin

Melvin E. Keener, *Coalition of Responsible Waste Incineration, Washington, DC,* Hazardous Waste Incineration

Steffen Kelch, *Deutsches Wollforschungsinstitut, Aachen, Germany,* Shape-Memory Polymers

Jay O. Keller, *Sandia National Laboratories, Livermore, CA,* Hydrogen Energy

Robert J. Keller, *Vinings Industries, Inc. Atlanta, GA,* Aluminates

J. Robert Kelly, *National Institute of Standards and Technology, Gaithersburg, MD,* Dental Materials

Corey J. Kenneally, *The Procter & Gamble Company, Cincinnati, OH,* Alcohols, Higher Aliphatic, Survey

D. Webster Keogh, *Los Alamos National Laboratory, Los Alamos, NM,* Thorium and Thorium Compounds; Uranium and Uranium Compounds

Rachid Kerboua, *GE Global Research Center, Niskayuna, NY,* Silicones

Pravin Khandare, *Occidental Chemical Corporation, Dallas, TX,* **Chlorotoluenes, Ring**

Yuri L. Khmelnitsky, *Albany Molecular Research, Albany, NY,* Microbial Transformations

Richard Kieffer, *Technical University of Vienna, Vienna, Austria,* Carbides, Survey

Barry T. Kilbourn, *Molycorp Inc., Brea, CA,* Cerium and Cerium Compounds

Jong Kyu Kim, *Rensselaer Polytechnic Institute, Troy, NY,* Light Emitting Diodes

Joon-Seop Kim, *Chosun University, Kwangju, Korea,* Ionomers

Glenn E. Kinard, *Air Products and Chemicals, Inc., Allentown, PA,* Cryogenic Technology

Desmond F. King, *Chevron Texaco, Tiburon, CA,* Fluidization

Michael G. King, *Selenium–Tellurium Development Association, Salt Lake City, UT,* Lead and Lead Alloys; Selenium and Selenium Compounds; Tellurium and Tellurium Compounds

R. E. King III, *Ciba Specialty Chemicals, Tarrytown, NY,* Antioxidants, Polymers

Ole Kirk, *Novozymes A/S, Bagsvaerd, Denmark,* Enzyme Applications, Industrial

Herbert A. Kirst, *Indianapolis, IN,* Macrolide Antibiotics

Ganesh M. Kishore, *Monsanto Company, St. Louis, MO,* Genetic Engineering, Plants

Yury V. Kissin, *Rutgers University, Piscataway, NJ,* Olefin Polymers, Introduction; Polyethylene, Linear Low Density (LLDPE); Polymers, Higher Olefins

John Klier, *The Dow Chemical Company, Midland, MI,* Microemulsions

William Klingensmith, *Akron Consulting Company, Akron, OH,* Rubber Compounding

Jerome M. Klosowski, *Klosowski Scientific, Inc., Bay City, MI,* Sealants

Edward A. Knaggs, *Consultant, Deerfield, IL,* Sulfonation and Sulfation

Jeffrey P. Knapp, *E.I. du Pont de Nemours & Co., Wilmington, DE,* Distillation, Azeotropic, and Extractive

Ted M. Knowlton, *Particulate Solids Research, Inc., Chicago, IL,* Fluidization

Edmond I. Ko, *Carnegie Mellon University, Pittsburgh, PA,* Aerogels

Joseph A. Kocal, *UOP LLC, Des Plaines, IL,* Alkylation

Dianna S. Kocurek, *Tischler/Kocurek, Round Rock, TX,* Waste Management, Hazardous

I. Fred Koenigsberg, *White & Case LLP, New York, NY,* Copyrights

Jan Kolařík, *Academy of Sciences of the Czech Republic, Prague, Czech Republic,* Polymer Blends

Daniel D. Koleske, *Sandia National Laboratories, Albuquerque, NM,* Semiconductors, Compound

Frans Kools, *Technological University, Eindhoven, Eindhoven, The Netherlands,* Ferrites

Peter W. Kopf, *TIAX LLC, Cambridge, MA,* Phenolic Resins

Sandra Kosinski, *New Jersey Institute of Technology, Newark, NJ,* Fiber Optics

Steven Kosmatka, *Portland Cement Association, Skokie, IL,* Cement

Boris Kosoy, *Odessa State Academy of Refrigeration, Odessa, Ukraine,* Heat Pipes

Joseph Kost, *Ben-Gurion University of the Negev, Beer-Sheva, Israel,* Drug Delivery Systems

Ed Kostansek, *Rohm and Haas Company, Spring House, PA,* Emulsions

R. H. Kottke, *Great Lakes Chemical Corporation, West Lafayette, IN,* Furan Derivatives

Matthew Koval, *NOAH Technologies Corporation, San Antonio, TX,* Thallium and Thallium Compounds

Deborah A. Kramer, *U.S. Geological Survey, Tucson, AZ,* Gallium and Gallium Compounds; Magnesium and Magnesium Alloys; Magnesium Compounds

Charles T. Kresge, *The Dow Chemical Company, Midland, MI,* Molecular Sieves

Charles B. Kreutzberger, *PPG Industries, Inc., Monroeville, PA,* Chloroformates and Carbonates

M. A. Krevalis, *Exxon Chemical Company, Baton Rouge, LA,* Carboxylic Acids

Ramesh Krishnamurti, *Occidental Chemical Corporation, Dallas, TX,* Chlorobenzenes

Michael Kroupa, *Dow Corning Corporation, Midland, MI,* Silicon and Silicon Alloys, Chemical and Metallurgical

Petr Kuban, *Texas Tech University, Lubbock, TX,* Capillary Separations

Volker Kuellmer, *ADM Nutraceutical, Decatur, IL,* Vitamin C (Ascorbic Acid)

Donald Kulich, *GE Plastics Technology Center, Pittsfield, MA,* Acrylonitrile-Butadiene-Styrene (ABS) Polymers

Anna Kultys, *Maria Curie-Sklodowska University, Lublin, Poland,* Sulfur-Containing Polymers

M. N. V. Ravi Kumar, *National Institute of Pharmaceutical Education & Research, Punjab, India,* Drug Delivery Systems

Neeraj Kumar, *National Institute of Pharmaceutical Education & Research, Punjab, India,* Drug Delivery Systems

Satish Kumar, *The Georgia Institute of Technology, Atlanta, GA,* Carbon Fibers

K. J. A. Kundig, *Metallurgical Consultant, Randolph, NJ,* Copper

T. Kundu, *The University of Arizona, Tucson, AZ,* Nondestructive Evaluation

Yeong-Jen Kuo, *Huntsman Performance Products, The Woodlands, TX,* Maleic Anhydride, Maleic Acid, and Fumaric Acid

Roger N. Kust, *Tetra Technologies, Inc., The Woodlands, TX,* Sodium Halides

Yakov Kutsovsky, *Cabot Corporation, Billerica, MA,* Carbon Black

Michael Ladisch, *Purdue University, West Lafayette, IN,* Bioseparations

Peter R. Lamb, *CSIRO Textile and Fibre Technology, Belmont, Australia,* Wool

John B. Lambert, *JBL Consulting, Lake Forest, IL,* Tantalum and Tantalum Compounds

Amedeo Lancia, *University of Napoli, Naples, Italy,* Calcium Sulfate

Smadar A. Lapidot, *Ben-Gurion University of the Negev, Beer-Sheva, Israel,* Drug Delivery Systems

George R. Lappin, *Albemarle Corporation, Baton Rouge, LA,* Alcohols, Higher Aliphatic, Synthetic Processes; Olefins, Higher

James Larminie, *Oxford Brookes University, Oxford, United Kingdom,* Fuel Cells

Serena Laschi, *University of Florence, Florence, Italy,* Biosensors

Manuel Laso, *Universidad Politecnica, ETSII, Madrid, Spain,* Absorption

Armin Lauterbach, *SQM Iodine Corporation, Norfolk, VA,* Iodine and Iodine Compounds

David P. Lawrence, *Lawrence Environmental, British Columbia, Canada,* Environmental Impact Assessment

Tom Leahy, *Mettler-Toledo, Inc., Inman, SC,* Weighing and Proportioning

Stuart E. Lebo, Jr., *LignoTech USA, Inc., Rothschild, WI,* Lignin

Diana M. Lee, *Lawrence Berkeley National Laboratory, Berkeley, CA,* Actinides and Transactinides

Stephanie P. Lee, *Columbia University, New York, NY,* Microfluidics

Thomas D. Lee, *PepsiCo, Valhalla, NY,* Sweeteners

H. David Leigh III, *Clemson University, Clemson, SC,* Refractories

Robert D. Lein, *The BOC Group, Inc., Murray Hill, NJ,* Noble Gases

José R. Leiza, *The University of the Basque Country, San Sebastian, Spain,* Copolymers

John Leman, *GE Global Research Center, Niskayuna, NY,* Silicones

Charles E. Lemke, *E. I. du Pont de Nemours & Co., Inc., Niagara Falls, NY,* Sodium and Sodium Alloys

Joseph J. Len, *Vinings Industries, Inc. Atlanta, GA,* Aluminates

Andreas Lendlein, *Deutsches Wollforschungsinstitut, Aachen, Germany,* Shape-Memory Polymers

Marguerite L. Leng, *Leng Associates, Midland, MI,* Pesticides

Gadi Lenz, *Kodeos Communications, Inc., South Plainfield, NJ,* Fiber Optics

J. A. Lepinski, *PT Perkasa Indobaja, Jakarta, Indonesia,* Iron

Claude Leray, *Centre National de la Recherche Scientifique, Montpellier, France,* Waxes

Alan Letki, *Alfa Laval Inc., Warminster, PA,* Centrifugal Separation

A. Leveque, *Rhodia Rare Earth, Hertfordshire, United Kingdom,* Lanthanides

Joel D. Levitt, *Springfield Resources, Inc., Plymouth Meeting, PA,* Maintenance

D. M. Lewis, *University of Leeds, Leeds, United Kingdom,* Dyes, Reactive

I. C. Lewis, *UCAR Carbon Company, Inc., Cleveland, OH,* Graphite, Artificial

Larry Lewis, *GE Global Research Center, Niskayuna, NY,* Silicones

Michael Lewis, *University of California, Davis, CA,* Beer and Brewing

Jun Li, *Georgia Institute of Technology, Atlanta, GA,* Self-Cleaning Materials—Lotus Effect Surfaces

T. Li, *ASARCO Inc., Phoenix, AZ,* Antimony and Antimony Alloys

Richard B. Lieberman, *Montell Polyolefins, Elkton, MD,* Polypropylene

JoAnn S. Lighty, *University of Utah, Salt Lake City, UT,* Hazardous Waste Incineration

Jerry Lii, *Columbia University, New York, NY,* Microfluidics

Charles B. Lindahl, *Elf Atochem North America, Inc., Tulsa, OK,* Antimony Compounds; Barium Compounds; Fluorine Compounds, Inorganic; Germanium and Germanium Compounds

Bruce F. Lipin, *U.S. Geological Survey, Reston, VA,* Chromium and Chromium Alloys

Charles W. Lipp, *The Dow Chemical Company, Freeport, TX,* Sprays

David W. Lipp, *Cytec Technology Corporation, Stamford, CT,* Acrylamide Polymers

Monica Lira-Cantú, *Materials Science Institute, Barcelona, Spain,* Hybrid Nanocomposite Materials

Zhenyu Liu, *Institute of Coal Chemistry, Shanxi, China,* Coal Liquefaction

Zhong Liu, *Tianjin Institute of Light Industry, Tianjin, China,* Pulp Bleaching

Teh C. Lo, *T. C. Lo & Associates, Wayne, NJ,* Extraction, Liquid–Liquid

Gary W. Loar, *McGean-Rohco, Inc., Cleveland, OH,* Chromium Compounds

J. Fred Lochary, *The Dow Chemical Company, Freeport, TX,* Alkanolamines from Olefin Oxides and Ammonia

David J. Locker, *Kodak Manufacturing Research and Engineering, Rochester, NY,* Photography

John E. Logsdon, *Union Carbide Corporation, Texas City, TX,* Ethanol

G. Gilbert Long, *North Carolina State University, Raleigh, NC,* Antimony Compounds

J. C. Long, *UCAR Carbon Company Inc., Columbia, TN,* Carbon

Rebecca Lopez-Garcia, *Tate & Lyle, Mexico,* Citric Acid

John F. Lorenc, *Schenectady Chemicals, Inc. Schenectady, NY,* Alkylphenols

Andrew B. Lowe, *The University of Southern Mississippi, Hattiesburg, MS,* Polymers, Water-Soluble

Steven Lowenkron, *The Dow Chemical Corporation, LaPorte, TX,* Aniline and Its Derivatives

V. Lowry, *GE Plastics Technology Center, Pittsfield, MA,* Acrylonitrile-Butadiene-Styrene (ABS) Polymers

Peter Luckie, *Penn State University, University Park, PA,* Separation, Size

Matthew C. Lucy, *University of Missouri-Columbia, Columbia, MO,* Genetic Engineering, Animals

Timothy B. Lueder, *Dow Corning Corporation, Midland, MI,* Sealants

Benno Lüke, *Krupp Uhde GmbH, Dortmund, Germany,* Chlorine

Henrik Lund, *Novozymes A/S, Bagsvaerd, Denmark,* Enzyme Applications, Industrial

George Lunn, *Center for Drug Evaluation and Research, Food and Drug Administration, Rockville, MD,* Chromatography, Liquid

John Lydon, *U.S. Department of Agriculture, ARS, Beltsville, MD,* Herbicides, Biotechnology

Jeremiah Lynch, *Exxon Chemical Company (Retired), Rumson, NJ,* Industrial Hygiene

Jesse L. Lynn, Jr., *Lever Brothers Company, Edgewater, NJ,* Detergency and Detergents

Mark J. Macielag, *Johnson & Johnson Pharmaceutical Research & Development, L.L.C., Raritan, NJ,* Antibacterial Agents, Overview; Sulfonamides

Arthur G. Mack, *Albemarle Corporation, Baton Rouge, LA,* Flame Retardants, Halogenated

Duncan J. Macquarrie, *University of York, York, United Kingdom,* Catalysts, Supported

Ian G. Macreadie, *CSIRO Molecular and Health Technologies, Parkville, Australia,* Yeasts; Yeasts, Molecular and Therapeutic Applications

Linda M. Macsavage, *Arcadia University,* Alkaloids

George MacZura, *Aluminum Company of America, Pittsburgh, PA,* Aluminum Oxide (Alumina), Calcined, Tabular, and Aluminate Cements

Arun Madan, *MVSystems, Inc., Golden, CO,* Semiconductors, Amorphous

Mauro Maestri, *Università di Bologna, Bologna, Italy,* Dendrimers, Luminescent

P. Maestro, *Rhodia SA, Aubervilliers, France,* Lanthanides

Doug Magde, *University of California, San Diego, La Jolla, CA,* Kinetic Measurements

Tariq Mahmood, *Elf Atochem North America, Inc., Tulsa, OK,* Antimony Compounds; Barium Compounds; Fluorine Compounds, Inorganic; Germanium and Germanium Compounds

Khaled Mahmud, *Cabot Corporation, Billerica, MA,* Carbon Black

Bernard Maisonneuve, *Akzo Chemicals, Inc., Dobbs Ferry, NY,* Amine Oxides

N. V. Majeti, *University of Saarland, Saarbrcken, Germany,* Drug Delivery Systems

Thomas G. Majewicz, *Aqualon Company, Palatine, IL,* Cellulose Ethers

Harry V. Makar, *Consultant, Ellicott City, MD,* Recycling, Metals

James Manganaro, *FMC Corporation, Princeton, NJ,* Barium Compounds

Ganpat Mani, *Honeywell International, Morristown, NJ,* Boron Halides

Uzi Mann, *Texas Tech University, Lubbock, TX,* Reactor Technology

Robert M. Manyik, *Union Carbide Corporation, South Charleston, WV,* Acetylene

Norma J. Maraschin, *Equistar Chemicals, Cincinnati, OH,* Polyethylene, Low Density (LDPE)

Vernon H. Markant, *E. I. du Pont de Nemours & Co., Inc., Niagara Falls, NY,* Sodium and Sodium Alloys

Kenric A. Marshall, *The Dow Chemical Company, Freeport, TX,* Chlorocarbons and Chlorohydrocarbons, Survey

F. Lennart Marten, *Air Products and Chemicals, Inc., Allentown, PA,* Vinyl Alcohol Polymers

A. E. Martin, *The Dow Chemical Company, Freeport, TX,* Glycols

S. J. Martin, *Sandia National Laboratories, Albuquerque, NM,* Sensors

Marco Mascini, *University of Florence, Florence, Italy,* Biosensors

Robert T. Mason, *Koppers Industries, Inc. (Retired), Pittsburgh, PA,* Naphthalene

James C. Masson, *JCM Consulting, Mooresville, NC,* Fibers, Acrylic

Louis R. Matricardi, *Consultant, Tonawanda, NY,* Manganese and Manganese Alloys

Michael A. Matthews, *University of South Carolina, Columbia, SC,* Green Chemistry

Ignacio Maturana, *SQM Nitratos SA, Santiago, Chile,* Sodium Nitrate and Nitrite

Ernest Mayer, *Consultant, Newark, DE,* Filtration

James W. Mayer, *Arizona State University, Tempe, AZ,* Ion Implantation

William J. Mazzafro, *Air Products and Chemicals, Inc., Allentown, PA,* Nitric Acid

Edward McBride, *DuPont Company, Wilmington, DE,* Ethylene-Acrylic Elastomers

Francis X. McConville, *FXM Engineering & Design, Worcester, MA,* Pilot Plants

Charles L. McCormick, *The University of Southern Mississippi, Hattiesburg, MS,* Polymers, Water-Soluble

Max P. McDaniel, *Chevron-Phillips Chemical Company, Bartlesville, OK,* Polyethylene, High Density (HDPE)

Daniel P. McDonald, *Catawba Valley Community College, Hickory, NC,* Lithium and Lithium Compounds

Joseph A. McGeough, *The University of Edinburgh, Edinburgh, United Kingdom,* Electrochemical Machining

J. Scott McIndoe, *University of Cambridge, Cambridge, United Kingdom,* Metal Carbonyls

David B. McKeever, *USDA Forest Service, Madison, WI,* Wood

Ronald J. McKinney, *E. I. du Pont de Nemours & Company, Inc., Wilmington, DE,* Nitriles

Hamish McNab, *University of Edinburgh, Edinburgh, United Kingdom,* Pyrolysis, Flash Vacuum

Timothy J. McNally, *LignoTech USA, Inc., Rothschild, WI,* Lignin

N. A. Meanwell, *Bristol-Myers Squibb, Wallingford, CT,* Quinolone Antibacterials

David E. Mears, *Unocal, Brea, CA,* Hydrocarbons

Gabriele Mei, *Basell Polyolefins, Ferrara, Italy,* Polypropylene

Roland E. Meissner III, *Meissner Engineering Company, La Canada Flintridge, CA,* Plant Layout; Plant Location

Sudhir K. Mendiratta, *Olin Corporation, Charleston, TN,* Chloric Acid and Chlorates; Perchloric Acid and Perchlorates

Philip H. Merrell, *Mallinckrodt, Inc., St. Louis, MO,* Sodium Halides

James A. Mertens, *The Dow Chemical Company, Freeport, TX,* Chloroethylenes and Chloroethanes

Stanley H. Mervis, *Polaroid Corporation (Retired), MA,* Photography, Instant

Dayal T. Meshri, *Advance Research Chemicals, Inc., Catoosa, OK,* Cobalt Compounds; Copper Compounds; Lead Compounds; Nickel Compounds

Jürgen O. Metzger, *Carl von Ossietzky Universität Oldenburg, Oldenburg, Germany,* Sustainable Development and Chemistry

D. E. Meyer, *University of Kentucky, Lexington, KY,* Reverse Osmosis

Peter Michels, *Albany Molecular Research, Albany, NY,* Microbial Transformations

Adriyan S. Milev, *University of Western Sydney, New South Wales, Australia,* Nanotechnology

Tom Millensifer, *Consultant, Littleton, CO,* Rhenium and Rhenium Compounds

Carol J. Miller, *Wayne State University, Detroit, MI,* Groundwater Monitoring

Charles E. Miller, *Westvaco Corporation, Charleston, SC,* Carbon, Activated

David J. Miller, *Union Carbide Corporation, South Charleston, WV,* Aldehydes

Matt C. Miller, *Dow Chemical Company, Freeport, TX,* Ethyl Chloride

Regis B. Miller, *USDA Forest Service, Madison, WI,* Wood

Keith R. Millington, *CSIRO Textile and Fibre Technology, Belmont, Australia,* Wool

Roy Mink, *U.S. Department of Energy, Washington, DC,* Geothermal Energy

Marilyn L. Minus, *The Georgia Institute of Technology, Atlanta, GA,* Carbon Fibers

Chanakya Misra, *Aluminum Company of America, Alcoa Center, PA,* Aluminum Oxide (Alumina), Hydrated

Stephen C. Mitchell, *Imperial College London, London, UK,* Aminophenols

Shahriar Mobashery, *Wayne State University, Detroit, MI,* Antibiotic Resistance

Irving Moch, Jr., *I. Moch & Associates, Inc., Wilmington, DE,* Membranes, Hollow-Fiber

A. Moeller, *Degussa AG, Hanan, Germany,* Hydrogen Peroxide

G. D. Moggridge, *University of Cambridge, Cambridge, United Kingdom,* Chemical Product Design

Jillian R. Moncarz, *E. I. du Pont de Nemours & Company, Inc., Deepwater, NJ,* Titanium Compounds, Organic

Victor M. Monroy, *Monroy Technology Concepts, LLC, Charlotte, NC,* Initiators, Anionic

John J. Mooney, *Environmental and Energy Technology and Policy Institute, Wyckoff, NJ,* Emission Control, Automotive

Amram Mor, *The Hebrew University of Jerusalem, Jerusalem, Israel,* Peptides, Antimicrobial

M. K. Moran, *M&T Chemicals, Inc., Rahway, NJ,* Tin Compounds

Matthew D. Moran, *McMaster University, Ontario, Canada,* Noble-Gas Compounds

Patrick Moran, *U.S. Naval Academy, Annapolis, MD,* Corrosion and Corrosion Control

Jeffrey O. Moreno, *Thompson Hine LLP, Washington, DC,* Transportation

Daniel A. Morgan, *The BOC Group, Inc., Murray Hill, NJ,* Noble Gases

Don Morgan, *O. S. Walker Company, Milwaukee, WI,* Magnetic Separation

James J. Morgan, *California Institute of Technology, Pasadena, CA,* Water

Giampiero Morini, *Basell Polyolefins Italia, Ferrara, Italy,* Ziegler-Natta Catalysts

Earl D. Morris, *The Dow Chemical Company, Midland, MI,* Acetic Acid, Halogenated Derivatives

Ian Morrison, *Cabot Corporation, Billerica, MA,* Dispersion

Hugh Morrow, *International Cadmium Council, Great Falls, VA,* Cadmium and Cadmium Alloys

Maurice Morton, *The University of Akron, Akron, Ohio,* Elastomers, Synthetic

Sanford L. Moskowitz, *American Economics Group and Chemical Heritage Foundation, Abington, PA,* Advanced Materials, Economic Evaluation; Synthetic Organic Chemicals, Economic Evaluation

Werner H. Mueller, *Hoechst-Celanese Corporation, Charlotte, NC,* Sodium Halides

Arun S. Mujumdar, *National University of Singapore, Singapore,* Drying

Rajiv Mukherjee, *Consultant, New Delhi, India,* Heat Exchanger Network Design

Chris J. Mulder, *International Malting Company, Milwaukee, WI,* Malts and Malting

Thomas L. Muller, *E. I. du Pont de Nemours & Company, Inc., Wilmington, DE,* Sulfuric Acid and Sulfur Trioxide

Michael J. Mummey, *Huntsman Performance Products, The Woodlands, TX,* Maleic Anhydride, Maleic Acid, and Fumaric Acid

T. F. Munday, *FMC Corporation, Princeton, NJ,* Phosphorus

Donald P. Murphy, *Parker Amchem, Madison Heights, MI,* Metal Surface Treatments

F. H. Murphy, *The Dow Chemical Company, Freeport, TX,* Glycols

Haydn H. Murray, *Indiana University, Bloomington, IN,* Clays, Survey; Clays, Uses

Raymond L. Murray, *Consultant, Raleigh, NC,* Nuclear Reactors; Waste Management, Radioactive

D. N. Prabhakar Murthy, *The University of Queensland, Brisbane, Australia,* Reliability

Ramiah Murugan, *Reilly Industries, Inc., Indianapolis, IN,* Pyridine and Pyridine Derivatives

Dino Musmarra, *University of Napoli, Naples, Italy,* Calcium Sulfate

Durai Muthusamy, *Shell Oil Company, Houston, TX,* Methyl Isobutyl Ketone

Jeffrey C. Myers, *Midrex Direct Reduction Corporation, Charlotte, NC,* Iron

Philip Myers, *Chevron Research and Technical Company, Orinda, CA,* Tanks and Pressure Vessels

Terry N. Myers, *Atofina Chemicals, Inc., King of Prussia, PA,* Initiators, Free-Radical; Peroxides, Organic

D. R. Nagaraj, *Cytec Industries, Inc., Stamford, CT,* Minerals Recovery and Processing

Vasantha Nagarajan, *DuPont Central Research and Development, Wilmington, DE,* Genetic Engineering, Microbes

Nobuyki Nagato, *Showa Denko K.K., Tokyo, Japan,* Allyl Alcohol and Its Derivatives

Vijay Naik, *Unilever Research India, Bangalore, India,* Soap

Kenichiro Nakashima, *Nagasaki University, Nagasaki, Japan,* Chemiluminescence, Analytical Applications

Raman Nambudripad, *Consultant, West Newton, MA,* Proteins

Kurt Nassau, *Consultant, Lebanon, NJ,* Color

Michael Nastasi, *Los Alamos National Laboratory, Los Alamos, NM,* Ion Implantation

Paul Natishan, *Naval Research Laboratory, Washington, DC,* Corrosion and Corrosion Control

Hildeberto Nava, *Reichhold Chemicals, Inc., Research Triangle Park, NC,* Polyesters, Unsaturated

J. R. D. Nee, *WR Grace & Company, Columbia, MD,* Fluid Catalytic Cracking (FCC) Catalysts and Additives

Lev Nelik, *Liquiflo Equipment Company, Garwood, NJ,* Pumps

L. H. Nemec, *Albemarle Corporation, Baton Rouge, LA,* Olefins, Higher

Marshall J. Nepras, *Stepan Company, Northfield, IL,* Sulfonation and Sulfation

Mary P. Neu, *Los Alamos National Laboratory, Los Alamos, NM,* Plutonium and Plutonium Compounds; Thorium and Thorium Compounds; Uranium and Uranium Compounds

John Newman, *University of California, Berkeley, Berkeley, CA,* Electrochemical Processing

William E. Newton, *Virginia Polytechnic Institute and State University, Blacksburg, VA,* Nitrogen Fixation

Yonghao Ni, *Limerick Pulp & Paper, New Brunswick, Canada,* Pulp Bleaching

Henry Nielsen, *Teledyne Wah Chang Albany, Albany, OR,* Zirconium and Zirconium Compounds

Lone Kierstein Nielsen, *Novozymes A/S, Bagsvaerd, Denmark,* Enzyme Applications, Industrial

Ralph H. Nielsen, *Teledyne Wah Chang Corporation, Albany, OR,* Hafnium and Hafnium Compounds; Zirconium and Zirconium Compounds

Alvin W. Nienow, *University of Birmingham, Edgbaston, Birmingham, United Kingdom,* Aeration, Biotechnology

Alfred F. Noels, *Université de Liège, Liège, Belgium,* Metathesis

Jacques W. M. Noordermeer, *DSM Elastomers, R&D, Geleen, The Netherlands,* Ethylene-Propylene Polymers

John Oakes, *John Oakes Associates, Cheshire, United Kingdom,* Dyes, Azo

J. A. H. Oates, *Limetec Consultancy Services, Derbyshire, United Kingdom,* Lime and Limestone

Gustavo Ober, *SQM Iodine Corporation, Norfolk, VA,* Iodine and Iodine Compounds

B. E. Obi, *The Dow Chemical Company, Midland, MI,* Vinylidene Chloride Polymers

Thomas F. O'Brien, *Consultant, Media, PA,* Chlorine

Julia I. O'Farrelly, *Johnson Matthey PLC, Reading, United Kingdom,* Platinum-Group Metals

Norma J. Ofsthun, *Fresenius Medical Care North America, Lexington, MA,* Hemodialysis

K. A. Ohemeng, *Paratek Pharmaceuticals, Inc., Boston, MA,* Quinolone Antibacterials

George A. Olah, *University of Southern California, Los Angeles, CA,* Friedel-Crafts Reactions

J. E. Oldfield, *Oregon State University, Corvallis, OR,* Tellurium and Tellurium Compounds

Anna Oliva, *State University of New York, Stony Brook, NY,* Herbicides, Biotechnology

Charles W. Olsen, Jr., *Dow Corning Corporation, Midland, MI,* Poly(fluorosilicones)

Hans Sejr Olsen, *Novozymes A/S, Bagsvaerd, Denmark,* Enzyme Applications, Industrial

David L. Olsson, *Rochester Institute of Technology, Rochester, NY,* Packaging, Containers for Industrial Materials

E. F. Olszewski, *ABB Lummus Global, Inc., Bloomfield, NJ,* Ethylene

Suzan Onel, *McKenna & Cuneo, LLP, Washington, DC,* Regulatory Agencies

Jarl Opgrande, *Noveon Kalama, Inc., Kalama, WA,* Benzaldehyde; Benzoic Acid

Judith P. Oppenheim, *Solutia Inc, Cantonment, FL,* Adipic Acid

Rodrigo Orefice, *University of Florida, Gainesville, FL,* Sol–Gel Technology

Anil Oroskar, *UOP LLC, Des Plaines, IL,* Adsorption, Liquid Separation

John Osepchuk, *Raytheon Company, Lexington, MA,* Microwave Technology

Michael J. Owen, *Dow Corning Corporation, Midland, MI,* Defoamers; Release Agents

John G. Owens, *3M Company, St. Paul, MN,* Fluoroethers and Fluoroamines

S. Ted Oyama, *Virginia Polytechnic Institute & State University, Blacksburg, VA,* Carbides, Survey

Toshitsugu Ozeki, *Kyowa Hakko Kogyo Company, Yokkaichi, Japan,* Amino Acids

E. Dickson Ozokwelu, *Amoco Chemical Company, Naperville, IL,* Toluene

Catherine E. Grégoire Padró, *Los Alamos National Laboratory, Los Alamos, NM,* Hydrogen Energy

Billie J. Page, *McGean-Rohco, Inc., Cleveland, OH,* Chromium Compounds

D. J. Page, *UCAR Carbon Company, Inc., Cleveland, OH,* Graphite, Artificial

Mark A. Paisley, *FERCO Enterprises, Inc., Columbus, OH,* Biomass Energy

P. Palmas, *UOP LLC, Des Plaines, IL,* Fluid Catalytic Cracking (FCC) Units, Regeneration

Robert J. Palmer, *Du Pont de Nemours International SA, Geneva, Switzerland,* Polyamides, Plastics

Sachin Pannuri, *Cambrex Technical Center, North Brunswick, NJ,* Biocatalysis

Anthony J. Papa, *Union Carbide Corporation, South Charleston, WV,* Amyl Alcohols

John R. Papcun, *Atotech, Cleveland, OH,* Ammonium Compounds; Boron Halides; Lithium and Lithium Compounds

Peter G. Pape, *Peter G. Pape Consulting, Saginaw, MI,* Silylating Agents

John F. Papp, *U.S. Geological Survey, Reston, VA,* Chromium and Chromium Alloys

Uday N. Parekh, *Air Products and Chemicals, Inc., Allentown, PA,* Hydrogen

Maurice J. Parks, *Dow Chemical Company, Freeport, TX,* Epoxy Resins

Barbara J. Parry, *Newalta Corporation, British Columbia, Canada,* Recycling, Oil

Edward E. Parry, *National Institute of Standards and Technology, Gaithersburg, MD,* Dental Materials

T. E. Parsons, *Eastman Chemical Company, Kingsport, TN,* Glycols

Nancy R. Passow, *Write For You!/NRP Associates, Inc., Englewood, NJ,* Regulatory Agencies

Angela N. Patterson, *General Electric, Blacksburg, VA,* Design of Experiments

John P. Paul, *Carter & Burgess, Inc., Fort Worth, TX,* Recycling, Rubber

Harold W. Paxton, *Carnegie Mellon University, Pittsburgh, PA,* Steel

Alan Pearson, *Aluminum Company of America, Alcoa Center, PA,* Aluminum Oxide (Alumina), Activated

Michael Pecht, *University of Maryland, College Park, MD,* Packaging of Electronic Materials

S. E. Pederson, *Union Carbide Corporation, South Charleston, WV,* Acrolein and Derivatives

Mel Pell, *ESD Consulting Services, Wilmington, DE,* Fluidization

Thomas W. Penrice, *Consultant, Mt. Juliet, TN,* Tungsten and Tungsten Alloys; Tungsten Compounds

Simon Penson, *Kraft Foods Corporation, Warwick, United Kingdom,* Coffee

Giuseppe Penzo, *Basell Polyolefins, Ferrara, Italy,* Polypropylene

Tilden Wayne Perry, *Purdue University, Van Buren, AR,* Feeds and Feed Additives, Ruminant

Lawrence D. Pesce, *E.I. du Pont de Nemours & Co., Memphis, TN,* Cyanides

Elizabeth M. Peters, *Mallinckrodt, Inc., St. Louis, MO,* Sodium Halides

Richard L. Petersen, *The University of Memphis, Memphis, TN,* Potassium Compounds

Charles Anthony Peterson, *Intel Corporation,* Atomic Force Microscopy—AFM

Francis P. Petrocelli, *Air Products and Chemicals, Inc., Allentown, PA,* Vinyl Acetate Polymers

Michael Petschel, *Parker Amchem, Madison Heights, MI,* Metal Surface Treatments

Roger C. Pettersen, *USDA Forest Service, Madison, WI,* Wood

Ha Q. Pham, *Dow Chemical Company, Freeport, TX,* Epoxy Resins

David G. Phillips, *CSIRO Textile and Fibre Technology, Belmont, Australia,* Wool

W. J. Piel, *ARCO Chemical Company, Newtown Square, PA,* Ethers

Fabrizio Piemontesi, *Basell Polyolefins Italia, Ferrara, Italy,* Ziegler-Natta Catalysts

Ronald Pierantozzi, *Air Products and Chemicals, Inc., Allentown, PA,* Carbon Dioxide

Anthony P. Pierlot, *CSIRO Textile and Fibre Technology, Belmont, Australia,* Wool

John R. Pierson, *Johnson Controls, Inc., Milwaukee, WI,* Batteries, Secondary Cells

Linda R. Pinckney, *Corning, Inc., Corning, NY,* Glass-Ceramics

Kenneth Pisarczyk, *Carus Chemical Company, LaSalle, IL,* Manganese Compounds

Sarma V. Pisupati, *Pennsylvania State University, University Park, PA,* Combustion Science and Technology

Jack R. Plimmer, *Consultant, Tampa, FL,* Herbicides; Insecticides

Riccardo Po, *Eni Chem Research Center, Novara, Italy,* Engineering Thermoplastics

Alphonsus V. Pocius, *3M Adhesive Technologies Center, St. Paul, MN,* Adhesion

Thomas J. Podlas, *Aqualon Company, Palatine, IL,* Cellulose Ethers

Ludwik Pokorny, *SQM Nitratos SA, Santiago, Chile,* Sodium Nitrate and Nitrite

Malcolm B. Polk, *Georgia Institute of Technology (Emeritus), Decatur, GA,* High Performance Fibers

Peter Pollak, *Fine Chemicals Business Consultant, Reinach, Switzerland,* Fine Chemicals

Graham Polley, *Consultant, Ulverston, United Kingdom,* Heat Exchanger Network Design

Warren H. Powell, *Consultant, Columbus, OH,* Nomenclature

Asohk Prabhu, *Nitto Denko America, Inc., San Jose, CA,* Packaging of Electronic Materials

G. K. Surya Prakash, *University of Southern California, Los Angeles, CA,* Friedel-Crafts Reactions

Laurence W. Prange, *Thompson Hine LLP, Washington, DC,* Transportation

R. David Prengaman, *RSR Corporation, Dallas, TX,* Lead and Lead Alloys

Duane B. Priddy, *The Dow Chemical Company, Midland, MI,* Styrene Plastics

Roger C. Prince, *Exxon Research and Engineering Company, Annandale, NJ,* Bioremediation

Marina Prisciandaro, *University of Napoli, Naples, Italy,* Calcium Sulfate

Salvatore Profeta, Jr., *University of South Carolina, Columbia, SC,* Molecular Modeling

Richard W. Prugh, *Chilworth Technology, Inc., Plainsboro, NJ,* Safety

R. D. Putnam, *Putnam Environmental Services, Research Triangle Park, NC,* Tellurium and Tellurium Compounds

Donald E. Putzig, *E. I. du Pont de Nemours & Company, Inc., Deepwater, NJ,* Titanium Compounds, Organic

Jinhao Qiu, *Tohoku University, Sendai, Japan,* Biomaterials, Prosthetics, and Biomedical Devices

Roderic P. Quirk, *The University of Akron, Akron, OH,* Initiators, Anionic

Daniel Raederstorff, *Roche Vitamins Ltd, Basel, Switzerland,* Nutraceuticals

Suresh Rajaraman, *GE Silicones, Waterford, NY,* Silicones

Richard B. Rajendren, *American Aerators, Inc., Monticello, MN,* Water Treatment, Aeration

Philip E. Rakita, *Armour Associates, Ltd., Philadelphia, PA,* Grignard Reactions

Venkoba Ramachandran, *ASARCO Inc., Salt Lake City, UT,* Lead and Lead Alloys

Venkat S. Raman, *Air Products and Chemicals, Inc., Allentown, PA,* Hydrogen

W. J. Reagan, *UOP LLC, Des Plaines, IL,* Fluid Catalytic Cracking (FCC) Units, Regeneration

R. L. Reddy, *UCAR Carbon Company, Inc., Cleveland, OH,* Graphite, Artificial

V. Prakash Reddy, *University of Southern California, Los Angeles, CA,* Friedel-Crafts Reactions

Jill Rehmann, *St. Joseph's College, Brooklyn, NY,* Nucleic Acids

Austin H. Reid, Jr., *E. I. du Pont de Nemours & Company, Wilmington, DE,* Technical Service

Abraham Reife, *Environmental Consultant, Toms River, NJ,* Dyes, Environmental Chemistry

Gary Reineccius, *University of Minnesota, St. Paul, MN,* Flavor Characterization

Signo T. Reis, *University of Missouri-Rolla, Rolla, MO,* Glass

James Rekoske, *UOP LLC, Des Plaines, IL,* Adsorption, Liquid Separation

Albert J. Repik, *Westvaco Corporation, Charleston, SC,* Carbon, Activated

Steve R. Reznek, *Cabot Corporation, Billerica, MA,* Carbon Black

A. J. Ricco, *Sandia National Laboratories, Albuquerque, NM,* Sensors

Joseph O. Rich, *Northwestern University, Evanston, IL,* Microbial Transformations

H. Wayne Richardson, *Phibro-Tech, Inc., Sumter, SC,* Cobalt Compounds; Copper Compounds; Recycling, Metals

Douglas S. Richart, *D.S. Richart Associates, Reading, PA,* Coating Processes, Powder

Martin M. Rieger, *M & A Rieger, Associates, Morris Plains, NJ,* Cosmetics

Christoph Riegger, *Roche Vitamins Ltd, Basel, Switzerland,* Nutraceuticals

J. R. Riley, *WR Grace & Company, Columbia, MD,* Fluid Catalytic Cracking (FCC) Catalysts and Additives

Riccardo Rinaldi, *Basell Polyolefins, Ferrara, Italy,* Polypropylene

Marguerite Rinaudo, *Centre National de la Recherche Scientifique, Grenoble, France,* Polysaccharides

John A. Rippon, *CSIRO Textile and Fibre Technology, Belmont, Australia,* Wool

Winston K. Robbins, *Exxon Research and Engineering Company, Annandale, NJ,* Petroleum

Kenneth L. Roberts, *North Carolina Agricultural and Technical State University, Greensboro, NC,* Nitrides

James Robinson, *Betz Dearborn, Trevose, PA,* Water Treatment

Peter W. Robinson, *Olin Corporation, Glen Carbon, IL,* Copper Alloys, Wrought

J. H. Robson, *Union Carbide Corporation, South Charleston, WV,* Glycols

Brendan Rodgers, *ExxonMobil Chemical Company, Baytown, TX,* Rubber Compounding

David J. Romenesko, *TempoInvestment Products, Inc., Midland, MI,* Poly(fluorosilicones)

Kenneth Rose, *Rensselaer Polytechnic Institute, Troy, NY,* Semiconductors, Silicon Based

Rodney D. Roseman, *University of Cincinnati, Cincinnati, OH,* Ceramics as Electrical Materials

Stephen L. Rosen, *University of Missouri–Rolla, Rolla, MO,* Polymers

Jack L. Rosette, *Forensic Packaging Concepts, Inc., Fort Mill, SC,* Packaging, Cosmetics and Pharmaceuticals

C. Philip Ross, *Glass Industry Consulting, Laguna Niguel, CA,* Recycling, Glass

F. S. Rosser, *UOP LLC, Des Plaines, IL,* Fluid Catalytic Cracking (FCC) Units, Regeneration

Alan Rossiter, *Rossiter & Associates, Bellaire, TX,* Energy Management

Eugene F. Rothgery, *Consultant, North Branford, CT,* Hydrazine and Its Derivatives

Jerry Rovner, *John Brown E&C, Houston, TX,* Methanol

Walter F. Rowe, *The George Washington University, Washington, D.C.,* Forensic Chemistry

Roger M. Rowell, *USDA Forest Service, Madison, WI,* Wood

Howard C. Rowles, *Air Products and Chemicals, Inc., Allentown, PA,* Cryogenic Technology

Slawomir Rubinsztajn, *GE Global Research Center, Niskayuna, NY,* Silicones

Scott Rudge, *FeRx Incorporated, Aurora, CO,* Electrophoresis

Abdul K. Rumaiz, *University of Delaware, Newark, DE,* Thin Film Formation Techniques

Wolfgang Runde, *Los Alamos National Laboratory, Los Alamos, NM,* Thorium and Thorium Compounds; Uranium and Uranium Compounds

W. R. Runyan, *Texas Instruments, Inc., Dallas, TX,* Silicon

Nelson W. Rupp, *National Institute of Standards and Technology, Gaithersburg, MD,* Dental Materials

Ian M. Russell, *CSIRO Textile and Fibre Technology, Belmont, Australia,* Wool

Douglas M. Ruthven, *University of Maine, Orono, ME,* Adsorption

B. Ryan, *Bristol-Myers Squibb, Wallingford, CT,* Quinolone Antibacterials

J. L. Ryans, *Eastman Chemical Company, Kingsport, TN,* Pressure Measurement

W. Janusz Rzeszotarski, *U.S. Food and Drug Administration, Rockville, MD,* Pharmaceuticals

J. L. Sabot, *Rhodia Rare Earths, Aubervilliers, France,* Lanthanides

Stephen H. Safe, *Texas A&M University, College Station, TX,* Halogenated Hydrocarbons, Toxicity and Environmental Impact

M. R. V. Sahyun, *3M Center, St. Paul, MN,* Photochemical Technology

Arnaud Saint-Jalmes, *Université Paris-Sud, Orsay, France,* Foams

Raj Sakamuri, *Clariant Corporation, Somerville, NJ,* Esters, Organic

Alvin Salkind, *Rutgers University, New Brunswick, NJ,* Batteries, Secondary Cells

Upasiri Samaraweera, *American Crystal Sugar Company, Moorhead, MN,* Sugar

A. Samat, *Université de la Méditerrannée, Marseille, France,* Chromogenic Materials, Electrochromic; Chromogenic Materials, Photochromic; Chromogenic Materials, Piezochromic; Chromogenic Materials, Thermochromic

Jose Sanchez, *Elf Atochem North America, Inc., Buffalo, NY,* Peroxides, Organic

Kathryn D. Sandefur, *University of Missouri–Rolla Coatings Institute, Rolla, MO,* Paint

Robert E. Sanders, Jr., *Aluminum Company of America, Alcoa Center, PA,* Aluminum and Aluminum Alloys

James R. Sandifer, *Eastman Kodak Company (Retired), Rochester, NY,* Electroanalytical Techniques

Stanley I. Sandler, *University of Delaware, Newark, DE,* Thermodynamics

Stanley R. Sandler, *Delaware County Community College, Springfield, PA,* Sulfur Compounds

A. T. Santhanam, *Kennametal, Inc., Latrobe, PA,* Carbides, Cemented

Alice Sapienza, *Simmons College, Boston, MA,* Research and Development Management

H. B. Sargent, *Consultant,* Acetylene

Dror Sarid, *University of Arizona, Tucson, AZ,* Atomic Force Microscopy—AFM

E. T. Sauer, *The Procter & Gamble Company, Cincinnati, OH,* Carboxylic Acids

J. D. Sauer, *Albemarle Corporation, Baton Rouge, LA,* Olefins, Higher

Alan W. Scaroni, *Pennsylvania State University, University Park, PA,* Combustion Science and Technology

R. P. Schaffer, *Consultant,* Acetylene

Wolfgang Schalch, *Roche Vitamins Ltd, Basel, Switzerland,* Nutraceuticals

Florian Schattenmann, *GE Global Research Center, Niskayuna, NY,* Silicones

William Scheffer, *Schenectady Chemicals, Inc. Schenectady, NY,* Alkylphenols

Brian E. Scheffler, *U.S. Department of Agriculture, ARS, Stoneville, MS,* Herbicides, Biotechnology

L. McDonald Schetky, *Memry Corporation, Brookfield, CT,* Shape-Memory Alloys

Hans Erik Schiff, *Novozymes A/S, Bagsvaerd, Denmark,* Enzyme Applications, Industrial

Steven L. Schilling, *Mobay Corporation, New Martinsville, WV,* Amines by Reduction

Mordechay Schlesinger, *University of Windsor, Ontario, Canada,* Electroplating

James H. Schlewitz, *Teledyne Wah Chang Albany, OR,* Niobium and Niobium Compounds; Zirconium and Zirconium Compounds

Eberhard Schmidt, *Carl von Ossietzky Universität Oldenburg, Oldenburg, Germany,* Sustainable Development and Chemistry

Frank J. Schmidt, *University of Missouri-Columbia, Columbia, MO,* Genetic Engineering, Procedures

Robert J. Schmidt, *UOP LLC, Des Plaines, IL,* Alkylation

Robert L. Schmitt, *Dow Chemical Company, Piscataway, NJ,* Ethylene Oxide Polymers

Uwe Schneidewind, *Carl von Ossietzky Universität Oldenburg, Oldenburg, Germany,* Sustainable Development and Chemistry

Rosalie A. Schnick, *Consultant, Lacrosse, WI,* Aquaculture Chemicals

Clifford J. Schoff, *Schoff Associates, Allison Park, PA,* Rheology and Rheological Measurements

Annemarie Schoonman, *Nestlé Research Center, Lausanne, Switzerland,* Flavor Delivery Systems

Laurier L. Schramm, *Saskatchewan Research Council, Saskatoon, Saskatchewan, Canada,* Colloids

William L. Schreiber, *Monmouth University, West Long Branch, NJ,* Perfumes

Gary J. Schrobilgen, *McMaster University, Ontario, Canada,* Noble-Gas Compounds

David M. Schubert, *Rio Tinto Minerals, Denver, CO,* Boron Hydrides, Heteroboranes, and their Metalla Derivatives

E. Fred Schubert, *Rensselaer Polytechnic Institute, Troy, NY,* Light Emitting Diodes

Gary E. Schumacher, *National Institute of Standards and Technology, Gaithersburg, MD,* Dental Materials

Christopher J. Sciarra, *Sciarra Laboratories, Inc., Hicksville, NY,* Aerosols

John J. Sciarra, *Sciarra Laboratories, Inc., Hicksville, NY,* Aerosols

Eric F. V. Scriven, *University of Florida, Gainesville, FL,* Pyridine and Pyridine Derivatives

Glenn T. Seaborg, *University of California, Berkeley, Oakland, CA,* Actinides and Transactinides

George C. Seaman, *Hoechst-Celanese Corporation, Corpus Christi, TX,* Formaldehyde

Kenneth R. Seddon, *The Queen's University of Belfast, Belfast, United Kingdom,* Ionic Liquids

Charles S. Sell, *Givaudan, Kent, United Kingdom,* Terpenoids

Kelly Sellers, *NOAH Technologies Corporation, San Antonio, TX,* Thallium and Thallium Compounds

Karl W. Seper, *Occidental Chemical Corporation, Grand Island, NY,* Chlorotoluenes, Benzyl Chloride, Benzal Chloride and Benzotrichloride

Richard J. Seymour, *Johnson Matthey PLC, Reading, United Kingdom,* Platinum-Group Metals

Lawrence J. Shadle, *U.S. Department of Energy, Morgantown, WV,* Coal Gasification

Timothy D. Shaffer, *ExxonMobil, Baytown, TX,* Butyl Rubber

S. Ismat Shah, *University of Delaware, Newark, DE,* Thin Film Formation Techniques

T. Shaikh, *University of Kentucky, Lexington, KY,* Aluminum Halides and Aluminum Nitrate

Reza Sharifi, *Pennsylvania State University, University Park, PA,* Combustion Science and Technology

Michael C. Shelton, *Eastman Chemical Company, Kingsport, TN,* Cellulose Esters, Inorganic Esters

George Shia, *Honeywell Specialty Chemicals, Buffalo, NY,* Fluorine

James E. Shigley, *Gemological Institute of America, Carlsbad, CA,* Diamond, Natural

M. M. Shreehan, *ABB Lummus Global, Inc., Bloomfield, NJ,* Ethylene

Samuel K. Sia, *Columbia University, New York, NY,* Microfluidics

Scott P. Sibley, *Goucher College, Baltimore, MD,* Semiconductors, Organic

Jeffrey J. Siirola, *Eastman Chemical Company, Kingsport, TN,* Separations Process Synthesis

Antonín Sikora, *Academy of Sciences of the Czech Republic, Prague, Czech Republic,* Polymer Blends

Geoffrey D. Silcox, *University of Utah, Salt Lake City, UT,* Hazardous Waste Incineration

Irwin Silverstein, *IBS Consulting in Quality LLC, Piscataway, NJ,* Quality

Edlyn S. Simmons, *The Procter & Gamble Company, Cincinnati, OH,* Patents, Literature

William T. Simpson, *USDA Forest Service, Madison, WI,* Wood

Vernon L. Singleton, *University of California, Davis, Davis, CA,* Wine

J. E. Singley, *Environmental Science & Engineering, Inc., Gainesville, FL,* Water Treatment

Kenneth E. Skog, *USDA Forest Service, Madison, WI,* Wood

Gregory C. Slack, *Clarkson University, Potsdam, NY,* Chromatography, Gas

William C. Sleppy, *Aluminum Company of America, Alcoa Center, PA,* Aluminum Compounds, Survey

Robert V. Slone, *Rohm and Haas Company, Spring House, PA,* Acrylic Ester Polymers; Methacrylic Ester Polymers

Charlene M. Smith, *Corning Incorporated, Corning, NY,* Silica, Vitreous

David E. Smith, *FMC Corporation, Philadelphia, PA,* Carbon Disulfide

Robert A. Smith, *AlliedSignal, Inc., Morristown, NJ,* Hydrogen Fluoride

Robin Smith, *University of Manchester Institute of Science and Technology, Manchester, United Kingdom,* Process Design

W. Ewen Smith, *University of Strathclyde, Glasgow, United Kingdom,* Raman Scattering

William L. Smith, *The Clorox Company, Pleasanton, CA,* Bleaching Agents

Ronald L. Smorada, *VersaCore Industrial Corporation, Kennett Square, PA,* Nonwoven Fabrics, Spunbonded

Thomas E. Snead, *Union Carbide Corporation, South Charleston, WV,* Acrolein and Derivatives

Gayle Snedecor, *The Dow Chemical Company, Freeport, TX,* Chloroethylenes and Chloroethanes

W. M. Snellings, *Union Carbide Corporation, South Charleston, WV,* Glycols

Nicholas H. Snow, *Seton Hall University, South Orange, NJ,* Chromatography, Gas

Scott Solis, *ExxonMobil Chemical Company, Baytown, TX,* Rubber Compounding

Gopalam Somasekhar, *Wyeth Pharmaceuticals, Collegeville, PA,* Vaccine Technology

Richard A. Sommer, *Wellman Furnaces, Inc., Shelbyville, IN,* Furnaces, Electric

Laszlo P. Somogyi, *Food Industry Consultant, Kensington, CA,* Food Additives

Hwiali Soo, *Union Carbide Technical Center, South Charleston, WV,* Ethylene Oxide

William C. Spangenberg, *Hammond Lead Products, Inc., Philadelphia, PA,* Lead Compounds

James G. Speight, *Consultant, CD&W, Inc., Laramie, WY,* Petroleum, Refinery Processes

Ron Spohn, *Occidental Chemical Corporation, Dallas, TX,* Chlorotoluenes, Ring

Giuseppe Spoto, *Universita di Catania, Catania, Italy,* Chemical Methods in Archaeology

Srivasan Sridhar, *Union Carbide, The Dow Chemical Co., South Charleston, WV,* Diamines and Higher Amines, Aliphatic

Apryll M. Stalcup, *University of Cincinnati, Cincinnati, OH,* Chiral Separations

Jeffrey W. Stansbury, *National Institute of Standards and Technology, Gaithersburg, MD,* Dental Materials

Annegret Stark, *Friedrich-Schiller-University of Jena, Jena, Germany,* Ionic Liquids

David M. Stark, *Monsanto Company, St. Louis, MO,* Genetic Engineering, Plants

Marshall W. Stark, *FMC Corporation, Bessemer City, NC,* Lithium and Lithium Compounds

Dale Steichen, *The Clorox Company, Pleasanton, CA,* Bleaching Agents

Judith Stein, *GE Global Research Center, Niskayuna, NY,* Silicones

Carl Steinecker, *MacDermid, Inc., New Hudson, MI,* Electroless Deposition

Norbert Steiner, *Degussa AG, Hanan, Germany,* Hydrogen Peroxide; Peroxides, Inorganic

R. Stepien, *GE Plastics Technology Center, Pittsfield, MA,* Acrylonitrile-Butadiene-Styrene (ABS) Polymers

Robert R. Stickney, *Texas A & M University, Bryan, TX,* Aquaculture

Edward I. Stiefel, *Exxon Research and Engineering Company, Florham Park, NJ,* Molybdenum and Molybdenum Alloys; Molybdenum Compounds

Elisabeth Stöcklin, *Roche Vitamins Ltd, Basel, Switzerland,* Nutraceuticals

William M. Stoll, *Consultant, Ligonier, PA,* Carbides, Industrial Hard

Alan M. Stolzenberg, *West Virginia University, Morgantown, WV,* Iron Compounds

Barbara Stuart, *University of Technology, Sydney, Broadway, Australia,* Infrared Spectroscopy

Werner Stumm, *Swiss Federal Institute of Technology (Deceased), Dübendorf, Switzerland,* Water

David M. Sturmer, *Eastman Kodak Company (Retired), Pittsford, NY,* Dyes, Sensitizing

The Sulphur Institute, *Washington, DC,* Sulfur

Phaik-Eng Sum, *Wyeth Research, Pearl River, NY,* Tetracyclines

James W. Summers, *The Geon Company, Avon Lake, OH,* Vinyl Chloride Polymers

H. N. Sun, *ExxonMobil, Baytown, TX,* Butadiene

K. M. Sundaram, *ABB Lummus Global, Inc., Bloomfield, NJ,* Ethylene

Ladislav Svarovsky, *FPS Institute, East Sussex, United Kingdom,* Sedimentation

Sönke Svenson, *Dendritic NanoTechnologies, Inc., Mount Pleasant, MI,* Dendrimers

Daniel R. Swiler, *Ferro Corporation, Washington, PA,* Pigments, Inorganic

Madhava Syamlal, *U.S. Department of Energy, Morgantown, WV,* Coal Gasification

Michael Szycher, *CardioTech International, Inc., Woburn, MA,* Biomaterials, Prosthetics, and Biomedical Devices

Tharwat Tadros, *Consultant, Berkshire, United Kingdom,* Surfactants

Koichi Takamura, *BASF Corporation, Charlotte, NC,* Polymer Colloids

Irene K. P. Tan, *University of Malaya, Kuala Lumpur, Malaysia,* Polyhydroxyalkanoates

Mami Tanaka, *Tohoku University, Sendai, Japan,* Biomaterials, Prosthetics, and Biomedical Devices

Rajan Tandon, *Sandia National Laboratories, Albuquerque, NM,* Ceramics, Mechanical Properties

Stavros Tavoularis, *University of Ottawa, Ontario, Canada,* Fluid Mechanics

Alan W. Taylor, *USDA Agricultural Research Service, College Park, MD,* Fertilizers

Harold A. Taylor, Jr., *Basics Mines, Summit Point, WV,* Graphite, Natural

John J. Taylor, *EPRI, Palo Alto, CA,* Nuclear Power Facilities, Safety

Paul Taylor, *GAF Corporation, Wayne, NJ,* Acetylene-Derived Chemicals

Richard F. Taylor, *TC Associates, Inc., West Boxford, MA,* Immunoassay; Microarrays

Roger Taylor, *Sussex University, Sussex, United Kingdom,* Fullerenes

Gijs Ten Berge, *Basell Polyolefins, Ferrara, Italy,* Polypropylene

John A. Tesk, *National Institute of Standards and Technology, Gaithersburg, MD,* Dental Materials

Curt Thies, *Thies Technology, Inc., Henderson, NV,* Microencapsulation

O. S. Thirunavukkarasu, *University of Regina, Saskatchewan, Canada,* Arsenic—Environmental Impact, Health Effects, and Treatment Methods

Dennis W. Thomas, *Eagle-Picher Technologies, LLC, Quapaw, OK,* Germanium and Germanium Compounds

Mary R. Thomas, *The Dow Chemical Company, Midland, MI,* Salicylic Acid and Related Compounds

Richard Thomas, *Ciba Specialty Chemicals, Tarrytown, NY,* Antioxidants, Polymers

R. O. Thribolet, *Consultant,* Acetylene

John K. Tien, *Columbia University, New York, NY,* Nickel and Nickel Alloys

Jefferson Tilley, *Hoffmann LaRoche, Inc., Nutley, NJ,* Antiobesity Drugs

Robert W. Timmerman, *FMC Corporation, Philadelphia, PA,* Carbon Disulfide

D. B. Todd, *Polymer Processing Institute, Newark, NJ,* Plastics Processing

E. Donald Tolles, *Westvaco Corporation, Charleston, SC,* Carbon, Activated

Aleksei I. Tolmachev, *National Academy of Sciences of Ukraine, Kyiv, Ukraine,* Polymethine Dyes

David L. Tomasko, *The Ohio State University, Columbus, OH,* Supercritical Fluids

Lauren M. Tonge, *Dow Corning Corporation, Midland, MI,* Poly(fluorosilicones)

G. Paull Torrence, *Hoechst-Celanese Corporation, Corpus Christi, TX,* Esterification

David L. Trent, *The Dow Chemical Company, Freeport, TX,* Propylene Oxide

Remi Trottier, *The Dow Chemical Company, Freeport, TX,* Particle Size Measurement

Andy H. Tsou, *ExxonMobil Chemical Company, Baytown, TX,* Butyl Rubber; Fillers

Kouichi Tsuji, *Osaka City University, Osaka, Japan,* X-Ray Technology

W. Gene Tucker, *James Madison University, Harrisonburg, VA,* Air Pollution and Control, Indoor

John H. Tundermann, *Inco Alloys International, Inc., Huntington, WV,* Nickel and Nickel Alloys

Albin F. Turbak, *Falcon Consultants, Inc., Sandy Springs, GA,* High Performance Fibers

Michael G. Turcotte, *TAMINCO Methylamines, Inc., Allentown, PA,* Amines, Lower Aliphatic Amines; Methylamines

D. L. Turk, *UCAR Carbon Company, Inc., Cleveland, OH,* Graphite, Artificial

Anthony P. F. Turner, *Cranfield University, Bedfordshire, United Kingdom,* Biosensors

Tzanko Tzanov, *University of Minito, Guimaraes, Portugal,* Bleaching Agents

Johan B. Ubbink, *Nestlé Research Center, Lausanne, Switzerland,* Flavor Delivery Systems

Eric Udd, *Blue Road Research, Gresham, OR,* Fiber Optics

Stefan Uhrlandt, *Degussa Corporation, Piscataway, NJ,* Silica

Shmuel D. Ukeles, *IMI (TAMI) Institute for Research and Development, Haifa Bay, Israel,* Bromine, Inorganic Compounds; Drilling Fluids

Saadet Ulas, *Vishwamitra Research Institute, Westmont, IL,* Sampling Techniques

Henri Ulrich, *Consultant, Guilford, CT,* Urethane Polymers

Joachim Ulrich, *Martin Luther University, Halle-Wittenberg, Halle (Saale), Germany,* Crystallization

L. L. Upson, *UOP LLC, Des Plaines, IL,* Fluid Catalytic Cracking (FCC) Units, Regeneration

Rajan Vaidyanathan, *University of Central Florida, Orlando, FL,* Shape-Memory Alloys

Michael R. Van De Mark, *University of Missouri–Rolla Coatings Institute, Rolla, MO,* Paint

Henk J. W. van den Haak, *Akzo Nobel, Arnhem, The Netherlands,* Dispersants

Pieter J. van der Valk, *Ferroxcube Nederland B.V., Eindhoven, The Netherlands,* Ferrites

Rudi van Eldik, *University of Erlangen-Nürnberg, Erlangen, Germany,* High Pressure Chemistry

Patrick Vandereecken, *Dow Corning Corporation, Seneffe, Belgium,* Sealants

Rajender S. Varma, *U.S. Environmental Protection Agency, Cincinnati, OH,* Microwave Technology—Chemical Synthesis Applications

Paul Vaughan, *CSIRO Molecular and Health Technologies, Parkville, Australia,* Yeasts

Mariano Velez, *University of Missouri-Rolla, Rolla, MO,* Glass

Zata M. Vickers, *University of Minnesota, St. Paul, MN,* Flavor Characterization

Jean-Paul Vidal, *Rhodia, Saint-Fons, France,* Vanillin

Tyrone L. Vigo, *U.S. Department of Agriculture, New Orleans, LA,* High Performance Fibers

Hugo O. Villar, *Triad Therapeutics, Inc., San Diego, CA,* Chemoinformatics

T. Viraraghavan, *University of Regina, Saskatchewan, Canada,* Arsenic—Environmental Impact, Health Effects, and Treatment Methods

Robert L. Virta, *U.S. Geological Survey, Reston, VA,* Asbestos

Kenneth Visek, *Akzo Chemicals, Inc., McCook, IL,* Amines, Fatty

Frank Vogt, *Arizona State University, Tempe, AZ,* Chemometrics

Urs von Stockar, *Laboratoire de Genie Chimique et Biologique, Lausanne, Switzerland,* Absorption

Bipin V. Vora, *UOP LLC, Des Plaines, IL,* Alkylation

Lisa Vrana, *Consultant, Columbus, OH,* Calcium and Calcium Alloys; Calcium Chloride; Calcium Fluoride

Walter H. Waddell, *ExxonMobil Chemical Company, Baytown, TX,* Fillers; Rubber Compounding; Silica, Amorphous

Mark Wadsworth, *Jet Propulsion Laboratory, Pasadena, CA,* Photodetectors

Frank S. Wagner, *Strem Chemicals, Inc., Newburyport, MA,* Acetic Acid; Acetic Anhydride; Rubidium and Rubidium Compounds

John D. Wagner, *Albemarle Corporation, Baton Rouge, LA,* Alcohols, Higher Aliphatic, Synthetic Processes; Olefins, Higher

Phillip J. Wakelyn, *The National Cotton Council of America, Washington, DC,* Cotton

Park W. Waldroup, *University of Arkansas, Fayetteville, AR,* Feeds and Feed Additives, Nonruminant

Jeanine Walenga, *Loyola University Medical Center, Maywood, IL,* Blood Coagulation and Anticoagulant Drugs

Jim Wallace, *M. W. Kellogg Company, Houston, TX,* Phenol

G. M. Wallraff, *IBM Research Division, San Jose, CA,* Lithographic Resists

Kevin C. Walter, *The Essex Technology Group, LLC, Aliso Viejo, CA,* Ion Implantation

Robert R. Walton, *Wellman Furnaces, Inc., Shelbyville, IN,* Furnaces, Electric

Vivian K. Walworth, *Jasper Associates, Concord, MA,* Photography, Instant

Guijun Wang, *University of New Orleans, New Orleans, LA,* Oxazolidinones, Antibacterial

Meng-Jiao Wang, *Cabot Corporation, Billerica, MA,* Carbon Black

Thomas J. Ward, *Clarkson University, Potsdam, NY,* Economic Evaluation

Roger H. Wardman, *Heriot-Watt University, Galashiels, United Kingdom,* Textiles

Rosemary Waring, *University of Birmingham, Birmingham, UK,* Aminophenols

Gerald S. Wasserman, *Kraft Foods Corporation, Tarrytown, NY,* Coffee

Richard W. Waterstrat, *National Institute of Standards and Technology, Gaithersburg, MD,* Dental Materials

Jeremiah J. Way, *Colorado State University, Fort Collins, CO,* Transport Phenomena for Chemical Reactor Design

Rob Weaver, *McCrone Research Institute, Chicago, IL,* Microscopy

Robert N. Webb, *ExxonMobil, Baytown, TX,* Butyl Rubber

Edwin Weber, *Technische Universität Bergakademie Freiberg, Freiberg, Germany,* Inclusion Compounds; Molecular Recognition

Peter Weber, *Roche Vitamins Ltd, Basel, Switzerland,* Nutraceuticals

Michele A. Weidner-Wells, *Johnson & Johnson Pharmaceutical Research & Development, L.L.C., Raritan, NJ,* Antibacterial Agents, Overview; Sulfonamides

Edward D. Weil, *Polytechnic University of New York, Brooklyn, NY,* Flame Retardants, Phosphorus; Sulfur Compounds

David A. Weitz, *Harvard University, Cambridge, MA,* Foams

Jeffrey Wengrovius, *GE Silicones, Waterford, NY,* Silicones

Robert H. Wentorf, Jr., *Greenwich, NY,* Diamond, Synthetic

R. A. Wessling, *The Dow Chemical Company, Midland, MI,* Vinylidene Chloride Polymers

Jack H. Westbrook, *Brookline Technologies, Ballston Spa, NY,* Materials Standards and Specifications

Charles W. Weston, *Freeport Research and Engineering Company, Bell Chase, LA,* Ammonium Compounds

David R. Whitcomb, *Eastman Kodak Company, Oakdale, MN,* Photothermographic and Thermographic Imaging Materials

D. R. White, *Measurement Standards Laboratory of New Zealand, Lower Hutt, New Zealand,* Temperature Measurement

David L. White, *Restoration Chemicals, Inc., Vincennes, IN,* Paint and Finish Removers

J. S. White, *White Technical Research Group, Argenta, IL,* Sugar

Robert H. White, *USDA Forest Service, Madison, WI,* Wood

Chris Whiteley, *Rhodes University, Grahamstown, South Africa,* Enzyme Inhibitors

Walter H. Whitlock, *The BOC Group, Inc., Murray Hill, NJ,* High Purity Gases

Denyce Wicht, *GE Global Research Center, Niskayuna, NY,* Silicones

Zeno W. Wicks, *Consultant, Louisville, KY,* Alkyd Resins; Coatings; Coatings for Corrosion Control, Organic; Drying Oils

Robert Wilczynski, *Rohm and Haas Company, Spring House, PA,* Methacrylic Acid and Derivatives

Mark Wilf, *Hydranautics, Oceanside, CA,* Water Desalination

Richard M. Wilkins, *Newcastle University, Newcastle upon Tyne, United Kingdom,* Controlled Release Technology, Agricultural

Laurence L. Williams, *Consultant, Stamford, CT,* Amino Resins and Plastics

M. Williams, *University of Kentucky, Lexington, KY,* Reverse Osmosis

Richard A. Wilsak, *Amoco Chemical Company, Naperville, IL,* Butylenes

Alan R. Wilson, *Defence Science and Technology Organisation, Melbourne, Australia,* Smart Materials

David Wilson, *The Dow Chemical Company, Freeport, TX,* Chelating Agents

Michael Wilson, *Wilson Technologies, Ltd.,* **Alkaloids**

Michael A. Wilson, *University of Western Sydney, New South Wales, Australia,* Nanotechnology

Jerrold E. Winandy, *USDA Forest Service, Madison, WI,* Wood

Robert M. Winslow, *Sangart, Inc., and University of California, San Diego, San Diego,CA,* Blood Substitutes

John Wohlgemuth, *BP Solar, Frederick, MD,* Solar Energy, Photovoltaic Cells

Suhad Wojkowski, *U.S. Department of Agriculture, New Orleans, LA,* Herbicides

John A. Wojtowicz, *Consultant, Goodyear, AZ,* Cyanuric and Isocyanuric Acids; Dichlorine Monoxide, Hypochlorous Acid, and Hypochlorites; N-Halamines; Ozone; Water Treatment of Swimming Pools, Spas, and Hot Tubs

C. P. Wong, *Georgia Institute of Technology, Atlanta, GA,* Embedding; Self-Cleaning Materials—Lotus Effect Surfaces

Christina Darkangelo Wood, *GE Global Research Center, Niskayuna, NY,* Silicones

Stewart Wood, *The Dow Chemical Company, Midland, MI,* Particle Size Measurement

Kermit E. Woodcock, *Gas Research Institute, Chicago, IL,* Gas, Natural

Calvin Woodings, *Calvin Woodings Consulting Ltd., Warwickshire, United Kingdom,* Fibers, Regenerated Cellulose

Gayle Woodside, *IBM Corporation, Austin, TX,* Waste Management, Hazardous

Mike Woolery, *Stratcor, Hot Springs, AR,* Vanadium Compounds

S. Davis Worley, *Auburn University, Auburn, AL,* N-Halamines

Andrew J. Woytek, *Air Products and Chemicals, Inc., Allentown, PA,* Halogen Fluorides

Nicholas G. Wright, *University of Newcastle upon Tyne, Newcastle upon Tyne, United Kingdom,* Silicon Carbide

C. W. Wrigley, *Food Science Australia, North Ryde, Australia,* Wheat and Other Cereal Grains

J. P. Wristers, *Exxon Chemical Company, Baytown, TX,* Butadiene

Mike Wu, *BP Amoco Chemicals, Naperville, IL,* Acrylonitrile Polymers, Survey and Styrene-Acrylonitrile (SAN)

Yun-Tai Wu, *DuPont Company, Wilmington, DE,* Ethylene-Acrylic Elastomers

Carl Wust, *FiberVisions, Covington, GA,* Fibers, Olefin

Nicholas Patrick Wynn, *Sulzer Chemtech GmbH, Neunkirchen, Germany,* Pervaporation

George Wypych, *ChemTec Laboratories, Inc., Ontario, Canada,* Solvents, Industrial

Marino Xanthos, *Polymer Processing Institute, Newark, NJ,* Plastics Processing

Jianwen Xu, *Georgia Institute of Technology, Atlanta, GA,* Embedding; Self-Cleaning Materials—Lotus Effect Surfaces

Eli Yablonovitch, *University of California, Los Angeles, Los Angeles, CA,* Electronic Materials

G. Yaluris, *WR Grace & Company, Columbia, MD,* Fluid Catalytic Cracking (FCC) Catalysts and Additives

Chen-Hsyong Yang, *Monsanto Company, St. Louis, MO,* Phosphorus Compounds

Gary L. Yingling, *McKenna & Cuneo, LLP, Washington, DC,* Regulatory Agencies

David Yoffe, *IMI (TAMI) Institute for Research and Development, Haifa Bay, Israel,* Bromine, Organic Compounds

Carmen M. Yon, *UOP, Fitzwilliam, NH,* Adsorption, Gas Separation

Elaine M. Yorkgitis, *3M Company, Automotive Division, Mendota Heights, MN,* Adhesives

Raymond A. Young, *The University of Wisconsin-Madison, Madison, WI,* Fibers, Vegetable

Marek Zaidlewicz, *Nicolaus Copernicus University, Torun, Poland,* Hydroboration

John I. Zerbe, *USDA Forest Service, Madison, WI,* Wood

Edward G. Zey, *Hoechst-Celanese Corporation, Corpus Christi, TX,* Esterification

Li Zhang, *Colorado State University, Fort Collins, CO,* Transport Phenomena for Chemical Reactor Design

Zhuqing Zhang, *Georgia Institute of Technology, Atlanta, GA,* Self-Cleaning Materials—Lotus Effect Surfaces

X. Zhao, *WR Grace & Company, Columbia, MD,* Fluid Catalytic Cracking (FCC) Catalysts and Additives

Shiping Zhu, *Unilever R&D, Bedford, United Kingdom,* Soap

J. Richard Zietz, *Ethyl Corporation, Baton Rouge, LA,* Alcohols, Higher Aliphatic, Synthetic Processes

B. L. Zoumas, *The Pennsylvania State University, Hershey Foods Corporation, Hershey, PA,* Chocolate and Cocoa

Michael Zviely, *Frutarom, Ltd. Haifa, Israel,* Aroma Chemicals

ABRASIVES

An abrasive is a substance used to abrade, smooth, or polish an object. Abrasive connotes very hard substances ranging from naturally occuring sands to the hardest material known, diamond.

There are four major forms of abrasive articles. A bonded abrasive is a three-dimensional composite of abrasive grains dispersed in a bond system. This bond system may be organic (eg, resinoid wheels), glassy inorganic bond (vitrified wheels), or metallic. Bonded abrasives are commercially available in a wide variety of forms, including wheels (most popular), stones, mounted points, saws, segments, and the like. Coated abrasives are generally described as a plurality of abrasive grains bonded to a backing. Nonwoven abrasives comprise a plurality of abrasive grains bonded into and onto a porous nonwoven web substrate. Nonwoven and coated abrasives are also available in a wide variety of converted forms of belts, sheets, disks, cones, flap wheels, etc. Loose abrasive slurries comprise a plurality of abrasive grains dispersed in a liquid medium, such as water. Loose abrasive slurries are typically employed in polishing-type applications where a very fine surface finish is desired.

These abrasive articles are used in a plethora of different refining processes including metal degating, grinding, shaping, cutting, deburring, finishing, sanding, cleaning, polishing, and planarizing. Today abrasive articles are employed in some aspect in many manufactured goods sold.

ABRASIVE MATERIALS

There are seven major properties of abrasive materials: hardness, toughness, refractoriness (melting temperature), chemical reactivity, thermal conductivity, fracture, and microstructure. Hardness is measured by the Mohs' scale, which is based on the relative scratch hardness of one mineral compared to another. This scale has two limitations: it is not linear and there is insufficient delineation. To ameliorate these deficiencies, Knoop devised a method in which a diamond indenter of pyramidal shape is forced into the material to be evaluated, and the depth of penetration is then determined from the length and width of the indentation produced.

An abrasive's toughness is often measured and expressed as the degree of friability, the ability of an abrasive grit to withstand impact without cracking, spalling, or shattering. Toughness is often considered a measure of resistance to fracture and given the symbol K_c.

Fracture characteristics of abrasive materials are important, as are the resulting grain shapes. Equiaxed shaped grains are generally preferred for bonded abrasive products and sharp, acicular shaped grains are typically preferred for coated abrasives. How the grains fracture in the grinding process determines the wear resistance and self-sharpening characteristics of the wheel or belt.

The first artificial abrasive was silicon carbide, which is produced from quartz sand and carbon in a large electric furnace. Reaction temperature range from 1800 to 2200°C. There are two basic types of silicon carbide: one is gray or black in color and the other, somewhat purer, slightly harder, but more friable form is green.

Fused aluminum oxide was manufactured in the Higgins furnace, which used a water-cooled steel shell instead of refractory lining. After crushing, further heat treatment in rotary furnaces is used, producing an exsolved dispersed phase that affects the impact strength of the resulting product. Most aluminum oxide is now fused in tilting furnaces and poured into ingots of sizes suitable for the desired rapid rate of cooling.

Bayer alumina (see ALUMINUM COMPOUNDS) is also the starting material for the production of fused white aluminum oxide abrasive. This white abrasive is widely used in tool grinding as well as in other applications requiring cool cutting, self-sharpening, or a damage-free workpiece. Special pink or ruby variations of the white abrasive are produced by adding small amounts of chromium compounds to the melt.

Sol–gel technology is used to improve the performance of aluminum oxide abrasives. Sol–gel processing permits the microstructure of the aluminum oxide to be controlled to a much greater extent than is possible by the fusion process, resulting in a crystal size several orders of magnitude smaller than that of the fused abrasives, with a corresponding increase in toughness.

Diamond is the hardest substance known. Abrasive applications for industrial diamonds include their use in rock drilling, as tools for dressing and trueing abrasive wheels, in polishing and cutting operations (as a loose powder), and as abrasive grits in bonded wheels and coated abrasive products.

Cubic boron nitride (CBN) is a synthetic mineral not found in nature. It is nearly as hard as diamond, yet it does not perform as well in the usual diamond grinding applications. However, CBN is an extremely efficient abrasive for grinding steel. CBN improves grinding wheel life by as much as 100 times over that of alumina, thus increasing productivity, reducing downtime for the wheel changes and dressing, and improving the quality of parts.

Boron carbide (B_4C) is produced by the reaction of boron oxide and coke in an electric arc furnace (70% B_4C) or by that of carbon and boric anhydride in a carbon resistance furnace (80% B_4C) (see BORON COMPOUNDS; REFRACTORY BORON COMPOUNDS). It is primarily used as a loose abrasive for grinding and lapping hard metals, gems, and optics.

Metallic abrasives are most commonly used as a blast medium to clean or to improve the properties of metallic surfaces.

Garnet is the name given to a group of silicate minerals possessing similar physical properties and crystal forms but differing in chemical composition. Of the seven existing, the two most important are pyrope.

Tripoli is a fine grained, porous, decomposed siliceous rock. Since Tripoli particles are rounded, not sharp, it has a mild abrasive action particularly suited for polishing. Rottenstone and amorphous silica are similar to Tripoli and find the same uses.

SIZING, SHAPING, AND TESTING OF ABRASIVE GRAINS

Sizing

Manufactured abrasives are produced in a variety of sizes that range from a pea-sized grit of 4 (5.2 mm) to submicron diameters. It is almost impossible to produce an abrasive grit that will just pass through one sieve size yet be 100% retained on the next smaller sieve. Thus a standard range was adopted in the United States that specifies a screen size through which 99.9% of the grit must pass, maximum oversize, minimum on-size, maximum through-size, and fines.

Shaping

Screening is a two-dimensional (2D) process and cannot give information about the shape of the abrasive particle. Desired shapes are obtained by controlling the method of crushing and by impacting or milling. Shape determinations are made optically and by measuring the loose-packed density of the abrasive particles; cubical-shaped particles pack more efficiently.

Testing

Chemical analyses are done on all manufactured abrasives, as well as physical tests such as sieve analysis, specific gravity, impact strength, and loose poured density (a rough measure of particle shape). Special abrasives such as sintered sol−gel aluminas require more sophisticated tests such as electron microscope measurement of alpha alumina crystal size, and indentation microhardness.

TYPES OF FLEXIBLE ABRASIVES

There are three types of flexible abrasive: coated, structured and nonwoven. A coated abrasive is defined as a plurality of abrasive particles adhered to a substrate, commonly called a backing. There are several typical constructions in this coated abrasive family, namely, conventional coated abrasives, lapping film, and structured abrasives. Over and in between the abrasive particles is a second binder, called a size coat; the size coat reinforces the abrasive particles. There may optionally be a third coating, called a supersize coating applied over the size coating. A second coated abrasive construction is commonly referred to as "lapping film," where a plurality of abrasive particles are randomly dispersed in a binder. This abrasive particle/binder composite is applied over

the front surface of the backing. A third coated abrasive construction, named a structured abrasive, comprises a plurality of shaped abrasive composites that are bonded to a backing.

The abrasive article binder system may contain additives that modify the polymer physical properties and/or positively affect the abrading performance of the resulting abrasive. Grinding aids are another class of additives generally preferred in dry metal grinding applications. Where grinding aids are generally preferred in dry metal grinding, antiloading materials are sometimes preferred in dry paint, wood sanding.

In recent years, a new coated abrasive construction has emerged, known as a structured abrasive. The structured abrasive comprises a plurality of shaped abrasive composites adhered to a backing; these shaped abrasive composites may be precisely or irregularly shaped. These shapes may be any geometric shape such as pyramidal, ridge-like, hemisphere, cube-like, and block-like. To make a structured abrasives article, a slurry is first prepared comprising a plurality of abrasive grains dispersed in a resin, along with optional additives. A production tool is generated comprising cavities having the desired shape, density, and size of abrasive composites. For precisely shaped abrasive composites this abrasive slurry is coated onto the backing and the resulting construction is brought into contact with the production tool. The abrasive slurry flows into the cavities of a production tool. Next, an energy source, such as uv light, is transmitted through the production tool and into the resin. The resin is at least partially cross-linked and the resulting abrasive slurry is solidified to form a plurality of abrasive composites bonded to a backing. For nonprecisely shaped abrasive composites, the abrasive slurry is coated into the cavities of a production tool, such as a rotogravure roll. The backing is brought into contact with the abrasive slurry and the rotogravure roll imparts a pattern to the abrasive slurry. The abrasive slurry is removed and then exposed to an energy source to at least partially cure the resin to form the structured abrasive.

Nonwoven abrasives are unique forms of abrasives that find use in many aspects of material finishing and surface cleaning. This category of abrasives provides a different interaction with the workpiece than coated abrasives or grinding wheels, and are, therefore, commonly called surface conditioning abrasives. Nonwoven abrasives are so named because of the random fibrous matrix on which they are based.

The first step in the manufacture of nonwoven abrasives is the creation of the carrier web. A mechanical carding, melt bonded, or air laid process in which crimped polymeric fibers are laid down on a carrier belt and passed through a variety of resin baths to give desired characteristics forms this web. The coated web is then passed through a drying oven to provide structural integrity. In some cases, an additional scrim layer may be mechanically bonded to provide even greater structural integrity.

In the second step, the "make coat" is applied, which can be a two-step resin/abrasive process or can be applied as a one-step resin/abrasive slurry. In the former, the pre-bond web is passed through a resin bath and the abrasive can be gravity applied or blown through the web using an air stream.

In order to create wheels, another resin coat, called the "size coat", is added. In this case, a "make coated" web is passed through a resin bath and lightly oven cured to produce a tacky web. This web is then treated in one of two ways to make wheels. In one process, winding the size-coated web into a roll that is cured before wheels are sliced from it makes "convolute" wheels. In the second process, "unitized" wheels are cut from slabs that have been compacted and cured under pressure.

There are many specialized forms and uses of bonded abrasives, but we list only four: honing and superfinishing, pulpstone wheels, crush-form grinding, and creep feed wheels.

J. Byers, ed., *Metalworking Fluids*, Marcel Dekker Inc., New York, 1994.

L. Coes, Jr., *Abrasives*, Springer-Verlag, New York, 1971, p. 2.

D. W. Olson, "Manufactured Abrasives", in *Minerals Yearbook 2000*, Vol. 1, *Metals and Minerals*, U.S. Dept. of the Interior, p. 5.2.

A Review of Diamond Sizing and Standards, IDA Bulletin, Industrial Diamond Association of America, Columbia, S.C., 1985.

Donna Bange
3M Company
Roger Gary
Milacron Inc.

ABSORPTION

Absorption, or gas absorption, is a unit operation used in the chemical industry and increasingly in environmental applications to separate gases by washing or scrubbing a gas mixture with a suitable liquid. One or more of the constituents of the gas mixture dissolves or is absorbed in the liquid and can thus be removed from the mixture. In some systems, this gaseous constituent forms a physical solution with the liquid or the solvent. In other cases, it undergoes a chemical reaction with one or more components of the liquid.

The purpose of such scrubbing operations may be any of the following: gas purification (eg, removal of air pollutants from exhaust gases or contaminants from gases that will be further processed), product recovery, or production of solutions of gases for various purposes. Several examples of applied absorption processes are shown in Table 1.

Gas absorption is usually carried out in vertical countercurrent columns as shown in Figure 1. The solvent is fed at the top of the absorber, whereas the gas mixture enters from the bottom. The absorbed substance is washed out by the solvent and leaves the absorber at the bottom as a liquid solution. The solvent is often recovered in a subsequent stripping or desorption operation. This second step is essentially the reverse of absorption and involves countercurrent contacting of the liquid loaded with solute using an inert gas or water vapor. Desorption is frequently carried out at higher temperatures and/or at lower pressure than the absorption step. The absorber may be a packed column, plate tower, or simple spray column, or a bubble column.

The fundamental physical principles underlying the process of gas absorption are the solubility of the absorbed gas and the rate of mass transfer. Information on both must be available when sizing equipment for a given application. Additionally, in the very frequent case of the design of countercurrent columns, it is also necessary to have information on the hydraulic capacity (eg, entrainment, loading, flooding) of the equipment. In addition to the fundamental design concepts based on solubility and mass transfer, many other practical details have to be considered during actual plant design and construction which may affect the performance of the absorber significantly. These details have been described in reviews and in some of the more comprehensive treatments of gas absorption and absorbers (see also DISTILLATION; HEAT EXCHANGE TECHNOLOGY).

Table 1. Typical Commercial Gas Absorption Processes

Treated gas	Absorbed gas, solute	Solvent	Function
coke oven gas coke	ammonia	water	by-product recovery
oven gas	benzene and toluene	straw oil	by-product recovery
reactor gases in manufacture of formaldehyde from methanol	formaldehyde	water	product recovery
drying gases in cellulose acetate fiber production	acetone	water	solvent recovery
natural and refinery gases	hydrogen sulfide	amine solutions	pollutant removal
flue gases	sulfur dioxide	water	pollutant removal
	carbon dioxide	amine solutions	by-product recovery
wet well gas	propane and butane	kerosene	gas separation
wet well gas	water	triethylene glycol	gas drying
ammonia synthesis gas	carbon monoxide	ammoniacal cuprous chloride solution	contaminant removal
roast gases	sulfur dioxide	water	production of calcium sulfite solution for pulping

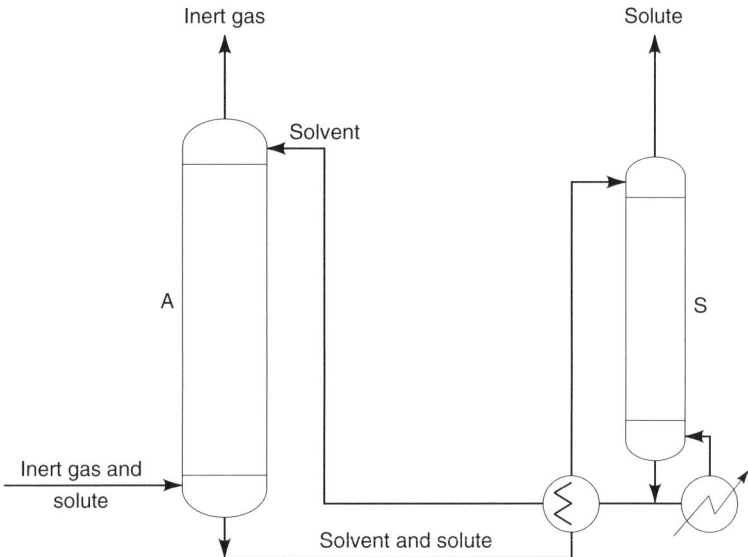

Figure 1. Absorption column arrangement with a gas absorber A and a stripper S to recover solvent.

MASS TRANSFER CONCEPTS

In order to determine the size of the equipment necessary to absorb a given amount of solvent per unit time, one must know not only the equilibrium solubility of the solute in the solvent, but also the rate at which the equilibrium is established; ie, the rate at which the solute is transferred from the gas to the liquid phase must be determined.

Mass Transfer Coefficients and Convection

Many theories have been developed in attempts to model mass transfer rates under the combined effects of molecular diffusion and turbulent convection. The classical model has been the film theory effectively assuming completely stagnant layers of a given thickness z_0 adjacent to the interface and a sudden change to the completely turbulent conditions prevailing in the bulk of the phase. Mass transfer is thus assumed to occur through these films only by molecular diffusion at steady state.

Absorption and Chemical Reaction

In instances where the solute gas is absorbed into a liquid or a solution where it is able to undergo chemical reaction, the driving forces of absorption become far more complex. The solute not only diffuses through the liquid film at a rate determined by the gradient of the concentration, but at the same time also reacts with the liquid at a rate determined by the concentrations of both the solute and the solvent at the point of interest. Calculating the concentration profiles through the liquid film requires formulating a differential mass balance over an infinitesimal control volume in the film which accounts for both diffusion and reaction of the solute gas and subsequently integrating it. The calculations show that these profiles are steeper and the rate of mass transfer higher than without chemical reaction.

DESIGN OF PACKED ABSORPTION COLUMNS

Discussion of the concepts and procedures involved in designing packed gas absorption systems shall first be confined to simple gas absorption processes without complications: isothermal absorption of a solute from a mixture containing an inert gas into a nonvolatile solvent without chemical reaction. Gas and liquid are assumed to move through the packing in a plug-flow fashion. Deviations such as nonisothermal operation, multicomponent mass transfer effects, and departure from plug flow are treated in later sections.

Standard Absorber Design Methods

Operating Line. As a gas mixture travels up through a gas absorption tower, the solute is transferred to the liquid phase and thus gradually removed from the gas. The liquid accumulates solute on its way down through the column so x increases from the top to the bottom of the column. The steady-state concentrations y and x at any given point in the column are interrelated through a mass balance around either the upper or lower part of the column (eq. 2), whereas the four concentrations in the streams entering and leaving the system are interrelated by the overall material balance.

Since the total gas and liquid flow rates per unit cross-sectional area vary throughout the tower, rigorous material balances should be based on the constant inert gas and solvent flow rates G'_M and L'_M, respectively, and expressed in terms of mole ratios Y' and X'. A balance around the upper part of the tower yields

$$G'_M Y' + L'_M X'_2 = G'_M Y'_2 + L'_M X' \qquad (1)$$

which may be rearranged to give

$$Y' = L'_M / G'_M \, (X' - X'_{A,2}) + Y'_{A,2} \qquad (2)$$

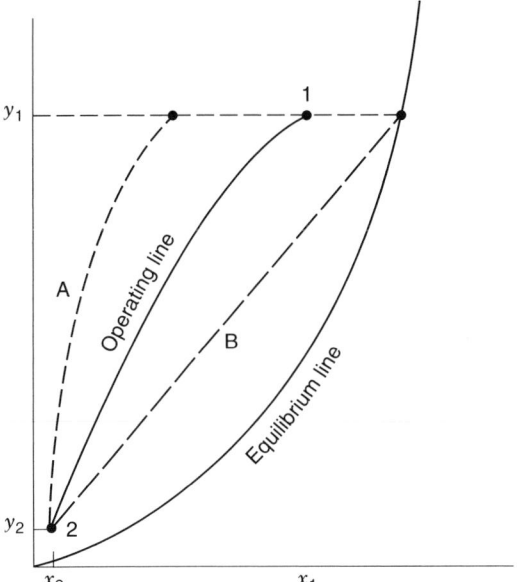

Figure 2. Operating lines for an absorption system: line A, high L_M/G_M ratio; solid line, medium L_M/G_M ratio; line B, L_M/G_M ratio at theoretical minimum necessary for the removal of the specified quantity of solute. Subscript 1 represents the bottom of tower, 2, the top of tower.

where G'_M and L'_M are in kg·mol/(h·m^2) [lb·mol/(h·ft^2)] and $Y' = y/(1-y)$ and $X' = x/(1-x)$. The overall material balance is obtained by substituting $Y' = Y'_1$ and $X' = X'_1$. For dilute gases the total molar gas and liquid flows may be assumed constant and a similar mass balance yields

$$y = L_M/G_M (x - x_2) + y_2 \qquad (3)$$

A plot of either equation 1 or 2 is called the operating line of the process as shown in Figure 2.

Design Procedure

The packed height of the tower required to reduce the concentration of the solute in the gas stream from $y_{A,1}$ to an acceptable residual level of $y_{A,2}$ may be calculated

by combining point values of the mass transfer rate and a differential material balance for the absorbed component.

Nonisothermal Gas Absorption

Nonvolatile Solvents. In practice, some gases tend to liberate such large amounts of heat when they are absorbed into a solvent that the operation cannot be assumed to be isothermal, as is usually done. The resulting temperature variations over the tower will displace the equilibrium line on a $y-x$ diagram considerably because the solubility usually depends strongly on temperature. Thus nonisothermal operation affects column performance drastically.

Axial Dispersion Effects

Effect of Axial Dispersion on Column Performance. Another assumption underlying standard design methods is that the gas and the liquid phases move in plug-flow fashion through the column. In reality, considerable departure from this ideal flow assumption exists and different fluid particles travel through the packing at varying velocities. This effect, usually called axial dispersion, counteracts the countercurrent contacting scheme for which the column is designed and thus lowers the driving forces throughout the packed bed. Neglect of axial dispersion results in an overestimation of the driving forces and in an underestimation of the number of transfer units needed. It may therefore lead to an unsafe design.

BUBBLE TRAY ABSORPTION COLUMNS

General Design Procedure

Bubble tray absorbers may be designed graphically based on a so-called McCabe-Thiele diagram. An operating line and an equilibrium line are plotted in y-x, Y'-X', or Y^0-X^0 coordinates using the principles for packed adsorbers outlined above (see Fig. 3). The minimum number of plates required for a specified recovery may be computed by assuming that equilibrium is reached between the two phases on each bubble tray. Thus the gas and the liquid leaving a tray are at equilibrium and a hypothetical tray

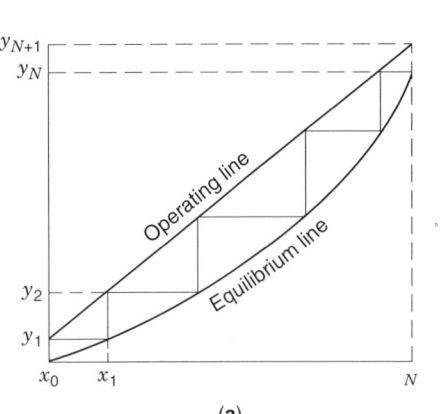

(a)

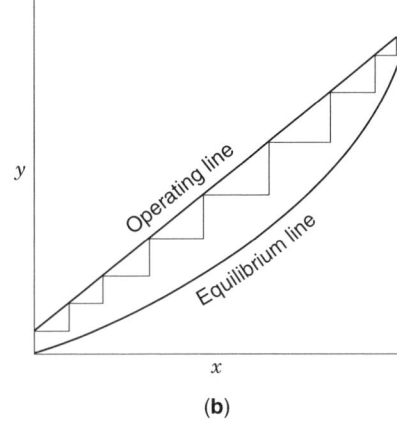

(b)

Figure 3. McCabe-Thiele diagram. (**a**) Number of theoretical plates, 5; (**b**) number of actual plates, 8.

capable of equilibrating the phase streams is termed a theoretical plate. Starting the calculation at the bottom of the tower, where the concentrations are y_{N+1} and x_N, the concentration leaving the lowest theoretical plate y_N may be found on the design diagram (Fig. 3**a**) by moving from the operating line vertically to the equilibrium line, because y_N is at equilibrium with x_N. Since the concentrations between two plates are always related by the operating line, x_{N-1} may be found from y_N by moving horizontally to the operating line. By repeating this sequence of steps until the desired residual gas concentration y_1 is reached, the number of theoretical plates can be counted.

The required number of actual plates, N_P, is larger than the number of theoretical plates, N_{TP}, because it would take an infinite contacting time at each stage to establish equilibrium. The ratio $N_{TP} \cdot N_P$ is called the overall column efficiency. This parameter is difficult to predict from theoretical considerations, however, or to correct for new systems and operating conditions. It is therefore customary to characterize the single plate by the so-called Murphree vapor plate efficiency, E_{MV}:

$$E_{MV} \equiv (y_n - y_{n+1})/(y_n^* - y_{n+1}) \qquad (4)$$

which indicates the fractional approach to equilibrium achieved by the plate. An efficiency of 80% means that the reduction in solute gas concentration effected by the plate is 80% of the reduction obtained from a theoretical plate. Corresponding actual plates may therefore be stepped off by moving from the operating line vertically only 80% of the distance between operating and equilibrium line (Fig. 3**b**). In some special cases having negligible resistance in the gas phase, E_{MV} values may become unreasonably small. It is then more logical to define a Murphree liquid plate efficiency, E_{ML}, simply by reversing the role of liquid and gas and by focusing on the change in liquid composition across the plate with respect to an equilibrium given by the leaving vapor.

Nonisothermal Gas Absorption

The computation of nonisothermal gas absorption processes is difficult because of all the interactions involved as described for packed columns. An very large number of plate calculations is normally required to establish the correct concentration and temperature profiles through the tower. Suitable algorithms have been developed and nonisothermal gas absorption in plate columns has been studied experimentally and the measured profiles compared to the calculated results. The close agreement between the calculated and observed pro-files was obtained without adjusting parameters. The plate efficiencies required for the calculations were measured independently on a single exact copy of the bubble cap plates installed in the five-tray absorber.

A general, approximate, short-cut design procedure for adiabatic bubble tray absorbers has not been developed, although work has been done in the field of nonisothermal and multicomponent hydrocarbon absorbers. An analytical expression has been developed which will predict the recovery of each component provided the stripping factor, ie, the group mG_M/L_M, is known for each component on each tray of the column. This requires knowledge of the temperature and total flow (G_M and L_M) profiles through the tower.

Capacity Limitations

The fluid flow capacity of a bubble tray may be limited by any of three principal factors: flooding, often the most restrictive of the limitations, occurs when the clear liquid height in the downcomer, H_{dc}, exceeds a certain fraction of the tray spacing; entrainment occurs when spray or froth formed on one tray enters the gas passages in the tray above; large hydraulic gradient at high liquid flow, which causes the caps near the liquid feed point will suffer.

H. S. Fogler, *Elements of Chemical Reaction Engineering*, 3rd ed., Prentice Hall, 1998.

D. Frenkel and B. Smit, *Understanding of Molecular Simulation From Algorithms to Applications*, Academic Press, 1999.

R. H. Perry, D. Green and J. O. Maloney, eds., *Perry's Chemical Engineer's Handbook*, 7th ed., McGraw-Hill Book Co., Inc., New York, 1997.

R. J. Sadus, *Molecular Simulation of Fluids: Theory, Algorithms and Object-Orientation*, Elsevier, New York, 1999.

MANUEL LASO
Universidad Politécnica de
Madrid, ETSII (Spain)
URS VON STOCKAR
École Polytechnique Fédérale,
Lausanne

ACETALDEHYDE

Acetaldehyde is a product of most hydrocarbon oxidations. It is an intermediate product in the respiration of higher plants and occurs in trace amounts in all ripe fruits that have a tart taste before ripening. The aldehyde content of volatiles has been suggested as a chemical index of ripening during cold storage of apples. Acetaldehyde is also an intermediate product of fermentation (qv), but it is reduced almost immediately to ethanol. It may form in wine (qv) and other alcoholic beverages after exposure to air imparting an unpleasant taste; the aldehyde reacts to form diethyl acetal and ethyl acetate. Acetaldehyde is an intermediate product in the decomposition of sugars in the it body and hence occurs in trace quantities in blood.

PROPERTIES

Physical Properties

Acetaldehyde is a colorless, mobile liquid having a pungent, suffocating odor that is somewhat fruity and quite pleasant in dilute concentrations. Its physical properties are given in Table 1.

Table 1. Physical Properties of Acetaldehyde

Properties	Values
formula weight	44.053
melting point, °C	−123.5
boiling point at 101.3 kPa[a] (1 atm), °C	20.16
density, g/mL	
d^0_4	0.8045
d^{15}_4	0.7846
coefficient of expansion per °C (0–30°C)	0.00169
refractive index, n^{20}_D	1.33113
vapor density (air = 1)	1.52
surface tension at 20°C, mN/m (= dyn/cm)	21.2
absolute viscosity at 15°C, mPa·s (= cP)	0.02456
specific heat at 0°C, J/(g·K)[b]	
15°C	2.18
25°C	1.41
$\alpha = C_p/C_v$ at 30°C and 101.3 kPa[a] (1 atm)	1.145

[a] To convert kPa to psi, multiply by 0.14503.
[b] To convert J to cal, divide by 4.187.

Acetaldehyde is miscible in all proportions with water and most common organic solvents, eg, acetone, benzene, ethyl alcohol, ethyl ether, gasoline, paraldehyde, toluene, xylenes, turpentine, and acetic acid.

Chemical Properties

Acetaldehyde is a highly reactive compound exhibiting the general reactivity of aldehydes (qv). Acetaldehyde undergoes numerous condensation, addition, and polymerization reactions; under suitable conditions, the oxygen or any of the hydrogens can be replaced.

Decomposition. Acetaldehyde decomposes at temperatures above 400°C, forming principally methane and carbon monoxide.

MANUFACTURE

Since 1960, the liquid-phase oxidation of ethylene has been the process of choice for the manufacture of acetaldehyde. There is, however, still some commercial production by the partial oxidation of ethyl alcohol and hydration of acetylene. The economics of the various processes are strongly dependent on the prices of the feedstocks. Acetaldehyde is also formed as a coproduct in the high temperature oxidation of butane. A more recently developed rhodium catalyzed process produces acetaldehyde from synthesis gas as a coproduct with ethyl alcohol and acetic acid. There are two variations for this commercial production: the two-stage process and the one-stage process. In the two-stage process ethylene is almost completely oxidized by air to acetaldehyde in one pass in a tubular plug-flow reactor made of titanium. Acetaldehyde produced in the first reactor is removed from the reaction loop by adiabatic flashing in a tower. The catalyst solution is recycled from the flash-tower base to the second stage (or oxidation reactor) where the cuprous salt is oxidized to the cupric state with air. In the one-stage process, ethylene, oxygen, and recycle gas are directed to a vertical reactor for contact with the catalyst solution under slight pressure. The gases are water-scrubbed and the resulting acetaldehyde solution is fed to a distillation column. Inert materials are eliminated from the recycle gas in a bleed-stream which flows to an auxiliary reactor for additional ethylene conversion.

HEALTH AND SAFETY FACTORS

Acetaldehyde appears to paralyze respiratory muscles, causing panic. It has a general narcotic action which prevents coughing, causes irritation of the eyes and mucous membranes, and accelerates heart action. When breathed in high concentration, it causes headache and sore throat. Carbon dioxide solutions in acetaldehyde are particularly pernicious because the acetaldehyde odor is weakened by the carbon dioxide. Prolonged exposure causes a decrease of both red and white blood cells; there is also a sustained rise in blood pressure. Mixtures of acetaldehyde vapor and air are flammable; they are explosive if the concentrations of aldehyde and oxygen rise above 4 and 9%, respectively. The threshold limit value (TLV) of acetaldehyde in air is 100 ppm.

USES

Acetaldehyde production is linked with the demand for acetic acid, acetic anhydride, cellulose acetate, vinyl acetate resins, acetate esters, pentaerythritol, synthetic pyridine derivatives, terephthalic acid, and peracetic acid. In 1996 acetic acid production represented 60% of the acetaldehyde demand. That demand has diminished as a result of the rising cost of ethylene as feedstock and methanol carbonylation as the preferred route to acetic acid (qv).

"Acetaldehyde," *Chemical Economics Handbook*, SRI International, Menlo Park, Calif., 2001.

R. Jira, in S. A. Miller, ed., *Ethylene and Its Industrial Derivatives*, Ernest Benn Ltd., London, 1969, pp. 639–553.

S. L. Levy and D. F. Othmer, *Ind. Eng. Chem.* **47**, 789 (1955).

R. L. Melnick, in E. Bingham, B. Cohrssen, and C. H. Powell, eds., *Patty's Toxicology*, 5th ed., John Wiley & Sons, Inc., New York, 2001.

H. J. HAGEMEYER
Texas Eastman Company

ACETIC ACID

Acetic acid, CH_3COOH, is a corrosive organic acid having a sharp odor, burning taste, and pernicious blistering properties. It is found in ocean water, oilfield brines, rain, and at trace concentrations in many plant and animal liquids. It is central to all biological energy pathways. Fermentation of fruit and vegetable juices yields 2–12% acetic acid solutions, usually called vinegar (qv). Any sugar-containing sap or juice can be transformed by bacterial or fungal processes to dilute acetic acid.

Most of the acetic acid is produced in the United States, Germany, Great Britain, Japan, France, Canada, and Mexico. Total annual production in these countries is close to four million tons. Uses include the manufacture of vinyl acetate and acetic anhydride. Vinyl acetate is used to make latex emulsion resins for paints, adhesives, paper coatings, and textile finishing agents. Acetic anhydride is used in making cellulose acetate fibers, cigarette filter tow, and cellulosic plastics.

PHYSICAL PROPERTIES

Acetic acid, fp 16.635°C, bp 117.87°C at 101.3 kPa, is a clear, colorless liquid. Water is the chief impurity in acetic acid although other materials such as acetaldehyde, acetic anhydride, formic acid, biacetyl, methyl acetate, ethyl acetoacetate, iron, and mercury are also sometimes found. Water significantly lowers the freezing point of glacial acetic acid as do acetic anhydride and methyl acetate. The presence of acetaldehyde or formic acid is commonly revealed by permanganate tests; biacetyl and iron are indicated by color.

The vapor density of acetic acid suggests a molecular weight much higher than the formula weight, 60.06. Indeed, the acid normally exists as a dimer, both in the vapor phase and in solution. This vapor density anomaly has important consequences in engineering computations, particularly in distillations.

Acetic acid containing <1% water is called glacial. A summary of the physical properties of glacial acetic acid is given in Table 1.

Table 1. Properties of Glacial Acetic Acid

Property	Value
freezing point, °C	16.635
boiling point, °C	117.87
density, g/mL at 20°C	1.0495
refractive index, n^{25}D	1.36965
heat of vaporization ΔH_v, J/g^a at bp	394.5
specific heat (vapor), J/(g·K)^a at 124°C	5.029
critical temperature, K	592.71
critical pressure, MPa^b	4.53
enthalpy of formation, kJ/mol^a at 25°C	
liquid	−484.50
gas	−432.25
normal entropy, J/(mol·K)^a at 25°C	
liquid	159.8
gas	282.5
liquid viscosity, mPa (=cP)	
20°C	11.83
40°C	8.18
surface tension, mN/m (=dyn/cm) at 20.1°C	27.57
flammability limits, vol % in air	4.0 to 16.0
autoignition temperature, °C	465
flash point, °C	
closed cup	43
open cup	57

^{a}To convert J to cal, divide by 4.184.
^{b}To convert MPa to psi, multiply by 145.

CHEMICAL PROPERTIES

Decomposition Reactions

Minute traces of acetic anhydride are formed when very dry acetic acid is distilled. Without a catalyst, equilibrium is reached after ~7 h of boiling, but a trace of acid catalyst produces equilibrium in 20 min. At equilibrium, ~4.2 mmol of anhydride is present per liter of acetic acid, even at temperatures as low as 80°C.

Acid–Base Chemistry

Acetic acid dissociates in water, $pK_a = 4 : 76$ at 25°C. It is a mild acid that can be used for analysis of bases too weak to detect in water. It readily neutralizes the ordinary hydroxides of the alkali metals and the alkaline earths to form the corresponding acetates. Other acids exhibit very powerful, superacid properties in acetic acid solutions and are thus useful catalysts for esterifications of olefins and alcohols.

Acetylation Reactions

Alcohols may be acetylated without catalysts by using a large excess of acetic acid.

$$CH_3COOH + ROH \longrightarrow CH_3COOR + H_2O$$

The reaction rate is increased by using an entraining agent such as hexane, benzene, toluene, or cyclohexane, depending on the reactant alcohol, to remove the water formed.

Nearly all commercial acetylations are realized using acid catalysts. Catalytic acetylation of alcohols can be carried out using mineral acids, eg, perchloric acid, phosphoric acid, sulfuric acid, benzenesulfonic acid, or methanesulfonic acid, as the catalyst.

MANUFACTURE

Commercial production of acetic acid was revolutionized in the decade 1978–1988. Most commercial production of virgin synthetic acetic acid is based on methanol carbonylation. By-product acetic acid recovery in other hydrocarbon oxidations, eg, in xylene oxidation to terephthalic acid and propylene conversion to acrylic acid, has also grown. Production from synthesis gas is increasing and the development of alternative raw materials is under serious consideration following widespread dislocations in the cost of raw material (see CHEMURGY).

Ethanol fermentation is still used in vinegar production.

Currently, almost all acetic acid produced commercially comes from acetaldehyde oxidation, methanol or methyl acetate carbonylation, or light hydrocarbon liquid-phase oxidation. Comparatively small amounts are generated by butane liquid-phase oxidation, direct ethanol oxidation, and synthesis gas.

Acetic acid has a place in organic processes comparable to sulfuric acid in the mineral chemical industries. About half of the world production comes from methanol carbonylation and about one-third from acetaldehyde oxidation. Another tenth of the world capacity can be attributed to butane–naphtha liquid-phase oxidation. Appreciable quantities of acetic acid are recovered from reactions involving peracetic acid.

HEALTH AND SAFETY FACTORS

Acetic acid has a sharp odor and the glacial acid has a fiery taste and will penetrate unbroken skin to make blisters. Prolonged exposure to air containing $5–10\,mg/m^3$ does not seem to be seriously harmful, but there are pronounced, undesirable effects from constant exposure to as high as $26\,mg/m^3$ over a 10-day period.

Care ought to be taken in handling acetic acid to avoid spillage or otherwise breathing vapors. Wash any exposed areas with large amounts of water. Once the odor of acetic acid vapors is noticeable, the area should be abandoned immediately.

Glacial acetic acid is dangerous, but its precise toxic dose is not known for humans. The LD_{50} for rats is said to be 3310 mg/kg, and for rabbits 1200 mg/kg. Ingestion of 80–90 g must be considered extraordinarily dangerous for humans. Vinegar, on the other hand, which is dilute acetic acid, has been used in foods and beverages since the most ancient of times.

"Acetic Acid", *Chemical News and Data, Chemical Profile*, http: www.chemexpo.com/news/Profile0102226.cfm, updated Feb. 26, 2001.

"Kohlenstoff," in K. von Baczko, ed., *Gmelins Handbuch der Anorganischen Chemie*, 8th ed., Teil C4, Frankfurt, 1975, 141–197.

R. J. Lewis, Sr., *Sax's Dangerous Properties of Industrial Materials*, 10th ed., John Wiley & Sons, Inc., New York, 2000, p. 15.

A. Popoff, J. J. Lagowski, ed., *Chemistry of Nonaqueous Solvents*, Vol. 3, Academic Press, New York, 1970.

FRANK S. WAGNER, JR.
Nandina Corporation

ACETIC ACID, HALOGENATED DERIVATIVES

The most important of the halogenated derivatives of acetic acid is chloroacetic acid. Fluorine, chlorine, bromine, and iodine derivatives are all known, as are mixed halogenated acids. For a discussion of the fluorine derivatives see FLUORINE COMPOUNDS, ORGANIC.

CHLOROACETIC ACID

Physical Properties

Pure chloroacetic acid ($ClCH_2COOH$), mol wt 94.50, $C_2H_3ClO_2$, is a colorless, white deliquescent solid. It has been isolated in three crystal modifications: α, mp 63°C,

β, mp 56.2°C, and γ, mp 52.5°C. Commercial chloroacetic acid consists of the α form.

Chemical Properties

Chloroacetic acid has wide applications as an industrial chemical intermediate. Both the carboxylic acid group and the α-chlorine are very reactive. It readily forms esters and amides, and can undergo a variety of α-chlorine substitutions. Electron withdrawing effects of the α-chlorine give chloroacetic acid a higher dissociation constant than that of acetic acid.

Manufacture

Most chloroacetic acid is produced by the chlorination of acetic acid using either a sulfur or phosphorus catalyst. The remainder is produced by the hydrolysis of trichloroethylene with sulfuric acid or by reaction of chloroacetyl chloride with water.

Health and Safety Aspects

Chloroacetic acid is extremely corrosive and will cause serious chemical burns. It also is readily absorbed through the skin in toxic amounts. Contamination of 5–10% of the skin area is usually fatal. The symptoms are often delayed for several hours. Single exposure to accidental spillage on the skin has caused human fatalities. The toxic mechanism appears to be blocking of metabolic cycles. Chloroacetic acid is 30–40 times more toxic than acetic, dichloroacetic, or trichloroacetic acid. When handling chloroacetic acid and its derivatives, rubber gloves, boots, and protective clothing must be worn. In case of skin exposure, the area should immediately be washed with large amounts of water and medical help should be obtained at once.

Uses

Major industrial uses for chloroacetic acid are in the manufacture of thioglycolic acid (29%), cellulose ethers (mainly carboxymethylcellulose,CMC) (24%) and herbicides (18%). Other industrial uses (29%) include manufacture of glycine, amphoteric surfactants, cyanoacetic acid, phenoxyacetic acid, and chloroacetic acid esters.

OTHER DERIVATIVES

Sodium Chloroacetate

Sodium chloroacetate, mol wt 116.5, $C_2H_2ClO_2Na$, is produced by reaction of chloroacetic acid with sodium hydroxide or sodium carbonate. In many applications, chloroacetic acid or the sodium salt can be used interchangeably.

Dichloroacetic Acid

Dichloroacetic acid ($Cl_2CHCOOH$), mol wt 128.94, $C_2H_2Cl_2O_2$, is a reactive intermediate in organic synthesis.

The liquid is totally miscible in water, ethyl alcohol, and ether. Dichloroacetic acid ($K_a = 5.14 \times 10^{-2}$) is a stronger acid than chloroacetic acid. Most chemical reactions are similar to those of chloroacetic acid, although both chlorine atoms are susceptible to reaction. Dichloroacetic acid is much more stable to hydrolysis than chloroacetic acid.

Trichloroacetic Acid

Trichloroacetic acid (Cl_3CCOOH), mol wt 163.39, $C_2HCl_3O_2$, forms white deliquescent crystals and has a characteristic odor. Trichloroacetic acid ($K_a = 0.2159$) is as strong an acid as hydrochloric acid. Esters and amides are readily formed. Trichloroacetic acid undergoes decarboxylation when heated with caustic or amines to yield chloroform. As with dichloroacetic acid, trichloroacetic acid can be converted to chloroacetic acid by the action of hydrogen and palladium on carbon.

Chloroacetyl Chloride

Chloroacetyl chloride ($ClCH_2COCl$) mol wt 112.94, $C_2H_2Cl_2O$, is the corresponding acid chloride of chloroacetic acid (see ACETYL CHLORIDE). Chloroacetyl chloride has a sharp, pungent, irritating odor. It is miscible with acetone and benzene and is initially insoluble in water. A slow reaction at the water–chloroactyl chloride interface, however, produces chloroacetic acid. When sufficient acid is formed to solubilize the two phases, a violent reaction forming chloroacetic acid and HCl occurs.

Since chloroacetyl chloride can react with water in the skin or eyes to form chloroacetic acid, its toxicity parallels that of the parent acid. Chloroacetyl chloride can be absorbed through the skin in lethal amounts.

Chloroacetate Esters

Two chloroacetate esters of industrial importance are methyl chloroacetate, $C_3H_5ClO_2$, and ethyl chloroacetate, $C_4H_7ClO_2$. Both esters have a sweet pungent odor and present a vapor inhalation hazard. They are rapidly absorbed through the skin and hydrolyzed to chloroacetic acid.

Bromoacetic Acid

Bromoacetic acid ($BrCH_2COOH$), mol wt 138.96, $C_2H_3BrO_2$, occurs as hexagonal or rhomboidal hygroscopic crystals. It is soluble in water, methanol, and ethyl ether. Bromoacetic acid undergoes many of the same reactions as chloroacetic acid under milder conditions, but is not often used because of its greater cost. Bromoacetic acid must be protected from air and moisture, since it is readily hydrolyzed to glycolic acid.

Dibromoacetic Acid

Dibromoacetic acid ($Br_2CHCOOH$), mol wt 217.8, $C_2H_2Br_2O_2$, is soluble in water and ethyl alcohol.

Tribromoacetic Acid

Tribromoacetic acid (Br_3CCOOH), mol wt 296.74, $C_2HBr_3O_2$, is soluble in water, ethyl alcohol, and diethyl ether. This acid is relatively unstable to hydrolytic conditions and can be decomposed to bromoform in boiling water.

Iodoacetic Acid

Iodoacetic acid (ICH_2COOH), mol wt 185.95, $C_2H_3IO_2$, is commercially available. The colorless, white crystals (mp 83°C) are unstable upon heating. Iodoacetic acid is soluble in hot water and alcohol, and slightly soluble in ethyl ether. Iodoacetic acid can be reduced with hydroiodic acid at 85°C to give acetic acid and iodine.

Diiodoacetic Acid

Diiodoacetic acid ($I_2CHCOOH$), mol wt 311.85, $C_2H_2I_2O_2$, occurs as white needles and is soluble in water, ethyl alcohol, and benzene.

Triiodoacetic Acid

Triiodoacetic acid (I_3CCOOH), mol wt 437.74, $C_2HO_2I_3$, is soluble in water, ethyl alcohol, and ethyl ether. Solutions of triiodoacetic acid are unstable as evidenced by the formation of iodine. Triiodoacetic acid decomposes when heated above room temperature to give iodine, iodoform, and carbon dioxide.

"Chloroacetic Acid," *ChemExpo, Chemical Profiles, http://63.23666.84.14/news/PROFILE00010101.cfm.*

Herbicide Handbook, 5th ed., Weed Science Society of America, Champaign, Ill., 1983, p. 128.

M. Sittig, ed., *Pesticide Manufacturing and Toxic Materials Control Encyclopedia*, Noyes Data Corp., Park Ridge, N.J., 1980.

Threshold Limit Values and Biological Exposure Indicies, 5th ed., American Conference of Government Industrial Hygienists, Cincinnati, Ohio, 1986, p. 122.

EARL D. MORRIS
JOHN C. BOST
The Dow Chemical Company

ACETIC ANHYDRIDE

Acetic anhydride, $(CH_3CO)_2O$, is a mobile, colorless liquid that has an acrid odor and is a more piercing lacrimator than acetic acid. It is the largest commercially produced carboxylic acid anhydride: U.S. production capacity is over 1×10^6 t yearly. Its chief industrial application is for acetylation reactions; it is also used in many other applications in organic synthesis, and it has some utility as a solvent in chemical analysis.

PHYSICAL AND CHEMICAL PROPERTIES

No dimerization of acetic anhydride has been observed in either the liquid or solid state. Decomposition, accelerated by heat and catalysts such as mineral acids, leads slowly to acetic acid. Acetic anhydride is soluble

Table 1. Physical Properties of Acetic Anhydride

Property	Value
freezing point, °C	−73.13
boiling point, °C at 101.3 kPa[a]	139.5
density, d_4^{20}, g/cm^3	1.0820
refractive index, n_D^{20}	1.39038
vapor pressure (Antoine equation), P in kPa[a] and T in K	$\ln(P) = \dfrac{14.6497 - 3467.76}{T - 67.0}$
heat of vaporization, ΔH_v, at bp, J/g[b]	406.6
specific heat, J/kg[c] at 20°C	1817
surface tension, mN/m (= dyn/cm)	
25°C	32.16
40°C	30.20
viscosity, mPa·s (= cP)	
15°C	0.971
30°C	0.783
heat conductivity, mW/(m·K)[d] at 30°C	136
electric conductivity, S/cm	2.3×10^{-8}

[a] To convert kPa to mm Hg, multiply by 7.5.
[b] To convert J/g to Btu/lb, multiply by 0.4302.
[c] To convert J to cal divide by 4.184.
[d] To convert mW/(m·K) to (Btu·ft)/(h·ft^2·°F), multiply by 578.

in many common solvents, including cold water. Although aqueous solutions are initially neutral to litmus, they show acid properties once hydrolysis appreciably progresses. Acetic anhydride ionizes to acetylium, CH_3CO^+, and acetate, $CH_3CO_2^-$, ions in the presence of salts or acids. Acetate ions promote anhydride hydrolysis. A summary of acetic anhydride's physical properties is given in Table 1.

Acetic anhydride acetylates free hydroxyl groups without a catalyst, but esterification is smoother and more complete in the presence of acids. Acetic anhydride can be used to synthesize methyl ketones in Friedel-Crafts reactions.

MANUFACTURE

The Acetic Acid Process

Prior to the energy crisis of the 1970s, acetic anhydride was manufactured by thermal decomposition of acetic acid at pressures of 15–20 kPa.

The Acetaldehyde Oxidation Process

Liquid-phase catalytic oxidation of acetaldehyde (qv) can be directed by appropriate catalysts, such as transition metal salts of cobalt or manganese, to produce anhydride. Either ethyl acetate or acetic acid may be used as reaction solvent. Acetaldehyde oxidation generates peroxyacetic acid, which then reacts with more acetaldehyde to yield acetaldehyde monoperoxyacetate. Subsequently, parallel reactions lead to formation of acetic acid and anhydride plus water.

Methyl Acetate Carbonylation

Anhydride can be made by carbonylation of methyl acetate in a manner analogous to methanol carbonylation to acetic acid. Methanol acetylation is an essential first step in anhydride manufacture by carbonylation. Surprisingly, there is limited nonproprietary experimental data on methanol esterification with acetic acid.

Prospective Routes to Acetic Anhydride

Methyl acetate–dimethyl ether carbonylation seems to be the leading new route to acetic anhydride production. The high energy costs of older routes proceeding through ketene preclude their reintroduction. Thermolysis of acetone, methyl acetate, and ethylidene diacetate suffers from the same costly energy consumption. Acetic acid cracking, under vacuum and at atmospheric pressure, continues to be used for anhydride manufacture in spite of the clear obsolescence of the processes.

Health and Safety Aspects

Acetic anhydride penetrates the skin quickly and painfully forming burns and blisters that are slow to heal. Anhydride is especially dangerous to the delicate tissues of the eyes, ears, nose, and mouth. The odor threshold is 0.49 mg/m^3, but the eyes are affected by as little as 0.36 mg/m^3 and electroencephalogram patterns are altered by only 0.18 mg/m^3. When handling acetic anhydride, rubber gloves that are free of pinholes are recommended for the hands, as well as plastic goggles for the eyes, and facemasks to cover the face and ears.

Acetic anhydride is dangerous in combination with various oxidizing substances and strong acids. Thermal decomposition of nitric acid in acetic acid solutions is accelerated by the presence of anhydride. Acetyl perchlorate is probably present in such solutions. These perchloric acid solutions are useful in metal finishing, but the risks in using them must be recognized.

Uses

The biggest use of acetic anhydride is in the preparation of cellulose acetates (86%), while the remaining 14% is consumed in miscellaneous uses including coatings, pesticides, aspirin, and acetaminophen.

Acetic anhydride is used in acetylation processes. There has been some diversification of anhydride usage in recent years. Acetic anhydride is used to acetylate various fragrance alcohols to transform them into esters having much higher unit value and vitamins are metabolically enhanced by acetylation. Anhydride is also extensively employed in metallography, etching, and polishing of metals, and in semiconductor manufacture. Starch acetylation furnishes textile sizing agents.

Acetic anhydride is a useful solvent in certain nitrations, acetylation of amines and organosulfur compounds for rubber processing, and in pesticides. Although acetic acid is unexceptional as a fungicide, small percentages of anhydride in acetic acid, or in cold water solutions are powerful fungicides and bactericides.

G. V. Jeffreys, *The Manufacture of Acetic Anhydride*, 2nd ed., The Institution of Chemical Engineers, London, 1964.

"Acetic Anhydride Design Problem" in J. J. McKetta, and W. A. Cunningham, eds., *Encyclopedia of Chemical Processing and Design*, Marcel Dekker, Inc., New York, Vol. 1, p. 271.

"Acetic Anhydride," Chemical Profiles, *http://www.chemexpo. com/news/PROFILE 010219.cfm*, revised, Feb. 19, 2001.

K. Boyes, in E. Bingham, B. Cohrssen, and C. H. Powell, eds., *Patty's Toxicology*, 5th ed., Vol. 2, John Wiley & Sons, Inc., New York, 2001, Chapt. 25.

FRANK S. WAGNER, JR.
Nandina Corporation

ACETONE

Acetone (2-propanone, dimethyl ketone, CH_3COCH_3), molecular weight 58.08 (C_3H_6O), is the simplest and most important of the ketones. It is a colorless, mobile, flammable liquid with a mildly pungent, somewhat aromatic odor, and is miscible in all proportions with water and most organic solvents. Acetone is an excellent solvent for a wide range of gums, waxes, resins, fats, greases, oils, dyestuffs, and cellulosics. It is used as a carrier for acetylene, in the manufacture of a variety of coatings and plastics, and as a raw material for the chemical synthesis of a wide range of products such as ketene, methyl methacrylate, bisphenol A, diacetone alcohol, methyl isobutyl ketone, hexylene glycol (2-methyl-2,4-pentanediol), and isophorone. Most of the world's manufactured acetone is obtained as a coproduct in the process for phenol from cumene and most of the remainder from the dehydrogenation of isopropyl alcohol. Numerous natural sources of acetone make it a normal constituent of the environment. It is readily biodegradable.

PROPERTIES

Selected physical properties are given in Table 1 and some thermodynamic properties in Table 2.

As for chemical properties, the closed cup flash point of acetone is −18°C and open cup is −9°C. The autoignition temperature is 538°C, and the flammability limits are 2.6−12.8 vol% in air at 25°C. Acetone shows the typical reactions of saturated aliphatic ketones.

MANUFACTURE AND SHIPMENT

Production of acetone by dehydrogenation of isopropyl alcohol began in the early 1920s and remained the dominant production method through the 1960s. In the mid-1960s, virtually all United States acetone was produced from propylene. However, by the mid-1970s 60% of United States acetone capacity was based on cumene hydroperoxide, which accounted for ~65% of the acetone produced.

Most of the world's acetone is now obtained as a coproduct of phenol by the cumene process. More then

90% of U.S. acetone is produced by this process. Cumene is oxidized to the hydroperoxide and cleaved to acetone and phenol. Dehydrogenation of isopropyl alcohol accounts for most of the acetone production not obtained from cumene. Minor amounts of acetone are made by other processes.

Acetone is produced in large quantities and is usually shipped by producers to consumers and distributors in drums and larger containers. Distributors repackage the acetone into containers ranging in size from small bottles to drums or even tank trucks. Specialty processors make available various grades and forms of acetone such as high purity, specially analyzed, analytical reagent grade, chromatography and spectrophotometric grades, and isotopically labeled forms, and ship them in ampoules, vials,

Table 1. Physical Properties

Property	Value
melting point, °C	−94.6
boiling point at 101.3 kPa[a], °C	56.29
refractive index, n_D	
at 20°C	1.3588
at 25°C	1.35596
electrical conductivity at 298.15 K, S/cm	5.5×10^{-8}
critical temperature, °C	235.05
critical pressure, kPa[a]	4701
critical volume, L/mol	0.209
critical compressibility	0.233
triple point temperature, °C	−94.7
triple point pressure, Pa[a]	2.59375
acentric factor	0.306416
solubility parameter at 298.15 K, $(J/m^3)^{1/2}$ [b]	19773.5
dipole moment, C·m[c]	9.61×10^{-30}
molar volume at 298.15 K, L/mol	0.0739
molar density, mol/L	
solid at −99°C	16.677
liquid at 298.15 K	13.506

[a]To convert kPa to mmHg, multiply by 7.501.
[b]To convert $(J/m^3)^{1/2}$ to $(cal/m^3)^{1/2}$, divide by 2.045.
[c]To convert C·m to debyes, divide by 3.336×10^{-30}.

Table 2. Thermodynamic Properties[a,b]

Property	Value
specific heat of liquid at 20°C, J/g	2.6
specific heat of vapor at 102°C, J/(mol·K)	92.1
heat of vaporization at 56.1°C, kJ/mol	29.1
enthalpy of vaporization, kJ/mol	30.836
enthalpy of fusion at melting point, J/mol	5691.22
heat of combustion of liquid, kJ/mol	1787
enthalpy of combustion, kJ/mol	−1659.17
entropy of liquid, J/(mol·K)	200.1
entropy of ideal gas, J/(mol·K)	295.349
Gibbs energy of formation, kJ/mol	−152.716
enthalpy of formation, kJ/mol	
ideal gas	−217.15
gas	−216.5
liquid	−248

[a]At 298.15 K unless otherwise noted.
[b]To convert J to cal, divide by 4.184.

bottles, or other containers convenient for the buyers. Barges and ships are usually steel, but may have special inner or deck-mounted tanks. Increasing in use, especially for international shipments, are intermodal (IM) portable containers, tanks suspended in frameworks suitable for interchanging among truck, rail, and ship modes of transportation.

STANDARDS AND QUALITY CONTROL

Higher or lower quality at more or less cost will meet the needs of some consumers. Acetone is often produced under contract to meet customer specifications that are different from those of ASTM D329. Specification tests are performed on plant streams once or twice per worker shift, or even more often if necessary, to assure the continuing quality of the product. The tests are also performed on a sample from an outgoing shipment, and a sample of the shipment is usually retained for checking on possible subsequent contamination. Tests on specialty types of acetone may require sophisticated instruments, eg, mass spectrometry for isotopically labeled acetone.

HEALTH AND SAFETY FACTORS

Acetone is among the solvents of comparatively low acute and chronic toxicity. High vapor concentrations produce anesthesia, and such levels may be irritating to the eyes, nose, and throat, and the odor may be disagreeable. Acetone does not have sufficient warning properties to prevent repeated exposures to concentrations that may cause adverse effects.

Acetone can be handled safely if common sense precautions are taken. It should be used in a well-ventilated area, and because of its low flash point, ignition sources should be absent. Flame will travel from an ignition source along vapor flows on floors or bench tops to the point of use. Sinks should be rinsed with water while acetone is being used to clean glassware, to prevent the accumulation of vapors. If prolonged or repeated skin contact with acetone could occur, impermeable protective equipment such as gloves and aprons should be worn.

Compatibility of acetone with other materials should be carefully considered, especially in disposal of wastes. It reacts with chlorinating substances to form toxic chloroketones, and potentially explosively with some peroxy compounds and a number of oxidizing mixtures. Mixed with chloroform, acetone will react violently in the presence of bases. Other incompatibilities are listed in the Sax handbook.

USES

Acetone is used as a solvent and as a reaction intermediate for the production of other compounds that are mainly used as solvents and/or intermediates for consumer products: direct solvent use, acrylics, bisphenol A, and aldol chemicals.

Also more than 70 thousand metric tons of acetone are used in small-volume applications, some of which are to make functional compounds such as antioxidants, herbicides, higher ketones, condensates with formaldehyde or diphenylamine, and vitamin intermediates.

"Acetone, Chemical Profile," Chem Expo, revised April 5, 1999, www.chemexpo.com, searched Nov. 29, 2001.

D. A. Morgott, "Acetone" in E. Bingham, B. Cohrssen, and C. H. Powell, eds., *Patty's Toxicology*, 5th ed., Vol. 6, Wiley-Interscience, New York, 2001, pp. 1–116.

J. A. Riddick, W. B. Bunger, and T. K. Sakano, "Organic Solvents, Physical Properties, and Methods of Purification," in *Techniques of Organic Chemistry*, Vol. 2, John Wiley & Sons, Inc., New York, 1986.

R. C. Weast and J. G. Grasselli, eds., *Handbook of Data on Organic Compounds*, 2nd ed., Vol. 6, CRC Press, Inc., Boca Raton, Fla., Compound No. 21433, p. 3731.

WILLIAM L. HOWARD
The Dow Chemical Company

ACETYLENE

Acetylene, C_2H_2, is a highly reactive, commercially important hydrocarbon. It is used in metalworking (cutting and welding) and in chemical manufacture. Chemical usage has been shrinking due to the development of alternative routes to the same products based on cheaper raw materials. The reactivity of acetylene is related to its triple bond between carbon atoms and, as a consequence, its high positive free energy of formation. Because of its explosive nature, acetylene is generally used as it is produced without shipping or storage.

PHYSICAL AND CHEMICAL PROPERTIES

The physical properties of acetylene are: the triple point is at $-80.55°C$ and 128 kPa (1.26 atm); the temperature of the solid under its vapor at 101 kPa (1 atm) is $-83.8°C$; the vapor pressure of the liquid at 20°C is 4406 kPa (43.5 atm); the critical temperature and pressure are 35.2°C and 6190 kPa (61.1 atm); the density of the gas at 20°C and 101 kPa is 1.0896 g/L; the specific heats of the gas, C_p and C_v (at 20°C and 101 kPa) are 43.91 and 35.45 J/mol·°C (10.49 and 8.47 cal/mol, °C), respectively; and the heat of formation ΔH_f at 0°C is 227.1 kJ/mol (54.3 kcal/mol). The dissolving powers of some of the better solvents are compared in Table 1.

Acetylene is highly reactive due to its triple bond and high positive free energy of formation. Important reactions involving acetylene are hydrogen replacements, additions to the triple bond, and additions by acetylene to other unsaturated systems. Moreover, acetylene undergoes polymerization and cyclization reactions. The formation of a metal acetylide is an example of hydrogen replacement, and hydrogenation, halogenation, hydrohalogenation,

Table 1. Solubility of Acetylene in Some Organic Liquids

Solvent	CAS Registry Number	bp, °C	Acetylene solubility[a]
acetone	[67-64-1]	56.5	237
acetonitrile	[75-05-8]	81.6	238
N,N-dimethylformamide (DMF)	[68-12-2]	153	278
dimethyl sulfoxide	[67-68-5]	189	269
N-methyl-2-pyrroli-dinone	[872-50-4]	202	213
γ-butyrolactone[b]	[96-48-0]	206	203

[a] g/L of solution at 15°C and 1520 kPa (15.0 atm) total pressure.
[b] Butanoic acid, 4-hydroxy-, lactone.

hydration, and vinylation are important addition reactions. In the ethynylation reaction, acetylene adds to a carbonyl group (see ACETYLENE-DERIVED CHEMICALS).

Many of the reactions in which acetylene participates, as well as many properties of acetylene, can be understood in terms of the structure and bonding of acetylene. Acetylene is a linear molecule in which two of the atomic orbitals on the carbon are *sp* hybridized and two are involved in π bonds. The lengths and energies of the C—H σ bonds and C≡Cσ + 2π bonds are as follows:

Bond	Bond length, nm	Energy, kJ/mol
≡C—H	0.1059	506
—C≡C—	0.1205	837

Metal Acetylides

The replacement of a hydrogen atom on acetylene by a metal atom under basic conditions results in the formation of metal acetylides that react with water in a highly exothermic manner to yield acetylene and the corresponding metal hydroxide. Certain metal acetylides can be prepared by reaction of the finely divided metal with acetylene in inert solvents such as xylene, dioxane, or tetrahydrofuran (THF) at temperatures of 38–45°C.

Acetylides of the alkali and alkaline-earth metals are formed by reaction of acetylene with the metal amide in anhydrous liquid ammonia:

$$C_2H_2 + MNH_2 \longrightarrow MC\equiv CH + NH_3$$

Aluminum triacetylide is formed from $AlCl_3$ and sodium acetylide in a mixture of dioxane and ethylbenzene at 70–75°C:

$$3\ NaC\equiv CH + AlCl_3 \longrightarrow Al(C\equiv CH)_3 + 3\ NaCl$$

Copper acetylides form under a variety of conditions. Cuprous acetylides are generally explosive, but their explosiveness is a function of the formation conditions and increases with the acidity of the starting cuprous solution. They are prepared by the reaction of cuprous salts with acetylene in liquid ammonia or by the reaction of cupric salts with acetylene in basic solution in the presence of a reducing agent such as hydroxylamine. Acetylides also form from copper oxides and salts produced by

exposing copper to air, moisture, and acidic or basic conditions. For this reason, copper or brasses containing >66% copper or brazing materials containing silver or copper should not be used in an acetylene system. Silver and mercury form acetylides in a manner similar to copper.

Acetylene Grignard reagents, which are useful for further synthesis, are formed by the reaction of acetylene with an alkylmagnesium bromide.

$$C_2H_2 + 2\ RMgBr \longrightarrow BrMgC\equiv CMgBr + 2\ RH$$

With care, the monosubstituted Grignard reagent can be formed and it reacts with aldehydes and ketones to produce carbinols (see GRIGNARD REACTIONS).

Hydrogenation

Acetylene can be hydrogenated to ethylene and ethane. The reduction of acetylene occurs in an ammoniacal solution of chromous chloride or in a solution of chromous salts in H_2SO_4. The selective catalytic hydrogenation of acetylene to ethylene, which proceeds over supported Group 8–10% (VIII) metal catalysts, is of great industrial importance in the manufacture of ethylene by thermal pyrolysis of hydrocarbons. Nickel and palladium are the most commonly used catalysts. Partial hydrogenation to ethylene is possible because acetylene is adsorbed on the catalyst in preference to ethylene.

Halogenation and Hydrohalogenation

Halogens add to the triple bond of acetylene. $FeCl_3$ catalyzes the addition of Cl_2 to acetylene to form 1,1,2,2-tetrachloroethane, which is an intermediate in the production of the industrial solvents 1,2-dichloroethylene, trichloroethylene, and perchloroethylene (see CHLOROCARBONS AND CHLOROHYDROCARBONS). Acetylene can be chlorinated to 1,2-dichloroethylene directly using $FeCl_3$ as a catalyst and a large excess of acetylene. The compound *trans*-$C_2H_2Cl_2$ is formed from acetylene in solutions of $CuCl_2$, CuCl, and HCl (24–26). Bromine in solution or as a liquid adds to acetylene to form first 1,2-dibromoethylene and finally tetrabromoethylene. Iodine adds less readily and the reaction stops at 1,2-diiodoethylene. Hydrogen halides react with acetylene to form the corresponding vinyl halides. An example is the formation of vinyl chloride that is catalyzed by mercuric salts.

Hydration

Water adds to the triple bond to yield acetaldehyde via the formation of the unstable enol (see ACETALDEHYDE). The reaction has been carried out on a commercial scale using a solution process with $HgSO_4/H_2SO_4$ catalyst. The vapor-phase reaction has been reported at 250–400°C using a wide variety of catalysts and even with no catalyst.

Addition of Hydrogen Cyanide

At one time the predominant commercial route to acrylonitrile was the addition of hydrogen cyanide to acetylene. This route has been completely replaced by the ammoxidation of propylene (SOHIO process) (see ACRYLONITRILE).

Vinylation

Acetylene adds weak acids across the triple bond to give a wide variety of vinyl derivatives. Alcohols or phenols give vinyl ethers and carboxylic acids yield vinyl esters (see VINYL POLYMERS). Vinyl ethers are prepared in a solution process at 150–200°C with alkali metal hydroxide catalysts, although a vapor-phase process has been reported. A wide variety of vinyl ethers are produced commercially. Vinyl acetate has been manufactured from acetic acid and acetylene in a vapor-phase process using zinc acetate catalyst, but ethylene is the currently preferred raw material. Vinyl derivatives of amines, amides, and mercaptans can be made similarly.

Ethynylation

Base-catalyzed addition of acetylene to carbonyl compounds to form -yn-ols and -yn-glycols (see ACETYLENE-DERIVED CHEMICALS) is a general and versatile reaction for the production of many commercially useful products.

Polymerization and Cyclization

Acetylene polymerizes at elevated temperatures and pressures that do not exceed the explosive decomposition point. Beyond this point, acetylene explosively decomposes to carbon and hydrogen. At 600–700°C and atmospheric pressure, benzene and other aromatics are formed from acetylene on heavy-metal catalysts.

Cuprous salts catalyze the oligomerization of acetylene to vinylacetylene and divinylacetylene. The former compound is the raw material for the production of chloroprene monomer and polymers derived from it. Nickel catalysts with the appropriate ligands smoothly convert acetylene to benzene or 1,3,5,7-cyclooctatetraene. Polymer formation accompanies these transition-metal catalyzed syntheses.

EXPLOSIVE BEHAVIOR

Gaseous Acetylene

Commercially pure acetylene can decompose explosively (principally into carbon and hydrogen) under certain conditions of pressure and container size. It can be ignited, ie, a self-propagating decomposition flame can be established, by contact with a hot body, by an electrostatic spark, or by compression (shock) heating. Ignition is generally more likely the higher the pressure and the larger the cross-section of the container.

When the wall of the container is heated, ignition occurs at a temperature that depends on the material of the wall and the composition of any foreign particles that may be present. In clean steel pipe, acetylene at 235–2530 kPa (2.3–25 atm) ignites at 425–450°C. In rusted steel pipe, acetylene at 100–300 kPa ignites at 370°C. In steel pipe containing particles of rust, charcoal, alumina, or silica, acetylene at 200–2500 kPa ignites at 280–300°C. Copper oxide causes ignition at 250°C and solid potassium hydroxide causes ignition at 170°C.

The predetonation distance (the distance the decomposition flame travels before it becomes a detonation) depends primarily on the pressure and pipe diameter when acetylene in a long pipe is ignited by a thermal, nonshock source.

The pressure developed by decomposition of acetylene in a closed container depends not only on the initial pressure (or more precisely, density), but also on whether the flame propagates as a deflagration or a detonation, and on the length of the container.

Flame Arresters. Propagation of a decomposition flame through acetylene in a piping system (by either deflagration or detonation) can be stopped by a hydraulic back pressure valve in which the acetylene is bubbled through water. It can also be stopped by filling the pipe with parallel tubes of smaller diameter, or randomly oriented Raschig rings. The presence of water or oil (on the walls or as mist) increases the effectiveness of the arrangement. Beds of granular ceramic material are effective with acetylene at cylinder pressure.

Ignition of Gaseous Acetylene Mixtures. Air in concentrations of less than ~13% inhibits ignition, but oxygen in any concentration promotes it. The data were obtained with relatively small containers and low ignition energies. With larger containers and higher ignition energies, the minimum pressure for ignition may be somewhat lower.

Acetylene—Air and Acetylene—Oxygen Mixtures

In acetylene–air mixtures, the normal mode of burning is deflagration in relatively short containers and detonation in pipes. In oxygen mixtures, detonation easily develops in both short and long containers.

Liquid and Solid Acetylene

Both the liquid and the solid have the properties of a high explosive when initiated by detonators or by detonation of adjoining gaseous acetylene. At temperatures near the freezing point neither form is easily made to explode by heat, impact, or friction, but initiation becomes easier as the temperature of the liquid is raised. Violent explosions result from exposure to mild thermal sources at temperatures approaching room temperature.

The minimum ignition energy of liquid acetylene under its vapor, when subjected to electrostatic sparks, has been found to depend on the temperature as indicated in Table 2. Ignition appears to start in gas bubbles within the liquid.

Table 2. Minimum Ignition Energy of Liquid Acetylene

temperature, °C	−78	−50	−40	−35	−30	−27
vapor pressure, kPa[a]	145	537	779	931	1103	1234
minimum ignition energy, J[b]	>11	1.5	0.98–4.1	0.98	0.68	0.13

[a]To convert kPa to atm, divide by 101.3.
[b]To convert J to cal, divide by 4.184.

MANUFACTURE FROM HYDROCARBONS

Although acetylene production in Japan and Eastern Europe is still based on the calcium carbide process, the large producers in the United States and Western Europe now rely on hydrocarbons as the feedstock. Now >80% of the acetylene produced in the United States and Western Europe is derived from hydrocarbons, mainly natural gas or as a coproduct in the production of ethylene. In Russia ~40% of the acetylene produced is from natural gas.

The hydrocarbon-to-acetylene processes that have been developed to commercial or pilot-plant scale must recognize and take advantage of the unique thermodynamic properties of acetylene. The common paraffinic and olefinic hydrocarbons are more stable than acetylene at ordinary temperatures. As the temperature is increased, the free energy of the paraffins and olefins become positive, while that of the acetylene decreases, until at >1400 K acetylene is the most stable of the common hydrocarbons. However, it is also evident that, although it has the lowest free energy of the hydrocarbons at high temperature, it is still unstable in relation to its elements C and H_2. Thus it is necessary to heat the feedstock extremely fast to minimize its decomposition to its elements and, for a similar reason, the quench must be extremely rapid to avoid the decomposition of the acetylene product.

Examination of the equilibrium composition of the product gas mixture under relevant reactor conditions indicates the restrictive process conditions required to optimize the production process.

Addressing the quench step of the reaction, it is most important to quench the equilibrium mixture as quickly as possible in order to preserve the high acetylene concentration. The effect of various gases injected into the stream not only indicated the effectiveness of the quenching medium, but also revealed a great deal about the dynamics of the high temperature equilibrium composition. Hydrogen is much more effective in preserving the acetylene than the inert gases argon, helium, or nitrogen. Thus hydrogen injection allows recovery of as much as 90% of the acetylene, whereas with the other gases <50% of the acetylene was recovered.

Process Technology

The processes designed to produce acetylene as the main product of a hydrocarbon feedstock are generally classified according to their energy source, ie, electricity or combustion. Two of these special cases are the production of acetylene by steam hydrocracking in oil refineries, and the potentially commercial process of producing acetylene from coal.

Hydrocracking is accomplished by electric discharge processes, the Hüls arc process, the Hoechst WLP process, the Hüls plasma process, the flame or partial combustion processes, the BASF process, the SBA process, the Montecatini process, the Hoechst HTP Process, and the BASF Submerged-Flame Process.

In summary, the bad features of partial combustion processes are the cost of oxygen and the dilution of the cracked gases with combustion products. Flame stability is always a potential problem. These features are more than offset by the inherent simplicity of the operation, which is the reason that partial combustion is the predominant process for manufacturing acetylene from hydrocarbons.

In addition, there are three more cracking processes: the Regenerative Furnace Processes, the Wulff Process, and the pyrolysis by direct firing.

The pyrolysis of methane results in a cracked gas that is relatively low in acetylene content and that contains predominantly a mixture of hydrogen, nitrogen, carbon monoxide, carbon dioxide, unreacted hydrocarbons, acetylene, and higher homologues of acetylene. In cases where a higher hydrocarbon than methane is used as feedstock, the converter effluent also contains olefins (ethylene, propylene, propadiene, butadiene), aromatics (benzene, naphthalene), and miscellaneous higher hydrocarbons. Most acetylene processes produce significant amounts of carbon black and tars that have to be removed before the separation of acetylene from the gas mixture.

The isolation of the acetylene from the various converters presents a complicated problem. The unstable, explosive nature of acetylene imposes certain restrictions on the use of the efficient separation techniques developed for other hydrocarbon systems. The results of decomposition and detonation studies on acetylene and its mixtures with other gases indicate that operating conditions where the partial pressure of acetylene exceeds 103–207 kPa (15–30 psi) should be avoided. Similar limitations apply to the operating temperatures that should not exceed 95–105°C. Low temperatures may lead to the appearance of liquid or solid acetylene or its homologues with concomitant danger of unexpected decompositions. In view of these severe operating restrictions, it is not surprising that all commercial processes for the recovery of hydrocarbon-derived acetylene are based on absorption–desorption techniques using one or more selective solvents.

The gases leaving the purification system are scrubbed with water to recover solvent and a continuous small purge of solvent gets rid of polymers. The acetylene purity resulting from this system is 99%. The main impurities in the acetylene are carbon dioxide, propadiene, and a very small amount of vinylacetylene.

MANUFACTURE FROM CALCIUM CARBIDE

Acetylene is generated by the chemical reaction between calcium carbide and water with the release of 134 kJ/mol (900 Btu/lb of pure calcium carbide).

$$CaC_2 + 2 H_2O \longrightarrow Ca(OH)_2 + C_2H_2$$

Because of the exothermic reaction and the evolution of gas, the most important safety considerations in the design of acetylene generators are the avoidance of excessively high temperatures and high pressures. The heat of reaction must be dissipated rapidly and efficiently in order to avoid local overheating of the calcium carbide which, in the absence of sufficient water, may become incandescent and cause progressive decomposition of the acetylene and the development of explosive pressures. Maintaining temperatures <150°C also minimizes polymerization of acetylene and other side reactions that may form undesirable contaminants. For protection against high pressures, industrial acetylene generators are equipped with pressure relief devices which do not allow the pressure to exceed 204.7 kPa (15 psig). This pressure is commonly accepted as a safe upper limit for operating the generator.

Most carbide acetylene processes are wet processes from which hydrated lime, $Ca(OH)_2$, is a by-product. The hydrated lime slurry is allowed to settle in a pond or tank after which the supernatant lime-water can be decanted and reused in the generator. Federal, state, and local legislation restrict the methods of storage and disposal of carbide lime hydrate and it has become increasingly important to find consumers for the by-product.

Carbide-to-Water Generation

This process is the one most widely used in the United States for generating acetylene from calcium carbide. Standards for the design and construction of acetylene-generating equipment using this technique have been developed over the years by the acetylene industry. Underwriters Laboratories, Inc. have generally accepted design criteria for acetylene generating equipment. A water capacity of 3.78 L (1 gal) per 0.454 kg of carbide and a gas-generating rate of 0.028 m^3 (1 ft^3) per hour per 0.454 kg of carbide hopper capacity is considered normal. These design criteria apply to gravity feed generators where it is possible to have an uncontrolled release of the entire carbide hopper contents into the generating chamber. Other high capacity generators (up to 283 m^3/h) are designed so that it is impossible to have uncontrolled feed of carbide (screw-feed type); therefore, the chamber water capacity can be reduced, the carbide hopper capacity can be increased, and the gas production capacity can be raised. These high capacity generators also must pass prescribed safety tests before sale.

There are two classes of acetylene generators: the low pressure generator which operates <108.2 kPa (15.7 psi), and the medium pressure generator which operates between 108.2 and 204.7 kPa (29.7 psi). The latter is more prevalent in the United States.

There are numerous variations in the design of commercially available carbide-to-water acetylene generators. Basically, however, they are practically identical in that they consist of a water vessel or reaction chamber, a carbide feed mechanism, and a carbide storage container that empties into the feed mechanism. The water vessel is equipped with a means of filling with water and draining the lime slurry.

Water-to-Carbide Generation

This method of acetylene production has found only limited acceptance in the United States and Canada but has been used frequently in Europe for small-scale generation. The rate of generation is regulated by the rate of water flow to the carbide. Hazardous hot spots may occur and overheating may lead to the formation of undesirable polymer by-products. This method is, therefore, used mainly in small acetylene generators such as portable lights or lamps where the generation rate is slow and the mass of carbide is small.

Dry Generator

This water-to-carbide acetylene generation method is used in certain large-scale operations. Absolute control of the addition of water is critical and the reacting mass of dry lime and unreacted carbide must be continuously mixed to prevent hazardous localized overheating and formation of undesirable polymer by-products. The gas stream is filtered to remove lime dust. The dry lime by-product is considered to be advantageous compared to the wet lime by-product.

Purification of Carbide Acetylene

The purity of carbide acetylene depends largely on the quality of carbide employed and, to a much lesser degree, on the type of generator and its operation. Carbide quality in turn is affected by the impurities in the raw materials used in carbide production, specifically, the purity of the metallurgical coke and the limestone from which the lime is produced.

The most commonly used dry methods employ oxidizing agents such as chromic acid or chromates, hypochlorite, permanganate, and ferric salts deposited on solid carriers such as diatomaceous earth arranged in beds or layers through which the gas is passed at ambient temperature. Some of the purifying media can be regenerated several times with diminishing effectiveness until they eventually lose their activity. Because of the high material and labor requirements, dry purification of acetylene is not practiced where large volumes of gas have to be treated. Large-scale acetylene installations exclusively employ continuous, wet purification processes. Such an intensive purification of acetylene is beneficial in cases where the gas is to be used in processes employing sensitive catalytic systems.

OTHER PROCESSES

To other processes that yield acetelene are coproduct acetylene from steam cracking, in which the steam cracking of petroleum liquids to produce olefine, mainly ethylene, small concentrations of acetylene are produced, and acetylene from coal, in which acetylene traditionally has been made from coal (coke) via the calcium carbide process. However, laboratory and bench-scale experiments have demonstrated the technical feasibility of producing the acetylene by the direct pyrolysis of coal.

SHIPMENT AND HANDLING

The design of equipment for the handling and use of acetylene must take into consideration the possibility of acetylene decompositions. The design parameters must consider various factors, namely, pressure, temperature, source of ignitions, and ultimate pressures which may result from a decomposition. Decompositions do not occur spontaneously but must have a source of ignition.

Exact design criteria on equipment for handling acetylene is not readily available because of the great number of factors involved. However, recommendations have been made concerning the equipment, piping, compressors, flash arresters, and proper materials.

Acetylene cylinders are constructed to stabilize acetylene and, thereby, safely avoid the hazard of a detonation. Cylinders constructed for other gases do not have the same features and it is extremely important that such cylinders not be charged with acetylene. Likewise, acetylene cylinders should not be charged with other gases, even though they are capable of containing those gases up to the service pressure of the cylinder. The basic feature of an acetylene cylinder that is different from all other cylinders is that it is entirely filled with a monolithic porous mass. It is this monolithic mass that stabilizes the acetylene and permits its safe shipment. Acetylene cylinders are fitted with safety devices to release the acetylene in the event of fire.

The filling and shipping of acetylene cylinders are subject to the regulations of the U.S. Department of Transportation. To completely charge acetylene cylinders in a reasonable period of time requires compression of acetylene to pressures >1833 kPa, usually in the range of 2074–2419 kPa (300–350 psi). Because acetylene at these pressures detonates, if a source of ignition is present, the cylinder charging plant must be carefully designed and constructed taking into account all of the safety hazards. The acetylene industry has prepared a basic set of guidelines.

Many factors must be taken into account in the charging operation, such as the rate of charging, cooling during charging, and interstage cooling of the gas to remove as much water as possible prior to the final drying. Water reduces the quantity of acetylene which may safely be carried, because the solubility of acetylene in water is less than in acetone. Other important factors include the mechanical reliability of the cylinder, valves and safety devices, the residual acetylene, and the presence of sufficient acetone to maintain the 0.58 ratio of acetylene/solvent. The charging of acetylene cylinders can be hazardous and should only be undertaken with the consent of the owner and by persons having full knowledge of the subject.

USES

Acetylene is used primarily as a raw material for the synthesis of a variety of organic chemicals (see ACETYLENE-DERIVED CHEMICALS). In the United States, this accounts for ~90% of acetylene usage and most of the remainder is used for metal welding or cutting. The chemical markets for acetylene are shrinking as ways are found to substitute lower cost olefins and paraffins for the acetylene, with some products now completely derived from olefinic starting materials.

Chemical Uses

At present the principal chemical markets for acetylene are its uses in the preparation of vinyl chloride, and 1,4-butanediol. Polymers from these monomers reach the consumer in the form of surface coatings (paints,

film, sheets, or textiles), containers, pipe, electrical wire insulation, adhesives, and many other products that total billions of kilograms. The acetylene routes to these monomers were once dominant but have been largely displaced by newer processes based on olefinic starting materials.

Fuel Uses

At one time acetylene was widely used for home, street, and industrial lighting. These applications disappeared with the advent of electrical lighting during the 1920s. However, one of the first fuel uses for acetylen, metalworking with the oxyacetylene flame, continues to consume a significant amount of acetylene.

T. Carreon, in E. Bingham, B. Cohrssen, and C. H. Powell, eds., *Patty's Toxicology*, 5th ed., Vol. 4, John Wiley & Sons, Inc., New York, 2001, pp. 119–120.

J. Larson, U. Loechner, and G. Tok, *Chemical Economics Handbook*, SRI International, Menlo Park, Calif., Aug. 2001.

S. A. Miller, *Acetylene—Its Properties, Manufacture and Uses*, Vol. 1 and Vol. 2, Academic Press, Inc., New York, 1965, 1966.

H. G. Viehe, *Chemistry of Acetylenes*, Marcel Dekker, Inc., New York, 1969.

RICHARD E. GANNON
Textron Defense Systems
ROBERT M. MANYIK
Union Carbide Corporation
C. M. DIETZ
H. B. SARGENT
R. O. THRIBOLET
R. P. SCHAFFER
Consultants

ACETYLENE-DERIVED CHEMICALS

Acetylene, C_2H_2, is an extremely reactive hydrocarbon, principally used as a chemical intermediate (see ACETYLENE). Because of its thermodynamic instability, it cannot easily or economically be transported for long distances. To avoid large free volumes or high pressures, acetylene cylinders contain a porous solid packing and an organic solvent. Acetylene pipelines are severely restricted in size and must be used at relatively low pressures. Hence, for large-scale operations, the acetylene consumer must be near the place of acetylene manufacture.

Because of its relatively high price, there have been continuing efforts to replace acetylene in its major applications with cheaper raw materials. Such efforts have been successful, particularly in the United States, where ethylene has displaced acetylene as raw material for acetaldehyde, acetic acid, vinyl acetate, and chlorinated solvents. Only a few percent of U.S. vinyl chloride production is still based on acetylene. Propylene has replaced acetylene as feed for acrylates and acrylonitrile. Even some recent production of traditional Reppe acetylene chemicals, such as butanediol and butyrolactone, is based on new raw materials.

REACTION PRODUCTS

Acetaldehyde

Acetaldehyde, C_2H_4O, (qv) was formerly manufactured principally by hydration of acetylene.

$$HC\equiv CH + H_2O \longrightarrow CH_3CHO$$

Many catalytic systems have been described, but acidic solutions of mercuric salts are the most generally used. This process has long been superseded by more economical routes involving oxidation of ethylene or other hydrocarbons.

Acrylic Acid, Acrylates, and Acrylonitrile

Acrylic acid, $C_3H_4O_2$, and acrylates were once prepared by reaction of acetylene and carbon monoxide with water or an alcohol, using nickel carbonyl as catalyst. In recent years this process has been completely superseded in the United States by newer processes involving oxidation of propylene. In western Europe, however, acetylene is still important in acrylate manufacture (see ACRYLIC ACID AND DERIVATIVES; ACRYLIC ESTER POLYMERS).

Cyclooctatetraene (COT)

Tetramerization of acetylene to cyclooctatetraene, although interesting, does not seem to have been used commercially. Nickel salts serve as catalysts. Other catalysts give benzene.

Ethylene

Although ethylene is produced by the cracking of hydrocarbons, hydrogenation of acetylene, using palladium on silica gel as catalyst is used for purification of ethylene containing small amounts of acetylene as contaminant (see ETHYLENE).

Vinyl Acetate

Vinyl acetate, $C_4H_6O_2$, used to be manufactured by addition of acetic acid to acetylene:

$$HC\equiv CH + CH_3COOH \longrightarrow CH_2=CHOOCCH_3$$

Liquid- and vapor-phase processes have been described; the latter appear to be advantageous, but it has been completely replaced since ~1982 by newer technology using oxidative addition of acetic acid to ethylene (see VINYL POLYMERS). In western Europe production of vinyl acetate from acetylene still remains a significant commercial route.

Vinylacetylene and Chloroprene

In the presence of cuprous salt solutions, acetylene dimerizes to vinylacetylene, Yields of 87% monovinylacetylene, together with 10% of divinylacetylene, have been described. Using cuprous chloride as catalyst, hydrogen chloride adds to acetylene, giving 2-chloro-1,3-butadiene,

chloroprene, C_4H_5Cl, the monomer for neoprene rubber. Manufacture via this process has been completely replaced by chlorination of butadiene (see CHLOROPRENE; POLYCHLOROPRENE).

Vinyl Chloride and Vinylidene Chloride

In the presence of mercuric salts, hydrogen chloride adds to acetylene giving vinyl chloride, C_2H_3Cl. As a route to vinyl chloride, this has been replaced by dehydrochlorination of ethylene dichloride in all but a few percent of current U.S. capacity. A combined process in which hydrogen chloride cracked from ethylene dichloride was added to acetylene was advantageous, but it is rarely used because processes to oxidize hydrogen chloride to chlorine with air or oxygen are cheaper (see VINYL POLYMERS).

In similar fashion, vinylidene chloride, $C_2H_2Cl_2$, has been prepared by successive chlorination and dehydrochlorination of vinyl chloride (see VINYLIDENE CHLORIDE MONOMER AND POLYMERS).

Vinyl Fluoride

Vinyl fluoride, C_2H_3F, the monomer for poly(vinyl fluoride), is manufactured by addition of hydrogen fluoride to acetylene (see FLUORINE CONTAINING COMPOUNDS, POLY(VINYL FLUORIDE)).

ETHYNYLATION REACTION PRODUCTS

The name ethynylation was coined by Reppe to describe the addition of acetylene to carbonyl compounds:

$$HC\equiv CH + RCOR' \longrightarrow HC\equiv CC(OH)RR'$$

Although stoichiometric ethynylation of carbonyl compounds with metal acetylides was known as early as 1899, Reppe's contribution was the development of catalytic ethynylation. Heavy metal acetylides, particularly cuprous acetylide, were found to catalyze the addition of acetylene to aldehydes. Although ethynylation of many aldehydes has been described, only formaldehyde has been catalytically ethynylated on a commercial scale. Copper acetylide is not effective as catalyst for ethynylation of ketones. For these, and for higher aldehydes, alkaline promoters have been used.

Propargyl Alcohol

Propargyl alcohol, 2-propyn-1-ol, C_3H_4O, is a colorless, volatile liquid, with an unpleasant odor that has been described as "mild geranium." Propargyl alcohol is miscible with water and with many organic solvents.

Manufacture. Propargyl alcohol is a by-product of butynediol manufacture. The original high pressure butynediol processes gave ~5% of the by-product; newer lower pressure processes give much less. Processes have been described that give much higher proportions of propargyl alcohol.

Shipment and Storage. Propargyl alcohol is available in tank cars, tank trailers, and drums. It is usually shipped in unlined steel containers and transferred through standard steel pipes or braided steel hoses; rubber is not recommended. Clean, rust-free steel is acceptable for short-term storage. For longer storage, stainless steel (types 304 and 316), glass lining, or phenolic linings (Lithcote LC-19 and LC-24, Unichrome B-124, and Heresite) are suitable. Aluminum, epoxies, and epoxy-phenolics should be avoided.

Health and Safety Factors. Although propargyl alcohol is stable, violent reactions can occur in the presence of contaminants, particularly at elevated temperatures. Heating in undiluted form with bases or strong acids should be avoided. Weak acids have been used to stabilize propargyl alcohol prior to distillation. Since its flash point is low, the usual precautions against ignition of vapors should be observed.

Propargyl alcohol is a primary skin irritant and a severe eye irritant and is toxic by all means of ingestion; all necessary precautions must be taken to avoid contact with liquid or vapors. The LD_{50} is 0.07 mL/kg for white rats and 0.06 mL/kg for guinea pigs.

Uses. Propargyl alcohol is a component of oil-well acidizing compositions, inhibiting the attack of mineral acids on steel (see CORROSION AND CORROSION CONTROL). It is also employed in the pickling and plating of metals.

Butynediol

Butynediol, 2-butyne-1,4-diol, is available commercially as a crystalline solid or a 35% aqueous solution manufactured by ethynylation of formaldehyde.

Manufacture. All manufacturers of butynediol use formaldehyde ethynylation processes. Variations of the original high pressure, fixed-bed process are still in use. However, all of the recent plants use low pressures and suspended catalysts.

The hazards of handling acetylene under pressure must be considered in plant design and construction. Although means of completely preventing acetylene decomposition have not been found, techniques have been developed that prevent acetylene decompositions from becoming explosive.

Shipment and Storage. Butynediol, 35% solution, is available in tank cars, tank trailers, and drums. Stainless steel, nickel, aluminum, glass, and various plastic and epoxy or phenolic liners have all been found satisfactory. Rubber hose is suitable for transferring. The solution is nonflammable and freezes at about −5°C.

Butynediol solid flakes are packed in polyethylene bags inside drums. The product is hygroscopic and must be protected from moisture.

Health and Safety Factors. Although butynediol is stable, violent reactions can take place in the presence of certain contaminants, particularly at elevated temperatures. In the presence of certain heavy metal salts, such as mercuric chloride, dry butynediol can decompose violently. Heating with strongly alkaline materials should be avoided.

Butynediol is a primary skin irritant and sensitizer, requiring appropriate precautions. Acute oral toxicity is relatively high.

Uses. Most butynediol produced is consumed by the manufacturers in manufacture of butanediol and butenediol. Small amounts are converted to ethers with ethylene oxide.

Butynediol is principally used in pickling and plating baths. Small amounts are used in the manufacture of brominated derivatives, useful as flame retardants.

Butenediol

2-Butene-1,4-diol is the only commercially available olefinic diol with primary hydroxyl groups. The commercial product consists almost entirely of the cis isomer.

trans-2-Butene-1,4-diol diacetate was prepared from 1,4-dibromo-2-butene in 1893 and hydrolyzed to the diol in 1926. The original preparation of the cis diol utilized the present commercial route, partial hydrogenation of butynediol.

Butenediol is very soluble in water, lower alcohols, and acetone. It is nearly insoluble in aliphatic or aromatic hydrocarbons.

Manufacture. Butenediol is manufactured by partial hydrogenation of butynediol. Although suitable conditions can lead to either cis or trans isomers, the commercial product contains almost exclusively cis-2-butene-1,4-diol. Trans isomer, available at one time by hydrolysis of 1,4-dichloro-2-butene, is unsuitable for the major uses of butenediol involving Diels-Alder reactions. The liquid-phase heat of hydrogenation of butynediol to butenediol is 156 kJ/mol (37.28 kcal/mol).

Newer, more selective processes use more active catalysts at lower pressures. In particular, supported palladium, alone or with promoters, has been found useful.

Shipment and Storage. Butenediol is available in unlined steel tank cars, tank trailers, and various sized drums. Because of its relatively high freezing point, tank cars are fitted with heating coils.

Health and Safety Factors. Butenediol is noncorrosive and stable under normal handling conditions. It is a primary skin irritant but not a sensitizer; contact with skin and eyes should be avoided. It is much less toxic than butynediol.

Uses. Butanediol is used to manufacture the insecticide Endosulfan, other agricultural chemicals, and pyridoxine (vitamin B_6) (see VITAMINS). Small amounts are consumed as a diol by the polymer industry.

Butanediol

1,4-Butanediol, tetramethylene glycol, 1,4-butylene glycol is produced commercially by catalytic hydrogenation of butynediol. Other processes used for commercial manufacture are described in the section on Manufacture.

Manufacture. Most butanediol is manufactured in Reppe plants via hydrogenation of butynediol. Recently, an alternative route involving acetoxylation of butadiene has come on stream and, more recently, a route based upon hydroformylation of allyl alcohol.

Shipment and Storage. Tank cars and tank trailers, selected to prevent color formation, are of aluminum or stainless steel, or lined with epoxy or phenolic resins; drums are lined with phenolic resins. Flexible stainless steel hose is used for transfer. Because of butanediol's high freezing point (~20°C) tank car coil heaters are provided.

Health and Safety Factors. Butanediol is much less toxic than its unsaturated analogues. It is neither a primary skin irritant nor a sensitizer. Because of its low vapor pressure, there is ordinarily no inhalation problem. As with all chemicals, unnecessary exposure should be avoided.

Uses. The largest uses of butanediol are internal consumption in manufacture of tetrahydrofuran (45%) and butyrolactone (22%). The largest merchant uses are for poly(butylene terephthalate) resins 24% (see POLYESTERS, THERMOPLASTIC) and in polyurethanes 5%, both as a chain extender and as an ingredient in a hydroxyl-terminated polyester used as a macroglycol. miscellaneous uses account for 4% of consumption and include uses as a solvent, as a coating resin raw material, and as an intermediate in the manufacture of other chemicals and pharmaceuticals.

Butyrolactone

γ-Butyrolactone, dihydro-2(3H)-furanone, was first synthesized in 1884 via internal esterification of 4-hydroxybutyric acid. In 1991 the principal commercial source of this material was dehydrogenation of butanediol. Manufacture by hydrogenation of maleic anhydride was discontinued in the early 1980s and resumed in the late 1980s.

Butyrolactone is completely miscible with water and most organic solvents. It is only slightly soluble in aliphatic hydrocarbons. It is a good solvent for many gases, for most organic compounds, and for a wide variety of polymers.

Manufacture. Butyrolactone is manufactured by dehydrogenation of butanediol. The old butyrolactone plant and process in Germany approximates the processes presently used. The dehydrogenation was carried out with preheated butanediol vapor in a hydrogen carrier over a supported copper catalyst at 230–250°C. The yield of butyrolactone after purification by distillation was ~90%. γ-Butyrolactone can also be prepared by catalytic hydrogenation of maleic anhydride.

Shipment and Storage. Butyrolactone is shipped in unlined steel tank cars and plain steel drums. Plain steel, stainless steel, aluminum, and nickel are suitable for storage and handling; rubber, phenolics, and epoxy resins are not suitable. Butyrolactone is hygroscopic and should be protected from moisture. Because of its low freezing point (−44°C), no provision for heating storage vessels is needed.

Health and Safety Factors. Butyrolactone is neither a skin irritant nor a sensitizer; however, it is judged to be a severe eye irritant in white rabbits. Because of its high boiling point (204°C), it does not ordinarily represent a vapor hazard.

Uses. Butyrolactone is principally consumed by the manufacturers by reaction with methylamine or ammonia to produce N-methyl-2-pyrrolidinone and 2-pyrrolidinone, C_4H_7NO, respectively. Considerable amounts are used as a solvent for agricultural chemicals and polymers, in dyeing and printing, and as an intermediate for various chemical syntheses.

OTHER ALCOHOLS AND DIOLS

Secondary acetylenic alcohols are prepared by ethynylation of aldehydes higher than formaldehyde. Although copper acetylide complexes will catalyze this reaction, the rates are slow and the equilibria unfavorable. The commercial products are prepared with alkaline catalysts, usually used in stoichiometric amounts.

Ethynylation of ketones is not catalyzed by copper acetylide, but potassium hydroxide has been found to be effective. In general, alcohols are obtained at lower temperatures and glycols at higher temperatures. Most processes use stoichiometric amounts of alkali, but true catalytic processes for manufacture of the alcohols have been described; the glycols appear to be products of stoichiometric ethynylation only.

Methylbutynol

2-Methyl-3-butyn-2-ol, prepared by ethynylation of acetone, is the simplest of the tertiary ethynols, and serves as a prototype to illustrate their versatile reactions. There are three reactive sites, ie, hydroxyl group, triple bond, and acetylenic hydrogen. Although the triple bonds and acetylenic hydrogens behave similarly in methylbutynol and in propargyl alcohol, the reactivity of the hydroxyl groups is very different.

Manufacture

In general, manufacture is carried out in batch reactors at close to atmospheric pressure. A moderate excess of finely divided potassium hydroxide is suspended in a solvent such as 1,2-dimethoxyethane. The carbonyl compound is added, followed by acetylene. The reaction is rapid and exothermic. At temperatures <5°C the product is almost exclusively the alcohol. At 25–30°C the glycol predominates. Such synthesis also proceeds well with non-complexing solvents such as aromatic hydrocarbons, although the conversion is usually lower.

Continuous processes have been developed for the alcohols, operating under pressure with liquid ammonia as solvent. Potassium hydroxide or anion exchange resins are suitable catalysts. However, the relatively small manufacturing volumes militate against continuous production.

Health and Safety Factors

Under normal conditions acetylenic alcohols are stable and free of decomposition hazard. The more volatile alcohols present a fire hazard.

The alcohols are toxic orally, through skin absorption, and through inhalation. The secondary alcohols are more toxic than the tertiary. The glycols are relatively low in toxicity.

Uses

The secondary acetylenic alcohols hexynol and ethyloctynol are used as corrosion inhibitors in oil-well acidizing compositions (see CORROSION AND CORROSION CONTROL). The tertiary alcohols methylbutynol and methylpentynol are used as chemical intermediates, for manufacture of Vitamin A and other products, and in metal plating and pickling operations. Dimethylhexynediol can be used in manufacture of fragrance chemicals and peroxide catalysts. Higher acetylenic glycols and ethoxylated acetylene glycols are useful as surfactants and electroplating additives.

VINYLATION REACTION PRODUCTS

Unlike ethynylation, in which acetylene adds across a carbonyl group and the triple bond is retained, in vinylation a labile hydrogen compound adds to acetylene, forming a double bond.

Catalytic vinylation has been applied to a wide range of alcohols, phenols, thiols, carboxylic acids, and certain amines and amides. Vinyl acetate is no longer prepared this way in the United States, although some minor vinyl esters such as stearates may still be prepared this way. However, the manufacture of vinyl-pyrrolidinone and vinyl ethers still depends on acetylene.

N-Vinylcarbazole

Vinylation of carbazole proceeds in high yields with alkaline catalysts. The product, 9-ethenylcarbazole, $C_{14}H_{11}N$, forms rigid high melting polymers with outstanding electrical properties.

Neurine

Neurine is trimethylvinylammonium hydroxide, $C_5H_{13}NO$. Tertiary amines and their salts vinylate readily at low temperatures with catalysis by free tertiary amines. Above ~50°C tetramethylammonium hydroxide is formed as a by-product; it is the sole product above 100°C.

N-Vinyl-2-pyrrolidinone

The major use of vinylpyrrolidinone is as a monomer in manufacture of poly(vinylpyrrolidinone) (PVP) homopolymer and in various copolymers, where it frequently imparts hydrophilic properties. These polymers are used in pharmaceutical and cosmetic applications, soft contact lenses, and viscosity index improvers. The monomer serves as a component in radiation-cured polymer compositions, serving as a reactive diluent that reduces viscosity and increases cross-linking rates (see VINYL POLYMERS, N-VINYLAMIDE POLYMERS).

Vinyl Ethers

The principal commercial vinyl ethers are methyl vinyl ether (methoxyethene, C_3H_6O); ethyl vinyl ether (ethoxy-ethene, C_4H_8O); and butyl vinyl ether (1-ethenyloxy-butane, $C_6H_{12}O$). Others, such as the isopropyl, isobutyl, hydroxybutyl, decyl, hexadecyl, and octadecyl ethers, as well as the divinyl ethers of butanediol and of triethylene glycol, have been offered as development chemicals (see ETHERS).

Manufacture. Variations of the old German vilyl elter plants are still in use. Vinylation of alcohols from methyl to butyl was carried out under pressure: typically 2–2.3 MPa (20–22 atm) and 160–165°C for methyl, and 0.4–0.5 MPa (4–5 atm) and 150–155°C for isobutyl.

High boiling alcohols were vinylated at atmospheric pressure. A tower packed with Raschig rings and filled with an alcohol containing 1–5% of KOH at 160–180°C was used. Acetylene was recycled continuously up through the tower. The heat of reaction, ~125 kJ/mol (30 kcal/mol), was removed by cooling coils. Fresh alcohol and catalyst were added continuously at the top and withdrawn at the bottom. Yields of purified, distilled product were described as quantitative.

Shipment and Storage. Methyl vinyl ether is available in tank cars or cylinders, while the other vinyl ethers are available in tank cars, tank wagons, or drums. Mild steel, stainless steel, and phenolic-coated steel are suitable for shipment and storage. If protected from air, moisture, and acidic contamination, vinyl ethers are stable for years.

Health and Safety Factors. Because of their high vapor pressures (methyl vinyl ether is a gas at ambient conditions), the lower vinyl ethers represent a severe fire hazard and must be handled accordingly. Contact with acids can initiate violent polymerization and must be avoided. Although vinyl ethers form peroxides more slowly than saturated ethers, distillation residues must be handled with caution.

Inhalation should be avoided.

The lower vinyl ethers do not appear to be skin irritants or sensitizers. Oral toxicity is very low.

Uses. Union Carbide consumes its vinyl ether production in the manufacture of glutaraldehyde. BASF and GAF consume most of their production as monomers (see VINYL POLYMERS). In addition to the homopolymers, the copolymer of methyl vinyl ether with maleic anhydride is of particular interest.

"1,4-Butandiol, Chemical Profile," *Chemical Market Reporter*, June 2000.

J. W. Copenhaver and M. H. Bigelow, *Acetylene and Carbon Monoxide Chemistry*, Reinhold Publishing Corp., New York, 1949, pp. 37–38.

S. A. Miller, *Acetylene, Its Properties, Manufacture and Uses*, Vol. 1, Academic Press, Inc., New York, 1965, pp. 24–28, 42–44.

W. Reppe and co-workers, *Ann.* **596**, 2 (1955).

EUGENE V. HORT
PAUL TAYLOR
GAF Corporation

ACROLEIN AND DERIVATIVES

Acrolein (2-propenal) (C_3H_4O), is the simplest unsaturated aldehyde ($CH_2=CHCHO$). The primary characteristic of acrolein is its high reactivity due to conjugation of the carbonyl group with a vinyl group. Controlling this reactivity to give the desired derivative is the key to its usefulness. Acrolein now finds commercial utility in several major products as well as a number of smaller volume products.

Acrolein is a highly toxic material with extreme lacrimatory properties. At room temperature, acrolein is a liquid with volatility and flammability somewhat similar to acetone; but unlike acetone, its solubility in water is limited. Commercially, acrolein is always stored with hydroquinone and acetic acid as inhibitors. Special care in handling is required because of the flammability, reactivity, and toxicity of acrolein.

MANUFACTURE

Acrolein was first produced commercially until the late 1930s. In 1957, bismuth molybdate catalysts capable of producing high yields of acrolein at high propylene conversions (>90%) and at low pressures were discovered. Over the next several decades, much industrial and academic research and development was devoted to improving these catalysts, which are used in the production processes for acrolein, acrylic acid, and acrylonitrile. All commercial acrolein manufacturing processes known today are based on propylene oxidation and use bismuth molybdate-based catalysts.

Many key improvements and enhancements to the bismuth molybdate based propylene oxidation catalysts have occurred since its discovery in 1957.

The most efficient catalysts are complex mixed-metal oxides that consist largely of Bi, Mo, Fe, Ni, and/or Co, K, and either P, B, W, or Sb. Many additional combinations of metals have been patented, along with specific catalyst preparation methods to adjust specific surface area, pore volume, and pore size distribution. Most catalysts used commercially today are extruded neat metal oxides as opposed to supported (coated) catalysts. Propylene conversions are generally >93%. The acrolein yields

Table 1. Properties of Acrolein

Property	Value
Physical properties	
molecular formula	C_3H_4O
molecular weight	56.06
specific gravity at 20/20°C	0.8427
boiling point, °C	52.69
at 101.3 kPa[a]	
Chemical properties	
autoignition temperature in air, °C	234
heat of combustion of 25°C, kJ/kg[b]	−27,589

[a]To convert kPa to mm Hg, multiply by 7.5.
[b]To convert kJ to kcal, divide by 4.184.

depend not only on the chemical composition of the catalyst, but also on the shape of the catalyst and catalyst loading configurations.

REACTIONS AND DERIVATIVES

Acrolein is a highly reactive compound because both the double bond and aldehydic moieties participate in a variety of reactions, including oxidation, reduction.

Acrolein is readily oxidized to acrylic acid, ($C_3H_4O_2$), by passing a gaseous mixture of acrolein, air, and steam over a catalyst composed primarily of molybdenum and vanadium oxides (see ACRYLIC ACID AND DERIVATIVES). Virtually all of the acrylic acid produced in the United States is made by the oxidation of propylene via the intermediacy of acrolein.

Direct formation of acrylic acid esters by oxidation of acrolein in the presence of lower alcohols has been studied. The intermediacy of acrylic acid is thereby avoided in the manufacture of these important acrylic acid derivatives.

The vapor-phase reduction of acrolein with isopropyl alcohol in the presence of a mixed-metal oxide catalyst yields allyl alcohol in a one-pass yield of 90.4%, with a selectivity to the alcohol of 96.4%.

The addition of alcohols to acrolein may be catalyzed by acids or bases. By the judicious choice of reaction conditions the regioselectivity of the addition may be controlled and alkoxy-propionaldehydes, acrolein acetals, or alkoxy-propionaldehyde acetals may be produced in high yields.

Reactions of acrolein with alcohols producing high yields of alkoxypropionaldehyde acetals are also known. The alkoxypropionaldehyde acetals may be useful as solvents or as intermediates in the synthesis of other useful compounds.

A new and potentially significant use of acrolein is the manufacture of 1,3-propanediol ($C_3H_6O_2$). Addition of water to acrolein forms 3-hydroxypropionaldehyde ($C_3H_6O_2$). Hydrogenation of 3-hydroxypropionaldehyde forms 1,3-propanediol.

Competitive routes to 1,3-propanediol are ethylene oxide hydroformylation and biofermentation of corn. The largest anticipated use of 1,3-propanediol is in the manufacture of polytrimethylene terephthalate (PTT).

One of the largest uses of acrolein is the production of 3-methylmercaptopropionaldehyde, (C_4H_8OS), which is an intermediate in the synthesis of D,L-methionine ($C_5H_{11}NO_2S$), an important chicken feed supplement.

An industrially useful reaction in which acrolein participates as the diene is that with methyl vinyl ether. The product, methoxydihydropyran, ($C_6H_{10}O_2$) is an intermediate in the synthesis of glutaraldehyde, ($C_5H_8O_2$).

In addition to its principal use in biocide formulations, glutaraldehyde has been used in the film development and leather tanning industries. It may be converted to 1,5-pentanediol, ($C_5H_{12}O_2$) or glutaric acid ($C_5H_8O_4$).

In the absence of inhibitors, acrolein polymerizes readily in the presence of anionic, cationic, or free-radical agents. The resulting polymers are insoluble, highly cross-linked solids with no known commercial use.

Copolymers, including one obtained by the oxidative copolymerization of acrolein with acrylic acid, a product of commercial interest, are known. There is a great variety of potential acrolein copolymers; however, significant commercial uses have not been developed. The possible application of polyacroleins or copolymers as polymeric reagents, polymeric complexing agents, and polymeric carriers has been recognized.

DIRECT USES OF ACROLEIN

Because of its antimicrobial activity, acrolein has found use as an agent to control the growth of microbes in process feed lines, thereby controlling the rates of plugging and corrosion (see WASTES, INDUSTRIAL).

Acrolein at a concentration of <500 ppm is also used to protect liquid fuels against microorganisms. In recent years, several acrolein derivatives have been proposed to provide a safer means of transport of acrolein to an application site.

HEALTH AND SAFETY FACTORS

The most frequently encountered hazards of acrolein are acute toxicity from inhalation and ocular irritation. Because of its high volatility, even a small spill can lead to a dangerous situation. Acrolein is highly irritating and a potent lacrimator.

Concentrations of acrolein vapor as low as 0.6 mg/m^3 (0.25 ppm) may irritate the respiratory tract, causing coughing, nasal discharge, chest discomfort or pain, and difficulty with breathing.

Acrolein vapor is highly irritating to the eyes, causing pain or discomfort in the eye, profuse lacrimation, involuntary blinking, and marked reddening of the conjunctiva. A small amount of acrolein may be fatal if swallowed. Acrolein is highly toxic by skin absorption. Brief contact may result in the absorption of harmful and possibly fatal amounts of material. There is no specific antidote for acrolein exposure.

Chronic human exposure is unlikely due to the lack of tolerance to acrolein.

Acrolein is very flammable. Acrolein is only partly soluble in water. The vapors are heavier than air and can travel along the ground and flash back from an ignition source.

Acrolein is a highly reactive chemical, and contamination of all types must be avoided.

Acrolein reacts slowly in water to form 3-hydroxypropionaldehyde and then other condensation products from aldol and Michael reactions.

Dimerization of acrolein is very slow at ambient temperatures but it can become a runaway reaction at elevated temperature ($\sim$90°C), a consideration in developing protection against fire exposure of stored acrolein.

STORAGE AND HANDLING

The following cautions should be observed: Do not destroy or remove inhibitor. Do not contaminate with alkaline or strongly acidic materials. Do not store in the presence of a water layer. In the event of spillage or misuse that cause a release of product vapor to the atmosphere, thoroughly ventilate the area, especially near floor levels where vapors will collect.

Suitable materials of construction are steel, stainless steel, and aluminum 3003. Galvanized steel should not be used. Plastic tanks and lines are not recommended.

Storage tanks should have temperature monitoring with alarms to detect the onset of reactions. Storage should be under an atmosphere of dry nitrogen and should vent vapors from the tank to a scrubber or flare.

In treatment of spills or wastes, the suppression of vapors is the first concern and the aquatic toxicity to plants, fish, and microorganisms is the second. Normal procedures for flammable liquids should also be carried out.

Acrolein, in *IARC Monographs on the Evaluation of Carcinogenic Risk of Chemicals to Humans. Some Monomers, Plastics and Synthetic Elastomers and Acrolein*, Vol. 19, International Agency for Research on Cancer, Lyon, France, 1979, pp. 479–494.

Dictionary of Organic Compounds, 5th ed., Vol. 5, Chapman and Hall, New York, 1982, p. 4784.

R. C. Schulz in J. I. Kroschwitz, ed., *Encyclopedia of Polymer Science and Engineering*, 2nd ed., Vol. 1, Wiley-Interscience, New York, 1985, pp. 160–169.

N. Yamashita in J. C. Salamone, ed., *Polymeric Materials Encyclopedia*, Vol. 1, CRC Press, Boca Raton, 1996, pp. 40–47.

W. G. ETZKORN
S. E. PEDERSEN
T. E. SNEAD
The Dow Company

ACRYLAMIDE

Acrylamide (NIOSH No: A533250) has been commercially available since the mid-1950s and has shown steady growth since that time, but is still considered a small volume commodity. Its formula, H_2C=$CHCONH_2$ (2-propeneamide), indicates a simple chemical, but it is by far the most important member of the series of acrylic and methacrylic amides. Water soluble polyacrylamides represent the most important applications. The largest use in this catagory is as a dewatering aid for sludges in the treatment of effluent from municipal wastewater treatment plants and industrial processes.

Other uses include flocculants in feed water treatment for industrial purposes, the mining industry and various other process industries, soil stabilization, papermaking aids, and thickeners. Smaller but none the less important uses include dye acceptors; polymers for promoting adhesion; additives for textiles, paints, and cement; increasing the softening point and solvent resistance of resins; components of photopolymerizable systems; and cross-linking agents in vinyl polymers.

Table 1. Physical Properties of Solid Acrylamide Monomer

Property	Value
molecular weight	71.08
melting point, °C	84.5 ±
boiling point, °C 0.67 kPa[a]	103

[a] To convert kPa to mm Hg, multiply by 7.5.

PHYSICAL AND CHEMICAL PROPERTIES

Acrylamide is a white crystalline solid that is quite stable at ambient conditions, and, even at temperatures as high as its melting point (for 1 day in the absence of light), no significant polymer formation is observed. Above its melting point, however, liquid acrylamide may polymerize rapidly with significant heat evolution. Precautions should be taken when handling even small quantities of molten material. In addition to the solid form, a 50% aqueous solution of acrylamide is a popular commercial product today. This solution is stabilized by small amounts of cupric ion (25–30 ppm based on monomer) and soluble oxygen. Several other stabilizers are also available for the aqueous monomer solution, such as ethylenediaminetetraacetic acid (EDTA), ferric ion, and nitrite. The only effect of oxygen is to increase the induction period for polymerization. The physical properties of solid acrylamide monomer are summarized in Table 1, and typical physical properties of a 50% solution in water appear in Table 2.

Acrylamide, C_3H_5NO, is an interesting difunctional monomer containing a reactive electron-deficient double bond and an amide group, and it undergoes reactions typical of those two functionalities. It exhibits both weak acidic and basic properties. The electron-withdrawing carboxamide group activates the double bond, that consequently reacts readily with nucleophilic reagents, eg, by addition.

Many of these reactions are reversible, and for the stronger nucleophiles they usually proceed the fastest.

MANUFACTURE

The current routes to acrylamide are based on the hydration of inexpensive and readily available acrylonitrile (C_3H_3N, 2-propenenitrile, vinyl cyanide, VCN, or cyanoethene) (see ACRYLONITRILE). For many years, the principal process for making acrylamide was a reaction of

Table 2. Physical Properties of 50% Aqueous Acrylamide Solution

Property	Value
pH	5.0–6.5
refractive index range, 25°C (48–52%)	1.4085–1.4148
viscosity, mPa (= cP) at 25°C	2.71
specific gravity, at 25°C	1.0412
boiling point at 101.3 kPa,[a] °C	99–104

[a] To convert kPa to mm Hg, multiply by 7.5.

acrylonitrile with $H_2SO_4 \cdot H_2O$ followed by separation of the product from its sulfate salt using a base neutralization or an ion exclusion column.

$$CH_2=CHCN + H_2SO_4 \cdot H_2O \longrightarrow CH_2=CHCONH_2 \cdot H_2SO_4$$

This process yields satisfactory monomer, either as crystals or in solution, but it also produces unwanted sulfates and waste streams. The reaction was usually run in glass-lined equipment at 90–100°C with a residence time of 1 h. Long residence time and high reaction temperatures increase the selectivity to impurities, especially polymers and acrylic acid, which controls the properties of subsequent polymer products.

The ratio of reactants had to be controlled very closely to suppress these impurities. Recovery of the acrylamide product from the acid process was the most expensive and difficult part of the process. Large scale production depended on two different methods. If solid crystalline monomer was desired, the acrylamide sulfate was neutralized with ammonia to yield ammonium sulfate. The acrylamide crystallized on cooling, leaving ammonium sulfate, which had to be disposed of in some way.

Acrylamide and its derivatives have been prepared by many other routes. The reactions of acryloyl chloride and acrylic anhydride with ammonia are classical methods.

A process for manufacturing acrylamide microemulsified homopolymer has been disclosed.

HEALTH AND SAFETY FACTORS

Contact with acrylamide can be hazardous and should be avoided. The most serious toxicological effect of exposure to acrylamide monomer is as a neurotoxin. In contrast, polymers of acrylamide exhibit very low toxicity. Since the solid form sublimes, the solid or powder form of acrylamide is more likely to be a problem than the aqueous form because of possible exposure to dusts and vapors. An important characteristic of the toxicity of acrylamide monomer is that the signs and symptoms of exposure to toxic levels may be slow in developing and can occur after ingestion of small amounts over a period of several days or weeks. It is therefore important that people who have been exposed to acrylamide be monitored by a qualified physician. Signs and symptoms include increased sweating of hands and feet, numbness or tingling of the extremities, or even paralysis of the arms and legs. Acrylamide is readily absorbed through unbroken skin, and the signs are the same as with ingestion. Eye contact can produce conjunctival irritation and slight corneal injury and can lead to systemic exposure if contact is prolonged and/or repeated. Inhalation of vapors, dusts, and/or mists can result in serious injury to the nervous system, but again, symptoms may be slow in developing.

Handling of dry acrylamide is hazardous primarily from its dust and vapor, and this is a significant problem, especially in the course of emptying bags and drums. This operation should be carried out in an exhaust hood with the operator wearing respiratory and dermal protection. Waste air from the above mentioned ventilation should be treated by a wet scrubber before purging to the open air, and the waste water should be fed to an activated sludge plant or chemical treatment facility. Solid acrylamide may polymerize violently when melted or brought into contact with oxidizing agents. Storage areas for solid acrylamide monomer should be clean and dry and the temperature maintained at 10–25°C, with a maximum of 30°C.

The 50% aqueous product is the most desirable where water can be tolerated in the process. Employees should not be permitted to work with acrylamide until thoroughly instructed and until they can practice the required precautions and safety procedures. Anyone handling acrylamide should practice strict personal cleanliness and strict housekeeping at all times. If contact is made, the affected skin area should be washed thoroughly with soap and water and contaminated clothing should be replaced. The need for good personal hygiene and housekeeping to prevent exposure cannot be overemphasized.

Aqueous solutions of 50% acrylamide should be kept between 15.5 and 38°C with a maximum of 49°C. Suitable materials of construction for containers include stainless steel (304 and 316) and steel lined with plastic resin (polypropylene, phenolic, or epoxy). Avoid contact with copper, aluminum, their alloys, or ordinary iron and steel.

Disposal of small amounts of acrylamide may be done by biodegradation in a conventional secondary sewage treatment plant, but any significant amounts should be avoided. Such waste material should not be allowed to get into a municipal waste treatment or landfill operation unless all appropriate precautions have been taken.

USES

The largest use of acrylamide in the United States is for the production of polyacrylamides and consumes 94% of the total. In this category, the largest use of polyacrylamides is in water treatment, which accounts for 56%. This includes use as a dewatering aid for sludge in the treatment of effluent from municipal wastewater treatment plants (eg, sewage) and industrial processes (pulp and paper plant wastewater). Polyacrylamides are also used as flocculents for feed water treatment for industrial purposes. Other uses include in pulp and paper production (24%), mineral processing (10%), *N*-methylacrylamide and other monomers (6%), and miscellaneous (4%).

"Acrylamide," *Chemical Profiles, Chem Expo, Http://63.236.84.14/news/profile.cfm.* May 6, 2002.

Aqueous Acrylamide, Forms 260-951-88, Analytical Method PAA 44, Chemical and Metals Department, The Dow Chemical Company, Midland, Mich., 1976.

E. J. Conway, R. J. Petersen, R. F. Colingsworth, J. G. Craca, and J. W. Carter, *Assessment of the Need for a Character of Limitations on Acrylamide and Its Components,* EPA MRI Project No. 4308-N, 1979.

R. L. Melnick, in E. Bingham, B. Cohrssen, and C. H. Powell, eds., *Patty's Toxicology,* 5th ed., Vol. 1, John Wiley & Sons, Inc., New York, 2001, p. 143.

C. E. HABERMANN
Dow Chemical

ACRYLAMIDE POLYMERS

The terminology used to describe acrylamide-containing polymers in the technical literature varies in its precision. In order to avoid confusion, throughout this article the term "poly(acrylamide)" will be reserved for the nonionic homopolymer of acrylamide, whereas the term "polyacrylamides" or "acrylamide polymers" will refer to acrylamide-containing polymers, including the homopolymer and copolymers. Specific nomenclature will be used for particular copolymers, for example, poly(acrylamide-co-sodium acrylate).

The diverse class of water-soluble and water-swellable polymers comprising polyacrylamides contains some of the most important synthetic polymeric materials used to improve the quality of life in our modern society. Acrylamide-containing polymers fall into three main categories: nonionic, anionic, and cationic.

Poly(acrylamide) is made by the free radical polymerization of acrylamide, which is derived from acrylonitrile by either catalytic hydrolysis or bioconversion. The unique chemistry of acrylamide, its favorable reactivity ratios with many comonomers, and the ability of poly(acrylamide) to be derivatized allows for a substantial variety of polymers to be tailor-made over a wide range of molecular weights (approximately 10^3–50×10^6 daltons), charge densities, and chemical functionalities.

One major application area for polyacrylamides is in solid–liquid separations. The largest market segments therein are for use as flocculants and dewatering aids for municipal wastewater, thickening aids for industrial wastewater, secondary clarification and clarification of potable water, solids removal from biological broths, and animal feed recovery from waste. Because of major concern for the environment, the allowable suspended solids in most effluent streams are becoming more restricted by government regulations. New technologies for producing cationic polymers with a wide range of charge levels, novel structures, and very high molecular weights have addressed this need. These polymers have greatly improved the dewatering performances of centrifuges, screw presses, and belt presses used for such purposes. This has resulted in drier dewatered solids, which has translated into lower costs to either landfill or incinerate the solids.

The largest volume applications for polyacrylamides in paper mills are in on-machine wet-end processes. Paper retention aids and drainage aids are used to flocculate or bind fillers, fibers, and pigments. Glyoxalated cationic polyacrylamides are used as strengthening agents and promoters for paper sizing. Other papermaking applications include off-machine processes for recovering fiber from recycled paper waste and for deinking.

High-molecular-weight polyacrylamides have also traditionally been used in the minerals processing industry. Recent polymer technology developments, including-ultra-high-molecular-weight and novel anionic polyacrylamides, have yielded important materials. These products are used as flocculants in coal mining, the Bayer process for alumina recovery (red mud flocculants), precious metals recovery, and the solid–liquid separation of underflow streams in a variety of mining processes. Novel chemical modifications of low molecular weight polyacrylamides have resulted in materials that are used as modifiers in the selective separation of metal sulfides and magnetite and as depressants and flotation aids.

One large market segment for anionic polyacrylamides had traditionally been in enhanced oil recovery. However, low oil prices have resulted in a large decline in such applications. Since 1990, polymer flooding has virtually disappeared in the United States. However, during 1999 crude oil prices started to increase.

Other significant application areas for polyacrylamides include soil conditioning and erosion control, drag reduction, sugar processing, additives in cosmetics, and super-absorbents.

PHYSICAL PROPERTIES

Solid Polyacrylamides

Completely dry poly(acrylamide) is a brittle white solid. It is nontoxic, unlike the monomer. Dry polyacrylamides (including copolymers) are commercially available as non-dusting powders and as spherical beads. These products can contain small amounts of additives that aid in both the stability and dissolution of the polymers in water. Commercially available acrylamide copolymer powders, which are typically dried under mild conditions, will usually contain about 5–15% water depending on their ionicity. The powders are hygroscopic, and generally become increasingly hygroscopic as the ionic character of the polymer increases. Cationic polymers are particularly hygroscopic.

Some physical properties of nonionic polyacrylamide are listed in Table 1.

Table 1. Physical Properties of Solid Polyacrylamide

Property	Value
density	1.302 g/cm^3 (23°C)
glass-transition temperature (T_g)	195°C
critical surface tension (γ_c)	52.3 mN/m (20°C)
chain structure	mainly heterotactic linear or branched, some head-to-head addition
crystallinity	amorphous (high molecular weight)
solvents	water, ethylene glycol, formamide
nonsolvents	ketones, hydrocarbons, ethers, alcohols
fractionation solvents	water–methanol
gases evolved on combustion in air	H_2, CO, CO_2, NH_3, nitrogen oxides

Table 2. Physical Properties of Polyacrylamide in Solution

Property	Value	Conditions
steric hindrance parameter (σ)	2.72	water @ 30°C
characteristic ratio (C_∞)	14.8	water @ 30°C
partial specific volume (v)	0.693 cm^3/g	water @ 20°C
theta temperature (Θ)	−8°C	water @ 25°C
theta conditions	0.40 v/v methanol/water	water @ 25°C
Flory χ parameter	0.48 ± 0.01	water @ 30°C
refractive index increment (dn/dc)	0.187 cm^3/g	$\lambda = 546.1$ nm
	0.185 cm^3/g	$\lambda = 632.8$ nm

Solution Properties

The amide group (—CONH$_2$) in polyacrylamide provides for its solubility in water and in a few other polar solvents such as glycerol, ethylene glycol, and formamide. A sense of poly(acrylamide)'s affinity for water can be acquired by examining a few characteristic parameters. Some physical properties of poly(acrylamide) in solution are collected in Table 2.

Solution Rheology

Solutions of polyacrylamides tend to behave as pseudo-plastic fluids in viscometric flows. Dilute solutions are Newtonian (viscosity is independent of shear rate) at low shear rates and transition to pseudoplastic, shear thinning behavior above a critical value of the shear rate. This critical shear rate decreases with the polymer molecular weight, polymer concentration, and the thermodynamic quality of the solvent. A second Newtonian plateau at high shear rates is not readily seen, probably due to mechanical degradation of the chains. Viscometric data for dilute and semidilute polyacrylamide solutions can often be fit to a Carreau model. It is wise to remember the cautions that were cited previously about mechanical degradation of the high-molecular-weight components of a polyacrylamide sample when analyzing rheological data.

The viscosities of fully dissolved, high-molecular-weight poly(acrylamide)s in aqueous solutions have often, but not always, been seen to change with time over the periods of days to weeks. Typically, the solution viscosity decreases with time.

ACRYLAMIDE POLYMERIZATION AND STRUCTURAL MODIFICATIONS OF POLY(ACRYLAMIDE)

Acrylamide (2-propenamide, C$_3$H$_5$ON) readily undergoes free-radical polymerization to high-molecular-weight poly(acrylamide). Free-radical initiation can be accomplished using organic peroxides, azo compounds, inorganic peroxides including persulfates, redox pairs, photoinduction, radiation-induction, electroinitiation, or ultrasonication.

The large amount of heat (82.8 kJ/mol) that evolves during polymerization can result in a rapid temperature rise. One way in which this exotherm problem has been addressed in commercial high-solids and high-molecular-weight processes has been through the use of an adiabatic gel process in which the initiation temperature is 0°C. In another approach, controllable-rate redox polymerization of aqueous acrylamide-in-oil emulsions can be carried out at moderate temperatures of 40–60°C in order to accommodate the exotherm and to achieve very high molecular weights. Additives greatly affect rate and the kinetics of polymerization. Chain transfer agents have been used purposely to control molecular weight, minimize insoluble polymer, and control cross-linking and the degree of branching in commercial preparations.

Polyacrylamide (PAM) is a relatively stable organic polymer. However, PAM can be degraded (eg, molecular weight decreases) under certain conditions. The amide functionality is acidic in nature and is capable of undergoing most of the chemical reactions of primary amides. Consequently, acrylamide polymers can be functionalized by post-polymerization chemical reactions. To obtain anionic derivatives, PAM can be hydrolyzed with caustic. Dry poly(acrylamide) is relatively stable. The onset of dry PAM decomposition occurs at 180°C. Inter-or intra-amide condensation to an imide can occur in acidic media at high temperatures (140–160°C). At temperatures above 160°C, thermal degradation, imidization, nitrile formation, and dehydration take place. Polymer stability is very important in actual applications in order to maintain consistent and excellent performance. In most applications, polymer solutions are prepared and used at moderate temperatures; however, there are exceptions such as in the harsh reservoir conditions (high temperature and high salinity) found in some enhanced oil recovery operations. Impurities such as residual persulfate from batch manufacturing can degrade the polymer.

Numerous types of oxygen scavengers are used to inhibit and prevent oxidative degradation. Effective compounds are thio compounds, hydroquinone, bisulfite, phenolic compounds, hydroxylamine, hydrazine, and others.

Hydrolyzed Polyacrylamide

Hydrolysis of polyacrylamide proceeds smoothly over a wide range of pH. At alkaline pH, three reaction kinetics constants have been described, k_0, k_1 and k_2. The subscripts characterize the number of neighboring carboxylate groups next to the amide group being hydrolyzed. Indirect evidence has shown that $k_0 > k_1 > k_2$. Under alkaline conditions, the rate of hydrolysis of polyacrylamide decreases with increasing conversion. The electrostatic repulsion from the increasing number of carboxylate groups in the backbone polymer opposes the approaching hydroxyl ion. Consequently, further hydrolysis will be severely retarded.

Hydrolysis of polyacrylamide proceeds slowly under acidic conditions. The undissociated carboxylic acid groups are protonated, neutral species under those conditions. An imide structure has been proposed to be an intermediate in the low-pH hydrolysis of polyacrylamide, yielding short blocks of carboxyl groups distributed along the polymer chain (see scheme below). To date, there has been limited application of these block copolymer structures, and ones with high molecular weight have not been commercialized.

Cationic Carbamoyl Polymers

Polyacrylamide reacts with formaldehyde, CH_2O, and dimethylamine, C_2H_7N, to produce aminomethylated polyacrylamide. A wide range of substitution can be produced in solution or in water-in-oil emulsion. ^{13}C-nmr studies have verified that the Mannich substitution reaction follows second-order kinetics. Several disadvantages of solution Mannich PAMs are the problem of handling high solution viscosities, the added expense of shipping low-solids formulations, and the limitations to applications with low-pH substrates due to the decrease in cationic charge with increasing pH. Quaternized aminomethylated products in water-in-oil emulsions with greater than 20% solids have been developed. The charges in both high- and low-charge products were nearly independent of pH. Microemulsion formulations have been developed and now replace certain polymer macroemulsions.

Sulfomethylation

The reaction of formaldehyde and sodium bisulfite with polyacrylamide under strongly alkaline conditions at low temperature to produce sulfomethylated polyacrylamides has been reported many times. It has been recently suggested that the expected sulfomethyl substitution is not obtained under the previously described strongly alkaline conditions of pH 10–12. This nmr study indicates that hydrolysis of polyacrylamide occurs and the resulting ammonia reacts with the sodium bisulfite and formaldehyde to form sulfomethyl amines and hexamethylenetetramine.

Reaction with Other Aldehydes

Polyacrylamide reacts with glyoxal, $C_2H_2O_2$, under mild alkaline conditions to yield a polymer with pendant aldehyde functionality. The rate of this reaction can be controlled by varying the pH and reaction temperature. Cross-linking is a competing reaction. The reaction rate increases rapidly with increasing pH and with increasing polymer concentration.

Transamidation

Polyacrylamide reacts with hydroxylamine, H_2NOH, to form hydroxamated polyacrylamides with loss of ammonia. This hydroxamation reaction occurs under alkaline conditions.

Hofmann Reaction

Polyacrylamide reacts with alkaline sodium hypochlorite, NaOCl, or calcium hypochlorite, $Ca(OCl)_2$, to form a polymer with primary amine groups. Optimum conditions for the reaction include addition of a slight molar excess of sodium hypochlorite followed by addition of concentrated sodium hydroxide at low temperature. A two-stage addition of sodium hydroxide minimizes a side reaction between the pendant amine groups and isocyanate groups formed by the Hofmann rearrangement. Cross-linking sometimes occurs if the polymer concentration is high. High temperatures can result in chain scission.

Reaction with Chlorine

Polyacrylamide reacts with chlorine under acid conditions or with NaOCl under mild alkaline conditions at low temperature to form reasonably stable N-chloropolyacrylamides. The polymers are water soluble and can provide good dry strength, wet strength, and wet web strength in paper.

CHEMISTRY OF ACRYLAMIDE COPOLYMERS

Solution polymerization, the inverse emulsion process, polymerization on moving belts, the dry bead process, microemulsion polymerization, environmentally friendly, polyacrylamides, disperson polymerization, inverse emulsions and biodegradable oils, and inverse emulsion polymerization acrylamide in near-critical and supercritical fluid conditions.

Cationic Copolymers

The largest segment of the acrylamide polymer market has been dominated by cationic copolymers. The copolymers of acrylamide (AMD) and cationic quaternary ammonium monomers are manufactured by various commercial processes, such as: The most widely used of these cationic comonomers are cationic quaternary amino derivatives of (meth)acrylic acid esters or (meth)acrylamides, and diallydimethylammonium chloride. This polymer is often used along with a high-molecular-weight anionic polyacrylamide in process-water clarification in paper deinking mills.

Anionic Copolymers

Anionic acrylamide copolymers such as poly(acrylamide-co-sodium acrylate), poly(acrylamide-co-ammonium acrylate), poly(acrylamide-co-sodium-2-acrylamido-2-methylpropanesulfonate (AMD/NaAMPS), and poly(acrylamide-co-2-acrylamido-2-methyl-1-propanesulfonic acid) (AMD/AMPS) have considerable practical importance. They can be prepared in solution, inverse emulsion, and inverse microemulsion.

AMD/AMPS copolymers and AMD/NaAMPS copolymers maintain their anionic charge at low pH and have a high tolerance to many divalent cations. They are used as flocculants for phosphate slimes, uranium leach residues, and coal refuse. There are also many oilfield applications.

SPECIFICATIONS, SHIPPING, AND STORAGE

The amount of residual acrylamide is usually determined for commercial polyacrylamides. In one method, the monomer is extracted from the polymer and the acrylamide content is determined by HPLC. A second method is based on analysis by cationic exchange chromatography.

For dry products the particle size distribution can be quickly determined by use of a shaker and a series of test sieves. Batches with small particles can present a dust hazard. The percentage of insoluble material is determined in both dry and emulsion products.

Polyacrylamide powders are typically shipped in moisture-resistant bags or fiber packs. Emulsion and solution polymers are sold in drums, tote bins, tank trucks, and tank cars. The transportation of dry and solution products is not regulated in the United States by the Department of Transportation, but emulsions require a DOT NA 1693 label.

Under normal conditions, dry polymers are stable for 1 year or more. The emulsion and solution products have somewhat shorter shelf lives.

HEALTH AND SAFETY FACTORS

Commercial Polyacrylamides

Dry cationic polyacrylamides have been tested in subchronic and developmental toxicity studies in rats. No adverse effects were observed in either study. Chronic studies of polyacrylamides in rats and dogs indicated no chronic toxicity or carcinogenicity. Dry nonionic and cationic material caused no skin and minimal eye irritation during primary irritation studies with rabbits. Dry anionic polyacrylamide did not produce any eye or skin irritation in laboratory animals. Emulsion nonionic polyacrylamide produced eye irritation in rabbits, while anionic and cationic material produced minimal eye irritation in rabbits. Emulsion nonionic, anionic, and cationic polyacrylamide produced severe, irreversible skin irritation when tested in rabbits that had the test material held in skin contact by a bandage for 24 h. This represents an exaggeration of spilling the product in a boot for several hours. When emulsion nonionic, cationic, and anionic polyacrylamides were tested under conditions representing spilling of product on clothing, only mild skin irritation was noted. Polyacrylamides are used safely for numerous indirect food packaging applications, potable water, and direct food applications.

Experimental Polyacrylamides

It is wise to treat any laboratory-prepared "experimental" polyacrylamide as if it contains substantial amounts of unreacted monomer unless it has been isolated and purified as described above. In the interest of safety, acrylamide solutions should be stored under the following conditions:

1. Maintain the storage temperature below 32°C (90°F) and above the solubility point.
2. Keep the solution free of contaminants.
3. Maintain the proper level of oxygen and Cu^{2+} inhibitors.
4. Maintain the pH at 5.2–6.0.
5. Store the solution in a container that is opaque to light.

It is recommended that these solutions be stored for no more than 3 mo due to depletion of the dissolved oxygen. All containers must be dated and no more than 93% full. Packaged acrylamide solutions should be consumed on a first-in, first-out basis.

J. Brandup and E. H. Immergut, eds., *Polymer Handbook*, 3rd ed., John Wiley & Sons, New York, 1989.

R. S. Farinato, S.-Y. Huang, and P. Hawkins in R. S. Farinato and P. L. Dubin, eds., *Colloid–Polymer Interactions: From Fundamentals to Practice*, John Wiley & Sons, New York, 1999.

P. J. Flory, *Principles of Polymer Chemistry*, Cornell University Press, Ithaca, NY, 1953.

S.-Y. Huang and D. W. Lipp in J. C. Salamone, ed., *Polymeric Materials Encyclopedia*, CRC Press, Inc., Boca Raton, Fla., 1996, p. 2427.

SUN-YI HUANG
DAVID W. LIPP
RAYMOND S. FARINATO
Cytec Industries

ACRYLIC ACID AND DERIVATIVES

The term acrylates includes derivatives of both acrylic (CH_2=CHCOOH) and methacrylic acids (CH_2=C(CH_3) COOH). Acrylic acid (propenoic acid) was first prepared in 1847 by air oxidation of acrolein. Interestingly, after use of several other routes over the past half century, it is this route, using acrolein from the catalytic oxidation of propylene, that is currently the most favored industrial process. Polymerization of acrylic esters has been known for just over a century, but it was not until 1930 that the technical difficulties of their manufacture and polymerization were overcome.

Acrylates are primarily used to prepare emulsion and solution polymers. The emulsion polymerization process provides high yields of polymers in a form suitable for a variety of applications. Acrylate emulsions are used in the preparation of both interior and exterior paints, floor polishes, and adhesives. Solution polymers of acrylates, frequently with minor concentrations of other monomers, are employed in the preparation of industrial coatings. Polymers of acrylic acid can be used as superabsorbents in disposable diapers, as well as in formulation of superior, reduced-phosphate-level detergents.

The polymeric products can be made to vary widely in physical properties through controlled variation in the ratios of monomers employed in their preparation, crosslinking, and control of molecular weight. They share common qualities of high resistance to chemical and environmental attack, excellent clarity, and attractive strength properties (see ACRYLIC ESTER POLYMERS).

PHYSICAL PROPERTIES

Physical properties of acrylic acid and representative derivatives appear in Table 1. Acrylic acid is a moderately strong carboxylic acid.

REACTIONS

Acrylic acid and its esters may be viewed as derivatives of ethylene, in which one of the hydrogen atoms has been replaced by a carboxyl or carboalkoxyl group. This functional group may display electron-withdrawing ability through inductive effects of the electron-deficient carbonyl carbon atom, and electron-releasing effects by resonance involving the electrons of the carbon–oxygen double bond. Therefore, these compounds react readily with electrophilic, free-radical, and nucleophilic agents.

Carboxylic Acid Functional Group Reactions

Polymerization is avoided by conducting the desired reaction under mild conditions and in the presence of polymerization inhibitors. Acrylic acid undergoes the reactions of carboxylic acids and can be easily converted to salts, acrylic anhydride, acryloyl chloride, and esters.

Salts are made by reaction of acrylic acid with an appropriate base in aqueous medium. They can serve as monomers and comonomers in water-soluble or water-dispersible polymers for floor polishes and flocculants.

Acrylic anhydride is formed by treatment of the acid with acetic anhydride or by reaction of acrylate salts with acryloyl chloride. *Acryloyl chloride* is made by reaction of acrylic acid with phosphorous oxychloride, or benzoyl or thionyl chloride. Neither the anhydride nor the acid chloride is of commercial interest.

Esters. Most acrylic acid is used in the form of its methyl, ethyl, and butyl esters. Specialty monomeric esters with a hydroxyl, amino, or other functional group

Table 1. Physical Properties of Acrylic Acid Derivatives

Property	Acrylic acid	Acrolein	Acrylic anhydride	Acryloyl chloride	Acrylamide
molecular formula	$C_3H_4O_2$	C_3H_4O	$C_6H_6O_3$	C_3H_3OCl	C_3H_5ON
melting point, °C	13.5	−88			84.5
boiling point[a], °C	141	52.5	38[b]	75	125[c]
refractive index[d], n_D	1.4185[e]	1.4017	1.4487	1.4337	

[a] At 101.3 kPa = 1 atm unless otherwise noted.
[b] At 0.27 kPa.
[c] At 16.6 kPa.
[d] At 20°C, unless otherwise noted.
[e] At 25°C.

are used to provide adhesion, latent cross-linking capability, or different solubility characteristics. The principal routes to esters are direct esterification with alcohols in the presence of a strong acid catalyst such as sulfuric acid, a soluble sulfonic acid, or sulfonic acid resins; addition to alkylene oxides to give hydroxyalkyl acrylic esters; and addition to the double bond of olefins in the presence of strong acid catalyst to give ethyl or secondary alkyl acrylates.

Amides. Reaction of acrylic acid with ammonia or primary or secondary amines forms amides. However, acrylamide (qv) is better prepared by controlled hydrolysis of acrylonitrile (qv). Esters can be obtained by carrying out the nitrile hydrolysis in the presence of alcohol.

Unsaturated Group Reactions

Free-radical-initiated polymerization of the double bond is the most common reaction and presents one of the more troublesome aspects of monomer manufacture and purification.

MANUFACTURE

For a method for the manufacture of acrylates to be commercially attractive, the raw material costs and utilization must be low, plant investment and operating costs not excessive, and waste disposal charges minimal.

In the 1980s cost and availability of acetylene made it an unattractive raw material for acrylate manufacture as compared to propylene, which has been readily available at attractive cost (see ACETYLENE-DERIVED CHEMICALS). As a consequence, essentially all commercial units based on acetylene, with the exception of BASF's plant at Ludwigshafen, have been shut down. All new capacity recently brought on stream or announced for construction uses the propylene route.

Propylene requirements for acrylates remain small compared to other chemical uses (polypropylene, acrylonitrile, propylene oxide, 2-propanol, and cumene for acetone and phenol). Hence, cost and availability are expected to remain attractive and new acrylate capacity should continue to be propylene-based for a few more years.

Propylene Oxidation

The propylene oxidation process is attractive because of the availability of highly active and selective catalysts and the relatively low cost of propylene. The process proceeds in two stages giving first acrolein and then acrylic acid (see ACROLEIN AND DERIVATIVES).

Single-reaction-step processes have been studied. However, higher selectivity is possible by optimizing catalyst composition and reaction conditions for each of these two steps. This more efficient utilization of raw material has led to two separate oxidation stages in all commercial facilities.

Catalysts. Catalyst performance is the most important factor in the economics of an oxidation process. It is measured by activity (conversion of reactant), selectivity (conversion of reactant to desired product), rate of production (production of desired product per unit of reactor volume per unit of time), and catalyst life (effective time on-stream before significant loss of activity or selectivity).

Catalyst performance depends on composition, the method of preparation, support, and calcination conditions. Other key properties include, in addition to chemical performance requirements, surface area, porosity, density, pore size distribution, hardness, strength, and resistance to mechanical attrition.

Oxidation Step. In the typical oxidation process the reactors are of the fixed-bed shell-and-tube type (about 3–5 m long and 2.5 cm in diameter) with a molten salt coolant on the shell side. The tubes are packed with catalyst, a small amount of inert material at the top serving as a preheater section for the feed gases. Vaporized propylene is mixed with steam and air and fed to the first-stage reactor. The feed composition is typically 5–7% propylene, 10–30% steam, and the remainder air (or a mixture of air and absorber off-gas).

Acrylic Acid Recovery. In the separations step, the acrylic acid is extracted from the absorber effluent with a solvent, such as butyl acetate, xylene, diisobutyl ketone, or mixtures, chosen for high selectivity for acrylic acid and low solubility for water and by-products. The extraction is performed using 5–10 theoretical stages in a tower or centrifugal extractor.

The extract is vacuum-distilled in the solvent recovery column, which is operated at low bottom temperatures to minimize the formation of polymer and dimer and is designed to provide acrylic acid-free overheads for recycle as the extraction solvent. A small aqueous phase in the overheads is mixed with the raffinate from the extraction step. This aqueous material is stripped before disposal both to recover extraction solvent values and minimize waste organic disposal loads.

Esterification. In this process acrylic acid, alcohol, and the catalyst, eg, sulfuric acid, together with the recycle streams are fed to the glass-lined ester reactor fitted with an external reboiler and a distillation column. Acrylate ester, excess alcohol, and water of esterification are taken overhead from the distillation column. The process is operated to give only traces of acrylic acid in the distillate. The bulk of the organic distillate is sent to the wash column for removal of alcohol and acrylic acid; a portion is returned to the top of the distillation column. If required, some base may be added during the washing operation to remove traces of acrylic acid.

Acetylene-Based Routes

Walter Reppe, the father of modern acetylene chemistry, discovered the reaction of nickel carbonyl with acetylene and water or alcohols to give acrylic acid or esters. This discovery led to several processes which have been in commercial use. The original Reppe reaction requires a stoichiometric ratio of nickel carbonyl to acetylene. The Rohm and Haas modified or semicatalytic process provides 60–80% of the carbon monoxide from a separate carbon monoxide feed and the remainder from nickel carbonyl.

Reppe's work also resulted in the high pressure route. In this process, acetylene, carbon monoxide, water, and a nickel catalyst react at about 200°C and 13.9 MPa (2016 psi) to give acrylic acid. Safety problems caused by handling of acetylene are alleviated by the use of tetrahydrofuran as an inert solvent. In this process, the catalyst is a mixture of nickel bromide with a cupric bromide promotor. The liquid reactor effluent is degassed and extracted. The acrylic acid is obtained by distillation of the extract and subsequently esterified to the desired acrylic ester. The BASF process gives acrylic acid, whereas the Rohm and Haas process provides the esters directly.

Acrylonitrile Route

This process, based on the hydrolysis of acrylonitrile, is also a propylene route since acrylonitrile (qv) is produced by the catalytic vapor-phase ammoxidation of propylene.

Important side reactions are the formation of either and addition of alcohol to the acrylate to give 3-alkoxypropionates. In addition to high raw material costs, this route is unattractive because of large amounts of sulfuric acid–ammonium sulfate wastes.

Ketene Process

The ketene process based on acetic acid or acetone as the raw material is no longer used commercially because the intermediate β-propiolactone is suspected to be a carcinogen. In addition, it cannot compete with the improved propylene oxidation process (see KETENES, KETENE DIMERS, AND RELATED SUBSTANCES).

Ethylene Cyanohydrin Process

This process, the first for the manufacture of acrylic acid and esters, has been replaced by more economical ones.

Other Syntheses

Acrylic acid and other unsaturated compounds can also be made by a number of classic elimination reactions. Acrylates have been obtained from the thermal dehydration of hydracrylic acid (3-hydroxypropanoic acid, from the dehydrohalogenation of 3-halopropionic acid derivatives, and from the reduction of dihalopropionates. Metallic oxide catalysts can produce acrylic acid by vapor phase catalytic oxidation of propane in high yield.

Vapor-Phase Condensations of Acetic Acid or Esters with Formaldehyde. Addition of a methylol group to the α-carbon of acetic acid or esters, followed by dehydration, gives the acrylates.

The procedure is technically feasible, but high recovery of unconverted raw materials is required for the route to be practical. Its development depends on the improvement of catalysts and separation methods and on the availability of low cost acetic acid and formaldehyde. Both raw materials are dependent on ample supply of low cost methanol.

Oxidative Carbonylation of Ethylene—Elimination of Alcohol from β-Alkoxypropionates. The procedure is based on the palladium catalyzed carbonylation of ethylene in the liquid phase at temperatures of 50–200°C. Esters are formed when alcohols are included. Anhydrous conditions are desirable to minimize the formation of by-products including acetaldehyde and carbon dioxide (see ACETALDEHYDE).

Although yields are excellent, the reaction medium is extremely corrosive, so high cost materials of construction are necessary. In addition, the high cost of catalyst and potential toxicity of mercury require that the inorganic materials be recovered quantitatively from any waste stream.

Dehydrogenation of Propionates. Oxidative dehydrogenation of propionates to acrylates employing vapor-phase reactions at high temperatures (400–700°C) and short contact times is possible. However, this route to acrylates is not currently of commercial interest because of the combination of low selectivity, high raw material costs, and purification difficulties.

Liquid-Phase Oxidation of Acrolein. As discussed before, the most attractive process for the manufacture of acrylates is based on the two-stage, vapor-phase oxidation of propylene. The second stage involves the oxidation of acrolein. Considerable art on the liquid-phase oxidation of acrolein is available, but this route cannot compete with the vapor-phase technology.

SPECIALTY ACRYLIC ESTERS

Higher alkyl acrylates and alkyl-functional esters are important in copolymer products, in conventional emulsion applications for coatings and adhesives, and as reactants in radiation-cured coatings and inks. In general, they are produced in direct or transesterification batch processes because of their relatively low volume.

Direct, acid catalyzed esterification of acrylic acid is the main route for the manufacture of higher alkyl esters. The most important higher alkyl acrylate is 2-ethylhexyl acrylate prepared from the available oxo alcohol 2-ethyl-1-hexanol (see ALCOHOLS, HIGHER ALIPHATIC). The most common catalysts are sulfuric or toluenesulfonic acid and sulfonic acid functional cation-exchange resins. Solvents are used as entraining agents for the removal of water of reaction. The product is washed with base to remove unreacted acrylic acid and catalyst and then purified by distillation.

HEALTH AND SAFETY FACTORS

The toxicity of common acrylic monomers has been characterized in animal studies using a variety of exposure routes. Toxicity varies with level, frequency, duration, and route of exposure. The simple higher esters of acrylic acid are usually less absorbed and less toxic than lower esters. In general, acrylates are more toxic than methacrylates.

Mucous membranes of the eyes, nose, throat, and gastrointestinal tract are particularly sensitive to irritation. Acrylates can produce a range of eye and skin irritations from slight to corrosive depending on the monomer.

Full eye protection should be worn whenever handling acrylic monomers; contact lenses must never be worn. Prolonged exposure to liquid or vapor can result in permanent eye damage or blindness. Excessive exposure to vapors causes nose and throat irritation, headaches, nausea, vomiting, and dizziness or drowsiness (solvent narcosis). Overexposure may cause central nervous system depression. Both proper respiratory protection and good ventilation are necessary wherever the possibility of high vapor concentration arises. Current TLV/TWA values are provided in *Material Safety Data Sheets* provided by manufacturers upon request.

Acrolein, acrylamide, hydroxyalkyl acrylates, and other functional derivatives can be more hazardous from a health standpoint than acrylic acid and its simple alkyl esters. Furthermore, some derivatives, such as the alkyl 2-chloroacrylates, are powerful vesicants and can cause serious eye injuries. Thus, although the hazards of acrylic acid and the normal alkyl acrylates are moderate and they can be handled safely with ordinary care to industrial hygiene, this should not be assumed to be the case for compounds with chemically different functional groups (see INDUSTRIAL HYGIENE; PLANT SAFETY; TOXICOLOGY).

USES

Most acrylic acid is consumed in the form of the polymer. The dominant share of acrylic acid is converted to esters. Today growth is in the demand for superabsorbents (SAPs) for use in diapers and hygienic products.

"Acrylic Acid" Chemical Profile, *Chemical Market Reporter* (April 1, 2002).

E. I. Becker and M. Tsutsui, eds., *Organometallic Reactions*, Vol. 3, Wiley-Interscience, New York, 1972.

T. O'Hara and co-workers, "Acrylic Acid and Derivatives" in *Ullmanns Encyclopedia of Industrial Chemistry*, 5th ed., Vol. A1, VCH, Verlagsgesellschaft mbH, Weinheim 1985, 161–176.

Storage and Handling of Acrylic and Methacrylic Esters and Acids, Bulletin 84C7, Rohm and Haas Co., Philadelphia, Pa., 1987; *Acrylic and Methacrylic Monomers—Specifications and Typical Properties, Bulletin 84C2*, Rohm and Haas Co., Philadelphia, Pa., 1986; *Rocryl Specialty Monomers—Specifications and Typical Properties, Bulletin 77S2*, Rohm and Haas Co., Philadelphia, Pa., 1989.

WILLIAM BAUER, JR.
Rohm and Haas Company

ACRYLIC ESTER POLYMERS

The first recorded preparation of the basic building block for acrylic ester polymers, acrylic acid relied on the air oxidation of acrolein. The first acrylic acid derivatives to be made were methyl acrylate and ethyl acrylate. Although these two monomers were synthesized in 1873, their utility in the polymer area was not discovered until 1880 when Kahlbaum polymerized methyl acrylate and tested its thermal stability. To his surprise, the polymerized methyl acrylate did not depolymerize at temperatures up to $320°C$. Despite this finding of incredibly high thermal stability, the industrial production of acrylic ester polymers did not take place for almost another 50 years.

PHYSICAL AND CHEMICAL PROPERTIES

Physical Properties

The structure of the acrylic ester monomers is represented by the following:

$$\begin{array}{c} H \\ \diagdown \\ H \diagup C = C \diagup H \\ \diagdown COOR \end{array}$$

The R ester group dominates the properties of the polymers formed. This R side-chain group conveys such a wide range of properties that acrylic ester polymers are used in applications varying from paints to adhesives and concrete modifiers and thickeners. The glass-transition range for a polymer describes the temperature range below which segmental pinning takes place and the polymer takes on a stiff, rigid, inflexible nature. Film properties are dramatically influenced by this changing of the polymer flexibility.

When copolymerized, the actylic ester monomers typically randomly incorporate themselves into the polymer chains according to the percentage concentration of each monomer in the reactor initial charge. Alternatively, acrylic ester monomers can be copolymerized with styrene, methacrylic ester monomers, acrylonitrile, and vinyl acetate to produce commercially significant polymers.

Acrylic ester monomers are typically synthesized from the combination of acrylic acid and an alcohol. The properties of the polymers they form are dominated by the nature of the ester side chain as well as the molecular weight of the product. Acrylic ester polymers are similar to others in that they show an improvement in properties as a function of molecular weight until a certain threshold beyond which no further improvement is observed.

Glass-Transition Temperature

The glass-transition temperature (T_g) (qv) describes the approximate temperature below which segmental rigidity (ie, loss of rotational and translational motion) sets in. Although a single value is often cited, in reality a polymer film undergoes the transition over a range of temperatures.

The rigidity upon cooling below T_g is manifested as an embrittlement of the polymer to the point where films are glass-like and incapable of handling significant mechanical stress without cracking. If, on the other hand, one raises the temperature to which a film is exposed above the glass-transition range, the polymer film becomes stretchable, soft, and elastic. For amorphous acrylic polymers, many physical properties show dramatic changes after passing through the glass-transition temperature range.

The most common thermal analyses used to determine the glass-transition temperature are dynamic mechanical analysis (dma) and differential scanning calorimetry (dsc).

The most common way of tailoring acrylic ester polymer properties is to copolymerize two or more monomers.

Molecular Weight

The properties of acrylic ester polymers (and most other types of polymers for that matter) improve as molecular weight increases. Beyond a certain level (100,000–200,000 for acrylic ester polymers) this improvement in polymer properties reaches a plateau.

Mechanical and Thermal Properties

The mechanical and thermal properties of a polymer are strongly dependent on the nature of the ester side-chain groups of its composite monomers.

Acrylic ester polymers are quite resilient to extreme conditions. This resilience gives finished products the durability that has earned acrylic polymers their reputation for value over time. In contrast to polymers of methacrylic esters, acrylic esters are stable when heated to high temperatures. Acrylic ester polymers are also resistant to oxidation.

Solublilty

Like most other properties, the side chain of acrylic ester polymers determines their solubility in organic solvents. Shorter side-chain polymers are relatively polar and will dissolve in polar solvents such as ether alcohols, ketones, and esters. With longer side-chain polymers, the solubility of a polymer shifts to the more hydrophobic solvents such as aromatic or aliphatic hydrocarbons. If a polymer is soluble in a given solvent, typically it is soluble in all proportions. Film formation occurs with the evaporation of the solvent, increase in solution viscosity, and the entanglement of the polymer chains. Phase separation and precipitation are not usually observed for solution polymers.

CHEMICAL PROPERTIES

Acrylic polymers and copolymers are highly resistant to hydrolysis. This property differentiates acrylic polymers from poly(vinyl acetate) and vinyl acetate copolymers.

Ultraviolet radiation is the other main stress encountered by polymers in the coatings arena. One hundred percent acrylic polymers are highly resistant to photodegradation because they are transparent to the vast majority of the solar spectrum. When uv-absorbing monomers, such as styrene, are incorporated into the polymer backbone, the uv-resistance of the resulting polymer decreases dramatically and a more rapid deterioration in polymer/coating properties is observed. On the other hand, a noncovalently bound uv absorber, such as hydroxybenzophenone further improves the uv stability of 100% acrylic polymers.

ACRYLIC ESTER MONOMERS AND POLYMERIZATION

Acrylic Ester Monomers

A wide variety of properties are encountered in the acrylic monomers area. A more complete listing of both monomers and their properties is found in the article ACRYLIC ACID AND DERIVATIVES.

The two most common methods for production of acrylic ester monomers are (1) the semicatalytic Reppe process, which utilizes a highly toxic nickel carbonyl catalyst, and (2) the propylene oxidation process, which primarily employs molybdenum catalyst. Because of its decreased cost and increased level of safety, the propylene oxidation process accounts for most of the acrylic ester production currently. In this process, acrolein is formed by the catalytic oxidation of propylene vapor at high temperature in the presence of steam.

A variety of methods are available for determining the purity of monomers by the measurement of their saponification equivalent and bromine number, specific gravity, refractive index, and color. Gas-liquid chromatography is useful in both the general measurement of monomer purity as well as the identification of minor species within a monomer solution.

POLYMERIZATION

Radical Polymerization

Free-radical initiators such as azo compounds, peroxides, or hydroperoxides are commonly used to initiate the polymerization of acrylic ester monomers. Photochemical and radiation-initiated polymerization are also possible. At constant temperature, the initial rate of polymerization is first order in monomer and one-half order in initiator. In addition to the standard side-chain variation, special functionality can be added to acrylic ester monomers by use of the appropriate functional alcohol.

Bulk Polymerization

Bulk polymerizations of acrylic ester monomers are characterized by the rapid formation of an insoluble network of polymers at low conversion with a concomitant rapid increase in reaction viscosity. These properties are thought to come from the chain transfer of the active radical via hydrogen abstraction from the polymer backbone. When two of these backbone radical sites propagate toward one another and terminate, a cross-link is formed.

Solution Polymerization

Of far greater commercial value than that of simple bulk polymerizations, solution polymerizations employ a co-solvent to aid in minimizing reaction viscosity as well as controlling polymer molecular weight and architecture. Lower polyacrylates are, in general, soluble in aromatic hydrocarbons, esters, ketones, and chlorohydrocarbons. Solubilities in aliphatic hydrocarbons, ethers, and alcohols are somewhat lower. As one moves to longer alcohol side-chain lengths, acrylics become insoluble in oxygenated organic solvents and soluble in aliphatic and aromatic hydrocarbons and chlorohydrocarbons.

Storage and handling equipment are typically made from steel. In order to prevent corrosion and the transfer of rust to product, moisture is typically excluded from solution polymer handling and storage systems. Because of the temperature-sensitive nature of the viscosity of solution polymers, the temperature of the storage tanks and tranfer lines is regulated either through prudent location of these facilities or through the use of insulation, heating, and cooling equipment.

Emulsion Polymerization

Emulsion polymerization is the most industrially important method of polymerizing acrylic ester monomers. The principal ingredients within this type of polymerization are water, monomer, surfactant, and water-soluble initiator. Products generated by emulsion polymerization find usage as coatings or binders in paints, paper, adhesives, textile, floor care, and leather goods markets. Because of their film-forming properties at room temperature, most commercial acrylic ester Polymers are copolymers of ethyl acrylate and butyl acrylate with methyl methacrylate.

Once packaged, the storage of acrylic latices is a nontrivial matter. Exposure of the material to extremes in temperature is avoided through prudent location of these facilities or the use of insulation, heating, and cooling equipment. Acrylic emulsion polymers, like many other types of polymers, are subject to bacterial attack. Proper adjustment of pH, addition of bactericides, and good housekeeping practices can alleviate the problems associated with bacterial growth.

Suspension Polymerization

Suspension polymers of acrylic esters are industrially used as molding powders and ion-exchange resins. In contrast to emulsion polymerization, initiation is accomplished by means of a monomer-soluble agent and occurs within the suspended monomer droplet. Water serves the same dual purpose as in emulsion (heat removal and polymer dispersion). The particle size of the final material is controlled through the control of agitation levels as well as the nature and level of the suspending agent.

Graft Copolymerization

Polymer chains can be attached to a preexisting polymer backbone of a similar or completely different composition to form what is termed a graft copolymer. Acrylic branches can be added to either synthetic or natural backbones. Attachment of graft polymer branches to preformed backbones is accomplished by chemical, photochemical, radiation, and mechanical means.

Living Polymerization

One of the most exciting areas currently in the radical polymerization of acrylic ester monomers is the field of living polymerization. Living polymers are defined as "polymers that retain their ability to propagate for a long time and grow to a desired maximum size while their degree of termination or chain transfer is still negligible."

Atom-transfer radical polymerization (ATRP) and nitroxide-mediated polymerization both show promise in terms of the ability to fine tune polymer architecture using living radical methods. The main drawbacks to the ATRP method of creating acrylic ester homo- and copolymers are the relatively long reaction times and the high levels of metal-containing initiator required.

Radiation-Induced Polymerization

Coatings can be formed through the application of high energy radiation to either monomer or oligomer mixture. Ultraviolet curing is the most widely practiced method of radiation-based initiation; this method finds its main industrial applications in the areas of coatings, printing ink, and photoresists for computer chip manufacturing. The main disadvantage of the method is that uv radiation is incapable of penetrating highly pigmented systems.

In order to avoid the problems associated with more highly pigmented systems, electron beam curing is employed. This high energy form of radiation is capable of penetrating through the entire coating regardless of the coating's pigment loading level.

Anionic Polymerization

The anionic polymerization of acrylic ester monomers, is accomplished by use of organometallic initiators in organic solvents. The main advantage to the use of anionic polymerization as opposed to other methods is its ability to generate stereoregular or block copolymers.

HEALTH AND SAFETY FACTORS

Acrylic polymers are categorized as nontoxic and have been approved for the handling and packaging by the FDA. The main concerns with acrylic polymers deal with the levels of residual monomers and the presence of non-acrylic additives (primarily surfactants) which contribute to the overall toxicity of a material. As a result, some acrylic latex dispersions can be mild skin or eye irritants.

During the manufacture of an acrylic polymer, precautions are taken to maintain temperature control. In addition to these measures, polymerizations are run under conditions wherein the reactor is closed to the outside environment to prevent the release of monomer vapor into the local environment. As for final product properties, acrylic latices are classified as nonflammable substances and solution polymers are classified as flammable mixtures.

USES

Because of their wide property range, clarity, and resistance to degradation by environmental forces, acrylic polymers are used in an astounding variety of applications: coating, textiles, adhesives, paper, leather finishing to impart impact strength and better substrate adhesion to cement, the manufacture of aqueous and solvent-based caulks and sealants, and as alternatives to nitrile rubbers in some hydraulic and gasket applications, transmission seals, vibration dampeners, dust boots, and steering and suspension seals.

J. Brandrup and E. H. Immergut, *Polymer Handbook*, 2nd ed., Wiley-Interscience, New York, 1975.

R. G. Gilbert, *Emulsion Polymerization: A Mechanistic Approach*, Academic Press. New York, 1995.

P. A. Lovell and M. S. El-Aasser *Emulsion Polymerization and Emulsion Polymers*, John Wiley & Sons, Inc., New York, 1997.

L. E. Nielsen, *Mechanical Properties of Polymers*, Van Nostrand Reinhold Co., Inc., New York, 1962, p. 122.

ROBERT V. SLONE
Rohm and Haas Company

ACRYLONITRILE

Prior to 1960, acrylonitrile (also called acrylic acid nitrile, propylene nitrile, vinyl cyanide, propenoic acid nitrile) was produced commercially by processes based on either ethylene oxide and hydrogen cyanide or acetylene and hydrogen cyanide. Today over 90% of the more than 4,000,000 metric tons produced worldwide each year are made using the Sohio-developed ammoxidation process. Acrylonitrile is among the top 50 chemicals produced in the United States as a result of the tremendous growth in its use as a starting material for a wide range of chemical and polymer products. Acrylic fibers remain the largest use of acrylonitrile; other significant uses are in resins and nitrile elastomers and as an intermediate in the production of adiponitrile and acrylamide.

PHYSICAL AND CHEMICAL PROPERTIES

Physical Properties

Acrylonitrile (C_3H_3N, mol wt = 53.064) is an unsaturated molecule having a carbon–carbon double bond conjugated with a nitrile group. It is a polar molecule because of the presence of the nitrogen heteroatom. Tables 1 and 2 list some physical properties and thermodynamic information, respectively, for acrylonitrile.

Table 1. Physical Properties of Acrylonitrile

Property	Value
appearance/odor	clear, colorless liquid with faintly pungent odor
boiling point, °C	77.3
freezing point, °C	−83.5
density, 20°C, g/cm^3	0.806
volatility, 78°C, %	>99
vapor density (air = 1)	1.8
pH (5% aqueous solution)	6.0–7.5
viscosity, 25°C, mPa·s (=cP)	0.34

Table 2. Thermodynamic Data^a

Property	Value
flash point, °C	0
autoignition temperature, °C	481
heat of combustion, liquid, 25°C, kJ/mol	1761.5
heat of vaporization, 25°C, kJ/mol	32.65

^{a} To convert kJ to kcal divide by 4.184.

Acrylonitrile is miscible in a wide range of organic solvents, including acetone, benzene, carbon tetrachloride, diethyl ether, ethyl acetate, ethylene cyanohydrin, petroleum ether, toluene, some kerosenes, and methanol.

Acrylonitrile has been characterized using infrared, Raman, and ultraviolet spectroscopies, electron diffraction, and mass spectroscopy.

CHEMICAL PROPERTIES

Acrylonitrile undergoes a wide range of reactions at its two chemically active sites, the nitrile group and the carbon–carbon double bond. Acrylonitrile polymerizes readily in the absence of a hydroquinone inhibitor, especially when exposed to light. Polymerization is initiated by free radicals, redox catalysts, or bases and can be carried out in the liquid, solid, or gas phase. Homopolymers and copolymers are most easily produced using liquid-phase polymerization (see ACRYLONITRILE POLYMERS). Acrylonitrile undergoes the reactions typical of nitriles, including hydration with sulfuric acid to form acrylamide sulfate ($C_3H_5NO \cdot H_2SO_4$, which can be converted to acrylamide (C_3H_5NO) by neutralization with a base; and complete hydrolysis to give acrylic acid ($C_3H_4O_2$). Acrylamide (qv) is also formed directly from acrylonitrile by partial hydrolysis using copper-based catalysts; this has become the preferred commercial route for acrylamide production. Industrially important acrylic esters can be formed by reaction of acrylamide sulfate with organic alcohols. Other reactions include addition of halogens across the double bond to produce dihalopropionitriles, and cyanoethylation by acrylonitrile of alcohols, aldehydes, esters, amides, nitriles, amines, sulfides, sulfones, and halides.

MANUFACTURING, PROCESSING, STORAGE, AND TRANSPORT

Acrylonitrile is produced in commercial quantities almost exclusively by the vapor-phase catalytic propylene ammoxidation process developed by Sohio:

$$C_3H_6 + NH_3 + \tfrac{3}{2} O_2 \xrightarrow{\text{catalyst}} C_3H_3N + 3 H_2O$$

The commercial process uses a fluid-bed reactor in which propylene, ammonia, and air contact a solid catalyst at 400–510°C and 49–196 kPa (0.5–2.0 kg/cm^2) gauge. It is a single-pass process with about 98% conversion of propylene, and uses about 1.1 kg propylene per kg of acrylonitrile produced. Useful by-products from the process are

HCN (about 0.1 kg per kg of acrylonitrile), which is used primarily in the manufacture of methyl methacrylate, and acetonitrile (about 0.03 kg per kg of acrylonitrile), a common industrial solvent. In the commercial operation the hot reactor effluent is quenched with water in a countercurrent absorber and any unreacted ammonia is neutralized with sulfuric acid. The resulting ammonium sulfate can be recovered and used as a fertilizer. Disposal of the process impurities has become an increasingly important aspect of the overall process, with significant attention being given to developing cost-effective and environmentally acceptable methods for treatment of the process waste streams. Current methods include deep-well disposal, wet air oxidation, ammonium sulfate separation, biological treatment, and incineration.

The active site on the surface of selective propylene ammoxidation catalyst contains three critical functionalities associated with the specific metal components of the catalyst: an α-H abstraction component such as Bi^{3+}, Sb^{3+}, or Te^{4+}; an olefin chemisorption and oxygen or nitrogen insertion component such as Mo^{6+} or Sb^{5+}; and a redox couple such as Fe^{2+}/Fe^{3+} or Ce^{3+}/Ce^{4+} to enhance transfer of lattice oxygen between the bulk and surface of the catalyst. The surface and solid-state mechanisms of propylene ammoxidation catalysis have been determined using Raman spectroscopy, neutron diffraction, x-ray absorption spectroscopy, x-ray diffraction, pulse kinetic studies, and probe molecule investigations.

Acrylonitrile must be stored in tightly closed containers in cool, dry, well-ventilated areas away from heat, sources of ignition, and incompatible chemicals. Storage vessels, such as steel drums, must be protected against physical damage, with outside detached storage preferred. Storage tanks and equipment used for transferring acrylonitrile should be electrically grounded to reduce the possibility of static spark-initiated fire or explosion.

Acrylonitrile is transported by rail car, barge, and pipeline. Department of Transportation (DOT) regulations require labeling acrylonitrile as a flammable liquid and poison.

HEALTH AND SAFETY FACTORS

Acrylonitrile is absorbed rapidly and distributed widely throughout the body following exposure by inhalation, skin contact or ingestion. However, there is little potential for significant accumulation in any organ, with most of the compound being excreted primarily as metabolites in urine.

The acute toxicity of acrylonitrile is relatively high. Signs of acute toxicity observed in animals include respiratory tract irritation and two phases of neurotoxicity, the first characterized by signs consistent with cholinergic over-stimulation and the second being CNS dysfunction, resembling cyanide poisoning. In cases of acute human intoxication, effects on the central nervous system characteristic of cyanide poisoning and effects on the liver, manifested as increased enzyme levels in the blood, have been observed.

Acrylonitrile is a severe irritant to the skin, eyes, respiratory tract and mucous membranes. It is also a skin sensitizer.

Acrylonitrile is a potent tumorigen in the rat. Tumors of the central nervous system, ear canal, and gastrointestinal tract have been observed in several studies following oral or inhalation exposure.

There is extensive occupational epidemiology data on acrylonitrile workers. These investigations have not produced consistent, convincing evidence of an increase in cancer risk, although questions remain about the power of the database to detect small excesses of rare tumors.

Experimental evaluations of acrylonitrile have not produced any clear evidence of adverse effects on reproductive function or development of offspring at doses below those producing paternal toxicity.

Acrylonitrile will polymerize violently in the absence of oxygen if initiated by heat, light, pressure, peroxide, or strong acids and bases. It is unstable in the presence of bromine, ammonia, amines, and copper or copper alloys. Neat acrylonitrile is generally stabilized against polymerization with trace levels of hydroquinone monomethyl ether and water.

Acrylonitrile is combustible and ignites readily, producing toxic combustion products such as hydrogen cyanide, nitrogen oxides, and carbon monoxide. It forms explosive mixtures with air and must be handled in well-ventilated areas and kept away from any source of ignition, since the vapor can spread to distant ignition sources and flash back.

Federal regulations (40 *CFR* 261) classify acrylonitrile as a hazardous waste and it is listed as Hazardous Waste Number U009. Disposal must be in accordance with federal (40 *CFR* 262, 263, 264, 268, 270), state, and local regulations only at properly permitted facilities. It is listed as a toxic pollutant (40 *CFR* 122.21) and introduction into process streams, storm water, or waste water systems is in violation of federal law. Federal notification regulations require that spills or leaks in excess of 100 lb (45.5 kg) be reported to the National Response Center.

USES

Acrylic fibers are used primarily for the manufacture of apparel, including sweaters, fleece wear, and sportswear, as well as for home furnishings, including carpets, upholstery, and draperies. Acrylic fibers consume about 57% of the acrylonitrile produced worldwide.

ABS resins and adiponitrile are the fastest growing uses for acrylonitrile (see ACRYLAMIDE POLYMERS). These resins normally contain about 25% acrylonitrile and are characterized by their chemical resistance, mechanical strength, and ease of manufacture.

Acrylamide is used primarily in the form of a polymer, polyacrylamide, in the paper and pulp industry and in waste water treatment as a flocculant to separate solid material from waste water streams (see ACRYLONITRILE POLYMERS). Other applications include mineral processing, coal processing, and enhanced oil recovery in which polyacrylamide solutions were found effective for displacing oil from rock.

Nitrile rubber finds broad application in industry because of its excellent resistance to oil and chemicals, its good flexibility at low temperatures, high abrasion and heat resistance (up to 120°C), and good mechanical properties. Nitrile rubber consists of butadiene–acrylonitrile copolymers with an acrylonitrile content ranging from 15 to 45% (see ELASTOMERS, SYNTHETIC, NITRILE RUBBER). In addition to the traditional applications of nitrile rubber for hoses, gaskets, seals, and oil well equipment, new applications have emerged with the development of nitrile rubber blends with poly(vinyl chloride) (PVC). These blends combine the chemical resistance and low temperature flexibility characteristics of nitrile rubber with the stability and ozone resistance of PVC. This has greatly expanded the use of nitrile rubber in outdoor applications for hoses, belts, and cable jackets, where ozone resistance is necessary.

Other acrylonitrile copolymers have found specialty applications where good gas-barrier properties are required along with strength and high impact resistance. Other applications include food, agricultural chemicals, and medical packaging.

A growing specialty application for acrylonitrile is in the manufacture of carbon fibers. They are used to reinforce composites (qv) for high performance applications in the aircraft, defense, and aerospace industries. These applications include rocket engine nozzles, rocket nose cones, and structural components for aircraft and orbital vehicles where light weight and high strength are needed. Other small specialty applications of acrylonitrile are in the production of fatty amines, ion-exchange resins, and fatty amine amides used in cosmetics, adhesives, corrosion inhibitors, and water treatment resins.

M. R. Antonio, R. G. Teller, D. R. Sandstrom, M. Mehicic, and J. F. Brazdil, *J. Phys. Chem.* **92**, 2939 (1988).

J. F. Brazdil, L. C. Glaeser, and R. K. Grasselli, *J. Phys. Chem.* **87**, 5485 (1983).

M. A. Dalin, I. K. Kolchin, and B. R. Serebryakov, *Acrylonitrile*, Technomic, Westport, Conn., 1971, 161–162.

JAMES F. BRAZDIL
BP, Nitriles Catalysis Research

ACRYLONITRILE–BUTADIENE–STYRENE (ABS) POLYMERS

Acrylonitrile–butadiene–styrene (ABS) polymers are composed of elastomer dispersed as a grafted particulate phase in a thermoplastic matrix of styrene and acrylonitrile copolymer (SAN). The presence of SAN grafted onto the elastomeric component, usually polybutadiene or a butadiene copolymer, compatabilizes the rubber with the SAN component. Property advantages provided by this graft terpolymer include excellent toughness, good dimensional stability, good processability, and chemical resistance. Property balances are controlled and optimized by adjusting elastomer particle size, morphology, microstructure, graft structure, and SAN composition and molecular weight. Therefore, although the polymer is a relatively

low cost engineering thermoplastic, the system is structurally complex. This complexity is advantageous in that altering these structural and compositional parameters allows considerable versatility in the tailoring of properties to meet specific product requirements. This versatility may be even further enhanced by adding various monomers to raise the heat deflection temperature, impart transparency, confer flame retardancy, and, through alloying with other polymers, obtain special product features.

PHYSICAL PROPERTIES

The range of properties typically available for general purpose ABS is illustrated in Table 1. Numerous grades of ABS are available including new alloys and specialty grades for high heat, plating, flaming-retardant, or static dissipative product requirements.

Impact Resistance

Toughness is a primary consideration in the selection of ABS for many applications. ABS is structured to dissipate the energy of an impact blow through shear and dilational modes of deformation. The inherent ductility of the matrix phase depends on the composition of the SAN copolymer and is reported to increase with increasing acrylonitrile content. Controlling rubber particle size, distribution, and microstructure are important in optimizing impact strength. Good adhesion between the rubber and the matrix phase is also essential and is achieved by an optimized graft structure. Typically, toughness is increased by increasing the rubber content and the molecular weight of the ungrafted SAN.

Table 1. Material Properties of General Purpose and Heat Distortion Resistant ABS

Properties	ASTM Method	High impact	Medium impact	Heat resistant
notched Izod impact at RT, J/m[b]	D256	347–534	134–320	107–347
tensile strength, MPa[c]	D638	33–43	30–52	41–52
elongation to yield, %	D638	2.8–3.5	2.3–3.5	2.8–3.5
Rockwell hardness	D785	80–105	105–112	100–111
heat deflection[d], °C at 1820 kPa[e]	D648	96–102	93–104	104–116
heat deflection[d], °C at 455 kPa[e]	D648	99–107	102–107	110–118
Vicat softening pt, °C	D1525	91–106	94–107	104–118
dielectric strength, kV/mm	D149	16–31	16–31	14–35
dielectric constant, $\times 10^6$ Hz	D150	2.4–3.8	2.4–3.8	2.4–3.8

[a] Material taken from Rubin ed., *Handbook of Plastic Materials and Technology*.
[b] To convert J/m to ft·lb/in. divide by 53.4.
[c] To convert MPa to psi multiply by 145.
[d] Annealed.
[e] To convert kPa to psi multiply by 0.145.

Rheology

Effects of structure of ABS on viscosity functions can be distinguished by considering effects at lower shear rates (<10/s) vs higher shear rates. At higher shear rates melt viscosity is primarily determined by ungrafted SAN structure and the percentage of graft phase.

By contrast, the graft phase structure has a marked effect on viscosity at small deformation rates. The long time relaxation spectra are affected by rubber particle–particle interactions, which are strongly dependent on particle size, grafting, morphology, and rubber content. Depending on particle surface area, a minimum amount of graft is needed to prevent the formation of three-dimensional networks of associated rubber particles.

Gloss

Surface gloss values can be achieved ranging from a very low matte finish at <10% (60° Gardner) to high gloss in excess of 95%. Gloss is dependent on the specific grade and the mold or polishing roll surface.

Electrical Properties

A new family of ABS products exhibiting electrostatic dissipative properties without the need for nonpolymeric additives or fillers (carbon black, metal) is now also commercially available. (See Table 1.)

Thermal Properties

ABS is also used as a base polymer in high performance alloys.

Color

ABS is sold as an unpigmented powder, unpigmented pellets, precolored pellets matched to exacting requirements, and "salt-and-pepper" blends of ABS and color concentrate. Color concentrates can also be used for online coloring during molding.

CHEMICAL PROPERTIES

Chemical Resistance

The term chemical resistance is generally used in an applications context and refers to resistance to the action of solvents in causing swelling or stress cracking as well as to chemical reactivity. In ABS the polar character of the nitrile group reduces interaction of the polymer with hydrocarbon solvents, mineral and vegetable oils, waxes, and related household and commercial materials. Good chemical resistance provided by the presence of acrylonitrile as a comonomer combined with relatively low water absorptivity (<1%) results in high resistance to staining agents (eg, coffee, grape juice, beef blood) typically encountered in household applications.

Processing Stability

Processing can influence resultant properties by chemical and physical means. Degradation of the rubber and matrix phases has been reported under very severe conditions. Thus the proper selection and control of process variables are important to maintain optimum performance in molded parts. Antioxidants (qv) added at the compounding step have been shown to help retention of physical properties upon processing.

Thermal Oxidative Stability

ABS undergoes autoxidation and the kinetic features of the oxygen consumption reaction are consistent with an autocatalytic free-radical chain mechanism. Comparisons of the rate of oxidation of ABS with that of polybutadiene and styrene–acrylonitrile copolymer indicate that the polybutadiene component is significantly more sensitive to oxidation than the thermoplastic component. Oxidation of polybutadiene under these conditions results in embrittlement of the rubber because of cross-linking; such embrittlement of the elastomer in ABS results in the loss of impact resistance.

Photooxidative Stability

Unsaturation present as a structural feature in the polybutadiene component of ABS (also in high impact polystyrene, rubber-modified PVC, and butadiene-containing elastomers) also increases liability with regard to photooxidative degradation. Such degradation only occurs in the outermost layer, and impact loss upon irradiation can be attributed to embrittlement of the rubber and possibly to scission of the grafted styrene–acrylonitrile copolymer. Appearance changes such as yellowing are also induced by irradiation and caused by chromophore formation in both the polybutadiene and styrene–acrylonitrile copolymer components.

Flammability

Flame retardancy is achieved by utilizing halogen in combination with antimony oxide or by alloys with PVC or PC. A new FR grade utilizing polymer-bound bromine has been developed to avoid additive bloom and toxicity.

MANUFACTURE AND PROCESSING

Manufacture

All manufacturing processes for ABS involve the polymerization of styrene and acrylonitrile monomers in the presence of an elastomer (typically polybutadiene or a butadiene copolymer) to produce SAN that has been chemically bonder or "grafted" to the rubber component termed the "substrate." These processes include rubber chemistry, graft chemistry, *emulsion process*, rubber substrate process, graft process, resin recovery process, air and water treatment, mass polymerization process, suspension process, and compounding.

PROCESSING

Good thermal stability plus shear thinning allow wide flexibility in viscosity control for a variety of processing methods. ABS exhibits non-Newtonian viscosity behavior. Viscosity can also be reduced by raising melt temperature; typically increasing the melt temperature 20 to 30°C within the allowable processing range reduces the melt viscosity by about 30%. ABS can be processed by all the techniques used for other thermoplastics: compression and injection molding, extrusion, calendering, and blow-molding (see PLASTICS PROCESSING).

Material Handling and Drying

Although uncompounded powders are available from some suppliers, most ABS is sold in compounded pellet form. The pellets are either precolored or natural to be used for in-house coloring using dry or liquid colorants or color concentrates. These pellets have a variety of shapes including diced cubes, square and cylindrical strands, and spheroids. The shape and size affect several aspects of material handling such as bulk density, feeding of screws, and drying (qv).

Secondary Operations

Thermoforming. ABS is a versatile thermoforming material. Forming techniques in use are positive and negative mold vacuum forming, bubble and plug assist, snapback and single- or twin-sheet pressure forming. It is easy to thermoform ABS over the wide temperature range of 120 to 190°C. As-extruded sheet should be wrapped to prevent scuffing and moisture pickup. Predrying sheet that has been exposed to humid air prevents surface defects; usually 1 to 3 h at 70–80°C suffices. Thick sheet should be heated slowly to prevent surface degradation and provide time for the core temperature to reach the value needed for good formability.

Cold Forming. Some ABS grades have ductility and toughness such that sheet can be cold formed from blanks 0.13–6.4 mm thick using standard metal-working techniques

Other Operations

Metallizing. ABS can be metallized by electroplating, vacuum deposition, and sputtering. Electroplating (qv) produces the most robust coating; progress is being made on some of the environmental concerns associated with the chemicals involved by the development of a modified chemistry. Attention must be paid to the molding and handling of the ABS parts since contamination can affect plate adhesion, and surface defects are magnified after plating. Also, certain aspects of part design become more important with plating (see ELECTROLESS PLATING; METALLIC COATINGS).

Fastening, Bonding, and Joining. Often parts can be molded with various snap-fit designs and bosses to receive rivets or self-tapping screws. Thermal-welding techniques that are easily adaptable to ABS are spin welding, hot plate welding, hot gas welding, induction welding, ultrasonic welding, and vibrational welding. ABS can also be nailed, stapled, and riveted. There are a variety of adhesives and solvent cements for bonding ABS to itself or other materials such as wood, glass, and metals.

USES

Its broad property balance and wide processing window has allowed ABS to become the largest selling engineering thermoplastic.

The largest market for ABS resins worldwide is for appliances. The majority of this consumption was for major appliances; extruded/thermoformed door and tank liners lead the way. Transparent ABS grades are also used in refrigerator crisper trays. Other applications in the appliance market include injection-molded housings for kitchen appliances, power tools, vacuum sweepers, sewing machines, and hair dryers. Transportation was the second largest market. Uses are numerous and include both interior and exterior applications.

Pipe and fittings remain a significant market for ABS at 13%, particularly in North America.

A large "value-added" market for ABS is business machines and other electrical and electronic equipment. Although general purpose injection-molding grades meet the needs of applications such as telephones and micro floppy disk covers, significant growth exists in more demanding flame-retardant applications such as computer housings and consoles.

Another use is medical application. Miscellaneous applications included toys, luggage, lawn and garden products, shower stalls, furniture and ABS resin blends with other polymers.

"ABS Resins, Chemical Profile," *Chemical Market Reporter* **263**, 27 (Jan. 13, 2003).

R. D. Deanin, I. S. Rabinovic, and A. Llompart, in *Multicomponent Polymer Systems* (Adv. in Chem. Ser. No. 99), American Chemical Society, Washington, D.C., 1971, p. 229.

M. G. Huguet and T. R. Paxton, *Colloidal and Morphological Behavior of Block and Graft Copolymers*, Plenum, New York, 1971, pp. 183–192.

C. T. Pillichody and P. D. Kelley, in I. I. Rubin, ed., *Handbook of Plastic Materials and Technology*, John Wiley & Sons, Inc., New York, 1990, Chapt. 3.

DONALD M. KULICH
S. K. GAGGAR
V. LOWRY
R. STEPIEN
GE Plastics, Technology Center

ACRYLONITRILE POLYMERS, SURVEY AND STYRENE–ACRYLONITRILE (SAN)

Acrylonitrile (AN), C_3H_3N, is a versatile and reactive monomer that can be polymerized under a wide variety of conditions and copolymerized with an extensive range of other vinyl monomers. Because of the difficulty of melt processing the homopolymer, acrylonitrile is usually copolymerized to achieve a desirable thermal stability, melt flow, and physical properties. As a comonomer, acrylonitrile (qv) contributes hardness, rigidity, solvent and light resistance, gas impermeability, and the ability to orient.

The utility of acrylonitrile in thermoplastics was first realized in its copolymer with styrene, C_8H_8. Styrene is the largest volume of comonomer for acrylonitrile in thermoplastic applications. Styrene–acrylonitrile (SAN) copolymers are inherently transparent plastics with high heat resistance and excellent gloss and chemical resistance. They are also characterized by good hardness, rigidity, dimensional stability, and load-bearing strength (due to relatively high tensile and flexural strengths). Because of their inherent transparency, SAN copolymers are most frequently used in clear application. These optically clear materials can be readily processed by extrusion and injection molding, but they lack real impact resistance.

The development of acrylonitrile–butadiene–styrene (ABS) resins, which contain an elastomeric component within a SAN matrix to provide toughness and impact strength, further boosted commercial application of the basic SAN copolymer as a portion of these rubber-toughened thermoplastics (see ACRYLONITRILE POLYMERS, ABS RESINS). When SAN is grafted onto a butadiene-based rubber, and optionally blended with additional SAN, the two-phase thermoplastic ABS is produced. ABS has the useful SAN properties of rigidity and resistance to chemicals and solvents, while the elastomeric component contributes real impact resistance. Because ABS is a two-phase system and each phase has a different refractive index, the final ABS is normally opaque. A clear ABS can be made by adjusting the refractive indexes through the inclusion of another monomer such as methyl methacrylate. ABS is a versatile material and modifications have brought out many specialty grades such as clear ABS and high-temperature and flame-retardant grades.

Another class of AN copolymers and multipolymers contains more than 60% acrylonitrile. These are commonly known as barrier resins and have found their greatest acceptance where excellent barrier properties toward gases, chemicals, and solvents are needed. They may be processed into bottles, sheets, films, and various laminates, and have found wide usage in the packaging industry (see BARRIER POLYMERS).

Acrylonitrile has found its way into a great variety of other polymeric compositions based on its polar nature and reactivity, imparting to other systems some or all of the properties noted above. Some of these areas include adhesives and binders, antioxidants, medicines, dyes, electrical insulations, emulsifying agents, graphic arts, insecticides, leather, paper, plasticizers, soil-modifying agents, solvents, surface coatings, textile treatments, viscosity modifiers, azeotropic distillations, artificial organs, lubricants, asphalt additives, water-soluble polymers, hollow spheres, cross-linking agents, and catalyst treatments.

SAN PHYSICAL PROPERTIES AND TEST METHODS; CHEMICAL PROPERTIES

Physical Properties

SAN resins possess many physical properties desired for thermoplastic applications. They are characteristically

Table 1. Physical/Mechanical Properties of Commercial Injection-Molded SAN Resins[a]

	Bayer Lustran 31-2060	Dow Tyril 100	ASTM Method
specific gravity (23/23°C)	1.07	1.07	D 792
Vicat softening point (°C)	110	108	D 1525
tensile strength, MPa[b]	72.4	71.7	D 638
ultimate elongation at breakage (%)	3.0	2.5	D 638
flexural modulus, GPa[c]	3.45	3.87	D 790
impact strength, notched Izod (J/m[d])	21.4 @ 0.125 in.	16.0 @ 0.125 in.	D 256
melt flow rate (g/10 min)	8.0	8.0	D 1238, cond. 1
refractive index, n_D	1.570	1.570	D 542
mold shrinkage (in./in.)	0.003–0.004	0.004–0.005	D 955
transmittance at 0.125-in. thickness (%)	89.0	89.0	D 1003
haze at 0.125-in. thickness (%)	0.8	0.6	D 1003

[a] Product literature.
[b] To convert MPa to psi, multiply by 145.
[c] To convert GPa to psi, multiply by 145,000.
[d] To convert J/m to ft lb/in., divide by 53.39.

hard, rigid, and dimensionally stable with load-bearing capabilities. They are also transparent, have high heat distortion temperatures, possess excellent gloss and chemical resistance, and adapt easily to conventional thermoplastic fabrication techniques.

SAN polymers are random linear amorphous copolymers. Physical properties are dependent on molecular weight and the percentage of acrylonitrile. An increase of either generally improves physical properties, but may cause a loss of processability or an increase in yellowness. Various processing aids and modifiers can be used to achieve a specific set of properties. Modifiers may include mold release agents, UV stabilizers, antistatic aids, elastomers, flow and processing aids, and reinforcing agents such as fillers and fibers. Methods for testing and some typical physical properties are listed in Table 1.

The properties of SAN resins depend on their acrylonitrile content. Both melt viscosity and hardness increase with increasing acrylonitrile level.

SAN CHEMICAL PROPERTIES

Chemical Properties

SAN resins show considerable resistance to solvents and are insoluble in carbon tetrachloride, ethyl alcohol, gasoline, and hydrocarbon solvents. They are swelled by solvents such as benzene, ether, and toluene. Polar solvents such as acetone, chloroform, dioxane, methyl ethyl ketone, and pyridine will dissolve SAN.

The properties of SAN are significantly altered by water absorption. The equilibrium water content increases with temperature while the time required decreases. A large decrease in T_g can result. Strong aqueous bases can degrade SAN by hydrolysis of the nitrile groups.

The molecular weight of SAN can be easily determined by either intrinsic viscosity or size-exclusion chromatography (SEC). Relationships for both multipoint and single point viscosity methods are available.

Residual monomers in SAN have been a growing environmental concern and can be determined by a variety of methods. Monomer analysis can be achieved by polymer solution or directly from SAN emulsions followed by "head space" gas chromatography (GC). Liquid chromatography (LC) is also effective.

SAN MANUFACTURE

The reactivities of acrylonitrile and styrene radicals toward their monomers are quite different, resulting in SAN copolymer compositions that vary from their monomer compositions. Further complicating the reaction is the fact that acrylonitrile is soluble in water (see ACRYLONITRILE) and slightly different behavior is observed between water-based emulsion and suspension systems and bulk or mass polymerizations. SAN copolymer compositions can be calculated from copolymerization equations and published reactivity ratios. The difference in radical reactivity causes the copolymer composition to drift as polymerization proceeds, except at the azeotrope composition where copolymer composition matches monomer composition. When SAN copolymer compositions vary significantly, incompatibility results, causing loss of optical clarity, mechanical strength, and moldability, as well as heat, solvent, and chemical resistance. The termination step has been found to be controlled by diffusion even at low conversions, and the termination rate constant varies with acrylonitrile content. The average half-life of the radicals increases with styrene concentration from 0.3 s at 20 mol% to 6.31 s with pure styrene. Further complicating SAN manufacture is the fact that both the heat and rate of copolymerization vary with monomer composition.

An emulsion model that assumes the locus of reaction to be inside the particles and considers the partition of AN between the aqueous and oil phases has been developed. The model predicts copolymerization results very well when bulk reactivity ratios of 0.32 and 0.12 for styrene and acrylonitrile, respectively, are used.

Commercially, SAN is manufactured by three processes: emulsion, suspension, and continuous mass (or bulk).

Emulsion Process

The emulsion polymerization process utilizes water as a continuous phase with the reactants suspended as microscopic particles. This low-viscosity system allows facile mixing and heat transfer for control purposes. An emulsifier is generally employed to stabilize the water-insoluble monomers and other reactants, and to prevent reactor fouling. With SAN the system is composed of water, monomers, chain-transfer agents for molecular weight control, emulsifiers, and initiators. Both batch and semibatch processes are employed. Copolymerization is normally carried out at 60 to 100°C to conversions of ~97%. Lower-temperature polymerization can be achieved with redox-initiator systems.

Compositional control for other than azeotropic compositions can be achieved with both batch and semibatch emulsion processes. Continuous addition of the faster reacting monomer, styrene, can be practiced for batch systems, with the feed rate adjusted by computer through gas chromatographic monitoring during the course of the reaction. For semibatch processes, adding the monomers at a rate that is slower than copolymerization can achieve equilibrium. It has been found that constant composition in the emulsion can be achieved after ca 20% of the monomers have been charged.

Suspension Process

Like the emulsion process, water is the continuous phase for suspension polymerization, but the resultant particle size is larger, well above the microscopic range. The suspension medium contains water, monomers, molecular weight control agents, initiators, and suspending aids. Stirred reactors are used in either batch or semibatch mode. The components are charged into a pressure vessel and purged with nitrogen. Copolymerization is carried out at 128°C for 3 h and then at 150°C for 2 h. Steam stripping removes residual monomers, and the polymer beads are separated by centrifugation for washing and final dewatering.

Continuous Mass Process

The continuous mass process has several advantages, including high space–time yield and good-quality products uncontaminated with residual ingredients such as emulsifiers or suspending agents. SAN manufactured by this method generally has superior color and transparency and is preferred for applications requiring good optical properties. It is a self-contained operation without waste treatment or environmental problems since the products are either polymer or recycled back to the process.

In practice, the continuous mass polymerization is rather complicated. Because of the high viscosity of the copolymerizing mixture, complex machinery is required to handle mixing, heat transfer, melt transport, and devolatilization. In addition, considerable time is required to establish steady-state conditions in both a stirred tank reactor and a linear flow reactor. Copolymerization is normally carried out between 100 and 200°C. Solvents are used to reduce viscosity or the conversion is kept to 40–70%, followed by devolatilization to remove solvents and monomers. Devolatilization is carried out from 120 to 260°C under vacuum at less than 20 kPa (2.9 psi). The devolatilized melt is then fed through a strand die, cooled, and pelletized.

Processing

SAN copolymers may be processed using the conventional fabrication methods of extrusion, blow molding, injection molding, thermoforming, and casting. Small amounts of additives, such as antioxidants, lubricants, and colorants, may also be used. Typical temperature profiles for injection molding and extrusion of predried SAN resins are as follows.

HEALTH AND SAFETY FACTORS

SAN resins themselves appear to pose few health problems in that SAN resins are allowed by FDA to be used by the food and medical for certain applications under prescribed conditions. The main concern over SAN resin use is that of toxic residuals, eg, acrylonitrile, styrene, or other polymerization components such as emulsifiers, stabilizers, or solvents. Each component must be treated individually for toxic effects and safe exposure level.

Acrylonitrile is believed to behave as an enzyme inhibitor of cellular metabolism, and it is classified as a probable human carcinogen of medium carcinogenic hazard and can affect the cardiovascular system and kidney and liver functions. Direct potential consumer exposure to acrylonitrile through consumer product usage is low because of little migration of the monomer from such products; the concentrations of acrylonitrile in consumer products are estimated to be less than 15 ppm in SAN resins.

Styrene, a main ingredient of SAN resins, is a possible human carcinogen (IARC Group 2B/EPA-ORD Group C). It is an irritant to the eyes and respiratory tract, and while prolonged exposure to the skin may cause irritation and central nervous system effects such as headache, weakness, and depression, harmful amounts are not likely to absorbed through the skin.

USES

Acrylonitrile copolymers offer useful properties, such as rigidity, gas barrier, chemical and solvent resistance, and toughness. These properties are dependent upon the acrylonitrile content in the copolymers. SAN copolymers offer low cost, rigidity, processability, chemical and solvent resistance, transparency, and heat resistance in which the properties provide the advantages over other competing transparent/clear resins, such as: polymethyl methacrylate, polystyrene, polycarbonate, and styrene–butadiene copolymers. SAN copolymers are widely used in goods such as housewares, packaging, appliances, interior automotive lenses, industrial battery cases, and medical parts.

Acrylonitrile copolymers have been widely used in films and laminates for packaging due to their excellent barrier properties. In addition to laminates, SAN copolymers are used in membranes, controlled-release formulations, polymeric foams, fire-resistant compositions, ion-exchange resins, reinforced paper, concrete and mortar

compositions, safety glasses, solid ionic conductors, negative resist materials, electrophotographic toners, and optical recording as well. SAN copolymers are also used as coatings, dispersing agents for colorants, carbon-fiber coatings for improved adhesion, and synthetic wood pulp. SAN copolymers have been blended with aromatic polyesters to improve hydrolytic stability, with methyl methacrylate polymers to form highly transparent resins, and with polycarbonate to form toughened compositions with good impact strength.

Acrylonitrile has contributed the desirable properties of rigidity, high temperature resistance, clarity, solvent resistance, and gas impermeability to many polymeric systems. Its availability, reactivity, and low cost ensure a continuing market presence and provide potential for many new applications.

D. M. Bennett, "Acrylic–Styrene–Acrylonitrile," in R. Juran, ed., *Modern Plastics Encyclopedia*, McGraw-Hill, Inc., New York, 1989, p. 96.

J. Brandup and E. H. Immergut, eds., *Polymer Handbook*, 3rd ed., Wiley-Interscience, New York, 1989, pp. II-165–II-171.

J. Santodonato and co-workers, *Monograph on Human Exposure to Chemicals in the Work Place; Styrene, PB86-155132*, Syracuse, N.Y., July 1985.

F. L. Reithel, "Styrene–Acrylonitrile (SAN)," in R. Juran, ed., *Modern Plastics Encyclopedia*, McGraw-Hill, Inc., New York, 1989, p. 105.

MICHAEL M WU
BP Amoco Chemicals

ACTINIDES AND TRANSACTINIDES

ACTINIDES

The actinide elements are a group of chemically similar elements with atomic numbers 89 through 103 and their names, symbols, and atomic numbers, are given in Table 1 (see THORIUM AND THORIUM COMPOUNDS; URANIUM AND URANIUM COMPOUNDS; PLUTONIUM AND PLUTONIUM COMPOUNDS; NUCLEAR REACTORS; and RADIOISOTOPES).

Each of the elements has a number of isotopes, all radioactive and some of which can be obtained in isotopically pure form. More than 200 in number and mostly synthetic in origin, they are produced by neutron or charged-particle-induced transmutations.

Thorium and uranium have long been known, and uses dependent on their physical or chemical, not on their nuclear, properties were developed prior to the discovery of nuclear fission.

Thorium, uranium, and plutonium are well known for their role as the basic fuels (or sources of fuel) for the release of nuclear energy. The importance of the remainder of the actinide group lies at present, for the most part, in the realm of pure research, but a number of practical applications are also known. The actinides present a

Table 1. The Actinide Elements

Atomic number	Element	Symbol	Mass number[a]
89	actinium	Ac	227
90	thorium	Th	232
91	protactinium	Pa	231
92	uranium	U	238
93	neptunium	Np	237
94	plutonium	Pu	242
95	americium	Am	243
96	curium	Cm	248
97	berkelium	Bk	249
98	californium	Cf	249
99	einsteinium	Es	254
100	fermium	Fm	257
101	mendelevium	Md	258
102	nobelium	No	259
103	lawrencium	Lr	262

[a] Mass number of longest lived or most available isotope.

storage-life problem in nuclear waste disposal and consideration is being given to separation methods for their recovery prior to disposal (see WASTE TREATMENT, HAZARDOUS WASTE; NUCLEAR REACTORS, WASTE MANAGEMENT).

Source

Only the members of the actinide group through Pu have been found to occur in nature. Thorium and uranium occur widely in the earth's crust in combination with other elements, and, in the case of uranium, in significant concentrations in the oceans. The extraction of these two elements from their ores has been studied intensively and forms the basis of an extensive technology. With the exception of uranium and thorium, the actinide elements are synthetic in origin for practical purposes; ie, they are products of nuclear reactions. High neutron fluxes are available in modern nuclear reactors, and the most feasible method for preparing actinium, protactinium, and most of the actinide elements is through the neutron irradiation of elements of high atomic number.

Experimental Methods of Investigation

All of the actinide elements are radioactive and, except for thorium and uranium, special equipment and shielded facilities are usually necessary for their manipulation. On a laboratory scale, enclosed containers (gloved boxes) are generally used for safe handling of these substances. In some work, all operations are performed by remote control.

The study of the chemical behavior of concentrated preparations of short-lived isotopes is complicated by the rapid production of hydrogen peroxide in aqueous solutions and the destruction of crystal lattices in solid compounds. These effects are brought about by heavy recoils of high energy alpha particles released in the decay process.

Special techniques for experimentation with the actinide elements other than Th and U have been devised because of the potential health hazard to the experimenter and the small amounts available. In addition, investigations are frequently carried out with the substance present in very low concentration as a radioactive tracer. Such procedures continue to be used to some extent with the heaviest actinide elements, where only a few score atoms may be available. Tracer studies offer a method for obtaining knowledge of oxidation states, formation of complex ions, and the solubility of various compounds. These techniques are not applicable to crystallography, metallurgy, and spectroscopic studies.

Microchemical or ultramicrochemical techniques are used extensively in chemical studies of actinide elements. If extremely small volumes are used, microgram or lesser quantities of material can give relatively high concentrations in solution. Balances of sufficient sensitivity have been developed for quantitative measurements with these minute quantities of material. Since the amounts of material involved are too small to be seen with the unaided eye, the actual chemical work is usually done on the mechanical stage of a microscope, where all of the essential apparatus is in view. Compounds prepared on such a small scale are often identified by x-ray crystallographic methods.

Properties

The close chemical resemblance among many of the actinide elements permits their chemistry to be described for the most part in a correlative way.

Oxidation States. The oxidation states of the actinide elements are summarized in Table 2. The most stable states are designated by bold face type and those which are very unstable are indicated by parentheses. These latter states do not exist in aqueous solutions and have been produced only in solid compounds.

The actinide elements exhibit uniformity in ionic types. In acidic aqueous solution, there are four types of cations.

Corresponding ionic types are similar in chemical behavior, although the oxidation–reduction relationships and therefore the relative stabilities differ from element to element.

Table 2. The Oxidation States of the Actinide Elements

						Atomic number and element								
89 Ac	90 Th	91 Pa	92 U	93 Np	94 Pu	95 Am	96 Cm	97 Bk	98 Cf	99 Es	100 Fm	101 Md	102 No	103 Lr
						(2)			(2)	(2)	2	2	**2**	
3	(3)	(3)	3	3	3	**3**	**3**	**3**	**3**	**3**	**3**	3	3	**3**
	4	4	4	4	**4**	4	4	4	(4)					
		5	5	**5**	5	5								
			6	6	6	6								
				7	(7)									

Hydrolysis and Complex Ion Formation. Hydrolysis and complex ion formation are closely related phenomena. Of the actinide ions, the small, highly charged M^{4+} ions exhibit the greatest degree of hydrolysis and complex ion formation. The degree of hydrolysis or complex ion formation decreases in the order $M^{4+} > MO_2^{2+} > M^{3+} > MO_2^+$. Presumably the relatively high tendency toward hydrolysis and complex ion formation of MO_2^{2+} ions is related to the high concentration of charge on the metal atom. On the basis of increasing charge and decreasing ionic size, it could be expected that the degree of hydrolysis for each ionic type would increase with increasing atomic number.

The tendency toward complex ion formation of the actinide ions is determined largely by the factors of ionic size and charge.

Actinide ions form complex ions with a large number of organic substances. Their extractability by these substances varies from element to element and depends markedly on oxidation state. A number of important separation procedures are based on this property.

Metallic State. The actinide metals, like the lanthanide metals, are highly electropositive. They can be prepared by the electrolysis of molten salts or by the reduction of a halide with an electropositive metal, such as calcium or barium.

Solid Compounds. The tripositive actinide ions resemble tripositive lanthanide ions in their precipitation reactions. Tetrapositive actinide ions are similar in this respect to Ce^{4+}. Thus the fluorides and oxalates are insoluble in acid solution, and the nitrates, sulfates, perchlorates, and sulfides are all soluble.

Thousands of compounds of the actinide elements have been prepared, and the properties of some of the important binary compounds are summarized in Table 3. The binary compounds with carbon, boron, nitrogen, silicon, and sulfur are of interest, because of their stability at high temperatures. A large number of ternary compounds, including numerous oxyhalides, and more complicated compounds have been synthesized and characterized. Also, hundreds of actinide organic derivatives, including

Table 3. Properties and Crystal Structure Data for Important Actinide Binary Compounds

Compound	Color	Melting point, °C	Symmetry	Space group or structure type	Density, g/mL
AcH_2	black		cubic	fluorite ($Fm3m$)	8.35
ThH_2	black		tetragonal	$F4/mmm$	9.50
Th_4H_{15}	black		cubic	$I43d$	8.25
α-PaH_3	gray		cubic	$Pm3n$	10.87
β-PaH_3	black		cubic	β-W	10.58
α-UH_3	?		cubic	$Pm3n$	11.12
β-UH_3	black		cubic	β-W ($Pm3n$)	10.92
NpH_2	black		cubic	fluorite	10.41
NpH_3	black		trigonal	$P3c1$	9.64
PuH_2	black		cubic	fluorite	10.40
PuH_3	black		trigonal	$P3c1$	9.61
AmH_2	black		cubic	fluorite	10.64
AmH_3	black		trigonal	$P3c1$	9.76
CmH_2	black		cubic	fluorite	10.84
CmH_3	black		trigonal	$P\bar{3}c1$	10.06
BkH_2	black		cubic	fluorite	11.57
BkH_3	black		trigonal	$P\bar{3}c1$	10.44
Ac_2O_3	white		hexagonal	La_2O_3 ($P\bar{3}m1$)	9.19
Pu_2O_3	?		cubic	$Ia3$ (Mn_2O_3)	10.20
Pu_2O_3	black	2085	hexagonal	La_2O_3	11.47
Am_2O_3	tan		hexagonal	La_2O_3	11.77
Am_2O_3	reddish brown		cubic	$Ia3$	10.57
Cm_2O_3	white to faint tan	2260	hexagonal	La_2O_3	12.17
Cm_2O_3			monoclinic	$C2/m$ (Sm_2O_3)	11.90
Cm_2O_3	white		cubic	$Ia3$	10.80
Bk_2O_3	light green		hexagonal	La_2O_3	12.47
Bk_2O_3	yellow-green		monoclinic	$C2/m$	12.20
Bk_2O_3	yellowish brown		cubic	$Ia3$	11.66
Cf_2O_3	pale green		hexagonal	La_2O_3	12.69
Cf_2O_3	lime green		monoclinic	$C2/m$	12.37
Cf_2O_3	pale green		cubic	$Ia3$	11.39
Es_2O_3	white		hexagonal	La_2O_3	12.7
Es_2O_3	white		monoclinic	$C2/m$	12.4
Es_2O_3	white		cubic	$Ia3$	11.79
ThO_2	white	ca 3050	cubic	fluorite	10.00
PaO_2	black		cubic	fluorite	10.45
UO_2	brown to black	2875	cubic	fluorite	10.95

Table 3. (*Continued*)

Compound	Color	Melting point, °C	Symmetry	Space group or structure type	Density, g/mL
NpO_2	apple green		cubic	fluorite	11.14
PuO_2	yellow-green to brown	2400	cubic	fluorite	11.46
AmO_2	black		cubic	fluorite	11.68
CmO_2	black		cubic	fluorite	11.92
BkO_2	yellowish-brown		cubic	fluorite	12.31
CfO_2	black		cubic	fluorite	12.46
Pa_2O_5	white		cubic	fluorite-related	11.14
Np_2O_5	dark brown		monoclinic	$P2_1/c$	8.18
α-U_3O_8	black-green	1150 (dec)	orthorhombic	$C2mm$	8.39
β-U_3O_8	black-green		orthorhombic	$Cmcm$	8.32
γ-UO_3	orange	650 (dec)	orthorhombic	$Fddd$	7.80
$AmCl_2$	black		orthorhombic	$Pbnm$ ($PbCl_2$)	6.78
$CfCl_2$	red-amber		?		
$AmBr_2$	black		tetragonal	$SrBr_2$ (P4/n)	7.00
$CfBr_2$	amber		tetragonal	$SrBr_2$	7.22
ThI_2	gold		hexagonal	$P6_3/mmc$	7.45
AmI_2	black	ca 700	monoclinic	EuI_2 ($P2_1/c$)	6.60
CfI_2	violet		hexagonal	CdI_2 ($P\bar{3}m1$)	6.63
CfI_2	violet		rhombohedral	$CdCl_2$ ($R\bar{3}m$)	6.58
AcF_3	white		trigonal	LaF_3 (P3c1)	7.88
UF_3	black	>1140 (dec)	trigonal	LaF_3	8.95
NpF_3	purple		trigonal	LaF_3	9.12
PuF_3	purple	1425	trigonal	LaF_3	9.33
AmF_3	pink	1393	trigonal	LaF_3	9.53
CmF_3	white	1406	trigonal	LaF_3	9.85
BkF_3	yellow-green		orthorhombic	YF_3 (Pnma)	9.70
BkF_3	yellow-green		trigonal	LaF_3	10.15
CfF_3	light green		orthorhombic	YF_3	9.88
CfF_3	light green		trigonal	LaF_3	10.28
$AcCl_3$	white		hexagonal	UCl_3 ($P6_3/m$)	4.81
UCl_3	green	835	hexagonal	$P6_3/m$	5.50
$NpCl_3$	green	ca 800	hexagonal	UCl_3	5.60
$PuCl_3$	emerald green	760	hexagonal	UCl_3	5.71
$AmCl_3$	pink or yellow	715	hexagonal	UCl_3	5.87
$CmCl_3$	white	695	hexagonal	UCl_3	5.95
$BkCl_3$	green	603	hexagonal	UCl_3	6.02
α-$CfCl_3$	green	545	orthorhombic	$TbCl_3$ ($Cmcm$)	6.07
β-$CfCl_3$	green		hexagonal	UCl_3	6.12
$EsCl_3$	white to orange		hexagonal	UCl_3	6.20
$AcBr_3$	white		hexagonal	UBr_3 ($P6_3/m$)	5.85
UBr_3	red	730	hexagonal	$P6_3/m$	6.55
$NpBr_3$	green		hexagonal	UBr_3	6.65
$NpBr_3$	green		orthorhombic	$TbCl_3$ ($Cmcm$)	6.67
$PuBr_3$	green	681	orthorhombic	$TbCl_3$	6.72
$AmBr_3$	white to pale yellow		orthorhombic	$TbCl_3$	6.85
$CmBr_3$	pale yellow-green	625±5	orthorhombic	$TbCl_3$	6.85
$BkBr_3$	light green		monoclinic	$AlCl_3$ (C2/m)	5.604
$BkBr_3$	light green		orthorhombic	$TbCl_3$	6.95
$BkBr_3$	yellow green		rhombohedral	$FeCl_3$ (R3)	5.54
$CfBr_3$	green		monoclinic	$AlCl_3$	5.673
$CfBr_3$	green		rhombohedral	$FeCl_3$	5.77
$EsBr_3$	straw		monoclinic	$AlCl_3$	5.62
PaI_3	black		orthorhombic	$TbCl_3$ ($Cmcm$)	6.69
UI_3	black		orthorhombic	$TbCl_3$	6.76
NpI_3	brown		orthorhombic	$TbCl_3$	6.82
PuI_3	green		orthorhombic	$TbCl_3$	6.92
AmI_3	pale yellow	ca 950	hexagonal	BiI_3 (R3)	6.35
AmI_3	yellow		orthorhombic	$PuBr_3$	6.95
CmI_3	white		hexagonal	BiI_3	6.40
BkI_3	yellow		hexagonal	BiI_3	6.02
CfI_3	red-orange		hexagonal	BiI_3	6.05
EsI_3	amber to light yellow		hexagonal	BiI_3	6.18
ThF_4	white	1068	monoclinic	UF_4 (C2/c)	6.20

Table 3. (*Continued*)

Compound	Color	Melting point, °C	Symmetry	Space group or structure type	Density, g/mL
PaF_4	reddish-brown		monoclinic	UF_4	6.38
UF_4	green	960	monoclinic	C2/c	6.73
NpF_4	green		monoclinic	UF_4	6.86
PuF_4	brown	1037	monoclinic	UF_4	7.05
AmF_4	tan		monoclinic	UF_4	7.23
CmF_4	light gray-green		monoclinic	UF_4	7.36
BkF_4	pale yellow-green		monoclinic	UF_4	7.55
CfF_4	light green		monoclinic	UF_4	7.57
α-$ThCl_4$	white		orthorhombic		4.12
β-$ThCl_4$	white	770	tetragonal	UCl_4 ($I4_1/amd$)	4.60
$PaCl_4$	greenish-yellow		tetragonal	UCl_4	4.72
UCl_4	green	590	tetragonal	$I4_1/amd$	4.89
$NpCl_4$	red-brown	518	tetragonal	UCl_4	4.96
α-$ThBr_4$	white		tetragonal	$I4_1/a$	5.94
β-$ThBr_4$	white		tetragonal	UCl_4	5.77
$PaBr_4$	orange-red		tetragonal	UCl_4	5.90
UBr_4	brown	519	monoclinic	2/c-/-	
$NpBr_4$	dark red	464	monoclinic	2/c-/-	
ThI_4	yellow	556	monoclinic	$P2_1/n$	6.00
PaI_4	black				
UI_4	black				
PaF_5	white		tetragonal	$I42d$	
α-UF_5	grayish white		tetragonal	$I4/m$	5.81
β-UF_5	pale yellow		tetragonal	$I42d$	6.47
NpF_5			tetragonal	$I4/m$	
$PaCl_5$	yellow	306	monoclinic	C2/c	
α-UCl_5	brown		monoclinic	$P2_1/n$	3.81
β-UCl_5	red-brown		triclinic	$P\bar{1}$	
α-$PaBr_5$			monoclinic	$P2_1/c$	
β-$PaBr_5$	orange-brown		monoclinic	$P2_1/n$	
UBr_5	brown		monoclinic	$P2_1/n$	
PaI_5	black		orthorhombic		
UF_6	white	64.02[a]	orthorhombic	$Pnma$	5.060
NpF_6	orange	55	orthorhombic	$Pnma$	5.026
PuF_6	reddish-brown	52	orthorhombic	$Pnma$	4.86
UCl_6	dark green	178	hexagonal	$P\bar{3}m1$	3.62

[a]At 151.6 kPa, to convert kPa to atm, divide by 101.3.

organometallic compounds, are known. A number of interesting actinide organometallic compounds of the π-bonded type have been synthesized and characterized. The triscyclopentadienyl compounds, although more covalent than the analogous lanthanide compounds, are highly ionic.

Crystal Structure and Ionic Radii. Crystal structure data have provided the basis for the ionic radii (coordination number = CN = 6). For both M^{3+} and M^{4+} ions there is an actinide contraction, analogous to the lanthanide contraction, with increasing positive charge on the nucleus. As a consequence of the ionic character of most actinide compounds and of the similarity of the ionic radii for a given oxidation state, analogous compounds are generally isostructural.

Absorption and Fluorescence Spectra. The absorption spectra of actinide and lanthanide ions in aqueous solution and in crystalline form contain narrow bands in the visible, near-ultraviolet, and near-infrared regions of the spectrum.

TRANSACTINIDES AND SUPERHEAVY ELEMENTS

The elements beyond the actinides in the Periodic Table can be termed the transactinides. These begin with the element having atomic number 104 and extend, in principle, indefinitely. Nine such elements were definitely known by 1996. Discovery of eight such elements has been confirmed and names approved by IUPAC/IUPAP as of the end of 2004. They are: Rutherfordium (104), Hahnium (105), Seaborgium (106), Bohrium (107), Hassium (108), Meitnerium (109), Darmstadtium (110), and Roentgenium (111).

Fully relativistic calculations have indicated that the electronic configurations of the transactinide elements will be different than those based on simple extrapolation from their lighter homologues in the same group in the periodic table. There may be changes in the valence electron configurations, differences in ionic radii, complexing ability, and other chemical properties. In agreement with relativistic calculations, reversals in trends in properties have been observed. The behavior of the transactinides cannot be simply extrapolated from their lighter homologues, and in some cases they behave more like actinides of the same oxidation state.It should be pointed out that although these atomic calculations give some general guidance for experimental research, they do not predict the behavior of molecular species under actual experimental conditions.

On the basis of simple extrapolation of known half-lives, it would appear that the half-lives of the elements beyond element 112 would become ever shorter as the atomic number increases, even for the isotopes with the longest half-life for each element. This would make future prospects for the existence of heavier transuranium elements appear extremely unlikely, but new theoretical calculations and experimental observations have changed this outlook and led to optimism concerning the prospects for the synthesis and identification of elements beyond the observed upper limit of the periodic table, elements that have come to be referred to as SuperHeavy Elements.

The postulated current synthesis of a broad range of chemical elements, possibly even including superheavy elements, in stars might enhance the prospects for finding even shorter-lived superheavy elements in cosmic rays; elements as heavy as uranium have apparently been found in cosmic rays emanating from such stars.

Although it now appears that many long-lived superheavy elements can exist, new imaginative production reactions and techniques for increasing the overall yields and provision for "stockpiling" long-lived products for future studies must be developed in order to explore this exciting new landscape.

J. J. Katz, G. T. Seaborg, and L. R. Morss, eds., *The Chemistry of the Actinide Elements*, 2nd ed., Chapman and Hall, London, 1986.

G. T. Seaborg, *Man-Made Transuranium Elements*, Prentice-Hall, Inc., Englewood Cliffs, N.J., 1963.

G. T. Seaborg, ed., *Transuranium Elements: Products of Modern Alchemy*, Benchmark Papers, Dowden, Hutchinson & Ross, Inc., Stroudsburg, Pa., 1978.

G. T. Seaborg and J. J. Katz, eds., *The Actinide Elements, National Nuclear Energy Series*, Div. IV, 14A, McGraw-Hill Book Co., Inc., New York, 1954.

Glenn T. Seaborg
Darleance C. Hoffman
University of California,
    Berkeley
Diana M. Lee
Lawrence Berkeley National
    Laboratory

ADHESION

Adhesion and adhesives play a role in many aspects of our daily lives. Secondary load-bearing structure in military and commercial aircraft is adhesively bonded to a large extent. The attainment of durable adhesion in these large structures is obviously tantamount in such applications. At the other end of the size scale, adhesion plays a role in the generation of modern electronics. Silicon chips are attached to lead frames by means of adhesives and electronic packages are often attached to circuit boards by means of adhesives. Indeed, the process of photolithography to generate microelectronic circuitry itself depends on adhesion. Nothing on this earth is independent of adhesion or the intermolecular forces that cause adhesion.

Adhesion is the physical attraction of the surface of one material for the surface of another. As such, it is dependent on the character of the physical forces that hold atoms and molecules together in each of the phases that are in contact and it is also dependent on how those forces match each other in the contact zone. An *adhesive* is a material that uses adhesion to effect an assembly between two other materials, which we call *adherends*. The assembly is known as an *adhesive bond* or an *adhesive joint*. The physical force necessary to break an adhesive joint is called practical *adhesion*. It is found that practical adhesion is dependent on adhesion, but also that practical adhesion is primarily determined by the physical properties of the adhesive and adherends. When a polymeric adhesive is used, it is often found that practical adhesion exceeds adhesion by one or many orders of magnitude. Much of the science of adhesion concerns itself with the attempt to predict practical adhesion from fundamental forces at interfaces combined with the physical properties of the adhesive and the adherends. Another part of the science of adhesion is concerned with the improvement of practical adhesion by providing proper surfaces to which appropriately designed adhesives will display increased attraction. The final part of adhesion science concerns itself with the generation of improved adhesives and primers.

ADHESION: BONDS AND FRACTURES

Wetting and Adhesion

To form an adhesive bond, a liquid adhesive must come into intimate contact with the surface of the adherend. If a drop of a liquid is placed on a smooth surface, it assumes a shape that is characteristic of the interaction of the liquid with the solid surface. When a liquid spreads on a surface completely (no contact angle), the liquid has *wetted the surface*. The first criterion for good adhesion is: *choose an adhesive and adherend such that the adhesive spontaneously completely wets the surface.*

ADHESION AND INTERDIFFUSION

The ultimate in intimate contact occurs when the adherend and the adhesive are so compatible that they dissolve into one another. In molecular parlance, the molecules in the adhesive diffuse into the adherend and vice versa. One means of gauging solubility between materials is the *solubility parameter*. The solubility parameter is based upon the heat of vaporization of a material. The heat of vaporization is a measure of intermolecular interactions. For two polymeric materials to be soluble in one another, the solubility parameters of the two materials must be almost identical, thus the intermolecular interactions may be closely matched. Alternatively, the enthalpy of solution of the two materials should be exothermic. Thus, one can formulate another criterion for adhesion, *for diffusive adhesive bonding, the solubility parameters of the two materials must be close to identical or their enthalpy of solution should be negative.*

LINEAR ELASTIC FRACTURE MECHANICS

In most cases, materials available to us contain flaws and/or cracks. Fracture mechanics allows us to examine the strength of materials when flaws are present. Linear elastic fracture mechanics assumes that the material or interface in question follows Hooke's law (proportionality of stress and strain) for all extensions. But materials, in general, are neither linear nor elastic and have many ways of dissipating mechanical energy other than cracking to generate surface. So in most cases,

$$\mathcal{G}_C >>>> W_A \qquad (1)$$

where $\mathcal{G}_c$ is the critical strain energy release rate and W_A is the work of adhesion. This means that $\mathcal{G}_C$ for most adhesive bonds is much larger than the energy necessary to break the interface. It also brings up the fact that much of the stress that can be placed on an adhesively bonded assembly could be dissipated by various mechanisms in the adherends or in the adhesive before that stress can be transferred to the interface.

COVALENT BONDING AND ADHESION

The interfacial forces discussed thus far have been those involving the interactions between electron distributions surrounding atoms or molecules. With the exception of ionic bonds, these interactions typically have a very small potential energy. Significantly higher potential energy of interaction can occur when protons or electron pairs are shared between atoms or molecules. The potential for significantly higher levels of interfacial interaction due to hydrogen or covalent bonding at interfaces has become a technology for generating interfacial chemistry. The best example of interfacial chemistry for improvement of adhesion phenomena is that of silane coupling agents. These silanes can be dissolved in water or water–alcohol mixtures in the presence of acidic or basic catalysts and applied to inorganic surfaces. Although the primary reason for use of silanes is improvement in the durability of adhesive bonds, often one is able to find that initial bond strength can be improved over that of untreated interfaces. Another criterion for adhesion is: *if possible, generate covalent bonds at interfaces.*

MECHANICAL ROUGHNESS AND ADHESION

In the absence of anything but van der Waals interactions at interfaces, an adhesive bond between an adhesive and a flat adherend will yield an adhesive joint that is likely to be mechanically weak. The reason for this situation becomes apparent when one considers the application of

a force to the edge of such an adhesive bond. The only interaction holding the two surfaces together being van der Waals interactions, the work to open the bond will be very low. If there is no other means of absorbing mechanical energy, the interfacial crack will propagate easily between the two surfaces.

If instead of a planar interface, a mechanically rough interface exists between the adhesive and the adherend, there could exist two new means of absorbing mechanical energy. As a crack is propagating from the edge of the adhesive bond, the crack can no longer follow a sharp interface. Instead, the crack must every so often encounter either the adherend or the adhesive. If there are overhangs under which the adhesive can flow, then once it hardens, the adhesive cannot pass those overhangs without significant plastic deformation.

If a rough surface is to be bonded, there is another important consideration. Adhesives have nonzero viscosity and, in many cases, adhesive viscosity will increase with time after application. With a rough surface, the adhesive must penetrate into the pores of that surface during the application time. The final criterion for good adhesion is: *the adherend should be microscopically rough and the adhesive should be low enough in viscosity to full (wet) the roughness on the surface.*

Adhesive Bond Breaking

Adhesive joints are mechanical systems. To evaluate the utility of an adhesive for a particular application, the adhesive bond must be examined mechanically. Specific tests have been developed to examine adhesives in their primary modes of loading: tensile tests, shear tests, fracture tests, peel tests, cleavage tests, and contact mechanics.

ADHESION AND PRACTICAL STRENGTH OF ADHESIVE BONDS

Engineering has made significant advances in the analysis of structures through the use of computer-based models based in finite element analysis (35). It is tempting to apply such models to adhesive joints. Such analysis can be done for adhesive bonds if one makes the assumption that adhesion is perfect and that the forces in the interphase transfer forces perfectly (36). We know that in many cases, such transfer is not perfect and the models are thus limited. The previous two sections have described the surface aspects of adhesion and some mechanical tests of adhesive bonds. For adhesion science to become predictive and useful in modeling of adhesive bonds, a connection must be made between the mechanical strength of adhesive bonds and the fundamental forces of adhesion.

Polymeric materials, including those used as adhesives, exhibit sensitivity to the rate of application of a mechanical stress. This phenomenon is known as time–temperature superposition. At low rates of application of mechanical stress, polymeric materials tend to be fully relaxed and act more liquid-like but at high rates of application of mechanical stress, polymeric materials tend to behave more like solids. The inverse is true as a function of temperature. One can measure the work necessary to break an adhesive bond as a function of the rate of application of mechanical stress. At low enough reduced rates (ie, high temperatures and low rates of application of mechanical stress), it is possible that the work to break an adhesive bond becomes independent of rate. That is, the polymer can comply completely with the applied mechanical stress and the only force retaining the adhesive on the surface is adhesion.

SUMMARY

Adhesion is a complex phenomenon whose understanding is addressed by three scientific disciplines: surface science, engineering mechanics, and polymer chemistry and physics. The phenomenon of adhesion is determined by the same intermolecular forces that provide the strength of all materials mitigated by the ability of adhesives to come into intimate contact with adherends. The strength of an adhesive bond is determined by a number of methods, most of which rely on destructive evaluation of an adhesive joint tailored to emphasize one or more modes of loading the joint. The measured strength of an adhesive bond results from a complex interweaving of adhesion with the physics of failure of the adhesive and the adherend.

R. D. Adams and W. C. Wake, *Structural Adhesives Joints in Engineering*, Elsevier Science, Inc., New York, 1984.

J. Israelachvili, *Intermolecular and Surface Forces*, Academic Press, London, 1992.

A. V. Pocius, *Adhesion and Adhesives Technology: An Introduction*, Carl Hanser Verlag, Munich, January, 1997.

I. M. Skeist, ed., *Handbook of Adhesives*, 3rd ed., Van Nostrand-Reinhold, New York, 1990.

ALPHONSUS V. POCIUS
3M Company

ADHESIVES

An adhesive is a material that is used to join two objects through nonmechanical means. It is placed between the objects, which usually are called *adherends* when it is part of a test piece or *substrates* when part of an assembly, to create an adhesive joint. Although some adhesives form joints that almost immediately are as strong as they will be in actual use, other adhesives require further operations for the adhesive joint to reach its full strength. Adhesives can be made in several different physical forms, and the form of a given adhesive will define the possible methods of its application to the substrate.

An adhesive is composed of a base chemical or a combination of chemicals that define its general chemical class. Most adhesives contain a curing agent or catalyst that will cause an increase in the molecular weight of the system and frequently the formation of a polymeric network. Nearly all adhesives also contain additives or modifiers that fine tune the adhesive and may significantly influence its behavior before and after formation of the adhesive joint. Additives or modifiers increasingly are chosen for their ability to provide more than one benefit, eg, a pigment may not only color but also may reinforce an adhesive. In some cases, the process used to combine these diverse ingredients will strongly influence the properties of an adhesive. Although inorganic adhesives do exist, this article will be restricted to organic polymeric adhesives.

Where an adhesive is the obvious choice, it is often the least expensive choice as well. In industrial situations, where the performance expected of the adhesive is high and broad and its cost is that of a specialty rather than a commodity material, it is common to see users take a systems approach to make the best choice of joining method or the best choice of adhesive, if adhesive bonding is seen to be the best joining method. The systems approach to choosing adhesives goes well beyond comparing the cost per gallon of adhesives. It considers the number of parts to be joined, the time and cost constraints of assembly, spatial limitations, the need for substrate surface cleaning or preparation, the cost of all application, fixturing, and curing equipment, environmental and safety requirements, disposal costs, and, finally, part performance, and lifetime.

FORMS AND TYPES OF ADHESIVES

As supplied, adhesives can be found in the form of low-viscosity liquids, viscous pastes, thin or thick films, semisolids, or solids. Before application to a substrate, an adhesive need not be sticky or otherwise particularly adherent. A distinct exception is the pressure-sensitive adhesive (PSA), which is inherently tacky when first made. The PSA remains throughout its useful lifetime essentially the same material it was when first made. All other forms and types of adhesives undergo a transformation that is central to their function as an adhesive. This transformation is usually carried out through imposition of time, heat, or radiation, either actively or passively.

An adhesive applied as a true solution or a dispersion of solids will dry through loss of water or another solvent, leaving behind a film of adhesive. A reactive adhesive system will form internal chemical bonds through the processes of cross-linking, chemical reaction that joins dissimilar long-chain molecules, or polymerization, chemical reaction that joins similar monomer units. Solid adhesives are heated in order to be applied and then on cooling become functional adhesives. The transformation from a liquid, paste, or semisolid to a functional adhesive is loosely termed *curing*. Additional general terms that refer to this transformation include *setting up* and *hardening*. Adhesives may also cure in stages. The first stage of curing is sometimes referred to as the B stage, and adhesives that have undergone some level of precure in their manufacture are often said to have been *B-staged*. For many adhesive applications, the ability of an adhesive to gel, precure, or develop green strength or handling strength is a key characteristic, being most important for parts that will be bonded and then transported to the next step in their processing. *Adhesives* are referred to as such before and after cure.

SYNTHETIC ADHESIVES

Pressure-Sensitive Adhesives

PSAs are inherently and permanently soft, sticky materials that exhibit instant adhesion or tack with very little pressure to surfaces to which they are applied. The level of adhesion may build with time and be surprisingly high. PSAs generally have a high cohesive strength and often can be removed from substrates without leaving a residue. Some applications take advantage of a PSA's ability to quickly form a strong bond and under stress force failure elsewhere in a system, an attribute used to advantage in tamper-proof packaging and price stickers. At the other end of the spectrum lie PSAs that can be repeatedly repositioned. The primary characteristics used to describe the performance of PSAs are tack, adhesion strength in peel, and resistance to shear forces.

Many PSA compositions contain a base elastomeric resin and a tackifier, which enhances the ability of the adhesive to instantly bond as well as its bond strength. The elastomer may be useful without cross-linking but will often require either chemical or physical cross-linking for establishment of sufficient cohesive strength. Heat, ultraviolet (uv), or radiation are usually the activators of the cross-linking, and suitable catalysts are used, with their choice depending on the base resin. Small amounts of epoxy or hydroxy functionality are sometimes added to allow uv cures if the base resins are not themselves uv-curable. Electron beam curing has received attention but tends to be more costly than uv curing.

The large bulk of PSAs are coated onto continuous webs or films to make pressure-sensitive tapes, labels, etc. While many PSAs continue to be coated out of organic solvents, many have been converted to water-based formulations or are extruded as hot-melt adhesives that upon cooling retain their tack. Aqueous emulsions of carboxylated styrene–butadiene and various acrylate copolymers are among the most useful as bases for water-based PSAs. Reinforcing agents such as phenolics and higher molecular weight relatives of the tackifiers are sometimes added to improve cohesive strength.

Hot-Melt Adhesives

Hot-melt adhesives are solid adhesives that are heated to a molten liquid state for application to substrates, applied hot, and then cooled, quickly setting up a bond. The largest uses of hot-melt adhesives are in packaging, bookbinding, disposable paper products, wood-bonding, shoemaking, and textile binding. The advantages of hot-melt adhesives include their easy handling in solid form, almost indefinite shelf life, generally nonvolatile nature, and, most importantly, ability to form bonds quickly without supplementary processing. They are considered friendly to the environment and are expected to see expanded use on a worldwide basis as the market continues to move away from solvent-based adhesives. The disadvantages of hot-melts lie in their tendency to damage substrates that cannot withstand their application temperatures, limited high temperature properties, and only moderate strength.

When there is some lack of cohesiveness in blends of base resins, compatibilizers may be used to improve the apparent miscibility of these resins. Hot-melts can be based on either amorphous or semicrystalline resins. Particularly in the case of semicrystalline resins, the rate of cooling can dramatically affect adhesion to a substrate.

Solution Adhesives

Adhesives delivered out of solutions are typically used for joining large areas destined for nonstructural or semistructural service. The solution may be made with an organic solvent or with water or may be an aqueous dispersion. It is important that the liquid carrier have some means of escaping from the bondline in order for the proper bond strength to develop.

Solvent-Based Solution Adhesives. Contact adhesives, activatable dry film adhesives, and solvent-weld adhesives make up the solvent-based adhesives.

The most widely used contact adhesive is a solution of polychloroprene or modified polychloroprene in solvent blends of aromatic hydrocarbons, aliphatic hydrocarbons, esters, or ketones, eg, toluene–hexane–acetone.

Water-Based Solution Adhesives. Solution adhesives based on water dispersions and aqueous emulsions are steadily gaining in use largely at the expense of solvent-based adhesives. These are rarely true solutions, with the exception of the viscosity modifiers often used to adjust flow characteristics. Dispersions of polyurethanes in water find use in bonding of plastic sheets and films, cloth, shoe parts, foams, PVC veneers, and carpets. Other water-dispersible resins can be added to the polyurethane dispersion to lower costs and modify performance characteristics. The largest group of water-dispersed or water-dissolved adhesives are made of natural products.

Poly(vinyl acetate) (PVAc) emulsions, the basis of the ubiquitous household white glues, are among the most familiar water-based adhesives. These are widely used for paper and wood bonding.

Structural Adhesives

Structural adhesives are designed to bond structural materials. Most any adhesive giving shear strengths in excess of ~7 MPa (~1000 psi) may be called a structural adhesive. Structural adhesives are generally the first choice when bonding metal, wood, and high-strength composites to construct a load-bearing structure. Bonds formed with structural adhesives cannot be reversed without damaging one or the other substrate. They are the only kind of adhesive that might be expected to be able to sustain a significant percentage of its initial failure load in a hot and humid or hot and dry environment. Any one of these descriptors names structural adhesives the strongest and most permanent type of adhesive. For good reason, they are sometimes referred to as engineering adhesives.

Epoxy Resins. Most epoxy adhesives are resins based on what is commonly known as the diglycidyl ether of bisphenol A (DGEBPA). A wide variety of epoxy resins are commercially available: monofunctional or polyfunctional, aliphatic, cyclic, or aromatic. Brominated epoxies may be useful where flammability is a concern. An oxirane functionality is all that is needed to make an epoxy resin, and structural adhesives are only one of over a dozen different uses for epoxy resins. Many epoxy resins on the market will not necessarily be suitable for adhesives, but their availability does expand the choices available for adhesive formulators. The specialty epoxy resins developed specifically for adhesive use sometimes will be more costly than the DGEBPA resins but may provide the basis for a specialty adhesive that can meet a unique need and therefore command a proportionally higher price.

Acrylics. The most well known characteristic of acrylics is their relatively high speed of reaction via free radical polymerization. Oxygen inhibits the polymerization of acrylic monomers to a useful extent, and its exclusion kicks off polymerization of monomeric acrylates.

A key ingredient in anaerobic acrylic adhesives is the acrylate monomer or monomers. In addition to the monomer acrylates, there generally is also present a diacrylate that acts as a cross-linker. Other ingredients used in these adhesives include stabilizers or polymerization inhibitors such as phenols or quinones; chelating agents that snatch up trace metals to prolong shelf life; and various modifiers such as inert fillers, inorganic and polymeric thickeners, elastomers to improve toughness, and bismaleimides that improve high temperature performance. The low viscosities and good wetting properties of these adhesives allow them to penetrate and flow in tight spaces, which is taken advantage of in many of their uses. Threadlocking and sealing are primary applications.

Urethanes. The core of a urethane adhesive is an isocyanate compound. Isocyanates react with a variety of functional groups having active hydrogens to generate a variety of linkages that give the resulting polymers their names.

Most polyurethane structural adhesives are two-part systems based on the reactions of isocyanates and polyisocyanates with oligomers or polymers having at least two hydroxyl groups, which are generically referred to as diols or polyols.

As a group, polyurethane structural adhesives produce bond strengths on the lower end of the strength scale for structural adhesives, but their high flexibility, usually strong peel strength, and generally good impact and fatigue resistance recommend their use when these characteristics are important. A variety of adhesives have been developed that incorporate polyurethanes into acrylic or epoxy structural adhesives.

Phenolics. Phenolic resins were the basis of the first synthetic structural adhesives. They are formed by the reaction of phenol (C_6H_6O), and formaldehyde (CH_2O). There are two types of phenolic resins, resoles and novolaks (or novolacs), the former being comprised of methylol-terminated resins and the latter, of phenol-terminated resins. Formulators can choose from a variety of commercially available phenolic compounds.

Epoxy–phenolics are important hybrid adhesives and offer an immensely useful combination of strength, toughness, durability, and heat resistance. Phenolic structural adhesives as a class of materials are highly resistant to most chemicals. Phenolic adhesives are found as powders, liquids, pastes, and supported and unsupported films. Phenolics are widely used in wood bonding and as foundry resins for making sand-shell molds.

Urea–Formaldehyde and Related Adhesives. Urea–formaldehydes (UF) are the most significant members of the class of materials known as the amino resins or aminopolymers. These are the polymeric condensation products of the reaction of aldehydes with amines or amides. These adhesives are widely used to make plywood and particleboard in processes utilizing heated hydraulic presses with multiple outlets for water vapor release.

High Performance Adhesives. A number of adhesive needs exist that require resistance to very high temperatures and other environmental stressors such as certain gases, solvents, radiation and mechanical loads. Heterocyclic polymers such as polyimides and polyquinoxalines have been the basis of most heat-resistant adhesives. Microelectronics adhesives often must not only withstand high temperature, but also conduct heat away from heat-sensitive parts. This has been the inevitable result of increasing miniaturization. Epoxies continue to be the basis of many microelectronics adhesives, but adhesives based on stiff-chained thermoplastic resins such as polyethersulfone and polyetheretherketone have made some inroads.

ADHESIVES MADE FROM NATURAL PRODUCTS

The first adhesives developed by humans were based on naturally available materials such as bone, blood, milk, minerals, and vegetable matter. Although replaced by synthetic adhesives for some applications, in certain industries, among them furniture, food, bookbinding, and textiles, adhesives based on natural products continue to be used to a significant extent. These adhesives can be divided into those based on proteins, carbohydrates, and natural rubbers or oils. Historically, *glue* is a term used to refer to adhesives made from animal matter or vegetable-based protein.

Protein-Based Adhesives

The protein sources for these adhesives include mammals, fish, milk, soybeans, and blood. Animal and fish parts that yield useful proteins include hides, skins, bones, and collagen from cartilage and connective tissues.

Carbohydrate-Based Adhesives

Carbohydrates are available from a wide variety of plants, the shells of marine crustaceans, and bacteria. The raw adhesive materials obtained from these sources include cellulose, starch, and gum. Cellulose is a semicrystalline polymeric form of glucose having a molecular weight of <1000 to nearly 30,000. The cellulose adhesives are film formers having a thermoplastic nature. A typical adhesive formulation includes a few percent of the cellulose, less than a percent each of a plasticizer and a natural protein, and the great balance of water or another solvent.

Other Nature-Based Adhesives

Natural rubber is an important adhesive component obtained from the rubber tree. Tannins are polymeric polyphenols isolated as one of two products from the bark of conifers and deciduous trees. Lignin is widely available as a waste material from pulp mills and has a complex structure. Tannin-based adhesives have attained some level of success in the marketplace. Despite considerable interest in and work toward more commercial use of lignins in adhesives for wood bonding, they have not yet succeeded in capturing market share. A vinyl-functionalized sugar has been developed for use in products including, most prominently, adhesives. Use of whey and whey by-products as adhesive components has been investigated. Modification of natural materials to make polyols and diisocyanates has been pursued in both the United States and the United Kingdom. It can be expected that additional plant-based monomers and polymers will be developed as the chemical industry comes to terms with the limited supply and rising costs of petrochemicals, making "green adhesives" a not-uncommon reality in the not-too-distant future.

ADHESIVE FORMULATION AND DESIGN

Formulating adhesives is both a skill and an art. There is more to adhesive formulation than the combining of various raw materials. The formulator must be a multidimensional technical professional able to juggle several different fields of science and engineering, legal issues, environmental considerations, computer hardware and software, and business concerns. It is not unusual to create a remarkable adhesive only to find that a key ingredient is unstable or too expensive for the intended market or poses unacceptable health and safety risks.

Better tools for adhesive formulation have been developed with the onset of the personal computer and computer workstations. On-line searching of and access to the scientific and patent literature as well as the information on business trends and supplier's products available on the Internet have made information gathering easier. Adhesive development accelerates more each year, and the savvy formulator must keep pace.

M. Ash and I. Ash, *Handbook of Adhesive Chemicals and Compounding Ingredients*, Synapse Information Resources, Endicott, N.Y., 1999.

I. Benedek and L. J. Heymans, *Pressure-Sensitive Adhesives Technology*, Marcel Dekker Inc., New York, 1997.

I. Benedek, *Development and Manufacture of Pressure-Sensitive Products*, Marcel Dekker, Inc., New York, 1999.

A. J. DeFusco, K. C. Sehgal, and D. R. Bassett, in J. M. Asua, ed., *Polymeric Dispersions: Principles and Applications*, Kluwer Academic Publishers, the Netherlands, 1997, pp. 379–396.

ELAINE M. YORKGITIS
3M Company

ADIPIC ACID

Adipic acid, hexanedioic acid, 1,4-butanedicarboxylic acid, mol wt 146.14, $HOOCCH_2CH_2CH_2CH_2COOH$, is a white crystalline solid with a melting point of ~152°C. Little of this dicarboxylic acid occurs naturally, but it is produced on a very large scale at several locations around the world. The majority of this material is used in the manufacture of nylon-6,6 polyamide, which is prepared by reaction with 1,6-hexanediamine. The large scale availability, coupled with the high purity demanded by the polyamide process, has led to the discovery of a wide variety of applications for the acid.

CHEMICAL AND PHYSICAL PROPERTIES

Adipic acid is a colorless, odorless, sour tasting crystalline solid. Its fundamental chemical and physical properties are listed in Table 1. The crystal morphology is monoclinic prisms strongly influenced by impurities. Both process parameters and additives profoundly affect crystal morphology in the crystallization of adipic acid, an industrially significant process. Aqueous solutions of the acid are corrosive. Generally, austenitic stainless steels containing nickel and molybdenum and >18% chromium are resistant.

CHEMICAL REACTIONS

Adipic acid undergoes the usual reactions of carboxylic acids, including esterification, amidation, reduction, halogenation, salt formation, and dehydration. Because of its bifunctional nature, it also undergoes several industrially significant polymerization reactions.

Table 1. Physical and Chemical Properties of Adipic Acid

Property	Value
molecular formula	$C_6H_{10}O_4$
molecular weight	146.14
melting point, °C	152.1 ± 0.3
specific gravity	1.344 at 18°C (sol)
	1.07 at 170°C (liq)
vapor density, air = 1	5.04
vapor pressure, Pa[a]	
solid at °C	
18.5	9.7
47.0	38.0
liquid at °C	
205.5	1,300
244.5	6,700
specific heat, kJ/kg K[b]	1.590 (solid state)
	2.253 (liquid state)
	1.680 (vapor, 300°C)
heat of fusion, kJ/kg[b]	115
melt viscosity, mPa s(= cP)	4.54 at 160°C
	2.64 at 193°C
heat of combustion, kJ/mol[b]	2,800

[a] To convert Pa to mm Hg divide by 133.3.
[b] To convert J to cal divide by 4.184.

Esters and polyesters comprise the second most important class of adipic acid derivatives, next to polyamides. The acid readily reacts with alcohols to form either the mono- or diester. Although the reaction usually is acid catalyzed, conversion may be enhanced by removal of water as it is produced. Recent modifications of adipic acid manufacturing processes have included methanol esterification of the dicarboxylic acid by-product mixture.

MANUFACTURE AND PROCESSING

Adipic acid historically has been manufactured predominantly from cyclohexane and, to a lesser extent, phenol. During the 1970s and 1980s, however, much research was directed to alternative feedstocks, especially butadiene and cyclohexene, as dictated by shifts in hydrocarbon pricing. All current industrial processes use nitric acid in the final oxidation stage. Growing concern with air quality may exert further pressure for alternative routes as manufacturers seek to avoid NO_x abatement costs, a necessary part of processes that use nitric acid.

Since adipic acid has been produced in commercial quantities for almost 50 years, it is not surprising that many variations and improvements have been made to the basic cyclohexane process. In general, however, the commercially important processes still employ two major reaction stages. The first reaction stage is the production of the intermediates cyclohexanone and cyclohexanol, usually abbreviated as KA, KA oil, ol-one, or anone-anol. The KA (ketone, alcohol), after separation from unreacted cyclohexane (which is recycled) and reaction by-products, is then converted to adipic acid by oxidation with nitric acid. An important alternative to this use of KA is its use as an intermediate in the manufacture of caprolactam, the monomer for production of nylon-6. The latter use of KA predominates by a substantial margin on a worldwide basis, but not in the United States.

Preparation of KA by Oxidation of Cyclohexane

There are three main variations to the basic cyclohexane oxidation process pioneered by DuPont in the 1940s. The first, which can be termed metal-catalyzed oxidation, is the oldest process still in use and forms the base for the other two. It employs a cyclohexane-soluble catalyst, usually cobalt naphthenate or cobalt octoate, and moderate temperatures (150–175°C) and pressures (800–1200 kPa).

Regardless of the techniques used to purify the KA oil, several waste streams are generated during the overall oxidation–separation processes and must be disposed of. The spent oxidation gas stream must be scrubbed to remove residual cyclohexane, but afterwards will still contain CO, CO_2, and volatile hydrocarbons (especially propane, butane, and pentane). This gas stream is either burned and the energy recovered, or it is catalytically abated. There are usually several aqueous waste streams arising from both water generated by the oxidation reactions and wash water.

An alternative to maximizing selectivity to KA in the cyclohexane oxidation step is a process which seeks to maximize cyclohexylhydroperoxide, also called P or CHHP. This peroxide is one of the first intermediates produced in the oxidation of cyclohexane. It is produced when a cyclohexyl radical reacts with an oxygen molecule to form the cyclohexylhydroperoxy radical. This radical can extract a hydrogen atom from a cyclohexane molecule, to produce CHHP and another cyclohexyl radical, which extends the free-radical reaction chain.

Another alternative to the basic cyclohexane oxidation process is one which maximizes only the yield of A. This process uses boric acid as an additive to the cyclohexane stream as both a promoter and an esterifying agent for the A that is produced. This is an energy-intensive step and can be quite a mechanical nuisance because of the requirement for handling boric acid solids. Without careful attention to energy conservation and engineering, the savings that accrue from the high yield can be more than offset.

Preparation of KA from Phenol

In past years, economics has dictated against the preparation of KA from phenol because of the relatively high cost of this material compared to cyclohexane. However, given new routes to phenol and occasional periods of overcapacity for this commodity chemical, such technology has been revisited. For example, the Solutia Benzene to Phenol process has been touted as a step-change in adipic acid manufacturing technology.

In both liquid- and vapor-phase phenol to KA processes, one can obtain varying ratios of K to A. If the desired product is to be further oxidized with nitric acid to make adipic acid, then there is an optimal concentration of K that maximizes the tradeoff between adipic yield and nitric acid usage. If the desired product is caprolactam, then a high K concentration would be required.

Nitric Acid Oxidation of Cyclohexanol(One)

Although many variations of the cyclohexane oxidation step have been developed or evaluated, technology for conversion of the intermediate ketone–alcohol mixture to adipic acid is fundamentally the same as originally developed by Du Pont in the early 1940s. This step is accomplished by oxidation with 40–60% nitric acid in the presence of copper and vanadium catalysts. The reaction proceeds at high rate, and is quite exothermic. Yield of adipic acid is 92–96%, the major by-products being the shorter chain dicarboxylic and succinic acids, and CO_2.

Because of the highly corrosive nature of the nitric acid streams, adipic acid plants are constructed of stainless steel, or titanium in the more corrosive areas, and thus have high investment costs.

Nitric acid oxidation may be used to recover value from waste streams generated in the cyclohexane oxidation portion of the process, such as the water wash and nonvolatile residue streams. The nitric acid oxidation step produces three major waste streams: an off-gas containing oxides of nitrogen and CO_2; water containing traces of nitric acid and organics from the water removal column; and a dibasic acid purge stream containing adipic, glutaric, and succinic acids. The fate of these waste streams varies widely, subject to the usually very complex environmental and regulatory situations at each individual manufacturing site. These issues are now a prime consideration, equal to economics, in the design of chemical processing systems in the petrochemical industry.

Other Routes to Adipic Acid

A number of adipic acid processes rely on feedstocks other than cyclohexane and phenol and produce neither K nor A as intermediates. Although these have been investigated, none has been employed at a commercial scale.

STORAGE, HANDLING, AND SHIPPING

When dispersed as a dust, adipic acid is subject to normal dust explosion hazards. The material is an irritant, especially upon contact with the mucous membranes. Thus protective goggles or face shields should be worn when handling the material. Prolonged contact with the skin should also be avoided.

The material should be stored in corrosion-resistant containers, away from alkaline or strong oxidizing materials. In the event of a spill or leak, nonsparking equipment should be used, and dusty conditions should be avoided. Spills should be covered with soda ash, then flushed to drain with large amounts of water.

SPECIFICATIONS AND ANALYSIS

Because of the extreme sensitivity of polyamide synthesis to impurities in the ingredients (eg, for molecular-weight control, dye receptivity), adipic acid is one of the purest materials produced on a large scale. In addition to food-additive and polyamide specifications, other special requirements arise from the variety of other applications. Typical impurities include monobasic acids arising from the air oxidation step in synthesis, and lower dibasic acids and nitrogenous materials from the nitric acid oxidation step. Trace metals, water, color, and oils round out the usual specification lists.

Standard methods for analysis of food-grade adipic acid are described in the Food Chemicals Codex. Classic methods are used for assay (titration), trace metals (As, heavy metals as Pb), and total ash.

Monobasic acids are determined by gas chromatographic analysis of the free acids; dibasic acids usually are derivatized by one of several methods prior to chromatographing. Methyl esters are prepared by treatment of the sample with BF_3–methanol, H_2SO_4–methanol, or tetramethylammonium hydroxide. Gas chromatographic analysis of silylation products also has been used extensively. Liquid chromatographic analysis of free acids or of derivatives also has been used. More sophisticated high-performance liquid chromatography (hplc) methods have been developed recently to meet the needs for trace analyses in the environment, in biological fluids, and other sources. Mass spectral identification of both dibasic and monobasic acids usually is done on gas chromatographically resolved derivatives.

HEALTH AND SAFETY FACTORS

Adipic acid is relatively nontoxic; no OSHA PEL or NIOSH REL have been established for the material.

Adipic acid is an irritant to the mucous membranes. In case of contact with the eyes, they should be flushed with water. It emits acrid smoke and fumes on heating to decomposition. It can react with oxidizing materials, and the dust can explode in admixture with air. Fires may be extinguished with water, CO_2, foam, or dry chemicals.

Airborne particulate matter and aerosol samples from around the world have been found to contain a variety of organic monocarboxylic and dicarboxylic acids, including adipic acid.

N_2O Abatement Technology

During the 1990s, the adipic acid industry updated its offgas control technology. These process modifications were based on the disclosure that nitrous oxide (N_2O), a gas-phase byproduct of adipic acid manufacture, is the major source of strasospheric nitric oxide (NO) and thus has a global warming potential many times more than CO_2.

Reducing flame burner technology represents the high-temperature option to N_2O abatement (1200–1500°C). Here, natural gas reduces N_2O to nitrogen, CO_2 and water. Other options commercially practiced depend on the use of a catalyst. An intermediate temperature process (1000–1500°C) developed by DuPont and Rhone Poulenc is based on the catalytic reaction of N_2O to NO. The low-temperature catalytic process (400–700°C) is designed to destroy N_2O without the formation of NO_x. Such facilities can be installed with or without heat recovery depending on the value of steam.

USES

About 86% of U.S. adipic acid production is used almost totally in the manufacture of nylon-6,6. The remaining 14% is sold in the merchant market for a large number of applications. These have been developed as a result of the large scale availability of this synthetic petrochemical commodity.

About 2.4% of U.S. consumption in 1999 was distributed among several other applications, amounting to several thousand tons each. Wet-strength resins based on polyamide–epichlorohydrin products consumed about 16,000–18,000 t in 1998. Unsaturated polyester resins (4000 t in 1998) are used in surface coatings, flexible alkyd resins (qv), coil coatings, and other coatings because of their curing properties. Adipic acid also is used as a food acidulant in jams, jellies, and gelatins. The synthetic lubricant market consumed about 7000 t as the C_{8-13} adipate esters in 1998, for gas turbines, compressors, and military jet engines. An environmentally significant use of the acid, and especially its dibasic acid by-products, is as a buffer in the scrubbing operation of power plant flue gas desulfurization. Adipoyl chloride is occasionally used as a softening agent for leather.

Adipic Acid, Product Bulletin E-99079-1, E. I. du Pont de Nemours & Co., Inc., 1989.

R. Keller in F. Snell and C. Hilton, eds., *Encyclopedia of Industrial Chemicals Analysis*, Vol. 4, Wiley-Interscience, New York, 1967, pp. 408–423.

R. J. Lewis, ed., *Sax's Dangerous Properties of Industrial Materials*, Willy, Hoboken, NJ, 2005.

Y. C. Yen and S. Y. Wu, *Nylon-6,6, Report No. 54B, Process Economics Program*, SRI International, Menlo Park, Calif., Jan. 1987, pp. 1–148.

JUDITH P. OPPENHEIM
GARY L. DICKERSON
Solutia Inc.

ADSORPTION

Adsorption is the term used to describe the tendency of molecules from an ambient fluid phase to adhere to the surface of a solid. This is a fundamental property of matter, having its origin in the attractive forces between molecules. The force field creates a region of low potential energy near the solid surface and, as a result, the molecular density close to the surface is generally greater than in the bulk gas. Furthermore, and perhaps more importantly, in a multicomponent system the composition of this surface layer generally differs from that of the bulk gas since the surface adsorbs the various components with different affinities. Adsorption may also occur from the liquid phase and is accompanied by a similar change in composition, although, in this case, there is generally little difference in molecular density between the adsorbed and fluid phases.

The enhanced concentration at the surface accounts, in part, for the catalytic activity shown by many solid surfaces, and it is also the basis of the application of adsorbents for low pressure storage of permanent gases such as methane. However, most of the important applications of adsorption depend on the selectivity, ie, the difference in the affinity of the surface for different components. As a result of this selectivity, adsorption offers, at least in principle, a relatively straightforward means of purification (removal of an undesirable trace component from a fluid mixture) and a potentially useful means of bulk separation.

FUNDAMENTAL PRINCIPLES

Forces of Adsorption

Adsorption can be classified as chemisorption or physical adsorption, depending on the nature of the surface forces. In physical adsorption the forces are relatively weak, involving mainly van der Waals (induced dipole–induced dipole) interactions, supplemented in many cases by electrostatic contributions from field gradient–dipole or –quadrupole interactions. By contrast, in chemisorption there is significant electron transfer, equivalent to the formation of a chemical bond between the sorbate and the solid surface. Such interactions are both stronger and more specific than the forces of physical adsorption and are obviously limited to monolayer coverage.

Heterogeneous catalysis generally involves chemisorption of the reactants, but most applications of adsorption in separation and purification processes depend on physical adsorption. Chemisorption is sometimes used in trace impurity removal since very high selectivities can be achieved. However, in most situations the low capacity imposed by the monolayer limit and the difficulty of regenerating the spent adsorbent more than outweigh this advantage. The higher capacities achievable in physical adsorption result from multilayer formation and this is obviously critical in such applications as gas storage, but it is also an important consideration in most adsorption separation processes since the process cost is directly related to the adsorbent capacity.

Selectivity

Selectivity in a physical adsorption system may depend on differences in either equilibrium or kinetics, but the great majority of adsorption separation processes depend on equilibrium-based selectivity. Significant kinetic selectivity is in general restricted to molecular sieve adsorbents—carbon molecular sieves, zeolites, or zeolite analogues. In these materials the pore size is of molecular dimensions, so that diffusion is sterically restricted. In this regime small differences in the size or shape of the diffusing molecule can lead to very large differences in diffusivity.

A degree of control over the kinetic selectivity of molecular sieve adsorbents can be achieved by controlled adjustment of the pore size. Control of equilibrium selectivity is generally achieved by adjusting the balance between electrostatic and van der Waals forces.

Hydrophilic and Hydrophobic Surfaces

Water is a small, highly polar molecular and it is therefore strongly adsorbed on a polar surface as a result of the large contribution from the electrostatic forces. Polar adsorbents such as most zeolites, silica gel, or activated alumina therefore adsorb water more strongly than they adsorb organic species, and, as a result, such adsorbents are commonly called hydrophilic. In contrast, on a nonpolar surface where there is no electrostatic interaction water is held only very weakly and is easily displaced by organics. Such adsorbents, which are the only practical choice for adsorption of organics from aqueous solutions, are termed hydrophobic.

The most common hydrophobic adsorbents are activated carbon and silicalite. The latter is of particular interest, since the affinity for water is very low indeed; the heat of adsorption is even smaller than the latent heat of vaporization.

Capillary Condensation

The equilibrium vapor pressure in a pore or capillary is reduced by the effect of surface tension. As a result, liquid sorbate condenses in a small pore at a vapor pressure that is somewhat lower than the saturation vapor pressure. In a porous adsorbent the region of multilayer physical adsorption merges gradually with the capillary condensation regime, leading to upward curvature of the equilibrium isotherm at higher relative pressure. In the capillary condensation region the intrinsic selectivity of the adsorbent is lost, so in separation processes it is generally advisable to avoid these conditions. However, this effect is largely responsible for the enhanced capacity of macroporous desiccants such as silica gel or alumina at higher humidities.

PRACTICAL ADSORBENTS

To achieve a significant adsorptive capacity an adsorbent must have a high specific area, which implies a highly porous structure with very small micropores. Such microporous solids can be produced in several different ways. Adsorbents such as silica gel and activated alumina are made by precipitation of colloidal particles, followed by dehydration (see ALUMINUM COMPOUNDS, ALUMINUM OXIDE (ALUMINA); SILICA, AMORPHOUS SILICA). Carbon adsorbents are prepared by controlled burn-out of carbonaceous materials such as coal lignite, and coconut shells (see CARBON, ACTIVATED CARBON). These procedures generally yield a fairly wide distribution of pore size. The crystalline adsorbents (zeolite and zeolite analogues) are different in that the dimensions of the micropores are determined by the crystal structure and there is therefore virtually no distribution of micropore size (see MOLECULAR SIEVES). Although structurally very different from the crystalline adsorbents, carbon molecular sieves also have a very narrow distribution of pore size. The adsorptive properties depend on the pore size and the pore size distribution as well as on the nature of the solid surface.

Amorphous Adsorbents

The amorphous adsorbents (silica gel, activated alumina, and activated carbon) typically have specific areas in the $200-1000$-m^2/g range, but for some activated carbons much higher values have been achieved (~1500 m^2/g). The difficulty is that these very high area carbons tend to lack physical strength and this limits their usefulness in many practical applications.

Crystalline Adsorbents

In the crystalline adsorbents, zeolites and zeolite analogues such as silicalite and the microporous aluminum phosphates, the dimensions of the micropores are determined by the crystal framework and there is therefore virtually no distribution of pore size. However, a degree of control can sometimes be exerted by ion exchange, since, in some zeolites, the exchangeable cations occupy sites within the structure which partially obstruct the pores.

Desiccants

A solid desiccant is simply an adsorbent that has a high affinity and capacity for adsorption of moisture so that it can be used for selective adsorption of moisture from a gas (or liquid) stream. The main requirements for an efficient desiccant are therefore a highly polar surface and a high specific area (small pores). The most widely used desiccants (qv) are silica gel, activated alumina, and the aluminum-rich zeolites (4A or 13X).

Loaded Adsorbents

Where highly efficient removal of a trace impurity is required it is sometimes effective to use an adsorbent preloaded with a reactant rather than rely on the forces of adsorption.

ADSORPTION EQUILIBRIUM

Henry's Law

Like any other phase equilibrium, the distribution of a sorbate between fluid and adsorbed phases is governed by the principles of thermodynamics. Equilibrium data are commonly reported in the form of an isotherm, which is a diagram showing the variation of the equilibrium adsorbed-phase concentration or loading with the fluid-phase concentration or partial pressure at a fixed temperature. In general, for physical adsorption on a homogeneous surface at sufficiently low concentrations, the isotherm should approach a linear form, and the limiting slope in the low concentration region is commonly known as the Henry's law constant. The Henry constant is simply a thermodynamic equilibrium constant and the temperature dependence therefore follows the usual van't Hoff equation:

$$\lim_{p \to 0} \left(\frac{\partial q}{\partial p}\right)_T \equiv K' = K_0' e^{-\Delta H_0/RT} \qquad (1)$$

in which $-\Delta H_0$ is the limiting heat of adsorption at zero coverage. Since adsorption, particularly from the vapor phase, is usually exothermic, $-\Delta H_0$ is a positive quantity and K' therefore decreases with increasing temperature.

Henry's law corresponds physically to the situation in which the adsorbed phase is so dilute that there is neither competition for surface sites nor any significant interaction between adsorbed molecules. At higher concentrations both of these effects become important and the form of the isotherm becomes more complex. The isotherms have been classified into five different types. Isotherms for a microporous adsorbent are generally of type I; the more complex forms are associated with multilayer adsorption and capillary condensation.

Langmuir Isotherm

Type I isotherms are commonly represented by the ideal Langmuir model:

$$\frac{q}{q_s} = \frac{bp}{1 + bp} \qquad (2)$$

where q_s is the saturation limit and b is an equilibrium constant which is directly related to the Henry constant ($K' = bq_s$).

Freundlich Isotherm

The isotherms for some systems, notably hydrocarbons on activated carbon, conform more closely to the Freundlich equation:

$$q = bp^{1/n} \qquad (n > 1.0) \tag{3}$$

Adsorption of Mixtures

The Langmuir model can be easily extended to binary or multicomponent systems:

$$\frac{q_1}{q_{s1}} = \frac{b_1 p_1}{1 + b_1 p_1 + b_2 p_2 + \cdots}; \quad \frac{q_2}{q_{s2}} = \frac{b_2 p_2}{1 + b_1 p_1 + b_2 p_2; \; + \cdots} \tag{4}$$

Thermodynamic consistency requires $q_{s1} = q_{s2}$, but this requirement can cause difficulties when attempts are made to correlate data for sorbates of very different molecular size. For such systems it is common practice to ignore this requirement, thereby introducing an additional model parameter. This facilitates data fitting but it must be recognized that the equations are then being used purely as a convenient empirical form with no theoretical foundation.

Ideal Adsorbed Solution Theory

Perhaps the most successful approach to the prediction of multicomponent equilibria from single-component isotherm data is ideal adsorbed solution theory. In essence, the theory is based on the assumption that the adsorbed phase is thermodynamically ideal in the sense that the equilibrium pressure for each component is simply the product of its mole fraction in the adsorbed phase and the equilibrium pressure for the pure component *at the same spreading pressure*. The theoretical basis for this assumption and the details of the calculations required to predict the mixture isotherm are given in standard texts on adsorption. Whereas the theory has been shown to work well for several systems, notably for mixtures of hydrocarbons on carbon adsorbents, there are a number of systems which do not obey this model. Azeotrope formation and selectivity reversal, which are observed quite commonly in real systems, are not consistent with an ideal adsorbed phase and there is no way of knowing a priori whether or not a given system will show ideal behavior.

ADSORPTION KINETICS

Intrinsic Kinetics

Chemisorption may be regarded as a chemical reaction between the sorbate and the solid surface, and, as such, it is an activated process for which the rate constant (k)

follows the familiar Arrhenius rate law:

$$k = k_0 e^{-E/RT} \tag{5}$$

Depending on the temperature and the activation energy (E), the rate constant may vary over many orders of magnitude.

In practice the kinetics are usually more complex than might be expected on this basis, since the activation energy generally varies with surface coverage as a result of energetic heterogeneity and/or sorbate-sorbate interaction. As a result, the adsorption rate is commonly given by the Elovich equation (1):

$$q = \frac{1}{k'} \ln(1 + k''t) \tag{6}$$

where k' and k'' are temperature-dependent constants.

In contrast, physical adsorption is a very rapid process, so the rate is always controlled by mass transfer resistance rather than by the intrinsic adsorption kinetics. However, under certain conditions the combination of a diffusion-controlled process with an adsorption equilibrium constant that varies according to equation 1 can give the appearance of activated adsorption.

A porous adsorbent in contact with a fluid phase offers at least two and often three distinct resistances to mass transfer: external film resistance and intraparticle diffusional resistance. When the pore size distribution has a well-defined bimodal form, the latter may be divided into macropore and micropore diffusional resistances. Depending on the particular system and the conditions, any one of these resistances may be dominant or the overall rate of mass transfer may be determined by the combined effects of more than one resistance.

ADSORPTION COLUMN DYNAMICS

In most adsorption processes the adsorbent is contacted with fluid in a packed bed. An understanding of the dynamic behavior of such systems is therefore needed for rational process design and optimization. What is required is a mathematical model which allows the effluent concentration to be predicted for any defined change in the feed concentration or flow rate to the bed. The flow pattern can generally be represented adequately by the axial dispersed plug-flow model, according to which a mass balance for an element of the column yields, for the basic differential equation governing the dynamic behavior,

$$-D_L \frac{\partial^2 c_i}{\partial z^2} + \frac{\partial}{\partial z}(vc_i) + \frac{\partial c_i}{\partial t} + \left(\frac{1-\epsilon}{\epsilon}\right)\frac{\partial \bar{q}_i}{\partial t} = 0 \tag{7}$$

The term $\partial \bar{q}_i/\partial t$ represents the overall rate of mass transfer for component i (at time t and distance z) averaged over a particle. This is governed by a mass transfer rate expression which may be thought of as a general functional relationship of the form

$$\frac{\partial \bar{q}}{\partial t} = f(c_i, c_j, \ldots, q_i, q_j, \ldots) \tag{8}$$

This rate equation must satisfy the boundary conditions imposed by the equilibrium isotherm and it must be thermodynamically consistent so that the mass transfer rate falls to zero at equilibrium.

Equilibrium Theory

The general features of the dynamic behavior may be understood without recourse to detailed calculations since the overall pattern of the response is governed by the form of the equilibrium relationship rather than by kinetics. Kinetic limitations may modify the form of the concentration profile but they do not change the general pattern.

Constant Pattern Behavior

In a real system the finite resistance to mass transfer and axial mixing in the column lead to departures from the idealized response predicted by equilibrium theory. In the case of a favorable isotherm the shock wave solution is replaced by a constant pattern solution. The concentration profile spreads in the initial region until a stable situation is reached in which the mass transfer rate is the same at all points along the wave front and exactly matches the shock velocity. In this situation the fluid-phase and adsorbed-phase profiles become coincident. This represents a stable situation and the profile propagates without further change in shape—hence the term constant pattern.

Length of Unused Bed

The constant pattern approximation provides the basis for a very useful and widely used design method based on the concept of the length of unused bed (LUB). In the design of a typical adsorption process the basic problem is to estimate the size of the adsorber bed needed to remove a certain quantity of the adsorbable species from the feed stream, subject to a specified limit (c') on the effluent concentration. The length of unused bed, which measures the capacity of the adsorber which is lost as a result of the spread of the concentration profile, is defined by

$$\text{LUB} = (1 - q'/q_0)L = (1 - t'/\bar{t})L \quad (9)$$

where q' is the capacity at the break time t' and $\bar{t}$; is the stoichiometric time (see Fig. 1). The values of t', $\bar{t}$, and

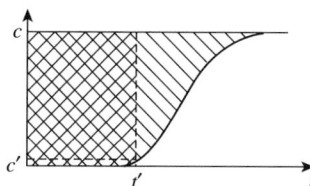

Figure 1. Sketch of breakthrough curve showing break time t' and the method of calculation of the stoichiometric time $\bar{t}$; and LUB. ▨ = the integral of equation 10; ▩ = the integral of equation 11.

hence the LUB are easily determined from an experimental breakthrough curve since, by overall mass balance,

$$\bar{t} = \frac{L}{v}\left[1 + \left(\frac{1-\epsilon}{\epsilon}\right)\left(\frac{q_0}{c_0}\right)\right] = \int_0^\infty \left(1 - \frac{c}{c_0}\right)dt \quad (10)$$

$$t' = \frac{L}{v}\left[1 + \left(\frac{1-\epsilon}{\epsilon}\right)\left(\frac{q'}{c_0}\right)\right] = \int_0^{t'} \left(1 - \frac{c}{c_0}\right)dt \quad (11)$$

The length of column needed can then be found simply by adding the LUB to the length calculated from equilibrium considerations, assuming a shock concentration front.

Proportionate Pattern Behavior

If the isotherm is unfavorable, the stable dynamic situation leading to constant pattern behavior can never be achieved. The equilibrium adsorbed-phase concentration lies above rather than below the actual adsorbed-phase profile. As the mass transfer zone progresses through the column it broadens, but the limiting situation, which is approached in a long column, is simply local equilibrium at all points ($c = c^*$) and the profile therefore continues to spread in proportion to the length of the column. This difference in behavior is important since the LUB approach to design is clearly inapplicable under these conditions.

Adsorption Chromatography

In a linear multicomponent system (several sorbates at low concentration in an inert carrier) the wave velocity for each component depends on its adsorption equilibrium constant. Thus, if a pulse of the mixed sorbate is injected at the column inlet, the different species separate into bands which travel through the column at their characteristic velocities, and at the outlet of the column a sequence of peaks corresponding to the different species is detected. Measurement of the retention time ($\bar{t}$) under known flow conditions thus provides a simple means of determining the equilibrium constant (Henry constant).

In an ideal system with no axial mixing or mass transfer resistance the peaks for the various components propagate without spreading. However, in any real system the peak broadens as it propagates and the extent of this broadening is directly related to the mass transfer and axial dispersion characteristics of the column. Measurement of the peak broadening therefore provides a convenient way of measuring mass transfer coefficients and intraparticle diffusivities.

APPLICATIONS

The applications of adsorbents are many and varied and may be classified as nonregenerative uses, in which the adsorbent is used once and discarded, and regenerative applications, in which the adsorbent is used repeatedly in a cyclic manner involving sequential adsorption and

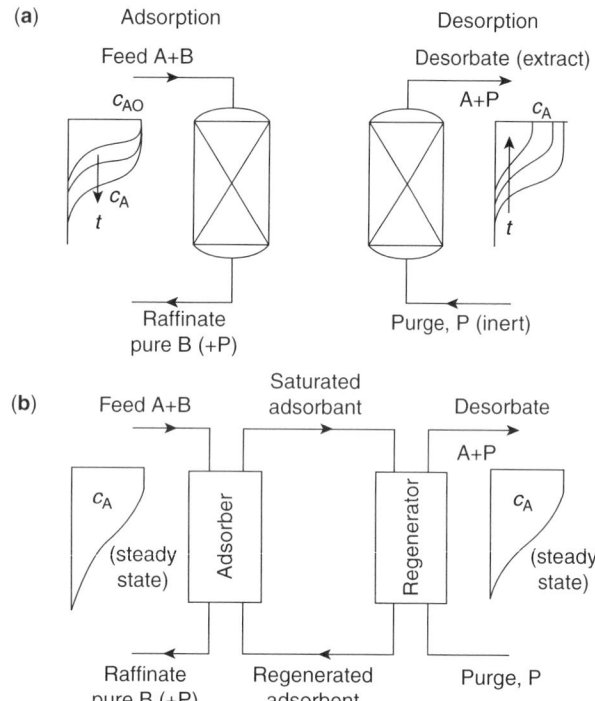

(a)

Adsorption

Feed A+B

c_{AO}

c_A

t

Raffinate
pure B (+P)

Desorption

Desorbate (extract)

A+P

c_A

t

Purge, P (inert)

(b)

Feed A+B

Saturated
adsorbant

Desorbate

A+P

c_A

Adsorber

Regenerator

c_A

(steady
state)

(steady
state)

Raffinate
pure B (+P)

Regenerated
adsorbent

Purge, P

Figure 2. The two basic modes of operation for an adsorption process: (**a**) cyclic batch system; (**b**) continuous countercurrent system with adsorbent recirculation. From Ruthven, 1984.

regeneration steps. Nonregenerative uses are desiccant in dual pane windows, odor removal in health care products, desiccant in refrigeration and air conditioning systems, and cigarette filters, while regenerative uses are water purification (some systems), removal of trace impurities from gases or liquid streams, bulk separations (gas or liquid), low pressure storage of methane, desiccant cooling (open-cycle air conditioning).

Adsorption Separation and Purification Processes

The main area of current application of adsorption is in separation and purification processes. Many different ways of operating such processes have been devised and it is helpful to consider the various systems according to the mode of fluid-solid contact (see Fig. 2).

The other major difference between adsorption processes lies in the method by which the adsorbent bed is regenerated. The three different methods are temperature swing, pressure swing, and displacement. For efficient removal of trace impurities it is normally essential to use a highly selective adsorbent on which the sorbate is strongly held. Temperature swing regeneration is therefore generally used in such applications. However, in bulk separations all three regeneration methods are widely used.

P. G. Ashmore, *Catalysis and Inhibition of Chemical Reactions*, Butterworths, London, 1963, p. 164.

A. E. Rodrigues, M. D. Le Van, and D. Tondeur, *Adsorption: Science and Technology*, NATO ASI E158, Kluwer, Amsterdam, 1989.

D. D. Do *Adsorption Analysis; Kinetics and Equilibria* Imperial College Press, London (1998).

D. M. Ruthven, *Principles of Adsorption and Adsorption Processes*, Wiley-Interscience, New York, 1984.

DOUGLAS M. RUTHVEN
University of Maine

ADSORPTION, GAS SEPARATION

Gas-phase adsorption is widely employed for the large-scale purification or bulk separation of air, natural gas, chemicals, and petrochemicals (Table 1). In these uses, it is often a preferred alternative to the older unit operations of distillation and absorption.

An adsorbent attracts molecules from the gas, the molecules become concentrated on the surface of the adsorbent, and are removed from the gas phase. Many process concepts have been developed to allow the efficient contact of feed gas mixtures with adsorbents to carry out desired separations and to allow efficient regeneration of the adsorbent for subsequent reuse. In nonregenerative applications, the adsorbent is used only once and is not regenerated.

Most commercial adsorbents for gas-phase applications are employed in the form of pellets, beads, or other granular shapes, typically ~1.5–3.2 mm in diameter. Most commonly, these adsorbents are packed into fixed beds through which the gaseous feed mixtures are passed.

Table 1. Commercial Adsorption Separations

Separation	Adsorbent
Gas bulk separations	
normal paraffins, isoparaffins, aromatics	zeolite
N_2/O_2	zeolite
O_2/N_2	carbon molecular sieve
CO, CH_4, CO_2, N_2, Ar, NH_3/H_2	zeolite, activated carbon
acetone/vent streams	activated carbon
C_2H_4/vent streams	activated carbon
H_2O/ethanol	zeolite
Gas purifications	
H_2O/olefin-containing cracked gas, natural gas, air, synthesis gas, etc.	silica, alumina, zeolite
CO_2/C_2H_4, natural gas, etc.	zeolite
organics/vent streams	activated carbon, others
sulfur compounds/natural gas, hydrogen, liquefied petroleum gas (LPG), etc.	zeolite
solvents/air	activated carbon
odors/air	activated carbon
NO_x/N_2	zeolite
SO_2/vent streams	zeolite
Hg/chlor–alkali cell gas effluent	zeolite

Normally, the process is conducted in a cyclic manner. When the capacity of the bed is exhausted, the feed flow is stopped to terminate the loading step of the process, the bed is treated to remove the adsorbed molecules in a separate regeneration step, and the cycle is then repeated.

The growth in both variety and scale of gas-phase adsorption separation processes, particularly since 1970, is due in part to continuing discoveries of new, porous, high-surface area adsorbent materials (particularly molecular sieve zeolites) and, especially, to improvements in the design and modification of adsorbents. These advances have encouraged parallel inventions of new process concepts. Increasingly, the development of new applications requires close cooperation in adsorbent design and process cycle development and optimization.

ADSORPTION PRINCIPLES

The design and manufacture of adsorbents for specific applications involves manipulation of the structure and chemistry of the adsorbent to provide greater attractive forces for one molecule compared to another, or, by adjusting the size of the pores, to control access to the adsorbent surface on the basis of molecular size. Adsorbent manufacturers have developed many technologies for these manipulations, but they are considered proprietary and are not openly communicated. Nevertheless, the broad principles are well known.

Adsorption Forces

Coulomb's law allows calculations of the electrostatic potential resulting from a charge distribution, and of the potential energy of interaction between different charge distributions. Various elaborate computations are possible to calculate the potential energy of interaction between point charges, distributed charges, etc.

Adsorption Selectivities

For a given adsorbent, the relative strength of adsorption of different adsorbate molecules depends on the relative magnitudes of the polarizability α, dipole moment μ, and quadrupole moment Q of each. Often, just the consideration of the values of α, μ, and Q allows accurate qualitative predictions to be made of the relative strengths of adsorption of given molecules on an adsorbent or of the best adsorbent type (polar or nonpolar) for a particular separation.

For example, the strength of the electric field F and field gradient ($\delta F = dF/dr$) of the highly polar cationic zeolites is strong. For this reason, nitrogen is more strongly adsorbed than is oxygen on such adsorbents, primarily because of the stronger quadrupole of N_2 compared to O_2.

In contrast, nonpolar activated carbon adsorbents lack strong electric fields and field gradients. Such adsorbents adsorb O_2 slightly more strongly than N_2, because of the slightly higher polarizability of O_2. Relative selectivities on nonpolar adsorbents often parallel the relative volatilities of the same compounds. Compounds with higher boiling points are more strongly adsorbed.

For a given adsorbate molecule, the relative strength of adsorption on different adsorbents depends largely on the relative polarizability and electric field strengths of adsorbent surfaces. On the one hand, water molecules, with relatively low polarizability but a strong dipole and moderately strong quadrupole moment, are strongly adsorbed by polar adsorbents (eg, cationic zeolites), but only weakly adsorbed by nonpolar adsorbents (eg, silicalite or nonoxidized forms of activated carbon). On the other hand, saturated hydrocarbons with low molecular weight have greater polarizabilities than does water, but no dipoles and only weak quadrupoles. These molecules are adsorbed less strongly than water on polar adsorbents, but more strongly than water on nonpolar adsorbents. Therefore, polar adsorbents are often called hydrophilic adsorbents and nonpolar adsorbents are called hydrophobic adsorbents.

Isotherms and Isobars

The equilibrium adsorbate loading vs adsorbate pressure (or concentration) at constant temperature is an adsorption isotherm, and the adsorbate loading vs temperature at constant adsorbate pressure is an adsorption isobar. The greater the strength of adsorption, the greater is the adsorbate loading at a given temperature and partial pressure of the adsorbate up to the point where the maximum adsorption capacity of the adsorbent has been attained.

The strength of adsorption of unsaturated hydrocarbons by a polar adsorbent (zeolite) is much greater than for saturated hydrocarbons, and increases with increasing carbon number. This observation may be understood as a consequence of the increasing polarizability of molecules with increasing numbers of bonds and the presence of dipole and stronger quadrupole moments in the unsaturated hydrocarbons compared to the saturated hydrocarbons.

Heats of Adsorption

Physical adsorption processes are exothermic, ie, they release heat. Because the entropy change ΔS on adsorption is negative (adsorbed molecules are more ordered than in the gas phase) and the free energy change ΔG must be negative for adsorption to be favored, thermodynamics ($\Delta G = \Delta H - T\Delta S$) requires the enthalpy change ΔH on adsorption (heat of adsorption) to be negative (exothermic). Adsorption strengths thus decrease with increasing temperature.

The integral heat of adsorption is the total heat released when the adsorbate loading is increased from zero to some final value at isothermal conditions. The differential heat of adsorption δH_{iso} is the incremental change in heat of adsorption with a differential change in adsorbate loading. This heat of adsorption δH_{iso} may be determined from the slopes of adsorption isosteres (lines of constant adsorbate loading) on graphs of $\ln P$ vs $1/T$ (Fig. 1) through the Clausius-Clapeyron relationship:

$$\frac{d \ln P}{d\,(1/T)} = -\frac{\delta H_{iso}}{R}$$

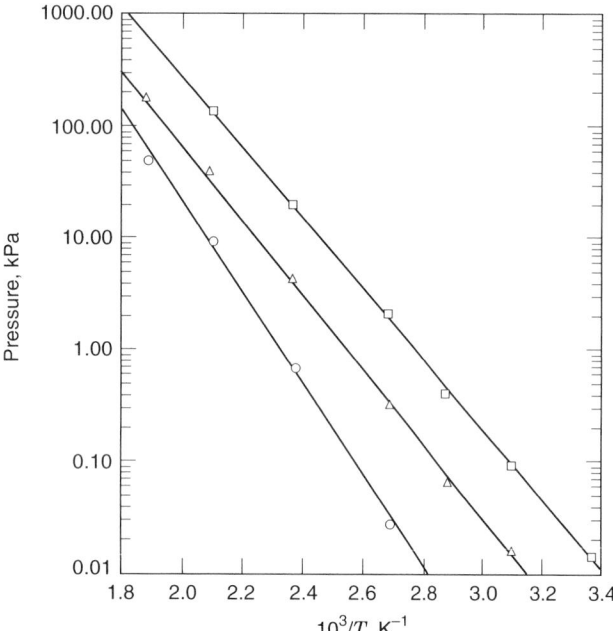

Figure 1. Adsorption isosteres, water vapor on 4A (NaA) zeolite pellets. H_2O loading: □, 15 kg/100 kg zeolite; △, 10 kg/100 kg, ○, 5 kg/100 kg. To convert kPa to mm Hg, multiply by 7.5. (Courtesy of Union Carbide.)

where R is the gas constant, P the adsorbate absolute pressure, and T the absolute temperature.

Isotherm Models

Many efforts have been made over the years to develop isotherm models for data correlation and design predictions for both single component and multicomponent adsorption. Unfortunately, no single model is accurate over broad ranges of adsorbent and adsorbate types, pressures, temperatures, and loadings, especially for multicomponent systems. The reason is probably due to deficiencies in the models in adequately describing both the heterogeneities of the surface and the effects of the adsorbate on the properties of the adsorbent itself. Most models assume the adsorbent is inert, ie, not altered by the presence of the adsorbate molecules; however, partial changes in some adsorbent properties are commonly observed.

Nevertheless, each of the more popular isotherm models have been found useful for modeling adsorption behavior in particular circumstances. The following outlines many of the isotherm models presently available.

Thermodynamically Consistent Isotherm Models. These models include both the statistical thermodynamic models and the models that can be derived from an assumed equation of state for the adsorbed phase plus the thermodynamics of the adsorbed phase.

Statistical Thermodynamic Isotherm Models. These approaches were pioneered by Fowler and Guggenheim and Hill, has been applied and this approach has been applied to modeling of adsorption in microporous adsorbents.

Semiempirical Isotherm Models. Some of these models—the Polanyi adsorption potential, the Radke-Prausnitz, the Toth, the UNILAN, the BET—have been shown to have some thermodynamic inconsistencies and should be used with due care. Nevertheless, they have each been found to be useful for data correlation and interpolation, as well as for the calculation of some thermodynamic properties.

Isotherm Models for Adsorption of Mixtures. Of the following models, all but the ideal adsorbed solution theory (IAST) and the related heterogeneous ideal adsorbed solution theory (HIAST) have been shown to contain some thermodynamic inconsistencies. They include Markham and Benton the Leavitt loading ratio correlation (LRC) method, the ideal adsorbed solution (IAS) model, the heterogeneous ideal adsorbed solution theory (HIAST), and the vacancy solution model (VSM).

Molecular Modeling

Since the 1980s, the availability of high speed computers has made it practical to model adsorbate–adsorbent interactions in micro- and meso-pores using statistical thermodynamic principles. Adsorbent pores have been modeled in various ways to predict adsorption equilibrium and diffusivity phenomenon—as flat surfaces, as narrow slits, and as the regular crystalline structures of zeolites. The results vary greatly depending on the choice of software used. Molecular modeling has been used to study pore size distribution by varying slit width and to interpret nmr data.

Adsorption Dynamics

An outline of approaches that have been taken to model mass-transfer rates in adsorbents has been given (see ADSORPTION). Extensive literature exists on the interrelated topics of modeling of mass-transfer rate processes in fixed-bed adsorbers, bed concentration profiles, and breakthrough curves and the related simple design concepts of WES, WUB, and LUB for constant-pattern adsorption.

Reactions on Adsorbents

To permit the recovery of pure products and to extend the adsorbent's useful life, adsorbents should generally be inert and not react with or catalyze reactions of adsorbate molecules. These considerations often affect adsorbent selection and/or require limits be placed upon the severity of operating conditions to minimize reactions of the adsorbate molecules or damage to the adsorbents.

ADSORBENT PRINCIPLES

Principal Adsorbent Types

Commercially useful adsorbents can be classified by the nature of their structure (amorphous or crystalline), by the sizes of their pores (micropores, mesopores, and macropores), by the nature of their surfaces (polar, nonpolar, or intermediate), or by their chemical composition. All of these characteristics are important in the selection of the best adsorbent for any particular application.

However, the size of the pores is the most important initial consideration because, if a molecule is to be adsorbed, it must not be larger than the pores of the adsorbent. Conversely, by selecting an adsorbent with a particular pore diameter, molecules larger than the pores may be selectively excluded, and smaller molecules can be allowed to adsorb.

Pore size is also related to surface area and thus to adsorbent capacity, particularly for gas-phase adsorption. Because the total surface area of a given mass of adsorbent increases with decreasing pore size, only materials containing micropores and small mesopores (nanometer diameters) have sufficient capacity to be useful as practical adsorbents for gas-phase applications.

The practical adsorbents used in most gas-phase applications are limited to the following types, classified by their amorphous or crystalline nature.

- *Amorphous*: silica gel, activated alumina, activated carbon, molecular sieve carbons and macroreticular resins.
- *Crystalline*: molecular sieve zeolites, and related molecular sieve materials that are not technically zeolites, eg, silicalite, $AlPO_4$s, and SAPOs, as well as mesoporous silicates/aluminosilicates, eg, MCM-41.

Assuming the pores are large enough to admit the molecules of interest, the most important consideration is the nature of the adsorbent surface, because this characteristic controls adsorption selectivity.

Practical adsorbents may also be classified according to the nature of their surfaces.

- *Highly polar*: molecular sieve zeolites with high aluminum and cation contents.
- *Moderately polar*: crystalline molecular sieves with low aluminum and low cation contents, silica gel, activated alumina, activated carbons with highly oxidized surfaces, crystalline molecular sieve $AlPO_4$'s.
- *Nonpolar*: silicalite, F-silicalite, other high silica content crystalline molecular sieves, activated carbons with reduced surfaces.

Adsorption Properties

Not only do the more highly polar molecular sieve zeolites adsorb more water at lower pressures than do the moderately polar silica gel and alumina gel, but they also hold onto the water more strongly at higher temperatures. For the same reason, temperatures required for thermal regeneration of water-loaded zeolites is higher than for less highly polar adsorbents.

Physical Properties

Physical properties of importance include particle size, density, volume fraction of intraparticle and extraparticle voids when packed into adsorbent beds, strength, attrition resistance, and dustiness. Any of these properties can be varied intentionally to tailor adsorbents to specific applications (See ADSORPTION LIQUID SEPARATION;

CARBON, ACTIVATED CARBON; ION EXCHANGE; MOLECULAR SIEVES; AND SILICON COMPOUNDS, SYNTHETIC INORGANIC SILICATES). Most commercial adsorbents for gas-phase separations are employed in the form of pellets, beads, or granular shapes, typically ~1.5–3.2 mm in size. However, a growing number of atmospheric presssure applications are employing adsorbents in the form of honeycomb monoliths and fabrics. These configurations have been introduced to reduce the pressure drop and thus utility consumption.

Deactivation

All adsorbents, no matter how inert, will be deactivated during extended usage by reaction with inpurities, reaction of adsorbates, or thermal damage. To compensate, adsorbent beds are sized to account for the gradual loss in capacity and to allow their use for a given period of time. Most commonly, at the end of its useful life, the adsorbent is dumped from the beds and replaced with fresh adsorbent.

ADSORPTION PROCESSES

Adsorption processes are often identified by their method of regeneration. Pressure-swing adsorption (PSA) and TSA are the most frequently applied process cycles for gas separation. Purge-swing cycles and nonregenerative approaches are also applied to the separation of gases. Special applications exist in the nuclear industry. Others take advantage of reactive sorption. Most adsorption processes use fixed beds, but some use moving or fluidized beds.

Temperature Swing

A temperature-swing or TSA cycle is one in which desorption takes place at a temperature much higher than adsorption. The principal application is for separations in which contaminants are present at low concentration, ie, for purification. The TSA cycles are characterized by low residual loadings and high operating loadings.

Pressure Swing

A PSA cycle is one in which desorption takes place at a pressure much lower than adsorption. Its principal application is for bulk separations where contaminants are present at high concentration. The PSA cycles are characterized by high residual loadings and low operating loadings. These low adsorption capacities for high concentrations mean that cycle times must be short, seconds to minutes, for reasonably sized beds. Fortunately, packed beds of adsorbent respond rapidly to changes in pressure. A purge usually removes the desorbed components from the bed, and the bed is returned to adsorption condition by repressurization. Applications may require additional steps. Systems with weakly adsorbed species are especially suited to PSA adsorption. The applications of PSA include drying, upgrading of H_2 and fuel gases, and air separation.

Purge Swing

A purge-swing adsorption cycle is one in which desorption takes place at the same temperature and total pressure as adsorption. Regeneration is accomplished either by partial-pressure reduction by an inert gas purge or by adsorbate displacement by an adsorbable gas. Its major application is for bulk separations when contaminants are at high concentration. Like PSA, purge cycles are characterized by high residual loadings, low operating loadings, and short cycle times (minutes). Mixtures of weakly adsorbed components are especially suited to purge-swing adsorption. Applications include the separation of normal from branched and cyclic hydrocarbons, and for gasoline vapor recovery.

Nonregenerative Processing

Gas-phase adsorption can also be used when regenerating the adsorbent is not practical. Most of these applications are used where the facilities to effect a regeneration are not justified by the small amount of adsorbent in a single unit. Nonregenerative adsorbents are used in packaging, dual-pane windows, odor removal, and toxic chemical protection.

Reactive Adsorption

Although chemisorbents are not used as extensively as physical adsorbents, a number of commercially significant processes employ chemisorption for gas purification.

Moving/Fluidized Beds and Wheels

Most adsorption systems use stationary-bed adsorbers. However, efforts have been made over the years to develop moving-bed adsorption processes in which the adsorbent is moved from an adsorption chamber to another chamber for regeneration, with countercurrent contacting of gases with the adsorbents in each chamber.

One use of fluidized-bed adsorption and moving-bed desorption is for removal of volatile organic carbon compounds from air. Another application for moving beds is in the treatment of flue or exhaust gases. Here the adsorbent flows downward in a cross-flow mode to minimize pressure drop. Also wheels have been commercialized for ambient pressure applications in order to reduce costly pressure drop. Their major application has been TSA cycles for drying and for VOC removal (see above).

DESIGN METHODS AND FUTURE DIRECTIONS

Design techniques for gas-phase adsorption range from empirical to theoretical, from simple to computationally intensive. Methods have been developed for equilibrium, mass transfer, and combined dynamic performance. Approaches are available for the regeneration methods of heating, purging, steaming, and pressure swing.

Adsorption

In the design of the adsorption step of gas-phase processes, two phenomena must be considered: equilibrium and mass transfer. Sometimes adsorption equilibrium can be regarded as that of a single component, but more often several components and their interactions must be accounted for. Design techniques for each phenomenon exist as well as some combined models for dynamic performance.

Equilibrium. Among the aspects of adsorption, equilibrium is the most studied and published. Many different adsorption equilibrium equations are used for the gas phase. Equally important is the adsorbed phase mixing rule that is used with these other models to predict multicomponent behavior.

Many simple systems that could be expected to form ideal liquid mixtures are reasonably predicted by extending pure-species adsorption equilibrium data to a multicomponent equation. The potential theory has been extended to binary mixtures of several hydrocarbons on activated carbon by assuming an ideal mixture, and to O_2 and N_2 on 5A and 10X zeolites. For most models of adsorptive equilibrium, however, the coefficients derived from pure species are not adequate to predict multicomponent equilibrium for nonideal mixtures.

Mass Transfer. The degree of approach to equilibrium that can be achieved in adsorption is determined by the mass-transfer rates. One useful design concept is the mass-transfer zone (MTZ), an extension of the ion-exchange zone method.

Most dynamic adsorption data are obtained in the form of outlet concentrations as a function of time. This LUB concept is commonly used for drying and desulfurization design in the natural gas industry and for air prepurification before cryogenic distillation.

Dynamic Performance. More complex models do not attempt to separate the equilibrium behavior from the mass-transfer behavior. Rather they treat adsorption as one dynamic process with an overall dynamic response of the adsorbent bed to the feed stream. Although numerical solutions can be attempted for the rigorous partial differential equations, simplifying assumptions are often made to yield more manageable calculating techniques. For systems with a large number of components, the design can be simplified by combining the adsorbates into pseudospecies based on Freundlich exponent and on mass-transfer coefficients.

Regeneration

In recent years, considerable effort has been expended to better understand and quantify the process of regeneration. Methods are available to predict thermal, purge, and steaming requirements. Models are available to simulate all of the regeneration types, temperature, pressure, and purge swings.

Thermal Requirements. When a TSA cycle is heating limited, the regeneration design is only concerned with transferring energy to the system.

Purge Requirements. The amount of purge gas needed in stripping-limited regeneration is similar to that for purge regeneration, but it differs primarily in the temperature at which the isothermal desorption occurs. For a pressure-swing process, the theoretical minimum volumetric purge/feed ratio is the ratio of the purge pressure to the feed pressure. For a thermal-swing process, the minimum purge/adsorbent ratio is the ratio of the heat capacity of the solid to that of the gas. Specific design purge data have been published for purge-swing activated-carbon automotive evaporative emissions control and for pressure-swing drying of pneumatic system air. The pneumatic system process exhibits an optimum purge ratio for maximizing the attainable dewpoint depression. An isothermal purge-swing model that uses Langmuir equilibrium to simulate adsorbent performance also exists.

Steaming Requirements. The steaming of fixed beds of activated carbon is a combination of thermal swing and displacement purge swing. The exothermic heat released when the water adsorbs from the vapor phase is much higher than is possible with heated gas purging.

Temperature Swing. Many thermal swing adsorption models have been used to carry out parametric analyses with a goal of energy minimization. A nonisothermal model for single components using equilibrium theory demonstrated that efficiency improves with increased purge contact time and high heat capacity purge gas but is minimally affected by initial bed loading, and the model defined conditions under which the desorption can be continued with a cold stream without additional overall purge gas.

Pressure Swing. Design equations have been developed to predict temperature rise, minimum bed length to retain the heat front, minimum purge rate, and effluent composition. A nonequilibrium, nonisothermal simulation program with a Freundlich isotherm equation was found to agree with data for drying with silica gel. A somewhat simpler isothermal model using an isotherm approximated by two straight lines successfully calculated the volumetric purge/feed ratio needed to achieve varying product dryness using silica gel. An adiabatic equilibrium model with a Langmuir isotherm was used to study the blowdown step of a cycle removing CO_2 on activated carbon and 5A zeolite. Changing to an isothermal assumption introduced significant errors into the results. The countercurrent pressurization step was investigated with an isothermal equilibrium model using a Langmuir isotherm for O_2 production from air with 5A zeolite. The model predicted the dependence of O_2 concentration on countercurrent pressure and was used to study other parameters. An isothermal model with linear isotherms and component-specific pore diffusivity was used and compared to data for the kinetic-limited separation of air by RS-10 zeolite. The simulations agreed well with the experimental parametric studies of time and pressure of feed, blowdown, purge, and pressurization.

Pressure Drop

The prediction of pressure drop in fixed beds of adsorbent particles is important. When the pressure loss is too high, costly compression may be increased, adsorbent may be fluidized and subject to attrition, or the excessive force may crush the particles. As discussed previously, RPSA relies on pressure drop for separation. Because of the cyclic nature of adsorption processes, pressure drop must be calculated for each of the steps of the cycle.

As far as future directions are concerned, advances in fundamental knowledge of adsorption equilibrium and mass transfer will enable further optimization of the performance of existing adsorbent types. Continuing discoveries of new adsorbent materials will also provide adsorbents with new combinations of useful properties. New adsorbents and adsorption processes will be developed to provide needed improvements in pollution control, energy conservation, and the separation of high value chemicals. New process cycles and new hybrid processes linking adsorption with other unit operations will continue to be developed.

Fundamentals

Marked improvements in the prediction of multicomponent equilibrium from single-component data will be achieved by developing more realistic theoretical models that provide for nonideal adsorbate phases and heterogeneities of surface energetics and geometries, and that allow for the effect of adsorbates on adsorbent properties. Molecular modeling and molecular-dynamic simulations of adsorption phenomena on high speed computers will enable better prediction of multicomponent adsorption behavior and design of adsorbents with desired properties.

New Adsorbent Materials

Hydrophobic molecular sieves, mesoporous molecular sieves, macroreticular resins, and new carbon molecular sieves will continue to find new application. Carbon nanotubes and pillared interlayer clays (PILCS) will become more available for commercial applications, including adsorption. Adsorbents with enhanced performance, both highly selective physical adsorbents and easily regenerated, weak chemisorbents will be developed.

Process Concepts

More hybrid systems involving gas-phase adsorption coupled with catalytic processes and with other separations processes (especially distillation and membrane systems) will be developed to take advantage of the unique features of each.

Design Methods

Improvements in the ability to predict multicomponent equilibrium and mass-transfer rate performance will allow continued improvements in the design of new adsorption systems and in the energy efficiency of existing systems.

Computer Systems

Improved "smart" control systems based on new computer capabilities and control algorithms will be used increasingly in adsorption systems to provide more efficient operation. Enhanced computer capabilities will also allow coupling of more sophisticated equilibrium models with more exact models for adsorption dynamics to provide improved design tools.

J. T. Collins, M. J. Bell, and W. M. Hewitt, in A. A. Moghissi and co-workers, eds., *Nuclear Power Waste Technology*, American Society of Mechanical Engineers, New York, 1978, Chapt. 4.

G. E. Keller, II, R. A. Anderson, and C. M. Yon, in R. W. Rousseau, ed., *Handbook of Separation Process Technology*, John Wiley & Sons, Inc., New York, 1987, pp. 644–696.

D. M. Ruthven, *Principles of Adsorption and Adsorption Processes*, John Wiley & Sons, Inc., New York, 1984.

R. T. Yang, *Gas Separation by Adsorption Processes*, Butterworths, Stoneham, Mass., 1987.

CARMEN M. YON
JOHN D. SHERMAN
UOP

ADSORPTION, LIQUID SEPARATION

Recovery and purification of the desired product are generally as important as the synthesis of the product itself. Although most of the value in chemical conversion is added via reaction, it is the separation that largely determines the capital cost of production. Nearly every chemical manufacturing operation requires the use of separation processes to recover and purify the desired product. In most circumstances, the efficiency of the separation process has a significant impact on both the quality and the cost of the product. Liquid-phase adsorption has long been used for the removal of contaminants present at low concentrations in process streams. In most cases, the objective is to remove a specific feed component; alternatively, the contaminants are not well defined, and the objective is to improve feed quality as defined by color, taste, odor, and storage stability. Deodorization of water, decolorization of sugar, ion exchange of fermentation broths are a few examples of processes in which trace impurities are removed. More recently the simulated moving bed (SMB) processes are finding applications in biotechnology semi batch and in protein purification.

While most of the strategies to remove trace impurities are batch processes, bulk adsorptive separation processes are continuous or semicontinuous in operation, because in bulk separation processes, where the feed component may be present in large enough concentration, it is imperative to maximize utilization of the adsorbent.

BATCH VERSUS CONTINUOUS OPERATION

Industrial-scale adsorption processes can be classified as batch or continuous. In a batch process, the adsorbent bed is saturated and regenerated in cyclic operation. In a continuous process, a countercurrent staged contact between the adsorbent and the feed and desorbent is established by either a true or a simulated recirculation of the adsorbent. The efficiency of an adsorption process is significantly higher in a continuous mode of operation than in a cyclic batch mode. In a batch chromatographic operation, the liquid composition at a given level in the bed undergoes a cyclic change with time, and large portions of the bed do not perform any useful function at a given time. In continuous operation, the composition at a given level is invariant with time, and every part of the bed performs a useful function at all times. The height equivalent of a theoretical plate (HETP) in a batch operation is roughly three times that in a continuous mode. For difficult separations, batch operation may require 25 times more adsorbent inventory and twice the desorbent circulation rate than does a continuous operation. In addition, in a batch mode, the four functions of adsorption, purification, desorption, and displacement of the desorbent from the adsorbent are inflexibly linked, whereas a continuous mode allows more degrees of freedom with respect to these functions, and thus a better overall operation.

PROCESSES AND OPERATIONS

Continuous Countercurrent Processes

The need for a continuous countercurrent process arises because the selectivity of available adsorbents in a number of commercially important separations is not high. Therefore, one stage of contacting cannot provide a good separation, and multistage contacting must be provided in the same way that multiple trays are required in fractionating materials with relatively low volatilities.

Moving-Bed Operation

In moving-bed system and a liquid-phase composition profile the adsorbent circulates continuously as a dense bed in a closed cycle and moves up the adsorbent chamber from bottom to top. Liquid streams flow down through the bed countercurrently to the solid.

Difficulties of Moving-Bed Operation. The use of a moving bed introduces the problem of mechanical erosion of the adsorbent. Obtaining uniform flow of both solid and liquid in beds of large diameter is also difficult. The performance of this type of operation can be greatly impaired by nonuniform flow of either phase.

Simulated Moving Bed Operation. In a moving-bed system, solid moves continuously in a closed circuit past fixed points of introduction and withdrawal of liquid. The same results can be obtained by holding the bed stationary and periodically moving the positions at which the various streams enter and leave. A shift in the positions of the introduction of the liquid feed and the withdrawal in the direction of fluid flow through the bed simulates the movement of solid in the opposite direction.

Of course, moving the liquid feed and withdrawal positions continuously is impractical. However, approximately the same effect can be produced by providing multiple liquid-access lines to the bed and periodically switching each stream to the adjacent line. Functionally, the adsorbent bed has no top or bottom and is equivalent to a toroidal bed. Therefore, the four liquid-access positions can be moved around the bed continually, always maintaining the same distance between the various streams.

Adsorbate–Adsorbent Interactions

An adsorbent can be visualized as a porous solid having certain characteristics. When the solid is immersed in a liquid mixture, the pores fill with liquid, which at equilibrium differs in composition from that of the liquid surrounding the particles of the adsorbent. These compositions can then be related to each other by enrichment factors that are analogous to relative volatility in distillation. The adsorbent is selective for the component that is more concentrated in the pores than in the surrounding liquid.

The choice of separation method to be applied to a particular system depends largely on the phase relations that can be developed by using various separative agents. Adsorption is usually considered to be a more complex operation than is the use of selective solvents in liquid–liquid extraction (see EXTRACTION, LIQUID–LIQUID), extractive distillation, or azeotropic distillation (see DISTILLATION, AZEOTROPIC AND EXTRACTIVE). Consequently, adsorption is employed when it achieves higher selectivities than those obtained with solvents.

ADSORBENTS

Practical Adsorbents

The search for a suitable adsorbent is generally the first step in the development of an adsorption process. A practical adsorbent has four primary requirements: selectivity, capacity, mass-transfer rate, and long-term stability. The requirement for adequate adsorptive capacity restricts the choice of adsorbents to microporous solids with pore diameters ranging from a few tenths to a few tens of nanometers.

Traditional adsorbents such as silica, SiO_2, activated alumina, $A1_2O_3$; and activated carbon, C, exhibit large surface areas and micropore volumes. The surface chemical properties of these adsorbents make them potentially useful for separations by molecular class. However, the micropore size distribution is fairly broad for these materials. This characteristic makes them unsuitable for use in separations in which steric hindrance can potentially be exploited (see ALUMINUM COMPOUNDS, ALUMINA; SILICON COMPOUNDS, SYNTHETIC INORGANIC SILICATES).

Typical polar adsorbents are silica gel and activated alumina. In general, the selectivities are parallel to those obtained by the use of selective polar solvents; in hydrocarbon systems, even the magnitudes are similar. Consequently, the commercial use of these adsorbents must compete with solvent-extraction techniques.

The principal nonpolar-type adsorbent is activated carbon. With some exceptions, the least polar component of a mixture is selectively adsorbed.

Polymeric resins are widely used in the food and pharmaceutical industries as cation–anion exchangers for the removal of trace components and for some bulk separations, such as fructose from glucose. These resins are primarily attractive for aqueous-phase separations and offer a fairly wide potential range of surface chemistries to fit a number of separation needs.

In contrast to these adsorbents, zeolites offer increased possibilities for exploiting molecular-level differences among adsorbates. Zeolites are crystalline aluminosilicates containing an assemblage of SiO_4 and $A1O_4$ tetrahedral joined together by oxygen atoms to form a microporous solid, which has a precise pore structure. Nearly 40 distinct framework structures have been identified to date. The versatility of zeolites lies in the fact that widely different adsorptive properties may be realized by the appropriate control of the framework structure, the silica-to-alumina ratio (Si/Al), and the cation form.

Desorbent

In addition to adsorbent, the desorbent or the eluant plays an important role in the commercial viability of the SMB process. Desorbent is usually physically separable from the product, ie, its boiling point must be either higher or lower by sufficient degrees. Also desorbent selectivity must fall between the two key components which one wants to separate in an SMB mode. The third and equally important property is for the desorbent to not hinder mass transfer. This can be very important in sterically hindered transfer processes such as in zeolites.

LIQUID-PHASE ADSORPTION AND SEPARATION OF PHARMACEUTICALS AND RELATED COMPOUNDS

Various modes of adsorptive separations have acquired a key role in processes for the production of numerous biologically significant compounds, including:

- Sugars (monosaccharide resolution, oligio-, and disaccharide purification, and recovery from molasses.
- Amino acids and peptides.
- Proteins, nucleic acids, and oligonucleotides.
- Antibodies [both monoclonal and recombinant].
- Pharmaceutically significant small organic molecules, including resolution of enantiomeric and diastereomeric compounds.

Although many similarities exist with petrochemical applications of adsorptive separations, there are two significant differences in liquid-phase adsorption processes for pharmaceuticals and related compounds. First, the sorbents, which are commonly referred to as "stationary phases", are generally composed of different materials and take slightly different forms compared to the sorbents used in petrochemical applications. Second, the factors that influence the selection of a mode of operation are significantly different in the purification of pharmaceutical and biologically active materials. Therefore, while the available modes of operation are not different, the frequency of application of the various modes differs between the two areas.

Sorbents and Eluents in Pharmaceutical Separations

As in any separation, the liquid-phase adsorptive separation of two or more compounds requires at least one physical property difference that can be exploited. In this manner, a mechanism can be chosen to exploit the desired physical property difference in the molecules to be separated and control the partitioning of the molecules between the liquid and adsorbed phases. The combination of the physical property being exploited and the mechanism through which partitioning occurs, in general, dictates the type of sorbent used to effect the separation.

Generally speaking, sorbents used in separation of pharmaceutical and biological substances fall into two broad categories: solids and gels. Solid packings are based on a polar or inert solid material, usually a metal oxide (eg, silica gel, zirconia, etc), which is typically spherical in shape and between 1 and 100 μ in diameter.

Gel-type packings are typically composed of carbohydrate matrices, with or without cross-linking with agarose or acrylamide. These gels are quite soft and can only be used in low- or moderate-pressure chromatography. Gel-type packings are often chosen for the separation of biologics, such as size exclusion-based purification of proteins.

Modes of Adsorptive Separation

While the applications of liquid-phase adsorptive separations in the petrochemical industry have tended toward the use of continuous processes, the same cannot be said for pharmaceutical and biomolecule separations. The potential reasons for the slow progress of continuous adsorptive separation processes (indeed, continuous process *in general*) in the pharmaceutical industry are potentially many. However, three particular hurdles to application can be readily identified.

First, and probably foremost, is the tremendous difference between the petrochemical and pharmaceutical applications "objective functions" that are used to determine the optimum process choice. Petrochemical process options tend toward minimization of the cost of manufacture (COM), while COM is rarely a substantial portion of the overall cost of development of a pharmaceutically important substance. In addition to cost differences, different valuations exist between the two sectors for such fundamental parameters as time-to-market and risk-avoidance. Finally, while petrochemical processes tend to be carried out at high volume in dedicated facilities, the low volumes

and product turn-over experienced in the pharmaceutical industry favor manufacturing through general rather than specialized pieces of equipment.

A second governing issue is the generally greater complexity of the most common adsorptive separations in the pharmaceutical and bioprocess industries as compared to the general chemical industry. A third seminal hurdle to be overcome by continuous adsorptive separations in pharmaceutical and biomolecule applications are the issues associated with validation and regulation, particularly in the application of current good manufacturing practices (cGMPs) to continuous separations.

These and other critical issues have led to the present situation in which batch adsorptive separation is preferred in pharmaceutical and biomolecule processes. This situation is, however, changing rapidly with a noticeable acceleration in the past 5 years. While batch adsorption processes will continue to be important for manufacturing of biologically important materials, there will continue to be cases which are particularly well suited to continuous processes, and these instances are likely to increase in frequency with each successful application.

Batch Processes

Batch processes are relatively abundant in the purification and separation of pharmaceutically important substances, particularly in the areas of peptide, proteins, and oligonucelotides. One commercial example is the purification of synthetic oligonucleotides for use as antisense drugs.

Cyclic Processes

Similar to batch separation processes, the pharmaceutical and biomolecule separations industry defines cyclic processes as those in which solute material is recycled, with or without fresh solute material, back to the sorbent bed for further processing. Preparative and process separations have used eluent recycle closed-loop recycle to improve eluent efficiency and/or the number of theoretical plates available for separation.

Continuous Processes

Although it had long been established as a viable, practical, and cost-effective liquid-phase adsorptive separation technique, the pharmaceutical and biomolecule separations community did not show considerable interest in SMB technology until the mid-1990s. Though resolution of enantiomers is only one such application of SMB in pharmaceuticals and biopharmaceuticals, it has quickly become an area of significant focus.

Enantiomeric resolutions are only one area of potential application of SMB in pharmaceutical and biotechnology manufacture. Indeed, the voluminous literature is filled with other applications, including the purification of paclitaxel, betaine recovery from beet molasses, antibiotics, and azeotropic protein separations. The later two cases required the use of SMB in a relatively new mode, that of providing an eluent composition gradient.

OUTLOOK

Liquid adsorption processes hold a prominent position in several applications for the production of high purity chemicals on a commodity scale. Many of these processes were attractive when they were first introduced to the industry and continue to increase in value as improvements in adsorbents, desorbents, and process designs are made. The Parex process alone has seen three generations of adsorbent and four generations of desorbent.

A surprisingly large number of important industrial-scale separations can be accomplished with the relatively small number of zeolites that are commercially available. The discovery, characterization, and commercial availability of new zeolites and molecular sieves are likely to multiply the number of potential solutions to separation problems. A wider variety of pore diameters, pore geometries, and hydrophobicity in new zeolites and molecular sieves as well as more precise control of composition and crystallinity in existing zeolites will help to broaden the applications for adsorptive separations and likely lead to improvements in separations that are currently in commercial practice.

J. E. Bauer, A. K. Chandhok, B. W. Scanlon, and S. A. Wilcher *A Comprehensive Look at Scaling-up SMB Chiral Separations,* Proceedings of Chiratech '97, Philadelphia, Catalyst Consultants Publishing, Inc., 1997.

R. R. Deshmukh, W. E. Leitch II, Y. S. Sanghvi, and D. L. Cole, in S. Ahuja, ed., *Separation Science and Technology,* Vol. 2 (Handbook of Bioseparations, Academic Press, San Diego, Calif., 2000.

J. J. Kipling, *Adsorption from Solutions of Non-Electrolytes,* Academic Press, Inc., New York, 1965.

D. M. Ruthven, *Principles of Adsorption and Adsorption Processes,* John Wiley & Sons, Inc., New York, 1984.

STANLEY A. GEMBICKI
UOP LLC

ADVANCED MATERIALS, ECONOMIC EVALUATION

The emergence of the advanced materials industry, beginning in the late 1970s, represents one of the most important and dynamic chapters in U.S. and international technological development. These materials possess new and different types of internal structures and exhibit a variety of novel physical and chemical properties that have a wide range of industrial and commercial applications.

The advanced materials industry encompasses such product areas as: biochemicals (including genetic-based materials); bioengineered materials; catalysts; ceramics and clays; coatings; composites; crystal materials; fuels; fullerenes; metal alloys; nanometarials (eg, nanotubes, nanopowders, nanospheres, nanofibers); optical and photonic materials; polymers (eg, plastics, rubber, fibers) and polymer matrices; powdered metals; sensor materials; superconducting materials; and thin films.

While advanced materials are highly diverse with respect to structure, physical and chemical characteristics, and applications, they form a coherent industry due to a number of criteria, including common processes and technical and economic interrelationships. The importance of the industry resides in the fact that its materials diffuse into and impact virtually all of the major industrial sectors, including aerospace, automotive, biomedical, construction, consumer electronics, defense, energy, food processing, healthcare, materials processing, mining, packaging, petrochemical, security, telecommunications and utilities.

The advanced materials field is fundamentally altering how, and even what, the world's leading industries produce, and how they structure themselves and perform their basic operations. Indeed, very few major new technologies can emerge without the application of advanced materials.

PROCESS FLOW AND TECHNOLOGY

While the advanced materials industry creates a wide range of products, common patterns emerge that characterize the process flow of the industry. First, the manufacturing processes that transform raw materials into advanced materials are generally chemical in nature. Typically, the processes employ thermal energy through various means—eg, thermal furnaces, laser technology, and high temperature reactors—to effect the requisite chemical reactions. Then too, an advanced material, once produced, generally requires subsequent processes to modify and prepare the material for market. Depending on the industry involved, additional processing steps may include assembling components or parts containing an advanced material into a final product to be distributed and sold to an original equipment manufacturer (OEM).

INDUSTRY STRUCTURE: COMPETITION, DIVERSITY, AND GEOGRAPHY

The advanced materials industry exhibits a high degree of competitiveness. While one company may dominate a new material for a short time, companies offering competing technology—whether a new product, process, or both—emerge rapidly thereafter. Also, product life cycles tend to be short as new companies and products regularly enter the market. New materials must also compete against older products that serve the same markets. The inevitable push of new technology is not always the case since markets, generally conservative in nature, do not readily accept novel materials over more familiar and tested products. Consumers of the older materials understand the advantages and limits of the established products and have adjusted their processes accordingly.

Within the industry, there are "captive" producers and "open" producers. The former produces advanced materials—eg, nanotubes—for internal use. They may also purchase advanced materials on the open market. Captive producers include universities and government and industrial laboratories. They may also include the larger, multinational corporations, such as Honeywell. On the other hand, "open" companies produce and sell their materials on the open market. Suppliers of advanced materials to the market may be small start up firms, large corporations, or academic and government laboratories.

The advanced materials industry encompasses a wide variety of companies in terms of size, diversity, and the degree of integration. These companies range from the large, integrated chemical, petrochemical, and process engineering firms.

While certain of the large corporations, continue to develop new material technologies, the established corporation typically has turned away from developing the most radical technologies. This is so for a variety of reasons including increasing development costs; decreasing returns on investment in R&D (due in part to the need to utilize an extensive R&D department and the high costs of retooling and restructuring existing, large-scale plants); the growing likelihood that new technology must compete against the innovating company's own existing products; and the difficulties and costs involved in establishing new supply lines and distribution networks (and the possible alienation of existing suppliers and customers). Moreover, the growing professionalization of corporate management blunts the desire and capability of companies to undertake new technological development. This is, in part, due to the rise of intellectual and cultural boundaries between managers that reflect the greater degree of specialization that characterizes corporate management professionals. Such specialists, embracing their own particular professional goals, problems, and even language, hinder the close cooperation and free flow of information between different departments, which is often required in successfully undertaking major technological projects.

In contrast, the small, start-up and "spin-off" firms enjoy a greater degree of flexibility in pursuing new technology. They are not burdened with an existing system of plants nor do they have to support an extensive supply, distribution, and R&D infrastructure. They do not have established products that could compete with new material technologies nor do they need to cosset an established network of suppliers, distributors, or customers. Indeed, commercializing radically new materials is absolutely critical to their existence since doing so distinguishes them from their competitors, both large corporations and other smaller newcomers, and is the key to capturing new markets. In addition, the organizational structure in these firms is more loosely organized. Often, individual executive managers undertake a wide range of functions, including engineering, procurement, and marketing. This informal organizational structure facilitates a free flow of information, know-how, and insight throughout the company and permits decisions involving new technology development to be made quickly and efficiently.

THE EMERGING ADVANCED MATERIALS

Industry specialists consider the following products as the most important advanced materials that have recently entered the market, are on the verge of commercialization, or may be commercialized over the next few years.

Bioengineered Materials

Biochemicals play an increasingly critical role in the advanced materials industry. An important biochemical technology that is just beginning commercialization is the so-called bioengineered materials. These materials bridge the biochemical and synthetic organic fields and promise to provide large volumes of synthetic materials over the next few years.

Bioengineering technology involves the biochemical transformation—in so-called "biorefineries"—of agricultural feedstock, by-products, and wastes into useful synthetic materials. Biorefineries are directly comparable to the petrochemical plant in that both petroleum and biomass refineries use a particular raw material to produce a wide range of synthetic products varying greatly in their chemical properties and physical characteristics and serving a diversity of markets. These products include synthetic plastics and packaging, clothing, fuel additives, chemicals (eg, alcohols, polymers, ethylene, phenolics, acetic acid, etc), biologics, food products, adhesives and sealants, and a variety of commodity and industrial products.

An important advantage of bioengineered materials is that they conserve on energy and provide additional markets for the products and wastes of the farming industry. Also, processing costs are low because the technology uses known methods and does not require complex and expensive bioreactors and associated facilities for upstream production. Scale-up of biorefineries appears to be rapid and relatively inexpensive. From the environmental viewpoint, these materials degrade more readily than traditional synthetics. They also generate fewer greenhouse gases and require less energy, water, and raw materials to produce compared to petroleum-based materials.

In general, the production of bioengineered materials occurs near to agricultural areas. Recent advances in bioengineering technology involve cellulosic- based feedstocks, including vegetable crops, starch-producing crops, oil seeds, wood and other lignocellulosic biomass. Current research in the field focuses on three major production technologies: fermentation (including enzyme processing), pyrolysis, and low temperature technology. Of these possible processes, fermentation, in conjunction with such operations as distillation and polymerization, appears to offer the most promising commercial method.

Advanced Ceramics

Ceramics are inorganic and nonmetallic materials. There are three major forms of ceramics: amorphous glasses, polycrystalline materials, and single crystals. Ceramics are generally made from powders and additives under high temperatures. Traditional types of ceramics includes bricks, tile, enamels, refractories, glassware, and porcelain. Advanced ceramic materials, developed more recently, possess superior physical, mechanical, and electrical properties. They are made from metal powders that undergo innovative processing methods.

Advanced ceramics increasingly enter into a wide variety of applications. In general, advanced ceramics extend equipment life, decrease fuel costs, and increase power and performance. Important ceramic products in electronics applications include both the pure and mixed oxides—alumina, zirconia, silica, ferrites—and doped barium and lead titanates. The important electronic application of these materials includes their use in substrates and packaging, capacitors, transformers, inductors, and piezoelectric devices and sensors.

Advanced ceramics are also used in structural applications due to their resistance to corrosion and high temperatures. Ceramics perform very well as a material for equipment components or as an industrial coating. Such ceramics are important materials for infrastructural applications, such as power plants, construction, and bridges as well as in industrial equipment, eg, bearings, seals, cutting tools. The automotive industry employs advanced structural ceramics in catalytic converters and for certain under-the-hood components. Important advanced structural ceramics include various forms of aluminum oxide, zirconia, silicon carbide, and silicon nitride.

Recent developments involve the creation of new types of advanced ceramics and production processes. One of the most important lines of research within the United States and internationally includes the development of ceramic metal matrix composites incorporating reinforcing materials such as carbon fibers. These materials possess superior mechanical properties, excellent thermal stability and a low friction coefficient (to serve as a superior lubricant). Research in the field continues to find a wider range of new composites and to reduce unit costs.

A second major area of advanced ceramics research involves new techniques to make ceramic powders. For example, the use of thermal plasmas may prove a superior process to generate very fine powders. The process produces high purity powders as well as eliminating intermediate production steps currently employed in traditional manufacturing methods.

A third line of research centers on the development of advanced ceramic coatings technology that allows the deposition of a thin layer of ceramic on complex surfaces at low costs for improved resistance to corrosion, mechanical wear, and thermal shocks. These processes impart superior properties without industry needing to go through the time and expense to make entirely new parts and components. Research in advanced ceramic coatings focuses on improving adhesion of the coating to surfaces, increasing the properties of the coatings, and reducing the costs of the coating process. The various coating processes currently employed or being investigated include plasma and flame spray, high velocity oxy-fuel deposition, and electron beam techniques.

Two advanced ceramic material areas appear particularly promising: nanoceramics and piezoelectric ceramics. While these materials currently account for a relatively small percentage of the total advanced ceramics market, they attract a disproportionate amount of research and development activity. As a result, they represent the cutting edge in the advanced ceramics field and promise further development, increased production, and a growing range of applications through 2012 and beyond.

Advanced Coatings

Advanced coatings provide a vital and growing area in the new materials field. These advances emerge from recent research in surface chemistry and solid state physics.

The advanced coatings field consists of an increasing number of smaller firms specializing in manufacturing particular types of coatings and coating application technology. Many of these firms are start-up operations that license technologies from the university and government (eg, NASA, DOE, DOD) and who carry out their own R&D to further commercialization. These firms create novel coating materials that provide new ways to protect surfaces from the environment—ie, heat, impacts, erosion, and chemical degradation—thus increasing the life and performance of components, equipment, and systems across a wide range of industries and technologies. The more advanced coatings impart to surfaces and objects heightened ability to sense and respond to the full range of changes in the environment.

Nanopowders and Nanocomposites

In recent years, powder metal technology has emerged as a major segment of the metals industry. It is a growing presence in a variety of commercial and industrial applications. Most recently, this field has been advancing into the still new area of nanotechnology. Nanopowders are typically composed of metals or metal mixtures and complexes with particulate sizes in the micron ranges. The potential markets for these materials depend on the fact that they can be formed into diverse shapes and forms possessing a variety of important mechanical, electrical, and chemical characteristics. Nanopowders are composed of a broad range of metals and their compounds. These include (but are not limited to) the oxides of aluminum, magnesium, iron, zinc, cerium, silver, titanium, yttrium, vanadium, manganese, and lithium; the carbides and nitrides of such metals as tungsten and silicon; and metal mixtures, such as lithium/titanium, lithium manganese, silver/zinc, copper/tungsten, indium/tin, antimony/tin, and lithium vanadium.

Nanopowder producers are either captive, ie, making powders for their own internal research and commercial use, or "open," ie, producing powders for sale to research organizations and commercial facilities.

Companies produce nanopowders through a number of processes, including furnace and laser-based technologies. Plasma chemical synthesis (PCS), eg, employs microwave methods to produce nanoparticulates through the creation and consequent rapid quenching of hot ionized gas plasma. Another process, a modification of the Xerox "emulsion aggregation" technology, utilizes emulsion polymerization technique. One of the more active areas in the nanopowder field is nano-based coatings. These materials, when added to a resin base, produces superior paints and varnishes.

One of the most important applications of nanopowders is in the manufacture of nanopowder–plastic composites. Typically, nanopowder composites contain under 6% by weight of nanometer-sized mineral particles embedded in resins. More recently, other plastics have come to the fore, such as polypropylene and polyester resins. A new generation of plastics with superior properties is currently being developed for future application.

Nanocomposites offer a range of beneficial properties including great strength and durability, shock resistance, electrical conductivity, thermal protection, gas impermeability, and flame retardancy. New and more sophisticated processes can manufacture composite powders with a uniform, nanolayer thick metallic or ceramic coating for high density parts with superior thermal, mechanical and electrical properties.

In additional there are many potential applications for nanopowders and nanopowder composites currently under investigation.

Nanocarbon Materials

Nanocarbon materials contain molecular-sized clusters composed of a number of carbon atoms arranged in various configurations. One such group falls into the category of fullerenes. In this case, a series of carbon atoms arranged spherically enclose one or more metal atoms. In the second type of material, the carbon atoms join together to form a tubular-like structure. These structures may or may not enclose metal atoms. These materials, known as nanotubes, have important applications in the advanced composites area.

Metal Fullerenes. Fullerenes in general refer to a group of materials composed of carbon structures of 60–90 carbon atoms, each enveloping a single metal atom. Two viable thermal processes produce advanced fullerenes. One technology involves application of the electric arc, using graphite to provide the carbon atoms. The second approach, referred to as the "soot-flame" process, is in fact currently utilized to manufacture certain traditional fullerenes. The advantage of the soot-flame process is that it is relatively cost efficient and permits carefully controlled production, and therefore more precise product design.

Nanotubes. Nanotubes are carbon-based structures with cylindrical shapes and diameters between 0.8 and 300 nm. Nanotubes resemble small, rolled tubes of graphite. As such they possess high tensile strength and can act as an excellent conductor or semiconductor material. There are two main varieties of nanotubes: single-walled and multiwalled. Multiwalled structures are the less pure form of nanotubes and offer only a limited num-

ber of applications. The more advanced, purer form of nanotube, ie, defined by a single-walled structure, is the more promising material commercially, especially for incorporation into polymer materials in the synthesis of composites with superior structural, thermal, and electrical characteristics.

Nanofibers

Nanofiber technology refers to the synthesis by various means of fiber materials with diameters less than 100 nm. Nanofiber technology remains a new but growing field with promising applications. In general, advantages of nanofibers depend on their high flexibility and therefore their ability to conform to a large number of three-dimensional configurations. They also have a very high surface area allowing a myriad of interactions with chemical and physical environments. Recent research suggests possible industrial applications as ceramic ultrafilters, gas separator membranes, electronic substrates, medical and dental composites, fiber reinforced plastics, electrical and thermal insulation, structural aerospace materials, and catalyst substrates for petrochemical synthesis. Nanofibers also may be applied in advanced optical systems, according to the shape, number, and composition of the fibers.

One of the most promising areas of nanofiber technology involves applications in the biomedical area. Nanofibers potentially can be integrated into advanced drug delivery systems. Even more importantly, nanofibers can produce three-dimensional collagen-based matrices or "scaffolds." When these scaffolds are "seeded" with specific types of human cells, blood vessels of small diameter are formed. These vessels can then be transplanted into a patient.

Thin Films

Advanced thin film materials represent one of the newest and most promising of the emerging material technologies. In general, thin film materials are composed of different advanced materials—polymers, metals, and polycrystals—layered a few tenths of an Angstrom deep onto a foundation or substrate, such as glass, acrylic, steel, ceramics, silica, and plastics. Whereas coatings are applied to surfaces, thin films often operate as stand alone components in a variety of products and systems including consumer electronics and electronic components, telecommunications devices, optical systems (eg, reflective, antireflective, polarizing, and beam splitter coatings), biomedical technology, sensor systems, electromagnetic and microwave systems, and energy sources and products (eg, batteries, photovoltaic cells).

The future success of thin-film technology depends to a large extent on the viability of the production process. One possible approach involves a thermal laser-based deposition technique. Also known as pulsed laser deposition (PLD), this method involves hitting a target composed of the desired film material with a laser beam of short pulse. This process avoids the formation of unwanted particulates that can cause defects in, and hinder performance of, the final film. The PLD process is often associated with metallic thin film materials. A similar process, called chemical vapor deposition, takes place in a vacuum and involves the diffusion and adsorption of the film material in the form of vapors onto the surface of the substrate. Variations of this process have yet to be fully developed. These processes include plasma enhanced chemical vapor deposition, ultraviolet injection liquid source chemical vapor deposition, and metallorganic chemical vapor deposition.

The second general type of process in the manufacture of thin films, known as electrostatic (or ionic) self-assembly (ESA), offers a superior technique in the production of organic, as well as metallic, thin-film materials. The ESA process conserves on costs because it does not require an ultraclean environment and is not energy intensive. The process also permits scale up for mass production manufacture through the use of automatic dipping machines and robot-controlled fabrication stations.

Thin-film technology brings together different advanced materials as the primary film substance. In general, these film materials can be organic polymers (eg, organic polymer electronic—OPE—synthetic resins), metals and alloys, or crystals of various sorts (eg, titanium dioxide, magnesium fluoride).

Metal-based thin films also appear close to achieving a reduction in the size of circuits and circuit components for electronic applications and may, in fact, compete against the polymer thin films in these markets. Metal-based thin films offer greater purity and durable interconnections between microcircuit components. These advantages allow future computers to be made much smaller and to operate faster than current technology. Metal-based thin films promise to advance a new generation of microelectronic and electromagnetic components including capacitors, resisters, thermistors, transducers, inductors, and related elements.

Both polymer- and metal-based thin films also provide a route to "printed" low-cost antennas for attachment onto different surfaces. These antennas possess large surface areas for capacitive coupling and may compete against certain types of metallic conductive coatings.

Despite a wide range of possible applications, a number of technical and economic risks exist. These problems include uncertain interface control; physical degradation of the polymer material in the presence of high temperatures, high electric fields, and exposure to solvents used in the circuit printing process (which limits the types of circuits that can be designed); uncontrolled charge leakage between thin film-based devices and circuit elements resulting in lower operating life and increased signal interference; reduced electrical performance and mechanical degradation due to impurities in the polymer or metal.

"Horizons in Advanced Materials," *The Update (Online)*, www. acq.osd.mil/bmdo/bmdolink/html/update/sum01/updhor1.htm (Summer 2001).

S. Moskowitz, "History of Refining," in J. Zumerchik, ed., *Macmillan Encyclopedia of Energy*, Vol. 3, Macmillan Reference U.S.A., New York, 2001.

U.S. Environmental Protection Agency, "Environmentally Friendly Anti-Corrosion Coatings," *Final Report* (March 2001).

J. Utterback, "The Dynamics of Product and Process Innovation in Industry," in C. Hill and J. Utterback, eds., *Technological Innovation for a Dynamic Economy*, Pergamon Press, New York, 1979, pp. 40–65.

SANFORD L. MOSKOWITZ
Villanova University

AERATION, BIOTECHNOLOGY

The supply of oxygen to a growing biological species, aeration, in aerobic bioreactors is one of the most critical requirements in biotechnology. It was one of the biggest hurdles that had to be overcome in designing bioreactors (fermenters) capable of turning penicillin from a scientific curiosity to the first major antibiotic. Aeration is usually accomplished by transferring oxygen from the air into the fluid surrounding the biological species, from where it is in turn transferred to the biological species itself. The rate at which oxygen is demanded by the biological species in a bioreactor depends very significantly on the species, on its concentration, and on the concentration of the other nutrients in the surrounding fluid (see CELL CULTURE TECHNOLOGY). There is no unique set of units used to define this rate requirement, but some typical figures are given in Table 1. The very wide range is noteworthy; during the course of a batch bioreaction, oxygen demand often passes through a marked maximum when the species is most biologically active.

The main reason for the importance of aeration lies in the limited solubility of oxygen in water, a value that decreases in the presence of electrolytes and other solutes and as temperature increases.

In addition to each bioreaction demanding oxygen at a different rate, there is a unique relationship for each between the rate of reaction and the level of dissolved oxygen. A typical generalized relationship is shown in Figure 1 for a particular species, eg, *Penicillium chrysogenum* or yeast. The shape of the curve is such that a critical oxygen concentration, C_{crit}, can be defined above which the rate of the bioreaction is independent of oxygen concentration, ie, zero order with respect to oxygen.

Table 1. Oxygen Demands of Biological Species

Biological species	kg O_2/($m^3 \cdot$h)
bacteria/yeasts	1–7
plant cells	0.03–0.3
seed priming[a]	$1–8 \times 10^{-2}$
mammalian cells[b]	$2–10 \times 10^{-3}$

[a]Based on a seed density of 100 kg/m^3.
[b]Based on a cell density of 10^{12} cells/m^3.

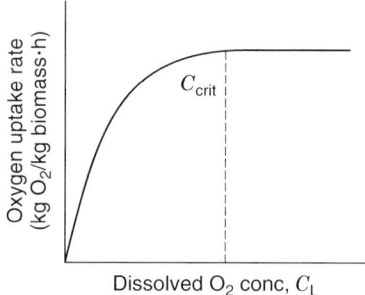

Figure 1. The relationship between rate of oxygen uptake and dissolved oxygen, concentration where C_{crit} is the critical oxygen concentration.

PRINCIPLES OF OXYGEN TRANSFER

The Basic Mass-Transfer Steps

The steps through which oxygen must pass in moving from air (or oxygen-enriched air) to the reaction site in a biological species consist of transport through the gas film inside the bubble, across the bubble–liquid interface, through the liquid film around the bubble, across the well-mixed bulk liquid (broth), through the liquid film around the biological species, and finally transport within the species (eg, cell, seed, microbial floc) to the bioreaction site. Each step offers a resistance to oxygen transfer. In the last step, the resistance to transport is negligible for freely suspended bacteria or cells that are extremely small, but for immobilized cells and biological flocs and pellets the rate of oxygen diffusion through their structure may be rate limiting. In the more complex situation, the rate of oxygen diffusion to the active sites depends on mass transfer through the external boundary layer followed by diffusion through the solid as governed by Fick's law.

The Basic Mass Transfer Relationship

The basic principles that underlie oxygenation (aeration) are exactly the same as those that determine the rate of transfer of any sparingly soluble gas (oxygen) from the gas stream (air) to the unsaturated liquid (broth). The rate at which this transfer takes place is dependent on four principal parameters. The first is the area of contact between the gas and the liquid. The other three mass-transfer rate parameters are the driving force available (ie, the difference in concentration of oxygen in the two phases); the two-phase fluid dynamics (including the effect of viscosity); and the chemical composition of the liquid.

AERATION IN BIOREACTORS

A huge variety of bioreactors has been developed and a thorough review is available. It is not feasible to consider them all and large numbers are only curiosities. A useful subdivision has been made into three generic types involving the way in which air is dispersed to give the desired specific surface area. These are bioreactors driven by rotating agitators (stirred tanks), bioreactors driven by gas compression (bubble columns/loop fermenters), and bioreactors driven by circulating liquid (jet loop reactors). The first two are the most important.

Stirred Tank Bioreactors

Traditionally, stirred tanks have been the most common types of bioreactors for aerobic processes and they remain so even in the face of newer designs. One of the main reasons is their extreme flexibility. Operational designs using controlled air flow rates up to ~1.5 vvm (volume of air/min per unit volume of fermentation fluid) and variable speed motors capable of transmitting powers up to about 5 W/kg with control down to close to zero are suitable for almost any bioreaction. These tanks are also relatively insensitive to fill, ie, to the proportion of liquid added to the bioreactor, and are therefore quite satisfactory for fed batch operations.

Bubble Column and Loop Bioreactors

Air driven bioreactors are said to offer these advantages: No opening for a shaft is required and therefore they are less likely to become contaminated; and they are very simple to operate on the very small scale and more economic on the very large scale where huge agitators and motors would otherwise be required.

Examples of air driven bioreactors are bubble column and loop bioreactors. The bubble column is clearly the simplest of these bioreactors to construct. However, because of its rather ill-defined liquid circulation, air-lift reactors having either internal (draught tube) or external (loop) circulation of broth have been introduced. The major disadvantages of all three types are the poor capability of handling very viscous fermentations, especially those having a yield stress; the inflexibility, especially of the airlift types, which only work well using a fill closely matched to the size of the bioreactor and its internals (this match affects both circulation rates, mixing and mass transfer, and bubble disengagement); and lack of independent control of dO_2 and mixing, since both are closely linked to the aeration rate. In contrast to stirred tank bioreactors, the bubble size may in certain cases be very dependent on the way the air is introduced, ie, on the type of sparger employed.

APPLICATIONS TO DIFFERENT BIOLOGICAL SPECIES

Mycelial Fermentations

Mycelial fermentations typically become viscous and shear thinning and difficult to mix. Agitation levels link with mycelial structure and if pelleted growth can be encouraged, for example, using *P. chrysogenum*, viscosity does not significantly increase. Difficulties may arise, however, if the pellets are too large, as a result of diffusion resistance within them, leading to oxygen starvation of the potentially active sites at the center of the pellet. Because of the high viscosity, adequate bulk blending, oxygen transfer, and cooling often require high specific power inputs.

Xanthan Gum Fermentations

Xanthan gum (and other polysaccharide) fermentations become very viscous toward the end whether batch or fed-batch. Therefore, satisfying the oxygen demand at this stage is very demanding of agitation power. Stirred bioreactors are therefore preferable using large impeller/ tank diameter ratios and more closely placed impellers.

High Oxygen Demanding Fermentations

High oxygen demanding fermentations often require a higher level of oxygen over relatively short periods of time, which means the flexibility of the stirred bioreactor is a distinct advantage.

Animal Cell Culture

Airlift and bubble column bioreactors have been considered necessary for handling fragile animal cells. However, more recent work has shown that bursting bubbles are much more damaging to these cells than agitation unless protective agents such as Pluronic F68 are included in the media.

Plant Cell Culture

Airlift bioreactors have been favored for plant cell systems since these cultures were first studied. However, they can give rise to problems resulting from flotation of the cells to form a "meringue" on the top and they are often rather viscous.

Seed Priming Bioreactors

Seed priming is a relatively new technique enabling seeds suspended in an osmotica to imbibe moisture and thus be brought to the point of germination. Both bubble column and stirred bioreactors have been used successfully, although the former requires high air rates to keep seeds in suspension.

Single-Cell Protein

Systems involving single-cell proteins are often very large throughput, continuous processing operations, and are ideal for airlift bioreactors of which the pressure cycle fermenter is a special case.

Biological Aerobic Wastewater Treatment

Biological aerobic wastewater treatment is a rather specialized biotechnical application, for which either aerated agitators or air spargers (diffusers) are used.

J. E. Bailey and D. F. Ollis, *Biochemical Engineering Fundamentals*, 2nd ed., McGraw-Hill Book Co., New York, 1986.

K. Kargi and M. Moo-Young, in M. Moo-Young, ed., *Comprehensive Biotechnology*, Vol. 2, Pergamon Press, Oxford, 1985, Chapt. 2.

A. W. Nienow, in C. F. Forster and D. A. J. Wase, eds., *Environmental Biotechnology*, Ellis Horwood, Chichester, 1987.

A. G. Pedersen, *Bioreactor and Bioprocess Fluid Dynamics*, (Ed. A.W. Nienow), BHR Group/MEP, London, U.K., 1997, p. 263.

ALVIN W. NIENOW
University of Birmingham

AEROGELS

Aerogels are solid materials that are so porous that they contain mostly air. Almost all applications of aerogels are based on the unique properties associated with a highly porous network. Envision an aerogel as a sponge consisting of many interconnecting particles which are so small and so loosely connected that the void space in the sponge, the pores, can make up for over 90% of its volume. The ability to prepare materials of such low density, and perhaps more importantly, to vary the density in a controlled manner, is indeed what make aerogels attractive in many applications.

SOL–GEL CHEMISTRY

Inorganic Materials

Sol–gel chemistry involves first the formation of a sol, which is a suspension of solid particles in a liquid, then of a gel, which is a diphasic material with a solid encapsulating a solvent. A detailed description of the fundamental chemistry is available in the literature. The chemistry involving the most commonly used precursors, the alkoxides $(M(OR)_m)$, can be described in terms of two classes of reactions:

Hydrolysis $-M-OR+H_2O \rightarrow -M-OH+ROH$
Condensation $-M-OH+XO-M- \rightarrow -M-O-M-$
$$+XOH$$
where X can either be H or R, an alkyl group

The important feature is that a three-dimensional gel network comes from the condensation of partially hydrolyzed species. Thus, the microstructure of a gel is governed by the rate of particle (cluster) growth and their extent of crosslinking or, more specifically, by the *relative* rates of hydrolysis and condensation.

Acid- and base-catalyzed gels yield micro- (pore width less than 2 nm) and meso-porous (2–50 nm) materials, respectively, upon heating. An acid-catalyzed gel which is weakly branched and contains surface functionalities that promote further condensation collapses to give micropores. This example highlights a crucial point: *the initial microstructure and surface functionality of a gel dictates the properties of the heat-treated product.*

Besides pH, other preparative variables that can affect the microstructure of a gel, and consequently, the properties of the dried and heat-treated product include water content, solvent, precursor type and concentration, and temperature.

In the preparation of a two-component systems, the minor component can either be a network modifier or a network former. In the latter case, the distribution of the two components, or mixing, at a molecular level is governed by the *relative* precursor reactivity. Qualitatively good mixing is achieved when two precursors have similar reactivities. When two precursors have dissimilar reactivities, the sol–gel technique offers several strategies to prepare well-mixed two-component gels. Two such strategies are prehydrolysis, which involves prereacting a less reactive precursor and chemical modification, which involves slowing down a more reactive precursor. The ability to control microstructure *and* component mixing is what sets sol–gel apart from other methods in preparing multicomponent solids.

Organic Materials

The sol–gel chemistry of organic materials is similar to that of inorganic materials. The first organic aerogel was prepared by the aqueous poly- condensation of resorcinol with formaldehyde using sodium carbonate as a base catalyst.

Resorcinol–formaldehyde gels are dark red in color and do not transmit light. The preparation of melamine–formaldehyde gels, which are colorless and transparent, is also aqueous-based. Since water is deleterious to a gel's structure at high temperatures and immiscible with carbon dioxide (a commonly used supercritical drying agent), these gels cannot be supercritically dried without a tedious solvent-exchange step. In order to circumvent this problem, an alternative synthetic route of organic gels that is based upon a phenolic–furfural reaction using an acid catalyst has been developed. The solvent-exchange step is eliminated by using alcohol as a solvent. The phenolic–furfural gels are dark brown in color.

Carbon aerogels can be prepared from the organic gels mentioned above by supercritical drying with carbon dioxide and a subsequent heat-treating step in an inert atmosphere.

Despite these changes, the carbon aerogels are similar in morphology to their organic precursors, underscoring again the importance of structural control in the gelation step. Furthermore, changing the sol–gel conditions can lead to aerogels that have a wide range of physical properties.

Inorganic–Organic Hybrids

One of the fastest growing areas in sol–gel processing is the preparation of materials containing both inorganic and organic components, because many applications demand special properties that only a combination of inorganic and organic materials can provide. In this regard, sol–gel chemistry offers a real advantage because its mild preparation conditions do not degrade organic polymers, as would the high temperatures that are associated with conventional ceramic processing techniques. The voluminous literature on the sol–gel preparation of inorganic–organic hybrids can be found in several recent reviews and the references therein.

PREPARATION AND MANUFACTURING

Supercritical Drying

The development of aerogel technology from the original work of Kistler to about late 1980s has been reviewed. Over this period, supercritical drying was the dominant method in preparing aerogels. Several advances, summarized in Table 1, have made possible the relatively safe supercritical drying of aerogels in a matter of hours. In recent years, the challenge has been to produce aerogel-like materials without using supercritical drying at all in an attempt to deliver economically competitive products.

Supercritical drying should be considered as part of the aging process, during which events such as condensation, dissolution, and reprecipitation can occur. The extent to which a gel undergoes aging during supercritical drying depends on the structure of the initial gel network. A higher drying temperature changes the particle structure of base-catalyzed silica aerogels but not that of acid-catalyzed ones. Gels that have uniform-sized pores can withstand the capillary forces during drying better because of a more uniform stress distribution. Such gels can be prepared by a careful manipulation of sol–gel parameters such as pH and solvent or by the use of so-called drying control chemical additives (DCCA).

Carbon dioxide is the drying agent of choice if the goal is to stabilize kinetically constrained structure, and materials prepared by this low-temperature route are referred to by some people as *carbogels*. In general, carbogels are also different from aerogels in surface functionality, in particular hydrophilicity.

However, even with carbon dioxide as a drying agent, the supercritical drying conditions can affect the properties of a product. Other important drying variables include the path to the critical point, composition of the drying medium, and depressurization.

For some applications it is desirable to prepare aerogels as thin films that are either self-supporting or supported on another substrate. All common coating

Table 1. Important Developments in the Preparation of Aerogels

Decade	Developments
1930	Using inorganic salts as precursors, alcohol as the supercritical drying agent, and a batch process; a solvent-exchange step was necessary to remove water from the gel.
1960	Using alkoxides as precursors, alcohol as the supercritical drying agent, and a batch process; the solvent exchange step was eliminated.
1980	Using alkoxides as precursor, carbon dioxide as the drying agent, and a semicontinuous process; the drying procedure became safer and faster. Introduction of organic aerogels.
1990	Producing aerogel-like materials without supercritical drying at all; preparation of inorganic–organic hybrid materials.

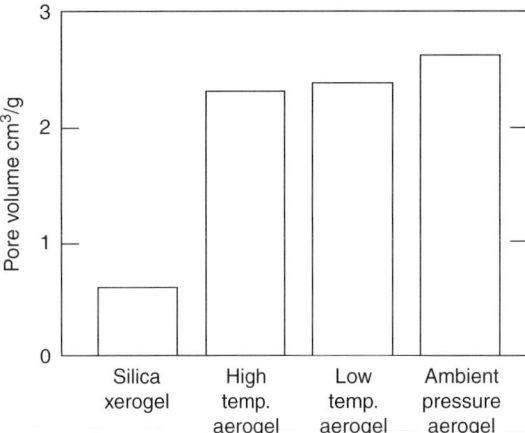

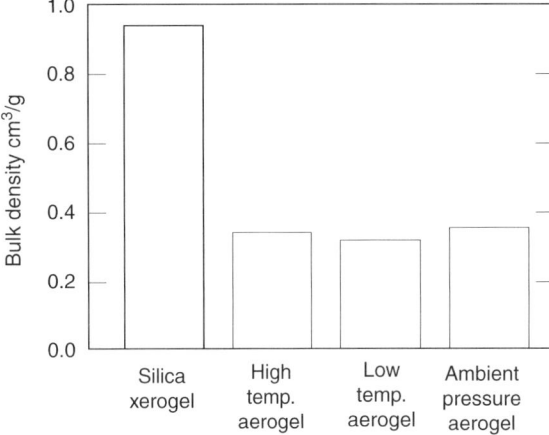

methods such as dip coating, spin coating, and spray coating can be used to prepare gel films.

In all the processes discussed above, the gelation and supercritical drying steps are done sequentially. Recently a process that involves the direct injection of the precursor into a strong mold body followed by rapid heating for gelation and supercritical drying to take place was reported. By eliminating the need of forming a gel first, this entire process can be done in less than three hours per cycle. Besides saving time, gel containment minimizes some stresses and makes it possible to produce near net-shape aerogels and precision surfaces. The optical and thermal properties of silica aerogels thus prepared are comparable to those prepared with conventional methods.

Ambient Preparations

Economic and safety considerations have provided a strong motivation for the development of techniques that can produce aerogel-like materials at ambient conditions, ie, without supercritical drying. The strategy is to minimize the deleterious effect of capillary pressure which is given by:

$$P = 2\sigma\cos(\theta)/r$$

where P is capillary pressure, σ is surface tension, θ is the contact angle between liquid and solid, and r is pore radius.

The equation above suggests that one approach would be to use a pore liquid that has a low surface tension. In fact, with a pore liquid that has a sufficiently small surface tension, ambient pressure acid catalyzed aerogels with comparable pore volume and with bulk density to those prepared with supercritical drying (see Fig. 1) have been produced.

For base-catalyzed silica gels, it has been shown that modifying the surface functionality is an effective way to minimize drying shrinkage. In particular, surface hydroxyl groups, the condensation of which leads to pore collapse, can be "capped off" via reactions with organic groups such as tetraethoxysilane and trimethylchlorosilane. This surface modification approach (also referred to as surface derivatization), initially developed for bulk specimens, has recently been applied to the preparation of thin films.

In changing surface hydroxyls into organosilicon groups, surface modification has an additional advantage of producing hydrophobic gels. This feature, namely the immiscibility of surface-modified gel with water, has led to the development of a rapid extractive drying process shown in Figure 2. This ambient pressure process offers improved heat transfer rates and, in turn, greater energy efficiency without compromising desirable aerogel properties.

Another approach to produce aerogels without supercritical drying is freeze drying, in which the liquid–vapor interface is eliminated by freezing a wet gel into a solid and then subliming the solvent to form what is known as a *cryogel*. The limited data available on freeze drying suggest that it might not be as

Figure. 1. Comparison of physical properties of silica xerogels and aerogels. Note the similar properties of the aerogels prepared with and without supercritical drying. Reproduced from C. J. Brinker and co-workers, *Mat. Res. Soc. Symp. Proc.* **271**, 567 (1992). Courtesy of the Materials Research Society.

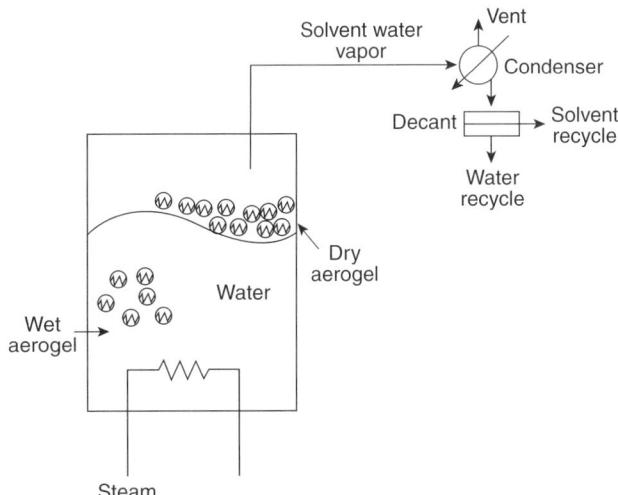

Figure. 2. Schematic diagram of an extractive drying process that produces aerogels at ambient pressure. Reproduced from D. M. Smith and co-workers, *Mat. Res. Soc. Symp. Proc.* **431**, 291 (1996). Courtesy of the Materials Research Society.

Table 2. Typical Values of Physical Properties of Silica Aerogels

Property	Values
density, kg/m^3	3–500
surface area, m^2/g	800–1000
pore sizes, nm	1–100
pore volume, cm^3/g	3–9
porosity, %	75–99.9
thermal conductivity, W/(m·K)	0.01–0.02
longitudinal sound velocity, m/s	100–300
acoustic impedance, kg/(m^2·s)	10^3–10^6
dielectric constant	1–2
Young's modulus, N/m^2	10^6–10^7

attractive as the above ambient approaches in producing aerogels on a commercial scale.

PROPERTIES

Table 2 summarizes the key physical properties of silica aerogels. A range of values is given for each property because the exact value is dependent on the preparative conditions and, in particular, on density.

APPLICATIONS

Aerogels are used in thermal insulation, catalystic, detection of high energy particles, piezoceramic, ultrasound transducers, integrated circuits, and as dehydrating agents.

SUMMARY

Aerogels have the potential of being marketable both as a commodity chemical (eg, in thermal insulation) and as a specialty chemical (eg, in electronic applications) because of their unique and tailorable properties. The next few years will be critical in assessing whether aerogels can penetrate and grow in either end of the market, as the field is changing rapidly with the development of cost-competitive technologies and novel applications.

C. J. Brinker and G. W. Scherer, *Sol-Gel Science: The Physics and Chemistry of Sol-Gel Processing*, Academic Press, New York, 1990.

J. Fricke, *Sci. Amer.* **256**(5), 92 (1988).

J. Livage, M. Henry, and C. Sanchez, *Prog. Solid State Chem.* **18**, 259 (1988).

M. Schneider and A. Baiker, *Catal. Rev.-Sci. Eng.* **37**(4), 515 (1995).

EDMOND I. KO
Carnegie Mellon University

AEROSOLS

Classically, aerosols are particles or droplets that range from ~0.15 to 5 µm in size and are suspended or dispersed in a gaseous medium such as air. However, the term aerosol, as used in this discussion, identifies a large number of products which are pressure-dispensed as a liquid or semisolid stream, a mist, a fairly dry to wet spray, a powder, or even a foam. This definition of aerosol focuses on the container and the method of dispensing, rather than on the size of the particles.

The aerosol container has enjoyed commercial success in a wide variety of product categories. Insecticide aerosols were introduced in the late 1940s. Additional commodities, including shave foams, hair sprays, antiperspirants, deodorants, paints, spray starch, colognes, perfumes, whipped cream, and automotive products, followed in the 1950s. Medicinal metered-dose aerosol products have also been developed for use in the treatment of asthma, migraine headaches, angina and diabetes. Food aerosols included whipped toppings and creams, cheese spreads, hors d'oeuvres, flavored syrups, and a host of similar products. Pharmaceutical topical aerosols include antibiotics, steroids, local anesthesia, etc.

Aerosol technology may be defined as involving the development, preparation, manufacture, and testing of products that depend on the power of a liquefied or compressed gas to expel the contents from a container. This definition can be extended to include the physical, chemical, and toxicological properties of both the finished aerosol system and the propellants.

Personal products were the fastest growing segment of the aerosol industry and still represent the largest of the categories. Other areas of growth in this category occurred in underarm deodorants, antiperspirants, pharmaceuticals, and industrial products, which should keep this category as the largest segment of the aerosol industry.

ADVANTAGES OF AEROSOL PACKAGING

Aerosol products are hermetically sealed, ensuring that the contents cannot leak, spill, or be contaminated. The aerosol packages can be considered to be tamper-proof. Aerosols deliver the product in an efficient manner generating little waste, often to sites of difficult access. By control of particle size, spray pattern, and volume delivered per second, the product can be applied directly without contact by the user.

The use of metered-dose valves in aerosol medical applications permits an exact dosage of an active drug to be delivered to the respiratory system where it can act locally or be systemically absorbed. Recent developments include the administration of insulin and other hormones by oral inhalation, thereby eliminating an injection.

FORMULATION OF AEROSOLS

All aerosols consist of product concentrate, propellant, container, and valve (including an actuator and dip tube). A typical aerosol system is shown in Figure 1.

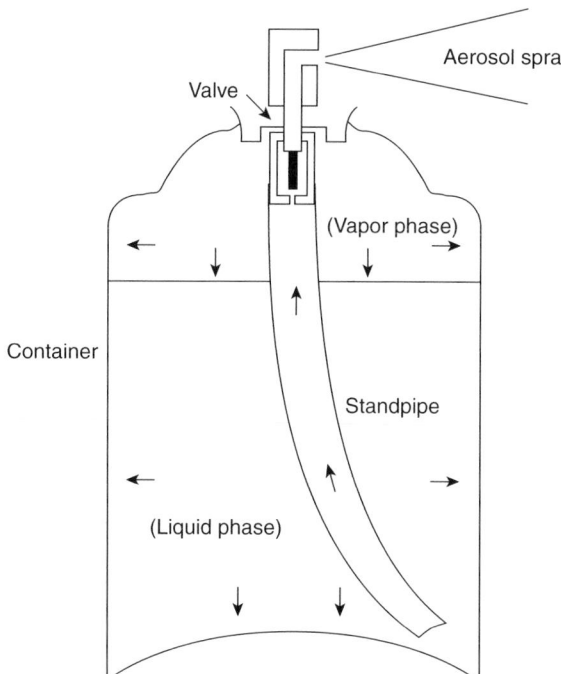

Figure 1. Solution-type aerosol system in which internal pressure is typically (35–40 psi) at 21°C. To convert kPa to psi, multiply by 0.145.

Product Concentrate

The product concentrate for an aerosol contains the active ingredient and any solvent or filler necessary. Various propellent and valve systems, which must consider the solvency and viscosity of the concentrate–propellent blend, may be used to deliver the product from the aerosol container. Systems can be formulated as solutions, emulsions, dispersions, dry powders, and pastes.

Propellants

The propellant, said to be the heart of an aerosol system, maintains a suitable pressure within the container and expels the product once the valve is opened. Propellants may be either a liquefied halocarbon, hydrocarbon, or halocarbon–hydrocarbon blend, or a compressed gas such as carbon dioxide (qv), nitrogen (qv), or nitrous oxide. Chlorofluorocarbons (CFC's) were the mainstay of propellants until 1986 when their use was curtailed and eventually eliminated due to their environmental impact. They were mainly replaced by hydrocarbons (butane, propane, and isobutane) and hydrofluorocarbons (HFC's). Most aerosol products currently utilize a hydrocarbon propellant. Pharmaceutical topical aerosols utilize a hydrocarbon or HFA propellant while metered dose inhales are made with only HFA propellants.

COMPONENTS

Containers

Aerosol containers, made to withstand a certain amount of pressure, vary in both size and materials of construction.

They are manufactured from tin-plated steel, aluminum, and glass. The most popular aerosol container is the three-piece tin-plated steel container. Glass containers, which are usually plastic coated, generally have thicker walls than conventional glass bottles.

Valves

The dispensing valve and actuator serve to close the opening through which the product and frequently the propellant entered the container, to retain the pressure within the container and to dispense the product in the precise form and dosage intended by the manufacturer and expected by the consumer. An aerosol valve, shown in Figure 2, consists of seven components. Many variations exist both for special purposes and to avoid existing patents.

Barrier–Type Systems

These systems separate the propellant from the product itself. The pressure on the outside of the barrier serves to push the contents from the container. The following types are available.

Piston Type. Since it is difficult to empty the contents of a semisolid from an aerosol container completely, a piston-type aerosol system has been developed. This systems utilizes a polyethylene piston fitted into an aluminum container. This sytem has been used successfully to package cheese spreads, cake decorating icings, and ointments. Since the products that use this system are semisolid and viscous, they are dispensed as a lazy stream

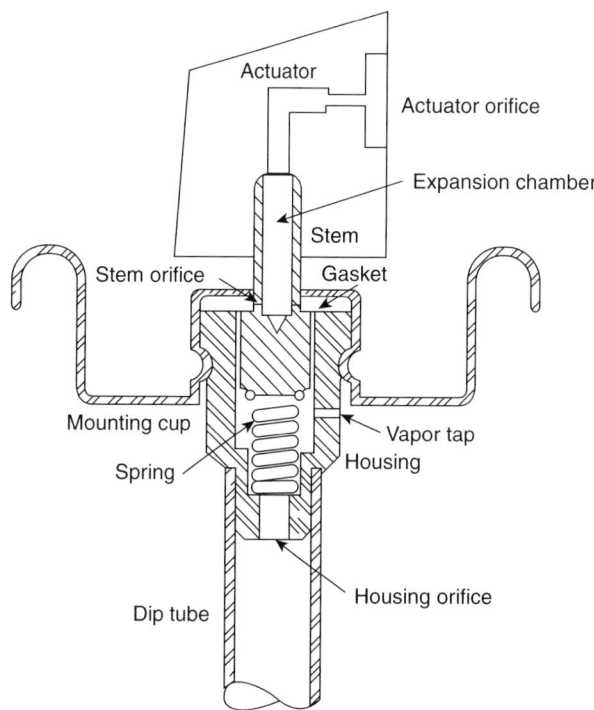

Figure 2. Aerosol valve components.

rather than as a foam or spray. This system is limited to viscous materials since limpid liquids, such as water or alcohol, will pass between the wall of the container and the piston.

Piston Bag Type. This system consists of a collapsible plastic bag fitted into a standard, three-piece, tin plate container. Since the product is placed into a plastic bag, there is no contact between the product and the container wall except for any product that may escape by permeation through the plastic bag.

Limpid liquids, such as water, can be dispensed either as a stream or fine mist depending on the type of valve used, while semisolid substances are dispensed as a stream. In order to prevent the gas from pinching the bag and preventing the dispensing of product, the inner plastic bag is accordion pleated. This system can be used for a variety of different pharmaceutical and nonpharmaceutical systems, including topical pharmaceutical products as a cream, ointment or gel.

A modification of this system dispenses the product as a gel that will then foam. This system, as well as the piston system, is used in postfoaming shave gels.

Other variations of these systems include using a laminated pouch that has been sealed onto a 1 in. valve. Another system includes filling the product into a latex bag that then expands. These systems have been used to dispense a variety of personal care products including some pharmaceutical gel products.

FILLING OF AEROSOLS

All aerosols are produced by either a cold or pressure-filling process. The cold fill process has been used for some aerosols that contain a metered-dose valve, although pressure filling is now the preferred method. For the most part, pressure filling is carried out either by an under-the-cup filler or through the valve. If an under-the-cup filler is used, a vacuum is drawn, the propellant is added (under the valve cup), and then the valve is sealed in place. Where filling is done through the valve stem, the product is first filled into the container, a valve is crimped into place, and, at the same time, a vacuum is drawn in the can.

The plastic bag type system consists of a collapsible plastic bag fitted into a standard three-piece, tin-plated container .There is no limitation on the viscosity of the product but compatibility with the plastic bag must be considered. This system has been used for caulking compounds, postfoaming gels, and depilatories.

Aerosol Guide, 8th ed., Aerosol Division, Chemical Specialties Manufacturers Association, Washington, D.C., 1995.

M. A. Johnson, *The Aerosol Handbook*, Wayne Dorland Company, Mendham, N.J., 1982. *Handbook of Pharmaceutical Excipients, Propellants*, American Pharmaceutical Association, Washington, D.C.; Pharmaceutical Press, London, United Kingdom, 2000, pp. 132–137, 184–187, 234–237, 355–352, 560.

P. A. Sanders, *Handbook of Aerosol Technology*, 2nd ed., Van Nostrand Reinhold Co., Inc., New York, 1979.

J. J. Sciarra, C. J. Sciarra, "Aerosols", in A. R. Gennaro, *Remington The Science and Practice of Pharmacy*, 20th ed., Lippincott Williams and Williams, 2000, pp. 963–979.

JOHN J. SCIARRA
CHRISTOPHER J. SCIARRA
Sciarra Laboratories, Inc.

AIR POLLUTION

Air pollution is the presence of any substance in the atmosphere at a concentration high enough to produce an undesirable effect on humans, animals, vegetation, or materials, or to significantly alter the natural balance of any ecosystem. Air pollutants can be solids, liquids, or gases, and can be produced by anthropogenic activities or natural sources. In this article, only nonbiological material is considered and the discussion of airborne radioactive contaminants is limited to radon, which is discussed in the context of indoor air pollution.

Over the past several decades concern about air pollutants has evolved; the current focus is on the effects of long term, chronic exposures to nonlethal concentrations of air pollutants, the effects of air pollution on global and regional climate, and the effects of air pollutants on global and regional atmospheric cycles (eg, stratospheric ozone depletion and acid deposition).

Health effects associated with chronic exposure to air pollution is a worldwide problem. Among the air pollutants of greatest concern are ozone, suspended particulate matter, nitrogen dioxide, sulfur dioxide, carbon monoxide, lead, and other toxins (detailed descriptions of these air pollutants are provided in subsequent sections). Of these pollutants, ozone is one of the most prevalent air pollutants in large cities and has been associated with increased respiratory illness and decreased lung function, particularly in children. Suspended particulate matter is the air pollutant most responsible for mortality worldwide. In addition to concerns about localized health impacts of urban air pollution, there has been a growing recognition of regional and global impacts of air pollutants on the natural balances of the earth's systems and on climate.

AIR POLLUTANTS

Air pollution is a complex mixture of many chemical species, and any description of air pollution must account for that heterogeneity.

Photochemical Smog

Photochemical smog is a complex mixture of constituents that are emitted directly to the atmosphere (primary pollutants) and constituents that are formed by chemical and physical transformations that occur in the atmosphere (secondary pollutants). Ozone (O_3), is generally the most abundant species formed in photochemical smog. Ozone is a secondary pollutant formed by the reactions of hydrocarbons and NO_x. Extensive studies have shown that O_3 is both a lung irritant and a phytotoxin. It is responsible for crop damage and is suspected of being a contributor to forest decline in Europe and in parts of the United States. There are, however, a multitude of other photochemical smog species that also have significant environmental consequences.

Since many components of photochemical smog are secondary, government regulatory agencies have attempted to reduce the incidence of smog episodes by controlling the emissions of precursor species. In the case of ozone, these precursor species are reactive hydrocarbons and nitrogen oxides.

Although photochemical smog is a complex mixture of many primary and secondary pollutants and involves a myriad of atmospheric reactions, there are characteristic pollutant concentration versus time profiles that are generally observed within and downwind of an urban area during a photochemical smog episode. In particular, the highest O_3 concentrations are generally found 10–100 km downwind of the major emission sources, unless the air is completely stagnant. This fact, in conjunction with the long lifetime of O_3 in the absence of high concentrations of NO, means that O_3 is a regional air pollution problem.

Particles are the major cause of the haze that is often associated with smog. The three most important components of particles produced in smog are organics, sulfates, and nitrates. Organic particles are formed when large VOC molecules, especially aromatics and cyclic alkenes, react and form condensable products. Sulfate particles are formed by a series of reactions initiated by the attack of OH on SO_2 in the gas phase or by liquid-phase reactions. Nitrate particles are formed by

$$HNO_{3(g)} + NH_3 \longleftrightarrow NH_4NO_{3(s)} \qquad (1)$$

or by the reactions of HNO_3 with NaCl or alkaline soil dust.

Volatile Organic Compounds (VOCs)

VOCs include any organic carbon compound that exists in the gaseous state in the ambient air. Reactive organic gases is also a term that is sometimes used to refer to a subset of VOCs that are reactive with hydroxl radical. VOC sources may be any process or activity utilizing organic solvents, coatings, or fuel. Emissions of VOCs are important: some are toxic by themselves, and most are precursors of O_3 and other species associated with photochemical smog.

Nitrogen Oxides (NO$_x$)

Most of the NO_x is emitted as NO, which is then oxidized to NO_2 in the atmosphere. All combustion processes are sources of NO_x. At the high temperatures generated during combustion, some N_2 is converted to NO in the

presence of O_2 and, in general, the higher the combustion temperature, the more NO_x produced.

Sulfur Oxides (SO$_x$)

The combustion of sulfur-containing fossil fuels, especially coal, is the major source of SO_x. Between 97 and 99 of the SO_x emitted from combustion sources is in the form of SO_2. The remainder is mostly sulfur trioxide (SO_3), which in the presence of atmospheric water vapor is immediately transformed into H_2SO_4, a liquid particulate. Both SO_2 and H_2SO_4 at sufficient concentrations produce deleterious effects on the respiratory system. In addition, SO_2 is a phytotoxin. Control strategies designed to reduce the ambient levels of SO_2 have been highly successful.

Carbon Monoxide

Carbon monoxide is emitted during any combustion process. Transportation sources account for about two-thirds of the CO emissions nationally, but, in certain areas, significant quantities of CO come from woodburning fireplaces and stoves.

Particulate Matter

Solid- and liquid-phase material in the atmosphere is variously referred to as particulate matter, particulates, particles, and aerosols. These terms are often used interchangeably. The original air quality standards in the United States were for total suspended particulates, (TSP), the weight of any particulate matter collected on the filter of a high volume air sampler. On the average, these samplers collect particles that are less than about 30–40 μm in diameter, but collection efficiencies vary according to both wind direction and speed. In 1987, the term PM_{10}, particulate matter having an aerodynamic diameter of 10 μm or less, was introduced.

Atmospheric particulate matter can be classified into three size modes: nuclei, accumulation, and large or coarse-particle modes. The bulk of the aerosol mass usually occurs in the 0.1–10-μm size range, which encompasses most of the accumulation mode and part of the large-particle mode. The nuclei mode is transient as nuclei, formed by combustion, nucleation, and chemical reactions. The nulceation mode particles coagulate and grow into the accumulation mode. Particles in the accumulation mode are relatively stable because they exceed the size range where coagulation is important, and they are too small to deposit out of the atmosphere quickly. Consequently, particles "accumulate" in this mode. The sources of large particles are mostly mechanical processes, such as dust entrainment by wind.

Lead

Lead is of concern because of its tendency to be retained by living organisms. When excessive amounts accumulate in humans, lead can inhibit the formation of hemoglobin and produce life-threatening lead poisoning. In smaller doses, lead is also suspected of causing learning disabilities in children.

Air Toxics

There are thousands of commercial chemicals used in the United States. Hundreds are emitted into the atmosphere and have some potential to adversely affect human health at certain concentrations; some are known or suspected carcinogens. Consequently, toxic air pollutants (TAPs) need to be prioritized based on risk analysis, so that those posing the greatest threats to health can be regulated.

Indoor Air Pollution

Indoor air pollution, the presence of air pollutants in indoor air, is of growing concern in offices and residential buildings. Numerous sources and types of pollutants found indoors can be classified into eight categories: tobacco smoke, radon, emissions from building materials, combustion products from inside the building, pollutants which infiltrate from outside the building, emissions from products used within the home, pollutants formed by reactions indoors, and biological pollutants. Concentrations of the pollutants depend on strength of the indoor sources, the ventilation rate of the building, and the outdoor pollutant concentration.

REGIONAL AND GLOBAL IMPACTS OF AIR POLLUTION

Photochemical smog is most severe in urban areas and has some impact at regional scales. There has been a growing recognition, however, that some air pollutants have impacts at regional to global scales. Three examples of impacts of air pollutants on natural balances of the earth's systems at regional and global scales are acid deposition, stratospheric ozone depletion, and global climate change.

Acid Deposition

Acid deposition, the deposition of acids from the atmosphere to the surface of the earth, can be dry or wet. Dry deposition involves acid gases or their precursors or acid particles coming in contact with the earth's surface and then being retained. The principal species associated with dry acid deposition are $SO_2(g)$, and acid sulfate particles (H_2SO_4 and NH_4HSO_4), and $HNO_3(g)$. Measurements of dry deposition are quite sparse. On the other hand, there are abundant data on wet acid deposition. Wet acid deposition, acid precipitation, is the process by which acids are deposited by rain or snow. The principal dissolved acids are H_2SO_4 and HNO_3. Other acids, such as HCl and organic acids, usually account for only a minor part of the acidity although organic acids can be significant contributors in remote areas.

Global Warming (The Greenhouse Effect)

The atmosphere allows solar radiation from the sun to pass through without significant absorption of energy.

Some of the solar radiation reaching the surface of the earth is absorbed, heating the land and water. Infrared radiation is emitted from the earth's surface, but certain gases in the atmosphere absorb this infrared (ir) radiation, and redirect a portion back to the surface, thus warming the planet and making life, as we know it, possible. This process is often referred to as the *greenhouse effect*. The surface temperature of the earth will rise until a radiative equilibrium is achieved between the rate of solar radiation absorption and the rate of ir radiation emission. Human activities, such as fossil fuel combustion, deforestation, agriculture and large-scale chemical production, have measurably altered the composition of gases in the atmosphere.

Table 1 is a list of the most important greenhouse gases along with their anthropogenic sources, emission rates, concentrations, residence times in the atmosphere, relative radiative forcing efficiencies, and estimated contribution to global warming. The primary greenhouse gases are water vapor, carbon dioxide, methane, nitrous oxide, chlorofluorocarbons, and tropospheric ozone.

Stratospheric Ozone Depletion

In the stratosphere, O_3 is formed naturally when O_2 is dissociated by (uv) solar radiation in the region 180–240 nm:

$$O_2 + uv \longrightarrow O + O \qquad (2)$$

and the atomic oxygen then reacts with molecular oxygen according to equation 2. Ultraviolet radiation in the 200–300 nm region can also dissociate O_3:

$$O_3 + uv \longrightarrow O_2 + O \qquad (3)$$

The strongest evidence that stratospheric O_3 depletion is occurring comes from the discovery of the Antarctic ozone hole. In recent years during the spring, O_3 depletions of 60% or more integrated over all altitudes and 95–100% in some layers have been observed over Antarctica. During winter in the southern hemisphere, a polar vortex develops that prevents the air from outside of the vortex from mixing with air inside the vortex. The depletion begins in August, as the approaching spring sun penetrates into the polar atmosphere, and extends into October. When the hole was first observed, existing chemical models could not account for the rapid O_3 loss, but attention was soon focused on stable reservoir species for chlorine. These compounds, namely, HCl and $ClNO_3$, are formed in competing reactions involving Cl and ClO that temporarily or permanently remove Cl and ClO from participating in the O_3 destruction reactions.

AIR QUALITY MANAGEMENT

In the United States, the framework for air quality management is the Clean Air Act (CAA), which defines two categories of pollutants: criteria and hazardous. For the criteria pollutants, the CAA charges the EPA with identifying those air pollutants that most affect public health and welfare, and setting maximum allowable ambient

Table 1. Greenhouse Gases and Global Warming Contribution[a]

Gas	Source (natural and anthropogenic)	Estimated anthropogenic emission rate	Preindustrial global concentration	Approximate current concentration	Estimated residence time in the atmosphere	Radiative forcing efficiency (absorptivity capacity) ($CO_2 = 1$)	Estimated contribution to global warming
carbon dioxide (CO_2)	fossil fuel combustion; deforestation	6000 M t/yr	280 ppm	355 ppm	50–200 yr	1	50%
methane (CH_4)	anaerobic decay (wetlands, landfills, rice patties) ruminants, termites, natural gas, coal mining, biomass burning	300–400 M t/yr	0.8 ppm	1.7 ppm	10 yr	58	12–19%
nitrous oxide (N_2O)	estuaries and tropical forests; agricultural practices, deforestation, land clearing, low temperature fuel combustion	4–6 M t/yr	0.385 ppm	0.31 ppm	140–190 yr	206	4–6%
chlorofluorocarbons (CFC-11 and CFC-12)	refrigerants, air conditioners, foam blowing agents, aerosol cans, solvents	1 M t/yr	0	0.0004–0.001 ppm	65–110 yr	4860	17–21%
tropospheric ozone (O_3)	photochemical reactions between VOCs and NO_x from transportation and industrial sources	not emitted directly	NA[b]	0.022 ppm	hours–days	2000	8%

[a] from ref. 27, M = million.
[b] Not applicable = NA.

83

air concentrations for these air pollutants. The air pollutants for which national ambient air quality standards are set are referred to as criteria pollutants. Six chemical species (CO, Pb, NO_2, O_3, PM, SO_2) have both primary and secondary National Ambient Air Quality Standards (NAAQS). The primary standards are intended to protect the public health with an adequate margin of safety. The secondary standards are meant to protect public welfare, such as damage to crops, vegetation, and ecosystems or reductions in visibility.

The NAAQS apply uniformly across the United States, whereas emissions standards for criteria pollutants depend on the severity of the local air pollution problem and whether an affected source already exists or is proposed. In addition, individual states have the right to set their own ambient air quality and emissions standards (which must be at least as stringent as the federal standards) for all pollutants and all sources except motor vehicles. With respect to motor vehicles, the CAA allows the states to choose between two sets of emissions standards: the Federal standards or the more stringent California ones.

To determine if NAAQS are met, states are required to monitor the criteria pollutants' concentrations in areas that are likely to be near or to exceed the standards. If an area exceeds a NAAQS for a given pollutant, it is designated as a nonattainment area for that pollutant, and the state is required to establish an SIP.

The SIP is a strategy designed to achieve emissions reductions sufficient to meet the NAAQS within a deadline that is determined by the severity of the local pollution problem. Areas that receive long (6 years or more) deadlines must show continuous progress by reducing emissions by a specified percentage each year. For SO_2 and NO_2, the initial SIPs were very successful in achieving the NAAQS. If a state misses an attainment deadline, fails to revise an inadequate SIP, or fails to implement SIP requirements, EPA has the authority to enforce sanctions such as banning construction of new stationary sources and withholding federal grants for highways.

In nonattainment areas, the degree of control on small sources is left to the discretion of the state and is largely determined by the degree of required emissions reductions. Large existing sources must be retrofitted with reasonable available control technology (RACT) to minimize emissions. All large new sources and existing sources that undergo major modifications must meet EPAs new source performance standards at a minimum. Additionally, in nonattainment areas, they must be designed using lowest achievable emission rate (LAER) technology, and emissions offsets must be obtained. Offsets require that emissions from existing sources within the area be reduced below legally allowable levels so that the amount of the reduction is greater than or equal to the emissions expected from the new source.

B. J. Finlayson-Pitts and J. N. Pitts, Jr., *Chemistry of the Upper and Lower Atmosphere*, Academy Press, New York, 2000.

T. E. Graedel, D. T. Hawkins, and L. D. Claxton, *Atmospheric Chemical Compounds Sources, Occurrence and Bioassay*, Academic Press, New York, 1986.

National Research Council, *Climate Change Science*, National Academy Press, Washington, D.C., 2001.

J. H. Seinfeld and S. N. Pandis, *Atmospheric Chemistry and Physics*, Wiley-Interscience, New York, 1998.

DAVID T. ALLEN
University of Texas

AIR POLLUTION AND CONTROL, INDOOR

Indoor air quality (IAQ) has become a growing environmental issue over the past 20 years. An increasing number of health and comfort problems have been reported in office buildings, schools, residences, and similar non-industrial settings.

Because the human body processes far more air than other environmental media, relatively small concentrations of contaminants can be of concern. Whereas we often think of water and food contamination in parts per million (ppm), many air contaminants are of concern at concentrations in micrograms per cubic meter ($\mu g/m^3$), which is roughly equivalent to parts per billion. The quality of the air we breathe while indoors is especially important, since on average we spend ~90% of our time inside buildings, not including industrial workplaces. Unfortunately, studies have shown that indoor concentrations of many contaminants are higher than outdoor concentrations.

CONCERNS

Historical Concerns

Bad odors, smoke, dampness, and infectious diseases have been indicators of poor indoor air quality for centuries. Controlling body odors (bioeffluents) has been the basis of ventilation standards since the eighteenth century. While these contaminants continue to be problems, the range of concerns broadened considerably in the 1970s—first in Europe, then in North America—as buildings were made tighter and operated with less ventilation to conserve energy.

Current Concerns

As more sophisticated measurements were made in the 1980s and 1990s, it became apparent that there are hundreds of measurable organic compounds in indoor air, present either as gases or associated with particles. Microbial contaminants and their associated gas-phase organic products of respiration and decomposition have also been investigated increasingly in recent years, but sampling and analytical methods are more cumbersome, so data are less complete.

The health concerns themselves are numerous. They range from vague dissatisfaction to frank irritation to chronic disease. Common terms found in the literature to classify the health and comfort effects or symptoms are:

- **Odor**.
- **Irritation** of eyes, nose, upper airways, throat, and skin.
- **Respiratory function decreases** in nonasthmatics including wheezing, cough, chest tightness, and shortness of breath.
- **Neurological symptoms** including nausea, dizziness, headache, loss of coordination, tiredness, and loss of concentration.
- **Immunological reactions** including inflammatory reactions, delayed hypersensitivity, and immediate hypersensitivity (allergic) reactions.
- Aggravation of **asthma**.
- **Cancer**.
- **Respiratory infections**.
- **Increased susceptibility** to infections or adverse responses to chemical substances.

Standards and Guidelines for Indoor Air Quality

Governmental regulatory bodies have set very few indoor air quality standards. The U.S. Food and Drug Administration (FDA) has set an indoor limit of $100\,\mu g/m^3$ of ozone for spaces where ozone is being generated, and the EPA guideline of 4 pCi/L for radon has become a *de facto* standard, but there are few others.

Lacking definitive IAQ standards, outdoor (ambient) air quality standards serve as a starting point. Since indoor concentrations are often higher than outdoor concentrations, indoor exposure (concentration × time) can be 10 or more times outdoor exposure.

Industrial workplace guidelines such as Threshold Limit Values (TLVs) are generally not applicable to residential, commercial, or institutional settings. They allow much higher concentrations than would be acceptable to the broad range of people who occupy typical "indoor" spaces.

Minimum ventilation rates required by building codes, many of which are based on ASHRAE Standard 62, Ventilation for Acceptable Indoor Air Quality, have historically been based primarily on the amount of outdoor air required to maintain body odor levels acceptable to at least 80% of visitors to the space.

Perhaps the primary way indoor air quality will be regulated in the future is through emission standards— established by regulatory bodies or private sector organizations—for materials and products. Outdoor air is regulated to a large extent by emissions standards, so it will not be surprising to see indoor air take a similar path. The main difference may be the greater involvement of the private sector.

Methods of Providing Good Indoor Air Quality

There are three basic ways to reduce exposures to indoor contaminants:

- Manage **sources** to prevent or reduce emissions.
- Provide **ventilation** to dilute and exhaust contaminants effectively.
- Remove contaminants by **air cleaners**.

CONTAMINANTS IN MODERN BUILDINGS

We Breathe a Complex Mixture

There is a wide range of natural and synthetic substances in both the gaseous and particulate phases. The comparison with industrial exposures is less similar. Indoor contaminants are more numerous but generally at lower concentrations than in industrial workplaces.

Sources

Many different types of materials are used in the construction, furnishing, maintenance, and operation of a building. In addition, there are various activities by occupants. As potential sources of indoor air quality problems, these items can be grouped into five categories: combustion sources, materials, activities, outdoor sources, and indoor chemical reactions.

VENTILATION AS A CONTROL METHOD

Types of Ventilation

Ventilation, as defined by ASHRAE, is "the process of supplying air to or removing air from a space for the purpose of controlling air contaminant levels, humidity, or temperature within the space." Ventilation can be either natural or mechanical. Natural ventilation is "ventilation provided by thermal, wind, or diffusion effects through doors, windows, or other intentional openings in the building." Mechanical ventilation is defined as "ventilation provided by mechanically powered equipment, such as motor-driven fans and blowers, but not by devices such as wind-driven turbine ventilators and mechanically operated windows." Infiltration and exfiltration, defined as air leakage inward or outward "through cracks and interstices and through ceilings, floors, and walls of a space or building", occurs in addition to intentional ventilation. While infiltration/exfiltration increases the air exchange of a space, it is so irregular and unpredictable that it is not normally considered in assessing the effectiveness of air quality control.

ASHRAE Ventilation Standard

ASHRAE Standard 62, Ventilation for Acceptable Indoor Air Quality, provides state-of-knowledge guidance from the scientific and technical communities on ventilation system design and operation practices that will help provide good air quality in commercial and residential buildings.

The Relationship With Energy

Based on the U.S. Department of Energy data, ~36% of total primary energy consumed in the United States is used to heat, cool, light and operate equipment in residential and commercial buildings. During the short-lived energy cost spikes in the 1970s, energy conservation was increased by tightening buildings to reduce infiltration and by reducing ventilation rates. Those energy conservation efforts tended to increase indoor relative humidity and contaminant levels, and highlighted the important relationship between energy conservation and indoor air quality. Given what we now know about indoor air quality, any future reductions in ventilation rates will have to be accompanied by either reduction in emissions from indoor sources, or increased use of air cleaning, or both.

Effectiveness of Ventilation

The effectiveness of ventilation in controlling indoor air quality has limitations. For example, of exhaust ventilation can be quite effective if the capture efficiency is high. Dilution ventilation generally works best if airflow is from low concentration areas toward high concentration areas, but this may not be practical, especially if the locations of major sources change with time. Furthermore, since much of ventilation air is recirculated, contaminants tend to get distributed throughout the building (or at least the portion of the building served by a given air handling system).

SOURCE MANAGEMENT AS A CONTROL METHOD

Prevention of Emissions

In principle, preventing IAQ problems by managing indoor sources can be very effective. The objective is to use materials and products with low (or no) emissions of substances that might cause odor, sensory irritation, or other health problems.

General Selection Criteria for Indoor Materials

Any building materials, furnishings, maintenance materials, or other contents of a building are selected with various physical, aesthetic, economic and environmental criteria in mind.

Emissions Criteria for Indoor Materials

From an indoor environmental point of view, there are several characteristics that could be considered in describing an "ideal" material. Not many materials will have all of these characteristics, of course. Compromises will usually have to be made, and many that at first seem undesirable may turn out to be quite acceptable.

Emissions Testing

Proper testing involves the use of environmental chambers with carefully controlled airflows and environmental conditions.

Emission rate testing and exposure modeling have become a significant steps in the design of some new and renovated office buildings. When coupled with appropriate attention to ventilation system design and operation, aesthetic and ergonomic factors, and building

maintenance, occupant complaints that characterize "sick buildings" are almost certain to be reduced. However, this process is currently hampered by the lack of data on emissions and clear understanding of relationships between exposure to emitted substances and health or sensory irritation.

Building Design, Operation, and Maintenance

Many indoor air quality (IAQ) problems can be prevented during building design. For example of selection of materials and products is now receiving increased attention. It is important to avoid sources with emissions that are too great to be diluted and exhausted by ventilation, or removed by affordable air cleaning devices. This is especially true for sources that are close to occupants, such as office furniture, furnishings, office supplies, and personal care products.

AIR CLEANING AS A CONTROL METHOD

General Comments

The third approach to providing good indoor air quality is air cleaning. Air cleaners can either be stand-alone units for treating air in a single room or a portion of it, or central-system units that are built into the heating, ventilating, and air-conditioning (HVAC) system. For the indoor situations covered by this article, mask-type filters are not considered.

If ventilation conditions meet standards and indoor sources are being managed to the extent practical, air cleaning may provide additional health benefit. In general, however, air cleaning has many of the same types of limitations as ventilation: Large volumes of air need to be processed to remove or destroy very small concentrations of contaminants. To be effective, air cleaner intakes need to be close to sources of key contaminants, the device needs to process a sufficient amount of air relative to the volume of the space being treated, and it must have reasonable collection or destruction efficiency and capacity.

Particle Filtration

There are two general types of particle removal devices used in buildings: fabric filters and electrostatic precipitators. The most common form for filters is the pleated type, although extended-area bag filters are used in some large buildings. Electrostatic precipitators, sometimes referred to as "electronic" air cleaners when marketed for residential applications, are physically similar to, but lighter in construction than industrial units; they also operate at lower voltages.

Control of Infectious Particles

Person-to-person transmission of bacterial and viral respiratory diseases occurs mainly via airborne particles that are evaporative residuals of droplets created by coughing and sneezing. These "droplet nuclei" are typically $1-3\,\mu m$ in diameter. Like other particles, their concentrations can be controlled by ventilation, source management (ie, care of infected persons), and particle air cleaners. Particle air cleaners with high removal efficiencies and recirculation rates can reduce droplet nuclei concentrations if located and operated so as to properly distribute the clean exhaust air. Another type of control system is ultraviolet germicidal irradiation (UVGI).

Gas/Vapor Removal

The possibilities for removing gases or vapors are much more limited than for particles. The most frequently considered (and used) approach is adsorption, particularly by activated carbon. However, the capacity of activated carbon for gases and vapors in the low ppm to high ppb range is too limited to make it a practical option for most applications.

Activated carbon has been used successfully to reduce ozone emissions from copy machines. Chemisorption has been considered for many years as a possible technology. In theory, adsorbed contaminants are chemically converted to substances that do not desorb. One of the major barriers to this technology is developing an adsorbent and reactive surface for the wide range of contaminants found in indoor air. Room temperature catalysis is another, conceptually similar, technology that has been investigated for some time as a way to reduce indoor concentrations of vapor-phase organic compounds. The version of this technique that has gotten most attention is photocatalytic oxidation.

R. B. Gammage and B. A. Berven, eds., *Indoor Air and Human Health*, 2nd ed., CRC/ Lewis Publishers, Boca Raton, Fla., 1996.

National Research Council, *Indoor Pollutants*, National Academy Press, Washington, D.C., 1981.

J. D. Spengler, J. M. Samet, and J. F. McCarthy, eds., *Indoor Air Quality Handbook*, McGraw-Hill, New York, 2001.

W. G. Tucker, B. P. Leaderer, L. Mølhave, and W. S. Cain, eds. "Sources of Indoor Air Contaminants: Characterizing Emissions and Health Impacts," *Annals of The New York Academy of Sciences*, New York, Vol. 641, 1992.

W. Gene Tucker
James Madison University

AIR POLLUTION CONTROL METHODS

Air pollution is the presence in the outdoor atmosphere (ambient air) of one or more contaminants in such quantities and for such duration as to be harmful or injurious to human health or welfare, animal or plant life, or property, or may unreasonably interfere with the enjoyment of life or property. Some of the most common air pollutants include particulate matter (PM), sulfur oxides (SO_x), nitrogen oxides (NO_x), volatile organic compounds (VOCs), and carbon monoxide (CO). Another major pollutant is ground-level ozone (O_3). Other important pollutants include lead (as leaded gasoline was phased out in the 1980s, the U.S. emissions of lead into the atmosphere dropped by 95% or more); hazardous air pollutants (HAPs), including lead, mercury, formaldehyde, benzene; and many others; several ozone-depleting compounds (such as the chlorofluorocarbons); and greenhouse gases, such as carbon dioxide (CO_2) and methane (CH_4).

Odors are not listed in federal law as pollutants but are identified in many state statutes as nuisance pollutants. Indoor air pollutants can include any of the above-mentioned pollutants, but because they are indoors, federal ambient standards do not apply.

NATIONAL AIR POLLUTION STANDARDS

There are two types of standards for air pollution – ambient (outdoors) air standards, and source performance standards (emission limits). National Ambient Air Quality Standards (NAAQSs) have been promulgated by the U.S. Environmental Protection Agency and adopted by many states. NAAQSs establish limits on the maximum allowable concentrations of several common pollutants in the outdoor atmosphere. There are many different types of sources and many different source performance standards, usually stated as emission limits. These limits can be listed as the mass of pollutant emitted per unit of time, or per unit amount of a certain input into the process, or per unit amount of a product or output from the process. The two types of standards work together in establishing the level of control that must be provided for any given source. Source emission standards set a baseline of control, but even if a source is designed to meet that level, if ambient standards are shown to be threatened, the source can be required to provide additional safeguards.

Standards and regulations for HAPs and greenhouse gases are still evolving. Individual HAPs coming from specific sources have very stringent limits. Any emission limits for greenhouse gases require international treaties and cooperation, and so far the U.S. has not agreed to any specific federal limits. However, a number of international companies with headquarters in the U.S. are making strides in limiting their emissions of CO_2.

In addition to the federal regulations of ambient air, and the setting of federal emission standards for major sources and for motor vehicles, local and state programs and efforts are crucial for the successful control of air pollution. Individual states are often authorized by EPA to administer and enforce air regulations. States administer permitting programs, review programs, and inspection/enforcement programs for new and existing sources to ensure that industries comply with the regulations. They review the annual stack testing that companies must conduct to prove that their emissions are within the limits set by their permits. States and local programs monitor ambient air quality and track whether counties are in attainment of the air quality standards. They conduct emission inventories and coordinate emissions reductions for one or more specific pollutants emitted from vastly different sources (eg, NO_x from motor vehicles and power plants) to comply with nonattainment rules. The need for air pollution control is often driven by compliance with regulations. Whether a region is meeting air quality regulations is determined by actual measurement or monitoring.

MEASUREMENT OF AIR POLLUTION

Ambient air sampling often requires continuous measurements for weeks, months, or years. The instruments must be able to detect and quantify pollutant concentrations in the ppm to ppb range for gases. Source sampling typically deals with gaseous pollutant concentrations in the ranges of tenths of a volume percent down to tens to hundreds of ppm. However, for dioxins and furans (and for certain other HAPs), very low concentrations must be measured in order to show compliance with stringent emissions limits.

Federal regulations require ambient air monitoring at strategic locations in every designated air quality control region. The number of required locations and complexity of monitoring increases with the population in the region and the level of pollution. Continuous monitoring is preferable, but for PM one 24-hour sample every sixth day is often acceptable. Continuous monitors are sent electronically to a computer, where they can be processed and placed onto the web. Special problems have been investigated using portable, vehicle-carried, or airborne ambient sampling equipment.

Ambient sampling may fulfill one or more of the following objectives: (1) establishing and operating a pollution alert network; (2) monitoring the effect of an emission source; (3) establishing a baseline prior to a proposed installation of a large source; (4) establishing seasonal or yearly trends; (5) pinpointing the source of an undesirable pollutant; (6) checking for hotspots in a city's transportation network; (7) obtaining permanent sampling records for legal action or for modifying regulations, and (8) correlating pollutant dispersion with meteorological, climatological, or topographic data, and with changes in societal activities.

The problems of source sampling are distinct from those of ambient sampling. Depending on the objectives or regulations, source sampling may be occasional or continuous. Typical objectives are (1) demonstrating legal compliance with regulations; (2) obtaining emission data; (3) measuring product loss or optimizing process operating variables; (4) obtaining data for engineering design; (5) determining collector efficiency for acceptance of purchased equipment; and (6) determining the need for the maintenance of process or control equipment.

Source exhaust gases may be at a high temperature or may contain high concentrations of water vapor or entrained mist, dust, or other interfering substances. Contaminants may be deposited onto or absorbed into the structure of the gas-extractive sampling probes or the adsorbents used to concentrate certain pollutants.

PREVENTION AND CONTROL OF AIR POLLUTION

The U.S. EPA has endorsed a hierarchical approach to solving pollution problems. At the base is pollution prevention/waste minimization, the most preferred approach. Next comes recycling and reuse of waste materials. Third comes treatment, fourth, disposal.

The preferred approach is to first try to prevent and/or minimize pollution. However, the addition of a control device (treatment) is often not the environmentally best or least costly approach. Process examination may reveal changes or alternatives that can eliminate or reduce pollutants, decrease the gas quantity to be treated, or render pollutants more amenable to collection. Listed below are some considerations for controlling pollutants without the addition of specific treatment devices.

Eliminate leaks or vents of the pollutant

Seal the system to prevent interchanges between system and atmosphere.

Use pressure vessels.

Interconnect vents on receiving and discharging containers.

Provide seals on rotating shafts and other necessary openings.

Recycle the exhaust stream rather than using fresh air or venting.

Change raw materials, fuels, or processing steps to reduce or eliminate the pollutant

Switch to a nonhydrocarbon-based cleaner.

Switch to a lower sulfur fuel.

Change the manner of process operation to prevent or reduce the formation of (or the air entrainment of) a pollutant.

Change the process itself to eliminate the step that produces the pollutant.

Reduce the quantity of pollutant released or the quantity of carrier gas to be treated

Minimize the entrainment of pollutants into a gas stream.

Reduce the number of points in the system in which materials can become airborne.

Recycle a portion of the process gas.

Design hoods to exhaust the minimum quantity of air necessary to ensure pollutant capture.

Use equipment for dual purposes

Use a fuel combustion furnace to serve as a pollutant incinerator (eg, design a larger volume furnace, or reduce gas flow to increase residence time).

Specific steps to illustrate these principles include the substitution of a low-sulfur coal at a power plant to reduce SO_2, using a water-based cleaner to rinse printed circuit boards to eliminate VOCs, changing raw materials (eg, eliminating a mercury containing metal ore), reducing operating temperatures to reduce NO_x formation, and installing well-designed hoods at emission points to effectively reduce the air quantity needed for pollutant capture.

Table 1. Checklist of Applicable Air Pollution Control Devices

APC equipment	Pollutant characteristics or control efficiency provided						
	Gaseous		Particulate		Control efficiency		
	Odors	Others	Liquid	Solid	Low	Medium	High
Gases							
absorption	•	•	•			•	•
adsorption							
one-use cannisters	•	•					•
regenerative beds	•	•					•
biofiltration	•	•					•
condensation	•	•			•		
chemical reaction	•	•			•	•	
Particulates							
cyclones			•	•	•	•	
electrostatic			•	•		•	•
precipitation							
filtration							
baghouses			•			•	
granular beds			•			•	
gravitational settling			•	•			
impingement			•		•		
wet scrubbers			•			•	•
Both							
oxidation							
catalytic oxidizers	•	•					•
incinerators	•	•	•	•			•
RTOs	•	•					•
thermal oxidizers	•	•		•			•
tall stacks[a] (dispersion)	•	•	•	•			

[a]Tall stacks are not a control device but are useful to reduce ground-level concentrations near the source. For certain pollutants, they may be used alone or in conjunction with other control devices.

SELECTION OF CONTROL EQUIPMENT

Engineering considerations include (1) the properties of the exhaust gas, (2) the properties of the pollutants, (3) knowledge of the regulations governing emissions, (4) the plant space or land available for construction of a new an (APC) system, pollution control and (5) the geography, climate, and other characteristics of the plant location.

The physical state of matter of a pollutant and its chemical characteristics are very important to the selection of an appropriate APC system. Starting with the checklist in Table 1, devices can be eliminated that are too inefficient or that simply are not physically capable of doing the job.

CONTROL OF GASEOUS EMISSIONS

Six methods are widely used for controlling gaseous emissions: absorption, adsorption, biofiltration, condensation, chemical reaction, and incineration. Atmospheric dispersion from a tall stack, considered as an alternative in the past, is not really a control method but is still used to help reduce final ambient concentrations to acceptable levels. Adsorption is applicable for many organic pollutants within a certain molecular weight range and can achieve contaminant removal down to extremely low levels (less than 1 ppm). It is often used for controlling organic vapors in relatively cool streams of air. Biofiltration is preferred for handling large gas volumes that have high humidity and quite dilute contaminant levels. Condensation is best for substances with rather low vapor pressures but that are present in relatively high concentrations in the air stream. Where refrigeration is needed for the final condensing step, elimination of noncondensible diluents is beneficial. Incineration, suitable only for combustibles, is used to destroy toxic, odorous, and other organic pollutants, and small concentrations of H_2S or CO. Specific gases such as sulfur oxides or nitrogen oxides often require specialized methods and are discussed later.

Volatile Organic Compounds (VOCs)

Volatile organic compounds (VOCs) are any organic compound with significant vapor pressure so that it can exist as a vapor in air. Emissions of VOCs come from many sources, including petroleum processing, painting, solvent cleaning, incomplete combustion of fuel, and others. Good attention to VOC control has resulted in significant decreases in U.S. emissions since the Clean Air Act Amendments of 1970: from 14.3 million tons then to 8.0 million in 1999. There are many techniques that can be applied to VOC control, including condensation, adsorption, incineration, and biofiltration.

Control of Acids and Other Gases

Absorption is particularly attractive for water-soluble pollutants in appreciable concentration; it is also applicable to dilute concentrations of gases having high solubility. These include HCl, Cl_2, HF, SO_2, H_2S, NH_3, and others. In many cases, absorption is enhanced by adding chemicals that control pH or that react with the pollutants once they are absorbed. In some instances, a nonaqueous scrubbing liquid may be used, but in most cases the absorbent liquid is water.

Water is the most common absorption liquid. It is used for removing highly soluble gases such as HCl and ammonia, and other gases, such as H_2S, HF, and SiF_4, especially if a caustic is added to the water. NH_3 can also be absorbed in water if the final contact is acidic. Gases such as SO_2, Cl_2, and H_2S can be absorbed more readily in alkaline solutions. The most common absorption processes use lime or limestone slurry, though sodium-based caustics are also used.

Disposal of recovered gaseous pollutants can be a problem. Precipitation of certain acid gases (especially SO_2) as insoluble sludges may be possible through the addition of lime, limestone, or other reagents. The sludge may be thickened by settling, and dewatered by centrifugation or filtration; however, sludges containing 70% water are not uncommon. Disposal to streams is not feasible and impounding in landfills or tailing ponds is becoming less acceptable. However, large $CaSO_3/CaSO_4$ sludge ponds can still be found at many coal-fired power plants. Conversion of the pollutant to a usable form is preferable, but usually involves added expense, even when selling the recovered material.

Sulfur dioxide is formed whenever any material containing sulfur is burned. Hence, SO_2 is prevalent in numerous industrial exhausts, including power plants, petroleum refineries, pulp and paper plants, phosphate fertilizer plants, nonferrous metal smelters, and others. For small sources (and some large ones) the best solution is to remove the sulfur prior to combustion, or simply buy a low-sulfur fuel.

There are two basic approaches to sulfur dioxide scrubbing: regenerative and throwaway. Within those two broad categories, there are numerous processes. For exhaust gases that have high concentrations of SO_2, as from copper or nickel smelters, it is possible to scrub with water and produce a sulfuric acid stream that can be sold. There are regenerative processes that can be applied at coal-fired power plants, in Japan and Germany much sulfur is being recovered this way. However, in the U.S., coal-fired power plants (this country's largest source of SO_2) mostly use a limestone-based throwaway process.

Nitric oxide (NO) and nitrogen dioxide (NO_2), commonly called NO_x, are formed whenever there is high-temperature combustion of anything using air as the oxygen source. At high temperatures (3000 to 3600°F), the nitrogen and oxygen molecules each split into highly reactive atomic forms and the atoms recombine as NO_x. The

major sources in the U.S. are power plants, large industrial furnaces, and motor vehicles. NO_x has been one of the most difficult pollutants to control—and emissions in the U.S. have actually risen over the last 30 years. More than half of all NO_x emissions in the United States comes from mobile sources.

Industrially, there are several methods for controlling NO_x, classifie as either combustion modifications or flue gas treatments. Because NO_x formation depends critically on the temperature and oxygen content in the flame zone, and on the time of exposure to these conditions, combustion controls attempt to reduce NO_x formation by various strategies. Some of the tactics are using low NO_x burners; low excess air firing; flue gas recirculation; off-stoichiometric combustion; gas reburning; reduced air preheating; and water injection. Flue gas treatment technologies focus mostly on chemically reducing the NO_x to N_2 and H_2O. The most widely used technology is selective catalytic reduction (SCR). This capital intensive process has been applied mostly at large power plants, with reasonable success. In this process, ammonia (or an other reducing agent) is injected into the hot flue gases, and then the mixture is passed over a catalyst.

Carbon dioxide (CO_2) is the thermodynamically stable end product of the combustion of any carbonaceous fuel and is the compound most responsible for global climate change (GCC). Excessive emission of CO_2 into the air is resulting in a steady increase in CO_2 concentrations in the atmosphere, and the acceleration of the greenhouse effect. CO_2 levels have risen to unprecedented levels, and temperatures are rapidly rising above historical norms. Because the burning of fossil fuels is so entrenched in the world's economy, there are only two ways to approach the CO_2 emissions problem. The first is to replace the use of fossil fuels with other energy sources. Biofuels, although still carbon based, are derived from crops that were grown recently (the plants utilize CO_2 from our current atmosphere) and simply replace that carbon back into the atmosphere when burned. Thus, these biofuels are carbon neutral, whereas fossil fuels add "new" carbon to the atmosphere (carbon that had been stored underground for millions of years). The other approach is to continue to burn fossil fuels and to capture the CO_2 before it is emitted. Scrubbing has been proposed as a way to reduce the CO_2 emissions from power plants. Because CO_2 is even less soluble than SO_2, such scrubbing would be very expensive. In addition, the problem of what to do with all that CO_2 would remain. Suggestions have been made for disposal including sequestering it in deep coal mines, old oil fields, or in the deep oceans.

CONTROL OF PARTICULATE MATTER EMISSIONS

Particulate matter (PM) is a term used to describe the many different types, sizes, and shapes of particles emitted from the myriad of industrial and other sources. The concentration of PM in a gas is expressed as a total mass concentration ($\mu g/m^3$), which accounts for particles of all types, sizes, and shapes. Because even the smallest particles are significantly larger and heavier than gas molecules, the control of PM often depends simply on separating and removing particles from the exhaust gas stream. Such separation usually makes use of the size and mass differences of PM compared with gases, but the selection of the best separation device often depends on a number of physical and chemical properties of the particles (eg, size, shape, density, electrical properties) as well as characteristics of the gas (eg, temperature, acidity, moisture content).

Particle sizes greatly influence the choice of control equipment. Large, heavy particles can be captured either by gravity settling or by centrifugal separation. Smaller particles may be caught efficiently by wet scrubbers, electrostatic precipitators, or fabric filters. The aerodynamic diameter of a specific particle is defined as the diameter of a unit density sphere that will settle in still air at the same velocity as the particle in question.

Sometimes a large number of small particles may escape collection, but the overall efficiency is still high because of the overwhelming mass in the larger sizes. Particle size distributions can be determined from a device called a cascade impactor. This device has various stages, each with an opening (to admit the air sample) that is immediately followed by an impaction plate. A sample of the air stream carrying the particles is directed through the cascade impactor, and as the air passes through the stages with openings of smaller and smaller sizes, the velocity increases. Large particles that have significant inertia and cannot change direction to avoid the plate are caught in the early stages of the device, smaller particles in the later stages. Each stage captures particles of a characteristic diameter.

A cyclone is a stationary tube or set of tubes in which the gas flows in a vortex, creating a centrifugal force that moves the particles to the walls of the tube(s). The vortex can be created by the gas entering tangentially or by spin vanes positioned in the cyclone inlet. The particles hit the walls and slide down to eventually exit by gravity out the bottom of the cyclone. The cleaned (but not perfectly clean) gas exits out the top. A typical tangential entry cyclone is shown in Figure 1. The advantages of cyclones are that they are low capital cost, have no moving parts, and can operate at high temperatures and under corrosive and erosive conditions. The disadvantages are that they tend to have low efficiencies (especially for small particles) and have high pressure drops. Depending on the design, cyclones trade off efficiency for throughput.

The process by which an electrostatic precipitator (ESP) removes particulate matter from a gas stream involves the following steps: (1) the creation of a high voltage drop between electrodes; (2) distributing the flow of gas between all plates uniformly; (3) the charging, migration, and collection of particles on oppositely charged plates, and (4) the removal of the bulk dust from the plates. Two major advantages of ESPs are that they collect particles with very high efficiencies and present very little resistance to gas flow (therefore causing only a slight pressure drop even when treating very large gas flows). Also, despite the very high voltage drop in an ESP there is very little current flow. So, with a low pressure

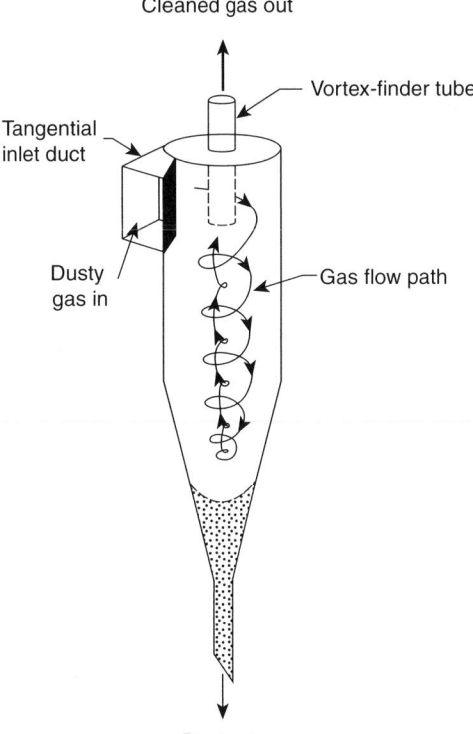

Figure 1. Schematic diagram of a cyclone.

drop and only slight electricity usage, the operating cost of an ESP is not as great as one might expect. Other advantages include the dry collection of dry materials or wet collection of wet mists. Disadvantages include high capital costs, relative inflexibility to operating changes, and limited application if the particles lack "good" electrical properties.

ESPs are used in a large number of coal-fired power plants but can also be found in numerous other industrial applications. Typically, large industrial ESPs are of the plate-and-wire type, composed of many steel plates in parallel with wires hung between the plates. The gases flow through the spaces between the plates, the particles are charged by the electric fields between the wires and the plates, and then they drift over to the plates to be collected. However, there are a number of other configurations, including tubular ESPs (more useful for collecting mists), and two-stage ESPs (often used for ventilation and indoor air cleaning). A complete electrostatic precipitator consists of discharge electrodes, collecting surfaces (plates or tubes), a suspension and tensioning system for discharge electrodes, a rapping system to remove dust from tubes, dust hoppers and a dust-removal system, a gas-distribution system and precipitator housing, and a power supply and control system.

There are thousands of large ESPs at power plants and numerous other industries, but ESPs do experience some operating problems. These usually are traceable to poor gas flow distribution, improper voltage control, incorrect rapping procedures, or too frequent or infrequent rapping. Nevertheless, even though the technology is nearly one hundred years old, ESPs remain reliable air pollution control devices for large and small industries throughout the world.

Fabric filtration is an ancient, well-accepted process for separating dry particulate matter from a gas stream. On an industrial scale, a number of fabric bags are installed in parallel inside a housing (hence the common term "baghouse") and the dusty gas flows slowly through the bags, leaving the dust behind on the fabric. The fabric is cleaned periodically, so the baghouse can continue to operate for long periods of time. There are many different types of fabrics, different ways of weaving them into bags, different ways of placing the bags in the house, different ways of passing the air into the compartments, and different methods for cleaning the dust off the filters. Baghouses are commonly classified by the cleaning method, namely reverseair, shaker, or pulse-jet.

Baghouses have several advantages: they are the most efficient of all the PM control technologies, they can operate on a wide variety of dusts, they are modular in design and manufacture, they operate with reasonable pressure drops, and they can collect the particles in a dry and reusable form. Disadvantages include that they require large floor areas, fabrics can be damaged by high temperatures or corrosive gases, they cannot operate in moist environments, and they have the potential for fire or explosion. Baghouse efficiency for a well-designed system is almost always greater than 99%; indeed, the most common source of particulate loss is through bag leaks, defects, or gaps where the bags attach rather than through direct penetration of the fabric.

Devices that collect particles by contacting the dusty gas with water are called wet scrubbers. Although they can act like gas absorbers and remove a substantial fraction of soluble gaseous pollutants, their main function is to remove particulate matter. These devices can be classified by their contacting mechanism and power consumption. Wet scrubber efficiency can vary, but high-energy venturis, for example, can be highly efficient even on small particles. Capital costs can be very reasonable, but the cost of water and sludge treatment can be substantial in some areas.

The advantages of wet scrubbers are that they can handle flammable and explosive dusts, provide some degree of gaseous pollutant control, collect wet mists, neutralize corrosive gases and dusts, and provide cooling of hot gases. Disadvantages include the potential for corrosion of the metal, potential water pollution problems, may freeze in cold climates, may produce a visible plume due to condensation, and may require expensive disposal of the waste sludge.

CONTROL OF MOBILE SOURCE EMISSIONS

Emissions from mobile sources include all the usual pollutants, but the three that are emitted in the largest quantities are carbon monoxide, nitrogen oxides, and volatile organic compounds. Particulate matter is an important pollutant from diesel engines and so can be of serious concern for on-road large trucks and buses and for nonroad construction equipment. Mobile source pollution is a world-wide problem; every large city in the world has traffic congestion and serious air pollution problems. In the United States, over the last 30 years, improvements in individual vehicle emission factors (EFs) has been so great that it has offset the growth in vehicle miles traveled (VMT), and total emissions (the product of EF and VMT) have gone down. However, in recent years, the improvements have been leveling out and total emissions are forecast to start to rise. In many other countries, emissions controls are not as strict as in the U.S., and motor vehicle emissions per VMT are much larger.

Gasoline engines can emit CO, VOCs, and NO_x simultaneously. Diesel engines operate at higher compression ratios, and much higher air/fuel ratios, and their CO and hydrocarbon emission factors (EFs) are relatively low, but their NO_x EFs are significantly higher compared with gasoline engines.

The best-known automotive pollution control device is the catalytic converter. This device is an exhaust gas clean-up device that simultaneously oxidizes CO and hydrocarbons while it reduces NO_x. The catalytic converter is designed to allow the exhaust gas to pass through small channels with very little pressure drop. The pollutants adsorb momentarily on a catalyst (platinum and palladium supported on aluminum oxide) and react. The diesel engine equivalent is called a trap-oxidizer, since there is a much higher concentration of PM in diesel exhaust, which takes longer to oxidize.

Fuels can have a big effect on emissions, and switching fuels is a viable pollution control measure. Gasoline is a blend of many different hydrocarbon fuels, as is diesel fuel (although diesel components are slightly heavier than gasoline components). Different components burn differently, and fuels are blended to meet industry specifications. Reid Vapor Pressure (RVP) is a measure of the volatility of gasoline. A higher RVP is required for good engine start-up in the winter. However, with higher RVP, gasoline has more of a tendency to evaporate, releasing hydrocarbon vapors into the air. One regulatory control measure used by the U.S. EPA to help control ozone is to require a much lower RVP in summer than winter. This reduces the emissions of VOCs, both from evaporation and from engine emissions. For areas that have a winter-time CO problem, the EPA has required that some oxygenates (such as ethanol or MTBE) be blended in with the gasoline. Having an oxygen molecule in the fuel helps complete the combustion and reduce CO formation.

Sulfur in fuel leads directly to more SO_2 emissions. But in addition, sulfur can temporarily poison catalytic converters and lead indirectly to more PM emissions. In the U.S., gasoline is highly refined and no longer has much sulfur in it. However, diesel fuel does. The U.S. EPA has promulgated new regulations for diesel fuel that will drop the sulfur content from about 500 ppm to 15 ppm in 2007 for on-road diesel and in 2010 for nonroad diesel fuel. This should greatly reduce both the sulfur oxides and the particulate matter from diesels.

High prices recently for petroleum fuels have spurred great interest in so-called alternative fuels and alternate technologies. The "new" fuels include ethanol, biodiesel, and hydrogen, and the new technologies include fuel cells, hybrids, and pure electric vehicles. Electric vehicles have batteries that must be recharged with electricity produced at power plants, so one might argue that one is simply displacing pollution. But often, power plants are well-controlled, so that the net result is a reduction in total air pollution. Fuel cells use a fuel (such as hydrogen) to combine at low temperature with oxygen to produce electricity on board the vehicle. Hybrid vehicles have both gasoline and electric motors and use regenerative braking and other techniques to recharge their batteries. Although still dependent on gasoline, they get substantially better gas mileage than regular engine cars, and thus reduce pollution emissions per mile. Fuel cell technology has still not been commercialined.

Ethanol comes from corn or another crop that can be grown domestically, reducing dependence on foreign oil. In addition, it is "carbon-neutral," meaning that the CO_2 emissions that come from burning ethanol simply replace the CO_2 that was absorbed in growing the corn to make the ethanol. Ethanol can be mixed with gasoline in low percentages and used in all cars today. E-85 is a blend of 85% ethanol and 15% gasoline that can be run in so-called flex-fuel vehicles, of which there are dozens of models being manufactured today. Brazil utilizes mostly ethanol in all its cars and has successfully weaned itself from dependence on foreign oil. However, there is still much debate over the efficacy of ethanol (its net energy effects), and its contributions to other air pollutants.

Biodiesel is a domestic, renewable fuel for diesel engines derived from natural oils like soybean oil or used cooking oils which meets certain fuel specifications. Like ethanol, it is carbon neutral. Biodiesel can be mixed in any concentration with petroleum-based diesel fuel for use in existing diesel engines with little or no modification to the engine. It is produced by a chemical process (called trans-esterification) which is a reaction of the vegetable oil (or animal fats) with methanol in the presence of sodium hydroxide to produce mono-alkyl esters and glycerin. The glycerin is removed from the oil, and can be sold as a by-product. Biodiesel is not the same thing as raw vegetable oil. Biodiesel can be produced in large quantities from current soybean or other crops and can displace petroleum diesel gallon for gallon with no ill effects on diesel equipment. The economics of biodiesel can be quite compelling. In addition, biodesel emits substantially less pollution than petroleum diesel, and there are no arguments about its overall net energy effects.

ODOR CONTROL

Odor control involves any process that results in a more acceptable perception of smell, whether as a result of dilution, removal or destruction of the offending substance, counteraction, or masking.

Odor Measurement

Both static and dynamic measurement techniques exist for odor. The objective is to measure odor intensity by determining the dilution necessary so that the odor is imperceptible or doubtful to a human test panel. That is, a given sample of odorous air is diluted with enough odor-free air to reach the detection threshold. An odor unit (o.u.) has been widely defined as $0.0283\,m^3$ ($1\,ft^3$) of air at the odor detection threshold. It is a dimensionless unit reprsenting the quantity of odor which when dispersed in 28.3 L ($1\,ft^3$) of odor-free air produces a positive response by 50% of panel members.

With the static dilution method, a known volume of an odorous sample is diluted with a known amount of nonodorous air, mixed, and presented statically (quiescently) to the test panel. The ASTM D1391 syringe dilution technique, the best known of these methods, involves preparation of a 100-mL glass syringe of diluted odorous air which is allowed to stand 15 min to assure uniformity. The test panel judge suspends breathing for a few seconds and slowly expels the 100-mL sample into one nostril. The test is made in an odor-free room with a minimum of 15 min between tests to avoid olfactory fatigue.

In the dynamic dilution method, odor dilution is achieved by continuous flow. Its advantages are more accurate results, simplicity, reproducibility, and speed. Devices known as dynamic olfactometers control the flow of both odorous and pure diluent air, provide for ratio adjustment to give desired dilutions, and present multiple, continuous samples for test panel observers at ports beneath ventilation hoods.

Odor Control Methods

Absorption, adsorption, and incineration are all typical control methods for gaseous odors; odorous particulates are controlled by the usual particulate control methods. However, a gas carrying odorous particulates may still require gaseous odor control treatment even after the particulates have been removed. For oxidizable odors, wet absorption combined with oxidation of the odorous compounds using such oxidants as hydrogen peroxide, ozone, chlorine, and $KMnO_4$ has been practiced; vapor phase oxidation (either thermal or catalytic) has also been employed.

R. L. Aldrich, "Environmental Laws and Regulations Related to Odor and Waste Gas Contaminants," *Biotechnology for Odor and Air Pollution Control*, Springer, New York, 2005.

S. Calvert and H. M. Englund, eds., *Handbook of Air Pollution Technology*, John Wiley & Sons, Inc., New York, 1984.

C. D. Cooper and F. C. Alley, *Air Pollution Control – A Design Approach*, 3rd ed., Waveland Press, Long Grove, Ill., 2002.

J. S. Devinny, M. A. Deshusses, and T. S. Webster, *Biofiltration for Air Pollution Control*, Lewis Publishers, Boca Raton, Fla., 1999.

C. David Cooper
University of Central
Florida

ALCOHOLS, HIGHER ALIPHATIC, SURVEY

SURVEY AND NATURAL ALCOHOLS MANUFACTURE

Monohydric, aliphatic alcohols with a hydrocarbon chain length of C6 and above are referred to as higher alcohols. For commercial products, the alcohol group is usually found in the primary, although secondary alcohols are occasionally seen. The hydrocarbon portion of the molecule is hydrophobic, while the hydroxyl group provides a reactive site for attaching a strong hydrophilic species. The combination of hydrophilic and hydrophobic properties in the same molecule yields an extremely good surfactant, readily biodegradable. Detergent alcohols generally have a hydrocarbon chain length of C12 and above with at least 35% linear chains. Plasticizer alcohols generally have a chain length of C6 to C13 with a linear or highly branched structure.

Both natural (oleochemical) and synthetic (petrochemical) routes are used to make the higher alcohols. Plasticizer alcohols are made primarily by the synthetic route, with only minor quantities obtained from natural feedstocks. Detergent alcohols are made by both routes, with global capacity currently split 50/50 synthetic/natural. The natural route is based on vegetable and animal fats, principally coconut, palm kernel, and tallow. The products are essentially all even chain length (ie, C12, C14, ...) and 100% linear. The synthetic route is classified as either Ziegler or oxo. The Ziegler process produces linear, even-chain products, while the oxo process produces linear, even- and odd-chain products.

Primary end uses of plasticizer alcohols are for flexible PVC and surface coatings applications in construction or automobile manufacturing.

PHYSICAL PROPERTIES

Specific gravity, boiling point, melt point, and viscosity are all a linear or a polynomial function of chain length. The branched alcohols have properties similar to the linears for equal carbon number, except that melt point is significantly lower for the branched molecules. The plasticizer alcohols are slightly soluble with or in water, but the detergent alcohols are generally insoluble. Branching does not have a significant impact on water solubility.

Heat of fusion, thermal conductivity, critical temperature, and flammability generally increase with chain length. Critical pressure and heat of vaporization are an inverse function of chain length.

CHEMICAL PROPERTIES

There are three principal types of reactions involving alcohols. These are the reactions involving the O–H bond, the C–O bond, and reactions with the alkyl portion of the molecule (ie, the C–H bond of the α carbon). Alcohols, like water, can act both as weak acids or weak bases. Reactions are generally done in the liquid phase, using either homogeneous or heterogeneous catalysts.

MANUFACTURE FROM FATS AND OILS

Both methyl esters and fatty acids are suitable intermediates for the manufacture of higher natural alcohols. Triglycerides from either vegetable or animal sources are the feedstocks used to produce these intermediates. The most useful chain lengths for detergent applications are C12 and C14. Health and personal care applications are more oriented toward C16 and C18. Common triglyceride feedstocks are coconut oil and palm kernel oil for detergent applications, and tallow, palm, or soybean oil for cosmetic applications. Less common is rapeseed oil as a feedstock for behenyl or eurucyl alcohol and castor oil as a feedstock of hydroxy-stearyl alcohol.

As the more widely used intermediate, methyl esters can be made by transesterification of fatty triglycerides or by direct esterification of fatty acids.

Transesterification of triglycerides with methanol is the predominant process for the manufacture of methyl esters.

Direct esterification of fatty acids to produce methyl esters is done in either a batch or a continuous process. Advantages of the continuous process are that a lower excess of methanol is required and the reaction time is much shorter. The direct esterification process is advantageous for making esters with a higher chain length purity than the parent triglycerides. Fatty acids are less widely used as an intermediate than methyl esters. On the other hand, fatty acids are more widely available than methyl esters for those who are purchasing, rather than producing, the intermediate material. In addition, the fatty acid process avoids the use of methanol, a flammable solvent subject to local environmental regulations.

There are two major porcesses in use: slurry based and fixed bed. In the methyl ester, slurry-based catalyst process, dry methyl ester, hydrogen, and catalyst slurry (typically copper chromite) are fed to a series of vertical reactors. The product stream is cooled and separated into liquid and gaseous phases, with the gaseous phase recycled. The catalyst is removed, largely recycled, and the liquid phase is stripped of methanol, then distilled. A similar flowsheet is used for the fatty acid based, wax ester process.

On paper, the process of direct hydrogenation of triglycerides should be less costly than either the fatty acid or the methyl ester route, since there are fewer processing steps and hence lower capital investment as well as lower labor and utilities. However, propylene glycol, not glycerine, is the main co-product generated, making the cost of fatty alcohol approximately equal for both processes.

SHIPMENT AND STORAGE

Higher alcohols are available in bulk quantities in 208-L (55-gal) drums, 23,000-L (6000-gal) tank trucks, 75,000-L (20,000-gal) tank cars, and marine barges. Some of the plasticizer alcohols are available in bottles and cans, eg, for perfume applications. Higher-melting alcohols (C16 and greater) are also generally available in a flaked form in 22.7-kg (50-lb) polyethylene or paper lined bags. Linear and branched alcohols of 6–9 carbon atoms are classified as combustible for shipment by the U.S. DOT due to their low flash points. Alcohols of C10 and above are classified as nonhazardous. The higher alcohols in anhydrous form do not attack common metals and thus may be stored in mild steel, but to avoid iron contamination a liner of zinc silicate, epoxy phenolic, or high baked phenolic may be used. For high melting alcohols, and for low melting alcohols under cold climate conditions, insulated tanks and heating coils of stainless steel are required. Avoid high temperatures during storage to maintain product quality. Carbonyl formation during storage may be minimized with a nitrogen blanket. Moisture in the atmosphere may be excluded by storing the product under a blanket of inert gas or by installing a dehumidifier. In order to minimize danger of fire in the handling of plasticizer range alcohols, tanks should be grounded, have no interior sources of ignition, be filled from the bottom of the tank to prevent static sparks, and be equipped with flame arrestors.

SPECIFICATIONS AND STANDARDS

Commercially available materials are found as both pure components and mixtures. Even-chain alcohols are produced from natural fats and oils and from the Ziegler process, and are highly linear. Odd-chain alcohols are produced by oxo-chemistry, and have some branched chains. Linear, detergent-range alcohols are marketed in the U.S. by P&G, Cognis, and Condea. Condea is the only company to produce C20 and C22 alcohols for health and personal care applications, using both the oleochemical and petrochemical routes. Branched, detergent-range alcohols are marketed in the U.S. by Shell.

Linear, plasticizer alcohols are marketed in the U.S. by P&G, Cognis, Condea, and BPAmoco. P&G and Cognis utilize natural fats and oils exclusively, while Condea and BPAmoco use petroleum-based feedstocks.

A range of branched, plasticizer range alcohols are marketed in the U.S. by Shell and Exxon. Both pure materials and mixtures are available. Shell produces odd-chain C9–C11 alcohols with 80% linearity. Exxon produces highly branched even- and odd-chain C6–C13 alcohols. 2-Ethyl hexanol is produced and marketed in the U.S. by BASF, Union Carbide, and Eastman.

HEALTH AND SAFETY FACTORS

The higher alcohols are among the less toxic of commonly used industrial or household chemicals. (Toxic effects generally decline as chain length increases.) The acute oral toxicity data on the higher alcohols indicate a low order of toxicity by the oral, dermal, or inhalation routes of exposure. Higher alcohols are poorly absorbed through the skin. The rate of dermal uptake for the neat material was found to decrease with increasing carbon number. Human repeated skin patch tests with a 1% alcoholic solution of C12–C15 alcohols indicate very slight to mild irritation which is nonfatiguing or sensitizing. Primary human skin irritation of C16 and C18 alcohols is nil; as these products have been historically used in cosmetic and personal care products. Inhalation hazard is slight; however, avoid sustained breathing of alcohol vapor or mist, to minimize any aspiration hazard.

USES

Detergent-Range Alcohols

Major classes of applications include surfactants, lubricants, cosmetics and personal care, and pharmaceutical and medical products. Surfactants and lubricants tend to use alcohol derivatives, while the other applications tend to use the alcohol as is. Alcohol ethoxylates have better compatibility with complex enzymes in liquid laundry products and are superior for cleaning body oil stains. The ether sulfates are mild for dish applications. Both surfactants have better tolerance for hard-water ions than any anionic surfactant available.

Very low levels of specialty alcohol ethoxylates are used as emulsifiers in cleansing cremes and a few other personal care products. Alcohol ethoxylates are also used in textile processing and metal degreasing. Alcohol sulfates and ether sulfates are used in hard surface, rug and upholstery cleaners, shampoos, bubble baths, toilet soaps, and other personal care products. Industrial uses include emulsion polymerization and agricultural emulsifiers. Alkyl glyceryl ether sulfonates and alkylpolyglucosides are small-volume, specialty surfactants. The former is used as a foam-boosting surfactant in light-duty liquid, shampoos, and combination soap-synthetic toilet bars. The latter is a nonionic surfactant with good solubility, foaming, and mildness and can be used in laundry, light-duty liquids, and personal care products. Fatty nitrogen derivatives include fatty amine oxides, ether amines, dialkyldimethylammonium quaternaries, and alkylbenzyldimethylammonium chlorides. Fatty amine oxides are surfactants used in light-duty dishwashing liquids, household cleaners, personal care products, and a few specialized industrial applications.

Polymethacrylate esters are employed in automotive and aircraft lubricating oils, as well as transmission and hydraulic fluids. Oligoesters of fatty alcohol and a pyrometallic anhydride are useful for waterproofing leather An aqueous belt lubricant composition is based on fatty alcohol polyglycerol ethers. A lubricant for cold working of metals is based on monoalkyl ethers of polyethylene glycol and fatty alcohols. Lubricating and antifriction characteristics of water-based drilling fluids are improved by additives such as fatty alcohol or esterified or ethoxylated fatty alcohols. A corrosion inhibitor for gas pipelines and other steel surfaces is made by adding higher aliphatic alcohols to a mixture containing butanol, urea, various surfactants, and amines.

C12, C14, and C16 alcohols are used in perfumes and fragrances. C18 alcohol is used in USP ointments and has been approved as a direct and indirect food ingredient and in over-the-counter drugs. In general, short-chain alcoholsunder about C16 tend to be irritants. Guerbet alcohols (2 alkyl-alcohols) and alkyl alkanoates are commonly uses for personal care applications. The longer-chain alcohols. generally greater than C24, have benefits as anti-inflammatories and antiviral agents.

Plasticizer Alcohols

Plasticizer alcohols are either linear or branched C6 to C13 molecules (excluding C12 and C13 linears, which are detergent range). The plasticizer alcohols are used primarily in plasticizers, but they also have applications in a wide range of industrial and consumer products. As with the detergent range alcohols, the plasticizer alcohols are mainly consumed as derivatives. A number of surfactants are made from the plasticizer-range alcohols. Ethoxylated mixtures of C9–C11 linear alcohols are used in hard surface cleaners and commercial detergents. Increased solubility and liquidity provide advantages over longer chain alcohols. Ethoxylates are also used for processing textiles, leather, pulp, and paper. C6–C10 and C8–C10 blends are used for ethoxylated/propoxylated/phosphated surfactants and C6–C10 sulfates and betaines. Automatic dishwasher formulations use propoxylated and ethoxylated linear alcohols. Ethoxylated tridecyl alcohol is used in nonionic surfactants, and the ether sulfate is used in a leading children's shampoo. 2-Ethyl hexanol is used as an additive in dispersing and wetting agents for pigment pastes, and as a sulfosuccinate molecule for wetting and scouring of textiles.

Plasticizer alcohols are used as solvents or as intermediates in the manufacture of insecticides and herbicides.

2-Ethyl hexanol and various amine derivatives are used as a feedstock in the manufacture of extractants for heavy metals. 2-Ethyl hexanol is also used as a direct solvent for defoaming in the paper, textile, and oil field industries; as low-volatility ingredient in solvent blends for dyestuffs and coatings; as an ingredient in solvent compositions to clean soil from dirty articles; as a flow and gloss improver in baking finishes; and its esters (with salicylic acid) have a variety of applications in the cosmetic, pharmaceutical, and food industries as flavor and fragrance chemicals.

M. Bockisch, *Fat & Oils Handbook*, AOCS Press, Champaign, 1998, pp. 803–808.

R. Modler, *CEH Marketing Research Report: Detergent Alcohols*, SRI International, Menlo Park, Calif., 2000.

J. A. Monick, *Alcohols, Their Chemistry, Properties and Manufacture*, Reinhold Book Corp., New York, 1968, pp. 189–193.

J. Roberts and M. Caserio, *Basic Principles of Organic Chemistry*, 2nd ed., Benjamin, New York, 1977, pp. 612–645.

Corey J. Kenneally
Procter & Gamble Co.

ALCOHOLS, HIGHER ALIPHATIC, SYNTHETIC PROCESSES

Higher aliphatic alcohols (C_6–C_{18}) are produced in a number of important industrial processes using petroleum-based raw materials. eg, These processes are summarized in Table 1, as are the principal synthetic products and most important feedstocks (qv).

By far the largest volume synthetic alcohol is 2-ethylexanol, $C_8H_{18}O$, used mainly in production of the poly(vinyl chloride) plasticizer bis(2-ethylhexyl) phthalate, $C_{24}H_{38}O_4$, commonly called dioctyl phthalate or DOP (see PLASTICIZERS). A number of other plasticizer primary alcohols in the C_6–C_{11} range are produced, as are large volumes of C_{10}–C_{18} synthetic, mainly primary, alcohols used as intermediates to surfactants (qv) for detergents. Other lower volume synthetic alcohol application areas include solvents and specialty esters.

E. R. Freitas and C. R. Gum, *Chem. Eng. Prog.*, 73 (Jan. 1979).

J. A. Monick, *Alcohols, Their Chemistry, Properties, and Manufacture*, Reinhold, New York, 1968.

K. Noweck and H. Ridder, *Ullmann's Encyclopedia of Industrial Chemistry*, 5th ed., VCH Verlagsgesellschaft mbh, Weinheim, Germany, 1987.

E. J. Wickson and H. P. Dengler, *Hydrocarbon Process.* **51**(11), 69 (1972).

John D. Wagner
George R. Lappin
J. Richard Zietz
Ethyl Corporation

Table 1. Synthetic Industrial Processes for Higher Aliphatic Alcohols

Process	Feedstock(s)	Principal products	Worldwide capacity, millions of tons
Ziegler (organoaluminum)	ethylene, triethylaluminum	primary C_6–C_{18} linear alcohols	0.3
oxo (hydroformylation)	olefins based on ethylene, propylene, butylene, or paraffins	primary alcohols	4.2
aldol	*n*-butyraldehyde	2-ethylhexanol	[a]
paraffin oxidation	paraffin hydrocarbons	secondary alcohols	0.2
Guerbet	lower primary alcohols	branched primary alcohols	[b]
Total			4.7

[a] Included in oxo process total.
[b] Less than 0.05.

ALCOHOLS, POLYHYDRIC

Polyhydric alcohols or polyols contain three or more CH_2OH functional groups. The monomeric compounds have the general formula $R(CH_2OH)_n$, where $n = 3$ and R is an alkyl group or $C.CH_2OH$; the dimers and trimers are also commercially significant.

Each polyhydric alcohol is a white solid, ranging from the crystalline pentaerythritols to the waxy trimethylol alkyls. The trihydric alcohols are very soluble in water, as is ditrimethylolpropane. Pentaerythritol is moderately soluble and dipentaerythritol and tripentaerythritol are less soluble. Table 1 lists the physical properties of these alcohols. Pentaerythritol and trimethylolpropane have no known toxic or irritating effects. Finely powdered pentaerythritol, however, may form explosive dust clouds at concentrations above 30 g/m^3 in air. The minimum ignition temperature is 450°C.

REACTIONS

Direct acetylation of pentaerythritol using acetic acid in aqueous solution or in toluene produces a mixture of acetates which can be fairly readily separated by chromatographic methods or distillation.

Long-chain esters of pentaerythritol have been prepared by a variety of methods.

Polyhydric alcohol mercaptoalkanoate esters are prepared by reaction of the appropriate alcohols and thioester using p-toluenesulfonic acid catalyst.

Pentaerythritol can be oxidized to 2,2-bis(hydroxymethyl) hydracrylic acid, $C_5H_{10}O_5$, Bromohydrins can be prepared directly from polyhydric alcohols. Borolane products of mixed composition can be synthesized by direct addition of boric acid to pentaerythritol.

Reaction between pentaerythritol and phosphorous trichloride yields the spirophosphite, 3,9-dichloro-2,4, 8,10-tetraoxa-3,9,-diphosphaspiro[5,5]-undecane, $C_5H_8Cl_2O_4P_2$.

The commercially important explosive pentaerythritol tetranitrate (PETN), $C_5H_8N_4O_{12}$, is produced by direct reaction of pentaerythritol in nitric or nitric–sulfuric acid media.

Aminoalkoxy pentaerythritols are obtained by reduction of the cyanoethoxy species obtained from the reaction between acrylonitrile, pentaerythritol, and lithium hydroxide in aqueous solution.

Tosylates of pentaerythritol and the higher homologues can be converted to their corresponding tetra-, hexa-, or octaazides by direct reaction of sodium azide, and azidobenzoates of trimethylolpropane and dipentaerythritol are prepared by reaction of azidobenzoyl chloride and the alcohols in pyridine medium.

Pentaerythritol can be converted to the biscyclic formal, 2,4,8,10-tetra-oxaspiro[5,5]undecane, $C_7H_{12}O_4$, by heating in the presence of formaldehyde or paraformaldehyde and an acid catalyst.

Simple alkyl and alkenyl ethers of pentaerythritol are produced on direct reaction of the polyol and the required alkyl or alkenyl chloride in the presence of quaternary alkylamine bromide.

MANUFACTURE

Pentaerythritol is produced by reaction of formaldehyde and acetaldehyde in the presence of a basic catalyst, generally an alkali or alkaline-earth hydroxide.

Dipentaerythritol and tripentaerythritol are obtained as by-products of the pentaerythritol process and may be further purified by fractional crystallization or extraction. Trimethylolethane and trimethylolpropane may be prepared using appropriate aldehyde in place of acetaldehyde. Ditrimethylolpropane is obtained as a by-product of the trimethylolpropane synthesis.

HEALTH AND SAFETY FACTORS

Pentaerythritol and trimethylolpropane are classified as nuisance particulate and dust, respectively. They are both nontoxic to animals by ingestion or inhalation and are essentially nonirritating to the skin or eyes.

USES

The most important industrial use of pentaerythritol is in a wide variety of paints, coatings, and varnishes, where the cross-linking capability of the four hydroxy groups is critical.

Table 1. Physical Properties of Polyhydric Alcohols

Property	Pentaerythritol	Dipentaerythritol	Tripentaerythritol	Trimethylolethane	Trimethylolpropane	Ditrimethylolpropane[a]
molecular formula	$C_5H_{12}O_4$	$C_{10}H_{22}O_7$	$C_{15}H_{32}O_{10}$	$C_5H_{12}O_3$	$C_6H_{14}O_3$	$C_{12}H_{26}O_5$
melting point, °C	261–262	221–222.5	248–250	202	58.8	112–114
boiling point, °C	276 (4 kPa)			283	289	210 (0.12 kPa)
solubility, g/100 g water at 25°C	7.23	0.28[b]	0.018[b]	soluble	soluble	2.6
density, g/mL	1.396	1.369	1.30		1.09	1.18

[a] Data supplied by Perstorp AB.
[b] Estimated value.

The explosives and rocket fuels formed by nitration of pentaerythritol to the tetranitrate using concentrated nitric acid are generally used as a filling in detonator fuses.

Pentaerythritol is used in self-extinguishing, nondripping, flame-retardant compositions with a variety of polymers, including olefins, vinyl acetate and alcohols, methyl methacrylate, and urethanes. Polymer compositions containing pentaerythritol are also used as secondary heat-, light-, and weather-resistant stabilizers with calcium, zinc, or barium salts, usually as the stearate, as the prime stabilizer. The polymers may be in plastic or fiber form.

Pentaerythritol in rosin ester form is used in hot-melt adhesive formulations, especially ethylene–vinyl acetate (EVA) copolymers, as a tackifier.

E. Berlow, R. H. Barth, and E. J. Snow, *The Pentaerythritols*, Reinhold Publishing Corp., New York, 1958.

WILLIAM N. HUNTER
Celanese Canada Inc.

ALDEHYDES

Aldehydes are carbonyl-containing organic compounds of the general formula RCH=O. The R group represents an aliphatic, aromatic, or heterosubstituted radical, except in formaldehyde, where R represents hydrogen. Aldehydes are inherently reactive compounds. The carbonyl group is susceptible to both oxidation and reduction, yielding acids and alcohols, respectively. Additionally, the carbonyl group is susceptible to nucleophilic addition, providing a means by which to form new chemical bonds. Furthermore, the presence of the carbonyl activates the hydrogens bound to the alpha carbon and thus provides an additional site of reactivity. Ketones are a related class of compounds having two alkyl groups attached to the carbonyl group $R_1R_2C=O$.

NOMENCLATURE

The common method of naming aldehydes corresponds very closely to that of the related acids, in the sense that the term "aldehyde" is added to the base name of the acid. eg, formaldehyde comes from formic acid, acetaldehyde from acetic acid, and butyraldehyde from butyric acid. If the compound contains more than two aldehyde groups, or is cyclic, the name is formed using "carbaldehyde". The IUPAC system of aliphatic aldehyde nomenclature is derived by replacing the final -e from the name of the parent acyclic hydrocarbon by the suffix -al. If two aldehyde functional groups are present, the suffix -dial is used. The prefix formyl is used with polyfunctional compounds.

PHYSICAL AND CHEMICAL PROPERTIES

The C_1 and C_2 carbon aliphatic aldehydes, formaldehyde and acetaldehyde are gases at ambient conditions whereas the C_3 (propanal) through C_{11} (undecanal) aldehydes are liquids, and higher aldehydes are solids at room temperature. The presence of hydrocarbon branching tends to lower the boiling or melting point, as does unsaturation in the carbon skeleton. Generally, an aldehyde has a boiling point between those of the corresponding alkane and alcohol. Aldehydes are usually soluble in common organic solvents and, except for the C_1 to C_5 aldehydes, are only sparingly soluble in water. The lower, C_1 to C_8, aldehydes have pungent, penetrating, unpleasant odors, some of which may be attributed to the presence of the corresponding acids that readily form by air oxidation. Above C_8, aldehydes have more pleasant odors in their diluted state, and some higher aldehydes are used in the perfume and flavoring industries. The C_9 aldehyde, nonanal, is reported to possibly be a human sex pheromone. Aldehydes must be kept from contact with air to retain purity.

Aldehydes are very reactive compounds. Reactions generally fall into two classes: those directly involving the carbonyl group and those occurring at the adjacent carbon atom. The polar nature of the carbonyl group (RCH—O) lends itself to nucleophilic addition, reduction and oxidation, and also affects the reactivity of its adjacent carbon atom by rendering its hydrogens relatively acidic. Aldehydes having acidic hydrogens must be protected against inadvertent contact with bases, as such contact may result in an exothermic condensation reaction that may become dangerous. Aldehydes can be reduced to the corresponding alcohols by catalytic hydrogenation using heterogeneous as well as homogeneous catalysts.

MANUFACTURE

Only a few are used on industrial scale. One important industrial process is by hydroformylation of olefins using synthesis gas and transition metal catalysts (oxo synthesis). Current commercial processes employ ligand-modified metal catalysts, which operate under milder conditions and provide better control of linear and branched aldehyde selectivities.

The direct oxidation of ethylene is used to produce acetaldehyde in the Wacker-Hoechst process. Another commercial aldehyde synthesis is the catalytic dehydrogenation of primary alcohols at high temperature in the presence of a copper or a copper-chromite catalyst. Although there are several other synthetic processes employedtend to be smaller-scale reactions.

Formaldehyde is produced in the largest volume and has comparable economic value to that of butyraldehyde.

CHARACTERIZATION

Aldehydes can be characterized qualitatively through the use of Tollens's or Fehling's reagents as well as by spectroscopic means. Additionally, aldehyde carbonyl groups can be derivatized. Hydrazone derivatives are often useful in isolating the aldehyde as a solid, crystalline material. The carbonyl group may also be oxidized to an acid.

HEALTH AND SAFETY FACTORS

Interest in the toxicity of aldehydes has focused primarily on formaldehyde, acetaldehyde, and acrolein. Little evidence exists to suggest that occupational levels of exposure to aldehydes would result in mutations, although some aldehydes are clearly mutagenic. There are, however, acute effects of aldehydes. Low-molecular-weight aldehydes, the halogenated aliphatic aldehydes, and unsaturated aldehydes are particularly irritating to the eyes, skin, and respiratory tract. The mucous membranes of nasal and oral passages and the upper respiratory tract can be affected, producing a burning sensation, an increased ventilation rate, bronchial constriction, choking, and coughing. If exposures are low, the initial discomfort may abate after 5–10 min but will recur if exposure is resumed. Furfural, the acetals, and aromatic aldehydes are much less irritating than formaldehyde and acrolein. Reports of sensitization reactions to formaldehyde are numerous.

Materials that have unquestionable anesthetic properties are chloral hydrate, paraldehyde, dimethoxymethane, and acetaldehyde diethyl acetal. The small quantities that can be tolerated by inhalation are usually metabolized so rapidly that no anesthetic symptoms occur.

The principal pathology experimentally produced in animals exposed to aldehyde vapors is that of damage to the respiratory tract and pulmonary edema. In general, the aldehydes are remarkably free of actions that lead to definite cumulative organic damage to tissues. Thus the aldehydes cannot generally be regarded as potent carcinogens. There is a significantly increased toxicological effect between saturated and unsaturated aliphatic aldehydes. The precautions for safely handling reactive unsaturated aldehydes are the same as for other highly active eye and pulmonary irritants.

Consult the Material Safety Data Sheets for details about individual compounds. Precautions for the higher aldehydes are essentially those for most other reactive organic compounds: adequate ventilation, fire and explosion precautions, and proper instruction in use of respiratory, eye, and skin protection.

USES

Aldehydes find the most widespread use as chemical intermediates. The production of acetaldehyde, propionaldehyde, and butyraldehyde as precursors of the corresponding alcohols and acids are examples. The aldehydes of low molecular weight are also condensed in an aldol reaction to form derivatives that are important intermediates for the plasticizer industry. 2-ethylhexanol, produced from butyraldehyde, is used in the manufacture of di (2-ethylhexyl) phthalate. Aldehydes are also used as intermediates for the manufacture of alcohols and ethers, resins, and dyes. Isobutyraldehyde is used as an intermediate for production of primary solvents and rubber antioxidants. Fatty aldehydes C_8 to C_{13} are used in nearly all perfume types and aromas. Polymers and copolymers of aldehydes exist and are of commercial significance.

The class of compounds known as acetals has been used in racing car fuels, gasoline additives, and paint and varnish solvents and hemiacetal acetal strippers. Because they are not as sensitive to alkalies or autoxidation, acetals find use as fragrances for alkaline formulations such as soaps, shampoos and heavy-duty detergents.

B. Cornils and W. A. Herrmann, eds., *Applied Homogeneous Catalysis with Organometallic Compounds*, Vol. 1. VCH Publishers, New York, 1996, pp. 29–90.

G. Hilgetag and A. Martini, eds., *Preparative Organic Chemistry*, 4th ed., Wiley-Interscience, New York, 1972, pp. 301–400.

H. O. House, *Modern Synthetic Reactions*, 2nd ed., W. A. Benjamin, Inc., Menlo Park, Calif., 1972.

S. Patai, *The Chemistry of the Carbonyl Group*, Wiley-Interscience, New York, 1966 and 1970.

ANTHONY G. ABATJOGLOU
DAVID J. MILLER
Union Carbide Corporation

ALKALOIDS

There is currently no simple definition of what is meant by alkaloid. Most practicing chemists working in the field would agree that most alkaloids, in addition to being products of secondary metabolism, are organic nitrogen-containing bases of complex structue, occuring for the most part in seed-bearing plants and having some physiological activity. The most recent catalog, listing nearly 10,000 alkaloids, contains compounds including not only nonbasic nitrogen-containing materials from plants but also substances occurring in animals. Because of their widespread distribution across all forms of life, alkaloids are intimately interwoven into the fabric of existence. Both our understanding of the roles these substances play in their respective sources and the possibility of genomic modification to adjust alkaloid production are just now being pursued.

OCCURRENCE, DETECTION, AND ISOLATION

The most recent compendium of alkaloids indicates that most of those so far detected occur in flowering plants and that the highest concentrations of them are probably to be found there, but it is almost certain that some concentration of alkaloids will be found almost everywhere. In the higher plant orders, somewhat more than one-half contain alkaloids in easily detected concentrations. Major alkaloid-bearing orders are *Campanulales, Centrospermae, Gentianales, Geraniales, Liliflorae, Ranales, Rhoedales, Rosales, Rubiales, Sapindales*, and *Tubiflorae*. Alkaloids have also been found in butterflies, beetles, millipedes, and algae and are known to be present in fungi, eg, agroclavine. They are found in toads (eg, bufotenine, an established hallucinogen in humans); in frogs, and in the musk deer.

The concentration of alkaloids, as well as the specific area of occurrence or localization within the plant or animal, can vary enormously. Thus the amount of nicotine apparently synthesized in the roots of various species of Nicotiana and subsequently translocated to the leaves varies with soil conditions, moisture, extent of cultivation, season of harvest, and other factors and may be as high as 8% of the dry leaf, whereas the amount of morphine in cerebrospinal fluid is of the order of 2–339 fmol/mL.

After detection of a presumed alkaloid, large quantities of the specific plant material are collected, dried, and defatted by petroleum ether extraction if the seed or leaf is investigated. This process usually leaves polar alkaloidal material but removes neutrals. The residue is extracted with dilute acid and filtered, and the acidic solution is made basic. Crystallization can occasionally be effected by adjustment of the pH. Once alkaloidal material has been found, taxonomically related plant material is also examined.

With the advent of, first, column, then preparative thin layer, and now high-pressure liquid chromatography, even very low concentrations of materials of physiological significance can be obtained in commercial quantities.

Most recently, with the advent of enzyme assay and genomic manipulation, the possiblity of utilization of callous or root tissue or even isolated enzymes along with genetic engineering techniques can be employed to enhance or modify production of specific alkaloids.

PROPERTIES

Most alkaloids are basic and they are thus generally separated from accompanying neutrals and acids by dilute mineral acid extraction. The physical properties of most alkaloids, once purified, are similar. Thus they tend to be colorless, crystalline, with definite melting points. However, among >10,000 individual compounds, these descriptions are overgeneralizations and some alkaloids are not basic, some are liquid, some brightly colored, some achiral, and in a few cases both enantiomers have been isolated in equal amounts; ie, the material as derived from the plant is racemic (or racemization has occurred during isolation).

ORGANIZATION AND USES

The building blocks of primary metabolism, from which biosynthetic studies have shown the large majority of

alkaloids to be built, are few and include the common amino acids ornithine, lysine; phenylalanine, tyrosine and tryptophan. Others are nicotinic acid, anthranilic acid, and histidine, and the nonnitrogenous acetate-derived fragment mevalonic acid. Mevalonic acid is the progenitor of isopentenyl pyrophosphate; its isomer 3,3-dimethylallyl pyrophosphate. A dimeric C_5 fragment (the C_{10} fragment) gives rise to the iridoid loganin and the trimer farnesyl pyrophosphate. The C_{15} fragment is also considered the precursor to the C_{30} steroid.

The physiologically and commercially important alkaloids of the ornithine-derived group of compounds, occurring widely in the Solanaceae and Convolulaceae as well as the Erythroxylaceae, include not only cocaine but also atropine and scopolamine.

Atropine, isolated from the deadly nightshade (*Atropa belladonna* L.) is the racemic form. Atropine is used to dilate the pupil of the eye in ocular inflammations and is available both as a parasympatholytic agent for relaxation of the intestinal tract and to suppress secretions of the salivary, gastric, and respiratory tracts. In conjunction with other agents, it is used as part of an antidote mixture for organophosphorus poisons. Scopolamine, an optically active, viscous liquid, also isolated from Solanaceae (eg, *Datura metel* L.) decomposes on standing and is thus usually both used and stored as its hydrobromide salt. The salt is employed as a sedative or, less commonly, as a prophylactic for motion sickness. It also has some history of use in conjunction with narcotics, as it appears to enhance their analgesic effects. Biogenetically, scopolamine is clearly an oxidation product of atropine, or, more precisely, because it is optically active, of ($\sim$)-hyoscyamine.

Cocaine had apparently been used by the natives of Peru prior to the European exploration of South America. Early explorers suggested that the leaves of, for example, *Erythroxylon coca* Lam. were chewed without apparent addiction by the indigenous peoples and with only mild numbing of the lips and tongue in return for increased endurance. Although it appears that the native populations, the practice of leaf chewing, the purified base obtained by simple extraction of the leaves has become a substance of abuse in the more civilized world. It is now recognized that the alkaloid itself is too toxic to be used as an anesthetic by injection.

In addition to the alkaloids in *Senecio* spp. (including asters and ragworts), members of this widely spread group of compounds are found in different genera (*Heliotropium, Trachelanthus*, and *Trichodesma*) within cosmopolitan families (eg, Boraginaceae and Leguminoseae). Most of these alkaloids are toxic, affecting the liver, and their ingestion is manifested in animals with the onset of symptoms associated with names such as the "horse staggers" or "walking disease". Isolates from Indian tobacco (Lobelia inflata L.), as a crude mixture of bases, have been recognized as expectorants. The same (or similar) fractions were also used both in the treatment of asthma and as emetics. The spores of *Lycopodium calvatum* L. (a club moss), sometimes called vegetable sulfur, have been used medicinally as an absorbent dusting powder; other uses as diverse as additives to gunpowder and suppository coatings have also been recorded. Only one pelletierine and, separately, a second acetoacetate and a second 1-dehydropiperidine, which could otherwise be combined to a second pelletierine, are used to generate both of the lycopodine and annotinine alkaloids.

The relatively small number of alkaloids derived from nicotinic acid (the tobacco alkaloids) are obtained from plants of significant commercial value and have been extensively studied. They are distinguished from the bases derived from ornithine and, in particular, lysine. These alkaloids include the substituted pyridone ricinine, $C_8H_8N_2O_2$, which is easily isolated in high yield as the only alkaloid from the castor bean (*Ricinus communis* L.). The castor bean is also the source of castor oil, which is obtained by pressing the castor bean and, rich in fatty acids, has served as a gentle cathartic.

The highly toxic alkaloid S-(−)-nicotine and related tobacco bases, including such materials as (−)anabasine, asine, $C_{10}H_{14}N_2$, are obtained from commercially grown tobacco plants. Various tobaccos have differing amounts of these and other bases, as well as different flavoring constituents, some of which are apparently habituating to some individuals. Currently, the assay of the (−)-nicotine content of tobacco, the annual world production of which is in excess of 7 million tons, is desirable and in some countries mandatory, although the toxicity of the unassayed plant bases may be as high as or higher than that of (−)-nicotine. There appears to be some evidence that cultivation of tobacco increases its alkaloid content, from which it can be argued that increased alkaloid content has insured survival of a particular cultivar. Millions in the Far East are apparently addicted to chewing ground betel nut. Among the alkaloids found in betel nut is arecoline, $C_8H_{13}NO_2$, an optically inactive, steam-volatile base that is used commercially as a vermifuge in dogs and is also a potent muscarinic agent. Arecoline may be derived from nicotinic acid by a rare reductive mechanism.

The importance of morphine as an analgesic, despite the danger of addiction and side effects that include depression of the central nervous system, slowing of respiration, nausea, and constipation, cannot be underestimated, and significant efforts have been expended to improve isolation techniques from crude dried opium extract. Depending on its source, the morphine content of poppy straw or dried exudate may be as high as ~20%. Although the details of current manufacturing processes are closely held secrets, early work has probably not been modified extensively. Usually, the crude opium is extracted with water and filtered, and the aqueous extract concentrated, mixed with ethanol, and made strongly basic with ammonium hydroxide. Morphine usually precipitates, while the other bases remain in solution, and is further purified by crystallization as its sulfate.

Codeine occurs in the opium poppy along with morphine but usually in much lower concentration. Because it is less toxic than morphine and, because its side effects are less marked, it has found widespread use in the treatment of minor pain. Much of the morphine found in crude opium is converted to codeine. The commercial conversion of morphine to codeine makes use of a variety of methylating agents, among which the most common are trimethylphenylammonium salts. In excess of 200 tons of codeine are consumed annually from production facilities scattered around the world.

The structures of the brightly colored (red-violet and yellow) alkaloids found in the order *Centrospermae* (cacti, red beet, etc), called betacyanins or betaxanthins, are relatively unstable water-soluble zwitterions. All betacyanins or betaxanthins may simply be imine derivatives (with the appropriate amino acid) of betalamic acid. There are a few simple indole derivatives derived, from tryptophan. Serotonin (5-hydroxytryptamine, was first isolated as a vasoconstrictor substance from beef serum and shown to be derived from tryptophan. The only slightly more complicated base harmine is found widely distributed in the *Leguminoseae* and *Rubiaceae*, extracts of which were at one time used therapeutically against tremors in Parkinson's disease. The seeds of the African rue, *Peganum harmala* L., which are rich in harmine and related alkaloids, have also been used as a tapeworm remedy.

Ingestion of contaminated flour infected by the parasitic fungus *Claviceps purpurea* results in the disease called ergotism (St. Anthony's Fire). Convulsive ergotism causes violent muscle spasms that bend the sufferer into otherwise unattainable positions and frequently leaves physical and mental scars; agroclavine and derivatives of lysergic acid are considered responsible. There are currently two medicinally valuable alkaloids of commercial import obtained from ergot. Commercial production involves generation parasitically on rye in the field or production in culture because a commercially useful synthesis is unavailable. The common technique today is to grow the fungus in submerged cult. Destruction of the aluminum complex with ammonia then permits hydrocarbon extraction of the alkaloid. Ergotamine is obtained from crude extract by formation of an aluminum complex. Reserpine is currently used as a hypotensive and acts as a sedative that reduces aggressiveness. At higher doses, reserpine has been reported to cause depression as well as peptic ulceration. There is some evidence that chronic administration in women results in an increased incidence of breast cance. The total synthesis of reserpine was a landmark synthesis.

The synthesis of strychnine was a truly monumental undertaking. Strychnine, although only moderately toxic when compared to other poisons, both naturally occurring and produced synthetically, probably owes its reputation to its literary use and its having had some use as a rodenticide. Poisoning is manifested by convulsions, and death apparently results from asphyxia. As little as 30–60 mg has been reported as fatal to humans, although at lower dosage it has received some medical use as an antidote for poisoning by central nervous system depressants, as a circulatory stimulant, and in treatment of delirium tremens. The useful medicinal dosage is normally <4 mg.

Recently, quinine has again become the treatment of choice for malaria as Plasmodium falciparum resistant to other drugs developed. Apperently, resistance to quinine is more difficult for the rapidly changing parasite population to acquire. Nonetheless, because quinine is not a prophylactic drug but rather a material which suppresses the overt manifestations of malaria, work continues on better treatment.

In addition to the alkaloids such as cyclobuxine-D and solanidine where the structural similarities to steroids are clear, there are the less obvious (but nonetheless also clearly related) Veratrum alkaloids. These compounds, of which protoveratrine A, $C_{41}H_{63}NO_{14}$, obtained from the rhizome of *Veratrum album* L. (Liliaceae), is a typical example, produce dramatic declines in blood pressure on administration and have been received by the medical community as good antihypertensive agents. Generally, however, the dosage must be individualized (slowly) from ~2 mg in 200 mL of saline upward. Because the therapeutically valuable dosage is similar to the toxic dose, and even nonlethal large doses may cause cardiac arrhythmias and peripheral vascular collapse, use of these compounds has frequently been limited to extreme cases where close attention can be accorded the patient.

In purine alkaloids the purine skeleton is not derived from histidine, as might be imagined, nor is it derived from any obvious amino acid progenitor. The nucleus common to xanthine, $C_5H_9N_4O_2$, and found in the bases of caffeine, theophylline, and theobromine $C_7H_8N_4O_2$, is created from small fragments that are attached to a ribosyl unit during synthesis and can presumably be utilized in a nucleic acid backbone subsequently. All three alkaloids—caffeine, theophylline, and theobromine occur widely in beverages commonly used worldwide.

The leaf and leaf buds of *Cammelia sinensis* (L.) O Kuntze and other related plants and most teas contain, depending on climate, specific variety, time of harvest, etc., somewhat <5% caffeine and smaller amounts of theophylline and theobromine. Coffee consists of various members of the genus *Coffea*, although the seeds of *Coffea arabica* L., believed to be indigenous to East Africa, are thought to have been the modern progenitor of the varieties of coffees currently available and generally cultivated in Indonesia and South America. The seeds contain less than ~3% caffeine which, bound to other agents, is set free during the roasting process. The caffeine may be sublimed from the roast or extracted with a variety of agents, such as methylene chloride, ethyl acetate, or dilute acid (eg, an aqueous solution of carbon dioxide) to generate decaffinated material. Supercritical extraction with carbon dioxide has also found to be useful. Two other commonly found sources of caffeine are kola (Cola) from the seeds of, eg, *Cola nitida* (Vent.) Schott and Engl., which contains 1–4% of the alkaloid, but little theophylline or theobromine, and cocoa (from the seeds of *Theobroma cacao* L.), which generally contains ~3% theobromine and significantly less caffeine.

All three of these materials are apparently central nervous system stimulants. It is believed that for most individuals caffeine causes greater stimulation than does theophylline. Theobromine apparently causes the least stimulation. There is some evidence that caffeine acts on the cortex and reduces drowsiness and fatigue, although habituation can reduce these effects.

Shikimic acid is a precursor of anthranilic acid and, in yeasts and *Escherichia coli* (a bacterium), anthranilic acid (o-aminobenzoic acid) is known to serve as a precursor of tryptophan. A similar but yet unknown path is presumed to operate in higher plants. Nonetheless, anthranilic acid itself is recognized as a precursor to a number of alkaloids. Thus damascenine, $(C_{10}H_{13}NO_3)$, from the seed coats of *Nigella damascena* has been shown to incorporate labeled anthranilic acid when unripe seeds of the plant are incubated with labeled precursor.

Coniine, implicated by Plato in the death of Socrates, is the major toxic constituent of *Conium maculatum* L. (poison hemlock) and apparently the first alkaloid to be synthesized. For years it was thought that coniine was derived from lysine, as were many of its obvious relatives containing reduced piperidine nuclei and a side chain, eg, pelletierine. However, it is now known that coniine is derived from a polyketooctanoic acid, $C_8H_{10}O_5$, or some other similar straight-chain analogue.

ECONOMIC ASPECTS

There are four broad classes of alkaloids whose general economic aspects are important: (1) the opiates such as morphine and codeine (2) cocaine (both licit and illicit); (3) caffeine and related bases in coffee and tea, and (4) the tobacco alkaloids such as nicotine.

N. C. Bruce in G. A. Cordell, ed., *The Alkaloids, Chemistry and Biology*, Vol. 57, Academic Press, New York, 2001, pp. 1–74.

J. S. Glasby, *Encyclopedia of the Alkaloids*, Vols. 1–4, Plenum Press, New York, 1975.

R. H. F. Manske and H. L. Holmes, eds., *The Alkaloids: Chemistry and Physiology*, Vol. 1, Academic Press, Inc., New York, 1950. This series gives a detailed exposition of the chemistry and pharmacology of the alkaloids, by structural class. Vol. 57, G. A. Cordell, ed. was published in 2001.

M. F. Roberts, in M. F. Roberts and M. Wink, eds., *Alkaloids, Biochemistry, Ecology and Medicinal Applications*, Plenum Press, New York, 1998, pp. 109–146.

DAVID R. DALTON
Temple University
LINDA M. MASCAVAGE
Arcadia University
MICHAEL WILSON
Wilson Technologies, Ltd.

ALKANOLAMINES FROM NITRO ALCOHOLS

The nitro alcohols, obtained by the condensing nitroparaffins with formaldehyde, may be reduced to a unique series of alkanolamines (β-amino alcohols). The condensation may occur one to three times, depending on the number of available hydrogen atoms on the α-carbon of the nitroparaffin, giving rise to amino alcohols with one to three hydroxyl groups.

Many members of this series are known, based on nitroparaffin condensations with aldehydes of longer chain length than formaldehyde. However, only the five primary amino alcohols discussed in the following are manufactured on a commercially-significant scale. N-Substituted derivatives of these compounds also have been prepared, but only 2-dimethylamino-2-methyl-1-propanol is available in commercial quantities.

PHYSICAL PROPERTIES

Because 2-amino-2-methyl-1-propanol (AMP) and 2-amino-2-ethyl-1,3-propanediol (AEPD) melt near room temperature and usually contain some water, they may be semisolid pastes, rather than crystalline solids. Water-diluted forms of both these alkanolamines are marketed because such solutions remain liquid at lower temperatures than do the pure compounds. These compounds are highly soluble in water.

AMP, AEPD, and others are very soluble in alchols, slightly soluble in aromatic hydrocarbons, and nearly insoluble in aliphatic hydrocarbons; TRIS(hydroxymethyl) aminomethane is appreciably soluble only in water.

Alkanolamines have high boiling points; under normal ambient conditions their vapor pressures are low. Current DOT regulations classify AMP, AMP-95, DMAMP, DMAMP-80, AEPD, and AB all as combustible liquids.

DMAMP and AMP are among the most strongly basic commercially-available amines. All alkanolamines have slight amine odors in the liquid state; the solid products are nearly odorless.

CHEMICAL PROPERTIES

The alkanolamines discussed here typically attack copper, brass, and aluminum but not steel or iron. Alkanolamines are useful as amination agents; however, the reactivity of both the amino and alcohol group must be considered in attempting any specific synthetic scheme with them.

With mineral acids, the alkanolamines form ammonium salts which hydrolyze readily in the presence of water and dissociate upon heating. Fatty acids give soaps which are highly efficient emulsifying agents with important industrial uses, particularly the soaps of AMP.

On heating, an alkanolamine soap first dehydrates to the amide; at higher temperature, the amide is dehydrated to an oxazoline.

These oxazolines have cationic surface-active properties and are emulsifying agents of the water-in-oil type. They are acid acceptors and, in some cases, corrosion inhibitors. Reactions of AMP with organic acids to create oxazoline functionality are useful as a tool for determination of double-bond location in fatty acids or for use as a carboxylic-acid protective group in synthesis. The oxazolines from AMPD, AEPD, and TRIS AMINO contain hydroxyl groups that can be esterified easily, giving waxes with saturated acids and drying oils with unsaturated acids.

Formaldehyde reacts with the hydrogen on the α-carbon of the fatty acid from which the oxazoline was formed to yield a vinyl monomer that can be polymerized or utilized for synthesis. These products are useful for modification of alkyd resins, preparation of paint vehicles, and copolymerization with other monomers.

Substitution on the amino group occurs readily, giving bases stronger than the parent amines. Alkanolamines react with nitro alcohols to form nitrohydroxylamines. Some of these compounds show antibacterial activity.

MANUFACTURE

The reduction of nitro alcohols to alkanolamines is readily accomplished by hydrogenation in the presence of Raney nickel catalyst. AMP, AEPD, and AB are purified by distillation, TRIS AMINO AND AMPD by crystallization. 2-Dimethylamino-2-methyl-1-propanol(markted primarily as DMAMP-80) is manufactured from AMP by hydrogenation in the presence of formaldehyde and purified by distillation.

Angus Chemical Company is the basic manufacturer of technical-grade TRIS AMINO. However, Angus and numerous processors offer recrystallized, higher purity grades of this alkanolamine for specialized applications.

HEALTH AND SAFETY FACTORS

Alkanolamines are only slightly toxic by ingestion. Undiluted DMAMP, AMP, and AB cause eye burns and permanent damage, if not washed out of the eye immediately. They are also severely irritating to the skin, causing burns upon prolonged or repeated contact. Of these three only AMP has been studied in subchronic and chronic oral studies. The principal effect noted was the action of AMP on the stomach as a result of its alkalinity. In general, the low volatility, and the applications for which these products are used, preclude the likelihood of exposure by inhalation.

AEPD is severely irritating to the eyes and should be washed out immediately but is only mildly irritating to the skin.

AMPD and TRIS AMINO, normally crystalline solids, are of less concern in terms of irritancy to skin and eyes.

In general, the toxicology of the alkanolamines is typical of alkaline materials, ie, the greater the base strength, the greater the effect. Neutralized alkanolamines are much less toxic.

Environmentally, these alkanolamines present little problem. AMP has been found to be degradable according to OECD guidelines, to be of low toxicity to fish and microorganisms, and to be nonaccumulative. TRIS AMINO is actually added to the water used for shiping living fish.

USES

Because they are closely related, the alkanolamines can sometimes be used interchangeably, but cost/performance considerations generally dictate a best choice for specific applications.

The fatty acid soaps of alkanolamines are excellent emulsification agents for use in functional fluids such as hydraulic and metalworking fluids. AMP, which enhances the chemical and biostablity of metal-working fluids, is widely used as a pigment co-dispersant in water-based paints and paper coatings. AMP is also the neutralizer/ solubilizer of the choice for use with acid-functional hair-fixative resins. TiO_2 and clay slurries utilizing AMP as part of the dispersant system are available in bulk for the paint and paper industries.

In general, water-soluble resins are amine salts of acidic polymers. Water-soluable coatings formulated with AMP-95 or DMAMP-80 exhibit superior performance.

The alkanolamines continue to find use in blocked-catalyst system for textile resins, coating resins, and adhesives.

When used in boiler water treatment, AMP provides excellent protection to condensate returns lines. The rapid formation at room temperature of oxazolidines from alkanolamines and formaldhyde provides a method for eliminating excess free formaldehyde from products such as urea–formaldehyde resins. AMP is useful for removing acid-gas contaminants such as CO_2 or H_2S from gas streams and has also been utilized in tertiary oil recovery to enhance removal of petroleum from marginal wells and been valuable in purification systems for fluid streams.

TRIS AMINO is used for a number of biomedical purpose. Together with ethylenediaminetetraacetic acid (EDTA) it can enhance the effectiveness of antimicrobial agents and antibiotics. AB, the only optically active alkanolamine in this series, is used as a raw material to produce the antituberclosis drug ethambutol.

Syntheticoxazolines has been utilized in many different applications. Oxazaline alkyd films are characterized by improved performance, particularly salt-spray resistance and gloss. Other oxazolines produced from alkanolamines are useful as oil-soluble surface-active agents and corrosion inhibitors. Synthetic oxazoline waxes promote lubricity and mar-resistance of coatings.

Oxazolidines, formed by reaction of alkanolamines with aldehydes, are useful as leather tanning agents and are effective curing agents for proteins, phenolic resins, moisture-cure urethanes, and as antimicrobial agents.

Applications of ANGUS AB in Chiral-Drug Synthesis, Benchmark No. 54, ANGUS Chemical Company, Buffalo Grove, Ill., August 1997.

Biotechnology Applications of TRIS AMINO (Tris Buffer), Technology Review No. 10, ANGUS Chemical Company, Buffalo Grove, Ill., March 1994.

M. J. Fountain and co-workers, *Proceedings of the International Water Conference, Engineering Society of Western Pennsylvania*, 45th, 1984, p. 393.

ALAN C. EACHUS
ALLEN F. BOLLMEIER
Angus Chemical Co.

ALKANOLAMINES FROM OLEFIN OXIDES AND AMMONIA

Ethylene oxide, propylene oxide, or butylene oxide react with ammonia to produce alkanolamines. Ethanolamines, $NH_{3-n}(C_2H_4OH)_n(C_2H_{OH})_n$ (n = 1,2,3, mono-, di-, and tri-), are derived from the reaction of ammonia with ethylene oxide. Isopropanolamines, $NH_{3-n}(CH_2CHOHCH_3)_n$ (mono-, di-, and tri-), result from the reaction of ammonia with propylene oxide. Secondary butanolamines, $NH_{3-n}(CH_2CHOHCH_2CH_3)_n$ (mono-, di-, and tri-), are the result of the reaction of ammonia with butylene oxide. Mixed alkanolamines can be produced from a mixture of oxides reacting with ammonia.

Ethanolamines have been commercially available for over 50 years and isopropanolamines, for over 40 years. *sec*-Butanolamines have been prepared in research quantities, but are not available commercially. Primary butanolamines, eg, 2-amino-1-butanol are made by a different chemical route.

A variety of substituted alkanolamines, shown in Table 1, are also available commercially, but have not reached the volume popularity of the ethanolamines and isopropanolamines.

PHYSICAL PROPERTIES

The freezing points of alkanolamines are moderately high, as shown in Table 2. The ethanolamines, monoisopropanolamine and mono-*sec*-butanolamine, are colorless liquids at or near room temperature. Di- and triisopropanolamine and di- and tri-*sec*-butanolamine are white solids at room temperature.

All the ethanolamines and isopropanolamines except monoisopropanolamine are available in low freezing grades, to provide liquid handling at room temperature.

Table 1. Some Physical Properties of Substituted Alkanolamines

Common name	Freezing point °C	Boiling point[a], °C
dimethylethanolamine	−59	135
diethylethanolamine		162
aminoethylethanolamine (AEEA)	−38[b]	244
methylethanolamine	−4.5	160
butylethanolamine	−2	199
N-acetylethanolamine	16	decompn
phenylethanolamine	11	285
dibutylethanolamine	−75[c]	229
diisopropylethanolamine	−39	191
phenylethylethanolamine	37[d]	decompn
methyldiethanolamine	−21	247
ethyldiethanolamine	<−44[c]	253
phenyldiethanolamine	57[b]	
dimethylisopropanolamine	−85[c]	126
N-(2-hydroxypropyl)ethylenediamine	−50[c]	155[e]

[a] At 101.3 kPa = atm unless otherwise noted.
[b] Pour point.
[c] Sets to a glass-like solid below this temperature.
[d] Melting point.
[e] At 8 kPa (60 mm Hg).

Table 2. Physical Properties of Alkanolamines Prepared from Ammonia and Olefin Oxides

Common name	Molecular formula	Freezing point, °C	Boiling point[a], °C	Water solubility[b], g/100 g	Viscosity[b], mPa · s (= cP)
monoethanolamine (MEA)	C_2H_7NO	10	171	∞	19
diethanolamine (DEA)	$C_4H_{11}NO_2$	28	268	∞	54 (60°C)
triethanolamine (TEA)	$C_6H_{15}NO_3$	21	340	∞	600
monoisopropanolamine (MIPA)	C_3H_9NO	3[c]	159	∞	23
diisopropanolamine (DIPA)	$C_6H_{15}NO_2$	44[c]	249	1200	86 (54°C)
triisopropanolamine (TIPA)	$C_9H_{21}NO_3$	44[c]	306	> 500	100 (60°C)
mono-sec-butanolamine	$C_4H_{11}NO$	3	169	∞	29
di-sec-butanolamine	$C_8H_{19}NO_2$	68–70	256	∞	890
tri-sec-butanolamine	$C_{12}H_{27}NO_3$	41–47	310	ca 7	ca 6000

[a] At 101.3 kPa = atm
[b] Approximate, at 25°C unless otherwise noted.
[c] Supercools; freezing points may show variation.

Alkanolamines have a mild ammoniacal odor and are extremely hygroscopic. The mono- and dialkanolamines have a basicity similar to aqueous ammonia; the trialkanolamines are slightly weaker bases.

CHEMICAL REACTIONS

Alkanolamines are bifunctional molecules because of the alcohol and the amine functional groups in the same compound. This allows them to react in a wide variety of ways, with similarities to primary, secondary, and tertiary amines, and primary and secondary alcohols.

ANALYTICAL TEST METHODS

Generally, alkanolamines are analyzed by gas chromatography or wet test methods.

STORAGE AND HANDLING

Stainless steel, 315L and 304L, is the preferred material of construction for shipment and storage of alkanolamines, if product quality is of importance.

Storage tanks, lines, and pumps should be heat traced and insulated to enable product handling. Temperature control is required to prevent product degradation because of color.

HEALTH AND SAFETY

Oral Toxicity

Alkanolamines generally have low acute oral toxicity, but swallowing substantial quantities could have serious toxic effects. including injury to mouth, throat, and digestive tract.

Vapor Toxicity

Laboratory exposure data indicate that vapor inhalation of alkanolamines presents low hazards at ordinary temperatures (generally, alkanolamines have low vapor pressures). Heated material may cause generation of sufficient vapors to cause adverse effects, including eye and nose irritation.

Eye Irritation

Exposure of the eye to undiluted alkanolamines can cause serious injury.

Skin Irriation

Monoethanolamine and monoisopro-panolamine, being strongly alkaline, are skin irritants, capable of producing serious injury in concentrations of 10% or higher upon repeated or prolonged contact.

Special Precautions

Use of sodium nitrite or other nitrosating agents in formulations containing alkanolamines could lead to formation of suspected cancer-causing nitrosamines.

Strong oxidizers and strong acids are incompatible with alkanolamines.

USES

Alkanolamines and their derivatives are used in a wide variety of household and industrial applications. Nonionic surfactants (alkanolamides) can be formed by the reaction of alkanolamines with fatty acids, at elevated temperatures. The amides can be liquid, water-soluble materials as produced from a 2:1 ratio, or solid, poorly water-soluble materials, or "super" amides produced from a 1:1 ratio of reactants. These products are useful as foam stabilizers and aid cleaning in laundry detergents, dishwashing liquids, shampoos, and cosmetics. They are also used as antistatic agents, glass coatings, fuel gelling agents, drilling mud stabilizers, demulsifiers, and in mining flotation. Reaction of alkanolamines with fatty acid at room temperature produces neutral alkanolamine soaps. Alkanolamine soaps are found in cosmetics, polishes, metalworking fluids, textile applications, agricultural products, household cleaners, and pharmaceuticals.

Alkanolamine salts are anionic surfactants formed from the reaction of alkanolamines and the acids of synthetic detergents, such as alkylarylsulfonates, alcohol sulfates, and alcohol ether sulfates. These add to the surfactants line used in detergents, cosmetics, textiles, polishes, agricultural sprays, household cleaners, pharmaceutical ointments, and metalworking compounds.

Salts of alkanolamines and inorganic acids are useful chemical intermediates, and are also used in corrosion inhibitors, antistatic agents, glass coatings, electroplating, high octane fuels, inks, metalworking, dust control in mining, and in textiles.

The Alkanolamines Handbook, The Dow Chemical Company, Midland, Mich., 1988.

Expert Panel of the Cosmetic Ingredient Review, *Final Report on the Safety Assessment for Diisopropanolamine, Triisopropanolamine, Isopropanolamine, Mixed Isopropanolamines*, Sept. 26, 1986; *Final Report for the Safety Assessment for Triethanolamine, Diethanolamine, Monoethanolamine*, May 19, 1983.

MARTHA R. EDENS
J. FREDLOCHARY
The Dow Chemical Company

ALKYD RESINS

While no longer the largest-volume vehicles in coatings, alkyds still are of major importance. Alkyds are prepared from polyols, dibasic acids, and fatty acids. They are polyesters, but in the coatings field the term "polyester" is reserved for "oil-free polyesters". The term "alkyd" is derived from alcohol and acid. Alkyds tend to be lower in cost than most other vehicles and tend to give coatings that exhibit fewer film defects during application. However, durability of alkyd films, especially outdoors, tends to be poorer than films from acrylics, polyesters, and polyurethanes. In a comparison of resistance to acid rain among coconut alkyd-MF, polyester-MF, and silicone-modified polyester-MF (MF—melamine—formaldehyde) coatings at five locations in the United States and Canada, the alkyd coating showed the poorest resistance.

TYPES OF ALKYDS

There are several types of alkyds. One classification is into oxidizing and nonoxidizing types. Oxidizing alkyds cross-link by the same mechanism as drying oils. Non-oxidizing alkyds are used as polymeric plasticizers or as hydroxyfunctional resins. A second classification is based on the ratio of monobasic fatty acids to dibasic acids utilized in their preparation. The terminology used was adapted from terminology used to classify varnishes. Varnishes with high ratios of oil to resin were called long oil varnishes; those with a lower ratio, medium oil varnishes; and those with an even lower ratio, short oil varnishes. Oil length of an alkyd is calculated by dividing the amount of "oil" in the final alkyd by the total weight of the alkyd solids, expressed as a percentage

Another classification is unmodified or modified alkyds. Modified alkyds contain other monomers in addition to polyols, polybasic acids, and fatty acids. Examples are styrenated alkyds and silicone alkyds. Closely related to alkyd resins, uralkyds and epoxy esters are also discussed.

OXIDIZING ALKYDS

Oxidizing alkyds can be considered synthetic drying oils: polyesters of one or more polyols, one or more dibasic acids, and fatty acids from one or more drying or semidrying oils.

When a film is applied, the initially naturally present hydroperoxides decompose to form free radicals in a process that establishes a chain reaction resulting in autooxidation. The characteristic undesirable odor of oil and alkyd paints during drying has been a factor motivating replacement of oil and alkyds in paints with latex, particularly for interior applications.

Dried films, especially of alkyds with three double-bond fatty acids, yellow with aging. The yellow color bleaches significantly when exposed to light; hence, yellowing is most severe when films are covered, such as by a picture hanging on a wall. Yellowing has been shown to result from incorporation of nitrogen compounds and is markedly increased by exposure to ammonia. The autoxidation rates of uncatalyzed nonconjugated oxidizing alkyds dry are slow, but metal salts (driers) catalyze drying. Mixtures of lead with cobalt and/or manganese are particularly effective, but lead driers cannot be used in consumer paints sold in interstate commerce in the United States. Combinations of cobalt and/or manganese with zirconium, frequently with calcium, which may promote drying, are commonly used. The amounts of driers needed are system specific. Their use should be kept to the minimum possible level, since they not only catalyze drying but also the reactions that cause postdrying embrittlement, discoloration, and cleavage.

Of the drying alkyds made by the monoglyceride process, soybean oil is used in the largest volume because it is economical and supplies are dependable. for alkyds made by the fatty acid process, tall order fatty acids (TOFAs) are more economical than soybean fatty acids. Both soybean oil and TOFAs contain significant amounts of linolenic acid. Premium cost "nonyellowing" alkyds are made with safflower or sunflower oils.

Applications in which fast drying and high cross-link density are important require alkyds made with drying oils. The rate of oxidative cross-linking is affected by the functionality of the drying oils used.

A critical factor involved in the choice of fatty acid is cost. Drying oils are agricultural products and, hence, tend to be volatile in price. Depending on relative prices, one drying oil is often substituted for another in certain alkyds. By adjusting for functionality differences, substitutions can frequently be made without significant changes in properties.

The dibasic acids used to prepare alkyds are usually aromatic. By far the most widely used dibasic acid is PA. The next most widely used dibasic acid is isophthalic acid. Since xylene is on the HAP list, its use is being reduced.

High solids alkyds tend to have lower functionality for cross-linking and a lower ratio of aromatic to aliphatic chains. Both changes increase the time for drying. There is also a decrease branching with the higher hydroxyl excess.

The effect of longer oil length on functionality can be minimized by using drying oils with higher average functionality. Use of oils containing linolenic or α-eleos-tearic acid is limited by their tendency to discolor. Saf-flower oil has a higher linoleic acid content and less linolenic acid than soybean oil. Proprietary fatty acids with 78% linoleic acid are commercially available. Early hardness of the films can be improved by using benzoic acid to esterify part of the free hydroxy groups.

Waterborne Alkyds. Work has been done to make alkyd resins for coatings that can be reduced with water, as the use of alkyd emulsions. The emulsions are stabilized with surfactants and can be prepared with little, if any, volatile solvent.

It is common to add a few percent of an alkyd–surfactant blend to latex paints to improve adhesion to chalky surfaces and, in some cases, to improve adhesion to metals. Another approach has been to make alkyds with an acid number in the range of 50, using secondary alcohols or ether alcohols as solvents.

Modified Alkyds. Oxidizing alkyds have been modified by reacting with a variety of other components; vinyl-, silicone-, phenolic-, and polyamide-modified alkyds are the most common examples.

Oxidizing alkyds can be modified by reaction with vinyl monomers. Essentially any vinyl monomer can be reacted in the presence of an alkyd to give a modified alkyd. Methyl methacrylate imparts better heat resistance than styrene but at higher cost.

Styrenated alkyd vehicles are often used for air dry primers. Application of topcoat without care to timing is likely to cause nonuniform swelling of the primer, leading to lifting of the primer and the development of wrinkled areas in the surface of the dried film. End users accustomed to using alkyd primers are particularly likely to encounter problems of lifting if they switch to styrenated alkyd primers.

Silicone resins have exceptional exterior durability but are expensive. Silicone modification of alkyd resins improves their exterior curability. The exterior durability of silicone-modified alkyd coatings is significantly better than that of unmodified alkyd coatings. Alkyd coatings modified with high-phenyl silicone resins have greater thermoplasticity, faster air drying, and higher solubility than high methyl silicone-modified alkyds. Silicone-modified alkyds are used mainly in outdoor air dry coatings for which application is expensive, as in a topcoat for steel petroleum storage tanks.

Phenolic-modified alkyds are made by heating the alkyd with a low molecular weight resole phenolic resin based on p-alkylphenols. The resins give harder films with improved water and chemical resistance as compared to the unmodified alkyd.

Ceramer (organic–inorganic hybrid) coatings prepared with long oil linseed alkyds and titanium tetraisopropoxide gave films with excellent hardness, tensile strength, flexibility, and impact resistance.

Polyamide-modified alkyds are used as thixotropic agents to increase the low shear viscosity of alkyd resin based paints. High solids thixotropic alkyds based on poly-amides made with aromatic diamines give superior performance in high solids alkyd coatings.

Nonoxidizing Alkyds

Certain low molecular weight short-medium and short oil alkyds are compatible with such polymers as nitro-cellulose and thermoplastic polyacrylates and can thus be used as plasticizers for these polymers. They have the advantage over monomeric plasticizers in that they do not volatilize appreciably when films are baked.

All alkyds, particularly the short-medium oil and short oil ones, are made with a large excess of hydroxyl groups, to avoid gelation. Butylated MF resins used in coatings provide somewhat better durability and faster curing than alkyd resins alone, with little increase in cost. The important advantage of relative freedom from film defects common to alkyd coatings can be retained. However, the high levels of unsaturation remaining in the cured films reduce resistance to discoloration on overbake and exterior exposure and cause loss of gloss and embrittlement on exterior exposure. These difficulties can be reduced by using nondrying oils like coconut oil with minimal levels of unsaturated fatty acids.

SYNTHESIS OF ALKYD RESINS

Various synthetic procedures are used to produce alkyd resins, either directly from oils or by using free fatty acids as raw materials.

With glycerol alkyds, to avoid first saponifying an oil to obtain fatty acids and glycerol and then reesterify the same groups in a different combination, the oil is first reacted with sufficient glycerol to give the total desired glycerol content, including the glycerol in the oil. Since PA is not soluble in the oil but is in the glycerol, transesterification of oil with glycerol must be carried out as a seperate step before the PA is added. This two-stage procedure is often called the monoglyceride process. The reaction is run under an inert atmosphere such as CO_2 or N_2 to minimize discoloration and dimerization.

It is often desirable to base an alkyd on a polyol other than glycerol. In this case, fatty acids must be used instead of oils, and the process can be performed in a single step with reduced time in the reactor. Any drying, semidrying, or nondrying oil can be saponified to yield fatty acids, but the cost of separating fatty acids from the reaction mixture increases the cost of the alkyd. A more economical alternative is to use TOFAs.

Most alkyds are produced using a reflux solvent such as xylene to promote the removal of water by azeotroping. The presence of solvent is desirable for other reasons: vapor serves as an inert atmosphere, reducing the amount of inert gas needed, and the solvent serves to avoid accumulation of sublimed solid monomers, mainly PA, in the reflux condenser.

Higher temperatures during the reaction obviously accelerate the reaction. If carried too far, there is a major risk of gelation. There are economic advantages to short reaction times, so it is desirable to operate at as high a temperature as possible without risking gelation.

Many variables affect the acid number and viscosity of alkyds. The composition of the fatty acids is a major factor affecting viscosity, and compositions of an oil or grade of TOFA can be expected to vary somewhat from lot to lot. Side reactions can also affect the viscosity–acid number relationship.

Urethane Derivatives

Uralkyds (urethane alkyds) are alkyd resins in which a diisocyanate has fully or partly replaced the PA usually used in the preparation of alkyds. One transesterifies a drying oil with a polyol such as glycerol or PE to make a "monoglyceride" and reacts it with some PA (if desired) and then with somewhat less diisocyanate than the equivalent amount of N=C=O based on the free OH content. Just like alkyds, uralkyds dry faster than the drying oil from which they were made, since they have a higher average functionality.

Two principal advantages of uralkyd over alkyd coatings are superior abrasion resistance and resistance to hydrolysis. Disadvantages are inferior color retention when aromatic isocyanates are used, higher viscosity of resin solutions at the same percent solids, and higher cost. Uralkyds made with aliphatic diisocyanates have better color retention, but are more expensive and have lower T_g. The largest use of uralkyds is in architectural coatings. Many so-called varnishes sold to the consumer today are based on uralkyds used as transparent coatings for furniture, woodwork, and floors, applications in which good abrasion resistance is important. Since they are generally made with aromatic isocyanates, they tend to turn yellow and then light brown with age.

Water-reducible polyunsaturated acid substituted aqueous polyurethane dispersions are also used. If aliphatic isocyanates are used, good color retention can be obtained. They are much more resistant to hydrolysis than conventional alkyd resins. Films also have excellent abrasion resistance.

Epoxy Esters

Bisphenol A (BPA) [4,4′-(1-methylethylidene)bisphenol] epoxy resins can be converted to what are commonly called epoxy esters by reacting with fatty acids. Drying or semidrying oil fatty acids are used so that the products cross-link by autoxidation. The epoxy groups undergo a ring-opening reaction with carboxylic acids to generate an ester and a hydroxyl group. These hydroxyl groups, as well as the hydroxyl groups originally present on the epoxy resin, can esterify with fatty acids.

Tall oil fatty acids are commonly used because of their low cost. Linseed fatty acids give faster cross-linking coatings because of higher average functionality. However, their viscosity is higher because of the greater extent of dimerization during esterification, and their cost is higher. For still faster cross-linking, part of the linseed fatty acids can be replaced with tung fatty acids, but the viscosity and cost are still higher. The color of epoxy esters

from linseed and linseed–tung fatty acids is darker than the tall oil esters. Dehydrated castor oil fatty acids give faster curing epoxy esters for baked coatings.

Epoxy esters are used in coatings in which adhesion to metal is important. It is common for epoxy coatings, including epoxy esters, to have good adhesion to metals and to retain adhesion after exposure of the coated metal to high humidity, a critical factor in corrosion protection. A distinct advantage of epoxy esters over alkyd resins is their greater resistance to hydrolysis and saponification.

The major uses for epoxy resins are in primers for metal and in can coatings, such as for bottle caps, in which the important requirements are adhesion and hydrolytic stability. In baking primers, it is sometimes desirable to supplement the cross-linking through oxidation by including a small amount of MF resin in the formulation to cross-link with part of the free hydroxyl groups on the epoxy ester.

Epoxy ester resins with good exterior durability (better than alkyds) can be prepared by reacting epoxy-functional acrylic copolymers with fatty acids. The product is an acrylic resin with multiple fatty acid ester side chains. Epoxy esters can also be made water reducible. The most widely used water-reducible epoxy esters have been made by reacting maleic anhydride with epoxy esters prepared from dehydrated castor oil fatty acids. Water-reducible epoxy esters are still used in spray-applied baking primers and primer surfacers. They are also used in dip coating primers in which nonflammability is an advantage. Their performance equals that of solvent–soluble epoxy ester primers.

USES AND ADVANTAGES

Coatings are the largest market for alkyds, and the largest use in coatings is in architectural paints, particularly in gloss enamels for application by contractors. They tend to prefer alkyd enamels over latex ones because coverage can be achieved with a single coat. Also alkyd paints can be applied at low temperatures, whereas latex paints can be applied only at temperatures above $\sim5\,^\circ$C. The do-it-yourself market is served primarily with latex paints because of ease of cleanup and lower odor. While initial gloss of alkyd enamels is higher than of latex enamels, the latex enamels exhibit far superior gloss retention, especially in exterior applications. Alkyd primers provide better adhesion to chalky surfaces than most latex paints.

Use of alkyds has been declining as they are replaced with resins having higher performance and lower volatile emissions. Higher solids alkyds have been replacing conventional solids alkyds.

The principal advantages of alkyds are low cost, low toxicity, and low surface tension, the latter of which permits wetting of most surfaces, including oily steel. Also, the low surface tension minimizes application defects such as cratering. The principal limitations are generally poorer exterior durability and corrosion protection than alternative coating resins. While high solids and waterborne alkyd resins are manufactured, their properties are generally somewhat inferior to conventional solvent borne alkyds.

The largest uses of alkyds in industrial applications is in general industrial coatings for such applications as machinery and metal-furniture. Significant amounts are used with UF resins in coatings for wood furniture. Alkyd resin-chlorinated rubber-based coatings are used in traffic paints. Some alkyds are still used in refinish paints for automobiles, since they give high-gloss coatings with a minimum of polishing. Higher grades of coatings are urethane coatings. Some nitrocellulose primers with nonoxidizing alkyd plasticizers and some alkyd underbody sealers are still used.

The largest use for uralkyds is as the vehicle for so-called urethane varnishes for the do-it-yourself market. The abrasion resistance of such coatings is greatly superior to that obtained with conventional varnishes or alkyd resins. Epoxy esters give coatings with markedly superior corrosion protection. Maleated epoxy esters give primers with equivalent properties of solvent borne epoxy ester coatings and are widely used in formulating waterborne primers for steel. Noncoatings applications include foundry core binders and printing inks, especially lithographic inks.

P. J. Bakker and co-workers, *Water-borne High-Solids, Powder Coating Symposium*, New Orleans, La, 2001, pp. 439–453.

A. Hofland, in J. E. Glass, ed., *Technology for Waterborne Coatings*, American Chemical Society, Washington, D.C. 1997, p. 183.

T. C. Patton, *Alkyd Resin Technology*, John Wiley & Sons, New York, 1962.

W. S. Sisson and R. J. Shah, *Proc. Waterborne, High Solids, Powder Coat. Symp.*, New Orleans, La., 2001, pp. 329–336.

ZENO W. WICKS, JR.
Consultant

ALKYLATION

Alkylation is the substitution of a hydrogen atom bonded to the carbon atom of a paraffin or aromatic ring by an alkyl group. The alkylations of nitrogen, oxygen, and sulfur are described in separate articles. This article covers important industrial technologies and the direction of future technological development.

Significant technological development has been made in the area of alkylation in recent years. Environmental concerns associated with mineral acid catalysts have encouraged process changes and the development of solid-bed alkylation processes. The application of heterogenous catalysts, especially zeolite catalysts, has led to new alkylation technologies. Research efforts to develop environmentally acceptable, economical technologies by applying new materials as alkylation catalysts will continue, and more new technologies are expected to be commercialized.

ALKYLATION OF PARAFFINIC HYDROCARBONS

Paraffin alkylation refers to the addition reaction of an isoparaffin and an olefin. The desired product is a higher molecular weight paraffin that exhibits a greater degree of branching than either of the reactants.

The principal industrial application of paraffin alkylation is in the production of premium-quality fuels for spark-ignition engines. Alkylation is now primarily used to provide a high octane blending component for automotive fuels. Future gasoline specifications will continue to favor the clean-burning characteristics and the low emissions typical of alkylate. Alkylate is an ideal gasoline blend stock because of its high octane and paraffinic nature. Alkylate production capacity is expected to grow as worldwide gasoline specifications become more stringent.

CATALYSTS AND REACTIONS

Although the alkylation of paraffins can be carried out thermally, catalytic alkylation is the basis of all processes in commercial use. Early studies of catalytic alkylation led to the formulation of a proposed mechanism based on a chain of ionic reactions. The reaction steps include the formation of a light tertiary cation, the addition of the cation to an olefin to form a heavier cation, and the production of a heavier paraffin (alkylate) by a hydride transfer from a light isoparaffin. This last step generates another light tertiary cation to continue the chain.

The catalysts used in the industrial alkylation processes are strong liquid acids, either sulfuric acid (H_2SO_4) or hydrofluoric acid (HF).

FEEDSTOCK AND PRODUCTS

Isobutane

Although other isoparaffins can be alkylated, isobutane is the only paraffin commonly used as a commercial feedstock.

Butylenes

Butylenes are the primary olefin feedstock to alkylation and produce a product high in trimethylpentanes. The research octane number, which is typically in the range of 94–98, depends on isomer distribution, catalyst, and operating conditions.

Propylene

Propylene alkylation produces a product that is rich in dimethylpentane and has a research octane typically in the range of 89–92.

Amylenes

Amylenes (C_5 monoolefins) produce alkylates with a research octane in the range of 90–93.

ALKYLATION OF AROMATIC HYDROCARBONS

Most of the industrially important alkyl aromatics used for petrochemical intermediates are produced by alkylating benzene with monoolefins. The most important monoolefins for the production of ethylbenzene, cumene, and detergent alkylate are ethylene, propylene, and olefins with 10–18 carbons, respectively.

ACID CATALYSTS AND REACTION MECHANISM

Acid catalysts promote the addition of alkyl groups to aromatic rings. Olefins, alcohols, ethers, halides, and other olefin-producing compounds can be used as alkylating reagents. In addition to traditional protonic acid catalysts (H_2SO_4, HF, phosphoric acid) and Friedel-Crafts-type catalysts ($AlCl_3$, boron fluoride), any solid acid catalyst having a comparable acid strength is effective for aromatic alkylation. Typical solid acid catalysts are amorphous and crystalline alumino-silicates, clays, ion-exchange resins, mixed oxides, and supported acids. Among these solid acid catalysts, ZSM-5, Y-type zeolites, and more recently MCM-22 and beta-zeolite have become the new commercial catalysts for aromatic alkylation.

BASE CATALYSTS AND REACTION MECHANISM

Alkali metals and their derivatives can catalyze the alkylation of aromatics with olefins. In contrast to acid-catalyzed alkylation, in which the aromatic ring is alkylated, an olefin is added to the alkyl group of aromatics over a base catalyst through a carbanion intermediate. The carbanion intermediate is produced from an aromatic compound by the abstraction of benzylic hydrogen as a proton by a base. The carbanion reacts with an olefin to grow the side chain of the aromatic compound.

INDUSTRIAL APPLICATIONS

Ethylbenzene

This alkylbenzene is almost exclusively used as an intermediate for the manufacture of styrene monomer. A small amount ($<1\%$) is used as a solvent and as an intermediate in dye manufacture.

Ethylbenzene is primarily produced by the alkylation of benzene with ethylene, although a small percentage of the world's ethylbenzene capacity is based on the superfractionation of ethylbenzene from mixed xylene streams.

Cumene

The demand for cumene has been risen at an average rate of 2–3% per year since 1970.

Currently, almost all cumene is produced commercially by two processes: a fixed-bed, kieselguhr-supported phosphoric acid catalyst system developed by UOP and a homogeneous $AlCl_3$ and hydrogen chloride catalyst system developed by Monsanto.

Two new processes using zeolite-based catalyst systems were developed in the late 1980s. Unocal's technology is based on a conventional fixed-bed system. CR&L has developed a catalytic distillation system based on an extension of the CR&L MTBE technology.

Cymene

Methylisopropylbenzene can be produced over a number of different acid catalysts by alkylation of toluene with propylene.

Detergent Alkylate

The synthetic detergent industry has become one of the largest chemical process industries. The most recent advance in detergent alkylation is the development of a solid catalyst system.

Industrial Processes. A variety of acid catalysts have been used for the production of alkylbenzenes by the alkylation of benzene with higher olefins (C_{10}–C_{15} detergent-range olefins). HF and $AlCl_3$ have been used since the 1960s and H_2SO_4 was used in some earlier units. In 1995, the first detergent alkylation unit, using a solid acid catalyst developed by UOP and CEPSA, was started-up. The Detal process offers superior LAB product quality and lower capital costs due to simplified catalyst handling and downstream product clean-up compared with either HF or $AlCl_3$.

The main reaction in detergent alkylation is the alkylation of benzene with the straight-chain olefins to yield a linear alkylbenzene:

$$R-CH=CH-R' \ + \ \bigcirc \ \longrightarrow \ R-CH-CH_2-R'$$

The most recent advance in detergent alkylation is the development of a solid catalyst system. UOP and Compania Espanola de Petroleos SA (CEPSA) have jointly developed the Detal process, which uses a fixed-bed heterogeneous aromatic alkylation catalyst system for the production of LAB.

Xylenes

The main application of xylene isomers, primarily *p*- and *o*-xylenes, is in the manufacture of polyester fibers, films, resins and plasticizers. Demands for xylene isomers and other aromatics such as benzene have steadily been increasing over the last two decades. Food packaging applications, which use polyester blends derived from *m*-xylene, increased 10–15% per year during the 1990s. This led to a significant increase in the capacity for the *m*-xylene isomer. The major source of xylenes is catalytic reforming of naphtha and pyrolysis of naphtha and gas oils.

Polynuclear Aromatics

The alkylation of polynuclear aromatics with olefins and olefin-producing reagents is effected by acid catalysts. The alkylated products are more complicated than are those produced by the alkylation of benzene because polynuclear aromatics have more than one position for substitution.

FUTURE TECHNOLOGY TRENDS

Over the years, improvements in aromatic alkylation technology have come in the form of both improved catalysts and improved processes. This trend is expected to continue into the future.

Catalysts

Nearly all of the industrially significant aromatic alkylation processes of the past have been carried out in the liquid phase with unsupported acid catalysts.

Since 1976, these forms of acids have become a significant environmental concern from both a physical handling and disposal perspective. This concern has fueled much development work toward solid acid catalysts, including zeolites, silica–aluminas, and clays.

Process

As solid acid catalysts have replaced liquid acid catalysts, they have typically been placed in conventional fixed-bed reactors. An extension of fixed-bed reactor technology is the concept of catalytic distillation being offered by CR&L. In catalytic distillation, the catalytic reaction and separation of products occurs in the same vessel. The concept has been applied commercially for the production of MTBE and is also being offered for the production of ethylbenzene and cumene.

A new alkylation method, the Alkymax process, was introduced by UOP in 1990. In addition to lowering the benzene content, the alkylate formed has a high octane value and can typically boost the octane of the gasoline pool by 0.5 RON.

OTHER ALKYLATIONS

Alkylation of Phenol

The hydroxyl group activates the alkylation of the benzene ring because it is a strong electron-donating group; therefore, the alkylation of phenol can be achieved with olefins and olefin-producing reagents under milder conditions than the alkylation of aromatic hydrocarbons. Alkylated phenol derivatives are used as raw materials for the production of resins, novolaks (alcohol-soluble resins of the phenol–formaldehyde type), herbicides, insecticides, antioxidants, and other chemicals.

Alkylation of Aromatic Amines and Pyridines

Commercially important aromatic amines are aniline, toluidine, phenylenediamines, and toluenediamines. The ortho alkylation of these aromatic amines with olefins, alcohols, and dienes to produce more valuable derivatives can be achieved with solid acid catalysts.

The alkylation of pyridine takes place through nucleophilic or homolytic substitution because the π-electron-deficient pyridine nucleus does not allow electrophilic substitution, eg, Friedel-Crafts alkylation.

HEALTH AND SAFETY FACTORS

Generally, specific health and safety factors relating to feedstock and products must be addressed for each particular industrial alkylation process. In addition, the properties of the catalyst systems employed in alkylation must be considered. In industrial applications, specialized procedures are required to ensure the safe handling of these materials. Replacing these materials with solid acid catalysts will become more important in the future.

The solid acid catalysts themselves present a disposal problem that favors the development of regenerable catalysts or the implementation of recycling procedures.

J. Branzaru, *Introduction to Sulfuric Acid Alkylation Unit Process Design*, Stratco Technology Conference, November, 2001.

T. Hutson and G. E. Hays, in L. F. Albright and A. R. Goldsby, eds., *Industrial and Laboratory Alkylations* (ACS Symposium Series) American Chemical Society, Washington, D.C., 1977, pp. 27–56.

R. A. Myers, ed., *Petroleum Refining Processes*, McGraw-Hill, New York, 1996.

G. Stefanidakis and J. E. Gwyn, in J. J. McKetta and W. A. Cunningham, eds., *Encyclopedia of Chemical Processing and Design*, Vol. 2, Marcel Dekker, New York, 1977, p. 357.

BIPIN V. VORA
JOSEPH A. KOCAL
PAUL T. BARGER
ROBERT J. SCHMIDT
JAMES A. JOHNSON
UOP LLC

ALKYLPHENOLS

Alkylphenols of greatest commercial importance have alkyl groups ranging in size from one to twelve carbons. The direct use of alkylphenols is limited to a few minor applications such as epoxy-curing catalysts and biocides. The vast majority of alkylphenols are used to synthesize derivatives which have applications ranging from surfactants to pharmaceuticals. The four principal markets are nonionic surfactants, phenolic resins, polymer additives, and agrochemicals.

Nonionic surfactants and phenolic resins based on alkylphenols are mature markets and only moderate growth in these derivatives is expected. Concerns over the biodegradability and toxicity of these alkylphenol derivatives to aquatic species may limit their use in the future. The use of alkylphenols in the production of both polymer additives and monomers for engineering plastics is expected to show above average growth as plastics continue to replace traditional building materials.

Alkylphenols containing 3–12-carbon alkyl groups are produced from the corresponding alkenes under acid catalysis. Alkylphenols containing the methyl group were traditionally extracted from coal tar. Today they are produced by the alkylation of phenol with methanol.

NOMENCLATURE

An alkylphenol is a phenol derivative wherein one or more of the ring hydrogens has been replaced by an alkyl group or groups. Phenol is a heading parent in the CAS indexing system. Appropriate names of alkylphenols for abstract citations can be derived by using the appropriate aids. The names generated in this manner are unambiguous and refer to a specific compound, but are lengthy and cumbersome to use. Common names are used on a daily basis and are especially prevalent for alkylphenols that have gained commercial importance.

PHYSICAL PROPERTIES

The physical properties of alkylphenols are comparable to phenol. The properties are strongly influenced by the type of alkyl substituent and its position on the ring. Alkylphenols, like phenol, are typically solids at 25°C. Their form is affected by the size and configuration of the alkyl group, its position on the ring, and purity. They appear colorless, or white, to a pale yellow when pure (Table 1).

The solubility of alkylphenols in water falls off precipitously as the number of carbons attached to the ring increases. They are generally soluble in common organic solvents: acetone, alcohols, hydrocarbons, toluene. Solubility in alcohols or heptane follows the generalization that "like dissolves like."

SYNTHESIS OF ALKYLPHENOLS

Alkyphenols can be synthesized by several approaches, including alkylation of a phenol, hydroxylation of an alkylbenzene, dehydrogenation of an alkyl-cyclohexanol, or ring closure of an appropriately substituted acyclic compound. The choice of an approach depends on the target alkylphenol, availability of the starting materials, and cost of processing.

CHEMICAL PROPERTIES

Alkylphenols undergo a variety of chemical transformations, involving the hydroxyl group or the aromatic nucleus that convert them to value-added products.

The Hydroxyl Group

The unshared pairs of electrons on hydroxyl oxygens seek electron deficient centers. Alkylphenols tend to be less nucleophilic than aliphatic alcohols as a direct result of the attraction of the electron density by the aromatic nucleus. The reactivity of the hydroxyl group can be enhanced in spite of the attraction of the ring current by use of a basic catalyst which removes the acidic proton from the hydroxyl group leaving the more nucleophilic alkylphenoxide.

Reactions Involving the Ring

The aromatic nucleus of alkylphenols can undergo a variety of aromatic electrophilic substitutions. Electron density from the hydroxyl group is fed into the ring. Besides activating the aromatic nucleus, the hydroxyl group controls the orientation of the incoming electrophile.

MANUFACTURE AND PROCESSING

Alkylphenols of commercial importance are generally manufactured by the reaction of an alkene with phenol in the presence of an acid catalyst. The alkenes used vary from

Table 1. Commercially Important Alkylphenols

Name	Molecular formula	Molecular weight	Physical form at 25°C	Boiling point, °C[a]	Freezing point, °C	Density[b], g/mL	Typical assay	Flash point, °C	Molten color APHA
4-tert-amylphenol	$C_{11}H_{16}O$	164.0	solid	249	90.0	0.915^{107}	99	121	200
4-tert-butylphenol	$C_{10}H_{14}O$	150.2	solid	237	97.5	0.890^{107}	98–99	117	100
2-sec-butylphenol	$C_{10}H_{14}O$	150.2	liquid	224	20.0	0.938^{43}	98	>93	100
4-cumylphenol	$C_{15}H_{16}O$	212.0	solid	335	70.0	1.029^{93}	99	188	100
4-dodecylphenol	$C_{18}H_{30}O$	262.0	liquid	334		0.914^{20}	89–95	>100	500
4-nonylphenol	$C_{15}H_{24}O$	220.3	liquid	310		0.933^{43}	90–95	146	100
4-tert-octylphenol	$C_{14}H_{22}O$	220.3	solid	290	81.0	0.940^{25}	90–98	132	200
2,4-di-tert-amylphenol	$C_{16}H_{26}O$	234.4	liquid	275	23.0	0.900^{49}	99	104	100
2,4-di-tert-butylphenol	$C_{14}H_{22}O$	206.3	solid	263	52.0	0.867^{82}	99	115	100
2,6-di-tert-butylphenol	$C_{14}H_{22}O$	206.3	solid	253	36.0	0.898^{43}	99	>99	100
di-sec-butylphenol	$C_{14}H_{22}O$	206.3	liquid			0.902^{66}	90	127	500
2,4-dicumylphenol	$C_{24}H_{26}O$	330.0	solid		65.0	1.030^{66}	99	462	100
2-methylphenol	C_7H_8O	108.1	solid	191	30.0	$1.049^{15.5}$	99	81	25
3-methylphenol	C_7H_8O	108.1	liquid	202	10.0	$1.042^{15.5}$	97	86	
4-methylphenol	C_7H_8O	108.1	solid	202	34.0	1.022^{25}	99	86	25
2,6-dimethylphenol	$C_8H_{10}O$	122.1	solid	203	48.0	1.020^{25}	99	88	

[a] At 101.3 kPa = 1 atm.
[b] At the temperature indicated by the superscript, °C.
[c] Mixture, branched chains.

single species, such as isobutylene, to complicated mixtures, such as propylene tetramer (dodecene). The alkene reacts with phenol to produce monoalkylphenols, dialkylphenols, and trialkylphenols. The monoalkylphenols comprise ~85% of all alkylphenol production.

The choice of catalyst is based primarily on economic effects and product purity requirements. More recently, the handling of waste associated with the choice of catalyst has become an important factor in the economic evaluation. Catalysts that produce less waste and more easily handled waste by-products are strongly preferred by alkylphenol producers. Some commonly used catalysts are sulfuric acid, boron trifluoride, aluminum phenoxide, methanesulfonic acid, toluene–xylene sulfonic acid, cationic-exchange resin, acidic clays, and modified zeolites.

The approach used to synthesize commercially available alkylphenols is Friedel-Crafts alkylation. The specific procedure typically uses an alkine as the alkylating agent and an acid catalyst, generally a sulfonic acid.

Reactors

Reactors used to produce alkylphenols are simple batch reactors, complex batch reactors, and continuous reactors. All of these reactors have good mixing and heat removal capability. Good mixing is required for contacting the alkene and catalyst with the phenol. Typically, alkene–alkene reactions compete with phenol–alkene reactions at operating conditions. Good mixing minimizes locally high alkene concentrations and thus favors the desired reactions relative to the undesired ones. Good heat removal capability is needed to maintain controlled temperatures because of the highly exothermic nature of these reactions. The selectivity of alkylation is greatly affected by temperature.

Purification

The method used to recover the desired alkylphenol product from the reactor output is highly dependent on the downstream use of the product and the physical properties of the alkylphenol. Some alkylphenol applications can tolerate "as is" reactor products, most significantly in the production of alkylphenol–formaldehyde resins. Most alkylphenols sold today require refinement. Distillation is by far the most common separation route.

SHIPMENT

Most commercially important alkylphenol production is of three types, unrefined alkylphenols, monoalkylphenols, and dialkylphenols. Together, these processes comprise over 95% of all alkylphenol production in the United States.

Large volumes of monoalkylphenols are shipped in liquid form by railcar, tank wagon, or export container. These shipping vessels must be stainless steel or phenolic resin lined carbon steel. For smaller volumes, drums and tote-tanks are used. For high freezing point alkylphenols, such as PTBP, the product is flaked and shipped in either bags or supersacs. For low freezing point products, such as p-nonylphenol (PNP) (fp < 20°C), the product is shipped in drums or tote-tanks.

ECONOMIC ASPECTS

Among the key variables in strategic alkylphenol planning are feedstock quality and availability, equipment capability, environmental needs, and product quality. In the past decade, environmental needs have grown enormously in their effect on economic decisions. The manufacturing cost of alkylphenols includes raw-material cost, nonraw-material variable cost, fixed cost, and depreciation.

Raw-material costs are the largest cost items over the lifetime of a plant and typically make up between 40 and 90% of the total manufacturing cost. The placement of plants near production facilities making alkenes and/or phenol is important to producers of alkylphenols. The raw-material costs are so important that a large fluctuation in a raw material price can drive a product from a reasonably profitable situation to a clearly unprofitable one.

HEALTH AND SAFETY FACTORS

The toxicity of alkylphenols as a class of compounds ranges from moderately toxic (oral rat LD_{50} 50–500 mg/kg) to practically nontoxic (oral rat LD_{50} 5,000–15,000 mg/kg) and most are irritants or corrosive towards skin.

In general, precautions should be taken when handling alkylphenols to avoid contact with the skin by wearing appropriate protective gloves, clothing, and a face shield or goggles. Most of the alkylphenols are combustible when heated and emit irritating vapors upon decomposition. Keep away from food. Alkylphenols should only be stored and handled in well-ventilated areas, and appropriate respiration equipment with carbon filters should be worn if PEL or TLV limits are exceeded.

COMMERCIAL USES AND DERIVATIVES OF ALKYLPHENOLS

4-tert-Amylphenol

4-tert-Amylphenol is employed as a germicide in cleaning solutions, but it is being replaced by environmentally safer quaternary ammonium salts. Another commercial application is in phenolic resins (novolaks and resoles).

4-tert-Butylphenol

Phenolic resin applications account for 60–70% of all 4-tert-butylphenol consumed worldwide.

2-sec-Butylphenol

A significant volume of 2-sec-butyl-4,6-dinitrophenol is used worldwide as a polymerization inhibitor in the production of styrene, where it is added to the reboiler of the styrene distillation tower to prevent the formation of polystyrene. Because of environmental concerns about 2-sec-butylphenol-based derivatives, the market growth is expected to be negative in the future, with the exception of possible significant growth in the use of the carbamate insecticide.

4-Cumylphenol

The major use of 4-cumylphenol is as a chain terminator for polycarbonates.

4-Dodecylphenol

The major use of technical grade 4-dodecylphenol is in lube oil additives. High-purity 4-dodecylphenol is used to produce specialty surfactants by its reaction with ehtylene oxide.

2-Methylphenol

The majority of 2-methylphenol is used in the production of novolak phenolic resins.

3-Methylphenol

A major use of 3-methylphenol is in the production of phenolic based antioxidants, which are particularly good at stabilizing polymers in contact with copper against thermal oxidative degradation. Another significant use of 3-methylphenol is in the production of herbicides and insecticides.

4-Methylphenol

The bulk of 4-methylphenol is used in the production of phenolic antioxidants.

4-Nonylphenol

The major use for 4-nonylphenol is in the production of nonionic surfactants. Another significant use of 4-nonylphenol is in the production of tris(4-nonylphenyl) phosphite (TNPP), a secondary antioxidant which protects organic materials against oxidative degradation by decomposing hydroperoxides.

4-tert-Octylphenol

4-tert-Octylphenol reacts with ethylene oxide under base catalysis and the resulting ethoxylates are used in many of the same applications as the 4-nonylphenol-based surfactants.

Another important application for 4-tert-octylphenol is in the production of phenolic resins. Other applications for 4-tert-octylphenol include chain termination of polycarbonates and the production of uv stabilizers.

Dialkylated Phenols

A major use for 2,4-di-tert-amylphenol is in the production of uv stabilizers; the principal one is a benzotriazole-based uv absorber, 2-(2⁰-hydroxy-3⁰,5⁰-di-tert-amylphenyl)-5-chlorobenzotriazole, which is widely used in polyolefin films, outdoor furniture, and clear coat automotive finishes. Another significant use for 2,4-di-tert-amylphenol is in the photographic industry.

The primary use for 2,4-di-tert-butylphenol is in the production of substituted triayl phosphites. The principal use for 2,6-di-tert-butylphenol is in the production of hindered phenoic antixidants, which accounts for 80–90% of all of this compound produced.

The only significant use for di-sec-butylphenol is as a specialty nonionic surfactant produced by reaction with ethylene oxide under base catalysis. This surfactant is registered with the EPA for use in emulsifying agrochemicals.

The largest use for 2,4-dicumylphenol is in the production of a uv stabilizer of the benzotriazole class, 2-(2'-hydroxy-3',5'-dicumylphenyl) benzotriazole, which is used in engineering thermoplastics where high molding temperatures are encountered.

The oxidative coupling of 2,6-dimethylphenol to yield poly (phenylene oxide) represents 90–95% of the consumption of 2,6-dimethylphenol.

P. R. Dean, *Index of Commercial Antioxidants and Antiozonants*, Technical Bulletin, Goodyear Chemicals, Akron, Ohio, 1983.

J. March, *Advanced Organic Chemistry Reactions, Mechanisms, and Structure*, 3rd ed., John Wiley & Sons, Inc., New York, 1985, p. 448.

"Nonylphenol, Chemical Profile, "*Chemical Market Reporter* (July 2, 2001).

R. L. Shriner, R. C. Fuson, D. Y. Curtin, and T. C. Morrill, *The Systematic Identification of Organic Compounds*, 6th ed., John Wiley & Sons, Inc., New York, 1980, p. 102.

JOHN F. LORENC
GREGORY LAMBETH
WILLIAM SCHEFFER
Schenectady Chemicals, Inc.

ALLYL ALCOHOL AND MONOALLYL DERIVATIVES

The technology of introducing a new functional group to the double bond of allyl alcohol was developed in the mid-1980s. Allyl alcohol is accordingly used as an intermediate compound for synthesizing raw materials such as epichlorohydrin and 1,4-butanediol, and this development is bringing about expansion of the range of uses of allyl alcohol.

PHYSICAL PROPERTIES

Allyl alcohol is a colorless liquid having a pungent odor; Its vapor may cause severe irritation and injury to eyes, nose, throat, and lungs. It is also corrosive. Allyl alcohol is freely miscible with water and miscible with many polar organic solvents and aromatic hydrocarbons, but is not miscible with n-hexane. It forms an azeotropic mixture with water and a ternary azeotropic mixture with water and organic solvents. Allyl alcohol has both bacterial and fungicidal effects. Properties of allyl alcohol are shown in Table 1.

Table 1. Properties of Allyl Alcohol

Property	Value
molecular formula	C_3H_6O
molecular weight	58.08
boiling point, °C	96.90
freezing point, °C	−129.00
density, d_4^{20}	0.8520
refractive index, n_D^{20}	1.413
viscosity at 20°C, mPa·s(= cP)	1.37
flash point, °C	25
solubility in water at 20°C, wt%	infinity

[a] Closed cup.

CHEMICAL PROPERTIES

Addition Reactions

The C=C double bond of ally alcohol undergoes addition reactions typical of olefinic double bonds.

Hydroformylation

Hydroformylation of allyl alcohol is a synthetic route for producing 1,4-butanediol, a raw material for poly(butylene terephthalate), an engineering plastic.

Substitution of Hydroxyl Group

The substitution activity of the hydroxyl group of allyl alcohol is lower than that of the chloride group of allyl chloride and the acetate group of allyl acetate. However, allyl alcohol undergoes substitution reactions under conditions in which satura- ted alcohols do not react. Reactions proceed in catalytic systems in which a π-allyl complex is considered as an intermediate.

Oxidation

The C=C double bond of allyl alcohol undergoes epoxidation by peroxide, yielding glycidol. This epoxidation reaction is applied in manufacturing glycidol as an intermediate for industrial production of glycerol.

INDUSTRIAL MANUFACTURING PROCESSES FOR ALLYL ALCOHOL

There are four processes for industrial production of allyl alcohol. One is alkaline hydrolysis of allyl chloride. A second process has two steps. The first step is oxidation of propylene to acrolein and the second step is reduction of acrolein to allyl alcohol by a hydrogen transfer reaction, using isopropyl alcohol. At present, neither of these two processes is being used industrially. Another process is isomerization of propylene oxide. Until 1984, all allyl alcohol manufacturers were using this process. Since 1985 Showa Denko K.K. has produced allyl alcohol industrially by a new process that they developed. This process, which was developed partly for the purpose of producing epichlorohydrin via allyl alcohol as the intermediate, has the potential to be the main process for production of allyl alcohol. The reaction scheme is as follows:

$$CH_2{=}CHCH_3 + CH_3COOH + 1/2\,O_2 \xrightarrow{Pd} CH_2{=}CHCH_2O\overset{\displaystyle O}{\overset{\displaystyle \|}{C}}CH_3$$

$$CH_2{=}CHCH_2O\overset{\displaystyle O}{\overset{\displaystyle \|}{C}}CH_3 + H_2O \underset{\longleftarrow}{\overset{H^+}{\longrightarrow}} CH_2{=}CHCH_2OH + CH_3COOH$$

MONOALLYL DERIVATIVES

In this article, mainly monoallyl compounds are described. Diallyl and triallyl compounds used as monomers are covered elsewhere.

REACTIVITY OF ALLYL COMPOUNDS

Hydrosilylation

The addition reaction of silane

to the C=C double bond of allyl compounds is applied in the industrial synthesis of silane coupling agents.

π-Allyl Complex Formation

Allyl halide, allyl ester, and other allyl compounds undergo oxidative addition reactions with low atomic valent metal complexes to form π-allyl complexs.

PHYSICAL PROPERTIES OF DERIVATIVES

The physical properties of some important monallyl compounds are summarized in Table 2.

Table 2. Properties of Important Allyl Compounds

Property	Allyl chloride	Allyl acetate	Allyl methacrylate	AGE[a]	Allyl amine	DMAA[b]
molecular formula	C_3H_5Cl	$C_5H_8O_2$	$C_7H_{10}O_2$	$C_6H_{10}O_2$	C_3H_7N	$C_5H_{11}N$
molecular weight	76.53	100.12	126.16	114.14	57.10	85.15
boiling point, °C	44.69	104	150	153.9	52.9	64.5
freezing point, °C	−134.5	−96	−60	−100	−88.2	
density, d_4^{20}	0.9382	0.9276	0.934	0.9698	0.7627	0.72
viscosity at 20°C, mPa·s(= cP)	3.36	0.52	13	1.20		0.44
flash point, °C	−31.7	6	33	57.2	−29	−23

[a] Allyl glycidyl ether.
[b] Dimethylallylamine.

ALLYL CHLORIDE

This derivative, abbreviated AC, is a transparent, mobile, and irritative liquid. It can be easily synthesized from allyl alcohol and hydrogen chloride. However, it is industrially produced by chlorination of propylene at high temperature.

Uses

Allyl chloride is industrially the most important allyl compound among all the allyl compounds. It is used mostly as an intermediate compound for producing epichlorohydrin, which is consumed as a raw material for epoxy resins.

ALLYL ESTERS

Allyl Acelate

Allyl acetate is produced mostly for manufacturing allyl alcohol.

Allyl Methacrylate

At present, allyl methacrylate, AMA, is used mostly as a raw material for silane coupling agents.

ALLYL ETHERS

The C—H bond of the allyl position easily undergoes radical fission, especially in the case of allyl ethers, reacting with the oxygen in the air to form peroxide compounds.

Therefore, in order to keep allyl ether for a long time, it must be stored in an air-tight container under nitrogen.

Allyl Glycidyl Ether

This ether is used mainly as a raw material for silane coupling agents and epichlorohydrin rubber.

ALLYL AMINES

Allylamine

This amine can be synthesized by reaction of allyl chloride with ammonia at the comparatively high temperature of 50–100°C, or at lower temperatures using $CuCl_2$ or CuCl as the catalyst.

Dimethylallylamine

Dimethylallylamine is used in the production of insecticides and pesticides.

SAFETY AND HANDLING

Most allyl compounds are toxic and many are irritants. Those with a low boiling point are lachrymators. Precautions should be taken at all times to ensure safe handling.

Allyl Alcohol, Technical Publication SC: 46–32, Shell Chemical Corp., San Francisco, Calif., Nov. 1, 1946.

H. Raech, Jr. *Allylic Resins and Monomers*, Reinhold Publishing Corporation, New York, 1965.

C. E. Schildknecht, *Allylic Compounds and Their Polymers*, John Wiley & Sons, Inc., New York, 1973.

NOBUYUKI NAGATO
Showa Denko K.K.

ALLYL MONOMERS AND POLYMERS

Allyl compounds comprise a large group of ethylenic compounds having unique reactivities and uses often contrasting with those of typical vinyl-type compounds (styrenes, acrylics, vinyl esters and ethers, and related compounds). In allyl compounds the double bond is not substituted by a strong activating group to promote polymerization but is attached to a carbon which generally bears one or more reactive hydrogen atoms. Unlike monovinyl compounds, monoallyl compounds do not form homopolymers of high molecular weight by free-radical or conventional ionic mechanisms; in general, only viscous liquid homopolymers of limited use have been obtained. This is explained by the low reactivity of the ethylenic double bond together with the high reactivity of hydrogen atoms on the allylic carbon in reducing the molecular weight by degradative chain transfer.

In contrast, many allyl compounds containing two or more reactive double bonds yield solid, high molecular weight polymers by initiation with suitable free-radical catalysts. A number of polyfunctional allyl esters have achieved importance in polymerization and copolymerization especially to obtain heat-resistant cast sheets and thermoset moldings. Another use is of minor proportions of polyfunctional allyl esters, eg, diallyl maleate, triallyl cyanurate, and triallyl isocyanurate, for crosslinking or curing preformed vinyl-type polymers such as polyethylene and vinyl chloride copolymers. These reactions are examples of graft copolymerization in which specific added peroxides or high energy radiation achieve optimum cross-linking. Small proportions of mono- or polyfunctional allylic monomers also may be added as regulators or modifiers of vinyl polymerization for controlling molecular weight and polymer properties. Polyfunctional allylic compounds of high boiling point and compatibility are employed as stabilizers against oxidative degradation and heat discoloration of polymers.

DIALLYL CARBONATE CAST PLASTICS

From a number of diallyl esters investigated, diallyl diglycol carbonate or diethylene glycol bis(allyl carbonate), DADC, was developed to produce by bulk polymerization cast sheets, lenses, and other shapes of outstanding scratch resistance, and optical and mechanical properties.

DADC Monomers

Reaction of allyl alcohol in the presence of alkali with diethylene glycol bis(chloroformate), obtained from the glycol and phosgene, gives the DADC monomer.

DADC monomer is a colorless liquid of mild odor. It is low in toxicity but can produce skin irritation. It is

Table 1. Typical Properties of Commercial DADC Monomer

Property	Value
appearance	clear, colorless liquid
color, APHA	10
odor	none to slight
specific gravity	1.15^{20}_{4}
refractive index, n_{D}^{20}	1.452
boiling point at 266 Pa[a], °C	166
melting point (supercooled), °C	−4 to 0
viscosity at 25°C, mm^2/s (=cSt)	15
flash point	
Seta closed cup, °C	173
Cleveland open cup, °C	186
water content, slightly hygroscopic, %	0.1

[a] To convert Pa to mm Hg, multiply by 0.0075.

fairly resistant to saponification by dilute alkali. Contact with strong alkali at higher temperature produces the more toxic allyl alcohol. Properties are given in Table 1. DADC is soluble in common organic solvents and in methyl methacrylate, styrene, and vinyl acetate. It is partially soluble in amyl alcohol, gasoline, and ligroin. It is insoluble in ethylene glycol, glycerol, and water.

The DADC monomer is available from several manufacturers, such as PPG Industries (CR-39), Akzo (Nouryset 200), Enichem (RAV), Rhône-Poulene (XR-80), Tokuyama Soda (TS-16), and Mitsui Toatsu (MR-3).

DADC Homopolymerization

Bulk polymerization of CR-39 monomer gives clear, colorless, abrasion-resistant polymer castings that offer advantages over glass and acrylic plastics in optical applications. Free-radical initiators are required for thermal or photochemical polymerization.

Casting of DADC

Sheets, rods, and lens preforms are cast from CR-39 or prepolymer syrup by methods similar to those used for methacrylate ester syrups. Usage in impact-resistant, lightweight eyewear lenses has grown rapidly and is now the principal application.

Coatings

In recent years methods have been developed to improve abrasion resistance of DADC polymer surfaces in optical devices and glazings by means of special coatings. Hard or glasslike coatings may be applied by near-vacuum vapor deposition of quartz (silica) or by hydrolysis of alkoxysilanes.

Modified Polymers and Copolymers

DADC pure monomer and mixtures with small amounts of comonomers or other additions are commercially available for casting. Monomer formulations are available including agents for protecting the eyes against uv light. Another grade is designed to absorb infrared radiation, and several modified monomers give copolymers of increased heat resistance and hardness.

Polymeric Nuclear-Track Detectors

DADC polymer is used in solid-state track detectors (SSTD) of nuclear particles, including alpha-particles, fast neutrons, cosmic rays, and ions of elements of atomic number 10 and above.

OTHER ALLYL CARBONATE POLYMERS

In bulk polymerization, triallyl carbonates show less than the 13% shrinkage of CR-39. For example, a trimethylolpropane derivative of average molecular weight 300 was treated with phosgene, and the resulting chloroformate, treated with allyl alcohol, gave a polyfunctional allyl carbonate monomer. The purified monomer was heated with a percarbonate initiator to form a polymer lens.

DIALLYL PHTHALATES AND THEIR POLYMERIZATION

The three isomeric diallyl phthalates are colorless liquids of mild odor, low volatility, and relatively slow polymerization in the early stages. At ca 25% conversion, the viscous liquid undergoes gelation and polymerization accelerates; however, the last monomer disappears at a slow rate. The monomers are prepared by conventional esterification.

Properties of two diallyl phthalate monomers, $C_{14}H_{14}O_4$, are given in Table 2.

DAP Copolymerization

The diallyl phthalates copolymerize readily with monomers bearing strong electron-attracting groups attached to the ethylenic group. These include maleic anhydride, maleate and fumarate esters, and unsaturated polyesters.

Diallyl Isophthalate

DAIP polymerizes faster than DAP, undergoes less cyclization, and yields cured polymers of better heat resistance, eg, up to ca 200°C. Besides application as heat-resistant molding powders for electronic and other applications, DAIP copolymers have been proposed for optical applications.

Table 2. Properties of Commercial Diallyl Esters[a]

Property	DAP[b]	DAIP[c]
CAS Registry Number	[131-17-9]	[1087-21-4]
boiling point, °C at 0.53 kPa[d]	161	181
density, g/mL	1.117^{25}	1.124^{20}
refractive index, n_{D}^{25}	1.518	1.5212
surface tension at 20°C, Pa[e]	3.9	3.54
viscosity at 20°C, mPa s(=cP)	12	17
freezing point, °C	below −70	−3
flash point, °C		171
solubility in gasoline at 25°C, %	24	miscible

[a] Sources: Osaka Soda Company, Hardwick Chemical Company, and FMC Corporation.
[b] Diallyl phthalate.
[c] Diallyl isophthalate.
[d] To convert kPa to mm Hg, multiply by 7.5.
[e] To convert Pa to dyn/cm^2, multiply by 10.

Telomerization

Polymerization of DAP is accelerated by telogens such as CBr_4, which are more effective chain-transfer agents than the monomer itself; gelation is delayed.

Uses

The largest use of diallyl phthalate thermoset polymers is in moldings and coatings for electronic devices requiring high reliability under long-term adverse environmental conditions. Diallyl phthalates are used with glass cloth and roving in tubular ducts, radomes, aircraft, and missile parts of high heat resistance. They offer the advantages of low volatility, little odor, and high heat resistance, DAP and DAIP can be polymerized by high energy radiation in lens molds. Coatings of silica and alumina by vaporization give antiglare, scratch-resistant lenses.

OTHER DIALLYL ESTERS

Tables 3 and 4 give properties of some diallyl esters. Dimethallyl phthalate has been copolymerized with vinyl acetate and benzoyl peroxide, and reactivity ratios have been reported.

ALLYL–VINYL COMPOUNDS

Monomers such as allyl methacrylate and diallyl maleate have applications as cross-linking and branching agents selected especially for the different reactivities of their double bonds; some physical properties are given in Table 4. These esters are colorless liquids soluble in most organic liquids but little soluble in water; DAM and DAF have pungent odors and are skin irritants.

Allyl Methacrylate (AMA)

Of the compounds containing both allyl and vinyl-type double bonds, allyl methacrylate is the most important. AMA is used as cross-linking agent with methacrylate esters in contact lenses. AMA is also used in low concentrations in curable acrylic coatings.

Table 3. Properties of Some Diallyl Esters

Diallyl ester	Molecular formula	CAS Registry Number	$Bp_a{}^a$	n_D^{20}	d_4^{20}
oxalate	$C_8H_{10}O_4$	[615-99-6]	$107_{1.9}$	1.4460	1.0081
malonate	$C_9H_{12}O_4$	[1797-75-7]	$119_{1.9}$	1.4489	1.060
succinate	$C_{10}H_{14}O_4$	[925-16-6]	$94_{0.13}$	1.4507	1.056
adipate	$C_{12}H_{18}O_4$	[2998-04-1]	$115_{0.13}$	1.4542	1.023
sebacate	$C_{16}H_{26}O_4$	[3137-00-6]	$164_{0.26}$	1.4550	0.978
tartrate	$C_{10}H_{14}O_6$	[57833-54-2]	$171_{1.3}$		1.187

^{a} To convert Pa to mm Hg, multiply by 0.0075.

POLYFUNCTIONAL ALLYL NITROGEN MONOMERS

Triallyl Cyanurate as Cross-linking Agent

Triallyl cyanurate (TAC), 2,4,6-tris(allyloxy)-s-triazine, 2,4,6-Tris(allyloxy)-s-triazine, and its isomer triallyl isocyanurate (TAIC) are used as cross-linking agents with comonomers and for aftercuring preformed polymers such as olefin copolymers in electrical insulations. TAC monomer melts at 20–25°C. It is prepared by gradual addition of cyanuric chloride to an excess of allyl alcohol in the presence of aqueous alkali. Properties of TAC and TAIC are given in Table 5.

Triallyl Cyanurate Cure of Preformed Polymers

TAC and TAIC are often used in small amounts with vinyl-type and condensation polymers for cured plastics, rubber and adhesive products of high strength, and heat and solvent resistance. In some cases, chemical stability is also improved. TAC has been applied to curing PVC elastomers. The use of TAC as a curing agent continues to grow for polyolefins and olefin copolymer plastics and rubbers.

TAIC as Curing Agent

Triallyl isocyanurate homopolymers are brittle, intractable, and of little use. Small amounts of TAIC together with DAP have been used to cure unsaturated polyesters in glass-reinforced thermosets.

Table 4. Properties of Allyl–Vinyl Monomers

Property	Allyl methacrylate, AMA	Diallyl maleate, DAM	Diallyl fumarate, DAF
structure	$CH_2{=}C(CH_3){-}COOA^a$	HC—COOA ‖ HC—COOA^a	AOOCCH ‖ HC—COOA^a
molecular formula	$C_7H_{10}O_2$	$C_{10}H_{12}O_4$	$C_{10}H_{12}O_4$
CAS Registry Number	[96-05-9]	[999-21-3]	[2807-54-7]
boiling point, $°C_{kPa}{}^b$	55_4	$112_{0.53}$	$140_{0.40}$
density at 25°C, g/cm³	0.930	1.070	1.0516
refractive index, n_D^{25}	1.453	1.4664^c	1.4669
viscosity mPa·s(=cP)	13	4.3	3.0
flash point, open cup, °C		123	74

^{a} A = allyl = $-CH_2CH{=}CH_2$
^{b} To convert kPa to mm Hg, multiply by 7.5.
^{c} At 20°C.

Table 5. Properties of TAC and TAIC

Property	TAC	TAIC
molecular formula	$C_{12}H_{15}N_3O_3$	$C_{12}H_{15}N_3O_3$
CAS Registry Number	[101-37-1]	[1025-15-6]
melting point, °C	31	24
boiling point, °C	$140_{67\ Pa}$[a]	$126_{40\ Pa}$[a]
density at 30°C, g/cm^3	1.1133	1.1720
refractive index n_D^{25}	1.5049	1.5115

[a] To convert Pa to mm Hg, multiply by 0.0075.

Diallyl Ammonium Polymers

N,N-Diallyldimethyl (DADM) ammonium salts are used for the preparation of polyelectrolytes used for aqueous coal flotation. DADM ammonium polyelectrolyte is used in coatings for copy paper. Copolymers of diallyldimethylammonium chloride with acrylamide have been used in electroconductive coatings. Molded polyamide surfaces can be hardened by grafting with N,N-diallylacrylamide monomer under exposure to electron beam. N,N-Diallyltartardiamide is a cross-linking agent for acrylamide reversible gels in electrophoresis.

OTHER ALLYL COMPOUNDS

Although much research has been carried out with allyl ethers there has been only limited commercial use. Multifunctional allyl glycidyl ether, a toxic liquid, has been used as an additive to epoxy resins, in copolymers with ethylene oxide and derivatives, and in copolymers with vinyl comonomers.

A great number of other allyl compounds have been prepared, especially allyl ethers and allyl ether derivatives of carbohydrates and other polymers. These compounds are miscible with most organic solvents. Much research has been directed toward the preparation of air-drying prepolymers or oligomers. Thus allyl groups have been introduced into low molecular weight polyesters, polyurethanes, and formaldehyde condensates, but curing rates have been low.

L. S. Luskin, *Modern Plastics Encyclopedia*, McGraw-Hill, New York, 1984–1985.

J. J. Mauer in E. A. Turi, ed., *Thermal Characteristics of Polymeric Materials*, Academic Press, Inc., New York, 1981, 571–708.

C. E. Schildknecht, *Allyl Compounds and Their Polymers*, Wiley-Interscience, New York, 1973.

R. DOWBENKO
PPG Industries

ALUMINATES

Among industrial users of sodium aluminate are producers of paper, paint pigments, silica, alumina or alumina-based catalysts, dishwasher detergents, molecular sieves, concrete, antacids, and others. Sodium aluminate is used in removal of phosphates from municipal and industrial waste waters and for clarification of industrial process and potable water. Commercial sodium aluminate products are available as liquids, and to a lesser degree, in solid form. The formula of anhydrous sodium aluminate is variously given as $NaAlO_2$ (aluminum sodium oxide), $Na_2O \cdot Al_2O_3$, or $Na_2Al_2O_4$. Commercial sodium aluminates are not accurately represented by these formulas because the products contain more than the stoichiometric amount of sodium oxide, Na_2O. The amount of excess caustic in commercial products is indicated by ratios of Na_2O/Al_2O_3 that are typically between 1.05 and 1.15 for dry products, and 1.26 and 1.5 for liquids.

PHYSICAL AND CHEMICAL PROPERTIES

Commercial grades of sodium aluminate contain both waters of hydration and excess sodium hydroxide. In solution, a high pH retards the reversion of sodium aluminate to insoluble aluminum hydroxide. The chemical identity of the soluble species in sodium aluminate solutions has been the focus of much work. Solutions of sodium aluminate appear to be totally ionic. The aluminate ion is monovalent and the predominant species present is determined by the Na_2O concentration.

MANUFACTURE

Small amounts of sodium aluminate are prepared in the lab by fusion of equimolar quantities of sodium carbonate and aluminum acetate, $Al(C_2H_3O_2)_3$, at 800°C. Other methods involve reaction of sodium hydroxide with amorphous alumina or aluminum metal. Commercial quantities of sodium aluminate are made from hydrated alumina, in the form of aluminum hydroxy oxide, $AlO(OH)$, or aluminum hydroxide, $Al(OH)_3$, a product of the Bayer process which is used to refine bauxite, the principal aluminum ore.

Commercial grades of sodium aluminate are obtained by digestion of aluminum trihydroxide in aqueous caustic at atmospheric pressure and near the boiling temperature.

USES

Sodium aluminate is used in the treatment of industrial and municipal water supplies. It is also an effective precipitant for soluble phosphate in sewage and is especially useful in wastewater having low alkalinity.

Large quantities of sodium aluminate are used in papermaking where it improves sizing, filler retention, and pitch deposition. The addition of sodium aluminate to titanium dioxide paint pigment improves the nonchalking performance of outdoor paints.

Sodium aluminate is widely used in the preparation of alumina-based catalysts. Aluminosilicate can be prepared by impregnating silica gel with alumina obtained from sodium aluminate and aluminum sulfate. Reaction of sodium aluminate with silica or silicates has produced porous crystalline aluminosilicates which are useful as adsorbents and catalyst support materials, ie, molecular sieves.

J. R. Glastonbury, *Chem. Ind.* (London), **121** (Feb. 1969).
U.S. Pat. 382,505 (May 8, 1888), K. J. Bayer.
U.S. Pat. 515,859 (Mar. 6, 1894), K. J. Bayer.

ROBERT J. KELLER
JOSEPH J. LEN
Vinings Industries, Inc.

ALUMINUM AND ALUMINUM ALLOYS

Aluminum, Al, is a silver-white metallic element in group III of the periodic table having an electronic configuration of $1s^2 2s^2 2p^6 3s^2 3p^1$. Aluminum exhibits a valence of $+3$ in all compounds except for a few high temperature gaseous species in which the aluminum may be monovalent or divalent. Aluminum is the most abundant metallic element on the surfaces of the earth and moon, comprising 8.8% by weight (6.6 atomic %) of the earth's crust. However, it is rarely found free in nature. Nearly all rocks, particularly igneous rocks, contain aluminum as aluminosilicate minerals.

Aluminum reflects radiant energy throughout the spectrum. It is odorless, tasteless, nontoxic, and nonmagnetic. Because of its many desirable physical, chemical, and metallurgical properties, aluminum is the most widely used nonferrous metal. The utility of the metal is enhanced by the formation of a stable adherent oxide surface that resists corrosion. Because of high electrical conductivity and lightness, aluminum is used extensively in electrical transmission lines. High purity aluminum is soft and lacks strength but its alloys, containing small amounts of other elements, have high strength-to-weight ratios. Alloys of aluminum are readily formable by many metalworking processes; they can be joined, cast, or machined and accept a wide variety of finishes. Aluminum, having a density about one-third that of ferrous alloys, is used in transportation and structural applications where weight saving is important.

PHYSICAL PROPERTIES

The properties of aluminum vary significantly according to purity and alloying. Physical properties for aluminum of a minimum of 99.99% purity are summarized in Table 1.

CHEMICAL PROPERTIES

Reactions with Elements and Inorganic Compounds

Aluminum reacts with oxygen O_2, having a heat of reaction of -1675.7 kJ/mol (-400.5 kcal/mol) Al_2O_3 produced.

Table 1. Physical Properties of Aluminum

Property	Value
atomic number	13
atomic weight	26.9815
density at 25°C, kg/m^3	2698
melting point, °C	660.2
boiling point, °C	2494
thermal conductivity at 25°C, W/m · K)	234.3
latent heat of fusion, J/g^a	395
latent heat of vaporization at bp ΔH_v, kJ/g^a	10,777
electrical conductivity	65% IACS^b
electrical resistivity at 20°C, Ω·m	2.6548×10^{-8}
temperature coefficient of electrical resistivity, Ω·m/°C	0.0043
electrochemical equivalent, mg/°C	0.0932
electrode potential, V	-1.66
magnetic susceptibility, g^{-1}	0.6276×10^{-6}
Young's modulus, MPa^c	65,000
tensile strength, MPa^c	50

^{a} To convert J to cal, divide by 4.184.
^{b} International Annealed Copper Standard.
^{c} To convert MPa to psi, multiply by 145.

Aluminum does not combine directly with hydrogen, but it does react with nitrogen sulfur and carbon in oxygen-free atmospheres at high temperatures. Very high purity aluminum, resistant to attack by most acids, is used in the storage of nitric acid, concentrated sulfuric acid, organic acids, and other chemical reagents. Aluminum is, however, dissolved by aqua regia.

Aluminum is attacked by salts of more noble metals. In particular, aluminum and its alloys should not be used in contact with mercury or mercury compounds.

Reaction with Organic Compounds

Aluminum is not attacked by saturated or unsaturated, aliphatic or aromatic hydrocarbons. Halogenated derivatives of hydrocarbons do not generally react with aluminum except in the presence of water, which leads to the formation of halogen acids. The chemical stability of aluminum in the presence of alcohols is very good and stability is excellent in the presence of aldehydes, ketones, and quinones.

MANUFACTURE AND PROCESSING

Raw Materials

Aluminum, the third most abundant element in the earth's crust, is usually combined with silicon and oxygen in rock. When aluminum silicate minerals are subjected to tropical weathering, aluminum hydroxide may be formed. Rock that contains high concentrations of aluminum hydroxide minerals is called bauxite. Although bauxite is, with rare exception, the starting material for the production of aluminum, the industry generally refers to metallurgical grade alumina, Al_2O_3, extracted from bauxite by the Bayer Process, as the ore. Aluminum is obtained by electrolysis of this purified ore Fig. 1.

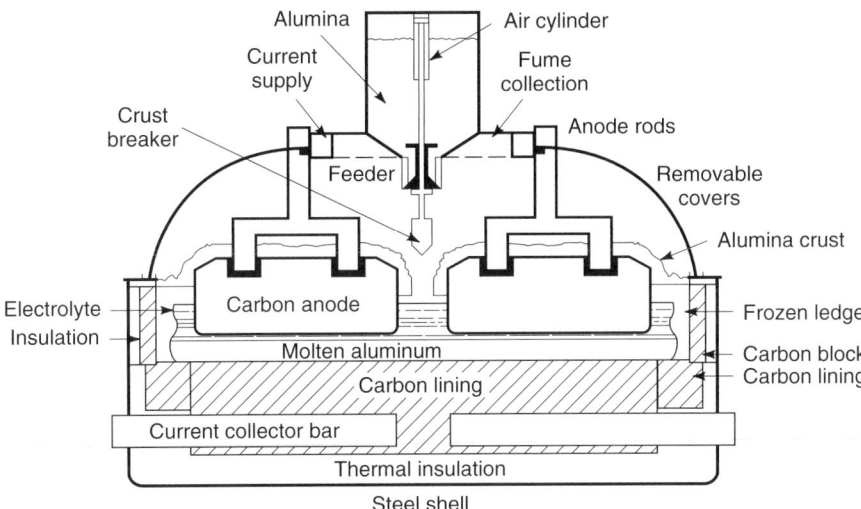

Figure 1. Aluminum electrolyzing cell with prebaked anode.

Energy Consideration

Table 2 gives a breakdown of the energy required to produce aluminum. Note that smelting consumes a about 65% of the required energy. In the United States most of this energy comes from fossil fuels. In other parts of the world, hydropower is a significant source of power for melting aluminum.

Energy Conservation. The U.S. Department of Energy (DOE) has sponsored research on inert anodes and refractory hard metal (RHM) composite cathodes. Success in these developments could significantly lower the energy required to reduce alumina. Inert anodes would eliminate the consumption of anode carbon and allow improved sealing of the cell for reduced heat loss and reduced fluoride emission. RHM cathodes would be wetted with a thin film of aluminum which would drain to a sump and provide stable cathodic surface rather than the present aluminum pool that sloshes about, owing to electrohydrodynamic effects. The stable cathode should improve current efficiency and also allow closer interelectrode spacing for reduced power consumption. The most promising route to reduced energy requirement, however, is through recycling of scrap aluminum. Recycling scrap requires less than 5% of the energy required to produce new metal. The recycling rate for aluminum beverage cans in the United States has typically been in the 62–67% range for the decade of the 1990's and into the 2005. Where possible, these recycling approaches should be extended to other aluminum scrap.

High Purity Aluminum

The Hall-Héroult process cannot ensure aluminum purity higher than 99.9%. Techniques such as electrolytic refining and fractional crystallization are required to produce metal of higher purity.

PRODUCTION

World smelter production and capacity are given in Table 3. In 2000, U.S. production of aluminum was 3.7×10^6 t. U.S. metal recovered from new and old scrap decreased by 7% to 3.45×10^6 t.

ECONOMIC ASPECTS

Aluminum prices have historically been more stable than other nonferrous metals. Since the 1970s, however,

Table 2. Energy Consumption Per Metric Ton of Aluminum Produced[a]

Operation	Thermal, MJ[b]	Electric, kW·h	Total energy	
			Fossil and hydro, MJ[b]	If all fossil, MJ[b]
mining and refining	30,000	480	35,200	35,200
smelting	19,000[c]	15,000	146,400	182,500
mill processing	19,000	1,830	33,700	38,900
Total	*68,000*	*17,310*	*215,300*	*256,600*

[a] Values are approximate. Actual energy consumption depends upon the particular plant, alloy produced, and product formed.
[b] To convert MJ to Mcal, divide by 4.184.
[c] Includes forming, baking, and fuel value of anodes.

Table 3. Growth of Aluminum Production Compared to Other Metals, $\times 10^3$t/yr

Year	Al[a]	Cu[b]	Mg[c]	Pb[c]	Zn
1900	5.7	449	0.01	877	479
1950	1,516	2,791	21	1,752	1,985
1960	4,732	4,631	93	2,436	3,019
1970	9,780	6,885	223	3,660	5,022
1980	16,043	7,984	317	5,456	6,115
1990	18,174	9,668	368	5,699	7,086
1999	23,074	11,337	393	6,120	8,406

[a] Al - primary production only.
[b] Cu - smelter production only.
[c] Mg, Pb - include both primary and secondary.

aluminum prices have fluctuated. These fluctuations reflect increased energy costs as well as increased costs of raw materials. Improvements in production processes as well as a rebalancing of demand and supply are expected to stabilize aluminum prices in the future.

U.S. imports for consumption decreased in 2000. Total exports increased 7% in 2000. Canada remains the largest shipper to the United States.

ENVIRONMENTAL CONSIDERATIONS

Fluoride emission from aluminum smelting cells has long been an area of great concern. Treatment consists of highly (over 99%) efficient dry scrubbers that catch particulates and adsorb HF on alumina that is subsequently fed to the cells. Hence, nearly all the fluoride evolved is fed back into the cell.

Hydrocarbon fumes evolved during anode baking are generally disposed of by burning. This treatment also catches the fluoride evolved during anode baking. Handling of alumina and coke presents dusting problems. Hoods and exhaust systems collect the dust, which is then separated from the exhaust air either by cyclones, electrostatic precipitators, filter bags, or a combination of these methods, and recycled to the process. Fumeless fluxing procedures remove hydrogen and undesirable metallic impurities.

The linings of aluminum reduction cells must be replaced periodically. These spent linings represent the largest volume of waste associated with the smelting process. Because they contain fluorides and cyanide, they must be either stored under roof or buried in landfills lined with impervious materials to prevent leaching and contamination of the environment.

Aluminum recovered in 2001 from purchased scrap was about 3.2×10^6 t, of which 60% came from new (manufacturing) scrap and 40% from old scrap (discarded aluminum products). Aluminum recovered from old scap was equivalent to approximately 20% apparent consumption.

ALUMINUM ALLOYS

Many of the properties of aluminum alloy products depend on metallurgical structure which is controlled both by the chemical composition and by processing. In addition to features such as voids, inclusions, grains, subgrains, dislocations, and vacancies which are present in virtually all metallic products, the structure of aluminum alloys is characterized by three types of intermetallic particles. Aluminum metallurgists refer to these as constituent particles, dispersoid particles, and precipitate particles. Constituent particles are formed during solidification, generally as a by-product of a divorced eutectic reaction, and range in size in the final product from about 1–20 micrometers. These negatively affect toughness of high strength alloy products. Dispersoids and precipitates both form by a solid-state reaction. The particles known as dispersoids characteristically form during thermal treatment of an ingot by precipitation of solid solution which exceeds maximum solid solubility because of nonequilibrium conditions during ingot solidification. Dispersoids, about 10–200 nm in the largest dimension, are present in most aluminum alloy products. Their primary function is to control grain size, grain orientation (texture), and degree of recrystallization. Particles classified as precipitates from during heat treatment of the final mill product by precipitation from a supersaturated solid solution that does not exceed the maximum equilibrium solid solubility. In the final product, their size may range from disks a few atoms thick by a few nm in diameter up to needlelike or platelike particles which may exceed 1 micrometer in the largest dimension. Precipitates may confer high strength. The nature of the constituent, dispersoid, and precipitate particles depends strongly on the phase diagrams of the particular alloy.

Binary Alloys

Aluminum-rich binary phase diagrams show three types of reaction between liquid alloy, aluminum solid solution, and other phases: eutectic, peritectic, and monotectic.

Al–Fe. The Al–Fe system, is important because virtually all commercial aluminum alloys contain some iron, Fe. The system has a eutectic at 1.9% Fe, but solid solubility of only 0.05% Fe.

Al–Mn. The Al–Mn system, the basis for the oldest yet most widely used aluminum alloys, is characterized by a eutectic at 1.95% Mn and 658°C. Maximum solid solubility is 1.76% manganese, Mn, and the intermetallic phase in aluminum-rich alloys is Al_6Mn.

Al–Cu. Many structural aluminum alloys contain significant amounts of copper, Cu. There is a eutectic in the Al–Cu system at 33.2% Cu and 548°C, but the important feature is the maximum solubility of 5.7% Cu at 548°C, which decreases drastically at lower temperatures. This decreasing solubility with decreasing temperature is necessary for the phenomenon known as age hardening or precipitation strengthening.

Al–Mg. Almost every commercial structural aluminum alloy contains magnesium as an alloying element. The Al–Mg system has a eutectic at 35% magnesium Mg, and 451°C. Maximum solid solubility is 14.9% Mg, and solubility decreases to about 0.8% Mg at room temperature. Despite this decreased solubility, precipitation strengthening by the mestable β′-phase precursor to the equilibrium β-phase Al_3Mg_2 precipitates is observed only at very high magnesium levels.

Al–Si. Al–Si alloys possess high fluidity and castability and are consequently used for weld wire, brazing, and as casting alloys. This system has a eutectic at 12.6% Si and 577°C; maximum solid solubility of Si is 1.65%.

Al–Li. Alloys containing about 2 to 3 percent lithium Li, earlier received much attention because of their low density and high elastic modulus. Each weight percent of lithium in aluminum alloys decreases density by about 3 percent and increases elastic modulus by about 6 percent. The system is characterized by a eutectic reaction at 8.1% Li at 579°C. The maximum solid solubility is 4.7% Li.

Al–Cr. Although no commercial alloys are based on this system, chromium, Cr, is an ingredient of several complex and commercially significant alloys. The Cr is added for control of grain structure. The Al–Cr system has a peritectic portion at 661°C where solid solubility is 0.7% Cr and liquid solubility is 0.4% Cr.

Al–Pb. Both lead, Pb, and bismuth, Bi, which form similar systems, are added to aluminum alloys to promote machinability by providing particles to act as chip breakers.

Al–Zn. Aluminum-rich binary alloys are not age hardenable to any commercial significance, and zinc, Zn, additions do not significantly increase the ability of aluminum to strain harden. Al–Zn alloys find commercial use as sacrificial claddings on high strength aircraft sheet or as sacrificial components in heat exchangers.

Al–Zr. This system has a peritectic reaction at 660.8°C at which solubility is 0.28% zirconium Zr, solid and 0.11% Zr liquid.

Ternary Alloys

Almost all commercial alloys are of ternary or higher complexity. Alloy type is defined by the nature of the principal alloying additions, and phase reactions in several classes of alloys can be described by reference to ternary phase diagrams. Minor alloying additions may have a powerful influence on properties of the product because of the influence on the morphology and distribution of constituents, dispersoids, and precipitates.

Al–Fe–Si

Iron and silicon, present in primary aluminum, may also be added to produce enriched alloys for specific purposes.

Al–Mg–Si

An important class of commercial alloys is based on the Al–Mg–Si system because of its precipitation hardening capabilities and good corrosion resistance.

Al–Mg–Mn

The basis for the alloys used as bodies, ends, and tabs of the cans used for beer and carbonated beverages is the Al–Mg–Mn alloy system. It is also used in other applications that require excellent weldability and corrosion resistance. These alloys have the unique ability to be highly strain hardened yet retain a high degree of ductility.

Al–Cu–Mg

The first precipitation hardenable alloy was an Al–Cu–Mg alloy. There is a ternary eutectic at 508°C, and there

are nine binary and five ternary intermetallic phases. For aluminum-rich alloys, only four phases are encountered in addition to the aluminum solid solution. Several commercial alloys are based on the age hardening characteristics of the metastable precursors of θ or S-phase, principally θ′ or S′.

Al–Mg–Zn

Although neither aluminum-rich binary Al–Zn nor Al–Mg alloys are precipitation hardenable, the ternary system is a source of alloys strengthened in this manner. Alloy compositions are selected for precipitation of an M- or η-phase precursor because T-phase is less effective as a strengthener. Commercially important alloys always contain more zinc than magnesium to provide attractive combinations of strength, extrudability, and weldability.

Al–Cu–Li

Although the addition of Cu to Al–Li alloys increases density, the boost in strength more than offsets the density increase so that Al–Cu–Li alloy products develop higher specific strengths (strength/density) than do binary Al–Li alloy products. Furthermore, the fracture toughness and corrosion resistance of products manufactured from Al–Cu–Li alloys are higher than these properties in binary Al–Li alloy products.

Al–Li–Mg

In aluminum-rich alloys the ternary phase, T, sometimes designated as Al_2LiMg, is encountered in addition to AlLi (δ), Al_3Mg_2 (β), and $Al_{12}Mg_{17}$ (γ). Assessment of the composition of the T-phase indicates that it contains 15.5 atomic % Mg and 32 atomic % Li.

Quaternary and Higher Alloys

Further additions to commercial aluminum alloys usually are made either to modify the metastable strengthening precipitates or to produce dispersoids.

Modifications to Precipitates. Silicon is sometimes added to Al–Cu–Mg alloys to help nucleate S′ precipitates without the need for cold work prior to the elevated temperature aging treatments. Additions of elements such as tin, Sn, cadmium, Cd, and indium In, to Al–Cu alloys serve a similar purpose for θ′ precipitates. Copper is often added to Al–Mg–Si alloys in the range of about 0.25% to 1.0% Cu to modify the metastable precursor to Mg_2Si. The copper additions provide a substantial strength increase. When the copper addition is high, the quaternary $Al_4CuMg_5Si_4$ Q-phase must be considered and dissolved during solution heat treatment.

The highest strength aluminum alloy products are based on the Al–Cu–Mg–Zn system and all are strengthened by precursors to the η-phase.

When combined with magnesium, silver has found commercial use as an alloying element in several aluminum alloys for specialized applications. It was added to an Al–Cu–Mg–Zn forging alloy to increase the resistance to stress-corrosion cracking and to an Al–Cu–Mg casting alloy to increase strength.

Dispersoid Formers. The three elements commonly added to precipitation hardenable alloys to form dispersoids are manganese, chromium, and zirconium. The amounts customarily used (0.5% Mn, 0.2% Cr, and 0.1% Zr) remain in supersaturated solid solution during ingot casting and precipitate as dispersoids during thermal treatment of the ingot. These dispersoids serve to minimize recrystallization during solution heat treatment of products such as plate, forgings, and extrusions which are hot-worked, and to maintain a fine recrystallized grain size in sheet and tubing which is cold-worked.

Foundry Alloys and Their Characteristics

Unalloyed aluminum does not have either mechanical properties or casting characteristics suitable for general foundry use, yet both can be greatly improved by the addition of other elements. The most common addition is silicon, which enhances fluidity, increases resistance to hot-cracking, and improves pressure tightness. Because binary Al–Si alloys have relatively low strengths and ductility, other elements such as copper and magnesium are added to obtain higher strengths through heat treatment. The compositions of representative foundry alloys are shown in Table 4.

Wrought Alloys and Their Characteristics

Alloys for the production of wrought products are selected for fabricability as well as their physical, chemical, and mechanical properties. Usually, these alloys are less highly alloyed than those for foundry use and contain less iron and silicon. A series of alloys based on the eutectic Al–Fe–Si composition, however, has been developed to provide good combinations of strength and formability in thin sheet products.

THERMAL TREATMENT OF ALLOYS

Aluminum alloys are subjected during manufacture to a variety of thermal treatments that range from heating to assist fabrication, to heating for control of final properties. Although the natural oxide film on aluminum provides good protection against surface oxidation and deterioration during such treatments, controlled atmospheres are sometimes employed for products requiring minimum surface oxide such as foil and sheet for reflectors.

Homogenization

Ingots are usually preheated prior to rolling, forging, or extrusion to increase workability. The process is commonly referred to as homogenization because chemical segregation of the major alloying elements that are completely soluble in the solid-state is reduced.

Annealing

The resistance to further deformation of aluminum alloy products at elevated temperatures reaches a steady value after a modest strain when the rate of formation of fresh dislocations is balanced by the rate of annihilation. This process is known as dynamic recovery. In-process annealing is employed to decrease the dislocation density thereby increasing the plasticity of the hot-rolled metal prior to cold rolling. This thermal process also

Table 4. Compositions of Aluminum Foundry Alloys

Aluminum Association designation	Casting process[a]	Alloying elements, %				Applications
		Si	Cu	Mg	Others[b]	
208.0	S	3.0	4.0			general purpose
213.0	P	2.0	7.0			cylinder heads, timing gears
242.0	S,P		4.0	1.5	2.0 Ni	cylinder heads, pistons
295.0	S	1.1	4.5			general purpose
B295.0	P	2.5	4.5			general purpose
308.0	P	5.5	4.5			general purpose
319.0	S,P	6.0	3.5			engine parts, piano plates
A332.0	P	12.0	1.0	1.0	2.5 Ni	pistons, sheaves
F332.0	P	9.5	3.0	1.0		pistons, elevated temperatures
333.0	P	9.0	3.5	0.3		engine parts, meter housings
355.0	S,P	5.0	1.3	0.5		general: high strength, pressure tightness
356.0	S,P	7.0		0.3		intricate castings: good strength, ductility
360.0	D	9.5		0.5	2.0 Fe max	marine parts, general purpose
380.0	D	8.5	3.5		2.5 Fe max	general purpose
A413.0	D	12.0				large intricate parts
443.0	D	5.3			2.0 Fe max	carburetors, fittings, cooking utensils
B443.0	S,P	5.3			0.8 Fe max	general purpose
514.0	S			4.0		hardware, tire molds, cooking utensils
520.0	S			10.0		aircraft fittings
A712.0	S		0.5	0.7	6.5 Zn	general purpose

[a] S, sand cast; P, permanent mold cast; D, pressure die cast.
[b] Aluminum and impurities constitute remainder.

modifies the crystallographic texture, a very important consideration in producing products requiring control of anisotropy.

Annealing is also employed as a final mill operation to produce a material having high formability for subsequent customer shaping or forming operations.

Solution Heat Treatment

Solution heat treatment is the first stage of a series of operations to achieve precipitation hardening.

Quenching

After solution treatment, the product is generally cooled to room temperature at such a rate to retain essentially all of the solute in solution.

Precipitation Heat Treatment

The supersaturated solution produced by the quench from the solution temperature is unstable, and the alloys tend to approach equilibrium by precipitation of solute. Because the activation energies required to form equilibrium precipitate phases are higher than those to form metastable phases, the solid solution decomposes to form G-P zones at room temperature (natural aging). Metastable precursors to the equilibrium phases are formed at the temperatures employed for commercial precipitation heat treatments (artificial aging).

SHAPING AND FABRICATING OF ALLOYS

Aluminum alloys are commercially available in a wide variety of cast forms and in wrought mill products produced by rolling, extrusion, drawing, or forging. The mill products may be further shaped by a variety of metal working and forming processes and assembled by conventional joining procedures into more complex components and structures.

CORROSION

Aluminum and aluminum alloys are employed in many applications because of the ability to resist corrosion. Corrosion resistance is attributable to the tightly adherent, protective oxide film present on the surface of the products. This film is 5–10 nm thick when formed in air; if disrupted it begins to form immediately in most environments. The loss in strength as a result of atmospheric weathering and corrosion is small, and the rate decreases with time. The amount of corrosion that occurs is a function of the alloy as well as the severity of the corrosive environment. Wrought alloys of the Al, Al–Mn, Al–Mg, and Al–Mg–Si types have excellent corrosion resistance in most weathering exposures including industrial and seacoast atmospheres. Alloys based on additions of copper, or copper, magnesium, and zinc, have significantly lower resistance to corrosion.

USES

Packaging has replaced the building and construction industry as the largest consumer of aluminum in the United State because aluminum is impermeable to gas, resistant to corrosion, and recyclable. The most prominent use of aluminum has been in containers for beer and carbonated beverages.

The largest market worldwide for aluminum products is in the building and construction industry.

Because of aluminum's low density, the field of transporation is another large market for aluminum alloys.

Aluminum is used in the home as household foil (0.18 mm thick), cooking utensils (the first commercial use of aluminum), refrigerators, air conditioners, appliances, insect screening, and hardware. It is also used for toys, sporting equipment, lawn furniture, lawn mowers, and portable tools.

Aluminum is an excellent conductor of electricity, having a volume conductivity 62% of that of copper. It also has many applications in the chemical and petrochemical industries such as for piping and tanks in alloys 1100, 3003, 6061, 6063, and the Al-Mg alloys.

J. Hatch, ed., *Aluminum: Properties and Physical Metallurgy*, American Society for Metals, Metals Park, Ohio, 1984.

T. G. Pearson, *The Chemical Background of the Aluminum Industry*, Monogr. 3, The Royal Institute of Chemistry, 1955.

P. A. Plunkert, "Aluminum," *Mineral Commodity Summaries*, U.S. Geological Survey, Reston, Va., Jan. 2002.

J. T. Staley, R. J. Rioja, R. K. Wyss, and J. Liu, *Processing to Improve High Strength Aluminum Alloy Products*, 9th Int. Conf. on Production Research, Cincinnati, Ohio, 1987.

ROBERT E. SANDERS, JR.
Alcoa Technical Center

ALUMINUM COMPOUNDS, SURVEY

The CAS registry lists 5,037 aluminum-containing compounds exclusive of alloys and intermetallics. Some of these are listed in Table 1.

CHEMICAL PROPERTIES

Aluminum, although highly electropositive, does not react with water under ordinary conditions because it is protected by a thin (2–3 nm) impervious oxide film that rapidly forms even at room temperature on nascent aluminum surfaces exposed to oxygen. If the protective film is overcome by amalgamation or scratching, water rapidly attacks to form hydrous aluminum oxide. Because of the tendency to amalgamate, aluminum and its alloys should not be used in contact with mercury or its compounds. Molten aluminum (mp 660°C) is known to react explosively with water. Thus the molten metal should not be allowed to touch damp tools or containers.

Table 1. Selected Aluminum Compounds from the CAS Registry.

Compounds	CAS Registry Number	Molecular formula
alum	[7784-24-9]	$KAl(SO_4)_2 \cdot 12\,H_2O$
alumina	[1344-28-1]	Al_2O_3
aluminum bromide	[77727-15-3]	$AlBr_3$
aluminum chlorhydroxide (ACH)	[12042-91-0]	$Al_2Cl(OH)_5$
aluminum(I) chloride	[13595-81-8]	$AlCl$
aluminum(III) chloride	[7446-70-6]	$AlCl_3$
aluminumchloride hexahydrate	[7784-13-6]	$AlCl_3 \cdot 6\,H_2O$
aluminum(I) fluoride	[13595-82-9]	AlF
aluminum(III) fluoride	[7784-18-1]	AlF_3
aluminum hydroxide	[21645-51-2]	$Al(OH)_3$
aluminum iodide	[7784-23-8]	AlI_3
aluminum(II) oxide	[14457-64-8]	AlO
aluminum silicate	[12141-46-7]	$Al_2(SiO_3)_3$
aluminum sulfate	[10043-01-3]	$Al_2(SO_4)_3$
aluminum sulfate octadecahydrate	[7784-31-8]	$Al_2(SO_4)_3 \cdot 18H_2O$
alunite	[12588-67-9]	$K_2Al_6(SO_4)_4(OH)_{12}$
anorthite	[1302-54-1]	$CaO \cdot Al_2O_3 \cdot 2\,SiO_2$
bauxite	[1318-16-7]	
boehmite	[1318-23-6]	$AlO(OH)$
calcium aluminate	[12042-78-3]	$Al_2O_3 \cdot 3\,CaO$
corundum	[1302-74-5]	$\alpha\text{-}Al_2O_3$
diaspore	[14457-84-2]	$\alpha\text{-}AlO(OH)$
gibbsite	[14762-49-3]	$\alpha\text{-}Al(OH)_3$
halloysite	[12244-16-5]	$Al_2Si_2O_5(OH)_4 \cdot 2\,H_2O$
kaolin	[1332-58-7]	$H_2Al_2Si_2O_8 \cdot H_2O$
kaolinite	[1318-74-7]	$Al_2O_3 \cdot 2\,SiO_2 \cdot 2\,H_2O$
kyanite	[1302-76-7]	$H_6O_5Si \cdot 2\,Al$
montmorillonite	[1318-93-0]	
nepheline	[12251-27-3]	$NaAl(OH)SiO_3$
nepheline	[12251-28-4]	$NaAl_2(OH)_2(SiO_3)_2 \cdot H_2O$
nepheline	[14797-52-5]	AlH_4O_4Si
sapphire	[1317-82-4]	Al_2O_3
sodium aluminate	[1302-42-7]	$NaAlO_2$
triethylaluminum	[97-93-8]	$(C_2H_5)_3Al$
triisobutylaluminum	[100-99-2]	$(C_4H_9)_3Al$
zeolite A	[1318-02-1]	$Na_{12}[(Al_{12}Si_{12})O_{48}] \cdot 27\,H_2O$
zeolite Y		$Na_{56}[(AlO_2)_{56}$ $(SiO_2)_{136}] \cdot 250\,H_2O$
zeolite X		$Na_{86}[(AlO_2)_{86}$ $(SiO_2)_{106}] \cdot 264\,H_2O$

COMMERCIALLY SIGNIFICANT COMPOUNDS

The aluminum containing compound having the largest worldwide market is metal grade alumina. Second is aluminum hydroxide. The split between additive and feedstock applications for $Al(OH)_3$ is roughly 50:50. Additive applications include those as flame retardants in products such as carpets and to enhance the properties of paper, plastic, polymer, and rubber products. Significant quantities are also used in pharmaceuticals, cosmetics, adhesives, polishes, dentifrices, and glass.

Feedstock applications of $Al(OH)_3$ for production of other chemicals include almost all of the 5000 plus compounds listed in the CAS registry.

Aluminum Sulfate (Alum)

Aluminum sulfate, $Al_2(SO_4)_3 \cdot 18\,H_2O$, also known as alum cake, is industrially produced by reaction of $Al(OH)_3$ and sulfuric acid, H_2SO_4, in agitated pressure vessels at about $170°C$. The commercial product has about 10% less water of hydration than the theoretical amount. Aluminum sulfate has largely replaced alums for the major applications as a sizing agent in the paper industry and as a coagulant to clarify municipal and industrial water supplies. In terms of worldwide production, it ranks third behind alumina and aluminum hydroxide, The U.S. exported 7,690 t and imported 23,500 t of aluminum sulfate in 2000.

Aluminum Halides

All the halogens form covalent aluminum compounds having the formula AlX_3. The commercially most important are the anhydrous chloride and fluoride, and aluminum chloride hexahydrate.

Anhydrous aluminum chloride, $AlCl_3$, is manufactured primarily by reaction of chlorine vapor with molten aluminum and used mainly as a catalyst in organic chemistry; Aluminum chloride hexahydrate, $AlCl_3 \cdot 6\,H_2O$, manufactured from aluminum hydroxide and hydrochloric acid, HCl, is used in pharmaceuticals and cosmetics as a flocculant and for impregnating textiles. Conversion of solutions of hydrated aluminum chloride with aluminum to the aluminum chlorohydroxy complexes serve as the basis of the most widely used antiperspirant ingredients.

Another cosmetic application of aluminum compounds is as lakes for lipstick manufacture. A water-soluble dye can become a lipstick ingredient if combined with compounds that are colorless and insoluble. The result, called a lake, is insoluble in both oil and water. Some dyes are laked with alumina; others are dissolved in water and treated with solutions that precipitate $Al(OH)_3$ with the dye molecules occluded in the precipitate. These lakes are mixed with castor oil, finely ground, and used as lipstick ingredients.

Organoaluminum Compounds

The alkyls and aryls, R_3Al (in monomer form), are colorless liquids or low melting solids easily oxidized and hydrolyzed when exposed to the atmosphere. Triethylaluminum (TEA), one of the most commercially important members of this family of chemicals, is so reactive it bursts into flame on contact with air; ie, it is pyrophoric, and it reacts violently with water. This behavior is typical and special techniques are necessary for the safe handling and use of organoaluminum compounds.

The alkylaluminum halides, R_nAlX_{3-n}, where X is Cl, Br, I, and R is methyl, ethyl, propyl, iso-butyl, etc, in monomer form, and $n = 1$ or 2, are less easily oxidized and hydrolyzed than the trialkyls. Organoaluminum hydrides such as diisobutylaluminum hydride [1191-15-7], (iso-$C_4H_9)_2$AlH, are also available. Organoaluminum compounds are used commercially in multimillion kg/yr quantities as catalysts or starting materials for the manufacture of organic compounds such as plastics, elastomers biodegradable detergents, and organometallics containing zinc, phosphorus, or tin.

Sodium Aluminate

Sodium aluminate is manufactured by dissolving high purity Al(OH)$_3$ in 50% sodium hydroxide solution. Sodium aluminate is used in water purification, in the paper industry, for the after treatment of TiO$_2$ pigment, and in the manufacture of aluminum containing catalysts and zeolite.

Zeolites

A large and growing industrial use of aluminum hydroxide and sodium aluminate is the manufacture of synthetic zeolites. Zeolites are aluminosilicates with Si/Al ratios between 1 and infinity. There are 40 natural, and over 100 synthetic, zeolites. All the synthetic structures are made by relatively low (100–150°C) temperature, high pH hydrothermal synthesis.

Zeolite-based materials are extremely versatile: uses include detergent manufacture, ion-exchange resins (ie, water softeners), catalytic applications in the petroleum industry, separation processes (ie, molecular sieves), and as an adsorbent for water, carbon dioxide, mercaptans, and hydrogen sulfide.

J. J. Eisch, *Comprehensive Organometallic Chemistry*, Vol. 1, Pergamon Press, Oxford, UK, 1982, Chapt. 6.

W. Gerhartz, ed., *Ullmann's Encyclopedia of Industrial Chemistry*, 5th ed., VCH, Weinheim, Germany, 1985.

L. D. Hart, ed., *Aluminum Chemicals: Science and Technology Handbook*, American Ceramics Society, Columbus, Ohio, 1990.

P. A. Plunkett, "Bauxite and Alumina", *Mineral Commodity Summaries*, U.S. Geological Survey, 2002.

WILLIAM C. SLEPPY
Aluminum Company of America

ALUMINUM HALIDES AND ALUMINUM NITRATE

The aluminum halides and aluminum nitrates have similar properties, with the exception of the family of aluminum fluoride compounds. Of the remaining members in this aluminum halide family, chloride derivatives are the most commercially important; aluminum bromide AlBr$_3$, aluminum iodide, AlI$_3$, and aluminum nitrate, Al(NO$_3$)$_3$ are of only minor commercial interest.

ALUMINUM CHLORIDE

The chemistry of aluminum chloride is influenced significantly by hydration. Aluminum chloride hexahydrate, AlCl$_3$·6H$_2$O, is a crystalline solid that dissolves easily in water, forming ionic species. Heating the hydrate results in the loss of hydrogen chloride, HCl, and formation of aluminum oxide, Al$_2$O$_3$. On the other hand, anhydrous aluminum chloride reacts violently with water, evolving heat, a gas consisting of hydrogen chloride and steam, and aluminum oxide particulates. Anhydrous aluminum chloride sublimes at 180°C, leaving no residue. The uses of anhydrous aluminum chloride and the hydrated form are also very different. The anhydrous material is a Lewis acid used as an alkylation catalyst. The hydrate is used principally as a flocculating aid.

Commercially, aluminum chloride is available as the anhydrous AlCl$_3$, as the hexahydrate, AlCl$_3$·6H$_2$O, or as a 28% aqueous solution designated 32°Be'. Polyaluminum chloride, or poly(aluminum hydroxy) chloride is a member of the family of basic aluminum chlorides. These are partially neutralized hydrates having the formula Al$_2$Cl$_{6-x}$(OH)$_x$·6H$_2$O, where $x = 1 - 5$.

ANHYDROUS ALUMINUM CHLORIDE

Properties

Anhydrous aluminum chloride is a hygroscopic, white solid that reacts with moisture in air. Properties are shown in Table 1.

Manufacture

In the United States anhydrous aluminum chloride is manufactured by the exothermic reaction of chlorine, Cl$_2$, vapor with molten aluminum.

Specifications and Packaging

Aluminum chloride's catalytic activity depends on its purity and particle size. Moisture contamination is an important concern; and exposure to humid air must be prevented to preserve product integrity. Moisture contamination can be determined by a sample's nonvolatile material content. After subliming, the material remaining is

Table 1. Physical Properties of Anhydrous Aluminum Chloride[a]

Property	Value
molecular weight	133.34054
density at 25°C, g/mL	2.46
sublimation temperature° C	180.2
triple point, C, 233 kPa[a]	192:5 ± 0:2
heat of formation, 25°C, kJ/mol[b]	−705:63 ± 0:84
heat of sublimation of dimer, 25°C, kJ/mol[b]	115:52 ± 2:3
heat of solution, 20°C, kJ/mol[b]	−329:1
heat of fusion, kJ/mol[b]	35:35 ± 0:84
entropy, 25°C, J/(K·mol)[b]	109:29 ± 0:42
heat capacity, 25°C, J/(K·mol)[b]	91.128

[a] To convert kPa to psi, multiply by 0.145.
[b] To convert J to cal, divide by 4.184.

principally nonvolatile aluminum oxide. Water contamination leads to a higher content of nonvolatile material.

In many chemical processes the catalyst particle size is important. The smaller the aluminum chloride particles, the faster it dissolves in reaction solvents. Particle-size distribution is controlled in the manufacturer's screening.

Aluminum chloride is available in a wide variety of moisture-free packages. Pails and drums are often used when fixed amounts of aluminum chloride are required for batch operations. For small operations, bags having a specially designed liner to maintain moisture-free product are available. For shipments from 200 to 1,200 kilograms net, suppliers offer 37.8, 75.7, 113.6, and 208-L drums. Semibulk bins hold up to 11,000 kilograms net. These returnable containers are constructed of fiberglass to make shipping, storage, and handling of aluminum chloride more convenient. Aluminum chloride can also be purchased in bulk truck trailers in quantity up to 90,000 kg net.

Safety and Handling

In storage, some reaction with moisture may occur and over time can lead to a pressure build-up from HCl in the container. Containers should be carefully vented before being opened. Safety goggles or face shields, rubber gloves, rubber shoes, and coveralls made of acid-resistant material should be used in handling. A NIOSH/OSHA-certified respirator is also required to prevent breathing fumes and dust. Aluminum chloride reacts with moisture in the skin, in the eyes, ears, nose, and throat.

Environmental Protection

Fumes resulting from exposure of anhydrous aluminum chloride to moisture are corrosive and acidic. Collection systems should be provided to conduct aluminum chloride dusts or gases to a scrubbing device. The choice of equipment, usually one of economics, ranges from simple packed-tower scrubbers to sophisticated high energy devices such as those of a Venturi design. Spills should be picked up before flushing thoroughly with water and neutralizing with soda ash or lime.

ALUMINUM CHLORIDE HEXAHYDRATE

The hexahydrate of aluminum chloride is a deliquescent, crystalline solid soluble in water and alcohol and usually made by dissolving aluminum hydroxide, $Al(OH)_3$, in concentrated hydrochloric acid.

Roofing granules and mineral aggregate for bituminous products are treated with aluminum chloride solution to improve adhesion of the asphalt. Pigmented coatings, containing sodium silicate, Na_2SiO_3, and used to color roofing granules, are insolubilized by spraying with aluminum chloride solution and then heating. Aluminum chloride hydrates are the alumina sources used in the manufacture of special forms of alumina and alumina-silica refractories. Aluminum chloride hydrate is used in textile finishing to impart crease recovery and

nonyellowing properties to cotton fabrics, antistatic characteristics to polyester, polymide, and acrylic fabrics, and to improve the flammability rating of nylon. Dye-bleeding of printed textile may be blocked by treatment with aluminum chloride and zinc acetate, $Zn(O_2CCH_3)_2$, followed by solubilizing with ethylenediamine tetraacetic acid, and washing from the fabric.

BASIC ALUMINUM CHLORIDES

The class of compounds identified as basic aluminum chlorides is used primarily in deodorant, antiperspirant, and fungicidal preparations. They have the formula $Al_2(OH)_{6-x}Cl_x$, where $x = 1-5$, and are prepared by the reaction of an excess of aluminum with 5–15% hydrochloric acid at a temperature of 67–97°C.

Hydrates of aluminum chloride and basic aluminum chlorides are also efective in a number of difficult water treatment problems.

ALUMINUM BROMIDE

Anhydrous aluminum bromide, $AlBr_3$, forms colorless trigonal crystals and exists in dimeric form, Al_2Br_6, in the crystal and liquid phases. Dissociation of the dimer to the monomer occurs in the gas phase. The bromide is produced commercially only in small quantities. This product melts at 97.45°C, boils at 256°C, and has a specific gravity at 25°C of 3.01.

Aluminum halides change from ionic to covalent character as the electronegativity of the halogen decreases (F > Cl > Br > I). Aluminum bromide, because of its covalent nature, is more soluble in many organic solvents than anhydrous aluminum chloride. Although its catalytic activity is moderate, it can be used in Friedel-Crafts reactions where selectivity is important. Anhydrous aluminum bromide, prepared from bromine and metallic aluminum, decomposes upon heating in air to bromine and alumina. Caution should be exercised in handling this hazardous compound because of its reactivity with water. Aluminum bromide may cause tissue burns, and both the anhydrous and the hydrate forms may be toxic upon ingestion.

Aluminum bromide hexahydrate, $AlBr_3·6H_2O$, may be made by dissolving aluminum or aluminum hydroxide in hydrobromic acid, HBr. This white, crystalline solid is precipitated from aqueous solution.

ALUMINUM IODIDE

Aluminum iodide, AlI_3, is a crystalline solid with a melting point of 191°C. The presence of free iodine in the anhydrous form causes the platelets to be yellow or brown. The specific gravity of this solid is 3.98 at 25°C. Aluminum iodide hexahydrate, $AlI_3·6H_2O$, and aluminum iodide pentadecahydrate, $AlI_3·15H_2O$, are precipitated from aqueous solution. They may be prepared by the reaction of hydroiodic acid, HI, with aluminum or aluminum hydroxide.

ALUMINUM NITRATE

Aluminum nitrate is available commercially as aluminum nitrate nonahydrate, $Al(NO_3)_3 \cdot 9 H_2O$. It is a white, crystalline material with a melting point of 73.5°C, that is soluble in cold water, alcohols, and acetone. Aluminum nitrate nonahydrate is prepared by dissolving aluminum or aluminum hydroxide in dilute nitric acid and crystallizing the product from the resulting aqueous solution. It is made commercially from aluminous materials such as bauxite.

Anhydrous aluminum nitrate is covalent in character, easily volatilized, and decomposes on heating. Hydrated aluminum nitrate is used in the preparation of insulating papers, on transformer core laminates, and in cathode-ray tube heating elements.

"Aluminum Chemicals," *Chemical Economics Handbook*, SRI International, Menlo Park, Calif., 2001.

P. J. Durrant and B. Durrant, *Introduction to Advanced Inorganic Chemistry*, John Wiley & Sons, Inc., New York, 1970, p. 570.

P. A. Plunkert, "Bauxite and Alumina," *Minerals Yearbook*, U.S. Geological Survey, Reston, Va, 2001.

K. Wade and A. J. Banister, *Comprehensive Inorganic Chemistry*, 1st ed., Pergamon Press, 1973.

G. W. GRAMS
Witco Corporation
B. CONLEY
T. SHAIKH
D. A. ATWOOD
University of Kentucky

ALUMINUM OXIDE (ALUMINA), ACTIVATED

The activated aluminas comprise a series of nonequilibrium forms of partially hydroxylated aluminum oxide, Al_2O_3. The chemical composition can be represented by $Al_2O_{(3-x)}(OH)_{2x}$ where x ranges from about 0 to 0.8. They are porous solids made by thermal treatment of aluminum hydroxide precursors and find application mainly as adsorbents, catalysts, and catalyst supports. Activated alumina, for purposes of this discussion, refers to thermal decomposition products (excluding α-alumina of aluminum trihydroxides, oxide hydroxides, and nonstoichiometric gelatinous hydroxides). The term "activation" is used in this article to indicate a change in properties resulting from heating (calcining).

Other names for these products are active alumina, gamma alumina, catalytic alumina, and transition alumina. Transition alumina is probably the most accurate because the various phases identified by x-ray diffraction are really stages in a continuous transition between the disordered structures immediately following decomposition of the hydrous precursors and the stable α-alumina which is the product of high temperature calcination.

PHYSICAL AND CHEMICAL PROPERTIES

In general, as a hydrous alumina precursor is heated, hydroxyl groups are driven off leaving a porous solid structure of activated alumina. The transformation is topotactic and little change in size or shape of the material is observed at low magnifications. At magnifications higher than about 10,000, changes in texture resulting from recrystallization can be seen. The physical properties of the material are set by the choice of precursor, the forming process, and the activation conditions.

Decomposition of Boehmite

Boehmite, AlO(OH), can be synthesized having surface areas ranging from about 1 to over 800 m^2/g, depending upon the method of preparation. The properties of activated boehmite products are strongly influenced by the crystallite size of the precursor material.

Activation Products of Aluminum Hydroxide

As gibbsite, α-Al(OH)$_3$, is heated, the surface area reaches a maximum of 300 m^2/g or more at about 650 K. As temperature is increased further, surface area decreases and the skeletal structure becomes more dense reflecting increased ordering of the crystalline structure during the progression from chi to kappa to alpha. At about 1450 K conversion to alpha alumina occurs, with a major rearrangement of crystal structure and corresponding decrease in surface area to about 5 m^2/g. These trend vary somewhat according to precursor crystal size, purity, and the atmosphere of heating.

MANUFACTURING PROCESSES

The large majority of activated alumina products are derived from activation of aluminum hydroxide, rehydrated alumina, or pseudoboehmite gel. Other commercial methods to produce specialty activated aluminas are roasting of aluminum chloride, $AlCl_3$, and calcination of precursors such as ammonium alum, $AlH_7NO_8S_2$. Processing is tailored to optimize one or more of the product properties such as surface area, purity, pore size distribution, particle size, shape, or strength. A process for the production of β-alumina solid electrolyte without calcination has been reported.

ECONOMIC ASPECTS

The least expensive products are those derived directly from Bayer-process gibbsite, and powders are generally less expensive than formed products. The soda content (0.2–0.3% Na_2O) of Bayer gibbsite makes it unattractive for many catalytic applications. Gel-based products are normally used where low soda level is required. Soda content of gels prepared from inorganic salts or aluminate solutions is typically about 0.03%, whereas soda in alkoxide-based gels is much lower. Specialty activated aluminas having purity as high as 99.99% are also available, at a much higher price. Shaped products used for adsorbent purposes are generally less sophisticated and therefore less expensive than catalytic products.

SAFETY AND HANDLING

Activated alumina is a relatively innocuous material from a health and safety standpoint. It is nonflammable and nontoxic. Fine dusts can cause eye irritation, and there is some record of lung damage because of inhalation of activated alumina dust mixed with silica and iron oxide. Normal precautions associated with handling of nuisance dusts should be taken. Activated alumina is normally shipped in moistureproof containers (bags, drums, sling bins) because of its strong desiccating action.

USES

Activated alumina is used commercially in catalytic processes as a catalyst, catalyst substrate, or as a modifying additive. Activated alumina serves as the catalyst in the Claus process for recovering sulfur from H_2S that originates from natural gas processing or petroleum refinery operations. Another catalytic application for promoted alumina is in automotive exhaust catalysts which enhance oxidation of hydrocarbons, carbon monoxide, and nitrogen oxide in exhaust gas. The largest tonnage single application for catalyst particles is in fluid cracking. These materials are typically made from zeolite having a clay or alumina–silica binder system to provide the necessary mechanical strength for fluid bed handling. A number of smaller, but nevertheless important, applications in which activated alumina is used as the catalyst substrate include alcohol dehydration, olefin isomerization, hydrogenation, oxidation, and polymerization. Activated alumina has been used for many years in the separation of various organic compounds by normal phase chromatography because of its natural hydrophilic surface characteristics. More recently, stable surface coatings have been developed which impart hydrophobic properties to the particle surface. One of the earliest uses for activated alumina was removal of water vapor from gases, and this remains an important application.

E. Bingham, B. Cohrssen, and C. H. Powell, eds., *Patty's Toxicology*, John Wiley & Sons, Inc., New York, Vol. 8, 2001, p. 1120.

J. C. Downing and K. P. Goodboy, "Claus Catalysts and Alumina Catalyst Materials and Their Application," in L. D. Hart, ed., *Alumina Chemicals Handbook*, American Ceramic Society, Westerville, Ohio, 1990.

B. C. Gates, J. R. Katzer, and G. C. A. Schuit, *Chemistry of Catalytic Processes*, McGraw-Hill, New York, 1979.

K. Wefers and C. Misra, *Oxides and Hydroxides of Aluminum, Alcoa Technical Paper 19*, revised, Alcoa Laboratories, Aluminum Company of America, Pittsburgh, Pa., 1987, p. 52.

ALAN PEARSON
Aluminum Company of America

ALUMINUM OXIDE (ALUMINA), CALCINED, TABULAR, AND ALUMINATE CEMENTS

CALCINED ALUMINA

Calcined aluminas are generally obtained from Bayer process gibbsite, α-Al(OH)$_3$, thermal decomposition of which follows the transition through the generic gamma alumina phases to α-alumina (corundum), α-Al$_2$O$_3$. Nonmineralized metal-grade or smelter-grade alumina (SGA) for aluminum production is calcined at lower temperatures and usually contains about 20 to 50% α-Al$_2$O$_3$. The remainder consists of higher temperature transition aluminas, usually theta, kappa, delta, and gamma, depending upon the consolidation of the original gibbsite structure, impurities, heating rate, and furnace atmosphere.

Preparation

Calcination of gibbsite has been done in rotary kilns for many years. Specialty calcined aluminas can also be prepared in stationary or fluid bed calciners similar to those used for producing SGA.

Ground, Calcined, and Reactive Aluminas. Most ceramic grade aluminas are supplied dry ground to about 95% −325 mesh (44 μm) using 85–90% Al$_2$O$_3$ ceramic ball, attrition, vibro-energy, or fluid-energy milling. Particles larger than 44 μm can be removed by air classification during continuous milling to produce 99+% −325 mesh product. More fully ground, or superground, calcined aluminas having particle size distributions that approximate the natural or ultimate crystal size of the Bayer grain as calcined are often desired.

Thermally reactive aluminas contain submicrometer crystals. These must be separated from the Bayer agglomerate during grinding to permit dense compaction upon ceramic forming, and thus, enhance densification upon sintering at lower temperatures. Such superground, thermally reactive aluminas exhibit higher densification rates when compacted and sintered into ceramic products, and complete densification is obtained about 200°C lower than using the coarser, continuously ground aluminas.

Specialty Aluminas. Process control techniques permit production of calcined specialty aluminas having controlled median particle sizes differentiated by about 0.5 μm. This broad selection enables closer shrinkage control of high tech ceramic parts. Production of pure 99.99% Al$_2$O$_3$ powder from alkoxide precursors, apparently in spherical form, offers the potential of satisfying the most advanced applications for calcined aluminas requiring tolerances of $\pm 0.1\%$ shrinkage.

USES

Calcined alumina markets consume slightly less than 50% of the specialty alumina chemicals production. Worldwide usage is estimated to be about 50% for refractories, 20% for abrasives, and 25% for ceramics. Calcined aluminas are also used in the manufacture of tabular alumina and calcium aluminate cements (CAC). Quantities are estimated to be over 200,000 and 100,000 t respectively.

World output of alumina increase 5% in 2000. Principal producing countries were Australia, the United States, China, and Jamaica. They accounted for 60% of world production.

Calcined aluminas are used in both electronic and structural ceramics. Electronic applications are dominant in the United States and Japan whereas mechanical applications are predominant in Europe. Specialty electronic integrated circuit packages generally use the low soda and thermally reactive aluminas.

Enamels, glass, chinaware glazes, china and hotel ware, and electrical porcelain insulators usually contain 5 to 25% alumina additions to increase strength and chip resistance, whereas electronic and mechanical alumina ceramics contain greater than 85% Al_2O_3 as calcined alumina. Coarse crystalline (2 to 10 μm) aluminas having 0.05 to 0.20% Na_2O are used in spark plug insulators, which is the largest use of alumina in the electronics field. High purity 99.99% Al_2O_3 is used to make translucent polycrystalline alumina tubes for sodium vapor lamps. Traditional glass tubes allow significant sodium diffusion at the operating temperatures of the sodium vapor lamps. All varieties of calcined aluminas are used in mechanical and technical applications. But when optimum hardness, density, and wear resistance are required, the thermally reactive aluminas are used in 95% and higher Al_2O_3 compositions. Lower price, normal soda calcined aluminas are used in compositions as low as 85% Al_2O_3 whenever lower performance can be tolerated.

Cutting tools of thermally reactive, high purity aluminas in combination with zirconia, titanium carbide or titanium nitride, the SIALONS, and boron, nitride have high mechanical strength, fracture toughness, and cutting behavior for high speed cutting of hard steel and cast iron. High mechanical strength, fine surface finish, high density, and high purity are also the requirements for alumina ceramics used in prosthetics such as hip joints and dental implants.

Other alumina ceramic applications include ceramic armor for bullet-proof vests, balls and rods for grinding media, abrasion-resistant tiles for lining coal and ash transfer lines in power stations, electrical high tension insulators, bioceramics, integrated electronic circuits, vacuum tube envelopes, r-f windows, rectifier housing, integrated circuit packages, and thick and thin film substrates.

TABULAR ALUMINA

Tabular alumina is a high density, high strength form of α-Al_2O_3 made by sintering an agglomerated shape of ground, calcined alumina. It is available in the form of smooth balls having diameters from 3 to 25 mm and imperfect 19 mm diameter spheres, which are crushed, screened, and ground to obtain a wide variety of graded, granular, and powdered products having various particle size distributions ranging from a top size of 12.7 mm to −325 mesh (44 μm).

Uses

The large α-Al_2O_3 crystals containing closed round pores make tabular alumina an excellent refractory raw material. Tabular alumina is the ideal base material for high alumina brick and monolith liners in the metal, ceramic, and petrochemical industries.

Tabular alumina also offers advantages over other materials as an aggregate in castables made from calcium aluminate cement. Other applications include their use in electrical insulators, electronic components, and kiln furniture.

ALUMINATE CEMENT

Refined calcined alumina is commonly used in combination with high purity limestone to produce high purity calcium aluminate cement (CAC). High purity CAC sinters readily in gas-fired rotary kiln calcinations at 1600–1700 K. CAC reactions are considered practically complete when content of free CaO is less than 0.15% and loss on ignition is less than 0.5% at 1373 K.

Uses

High purity CA cements are primarily used as binders for high strength refractory castables to form linings up to about 1.0 m thick, as, for example, in iron blast furnaces.

The high purity CAC finds extensive use as an efficient binder for other aggregates such as fire clays, kaolin, and alusite, kyanite, pyrophyllite, sillimanite, mullite, and refractory grade bauxite, having the added advantage of increasing the refractoriness of some of these aggregates. The many applications cited for tabular alumina in refractories are also common for high purity CAC.

High purity CAC is also used as a steel slag conditioner during ladle refining of steel.

Advances in Ceramics, Vol. 13, The American Ceramic Society, Columbus, Ohio, 1985.

T. J. Carbone, "Production Processes, Properties, and Applications for Calcined in High-Purity Aluminas," in L. D. Hart, ed., *Alumina Chemicals: Science and Technology Handbook*, The American Ceramic Society, Columbus, Ohio, 1990.

G. MacZura, "Production Processes, Properties, and Applications for Tabular Alumina Refractory Aggregates," in L. D. Hart, ed., *Alumina Chemicals: Science and Technology Handbook*, The American Ceramic Society, Columbus, Ohio, 1990.

T. D. Robson, *High-Alumina Cements and Concretes*, John Wiley & Sons, Inc., New York, 1962.

GEORGE MACZURA
Aluminum Company of America

ALUMINUM OXIDE (ALUMINA), HYDRATED

The terms "alumina hydrates" or "hydrated aluminas" are used in industry and commerce to designate aluminum hydroxides. These compounds are true hydroxides and do not contain water of hydration. Several forms are known; the most well-defined crystalline forms are the trihydroxides, Al(OH)$_3$: gibbsite, bayerite, and nordstrandite. In addition, two aluminum oxide–hydroxides, AlO(OH), boelimite and diaspore, have been clearly defined.

The terms "gelatinous alumina" or "alumina gel" cover a range of products in which colloidal hydrated alumina is the predominant solid phase. Structural order varies from x-ray indifferent (amorphous) to some degree of crystallinity. The latter product has been named pseudoboehmite or gelatinous boehmite. Its x-ray diffraction pattern shows broad bands that coincide with the strong reflections of the well-crystallized boehmite.

CRYSTALLINE ALUMINA HYDRATES

The mineralogical, structural, physical, and thermodynamic properties of the various crystalline alumina hydrates are listed in Tables 1, 2, and 3, respectively. X-ray diffraction methods are commonly used to differentiate between materials.

GELATINOUS ALUMINIUM HYDROXIDES

Apart from the crystalline forms, aluminum hydroxide often forms a gel. Fresh gels are usually amorphous, but crystallize on aging, and gel composition and properties depend largely on the method of preparation. Gel products have considerable technical use.

PHASE RELATIONS IN THE Al$_2$O$_3$–H$_2$O SYSTEM

Under equilibrium vapor pressure of water, the crystalline trihydroxides, Al(OH)$_3$ convert to oxide–hydroxides at above 100°C. Below 280–300°C, boehmite is the prevailing phase, unless diaspore seed is present. Although spontaneous nucleation of diaspore requires temperatures in excess of 300°C and 20 MPa (200 bar) pressure, growth on seed crystals occurs at temperatures as low as 180°C. For this reason it has been suggested that boehmite is the metastable phase, although its formation is kinetically favored at lower temperatures and pressures. The ultimate conversion of the hydroxides to corundum,

Table 1. Mineralogical Properties of Aluminum Hydroxides

Material	Index of refraction[a]			Cleavage	Brittleness	Mohs' hardness	Luster
	α	β	γ				
gibbsite	1.568	1.568	1.587	(001) perfect	tough	2½ – 3½	pearly vitreous
boehmite	1.649	1.659	1.665	(010)		3½ – 4	
diaspore	1.702	1.722	1.750	(010) perfect	brittle	6½ – 7	brilliant pearly

[a]The average index of refraction for bayerite is 1.583.

Table 2. Structural Properties of Aluminum Hydroxides

Material	Crystal system[a]	Space group	Unit axis length, nm			Angle	Density, g/cm^3
			a	b	c		
Al(OH)$_3$							
gibbsite	monoclinic[b]	C$_{2h}^5$	0.8684	0.5078	0.9136	94°34′	2.42
bayerite	monoclinic	C$_{2h}^5$	0.5062	0.8671	0.4713	90°27′	2.53
nordstrandite	triclinic	C$_1^1$	0.5114	0.5082	0.5127	70°16′ 74°0′ 58°28′	
AlO(OH)							
boehmite	orthorhombic	D$_{2h}^{17}$	0.2868	0.1223	0.3692		3.01
diaspore	orthorhombic	D$_{2h}^{16}$	0.4396	0.9426	0.2844		3.44

[a]Unit cell contains two molecules unless otherwise indicated.
[b]Unit cell contains four molecules.

Table 3. Thermodynamic Data for Crystalline Aluminum Hydroxides at 298.15 K and 0.1 MPa[a]

Substance	Molecular weight	Molar vol, cm^3/mol	ΔH_f, kJ/mol[b]	ΔG_f, kJ/mol[b]	$S°$, J/(mol·K)[b]	C_p, J/(mol·K)[b]
gibbsite	78.004	31.956	−1293.2	−1155.0	68.44	91.7
bayerite	78.004		−1288.2	−1153.0		
boehmite	59.989	19.55	−990.4	−915.9	48.43	65.6
diaspore	59.989	17.76	−999.8	−921.0	35.33	53.3

[a]To convert MPa to psi, multiply by 145.
[b]To convert J to cal, divide by 4.184.

Table 4. Properties of Commercial Grade Aluminum Hydroxides

Property	Normal coarse grade[a]	Normal white grade[b]	Ground[c]	Fine precipitated[d]
Al_2O_3, wt %	65.0	65.0	65.0	64.7
SiO_2, wt %	0.012	0.01	0.02	0.04
Fe_2O_3, wt %	0.015	0.004	0.03	0.01
Na_2O (total), wt %	0.40	0.15	0.30	0.45
Na_2O (soluble), wt %	0.05	0.05	0.05	0.1–0.25
LOI at 1200°C, wt %[e]	34.5	34.5	34.5	34.5
moisture at 100°C, wt %	0.1	0.1	0.4	0.3–1.0
specific gravity	2.42	2.42	2.42	2.42
bulk density (loose), g/cm^3	1.2–1.4	1.0–1.1	0.7–1.25	0.13–0.22
surface area, m^2/g	0.1	0.15	2–4	6–8
color	off-white	white	off-white	white
refractive index	1.57	1.57	1.57	1.57
Mohs' hardness	2.5–3.5	2.5–3.5	2.5–3.5	2.5–3.5

Particle size, cumulative wt%

retained 100 mesh = 149 μm	5–20	0–1		
retained 200 mesh = 74 μm	65–90	5–15		
retained 325 mesh = 44 μm	90–98	30–65		
passing 325 mesh = 44 μm	2–10	35–70	1–2	0.1–0.2
			98–99	99.8
median particle size, μm			6.5–9.5	0.6

[a] Alcoa C-30.
[b] Alcoa C-31.
[c] Alcoa C-330.
[d] Alcoa Aydral 710.
[e] Loss on ignition.

Al_2O_3, the final oxide form, occurs above 360°C and 20 MPa.

PRODUCTION

Aluminum hydroxides are technically the most widely used members of the alumina chemicals family. The most important source of aluminum hydroxides is the bauxite refining plant for alumina production. A small amount of somewhat purer aluminum hydroxide is produced by the Sinter process.

Several commercial grades of aluminium hydroxide are produced. The properties of some grades are given in Table 4.

Hydroxide grades can be surface-treated to modify dispersion behavior and rheological properties. The most widely used surface coating agents are stearic acid and stearates. Additionally, compounds from the silane group have been used as coupling agents to give improved adhesion to polymers when the hydroxide is used as a filler.

SHIPPING AND ANALYSIS

Shipping of aluminum hydroxide powders is usually in paper bags of 10 to 25 kg size. Bulk shipment by road or rail wagons is also common. Aluminum hydroxides are not hygroscopic but could be dusty; precautions against dust inhalation should be taken during handling.

ECONOMIC ASPECTS

U.S aluminum hydroxide production capacity was nearly 940×10^3t in 1997. About 90% of world production came from the Bayer process; the remaining came from Sinter, Ziegler, and gel processes. In 2000, 1×10^3t of aluminum hydroxide were consumed in the U.S.; 960×10^3t were consumed in Western Europe, and 430×10^3t were used in Japan.

HEALTH AND SAFETY FACTORS

Aluminum hydroxides are minimally absorbed by the body, and LD_{50} values for ingestion are unavailable. Death upon ingestion occurs from intestinal blockage rather than systemic aluminum toxicity. It is only as a fine particulate suspended in air that aluminum hydroxides may gain entry (via the lungs) into the body in amounts of physiological significance. Evidence collected among workers in the alumina refining industry has failed to show any effect of aluminum hydroxide dust on the lungs. However, in recognition of the possible adverse effects of long term exposure to alumina dusts, threshold limit values have been established by the ACGIH as follows: 10 mg/m^3 TLV–TWA and 20 mg/m^3 TLV–STEL. Aluminum hydroxide and aluminum hydroxide oxide are reported in EPA TSCA inventory.

USES

Aluminum hydrate used in flame retardants, reinforcement fillers in plastics, elastomers, and adhesives, filler pigments, coatings in papermaking, precursors for the production of activated alumina and other specialty aluminas, and as a raw material for the production of aluminum.

R. L. Bertholf, M. R. Wills, and J. Savory, "Aluminum" in H. G. Seiter and H. Sigel, eds., *Handbook on Toxicity of Inorganic Compounds*, Marcel Dekker, New York, 1988, Chapt. 6.

H. Ginsberg and K. Wefers, *Aluminum and Magnesium*, Vol. 15, Die Metallischen Rohstoffe, Enke Verlag, Stuttgart, Germany, 1971.

L. D. Hart, ed., *Alumina Chemicals Science and Technology Handbook*, The American Ceramic Society, Westerville, Ohio, 1990.

C. Misra, *Industrial Alumina Chemicals, ACS Monogr. 184*, American Chemical Society, Washington, D.C., 1986.

CHANAKYA MISRA
Aluminum Company of America

ALUMINUM SULFATE AND ALUMS

Aluminum sulfate octadecahydrate, $Al_2(SO_4)_3 \cdot 18H_2O$, and its aqueous solutions are used primarily in the paper industry for sizing and as a flocculating agent in water and wastewater treatment. This material is often called papermakers' alum or alum. Because this salt is precipitated from aqueous solution, aluminum sulfate hydrate, $Al_2(SO_4)_3 \cdot nH_2O$, can have variable composition and is sometimes referred to as cake alum or patent alum. The solid commercial hydrate, generally written as the 18-hydrate, is typically dehydrated to correspond to from 17.0–17.5% Al_2O_3 where n = 13–14. This dehydrated form is called dry alum, ground or lump. Aluminum sulfate solutions are typically 7.5–8.5% Al_2O_3 and are known as liquid alum.

Anhydrous aluminum sulfate, $Al_2(SO_4)_3$, is a specialty item used in food applications.

PROPERTIES

Over 50 acidic, basic, and neutral aluminum sulfate hydrates have been reported. Only a few of these are well-characterized because the exact compositions depend on conditions of precipitation from solution. Variables such as supersaturation, nucleation and crystal growth rates, occlusion, nonequilibrium conditions, and hydrolysis can each play a role in the final composition. Commercial dry alum is likely not a single crystalline hydrate, but rather it contains significant amounts of amorphous material.

MANUFACTURE

In the United States, aluminum sulfate is usually produced by the reaction of bauxite or clay with sulfuric acid.

OTHER ALUMS

The word alum is derived from the Latin *alumen*, which was applied to several astringent substances, most of which contained aluminum sulfate. Unfortunately, the term alum is now used for several different materials. Papermakers' alum or simply alum refers to commercial aluminum sulfate. Common alum or ordinary alum usually refers to potash alum which can be written in the form $K_2SO_4 \cdot Al_2(SO_4)_3 \cdot 24H_2O$, or it can refer to ammonium alum, ammonium aluminum sulfate. The term is also applied to a whole series of crystallized double sulfates $[M(I)M'(III)(SO_4)_2 \cdot 12H_2O]$ having the same crystal structure as the common alums, in which sodium and other univalent metals may replace the potassium or ammonium, and other metals may replace the aluminum. Even the sulfate radical may be replaced, by selenate, for example. Some examples of alums are cesium alum, $CsAl(SO_4)_2 \cdot 12H_2O$; iron alum, $KFe(SO_4)_2 \cdot 12H_2O$; chrome alum, $KCr(SO_4)_2 \cdot 12H_2O$; and chromoselenic alum, $KCr(SeO_4)_2 \cdot 12H_2O$.

Pseudoalums are a series of double sulfates, such as iron(II) aluminum sulfate, $FeSO_4 \cdot Al_2(SO_4)_3 \cdot 24H_2O$, containing a bivalent metal ion in place of the univalent element of ordinary alums. These pseudoalums have different crystal structures from those of the ordinary alums.

In industrial practice it is generally the aluminum content of alums that is important. Because aluminum sulfate is widely available, other alums are more in the nature of specialty items and are no longer produced in quantities comparable to those of aluminum sulfate.

W. Gerhartz, *Ullmann's Encyclopedia of Industrial Chemistry*, 5th ed., vol. **A1**, VCH, Deerfield Beach, Fla., 1985, pp. 527–534.

J. W. Mellor, *A Comprehensive Treatise on Inorganic and Theoretical Chemistry*, Vol. **5**, Longmans, Green and Co. Ltd., London, U.K., pp. 332–357.

Chemical Economics Handbook, Standord Research Institure, Menlo Park, Calif., Sept. 1991, parts 702.1000H, 702.1003U–1004P.

K. V. DARRAGH
C. A. GREEN
Rhône-Poulenc, Inc.

AMIDES, FATTY ACID

Fatty acid amides are of the general formula

$$R-\overset{\overset{\displaystyle O}{\|}}{C}-N\overset{\displaystyle R'}{\underset{\displaystyle R''}{<}}$$

in which R may be a saturated or unsaturated alkyl chain derived from a fatty acid. They can be divided into three categories. The first is primary monoamides in which R is a fatty alkyl or alkenyl chain of C_5–C_{23} and R' = R" = H. The second, and by far the largest category, is substituted monoamides, including secondary, tertiary, and alkanolamides in which R is a fatty alkyl or alkenyl chain of C_5–C_{23}; R′ and R″ may be a hydrogen, fatty alkyl, aryl, or alkylene oxide condensation groups with at least one alkyl, aryl, or alkylene oxide group. The third category is bis(amides) of the general formula

$$R-\overset{\overset{\displaystyle O}{\|}}{C}-\underset{\underset{\displaystyle R'}{|}}{\overset{\overset{\displaystyle R''}{|}}{N}}-(CH)_x-\underset{\underset{\displaystyle R'}{|}}{N}-\overset{\overset{\displaystyle O}{\|}}{C}-R$$

where R groups are fatty alkyl or alkenyl chains. R′ and R″ may be hydrogen, fatty alkyl, aryl, or alkylene oxide condensation groups. Other amides include halogenated amides and multifunctional amides such as amidoamines and polyamides.

PHYSICAL PROPERTIES

Many of the physical properties of fatty acid amides have been explained on the basis of the tautomeric

structures:

$$R-\overset{\overset{\displaystyle O}{\|}}{C}-\underset{\underset{\displaystyle H}{|}}{N}-H \rightleftharpoons R-\overset{\overset{\displaystyle O^-}{|}}{C}-\underset{\underset{\displaystyle H}{|}}{N^+}-H$$

Primary and secondary amides show strong hydrogen bonding that accounts for their high melting points and low solubilities in most solvents. With tertiary amides (disubstituted amides), hydrogen bonding is not possible as exhibited by their increased solubility and lower melting points.

Amides have a strong tendency to reduce friction by adsorption on surfaces. This coating action may be attributed to their hydrophobic character and strong hydrogen bonding.

CHEMICAL PROPERTIES

Amides in general are stable to elevated processing temperatures, air oxidation, and dilute acids and bases.

Table 1. Fatty Amides, Producers, and Trade Names

Product category, trade name	Company	Web site
	Alkyl amides	
Armid	Akzo Nobel Surface Chemistry AB	www.surface.akzonobel.com
Petrac Vyn-Eze Addit.	Ferro Corp./Polymer Additives Divison	www.ferro.com
	Ethylene bis(stearamide)	
Alkamide STEDA	Rhodia Home, Personal Care, Industrial Ingreds. (HPCII),	www.rpsurfactants.com
Glycowax 765	Lonza Inc.	www.lonza.com
Advawax	Rohm and Haas Co.	www.rohmhaas.com
Kemamide W-39	Crompton Corp./Olefins & Styrenics	www.uniroyalchemical.com
	Amido amines	
Indulin QTS	Westvaco Corp., Chemical Division	www.westvaco.com
	Ethoxylated amides	
Schercoterge 140	Scher Chemicals, Inc.	
Bermodol Amadol	Akzo Nobel Surface Chemistry AB	www.bermodol.com www.surface.akzonobel.com
	Alkanolamides	
Ablumide	Taiwan Surf.	www.taiwansurfactant.com.tw
Amidex	Chemron Corp.	www.chemron.com
Alkamide	Rhodia Home, Personal Care, Industrial Ingreds. (HPCII)	www.rpsurfactants.com
Mackamide	McIntyre	www.mcintyregroup.com
NINOL Manromid STEPANOL	Stepan Co Stepan UK Ltd.	www.stepan.com
Monamid	Uniqema	www.uniqema.com
Surfonamide Empilan	Huntsman Corp.	www.huntsman.com
Aminol Amidet	Kao Corp.	www.kao.co.jp
Calamide	Pilot Chemical Co.	www.pilotchemical.com
Chimipal Rolamid	Cesalpinia Chemicals SpA	www.cesalpinia.com
Comperlan	Cognis Deutschland GmbH	www.es.cognis.com
Witcamide	Akzo Nobel Surface Chemistry AB	www.surface.akzonobel.com
Foamid	Alzo International Inc.	www.alzointernational.com
Incromide	Croda Chemicals (SA) (Pty) Ltd	www.croda.com
Mazamide.	BASF Corp./Performance Chemicals	www.basf.com/businesses/chemicals/performance

Source: Industrial Surfactants Electronic Handbook, 2002 ed.

Stability is reduced in amides containing unsaturated alkyl chains; unsaturation offers reactive sites for many reactions. Hydrolysis of primary amides catalyzed by acids or bases is rather slow compared to other fatty acid derivatives. Even more difficult is the hydrolysis of substituted amides. The dehydration of amides that produce nitriles is of great commercial value.

SYNTHESIS AND MANUFACTURE

Unsubstituted Amides

The most widely used synthetic route for primary amides is the reaction of fatty acid with anhydrous ammonia.

Substituted Amides

Most monosubstituted and disubstituted amides can be synthesized with or without solvents from fatty acids and alkylamines.

Bisamides

Most bisamides are propared by the reaction of the primary fatty amide and formaldehyde in the presence of an acid catalyst, or by the reaction of ethylene diamine with fatty acid.

COMMERCIAL ASPECTS

Many fatty amides which are available from various manufactures are listed in Table 1.

R. Beck, in D. R. Karsa, ed., *Industrial Applications of Surfactants IV*, Royal Society of Chemistry Special Publication No. 230, 1999, pp. 115–129.

A. Lif and M. Hellsten, *Nonionic Surfactants*, Vol. 72, Marcel Dekker, New York, 1998, pp. 177.

A. L. McKenna, *Fatty Amides*, Witco Chemical Corporation, Tenn., 1982, p. 1.

S. H. Shapiro, in E. S. Pattison, ed., *Fatty Acids and Their Industrial Application*, Vol. 5, Marcel Dekker, New York, 1968, p. 77.

INGEGÄRD JOHANSSON
Akzo Nobel Surface Chemistry
AB

AMINE OXIDES

Amine oxides, known as *N*-oxides of tertiary amines, are classified as aromatic or aliphatic, depending on whether the nitrogen is part of an aromatic ring system or not. This structural difference accounts for the difference in chemical and physical properties between the two types.

The higher aliphatic amine oxides are commercially important because of their surfactant properties and are used extensively in detergents. Amine oxides that have surface-acting properties can be further categorized as nonionic surfactants; however, because under acidic conditions they become protonated and show cationic properties, they have also been called cationic surfactants. Typical commercial amine oxides include the types shown in Table 1.

Aromatic amine oxides, produced on a much smaller scale and having some pharmaceutical importance, do not demonstrate the surface-acting properties that the aliphatic amine oxides do.

CHEMICAL PROPERTIES

Decomposition

Most amine oxides undergo thermal decomposition between 90 and 200°C. Aromatic amine oxides generally decompose at higher temperatures than aliphatic amine oxides and yield the parent amine.

Reduction

Just as aromatic amine oxides are resistant to the foregoing decomposition reactions, they are more resistant than aliphatic amine oxides to reduction.

Table 1. Commercial Amine Oxides

Name	Molecular formula	Structural formula
dimethyl dodecyl amine oxide	$C_{14}H_{31}NO$	$CH_3(CH_2)_{11}\overset{\displaystyle CH_3}{\underset{\displaystyle CH_3}{N}}\!\rightarrow\!O$
dihydroxyethyl-dodecylamine oxide	$C_{16}H_{35}NO_3$	$CH_3(CH_2)_{11}\overset{\displaystyle CH_2CH_2OH}{\underset{\displaystyle CH_2CH_2OH}{N}}\!\rightarrow\!O$
dimethyltetra-decyl-amidopropyl amine oxide	$C_{20}H_{40}NO_2$	$CH_3(CH_2)_{13}\overset{\displaystyle O}{\overset{\|}{C}}NHCH_2CH_2CH_2\overset{\displaystyle CH_3}{\underset{\displaystyle CH_3}{N}}\!\rightarrow\!O$
N-dodecylmor-pholine *N*-oxide	$C_{16}H_{33}NO_2$	$CH_3(CH_2)_{11}\overset{\displaystyle CH_3}{\underset{\displaystyle CH_3}{N}}\!\rightarrow\!O$
1-hydroxyethyl-2-octadecyl imidazoline oxide	$C_{23}H_{46}N_2O_2$	$CH_3(CH_2)_{11}\overset{\displaystyle CH_2CH_2OH}{\underset{\displaystyle CH_2CH_2OH}{N}}\!\rightarrow\!O$
N,*N'*,*N'*-hydroxy ethyl-*N*-octadecyl-1,3-propylene-diamine oxide	$C_{27}H_{58}N_2O_5$	

Alkylation

Alkylating agents such as dialkyl sulfates and alkyl halides react with aliphatic amine oxides to form trialkylalkoxyammonium quaternaries.

Acylation

Aliphatic amine oxides react with acylating agents such as acetic anhydride and acetyl chloride to form either *N,N*-dialkylamides and aldehyde, the Polonovski reaction, or an ester, depending upon the polarity of the solvent used.

MANUFACTURING AND PROCESSING

Linear alpha-olefins are the source of the largest volume of aliphatic amine oxides. The olefin reacts with hydrogen bromide in the presence of peroxide catalyst, to yield primary alkyl bromide, which then reacts with dimethylamine to yield the corresponding alkyldimethylamine. Fatty alcohols and fatty acids are also used to produce amine oxides.

Amine oxides used in industry are prepared by oxidation of tertiary amines with hydrogen peroxide solution using either water or water and alcohol solution as a solvent.

ECONOMIC ASPECTS

Demand for amines in the United States is expected to grow to $\$ 1.9 \times 10^9$ in 2004. Specialty amines, the group in which amine oxides are categorized, lead the demand because of strong performance characteristics. A major use for amine oxides is as surfactants in a variety of soaps, detergents and personal care products.

HEALTH AND SAFETY FACTORS

Aliphatic amine oxides such as alkyldimethylamine oxides and alkylbis (2-hydroxylethyl)amine oxides range from practically nontoxic to slightly toxic.

Among the aromatics, 4-nitroquinoline *N*-oxide is a powerful carcinogen producing malignant tumors when painted on the skin of mice. The 2-methyl, 2-ethyl, and 6-chloro derivatives of 4-nitroquinoline oxide are also carcinogens.

USES

Aliphatic amine oxides find wide use in the detergent and personal care industries. Other uses for amine oxides are found in paper and textile production, electroplating, oil and petroleum, plastics and rubber, metal and mining, polymerization, and photographic industries.

Amine oxides compete with alkanolamides as foam boosters in the detergent and personal care industry. Although amine oxides are more expensive than alkanolamides they have the advantage of being milder to the skin and eyes and are more effective surfactants, so that on a cost performance basis they are a better buy than alkanolamides in many cases.

Alkyl amine oxides also increase viscosity, emolliency, detergency, and antistatic properties in many detergent and cosmetic formulas.

Amine oxides are used in synthetic organic chemistry in the preparation of olefins, or phase-transfer catalysts, in alkoxylation reactions, in polymerization, and as oxidizing agents.

In the area of textile and synthetic fiber processing, amine oxides have been used as dyeing auxiliaries as well as wetting agents, as antistatic agents, and as bleaching agents.

The biochemistry of heteroaromatic amine oxides has led to the synthesis of many biochemically and pharmaceutically important compounds. Aromatic amine oxides are useful as analgesics, antihistamines, antitussives, diuretics, tranquilizers, and drug potentiators. In many cases, the *N*-oxides of pharmacologically active tertiary amines have added benefits, ranging from lower toxicity and better solubility to enhanced therapeutic behavior. The biological activity of these materials has led to patented uses as bactericides, fungicides, insecticides, nematocides, filaricides, amoebicides, anthelmintics, antiparasitics, and disinfectants.

Other uses of aliphatic amine oxides are as corrosion inhibitors for nonferrous metals and in aqueous systems, as fuel oil anti-icing and pour-point additives that also depress combustion chamber fouling, in the plastic industry as molecular weight regulators in ethylene and propylene copolymerization, in photography to prevent waterspots in drying photographic films, as complexing developers and dyes, and as asphalt emulsifiers.

L. W. Burnette, in M.J. Shick, ed., *Nonionic Surfactants*, Vol. I, Marcel Dekker, Inc., New York, 1967, pp. 403–410.

E. Ochiai, *Aromatic Amine Oxides*, Elesevier Publishing Co., Amsterdam, 1967.

J. D. Sauer, in J.M. Richmond, ed., *Surfactant Science Series*, Vol. 34, Marcel Dekker, New York, 1990, pp. 275–295.

P. A. S. Smith, *The Chemistry of Open-Chain Organic Nitrogen Compounds*, Vol. II, W.A. Benjamin, Inc., New York, 1966, pp. 21–28.

B. Maisonneuve
Akzo Chemicals, Inc.

AMINES BY REDUCTION

Amines are derivatives of ammonia in which one or more of the hydrogens is replaced with an alkyl, aryl, cycloalkyl, or heterocyclic group. When more than one hydrogen has been replaced, the substituents can either be the same or different. Amines are classified as primary, secondary, or tertiary, depending on the number of hydrogens that have been replaced. General structures for ammonia as well as primary, secondary, and tertiary amines are shown below.

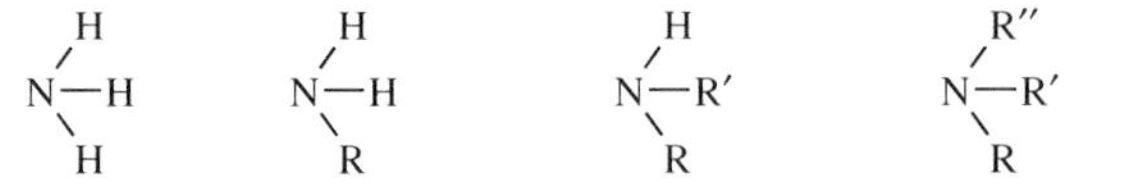

ammonia   primary amine   secondary amine   tertiary amine

In reductive methods of making amines, the nitrogen is already incorporated in the molecule, and the amine is formed by reducing the oxidation state of the compound with the addition of hydrogen. In theory, many different types of nitrogen-containing compounds can be reduced to amines. In practice, however, nitriles or nitro compounds are usually used because they are the most easily obtained starting materials.

There are several commercial processes for reducing nitro or nitrile groups to amines. Most large-volume aromatic and aliphatic amines are made by continuous high-pressure catalytic hydrogenation. Nitro compounds can also be reduced in good yields with iron and hydrochloric acid in the Béchamp process. Other more specialized, methods used for making amines by reduction are also used on occasion.

CATALYTIC HYDROGENATION

In catalytic hydrogenation, a compound is reduced with molecular hydrogen in the presence of a catalyst. Some examples of these reactions follow:

Nitro $R=NO_2 + 3 H_2 \longrightarrow R=NH_2 + 2 H_2O$

Nitrile $R=CN + 2 H_2 \longrightarrow R=CH_2NH_2$

Amide
$$\underset{\text{}}{R-\overset{\overset{O}{\|}}{C}-NH_2} + 2 H_2 \longrightarrow R-CH_2NH_2 + H_2O$$

Thioamide
$$\underset{\text{}}{R-\overset{\overset{S}{\|}}{C}-NH_2} + 2 H_2 \longrightarrow R-CH_2NH_2 + H_2S$$

Azo $R-N=N-R' + 2 H_2 \longrightarrow R-NH_2 + R'-NH_2$

Catalytic hydrogenation is the most efficient method for the large scale manufacture of many aromatic and aliphatic amines. Aromatic amines are usually made by hydrogenating the corresponding nitro compound, whereas the aliphatic amines generally start with the corresponding nitrile. Certain aliphatic amines can be prepared by reduction of corresponding aromatic amines using catalytic hydrogenation.

BÉCHAMP PROCESS

In the Béchamp process, nitro compounds are reduced to amines in the presence of iron and an acid. This is the oldest commercial process for preparing amines, still used in the dyestuff industry for the production of small volume amines and for the manufacture of iron oxide pigments where aniline is produced as a by-product.

The overall reaction in the Béchamp process is as follows

$$4 RNO_2 + 9 Fe + 4 H_2O \xrightarrow{FeCl_2} 4 RNH_2 + 3 Fe_3O_4$$

MISCELLANEOUS REDUCTIONS

The method of reducing aromatic nitro compounds with divalent sulfur is known as the Zinin reduction. This reaction can be carried out in a basic media using sulfides, polysulfides, or hydrosulfides as the reducing agent.

Sodium bisulfite, $NaHSO_3$, is occasionally used to perform simultaneous reduction of a nitro group to an amine and the addition of a sulfonic acid group.

Both nitro compounds and nitriles can be reduced electrochemically. One advantage of electrochemical reduction is the cleanliness of the operation, which results in a minimum of by-products.

Metal hydrides and amalgams are sometimes the preferred method of reducing various functional groups in the laboratory, especially when the necessary equipment for catalytic hydrogenations is unavailable. However, these reagents are usually too expensive to make their use on a large commercial scale feasible.

ENVIRONMENTAL AND SAFETY ASPECTS

Amines, nitro compounds, nitriles, and the various solvents and reagents used in the preparation of amines by reduction vary widely in the hazards they may pose. Some of these materials are acutely toxic by ingestion, inhalation, or absorption through the skin. Others are skin irritants or sensitizers. Still others may cause damage by chronic exposure to organs, such as the liver, or may be carcinogenic. Since amines vary so widely in their potential danger, no general rules can govern their safe use in all cases. The Material Safety Data Sheet (MSDS) for the material in question should be consulted before the chemicals are used.

R. L. Augustine, *Catalytic Hydrogenation*, Marcel Dekker, Inc., New York, 1965.

P. N. Rylander, *Catalytic Hydrogenation in Organic Synthesis*, Academic Press, New York, 1979.

K. Schofield, *Aromatic Nitration*, Cambridge University Press, Cambridge, UK, 1980.

T. Urbanski, *Chemistry and Technology of Explosives*, Vol. 1, MacMillan Company, New York, 1964.

STEVEN L. SCHILLING
Mobay Corporation

AMINES, CYCLOALIPHATIC

Cycloaliphatic amines are comprised of a cyclic hydrocarbon structural component and an amine functional group external to that ring. Included in an extended cycloaliphatic amine definition are aminomethyl cycloaliphatics.

Table 1. Properties of Primary Aminocycloalkanes

Cycloaliphatic amine	Molecular formula	Boiling point, °C	Flash point, °C	Specific gravity, g/mL	Refractive index, n_D
cyclopropylamine	C_3H_7N	49	−26	0.824	1.4210
cyclobutylamine	C_4H_9N	82	−4	0.833	1.4363
cyclopentylamine	$C_5H_{11}N$	108	17	0.863	1.4478
cyclohexylamine	$C_6H_{13}N$	134	32	0.868	1.4565
cycloheptylamine	$C_7H_{15}N$	169	42		1.4724
cyclooctylamine	$C_8H_{17}N$	190	80	0.928	1.4804
cyclododecylamine	$C_{12}H_{25}N$	280^a	121		

^{a} Melting point 27°C.

Although some cycloaliphatic amine and diamine products have direct end-use applications, their major function is as low-cost organic intermediates sold as moderate-volume specification products.

PHYSICAL PROPERTIES

For simple primary amines directly bonded to a cycloalkane by a single C−N bond to a secondary carbon the homologous series is given in Table 1.

Up through C_8 each is a colorless liquid at room temperature. The ammoniacal or fishy odor and high degree of water solubility decrease with increased molecular weight and boiling point for these corrosive, hygroscopic mobile fluids.

Table 2 lists properties of the cycloaliphatic diamines.

CHEMICAL PROPERTIES

Cycloaliphatic amines are strong bases with chemistry similar to that of simpler primary, secondary, or tertiary amines. Upon reaction with nitrous acid, primary amines evolve nitrogen and generate alcohols; secondary amines form mutagenic nitrosamines.

Salt formation with Brønsted and Lewis acids and exhaustive alkylation to form quaternary ammonium cations are part of the rich derivatization chemistry of these amines.

Primary cycloaliphatic amines react with phosgene to form isocyanates.

Cycloaliphatic diamines react with dicarboxylic acids or their chlorides, dianhydrides, diisocyanates and di-(or poly-)epoxides as comonomers to form high molecular weight polyamides, polyimides, polyureas, and epoxies.

Table 2. Properties of Cycloaliphatic Diamines

Diamine	Molecular formula	Boiling point^a, °C	Flash point, °C
cis,trans-1,2-cyclohexanediamine	$C_6H_{14}N_2$	183	75
cis-1,2-cyclohexanediamine	$C_6H_{14}N_2$	182	72
(±)trans-1,2-cyclohexanediamine	$C_6H_{14}N_2$		
(+)trans-1,2-cyclohexanediamine	$C_6H_{14}N_2$		
(−)trans-1,2-cyclohexanediamine	$C_6H_{14}N_2$		
cis,trans-1,3-cyclohexanediamine	$C_6H_{14}N_2$		91
cis-1,3-cyclohexanediamine	$C_6H_{14}N_2$	198	
trans-1,3-cyclohexanediamine	$C_6H_{14}N_2$	203	
methylcyclohexanediamine	$C_7H_{16}N_2$	99 (1.66)	83
cis,trans-1,3-cyclohexanediamine,2-methyl			
cis,trans-1,3-cyclohexanediamine,4-methyl			
cis,trans-1,4-cyclohexanediamine	$C_6H_{14}N_2$	181	80
cis-1,4-cyclohexanediamine	$C_6H_{14}N_2$		
trans-1,4-cyclohexanediamine	$C_6H_{14}N_2$	197	71
cis,trans-1,8-menthanediamine	$C_{10}H_{22}N_2$	210	102
cis,trans-1,3-di(aminomethyl)cyclohexane	$C_8H_{18}N_2$		106
cis-1,3-di(aminomethyl)cyclohexane		114 (1.07)	
trans-1,3-di(aminomethyl)cyclohexane			
cis,trans-1,4-di(aminomethyl)cyclohexane	$C_8H_{18}N_2$	245	107
cis-1,4-di(aminomethyl)cyclohexane	$C_8H_{18}N_2$		
trans-1,4-di(aminomethyl)cyclohexane			
cis,trans-isophoronediamine	$C_{10}H_{22}N_2$	252	112
methylenedi(cyclohexylamine)	$C_{13}H_{26}N_2$	162 (2.40)	>110
isopropylidenedi(cyclohexylamine)	$C_{15}H_{30}N_2$	182 (1.32)	>110
3,3'-dimethylmethylene-di(cyclohexylamine)	$C_{15}H_{30}N_2$	160 (0.27)	174
cis,trans-tricyclodecanediamine^b	$C_{12}H_{22}N_2$	~314	165

^{a} At 101.3 kPa unless otherwise indicated by the value (in kPa) in parentheses. To convert kPa to mm Hg, multiply by 7.5.
^{b} (4,7-Methano-1H-indene-dimethaneamine, octahydro).

MANUFACTURE AND PROCESSING

Cycloaliphatic amine synthesis routes may be described as distinct synthetic methods, though practice often combines, or hybridizes, the steps that occur: amination of cycloalkanols, reductive amination of cyclic ketones, ring reduction of cycloalkenylamines, nitrile addition to alicyclic carbocations, reduction of cyanocycloalkanes to aminomethylcycloalkanes, and reduction of nitrocycloalkanes or cyclic ketoximes.

SHIPMENT

Shipment of these liquid products is by nitrogen-blanketed tank truck or tank car. Drum shipments are usually in carbon steel, DOT-17E.

ECONOMIC ASPECTS

The production economics of cycloaliphatic amines are dominated by raw material charges and process equipment capital costs. Reductive alkylations and aminations require pressure-rated reaction vessels and fully contained and blanketed support equipment. Nitrile hydrogenations are similar in their requirements. Arylamine hydrogenations have historically required very high pressure vessel materials of construction.

SPECIFICATIONS, STANDARDS, AND QUALITY CONTROL

Liquid cycloaliphatic amines and diamines have exacting purity and color standards. Almost all are sold to specification, not performance, standards. Use as isocyanate precursors requires low-water content criteria for these hygroscopic fluids; hence, nitrogen blanketing is often specified for product sampling as well as storage and transport.

HEALTH AND SAFETY FACTORS

Cycloaliphatic amines and diamines are extreme lung, skin, and eye irritants. MSD sheets universally carry severe personal protective equipment use warnings due to the risk of irreversible eye damage.

DERIVATIVES

Before a jaunary 1, 1970 FDA ban, cyclamate noncaloric sweeteners were the major derivatives driving cyclohexylamine production. Cyclohexylamine condensed with mercaptobenzothiazole produces the large volume moderated rubber accelerator N-cyclohexyl-2-benzothiazolesulfenamide.

1,3-Dicyclohexylcarbodiimide is an important peptide condensing agent and analytical reagent.

trans-1,2-Cyclohexanediamine is derivatized by Mannich reaction of formaldehyde and HCN, then hydrolyzed to the tetraacetate and sold as a chelating agent.

Methylene di(cyclohexylisocyanate) (MDCHI, Desmodur W) is the dominant derivative of MDCHA and is used in light-stable urethanes.

Isophoronediisocyanate, made by phosgenation of IPD, competes effectively in this same polyurethane market, predominantly coatings, and is the major commercial application of isophoronediamine.

1,4-Cyclohexanediamine, from hydrogenation of *p*-phenylenediamine may be easily phosgenated, unlike the corresponding 1,2- and 1,3- isomers to produce a useful diisocyanate for performance polyurethanes efficiently, particularly *trans*-1,4-cyclohexanediisocyanate (CHDI).

A representative agrochemical application of cycloaliphatic amines is the reaction of the commercial 30/70 *cis/trans* isomer mixture of 2-methylcyclohexylamine with phenylisocyanate to give the crabgrass and weed control agent Siduron (1-(2-methylcyclohexyl)-3-phenylurea). The preplant herbicide Cycloate, used for sugar beets, vegetable beets, and spinach (*S*-ethyl-*N*-ethyl-*N*-cyclohexylthiocarbamate, incorporates *N*-ethylcyclohexylamine. The herbicide Hexazinone (3-cyclohexyl-6-dimethylamino-1-methyl-1,3,5-triazine-2,4-dione) is prepared from cyclohexylisocyanate.

"Cyclohexylamine, Chemical Profile", *Chemical Market Reporter*, (May 28, 2001).

R. J. Lewis, Sr., *Sax's Dangerous Properties of Industrial Materials*, 10th ed., Vol. 2, John Wiley & Sons, Inc., New York, 2000.

J. March, *Advanced Organic Chemistry, Reactions, Mechanisms and Structure*, 3rd ed., John Wiley & Sons, Inc., New York, 1985.

T. F. Mika and R. S. Bauer, in C. A. May, ed., *Epoxy Resins Chemistry and Technology*, 2nd ed., Marcel Dekker, New York, 1988, 465–550.

JEREMIAH P. CASEY
Air Products and Chemicals

AMINES, FATTY

Fatty amines are nitrogen derivatives of fatty acids, olefins, or alcohols prepared from natural sources, fats and oils, or petrochemical raw materials. Commercially available fatty amines consist of either a mixture of carbon chains or a specific chain length from C_8–C_{22}. The amines are classified as primary, secondary, or tertiary depending on the number of hydrogen atoms of an ammonia molecule replaced by fatty alkyl or methyl groups (Fig. 1).

The amino nitrogen is most frequently found on a primary carbon atom, but secondary and tertiary carbon substitution derivatives have been made and are commercially available. Fatty amines are cationic surface-active compounds, which strongly adhere to surfaces by either physical or chemical bonding, thus modifying surface properties. Important commercial products are prepared using fatty amines as reactive intermediates.

Commercially available fatty amines are most frequently prepared from naturally occurring materials by

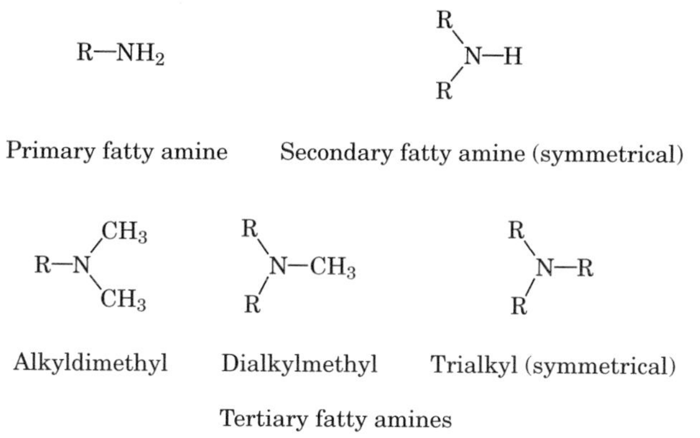

Figure 1. Types of commercially available fatty amines. R = $C_8 - C_{22}$.

hydrogenation of a fatty nitrile intermediate using a variety of catalysts.

Fatty amines derived from fats and oils, containing several carbon-chain-length moieties, are designated as such by common names which describe these mixtures: tallowalkylamines, cocoalkylamines, and soya alkylamines, for example. High purity fatty amines are also commercially available. These amines are prepared by distillation of either the precursor fatty acid or amine product mixture.

Trade names are commonly used for commercial products.

PHYSICAL PROPERTIES

Data on physical properties of fatty amines have been well documented and summarized in many reference works on fatty acids and nitrogen derivatives. It is evident that (1) melting points within a homologous series of single-chain-length fatty amines increase with molecular weight; (2) symmetrical secondary amines have a higher melting point than the primary amine of the same alkyl group, but are lower melting than a primary amine with the same number of carbon atoms (hydrogen bonding); (3) symmetrical tertiary amines are lower melting than a symmetrical secondary amine of the same alkyl group; (4) symmetrical tertiary amines are lower melting than a primary or secondary amine containing the same number of carbon atoms; and (5) unsaturation lowers the melting point of the fatty amine, eg, oleyl amine versus 1-octadecylamine and ditallowalkylamines versus dihydrogenated tallowalkylamines.

Boiling points of fatty amines have been reported. A direct correlation between molecular weight and boiling point is observed. Mixtures of primary fatty amines prepared from fats and oils can be separated into component amines by fractional distillation; an approximately 10°C increment in boiling point per carbon in the chain length is maintained throughout the series.

Fatty amines are insoluble in water, but soluble in organic solvents to varying degrees. Water, however, is soluble in the amines, and hydrates are formed.

CHEMICAL PROPERTIES

General amine chemistry is applicable to fatty amines. Many chemical reactions using fatty amines as reactive intermediates are run on an industrial scale to produce a wide range of important products. Important industrial reactions are salt formation, methylation of primary and secondary fatty amines, quaternization, ethoxylation and propoxylation, oxidation by hydrogen peroxide, and cyanoethylation.

MANUFACTURE

The principal industrial production route used to prepare fatty amines is the hydrogenation of nitriles, a route which has been used since the 1940s. Commercial preparation of fatty amines from fatty alcohols is a fairly new process, created around 1970, which utilizes petrochemical technology, Ziegler or Oxo processes, and feedstock.

Fatty amine products are normally shipped in 55-gal (208-L), lined and unlined, steel drums or in tank cars or tank trucks for bulk shipments. High melting amines can be flaked and shipped in cardboard cartons or paper bags. The amines are corrosive to skin and eyes. Protective splash goggles and gloves should be worn when handling these materials.

ECONOMIC ASPECTS

Demand for amines in the United States will grow. Advances will be led by the specialty amines. The fatty amines will show more modest gains, since it is a mature market. Growth will exist in water treatment and plastics. The set markets, detergents, cleaning products, personal care products, and agricultural products will advance moderately. Fatty amines and derivatives are used in fabric softeners, dishwashing liquids, car wash detergents, and carpet cleaners.

Fatty amines and ethanolamines are the largest amine types used in the detergent and cleaner markets.

The top U.S. producers of amines are Union Carbide, Air Products and Chemicals, Huntsman, and Dow Chemical.

The major source of raw materials for the preparation of fatty amines is fats and oils such as tallow and the coconut, soya, and palm oils. Cost of the amines can vary, owing to supply of raw materials.

HEALTH AND SAFETY FACTORS

Skin and Eye Irritation

Fatty alkylamines are generally considered to be irritating to both the skin and eyes.

Oral Toxicity

Depending on the chemical class, most fatty amines range from moderately toxic to practically nontoxic by acute oral ingestion.

Dermal Toxicity

Fatty alkylamines are not considered especially toxic with regard to skin penetration and systemic absorption into the body; certain polyamines may be absorbed through the skin to a much greater degree.

Inhalation

Long-chain amines are not considered an inhalation hazard at ambient conditions because of their relatively low volatility.

USES

Fatty amines and chemical products derived from the amines are used in many industries. Used for the nitrogen derivatives may be broken down as follows as a percentage of the total market: fabric softeners (46%), oil-field chemicals (15%), asphalt emulsifiers (10%), petroleum additives (10%), mining (4%), and others (15%).

Amine salts, especially acetate salts prepared by neutralization of a fatty amine with acetic acid, are useful as flotation agents (collectors), corrosion inhibitors, and lubricants.

The single largest market use for quaternary fatty amines is in fabric softeners. Monoalkyl quaternaries (chloride) have been used in liquid detergent softener antistat formulations (LDSAs), dialkyldimethyl quaternaries (chloride) in the rinse cycle, and dialkyldimethyl quaternaries (sulfate) as dryer softeners.

Another significant use for dialkyldimethyl quaternary ammonium salts and alkylbenzyldimethylammonium salts is in preparing organoclays for use as drilling muds, paint thickeners, and lubricants.

Betaines, or specialty quaternaries, are used in the personal care industry in shampoos, conditioners, foaming, and wetting agents.

Examples of uses for amine oxides include detergent and personal care areas as a foam booster and stabilizer, as a dispersant for glass fibers, and as a foaming component in gas recovery systems.

Important uses for the diamines include corrosion inhibitors, gasoline and fuel oil additives, flotation agents, pigment wetting agents, epoxy curing agents, herbicides, and asphalt emulsifiers. Fatty amines and derivatives are widely used in the oil field, as corrosion inhibitors, surfactants, emulsifying/deemulsifying and gelling agents. In the mining industry, amines and diamines are used in the recovery and purification of minerals, flotation, and benefication. A significant use of fatty diamines is as asphalt emulsifiers for preparing asphalt emulsions. Diamines have also been used as epoxy curing agents, corrosion inhibitors, gasoline and fuel oil additives, and pigment wetting agents. Oleylamine is a petroleum additive useful as a detergent in gasoline. In addition, derivatives of the amines, amphoterics, and long-chain alkylamines are used as anionic and cationic surfactants in the personal care industry.

R. A. Reck, "Polyoxyethylene Alkylamines," in M. Schick, ed., *Nonionic Surfactants, Surfactant Science Series*, Vol. 1, Marcel Dekker, Inc., New York, 1967.

R. A. Reck, in R. W. Johnson and E. Fritz, ed., *Fatty Acids in Industry*, Marcel Dekker, Inc., New York, 1989.

S. H. Shapiro, "Commercial Nitrogen Derivatives of Fatty Acids," in E. Pattison, ed., *Fatty Acids and Their Industrial Applications*, Marcel Dekker, Inc., New York, 1968, 77–154.

N. O. V. Sonntag, "Nitrogen Derivatives," in K. S. Markley, ed., *Fatty Acids*, Part 3, John Wiley & Sons, Inc., New York, 1964, 1551–1715.

K. Visek
Akzo Chemicals Inc.

AMINES, LOWER ALIPHATIC

Lower aliphatic amines are derivatives of ammonia with one, two, or all three of the hydrogen atoms replaced by alkyl groups of five carbons or less. Amines with higher alkyl groups are known as fatty amines. Amines are toxic, colorless gases or liquids, highly flammable, and have strong odors. Lower molecular-weight amines are water soluble and are sold as aqueous solutions and in pure form. Amines react with water and acids to form alkylammonium compounds analogous to ammonia. The base strengths in water of the primary, secondary, and tertiary amines and ammonia are essentially the same.

Primary and secondary amines can also act as very weak acids ($K_a = 10^{-33}$). They react with acyl halides, anhydrides, and esters with rates depending on the size of the alkyl groups. The lower aliphatic amines are widely used in the manufacture of pharmaceutical, agricultural, textile, rubber, and plastic chemicals.

PHYSICAL PROPERTIES

Table 1 lists the names, molecular formulas, molecular weights, and common names or synonyms of the commercially important alkylamines. Thermodynamic data are available for only the lower alkylamines and are mainly estimates based on a few experimental determinations. Recently, quantum-mechanics-based computational methods have been used to calculate equilibrium constants for amines systems. This methodology may find practical application as the product selectivities from many manufacturing processes appear to be limited by thermodynamic equilibria.

CHEMICAL PROPERTIES

The chemistry of the lower aliphatic amines is dominated by their basicity and their nucleophilic character, which result from the presence of an unshared pair of electrons on the nitrogen atom. Due to their basicity, amines are often used as neutralization agents or pH adjusters.

Amines react with a variety of substrates such as epoxides, aldehydes and ketones, alkyl halides, carboxylic acids/halides/esters/anhydrides, and carbon disulfide to produce products used in agricultural, pharmaceutical, textile, polymer, and rubber chemical applications. Allylamines are somewhat unique in that both amine and olefin functionalities are available.

Table 1. Commercial Lower Aliphatic Amines

Alkylamine	Molecular Formula	Molecular Weight	Synonym or common abbreviation
Ethylamines			
ethylamine	C_2H_7N	45.08	monomethylamine, aminomethane, MEA
diethylamine	$C_4H_{11}N$	73.14	N-ethylethanamine, DEA
triethylamine	$C_6H_{15}N$	101.19	N,N-diethylethanamine, TEA
n-Propylamines			
n-propylamine	C_3H_9N	59.11	mono-n-propylamine, 1-aminopropane,1-propanamine, MNPA
di-n-propyl-amine	$C_6H_{15}N$	101.19	N-propyl-1-propanamine, DNPA
tri-n-propyl-amine	$C_9H_{21}N$	143.27	N,N-dipropyl-1-propanamine, TNPA
iso-Propylamines			
isopropylamine	C_3H_9N	59.11	2-aminopropane, 2-propanamine, MIPA
diisopropyl-amine	$C_6H_{15}N$	101.19	N-(1-methylethyl)-2-propanamine, DIPA
Allylamines			
allylamine	C_3H_7N	57.10	monoallylamine, 2-propenamine, 3-aminopropene
diallylamine	$C_6H_{11}N$	97.16	N-2-propenyl-2-propenamine, di-2-propenylamine
triallylamine	$C_9H_{15}N$	137.23	N,N-di-2-propenyl-2-propenamine,tris(2-propenyl)amine
n-Butylamines			
n-butylamine	$C_4H_{11}N$	73.14	mono-n-butylamine, 1-aminobutane, MNBA
di-n-butylamine	$C_8H_{19}N$	129.25	N-butyl-1-butanamine, DNBA
tri-n-butylamine	$C_{12}H_{27}N$	185.36	N,N-dibutyl-1-butanamine, TNBA
Isobutylamines			
isobutylamine	$C_4H_{11}N$	73.14	monoisobutylamine, 2-methyl-1-propanamine, 1-amino-2-methylpropane,1-aminobutane, MIBA
diisobutylamine	$C_8H_{19}N$	129.25	2-methyl-N-(2-methylpropyl)-1-propanamine, DIBA
triisobutylamine	$C_{12}H_{27}N$	185.36	2-methyl-N,N-bis(2-methylpropyl)-1-propanamine, TIBA
sec-Butylamine			
sec-butylamine	$C_4H_{11}N$	73.14	2-aminobutane, 2-butanamine, 1-methylpropanamine
tert-Butylamine			
tert-butylamine	$C_4H_{11}N$	73.14	2-methyl-2-propanamine, 2-aminoisobutane, 1,1-dimethylethanamine, trimethylaminomethane
Amylamines			
amylamine	$C_5H_{13}N$	87.17	mixture of 1-pentylamine and 2-methyl-1-butylamine
diamylamine	$C_{10}H_{23}N$	157.30	mixture of linear and branched isomers
triamylamine	$C_{15}H_{33}N$	227.44	mixture of linear and branched isomers
Mixed Amines			
dimethylethylamine	$C_4H_{11}N$	73.14	N,N-dimethylethanamine, N-ethyldimethylamine
dimethyl-n-propylamine	$C_5H_{13}N$	87.17	N,N-dimethyl-1-propanamine, propyldimethylamine
ethyl-n-butylamine	$C_6H_{15}N$	101.19	N-ethyl-1-butanamine, butylethylamine, EBA
dimethyl-n-butylamine	$C_6H_{15}N$	101.19	N,N-dimethyl-1-butanamine, butyldimethylamine, DMBA

Alkylamines are corrosive to copper, copper-containing alloys (brass), aluminum, zinc, zinc alloys, and galvanized surfaces.

MANUFACTURE

Lower aliphatic amines can be prepared by a variety of methods, and from many different types of raw materials. By far the largest commercial applications involve the reaction of alcohol with ammonia to form the corresponding amines. Other methods are employed depending on the particular amine desired, raw material availability, plant economics, and the ability to sell co-products. The following manufacturing methods are used commercially to produce the lower alkylamines:

Method 1. Alcohol amination: amination of an alcohol over a metal catalyst under reducing conditions or over a solid acid catalyst at high temperature.

Method 2. Reductive alkylation: reaction of an amine or ammonia and hydrogen with an aldehyde or ketone over a hydrogenation catalyst.

Method 3. Ritter reaction: reaction of hydrogen cyanide with an olefin in an acidic medium to produce a primary amine.

Method 4. Nitrile reduction: reaction of a nitrile with hydrogen over a hydrogenation catalyst.

Method 5. Olefin amination: reaction of an olefin with ammonia.

Method 6. Alkyl halide amination: reaction of ammonia or alkylamine with an alkyl halide.

SHIPMENT AND HANDLING

All of the lower alkylamines are classified as either flammable or combustible liquids at normal temperatures and pressure with the exception of monoethylamine, which is a flammable gas under these conditions. Anhydrous monoethylamine therefore is shipped under pressure in bulk tank trucks and railcars. Both monoethylamine and monoisopropylamine are available as 70% solutions in water and are shipped in this form as flammable liquids. The liquid amines are available in drums and isocontainers as well as tank trucks and railcars.

The lower alkylamines are toxic and have strong odors. Labeling and packaging of amines must conform with Department of Transportation (DOT) requirements. Amine shipments are regulated by the Coast Guard, the DOT, the International Air Transport Association (IATA), and in some cases, the Drug Enforcement Administration (DEA).

HEALTH AND SAFETY FACTORS

The lower alkylamines all have strong fishy or ammoniacal odors. As a general practice, exposure to all alkylamines should be limited, and therefore they should be handled only in well-ventilated areas. A full face shield with goggles underneath, neoprene, nitrile, or butyl rubber gloves and impervious clothing should be worn when working with alkylamines.

The lower alkylamines are toxic by ingestion, inhalation, and/or skin absorption. Alkylamine vapors in low concentrations can cause lacrimation, conjunctivitis, and corneal edema when absorbed into the tissue of the eye from the atmosphere. Inhalation of vapors may cause irritation in the respiratory tract. Contact of undiluted product with the eyes or skin quickly causes severe irritation and pain and may cause burns, necrosis, and permanent injury. Repeated exposure may result in adverse respiratory, eye, and/or skin effects. If contact with the eyes or skin occurs, the affected area should be washed with water for at least 15 min. If these products are inhaled, the patient should be moved to fresh air and assisted with respiration if required.

B. C. Challis and A. R. Butler in S. Patai, ed., *The Chemistry of the Amino Group*, Interscience, London, 1968, pp. 277–347.

J. A. Dean, ed. *Lange's Handbook of Chemistry* McGraw-Hill, Inc., New York, 1992.

P. A. S. Smith, *The Chemistry of Open-Chain Organic Nitrogen Compounds: Vol. 1*, W. A. Benjamin, Inc., New York, 1965.

J. Volf and J. Pasek in L. Cerveny, ed., *Studies in Surface Science and Catalysis, Vol. 27: Catalytic Hydrogenation*, Elsevier, Amsterdam, 1986, pp. 105–144.

MICHAEL G. TURCOTTE
KATHRYN S. HAYES
Air Products and Chemicals, Inc.

AMINO ACIDS

SURVEY

Amino acids are the main components of proteins. Approximately twenty amino acids are common constitu-

ents of proteins and are called protein amino acids or primary protein amino acids, because they are found in proteins as they emerge from the ribosome in the translation process of protein synthesis, or natural amino acids.

Hydroxylated amino acids (eg, 4-hydroxyproline, 5-hydroxylysine) and *N*-methylated amino acids (eg, *N*-methylhistidine) are obtained by the acid hydrolysis of proteins. γ-Carboxyglutamic acid occurs as a component of some sections of protein molecules; it decarboxylates spontaneously to L-glutamate at low pH. These examples are formed upon the nontranslational modification of protein and are often called secondary protein amino acids.

The presence of many nonprotein amino acids has been reported in various living metabolites, such as in antibiotics, some other microbial products, and in nonproteinaceous substances of animals and plants. Plant amino acids and seleno amino acids have been reviewed.

The general formula of an α-amino acid may be written:

$$R-\overset{*}{C}H-COOH$$
$$|$$
$$NH_2$$

The asterisk signifies an asymmetric carbon. All of the amino acids, except glycine, have two optically active isomers designated D- or L-. Isoleucine and threonine also have centers of asymmetry at their β-carbon atoms. Protein amino acids are of the L-α-form as illustrated in Table 1.

Amino acids are important components of the elementary nutrients of living organisms. For humans, ten amino acids are essential for existence and must be ingested in food. The nutritional value of proteins is governed by the quantitative and qualitative balance of individual essential amino acids.

The nutritional value of a protein can be improved by the addition of amino acids of low abundance in that protein. Thus the fortification of plant proteins such as wheat, corn, and soybean with L-lysine, DL-methionine, or other essential amino acids (L-tryptophan and L-threonine) is expected to alleviate some food problems. Such fortification has been widespread in the feedstuff of domestic animals.

Proteins are metabolized continuously by all living organisms, and are in dynamic equilibrium in living cells. The role of amino acids in protein biosynthesis has been described. Most of the amino acids absorbed through the digestion of proteins are used to replace body proteins. The remaining portion is metabolized into various bioactive substances such as hormones and purine and pyrimidine nucleotides, (the precursors of DNA and RNA) or is consumed as an energy source.

All of the protein amino acids are currently available commercially and their uses are growing. Amino acids and their analogues have their own characteristic effects in flavoring, nutrition, and pharmacology.

In the food industries a number of amino acids have been widely used as flavor enhancers and flavor modifiers. For example, monosodium L-glutamate is well-known as a meat flavor-enhancer; an enormous quantity of it is now used in various food applications. Protein, hydrolyzed by

Table 1. α-Amino Acids

Common name	Abbreviation	Systematic name	Formula	Molecular weight
		Monocarboxylic		
Aliphatic				
glycine	Gly	aminoacetic acid	H_2NCH_2COOH	75.07
alanine	Ala	2-aminopropanoic acid	$CH_3\underset{\underset{NH_2}{\vert}}{C}HCOOH$	89.09
L-alanine				
D-alanine				
DL-alanine				
valine[a]	Val	2-amino-3-methylbutanoic acid	$(CH_3)_2\underset{\underset{NH_2}{\vert}}{C}HCHCOOH$	117.15
L-valine				
D-valine				
DL-valine				
leucine[a]	Leu	2-amino-4-methylpentanoic acid	$(CH_3)_2CHCH_2\underset{\underset{NH_2}{\vert}}{C}HCOOH$	131.17
L-leucine				
D-leucine				
DL-leucine				
isoleucine[a]	Ileu	2-amino-3-methylpentanoic acid	$CH_3CH_2\underset{\underset{CH_3}{\vert}}{C}H{-}\underset{\underset{NH_2}{\vert}}{C}HCOOH$	131.17
L-isoleucine				
D-isoleucine				
DL-isoleucine				
Aliphatic containing —OH, —S—, —NH— group				
serine	Ser	2-amino-3-hydroxypropanoic acid	$HOCH_2\underset{\underset{NH_2}{\vert}}{C}HCOOH$	105.09
L-serine				
D-serine				
DL-serine				
threonine[a]	Thr	2-amino-3-hydroxybutanoic acid	$CH_3\underset{\underset{OH}{\vert}}{C}H{-}\underset{\underset{NH_2}{\vert}}{C}HCOOH$	119.12
L-threonine				
D-threonine				
DL-threonine				
cysteine	Cys	2-amino-3-mercaptopropanoic acid	$HSCH_2\underset{\underset{NH_2}{\vert}}{C}HCOOH$	121.16
L-cysteine				
D-cysteine				
DL-cysteine				
cystine	(Cys)$_2$	3,3'-dithio-bis-(2-aminopropanoic acid)	$\underset{\underset{NH_2}{\vert}}{S}CH_2CHCOOH$ $\underset{\underset{NH_2}{\vert}}{S}CH_2CHCOOH$	240.30
L-cystine				
D-cystine				
DL-cystine				
methionine[a]	Met	2-amino-4-methylthiobutanoic acid	$CH_3SCH_2CH_2\underset{\underset{NH_2}{\vert}}{C}HCOOH$	149.21
L-methionine				
D-methionine				
DL-methionine				
lysine[a]	Lys	2,6-diaminohexanoic acid	$H_2N(CH_2)_4\underset{\underset{NH_2}{\vert}}{C}HCOOH$	146.19

Table 1. (*Continued*)

Common name	Abbreviation	Systematic name	Formula	Molecular weight
DL-lysine D-lysine L-lysine arginine[b] L-arginine D-arginine DL-arginine	Arg	2-amino-5-guanidopentanoic acid	$HN{=}CNH(CH_2)_3CHCOOH$ $\quad\vert\qquad\qquad\quad\vert$ $H_2N\qquad\qquad\quad NH_2$	174.20
Aromatic phenylalanine[a]	Phe	2-amino-3-phenylpropanoic acid	$C_6H_5CH_2CHCOOH$ $\qquad\qquad\vert$ $\qquad\quad NH_2$	165.19
L-phenylalanine D-phenylalanine DL-phenylalanine tyrosine	Tyr	2-amino-3-(4-hydroxy-phenyl)- propanoic acid	$HO{-}\bigcirc{-}CH_2CHCOOH$ $\qquad\qquad\qquad\quad\vert$ $\qquad\qquad\qquad NH_2$	181.19
L-tyrosine D-tyrosine DL-tyrosine				
Heterocyclic proline	Pro	2-pyrrolidinecarboxylic acid		115.13
L-proline D-proline DL-proline hydroxyproline	Hypro	4-hydroxy-2-pyrrolidine carboxylic acid		131.13
L-hydroxyproline D-hydroxyproline DL-hydroxyproline hydroxyproline L-hydroxyproline (*trans*) D-hydroxyproline (*trans*) DL-hydroxyproline (*trans*) histidine[b]	His	2-amino-3-imidazolepropanoic acid		155.16
L-histidine D-histidine DL-histidine tryptophan[a]	Trp	2-amino-3-indoylpropanoic acid		204.22
L-tryptophan D-tryptophan DL-tryptophan				
		Dicarboxylic		
aspartic acid	Asp	2-amino-butanedioic acid	$HOOCCH_2CHCOOH$ $\qquad\qquad\quad\vert$ $\qquad\qquad NH_2$	133.10
L-aspartic acid D-aspartic acid DL-aspartic acid glutamic acid	Glu	2-amino-pentanedioic acid	$HOOCCH_2CH_2CHCOOH$ $\qquad\qquad\qquad\vert$ $\qquad\qquad\quad NH_2$	147.13
L-glutamic acid D-glutamic acid				

Table 1. (*Continued*)

Common name	Abbreviation	Systematic name	Formula	Molecular weight
asparagine	Asn	2-amino-3-carbamoylpropanoic acid	$H_2NCOCH_2CHCOOH$ $\underset{NH_2}{|}$	132.12
L-asparagine				
D-asparagine				
DL-asparagine				
glutamine	Gln	2-amino-4-carbamoylbutanoic acid	$H_2NCOCH_2CH_2CHCOOH$ $\underset{NH_2}{|}$	146.15
L-glutamine				
D-glutamine				
DL-glutamine				

[a] Essential amino acid.

[b] Arginine and histidine are also essential for children.

acid or enzyme to be palatable, has been used for a long time in flavoring agents. The addition of L-glutamate, L-aspartate, glycine, DL-alanine, and other palatable amino acids can improve flavoring by these protein hydrolyzates. In addition, some nucleotides, such as 5′-inosinic acid and 5′-guanylic acid, have a synergistic effect on the meat flavor enhancing effects of L-glutamate and L-aspartate. Tricholomic acid and ibotenic acid, nonprotein amino acids found in mushrooms, have 4 to 25 times stronger umami taste than L-glutamic acid. However, they have not been used in food.

Some peptides have special tastes. L-Aspartyl phenylalanine methyl ester is very sweet and is used as an artificial sweetener. In contrast, some oligopeptides (such as L-ornithinyltaurine·HCl and L-ornithinyl-β-alanine· HCl), and glycine methyl or ethyl ester·HCl have been found to have a very salty taste.

Amino acids are also used in medicine. Amino acid infusions prepared from crystalline amino acids are used as nutritional supplements for patients before and after surgery. Some amino acids and their analogues are used for treatment of major diseases. L-DOPA, L-3-(3,4-dihydroxyphenyl) alanine, is an important drug in the treatment of Parkinson's disease, and L-glutamine and its derivatives are used for treatment of stomach ulcers. α-Methyl-DOPA is an effective antidepressant. Some peptides, eg, oxytocin, angiotensin, gastrin, and cerulein, have hormonal effects which have medical utility. The physiological effect of glutathione (L-glutamyl-L-cysteinyl glycine) has been reviewed.

Amino acid polymers like poly(γ-methyl-L-glutamate) have been developed as raw materials for artificial leathers. Derivatives of amino acids are now finding new applications in industry and agriculture.

PHYSICAL PROPERTIES

Melting Point

Amino acids are solids, even the lower carbon-number amino acids such as glycine and alanine. The melting points of amino acids generally lie between 200 and 300°C. Frequently amino acids decompose before reaching their melting points (Table 2).

Crystalline Structures

The crystal shape of amino acids varies widely, eg, monoclinic prisms in glycine and orthorhombic needles in L-alanine. X-ray crystallographic analyses of 23 amino acids have been described.

Dipole

Every amino acid molecule has two equal electric charges of opposite sign caused by the amino and carboxyl groups on the α-carbon atom.

$$R\!-\!CH\!-\!COO^-$$
$$\underset{NH_3^+}{|}$$

The dielectric constants of amino acid solutions are very high. Their ionic dipolar structures confer special vibrational spectra (Raman, ir), as well as characteristic properties (specific volumes, specific heats, electrostriction).

Optical Configuration

With the exception of glycine, all α-amino acids contain at least one asymmetric carbon atom and may be characterized by their ability to rotate light to the right (+) or to the left (−), depending on the solvent and the degree of ionization. Specific rotations are given in Table 2. They are also characterized by the stereochemical configuration of the asymmetric carbon based on the configuration of glyceraldehyde; D,L-notation is popular for amino acids, but *R,S*-notation is a more precise designation of chirality.

Solubility

In all instances there are at least two polar groups, acting synergistically on the solubility in water. The solubility of amino acids having additional polar groups, eg, −OH, −SH, −COOH, −NH$_2$, is even more enhanced.

Table 2. Physical Constants of Amino Acids

Amino acid		Melting point, °C	Density, d_{t1}^{t2}	Specific rotation			
				$[\alpha]_D$	t, °C	c, %	Solvent
Ala	L-	297 (dec)	1.401	+2.8	25	6	H$_2$O
		314 (dec)	1.432[23]	+2.8	25	6	H$_2$O
	L-·HCl	204 (dec)		+8.5	26	9.3	
	D-	314 (dec)		−13.6	25	1	6 N HCl
	DL-	264 (dec)	1.424				
		295 (dec)	1.424				
Arg	L-	244 (dec)		+12.5	20	3.5	H$_2$O
	L-·HCl	235 (dec)		+12.0	20	4	
	DL-	217–218					
Asn	L-·H$_2$O	234–235	1.543[15]$_4$	−5.42	20	1.3	
	D-·H$_2$O	215		+5.41	20	1.3	
		234.5	1.543[15]$_4$	+5.41	20	1.3	
	DL-·H$_2$O	182–183	1.4540[15]$_4$				
Asp	L-	270–271	1.661[12.5]	+25.0	20	1.97	6 N HCl
		324 (dec)	1.6613[13]$_{13}$	+24.6	24	2	6 N HCl
	D-			−23.0	27	2.30	6 N HCl
		269–271	1.6613[13]$_{13}$	−25.5	20		HCl
	DL-	338–339	1.6632[13]$_{13}$				
Cys	L-			+6.5	25		5 N HCl
		240 (dec)		+9.8	30	1.3	H$_2$O
	L-·HCl	175–178		+5.0	25		5 N HCl
(Cys)$_2$	L-	260–261 (dec)	1.677	−223.4	20	1	1 N HCl
	D-			+223	20		1 N HCl
		247–249		+224	20	1	1 N HCl
	DL-	260					
Glu	L-	247–249 (dec)	1.538[20]$_4$	+31.4	22.4		6 N HCl
		224–225 (dec)	1.538[20]$_4$	+31.4	22	1	6 N HCl
	L-·HCl	214 (dec)		+24.4	22	6	
	D-			−30.5	20	1.0	6 N HCl
		213 (dec)	1.538[20]$_4$	−31.7	25		1.7 N HCl
	DL-	225–227 (dec)	1.4601[20]				
		199 (dec)	1.4601[20]$_4$				
Gly		233 (dec)	1.1607				
		262 (dec)	0.828[17]				
His	L-	287 (dec)		−39.74	20	1.13	
		287 (dec)		−39.7	20	1.13	H$_2$O
	L-·HCl·H$_2$O	259 (dec)		+8.0	26	2	3 N HCl
	D-	287 (dec)		+40.2	20		H$_2$O
	DL-	285 (dec)					
Ileu	L-	284 (dec)		+11.29	20	3	
				+40.61	20	4.6	6.1 N HCl
		285–286 (dec)		+12.2	25	3.2	H$_2$O
				+36.7		4	1 N HCl
	D-	283–284 (dec)		−12.2	20	3.2	H$_2$O
				−40.7		1	5 N HCl
	DL-	280 (dec)					
Leu	L-	293–295 (dec)		−10.8	25	2.2	
		293–295	1.293[18]$_4$	−10.42	25	22	H$_2$O
	D-	293		+10.34	20		
	DL-	332 (dec)					
		293–295	1.293[18]$_4$				
Lys	L-	224.5 (dec)		+25.9	23	2	6 N HCl
		224–245 (dec)		+14.6	20	6	H$_2$O
	L-·HCl	263–264		+14.6	25	2	0.6 N HCl
	L-·2HCl	193		+15.3	20	2	
		201–202		+15.29	20		H$_2$O
	DL-·HCl	260–263					
	DL-·2 HCl	187–189					

Table 2. (*Continued*)

Amino acid		Melting point, °C	Density, d_{t1}^{t2}	Specific rotation [α]$_D$	t, °C	c, %	Solvent
Met	L-	280–282 (dec)		−8.2	25		
		283 (dec)		−8.2	25	1	H$_2$O
	DL-	281 (dec)	1.340				
Phe	L-	283 (dec)		−35.1	20	1.94	
	D-	285 (dec)		+35.0	20	2.04	
	DL-	271–273 (dec)					
Pro	L-	220–222 (dec)		−52.6	20	0.58	0.5 N HCl
		220–222 (dec)		−80.9	20	1	H$_2$O
	DL-	205 (dec)					
Hyp	L-	274		−76.5		2.5	H$_2$O
Ser	L-	228 (dec)		−6.83	20	10	H$_2$O
	D-	228 (dec)		+6.87	20	10	H$_2$O
	DL-	246 (dec)	1.537				
		246 (dec)	1.603$^{22.5}$				
Thr	L-	255–257 (dec)		−28.3	26	1.1	
	DL-	229–230 (dec)					
Trp	L-	289 (dec)		−31.5	23	1	
				+2.4	20		0.5 N HCl
				+0.15	20	2.43	0.5 N NaOH
		290–292 (dec)		−31.5	20	0.5	H$_2$O
				+6.1	20	11	1 N NaOH
	D-	281–282		+33	20		H$_2$O
	DL-	282					
Tyr	L-	342–344 (dec)	1.456	−10.6	22	4	1 N HCl
				−13.2	18	4	3 N NaOH
	D-	310–314 (dec)		+10.3	25	4	1 N HCl
	DL-	316 (dec)					
		340 (dec)					
Val	L-	315	1.230	+22.9	23	0.8	20% HCl
		93–96(?)	1.230	+22.9	23	0.8	20% alc
	D-	156–157.5		−29.4	20		20% alc
	DL-	298 (dec)	1.310				

Dissociation

In aqueous solution, amino acids undergo a pH-dependent dissociation:

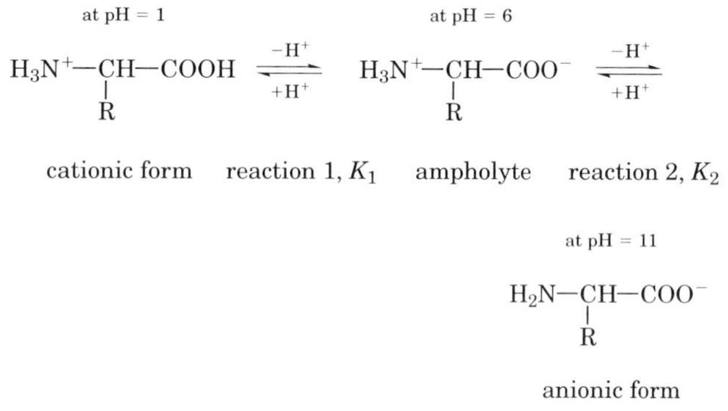

at pH = 1   at pH = 6

$$H_3N^+\!-\!CH\!-\!COOH \underset{+H^+}{\overset{-H^+}{\rightleftharpoons}} H_3N^+\!-\!CH\!-\!COO^- \underset{+H^+}{\overset{-H^+}{\rightleftharpoons}}$$
$$\qquad\quad | \qquad\qquad\qquad\qquad\qquad |$$
$$\qquad\quad R \qquad\qquad\qquad\qquad\qquad R$$

cationic form   reaction 1, K_1 ampholyte   reaction 2, K_2

at pH = 11

$$H_2N\!-\!CH\!-\!COO^-$$
$$\qquad\quad |$$
$$\qquad\quad R$$

anionic form

SYNTHESIS OF α-AMINO ACIDS

Many methods for chemical synthesis of α-amino acids have been established. Because excellent reviews have been published, well-known reactions are introduced here only by their names and synthetic pathways.

Synthetic Pathways: Strecker Synthesis

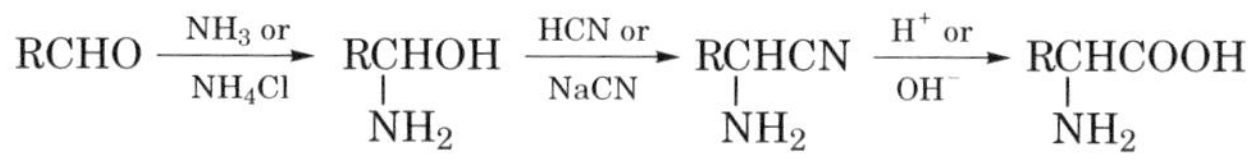

$$RCHO \xrightarrow[NH_4Cl]{NH_3 \text{ or}} \underset{\underset{NH_2}{|}}{RCHOH} \xrightarrow[NaCN]{HCN \text{ or}} \underset{\underset{NH_2}{|}}{RCHCN} \xrightarrow[OH^-]{H^+ \text{ or}} \underset{\underset{NH_2}{|}}{RCHCOOH}$$

Bucherer Synthesis

$$RCHO \xrightarrow[NaCN]{(NH_4)_2CO_3} \begin{matrix} RCH\!-\!C{\diagup}^O \\ |\qquad\quad NH \\ NH\!-\!C{\diagdown}_O \end{matrix} \xrightarrow{OH^-} \underset{\underset{NH_2}{|}}{RCHCOOH}$$

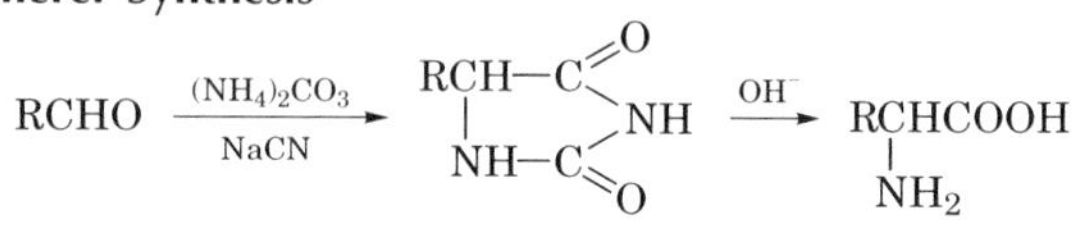

These two methods are popular for α-amino acid synthesis, and used in the industrial production of some amino acids since raw materials are readily available.

Amination of α-Halogeno Carboxylic Acids

Original Method

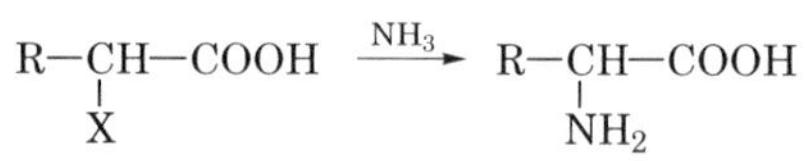

$$\underset{\underset{X}{|}}{R\!-\!CH\!-\!COOH} \xrightarrow{NH_3} \underset{\underset{NH_2}{|}}{R\!-\!CH\!-\!COOH}$$

Gabriel's Modification

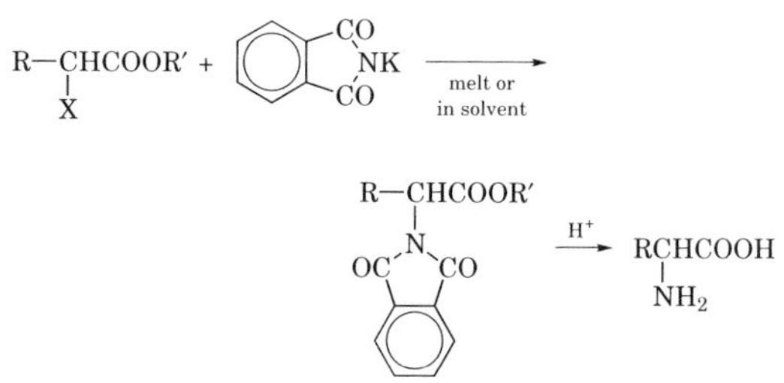

$$\underset{\underset{X}{|}}{R\!-\!CHCOOR'} + \text{(phthalimide)NK} \xrightarrow[\text{in solvent}]{\text{melt or}}$$

$$\begin{matrix} R\!-\!CHCOOR' \\ | \\ N \\ OC{\diagup}\quad{\diagdown}CO \end{matrix} \xrightarrow{H^+} \underset{\underset{NH_2}{|}}{RCHCOOH}$$

Alkylation of Active Methylene Compounds

Erlenmeyer Synthesis and Others. Hydantoin, azlactone, diketopiperazine, etc., are readily available, so that these methods are often utilized.

Amination of α-Keto Acids

α-Keto acids are catalytically reduced

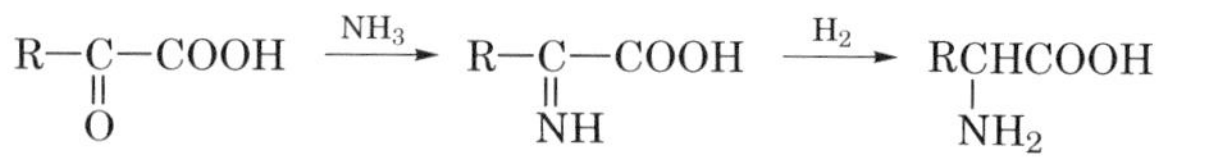

$$\underset{\underset{O}{\|}}{R\!-\!C\!-\!COOH} \xrightarrow{NH_3} \underset{\underset{NH}{\|}}{R\!-\!C\!-\!COOH} \xrightarrow{H_2} \underset{\underset{NH_2}{|}}{RCHCOOH}$$

in the presence of ammonia. α-Keto acids are readily prepared by hydrolysis of substituted hydantoins or double carbonylation of benzyl halide in the case of phenylpyruvic acid. Enzymatic amination of α-keto acids has been developed by many research groups.

Reduction of α-Ketoxime

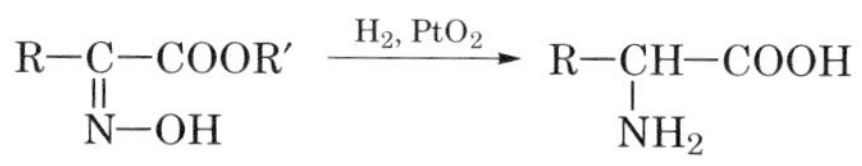

$$\underset{\underset{N-OH}{\|}}{R\!-\!C\!-\!COOR'} \xrightarrow{H_2, PtO_2} \underset{\underset{NH_2}{|}}{R\!-\!CH\!-\!COOH}$$

Reduction of α-Nitro Carboxylic Acid

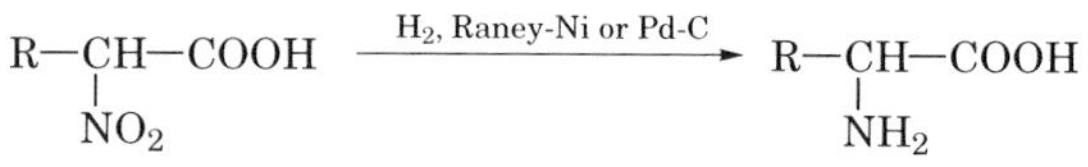

$$\underset{\underset{NO_2}{|}}{R\!-\!CH\!-\!COOH} \xrightarrow{H_2, \text{Raney-Ni or Pd-C}} \underset{\underset{NH_2}{|}}{R\!-\!CH\!-\!COOH}$$

Hofmann Degradation

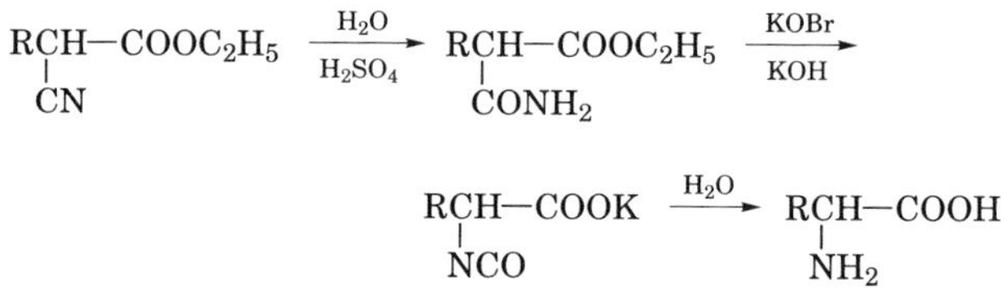

$$\underset{\underset{CN}{|}}{RCH\!-\!COOC_2H_5} \xrightarrow[H_2SO_4]{H_2O} \underset{\underset{CONH_2}{|}}{RCH\!-\!COOC_2H_5} \xrightarrow[KOH]{KOBr}$$

$$\underset{\underset{NCO}{|}}{RCH\!-\!COOK} \xrightarrow{H_2O} \underset{\underset{NH_2}{|}}{RCH\!-\!COOH}$$

Schmidt Reaction

$$\underset{\underset{COCH_3}{|}}{RCH\!-\!COOC_2H_5} + HN_3 \xrightarrow{H_2SO_4}$$

$$\underset{\underset{NHCOCH_3}{|}}{RCHCOOC_2H_5} \xrightarrow{H_2O} \underset{\underset{NH_2}{|}}{RCHCOOH}$$

$$RCH(COOH)_2 + HN_3 \xrightarrow{H_2SO_4} \underset{\underset{NH_2}{|}}{RCHCOOH}$$

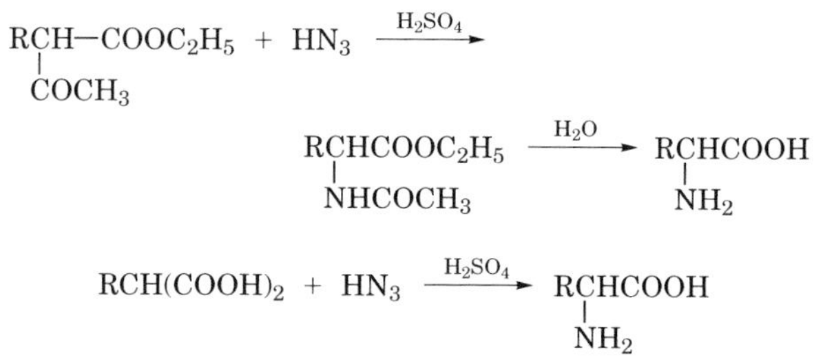

Curtius Degradation

$$RCH{-}COOK \xrightarrow{NH_2NH_2} RCH{-}COOK \xrightarrow{HNO_2}$$
$$\underset{COOC_2H_5}{} \qquad \underset{CONHNH_2}{}$$

$$RCH{-}COOK \xrightarrow[H_2O]{H^+} RCHCOOH$$
$$\underset{CON_3}{} \qquad \underset{NH_2}{}$$

$$RCHCN \xrightarrow{NH_2NH_2} RCHCN \xrightarrow{HNO_2}$$
$$\underset{COOC_2H_5}{} \qquad \underset{CONHNH_2}{}$$

$$RCHCN \xrightarrow[H_2O]{H^+} RCHCOOH$$
$$\underset{CON_3}{} \qquad \underset{NH_2}{}$$

Amine Addition to Double Bond

Production of D,L-aspartic acid from maleic acid ester or fumaric acid ester is a typical example.

$$C_2H_5OOCCH{=}CHCOOC_2H_5 \xrightarrow{NH_3}$$

$$H_2NCOCH_2CH \underset{NH-CO}{\overset{CO-NH}{<}} CHCH_2CONH_2 \xrightarrow{H^+}$$

$$HOOCCH_2CHCOOH$$
$$\underset{NH_2}{}$$

Carbonylation of Aldehyde

This method is noteworthy as an efficient one-step synthesis.

Wakamatsu Reaction

$$R{-}CHO + R'CONH_2 + CO \xrightarrow{H_2}{Co_2(CO)_8}$$

$$R{-}CH{-}COOH \longrightarrow R{-}CH{-}COOH$$
$$\underset{NHCOR'}{} \qquad \underset{NH_2}{}$$

Modified Method

$$C_6H_5CH{-}CH_2 + CH_3CONH_2 + CO \xrightarrow{H_2}{Co_2(CO)_8, Ti(O\text{-}iC_3H_7)_4}$$
$$\underset{O}{}$$

$$C_6H_5CH_2CHCOOH$$
$$\underset{NHCOCH_3}{}$$

Optical Resolution

In many cases only the racemic mixtures of α-amino acids can be obtained through chemical synthesis. Therefore, optical resolution is indispensable to get the optically active L- or D-forms in the production of expensive or uncommon amino acids. The optical resolution of amino acids can be done in two general ways: physical or chemical methods which apply the stereospecific properties of amino acids, and biological or enzymatic methods which are based on the characteristic behavior of amino acids in living cells in the presence of enzymes.

ASYMMETRIC SYNTHESIS

Asymmetric synthesis is a method for direct synthesis of optically active amino acids and finding efficient catalysts is a great target for researchers. Asymmetric syntheses are classified as either enantioselective or diastereoselective reactions. Asymmetric hydrogenation has been applied for practical manufacturing of L-DOPA and L-phenylalanine, but conventional methods have not been exceeded because of the short life of catalysts.

MANUFACTURE AND PROCESSING

Since the discovery of amino acids in animal and plant proteins in the nineteenth century, most amino acids have been produced by extraction from protein hydrolyzates. However, there are many problems in the efficient isolation of the desired amino acid in the pure form.

The rapid development of fermentative production and enzymatic production have contributed to the lower costs of many protein amino acids and to their availability in many fields as economical raw materials.

Reaction of α-Amino Acids

α-Amino acids are ampholytic compounds. The chemical reactions of amino acids can be classified according to their carboxyl, amino, and side-chain groups. Most of the reactions have been well known for a long time; the details of these reactions have been reviewed.

Reactions of the amono group include n-acylation, reaction with phosgene, formation of Schiff bases, the Maillard reaction (nonenzymatic glycation), and substitution reactions.

Reactions of the carboxyl group include esterification, amidation, acid chloride formation, reduction to amino alcohols, and anhydride formation.

Reactions depending on both amino and carboxyl groups include formation of diketopiperazines, formation of hydantoin, Strecker degradation (oxidative deamination), formation of n-carboxy-α-amino acid anhydride (NCA), and ninhydrin-color reaction.

Other reactions include salt formation and metal chelation, synthesis of peptide, and induction of asymmetry by amino acids.

HEALTH AND SAFETY FACTORS

Nutrition

Protein amino acids, which are not synthesized by the body and should be supplied as nutrients to maintain life, are called essential amino acids. For humans, L-arginine, L-histidine, L-isoleucine, L-leucine, L-lysine,

L-methionine, L-phenylalanine, L-valine, L-threonine, and L-tryptophan are essential amino acids. However, in adults, L-arginine and L-histidine are somewhat synthesized in cells. For histidine, there is evidence that it is dietetically essential for the maintenance of nitrogen balance. On the other hand, those amino acids which are synthesized in apparently adequate amounts are nonessential amino acids: L-alanine, L-asparagine, L-aspartic acid, L-cysteine, L-glutamic acid, L-glutamine, glycine, L-proline, L-serine, and L-tyrosine. Of these, L-tyrosine and L-cysteine are essential for children. Recent advances in nutritional studies of amino acids have led to development of amino acid transfusion.

Biosynthesis of Protein

The human body is maintained by a continuous equilibrium between the biosynthesis of proteins and their degradative metabolism where the nitrogen lost as urea (about 85% of total excreted nitrogen) and other nitrogen compounds is about 12 g/d under ordinary conditions.

Toxicity of α-Amino Acid

LD_{50} values of α-amino acids are listed in Table 3. L-Lysine and L-arginine are mutually antagonistic.

Metabolism of Amino Acids

The amino acids are metabolized principally in the liver to a variety of physiologically important metabolites, eg, creatine (creatinine), purines, pyrimidines, hormones, lipids, amino sugars, urea, ammonia, carbon dioxide, and energy sources.

As Neurotransmitters

Several amino acids serve as specialized neurotransmitters in both vertebrate and invertebrate nervous systems.

Table 3. Toxicity[a] of Amino Acids

Amino acid	LD_{50}	
	Oral	Intraperitoneal
L-Arg·HCl	12 g	
L-Cys	5580 mg	1620 mg
L-Cys·HCl		1250 mg[b]
L-(Cys)₂	25 g	
L-His	7930 mg	
L-Ileu		6822 mg
L-Leu		5379 mg
D-Leu		6429 mg
L-Lys·HCl	10 g	4019 mg
L-Met	36 g	4328 mg
L-Phe		5287 mg
D-Phe		5452 mg
DL-Thr		3098 mg
L-Trp		1634 mg
D-Trp		4289 mg
L-Val		5390 mg
D-Val		6093

[a] Rat, unless otherwise noted.
[b] Mouse.

These amino acids can be classified as inhibitory transmitters, such as γ-aminobutyric acid (GABA) and glycine, and excitatory amino acids, examples of which are L-glutamic acid and L-aspartic acid.

Modification of Amino Acid in Protein Molecules

Protein kinases, whose activities are regulated by secondary messengers, such as cyclic nucleotide and Ca^{2+}, modify physiologically important proteins by phosphorylating the hydroxy moiety of serine, threonine, and tyrosine in protein molecules. Consequently, various cellular functions, cell growth, and cell differentiation are seriously affected.

USES

Amino acids are used in feeds, food, parenteral and enteral nutrition, medicine, cosmetics, and raw materials for the chemical industry.

The agricultural products used as feedstuffs for domestic animals are different, depending on the areas where they are used. These feedstuffs do not always meet the essential amino acid requirements for the economical growth of the animals and usually require DL-methionine and L-lysine supplements as the first and/or second limiting amino acid. The addition of these amino acids to the feeds saves the use of feed protein without affecting the growth response of animals.

Each amino acid has its characteristic taste of sweetness, sourness, saltiness, bitterness, or "umami". Umami taste, which is typically represented by L-glutamic acid salt (and some 5'-nucleotide salts), makes food more palatable and is recognized as a basic taste, independent of the four other classical basic tastes of sweet, sour, salty, and bitter.

Monosodium L-glutamate MSG is utilized as a food flavor enhancer in various seasonings and processed foods. D-Glutamate is tasteless. L-Aspartic acid salt has a weaker taste of umami. Glycine and L-alanine are slightly sweet.

Aspartame (L-aspartyl-L-phenylalanine methyl ester) is about 200 times sweeter than sucrose. The Acceptable Daily Intake (ADI) has been established by JECFA as 40 mg/kg/day. Demand for L-phenylalanine and L-aspartic acid as the raw materials for the synthesis of aspartame has been increasing. Derivatives of aspartame are also described as flavor modifiers and sweeteners in chewing gum.

In traditional cooking of proteinaceous foods, the fundamental difference between Western and Asiatic cultures is that the former cooks proteins with unseasoned fats and the latter cooks with many kinds of traditional seasonings that have tastes of amino acids. Western cultures have some traditional foods with amino acid taste such as cheese. Protein hydrolysates are popular as seasonings.

The enzymatic hydrolysates of milk casein and soy protein sometimes have a strong bitter taste. The bitter taste is frequently developed by pepsin, chymotrypsin, and some neutral proteases and accounted for by the existence of peptides that have a hydrophobic amino acid in the carboxylic terminal. Amino acids play a role in food processing in the development of a cooked flavor as the result of a chemical reaction called the nonenzymatic browning reaction.

Currently available proteins are all deficient to greater or lesser extent in one or more of the essential amino acids.

Amino acid transfusion has been widely used since early times to maintain basic nitrogen metabolism when proteinaceous food cannot be eaten. Special amino acid mixtures (eg, branched chain amino acids-enriched solution) have been developed for the treatment of several diseases.

Many amino acids have been used or studied for pharmaceutical purposes. L-Glutamine has been used as a remedy for gastric and duodenal ulcers. L-DOPA [L-3-(3,4-dihydroxyphenyl)alanine] has been widely employed as an antiparkinsonism agent. L-α-MethylDOPA is an effective antihypertensive drug. L-Tryptophan and 5-hydroxy-L-tryptophan are effective as antidepressants. In animal experiments it was demonstrated that L-tryptophan induces sleep. Potassium aspartate is widely used for improving disturbances in electrolyte metabolism. Calcium aspartate is known as a calcium supplement. Glutamic acid hydrochloride is a gastric acidifier which acts to counterbalance a deficiency of hydrochloric acid in the gastric juice. L-Arginine and L-ornithine are used for ammonia detoxification. p-Hydroxy-D-phenylglycine, D-phenylglycine, D-cysteine, D-aspartic acid are important as the side chains of β-lactam antibiotics (see ANTIBIOTICS, β-LACTAMS). D-Homophenylalanine is a raw material for chemical synthesis of enalapril, an inhibitor of angiotensin converting enzymes. D-Valine is the raw material for the chemical synthesis of pyrethroid agricultural chemicals.

Some of the recently reported medical applications include: glutamic acid decarboxylase for treating type I diabetes; compositions based on proline, glycine, and lysine for treating lesions and wounds; modulation of Bruton's tyrosine kinease and intermediates for the treatment of osteoporosis; and the regulation of T-cell mediated immunity by tryptophan.

Amino acids and their derivatives occur in skin protein, and they exhibit a controlling or buffering effect of pH variation in skin and a bactericidal effect. Serine is one component of skin care cream or lotion. N-Acylglutamic acid triethanolamine monosalt is used for shampoo. Glucose glutamate is a moisturizing compound for hair and skin. New histidine derivatives as free antiradical agents in cosmetics has been described.

Cysteine is used as a reductant for cold wave treatment in place of thioglycolic acid. N-Lauroylarginine ethyl ester is applied as the hydrochloride as a preservative. Urocanic acid which is derived from histidine is used in skin cream as a uv absorber.

Recently, as some amino acids (eg, L-glutamic acid, L-lysine, glycine, DL-alanine, DL-methionine) have become less expensive chemical materials, they have been employed in various application fields. Poly (amino acid)s are attracting attention as biodegradable polymers in connection with environmental protection.

N-Acylglutamates, sodium N-lauroyl sarcosinate, and N-acyl-β-alanine Na salt are used in the cosmetic field as nontoxic surfactants. Some of them (eg, N-acylglutamic acid dibutylamide) are used as oil gellating agents to recover effluent oil in seas and rivers.

Ferroelectric liquid crystals have been applied to LCDs (liquid crystal displays) because of their quick response. Ferroelectric liquid crystals have chiral components in their molecules, some of which are derived from amino acids. Concentrated solutions (10–30%) of α-helix poly (amino acid)s show a lyotropic cholesteric liquid crystalline phase, and poly(glutamic acid ester) films display a thermotropic phase. Their practical applications have not been determined.

Poly(γ-methyl glutamate) that has excellent weatherability, is nonyellowing, and has high moisture permeability and heat resistance was developed as the original coating agent for artificial leather. To improve flexibility and stretch, a block copolymer with polyurethane was developed. Poly(L-leucine) is being tested as artificial skin or as a wound dressing.

Various types of protected amino acids for peptide synthesis are available commercially.

As hardeners and vulcanizing agents, acylhydrazide derivatives of amino acids are used for epoxy resins.

As vulcanizing agents, amino acids with or without sulfur are used for nipple rubber of babies' bottles and rubbers used in medical applications.

K. Aida and co-eds., Biotechnology of Amino Acid Production, Elsevier, Amsterdam, The Netherlands, 1986.

B. Alberts and co-eds., Molecular Biology of the Cell, 2nd ed., Garland Publishing, New York, 1989.

R. Funabiki in A. Yoshida and co-eds., Nutrition: Proteins and Amino Acids, Japan Science Society Press, Tokyo, Japan, 1990.

P. Hardy in G. C. Barrett, ed., Chemistry and Biochemistry of the Amino Acids, Chapman and Hall, London, 1985.

KAZUMI ARAKI
University of East Asia
TOSHITSUGU OZEKI
Kyowa Hakko Kogyo Company

AMINO RESINS AND PLASTICS

Amino resins are thermosetting polymers made by combining an aldehyde with a compound containing an amino ($-NH_2$) group. Urea–formaldehyde (U/F) accounts for over 80% of amino resins; melamine–formaldehyde accounts for most of the rest. Other aldehydes and other amino compounds are used to a very minor extent.

The principal attractions of amino resins and plastics are water solubility before curing, which allows easy application to and with many other materials; colorlessness, which allows unlimited colorability with dyes and pigments, excellent solvent resistance in the cured state; outstanding hardness and abrasion resistance; and good heat resistance. Limitations of these materials include release of formaldehyde during cure and, in some cases, such as in foamed insulation, after cure, and poor outdoor weatherability for urea moldings. Repeated cycling of wet and dry conditions causes surface cracks. Melamine moldings have relatively good outdoor weatherability.

Amino resins are manufactured throughout the industrialized world to provide a wide variety of useful products. Adhesives, representing the largest single market, are used to make plywood, chipboard, and sawdust board. Other types are used to make laminated wood beams, parquet flooring, and for furniture assembly. Some amino resins are used as additives to modify the properties of other materials. Amino resins are also often used for the cure of other resins such as alkyds and reactive acrylic polymers.

The term "amino resin" is usually applied to the broad class of materials regardless of application, whereas the term "aminoplast" or sometimes "amino plastic" is more commonly applied to thermosetting molding compounds based on amino resins. Amino plastics and resins have been in use since the 1920s. Compared to other segments of the plastics industry, they are mature products, and their growth rate is only about one-half of that of the plastics industry as a whole. They account for ~3% of the U.S. plastics and resins production.

Aminoplasts and other thermosetting plastics are molded by an automatic injection molding process similar to that used for thermoplastics, but with an important difference. Instead of being plasticized in a hot cylinder and then injected into a much cooler mold cavity, the thermosets are plasticized in a warm cylinder and then injected into a hot mold cavity where the chemical reaction of cure sets the resin to the solid state. The process is best applied to relatively small moldings. Melamine plastic dinnerware is still molded by standard compression-molding techniques. The great advantage of injection molding is that it reduces costs by eliminating manual labor, thereby placing the amino resins in a better position to compete with thermoplastics.

The future for amino resins and plastics seems secure because they can provide qualities that are not easily obtained in other ways. New developments will probably be in the areas of more highly specialized materials for treating textiles, paper, etc, and for use with other resins in the formulation of surface coatings, where a small amount of an amino resin can significantly increase the value of a more basic material. Additionally, since amino resins contain a large proportion of nitrogen, a widely abundant element, they may be in a better position to compete with other plastics as raw materials based on carbon compounds become more costly.

RAW MATERIALS

Most amino resins are based on the reaction of formaldehyde with urea or melamine.

Since melamine resins are derived from urea, they are more costly and are therefore restricted to applications requiring superior performance. Essentially all of the melamine produced is used for making amino resins and plastics.

CHEMISTRY OF RESIN FORMATION

The first step in the formation of resins and plastics from formaldehyde and amino compounds is the addition of formaldehyde to introduce the hydroxymethyl group, known as methylolation or hydroxymethylation:

$$R-NH_2 + HCHO \longrightarrow R-NH-CH_2OH$$

The second step is a condensation reaction that involves the linking together of monomer units with the liberation of water to form a dimer, a polymer chain, or a vast network. This reaction is usually referred to as methylene bridge formation, polymerization, resinification, or simply cure, and is illustrated in the following equation:

$$RNH-CH_2OH + H_2NR \longrightarrow RNH-CH_2-NHR + H_2O$$

Success in making and using amino resins largely depends on the precise control of these two chemical reactions. Consequently, these reactions have been much studied.

MANUFACTURE

Precise control of the course, speed, and extent of the reaction is essential for successful manufacture. Important factors are mole ratio of reactants; catalyst (pH of reaction mixture); and reaction time and temperature. Amino resins are usually made by a batch process. In the manufacture of amino resins, every effort is made to recover and recycle the raw materials. However, there may be some loss of formaldehyde, methanol, or other solvent as tanks and reactors are vented. Some formaldehyde, solvents, and alcohols are also evolved in the curing of paint films and the curing of adhesives and resins applied to textiles and paper. The amounts of material evolved in curing the resins may be small so that it may be difficult to justify the installation of complex recovery equipment. However, in the development of new resins for coatings and for treating textiles and paper, emphasis is being placed on those compositions that evolve a minimum of by-products on curing.

Urea–formaldehyde adhesives are used in the manufacture of plywood, in the fortification of starch adhesives for manufacture of paper bags and corrugated box boards, the production of "floral" and insulating foams, high quality sandpaper, and parquet flooring. Large volumes of urea resins are also sold in the United States for particle board bonding and other uses, at relatively low prices.

Melamine or melamine–ureas are used in the manufacture of truck and railroad flooring, laminated lumber, beams, exterior doors, marine plywood, toilet seats, and school furniture. The bonds in these products meet a variety of commercial, military, and federal specifications for exterior waterproof adhesives.

Urea molding compound has found wide use and acceptance in the electrical surface wiring device industry. Typical applications are circuit breakers, switches, wall plates, and duplex outlets. Urea is also used in closures, stove hardware, buttons, and small housings.

Melamine molding compound is used primarily in dinnerware applications for both domestic and institutional use. It is also used in electrical wiring devices, ashtrays, buttons, and housings.

The emergence of a new amino application is rare at this point in its relatively long life, but one such has appeared and is growing rapidly. Because of the relative hardness of both urea and melamine moldings, a unique use has been developed for small, granular sized particles of cut up molded articles. It is the employment of a pressurized stream of plastic particles to remove paint without damaging the surface beneath, and can be compared to a sandblasting operation. This procedure is gaining wide acceptance by both commercial airlines and the military for the refinishing of painted surfaces. It does not harm the substrate and eliminates the use of chemicals formerly used in stripping paint.

Production of decorated melamine plastic dinner plates makes use of molding and laminating techniques. The pattern is printed on the same type of paper used for the protective overlay of decorative laminates, treated with melamine resin and dried, and then cut into disks of the appropriate size.

The excellent electrical properties, hardness, heat resistance, and strength of melamine resins makes them useful for a variety of industrial applications. Cured amino resins are far too brittle to be used alone as surface coatings for metal or wood substrates, but in combination with other film formers (alkyds, polyesters, acrylics, epoxies) a wide range of acceptable performance properties can be achieved. A wide selection of amino resin compositions is commercially available. Amino resins based on urea have advantages in low temperature cure response and low cost. However, they are not as stable to ultraviolet (uv) radiation as melamine resins, and have poorer heat resistance; therefore, they have been successful primarily in interior wood finishes. Melamine resins, on the other hand, are uv stable, have excellent heat resistance, film hardness, and chemical resistance. They therefore dominate amino resin usage in (OEM) automotive coatings, general metals finishes, container coatings (both interior and exterior), and prefinished metal applications. Glycoluril resins have also found use in prefinished metal, primarily because of their high film flexibility properties. Unalkylated glycoluril resins are unique in that they are stable under slightly acidic conditions and have therefore found use in low temperature cure waterborne finishes. Benzoguanamine resins have historically been successful in appliance finishes because of their superior chemical resistance and specifically their detergent resistance. However, they have both poor uv resistance and economics, which have limited their use in other application areas.

The principal problems facing amino resins in recent industrial coatings have been their formaldehyde emission and low temperature cure performance. Significant progress has been made in reducing the residual free formaldehyde in the amino resin, but formaldehyde generation on baking must still be addressed. Concerning low temperature cure performance, emphasis is being placed on catalyst selection. The development of improved catalysts is the most promising solution to low temperature cure performance enhancement.

Most amino resins used commercially for finishing textile fabrics are methylolated derivatives of urea or melamine. Although these products are usually monomeric, they may contain some polymer byproduct.

Amino resins react with cellulosic fibers and change their physical properties. They do not react with synthetic fibers, such as nylon, polyester, or acrylics, but may self-condense on the surface. This results in a change in the stiffness or resiliency of the fiber. Partially polymerized amino resins of such molecular size that prevents them from penetrating the amorphous portion of cellulose also tend to increase the stiffness or resiliency of cellulose fibers.

The most versatile textile-finishing resins are the melamine–formaldehyde resins. They provide wash-and-wear properties to cellulosic fabrics, and enhance the wash durability of flame-retardant finishes. Butylated melamine–formaldehyde resins of the type used in surface coatings may be used in textile printing-ink formulations.

Much less important than the melamine–formaldehyde and urea–formaldehyde resins are the methylol carbamates. They are urea derivatives since they are made from urea and an alcohol (R can vary from methyl to a monoalkyl ether of ethylene glycol).

Other amino resins used in the textile industry for rather specific properties have included the methylol derivatives of acrylamide, hydantoin, and dicyandiamide.

Melamine resins are also used to improve the adhesion of rubber to reinforcing cord in tires. Amino resins are used by the paper industry in large volume for a variety of applications. The resins are divided into two classes according to the mode of application. Resins added to the fiber slurry before the sheet is formed are called wet-end additives and are used to improve wet and dry strength and stiffness. Resins applied to the surface of formed paper or board, almost invariably together with other additives, are used to improve the water resistance of coatings, the sag resistance in ceiling tiles, and the scuff resistance in cartons and labels.

The integrity of a paper sheet is dependent on the hydrogen bonds that form between the fine structures of cellulose fibers during the pressing and drying operations. The bonds between hydroxyl groups of neighboring fibers are very strong when the paper is dry but are severely weakened as soon as the paper becomes wet. Bonding between the hydroxyls of cellulose and water is as energetic as bonding between two cellulose hydroxyl groups. Consequently, ordinary paper loses most of its strength when it is wet or exposed to very high humidity.

Many materials have been used over the years in an effort to correct this weakness in paper. If water can be prevented from reaching the sites of the bonding by sizing or coating the sheet, then a measure of wet strength may be attained. Formaldehyde, glyoxal, polyethylenimine, and, more recently, derivatized starch and derivatized cationic polyacrylamide resins have been used to provide temporary wet strength. The first two materials must be applied to the formed paper but the other materials are substantive to the fiber and may be used as wet-end additives.

Today three major types of wet-strength resins are used in papermaking:polyamide–polyamine resins cross-linked with epichlorohydrin are used in neutral to alkaline papers; cationic polyacrylamide resins cross-linked with glyoxal are used for acid to neutral papers; and melamine–formaldehyde resins are used for acid papers.

Wet-strength applications account for the majority of amino resin sales to the paper industry, but substantial volumes are sold for coating applications. The largest use is to improve the resistance of starch-clay coatings to dampness.

Closely allied to resins for treating paper are the resins used to treat regenerated cellulose film (cellophane) that does not have good water resistance unless it is coated with nitrocellulose or poly(vinylidene chloride). Water-soluble melamine–formaldehyde resins are used in the tanning of leather in combination with the usual tanning agents. By first treating the hides with a melamine–formaldehyde resin, the leather is made more receptive to other tanning agents and the finished product has a lighter color. The amino resin is often referred to as a plumping agent because it makes the finished leather firmer and fuller.

Urea–formaldehyde resins are used in the manufacture of foams. They are also used as the binder for the sand cores used in the molds for casting hollow metal shapes.

REGULATORY CONCERNS

Both urea– and melamine–formaldehyde resins are of low toxicity. Melamine–formaldehyde resins may be used in paper that contacts aqueous and fatty foods according to 21 *CFR* 121.181.30. However, because a lower PEL has been established by OSHA, some mills are looking for alternatives. Approaches toward achieving lower formaldehyde levels in the resins have been reported; the efficacy of these systems needs to be established. Although alternative resins are available, significant changes in the papermaking operation would be required in order for them to be used effectively.

W. Lindlaw, *The Preparation of Butylated Urea–Formaldehyde and Butylated Melamine Formaldehyde Resins Using Celanese Formcel and Celanese Paraformaldehyde,* Technical Bulletin, Celanese Chemical Co., New York, Table XIIA.

U.F. Concentrate-85, Technical Bulletin, Allied Chemical Corp., New York, 1985.

J. F. Walker, *Formaldehyde, American Chemical Society Monograph, No. 159,* 3rd ed., Reinhold Publishing Corp., New York, 1964.

J. P. Weidner, ed., *Wet Strength in Paper and Paper Board, Monograph Series, No. 29,* Technical Association of Pulp and Paper Industry, New York, 1965.

LAURENCE L. WILLIAMS
American Cyanamid Company

AMINOPHENOLS

Aminophenols and their derivatives are of commercial importance, both in their own right and as intermediates in the photographic, pharmaceutical, and chemical dye industries. They are amphoteric and can behave either as weak acids or weak bases, but the basic character usually predominates. 3-Aminophenol (**1**) is fairly stable in air, unlike 2-aminophenol (**2**) and 4-aminophenol (**3**), which easily undergo oxidation to colored products. The former are generally converted to their acid salts, whereas 4-aminophenol is usually formulated with low concentrations of antioxidants which act as inhibitors against undesired oxidation.

(1)   (2)   (3)

PHYSICAL PROPERTIES

The simple aminophenols exist in three isomeric forms, depending on the relative positions of the amino and hydroxyl groups around the benzene ring. At room temperature they are solid crystalline compounds. In the past, the commercial grade materials were usually impure and colored because of contamination with oxidation products, but now virtually colorless, high purity commercial grades are available. The solubilities of these compounds in common solvents of differing polarities (dielectric constants) are given in Table 1.

2-Aminophenol

This compound forms white orthorhombic bipyramidal needles when crystallized from water or benzene, which readily become yellow-brown on exposure to air and light.

3-Aminophenol

This is the most stable of the isomers under atmospheric conditions. It forms white prisms when crystallized from water or toluene.

Table 1. General Properties of Aminophenols

Property	2-Aminophenol	3-Aminophenol	4-Aminophenol
alternative names	2-hydroxyaniline	3-hydroxyaniline	4-hydroxyaniline
	2-amino-1-hydroxybenzene	3-amino-1-hydroxybenzene	4-hydroxy-1-aminobenzene
C.I. designation	76,520		
CAS Registry Number	[95-55-6]	[591-27-5]	[123-30-8]
molecular formula	C_6H_7NO	C_6H_7NO	C_6H_7NO
molecular weight	109.13	109.13	109.13
melting point, °C	174	122–123	189–190[a]
boiling point, °C			
0.04 kPa			130[a], 110[b]
0.4 kPa			150
1.07 kPa			167
1.47 kPa	153[b,c]		174
101.3 kPa		164	284
ΔH_f, kJ/mol[c]	-191.0 ± 0.9	-194.1 ± 1.0	-190.6 ± 0.9[d]

[a] Decomposes.
[b] Sublimes. To convert kPa to mm Hg, multiply by 7.5.
[c] In the crystalline state. To convert kJ to kcal, divide by 4.184.
[d] −179.1 is also quoted.

CHEMICAL PROPERTIES

The chemical properties and reactions of the aminophenols and their derivatives are to be found in detail in many standard chemical texts. The acidity of the hydroxyl function is depressed by the presence of an amino group on the benzene ring; this phenomenon is most pronounced with 4-aminophenol. The amino group behaves as a weak base, giving salts with both mineral and organic acids. The aminophenols are true ampholytes, with no zwitterion structure; hence they exist either as neutral molecules (4), or as ammonium cations (5), or phenolate ions (6), depending on the pH value of the solution.

(5)   (4)   (6)

The aminophenols are chemically reactive, undergoing reactions involving both the aromatic amino group and the phenolic hydroxyl moiety, as well as substitution on the benzene ring. Oxidation leads to the formation of highly colored polymeric quinoid structures. 2-Aminophenol undergoes a variety of cyclization reactions. Important reactions include alkkylation, acylation, diazonium salt formation, cyclization rections, condensation reactions, and reactions of the benzene ring.

MANUFACTURE AND PROCESSING

Aminophenols are either made by reduction of nitrophenols or by substitution. Reduction is accomplished with iron or hydrogen in the presence of a catalyst. Catalytic reduction is the method of choice for the production of 2- and 4-aminophenol. Electrolytic reduction is also under industrial consideration, and substitution reactions provide the major source of 3-aminophenol.

PURIFICATION

Contaminants and by-products that are usually present in 2- and 4-aminophenol made by catalytic reduction can be reduced or even removed completely by a variety of procedures. These include treatment with 2-propanol, with aliphatic, cycloaliphatic, or aromatic ketones, with aromatic amines, with toluene or low mass alkyl acetates, or with phosphoric acid, hydroxyacetic acid, hydroxypropionic acid, or citric acid. In addition, purity may be enhanced by extraction with methylene chloride, chloroform, or nitrobenzene. Another method employed is the treatment of aqueous solutions of aminophenols with activated carbon.

ECONOMIC ASPECTS

Production figures for the aminophenols are scarce, the compounds usually being classified along with many other aniline derivatives. Most production of the technical grade materials (95% purity) occurs on-site as they are chiefly used as intermediate reactants in continuous chemical syntheses. World production of the fine chemicals (99% purity) is probably no more than a few hundred metric tons yearly with 4-aminophenol being the least expensive.

STORAGE

Under atmospheric conditions, 3-aminophenol is the most stable of the three isomers. Both 2- and 4-aminophenol are unstable; they darken on exposure to air and light and should be stored in brown glass containers, preferably in an atmosphere of nitrogen. The use of activated iron oxide in a separate cellophane bag inside the storage container, or the addition of stannous chloride, or sodium bisulfite inhibits the discoloration of aminophenols. The salts, especially the hydrochlorides, are more resistant to oxidation and should be used where possible.

HEALTH AND SAFETY FACTORS

In general, aminophenols are irritants. Their toxic hazard rating is slight to moderate and their acute oral toxicities in the rat (LD_{50}) are quoted as 1.3, 1.0, and 0.375 g/kg body weight for the 2-, 3-, and 4-isomer, respectively. Repeated contamination may cause general itching, skin sensitization, dermatitis, and allergic reactions. Immunogenic conjugates are spontaneously produced upon exposure to 2- and 4-aminophenol. Methemoglobin formation with subsequent cyanosis is another possible complication. Inhalation of aminophenols causes irritation of the mucosal membranes and may precipitate allergic bronchial asthma. Thermal decomposition will release toxic fumes of carbon monoxide and nitrogen oxides.

4-Aminophenol is a selective nephrotoxic agent and interrupts proximal tubular function. Teratogenic effects have been noted with 2- and 4-aminophenol in the hamster, but 3-aminophenol was without effect in the hamster and rat. Obviously, care should be taken in handling these compounds with the wearing of chemical-resistant gloves and safety goggles; prolonged exposure should be avoided. Contaminated clothing should be removed immediately and the affected area washed thoroughly with running water for at least 10 minutes.

The addition of slaked lime and the initiation of polymerization reactions with H_2O_2 and ferric or stannous salts are techniques employed to remove aminophenols from waste waters.

USES

The aminophenols are versatile intermediates and their principal use is as synthesis precursors; their products are represented among virtually every class of stain and dye.

Both 2- and 4-aminophenols are strong reducing agents and are employed as photographic developers under the trade names of Atomal and Ortol (2-aminophenol); Activol, Azol, Certinal, Citol, Paranol, Rodinol, Unal, and Ursol P (4-aminophenol); 2-aminophenol is a principal intermediate in the synthesis of such heterocyclic systems as oxyquinolines, phenoxazines, and benzoxazoles. The last-named compounds have been used as inflammation inhibitors, and other derivatives have potential as antiallergic agents. In addition, 2-aminophenol is specifically used for shading leather, fur, and hair from grays to yellowish brown. It has also found application in the determination and extraction of certain precious metals.

3-Aminophenol has been used as a stabilizer of chlorine-containing thermoplastics, although its principal use is as an intermediate in the production of 4-amino-2-hydroxybenzoic acid, a tuberculostat. This isomer is also employed as a hair colorant and as a coupler molecule in hair dyes.

Nitrogen-substituted 4-aminophenols have long been known as antipyretics and analgesics, and the production of these derivatives represents significant use of this compound. 4-Aminophenol is also used as a wood stain,

Table 2. Derivatives of Aminophenols

Common name	CAS Registry Number	Molecular formula	Molecular weight	Melting point, °C	Boiling point[a], °C
Derivatives of 2-aminophenol					
2-amino-4-nitrophenol	[99-57-0]	$C_6H_6N_2O_3$	154.13	80–90	
2-amino-4,6-dinitrophenol	[96-91-3]	$C_6H_5N_3O_5$	199.13	169–170	
2-amino-4,6-dichlorophenol	[527-62-8]	$C_6H_5Cl_2NO$	168.15	95–96	
2,4-diaminophenol	[95-86-3]	$C_6H_8N_2O$	124.14	78–80[b]	
acetarsone	[97-44-9]	$C_8H_{10}AsNO_5$	275.08	240–250[b]	
Derivatives of 3-aminophenol					
3-(N,N-dimethylamino)phenol	[99-07-0]	$C_8H_{11}NO$	137.18	87	265–268 206 (13.3) 194 (6.7) 153 (0.7)
3-(N,N-methylamino)phenol	[14703-69-6]	C_7H_9NO	123.15		170 (1.6)
3-(N,N-diethylamino)phenol	[91-68-9]	$C_{10}H_{15}NO$	165.23	78	276–280 209–211 (1.6)
3-(N-phenylamino)phenol	[101-18-8]	$C_{12}H_{11}NO$	185.22	81.5–82	340
4-amino-2-hydroxybenzoic acid	[65-49-6]	$C_7H_7NO_3$	153.13	150–151	
Derivatives of 4-aminophenol					
4-(N-methylamino)phenol	[150-75-4]	C_7H_9NO	123.15	87	168–169 (2)
4-(N,N-dimethylamino)phenol	[619-60-3]	$C_8H_{11}NO$	137.18	75–76	101–103 (0.067)
4-hydroxyacetanilide	[103-90-2]	$C_8H_9NO_2$	151.15	169–171	
4-ethoxyacetanilide	[62-44-2]	$C_{10}H_{13}NO_2$	179.21	134–135	
N-(4-hydroxyphenyl)glycine	[122-87-2]	$C_8H_9NO_3$	167.16	245–247[b]	

[a] At 101.3 kPa = 760 mm Hg unless otherwise noted in parentheses. Values in parentheses are in the kPa. To convert kPa to mm Hg, multiply by 7.5.
[b] Decomposes.

imparting a roselike color to timber, and as a dyeing agent for fur and feathers.

DERIVATIVES

The derivatives of the aminophenols have important uses both in the photographic and the pharmaceutical industries. They are also extensively employed as precursors and intermediates in the synthesis of more complicated molecules, especially those used in the staining and dye industry. All of the major classes of dyes have representatives that incorporate substituted aminophenols; these compounds produced commercially as dye intermediates have been reviewed.

Details of the more commonly encountered derivatives of the aminophenols can be found in standard organic chemistry texts. Table 2 shows the major aminophenol derviatives.

Beilstein's Handbuch der Organischen Chemie, Julius Springer, Berlin, 1918, Section 13, and Section 13(2).

S. Coffey, ed., *Rodd's Chemistry of Carbon Compounds*, 2nd ed., Vol. 3A, Elsevier Publishing Co., Amsterdam, The Netherlands, 1971.

Colour Index, 3rd ed., Society of Dyers and Colorists, Bradford, England, Vol. 4, 1971, pp. 4001–4863 and Vol. 6, 1975.

H. Zollinger, *Azo and Diazo Chemistry*, Interscience, New York, 1961.

STEPHEN C. MITCHELL
PAUL CARMICHAEL
Imperial College London

ROSEMARY WARING
University of Birmingham

AMMONIA

Ammonia, NH_3, a colorless alkaline gas, is lighter than air and possesses a unique, penetrating odor.

The synthesis of ammonia directly from hydrogen and nitrogen on a commercial scale was pioneered by Haber and Bosch in 1913, for which they won Nobel prizes. Further developments in economical, large-scale ammonia production for fertilizers have made as significant impact on increases in the world's food suply: today as much as 40% of human's protein needs are derived indirectly from atmospheric nitrogen fixed by the Haber-Bosch process and its successors.

Nitrogen, in various forms, is an important element to life. It is a building block for amino and nucleic acids. Bound nitrogen cycles through the biosphere among plants, animals, and micro-organisms. The ultimate source of all this nitrogen is the Earth's atmosphere. Atmospheric nitrogen can only be fixed to ammonia by only a relatively number of species of micro-organisms. Some are free living, such as the cyanobacteria; others are symbiotic with certain plants such as legumes.

Many important crops tend to take up nitrogen mainly in the nitrate form. This form of nitrogen is produced by soil organisms from various bound nitrogen sources, or is intentionally applied as a nitrate fertilizer. Unfortunately, the nitrate ion is prone to groundwater leaching. Nitrate contamination of groundwater became a significant environmental issue in the 1980s. Improved farm management techniques have been the main tools used to deal with this issue.

PHYSICAL PROPERTIES

Table 1 lists the important physical properties of ammonia. The flammable limits of ammonia in air are 16 to 25% by volume; in oxygen the range is 15 to 79%. Such mixtures can explode although ammonia–air mixtures are quite difficult to ignite. The ignition temperature is about 650°C.

Ammonia is readily absorbed in water to make ammonia liquor. Additional thermodynamic properties may be found in the literature. Considerable heat is evolved during the solution of ammonia in water: approximately 2180 kJ (520 kcal) of heat is evolved upon the dissolution of 1 kg of ammonia gas.

Ammonia is an excellent solvent for salts, and has an exceptional capacity to ionize electrolytes. The alkali metals and alkaline earth metals (except beryllium) are readily soluble in ammonia. Iodine, sulfur, and phosphorus dissolve in ammonia. In the presence of oxygen, copper is readily attacked by ammonia. Potassium, silver, and uranium are only slightly soluble. Both ammonium and beryllium chloride are very soluble, whereas most other metallic chlorides are slightly soluble or insoluble. Bromides are in general more soluble in ammonia than

Table 1. Physical Properties of Anhydrous Ammonia

Property	Value
molecular weight	17.03
boiling point, °C	−33.35
freezing point, °C	−77.7
critical temp, °C	133.0
critical pressure, kPa[a]	11,425
specific heat, J/(kg·K)[b]	
0°C	2097.2
100°C	2226.2
200°C	2105.6
heat of formation of gas, ΔH_f, kJ/mol[b]	
0 K	−39,222
298 K	−46,222
solubility in water, wt %	
0°C	42.8
20°C	33.1
40°C	23.4
60°C	14.1
specific gravity	
−40°C	0.690
0°C	0.639
40°C	0.580

[a] To convert kPa to psi, multiply by 0.145.
[b] To convert J to cal, divide by 4.184.

chlorides, and most of the iodides are more or less soluble. Oxides, fluorides, hydroxides, sulfates, sulfites, and carbonates are insoluble. Nitrates (eg, ammonium nitrate) and urea are soluble in both anhydrous and aqueous ammonia making the production of certain types of fertilizer nitrogen solutions possible. Many organic compounds such as amines, nitro compounds, and aromatic sulfonic acids, also dissolve in liquid ammonia. Ammonia is superior to water in solvating organic compounds such as benzene, carbon tetrachloride, and hexane.

CHEMICAL PROPERTIES

Ammonia is comparatively stable at ordinary temperatures, but decomposes into hydrogen and nitrogen at elevated temperatures. The rate of decomposition is greatly affected by the nature of the surfaces with which the gas comes into contact: glass is very inactive; porcelain and pumice have a distinct accelerating effect; and metals such as iron, nickel, osmium, zinc, and uranium have even more of an effect.

Ammonia reacts readily with a large variety of substances. Oxidation at a high temperature is one of the more important reactions, giving nitrogen and water.

Of major industrial importance is the reaction of ammonia and carbon dioxide, giving ammonium carbamate, $CH_6N_2O_2$:

$$2\,NH_3 + CO_2 \longrightarrow NH_2CO_2NH_4$$

which then decomposes to urea and water:

$$NH_2CO_2NH_4 \longrightarrow NH_2CONH_2 + H_2O$$

SOURCE AND SUPPLIES

Ammonia is a world-class commodity, manufactured in more than 80 countries. China and the former U.S.S.R. are currently the largest producers, after the United States.

MANUFACTURE

The ammonia synthesis reaction is deceptively simple: nitrogen is combined with hydrogen in a 1:3 stoichiometric ratio to give ammonia with no by-products. The difficulty lies in how to obtain the hydrogen needed for the reaction. The hydrogen production method is the main source of distinction between the various ammonia production routes. In fact, the majority of the equipment in a typical ammonia plant is devoted to hydrogen production rather than ammonia synthesis.

The bulk of world ammonia production is based on steam reforming. Any hydrocarbon feed that can be completely vaporized can be used. Natural gas and naphtha are common. The bulk of existing and nearly all new steam reforming facilities use natural gas for feedstock. Economics and availability influence the decision on feedstock and production technology. The recent trend toward using ammonia and other synthesis gas products to monetize remote natural gas suggests that steam reforming of

natural gas will maintain its dominant position in the foreseeable future.

The exact technology used within a facility depends upon the age of the plant, economics, and the overall fit of the ammonia facility within the entire nitrogen or syngas complex. A standard steam methane reforming plant uses conventional gas purification, but there are other technology routes, based on partial oxidation, water electrolysis plus predictions on future technologies.

BY-PRODUCT HYDROGEN

By-product hydrogen is generated in units that were designed for other purposes but happen to produce hydrogen as a by-product. Examples include many petroleum refining operations such as catalytic reforming, catalytic crackers, thermal crackers, and cokers. Other major by-product sources are offgas from ethylene crackers, offgas from industrial electrolysis processes such as those used for caustic/chlorine manufacture, and tail gases from other synthesis gas processes such methanol production.

A very small portion of on-purpose hydrogen production comes from electrolysis. The electrolysis route is not currently used for ammonia production. For hydrogen, it is currently only competitive for small-scale production and/or for high purity hydrogen applications such as hydrogen for use in semiconductor processing.

The electrolysis route fits well with the vision of using either hydrogen or ammonia as an energy carrier. In this view, renewable-based electricity generated is generated from hydroelectric, solar, geothermal, ocean thermal, or other sources. The electricity is used to split water, the hydrogen is stored and/or transported, and then electricity is regenerated from the hydrogen using a fuel cell. Temporary conversion of the hydrogen to ammonia followed by back-conversion to hydrogen is envisioned as one way to simplify storage and transport. In either case, the hydrogen or the ammonia is merely being used as an energy carrier rather than as an ultimate source of energy.

At the time of this writing, the hydrogen economy is still in the future. Production costs and lack of infrastructure remain as significant hurdles. Continued availability of low-cost nonrenewable hydrocarbons and their associated existing and largely depreciated infrastructures will slow commercialization progress in this area.

Biological processes based on genetic engineering have the potential to eliminate the synthetic nitrogen fertilizer industry completely. The genes responsible for nitrogen fixation could potentially be isolated from symbiotic soil organisms and then transferred and expressed in nonleguminous crops. Alternatively, nitrogen fixation could be transferred and expressed in a wider range of natural soil bacteria. At this point in time, these are rather futuristic scenarios.

STORAGE AND SHIPMENT

Storage

Anhydrous ammonia is ordinarily stored in refrigerated tanks at the plant site and major distribution points atmospheric pressure and at temperature of ca −33°C.

Shipping

Ammonia is usually transported for long distances by barge, pipeline, rail, and for short distances by truck. Factors that govern the type of carrier used in anhydrous ammonia transportation systems are distance, location of plant site in relation to consuming area, availability of transportation equipment, and relative cost of available carriers.

ECONOMIC ASPECTS

Examination of typical ammonia production economics explains the recent trend in new plant location.

Over the past decade inflation rates have been quite low. Natural gas prices have been more of a function of geography rather than time.

Whereas manufacturing costs are strongly influenced by energy prices and capital investment, ammonia selling prices are determined by supply and demand. Most ammonia production is processed or used in the countries where it is produced: world trade of ammonia accounts for only 11% of world production.

ANALYTICAL METHODS

Anhydrous ammonia is normally analyzed for moisture, oil and residue. The ammonia is first evaporated from the sample and the residue tested. In most instances, the amount of oil and sediment in the samples are insignificant and the entire residue is assumed to be water. For more accurate moisture determinations, the ammonia can be dissociated into nitrogen and hydrogen and the dewpoint of the dissociated gas obtained. This procedure works well where the concentration of water is in the ppm range. Where the amount of water is in the range of a few hundredths of a percent, acetic acid and methanol can be added to the residue and a Karl Fischer titration performed to an electrometrically detected end point.

ENVIRONMENTAL CONCERNS

Ammonia production by steam reforming of natural gas is a relatively clean operation and presents no unique environmental problems. NO_x emissions from the flue gas of the primary reformer can be suitably treated by a combination of conventional control techniques including low NO_x burners, selective catalytic reduction and flue gas scrubbing.

Ammonia production using heavy feedstocks raises additional environmental issues. Particulate emissions from solids handling of the feeds must be controlled. The soot, ash and slag produced from these routes must also be disposed of in an environmentally acceptable manner. These feeds are also more likely to create liquid and gases by-products such as tars, phenols, sulfur, cyanides, etc, which must be handled properly.

HEALTH AND SAFETY FACTORS

Fire is the most frequent cause of safety incidents in ammonia production. The most common fuel sources were flanges leaking hydrogen rich gases and oil leaks associated with compressor and pump lubricating–sealing systems.

Ammonia is a strong local irritant that also has a corrosive effect on the eyes and the membranes of the pulmonary system. Current OSHA standards specify the threshold limit value (TLV) 8-h exposure to ammonia as 50 ppm (35 mg/m^3). However, the ACGIH recommends a TLV of 25 ppm. Respiratory protection should be provided for workers exposed to ammonia. Protective clothing such as rubber aprons, boots, gloves and goggles should be worn when handling ammonia.

USES

Non-fertilizer use accounts for about 20% of U.S. annual consumption. Globally, the chemical markets for ammonia are smaller, accounting for only about 15% of consumption. The vast majority of ammonia is used for nitrogen fertilizers. Three principal fertilizers, ammonium nitrate, NH_4NO_3, ammonia sulfate, $(NH_4)SO_4$, and ammonium phosphate, $(NH_4)_3PO_4$, are made by reaction of the respective acids with ammonia.

The catalytic oxidation of ammonia in the presence of methane is commercially used for the production of hydrogen cyanide.

Acrylonitrile, $CH_2{=}CHC{\equiv}N$, is produced in commercial quantities almost exclusively by the ammoxidation of propylene.

Urea, formed from the reaction of ammonia with carbon dioxide, has enjoyed a long-term steady rise in market share to point where today it is the most common nitrogen fertilizer. This increase in market share for urea has been at the expense of ammonium nitrate, ammonium sulfate and direct application of ammonia. Caprolactam is made from ammonia and cyclohexanone. Caprolactam is the monomer for nylon 6. By-product ammonium sulfate from this route is a major source of this fertilizer ingredient. Ammonia is also used in the production of pyridines, amines and amides.

Chemical Profiles - Ammonia, Nov. 29, 1999, Schnell Publishing Company, Web site: www.chemexpo.com.

I. Dybkjaer, "Design and Operating Experience of Large Ammonia Plants", *Ammonia Plant Safety and Related Facilities*, American Institute of Chemical Engineers, Vol. 34, 1994, pp. 199–209.

International Fertilizer Industry Association, Web site: www.fertilizer.org.

Occupational Health Guideline for Ammonia, U.S. Dept. of Health and Human Services, Washington, D.C., 1978.

TIM EGGEMAN
Neoterics International

AMMONIUM COMPOUNDS

There are a considerable number of stable crystalline salts of the ammonium ion, NH_4^+. Several are of commercial importance because of large-scale consumption in fertilizer and industrial markets. The ammonium ion is about the same size as the potassium and rubidium ions, so these salts are often isomorphous and have similar solubility in water. Compounds in which the ammonium ion is combined with a large, uninegative anion are usually the most stable. Ammonium salts containing a small, highly charged anion generally dissociate easily into ammonia and the free acid.

AMMONIUM ACETATES

Both normal or neutral ammonium acetate, $NH_4C_2H_3O_2$, and the acid salt are known. The normal salt results from exact neutralization of acetic acid using ammonia; the acid salt is composed of the neutral salt and acetic acid. The normal salt, CH_3COONH_4, is a white, deliquescent, crystalline solid, formula wt 77.08, having a specific gravity of 1.073. It is quite soluble in water or ethanol: 148 g dissolve in 100 g of water at 4°C.

AMMONIUM CARBONATES

Ammonium Bicarbonate

Ammonium bicarbonate, also known as ammonium hydrogen carbonate or ammonium acid carbonate, is easily formed. However, it decomposes below its melting point, dissociating into ammonia, carbon dioxide, and water.

Ammonium bicarbonate is produced as both food and standard grade and the available products are normally very pure.

Ammonium Carbonate

Normal ammonium carbonate, mp 43°C, formula wt 96.09, is a crystalline solid. The commercial product may be produced by passing carbon dioxide into an absorption column containing aqueous ammonia solution and causing distillation. Vapors containing ammonia, carbon dioxide, and water condense to give a solid mass of crystals. Ammonium carbonate is the principal ingredient of smelling salts because of its characteristic strong ammonia odor. It is also used for other medicinal purposes and as a leavening agent.

AMMONIUM CITRATE

Diammonium citrate, $(NH_4)_2C_6H_6O_7$, mol wt 226.19, is soluble in an equal weight of water but is only slightly soluble in ethanol. The pH of a 0.1 M solution is 4.3. It is made by neutralization of citric acid with ammonia; the crystalline or granular product is used as a chemical reagent and pharmaceutically as a diuretic.

AMMONIUM HALIDES

Ammonium chloride, NH_4Cl, ammonium bromide, NH_4Br, ammonium fluoride and ammonium iodide, NH_4I, are crystalline, ionic compounds of formula wts 53.49, 97.94, and 144.94, respectively. Their densities d_4^{20} systematically follow the increase in formula weight: 1.53, 2.40, and 2.52. All three exist in two crystal modifications: the chloride, bromide, and iodide have the CsCl structure below temperatures of 184.5, 137.8, and −17.6°C, respectively; each reversibly transforms to the NaCl structure at higher temperatures.

The solubility of the ammonium halides in water also increases with increasing formula weight. All ammonium halides exhibit high vapor pressures at elevated temperatures, and thus, sublime readily. The vapor formed on sublimation consists not of discrete ammonium halide molecules, but is composed primarily of equal volumes of ammonia and hydrogen halide. Aqueous solutions of ammonium halides, like the other ammonium salts of strong acids, are acidic.

Ammonium Chloride

Manufacture. Production by direct reaction of ammonia and hydrochloric acid is simple but usually economically unattractive; a process based on metathesis or double decomposition is generally preferred.

Several commercial grades are available: fine crystals of 99 to 100% purity, large crystals, pressed lumps, rods, and granular material.

Uses. Ammonium chloride is used as a nitrogen source for fertilization of rice, wheat, and other crops in Japan, China, India, and Southeast Asia. Ammonium chloride has a number of industrial uses, most importantly in the manufacture of dry-cell batteries, where it serves as an electrolyte. It is also used to make quarrying explosives, as a hardener for formaldehyde-based adhesives, as a flame suppressant, and in etching solutions in the manufacture of printed circuit boards.

Ammonium Bromide and Iodide

Manufacture. Ammonium bromide and ammonium iodide are manufactured either by the reaction of ammonia with the corresponding hydrohalic acid or, more economically, by the reaction of ammonia with elemental bromine or iodine.

Uses. Ammonium bromide is available as a dry technical grade or as 38 to 45% solutions. It is used to manufacture chemical intermediates and in photographic chemicals; it also has some flame retardant applications.

AMMONIUM NITRATE

Ammonium nitrate, NH_4NO_3, formula wt 80.04, is the most commercially important ammonium compound both in terms of production volume and usage. It is the principal component of most industrial explosives and nonmilitary blasting compositions; however, it is used primarily as a nitrogen fertilizer.

One general disadvantage of nitrogen fertilizers, and ammonium nitrate in particular, is that the nitrate ion is more prone to leach through the soil profile and enter the groundwater.

Physical and Chemical Properties

Ammonium nitrate is a white, crystalline salt, $d_4^{20} = 1.725$, that is highly soluble in water. Although it is very hygroscopic, it does not form hydrates.

Solid ammonium nitrate occurs in five different crystalline forms (Table 1), detectable by time–temperature cooling curves.

The specific heat of solid β-phase ammonium nitrate is 1.70 J/g (0.406 cal/g) between 0 and 31°C. Ammonium nitrate has a negative heat of solution in water, and can therefore be used to prepare freezing mixtures.

Decomposition and Detonation Hazard

Ammonium nitrate is considered a very stable salt, When the salt is heated to temperatures from 200 to 230°C, exothermic decomposition occurs. The reaction is rapid, but it can be controlled, and it is the basis for the commercial preparation of nitrous oxide.

Above 230°C, exothermic elimination of N_2 and NO_2 begin. The final violent exothermic reaction occurs with great rapidity when ammonium nitrate detonates. When used in blasting, ammonium nitrate is mixed with fuel oil and sometimes sensitizers such as powdered aluminum.

Manufacture

Modern commercial processes, rely almost exclusively on the neutralization of nitric acid, produced from ammonia through catalyzed oxidation, with ammonia.

Health and Safety Factors

Ammonium nitrate can be considered a safe material if treated and handled properly. Potential hazards include those associated with fire, decomposition accompanied by generation of toxic fumes, and explosion.

AMMONIUM NITRITE

Ammonium nitrite, NH_4NO_2, a compound of questionable stability, can be prepared by reaction of barium nitrite and aqueous ammonium sulfate.

Table 1. Crystalline Forms of Ammonium Nitrate

Designation	Temperature range, °C	Crystal system
α	< -18	tetragonal
β	$-18 - 32.1$	rhombic
γ	$32.1 - 84.2$	rhombic
δ	$84.2 - 125.2$	tetragonal
ε	$125.2 - 169.6$	cubic

AMMONIUM SULFATE

Ammonium sulfate, $(NH_4)_2SO_4$, is a white, soluble, crystalline salt having a formula wt of 132.14. The crystals have a rhombic structure; d_4^{20} is 1.769.

The solubility of ammonium sulfate in 100 g of water is 70.6 g at 0°C and 103.8 g at 100°C. It is insoluble in ethanol and acetone, does not form hydrates, and deliquesces at only about 80% relative humidity.

Manufacture

Ammonium sulfate is produced from the direct neutralization of sulfuric acid with ammonia.

Uses

Almost all ammonium sulfate is used as a fertilizer; for this purpose it is valued both for its nitrogen content and for its readily available sulfur content.

AMMONIUM SULFIDES

Ammonia combines with hydrogen sulfide, sulfur, or both, to form various ammonium sulfides and polysulfides. Generally these materials are somewhat unstable, tending to change in composition on standing. Ammonium sulfides are used by the textile industry. Their include ammonium sulfide and ammonium hydrosulfide.

Ammonium Sulfide

Ammonium sulfide, $(NH_4)_2S$, can be produced by the reaction of hydrogen sulfide with excess ammonia.

Solid ammonium sulfide is normally marketed as a 40–44% aqueous solution.

Ammonium Hydrosulfide

The reaction of equimolar amounts of ammonia and hydrogen sulfide results in the formation of ammonium hydrosulfide, NH_4HS, which is also produced by the loss of ammonia from ammonium sulfide. The hydrosulfide is very soluble in water, liquid ammonia, liquid hydrogen sulfide, and alcohol. Vapors from the hydrosulfide, composed of ammonia and hydrogen sulfide, are very toxic.

QUATERNARY AMMONIUM COMPOUNDS

There are a vast number of quaternary ammonium compounds or quaternaries. Many are naturally occurring and have been found to be crucial in biochemical reactions necessary for sustaining life. A wide range of quaternaries are also produced synthetically and are commercially available. Over 344,000 metric tons of quaternary ammonium compounds are produced annually in the United States. The economic value is estimated at $810 MM. These have many diverse applications (see Table 2). Most are eventually formulated and make their way to the marketplace to be sold in consumer products.

Table 2. Selected Quaternary Ammonium Compounds and Their Applications

Quaternary	Industry	Application and function	Comments
many compounds claimed, as an example: polypropoxylated (6) choline chloride	agricultural	surfactant	key component in glyphosphate composition
tetrabutylammonium hydroxide	chemical	phase-transfer catalyst	used as a catalyst for the production of gem-dichloro compounds
alkyltrimethyl or dialkydimethyl type quaternaries, as an example dicocoalkyldimethylammonium chloride	chemical	emulsifier	emulsifier for silanes useful as masonry water repellents
many compounds claimed, as an example: dicocoalkyldimethylammonium chloride	chemical	complex agent	complexed with anionic dyes to produce a formulation free of inorganic salts
many perfluoroalkyl quaternary ammonium compounds claimed, as an example: di(4,4,5,5,6,6,7,7,7-nonafluoroheptyl)dimethylammonium chloride	chemical	phase-transfer catalyst	a family of novel quaternaries useful as phase-transfer catalysts especially in basic media
many diquaternary ammonium compounds claimed, as an example: 1,3-bis(dipentylethylammonium)propane dibromide	chemical	catalysts	catalyst for the interfacial polymerization polycarbonate preparation
many compounds claimed, as an example: tridecylmethylammonium chloride	chemical	component in catalyst	key component in a catalyst composition containing a zirconium compound
di(hydrogenated tallowalkyl)dimethyl ammonium chloride and others	chemical	nonvolatile compositions	compositions containing vegetable oils as diluents useful in organoclays
many compounds claimed, as an example: diallyldimethylammonium chloride	chemical	antimicrobial	antibacterials for polymer latexes and resins
benzyltrimethylammonium chloride and others	defense	surfactant	component in a formulation to neutralize chemical and biological warfare agents
many compounds claimed, as an example: benzyldimethyl[2-(3,5-di-*tert*-butyl 4-hyroxybenzoyloxy)ethyl]ammonium *m*-nitrobenzenesulfonate	electronics	charge control agent	component in toner compositions
benzylacetyldimethylammonium chloride	electronics	dispersant	component in electrostatic liquid developer formulation useful in color copying
mixtures of quaternary ammonium hydroxides and halides	electronics	dissolution agent	components in positive photoresist formulation to improve the dissolution selectivity between exposed and unexposed portion of photoresist
polyquaternary ammonium compounds	electronics	binder	key component in information storage layer of electronic recording medium
many compounds claimed, as example: methyltrioctylammonium chloride and tridecylmethylammonium chloride	electronics	surfactant	component in blocking layer of an electrographic photosensitive material
mixtures of quaternary ammonium hydroxides and carbonates, as an example: tetramethyl ammonium hydroxide and tetramethylammonium hydrogen carbonate	electronics	surfactant/buffering agent	components in a developing solution for producing printed circuit boards
as an example: tetrabutylammonium tetrafluoroborate	electronics	charge control agent	additive for the preparation of phase change inks with increased specific conductance
many compounds claimed, as an example: tetradecyltrimethylammonium bromide	electronics	biocides	biocides for color reversal photographic film and photographic reversal bath
tetramethylammonium hydroxide	electronics	surface active agent	in formulation for polishing semiconductor wafers
quaternary ammonium hydroxide	electronics	surfactant	key component in removing agent formulation for producing semiconductor integrated circuits

Table 2. (*Continued*)

Quaternary	Industry	Application and function	Comments
quaternary ammonium hydroxide	electronics	surfactant	components in cleaning formulations for semiconductor devices
quaternary ammonium hydroxide	electronics	surfactant	component in cleaning composition for removing plasma etching residues
tetraalkylammonium halide	food	antimicrobial	compositions for removal and prevention of microbial contamination
many compounds claimed	household	antimicrobial	disenfecting component in cleaning composition
hexadecylpyridinium chloride	household	antimicrobial	antimicrobial is dispersed throughout plastic toothbrush
dimethyldialkylquaternary ammonium	household	antimicrobial	claimed synergy when used in combination with water soluble anionic surfactant
pentamethyltallowalkyl-1,3-propanediammonium chloride	household	antimicrobial	used in a composition for deodorizing footware
hexadecyltrimethylammonium chloride	household	surfactant	component in vicoselastic thickening system for opening drains
ethoxylated quaternary ammonium compounds	household	surfactant	used in a formulation to remove road-film
trialkylammoniumacetylpyrrolidone chloride	household	bleach activator	in a detergent or cleaning formulation
ester based quaternary ammonium compounds	household	fabric softening	for the preparation of rinse cycle fabric softening formulations
amidoamine and branched quaternary ammonium compounds	household	fabric softening	for the preparation of rinse cycle fabric softening formulations
ester based quaternary ammonium compounds	household	antistatic	for use in dryer-activated fabric conditioning and antistatic compositions
alkoxylated quaternary ammonium compounds	household	detergents	for improved performance in detergent formulations such as optimum grease and soil removal, enhancement of bleach efficacy, and better cold temperature performance
many compounds claimed, as an example: hexadecyltrimethylammonium bromide	household	antimicrobial	for use in liquid laundry detergent composition
ester based quaternary ammonium compounds	household	fabric softening	for use in liquid detergent formulations that soften fabric
2-hydroxyethyltrialkyl ammonium halide	mining	complexing agent for gold anions	quat is adsorbed onto porous polymer resin and forms ion pairs with gold anions
quaternary ammonium salts and mixtures	mining	froth flotation	recovery of petalite free of feldspar
many compounds claimed, as an example: didecyldimethylammonium chloride	other	antimicrobial	active component in antimicrobial formulations
pyridinium halides such as decylpyridinium bromide	other	anticorrosion	functions as corrosion inhibitor to protect metal surfaces from acid
polyquaternary ammonium	other	superabsorbent	cationic copolymer with improved water absorbing properties
unsaturated quaternary ammonium compounds	other	surfactant	component in a photo-curable antifogging composition for glass
polyquaternary ammonium	other	pesticide	to control the infestation of snails in aqueous systems
2-ethylhexylhydrogenated-*t*-allowalkyl-dimethyl ammonium methosulfate	organoclay	modify smectite-type clay	organoclay product containing branched chain quaternary ammonium compounds

Table 2. (*Continued*)

Quaternary	Industry	Application and function	Comments
alkyl quaternary ammonium salt	organoclay	modify mineral clay	organoclay composition comprising a mineral clay treated with an alkyl quaternary ammonium salt
ester based quaternary ammonium compounds	organoclay	modify smectite-type clay	organoclays useful for nonaqueous systems like paints, inks and coatings
preferred is dimethyldi(hydrogenated tallowalkyl)ammonium chloride	organoclay	modify smectite-type clay	organoclays useful for nanocomposites and rheological additives
many compounds claimed, as an example: octadecyltrimethylammonium chloride	paper	active in paper softening composition	key component to prepare tissue with a soothing feeling
many compounds claimed, as an example: dimethyldi(hydrogenated tallowalkylalkyl) ammonium methosulfate	paper	component in tissue paper web	acts as an antimigration material for emollient lotion
ester based quaternary ammonium compounds	paper	softening	for the manufacture of soft absorbent paper products such as paper towels, facial tissue and toilet tissue
polyquaternary ammonium	personal care	thickeners and dispersants	quaternary ammonium copolymers useful in cosmetic compositions
many compounds claimed, as an example: N-methyl-N,N-bis(2-($C_{16/18}$-acyloxy)ethyl)-N-hydroxyethylammonium methosulfate	personal care	conditioning shampoo	conditioning component
as an example: 1,2-ditallowalkyloxy-3-trimethylammonium-propane chloride	personal care	hair conditioner	active ingredient in formulation
unsaturated quaternary ammonium compounds	personal care	delivery agent	used in a formulation to deliver the active component to the hair or skin
polyquaternary ammonium	personal care	antimicrobial	used as a disinfectant for ophthalmic compositions
as an example: dimethyldi(hydrogenated tallowalkyl)ammonium chloride	petroleum	drilling fluids	drilling fluid compositions having special rheological properties
quaternary ammonium hydroxides	petroleum	key component of a process	method for scavenging mercaptans in hydrocarbon fluid
dicocoalkyldimethylammonium chloride	petroleum	oil spill recovery	oil spill rediation agent containing organoclay and waste paper
eg, tetrabutylammonium bromide	petroleum	phase-transfer catalyst	method of removing contaminants from petroleum distillates and contaminants from used oils
ester based quaternary ammonium compounds	petroleum	surfactant	used in a composition for enhanced recovery of crude oils
chloromethyl(8-hexadecenyl)dimethyl-ammonium chloride	pharmaceutical	antitumorous	active component in composition
cationic lipids	pharmaceutical	receptor	improved cell targeting ability for the delivery of molecules into cells
polyquaternary ammonium	pharmaceutical		used to lower cholesterol levels
hydrogenatedtallowalkyltrimethyl ammonium chloride	remediation	component in an organoclay	organoclay adsorbs dissolved heavy metals including lead and radiactive contaminants from aqueous solutions
many compounds claimed, as an example: benzyltrimethylammonium chloride	remediation	component in an organoclay	removal of aromatic petroleum-based contaminants from water
trimethylbetahydroxyethyl ammonium hydroxide and others	rubber	accelerator	for vulcanization of rubber with nontoxic material
dialkyldimethyl quaterary ammonium compounds	wood	biocides	biocidal component in a formulation to waterproof and preserve wood

Applications range from cosmetics to hair preparations to clothes softeners, sanitizers for eating utensils, and asphalt emulsions.

Physical Properties

Most quaternary compounds are solid materials that have indefinite melting points and decompose on heating. Physical properties are determined by the chemical structure of the quaternary ammonium compound as well as any additives such as solvents. The simplest quaternary ammonium compound, tetramethylammonium chloride, is very soluble in water insoluble in nonpolar solvents. As the molecular weight of the quaternary compound increases, solubility in polar solvents decreases and solubility in nonpolar solvents increases. The ability to form aqueous dispersions is a property that gives many quaternary compounds useful applications. Placement of polar groups, eg, hydroxy or ethyl ether, in the quaternary structure can increase solubility in polar solvents.

Higher order aliphatic quaternary compounds, where one of the alkyl groups contains ~10 carbon atoms, exhibit surface-active properties. These compounds compose a subclass of a more general class of compounds known as cationic surfactants. These have physical properties such as substantivity and aggregation in polar media that give rise to many practical applications. In some cases the ammonium compounds are referred to as inverse soaps, because the charge on the organic portion of the molecule is cationic rather than anionic.

Chemical Properties

Reactions of quaternaries can be categorized into three types: Hofmann eliminations, displacements, and rearrangements. Thermal decomposition of a quaternary ammonium hydroxide to an alkene, tertiary amine, and water is known as the Hofmann elimination. This reaction has not been used extensively to prepare olefins. Some cyclic olefins, however, are best prepared this way. Exhaustive methylation, followed by elimination, is known as the Hofmann degradation and is important in the structural determination of unknown amines, especially for alkaloids.

Naturally Occurring Quaternaries

Many types of aliphatic, heterocyclic, and aromatic derived quaternary ammonium compounds are produced both in plants and invertebrates. Examples include thiamine (vitamin B_1) choline and acetylcholine. These have numerous biochemical functions. Several quaternaries are precursors for active metabolites.

Biochemically, most quaternary ammonium compounds function as receptor-specific mediators. Because of their hydrophilic nature, small molecule quaternaries cannot penetrate the alkyl region of bilayer membranes and must activate receptors located at the cell surface. Quaternary ammonium compounds also function biochemically as messengers, which are generated at the inner surface of a plasma membrane or in a cytoplasm in response to a signal. They may also be transferred through the membrane by an active transport system.

General types of physiological functions attributed to quaternary ammonium compounds are curare action, muscarinic–nicotinic action, and ganglia blocking action. The active substance of curare is a quaternary that can produce muscular paralysis without affecting the central nervous system or the heart. Muscarinic action is the stimulation of smooth-muscle tissue. Nicotinic action is primary transient stimulation and secondary persistent depression of sympathetic and parasympathetic ganglia.

Synthesis and Manufacture

A wide variety of methods are available for the preparation of quaternary ammonium compounds. Significantly fewer can be used on a commercial scale.

Quaternary ammonium compounds are usually prepared by reaction of a tertiary amine and an alkylating agent.

Primary and secondary amines are usually converted to tertiary amines using formaldehyde and hydrogen in the presence of a catalyst. This process, known as reductive alkylation, and is attractive commercially. The desired amines are produced in high yields and without significant by-product formation. Quaternization by reaction of an appropriate alkylating reagent then follows.

Synthesis and Manufacture of Amines. The chemical and business segments of amines and quaternaries are so closely linked that it is difficult to consider these separately. The majority of commercially produced amines originate from three amine raw materials: natural fats and oils, α-olefins, and fatty alcohols. Most large commercial manufacturers of quaternary ammonium compounds are fully back-integrated to at least one of these three sources of amines. The amines are then used to produce a wide array of commercially available quaternary ammonium compounds. Some individual quaternary ammonium compounds can be produced by more than one synthetic route.

Nitrile Intermediates. Most quaternary ammonium compounds are produced from fatty nitriles, which are in turn made from a natural fat or oil-derived fatty acid and ammonia. The nitriles are then reduced to the amines. A variety of reducing agents may be used.

Fats, Oils, or Fatty Acids. The primary products produced directly from fats, oils, or fatty acids without a nitrile intermediate are the quaternized amidoamines, imidazolines, and ethoxylated derivatives. Reaction of fatty acids or tallow with various polyamines produces the intermediate dialkylamidoamine. By controlling reaction conditions, dehydration can be continued until the imidazoline is produced. Quaternaries are produced from both amidoamines and imidazolines by reaction with methyl chloride or dimethyl sulfate. The amidoamines can also react with ethylene oxide to produce ethoxylated amidoamines that are then quaternized.

These compounds and their derivatives can be manufactured using relatively simple equipment compared to that required for the fatty nitrile derivatives. Cyclization of amidoamines to imidazolines requires higher reaction temperatures and reduced pressures. Prices of imidazolines are therefore high.

Olefins and Fatty Alcohols. Alkylbenzyldimethylammonium (ABDM) quaternaries are usually prepared from α-olefin or fatty alcohol precursors. Manufacturers that start from the fatty alcohol usually prefer to prepare the intermediate alkyldimethylamine directly by using dimethylamine and a catalyst rather than from fatty alkyl chloride.

Quaternized Esteramines. Esterquaternary ammonium compounds or esterquats can be formulated into products that have good shelf stability. Many examples of this type of molecule have been developed. Quaternized esteramines are usually derived from fat or fatty acid that reacts with an alcoholamine to give an intermediate esteramine. The esteramines are then quaternized.

Analytical Methods

There are no universally accepted wet analytical methods for the characterization of quaternary ammonium compounds. The American Oil Chemists' Society (AOCS) has established, however, a number of applicable tests, including sampling, color, moisture, amine value, ash, iodine value, average molecular weight, pH, and flash point.

Numerous "wet chemical" methods have been developed for the determination of the activity of quaternary ammonium samples.

The chain length composition of quaternaries can be determined by gas chromatography.

Liquid chromatography has been widely applied for analysis of quaternaries.

Nuclear magnetic resonance (nmr) spectroscopy is useful for determining quaternary structure.

Health and Safety Factors

Acute oral toxicity data show most structures to have an LD_{50} in the range of 100–5000 mg kg. Many quaternaries are considered to be moderately to severely irritating to the skin and eyes.

Some quaternary ammonium compounds are potent germicides, toxic in small (1 mg L range) quantities to a wide range of microorganisms. Bactericidal, algicidal, and fungicidal properties are exhibited. Ten-minute-contact kills of bacteria are typically produced by quaternaries in concentration ranges of 50–333 mg L. Acute toxicity at low (1 mg L) levels has been reported in invertebrates, snails, and fish. In plant systems, growth inhibition of green algae and great duckweed occurs at 3–5 mg L.

Over the last decade, considerable advances have been made in understanding the metabolic pathway of quaternary ammonium compounds. Numerous internationally recognized standardized methods are available for assessing the biodegradation of chemicals under aerobic conditions. There are three testing method levels: ready biodegradeability, inherent biodegradeability, and simulation of biological treatment systems.

Ready biodegradeability tests are used primarily for regulatory purposes. These include the Closed Bottle test, MITI I test, and the Sturm test. There are three standard test methods to assess inherent biodegradeability. These are the MITI II test, Zahn-Wellens and SCAS test. Sludge (CAS) and Semi-Continuous Activated Sludge (SCAS) tests are used to simulate behavior of materials in biological treatment systems. The CAS test is believed to give realistic results similar to full-scale treatment.

A dynamic equilibrium exists between quaternary ammonium species in the aqueous phase and those existing as a soild after absorption. Thus, only the fraction in the aqueous phase is "available" at any given time to be biodegraded. Bioavailability is defined as that fraction of material that is readily accessible to microbial degradation.

Most uses of quaternary ammonium compounds can be expected to lead to these compounds' eventual release into wastewater treatment systems except for those used in drilling muds. Useful properties of the quaternaries as germicides can make these compounds potentially toxic to sewer treatment systems. It appears, however, that quaternary ammonium compounds are rapidly degraded in the environment and strongly sorbed by a wide variety of materials. Quaternaries appear to bind anionic compounds and thus are effectively removed from wastewater by producing stable, lower toxicity compounds. Under normal circumstances these compounds are unlikely to pose a significant risk to microorganisms in wastewater treatment systems. Microbial populations acclimate readily to low levels of quaternary compounds and biodegrade them.

Newer classes of quaternaries, eg, esters and betaine esters, have been developed. These materials are more readily biodegraded. The mechanisms of antimicrobial activity and hydrolysis of these compounds have been studied. Applications as surface disinfectants, antimicrobials, and in vitro microbiocidals have also been reported.

Uses

Uses of quaternary ammonium compounds range from surfactants to germicides and encompass a number of diverse industries.

Fabric Softening. The use of quaternary surfactants as fabric softeners and static control agents can be broken down into three main household product types: rinse cycle softeners; tumble dryer sheets; and detergents containing softeners, also known as softergents. Rinse cycle softeners are aqueous dispersions of quaternary ammonium compounds designed to be added to the wash during the last rinse cycle. Original products contained from 3–8% quaternary ammonium compound, typically di(hydrogenated tallow)alkyldimethylammonium chloride (DHTDMAC).

Tumble dryer sheets contain a quaternary ammonium compound formulation applied to a nonwoven sheet typically made of polyester or rayon. These sheets are added with wet clothes to the tumble dryer. Although these products afford some softening to the clothes, their greatest strength is in preventing static charge buildup on clothes during the drying cycle and during wear. A nonionic surfactant, such as an ethoxylated alcohol or fatty acid, is typically used in combination with the quaternary ammonium compound. The nonionics are known as release agents or distribution agents. More efficient transfer of the quaternary from the substrate to the drying fabric can be obtained.

Detergents containing softeners are also produced. These softergents are made from complex formulas in order to accomplish both detergency and softening during the wash cycle. These formulations typically contain quaternary ammonium compounds mixed with other materials such as clays for softening, in conjunction with the typical nonionic and anionic cleaning surfactant. The consumer benefits of quaternaries are fabric softening, antistatic properties, ease of ironing, and reduction in energy required for drying.

Hair Care. Quaternary ammonium compounds are the active ingredients in hair conditioners. Quaternaries are highly substantive to human hair because the hair fiber has anionic binding sites at normal pH ranges. Surface analysis by X-ray photoelectron spectroscopy (XPS) has shown specific 1:1 (ionic) interaction between cationic alkyl quaternary surfactant molecules and the anionic sulfonate groups present on the hair surface. The use of quaternaries as hair conditioners can be broken down into creme rinses and shampoo conditioners.

Creme rinses are applied to the hair after washing. Frequently used quaternaries in creme rinses are dodecyltrimethylammonium chloride, dimethyloctadecyl (pentaethoxy)ammonium chloride, benzyldimethyloctadecylammonium chloride, and dimethyldioctadecylammonium chloride.

Conditioning shampoos are formulations that contain anionic surfactants for cleaning hair and cationic surfactants for conditioning. The quaternary ammonium compounds most often used are either trihexadecylmethylammonium chloride, ethoxylated quaternaries, or one of the polymeric quaternaries. The polymeric quaternaries have either a natural or a synthetic backbone and numerous quaternary side functions. The polymer may offer an advantage by showing a high degree of affinity to the human hair surface and providing better compatibility with the other ingredients of conditioner shampoos.

Regardless of how the conditioner is applied or what the structure of the quaternary is, benefits provided to conditioned hair include the reduction of combing forces, increased luster, and improved antistatic properties.

Germicides. The third largest market for quaternaries is sanitation. Generally, quaternaries offer several advantages over other classes of sanitizing chemicals, such as phenols, organohalides, and organomercurials, in that quaternaries are less irritating, low in odor, and have relatively long activity. The first use of quaternaries in the food industry occurred in the dairy industry for the sanitization of processing equipment. Quaternaries find use as disinfectants and sanitizers in hospitals, building maintenance, and food processing in secondary oil recovery for drilling fluids and in cooling water applications.

Quaternaries have also received extensive attention for use as a general medicinal antiseptics and in the pharmaceutical area as skin disinfectants and surgical antiseptics. In addition, quaternaries have been used in the treatment of eczema and other dermatological disorders as well as in contraceptive formulations and ocular solutions for contact lenses.

Alkylbenzyldimethyl quaternaries (ABDM) are used as disinfectants and preservatives. The most effective alkyl chain length for these compounds is between 10 and 18 carbon atoms. Alkyltrimethyl types, alkyldimethylbenzyl types, and didodecyldimethylammonium chloride exhibit excellent germicidal activity. Dialkyldimethyl types are effective against anaerobic bacteria such as those found in oil wells. One of the most effective and widely used biocides is didecyldimethylammonium chloride.

Organoclays. Another large market for quaternary ammonium salts is the manufacture of organoclays, ie, organomodified clays. Clay particles are silicate minerals that have charged surfaces and that attract cations or anions electrostatically. Organoclays are produced by ion-exchange reaction between the quaternary ammonium salt and the surface of the clay particles. The quaternary ammonium salt displaces the adsorbed cations, usually sodium or potassium, producing an organomodified clay. The new modified clay exhibits different behavior from that of the initial clay. Most importantly, it is preferentially wet by organic liquids and not by water.

The main use of these clays is to control, or adjust, viscosity in nonaqueous systems. Organoclays can be dispersed in nonaqueous fluids to modify the viscosity of the fluid so that the fluid exhibits non-Newtonian thixotropic behavior. Important segments of this area are drilling fluids, greases, lubricants, and oil-based paints. Quaternaries used to produce organoclays are dimethyldi (hydrogenated tallow)alkylammonium chloride, dimethyl (hydrogenated tallow)alkylbenzylammonium chloride, and methyldi(hydrogenated tallow)alkylbenzylammonium chloride.

Miscellaneous Uses. Many quaternaries have been used as phase-transfer catalysts (PTCs). A PTC increases the rate of reaction between reactants in different solvent phases. Usually, water is one phase and a water-immiscible organic solvent is the other. Common quaternaries employed as phase-transfer agents include benzyltriethylammonium chloride, tetrabutylammonium bromide, tributylmethylammonium chloride, and hexadecylpyridinium chloride.

Polyamine-Based Quaternaries. Another important class of quaternaries are the polyamine based or polyquats. Generally, polyamine-based quaternaries have been used in the same applications as their monomeric counterparts.

Perfluorinated Quaternaries. Perfluorinated quaternaries are another important, but smaller, class of quaternary ammonium compounds. In general, these are similar to their hydrocarbon counterparts but have at least one of the hydrocarbon chains replaced with a perfluoroalkyl group. These compounds are generally much more expensive than hydrocarbon-based quaternaries, so they must offer a significant performance advantage if they are to be used. Production volumes of perfluorinated quaternary ammonium compounds are significantly smaller than those of other classes. Many of these quaternaries have proprietary chemical structures. They are used in water-based coating applications to promote leveling, spreading, wetting, and flow control.

W. L. Jolly, *The Inorganic Chemistry of Nitrogen*, W. A. Benjamin, Inc., 1964.

D. H. Lauriente, "Ammonium Nitrate," *Chemical Economics Handbook*, SRI, Menlo Park, CA, Oct. 2000.

R. J. Lewis, Sr., *Dangerous Properties of Industrial Materials*, Vol. 2, John Wiley & Sons, Inc., New York, 2000.

A. F. Wells, *Structural Inorganic Chemistry*, 5th ed., Clarendon Press, Oxford, UK, 1984, pp. 362–363.

Charles W. Weston
Freeport Research and
Engineering Company

John R. Papcun
Atotech

Maurice Dery
Akzo Nobel Chemicals, Inc.

AMYL ALCOHOLS

Amyl alcohol describes any saturated aliphatic alcohol containing five carbon atoms. This class consists of three pentanols, four substituted butanols, and a disubstituted propanol, ie, eight structural isomers $C_5H_{12}O$: four primary, three secondary, and one tertiary alcohol. In addition, 2-pentanol, 2-methyl-1-butanol, and 3-methyl-2-butanol have chiral centers and hence two enantiomeric forms.

The odd-carbon structure and the extent of branching provide amyl alcohols with unique physical and solubility properties and often offer ideal properties for solvent, surfactant, extraction, gasoline additive, and fragrance applications. Amyl alcohols have been produced by various commercial processes in past years. Today the most important industrial process is low-pressure rhodium-catalyzed hydroformylation (oxo process) of butenes.

Mixtures of isomeric amyl alcohols (1-pentanol and 2-methyl-1-butanol) are often preferred because the different degree of branching imparts a more desirable combination of properties; they are also less expensive to produce commercially.

PHYSICAL PROPERTIES

With the exception of neopentyl alcohol (mp 53°C), the amyl alcohols are clear, colorless liquids under atmospheric conditions, with characteristic, slightly pungent and penetrating odors. They have relatively higher boiling points than ketonic or hydrocarbon counterparts and are considered intermediate boiling solvents for coating systems (Table 1).

Commercial primary amyl alcohol is a mixture of 1-pentanol and 2-methyl-1-butanol, in a ratio of ca. 65 to 35. Typical physical properties of this amyl alcohol mixture are listed in Table 2.

Like the lower alcohols, amyl alcohols are completely miscible with numerous organic solvents and are excellent solvents for nitrocellulose, resin lacquers, higher esters, and various natural and synthetic gums and resins. However, in contrast to the lower alcohols, they are only slightly soluble in water. Only 2-methyl-2-butanol exhibits significant water solubility.

CHEMICAL PROPERTIES

The amyl alcohols undergo the typical reactions of alcohols which are characterized by cleavage at either the oxygen–hydrogen or carbon–oxygen bonds. Important reactions include dehydration, esterification, oxidation, amination, etherification, and condensation.

MANUFACTURE

Three significant commercial processes for the production of amyl alcohols include separation from fusel oils, chlorination of C-5 alkanes with subsequent hydrolysis to produce a mixture of seven of the eight isomers Pennsalt, and a low-pressure oxo process, or hydroformylation, of C-4 olefins followed by hydrogenation of the resultant C-5 aldehydes.

The oxo process is the principal one in practice today; only minor quantities, mainly in Europe, are obtained from separation from fusel oil. *tert*-Amyl alcohol is produced on a commercial scale in lower volume by hydration of amylenes.

SHIPPING AND STORAGE

Amyl alcohols are best stored or shipped in either aluminum, lined steel, or stainless-steel tanks. Baked phenolic is a suitable lining for steel tanks. Plain steel tanks can also be used for storage or shipping. However, storage of aqueous solutions can cause rusting. Also, the alcohols are sufficiently hygroscopic so that moisture pick-up can cause rusting of plain steel storage tanks. Storage and transfer under dry nitrogen is recommended. Storage and handling facilities should be in compliance with the OSHA "Flammable and Combustible Liquids" regulations. Piping and pumps can be made from the same metals as used for storage tanks. The freezing points of amyl alcohols are low and they remain fluid at cold outside temperatures, thus allowing storage facilities above or below ground.

ECONOMIC ASPECTS

All eight amyl alcohol isomers are available from fine-chemical-supply firms in the United States. Five of them,

Table 1. The Amyl Alcohols and Some of Their Physical Properties

Properties	1-Pentanol	2-Pentanol	3-Pentanol	2-Methyl-1-butanol	3-Methyl-1-butanol	2-Methyl-2-butanol	3-Methyl-2-butanol	2,2-Dimethyl-1-propanol
common name	n-amyl alcohol	sec-amyl alcohol			isoamyl alcohol	$tert$-amyl alcohol		neopentyl alcohol
critical temperature, °C	315.35	287.25	286.45	291.85	306.3	272.0	300.85	276.85
critical pressure, kPa[a]	3868.	3710.	3880.	3880.	3880.	3880.	3960.	3880.
critical specific volume, mL/mol	326.5	328.9	325.3	327	327	327	327	327
critical compressibility	0.25810	0.26188	0.27128	0.27009	0.26335	0.27992	0.27133	0.27745
boiling point at pressure, °C								
101.3 kPa[a]	137.8	119.3	115.3	128.7	130.5	102.0	111.5	113.1
40 kPa	111.5	93.8	90.9	103.5	105.6	78.3	87.2	89.0
1.33 kPa	44.6	32.0	27.7	40.2	43.0	21.0	26.0	25.0
vapor pressure[b] kPa[a]	0.218	0.547	0.761	0.274	0.200	1.215	0.810	0.929
melting point, °C	−77.6	−73.2	−69.0	<−70	−117.2	−8.8 / −88	forms glass	54.0
heat of vaporization at normal boiling point, kJ/mol[c]	44.83	43.41	42.33	44.75	43.84	40.11	41.10	41.35
ideal gas heat of formation[d], kJ/mol[c]	−298.74	−313.80	−316.73	−302.08	−302.08	−329.70	−314.22	−319.07
liquid density[b], kg/m^3	815.1	809.4	820.3	819.1	810.4	809.6	818.4	851.5[e]
liquid viscosity[b], mPa·s(=cP)	4.06	4.29	6.67	5.11	4.37	4.38	3.51d	2.5[e]
surface tension[b], mN/m(=dyn/cm)	25.5	24.2	24.6	25.1^d	24.12	22.7	23.0d	14.87[e]
refractive index[d]	1.4080	1.4044	1.4079	1.4086	1.4052	1.4024	1.4075	1.3915
solubility parameter[d] (MJ/m^3)$^{0.5f}$	22.576	21.670	21.150	22.274	22.322	20.758	21.607	19.265[e]
solubility in water[b], wt %	1.88	4.84	5.61	3.18	2.69	12.15	6.07	3.74
solubility of water in[b], wt %	9.33	11.68	8.19	8.95	9.45	24.26	11.88	8.23

[a]To convert kPa to mm Hg, multiply by 7.5.
[b]At 20°C unless otherwise noted.
[c]To convert kJ/mol to cal/mol, multiply by 239.
[d]At 25°C.
[e]At the melting point.
[f]To convert (MJ/m^3) to (cal)$^{0.5}$, divide by 2.045.

Table 2. Physical Properties of Primary Amyl Alcohol, Mixed Isomers[a]

Property	Value
molecular weight	88.15
boiling point at 101.13 kPa[b], °C	133.2
freezing point, °C	−90[c]
specific gravity 20/20 °C	0.8155
absolute viscosity at 20 °C, mPa·s(=cP)	4.3
vapor pressure at 20 °C, kPa[b]	0.27
flash point (closed cup), °C	45
solubility at 20 °C, by wt %	
in water	1.7
water in	9.2

[a] 65/35 blend, ie, a mixture of 1-pentanol and 2-methyl-1-butanol, 65/35 wt %, respectively.
[b] To convert kPa to mm Hg, multiply by 7.5.
[c] Sets to glass below this temperature.

1-pentanol, 2-pentanol, 2-methyl-1-butanol, 3-methyl-1-butanol, and 2-methyl-2-butanol (tert-amyl alcohols) are available in bulk in the United States; in Europe all but neopentyl alcohol are produced. In 2001, 8.4×10^6 t of oxo chemicals were produced. Oxo chemicals are expected to grow at a rate of 1.4% per year through 2007.

HEALTH AND SAFETY FACTORS

The main effects of prolonged exposure to amyl alcohols are irritation to mucous membranes and upper respiratory tract, significant depression of the central nervous system, and narcotic effects from vapor inhalation or oral absorption. All the alcohols are harmful if inhaled or swallowed, appreciably irritating to the eyes and somewhat irritating to uncovered skin on repeated exposure. Prolonged exposure causes nausea, coughing, diarrhea, vertigo, drowsiness, headache, and vomiting. The toxicity of 3-methyl-2-butanol and 2,2-dimethyl-1-propanol has not been thoroughly investigated.

All of the amyl alcohols are TSCA and EINECS (European Inventory of Existing Commercial Chemical Substances) registered.

The amyl alcohols are readily flammable substances; tert-amyl alcohol is the most flammable (closed cup flash point, 19 °C). Their vapors can form explosive mixtures with air.

USES

Solvents and coatings are the biggest market for C5 alcohols. 1-pentanol and 2-methyl-1-butanol is used for zinc diamyldithiophosphate lubrication oil additives as important corrosion inhibitors and antiwear additives. Amyl xanthate salts are useful as frothers in the flotation of metal ores because of their low water solubility and miscibility with phenolics and natural oils. Potassium amyl xanthate, a collector in flotation of copper, lead, and zinc ores, is no longer produced in the United States.

Another significant application for amyl alcohols is for production of amyl acetates.

As solvents, the amyl alcohols are good solvents and diluents for lacquers, hydrolytic fluids, dispersing agents in textile printing inks, industrial cleaning compounds, natural oils such as linseed and castor, synthetic resins such as alkyds, phenolics, urea–formaldehyde maleics, and adipates, and naturally occurring gums, such as shellac, paraffin waxes, rosin, and manila. In solvent mixtures they dissolve cellulose acetate, nitrocellulose, and cellulosic ethers.

The principal component of primary amyl alcohol, 1-pentanol, although itself a good solvent, is useful for the preparation of specific chemicals such as pharmaceuticals and other synthetics.

Growth applications for amyl alcohols appear to be shifting toward higher boiling esters as plasticizers, perfumes, fragrances, and production of fine chemicals.

tert-Amyl alcohol is employed in formulations for stabilizing 1,1,1-trichloroethane (a replacement for trichloroethylene) which is used for degreasing metals, especially aluminum, copper, zinc, and iron and their alloys. The tert-amyl alcohol in stabilizing formulations allowed only negligible reaction between the 1,1,1-trichloroethane and metal. tert-Amyl alcohol is also used for stabilizing 1,1,1-trichloroethane mixtures for rosin flux removal compositions, eg, ionic and nonionic fluxes from circuit boards and in stabilizer compositions for 1,1,1-trichloroethane for dry cleaning applications where it is durable against repeated use, without causing corrosion of the dry cleaner metal components and conforms with environmental and health standards enacted by the Occupational Safety and Health Act. tert-Amyl alcohol also has solvent use in the preparation of epoxy-containing novolak resins with low chloride content and in mixtures with surfactants in enhanced petroleum recovery by flooding.

Other applications of amyl alcohols include their use as flavor and fragrance chemicals. Isoamyl salicylate is used to a large extent in soap and cosmetic fragrances because of its cost effectiveness. Isoamyl alcohol is used as the extracting solvent for purification of wet process phosphoric acid. t-Amyl methyl ether (TAME) is a useful gasoline additive as an octane booster. Amyl cinnamic aldehyde is an important ester of amyl alcohol. 1-Pentanol is used as an alcohol cosurfactant in a variety of applications. Amyl alcohols are used for the preparation of a variety of herbicides, fungicides, and pesticides. Amyl alcohols are a superior medium, compared to either benzyl alcohols or dichloromethane, for preparation of magnesia suspensions for electrophoretic deposition. Isoamyl alcohol is an intermediate in the synthesis of pyrethroids.

Amyl Alcohols or Pentanols (C5H12O). Toxicology Card No. 206, Cahiers de Notes Documentaires, No. 118, 1st quarter, National Institute for Research and Safety, Paris, 1985, 143–146.

Chemical Economics Handbook, SRI International, Menlo Park, Calif., Nov. 2002.

J. S. Riddick and W. B. Bunger, Organic Solvents: Physical Properties and Methods of Purification, 3rd ed., Wiley-Interscience, New York, 1970.

J. A. Riddick, W. B. Bunger, and T. K. Sakano, *Techniques of Chemistry: Organic Solvents*, Vol. 2, 4th ed., John Wiley & Sons, Inc., New York, 1986.

ANTHONY J. PAPA
Union Carbide Chemicals
and Plastics Company

ANILINE AND ITS DERIVATIVES

Aniline (benzenamine) is the simplest of the primary aromatic amines. Aromatic amines can be produced by reduction of the corresponding nitro compound, the ammonolysis of an aromatic halide or phenol, and by direct amination of the aromatic ring. At present, the catalytic reduction of nitrobenzene is the predominant process for manufacture of aniline. To a smaller extent aniline is also produced by ammonolysis of phenol.

Important analogs of aniline include the toluidines, xylidines, anisidines, phenetidines, and its chloro-, nitro-, *N*-acetyl, *N*-alkyl, *N*-aryl, *N*-acyl, and sulfonic acid derivatives.

PHYSICAL PROPERTIES

Pure, freshly distilled aniline is a colorless, oily liquid that darkens on exposure to light and air. It has a characteristic sweet, aminelike aromatic odor. Aniline is miscible with acetone, ethanol, diethyl ether, and benzene, and is soluble in most organic solvents. The physical properties of aniline are given in Table 1.

CHEMICAL PROPERTIES

Aromatic amines are usually weaker bases than aliphatic amines.

Aromatic amines form additional compounds and complexes with many inorganic substances, such as zinc chloride, copper chloride, uranium tetrachloride, or boron trifluoride. Various metals react with the amino group to form metal anilides; and hydrochloric, sulfuric, or phosphoric acid salts of aniline are important intermediates in the dye industry. Important reactions include *n*-alkylation, ring alkylation, acylation, condensation, crylization, reaction with nitrous acid, oxidation, halogenation, sulfonation, nitration, and reduction.

MANUFACTURING AND PROCESSING

The predominant process for manufacture of aniline is the catalytic reduction of nitrobenzene with hydrogen.

DuPont uses a liquid-phase hydrogenation process that employs a palladium–platinum-on-carbon catalyst. The process uses a plug-flow reactor that achieves essentially quantitative yields, and the product exiting the reactor is virtually free of nitrobenzene.

Table 1. Physical Properties of Aniline

Property	Value
molecular formula	C_6H_7N
molecular weight	93.129
boiling point, °C	
101.3 kPa[a]	184.4
4.4 kPa[a]	92
1.2 kPa[a]	71
freezing point, °C	−6.03
density, liquid, g/mL	
20/4°C	1.02173
20/20°C	1.022
density, vapor (at bp, air = 1)	3.30
refractive index, n^{20}_D	1.5863
viscosity, mPa·s(=cP)	
20°C	4.35
60°C	1.62
enthalpy of dissociation, kJ/mo[b]	21.7
heat of combustion, kJ/mol[b]	3394
ionization potential, eV	7.70
dielectric constant, at 25°C	6.89
dipole moment at 25°C (calcd), C·m[c]	5.20×10^{-30}
specific heat at 25°C, J/(g·K)[b]	2.06
heat of vaporization, J/g[b]	478.5
flash point, °C	
closed cup	70
open cup	75.5
ignition temperature, °C	615
lower flammable limit, vol %	1.3

[a]To convert kPa to mm Hg, multiply by 7.5.
[b]To convert J to cal, divide by 4.184.
[c]To convert C·m to debye, multiply by 3×10^{29}.

Demand in 2000 was 823×10^6 kg (1815×10^6 lb). Projected demand for 2004 is 959×10^6 kg (2115×10^6 lb). Demand equals production plus imports less exports. Although somewhat depressed by comparison to its performance in recent years, aniline should continue to produce better than GDP growth.

Prices remained stable over the period 1995–2000. List price was $0.20–0.23/kg ($0.45–0.50/lb) tanks, fob, Current price $0.37–0.39/lb tanks, fob.

ANALYTICAL METHODS

The typical analytical method used are spectroscopy, examination of the ultraviolet spectrum, nuclear magnetic resonance imaging, and gas chromatography. The latter method offers a rapid and accurate for determination of aniline in mixtures and in the method of choice for quality control used by producers of anailine.

STORAGE AND HANDLING

The flash point of aniline (70°C) is well above its normal storage temperature, but, aniline should be stored and used in areas with minimum fire hazard. Air should not be allowed to enter equipment containing aniline liquid or vapor at temperatures equal to or above its flash point.

Strong oxidizing agents, such as nitric acid, perchloric acid, or ozone may cause aniline to oxidize spontaneously.

Hexachloromelamine and trichloromelamine react violently with aniline, and in confined conditions the mixtures will explore or catch fire.

Aniline is slightly corrosive to some metals. It attacks copper, brass, and other copper alloys, and use of these metals should be avoided in equipment that is used to handle aniline. For applications in which color retention is critical, the use of 400-series stainless steels is recommended.

Aniline is shipped in tank truck and tank car quantities and is classified by the U.S. Department of Transportation (DOT) as a Class B poison (UN 1547). It must carry a poison label.

Wastes contaminated with aniline may be listed as RCRA Hazardous Waste, and if disposal is necessary the waste disposal methods used must comply with U.S. federal, state, and local water pollution regulations. The aniline content of wastes containing high concentrations of aniline can be recovered by conventional distillation. Biological disposal of dilute aqueous aniline waste streams is feasible if the bacteria are acclimated to aniline. Aniline has a 5-day BOD of 1.89 g of oxygen per gram of aniline.

Aniline can be safely incinerated in properly designed facilities. It should be mixed with other combustibles such as No. 2 fuel oil to ensure that sufficient heating values are available for complete combustion of aniline to carbon dioxide, water, and various oxides of nitrogen. Abatement of nitrogen oxides may be required to comply with air pollution standards of the region.

HEALTH AND SAFETY FACTORS

Aniline is highly toxic and may be fatal if swallowed, inhaled, or absorbed through the skin. Aniline vapor is mildly irritating to the eye, and in liquid form it can be a severe eye irritant and cause corneal damage. The first sign of aniline poisoning is cyanosis, a bluish tinge to the lips and tongue, caused by conversion of the blood hemoglobin to methemoglobin. As methemoglobin concentration of the blood rises above a certain level, death may result from anoxia.

The U.S. Department of Labor (OSHA) has ruled that an employee's exposure to aniline in an 8-h work shift of a 40-h work week shall not exceed an 8-h time-weighted average (TWA) of 5 ppm vapor in air, 2 ppm skin. The American Conference of Governmental Industrial Hygienists (ACGIH) recommends a threshold limit value (TLV) of 2 ppm aniline vapor in air, TWA for an 8-h work day.

Table 2. Aniline Derivatives

Class of compound and common name	Molecular formula	Condensed structural formula	Appearance	Melting point, °C	Boiling point, °C	Commercial derivatives and uses
salts						
aniline hydrochloride	$C_6H_7N \cdot ClH$	$C_6H_5NH_2 \cdot HCl$	white solid	198	245	aniline black
aniline sulfate	$C_6H_7N \cdot 1/2H_2O_4S$	$(C_6H_5NH_2)_2 \cdot H_2SO_4$	white crystals			sulfanilic acid
N-alkyl, N-aryl						
N-methylaniline	C_7H_9N	$C_6H_5NHCH_3$	yellow liquid	−57	194.6	
N,N-dimethylaniline	$C_8H_{11}N$	$C_6H_5N(CH_3)_2$	yellow liquid (darkens in air)	2	193−194	vanillin; Michler's ketone; alkylating agents; dyes
N-ethylaniline	$C_8H_{11}N$	$C_6H_5NHC_2H_5$	colorless liquid (darkens in air)	−63.5	204.7	explosive stabilizer; dyes
N,N-diethylaniline	$C_{10}H_{15}N$	$C_6H_5N(C_2H_5)_2$	pale yellow liquid	−38.8	215−216	alkylating agent
N-benzyl-N-ethylaniline $C_6H_5N(C_2H_5)$ $CH_2C_6H_5$	$C_{15}H_{17}N$ light yellow oil	314	triphenylmethane dyes			
diphenylamine	$C_{12}H_{11}N$	$C_6H_5NHC_6H_5$	white crystals, floral odor	54−55	302	rubber antioxidants; phenothiazine
C-alkyl						
o-toluidine	C_7H_9N	$H_3CC_6H_4NH_2$	yellow liquid (darkens in air)		200−202	triphenylmethane dyes; safranine colors
m-toluidine			colorless liquid	−30.4	203−204	dyes
p-toluidine			white crystals	44−45	200−201	Basic Red 9; Acid Green 25
2,3-xylidine	$C_8H_{11}N$	$(H_3C)_2C_6H_3$ NH_2	liquid		221−222	
2,4-xylidine			liquid	16	214	Solvent Orange 7; Direct Violet 14
2,5-xylidine			oily liquid	15.5	213.5	p-xyloquinone; Red 26; Direct Violet 7
2,6-xylidine			colorless liquid	11−12	216−217	formerly in dyes

Table 2. (*Continued*)

Class of compound and common name	Molecular formula	Condensed structural formula	Appearance	Melting point, °C	Boiling point, °C	Commercial derivatives and uses
3,4-xylidine			solid	51	226	synthetic riboflavin
3,5-xylidine			oil	9.8	220–221	azo dyes
C-alkoxy						
o-anisidine	C_7H_9NO	$H_3COC_6H_4NH_2$	yellow liquid (darkens in air)	5–6	225	guaiacol synthesis; Direct Red 24; Solvent Red 1
m-anisidine			oily liquid	−1 – 1	251	
p-anisidine			white solid	57	243	dyes
o-phenetidine	$C_8H_{11}NO$	$H_5C_2OC_6H_4NH_2$	oily liquid		231–233	dyes
p-phenetidine			liquid (darkens in air)	3–4	254–255	phenacetin; phenocoll; rubber antioxidant; dyes
p-cresidine	$C_8H_{11}NO$	$H_3CO(CH_3)$-$C_6H_3NH_2$	white crystals	52–54	235	FD&C Red 40
N-acyl						
formanilide	C_7H_7NO	$HCONHC_6H_5$	white crystals	50	271	analgesic and antipyretic
acetanilide	C_8H_9NO	$CH_3CONHC_6H_5$	colorless crystals	114.3	304	intermediate for sulfa drugs; hydrogen peroxide stabilizer; azo dyes
acetoacetanilide	$C_{10}H_{11}NO_2$	CH_3COCH_2-$CONHC_6H_5$	white crystals	86		intermediate for pyrazolones and pyrimidines; Hansa yellows; benzidine yellow pigments
chloroanilines						
2-chloroaniline	C_6H_6ClN	$ClC_6H_4NH_2$	colorless liquid	−14	208–210	dyes
3-chloroaniline			colorless liquid	−10	230–231	dyes
4-chloroaniline			colorless liquid	72.5	232	azoic dye coupling Component 10 and 15
2,5-dichloroaniline	$C_6H_5Cl_2N$	$Cl_2C_6H_3NH_2$	needle crystals	51	251	dyes
3,4-dichloroaniline			white crystals	71.5	272	herbicides; dyes
sulfonated anilines						
orthanilic acid	$C_6H_7NO_3S$	$H_2NC_6H_4SO_3H$	colorless crystals	>320 dec		dyes
metanilic acid			white crystals	dec		Acid Yellow 36, Direct Yellow 44 dyes
sulfanilic acid			white crystals	288		Acid Orange 1; Food Yellow 3 dyes
nitroanilines						
2-nitroaniline	$C_6H_6N_2O_2$	$O_2NC_6H_4NH_2$	golden crystals	71–72	284	Vat Orange 7 and Red 14 dyes
3-nitroaniline			yellow crystals	114	305–307 dec	synthetic intermediate; dyes
4-nitroaniline			pale yellow crystals	148–149	332	dyes; intermediate for 1,4-phenylenediamine
2,4-dinitroaniline	$C_6H_5N_3O_4$	$(O_2N)_2C_6H_3NH_2$	yellow crystals	187–188		Pigment Orange 5; dyes
2,4,6-trinitroaniline	$C_6H_4N_4O_6$	$(O_2N)_3C_6H_2NH_2$	yellow solid	192–195	explodes	explosives; detonators

Based on tests with laboratory animals, aniline may cause cancer. In view of the above, aniline should be handled in areas with adequate ventilation and skin exposure should be avoided by wearing the proper safety equipment. Recommended personal protective equipment includes hard hat with brim, chemical safety goggles, full length face shield, rubber gauntlet gloves, rubber apron, and rubber safety shoes or rubber boots worn over leather shoes.

USES

The major uses of aniline are in the manufacture of polymers, rubber, agricultural chemicals, dyes and pigments, pharmaceuticals, and photographic chemicals. Production of MDI (4,4-methylene diphenyl diisoyanate) accounts for 85% of aniline use. Other uses: rubber processing chemicals, 9%; herbicides, 2%; dyes and pigments, 2%; speciality fibers, 1%; miscellaneous including explosives, epoxy curing agents, and pharmaceuticals, 1%.

New uses for aniline described in recent patents include; aniline disulfide derivatives for treating allergic diseases, aniline compound in a hair dye composition and method of dyeing hair, and fluorine-containing aniline compounds as a starting material for insecticides.

The major consuming use MDI is tied to depressed economic conditions, but MDI growth continues to expand as new uses of polyurethanes are promoted outside traditional construction and refrigeration areas.

OTHER DERIVATIVES

Most derivatives of aniline are not obtained from aniline itself, but are prepared by hydrogenation of their nitroaromatic precursors. The exceptions—for example, N-alkylanilines, N-arylanilines, sulfonated anilines, or the N-acyl derivatives—can be prepared from aniline. Nitroanilines are usually prepared by ammonolysis of the corresponding chloronitrobenzene. Special isolation methods may be required for some derivatives if the boiling points are close and separation by distillation is not feasible. Table 2 lists some of the derivatives of aniline that are produced commercially.

"Aniline, Chem Profile," *Chemical Week*, Jan. 21, 2002.

Du Pont Aniline Properties, Uses, Storage, and Handling Bulletin, E. I. du Pont de Nemours & Co., Inc., 1983.

R. J. Lewis, Sr., *Sax's Dangerous Properties of Industrial Materials*, 10th ed., Vol. 3, John Wiley & Sons, Inc., New York, 2000.

J. W. McDowell and J. Northcott, in C. L. Hilton, ed., *Encyclopedia of Industrial Chemical Analysis*, Vol. 5, Wiley-Interscience, New York, 1967, 421–459.

BIJAN AMINI
E. I. du Pont de Nemours & Co.,
Inc.

ANTIAGING AGENTS

Aging in humans is associated with a decline in physical vigor and function, with progressive deterioration in most major organ systems, including the central nervous system and immune system functions. The current view is that molecular damage and disorder that occur with age in macromolecules are largely responsible for the age-related changes observed at the organism level. Such damage to macromolecules may be caused by free radicals and by the formation of advanced glycation end-products (AGEs). It is still unclear, however, whether AGEs, which accumulate to high levels in many age-related chronic diseases, are the cause or the consequences of the diseases.

Some inhibitors of glycation and free radical formation are showing promise against chronic conditions, particularly diabetes and its related complications.

Aging is associated with an increase in oxidative damage to cellular macromolecules, probably arising from the electron transport system as part of the day-to-day metabolic process. Agents that can lower oxidative stress and damage are beginning to show potential as calorie restriction mimics. Several hormones have been shown to improve certain changes associated with human aging, such as body mass composition. For skin aging, basic cell functions are reduced with advancing age. Therefore, an effective antiaging agent should provide acceleration of mitochondrial activity, enhanced cell proliferation, and increased matrix component synthesis in dermal fibroblasts.

The central nervous system suffers from a gradual decline in cognition, behavior, and function with advancing age. Alzheimer's disease (AD) is the most common form of dementia in the elderly. Currently, available treatments for AD only diminish certain symptoms but cannot halt the dementing process. New therapies currently being developed for AD include agents that target amyloid β-peptide and downstream pathological changes as well as agents that increase the activity of the cholinergic transmitter system. Newer inhibitors of cholinesterase are more selective and show fewer side effects than the first generation series of inhibitors.

Age is a major risk factor for the development of chronic diseases of the cardiovascular system, and such disease are a direct result of atherosclerosis. Since atherosclerosis has both an autoimmune and an inflammatory component, new approaches for the treatment and prevention of heart disease are beginning to focus on these areas as well as the traditional risk factors such as dyslipidemia.

It is believed that pharmacologic intervention to restore immune function in the elderly will provide widespread benefits in helping to maintain health in advanced age. Age-related conditions that involve the musculoskeletal system include osteroarthritis and rheumatoid arthritis as well as osteoporosis. Several biologicals are now in use for treatment of rheumatoid arthritis. Future treatments will likely involve the inhibition of proinflammatory cytokines by small molecules.

Finally, age is a major risk factor for developing cancer in humans. Many agents are being investigated for their ability to inhibit the formation and progression of various types of cancer, including breast, prostate, and colorectal cancers. No treatment on the market today has been proven to slow human aging. Medical interventions for age-related diseases do result in an increase in life expectancy, but none have been proven to modify the underlying processes of aging.

The current view is that random damage that occurs within cells and among extracellular molecules is responsible for many of the age-related changes that are observed in organisms. Molecular disorder occurs and accumulates within cells and their products because this occurrence outpaces the cell's repair mechanisms.

ORGAN SYSTEMS

Skin and Hair

Skin aging, which includes photoaging and intrinsic aging, causes the formation of wrinkles and sagging. Dermal matrix components change qualitatively and quantitatively over time in aging skin. In addition, basic cell functions such as proliferation, mitochondrial respiration, and production of matrix components in dermal fibroblasts are reduced with aging. Thus, an effective antiaging agent for skin should provide acceleration of mitochondrial activity, enhanced cell proliferation, increased matrix component synthesis and improvement of collagen bundle fiber. Many cosmetic formulations have been marketed, with claims of preventing or reducing wrinkles and lines, containing a variety of agents.

Clostridium botulinum toxin (Botox) type A has been widely used aesthetically for the past 15 years for facial skin rejuvenation. Botox works by clinically paralyzing the facial muscles underlying the lines and wrinkles on the surface with restoration of muscle activity usually commencing between three and four months after injection.

Hair loss (androgenetic alopecia) occurs in men and women, and is characterized by the loss of hair from the scalp in a defined pattern. The involvement of androgens in androgenetic alopecia has been established for some time, and is well accepted. Eunuchs, who lack androgens, do not go bald. Individuals who lack a functional androgen receptor are androgen insensitive and develop as females; again, these individuals do not bald. Likewise, no baldness is seen in individuals who lack 5α-reductase, the enzyme that converts testosterone to the potent androgen dihydrotestosterone (DHT). The exact mechanism(s) through which androgens act to cause baldness remain unclear; however, given that the complex formed between the androgen receptor (AR) and androgen acts as a transcription factor, it is likely that genes controlling hair follicle cycling are regulated by androgen. Without treatment, androgenetic alopecia is a progressive condition. Only two pharmaceutical agents are approved for the treatment of androgenetic alopecia in males: topical minoxidil and oral finasteride.

Minoxidil is a vasodilator originally used to treat high blood pressure, but a topical formulation was developed when patients treated with the drug showed increased hair growth. Finasteride is a synthetic azo-steroid and is a highly selective and potent 5α-reductase type-2 inhibitor, thus lowering DHT levels in scalp and serum by >60% at a daily dose of only 1 mg. Future therapies may include the use of androgen-receptor blockers.

CENTRAL NERVOUS SYSTEM

Alzheimer's disease (AD) is becoming the most common form of dementia in the elderly worldwide. It is characterized by a gradual decline in three domains: cognition, behavior, and function. Available treatments for AD diminish only certain symptoms and cannot halt the dementing process. The AGE accumulation in the centerl nervous system (CNS), eg, may be related to the aging process and the degenerating process of AD neurons. New therapies currently being developed include therapeutic agents that target amyloid β-peptide and downstream pathological changes. Inhibition of the formation of amyloid peptide from its precursor protein is an attractive target for blocking the cascade process leading to the development of neurodegenerative disease. Bafilomycin A, eg, and its analogues are of interest, since they very effectively and selectively block the formation of amyloid peptide by an indirect inhibition of β-secretase activity.

Currently, there are only a few therapeutic agents on the market for the treatment of AD. The main pharmacological effect of most of the agents is to improve the cognitive functions decreased in AD due to hypofunction of the cholinergic transmitter system.

Newer inhibitors, which are more selective, include Aricept, Galanthamine, and Eptastigmine. One drug, Rivastigmine, successfully targets acetylcholinesterase in the brain as opposed to peripheral forms. These second-generation inhibitors show considerably fewer side effects than their first-generation predecessors.

One of the promising approaches in the development of preventive therapies of AD is a design of agents based on derivatives and analogs of melatonin. Melatonin is an endogenous hormone that has been shown to be effective against oxidative stress in the CNS.

Antiinflammatory agents are predicted to be of use in AD therapy because neurodegenerative changes in the AD brain are accompanied by inflammatory reactions of the CNS. Nonsteroidal anti-inflammatory agents (NSAIDS) have been shown to decrease the risk of developing AD in epidemiological studies. Ibuprofen was the first in the series of NSAIDS to be suggested for AD therapy and the activity of these types of compounds is thought to be due primarily to the nonspecific inhibition of the cyclooxygenases (COXs). Other promising NSAIDS being studied in AD are Naproxen and Rofecoxib (Vioxx).

CARDIOVASCULAR SYSTEM

Aging of the population will undoubtedly result in a concomitant increase in the incidence of the most common chronic cardiovascular diseases, including coronary artery disease, heart failure, myocardial infarction, and stroke. These diseases are direct consequences of atherosclerosis (AS), a multifactorial process that is both an autoimmune and an inflammatory condition. The immune system plays a major role in the development and progression of AS involving macrophages and activated lymphocytes. New guidelines from the Adult Treatment Panel III (ATP III) of the National Cholesterol Education Program recommend blood lipid management beyond low density lipoprotein (LDL) lowering, including aggressive treatment of elevated triglycerides, since recent studies show that elevated triglycerides significantly increase cardiovascular disease risk. High density cholesterol (HDL), on the other hand, appears to play a protective role against development of AS by several mechanisms, including "reverse cholesterol transport," inhibition of oxidation or aggregation of LDL, and modulation of inflammatory responses to favor vasoprotection. Thus, raising HDL while lowering LDL levels would be beneficial for prevention and treatment of AS.

The beneficial effects of 3-hydroxy-3-methylglutaryl CoA (HMG-CoA) reductase inhibitors (statins) in the treatment and prevention of cardiovascular disease have generally been attributed to their ability to lower cholesterol biosynthesis. The three most studied and widely used statins include atorvastatin, simvastatin, and pravastatin.

IMMUNE SYSTEM

The functional capacity of the immune system gradually declines with age. The age-related alterations that occur in the immune system may be referred to as immunosenescence and involve both the innate and the adaptive immune responses. These alterations account for the increased susceptibility to certain microbial infections, autoimmune diseases, or malignancies in the elderly and contribute to increased morbidity and mortality with age. Hence, the restoration of immunological function is expected to have a beneficial effect in reducing pathology and maintaining a healthy condition in advanced age.

MUSCULOSKELETAL SYSTEM

One of the most effective agents for treatment of rheumatoid arthritis (RA) is etanercept (Enbrel), which is a biological disease-modifying antirheumatic drug (DMARD) that works by blocking the proinflammatory cytokine tumor necrosis factor-alpha (TNF-α).

Another biological that blocks TNF-α is infliximab (Remicade), a monoclonal antibody to TNF used effectively in Crohn's disease as well as RA. Whether or not a diet that includes omega-3 fatty acids is useful in prevention of RA or bone loss is not yet known. Glucosamine, or its sulfate, has been widely used to treatosteoarthritis in humans and is thought to work by suppressing neutrophil function and activation.

In addition, glucosamine also has been shown to inhibit inducible nitric oxide production and shows anti-inflammatory activity.

CANCER PREVENTION

Age is clearly the single most important risk factor for development of prostate cancer in men and breast cancer in women. Management options for women at high risk for breast cancer include close surveillance, chemoprevention, and prophylactic mastectomy. Chemoprevention refers to the use of specific natural or synthetic chemical agents to reverse, suppress, or prevent the progression to invasive cancer. The ideal chemopreventive agent is safe and nontoxic over the long term. Prevention of breast cancer is still under clinical investigation with only one drug, tamoxifen, showing benefit in high risk patients. Raloxifene is another nonsteroidal antiestrogen being studied for potential chemoprevention of breast cancer.

Prostate cancer is the second leading cause of cancer death in the United States. There is evidence suggesting that androgenic influences over a period of time encourage the process of prostate carcinogenesis. Moreover, early prostate tumors are often androgen dependent, but androgen insensitive tumors inevitably develop that then have a very poor prognosis. This fact underscores the need for prevention strategies such as chemoprevention. Antiandrogens are among the promising chemopreventive agents for prostate cancer, since prostate epithelium is androgen dependent.

Lycopene is a carotenoid derived largely from tomato-based products. Recent epidemiological studies have suggested a potential benefit of this natural product against the risk of prostate cancer, especially the more advanced and aggressive form.

An impressive body of epidemiological data suggests an inverse relationship between colorectal cancer risk and regular use of nonsteroidal antiinflammatory drugs (NSAIDs), including aspirin. Clinical trials with NSAIDs have demonstrated that NSAID treatment caused regression of preexisting colon adenomas in patients with familial adenomatous polyposis. In addition, several phytochemicals with anti-inflammatory activity and NSAIDs act to retard, block or reverse colon carcinogenesis.

D. Aronson, *J. Hypertension* **21**, 3 (2003).

G. Fernandes (2003) *Abstracts of Papers, 225th ACS National Meeting, New Orleans, LA, United States, March 23-27, 2003,* AGFD-019.

G. Miltiadous, J. Papakostas, G. Chasiotis, K. Seferiadis, and M. Elisaf, *Atherosclerosis (Shannon, Ireland)* **166**, 199 (2003).

S. Rahbar, and J. L. Figarola, *Curr. Med. Chem.: Immunol., Endocrine Metabolic Agents* **2**, 135 (2002).

HOWARD B. COTTAM
University of California

ANTIBACTERIAL AGENTS, OVERVIEW

Antibacterial agents are synthetic compounds derived from petrochemical sources and other small chemical building blocks that either kill or prevent the growth of bacteria. For the purposes of this survey, antibacterial agents are distinguished from antibiotics, antiseptics, disinfectants, and preservatives. Antibiotics are chemical substances isolated from natural sources, or their semisynthetic derivatives, that kill microorganisms or inhibit their growth (see also ANTIBIOTICS). Antiseptics are chemical substances with antimicrobial properties that are used on the surface of living tissues, such as the skin or mucous membranes. In contrast to antibacterial agents and antibiotics, antiseptics do not necessarily exhibit selective toxicity for the microbial cell relative to the host cell. Disinfectants are chemical substances that kill microorganisms when applied to inanimate objects. Preservatives are generally static agents that slow the decomposition of organic substances by inhibiting the growth of microorganisms.

Antibacterial agents are commonly used to treat and/or prevent infections due to pathogenic bacteria in humans and animals. Although injectable dosage forms of some antibacterial agents have been developed, most of the drugs used in modern antibacterial chemotherapy were designed to achieve high systemic blood levels following oral administration. Given their synthetic origin, the antibacterial agents also contain few or no chiral centers, in contrast to antibiotics derived from natural sources, which frequently contain multiple contiguous stereocenters. Thousands of analogues of antibacterial agents have been prepared in an effort to identify compounds with an enhanced spectrum of activity, improved pharmacokinetics, or a greater safety margin. Nevertheless, a relatively small number (~100) of antibacterial agents have been marketed for clinical or veterinary use.

The mechanism of action of antibacterial agents varies depending on the structural class. Some agents interfere with bacterial deoxyribonucleic acid (DNA) or protein synthesis (quinolones, oxazolidinones); others inhibit the activity of an enzyme or enzymes involved in bacterial cell metabolism (sulfonamides, diaminopyrimidines, nitrofurans, isoniazid, ethionamide). Some classes of antibacterial agents, such as the sulfonamides, nitrofurans, and oxazolidinones are bacteriostatic (ie, bacterial cell growth is inhibited). Others, such as the quinolones (against gram-positive and gram-negative bacteria) and isoniazid (against mycobacteria), are bactericidal (ie, bacteria are killed).

NOMENCLATURE

Antibacterial agents are identified by three different types of names:

1. The chemical name is usually long and cumbersome and is based on conventional chemical nomenclature rules.
2. The generic name frequently has a common stem for a specific class of agents. For example, the generic

names for the quinolone family end in "-oxacin." Since this is a nonproprietary name, more than one brand name drug can have the same generic name.
3. The brand (trade) name is a proprietary name given by the manufacturer and is often based on commercial considerations.

The following example shows the difference between the three types of names for the same compound.

1. Chemical name—(S)-9-fluoro-2,3-dihydro-3-methyl-10-(4-methyl-1-piperazinyl)-7-oxo-7H-pyrido[1,2,3-de]-1,4-benzoxazine-6-carboxylic acid
2. Generic name—levofloxacin
3. Trade name—Levaquin

Generic names are usually preferred in scientific communications.

CLASSIFICATION OF ANTIBACTERIAL AGENTS

Antibacterial agents can be classified according to their molecular features. Agents within the same chemical family usually act by the same mechanism of action. However, since several chemical classes may exert the same, or closely related, mode of action, antibacterial agents may also be broadly classified according to the bacterial target affected. It is also possible to classify agents according to the therapeutic indication. For the purposes of this survey, antibacterial agents are classified either according to the clinical indication for which the drug is used (eg, antitubercular agents), or according to the salient molecular features (eg, oxazolidinones).

Antitubercular Agents

The synthetic first-line antitubercular agents can be subdivided into two broad categories according to structure. The first group, consisting of isoniazid, pyrazinamide, and ethionamide contain a heteroaryl hydrazide, amide, or thioamide. Isoniazid and ethionamide are metabolized by mycobacteria to electrophilic intermediates, which then inhibit the synthesis of mycolic acids essential to bacterial viability. Pyrazinamide has been shown to inhibit fatty acid synthesis by preventing the formation of precursors needed for the synthesis of mycolic acids. These agents all exhibit activity against *Mycobacterium tuberculosis*. Ethionamide also inhibits the growth of other slowly growing mycobacteria. Despite chemical similarities, these three agents do not always exhibit cross-resistance.

The second structural type of antitubercular agents is represented by ethambutol, which contains a symmetrical diamino-dihydroxy aliphatic chain. It has been proposed that ethambutol prevents mycobacterial cell wall synthesis by inhibiting the production of arabinan.

Nitrofurans

The nitrofuran class of antibacterial agents contain a 5-nitro-2-furanyl moiety. Compounds in this family are usually hydrazone derivatives of 5-nitro-2-furancarbox-aldehyde. However, several compounds are olefinic deriva-tives. It has been shown that nitrofurans are converted by bacterial reductases to reactive intermediates that can inhibit a number of bacterial enzymes, including those responsible for DNA and ribonucleic acid (RNA) synthesis and carbohydrate metabolism. This class is active against a wide spectrum of gram-positive and gram-negative organisms, including enterococci and *Escherichia coli*, respectively, but has dropped out of widespread use in the United States for severe infections due to the dis-covery and development of new classes of agents.

Oxazolidinones

The oxazolidinone class of antibacterial agents, exemplified by linezolid, contains a 3-aryl-5-acetamidomethyloxazolidin-2-one pharmacophore essential for biological activity. These agents selectively bind to the P site of the 23S RNA compo-nent of the 50S ribosomal subunit, thus inhibiting protein synthesis at an early stage of translation, possibly by inhi-biting translocation of fMet-tRNA. Oxazolidinones have microbiological activity against a variety of susceptible and multidrug-resistant gram-positive organisms.

Quinolones

The 4-quinolone class of antibacterial agents contains a 3-carboxylic acid attached to the core quinolone or naphthyridone nucleus. In addition, the N-1 nitrogen is arylated or alkylated. The quinolones target the essen-tial bacterial type II topoisomerases, DNA gyrase and topoisomerase IV, the relative potency depending on the organism and the specific compound. Quinolones are broad-spectrum bactericidal agents against a variety of gram-positive and gram-negative pathogens, including some anaerobic bacteria and intracellular pathogens.

Sulfonamides and 2,4-Diaminopyrimidines

The sulfonamide class of antibacterial agents includes N-1 derivatives of *para*-aminobenzenesulfonamide. Sulfona-mides compete with *para*-aminobenzoic acid (PABA) for incorporation into folic acid in a reaction catalyzed by dihydropteroate synthase. The 5-substituted-2,4-diamino-pyrimidines inhibit the enzyme dihydrofolate reductase, the next step in the biosynthesis of tetrahydrofolic acid. In particular, the drug combinations, sulfadoxine-pyrimethamine and trimethoprim-sulfamethoxazole have been used to treat uncomplicated malaria and bacterial urinary tract infections, bronchitis and otitis media, respectively. These agents are bacteriostatic. Emergence of resistance to sulfonamides as well as the introduction of new classes of more potent antibacterial agents has diminished the clinical usefulness of this class.

PREPARATION AND MANUFACTURE

Antitubercular Agents

Isoniazid, pyrazinamide and ethionamide, related in structure, are manufactured by similar routes. Isoniazid is prepared by condensation of ethyl isonicotinate with hydrazine hydrate, or alternatively by heating 4-cyano-pyridine with hydrazine hydrate in aqueous alkaline solution. A variation of this procedure involves the reac-tion of isonicotinic acid with hydrazine hydrate in the presence of a catalyst, such as alumina, titanium tetra-butoxide, or a sulfonic acid cation exchanger.

Pyrazinamide is produced by ammonolysis of an alkyl ester of pyrazinoic acid. Alternatively, pyrazinamide can be obtained from the reaction of pyrazinecarbonitrile with aqueous ammonia or by hydrolysis of pyrazinecarbo-nitrile under acidic or alkaline conditions. The alkyl pyr-azinoates are produced by acid-catalyzed esterification of pyrazinoic acid in the presence of a lower alkanol. Pyrazi-noic acid, in turn, is prepared by two general methods: (1) potassium permanganate oxidation of quinoxaline, fol-lowed by decarboxylation of the intermediate pyrazine-2,3-dicarboxylic acid, and (2) oxidation of methylpyrazine with selenious acid in pyridine or, alternatively, reaction of ethylpyrazine with potassium permanganate. Pyrazine-carbonitrile is readily prepared by ammonoxidation of methylpyrazine.

Two general methods are available for the preparation of ethionamide. Both routes converge at the key inter-mediate, 4-cyano-2-ethylpyridine, which is converted to ethionamide by treatment with hydrogen sulfide gas. In the first method, radical alkylation of methyl isonicotinate with dipropionyl peroxide solution affords a mixture of 3-ethyl and 2-ethyl isomers, which can be converted to the corresponding 4-cyanopyridine derivatives by con-densation with ammonia, followed by dehydration of the resulting amide in the presence of alumina. Alternatively, 4-cyano-2-ethylpyridine can be prepared directly by radi-cal alkylation of 4-cyanopyridine.

In the second general method, 2-ethylpyridine is converted to 4-cyano-2-ethylpyridine through a series of steps, including oxidation to 2-ethylpyridine-*N*-oxide, chlorination to give 4-chloro-2-ethylpyridine, treatment with an alkali metal bisulfite or pyrosulfite to give the intermediate 2-ethylpyridine-4-sulfonic acid, and finally reaction with an alkali metal cyanide.

The most common method for the manufacture of ethambutol begins with (+)-2-amino-1-butanol, obtained from resolution of racemic 2-amino-1-butanol with L-glutamic acid. Another approach to the preparation of ethambutol uses butadiene monoepoxide as an inexpen-sive source of the carbon atoms of the molecule.

Nitrofurans

The majority of the nitrofurans are commercially pre-pared by the condensation of either 5-nitro-2-furancarbox-aldehyde or 5-nitro-2-furancarboxaldehyde diacetate with the appropriate hydrazine derivative.

Oxazolidinones

In the typical drug discovery route for the synthesis of the oxazolidinone class of antibacterial agents, the oxazolidinone ring is formed by reaction of the anion of the appropriately substituted Cbz-protected aniline with (R)-glycidyl butyrate. Several recent process patent applications have disclosed the condensation of the above carbamate with nitrogen-containing three carbon reagents in place of (R)-glycidyl butyrate.

Quinolones

There are two common methods for the synthesis of the quinolone core structure. The earlier method relies on a Gould-Jacobs cyclization reaction between the appropriate aniline and diethyl ethoxymethylenemalonate. Norfloxacin and other analogs of this type have been synthesized in this manner. However, this method is unsatisfactory for N-1 aryl substituents or for substituents that would be derived from an unreactive alkyl halide, such as cyclopropyl. In addition, there is the potential for the formation of regio-isomers during the cyclization of unsymmetrical anilines.

The most widely utilized methodology allows for greater flexibility and produces a wide variety of quinolones including the tricyclic analogues. The ring closure occurs by an intramolecular nucleophilic aromatic substitution reaction of an appropriately substituted enamine. Ciprofloxacin, levofloxacin and numerous other quinolones have been synthesized by this method.

Sulfonamides and 2,4-Diaminopyrimidines

The sulfonamides are usually prepared by the reaction of N-acetylbenzenesulfonyl chloride with the appropriate amine and an equivalent of base (or 2 equiv of the appropriate amine), followed by basic hydrolysis of the acetamide functionality. The sulfonyl chloride is synthesized by chlorosulfonation of acetanilide.

THERAPEUTIC UTILITY

Antitubercular Agents

Treatment of infections caused by M. tuberculosis is most effective when multiple drugs are used for at least 4 months, due to the slow growth of mycobacteria and their propensity to develop resistance during monotherapy. Regimens recommended by The American Thoracic Society Medical Section of the American Lung Association and The Tuberculosis Committee of the Infectious Disease Society of America in conjunction with the Division of Tuberculosis Elimination of the Centers for Disease Control and Prevention employ the use of three drug combinations unless that particular geographic region reports more than 4% of TB isolates resistant to isoniazid, in which case a four drug regimen is recommended. The most common combinations are isoniazid, rifampicin, and pyrazinamide, with the addition of either ethambutol or streptomycin depending upon geographical resistance profiles. Note that rifampicin and streptomycin are considered to be antibiotics, and, as such, are discussed in detail elsewhere. If a fully drug-susceptible strain is involved, a two drug regimen may be employed, such as rifampicin and isoniazid. Isoniazid alone for up to 12 months, or a rifampicin–pyrazinamide regimen for 2 months, has been recommended for prophylactic treatment of TB-infected asymptomatic HIV-infected patients.

Nitrofurans

Urinary tract infections are the major area in which nitrofurans are used. Nitrofurantoin is not indicated for the treatment of more complicated renal infections, such as pyelonephritis or perinephric abscesses. Certain nitrofuran derivatives have been used in veterinary practice to treat or prevent protozoal and bacterial infections in both nonfood and food-producing animals. Use of these carcinogenic and teratogenic drugs, however, results in unacceptably elevated residues in edible tissues, leading to an FDA ban on the use of nitrofurans in food-producing animals in May 2002.

Oxazolidinones

Linezolid, the first and only oxazolidinone currently approved for therapeutic use by the regulatory agencies, represents the first new structural class of antibacterial agent in 35 years. Because it targets a stage of protein synthesis different from other agents, it does not demonstrate cross-resistance with any other antibiotic or antibacterial agent. Linezolid is specifically indicated for the treatment of adult patients with infections caused by linezolid-susceptible strains of vancomycin-resistant Enterococcus faecium, including bacteremia. Linezolid can be used in both nosocomial and community-acquired pneumonia caused by S. aureus (including methicillin-resistant strains for nosocomial pneumonia) or S. pneumoniae, with combination therapy indicated if gram-negative organisms are present. It has been approved for treatment of both complicated and uncomplicated skin and skin structure infections caused by S. aureus or Streptococcus pyogenes.

Quinolones

Quinolones are broad-spectrum agents with antibacterial activity against both gram-positive and gram-negative bacteria, including anaerobic and intracellular pathogens. Their target of bacterial DNA topoisomerase means that they exhibit minimal cross-resistance with most other antibacterial agents, and generally retain activity against the penicillin-resistant streptococci that are becoming highly prevalent in community-acquired infections. Quinolones have been approved for multiple therapeutic indications, dependent upon the individual agent. Some quinolones have been approved for treatment of serious infections including both complicated and uncomplicated skin and skin structure infections and nosocomial pneumonia. They may also be used to treat genitourinary tract infections including complicated and uncomplicated urinary tract infections (mild to moderate), acute pyelonephritis, prostatitis, various urethral and cervical infections, pelvic inflammatory disease, and uncomplicated cystitis.

Additionally, six quinolones are marketed exclusively for use in veterinary medicine: danofloxacin, difloxacin, enrofloxacin, marbofloxacin, orbifloxacin, and sarafloxacin. The human drugs, ciprofloxacin, ofloxacin, and trovafloxacin, are occasionally used in companion animal medicine.

Sulfonamides and 2,4-Diaminopyrimidines

These agents have been surpassed as first line agents for most bacterial infections. However, sulfonamides alone are still utilized for the treatment of urinary tract infections due to susceptible enteric bacteria. The combination of a sulfonamide with trimethoprim, a dihydrofolate reductase inhibitor, can be used for treatment of a number of less serious microbial infections including urinary tract infections, acute otitis media due to susceptible strains of *S. pneumoniae* or *Haemophilus influenzae*, and acute exacerbations of chronic bronchitis in adults due to susceptible strains of *S. pneumoniae* or *H. influenzae*. This combination is also used for treatment of travelers' diarrhea in adults due to susceptible strains of enterotoxigenic *E. coli*, for shigellosis, and for enteritis caused by susceptible strains of *Shigella flexneri* and *Shigella sonnei*. The combination of trimethoprim and sulfamethoxazole is the treatment of choice for *Pneumocystis carinii* pneumonia, and is also used as prophylaxis against this common fungal infection in immunocompromised or acquired immune deficiency syndrome (AIDS) patients.

Sulfonamide–trimethoprim combinations have been utilized for the treatment of bacterial disease in a variety of animal species.

A. Albert, *Selective Toxicity: The Physico-Chemical Basis of Therapy*, 5th Ed., Chapman and Hall, London, 1973, p. 134.

J. Grange, in F. O'Grady, H. Lambert, R. G. Finch, and D. Greenwood, eds., *Antibiotic and Chemotherapy: Anti-infective Agents and Their Use in Therapy*, Churchill Livingstone, New York, 1997, pp. 499–512.

M. Marketos, *Top 200 Drugs by Retail Sales in 2001*, Drugtopics. com, http://www.drugtopics.com/be_core/content/journals/d/data/2002/0218 (Feb. 3, 2003).

J. W. A. Petri, in J. G. Hardman, L. E. Limbird, and A. Goodman Gilman, eds., *Goodman and Gilman's The Pharmacological Basis of Therapeutics"*, McGraw-Hill, New York, 2001, pp. 1171–1187.

MARK J. MACIELAG
KAREN BUSH
MICHELE A. WEIDNER-WELLS
Johnson & Johnson
    Pharmaceutical Research &
    Development, L.L.C.

ANTIBIOTIC RESISTANCE

It is widely accepted that bacteria as living organisms came to existence over 3.5 billion years ago. As these organisms became compelled to interact with other living entities, they became more complex and evolved the biochemical means for influencing the existence of each other. One of these evolutionary developments was the advent of biochemical pathways for production of antiobiotics. In essence, if growth of a competitor were to be influence, more resources would be available for growth of the original organism. As such, multiple pathways for generation of "secondary metabolite", which include molecules that have antibacterial properties, have evolved. A number of these secondary metabolites with antibacterial properties have been discovered over the past few decades. The structures of many of them have been altered by chemists to expand their properties or to impart desirable chemical traits to them. Many of these molecules have found clinical use over the years.

MOLECULAR TARGETS FOR ANTIBIOTICS

Many of the first antibiotics discovered in the past 60 years have been natural products from microbial systems. To date, antibiotics that trace their origins to natural products dominate the armamentarium of clinically useful antibiotics. These are molecules that interfere with the biochemical processes of bacteria with some specificity, hence they are useful in mammalian hosts.

A total of over 70 bacterial genomes have been sequenced to date. It has been proposed that somewhere between 20 and 200 or so genes are critical for survival of a broad spectrum of bacteria. Known antibiotics interfere with a small number of biochemical processes coinciding with these critical genes. These processes include metabolic pathways, disruption of the integrity of the cytoplasmic membrane, inhibition of protein biosynthesis, inhibition of DNA biosynthesis, and disruption of the biosynthesis of the cell wall, of which the last three targets are especially important.

BIOCHEMICAL STRATEGIES FOR RESISTANCE TO ANTIBIOTICS

Development of antibiotic resistance is very complex. It is the result of a series of genotypic and phenotypic interactions of the biological systems of the host, pathogen, and antibiotic. Mutagenesis and gene acquisition are two important mechanisms in bacterial survival in the face of antibiotic or other life threatening challenges. There are many factors that effect the appearance and spread of acquired antibiotic resistance. Among these, the mutation frequency and the biological cost of resistance have become of increasing importance in understanding antibiotic resistance. The mutation frequency measures all the mutations present in a given population regardless of the status of the bacterial growth at which the mutation appears. Mutations happen randomly throughout the genome, and the rate by which the resistant mutants form will depend on the size of the genome and the bacterial population. When the mutation impairs a given gene product, the organism may die. However, should the mutation not be lethal, then it creates an incremental change in the organism.

Bacteria have evolved many mechanisms for acquiring resistance genes. These mechanisms enable bacteria to move DNA sequences from cell to cell via conjugation and transformation, or from one genome to another via classical recombination, transposition, and site-specific recombination. The site-specific recombination mechanism is important in acquisition and spread of the bacterial resistance.

The first β-lactam antibiotic to be used clinically was penicillin G (mid-1940s). This molecular class, including other β-lactam antibiotics, has enjoyed exceptional success clinically because it inhibits a step such as cross-linking of the cell wall, that is unique to bacteria. Barring the allergic response by a small fraction of the population to these antibiotics, these molecules generally are not toxic to the host. It is important to note that at least four distinct mechanisms for resistance to β-lactam antibiotics have been documented.

The most common mechanisms of resistance to β-lactams is through the expression of β-lactamases. These enzymes hydrolyze the β-lactam moiety of the drug, rendering it inactive. The success of this strategy is underscored by the fact that over 350 such enzymes have been identified from clinical strains. These enzymes fall into four structural classes, all of which appear to follow a distinct catalytic mechanism.

The second mechanism of resistance to β-lactam antibiotic is the evolutionary acquisition of DD-transpeptidases—the target enzymes—with reduced affinity for these drugs.

β-Lactam antibiotics must reach the outer surface of the cytoplasmic membrane to inhibit the PBPs. Hence, in gram-negative bacteria, β-lactam antibiotic has to penetrate the outer membrane to reach its target. This penetration takes place through the channel-forming proteins, namely, porins. These proteins transverse the outer membrane and are the portals through which the nutrients enter the cell. This is a means for resistance to imipenem, a member of the carbapenem class of β-lactam antibiotics.

The fourth mechanism of resistance to β-lactam antibiotics was discovered only recently. It has been reported that there exists an LD-transpeptidase that is capable of carrying out the cross-linking reaction not with the penultimate D-Ala residue, but rather with the third amino acid.

Vancomycin has been considered to be the antibiotics of last resort against gram-positive infections, especially the ones caused by the methicillin-resistant S.aureus. Over the past 10 years, a number of Enterococcus strains with high-level inducible resistance to vancomycin and its analogues have been identified. Vancomycin resistance is also seen in methicillin-resistant S. aureus clinical strains (VRSA). Such strains would appear to lack the enterococcal van genes, which suggest the possibility for other mechanisms in resistance to vancomycin.

Aminoglycosides are another class of antibiotics used against the infections caused by gram-positive and gram-negative bacteria. As with any other class of antibiotics, their antimicrobial properties are compromised by bacterial resistance. Methylation of the ribosomal binding site is known to cause resistance to gentamicin, as an example of altered target. This mechanism is observed only in aminoglycoside-produceding organisms.

Erythromycin, a macrolide antibiotic, is commonly used against gram-positive bacteria. Macrolide antibiotics inhibit bacterial protein synthesis by binding to the 23S rRNA of the 50S ribosomal subunit. The first resistant clinical isolates to macrolides were S. aureus, but subsequently resistance transferred to other organisms. Bacteria have developed three mechanisms that protect them from the action of the macrolides: target site alteration, antibiotic modification, and altered antibiotic transport. Target site alteration is the most common mechanism of resistance in the organisms that produce this antibiotic.

Bacteria have an intrinsic mechanism for protection from any toxic compounds in their environment. The gram-negative bacteria and gram-positive mycobacteria combine two mechanisms of resistance. First, the outer membrane and the mycolate-containing cell wall, respectively, produce effective permeability barriers. Second, the antibiotics that make it through the first outer membrane barrier are pumped out by the multidrug resistance efflux pumps (MDR). In gram-negative bacteria, MDR pumps interact with outer membrane channels and accessory proteins, forming multisubunit complexes that extrude antibiotics directly into the medium, bypassing the outer-membrane barrier.

Microorganisms have the ability to irreversibly attach to and grow on a surface and produce extracellular polysaccharides that facilitate attachment and matrix formation. Such matrix association of cells is known as biofilms. Biofilms may form on any surface, but their formation on the surface of indwelling medical devices, tooth enamel, heart valves or the lung, and middle ear is of biomedical concern.

Biofilm-associated organisms have altered phenotypes with respect to growth rate and gene transcription due to biofilm composition and structure. Biofilm consist of microcolonies held together by an extracellular matrix (polysaccharides). Its structure is hetrogeneous, with water channels that allow transport of essential nutrients and oxygen to the cells within the biofilm. Furthermore, biofilm-associated organisms have reduced growth rates that might as well minimize antimicrobial intake rate. Clearly, resistance in biofilms is more complicated: Multiple resistance mechanisms can act in concert. Adherence of bacteria to implanted medical devices or damaged tissues in the form of biofilm and inherent resistance contribute to duration of bacterial infections.

R. M. Hall, in D. J. Chadwick, and J. Goode, eds., *Antibiotic Resistance: Origins, Evolution, Selection and Spread*, Ciba Foundation Symposium 270, John Wiley & Sons, Inc., Chichester, 1996.

L. P. Kotra, S. Vakulenko, and S. Mobashery, *Microbes Infect.* **2**, 651 (2000).

K. Lewis, A. A. Salyers, H. W. Taber, and R. G. Wax, eds., *Bacterial Resistance to Antimicrobials*, Dekker, New York, 2001.

S. Mobashery and E. Azucena, *Encyclopedia Life Science London*, Nature Publishing Group, UK, 2000.

DASANTILA GOLEMI-KOTRA
SHAHRIAR MOBASHERY
Wayne State University

ANTIMONY AND ANTIMONY ALLOYS

Antimony, Sb, belongs to Group 15 (VA) of the Periodic Table which also includes the elements arsenic and bismuth. It is in the second long period of the table between tin and tellurium. Antimony, which may exhibit a valence of +5, +3, 0, or −3 (see ANTIMONY COMPOUNDS), is classified as a nonmetal or metalloid, although it has metallic characteristics in the trivalent state. There are two stable antimony isotopes that are both abundant and have masses of 121 (57.25%) and 123 (42.75%).

HISTORY AND OCCURRENCE

The word antimony (from the Greek *anti* plus *monos*) means "a metal not found alone" and, in fact, native antimony is seldom found in nature because of its high affinity for sulfur and metallic elements such as copper, lead, and silver. Over 100 naturally occurring minerals of antimony have been identified. Occasionally native metallic antimony is found; however, the most important source of the metal is the mineral stibnite (antimony trisulfide), Sb_2S_3. Most of the antimony produced in the United States is from the complex antimony deposits found in Idaho, Nevada, Alaska, and Montana. Ores of the complex deposits are mined primarily for lead, copper, zinc, or precious metals; antimony is a by-product of the treatment of these ores.

PROPERTIES

Physical properties of antimony are given in Table 1. Antimony, a silvery white, brittle, crystalline solid, is a poor conductor of electricity and heat. On solidification, pure antimony contracts 0.79 ± 0.14 vol% (10).

Antimony is ordinarily quite stable and not readily attacked by air or moisture. Under controlled conditions antimony reacts with oxygen to form the oxides Sb_2O_3, Sb_2O_4, and Sb_2O_5.

Table 1. Physical Properties of Antimony

Property	Value
at wt	121.75
mp, °C	630.8
bp, °C	1753
density at 25°C, kg/m³	6684
crystal system	hexagonal (rhombohedral)
hardness, Mohs' scale	3.0–3.5
latent heat of fusion, J/mol[a]	19,874
latent heat of vaporization, J/mol[a]	195,250
coefficient of linear expansion at 20°C, μm/(m·°C)	8–11
electrical resistivity at 0°C, μΩ·cm	39
magnetic susceptibility at 20°C, cgs	-99.0×10^{-6}
specific heat at 25°C, J/(mol·K)[1]	25.2
thermal conductivity at 0°C, W/(m·K)	25.5

[a]To convert J to cal, divide by 4.184

PROCESS METALLURGY

The antimony content of commercial ores ranges from 5 to 60%, and determines the method of treatment, either pyrometallurgical or hydrometallurgical. In general, the lowest grades of sulfide ores, 5–25% antimony, are volatilized as oxides; 25–40% antimony ores are smelted in a blast furnace; and 45–60% antimony ores are liquated or treated by iron precipitation. The blast furnace is generally used for mixed sulfide and oxide ores, and for oxidized ores containing up to ∼40% antimony; direct reduction is used for rich oxide ores. Some antimony ores are treated by leaching and electrowinning to recover the antimony. The concentrates may be leached directly or converted into a complex matte first. The most successful processes use an alkali hydroxide or sulfide as the solvent for antimony (see METALLURY, SURVEY).

By-Product and Secondary Antimony

Antimony is often found associated with lead ores. The smelting and refining of these ores yield antimony-bearing flue, baghouse, and Cottrell dusts, drosses, and slags. These materials may be treated to recover elemental antimony or antimonial lead from which antimony oxide or sodium antimonate may be produced.

Recycling of antimony provides a large proportion of the domestic supply of antimony. Secondary antimony is obtained from the treatment of antimony-bearing lead and tin scrap such as battery plates, type metal, bearing metal, antimonial lead, etc.

Refining

The metal produced by a simple pyrometallurgical reduction is normally not pure enough for a commercial product and must be refined. Impurities present are usually lead, arsenic, sulfur, iron, and copper. The iron and copper concentrations may be lowered by treating the metal with stibnite or a mixture of sodium sulfate and charcoal to form an iron-bearing matte that is skimmed from the surface of the molten metal. The metal is then treated with an oxidizing flux consisting of caustic soda or sodium carbonate and niter (sodium nitrate) to remove the arsenic and sulfur. Lead cannot be readily removed from antimony, but material high in lead may be used in the production of antimony-bearing, lead-based alloys. The yield of refined antimony from the matting and fluxing technique is 85–90%. Impure metal may be refined by electrolysis, although this procedure is not as economical as the pyrometallurgical treatment.

ENVIRONMENTAL CONCERNS

Antimony is a common air pollutant that occurs at an average concentration of 0.001 μg/m³. Antimony is released into the environment from buring fossil fuels and from industry. In the air, antimony is rapidly attached to suspended particles and thought to stay in the air for 30–40 days. The impact of antimony and antimony compounds on the environment has not been extensively studied to date.

HEALTH AND SAFETY FACTORS

Although metallic antimony may be handled freely without danger, it is recommended that direct skin contact with antimony and its alloys be avoided. Properly designed exhaust ventilation systems and/or approved respirators are required for operations that create dusts or fumes. As with other heavy metals, orderly housekeeping practice and good personal hygiene are necessary to prevent ingestion of (or exposure to) antimony.

USES

Antimony in the unalloyed state is extremely brittle and is not easily fabricated. For this reason, the use of the pure metal is restricted to ornamental applications.

Antimony Alloys

Approximately one-half of the total antimony demand is for metal used in antimony alloys. Antimonial lead is a term used to describe lead alloys containing antimony in proportions of up to 25%. Most commercial lead–antimony alloys have antimony contents <11%.

The largest application for antimonial lead is its use as a grid metal alloy in the lead acid storage battery (see BATTERIES, SECONDARY, LEAD–ACID). Demand for high performance SLI batteries has led to the development of smaller, lighter batteries that require less maintenance. The level of antimony is being decreased from the conventional 3–5% to 1.75–2.75% to minimize the detrimental effects.

Tin–antimony–copper and lead–antimony–tin white bearing alloys, commonly referred to as babbitt, are used to reduce friction and wear in machinery and help prevent failure by seizure or fatigue. Addition of antimony to babbitts increases strength and hardness. The choice of babbitt depends on the application and resultant desired properties.

Semiconductor and Solar Cells

High purity (up to 99.9%) antimony has a limited but important application in the manufacture of semiconductor devices and is utilized in such applications as infrared in devices, diodes, and Hall-effect components. High efficiency solar cells have been produced that comprise two layers: one of gallium arsenide and the other of gallium antimonide (see SOLAR ENERGY).

Antimony with a purity as low as 99.9 + % is an important alloying ingredient in the bismuth telluride Bi_2Te_3, class of alloys, which are used for thermoelectric cooling.

Antimony Compounds

The greatest use of antimony compounds is in flame retardants (qv) for plastics, paints, textiles, and rubber. Antimony trioxide and sodium antimonate are added to specialty glasses as decolorizing and fining agents, and

are used as opacifiers in porcelain enamels. Antimony oxides are used as white pigments in paints, whereas antimony trisulfide and pentasulfide yield black, vermillion, yellow, and orange pigments. Camouflage paints contain antimony trisulfide, which reflects infrared radiation. In the production of red rubber, antimony pentasulfide is used as a vulcanizing agent. Antimony compounds are also used in catalysts, pesticides, ammunition, and medicines (see ANTIMONY COMPOUNDS).

J. F. Carlin, Jr., "Antimony," in *Mineral Commodity Summaries*, U.S. Geological Survey, Reston, Va., Jan. 2002.

R. Fairbridge, ed., *Encyclopedia of Geochemistry and Environmental Sciences*, Vol. IV, Van Nostrand Reinhold Co., New York, 1972.

L. Gallicchio, B. A. Fowler, and E. F. Madden, "Arsenic, Antimony, and Bismuth," in E. Bingham, B. Cohrssen, and C. H. Powell, eds., *Patty's Toxicology*, 5th ed., Vol. 2, John Wiley & Sons, Inc., New York, 2001, Chapt. 36.

J. W. Mellor, *Comprehensive Treatise on Inorganic and Theoretical Chemistry*, Vol. 9, Longmans, Green, & Co., Inc., New York, 1929.

T. LI
ASARCO Inc.

ANTIMONY COMPOUNDS

Antimony is the fourth member of the nitrogen family and has a valence shell configuration of $5s^25p^3$. The utilization of these orbitals and, in some cases, of one or two $5d$ orbitals, permits the existence of compounds in which the antimony atom forms three, four, five, or six covalent bonds.

The valence bond theory in its most elementary form predicts that trivalent compounds of antimony should have pyramidal structures derived from the $5p$ orbitals and that the $5s$ electrons should act as an inert pair. Many trivalent derivatives of antimony, however, have intervalency angles significantly larger than the 90° angle predicted by this model. The fact that the bond angles are often considerably less than the regular tetrahedral value of 109.5° may be ascribed to repulsion by the lone pair.

INORGANIC COMPOUNDS OF ANTIMONY

Stibine

Stibine SbH_3, mp, is a colorless, poisonous gas having a disagreeable odor. It is the only well-characterized binary compound of antimony and hydrogen, although distibine Sb_2H_4, has been reported. Stibine is readily soluble in organic solvents such as carbon disulfide or ethanol, and is slightly soluble in water.

Metallic Antimonides

The most important binary compounds of antimony with metallic elements are indium antimonide, InSb, gallium antimonide, GaSb, and aluminum antimonide, AlSb, which find extensive use as semiconductors. The alkali metal antimonides, such as lithium antimonide and sodium antimonide, do not consist of simple ions. Rather, there is appreciable covalent bonding between the alkali metal and the Sb as well as between pairs of Na atoms.

Antimony Trioxide

Antimony(III) oxide (antimony sesquioxide), Sb_2O_3, is dimorphic, existing in an orthorhombic modification; valentinite is colorless (sp gr 5.67) and exists in a cubic form; and senarmontite, Sb_4O_6, is also colorless (sp gr 5.2). The cubic modification is stable at temperatures below $570°C$ and consists of discrete Sb_4O_6 molecules. This solid crystallizes in a diamond lattice with an Sb_4O_6 molecule at each carbon position.

Antimony Tetroxide

Antimony(III,V) oxide, antimony dioxide, SbO_2 and Sb_2O_4, occurs in two modifications. Orthorhombic antimony tetroxide has long been known as the mineral cervantite, α-Sb_2O_4, (colorless, sp gr 4.07). More recently a monoclinic modification, β-Sb_2O_4, has been recognized. In both dimorphs half of the antimony is in the $+3$ oxidation state, half in the $+5$ state. The antimony environments are quite similar in both modifications.

Antimony Pentoxide Hydrates

Antimonic acid (antimony(V) acid), and antimony(V) oxide, $Sb_2O_5 \cdot nH_2O$, are both hydrates of Sb_2O_5. Commercial antimony pentoxide is either hydrated Sb_2O_5 or at times β-Sb_2O_4. Material having the approximate composition $Sb_2O_5 \cdot 3.5H_2O$ may be prepared by hydrolysis of antimony pentachloride or by acidification of potassium hexahydroxoantimonate(V), $KSb(OH)_6$, followed by filtration and drying to constant weight in air at room temperature. This substance is a white solid which loses water upon

heating and becomes yellow in color. This loss of water fails to correspond to definitive ratios of $H_2O:Sb_2O_5$, nor is the composition Sb_2O_5 attained. At about $700°C$ the material is anhydrous and white in color; this is an antimony oxide, Sb_6O_{13}, containing both Sb(III) and Sb(V) and having a cubic pyrochlore-type structure. Hydrated antimony pentoxide (antimonic acid) is essentially insoluble in nitric acid solutions, only very slightly soluble in water, but dissolves in aqueous KOH.

Antimony Trifluoride

Antimony(III) fluoride, SbF_3, is a white, crystalline, orthorhombic solid. The molecule shows a very distorted octahedral arrangement. Antimony trifluoride is extremely soluble in water, the solubility being increased by the presence of hydrofluoric acid. It is also very soluble in polar solvents such as methanol, 154 g/100 mL, and acetone. Table 1 lists physical constants for the antimony halides.

Antimony Trichloride

Antimony(III) chloride, $SbCl_3$, is a colorless, crystalline solid, readily soluble in hydrochloric acid; water, ca 9% at $25°C$, increasing with temperature; $CHCl_3$, 22%; CCl_4, 13%; benzene; CS_2; and dioxane.

Antimony Tribromide and Triiodide

Antimony(III) bromide, $SbBr_3$, is a colorless, crystalline solid having a pyramidal dimorphic molecular structure and an acicular (α-$SbBr_3$) and a bipyramidal (β-$SbBr_3$) habit.

Antimony(III) iodide, SbI_3, forms red rhombohedral crystals, intermediate in structure between a molecular and an ionic crystal. In SbI_3 vapor there is no indication of association.

Both antimony tribromide and antimony triiodide are prepared by reaction of the elements. Their chemistry is similar to that of $SbCl_3$ in that they readily hydrolyze,

Table 1. Physical Constants of the Antimony Halides

Parameter	Antimony trifluoride	Antimony trichloride	Antimony tribromide	Antimony triiodide[a]	Antimony pentafluoride	Antimony pentachloride
formula	SbF_3	$SbCl_3$	$SbBr_3$	SbI_3	SbF_5	$SbCl_5$
mp, °C	291 ± 1	73.2	96.0 ± 0.5	170.5	6	3.2 ± 0.1
bp, °C	346 ± 10	222.6	287	401	150	68^b, 140^c
$\Delta_f^°$ at 298°C, kJ/mol[d]	-915.5	-382.2	-259.4	-100.4		-450.8 ± 6.2
$S°$ at 298°C, J/(mol·K)[d]	127	184	207	216 ± 1		263 ± 12
ΔH_{fusion}, kJ/mol[d,e]	21.4			$22.7_{444} \pm 0.2$		
ΔS_{fusion}, J/(mol·K)[d,e]	38.2			$51.5_{444} \pm 0.4$		
ΔH_{vap}, kJ/mol[d,e]	$102.8_{298} \pm 1.3$	46.72_{496}	53.2_{560}			43.45_{449}
ΔS_{vap}, J/(mol·K)[d,e]	$175.8_{298} \pm 2.5$	93.3_{496}	94.9_{560}			95.44_{449}
C_p, J/mol·K[d]		108^f		96^f, 144^g		

[a] The ΔH_{subl} at 298°C is 101.6 ± 0.4 kJ/mol (24.3 ± 0.1 kcal/mol).
[b] At a pressure of 1.82 kPa.
[c] Decomposes at atmospheric pressure, 101.3 kPa.
[d] To convert from J to cal, divide by 4.184.
[e] At the temperature in °C indicated by the subscript.
[f] Value given is for solid.
[g] Value given is for liquid.

form complex halide ions, and form a wide variety of adducts with ethers, aldehydes, mercaptans, etc. They are soluble in carbon disulfide, acetone, and chloroform.

Antimony Pentafluoride

Antimony(V) fluoride, SbF_5, is a colorless, hygroscopic, viscous liquid that has SbF_6 units with cis-fluorines bridging to form polymeric units. ^{19}F nmr shows that at low temperatures there are three different types of F atoms. Contamination with a small amount of HF markedly decreases the extent of polymerization. The vapor density at 150°C corresponds to the trimer. The solid is a cis-fluorine-bridged tetramer.

Antimony Pentachloride

Antimony(V) chloride, $SbCl_5$, is a colorless, hygroscopic, oily liquid that is frequently yellow because of the presence of dissolved chlorine; it cannot be distilled at atmospheric pressure without decomposition, but the extrapolated normal boiling point is 176°C. In the solid, liquid, and gaseous states it consists of trigonal bipyramidal molecules with the apical chlorines being somewhat further away than the equatorial chlorines.

Antimony Trisulfide

Antimony(III) sulfide (antimony sesquisulfide), SbS_3, exists as a black crystalline solid, stibnite, and as an amorphous red to yellow-orange powder. The crystal structure of stibnite contains two distinctly different antimony sites and consists of two parallel Sb_4S_6 chains that are linked together to form crumpled sheets (two per unit cell).

Antimony Pentasulfide

Antimony pentasulfide, Sb_2S_5, is a yellow to orange amorphous solid of indefinite composition. It is frequently given the formula Sb_2S_5, but actually consists of Sb(III) with a variable quantity of sulfur.

Antimony(III) Salts

Concentrated acids dissolve trivalent antimony compounds. From the resulting solutions it is possible to crystallize normal and basic salts, eg, antimony(III) sulfate, $Sb_2(SO_4)_3$; antimonyl sulfate, $(SbO)_2SO_4$; antimony(III) phosphate, $SbPO_4$; antimony(III) acetate, $Sb(C_2H_3O_2)_3$; antimony(III) nitrate, $Sb(NO_3)_3$; and antimony(III) perchlorate trihydrate, $Sb(ClO_4)\cdot 3H_2O$. The normal salts all hydrolyze readily.

Hexafluoroantimonates

Hexafluoroantimonic acid, $HSbF_6\cdot 6H_2O$, is prepared by dissolving freshly prepared hydrous antimony pentoxide in hydrofluoric acid or adding the stoichiometric amount of 70% HF to SbF_5. Both of these reactions are exothermic and must be carried out carefully.

Compounds Containing Sb–O–C or Sb–S–C Linkages

A large number of compounds have been prepared in which the antimony atom is linked to carbon through an oxygen or sulfur atom. The simplest of these compounds are esters of the hypothetical antimonic acid, H_3SbO_3, or thioantimonic acid, H_3SbS_3.

By far the largest group of compounds containing the Sb–O–C linkage are those obtained by reaction of an antimony oxide with an α-hydroxy acid, o-dihydric phenol, sugar alcohol, or some other polyhydroxy compound containing at least two adjacent hydroxyl groups. The best known compound of this type is antimony potassium tartrate (tartar emetic) prepared by refluxing potassium hydrogen tartrate with freshly precipitated antimony trioxide.

ORGANOANTIMONY COMPOUNDS

A wide variety of compounds containing the Sb–C bond are known. Organoantimony compounds can be broadly divided into Sb(III) and Sb(V) compounds. The former may contain from one to four organic groups, and the Sb(V) compounds from one to six organic groups. With a few exceptions, the nomenclature used here is that proposed by the International Union of Pure and Applied Chemistry. There are a number of heterocyclic compounds in which one or more antimony atoms are members of the heterocycle.

Primary and Secondary Stibines

Relatively few primary ($RSbH_2$) and secondary (R_2SbH) stibines are known. Methylstibine, CH_5Sb, ethylstibine, C_2H_7Sb, isopropylstibine, C_3H_9Sb, and butylstibine, $C_4H_{11}Sb$, have been prepared by the reduction of the corresponding alkyldichlorostibines using lithium aluminum hydride or sodium borohydride. All of the alkylstibines are thermally unstable, easily oxidizable, colorless liquids with strong alliaceous odors. Decomposition products include hydrogen and nonvolatile black solids analyzing for $(RSb)_x$. Reaction of the alkylstibines with hydrogen chloride produces lustrous, pale green polymeric solids also analyzing for $(RSb)_x$.

Tertiary Stibines

A large number of trialkyl- and triarylstibines are known. They are usually prepared by the interaction of a reactive organometallic compound and an antimony trihalide, a halostibine, or a dihalostibine. The type of organometallic compound most widely employed in these syntheses is the Grignard reagent. Organolithium, organocadmium, organoaluminum, and organomercury compounds have also been used. Triarylstibines can be readily prepared from an aryl halide, an antimony trihalide, and sodium.

Trialkylstibines are sensitive to oxygen, and in some cases they ignite spontaneously in air. Trimethylstibine, C_3H_9Sb, may explode on contact with atmospheric oxygen. Triarylstibines usually do not react with air, and they are quite stable thermally. Trialkylstibines are powerful reducing agents and can convert halides of phosphorus or antimony to the corresponding elements. Triarylstibines are much less reactive as reducing agents, but they are readily oxidized by halogens, interhalogens, pseudohalogens, sulfur, and fuming nitric acid.

Halostibines, Dihalostibines, and Related Compounds

Alkyldihalo- and dialkylhalostibines are highly reactive substances which are rapidly oxidized in air. Some are spontaneously inflammable. The aromatic counterparts are less susceptible to air oxidation but are readily oxidized by halogens. Alkaline hydrolysis of the dihalo- and halostibines yields compounds of the types $(RSbO)_x$ and $R_2SbOSbR_2$, respectively, whereas reaction of the stibine with sodium sulfide gives the analogous sulfur compounds.

Distibines and Distibenes

A considerable number of tetraalkyl- and tetraaryldistibines have been investigated. These are usually obtained by the reduction of a dialkyl- or diarylhalostibine with sodium hypophosphite or magnesium. Distibines undergo a variety of interesting reactions and have also attracted attention because a number of these substances are thermochromic.

Cyclic and Polymeric Substances Containing Antimony–Antimony Bonds

A number of organoantimony compounds containing rings of four, five, or six antimony atoms have been prepared. Polymeric substances, $(RSb)_x$ or $(ArSb)_x$, have been obtained by the decomposition of primary stibines, the reaction of primary stibines with hydrogen chloride, the treatment of primary stibines with dibenzylmercury, or the reduction of dihalostibines. Most of these polymers have not been well characterized.

Antimonin and its Derivatives

Antimonin(stibabenzene), C_5H_5Sb, the antimony analogue of pyridine, can be prepared by the dehydrohalogenation of a cyclic chlorostibine using 1,5-diazabicyclo[4.3.0]non-5-ene. A number of derivatives of antimonin are also known. The potential aromaticity of this ring system has aroused considerable interest and has been investigated with the aid of spectroscopy as well as *ab initio* molecular orbital calculations. There seems to be no doubt that antimonin does possess considerable aromatic character.

Stibonic and Stibinic Acids

The stibonic acids, $RSbO(OH)_2$, and stibinic acids, $R_2SbO(OH)$, are quite different in structure from their phosphorus and arsenic analogues. The stibonic and stibinic acids are polymeric compounds of unknown structure and are very weak acids.

Stibine Oxides and Related Compounds

Both aliphatic and aromatic stibine oxides, R_3SbO, or their hydrates, $R_3Sb(OH)_2$, are known. Thus both dihydroxotrimethylantimony, $C_3H_{11}O_2Sb$, and trimethylstibine oxide, C_3H_9OSb, have been prepared.

The structure of triphenylstibine oxide, $C_{18}H_{15}OSb$, has been the subject of considerable controversy. Apparently it can exist in two different forms; as prismatic crystals, mp 221–222°C, and as an amorphous powder. The structure of the crystalline form was shown by x-ray diffraction to be a dimer containing a planar four-membered ring with Sb–O–Sb bonds.

Pentacovalent Antimony Halides and Related Compounds

Antimony halides of the types $RSbX_4$, R_2SbX_3, R_3SbX_2, and R_4SbX, where X is a halogen, are known, but compounds of the first type have only been isolated and characterized where R is aryl.

Both dialkyl- and diaryltrihaloantimony compounds are known, although only a few dialkyl compounds have been described. The trichlorides have been obtained by the chlorination of either dialkylchlorostibines or tetraalkyldistibines with sulfuryl chloride. Dimethyltrichloroantimony, $C_2H_6Cl_3Sb$, is dimeric in the solid state but is monomeric in solution. The dimer exists in two different forms, covalent and ionic. The covalent form contains bridging chlorine atoms; the ionic form possesses the structure $[(CH_3)_4Sb] [SbCl_6]$.

Stibonium Ylids and Related Compounds

In contrast to phosphorus and arsenic, only a few antimony ylids have been prepared. Until quite recently triphenylstibonium tetraphenylcyclopentadienylide, $C_{47}H_{35}Sb$, was the only antimony ylid that had been isolated and adequately characterized. A new method, utilizing an organic copper compound as a catalyst, has resulted in the synthesis of a number of new antimony ylids. Among the ylids prepared by this method are those in which X and Y are $C_6H_5SO_2$, $4-CH_3C_6H_4SO_2$, or CH_3CO or where X is CH_3CO and Y is C_6H_5CO. These ylids are solids, stable in a dry atmosphere, but readily hydrolyzed in protic solids by traces of moisture. Attempts to carry out the Wittig reaction with the stibonium ylids containing sulfonyl or carbonyl substituents, using highly reactive 2,4-dinitrobenzaldehyde as the substrate, were unsuccessful (188). Closely related to the ylids are imines of the type $R_3Sb=NR'$, where R is either an alkyl or an aryl group.

Organoantimony Compounds with Five Sb–C Bonds

A number of pentaalkyl- and pentaalkenylantimony compounds have been prepared from tetraalkyl- or tetraalkenylstibonium halides and alkyl or alkenyllithium or Grignard reagents.

HEALTH AND SAFETY FACTORS

OSHA has a TWA standard on a weight of Sb basis of 0.5 mg/m^3 for antimony in addition to a standard TWA of 2.5 mg/m^3 for fluoride. Most antimony compounds are poisonous by ingestion, inhalation, and intraperitoneal routes locally antimony compounds irritate the skin and mucous membranes. NIOSH has issued a criteria document on occupational exposure to inorganic fluorides. Antimony pentafluoride is considered by the EPA to be an extremely hazardous substance and releases of 0.45 kg or more reportable quantity (RQ) must be reported. Antimony trifluoride is on the CERCLA list and releasing of 450 kg or more RQ must be reported.

ENVIRONMENTAL IMPACT

Antimony is a common air pollutant that occurs at an average concentration of 0.001 µg/m^3. Antimony is released into the environment from burning fossil fuels and from industry. In the air, antimony is rapidly attached to suspended particles and thought to stay in the air for 30 to 40 days. Antimony is found at low levels in some lakes, rivers, and streams, and may accumulate in sediments. Although antimony concentrations have been found in some freshwater and marine invertebrates, it does not biomagnify in the environment. The impact of antimony and antimony compounds on the environment has not been extensively studied to date.

J. F. Carlin, Jr., *Minerals Yearbook*, U.S. Geological Survey, Reston, Va., 2001.

G. O. Doak and L. D. Freedman, *Organometallic Compounds of Arsenic, Antimony, and Bismuth*, John Wiley & Sons, Inc., New York, 1970.

R. J. Lewis, Sr., "*Sax's Dangerous Properties of Industrial Materials*, Vol. 2, John Wiley & Sons, Inc., New York, 2000, p. 280.

F. G. Mann, *The Heterocyclic Derivatives of Phosphorus, Arsenic, Antimony and Bismuth*, 2nd ed., Wiley-Interscience, New York, 1970.

Leon D. Freedman
G. O. Doak
G. Gilbert Long
North Carolina State University
Tariq Mahmood
Charles B. Lindhal
Elf Atochem North America, Inc.

ANTIOBESITY DRUGS

MEDICAL ASPECTS OF OBESITY

Obesity is an increasingly prevalent, complex disease with multiple etiologies and profound medical consequences. In addition to being a cosmetic problem, overweight and obesity are associated with an enhanced likelihood of developing chronic conditions including hypertension, hyperlipidemia, coronary heart disease, diabetes, cancer, gall bladder disease, and arthritis.

To define the obese state in a clinical setting, it is necessary to have a means of estimating the amount of adipose (fat) tissue relative to lean body mass. Large clinical studies typically employ measures of skin-fold thickness, waist/hip ratio, waist circumference, or more commonly, BMI= (weight in kg)/(height in m)^2] as a quantitative measure

Table 1. Weight Classification by BMI[a]

NHLBI[b] Terminology	BMI, kg/m^2	WHO[c] classification
underweight	<18.5	underweight
normal	18.5–24.9	normal range
overweight	25.0–29.9	preobese
obesity class 1	30.0–34.9	obese class 1
obesity class 2	35.0–39.9	obese class 2
obesity class 3	>40.0	obese class 3

[a]Reproduced with permission from Kuczmarski and co-workers.
[b]National Heart, Lung and Blood Insitute.
[c]World Health Organization.

of obesity. Commonly accepted classifications for stages of obesity based on BMI are summarized in Table 1.

In addition to the obvious role of the environment, there is a significant predisposing genetic component to obesity, as indicated in a number of twin and adoption studies. Although there are rare cases of extreme obesity that can be ascribed to a defect in a single gene, a considerable body of evidence supports a polygenic contribution to most forms of obesity. Intense efforts are under way to identify candidate genes, since it is estimated that genetic factors account for 70% of the variation in weight between individuals. Obesity is rare in many third world populations until the people become westernized, adopting energy-rich diets and sedentary lifestyles.

The balance between energy intake and energy expenditure is finely regulated in most people, and they maintain relatively consistent body weights for long periods despite variations in their day-to-day nutrient consumption. Although the mechanism is not understood at present, individuals may have a metabolic set point that provides a homeostatic drive toward weight maintenance. For many obese patients, this set point is inappropriately high and favors an increase in metabolic efficiency and restoration of body mass after weight loss. One attractive approach to treating obesity would be to find a way to reset this metabolic set point to a lower BMI.

TREATMENT OF OBESITY

Introduction

Obesity is a difficult condition to treat. Dietary restriction of caloric intake is the first line therapy and is optimally combined with an exercise program to promote loss of fat relative to lean body mass. Drug treatments that help to suppress appetite, increase energy expenditure, or decrease fat absorption are available and may be beneficially used in conjunction with a comprehensive weight loss program. The majority of formerly obese patients eventually regain their excess weight lost through diet, and thus a truly successful program must include long-term behavior modification.

Anorectics (Appetite Suppressants)

Appetite suppressants are widely used as an adjunct to dietary restriction, and sympathomimetic amines have traditionally been used for this purpose. The sympathetic or adrenergic nervous system operates in juxtaposition to the parasympathetic nervous system to maintain homeostasis in response to physical activity and physical or psychological stress.

Compounds structurally related to the endogenous sympathomimetic amines have classically been employed as appetite suppressants.

Lipid Adsorption Inhibitors

Orlistat. Dietary fat occurs largely in the form of triglycerides that must be hydrolyzed through the action of lipases, primarily pancreatic lipase, in the digestive tract prior to absorption. When maintenance or induction of weight loss is desired with a minimal impact on meal composition, an attractive approach to limiting caloric intake is to minimize the absorption of fat by inhibiting pancreatic lipase.

The drug acts locally in the gastrointestinal (GI) tract to reversibly inhibit lipases through attack of a lipase serine hydroxyl group on the γ-lactone carbonyl to give an inactivated acyl enzyme. Less than 2% of an oral dose is absorbed, and systemic side effects are virtually unknown. The side effects, which do occur, are a consequence of its mechanism of action and relate to the presence of excess fat in the GI tract.

FUTURE DEVELOPMENTS

As noted, there is a significant unmet medical need for better modalities for the management of overweight and obesity together with their consequent morbidity. The size of the patient population in developed countries, and the staggering direct and indirect costs, have prompted an intense effort on the part of academic as well as pharmaceutical company laboratories to understand the driving forces governing nutrient absorption, satiety, and energy utilization.

Centrally Acting Drugs

Concentrated in the arcuate nucleus in the hypothalamus, leptin receptors mediate both inhibitory and stimulatory signals, projecting into the lateral hypothalamic/perifornical areas and paraventricular nucleus, respectively, to inhibit food intake. Clinical trials of leptin itself as an anorectic agent were disappointing; however, investigation of neurons expressing leptin receptors has led to the identification of several potential drug targets. All of which are peptidic neurotransmitters that interact with specific G-protein-coupled receptors (GPCR) to either stimulate (NPY, AGRP, orexins, MCH) or inhibit (α-MSH) food intake.

Peripherally Acting Drugs

?tlsb=-0.03w>Drugs that stimulate or block peripheral adrenergic receptors have been in widespread use for some time, eg, as antihypertensives, antianginal agents,

and bronchodilators. Detailed investigations with some of these led to the identification of the β_3-adrenergic receptor, which mediates catecholamine-induced lipolysis in brown adipose tissue leading to increases in thermogenesis. Compounds tested clinically to date were insufficiently selective for the human β_3-adrenergic receptor, but newer, more promising compounds are being evaluated.

Protein tyrosine phosphatase-1b (PTP-1b) has become an exciting target for drug discovery based on a paper describing a mouse model in which the gene coding for PTP-1b was knocked out. These findings have prompted an intense search within the pharmaceutical industry for a small-molecule PTP-1b inhibitor that might serve to effectively treat both type 2 diabetes and obesity.

Locally released gut neuropeptides such as cholecystokinin (CCK) and glucagon-like peptide 1 (GLP-1) regulate food intake, probably through stimulation of vagal afferent fibers.

Finally, compounds that inhibit fat absorption or utilization may be of interest. In addition to gut lipases, targets of such drugs could include the intestinal fatty acid transport protein FATP4; the enzyme acyl coenzyme (Co-A): diacylglycerol transferase (DGAT), which catalyzes a key step in triglyceride synthesis; and the enzyme complex fatty acid synthetase, which is responsible for the conversion of malonyl Co-A to palmitoyl coenzyme-A.

SUMMARY

The available appetite suppressants based on stimulation of the sympathetic nervous system have fallen out of favor for the treatment of morbid obesity with the exception of sibutramine. The introduction of the fat absorption inhibitor orlistat has offered a novel approach, free of the side effects and tolerance development associated with present centrally acting agents. Although drug treatment is not likely to replace diet and behavior modification for the control of obesity, the profound medical need combined with intensive research on both peripheral and central pathways involved in regulation of nutrient absorption, meal size, and energy utilization is certain to lead to new opportunities for the control and maintenance of desirable body weight. Ultimately, the measure of success in the overall treatment of obesity will be a reduction in the associated morbidity.

B. B. Hoffman and R. J. Lefkowitz in J. G. Hardman, L. E. Limbird, P. B. Molinoff, R. W. Ruddon, and A. G. Gilman, ed, *Goodman & Gilman The Pharmacological Basis of Theraputics*, 9th ed., McGraw-Hill, New York, 1996.

P. G. Kopelman, *Nature (London)* **404**, 635–643 (2000).

R. J. Kuczmarski, M. D. Carroll, K. M. Flegal, and R. P. Troiano, *Obesity Res.* **5**, 542–548 (1997).

JEFFERSON W. TILLEY
Hoffmann-La Roche Inc.

ANTIOXIDANTS, POLYMERS

Antioxidants are used to retard the reaction of organic materials, such as synthetic polymers, with atmospheric oxygen. Such reaction can cause degradation of the mechanical, aesthetic, and electrical properties of polymers; loss of flavor and development of rancidity in foods; and an increase in the viscosity, acidity, and formation of insolubles in lubricants. The need for antioxidants depends upon the chemical composition of the substrate and the conditions of exposure. Relatively high concentrations of antioxidants are used to stabilize polymers such as natural rubber and polyunsaturated oils. Saturated polymers have greater oxidative stability and require relatively low concentrations of stabilizers.

MECHANISM OF UNINHIBITED AUTOXIDATION

The mechanism by which an organic material (RH) undergoes autoxidation involves a free-radical chain reaction is shown below:

Initiation

$$RH \longrightarrow \text{free radicals, eg, } R\cdot, ROO\cdot, RO\cdot, HO\cdot \quad (1)$$
$$ROOH \longrightarrow RO\cdot + OH\cdot \quad (2)$$
$$2\,ROOH \longrightarrow RO\cdot + ROO\cdot + H_2O \quad (3)$$
$$ROOR \longrightarrow 2\,RO\cdot \quad (4)$$

Propagation

$$R\cdot + O_2 \longrightarrow ROO\cdot \quad (5)$$
$$ROO\cdot + RH \longrightarrow ROOH + R\cdot \quad (6)$$

Termination

$$2\,R\cdot \longrightarrow R\!-\!R \quad (7)$$
$$ROO\cdot + R\cdot \longrightarrow ROOR \quad (8)$$
$$2\,ROO\cdot \longrightarrow \text{nonradical products} \quad (9)$$

Initiation

Free-radical initiators are produced by several processes. The high temperatures and shearing stresses required for compounding, extrusion, and molding of polymeric materials can produce alkyl radicals by homolytic chain cleavage. Oxidatively sensitive substrates can react directly with oxygen, particularly at elevated temperatures, to yield radicals.

Propagation

Propagation reactions (eqs. 5 and 6) can be repeated many times before termination by conversion of an alkyl or peroxy radical to a nonradical species. Homolytic decomposition of hydroperoxides produced by propagation reactions increases the rate of initiation by the production of radicals.

RADICAL SCAVENGERS

Hydrogen-donating antioxidants (AH), such as hindered phenols and secondary aromatic amines, inhibit oxidation by competing with the organic substrate (RH) for peroxy radicals. This shortens the kinetic chain length of the propagation reactions.

$$ROO\cdot + AH \xrightarrow{k_{17}} ROOH + A\cdot \quad (10)$$
$$ROO\cdot + RH \xrightarrow{k_{6}} ROOH + R\cdot \quad (11)$$

Because k_{17} is $\gg k_6$, hydrogen-donating antioxidants generally can be used at low concentrations. The usual concentrations in saturated thermoplastic polymers range from 0.01 to 0.05%, based on the weight of the polymer. Higher concentrations, ie, ~ 0.5–2%, are required in substrates that are highly sensitive to oxidation, such as unsaturated elastomers and acrylonitrile–butadiene–styrene (ABS).

Hindered Phenols

Even a simple monophenolic antioxidant has a complex chemistry in an autooxidizing substrate. Stilbenequinones absorb visible light and cause some discoloration. However, upon oxidation phenolic antioxidants impart much less color than aromatic amine antioxidants and are considered to be nondiscoloring and nonstaining.

The effect substitution on the phenolic ring has on activity has been the subject of several studies. Hindering the phenolic hydroxyl group with at least one bulky alkyl group in the ortho position appears necessary for high antioxidant activity. Nearly all commercial antioxidants are hindered in this manner.

Aromatic Amines

Antioxidants derived from *p*-phenylenediamine and diphenylamine are highly effective peroxy radical scavengers. They are more effective than phenolic antioxidants for the stabilization of easily oxidized organic materials, such as unsaturated elastomers. Because of their intense staining effect, derivatives of *p*-phenylenediamine are used primarily for elastomers containing carbon black (qv). *N,N′*-Disubstituted-*p*-phenylenediamines are used in greater quantities than other classes of antioxidants.

Hindered Amines

Hindered amines are extremely effective in protecting polyolefins and other polymeric materials against photodegradation. They usually are classified as light stabilizers rather than antioxidants.

Most of the commercial hindered-amine light stabilizers (HALS) are derivatives of 2,2,6,6-tetramethylpiperidine. These stabilizers funtion as light-stable antioxidants to project polymers.

Hydroxylamines

A relatively new stabilizer chemistry, commercially introduced in 1996, based on the hydroxylamine functionality, can serve as a very powerful hydrogen atom donor and free-radical scavenger. This hydroxylamine chemistry is extremely powerful on an equivalent weight basis in comparison to conventional phenolic antioxidants and phosphite melt-processing stabilizers. In terms of its scavenger capability, however, it is more effective during melt processing of the polymer than during long term thermal stability (ie, below the melting point of the polymer).

Benzofuranones

In 1997, a fundamentally new chemistry was introduced, that not only inhibits the autoxidation cycle but attempts to shut it down as soon as it starts. The exceptional stabilizer activity of the class of benzofuranones is due to the ready formation of a stable benzofuranyl radical by donation of the weakly bonded benzylic hydrogen atom.

Benzofuranones are similar to hydroxylamines in that on an equivalent weight basis, they are more powerful than conventional phenolic antioxidants or phosphite-based melt-processing stabilizers. Once again, note that even though benzofuranones are capable of providing free-radical scavenging chemistry similar to phenolic anti-oxidants, the effective temperature domain is typically above the melting point of the polymer; eg, during melt processing (similar to hydroxylamines).

PEROXIDE DECOMPOSERS

Thermally induced homolytic decomposition of peroxides and hydroperoxides to free radicals (eqs. 2–4) increases the rate of oxidation. Decomposition to nonradical species removes hydroperoxides as potential sources of oxidation initiators. Most peroxide decomposers are derived from divalent sulfur and trivalent phosphorus.

EFFECT OF TEMPERATURE

As mentioned above, certain types of antioxidants provide free-radical scavenging capability; albeit over different temperature ranges. Figure 1 illustrates this in a general fashion for representative classes of stabilizers, over the temperature range of 0–300°C.

As a representative example, hindered phenols are capable of providing long term thermal stability below the melting point of the polymer, as well as melt-processing stability above the melting point of the polymer. As such most (if not all) hindered phenols are useful across the entire temperature range.

Thiosynergists, in combination with a hindered phenol, contribute to long term thermal stability, primarily below the melting point of the polymer. In extreme cases, where peroxides have built up in the polymer, thiosynergist can be shown to have a positive impact during melt processing. This finding, however, is not the norm, and this type of melt processing efficacy has been left out of the figure.

Hindered amines, commonly thought of as being useful for uv stabilization, are also useful for long term thermal stability below the melting point of the polymer. This effectiveness is due to the fact that hindered amines work by a free-radical scavenging mechanism, but they are virtually ineffective at temperatures >150°C. Therefore, hindered amines, when used as a reagent for providing long term thermal stability, should always be used in combination with an effective melt-processing stabilizer.

Phosphites, hydroxylamines, and lactones, are most effective during melt processing; either through free-radical scavenging or hydroperoxide decomposition.

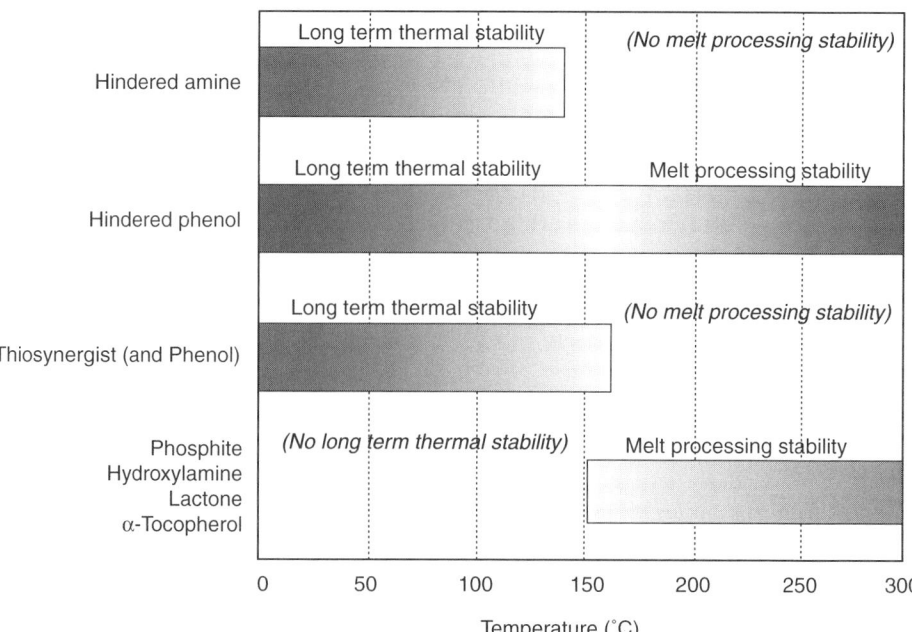

Figure 1. General representation of effective temperature ranges for selected types of antioxidants.

They are not effective as long term thermal stabilizers. These type of stabilizers also help with long term thermal stability.

One anomaly that should be pointed out are the hindered phenols based on tocopherols. Even though tocopherols, such as Vitamin E, fall into the general class of hindered phenols, they behave more as melt processing stabilizers, and less as reagents for providing long term thermal stability, at least with regard to polymer stabilization.

ANTIOXIDANT BLENDS

In practical application, it is reasonable to use more than one type of antioxidant in order to meet the requirements of the application, such as melt-processing stability as well as long term thermal stability. The most common combination of stabilizers used, particularly in polyolefins, are blends of a phenolic antioxidant and a phosphite melt-processing stabilizer. Another common combination is a blend of a phenolic antioxidant and a thioester; especially for applications that require long term thermal stability. These common phenol-based blends have been used successfully in many different types of end-use applications. The combination of phenolic, phosphite, and lactone moieties represents an extremely efficient stabilization system since all three components provide a specific function.

For color critical applications requiring "phenol free" stabilization, synergistic mixtures of hindered amines (for both uv stability as well as long term thermal stability) with a hydroxylamine or benzofuranone (for melt processing), with or without a phosphite, can be used to avoid discoloration typically associated with the overoxidation of the phenolic antioxidant.

APPLICATION OF ANTIOXIDANTS IN POLYMERS

Nearly all polymeric materials require the addition of antioxidants to retain physical properties and to ensure an adequate service life. The selection of an antioxidant system is dependent upon the polymer and the anticipated end use.

Polyolefins

Low concentrations of stabilizers ($< 0.01\%$) are often added to polyethylene and polypropylene after synthesis and prior to isolation to retard oxidation of the polymer before they are exposed to sources of oxygen or air. Higher concentrations are added downstream during the conversion of the reactor product to a pelletized form. The antioxidant components and concentrations are selected by the manufacturer to yield general purpose grades, or can be optimized to meet a specific end-use application.

These polymers can be subjected to temperatures as high as $300°C$, during cast film extrusion and thin wall injection molding. In these type of demanding applications, processing stabilizers are used to decrease both the change in viscosity (molecular weight) of the polymer

melt and the development of color. A phosphite, such as tris(2,4-di-*tert*-butylphenyl)phosphite or bis(2,4-di-*tert*-butylphenyl)pentaerythritol diphosphite, in combination with a phenolic antioxidant such as octadecyl 3,5-di-*tert*-butyl-4-hydroxyhydrocinnamate may be used. Concentrations usually range from 0.01 to 0.5%, depending on the polymer and the severity of the processing conditions. For long term exposure, a persistent antioxidant like tetrakis[methylene(3,5-di-*tert*-butyl-4-hydroxyhydrocinnamate)]methane, at a concentration of 0.1–0.5%, may be added to the base stabilization package.

Polyamides

Due to their excellent mechanical properties at high temperatures, polyamides, particularly mineral and glass-filled grades, are finding increased usage in demanding applications such as automotive under-the-hood application.

The traditional stabilization system for aliphatic polyamides are copper salts. Typical systems are based on low levels of copper (< 50 ppm) and iodide or bromide salts. The mechanism of stabilization is not well understood but may be due to hydroperoxide decomposition initiated by metal ions. These systems are effective in polyamides whereas in other polymers, such as polylefins, small amounts of oxidized copper can often as prodegradants. Good dispersion of the copper is critical to good performance.

Styrenics

Unmodified styrenics such as crystal polystyrene are relatively stable and under most end-use conditions it is not necessary to add antioxidants. Low levels (0.1%) of a hindered phenolic antioxidant are added to protect the polymer during repeated processing of scrap.

Polyesters

Poly(ethylene terephthalate) (PET) requires little to no antioxidant during thermal processing. In some cases, phosphites are added to improve the color of regrind. the other hand, poly(butylene terephthalate) (PBT), because of its higher hydrocarbon content, is more susceptible to oxidative degradation than PET.

Polycarbonate

Polycarbonate (PC) is susceptible to photooxidation, and antioxidants are necessary to maintain the low color and high transparency critical to its end-use applications.

Polyacetal

Polyacetals thermal decompose by an acid-catalyzed depolymerization process starting at the chain ends. The polymer structure is stabilized by end capping of the polymer and introducing comonomers to interrupt the unzipping.

Polyurethanes

The oxidative stability of polyurethanes (PURs) is highly dependent on the chemical nature of both the polyol component and the isocyanate. Thus, PUR derived from a polyester polyol is typically more stable than one derived from a polyether polyol.

Elastomers

Polyunsaturated elastomers are sensitive to oxidation. Stabilizers are added to the elastomers prior to vulcanization to protect the rubber during drying and storage. Nonstaining antioxidants such as butylated hydroxytoluene, 2,4-bis(octylthiomethyl)-6-methylphenol, 4,4′-bis(α,α-dimethylbenzyl)diphenylamine, or a phosphite such as tris(nonylphenyl)phosphite may be used in concentrations ranging from 0.01 to 0.5%. Staining antioxidants such as N-isopropyl-N′-phenyl-p-phenylenediamine are preferred for the manufacture of tires. These potent antioxidants also have antiozonant activity and retard stress cracking of the vulcanized rubber.

Poly(vinyl chloride)

While reasonably stable with respect to oxidative degradation, poly(vinyl chloride) (PVC) is very susceptible to thermal degradation. Protection of PVC from thermal degradation leading primarily to dehydrohalogenation reactions is out of the scope of this article.

Fuels and Lubricants

Gasoline and jet engine fuels contain unsaturated compounds that oxidize on storage, darken, and form gums and deposits. Radical scavengers such as 2,4-dimethyl-6-tert-butylphenol, 2,6-di-tert-butyl-p-cresol, 2,6-di-tert-butylphenol, and alkylated paraphenylene diamines are used in concentrations of ~5–10 ppm as stabilizers.

Lubricants for gasoline engines are required to withstand harsh conditions. The thin films of lubricants coating piston walls are exposed to heat, oxygen, oxides of nitrogen, and shearing stress. Relatively high concentrations of primary antioxidants and synergists are used to stabilize lubricating oils. Up to 1% of a mixture of hindered phenols, of the type used for gasoline, and secondary aromatic amines, such as alkylated diphenylamine and alkylated phenyl-α-napththylamine, are used as the primary antioxidants. About 1% of a synergist, zinc dialkyldithiophosphonate, is added as a peroxide decomposer. Zinc dialkyldithiophosphates are cost effective multifunctional additives.

ANCILLARY PROPERTIES

In reality, there is more to antioxidants than providing stability to the polymer by quenching free radicals and decomposing hydroperoxides. Other key issues besides rates of reactivity and efficiency include performance parameters such as volatility, compatibility, color stability, physical form, taste or odor, regulatory issues associated with food contact applications, and polymer performance versus cost.

HEALTH AND SAFETY FACTORS

Safety is assessed by subjecting the antioxidant to a series of animal toxicity tests, eg, oral, inhalation, eye, and skin tests. Mutagenicity tests are also carried out to determine possible or potential carcinogenicity. Granulated and liquid forms of antioxidants are receiving greater acceptance to minimize the inhalation of dust and to improve flow characteristics.

A number of antioxidants have been regulated by the U.S. Food and Drug Administration as indirect additives for polymers used in food contact applications (primarily food packaging). Acceptance is determined by subchronic or chronic toxicity in more than one animal species and by the concentration expected in the diet, based on the amount of the additive extracted from the polymer by solvents that simulate food in their extractive effects.

J. Galbo, "Light Stabilizers" in Encyclopedia of Polymeric Materials, CRC Press, 1996, pp. 3616–3623.

Plastics Additives, Chemical Marketing Reporter, 7 (June 19, 2000) search date March 14, 2001.

Pospisil and P. P. Klemchuk, eds., Oxidation Inhibition in Organic Materials, Vols. I and II, CRC Press Inc., Boca Raton, Fla., 1990.

K. Schwarzenhach, in H. Zweifel, ed. Plastics Additives Handbook, Hanser, Munich, 2000, pp. 1–137.

RICHARD THOMAS
Ciba Specialty Chemicals

MARTIN DEXTER
Consultant

R. E. KING III
Ciba Specialty Chemicals

ANTIVIRAL AGENTS

In the 10 years since the last publication of an article on antiviral agents (1), research on this topic has taken on an explosive course, largely because of the growing threat of the epidemic of AIDS (acquired immunodeficiency syndrome) in the western hemisphere, which not only intensified research on HIV (human immunodeficiency virus), but also on other opportunistic viral infections associated with HIV, such as HCV (hepatitis C virus), HBV (hepatitis B virus), and CMV (cytomegalovirus). Thanks to many rapid advances made from chemical, biochemical, as well as molecular biological fronts that led to effective anti-HIV therapies including the most successful combination drug regimen, the threat from HIV infection has been somewhat downplayed in recent years from the edict of a "death sentence" to that of a "manageable illness". On the other hand, a relatively less known virus such as the West Nile virus (WNV), or the less heeded viruses such as HCV and HBV, have suddenly taken up the center stage. The focus here is on four major viruses, including HIV, HBV, HCV, and WNV. Also, since nucleoside analogues have played a major role as therapeutics in combating these viruses, a special emphasis has been placed on this class of drugs.

CLASSIFICATION OF VIRUSES

In the early twentieth century, viruses were classified based on the hosts they infected: (a) plant viruses, (b) animal viruses, and (c) bacteriophages. The present-day broad classification of viruses is based on the genetic material they contain: DNA or RNA viruses. They may contain single-stranded DNA (parvoviruses), double-stranded DNA (herpesviruses), single-stranded RNA (poliovirus), or double-stranded RNA (reoviruses). The RNA viruses are unique in that they are the only living organisms that use RNA to store their genetic information. All other reproducing forms of life employ DNA. The more subtle classification of viruses, however, would include their hosts, chemical composition (including nucleic acid, protein, presence or absence of lipid envelope), shape, size, and symmetry. The major *animal viruses* can thus be subdivided into 18 categories: (a) herpesviruses, (b) papovaviruses, (c) adenoviruses, (d) poxviruses, (e) hepadnaviruses, (f) retroviruses, (g) orthomyxoviruses, (h) picornaviruses, (i) togaviruses, (j) rhabdoviruses, (k) paramyxoviruses, (l) reoviruses, (m) parvoviruses, (n) arenaviruses, (o) bunyaviruses, (p) filoviruses, and (q) coronaviruses.

THE GENERAL PROCESS OF VIRAL INFECTION AND THE AVAILABLE REMEDIES

In order to discover site- or process-specific antiviral agents, it is important to understand the specific biochemical processes that occur during viral infection. In all, there are seven stages in a typical viral infection process: (a) *adsorption*: The attachment of the virus to specific receptors on the cell surface; (b) *penetration*: The viral entrance into the cell by penetration through plasma membrane; (c) *uncoating*: The release of viral nucleic acid from the covering proteins; (d) *transcription*: The production of viral mRNA from the viral genome; (e) *translation*: The synthesis of viral proteins, including coat proteins and enzymes necessary for viral replication, as well as replication of viral nucleic acid (ie, the parental genome or complimentary strand); (f) *virion assembly*: The assembly of individual components of the viron (nucleic acid and structural proteins synthesized in stage (e), and transportation to the site of nucleocapsid assembly, followed by autocatalytic assembly; (g) *release*: For viruses with icosahedral symmetry that do not have an envelope, this stage comes after disintegration of host cell as a result of the killing action of the infecting virus; for enveloped viruses, the assembled nucleocapsids move toward the modified membrane areas where the synthesized viral matrix protein replaces the cellular membrane proteins, and then nucleocapsids bud through the modified membrane, wrapping themselves into a portion of membrane in the process.

The preferred approach to combat viral diseases is the prevention of infection by active immunization. There are a number of successful vaccines for prophylaxis of some viruses such as polio, mumps, measles, influenza, encephalitis, hepatitis, and smallpox. On the other hand, there has been less success in the prevention of viruses such as the HIV, HSV, and RSV.

Two major virus-specific processes are normally targeted in order to develop selective antiviral agents: (a) early events, including adsorption, penetration and uncoating, and (b) later synthetic events that concern intracellular replication of the virus. Other examples of antiviral agents targeted at the early events of viral activity include DIQA (3,4-dihydro-1-isoquinolineacetamide HCl), which has a broad-spectrum antiviral activity against lethal influenza A virus, echovirus, Columbia SK virus, herpes simplex virus, and rhino virus.

There exists a much larger pool of synthetic drugs that target the later events in a virus life cycle as compared to only a few synthetic or natural products that target the early events. These events are known to be virus specific. The viruses synthesize and utilize specific enzymes and proteins, and more importantly, the replication of viral genetic codes is also virus specific. The specificity in the synthesis of viral DNA or RNA is conferred by the virus-specific enzymes such as kinases, helicases, polymerases, transcriptases, reductases, etc. Viruses are more prone to mutations as compared with other microorganisms. The mutation rate of a virus is much higher than that of its host cell. Its mutants possess an excellent chance to be accommodated in the new host cell and escape from the host immune responses. On the other hand, high mutation rate means less selectivity toward substrates for the enzymes involved in DNA/RNA replication process. Once the potential drug candidate (an unnatural nucleotide analogue, for example) enters the catalytic site, it may disrupt or terminate the activity of enzymes. The unnatural nucleotides can be incorporated into DNA double helix, distort the DNA structure, and utimately stop the virus replication.

There are also many compounds that fall outside the nucleoside family. A distinct example comes from the ever-growing fight against AIDS. In the late stage of a virus life cycle, the virus-specific processing of certain viral proteins by viral or cellular proteases is crucial. It was revealed that HIV expresses three genes as precursor polypepteins. Two of these gene products (designated as P55gag and p160 gag-pol proteins) undergo cleavage at several sites by a virally encoded protease to form structural proteins and enzymes required for replication. This fact has stimulated the research efforts to find safe and effective inhibitors for the viral protease. It is believed that the inhibitors should resemble a small portion of the substrate polyprotein structure but contain an isosteric replacement for the scissible (hydrolyzable) peptide bond that mimics the transition state for the hydrolysis of that bond, which is stable against cleavage. This has led to the discovery of several successful clinical candidates for HIV infection.

NUCLEOSIDE ANALOGUES AS ANTIVIRAL AGENTS

Initially, the term "nucleoside" was referred to the purine and pyrimidine N-glycosides derived from nucleic acid. However, after the discovery of pseudouridine (5-β-D-ribofuranosyluracil), a natural constituent of tRNA, it became a common practice to consider even those molecules whose heterocyclic rings are connected to the sugar moieties at the anomeric junctions through carbon–carbon single bonds. Such compounds are classified as *C-nucleosides*. Another interesting class of nucleosides, called *L-nucleosides*, are lately emerging as powerful antiviral compounds. The sugar parts of these nucleoside analogues possess the L- instead of the natural D-configuration. Furthermore, considering the possibility of existence of the carbon linking the base to the heterocycle into α- or β- anomeric form (β being the natural form), there exists an additional category of *α-nucleosides*. These different classes of nucleosides are contrasted below, using uridine as an example.

D-Uridine

L-Uridine

α-Uridine

C-Uridine
(pseudouridine)

In contrast to the limited number of natural nucleosides, numerous synthetic nucleosides are now available for treating viral infections. This is because of the unlimited possibilities for modifications both at the carbohydrate and the base sites.

THE VIRUSES OF CURRENT HEALTH CONCERN AND THE RELATED ANTIVIRAL THERAPY

As mentioned earlier, the following four viruses are currently of prime health concern worldwide: HIV, HBV, HCV and WNV.

Human Immunodeficiency Virus (HIV)

Perhaps no other virus in recent history has stirred more global panic and paranoia than HIV, an etiological agent causing the acquired immunodeficiency syndrome (AIDS). Despite intense efforts from several research fronts including chemistry, biochemistry, biology, and biotechnology, and not to mention epidemiology and prevention measures, the fight to conquer HIV altogether still remains largely elusive. As proven techniques of viral attack seem inadequate against HIV, and chances for a suitable vaccine continue to be disappointing, the current research trend is to focus on the complete viral life cycle and the replication process for new targets. The ultimate success may lie in the power of modern molecular biology to explore every aspect of the HIV life cycle and every response of the human body toward viral invasion. The tools of biotechnology have greatly aided in sequencing the viral genome as well as the proteins that are associated with it. So, it is important to review the current status of knowledge on the viral structure and its life cycle before delving into what is being targeted for antiviral therapy.

As classified earlier, HIV is a retrovirus consisting of two copies of a single-stranded RNA genome and a few replicative and accessory proteins within the boundaries of a lipoprotein shell (see Figure 1), known as the viral envelope. Embedded in the viral envelope is a complex protein known as *env*, which consists of an outer protruding cap glycoprotein (gp) 120, and a stem gp41. Within the viral envelope is an HIV protein called p17 (matrix), and within this is the viral core or capsid, which is made of another viral protein p24 (core antigen). The major elements contained within the viral core are two single strands of HIV–RNA, a protein p7 (nucleocapsid), and three enzyme proteins, p66 (reverse transcriptase), p11 (protease), and p31 (integrase).

In addition to HIV *reverse transcriptase* and *integrase*, the two key enzymes involved in the viral replication process as described above, a third enzyme called HIV *protease* is also a viable target for antiviral therapy. After replication within a host cell, when new viral particles are ready to break off to infect other cells, protease plays a vital role in cutting longer protein strands into smaller parts needed to assemble a mature virus.

HIV protease plays a critical function in the HIV life cycle. Consequently, the detailed structural analysis of HIV protease has led to the discovery of protease inhibitors, one of the important components in the *highly active antiretroviral therapy*, commonly referred to as HAART Therapy that consists of multiple drug regimen aimed at different targets in the viral life cycle. The "cocktail" regimen normally includes a protease inhibitor along with two HIV RT inhibitors, one a nucleoside and the other a nonnucleoside.

Antiviral Therapy for HIV Infections

Vaccines. Clinical trials of candidate HIV vaccines have so far been only informative. In the absence of validated correlates of immune protection, larger trials of the most promising candidates will be needed. Furthermore, as promising candidates advance to efficacy trials, there does appear to be room for improvement. There is at least as much if not more known about the HIV genome than other pathogens for which vaccines have successfully been made. Advances in genomics and micro-array technologies will likely have multiple applications in the field of HIV vaccine development.

Chemotherapy. Currently, there are a total of 16 drugs that have been approved by the U.S. Food and Drug Administration (FDA) for the treatment of AIDS. Seven out of the 16 are nucleoside-based reverse transcriptase inhibitors (NRTI), three are nonnucleoside reverse

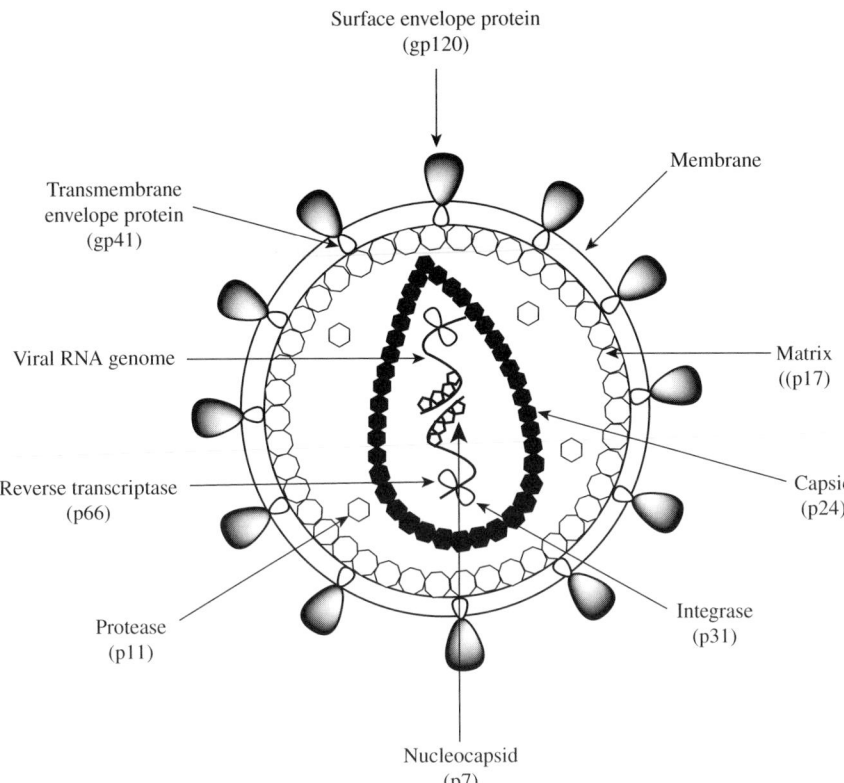

Surface envelope protein
(gp120)

Transmembrane
envelope protein
(gp41)

Membrane

Viral RNA genome

Matrix
((p17)

Reverse transcriptase
(p66)

Capsid
(p24)

Protease
(p11)

Integrase
(p31)

Nucleocapsid
(p7)

Figure 1. The molecular structure of HIV.

transcriptase inhibitors (NNRTI), and 6 are protease inhibitors.

Two other classes of drugs against HIV that are currently under clinical trials are *Integrase* and *Fusion* Inhibitors. HIV integrase is the third key enzyme in HIV replication besides protease and reverse transcriptase. As described earlier, the integration of provirus into the host genome is catalyzed by virally encoded integrase which has multiple functions. First, it acts as an exonuclease to cut the complementary viral DNA produced by HIV–RT to the appropriate size. Second, it serves as an endonuclease to cut the host DNA so as to facilitate insertion of the provirus. Finally, it acts as a ligase to fuse the host and viral DNAs into a seamless whole. Currently, a few drugs are being developed for inhibition of this integration step. *HIV Fusion Inhibitors* are a new class of drugs that bind to the viral protein gp120 and prevent HIV from infecting host cells. The virus is basically frozen on the surface of the cell preventing it from entering the cell, and therefore cannot propagate in an HIV-infected person.

In summary, despite enormous progress made in understanding its life cycle as well as its potentially viable targets for the development of antiviral therapies, HIV remains an elusive virus even in the face of the successful HAART therapy. The major obstacle in conquering HIV through therapy concerns the high level of viral mutagenicity and the consequent drug resistance. Its eerie ability to integrate into the host to kill the very cells that are normally mobilized to confront the invader makes HIV a formidable virus. Neither the efficacious long-term

therapies nor the uniformly effective vaccines against HIV are yet close in sight, but the international research to fight the virus continues unabated.

Hepatitis B Virus (HBV)

Mother to infant transmission accounts for most cases of HBV in the undeveloped countries, whereas unsafe sex and body fluid contacts are the major forms of transmission in the developed countries. The viral transmission occurs primarily through blood and/or sexual contact, though other methods of transmission have also been suggested. Transmission is most efficient via percutaneous mode, whereas sexual transmission is somewhat inefficient. The virus is primarily found in the blood of infected individuals. HBV has also been detected in other body fluids including urine, saliva, nasopharyngeal fluids, semen, and menstrual fluids.

HBV is responsible for both acute and chronic hepatitis. Individuals infected with acute HBV show no apparent clinical signs of the disease, but at the end of the incubation period, flu-like symptoms, such as fever, fatigue, and general discomfort, and in some cases jaundice, will occur. About 2–10% of the adult acute HBV carriers will become chronic carriers of the disease, but in infants this percentage is >90% via neonatal exposure. In the last few decades, the correlation between HBV and the development of HCC has been well established, although the mechanism by which HBV transforms hepatocytes still remains elusive. Before HBV can transform a cell, the virus must first infect it. However, the mechanism through which HBV enters hepatocytes has not been resolved despite further understanding of the viral proteins involved. Vaccines are available against HBV, but they may not be 100% effective against all variants of HBV. Furthermore, there is no cure for individuals already infected.

Antiviral Therapy for HBV Infections

Vaccine. HBV vaccine is the first successful *recombinant vaccine* against a human infectious disease, in particular, against a mucosal virus. Although the vaccine can help prevent the spread of HBV, it is not useful for those 350 million people who have already been chronically infected with the virus.

Anti-HBV Therapy. (*a*) *Interferons:* Interferons are a family of proteins—α, β, ω, and γ—that are induced in response to viral infections or double-stranded RNA. Interferon α is the most effective against HBV. (*b*) *Nucleoside Analogues*: Nucleoside analogues are currently the most intensely studied anti-HBV agents. Some of the leading candidates include 3TC (lamivudine), BMS-200475, lobucavir PMEA (adefovir), adefovir dipivoxil, penciclovir (famciclovir) and L-FMAU. Lamivudine (3TC) is the first, effective, and reasonably well tolerated, oral treatment for chronic HBV infection, approved by FDA. The other approved drug for chronic HBV infection is adefovir

dipivoxil. Lamivudine is an inhibitor of RT, and is in clinical use in HIV-infected individuals. The use of lamivudine on patients with HBV infection clearly shows the development of resistance arising from base-pair substitutions at a specific locus called YMDD of the viral DNA polymerase, resulting in significant clinical problem. So, the future of HBV chemotherapy may reside in combination drug therapy with newer, less toxic nucleoside analogues, along with other classes of agents including immunomodulators. With the discovery of 3TC, a nucleoside with a sugar moiety in the unnatural L-configuration, and its potent dual activity against both HIV and HBV, the interest in L-nucleoside analogues has taken on an explosive course. A number of L-nucleosides are currently undergoing preclinical and clinical trials against both HIV and HBV as well as other viral infections, and are listed in Figure 2. Beneficial features of L-nucleosides include an antiviral activity comparable or sometimes greater than their natural D-counterparts, a more favorable toxicological profile, and more importantly, a greater metabolic stability due to their lower susceptibility to catabolic and hydrolytic enzymes.

The ring-expanded nucleosides **REN-1** and **REN-2** containing the imidazo[4,5-*e*][1,3]diazepine ring system, along with nucleoside **REN-3**, containing the imidazo[4,5-*e*][1,2,4]triazepine ring system (Figure 3), exhibit potent and selective *in vitro* anti-hepatitis B virus (anti-HBV) activity in cultured human hepatoblastoma 2.2.15 cells.

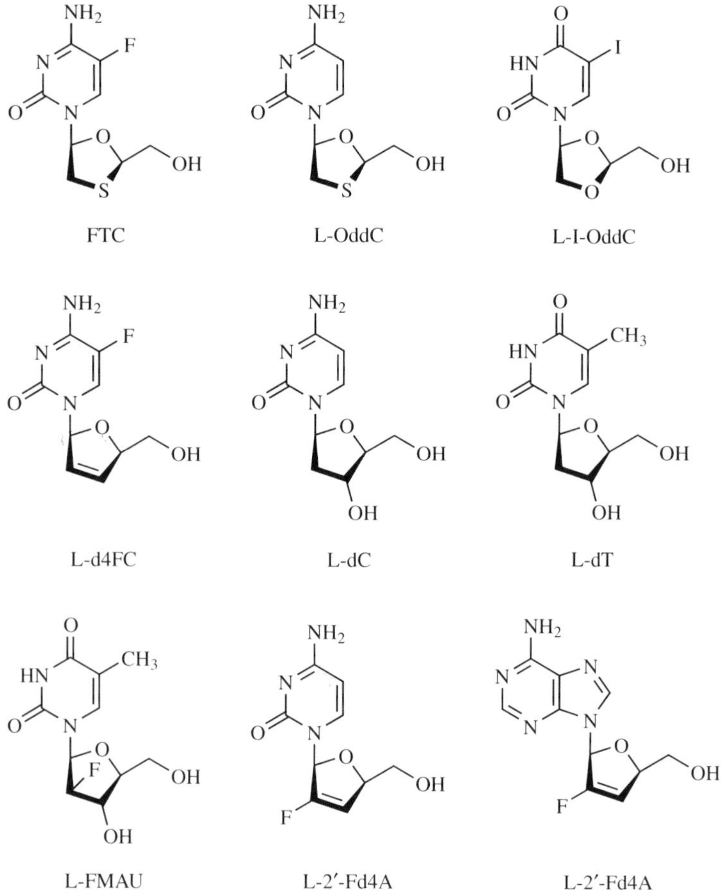

Figure 2. L-Nucleoside analogues that are currently undergoing preclinical and clinical trials against HIV, HBV, and other viral infections.

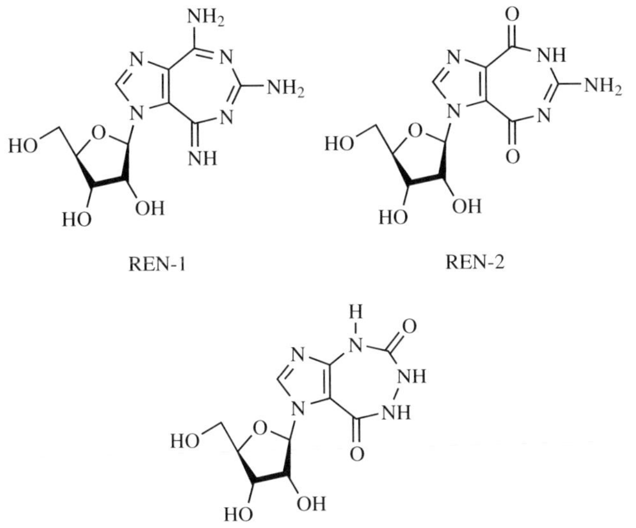

Figure 3. Ring-expanded nucleosides with potent *in vitro* anti-hepatitis B virus (anti-HBV) activity with little, if any, toxicity.

Hepatitis C Virus (HCV)

The hepatitis C virus (HCV) is one of the most dreadful infectious diseases of modern times. What makes it so dreadful is that most people do not even know that they have been infected with the virus as it can remain dormant for scores of years in the infected individual without revealing any signs or symptoms of the disease. Some estimates say the number of HCV-infected individuals may be four times the number of those infected with the AIDS virus, the main differences being that hepatitis C does not kill as quickly as AIDS. Until 1989, HCV was known by the name non-A, non-B hepatitis, when scientists at Chiron, Inc. succeeded in isolating portions of the HCV genome and conclusively demonstrated that the virus was indeed responsible for the noted pathogenicity that did not fit the category of either the A or B type hepatitis, and so classified it as type C. Subsequently, the complete genomes of various HCV isolates were cloned and sequenced by several research groups.

HCV is one of the major causes of chronic liver disease in the United States. It accounts for ~15% of acute viral hepatitis, 60–70% of chronic hepatitis, and up to 50% of cirrhosis, end-stage liver disease, and liver cancer. A conspicuous characteristic of hepatitis C is its tendency to cause chronic liver disease. At least 75% of patients with acute hepatitis C ultimately develop chronic infection, and most of these patients have accompanying chronic liver disease. But chronic hepatitis C varies greatly in its course and outcome. At one end of the spectrum are patients who have no signs or symptoms of liver disease and completely normal levels of serum liver enzymes. At the other end of the spectrum are patients with severe hepatitis C who have symptoms, HCV–RNA in serum, and elevated serum liver enzymes, and who ultimately develop cirrhosis and end-stage liver disease. In the middle of the spectrum are many patients who have few or no symptoms, mild-to-moderate elevations in liver enzymes, and an uncertain prognosis.

The virus is transmitted primarily by blood and blood products. Sexual transmission between monogamous couples is rare but HCV infection is more common in sexually promiscuous individuals. Perinatal transmission from mother to fetus or infant is also relatively low but possible. Many individuals infected with HCV have no obvious risk factors. Most of these persons have probably been inadvertently exposed to contaminated blood or blood products.

HCV is also considered an opportunistic infection in HIV-infected individuals, and about one quarter of them are also infected with HCV. Since HCV is transmitted primarily by large or repeated direct percutaneous (ie, passage through the skin by puncture) exposures to contaminated blood, coinfection with HCV is common (50–90%) especially among HIV-infected injection drug users. Also, HCV infection progresses more rapidly to liver damage in HIV-infected persons. HCV infection may also impact the course and management of HIV infection. Prevention of HCV infection for those not already infected and reducing chronic liver disease in those who are infected are important concerns for HIV-infected individuals and their health care providers.

HCV is a member of the family of RNA viruses called *Flaviviridae* to which also belongs the West Nile virus, another frightful virus. The viruses of the *flaviviridae* family are small, enveloped, spherical particles of 40–50 nm in diameter with single-stranded, positive sense RNA genomes. They are known to be the cause of severe encephalitic, hemorrhagic, hepatic, and febrile illnesses in humans.

Antiviral Therapy for HCV Infections

Vaccines. There is no vaccine for HCV and vaccines for hepatitis A and B do not provide immunity against hepatitis C. There are various strains of HCV and the virus undergoes mutations. Consequently, it will be difficult to develop a vaccine. Also, there is no effective immune globulin preparation. Furthermore, despite the discovery of HCV by molecular biological methods and the successful sequencing of the entire genome, a permissive cell culture system for propagating HCV has yet to be established.

Currently Available Treatments. *(a) Interferons:* All current treatment protocols for hepatitis C are based on the use of various preparations of interferon alpha, which are administered by intramuscular or subcutaneous injection. Interferon alpha is a naturally occurring glycoprotein that is secreted by cells in response to viral infections. It exerts its effects by binding to a membrane receptor. *(b) Nucleoside Analogue Ribavirin and Combination Therapy:* As mentioned earlier, ribavirin is a synthetic nucleoside containing a five-membered triazole ring, which has shown activity against a broad spectrum of viruses. In several studies, oral ribavirin was examined as a single agent for the treatment of adults with chronic hepatitis C. Although decreases in serum alanine transaminase (ALT) activities were seen with treatment, the overall results of these studies were discouraging as sustained-responses were rarely achieved. Because of its partial effectiveness, ribavirin was studied in subsequent trials in combination with interferon alpha. It was discovered that the addition of ribavirin to interferon alpha-2b is superior to interferon alpha-2b alone in the treatment of chronic hepatitis C, especially in achieving a sustained response in patients not previously treated with interferon. For eligible patients with chronic hepatitis C, a peginterferon alpha plus ribavirin is likely to be the best treatment option for the near future.

The treatment using interferon alpha with or without ribavirin is, nevertheless, associated with may side effects. During treatment, patients must be monitored carefully for side effects including flu-like symptoms, depression, rashes, other unusual reactions and abnormal blood counts.

Current Research Trends in Mechanism-Based HCV Inhibitors. The current intensive effort to discover novel therapies to treat HCV infection is aimed primarily at specific processes that are essential to HCV replication. Another class of compounds being developed as HCV inhibitors are ribozymes, which inhibit viral replication by cleavage of the target HCV genomic RNA.

West Nile Virus (WNV)

As noted above, both WNV and HCV belong to the same family of viruses called *Flaviviridae*. However, unlike HCV, WNV can be isolated from clinical specimens by tissue culture methods, and therefore, is used as a close mimic of HCV in experimental models.

Antiviral Therapy for WNV Infections

There are no currently approved drugs or vaccines for treating or preventing the disease in humans, although a vaccine has recently been approved for horses. Although ribavirin was initially reported to halt the viral replication, the need to use very high doses of the drug proved too toxic to be clinically useful. Furthermore, since WNV is still a rare virus affecting humans, there is not enough incentive for drug companies to develop anti-WNV drugs, but this scenario is likely to change as more and more cases of infection emerge.

CONCLUSION

The molecular structure, life cycle, mode of infection, and replication process of four major viruses of current health scare, including HIV, HBV, HCV, and WNV, have been discussed at length with cursory references to other human viruses. While no vaccine nor total cure is yet available against HIV infection, great strides have been made in antiviral therapy to enable classification of AIDS as a manageable illness from that of an "absolute death sentence" only a few years ago. With regard to HBV infection, a vaccine is now available, but the initial clinical trials with a number of nucleoside analogues as

anti-HBV agents were disappointing in light of severe toxicities associated with them. The ultimate success in treating HBV may lie in the combination drug therapy similar to the successful HAART therapy applied against HIV infection. Unfortunately, neither vaccines nor good drugs are currently available for treating HCV or WNV, the two viruses belonging to the family of *flaviviridae*, but a vast array of information is being rapidly accumulated on the structural biology and molecular virology of the two viruses to afford development of suitable antiviral therapies against them in the near future.

J. N. Delgado and W. A. Remers, *Textbook of Organic Medicinal and Pharmaceutical Chemistry*; 10th ed., Lippincott-Raven Publishers, Philadelphia, 1998, p. 329.

G. Gumina, Y. Chong, H. Choo, G.-Y. Song, and C. K. Chu, "L-Nucleosides: Antiviral Activity and Molecular Mechanism," in R. S. Hosmane, ed., *Current Topics in Medicinal Chemistry: Recent Developments in Antiviral Nucleosides, Nucleotides and Oligonucleotides*, Bentham Science Publishers Ltd., Karachi, 2002, pp. 1065–1086.

J. S. Oxford and S. Patterson, *Developments in Antiviral Therapy*, Academic Press, London, 1980, pp. 119–131.

S. Ren and E. J. Lien, "Development of HIV Protease Inhibitors: A Survey." *Fortschritte der Arzneimittelforschung-Progress in Drug Research-Progres des Recherches Pharmaceutiques.* 2001, Spec, pp. 1–34.

M. M. Mangos and M. J. Damha, in R. S. Hosmane, ed., *Current Topics in Medicinal Chemistry: Recent Developments in Antiviral Nucleosides, Nucleotides and Oligonucleotides*; Bentham Science Publishers, Ltd., Karachi, 2002, pp. 1147–1171.

RAMACHANDRA S. HOSMANE
University of Maryland

AQUACULTURE

One definition of aquaculture is the rearing of aquatic organisms under controlled or semicontrolled conditions. Another, used by the Food and Agriculture Organization (FAO) of the United Nations, is that aquaculture is, "the farming of aquatic organisms, including fish, molluscs, crustaceans, and aquatic plants." Included within those broad definitions are activities in fresh, brackish, marine, and even hypersaline waters. The term *mariculture* is often used in conjunction with aquaculture in the marine environment.

Public sector aquaculture involves production of aquatic animals to augment or establish recreational and commercial fisheries. The FAO definition of aquaculture also indicates that farming implies ownership of the organisms being cultured, which would seem to exclude public sector aquaculture.

In recent years, aquaculture has been increasingly used as a means of aiding in the recovery of threatened and endangered species. Those efforts are currently public sector activities, although there is interest in the private sector to become involved. Going hand in hand with attempts to recover endangered species are enhancement stocking programs aimed at releasing juvenile animals to rebuild stocks of aquatic animals that have been reduced due to overfishing.

The bulk of global production from aquaculture is utilized directly as human food, with public aquaculture playing a minor role in many nations or being absent. Private aquaculture is not only about human food production, however. In some regions, well-developed private sector aquaculture is involved in the production of bait and ornamental fishes and invertebrates.

Aquatic plants are cultured in many regions of the world. Private aquaculture existed as a minor industry for many decades, coming into prominence in the 1960s. Since then the United States has become one of the leaders in aquaculture research and development, although not in production.

Aquaculture production continues to grow annually, but increasing competition for suitable land and water, problems associated with wastewater from aquaculture facilities, disease outbreaks, and potential shortages of animal protein for aquatic animal feeds are having, or may have, negative effects on future growth. New technology, including the application of genetic engineering approaches to improving performance and disease resistance in aquatic species, along with the development of water reuse (recirculating) systems and the establishment of offshore facilities, may provide the impetus for a resurgence of growth in the industry.

REGULATION

The extent to which governments regulate aquaculture varies greatly from one nation to another. In some parts of the world, particularly in developing nations, there has historically been little or no regulation. Unregulated expansion of aquaculture in some countries has led to pollution problems, destruction of valuable habitats such as mangrove swamps, and has enhanced the spread of disease from one farm to another. The need for imposing regulations is now becoming evident around the world.

In developed countries there may or may not be a standardized set of national regulations. The United States is an example of a mixture of local, state, and federal regulations.

SPECIES UNDER CULTIVATION

This article emphasizes aquatic animal production, but many hundreds of thousands of people are involved, worldwide, in aquatic plant production. The quantity of brown seaweeds, red seaweeds, green seaweeds, and other algae produced in 1996 was estimated at over 7.7 million metric tons.

Animal aquaculture is concentrated on finfish, molluscs, and crustaceans. Sponges, echinoderms, tunicates, turtles, frogs, and alligators are also being cultured, but production is insignificant in comparison with the three principal groups.

CULTURE SYSTEMS

At one extreme aquaculture can be conducted with a small amount of intervention from humans and the employment of little technology. At the other is total environmental control and the use of computers, molecular genetics, and complex modern technology. Many aquaculturists operate between the extremes. In general, as the level of culture intensity increases, stocking density, and as a consequence, production per unit area of culture system or volume of water, increases.

The most extensive types of aquaculture involve minimal human intervention to promote increases in natural productivity. One of the most extensive forms of culture involves placing oyster, clam, or other types of shell (cultch) on the bottom in intertidal areas that are known to have good oyster reproduction. The next level of intensity might involve placing bags of cultch out in nature to collect spat in a productive area that already has sufficient quantities of natural cultch. The next step in increasing oyster culture intensity might involve hatchery production and settling of spat on cultch. Control of predators such as starfish and oyster drills could easily be a part of culture at all levels.

The stocking of ponds, lakes, and reservoirs to increase the production of desirable fishes that depend on natural productivity for their food supply and are ultimately captured by recreational fishermen or for subsistence is another example of extensive aquaculture.

Most of the aquaculture practiced around the world is conducted in static ponds. Fertilization of ponds to increase productivity is the next level of intensity with respect to fish culture.

With the application of increased technology and control over the culture system, intensity continues to increase. Utilization of specific pathogen-free animals, provision of nutritionally complete feeds, careful monitoring and control of water quality, and the use of animals bred for good performance, can lead to impressive production levels.

Where water is plentiful and inexpensive, raceway culture is an attractive option and one which allows for production levels well in excess of what is possible in ponds. Linear raceways are essentially channels that are longer than they are wide, and are usually no deeper than 1–2 m. High density raceways used in production facilities are commonly constructed of poured concrete. Small raceways of the type used in hatcheries and research facilities may be constructed of fiberglass or other resilient materials. Circular raceways, called tanks, are also used by aquaculturists.

Salmon, steelhead trout, and a variety of marine fishes are currently being reared in net-pens. Most net-pens are located in protected waters since they are easily damaged or destroyed by storms.

Competition by various user groups for space in protected coastal waters in much of the world has led to strict controls and in some cases prohibitions against the establishment of inshore net-pen facilities. As a result, there is growing interest in developing the technology to move offshore.

The highest levels of intensity that can be found in aquaculture systems are associated with totally closed systems, often called recirculating systems. In these systems, all water passing through the chambers in which the finfish or shellfish are held is continuously treated and reused. It is necessary to add some water to such systems to make up for that lost to evaporation, splashout, and in conjunction with solids removal. Most of the recirculating systems in use today are operated in a mode between entirely closed and completely open. In many a significant percentage of replacement water is added either continuously or intermittently on a daily basis.

Recirculating systems often feature other types of apparatus, such as foam strippers and supplemental aeration. The technology for denitrifying nitrate to nitrogen gas has developed to the point that it may find a place in commercial culture systems in the near future. Computerized water-quality monitoring systems that will sound alarms and call emergency telephone numbers to report system failures to the culturists are also finding increased use.

The technology involved makes recirculating systems expensive to construct and operate. However, recirculating systems can make aquaculture feasible in locations where conditions would not otherwise be conducive to successful operations. Another approach to aquaculture is enhancement, which involves spawning and rearing aquatic organisms to a size large enough that the organisms will have a good chance of survival in nature.

WATER SOURCES AND QUALITY

Sources of water for aquaculture include municipal supplies, wells, springs, streams, lakes, reservoirs, estuaries, and the ocean. The water may be used directly from the source or it may be treated in some fashion prior to use (see WATER).

Many aquaculture facilities that utilize surface waters and those that obtain their water from wells other than artesian wells are required to pump the water into their facilities. Pumping costs can be a major expense, particularly when the facility requires continuous inflow.

Surface water can sometimes be obtained through gravity flow by locating aquaculture facilities at elevations below those of adjacent springs, streams, lakes, or reservoirs. Coastal facilities may be able to obtain water through tidal flow.

For many freshwater species that can be characterized as warmwater (such as channel catfish and tilapia) or coldwater (such as trout), the conditions outlined in Table 1 should provide an acceptable environment. So-called midrange species are those with an optimum temperature for growth of about 25°C. Typically they do well under the conditions, other than temperature, specified in Table 1 for coldwater species.

The water quality criteria for each species should be determined from the literature or through experimentation when literature information is unavailable. Synergistic effects that occur among water quality variables can have an influence on the tolerance a species has under

Table 1. General Water Quality Requirements for Cold- and Warmwater Aquatic Animals in Fresh Water

Variable	Acceptable level or range	
	Coldwater	Warmwater
temperature, °C	<20	26–30
alkalinity, mg/L	10–400	50–400
dissolved oxygen, mg/L	>5	≥5
hardness, mg/L	10–400	50–400
pH	6.5–8.5	6.5–8.5
total ammonia, mg/L	<0.1	<1.0
ferrous iron, mg/L	0	0
ferric iron, mg/L	0.5	0–0.5
carbon dioxide, mg/L	0–10	0–15
hydrogen sulfide, mg/L	0	0
cadmium, μg/L	<10	<10
chromium, μg/L	<100	<100
copper, μg/L	<25	<25
lead, μg/L	<100	<100
mercury, μg/L	<0.1	<0.1
zinc, μg/L	<100	<100

any given set of circumstances. Biocides should not be present in water used for aquaculture.

Most aquaculture facilities release water constantly or periodically into the environment without passing it through a municipal sewage treatment plant. The effects of those effluents on natural systems have become a subject of intense scrutiny. There have even been demands that some existing operations should be shut down. Research is currently underway to develop feeds containing reduced levels of nutrients or to provide nutrients in forms that can better be utilized by the culture animals. The goal in both approaches is to reduce discharges of nutrients to the environment through excretion.

NUTRITION AND FEEDING

There are cases in which intentional fertilization is commonly used by aquaculturists in order to produce desirable types of natural food for the species under culture. Provision of live foods is currently necessary for the early stages of many aquaculture species because acceptable prepared feeds have yet to be developed. However, some of the most popular aquaculture species accept prepared feeds from first feeding. Included are catfish, tilapia, salmon, and trout.

Requirements for energy, protein, carbohydrates, lipids, vitamins and minerals have been determined for the species commonly cultured. Since feeds contain other substances than those required by the animals of interest, knowledge is also required of antinutritional factors in feedstuffs and on the use of additives. Certain feed ingredients contain chemicals that retard growth or may actually be toxic. Restriction on the amount of the feedstuffs used is one way to avoid problems. In some cases, as is true of trypsin inhibitor, proper processing can destroy the antinutritional factor in the feed ingredient.

When color development is an important consideration whether external or of the flesh, it can be achieved by incorporating ingredients that contain pigments or by adding extracts or synthetic compounds. One class of additives used to impart color is the carotenoids.

Prepared feeds are marketed in various sizes from very small particles (fines) through crumbles, flakes, and pellets. Pelleted rations may be hard, semimoist, or moist. The most widely used types of prepared feeds are produced by pressure pelleting or extrusion.

Nearly all aquaculture feeds contain at least some animal protein since the amino acid levels in plant proteins typically cannot meet the requirements of most aquatic animals. Fish meal is the most commonly used source of animal protein in aquaculture feeds. Most formulations contain a few percent of added fat from such sources as fish oil, tallow, or more commonly, oilseed oils such as corn oil and soybean oil. Complete rations contain added vitamins, and minerals. Purified amino acids, binders, carotenoids, and antioxidants are other components found in many feeds. Growth hormone and antibiotics are sometimes used. Regulations on the incorporation of hormones along with other chemicals and drugs into aquatic animal feeds are in place in the United States and some other countries. Few such regulations have been promulgated in developing nations.

Feeding practices vary from species to species. It is important not to overfeed since waste feed can also lead to degradation of water quality. Most species require only three to four percent of body weight in dry feed daily for optimum growth. Very young animals are an exception. They are fed at a higher rate because they are growing rapidly and consume a greater daily percentage of body weight than older animals.

REPRODUCTION AND GENETICS

Selective breeding has long been practiced as a mean of improving aquaculture stocks. Most of the species that are being reared in significant quantities around the world are produced in hatcheries using either captured or cultured broodstock.

Fish breeders have worked with varying degrees of success to improve growth and disease resistance in a number of species. As genetic engineering techniques are adapted to aquatic animals, dramatic and rapid changes in the genetic makeup of aquaculture species may be expected. However, since it is virtually impossible to prevent escapement of aquacultured animals into the natural environment, potential negative impacts of such organisms on wild populations cannot be ignored.

DISEASES AND THEIR CONTROL

Aquatic animals are susceptible to a variety of diseases including those caused by viruses, bacteria, fungi, and parasites. A range of chemicals and vaccines has been developed for treating the known diseases, although some conditions have resisted all control attempts to date. In some nations, severe restrictions on the use of therapeutants has impaired that ability of aquaculturists to control disease

outbreaks. The United States is a good example of a nation in which the variety of treatment chemicals is limited by government regulators. Managing conditions in the culture environment to keep stress to a minimum is one of the best methods of avoiding diseases.

HARVESTING, PROCESSING, AND MARKETING

Harvesting techniques vary depending on the type of culture system involved. Seines are often used to capture fish from ponds, or the majority of the animals can be collected by draining the pond through netting. Fish pumps are available that can physically transfer aquatic animals directly onto hauling trucks from ponds, raceways, cages and net-pens without causing skin abrasions, broken fins, or other damage.

Aquaculturists may harvest, and even process their own crops, although custom harvesting and hauling companies are often available. Some processing plants also provide harvesting and live-hauling services.

Centralized processing plants specifically designed to handle regional aquaculture crops are established in areas where production is sufficiently high. In coastal regions, aquacultured animals are often processed in plants that also service capture fisheries.

Marketing can be done by aquaculturists who operate their own processing facilities. Most aquaculture operations depend on a regional processing plant to market the final product.

C. E. Boyd, *Bottom Soils, Sediment, and Pond Aquaculture*, Chapman and Hall, New York, 1995.

J. Huegenin and J. Colt, *Design and Operating Guide for Aquaculture Seawater Systems*, Elsevier, New York, 1989.

National Research Council, *Nutrient Requirements of Fish*, National Academy Press, Washington, D.C., 1993.

R. R. Stickney, *Principles of Aquaculture*, John Wiley & Sons, Inc., New York, 1994.

ROBERT R. STICKNEY
Texas A&M University

AQUACULTURE CHEMICALS

Intensive or extensive culture of aquatic animals requires chemicals that control disease, enhance the growth of cultured species, reduce handling trauma to organisms, improve water quality, disinfect water, and control aquatic vegetation, predaceous insects, or other nuisance organisms. The aquaculture chemical needs for various species have been described for rainbow trout, *Oncorhynchus mykiss* (1); Atlantic and Pacific salmon, *Salmo salar* and *Oncorhynchus* sp. (2); channel catfish, *Ictalurus punctatus* (3); striped bass, *Morone saxatilis* (4); milkfish, *Chanos chanos* (5); mollusks (6); penaeid *Penaeus* shrimp (7); and a variety of other freshwater and marine species (8).

Laws and regulations on the use of chemicals in aquaculture vary by country and serve to ensure safe and effective use and protection of humans and the environment. Regulations and therapeutants or other chemicals that are approved or allowed for use in the United States, Canada, Europe, Japan, Chile, and Australia are presented below.

REGULATION OF AQUACULTURE CHEMICALS IN THE UNITED STATES

In the United States, the U.S. Food and Drug Administration (FDA) and the U.S. Environmental Protection Agency (EPA) regulate the application of chemicals to organisms or to their environments. FDA controls the use of drugs and anesthetics and EPA controls the application of chemicals and pesticides to the environment. In cases that involve treatments to control pathogens that are present in the water, the jurisdiction becomes unclear and has been changed over time. Each agency develops appropriate guidelines and policies to implement the laws for its field of responsibility.

REGISTERED AQUACULTURE CHEMICALS IN THE UNITED STATES

Antibacterials

Few therapeutants are registered in the United States for use on any cultured aquatic species. In the most critical area of antibacterials, only two (Terramycin for Fish and Romet-30) are approved and available.

Fungicides

Formalin is the only fungicide approved by FDA for use on eggs of all fish at 1,000–2,000 mg/L for 15 min. Delivery apparatus has been developed to reduce human exposure toformalin.

Parasiticides

Formalin is the only parasiticide currently approved for use on all fish and penaeid shrimp. It is registered for use on all fish at concentrations up to 250 mg/L for 1 h in tanks and raceways and 15 to 25 mg/L for an indefinite period in ponds and for penaeid shrimp at 50–100 mg/L for up to 4 h in tanks and raceways and 25 mg/L in ponds. A second chemical, trichlorfon (Masoten) was registered for use on nonfood fishes by EPA but is not currently available. Vinegar (glacial acetic acid) and salt (sodium chloride) are also used to control external parasites on fishes and CVM classifies these compounds as unapproved drugs of Low Regulatory Priority.

Disinfectants

Several disinfecting agents can be used in hatcheries and two are of particular interest. Because they are considered as unapproved drugs of Low Regulatory Priority by FDA, povidone-iodine compounds can be used to disinfect the surface of eggs. Benzalkonium chloride and benzethonium chloride (quaternary ammonium compounds) are allowed at 2 mg/L by FDA to disinfect water.

Water Treatment Compounds

Of particular interest ispotassium permanganate which is exempted from registration by EPA when used as an oxidizer or detoxifier and can control certain parasites, external bacteria, and possibly fungi.

Spawning Aids

One spawning aid is approved in the United States, human chorionic gonadotropin (Chorulon).

Anesthetics

Tricaine methanesulfonate (MS-222) is the only currently approved anesthetic and requires a 21-day withdrawal time. Both carbon dioxide and sodium bicarbonate have also been used as anesthetics and are classified as unapproved drugs of Low Regulatory Priority by FDA; however, both chemicals are difficult to use with consistent results and involve long induction and recovery periods.

Herbicides

An array of herbicides is registered for use in aquatic sites, but copper sulfate and diquat dibromide are of particular interest because they also have therapeutic properties.

Piscicides

The two piscicides, antimycin androtenone, are both used in ponds to control nuisance fish.

REGULATION AND REGISTRATION OF AQUACULTURAL CHEMICALS OUTSIDE THE UNITED STATES

The control of aquaculture drugs varies among countries from no regulation to restrictive regulations. Generally, few requirements are needed for a therapeutant to be licensed or registered in South America, Africa, and most of Asia. Seafood-exporting countries are increasingly concerned because importing countries may no longer accept products without a guarantee that the products contain no chemical residues of concern.

Canada

Except for environmental studies, requirements for registration data in Canada are similar to requirements in the United States. However, Canada has significantly different regulations and approval processes. Canadian aquaculturalists use drugs that are either licensed for other food animals and prescribed by veterinarians or used in an emergency under the direction of the Canadian Bureau of Veterinary Drugs (BVD). The BVD is concerned about the lack of data on the pharmacokinetics of fishes, especially the difference in uptake of drugs at a range of temperatures. Chloramphenicol and tributyltin compounds are two classes of compounds that cannot be used in Canadian aquaculture. Canadian regulations also differ from the

United States in that they have no minor-use policy or classifications such as Low Regulatory Priority drugs.

Europe

The European Agency for the Evaluation of Medicinal Products (EMEA) regulates the approvals of all Veterinary Medicinal Products in Europe and establishes the Maximum Residue Limit (MRL) for each animal drug. Those chemicals that should have established MRLs were banned from use if they were not established by January 1, 2000. Requirements for MRLs for drugs are divided into four groups or annexes: Annex I = fixed MRL, Annex II = No MRL needed, Annex III = temporary MRL, and Annex IV = No MRL can be established. Banned from use are drugs such as the nitrofurans and chloramphenicol that have been placed in Annex IV. A MRL is required before a member country can evaluate a drug for approval. Under current regulations, veterinary medicines allowed for use in various European countries are either fully licensed for aquacultural use (oxytetracycline, oxolinic acid) or can be prescribed by veterinarians if (1) the drugs are licensed for use on other food animals or in humans, (2) only a limited number of animals are treated, and (3) a 500 degree day withdrawal time is observed. Fish specific MRL approvals are available only for amoxicillin, potentiated sulfonamides, oxolinic acid, flumequine, sarafloxacin, oxytetracycline, and thiamphenicol.

Japan

In Japan, registration of drugs for aquatic species requires the same data as those required for drugs used on terrestrial animals. The Ministry of Agriculture, Forests, and Fisheries and the Ministry of Welfare control the use of chemicals in aquaculture in Japan. As of April 2001, more chemicals were registered for aquacultural use in Japan than in any other country.

Chile

The Servico Agricola y Ganadero has recently increased its scrutiny of drugs used in aquaculture. New approvals have and will become even more difficult. The agency will accept foreign data but some data are required to be generated in Chile. There are five drugs—flortenicol, flumequine, oxolinic acid, dxytetacycline, and savafloxacin—currently approved in Chile. Other drugs being used or under consideration for approval include amoxicillin, benzocaine, chloramine-T, and MS-222.

Australia

In the past ten years, Australia has increased its aquaculture production and as a result has begun to register drugs and chemicals for that use. The National Registration Authority for Agricultural and Veterinary Chemicals has registered the following: benzocaine as a sedative and anesthetic for finfish and abalone, formalin to control protozoan and metazoan ectoparasites on fish and epicommensal ciliates on shrimp, flubendazole to control gill flukes on ornamental fish, leutinizing hormone releasing hormone analogue to induce spawning in finfish broodstock, methyltestosterone to produce female salmonid fish stocks, and trifluralin as a selective herbicide for prawn larvae mycosis. Certain chemicals have been exempted from the need for registration: calcium carbonate, $CaCO_3$; calcium hydroxide, $Ca(OH)_2$; calcium oxide, CaO; magnesium carbonate; calcium sulfate, CaO_4S; zeolite, $Na_2O \cdot Al_2O_3 \cdot (SiO_2)X.(H_2O)Y$; aluminum sulfate, $Al_2O_{12}S_3$; ferric chloride, Cl_3Fe; and inorganic and organic fertilizers.

PROMISING CHEMICALS FOR REGISTRATION FOR AQUACULTURE

More therapeutants and vaccines may soon be added to the medicine chest of fish farmers. A variety of chemicals have potential for registration and use in aquaculture.

Antibacterials

Research has been conducted on three important external and systemic antibacterial compounds in the United States: chloramine-T, sarafloxacin, and erythromycin. In addition, florfenicol, an oral antibacterial is being considered for development in the United States by the sponsor for use on catfish and salmonids. Amoxicillin is another antibacterial that has potential for development for control of streptococcal infections in tilapia and hybrid striped bass.

Fungicides and Parasiticides

UMESC is working on several fronts to improve the availability of fungicides and parasiticides. UMESC, with funding from Bonneville Power Administration screened and tested promising candidates for replacement of malachite green as both fungicides and parasiticides. Although several compounds show promise for controlling fungi, the best fungicide candidate was identified as hydrogen peroxide.

The approval of formalin as a fungicide now extends to all fish eggs and may soon extend to all fish.

The Harry K. Dupree Stuttgart National Aquaculture Research Center, Stuttgart, Arkansas (SNARC) has developed data on copper sulfate with funds from the IAFWA Project for the control of *Ichthophthirius*, an external protozoan that causes significant losses in the catfish industry. Other promising parasiticides include praziquantel, fumagillin, and sea lice control agents (eg, azamethiphos, cypermethrin, emamectin, and hydrogen peroxide).

Disinfectants

Promising disinfectants include ultraviolet (uv) light andozone, O_3.

Anesthetics

Ethyl aminobenzoate (benzocaine), $C_9H_{11}NO_2$, was a candidate anesthetic but it would have required additional mammalian safety studies, did not have a sponsor, and probably would not have allowed the use of spawned-out broodstock carcasses to be used for pet or human food.

AQUI-S, an anesthetic developed in New Zealand has great potential as a zero withdrawal drug and is being developed by its sponsor for worldwide approval. In the United States, UMESC with funds from the IAFWA Project is developing the anesthetic for use on all fish. Electronarcosis is an alternative to chemical anesthesia that uses varying electrical frequencies to rapidly anesthetize fishes and allow gentle recovery.

D. J. Alderman in J. F. Muir and R. J. Roberts, eds., *Recent Advances in Aquaculture*, Vol. 3, Timber Press, Portland, Oreg., 1988.

Joint Subcommittee on Aquaculture, *Guide to Drug, Vaccine, and Pesticide Use in Aquaculture*, Texas Agricultural Extension Service, College Station, Tex., 1994.

R. A. Schnick, *Use of Chemicals in Fish Culture: Past and Future*, in D. J. Smith, W. H. Gingerich, and M. G. Beconi-Barker, eds., Xenobiotics in Fish, Kluwer Academic/ Plenum Publishing Corp., New York, Chapt. 1 1999.

C. J. Sindermann and D. V. Lightner, *Disease Diagnosis and Control in North American Marine Aquaculture*, Elsevier, New York, 1988.

ROSALIE A. SCHNICK
Aquaculture Chemicals

AROMA CHEMICALS

Aroma chemicals are an important group of organic molecules used as ingredients in flavor and fragrance compositions (see FLAVORS; PERFUMES). Aroma chemicals consist of natural, nature-identical, and artificial molecules. Natural products are obtained directly from the plant or animal sources by physical procedures. Nature-identical compounds are produced synthetically, but are chemically identical to their natural counterparts. Artificial flavor substances are compounds that have not yet been identified in plant or animal products for human consumption.

There are ca. 3000 different molecules that find use in the production of flavor and fragrance compositions. Synthetic ingredients play a major part as components due to their convenient availability and the relatively lower costs compared to natural molecules from isolation of relatively limited natural sources.

ODORS DESCRIPTORS

The odors of single chemical compounds (aroma chemicals) are very difficult to describe unequivocally. The odors of complex mixtures called compounds are often impossible to describe unless one of the components is so characteristic that it determines the odor or flavor of the composition. Although an objective classification is not possible, an odor can be described by adjectives such as flowery, fruity, woody, or hay-like, which will relate to natural occurring or other well-known products with such odors characteristics.

A few terms used to describe odors are listed in Table 1, with a few examples.

GENERAL PRODUCTION ROUTES

Aroma chemicals are specific molecules of particular aroma, which can be obtained by isolation from natural sources, with or without chemical modifications, using natural molecules as precursors for many aroma chemicals (partial synthesis); from petrochemical raw materials; or by synthesis from cyclic and aromatic precursors.

The Use of Natural Molecules as Precursors

One of the most useful sources for natural molecules as chemical precursors is turpentine oil, originated from *Pinus* sp. The oil contains 60–70% of α-pinene and β-pinene, along with other natural molecules, i.e, α-phellandrene, γ-terpinene, anethole, caryophyllene, 3-carene, and camphene (see Figs. 1 and 2).

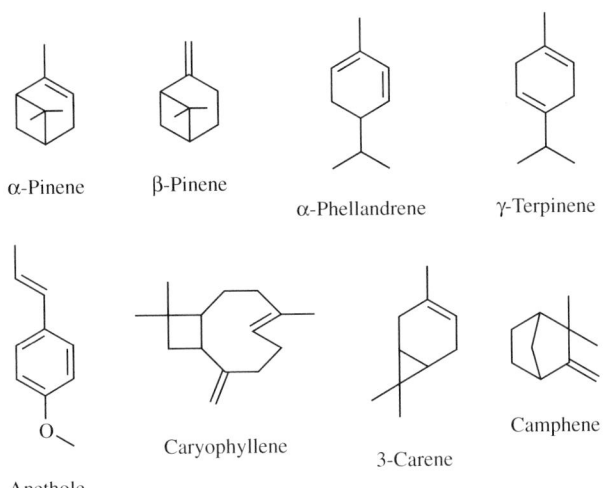

α-Pinene   β-Pinene

α-Phellandrene   γ-Terpinene

Anethole   Caryophyllene   3-Carene   Camphene

The Use of Petrochemicals as Precursors

Synthesis from petrochemical precursors of one-to-five carbon atoms, ie, carbon monoxide/formaldehyde, acetylene, isobutylene, and isoprene, represents one of the most important routes to produce aroma chemicals.

Aromatic molecules, eg benzene, toluene, xylenes, phenol, cresols, and naphthalene, are also important precursors for aroma chemicals.

FUNCTIONAL GROUPS OF AROMA CHEMICALS

As mentioned before, over 3000 specific chemical molecules are used in the F&F industry, but only a few hundreds are produced in a scale between 20 and 50 mt year. These molecules include most of the functional groups,

Table 1. Terms to Describe Odors

Odor	Description	Examples
aldehydic	note of the long-chain fatty aldehydes, eg, fatty–sweaty, ironed laundry, seawater	*n*-Decanal *n*-Octanal
animalic	typical notes from the animal kingdom, eg, musk, castoreum, skatol, civet, ambergis	Indole Ambrox
balsamic	heavy, sweet odors, eg, cocoa, vanilla, cinnamon	Vanillin isobutyrate Cinnamaldehyde
camphoraceous	reminiscent of camphor	2-Adamantanone (+)-Isoborneol
citrus	fresh, stimulating odor of citrus fruits such as lemon or orange	Citral Citronellal
earthy	humus-like, reminiscent of humid earth	2- Ethylfenchol 6-Isobutylquinoline
fatty	reminiscent of animal fat and tallow	$CH_3(CH_2)_8CO_2(CH_2)_4CH_3$ Amyl decanote (*E,E*)-2,4-Decadienal
floral, flowery	generic terms for odors of various flowers	Tetrahydrolinalool Geraniol

Table 1. (*Continued*)

Odor	Description	Examples
fruity	generic terms for odors of various fruits	$CH_3CO_2(CH_2)_7CH_3$ *n*-Octyl acetate *trans*-2-Hexenylacetate
green	typical odor for freshly cut grass and leaves	*cis*-3-Hexenol 2-(Cyclohexyl)-propanal
herbaceous	noncharacteristic, complex odor of green herbs with, eg, sage, minty, eucalyptus-like, or earthy nuances	Estragole Citronellylethyl ether
medicinal	odor reminiscent of disinfectants, eg, phenol, lysol, methyl salycilate	Phenol Methyl salicylate
metallic	typical odor observed near metal surfaces, eg, brass or steel	2,5-Dimethyl-2-vinyl-4-hexenenitrile Benzyl methyl disulfide
minty	peppermint-like odor	(−)-Menthol Menthone

Table 1. (*Continued*)

Odor	Description	Examples
mossy	typical note reminiscent of forests and seaweed	Ethyl 2-hydroxy-4-methoxy-6-methyl-benzoate 3-Methoxy-5-methyl phenol
powdery	odor identified with toilet powders, sweet-diffusive	2-(1-Cyclohexenyl)-cyclohexanone Methyl-β-naphthyl ketone
resinous	aromatic odor of tree exudates	2-Isopropyl-5-methyl-2-hexenal 3-Phenylpropionic acid
spicy	generic term for odors of various spices	Carvacrol 2,4-Dimethyl-acetophenone
waxy	odor resembling that of candle wax	*n*-Decanal Citronellyl-isobutyrate
woody	generic term for the odor of wood, eg, cedarwood, sandalwood	α-Cedrene *cis-p-tert*-Butyl-cyclohexylacetate

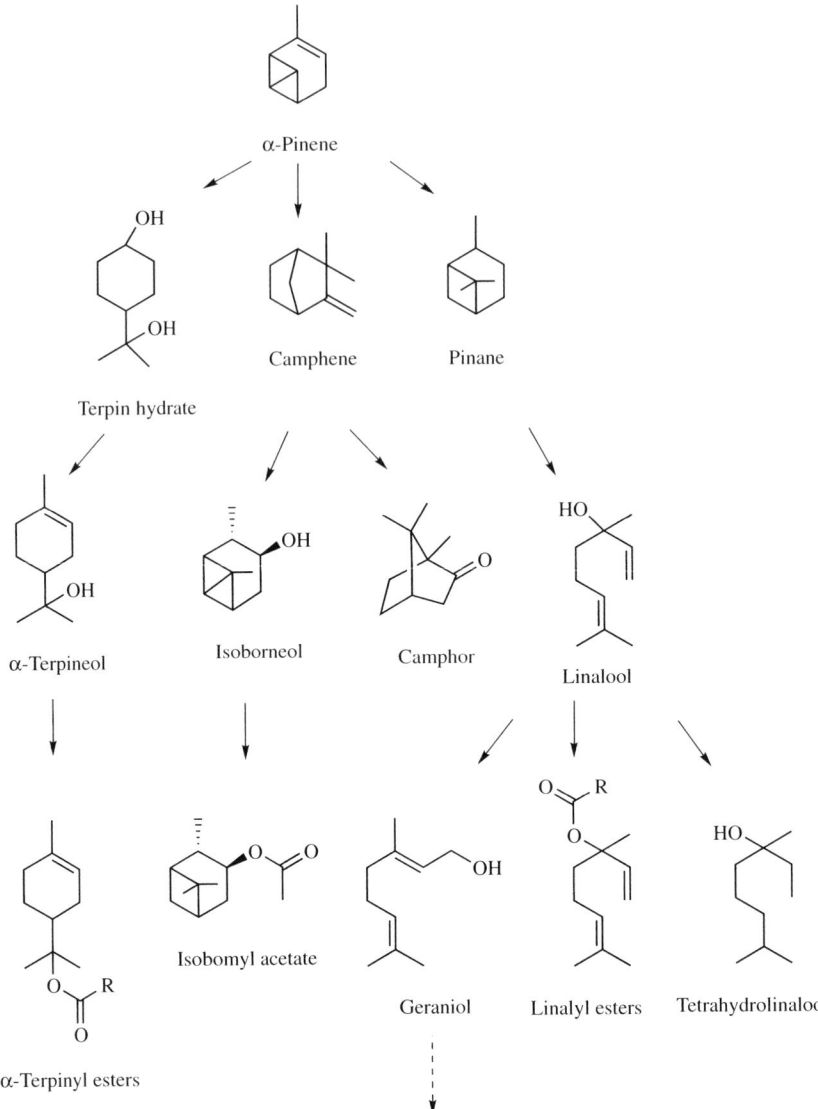

Figure 1. α-Pinene as a natural precursor for aroma chemicals.

from aliphatic molecules to heterocyclic ones, according to the following list:

- Hydrocarbons (aliphatic, acyclic terpenes, cyclic terpenes, benzenoids)
- Alcohols (aliphatic, alicyclic, cyclic)
- Ethers
- Aldehydes and ketones (including acetals and ketals)
- Carboxylic acids
- Esters and lactones
- Nitriles
- Amines
- Nitroaromatic compounds
- Thio compounds
- Heterocyclic molecules

The following sections contain selected examples from the functional group.

Hydrocarbons

Hydrocarbons include simple aliphatic molecules, terpenes—both acyclic and cyclic, and benzene rings.

Terpenes. Terpenes (see TERPENOIDES; OILS, ESSENTIAL; FLAVORS; PERFUMES) are a group of plant originated natural products, which are usually composed of usually two, three, four, five, six or eight units of C_5 atoms. These units are formally derived from 2-methyl-1,3-butadiene (isoprene).

2-Methyl-1,3-butadiene (Isoprene)

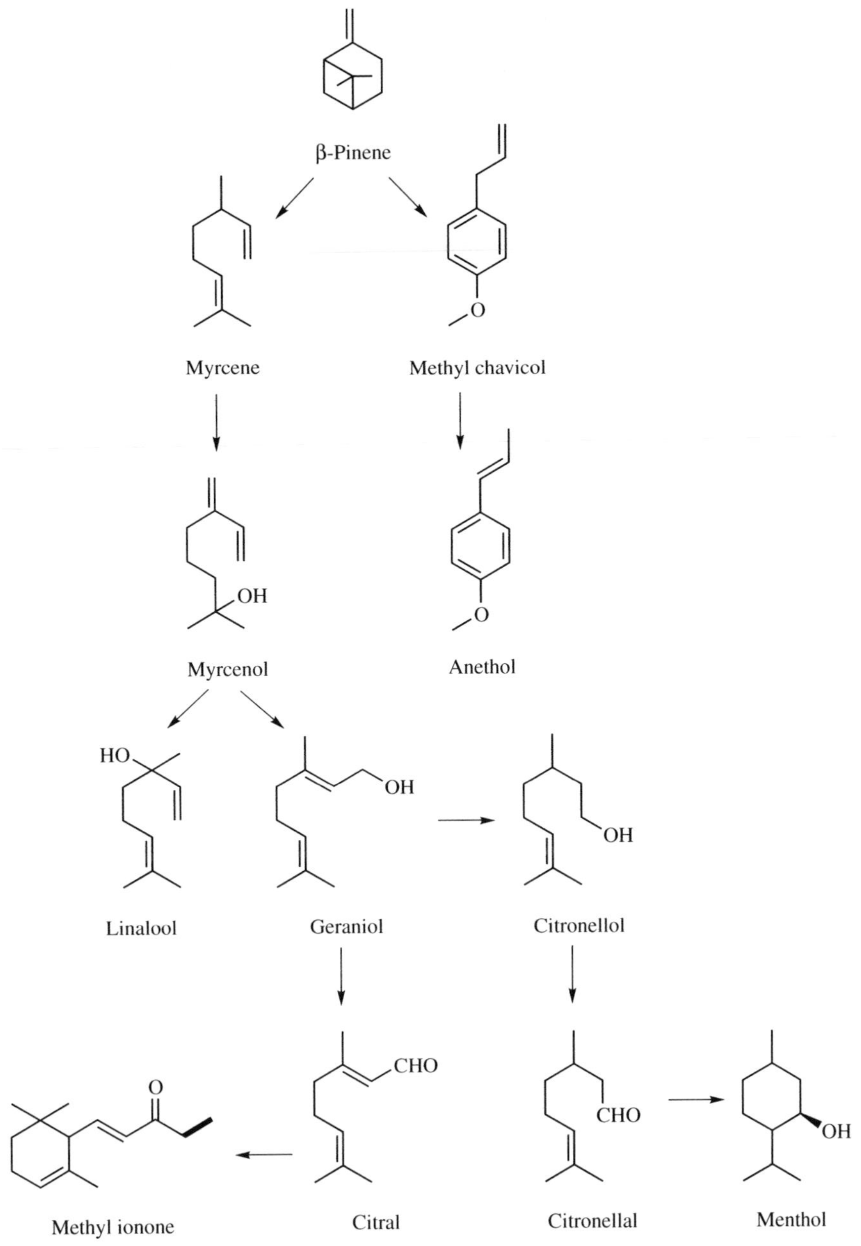

Figure 2. β-Pinene as a natural precursor for aroma chemicals.

These molecules are named as follows:

Name	Number of isoprene units	Number of carbon atoms
monoterpenes	2	10
sesquiterpenes	3	15
diterpenes	4	20
sesterterpenes	5	25
triterpenes	6	30
tetraterpenes	8	40

Terpenes are formed in nature via the "two carbons metabolism", a process enabled by acetyl coenzyme A (CoA), which is produced from pyruvic acid. Acetyl—CoA forms mevalonic acid, which loses one carbon atom by decarboxylation to yield a C_5 unit—isopentenyl pyrophosphate.

Alcohols

The alcohol function is found in simple aliphatic molecules, in acyclic and cyclic terpenes, and in molecules containing benzene rings. Phenols are also contained in this group of aroma chemicals.

Ketones. 1,3,4,6,7,8-Hexahydro-4,6,6,7,8,8-hexamethyl-cyclopenta-(g)-2-benzopyran, also commercially as eg, Galaxolide, Abbalide, is synthesized as following: There is a condensation–cyclization stage of tert-amyl alcohol and α-methyl styrene in acidic conditions to obtain the indane system, followed by a Friedel–Crafts reaction

Table 2. Saturated Carboxylic Acids

Name	Organoleptic characteristics	Structure
formic acid	pungent, acidic, sour, astringent with a fruity depth	HCO_2H
acetic acid	sour, vinegar-like	CH_3CO_2H
propionic acid	sour, fruity on dilution	$CH_3CH_2CO_2H$
butyric acid	penetrating, reminiscent of rancid butter	$CH_3(CH_2)_2CO_2H$
valeric acid	strongly acidic, caprylic, cheese-like	$CH_3(CH_2)_3CO_2H$
caproic acid	acidic, caprylic, fatty	$CH_3(CH_2)_4CO_2H$
oenanthic acid	caprylic, fatty, green	$CH_3(CH_2)_5CO_2H$
caprylic acid	caprylic, fatty, oily	$CH_3(CH_2)_6CO_2H$
pelargonic acid	oily, fatty, caprylic; cheesy with a mild creamy background	$CH_3(CH_2)_7CO_2H$
capric acid	sour, fatty aroma	$CH_3(CH_2)_8CO_2H$
undecylic acid	fatty, fruity aspects	$CH_3(CH_2)_9CO_2H$
lauric acid	mild fatty	$CH_3(CH_2)_{10}CO_2H$
myristic acid	faint oily, fatty	$CH_3(CH_2)_{12}CO_2H$
palmitic acid	faint oily aroma	$CH_3(CH_2)_{14}CO_2H$
stearic acid	fatty, stearinic	$CH_3(CH_2)_{16}CO_2H$

with propylene oxide to get the side chain. The side chain is finally closed to the isochromanic system using formaldehyde.

Methyl dihydrojasmonate is also known commercially by the names Hedione, Claigeon. It is synthesized by the Michael addition of diethyl malonate to the pentyl cyclopentenone to obtain the second side chain, followed by hydrolysis and decarboxylation, and finaly esterification.

Carboxylic Acids

Saturated Carboxylic Acids (see Table 2)

K. Bauer, D. Garbe, and H. Surburg, *Common Fragrance and Flavor Materials*, 3rd revised ed., VCH, Germany, 1997.
Flavors and Fragrances Report, SRI International, 1992

MICHAEL ZVIELY
Frutarom, Ltd.

ARSENIC AND ARSENIC ALLOYS

Arsenic, although often referred to as a metal, is classified chemically as a nonmetal or metalloid and belongs to Group 15 (VA) of the Periodic Table (as does antimony). The principal valences of arsenic are +3, +5, and −3. Only one stable isotope of arsenic having mass 75 (100% natural abundance) has been observed.

PROPERTIES

Physical properties of α-crystalline metallic arsenic are given in Table 1. The properties of β-arsenic are not completely defined.

Metallic arsenic is stable in dry air, but when exposed to humid air the surface oxidizes, giving a superficial golden bronze tarnish that turns black upon further exposure. The amorphous form is more stable to atmospheric oxidation. Upon heating in air, both forms sublime and the vapor oxidizes to arsenic trioxide, As_2O_3. Although As_4O_6 represents its crystalline makeup, the oxide is more commonly referred to as arsenic trioxide.

Elemental arsenic combines with many metals to form arsenides.

Arsenic vapor, As_4, does not combine directly with hydrogen to form hydrides. However, arsine (arsenic hydride), AsH_3, a highly poisonous gas, forms if an intermetallic compound such as AlAs is hydrolyzed or treated with HCl. Arsine may also be formed when arsenic compounds are reduced using zinc in hydrochloric acid. Heating to 250°C decomposes arsine into its elements.

Metallic arsenic is not readily attacked by water, alkaline solutions, or nonoxidizing acids. It reacts with concentrated nitric acid to form orthoarsenic acid, H_3AsO_4. Hydrochloric acid attacks arsenic only in the presence of an oxidant.

Arsenic may be detected qualitatively as a yellow sulfide, As_2S_3, by precipitation from a strongly acidic HCl solution. Other members of this group that are normally precipitated with hydrogen sulfide do not interfere if the solution contains 25% or more hydrochloric acid. Trace quantities of arsenic may be detected by first converting to arsine.

HEALTH AND SAFETY FACTORS

The toxicity of arsenic ranges from very low to extremely high depending on the chemical state. Metallic arsenic and arsenious sulfide, As_2S_3, have low toxicity. Arsine is extremely toxic. The toxicity of other organic and inorganic arsenic compounds varies.

Arsenic is classified as a carcinogen by the International Agency for Research on Cancer (IARC). The handling of arsenic in the workplace should be in compliance with the Occupational Safety and Health Administration

Table 1. Physical Properties of Arsenic

Property	Value
atomic weight	74.9216
mp at 39.1 MPa[a], °C	816
bp, °C	615[b]
density at 26°C, kg/m³	5,778
specific heat at 25°C, J/(mol·K)[c]	24.6
linear coefficient of thermal expansion at 20°C, μm/(m·°C)	5.6
electrical resistivity at 0°C, μΩ·cm	26
hardness, Mohs' scale	3.5

[a]To convert MPa to psi, multiply by 145.
[b]Sublimes.
[c]To convert to cal/(mol·K), divide by 4.184.

(OSHA) regulations: the maximum permissible exposure limit for arsenic in the workplace is 10 µg/m^3 of air as determined as an average over an 8-h period. Precaution should be taken to avoid accidental generation of arsine gas; the maximum permitted exposure is 0.05 ppm in air per 8-h period five days per week. Disposal of arsenical products should be in compliance with federal and local government environmental regulations.

ENVIRONMENTAL CONCERNS

The location and extent of arsenic in ground water was the subject of a U.S. Geological Survey study in 2000. The presence of arsenic in ground water is due largely to minerals dissolving. Data on 19,000 samples of potable water showed that the arsenic concentration was lower that the 50 µ/L, which was the EPA standard at that time.

For a detailed discussion of this topic see ARSENIC, ENVIRONMENTAL IMPACT, HEALTH EFFECTS, AND TREATMENT METHODS.

USES

The use of many arsenical chemicals are subject to registration and must comply with federal and local government environmental regulations.

Wood Preservative

The largest use for arsenic (as arsenic trioxide) is in the production of wood preservatives.

Semiconductor Applications

A limited but important demand for metallic arsenic of 99.99% and greater (exceeding 99.999+%) purities exists in semiconductor applications (see SEMICONDUCTORS).

HP arsenic is used in the manufacture of photoreceptor arsenic-selenium alloys for xerographic plain paper copiers (see ELECTROPHOTOGRAPHY).

Arsenic from the decomposition of high purity arsine gas may be used to produce epitaxial layers of III–V compounds, such as InAs, GaAs, AlAs, and as an n-type dopant in the production of germanium and silicon semiconductor devices. A group of low melting glasses based on the use of high purity arsenic were developed for semiconductor and ir applications.

Other

Other uses for arsenic metal are as an additive to improve corrosion resistance and tensile strength in copper alloys and as a minor additive (0.01–0.5% 0 to increase strength of posts and grids in lead storage batteries. Arsenic acid is used by the glass industry as a fining agent to disperse air bubbles. Arsenic is also used in some herbicides.

ALLOYS

Arsenic metal is used primarily in alloys in combination with lead and, to a lesser extent, copper.

Trace quantities of arsenic are added to lead-antimony grid alloys used in lead–acid batteries (see BATTERIES, LEAD ACID).

Minor additions of arsenic (0.02–0.5%) to copper (qv) and copper alloys (qv) raise the recrystallization temperature and improve corrosion resistance. In some brass alloys, small amounts of arsenic inhibit dezincification, and minimize season cracking.

Phosphorized deoxidized arsenical copper (alloy 142) is used for heat exchangers and condenser tubes.

H. C. Beard, *The Radiochemistry of Arsenic*, NAS-NS 3002, U.S. Atomic Energy Commission, Washington D.C., Jan. 1960.

W. H. Lederer and R. J. Fensterheim, eds., *Arsenic: Industrial, Biomedical, Environmental Perspectives*, Van Nostrand Reinhold Co., Inc., N.Y., 1983.

R. Reddy, ed., *Arsenic Metallurgy-Fundamental and Applications*, The Metallurgical Society (AIME), Warrendale, Pa., 1988.

R. G. Reese, Jr., Arsenic *Minerals Yearbook 2000*, U.S. Geological Survey, Reston, Va., Jan. 2002.

S. C. CARAPELLA, JR.
Consultant

ARSENIC—ENVIRONMENTAL IMPACT, HEALTH EFFECTS, AND TREATMENT METHODS

Arsenic, a cancer causing substance, is present in a variety of forms in soil, water, air, and food. As a naturally occurring element in the earth's crust, arsenic enters into aquifers and wells through natural activities, and to the water cycle as a result of anthropogenic activities. The four arsenic species commonly reported are arsenite [As(III)], arsenate [As(V)], monomethyl arsenic acid (MMA), and dimethyl arsenic acid (DMA). It is generally known that As(III) is more toxic than As(V) and inorganic arsenicals are more toxic than organic derivatives.

Extensive arsenic contamination of surface and subsurface waters has been reported in many parts of the world, thereby threatening the health of a number of people in the affected areas. Due to human health concerns, arsenic content standard for drinking water has been lowered in many countries.

OCCURRENCE OF ARSENIC IN THE ENVIRONMENT

Arsenic is mainly transported to the environment by water. Arsenic contamination of subsurface waters is believed to be geological, and high arsenic concentrations in groundwater may result from dissolution of, or desorption from iron oxide, and oxidation of arsenic pyrites. In addition, the occurrence of arsenic in groundwater depends on factors such as redox conditions, ion exchange, precipitation, grain size, organic content, biological activity, and characteristics of the aquifer.

Natural Sources

The natural weathering processes contribute ~40,000 tons of arsenic to the global environment annually, while twice this amount is being released by human activities. The primary natural sources are weathering of rocks, geothermal, and volcanic activity; rocks are the major reservoirs for arsenic, and soils and oceans are the remaining natural sources of arsenic.

Anthropogenic Sources

Anthropogenic activities such as mining and smelting activities, and the use of pesticides and fossil fuels have resulted in a dramatic effect on natural environmental arsenic levels. In addition, arsenic and arsenic compounds are used in pigments and dyes, preservatives of animal hides and wood, pulp and paper production, electroplating, battery plates, dye and soaps, ceramics and in the manufacture of semiconductors, glass, and various pharmaceutical substances. Chromated copper arsenate (CCA), an inorganic arsenic compound that is used to treat lumber, accounts for ~90% of the arsenic used annually by industry in the United States.

ARSENIC EXPOSURE AND HEALTH EFFECTS

Arsenic Exposure

All humans are exposed to low levels of arsenic through drinking water, air, food, and beverages. Consumption of food and water are the major sources of arsenic exposure for the majority of the affected people. In addition, workers involved in the operations of mining and smelting of metals, pesticide production and application, production of pharmaceutical substances, and glass manufacturing have a high level of occupational exposure to arsenic.

Arsenic Toxicity and Health Effects

Arsenic is considered as a notorious poison because of its toxicity. The toxicity of arsenic depends on its speciation. The toxicity of arsenite is 25–60 times higher than that of arsenate, and the toxicity decreases in the order of arsine > inorganic As(III) > organic As(III) > inorganic As(V) > organic As(V) > arsonium compounds and elemental arsenic. Recent studies showed that arsenite was more prevalent in groundwater than arsenate.

Cancer Effects

The association of arsenic in drinking water and skin cancer was first reported in Taiwanese people. Based on the Taiwanese data, the USEPA estimated the lifetime risk of developing skin cancer as 1 or 2 per 1000 people for each microgram of inorganic arsenic per liter of drinking water.

An additional strong evidence that drinking arsenic-contaminated water causes cancer is from Chile, where the population studied was nearly 10 times larger than that of the Taiwanese study population. In both the Taiwanese and Chilean studies, the people were exposed to a high level of arsenic (>500 µg/L) in drinking water.

Non-Cancerous Effects

Non-cancerous effects have been reported in humans after exposure to drinking inorganic arsenic contaminated water. Inorganic arsenic in drinking water may affect many organs including central and peripheral nervous systems, dermal, cardiovascular, gastrointestinal, and respiratory systems.

ARSENIC DETERMINATION

In the past, measurement of total elemental concentrations was considered to be sufficient for environmental considerations. Since the element occurs in different species and the species have different properties, a determination of total concentration of an element alone may not provide adequate information about the physical/chemical forms of the element and its toxicological properties. Therefore, it is essential to determine the individual species of an element, enabling one to obtain realistic information about the toxicity and transformation of the species. The term speciation refers to the determination of different oxidation states of an element that prevail in a certain specimen or to the identification and quantification of the biologically active compounds to which the element is bound.

TREATMENT TECHNOLOGIES FOR ARSENIC REMOVAL

Various treatment methods have been reported in the literature to remove arsenic effectively from the drinking water. Such treatment methods include coagulation/filtration, adsorption on activated alumina, adsorption on activated carbon, adsorption on ion-exchange resin, adsorption on hydrous ferric oxides, adsorption on various iron oxides, and adsorption/filtration by manganese greensand. After careful review, the USEPA suggested ion-exchange, activated alumina, reverse osmosis, modified coagulation/filtration, and modified lime softening as the best available technologies (BAT) based on arsenate removal.

Arsenic Removal by the Coagulation/Filtration Process

Coagulation/filtration processes are mainly used in large-scale water utilities. Based on the type and initial concentration of the contaminant, either precipitation or coprecipitation or both play an important role in the removal during coagulation. In a coagulation process, arsenic removal is dependent on adsorption and coprecipitation of arsenic onto metal hydroxides.

Arsenic Removal by Activated Alumina

Activated alumina (AA) treatment is a physical/chemical process by which ions in the drinking water are removed by the oxidized AA surface. Activated alumina treatment is considered to be an adsorption process, even though the reactions involved in the process involve actually an exchange of ions.

Arsenic Removal by Ion Exchange

Although ion exchange is an efficient treatment system for arsenic removal from drinking water, its application is limited to small and medium scale point-of-entry (POE) systems because of its high treatment cost as compared to other treatment technologies. Ion exchange is an ion-selective process, which removes As(V) significantly but does not remove As(III). It is an effective process for arsenic removal, if the source water contains <500-mg/L total dissolved solids and <150-mg/L sulfate; preoxidation of As(III) to As(V) is necessary.

The Role of Iron Oxides in Arsenic Removal

Iron oxides, oxyhydroxides, and hydroxides (all are called iron oxides) consist of Fe in association with O and/or OH. They differ in composition, in the nature of Fe, and in crystal structure. There are 16 iron oxides, and these iron oxides play an important role in a variety of industrial applications, including pigments for the paint industry, catalyst for industrial synthesis, and raw material for iron and steel industry. The application of iron oxide has been extended to remove metals from water and wastewater.

Recent studies showed that iron-based materials are effective in reducing arsenic to a low level in drinking water, which indicated that filtration systems containing iron-based materials offer an excellent choice among the treatment systems available for arsenic removal in small water facilities.

J. M. Azcue and J. O. Nriagu, in J. O. Nriagu, ed., *Arsenic in the Environment, Part 1: Cycling and Charecterization*, John Wiley & Sons, Inc., New York, 1994.

S. Niu, S. Cao, and E. Shen, in C. O. Abernathy, R. L. Calderon, and W. R. Chappell, eds., *Arsenic Exposure and Health Effects*, Chapmann and Hall, London, 1997.

USEPA, http://www.epa.gov/safewater/arsenic.html (October 31, 2001).

E. A. Woolson, in B. A. Fowler, ed., *Biological and Environmental Effects of Arsenic*, Elsevier Science, Elmsford, N. Y., 1983.

O. S. Thirunavukkarasu
T. Viraraghavan
University of Regina

ASBESTOS

Asbestos is a generic term referring to six types of naturally occurring mineral fibers that are or have been commercially exploited. These fibers belong to two mineral groups: serpentines and amphiboles. The serpentine group contains a single asbestiform variety: chrysotile. There are five asbestiform varieties of amphiboles: anthophyllite asbestos, cummingtonite, grunerite asbestos (amosite), riebeckite asbestos (crocidolite), tremolite asbestos, and actinolite asbestos. Usually, the term asbestos is applied only to those varieties that have been commercially exploited. That does not preclude the occurrence of other asbestos-like minerals, however.

The asbestos varieties share several properties: (1) they occur as bundles of fibers that can be easily separated from the host matrix or cleaved into thinner fibers; (2) the fibers exhibit high tensile strengths; (3) they show high length: diameter (aspect) ratios, with a minimum of 20 and up to 1000; (4) they are sufficiently flexible to be spun; and (5) macroscopically, they resemble organic fibers such as cellulose. Since asbestos fibers are all silicates, they exhibit several other common properties, such as incombustibility, thermal stability, resistance to biodegradation, chemical inertia toward most chemicals, and low electrical conductivity. The usual definition of asbestos fiber excludes numerous other fibrous minerals that may possess an asbestiform habit but do not exhibit all of the properties of asbestos.

The fractional breakdown of the recent world production of the various fiber types shows that the industrial applications of asbestos fibers have now shifted almost exclusively to chrysotile. Amosite and crocidolite are no longer being mined although some probably is still being sold from stock. Current use of amosite and crocidolite is estimated to be less than a few hundred tons annually. Actinolite asbestos, anthophyllite asbestos, and tremolite asbestos may be still mined in small amounts for local use; production probably is <100 tons annually.

GEOLOGY AND FIBER MORPHOLOGY

The genesis of asbestos fibers as mineral deposits required certain conditions with regard to chemical composition, nucleation, and fiber growth; such conditions must have prevailed over a period sufficiently long and perturbation-free to allow a continuous growth of the silicate chains into fibrous structures. Some of the important geological or mineralogical features of the industrially significant asbestos fibers are summarized in Table 1. More emphasis is given to chrysotile in the following section owing to its total dominance in the industry over the years.

Only three varieties of amphibole fibers will be discussed because (1) crocidolite and amosite were the only amphiboles with significant industrial uses in recent years; and (2) tremolite, although having essentially no industrial application, may be found as a contaminant in other fibers or in other industrial minerals (eg, chrysotile and talc).

Chrysotile

Chrysotile belongs to the serpentine group of minerals, varieties of which are found in ultra basic rock formations located in many places in the world.

Growth of chrysotile fibers at right angles to the walls of cracks (cross-vein) in massive serpentine formations led to the most common type of chrysotile deposit. Most of the industrial chrysotile fibers are extracted from deposits where fiber lengths can reach several centimeters, but most often do not exceed 1 cm.

Chrysotile is a hydrated magnesium silicate and its stoichiometric chemical composition may be given as

Table 1. Geological Occurence of Asbestos Fibers

	Chrysotile	Amosite	Crocidolite	Tremolite
mineral species	chrysotile	cummingtonite-grunerite	riebeckite	tremolite
structure	as veins in serpentine and mass fiber deposits	lamellar, coarse to fine, fibrous and asbestiform	fibrous in ironstones	long, prismatic, and fibrous aggregates
origin	alteration and metamorphism of basic igneous rocks rich in magnesium silicates	metamorphic	regional metamorphism	metamorphic
essential composition	hydrous silicates of magnesia	hydroxy silicate of Fe and Mg	hydroxy silicate of Na, Mg, and Fe	hydroxy silicate of Ca and Mg

$Mg_3Si_2O_5(OH)_4$. However, the geothermal processes that yield the chrysotile fiber formations usually involve the codeposition of various other minerals.

Chrysotile fibers can be extremely thin, the unit fiber having an average diameter of ~25 nm (0.025 μm). Industrial chrysotile fibers are aggregates of these unit fibers that usually exhibit diameters from 0.1 to 100 μm; their lengths range from a fraction of a millimeter to several centimeters, though most of the chrysotile fibers used are <1 cm.

Amphiboles

The amphibole group of minerals is widely found throughout the earth's crust. Their chemical composition can vary widely. Of the amphiboles, only a few varieties have an asbestiform habit and the latter occur in relatively low quantities.

The chemical composition of amphiboles readily reflects the complexity of the environment in which they formed.

From their respective compositions, the amphibole fibers can be viewed as a series of minerals in which one cation is progressively replaced by another at a given site.

The two most important amphibole asbestos minerals are amosite and crocidolite, and both are hydrated silicates of iron, magnesium, and sodium (crocidolite only). Although the macroscopic visual aspect of clusters of various types of asbestos fibers is similar, significant differences between chrysotile and amphiboles appear at the microscopic level. Under the electron microscope, chrysotile fibers are seen as clusters of fibrils, often entangled, suggesting loosely bonded, flexible fibrils. Amphibole fibers, on the other hand, often appear individually, rather than in fiber bundles.

CRYSTAL STRUCTURE OF ASBESTOS FIBERS

The microscopic and macroscopic properties of asbestos fibers stem from their intrinsic, and sometimes unique, crystalline features. As with all silicate minerals, the basic building blocks of asbestos fibers are the silicate tetrahedra that may occur as double chains $(Si_4O_{11})^{6-}$, as in the amphiboles, or in sheets $(Si_4O_{10})^{4-}$, as in chrysotile.

Chrysotile

In the case of chrysotile, an octahedral brucite layer having the formula $[Mg_6O_4(OH)_8]^{4-}$ is intercalated between each silicate tetrahedra sheet. This arrangement results in curvature of the sheets to form the tubular structure of chrysotile.

Amphiboles

The crystalline structure common to amphibole minerals consists of two ribbons of silicate tetrahedra placed back to back. Various cations are found between the ribbon structures.

PROPERTIES OF ASBESTOS FIBERS

Asbestos fibers used in most industrial applications consist of aggregates of smaller units (fibrils), which is most evident in chrysotile that exhibits an inherent, well-defined unit fiber. Diameters of fiber bundles in bulk industrial samples may be in the millimeter range in some cases; fiber bundle lengths may be several millimeters to 10 cm or more.

The mechanical processes employed to extract the fibers from the host matrix, or to further separate (defiberize, open) the aggregates, can impart significant morphological alterations to the resulting fibers. Typically, microscopic observations on mechanically opened fibers reveal fiber bends and kinks, partial separation of aggregates, fiber end-splitting, etc. The resulting product thus exhibits a wide variety of morphological features.

Morphological variances occur more frequently with chrysotile than amphiboles. The crystal structure of chrysotile, its higher flexibility, and interfibril adhesion allow for a variety of intermediate shapes when fiber aggregates are subjected to mechanical shear. Amphibole fibers are generally more brittle and accommodate less morphological deformation during mechanical treatment.

Fiber Length Distribution

For industrial applications, the fiber length and length distribution are of primary importance because they are closely related to the performance of the fibers in matrix reinforcement. Representative distributions of fiber lengths and diameters can be obtained through measurement and statistical analysis of microphotographs; fiber length distributions have also been obtained from automated optical analyzers.

Physicochemical Properties

The industrial applications of chrysotile fibers take advantage of a combination of properties: fibrous morphology, high tensile strength, resistance to heat and corrosion, low electrical conductivity, and high friction

Table 2. Physical and Chemical Properties of Asbestos Fibers

Property	Chrysotile	Amosite	Crocidolite	Tremolite
color	usually white to grayish green; may have tan coloration	yellowish gray to dark brown	cobalt blue to lavender blue	gray-white, green, yellow, blue
luster	silky	vitreous to pearly	silky to dull	silky
hardness, Mohs	2.5–4.0	5.5–6.0	4.0	5.5
specific gravity	2.4–2.6	3.1–3.25	3.2–3.3	2.9–3.2
optical properties	biaxial positive parallel extinction	biaxial positive parallel extinction	biaxial negative oblique extinction	biaxial negative oblique extinction
refractive index	1.53–1.56	1.63–1.73	1.65–1.72	1.60–1.64
flexibility	high	fair	fair to good	poor, generally brittle
texture	silky, soft to harsh	coarse but somewhat pliable	soft to harsh	generally harsh
spinnability	very good	fair	fair	poor
tensile strength, MPa[a]	1100–4400	1500–2600	1400–4600	<500
resistance to:				
acids	weak, undergoes fairly rapid attack	fair, slowly attacked	good	good
alkalies	very good	good	good	good
surface charge, mV (zeta potential)	+13.6 to +54[b]	−20 to −40	−32	
decomposition temperature, °C	600–850	600–900	400–900	950–1040
residual products	forsterite, silica, eventually enstatite	Fe and Mg pyroxenes, magnetite, hematite, silica	Na and Fe pyroxenes, hematite, silica	Ca, Mg, and Fe pyroxenes, silica

[a]To convert MPa to psi, multiply by 145.
[b]Chrysotile fibers tend to become negative after weathering and/or leaching.

coefficient. In many applications, the surface properties of the fibers also play an important role; in such cases, a distinction between chrysotile and amphiboles can be observed because of differences in their chemical composition and surface microstructure. Technologically relevant physical and chemical properties of asbestos fibers are given in Table 2.

Thermal Behavior. Asbestos fiber minerals are hydrated silicates so their behavior as a function of temperature is related first to dehydration (or dehydroxylation) reactions. In the case of chrysotile, the crystalline structure is stable up to ~550°C [depending on the heating period], where the dehydroxylation of the brucite layer begins. This process is completed near 750°C and is characterized by a total weight loss of 13%. The resulting magnesium silicate recrystallizes to form forsterite and silica in the temperature range 800–850°C, as an exothermic process.

Tensile Strength. The inherent tensile strength of a single asbestos fiber, based on the strength of Si–O–Si bonds in the silicate chain, should be near 10 GPa (1.45×10^6 psi). However, industrial fibers exhibit substantially lower values, because of the presence of various types of structural or chemical defects.

Asbestos Fibers in Aqueous Media. Although asbestos fibers cannot be viewed as water-soluble silicates, prolonged exposure of chrysotile or amphiboles to water (especially at high temperature) leads to slow progressive leaching of both their metal and silicate components. In the case of chrysotile fibers (in a given amount of water), the brucite layer will, fairly rapidly, dissolve in part, with concomitant increase in the pH of the solution.

Other Bulk Physical Properties. The hardness of asbestos fibers is comparable to that of other crystalline or glassy silicates. Compared to glass fibers, amphiboles have similar hardness values, while chrysotile shows lower hardness values.

The high electrical resistivity of asbestos fibers is well known and has been widely exploited in electrical insulation applications. In general, the resistivity of chrysotile is lower than that of the amphiboles, particularly in high humidity environments because of the availability of soluble ions.

With respect to magnetic properties, the intrinsic magnetic susceptibility of pure chrysotile is very weak. However, the presence of associated minerals such as magnetite, as well as substitution ions (Fe, Mn), increases the magnetic susceptibility to values ~ 1.9–3.5×10^{-6}/g Oe. With amphiboles, the magnetic susceptibility is much higher, mainly because of the high iron content.

Surface Properties

Surface Area. The specific surface area of industrial asbestos fibers obviously depends on the extent of their defiberization (opening), and is usually between 1 and 30 m²/g. With regard to amphibole fibers, surface areas of 1.8–9 m²/g have been reported for crocidolite and 1.3–5.5 m²/g for amosite.

Surface Charge in Aqueous Media. Because of dissolution–ionization effects, the surface of asbestos fibers in aqueous dispersions adopts an electrostatic charge. In the case of chrysotile, partial dissolution of the brucite layer leads to a positive surface charge (or potential), which is strongly influenced by the solution pH. In the case of amphibole fibers, the surface charge seems dominated by the silica component, and is generally observed to be negative, increasing toward 0 as the pH is decreased. Since the progressive leaching of magnesium from the external brucite layer of chrysotile gradually exposes silica, the surface potential rapidly decreases early in the leaching reaction.

Adsorption and Surface Chemical Grafting. As with silica and many other silicate minerals, the surface of asbestos fibers exhibits a significant chemical reactivity. In particular, the highly polar surface of chrysotile fibers promotes adsorption (physi- or chemisorption) of various types of organic or inorganic substances.

MINING AND MILLING TECHNOLOGIES

The finding and mapping of chrysotile asbestos ore deposits usually relies on magnetometric surveys largely because magnetite is associated with asbestos deposits, except in the case of ore bodies located in sedimentary formations. As in other mining operations, core drilling is used for a precise evaluation of the grade and volume of the ore body.

The choice of a particular mining method depends on a number of parameters, typically the physical properties of the host matrix, the fiber content of the ore, the amount of sterile materials, the presence of contaminants, and the extent of potential fiber degradation during the various mining operations. Most of the asbestos mining operations are of the open pit type, using bench drilling techniques.

The fiber extraction (milling) process must be chosen so as to optimize recovery of the fibers in the ore, while minimizing reduction of fiber length. Since the asbestos fibers have a chemical composition similar to that of the host rock, the separation processes must rely on differences in the physical properties between the fibers and the host rock rather than on differences in their chemical properties. Dry milling operatons are currently the most widely used.

Wet milling operations, where the asbestos is dispersed in water and not dried until after the final separation process is completed, offer advantages in dust control and the separation of mineral contaminants from the fiber product. However, wet process technology currently is used in only a few small-scale milling operations.

FIBER CLASSIFICATION AND STANDARD TESTING METHODS

In the beneficiation of asbestos fibers, several parameters are considered critically important and are used as standard evaluation criteria: length (or length distribution),

degree of opening and surface area, performance in cement reinforcement, and dust and granule content. The measurement of fiber length is important since the length determines the product category in which the fibers will be used and, to a large extent, their commercial value.

Dry Classification Method

The most widely accepted method for chrysotile fiber length characterization in the industry is the Quebec Standard (QS) test, which is a dry sieving method (on vibrating screens) that enables the fractionation of an asbestos fiber sample into four fractions of decreasing sizes (>2 mesh, <2 and ≥ 4 mesh, <4 and ≥ 10 mesh, and <10 mesh, with mesh defined as the number of openings per linear inch).

Wet Classification Method

A second industrially important fiber length evaluation technique is the Bauer–McNett (BMN) classification. In this method, a fiber slurry is circulated through a series of four grids with decreasing opening size (positioned vertically), thus yielding five fractions ($+4$, $+14$, $+35$, $+200$, and -200 mesh). A similar method, using smaller samples and horizontal grids has also been developed and referred to as the Turner–Newall classification. Other classification techniques that provide some insight on fiber lengths are the Ro–Tap test, the Suter–Webb Comb, and the Wash test.

INDUSTRIAL APPLICATIONS

Asbestos fibers have been used in a broad variety of industrial applications. In the peak period of asbestos consumption in industrialized countries, some 3000 applications, or types of products, have been listed. Because of recent restrictions and changes in end use-markets, most of these applications have been abandoned and the remainder is pursued under strictly regulated conditions.

The main properties of asbestos fibers that can be exploited in industrial applications are their thermal, electrical, and sound insulation; nonflammability; matrix reinforcement (cement, plastic, and resins); adsorption capacity (filtration, liquid sterilization); wear and friction properties (friction materials); and chemical inertia (except in acids). These properties have led to several main classes of industrial products or applications: cement, brake drums and pads, roofing, flooring, and gaskets; loose fiber as in heat, electrical or sound insulations; and textiles as in cloth, tubing, and jointing. The first category encompasses more than 98% of currently manufactured products. Asbestos-cement products alone account for about 85% of world use of asbestos.

ALTERNATIVE INDUSTRIAL FIBERS AND MATERIALS

Table 3 lists some of the materials and fibers that have been suggested or used in the development of

Table 3. Asbestos Substitutes and Relative Costs[a]

Minerals	Synthetic mineral fibers	Synthetic organic fibers
	< 2 $/kg[b]	
attapulgite	mineral wool	
diatomite	glass wool	
mica		
perlite		
sepiolite		
talc		
vermiculite		
wollastonite		
asbestos, grades 3–7		
	2–10 $/kg	
	steel fibers	polypropylene (PP)
	continuous filament glass	poly(vinyl alcohol) (PVA)
	alkali-resistant glass	polyacrylonitrile (PAN)
	aluminosilicates	
	10–20 $/kg	
	continuous filament glass	polytetrafluoroethylene (PTFE)
	>20 $/kg	
	alumina fibers	polybenzimidazole (PBI)
	silica fibers	aramid fibers
	graphite fibers	pitch and PAN carbon fibers

[a]In U.S. $, 1989.
[b]The natural organic fiber, cellulose (pulp), also falls in the < $2/kg range.

asbestos-free products along with an estimate of the cost ranges of asbestos fibers and several types of substitution materials.

HEALTH AND SAFETY FACTORS

The relationship between workplace exposure to airborne asbestos fibers and respiratory diseases is one of the most widely studied subjects of modern epidemiology.

The research efforts resulted in a consensus in some areas, although controversy still remains in other areas. It is widely recognized that the inhalation of long (considered usually as >5 μm), thin, and durable fibers in high concentrations over a long period of time can induce or promote lung cancer. It is also widely accepted that asbestos fibers can be associated with three types of diseases: asbestosis.

A further consensus developed within the scientific community regarding the relative carcinogenicity of the different types of asbestos fibers. There is strong evidence that the genotoxic and carcinogenic potentials

of asbestos fibers are not identical; in particular mesothelial cancer is most strongly associated with amphibole fibers.

Regulation

The identification of health risks associated with long-term, high level exposure to asbestos fibers, together with the fact that large quantities of these minerals were used (several million tons annually) in a variety of applications, prompted the enactment of regulations to limit the maximum exposure of airborne fibers in workplace environments.

M. A. Bernarde, ed., *Asbestos: The Hazardous Fiber*, CRC Press, Boca Raton, Fla., 1990, pp. 4, 30–38, 41, 80.

J. Harrod and V. Thorpe, *Asbestos, Politics and Economics of a Lethal Product*, International Federation of Chemical, Energy and General Workers' Unions, Geneva, Switzerland, 1984.

A. A. Hodgson, in L. Michaels and S. S. Chissick, eds., *Asbestos: Properties, Applications and Hazards*, Vol. 1, John Wiley & Sons, Inc., New York, 1979, pp. 89–90, 93.

R. L. Virta, *Asbestos*, U.S. Geological Survey Minerals Yearbook 2000 preprint, 2001, p. 7.

ROBERT L. VIRTA
U.S. Geological Survey

ATOMIC FORCE MICROSCOPY—AFM

PRINCIPLES

The atomic force microscope (AFM) was invented in 1986 at Stanford University by Binnig, Quate, and Gerber as a means of measuring interatomic forces. Their invention was based on a commonly used tool called a profilometer. In this device, a stylus attached to a spring is placed in contact with any relatively flat sample to be examined. The stylus is dragged along the surface of the sample and the deflection of the spring, measured with a variety of techniques, is translated into an image of the sample surface.

Contact Mode of Operation

Among the methods for monitoring the displacement of the cantilever in modern AFM systems, one finds optical deflection, optical interference, and cantilever-mounted strain gauges. Most current systems, however, are based on the optical deflection technique as shown in Figure 1.

Cantilever Assemblies

Modern AFM cantilever assemblies are made from a variety of materials. Most common today are microfabricated silicon or silicon-nitride cantilevers.

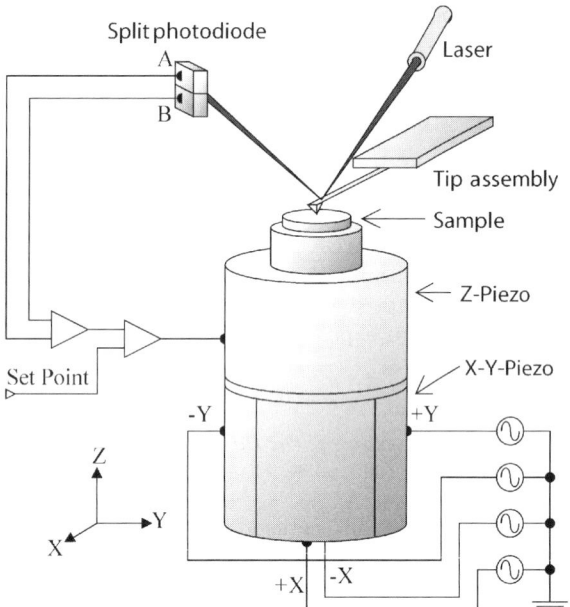

Figure. 1. A schematic diagram of a typical AFM system showing the piezoelectric tube scanner with its electric contacts, the sample mounted on top of the scanner, the tip assembly, and the optical elements.

Tip-Sample Interactions

To fully understand the operation of the AFM, it is important to consider the forces acting between the tip and the sample. However, the flexibility of the cantilever complicates a direct measurement of these forces. To simplify the analysis, assume a tip attached to a cantilever with infinite stiffness such that the tip–sample separation is controlled directly by the substrate–sample separation. Later, this restriction will be relaxed.

Resolution

A scanning tunneling microscope (STM), a predecessor of the AFM that requires a conducting sample, operates via a tunneling current that is exponentially dependant on the tip–sample separation distance. Therefore, the predominant contributor of current to the signal is from the very bottom of the lowest atom on the tip, with no contributions coming from atoms farther away from the sample. This tool is therefore capable of providing true atomic resolution. The tip–sample interaction of the AFM, however, is not as sensitive to distance, so atoms other then the end-most atom can contribute to the image, the number of which is dependent on the sharpness of the tip. The AFM, in general, cannot detect the presence or absence of individual atoms. Only under highly controlled conditions is the AFM capable of true atomic resolution. For large structures, in general, the resolution is mostly limited by tip–sample topographic convolution effects.

Noncontact Mode of Operation

The contact mode AFM operates solely within the repulsive region of the tip–sample force curve. This mode of operation can be highly destructive to the sample, particularly in the case of soft samples such as organic or biological material.

In the noncontact mode of operation, the tip is within the attractive region of the tip–sample force curve and the interaction forces are much smaller and therefore more difficult to measure. Such an operation requires a phase- sensitive method. To that end, the tip is oscillated with an amplitude of several nanometers near its resonance frequency.

Along with variations in amplitude, it is also possible to produce images based on variations in the phase lag between the cantilever driving force and the response of the cantilever.

Intermittent-Contact Mode Atomic Force Microscopy (FM)

Although the noncontact mode of operation significantly reduces the sample damage associated with the contact mode of operation, the repulsive interaction of the latter provides a higher force and therefore higher speed. In the case of samples with large topographic features, it may be advantageous to operate in the intermittent-contact mode of operation. In this mode of operation, the tip is still vibrated. However, here the tip is repeatedly brought from the attractive region into the repulsive region of the tip–sample force curve. Therefore, while increasing the interaction force to that near the contact mode of operation, the shear forces of the contact mode of operation are still avoided. This mode is commonly called the Tapping Mode by Digital Instruments or Intermittent Mode by other companies.

VARIANTS OF ATOMIC FORCE MICROSCOPY

The AFM has proven to be a useful base for a wide range of related techniques for mapping other qualities of a sample including, eg, electrical properties, thermal conductivity, elasticity, and friction.

Scanning Capacitance Microscopy

One technique used for electrical characterization of semiconductor materials is scanning capacitance microscopy (SCM). This technique maps the tip–sample capacitance at a given voltage at every point in the scan using the contact mode of operation. Such a measurement can yield the local doping level of semiconductor substrate.

Tunneling AFM

A standard technique for characterizing MOS devices on large-scale areas consists of obtaining $I–V$ curves across meter-size metal disks fabricated on the surface of the sample. These $I–V$ curves provide information on charge trapping, dielectric strength, and oxide degradation. In its standard form, however, this technique lacks the required resolution. Therefore, a new form of AFM has been developed that uses a conducting tip and a constant tip–sample bias that gives rise to a varying tunneling current through a thin dielectric film. Such an AFM is called a tunneling atomic force microscope (TAFM) or TUNA by Digital

Instruments. Here the tunneling current depends on the local thickness of the probed dielectric, making it possible to obtain dual maps of the topography of the top and bottom surfaces of the dielectric, as well as a map of the thickness or electrical quality of the dielectric, all at a resolution equaling approximately the thickness of the dielectric.

Electron Tunneling. MOS devices are typically fabricated on doped silicon that is thermally oxidized to produce a few nanometers-thick oxide. To control the generation of a conducting channel in the silicon beneath the oxide, a metal or polysilicon gate is fabricated above the oxide. If a large enough bias is placed across the thin barrier layer, such as the oxide in the MOS structure, quantum mechanical tunneling of electrons occurs through the layer.

Image Interpretation. The TAFM uses a conducting probe scanned over a dielectric of interest to produce images containing features that are interpreted as variations in one or more of the parameters mentioned earlier. In most cases, the dominant variation is assumed to be the change in the oxide thickness.

Current Feedback. Although a constant bias applied between tip and sample allows some imaging to be performed, it is easy to conceive of cases where this mode is not desirable. The reason is that since the tunneling equations are depend exponentially on the ratio of the applied bias and the oxide thickness, a sample containing both thin and thick regions of the oxide can cause the current to rapidly swing from below the noise level to above the range of the current detector, producing only a small range of sensitivity. On the other hand, maintaining a constant tunneling current across the oxide yields a direct, linear map of the thickness of the oxide. Note that such an interpretation breaks down at extremely thin oxides. Note, on the other hand, that forcing a current through an oxide can create charge-trapping centers that affect the tunneling current. Constant current imaging, however, makes this effect uniform over the surface of the sample, avoiding the large currents associated with a constant-voltage imaging of thin oxides. Care must be taken in controlling the current density during imaging while maintaining a reasonable scan speed.

Experimental System TAFM. TAFM system uses a conventional AFM to map the surface of a dielectric with areas that can be as large as 100×100 μm^2. For obtaining TAFM maps, a standard silicon or silicon–nitride cantilever is replaced with a conducting cantilever and the area is scanned while applying a tip–sample bias. A second, independent, computer-controlled feedback system simultaneously monitors the tip–sample tunneling current through the dielectric and adjusts the applied positive sample bias to maintain a constant 100-fA set-point current.

This system is also capable of producing I–V curves at a chosen location by using the AFM to position the tip at a particular location, ramping the applied bias and recording the current.

TAFM Tip Convolution. In the same manner as AFM imaging, TAFM imaging can suffer from tip–sample topographic convolution. However, where AFM convolution is a relatively simple addition of the radii of the probed surface feature and the tip in a given direction, the TAFM convolution is more complex. When the tip is in direct contact with a flat surface surrounding a feature, the current flows in a continuum of paths directly down through the dielectric. Upon contacting the feature, the tip lifts off the flat surface. At this point, one might imagine that if the tip lifts 1.5 nm above a 5-nm thick feature, then the resulting current would indicate a 6.5-nm thick feature. There are two complications that prevent this from being the case. One is the addition of a vacuum barrier in the original path and the other is the addition of a second continuum of paths contributing to the total current.

Conclusion. The TAFM has been developed as a method for locating and characterizing defects in thin dielectrics. Its ability to do so has been demonstrated on both a manufacturing defect, and on contaminated samples. Although currently useful as a tool for qualitative analysis, continued development would help to improve quantitative analysis. In particular, the issue of sample charging needs to be addressed further as this is a significant effect on the imaging process.

Scanning Spreading Resistance Microscopy

Scanning spreading resistance microscopy is another technique for profiling doped semiconductor structures using a conductive-tip AFM in the contact mode of operation. Typically, this method is applied to the cross-section of semiconductor device structures.

Lateral Force Microscopy

Lateral force microscopy scans the sample such that twisting is along the axis of the cantilever. Using an additional set of photodetectors, namely, a quadrant structure, yields a measure of changes in friction due to the sample material.

Electric/Magnetic Force Microscopy

Electric force microscopy is based on the use of a conducting-tip AFM to map the tip–sample electric field, while the magnetic counterpart uses a tip whose apex is magnetized.

Kelvin Probe Microscopy

Kelvin probe microscopy is a technique for measuring the relative work function of a sample and tip.

Force Modulation Microscopy

This AFM variant is operated in a constant-force contact mode with a small oscillation that is applied to the tip at a frequency higher than the frequency response of the feedback electronics of the AFM.

Scanning Thermal Microscopy

Scanning thermal microscopy is a method for measuring the thermal conductance of a sample that exhibits usually a lower resolution than obtained by other AFM techniques.

EXAMPLES OF ATOMIC FORCE MICROSCOPY APPLIED TO SEMICONDUCTORS

Dielectric breakdown of thin gate oxides fabricated on silicon wafers is, in general, a result of a large variety of defects including, eg, particulate contamination, pinholes, and surface roughness. One particular limiting factor affecting the quality of these oxides results from the presence of a variety of metallic islands introduced during wafer processing. Low concentrations of copper in particular have been observed to cause defects when the copper was deposited from hydrofluoric acid (HF) solutions in ultrathin oxides.

AFM Imaging of Silicon Etching

Researchers have observed that silicon dissolution occurs primarily next to copper precipitation. Pits result from significant amounts of silicon being removed. AFM examinations of contaminated samples that have been cleaned in either HF or HCl to remove copper show that the surface features remain intact. This result indicates that silicon dissolution occurs not directly beneath the copper deposit but in the area surrounding it, while the silicon beneath the copper is protected from etching.

TAFM Characterization of Copper Contaminated Gate Oxides

The TAFM system was used to electrically examine several samples of silicon contaminated with large quantities of copper. In this experiment, copper-contaminated wafers were thermally oxidized to 4–20 nm in thickness.

SUMMARY

This article covered three topics: The principles of AFM, variants of AFM, and examples of AFM applied to semiconductors. The first topic discussed the three main modes of operation of AFM, namely, the contact, noncontact, and intermittent-contact (tapping) modes. Also discussed were cantilever assemblies, tip–sample interactions, and resolution and manufacturers of AFMs.

The second topic described scanning spreading resistance microscopy, lateral force microscopy, electric and magnetic force microscopy, kelvin probe microscopy, force modulation microscopy, and scanning thermal microscopy. Again, these are the most prevalent variants of AFM that can be found in every well-equipped laboratory.

The third topic deals with examples of AFM applied to semiconductors. Here, this particular choice was made because of two reasons. The first reason is that the list of examples of results obtained with AFM, even if one were to collect only the most obvious ones, is remarkably rich and could not be fit into this article. The second reason for presenting this particular choice is that the utilization of variants of AFM in new fields of technology appears at an ever growing pace.

U. Hartman, "Theory of Non-contact Force Microscopy", *Scanning Tunneling Microscopy III*, Springer-Verlag, 1993, p. 293.

T. G. Ruskell, R. K. Workman, D. Chen, D. Sarid, S. Dahl, and S. Gilbert, *Appl. Phys. Lett.* **68**, 93 (1996).

D. Sarid, *Exploring Scanning Probe Microscopy using Mathematica*, John Wiley & Sons, Inc., Interscience, 1997.

CHARLES ANTHONY PETERSON
Intel Corporation
DROR SARID
College of Optical Sciences,
    University of Arizona

B

BARIUM

Barium, Ba, belongs to Group 2. (IIA) of the Periodic Table. Calcium, strontium, and barium belong to the alkaline earth metals and form a closely allied series in which the chemical and physical properties of the elements and their compounds vary systematically with increasing size, ionic and electropositive nature, and specific density. The properties are greatest for barium.

OCURRENCE

In its natural form, barium never occurs as the metal because of reactivity, but is almost always found as the ore barite, $BaSO_4$, which is also known as heavy spar. A smaller deposit is found as barium carbonate, $BaCO_3$ (witherite) barium carbonate can easily be decomposed by heating (calcination) to BaO. Barium oxide is used commercially for the production of barium metal.

PHYSICAL PROPERTIES

Pure barium is a silvery white metal, although contamination with nitrogen lead to a yellowish color. The metal is relatively soft and ductile and may be worked readily. It is fairly volatile (though less than magnesium) and this property is used to advantage in commercial production. Barium has a body-centered entered cubic (bcc) crystal structure at atmospheric pressure, but undergoes structural phase transitions at high pressure. Barium also exhibits hig ressure induced superconductivity at low temperatures and is an essential component of several high temperature superconductors eg, $YBa_2Cu_3O_7$.

Physical properties of barium are listed in Table 1.

CHEMICAL PROPERTIES

Barium has a valence electron configuration of $6s_2$ and characteristically forms divalent compounds. It is an extremely reactive metal and its compounds possess large free energies of formation.

ENVIRONMENTAL CONCERNS

Barium occurs in seawater in a concentration of 6 μg/L. ln fresh water, the barium content depends on the occurence of barium and the concentration of anions that form barium salts of low solubility such as sulfate and carbonate ions. Values in drinking water vary with location.

Barium levels in the air are not well documented. There is no correlation between the degree of industrilization and the barium concentration in the air. Higher levels are found in areas with high natural dust levels. Anthropogenic emissions are primarily industrial. Other atmospheric emissions result from the handling of barium

Table 1. Physical Properties of Barium

Physical properties						Value
atomic number						56
relative atomic mass A_r						137.34
mass number (natural abundance, %) of stable isotopes:						130 (0.101), 132 (0.097), 134 (2.42), 135 (6.59), 136 (7.81), 137 (11.3), 138 (71.7)
density at 20°C, g/cm^3						3.74
melting point °C						726.2
boiling point, at 101.3 kPa, °C^a						1637
hardness (mohs scale)						1.25
coefficient of thermal expansion, α_1, (mean, 0–100°C)						$1.8 \times 10^{-5} K^{-1}$
heat of fusion, ΔH_m, kJ/mol^b						7.98
heat of vaporization, ΔH_v, kJ/mol						140.3 kJ/mol
specific heat capacity c at 20°C J/hg·K						192 J
at 900°C, J/hg·K						230 J kg
vapor pressure at						
temperature, °C	630	730	1050	1300	1520	1637
pressure, kPa	0.00133	0.0133	0.133	13.3	53.3	101.3
electrical resistivity, Ω cm						40×10^{-6}
extra high purity						30×10^{-6}
liquid barium at mp						314×10^{-6}

^{a} To convert kPa to mm Hg, multiply by 7.5.
^{b} To convert J to cal, divide by 4.18.

Co or materials containing barium compounds, such as welding wire.

HEALTH AND SAFETY FACTORS

If barium metal contacts moisture in the eyes, on the skin, or in the respiratory tract, severe corrosive irritation may result. Inhalation of dust or fume may cause severe respiratory irritation, cough, difficulty in breathing, and chemical pnemonitis. Contact with skin causes irritation and possible corrosive damage.

Barium metal poisoning is virtually unknown in industry, although the potential exists when the soluble barium compound forms are used. When ingested or given orally, the soluble, ionized barium compounds exert a profound effect on all muscles and especially smooth muscles, markedly increasing their contractility.

USES

The major use of barium is the production of barium–aluminum alloy—evaporation getters (gas absorbers) in CRTs (cathode ray tubes) for television sets and computer monitors to generate and to maintain high vacuum by reaction with detrimental gases. Barium is used as getter material in X-ray and emitter tubes and in sodium vapor lamps.

C. L. Mantell, in C. A. Hampel, ed., *Rare Metals Handbook*, 2nd ed., Reinhold Publishing Corp., London, 1961, pp. 15–31.

J. W. Mellor, *Comprehensive Treatise on Inorganic and Theoretical Chemistry*, Vol. 3, Longmans, Green & Co., Inc., New York, 1923, pp. 619–652.

A. L. Reeves, "Barium", in L. Friberg, G. F. Nordberg, and B. Velimir, eds., *Handbook on the Toxicology of Metals— Volume II: Specific Metals*, Elsevier Science Publishers, Amsterdam, 1986, pp. 84–93.

J. P. Searls, "Barite" *Mineral Commodity Summaries*, U. S. Geological Survey, Jan. 2002.

CLAUDIO BOFFITO
Saes Getters SpA

BARIUM COMPOUNDS

In its natural form, barium, Ba, never occurs as the metal but is almost always found as the ore barite, $BaSO_4$. More than 90% of all barium is actually used as the ore, albeit after preliminary beneficiation. Witherite, the only other significant natural barium ore, is not mined commercially.

Barium is a member of the alkaline-earth group of elements in Group 2 (IIA) of the Period Table.

In metallic form, barium is very reactive, reacting readily with water to release hydrogen. In aqueous solution it is present as an ion with a +2 charge. Barium acetate, chloride, hydroxide, and nitrate are water-soluble, whereas barium arsenate, chromate, fluoride, oxalate, and sulfate are not.

BARITE

Barite, natural barium sulfate, $BaSO_4$, commonly known as barytes, and sometimes as heavy spar, till, or cawk, occurs in many geological environments in sedimentary, igneous, and metamorphic rocks. Commercial deposits are of three types: vein and cavity filling deposits; residual deposits; and bedded deposits. Barite is widely distributed and has minable deposits in many countries.

Mineralogically, barite crystallizes in the dipyramidal class of the orthorhombic system. Barite is most commonly associated with quartz, chert, jasperoid, calcite, dolomite, siderite, rhodochrosite, celestite, gluorite, various sulfide minerals, and their oxidation products.

Barite is a moderately soft crystalline mineral, Mohs' hardness 3–3.5; sp gr 4.3–4.6; n_D 1.64. The ore is white opaque to transparent, but impurities can produce pale shades of yellow, green, blue, brown, red, or gray-black. The most important impurities are Fe_2O_3, Al_2O_3, SiO_2, and $SrSO_4$, all of which are undesirable in chemical-grade barite. When the barite is used for drilling mud, the iron content can be permitted to be much higher than for other uses.

Uses

Its largest use is in drilling muds. Finely ground barite which may be bleached, usually by sulfuric acid, or unbleached, is used as a filler or extender in paints (qv), especially in automotive undercoats, where its low oil absorption, easy wettability in oils, and good sanding properties are advantageous (see FILLERS). It is also used as a filler in plastics and rubber products.

In the glass (qv) and ceramic industry (see CERAMICS), barite can be used both as a flux, to promote melting at a lower temperature or to increase the production rate, and as an additive to increase the refractive index of glass.

BARIUM ACETATE

Barium acetate, $Ba(C_2H_3O_2)_2$, crystallizes from an aqueous solution of acetic acid and barium carbonate or barium hydroxide. The level of hydration depends on crystallization temperature.

BARIUM BROMIDE

Barium bromide, $BaBr_2$, mp 854°C, density 4.781 g/mL, also exists as barium bromide dihydrate, $BaBr_2 \cdot 2H_2O$, dehydration temperature 120°C, density 3.58 g/mL. Barium bromide is very soluble in methanol, yet almost insoluble in ethanol.

BARIUM CARBONATE

Most barium compounds are prepared from reactions of barium carbonate, $BaCO_3$, which is commercially manufactured by the "black ash" process from barite and coke in a process identical to that for strontium carbonate production.

Precipitated or synthetic barium carbonate is the most commercially important of all the barium chemicals except for barite.

Uses

There are several different grades of barium carbonate manufactured to fit the specific needs of a wide variety of applications: very fine, highly reactive grades are made for the chemical industry; coarser and more readily handleable grades are mainly supplied to the glass industry.

The main use for barium carbonate is in the manufacture of glass.

In 2000, 50% of the barium carbonate produced was used in glass manufacturing. Other uses include manufacturing of brick and clay products, barium chemicals, barium ferrites, and in the production of photographic papers.

BARIUM CHLORIDE

Both anhydrous barium chloride, $BaCl_2$, mol wt 208.25, density 3.856 g/mL, and barium chloride dihydrate, $BaCl_2 \cdot 2 H_2O$, mol wt 244.28, density 3.097 g/mL, are produced from a filtered aqueous solution formed by the reaction of hydrochloric acid and $BaCO_3$ or BaS. If BaS is used, the H_2S generated must be appropriately handled.

$BaCl_2$ is used in heat treating baths because of the eutectic mixtures it readily forms with other chlorides. $BaCl_2$ is also used to set up porcelain enamels for sheet steel (see ENAMELS, PORCELAIN OR VITREOUS; STEEL), and it is used to produce blanc fixe.

BARIUM 2-ETHYLHEXANOATE

Barium 2-ethylhexanoate, $Ba(C_8H_{16}O_2)_2$, also known as barium octanoate or barium octoate is usually used in synergistic combination with cadmium or zinc organic salts as a thermal stabilizer for PVC.

BARIUM FLUORIDE

Barium fluoride, BaF_2, is a white crystal or powder. Under the microscope crystals may be clear and colorless.

Barium fluoride is used commercially in combination with other fluorides for arc welding (qv) electrode fluxes. Other reported uses of barium fluoride include the manufacture of fluorophosphate glass; stable fluoride glass; fluoroaluminate glass; fluorozirconate glass; infared transmitting glass; in oxidation-resistant ceramic coatings; in the manufacture of electric resistors; as a superconductor with copper oxide; as a fluoride optical fiber (see FIBER OPTICS; GLASS); and in a high repetition rate uv excimer laser.

BARIUM HYDROSULFIDE

Barium hydrosulfide, $Ba(HS)_2$, is formed by absorption of hydrogen sulfide into barium sulfide solution. On addition of alcohol, barium hydrosulfide tetrahydrate, $Ba(HS)_2 \cdot 4 H_2O$, crystallizes as yellow rhombic crystals that decompose at 50°C. Solid barium hydrosulfide is very unstable.

BARIUM HYDROXIDE

Barium hydroxide is the strongest base and has the greatest water-solubility of the alkaline-earth elements. Barium hydroxide (barium hydrate, caustic baryta) exists as the octahydrate, $Ba(OH)_2 \cdot 8 H_2O$, the monohydrate, $Ba(OH)_2 \cdot H_2O$, or as the anhydrous material, $Ba(OH)_2$. The octahydrate and monohydrate have sp gr 2.18 and 3.74, respectively. The mp of the octahydrate and anhydrous are 77.9°C and 407°C, respectively.

Barium hydroxide is used in the manufacture of barium greases and plastic stabilizers such as barium 2-ethylhexanoate, in papermaking, in sealing compositions (see SEALANTS), vulcanization accelerators, water purification, pigment dispersion, in a formula for self-extinguishing polyurethane foams, and in the protection of objects made of limestone from deterioration (see FINE ART EXAMINATION AND CONSERVATION; LIME AND LIMESTONE).

BARIUM IODIDE

Barium iodide dihydrate, $BaI_2 \cdot 2 H_2O$, crystallizes from hot aqueous solution. BaI_2 is useful in making other iodides. BaI_2 has been cited for producing ir transparent glasses that are useful in power transmission from CO and CO_2 lasers (qv); as a catalyst promoter in carbonylation reactions; as being useful in chemical vapor depositionas a precursor in forming the superconducting composition $YBa_2Cu_3O_{7-x}$; as a sintering aid for aluminum nitride, and in phosphor formulations for cathode-ray tubes.

BARIUM METABORATE

Barium metaborate monohydrate, $Ba(BO_2)_2 \cdot H_2O$, has a sp gr of 3.25–3.35, and can be prepared from the reaction of a solution of BaS and sodium tetraborate.

$Ba(BO_2)_2 \cdot H_2O$, used in flame retardant plastic formulations and as a partial or complete replacement for antimony oxide (see FLAME RETARDANTS), is excellent as an afterglow suppressant. The low refractive index of $Ba(BO_2)_2$ results in greater transparency and brighter colors in formulated plastics. Barium metaborate has been reported in paint formulations to convey insecticidal properties (see INSECT CONTROL TECHNOLOGY). $Ba(BO_2)_2$ has been reported to be used in antibacterial coatings for aluminum heat exchanger surfaces of air conditioners.

BARIUM NITRATE

Barium nitrate, $Ba(NO_3)_2$, occurs as colorless crystals; mp 592°C; sp gr 3.24. It is used in pyrotechnic green flares, fracer bullets, primers, and in detonators.

BARIUM NITRITE

Barium nitrite, $Ba(NO_2)_2$, crystallizes from aqueous solution as barium nitrite monohydrate, $Ba(NO_2)_2 \cdot H_2O$, which has yellowish hexagonal crystals, sp gr 3.173. The monohydrate loses its water of crystallization at 116°C. It has been used in diazotization reactions.

BARIUM OXIDE AND PEROXIDE

Barium oxide, BaO, occurs as colorless cubic or hexagonal crystals; mp 1923°C; sublimation ca 2000°C; bp ca 3088°C; sp gr (cubic) 5.72, (hexagonal) 5.32.

Of the alkaline-earth carbonates, $BaCO_3$ requires the greatest amount of heat to undergo decomposition to the oxide. Thus carbon in the form of coke, tar, or carbon black, is added to the carbonate to lower reaction temperature from about 1300°C in the absence of carbon to about 1050°C.

BaO is used to impart improved strength to porcelain, as a solid base catalyst, in specialty cements, and for drying gases.

When heated in air or oxygen to 500°C, barium oxide is converted readily to barium peroxide. BaO_2.

Reported uses of BaO_2 include in the cathodes of fluorescent lamps, formation of $YBa_2Cu_3O_{7-x}$ superconducting phase from CuN_3, BaO_2, and Y_2O_3, and as a drying agent forlithographic inks.

BARIUM SODIUM NIOBIUM OXIDE

Barium sodium niobium oxide, $Ba_2NaNb_5O_{15}$, finds application for its dielectric, piezoelectric, nonlinear crystal and electro-optic properties. It has been used in conjunction with lasers for second harmonic generation and frequency doubling. The crystalline material can be grown at high temperature, mp ca 1450°C.

BARIUM SULFATE

Barium sulfate, $BaSO_4$, occurs as colorless rhombic crystals, mp 1580°C (dec); sp gr 4.50. It is soluble in concentrated sulfuric acid, forming an acid sulfate; dilution with water reprecipitates barium sulfate. Precipitated $BaSO_4$ is known as blanc fixe.

Because of its extreme insolubility, barium sulfate is not toxic. In medicine, barium sulfate is widely used as an x-ray contrast medium (see IMAGING TECHNOLOGY; X-RAY TECHNOLOGY). It is also used in photographic papers, filler for plastics, and in concrete as a radiation shield.

BARIUM SULFIDE

Impure barium sulfide with 20–35% contaminants is produced in large volume by the black ash kiln. Pure barium sulfide, BaS, occurs as colorless cubic crystals, sp gr 4.25 and as hexagonal plates of barium sulfide hexahydrate, $BaS \cdot 6H_2O$. BaS melts at 2227°C.

BaS is used in the manufacture of lithophone, useful as a white pigment in paints, and has been used in the production of thin-film electroluminescent phosphors.

BARIUM TITANATE

The basic crystal structure of barium titanate, $BaTiO_3$, the so-called perovskite structure, after the mineral, $CaTiO_3$, leads to unique, outstanding dielectric properties.

Barium titanate has widespread use in the electronics industry, ie, miniature capacitors (see CERAMICS AS ELECTRICAL MATERIALS).

YTTRIUM–BARIUM–COPPER OXIDE

Yttrium–barium–copper oxide, $YBa_2Cu_3O_{7-x}$, is a high T_c material which has been found to be fully superconductive at temperatures above 90 K, a temperature that can be maintained during practical operation. Ultrapure powders of yttrium–barium–copper oxide that are sinterable into single-phase superconducting material at low temperatures are required, creating a worldwide interest in high purity barium chemicals.

HEALTH AND SAFETY ASPECTS

Environmental Levels and Exposures

Agricultural soils contain Ba^{2+} in the range of several micrograms per gram. The Environmental Protection Agency, under the Safe Drinking Water Act, has set a limit for barium of 1 mg/L for municipal waters in the United States.

Toxicity

The toxicity of barium compounds depends on solubility.

Soluble Compounds

The mechanism of barium toxicity is related to its ability to substitute for calcium in muscle contraction. Toxicity results from stimulation of smooth muscles of the gastrointestinal tract, the cardiac muscle, and the voluntary muscles, resulting in paralysis. Skeletal, arterial, intestinal, and bronchial muscle all seem to be affected by barium.

F. A. Cotton and G. Wilkinson, *Advanced Inorganic Chemistry*, John Wiley & Sons, Inc., New York, 1980, pp. 271–273.

Industrial Minerals and Rocks, 4th ed., American Institute of Mining, Metallurgical, and Petroleum Engineers, Inc., New York, 1975, pp. 427–442.

A. L. Reeves, in L. Friberg, G. F. Nordberg, and V. B. Vouk, eds., *Handbook on the Toxicology of Metals*, Vol. 2, 2nd ed., Elsevier, New York, 1986, 84–94.

J. P. Searls, "Barite", *Mineral Commodity Summaries*, U.S. Geological Survey, Reston, Va., Jan. 2002.

PATRICK M. DIBELLO
JAMES L. MANGANARO
ELIZABETH R. AGUINALDO
FMC Corporation
TARIQ MAHMOOD
CHARLES B. LINDAHL
Elf Atochem North America Inc.

BARRIER POLYMERS

Barrier polymers are used for many packaging and protective applications. As barriers they separate a system, such as an article of food or an electronic component, from an environment. That is, they limit the introduction of matter from the environment into the system or limit the loss of matter from the system or both.

All polymers are barriers to some degree; however, no polymer is a perfect barrier. Polymers only limit the movement of substances. When a polymer limits the movement enough to satisfy the requirements of a particular application, it is a barrier polymer for that application. Hence, barrier polymer finds definition in the application.

THE PERMEATION PROCESS

Barrier polymers limit movement of substances, hereafter called permeants. The movement can be through the polymer or, in some cases, merely into the polymer. The overall movement of permeants through a polymer is called permeation, which is a multistep process. First, the permeant molecule collides with the polymer. Then, it must adsorb to the polymer surface and dissolve into the polymer bulk. In the polymer, the permeant "hops" or diffuses randomly as its own thermal kinetic energy keeps it moving from vacancy to vacancy while the polymer chains move. The random diffusion yields a net movement from the side of the barrier polymer that is in contact with a high concentration or partial pressure of the permeant to the side that is in contact with a low concentration of permeant. After crossing the barrier polymer, the permeant moves to the polymer surface, desorbs, and moves away.

Permeant movement is a physical process that has both a thermodynamic and a kinetic component. For polymers without special surface treatments, the thermodynamic contribution is in the solution step. The permeant partitions between the environment and the polymer according to thermodynamic rules of solution. The kinetic contribution is in the diffusion. The net rate of movement is dependent on the speed of permeant movement and the availability of new vacancies in the polymer.

SMALL MOLECULE PERMEATION

Permanent Gases

Table 1 lists the permeabilities of oxygen, nitrogen, and carbon dioxide for selected barrier and nonbarrier polymers at 20°C and 75% rh. The effect of temperature and humidity are discussed later. For many polymers, the permeabilities of nitrogen, oxygen, and carbon dioxide are in the ratio 1:4:14.

The traditional definition of a barrier polymer required an oxygen permeability <2 nmol/(m · s · GPa) [originally, <(cc · mil)/(100 in.^{2} · d · atm)] at room temperature. This definition was based partly on function and partly on conforming to the old commercial unit of permeability. The old commercial unit of permeability was created so that

Table 1. Permeabilities of Selected Polymers

| Polymer | Gas permeability nmol/(m · s · GPa) | | |
	Oxygen	Nitrogen	Carbon dioxide
vinylidene chloride copolymers	0.02–0.30	0.005–0.07	0.1–1.5
ethylene–vinyl alcohol (EVOH) copolymers, dry at 100% rh	0.014–0.095, 2.2–1.1		
nylon-MXD6^a	0.30		
nitrile barrier polymers	1.8–2.0		6–8
nylon-6	4–6		20–24
amorphous nylon (Selar^b PA 3426)	5–6		
poly(ethylene terephthalate) (PET)	6–8	1.4–1.9	30–50
poly(vinyl chloride) (PVC)	10–40		40–100
high density polyethylene	200–400	80–120	1200–1400
polypropylene	300–500	60–100	1000–1600
low density polyethylene	500–700	200–400	2000–4000
polystyrene	500–800	80–120	1400–3000

^{a}Trademark of Mitsubishi Gas Chemical Co.
^{b}Trademark of Du Pont.

the oxygen permeability of Saran Wrap brand plastic film, a trademark of The Dow Chemical Company, would have a numerical value of 1. However, the traditional definition of a barrier polymer is a good starting point for food packaging.

Poly(ethylene terephthalate), with an oxygen permeability of 8 nmol/(m · s · GPa), is not considered a barrier polymer by the old definition; however, it is an adequate barrier polymer for holding carbon dioxide in a 2-L bottle for carbonated soft drinks. The solubility coefficients for carbon dioxide are much larger than for oxygen. For the case of the PET soft drink bottle, the principal mechanism for loss of carbon dioxide is by sorption in the bottle walls as 500 kPa (5 atm) of carbon dioxide equilibrates with the polymer. For an average wall thickness of 370 μm (14.5 mil) and a permeability of 40 nmol/(m · s · GPa), many months are required to lose enough carbon dioxide (15% of initial) to be objectionable.

The diffusion and solubility coefficients for oxygen and carbon dioxide in selected polymers have been collected in Table 2. Determination of these coefficients is neither common, nor difficult. Methods are discussed later. The values of S for a permeant gas do not vary much from polymer to polymer. The large differences that are found for permeability are due almost entirely to differences in D.

Polymers with Good Barrier-to-Permanent Gases

Those polymers that are good barriers to permanent gases, especially oxygen, have important commercial significance.

Table 2. Diffusion and Solubility Coefficients for Oxygen and Carbon Dioxide in Selected Polymers at 23°C, Dry

Polymer	Oxygen		Carbon dioxide	
	D, m^2/s	S, nmol/(m^3·GPa)^a	D, m^2/s	S, nmol/(m^3·GPa)^a
vinylidene chloride copolymer	1.2×10^{-14}	1.01×10^{13}	1.3×10^{-14}	3.2×10^{13}
EVOH copolymer^b	7.2×10^{-14}	2.4×10^{12}		
acrylonitrile barrier polymer	1.0×10^{-13}	1.0×10^{13}	9.0×10^{-14}	4.4×10^{13}
PET	2.7×10^{-13}	2.8×10^{13}	6.2×10^{-14}	8.1×10^{14}
PVC	1.2×10^{-12}	1.2×10^{13}	8.0×10^{-13}	9.7×10^{13}
polypropylene	2.9×10^{-12}	1.1×10^{14}	3.2×10^{-12}	3.4×10^{14}
high density polyethylene	1.6×10^{-11}	7.2×10^{12}	1.1×10^{-11}	4.3×10^{13}
low density polyethylene	4.5×10^{-11}	2.0×10^{13}	3.2×10^{-11}	1.2×10^{14}

^{a}For unit conversion, see equation 5.
^{b}42 mol % ethylene.

Vinylidene chloride copolymers are available as resins for extrusion, latices for coating, and resins for solvent coating. Comonomer levels range from 5 to 20 wt%. Common comonomers are vinyl chloride, acrylonitrile, and alkyl acrylates. The permeability of the polymer is a function of type and amount of comonomer. As the comonomer fraction of these semicrystalline copolymers is increased, the melting temperature decreases and the permeability increases. The permeability of vinylidene chloride homopolymer has not been measured.

Hydrolyzed ethylene–vinyl acetate copolymers, commonly known as EVOH copolymers, are usually used as extrusion resins, although some may be used in solvent-coating applications.

Copolymers of acrylonitrile are used in extrusion and molding applications. Commercially important comonomers for barrier applications include styrene and methyl acrylate.

Polyamide polymers can provide good-to-moderate barrier-to-permeation by permanent gases.

Often, used polymers have adequate properties for some applications. Poly(ethylene terephthalate) is used to make films and bottles. This polymer is commonly made from ethylene glycol and dimethyl terephthalate or from ethylene glycol and terephthalic acid. PET is a moderate barrier-to-permanent gases; however, it is an excellent barrier to flavors and aromas. The oxygen permeability decreases slightly with increasing humidity. Poly(vinyl chloride) is a moderate barrier-to-permanent gases. Plasticized PVC is used as a household wrapping film. The plasticizers greatly increase the permeabilities.

Water–Vapor Transmission. Table 3 lists WVTR values for selected polymers. Comparison of Tables 1 and 3 shows that often there is a reversal of roles. Those polymers that are good oxygen barriers are often poor water–vapor barriers and vice versa, which can be rationalized as follows. Barrier polymers often rely on dipole–dipole interactions to reduce chain mobility and, hence, diffusional movement of permeants. These dipoles can be good sites for hydrogen bonding. Water molecules are attracted to these sites, leading to high values of S. Furthermore, the water molecules enhance D by interrupting the attractions and chain packing. Polymer molecules without dipole–dipole interactions, such as polyolefins, dissolve very little water

and have low WVTR and permeability values. The low values of S more than compensate for the naturally higher values of D.

Large Molecule Permeation

The permeation of flavor, aroma, and solvent molecules in polymers follows the same physics as the permeation of small molecules. However, there are two significant differences. For these larger molecules, the diffusion coefficients are much lower and the solubility coefficients are much higher. This means that steady-state permeation may not be reached during the storage time of some packaging situations. Hence, large molecules from the environment might not enter the contents, or loss of flavor molecules would be limited to sorption into the polymer. However, since the solubility coefficient is large, the loss of flavor could be important solely from sorption in the polymer. Furthermore, the large solubility coefficient can lead to enough sorption of the large molecule that plasticization occurs in the polymer, which can increase the diffusion coefficient. Generally, vinylidene chloride copolymers and glassy polymers such as polyamides and EVOH are good barriers to flavor and aroma permeation whereas the polyolefins are poor barriers.

Table 3. WVTR of Selected Polymers^a

Polymer	WVTR, nmol/(m·s)
vinylidene chloride copolymers	0.005–0.05
HDPE	0.095
polypropylene	0.16
LDPE	0.35
EVA, 44 mol% ethylene^b	0.35
PET	0.45
PVC	0.55
EVA, 32 mol% ethylene^b	0.95
nylon-6,6, nylon-11	0.95
nitrile barrier resins	1.5
polystyrene	1.8
nylon-6	2.7
polycarbonate	2.8
nylon-12	15.9

^{a}At 38°C and 90% rh unless otherwise noted (13).
^{b}Measured at 40°C.

PHYSICAL FACTORS AFFECTING PERMEABILITY

Several physical factors can affect the barrier properties of a polymer. These include temperature, humidity, orientation, and cross-linking.

Temperature

The temperature dependence of the permeability arises from the temperature dependencies of the diffusion coefficient and the solubility coefficient. The permeabilities increase ~5% per degree in polymers that are below their T_g such as acrylonitrile copolymers, EVOH, and PET.

Humidity

When a polymer equilibrates with a humid environment, it absorbs water.

Orientation

The effect of orientation on the permeability of polymers is difficult to assess because the words orientation and elongation or strain have been used interchangeably in the literature. Diffusion in some polymers is unaffected by orientation; in others, increases or decreases are observed.

Cross-Linking

Cross-linking has been shown to decrease the diffusion coefficient.

BARRIER STRUCTURES

Barrier polymers are often used in combination with other polymers or substances. The combinations may result in a layered structure either by coextrusion, lamination, or coating. The combinations may be blends that are either miscible or immiscible. In each case, the blend seeks to combine the best properties of two or more materials to enhance the value of a final structure.

Layered Structures

Whenever a barrier polymer lacks the necessary mechanical properties for an application or the barrier would be adequate with only a small amount of the more expensive barrier polymer, a multilayer structure via coextrusion or lamination is appropriate. Whenever the barrier polymer is difficult to melt process or a particular traditional substrate such as paper or cellophane is necessary, a coating either from latex or a solvent is appropriate. A layered structure uses the barrier polymer most efficiently since permeation must occur through the barrier polymer and not around the barrier polymer. No short cuts are allowed for a permeant.

Immiscible Blends

When two polymers are blended, the most common result is a two-phase composite. The most interesting blends have good adhesion between the phases, either naturally or with the help of an additive. The barrier properties of an immiscible blend depend on the permeabilities of the polymers, the volume fraction of each, phase continuity, and the aspect ratio of the discontinuous phase. Phase continuity refers to which phase is continuous in the composite. Continuous for barrier applications means that a phase connects the two surfaces of the composite. Typically, only one of the two polymer phases is continuous, with the other polymer phase existing as islands. It is possible to have both polymers be continuous.

Miscible Blends

Sometimes a miscible blend results when two polymers are combined. A miscible blend has only one amorphous phase because the polymers are soluble in each other. There may also be one or more crystal phases. Simple theory has supported the empirical relation for the permeability of a miscible blend.

PREDICTING PERMEABILITIES

Reasonable prediction can be made of the permeabilities of low molecular weight gases such as oxygen, nitrogen, and carbon dioxide in many polymers. The diffusion coefficients are not complicated by the shape of the permeant, and the solubility coefficients of each of these molecules do not vary much from polymer to polymer. Hence, all that is required is some correlation of the permeant size and the size of holes in the polymer matrix. Reasonable predictions of the permeabilities of larger molecules such as flavors, aromas, and solvents are not easily made. The diffusion coefficients are complicated by the shape of the permeant, and the solubility coefficients for a specific permeant can vary widely from polymer to polymer.

The permachor method is an empirical method for predicting the permeabilities of oxygen, nitrogen, and carbon dioxide in polymers. In this method, a numerical value is assigned to each constituent part of the polymer. An average number is derived for the polymer, and a simple equation converts the value into a permeability. This method has been shown to be related to the cohesive energy density and the free volume of the polymer. The model has been modified to liquid permeation with some success.

For larger molecules, independent predictions of the diffusion coefficients and the solubility coefficients are required.

MEASURING BARRIER PROPERTIES

Measuring the barrier properties of polymers is important for several reasons. The effects of formulation or process changes need to be known, new polymers need to be evaluated, data are needed for a new application before a large investment has been made, and fabricated products need to have performance verified.

Oxygen Transport

The most widely used methods for measuring oxygen transport are based upon the Ox-Tran instrument (Modern

Controls, Inc.). Several models exist, but they all work on the same principle. The most common application is to measure the permeability of a film sample.

Water Transport

Two methods of measuring WVTR are commonly used. The newer method uses a Permatran-W (Modern Controls, Inc.). The other method is the ASTM cup method.

Carbon Dioxide Transport

Measuring the permeation of carbon dioxide occurs far less often than measuring the permeation of oxygen or water. A variety of methods are used; however, the simplest method uses the Permatran-C instrument (Modern Controls, Inc.).

Flavor and Aroma Transport

Many methods are used to characterize the transport of flavor, aroma, and solvent molecules in polymers. Each has some value, and no one method is suitable for all situations. Any experiment should obtain the permeability, the diffusion coefficient, and the solubility coefficient. Furthermore, experimental variables might include the temperature, the humidity, the flavor concentration, and the effect of competing flavors.

SAFETY AND HEALTH FACTORS

The use of safe materials is vital for barrier applications, particularly for food, medical, and cosmetics packaging. Suppliers of specific barrier polymers can provide the necessary details, such as material safety data sheets, to ensure safe processing and use of barrier polymers.

J. Comyn, ed., *Polymer Permeability*, Elsevier Applied Science Publishers, Ltd., Barking, UK, 1985.

J. Crank and G. S. Park, eds., *Diffusion in Polymers*, Academic Press, London, 1968.

R. M. Felder and G. S. Huvard, in R. A. Fava, ed., *Methods of Experimental Physics*, Vol. 16, Part C, Academic Press, New York, 1980, Chapt. 17, pp. 315–377.

L. E. Gerlowski, in W. J. Koros, ed., *Barrier Polymers and Structures*, ACS Symposium Series No. 423, American Chemical Society, Washington, D.C., 1990, Chapt. 8, pp. 177–191.

PHILLIP DeLASSUS
The University of Texas—Pan American

BATTERIES, INTRODUCTION

Batteries are storehouses for electrical energy "on demand". They range in size from large house-sized batteries for utility storage, cubic foot-sized batteries for automotive starting, lighting, and ignition, down to tablet-sized batteries for hearing aids and paper-thin batteries for memory protection in electronic devices.

In bulk chemical reactions, an oxidizer (electron acceptor) and fuel (electron donor) react to form products resulting in direct electron transfer and the release or absorption of energy as heat. By special arrangements of reactants in devices called batteries, it is possible to control the rate of reaction and to accomplish the direct release of chemical energy in the form of electricity on demand without intermediate processes.

Figure 1 schematically depicts an electrochemical reactor in which the chemical energy stored in the electrodes is manifested directly as a voltage and current flow. The electrons involved in the chemical reactions are transferred from the active materials undergoing oxidation to the oxidizing agent by means of an external circuit. The passage of electrons through this external circuit generates an electric current, thus providing a direct means for energy utilization without going through heat as an intermediate step. As a result, electrochemical reactors can be significantly more efficient than Carnot cycle heat engines.

The three main types of batteries are primary, secondary, and reserve. A primary battery is used or discharged once and discarded. Secondary or rechargeable batteries can be discharged, recharged, and used again. Reserve batteries are normally special constructions of primary battery systems that store the electrolyte apart from the electrodes, until put into use. They are designed for long-term storage before use. Fuel cells (qv) are not discussed herein.

THERMODYNAMICS

Batteries can be thought of as miniature chemical reactors that convert chemical energy into electrical energy on demand. The thermodynamics of battery systems follow directly from that for bulk chemical reactions. For the general reaction

$$a\mathrm{A} + b\mathrm{B} \rightleftharpoons c\mathrm{C} + d\mathrm{D} \qquad (1)$$

the basic thermodynamic equations for a reversible electrochemical transformation are given as

$$\Delta G = \Delta H - T\Delta S \qquad (2)$$

$$\Delta G^\circ = \Delta H^\circ - T\Delta S^\circ \qquad (3)$$

where ΔG is the Gibbs free energy, or the energy of a reaction available for useful work, ΔH is the enthalpy, or the energy released by the reaction, ΔS is the entropy, or the heat associated with the organization of material, and T is the absolute temperature. The superscript $^\circ$ is used to indicate that the value of the function is for the material in the standard state at 25°C and unit activity. Although the Helmholtz free energy ΔA is used to describe constant volume situations found in battery systems, the use of the Gibbs free energy ΔG is adequate to describe practical battery systems.

The terms ΔG, ΔH, and ΔS are state functions and depend only on the identity of the materials and the initial and final states of the reaction.

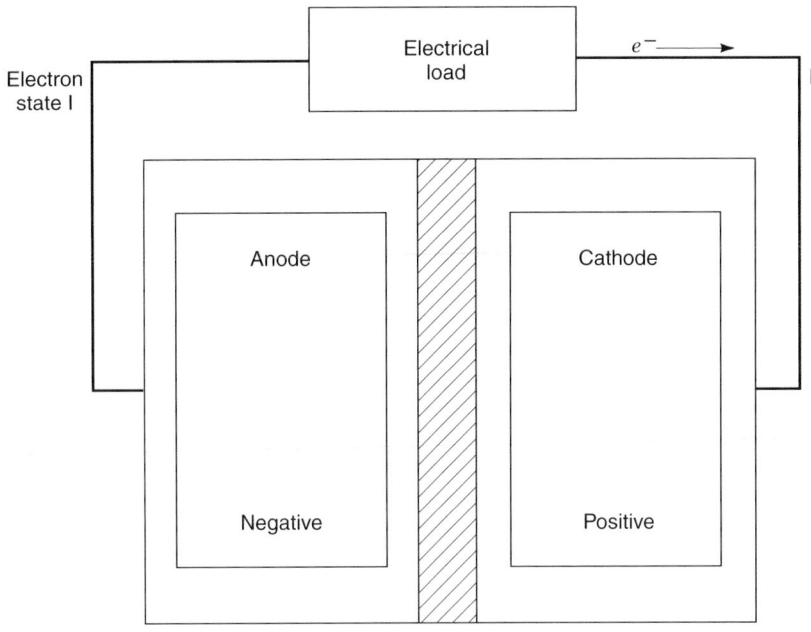

Figure 1. Schematic representation of a battery system also known as an electrochemical transducer where the anode, also known as electron state I, may be comprised of lithium, magnesium, zinc, cadmium, lead, or hydrogen, and the cathode, or electron state II, depending on the composition of the anode, may be lead dioxide, manganese dioxide, nickel oxide, iron disulfide, oxygen, silver oxide, or iodine.

Because ΔG is the net useful energy available from a given reaction, in electrical terms, the net available electrical energy from a reaction is given by

$$-\Delta G = nFE \tag{4}$$

and

$$-\Delta G^{\circ} = nFE^{\circ} \tag{5}$$

where n is the number of electrons transferred in the reaction, F is Faraday's constant, E is the voltage or electromotive force (emf) of the cell, and E° is the voltage at 25°C and at unit activity. The voltage is unique for each group of reactants comprising the battery system. The amount of electricity produced is determined by the total amount of materials involved in the reaction. The voltage may be thought of as an intensity factor, and the term nF may be considered a capacity factor.

The more negative the value of ΔG, the more energy or useful work can be obtained from the reaction. Reversible processes yield the maximum output. In irreversible processes, a portion of the useful work or energy is used to help carry out the reaction. The cell voltage or emf also has a sign and direction. Spontaneous processes have a negative free energy and a positive emf; the reaction, written in a reversible fashion, goes in the forward direction.

ELECTROLYTES

Electrolytes are a key component of electrochemical cells and batteries. Electrolytes are formed by dissolving an ionogen into a solvent. When salts are dissolved in a solvent such as water, the salt dissociates into ions through the action of the dielectric solvent. Strong electrolytes, ie,

salts of strong acids and bases, are completely dissociated in solution into positive and negative ions. The ions are solvated but positive ions tend to interact more strongly with the solvent than do the anions. The ions of the electrolyte provide the path for the conduction of electricity by movement of charged particles through the solution. The electrolyte also provides the physical separation of the positive and negative electrodes needed for electrochemical cell operation.

Transport properties of the electrolyte, as well as electrode reactions, have a significant impact on battery operation. The electrode reactions and ionic transference that occur during discharge result in considerable modifications to the solution composition at each electrode compartment. The negative and positive electrode compartments can lose or gain electrolyte and solvent depending on the transference numbers of the ions and the electrode reactions. The composition of the electrolyte in the separator between the two compartments generally remains unchanged.

Each electrolyte is stable only within certain voltage ranges. Exceeding these limits results in decomposition. The stable range depends on the solvent, electrolyte composition, and purity level. In aqueous systems, hydrogen and oxygen form when the voltage limit is exceeded. In the nonaqueous organic solvent-based systems used for lithium batteries, exceeding the voltage limit can result in polymerization or decomposition of the solvent system. It is especially important to remove traces of water from the nonaqueous electrolytes, as water can catalyze the electrolytic decomposition of the organic solvent.

In addition to the liquid conductors described above, two types of solid-state ionic conductors have been developed; one involves inorganic compounds and the other is based on polymeric materials. Several inorganic solids have been found to have excellent conductivity resulting wholly from ionic motion in the crystal lattice. Conductivity is related to specific crystal structures in which one ion, usually the cation, can move freely through the lattice.

A second type of solid ionic conductor, which is based around polyether compounds such as poly(ethylene oxide) (PEO), has been discovered and characterized. The polyethers can complex and stabilize lithium ions in organic media. They also dissolve salts such as $LiClO_4$ to produce conducting solid solutions.

A third type of "solid" ionic conductor is based on the ability of some polymers to absorb organic electrolytes, while maintaining a solid physical dimension. These are called "plasticized" or "gel" polymer electrolytes. These electrolytes have good conductivity at room temperature in contrast to the pure polymer electrolytes that have good conductivity only at $>60°C$.

ELECTRICAL DOUBLE LAYER

When two conducting phases come into contact with each other, a redistribution of charge occurs as a result of any electron energy level difference between the phases. If the two phases are metals, electrons flow from one metal to the other until the electron levels equilibrate. When an electrode, ie, electronic conductor, is immersed in an electrolyte, ie, ionic conductor, an electrical double layer forms at the electrode–solution interface resulting from the unequal tendency for distribution of electrical charges in the two phases. Because overall electrical neutrality must be maintained, this separation of charge between the electrode and solution gives rise to a potential difference between the two phases, equal to that needed to ensure equilibrium.

On the electrode side of the double layer, the excess charges are concentrated in the plane of the surface of the electronic conductor. On the electrolyte side of the double layer, the charge distribution is quite complex. The potential drop occurs over several atomic dimensions and depends on the specific reactivity and atomic structure of the electrode surface and the electrolyte composition. The electrical double layer strongly influences the rate and pathway of electrode reactions.

The region of the gradual potential drop from the Helmholtz layer into the bulk of the solution is called the Gouy or diffuse layer. The Gouy layer has similar characteristics to the ion atmosphere from electrolyte theory. This layer has an almost exponential decay of potential with increasing distance. The thickness of the diffuse layer may be approximated by the Debye length of the electrolyte.

Electrical double layers are not confined to the interface between conducting phases. Solid particles of active mass, or of conductive additives of colloidal size, can acquire an electric charge by specific adsorption of cations or anions from the electrolyte or by reaction of surface moieties with components of the solution. The resulting excess charge on the particle is neutralized by a diffuse or Gouy layer in the solution. The electrokinetic properties of the interface and zeta potential concepts are based on the characteristics of the Gouy layer. Migration of the colloidal-sized solid particles can occur under the influence of an applied electric field during battery operations.

Electrically, the electrical double layer may be viewed as a capacitor with the charges separated by a distance of the order of molecular dimensions. The measured capacitance ranges from about two to several hundred microfarads per square centimeter depending on the structure of the double layer, the potential, and the composition of the electrode materials.

KINETICS AND TRANSPORT

Activation Processes

Reactions must occur at a reasonable rate to be useful in battery applications. The rate or ability of battery electrodes to produce current is determined by the kinetic processes of electrode operations, not by thermodynamics. Thermodynamics describes the characteristics of reactions at equilibrium when the forward and reverse reaction rates are equal. Electrochemical reaction kinetics follow the same general considerations as those of bulk chemical reactions. Two differences are a potential drop that exists between the electrode and the solution because of the electrical double layer at the electrode interface, and the reaction that occurs at interfaces that are two-dimensional (2D) rather than in the three-dimensional (3D) bulk.

Transport Processes

The velocity of electrode reactions is controlled by the charge-transfer rate of the electrode process or by the velocity of the approach of the reactants to the reaction site. The movement or transport of reactants to and from the reaction site at the electrode interface is a common feature of all electrode reactions. Transport of reactants and products occurs by diffusion, by migration under a potential field, and by convection. The complete description of transport requires a solution to the transport equations. Molecular diffusion in electrolytes is relatively slow. Although the process can be accelerated by stirring, enhanced mass transfer (qv) by stirring or convection is not possible in most battery designs. Natural convection from density changes does occur but does not greatly enhance transport in battery operation. Lead acid batteries, used for motive power and stationary applications, are given a gassing overcharge on a regular basis. The gas evolution stirs up the electrolyte and equalizes the sulfuric acid concentration in the electrolyte.

PRACTICAL BATTERY SYSTEMS

Most battery electrodes are porous structures in which an interconnected matrix of solid particles, consisting of both nonconductive and electronically conductive materials, is filled with electrolyte. When the active mass is nonconducting, conductive materials, usually carbon or metallic powders, are added to provide electronic contact to the active mass. The solids occupy 60–80% of the volume of a typical porous battery electrode. Most battery electrode structures do not have a well-defined planar surface but have a complex surface extending throughout the volume of the porous electrode. Macroscopically, the porous electrode behaves as a homogeneous unit.

When a battery produces current, the sites of current production are not uniformly distributed on the electrodes. The nonuniform current distribution lowers the expected performance from a battery system and causes excessive heat evolution and low utilization of active materials. Two types of current distribution, primary and secondary, can be distinguished. The primary distribution is related to the current production based on the geometric surface area of the battery construction. Secondary current distribution is related to current production sites inside the porous electrode itself. Most practical battery constructions have nonuniform current distribution across the surface of the electrodes. This primary current distribution is governed by geometric factors such as height (or length) of the electrodes, the distance between the electrodes, the resistance of the anode and cathode structures, by the resistance of the electrolyte, and by the polarization resistance or hinderance of the electrode reaction processes.

Cell geometry, such as tab/terminal positioning and battery configuration, strongly influence primary current distribution. The monopolar construction is most common. Several electrodes of the same polarity may be connected in parallel to increase capacity. The current production concentrates near the tab connections unless special care is exercised in designing the current collector. Bipolar construction, wherein the terminal or collector of one cell serves as the anode and cathode of the next cell in pile formation, leads to greatly improved uniformity of current distribution.

Whereas, current producing reactions occur at the electrode surface, they also occur at considerable depth below the surface in porous electrodes. Porous electrodes offer enhanced performance through increased surface area for the electrode reaction and through increased mass-transfer rates from shorter diffusion path lengths. The key parameters in determining the reaction distribution include the ratio of the volume conductivity of the electrolyte to the volume conductivity of the electrode matrix, the exchange current, the diffusion characteristics of reactants and products, and the total current flow. The porosity, pore size, and tortuosity of the electrode all play a role.

The positive electrode in a battery system is most often a metal oxide, but it may also be a metal sulfide or halide. Generally, these materials are relatively poor electrical conductors and exhibit extremely high ohmic polarizations (impedances) if not combined with supporting electronic conductors such as graphite, lead, silver, copper, or nickel in the form of powder, rod, mesh, wire, grid, or other configurations. In almost all cases, the negative electrode is a metallic element of sufficient conductivity to require only minimal supporting conductive structures. Exceptions are the oxygen (air, positive) and hydrogen gas (negative) electrodes that require a substantial conductive, catalytically active, surface support that also serves as current collector.

Although there are a multitude of chemical reactions that can store and release energy, only a few have the characteristics requisite for use in commercial batteries. A set of criteria can be established to characterize reactions suitable for battery development. The principal features necessary for battery reactions include (1) Mechanical and chemical stability, ie, the reactants or active masses and cell components must be stable over time (5 years or more) in the operating environment and must reform in their original condition on recharge; (2) energy content, ie, the reactants must have sufficient energy content to provide a useful voltage and current level; (3) power density, ie, the reactants must be capable of reacting at rates sufficient to deliver useful rates of electricity; (4) temperature range, ie, the reactants must be able to maintain energy, power, and stability over a normal operating environment; (5) safety, ie, the battery must be safe in the normal operating environment as well as under mild abusive conditions; and (6) cost, ie, the reactants and the materials of construction should be inexpensive and in good supply.

R. J. Brodd and A. Kozawa, in E. B. Yeager and A. J. Salkind, eds., *Techniques of Electrochemistry*, John Wiley & Sons., Inc., New York, 1978, Vol. 3.

G. Pistoia, ed., *Lithium Batteries*, Elsevier Science, New York, 1993.

C. A. Vincent, B. Scrosati, M. Lazzari, and F. Bonino, *Modern Batteries*, 2nd ed., Edward Arnold, Ltd., London, 1997.

D. Linden, ed., *Handbook of Batteries and Fuel Cells*, 2nd ed., McGraw-Hill Book Co., New York, 1995.

RALPH BRODD
Broddarp of Nevada, Inc.

BATTERIES, PRIMARY CELLS

Primary cells are galvanic cells designed to be discharged only once, and attempts to recharge them can present possible safety hazards. The cells are designed to have the maximum possible energy in each cell size because of the single discharge. Thus, comparison between battery types is usually made on the basis of the energy density in $W \cdot h/cm^3$. The specific energy, $W \cdot h/kg$, is often used as a secondary criterion for primary cells, especially when the application is weight-sensitive, as in space applications. The main categories of primary cells are carbon–zinc, known as heavy-duty and general purpose; alkaline, cylindrical, and miniature; lithium; and reserve or specialty cells.

CARBON–ZINC CELLS

Carbon–zinc batteries are the most commonly found primary cells worldwide and are produced in almost every country. Traditionally there are a carbon rod, for cylindrical cells, or a carbon-coated plate, for flat cells, to collect the current at the cathode and a zinc anode. There are two basic versions of carbon–zinc cells: the Leclanché cell and the zinc chloride, $ZnCl_2$, or heavy-duty cell. Both have zinc anodes, manganese dioxide, MnO_2, cathodes, and include zinc chloride in the electrolyte. The Leclanché cell also has an electrolyte saturated with ammonium chloride, NH_4Cl. Additional undissolved ammonium chloride is usually added to the cathode, whereas the zinc chloride cell has at most a small amount of ammonium chloride added to the electrolyte. Both types are dry cells in the sense that there is no excess liquid electrolyte in the system. The zinc chloride cell is often made using synthetic manganese dioxide and gives higher capacity than the Leclanché cell, which uses inexpensive natural manganese dioxide for the active cathode material. The MnO_2 is only a modest conductor. Thus the cathodes in both types of cell contain 10–30% carbon black in order to distribute the current. Because of the ease of manufacture and the long history of the cell, this battery system can be found in many sizes and shapes.

Performance

Carbon–zinc cells perform best under conditions of intermittent use, and many standardized tests have been devised that are appropriate to such applications as light and heavy flashlight usage, radios, cassettes, and motors (toys). The most frequently used tests are American National Standards Institute (ANSI) tests. The tests are carried out at constant resistance and the results reported in minutes or hours of service. Figure 1 shows typical results under a light load for different size cells.

To compare one battery with another, it is useful to compute the energy density from these data. Because the voltage declines with capacity, the average voltage during the discharge is used to compute an average current, which is then multiplied by the service in hours to give the ampere-hours of capacity. Watt-hours of energy can be obtained by multiplying again by the average voltage.

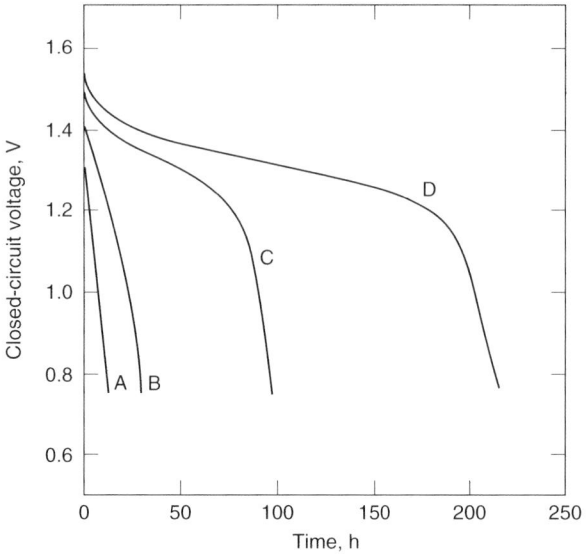

Figure 1. Hours of service on 40-Ω discharge for 4 h/d radio test at 21°C for A, RO3 "AAA"; B, R6 "AA"; C, R14 "C"; and D, R20 "D" paper-lined, heavy-duty zinc chloride cells.

CYLINDRICAL ALKALINE CELLS

Primary alkaline cells use sodium hydroxide or potassium hydroxide as the electrolyte. They can be made using a variety of chemistries and physical constructions. The alkaline cells are mostly of the limited electrolyte, dry cell type. Most primary alkaline cells are made using zinc as the anode material; a variety of cathode materials can be used. Primary alkaline cells are commonly divided into two classes, based on type of construction: the larger, cylindrically shaped batteries, and the miniature, button-type cells. Cylindrical alkaline batteries are mainly produced using zinc–manganese dioxide chemistry, although some cylindrical zinc–mercury oxide cells are made.

Cylindrical alkaline cells are zinc–manganese dioxide cells having an alkaline electrolyte, which are constructed in the standard cylindrical sizes, R20 "D", R14 "C", R6 "AA", RO3 "AAA", as well as a few other less common sizes. They can be used in the same types of devices as ordinary Leclanché and zinc chloride cells. Moreover, the high level of performance makes them ideally suited for applications such as toys, audio devices, and cameras.

Performance

Alkaline manganese dioxide batteries have relatively high energy density, as can be seen from Table 1. This results in part from the use of highly pure materials, formed into electrodes of near optimum density. Moreover, the cells are able to function well with a rather small amount of electrolyte. The result is a cell having relatively high capacity at a fairly reasonable cost.

MINIATURE ALKALINE CELLS

Miniature alkaline cells are small, button-shaped cells which use alkaline NaOH or KOH electrolyte and generally have zinc anodes, but may have a variety of cathode materials. They are used in watches, calculators, cameras, hearing aids, and other miniature devices.

Cylindrical alkaline cells are made in only a few standard sizes and have only one important chemistry. In contrast, miniature alkaline cells are made in a large number of different sizes, using many different chemical systems. Whereas the cylindrical alkaline batteries are multipurpose batteries, used for a wide variety of devices under a variety of discharge conditions, miniature alkaline batteries are highly specialized, with the cathode material, separator type, and electrolyte all chosen to match the particular application.

Zinc–Mercuric Oxide Batteries

Miniature zinc–mercuric oxide batteries have a zinc anode and a cathode containing mercuric oxide, HgO.

Miniature zinc–mercuric oxide batteries function efficiently over a wide range of temperatures and have good storage life.

Although the zinc–mercuric oxide battery has many excellent qualities, increasing environmental concerns have led to a de-emphasis in the use of this system. The main environmental difficulty is in the disposal of the

Table 1. Characteristics of Aqueous Primary Batteries

Parameter	Carbon–zinc (Zn/MnO$_2$)	Alkaline manganese dioxide(Zn–MnO$_2$)	Mercuric oxide (Zn–HgO)	Silver oxide(Zn–Ag$_2$O)	Zinc–air (Zn–O$_2$)
nominal voltage, V	1.5	1.5	1.35	1.5	1.25
working voltage, V	1.2	1.2	1.3	1.55	1.25
specific energy, W·h/kg	40–100	80–95	100	130	230–400
energy density, W·h/mL	0.07–0.17	0.15–0.25	0.40–0.60	0.49–0.52	0.70–0.80
temperature range, °C					
storage	−40–50	−40–50	−40–60	−40–60	−40–50
operating	−5–55	−20–55	−10–55	−10–55	−10–55

cell. Both the mercuric oxide hi the fresh cell and the mercury reduction product in the used cell have long-term toxic effects.

Zinc–Silver Oxide Batteries

Miniature zinc–silver oxide batteries have a zinc anode, and a cathode containing silver oxide, Ag$_2$O. Miniature zinc–silver oxide batteries are commonly used in electronic watches and in other applications where high energy density, a flat discharge profile, and a higher operating voltage than that of a mercury cell are needed. These batteries function efficiently over a wide range of temperatures and are comparable to mercury batteries in this respect. Miniature zinc–silver oxide batteries have good storage life.

Divalent Silver Oxide Batteries

It is possible to produce a silver oxide in which the silver has a higher oxidation state, approaching a composition of AgO. This material can provide both higher capacity and higher energy density than Ag$_2$O. Alternatively, a battery can be made with the same capacity as a monovalent silver cell, but with cost savings. However, some difficulties with regard to material stability and voltage regulation must be addressed.

Zinc–Manganese Dioxide Batteries

The combination of a zinc anode and manganese dioxide cathode, which is the dominant chemistry in large cylindrical alkaline cells, is used in some miniature alkaline cells as well. Overall, this type of cell does not account for a large share of the miniature cell market. It is used hi cases where an economical power source is wanted and where the devices can tolerate the sloping discharge curve shown in Figure 2.

Zinc–Air Batteries

Zinc–air batteries offer the possibility of obtaining extremely high energy densities. Instead of having a cathode material placed in the battery when manufactured, oxygen from the atmosphere is used as cathode material, allowing for a much more efficient design. The construction of a miniature air cell is shown in Figure 3. From the outside, the cell looks like any other miniature cell, except for the air access holes in the can. On the inside, however, the anode occupies much more of the internal volume of the cell. Rather than the thick cathode pellet, there is a thin

layer containing the cathode catalyst and air distribution passages. Air enters the cell through the holes in the can and the oxygen reacts at the surface of the cathode catalyst. The air access holes are often covered with a protective tape, which is removed when the cell is placed in service.

The performance level of air cells is exceptional, but these are not general-purpose cells. They must be used in applications where the usage is largely continuous, and where the discharge level is relatively constant and well-defined. The reasons for these limitations lies in the fact that the cell must be open to the atmosphere and the holes that allow oxygen into the cell also allow other gases to enter or leave the cell.

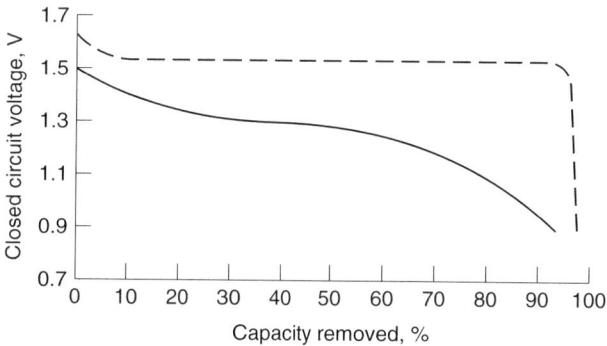

Figure 2. Discharge curves for miniature zinc–silver oxide batteries (–), and zinc–manganese dioxide batteries (—). Courtesy of Eveready Battery Co.

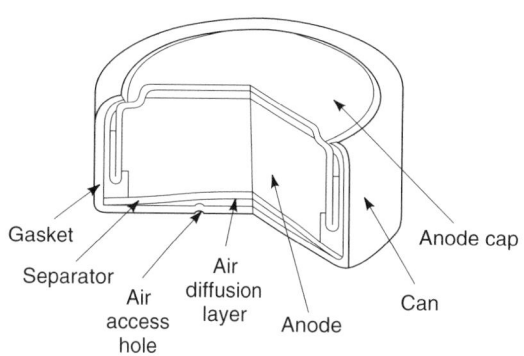

Figure 3. Cutaway view of a miniature air cell battery. Courtesy of Eveready Battery Co.

Miniature air cells are mainly used in hearing aids, where they are required to produce a relatively high current for a relatively short time period such as a few weeks. In this application they provide exceptional performance compared to other batteries.

LITHIUM CELLS

Cells having lithium anodes are generally called lithium cells regardless of the cathode. They can be conveniently separated into two types: cells having solid cathodes and cells having liquid cathodes. Cells having liquid cathodes also have liquid electrolytes and in fact, at least one component of the electrolyte solvent and the active cathode material are one and the same. Cells having solid cathodes may have liquid or solid electrolytes but, except for the lithium–iodine system, those having solid electrolytes are not yet commercial.

All of the cells take advantage of the inherently high energy of lithium metal and its unusual film-forming property.

Much analytical study has been required to establish the materials for use as solvents and solutes in lithium batteries. Among the best organic solvents are cyclic esters, such as propylene carbonate (PC), $C_4H_6O_3$, ethylene carbonate (EC), $C_3H_4O_3$, and butyrolactone, $C_4H_6O_2$, and ethers, such as dimethoxyethane (DME), $C_4H_{10}O_2$, the glymes, tetrahydrofuran (THF), and 1,3-dioxolane, $C_3H_6O_2$. Among the most useful electrolyte salts are lithium perchlorate, $LiClO_4$, lithium trifluoromethanesulfonate, $LiCF_3SO_3$, lithium tetrafluoroborate, $LiBF_4$, and lithium hexafluoroarsenate, $LiAsF_6$. A limitation of these organic electrolytes is the relatively low conductivity, compared to aqueous electrolytes. This limitation, combined with the generally slow kinetics of the cathode reactions, has forced the use of certain designs, such as thin electrodes and very thin separators, in all lithium batteries. This usage led to the development of coin cells rather than button cells for miniature batteries and jelly or Swiss roll designs rather than bobbin designs for cylindrical cells.

Many of the cylindrical cells have glass-to-metal hermetic seals, although this is becoming less common because of the high cost associated with this type of seal. Alternatively, cylindrical cells have compression seals carefully designed to minimize the ingress of water and oxygen and the egress of volatile solvent. These construction designs are costly and the high price of the lithium cell has limited its use. However, the energy densities are superior.

Solid Cathode Cells

Solid cathode cells include lithium–manganese dioxide cells, lithium–carbon monofluoride cells, lithium–iron disulfide cells, and lithium–iodine cells.

Liquid Cathode Cells

Liquid cathode cells include lithium–sulfur dioxide cells and lithium–thionyl chloride cells.

RESERVE BATTERIES

Reserve batteries have been developed for applications that require a long inactive shelf period followed by intense discharge during which high energy and power, and sometimes operation at low ambient temperature, are required. These batteries are usually classified by the mechanism of activation which is employed. There are water-activated batteries that utilize fresh or seawater; electrolyte-activated batteries, some using the complete electrolyte, some only the solvent; gasactivated batteries where the gas is used as either an active cathode material or part of the electrolyte; and heat-activated or thermal batteries which use a solid salt electrolyte activated by melting on application of heat.

Activation of these batteries involves adding the missing component which can be done in a simple way, such as pouring water into an opening in the cell, for water-activated cells, or in a more complicated way by using pistons, valves, or heat pellets activated by gravitational or electric signals for the case of the electrolyte- or thermal-activation types. Such batteries may be stored for 10–20 yr while awaiting use. Reserve batteries are usually manufactured under contract for various government agencies such as the U.S. Department of Defense although occasional industrial or safety uses have been found. Many of the electrochemical systems involved in these batteries are beyond the scope of this article.

The lithium-thionyl chloride, or the lithium sulfur dioxide, system is often used in a reserve battery configuration in which the electrolyte is stored in a sealed compartment which upon activation may be forced by a piston or inertial forces into the interelectrode space. Most applications for such batteries are in mines and fuse applications in military ordnance.

One variant of the liquid cathode reserve battery is the lithium–water cell in which water serves as both the liquid cathode and the electrolyte. A certain amount of corrosion occurs, but sufficient lithium is provided to compensate. These cells are mostly used in the marine environment where water is available or compatible with the cell reaction product. Common applications are for torpedo propulsion and to power sonobuoys and submersibles.

The last type of reserve cell is the thermally activated cell. The older designs use calcium or magnesium anodes; newer types use lithium alloys as anodes.

The heat pellet used for activation in these batteries is usually a mixture of a reactive metal such as iron or zirconium, and an oxidant such as potassium perchlorate. An electrical or mechanical signal ignites a primer, which then ignites the heat pellet which in turn melts the electrolyte. Sufficient heat is given off by the high current to sustain the necessary temperature during the lifetime of the application. Many millions of these batteries have been manufactured for military ordnance and employed in rockets, bombs, missiles, etc.

R. J. Brodd, *Batteries for Cordless Appliances*, John Wiley & Sons, Inc., New York, 1987.

T. R. Crompton, *Battery Reference Book*, Butterworths, London, 1990.

D. Linden, ed., *Handbook of Batteries and Fuel Cells*, McGraw-Hill, New York, 1984.

G. W. Heise and N. C. Cahoon, eds., *The Primary Battery*, Vol. 1 and 2, John Wiley & Sons, Inc., New York, 1971, 1976.

GEORGE BLOMGREN
JAMES HUNTER
Eveready Battery Company, Inc.

BATTERIES, SECONDARY CELLS

ALKALINE CELLS

Alkaline electrolyte storage battery systems are more suitable than others in applications where high currents are required, because of the high conductivity of the electrolyte. Additionally, in almost all of these battery systems, the electrolyte which is usually an aqueous solution containing 25–40% potassium hydroxide, KOH, does not enter into the chemical reaction. Thus concentration and cell resistance are invariant with state of discharge and these battery systems give high performance and have long cycle life. The annual battery supply for this application was forecasted to increase to over a million units by the year 2008.

Positive electrode active materials have been made from the oxides or hydroxides of nickel, silver, manganese, copper, mercury, and from oxygen. Negative electrode active materials have been fabricated from various geometric forms of cadmium, Cd, iron, Fe, and zinc, Zn, and from hydrogen. Two different types of hydrogen electrode designs are common: those used in space, which employ hydrogen as a gas, and those used in consumer batteries, where the hydrogen is used as a metallic hydride. As indicated in Table 1, nine electrode combinations exist in some scale of commercial production. Five system combinations are in the research/development stage, and two have been abandoned before or after commercial production for reasons such as short life, high cost, low voltage, low energy density, and excessive maintenance.

The annual production value of small-sealed nickel cadmium cells is approximately 1.0×10^9. Environmental considerations relating to cadmium have necessitated changes in the fabrication techniques, as well as recovery of failed cells. Battery system designers have switched to nickel–metal hydride (MH) cells for high power applications. However, the highest discharge/recharge rates are still achieved with nickel–cadmium cells typically in "AA"-size cells, to increase capacity in the same volume and avoid the use of cadmium.

Many of the most recent applications for alkaline storage batteries require higher energy density and lower cost designs than previously available. Materials such as foam and/or fiber nickel, Ni, mats as substrates, and new processing techniques including plastic bounded, pasted, or electroplated electrodes, have enabled the alkaline storage battery to meet these new requirements, while reducing environmental problems in the manufacturing plants. In addition, substantial technical efforts have been devoted to the recovery of used batteries. The most recent innovations in materials relate to the development

Table 1. Rechargeable Alkaline Storage Battery Systems

System[a]	Historical name	Voltage, V	Production[b]
nickel–cadmium	Jungner	1.30	vl
nickel–iron	Edison	1.37	vs
nickel–zinc	Drumm	1.70	vs
nickel–hydrogen (H$_2$ or MH)		1.30	vl
silver–cadmium		1.38 and 1.16[c]	vs
silver–iron	Jirsa	1.45 and 1.23[c]	vs
silver–zinc	Andre	1.86 and 1.60[c]	s
silver–hydrogen		1.38 and 1.16[c]	vs
manganese–zinc[d]		1.52	vs
mercury–cadmium		0.92	r
air(oxygen)–zinc		1.60	r
air(oxygen)–iron		1.40	r
air(oxygen)–aluminum			vs
copper–lead		1.20	r
copper–cadmium	Darrieus Waddell-Entz,	0.45	n
copper–zinc	Edison-LeLande Lelande-Chaperon	0.85	n

[a]The substance named first represents the positive electrode; the substance named second is the negative electrode. In all cases except for air(oxygen) systems, the active electrode material is the oxide or the hydroxide of the named species.
[b]vl = >100 × 10^6 A·h/yr product; l = >25 × 10^6 A·h/yr; s = >5 × 10^6 A·h/yr; vs = <5 × 10^6 A·h/yr; r = research and development phase; and n = no longer in production.
[c]Silver oxide electrodes have two voltage plateaus.
[d]Secondary system designs.

of metal–hydride alloys for the storage and electrochemical utilization of hydrogen. Modifications to the chemical structure and/or the cell design of manganese dioxide, MnO_2, electrodes have resulted in sufficient improvement to allow the reintroduction of the rechargeable MnO_2–zinc cell to the market as a lower cost, albeit lower performance, alternative to nickel–cadmium consumer size cells. Improvements in materials science and electrical circuits have lead to better separators, seals, welding techniques, feedthroughs, and charging equipment.

Nickel–Cadmium Cells

Electrodes. A number of different types of nickel oxide electrodes have been used. The term nickel oxide is common usage for the active materials that are actually hydrated hydroxides at nickel oxidation state 2+, in the discharged condition, and nickel oxide hydroxide, $NiO \cdot OH$, nickel oxidation state 3+, in the charged condition. Nickelous hydroxide, $Ni(OH)_2$, can be precipitated from acidic solutions of bivalent nickel either by the addition of sodium hydroxide or by cathodic processes to cause an increase in the interfacial pH at the solution–electrode surface (see NICKEL AND NICKEL ALLOYS; NICKEL COMPOUNDS).

The many varieties of practical nickel electrodes can be divided into two main categories. In the first, the active nickelous hydroxide is prepared in a separate chemical reactor and is subsequently blended, admixed, or layered with an electronically conductive material. This active material mixture is afterwards contained in a confining porous metallic structure or pasted onto a metallic mat or grid. Electrodes for pocket, tubular, pasted, and most button cells are made this way.

The other type of nickel electrode involves constructions in which the active material is deposited *in situ*. This includes the sintered-type electrode in which nickel hydroxide is chemically or electrochemically deposited in the pores of a 80–90% porous sintered nickel substrate that may also contain a reinforcing grid.

Almost all the methods described for the nickel electrode have been used to fabricate cadmium electrodes. However, because cadmium, cadmium oxide, CdO, and cadmium hydroxide, $Cd(OH)_2$, are more electrically conductive than the nickel hydroxides, it is possible to make simple pressed cadmium electrodes using less substrate (see CADMIUM AND CADMIUM ALLOYS; CADMIUM COMPOUNDS). These are commonly used in button cells.

Electrochemistry and Crystal Structure. The solid-state chemistry of the nickel electrode is complex. Nickel hydroxide in the discharged state has a hexagonal layered lattice, where planes of Ni^{2+} ions are sandwiched between planes of OH^-. This structure, similar to that of cadmium iodide, CdI_2, is common to seven metal hydroxides including those of cadmium and cobalt. There are various hydrated and nonhydrated nickel hydroxides that have slightly different crystal habitats and electrochemical potentials. The most common form of charged material observed in batteries is $NiOOH$, density = 4.6 g/mL. In comparison, $Ni(OH)_2$ has a density of 4.15 g/mL. Thus

the theoretical change in density on charge–discharge is less than 10%, and the kinetics involve only a proton transfer.

The chemistry, electrochemistry, and crystal structure of the cadmium electrode is much simpler than that of the nickel electrode. The overall reaction is generally recognized as:

$$Cd + 2\,OH^- \underset{\text{charge}}{\overset{\text{discharge}}{\rightleftharpoons}} Cd(OH)_2 + 2\,e^-$$

However, there is a strong likelihood of a soluble intermediate in the formation of $Cd(OH)_2$. Cadmium has an appreciable solubility in alkaline solutions: $\sim 2 \times 10^{-4}$ mol/L in $8\,M$ potassium hydroxide at room temperature. In general it is believed that the solution process consists of anodic dissolution of cadmium ions in the form of complex hydroxides (see CADMIUM COMPOUNDS).

In more recent studies involving cyclic line scan voltammetry of the nickel electrode, it was suggested that nickel can exist in the positive 2, 3, and 4 oxidation states. The structural parameters using transmission extended x-ray absorption fine structure (exafs) and *in situ* electrodes confirmed earlier x-ray data, showing that the presence of cobalt, Co, does not change crystal lattice parameters. The density and compressibility of nickel electrodes have been found to be highly variable. Cobalt additions appear to reduce the compressibility of nickel hydroxide, resulting in a firmer attachment of the active material to the substrate. However, a felt metal grid has been shown to reduce shear failures of the electrode structure and minimize the need for stabilizing additives, such as cobalt.

Sealed Cells. Most sealed cells are based on the principles appearing in patents of the early 1950s where the virtues of limiting electrolyte, a separator that would absorb and retain electrolyte, and leaving free passage for the oxygen from the positive to the negative plate were described. First, the negative electrode has a surplus of uncharged active material so that the positive plate starts to produce oxygen before the negative plate is fully charged. The oxygen reacts with the negative active material, so that the negative electrode never becomes fully charged and consequently never evolves hydrogen. Second, the amount of electrolyte used is generally lower than can normally be absorbed in the electrodes and separators, facilitating the transfer of oxygen from the positive to the negative plate. Oxygen transport, at least to a certain extent is carried out in the gaseous stage. Third, the separators generally used can pass oxygen to the gaseous state for rapid transfer to the negative plate. Although both pocket and sintered electrodes of the nickel–cadmium type have been used in sealed-cell construction, the preponderant majority of cells in commercial production use sintered positive (nickel) electrodes, and either sintered or pasted negative (cadmium) electrodes.

Cell Fabrication Methods   *Pocket Cells.* A view of a pocket electrode nickel–cadmium cell is shown in Figure 3. The essential steps of positive (nickel) electrode construction are (*1*) cold-rolled steel ribbon is cut to proper width and is perforated using either needles or rolls; (2) the perforated steel ribbon is nickel-plated and usually annealed in hydrogen. The ribbon is formed into a trough shape, is filled with active material by either a briquetting or a powder-filling technique; (3) a second strip is formed into a lid that covers and locks with the filled trough; (4) the filled strips are cut to length and are arranged to form an electrode sheet by interleaving. This operation, carried out by means of rollers in a forming roll, is often combined with the pressing of a pattern into the electrode sheet in order to ensure good contact between ribbon and active material and   to add mechanical strength to the construction; and   (5) the electrode sheet is then cut to pieces of appropriate size and side bedding and lugs attached to form a metallic frame. The frame material is usually also cold-rolled steel ribbon.

The pockets are usually arranged horizontally in the electrodes, but in a few cases vertical pockets are used. No significant difference has been observed between the two arrangements.

Pocket-type cadmium electrodes are made by a procedure similar to that described for the positive electrode. Because cadmium active material is more dense than nickel active material, and because cadmium has a 2+ valence, cadmium electrodes, when fabricated to equal thicknesses, have almost twice the working capacity of the nickel electrode. A cell having considerably greater negative capacity provides for loss of negative capacity during life and avoids generation of hydrogen during charging. Thus in actual practice plates of equal thickness are used.

After the individual pocket electrodes are fabricated, they are assembled into electrode groups. Electrodes of the same polarity are electrically and mechanically connected to each other and to a pole bolt.

To complete the assembly of a cell, the interleaved electrode groups are bolted to a cover and the cover is sealed to a container. Originally, nickel-plated steel was the predominant material for cell containers but, more recently plastic containers have been used for a considerable proportion of pocket nickel-cadmium cells. Polyethylene, high impact polystyrene, and a copolymer of propylene and ethylene have been the most widely used plastics.

Steel containers are mechanically stronger than plastic and easier to fabricate in large sizes. They dissipate heat better and tend to keep the electrodes cooler during high temperature or high rate operations. However, cells assembled in steel containers must not have contact with each other during assembly to prevent intercell shorts. Plastic containers are the better option for most small and medium-sized cells because they require no protection against corrosion, they permit visual observation of the electrolyte level, they are lighter than steel containers, they can be closely packed into a battery, and small cells can be cemented or taped into batteries, eliminating a tray.

Tubular Cells. Although the tubular nickel electrode invented by Edison is almost always combined with an iron negative electrode, a small quantity of cells is produced in which nickel in the tubular form is used with a pocket cadmium electrode. This type of cell construction is used for low operating temperature environments, where iron electrodes do not perform well or where charging current must be limited.

Sintered Cells. The fabrication of sintered electrode batteries can be divided into five principal operations: preparation of sintering-grade nickel powder; preparation of the sintered nickel plaque; impregnation of the plaque with active material; assembly of the impregnated plaques (often called plates) into electrode groups and into cells; and assembly of cells into batteries.

A good powder for sintering purpose should be very pure (Fe, Cu, and S should be especially avoided) and should have a very low apparent density, in the range of 0.5–0.89 g/mL. An excellent powder for this purpose is made by the decomposition of nickel carbonyl, $Ni(CO)_4$, (see CARBONYLS).

Other Cells. Other methods to fabricate nickel–cadmium cell electrodes include those for the button cell, used for calculators and other electronic devices. This cell, the construction of which is illustrated in Figure 1, is commonly made using a pressed powder nickel electrode mixed with graphite that is similar to a pocket electrode. The cadmium electrode is made in a similar manner. The active material, graphite blends for the nickel electrode, are almost the same as that used for pocket electrodes, ie, 18% graphite.

Lower cost and lower weight cylindrical cells have been made using plastic bound or pasted active material pressed into a metal screen. These cells suffer slightly in utilization at high rates compared to a sintered-plate cylindrical cell, but they may be adequate for most applications.

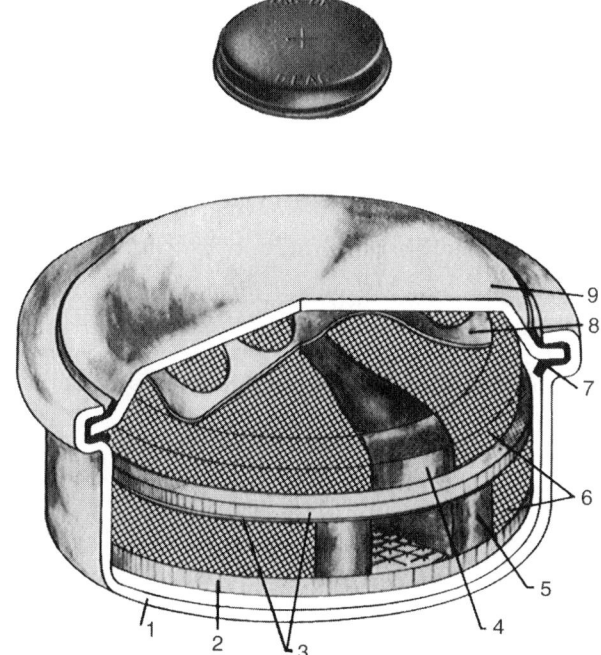

Figure 1. Section of disk-type cell where: 1, is the cell cup; 2, is the bottom insert; 3, is the separator; 4, is the negative electrode; 5, is the positive electrode; 6, is the nickel wire gauze; 7, is the sealing washer; 8, is the contact spring; and 9, is the cell cover.

Applications. In the U.S., rechargeable NiCd batteries provide power for about three-fourths of most common portable products. Uses are divided into three categories: pocket cells are used in emergency lighting, diesel starting, and stationary and traction applications where the reliability, long life, medium-high rate capability, and low temperature performance characteristics warrant the extra cost over lead–acid storage batteries; sintered, vented cells are used in extremely high rate applications, such as jet engine and large diesel engine starting; and sealed cells, both the sintered and button types, are used in computors, phones, cameras, portable tools, electronic devices, calculators, cordless razors, toothbrushes, carving knives, flashguns, and in space applications, where nickel–cadmium is optimum because it can be recharged a great number of cycles and given prolonged trickle overcharge. Cells of this category are generally made in sizes comparable to conventional dry cells, such as "D", "C", "AA", etc.

Charger Technology. Alkaline storage batteries are commonly charged from rectified d-c equipment, solar panels, or other d-c sources and have fairly good tolerance to ripple and transient pulses. Because the voltage of the nickel electrode is variable, the cutoff voltage is not a good control parameter for nickel–cadmium cells. It is, however, often used in vented cell chargers. For sealed nickel–cadmium cells, and other systems where a combination mechanism exists, a negative voltage slope detector is often incorporated into the charger control circuit. As the charge of a cell or battery progresses there is a slow rise in the unit voltage. When the nickel electrode approaches full charge, the oxygen evolved combines with the cadmium electrode reducing the overpotential on the cadmium electrode. This slight change in cell voltage can be detected by electronic voltage slope detectors and used to reduce the charge current or shut off the charge. A method for charging sterilizable rechargeable batteries has been reported.

Nickel–Iron Cells. In the 1980s–1990s there was a renewed interest in a high rate sintered electrode design for electric vehicle applications. However, when interest shifted to hybrid electric vehicles, in the early 2000s, with lower battery capacity requirements, interest shifted to Ni-MH because of superior performance.

Electrochemistry and Kinetics. The electrochemistry of the nickel–iron battery and the crystal structures of the active materials depends on the method of preparation of the material, degree of discharge, the age (life cycle), concentration of electrolyte, and type and degree of additives, particularly the presence of lithium and cobalt. A simplified equation representing the charge–discharge cycle can be given as:

$$2\,NiOOH^* + \alpha\text{–Fe} + 2\,H_2O \rightleftharpoons 2\,Ni(OH)_2^* + Fe(OH)_2$$

where the asterisks indicate adsorbed water and KOH.

Electrode Structures. The classical iron active material for pocket and pasted iron electrodes was formed by roasting recrystallized ferrous sulfate, $FeSO_4$, in an oxidizing atmosphere to ferric oxide, Fe_2O_3, and then reducing the latter in hydrogen. The α-iron formed was then heated to a mixture of Fe_3O_4 and Fe. As such it was pure enough to be used for pharmaceutical purposes. For battery use, a small amount of sulfur, as FeS, was added as were other additives which were believed to increase the cycle life by acting as depassivating agents, ie, helping to reduce the tendency of iron to evolve hydrogen upon standing in alkaline electrolyte.

A study of sintered iron electrodes claimed advantages of high rate capability, long life, and low hydrogen evolution.

Sintered nickel electrodes used in nickel iron cells are usually thicker than those used in Ni/Cd cells. These result in high energy density cells, because very high discharge rates are usually not required.

Performance Characteristics. The sintered nickel-sintered iron design battery has outstanding power characteristics at all states of discharge making them attractive to the design of electric vehicles (EV) which must accelerate with traffic even when almost completely discharged. Although the evolution of hydrogen is a problem preventing sealed cell design, introduction of automatic watering systems have ameliorated the maintenance time requirements.

Silver–Zinc Cells

The silver–zinc battery has the highest attainable energy density of any rechargeable system in use except for lithium-ion cells as of this writing. In addition, it has an extremely high rate capability coupled with a very flat voltage discharge characteristic. Its use, in the early 1990s, was limited almost exclusively to the military for various aerospace applications such as satellites and missiles, submarine and torpedo propulsion applications, and some limited portable communications applications. The main drawback of these cells is the rather limited lifetime of the silver–zinc system. Life is normally limited to less than 200 cycles with a total wet-life of no more than about two years. The silver–zinc system also carries a very high cost and applications are justified only where cost is a minor factor. The high cost of silver battery systems is attributable to the cost of the active silver material used in the positive electrodes.

Cellophane or its derivatives have been used as the basic separator for the silver–zinc cell since the 1940s. Cellophane is hydrated by the caustic electrolyte and expands to approximately three times its dry thickness inside the cell exerting a small internal pressure in the cell.

A second lifetime limitation is the zinc anode. In spite of the separator and cell designs, some zinc material is solubilized during the charge–discharge reaction. Over a period of cycling there is a shift of active material, originally distributed evenly over the face of the electrode, to the center and bottom areas of the electrode. This shape change limits the life of the cell as exemplified by a fading of the capacity and a build-up of internal pressure that may eventually lead to a short circuit.

Electrochemistry. Silver–zinc cells have some unusual thermodynamic properties. The equations indicate that the higher valence silver oxide is AgO, silver(II) oxide. However, in the crystallographic unit cell, which is monolithic, there are four silver atoms and four oxygen atoms, and none of the Ag–O bonds conforms to a silver(II) bond length. Instead there are two Ag–O bonds of 0.218 nm corresponding to silver(I) and two Ag–O bonds of 0.203 nm corresponding to silver(III). This structure has also been proposed on the basis of magnetic and semiconductor properties and confirmed using neutron diffraction.

Electrodes. All of the finished silver electrodes have certain common characteristics: the grids or substrates used in the electrodes are usually made of silver, although in some particular cases silver-plated copper is used. Material can be in the form of expanded silver sheet, silver wire mesh, or perforated silver sheet. In any case, the intent is to provide electronic contact of the external circuit of the battery or cell and the active material of the positive plate. Silver is necessary to avoid any possible oxidation at this junction and the increased resistance that would result.

There are three methods of silver electrode fabrication: (1) the slurry pasting of monovalent or divalent silver oxide to the grid, drying, reducing by exposure to heat, and then sintering to agglomerate the fine particles into an integral, strong structure; (2) the dry processing of fine silver powders by pressing in a mold or by a continuous rolling operation onto a silver grid followed by sintering; and, (3) the use of plastic-bonded active material formed by imbedding the active material (fine silver powder) in a plastic vehicle such as polyethylene, which can then be milled into flexible sheets. These sheets are cut to size, pressed in a mold on both sides of a conductive grid, and the pressed electrode subjected to sintering where the plastic material is fired off, leaving the metallic silver.

Silver electrodes prepared by any of the three methods are almost always subjected to a sintering operation prior to cell or battery assembly.

Zinc electrodes for secondary silver–zinc batteries are made by one of three general methods: the dry-powder process, the slurry-pasted process, or the electroformed process. The active material used in any of the processes for the manufacture of electrodes is a finely divided zinc oxide powder, USP grade 12.

Silver–Zinc Separators. The basic separator material is a regenerated cellulose (unplasticized cellophane) which acts as a semipermeable membrane allowing ionic conduction through the separator and preventing the migration of active materials from one electrode to the other. Usually, multiple layers are used in cell fabrication. A stronger separator is one made of sausage casing material (FSC), a regenerated cellulose similar to cellophane but including some fibrous material. Another method of extending the life of the cellulosic separators

has been to incorporate a silver organic compound, eg, silver xanthate, into the cellulosic separator.

Electrolyte. The electrolyte in silver–zinc cells is 30–45% KOH. The lower concentrations in this range have higher conductivities and are preferred for high rate cells. Higher concentrations have a less deleterious effect on cellulosic separators and are preferable for extended life characteristics.

Cell Hardware. Cell jars are constructed almost exclusively of injection-molded plastics, which are resistant to the strong alkali electrolyte. The most generally used materials are modified styrenes or copolymers of styrene and acrylonitrile (SAN). Another material that has been found to increase shock resistance of cells is ABS plastic (acrylonitrile–butadiene–styrene).

Performance. *Charging.* Charging of silver–zinc cells can be done by one of several methods. The constant-current method which is most common consists of a single rate of current usually equivalent to a full input within the 12–16-h period.

Discharge. Silver–zinc cells have one of the flattest voltage curves of any practical battery system known. However, there are two voltage plateaus.

Performance of silver–zinc cells is normally considered to be adequate in the temperature range of 10–38°C. If a wider temperature range is desired silver–zinc cells and batteries may be used in the range 0–71°C without any appreciable derating. Lower temperatures result in some reduction of cell voltage capacities at medium to high rates, and higher temperatures curtail life because of deterioration of the separator materials.

Cell Life. Silver–zinc cells are usually manufactured as either low or high rate cells. Low rate cells contain fewer and thicker electrodes and have many layers of separator (up to the equivalent of 10 layers of cellophane). High rate cells, on the other hand, contain many thinner electrodes and have separator systems of the equivalent of three to four layers of cellophane. Approximately 10–30 cycles can be expected for high rate cells depending on the temperature of use, the rate of discharge, and methods of charging. Low rate cells have been satisfactorily used for 100–300 cycles under the proper conditions. In general, the overall life of the silver–zinc cell with the separator systems normally in use is approximately 1–2 yr.

Other Silver Positive Electrode Systems

Silver–Cadmium Cells. In satellite applications the nonmagnetic property of the silver–cadmium battery was of utmost importance because magnetometers were used on satellites to measure radiation and the effects of magnetic fields of energetic particles. Satellites had to be constructed of nonmagnetic components in sealed batteries.

Silver–cadmium satellite batteries have been used in cyclic periods of five hours or more with discharge times of 30–60 min. Operational and test programs have shown cycle life periods of 3 yr at low temperatures. At temperatures of 40°C and 50°C, the cycle life is 1 yr and 0.2 yr, respectively. The cycle life at intermediate temperatures is 1.4–2.0 yr.

Another application for silver–cadmium batteries is propulsion power for submarine simulator-target drones.

Silver–Iron Cells. The silver–iron battery system combines the advantages of the high rate capability of the silver electrode and the cycling characteristics of the iron electrode. Development has been undertaken to solve problems associated with deep cycling of high power batteries for ocean systems operations.

Cells consisted of porous sintered silver electrodes and high rate iron electrodes. The latter were enclosed with a seven-layered, controlled-porosity polypropylene bag which serves as the separator. The electrolyte contains 30% KOH and 1.5% LiOH.

Applications have been found for these batteries in emergency power applications for telecommunications systems in tethered balloons. Unfortunately, the system is expensive because of the high cost of the silver electrode. Applications are, therefore, generally sought where recovery and reclamation of the raw materials can be made.

Nickel–Zinc Cells

Nickel–zinc cells offer some advantages over other rechargeable alkaline systems. The single-level discharge voltage, 1.60–1.65 V/cell is approximately 0.35–0.45 V/cell higher than nickel–cadmium or nickel–iron and approximately equal to that of silver–zinc. In addition, the use of zinc as the negative electrode should result in a higher energy density battery than either nickel–cadmium or nickel–iron and a lower cost than silver–zinc. In fact, nickel–zinc cells having energy densities in the range of 50–60 W·h/kg have been successfully demonstrated.

Some efforts toward sealed battery development were made. However, a third electrode, an oxygen recombination electrode was required to reduce the cost of the system. High rate applications such as torpedo propulsion were investigated and moderate success achieved using experimental nickel–zinc cells yielding energy densities of 35 W·h/kg at discharge rates of 8 C. A commercial nickel–zinc battery is considered to be a likely candidate for electric vehicle development. If the problems of limited life and high installation cost are solved, a nickel–zinc EV battery could provide twice the driving range for an equal weight lead–acid battery. Work is developmental; there is only limited production of nickel–zinc batteries.

Cell Construction. Nickel–zinc batteries are housed in molded plastic cell jars of styrene, SAN, or ABS material for maximum weight savings. Nickel electrodes can be of the sintered or pocket type, however, these types are not cost effective and several different types of plastic-bonded nickel electrodes have been developed.

Nickel hydrate, usually 5–10% cobalt added, serves as the active material and is mixed with a conductive carbon, eg, graphite. The active mass is mixed with an inert organic binder such as polyethylene or poly(tetrafluoroethylene) (TFE). The resultant mass is rolled into sheets on a compounding mill or pressed into electrodes as a dry powder on a nickel grid.

Negative electrodes are fabricated of zinc oxide by any of the methods (pasting, pressing, etc) described. Binders, usually TFE, are used to reduce the solubility of the electrode in KOH. In addition, other techniques such as extended edges, inert extenders, contouring, and variable density have been tried in an effort to reduce shape change of the negative electrode upon cycling. Separators are both of the organic and inorganic type.

Performance. The limited life of nickel–zinc batteries is the principal drawback to widespread use.

Nickel–Hydrogen Cells

There are two types of nickel–hydrogen cells; those that employ a gaseous H_2 electrode and those that utilize a metal hydride, MH.

Gaseous Hydrogen Systems. The nickel–hydrogen cell incorporating a gaseous hydrogen electrode is a hybrid consisting of one gaseous and one solid electrode. The nickel electrode is of the type used in a nickel–cadmium battery and the hydrogen electrode is a gas diffusion electrode of the type used in alkaline fuel cells (qv). These two electrodes are capable of extremely long, stable life.

During charge the nickel hydroxide is converted to NiOOH, the charged state of nickel, and on the surface of the hydrogen electrode, hydrogen gas is evolved. By placing the electrode stack in a sealed container, the hydrogen is captured for subsequent reuse. During discharge the same hydrogen is reconsumed on the same electrode surface and the nickel electrode is reduced to provide electric energy. Another desirable feature of this battery system is its capability of high rate of overcharge.

Metal Hydride Systems. The success of the gaseous nickel–hydrogen system led to the investigation of replacing the gaseous hydrogen with metal hydrides in order to reduce the cell pressure and the volume required for hydrogen storage. A number of metal hydrides were developed for reversible hydrogen storage. Of particular interest were $LaNi_5$ and $MmNi_5$.

Other Cell Systems Silver–Hydrogen Cells

With the development of the nickel–hydrogen system limited attention was directed to the development of a silver–hydrogen cell. The main characteristics of interest were the potential for a higher gravametric energy density based on the ligher weight of the silver electrode vs that of the nickel. The packaging approach utilized for this battery is similar to that for nickel–hydrogen single cylindrical cells as shown in Figure 2. The silver electrode is typically the sintered type used in rechargeable silver–zinc cells. The hydrogen electrode is a Teflon-bonded platinum black gas diffusion electrode.

Because the silver oxide electrode is slightly soluble in the potassium hydroxide electrolyte the separator is of a barrier type to minimize silver diffusion to the opposite electrode.

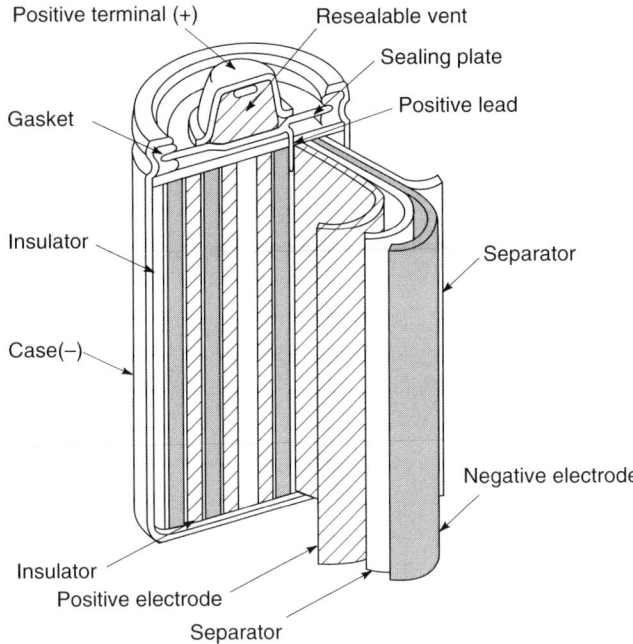

Figure 2. Schematic diagram of Ni–H(MH) cell.

Zinc–Oxygen Cells. On the basis of reactants the zinc–oxygen or air system is the highest energy density system of all the alkaline rechargeable systems with the exception of the $H_2 \cdot O_2$ one. The reactants are cheap and abundant and therefore a number of attempts have been made to develop a practical rechargeable system.

Iron–Air Cells. The iron–air system is a potentially low cost, high energy system being considered mainly for mobile applications. The iron electrode, similar to that employed in the nickel–iron cell, exhibits long life and therefore this system could be more cost effective than the zinc–air cell.

Hydrogen–Oxygen Cells. The hydrogen–oxygen cell can be adapted to function as a rechargeable battery, although this system is best known as a primary one (see FUEL CELLS).

Because of cross-gas leakage and other complexities, a single-cell was developed. In this configuration the electrodes were constructed in the form of a cylinder; the hydrogen gas was stored in the central compartment, and the oxygen gas in the space between the electrode core and the outer pressure vessel.

Primary alkaline fuel cells were developed for space applications.

Electrolyte

Potassium hydroxide is the principal electrolyte of choice for the above batteries because of its compatibility with the various electrodes, good conductivity, and low freezing point temperature.

Health and Safety Factors

The potassium hydroxide electrolyte used in alkaline batteries is a corrosive hazardous chemical. It is a poison and if ingested attacks the throat and stomach linings. Immediate medical attention is required. It slowly attacks skin if not rapidly washed away. Extreme care should be taken to avoid eye contact that can result in severe burns and blindness. Protective clothing and face shields or goggles should be worn when filling cells with water or electrolyte and performing other maintenance on vented batteries.

Alkaline batteries generate hydrogen and oxygen gases under various operating conditions. This can occur during charge, overcharge, open circuit stand, and reversal. In vented batteries free ventilation should be provided to avoid hydrogen accumulations surrounding the battery. A vented battery must never be placed in a sealed container for which it was not designed.

Alkaline batteries are capable of high current discharges and accidental short circuits should be avoided. Spontaneous low resistance internal short circuits can develop in silver–zinc and nickel–cadmium batteries.

LEAD ACID CELLS

The lead–acid battery is one of the most successful electrochemical systems and the most successful storage battery developed. About 87% of the lead (qv), Pb, consumption in the United States was for batteries in 2001.

The lead–acid battery consists of a number of cells in a container. These cells contain positive (PbO_2) and negative (Pb) electrodes or plates, separators to keep the plates apart, and sulfuric acid, H_2SO_4, electrolyte. The battery reactions are highly reversible, so that the battery can be discharged and charged repeatedly. The number of charge–discharge cycles that can be obtained depends strongly on the use mode and can vary from several hundred to thousands of cycles.

Each cell has a nominal voltage of 2 V and capacities typically vary from 1 to 2000 ampere-hours. Lead–acid cells can be operated with coulombic efficiencies as high as 95% and with energy efficiencies greater than 80%. The many cell designs available for a wide variety of uses can be divided into three main categories: automotive, industrial, and consumer. Industrial batteries are used for heavy-duty application such as motive and standby power. More recently, the use of batteries for utility peak shaving has been increasing. Consumer batteries for emergency lighting, security alarm systems, cordless convenience devices and power tools, and small engine starting is one of the fastest growing markets for the lead–acid battery.

In Figure 3, the cutaway view of the automotive battery shows the components used in its construction. This and industrial motive power batteries have the standard free electrolyte systems and operate only in the vertical position.

Two types of batteries having immobilized electrolyte systems are also made. They are most common in consumer

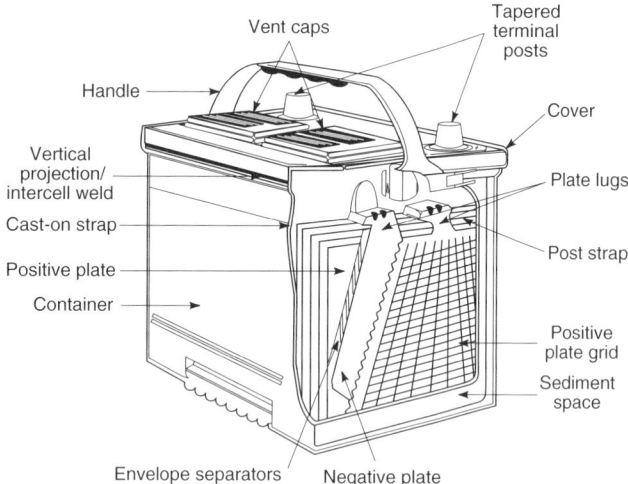

Figure 3. Cutaway view of an automotive SLI lead–acid battery container and cell element. Courtesy of Johnson Controls, Inc.

applications, but their use in industrial and SLI applications is increasing. Both types have low maintenance requirements and usually can be operated in any position. They are sometimes called valve regulated or recombinant batteries because they are equipped with a one-way pressure relief vent and normally operate in a sealed condition with an oxygen recombination cycle to reduce water loss.

In the gelled electrolyte battery, the sulfuric acid electrolyte has been immobilized by a thixotropic gel. This is made by mixing an inorganic powder such as silicon dioxide, SiO_2, with the acid. Other cells use a highly absorbent separator to immobilize the electrolyte.

Cell Thermodynamics

The chemical reaction of the lead–acid battery was explained as early as 1882. The double sulfate theory has been confirmed by a number of methods as the only reaction consistent with the thermodynamics of the system. The thermodynamics of the lead–acid. Other battery has been reviewed in great detail.

Lead sulfate is formed as the battery discharges, sulfuric acid is regenerated as the battery is charged. The open circuit voltage of the lead–acid battery is a function of the acid concentration and temperature. A review of this subject is available. The Nernst equation may be used to calculate the open circuit cell voltage. The battery voltage is then obtained by multiplying the cell voltage by the number of cells.

Lead Grid Corrosion

The corrosion of the lead grid at thelead dioxide electrode is one of the primary causes of lead–acid battery failure. The mechanisms of lead corrosion in sulfuric acid have been studied and good reviews of the literature are available.

Charge–Discharge Processes

An excellent review covers the charge and discharge processes in detail and ongoing research on lead–acid batteries may be found in two symposia proceedings. Detailed studies of the kinetics and mechanisms of lead–acid battery reactions are published continually.

At high discharge rates, such as those required for starting an engine, the voltage drops sharply primarily because of the resistance of the lead current collectors. This voltage drop increases with the cell height and becomes significant even at moderate discharge rates in large industrial cells. Researchers have measured this effect in industrial cells and have developed a model which has been used to improve grid designs for automotive batteries.

Self-Discharge Processes. The shelf life of the lead–acid battery is limited by self-discharge reactions, which proceed slowly at room temperature. High temperatures reduce shelf life significantly. The reactions which can occur are well defined and self-discharge rates in lead–acid batteries having immobilized electrolyte and limited acid volumes have been measured.

The lead current collector in the positive lead–dioxide plate corrodes and the compounds which form are a function of the acid concentration and positive electrode voltage. Other reactions which take place at the positive electrode are oxygen evolution, oxidation of organics, sulfation of PbO (in new cells), and oxidation of additives such as antimony, in the grid alloy. Similar reactions can be written for other metallic additives. At the negative electrode two more reactions can occur: hydrogen evolution and oxygen recombination.

Overcharge Reactions. Water electrolysis during overcharge is an irreversible process. Theoretically, water should decompose at a voltage below the voltage required to recharge a lead–acid battery. However, the rate of water electrolysis is much slower than the rate of the recharge reaction. Thus the lead–acid battery can operate with as little as 5% excess charge to compensate for water electrolysis. Use of lead–antimony alloys for the current collectors in lead–acid batteries increases water loss. Some of these batteries need regular maintenance by addition of water to replace the water lost on overcharge. Many newer designs, however, use either lower concentrations of antimony in the alloy or lead–calcium alloys to reduce water loss (see LEAD ALLOYS). This is the basis for the maintenance-free batteries.

Material Fabrication and Manufacturing Processes

The lead–acid battery is comprised of three primary components: the element, the container, and the electrolyte. The element consists of positive and negative plates connected in parallel and electrically insulating separators between them. The container is the package which holds the electrochemically active ingredients and houses the external connections or terminals of the battery. The electrolyte, which is the liquid active material and ionic conductor, is an aqueous solution of sulfuric acid.

Plates. Plates are the part of the cell that ultimately become the battery electrodes. The plates consist of an electrically conductive grid pasted with a lead oxide–lead sulfate paste which is the precursor to the electrode active materials which participate in the electrochemical charge–discharge reactions.

Paste Mixing. The active materials for both positive and negative plates are made from the identical base materials. Lead oxide, fibers, water, and a dilute solution of sulfuric acid are combined in an agitated batch mixer or reactor to form a pastelike mixture of lead sulfates, the normal, tribasic, and tetrabasic sulfates, plus PbO, water, and free lead. The positive and negative pastes differ only in additives to the base mixture. Organic expanders, barium sulfate, $BaSO_4$, carbon, and occasionally mineral oil are added to the negative paste. Red lead, or minium, Pb_3O_4, is sometimes added to the positive mix.

Grids. The grid in a battery plate performs two vital functions. It acts as the mechanical support framework of the plate during manufacturing, and provides uniform, efficient current flow to and from the plate during formation and use. Grids are designed to have a lug or current collection point, current carrying arms, structural frame, and usually feet.

The grid must possess sufficient stiffness to prevent damage or distortion during the casting, plate pasting, and battery assembly operations. During the life of the battery the grid must bear significant loads such as active material weight, its own weight, and stresses resulting from corrosion and volumetric changes caused by the sulfation of lead and lead dioxide. The most severe environment for the grid is in the positive plate.

Separators. The separator's purpose is to isolate the positive and negative electrodes electronically while allowing ionic exchange between the two.

Electrolyte

Sulfuric Acid. Sulfuric acid is a primary active material of the battery. It must be present to provide sufficient sulfate ions during discharge and to retain suitable conductivity. Lead–acid batteries generally use an aqueous solution of acid in either a free-flowing or in an immobilized state.

Battery Assembly

The cell element (Fig. 3) is normally constructed from groupings of positive and negative plates. The number and size of plates of each type is determined by the desired performance level for the battery. Positive and negative plates are alternately stacked using separators in between to form the proper plate count. When the element is assembled, this piece moves to the final assembly.

The elements are inserted into containers such as that shown in Figure 3. The voltage of each charged cell is approximately 2 V, thus three elements are connected in series in 6 V batteries, six in 12 V batteries, and so forth. Once in the container, the strap vertical projections are fused together in a series arrangement to produce an unformed battery. When the fusing operation is complete, the cover is sealed to the container.

OTHER CELLS

The proliferation of portable electronic devices has fueled rapid market growth for the rechargeable battery industry. Miniaturization of electronics coupled with consumer demand for lightweight batteries providing ever longer run times continues to spur interest in advanced battery systems. Interest also continues to run strong in electric vehicles (EVs) and the large auto manufacturers continue to develop prototype EVs. It is clear that advances in battery technology are required for a widely acceptable EV. Advanced batteries continue to play a strong role in other applications such as load leveling for the electric utility industry and satellite power systems for aerospace.

Ambient Temperature Lithium Systems

Traditionally, secondary battery systems have been based on aqueous electrolytes. Whereas these systems have excellent performance, the use of water imposes a fundamental limitation on battery voltage because of the electrolysis of water, either to hydrogen at cathodic potentials or to oxygen at anodic potentials. The application of nonaqueous electrolytes affords a significant advantage in terms of achievable battery voltages. By far the most actively researched field in nonaqueous battery systems has been the development of practical rechargeable lithium batteries. These are systems that are based on the use of lithium metal, Li, or a lithium alloy, as the negative electrode (see LITHIUM AND LITHIUM COMPOUNDS).

The use of lithium as a negative electrode for secondary batteries offers a number of advantages. Lithium has the lowest equivalent weight of any metal and affords very negative electrode potentials when in equilibrium with solvated lithium ions resulting in very high theoretical energy densities for battery couples. These high theoretical energy densities have prompted a wealth of research activity in a wide variety of experimental battery systems. However, realization of the technology to commercialize these systems has been slow.

A key technical problem in developing practical lithium batteries has been poor cycle life attributable to the lithium electrode. The highly reactive nature of freshly plated lithium leads to reactions with electrolyte and impurities to form passivating films that electrically isolate the lithium metal.

The choice of battery electrolyte is of paramount importance to achieving acceptable cycle life because of the high reducing power of the metallic lithium. The formation of surface films on the lithium electrode imparts the apparent stability of the electrolyte to the electrode. It is critical to determining lithium cycling efficiency. In addition to providing a stable film in the presence of lithium, the electrolyte must satisfy additional requirements including good conductivity, being in the liquid range over the battery operating temperature, and electrochemical stability over a wide voltage range. Solubility of the electrolyte salt in the solvent system is important in achieving good conductivity. In order to satisfy the various electrolyte system requirements, the use of mixed solvent electrolytes has become common in practical cells. Examples are tetrahydrofuran, C_4H_8O, -based electrolytes or ethylene carbonate, $C_2H_4O_3$, −propylene carbonate, $C_4H_6O_3$, mixed solvent systems.

A second class of important electrolytes for rechargeable lithium batteries are *solid* electrolytes. Of particular importance is the class known as solid polymer electrolytes (SPEs). SPEs are polymers capable of forming complexes with lithium salts to yield ionic conductivity. The best known of the SPEs are the lithium salt complexes of poly(ethylene oxide) (PEO), $-(CH_2CH_2O)_n-$, and poly(propylene oxide) (PPO).

The lithium or lithium alloy negative electrode systems employing a liquid electrolyte can be categorized as having either a solid positive electrode or a liquid positive electrode. Systems employing a solid electrolyte employ solid positive electrodes to provide a solid-state cell. Another class of lithium batteries are those based on conducting polymer electrodes. Several of these systems have reached advanced stages of development or initial commercialization such as the Seiko Bridgestone lithium polymer coin cell.

The most important rechargeable lithium batteries are those using a solid positive electrode within which the lithium ion is capable of intercalating. These intercalation, or insertion, electrodes function by allowing the interstitial introduction of the Li^+ ion into a host lattice. A large number of inorganic compounds have been investigated for their ability to function as a reversible positive electrode in a lithium battery. Intercalation electrodes have found wide application in systems employing both liquid or solid electrolytes.

Solid Electrolyte Systems. Whereas there has been considerable research into the development of solid electrolyte batteries, development of practical batteries has been slow because of problems relating to the low conductivity of the solid electrolyte. The development of an all solid-state battery would offer significant advantages. Such a battery would overcome problems of electrolyte leakage, dendrite formation, and corrosion that can be encountered with liquid electrolytes. The general configuration of one system that has reached an advanced stage of development is shown in Figure 4. The negative electrode consists of thin lithium foil. The composite cathode is composed of vanadium oxide, V_6O_{13}, mixed with polymer electrolyte.

A new all solid-state lithium battery employing a positive electrode comprised of organosulfur polymers, $-(SRS)_n-$, has been reported. During discharge of the battery, current is produced by cleavage of the sulfur−sulfur bonds in the polymer, depolymerizing the polymer. On

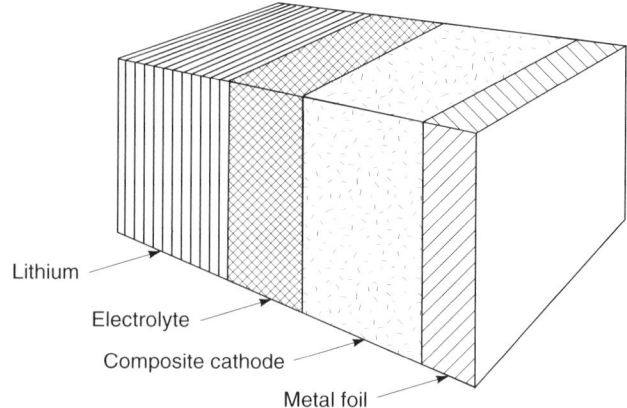

Figure 4. Configuration for a solid polymer electrolyte rechargeable lithium cell where the total thickness is 100 μm. Courtesy of Mead Corp.

charge, the process is reversed and the disulfides are polymerized back to their original form. This use of a polymerization−depolymerization reaction for a battery electrode is unique and this electrode is expected to offer significantly improved rate capability over intercalation electrodes. The organosulfur electrodes provide excellent stability and reversibility, which should result in long cycle life. Cells constructed to date have employed a PEO electrolyte and hence require operation at elevated temperature.

Advanced Systems. Applications for the coin and button secondary lithium cells is limited. However, researchers are working to develop practical "AA"-sized and larger cells. Several systems have reached advanced stages of development.

One of the most widely studied intercalation electrode materials is titanium disulfide, TiS_2. A number of factors make TiS_2 attractive for secondary lithium cells including good rate capability, high theoretical energy density, and a highly reversible intercalation reaction. Very high cycle lives have been demonstrated for TiS_2 electrodes. One of the earliest efforts to develop commercial lithium cells employed TiS_2.

A rechargeable lithium "AA" cell employing niobium triselenide, $NbSe_3$, as the positive electrode material has been developed. The key to this system was a method for thermally growing $NbSe_3$ fibers that can then be pressed onto a current collector to provide a very high energy density electrode. No binder or additional conductive material is required. The excellent properties of this electrode also allow for higher discharge currents than those typically available for rechargeable lithium systems.

Other solid cathode systems that have been widely investigated include those containing lithium cobalt oxide, $LiCoO_2$, vanadium pentoxide, V_2O_5, and higher vanadium oxides, eg, V_6O_{13}.

In addition to cells employing solid positive electrodes, a rechargeable lithium cell employing sulfur dioxide SO_2, as a liquid electrode has been developed.

High Temperature Systems

Lithium–Aluminum/Metal Sulfide Batteries. The use of high temperature lithium cells for electric vehicle applications has been under development since the 1970s. Advances in the development of lithium alloy–metal sulfide batteries have led to the Li–Al/FeS system, where the following cell reaction occurs.

$$2\,LiAl_x + FeS \rightleftharpoons Fe + Li_2S + 2x\,Al$$

The cell voltage is 1.33 V to give a theoretical energy density of 458 W·h/kg. The cell employs a molten salt electrolyte, most commonly a lithium chloride/potassium chloride, LiCl–KCl eutectic mixture. The cell is generally operated at 400–500°C. The negative electrode is composed of lithium–aluminum alloy, which operates at about 300 mV positive of pure lithium. The positive electrode is composed of iron sulfide mixed with a conductive agent such as carbon or graphite. Electrodes are constructed by cold pressing powder onto current collectors.

Development of practical and low cost separators has been an active area of cell development. Cell separators must be compatible with molten lithium, restricting the choice to ceramic materials. Early work employed boron nitride, BN, but a more desirable separator has been developed using magnesium oxide, MgO, or a composite of MgO powder–BN fibers.

Li–Al/FeS cells have demonstrated good performance under EV driving profiles and have delivered a specific energy of 115 W·h/kg for advanced cell designs. Cycle life expectancy for these cells is projected to be about 400 deep discharge cycles. This system shows considerable promise for use as a practical EV battery.

A similar system under development employs iron disulfide, FeS_2, as the positive electrode. Whereas this system offers a higher theoretical energy density than does Li–Al/FeS, the FeS_2 cell is at a lower stage of development.

Sodium–Sulfur. The best known of the high temperature batteries is the sodium–sulfur, Na–S, battery. The cell reaction is best represented by the equation:

$$2\,Na + 3\,S \rightleftharpoons Na_2S_3$$

occurring at a cell voltage of 1.74 V, to give a specific energy of 760 W·h/kg. The cell is constructed using a solid electrolyte typically consisting of β-alumina, β-Al_2O_3, ceramic, although borate glass fibers have also been used. These materials have high conductivities for the sodium ion. The negative electrode consists of molten sodium metal and the positive electrode of molten sulfur. Because sulfur is not conductive, a current collection network of graphite is required. The cell is operated at about 350°C.

The Na–S battery couple is a strong candidate for applications in both EVs and aerospace. Projected performance for a sodium–sulfur-powered EV van is shown in Table 2 for batteries having three different energies.

Table 2. Electric Vehicle Battery Performance

Parameter	Battery			
	Lead–acid	Sodium–sulfur		
battery energy, kW·h	40.0	40.0	60.0	85.0
range, km	84.0	113.0	169.0	242.0
max payload, t	0.9	1.7	1.6	1.6
battery weight, kg	1250.0	330.0	424.0	580.0

The Na–S system is expected to provide significant increases in energy density for satellite battery systems. In-house testing of Na–S cells designed to simulate mid-altitude (MAO) and geosynchronous orbits (GEO) demonstrated over 6450 and over 1400 cycles, respectively.

Difficulties with the Na–S system arise in part from the ceramic nature of the alumina separator: the specific β-alumina is expensive to prepare; and the material is brittle and quite fragile. Separator failure is the leading cause of early cell failure. Cell failure may also be related to performance problems caused by polarization at the sodium/solid electrolyte interface. Lastly, seal leakage can be a determinant of cycle life. In spite of these problems, however, the safety and reliability of the Na–S system has progressed to the point where pilot plant production of these batteries is anticipated for EV and aerospace applications.

A battery system closely related to Na–S is the Na-metal chloride cell. The cell design is similar to Na–S; however, in addition to the β-alumina electrolyte, the cell also employs a sodium chloroaluminate, $NaAlCl_4$, molten salt electrolyte. The positive electrode active material consists of a transition metal chloride such as iron(II) chloride, $FeCl_2$, or nickel chloride, $NiCl_2$, in lieu of molten sulfur. This technology is in a younger state of development than the Na–S.

"Cadmium," *Mineral Commodity Summaries*, U.S. Geological Survey, Reston, Va., Jan. 2003; *Minerals Yearbook*, 2001.

P. A. Nelson, *4th International Seminar on Lithium Battery Technology and Applications*, Deerfield Beach, Fla., Mar. 1989.

J. A. Ober, "Lithium," *Minerals Yearbook*, U.S. Geological Survey, Reston, Va., 2002.

H. Takeshita, *Presentation at the 20th International Seminar and Exhibit on Primary and Secondary Batteries*, March 17, 2003.

A. Hooper, in K. M. Abraham and B. B. Owens, eds., *Materials and Processes for Lithium Batteries*, Vol. 89–4, The Electrochemical Society, Inc., Cleveland, Ohio, 1980, 15–32.

ALVIN J. SALKIND
MARTIN KLEIN
Rutgers University

KATHRYN R. BULLOCK
JOHN R. PIERSON
Johnson Controls, Inc.

PAUL R. GIFFORD
Ovonic Battery Company

BEER AND BREWING

Beer (Latin: bibere, to drink. Old English, beor. Middle English, bere) may be defined as a mildly alcoholic beverage made by the fermentation of an aqueous extract of cereals (grains). Cereals contain carbohydrates, mainly in the form of starch, which brewers' yeasts cannot ferment, and so breakdown of starch to fermentable sugar is a central feature of beer-making processes. This is in contrast to wines in which fermentable sugars are preformed in the raw materials (eg, fruits such as grapes). Thus, sake, for eg, commonly called a rice wine, is in fact a beer. The grain mainly used for beer-making is barley, with rice and corn as adjunct, and some beers are made partly from malted wheat. Others grains such as sorghum can be used, especially in the manufacture of traditional beers, and oats in, eg, a few stouts.

Manufacture of beer has five main stages: (1) malting, (2) brewing, (3) fermentation, (4) finishing, and (5) packaging.

OVERVIEW OF THE PROCESS

Beer is a food product and subject to all the regulations concerning food production and distribution. Malthouses and breweries therefore are impeccably clean and sanitary places not only to meet the provisions of those regulations but also because the brewing process and beer itself are subject to attack by unwanted microorganisms. While these organisms pose no danger from the point of view of transmitting illness to the consumer, they can easily spoil beer flavor, eg, by causing sour tastes and unwanted aromas and by forming hazes.

Malting takes ~8 days. However, the malt spends a good deal more time than that in the malthouse before it is sold to brewers because it needs to be cleaned, matured, analyzed, and then blended to meet brewers' specification before sale. In the malting process, barley is first wetted (steeped) for 2 days and then put to germinate in an appropriate vessel where it is aerated and turned regularly for ~4 or 5 days. Then, heating in a kiln for up to 2 days dries the green malt. This imbues the product with intense malty flavors and color, both of which become part of the character of the beer made from it. *Brewing* follows malting. In the brewhouse, five distinct stages occur in a period of ~5–6 h: (1) the malt is milled (ground up) and then mixed with suitable water and (2) heated through a precise temperature program (mashed). This converts starch to fermentable sugar. The liquid mass is then transferred (3) to a device to separate the insoluble spent grains (mainly the husk of the malt and precipitated protein) from the sugary aqueous extract called wort. The wort is then boiled (4) with hops, to stabilize it and to impart bitterness. The spent hops are removed (5) and the wort is then cooled, which concludes the brewhouse operations. *Fermentation* follows. A desirable yeast culture is added to the wort and during the course of ~3–9 days (depending on temperature) the yeast converts the sugar present mainly into alcohol and carbon dioxide, but also forms a myriad of flavor compounds that, with the flavors from malt and hops,

combine to create the final beer flavor. After this primary fermentation the beer is *finished*, ie matured (aged) and carbonated by further treatments, eg, secondary fermentation. It is then filtered to make it brilliantly clear and stabilized to prevent changes in the market place. Finally this beer is *packaged* at high speed into bottles and cans of various convenient sizes and into kegs for sale with strict control of oxygen access.

MALTING

The objective of malting is to achieve "modification" of the barley grain. Barley lacks suitable enzymes for brewing, lacks friability (ie, it is a hard grain) and so is difficult to mill, it lacks suitable aroma and flavor and color, it lacks simple nitrogenous compounds suitable for yeast growth, and it contains β-glucans, which make aqueous barley extracts viscous and difficult to filter. Modification, which is the sum of changes as barley is converted into malt, and is achieved by partially germinating the grain, solves all of these problems when done well. Barley, unlike wheat, has an adherent husk that protects the grain during malting.

Malt is made from barleys that are approved or selected varieties known as malting barleys and that meet necessary specifications. Typically, malting barleys tend to germinate vigorously, modify evenly and completely, have a low nitrogen or protein content, and are relatively plump, ie, tend to have large kernels that contain a lot of starch relative to the amount of husk. There are two kinds of malting barleys, six- and two-row, and brewers use blends of both kinds. Generally, malts made from six-row barleys tend to yield rather more enzymes (expressed as diastatic power, DP, or starch-splitting ability) than malts from two-row barleys. On the other hand, malts made from two-row barleys yield a little more extract (more soluble solids to the wort) than six-row barleys.

The quality of malt is gauged by a number of laboratory measures. These include (1) determination of Diastatic Power (DP) that is a measure of the amylase enzymes present (especially β-amylase). (2) Ease of milling (using an instrument called the friabilmeter), or direct determination of the β-glucan content, or the ease of extraction (called the coarse/fine difference) is used to gauge the degree of modification. (3) FAN is used to measure the presence of potential yeast nutrients (amino acids, eg). The overall quality of the malt resides in its extract yield; ie, the amount of material that can be dissolved from it in a laboratory scale mashing process; the value is usually ~80%. By these measures and others maltsters and brewers determine the extent to which the necessary changes in barley have been achieved when producing a particular batch of malt. In practice, at the malthouse, many batches of malt are blended together to meet the brewers' specifications. The brewer is concerned with three things: (1) *kernel size* expressed here as assortment (by screening) and as 1000-kernel weight; (2) *modification* expressed as growth of shoot (length of kernel), coarse/fine difference % (difference in extract between fine and coarse grind malt), and mash soluble protein expressed as a percentage of soluble over total protein (S/T%), and possibly wort viscosity; (3) *enzyme content* expressed as DP (mainly β-amylase) and as α-Amylase (DU, dexrinizing units).

HOPS

Hops are a crucial component of beer although only ∼4–8 oz/barrel (120–240 g/hL) are used. Their primary role is to give bitterness to beer. Although humans do not usually like bitterness, a sufficient and balanced inclusion of bitter character in beers is necessary for a satisfactory product. How the brewer handles the bitter quality of hops in creating a beer is important in differentiating one beer from another, and meeting the needs of the target consumer population. Hops also can contribute delicate aromas to beers, that, in conjunction with those flavors arising from the yeast and malt, creates the overall impression of a beer. Hops are used in the kettle-boiling process in a brewery. The key chemical reaction of wort boiling is the conversion of relatively insoluble α-acids present in the hops to quite soluble iso-α-acids that persist into the beer.

WATER

Water makes up ∼95% by volume of most beers, but the quality of water used in brewing, beyond mere potability, can have an impact on its quality and flavor characteristics. In general, hardness is desirable for brewing purposes (ie, as a beer component) and alkalinity is not. It is a matter of pH or acidity.

The bulk of water used in breweries is for cleaning and sanitizing the plant and for raising steam for transporting energy about the brewery. Soft water (ie, lacking Ca^{2+} and Mg^{2+}) is preferred for these purposes because it does not react with cleaners or deposit "stone" on surfaces being cleaned and sludge in steam boilers. The most common cleaners are based on caustic soda or other strongly alkaline agents.

BREWING

A brewery is divided into three main parts: (1) the brewhouse where the malt is extracted with hot water to make "wort", (2) the cellars where fermentation by yeast takes place and the beer is matured and clarified, and (3) the packaging hall.

Brewhouse

In the brewhouse, the operations are (1) milling for crushing the malt, (2) mashing for extracting the malt with water, (3) filtration to separate spent grain solids from liquid wort, (4) kettle-boiling for stabilizing the wort and extracting the hops, and (5) wort clarification and cooling. The purpose of the brewhouse is to prepare the malt for extraction by *milling*, produce the extract in *mashing*, recover the extract by filtration (called *lautering/mash filtration*), and stabilize it by *boiling*.

Milling

The objective of milling is to crush the malt in such a way that the later processes can operate at maximum efficiency. Thus, finely milled malt will be easily extracted, but it will be difficult to separate the liquid (called "wort") from the insoluble material (called "spent grain"). Depending on the filtration device available therefore, the brewer decides on the most suitable milling strategy. Ideally, the endosperm of the malt is reduced to fine particles and the husk remains intact. Almost all mills in North America are dry mills. That is, the malt enters the mill dry and exits as a dry grist. These are roll mills. They might have three pairs of rolls as in a six-roll mill, although simpler ones are common, especially in small breweries. Wet mills are also used widely around the world. In such mills, the grain is wetted before passing through a single pair of rolls and exits the mill as a slurry of malt in water that is pumped directly to the mash vessel. Hammer mills that reduce malt almost to a powder can be used with some kinds of mash filters.

Mashing comprises extracting the milled malt with a predetermined volume of water (which establishes mash thickness).

Wort Separation

At the end of mashing, the spent grain must be separated from the dense solution (called wort) of sugar and other materials extracted from malt and other grains. This is done by filtration. Two alternative device may be used, a *lauter vessel* or a *mash filter*. The operating principle of both devices is the same. A lauter vessel is a broad flat vessel in which the mash is spread over a false bottom with slots in it. After the mash settles, the wort is drawn slowly through the settled spent grain, where it is clarified, and exits the vessel through the false bottom. In a mash filter, the mash is held in a quite shallow layer against a vertical filter cloth. In either case, the wort, substantially freed of suspended solids, is produced over a period of ∼1.5–2 hs and flows to the wort kettle.

Boiling

Clarified wort from the lauter or mash filter is unstable in several ways: it could possibly contain (1) some active enzymes and so be subject to further change, or (2) unwanted microorganisms that inevitably find their way to warm moist sugary environments, or (3) excessive proteins and polyphenols that could easily cause hazes in beers. By boiling the wort, remnant enzymes are inactivated, bacteria are killed, and much of the protein and polyphenol is precipitated. This precipitate is called "hot trub" or "hot break". In addition, (4) the kettle boil concentrates the wort by evaporation of water, (5) removes the unwanted volatile components of hops and malt, and, most importantly, (6) effects the isomerization of α-acids (which are insoluble in wort and beer) into the bitter and soluble iso-α-acids. There is also evidence that denaturation (loss of native structure) of proteins during boiling helps form polypeptides that have foam-stabilizing properties.

After boiling, the spent hop material and precipitated trub must be removed. Whole hops, if used, must be removed by a *hop strainer*, but a whirlpool separator best removes the particulate matter from pellets.

FERMENTATION

Following wort cooling and aeration or oxygenation, yeast is added and the wort–yeast mixture enters the fermentation cellar. Many beer characteristics, especially the alcohol content, are determined by the strength of the wort at the beginning of fermentation. This is expressed as the original specific gravity or O.G., a measure of density. Density is also expressed directly as specific gravity = 1.040–1.048, which is the ratio of the weight of wort to the weight of water). However, these days brewers commonly use *high gravity brewing* throughout the fermentation and finishing processes and then, just before packaging, dilute the beer to sales strength using carbonated water free of oxygen. This practice assures the most efficient use of brewery capacity.

Brewers recover yeast from a completed fermentation to start another, and in this way have nurtured certain yeasts for many centuries. Brewers' yeasts therefore can no longer be found in nature. Brewers have naturally selected yeasts that particularly meet their requirements. For example, *ale yeast*, when used in small traditional vessels, concentrates at the surface of the fermenting beer where it can be easily recovered by "skimming." These yeasts are named *Saccharomyces cerevisae*; this designation includes wine yeasts and baker's yeasts, too. *Lager yeast* also has an unusual property that assured the early brewers a reasonably constant yeast supply and a means to recover it: It can grow and ferment at low temperatures and settles readily (bottom yeast). In addition, low temperatures allowed more of the CO_2 evolved in fermentation to remain in solution and so lagers were much more easy to carbonate than ales. Lager yeasts are named *Saccharomyces carlsbergensis* or *Saccharomyces uvarum*. In practice, brewers use lager yeasts at lower temperatures (say 8–14°C) than ale yeasts (say 20°C), and this might well account for the differences in flavor between ales and lagers.

To initiate fermentation, yeast is added to cooled wort in a process called "pitching". Detectors monitor yeast addition to assure this addition is done accurately. As it exits the brewhouse, wort is cooled to ~17–20°C for ale fermentations and 8–10°C for lager fermentations. As a result ale fermentations are shorter (~3 days) than lager fermentations (~1 week).

The progress of fermentation is easily measured by the change in specific gravity of the wort as it becomes beer, or the production of alcohol or carbon dioxide. Fermentation is slow at first then, as the yeast begins to grow, becomes much more rapid and the liquid tends to warm up. Alcohol, carbon dioxide, and most flavor compounds are produced roughly in step with yeast growth. Fermentation then slows as the fermentable sugar is exhausted and the yeast begins to fall out of suspension (flocculate). When all the fermentable sugar is used up and there can be no further

change in specific gravity, the brewer cools the beer to encourage further yeast settlement. This brings the primary fermentation to an end. The green beer is now ready for the final stages of processing that are designed to (*1*) mature the green beer, (*2*) carbonate it, (*3*) clarify it, and (*4*) render it stable. These processes are *secondary fermentation* and *finishing*.

SECONDARY FERMENTATION AND FINISHING

There are three general strategies for maturing the green beer. The first is simply to cool the beer to low temperature (0°C or below) after primary fermentation and hold it for some period of time such as a week or two. This is called *aging*; carbon dioxide can be injected at some stage to achieve carbonation. Second, *krausening* is a widely used secondary fermentation strategy that involves mixing freshly fermenting wort containing yeast into the green beer. This mixture is held at ~8°C for ~3 weeks. During this time, the yeast slowly ferments the added sugar and the carbon dioxide formed is entrapped for carbonation of the beer. At the same time, undesirable flavor compounds such as diacetyl and acetaldehyde are reduced by the yeast action to more or less flavorless compounds. The third strategy of secondary fermentation is called *lagering*. In this process the beer, toward the end of primary fermentation, is cooled somewhat to flocculate much, but not all, of the yeast and is then moved to a new vessel before all the fermentable sugar is exhausted. The yeast ferments out the last few degree of gravity slowly at ~8°C, again with entrapment of CO_2 for carbonation, and reduction of diacetyl and acetaldehyde. A modern maturation strategy that considerably shortens the maturation time for many beers is called the *diacetyl rest*. At the end of the primary fermentation, the beer is simply held at fermentation temperature (~15°C at this stage for lagers, 20°C for ales) until the diacetyl is reduced to specification, as determined by measurement.

Finishing

These processes concern (*1*) *filtration* of the beer at low temperature such as minus 2°C, so that it is brilliantly clear and (*2*) treating the beer with *stabilizing agents* so that it remains brilliantly clear during its sojourn in the market place. Beer is often filtered twice: a rough filtration using diatomaceous earth to remove the vast bulk of the suspended particles and then a polish filtration using, eg, sheet filters to achieve brilliant clarity. Centrifugation sometimes replaces rough filtration, or *finings* (isinglass, ie, specially prepared collagen) can be used to coagulate and settle particles. After rough filtration, stabilizing agents are commonly added. Modern stabilizing agents are, for the most part, insoluble adsorbents. Beer that has gone through these finishing processes is now mature in flavor, properly carbonated, brilliantly clear and stable against chill-haze formation, and free of oxygen.

PACKAGING

In packaging into bottles two technical factors come into play. First, packaging must be very rapid, ~2000 units/

min, eg, so that the large volumes of beer produced by modern breweries can be broken down to consumer units in a reasonably short time. And second, oxygen (air) must be rigorously excluded from the package because it harms beer flavor.

Most beer in bottles or cans is *pasteurized*. That is, the beer is heated briefly to kill any microorganisms that might be present that could spoil beer flavor. A pasteurizer is a large tunnel through which the beer cans or bottles move on an endless belt. The containers are sprayed with increasingly hot water to raise their temperature to 60–62°C. They are held at this temperature as long as required, and then cooled by water sprays. There are two alternative techniques to tunnel pasteurization for dealing with the few microbes that might enter beer: "flash" pasteurization, in which the beer before packaging flows through a heat exchanger and is rapidly heated up and cooled down, and second, bacteria present can be filtered out of the beer by extremely tight membrane filtration.

W. A. Hardwick, ed., *Handbook of Brewing*, Marcel Dekker, New York, 1995.

M. Jackson, *The New World Guide to Beer*, Running Press, Philadelphia, London, 1988.

W. Kunze, *Technology Brewing and Malting*, VLB, Berlin, 1996.

L. C. Verhagen, ed., *Hops and Hop Products, Manual of Good Practice*, Getranke- Fachverlag Hans Carl, Nurnburg, 1997.

M. J. Lewis and T. W. Young, *Brewing*, 2nd ed., Kluwer Academic/Plenum Publishers, New York, 2002.

MICHAEL J. LEWIS
University of California

BENZALDEHYDE

Benzaldehyde, C_6H_5CHO, is the simplest and quite possibly the most industrially useful member of the family of aromatic aldehydes. Benzaldehyde exists in nature, primarily in combined forms such as a glycoside in almond, apricot, cherry, and peach seeds. The characteristic benzaldehyde odor of oil of bitter almond occurs because of trace amounts of free benzaldehyde formed by hydrolysis of the glycoside amygdalin.

PHYSICAL PROPERTIES

Physical properties of benzaldehyde are listed in Table 1.

MANUFACTURE

The only industrially important processes for the manufacturing of synthetic benzaldehyde involve the hydrolysis of benzal chloride and the air oxidation of toluene. The hydrolysis of benzal chloride, which is produced by the side-chain chlorination of toluene, is the older of the two processes. It is not utilized in the United States but is used in Europe, India, and China. Other processes,

Table 1. Physical Properties of Benzaldehyde

Property	Value
molecular formula	C_7H_6O
molecular weight	106.12
boiling point, °C at 101.3 kPa[a]	179
melting point, °C	−26
flash point, closed cup, °C	63
autoignition temperature, °C	192
refractive index, n^{20}	1.5455
viscosity, mPa·s (=cP) at 25°C	1.321
density, g/cm³ at 25°C	1.046
specific heat (liquid) at 25°C, J/g·K[b]	1.615
latent heat of vaporization[c], J/g[b]	362
standard heat of combustion, kJ/g[b]	−31.9
solubility in water at 20°C, wt %	~0.6
solubility of water in at 20°C, wt %	~1.5

[a]To convert kPa to atm, divide by 101.3.
[b]To convert J to cal, divide by 4.184.
[c]At the boiling point (179°C).

including the oxidation of benzyl alcohol, the reduction of benzoyl chloride, and the reaction of carbon monoxide and benzene, have been utilized in the past, but they no longer have any industrial application.

HEALTH AND SAFETY FACTORS

The oral LD_{50} for benzaldehyde is reported as 1300 mg/kg in rats and as 1000 mg/kg in guinea pigs. Based upon these values, benzaldehyde is considered a moderately toxic substance when ingested. Studies of the carcinogenic effects of benzaldehyde are currently in progress. In the industrial setting, exposure to benzaldehyde through eye and skin contact and inhalation is far more prevalent than ingestion incidence. Overexposure to benzaldehyde vapors is irritating to the upper respiratory tract and produces central nervous system depression with possible respiratory failure. Contact may cause eye and skin irritation.

HANDLING

The low autoignition temperature of benzaldehyde (192°C) presents safety problems since benzaldehyde can be ignited by exposure to low pressure steam piping, for example. Benzaldehyde may also spontaneously ignite when soaked into rags or clothing or adsorbed onto activated carbon. Bulk storage of benzaldehyde should be made under a nitrogen blanket, since benzaldehyde is easily oxidized to benzoic acid upon exposure to air. All storage tank openings should be easily accessible for cleaning, since they will have a tendency to plug with benzoic acid. Benzaldehyde is stored in noninsulated type 304 stainless steel storage tanks.

USES

Benzaldehyde is a synthetic flavoring substance, sanctioned by the U.S. Food and Drug Administration (FDA) to be generally recognized as safe (GRAS) for foods (21 CFR 182.60). Both "pure almond extract" and "imitation almond extract" are offered for sale. Each contains 2.0–2.5 wt% benzaldehyde in an aqueous solution containing approximately one-third ethyl alcohol.

Benzaldehyde is widely used in organic synthesis, where it is the raw material for a large number of products. In this regard, a considerable amount of benzaldehyde is utilized to produce various aldehydes, such as cinnamic and methyl, butyl, amyl, and hexyl cinnamic aldehydes. The single largest use for benzaldehyde, however, is the production of benzyl alcohol via hydrogenation.

Benzaldehyde, Product Information Bulletin, Noveon Kalama, Inc., Kalama, Wash., June 15, 2002.

L. F. Fieser and M. Fieser, *Organic Chemistry*, D. C. Heath and Co., Boston, 1944.

Food Chemicals Codex, 4th ed., National Academy Press, Washington, D.C., 1996.

R. H. Perry and co-workers, *Perry's Chemical Engineer's Handbook*, 6th ed., McGraw Hill Book Co., New York, 1984.

JARL L. OPGRANDE
EDWARD BROWN
MARTHA HESSER
JERRY ANDREWS
Noveon Kalama, Inc.

BENZENE

Benzene, C_6H_6, is a volatile, colorless, and flammable liquid aromatic hydrocarbon possessing a distinct, characteristic odor. Benzene is used as a chemical intermediate for the production of many important industrial compounds, such as styrene (polystyrene and synthetic rubber), phenol (phenolic resins), cyclohexane (nylon), aniline (dyes), alkylbenzenes (detergents), and chlorobenzenes. These intermediates, in turn, supply numerous sectors of the chemical industry producing pharmaceuticals, specialty chemicals, plastics, resins, dyes, and pesticides. In the past, benzene has been used in the shoe and garment industry as a solvent for natural rubber. Benzene has also found limited application in medicine for the treatment of certain blood disorders, such as polycythemia and malignant lymphoma, and further in veterinary medicine as a disinfectant. Benzene, along with other light high octane aromatic hydrocarbons such as toluene and xylene, is used as a component of motor gasoline. Although this use has been largely reduced in the United States, benzene is still used extensively in many countries for the production of commercial gasoline. Benzene is no longer used in appreciable quantity as a solvent because of the health hazards associated with it.

Since the 1950s, benzene production from petroleum feedstocks has been very successful and accounts for ~95% of all benzene obtained. Less than 5% of commercial benzene is derived from coke oven light oil.

Benzene is the simplest and most important member of the aromatic hydrocarbons and should not be confused with benzine, a low boiling petroleum fraction composed chiefly of aliphatic hydrocarbons.

PHYSICAL PROPERTIES

The physical and thermodynamic properties of benzene are shown in Table 1. Benzene forms minimum-boiling azeotropes with many alcohols and hydrocarbons. Benzene also forms ternary azeotropes.

Structure

The representation of the benzene molecule has evolved from the Kekulé ring formula (1) to the more electronically accurate (2), which indicates all carbon–carbon bonds are identical.

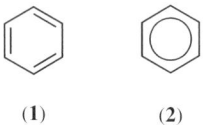

(1)   (2)

Resonance Stabilization. Benzene has great thermal stability. It has a lower heat of formation from the elements than the corresponding structure (1) possessing three fixed, ethylene-type double bonds. Similarly, when benzene is decomposed into carbon and hydrogen, it absorbs more energy than is predicted by the Kekulé formula.

CHEMICAL PROPERTIES

Benzene undergoes substitution, addition, and cleavage of the ring; substitution reactions are the most important for industrial applications.

Table 1. Physical and Thermodynamic Properties of Benzene[a]

Property	Value
mol wt	78.115
freezing point, °C in air at 101.3 kPa[b]	5.530
boiling point, °C at 101.3 kPa[b]	80.094
density, g/cm³	
20°C	0.8789
25°C	0.8736
vapor pressure, 25°C, kPa[c]	12.6
refractive index, n_D, 25°C	1.49792
surface tension, 25°C, mN/m (= dyn/cm)	28.20
viscosity, absolute, 25°C in mPa·s (= cP)	0.6010
critical temperature, °C	289.01
critical pressure, kPa[b]	4.898×10^3
critical volume, cm³/mol	259.0
heat of formation	
g, kJ/mol	82.93
L, kJ/mol	49.08
heat of combustion, kJ/mol[d,e]	3.2676×10^3
heat of fusion, kJ/mol	9.866
heat of vaporization, 25°C, kJ/mol	33.899
solubility in H_2O, 25°C, g/100 g H_2O	0.180

[a]Courtesy of the Thermodynamics Research Center, The Texas A&M University System.
[b]To convert kPa to atm, divide by 101.3.
[c]To convert kPa to mmHg, multiply by 7.5.
[d]To convert kJ to kcal, divide by 4.184.
[e]At 298.15 K and constant pressure to CO_2 and H_2O.

MANUFACTURE

Petroleum-derived benzene is commercially produced by reforming and separation, thermal or catalytic dealkylation of toluene, and disproportionation. Benzene is also obtained from pyrolysis gasoline formed in the steam cracking of olefins.

SPECIFICATIONS, STANDARDS, AND TEST METHODS

Several different grades of benzene are commercially available. The most common grades are benzene 535, benzene 485 (nitration grade), benzene 545, and thiophene-free benzene. Specifications and the corresponding American Society for Testing and Material (ASTM) test procedures for these various types are shown in Table 2.

ENVIRONMENTAL CONSIDERATIONS

Benzene is classified as a hazardous waste by the Environmental Protection Agency (EPA) under subtitle C of the Resource and Recovery Act (RCRA). Effective Sept. 25, 1990, solid wastes containing more than 0.5-mg/mL benzene must be treated in accordance with applicable RCRA regulations. Benzene is also subject to annual reporting of environmental releases as described in Section 313 of the Emergency and Community Right to Know Act of 1986. Benzene emissions and effluent streams from petroleum refineries or benzene processing plants are also subject to strict federal regulations. Federal waste management procedures must be complied with for any industrial process involving manufacture, transport, treatment, or disposal of benzene.

A complete description of the new EPA regulations concerning benzene and other hazardous wastes is found in the *Federal Register*. Further information regarding the handling and disposal of toxic or hazardous wastes is in the CFR, Vol. 40.

HEALTH AND SAFETY FACTORS

At room temperature and atmospheric pressure, benzene is sufficiently vaporized to pose an inhalation hazard. Benzene is a toxic substance that can produce both acute and chronic adverse health effects. It is generally recognized that prolonged or repeated exposure to benzene can result in serious damage to the blood-forming elements.

Inhalation of 3000 ppm benzene can be tolerated for 0.5–1 h; 7500 ppm causes toxic effects in 0.5–1 h; and 20,000 ppm is fatal in 5–10 min. The lethal oral dose for an adult is ~15 mL. Repeated skin contact is reported to cause drying, defatting, dermatitis, and the risk of secondary infection if fissuring occurs.

In chronic benzene intoxication, mild poisoning produces headache, dizziness, nausea, stomach pain, anorexia, and hypothermia. In severe cases, pale skin, weakness, blurred vision, and dyspnea occur on exertion. Hemorrhagic tendencies include petechia, easy bruising, and bleeding gums. Bone marrow depression produces a decrease in circulating peripheral erythrocytes and leucocytes. Fatalities from chronic exposure show at autopsy severe bone marrow aplasia, and necrosis or fatty degeneration of the heart, liver, and adrenals.

Acute benzene poisoning results in central nervous system (CNS) depression and is characterized by an initial

Table 2. Specifications for Commercial Grades of Benzene

ASTM test	Benzene 535[a]	Benzene 485[b]	Industrial-grade[c]
appearance	clear liquid, free from sediment or haze at 18–24°C	clear liquid, free from sediment or haze at 18–24°C	clear liquid, free from sediment at 18–24°C
relative density, 14.56–15.56°C, D3505	0.8820–0.8860	0.8820–0.8860	0.875–0.886
density, 20°C, g/cm^3, D4052	0.8780–0.8820	0.8780–0.8820	0.871–0.882
color pt-co scale, D1209	20 max	20 max	20 max
total distillation range, 101.3 kPa,[d] D850	1.0°C max, including the temperature of 80.1°C	1.0°C max, including the temperature of 80.1°C	2.0°C max, including the temperature of 80.1°C
solidification point, D852	5.35°C (anhydrous)	not lower than 4.85°C (anhydrous)	
acid wash color, D848	1 max	2 max	3 max
acidity, D847	none detected	none detected	none detected
H$_2$S and SO$_2$, D853	none detected	none detected	none detected
thiophene, D1685	1 mg/kg max		
copper corrosion, D849	pass	pass	copper strip shallnot show iridescence, a gray or black deposit, or discoloration
nonaromatics, D2360	0.15 wt% max		

[a]ASTM D2359-85a.
[b]Nitration-grade, ASTM D835-85.
[c]ASTM D836-84.
[d]To convert kPa to mmHg, multiply by 7.5.

Table 3. National Exposure Limits for Benzene

Country		Concentration	Status
Australia	TWA	5 ppm (16 mg/m^3)	carcinogen
Belgium	TWA	10 ppm (32 mg/m^3)	carcinogen
Czechoslovakia	TWA	10 mg/m^3; STEL 20 mg/m^3	
Denmark	TWA	5 ppm (16 mg/m^3)	skin, carcinogen
Finland	TWA	5 ppm (15 mg/m^3); STEL 10 ppm (30 mg/m^3)	skin
France	TWA	5 ppm (16 mg/m^3)	carcinogen
Germany (DFG MAK)	none		
Hungary	STEL	5 mg/m^3	skin, carcinogen
India	TWA	10 ppm (30 mg/m^3)	carcinogen
Ireland	TWA	5 ppm (16 mg/m^3)	carcinogen
Japan (JSOH)			carcinogen
The Netherlands	TWA	2.3 ppm (7.5 mg/m^3)	skin
The Philippines	TWA	25 ppm (80 mg/m^3)	skin
Poland	TWA	10 mg/m^3; STEL 40 mg/m^3	skin
Russia	TWA	10 ppm (5 mg/m^3); STEL 25 ppm (15 mg/m^3)	skin, carcinogen
Sweden	TWA	1 ppm (3 mg/m^3); STEL 5 ppm (16 mg/m^3),	skin, carcinogen
Switzerland	TWA	5 ppm (16 mg/m^3)	skin, carcinogen
Thailand	TWA	10 ppm (30 mg/m^3); STEL 25 ppm (75 mg/m^3)	skin
Turkey	TWA	20 ppm (64 mg/m^3)	skin
United Kingdom (HSE MEL)	TWA	5 ppm (16 mg/m^3)	

euphoria followed by staggered gait, stupor, coma, and convulsions.

Treatment for acute exposure to benzene vapor involves removing the subject from the affected area, followed by artificial respiration with oxygen; intubation and cardiac monitors may be necessary for severe acute exposures. Because of its low surface tension, benzene poses a significant aspiration hazard if the liquid enters the lungs. Emesis is indicated in alert patients if more than 1 mL of benzene per kg of body weight has been ingested and less than two hours have passed between ingestion and treatment.

Treatment for chronic benzene poisoning is supportive and symptomatic, with chemotherapy and bone marrow transplants as therapeutic agents for leukemia and aplastic anemia.

REGULATIONS

Because of the potential hazards associated with benzene, exposure to benzene in the workplace has been heavily regulated in the United States. Benzene is considered one of the ~40 known human carcinogens. Twenty countries have been reported to limit occupational exposure to benzene by regulation or recommended guideline. These occupational exposure limits are shown in Table 3.

USES

Benzene is now used primarily as an intermediate in the manufacture of industrial chemicals. Approximately 95% of U.S. benzene is consumed by industry for the preparation of polymers, detergents, pesticides, pharmaceuticals, and allied products.

Benzene is alkylated with ethylene to produce ethylbenzene, which is then dehydrogenated to styrene, the most important chemical intermediate derived from benzene.

Benzene is alkylated with propylene to yield cumene (qv). Cumene is catalytically oxidized in the presence of air to cumene hydroperoxide, which is decomposed into phenol and acetone (qv). Phenol is used to manufacture caprolactam (nylon) and phenolic resins such as bisphenol A.

Benzene is hydrogenated to cyclohexane. Cyclohexane is then oxidized to cyclohexanol, cyclohexanone, or adipic acid (qv). Adipic acid is used to produce nylon.

Nitration of benzene yields nitrobenzene, which is reduced to aniline, an important intermediate for dyes and pharmaceuticals. Benzene is chlorinated to produce chlorobenzene, which finds use in the preparation of pesticides, solvents, and dyes.

Some of benzene consumed was used for the manufacture of straight- or branched-chain detergent alkylate. Linear alkane sulfonates (LAS) are widely used as household and laundry detergents.

R. Hoag, "Benzene," *Chemical Economics Handbook*, SRI, Menlo Park, Calif., Oct. 2000.

W. E. Garwood, N. Y. Chen, and F. G. Dwyer, *Shape Selective Catalysis in Industrial Applications*, Marcel-Dekker, New York, 1989, pp. 205–218.

E. G. Hancock, *Benzene and Its Industrial Derivatives*, Halsted Press, a division of John Wiley & Sons, Inc., New York, 1975, p. 55.

D. Walsh, ed., *Chemical Safety Data Sheets*, Vol. 1, *Solvents, Benzene*, The Royal Society of Chemistry, Science Park, Cambridge, UK, 1988, p. 5.

WILLIAM FRUSCELLA
Unocal Corporation

BENZOIC ACID

Benzoic acid, C_6H_5COOH, the simplest member of the aromatic carboxylic acid family.

In the United States, virtually all benzoic acid is manufactured by the continuous liquid-phase air oxidation of toluene. Benzoic acid, its salts, and esters are very useful and find application in medicinals, food and industrial preservatives, cosmetics, resins, plasticizers, dyestuffs, and fibers.

OCCURRENCE

Benzoic acid in the free state, or in the form of simple derivatives such as salts, esters, and amides, is widely distributed in nature.

PROPERTIES

Selected physical properties of benzoic acid are given in Table 1.

In its chemical behavior benzoic acid shows few exceptional properties; the reactions of the carboxyl group are normal, and ring substitutions take place as would be predicted.

MANUFACTURE

Benzoic acid is almost exclusively manufactured by the cobalt catalyzed liquid-phase air oxidation of toluene. The basic process usually consists of a large reaction vessel in which air is bubbled through pressurized hot liquid toluene containing a soluble cobalt catalyst as well as the reaction products, a system to recover hydrocarbons from the reactor vent gases, and a purification system for the benzoic acid product.

Table 1. Physical Properties of Benzoic Acid

molecular formula	$C_7H_6O_2$
mp, °C	122.4
bp, at 101.3 kPa,^{a} °C	249.2
density	
solid, d_4^{24}	1.316
liquid, d_4^{180}	1.029
refractive index, n_D^{b}, liquid	1.504
viscosity at 130°C, mPa·s (= cP)	1.26
surface tension at 130°C, mN/m (= dyn/cm)	31
specific heat, J/g^c	
solid	1.1966
liquid	1.774
heat of fusion, J/g^c	147
heat of combustion, kJ/mol^c	3227
heat of formation at 26.16°C, kJ/mol^c, solid	−385
heat of vaporization, at 140°C, J/g^c	534
at 249°C, J/g^c	425
dissociation constant, K_a, at 25°C	6.339×10^{-5}
flash point, °C	121–131
autoignition temperature, °C, in air	573
pH of saturated aqueous solution at 25°C	2.8

^{a}To convert kPa to atm, divide by 101.3.
^{b}At 131.9°C.
^{c}To convert J to cal, divide by 4.184.

SPECIFICATIONS, ANALYSIS, PACKAGING, AND SHIPMENT

Benzoic acid is available as technical grade as well as grades meeting the specifications of the *United States Pharmacopeia (USP)*, the *Food Chemicals Codex (FCC)*, or the *British Pharmacopeia (BP)*.

Trace impurities typically present in technical grade benzoic acid include methyl diphenyls and phthalic acid. Gas chromatography (gc) and high pressure liquid chromatography (hplc) are useful for determining the concentrations of those impurities.

Technical grade benzoic acid is available in molten as well as solid forms (called flakes or chips). USP/FCC grade is available in solid form, either as crystals or powder. The solid forms of technical grade is usually packaged in 25-kg polylined bags and also in a flexible intermediate bulk container (FIBC), each FIBC containing from 500 to 1000 kg of product. USP/FCC grade is usually packaged in polylined fiber drums, each containing 100 lb (45.5 kg).

Molten technical benzoic acid may be transported in type 316 stainless steel tank cars, usually 76 m³ (20,000 gal) of product, or in ~5000 gal (19 m³) 316 stainless steel tank trucks.

HEALTH AND SAFETY ASPECTS

Benzoic acid's toxicity is rated as moderate based upon its LD_{50} (oral-rat) of 2530 mg/kg. Healthy individuals may tolerate small doses (<0.5 g of benzoates per day) mixed with food without ill effects. Manufacturer's product and information bulletins provide an excellent source for information regarding the safety and handling of benzoic acid.

The principal safety concern in handling molten benzoic acid is its elevated temperature. Thermal burns may result from improper handling of the molten product.

USES

Although the main uses for benzoic acid are as a chemical raw material, it also has numerous direct uses. Benzoic acid is used in substantial quantities to improve the properties of various alkyd resin coating formulations, where it tends to improve gloss, adhesion, hardness, and chemical resistance.

Benzoic acid is also used as a down-hole drilling mud additive where it functions as a temporary plugging agent in subterranean formations.

In medicine, its principal use is external; it is used in dermatology as an antiseptic stimulant and irritant. Combined with salicylic acid, benzoic acid is employed in the treatment of ringworm of the scalp and other skin diseases (Whitfield's ointment).

The largest use for benzoic acid is as a chemical raw material in the production of phenol, caprolactam, glycol dibenzoate esters, and sodium and/or potassium benzoate.

BENZOIC ACID DERIVATIVES

Benzoic acid derivatives include *benzoyl chloride* (C_6H_5 COCl, mp, −1°C; bp, 197.2°C at 101.3 kPa; d_4^{25}, 1.2070);

benzoic anhydride $(C_6H_5CO)_2O$, mp, 42°C; bp, 360°C at 101.3 kPa; benzoic acid salts [*ammonium benzoate* $(C_6H_5COONH_4$, mp, 198°C), *sodium benzoate* $(C_6H_5 COONa)$ and *lithium benzoate* $(LiC_7H_5O_2)$ and *potassium benzoate* (C_6H_5COOK)]; and benzoic acid esters [*benzyl benzoate* $(C_6H_5COOCH_2C_6H_5$, mp, 21°C, d_4^{25}, 1.118; bp, 323–324°C at 101.3 kPa), *butyl benzoate* $(C_6H_5COOC_4H_9$, mp, −22°C; bp, 250°C at 101.3 kPa), *ethyl benzoate* $(C_6H_5COOC_2H_5$, mp, −35°C; bp, 212°C at 101.3 kPa), *n-hexyl benzoate* $(C_6H_5COOC_6H_{13}$, bp, 272°C at 103.9 kPa), *methyl benzoate* $(C_6H_5COOCH_3$, bp, 198–200°C at 101.3 kPa), *phenyl benzoate* $(C_6H_5COOC_6H_5$, mp, 70–71°C; bp, 314°C at 101.3 kPa), and *alkyl (C12-15) benzoate* $(C_{20}H_{32}O_2$ (av), bp, 300°C at 101.3 kPa)].

R. H. Perry and co-workers, *Perry's Chemical Engineer's Handbook*, 6th ed., McGraw-Hill, New York, 1984, pp. 3–50.

Food Chemicals, Codex, 4th ed., National Academy Press, Washington, D.C.

GRAS (Generally Recognized as Safe) Food Ingredients: Benzoic Acid and Sodium Benzoate (PB-221, PB-228), National Technical Information Service, U.S. Department of Commerce, Washington, D.C., September 1972.

Benzoic Acid Material Safety Data Sheet, Noveon–Kalama Inc., Kalama, Wash.

JARL L. OPGRANDE
EDWARD E. BROWN
MARTHA HESSER
JERRY ANDREWS
Noveon Kalama, Inc.

BERYLLIUM, BERYLLIUM ALLOYS, AND COMPOSITES

BERYLLIUM

Beryllium, Be, specific gravity = 1.848 g/mL, and mp = 1287°C, is the only light metal having a high melting point. The majority of the beryllium commercially produced is used in alloys, principally copper–beryllium alloys (see CAST COPPER ALLOYS). The usage of unalloyed beryllium is based on its nuclear and thermal properties, and its uniquely high specific stiffness, ie, elastic modulus/density values. Beryllium oxide ceramics (qv) are important because of the very high thermal conductivity of the oxide while also serving as an electrical insulator.

Occurrence

The beryllium content of the earth's surface rocks has been estimated at 4–6 ppm. Although 45 beryllium-containing minerals have been identified, only beryl and bertrandite are of commercial significance.

Gemstone beryl (emerald, aquamarine, and beryl) approaches a pure beryllium–aluminum–silicate composition, $3BeO \cdot Al_2O_3 \cdot 6SiO_2$. Beryl is usually obtained as a by-product from mining zoned pegmatite deposits to recover feldspar, spodumene, or mica.

Bertrandite, $4BeO \cdot 2SiO_2 \cdot H_2O$, became of commercial importance in 1969 when the deposits of Spor Mountain in the Topaz district of Utah were opened. These deposits are believed to have been derived from fluorine-rich hydrothermal solutions at shallow depths. Although some beryl is processed, the majority of beryllium is now obtained from bertrandite.

Properties

A summary of physical and chemical constants for beryllium is compiled in Table 1. One of the more important characteristics of beryllium is its pronounced anisotropy resulting from the close-packed hexagonal crystal structure. This factor must be considered for any property that is known or suspected to be structure sensitive.

At ambient temperatures beryllium is quite resistant to oxidation; highly polished surfaces retain the brilliance for years.

Beryllium is susceptible to corrosion under aqueous conditions especially when exposed to solutions containing the chloride ion. It is rapidly attacked by seawater. Protective systems used for beryllium include chromic acid passivation, chromate conversion coatings, chromic acid anodizing, electroless plating (qv), and paints.

Beryllium reacts readily with sulfuric, hydrochloric, and hydrofluoric acids. It reacts with fused alkali halides releasing the alkali metal until equilibrium is established. It does not react with fused halides of the alkaline earth metals to release the alkaline earth metal.

Chemically, beryllium is closely related to aluminum from which complete separation is difficult.

Table 1. Physical and Chemical Properties of Beryllium

Parameter	Value
transformation pt, hcp to bcc, K	1527
mp, °C	1287
bp, °C	2472
density, g/mL	
at 298 K	1.8477
at 1773 K	1.42
heat of fusion, ΔH_{fus}, J/g[a]	1357
heat of sublimation, ΔH_s, kJ/g[a]	35.5–36.6
heat of vaporization, ΔH_v, kJ/g[a]	25.5–34.4
heat of transformation, J/g[a]	837
standard entropy, S°, J/(g·K)[a]	1.054
standard enthalpy, H°, J/g[a]	216
contraction on solidification, %	3
vapor pressure, MPa[b]	
specific heat, J/(g·K)[a]	
thermal conductivity at 298 K, W/(m·K)	220
linear coefficient of thermal expansion, 278–333 K[c]	11.4×10^{-6}
electrical resistivity at 298 K, $\Omega \cdot m$	4.31×10^{-8}
reflectivity, %	
white light	50–55
infrared (10.6 μm)	98
sound velocity, m/s	12,600

[a]To convert J to cal, divide by 4.184.
[b]To convert MPa to psi, multiply by 145.
[c]Value is for unworked, isostatically pressed powder metallurgy metal.

Ore Processing

Sulfate Extraction of Beryl. The Kjellgren-Sawyer sulfate process is used commercially for the extraction of beryl.

Extraction of Bertrandite. Bertrandite-containing tuff from the Spor Mountain deposits is wet milled to provide a thixotropic, pumpable slurry of below 840 μm (−20 mesh) particles. This slurry is leached with sulfuric acid at temperatures near the boiling point. The resulting beryllium sulfate solution is separated from unreacted solids by countercurrent decantation thickener operations. Water conservation practices are essential in semiarid Utah, so the wash water introduced in the countercurrent decantation separation of beryllium solutions from solids is utilized in the wet milling operation.

Production of Beryllium Metal

Reduction of Beryllium Fluoride with Magnesium. The Schwenzfeier process is used to prepare a purified, anhydrous beryllium fluoride, BeF_2, for reduction to the metal.

Electrolytic Processes. The electrolytic procedures for both electrowinning and electrorefining beryllium have primarily involved electrolysis of the beryllium chloride, $BeCl_2$, in a variety of fused-salt baths. The chloride readily hydrolyzes making the use of dry methods mandatory for its preparation (see BERYLLIUM COMPOUNDS). For both ecological and economic reasons there is no electrolytically derived beryllium available in the market place.

Vacuum Melting and Casting. A vacuum melting operation is required for beryllium regardless of its origin.

Because beryllium is primarily used as a powder metallurgy product or as an alloying agent, casting technology in the conventional metallurgical sense is not commonly utilized with the pure metal.

Fabrication

Most beryllium hardware is produced by powder metallurgy techniques achieving fine-grained microstructure having a nearly random crystallographic orientation thus providing a strong material with substantial ductility at room temperature. For some specialized applications, sheet and foil have been rolled from cast beryllium ingot. Such material exhibits an average grain size of 50–100 μm as compared to the typical 12 μm or less of the powder metallurgy products.

Beryllium powder is manufactured from vacuum-cast ingot using impact grinding or jet milling.

Hot-isostatic-pressing (HIP) is replacing the vacuum hot-pressing procedure for all but the largest shapes. Cold-isostatic-pressing followed by vacuum sintering or HIP is also used to manufacture smaller intricate shapes.

Beryllium sheet is produced by rolling powder metallurgy billets clad in steel cans at 750–790°C. Beryllium foil down to 12.5 μm (0.0005 in.) in gauge is commercially available. Extrusion is also carried out in this temperature region, again using steel cans to contain the powder metallurgy billet.

Safe Handling

Beryllium, beryllium-containing alloys and composites, and beryllium oxide ceramic in solid or massive form present no special health risk. However, like many industrial materials, beryllium, beryllium-containing alloys and composites, and beryllium oxide ceramic may present a health risk if handled improperly. Care must be taken in the fabrication and processing of beryllium products to avoid inhalation of airborne beryllium-containing particulate such as dust, mist, or fume in excess of the prescribed occupational exposure limits. Inhalation of fine airborne beryllium may cause chronic beryllium disease, a serious lung disorder, in certain sensitive individuals.

Occupational Exposure Limits. The U.S. Occupational Safety and Health Administration (OSHA) has set mandatory limits for occupational respiratory exposures.

RECYCLING

Beryllium, beryllium-containing alloys, and beryllium oxide ceramic can be recycled.

Uses

Beryllium is used extensively as a radiation window, both in source and detector applications, because of its ability to transmit radiation, particularly low energy X-rays.

Beryllium is used in the space shuttle orbiter as window frames, umbilical doors, and the navigation base assembly. An important application for beryllium is inertial guidance components for missiles and aircraft.

Beryllium is important as a sensor support material in advanced fire-control and navigation systems for military helicopters and fighter aircraft utilizing the low weight and high stiffness of the material to isolate instrumentation from vibration. It is also used for scanning mirrors in tank fire-control systems.

Beryllium is used in satellite structures in the form of both sheet and extruded tubing and is a very important material for all types of space optics. Beryllium oxide ceramic applications take advantage of high room temperature thermal conductivity, very low electrical conductivity, and high transparency to microwaves in microelectronic substrate applications.

BERYLLIUM COMPOSITES

Beryllium's reactivity limits its uses in the formulation of composite materials. There are only two composite formulations available. These are beryllium–aluminum and beryllium–beryllium oxides.

ALLOYS CONTAINING BERYLLIUM

A small beryllium addition produces strong effects in several base metals. In copper and nickel this alloying element promotes strengthening through precipitation hardening. In aluminum alloys a small addition improves oxidation resistance, castability, and workability. Other advantages are produced in magnesium, gold, and zinc. Many other alloying compositions have been researched, but no alloy with commercial importance approaching these dilute alloys has emerged.

Copper Beryllium Alloys

Wrought copper–beryllium alloys rank high among copper alloys in attainable strength and, at this high strength, useful levels of electrical and thermal conductivity are retained (see COPPER ALLOYS). Applications include uses in electronic components where their strength-formability-elastic modulus combination leads to use as electronic connector contacts; electrical equipment where fatigue strength, conductivity, and thermal relaxation resistance leads to use as switch and relay blades; control bearings where antigalling features are important; housings for magnetic sensing devices where low magnetic susceptibility is critical; and resistance welding systems where hot-hardness and conductivity are important in structural components.

Hardness, thermal conductivity, and castability are important in most casting alloy applications. For example, casting alloys are used in molds for plastic component production where fine cast-in detail such as wood or leather texture is desired. These alloys are also used for thermal management in welding equipment, waveguides, and mold components such as core pins. High strength alloys are used in sporting equipment such as investment cast golf club heads. Cast master alloys of beryllium in copper, nickel, and aluminum are used in preparing casting alloys or otherwise treating alloy melts.

Nickel–Beryllium Alloys. Dilute alloys of beryllium in nickel, like their copper–beryllium counterparts, are age hardenable. Nickel–beryllium alloys are distinguished by very high strength; good bend formability in strip; and high resistance to fatigue, elevated temperature softening, stress relaxation, and corrosion. Wrought nickel–beryllium is available as strip, rod, and wire and is used in mechanical, electrical, and electronic components that must exhibit good spring properties at elevated temperatures. Examples include thermostats, bellows, pressure sensing diaphragms, other high reliability mechanical springs, plus burn-in connectors and sockets.

A variety of nickel–beryllium casting alloys exhibit strengths nearly as high as the wrought products with castability advantages. Casting alloys are used in molds and cores for glass and plastic molding, and in jewelry and dental applications by virtue of their high replication of detail in the investment casting process.

Aluminum-Beryllium Alloys. Small additions of beryllium to aluminum systems are known to improve consistency. When as little as 0.005–0.05 wt% beryllium is added as a master alloy to an aluminum alloy during melting, a protective surface oxide film is formed. This film reduces drossing, increases cleanliness, and improves fluidity. Preferentially oxidizable alloy additions such

as magnesium and sodium are protected from oxidation during melting and casting. Hydrogen absorption is also reduced, as are mold reactions. Castings thus have improved surface finish, consistent strength, and higher ductility. Additional benefits cited include reduced tarnishing, improved buffing and polishing response, and consistency of aging response, particularly in alloys containing magnesium or silicon. Applications include aircraft skin panels and aircraft structural castings in alloy A357.

D. R. Floyd and J. N. Lowe, eds., *Beryllium Science and Technology*, Vol. 2, Plenum Press, New York, 1979.

H. H. Hausner, ed., *Beryllium, Its Metallurgy and Properties*, University of California Press, Berkeley, Calif., 1965.

M. D. Rossman, O. P. Preuss, and M. B. Powers, eds., *Beryllium—Biomedical and Environmental Aspects*, Williams & Wilkins, Baltimore, Md., 1991, p. 319.

D. W. White and J. E. Burke, eds., *The Metal Beryllium*, American Society for Metals, Novelty, Ohio, 1955.

DONALD J. KACZYNSKI
Brush Wellman Inc.

BERYLLIUM COMPOUNDS

BERYLLIUM CARBIDE

Beryllium carbide, Be_2C, may be prepared by heating a mixture of beryllium oxide and carbon to 1950–2000°C, or heating a blend of beryllium and carbon powders to 900°C under mechanical pressure of 3.5–6.9 MPa (500–1000 psi). The metal–carbon reaction is easier to carry out and is accompanied by a substantial exotherm.

The melting point is 2250–2400°C and the compound dissociates under vacuum at 2100°C. This compound is not used industrially, but Be_2C is a potential first-wall material for fusion reactors, one on the very limited list of possible candidates (see FUSION ENERGY).

BERYLLIUM CARBONATES

Beryllium carbonate tetrahydrate, $BeCO_3 \cdot 4\,H_2O$, has been prepared by passing carbon dioxide through an aqueous suspension of beryllium hydroxide. It is unstable and is obtained only when the solution is under carbon dioxide pressure. Beryllium oxide carbonate is precipitated when sodium carbonate is added to a beryllium salt solution.

Soluble beryllium, carbonate complexes are produced by dissolving beryllium oxide carbonate or hydroxide in ammonium carbonate. The solid beryllium oxide carbonate intermediates are obtained by a laboratory procedure for preparing pure beryllium salt solutions by reaction with aqueous mineral or organic acids.

BERYLLIUM CARBOXYLATES

The beryllium salts of organic acids can be divided into normal carboxylates, $Be(RCOO)_2$, and beryllium oxide carboxylates, $Be_4O(RCOO)_6$. The latter are prepared by dissolving beryllium oxide, hydroxide, or the oxide carbonate in an organic acid, followed by evaporation to give either a solid or an oily liquid.

BERYLLIUM HALIDES

The properties of the fluoride differ sharply from those of the chloride, bromide, and iodide. Beryllium fluoride is essentially an ionic compound, whereas the other three halides are largely covalent. The fluoroberyllate anion is very stable.

Beryllium fluoride, is produced commercially by the thermal decomposition of diammonium tetrafluoroberyllate, $(NH_4)_2BeF_4$.

Beryllium fluoride is hygroscopic and highly soluble in water, although its dissolution rate is slow. Compounds containing the BeF^{2-} ion are the most readily obtained, though compounds containing other fluoroberyllate ions can also be obtained, eg, NH_4BeF_3, depending on conditions.

Beryllium chloride, $BeCl_2$, is prepared by heating a mixture of beryllium oxide and carbon in chloride at 600–800°C.

Beryllium bromide, $BeBr_2$, and beryllium iodide, BeI_2, are prepared by the reaction of bromine or iodine vapors, respectively, with metallic beryllium at 500–700°C. They cannot be prepared by wet methods.

BERYLLIUM HYDRIDE

Beryllium hydride, BeH_2, is best prepared by the controlled pyrolysis of di-*tert*-butyl beryllium, $C_8H_{18}Be$, at 200°C.

BERYLLIUM HYDROXIDE

Beryllium hydroxide, $Be(OH)_2$, exists in three forms. On addition of alkali to a beryllium salt solution to obtain a slightly basic pH, a slimy, gelatinous beryllium hydroxide is produced. Aging this amorphous product results in a metastable tetragonal crystalline form, which after months of standing transforms into a stable orthorhombic crystalline form. The orthorhombic modification is also precipitated from a sodium beryllate solution containing >5 g/L db Be by hydrolysis near the boil.

BERYLLIUM INTERMETALLIC COMPOUNDS

Beryllium forms intermetallic compounds, referred to as beryllides, with most metals. They are usually prepared by a solid-state reaction of the blended powder constituents at ~1260°C. The properties exhibited by some beryllides include excellent oxidation resistance, high strength at elevated temperature, good thermal conductivity, and low densities as compared with refractory metals and ceramic materials (see CERAMICS; REFRACTORIES).

BERYLLIUM NITRATE

Beryllium nitrate tetrahydrate, $Be(NO_3)_2 \cdot 4H_2O$, is prepared by crystallization from a solution of beryllium hydroxide or beryllium oxide carbonate in a slight excess of dilute nitric acid.

BERYLLIUM NITRIDE

Beryllium nitride, Be_3N_2, is prepared by the reaction of metallic beryllium and ammonia gas at 1100°C.

BERYLLIUM OXALATE

Beryllium oxalate trihydrate, $BeC_2O_4 \cdot 3H_2O$, is obtained by evaporating a solution of beryllium hydroxide or oxide carbonate in a slight excess of oxalic acid. The compound is very soluble in water.

BERYLLIUM OXIDE

Beryllium oxide, BeO, is the most important high purity commercial beryllium chemical. In the primary industrial process, beryllium hydroxide extracted from ore is dissolved in sulfuric acid. The solution is filtered to remove insoluble oxide and sulfate impurities. The resulting clear filtrate is concentrated by evaporation and upon cooling high purity beryllium sulfate, $BeSO_4 \cdot 4H_2O$, crystallizes. This salt is calcined at carefully controlled temperatures between 1150 and 1450°C, selected to give tailored properties of the beryllium oxide powders as required by the individual beryllia ceramic fabricators.

High purity beryllium oxide powder is fabricated by classical ceramic-forming processes such as dry pressing, isostatic pressing, extrusion, tape casting, and slip casting.

BERYLLIUM SULFATE

Beryllium sulfate tetrahydrate, $BeSO_4 \cdot 4H_2O$, is produced commercially in a highly purified state by fractional crystallization from a beryllium sulfate solution obtained by the reaction of beryllium hydroxide and sulfuric acid.

HEALTH AND SAFETY FACTORS

Care must be taken in the fabrication and processing of beryllium products to avoid inhalation of airborne beryllium particulate matter such as dusts, mists, or fumes in excess of prescribed work place limits. Inhalation of fine airborne beryllium may cause chronic beryllium disease, a serious lung disorder, in certain sensitive individuals.

W. W. Beaver, in D. W. White and J. E. Burke, eds., *The Metal Beryllium, American Society for Metals*, Novelty, Ohio, 1955, pp. 570–598.

L. D. Cunningham, "Beryllium", *Mineral Commodity Summaries*, U.S. Geological Survey, Jan. 2002.

D. A. Everest, *The Chemistry of Beryllium*, Elsevier Publishing Company, Amsterdam/London/New York, 1964.

"Argnoberyllium Compounds" in *G. Melins Handbook of Inorganic Chemistry*, Part 1, 8th ed., 1987.

Donald J. Kaczynski
Brush Wellman Inc.

BIOCATALYSIS

Bioorganic catalysis can be defined as the use of biological systems (whole cells or pure enzymes) to produce organic compounds. The production of alcohol via fermentation, of vinegar via oxidation of ethanol by acetic acid bacteria, and the production of cheese via enzymatic breakdown of milk proteins are well-known examples. Biocatalysts are ever increasingly being exploited for the production of industrially important materials, in many cases competing with traditional chemical methods and in some instances performing reactions that traditional chemistry methods cannot. As a result, biocatalysis is now being considered as another implement in the chemist's arsenal for tackling chemical transformations.

Debunking the myth of biocatalysts is an ongoing task and there are many examples being presented to elucidate the many advantages of using biocatalysts for organic reactions. Even though most biocatalysts work in relatively moderate reaction conditions, advances in isolation and expression of extremophilic enzymes and genetic modifications of mesophilic enzymes have created biocatalysts that can tolerate substantially harsh conditions. Some enzymes isolated from hyperthermophilic organisms are active at temperature as high as 140°C and others isolated from psychrophilic organisms are active at temperatures as low as 4°C. Some enzymes from barophilic organisms isolated near deep-sea vents tolerate pressures as high as 100 bar. Even though enzymes are specific with respect to the type of reaction they catalyze, most of them have activity on a wide range of substrates. In addition, there are many instances where enzymes have been tailored using genetic engineering to suit the conditions in biotransformation processes. There are biologically catalyzed equivalents for almost all of the chemical reactions, even the Diels–Alder reaction and the Claisen rearrangement. Advances in genetic engineering and fermentation technology have enabled the production of many enzymes at high concentrations in high cell density fermentations, resulting in a substantial decrease in cost of manufacture of biocatalysts.

WHOLE CELLS VERSUS PURE ENZYMES

There are mainly two biological entities that can compete as bioorganic catalysts, and they are isolated enzymes and whole cells (microbial, plant, or animal). Several parameters that ultimately affect the cost of a process (product purity, throughput, stability) are important while selecting a particular form of biocatalyst in an industrial biotransformation process. An ideal process would be one where the culturing of the cells would also accomplish the biotransformation at a high product concentration. However, a lot of the substrates and products at any significant level are toxic to cell growth. In addition, separation of the product from the cell broth can be very difficult and there may be undesirable side reactions during cell growth. This necessitates the separation of cell growth and biotransformation. The only time whole cell biocatalysis is advantageous is when the biotransformation involves multiple enzymes, where cofactor regeneration is necessary, or when enzyme isolation is needed and is difficult.

Immobilized cells could be used instead of immobilized enzymes to eliminate the costly isolation step while retaining the ability to recycle the biocatalyst. Typical cell immobilization methods are adsorption, gel entrapment, and compartmentalization in polymer matrices. The advantages for immobilized cells remain the same as for immobilized enzymes, except that improved enzyme stability is rarely observed. The disadvantages may be exacerbated because of the immobilization method and the fact that there is another barrier for the substrate to get to the enzyme, namely the cell membrane.

Crude enzyme probably represents the simplest form of prepared biocatalyst. The advantage is that there is little preparatory cost compared to immobilized enzymes, cells, or purified enzymes. The main disadvantage is that the biocatalyst cannot be recycled. Crude enzyme can be prepared by spray drying or freeze drying the cell culture. Freeze drying which is prevalent at the laboratory scale cannot be practiced economically at the large scale.

BIOCATALYST PERFORMANCE

Biocatalyst performance parameters such as activity, selectivity, and stability can be altered by modifying the enzyme or the environment around it. The latter approach uses the principle of changing the solvent environment around the enzyme molecule (commonly referred to as solvent engineering) while the former approach uses the redesigning of enzyme to meet the desired goals. Solvent properties such as dipole, dielectric constant, hydrophobicity, and density have been shown to cause predictable effect on enzyme stability, activity, and enantioselectivity. The solvent engineering approach has been demonstrated mainly at small scale but has not been scaled up into commercial processes because enzyme catalytic efficiencies are 2–6 orders of magnitude lower in nonaqueous media than in aqueous solutions.

The other approach to improving enzyme performance parameters is by enzyme modification at the molecular level (commonly referred to as enzyme engineering). The enzyme can be altered to tackle the problem of production cost, poor activity, stereoselectivity, and stability. The earliest approach to engineering enzymes is through rational design on the basis of structure–function relationship. This requires knowledge regarding the protein structure either by x-ray crystallography or molecular modeling or a combination of both. This can be tedious in an industrial setting where the enzyme has to be improved within a matter of months if not weeks.

Error-prone polymerase chain reaction (PCR) is a powerful tool to introduce errors in the DNA sequence coding for the enzyme of interest. The advantage of this method is the speed at which mutations can be introduced and the fact that they can be targeted into the gene of interest. The result is a random mutation of the enzyme, resulting in many candidates in a short amount of time. A proper screening or selection method can then be used to pick the enzyme with the required property.

BIOORGANIC CATALYSIS

The enzymes catalyzing oxidation/reduction reactions constitute an important class from an industrial perspective since industrial chemistry involves many oxidation/reduction reactions. The use of oxidoreductases for organic synthesis has been under intense investigation for the past several decades and still continues.

Dehydrogenases, which constitute the largest class within the oxidoreductases, are enzymes that catalyze the reduction and oxidation of carbonyls and alcohols, respectively.

Chemical methods for cofactor regeneration using sodium dithionite, phenazine methosulfate, and flavin mononucleotide have been fairly successful. However the utility of chemical regeneration method depends on ease of separation of product and number of cofactor turnover in the reaction system.

Electrochemical methods have been studied as a means of regenerating cofactors. Electrochemical methods, although widely used in biosensors, need to demonstrate economic feasibility (high turnover number) before being accepted as method for regenerating cofactors. Another method is to use a second enzyme system to recycle the cofactor, and this has been successfully used in a small-scale process producing multi-kilogram quantities.

A glucose dehydrogenase (GDH) cofactor regenerating system has been used to recycle NADPH and NADH via the oxidation of glucose to gluconolactone. Gluconolactone then spontaneously hydrolyses to form gluconic acid, making the reaction scheme favorable for both NADH and NADPH generation. Other regenerating systems have been used such as glucose-6-phosphate dehydrogenase, alcohol dehydrogenases, and hydrogenase. The latter methods are less attractive than the FDH or GDH system for a variety of reasons.

The enzyme recycling principle can be applied using whole living cells instead of isolated enzymes. In this case, the main enzymatic reaction and the cofactor regeneration reaction are carried out during the growth of living cells.

Enolate reductases which reduce C–C unsaturated bonds are another class of enzymes that require cofactor recycling. Since the products of these reactions can result in a chirally pure product, they constitute an important class of reactions.

Biological oxidation reactions achieve heteroatom oxygenation, aromatic hydroxylation, Bayer–Villiger oxidation, double bond epoxidation, and nonactivated carbon atom hydroxylation of substrates, which is difficult via conventional chemistry.

The regioselective oxidation of polyols is of practical interest because biocatalysts can selectively oxidize one hydroxyl group without requiring any protection of the remaining hydroxyl groups. This is a feat that cannot be achieved by conventional chemical oxidants. Selective oxidation of hydroxyl groups in steroids is an important reaction carried by cholesterol oxidase. Since substrate solubility is a problem in aqueous systems, steroid oxidations can be carried out in organic solvents using PEG modification of the enzymes as a method to make the enzymes soluble in organic solvents.

Oxygenases are enzymes that incorporate molecular oxygen directly into the substrate. Oxygenase-catalyzed oxidations are important since direct addition of molecular oxygen into unactivated organic substrates is very difficult to accomplish using conventional chemistry. Monooxygenases incorporate one atom of oxygen whereas dioxygenases incorporate two atoms of oxygen into a substrate.

Another class of reactions that is very important is the Baeyer–Villiger reactions where ketones are oxidized into esters and lactones. The fact that there is chiral recognition by the enzymes sets them apart from conventional methods. Since flavin- and nicotinamide-dependent monooxygenases are usually involved in these reactions, whole-cell biocatalysis is utilized most of the time.

Aldolase-catalyzed asymmetric C–C bond formations which are carried out in a neutral pH aqueous environment are a very important class of reactions. Other commercially important reactions catalyzed by aldolases are for the synthesis of unusual sugars, polyhydroxylated alkaloids, novel C–C polymers, and analogues of N-acetylneuraminic acid.

The formation of cyanohydrins is catalyzed by oxynitrilase. Chiral cyanohydrins are important intermediates in the synthesis of pharmaceuticals, agrochemicals, or liquid crystals.

Hydrolases catalyze the hydrolysis of various bonds such as amides and esters. Among the hydrolases, lipases, esterases, and proteases are most widely used enzymes. Hydrolases are routinely used in organic synthesis since they do not require cofactors and a large number of them possessing relaxed substrate specificities are available from different sources. Lipases are the most widely used hydrolases and they catalyze the hydrolysis of triglycerides into fatty acids and glycerol. On the basis of triglyceride hydrolysis, microbial lipases can be classified into two groups. Lipases of the first group have no regiospecificity and release fatty acids from all three positions of glycerol. In contrast, lipases of second group release fatty acids regioselectively from the outer 1- and 3-positions of triglycerides. Lipases have enormous potential in chemical synthesis because of several reasons: (1) lipases are stable in organic environment; (2) lipases possess broad substrate specificity; (3) lipases exhibit high regio- and enantioselectivity.

K. Drauz and H. Waldmann, eds., *Enzyme Catalysis in Organic Synthesis*, Vol. VII, 1995, p. 598.

K. Faber, *Biotransformations in Organic Chemistry*, Springer-Verlag, Berlin, 1995, p. 3.

E. Schmidt, in J. M. S. Cabral, D. Best, L. Boross, and J. Tramper, eds., *Applied Biocatalysis*, Harwood Academic Publishers, Chur, Switzerland, 1994, p. 133.

D. I. Stirling, *Chirality in Industry*, John Wiley & Sons, Inc., New York, 1992, p. 209.

SACHIN PANNURI
ROCCO DISANTO
SANJAY KAMAT
Cambrex Technical Center

BIOMASS ENERGY

Concerns about the long-term stability of supply of fossil fuel resources coupled with concerns about the environmental impacts of the use of these fuels has led to the investigation of the use of more renewable resources. Primary among these renewable resources is biomass. Biomass currently provides a $\sim$10–11% of the worlds primary energy; however, the various forms of biomass can potentially supply a quantity of energy several times that of the current global demand.

Increased environmental pressures including concerns relative to air quality and to the disposal of biomass containing wastes such as construction and demolition wastes, municipal solid wastes, sewage sludges, and urban biomass (yard) wastes further expanded the interest in energy production from biomass materials.

WHAT IS BIOMASS?

Biomass quite simply is any organic material that is or was derived from plants or animals. A more legalistic definition has been proposed by researchers and several governments throughout the world. This definition describes biomass as "all nonfossil organic materials that have an intrinsic chemical energy content. This includes all water- and land-based vegetation and trees, or virgin biomass, and organic components of waste materials such as municipal solid waste (MSW), municipal biosolids (sewage) and animal wastes (manures), forestry and agricultural residues, and certain types of industrial wastes." The most common example of biomass is fuelwood.

WHY BIOMASS ENERGY?

Biomass derived energy is considered a "renewable" energy source. That is, it is a source of energy that can be utilized without depleting the reserves. As long as the conditions exist (sunlight, water, and the organic substrate) additional biomass can be produced to replenish that used for energy. Interest in renewable energy is increasing due to dwindling supplies of petroleum and natural gas, two fossil fuels, and to mitigate global climate change.

An increase in the use of biomass energy provides a means to both reduce the consumption of fossil fuels and to reclaim a portion of land that has been deforested. Because solar energy is an integral part of the production of most biomass, some consider biomass a form of stored solar energy. Biomass based energy, however, provides the advantage of being available at all times (dispatchable) regardless of atmospheric conditions.

Biomass as an energy source is not free of environmental issues. Biomass is renewable and carbon-neutral only if replacement through new growth occurs at, at least, the same rate at which it is harvested. Furthermore, in many undeveloped countries, cooking and heating using biomass is accomplished in relatively crude stoves or fireplaces. These devices can release carbon monoxide, unburned hydrocarbons, and particulates (as smoke). More advanced, high efficiency biomass-to-energy conversion systems introduce their own unique set of environmental and safety concerns. Intermediate products such as synthesis gas or biomass derived liquids can be hazardous, flammable, or toxic. Most, if not all, of these potential intermediates can be handled safely through proper application of industry accepted safety and health guidelines without any widespread expansion.

Biomass ash, from any conversion system, is generally recyclable to the soil unless the incoming biomass has been contaminated with heavy metals (as is the case with some municipal residues). However, biomass ash, particularly from combustion devices, is a very fine (sometimes <5 μ) material leading to potential emissions of fine particulate matter, PM-2.5.

HISTORICAL USES OF BIOMASS ENERGY

Biomass in the United States historically was used for simple heating by direct combustion in wood stoves or fireplaces. Such technologically simple uses continue today in developing countries throughout the world. Biomass, due to its ready availability, provides a simple and reliable energy supply that can be adapted for home and simple industrial purposes. In the United States, the primary form of biomass used for these simple applications is fuelwood. Fuelwood, in the form of logs or brush is burned in air to provide heat. The combustion of biomass in a fireplace or wood stove is accomplished by reacting the biomass feed material in air according to the following reactions:

$$C + O_2 \rightarrow CO_2 \tag{1}$$

and

$$2\,H + 1/2\,O_2 \rightarrow H_2O \tag{2}$$

Power Generation from Biomass

Boiler applications further provide the opportunity to generate electric power from the incoming biomass. Biomass is burned to generate steam. The steam is then used to turn a turbine for the generation of electric power.

Other Commercial Uses of Biomass Combustion

The pulp and paper industry is a primary user of biomass combustion for its energy needs. In addition to direct combustion of wood and woody residues, a significant portion of the energy in a modern pulp mill is derived from the combustion of black liquor, a by-product of the pulping operation. Black liquor contains the spent pulping chemicals and lignin content from the incoming wood pulp. The lignin content provides the energy content in the liquor.

Wood and wood residues (bark, twigs, etc) are collectively known as hog fuel. In a pulp mill, the hog fuel is burned in a power boiler and produces steam for power generation.

District Heating and Combined Heat and Power

Biomass based combined heat and power (CHP) systems typically combust biomass to produce high pressure steam that is used for power generation. Lower pressure steam is then extracted from the turbine system and used for district heating.

Charcoal

While not used extensively in the United States as an energy source, charcoal plays a major role in many countries as a primary source of fuel for heating, cooking, power production, and metal processing. Charcoal can be made from virtually any organic material. The primary source is fuelwood or coconut shells.

Environmental concerns regarding charcoal have recently become important. If charcoal is produced on a sustainable basis, ie, without deforestation, many of these concerns can be alleviated.

OTHER ENERGY RECOVERY SYSTEMS

Waste-to-Energy Facilities

Waste-to-energy facilities typically use the same type of stoker-grate combustor as described above for wood fired power generation.

Ethanol

Ethanol is a form of alcohol found in beverages such as wine and beer. It is readily produced through the natural fermentation of the starches and complex sugars present in many forms of biomass.

DEVELOPING METHODS FOR CONVERSION OF BIOMASS INTO ENERGY

Gasification

Gasification is quite simply the conversion of a solid or liquid material into a gaseous fuel. The resulting fuel gas has an energy content or "heating value" ranging from 10 to 50% of the heating value of natural gas. The wide variation in heating values is a direct result of both the reactor type used and the reactants chosen for the gasification reactions. Gasification is probably the most flexible conversion system for biomass materials as

the fuel gas produced can be used directly as a fuel for heating applications, be used for the production of power in gas turbines or fuel cells, or used as a synthesis gas for the production of liquid fuels, chemicals, or hydrogen.

Gasification Reactions. There are three main types of reactors used for biomass gasification. These are fixed, fluidized, and entrained bed. In addition, there are some hybrid reactor types such as circulating fluidized bed and steam reforming fluidized bed that utilize properties of two reactor types to enhance the conversion reactions.

Use of the Gas from Gasification. Gases derived from the gasification of biomass have a potentially wide range of end uses. In the simplest application, the gas may simply be used as a fuel for heating or in industrial furnaces such as lime kilns in the pulp and paper industry.

Higher overall efficiencies can be achieved by use of the gases in more demanding applications (those requiring a higher degree of cleanup of the gases). These include, use in an internal combustion engine or a gas combustion turbine for the direct production of power.

The diluent effect of nitrogen present in low calorific value gases limits their application for more advanced chemical synthesis applications. Medium calorific value gases, on the other hand, are well suited for synthesis applications. A simple and the most commonly considered synthesis application is the production of methanol or other alcohols from the biomass derived gas. Like ethanol, however, methanol (and higher alcohols) are not fungible fuels, therefore limiting their acceptance by the petroleum industry. Methanol as a fuel is even more susceptible to phase separation in gasoline blends than ethanol. Synthesis gas from biomass can be utilized to generate essentially any product that would be produced from a petrochemical based synthesis gas. This includes chemical intermediates, polymers, fuel additives, or hydrogen.

Pyrolysis

Pyrolysis is a similar technology to gasification in that biomass feedstocks are heated in an anaerobic (no air present) environment to break down the biomass into primarily liquid hydrocarbons. In pyrolysis systems currently under development, the production of liquids is enhanced by rapidly heating the biomass in a "fast pyrolysis" mode. Fast pyrolysis is a process that yields a liquid product that is referred to by many names including pyrolysis liquid, pyrolysis oil, bio-crude-oil, bio-oil, bio-fuel-oil, pyroligneous tar, pyroligneous acid, wood liquids, wood oil, wood distillates and liquid wood.

Uses of Bio-Oil. Bio-oil can be used in a variety of primarily industrial applications. The primary use is as a substitute or cofired fuel with fuel oil. Because the bio-oils are chemically similar to some organic chemical intermediates, other uses include resins (used in plywood manufacture and as adhesives), agri-chemicals, and other specialty chemicals.

Relatively small quantities of bio-oils are sold as food flavorings and additives. The liquid smoke products found in supermarkets are typical of this type of bio-oil product.

Biodiesel

Biodiesel is a term applied to a fuel derived from the transesterification of used vegetable oils or animal fats. In the production of biodiesel, the triglycerides in the fats and oils are reacted with methanol to make methyl esters and glycerine. The glycerine produced can be sold as a by-product, however, due to large supplies of glycerine produced during soap manufacture, the income from the sale is likely to be small.

Uses of Biodiesel. Much like ethanol, biodiesel is blended with traditional diesel fuel in varying proportions (typically 20% biodiesel to 80% petroleum diesel fuel). Unlike ethanol, however, much higher concentrations of biodiesel can be blended without requiring a modification to the engine.

Anaerobic Digestion

Anaerobic digestion is the biological degradation of organic material in the absence of oxygen. The product of such digestion is a gas containing primarily methane (CH_4) and carbon dioxide (CO_2). The digestion process occurs naturally in landfills, manure disposal sites, and other such residue disposal sites. It has also been applied in a more industrial setting in controlled digestors or "bio-gas" reactors.

Commercial Application of Anaerobic Digestion. The gas produced (landfill gas) from anaerobic digestion in MSW landfills is being collected and used as a fuel for industrial heating and power generation at over 330 MSW landfills in the U.S.

Advanced Ethanol Production Methods

Lignocellulosics (wood, straw, and grasses) are in abundant supply and can potentially supply a source for the production of ethanol. Effective utilization of this resource can greatly enhance the quantity of ethanol that is produced for energy. The primary chemical components of these materials, however, are cellulose (~40–50%) and hemicellulose (~25–30%). From a theoretical standpoint, the same quantity of ethanol could be realized from these materials as from high sugar containing biomass such as corn. A more complex conversion process is required to realize this potential as the cellulose and hemicellulose must first be converted into sugars so that fermentation can take place. Cellulose converts via hydrolysis into six-carbon sugars (primarily glucose) that are readily fermented with yeasts to produce ethanol.

Hemicellulose, on the other hand, is converted into mainly five-carbon sugar precursors with xylose as a major product. The C_5 sugars are not readily fermented into ethanol without additional conversion steps and they inhibit the hydrolysis reactions once they are formed.

ENVIRONMENTAL BENEFITS FROM BIOMASS ENERGY

The use of biomass as an energy source, superficially results in a net zero change in carbon dioxide. Carbon dioxide is the predominant greenhouse gas found in the atmosphere. However, if forests are cleared for agricultural applications or development, this balance is no longer valid. Other greenhouse gases, namely, methane and nitrous oxide, are produced both by fossil energy conversion systems and during the growing and harvesting of biomass. The balance of these emissions and their ultimate impacts on the environment can only be accurately determined by the use of life cycle assessment (LCA) techniques.

Reductions in greenhouse gas emissions are also realized when ethanol is used as a transportation fuel additive.

Other environmental benefits beyond the reduction of greenhouse gases can be realized by the use of biomass derived energy products. Cofiring of biomass with fossil fuels in boiler systems results in reductions of both sulfur and nitrogen oxides proportionate to the quantity of biomass being used.

In some boiler applications, medium calorific value gas from biomass has been used as a reburn fuel in fossil fuel boilers.

CONCLUSION

Biomass is one of the largest renewable resources and provides advantages when compared to other renewable resources. Furthermore, biomass is the only renewable carbon-based resource with minimal greenhouse gas emission potential. Due to the flexibility of biomass energy technologies, a wide range of products can be produced ranging from basic energy products to refined chemicals, pharmaceuticals and fertilizers.

American Biomass Association, "Biomass Clean Energy for America", internet site, www.biomass.org.

H. Chum and R. Overend, "Biomass and Bioenergy in the United States", *Advances in Solar Energy*, Vol. 15, American Solar Energy Society, 2002.

IEA *World Energy Outlook*, Paris, France, Organization for Economic Co-operation and Development, 2002.

R. Matthews and K. Robertson, "Answers to Ten Frequently Answered Questions about Bioenergy, Carbon Sinks, and Their Role in Global Climate Change," IEA Bioenergy Task 38, internet site, www.joanneum.at/iea-bioenergy-task38/, 2001.

MARK A. PAISLEY
FERCO Enterprises, Inc.

BIOMATERIALS, PROSTHETICS, AND BIOMEDICAL DEVICES

Prosthetics or biomedical devices are objects which serve as body replacement parts for humans and other animals or as tools for implantation of such parts. An implanted prosthetic or biomedical device is fabricated from a biomaterial and surgically inserted into the living body by a physician or other health care provider. Such implants are intended to function in the body for some period of time in order to perform a specific task. Prosthetics and biomedical devices are composed of biocompatible materials, or biomaterials. Polymers, metals, and ceramics originally designed for commercial applications have been adapted for prostheses, opening the way for implantable pacemakers, vascular grafts, diagnostic/therapeutic catheters, and a variety of other orthopedic devices. The term prosthesis encompasses both external and internal devices. This article concentrates on implantable prostheses.

BIOMATERIALS

A biomaterial is defined as a systemic, pharmacologically inert substance designed for implantation or incorporation within the human body. A biomaterial must be mechanically adaptable for its designated function and have the required shear, stress, strain, Young's modulus, compliance, tensile strength, and temperature-related properties for the application. Moreover, biomaterials ideally should be nontoxic, ie, neither teratogenic, carcinogenic, or mutagenic; nonimmunogenic; biocompatible; biodurable, unless designed as bioresorbable; sterilizable; readily available; and possess characteristics allowing easy fabrication. The traditional areas for biomaterials are plastic and reconstructive surgery, dentistry, and bone and tissue repair. A widening variety of materials are being used in these areas. Artificial organs play an important role in preventive medicine, especially in the early prevention of organ failure.

To be biocompatible is to interact with all tissues and organs of the body in a nontoxic manner, not destroying the cellular constituents of the body fluids with which the material interfaces. In some applications, interaction of an implant with the body is both desirable and necessary, as, for example, when a fibrous capsule forms and prevents implant movement.

MEDICAL DEVICES

Medical devices are officially classified into one of three classes. Class I devices are general controls that are primarily intended as devices that pose no potential risk to health, and thus can be adequately regulated without imposing standards or the need for premarket review.

Class II devices have performance standards and are applicable when general controls are not adequate to assure the safety and effectiveness of a device, based on the potential risk to health posed by the device.

Class III devices require premarket approval; they comprise life-supporting and/or life-sustaining devices, unless adequate justification is given for classifying it in another category. Class III also contains devices after 1976 that are not sufficiently similar to pre-1976 devices, and devices that were regulated as new drugs before 1976.

CARDIOVASCULAR DEVICES

Treatment of cardiovascular diseases is a vast and growing industry (see CARDIOVASCULAR AGENTS). Open-heart surgery, cardiac pacing, heart transplants, implantable valves, and coronary angioplasty and clot busters, have been developed, but none of the great advances in cardiovascular medicine is preventive.

Cardiovascular Problems

Plaque. A heart attack, or myocardial infarction, results from insufficient delivery of oxygen to parts of the heart muscle owing to restricted blood flow in the coronary arteries. The heart attack is often precipitated by a clot, or thrombus, which forms on a severely narrowed portion of a coronary artery. Silent ischemia is somewhat reduced blood supply from narrowing of the arteries. Plaque also causes other problems such as strokes and aneurysms, as well as complications of peripheral vascular disease.

Lethal Arrhythmias. Arrhythmias are a second significant source of cardiovascular problems. An arrhythmia is an abnormal or irregular heart rhythm. Bradyarrhythmias result in heart rates that are too slow; tachyarrhythmias cause abnormally fast rates.

Arrhythmias are caused by disturbances of the normal electrical conduction patterns synchronizing and controlling heartbeats.

Fibrillation is uncontrolled electrical activity.

Valvular Disease. Valve problems severely limit the efficiency of the heart's pumping action bringing forth definitive symptoms. There are two types of conditions, both of which may be present in the same valve. The first is narrowing, or stenosis, of the valve. The second condition is inability of the valve to close completely.

Cardiomyopathy. Cardiomyopathy, or diseased heart muscle, may reach a point at which the heart can no longer function. It arises from a combination of factors, including hypertension, arrhythmias, and valve disease. Other problems, such as congestive heart failure, cause the interrelated heart–lung system to break down. Because the heart can no longer adequately pump, fluid builds up in the lungs and other areas.

Device Solutions

The first big step in cardiovascular devices was the development of a heart–lung machine in 1953. The ability to shut down the operation of the heart and lungs and still maintain circulation of oxygenated blood throughout the body made open-heart surgery possible. The principal components of the heart–lung machine, the oxygenator and pump, take over the functions of the lungs and heart.

Atherosclerosis. The first solution to the problem of atherosclerosis was the coronary artery bypass graft (CABG) procedure, first performed in 1964.

The second step toward solving cardiovascular disease from atherosclerosis, ie, angioplasty, was preceded by the diagnostic tool of angiocardiography by nearly 20 years. Angiocardiography, or angiography, permits x-ray diagnosis using a fluoroscope. A radiopaque contrast medium is introduced into the arteries through a catheter (see RADIOPAQUES), and angiography allows accurate location of the plaque blockage. Percutaneous transluminal coronary angioplasty (PTCA), a nonsurgical procedure, emerged in the 1980s as a viable method for opening up blocked arteries.

Arrhythmias. The first solution to cardiovascular problems arising from arrhythmias came about as a result of a complication caused by open-heart surgery. During procedures to correct congenital defects in children's hearts, the electrical conduction system often became impaired, and until it healed, the heart could not contract sufficiently without outside electrical stimulation. The first implantable pacemaker, introduced in 1960, provided a permanent solution to a chronic bradyarrhythmia condition.

Valve Problems. The primary solution to valve problems has been implantable replacement valves. The introduction of these devices necessitates open-heart surgery. There are two types of valves available: tissue (porcine and bovine) and mechanical. The disadvantage of tissue valves is that these have a limited life of about seven years before they calcify, stiffen, and have to be replaced. The mechanical valves can last a lifetime, but require anticoagulant therapy.

Cardiomyopathy. The best available solution to cardiomyopathy may be one that is less sophisticated than transplant surgery or the artificial heart. The cardiomyoplasty-assist system combines earlier electrical stimulation technology with a new surgical technique of utilizing muscle from another part of the body to assist the heart.

Interventional Procedures

The emergence of angioplasty created a specialty called interventional cardiology. Interventional cardiologists not only implant pacemakers and clear arteries using balloon catheters, but they also use balloons to stretch valves (valvuloplasty). In addition, they work with various approaches and technologies to attack plaque, including laser (qv) energy, mechanical cutters and shavers, stents to shore up arterial walls and deliver drugs, and ultrasound to break up plaque or to visualize the inside of the artery. Typically, procedures have become less invasive as technology evolves.

Clinical evaluation is underway to test transvenous electrodes. Transvenous leads permit pacemakers to be implanted under local anesthesia while the patient is awake, greatly reducing recovery time and risk.

Coronary bypass surgery and angioplasty are vastly different procedures, but both procedures seek to revascularize and restore adequate blood flow to coronary arteries.

Other cardiovascular devices developed initially for use in open-heart surgery are used extensively in other parts of the hospital and, in many cases, outside the hospital: these include portable cardiopulmonary support systems and blood pumps and oxygenators.

Biomaterials for Cardiovascular Devices

Perhaps the most advanced field of biomaterials is that for cardiovascular devices. The development of implantable-grade synthetic polymers, such as silicones and polyurethanes, has made possible the development of advanced cardiac assist devices (see SILICON COMPOUNDS, SILICONES; URETHANE POLYMERS).

Dramatic developments and growth are also taking place in other areas such as the use of laser systems intended to ablate significant amounts of plaque. Mechanical or atherectomy devices to cut, shave, or pulverize plaque have been tested extensively in coronary arteries. One of the more intriguing cardiovascular developments is cardiomyoplasty where implantable technologies are blended with another part of the body to take over for a diseased heart. Cardiomyoplasty could greatly reduce the overwhelming need for heart transplants. It might also eliminate the need for immunosuppressive drugs.

Pacemakers. Implantable tachyrhythmia devices, available for some years, address less dangerous atrial tachyarrhythmias and fibrillation.

Surgical Devices. Surgical devices comprise the equipment and disposables to support surgery and to position implantable valves and a variety of vascular grafts. Central to open-heart surgery is the heart–lung machine and a supporting cast of disposable products. Two devices, the oxygenator and the centrifugal pump, amount to significant market segments in their own right. Other disposables include cardiotomy reservoirs, filters, tubing packs, and cardioplegia products to cool the heart.

Centrifugal pumps are increasingly being used as a safer and more effective alternative to the traditional roller pump in open-heart surgery and liver transplants. Implantable valves, particularly mechanical valves which continue to encroach on tissue valves, are unique. Methods such as valvuloplasty, mitral valve repair, or use of ultrasound are unlikely to reduce the number of valve replacements into the twenty-first century.

Vascular grafts are tubular devices implanted throughout the body to replace blood vessels which have become obstructed by plaque, atherosclerosis, or otherwise weakened by an aneurysm.

Cardiac-Assist Devices. The principal cardiac-assist device, the intra-aortic balloon pump (IABP), is used primarily to support patients before or after open-heart surgery, or patients who go into cardiogenic shock.

Other devices, which can completely take over the heart's pumping function, are the ventricular assist devices (VADs), supporting one or both ventricles. Considerable interest has emerged in devices providing cardiopulmonary support (CPS), ie, taking over the functions of both the heart and lungs without having to open up the chest.

Other specialized applications of cardiac arrest devices include extracorporeal membrane oxygenation (ECMO) which occurs when the lungs of a premature infant cannot function properly.

Artificial Hearts. In 1980, the National Heart, Lung and Blood Institute of NIH established goals and criteria for developing heart devices and support techniques in an effort to improve the treatment of heart disease. This research culminated in the development of both temporary and permanent left ventricular-assist devices that are tether-free, reliable over two years, and electrically powered.

In contrast, the total artificial heart (TAH) is designed to overtake the function of the diseased natural heart. One successful total artificial heart is ABIOMED's electric TAH.

Heart Valves. The most commonly used valves as of the mid-1990s include mechanical prostheses and tissue valves. Caged-ball, caged-disk, and tilting-disk heart valves are the types most widely used.

Blood Salvage. Surgical centers have a device, called the Cell Saver (Haemonetics), that allows blood lost during surgery to be reused within a matter of minutes, instead of being discarded.

Use of intraoperative autotransfusion (IAT) eliminates disease transmission, compatibility testing, and immunosuppression that may result from the use of homologous blood products, reduces net blood loss of the patient, and conserves the blood supply.

Blood Access Devices. An investigational device called the Osteoport system allows repeated access to the vascular system via an intraosseous infusion directly into the bone marrow.

Blood Oxygenators. The basic construction of an oxygenator involves any one of several types of units employing a bubble-type, membrane film-type, or hollow-fiber-type design. The most important advance in oxygenator development was the introduction of the membrane-type oxygenator. These employ conditions very close to the normal physiological conditions in which gas contacts occur indirectly via a gas-permeable membrane.

Polyurethanes as Biomaterials

Much of the progress in cardiovascular devices can be attributed to advances in preparing biostable polyurethanes (see URETHANE POLYMERS). Biostable polycarbonate-based polyurethane materials offer far-reaching capabilities to cardiovascular products. These and other polyurethane materials offer significant advantages for important

long-term products, such as implantable ports, hemodialysis, and peripheral catheters; pacemaker interfaces and leads; and vascular grafts.

Implantable Ports. The safest method of accessing the vascular system is by means of a vascular access device (VAD) or port. Vascular access ports typically consist of a self-sealing silicone septum within a rigid housing which is attached to a radiopaque catheter (see RADIOPAQUES). The catheter must be fabricated from a low modulus elastomeric polymer capable of interfacing with both soft tissue and the cardiovascular environment. A low modulus polyurethane-based elastomer is preferred to ensure minimal trauma to the fragile vein.

Vascular Grafts. The advent of newer polyurethane materials is expected to lead to a new generation of cardiovascular devices. The characteristics of polyurethanes, combined with newer manufacturing techniques, should translate into direct medical benefits for the physician, the hospital, and the patient.

ORTHOPEDIC DEVICES

Bone is formed through a highly complex process that begins with the creation of embryonic mesenchymal cells. These cells, found only in the mesoderm of the embryo, migrate throughout the human body to form all the types of skeletal tissues including bone, cartilage, muscle, tendon, and ligament.

A bone is classified according to shape as flat, long, short, or irregular. A living bone consists of three layers: the periosteum, the hard cortical bone, and the bone marrow or cancellous bone.

Joints are structurally unique. They permit bodily movement and are bound together by fibrous tissues known as ligaments. Most larger joints are encapsulated in a bursa sac and surrounded by synovial fluid which lubricates the joint continuously to reduce friction.

The bearing surface of each joint is cushioned by cartilage. This tissue minimizes friction. The cartilage also reduces force on the bone by absorbing shock.

Ligaments function to tie two bones together at a joint, maintain joints in position preventing dislocations, and restrain the joint's movements. Ligaments may be reattached to bone by the use of an orthopedic anchor.

The function of the tendon is to attach muscles to bones and other parts.

The meniscus is skeletal system fibrocartilage-like tissue. The function of the meniscus is to absorb shock by cushioning and distributing forces evenly throughout a joint, and provide a smooth articulating surface for the cartilages of the adjoining bones.

The body's frame or skeleton is constructed as a set of levers powered or operated by muscle tissue. A typical muscle consists of a central fibrous tissue portion, and tendons at either end. One end of the muscle, known as the head, is attached to tendon tissue, which is attached to bone that is fixed, and known as the point of origin. The other end of the muscle is attached to a tendon. This tendon is attached to bone that is the moving part of the joint. This end of the muscle is known as the insertion end.

Muscle tissue is unique in its ability to shorten or contract. The human body has three basic types of muscle tissue histologically classified into smooth, striated, and cardiac muscle tissues. Only the striated muscle tissue is found in all skeletal muscles.

Bones function as levers; joints function as fulcrums; muscle tissue, attached to the bones via tendons, exert force by converting electrochemical energy (nerve impulses) into tension and contraction, thereby facilitating motion.

Soft Tissue Injuries

Some of the more common soft tissue injuries are sprains, strains, contusions, tendonitis, bursitis, and stress injuries, caused by damaged tendons, muscles, and ligaments. A sprain is a soft tissue injury to the ligaments. A sprain is a soft tissue injury to the ligaments.

A strain is the result of an injury to either the muscle or a tendon, usually in the foot or leg. Strain is a soft tissue injury resulting from excessive use, violent contraction, or excessive forcible stretch.

A contusion is an injury to soft tissue in which the skin is not penetrated, but swelling of broken blood vessels causes a bruise. A bruise, also known as a hematoma, is caused when blood coagulates around the injury causing swelling and discoloring skin.

Tendonitis, an inflammation in the tendon or in the tendon covering, is usually caused by a series of small stresses that repeatedly aggravate the tendon, preventing it from healing properly, rather than from a single injury.

A bursa, a sac filled with fluid located around a principal joint, is lined with a synovial membrane and contains synovial fluid. Repeated small stresses and overuse can cause the bursa in the shoulder, hip, knee, or ankle to swell. This swelling and irritation is referred to as bursitis.

Bone Fractures

Bone fractures are classified into two categories: simple fractures and compound, complex, or open fractures. In the latter the skin is pierced and the flesh and bone are exposed to infection. A bone fracture begins to heal nearly as soon as it occurs. Therefore, it is important for a bone fracture to be set accurately as soon as possible.

Stress fractures occur when microfractures accumulate because muscle tissue becomes fatigued and no longer protects from shock or impact.

Fracture Treatment

The movement of a broken bone must be controlled because moving a broken or dislocated bone causes additional damage to the bone, nearby blood vessels, and nerves or other tissues surrounding the bone. Indeed, emergency treatment requires splinting or bracing a fracture injury before further medical treatment is given. All treatment forms for fractures follow one basic rule: the broken pieces must be repositioned and prevented from moving out of place until healed. Broken bone ends heal by growing back together, ie, new bone cells form around the edge of the broken pieces. Specific bone fracture treatment depends on the severity of the break and the bone involved, ie, a broken bone in the spine is treated differently from a broken rib or a bone in the arm.

Treatments used for various types of fractures are cast immobilization, traction, and internal fixation.

Traction is typically used to align a bone by a gentle, constant pulling action. The pulling force may be transmitted to the bone through skin tapes or a metal pin through a bone. Traction may be used as a preliminary treatment, before other forms of treatment or after cast immobilization.

In internal fixation, an orthopedist performs surgery on the bone. During this procedure, the bone fragments are repositioned (reduced) into their normal alignment and then held together with special screws or by attaching metal plates to the outer surface of the bone. The fragments may also be held together by inserting rods (intramedullary rods) down through the marrow space into the center of the bone.

Joint Replacement

The most frequent reason for performing a total joint replacement is to relieve the pain and disability caused by severe arthritis.

A total joint replacement is a radical surgical procedure performed under general anesthesia, in which the surgeon replaces the damaged parts of the joint with artificial materials. Whereas hips and knees are the joints most frequently replaced, because the scientific understanding of these is best, total joint replacement can be performed on other joints as well, including the ankle, shoulder, fingers, and elbow.

The materials used in a total joint replacement are designed to enable the joint to function normally. The artificial components are generally composed of a metal piece that fits closely into bone tissue. The metals are varied and include stainless steel or alloys of cobalt, chrome, and titanium. The plastic material used in implants is a polyethylene that is extremely durable and wear-resistant. Also, a bone cement, a methacrylate, is often used to anchor the artificial joint materials into the bone. Cementless joint replacements have more recently been developed. In these replacements, the prosthesis and the bone are made to fit together without the need for bone cement.

The principal complication for total joint replacement is infection, which may occur just in the area of the incision or more seriously deep around the prosthesis.

Loosening of the prosthesis is the most common biomechanical problem occurring after total joint replacement surgery. If loosening is significant, a second or revision total joint replacement may be necessary. Another complication which sometimes occurs after total joint replacement, generally right after the operation, is dislocation, the result of weakened ligaments. In most cases the dislocation can be relocated manually by the orthopedic surgeon. Very rarely is another operation necessary.

BIORESORBABLE POLYMERS

The concept of using biodegradable materials for implants which serve a temporary function is a relatively new one. This concept has gained acceptance as it has been realized that an implanted material does not have to be inert, but can be degraded and/or metabolized *in vivo* once its function has been accomplished. Resorbable polymers have been utilized successfully in the manufacture of sutures, small bone fixation devices, and drug delivery systems (qv).

One area in which predictable biodegradation is used is the area of degradable surgical sutures. One such suture, catgut, is infection-resistant. The biodegradation of catgut results in elimination of foreign material that otherwise could serve as a nidus for infection or, in the urinary tract, calcification. As a result, chromic catgut, which uses chromic acid as a cross-linking agent, is still preferred in some procedures.

Polylactic Acid

Polylactic acid (PLA) was introduced in 1966 for degradable surgical implants.

Polyglycolic Acid

Polyglycolic acid (PGA), is also known as polyglycolide. An important difference between polylactide and polyglycolide, is that polyglycolide (mp 220°C) is higher melting than poly-L-lactide (mp 170°C). Although the polymerization reaction in both cases is reversible at high temperature, melt processing of polyglycolide is more difficult because the melting temperature is close to its decomposition temperature.

Poly(lactide-*co*-glycolide)

Mixtures of lactide and glycolide monomers have been copolymerized in an effort to extend the range of polymer properties and rates of *in vivo* absorption.

Polydioxanone

Fibers made from polymers containing a high percentage of polyglycolide are considered too stiff for monofilament suture and thus are available only in braided form above the microsuture size range.

Poly(ethylene oxide)–Poly(ethylene terephthalate) Copolymers

The poly(ethylene oxide)–poly(ethylene terephthalate) (PEO/PET) copolymers were developed in an attempt to simultaneously reduce the crystallinity of PET, and increase its hydrophilicity to improve dyeability. PEO/PET copolymers with increased PEO contents produce surfaces that approach zero interfacial energy between the implant and the adjacent biological tissue. The collagenous capsule formed around the implant is thinner as the PEO contents increase.

Poly(glycolide-*co*-trimethylene carbonate)

Another successful approach to obtaining an absorbable polymer capable of producing flexible monofilaments has involved finding a new type of monomer for copolymerization with glycolide. This suture reportedly has excellent flexibility and superior *in vivo* tensile strength retention compared to polyglycolide. It has been absorbed without adverse reaction in about seven months.

Poly(ethylene carbonate)

Like polyesters, polycarbonates (qv) are bioabsorbable only if the hydrolyzable linkages are accessible to enzymes and/or water molecules. Because poly(ethylene carbonate) hydrolyzes more rapidly *in vivo* than *in vitro*, enzyme-catalyzed hydrolysis is postulated as a contributing factor in polymer absorption. Copolymers of polyethylene and polypropylene carbonate have been developed as an approach to achieving the desired physical and pharmacological properties of microsphere drug delivery systems.

Polycaprolactone

Polycaprolactone is absorbed very slowly *in vivo*, releasing ε-hydroxycaproic acid as the sole metabolite. Degradation occurs in two phases: nonenzymatic bulk hydrolysis of ester linkages followed by fragmentation, and release of oligomeric species. Polycaprolactone fragments ultimately are degraded in the phagosomes of macrophages and giant cells, a process that involves lysosome-derived enzymes.

Poly(ester–amides)

The rationale for designing poly(ester–amide) materials is to combine the absorbability of polyesters (qv) with the high performance of polyamides (qv). Two types have been reported. Both involve the polyesterification of diols that contain preformed amide linkages. Poly(ester–amides) obtained from bis-oxamidodiols have been reported to be absorbable only when oxalic acid is used to form the ester linkages. Poly(ester–amides) obtained from bis-hydroxyacetamides are absorbable regardless of the diacid employed, although succinic acid is preferred.

Poly(orthoesters)

The degradation of a bioresorbable polymer occurs in four stages: hydration, loss of strength, loss of integrity, and loss of mass.

Poly(orthoesters) represent the first class of bioerodible polymers designed specifically for drug delivery applications. *In vivo* degradation of the polyorthoester shown, known as the Alzamer degradation yields 1,4-cyclohexanedimethanol and 4-hydroxybutyric acid as hydrolysis products.

Poly(anhydrides)

Poly(anhydrides) are another class of synthetic polymers used for bioerodible matrix, drug delivery implant experiments.

SHAPE MEMORY ALLOYS

TiNi shape memory alloy (SMA) has attracted much attention for biomedical applications such as implants (bone plate and marrow needle) and for surgical and dental instruments, devices and fixtures, such as orthodontic fixtures and biopsy forceps (see SHAPE MEMORY ALLOYS). This is due to its excellent biocompatibility and mechanical characteristics. The first example of successful biomedical and dental applications of SMA are available and many new applications are being developed.

Shape Memory Effect

Pre-deformed SMAs have the ability to remember their original shape before deformation and are able to recover the shape when heated if the plastic deformation takes place in the martensite phase. The shape recovery is the result of transformation from the low-temperature martensite phase to the high-temperature austenite phase when it is heated. The shape memory effect makes it easy to deploy an SMA appliance in the body and makes it possible to create a pre-stress after deployment when necessary.

Superelasticity

SMAs exhibit superelasticity when they are in the austenite phase. An important feature of superelastic materials is that they exhibit constant loading and unloading stesses over a wide range of strain.

The superelasticity of SMAs makes it easy to depoly SMA stents. Stents made from stainless steel are expanded against the vessel wall by plastic deformation caused by the inflation of a balloon placed inside the stent. TiNi stents, on the other hand, are self-expanding.

Hysteresis of SMA

Superelastic SMA demonstrates a hysteretic stress–strain relationship. Hysteresis is usually regarded as a drawback for traditional engineering application, but it is a useful characteristic for biomedical applications. If the SMA is set at some stress–strain state, upon unloading during deployment, it should provide a light and constant chronic force against the organ wall even with a certian amount of further strain release. On the other hand, it would generate a large resistive force to crushing if it is compressed in the opposite direction.

Anti-kinking Properties

The stress of stainless steel remains nearly constant in the plastic region. This means that a small increase of stress in the plastic region could lead to a drastic increase of strain or the failure of the medical appliance made from stainless steel. The increase in stiffness would prevent the local strain in the high strain areas from further increasing and cause the strain to be partitioned in the areas of lower strain. Hence, strain localization is prevented by creating a more uniform strain than could not be realized with a conventional material.

C. P. Sharma and M. Szycher, *Blood Compatible Materials and Devices*, Technomic Publishing Co., Inc., Lancaster, Pa., 1991.

M. Szycher, *Introduction to Biomedical Polymers*, ACS Audio Courses, American Chemical Society, Washington, D.C., 1989.

M. Szycher, *High Performance Biomaterials*, Technomic Publishing Co., Inc., Lancaster, Pa., 1991.

M. Szycher, *Szycher's Dictionary of Biomaterials and Medical Devices*, Technomic Publishing Co., Inc., Lancaster, Pa., 1992.

MICHAEL SZYCHER
PolyMedica Industries, Inc.
JINHAO QIU
MAMI TANAKA
Institute of Fluid Science, Tohoku University

BIOREMEDIATION

Bioremediation is the process of judiciously exploiting biological processes to minimize an unwanted environmental impact; usually it is the removal of a contaminant from the biosphere.

Bioremediation is already a commercially viable technology. There are significant opportunities to enlarge upon this success. Bioremediation has applications in the gas phase, in water, and in soils and sediments. For water and soils, the process can be carried out *in situ*, or after the contaminated medium has been moved to some sort of contained reactor (*ex situ*). The former is generally rather cheaper, but the latter may result in such a significant increase in rate that the additional cost of manipulating the contaminated material is overshadowed by the time saved. Bioremediation may explicitly exploit bacteria, fungi, algae, or higher plants. Each, in turn, may be part of a complex food-web, and optimizing the local ecosystem may be as important as focusing solely on the primary degraders or accumulators.

Bioremediation usually competes with alternative approaches to achieving an environmental goal. It is typically among the least expensive options, but an additional important consideration is that in many cases bioremediation is a permanent solution to the contamination problem, since the contaminant is completely destroyed or collected. Some of the alternatives technologies, such as thermal desorption and destruction of organics, are also permanent, solutions, but the simplest, removing the contaminant to a dump site, merely moves the problem, and may well not eliminate the potential liability. Furthermore, by its very nature bioremediation addresses the bioavailable part of any contamination. The same cannot necessarily be said for nonbiological technologies, which may leave bioavailable contaminants at low levels.

Bioremediation also has the advantage that it can be relatively nonintrusive, and can sometimes be used in situations where other approaches would be severely disruptive. On the other hand, bioremediation is usually slower than most physical techniques, and may not always be able to meet some very strict cleanup standards.

GENERAL TECHNOLOGICAL ASPECTS

Table 1 lists some of the technologies in use today in bioremediation.

ORGANIC CONTAMINANTS

Hydrocarbons

Constituents.   Hydrocarbons get into the environment from biogenic and fossil sources. Methane is produced by

Table 1. Some Technological Definitions Relevant to Bioremediation

Technology	Description
air sparging; aquifer sparging; biosparging	injection of air to stimulate aerobic degradation; may also stimulate volatilization
air stripping	injection of air to stimulate volatilization
aquifer bioremediation	*in situ* bioremediation in an aquifer, usually by adding nutrients or co-substrates
aquifer sparging	injection of air into a contaminated aquifer to stimulate aerobic degradation, may also stimulate volatilization
batch reactor	a bioreactor loaded with contaminated material, and run until the contaminant has been consumed, then emptied, and the process is repeated
bioactive barrier; bioactive zone; biowall	a zone, usually subsurface, where biodegradation of a contaminant occurs so that no contaminant passes the barrier
bioaugmentation	addition of exogenous bacteria with defined degradation potential (or rarely indigenous bacteria cultivated in a reactor and reapplied)
biofilm reactor	a reactor where bacterial communities are encouraged on a high surface area support, biofilms often have a redox gradient so that the deepest layer is anaerobic while the outside is aerobic
biofiltration	usually an air filter with degrading organisms supported on a high surface area support such as granulated activated carbon
biofluffing	augering soil to increase porosity
bioleaching	extracting metallic contaminants at acid pH
biological fluidized bed; fluidized-bed bioreactor	bioreactor where the fluid phase is moving fast enough to suspend the solid phase as a fluid-like phase
biopile; soil heaping	an engineered pile of excavated contaminated soil, with engineering to optimize air, water, and nutrient control
bioslurping	vacuum extraction of the floating contaminant, water, and vapor from the vadose zone; the air flow stimulates biodegradation
biostimulation	optimizing conditions for the indigenous biota to degrade the contaminant
biotransformation	the biological conversion of a contaminant to some other form, but not to carbon dioxide and water
biotrickling filter	a reactor where a contaminated gas stream passes up a reactor with immobilized micro organisms on a solid support, while nutrient liquor trickles down the reactor
bioventing	vacuum extraction of contaminant vapors from the vadose zone, thereby drawing in air that stimulates the biodegradation of the remainder
borehole bioreactor; in-well bioreactor	the addition of nutrients and electron acceptor to stimulate the biodegradation *in situ* in a contaminated aquifer
closed-loop bioremediation	groundwater recovery, a bioreactor, and low-pressure reinjection to maximize nutrient use, and maintain temperature in cold climates
composting	addition of biodegradable bulking agent to stimulate microbial activity; optimal composting generally involves self-heating to 50–60°C
constructed wetland	artificial marsh for bioremediation of contaminated water
continuous stirred tank reactor (CSTR)	a completely mixed bioreactor
digester	usually an anaerobic bioreactor for digestion of solids and sludges that generates methane
ex-situ bioremediation	usually the bioremediation of excavated contaminated soil in a biopile, compost system or bioreactor
fixed-bed bioreactor	bioreactor with immobilized cells on a packed column matrix
land-farming; land treatment	application of a biodegradable sludge as a thin layer to a soil to encourage biodegradation; the soil is typically tilled regularly
natural attenuation; intrinsic bioremediation	unassisted biodegradation of a contaminant
phytoextraction	the use of plants to remove and accumulate contaminants from soil or water to harvestable biomass
phytofiltration	the use of completely immersed plant seedlings, to remove contaminants from water
phytoremediation	the use of plants to effect bioremediation
phytostabilization	the use of plants to stabilize soil against wind and water erosion
pump and treat	pumping groundwater to the surface, treating, and reinjection or disposal
rhizofiltration	the use of roots to immobilize contaminants from a water stream
rotating biological contactor	bioreactor with rotating device that moves a biofilm through the bulk water phase and the air phase to stimulate aerobic degradation
sequencing batch reactor	periodically aerated solid phase or slurry bioreactor operated in batch mode
soil-vapor extraction	vacuum-assisted vapor extraction

anaerobic bacteria in enormous quantities in soils, sediments, ruminants and termites, and it is consumed by methanotropmc bacteria on a similar scale. Submarine methane seeps support substantial oases of marine life, with a variety of invertebrates possessing symbiotic methanotrophic bacteria. Thus, methano-trophic bacteria are ubiquitous in aerobic environments. Plants generate large amounts of volatile hydrocarbons, including isoprene and a range of terpenes. These compounds provide an abundant substrate for hydrocarbon-degrading organisms.

Crude oil has been part of the biosphere for millennia, leaking from oil seeps on land and in the sea. Crude oils are very complex mixtures, primarily of hydrocarbons although some components do have heteroatoms such as nitrogen (eg, carbazole) or sulfur (eg, dibenzothiophene). Chemically, the principal components of crude oils and refined products can be classified as aliphatics, aromatics, naphthenics, and asphaltic molecules. When crude oils reach the surface environment the lighter molecules evaporate, and are either destroyed by atmospheric photooxidation or are washed out of the atmosphere in rain, and are biodegraded. Some molecules, such as the smaller aromatics (benzene, toluene, etc) have significant solubilities, and can be washed out of floating slicks, whether these are at sea, or on terrestrial water tables. Fortunately the majority of molecules in crude oils, and refined products made from them, are biodegradable, at least under aerobic conditions.

Biodegradation. Methane and the volatile plant terpenes are fully biodegradable by aerobic organisms, and most refined petroleum products are essentially completely biodegradable under aerobic conditions. The least biodegradable material, principally polar molecules and asphaltenes, lacks the "oily" feel and properties that are associated with oil. These are essentially impossible to distinguish from more recent organic material in soils and sediments, such as the humic and fulvic acids, and appear to be biologically inert.

Numerous bacterial and fungal genera have species able to degrade hydrocarbons aerobically and the pathways of degradation of representative aliphatic, naphthenic and aromatic molecules have been well characterized in at least some species. It is a truism that the hallmark of an oil-degrading organism is its ability to insert oxygen atoms into the hydrocarbon, and there are many ways in which this is achieved. Figures 1 and 2 show the most well-studied.

In recent years it has become clear that at least some hydrocarbons are oxidized by bacteria under completely anaerobic conditions, where the oxygen is probably coming from water. Limited hydrocarbon biodegradation has now been shown under sulfate-, nitrate-, carbon dioxide- and ferric iron-reducing conditions (Table 2). Figure 3 shows the intermediates identified in anaerobic toluene degradation in different organisms. It is noteworthy that while organisms capable of aerobic oil biodegradation seem to be ubiquitous, organisms capable of the anaerobic degradation of hydrocarbon have to date only been found in a few places.

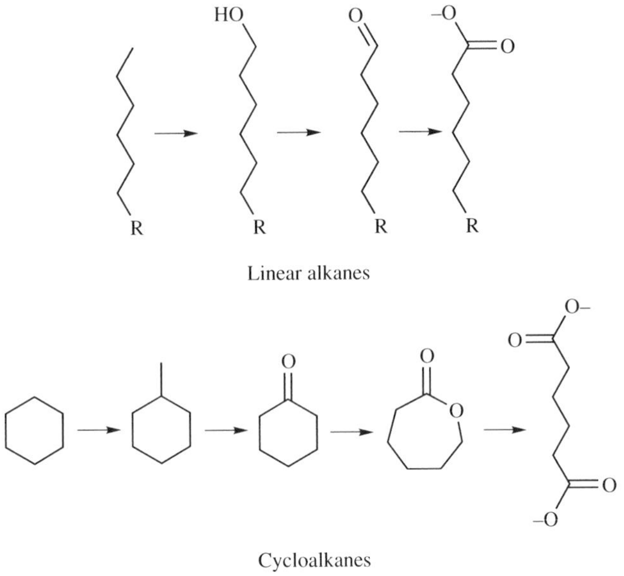

Figure 1. Initial steps in the biodegradation of linear and cyclic alkanes.

Although the majority of molecules in crude oils and refined products are hydrocarbons, the U.S. Clean Air Act amendment of 1990 mandated the addition of oxygenated compounds to gasoline in many parts of the United States. The requirement is usually that 2% (w/w) of the fuel be oxygen, which requires that 5–15% (v/v) of the gasoline be an oxygenated additive (eg, methanol, ethanol, methyl *tert*-butyl ether (MTBE), etc). Although methanol and ethanol are readily degraded under aerobic conditions, the degradability of MTBE remains something

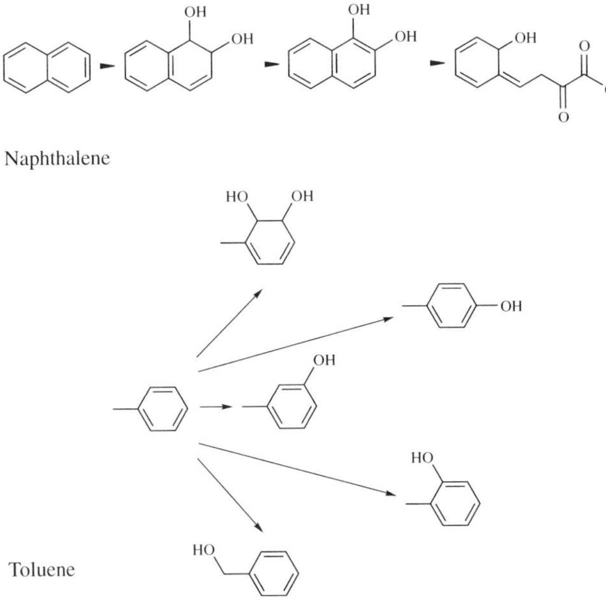

Figure 2. Initial steps in the aerobic degradation of naphthalene, as a representative multiringed aromatic, and toluene. The different initial steps of toluene degradation are examples of the diversity found in different organisms.

Table 2. Hydrocarbons that have been shown to be Biodegraded under Anaerobic Conditions

Electron acceptor	Substrate
nitrate (to nitrogen)	heptadecene
	toluene, ethylbenzene, xylene
	naphthalene
	terpenes
iron(III) (to iron(II))	toluene
manganese(IV) (to Mn(II))	toluene
sulfate (to sulfide)	hexadecane, alkylbenzenes
	benzene
	naphthalene, phenanthrene
CO_2 (to methane)	toluene, xylene

of an open question. The compound was previously very rare in the environment, but now it is one of the major chemicals in commerce. At first it seemed that the compound was completely resistant to biodegradation, but complete mineralization has now been reported. Whether biodegradation can be optimized for effective biore-mediation remains to be seen.

Bioremediation. Crude oil and refined products are readily biodegradable under aerobic conditions, but they are only incomplete foods since they lack any significant nitrogen, phosphorus, and essential trace elements. Bioremediation strategies for removing large quantities of hydrocarbon must therefore include the addition of fertilizers to provide these elements in a bioavailable form.

Air. Hydrocarbon vapors in air are readily treated with biofilters. These are typically rather large devices with a very large surface area provided by bulky material such as a bark or straw compost. The contaminated air, perhaps from a soil vapor-extraction treatment, or from a factory using hydrocarbon solvents, is blown through the filter, and organisms, usually indigenous to the filter material or provided by a soil or commercial inoculum, grow and consume the hydrocarbons.

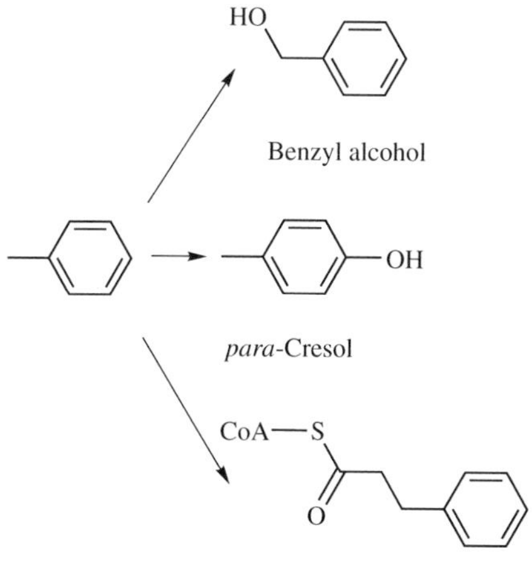

Figure 3. Proposed initial steps in the anaerobic biodegradation of toluene in different organisms.

Sea. Significantly more oil reaches the world's oceans from municipal sewers than widely covered crude oil spills. Physical collection of spilled oil is the preferred remediation option, but if skimming is unable to collect the oil, biodegradation and perhaps combustion or photooxidation are the only routes for eliminating of the spill. One approach to stimulating biodegradation is to disperse the oil with chemical dispersants. Modern dispersants and application protocols stimulate biodegradation by increasing the surface area of the oil available for microbial attachment, and perhaps providing nutrients to stimulate microbial growth.

Shorelines. The successful bioremediation of shorelines affected by the spill from the *Exxon Valdez* in Prince William Sound, Alaska, was perhaps the largest project to date. Bioremediation focused on the addition of nitrogen and phosphorus fertilizers to partially remove the nutrient-limitation on oil degradation. Of course the addition, of fertilizers was complicated by the fact that oiled shorelines were washed by tides twice a day. Two fertilizers were used in the full-scale applications; one, an oleophilic product known as Inipol EAP22 (trademark of CECA, Paris, France), was a microemulsion of a concentrated solution of urea in an oil phase of oleic acid and trilaurethphosphate, with butoxyethanol as a cosolvent. This product was designed to adhere to oil, and to release its nutrients to bacteria growing at the oil-water inter-face. The other fertilizer was a slow-release formulation of inorganic nutrients, primarily ammonium nitrate and ammonium phosphate, in a polymerized vegetable oil skin. This product, known as Customblen (trademark of Grace-Sierra, Milpitas, California), released nutrients with every tide, and these were distributed throughout the oiled zone as the tide fell. Fertilizer application rates were carefully monitored so that the nutrients would cause no harm, and the rate of oil biodegradation was stimulated between two- and five-fold.

Areas where there are currently few remediation options but where bioremediation may provide an option include oiled marshes, mangroves, and coral reefs. Bioremediation also offers options for dealing with oiled material, such as seaweed, that gets stranded on shorelines; composting has been shown to be effective.

Groundwater. Spills of refined petroleum product on land, and leaking underground storage tanks, sometimes contaminate groundwater. Stand-alone bioremediation is an option for these situations, but "pump and treat" is the more usual treatment. Contaminated water is brought to the surface, free product is removed by flotation, and the cleaned water re-injected into the aquifer or discarded. Adding a bioremediation component to the treatment, typically by adding oxygen and low levels of nutrients, is an appealing and cost-effective way of stimulating the degradation of the residual hydrocarbon not extracted by the pumping. This approach is becoming widely used.

Typically only small aromatic molecules, the infamous BTEX (benzene, toluene, ethylbenzene, and xylenes), are soluble enough to contaminate groundwater. With the advent of oxygenated gasolines, it is expected that these oxygenates [ethanol, methanol, MTBE (niethyl-*terf*-butyl ether) etc] will also be found in groundwater. These contaminants are biodegradable, and some biodegradation is probably already occurring when the contamination is discovered. The cheapest approach to remediation is, thus, to allow this intrinsic process to continue.

Intrinsic bioremediation is becoming an acceptable option in locations where the contaminated groundwater poses little threat to environmental health. Nevertheless, it may not be the lowest cost option if there are extensive monitoring and documentation costs involved for several years. In such cases it may well be more cost effective to optimize conditions for biodegradation.

One approach is to optimize the levels of electron acceptors. Slow release formulations of inorganic peroxides, such as magnesium peroxide, have recently been used with success. Nitrate may be added, although there are sometimes regulatory limitations on the amount of this material that may be added to ground-water. Ferric iron availability may be manipulated by adding ligands.

If there are significant amounts of both volatile and nonvolatile contaminants, remediation may be achievedby a combination of liquid and vapor extraction of the former, and bioremediation of the latter. This combination has been termed "bioslurping".

The majority of remediation operations include stopping the source of the contamination, but in some cases this is impossible, either because of the location of the spill, or because it is over a large area, and not a point source. In these situations it may be possible to intercept the flow of contaminated groundwater on-site, and ensure that no contamination passes. Approaches include biowall, trench biosparge, funnel and gate, bubble curtain, sparge curtain and engineered trenches and gates. Both aerobic and anaerobic designs have been successfully installed.

Where there are large volumes of contaminated water under a small site, it is sometimes most convenient to treat the contaminant in a biological reactor at the surface.

Of course the presence of a liquid phase of hydrocarbon in a soil gives rise to vapor contamination in the yadose zone above the water table. This can be treated by vacuum extraction, and the passage of the exhaust gases through a biofilter (see above) can be a cheap and effective way of destroying the contaminant permanently.

Soil. Spills, from production facilities and pipelines often involve both oil and brine. Successful bioremediation strategies must therefore include remediating the brine. In wet regions the salt is eventually diluted by rainfall, but in arid regions, and to speed the process in wetter regions, gypsum is often added to restore soil porosity.

Many hydrocarbons bind quite tightly to soil components, and are thereby less available to microbial degradation. Intrinsic biodegradation occurs, but it usually only removes the lightest refined products, such as gasoline, diesel and jet fuel. Active intervention is typically required. Usually the least expensive approach is *in situ* remediation, typically with the addition of nutrients, and the attempted optimization of moisture and oxygen by tilling.

Deeper contamination may be remedied with bioventing, where air is injected through some wells, and extracted through others to both strip volatiles and provide oxygen to indigenous organisms. Fertilizer nutrients may also be added. This is usually only a viable option with lighter refined products.

A recent suggestion has been to use plants to stimulate the microbial degradation of the hydrocarbon (hydrocarbon phytoremediation). The plants are proposed to help deliver air to the soil microbes, and to stimulate microbial growth in the rhizosphere by the release of nutrients from the roots. The esthetic appeal of an active phytoremediation project can be very great.

When soil contamination extends to some depth it may be prefeable to excavate the contaminated soil and put it into "biopiles" where oxygen, nutrient and moisture levels are more easily controlled. Composting by the addition of readily degradable bulking agents is also a useful option for relatively small volumes of excavated contaminated soil.

Slurry bioreactors offer the most aggressive approach to maximizring contact between the contaminated soil and the degrading organisms. Slurry bioreactors are usually the most expensive bioremediation option because of the large power requirements, but under some conditions this cost is offset by the rapid biodegradation that can occur.

In all these cases it is important to bear in mind that although the majority of hydrocarbons are readily biodegraded, some are very resistant to microbial attack. It is thus important to run laboratory studies to ensure that the contaminant is sufficiently biodegradable that clean-up targets can be met.

Halogenated Organic Solvents Constituents. Halogenated organic solvents are widely used in metal processing, electronics, dry cleaning and paint, paper and textile manufacturing and are fairly widespread contaminants. Unlike the hydrocarbons, the halogenated solvents typically have specific gravities greater than 1, and they generally sink to the bottom of any groundwater, and float on the bedrock. For this reason they are sometimes known as DNAPLs for dense nonaqueous phase liquids.

Biodegradation. Halogenated solvents are degraded under aerobic and anaerobic conditions. The anaerobic process is typically a reductive dechlorination that progressively removes one halide at a time.

The simplest chlorinated alkaries, alkenes, and alcohols (eg, chloromethane, dichloromethane, chloroethane, 1,2-dichloroethane, vinyl chloride, and 2-chloroethanol) serve as substrates for aerobic growth for some bacteria, but the majority of halogenated solvents cannot support growth. Nevertheless these compounds are mineralized under aerobic conditions.

Bioremediation

Air. Biofilters are an effective way of dealing with air from industrial processes that use halogenated solvents that support aerobic growth. Both compost-based dry systems and trickling filter wet systems are in use. Similar filters could be incorporated into pump-and-treat operations.

Groundwater and Soil. Pumping out the liquid phase is an obvious first step if the contaminant is likely to be mobile, but *in situ* bioremediation is a promising option. Thus, the U.S. Department of Energy is investigating the use of anaerobic *in situ* degradation of carbon tetrachloride with nitrate as electron acceptor, and acetate as electron donor.

Trichloroethylene, the most frequent target of remediation, is only metabolized co-metabolically. Remediation operations thus incorporate the addition of co-metabolized substrate.

Plants may have a role to play in enhancing microbial biodegradation of halogenated solvents, for it has recently been shown that mineralization of radiolabelled trichlorothylene is substantially greater in vegetated rather than unvegetated soils, indicating that the rhizo-sphere provides a favorable environment for microbial degradation of organic compounds.

Methane has been used in aerobic bioreactors that are part of a pump-and-treat operation, and toluene and phenol have been used as co-substrates at the pilot scale. Anaerobic reactors have also been developed for treating trichloroethylene.

Groundwater contaminated with other halogenated solvents can also be treated in aboveground reactors. Aerobic reactors are useful for those compounds that can support growth. Sequential anaerobic and aerobic reactors are capable of mineralizing tetrachloroethylene.

Halogenated Organic Compounds

Constituents. Complex halogenated organic compounds have been widely used in commerce in the last fifty years. Representative examples are pentachlorophenol, (2,4-dichlorophenoxy)acetate, DDT and polychlorinated biphenyls (PCBs). They may not seem a good target for bioremediation but some successful applications have been developed.

Biodegradation. An important characteristic of degradation is the cleavage of carbon–chlorine bonds, and the enzymes that catalyze these reactions, the dehalogenases, are being characterized. The reductive dechlorination seen with carbon tetrachloride and tetrachloroethylene seems to be a general phenomenon, and even compounds as persistent as DDT and the polychlorinated biphenyls are reductively dechlorinated under some conditions, particularly under methanogenic conditions. Some compounds, such as pentachlorophenol, can be completely mineralized under anaerobic conditions, but the more recalcitrant ones require aerobic degradation after reductive dehalogenation.

Bioremediation *Soil*. Pentachlorophenol has been the target of bioremediation at a number of wood-treatment facilities, and good success has been achieved in several applications. In *situ* degradation has been stimulated by bioventing. Just as with the halogenated solvents, it seems that plants stimulate microbial degradation of pentachlorophenol in the rhizosphere.

The kinetics of such *in situ* degradation are rather slow, however, and more active bioremediation is usually attempted. For example, contaminated soil at the Champion Superfund site in Libby, Montana, was placed into 1-acre land treatment units in 6-in. layers, and irrigated, tilled, and fertilized. Under these conditions, the half-lives of pentachlorophenol, pyrene, and several other polynuclear aromatic hydrocarbons, initially present at around 100–200 ppm, were on the order of 40 days. Composting, and bioremediation focusing on the use of white-rot fungi, has also met with success at the pilot scale. Others have used fed-batch or fluidized-bed bioreactors to stimulate the biodegradation of pentachlorophenol. This allows significant optimization of the process and increases in rates of degradation by tenfold.

A major concern when remediating wood-treatment sites is that pentachlorophenol was often used in combination with metal salts, and these compounds, such as chromated-copper-arsenate, are potent inhibitors of at least some pentachlorophenol degrading organisms. Sites with significant levels of such inorganics may not be suitable candidates for bioremediation.

The phenoxy-herbicide, 2,4-D, has been successfully bioremediated in a soil contaminated with such a high level of the compound (710 ppm) that it was toxic to microorganisms. Success relied on washing a significant fraction of the contaminant off the soil and adding bacteria enriched from a less contaminated site. Success was achieved in remediating both soil washwater and soil in a bioslurry reactor. 2,4-D is also effectively degraded in composting, with about half being completely mineralized, and the other half becoming incorporated in a nonextractable form in the residual soil organic matter.

Cultures are being found that can degrade both polychlorinated biphenyls and petroleum hydrocarbons. There is also interest in the role of rhizosphere organisms in polychlorinated biphenyl degradation, particularly since some plants exude phenolic compounds into the rhizosphere that can stimulate the aerobic degradation of the less chlorinated biphenyls.

Groundwater. A successful groundwater bioremediation of pentachlorophenol is being carried out at the Libby Superfund site described above. A shallow aquifer is present at 5.5 to 21 m below the surface, and a contaminant plume is nearly 1.6 km in length. Nutrients and hydrogen peroxide were added at the source area and approximately half way along the plume, and pentachlorophenol concentrations decreased from 420 ppm to 3 ppm where oxygen concentrations were successfully raised.

Pentachlorophenol is readily degraded in biofilm reactors, so bioremediation is a promising option for the treatment of contaminated groundwater brought to the surface as part of a pump-and-treat operation.

River and Pond Sediments. Much of the work on polychlorinated biphenyls has focused on the remediation of aquatic sediments, particularly from rivers, estuaries, and ponds. Harkness and co-workers have successfully stimulated aerobic biodegradation in large caissons in the Hudson River by adding inorganic nutrients, biphenyl, and hydrogen peroxide, but found that repeated addition of a polychlorinated-biphenyl degrading bacterium (*Alcaligenes eutrophus* H850) had no beneficial effect. Whether this approach can be scaled-up for large-scale use, with a net environmental benefit, remains to be seen.

Nonchlorinated Pesticides and Herbicides

Constituents. It is unusual for these compounds to become contaminants where they are applied correctly, but manufacturing facilities, storage depots and rural airfields where crop-dusters are based have had spills that can lead to long lasting contamination.

Biodegradation. The vast majority of pesticides, herbicides, fungi; cides, and insecticides in use today are biodegradable, although the intrinsic biodegradability of individual compounds is one of the variables used in deciding which compound to use for which task.

Compounds with organophosphate moieties, such as Diazinon, Methyl Parathion, Coumaphos and Glyphbsate are usually hydrolyzed at the phosphorus atom. Indeed several *Flavobacterium* isolates are able to grow using parathion and diazinon as sole sources of carbon.

Very few pure cultures of microorganisms are able to degrade triazines such as Atrazine, although some *Pseudomonads* are able to use the compound as sole source of nitrogen in the presence of citrate or other simple carbon substrates. The initial reactions seem to be the removal of the ethyl or isopropyl substituents on the ring, followed by complete mineralization of the triazine ring.

Nitroaromatic compounds, such as Dinoseb, are degraded under aerobic and anaerobic conditions. The nitro group may be cleaved from the molecule as nitrite, or reduced to an amino group under either aerobic or anaerobic conditions. Alternatively, the ring may be the subject of reductive attack. Recent work has isolated a *Clostridiurm bifermentans* able to anaerobically degrade dinoseb cometabolically in the presence of a fermentable substrate. The dinoseb was degraded to below detectable levels, although only a small fraction was actually mineralized to CO_2.

Carbamates such as Aldicarb undergo degradation under both aerobic and anaerobic conditions.

Bioremediation

Groundwater. Atrazine dominated the world herbicide market in the 1980s, and contamination of groundwater has been reported in several locations in the U.S., Europe, and South Africa. Successful biodegradation has been achieved with indigenous organisms in laboratory mesocosms after a lag phase, and once activity was found, it remained. It is clear that intrinsic remediation is likely to lead to the disappearance of atrazine from groundwaters.

If more active treatment is required, such as pump and treat, it is possible that biological reactors will be a cost-effective replacement for activated carbon filters.

Marsh and Pond Sediments. Herbicides and pesticides are detectable in marsh and pond sediments, but intrinsic biodegradation is usually found to be occurring.

Soil. Herbicides and pesticides are of course metabolized in the soil to which they are applied, and there are many reports of isolating degrading organisms from such sites. Little work has yet been presented where the biodegradation of these compounds has been successfully stimulated by a bioremediation approach.

It is a general observation that herbicide degradation occurs more readily in cultivated than fallow soil, suggesting that rhizosphere organisms are effective herbicide degraders. Whether this can be effectively exploited in a phytoremediation strategy remains to be seen.

Military Chemicals

Constituents. The military use a range of chemicals such as explosives and propellants that are sometimes termed "energetic molecules." Generally speaking, modern explosives are cyclic, often heterocyelic, composed of carbon, nitrogen and oxygen, eg; 2,4,6-trinitrotoluene; RDX (Royal Demolition eXplosive; hexahydro-l,3,5-trinitro-l,2,3-triazine); HMX (High Melting eXplosive; octahydro-l,3,5,7-tetranitro-l,3,5,7-tetrazocine); and *N,N*-dimethylhydrazine. These compounds are sometimes present at quite nigh levels in soils and groundwater on military bases and production sites. Bioremediation is a promising new technology for treating sites contaminated with such compounds.

Bioremediation may also be an appropriate tool for dealing with nemica agents such as the mustards and organophosphate neurotoxins, but little work on actual bioremediation has been published.

Biodegradation. Nitrosubstituted compounds are subject to a variety of degradative processes. Under anaerobic conditions TNT is readily reduced to the corresponding aromatic amines and subsequently deaminated to toluene. As shown in the section on hydrocarbons, the latter can be mineralized under anaerobic conditions, leading to the potentially complete mineralization of TNT in the absence of oxygen.

Under aerobic conditions TNT can be mineralized by a range of bacteria and fungi, often co-metabolically with the degradation of a more degradable substrate. There is even evidence that some plants are able to deaminate TNT reductively.

RDX and HMX are rather more recalcitrant, especially under aerobic conditions, but there are promising indications that biodegradation can occur under some conditions, especially composting.

Little work has been reported on the biodegradation of dimethylhydrazine.

Bioremediation

Groundwater. Nitrotoluenes have been detected in groundwater in some areas, and intrinsic remediation may be occurring at some sites by anaerobic degradation.

A commercial technology, the SABRE process, treats contaminated water and soil in a two-stage process by adding a readily degradable carbon and an inoculum of anaerobic bacteria able to degrade the contaminant. An initial aerobic fermentation removes oxygen so that the subsequent reduction of the contaminant is not accompanied by oxidative polymerization.

Soil. Composting of soils contaminated by high explosives is being carried out at the Umatilla Army Depot near Hermiston, Oregon. If this is successful, there are 30 similar sites on the National Priority List that could be treated in a similar way.

Other Organic Compounds

The majority of organic compounds in commerce are biodegradable, so bioremediation is a potential option for cleaning up after industrial and transportation accidents.

INORGANIC CONTAMINANTS

Nitrogen Compounds

Constituents. Nitrate levels are regulated in groundwater because of concerns for human and animal health. Ammonia is regulated in streams and effluents as a potential fish toxicant, and any nitrogenous contaminant is a potential problem in water because of its stimulatory effect on the growth of algae. Other nitrogenous contaminants include cyanides in mine waters. Fortunately, all are amenable to biological treatment.

Biodegradation

The biological mineralization of fixed nitrogen is well studied; ammonia is oxidized to nitrite, and nitrite to nitrate, by autotrophic bacteria, and nitrate is reduced to nitrogen by anaerobic bacteria. On the other hand, ammonia and nitrate are essential nutrients for plant and bacterial growth, so one option is to use these organisms to take up and use the contaminants.

Bioremediation

Surface Water. One example of exploiting biology to handle excess nitrogen in a surface water is the case of the Venice Lagoon in Italy. About a million tons of sea lettuce (*Ulva*) grows in the lagoon annually because of the high levels of nitrogenous nutrients in this relatively landlocked bay. This material is harvested, composted and sold as a low-cost remediation of this problem.

A more constrained opportunity for nitrate bioremediation arose at the U.S.-DoE Weldon Spring Site near St. Louis, Missouri which had been a uranium and thorium processing facility. Two pits had nitrate levels that required treatment before discharge. Bioremediation by the addition of Calcium acetate as a carbon source successfully treated more than 19 million liters of water at a reasonable cost.

Groundwater. One approach to minimizing the environmental impact of excess nitrogen in groundwater migrating into rivers and aquifers is to intercept the water with rapidly growing trees, such as poplars, that will use the contaminant as a fertilizer.

An alternative approach is to add a readily degradable substrate to the contaminated aquifer, in the absence of oxygen, to stimulate bacterial denitrification.

Metals and Metalloids. A wide range of metallic and nonmetallic elements are present as contaminants at industrial and agricultural sites throughout the world, both in ground and surface water, and in soils. They pose a quite different problem from that of organic contaminants, since they cannot be degraded so that they disappear. Some metal and metalloid elements have radically different bioavailabili-ties and toxicities de- pendingon their redox state, so one option is stabilization by converting them to their least toxic form. This can be a very effective way of minimizing the environmental impact of a contaminant, but if the contaminant is not removed from the environment, there is always the possibility that natural processes, biological or abiological, may reverse the process. Removing the contaminants from water phases is relatively straightforward, and the wastewater treatment industry practices this on an enormous scale. Pump-and-treat systems that mimic wastewater treatments are already being used for several contaminants, and less complex systems involving biological mats are a promising solution for less demanding situations.

In the past, removing metal and metalloid contaminants from soil has been impossible, and site clean-up has meant excavation and disposal in a secure landfill. An exciting new approach to this problem is phytoextraction, where plants are used to extract contaminants from the soil and harvested.

Bioremediation

Water. Groundwater can be treated in anaerobic bioreactors that encourage the growth of sulfate reducing bacteria, where the metals are reduced to insoluble sulfides, and concentrated in the sludge.

Phytoremediation is not yet being used commercially, but results at several field trial suggest that commercialization is not far away. Perhaps the biggest success to date is the successful rhizofiltration of radionuclides from a Department of Energy site at Ashtabula, Ohio, where uranium concentrations of 350 ppb were reduced to less than 5 ppb, well below groundwater standards, by Sunflower roots.

Mine Drainage. In recent years it has become clear that the environmental impact of acid mine drainage can be minimized by the construction of artificial wetlands that combine geochemistry and biological treatments. These systems are being designed for a range of wastewaters, most of which fall outside the scope of this article.

Soil. The results of the first reported field trial of the use of hy-peraccumulating plants to remove metals from a soil contaminated by sludge applications were positive. However, the rates of metal uptake suggest a time scale of decades for complete cleanup. Trials with higher biomass plants, such as *B. juncea*, are underway.

The bacterial reduction of Cr(VI) to Cr(III) is also being used to reduce the hazards of chromium in soils and water.

Table 3. Will Bioremediation be a Suitable Treatment for a Site Contaminated with Organic, Nitrogenous, or Organic Contaminants?

Organic

Is the contaminant biodegradable? If the contaminant is a complex mixture of components, are the individual chemical species biodegradable? If the contaminant has been at the site for some time, biodegradation of the most readily degradable components may have already occurred. Is the residual contamination biodegradable?
Are degrading organisms present at the site?
What is limiting their growth and activity? Can this be added effectively?
Are the levels of contaminant amenable to bioremediation? Are they toxic to microorganisms? Are they so abundant that even substantial microbial activity will take too long to clean the site?
Are the clean-up standards reasonable? Are biological processes known to degrade substrates down to the levels required?

Nitrogenous

Are appropriate microorganisms present at the site?
What is limiting their growth and activity? Can this be added effectively?
Are the levels of contaminant amenable to bioremediation? Are they toxic to microorganisms? Are they so abundant that even substantial microbial activity will take too long to clean the site?
Can the nitrogenous compound be used by plants?
Are the clean-up standards reasonable? Are biological processes known to degrade substrates down to the levels required?

Inorganic

Can the contaminant be made less hazardous by changing its redox state?
Can the contaminant be brought to a reactor or constructed wetland where biological systems, microbial or plant, can extract and immobilize the contaminant?
Can plants extract the contaminant from the soil matrix?
Are the clean-up standards reasonable? Are biological processes known to accumulate contaminants down to the levels required?

CONCLUSIONS

Bioremediation has many advantages over other technologies, both in cost and in effectively destroying or extracting the pollutant. An important issue is thus when to consider it, and a series of questions may lead to the appropriate answer (Table 3).

If the answers to the questions in Table 3 lead to the selection of bioremediation, it then becomes important to assess the success of the bioremediation strategy in achieving the cleanup criteria. A major disadvantage of bioremediation is that it is typically rather slower than competing technologies such as thermal treatments. How can regulators and responsible parties gain confidence during this time that success will indeed be achieved? The National Research Council has recently addressed this issue, and suggested a three-fold strategy for "proving" bioremediation: (1) a documented loss of contaminants from the site; (2) laboratory tests showing the potential of endogenous microbes to catalyze the reactions of interest; and (3) some evidence that this potential is achieved in the field.

Finally a caveat. Despite its documented success in many situations, bioremediation may not always be able to meet current clean up criteria for a particular site. Some standards are so tight that they are essentially "detection limit" standards, and it is not clear that biological processes will be able to remove contaminants to such low levels. Bioremediation will be more likely to fulfill its promise as an important tool in contaminated site remediation if there is progress towards standards based on bioavailability and net environmental benefit from the clean up, rather than on arbitrary absolute standards.

M. A. Alexander, *Biodegradation and Bioremediation*, Academic Press, New York, 1994.

Bioremediation; when does it work?, National Research Council, National Academy Press, 1993.

R. L. and D. L. Crawford, *Bioremediation: Principles and Practice*, Cambridge University Press, 1996.

L. Y. Young and C. E. Cerniglia, eds., Wiley-Liss, *Microbial Transformation and Degradation of Toxic Organic Compounds*, New York, 1995.

ROGER C. PRINCE
Exxon Research and Engineering
Company

BIOSENSORS

A chemical sensor is a device that transforms chemical information, ranging from the concentration of a specific sample component to total composition analysis, into an analytically useful signal. Chemical sensors usually contain two basic components connected in series: a chemical (molecular) recognition system (receptor) and a physico-chemical transducer. Biosensors are chemical sensors in which the recognition system utilizes a biochemical mechanism. The current consensus is that this term should be reserved for sensors incorporating a biological element such as an enzyme, antibody, nucleic acid, microorganism, or cell where the biological element is in intimate contact with the transducer.

Figure 1 presents a generalized scheme of a biosensor.

In a biosensor, the biological component and the transducer are in intimate contact; this distinguishes this class of device from an analytical system that incorporates additional processing steps like separation processes or incubating chambers (ie, as in flow injection analysis). Thus, a biosensor should be a reagentless analytical device, although the presence of ambient

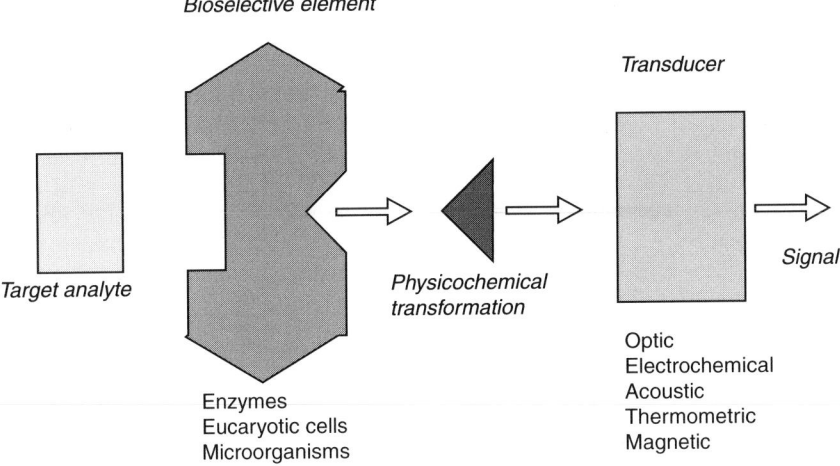

Bioselective element

Transducer

Signal

Physicochemical
transformation

Target analyte

Enzymes
Eucaryotic cells
Microorganisms
Antibodies

Optic
Electrochemical
Acoustic
Thermometric
Magnetic

Figure 1. Scheme of a generic biosensor.

cosubstrates (eg, oxygen for oxidoreductase) is often required for the detection of the analyte. Immobilization of the biological component can be performed using a variety of methods such as chemical or physical adsorption, physical entrapment within a membrane or gel, cross-linking of molecules, or covalent binding. The biological components used in biosensor construction can be divided into two broad categories: catalytic biosensors, where the primary sensing event results from a chemical transformation (catalyzed by an enzyme either isolated or retained in a microorganism or tissue); or affinity biosensors, in which the sensing event is dependent on an essentially irreversible binding of the target molecules (eg, affinity sensors based on antigen–antibody or nucleic acid interactions).

CATALYTIC BIOSENSORS

Catalytic biosensors are based on a reaction involving macromolecules that are present in their original biological environment, have been isolated previously, or have been manufactured. Biocatalytic elements normally used are enzymes (mono- or multisystems), whole cells from microorganisms such as bacteria or fungi, or tissues directly excised from the original organism.

The development of enzyme-based sensing devices, particularly those based on electrochemistry, continues to attract considerable attention. Of the electrochemical transducing systems available (amperometric, potentiometric, conductometric, and impedimetric), amperometric biosensors have dominated both research and commercial activity to date, largely because of their relative simplicity and flexibility. The most common example of an amperometric biosensor is the mediated amperometric glucose biosensor; this device involves the use of the enzyme glucose oxidase (GOD) as the biological recognition element.

AFFINITY BIOSENSORS: IMMUNOSENSORS

Affinity biosensors are analytical devices comprising a biological affinity element such as an antibody, receptor

protein, biomimetic material, or DNA, interfaced to a signal transducer, to convert the concentration of an analyte to a measurable electronic signal.

Because of their high affinity, versatility, and commercial availability, antibodies are the most widely reported biological recognition elements used in affinity-type biosensors; in this case, the affinity biosensor is known as an "immunosensor".

Antibody-based biosensors (immunosensors), where the antibody or antigen is immobilized to the transducer, have been constructed in a variety of ways, but generally fall into one of three basic configurations. These formats involve direct noncompetitive assays, competitive (direct or indirect) assays, or sandwich-type assay formats. In the case of the direct noncompetitive assay format, a unique optical or electrochemical property of the analyte of interest is observed as the target compound binds to the recognition site of the bioaffinity element and accumulates on the sensor surface.

The simplest biochemical immunosensor consists of an antibody or antigen immobilized to a transducer and results in a signal generated from the binding of an analyte to an antibody at the sensor surface.

Optical immunosensors based on fluorescence have also been realized. Optical immunosensors have been developed for many environmental applications including pesticide analysis for analytes such as parathion and triazine. Optical evanescent-wave technology has been used to streamline the design of affinity biosensors that contain a label or marker.

Piezoelectric biosensors have been reported for a range of applications in biochemistry and affinity biosensors construction. These sensors operate by observation of the frequency of oscillation or the propagation of an acoustic wave through the solid-state device. Sensing is usually achieved by correlating the acoustic wave-propagation variations to the amount of analyte captured at the surface, and then to the amount or concentration of analyte present in the sample exposed to the sensor. Alternatively, changes in the physical properties of interfacial thin films in response to analyte may be measured.

AFFINITY BIOSENSORS: DNA BIOSENSORS

DNA biosensors represent a very important class of affinity biosensor. In this developing aspect of the field, DNA strands are used to detect the binding of oligonucleotides (gene probes). Such devices are known as "DNA biosensors" or more generally as "DNA Chips". DNA biosensors involve the use of nucleic acids as biological recognition elements to explore novel hybridization probes, transduction strategies, and potential practical applications. Transduction strategies that have been reported include electrochemical, acoustic, and optical techniques. For each of these strategies, the biosensor format typically relies on immobilization of single stranded (ss) DNA to the sensor surface followed by hybridization of a complimentary sequence. Detection of the formation of double stranded (ds) DNA has been facilitated through the use of a variety of electrochemical and optical tracers that bind or intercalate in to ds DNA.

Electrochemical techniques have been used to differentiate between ss (prehybridized) and ds (hybridized) DNA using several approaches. The detection of small molecule binding to DNA and general DNA damage resulting form ionizing radiation, dimethyl sulfate, etc has been described by following variation in the electrochemical signal derived from guanine. These approaches include the use of redox active intercalators that accumulate into ds DNA, transition metal complexes that mediate oxidation of the guanine base, or the direct oxidation of guanine.

The mass change associated with DNA hybridization may also be detected by employing piezoelectric devices. A further example of piezoelectric DNA-based biosensors has been developed to detect bacteria toxicity in environmental samples. In this system, a biotinylated oligonucleotide probe was immobilized on a streptavidin coated gold disk. Streptavidin was covalently linked to the thiol/carboxylated dextran modified gold surface. The immobilized probe was then reacted with a solution of the target oligonucleotide. The interactions of the immobilized DNA strand with a complementary and a noncomplementary sequence in solution were studied. The hybridization reaction was also performed on real samples of DNA extracted from different *Aeromonas* strains isolated from water, vegetables, or human specimens and amplified by PCR. These experiments showed that it was possible to distinguish between strains that contain the aerolysin gene and those that do not, hence furnishing an assay for the toxicity of the bacterium.

SYNTHETIC RECEPTORS AND BIOMIMETIC SENSORS

In biosensor technology, the recognition element is conventionally isolated from living systems, measuring interactions such as antigen–antibody or utilizing the substrate specificity obtained with enzymes. However, for a number of potential applications, no biocomponent is available or requirements such as stability cannot be met using biological molecule. For this reason, synthetic bioreceptors have attracted considerable attention as a potential new avenue for biosensor development. Semisynthetic receptors have been created by modifying enzymes and antibodies at both the pre- and posttranslational level. The design of semisynthetic receptors in biosensors may be superseded by totally synthetic ligands produced with the aid of computational chemistry, combinatorial chemistry, molecular imprinting, self-assembly, rational design, or combinations of one or more of these.

Molecularly imprinted polymers are proposed for a number of applications including use as the stationary phase in high performance liquid chromatography (HPLC) and, in thin-layer chromatography (TLC) to separate chiral products and as a replacement for antibodies in biosensor construction.

Molecularly imprinted polymer (MIP) technology has also been applied for environmental applications.

Transducing the binding event in molecularly imprinted polymers into a detectable signal has proved quite a challenge. However, creative solutions have been achieved with both optical and electrochemical configurations.

MASS PRODUCTION OF BIOSENSORS: THICK-FILM TECHNOLOGY

Over the past two decades, there has been increasing interest in the application of simple, rapid, inexpensive, and disposable biosensors in fields such as clinical, environmental, or industrial analysis. The most common disposable biosensors are those produced by thick-film technology. A thick-film biosensor configuration is normally considered to be one that comprises layers of special inks or pastes deposited sequentially onto an insulating support or substrate. One of the key factors that distinguishes a thick-film technique is the method of film deposition, namely, screen printing, which is possibly one of the oldest forms of graphic art reproduction. To create a thick-film electrode, a conductive or dielectric film is applied to a substrate. The film is applied through a mask contacting the substrate and deposited films are obtained by pattern transfer from the mask. Conventionally, thick-film electrodes were baked at temperature ranging from 300 to 1200°C to drive off solvents from the applied paste and to cure the pattern paste. In commercial biosensors, this step is usually avoided since it would damage incorporated enzymes. Alternatives include cold-cured ink formulations and a photocured process using ultraviolet(uv) light.

More complex structures can be built up by repeating the print process using the materials appropriate to the specific design and a range of mask designs. A variety of screen-printed thick-film devices can thus be fabricated as a base for disposable electrochemical immunosensors, DNA sensors, and enzyme electrodes. These planar devices present many advantages including disposability, which is a very important characteristic when working with real samples, and small dimensions, which facilitates the design of portable measuring systems.

Screen-printed electrodes have also been fabricated to detect blood glucose. Screen-printed electrodes can also be designed to incorporate bioaffinity molecules like antibodies, in order to obtain disposable immunosensors for clinical and environmental applications. The electrode constitutes both the solid phase for the immunoassay and the electrochemical transducer.

APPLICATIONS OF BIOSENSORS

The vast majority of commercial activity to date, in the field of biosensors, has been focused on medical applications and a substantial market now exists for such devices especially for home blood glucose measurement. The second most reported application area for biosensors is for environmental analysis. Less reported, but of enormous current interest, is the application of biosensors for defence (detection of biological and chemical warfare agents) and security (detection of drugs, explosives, identity, etc). Biosensors are also expected to impact on the food and process industries.

Enzyme-based biosensors have generally been used to measure batch samples, but a number of attempts have been made to monitor processes on-line. In most cases, the analysis must be performed outside the reactor or process line (ex situ). Other disadvantages of biosensors for process monitoring are that they are invasive, temperature dependent, subject to fouling, and need to be recalibrated and regenerated frequently. Applications of on-line biosensors have included process optimization in pilot scale and control of animal cell culture where the product is very valuable. Bioluminescent sensors have also been developed for on-line applications and consist of a bioluminescent enzyme and an optical transducer. Optical biosensors provide a rapid and highly selective detection system.

Effective on-line sensing is arguably the most difficult technical hurdle facing biosensor technology. Progress in the related area of real-time in vivo has advanced significantly in recent years. Two new commercial devices have been launched for monitoring glucose in vivo. One offers a miniaturized, sterilizable, and implantable enzyme electrode produced using microfabrication technology, while the other is a transcutaneous device, which obtains a sample using reverse ionophoresis.

U. Bilitewski and A. P. F. Turner, *Biosensors for Environmental Monitoring*, Harwood Academic Publishers, 2000.

J. D. Newman and A. P. F. Turner. *Biosensors and Bioelectronics* **20**, 2435–2453 (2005).

S. Piletsky and A. P. F. Turner, *Molecular Imprinting*, Landes Bioscience, Georgetown, Texas, 2006.

A. P. F. Turner, *Science* **290**(5495), 1315–1317 (2000).

A. P. F. TURNER
Cranfield University, UK

BIOSEPARATIONS

The large-scale purification of proteins and other bioproducts is the final production step, prior to product packaging, in the manufacture of therapeutic proteins, specialty enzymes, diagnostic products, and value-added products from agriculture. These separation steps, taken to purify biological molecules or compounds obtained from biological sources, are referred to as bioseparations. Large-scale bioseparations combine art and science. Bioseparations often evolve from laboratory-scale techniques, adapted and scaled up to satisfy the need for larger amounts of extremely pure test quantities of the product. Uncompromising standards for product quality, driven by commercial competition, applications, and regulatory oversight, provide many challenges to the scale-up of protein purification. The rigorous quality control embodied in current good manufacturing practices, and the complexity and lability of the macromolecules being processed provide other practical issues to address.

Recovery and purification of new biotechnology products is the fastest growing area of bioseparations.

Manufacturing approaches for selected bioproducts of the new biotechnology impact product recovery and purification. The most prevalent bioseparations method is chromatography (qv). Thus the practical tools used to initiate scaleup of process liquid chromatographic separations starting from a minimum amount of laboratory data are given.

ECONOMIC ASPECTS

The development of biotechnology processes in the biopharmaceutical and bioproduct industries is driven by the precept of being first to market while achieving a defined product purity, and developing a reliable process to meet validation requirements. The economics of bioseparations are important, but are likely to be secondary to the goal of being first to market.

The three main sources of competitive advantage in the manufacture of high value protein products are first to market, high product quality, and low cost. The first company to market a new protein biopharmaceutical, and the first to gain patent protection, enjoys a substantial advantage. In the absence of patent protection, product differentiation becomes very important. Differentiation reflects a product that is purer, more active, or has a greater lot-to-lot consistency.

Biopharmaceuticals and Protein Products

Purification of proteins is a critical and expensive part of the production process, often accounting for ≥50% of total production costs. Hence, bioseparation processes have a significant impact on manufacturing costs. For small-volume, very high value biotherapeutics, however, these costs may be considered secondary to the first to market principle unless a lower cost competitor surfaces.

BIOPRODUCT SEPARATIONS

The task of quickly specifying, designing, and scaling-up a bioproduct separation is not simple. These separations are carried out in a liquid phase using macromolecules which are labile, and where conformation and heterogeneous chemical structure undergoing even subtle change during purification may result in an unacceptable product. A typical purification scheme for biopharmaceutical proteins involves the harvesting of protein-containing material or cells, concentration of protein using ultrafiltration (qv), initial chromatographic steps, viral clearance steps, additional chromatographic steps, again concentration of protein using ultrafiltration, and finally formulation.

Biosynthetic Human Insulin from *E. coli*

Human insulin was the first animal protein to be made in bacteria in a sequence identical to the human pancreatic peptide. Expression of separate insulin A and B chains were achieved in *Escherichia coli* K-12 using genes for the insulin A and B chains synthesized and cloned in frame with the β-galactosidase gene of plasmid pBR322.

Recovery and Purification. The production of Eli Lilly's human insulin requires an estimated 31 principal processing steps of which 27 are associated with product recovery and purification. The production process for human insulin, based on a fermentation which yields proinsulin. Whereas the exact sequence has not been published, the principle steps in the purification scheme are outlined in Figure 1.

Ion-exchange chromatography removes most of the impurities that form during processing and is followed by reversed-phase chromatography which separates insulin from structurally similar insulin-like components. Then size-exclusion (gel-permeation) chromatography is introduced to remove salts and other small molecules from the insulin. This sequence follows the principle of orthogonality of separation sequence, ie, each step is based on a different property, in this case charge, solubility, and size, respectively. Near the end of the chromatography sequence, the insulin may be concentrated by precipitation to form insulin zinc crystals.

Yield Losses

The numerous steps incur a built-in yield loss. For example, if only 2% yield loss were to be associated with each step, the overall yield for a purification sequence of 10 steps would be as in equation 1:

$$\eta = 100(1 - L/100)^n = 100(1 - 0.02)^{10} = 81.7\% \qquad (1)$$

where η denotes yield, L the percent yield loss at each step, and n the number of steps. Maximizing recovery at each step is important.

The purification of human insulin involves five separate alterations in the molecular structure, and hence, changes in physicochemical properties during its recovery and purification. The final purification steps rely on multiple properties of the insulin, such as size, hydrophobicity, ionic charge, and crystallizability. The final purity level is reported to be >99.99%.

Tissue Plasminogen Activator from Mammalian Cell Culture

In 1981 the Bowes melanoma cell line was found to secrete tissue plasminogen activator (known as mt-PA) at

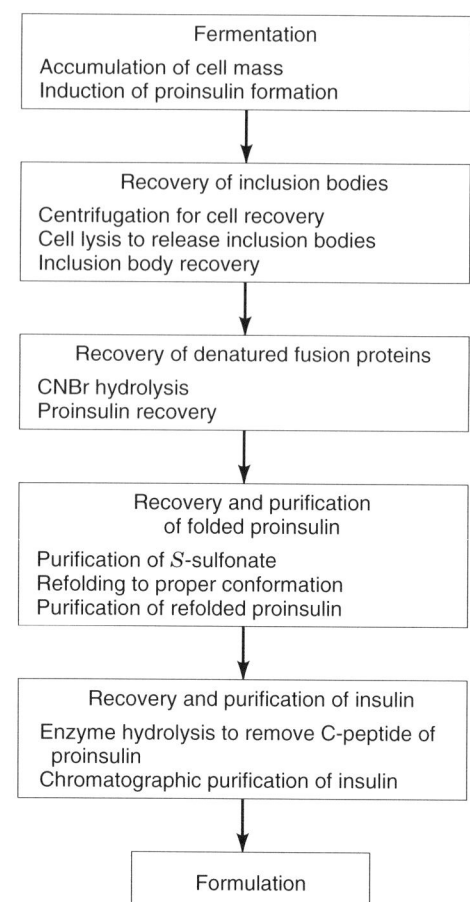

Figure 1. Process flow sheet for human insulin production, recovery, and purification.

100 × higher concentrations than normal, making possible the isolation and purification of this enzyme in sufficient quantities that antibodies could be generated and assays developed to lead to cloning of the gene for this enzyme and subsequent expression of the enzyme in both *E. coli* and a Chinese hamster ovary (CHO) cell line. Comparison of the melanoma and recombinant forms of the enzymes showed the same activity toward dissolution of blood clots.

Characteristics of t-PA. Tissue plasminogen activator, a proteolytic, hydrophobic enzyme, has a molecular weight of 66,000, 12 disulfide bonds, 4 possible glycosylation sites, and a bridge of 6 amino acids connecting the principal protein structures. Only three of the sites (Asn-117–118, −448) are actually glycosylated. When administered to heart attack victims it dissolves clots consisting of platelets in a fibrin protein matrix and acts by clipping plasminogen, an active precursor protein found in the blood, to form plasmin, a potent protease that degrades fibrin. Concentrations of plasminogen activator are low in blood and tissues.

t-PA Production. Recombinant technology provides the only practical means of rt-PA production. The amount of t-PA required per dose is on the order of 100 mg. Cell lines of transformed CHO cells, selected for high levels of rt-PA expression using methotrexate, are grown in large fermenters. The purification steps for rt-PA must therefore separate out cells, virus, and DNA. Recovering a protein derived from mammalian cells involves a number of steps. In one possible scheme, the culture medium is separated from the cell by sterile filtration (see MICROBIAL AND VIRAL FILTRACTION). This is followed by additional removal by cross-flow filtration, ultrafiltration, and chromatography to remove DNA and remaining viruses. The product protein then undergoes purification by chromatography.

The separation of cells from the culture media or fermentation broth is the first step in a bioproduct recovery sequence. Whereas centrifugation is common for recombinant bacterial cells (see CENTRIFUEAL SEPERATION), the final removal of CHO cells utilizes sterile-filtration techniques.

The possibility that DNA from recombinant immortal cell lines such as CHO cells could cause oncogenic (gene altering) events resulting in cancer was a concern during development of the rt-PA purification sequence. The goal for rt-PA purification is to reduce the DNA to <10 pg/dose (10^{-11} g/dose). A level of 0.1 pg was achieved. Special assays are required to detect and quantify these very low levels of DNA in the final product.

Ultrafiltration followed by ion-exchange (qv) chromatography (qv) and then a final round of ultrafiltration concentrate the dilute protein while purifying it.

Independent Assays for Proving Virus Removal. Retroviruses and viruses can also be present in culture fluids of mammalian cell lines. Certainly the absence of virus can be difficult to prove.

Viral clearance can be achieved by use of chaotropes such as urea or guanidine, pH extremes, detergents, heat, formaldehyde, proteases or DNA'ses organic solvents such as formaldehyde, or ion-exchange or size-exclusion chromatography. Because only the inactivation or removal which can be measured counts as validation, a sequence of orthoganol removal/fractionation steps must be used.

MANUFACTURE OF BIOLOGICS AND GOVERNMENT REGULATION

The definition of biologics versus drugs continues to evolve. Assignment is made on a case by case basis. Section 351 of the Public Health Service Act defines a biologic product as "any virus, therapeutic serum, toxin, antitoxin, vaccine, blood, blood component or derivative, allergenic product, or analogous product...applicable to the prevention, treatment, or cure of diseases or injuries in man." To compensate for the incomplete analytical capability to define biologics, regulatory agencies include parameters of the process used to make biologics in the control and monitoring. Changes in the process may yield a different product from that previously reviewed and approved. A different product requires a new license. Thus substantial barriers exist in terms of effort, money, and time to making significant changes in processes used to produce licensed biologies, although regulations are evolving.

Biologics are subject to licensing provisions that require that both the manufacturing facility and the product be approved. All licensed products are subject to specific requirements for lot release by the FDA.

The design of bioseparation unit operations is influenced by these governmental regulations. The constraints on process development grow as a recovery and purification scheme undergo licensing for commercial manufacture. Given changes in the products, their regulation, process, and analytical technology, regulations are continually evolving as well.

PROTEIN CHROMATOGRAPHY

Proteins and nucleotides are macromolecular biomolecules. Mixtures of biomolecules are fractionated based on differences in charge; molecular weight, shape, and size; solubility in organic solvents; surface hydrophobic character; and types of active sites using ion-exchange, size-exclusion (gel-permeation), and reversed-phase, hydrophobic interaction (surface hydrophobicity), and affinity chromatographies, respectively. The appropriate separation may be selected from these five basic classes of chromatography. More than 30% of the purification steps for laboratory-scale protein purification procedures use ion-exchange and/or gel filtration, and at least 20% use affinity chromatography. This pattern is likely to be consistent with industrial practice.

Reversed-phase chromatography is widely used as an analytical tool for protein chromatography, but it is not as commonly found on a process scale for protein purification because the solvents which make up the mobile phase, ie, acetonitrile, isopropanol, methanol, and ethanol, reversibly or irreversibly denature proteins. Hydrophobic interaction chromatography appears to be the least common process chromatography tool, possibly owing to the relatively high costs of the salts used to make up the mobile phases.

Liquid Chromatographs. The basic equipment for liquid chromatography is shown in Figure 2. The column is packed with an adsorbent, ie, the stationary phase. The mixture to be separated is pushed through the column by the eluent or mobile phase. Isocratic chromatography, carried out at a constant flow rate, buffer composition, and temperature, is usually associated with size-exclusion separations. Gradient chromatography typically uses a constant flow rate and temperature, but the composition of the element is altered by mixing two or more buffer reservoirs to achieve a steadily changing salt concentration or changes in pH. The gradients formed are reported in terms of concentration at the inlet of the chromatography column; protein is detected at the column outlet. A chromatogram of the type illustrated in Figure 3 results.

Ion-Exchange Chromatography. Ion-exchange chromatography is initiated by eluting an injected sample through a column using a buffer but no NaCl or other displacing salt. The protein, which has charged sites spread

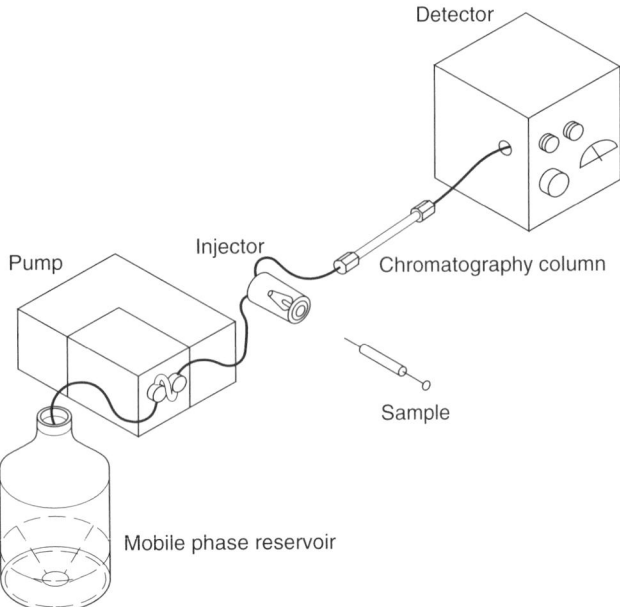

Figure 2. Schematic of a process liquid chromatography system. Courtesy of K. Hanaker.

over its surface, displaces anions or cations previously equilibrated on the stationary phase, ie, the protein sites exchange with the salt counterions associated with the ion-exchange stationary phase. A protein having a greater number and/or density of charged sites displaces or exchanges more ions and hence binds more strongly than a protein having a lower charge number or charge density.

After the column is loaded, proteins of similar size and shape are separated by differential desorption from the ion exchanger by using an increasing salt gradient of the mobile phase. The more weakly bound macromolecules elute first; the most tightly bound elute last, at the highest salt concentration. Figure 3 is an example of an anion-exchange separation.

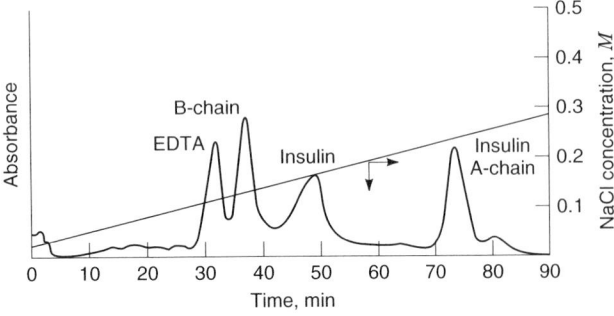

Figure 3. Anion-exchange separation of insulin, and insulin A- and B-chains, over diethylaminoethyl (DEAE) in a 10.9 × 200 mm column having a volume of 18.7 mL. Sample volume is 0.5 mL and protein concentration in 16.7 mM Tris buffer at pH 7.3 is 1 mg/mL for each component in the presence of EDTA. Eluent (also 16.7 mM Tris buffer, pH 7.3) flow rate is 1.27 mL/min, and protein detection is by uv absorbance at 280 nm. The straight line depicts the salt gradient. Courtesy of the American Chemical Society.

Size Exclusion (Gel-Permeation) Chromatography. Size-exclusion chromatography is often referred to as gel-permeation chromatography because the stationary phases are usually made up of soft spherical particles which resemble gels. Separation occurs by a molecular sieving effect (see MOLECULAR SILVERS; SIZE SEPERATION).

The apparatus utilized to carry out size-exclusion (gel-permeation) chromatography is analogous to that used for isocratic operating conditions. The column is packed with a gel-filtration stationary phase, selected according to the molecular weight of the protein of interest.

An example of a size-exclusion chromatogram is given in Figure 4, for both a bench-scale (23.5 mL column) separation. The stationary phase is Sepharose CL-6B, a cross-linked agarose with a nominal molecular weight range of $\sim 5000 - 2 \times 10^6$.

Buffer Exchange and Desalting. A primary use of size-exclusion chromatography (sec) is for removal of salt or buffer from the protein, ie, desalting and buffer exchange. The difference in molecular weights is large; salts generally have a mol wt<200, whereas mol wts of proteins are between 10,000 and 60,000.

Alternative methods of desalting and buffer exchange include continuous diafiltration, countercurrent dialysis (ccd), a membrane separation technique, and cross-flow filtration, which uses membranes (see MICROBIAL AND VIRAL FILTRATION). Buffer exchange, used to remove denaturing agents in order to induce refolding of proteins, to remove buffers between purification steps, or to remove buffers and other reagents from the final product, is usually carried out at later steps in a recovery sequence.

The use of sec is likely to continue to be a widely practiced technology in industry. Rapid size-exclusion columns for the purpose of buffer exchange have been developed which enable desalting to be achieved at linear velocities of 500–600 cm/h, significantly increasing

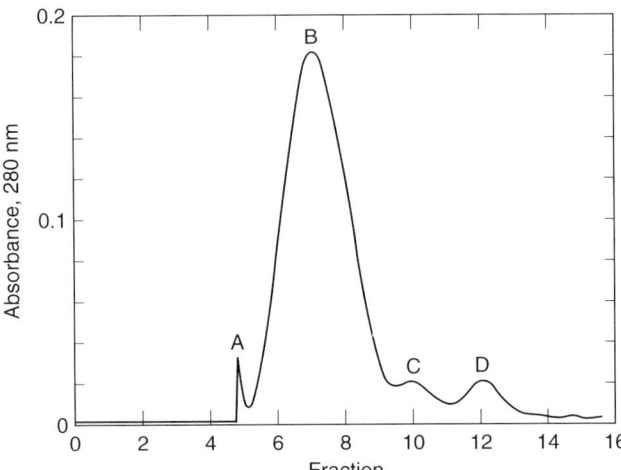

Figure 4. Chromatogram of size-exclusion separation of IgM (mol wt = 800,000) from albumin (69,000) where A–D correspond to IgM aggregates, IgM, monomer units, and albumin respectively, using FPLC Superose 6 in a 1 × 30 – cm long column. Courtesy of American Chemical Society.

throughput and reducing operating time and plant floor space. Further, sec using gel-filtration media on cellulosic-based material has a special niche for the partial and controlled separation of denaturing salts from recombinant proteins for purposes of refolding. The development of such rapid desalting techniques is important because of the larger volumes of proteins needing to be processed in industry.

Product Monitoring and Peptide Mapping. One purpose in monitoring a protein product is to detect the presence of a change in which as little as one amino acid has been chemically or biologically altered or replaced during the manufacturing process. Variant amino acid(s) in a protein may not affect protein retention during reversed-phase chromatography if the three-dimensional structure of the polypeptide shields the variant residue from the surface of the reversed-phase support. Reversed-phase chromatography discriminates between different molecules on the basis of hydrophobicity. Because large proteins may contain only small patches of hydrophobic residues, these patches may not correlate to the molecular modifications which a reversed-phase analytical method seeks to detect. The reversed-phase method must therefore be completely validated, and preferably combined with controlled chemical and/or proteolytic hydrolysis followed by chromatography or electrophoresis (see ELECTROSEPARATION) of the cleared protein to give a map of the resulting peptide fragments.

A peptide map is generated by cleaving a previously purified protein using chemicals or enzymes. Hydrolytic agents having known specificity are used to perform limited proteolysis followed by resolution and identification of all the peptide fragments formed. Identification of changes, and reconstruction of the protein's primary structure, is then possible. Reagents and enzymes which cleave specific bonds are discussed in the literature.

Reversed-Phase Process Chromatography. Polypeptides, peptides, antibiotics, alkaloids, and other low molecular weight compounds are amenable to process chromatography by reversed-phase methods. There are numerous examples of bioproducts which have been purified using reversed-phase chromatography. The manufacture of salmon calcitonin, a 32-residue peptide used for treatment of post-menopausal osteoporosis, hypercalcemia, and Paget's disease of the bone, includes reversed-phase chromatography. This peptide, commercially prepared on a kilogram scale by a solid-phase synthesis, is then purified by a multimodal purification train. Reversed-phase chromatography is the dominant technique used by Rhône-Poulenc Rorer.

Small Particle Silica Columns. Process-scale reversed-phase supports can have particle sizes as small as 5–25 μm. Unlike polymeric reversed-phase sorbents, these small-particle silica-based reversed-phase supports require high pressure equipment to be properly packed and operated. The introduction of axial compression columns has helped promote the use of high performance silica supports on a process scale. Resolution approaching

that of an analytical-scale separation can be achieved using these columns that can also be quickly packed.

Hydrophobic Interaction Chromatography. Hydrophobic interactions of solutes with a stationary phase result in their adsorption on neutral or mildly hydrophobic stationary phases. The solutes are adsorbed at a high salt concentration, and then desorbed in order of increasing surface hydrophobicity, in a decreasing kosmotrope gradient. This characteristic follows the order of the lyotropic series for the anions: phosphates > sulfates > acetates > chlorides > nitrates > thiocyanates. Anions which precipitate proteins less effectively than chloride (nitrates and thiocyanates) are chaotropes or water structure breakers, and have a randomizing effect on water's structure; the anions preceding chlorides, ie, phosphates, sulfates, and acetates, are polar kosmotropes or water structure makers. These promote precipitation of proteins. Kosmotropes also promote adsorption of proteins and other solutes onto a hydrophobic stationary phase. These kosmotropes have other beneficial characteristics which include increasing the thermal stability of enzymes, decreasing enzyme inactivation, protecting against proteolysis, increasing the association of protein subunits, and increasing the refolding rate of denatured proteins. Hence, utilization of hydrophobic interaction chromotography is attractive for purification of proteins where recovery of a purified protein in an active and stable conformation is desired.

Various types of proteins have been purified using hydrophobic interaction chromatography including alkaline phophatase, estrogen receptors, isolectins, strepavidin, calmodulin, epoxide hydrolase, proteoglycans, hemoglobins, and snake venom toxins. In the case of cobra venom toxins, the order of elution of the six cardiotoxins supports the hypothesis that the mechanism of action is related to hydrophobic interactions with the phospholipids in the membrane.

The recovery of recombinant chymosin from a yeast fermentation broth showed that large-scale hydrophobic interaction chromatography could produce an acceptable product in one step.

Affinity Chromatography. The concept of affinity chromatography, credited to the discovery of biospecific adsorption in 1910, was reintroduced as a means to purify enzymes in 1968. Substrates and substrate inhibitors diffuse into the active sites of enzymes irreversibly or reversibly binding there. Conversely, if the substrate or substrate inhibitor is immobilized through a covalent bond to a solid particle of stationary phase having large pores, the enzyme should be able to diffuse into the stationary phase and bind with the substrate or inhibitor. Because the substrate is small (mol wt<500) and the enzyme large (>15,000), the diffusion of the enzyme to its binding partner at a solid surface can be sterically hindered. The placement of the substrate at the end of an alkyl or glycol chain tethered to the stationary phase's surface reduces hindrance and forms the basis of affinity chromatography. This concept has also been applied to ion-exchange chromatography under the names of tentacle or fimbriated stationary phases.

The realization that enzymes could be selectively retained in a chromatography column packed with particles of immobilized substrates or substrate analogues led to experiments with other pairs of binding partners. Numerous applications of affinity chromatography developed, given the specific and reversible yet strong affinity of biological macromolecules for numerous specific ligands or effectors. These interactions have been exploited for purposes of highly selective, but often expensive protein purifications, recovery of messenger ribonucleic acid (mRNA) in some recombinant DNA applications, and study of mechanisms of protein binding with effector molecules.

Minimization of Nonspecific Binding. The purpose of affinity chromatography is the highly selective adsorption and subsequent recovery of the target biomolecule. Loss of specificity occurs when macromolecules, other than the targeted materials, adsorb onto the stationary phase owing to hydrophobic or ionic interactions.

The ideal matrices for anchoring binding ligands are nonionic, hydrophilic, chemically stable, and physically robust. The most popular matrices are polysaccharide based, principly owing to their hydrophilic character and history of use as size-exclusion or gel-permeation gels, although glass beads, polyacrylamide gels, cross-linked dextrans (Sephadex), and agarose synthesized into a bead form have all been used.

Multistep Processes. An excellent synopsis and industrial viewpoint of affinity chromatography and its fit with other bioseparations unit operations is available. Ligands range from the low molecular weight components, eg, arginine and benzamidine, which both bind trypsin-like proteases, triazine dyes, and metal chelates; to high molecular weight ligands, eg, protein A, immunoglobins, and monoclonal antibodies. The blood factor VIII, purified by monoclonal affinity techniques, was approved by the U.S. FDA in 1988. Limitations of affinity chromatography as an industrial separation technique can be due to leaching of bound ligands from the column into the product at ppm levels, nonspecific interactions resulting in contamination of the target molecule, and failure of the affinity ligand to differentiate all variant forms of a protein or polypeptide. Because many antibody preparations cannot differentiate between minor structural changes in proteins, affinity chromatography must be followed by other separation steps, and does not provide a one-step purification.

Receptor Affinity Chromatography. Receptor affinity chromatography is a selective form of immunoaffinity chromatography which is based on antigen-antibody interactions. Protein or polypeptide ligands used in preparing receptor affinity supports are themselves products of fermentation of recombinant microorganisms and are subjected to a separate sequence of purification steps, prior to being reacted with a functionalized stationary phase to form the affinity support. The resulting affinity chromatography columns are expensive when viewed on the basis of cost of support/unit volume of stationary phase. The cost/benefit ratio would still be attractive because process-scale columns can be small (volumes on the order of 1–10 L). Moreover, as with other types of affinity chromatography, purification of dilute but highly active protein is possible.

C. A. Bisbee, *GEN* **13**(14), 8–9 (1993).

S. E. Builder, R. van Reis, N. Paoni, and J. Ogez, Proc. 8th Int. Biotechnol. Symp., Paris, July, 1989.

Committee on Bioprocess Engineering, National Research Council, *Putting Biotechnology to Work: Bioprocess Engineering*, National Academy of Sciences, Washington, D.C., 1992, 2–22.

S. M. Wheelwright, *Protein Purification: Design and Scale-up of Downstream Processing*, Hanser Publishers, Munich, Germany, 1991, 1–9, 61, 213–217.

R. C. Willson and M. R. Ladisch, in M. R. Ladisch, R. C. Willson, C-C. Painton, and S. E. Builder, eds., *ACS Symp. Ser.* **427**, 1–13 (1990).

MICHAEL R. LADISCH
Purdue University

BISMUTH AND BISMUTH ALLOYS

Bismuth (Bi) is a very brittle, silvery metal having a high metallic luster. The element is found in the periodic table of the elements under Group 15 (Va). The atomic number is 83, and the atomic weight is 208.98. Bismuth is next to lead on the periodic table and exhibits many characteristics of lead except that bismuth is considered nontoxic despite its heavy-metal status.

OCCURRENCE

Bismuth is a minor metal, that is, it is a mining by-product and is therefore not mined for its own intrinsic value. Usually, bismuth is a by-product of lead or copper ores.

The concentration of bismuth in the earth's crust has not been clearly determined. Estimates range from 0.008 to 0.1 ppm.

PROPERTIES

Bismuth has unique properties that make it a valuable metal for certain industrial applications. These properties are characterized by a low melting point, a high density, and expansion on solidification. Bismuth is one of only two metals that expand on solidification. The solid metal floats on molten metal just as ice floats on water. Bismuth is a poor electrical conductor and is the most diamagnetic of the metals. The thermal conductivity of bismuth is lower than that of most other metals. Table 1 lists many of the properties of bismuth.

PRODUCTION

Bismuth is a mining by-product, primarily from the mining of ores such as copper and lead. Tungsten, tin, and molybdenum ores also may contain bismuth. Significant quantities of bismuth are mined in Australia, Bolivia, Canada, China, Japan, Mexico, Peru, and the United States.

MANUFACTURE AND PROCESSING

Four basic forms of bismuth are readily available commercially: ingot, needle, pellet, and powder. Bismuth ingots range from 4.5 to 20 kg each, depending on the producer. Ingots are used mostly in metallurgical applications and in making fusible alloys. Bismuth needle is typically 0.16 cm (0.0625 in.) in diameter by nominally 2.54 cm (1 in.) in length. Needle is primarily used in the production

Table 1. Physical Properties of Bismuth

Property	Value
boiling point °C	1,560
Bi–Bi bond length at 25°C, nm	0.309
crystal ionic radius, nm	
$\quad$ Bi$^+$	0.098
$\quad$ Bi^{3+}	0.096
$\quad$ Bi^{5+}	0.074
crystal structure	Rhombohedral
density, kg/m^3	
$\quad$ 20°C	9800
$\quad$ 271°C^a	9740
$\quad$ 271°C^b	10,070
$\quad$ 600°C	9660
electrical resistivity, Ω·cm	
$\quad$ 0°C	106×10^{-6}
$\quad$ 20°C	120×10^{-6}
expansion on freezing, % by volume	3.3
hardness, Mohs scale	2.5
magnetic susceptibility	
$\quad$ solid	-280×10^{-13}
$\quad$ liquid	-10.5×10^{-13}
melting point, °C	271.3
vapor pressure, kPa^c	
$\quad$ 400°C	1.013×10^{-4}
$\quad$ 600°C	1.013×10^{-1}
$\quad$ 880°C	1.013×10^{2}
$\quad$ 1420°C	1.013×10^{5}
viscosity, mPa·s	
$\quad$ 285°C	1.610
$\quad$ 304°C	1.662
$\quad$ 365°C	1.460
$\quad$ 451°C	1.280
$\quad$ 600°C	0.998

^{a}Solid.
^{b}Liquid.
^{c}To convert kPa to psi, divide by 6.895×10^3.

of bismuth compounds for pharmaceutical and catalyst applications. The high surface area of the needle makes it easy to dissolve in various acids. Bismuth powder is produced in varying mesh sizes for the electronics industry.

Recovery of Bismuth from Tin Concentrates

Bismuth is leached from roasted tin concentrates and other bismuth-bearing materials by means of hydrochloric acid.

The Sperry process for making white lead in an electrolytic cell recovers bismuth as a by-product in the anode slimes.

Refining

The alloy of bismuth and lead from the separation procedures is treated with molten caustic soda to remove traces of such acidic elements as arsenic and tellurium. It is then subjected to the Parkes desilverization process to remove the silver and gold present. This process is also used to remove these elements from lead.

The desilverized alloy now contains bismuth as well as lead and zinc. Removed of the lead and zinc is facilitated by the fact that the formation of zinc and lead chlorides precedes the formation of bismuth chloride ($BiCl_3$), when the alloy is treated at 500°C with chlorine gas. Zinc chloride ($ZnCl_2$), forms first, and after its removal lead chloride ($PbCl_2$) forms preferentially.

ECONOMIC ASPECTS

The supply of bismuth is dependent on the supply of the associated metals with which it is mined. To increase the supply of bismuth, it would be necessary to mine more copper or lead to get more bismuth from the ore, which would not make economic sense.

GRADES AND SPECIFICATIONS

The purity of bismuth ranges from 99.99 to 99.999%, depending on the use. Ingot for metallurgical and fusible alloy applications is commercially available at 99.99% pure. Needle for catalyst and other chemical applications is usually supplied as 99.99% pure. Pharmaceuticals require bismuth to be minimum 99.997% pure. This is typically supplied as needle. Needle of 99.999% purity is useful for the manufacture of high-purity bismuth compounds for medical, electronic and ceramic applications.

ANALYTICAL METHODS

Many of the methods available for determination of bismuth are not very selective; thus it is necessary to separate the bismuth from interfering substances. Titration of bismuth with EDTA (ethylenediamine tetraacetic acid) has been found to be one of the best general methods for determining both macroscale and semimicroscale quantities of bismuth.

Table 2. Alloy Compositions and Uses

Melting point, °C	Bismuth	Lead	Tin	Cadmium	Indium	Uses[a]
47	44.7	22.6	8.3	5.3	19.1	FSD, LB
70	50	26.7	13.3	10	0	FSD, LB, RS, T, W
95	52.5	32	15.5	0	0	FSD, RS
138	58	0	42	0	0	FC, FSD, SMF, W
138/170	40	0	60	0	0	FSD, IC

[a]Abbreviations: FC = fusible core; FSD = fusible safety devices; IC = investment casting; LB = lens blocking; RS = radiation shielding; SMF = sheetmetal forming; T = tube bending; W = work holding.

For the determination of trace amounts of bismuth, atomic absorption spectrometry is probably the most sensitive method. The low neutron cross section of bismuth virtually rules out any determination of bismuth based on neutron absorption or neutron activation.

ENVIRONMENTAL CONCERNS

Bismuth is considered both nontoxic and noncarcinogenic despite the fact that it is a heavy metal. It exhibits many of the characteristics of lead, which is next to it on the periodic table of the elements, yet it is not poisonous as lead.

Recycling and Disposal

Bismuth has been involved in recycling since long before recycling became mandated by the Environmental Protection Agency. Because bismuth alloys melt at relatively low temperatures, it is relatively easy to reclaim and reprocess them. Bismuth-containing materials should not be disposed of because, of the ease in refining and the inherent value of the bismuth contained in the material.

Health and Safety Factors

No industrial poisoning from bismuth has been reported. The use of bismuth compounds in the medical field for several hundred years indicates the safety of the material. However, precautions should be taken against the careless handling of bismuth and its compounds; ingestion and inhalation of dusts and fumes should be avoided.

USES

The uses of bismuth can be broken down into three primary categories: chemical, metallurgical additive, and fusible alloy.

The chemical category can be broken down into pharmaceuticals, cosmetics, catalysts, industrial pigments, and electronics, and the metallurgical additive category into steel, aluminum, and castiron additives. The fusible alloy category is divisible into more than a dozen subcategories dependent on a specific application.

BISMUTH ALLOYS

Because bismuth expands on solidification and because it alloys with certain other metals to give low-melting alloys, bismuth is particularly well suited for a number of uses. Alloys of bismuth can be made that expand, shrink, or remain dimensionally stable on solidification. So, bismuth alloys have lent themselves to a wide range of industrial applications. Composition and uses are summarized in Table 2.

Among the uses of bismuth alloys are to align and set punches in die plates; for radiation shielding in the medical industry; for electromagentic and radiofrequency shielding; tube bending; securing optical lenses for grinding; tempering steel; proofcasting of dyes and molds; work holding and work supporting; in sheetmetal forming dyes; and as an alternative to lead shot in hunting.

P. Koslosky, Jr., in *The Bulletin of the Bismuth Institute* Vol. 69, Bismuth Institute, Grimbergen, Belgium, 1996, pp. 1–2.

Mineral Industry Surveys, Bismuth Annual Review, U.S. Bureau of Mines, Washington, D.C., 1990.

M. D. Fickling, Review of Bismuth in 1996, *Mining Journal* Annual Review, 1997.

N. I. Sax, *Dangerous Properties of Industrial Materials*, 11th ed., Wiley, Hoboken, N.J., 2004.

MARK J. CHAGNON
Atlantic Metals & Alloys

BISMUTH COMPOUNDS

Bismuth is the fifth member of the nitrogen family of elements and, like its congeners, possesses five electrons in its outermost shell, $6s^2 6p^3$. In many compounds, the bismuth atom utilizes only the three $6p$ electrons in bond formation and retains the two $6s$ electrons as an inert pair. Compounds are also known where bismuth is bonded to four, five, or six other atoms. Many bismuth compounds do not have simple molecular structures and exist in the solid state as polymeric chains or sheets.

Technical information concerning bismuth and its compounds is distributed periodically by the Bismuth Institute, a nonprofit organization incorporated in La Paz, Bolivia, that has an information center in Brussels.

Information on the production and consumption of bismuth is available from the U.S. Geological Surveys *Minerals Yearbook* on the World Wide Web.

World production of bismuth in 2000 stood at ~4000 tons/year. Major applications of bismuth compounds include pharmaceuticals, additives to ceramics, plastics, catalysts for use in industrial organic chemistry, pigments in cosmetics, and paints. Metallic bismuth is widely used in metallurgy (steel additives, fusible alloys etc). Ecological considerations favor applications of bismuth because it is considered nontoxic and noncarcinogenic, notwithstanding its heavy-metal status.

INORGANIC COMPOUNDS OF BISMUTH

Inorganic compounds of bismuth include bismuthine, bismuthides, bismuth halides, bismuth oxide halides, bismuth oxides and bismuthates, higher oxides of bismuth and related compounds, sulfides are related compounds, and bismuth salts. Properties of important inorganic bismuth compounds are given in Table 1.

ORGANOBISMUTH COMPOUNDS

In a manner similar to phosphorus, arsenic, and antimony, the bismuth atom can be either tri- or pentacovalent. However, organobismuth compounds are less stable thermally than the corresponding phosphorus, arsenic, or antimony compounds, and there are fewer types of organobismuth compounds. The chemistry of organobismuth compounds has been described in several publications. The use of organobismuth compounds, as well as organoantimony ones, in organic synthesis has been exhaustively reviewed.

Primary and Secondary Bismuthines

Unstable methylbismuthine, CH_3BiH_2, and dimethylbismuthine, $(CH_3)_2BiH$ are prepared by the lithium aluminum hydride reduction of methyldichlorobismuthine, CH_3BiCl_2 and dimethylchlorobismuthine, respectively, in a nitrogen atmosphere at $-150°C$. Tertiary bismuthines appear to have a number of uses in synthetic organic chemistry, eg, they promote the formation of 1,1,2-trisubstituted cyclopropanes by the interaction of electron-deficient olefins and dialkyl dibromomalonates. They have also been employed for the preparation of thin films of superconducting bismuth strontium calcium copper oxide, as cocatalysts for the polymerization of alkynes, as inhibitors of the flammability of epoxy resins, and for a number of other industrial purposes.

Halobismuthines, Dihalobismuthines, and Related Compounds

Chloro-, dichloro-, bromo-, and dibromobismuthines are prepared by the reaction of a tertiary bismuthine and bismuth trichloride or tribromide. Iodo- and diiodobismuthines are easily obtained by the reaction of the corresponding chloro-, dichloro-, bromo-, or dibromobismuthine with sodium or potassium iodide.

Dibismuthines, Dibismuthenes, and Cyclo-Bismuthines

About a dozen dibismuthines are known. These compounds can be obtained by the reaction of a sodium dialkyl- or diarylbismuthide and a 1,2-dihaloethane.

Several dibismuthines have also been obtained by the addition of the stoichiometric amount of sodium to a solution of a halobismuthine in liquid ammonia.

The best method for the synthesis of tetraphenyldibismuthine, $(C_6H_5)_4Bi_2$, involves the reduction of diphenyliodobismuthine, $(C_6H_5)_2BiI$ using bis(cyclopentadienyl)cobalt(II), $(C_5H_5)_2Co$. dibismuthines tend to be thermally unstable and very senstive to oxidation.

Bismin and Its Derivatives

Bismin (bismabenzene), C_5H_5Bi, the bismuth analogue of pyridine, has never been isolated, but it can be formed in solution by the dehydrohalogenation of 1-chloro-1,4-dihydrobismin, C_5H_6BiCl, using 1,8-diazabicyclo[5.4.0]undec-7-ene (DBU) at low temperatures.

Diarylbismuthinic Acids and Their Esters

Although organobismuth (V) compounds containing three, four, or five Bi—C bonds have been well known, no compounds containing two such bonds had been prepared until a number of methyl diarylbismuthinates (diarylmethoxybismuth oxides) were reported in 1988.

Because organobismuth (V) compounds have found considerable use as oxidizing agents, the oxidizing ability of methyl di-1-napthylbismuthinate, $C_{21}H_{17}BiO_2$, was investigated.

Trialkyl- and Triarylbismuth Dihalides and Related Compounds

The triarylbismuth dihalides constitute an important class of organobismuth compounds. Only very few trialkylbismuth dihalides are kown. Trimethylbismuth dichloride, prepared by oxidative chlorination of Me_3Bi with SO_2Cl_2 at $-78°C$ is unstable at room temperature.

Triarylbismuthines are readily chlorinated or brominated to the corresponding dichlorides or dibromides, using chlorine or bromine. Triarylbismuth difluorides have been prepared from the dichlorides by metathesis with potassium fluoride or by direct fluorination of triarylbismuthines with fluorine diluted with argon. No triarylbismuth diiodides are known. However, the two compounds have been prepared from triphenylbismuthine and iodine azide or iodine isocyanate. The triarylbismuth dihalides are stable crystalline solids with high melting points. In recent years organobismuth(V) compounds have found increasing use as reagents in organic synthesis. Thus they have been used for the oxidation of primary and secondary alcohols to the corresponding aldehydes and ketones for the oxidative cleavage of vicinal glycols and for the O-, C-, and N-arylation of a wide variety of organic compounds. Because most of these reactions occur under relatively mild conditions, organobismuth(V) reagents have proved to be of particular value when the substrates are sensitive natural products.

Table 1. Physical Properties of Bismuth Compounds

Bismuth compound	Formula	Mp, °C	Bp, °C	$\Delta H^\circ_{f,298}$, kJ/mol[a]	S°_{298}, J/mol·K[a]	ΔH°_{fusion}, kJ/mol[a]	ΔS°_{fusion}, J/mol·K[a]	$\Delta H^\circ_{subl,298}$, kJ/mol[a]	$\Delta S^\circ_{subl,298}$, J/mol·K[a]	Bi–X bond energy, kJ/mol[a]
bismuth trifluoride	BiF_3	649[b]	900 ± 10	−900 ± 13	123 ± 4	21.6 ± 0.6	23.4 ± 0.8	201 ± 3	195 ± 3	381
bismuth pentafluoride	BiF_5	151	230							
bismuth trichloride	$BiCl_3$	233.5	44.7	−379	174 ± 6	23.9		114 ± 1	183 ± 2	279
bismuth monochloride	$BiCl$			−131	94.5					300 ± 4
bismuth oxychloride	$BiOCl$			−367	120					
bismuth tribromide	$BiBr_3$	219	441	−276 ± 2	190 ± 1	21.8		115 ± 1	182 ± 1	233.0 ± 1.4
bismuth triiodide	BiI_3	408.6	542[c]	−151 ± 4	224.8 ± 0.8	39.1 ± 0.3	57.3 ± 0.4	134.3 ± 0.5	183.4 ± 0.8	181 ± 5
bismuth trioxide	Bi_2O_3[d]	824		−574	151.5					
bismuth trisulfide	Bi_2S_3	850		−143	200					
bismuth tritelluride	Bi_2Te_3			−77.4	260.9					

[a] To convert J to cal, divide by 4.184.
[b] The mp frequently cited is 120 C higher than this and is, apparently, for material contaminated with oxyfluoride.
[c] The normal bp has been extrapolated from vapor-pressure data.
[d] Monoclinic.

290

In addition to use in organic synthesis, triarylbismuth dihalides and related compounds have found limited industrial use. A patent has been issued for the use of such compounds as antifungal agents on plastics or fibrous material.

Triphenylbismuth dichloride, $(C_6H_5)_3BiCl_2$ is active against bean rust. Triarylbismuth dihalides have been used as catalysts for the carbonation of epoxides to form cyclic carbonates.

Quaternary Bismuth Compounds

It was not until 1952 that tetraphenylbismuth bromide, $(C_6H_5)_4BiBr$ was obtained from pentaphenylbismuth, $(C_6H_5)_5Bi$ and 1 M equivalent of bromine at $-70°C$. In a similar manner, tetraphenylbismuth chloride, $(C_6H_5)_4BiCl$ and tetraphenylbismuthonium tetrafluoroborate, $(C_6H_5)_4$ $Bi[BF_4]$ are obtained from pentaphenylbismuth and hydrogen chloride or hydrogen tetrafluoroborate, respectively.

A number of other tetraarylbismuth compounds Ar_4BiY, where Y is a group, such as NO_3^-, ClO_4^-, OCN^-, N_3^-, etc, have been prepared from the chloride by metathesis.

Many quaternary bismuth compounds are unstable. They have not found extensive use in industry or in organic synthesis.

Bismuthonium Ylides

Prior to 1988 the only bismuthonium ylides known were (1) and (2). Compound (1) is an unstable blue solid that cannot be obtained in a pure state; structure (2), however, is stable. Compound (2) was obtained from triphenylbismuth carbonate and dimedone. More recently a number of bismuthonium ylides, eg, (3), have been prepared and their reactions studied.

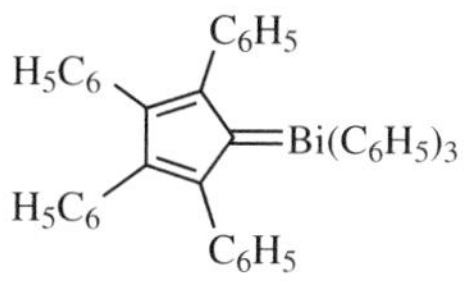

(1) 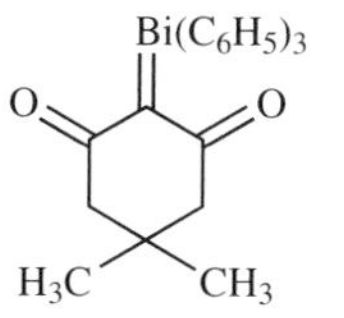   (2) 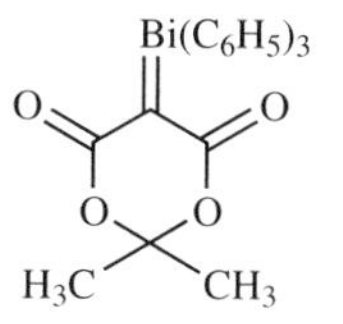   (3)

Qinquenary Bismuth Compounds

A number of pentaarylbismuth compounds and pentamethylbismuth are known. Pentaphenylbismuth, $C_{30}H_{25}Bi$ was first prepared by means of the following reaction:

$$(C_6H_5)_3BiCl_2 + 2\ C_6H_5Li \longrightarrow (C_6H_5)_5Bi + 2\ LiCl$$

It can also be prepared by the reaction of phenyllithium with tetraphenylbismuth chloride or the N-triphenylbismuth derivative of 4-toluenesulfonamide.

Pentaphenylbismuth is a violet-colored, crystalline compound that decomposes spontaneously after standing for several days in a dry nitrogen atmosphere. Pentaphenylbismuth has been studied as a reagent in organic synthesis where it can act either as an oxidizing or an arylating agent. Thus it can be used for the oxidation of primary or secondary alcohols to aldehydes or ketones, respectively.

BISMUTH COMPOUNDS USED IN MEDICINE

During the 1920s, it was shown that bismuth compounds were comparable in efficacy to the best antisyphilic drugs then available. During the next quarter of a century, bismuth compounds became widely used as adjuncts to the arsenical therapy of syphilis. Now however, antibiotics, especially penicillin, have made both arsenic and bismuth compounds completely obsolete for the treatment of syphilis. Bismuth compounds were employed for the treatment of amoebic dysentery, certain skin diseases, and several spirochetal diseases besides syphilis, but these substances are now seldom considered the drugs of choice. Various insoluble preparations of bismuth, especially the subcarbonate, subnitrate, subgallate, subcitrate, and subsalicylate, are employed for the treatment of gastrointestinal disorders. Eradication of *H. pyloris* is achieved by means of a combination therapy that uses bismuth compounds and other substances.

Bismuth subsalicylate, Pepto-Bismol, is a basic salt of varying composition, corresponding approximately to o-$HOC_6H_4CO_2(BiO)$. It does appear to be effective for the relief of mild diarrhea and for the prevention of travelers' diarrhea in the symptomatic treatment of isosporiasis, a disease caused by the intracellular parasite *Isospora belli*. Bismuth subcarbonate (basic bismuth carbonate) is a white or pale yellow powder that is prepared by interaction of bismuth nitrate and a water-soluble carbonate. It has been widely used as an antacid.

De-Nol tripotassium dicitratobismuthate (bismuth subcitrate), is a buffered aqueous suspension of a poorly defined, water-insoluble bismuth compound. It is said to be very effective for the treatment of gastric and duodenal ulcers.

Bismuth subnitrate (basic bismuth nitrate), can be prepared by the partial hydrolysis of the normal nitrate with boiling water. It has been used as an antacid and in combination with iodoform as a wound dressing.

Bismuth subgallate (basic bismuth gallate), Dermatol, is a bright yellow powder that can be prepared by the interaction of bismuth nitrate and gallic acid in an acetic acid medium. It has been employed as a dusting powder in some skin disorders and as an ingredient of suppositories for the treatment of hemorrhoids. It has been taken orally for many years by colostomy patients in order to control fecal odors, but the drug may cause serious neurological problems.

L. L. Brunton, in A. G. Gilman, T. W. Rall, A. S. Niews, and P. Taylor, eds., *Goodman and Gilman's The Pharmacological Basis of Therapeutics*, 8th ed., Pergamon Press, New York, 1990, pp. 910, 911.

G. O. Doak and L. D. Freedman, *Organometallic Compounds of Arsenic, Antimony, and Bismuth*, Wiley-Interscience, New York, 1970.

N. C. Norman, ed. in *Chemistry of Arsenic, Antimony and Bismuth*, Blackie Academic and Professional, Thomson Science; London, 1998.

S. Patai, ed., *The Chemistry of Organic Arsenic, Antimony and Bismuth Compounds*, John Wiley & Sons, Inc., Chichester, UK, 1994.

HANS JOACHIM BREUNIG
University of Bremen

BLEACHING AGENTS

A bleaching agent is a material that lightens or whitens a substrate through chemical reaction. The bleaching reactions usually involve oxidative or reductive processes that degrade color systems. These processes may involve the destruction or modification of chromophoric groups in the substrate as well as the degradation of color bodies into smaller, more soluble units that are more easily removed in the bleaching process. The most common bleaching agents generally fall into two categories: chlorine and its related compounds (such as sodium hypochlorite) and the peroxygen bleaching agents such as hydrogen peroxide and sodium perborate. Reducing bleaches represent another category. Bleaching agents are used for textile, paper, and pulp bleaching as well as for home laundering.

CHLORINE-CONTAINING BLEACHING AGENTS

Chlorine-containing bleaching agents are the most cost-effective bleaching agents known. They are also effective disinfectants, and water disinfection is often the largest use of many chlorine-containing bleaching agents. They may be divided into four classes: chlorine, hypochlorites, *N*-chloro compounds, and chlorine dioxide.

The first three classes are called available chlorine compounds and are related to chlorine by the equilibria in equations 1–4. These equilibria are rapidly established in aqueous solution, but the dissolution of some hypochlorite salts and *N*-chloro compounds can be quite slow.

$$Cl_2 \text{ (gas)} \rightleftharpoons Cl_2 \text{ (aq)} \qquad (1)$$

$$Cl_2 \text{ (aqueous)} + H_2O \rightleftharpoons HOCl + H^+ + Cl^- \qquad (2)$$

$$HOCl \rightleftharpoons H^+ + OCl^- \qquad (3)$$

$$RR'NCl + H_2O \rightleftharpoons HOCl + RR'NH \qquad (4)$$

The total concentration or amount of chlorine-based oxidants is often expressed as available chorine or less frequently as active chlorine. Available chlorine is the equivalent concentration or amount of Cl_2 needed to make the oxidant according to equations 1–4. Active chlorine is the equivalent concentration or amount of Cl atoms that can accept two electrons. This is a convention, not a description of the reaction mechanism of the oxidant. Because Cl_2 accepts only two electrons as do HOCl and monochloramines, it has only one active Cl atom according to the definition. Thus the active chlorine is always one-half of the available chlorine. The available chlorine is usually measured by iodometric titration. The weight of available chlorine can also be calculated by equation 5,

$$\text{weight available chlorine} = 70.9 \times \text{moles of oxidant} \\ \times \frac{\text{number active Cl atoms}}{\text{molecule}}$$

$$(5)$$

where 70.9 represents the mol wt of Cl_2 and moles of oxidant can be represented wt oxidant/mol wt of oxidant.

In solutions, the concentration of available chlorine in the form of hypochlorite or hypochlorous acid is called free-available chlorine. Commercially important solid available chlorine bleaches are usually more stable than concentrated hypochlorite solutions. They decompose very slowly in sealed containers. But most of them decompose quickly as they absorb moisture from air or from other ingredients in a formulation. This may release hypochlorite that destroys other ingredients as well.

Chlorine

Except to bleach wood pulp and flour, chlorine itself is rarely used as a bleaching agent.

Hypochlorites

The principal form of hypochlorite produced is sodium hypochlorite, NaOCl. Other hypocholorites include calcium hypochlorite, bleach liquor, bleaching powder and topical bleach, dibasic magnesium hypochlorite, lithium hypochlorite, chlorinated trisodium phosphate, and hypochlorous acid.

N-Chloro Compounds

Chlorinated Isocyanurates. The principal solid chlorine bleaching agents are the chlorinated isocyanurates. The one used most often for bleaching applications is sodium dichloroisocyanurate dihydrate with 56% available chlorine.

Other *N*-choloro compounds include halogenated hydantoins, sodium *N*-chlorobenzenesulfonamide (chloramine B), sodium *N*-chloro-*p*-toluenesulfonamide (chloramine T), *N*-chlorosuccinimide, and trichloromelamine.

Chlorine Dioxide

Chlorine dioxide, ClO_2, is a gas that is more toxic than chlorine. Large amounts for pulp bleaching are made by several processes in which sodium chlorate is reduced with chloride, methanol, or sulfur dioxide in highly acidic solutions by complex reactions. For most other purposes, chlorine dioxide is made from sodium chlorite.

PEROXYGEN COMPOUNDS

Peroxygen compounds contain the peroxide linkage $(-O-O-)$ in which one of the oxygen atoms is active.

Hydrogen Peroxide

Hydrogen peroxide is one of the most common bleaching agents. It is the primary bleaching agent in the textile industry, and is also used in pulp, paper, and home laundry applications. In textile bleaching, hydrogen peroxide is the most common bleaching agent for protein fibers and is also used extensively for cellulosic fibers.

Solid Peroxygen Compounds

Hydrogen peroxide reacts with many compounds, such as borates, carbonates, pyrophosphates, sulfates, silicates, and a variety of organic carboxylic acids, esters, and anhydrides to give peroxy compounds or peroxyhydrates. A number of these compounds are stable solids that hydrolyze readily to give hydrogen peroxide in solution.

Compounds include perborates, sodium carbonate peroxyhydrate, and peroxymonosulfate.

Peracids

Peracids are compounds containing the functional group $-OOH$, derived from an organic or inorganic acid functionality. Typical structures include $CH_3C(O)OOH$, derived from acetic acid and $HOS(O)_2OOH$ (peroxymonosulfuric acid), derived from sulfuric acid. Peracids have superior cold water bleaching capability versus hydrogen peroxide because of the greater electrophilicity of the peracid peroxygen moiety. Lower wash temperatures and phosphate reductions or bans in detergent systems account for the recent utilization and vast literature of peracids in textile bleaching.

Peracids can be introduced into the bleaching system by two methods. They can be manufactured separately and delivered to the bleaching bath with the other components or as a separate product. Peracids can also be formed *in situ*, utilizing the perhydrolysis reaction shown in equation 6.

$$\underset{\text{R–C–Z}}{\overset{\text{O}}{\parallel}} + {}^-OOH \longrightarrow RCOOH + Z^- \qquad (6)$$

Peracid Precursor Systems

Compounds that can form peracids by perhydrolysis are almost exclusively amide, imides, esters, or anhydrides. Tetraacetylethylenediamine is utilized in >50% of western European detergents.

Nonanoyloxybenzene sulfonate (NOBS) is used in detergent products in the United States and Japan.

Preformed Peracids

Peracids can be generated at a manufacturing site and directly incorporated into formulations without the need for *in situ* generation. Two primary methods are utilized for peracid manufacture. The first method uses the equilibrium shown in equation 7 to generate the peracid from the parent acid.

$$\underset{\text{RCOH}}{\overset{\text{O}}{\parallel}} + H_2O_2 \rightleftharpoons \underset{\text{RCOOH}}{\overset{\text{O}}{\parallel}} + H_2O \qquad (7)$$

The equilibrium is shifted by removal of the water or removal of the peracid by precipitation. Peracids can also be generated by treatment of an anhydride with hydrogen peroxide to generate the peracid and a carboxylic acid.

$$\underset{\text{RCOCR}}{\overset{\text{O} \quad \text{O}}{\parallel \quad \parallel}} + H_2O_2 \longrightarrow \underset{\text{RCOOH}}{\overset{\text{O}}{\parallel}} + \underset{\text{RCOH}}{\overset{\text{O}}{\parallel}} \qquad (8)$$

The latter method typically requires less severe conditions than the former because of the labile nature of the organic anhydride. Both of these reactions can result in explosions and significant precautions should be taken prior to any attempted synthesis of a peracid.

Peracid Decomposition

Peracids, whether preformed or formed *in situ* via the perhydrolysis reaction, are susceptible to decomposition in an aqueous bleaching bath. The decomposition is caused by the occurrence of one of four reactions. The peracid can decompose as a result of oxidation of the bleachable material. Transition metals present even at extremely low concentration in the bath from the incoming water can decompose the peracid catalytically. To minimize this effect, metal-sequestering agents have been proposed to prevent the degradation of the peracid in solution. Peracids can also hydrolyze to the parent acid and hydrogen peroxide because of the large excess of water present in the aqueous bleaching bath. This is generally a kinetically slow process. A final decomposition mechanism involves the reaction of 2 mol of peracid generating 2 mol of parent acid and 1 mol of oxygen.

REDUCING BLEACHES

The reducing agents generally used in bleaching include sulfur dioxide, sulfurous acid, bisulfites, sulfites, hydrosulfites (dithionites), sodium sulfoxylate formaldehyde, and sodium borohydride. These materials are used mainly in pulp and textile bleaching.

ENZYMES FOR BLEACHING IN THE TEXTILE INDUSTRY

Bleaching of Textile Substrates

Enzymes are used for bleaching purposes in the textile industry for decolorization of dyehouse effluents, bleaching of released dyestuff, and inhibiting dye transfer. Enzymatic systems suitable for bleaching of textile materials are further classified as enzymes for production of the bleaching agent and enzymes acting directly on the textile substrate.

Biogeneration of Hydrogen Peroxide for Bleaching

The most common bleaching agent, hydrogen peroxide, may be produced enzymatically by glucose oxidase, which catalyzes the conversion of β-D-glucose in aqueous solutions in the presence of oxygen as electron acceptor. Similar enzymatic systems are used as constituents in detergent formulations to generate controlled rates of hydrogen peroxide.

Hydrogen Peroxide Generation with Immobilized Glucose Oxidase

One reason that a bleaching process using enzymatically produced peroxide is not yet commercial is that the enzymes are still quite expensive. Immobilization of the enzymes provides long-term application of the enzyme at lower process cost. Immobilization accounts for easy recovery and recycling, reduced enzyme dosage, and continuous operations.

Bleaching of Textiles Chloroperoxidases

The haloperoxidases form a class of heme-containing enzymes capable of oxidizing halides in the presence of hydrogen peroxide to the corresponding hypohalous acid. If an appropriate nucleophile is present, a reaction occurs with formation of hypohalous acid, whereby bleaching takes place at lower than the conventional bleaching temperature and pH. This type of bleaching reduces damage to fiber and avoids the use of bleaching auxiliaries. A recent patent reported the use of haloperoxidases for textile treatment.

Chloroperoxidases application alone cannot replace chemical bleaching. The enzymatic pretreatment might be an alternative to the repeated conventional bleaching process or a way to reduce significantly the peroxide dosage in subsequent chemical bleaching. However, this enzymatic system might face problems with environmental restrictions regarding the absorbable organic halogens by-products in industrial effluents.

Bleaching of Textiles with Laccases

Laccases are multi-copper enzymes that catalyze the oxidation of a wide range of inorganic substances using oxygen as an electron acceptor. The oxidation of a reducing substrate typically involves formation of a free (cation) radical after the transfer of a single electron to laccase. Laccases have found various biotechnological and environmental applications; eg, removal of toxic compounds from polluted effluents through oxidative enzymatic coupling of the contaminants leading to insoluble complex structures, or as biosensors for phenols. Laccases have been extensively used in bleaching of craft pulp in delignification and demethylation. Capability of laccases to act on chromophore compounds suggested their application in industrial decolorization processes. However, these enzymes have not yet been used for bleaching of textiles, despite promising experimental results. Apparently, laccase pretreatment alone does not improve the whiteness of the textile material; moreover, it generates color. However, after hydrogen peroxide bleaching, the whiteness of the enzymatically pretreated fabrics reaches whiteness index enhancement comparable with the whiteness achieved in two consecutive peroxide bleaching runs. Laccase produces colored substances when suitable substrate is present.

Enzymes for Treatment of Textile Bleaching Effluents

The washing process after bleaching consumes large amounts of water, since any residual hydrogen peroxide has to be removed to avoid problems in subsequent dyeing processes. Minor modifications of the dye molecule can result in color loss. The demands to reduce water consumption by reducing or eliminating the washing cycle after bleaching while ensuring good reproducibility of dyeing can be met by application of catalases. Catalase is widely distributed enzyme in nature and well known for its ability to catalyze the conversion of hydrogen peroxide to water and gaseous oxygen. It has found numerous applications in food science, industrial food production, and medical any analytical fields. Commercial products containing catalase for textile applications are also available. These have been used to decompose residual hydrogen peroxide in fabric prior to dyeing, and are normally applied after draining the bleaching bath and refilling it with fresh water. A new, unconventional dyeing technique, ie, dyeing within the bleaching bath, is now implemented. In this technique, the bleaching bath, containing the fabric, is treated directly with catalase to destroy the residual hydrogen peroxide, and then being reused for reactive dyeing.

Removal of Residual Peroxide in Bleaching Effluents with Immobilized Catalases

Major problems in the use of catalases arise from the high temperature and alkalinity of the bleaching and washing liquors. The sensitivity of catalytically active protein structures to high temperatures, extreme pH, and other denaturing causes is one of the most important factors in the commercialization and industrial application of enzymes. In general, stabilization of the enzymes can be achieved in several ways: screening for stable ones such as thermophiles and extremophiles, chemical modification, protein engineering, immobilization, or stabilizing additives use. Interactions between dye and protein renders the use of soluble catalase inappropriate; alternatively, immobilized catalase can be used.

ECONOMIC ASPECTS

The chemicals used for bleaching have a variety of uses outside of bleaching technology. As a consequence, detailed information regarding production of these materials for bleach use is limited.

Sodium hypochlorite accounts for 92% of global use of the hypochlorite bleaches. Calcium hypochlorite accounts for the remaining 8%.

Residential use of sodium hypochlorite breaks down into the following categories: laundry bleaching (80%); sanitizers (18%); and residential pool and spa treatment (2%). Industrial uses are as follows: industrial and municipal water treatment (45%); commercial and municipal swimming pool treatment (33%); commercial laundry bleach (5%); liquid dishwashing detergent (5%); textile bleaching (4%), chemical (4%), and miscellaneous (4%).

Calcium hypochlorite is used as a shock treatment in swimming pools to boost the chlorine levels in pools. Swimming pool treatment accounts for 75% of calcium hypochlorite use; the remaining is used for municipal and industrial water treatment.

HEALTH AND SAFETY FACTORS

Much new information regarding the toxicities of sodium chlorite and sodium hypoclorite has become available, primarily as a result of safety concerns about chlorinated drinking water. In general, human population studies and animal bioassays have not found an association between exposure to these compounds and an increased risk of cancer, reproductive, or teratogenic effects.

The International Agency for Research on Cancer has concluded that there is inadequate evidence for the carcinogenicity of sodium hypochlorite in animals, and that sodium hypochlorite is not classifiable as to its carcinogenicity in humans.

The current OSHA PEL for perchlorates as nuisance dust is 15 mg/m^3. The California groundwater standard is 18 µg/L.

Hydrogen peroxide is a confirmed carcinogen and is moderately toxic by inhalation, ingestion, and skin contact. OSHA PEL TWA is ppm no, as is the ACGIH TLV.

Sulfur dioxide OSHA PEL TWA and ACGIH TLV TWA are both 2 ppm, STEL. ACGIH notes not classifiable as a human carcinogen.

USES

Laundering and Cleaning

Home and Institutional Laundering. The most widely used bleach in the United States is liquid chlorine bleach, an alkaline aqueous solution of sodium hypochlorite. This bleach is highly effective at whitening fabrics and also provides germicidal activity at usage concentrations. Liquid chlorine bleach is sold as a 5.25% solution and 1 cup provides 200 ppm of available chlorine in the wash. Liquid chlorine bleaches are not suitable for use on all fabrics. Dry and liquid bleaches that deliver hydrogen peroxide to the wash are used to enhance cleaning on fabrics. They are less efficacious than chlorine bleaches but are safe to use on more fabrics. The dry bleaches typically contain sodium perborate in an alkaline base whereas the liquid peroxide bleaches contain hydrogen peroxide in

an acidic solution. Detergents containing sodium perborate tetrahydrate are also available.

The worldwide decreasing wash temperatures, which decrease the effectiveness of hydrogen peroxide based bleaches, have stimulated research to identify activators to improve bleaching effectiveness. Tetraacetylethylenediamine is widely used in European detergents to compensate for the trend to use lower wash temperatures. TAED generates peracetic acid in the wash in combination with hydrogen peroxide. TAED has not been utilized in the United States, where one activator, nonanoyloxybenzene sulfonate (NOBS) has been commercialized and incorporated into several detergent products. NOBS produces pernonanoic acid when combined with hydrogen peroxide in the washwater and is claimed to provide superior cleaning to perborate bleaches.

In industrial and institutional bleaching either liquid or dry chlorine bleaches are used because of their effectiveness, low cost, and germicidal properties. Dry chlorine bleaches, particularly formulated chloroisocyanurates, are used in institutional laundries.

Hard Surface Cleaners and Cleansers. Bleaching agents are used in hard surface cleaners to remove stains caused by mildew, foods, etc, and to disinfect surfaces. Disinfection is especially important for many industrial uses. Alkaline solutions of 1–5% sodium hypochlorite that may contain surfactants and other auxiliaries are most often used for these purposes. These are sometimes thickened to increase contact times with vertical surfaces. A thick, alkaline cleaner with 5% hydrogen peroxide is also sold in Europe. Liquid abrasive cleansers with suspended solid abrasives are also available and contain ~1% sodium hypochlorite. Powdered cleansers often contain 0.1–1% available chlorine and they may contain abrasives. Sodium dichloroisocyanurate is the most common bleach used in powdered cleansers, having largely replaced chlorinated trisodium phosphate. Calcium hypochlorite is also used. Dichloroisocyanurates are also used in effervescent tablets that dissolve quickly to make cleaning solutions. In-tank toilet cleaners use calcium hypochlorite, dichloroisocyanurates, or N-chlorosuccinimide to release hypochlorite with each flush, to prevent stains.

One powdered toilet bowl cleaner uses potassium peroxymonosulfate and sodium chloride to generate hypochlorite in situ.

Automatic Dishwashing and Warewashing. The primary role of bleach in automatic dishwashing and warewashing is to reduce spotting and filming by breaking down and removing the last traces of adsorbed soils. They also remove various food stains such as tea. All automatic dishwashing and warewashing detergents contain alkaline metal salts or hydroxides.

Textile Bleaching

Many textiles are bleached to remove any remaining soil and colored compounds before dyeing and finishing. Bleaching is usually preceded by washing in hot alkali to remove most of the impurities in a process called scouring. Bleaching is usually done as part of a continuous process, but batch processes are still used. Bleaching conditions vary widely, depending on the equipment, the bleaching agent, the type of fiber, and the amount of whiteness required for the end use.

Cotton and Cotton–Polyester. Cotton is the principal fiber bleached today, and almost all cotton is bleached. About 80–90% of all cotton and cotton–polyester fabric is bleached with hydrogen peroxide.

Other Cellulosics. Rayon is bleached similarly to cotton but under milder conditions since the fibers are more easily damaged and since there is less colored material to bleach. Cellulose acetate and triacetate are not usually bleached. They can be bleached like rayon, except a slightly lower pH is used to prevent hydrolysis. The above fibers are most commonly bleached with hydrogen peroxide. Linen, flax, and jute require more bleaching and milder conditions than cotton, so multiple steps are usually used.

Synthetic Fibers. Most synthetic fibers are sufficiently white and do not require bleaching. For white fabrics, unbleached synthetic fibers with fluorescent whitening agents are usually used. When needed, synthetic fibers and many of their blends are bleached with sodium chlorite solutions at pH 2.5–4.5 for 30–90 min at concentrations and temperatures that depend on the type of fiber. Solutions of 0.1% peracetic acid are also used at pH 6–7 for 1 h at 80–85°C to bleach nylon.

Wool and Silk. Wool must be carefully bleached to avoid fiber damage. It is usually bleached with 1–5% hydrogen peroxide solutions at pH 8–9 for several hours at 40–55°C or at pH 5.5–8 for 20–60 min at 70–80°C. Silk is bleached similarly, but at slightly higher temperatures.

Bleaching of Other Materials

Hair. Hydrogen peroxide is the most satisfactory bleaching agent for human hair.

Fur. The coloring matter in fur is usually bleached using hydrogen peroxide stabilized with sodium silicate.

Foodstuffs, Oils. Sulfur dioxide is used to preserve grapes, wine, and apples; the process also results in a lighter color. During the refining of sugar, sulfur dioxide is added to remove the last traces of color. Flour can be bleached with a variety of chemicals, including chlorine, chlorine dioxide, oxides of nitrogen, and benzoyl peroxide. Bleaching agents such as chlorine dioxide or sodium dichromate are used in the processing of nonedible fats and fatty oils for the oxidation of pigments to colorless forms.

S. H. Batra, in M. Lewin and E. M. Pearce, eds., *Handbook of Fiber Service and Technology*, Vol. 4, Marcel Dekker, New York, 1985.

S. H. Higgins, *A History of Bleaching*, Longmans, Green and Co., 1924.

E. Linak, F. Dubas, and A. Kishi, *Chemical Economic Handbook*, Stanford Research Institute, Menlo Park, Calif, Feb. 2001.

T. Teitelbaum In E. Bingham, B. Cohrssen, and C. H. Powell, eds., *Patty's Toxicology*, 5th ed.,Vol. 3, John Wiley & Sons, Inc., New York, 2001, pp. 797–799.

JAMES P. FARR
WILLIAM L. SMITH
DALE S. STEICHEN
The Chlorox Company
TZANKO TZANOV
ARTHUR CAVACO-PAULO
University of Minito

BLOOD, COAGULANTS AND ANTICOAGULANTS

Cardiovascular disease, including intravascular clot formation, represents the primary cause of death in the Western world. Blood coagulation is essential to our health; however, when it proceeds abnormally, myocardial infarction (heart attack), stroke, or pulmonary embolism can result. Pharmacologic interventions to control and correct these thromboembolic disorders have recently made much progress as the mechanisms of blood clotting have become better understood.

The hemostatic system achieves a balance between the fluid and solid states of blood. When the integrity of the vascular system has been compromised, the blood clots to preserve the continuity of the vasculature and the blood supply. The initial response is the formation of the platelet aggregate. Platelets in the flowing blood rapidly adhere to the exposed subendothelial vessel wall matrix and become activated at the site where the endothelial cells have been damaged. During this activation process, products from the platelets are released, causing further platelet activation and platelet aggregation. The platelet plug initially arrests the loss of blood. This, however, is not a permanant block.

Simultaneous with platelet activation, the coagulation cascade is activated. The formation of a fibrin clot stabilizes the platelet plug. When a blood clot is no longer needed, it is broken down (lysed) by activated components of the fibrinolytic system.

Both the coagulation system and the fibrinolytic system are composed of several activators and inhibitors that provide for efficient physiological checks and balances. If any one component is over- or underactivated due to congenital or acquired abnormalities, pathologic blood clotting (thrombosis) occurs. As the components of hemostasis are many, there are multiple targets for therapeutic intervention. Bleeding complications can arise if the balance is pushed to the other extreme with drug treatment or due to physiologic abnormalities.

THE HEMOSTATIC SYSTEM

Vascular Endothelium

The vascular endothelium plays an important role in hemostasis in that quiescent endothelial cells act as a barrier separating the flowing blood from subendothelial components such as tissue factor (activation of coagulation) and collagen (activation of platelets). The functional interactions of endothelial cells can be either procoagulant or anticoagulant in nature. These actions can lead to either maintenance of normal hemostasis or to pathologic occlusive disorders (stenosis, thrombosis).

Platelets

Platelets are disk-shaped, anuclear cells that contain a contractile system, storage granules, and cell surface receptors. Platelets normally circulate in a nonactivated state in the blood but are extremely reactive to changes in their environment.

Upon activation the expression of cell receptors and procoagulant phospholipids on the platelet surface is upregulated. During the activation process, there is a morphologic shape change in the overall platelet structure as pseudopods are formed. This change facilitates the platelet aggregation process. Platelet aggregates serve to plug the damage to the vascular wall.

An increase in cytosolic calcium levels leads to activation of internal platelet enzymes with the subsequent release of platelet granule contents. The release of platelet granule contents leads to further platelet activation and aggregation as well as coagulation activation.

Of particular interest in the study of thrombosis and antithrombotic drugs is the acute coronary syndrome (ACS), which encompasses unstable angina, non-ST segment myocardial infarction (MI), and acute ST elevation (transmural) myocardial infarction (AMI). ACS stems from rupture of atherosclerotic plaque, leading to intravascular thrombosis. The role of platelets in ACS, in addition to the role of thrombin and the coagulation system, has been the focus of extensive drug development.

The Coagulation System

The plasma proteins that comprise the coagulation system are referred to as coagulation factors. Several of the coagulation factors are dependent on vitamin k for the structural formation required for activity.

The outcome of activation of the various factors that comprise the coagulation system is to generate thrombin. Excessive thrombin generation (a hypercoagulable state) results in unwanted blood clots that cause tissue ischemia. Depending on the location of the thrombus, skeletal muscle, heart, lung, brain, or other organs are affected. Inhibition of one or more of the coagulation factors that excessively reduces thrombin generation, such as by a congenital factor deficiency or overdose of anticoagulant treatment, may result in bleeding.

Natural Inhibitors of Coagulation

Antithrombin (AT) is a single-chain glycoprotein that is the primary inhibitor of coagulation and targets most coagulation factors. The efficient inhibition of proteases by AT requires heparin as a cofactor. A deficiency of AT, due to low protein levels or to functionally abnormal molecules, predisposes an individual to thrombotic complications.

Heparin cofactor II (HCII) is another plasma inhibitor that resembles AT but has a higher protease specificity than AT. Of the coagulation enzymes, it only inhibits thrombin.

Tissue factor pathway inhibitor (TFPI) serves an important function in the control of coagulation activation. It binds to the FVIIa-tissue factor complex and also to FXa thereby inhibiting fibrin formation.

Protein C is another important natural anticoagulant. Protein C is a vitamin K-dependent zymogen. Protein C derives its anticoagulant properties from its ability to cleave and inactivate membrane-bound forms of factors Va and VIIIa.

The Fibrinolytic System

The fibrinolytic system keeps the formation of blood clots in check. Like the coagulation cascade, this system consists of a number of serine protease activators and inhibitors. Plasminogenis converted to plasmin is the active enzyme that breaks down fibrin clots. Plasminogen activator inhibitor-1(PAI-1) and thrombin activatasie fibrinolytic inhibitor (TAFZ) are important inhibitors of fibrinolysis.

Inflammation

Inflammation is integregrated with hemostatic activation. Cell-cell binding (leukocyter, platelets, endothelial cells) and cytokines play a role in activation mechanisms.

THERAPEUTIC INTERVENTION OF THROMBOEMBOLIC DISORDERS

Thrombosis is associated with a high degree of morbidity and mortality. Anticoagulant drugs (heparin, and warfarin-derivatives) are not specific in mechanism; they target the inhibition of thrombin, thrombin generation and the initiation of coagulation, among other factors. Specific plasma and cellular sites within the hemostatic network are now targeted by a host of newly developed anticoagulant, antithrombotic, and antiplatelet, drugs.

Heparin

Heparin is a family of polysaccharide species whose chains are made up of alternating 1–4 linked and variously sulfated residues of uronic acid and D-glucosamine. It is a strongly anionic glycosaminoglycan that contains three functional side groups: $-OSO_3^-$, $-NHSO_3^-$, and $-COO^-$. Heparin is not one molecule but a heterogeneous mixture of different molecules. Owing to its structural heterogeneity, heparin exhibits a number of pharmacologic properties. Among these are antilipemic and antiviral properties; it can also inhibit tumor growth.

Foremost among the actions of heparin is its ability to inhibit blood clotting. Heparin produces little anticoagulant or antithrombotic effect directly. Rather, its effects are mediated through specific saccharide sequences that bind to one of several endogenous plasma proteins that include AT, HCII, and TFPI. The major antithrombotic activity of heparin is its ability to inhibit thrombin (antithrombin or anti-factor IIa activity) and factor Xa (anti-factor Xa activity). Administration of heparin causes an increase in the plasma levels of TFPI that adds to its antithrombotic action.

Heparin is administered either by intravenous infusion or subcutaneous injection. Heparin binds to a variety of plasma proteins in the blood, thereby lowering its bioavailability and producing a variable anticoagulant response.

Clinical Uses of Heparin. Heparin is the drug of choice for effective treatment of venous thrombosis and pulmonary embolism (PE). Mortality is reduced in patients receiving heparin for the treatment of PE. Heparin is also used for prophylaxis in patients at risk of developing deep venous thrombosis (DVT) and PE.

Heparin is also used to anticoagulate patients with ACS. Heparin can prevent AMI and recurrent refractory angina in patients with unstable angina. In patients with a previous MI, heparin significantly reduces reinfarction and death. Heparin is also used to treat thrombotic stroke.

Blood in contact with a foreign surface will clot within minutes if left without anticoagulant. Heparin has been used as a flush solution for most catheters inserted in hospitalized patients. Heparin-coated devices are now being produced that eliminate the need for heparin administration directly to the patient. Heparin is used with interventional cardiology procedures and renal dialysis. The most extreme case where the highest level of anticoagulation is needed is with the heart-lung machine used for cardiopulmonary bypass in cardiac surgery, where heparin is useful.

Heparin can be used safely in pregnancy because it does not cross the placental barrier and does not cause unwanted effects on the fetus. Heparin is effectively used in the pediatric population for the same indications as in the adult.

Part of heparin's attractiveness for use as an anticoagulant in surgical situations relates to the relative ease in which it can be neutralized upon completion of the procedure or in the event of an overdose with protamine.

Heparin has provided reliable thromboprophylaxis for many years and remains a useful, effective drug that is easily dosed, monitored and neutralized.

Side Effects of Heparin Therapy. The most common side effect of heparin therapy is hemorrhage, ranging from minor to life threatening. Heparin-induced thrombocytopenia (HIT), which occurs in ~3% of patients exposed to heparin, is perhaps the worst of all drug-induced allergic reactions. Long-term heparin therapy has been shown to produce osteoporosis. Heparin-induced skin necrosis is a rare complication of subcutaneously administered heparin.

Low Molecular Weight Heparin

The depolymerization of heparin (the original unfractionated heparin) either chemically (nitrous acid degradation, benzylation-alkaline hydrolysis, peroxidative cleavage), enzymatically (heparinase), or by physicochemical means (γ irradiation) results in the production of clinically useful drugs known as low molecular weight (LMW) heparins.

Clinical Uses of LMW Heparins. Four LMW heparins have been approved for use in the United States enoxaparin (Lovenox, Sanofi-Aventis), ardeparin (Normiflo, Wyeth-Ayerst), dalteparin (Fragmin, Pfizer) and tinzaparin (Innohep, Pharmion). The LMW heparins are used in prevention of venous thrombosis in patients undergoing abdominal surgery, hip/knee repair/replacement or medically ill patients with restricted mobility; treatment of existing venous thrombosis; and prevention of ischemic complications in patients with unstable angina/non-Q-wave MI.

LMW heparins can also be used as anticoagulants in patients with end-stage renal disease requiring extracorporeal hemodialysis treatment. They can be used in children and are the drug of choice in pregnant women requiring anticoagulation.

At-home dosing with LMW heparin is as safe and effective as in-hospital treatment by heparin infusion. Newer indications for possible use of LMW heparins are in the management of thrombotic stroke and in cancer patients. LMW heparins may not only decrease the incidence of cancer associated thrombosis but may also positively affect all-cause mortality.

LMW heparins are effective for the reduction of restenosis after interventional cardiologic procedures, maintenance of peripheral arterial and coronary graft patency, and as adjunct anticoagulants in stenting and other interventional cardiologic procedures. LMW heparins will not become the drug of choice in surgical settings where a short half-life anticoagulant is required. Reversal agents such as protamine do not completely block the antithrombotic activity of LMW heparin, and there are no commonly available devices/assays to effectively monitor the high drug levels required.

Side Effects of LMW Heparin Therapy. LMW heparins are less likely to cause hemorrhagic complications than unfractionated heparin during treatment of venous thrombosis.

LMW heparin should not be given to a patient suspected of having HIT.

Other side effects of heparin, such as osteoporosis are reduced with LMW heparins.

Synthetic Heparins

Pentasaccharide (fondaparinux, GSK) is the first synthetic heparin. It is composed of a specific saccharide sequence that allows for tight binding to AT and high anti-factor Xa activity. The chemically synthesized pentasaccharide is free of viral or other animal contaminants. Fondaparinux can be used to prevent venous thrombosis following orthopedic surgery.

The synthesis of pentasaccharide has opened the door for the possibility of synthesizing other heparin-like agents like idraparinux that exhibit a higher affinity to AT, more potent anti-factor Xa activity, and an extended half-life.

Oral Heparin

Recent attempts to produce heparin formulations that exhibit oral bioavailability have met with varying degrees of success.

Non-Heparin Glycosaminoglycans

Dermatans, heparans, and chondroitin sulfates represent nonheparin glycosaminoglycans (GAGs) that are used mainly in the intravenous management of DVT. These drugs can be given to patients who are heparin compromised.

Dermatan Sulfate. Dermatan sulfate is a glycosaminoglycan polymer of iduronic acid and N-acetylated galactosamine that is unable to interact with AT but rather complexes with HCII to inhibit thrombin generation. The advantage it has over heparin as an antithrombotic agent is a lower risk of bleeding complications.

Intimatan. Due to its unique structure, intimatan has higher anti-thrombin potency than the naturally occurring, parent dermatan sulfate. Intimatan has not been studied in humans yet.

Vitamin K Antagonists

Long-term prophylaxis against thrombosis is typically achieved using vitamin K antagonists (VKA). In the United States, warfarin (Coumadin) is most commonly used. Acenocoumarol and phenprocoumon, with shorter and longer half-lives than warfarin, respectively, are used in other countries. The VKAs interfere with blood coagulation by inhibiting vitamin K reductase and vitamin K epoxide reductase coagulation factors produced in the presence of warfarin lose their clotting activity.

The primary benefit of VKAs over the heparins is their ability to be administered orally. As such, VKAs are commonly used for the prophylaxis and treatment of DVT (in particular when long-term treatment is needed), for anticoagulation of patients with atrial fibrillation, and for anticoagulation of patients with mechanical heart valves. Patients experiencing AMI may also benefit from anticoagulant treatment with VKAs.

Hemorrhage is by far the most frequent complication of VKA therapy. As this class of drugs has a relatively narrow safety–efficacy margin, frequent monitoring of drug levels is necessary.

The anticoagulant actions of VKAs are not easily reversed, complicating the anticoagulant management of treated patients who require surgical intervention. Skin necrosis can occur during treatment in individuals with protein C deficiency, activated protein C resistance (factor V Leiden), and protein S deficiency. Teratogenic effects of VKAs on fetuses preclude its use in the anticoagulant management of pregnant patients.

Direct Thrombin Inhibitors

This new class of antithrombotic agents includes hirudin, (Lepirudin, Berlex) a leech-derived protein the synthetic peptide inhibtor hirulog (bivalirudin); and argatroban (GSK), the medicines co.

The importance of thrombin inhibitors is that they can be effectively used as anticoagulants in patients with HIT. Thrombin inhibitors offer other advantages over heparin in that they directly inhibit thrombin without the need for plasmatic cofactors. They can inhibit both fluid-phase and clot-bound thrombin. Because they are not neutralized by plasma proteins, they have predictable and consistent pharmacodynamics, which can translate into fast therapeutic control and fewer treatment failures.

Factor Xa Inhibitors

Factor Xa inhibitors are a diverse class of agents, each with distinct characteristics. Several agents have been identified that are able to either directly bind to and inhibit factor Xa or indirectly inhibit factor Xa. Thus, these agents derive their antithrombotic activity from their ability to inhibit thrombin generation.

Oral Anticoagulants

Today VKAs are the only oral antithrombotics. The high risk of bleeding with VKAs and the need for frequent monitoring, however, have prompted the development of new oral antithrombotic agents. Both thrombin and factor Xa inhibitors are being developed.

Other Inhibitors

Agents that specifically target either factor VIIa or tissue factor are in development. The natural inhibitors of the coagulation system are also targets for potential drug development. In addition to TFPI, AT, HCII, C_1-esterase inhibitor, and PAI-1 are under development. Protein Ca concentrate, which targets the inhibition of factor Va and factor VIIIa, has shown successful outcomes in patients with sepsis and disseminated intravascular coagulation (DIC).

Antiplatelet Drugs

A wide variety of antiplatelet agents that inhibit different aspects of the platelet activation response have proven clinical efficacy or are currently under development.

Aspirin, the most widely used antiplatelet agent, blocks platelet activation by inhibiting cyclooxygenase (COX) thereby limiting thromboxane (a potent platelet aggregation activator) generation. Aspirin offers clinical benefit in both the primary and secondary prevention of cardiovascular events, prevention of DVT and PE, and is used in the treatment of AMI, stable and unstable angina, carotid artery stenosis, ischemic stroke, and placental insufficiency. Aspirin is also used in combination with other antiplatelet drugs during PCI and in the prophylaxis of thrombotic complications following PCI. After coronary bypass grafting (CABG surgery) patients are put on life-long aspirin therapy. Asprin has been shown in vascular patients to reduce the risk of stroke, MI and death by ~25%.

New anti platelet drugs that forget for the treatment of cardiology patients. Clopidoge (plavix) blocks ADP acceptor. Several types of glycoprotein IIb/IIIa receptor blockers are also available. Other specific inhibitors of platelet activation available dipyridamole (Persantine), receptor and cilostazol (Pletaol).

Pharmacologic Considerations

The future holds promise for effective new antithrombotic drugs in individual and specific indications.

Today the LMW heparins are the drugs of choice of several thrombotic indications. From identified structure–activity relationships, such heparinomimetics as fondaparinux have been developed. Additionally, studies to develop drugs with antithrombotic activities but without anticoagulant aspects are in progress. Direct acting factor Xa inhibitors are currently in clinical development.

Synthetic agents have certain advantages over naturally derived products, not the least of which is their specific chemical design to target desired biological effects. How and where each drug is used clincially, and how each will compete with the standard heparin, warfarin, and aspirin treatments remains to be determined. The newly developed drugs are mostly monotherapeutic and do not mimic the polytherapeutic actions of heparins. Heparin and its derived and modified forms will therefore continue to play an important role in the management of thrombosis.

B. Casu, *Heparin: Chemical and Biological Properties, Clinical Applications*, Edward Arnold, London, 1989, pp. 25–50.

R. W. Colman, J. Hirsh, V. J. Marder, and E. W. Salzman, *Hemostasis and Thrombosis. Basic Principles and Clinical Practice*, 3rd ed., J.B. Lippincott, Philadelphia, Pa., 1994.

W. Jeske, H. L. Messmore, and J. Fareed, Pharmacology of Heparin and Oral Anticoagulants, in *Thrombosis and Hemorrhage*, 2nd ed., Williams and Wilkins, Baltimore, Chapt. 55, 1998, pp. 1193–1213.

P. W. Majerus, G. J. Broze, J. P. Miletich, and D. M. Tollefsen, *Anticoagulant, Thrombolytic, and Antiplatelet Drugs, in Goodman and Gilman's, The Pharmacological Basis of Therapeutics*, 8th ed., Pergamon Press, New York, Chapt. 55, 1990, pp. 1311–1331.

JEANINE M. WALENGA
WALTER P. JESKE
PETER BACHER
JAWED FAREED
Loyola University

BLOOD SUBSTITUTES

Artificial blood is herein defined as consisting of red cell substitutes. Red cell substitutes are solutions intended for use in patients whose red cells are either not available or their use is to be avoided for other reasons. Despite enormous effort, more than 100 years of research have not produced a solution that can be safely used in humans.

In 1983, the move to develop red cell substitutes intensified when recognition that the acquired immune deficiency syndrome (AIDS) could be transmitted by the blood-borne human immunodeficiency virus (HIV) produced grave concern for the nation's blood supply. Since that time, modernized blood bank methods have dramatically reduced the risk of transfused blood. Furthermore, indications for transfusion have been reevaluated, and the use of blood products has become much more efficient. More careful screening of donors, testing of all donated units, and a general awareness in the donor population have all contributed to a decreased risk from transfusion-contracted AIDS.

Historically, many blood substitutes have been tried: milk, normal saline, Ringer's solution (a potassium/calcium conentration), gum saline, blood plasma and serum, albumin, starches, perfluorocarbons, cell-free hemoglobin, and modified encapsulated hemoglobin.

HEMOGLOBIN MODIFICATIONS

Reactivity

Hemoglobin can exist in either of two structural conformations, corresponding to the oxy (R, relaxed) or deoxy (T, tense) states. The key differences between these two structures are that the constrained T state has a much lower oxygen affinity than the R state and, the T state has a lower tendency to dissociate into subunits that can be filtered in the kidneys. Therefore, stabilization of the T conformation would be expected to both reduce renal filtration and maintain oxygen affinity similar to that of red cells.

The transition between the T and R states of hemoglobin is also deeply involved in the Bohr (pH) effect and cooperativity. Therefore, stabilization of either of the two structures should diminish these effects, which might have impo rtant physiologic consequences.

Stabilization of the T conformation under normal conditions is illustrated by the reaction of 2,3-diphosphoglycerate, (2,3-DPG), as shown in Figure 1.

All of the reactions considered to be useful in the production of hemoglobin-based blood substitutes use chemical modification at one or more of the sites discussed above. Table 1 lists the different types of modifications with examples of the most common reactions for each. Differences in the reactions are determined by the dimensions and reactivity of the cross-linking reagents. Because the function of hemoglobin in binding and releasing oxygen is intricately connected to the transition between T and R conformations, it is not surprising that the p_{50} yields are highly variable. Even small differences among

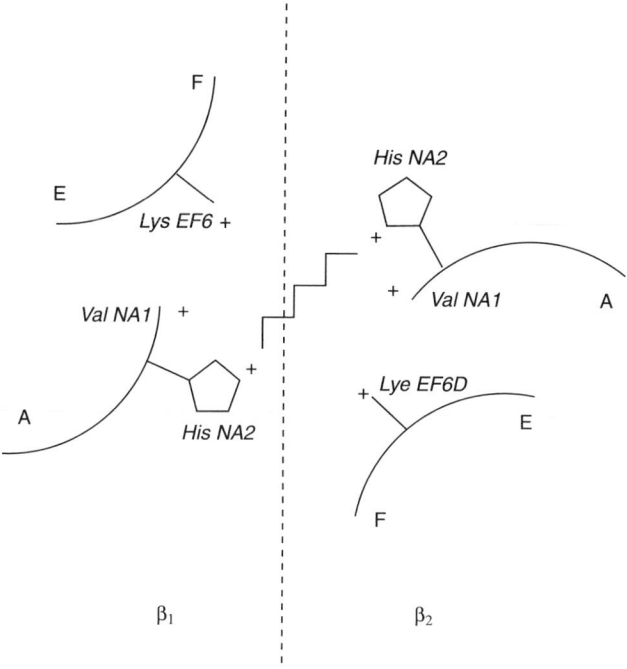

Figure 1. Reaction of 2,3-DPG and deoxyhemoglobin. The molecule fits into the central cavity of hemoglobin and forms salt bridges with valine $NA_1\beta$, histidines $NA_2\beta$, $H_{21}\beta$, and lysine $EF_6\beta$. A, E, and F refer to specific hemoglobin helices and NA is the sequence from the amino-terminals to the A helix.

structures of the reagents can yield products having very different properties. In addition, the conditions of the reaction are very important, not only in regard to the state of ligation, ie, oxygen saturation, but also in regard to the presence of agents or molecules that block or compete for certain reactive sites.

A further complication of these reactions is that many nonhemoglobin proteins also contain reactive groups that may also be co-modified. These molecules, if present at the

time of reaction, could affect the properties of the final solution. For this reason, derivatives prepared for studies of the hemoglobin molecule per se must start with highly purified stroma-free hemoglobin.

HEMOGLOBIN SOURCES

Purification

Hemoglobin is provided by the red blood cell in highly purified form. However, the red cell contains many enzymes and other proteins, and red cell membranes contain components, such as phospholipids, that could be toxic. Furthermore, plasma proteins and other components could trigger immune reactions in recipients. The chemical modification reactions discussed here are not specific for hemoglobin; they may modify other proteins as well. Indeed, multifunctional reagents could actually couple hemoglobin to nonhemoglobin proteins.

Rabiner's method for the filtration purification of hemoglobin was thought to be a significant advance over older centrifugation methods. However, hemoglobin prepared in this way still caused unwanted reactions in human recipients. The crystallization method showed fewer toxic effects in animals, but batch-to-batch reproducibility was uneven. Ultrapurification of hemoglobin using ion-exchange chromatographic technique is possible but is tedious and expensive.

Outdated Human Blood

If clinical efficacy and safety of hemoglobin solutions can be shown, the demand for product would soon outstrip the supply of outdated human blood. About 12 million units of blood (1 unit is ~480 mL) are used in the United States each year, and only about 1% outdate. The primary use of blood is in intraoperative and emergency settings. The quantity of blood available for use in production of blood substitutes depends on the willingness of donors who are qualified to donate, and on the efficient matching of donor blood to the recipients.

Bovine Hemoglobin

One solution to the hemoglobin supply problem is to use nonhuman sources of blood such as cows (bovine hemoglobin) as a starting material. The ultimate success of bovine, or any other hemoglobin, depends on demonstration of safety, not supply. One problem in using bovine hemoglobin is the fear of the bovine spongiform encephalitis (BSE) virus. This virus, related to the Scrapie organism, has been detected in cows in Europe as well as other mammals in North America. The variant Creuzfield-Jacob Disease (vCJD) has been associated with BSE in Europe, and it is known that BSE can be transmitted by blood in animals. Although there are no known cases of human transmission by blood transfusion at this time, the FDA has placed restrictions on the importation of blood from Europe into the United States. At this time, there is no adequate test for BSE in donated blood that could be implemented on a large scale.

Table 1. Classes of Hemoglobin Modification[a]

Class	Examples
amino-terminal modification	carbamylation
	carboxymethylation
	pyridoxylation
	acetaldehyde
Lys EF6(82)β modification	mono(3,5-dibromosalicyl)fumarate
Val NA1(1)β-Lys EF6(82)β cross-link	2-nor-2-formylpyridoxal 5'-phosphate (NFPLP)
	bis(pyridoxal) tetraphosphate (bis-PL)P4
Lys G6(99)α1-Lys G6(99)α2 cross-link	bis(3,5-dibromosalicyl)fumarate
2,3-DPG analogue	pyridoxal 5'-phosphate
surface, multisite	glutaraldehyde
	polyaldehydes
	ring-open dials
	diimidate esters
conjugated hemoglobin	dextran, starch aldehydes
	poly(ethylene glycol)

Recombinant Hemoglobin

An alternative and novel source of hemoglobin, which is used for modification, is from microorganisms the genome of which has been modified to contain globin genes for recombinant hemoglobin (rHb) production. Significant strides have been made in this approach, and it is possible to express both human α- and β-globin chains in E. coli.

Much of the development work with recombinant hemoglobin for commercial purposes has been done in commercial laboratories, so not all details of the process are available. However, it is likely that production on the scale needed for a viable red cell substitute product could be a problem. One unit of blood (500 mL) contains ~15 g/dL of hemoglobin, so a total of 75 g of hemoglobin would be needed to produce a unit. If the yield is 0.1 g/L of culture medium, 750 L of cell culture would be needed. In the future, it might be possible to express hemoglobin genes in higher organisms; synthesis of functional human hemoglobin has already been reported in yeast, transgenic mice, and pigs. However, these approaches present additional purification, logistic, and economic problems. Purification of rHb could also be a significant challenge, since it would need to be separated from media components and other microorganism products. Endotoxin contamination could be a serious problem for bacteria.

CURRENT STATUS OF ARTIFICIAL BLOOD

The magnitude of the undertaking to produce "Artificial Blood" cannot be overestimated. In the past 20 years, >$2 billion has been expended on attempts to do so. To date, only one product has been presented to the FDA for licensure approval. At the same time, several prominent attempts, fueled by significant investments from large pharmaceutical companies have, failed.

Some of the major problems that have plagued the development of hemoglobin solutions appear to be solved. For example, renal failure, a consequence of renal filtration of dissociated hemoglobin molecules, has been eliminated by cross-linking or polymerization chemistry and strict elimination of unreacted hemoglobin. Modern perfluorocarbon emulsions can be stored at refrigerator temperature and need not be reconstituted prior to use. Optimal sizing of the emulsion particles has decreased some of the side effects that have hampered development of earlier emulsions. However, some significant scientific problems remain to be solved or clarified in the field.

For hemoglobin-based products there is still no agreement on the mechanism of vasoconstriction. This problem does not affect all products. Most studies in the published literature are based on αα-Hb or DCL-Hb, mainly because these products have been readily available to researchers. Understandably, this has been a sensitive subject with commercial developers, and fewer basic studies are available with polymerized products that are in advanced clinical trials.

Both theories of vasoconstriction, NO binding and O_2 autoregulation, are being studied. To date it is not certain if they are mutually exclusive or what their physiological and clinical consequences are.

The mechanism of the frequently reported gastrointestinal (GI) side effects of hemoglobin solutions is not clear. Studies have been done in animals showing that some hemoglobin solutions interfere with esophageal motility and gastric emptying. However, how plasma hemoglobin can interfere with smooth muscle function is not established. Other possible concerns center around the effect of hemoglobin on cellular function. Some preparations stimulate monocytes and macrophages; others degranulate basophils and reduce eosinophil mobility. Still other reports suggest that neutrophil function and platelet-endothelial adhesion can be affected. Work is still in the early stages on these phenomena, and it is not yet established whether or not they have clinical implications.

The relatively short intravascular persistence of any of the products remains a problem for certain clinical applications such as chronic anemia, but should not impact use in many elective surgical procedures or in trauma, where the patient's own marrow should replenish red cells in a few days. This issue is complicated by the fact that detailed metabolic pathways for many of the products are still not completely defined, so the impact of the rate of breakdown is not known. On the positive side, clinical trials with all of the products studied so far have shown the ability to reduce the number of patients that receive allogeneic transfusions and the number of units transfused in elective surgical procedures. The U.S. FDA has stated publicly that such reduction or elimination of allogeneic blood could be a basis for approval of these products. Establishing reduced mortality as an end point is a more difficult task, and the FDA has stated that while it will not require such demonstration as a prerequisite to approval, it will require phase II clinical trials in trauma to establish safety in this application.

Finally, the cost of new products will be an important issue once regulatory approval has been won. Some commercial developers and financial analysts believe that a safe and effective product that can reduce the number of patients who receive transfusions or the number of units of allogeneic blood transfused, will command a price significantly higher than that of banked blood. However, as no products are as yet approved, this remains to be seen.

A. S. Acharya, B. N. Manjula, and P. Smith, *Hemoglobin Crosslinkers*, Albert Einstein College of Medicine of Yeshiva University, New York, 1996, pp. 1–16.

A. S. Rudolph, R. Rubinovici, and G. Z Feuerstein, eds., *Red Blood Substitutes. Basic Principles and Clinical Application*, Marcel Dekker, New York, 1998.

R. Winslow, *Hemoglobin-based Red Cell Substitutes*, Johns Hopkins University Press, Baltimore, M.D., 1992.

H. Wintrobe, ed., *Blood, Pure and Eloquent*, McGraw-Hill, New York, 1980.

ROBERT M. WINSLOW
Sangart, Inc., and University of
California, San Diego

BORON

Boron, B, is unique in that it is the only nonmetal in Group 13 (IIIA) of the Periodic Table. Boron, at wt 10.81, at no. 5, has more similarity to carbon and silicon than to the other elements in Group 13. There are two stable boron isotopes, ^{10}B and ^{11}B, which are naturally present at 19.10–20.31% and 79.69–80.90%, respectively. There is a very low cosmic abundance of boron, but its occurrence at all is surprising for two reasons. First, boron's isotopes are not involved in a star's normal chain of thermonuclear reactions, and second, boron should not survive a star's extreme thermal condition. The formation of boron has been proposed to arise predominantly from cosmic ray bombardment of interstellar gas in a process called spallation.

Boron is the 51st most common element present in the earth's crust at a concentration of three grams per metric ton.

PROPERTIES

Elemental boron has a diverse and complex chemistry, primarily influenced by three circumstances. First, boron has a high ionization energy, 8.296, 23.98, and 37.75 eV for first, second, and third ionization potentials, respectively. Second, boron has a small size. Third, the electronegativities of boron (2.0), carbon (2.5), and hydrogen (2.1) are all very similar, resulting in extensive and unusual covalent chemistry.

Boron has electronic structure $1s^2 2s^2 2p$ and an expected valence of three. Because of the high ionization energies there is no formation of univalent compounds as for the other Group 13 elements.

Boron also has a high affinity for oxygen-forming borates, polyborates, borosilicates, peroxoborates, etc. Boron reacts with water at temperatures above 100°C to form boric acid and other boron compounds.

The physical properties of elemental boron are significantly affected by its purity and crystal form. In addition to its being an amorphous powder, boron has four crystalline forms: α-rhombohedral, β-rhombohedral, α-tetragonal, and β-tetragonal. The α-rhombohedral form has mp 2180°C, sublimes at ~3650°C, and has a density of 2.45 g/mL. Amorphous boron, by comparison, has mp 2300°C, sublimes at ~2550°C, and has a density of 2.35 g/mL.

Boron is an extremely hard, refractory solid having a hardness of 9.3 on Mohs' scale and a very low (1.5×10^{-6} ohm^{-1} cm^{-1}) room temperature electrical conductivity so that boron is classified as a metalloid or semiconductor. These values are for the α-rhombohedral form.

The electron-deficient character of boron also affects its allotropic forms. Its high ionization energies and small size prevent boron from adopting metallic bonding to compensate for its electron deficiency and that of other hypoelectronic elements.

PREPARATION

Three methods—electrolytic reduction, chemical reduction, and thermal decomposition—are used on a laboratory scale.

A high purity (>99%) boron comes from the direct thermal decomposition of boron hydrides such as diborane, B_2H_6.

PRODUCTION

The Moissan process, the reduction of boric oxide with magnesium, is the most widely used commercial process for producing boron.

Another commercial process yields high purity boron of >99%. In this process boron hydrides, such as diborane, are thermally decomposed.

HEALTH AND SAFETY FACTORS

Boron is a trace element that is essential to human health and behavior. Evidence points to the fact that boron may reduce either the symptoms or incidence of arthritis. There have been no reports of its toxicity in humans.

USES

Elemental boron is used in very diverse industries from metallurgy to electronics. Other areas of application include ceramics, propulsion, pyrotechnics, and nuclear chemistry.

H. E. Boyer and T. L. Gall, eds., *Metal Handbook, Desk Edition*, American Society for Metals, Metals Park, Ohio, 1985, pp. 4–11.

P. A. Lyday, "Boron," *Minerals Yearbook*, U.S. Geological Survey, Reston, Va., 2001.

A. G. Schauus, "Boron," http://www.traceminerals.com/products/boron.html, undated, accessed May 13, 2002.

E. P. Wohlfarth and K. H. J. Buschow, eds., *Ferromagnetic Materials*, North-Holland, Amsterdam, The Netherlands, 1988, p. 1.

LINDA H. JANSEN
Callery Chemical Company

BORON HALIDES

The boron trihalides—boron trifluoride, BF_3, boron trichloride, BCl_3, and boron tribromide, BBr_3—are important industrial chemicals having increased usage as Lewis acid catalysts and in chemical vapor deposition (CVD) processes. Boron halides are widely used in the laboratory as catalysts and reagents in numerous types of organic reactions and as starting material for many organoboron and inorganic boron compounds.

BCl_3, BBr_3, and BI_3 undergo exchange reactions to yield mixed boron halides. Boron trihalides can be reduced to elemental boron by heating and presence of alkali metals, alkaline-earth metals, or H_2.

Reactions of boron trihalides that are of commercial importance are those of BCl_3 and, to a lesser extent, BBr_3, with gases in chemical vapor deposition (CVD).

Metal halides react with BF_3 when heated to form BX_3 and the metal fluoride.

MANUFACTURE

Boron Trifluoride

Boron trifluoride is prepared by the reaction of a boron-containing material and a fluorine-containing substance in the presence of an acid. The traditional method used borax, fluorspar, and sulfuric acid.

A process for recovering boron trifluoride from oligo-merization mix have been reported.

Boron Trichloride

Boron trichloride is prepared on a large scale by the reaction of Cl_2 and a heated mixture of borax, $Na_2BO_4O_7 \cdot 10\,H_2O$, and crude oil residue in a rotary kiln heated to $1038°C$. On a smaller scale, BCl_3 can be prepared by the reaction of Cl_2 and a mixture of boron oxide, B_2O_3, petroleum coke, and lampblack in a fluidized bed.

Boron Tribromide

Boron tribromide is produced on a large scale by the reaction of Br_2 and granulated B_4C at $850–1000°C$ or by the reaction of HBr with CaB_6 at high temperatures. Most of the methods for preparing BBr_3 are similar to those for preparation of BCl_3.

Boron Triiodide

Boron triiodide is not manufactured on a large scale.

SHIPMENT AND HANDLING

Boron trifluoride gas is nonflammable and is shipped in DOT 3A and 3AA steel cylinders at a pressure of approximately 12,410 kPa (1800 psi). Boron trifluoride is classified as a poison gas, both domestically and internationally. Cylinders must have a poison gas diamond and an inhalation hazard warning label. Tube trailers carry both a poison gas placard and an inhalation hazard warning. Cylinders containing 27.2 kg and tube trailers containing 4.5–10 metric tons are available. If boron trifluoride is compressed using oil as a compressor lubricant, it must not be used with oxygen under pressure nor with gauges, valves, or lines that are to be used with oxygen.

In as much as the gas hydrolyzes readily, all equipment should be purged repeatedly, using inert dry gas, before admitting boron trifluoride. Under anhydrous conditions, carbon steel equipment is satisfactory. Stainless steel and aluminum silicon bronze may also be used. Stainless steel tubing is recommended for both temporary and permanent connections.

In the presence of moisture, boron trifluoride may be handled in polytetrafluoroethylene (PTFE), polyethylene, Pyrex glass (limit to atmospheric pressure), or Hastelloy C containers. At $600°C$, stainless steel (304 L) and Hastelloy N are attacked by BF_3; Hastelloy C is more resistant. Kel

F and PTFE serve as satisfactory gasket and packing materials, whereas rubber, fiber, polymerizable materials, or organic oxygen- and nitrogen-containing compounds must be avoided. Because boron trifluoride is soluble in, and reacts with, many liquids, the gas must not be introduced into any liquid unless a vacuum break or similar safety device is employed.

BCl_3 is shipped in steel cylinders; BBr_3 is shipped in glass bottles (0.45- and 2.3-kg, net) and 91-kg (net) monel drums (10). Both BCl_3 and BBr_3 are classed as corrosive liquids and must be shipped by private carriers.

SPECIFICATIONS AND ANALYTICAL METHODS

Commercial boron trifluoride is usually approximately 99.5% pure. The common impurities are air, silicon tetrafluoride, and sulfur dioxide. An excellent procedure for sampling and making a complete analysis of gaseous boron trifluoride has been developed.

Analysis for boron, halide, free halogen, and silicon is carried out by standard methods following hydrolysis of BX_3.

HEALTH AND SAFETY FACTORS

Boron trifluoride is primarily a pulmonary irritant. The toxicity of the gas to humans has not been reported, but laboratory tests on animals gave results ranging from an increased pneumonitis to death. The TLV is 1 ppm. Inhalation toxicity studies in rats have shown that exposure to BF_3 at 17 mg/m^3 resulted in renal toxicity, whereas exposure at 6 mg/m^3 did not result in a toxic response. Prolonged inhalation produced dental fluorosis. High concentrations burn the skin similarly to acids such as HBF_4 and, if the skin is subject to prolonged exposure, the treatment should be the same as for fluoride exposure and hypocalcemia. No chronic effects have been observed in workers exposed to small quantities of the gas at frequent intervals over a period of years.

Boron trichlorides are highly reactive, toxic, and corrosive; these trihalides (BCl_3, BBr_3, BI_3) react vigorously, even explosively, with water. High temperature decomposition of BX_3 can yield toxic halogen-containing fumes. Boron trichloride is a poison by inhalation and a severe irritant to skin, eyes and mucous membranes. BCl_3, BBr_3, and BI_3 emit toxic fumes when heated to decomposition. Safe handling, especially of BCl_3, has been reviewed.

USES

Boron Trifluoride

Boron trifluoride is an excellent Lewis acid catalyst for numerous types of organic reactions. Its advantages are ease of handling as a gas and the absence of undesirable tarry by-products. As an electrophilic molecule, it is an excellent catalyst for Friedel-Crafts and many other types of reactions.

Boron trifluoride and some of its adducts have widespread application as curing agents for epoxy resins and in preparing alcohol-soluble phenolic resins.

Boron trifluoride catalyst is used under a great variety of conditions, either alone in the gas phase or in the presence of many types of promoters. Many boron trifluoride coordination compounds are also used.

Boron trifluoride is also employed in nuclear technology, by utilizing several nuclear characteristics of the boron atom. It is used for the preparation of boranes.

Diborane is obtained from reaction with alkali metal hydrides; organoboranes are obtained with a suitable Grignard reagent.

Boron trifluoride has been used in mixtures to prepare boride surfaces on steel and other metals and as a lubricant for casting steel.

Boron Trichloride

Much of the BCl_3 consumed in the United States is used to prepare boron filaments by CVD. These high performance fibers are used to reinforce composite materials made from epoxy resins and metals (Al, Ti). The principal markets for such composites are aerospace industries and sports equipment manufacturers.

Another important use of BCl_3 is as a Friedel-Crafts catalyst in various polymerization, alkylation, and acylation reactions, and in other organic syntheses.

BCl_3 is also used for the production of halosilanes, in the preparation of many boron compounds, in the production of optical wave guides, and for the prevention of solid polymer formation in liquid SO_3; for the removal of SiO_2 from SiC powders, carbochlorination of kaolinitic ores, and removal of impurities from molten Mg. It is also used as a critical solvent in metal recovery from chlorination processes, for the removal of potential catalyst poisons from hydrocarbon oils, and in the production of lithium–thionyl chloride batteries. Other than production of boron fibers, important CVD processes involving BCl_3 include production of boron carbide-coated carbon fiber; doping Si or Ge with B and for doping electric or photoconducting polymers, in the preparation of scratch-resistant silicon-based coatings, and in glass-fiber technology; production of boron nitride, and metal borides. BCl_3 is also used in reactive ion etching and plasma etching in the production of silicon-integrated circuits and devices, in the dry etching of boron nitride, gallium arsenide, and SnO_2, and Al–Si.

Boron Bromide

A large portion of the BBr_3 produced in the United States is consumed in the manufacture of proprietory pharmaceuticals (qv). BBr_3 is used in the manufacture of isotopically enriched crystalline boron, as a Friedel-Crafts catalyst in various polymerization, alkylation, and acylation reactions, and in semiconductor doping and etching. BBr_3 is a very useful reagent for cleaving ethers, esters, lactones, acetals, and glycosidic bonds; it is used to deoxygenate sulfoxides and in the preparation of image-providing materials for photography.

Boron Triiodide

There are no large-scale commercial uses of boron triiodide. It is used to clean equipment for handling UF_6 and in the manufacture of lithium batteries.

DERIVATIVES

Fluoroboric Acid and the Fluoroborate Ion

Fluoroboric acid, generally formulated as HBF_4, does not exist as a free, pure substance. The acid is stable only as a solvated ion pair, such as $H_3O^+BF^-_4$; the commercially available 48% HBF_4 solution approximates $H_3O^+BF^-_4 \cdot 4H_2O$. Other names used infrequently are hydrofluoroboric acid, hydroborofluoric acid, and tetrafluoroboric acid. Salts of the acid are named as fluoroborates or occasionally borofluorides. The acid and many transition-metal salts are used in the electroplating (qv) and metal finishing industries. Some of the alkali metal fluoroborates are used in fluxes.

Properties. Fluoroboric acid is stable in concentrated solutions, and hydrolyzes slowly in aqueous solution to hydroxyfluoroborates.

Manufacture, Shipping, and Waste Treatment. Fluoroboric acid (48%) is made commercially by direct reaction of 70% hydrofluoric acid and boric acid, H_3BO_3. The reaction is exothermic and must be controlled by cooling.

The commercial product is usually a 48–50% solution which contains up to a few percent excess boric acid to eliminate any HF fumes and to avoid HF burns. Reagent-grade solutions are usually 40%. Vessels and equipment must withstand the corrosive action of hydrofluoric acid. For a high quality product the preferred materials for handling HBF_4 solutions are polyethylene, polypropylene, or a resistant rubber such as neoprene (see ELASTOMERS, SYNTHETIC). Where metal must be used, ferrous alloys having high nickel and chromium content show good resistance to corrosion. Impregnated carbon (Carbate) or Teflon can be used in heat exchangers. Teflon-lined pumps and auxilliary equipment are also good choices. Working in glass equipment is not recommended for fluoroboric acid or any fluoroborate.

Fluoroboric acid and some fluoroborate solutions are shipped as corrosive material, generally in polyethylene-lined steel pails and drums or in rigid nonreturnable polyethylene containers. Acid spills should be neutralized with lime or soda ash.

Waste treatment of fluoroborate solutions includes a pretreatment with aluminum sulfate to facilitate hydrolysis, and final precipitation of fluoride with lime. The aluminum sulfate treatment can be avoided by hydrolyzing the fluoroborates at pH 2 in the presence of calcium chloride; at this pH, hydrolysis is most rapid at elevated temperature.

Analysis. Fluoroboric acid solutions and fluoroborates are analyzed gravimetrically using nitron or tetraphenylarsonsium chloride. A fluoroborate ion-selective electrode has been developed.

Health and Safety Factors. Fluoroborates are excreted mostly in the urine. Sodium fluoroborate is absorbed almost completely into the human bloodstream and over a 14-d experiment all of the $NaBF_4$ ingested was found in the urine. Although the fluoride ion is covalently bound to boron, the rate of absorption of the physiologically inert BF_4^- from the gastrointestinal tract of rats exceeds that of the physiologically active simple fluorides.

Uses. Printed circuit tin–lead plating is the main use of fluoroboric acid. However, the Alcoa Alzak process for electropolishing aluminum requires substantial quantities of fluoroboric acid. The high solubility of many metal oxides in HBF_4 is a decided advantage in metal finishing operations.

Fluoroboric acid is used as a stripping solution for the removal of solder and plated metals from less active substrates. A number of fluoroborate plating baths require pH adjustment with fluoroboric acid.

A low grade fluoroboric acid is used in the manufacture of cryolite for the electrolytic production of aluminum. As a strong acid, fluoroboric acid is frequently used as an acid catalyst; e.g, in synthesizing mixed polyol esters. This process provides an inexpensive route to confectioner's hard-butter compositions, which are substitutes for cocoa butter in chocolate candies.

Main Group

Properties. The alkali-metal and ammonium fluoroborate compounds differ chemically from the transition-metal fluoroborates, usually separating in anhydrous from. This group is very soluble in water, except for the K, Rb, and Cs salts.

These hydrated fluoroborates can be dehydrated completely to the anhydrous salts, which show decreasing stabilities: $Ba > Sr > Ca > Mg$.

The anhydrous magnesium salt is least stable thermally.

Manufacture. Fluoroborate salts are prepared commercially by several different combinations of boric acid and 70% hydrofluoric acid with oxides, hydroxides, carbonates, bicarbonates, fluorides, and bifluorides. Fluoroborate salts are substantially less corrosive than fluoroboric acid, but the possible presence of HF or free fluorides cannot be overlooked. Glass vessels and equipment should not be used.

Sodium Fluoroborate. Sodium fluoroborate is prepared by the reaction of NaOH or Na_2CO_3 with fluoroboric acid, or by treatment of disodium hexafluorosilicate with boric acid.

Potassium Fluoroborate. Potassium fluoroborate is produced as a gelatinous precipitate by mixing fluoroboric acid and KOH or K_2CO_3.

Ammonium and Lithium Fluoroborates. Ammonia reacts with fluoroboric acid to produce ammonium fluoroborate.

Magnesium Fluoroborate. Treatment of magnesium metal, magnesium oxide, or magnesium carbonate with HBF_4 gives magnesium fluoroborate. The MgF_2 is filtered and the product is sold as a 30% solution.

Uses. Alkali metal and ammonium fluoroborates are used mainly for the high temperature fluxing action required by the metals-processing industries. The tendency toward BF_3 dissociation at elevated temperatures inhibits oxidation in magnesium casting and aluminum alloy heat treatment.

The molten salts quickly dissolve the metal oxides at high temperatures to form a clean metal surface. Other uses are as catalysts and in fire-retardant formulations.

Potassium Fluoroborate. The addition of potassium fluoroborate to grinding wheel and disk formulations permits lower operating temperatures. Cooler action is desirable to reduce the burning of refractory materials such as titanium and stainless steels. Excellent results in grinding wheels are also obtained with $NaBF_4$. A process for boriding steel surfaces using B_4C and KBF_4 as an activator improves the hardness of the base steel. Fluxes for aluminum bronze and silver soldering and brazing contain KBF_4.

Fire retardance is imparted to acrylonitrile polymers by precipitating KBF_4 within the filaments during coagulation. In polyurethanes, KBF_4 and NH_4BF_4 reduce smoke and increase flame resistance. Both the potassium and ammonium salts improve the insulating efficiency of intumescent coatings. of up to 99.5% purity with KBF_4 containing the ^{10}B isotope is used in the nuclear energy field as a neutron absorber.

Sodium Fluoroborate. Sodium fluoroborate can be used in the transfer of boron to aluminum alloys, but the efficiency is lower than for KBF_4. Sodium fluoroborate in an etching solution with sulfamic acid and H_2O_2 aids in removing exposed lead in printed circuit manufacture. Sodium and lithium fluoroborates are effective flame retardants for cotton and rayon.

Ammonium Fluoroborate. Ammonium fluoroborate blends with antimony oxide give good results in flame-retarding polypropylene. The complete thermal vaporization makes ammonium fluoroborate an excellent gaseous flux for inert-atmosphere soldering. A soldering flux of zinc chloride and ammonium fluoroborate is used in joining dissimilar metals such as Al and Cu. Ammonium fluoroborate acts as a solid lubricant in cutting-oil emulsions for aluminum rolling and forming.

Lithium Fluoroborate. Lithium fluoroborate is used in a number of batteries as an electrolyte, for example in the lithium–sulfur battery.

Miscellaneous. Flame-resistant cross-linked polyethylene can be made with a number of fluoroborates and antimony oxide.

Transition-Metal and Other Heavy-Metal Fluoroborates

The physical and chemical properties are less well known for transition metals than for the alkali metal fluoroborates. Most transition-metal fluoroborates are strongly hydrated coordination compounds and are difficult to dry without decomposition. Decomposition frequently occurs during the concentration of solutions for crystallization. The stability of the metal fluorides accentuates this problem. Loss of HF because of hydrolysis makes the reaction proceed even more rapidly. Even with low temperature vacuum drying to partially solve the decomposition, the dry salt readily absorbs water. The crystalline solids are generally soluble in water, alcohols, and ketones but only poorly soluble in hydrocarbons and halocarbons.

Manufacture. The transition- and heavy-metal fluoroborates can be made from the metal, metal oxide, hydroxide, or carbonate with fluoroboric acid. Because of the difficulty in isolating pure crystalline solids, these fluoroborates are usually available as 40–50% solutions, $M(BF_4)_x$. Most of the solutions contain about 1–2% excess fluoroboric acid to prevent precipitation of basic metal complexes. The solutions are usually sold in 19 and 57 L polyethylene containers.

In some cases, particularly with inactive metals, electrolytic cells are the primary method of manufacture of the fluoroborate solution.

Anhydrous silver fluoroborate is made by the addition of BF_3 gas to a suspension of AgF in ethylbenzene.

Uses. Metal fluoroborate solutions are used primarily as plating solutions and as catalysts. The Sn, Cu, Zn, Ni, Pb, and Ag fluoroborates cure a wide range of epoxy resins at elevated or ambient room temperature. In the textile industry, zinc fluoroborate is used extensively as the curing agent in applying resins for crease-resistant finishes. The manufacture of linear polyester is catalyzed by Cd, Sn, Pb, Zn, or Mn fluoroborates. Metal fluoroborate electroplating baths are employed where speed and quality of deposition are important.

B. R. Gragg, in K. Niedenzu, K. C. Buschbeck, and P. Merlet, eds., *Gmelin Handbook of Inorganic Chemistry, Borverbindungen*, Vol. 19, Springer-Verlag, Berlin, Germany, 1978, pp. 109, 168.

E. J. Largent, "Metabolism of Inorganic Fluoride" in *Fluoridation as a Public Health Measure*, American Association for the Advancement of Science, Washington, D.C., 1954, 49–78.

A. Meller, "Boron Compounds," in *Gmelin Handbook of Inorganic Chemistry*, 2nd Suppl., Vol. 2, Springer-Verlag, Berlin, Germany, 1982, pp. 77, 154.

G. Urry, in E. Muetterties, ed., *The Chemistry of Boron and Its Compounds*, John Wiley & Sons, Inc., New York, 1967, p. 325.

FAZLUL ALAM
U.S. Borax Research Corporation
FRANCIS EVANS
GANPAT MANI
Allied Signal, Inc.
JOHN R. PAPCUN
Atotech

BORON HYDRIDES, HETEROBORANES, AND THEIR METALLA DERIVATIVES

The boron hydrides, including the polyhedral boranes, heteroboranes, and their metalla derivatives, encompass an amazingly diverse area of chemistry. This class contains the most extensive array of structurally characterized cluster compounds known. Included here are many unique clusters possessing idealized molecular geometries ranging over every point group symmetry from identity (C_1) to icosahedral (I_h). Because boron hydride clusters may be considered in some respects to be progenitorial models of metal clusters, their development has provided a framework for the development of cluster chemistry in general as well as for chemical bonding theory.

NOMENCLATURE

The nomenclature of boron hydride derivatives has been somewhat confusing and many inconsistencies exist in the literature. The structures of some reported boron hydride clusters are so complicated that only a structural drawing or graph, often accompanied by explanatory text, is used to describe them. Nomenclature systems often can be used to describe compounds unambiguously, but the resulting descriptions may be so unwieldy that they are of little use. The International Union of Pure and Applid Chemistry (IUPAC) and the Chemical Abstract Service (CAS) have made recommendations, and nomenclature methods have now been developed that can adequately handle nearly all cluster compounds; however, these methods have yet to be widely adopted. For the most part, the nomenclature used in the original literature is retained herein.

The neutral boron hydrides are termed "boranes". The molecule BH_3 is called borane or borane(3). For more complex polyboranes, the number of boron atoms is indicated by the common prefixes di-, tri-, tetra-, etc., and the number of hydrogens (substituents) is given by an arabic numeral in parentheses following the name. The position of the substituents can be designated precisely from framework-numbering conventions. Because numbering systems are used, it is advisable to refer to structural diagrams.

Borane polyhedra have both closed and open skeletons, and it has become common practice to include the appropriate structural classification in the compound's name. Closed polyhedra having only triangular faces are termed *closo*. Open structures are designated *nido*, *arachno*, or *hypho*. Boron hydride anions are generally termed "hydroborates" using prefixes to designate the number of hydrogens and borons; the charge follows the name in parentheses.

When a boron atom of a borane is replaced by a heteroelement, the compounds are called carbaboranes, phosphaboranes, thiaboranes, azaboranes, etc., by an adaptation of organic replacement nomenclature. The original term "carborane" has been widely adopted in preference to carbaborane, the more systematic name. The numbering of the skeleton in heteroboranes is such that the heteroelement is given the lowest possible number consistent with the conventions of the parent borane. When different heteroelements occur in combination in a polyhedron, *Chemical Abstracts* gives priority by descending group number and increasing atomic number within a group. However, the hierarchy in the original literature often gives the lowest number to the element of lowest atomic number. This convention carries over to the metallaboranes and metallaheteroboranes when a metal occupies a polyhedral vertex. The arabic numeral in parentheses following the name does not include exopolyhedral ligands bonded to the metal, only the total of the hydrogen atoms plus other substituents bonded to boron and main group heteroelements in the skeletal framework.

A variety of heteroboranes, metallaboranes, and metallaheteroboranes exist that contain more than one interconnected polyhedral cluster. These complex clusters are referred to as conjuncto-boranes. Conjuncto-boranes may be interconnected by sharing a single common boron atom, having a direct B—B bond between two clusters, sharing two boron atoms at a polyhedral edge or three boron atoms at a face, or more extensive polyhedral fusion by the sharing of four or more boron atoms between clusters.

The *commo* prefix is often used to indicate that the metal vertex is shared by two polyhedra. The *commo* nomenclature of metallaboranes and metallaheteroboranes is a widely used special case of the IUPAC recommended conjuncto nomenclature.

STRUCTURAL SYSTEMATICS

Because the polyhedral boron hydrides are cage molecules, which usually possess triangular faces, their idealized geometries can be described accurately as deltahedra or deltahedral fragments. The left-hand column of Figure 1 illustrates the deltahedra containing $n = 6–12$ vertices: the octahedron, pentagonal bipyramid, bisdisphenoid, symmetrically tricapped trigonal prism, bicapped square antiprism, octadecahedron, and icosahedron. These idealized structures are convex deltahedra except for the octahedron, which is not a regular polyhedron. The left-hand column of Figure 1 also represents the class of deltahedral *closo* molecules from which the other idealized structures (deltahedral fragments) can be generated systematically. Any *nido* or *arachno* cluster can be generated from the appropriate deltahedron by ascending a diagonal from left to right. This progression generates the *nido* structure (center column) by removing the most highly connected vertex of the deltahedron, and the *arachno* structure (right column) by removal of the most highly connected atom of the open (nontriangular) face of the *nido* cluster. The structural correlations shown in Figure 1 were formulated in 1971, and subsequently elaborated. The terms *closo, nido, arachno,* and *hypho* are derived from Greek and Latin and imply closed, nestlike, weblike, and netlike structures, respectively. These classifications apply equally well to boranes, heteroboranes, and their metalla analogues and are intimately connected to a quantity known as the framework, or skeletal,

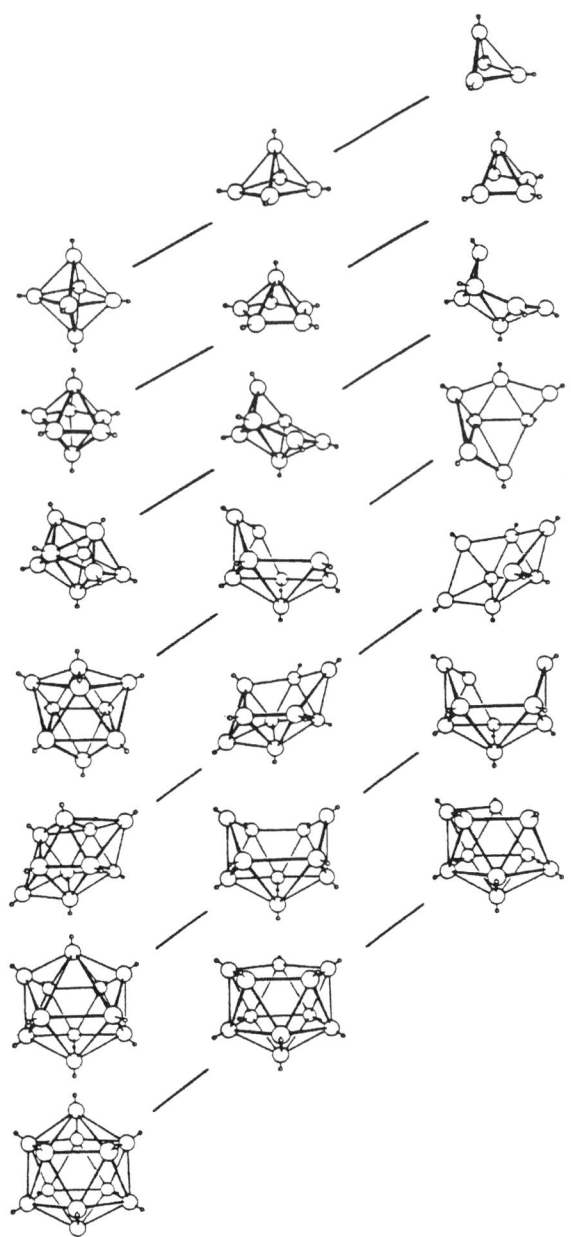

Figure 1. Idealized deltahedra and deltahedral fragments for *closo, nido,* and *arachno* boranes and heteroboranes. From left to right, the vertical columns give generic *closo, nido,* and *arachno* frameworks; bridge hydrogens and BH_2 groups are not shown, but when appropriate they are placed around the open (nontriangular) face of the framework.

electron count. The partitioning of electrons into framework and exopolyhedral classes allows for predictions of structures in most cases, even though these systematics are not concerned explicitly with the assignments of localized bonds within the polyhedral skeletons of these molecules. That is, the lines depicting the skeletons of the structures illustrated are not electron-pair, or "electron precise", bonds. The lines merely serve to join nearest neighbors and illustrate cluster geometries. However, exopolyhedral lines do represent the usual electron precise bonds.

Placement of Heteroatoms

Many of the deltahedra and deltahedral fragments of Figure 1 have two or more nonequivalent vertices. Nonequivalent vertices are recognized as having a different order; ie, a different number of nearest neighbor vertices within the framework. Heteroatoms generally exhibit a positional preference based on the order of the polyhedral vertex and the electron richness of the heteroatom relative to boron. Electron-rich heteroatom groupings contribute more framework electrons than a : B—H moiety, which has two framework electrons, and generally prefer low-order vertices, ie, those having fewer neighbors.

Placement of Extra Hydrogens

The placement of extra hydrogens plays a crucial role in determining the structures adopted by boranes and carboranes. However, the exact position of extra hydrogens sometimes depends on the physical state of the molecule, eg, the tridecahydrodecaborate(1−) anion, $[B_{10}H_{13}]^-$ exhibits different bridge hydrogen placements in the crystal and in solution as can be inferred from experimental evidence, but the solution data are also consistent with a dynamic process of bridge hydrogen tautomerism.

The placement of bridge hydrogens may be the most important variable in the determination of relative isomer stabilities, outranking placement of heteroatoms.

M—H—B Bridges

Numerous metallaboranes and metallaheteroboranes are known to contain hydrogens bridging between a metal atom and a skeletal boron atom, but complexes containing covalently bound tetrahydroborate(−1), $[BH_4]^-$, constitute the prototypical class. Metal tetrahydroborates have been reviewed.

Exceptions to Structural Systematics

When strong electron donating or withdrawing groups are present as substituents attached to boron in polyboranes, there is the possibility of structural anomalies. In some cases, electron deficiency of boron apparently can be ameliorated by back-donation instead of by the multicenter bonding afforded in a cage framework. Thus it has been suggested that exceptions to the electron-counting paradigms may occur where back-donation from the substituent to a cluster boron is possible. Some metallacarboranes present anomalies to the electron-counting formalisms.

BONDING

Localized Bonds

Because boron has more valence orbitals than valence electrons, its compounds are often called electron deficient. This electron deficiency is partly responsible for the great interest surrounding boron hydride chemistry and molecular structure.

The elucidation of the structure of diborane(6) led to the description of a new bond type, the three-center two-electron bond, in which one electron pair is shared by three atomic centers. The delocalization of a bonding pair over a three-center bond allows for the utilization of all the available orbitals in an electron-deficient system. This key point led to the formulation of a valence bond description of the bonding in boron hydrides, sometimes termed a topological description. The valence structures of this topological approach give localized bonding descriptions that include delocalized three-center bonds in the basis set of bond types. In addition to the B—H—B three-center bridge bond depicted, a B—B—B three-center bond was introduced to describe bonding in the framework.

The valence theory includes both types of three-center bonds shown as well as normal two-center, B—B and B—H, bonds.

BORANES

Nido and Arachno Boranes

These boranes are generally more reactive and less stable thermally than the corresponding *closo* boranes. The most extensively studied boranes include diborane(6), B_2H_6, tetraborane(10), B_4H_{10}, pentaborane(9), B_5H_9, and decaborane(14), $B_{10}H_{14}$. This subject has been reviewed. A great deal of early work in this area was associated with the government-sponsored high energy fuels programs. Some of this work is summarized. The *nido* and *arachno* boranes smaller than $B_{10}H_{14}$ are quite reactive toward oxygen and water. The properties of selected boranes are given in Table 1.

Table 1. Physical Properties of Boranes

Borane	Molecular formula	Mp, °C	Bp, °C	ΔH_f°, kJ/mol[a]	ΔG_f°, kJ/mol[a]	ΔS_{298}°, J/(K·mol)[a]
diborane(6)	B_2H_6	−164.9	−92.6	35.5	86.6	232.0
tetraborane(10)	B_4H_{10}	−120	18	66.1		
pentaborane(9)	B_5H_9	−46.6	48	73.2	174	275.8
pentaborane(11)	B_5H_{11}	−123	63	103.0		
hexaborane(10)	B_6H_{10}	−62.3	108	94.6		
decaborane(14)	$B_{10}H_{14}$	99.7	213	31.5	216.1	353.0

[a] To convert J to cal, divide by 4.184.

Molecular Orbital Descriptions

In addition to the localized bond descriptions, molecular orbital (MO) descriptions of bonding in boranes and carboranes have been developed. Early work on boranes helped one of the most widely applicable approximate methods, the extended Hückel method. Molecular orbital descriptions are particularly useful for *closo* molecules where localized bond descriptions become cumbersome because of the large number of resonance structures that do not accurately reflect molecular symmetry.

Reactions of Boranes with Lewis Bases

Boranes that contain a BH_2 moiety eg, B_2H_6, B_4H_{10}, B_5H_{11}, hexaborane, B_6H_{12}, and nonaborane, B_9H_{15}, can generally be cleaved by nucleophiles in two ways termed symmetrical and unsymmetrical bridge cleavage. By using neutral bases, the two modes of cleavage lead to molecular and ionic fragments, respectively. Certain base adducts of borane, BH_3, such as triethylamine borane, $(C_2H_5)_3N \cdot BH_3$, dimethylsulfide borane, $(CH_3)_2S \cdot BH_3$, and tetrahydrofuran borane, $C_4H_8O \cdot BH_3$, are more easily and safely handled than B_2H_6 and are commercially available. These compounds find wide use as reducing agents and in hydroboration reactions. Base displacement reactions can be used to convert one adduct to another.

Proton Abstraction

Although the exopolyhedral hydrogens of *nido* and *arachno* boranes are generally considered hydridic, the bridge hydrogens are acidic, as first demonstrated by titration of $B_{10}H_{14}$ and deuterium exchange.

Polyhedral Expansion

The term "polyhedral expansion" is used to describe a host of reactions in which the size of the polyhedron is increased by the addition of new vertex atoms whether boron, heteroelements, or metals.

Electrophilic Attack

A variety of boranes, heteroboranes, and metallaboranes undergo electrophilic substitution.

Closo Borane Anions

This group contains a homologous series of very stable polyhedral anions, $[closo-B_nH_n]^{2-}$, $n = 6-12$. Just as the previously known boron hydrides might be considered as analogues of aliphatic hydrocarbons, the *closo* borane anions are analogues of aromatic hydrocarbons. The stability of the *closo* anions is attributable to electron delocalization in a unique 3D aromaticity. Unlike their *nido* and *arachno* counterparts with bridging hydrogens, proton abstraction does not, for practical purposes, occur in *closo* borane chemistry. Instead, acid catalysis is important in their substitution chemistry. The best-known members of this series, $[closo-B_{10}H_{10}]^{2-}$ and $[closo-B_{12}H_{12}]^{2-}$, were first reported in 1959 and 1960 and were the subject of detailed studies.

Table 2. Properties of Alkali Metal Tetrahydroborates

Property	Compound				
	$LiBH_4$	$NaBH_4$	KBH_4	$RbBH_4$	$CsBH_4$
mp,°C	268	505	585		
decomp. temp.,°C	380	315	584	600	600
density, g/mL	0.68	1.08	1.17	1.71	2.40
refractive index	1.547	1.490	1.487	1.498	122
lattice energy, kJ/mol[a]	792.0	697.5	657	648	630.1
ΔH_f°, kJ/mol[a]	−184	−183	−243	−246	−264
ΔS_{298}°, J/(mol·K)[a]	−128.7	−126.3	−161	−179	−192

[a] To convert J to cal, divide by 4.184.

Tetrahydroborates

The tetrahydroborates constitute the most commercially important group of boron hydride compounds. Tetrahydroborates of most of the metals have been characterized and their preparations have been reviewed. The most important commercial tetrahydroborates are those of the alkali metals. Some properties are given in Table 2.

The use of tetrahydroborates, as well as the boranes and organoboranes, for organic transformations has proven to be even more significant because these reduction reactions are highly selective and nearly quantitative. The reducing characteristics of borohydrides may be varied by changing the associated cation and by changing the solvent. Borohydrides are often the reagents of choice for the reduction of aldehydes and ketones to the corresponding alcohols, especially when selective reduction in the presence of other functional groups is required. Many other functional groups, such as acid chlorides, imines, and peroxides, can also be reduced using borohydrides. There is also considerable interest in the use of Lithium and sodium tetrahydroborates for hydrogen storage.

HETEROBORANES

Heteroboranes contain heteroelements classified as nonmetals. The heteroatoms known to form part of a borane polyhedron include C, N, O, Si, P, Ge, As, S, Se, Sb, and Te either alone or in combination. In principle, heteroboranes containing a variety of heteroatoms could have a wide range of skeletal sizes. Of these, the carboranes have by far the greatest demonstrated scope of chemistry.

Carboranes

The term "carborane" is widely used as a contraction of the IUPAC approved nomenclature "carbaborane". The first carboranes, including isomers of $C_2B_3H_5$, $C_2B_4H_6$, and $C_2B_5H_7$, were prepared in the mid-1950s. These carboranes were obtained as mixtures in low yield from the reaction of smaller boranes such as pentaborane(9) with acetylene in a silent electric discharge. The discovery of the icosahedral closo-1,2-dicarbadodecaborane(12), 1,2-$C_2B_{10}H_{12}$, came soon after and led to a rapid development of carborane chemistry.

The discovery of the base-promoted degradation of the isomeric closo-$C_2B_{10}H_{12}$ cages provided one of the most important carborane anion systems, the isomeric [nido-$C_2B_9H_{12}$]$^-$ anions,

$$closo - C_2B_{10}H_{12} + RO^- + 2\,ROH \longrightarrow$$
$$[nido - C_2B_9H_{12}]^- + B(OR)_3 + H_2$$

where R = CH_3, C_2H_5, etc. The [nido–$C_2B_9H_{12}$]$^-$ cages, and their C-substituted derivatives, are commonly referred to as dicarbollide ions, derived from the Spanish olla, meaning a bowl. Aside from their extensive use in metallacarborane chemistry, the dicarbollide anions are important intermediates in the synthesis of other carborane compounds.

The arachno carboranes 1,3-$C_2B_7H_{13}$ and 6,9-$C_2B_8H_{14}$ are unusual in that two of the extra hydrogens occur in CH_2 groups. The other two extra hydrogens are present as B−H−B bridges.

As with the simple boranes, the closo carboranes are generally more thermally stable than the corresponding nido and arachno species.

Cage rearrangements in polyhedral carboranes are well known. Although most carborane cages are stable at room temperature, many undergo rearrangements at elevated temperatures. Carborane isomers obtained by conventional synthetic routes are often kinetic products and not the thermodynamically most stable isomers. When subjected to elevated temperatures below the ultimate decomposition temperatures, carboranes often undergo rearrangements to the more stable isomers.

A diversity of polyhedral carborane cage-containing polymers has been prepared. The best known of these are elastomeric polycarboranylsiloxanes which were developed by Olin Corp. and Union Carbide Corp. in the 1970s. These are based on m-carborane cages linked by polysiloxane groups with direct C−Si bonds. The properties of these materials can be varied by changing the length and substituents of the polysiloxane linkages as well as their overall molecular weights. Some of these materials have excellent thermal stabilities, chemical resistance, and high temperature elastomeric properties. Polymers of this type, known under the trade name Dexsil, were commercial materials, useful as high temperature stationary phases in gas chromatography among other applications. These compounds, however, have not been produced commercially for many years. The organic and organometallic chemistry of closo carborane derivatives has been reviewed.

Other Heteroboranes

Other well-documented families of heteroboranes include the azaboranes, thiaboranes, phosphaboranes, arsenaboranes, stibaboranes, selenaboranes, and telluraboranes.

METALLABORANES

Transition-Element Metallaboranes

The transition-metal hydroborate cluster, $HMn_3(CO)_{10}$ $(BH_3)_2$, containing a B_2H_6 moiety, which is multiplied bridging between three manganese carbonyl and manganese carbonyl hydride centers via M−H−B bridges, might be regarded as the first structurally characterized metallaborane cluster. This and similar clusters were isolated in the 1960s as by-products in the synthesis of transition-metal carbonyl hydrides by sodium borohydride reduction of metal carbonyls, a standard method for the preparation of transition-metal hydride complexes and clusters since the 1970s. Indeed, the [BH_4]$^-$ anion acts as a ligand in a wide variety of metal complexes in which from one to all four hydrogen atoms are involved in bonding to metals. However, the chemistry of stable metallaboranes that incorporate metals in vertex positions of polyhedral borane clusters was developed somewhat later. To date a great many metallaborane clusters have been characterized covering a wide range of metals, sizes, and polyhedral fragment geometries.

One of the most extensive classes of metallaboranes are those derived from decaborane, which in most cases produces 11-vertex metallaborane products.

The [$B_{10}H_{12}$]$^-$ anion can also be considered as a bidentate ligand that coordinates metals between boron atoms 2,11 and 3,8, the metal at position 7 such that the metal in effect occupies the position of a bridge hydrogen of the conjugate acid borane. Situations in which a metal vertex may be regarded as equivalent to an H$^+$, BH^{2+}, or the BH$_2^+$ moiety have also been discussed.

The first closo metallaborane complexes prepared were the nickelaboranes [closo–(η^5–C_5H_5)Ni($B_{11}H_{11}$)]$^-$ and closo-1,2–(η^5–C_5H_5)$_2$–1,2–$Ni_2B_{10}H_{10}$. These metallaboranes display remarkable hydrolytic, oxidative, and thermal stability.

Main Group Element Metallaboranes

A variety of metallaborane clusters, which incorporate main group metals in vertex positions of polyhedral metallaborane clusters, have been reported.

Exopolyhedral Metallaboranes

Polyboranes may bind exopolyhedral metals in a variety of ways. Most commonly, metals are bound via M−H−B interactions. In other cases, metals may formally replace bridging hydrogen atoms at edge positions to give B−M−B interactions. Metals may also be attached to polyborane cages by direct M−B−σ bonds.

Metallacarboranes

The isomeric [nido–$C_2B_9H_{11}$]$^{2-}$ anions and their C-substituted derivative, which are commonly known as dicarbollide ions, form stable complexes with most of the metallic elements. Indeed, nearly all metals can be

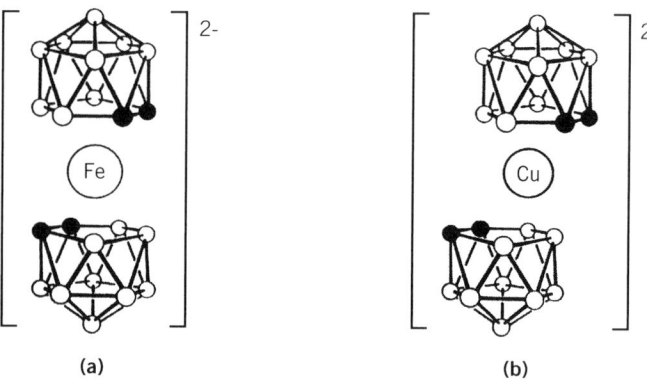

Figure 2. Exemplary structures of (**a**) unslipped and (**b**) slipped metallacarbollide dicarbollide sandwich derivatives where ○ represents BH; ● CH.

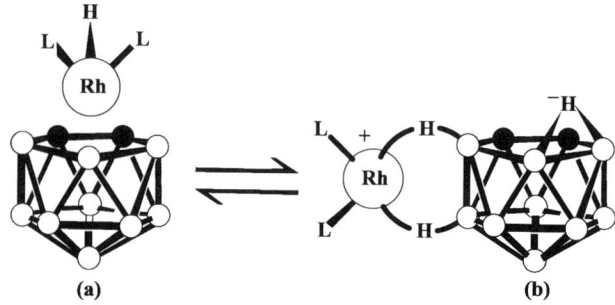

Figure 3. (**a**) The structure of *closo*-3,3-[(C$_6$H$_5$)$_3$P]$_2$-3-H-3,1,2-RhC$_2$B$_9$H$_{11}$, and (**b**) one isomer of its tautomer *exo–nido*-(L$_2$Rh)$_2$-(μ-H)$_2$-7,8-C$_2$B$_9$H$_{12}$, where L is (C$_6$H$_5$)$_3$P.

combined with polyboron hydride clusters to produce an apparently limitless variety of cluster compounds.

Transition-Metal Metallacarboranes

The first demonstration of the insertion of a transition metal into an open face of a carborane cage was with the iron sandwich compound [*commo*–Fe(C$_2$B$_9$H$_{11}$)$_{12}$]$^{2-}$. This product is readily air oxidized to [*commo*–(C$_2$B$_9$H$_{11}$)$_2$Fe]$^-$, a complex containing a formal Fe^{3+} center. These complexes, as well as those formed from a variety of formally d^3, d^5, d^6, and d^7 transition metals, have symmetrical sandwich structures of the type shown in Figure 1**a**. By contrast, d^8 and d^9 metals form slipped sandwich structures as shown in Figure 2**b**.

Exopolyhedral Metallacarboranes

Many metallacarboranes are known that exhibit exopolyhedral bonding to metals. Most commonly in these compounds metals are bound via M–H–B interactions in which the B–H group can be regarded as a two-electron donor to the metal center. In other cases, M–B, M–C, or M–M bonding may be involved.

Host–Guest Chemistry-Carborane Anticrowns

An extensive literature exists for compounds that complex cations, including the crown ethers and more complex host–guest chemistries. However, compounds that display selective anion complexation are more unusual. Anion complexation has received increasing attention recently because of its importance to biology and analytical chemistry.

Metallacarboranes in Catalysis

One of the most intensely studied metallacarborane complexes is the exopolyhedral metallacarborane *closo*–3,3–[P(C$_6$H$_5$)$_3$]$_2$–3–H–3,1,2–RhC$_2$B$_9$H$_{11}$, shown in Figure 3**a**, and its cage C–substituted derivatives. The three available isomers of *closo*–[P(C$_6$H$_5$)$_3$]$_2$(H)Rh–C$_2$ B$_9$H$_{11}$ are

synthesized in high yield by the oxidative addition of [P(C$_6$H$_5$)$_3$]$_2$RhCl with the appropriate [*nido*-C$_2$B$_9$H$_{12}$]$^-$ ion, which may also be made chiral by the attachment of a single-alkyl or -aryl group at a carbon position. The resulting hydridorhodacarboranes are quite robust and function as homogeneous-catalyst precursor for a number of reactions including the olefin hydrogenation the deuteration of B–H groups and alkenyl acetates.

Main Group Element Carborane Derivatives

Main group element carborane derivatives have been reviewed. Only a few alkaline-earth element metallacarborane derivatives have been characterized.

f-Block Element Metallacarborane Derivatives

The first actinide metallacarborane complex, *commo*-(C$_2$B$_9$H$_{11}$)UCl$_2$, was prepared in 1977. The coordination geometry of this complex can be described as distorted tetrahedral with the four positions occupied by two η^5–bound [7,8-C$_2$B$_9$H$_{11}$]$^{2-}$ cages and two chloride ions. Complexes of this type are often referred to as bent sandwiches because of the configuration of the two-dicarbollide cage about the metal center, which is analogous to the corresponding pentamethyl cyclopentadiene–metal complexes.

Boron Neutron Capture Therapy

There is great interest in the use of polyboron hydride compounds for boron neutron capture therapy (BNCT) for the treatment of cancers and other diseases. Boron-10 is unique among the light elements in that it possesses an unusually high neutron capture nuclear cross-section (3.8×10^{-25} m^2, 0.02–0.05 eV neutron). The nuclear reaction between ^{10}B and low-energy thermal neutrons yields alpha particles and recoiling lithium-7 nuclei.

Because the cytotoxic effects of the energetic lithium-7 and α particles are spacially limited to a range of only about one-cell diameter, the destructive effects are confined to only cells near the site of the event. Thus BNCT involves the selective delivery of sufficiently high concentrations of ^{10}B-containing compounds to tumor sites followed by the irradiation of these sites with a beam of relatively nondestructive thermal neutrons. The resulting cytotoxic reaction can then in theory destroy the tumor cells that are intimately associated with the ^{10}B target. The great advantage of BNCT is its cell level selectivity for destroying tumor cells without significant damage to healthy tissue. This attribute makes BNCT particularly valuable in the treatment of tumors that are difficult or impossible to remove by surgical methods.

The challenge of BNCT lies in the development of practical means for the selective delivery of approximately 10^9 ^{10}B atoms to each tumor cell for effective therapy using short neutron irradiation times.

To date, the most extensively studied polyboron hydride compounds in BNCT research have been the icosahedral mercaptoborane derivatives $Na_2[B_{12}H_{11}SH]$ (called BSH) and $Na_4[(B_{12}H_{11}S)_2]$, which have been used in human trials, with some success. New generations of tumor-localizing boronated compounds are being developed. The dose-selectivity problem of BNCT has been approached using boron hydride compounds in combination with a variety of delivery vehicles, including boronated polyclonal and monoclonal antibodies, porphyrins, amino acids, nucleotides, carbohydrates, and liposomes. BNCT has also been proposed as a treatment for other diseases such as arthritis.

Boranes as Pharmacophores

The unique properties of polyhedral boranes, such as hydrophobicity, steric bulk, stability under physiological conditions, and ease of functionalization, have been exploited in the design of new pharmaceutical agents. Carboranes have also been used as pharmacophores in retinoid antagonists and other biologically active molecules of therapeutic interest.

A related potential medical application of metallacarboranes is based on the highly favorable kinetic stability of many metallacarborane complexes under physiological conditions and ability to form stable complexes with a wide range of metals. These features make certain functionalized metallacarboranes containing radiometals ideal choices for use as medical imaging reagents. The use of antibody-conjugated bridged dicarbollide metallacarborane (venus fly trap) chelate complexes incorporating γ-emitting $^{57}Co^{3+}$ in the imaging tumors has been reported.

ECONOMIC ASPECTS

Despite the fact that many boron hydride compounds possess unique and useful chemical and physical properties, very few of these compounds have yet undergone significant commercial exploitation. This is largely owing to the extremely high cost of most boron hydride materials, which has discouraged development of all but the most exotic applications. Nevertheless, considerable commercial potential is foreseen for boron hydride materials if and when economical and reliable sources become available. Only the simplest of boron hydride compounds, most notably sodium tetrahydroborate, $NaBH_4$, diborane(6), B_2H_6, and some of the borane adducts, eg, amine boranes, are now produced in significant commercial quantities.

Sodium Tetrahydroborate, Na[BH₄]

This air-stable white powder, commonly referred to as sodium borohydride, is the most important commercial boron hydride material. It is used in a variety of industrial processes including bleaching of paper pulp and clays, preparation and purification of organic chemicals and pharmaceuticals, textile dye reduction, recovery of valuable metals, wastewater treatment, and production of dithionite compounds. More than 6 million lb of this material supplied as powder, pellets, and aqueous solution (called Borol solution), were produced in 2001.

Diborane(6), B₂H₆

This spontaneously flammable gas is consumed primarily by the electronics industry as a dopant in the production of silicon wafers for use in semiconductors. It is also used to produce amine boranes and the higher boron hydrides. Several hundred tons were manufactured worldwide in 2001.

Borane Adducts

Trialkylamine and dialkylamine boranes, such as tri-*tert*-butylamine borane and dimethylamine borane, are mainly used in electroless plating processes. Other borane adducts, such as THF–borane and dimethyl sulfide–borane are used for specialized reduction reactions.

Organoboron Hydrides

A variety of organoboron hydrides produced by hydroboration of olefins is commercially available. These are widely used in organic synthesis and the manufacture of pharmaceuticals.

Polyhedral Boron Hydrides

Although relatively large quantities of polyhedral boron hydrides and carboranes have been produced under various government contracts, these materials are not currently produced on large-scale commercial basis.

Polyhedral Boron Hydrides

These are used as experimental agents in neutron capture therapy of cancers, and as burn-rate modifiers (accelerants) in gun and rocket propellant compositions. A salt of the $[B_{12}H_{12}]^{2-}$ anion is used in the fuse system of the passenger-side automotive airbag.

Carboranes

These are used as experimental agents in neutron capture therapy, and as burn-rate modifiers in gun and rocket propellants. They have been used as in high-temperature elastomers and high-temperature gas–liquid chromatography stationary phases and have potential for use in other unique materials, optical switching devices, and gasoline additives.

Metallacarboranes

These have potential for use in homogeneous catalysis, including hydrogenation, hydrosilylation, isomerization, hydrosilanolysis, phase-transfer, burn rate modifiers in gun and rocket propellants, neutron capture therapy, medical imaging, processing of radioactive waste, analytical reagents, and as ceramic precursors.

S. Bresadola, in R. N. Grimes, ed., *Metal Interactions with Boron Clusters*, Plenum Press, New York, 1982.

R. T. Holtzmann, ed., *Production of Boranes and Related Research*, Academic Press, New York, 1967.

E. L. Muetterties and W. H. Knoth, *Polyhedral Boranes*, Marcel Dekker, Inc., New York, 1968.

S. G. Shore, in E. L. Muetterties, ed., *Boron Hydride Chemistry*, Academic Press, New York, 1975, Chapt. 3.

DAVID M. SCHUBERT
U.S. Borax Inc.

BORON OXIDES, BORIC ACID, AND BORATES

Boron, the fifth element in the periodic table, does not occur in nature in its elemental form. Rather, boron combines with oxygen as a salt or ester of boric acid. There are more than 200 minerals that contain boric oxide but relatively few that are of commercial significance. In fact, three minerals represent almost 90% of the borates used by industry: borax, a sodium borate; ulexite, a sodium–calcium borate; and colemanite, a calcium borate. These minerals are extracted in California and Turkey and to a lesser extent in Argentina, Bolivia, Chile, Peru, and China. China and Russia also have some commercial production from magnesium borates and calcium borosilicates. These deposits furnish nearly all of the world's borate supply at this time.

BORON OXIDES

Boric Oxide

Boric oxide, B_2O_3, formula wt 69.62, is the only commercially important oxide. It is also known as diboron trioxide, boric anhydride, or anhydrous boric acid. B_2O_3 is normally encountered in the vitreous state. This colorless, glassy solid has a Mohs' hardness of 4 and is usually prepared by dehydration of boric acid at elevated temperatures.

Boric oxide is an excellent Lewis acid.

The physical properties of vitreous boric oxide (Table 1) are somewhat dependent on moisture content and thermal history.

The uses of boric oxide relate to its behavior as a flux, an acid catalyst, or a chemical intermediate. The fluxing

Table 1. Physical Properties of Vitreous Boric Oxide

Property	Value	Reference
vapor pressure[a], 1331–1808 K	$\log P = 5.849 - \dfrac{16960}{T}$	20
heat of vaporization, ΔH_{vap}, kJ/mol[b]		
1500 K	390.4	21
298 K	431.4	21
boiling point, extrapolated	2316°C	17
viscosity, $\log \eta$, mPa·s(= cP)		
350°C	10.60	22
700°C	4.96	22
1000°C	4.00	22
density, g/mL		
0°C	1.8766	
18–25°C	1.844	
18–25°C[c]	1.81	
500°C[d]	1.648	22
1000°C[d]	1.528	22
index of refraction, 14.4°C	1.463	
heat capacity (specific), J/(kg·K)[b]		
298 K	62.969	17
500 K	87.027	17
700 K	132.63	17
1000 K	131.38	17
heat of formation[e], ΔH_f, kJ,[b] 298.15 K	-1252.2 ± 1.7	17

[a] P is in units of kPa; T is in K. To convert kPa to torr, multiply by 7.5.
[b] To convert J to cal, divide by 4.184.
[c] Well-annealed.
[d] Quenched.
[e] For $2\,B(s) + \frac{3}{2}O_2(g) \longrightarrow B_2O_3$ (glass).

action of B_2O_3 is important in preparing many types of glass, glazes, frits, ceramic coatings, and porcelain enamels.

Boric oxide is used as a catalyst in many organic reactions. It also serves as an intermediate in the production of boron halides, esters, carbide, nitride, and metallic borides.

Boron Monoxide and Dioxide

High-temperature vapor phases of BO, B_2O_3, and BO_2 have been the subject of a number of spectroscopic and mass spectrometric studies aimed at developing theories of bonding, electronic structures, and thermochemical data. Values for the principal thermodynamic functions have been calculated and compiled for these gases.

Lower Oxides

A number of hard, refractory suboxides have been prepared either as by-products of elemental boron production or by the reaction of boron and boric acid at high temperatures and pressures. It appears that the various oxides represented as B_6O, B_7O, $B_{12}O_2$, and $B_{13}O_2$ may all be the same material in varying degrees of purity.

BORIC ACID

The name "boric acid" is usually associated with orthoboric acid, which is the only commercially important form of

Table 2. Thermodynamic Properties of Crystalline Boric Acid, B(OH)$_3$

Temperature, K	C°_p,J/(kg·K)^a	S°, J/K^a	$H^\circ-H^\circ_{298}$, J/mol^a
0	0	0	−13393
100	35.92	28.98	−11636
200	58.74	61.13	−6866
298	81.34	88.74	0
400	100.21	115.39	9284

^{a}To convert J to cal, divide by 4.184.

boric acid and is found in nature as the mineral sassolite. Three crystalline modifications of metaboric acid also exist. All these forms of boric acid can be regarded as hydrates of boric oxide and formulated as B$_2$O$_3$·3H$_2$O for orthoboric acid and B$_2$O$_3$·H$_2$O for metaboric acid.

Properties

The standard heats of formation of crystalline orthoboric acid and the three forms of metaboric acid are $\Delta H^\circ = -1094.3$ kJ/mol (-261.54 kcal/mol) for B(OH)$_3$; -804.04 kJ/mol (-192.17 kcal/mol) for HBO$_2$-I; -794.25 kJ/mol (-189.83 kcal/mol) for HBO$_2$-II; and -788.77 kcal/mol (188.52 kcal/mol) for HBO$_2$-III. Values for the principal thermodynamic functions of B(OH)$_3$ are given in Table 2.

The solubility of boric acid in water increases rapidly with temperature. The heat of solution is somewhat concentration dependent.

The presence of inorganic salts may enhance or depress the aqueous solubility of boric acid: it is increased by potassium chloride as well as by potassium or sodium sulfate but decreased by lithium and sodium chlorides.

Boric acid is quite soluble in many organic solvents.

Manufacture

The majority of boric acid is produced by the reaction of inorganic borates with sulfuric acid in an aqueous medium. Sodium borates are the principal raw material in the United States. European manufacturers have generally used partially refined calcium borates, mainly colemanite from Turkey. Turkey uses both colemanite and tincal to make boric acid.

Uses

Boric acid has a surprising variety of applications in both industrial and consumer products. It serves as a source of B$_2$O$_3$ in many fused products, including textile fiber glass, optical and sealing glasses, heat-resistant borosilicate glass, ceramic glazes, and porcelain enamels. It also serves as a component of fluxes for welding and brazing.

A number of boron chemicals are prepared directly from boric acid. These include synthetic inorganic borate salts, boron phosphate, fluoborates, boron trihalides, borate esters, boron carbide, and metal alloys such as ferroboron.

Boric acid catalyzes the air oxidation of hydrocarbons and increases the yield of alcohols by forming esters that prevent further oxidation of hydroxyl groups to ketones and carboxylic acids.

The bacteriostatic and fungicidal properties of boric acid have led to its use as a preservative in natural products such as lumber, rubber latex emulsions, leather, and starch products.

NF-grade boric acid serves as a mild, nonirritating antiseptic in mouthwashes, hair rinse, talcum powder, eyewashes, and protective ointments. Although relatively nontoxic to mammals, boric acid powders are quite poisonous to some insects. With the addition of an anticaking agent, they have been used to control cockroaches and to protect wood against insect damage.

Inorganic boron compounds are generally good fire retardants.

Because boron compounds are good absorbers of thermal neutrons, owing to isotope [10]B, the nuclear industry has developed many applications. High-purity boric acid is added to the cooling water used in high-pressure water reactors.

SOLUTIONS OF BORIC ACID AND BORATES

Polyborates and pH Behavior

Whereas boric acid is essentially monomeric in dilute aqueous solutions, polymeric species may form at concentrations above 0.1 M. The conjugate base of boric acid in aqueous systems is the tetrahydroxyborate anion sometimes called the metaborate anion, B(OH)$_4^-$. This species is also the principal anion in solutions of alkali metal (1:1) borates such as sodium metaborate, Na$_2$O·B$_2$O$_3$·4H$_2$O. Mixtures of B(OH)$_3$ and B(OH)$_4^-$ appear to form classical buffer systems where the solution pH is governed primarily by the acid:salt ratio, i.e.,$[H^+] = K\alpha[BOH]_3 : B(OH)_4^- = 1$. This relationship is nearly correct for solutions of sodium or potassium (1:2) borates, eg, borax, where the ratio B(OH)$_3$:B(OH)$_4^-$ = 1 and the pH remains near 9 over a wide range of concentrations. However, for solutions that have pH values much greater or less than 9, the pH changes greatly on dilution as shown in Figure 1.

This anomalous pH behavior results from the presence of polyborates, which dissociate into B(OH)$_3$ and B(OH)$_4^-$ as the solutions are diluted.

Solubility Trends

Formation of polyborates greatly enhances the mutual solubilities of boric acid and alkali borates:

Sodium borate solutions near the Na$_2$O:B$_2$O$_3$ ratio of maximum solubility can be spray-dried to form an amorphous product with the approximate composition Na$_2$O·4B$_2$O$_3$·4H$_2$O commonly referred to as sodium octaborate. This material dissolves rapidly in water without any decrease in temperature to form supersaturated solutions.

The Polyborate Species

From a series of very rigorous pH studies, a series of equilibrium constants involving the species B(OH)$_3$, B(OH)$_4^-$, and the plyions B$_3$O$_3$(OH)$_5^{2-}$, B$_3$O$_3$(OH)$_4$, B$_5$O$_6$(OH)$_4^-$, and B$_4$O$_5$(OH)$_4^{2-}$ have been calculated. The relative

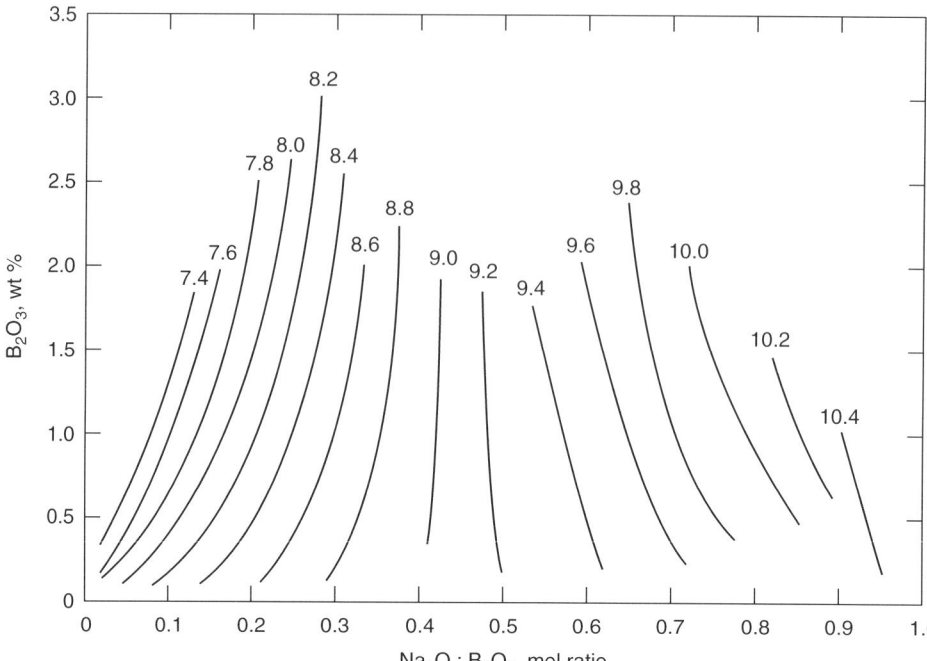

Figure 1. Values of pH in the system $Na_2O-B_2O_3-H_2O$ at 25°C.

populations of these species as functions of pH are shown in Figure 2.

Sources and Supplies

A limited number of geographic regions on earth contain borate deposits. The two main producers and regions are Borax in California and Etibank in Turkey. Other minor producers exist in the Andes region of northern Chile and Argentina, Bolivia, Peru, California, Russia, and China.

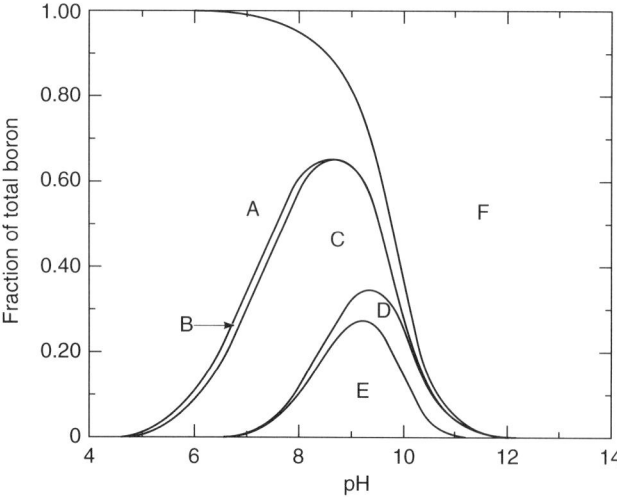

Figure 2. Distribution of boron in A, $B(OH)_3$; B, $B_5O_6(OH)_4^-$; C, $B_3O_3(OH)_4^-$; D, $B_3O_3(OH)_5^{2-}$; E, $B_4O_5(OH)_4^{2-}$; F, $B(OH)_4^-$; where total B_2O_3 concentration is 13.93 g/L. At a given pH, the fraction of the total boron in a given ion is represented by the portion of a vertical line falling within the corresponding range.

SODIUM BORATES

The solubility–temperature curves for the $Na_2O-B_2O_3-H_2O$ system are given in Figure 3 (Table 3).

Disodium Tetraborate Decahydrate (Borax Decahydrate)

Disodium tetraborate decahydrate, $Na_2B_4O_7 \cdot 10H_2O$ or $Na_2O \cdot 2B_2O_3 \cdot 10H_2O$, formula wt 381.36; monoclinic; sp gr, 1.71; specific heat 1.611 kJ/(kg·K) [0.385 kcal/(g°C] at 25–50°C; heat of formation −6.2643 MJ/mol (−1497 : 2 kcal/mol); exists in nature as the mineral borax. Its crystal habit, nucleation, and growth rate are sensitive to inorganic and surface active organic modifiers.

Disodium Tetraborate Pentahydrate (Borax Pentahydrate)

Although referred to as borax pentahydrate, well-formed crystals actually contain not five but 4.67 moles of water, $Na_2B_4O_7 \cdot 4.67H_2O$ or $Na_2O \cdot 2B_2O_3 \cdot 4.67H_2O$. The structural formula is best represented as $Na_2[B_4O_5(OH)_4] \cdot 2.67H_2O$; formula wt, 286.78; trigonal; rhombohedral crystal shape; sp gr; measured 1.880 crystallographic 1.912; specific heat, 1.32 kJ/(kg·K) [0.316 kcal/(g·°C)]; heat of formation: −4.7844 MJ/mol (−1143.5 kcal/mol). It is found in nature as a fine-grained mineral, tincalconite, formed by dehydration of borax. Solubility data in water are given in Table 3.

Disodium Tetraborate Tetrahydrate

Disodium tetraborate tetrahydrate, $Na_2B_4O_7 \cdot 4H_2O$ or $Na_2O \cdot 2B_2O_3 \cdot 4H_2O$, formula wt 273.27; monoclinic; sp gr 1.908; specific heat ca 1.2 kJ/(kg·K) [0.287 kcal/(g·°C)]; heat of formation −4.4890 MJ/mol (−1072.9 kcal/mol); exists in nature as the mineral kernite and has a structural formula $Na_2[B_4O_6(OH)_2] \cdot 3H_2O$. The crystals

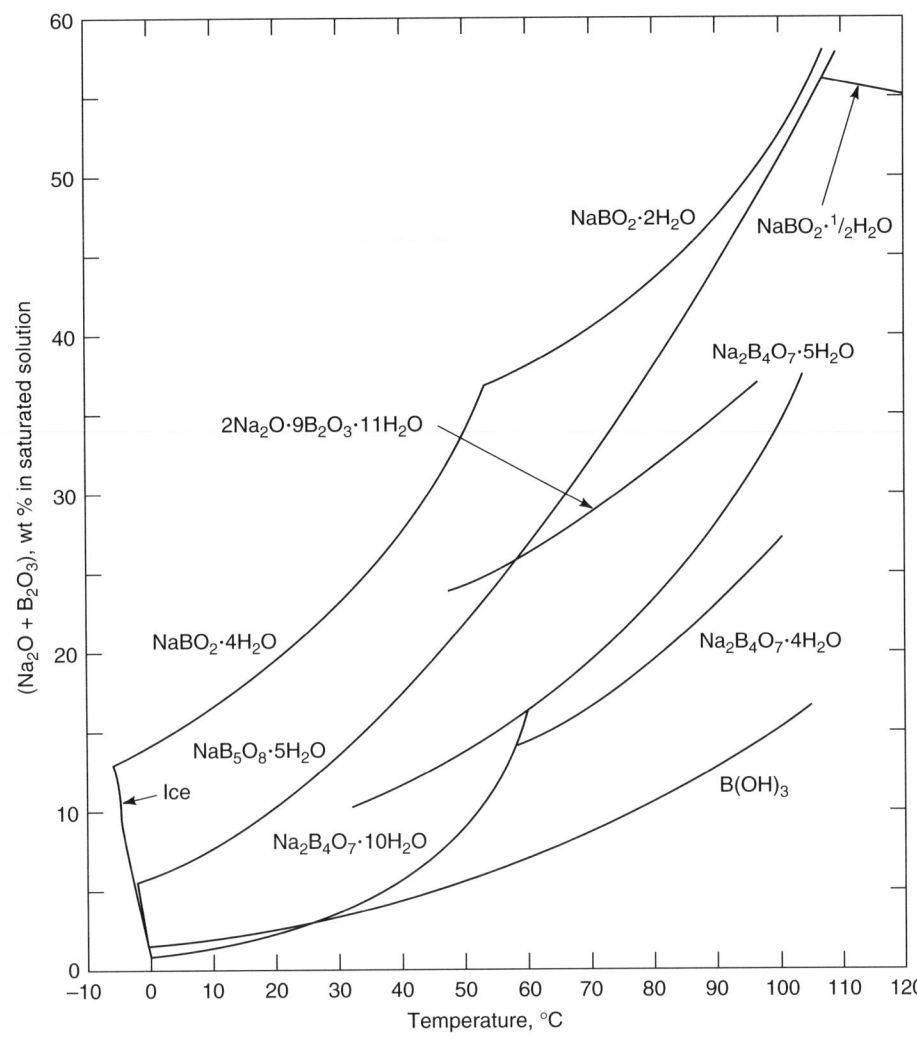

Figure 3. Solubility–temperature curves for boric acid, borax, sodium pentaborate, and sodium metaborate. Courtesy of The American Chemical Society.

have two perfect cleavages and, when ground, form elongated splinters. The water solubility of kernite is shown in Table 3.

Disodium Tetraborate (Anhydrous Borax)

Disodium tetraborate, $Na_2B_4O_7$ or $Na_2O \cdot 2B_2O_3$, formula wt, 201.21; sp gr (glass), 2.367, (α-crystalline form) 2.27; heat of formation (glass), -3.2566 MJ/mol (-778.34 kcal/mol), (α-crystalline form), -3.2767 MJ/mol (-783.2 kcal/mol); exists in several crystalline forms as well as a glassy form.

Disodium Octaborate Tetrahydrate

The composition of a commercially available sodium borate hydrate, 66.3 wt% B_2O_3, POLYBOR, corresponds quite closely to that of a hypothetical compound, disodium octaborate tetrahydrate, $Na_2B_8O_{13} \cdot 4H_2O$ or $Na_2O \cdot 4B_2O_3 \cdot 4H_2O$. This product dissolves rapidly in water without the temperature decrease, which occurs when the crystalline borates dissolve, and easily forms viscous supersaturated solutions at elevated temperatures. The solution pH decreases as the concentration increases.

Sodium Pentaborate Pentahydrate

Sodium pentaborate pentahydrate, $NaB_4O_8 \cdot 5H_2O$ or $Na_2O \cdot 5B_2O_3 \cdot 10H_2O$; formula wt 295.11; monoclinic; spgr, 1.713; exists in nature as the mineral sborgite. Heat capacity, entropy, and other thermal measurements have been made at 15–345 K.

Sodium pentaborate can easily be crystallized from a solution having a $Na_2O:B_2O_3$ mol ratio of 0.2. Its water solubility (Table 3) exceeds that of borax and boric acid. Its pH decreases with solution concentration.

Sodium Metaborate Tetrahydrate

Sodium metaborate tetrahydrate, $NaBO_2 \cdot 4H_2O$ or $Na_2O \cdot B_2O_3 \cdot 8H_2O$; formula wt 137.86; triclinic; sp gr 1.743; is easily formed by cooling a solution containing borax and an amount of sodium hydroxide just in excess of the theoretical value. It is the stable phase in contact with its saturated solution between 11.5 and 53.6°C. At temperatures above 53.6°C, the dihydrate, $NaBO_2 \cdot 2H_2O$, becomes the stable phase. The water solubility of sodium metaborate is given in Table 3.

Table 3. Aqueous Solubilities of Alkali Metal and Ammonium Borates at Various Temperatures

Compound	Solubility, wt% anhydrous salt, at °C											
	0	10	20	25	30	40	50	60	70	80	90	100
$Li_2O \cdot 5B_2O_3 \cdot 10H_2O^a$							20.88	24.34	27.98	31.79	36.2	41.2
$Li_2O \cdot 2B_2O_3 \cdot 4H_2O$	2.2–2.5	2.55	2.81	2.90	3.01	3.26	3.50	3.76	4.08	4.35	4.75	5.17
$Li_2O \cdot B_2O_3 \cdot 16H_2O^b$	0.88	1.42	2.51	3.34	4.63	9.40						
$Li_2O \cdot B_2O_3 \cdot 4H_2O$						7.40	7.84	8.43	9.43	10.58	11.8	13.4,
										9.75	9.7	9.70^c
$Na_2O \cdot 5B_2O_3 \cdot 10H_2O$	5.77	7.90	10.55	12.20	13.72	17.50	21.72	26.88	32.25	38.1	44.3	51.0
$Na_2O \cdot 2B_2O_3 \cdot 10H_2O$	1.18	1.76	2.58	3.13	3.85	6.00	9.55	15.90				
$Na_2O \cdot 2B_2O_3 \cdot 4.67H_2O^d$								16.40	19.49	23.38	28.37	34.63
$Na_2O \cdot 2B_2O_3 \cdot 4H_2O^e$								14.82	17.12	19.88	23.31	28.22
$Na_2O \cdot B_2O_3 \cdot 8H_2O^f$	14.5	17.0	20.0	21.7	23.6	27.9	34.1					
$Na_2O \cdot B_2O_3 \cdot 4H_2O$								38.3	40.7	43.7	47.4	52.4
$K_2O \cdot 5B_2O_3 \cdot 8H_2O$	1.56	2.11	2.82	3.28	3.80	5.12	6.88	9.05	11.7	14.7	18.3	22.3
$K_2O \cdot 2B_2O_3 \cdot 4H_2O$		9.02	12.1	13.6	15.6	19.4	24.0	28.4	33.3	38.2	43.2	48.4
$K_2O \cdot B_2O_3 \cdot 2.5H_2O$		42.3	43.0	43.3	44.0	45.0	46.1	47.2	48.2	49.3	50.3	
$Rb_2O \cdot 5B_2O_3 \cdot 8H_2O$	1.58	2.0	2.67	3.10	3.58	4.82	6.52	8.69	11.4	14.3	18.1	23.75^g
$Cs_2O \cdot 5B_2O_3 \cdot 8H_2O^h$	1.6	1.85	2.5	2.97	3.52	4.8	6.4	8.31	10.5	13.8	18.0	23.45^i
$(NH_4)_2O \cdot 2B_2O_3 \cdot 4H_2O$	3.75	5.26	7.63	9.00	10.8	15.8	21.2	27.2	34.4	43.1	52.7	
$(NH_4)_2O \cdot 5B_2O_3 \cdot 8H_2O$	4.00	5.38	7.07	8.03	9.10	11.4	14.4	18.2	22.4	26.4	30.3	

^{a} Incongruent solubility below 37.5 or 40.5°C.

^{b} Transition point to tetrahydrate, 36.9 or 40°C.

^{c} At 101.2°C.

^{d} Commonly known as the five hydrate, transition point to decahydrate, 60.7°C, 16.6% $Na_2B_4O_7$.

^{e} Transition point to decahydrate, 58.2°C, 14.55% $Na_2B_4O_7$.

^{f} Transition point to tetrahydrate, 53.6°C, 36.9% $Na_2B_2O_4$.

^{g} At 102°C.

^{h} Dicesium tetraborate pentahydrate [12228-83-0], $Cs_2O \cdot 2B_2O_3 \cdot 5H_2O$, anddicesium diborate heptahydrate [66634-85-3], $Cs_2O \cdot B_2O_3 \cdot 7H_2O$, also exist. The former has incongruent solubility; the latter has a solubility of 36.8 wt% anhydrous salt at 18°C.

^{i} At 101.65°C.

Sodium Metaborate Dihydrate

Sodium metaborate dihydrate, $NaBO_2 \cdot 2H_2O$ or $Na_2O \cdot B_2O_3 \cdot 4H_2O$; formula wt 101.83; triclinic; sp gr 1.909; can be prepared by heating a slurry of the tetrahydrate above 54°C, by crystallizing metaborate solutions at 54–80°C, or by dehydrating the tetrahydrate in vacuum. The water solubility for the dihydrate is shown in Table 3.

Sodium Perborate Hydrates

Peroxyborates are commonly known as perborates, written as if the perborate anion were BO_3. X-ray crystal structure has shown that they contain the dimeric anion $[(HO)_2B(O_2)2B(OH)_2]^{2-}$. Three sodium perborate hydrates, $NaBO_3 \cdot xH_2O$ ($x = 1, 3,$ and 4) are known. Only the mono- and tetrahydrate are of commercial importance, primarily as bleaching agents in laundry products.

$$\begin{bmatrix} HO & O-O & OH \\ & B \quad\quad B & \\ HO & O-O & OH \end{bmatrix}^{2-}$$

Sodium perborate tetrahydrate, $NaBO_3 \cdot 4H_2O$ or $Na_2B_2(O_2)2(OH)_4 \cdot 6H_2O$, is triclinic; heat of formation, −2112 kJ/mol (−504.8 kcal/mol) (crystal), −921 kJ/mol (−220.2 kcal/mol) ($1M$ soln) and contains 10.4 wt% active oxygen. It melts at 63°C by dissolving in its own water of hydration and on heating to 250°C decomposes rapidly and completely to oxygen and sodium metaborate. In water its decomposition, which is important in its use as a bleach, is accelerated by catalysts or elevated temperature.

Sodium perborate trihydrate, $NaBO_3 \cdot 3H_2O$ or $Na_2B_2(O_2)_2(OH)_4 \cdot 4H_2O$, triclinic, contains 11.8 wt% active oxygen. It has been claimed to have better thermal stability than the tetrahydrate but has not been used commercially. The trihydrate can be made by dehydration of the tetrahydrate or by crystallization from a sodium metaborate and hydrogen peroxide solution in the present of trihydrate seeds.

Sodium perborate monohydrate, $NaBO_3 \cdot H_2O$ or $Na_2B_2(O_2)_2(OH)_4$, 16.0 wt% active oxygen, is commercially prepared by dehydration of the tetrahydrate.

Manufacturing, Production and Processing

Both sedimentary and metamorphic borate mineral deposists are exploited, although sedimentary deposits are by far the larger sources. Ore extraction is typically accomplished using conventional surface and underground mining tehniques. Solution mining has also been employed, albeit on a small scale. Sedimentary sodium borate minerals borax and kernite, the sedimentary calcium mineral colemanite, the sedimentary calcium–sodium mineral ulexite, and the metamorphic mineral datolite are the principal ore types exploited around the world.

In nearly all end uses, borates are either dispersed in low concentrations (eg, detergents, agricultural products) or incorporated into products from which borate separation would be difficult (eg, ceramics, fiberglass). Because borates cannot easily be recycled, nearly all borates used in commerce are obtained from virgin sources located in remote desert regions.

Commercial borate ores fall into basic types: sodium borates, which are relatively soluble in water; and calcium or sodium–calcium borate ores, which are relatively insoluble in water. Because of these intrinsic properties, sodium borate ores are often used to produce refined borate products (various forms of sodium borates and boric acid) whereas calcium and sodium–calcium borate ores are often used to supply mineral borates.

Shipment

Today, borates are moved by virtually all available forms of transportation. Bulk, intercontinental shipments are made in ocean-going vesels with capacities of −40,000 tones. Barges, railcars, and trucks move borate products from sources or shipping ports to customers. Intermediate stock points are located around the world to provide working inventories and to guard against supply interruptions.

Health and Safety Factors

Cases of industrial intoxication on exposure to inorganic borates have not been reported. There is a large body of literature on the toxicology of boric acid and borax. Acute oral LD_{50} in the rat is 3000–4000 mg/kg for boric acid and 4500–6000 mg/kg for borax. These values are comparable to sodium chloride, LD_{50} 3750 mg/kg. Ingested boric acid is excreted rapidly in the urine with a half-life of 21 h. Chronic ingestion studies (high dosage level and repeated exposure) indicates some reproductive toxicity in animals, but adequate evidence for these effects in humans is lacking. Studies indicate no evidence of carcinogenic or mutagenic activity. Boric acid and borax are poorly absorbed through healthy skin and do not cause skin irritation.

Sodium metaborate hydrates are more alkaline than borax and greater care is required in handling. The metaborate material is harmful to the eyes and can cause skin irritation. Gloves, goggles, and a simple dust mask should be used when handling sodium metaborate powder.

Boron in the form of borate is an essential micronutrient for the healthy growth of plants and is present in the normal daily human diet at an estimated level of 3–40 mg as boron. It is not a proven essential micronutrient for animals.

The handling of boric acid and borax is generally not considered dangerous. There are no fire risks associated with the storage or use of inorganic borates, and they are not explosive.

Uses

The single largest end use for borate is insulation fiberglass or glass wool. Insulation fiberglass accounts for 20% of the total world B_2O_3 demand and >30% of the total demand in North America. The second largest end use is perborate and cleaning products, accounting for just under 20% of world demand. Much of this use category consists of perborate, which is included in the formulations of various soaps. Approximately 95% of the perborate demand originates in Europe, with the remainder arising mostly from North America. Conversely, the demand for cleaning products is approximately 90% from North America and the remainder from Europe.

The third largest end use for borates is textile fiberglass, accounting for roughly 15% of world borate demand. Another 10% of the world's borates end up in ceramics and frits. The final catchall category, "other uses," accounts for an additional 15% of global demand for borates, for relatively minor applications such as cellulose and adhesives. Borosilicate glass accounts for 10% of B_2O_3 world demand; >70% of this demand arises from North America and Europe.

OTHER ALKALI METAL AND AMMONIUM BORATES

Dipotassium Tetraborate Tetrahydrate

Dipotassium tetraborate tetrahydrate, $K_2B_4O_7 \cdot 4H_2O$ or $K_2O \cdot 2B_2O_3 \cdot 4H_2O$, formula wt, 305.49; orthorhombic; sp gr 1.919; is much more soluble than borax in water. Solubility data are given in Table 3.

Potassium Pentaborate Tetrahydrate

Potassium pentaborate tetrahydrate, $KB_5O_8 \cdot 4H_2O$ or $K_2O \cdot 5B_2O_3 \cdot 8H_2O$; formula wt 293.20; orthorhombic prisms; sp gr 1.74; heat capacity 329.0 J/(mol·K) [78.6 cal/.(mol·K)] at 296.6 K; is much less soluble than sodium pentaborate (Table 3).

Diammonium Tetraborate Tetrahydrate

Diammonium tetraborate tetrahydrate, $(NH_4)_2B_4O_7 \cdot 4H_2O$ or $(NH_4)_2O \cdot 2B_2O_3 \cdot H_2O$; formula wt 263.37; monoclinic; sp gr 1.58; is readily soluble in water (Table 3). The pH of solutions of diammonium tetraborate tetrahydrate is 8.8 and independent of concentration.

Ammonium Pentaborate Tetrahydrate

Ammonium pentaborate tetrahydrate, $NH_4B_5O_8 \cdot 4H_2O$ or $(NH_4)_2O \cdot 5B_2O_3 \cdot 8H_2O$; formula wt 272.13; sp gr, 1.567; heat capacity 359.4 J/(mol·K) [85.9 cal/(mol·K)] at 301.2 K; exists in two crystalline forms, orthorhombic (α) and monoclinic (β). Solubility data are given in Table 3.

Lithium Borates

Two lithium borates are of minor commercial importance, the tetraborate trihydrate and metaborate hydrates.

Manufacture

Potassium tetraborate tetrahydrate may be prepared from an aqueous solution of KOH and boric acid having a $B_2O_3:K_2O$ ratio of about 2 or by separation from a

KCl–borax solution. Potassium pentaborate is prepared in a manner analogous to that used for the tetraborate, but the strong liquor has a B_2O_3:K_2O ratio near 5.

Ammonium tetraborate tetrahydrate is prepared by crystallization from an aqueous solution of boric acid and ammonia having a B_2O_3:$(NH_4)_2O$ ratio of 1.8:2.1. Ammonium pentaborate is similarly produced from an aqueous solution of boric acid and ammonia having a B_2O_3:$(NH_4)_2O$ ratio of 5. Supersaturated solutions are easily formed, and the rate of crystallization is proportional to the extent of supersaturation. A process for the production of ammonium pentaborate by precipitation from an aqueous ammonium chloride–borax mixture has been patented.

Health and Safety Factors

Little toxicological data are available on borates other than boric acid and borax. Most water-soluble borates have the same toxicological effects as borax when adjusted to account for differences in B_2O_3 content.

Uses

Dipotassium tetraborate tetrahydrate is used to replace borax in applications where an alkali metal borate is needed but sodium salts cannot be used or where a more soluble form is required. The potassium compound is used as a solvent for casein, as a constituent in welding fluxes, and a component in diazotype developer solutions. Potassium pentaborate tetrahydrate is used in fluxes for welding and brazing of stainless steels for nonferrous metals. Diammonium tetraborate tetrahydrate is used when a highly soluble borate is desired but alkali metals cannot be tolerated. It is used mostly as a neutralizing agent in the manufacture of urea–formaldehyde resins and as an ingredient in flameproofing formulations. Ammonium pentaborate tetrahydrate is used as a component of electrolytes for electrolytic capacitors, as an ingredient in flameproofing formulations, and in paper coatings.

CALCIUM-CONTAINING BORATES

Dicalcium Hexaborate Pentahydrate

Dicalcium hexaborate pentahydrate, $Ca_2B_6O_{11} \cdot 5H_2O$ or $2CaO \cdot 3B_2O_3 \cdot 5H_2O$; formula wt 411.08; monoclinic; sp gr, 2.42; heat of formation -3.469 kJ/mol (-0.83 kcal/mol); exists in nature as the mineral colemanite. Its solubility in water is about 0.1% at $25°C$ and 0.38% at $100°C$.

Sodium Calcium Pentaborate Octahydrate

Sodium calcium pentaborate octahydrate, $NaCaB_5O_9 \cdot 8H_2O$ or $Na_2O \cdot 2CaO \cdot 5B_2O_3 \cdot 16H_2O$; formula wt 405.23; triclinic; sp gr, 1.95; exists in nature as the mineral ulexite.

Its solubility in water at $25°C$ is 0.5% as $NaCaB_5O_9$.

Sodium Calcium Pentaborate Pentahydrate

Sodium calcium pentaborate pentahydrate, $NaCaB_5O_9 \cdot 5H_2O$ or $Na_2O \cdot 2CaO \cdot 5B_2O_3 \cdot 10H_2O$; formula wt 351.19;

monoclinic; sp gr, 2.14; exists in nature as the mineral probertite.

Manufacture

The alkaline-earth metal borates of primary commercial importance are colemanite and ulexite. Both of these borates are sold as impure ore concentrates from Turkey, the principal world supplier.

Uses

Colemanite, $2CaO \cdot 3B_2O_3 \cdot 5H_2O$, is used in the production of boric acid and borax, as well as in several direct applications. It is a highly desirable material for the manufacture of the E-glass used in textile glass fibers and plastic reinforcement (where sodium cannot be tolerated). High As or Fe levels in the ore concentrate can limit its use in this application. Colemanite has seen limited application as a slagging material in steel manufacture. It is also used in some fire retardants and as a precursor to some boron alloys.

Ulexite, $NaCaB_5O_9 \cdot 8H_2O$, and probertite, $NaCaB_5O_9 \cdot 5H_2O$, have found application in the production of insulation fiber glass and borosilicate glass as well as in the manufacture of other borates.

OTHER METAL BORATES

Borate salts or complexes of virtually every metal have been prepared. For most metals, a series of hydrated and anhydrous compounds may be obtained by varying the starting materials and/or reaction conditions. Some have achieved commercial importance.

In general, hydrated borates of heavy metals are prepared by mixing aqueous solutions or suspensions of the metal oxides, sulfates, or halides and boric acid or alkali metal borates such as borax. The precipitates formed from basic solutions are often sparingly-soluble amorphous solids having variable compositions. Crystalline products are generally obtained from slightly acidic solutions.

Anhydrous metal borates may be prepared by heating the hydrated salts to $300-500°C$, or by direct fusion of the metal oxide with boric acid or B_2O_3. Many binary and tertiary anhydrous systems containing B_2O_3 form vitreous phases over certain ranges of composition.

Barium Metaborate

Three hydrates of barium metaborate, $BaO \cdot B_2O_3 \cdot xH_2O$, are known.

Barium metaborate is used as an additive to impart fire-retardant and mildew-resistant properties to latex paints, plastics, textiles, and paper products.

Copper, Manganese, and Cobalt Borates

Borate salts of copper, manganese, and cobalt are precipitated when borax is added to aqueous solutions of the metal(II) sulfates or chlorides. However, these materials are no longer produced commercially.

Zinc Borates

A series of hydrated zinc borates have been developed for use as fire-retardant additives in coatings and polymers.

BORON PHOSPHATE

Boron phosphate, BPO_4, is a white, infusible solid that vaporizes slowly above $1450°C$, without apparent decomposition. It is normally prepared by dehydrating mixtures of boric acid and phosphoric acid at temperatures up to $1200°C$.

The principal application of boron phosphate has been as a heterogeneous acid catalyst.

ENVIRONMENTAL CONCERNS, GENERAL

Boron is present everywhere in trace amounts, but does not exist by itself in nature. Rather, boron combines with oxygen and other elements to form boric acid, or inorganic salts called borates.

Trace amounts of boron are enviornmentally ubiquitous, essential to plant life and nutritionally important to humans. However, as with any substance, the dose makes the poison: overexposure to boron compounds can be hazardous to plants, animals, and even people.

"Boron", *Mineral Commodity Summaries*, U.S. Geological Survey, Reston, Va., 2006.

P. Harben and Bates, *Geology of the Nonmetallics*; R. Hartnoll Ltd., U.K. 1984.

P. Harben and Kuzvart, *Industrial Minerals, a Global Geology*, Industrial, Min. Information Ltd., Industrial Min. Division of Metal Bulletin, U.K., 1996.

R. Kistler and C. Helvaci in D. Carr, ed., *Industrial Minerals and Rocks*, 6th ed., Society of Minning Engineering, New York, 1994, pp. 171–186.

MICHAEL BRIGGS
U.S. Borax, Inc.

BROMINE

Bromine, Br_2, is the only nonmetallic element that is a liquid at standard conditions. The bromine (Br) atom has at no. 35, at wt 79.904, and belongs to Group 17 (VIIA) of the Periodic Table, the halogens. Its electronic configuration is $1s^2 2s^2 2p^6 3s^2 3p^6 3d^{10} 4s^2 4p^5$. The element's known isotopes range in mass number from 74 to 90. Isotopes usable as radioactive tracers are 77, 80, 80m (metastable), and 82. Bromine has two stable isotopes, ^{79}Br and ^{81}Br. The most common valence states are −1 and +5, but +1, +3, and +7 are also observed.

PHYSICAL PROPERTIES

Bromine is a dense, dark red, mobile liquid that vaporizes readily at room temperature to give a red vapor that

Table 1. Physical Properties of Bromine

Property	Value
stable isotope abundance, %	
^{79}Br	50.54%
^{81}Br	49.46%
mol wt	159.808
freezing point, °C	−7.25
bp, °C	58.8
density, g/mL	
15°C	3.1396
20°C	3.1226
25°C	3.1055
30°C	3.0879
vapor density, g/L, 0°C, 101.3 kPa[a]	7.139
refractive index	
20°C	1.6083
25°C	1.6475
viscosity, mm²/s (=cSt)	
20°C	3.14×10^{-1}
30°C	2.88×10^{-1}
40°C	2.64×10^{-1}
50°C	2.45×10^{-1}
surface tension, mN/m (=dyn/cm), 25°C	40.9
solubility parameter, 25°C, $(J/cm^3)^{1/2b}$	23.5
critical temperature, °C	311
critical pressure, MPa[c]	10.3
thermal conductivity, W/(m·K)	0.123
specific conductivity, $(\Omega \cdot cm)^{-1}$	9.10×10^{-12}
dielectric constant, 25°C, 10⁵ Hz	3.33
electrical resistivity, 25°C, $\Omega \cdot cm$	6.5×10^{10}
expansion coefficient from 20–30°C, per °C	0.0011
compressibility, 20°C from 0–10 MPa[c]	62.5×10^{-6}
heat of vaporization, 50°C, J/g[b]	187
heat of fusion, −7.25°C, J/g[b]	66.11
heat capacity, J/mol[b]	
15 K	7.217
30 K	22.443
60 K	36.33
240 K	57.94
265.9 K	61.64
265.9 K[d]	77.735
288.15 K[e]	78.66
electronegativity	3.0
electron affinity, kJ[b]	330.5

[a]To convert kPa to mmHg, multiply by 7.50.
[b]To convert J to cal, divide by 4.184.
[c]To convert MPa to bar, multiply by 10.
[d]Solid bromine.
[e]Liquid bromine.

is highly corrosive to many materials and human tissues. Bromine liquid and vapor, up to ~600°C, are diatomic (Br_2). Table 1 summarizes the physical properties of bromine.

Bromine is moderately soluble in water, 33.6 g/L at 25°C. Bromine is soluble in nonpolar solvents and in certain polar solvents such as alcohol and sulfuric acid. Bromine can function as a solvent.

CHEMICAL PROPERTIES

One of the central features of the chemistry of the halogens is the tendency to acquire an electron to form either a negative ion, X^-, or a single covalent bond, $-X$, and bromine is no exception.

Reaction with Hydrogen and Metals

Bromine combines directly with hydrogen at elevated temperatures and this is the basis for the commercial production of hydrogen bromide. Heated charcoal and finely divided platinum metals are catalysts for the reaction.

Bromine reacts with essentially all metals, except tantalum and niobium, although elevated temperatures are sometimes required.

Reaction with Other Halides

Bromide ion is oxidized by chlorine to bromine, which is the basic reaction in the production of bromine from brines, seawater, bitterns, or bromine-containing wastes.

Reaction with Nonmetals

Bromine oxidizes sulfur and a number of its compounds. Bromine also oxides red phosphorus and some phosphorus compounds. Ammonia, hydrazine, nitrites, and azides are oxidized by bromine. Nitrogen is often a product of such reactions.

Under certain circumstances bromine reacts with ammonia and amino compounds to form bromamide, NH_2Br, bromimide, $NHBr_2$, or nitrogen bromide, NBr_3,. Bromine oxidizes carbon and reacts with carbon monoxide to form carbonyl bromide.

Reactions in Water

The ionization potential for bromine is 11.8 eV and the electron affinity is 3.78 eV. The heat of dissociation of the Br_2 molecule is 192 kJ (46 kcal).

When bromine dissolves in water, it partially disproportionates, Br_2.

The equilibrium constant for this reaction at $25°C$ is $7.2 \times 10^9\ M^2$. Light catalyzes the decomposition of hypobromous acid to hydrogen bromide and oxygen.

In the dark, hypobromous acid decomposes to bromic acid and bromine. Bromic acid is relatively unstable and decomposes slowly to give bromine and oxygen.

Reactions with Organic Compounds

The addition of bromine to unsaturated carbon compounds occurs readily.

Bromine reacts directly with alkanes but this reaction has little value, because mixtures are obtained. However, photochemical bromination of alkyl bromides can be quite selective. The bromination of aromatic hydrocarbons can occur either in a side chain or on the ring, depending on conditions. In the presence of sunlight, alkylbenzenes are brominated predominately in the side chain.

In the presence of halogen Lewis acids, such as metal halides or iodine, aromatic hydrocarbons are halogenated on the ring.

Phenols and phenol ethers readily undergo mono-, di-, or tribromination in inert solvents, depending on the amount of bromine used.

Heterocyclic compounds range from those, such as furan, which is readily halogenated and tends to give polyhalogenated products, to pyridine, which forms a complex with aluminum chloride that can only be brominated to 50% reaction.

Aliphatic ketones are readily brominated in the α position.

Bromination of aldehydes is more complicated, because bromination can take place on the aldehyde carbon as well as the α-carbon.

Acids and esters are less easily brominated than aldehydes or ketones. Bromination of α-chloro ethers proceeds readily and often gives 90–95% yields. Bromine can replace sulfonic acid groups on aromatic rings that also contain activating groups.

Organometallic compounds can react with bromine to give bromides, but because organometallic compounds are frequently made from bromides the reaction with iodine to give iodides is of more synthetic significance.

Charge-Transfer Compounds

Similar to iodine and chlorine, bromine can form charge-transfer complexes with organic molecules that can serve as Lewis bases.

OCCURRENCE

Bromine is widely distributed in nature but in relatively small amounts. The only natural minerals that contain bromine are some silver halides, including bromyrite AgBr, embolite Ag (Cl, Br), and iodobromite, Ag (Cl, Br, I). The biggest source of commercial bromine is Dead Sea brines. They contain 5 g/L bromine in the open sea, 6.5 g/L in the southern basin, near Ein Bokek, and up to 12 g/L in the end brine of potash production, that is the raw material for production of bromine in Israel. Their quantity is practically unlimited. Other important sources are underground brines in Arkansas, which contain 3–5 g/L bromine, and in China, Russia, and the United Kingdom; bitterns from mined potash in France and Germany; seawater or seawater bitterns in France, India, Italy, Japan, and Spain.

An average of ∼7 ppm of bromine is found in terrestrial plants, and edible foods contain up to 20 ppm. Among animals the highest bromide contents are found in sea life, such as fish, sponges, and crustaceans. Animal tissues contain 1–9 ppm of bromide and blood 5–15 ppm. The World Health Organization has set a maximum acceptable bromide intake for humans at 1 mg/kg of body weight per day. In adult males, the bromine content in serum has been found to be 3.2–5.6 µg/mL, in urine 0.3–7.0 µg/mL, and in hair 1.1–49.0 µg/mL. Bromine may be an essential trace element as are the other halides.

Bromine compounds are found in the atmosphere in small amounts; the sea is a primary source. Rainfall over the Pacific and Indian Oceans has been found to contain 60–80 µg/L of bromine. Approximately 15–20 parts per trillion (ppt) (v/v) of bromine is found in the stratosphere. It is up from 10 ppt a decade ago, maybe due to the use of brominated fire suppressants (CF_3Br, etc).

The inorganic forms of Br in the stratosphere are likely involved in ozone destruction processes.

MANUFACTURE

Bromine occurs in the form of bromide in seawater and in natural brine deposits, always together with chloride. In all current methods of bromine production, chlorine, which has a higher reduction potential than bromine, is used to oxidize bromide to bromine.

There are four principal steps in bromine production: (1) oxidation of bromide to bromine; (2) stripping bromine from the aqueous solution; (3) separation of bromine from the vapor; and (4) purification of the bromine. Most of the differences between the various bromine manufacturing processes are in the stripping and purification step.

ECONOMIC ASPECTS

Facilities for manufacturing bromine are primarily located near sources of natural brines or bitterns containing usable levels of bromine. The biggest single bromine plant is erected by the Dead Sea in Israel.

The costs of building and maintaining a bromine plant are high because of the corrosiveness of brine solutions that contain chlorine and bromine and require special materials of construction. The principal operating expenses are for pumping, steam, environmental costs, energy, and chlorine. The plants are very capital intensive.

REQUIREMENTS AND SPECIFICATIONS

Typical specifications for bromine produced in a modern plant generally exceed the ACS requirements for bromine used as a reagent chemical.

HEALTH AND SAFETY ASPECTS AND HANDLING

Consequences of Exposure

Bromine has a sharp, penetrating odor. The Occupational Safety and Health Administration (OSHA) threshold limit value–time-weighted average for an 8-h workday and 40-h workweek is 0.1 ppm in air. Monitors are available for determining bromine concentrations in air. Concentrations of ∼1 ppm are unpleasant and cause eyes to water; 10 ppm are intolerable. Inhalation of 10 ppm and higher concentrations of bromine causes severe burns to the respiratory tract and is highly toxic. Symptoms of overexposure include coughing, nose bleed, feeling of oppression, dizziness, headache, and possibly delayed abdominal pain and diarrhea. Pneumonia may be a late complication of severe exposure.

Liquid bromine produces a mild cooling sensation on first contact with the skin, which is followed by a sensation of heat. If bromine is not removed immediately by flooding with water, the skin becomes red and finally brown, resulting in a deep burn that heals slowly. Contact with concentrated vapor can also cause burns and blisters. For very small areas of contact in the laboratory, a 10% solution of sodium thiosulfate in water can neutralize bromine and such a solution should be available when working with bromine. Bromine is especially hazardous to the tissues of the eyes where severely painful and destructive burns may result from contact with either liquid or concentrated vapor. Ingestion causes severe burns to the gastrointestinal tract.

Protective Equipment

Totally enclosed systems should be used for processes involving bromine. For handling bromine in the laboratory, the minimum safety equipment should include chemical goggles, rubber gloves (Buna-*N* or neoprene rubber), laboratory coat, and fume hood. For handling bromine in a plant, safety equipment should include hard hat, goggles, neoprene full coverage slicker, Buna-*N* or neoprene rubber gloves, and neoprene boots. For escaping from an area where a bromine release has occurred, a full face respirator with an organic vapor–acid gas canister is desirable. For emergency work in an area with bromine concentrations >0.1 ppm, a self-contained breathing apparatus can be used until the air supply gets low. For longer term work in elevated bromine concentrations, an air-line respirator is essential.

Reactivity

Bromine is nonflammable but may ignite combustibles, such as dry grass, on contact. Handling bromine in a wet atmosphere, extreme heat, and temperatures low enough to cause bromine to solidify (−6°C) should be avoided. Bromine should be stored in a cool, dry area away from heat. Materials that should not be permitted to contact bromine include combustibles, liquid ammonia, aluminum, titanium, mercury, sodium, potassium, and magnesium. Bromine attacks some forms of plastics, rubber, and coatings.

Spills and Disposal Procedures

If a spill occurs outdoors, personnel should stay upwind of it. Broine vapors may be neutralized by gaseous ammonia. Small spills may be neutralized with lime water slurry (most common procedure) or 10–30% of K_2CO_3 or Na_2CO_3 solution. No decontaminants other than water should be used on humans. Under the Comprehensive Environmental Response, Compensation, and Liability Act (CERCLA) regulations in effect at the end of 1986, bromine is regulated as a hazardous waste or material. Therefore, it must be disposed of in an approved hazardous waste facility in compliance with the Environmental Protection Agency (EPA) and/or other applicable local, state, and federal regulations and should be handled in a manner acceptable to good waste management practice.

Materials of Construction

Glass has excellent corrosion-resistance to wet or dry bromine. Lead is very useful for bromine service if water is <70 ppm. The bromine corrosion rate increases with concentrations of water and organics. Tantalum and niobium have excellent corrosion-resistance to wet or dry bromine. Nickel and nickel alloys such as Monel 400 and Hastelloys B and C have useful resistance for dry bromine but is rapidly attacked by wet bromine. Steel and stainless steel materials are not recommended. The fluoropolymers Kynar, Halar, and Teflon are highly resistant to bromine but are somewhat permeable. The rate depends on temperature, pressure, and structure (density) of the fluoropolymer. Other polymers are not recommended.

Storage and Transportation

Bromine in bulk quantities is shipped domestically in 7570 and 15,140 L lead-lined pressure tank cars or 6435–6813 L nickel-clad pressure tank trailers.

USES

An important use of bromine compounds is in the production of flame retardants. Brominated polymers are used in flame retardant applications and bromine-containing epoxy sealants are used in semiconductor devices.

Bromine has some use in swimming pools and in bleaching. It is also a disinfectant for cooling water and wastewater. Its main use is as a chemical reactant.

Zinc–bromine storage batteries are under development as load-leveling devices in electric utilities.

A. J. Downs and C. J. Adams, "Chlorine, Bromine, Iodine, and Astatine," in J. C. Bailar, Jr. and co-eds., *Comprehensive Inorganic Chemistry*, Vol. 2, Pergamon Press, Oxford, 1973, pp. 1107–1594.

V. Gutmann, ed., *Halogen Chemistry*, Academic Press, New York, 1967 (three volumes).

V. Gutmann, ed., *MTP Int. Rev. Sci.: Inorg. Chem., Ser. One,* **3** (1972).

Z. E. Jolles, ed., *Bromine and its Compounds*, Academic Press, New York, 1966.

Baruch Grinbaum
IMI(TAMI) Institute for
Research and Development

Mira Freiberg
Dead Sea Bromine Group

BROMINE, INORGANIC COMPOUNDS

The main classes of inorganic bromine compounds include (a) bromamines, (b) hydrogen bromide and hydrobromic acid, (c) metal and bromides, (d) nonmetal bromides, (e) bromine halides, (f) bromine oxides, (g) oxygen acids and their salts.

Table 1. Physical Characteristics of Hydrogen Bromide

mp $-86°C$
bp $-67°C$ (101.3 kPa)
liquid density 2.152 g/mL
heat of fusion at mp 29.8 kJ/kg (7.12 kcal/kg)
heat of vaporization at $-66.7°C$, 218 kJ/kg (52 kcal/kg)
heat capacity J/(kg·K) [cal/(kg·K)]
   (i) solid at $-91°C$, 636 [152]
   (ii) liquid 737 [176]
   (iii) gas at $27°C$, 356 [85]
critical temperature $89.8°C$
critical pressure 8510 kPa (84 atm)

BROMAMINES

The bromamines are highly unstable compounds, having a tendency to explode at low temperatures, if they are isolated. Traces of these compounds can be formed when water containing small amounts of bromide and ammonia is chlorinated.

BROMIDES

Hydrogen Bromide

Hydrogen bromide, HBr (hydrobromic acid), is a colorless, corrosive gas that fumes strongly in moist air. It is extremely irritating to the eyes, nose, and throat. Some of the physical properties of anhydrous hydrogen bromide gas are summarized in Table 1.

Hydrobromic acid is one of the strongest mineral acids known. It is considered a more effective leaching agent than hydrochloric acid for some mineral ores because of its higher boiling point and stronger reducing action. Certain higher oxides such as ceric oxides are readily dissolved in HBr. The acid forms complexes with the bromides of several metals. Bromine is highly soluble in concentrated aqueous hydrobromic acid.

Safety and Environmental Considerations. The liquids and vapors of HBr are highly corrosive to human tissue. The threshold limit value for hydrogen bromide gas in an 8-h day is 3 ppm time-weighted average. Inhalation of the vapor (when present at highly hazardous concentrations) is highly irritating to the nose and throat. Symptoms from HBr overexposure include coughing, choking, burning in the throat, wheezing, or asphyxiation. Ingestion of the HBr vapor causes severe burns of the mouth and stomach while skin contact can cause severe burns. In the case of liquid or vapor contact with the eyes, permanent damage may result. It is therefore imperative to employ the proper safety equipment when handling HBr including a safety shower and eye bath.

Most metals, concrete, and other construction materials are corroded by HBr. Suitable materials of construction include some fiber glass-reinforced plastics, some chemically resistant rubbers, polyvinyl chloride (PVC),

Teflon, polypropylene and ceramic-, rubber-, and glass-lined steel. Metals that are used include Hastelloy B, Hastelloy C, and titanium. The Hastelloys are only suitable at ambient temperatures. Hydrogen bromide under pressure in glass at or above room temperature can attack the glass, resulting in unexpected shattering.

Technical 48 and 62% acids range from colorless-to-light yellow liquids, which are available in drums or tank trailers and tank car quantities. They are classified under DOT regulations as corrosive materials. Anhydrous HBr is available in cylinders under its vapor pressure (~2.4 MPa or 350 psi) at 25°C. It is classified as a nonflammable gas.

Uses and Applications

1. Hydrogen bromide is used in the manufacture of inorganic bromides. Metal hydroxides or carbonates are used for the neutralization.
2. Hydrogen bromide is also a raw material in the synthesis of alkyl bromides from alcohols.
3. Hydrobromination of olefins. The addition can take place by an ionic mechanism, usually in a polar solvent according to Markovnikove's rule to yield a secondary alkyl bromide. By using a free-radical catalyst in aprotic, nonpolar solvents, dry HBr reacts with an olefin to produce a primary alkyl bromide as the predominant product. Primary alkyl bromides are used in synthesizing other compounds and are 40–60 times as reactive as the corresponding chlorides.
4. Hydrogen bromide adds to acetylene to form vinyl bromide or ethylidene bromide, depending on the stoichiometry.
5. Hydrogen bromide cleaves acyclic and cyclic ethers.
6. Hydrogen bromide adds to the cyclopropane group by ring opening.
7. Addition of hydrogen bromide to quinones produces bromohydroquinones.
8. Hydrogen bromide and aldehydes can be used to introduce bromoalkyl groups into various molecules, eg, bromoethylation of aromatic nuclei.
9. In the petroleum industry, HBr serves as an alkylation catalyst.
10. Hydrogen bromide is claimed as a catalyst in the controlled oxidation of aliphatic and alicyclic hydrocarbons to ketones, acids, and peroxides.
11. Applications of HBr with NH_4Br or with H_2S and HCl as promoters for the dehydrogenation of butene to butadiene.
12. Hydrogen bromide is used in the replacement of aliphatic chlorine by bromine in the presence of an aluminum catalyst.
13. Hydrogen bromide also finds use in the electronics industry.
14. Hydrogen bromide is used to make a catalyst for PTA production or is used as the catalyst itself.

Metal and Nonmetal Bromides

Some of the physical properties have been redetermined in recent years for some of the above metal bromide solutions.

Manufacture. The metal bromides are generally prepared by the reaction of hydrogen bromide with the metals, hydroxides, carbonates, or oxides of the metals. For the specific cases of alkali and alkaline earth bromides, the procedure involves the neutralization of the corresponding hydroxide or carbonate with hydrobromic acid, evaporation of excess water, and crystallization.

Metal bromide solutions can be prepared on a laboratory scale using the reaction of the appropriate carbonate or sulfate with barium bromide.

A more recent process suggested for the preparation of inorganic bromide salts is the reaction of the respective chloride salt with 1,2-dibromoethane (DBE).

The production of the nonmetal bromides is accomplished through the reaction of the nonmetal with hydrogen bromide or bromine at an elevated temperature.

Applications. The main applications of inorganic bromides are in drilling fluids, biocides, photography, pharmaceutics, catalysts, and brominating agents.

BROMINE HALIDES

Bromine is capable of forming a number of bromine halide compounds under varying process conditions. In general, these compounds have properties intermediate between those of the parent halogens.

Bromine chloride, BrCl, is formed when bromine and chlorine react reversibly in the liquid and vapor phases at room temperature. BrCl has uses in organic synthesis involving the addition across olefinic double bonds to produce bromochloro compounds, and for aromatic brominations, where an aromatic bromide and HCl are produced.

BrCl also finds use as a disinfectant in wastewater treatment, where it enjoys the following advantages over chlorine: activity maintained over a wider pH range, more rapid disinfection, effective at lower residual concentrations, and lower aquatic toxicity. In addition, bromine chloride has applications as a brominating agent in the preparation of fire-retardant chemicals, pharmaceuticals, high density brominated liquids, agricultural chemicals, dyes, and bleaching agents.

Bromine monofluoride, BrF, may be prepared by direct reaction between bromine and fluorine. Because of its high tendency to disproportionate, it has never been prepared in its pure form.

Bromine trifluoride, BrF_3 can be formed by the reaction between gaseous fluorine and liquid bromine at a temperature of 200°C.

Bromine trifluoride is a useful solvent for ionic reactions that need to be carried out under highly oxidizing conditions. As a strong fluorinating agent, it finds use in organic syntheses and in the formation of inorganic fluorides. Bromine trifluoride is highly toxic and corrosive to all tissues. Rail shipments of bromine trifluoride require "Oxidizer" and "Poison" labeling.

Bromine pentafluoride, BrF_5, can be formed either by (1) the reaction of bromine with excess fluorine above 150°C or by (2) the reaction of BrF_3 vapor and gaseous fluorine at 200°C. BrF_5 is used as a fluorinating agent in organic syntheses and as the oxidizer component of some rocket propellants.

BrF_5 is highly toxic and corrosive to the skin. It requires the "Oxidizer" label for rail shipments and the "Corrosive" label for air shipments.

The two bromine fluorides—the trifluoride and pentafluoride—are available commercially.

Iodine bromide, IBr, is a black solid (mp 41°C, subl 50°C) that is more stable than bromine monochloride. It is formed by the reaction between bromine and iodine. There is also evidence for the existence of a tribromo species, iodine tribromide IBr_3. These iodine–bromine compounds are soluble in carbon tetrachloride and acetic acid and are used as halogenating agents.

BROMINE OXIDES

None of the bromine oxides are stable at ordinary temperatures and none are considered to be of any practical importance.

OXYGEN ACIDS AND SALTS

The oxygen acids of bromine are unstable strong oxidizing agents, which exist only at ambient temperatures in solution.

Hypobromous acid is used as a strong bactericide and as a water disinfectant.

The hypobromites find use as bleaching and desizing agents in the textile industry.

Because, in general, halite ions and halous acids do not arise during the hydrolysis of halogens, the existence of bromous acid HOBrO is considered to be in doubt.

Sodium bromite, $NaBrO_2$, is available in the United States at a solution concentration of ~10%, is used as a desizing agent in the textile industry. Sodium bromite has been synthesized as 99.6% pure $NaBrO_2 \cdot 3\,H_2O$, and is available commercially in Europe. The sodium salt is stable and can be stored for long periods of time.

Barium bromite, $Ba(BrO_2)_2$ potassium bromite, $KBrO_2$; and lithium bromite $LiBrO_2$ can be prepared in a similar way to sodium salt under vacuum at 0°C. Anhydrous lithium and barium bromite have been prepared by heating bromate salts.

Such compounds are stable for at least 3 months. The bromites find industrial application as desizing agents for textiles and in water treatment.

Bromic acid, $HBrO_3$, compounds can be prepared by (1) the reaction of alkali bromates with dilute sulfuric acid or (2) by the electrolysis of bromine in bromic acid solution with platinum or lead dioxide anodes. $HBrO_3$ is a strong acid (pK < 0). It is also a strong oxidant, with a standard potential of 1.47 V for the reaction $HBrO_3 \rightarrow 0.5\ Br_2$ (in acid medium).

The industrially important bromates—sodium and potassium bromate—are commonly produced by the electrolytic oxidation of the corresponding aqueous bromide solution in the presence dichromate at a temperature of 65–70°C.

Sodium bromate, $NaBrO_3$, is a strong oxidizing agent, which can cause fires upon contact with organic materials. As the bromates are a source of active oxygen, when either heated, subjected to shock, or acidified, they represent potential fire and explosion hazards. The bromates are usually packed in polyethylene-lined filter drums. Metal drums have also been used for packaging. The bromates are considered quite stable in storage.

Some physical characteristics and uses are hereby listed for selected bromates:

1. Sodium bromate: Colorless crystal or powder. specific gravity 3.34, mp 381°C (decomposes with oxygen evolution). Water solubility at 25°C is 28.39 g/100 g solution while at 80°C, the solubility is 43.1 g/100 g solution. Uses: (a) Neutralizer–oxidizer in hair wave preparations. (b) Used in mixtures with sodium bromide in gold mining applications. (c) Used in applications in textile bleaching.

2. Potassium bromate, $KBrO_3$ White crystals or powder, specific gravity 3.27, decomposes at ~327°C. Water solubility at 25°C is 7.53 g/100-g solution while at 80°C, the solubility is 25.4 g/100-g solution. This bromate is a powerful oxidizing agent, either in the pure state or when blended with magnesium carbonate. Uses: (a) Flour treatment to improve baking characteristics. (b) Hair wave solutions. (c) Malting of beer. (d) Cheese manufacture. (e) Used as an analytic standard.

3. Barium bromate, $Ba(BrO_3)_2 \cdot H_2O$. Exists as white crystals. Specific gravity 4.0. Loses water of crystallization at 180–200°C and decomposes at ~270°C. Produced by reaction between bromine and barium hydroxide. This bromate is used as a corrosion inhibitor.

Some of the characteristic reactions of aqueous bromates include

1. Reaction with hydrogen peroxide to yield O_2.
2. Reaction with PH_3 to yield phosphoric acid.
3. Reaction with HBr to yield Br_2.
4. Reaction with XeF_2, F_2 (alkali) to yield BrO_4^-.
5. Reaction with OCl^- (slow) to yield ClO_4^-.
6. Reaction with nitrous acid to yield nitric acid.

Perbromic acid $HBrO_4$, is a strong acid, which is completely dissociated in aqueous solutions. This acid is generally prepared by passing sodium perbromate solution through a cation-exchange resin, which is in the hydrogen form. Perbromic acid is very similar to perchloric acid, having little oxidizing activity in dilute solution but being quite capable of reacting violently when concentrated.

Sodium perbromate, $NaBrO_4$, is highly soluble in water, while potassium perbromate, $KBrO_4$, is soluble only to the extent of ~0.2 M at room temperature.

F. A. Cotton and G. Wilkinson, *Advanced Inorganic Chemistry*, 5th ed., John Wiley & Sons, Inc., New York, 1988, pp. 572–573.

A. J. Downs and C. J. Adams in J. C. Bailar, Jr., and co-ed., *Comprehensive Inorganic Chemistry*, Vol. 2, Pergamon, Oxford, U.K. 1973, p. 1400.

N. N. Greenwood and A. Earnshaw, "The Halogens: Fluorine, Chlorine, Bromine, Iodine, and Astatine," *Chemistry of the Elements*, Pergamon Press, 1984, Chapt. 17.

D. Shriver and P. Atkins, *Inorganic Chemistry*, 3rd ed., W. H. Freeman & Co., New York 1999.

S. D. UKELES
OB IMI (TAMI)
M. FREIBERG
DSBG

BROMINE, ORGANIC COMPOUNDS

Organic bromine compounds, which are organic compounds in which the bromine is covalently bonded to carbon or (rarely) to nitrogen and oxygen, are a very important group of organic halogen compounds. Even naturally occurring bromine containing organic compounds produced by marine and terrestrial plants, bacteria, fungi, insects, marine animals, and some higher animals, number nearly 1500 compounds. Organic bromine compounds are the predominant industrial bromine compounds and, in terms of bromine consumption, account for ~80% of bromine production. The industrially produced organic bromine compounds can be divided into two main groups: (*1*) Organic bromine compounds in which the bromine atom is retained in the final molecular structure, and where its presence contributes to the properties of the desired products, the largest segment in terms of consumed volumes, (*2*) Organic bromine compounds, which have traditionally played an important role as intermediates in the production of agrochemicals, pharmaceuticals and dyes.

CHEMICAL PROPERTIES

Substitution of bromine by other groups proceeds as a nucleophilic, electrophilic, or radical process. Nucleophilic displacement of bromine, both in aliphatic and aromatic molecules, by neutral or anionic nucleophiles, is the leading process in the application of brominated intermediates. As a rule, the reactivity of bromine compounds in the nucleophilic substitution is greater than the corresponding chlorine derivatives, owing to the difference in the bond energies (C–Br 276 kJ/mol vs. C–Cl 328 kJ/mol). This reactivity is the main advantage of using bromine compounds as intermediates. The reaction of aliphatic bromides, either with water or with dilute aqueous solutions of bases, gives alcohols, the reaction with alcoholates gives ethers, the reaction with salts of carbonic acids gives esters, and the reaction with sodium cyanide gives nitriles. Interaction with ammonia, both in solution and in the gaseous phase, gives primary, secondary, or tertiary amines and quaternary ammonium salts, depending on the reaction conditions.

PREPARATION AND PRODUCTION

Organic bromine compounds can be produced by a great number of different chemical reactions; however, substitutive and additive bromination are the most common methods employed in industrial processes.

ALIPHATIC BROMINE COMPOUNDS

Methyl Bromide

Methyl bromide, CH_3Br, bromomethane, is a colorless liquid or gas with practically no odor. Its physical properties are mp $-93.7°C$; bp $3.56°C$; d_4^{20} 1.6755 kg/m^3; 3.974 kg/m^3; n_D^{20} 1.4218; vapor pressure at $20°C$, 189.3 kPa (1420 mmHg); viscosity at -20, 0, and $25°C$: 0.475, 0.397, and 0.324 mPa·s, respectively. Heat capacity of the liquid at $-13°C$ and of the vapor at $25°C$, 824 (197) and 448 (107) J/kg·K, (cal/kg·K), respectively; heat of vaporization at $3.6°C$, 252 J/g (60.2 cal/g); critical temperature (calculated) $194°C$; expansion coefficient -15 to $3°C$, 0.00163/K; dielectric constant at $0°C$ and 0.001–0.01 MHz, 9.77; dipole moment gas, 1.81D. Methyl bromide is miscible with most organic solvents and forms a bulky, crystalline hydrate below $4°C$. Its solubility in water varies with pressure: at normal pressure, methyl bromide plus water vapor, the solubility is 1.75 g/100-g solution ($20°C$).

Methyl bromide reacts with several nucleophiles and is a useful methylation agent for the preparation of ethers, sulfides, amines, etc. Tertiary amines are methylated by methyl bromide to form quaternary ammonium bromides.

Methyl bromide, when dry (<100 ppm water), is inert toward most materials of construction. Carbon steel is recommended for storage vessels, piping, pumps, valves, and fittings. Copper, brass, nickel, and their alloys are sometimes used. Aluminum, magnesium, zinc, and alloys of these metals should not be used, because under some conditions dangerous pyrophoric Grignard-type compounds may be formed. A severe explosion due to the ignition of a methyl bromide–air mixture by pyrophoric methylaluminum bromides produced by the corrosion of an aluminum fitting has been reported. Nylon and poly(vinyl chloride) (PVC) should also be avoided for handling methyl bromide.

Methyl bromide is nonflammable over a wide range of concentration in air at atmospheric pressure, and offers practically no fire hazards. With an intense source of ignition its explosive limits are from 13.5 to 14.5% by volume.

The commercial manufacture of methyl bromide is based on the reaction of hydrogen bromide with methanol. The hydrogen bromide used could be generated in situ from bromine and a reducing agent. The uses of sulfur or hydrogen sulfide as reducing agents are described, the latter process having the advantage. A new continuous process for the production of methyl bromide from methanol and aqueous HBr in the presence of a silica supported heteropolyacid catalyst has recently been described methyl bromide can also be coproduced with other organic bromine compounds by the reaction of the methanol solvent with hydrogen bromide formed as a by-product.

Worldwide, methyl bromide is used principally as a space fumigant used for killing soil parasites (nematodes, fungi, weeds, insects, and rodents) in agriculture and for the sanitation of cereal and other crops under storage and before shipment. Methyl bromide is also used as an intermediate for the manufacture of pharmaceuticals and chemical reagents. The current world consumption of methyl bromide as an intermediate is ~1000 t/a (~1.5% of total world consumption).

Due to its role in the depletion of the ozone layer, an international agreement has been reached calling for its reduced consumption and complete phasing out in the developed countries.

The nonagricultural uses of methyl bromide (its use in organic synthesis) are not restricted, provided that the compound is destroyed during the reaction.

Methyl bromide is a toxic compound. Repeated splashes on the skin result in severe skin lesions. In cases of lesser exposure, a severe itching dermatitis can develop. Overexposure to methyl bromide may cause dizziness, nausea, vomiting, headache, drowsiness, dimming of vision, and convulsions in the short term. Repeated and prolonged exposure to lower concentrations (30–100 ppm) causes severe nervous system effects. The time-weighted average limit for daily 8-h exposure to the vapor in air is 5 ppm by volume, or 19 mg/m^3; the short-time exposure limit is 15 ppm.

Bromochloromethane, CH_2BrCl, methylene chlorobromide, is a colorless liquid with a characteristic sweet odor. Its physical properties are bp 68.1°C; mp −86.5°C; d^{25}_4 1.9229 kg/m^3; n^{25}_D 1.4808; vapor pressure at 20°C 117 mmHg; heat of vaporization at bp 232 kJ/kg (55.4 kcal/kg). The liquid is completely miscible with common organic solvents. Its solubility in water is 0.9%. Common routes for its production involve the partial replacement of chlorine in dichloromethane by a halogen-exchange reaction using either bromine or hydrogen bromide. Both processes are carried out in the presence of aluminum or aluminum trihalide. Other patented processes to produce bromochlorometane include the gas-phase bromination of methyl chloride with a mixture of chlorine and HBr or bromine and chlorine, and the liquid-phase displacement reaction of dichloromethane with inorganic bromides. The major use of bromochloromethane is as a fire-extinguishing fluid, its effectiveness per unit weight makes it suitable for use in aircraft and portable systems. It is also used as an explosion suppression agent, as a solvent, and as an intermediate in the manufacture of some insecticides.

Dibromomethane, CH_2Br_2, methylene bromide, is a similar liquid, mp −52.7°C; bp 96.9°C; d^{20}_4 2.4956 kg/L; n^{25}_D 1.5419; vapor pressure 34.9 mmHg (20°C). Its solubility in water is 1.17 g/100 g at 15°C. The compound is produced by the same methods as bromochloromethane. The compound is used as a high-density solvent (mineral separation, gauge fluid) and as an intermediate. Dibromomethane is more toxic than bromochloromethane.

Both dibromomethane and bromochloromethane (or mixtures of these two compounds) are used as solvents for bromination reactions, especially for the production of polybrominated aromatic compounds and polymers.

Tribromomethane, $CHBr_3$, bromoform, as a pure liquid has a mp 8.3°C, bp 149.5°C, d^{20}_4, 2.8912 kg/L, n^{19}_D 1.5419, vapor pressure 5 mmHg (20°C). Its water solubility is ~0.3 g/100 g at 25°C. Bromoform is prepared by reaction of bromine with acetone or ethanol in the presence of sodium hydroxide. Uses have been found as high-density solvent for mineral separation, in gauge fluids and as an intermediate. Bromoform is a toxic and irritant compound, TLV 0.5 ppm (skin).

Tetrabromomethane, CBr_4, carbon tetrabromide, crystallize in two forms, α-form mp 48.4°C, β-form mp 92.5°C, bp 189.5°C; d^{20}_4 3.240 kg/m^3, $d^{99.5}$ 2.9609, $n^{99.5}_D$ 1.600. It is prepared by the replacement of chlorine in carbon tetrachloride using hydrogen bromide and an aluminum halide catalyst or by action of sodium hypobromite on bromoform by an extension of the haloform reaction. Tetrabromomethane is used as intermediate both in ionic and in homolytic reactions and telomerisation processes.

Bromotrifluoromethane, $CBrF_3$, bromochlorodifluoro methane, $CBrClF_2$, and 1,2-dibromotetrafluoroethane, $CBrF_2 CBrF_2$, are volatile bromine-containing halogenofluorocarbons, known under the technical name "halons." Halons are fire-extinguishing agents used to replace the more toxic methyl bromide and carbon tetrachloride.

Ethylene Dibromide

Ethylene dibromide CH_2BrCH_2Br (ethylene bromide, 1,2-dibromoethane), commonly abbreviated as EDB, is a clear, colorless liquid with a characteristic sweet odor. Its properties include mp 9.9°C; bp 131.4°C; d^{20}_4 2.1792 kg/L, n^{20}_D 1.5380, vapor pressure 1.13 (8.5), 15.98 (119.8), and 38.03 kPa·s (285.2 mmHg) at 20, 75, and 100°C, respectively; viscosity 1.727 mPa·s (20°C); heat capacity of the solid at 15.3°C, 519 J/kg·K (124 cal/kg·K) and of the liquid at 21.3°C, 724 J/kg·K (173 cal/kg·K); heat of fusion at 9.9°C, 53.4 J/g (12.76 cal/g); heat of vaporization at bp 191 J/g (45.7 cal/g); heat of transition at −23.6°C, 10.34 J/g (2.47 cal/g); critical temperature, 309.8°C; critical pressure 7154 kPa (70.6 atm); expansion coefficient at 15–30°C, 0.000958/K;

dielectric constant at 20.5°C (0.1 MHz), 4.77. The liquid is completely miscible with carbon tetrachloride, benzene, gasoline, and anhydrous alcohols at 25°C and its solubility in water at 20°C is 0.404 g/100-g solution.

EDB is nonflammable and quite stable under ordinary conditions. Ethylene glycol is produced by its high temperature hydrolysis under pressure. Reaction with metals (zinc, magnesium) yields ethylene, reaction with ammonia proceeds with explosion, yielding ethylenediamine and higher polymers.

The largest single application of EDB has traditionally been its use as a lead scavenger in leaded gasoline. The second-largest traditional use of EDB was as an insect fumigant and soil nematocide. In 1983, however, the EPA banned the use of EDB in most agricultural applications because of concerns about the chemical's toxicity. Currently, most EDB in the USA is produced for export.

Other uses of EDB are as an intermediate for pharmaceuticals and dyes, where it provides an "ethylene bridge" in the molecular structure. EDB is used as a nonflammable solvent for resins, gums, and waxes. Additionally, EDB can be used as a raw material in the synthesis of chemicals such as vinyl bromide and styrenic block copolymers.

EDB is an acutely toxic, severely irritating to skin, mutagenic, and carcinogenic compound. Its current time-weighted average limit is 20 ppm.

INDUSTRIAL CHEMICAL PRODUCTS

Flame Retardants

Brominated flame retardants are the more significant and voluminous part of all bromine derivatives. Bromine consumption in flame retardants has risen substantially from the early 1990s and at present forms, on average, one-half of organic bromine compounds consumption and ~40% of the total bromine consumption.

Brominated flame retardants can be divided into two groups according to their chemical structure: brominated aliphatic compounds and brominated aromatic compounds. The latter are much more stable and may be used in thermoplastics at fairly high temperatures without the use of stabilizers and at very high temperature with stabilizers.

Brominated flame retardants can also be divided into two general classes according to their relation to polymers—additive flame retardants and reactive flame retardants. Additives are mixed into the polymer in common polymer processing equipment. Reactive flame retardants literally become part of the polymer by either reacting into the polymer backbone or grafting onto it.

The use of brominated compounds will benefit from more exacting fire safety standards in consumer products, building products, automobile and aircraft components. The proliferation of computers and other consumer electronics is boosting demand for plastics that have enhanced flame retardancy characteristics. Because of the effectiveness and performance advantages of brominated flame retardants, smaller amounts can be used, which enhances the cost effectiveness of these products while maintaining the functional characteristics of the host material.

While chlorine controls the majority of the water treatment market, brominated compounds are becoming increasingly popular. In general, both industrial and consumer segments of the water treatment industry are increasingly replacing chlorine and chlorinated compounds as sanitizers and biocides with bromine-based products. Chlorine is now subject to a wide range of EPA limitations, and although bromine is itself a halogen, no restrictions have yet been placed on brominated biocides. In addition, the greater strength of bromine-based biocides allows treatment of a given amount of water with considerably less biocide. This not only reduces the amount of halogen released into the environment, but can also reduce costs for municipal and industrial water treatment plants. Benefits accrue through reduced chemical costs, avoidance of the dechlorination step common in chlorine water treatment, and reduced corrosion to condensers, tubing and other equipment.

Brominated biocides also perform better than chlorinated biocides in a number of industrial applications because of their higher tolerance to a wide range of pH levels, a concern in cooling towers and process waters. Brominated products are both more effective at higher pH ranges than chlorine, and are estimated to be three times more effective than chlorine at controlling algae blooms.

Apart from their increasing use in industrial and municipal water treatment, bromine derivatives are also registering gains in the consumer water conditioning market as biocides for spas and hot tubs. In these applications, DBDMH and Halobrom are displacing chlorinated biocides for the same reasons as in large-scale water treatment.

In addition, brominated biocides are more stable than chlorine in higher temperature waters, do not readily decompose on exposure to sunlight and are less irritating to the eyes and mucous membranes. This is particularly important in the recreational segment of the water treatment market.

Pharmaceuticals

Organic pharmaceuticals containing bromine can be divided into two groups. The main group includes actual organic bromine compounds, in which bromine is bonded with the carbon atom. The second group includes salts of hydrobromic acid and ammonium organic compounds.

Dyes and Indicators

The effect of bromine in dye or indicator molecules in place of hydrogen includes a shift of light absorption to longer wavelengths, increased dissociation of phenolic hydroxyl groups, and lower solubility. The first two effects probably result from increased polarization caused by bromine's electronegativity compared to that of hydrogen.

N. De Kimpe and R. Verhe in S. Patai and Z. Rappoport, eds., *The Chemistry of Functional Groups*, Update Vol., John Wiley Sons, Inc., New York, 1988.

C. Hill, *Activation and Functionalization of Alkanes*, John Wiley & Sons, Inc., New York, 1989.

W. Paulus, *Microbiocides for the Protection of Materials*, Chapman & Hall, London, 1993.

Y. Sasson, M. Weiss, and G. Barak in D. Price, B. Iddon, and B. J. Wakefield, eds., *Bromine Compounds Chemistry and Applications*, Elsevier, 1988, pp. 252–271.

DAVID IOFFE

IMI (TAMI)

ARIEH KAMPF

DSBG

BUTADIENE

Butadiene, C_4H_6, exists in two isomeric forms: 1,3-butadiene, $CH_2=CH-CH=CH_2$, and 1,2-butadiene, $CH_2=C=CH-CH_3$. 1,3-Butadiene is a commodity product of the petrochemical industry with a 2000 U.S. production of 4.4 billion pounds (2.0×10^9 kg). Elastomers consume the bulk of 1,3-butadiene, led by the manufacture of styrene–butadiene rubber (SBR). 1,3-Butadiene is manufactured primarily as a coproduct of steam cracking to produce ethylene in the United States, Western Europe, and Japan. However, in certain parts of the world it is still produced from ethanol.

The other isomer, 1,2-butadiene, a small by-product in 1,3-butadiene production, has no significant current commercial interests. However, there are a number of publications and patents on its recovery and applications, particularly in the specialty polymer area and as a gel inhibitor.

PROPERTIES

1,3-Butadiene is a noncorrosive, colorless, flammable gas at room temperature and atmospheric pressure. It has a mildly aromatic odor. It is sparingly soluble in water, slightly soluble in methanol and ethanol, and soluble in organic solvents like diethyl ether, benzene, and carbon tetrachloride. Its important physical properties are summarized in Table 1. 1,2-Butadiene is much less studied. It is a flammable gas at ambient conditions. Some of its properties are summarized in Table 2.

MANUFACTURE AND PROCESSING

The pattern of commercial production of 1,3-butadiene parallels the overall development of the petrochemical industry. Since its discovery via pyrolysis of various organic materials, butadiene has been manufactured from acetylene as well as ethanol, both via butanediols (1,3- and 1,4-) as intermediates. On a global basis, the importance of these processes has decreased substantially

Table 1. Physical Properties of 1,3-Butadiene[a]

Property	Value
RTECS accession number	EI9275000
UN number	1010
molecular formula	C_4H_6
molecular weight	54.092
boiling point at 101.325 kPa[a], °C	−4.411
freezing point, °C	−108.902
critical temperature, °C	152.0
critical pressure, MPa[b]	4.32
critical volume, cm^3/mol	221
critical density, g/mL	0.245
density (liquid), g/mL at	
0°C	0.6452
15°C	0.6274
20°C	0.6211
25°C	0.6194
50°C	0.5818
density (gas) (air = 1)	1.9
heat capacity at 25°C, J/(mol·K)[c]	79.538
refractive index, n_D at −25°C	1.4292
solubility in water at 25°C, ppm	735[e]
viscosity (liquid), mPa·s (=cP) at	
−40°C	0.33
0°C	0.25
40°C	0.20
heat of formation, gas, kJ/mol[c]	110.165
heat of formation, liquid, kJ/mol[c]	88.7
free energy of formation, kJ/mol[c]	150.66
heat of vaporization, J/g[d] at	
25°C	389
boiling point	418
flash point, °C	−85
autoignition temperature, °C	417.8
explosion limits in air, vol %	
lower	2.0
upper	11.5
minimum oxygen for combustion (MOC), %v/vO_2	
N_2–air	10
CO_2–air	13
absorption	
λ, cm^{-1}	217
log ε	4.32

[a] To convert kPa to mm Hg, multiply by 7.5.
[b] To convert MPa to psi, multiply by 145.
[c] To convert J to cal, divide by 4.184.
[d] 245 mol ppm.

Table 2. Physical Properties of 1,2-Butadiene[a]

Property	Value
molecular formula	C_4H_6
molecular weight	54.092
boiling point at 101.325 kPa,[a] °C	10.85
freezing point, °C	−136.19
density (liquid), g/mL at	
0°C	0.676
20°C	0.652
heat of formation at 25°C (gas), kJ/mol[b]	162.21
heat of vaporization at 25°C, kJ/mol[b]	23.426
refractive index at 1.3°C	1.4205

[a] To convert kPa to mm Hg, multiply by 7.5.
[b] To convert kJ to kcal, divide by 4.184.

because of the increasing production of butadiene from petroleum sources. China and India still convert ethanol to butadiene using the two-step process, while Poland and the former USSR use a one-step process. In the past, butadiene was also produced by the dehydrogenation of *n*-butane and oxydehydrogenation of *n*-butenes. However, butadiene is now primarily produced as a by-product in the steam cracking of hydrocarbon streams to produce ethylene. Except under market dislocation situations, butadiene is almost exclusively manufactured by this process in the United States, Western Europe, and Japan.

HANDLING, STORAGE, AND SHIPPING

Butadiene reacts with a large number of chemicals, has an inherent tendency to dimerize and polymerize, and is toxic. Therefore, specific handling, storage, and shipping procedures must be followed.

Butadiene is primarily shipped in pressurized containers via railroads or tankers. U.S. shipments of butadiene, which is classified as a flammable compressed gas, are regulated by the Department of Transportation. Most other countries have adopted their own regulations. Other information on the handling of butadiene is also available. As a result of the extensive emphasis on proper and timely responses to chemical spills, a comprehensive handbook from the National Fire Protection Association is available.

ECONOMIC ASPECTS

Since the bulk of butadiene is recovered from steam crackers, its economics is very sensitive to the selection of feedstocks, operating conditions, and demand patterns. Butadiene supply and, ultimately, its price are strongly influenced by the demand for ethylene, the primary product from steam cracking. Currently, there is a worldwide surplus of butadiene. Announcements of a number of new ethylene plants will likely result in additional butadiene production, more than enough to meet worldwide demand for polymers and other chemicals. When butadiene is in excess supply, ethylene manufacturers can recycle the butadiene as a feedstock for ethylene manufacture.

HEALTH AND SAFETY FACTORS

Short-term exposure to high concentrations of butadiene may cause irritation to the eyes, nose, and throat. Dermatitis and frostbite may result from exposure to the liquid and the evaporating gas. Long-term physiological reactions to 1,3-butadiene may vary individually.

Exposure studies using mice and rats have demonstrated species differences in butadiene toxicity and carcinogenicity.

In several epidemiological studies elevation in mortality were observed for small subgroups and tumor types. There have been many reviews published on the toxicity of butadine.

Table 3. Pattern of Butadiene Uses in the United States, 2000

End use	Percentage of total, %
synthetic elastomers	61
styrene–butadiene rubber (SBR)	29
polybutadiene (BR)	27
polychloroprene (Neoprene)	3
nitrile rubber (NR)	2
polymers and resins	18
acrylonitrile–butadiene–styrene (ABS)	6
styrene–butadiene copolymer (latex)	12
chemicals and other uses	21
adiponitrile	14
others	7

*Chemical News and Data, March 6, 2000.

USES

Butadiene is used primarily in polymers, including SBR, BR, ABS, SBL, and NR. In 2000 these uses accounted for about 89% of butadiene consumed in the United States. Styrene–butadiene rubber, the single largest user of butadiene, consumes 29% of the total, followed by polybutadiene rubber at 27%. Consumption for the other polymers, ABS, SBL, polychloroprene, and nitrile rubber are listed in Table 3.

Another significant butadiene use is for manufacturing adiponitrile, $NC(CH_2)_4CN$, a precursor for nylon-6,6 production. Other miscellaneous chemical uses, such as for ENB (ethylidene norbornene) production, account for 7% combined.

New butadiene applications are expected to grow from a small base, particularly liquid hydroxy-terminated homopolymers used in polyurethane for sealants, waterproofing, electrical encapsulation, and adhesive. A recent patent describes an epoxy resin composition for semiconductor encapsulation comprising an epoxy resin, phenolic resin, and butadiene rubber particles. An invention for improved polybutadiene composition suitable for use in molded golf ball cores has been described. A butadiene rubber adhesive composition with specified low syndiotactic butadiene has been patented.

"Butadiene", *ChemExpo Chemical Profile*, Chemical News and Data, March 6, 2000, http://www.chemexpo.com/Profile, searched Feb. 14, 2002.

T. Carreon, "Aliphatic Hydrocarbons" in E. Bingham, B. Cohrssen, C. H. Powell, eds., *Patty's Toxicology*, Vol. 4, John Wiley & Sons, Inc., New York, 2001, Chapt. 49.

R. W. Gallant, *Physical Properties of Hydrocarbons*, Gulf Publishing Co., Houston, Tex., 1968.

J. Larson, Butadiene, *Chemical Economics Handbook*, SRI, Menlo Park, Calif., Nov. 2000.

H. N. SUN
J. P. WRISTERS
Exxon Chemical Company

BUTYL ALCOHOLS

Butyl alcohols encompass the four structurally isomeric 4-carbon alcohols of empirical formula $C_4H_{10}O$. One of these, 2-butanol, can exist in either the optically active $R(-)$ or $S(+)$ configuration or as a racemic $(\pm)$ mixture.

PHYSICAL AND CHEMICAL PROPERTIES

The butanols are all colorless, clear liquids at room temperature and atmospheric pressure with the exception of t-butyl alcohol which is a low melting solid (mp 25.82°C); it also has a substantially higher water miscibility than the other three alcohols. Physical constants of the four butyl alcohols are given in Table 1.

Physical constants for the optically pure stereoisomers of 2-butanol have been reported as follows:

	d^{t}_{4}	n^{D}_{20}	$[\alpha]^{D}_{27}$
(S)-(+)-2-butanol	0.8025^{27}	1.3954	+13.52
(R)-(−)-2-butanol	0.8042^{25}	1.3970	−13.52

Butyl alcohol liquid vapor pressure/temperature responses, which are important parameters in direct solvent applications, are presented in Figure 1.

The butanols undergo the typical reactions of the simple lower chain aliphatic alcohols.

MANUFACTURE

The principal commercial source of 1-butanol is n-butyraldehyde, obtained from the Oxo reaction of propylene.

USES

The largest volume commercial derivatives of 1-butanol in the United States are n-butyl acrylate and methacrylate. These are used principally in emulsion polymers for latex paints, in textile applications and in impact modifiers for rigid poly(vinyl chloride).

Isobutyl alcohol has replaced n-butyl alcohol in some applications where the branched alcohol appears to have preferred properties and structure.

HEALTH, SAFETY, AND ENVIRONMENTAL CONSIDERATIONS

All four butanols are thought to have a generally low order of human toxicity. However, large dosages of the butanols generally serve as central nervous system depressants and mucous membrane irritants.

All four butanols are registered in the United States on the Environmental Protection Agency Toxic Substances Control Act (TSCA) Inventory, a prerequisite for the manufacture or importation for commercial sale of any chemical substance or mixture in quantities greater than a

Table 1. Physical Properties of the Butyl Alcohols (Butanols)

Common Name	n-Butyl alcohol	Isobutyl alcohol	sec-Butyl alcohol	t-Butyl alcohol
systematic name	1-butanol	2-methyl-1-propanol	2-butanol	2-methyl-2-propanol
formula	$CH_3(CH_2)_3OH$	$(CH_3)_2CHCH_2OH$	$CH_3CH(OH)C_2H_5$	$(CH_3)_3COH$
critical temperature, °C	289.90	274.63	262.90	233.06
critical pressure, kPa[a]	4423	4300	4179	3973
critical specific volume, m³/kg mol	0.275	0.273	0.269	0.275
normal boiling point, °C	117.66	107.66	99.55	82.42
melting point, °C	−89.3	−108.0	−114.7	25.82
ideal gas heat of formation at 25°C, kJ/mol[b]	−274.6	−283.2	−292.9	−312.4
heat of fusion, kJ/mol[b]	9.372	6.322	5.971	6.703
heat of vaporization at normal boiling point, kJ/g[b]	43.29	41.83	40.75	39.07
liquid density, kg/m³ at 25°C	809.7	801.6	806.9	786.6[c]
liquid heat capacity at 25°C, kJ/(mol·K)[b]	0.17706	0.18115	0.19689	0.2198 at mp
refractive index at 25°C	1.3971	1.3938	1.3949	1.3852
flash point, closed cup, °C	28.85	27.85	23.85	11.11
dielectric constant, ε	17.5^{25}	17.93^{25}	16.56^{20}	12.47^{30}
dipole moment × 10³⁰, C·m[d]	5.54	5.47	5.54	5.57
solubility parameter, (MJ/m³)^0.5 [e] at 25°C	23.354	22.909	22.541	21.603
solubility in water at 30°C, % by weight	7.85	8.58	19.41	miscible
solubility of water in alcohol at 30°C, % by weight	20.06	16.36	36.19	miscible

[a]To convert kPa to mm Hg, multiply by 7.50.
[b]To convert kJ to kcal, divide by 4.184.
[c]For the subcooled liquid below melting point.
[d]To convert C·m to debyes, divide by 3.336×10^{-30}.
[e]To convert (MJ/m³)^0.5 to (cal/cc)^0.5, multiply by $0.239^{0.5}$.

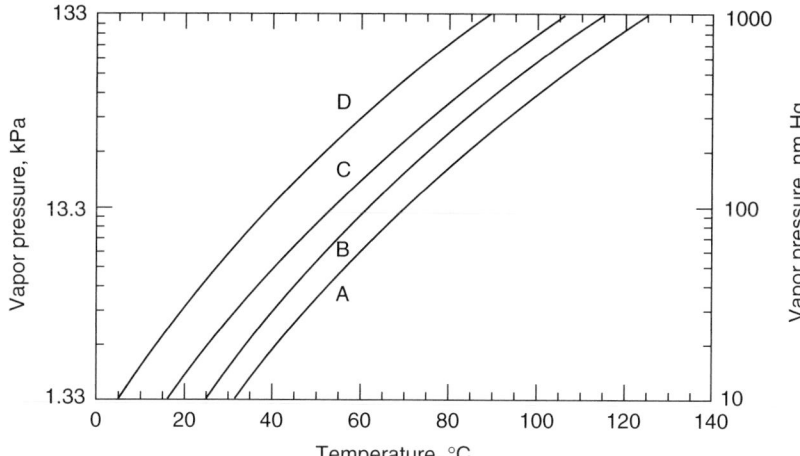

Figure 1. Vapor pressure of butyl alcohols: A, *n*-butyl; B, isobutyl; C, *sec*-butyl; D, *t*-butyl. To convert kPa to mm Hg, multiply by 7.5.

1000 pounds (454 kg). Additionally, the manufacture and distribution of the butanols in the United States are regulated under the Superfund Amendments and Reauthorization Act (SARA), Section 313, which requires that anyone handling at least 10,000 pounds (4545 kg) a year of a chemical substance report to both the EPA and the state any release of that substance to the environment.

STORAGE AND HANDLING

The C-4 alcohols are preferably stored in baked phenolic-lined steel tanks. However, plain steel tanks can also be employed provided a fine porosity filler is installed to remove any contaminating rust.

Storage under dry nitrogen is also recommended since it limits flammability hazards as well as minimizing water pickup.

AIChE Design Institute for Physical Property Data, Project 801 Source File Tape, Revision Dates Aug. 1989 and 1990.

Chemical Economics Handbook, SRI Consulting, Menlo Park, Calif.

R. H. Crabtree, *The Organometallic Chemistry of the Transition Metals*, John Wiley & Sons, Inc., New York, 1988, pp. 173–176.

ERNST BILLIG
Union Carbide Chemicals and
Plastics Company Inc.

BUTYLENES

Butylenes are C_4H_8 mono-olefin isomers: 1-butene, *cis*-2-butene, *trans*-2-butene, and isobutylene (2-methylpropene). These isomers are usually coproduced as a mixture and are commonly referred to as the C_4 fraction. These C_4 fractions are usually obtained as by-products from petroleum refinery and petrochemical complexes that crack petroleum fractions and natural gas liquids. Since the C_4 fractions almost always contain butanes, it is also known as the B–B stream. The linear isomers are referred to as butenes.

PHYSICAL PROPERTIES

For any industrial process involving vapors and liquids, the most important physical property is the vapor pressure. Table 1 presents values for the constants for a vapor-pressure equation and the temperature range over which the equation is valid for each butylene.

Table 2 presents other important physical properties for the butylenes. Thermodynamic and transport properties can also be obtained from other sources.

CHEMICAL PROPERTIES

The carbon–carbon double bond is the distinguishing feature of the butylenes and as such, controls their chemistry. The carbon–carbon bond, acting as a substitute, affects the reactivity of the carbon atoms at the alpha positions through the formation of the allylic resonance structure. This structure can stabilize both positive and negative charges. Thus allylic carbons are more reactive to substitution and addition reactions than alkane carbons. Therefore, reactions of butylenes can be divided into two broad categories: (*1*) those that take place at the double bond itself, destroying the double bond; and (*2*) those that take place at alpha carbons.

The electron-rich carbon–carbon double bond reacts with reagents that are deficient in electrons, eg, with electrophilic reagents in electrophilic addition, free radicals in free-radical addition, and under acidic conditions with another butylene (cation) in dimerization.

MANUFACTURE

The C_4 isomers are almost always produced commercially as by-products in a petroleum refinery/petrochemical process.

There are two important sources for the commercial production of butylenes: catalytic or thermal cracking,

Table 1. Vapor-Pressure Equation Constants for the Butanes, Butylenes, and Butadienes

	A	B	C	D	N	Temperature range, K
n-butane	61.5623	−4259.90	−6.20315	3.07575×10^{-7}	2.5	135–423
isobutane	66.7163	−4237.62	−7.08156	4.00506×10^{-7}	2.5	129–408
1–butene	78.8760	−4713.65	−9.05743	1.28654×10^{-5}	2.0	126–416
cis–2–butene	71.9534	−4681.34	−7.87527	1.00237×10^{-5}	2.0	203–358
trans–2–butene	74.3950	−4648.45	−8.33977	1.20897×10^{-5}	2.0	195–358
isobutylene	83.8683	−4822.95	−9.90214	1.51060×10^{-5}	2.0	194–359
1,2-butadiene	49.5031	−4021.95	−4.28893	5.13547×10^{-6}	2.0	200–284
1,3-butadiene	73.0016	−4547.77	−8.11105	1.14037×10^{-5}	2.0	164–425

and steam cracking. In these two processes, butylenes are always produced as by-products.

There are other commercial processes available for the production of butylenes. However, these are site or manufacturer specific, eg, the Oxirane process for the production of propylene oxide; the disproportionation of higher olefins; and the oligomerization of ethylene. Any of these processes can become an important source in the future.

More recently, the Coastal Isobutane process began commercialization to produce isobutylene from butanes for meeting the expected demand for methyl-tert-butyl ether.

New Technologies

Several technologies are emerging for the production of isobutylene to meet the expected demand for isobutylene:

Table 2. Physical Properties of the Butylenes

Property	Values			
	1-Butene	cis-2-Butene	trans-2-Butene	Isobutylene
molecular weight	56.11	56.11	56.11	56.11
melting point, K	87.80	134.23	167.62	132.79
boiling point, K	266.89	276.87	274.03	266.25
critical temperature, K	419.60	435.58	428.63	417.91
critical pressure, MPa[a]	4.023	4.205	4.104	4.000
critical volume, L/mol	0.240	0.234	0.238	0.239
critical compressibility factor	0.277	0.272	0.274	0.275
acentric factor	0.1914	0.2018	0.2186	0.1984
Ideal gas properties[b] at 298.15 K				
Hf, kJ/mol	−0.126	−6.99	−11.18	−16.91
Gf, kJ/mol	71.34	65.90	63.01	58.11
Cp, J/mol · K	85.8	79.4	88.3	90.2
Hvap, kJ/mol	20.31	22.17	21.37	20.27
Hcomb, kJ/mol	−2719	−2712	−2708	−2724
Saturated vapor at 298.15 K				
viscosity, mPa · s(= cP)	0.00776	0.00782	0.00763	0.00816
thermal conductivity, W/(m · K)	0.0151	0.0135	0.0144	0.0158
Saturated liquid				
density at 298.15 K, mol/L	10.47	11.00	10.69	10.49
surface tension at 298.15 K, mN/m(= dyn /cm)	0.0121	0.0140	0.0132	0.0117
Cp at 266 K, J/mol · K[b]	121.6	118.8	121.8	123.3
viscosity at 266 K, mPa · s(= cP)	0.186	0.214	0.214	0.228
thermal conductivity at 266 K, W/(m · K)	0.120	0.124	0.121	0.117
Flammability limits, vol % in air				
lower limit	1.6	1.6	1.8	1.8
upper limit	9.3	9.7	9.7	8.8
autoignition temperature, K	657	598	597	738

[a]To convert MPa to atm, multiply by 9.869.
[b]To convert kJ to kcal, divide by 4.184.

(*1*) deep catalytic cracking; (*2*) superflex catalytic cracking; (*3*) dehydrogenation of butanes; and (*4*) the Coastal process of thermal dehydrogenation of butanes.

Separation and Purification of C₄ Isomers

1-Butene and isobutylene cannot be economically separated into pure components by conventional distillation because they are close boiling isomers 2-Butene can be separated from the other two isomers by simple distillation. There are four types of separation methods available: (*1*) selective removal of isobutylene by polymerization and separation of 1-butene; (*2*) use of addition reactions with alcohol, acids, or water to selectively produce pure isobutylene and 1-butene; (*3*) selective extraction of isobutylene with a liquid solvent, usually an acid; and (*4*) physical separation of isobutylene from 1-butene by absorbents.

There are three important processes for the production of isobutylene: (*1*) the extraction process using an acid to separate isobutylene; (*2*) the dehydration of *tert*-butyl alcohol, formed in the Arco's Oxirane process; and (*3*) the cracking of MTBE.

SHIPMENT AND HANDLING

Storage and Transportation

Handling requirements are similar to liquefied petroleum gas (LPG). Storage conditions are much milder. Butylenes are stored as liquids at temperatures ranging from 0 to 40°C and at pressures from 100 to 400 kPa (1–4 atm). Their transportation is also similar to LPG; they are shipped in tank cars, transported in pipelines, or barged.

HEALTH AND SAFETY FACTORS

The effect of long-term exposure is not known, hence, they should be handled with care. They are volatile and asphyxiants. 1-Butene, *cis* 2-butene, and *trans* 2-butene are very dangerous fire hazards when exposed to heat or flame. *cis* 2-Butene and *trans*-2-butene are very likely to explode. 1-Butene is moderately explosive. 2-Butene emits acrid smoke and irritating vapors when heated to decomposition. Care should be taken to avoid spills because they are extremely flammable. Physical handling requires adequate ventilation to prevent high concentrations of butylenes in the air. Explosive limits in air are 1.6 to 9.7% of butylenes. Their flash points range from −80 to −73°C. Their autoignition is around 324 to 465°C (Table 2). Water and carbon dioxide extinguishers can be used in case of fire.

USES

Among the butylenes, isobutylene has become one of the important starting materials for the manufacture of polymers and chemicals. There are many patents that describe the use of isobutylene or its derivatives to produce insecticides, antioxidants, elastomers, additives for lubricating oils, adhesives, sealants, and caulking compounds.

D. F. Blaser, in J. J. McKetta, ed., *Encyclopedia of Chemical Processing and Design*, Vol. 10, Marcel Dekker, Inc., New York, 1979, p. 13.

"Butadiene Chemical Profile", *Chemical Market Reporter*, March 25, 2002.

L. F. Fieser and M. Fieser, *Advanced Organic Chemistry*, Reinhold Publishing Co., New York, 1961, p. 162.

W. A. Pryer, *Introduction to Free Radical Chemistry*, Prentice-Hall, Englewood Cliffs, N.J., 1965.

NARASIMHAN CALAMUR
MARTIN E. CARRERA
RICHARD A. WILSAK
Amoco Chemical Company

BUTYL RUBBER

Isobutylene has been of interest since the early days of synthetic polymer research when Friedel Crafts catalysts were used to prepare elastic materials. Isobutylene polymers of commercial importance include homopolymers and copolymers containing small amounts of isoprene or *p*-methylstyrene. Currently, chlorinated and brominated derivatives of butyl(poly[isobutylene-*co*-isoprene]) have the highest sales volume.

The first use for butyl rubber was as inner tubes, whose air-retention characteristics contributed significantly to the safety and convenience of tires. Good weathering, ozone resistance, and oxidative stability have led to applications in mechanical goods and elastomeric sheeting. Automobile tires were manufactured for a brief period from butyl rubber, but poor abrasion resistance curtailed this development.

Halogenated butyl rubber greatly extended the usefulness of butyl rubber by providing much higher vulcanization rates and improving the compatibility with highly unsaturated elastomers, such as natural rubber and styrene–butadiene rubber (SBR). These properties permitted the production of tubeless tires with chlorinated or brominated butyl inner liners. The retention of air pressure and low intercarcass pressure extended tire durability.

Polyisobutylene is produced in a number of molecular weight grades and each has found a variety of uses. The low molecular weight liquid polybutenes have applications as adhesives, sealants, coatings, lubricants, and plasticizers, and for the impregnation of electrical cables. Moderate molecular weight polyisobutylene was one of the first viscosity-index modifiers for lubricants. High molecular weight polyisobutylene is used to make uncured rubbery compounds and as an impact additive for thermoplastics.

PROCESS CHEMISTRY

Butyl rubber is prepared from 2-methylpropene (isobutylene) and 2-methyl-1,3-butadiene (isoprene). Isobutylene with a purity of >99.5 wt% and isoprene with a purity of

>98 wt% are used to prepare high molecular weight butyl rubber. Water and oxygenated organic compounds are minimized by feed purification systems because these impurities interfere with the cationic polymerization mechanism. Copolymers of isobutylene can also be prepared from mixed C_4 olefin containing streams that contain *n*-butene. These copolymers are generically known as polybutenes.

Isobutylene Polymerization Mechanism

The carbocationic polymerization of isobutylene and its copolymerization with viable comonomers like isoprene and *p*-methylstyrene is mechanistically complex.

Halobutyls. Chloro- and bromobutyls are commercially the most important derivatives. The halogenation reaction is carried out in hydrocarbon solution using elemental chloride and bromine (equimolar with the enchained isoprene).

Other Derivatives. Conjugated-diene butyl (CDB) is obtained by the controlled dehydrohalogenation of halogenated butyl rubber.

Isobutylene–Isoprene–Divinylbenzene Terpolymers. A partially cross-linked terpolymer of isobutylene, isoprene, and partially reacted divinyl benzene is commercially available. This material is used primarily in the manufacture of sealant tapes and caulking compounds.

Liquid Butyl Rubber. Degradation of high molecular weight butyl rubber by extrusion at high shear rates and temperatures produces a liquid rubber with a viscosity average molecular weight (M_v) between 20,000 and 30,000. The relatively low viscosity aids in formulating high solids compounds for use in sealants, caulks, potting compounds, and coatings. Resulting compounds can be poured, sprayed, and painted.

New Materials

Star-Branched (SB) Butyl. Butyl rubbers have unique processing characteristics because of their viscoelastic properties and lack of crystallization of compounds on extension. They exhibit both low green strength and low creep resistance as a consequence of high molecular weight between entanglements. To enhance the strength of uncured traditional butyl rubber a relatively high molecular weight is required. Increasing molecular weight also causes an increase in relaxation time along with high viscosity.

SB butyl rubbers offer a unique balance of viscoelastic properties, resulting in significant processability improvements. Dispersion in mixing and mixing rates are improved. Compound extrusion rates are higher, die swell is lower, shrinkage is reduced, and surface quality is improved. The balance between green strength and stress relaxation at ambient temperature is improved, making shaping operations such as tire building easier. Several grades of ExxonMobil SB butyl polymers including copolymer, chlorinated, and brominated copolymers are commercially available.

Brominated Poly(isobutylene-co-p-methylstyrene). *para*-Methylstyrene (PMS) can be readily copolymerized with isobutylene via classical carbocationic copolymerization using a strong Lewis acid, eg, $AlCl_3$ or alkyl aluminum in methyl chloride, at low temperature. These new high molecular weight copolymers encompass an enormous range of properties, from polyisobutylene-like elastomers to poly(*p*-methylstyrene)-like tough, hard plastic materials with T_g's above 100°C, depending on monomer ratio. The brominated copolymer can be cross-linked with a variety of cross-linking systems. The benzyl bromide in the brominated copolymer can also be easily converted by nucleophilic substitution reactions to a variety of other functional groups and graft copolymers as desired for specific properties and applications.

Thermoplastic Elastomers. With the structural control inherent in living polymerization, new block copolymers containing polyisobutylene are possible. As thermoplastic elastomers, these materials offer other advantages, owing to the intrinsic properties of polyisobutylene, namely low permeability and low dynamic modulus. Many more architecturally designed polyisobutylenes are possible through living polymerization techniques.

MANUFACTURING

Most of the butyl polymers made commercially are produced by copolymerizing isobutylene and isoprene in precipitation processes that use methyl chloride as the diluent and a catalyst system comprising a Lewis acid and an alkyl halide.

The manufacture of butyl rubber, poly(isobutylene-*co-p*-methylstyrene) (Exxpro backbone) and high molecular weight polyisobutylene (Vistanex) requires a complex manufacturing process consisting of feed purification, feed blending, polymerization, slurry stripping, and finishing.

The manufacture of halobutyl rubbers such as Bromobutyl, Chlorobutyl, and Exxpro [bromopoly(isobutylene-*co-p*-methylstyrene)] requires a second chemical reaction: the halogenation of the polymer backbone.

STRUCTURE, PROPERTIES, AND PRODUCT APPLICATIONS OF ISOBUTYLENE-BASED POLYMERS

Molecular Structure

Polyisobutylene. Isobutylene polymerizes in a regular head-to-tail sequence to produce a polymer having no asymmetric carbon atoms. The glass-transition temperature of PIB is about −70°C.

Polybutenes. Copolymerization of mixed isobutylene and 1-butene containing streams with a Lewis acid catalyst system yields low molecular weight copolymers, from several hundreds to a few thousand, that are clear, colorless, and viscous liquids. The chain ends are unsaturated and are often chemically modified to provide a certain functionality.

Butyl Rubber. In butyl rubber, isoprene is enchained predominantly (90–95%) by 1,4-addition and head-to-tail arrangement. The glass-transition temperature of butyl rubber is about $-65°C$.

Halogenated Butyl Rubber. Halogenation at the isoprene site in butyl rubber proceeds by a halonium ion mechanism leading to the formation of a predominated exomethylene alkyl halide structure in both chlorinated and brominated rubbers. Upon heating, the exo allylic halide rearranges to give an equilibrium distribution of exo and endo structures. Halogenation of the unsaturation has no apparent effects on the butyl backbone and on the butyl glass-transition temperature. However, cross-linked halobutyl rubbers do not crystallize on extension as a result of the backbone irregularities introduced by the halogenated isoprene units.

Star-Branched (SB) Butyl. Introduction of a branching agent of styrene–butadiene–styrene (SBS) block copolymer during cationic polymerization of butyl leads to a SB butyl that, in general, contain 10–15% of star polymers, with remaining linear butyl chains.

Brominated Poly(isobutylene-*co-p*-methylstyrene). Copolymerization of isobutylene with *para*-methylstyrene (PMS) produces a saturated copolymer backbone with randomly distributed pendant *para*-methyl-substituted aromatic rings.

Physical Properties

Permeability. Primary uses of polyisobutylene and isobutylene copolymers of butyl, halobutyl, SB butyl, and brominated poly(isobutylene-*co-p*-methylstyrene) in elastomeric vulcanized compounds rely on their properties of low air permeability and high damping. In comparison with many other common elastomers, polyisobutylene and its copolymers are notable for their low permeability to small-molecule diffusants as a result of their efficient intermolecular packing as evidenced by their relative high density of 0.917 g/cm^3.

Dynamic Damping. Polyisobutylene and isobutylene copolymers are high damping at $25°C$, with loss tangents covering more than eight decades of frequencies even though their T_gs are less than $-60°C$. This broad dispersion in PIB's dynamic mechanical loss modulus is unique among flexible-chain polymers and is related to its broad glass–rubber transition.

Chemical Properties. PIB is a saturated hydrocarbon where the unsaturated chain ends could undergo reactions typical of a hindered olefin. These end groups are used, particularly in low molecular weight materials

where plenty of chain ends are available, as a route to functionization, such as the introduction of amine groups to PIB for producing dispersants for lubricating oils.

Elastomeric Vulcanizates

As with almost all rubbers, the applications of isobutylene copolymers in rubber goods require compounding and subsequent vulcanization, or cross-linking. During compounding, various fillers, processing aids, plasticizers, tackifiers, curatives, and antidegradants are added.

Mineral fillers are used for light-colored rubber compounds. Petroleum-based oils are commonly used as plasticizers.

Vulcanization

Vulcanization or curing in isobutylene polymers introduces chemical cross-links via reactions involving either allylic hydrogen or allylic halogen in butyl or halobutyl, respectively, or benzylic bromine in brominated poly(isobutylene-*co-p*-methylstyrene) to form a polymer network.

HEALTH AND SAFETY FACTORS

Polyisobutylene, isobutylene–isoprene copolymers, and isobutylene-*p*-methylstyrene copolymers are considered to have no chronic hazards associated with exposure under normal industrial use. Although many vulcanizates are inert, specific recommendations should be sought from suppliers.

USES

The polyisobutylene portion of the isobutylene copolymers, such as butyl, halobutyl, and brominated poly(isobutylene-*co-p*-methylstyrene), imparts chemical and physical characteristics that make them highly useful in a wide variety of applications. The low degree of permeability to gases accounts for the largest uses of butyl and halobutyl rubbers, namely as inner tubes and tire inner-liners. These same properties are also of importance in air cushions, pneumatic springs, air bellows, accumulator bags, and pharmaceutical closures. The thermal stability of butyl rubber makes it ideal for rubber tire-curing bladders, high-temperature service hoses, and conveyor belts for hot material handling. With the added thermal stability in brominated poly(isobutylene-*co-p*-methylstyrene), it has made inroads in applications of bladders and curing envelopes.

Isobutylene-based polymers exhibit high damping and have uniquely broad damping and shock absorption ranges in both temperature and frequency. Molded rubber parts from butyl and halobutyl find wide applications in automobile suspension bumpers, exhaust hangers, and body mounts.

Blends of halobutyl or brominated poly(isobutylene-*co-p*-methylstyrene) with high diene rubbers are used in tire sidewalls and tread compounds.

Polybutenes enjoy extensive uses as adhesives, caulks, sealants, and glazing compounds. They are also used as plasticizers in rubber formulations with butyl, SBR, and natural rubber. In linear low density polyethylene (LLDPE) blends, they induce cling to stretch-wrap films. Polybutenes when modified at their chain-end unsaturations with polar functionality are widely employed in lubricants as dispersants. Blends of polybutene with polyolefins produce semisolid gels that can be used as potting and electrical-cable filling materials.

J. Bandrupt and E. H. Immergut, eds., *Polymer Handbook*, 3nd ed., John Wiley & Sons, Inc., New York, 1989.

B. Ivan and J. P. Kennedy, *Designed Polymers by Carbocationic Macromolecular Engineering: Theory and Practice*, Hanser Publishers, Munich, 1991.

M. Morton, ed., *Rubber Technology*, 3rd ed., Chapman and Hall, London, 1995.

E. G. M. Tornqvist, in J. P. Kennedy and E. G. M. Tornqvist, eds., *Polymer Chemistry of Synthetic Elastomers*, Part 1, John Wiley & Sons, Inc., New York, 1968, p. 76.

ROBERT N. WEBB
TIMOTHY D. SHAFFER
ADDY H. TSOU
ExxonMobil

BUTYRALDEHYDES

The two isomeric butanals, *n*- and isobutyraldehyde, C_4H_8O, are produced commercially almost exclusively by the Oxo Reaction of propylene. They also occur naturally in trace amounts in tea leaves, certain oils, coffee aroma, and tobacco smoke.

PHYSICAL PROPERTIES

The butanals are highly flammable, colorless liquids of pungent odor. Their physical properties are shown in Table 1.

These aldehydes are miscible with most organic solvents, eg, acetone, ether, ethanol, and toluene, but are only slightly soluble in water.

CHEMICAL PROPERTIES

The reactions of *n*- and isobutyraldehyde are characteristic aldehyde reactions of oxidation, reduction, and condensation.

ECONOMIC ASPECTS

The biggest markets for C_4–C_5 alcohols are solvents and coatings.

2-Ethylhexanol is used in the production of acrylate and methacrylic esters. Their principal markets are as acrylic emulsion polymers for pressure sensitive adhesives,

Table 1. Physical Properties of C-4 Aldehydes

	n-Butyraldehyde	Isobutyraldehyde
formula	$CH_3CH_2CH_2CHO$	$(CH_3)_2CHCHO$
systematic name	butanal	2-methyl propanal
critical temperature, °C	263.95	233.85
critical pressure, kPa[a]	4000	4100
critical specific volume, m³/(kg-mol)[b]	0.258	0.263
melting point, °C	−96.4	−65.0
normal boiling point, °C	74.8	64.1
coefficient of expansion1 at 20°C	0.00114	
refractive index at 25°C	1.3766	1.3698
liquid density at 20°C, kg/m³[c]	801.6	789.1
liquid heat capacity at 25°C, kJ/(mol·K)[d]	0.16333	0.15581
heat of vaporization at normal boiling point, kJ/mol[d]	30.72	31.23
ideal gas heat of formation at 25°C, kJ/mol[d]	−204.8	−215.8
heat of fusion, kJ/mol[d]	11.1	12.0
dipole moment, C·m[e]	9.07×10^{-30}	9.0×10^{-30}
dielectric constant ε at °C	13.426	
solubility parameter, at 25°C, (MJ/m³)^{0.5}[f]	18.666	18.446
solubility in water at 25°C, wt %	8.36	6.47
solubility of water in at 25°C, wt %	3.45	2.60

[a] To convert kPa to mm Hg, multiply by 7.50.
[b] To convert m³/kg-mol to mL/mol, multiply by 1000.
[c] To convert m³/kg-mol to mL/mol, multiply by 1000.
[d] To convert kJ to kcal, divide by 4.184.
[e] To convert C·m to Debye, D, divide by 3.36×10^{-30}.
[f] To convert (MJ/m³)^{0.5} to (cal/cc)^{0.5}, multiply by $0.239^{0.5}$.

textiles, and surface coatings. Demand for water-based acrylic products has increased because of the stringent air-emission regulations on solvent-based products.

The largest use of *n*-butanol is in the production of *n*-butyl acrylate and *n*-butyl methacrylate. These compounds are use in emulsified and solution polymers that are used in latex surface coatings and enamels.

STORAGE AND HANDLING

Stainless steel, baked phenolic lined steel, or aluminum are often used for storage and handling of *n*- and isobutyraldehyde. The butanals are flammable and reactive, are easily oxidized on exposure to air, and in contact with acid, bases, or certain metal ions (eg, iron), will undergo exothermic condensation reactions. Storage of the aldehydes under nitrogen will avoid these problems and preserve the integrity of the material. There is some evidence that water stabilizes aldehydes against certain types of exothermic condensation reactions, possibly by precipitating any soluble iron species as hydrous iron oxides.

HEALTH, SAFETY, AND ENVIRONMENTAL FACTORS

n-Butyraldehyde is moderately toxic by ingestion, inhalation skin contact, intraperitoneal, and subcutaneous routes. It is a severe skin and eye irritant. It is a highly flammable liquid. When heated to decomposition, it emits acid smoke and fumes. No threshold limit value has been assigned for either butyraldehyde or isobutyraldehyde. Isobutryaldehyde is moderately toxic by ingestion, skin contact and inhalation. It is a severe skin and eye irritant and is a highly flammable liquid that can react vigorously with reducing materials. When heated to decomposition, it emits smoke and fumes.

The flash points for the butanals are well below room temperature. Thus, precautions must be taken to avoid heat, sparks, or open flame.

Both butanals are on the United States Toxic Substances Control Act (TSCA) Inventory, a prerequisite for the manufacture or importation for commercial sale of any chemical substance or mixture in quantities greater than one thousand pounds (455 kg). Additionally, the manufacture and distribution of the butanals in the United States are regulated under the Superfund Amendments and Reauthorization Act (SARA), Section 313, which requires that anyone handling at least ten thousand pounds (4550 kg) a year of a chemical substance report to both the EPA and the state any release of that substance to the environment.

"*n*-Butanol," Chemical Profile, *Chemical Market Reporter*, (Oct. 28, 2002).

H. G. Hagemeyer and G. C. DeCroes, *The Chemistry of Isobutyraldehyde and its Derivatives*, Eastman Kodak Co., 1953.

R. J. Lewis, Sr., *Sax's Dangerous Properties of Industrial Materials*, 10th ed., John Wiley & Sons, Inc., New York, 2000.

Union Carbide Specification Method (Butyraldehyde) 1B-4A4-1i, June 1, 1990.

ERNST BILLIG
Union Carbide Chemicals and
Plastics Company Inc.

C

CADMIUM AND CADMIUM ALLOYS

Cadmium, Cd, in Group 12 (IIB) of the periodic table of elements, a soft, ductile, silver-white or bluish-white metal.

Cadmium metal may be recovered as a byproduct of zinc, lead, or copper smelting operations and is utilized commercially as a corrosion-resistant coating on steel, aluminum, and other nonferrous metals. Cadmium is also added to some nonferrous alloys to improve properties such as strength, hardness, wear resistance, castability, and electrochemical behavior or to other nonferrous alloys to obtain lower melting temperatures and improved brazing and soldering characteristics. Cadmium metal is also used as the basis to prepare cadmium compounds of commercial importance such as cadmium oxide, cadmium hydroxide, cadmium sulfide and cadmium nitrate.

PROPERTIES

The physical, thermal, electrical, magnetic, optical, and nuclear properties of cadmium metal are summarized in Table 1. The chemical properties of cadmimum metal in general resemble those of zinc, especially under reducing conditions and in covalent compounds. The crystal structure of the protective cadmium oxide film is hexagonal and exhibits a very low coefficient of friction, making cadmium coatings ideal for friction applications. In oxides, fluorides and carbonates, and under oxidizing conditions, cadmium may behave similarly to calcium. It also forms a relatively large number of complex ions with other ligand species such as ammonia, cyanide, and chloride. Cadmium is a fairly reactive metal. It dissolves slowly in dilute hydrochloric or sulfuric acids, but dissolves rapidly in hot dilute nitric acid. All of the halogens, phosphorus, sulfur, selenium, and tellurum also react readily with cadmium at elevated temperatures.

Unlike zinc, cadmium is not markedly amphoteric, and cadmium hydroxide is virtually insoluble in alkaline media. Like zinc, however, cadmium forms a protective oxide film that reduces its corrosion/oxidation rate in atmospheric service. Both metals exhibit low corrosion rates over the range of pH ~5–10. In more acidic and alkaline environments, their corrosion rates increase dramatically. Cadmium is generally preferred for marine or alkaline service, whereas zinc is often as good or better in heavy industrial exposures containing sulfur or ammonia.

MANUFACTURING AND PROCESSING

Cadmium metal is produced mainly as a by-product of the beneficiation and refining of zinc sulfide ore concentrates and, to a lesser degree, from the processing of complex zinc, lead, and copper ores and their concentrates. Cadmium is also increasingly being recovered through the recycling of nickel–cadmium batteries. It is estimated that 90–98% of the cadmium present in zinc ores is recovered in the mining and benefication stages of the extraction process.

Refining of zinc/cadmium concentrates to separate and purify the two metals can be accomplished by either hydrometallurgical (electrolytic) or pyrometallurgical (high temperature) techniques. Today, virtually all primary zinc and its cadmium by-product are produced by electrolytic production techniques, but some cadmium is also recovered from the smelting and refining of lead and/or copper from complex ores.

ECONOMIC ASPECTS

It is difficult to establish specific cadmium consumption patterns accurately on a country-by-country basis since so much of the cadmium supply and demand patterns are carried out on a global scale. Worldwide refined cadmium metal consumption has remained essentially constant since 1996, whereas primary cadmium metal production has decreased drastically, especially in Europe, and recycled cadmium metal production has increased during that same time period.

Prices for the two grades of cadmium metal (99.95% and 99.99% purity) are generally within $0.10 of each other, and large volume cadmium chemicals such as cadmium oxide command only a modest premium over the base metal. However, specialized, low-volume cadmium chemicals with specific chemical, particle size or other requirements may command much higher premiums.

SPECIFICATIONS FOR CADMIUM AND CADMIUM COMPOUNDS

There are many specific specifications for cadmium metals, compounds and products prepared by various materials and products manufacturers, as well as the national, military, regional and international standards covering cadmium materials such as ASTM in the United States, EN standards in Europe, and ISO standards internationally. Each NiCd battery producer has a specific requirement for the cadmium compounds such as cadmium oxide, cadmium hydroxide, or cadmium nitrate, which they may utilize in the production of their NiCd batteries. Similarly, pigment and stabilizer producers have specific compositional, particle size, and other requirements for the cadmium compounds used in their formulations. Many different zinc-, lead-, and copper-based materials also have maximum cadmium contents levels.

RECYCLING

There has been increasing emphasis on the collection and recycling of spent cadmium products to reduce any risk to

Table 1. Physical Properties of Cadmium

Property	Value
atomic weight	112.40
melting point, °C	321.1
boiling point,°C	767
latent heat of fusion, kJ/mol[a]	6.2
latent heat of vaporization, kJ/mol[a]	99.7
specific heat, J/(mol·K)[a]	
20°C	25.9
321–700°C	29.7
coefficient of linear expansion at 20°C, μm/(cm.°C)	0.313
electrical resistivity, μω·cm	
22°C	7.27
400°C	34.1
600°C	34.8
700°C	35.8
electrical conductivity, % IACS[b]	25
density, kg/m^3	
26°C	8642
330°C (liq>)	8020
400°C	7930
600°C	7720
volume change on fusion, % increase	4.74
thermal conductivity, W/(m·K)	
273 K	98
373 K	95
573 K	89
vapor pressure, kPa[c]	
382°C	0.1013
473°C	1.013
595°C	10.13
767°C	101.3
surface tension, mN/m (= dyn/cm)	
330°C	564
420°C	598
450°C	611
viscosity, mPa·s (= cP)	
340°C	2.37
400°C	2.16
500°C	1.84
600°C	1.54
molar magnetic susceptibility, cm^3/mol (= emu/mol)	-19.8×10^{-6}
Brinell hardness, kg/mm^2	16–23
tensile strength, MPa[d]	71
elongation, %	50
Poisson's ratio	0.33
modulus of elasticity, GPa[e]	49.9
shear modulus, GPa[e]	19.2
thermal neutron capture cross-section at 2200 m/s, m^2/atom	$2450 \pm 50 \times 10^{-28}$

[a]To convert J to cal, divide by 4.184.
[b]IACS = International Annealed Copper Standard.
[c]To convert kPa to mm Hg, multiply by 7.5.
[d]To convert MPa to psi, multiply by 145.
[e]To convert GPa to psi, multiply by 145,000.

human health and the environment from the disposal of cadmium products. Since NiCd batteries account for the vast majority (82%) of cadmium consumption, the main efforts have centered on the collection and recycling of NiCd batteries. However, cadmium-coated products, cadmium alloys, and CdTe photovoltaic modules are also recyclable and efforts have been made to collect and recycle these products as well.

ENVIRONMENTAL CONCERNS

Cadmium Emissions

Cadmium emissions arise from two principal source categories: natural sources and man-made or anthropogenic sources. Man-made cadmium emissions arise either from the manufacture, use and disposal of products intentionally utilizing cadmium, or the presence of cadmium as a natural but not functional impurity in noncadmium-containing products. Anthropogenic cadmium emissions have declined dramatically since the late 1960s and are still declining today. Considerable progress is now also being made in reduction of diffuse contamination from cadmium products through collection and recyling programs of cadmium-containing products. Cadmium emissions from products where cadmium is present as an impurity have not been reduced as significantly and appear to be the one remaining area where additional reductions might be achieved.

HEALTH AND SAFETY FACTORS

Excess cadmium exposure produces adverse health effects on human beings. For virtually all chemicals, adverse health effects are noted at sufficiently high total exposures. For certain elements such as copper and zinc, which are essential to human life, a deficiency as well as an excess can cause adverse health effects. Cadmium is not regarded as essential to human life. The relevant questions with regard to cadmium exposure are the total exposure levels and the principal factors that determine the levels of cadmium exposure and the adsorption rate of the ingested or inhaled cadmium by the individual, in other words, the pathways by which cadmium enters the food chain, which is the principal pathway of cadmium exposure for most human beings.

USES

Cadmium metal and cadmium compounds are utilized in five principal product areas that include NiCd batteries, pigments, stabilizers, coatings, and alloys and other miscellaneous products. The International Cadmium Association has for some years made estimates of cadmium consumption patterns by various end-use categories, most recently batteries, 82%; pigments, 10%; stabilizers, 1.5%; coating, 6%; and alloys, 0.5%. Stabilizers and alloys are rapidly disappearing as end-use applications for cadmium metal and compounds.

ALLOYS

Cadmium-containing alloys are generally considered in two groups: those where small amounts of cadmium are utilized to improve some feature of the base alloy and those where cadmium is part of a complex low melting point alloy where the presence of cadmium helps to lower the melting point. The low-melting, cadmium-containing alloys, however, are generally being replaced by cadmium-free compositions except in a few critical, safety-related applications.

Cadmium forms a number of compounds that exhibit semiconducting or electronic behavior that makes them useful for a wide variety of applications. The amounts utilized are usually quite small, but these minor uses often have major technological and social importance.

Cadmium mercury telluride is utilized in infrared imaging systems, and cadmium selenides are used in thin-film transistors for switching applications.

An important use for cadmium use in alloys is the copper-cadmium and copper-cadmium-titanium thermal and electrical conductivity alloys and the silver-cadmium oxide electrical contact alloys.

Silver–indium–cadmium alloys are utilized as control rods in some pressurized-water nuclear reactors, and cadmium sheet is used for nuclear shielding. These are specialized uses.

Applications that show promise for the future include NiCd battery–powered electric vehicles, hybrid electric vehicles and electric buses. Another area of promise for future cadmium applications is CdTe solar cells and NiCd batteries for energy storage of the output from those solar cells. This area has become one of increasing interest, especially in Third World nations in tropical zones, where solar power is a viable option and extensive power infrastructures are not available.

Whatever new applications are developed for cadmium for the future, they will have to be recyclable.

CRC Handbook of Chemistry and Physics, 77th ed. CRC Press, Boca Raton, Fla., 1996.

M. Farnsworth, *Cadmium Chemicals*, Internal Lead Zinc Research Organization, Inc., New York, 1980.

H. Morrow, "Cadmium (CD)," in *Metals Handbook*, 10th ed., Vol. **2**, ASM International, Metals Park, Ohio, 1990.

Organisation for Economic Co-operation and Development (OECD), report from Session F, "Sources of Cadmium in Waste," *Chairman's Report of the Cadmium Workshop*, ENV/MC/CHEM/RD(96)1, Stockholm, Sweden, Oct. 1995.

HUGH MORROW
International Cadmium
Association

CADMIUM COMPOUNDS

Naturally occurring cadmium compounds are limited to the rare minerals, greenockite, CdS, and otavite, an oxycarbonate, but neither is an economically important source of cadmium metal or its compounds. Instead, cadmium compounds are more usually derived from metallic cadmium, which is produced as a by-product of lead–zinc smelting or electrolysis. Typically, this cadmium metal is burnt as a vapor, to produce the brown-black cadmium oxide, CdO,

which then acts as a convenient starting material for most of the economically important compounds.

PROPERTIES

In general, cadmium compounds exhibit properties similar to the corresponding zinc compounds. Compounds and properties are listed in Table 1.

INORGANIC COMPOUNDS

The inorganic compounds include the cadmium arsenides, antimonides, and phosphides; cadmium borates; cadmium carbonate; cadmium complexes (Table 2); cadmium halides; cadmium hydroxide; cadmium nitrate; cadmium oxide; cadmium phosphates; cadmium selenide and telluride; cadmium silicates; cadmium sulfate; cadmium sulfide; and cadmium tungstate.

ORGANIC COMPOUNDS

Many organocadmium compounds are known, but few have been of commercial importance. They include dialkyl and diaryl cadmium compounds, cadmium acetate, and organocadmium soaps.

ECONOMIC ASPECTS

One of the most promising applications is the use of cadmium telluride solar cells to convert sunlight into electricity and the use of NiCd batteries to store that energy.

HEALTH AND SAFETY FACTORS

Cadmium, both as the free metal and in its compounds, is highly toxic and has been designated one of the 100 most hazardous substances under Section 110 of the Superfund Amendments and Reauthorization Act of 1986. The EPA maintains that cadmium is a probable human carcinogen (Group B1) but only by the inhalation route.

ENVIRONMENTAL CONCERNS

Cadmium discharges to air and water are decreasing. Most of the cadmium released to the environment is now in the form of solid wastes such as coal ash, sewage sludge, flue dust, and fertilizers. Cadmium compounds are employed mainly as the negative electrode materials in NiCd batteries, as ultraviolet light and weathering stabilizers for poly(vinyl chloride) (PVC), and as red, orange, and yellow pigments in plastics, ceramics, glasses, enamels, and artists' colors.

USES

Cadmium hydroxide is the anode material of Ag–Cd and Ni–Cd rechargeable storage batteries. Cadmium sulfide,

Table 1. Physical and Chemical Properties of Selected Cadmium Compounds

Compound	$\Delta H^\circ_{f,298}$, kJ/mol[a]	$\Delta G^\circ_{f,298}$, kJ/mol[a]	S°_{298}, J/mol·K[a]	Density, g/mL	C°_p, J/mol·K[a]	Mp, °C	ΔH_{fus}, kJ/mol[a]	Aqueous solubility, g/100 g H$_2$O[b]	Crystal[c] structure	Unit cell dimensions, nm
cadmium antimonide CdSb	−14.4	−13.0	92.9	6.92		452	32.05^a		ortho-rhomb	$a = 0.6471$, $b = 0.8253$, $c = 8.526$
cadmium bromide CdBr$_2$	−316	−296	137.2	5.192	76.7	568	20.92	95_{18}	hex	$a = 0.395$, $c = 1.867$
cadmium carbonate CdCO$_3$	−751	−669	92.5	4.26 ($a = 0.61306$)		332 dec		2.8×10^{-6}	rhomb	$a = 0.61306$
cadmium chloride CdCl$_2$	−391	−344	115.3	4.05	74.7	568	22.176	128.6_{30}	hex	$a = 0.3854$, $c = 1.746$
cadmium fluoride CdF$_2$	−700	−648	77.4	6.39		1048	22.594	3.45_{25}	cubic	$a = 0.53880$
cadmium hydroxide Cd(OH)$_2$	−561 (hex)	−474	96.2 ($a = 0.3475$)	4.79		150 dec		2.6×10^{-4}	hex	$a = 0.3475$, $c = 0.467$
cadmium iodide CdI$_2$, α-form	−203	−201	161.1	5.67	80.0	387	33.472	86_{25}	hex	$a = 0.424$, $c = 0.684$
cadmium nitrate Cd(NO$_3$)$_2$	−456	−255	197.9			350		109_0	(hex)[d] cubic	$(c = 1.367)^e$, $a = 0.756$
cadmium nitrate tetrahydrate Cd(NO$_3$)$_2$ · 4H$_2$O	−1649			2.455		59.4	32.636	215_0	ortho-rhomb	$a = 0.583$, $b = 2.575$, $c = 1.099$

343

Table 1. (*Continued*)

Compound	$\Delta H^\circ_{f,298}$, kJ/mol[a]	$\Delta G^\circ_{f,298}$, kJ/mol[a]	S°_{298}, J/mol·K[a]	Density, g/mL	C°_p, J/mol·K[a]	Mp, °C	ΔH_{fus}, kJ/mol[a]	Aqueous solubility, g/100 g H_2O[b]	Crystal[c] structure	Unit cell dimensions, nm
cadmium oxide CdO	−258	−228	54.8	8.2	43.4	1540 sub	243.509 sub	9.6×10^{-4}	cubic	$a = 0.46953$
cadmium selenide CdSe, α-form	−136	−100	96.2	5.81 ($a = 0.46953$)		1350 dissoc	305.307 dissoc		hex (cubic)e	$a = 0.4309$ $c = 0.7021$
cadmium *m*-silicate CdSiO$_3$	−1189	−1105	97.5	4.928	88.6	1242			monoclinic	$(a = 0.605)$e $a = 1.504$ $b = 0.710$ $c = 0.696$
cadmium sulfate CdSO$_4$	−933	−823	123.0	4.691	99.6	1000	20.084	76.6_{20}	ortho–rhomb	$a = 0.4717$ $b = 0.6559$ $c = 0.4701$
cadmium sulfate hydrate CdSO$_4 \cdot H_2O$	−1240	−1069	154.0	3.79	134.6	105 trans			monoclinic	$a = 7.607$ $b = 0.7541$ $c = 8.186$
cadmium sulfate hydrate 3CdSO$_4 \cdot 8H_2O$	−1729	−1465	229.6	3.09	213.3	80 trans		113_0	monoclinic	$a = 0.947$ $b = 1.184$ $c = 1.635$
cadmium sulfide CdS, α-form	−162	−156	64.9	4.82 (4.50)e		980 sub in N$_2$ 1045	201.669 subl	$1.3 \times 10^{-4}_{18}$	hex	$a = 0.41348$ $c = 0.6749$
cadmium telluride CdTe	−92	−92	100.4	6.20					(cubic)e hex	$(a = 0.5818)$e $a = 0.457$ $c = 0.747$ $(a = 0.6480)$e

[a]To convert J to cal, divide by 4.184.
[b]Subscript denotes temperature in °C.
[c]Ortho–rhomb is orthorhombic and hex is hexagonal.
[d]β-form.

344

Table 2. Thermodynamic and Stability Constant Data for Selected Aqueous Cadmium Complexes[a]

Complex ion	$\Delta H^\circ_{f,298}$ kJ/mol[b]	$\Delta G^\circ_{f,298}$ kJ/mol[b]	Stability constant
CdCl$^+$	−240.5	−224.4	log $K_1 = 1.32$
CdCl$^-_3$	−561.0	−487.0	log $K_3 = 0.09$
Cd(CN)$_4^{2-}$	428.0	507.5	log $K_4 = 3.58$
Cd(NH$_3$)$_2^{2+}$	−266.1	−159.0	log $K_2 = 2.24$
Cd(NH$_3$)$_4^{2+}$	−450.2	−226.4	log $K_4 = 1.18$
CdBr$^+$	−200.8	−193.9	log K$_1 = 1.97$
CdBr$^-_3$		−407.5	log $K_3 = 0.24$
CdI$^+$	−141.0	−141.4	log $K_1 = 2.08$
CdI$^-_3$		−259.4	log K$_3 = 2.09$
CdI$_4^{-2}$	−341.8	−315.9	log $K_4 = 1.59$
CdSCN$^+$		7.5	log $K_1 = 1.90$
Cd(SCN)$_4^{-2}$			log $K_4 = $ ca 0.1
Cd(N$_3$)$_4^{-2}$		1,295.0	log $K_4 = 0.76$

[a] Standard state $M = 1$.
[b] To convert kJ to kcal, divide by 4.184.

selenide, and especially telluride find utility in solar cells. Cadmium sulfide, lithopone, and sulfoselenide are used as colorants (orange, yellow, red) for plastics, glass, glazes, rubber, and fireworks.

A cadmium sulfide interface layer can improve III-V semiconductor device performance.

In flexible PVC, cadmium salts of long-chain organic acids, such as stearate and laurate, are used in combination with similar Ba^{2+} salts as heat and light stabilizers, but these uses are in decline since these compounds can now be replaced by less toxic compounds. Cadmium cyanide, acetate, fluoroborate, or sulfate is used as an electrolyte in coating a thin cadmium layer, ie, electroplating, onto other metals thereby imparting enhanced corrosion protection. Cadmium protective overlayers are also deposited by mechanical plating or vapor deposition.

The cadmium chalcogenide semiconductors have found numerous applications ranging from rectifiers to photoconductive detectors in smoke alarms. Many Cd compounds, eg, sulfide, tungstate, selenide, telluride, and oxide, are used as phosphors in luminescent screens and scintillation counters. Glass colored with cadmium sulfoselenides is used as a color filter in spectroscopy and has recently attracted attention as a third-order, nonlinear optical switching material. Dialkylcadmium compounds are polymerization catalysts for production of poly(vinyl chloride) (PVC), poly(vinyl acetate) (PVA), and poly(methyl methacrylate) (PMMA). Mixed with TiCl$_4$, they catalyze the polymerization of ethylene and propylene.

Dimethyl cadmium is toxic and can be replaced with a more stable cadmium oxide to form nanocrystals for use in electronics and optoelectronic devices. New forms of cadmium sulfide crystals have been developed for use in the manufacture of solar panels. The new crystals facilitate electron flow and could boost efficiency by more than 20%.

M. Farnsworth, *Cadmium Chemicals*, International Lead Zinc Research Org., New York, 1980.

M. Jakubowski, in E. Bingham, B. Cohrssen, and C. H. Powell, eds., *Patty's Toxicology*, 5th ed., Vol. 2, John Wiley & Sons, Inc., New York, 2001, p. 309.

K. Matsumoto and K. Fuwa, "Cadmium," in E. C. Foulkes, ed., *Handbook of Experimental Pharmacology*, Vol. 80, Springer-Verlag, Berlin, Heidelberg, 1986.

U.S. Bureau of Mines, *Mineral Industry Surveys*, Cadmium, U.S. Dept. of the Interior, Washington, D.C., 1989 and 1990.

NORMAN HERRON
E. I. du Pont de Nemours & Co., Inc.

CALCIUM AND CALCIUM ALLOYS

Calcium, Ca, a member of Group 2 (IIA) of the Periodic Table between magnesium and strontium, is classified, together with barium and strontium, as an alkaline-earth metal and is the lightest of the three. Calcium metal does not occur free in nature; however, in the form of numerous compounds it is the fifth most abundant element, constituting 3.63% of the earth's crust.

Calcium is mainly used as a reducing agent for many reactive, less common metals; to remove bismuth from lead; as a desulfurizer and deoxidizer for ferrous metals and alloys; and as an alloying agent for aluminum, silicon, and lead. Small amounts are used as a dehydrating agent for organic solvents and as a purifying agent for removal of nitrogen and other impurities from argon and other rare gases.

PHYSICAL PROPERTIES

Pure calcium is a bright, silvery-white metal, although under normal atmospheric conditions freshly exposed surfaces of calcium quickly become covered with an oxide layer. The metal is extremely soft and ductile, having a hardness between that of sodium and aluminum. It can be work-hardened to some degree by mechanical processing. Although its density is low, calcium's usefulness as a structural material is limited by its low tensile strength and high chemical reactivity.

Some of the more important physical properties of calcium are given in Table 1. Measurements of the physical properties of calcium are usually somewhat uncertain, owing to the effects that small levels of impurities can exert.

CHEMICAL PROPERTIES

Calcium has a valence electron configuration of $4s^2$ and characteristically forms divalent compounds. It is very reactive and reacts vigorously with water, liberating hydrogen and forming calcium hydroxide, Ca(OH)$_2$. Calcium does not readily oxidize in dry air at room temperature but is quickly oxidized in moist air or in dry oxygen at about 300°C. Calcium reacts with fluorine at room temperature and with the other halogens at 400°C. When heated to 900°C, calcium reacts with nitrogen to form calcium nitride Ca$_3$N$_2$.

Calcium is an excellent reducing agent and is widely used for this purpose.

Table 1. Physical Properties of Calcium[a]

Property	Value		
atomic weight	40.08		
electron configuration	$1s^2 2s^2 2p^6 3s^2 3p^6 4s^2$		
stable isotopes			
atomic weight 40 42 43	44	46	48
natural abundance, % 96.947 0.646 0.135 0.18	2.083	0.186	
specific gravity at 20°C, kg/m³	1.55×10^3		
melting point, °C	839 ± 2		
boiling point, °C	1484		
heat of fusion, ΔH_{fus}, kJ/mol[b]	9.2		
heat of vaporization, ΔH_{vap}, kJ/mol[b]	161.5		
heat of combustion, kJ/mol[b]	634.3		
vapor pressure			
pressure, kPa[c] 0.133 1.33	13.3	53.3	101.3
temperature, °C 800 970	1200	1390	1484
specific heat at 25°C, J/(g·K)[b]	0.653		
coefficient of thermal expansion, 0–400°C, m/(m·K)	22.3×10^{-6}		
electrical resistivity at 0°C, $\mu\Omega \cdot$cm	3.91		
electron work function, eV	2.24		
tensile strength (annealed), MPa[c]	48		
yield strength (annealed), MPa[c]	13.7		
modulus of elasticity, GPa[c]	22.1–26.2		
hardness (as cast)			
HB[d]	16–18		
HR B[e]	36–40		

[a]Refs. (4–6).
[b]To convert J to cal, divide by 4.184.
[c]To convert kPa to psi, multiply by 0.145.
[d]Brinell Hardness scale.
[e]Rockwell B Hardness scale.

Commercially produced calcium metal is analyzed for metallic impurities by emission spectroscopy. Carbon content is determined by combustion, whereas nitrogen is measured by Kjeldahl determination.

MANUFACTURE

Although in Western countries the aluminothermic process has now completely replaced the electrolytic method, electrolysis is believed to be the method used for calcium production in the People's Republic of China and the Commonwealth of Independent States (CIS). This process likely involves the production of a calcium–copper alloy, which is then redistilled to give calcium metal.

For certain applications, especially those involving reduction of other metal compounds, better than 99% purity is required. This can be achieved by redistillation.

SHIPMENT

Because of its extreme chemical reactivity, calcium metal must be carefully packaged for shipment and storage. The metal is packaged in sealed argon-filled containers. Calcium is classed as a flammable solid and is nonmailable. Sealed quantities of calcium should be stored in a dry, well-ventilated area so as to remove any hydrogen formed by reaction with moisture.

ECONOMIC ASPECTS

Calcium consumption is primarily for the production of maintenance-free and sealed lead–acid batteries, in the steel industry, and for permanent magnet manufacture. These markets are fairly stable and strong markets. Use as a reducing agent of rare-earth oxides for permanent magnet manufacture is expected to increase due to an expansion of this market.

GRADES AND SPECIFICATIONS

Calcium is usually sold as crowns, broken crown pieces, nodules, or billets. The purity of these forms is at least 98%. If a higher quality of the metal is required, it can be redistilled to remove additional impurities.

HEALTH AND SAFETY FACTORS

Inhalation of calcium metal produces damaging effects on the mucous membranes and upper respiratory tract. Symptoms may include irritation of the nose and throat and difficulty breathing. It may also cause lung edema, which is a medical emergency. If breathing is difficult, give oxygen and get medical attention. If ingested, caustic lime will form due to reaction with moisture. Large amounts can have a corrosive effect. Abdominal pain, nausea, vomiting, and diarrhea are symptoms. If swallowed, do not induce vomiting but give large amounts of water. As calcium metal is corrosive, contact with the skin can cause pain, redness, and a severe burn. Contact with the eyes will cause redness and pain with possible burns and damage to the eye tissues. If calcium metal comes in contact with the skin or eyes, flush with water for at least 15 min and get medical attention. Protective clothing should include boots, gloves, lab coat, apron, or coveralls in addition to chemical safety goggles. As calcium metal is a very water-reactive flammable sold, reaction with water, steam, and acids to release flammable/explosive hydrogen gas should be avoided. Water must not be used to extinguish a fire with calcium metal.

Calcium metal should be stored in a tightly closed container in a cool, dry, ventilated area under nitrogen. It should be kept away from water or locations where water may be needed for a fire. This material should be handled as a hazardous waste for diposal purposes.

In the case of a spill, the material should be collected quickly and transferred to a container of kerosene, light oil, or similar hydrocarbon fluid for recovery. Exposure to air should be minimized. Do not use water on the metal or where it spilled if significant quantities still remain. Waste calcium should be packaged under hydrocarbon fluid and sent to an approved waste disposal facility.

CALCIUM ALLOYS

Calcium alloys can be produced by various techniques. However, direct alloying of the pure metals is normally used in the production of 80% calcium–magnesium, 70% magnesium–calcium, and 75% calcium–aluminum alloys.

Lead alloys containing small amounts of calcium are formed by plunging a basket containing a 77 or 75% calcium–23–25% Al alloy into a molten lead bath or by stirring the Ca–Al alloy into a vortex created by a mixing impellor.

Alloys of calcium with silicon are used in ferrous metallurgy (qv) and are generally produced in an electric furnace from CaO (or CaC$_2$), SiO$_2$, and a carbonaceous reducing agent.

USES

The most significant use of calcium is for improvement of steel. It is used as an aid in removing bismuth in lead refining and as a desulfurizer and deoxidizer in steel refining. Addition of calcium causes inclusions in the steel to float out by modifying the melting point of these inclusions.

Calcium has multiple other uses: reducing oxides of the rare-earth neodymium and boron for alloying with metallic iron for use in neodymium–iron–boron permanent magnets and use as a reducing agent to recover hafnium, plutonium, thorium, tungsten, uranium, vanadium, and the rare-earth metals from their oxides and fluorides.

F. Emley, in D. M. Considine, ed., *Chemical and Process Technology Encyclopedia*, McGraw-Hill Book Co., Inc., New York, 1974.

C. L. Mantell, in C. A. Hampel, ed., *The Encyclopedia of the Chemical Elements*, Reinhold Publishing Corp., New York, 1968.

M. Miller, "Calcium and Calcium Compounds," *Minerals Yearbook*, Vol. 1, U.S. Bureau of Mines, U.S. Department of the Interior, Washington, D.C., 1991, pp. 317–324.

R. C. Weast, ed., *Handbook of Chemistry and Physics*, 57th ed., CRC Press, Cleveland, Ohio, 1976, pp. B1, B3, B12, B276, D62, D165, D211, E81, F170.

LISA M. VRANA
Consultant

CALCIUM CARBIDE

Chemically pure calcium carbide is a colorless solid; however, the pure material can be prepared only by very special techniques. Commercial calcium carbide is composed of calcium carbide, calcium oxide, CaO, and other impurities present in the raw materials. The commercial product's calcium carbide content varies and is sold on the basis of the acetylene yield. Industrial-grade carbide contains about 80% as CaC$_2$, 15% CaO, and 5% other impurities.

Annual worldwide production of calcium carbide reached a peak of 8000 10^3 t. in the 1960s and has declined steadily since then to about 4700 10^3 t today, the principal reason being the substitution of acetylene from petrochemical sources (from by-product ethylene production and thermal cracking of hydrocarbons). Calcium carbide is used extensively as a desulfurizing reagent in steel and ductile iron production, which allows these manufacturers to use high sulfur coke without the penalty of excessive sulfur in the resultant products. Many countries produce calcium carbide; the largest producer is China.

PROPERTIES

Table 1 list the more important physical properties of calcium carbide.

Reaction with Water

The exothermic reaction of calcium carbide and water yielding acetylene forms the basis of the most important industrial use of calcium carbide:

$$CaC_2 + 2\,H_2O \longrightarrow C_2H_2 + Ca(OH)_2 \quad (H = -130\ kJ/mol)$$

Table 1. Physical Properties of Calcium Carbide

Property	Value
mol wt	64.10
mp, °C	2300
crystal structure	
phase I, 25–447°C	face centered tetragonal
phase II,	triclinic
phase III[a]	monoclinic
phase IV, >450°C	fcc
commercial	grain structure, 7–120 µm
specific gravity,	
commercial-grade	
at 15°C	2.34
2000°C[b]	1.84
electrical conductivity,	
technical-grade,	
(ohm–cm)$^{-1}$	
at 25°C	3,000–10,000
1000°C	200–1,000
1700°C[b]	0.36–0.47
1900°C[b]	0.075–0.078
viscosity at 1900°C,	
MPa·s (=CP)	
50% CaC$_2$	6000
87% CaC$_2$	1700
specific heat, 0–2000°C,	74.9
J/mol·K[c]	
heat of formation, H_f, 298,	-59 ± 8
kJ/mol[c]	
latent heat of fusion, ΔH_{fus},	32
kJ/mol[c]	

[a] Phase III is metastable.
[b] Material is a liquid.
[c] To convert from J to cal, divide by 4.184.

where H_r = heat produced in the reaction. Wet and dry processes are in use for generating acetylene from calcium carbide.

Reaction with Sulfur

An important use of calcium carbide has developed in the iron and steel industy, where it was found to be a very effective desulfurizing agent for blast-furnace iron.

Reaction with Nitrogen

Calcium cyanamide is produced from calcium carbide

$$CaC_2 + N_2 \longrightarrow CaCN_2 + C \qquad (H_r = -295\,kJ/mol)$$

The reaction is carried out in a refractory oven by passing nitrogen gas through finely pulverized carbide at a temperature of 1000–1200°C.

MANUFACTURE AND PRODUCTION

Calcium carbide is produced commercially by reaction of high purity quicklime and a reducing agent such as metallurigical or petroleum coke in an electric furnace at 2000–2200°C.

$$CaO + 3\,C \longrightarrow CaC_2 + CO \qquad (H_r = 466\,kJ/mol)$$

Commercial calcium carbide, containing about 80% CaC_2, is formed in the liquid state. Impurities are mainly CaO and impurities present in raw materials. CO is usually collected and used as a fuel in lime production or drying of the coke used in the process. The liquid calcium carbide is tapped from the furnace into cooling molds.

SHIPMENT AND TRANSPORTATION

Since calcium carbide produces flammable acetylene gas in the presence of water, it is classified as a hazardous material. In general, calcium carbide is shipped in metal containers, either in bulk bins of up to 4-ton capacity, or in drums ranging in size from 50 to 230 kg. Under certain conditions shipment is allowed in reinforced bulk bags of up to 1-ton capacity. Calcium carbide containers are marked "Flammable solid, dangerous when wet" with the United Nations designation UN 1402.

Calcium carbide for desulfurization is usually sold on the basis of a specified minimum CaC_2 content, together with minimum levels of various additives, and a specified particle size distribution. This can vary considerably according to the reagent formulation. Domestic shipments are either in steel tote bins of 2–4-ton capacity, bulk railcars, or trucks with pneumatic unloading capabilities. Containers are usually pressurized slightly with an inert gas such as dry nitrogen to avoid reagent contact with atmospheric moisture and generation of acetylene.

ENVIRONMENTAL CONCERNS

The major environmental problem in carbide production is the prevention of particulate dust emission. Normally, cloth filtration equipment is used in handling of raw materials, at furnace tapping, and product crushing and screening. The carbide furnace CO gas steam can be treated by high temperature cloth filtration, ceramic filters, or wet scrubbers. The dust collected by either wet or dry methods may contain trace cyanide that must be treated before disposal.

The hollow electrode system allows for the use of small coke and lime particles, which could not otherwise be used in the furnace in the main furnace charge. A covered furnace collects all the CO gas generated during carbide manufacture. This fuel gas can provide energy for a lime kiln operation or drying carbon raw materials.

Acetylene generators produce a calcium hydroxide coproduct during acetylene manufacture. In the dry process the powder can be used for acid neutralization or soil stabilization. In the wet process the slurry can be used for flue-gas desulfurization at utilities. If the wet process is used, the hydrate is collected in settling lagoons for disposal. The dry process produces a powder hydrate that can be used for flue gas desulfurization or soil stabilization.

HEALTH AND SAFETY FACTORS

The usual precautions must be observed around high-tension electrical equipment supplying power.

Although acetylene is considered to be a material with a very low toxicity, a threshold value (TLV) of 2500 ppm has been established by NIOSH. In the presence of small amounts of water, calcium carbide may become incandescent and ignite the evolved air–acetylene mixture. Non-sparking tools should be used when working in the area of acetylene-generating equipment.

USES

Calcium carbide has three primary applications today. It is used to produce acetylene gas for heating, oxyacetylene welding, metal cutting, acetylene black, and acetylene-derived chemicals such as vinyl ethers and acetylenic chemicals and alcohols. It is also used as a desulfurizing reagent for iron and steel, and as an intermediate for calcium cyanamide manufacture.

Acetylene

Acetylene production accounts for about 75% of calcium carbide consumption. There are two main areas of use: industrial applications, which include heating, welding, and cutting; and chemical uses, which include calcium carbide–derived acetylene for various chemicals.

Since calcium carbide-derived acetylene competes with less expensive acetylene derived from the oxidation of natural gas and acetylene produced as a by-product from the manufacture of ethylene, it is not used on large scale for chemical production.

Calcium Cyanamide

The most important use of calcium cyanamide is as fertilizer, but it is also effective as a herbicide and defoliant. It was used as a starting material for ammonia until it was displaced by the Haber process. Several derivatives of calcium cyanamide have also been developed: hydrogen cyanamide (intermediate for insecticides, pharmaceuticals, soil sterilants), dicyandiamide (intermediate for flame and fire retardants, viscosity reducers for glues and adhesives, nitrification inhibitors in fertilizers), melamine (thermoset plastic), and calcium cyanide (gold extraction and fumigant).

Desulfurizing Reagent

Calcium carbide for metallurgical applications accounts for about 25% of its use. Carbide for desulfurization in steel mills has allowed steel mills to use high-sulfur coke in the blast furnace without the penalty of excessive sulfur in the resultant steel. The reagent generally consists of commercial carbide that has been ground to a powder, or the pulverized carbide mixed with other ingredients such as lime, limestone, graphite, coal, solid hydrocarbons, and silicone. The powdered regent is injected into molten iron using an inert-gas carrier. Calcium carbide competes with magnesium based desulfuring reagents, and switching costs between these reagents are low. Calcium carbide has the advantage of being less expensive; however, magnesium has a higher affinity for sulfur.

Calcium carbide and magnesium are also used as simple blends or in co-injection processes. With coinjection, the two reagents can be injected simultaneously or sequentially. The less expensive carbide removes 90% of the sulfur, and then the remaining sulfur is removed by magnesium injection.

For the foreseeable future it is expected that both reagents will be used in steel mills and demand will change in line with steel production. Carbide Graphite, Elkem-American, and Metaloides all produce powdered carbide for desulfurization. All producers market carbide of a granular size for desulfurizing ductile iron in foundries. This market is expected to be stable.

Current Industrial Report, Fuorgaric Chemicals Series MA28A, U.S. Dept. of Commerce, Bureau of Census.

M. Haley, *Chemical Economics Handbook, Calcium Carbide, United States*, SRI, Menlo Park, Calif., Dec. 1989.

D. W. K. Hardie, *Acetylene, Manufacture and Uses*, Oxford University Press, London, UK, 1965.

F. W. Kampmann and W. Portz, *Chemicals from Coal via the Carbide Route*, Hoechst A. G., Heurth-Knapsack D-5030, Germany, 1991.

WILLIAM L. CAMERON
William Cameron Consulting

CALCIUM CARBONATE

Calcium carbonate, $CaCO_3$, mol wt 100.09, occurs naturally as the principal constituent of limestone, marble,

and chalk. Powdered calcium carbonate is produced by two methods on the industrial scale. It is quarried and ground from naturally occurring deposits and in some cases beneficiated. It is also made by precipitation from dissolved calcium hydroxide and carbon dioxide. The natural ground calcium carbonate and the precipitated material compete industrially based primarily on particle size and the characteristics imparted to a product.

Calcium carbonate is one of the most versatile mineral fillers and is consumed in a wide range of products including paper, paint, plastics, rubber, textiles, caulks, sealants, and printing inks. High-purity grades of both natural and precipitated calcium carbonate meet the requirements of the *Food Chemicals Codex* and the *United States Pharmacopeia* and are used in dentifrices, cosmetics, foods, and pharmaceuticals.

PROPERTIES

Calcium carbonate occurs naturally in three crystal structures: calcite, aragonite, and, although rarely, vaterite.

The commercial grades of calcium carbonate from natural sources are either calcite, aragonite, or sedimentary chalk. In most precipitated grades aragonite is the predominant crystal structure. The essential properties of the two common crystal structures are shown in Table 1.

ECONOMIC ASPECTS

Consumption of fine ground calcium carbonate is six times that of the precipitated calcium carbonate. The United States market for fine ground calcium carbonate was expected to grow at the rate of 2% through 2004. In Western Europe the consumption of fine ground $CaCO_3$ is used in the paper industry and exceeds the use of precipitated $CaCO_3$.

HEALTH AND SAFETY FACTORS

Calcium carbonate is listed as a food additive and is not considered a toxic material. The exposure to dust is

Table 1. Properties of Calcium Carbonate

Property[a]	Calcite	Aragonite
specific gravity	2.60–2.75	2.92–2.94
hardness, Mohs'	3.0	3.5–4.0
solubility at 18°C, g/100 g H2O	0.0013	0.0019
melting point, °C	1339[a] dec 900	[b]
index of refraction		
α		1.530
β		1.680
γ		1.685
ω	1.658	
ε	1.486	

[a]At 10.38 MPa (102.5 atm).
[b]Decomposes to calcite at temperatures >400°C.

regulated and a Threshold Limit Value–Time-Weighted Average (TLV–TWA) of 10 mg/m³ is set. Both natural ground and precipitated calcium carbonates can contain low levels of impurities that are regulated.

USES

The use of calcium carbonate in paint, paper, and plastics make up the principal part of the market. In the paper industry calcium carbonate products find two uses: as a filler in the papermaking process and as a part of the coating on paper.

The plastics industry is a primary consumer of calcium carbonate products. Flexible and rigid PVC, polyolefins, thermosets, and elastomers, including rubber, utilize a wide variety of coated and uncoated grades.

Increased loadings of calcium carbonate in thermosets reduce cost and provide better surface characteristics.

Calcium carbonate is one of the most common filler/extenders used in the paint and coatings industry. The main function of calcium carbonate in paint is as a low cost extender.

It is also used in industrial finishes and powder coatings, and continues to be used in its original applications, putty, as well as caulks, sealants, adhesives, and printing inks. Large volumes are used in carpet backing and in joint cements.

Calcium carbonate is used in flue gas desulfurization.

Calcium carbonate is used in food and pharmaceutical applications, for both its chemical and physical properties. It is used as an antacid, as a calcium supplement in foods, as a mild abrasive in toothpaste, and in chewing gum. It can also be used as a builder in detergent compositions.

Food Chemicals Codex, 3rd ed., National Academy of Science, Washington, D.C., 1981, p. 46.

H. S. Katz and J. V. Milewski, *Handbook of Fillers for Plastics*, Van Nostrand Reinhold Co., New York, 1987, p. 123.

C. Klein and C. S. Hurlbut, Jr., *Manual of Mineralogy*, John Wiley & Sons, Inc., New York, 1985, pp. 328, 335.

R. J. Reeder, ed., *Carbonates, Mineralogy and Chemistry*, Mineralogical Society of America, Washington, D.C., 1990, p. 191.

F. Patrick Carr
David K. Frederick
OMYA, Inc.

CALCIUM CHLORIDE

Calcium chloride, $CaCl_2$, is a white, crystalline salt that is very soluble in water. In its anhydrous form it is 36.11% calcium and 63.89% chlorine. It forms mono-, di-, tetra-, and hexahydrates. Calcium chloride is found in small quantities, along with other salts, in seawater and in many mineral springs. It also occurs as a constituent of some natural mineral deposits. Natural brines account for 70–75% of the United States $CaCl_2$ production.

PROPERTIES

The properties of calcium chloride and its hydrates are summarized in Table 1.

Calcium Chloride Solutions

Because of its high solubility in water, calcium chloride is used to obtain solutions having relatively high densities.

Viscosity is an important property of calcium chloride solutions in terms of engineering design and in application of such solutions to flow through porous media. Data and equations for estimating viscosities of calcium chloride solutions over the temperature range of 20–50°C are available.

MANUFACTURE AND PRODUCTION

Calcium chloride is produced in commercial amounts using many different procedures: (*1*) refining of natural brines, (*2*) reaction of calcium hydroxide with ammonium chloride in Solvay soda and production, and (*3*) reaction of hydrochloric acid with calcium carbonate. The first two processes account for over 90% of the total calcium chloride production.

ECONOMIC ASPECTS

Calcium chloride consumption is very dependent on the weather. The deicing, dust control, and road stabilization markets are, thus, affected by these conditions. Industry sources believe that the use of calcium chloride as a growth-enhancing macronutrient may be a future market in the agricultural sector.

There is currently an excess of capacity in the calcium chloride industry, which is only expected to become more actue as additional synthetic or by-product capacity increases.

ENVIRONMENTAL CONCERNS

Calcium chloride is not harmful to the environment; calcium is essential for all organisms. At concentrations above 1000 ppm, calcium chloride has been found to retard plant growth and can damage plant foliage.

HEALTH AND SAFETY FACTORS

In general, calcium chloride is not considered to be toxic. Because calcium chloride is hygroscopic, common safety precautions should be used: wearing gloves, long-sleeved clothing, shoes, and safety glasses. Contact with skin may cause mild irritation on dry skin. Strong solutions or solid in contact with moist skin may cause severe

Table 1. Properties of Calcium Chloride Hydrates

Property	$CaCl_2 \cdot 6H_2O$	$CaCl_2 \cdot 4H_2O$	$CaCl_2 \cdot 2H_2O$	$CaCl_2 \cdot H_2O$	$CaCl_2$
mol wt	219.09	183.05	147.02	129.00	110.99
composition, wt % $CaCl_2$	50.66	60.63	75.49	86.03	100.00
mp, °C	30.08	45.13	176	187	772
sp gravity, d_4^{25}	1.71	1.83	1.85	2.24	2.16
heat of fusion or transition, kJ/mol[a]	43.4	30.6b	12.9	17.3	28.5
heat of solution in water[b], kJ/mol[a]	15.8	−10.8	−44.05	−52.16	−81.85
heat of formation, at 25°C, kJ/mol	−2608	−2010	−1403	−1109	−795.4
heat capacity, at 25°C, J/(g·°C)[a]	1.66	1.35	1.17	0.84	0.67

[a] To convert J to cal, divide by 4.184.

[b] To infinite dilution.

irritation and possibly burns. Calcium chloride can irritate and burn eyes from the heat of hydrolysis and chloride irritation. Inhalation may irritate the lungs, nose, and throat with symptoms of coughing and shortness of breath. Ingestion may cause irritation to the mucous membrane due to the heat of hydrolysis. Large amounts can cause gastrointestinal upset, vomiting, and abdominal pain.

Dry bulk calcium chloride can be stored in construction-grade bins. Care should be taken to minimize moisture. It should be kept in a tightly closed container, stored in a cool, dry, ventilated area.

USES

Calcium chloride, manufactured for over 100 years, has been used for a variety of purposes. The primary $CaCl_2$ markets have not changed since the 1950s. Significant markets in the United States are for deicing during winter, roadbed stabilization, and as a dust palliative during the summer. Use as an accelerator in the ready-mix concrete industry is sizable, but there is still concern about chlorine usage because of possible corrosion of steel in highways and buildings. Calcium chloride is also used in oil- and gas-well drilling.

Food

Calcium chloride is used in the food industry to increase firmness of fruits and vegetables, such as tomatoes, cucumbers, and jalapenos, and prevent spoilage during processing. Food-grade calcium chloride is used in cheese making to aid in rennet coagulation and to replace calcium lost in pasteurization. It also is used in the brewing industry both to control the mineral salt characteristics of the water and as a basic component of certain beers.

Calcium Chloride, 2nd ed., Standard B550-90, American Water Works Association, Denver, Colo., 1990.

Calcium Chloride MSDS, Mallinckrodt Baker, Inc., Phillipsburg, NH, 2000.

M. Miller, "Calcium and Calcium Compounds," *Minerals Yearbook*, Vol. 1, U.S. Bureau of Mines, U.S. Department of the Interior, Washington, D.C., 1991.

F. W. Pontius, ed., *Water Quality and Treatment; A Handbook of Community Water Supplies*, American Water Works Association, 4th ed., McGraw-Hill, Inc., New York, 1990.

LISA M. VRANA
Consultant

CALCIUM FLUORIDE

Fluorspar lowers the melting point of minerals and reduces the viscosity of slags. The term "fluorspar" describes ores containing substantial amounts of the mineral fluorite, CaF_2, but fluorspar is often used interchangeably with fluorite and calcium fluoride.

Calcium fluoride has the formula CaF_2 and a molecular weight of 78.07; it is 51.33% calcium and 48.67% fluorine. Calcium fluoride occurs in nature as the mineral fluorite or fluorspar. It is prepared from the reaction of $CaCO_3$ and HF.

Fluorine mineral deposits are closely associated with fault zones. In the United States, significant fluorspar deposits occur in the Appalachian Mountains and in the mountainous regions of the West. Worldwide, large deposits of fluorspar are found in China, Mongolia, France, Morocco, Mexico, Spain, South Africa, and countries of the former Soviet Union.

PROPERTIES

Some of the important physical properties of calcium fluoride are listed in Table 1. Pure calcium fluoride is without color. However, natural fluorite can vary from transparent and colorless to translucent and white, wine-yellow, green, greenish blue, violet-blue, and sometimes blue, deep purple, bluish black, and brown. These color variations are produced by impurities and by radiation damage (color centers).

Table 1. Physical Properties of Calcium Fluoride

Property	Value
formula weight	78.08
composition, wt %	
Ca	51.33
F	48.67
melting point, °C	1402
boiling point, °C	2513
heat of fusion, kJ/mol[a]	23.0
heat of vaporization at bp, kJ/mola	335
vapor pressure at 2100°C, Pa[b]	1013
heat capacity, C_p, kJ/(mol·K)[a]	
solid at 25°C	67.03
solid at mp	126
liquid at mp	100
entropy at 25°C, kJ/(mol·K)[a]	68.87
heat of formation, solid at 25°C, kJ/mol[a]	−1220
free energy of formation, solid at 25°C, kJ/mol[a]	−1167
thermal conductivity, crystal at 25°C, W/(m·K)	10.96
density, g/mL	
solid at 25°C	3.181
liquid at mp	2.52
thermal expansion, average 25 to 300°C, K^{-1}	22.3×10^{-6}
compressibility, at 25°C and 101.3 kPa (=1 atm)	1.22×10^{-8}
hardness	
Mohs' scale	4
Knoop, 500-g load	158
solubility in water, g/L at 25°C	0.146
refractive index at 24°C, 589.3 nm	1.43382
dielectric constant at 30°C	6.64
electrical conductivity of solid, (Ω·cm^{-1})	
at 20°C	1.3×10^{-18}
at 650°C	6×10^{-5}
at mp	3.45
optical transmission range, nm	150 to 8000

[a] To convert J to cal, divide by 4.184.
[b] To convert Pa to mm Hg, multiply by 7.5×10^{-3}.

MINING

Underground mining procedures are used for deep fluorspar deposits, and open-pit mines are used for shallow deposits or where conditions do not support underground mining techniques.

Beneficiation

Most fluorspar ores as mined must be concentrated or beneficiated to remove waste. Metallurgical-grade fluorspar is sometimes produced by hand-sorting lumps of high grade ore. In most cases the ore is beneficiated by gravity concentration with fluorspar and the waste minerals, having specific gravity values of >3 and <2.8, respectively.

PREPARATION

CaF_2 is manufactured by the interaction of H_2SiF_6 with an aqueous carbonate suspension; by the reaction of $CaSO_4$ with NH_4F; by the reaction of HF with $CaCO_3$ in the presence of NH_4F; by reaction of $CaCO_3$ and NH_4F at 300–350°C followed by calcining at 700–800°C; by reaction of NH_4F and $CaCO_3$; and from the thermal decomposition of calcium trifluoroacetate.

SHIPMENT

Truck, rail, barge, and ship are used to transport fluorspar. The different grades are shipped in different forms: metallurgical grade as a lump or gravel; acid grade as a damp filtercake containing 7–10% moisture to facilitate handling and reduce dust.

ECONOMIC ASPECTS

For many years the United States has relied on imports for more than 80% of its fluorspar needs.

GRADES AND SPECIFICATIONS

Ceramic-grade and acid-grade fluorspars (ceramic-spar and acid-spar) have the typical analysis shown in Table 2. Both types are usually finely ground.

Metallurgical fluorspar (met-spar) is sold as gravel, lump, or briquettes.

HEALTH AND SAFETY FACTORS

Fluorite is not classified as a hazardous material. However, every precaution should be taken to prevent contact of calcium fluoride with an acid since formation of hydrofluoric acid will result. Because of the low solubility of calcium fluoride, the potential problem of fluoride-related toxicity is reduced. However, ingesting large amounts may cause vomiting, abdominal pain, and diarrhea. Continued oral ingestion of calcium fluoride could produce symptoms of fluorosis.

Beneficiation facilities require air and water pollution control systems, including efficient control of dust emissions, treatment of process water, and proper disposal of tailings. Avoid breathing fluorspar dust and contacting fluorspar with acids.

Table 2. Analyses of Ceramic- and Acid-Grade Fluorspar, wt%

Assay	Ceramic	Acid
CaF_2	90.0–95.5	96.5–97.5
SiO_2	1.2–3.0	1.0
$CaCO_3$	1.5–3.4	1.0–1.5
MgO		0.15
B		0.02
Zn		0.02
Fe_2O_3	0.10	0.10
P_2O_5		0.03
$BaSO_4$		0.2–1.3
R_2O_3[a]	0.15–0.25	0.1–0.3

[a] R_2O_3 is any trivalent metal oxide, eg, Al_2O_3.

Shipping and storage containers, when empty, can be hazardous as they will contain residues. Calcium fluoride should be kept in a tightly closed container, stored in a cool, dry, ventilated area.

ENVIRONMENTAL CONCERNS

Plants sensitive to fluorides have been shown to show signs of injury at concentrations of 1.0–4.0 µg of fluoride/m^3 over a 24-h period of exposure or at 0.5–1.0 µg of fluoride over a one-month period of exposure. Cattle consuming a diet containing 40 ppm of fluoride or more developed symptoms of fluorosis: dental defects, bone lesions, lameness, and reduced appetite, with weight loss and diminished milk yield.

Fluorides used in pesticides may cause severe illness or death if ingested. Fluorocarbons are very stable but do pose a health hazard: Burning can result in the release of phosgene gas, which is toxic. Some fluorocarbons as gases can replace the normal air supply in confined spaces if more dense than air, resulting in suffocation.

USES

Fluorspar is considered to be a commodity of strategic and commercial importance as the United States import reliance is high and fluorspar is necessary in steel and aluminum production. Fluorspar is also the primary source of fluorine and its compounds. Over 80% of reported fluorspar consumption is used in the manufacture of hydrofluoric acid and 20% in the steel and iron industry. Fluorspar is the starting material for the production of HF.

An estimated 90% of reported fluorspar consumption has gone to the manufacture of hydrofluoric acid and aluminum fluoride. The remaining 10% was used as a flux in steel making, in iron and steel foundries, for primary aluminum production, glass manufacture, enamels, welding rod coatings, and other uses or products.

Fluorspar is marketed in several grades, the three principal ones being acid, ceramic, and metallurgical.

Acid-grade fluorspar is used primarily as a feedstock in the manufacture of hydrofluoric acid and to produce aluminum fluoride.

Ceramic-grade fluorspar is used in the production of glass and enamel to make welding rod coatings and as a flux in the steel industry.

Metallurgical-grade fluorspar is primarily used as a fluxing agent by the steel industry.

Calcium Fluoride MSDS, Mallinckrodt Baker, Inc., Phillipsburg, N.J., 2001.

J. J. McKetta, ed., *Encyclopedia of Chemical Processing and Design*, Vol. 23, Marcel Dekker, Inc., New York, 1985, pp. 270–295.

M. Miller, "Flurospar," *Mineral Commodity Summaries*, U.S. Geological Survey, U.S. Department of the Interior, Washington, D.C., Jan. 2003.

D. R. Shawe, ed., *Geology and Resources of Fluorine in the United States*, U.S. Geological Survey Professional Paper 933, Washington, D.C., 1976, 1–5, 18, 19, 82–87.

Lisa M. Vrana
Consultant

CALCIUM SULFATE

Calcium sulfate, CaSO$_4$, has several forms, ie, calcium sulfate dihydrate (commercially known as gypsum), calcium sulfate anhydrous (anhydrite), calcium sulfate hemihydrate, present in two different structures, α-hemihydrate and β-hemihydrate (commercial name of β-form: stucco or plaster of Paris). In natural deposits, the main form of calcium sulfate is the dihydrate. Some anhydrite is also present in most areas, although to a lesser extent. Mineral composition can be found in Table 1.

Stucco has the greatest commercial significance of these materials. It is the primary constituent used to produce boards and plasters as the primary wall claddingmaterials in modern building construction and in formulated plasters used in job- or shop-site applications.

About 20–25 million metric tons of calcium sulfate are consumed annually. About 80% is processed into the commercially usable hemihydrate. Gypsum and its dehydrated form have been used by builders and artists in ornamental and structural applications for >5000 years.

PROPERTIES

Table 2 lists the physical properties of calcium sulfate in its different forms.

SOURCES

The natural, or mineral, form of calcium sulfate is most widely extracted by mining or quarrying and is commercially used.

Gypsum is also obtained as a by-product of various chemical processes. The main sources are from processes involving scrubbing gases evolved in burning fuels that contain sulfur, such as coal used in electrical power generating plants, and the chemical synthesis of chemicals, such as sulfuric acid, phosphoric acid, titanium dioxide, citric acid, and organic polymers.

Table 1. Gypsum Forms and Composition

Common name	Molecular formula	Composition, wt %		
		CaO	SO$_3$	Combined H$_2$O
anhydrite	CaSO$_4$	41.2	58.8	
gypsum	CaSO$_4$·2 H$_2$O	32.6	46.5	20.9
stucco	CaSO$_4$·$^1/_2$ H$_2$O	38.6	55.2	6.2

Table 2. Physical Properties of Calcium Sulfate

Property	Dihydrate	Hemihydrate	Anhydrite
mol wt	172.17	145.15	136.14
transition point, °C	128[a]	163[b]	
mp[c], °C	1450	1450	1450
specific gravity	2.32		2.96
solubility at 25°C, g/100 g H_2O	0.24	0.30	0.20
hardness, Mohs'	1.5–2.0		3.0–3.5

[a]Hemihydrate is formed.
[b]Anhydrous material is formed.
[c]Compound decomposes.

THERMODYNAMIC AND KINETICS OF GYPSUM FORMATION–DECOMPOSITION

The thermodynamic properties of gypsum formation by precipitation can be evaluated considering the liquid–solid equilibrium between calcium and sulfate ions in solution and solid $CaSO_4 \cdot 2H_2O$:

$$Ca^{2+} + SO_4^{2-} + 2H_2O = CaSO_4 \cdot 2H_2O$$

MANUFACTURE

Natural Gypsum

Gypsum rock from the mine or quarry is crushed and sized to meet the requirements of future processing or extracted for direct marketing of the dihydrate as a cement retarder. Once subjected to a secondary crusher, calcining, and drying, the product is fine ground. Fine-ground dihydrate is commonly called land plaster, regardless of its intended use. The degree of fine grinding is dictated by the ultimate use. The majority of fine-ground dihydrate is used as feed to calcination processes for conversion to hemihydrate.

β-Hemihydrate

The dehydration of gypsum commonly referred to as calcination in the gypsum industry, is used to prepare hemihydrate, or anhydrite. Kettle calcination continues to be the most commonly used method of producing β-hemihydrate.

α-Hemihydrate

Three processing methods are used for the production of α-hemihydrate. One, developed in the 1930s, involves charging lump gypsum rock 1.3–5 cm in size into a vertical retort, sealing it, and applying steam at a pressure of 117 kPa (17 psi) and a temperature of ~123°C. After calcination under these conditions for 5–7 h, the hot moist rock is quickly dried and pulverized.

Another method, first reported in the 1950s, has lower water demand. The dihydrate is heated in a water solution containing a metallic salt, such as $CaCl_2$, at pressures not exceeding atmospheric. A third method, developed in 1967, prepares very low water demand α-hemihydrate by autoclaving powdered gypsum in a slurry. A crystal-modifying substance such as succinic acid or malic acid

is added to the slurry in the autoclave to produce large squat crystals.

Anhydrite. In addition to kettle calcination, soluble anhydrite is commercially manufactured in a variety of forms, from fine powders to granules 4.76 mm (4 mesh) in size, by low-temperature dehydration of gypsum.

SHIPMENT

Gypsum and gypsum products are bulky and relatively low in cost. In North America, factors of varying regional supply and demand for building products not withstanding, the normal economic overland shipping range is ~500 km. For overland shipments, there has been a steady shift from rail to motor transport.

Formulated plasters utilizing specially processed calcined gypsum are packaged in multiple paper bags having moisture vapor–resistant liners. This type of packaging protects the contents from airborne moisture keeping the plaster more stable with respect to setting time and mixing water demand over longer periods of warehousing. Manufactured board products are most often bundled, two pieces face to face, stacked in units for transport to dealers' yards, and reshipped to individual job sites as construction schedules dictate. Specialized, labor saving, power driven handling equipment has been developed for stocking boards on construction sites.

USES

Uncalcined Gypsum

Calcium sulfate, generally in the form of gypsum, is added to Portland cement (qv) clinker to stop the rapid reaction of calcium aluminates (flash set) and accelerate strength development. Another notable use of uncalcined gypsum is in agricultural soil treatment.

FGD (disulfogypsum) Gypsum

In countries with many years of gypsum tradition and with a well-developed market for gypsum-based building materials, the gypsum industry uses more raw gypsum than the amount produced by the FGD plants. This enables complete utilization of FGD gypsum by substitution of a part of the natural raw gypsum that occurs. For example, in Germany, the extra capacity of gypsum required between 1998 and 2009 that could be satisfied by using FGD gypsums is 31%.

Calcined Anhydrite

Soluble anhydrite has physical properties similar to those of gypsum plaster. Its outstanding property is its extreme affinity for any moisture, which makes it a very efficient drying agent.

Hemihydrate

The ability of plaster of Paris to readily revert to the dihydrate form and harden when mixed with water is the basis for its many uses. Of equal significance, is the ability to control the time of rehydration. Other favorable properties include its fire resistance, excellent thermal and hydrometric dimensional stability, good compressive strength, and neutral pH.

The largest single use of calcined gypsum in North America is in the production of gypsum board, which replaced plaster in the United States during the 1960s as the main wall cladding material.

Molding plasters have been used for centuries to form cornices, columns, decorative moldings, and other interior building features. A moderate amount of plaster is used in making impressions and casting molds for bridges, and by dental laboratories.

K. Rumph, *Proceedings of Gypsum 2000*, 6th International Conference of Natural and Synthetic Gypsum, May 16–19, 2000, Toronto, Canada, 2000.

O. Söhnel, and J. Garside, *Precipitation*, Butterworth-Heinemann Ltd, Oxford, (1992).

W. Stumm, and J. J. Morgan, *Aquatic Chemistry*, 3rd ed., John Wiley & Sons, Inc., New York, 1996.

Web site "The Mineral Gallery", http://209.51.193.54//minerals/sulfates/gypsum/gypsum.htm.

Amedeo Lancia
Dino Musmarra
Marina Prisciandaro
University of Napoli

CAPILLARY SEPARATIONS

Capillary scale separations, most typically implemented in chromatography and electrophoresis, generally offer higher performance indices than their larger-scale counterparts. Capillary chromatography and electrophoresis are typically used for the analyses of complex samples containing a large number of analytes. In chromatography, analytes are separated based on their differential partitioning between an immobile stationary phase and a mobile fluid phase. In electrophoresis, different charged analytes are separated based on the different velocities with which they move in an electric field. Micellar electrokinetic chromatography (MEKC) and capillary electrochromatography (CEC) are hybrid techniques where separation by differential partitioning and movement under the influence of an electric field are both involved.

Though chromatography and electrophoresis differ in how separation is achieved, many common principles apply to both. In both techniques, the sample is introduced at the head of the column as a narrow zone and the analytes separate as they move to the exit end. The effluent analyte bands are detected as peaks using suitable detectors. The parameters of the separation process, such as elution/migration time, efficiency (how narrow peak is while eluting at a given time, often given as the number of theoretical plates), resolution between adjacent peaks, etc., are all defined similarly. Some aspects are different between them, however.

BASIC CHROMATOGRAPHY PRINCIPLES

As the analyte is partitioned between the mobile and the stationary phase, an equilibrium or distribution constant K can be defined

$$K = \frac{C_S}{C_M}$$

where C_S and C_M are the concentration of the analyte in the stationary and the mobile phase, respectively. The longer time the analyte spends in the stationary phase, the longer is its retention time, t_r, the total time the analyte spends in the separation process. The mobile phase moves through the column with a constant velocity; all unretained analytes move at the same velocity, eluting at the so-called dead time, t_0. All other analytes, which are retained to any extent by the stationary phase, elute later, the time increasing with increasing K. Figure 1 shows a model chromatogram with two analyte peaks with corresponding retention times t_{r1} and t_{r2} and an unretained molecule (eluting at t_0). Since t_r is a function of the mobile phase flow rate, the retention volume, V_r, the volume of the mobile phase necessary to elute the analyte, which is independent of the flow rate, is sometimes used instead of t_r:

$$V_r = t_r Q$$

where Q is the volumetric flow rate of the mobile phase. Similarly, the dead volume or holdup volume for an unretained analyte, V_0, can be defined from this equation by substituting t_0 for t_r.

The retention factor, k', used to describe the movement of the analytes through the column, is also commonly used to describe retention and is readily computed

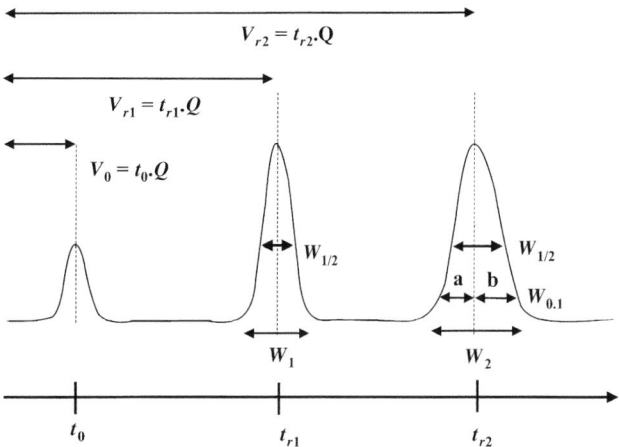

Figure 1. A model chromatogram.

as follows:

$$k' = \frac{t_r - t_0}{t_0} = \frac{V_r - V_0}{V_0}$$

Parameters to compute k' can be easily obtained from a chromatogram such as Figure 1. For practical reasons, the best chromatograms result with retention factors in the range of 2–20. If k' values are too low, analytes are not well separated; they all elute near the dead time; if k' values are too high, analysis time increases and the ability to detect an analyte also suffers because a broad wide peak of the same total area as an otherwise sharp narrow peak results.

The difference in retention factors or retention times is not the only factor that governs how well two or more analytes separate. Band broadening occurs during the separation process.

In chromatography, several processes contribute to the final zone width, such as resistance to mass transfer, eddy diffusion (in packed columns), longitudinal diffusion, and extra-column effects. The overall peak variance is the sum of the individual variances:

$$\sigma_{tot}^2 = \sigma_{res}^2 + \sigma_{eddy}^2 + \sigma_{long}^2 + \sigma_{extra}^2$$

For ideal (Gaussian) analyte peaks, the width of the peak is directly related to the variance, σ^2. The variance per unit column length is the plate height, H:

$$H = \frac{\sigma^2}{L}$$

where L is the column length. Capillary separation systems excel in their separation abilities relative to their larger counterparts in that small plate heights, typically expressed in micrometers (μm), are observed. The total number of theoretical plates in a column, N, which ultimately governs the quality of the separation, is hence given by

$$N = \frac{L}{H}$$

For idealized Gaussian peaks, experimentally N can be readily calculated from the width at half height of the peak, $w_{\frac{1}{2}}$:

$$N = 5.54 \left(\frac{t_r}{w_{1/2}} \right)^2$$

For capillary gas chromatography (GC), the most mature and perhaps the most widely practiced of all separation techniques, the number of theoretical plates achieved in a typical separation system can be in millions. True Gaussian peaks, are, however, rarely observed in practice. For asymmetric peaks, the ratio, b/a, of the front and the tail half-widths, $w_{0.1}$, of a peak measured at 10% of the peak height, is used. For a symmetric peak,

the value of b/a is unity, for a tailing peak the value will be >1, for a fronting peak the value will be <1. The parameter N is then calculated as:

$$N = \frac{41.7(t_r/w_{0.1})^2}{1.25 + (b/a)}$$

As band broadening increases, plate height, H, increases. In the classical van Deemter equation, H, is described as a function of mobile phase flow velocity:

$$H = A + B/v + Cv$$

where A, B, and C are coefficients pertaining to eddy diffusion, longitudinal diffusion and resistance to mass transfer, respectively, and v is the velocity of the mobile phase.

The third term in this equation relates to the rate of transport of the analyte molecules between the stationary and mobile phase. As the mobile phase velocity through the column is increased, it has less and less time to equilibrate with the stationary phase. Some molecules are left behind in the stationary phase that would not have been if equilibrium were achieved. This nonequilibrium condition persists along the column length and results in band broadening that increases with increasing mobile-phase velocity. The coefficient C is proportional to the square of diameter of the packed particles in packed columns or conversely, square of the column diameter for open tubular capillaries and the effective thickness of the active layer on the stationary phase. It is also inversely proportional to the diffusion coefficient of the analyte, both in the mobile and stationary phases.

Resolution

Resolution, R_s, provides a quantitative measure of separation between two analytes and is given by equation:

$$R_s = \frac{2(t_{r2} - t_{r1})}{w_1 + w_2}$$

where t_r and w are the retention times and base widths of respective analyte peaks.

BASIC CAPILLARY ELECTROPHORESIS (Ce) PRINCIPLES

Whereas chromatography is carried out both in larger bore columns and in the capillary format, for a variety of reasons there is no larger bore equivalent of ce. In ce, analytes are carried along a capillary in which an electric field is present along the length of the capillary. The analytes separate because they exhibit different mobilities (speeds) due to differences in their size and charge.

In chromatography, the movement of the sample constituents takes place due to the movement of the mobile phase, in turn accomplished by a pressure gradient. In ce, bulk flow of the separation medium (often called the background electrolyte, BGE) can occur due to electroosmosis but is not necessarily essential for the separation of the analyte components, because these can move without bulk electroosmotic flow. However, in most cases, electroosmosis is present and plays an important role in the separation process.

Concepts such as number of theoretical plates, N, or resolution, R_s, apply to CE in much the same way as in chromatography.

SAMPLE PREPARATION FOR CHROMATOGRAPHY AND CAPILLARY ELECTROPHORESIS

Sample preparation or extraction prior to the actual separation and analysis is sometimes more challenging than separation/analysis and is often tedious and time consuming. It is particularly important in trace analysis applications (pesticides in fruit/vegetables, polychlorinated biphenyls in soil or sediments) where the analyte must first be extracted/separated from the matrix. Currently, several methods are used for sample treatment prior to chromatographic analysis. For air samples, a variety of sample collection devices, including adsorbent tubes, filters, or denuders, are used.

In liquid–liquid extraction the analyte(s) of interest are extracted from the sample (usually aqueous) into an immiscible extractant phase. Liquid–liquid extraction or soxhlet extraction are the most commonly prescribed analyte extraction method in most United States Environmental Protection Agency (US EPA) and Association of Official Analytical Chemists (AOAC) methods.

For extraction of liquid samples, solid phase extraction (SPE) has become popular. A sample containing the analytes is passed through the solid packing material contained in a SPE-tube or embedded in a disk. In the more common mode of using SPE, analytes of interest are retained on the packing while most matrix components are unretained. SPE is a convenient, inexpensive, and time-saving sample preparation mode for many liquid samples; it significantly reduces the volume of organic solvents used. For complex samples, SPE is often used after SFE or ASE.

Solid phase microextraction (SPME) is a miniaturized form of SPE, particularly useful for the analysis of gaseous samples. SPME requires no solvent, the fiber is reusable and can be used for concentration of a variety of analytes and is thus particularly attractive.

Headspace and purge-trap methods are used for analyzing volatile analytes from a liquid, often an aqueous sample, most typically for analysis by capillary GC. Headspace sampling is a static method for the collection of a volatile analyte.

In a purge-and-trap procedure, an inert gas is bubbled through the (typically aqueous) sample causing the removal of the volatile compounds into the gaseous phase. The gas is made to flow through a sorbent trap that retains the analytes. Subsequently, the analytes are desorbed from the trap (typically by heating) and injected into a gas chromatograph (GC).

CAPILLARY GAS CHROMATOGRAPHY (Gc)

In gc, the sample is vaporized and injected onto the chromatographic column. Flowing gaseous mobile phase (carrier gas) transports the sample through the column and separation between analytes takes place based on the partitioning of the analytes between this mobile phase and the immobilized stationary phase on the capillary wall.

Instrumentation

A typical gc consists of a pressurized source of a carrier gas (typically a gas cylinder), an injector, a capillary column, a controlled temperature oven and a detector. A typical gc will also contain several pressure regulators and/or flow controllers, since it is vital to control the carrier gas flow rate.

Carrier Gas

The carrier gas transports the sample through the column. It does not interact with the analytes. Occasionally, the choice of the carrier gas is dictated by the detector used. Common carrier gases used are He, N_2, and H_2; sometimes Ar, CO_2, and air.

Sample Injection

Microsyringe Injection. Direct injection of a liquid sample with a microsyringe, although commonly used with packed columns, is less commonly used in capillary gc since overloading often occurs.

Split Injection. Split injection is one of the most common injection techniques in capillary gc, since it allows injection of samples independent of the solvent choice, at any column temperature and causes minimal band broadening.

Splitless Injection. This injection mode was devised to allow trace analysis, and allows injection of relatively large sample volumes (1–10 µL) on the columns with a bore ≥ 300 µm. Two types of splitless injection are cold trapping and the retention gap injection mode.

Programmed Temperature Vaporization (PTV). A PTV based injector can be used in different modes. Split injection minimizes both the pressure wave and discrimination against less volatile analytes. Solvent elimination mode can be used to inject large sample volumes if the analyte is much less volatile than the solvent. The cold solvent splitless mode relies on cold trapping and solvent effect. The precision and accuracy attainable by a PTV based injector is significantly superior to that attainable in classical split and splitless injection modes.

Cold On-Column Injection. This technique eliminates discrimination of less volatile analytes and sample decomposition. It is easy to implement and automate; small sample volumes, 0.2–2 µL, are typically used.

Controlled Temperature Oven

A thermostated oven is used in GC, which requires exact control of the column temperature. Modern gas chromatographs are typically equipped with a programmable temperature oven that quickly and reproducibly changes the temperature as a function of time. The upper temperature limit is at least 350 °C.

Detectors for Gas Chromatography

The detector is one of the most important components of a separation system; it may or may not be integrated with the rest of the instrument. An accurate and precise detection of analyte peaks is required for both qualitative and quantitative analysis with any separation system. Ideally the detector should be sensitive, have linear response to the analytes over a wide range, be stable and reproducible, and have a short response time. No detector meets all of the ideal criteria above. In reality, a detector is chosen based on the particular analytical problem. GC detectors, almost all of which are available for use with capillary systems, can be broadly subdivided into ionization detectors, detectors measuring bulk physical properties, optical detectors, and electrochemical detectors.

Columns

Columns represent perhaps the heart of a chromatographic system; after all, the actual separation takes place on the column. Columns used in capillary GC today are fused silica or glass capillaries ranging in inner diameter from 50 to 700 µm.

Stationary Phases

Stationary phases provide means for the analyte separation. The requirement for a suitable stationary phase is low volatility and thermal stability, chemical inertness and suitable values of k' and α for the analyte mixture to be separated.

The number of stationary phases in common use has decreased rapidly over the years. Only a limited number of stationary phases is now used for the vast majority of applications.

Hyphenated Techniques

Gc–Ms. Coupled or hyphenated techniques are among the most powerful in analytical chemistry. Capillary gc is often coupled to a selective detector such as a mass spectrometer or Fourier transform infrared spectroscopy (Ftir) that can provide additional, especially structural, information. Direct introduction of the effluent from a capillary gc into a ms without special interfaces is possible. Gc–ms systems have been used for the identification and quantitation of thousands of myriad components present in natural and biological systems, including odor and flavor components, water pollutants, or drug metabolites and pharmaceuticals.

Gc–Ftir. Like gc–ms, gc–ftir provides a potent means for separation and identification of analytes in complicated mixtures. In this technique, the end of the GC column is connected to a narrow bore gold-coated light pipe in which the infrared (ir) absorption spectrum is acquired.

Multidimensional Gc. In two-dimensional (2-D) gc or multidimensional gc, one or more peaks eluting from a first gc column is separated on a second gc column offering a different selectivity, resulting in very high resolving power.

This technique resembles the fast scanning mass spectra in gc–ms. It is particularly useful in cases where a number of structurally analogous compounds or isomers coexist that mass spectrometry cannot differentiate. Multidimensional gc has been widely applied in the analysis of petroleum products that often contain closely structurally related isomers, in the analysis of traces of structurally analogous chlorinated substances in the environment and the analysis of food and fragrances.

Applications

Capillary gc is applied for the analysis and characterization of an enormous number of compounds in various areas, with the primary focus on volatile, volatilizable and semivolatile compounds. The main application areas include (1) industrial chemical analysis, such as analysis of mixtures of acids, alcohols, aldehydes, amines, esters, phenols, and other aromatics (2) analysis of industrial solvents and their mixtures as well as analysis of permanent gases and hydrocarbons in various industries, natural and synthetic gases, refinery gases, sulfur containing gases, permanent and noble gases and freons, gasoline and naphtha, etc. (3) analysis of flavors, fragrances, antioxidants and preservatives in food, beverages, perfumes, and cosmetic products, and (4) analysis of pharmaceutical formulations, including antidepressants, antiepileptics, anticonvulsants, opiates, barbiturates, alkaloids, etc. Capillary gc is particularly useful for the analysis of volatile and semivolatile analytes in environmental samples.

LIQUID CHROMATOGRAPHY (Lc)

In lc, a liquid mobile phase is used. In contrast to gc, the choice of the mobile phase affects the analyte retention. There are also many more stationary phases available, permitting much greater latitude in developing a separation method. However, a greater choice can also sometimes mean greater complexity.

As in gc, capillary lc offers better separation performance than its larger-bore counterparts. However, it has not yet become as widely used as capillary gc. Capillary lc has established a niche for itself where the ability to analyze minute sample volumes, limited consumption of mobile phase, and high resolution are important.

The popularity of lc stems from its ability to accomplish almost any separation, given appropriate sample preparation. This is especially true for nonvolatile thermally unstable species, such as amino acids, proteins, carbohydrates, drugs, antibiotics, pesticides as well as myriad inorganic and ionic species.

Detectors

All detectors that are used in capillary scale lc can also be used in ce. With the exception of refractive index and conductivity detectors, bulk property detectors are not commonly used. Many detection principles allow on-column detection and thus eliminate extraneous dispersion in transit to the detector. Some gc detectors can be used with capillary lc, but the practice is not common.

The key lc detectors are absorbance detectors, fluorescence detectors, electrochemical detectors, conductometric detectors, refractive index (ri) detectors, and evaporative light scattering detectors.

Stationary and Mobile Phases

In lc both the stationary and mobile phase play an important role in the analyte partitioning mechanism. The type of mobile phase depends on the chromatographic mode used. In "normal-phase" lc, a nonpolar solvent such as hexane or isopropyl ether are used, in "reversed phase" (RP) lc, by far the most common mode in which lc is currently practiced, aqueous eluents containing methanol or acetonitrile (occasionally, tetrahydrofuran) as well as buffering salts/acids/bases are used.

Normal Phase Lc

In normal phase lc a column is packed with a polar stationary phase; the mobile phase is of a nonpolar or moderately polar nature. The polarity of the analytes plays crucial role in their retention on the stationary phase.

Reversed Phase Lc (Rplc)

Rplc accounts for the majority of present lc separations. In reversed-phase chromatography, the elution order of the analytes is opposite to the normal-phase chromatography, ie, the most polar analyte elutes first. A nonpolar stationary phase and a polar mobile phase is used.

Lc–Ms

The great potential in coupling liquid chromatography with mass spectrometry (lc–ms) has been recognized especially in recent years. Coupling of lc to ms provides an important tool in qualitative and quantitative analysis of various samples and is useful in several areas such as analysis of pharmaceuticals, biochemical and environmental samples.

Applications of Capillary Lc

LC is the most popular and widely used analytical separation method today. Thousands of Lc-based analyses are in routine use, for analysis of pharmaceutical formulations and drugs, food, and consumer products, large molecules of biological importance such as DNA, DNA-fragments or peptides, and proteins. The Lc analysis of mixtures of nonvolatile or semivolatile pollutants supplement the available gc methodologies.

Virtually all methods developed for conventional lc with large bore columns can be readily implemented to the capillary scale. While capillary lc is not at the moment poised to displace conventional scale lc (as happened with gc), its future is increasingly brighter.

Ion Chromatography (Ic)

Ic, an important subclass of lc, is used primarily to determine inorganic and organic anions and cations.

Applications of Ic

Ic is applied for separation and analysis of small charged analytes in various samples. It is used particularly for the analysis of different types of water samples, such as tap, drinking, rain or wastewater from industrial processes. Several industries, such as pulp and paper, chemical, pharmaceutical, power generation and the semiconductor industry depend on Ic analysis.

SUPERCRITICAL FLUID CHROMATOGRAPHY (Sfc)

In supercritical chromatography, the sample is injected onto the column and carried through it by a mobile phase consisting of a supercritical fluid, that is maintained in the separation system at a temperature higher than its critical temperature T_c and critical pressure P_c. Properties of supercritical fluids such as density, viscosity, and diffusion coefficient are intermediate between those of liquids and gases. This makes sfc an interesting alternative to both gc and lc. Sfc can be advantageously used for compounds not easily analyzed by gc; eg, nonvolatile, thermally labile, or polymeric compounds or compounds lacking chromophores or electrochemical properties that are not so easily detected by conventional lc detectors.

Instrumentation

A typical instrument for capillary sfc resembles the instrumentation for gc. The chromatograph consists of a gas tank (usually containing CO_2, with or without added modifiers, as the mobile phase), a syringe pump, an injector, a column placed in the thermostatic oven, a pressure restrictor and a detector. The instrumentation for capillary sfc differs slightly from the packed column sfc due to the different pressures generated in the two systems.

Mobile Phase

As in lc, the mobile phase affects the retention of the analytes. By changing the composition of the mobile phase the selectivity factor, α, can be changed and the selectivity of the separation altered.

Injectors

The type of injector and the size of the injected sample depend primarily on the columns used. For capillary columns, with inner diameter of several tens of micrometers, the sample volume should be sufficiently small so that the column overloading is minimized.

Detectors

Sfc benefits from being an intermediate methodology between gc and lc; typically, detection can be accomplished by either gc or lc-detectors.

Columns

The use of packed or capillary columns in sfc is dictated by the needs of the particular analysis.

Stationary Phases

Stationary phases used for open tubular sfc are those routinely used in gc, which are further cross-linked to increase their stability.

Applications

Supercritical fluid chromatography (often following supercritical fluid extraction, sfe) has been used, eg, for measuring caffeine in coffee or nicotine in tobacco. Sfc has been used for the analysis of fats, oils and food related samples, natural products such as terpenoids or lipids, low molecular weight polymers, fossil fuels and synthetic lubricants, thermally labile large molecules, including biomolecules and synthetic polymers. The interesting properties of the mobile phase makes sfc a very attractive technique in principle; however, thus far sfc has not enjoyed the degree of commercial success originally hoped for.

CAPILLARY ELECTROPHORESIS

Capillary electrophoresis differs in several important ways from the three chromatographic methods described above. The mode of injection, the mode of sample transport, and the mechanism of analyte separation are all different. Ce evolved from the planar electrophoretic methods performed on paper or a thin gel. In ce, a fused silica capillary is filled with a buffered background electrolyte (BGE) solution, which serves as the separation medium. Unlike the chromatographic methods, in ce there are no pumps or pressurized fluid sources. The two ends of the capillary are immersed in the reservoirs containing the BGE. Two noble metal electrodes, usually Pt, are placed in each vial as well. The movement of the electrolyte and sample through the capillary is accomplished by applying a high voltage (HV) (typ. 10–30 kV across a 40–75 cm long, 25–100-μm bore capillary) across the electrodes.

Capillary electrophoresis is characterized by the very high efficiency due to the pluglike flow profile of the electrophoretic flow.

Basic Instrumentation in Ce

A simple ce apparatus consists of a HV supply capable of providing a potential difference in the range of ±30 kV, two platinum electrodes, placed in the two electrolyte vials, a fused silica capillary, and a detector.

Injection Techniques in Ce

Although valve-based injection has been used in ce, this is not common. The two most used injection techniques are the hydrodynamic (HD) and the electrokinetic (EK) injection methods.

Detection in Ce

Detection methods and detectors in ce are essentially identical to those used in capillary lc with one caveat that modifications may need to be made for the high voltage present in the capillaries.

Counter versus Co-current Electrophoretic versus Electroosmotic Flow (EOF)

Manipulation of the EOF can be used to achieve additional resolution of closely migrating peaks.

Factors Affecting the Separation Selectivity in Ce

Separation selectivity in ce is mainly achieved by modification of the BGE. Several factors affect separation efficiency and migration time.

RELATED TECHNIQUES

Separation Modes in Ce

CE is typically carried out in the zone injection and separation mode, which is therefore often referred to as capillary zone electrophoresis (cze). There are other modes of capillary electrophoretic separations, such as isotachophoresis.

Electrophoresis in Sieving Media: DNA Analysis

Electrophoresis in sieving media provides a method for separation of high molecular weight molecules, such as proteins, nucleic acids, and their fragments based on their size. While proteins can usually be separated based on the differences in charge and size using ce, nucleic acids, synthetic nucleotides, DNA restriction fragments and higher DNA strains posses very similar charge densities and their separation by ce is difficult. Two modes are in use: capillary gel electrophoresis (cge) and dynamic size-sieving ce.

Micellar Electrokinetic Chromatography

Ce methods discussed thus far can only separate charged analytes. In mekc, the BGE consists of a surfactant above its critical micelle concentration (CMC). Surfactants form micelles, spherical aggregates of several molecules of surfactants, above their CMC. In a micelle, the hydrophobic tails (usually alkyl chains) face the interior of the sphere, while the ionic groups extend into the surrounding media. Most mekc applications use typical anionic surfactants like SDS or cationic surfactants like CTAB. The charged micelles move at a rate different from that of the bulk liquid. Neutral analytes can remain in the bulk liquid or partition to the micelle, the "pseudostationary phase". Differences in this partition behavior for different analytes lead to different effective velocities.

Thus, the differential partitioning of the nonpolar, uncharged, analytes between the BGE and the hydrophobic micellar interior is the principle of separation by mekc, analogous to reversed-phase lc.

MEKC extends the separation range of ce-like methods to uncharged molecules. Mekc can be applied for the analysis of variety of analytes such as amino acids and polypeptides, DNA adducts, flavonoids and steroids, drugs and environmental samples.

Capillary Electrochromatography

Capillary electrochromatography combines the high selectivity and versatility of the stationary phases from lc with the high efficiency of electrically driven flow in ce. The column is essentially the same as a packed lc column (although some aficionados believe that cec columns need to be packed electrokinetically). Since no pumps are involved for the mobile phase delivery, the column can be packed with the particles as small as 0.5 μm, thus providing very high column efficiencies. The eluent is driven through the column electrokinetically, by applying voltage. Extraordinarily high separation efficiencies (≥1 million plates) have been reported.

Will cec take up some of the applications of lc? Or will the fate of cec be more like SFC, which is yet to fulfill its originally envisioned market potential? The technique is still not fully mature and reproducibility is often limited. The technique is still in its infancy and only time will prove its value.

Ce–Ms

Like lc–ms, ce–ms provides identification and structural information. Compatibility of the ce BGE to ms, flow rate matching, termination of HV, etc. are important issues in a successful interface.

Applications

Due to its high resolving power, short analysis times and minute sample consumption, capillary electrophoresis plays an important role in analysis of small samples such as in the life sciences, notably analysis of single cells, human and plant tissues and for monitoring of in vivo processes. Ce can be applied for variety of samples and the range of analytes overlaps application areas of ic and lc. However, it has not significantly displaced either of these for routine quantitative analysis. Despite the advent of mekc, ce is not competitive in the marketplace with lc or gc for the analysis of neutral analytes. Main application areas of capillary electrophoresis and related techniques are DNA sequencing, analysis of amino acids, bases, nucleosides and nucleotides, proteins and peptides. High separation efficiency and little need for pretreatment makes it an excellent method for analysis of various enzymatic digests, peptide mapping, analysis of fermentation broths, biological fluids and food sample.

Another important area of ce is chiral separations, due to the decreased costs of the chiral selectors needed for ce-separation. Further application areas include analysis of drugs, carbohydrates, small organic molecules and inorganic ions.

M. Caude, D. Thiebaut, *Practical Supercritical Fluid Chromatography and Extraction*, Harwood Academic Publishers, Amsterdam, The Netherlands, 1999.

R. Kuhn and S. Hoffstetter-Kuhn, *Capillary Electrophoresis: Principles and Practice*, Springer Verlag, Berlin, 1993.

R. A. Meyers, *Encyclopedia of Analytical Chemistry*, Wiley-Interscience, John Wiley & Sons, Ltd., Chichester, U.K., 2000 Volumes 12 and 13.

D. A. Skoog, F. J. Holler, and T. A. Nieman, *Principles of Instrumental Analysis-5th ed.*, Harcourt Brace College Publishers, Philadephia, 1998.

Petr Kuban
Pumendu K. Dasgupta
Texas Tech University

CARBIDES

Carbon reacts with most elements of the Periodic Table to form a diverse group of compounds known as carbides, some of which are extremely important in technology. For example, calcium carbide, CaC_2, is a source of acetylene; silicon carbide, SiC, and boron carbide, B_4C, are used as abrasives; tungsten carbide, WC, titanium carbide, TiC, and tantalum (niobium) carbide, TaC(NbC) find use as structural materials at extremely high temperatures or in corrosive atmospheres. Cementite, Fe_3C, and the multimetallic complexes $(Co,W)_6C$, $(Cr,Fe,Mo)_{23}C_6$, and $(Cr,Fe)_7C_3$ are the components in tool steels and Stellite-type alloys responsible for their hardness, wear resistance, and excellent cutting performance. There are also emerging uses of these materials as catalysts.

Figure 1 provides a survey of the most important and well-known binary compounds of carbon, according to their position in the periodic system. They are divided into four main groups: the saltlike, metallic, diamondlike, and volatile compounds of carbon, which have ionic, metallic, semiconductor, or covalent character, but these and further subdivisions are not rigid and there are a number of transitional cases. Whereas the members of Groups 2 (IIA) and 3 (IIIB) are classified as saltlike carbides, some of their properties, eg, Be_2C, with a very high degree of hardness, correspond to diamondlike carbides. Conversely, some monocarbides of Group 3 (IIIB), eg, scandium carbide, ScC, and uranium carbide, UC, as well as thorium carbide, ThC, have pronounced metallic characteristics.

Carbides are generally stable at high temperatures and thus can be prepared by the direct reaction of carbon with metals or metal-like materials at high temperatures. This does not apply to the acetylides and the alkali metal–graphite compounds, which although being carbon compounds, fit only marginally into the category of carbides. Similarly, the large class of coordination compounds, ML_y, known as organometallics are not typically considered

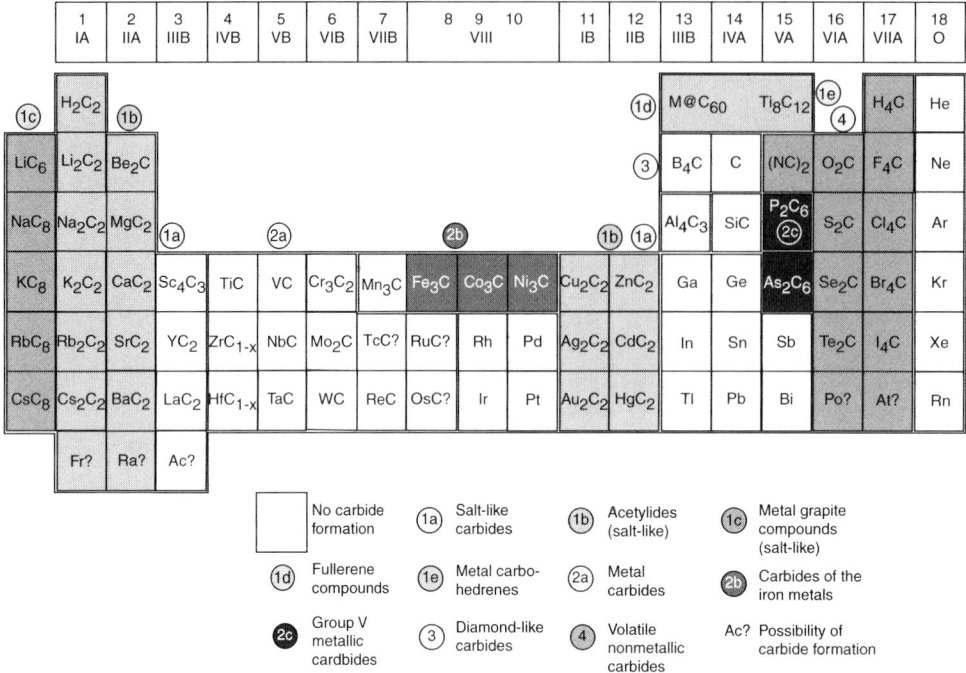

Figure 1. Principal binary compounds of carbon.

carbides, even though the ligands, L, are attached to the metal center by metal–carbon bonds. The same applies to compounds such as M_x at C_{60} or M_x at C_{70}, formed from diverse elements by association with fullerene structures and metallo-carbohedrene clusters such as Ti_8C_{12}, in which metals are part of the polyhedral cage. The volatile compounds of carbon are also excluded.

V. A. Gubanov, A. L. Ianovsky, and V. P. Zhukov, *Electronic Structure of Refractory Carbides and Nitrides*, Cambridge University Press, Cambridge, 1994.

S. T. Oyama, ed., *The Chemistry of Transition Metal Carbides and Nitrides*, Blackie Academic and Scientific, London, 1996.

L. Toth, *Transition Metal Carbides and Nitrides*, Academic Press, New York, 1971.

S. TED OYAMA
Virginia Polytechnic Institute
and State University

RICHARD KIEFFER
Technical University of Vienna

CARBIDES, CEMENTED

Cemented carbides belong to a class of hard, wear-resistant, refractory materials in which the hard carbides of Group 4–6 (IVB-VIB) metals are bound together or cemented by a soft and ductile metal binder, usually cobalt or nickel. Although the term "cemented carbide" is widely used in the United States, these materials are better known internationally as hard metals.

MANUFACTURE

Cemented carbides are manufactured by a powder metallurgy process, consisting of a sequence of carefully controlled steps designed to obtain a final product having specific properties, microstructure, and performance. The carbides or the carbide solid solution powders are prepared and blended with very fine binder metal powder, cobalt or nickel, in ball mills, vibratory mills, or attritors using carbide balls. The powder blends are compacted, presintered, and shaped, and the carbide subjected to sintering and postsintering operations. The sintered product may either be directly put to use or ground, polished, and coated.

Inhalation of extremely fine carbide, cobalt, and nickel powders should be avoided. Efficient exhaust devices, dust filters, and protective masks are essential when handling these powders.

RECYCLING OF SCRAP

Recycling of cemented carbide scrap is of growing importance. In one method, the sintered scrap is heated to 1700–1800°C in a vacuum furnace to vaporize some of the cobalt and embrittle the material. After removal from the furnace the material is crushed and screened. Chemical recycling and the zinc reclaim process are also used.

TOOL FAILURE MODES

In simple terms, a cutting tool must be harder than the material being cut. In addition to abrasive wear, high temperatures and stresses can cause blunting of the tool tip from plastic deformation, and high stresses can lead to catastrophic fracture. The workpiece may chemically interact with the tool material. The tool may also experience repeated impact loads during interrupted cuts. The useful life of the tool depends on its response to the conditions existing at the tool tip, including crater wear, flank wear, built-up edge, depth-of-cut notching, thermal cracks.

EVALUATION OF PROPERTIES

In addition to chemical analysis, a number of physical and mechanical property evaluations are required to determine cemented carbide quality. Standard test methods employed by the industry for abrasive wear resistance, apparent grain size, apparent porosity, coercive force, compressive strength, density, fracture toughness, hardness, linear thermal expansion, magnetic permeability, microstructure, Poisson's ratio, transverse rupture strength, and Young's modulus are set forth by ASTM/ANSI and the ISO.

Among the physical properties, density is very sensitive to composition and porosity of the cemented carbide and is widely used as a quality control test. Magnetic properties most often measured are magnetic saturation and coercive force.

The properties and performance of cemented carbide tools depend not only on the type and amount of carbide but also on carbide grain size and the amount of binder metal. Information on porosity, grain size and distribution of WC, solid solution cubic carbides, and the metallic binder phase is obtained from metallographically polished samples. Optical microscopy and scanning and transmission electron microscopy are employed for microstructural evaluation.

Hardness and transverse rupture strength (TRS) are often used as quality control tests.

METAL-CUTTING APPLICATIONS

Tools and Toolholding

Early carbide metal-cutting tools consisted of carbide blanks brazed to steel holders. Indexable inserts were introduced in the 1950s. In this configuration the so-called throwaway carbide insert is secured in the holder pocket by a clamp or some other holding device instead of a braze. When a cutting edge wears, a fresh edge is rotated or indexed into place.

Compositions

For machining purposes, alloys having 5–12 wt% Co and carbide grain sizes from 0.5 to >5 µm are commonly used. The straight WC–Co alloys have excellent resistance to simple abrasive wear and are widely used in machining materials that produce short chips, eg, gray cast-iron, nonferrous alloys, high temperature alloys, etc.

Ultrafine-grained WC—Co Alloys

Since the late 1980s there has been a tendency toward ultrafine-grained cemented carbides (WC grain sizes <0.5 µm) for woodworking tools, printed circuit board drills, and endmills.

Cemented Carbides for Steel Machining

Additions of cubic carbides such as TiC, TaC, and NbC impart chemical stability and crater wear resistance to WC–Co alloys.

Coated Carbide Tools

Chemical vapor deposited (CVD) TiC coatings were used in the late 1960s to combat wear on steel watch parts and cases. When applied to cutting tools, the relatively thin (~5 µm) TiC coatings extended tool life in steel and cast iron machining by a factor of two to three by suppressing the crater wear and flank wear. Hard coatings reduce frictional forces at the chip–tool interface, which in turn reduce the heat generated in the tool resulting in lower tool tip temperatures. Coatings also permit the use of higher cutting speeds, boosting machining productivity. Currently, coated carbides account for nearly 75% of all indexable metal-cutting inserts used in the United States and CVD accounts for ~70% of coated carbides.

In the mid-1980s physical vapor deposition (PVD) emerged as a commercially viable process for applying hard TiN coatings onto cemented carbide tools. A number of factors make PVD process attractive for use with cemented carbide tools: (1) lower deposition temperature (<550°C) prevents eta-phase formation and produces finer grain sizes in the coating layer; (2) PVD coatings are usually crack-free; (3) depending on the deposition technique, compressive residual stresses, which are beneficial in resisting crack propagation, may be introduced in the coating; (4) PVD coating preserves the transverse rupture strength of the carbide substrate, whereas the CVD process generally reduces the TRS by as much as 30%; and (5) PVD coatings can be applied uniformly over sharp cutting edges.

PVD-coated tools are thus successfully employed in operations where sharp edges are most beneficial, including milling, turning, boring, drilling, threading, and grooving. Newer PVD coatings are rapidly becoming commercially available. These include TiCN, TiAlN, Cr_3C_2, CrN, TiB_2, layers of TiN–TiAlN, and aluminum-rich AlTiN.

NONMETAL-CUTTING APPLICATIONS

Today almost one half of the total production of cemented carbides is used for nonmetal-cutting applications such as mining, oil and gas drilling, transportation and construction, metalforming, structural and fluid-handling components, and forestry tools. The majority of compositions used in these applications comprise straight WC–CO grades.

Metal-forming applications include drawing dies, rolls for hot and cold forming of strips and bars, cold heading dies, forward and back extrusion punches, swaging hammers and mandrels, and can body punches and dies.

The high abrasion resistance and edge strength of carbides make them ideal for use as slitter knives for trimming steel cans and stainless and carbon steel strips, cutting abrasive materials in the paper, cellophane and plastic industries, and for slitting magnetic tapes for audio, video, and computer applications.

Cold-forming equipment such as extrusion or heading punches and dies are made from cemented carbides to produce a variety of parts such as wrist pins, spark plug shells, bearing retainer cups, and propeller shaft ends.

The high elastic modulus, compressive strength, and wear resistance of cemented carbides make them ideal candidates for use in boring bars, long shafts, and plungers, where reduction in deflection, chatter, and vibration are concerns. Another application area for carbides is the synthetic diamond industry, where carbides are used for dies and pistons.

The rigidity, hardness, and dimensional stability of cemented carbides, coupled with their resistance to abrasion, corrosion, and extreme temperatures, provide superior performance in fluid-handling components such as seal rings, bearings, valve stems and seats, and nozzles. In the transportation and construction industry, steel tools having cemented carbide cutting tips are used for road planing, soil stabilization, asphalt reclamation, vertical and horizontal drilling, trenching, dredging, tunnel boring, forestry, and for snowplow blades, tire studs, and street sweeper skids.

Cemented carbides play a crucial role in the recovery of metallic ores and nonmetals by underground or open-pit mining practices, recovery of minerals such as coal, potash, and trona, and drilling for oil and gas. In the oil and gas drilling industry tungsten carbide buttons are used in steel drill bodies for deep penetration of metamorphic, igneous, and sedimentary rocks.

Cemented tungsten carbides also find use as a support for PCD cutting tips, or as a matrix alloy with cobalt, nickel, copper, and iron, in which diamond particles are embedded. These tools are employed in a variety of industries including mineral exploration and development; oil and gas exploration and production; and concrete, asphalt, and dimension stone cutting.

ECONOMIC ASPECTS

Cemented carbide inserts and tools for metal cutting and metal working have traditionally accounted for the largest percentage of carbide industry sales. However, carbide tool consumption in nonmetal-working fields, notably in the construction and transportation industries, has grown rapidly.

Developments in materials, coatings, and insert geometries have claimed an increasing share of research and development budgets in the cemented carbide industry. Important economic benefits of these effects have been an increase in tool performance and significant increase in metal-cutting productivity. Continuing developments in computer numerically controlled machining systems have placed a heavy emphasis on tool reliability and consistency, which in turn puts pressure on the industry to invest increasing amounts of capital in developing new materials and processes.

K. J. A. Brookes, *World Directory and Handbook of Hardmetals and Hard Materials*, 6th ed., International Carbide Data, East Barnet Hertfordshire, U.K., 1996.

A. T. Santhanam, P. Tierney, and J. L. Hunt, *Metals Handbook, Properties and Selection*, Vol. 2, 10th ed., 1990, pp. 950–977.

G. Schneider, Jr., *Principles of Tungsten Carbide Engineering*, 2nd ed., Society of Carbide and Tool Engineers, ASM International, Materials Park, Ohio, 1989.

E. M. Trent, *Metal Cutting*, 3rd ed., Butterworth-Heinemann Ltd., Oxford, U.K., 1991.

A. T. Santhanam
Kennametal Inc.

CARBIDES, INDUSTRIAL HARD

The four most important carbides for the production of cemented carbides are tungsten carbide, WC; titanium carbide, TiC; tantalum carbide, TaC; and niobium carbide, NbC. The binary and ternary solid solutions of these carbides such as WC–TiC and WC–TiC–TaC (NbC) are also of great importance. Chromium carbide (3:2), Cr_3C_2; molybdenum carbide, MoC; and molybdenum carbide (2:1), Mo_2C; vanadium carbide, VC; hafnium carbide, HfC; and zirconium carbide, ZrC; have minor significance. Carbides and their solid solutions are generally combined with cobalt and used in the form of cemented carbides.

PREPARATION

In general, the carbides of metals of Groups 4–6 (IVB–VIB) are prepared by reaction of elementary carbon or hydrocarbons and metals and metal compounds at high temperatures. The process may be carried out in the presence of a protective gas, under vacuum, or in the presence of an auxiliary metal (menstruum). Methods include carburization by fusion, carburization by thermal diffusion, carburization by menstruum process, carburization by exothermic thermochemical reaction, and reduction.

The physical properties of the primary carbides are listed in Table 1.

QUALITY CONTROL METHODS FOR PRIMARY CARBIDES

Analytical control of primary hard carbides for cemented carbides is normally done on a production-lot basis, with powder samples of uniformly blended lots being issued routinely to control laboratories, which, among major carbide producers commonly have widely diverse analytical capabilities, while more numerous smaller carbide producers rely on independent laboratories. The rapid growth of semi-automatic analytical equipment over the past decades has resulted in large cost savings as well as faster delivery of results.

Table 1. Physical Properties of Primary Carbides

Property	WC	TiC	TaC	NbC
mol wt	195.87	59.91	192.96	104.92
carbon, wt%	6.13	20.05	6.23	11.45
crystal structure	hex, Bh	fcc, B1	fcc, B1	fcc, B1
lattice constants, nm	$a = 0.29065$	0.43305	0.4454	0.4470
	$c = 0.28366$			
density, g/cm^3	15.7	4.93	14.48	7.78
mp, °C	2720	2940	3825	3613
microhardness, kg/mm^2	1200–2500	3000	1800	2000
modulus of elasticity, N/mm^{2a}	696,000	451,000	285,000	338,000
transverse rupture strength, N/mm^{2b}	550–600	240–400	350–450	300–400
coefficient of thermal expansion, K^{-1}	$a = 5.2 \times 10^{-6}$	7.74×10^{-6}	6.29×10^{-6}	6.65×10^{-6}
	$c = 7.3 \times 10^{-6}$			
thermal conductivity, W/(m · k)	121	21	22	14
heat of formation, $\Delta H_{f,\,298}$, kJ/mol^c	−40.2	−183.4	−146.5	−140.7
specific heat, J/(mol·)c	39.8	47.7	36.4	36.8
electrical resistivity, μΩ· cm	19	68	25	35
superconducting temperature, <K	1.28	1.15	9.7	11.1
Hall constant, cm^3/(A · s)	-21.8×10^{-4}	-15.0×10^{-4}	-1.1×10^{-4}	-1.3×10^{-4}
magnetic susceptibility	+10	+6.7	+9.3	+15.3

[a] Face-centered cubic = fcc and hexagonal = hex.
[b] To convert N/mm^2 (MPa) to psi, multiply by 145.
[c] To convert J to cal, divide by 4.184.

ECONOMIC ASPECTS

Three categories of metallurgical refining support manufacturers of cemented carbides. The first involves extraction from ore concentrates of tungsten minerals to produce tungstic oxide, WO_3, or ammonium paratungstate (APT), and the extraction from ore concentrates of Ti, Ta, and Nb minerals to produce refined oxides of titanium, tantalum, or niobium. The second step converts WO_3, APT, and oxides of titanium, tantalum, and niobium to the respective primary carbides or solid solutions of these metals. A third category of refining combines the first two steps to convert mineral concentrates of these metals directly into primary carbides. These refining categories also apply in general to the production of the auxiliary carbides.

Some refining, traditionally the conversion of the metal oxides to carbides, is carried out by the cemented carbide manufacturers themselves, chiefly by larger, more vertically integrated companies, while smaller cemented carbide manufacturers obtain their supplies either from independent refiners specializing in primary metal powders, carbides, nitrides, carbonitrides, and many lower volume accessory materials used in cemented carbide production or from the more integrated cemented carbide manufacturers. Consolidation within the cemented carbide industry is, however, an important ongoing evolution, one aspect of which is growth in cross-supply arrangements between larger companies of primary carbides and other precursor materials, in a mutual search for cost savings.

Recently, many of the advances in the preparation of primary carbides have favored metalcutting and metalworking components in the manufacturing industries rather than the resource industries, eg, the mining and processing of coal and hard minerals, exploration for and development of oil and gas fields, and agriculture. That unalloyed WC has been virtually the sole carbide used in these latter fields suggests that great tool opportunities may yet lie ahead in these fields. For metalcutting and metalworking fields, an ongoing development of production methods for carbides and carbonitrides in the form of chemical vapor- and physical vapor-deposited coatings on cemented carbide tool components stand out as a major leap forward for the cemented carbide tool industry, increasing productivity and profitability. A further important advance in carbide metallurgy has been seen in the development of submicron and ultrafine WC powders, which, in cemented carbide form, reach combinations of high strength and high hardness. Ongoing developments have motivated new independent enterprises specializing in this field, thus creating a potential for a new line of independent suppliers to the industry.

K. J. A. Brookes, *World Directory and Handbook of Hardmetals and Hard Materials*, 6th ed., International Carbide Data, East Barnet, Hertfordshire, U.K., 1996.

T. J. Kosolapova, *Carbides: Properties, Production and Applications*, Plenum Press, New York, 1971.

M. MacInnis and T. Kim, in T. C. Lo, M. H. I. Baird, and C. Hanson, eds., *Handbook for Solvent Extraction*, John Wiley & Sons, Inc., New York, 1983.

E. K. Storms, *The Refractory Carbides*, Academic Press, New York, 1967.

W. M. STOLL
Consulting Metallurgist

CARBOHYDRATES

Carbohydrates are found in all plant and animal cells. They are the most abundant of the organic compounds, so abundant that it is estimated that well over one-half of the organic carbon on earth exists in the form of carbohydrates. Most carbohydrates are produced and found in plants. Carbohydrate molecules make up about three-fourths of the dry weight of plants; most is found in cell walls as structural components. Carbohydrates also constitute important energy reserves in plants; one carbohydrate, starch, provides about three-fourths of the calories in the average human diet on a worldwide basis. But the nutritional aspects are only a part of the story of carbohydrates. They have many important commercial uses in such diverse areas as adhesive, agricultural chemicals, fermentation, food, pharmaceutical, textile and paper and related products, and in petroleum production. Because the basic carbohydrate molecule is functionalized at every carbon atom, and because carbohydrates seldom occur as simple sugars but rather combined with each other or other compounds, the variety of carbohydrates in nature is large, and the number of theoretical possibilities is almost limitless.

CLASSIFICATION

The basic carbohydrate molecule possesses an aldehyde or ketone group and a hydroxyl group on every carbon atom except the one of the carbonyl group. As a result, carbohydrates are defined as aldehyde or ketone derivatives of polyhydroxy alcohols and their reaction products. The formula for glucose and related sugars ($C_6H_{12}O_6$) contains hydrogen and oxygen atoms in the ratio in which they are found in water. The name carbohydrate (hydrate of carbon) is derived from the fact that the basic carbohydrate molecule has the formula $C_n(H_2O)_n$.

Numerical prefixes designate the number of carbon atoms are tri-, tetra-, penta-, hexa-, hepta-, etc. In systemic nomenclature, the suffix for the names of aldehyde sugars is -ose and for ketone sugars -ulose. The two classification systems can be joined in a single-word description. Common names are frequently used, creating exceptions to systematic nomenclature.

Monosaccharides are most often joined together in chains. Oligosaccharides are carbohydrate chains that yield 2–10 monosaccharide molecules upon hydrolysis.

Most carbohydrates exist in the form of polysaccharides. Polysaccharides give structure to the cell walls of land plants (cellulose), seaweeds, and some microorganisms and store energy (starch in plants, glycogen in animals). They are important in the human diet and in many commercial applications.

CHEMISTRY OF SACCHARIDES

Most carbohydrates have two kinds of reactive groups: the carbonyl group and primary and secondary hydroxyl groups.

Reactions of the Carbonyl Group

Ring Forms. Aldehydes and ketones react with compounds containing a hydroxyl group (alcohols) to form first hemiacetals and then acetals. Because aldose and ketose molecules have a carbonyl group and hydroxyl groups on the same carbon chain, they can form hemiacetal structures intramolecularly, as well as by reacting with another molecule. Such an intramolecular reaction forms a ring. The most common rings are the six-membered pyranose ring, a cyclic structure composed of five carbon atoms and one oxygen atom, and the five-membered furanose ring, a cyclic structure composed of four carbon atoms and one oxygen atom.

Glycosides, Oligosaccharides, and Polysaccharides. Few monosaccharides are found free in nature, and these few are usually present in only small amounts. Most monosaccharides occur in combinations, most often with either more of the same sugar or different sugars in the form of polymers (polysaccharides). Less frequently, except in the case of sucrose, they are joined together in oligosaccharide chains. Mono- and oligosaccharides may also be linked to nonsugar organic compounds. These combined forms of sugars are known as glycosides.

Oligo- and polysaccharides have reducing and nonreducing ends. A reducing sugar is a carbohydrate that contains an aldehyde or ketone group, either free or in a hemiacetal form, which in aqueous solution is always in equilibrium with the free form. The aldehyde group (and the ketone group, after isomerization to an aldehyde group under basic conditions) can be oxidized to a carboxyl group, ie, act as a reducing agent. The reducing end of an oligo- or polysaccharide is the one end not involved in a glycosidic linkage and can, therefore, react as an aldehyde or ketone. The sugar units constituting all other ends are attached through glycosidic (acetal) bonds and are, therefore, nonreducing ends. Reducing and nonreducing ends can be demonstrated with the structure of lactose (β-D-galactopyranosyl-α-D-glucopyranose), the reducing disaccharide of milk.

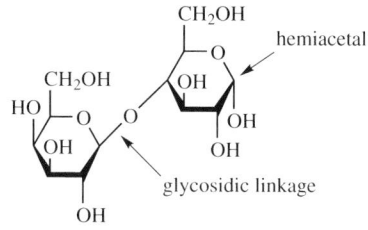

nonreducing end   reducing end

Additional sugar units added to either end of disaccharides form higher oligosaccharides.

Polysaccharides are naturally occurring polymers of monosaccharide (sugar) units. In precise chemical nomenclature, polysaccharides are glycans and are described as being composed of glycosyl units.

Synthetic Methods. Although mono- and oligosaccharides are most often made by depolymerization of polysaccharides, oligosaccharides and other compounds with glycosidic bonds can be made synthetically. The classic and still widely used reaction is the Koenigs-Knorr reaction; many modifications of it are known.

Glycoconjugates. Another class of carbohydrates is the glycoconjugates, which is composed of glycoproteins, proteoglycans, peptidoglycans, and glycolipids.

Sucrose and Derivatives of Sucrose. By far the most abundant of the naturally occurring oligosaccharides is the disaccharide sucrose, ordinary table sugar from sugarcane or sugar beets. The two monosaccharide units in sucrose are α-D-glucopyranosyl and β-D-fructofuranosyl units.

Oxidation to Sugar Acids and Lactones. When the aldehyde group of an aldose is oxidized, the resulting compound is an aldonic acid (salt form = aldonate). Some aldonic acids are products of carbohydrate metabolism.

Reduction. Mono- and oligosaccharides can be reduced to polyols (polyhydroxy alcohols) termed alditols. Common examples of compounds in this class are D-glucitol (sorbitol) made by reduction of D-glucose, and xylitol made by reduction of D-xylose.

Cyclitols. Cyclitols are polyhydroxycycloalkanes and -alkenes. They are widely distributed in nature, though never in large quantities. The most abundant of these carbocyclic compounds are the hexahydroxycyclohexanes, commonly called inositols, and their methyl ethers.

Reactions of Hydroxyl Groups

Reduction and Oxidation. Hydroxyl groups can be both oxidized to carbonyl groups and removed by reduction. Sugars that have the hydroxyl group missing from one or more of the carbon atoms are called deoxy sugars.

Esterification. The hydroxyl groups of sugars can react with organic and inorganic acids just as other alcohols do. Both natural and synthetic carbohydrate esters are important in various applications. Phosphate monoesters of sugars are important in metabolic reactions.

Etherification. Carbohydrates are involved in ether formation, both intramolecularly and intermolecularly.

Acetalation. As polyhydroxy compounds, carbohydrates react with aldehydes and ketones to form cyclic acetals.

Replacement of Hydroxyl Groups. Replacement of a hydroxyl group with an amino group at any position produces an aminodeoxysugar.

Isomerization. An aldose can be converted into another aldose and a ketose, and a ketose converted into two aldoses. It is for this reason that ketoses are reducing sugars.

Modifications of the Carbon Chain

Branched-chain sugars are found in nature. For example, cladinose (2,6-dideoxy-3-*C*-methyl-3-*O*-methyl-L-*ribo*-hexose is a component of erythromycin.

Unsaturated sugars are useful synthetic intermediates. The most commonly used are the so-called glycals (1,5- or 1,4-anhydroalditol-1-enes).

USES

Carbohydrates have widespread utilization, both as low cost, high volume commodities and as low volume specialty chemicals. Significant uses in terms of volume are surveyed here. Not covered are the lower volume uses involving carbohydrates either in the native state or in modified form; these are mainly pharmaceutical applications involving antibiotics, antigens, and synthetic drugs. In the case of drugs, monosaccharides are important as chiral synthons (chirons) as well as being used more directly to make products such as the nucleoside analogues AraA [9-(β-D-arabinofuranosyl)-9*H*-purin-6-amine], an antineoplastic and antiviral compound known by a number of trade names, and AZT (3′-azido-3′-deoxythymidine), an antiviral compound also known by a variety of trade names.

Neither are the considerable uses of carbohydrates as carbon sources for various fermentations or the uses of unrefined carbohydrates, flours, eg, described here.

Monosaccharides

In commerce, crystalline α-D-glucopyranose is generally known as dextrose. Glucose is also isomerized to D-fructose to produce high-fructose syrup (HFS). Crystalline D-fructose also finds use in the food industry. The HFS annual consumption, in the United States, is about 12,000,000 tons.

Oligosaccharides

Sucrose is widely used in the food industry to sweeten, control water activity, add body or bulk, provide crispness, give surface glaze or frost, form a glass, provide viscosity, and impart desirable texture. It is used in a wide variety of products from bread to medicinal syrups.

Oligo- and higher saccharides are produced extensively by acid- and/or enzyme-catalyzed hydrolysis of starch, generally in the form of syrups of mixtures. These products are classified by their dextrose equivalency (DE), which is inversely proportional to their molecular size and is a measure of their reducing power, with the DE value of anhydrous D-glucose defined as 100.

Another class of products are the cyclodextrins or cycloamyloses, a family of cyclic oligosaccharides containing α-D-glucopyranosyl units, most commonly seven (β-cyclodextrin, cycloheptaamylose, cyclomaltoheptaose). These stable complexes are useful in the food industry to provide stable flavors and fragrances in dry powder form, in the pharmaceutical industry, and in other applications where increased chemical and/or physical stability, solubility control, or controlled release is desired, eg, with agricultural chemicals.

Polysaccharides

Since polysaccharides are the most abundant of the carbohydrates, it is not surprising that they comprise the greatest part of industrial utilization. Most of the low molecular weight carbohydrates of commerce are produced by depolymerization of starch. Polysaccharide materials of commerce can be thought of as falling into three classes: cellulose, a water-insoluble material; starches, which are not water-soluble until cooked; and water-soluble gums.

Cellulose. Cellulose is the principal cell-wall component of higher plants and the most abundant polysaccharide. Approximately one-half the mass of perennial plants and one-third the mass of annual plants is cellulose. The greatest amount of cellulose used is the purified, but not highly purified, wood pulp that is used in the manufacture of paper, associated products, absorbents, rayons, and nonwovens. A number of derivatives of cellulose are also commercial entities.

Hemicelluloses and Related Polysaccharides. Hemicelluloses are a large group of polysaccharides that are associated with cellulose in the primary and secondary cell walls of all higher plants, but otherwise have no relationship to cellulose. They are also present in some other plants.

Hemicelluloses are heteroglycans. They do not comprise a distinct class of chemical structures. Constituent monosaccharides are D-xylose, D-mannose, D-glucose, D-galactose, L-galactose, L-arabinose, D-glucuronic acid, 4-O-methyl-D-glucuronic acid, D-galacturonic acid, and to a lesser extent L-rhamnose, L-fucose, and various methyl ethers of neutral sugars, with a limit of perhaps six different glycosyl units per polysaccharide molecule. Both woody and nonwoody tissues contain 20–35% hemicelluloses. Some are neutral polymers, but most are acidic.

Starches. Starch occurs in the form of granules, which must be cooked before they will release their water-soluble molecules. Starch use is widespread and permeates the entire economy because it (corn starch in particular) is abundantly available, inexpensive, and occurs in the form of granules that can be easily handled and reacted.

All green plants package and store carbohydrate (D-glucose) in the form of starch granules. Starch granules are quasi-crystalline, dense, insoluble in cold water, and only partially hydrated. The sizes and shapes of granules are specific for the plant of origin. Granules can be easily isolated from suspensions by filtration or centrifugation, resuspended, reacted, and recovered.

Normal corn starch is composed of ~28% of the linear polysaccharide amylose and ~72% of the branched polysaccharide amylopectin. Through genetic manipulation, corn cultivars with altered starch compositions have been developed. Various modified and derivatized starches are produced by treating a slurry of starch granules with chemicals or enzymes.

General Properties of Starches. Undamaged starch granules are not soluble in room temperature water. Heating a starch in water causes the granules to gelatinize.

The properties of starches are a reflection of the properties of their constituent amylose and amylopectin molecules.

In general, derivatization (etherification or esterification) increases solution and gel clarity, reduces the tendency to gel, improves water binding, increases freeze–thaw stability, reduces the gelatinization temperature, increases peak viscosity, and reduces the tendency to retrograde.

In general, all starches can be digested in the human small intestine, and the absorbed D-glucose is used for energy and a source of carbon. Cooked (pasted) starch is much, usually to a high degree, more digestible than is raw starch, and there are nondigestible (resistant) forms of starch.

Oxidized Starches. Alkaline hypochlorite treatment introduces carboxyl and carbonyl groups, effects some depolymerization, and produces whiter (bleached) products that produce softer, clearer gels. Much of the hypochlorite-oxidized starch and all the ammonium persulfate-oxidized starch is used in the paper industry. The low solution viscosity at high solids and good binding and adhesive properties of oxidized starches make them especially effective as textile and paper sizes.

Dextrins. Dextrins, like oxidized starches, are so-called converted starches. High-solids solutions of some of the more highly converted dextrins produce the tacky, quick-setting adhesives used in paper products.

Acid-Modified Starches. Acid-modified starches are prepared by treating a suspension of starch granules with dilute mineral acid. Acid-modified starches, also called thin-boiling and acid-thinned starches, are used in large quantities as textile warp sizes.

Starch Ethers. A large number of starch ethers has been prepared and patented; only a few are manufactured and used commercially. Commercially available starch ethers are the hydroxyalkyl ethers, hydroxyethylstarch and hydroxypropylstarch, and cationic starches.

Hydroxyethylstarch is widely used with synthetic latexes in the surface sizing of paper and as a coating binder. For these uses, the hydroxyethylstarch is acid-thinned, oxidized, or dextrinized. Hydroxypropylstarch is used in foods to provide viscosity stability and to ensure water-holding during low-temperature storage.

Starch Esters. As with the starch ethers, a large number of starch esters have been prepared and patented, but only a few are manufactured and used commercially. Starch acetates are used in foods to provide paste clarity and viscosity stability at low temperatures. Starch phosphate monoesters are used primarily in foods as pudding starches and in oil-in-water emulsions.

Cross-linked Starches. The polymer chains in starch granules can be cross-linked with difunctional reagents that form diethers or diesters. The properties imparted to the starch by such cross-linking are unique and, therefore, these derivatives are considered separately.

Food starches, especially those made from waxy maize, potato, and tapioca starch, are usually both cross-linked and phosphorylated, acetylated, or hydroxypropylated to provide appropriate cooking, viscosity, and textural properties. Examples of their application are their use in canned foods that are to be retort-sterilized and in the preparation of spoonable salad dressings, where products stable to high shear at low pH are required.

Cationic Starches. Commercial cationic starches are starch ethers that contain a tertiary amino or quaternary ammonium group. Cationic starches are used in papermaking.

Pregelatinized Starches. Suspensions of starches and starch derivatives can be pasted/cooked and dried to yield a variety of products that can be dispersed in cold water to yield pastes comparable to those obtained by cooking granular starch products. These products are made for convenience of use.

Starch Graft Copolymers. Graft copolymers can be made by forming radicals on a chain of a starch or a modified starch, particularly hydroxyethylstarch, most commonly with cerium(III) ions, then introducing a monomer. Commercial products that have been made in this way are starch–graft–styrene–butadiene latex copolymer and starch–graft–polyacrylonitrile copolymer, which was subsequently treated with alkali to convert the nitrile groups to a mixture of carbamoyl and carboxylate groups.

Cold-Water Swelling Starches. Special physical treatment produces starch granules that will swell in water without heating. Molecular dispersions can be formed by application of shear to the swollen granules.

Water-Soluble Gums/Hydrocolloids. Gums are polymeric substances that, in an appropriate solvent or swelling agent, form highly viscous dispersions or gels at low dry-substance content. Commonly, the term "industrial gums" refers to water-soluble polysaccharides (glycans in official carbohydrate nomenclature) or polysaccharide derivatives used industrially. They are classified both by structure, and by source (Table 1). Particularly in the food industry, the term "hydrocolloid" is often used interchangeably with "gum".

The usefulness of such industrial gums is based on their physical properties, in particular their capacity to thicken and/or gel aqueous solutions and otherwise to control water. Because all gums modify or control the flow of aqueous solutions, dispersions, and suspensions, the choice of which gum to use for a particular application often depends on its secondary characteristics. These characteristics are responsible for their utilization as adhesives, binders, bodying agents, bulking agents, crystallization inhibitors, clarifying agents, cloud agents, emulsifying agents, emulsification stabilizers, encapsulating agents, film formers, flocculating agents, foam stabilizers, gelling materials, mold release agents, protective colloids, suspending agents, suspension stabilizers, swelling agents, syneresis inhibitors, texturing agents, and whipping agents, in coatings, and for water absorption and binding.

Gums are tasteless, odorless, colorless, and nontoxic. None, except the starches and starch derivatives, are broken down by human digestive enzymes, but all are subject to microbiological attack. All can be depolymerized by acid- and enzyme-catalyzed hydrolysis of the glycosidic (acetal) linkages joining the monomeric (saccharide) units.

All native and modified polysaccharides have a range of molecular weights. The average composition and distribution of molecular weights in a gum sample can vary with the source, the conditions used for isolation or preparation, and any subsequent treatment(s). In all

Table 1. Classification of Commercial Polysaccharides by Source

Class	Examples
algae (seaweeds)	agars, algins, carrageenans, furcellarans, laminarans
higher plants	
insoluble	cellulose
extract	pectins
seeds	corn starches, rice starches, wheat starches, guar gum, locust bean gum, psyllium seed gum
tubers and roots	potato starch, tapioca/cassava starch, konjac glucomannan
exudates	gum arabics, gum karaya, gum tragacanth
microorganisms (fermentation gums)	curdlan, dextrans, gellan, pullulan, scleroglucan, welan, xanthans
animal	chitins/chitosans (also a cell-wall constituent of some fungi)
derived	
from cellulose	carboxymethylcelluloses, cellulose acetates, cellulose acetate butyrates, cellulose nitrates, ethylcellulose, hydroxyalkylcelluloses, hydroxyalkylalkylcelluloses, methylcelluloses
from starches[a]	starch acetates, starch adipates, starch 1-octenylsuccinates, starch phosphates, starch succinates, carboxymethylstarches, hydroxyethylstarches, hydroxypropylstarches, cationic starches, oxidized starches, dextrins
from guar gum	carboxymethylguar gum, carboxymethyl(hydroxypropyl)-guar gum, hydroxyethylguar gum, hydroxypropylguar gum, cationic guar gum
synthetic	polydextrose

[a]It is common for a commercial modified starch to have undergone two or more different reactions or treatments.

except bacterial polysaccharides, the percentage of individual monomeric unit types varies from molecule to molecule and from sample to sample. Because both molecular size and structure determine physical properties, various functional types of a given gum are produced by controlling the source and isolation procedure (in the case of natural gums) or derivatization method (in the case of derived gums) and subsequent treatment(s).

In general, gums do not form true solutions. Rather, because of their molecular weights and intermolecular interactions, they form dispersions, where the particles may be dispersed molecules and/or aggregated clusters of molecules. The rheology or flow characteristics and gel properties of gum solutions is a function of particle solvation, particle size, particle shape, particle flexibility and ease of deformation, and the presence and magnitude of charges. In general, the rheology of gum solutions is pseudoplastic or thixotropic, ie, they exhibit shear thinning. Most gums are available in a range of viscosity grades.

Polysaccharide gels in general are composed of 99.0–99.5% water and 0.5–1.0% gum. Important characteristics of gels are means of gelation (chemical gelation, thermogelation), reversibility, texture (brittle, elastic, plastic), rigidity (rigid or firm, soft or mushy), tendency for syneresis, and cutable or spreadable. Gels are composed of interconnected fringed micelles (junction zones).

Algins. Algins are salts (generally sodium, ammonium, or potassium or esters (propylene glycol) of alginic acid.

Advances in Carbohydrate Chemistry and Biochemistry, Academic Press, Inc., San Diego, Calif.

J. N. BeMiller and Whistler, *Carbohydrate Chemistry for Food Scientists*, 2nd ed., American Association of Cereal Chemists, St. Paul, Minn., 2003.

B. Ernst, G. W. Hart, and P. Sinaÿ, eds., *Carbohydrates in Chemistry and Biology*, Vol. 1, Wiley-VCH, Weinheim, Germany, 2000.

J. Lehmann, *Carbohydrates. Structure and Biology*, Thieme, Stuttgart, Germany, 1998.

James N. BeMiller
Purdue University

CARBON

Elemental carbon, atomic number six in the periodic table, at wt 12.011, occurs naturally throughout the world in either its crystalline, more ordered, or amorphous, less ordered, form. Carbonaceous materials such as soot or charcoal are examples of the amorphous form, whereas graphite and diamond are crystalline. Carbon atoms bond with other carbon atoms as well as with other elements, principally hydrogen, nitrogen, oxygen, and sulfur, to form carbon compounds, which are the subject of organic chemistry. In its many varying manufactured forms, carbon and graphite can exhibit a wide range of electrical, thermal, and chemical properties that are controlled by the selection of raw materials and thermal processing during manufacture.

CRYSTALLOGRAPHIC STRUCTURE

There are two allotropes of carbon: diamond and graphite. The diamond, or isotropic form, has a crystal structure that is face-centered cubic with interatomic distances of 0.154 nm. Each atom is covalently bonded to four other carbon atoms in the form of a tetrahedron. The nature of the bonding explains the differences in properties of the two allotropic forms. The hardness of diamond is derived from the regular three-dimensional network of σ-bonds. Graphite, or the anisotropic form, has a structure that is composed of infinite layers of carbon atoms arranged in the form of hexagons lying in planes.

Terminology

The electronic ground state of carbon is $1s^2$, $2s^2$, $2p^2$. In diamond, the $2s$ and $2p$ electrons mix to form four equivalent covalent σ-bonds. In graphite, three of the four electrons form strong covalent π-bonds with the adjacent in-plane carbon atoms. The fourth electron forms a less strong bond between the planes.

A wide variety and range of bulk carbon forms are available within the industry. In general, commercial forms are loosely characterized as carbon or graphite, but they are distinctly different. The term *manufactured carbon* refers to a bonded granular carbon body whose matrix has been subjected to a temperature typically between 900 and 2400°C. *Manufactured graphite* refers to a bonded granular carbon body whose matrix has been subjected to a temperature typically in excess of 2400°C.

OTHER FORMS OF CARBON AND GRAPHITE

The versatility and uniqueness of carbon and graphite attest to its widespread use for a variety of industrial applications.

Because of graphite's stability at high temperatures, flexible foil graphite is useful in applications requiring thermal stability in corrosive environments, eg, gaskets and valve packings, and is often used as a replacement for asbestos gaskets.

Carbon–graphite foam is a unique material whose low thermal conductivity, mechanical stability over a wide range of temperatures from room temperature to 3000°C, and light weight make it a prime candidate for thermal protection of new, emerging carbon–carbon aerospace reentry vehicles.

Pyrolytic graphite is used for missile components, rocket nozzles, and aircraft brakes for advanced high performance aircraft. Coated on surfaces or infiltrated into porous materials, it is also used in other applications, such as nuclear fuel particles, prosthetic devices, and high temperature thermal insulators.

Glassy, or vitreous, carbon is a black, shiny, dense, brittle material with a vitreous or glasslike appearance whose chemical inertness and low permeability have made it a useful material for chemical laboratory crucibles and other vessels.

Carbon and graphite paper is produced from carbon fibers by conventional papermaking methods. This form of carbon and graphite has outstanding electrical conductivity, corrosion resistance, and moderately high strength.

H. O. Pierson, *Handbook of Carbon, Graphite, Diamond and Fullerenes: Processes and Applications*, Noyes Publications, New York, 1994.

A. Swertka and E. Swertka, *A Guide to the Elements*, Oxford University Press, New York, 1998.

P. A. Thrower, ed., *Chemistry and Physics of Carbon; A Series of Advances*, Marcel Dekker, New York, 1999.

S. Yoshimura and R. P. Chang, *Supercarbons: Synthesis, Properties, and Applications*, Springer Verlag, New York, 2000.

J. C. Long
J. M. Criscione
UCAR Carbon Company Inc.

CARBON, ACTIVATED

Activated carbon is a predominantly amorphous solid that has an extraordinarily large internal surface area and pore volume. These unique characteristics are responsible for its adsorptive properties, which are exploited in many different liquid- and gas-phase applications. Through choice of precursor, method of activation, and control of processing conditions, the adsorptive properties of products are tailored for applications as diverse as the purification of potable water and the control of gasoline emissions from motor vehicles.

PHYSICAL AND CHEMICAL PROPERTIES

The structure of activated carbon is best described as a twisted network of defective carbon layer planes, cross-linked by aliphatic bridging groups. X-ray diffraction patterns of activated carbon reveal that it is nongraphitic, remaining amorphous because the randomly cross-linked network inhibits reordering of the structure even when heated to 3000°C. This property of activated carbon contributes to its most unique feature, namely, the highly developed and accessible internal pore structure. The surface area, dimensions, and distribution of the pores depend on the precursor and on the conditions of carbonization and activation.

Functional groups are formed during activation by interaction of free radicals on the carbon surface with atoms such as oxygen and nitrogen, both from within the precursor and from the atmosphere. The functional groups render the surface of activated carbon chemically reactive and influence its adsorptive properties. Activated carbon is generally considered to exhibit a low affinity for water, which is an important property with respect to the adsorption of gases in the presence of moisture. However, the functional groups on the carbon surface can interact with water, rendering the carbon surface more hydrophilic. Surface oxidation, which is an inherent feature of activated carbon production, results in hydroxyl, carbonyl, and carboxylic groups that impart an amphoteric character to the carbon, so that it can be either acidic or basic. The electrokinetic properties of an activated carbon product are, therefore, important with respect to its use as a catalyst support. As well as influencing the adsorption of many molecules, surface oxide groups contribute to the reactivity of activated carbons toward certain solvents in solvent recovery applications.

In addition to surface area, pore size distribution, and surface chemistry, other important properties of commercial activated carbon products include pore volume, particle size distribution, apparent or bulk density, particle density, abrasion resistance, hardness, and ash content.

MANUFACTURE AND PROCESSING

Commercial activated carbon products are produced from organic materials that are rich in carbon, particularly coal, lignite, wood, nut shells, peat, pitches, and cokes. The choice of precursor is largely dependent on its availability, cost, and purity, but the manufacturing process

and intended application of the product are also important considerations. Manufacturing processes fall into two categories, thermal activation and chemical activation. The effective porosity of activated carbon produced by thermal activation is the result of gasification of the carbon at relatively high temperatures, but the porosity of chemically activated products is generally created by chemical dehydration reactions occurring at significantly lower temperatures.

Forms of Activated Carbon Products

To meet the engineering requirements of specific applications, activated carbons are produced and classified as granular, powdered, or shaped products.

SHIPMENT AND STORAGE

Activated carbon products are shipped in bags, drums, and boxes in weights ranging from about 10 to 35 kg. Containers can be lined or covered with plastic and should be stored in a protected area both to prevent weather damage and to minimize contact with organic vapors that could reduce the adsorption performance of the product. Bulk quantities of activated carbon products are shipped in metal bins and bulk bags, typically $1-2$ m^3 in volume, and in railcars and tank trucks. Bulk carbon shipments are generally transferred by pneumatic conveyors and stored in tanks. However, in applications such as water treatment where water adsorption does not impact product performance, bulk carbon may be transferred and stored as a slurry in water.

ECONOMIC ASPECTS

U.S. producers of activated carbon are listed in Table 1.

Growth was expected at the rate of 4.5% through 2004. Demand is driven by environmental considerations. In many applications, adtivated carbon has best available technology status. However, some alternative systems may be more cost effective.

SPECIFICATIONS

Activated carbon producers furnish product bulletins that list specifications, usually expressed as a maximum

Table 1. U.S. Producers of Activated Carbon and Their Capacities

Producer	Capacity, $\times 10^6$ kg ($\times 10^6$ lb)	
Barnebey & Sutcliffe, Columbus, Ohio	13.6	(30)
Calgon Carbon, Catlettsburg, Ky.	63.5	(140)
Calgon Carbon, Pearlington, Miss.	18.1	(40)
Norit Americas, Pryor, Okla.	15.8	(35)
Norit Americas, Marshall, Texas	45.3	(100)
Royal Oak, Romeo, Fla.	9.1	(20)
Westvaco, Covington, Va.	22.7	(50)
Westvaco, Wickliffe, Ky.	22.7	(50)
Total	*210.8*	*(465)*

or minimum value, and typical properties for each grade produced. Standards helpful in setting purchasing specifications for granular and powdered activated carbon products have been published.

HEALTH AND SAFETY FACTORS

Activated carbon generally presents no particular health hazard as defined by NIOSH. However, it is a nuisance and mild irritant with respect to inhalation, skin contact, eye exposure, and ingestion. On the other hand, special consideration must be given to the handling of spent carbon that may contain a concentration of toxic compounds.

Activated carbon products used for decolorizing food products in liquid form must meet the requirements of the *Food Chemical Codex* as prepared by the Food & Nutrition Board of the National Research Council.

According to the National Board of Fire Underwriters, activated carbons normally used for water treatment pose no dust explosion hazard and are not subject to spontaneous combustion when confined to bags, drums, or storage bins. However, activated carbon burns when sufficient heat is applied; the ignition point varies between about 300 and 600°C.

Dust-tight electrical systems should be used in areas where activated carbon is present, particularly powdered products. When partially wet activated carbon comes into contact with unprotected metal, galvanic currents can be set up; these result in metal corrosion.

Manufacturer material safety data sheets (MSDS) indicate that the oxygen concentration in bulk storage bins or other enclosed vessels can be reduced by wet activated carbon to a level that will not support life. Therefore, self-contained air packs should be used by personnel entering enclosed vessels where activated carbon is present.

ENVIRONMENTAL CONCERNS

Activated carbon is a recyclable material that can be regenerated. In general, regeneration of spent carbon is considerably less expensive than the purchase of new activated carbon.

LIQUID-PHASE APPLICATIONS

Activated carbons for use in liquid-phase applications differ from gas-phase carbons primarily in pore size distribution. Liquid-phase carbons have significantly more pore volume in the macropore range, which permits liquids to diffuse more rapidly into the mesopores and micropores. The larger pores also promote greater adsorption of large molecules, either impurities or products, in many liquid-phase applications. Specific-grade choice is based on the isotherm and, in some cases, bench or pilot scale evaluations of candidate carbons. Liquid-phase activated carbon can be applied either as a powder, granular, or shaped form.

Batch-stirred vessels are most often used in treating material with powdered activated carbon. The type of carbon, contact time, and amount of carbon vary with the desired degree of purification. The efficiency of activated

carbon may be improved by applying continuous, countercurrent carbon–liquid flow with multiple stages.

Granular and shaped carbons are used generally in continuous systems where the liquid to be treated is passed through a fixed bed.

Liquid-phase applications of activated carbon include potable water treatment, industrial and municipal wastewater treatment, sweetener decolorization, groundwater remediation, food, beverages, and cooking oil, pharmaceuticals, mining, and miscellaneous uses.

GAS-PHASE APPLICATIONS

Gas-phase applications of activated carbon include separation, gas storage, and catalysis. Although only 20% of activated carbon production is used for gas-phase applications, these products are generally more expensive than liquid-phase carbons and account for about 40% of the total dollar value of shipments. Most of the activated carbon used in gas-phase applications is granular or shaped. Gas-phase applications account for 18% of total activated carbon. They include air purification, 42%; automotive emission control, 21%; solvent vapor recovery, 14%; cigarette filters medium, 2%; miscellaneous, 21% (33). Separation processes comprise the main gas-phase applications of activated carbon. These usually exploit the differences in the adsorptive behavior of gases and vapors on activated carbon on the basis of molecular weight and size. For example, organic molecules with a molecular weight greater than about 40 are readily removed from air by activated carbon.

P. N. Cheremisinoff and F. Ellerbusch, *Carbon Adsorption Handbook*, Ann Arbor Science Publishers, Inc., Ann Arbor, Mich., 1978, 539–626. An excellent reference book on activated carbon, ranging from theoretical to applied aspects.

W. Gerhartz, Y. S. Yamamoto, and F. Thomas Campbell, eds., *Ullmann's Encyclopedia of Industrial Chemistry*, 5th ed., Vol. A5, VCH Publishers, New York, 1986, pp. 124–140. Good descriptions of activation processes.

J. W. Hassler, *Activated Carbon*, Chemical Publishing Co., Inc., New York, 1963, pp. 1–14. A comprehensive account of the development and use of activated carbon products to about 1960.

B. McEnaney and T. J. Mays, in H. Marsh, ed., *Introduction to Carbon Science*, Butterworths, London, 1989, pp. 153–196. A good introduction to carbon science in general.

FREDERICK S. BAKER
CHARLES E. MILLER
ALBERT J. REPIK
E. DONALD TOLLES
Westvaco Corporation
    Charleston Research Center

CARBON BLACK

Carbon black is a generic term for an important family of products used principally for the reinforcement of rubber, as a black pigment, and for its electrically conductive properties. It is a fluffy powder of extreme fineness and high surface area, composed essentially of elemental carbon. Plants for the manufacture of carbon black are strategically located worldwide in order to supply the rubber tire industry, which consumes 70% of carbon black production. About 20% is used for other rubber products and 10% is used for special nonrubber applications. World capacity in 2001 was estimated at >8 million metric tons.

A number of processes have been used to produce carbon black, including the oil-furnace, impingement (channel), lampblack, the thermal decomposition of natural gas, and decomposition of acetylene. Since over 95% of the total output of carbon black is produced by the oil-furnace process, this article emphasizes this process.

PHYSICAL STRUCTURE OF CARBON BLACK

Molecular and Crystallite Structure

The chemical composition of carbon blacks is shown in Table 1. The arrangement of carbon atoms in carbon black has been well established by X-ray diffraction methods. The diffraction patterns show diffuse rings at the same positions as diffraction rings from pure graphite. The suggested relation to graphite is further emphasized as carbon black is heated to 3000°C. The diffuse reflections sharpen, but the pattern never approaches that of true graphite. Carbon black has a degenerated graphitic crystalline structure. Whereas graphite has three-dimensional order, carbon black has two-dimensional order. The X-ray data indicate that carbon black consists of well-developed graphite platelets stacked roughly parallel to one another but random in orientation with respect to adjacent layers.

Morphology

Morphologically, carbon blacks differ in primary "particle" or nodule size, surface area, aggregate size, aggregate shape, and in the distribution of each of these.

Primary "Particle" (Nodule) Size. The smallest discrete entity of carbon black is the aggregate, the "particle" size.

Aggregate Morphology (Structure). The term "structure" is widely used in the carbon black and rubber industries to describe the aggregate morphology.

Structure comparisons of grades with different surface areas cannot be made. It is now known that the properties associated with structure are associated principally with the bulkiness of individual aggregates. Aggregates of the same mass, surface area, and number of nodules have high structure in the open bulky and filamentous arrangement and a low structure in a more clustered compact arrangement. Therefore, the structure is now used to describe the relative void volume characteristics of grades of black of the same surface area. Structure is determined by aggregate size and shape, and their distribution. In composite systems, structure is a principal feature that determines the performance of carbon black as a reinforcing agent and as a pigment. In liquid media, structure affects rheological properties such a viscosity and yield point. In rubber, viscosity, extrusion die swell, modulus, abrasion resistance, dynamic properties, and electrical conductivity are affected by structure.

CLASSIFICATION

Carbon blacks have been classified by their production process, by their production feedstocks such as acetylene blacks, by their application field, such as rubber blacks, color blacks, electric conductive blacks, and by properties of end use products such as high abrasion furnace black (HAF) and fast extrusion furnace black (FEF). From their applications, carbon black is classified into two groups: one used for rubber products, and another for nonrubber applications, referred as special blacks. Generally, special carbon blacks cover a wider range of morphology and surface chemistry than rubber blacks.

CARBON BLACK FORMATION

Mechanisms of carbon black formation must account for the experimental observations of the unique morphology and microstructure of carbon black. These features include the presence of nodules, or particles, multiple growth centers within some nodules, the fusion of nodules into large aggregates, and the paracrystalline or concentric layer plane structure of the aggregates.

MANUFACTURE

The Oil-Furnace Process

The oil-furnace process accounts for >95% of all carbon black produced in the world. Over the 50 years since its invention, the oil-furnace process has undergone several cycles of improvements that have resulted in higher yields, larger process trains, better energy economy, and improved product performance.

Table 1. Chemical Composition of Carbon Blacks

Type	Carbon, %	Hydrogen, %	Oxygen, %	Sulfur, %	Nitrogen, %	Ash, %	Volatile, %
furnace rubber-grade	97.3–99.3	0.20–0.80	0.20–1.50	0.20–1.20	0.05–0.30	0.10–1.00	0.60–1.50
medium thermal	99.4	0.30–0.50	0.00–0.12	0.00–0.25	NA[a]	0.20–0.38	
acetylene black	99.8	0.05–0.10	0.10–0.15	0.02–0.05	NA[a]	0.00	<0.40

[a] Not available = NA.

The principal pieces of equipment are the air blower, process air and oil preheaters, reactors, quench tower, bag filter, pelletizer, and rotary dryer. The basic process consists of atomizing the preheated oil in a combustion gas stream formed by burning fuel in air preheated to 1400 to >1800°C. Leaving the formation zone, the carbon-black-containing gases are quenched by spraying water into the stream. The partially cooled smoke is then passed through a heat exchanger, where incoming air is preheated. Additional quench water is used to cool the smoke to a temperature consistent with the life of the bag material used in the bag filter. The bag filter separates the unagglomerated carbon black from the by-product tail gas, which contains nitrogen, hydrogen, carbon monoxide, carbon dioxide, and water vapor. It is mainly nitrogen and water vapor.

The fluffy black from the bag filter is mixed with water, typically in a pin mixer, to form wet granules. These are dried in a rotary dryer, and the dried product is conveyed to bulk storage tanks. For special purposes, dry pelletization in rotary drums is also practiced. Most carbon black is shipped by rail or in bulk trucks. Various semibulk containers are also used including IBCs and large semibulk bags. Some special purpose blacks are packed in paper or plastic bags.

Feedstocks. Feedstocks for the oil-furnace process are heavy fuel oils. Preferred oils have high aromaticity, are free of suspended solids, and have a minimum of asphaltenes. Suitable oils are catalytic cracker residue (once residual catalyst has been removed), ethylene cracker residues, and distilled heavy coal tar fractions. Other specifications of importance are freedom from solid materials, moderate to low sulfur, and low alkali metals. The ability to handle such oils in tanks, pumps, transfer lines, and spray nozzles is also a primary requirement.

HEALTH AND SAFETY FACTORS

There are two major areas of carcinogenic concern with carbon black. Of these, the one that has historically attracted the most attention is the few tens to hundreds of pasts per million (ppms) of polynuclear aromatic hydrocarbons (PAH) that are adsorbed on the surface of most blacks. An extract made by exhaustive extraction of these materials with aromatic solvent has been shown to cause skin tumors in rodents. The PAH on carbon black is very tightly adsorbed and is not liberated by biological fluids. Hence, this material is believed to have little or no bioavailability.

Carbon black inhalation is currently regulated by Occupational Safety and Health Administration (OSHA) in the United States by the Health and Safety Executive (HSE) in the United Kingdom, and by the MAK Commission in Germany.

Carbon blacks will burn in air if ignited and once ignited, are difficult to extinguish. In bulk storage, local hot spots can exist for very long periods. Great care needs to be exercised where a smoldering fire is suspected as there can be accumulations of carbon monoxide in enclosed spaces.

Carbon black dust clouds in air are not considered flammable. Carbon blacks have a high ignition energy requirement, and entrained dust clouds do not propagate flame, nor exhibit substantial overpressures. The reason is that they presumably have essentially no combustible volatile matter. Carbon black in air can be incinerated but only with difficulty, requiring long burn-out times.

ENVIRONMENTAL CONCERNS

The carbon black industry takes extreme efforts to confine the product during all stages of manufacturing and transport. Highly efficient bag filters are used to collect the product. After collection the fluffy carbon black is densified and pelletized to minimize dusting during shipment and use by customers. The process gas leaving the bag filter contains primarily water, nitrogen, carbon monoxide, carbon dioxide, and hydrogen. A portion of these gases is burned for internal plant fuel, and the residual gas is generally burned in a flare or incinerator. Where local conditions warrant, the remaining gas may be used to generate power or steam, either for the plant itself or for merchant sale.

Like all other operators of combustion equipment, carbon black plants are subject to the usual pressures for reduced sulfur and nitrogen oxide emissions. It appears that the use of lower sulfur feedstock is the most economic way of reducing sulfur emissions. Redesign of combustion equipment for nitrogen oxide reduction is showing some promise. The primary NO_x issues arise from the combustion of tail gas, since the carbon black production process is exceedingly fuel-rich.

USES

About 89% of total U.S. consumption of carbon black in 2000 was in the rubber industry and 70% for tires. About 10% is consumed for other automotive products and 9% for rubber products unrelated to the automotive industry. The automotive industry accounts for 80% of consumption, and 11% of the blacks is for nonrubber uses. Its main applications are related to pigmentation, ultraviolet (uv) absorption, and electrical conductivity of other products such as plastics, coating, and inks. These carbon blacks are also termed special blacks.

Carbon black is a major component in the manufacture of rubber products, with a consumption second only to rubber itself. It is by far the most active rubber reinforcing agent, owing to its unique ability to enhance the physical properties of rubbers. Table 2 lists the principal rubber grades by their N-number classification, general rubber properties, and typical uses.

The consumption of the various carbon black grades can be divided into tread grades for tire reinforcement and nontread grades for nontread tire use and other rubber applications. Table 3 shows the distribution of

Table 2. Application of Principal Rubber-Grade Carbon Blacks

Designation	General rubber properties	Typical uses
N110, N121	very high abrasion resistance	special tire treads, airplane, off-the-road, racing
N220, N299, N234	very high abrasion resistance, good processing	passenger, off-the-road, special service tire treads
N339, N347, N375, N330	high abrasion resistance, easy processing, good abrasion resistance	standard tire treads, rail pads, solid wheels, mats, tire belt, sidewall, carcass, retread compounds
N326	low modulus, good tear strength, good fatigue, good flex cracking resistance	tire belt, carcass, sidewall compounds, bushings, weather strips, hoses
N550	high modulus, high hardness, low die swell, smooth extrusion	tier innerliners, carcass, sidewall, innertubes, hose, extruded goods, v-belts
N650	high modulus, high hardness, low die swell, smooth extrusion	tire innerliners, carcass, belt, sidewall compounds, seals, friction compounds, sheeting
N660	medium modulus, good flex fatigue resistance, low heat buildup	carcass, sidewall, bead compounds, innerliners, seals, cable jackets, hose, soling, EPDM compounds
N762	high elongation and resilience, low compression set	mechanical goods, footwear, innertubes, innerliners, mats

production of types for these uses. A typical passenger tire has several compounds and uses five to seven different carbon black grades.

Besides reinforcement for rubber, the principal functions that carbon black imparts to a compound material are color, uv damage resistance, electrical conductivity, nondegradation of polymer physical properties, and ease of dispersion. The carbon blacks used for these

Table 3. Carbon Black Production by Grade in the United States for 2000 (K metric tons)

N330 high abrasion	0.623
N550 fast extruding	0.138
N762 semireinforcing	0.129
N660 general purpose	0.356
N110 super abrasion	0.061
N220 intermediate super abrasion	0.170
N990 thermal	0.014
Total	*1.493*

purposes are classified as special-grade blacks. Smaller volume applications exploit other principal attributes, such as chemical inertness, thermal stability, and an open porous structure. The secondary attributes include chemical and physical purity, low affinity for water adsorption, and ease of transportation and handling.

In 2000, 11% of the U.S. consumption of carbon black was special blacks. About 51% of special blacks are used in plastics, 32% in printing inks, 5% in paint, 3% in paper, and 9% in miscellaneous applications.

J. F. Aucher, T. Kälin, and Y. Sakuma, "CEH Marketing Research Report, Carbon Black," *Chemical Economic Handbook*, SRI International, January 2002.

J.-B. Donnet and A. Voet, *Carbon Black, Physics, Chemistry, and Elastomer Reinforcement*, Chapt. 4, Marcel Dekker, Inc., New York, 1976.

R. J. McCunney, H. J. Muranko, P. A. Valberg, in E. Bingham, B. Cohrssen, and C. H. Powell, eds., *Patty's Toxicology*, 5th ed., Vol. 8, John Wiley & Sons, Inc., New York, 2001.

N. Probst, in J.-B. Donnet, R. C. Bansal, and M.-J. Wang, eds., *Carbon Black, Science and Technology*, Chapt. 8, Marcel Dekker, Inc., New York, 1993.

MENG-JIAO WANG
CHARLES A. GRAY
STEVE A. REZNEK
KHALED MAHMUD
YAKOV KUTSOVSKY
Cabot Corporation

CARBON DIOXIDE

Carbon dioxide, CO_2, is a colorless gas with a faintly pungent odor and acid taste first recognized in the sixteenth century as a distinct gas through its presence as a by-product of both charcoal combustion and fermentation. Today carbon dioxide is a by-product of many commercial processes: synthetic ammonia production, hydrogen production, substitute natural gas production, fermentation, limestone calcination, certain chemical syntheses involving carbon monoxide, and reaction of sulfuric acid with dolomite. Generally present as one of a mixture of gases, carbon dioxide is separated, recovered, and prepared for commercial use as a solid (dry ice), liquid, or gas.

Carbon dioxide is also found in the products of combustion of all carbonaceous fuels, in naturally occurring gases, as a product of animal metabolism, and in small quantities, about 0.03 vol %, in the atmosphere. Its many applications include beverage carbonation, chemical manufacture, firefighting, food freezing, foundry-mold preparation, greenhouses, mining operations, oil well secondary recovery, rubber tumbling, therapeutical work, welding, and extraction processes. Although it is present in the atmosphere and the metabolic processes of animals and plants, carbon dioxide cannot be recovered economically from these sources.

Table 1. Properties of Carbon Dioxide

Property	Value
sublimation point at 101.3 kPa[a], °C	−78.5
triple point at 518 kPa[b] °C	−56.5
critical temperature, °C	31.1
critical pressure, kPa[b]	7383
critical density, g/L	467
latent heat of vaporization, J/g[c]	
at the triple point	353.4
at 0°C	231.3
gas density at 273 K and 101.3 kPa[a], g/L	1.976
liquid density	
at 273 K, g/L	928
at 298 K and 101.3 kPa[a] CO_2, vol/vol	0.712
viscosity at 298 K and 101.3 kPa[a], mPa·s(=cP)	0.015
heat of formation at 298 K, kJ/mol[d]	393.7

[a] 101.3 kPa = 1 atm.
[b] To convert kPa to psia, multiply by 0.145.
[c] To convert J/g to Btu/lb, multiply by 0.4302.
[d] To convert kJ/mol to Btu/mol, multiply by 0.9487.

PHYSICAL PROPERTIES

Some values of physical properties of CO_2 appear in Table 1.

CHEMICAL PROPERTIES

Carbon dioxide, the final oxidation product of carbon, is not very reactive at ordinary temperatures. However, in water solution it forms carbonic acid H_2CO_3, which forms salts and esters through the typical reactions of a weak acid.

Although carbon dioxide is very stable at ordinary temperatures, when it is heated above 1700°C the reaction forming CO proceeds to the right to an appreciable extent (15.8%) at 2500 K. This reaction also proceeds to the right to a limited extent in the presence of ultraviolet light and electrical discharges. Carbon dioxide may be reduced by several means. The most common of these is the reaction with hydrogen.

Carbon dioxide reacts with ammonia as the first stage of urea manufacture to form ammonium carbamate. The ammonium carbamate then loses a molecule of water to produce urea, $CO(NH_2)_2$. Commercially, this is probably the most important reaction of carbon dioxide, used worldwide in the production of urea for synthetic fertilizers and plastics.

Radioactive Carbon

In addition to the common stable carbon isotope of mass 12, traces of a radioactive carbon isotope of mass 14 with a half-life estimated at 5568 years are present in the atmosphere and in carbon compounds derived from atmospheric carbon dioxide.

Carbon dioxide containing known amounts of ^{14}C has been used as a tracer in studying botanical and biological problems involving carbon and carbon compounds and in organic chemistry to determine the course of various chemical reactions and rearrangements. It has also been used in testing gaseous diffusion theory with mixtures of CO_2 and [14]CO_2 at elevated pressures.

Environmental Chemistry

Carbon dioxide plays a vital role in the earth's environment. It is a constituent in the atmosphere and, as such, is a necessary ingredient in the life cycle of animals and plants.

The balance between animal and plant life cycles as affected by the solubility of carbon dioxide in the earth's water results in the carbon dioxide content in the atmosphere of about 0.03 vol %. However, carbon dioxide content of the atmosphere seems to be increasing as increased amounts of fossil fuels are burned. There is some evidence that the rate of release of carbon dioxide to the atmosphere may be greater than the earth's ability to assimilate it.

The effects of such an increase, if it occurs, are not known. It could result in a warmer temperature at the earth's surface by allowing the short heat waves from the sun to pass through the atmosphere while blocking larger waves that reflect back from the earth.

MANUFACTURE

Sources of carbon dioxide for commercial carbon dioxide recovery plants are (1) synthetic ammonia and hydrogen plants in which methane or other hydrocarbons are converted to carbon dioxide and hydrogen ($CH_4 + 2 H_2O \longrightarrow CO_2 + 4 H_2$); (2) flue gases resulting from the combustion of carbonaceous fuels; (3) fermentation in which a sugar such as dextrose is converted to ethyl alcohol and carbon dioxide ($C_6H_{12}O_6 \longrightarrow 2 C_2H_5OH + 2 CO_2$); (4) lime-kiln operation in which carbonates are thermally decomposed ($CaCO_3 \longrightarrow CaO + CO_2$); (5) sodium phosphate manufacture ($3 Na_2CO_3 + 2 H_3PO_4 \longrightarrow 2 Na_3PO_4 + 3 CO_2 + 3 H_2O$); and (6) natural carbon dioxide gas wells.

Methods of Purification

Although carbon dioxide produced and recovered by the methods outlined above has a high purity, it may contain traces of hydrogen sulfide and sulfur dioxide, which cause a slight odor or taste. The fermentation gas recovery processes include a purification stage, but carbon dioxide recovered by other methods must be further purified before it is acceptable for beverage, dry ice, or other uses. The most commonly used purification methods are treatments with potassium permanganate, potassium dichromate, or active carbon.

Methods of Liquefaction and Solidification

Carbon dioxide may be liquefied at any temperature between its triple point (216.6 K) and its critical point (304 K) by compressing it to the corresponding liquefaction pressure, and removing the heat of condensation.

Solidification. Liquid carbon dioxide from a cylinder may be converted to "snow" by allowing the liquid to expand to atmospheric pressure. This simple process is used only where very small amounts of solid carbon dioxide are required, because less than one-half of the liquid is recovered as solid. Solid carbon dioxide is produced in blocks by hydraulic presses.

ECONOMIC ASPECTS

In 1998, 35×10^6t of gaseous carbon dioxide was consumed. Liquid carbon dioxide consumption was 6.6×10^6t. Gaseous carbon dioxide is used for enhanced oil recovery and, thus, it exceeded liquid use.

In the United States, 27.9×10^6t of gaseous carbon dioxide was used in the oil fields and 5.8×10^6t was used in urea production. Western Europe consumed 3.5×10^6t. Growth in the U.S. and Western Europe was expected at the rate of 3–4% through 2003. In Japan, a 2% growth rate was expected.

Much more carbon dioxide is generated daily than is recovered. The decision whether or not to recover by-product carbon dioxide often depends on the distance and cost of transportation between the carbon dioxide producer and consumer. For example, it has become profitable to recover more and more carbon dioxide from CO_2-rich natural gas wells in Texas as the use of carbon dioxide in secondary oil recovery has increased.

HEALTH AND SAFETY FACTORS

Although carbon dioxide is a constituent of exhaled air, high concentrations are hazardous. Up to 0.5 vol% carbon dioxide in air is not considered harmful, but carbon dioxide concentrates in low spots because it is one and one-half times as heavy as air. Five vol% carbon dioxide in air causes a threefold increase in breathing rate, and prolonged exposure to concentrations higher than 5% may cause unconsciousness and death. Ventilation sufficient to prevent accumulation of dangerous percentages of carbon dioxide must be provided where carbon dioxide gas has been released or dry ice has been used for cooling.

USES

A large portion of the carbon dioxide recovered is used at or near the location where it is generated as an ingredient in a further processing step.

About 51% of the carbon dioxide consumed in the United States is used in the food industry. Approximately 18% of carbon dioxide output is used for beverage carbonation. Both soft drinks and beer production consume the largest quantity of CO_2, for carbonation. About 10% of the carbon dioxide produced is for chemical manufacturing. Refrigeration of foodstuffs, especially ice cream, meat products, and frozen foods, is the principal use for solid carbon dioxide. Liquid carbon dioxide provides the most readily available method of rapid refrigeration and is used for rapid chilling of loaded trucks and rail cars before shipment. In addition to chemical synthesis and enhanced oil recovery, gaseous carbon dioxide is used in the carbonated beverage industry. The addition of small amounts of carbon dioxide to the atmosphere in greenhouses greatly improves the growth rate of vegetables and flowers.

Chemical Economics Handbook, SRI International, Menlo Park, Calif., June 2000.

L. H. Chen, *Thermodynamic and Transport Properties of Gases, Liquids and Solids*, McGraw-Hill Book Co., Inc., New York, 1959, 358–369.

C. D. Leikauf in E. Bingham, B. Cohrssen, and C. H. Powell, Eds., *Patty's Toxicology,* 5th ed., Vol. 5, John Wiley & Sons, Inc., New York, 2001.

A. V. Slack and G. R. James, eds., *Ammonia Part II*, Vol. 2, Marcel Dekker, New York, 1974.

RONALD PIERANTOZZI
Air Products and Chemicals, Inc.

CARBON DISULFIDE

Carbon disulfide (carbon bisulfide, dithiocarbonic anhydride), CS_2, is a toxic, dense liquid of high volatility and flammability. It is an important industrial chemical and its properties are well established. Low concentrations of carbon disulfide naturally discharge into the atmosphere from certain soils, and carbon disulfide has been detected in mustard oil, volcanic gases, and crude petroleum. Carbon disulfide is an unintentional by-product of many combustion and high temperature industrial processes where sulfur compounds are present.

Carbon disulfide was first prepared nearly two hundred years ago, but commercial uses grew rapidly from about 1929 to 1970, when the principal applications included manufacturing viscose rayon fibers, cellophane, carbon tetrachloride, flotation aids, rubber vulcanization accelerators, fungicides, and pesticides. Production of carbon disulfide in the United States has declined in recent years. Other chemical fibers and films, as well as environmental and toxicity considerations related to carbon tetrachloride, have had significant impact on the demand for carbon disulfide.

PHYSICAL PROPERTIES

Pure carbon disulfide is a clear, colorless liquid with a delicate, etherlike odor. Carbon disulfide is slightly miscible with water, but it is a good solvent for many organic compounds. Thermodynamic constants, vapor pressure, spectral transmission, and other properties of carbon disulfide have been determined. Principal properties are listed in Table 1.

Carbon disulfide is completely miscible with many hydrocarbons, alcohols, and chlorinated hydrocarbons. Phosphorus and sulfur are very soluble in carbon disulfide.

CHEMICAL PROPERTIES

The low flash point temperature of $-30°C$ at atmospheric pressure and wide flammability range of carbon disulfide deserve special attention. The flash point is lowered if the pressure is decreased or the oxygen content enriched. The flammability limits or explosive ranges depend on conditions of temperature, pressure, and geometry of the enclosure. Flammability limits of 1.06–50.0 vol % carbon disulfide in air are reported for upward propagation and 1.91–35.0 vol % for downward propagation in a 75-mm diameter glass tube.

Carbon disulfide chemistry is thoroughly described in several publications.

Table 1. Properties of Carbon Disulfide

Property	Values		
General			
melting point, K	161.11		
latent heat of fusion, kJ/kg[a]	57.7		
boiling point at 101.3 kPa[b], °C	46.25		
flash point at 101.3 kPa[b], °C	−30		
ignition temperature in air, °C			
10-s lag time	120		
0.5-s lag time	156		
critical temperature, °C	273		
critical pressure, kPa[b]	7700		
critical density, kg/m^3	378		
solubility H_2O in CS^2			
at 10°C, ppm	86		
at 25°C, ppm	142		
dielectric constant	2.641		
Liquid at temperature, °C	0°C	20°C	46.25°C
density, kg/m^3	1293	1263	1224
specific heat, J/kg·K[a]	984	1005	1030
latent heat of vaporization, kJ/kg[a]	377	368	355
surface tension, mN/M (=dyn/cm)	35.3	32.3	28.5
thermal conductivity, W/m·KW/(m·K)		0.161	
viscosity, mPa·s(= cP)	0.429	0.367	0.305
refractive index, n^D	1.6436	1.6276	
solubility in water, g/kg soln	2.42	2.10	0.48
vapor pressure, kPa[b]	16.97	39.66	101.33
Gas at temperature, °C[c]	46.25	200	400°C
density, kg/m^3	2.97	1.96	1.37
specific heat, J/kg·KJ/(kg·K)[a,d]	611	679	730
viscosity, mPa·s(= cP)	0.0111	0.0164	0.0234
thermal conductivity, W/m·KW/(m·K)	0.0073		
Thermochemical data at 298 K[a]			
heat capacity, C^0p, J/mol·KJ/(mol·K)[a]	45.48		
entropy, S^0, J/mol·KJ/(mol·K)[a]	237.8		
heat of formation, $H^0{}_f$, kJ/mol[a]	117.1		
free energy of formation, $G^0{}_f$, kJ/mol[a]	66.9		

[a]To convert J to cal, divide by 4.184.
[b]To convert kPa to atm, divide by 101.3.
[c]At absolute pressure, 101.3 kPa.
[d]$C_p/C_v = 1.21$ at 100°C (2).

MANUFACTURE

The earliest method for manufacturing carbon disulfide involved synthesis from the elements by reaction of sulfur and carbon as hardwood charcoal in externally heated retorts. Safety concerns, short lives of the retorts, and low production capacities led to the development of an electric furnace process, also based on reaction of sulfur and charcoal. The commercial use of hydrocarbons as the source of carbon was developed in the 1950s, and it was still the predominant process worldwide through the end of the century. That route, using methane and sulfur as the feedstock, provides high capacity in an economical, continuous unit.

Processes

Sulfur vapor reacts with other hydrocarbon gases, such as acetylene or ethylene, to form carbon disulfide. Light gas oil was reported to be successful on a semiworks scale. In the reaction with hydrocarbons or carbon, pyrites can be the sulfur source. With methane and iron pyrite the reaction products are carbon disulfide, hydrogen sulfide, and iron or iron sulfide. Pyrite can be reduced with carbon monoxide to produce carbon disulfide.

HANDLING, SHIPMENT, AND STORAGE

Transportation of carbon disulfide is controlled by federal regulations. Acceptable shipping containers include drums, tank trucks, special portable tanks, and rail tank cars. Barges have been used in the past. The United States Department of Transportation classifies carbon disulfide as a flammable liquid and a poison. For ship transport, carbon disulfide must be marked as a marine pollutant. All air transport, cargo or passenger, is forbidden.

Carbon disulfide is normally stored and handled in mild steel equipment. Tanks and pipes are usually made from steel. Valves are typically cast-steel bodies with chrome steel trim. Lead is sometimes used, particularly for pressure relief disks. Copper and copper alloys are attacked by carbon disulfide and must be avoided. Carbon disulfide liquid and vapor become very corrosive to iron and steel at temperatures above about 250°C. High chromium stainless steels, glass, and ceramics may be suitable at elevated temperatures.

Contact of carbon disulfide with air should be avoided because the combination of high volatility, wide flammability range, and low ignition temperature results in a readily combustible mixture. Carbon disulfide must be stored in inert-blanketed, closed tanks.

Small carbon disulfide fires can be smothered with carbon dioxide. Large fires can be controlled with certain types of foams or by a fog or spray of water with attention to proper impoundment of the contaminated water runoff.

ECONOMIC ASPECTS

Depending on energy and raw material costs, the minimum economic carbon disulfide plant size is generally in the range of about 2000–5000 tons per year for an electric furnace process and 15,000–20,000 tons per year for a hydrocarbon-based process. A typical charcoal–sulfur facility produces approximately 5000 tons per year. Hydrocarbon–sulfur plants tend be on the scale of 50,000–200,000 tons per year.

Demand for carbon disulfide has apparently bottomed out at approx. 72.5×10^3 t over the last few years. Rayon is CS's largest market, but represents only 4% of the synthetic fiber market.

Carbon disulfide is used in the manufacture of rubber vulcanization accelerators. Production of these accelerators required 12.7×10^3 t in 2001. Demand in agricultural chemicals is increasing slightly because of the use of Metam sodium as a replacement for methyl bromide.

SPECIFICATIONS AND QUALITY CONTROL

Modern plants generally produce carbon disulfide of about 99.99% purity. High product quality is ensured by closely controlled continuous fractional distillation.

HEALTH AND SAFETY FACTORS

Care must be exercised in handling carbon disulfide because of both health concerns and the danger of fire or explosions. Occupational exposure potentially may involve as many as 20,000 workers in the United States. Ingestion is rare, but a 10 mL dose can prove fatal. Contact usually occurs by inhalation of vapor. However, vapor and liquid can be absorbed through intact skin, and poisoning may occur by the dermal route. Repeated contact of liquid carbon disulfide with the skin can cause inflammation and cracking, because carbon disulfide removes protective waxes and oils. Extended skin contact results in blistering and possibly second- and third-degree burns. Precautions should be taken to avoid breathing of vapors or mists that may contain carbon disulfide. Contact with skin or eyes should also be avoided, and adequate safety gear should be worn, including goggles, impervious gloves, and appropriate clothing.

The odor threshold of carbon disulfide is about 1 ppm in air but varies widely, depending on individual sensitivity and purity of the carbon disulfide. However, using the sense of smell to detect excessive concentrations of carbon disulfide is unreliable because of the frequent co-presence of hydrogen sulfide that dulls the olfactory sense.

Immediate effects of overexposure to carbon disulfide vapors range from headache, dizziness, nausea, and vomiting to life-threatening convulsions, unconsciousness, and respiratory paralysis. For an exposure time of 30 min, 1150 ppm carbon disulfide in air results in serious symptoms, 3210 ppm is dangerous to life, and 4815 ppm is fatal. Prolonged and repeated exposure to carbon disulfide vapor can affect both the central and peripheral nervous systems. Manifestations of long-term overexposure may include headache, vertigo, irritability, nervousness, depression, mental derangement, memory loss, muscular weakness, fatigue, insomnia, eating disorders, gastrointestinal disturbances, impaired vision, diminished reflexes, numbness, and difficulty walking. Certain workers exposed to a time-weighted average of 11 ppm experienced headaches and dizziness, and those with an average of 186 ppm had additional complaints of nervousness, fatigue, sleep problems, and weight loss. Repeated exposure to relatively high concentrations of carbon disulfide has long been known to cause serious neurological and psychological impairments. In recent years, previously unrecognized and more subtle toxic effects of repeated lower level exposures became evident. This led OSHA in 1989 to reduce permissible concentration limits to 4 ppm (12 mg/m^3) maximum time-weighted average for 8-h exposure and 12 ppm (36 mg/m^3) maximum for 15 min short-term exposure (130). OSHA states that the new limits should substantially reduce the risk of both cardiovascular disease and adverse reproductive effects associated with carbon disulfide. Analysis of urine specimens for carbon disulfide metabolites by an iodine–azide test and other methods can indicate overexposure.

Health hazards linked to carbon disulfide are extensively covered. Also available are epidemiological studies, general reviews containing many references, and a Material Safety Data Sheet.

USES

United States applications of carbon disulfide are rayon production (44%), agriculture and other chemicals (35%), rubber chemicals (18%), cellophane and other regenerated cellulosics (3%).

Carbon disulfide is used to make intermediates in the manufacture of rubber vulcanization accelerators including thiocarbanilide used in dyes. Thiophosgene (thiocarbonyl chloride), made from carbon disulfide, is a useful intermediate in the synthesis of many organic sulfur compounds. Xanthates (dithiocarbonates) produced from carbon disulfide are widely employed as flotation chemicals for metal sulfide ores in the mining industry. The solvent properties of carbon disulfide find a wide range of industrial uses, including various dehydration, extraction, reaction, and separation applications. A useful laboratory chemical, carbon disulfide is a reactant in the synthesis of many compounds and a solvent in Friedel-Crafts reactions. Its solvent properties are useful in spectroscopy and in solubilizing phosphorus and sulfur.

Pharmaceutical intermediates, such as thiocarbanilide and thiocyanates, are prepared from carbon disulfide. Methionine, an essential amino acid, is manufactured from carbon disulfide intermediates. Carbon disulfide is a starting material for several fungicides, soil fumigants, and insecticides or their intermediates.

E. Bingham, in E. Bingham, B. Cohrssen, and C. H. Powell eds., *Patty's Toxicology*, 5th ed., Vol. 3 John Wiley & Sons, Inc., New York, 2001, p. 505.

"Carbon Disulfide," Chemical Profile, *Chemical Market Reporter*, August 5, 2002.

W. L. Faith, D. B. Keyes, and R. L. Clark, *Industrial Chemicals*, 3rd ed., John Wiley & Sons, Inc., New York, 1965, 223–228.

M. D. S. Lay, M. W. Sauerhoff, and D. R. Saunder, in W. Gerhartz, ed., *Ullmann's Encyclopedia of Industrial Chemistry*, 5th ed., VCH Verlagsgesellschaft, Germany, 1986, pp. 185–195.

DAVID E. SMITH
ROBERT W. TIMMERMAN
FMC Corporation

CARBON FIBERS

Carbon fibers contain at least 90% carbon by weight. In graphite, one form of carbon, the elastic modulus is much higher parallel to the plane than it is perpendicular to it. Carbon fibers are formed when long, thin graphite planes are packed together. Alignment of the graphite planes parallel to the fiber axis leads to a high tensile modulus and electrical and thermal conductivity parallel to the fiber axis.

Polymeric materials that leave a carbon residue and do not melt upon pyrolysis in an inert atmosphere are generally considered candidates for carbon fiber production. The first carbon fibers were produced by Edison in the United States and Swan in England, respectively, more than a century ago. Although cellulose was the early precursor used for carbon fibers, today polyacrylonitrile (PAN) is the predominant carbon fiber precursor, followed by petroleum pitch precursor. Carbon fibers are also being produced by decomposing gaseous hydrocarbons at high temperatures.

PROCESSING OF CARBON FIBERS

PAN-based Carbon Fibers

The steps involved in producing carbon fibers from PAN include polymerization of the PAN, spinning of fibers, thermal stabilization, carbonization, and graphitization. The typical carbon yield from PAN-based precursors is 50–60%.

PAN fibers can be spun by wet, melt, dry, gel, and dry-jet wet-spinning, with wet spinning being the commonly used process. In the wet-spinning process, the polymer is directly extruded in the coagulation bath and the fiber is subsequently drawn at ~100°C. Wet-spun PAN precursor fibers typically have a circular or dog-bone-shaped cross section. The cross-section shape of the ultimate carbon fiber resembles the shape of the precursor fiber.

The typical diameter of the PAN precursor fiber is about 15 µm, which ultimately results in a carbon fiber of about 7 µm in diameter. When processed under comparable conditions, the tensile strength of the carbon increases with decreasing fiber diameter. Therefore, higher tensile strength carbon fibers can be produced by decreasing the diameter of the PAN precursor fiber. Small-diameter PAN-based carbon fibers (100 nm to about 2 µm) can be processed by electrospinning. Carbonization of such fibers has the potential to yield carbon fibers with a tensile strength close to the theoretical value.

Pitch-based Carbon Fibers

Both isotropic and mesophase pitches are used to produce carbon fibers with a low (100 GPa) and high moduli (up to 900 GPa), respectively. Pitch is produced from petroleum or coal tar made up of fused aromatic rings. Pitch-based carbon fibers are theoretically capable of a modulus equal to that of graphite at 1050 GPa, significantly higher than the highest modulus obtained from PAN-based carbon fibers of 650 GPa. Pitch-based carbon fibers also demonstrate better electrical and thermal properties

than PAN-based fibers. Mesophase pitch-based carbon fiber production is, however, an expensive process when compared with PAN-based carbon fiber production. Production of pitch-based carbon fibers involve melt spinning of pitch precursor fibers, stabilization (oxidation), carbonization, and graphitization.

Cellulose-based Carbon Fibers

Man-made cellulosic fibers, such as Rayon, are used to produce carbon fibers. There are three main stages for rayon-based carbon fibers: (1) low-temperature decomposition, (2) carbonization, and (3) graphitization. The mechanical properties show improvement after graphitization with Young's modulus ranging from 170 to 500 GPa and tensile strength from 1 to 2 GPa for some commercial fibers. The rayon precursor is also used for making activated carbon fiber. The production of rayon-based carbon fibers is now almost nonexistent.

Gas-Phase Grown Carbon Fibers

Gas-phase grown carbon fibers, or vapor-grown carbon fibers (VGCFs), are made by decomposing gaseous hydrocarbons at temperatures between 300°C and 2500°C in the presence of a metal catalyst like Fe or Ni that is either fixed to a substrate or fluidized in space. The properties of the carbon fibers are affected by the residence time of thermal decomposition and the temperature of the furnace.

Carbon Nanotubes

Carbon nanotubes (CNTs) have been the subject of intensive research, due to their remarkable mechanical, electrical, and thermal properties. Single-wall nanotubes (SWNTs, typical diameter 0.7 to 1.5 nm) can be thought of as the ultimate carbon fiber, because of their perfect graphitic structure, low density, and alignment with respect to each layer, giving them exceptional engineering properties and light weight. The combination of density, mechanical, thermal, and electrical properties of SWNTs is unmatched. The translation of these properties into macroscopic structures is the current challenge for materials scientists and engineers.

STRUCTURE, PROPERTIES, AND MORPHOLOGY OF CARBON FIBERS

Structure and Morphology

The fine structure of carbon fibers consists of basic structural units of turbostratic carbon planes. Carbon fibers typically exhibit a skin-core texture that can result from a higher preferred orientation and a higher density of material at the fiber surface. The formation of the skin is also associated with the coagulation conditions during PAN precursor fiber spinning.

Properties

Properties of some commercial carbon fibers are listed in Table 1. The axial compressive strength of the PAN-based

Table 1. Properties of Various High-Performance Fibers

Fiber	Tensile strength (GPa)	Tensile modulus (GPa)	Elongation to break (%)	Density, ρ (g/cm^3)	Thermal conductivity (W/mK)	Electrical conductivity (S/m)
Hexcel Magnamite® PAN-based						
AS4	4.27	228	1.87	1.79		6.54E + 04
AS4C	4.34	231	1.88	1.78		
IM4	4.79	276	1.74	1.78		
IM8	5.58	304	1.84	1.79		
PV42/850	5.76	292	1.97	1.79		
Cytec Thornel® PAN-based						
T300	3.75	231	1.4	1.76	8	5.56E + 04
T650/35	4.28	255	1.7	1.77	14	6.67E + 04
T300C	3.75	231	1.4	1.76	8	5.56E + 04
Toray Torayca® PAN-based						
T300	3.53	230	1.5	1.76		
T700SC	4.90	230	2.1	1.80		
M35JB	4.70	343	1.4	1.75		
M50JB	4.12	475	0.9	1.88		
M55J	4.02	540	0.8	1.91		
M30SC	5.49	294	1.9	1.73		
Cytec Thornel® Pitch-based						
P-25	1.38	159	0.9	1.90	22	7.69E + 04
P-55S	1.90	379	0.5	1.90	120	1.18E + 05
P-100S	2.41	758	0.3	2.16	520	4.00E + 05
P-120S	2.41	827	0.3	2.17	640	4.55E + 05
K-800X	2.34	896		2.20	900–1000	6.67E + 05 to 8.33E + 05
K-1100	3.10	965		2.20	900–1000	7.69E + 05 to 9.09E + 05

carbon fibers is higher than those of the pitch-based fibers and decreases with an increasing modulus in both cases. The higher orientation, higher graphitic order, and larger crystal size all contribute negatively to the compressive strength. PAN-based carbon fibers typically fail in the buckling mode, whereas pitch-based fibers fail by shearing mechanisms. This suggests that the compressive strength of intermediate modulus PAN-based carbon fibers may be higher than what is being realized in the composites. Changes in the fiber geometry, effective fiber aspect ratio, fiber/matrix interfacial strength, as well as matrix stiffness can result in fiber compressive strength increase until the failure mode changes from buckling to shear. High compressive strength fibers also exhibit a high shear modulus.

The electrical and thermal conductivities increase with the increasing fiber tensile modulus and the carbonization temperature. The electrical conductivity of PAN-based carbon fibers is in the range of 10^4 to 10^5 S/m, whereas that of the pitch-based carbon fibers is in the range of 10^5 to 10^6 S/m. The electrical conductivity increases with temperature because as the temperature is raised, the density and carrier (electrons and holes) mobility increases. Defects are known to cause carrier scattering. An increase in modulus is due to increased orientation of the carbon planes which decreases the concentration of defects and subsequently decreases carrier scattering. Carbon fiber resistance to oxidation increases with the degree of graphitization.

SURFACE TREATMENT

Carbon fibers used in composites are often coated or surface treated to improve interaction between the fiber surface and the matrix. Surface treatment usually results in development of specific polar groups and/or roughness on the surface, for enhanced interaction with the matrix. Carbon fibers can be sized by application of thin coating with epoxy resin or other polymers to make them compatible with a particular matrix. As carbon fibers degrade in the presence of oxygen above 400°C and are stable in an inert environment up to above 2000°C, they can be protected from oxidative degradation by application of a coating such as SiC, Si_3N_4, BN, and Al_2O_3.

APPLICATIONS

Carbon fiber costs have come down significantly over the last 20 years, and PAN-and pitch-based fiber production technology seems to have matured. Carbon fibers are used in aerospace, aircraft, nuclear, sporting goods, biomedical, and high-end automotives. The high strength, toughness, and low density of carbon fibers make them suitable for aerospace and sporting goods applications. Carbon fibers are also used for chemical protective clothing, electromagnetic shielding, and as fire-retardant nonwovens. Rayon-based carbon fibers are used for heat shielding.

Carbon fiber composites are made with polymer, metal, ceramic, and carbon matrices. Although the composite materials do not yield the same mechanical properties as the fibers alone, the matrix adds other important properties to the composite for specific applications and holds the fiber together. As compared with most matrices, carbon fiber's coefficient of thermal expansion is typically two orders of magnitude lower; therefore, they can improve the dimensional stability of the composite. Although the early development of carbon fibers was prompted by defense and NASA applications, carbon fiber usage in the civilian aerospace industry is increasing at a rapid rate.

HEALTH AND SAFETY FACTORS

Safety concerns in handling carbon fibers fall into three categories: dust inhalation, skin irritation, and electrical shorting of equipment. Additionally, the protective finish, or size, which is applied to the fiber may necessitate additional safety precautions. The most common sizes are epoxies that may contain cross-linking or curing agents that produce severe skin reactions. Groups such as SACMA provide general information on the safety of carbon fibers and composites. However, as carbon fiber sizes are specially formulated to match end-use requirements, it is best to consult with the Material Safety Data Sheet (MSDS) available from the suppliers for specific handling requirements.

D. D. L. Chung, *Carbon Fiber Composites*, Butterworth-Heineman, Boston, 1994, pp. 3–11.

J. B. Donnet and R. C. Bansal, *Carbon Fibers*, 2nd ed., Marcel Dekker, New York, 1990.

E. Fitzer, in J. L. Figueiredo, C. A. Bernardo, R. T. K. Baker, and K. J. Huttinger, eds., *Carbon Fibers Filaments and Composites*, Kluwer Academic, Dordrecht, 1990, pp. 3–4.

L. H. Peebles, *Carbon Fiber – Formation, Structure, and Properties*, CRC Press, Boca Raton, 1995.

MARILYN L. MINUS
SATISH KUMAR
The Georgia Institute of
Technology

CARBON MONOXIDE

Carbon monoxide, CO, gaseous in normal atmospheric conditions (15°C and 101.3 kPa), is a colorless, odorless, and highly toxic gas. Carbon monoxide is produced by the incomplete combustion of carbon in solid, liquid, and gaseous fuels. Carbon monoxide is also a by-product of highway vehicle exhaust, which contributes about 60% of all CO emissions in the U.S. Industrially produced carbon monoxide is used in the chemical and metallurgical industries, for the synthesis of several compounds (e.g., acetic acid, polycarbonates, polyketones, etc.), and the creation of reducing atmospheres, respectively.

Table 1. Physical Properties of Carbon Monoxide

Property	Value
mol wt	28.011
melting point, K	68.09
boiling point, K	81.65
ΔH, fusion at 68 K, kJ/mol[a]	0.837
ΔH, vaporization at 81 K, kJ/mol[a]	6.042
density at 273 K, 101.3 kPa[b], g/L	1.2501
sp gr[c], liquid, 79 K	0.814
sp gr[d], gas, 298 K	0.968
critical temperature, K	132.9
critical pressure, MPa[b]	3.496
critical density, g/cm^3	0.3010
triple point	
temperature, K	68.1
pressure, kPa[e]	15.39
$\Delta G°$ formation at 298 K, kJ/mol[a]	−137.16
$\Delta H°$ formation at 298 K, kJ/mol[a]	−110.53
$S°$ formation at 298 K, kJ/(mol·K)[a]	0.1975
$C°_p$ at 298 K, J/(mol·K)[a]	29.1
$C°_v$ at 298 K, J/(mol·K)[a]	20.8
autoignition temperature, K	882
bond length, nm	0.11282
bond energy, kJ/mol[a]	1070
force constant, mN/m = (dyn/cm)	1,902,000
dipole moment, C·m[f]	0.374×10^{-30}
ionization potential, eV	14.01
flammability limits in air[g]	
upper limit, %	74.2
lower limit, %	12.5

[a] To convert J to cal, divide by 4.184.
[b] 101.3 kPa = 1 atm; to convert MPa to atm, multiply by 9.87.
[c] With respect to water at 277 K.
[d] With respect to air at 298 K.
[e] To convert kPa to torr, multiply by 7.5.
[f] To convert C·m to debye, multiply by 2.99×10^{29}.
[g] Saturated with water vapor at 290 K.

PHYSICAL PROPERTIES

Gaseous carbon monoxide is colorless, odorless, tasteless, flammable and highly toxic. It becomes a liquid at 81.62 K. Carbon monoxide is flammable in air over a wide range of concentration: lower limit of 12.5% and upper limit of 74% at 20°C and 101.3 kPa. Carbon monoxide is moderately soluble in water at low temperatures but virtually insoluble above 70°C.

Selected physical properties are listed in Table 1. Solubility data are listed in Table 2.

CHEMICAL PROPERTIES

Carbon monoxide is a reducing agent that reacts with oxidizers and salts such as iodic anhydride, palladium salts, and red mercuric oxide. Catalytic reduction of carbon monoxide produces methane. Catalytic oxidation of carbon monoxide leads to carbon dioxide. Carbon monoxide reacts violently with oxygen difluoride, chlorine produced by phosgene decomposition, and barium peroxide. Hydrogenation of carbon monoxide yields products that vary with catalysts and conditions: methane, benzene, olefins,

Table 2. Aqueous Solubility of CO at STP, L/L

Temperature, °C	Bensen coefficient
0	0.03516
5	0.03122
10	0.02782
15	0.02501
20	0.02266
25	0.02076
30	0.01915
40	0.01647
50	0.01420
60	0.01197
70	0.00998
80	0.00762
90	0.00438

paraffin waxes, hydrocarbon high polymers, methanol, higher alcohols, ethylene glycol, and glycerol.

Industrially Significant Reactions of Carbon Monoxide

Chemicals including acetic acid, acetic anhydride, vinyl acetate monomer, formic acid, propionic acid, dimethyl carbonate, methyl methacrylate, trialkylacetic acids, adipic acid, phosgene, diisocyanates, polycarbonates, and polyketones are examples of industrially significant final products.

General Reactions of Carbon Monoxide

With Hydrogen. In a liquid-phase high-pressure reaction (60 MPa or 600 atm), a rhodium cluster complex catalyzes the direct formation of ethylene glycol, propylene glycol, and glycerol from synthesis gas.

With Alcohols, Ethers, and Esters. Carbon monoxide reacts with alcohols, ethers, and esters to give carboxylic acids.

With Unsaturated Compounds. The reaction of unsaturated organic compounds with carbon monoxide and molecules containing an active hydrogen atom leads to a variety of interesting organic products. The hydroformylation reaction is the most important member of this class of reactions.

Oxidative Carbonylation. Carbon monoxide is rapidly oxidized to carbon dioxide; however, under proper conditions, carbon monoxide and oxygen react with organic molecules to form carboxylic acids or esters. With olefins, unsaturated carboxylic acids are produced, whereas alcohols yield esters of carbonic or oxalic acid.

Isocyanate Synthesis. In the presence of a catalyst, nitroaromatic compounds can be converted into isocyanates, using carbon monoxide as a reducing agent.

Dimethylformamide. The industrial solvent dimethylformamide is manufactured by the reaction between carbon monoxide and dimethylamine.

Aromatic Aldehydes. Carbon monoxide reacts with aromatic hydrocarbons or aryl halides to yield aromatic aldehydes.

Metal Carbonyls. Carbon monoxide forms metal carbonyls or metal carbonyl derivatives with most transition metals.

MANUFACTURE

Commercial carbon monoxide is a co-product, along with hydrogen, of synthetic gas (syngas) production. Several technologies, based on steam re-forming or partial oxidation processes, are used to produce syngas. The principal components of the resulting syngas, hydrogen and carbon monoxide, are then separated and purified by pressure swing adsorption and/or cryogenic distillation. The purity of the final carbon monoxide product typically ranges from 97% to 99.9%. The nature and level of the impurities remaining in the final carbon monoxide product are usually more critical than the total purity for chemical synthesis applications.

PRODUCTION

Carbon monoxide and hydrogen are generally produced simultaneously by syngas plants. Carbon monoxide production for relatively large users typically falls into one of three cases. They are (1) a plant on the user's property owned and operated by the user; (2) a plant that is on or adjacent to the user's property, owned and operated by an industrial gas company per a long-term contract between the industrial gas supplier and the CO user; and (3) a plant owned and operated by an industrial gas company that supplies carbon monoxide to several users. In the first two categories, carbon monoxide is generally supplied to the user's site via a pipe. In the last case, carbon monoxide is distributed via a pipeline with branches to several users' sites. In most cases, pure hydrogen produced from a syngas plant is distributed similarly to CO. For most applications, an average capacity of a CO plant is approximately 4000 to 8000 Nm^3/h. Some of the largest plants built can produce up to around 25,000 Nm^3/h.

SHIPMENT

Gas by pipeline is a cost-effective way to manufacture and supply CO to the user. Losses and distribution costs are minimized. Carbon monoxide pipeline networks are found in heavily industrialized areas such as the Gulf Coast in the United States, and in the Rotterdam area in Europe.

For small-volume customers involved with applications such as CO-lasers, high-pressure cylinders are the supply mode of choice. Under present regulations, the cylinders authorized for carbon monoxide service, per TC/DOT specifications, must be requalified by hydrostatic test every five years.

Carbon monoxide can also be delivered via high pressure tube trailers, typically containing 50,000 to 100,000 scf each. Liquid carbon monoxide is available only from a very small number of suppliers due to the safety and health risks associated with handling and stocking the product under liquid form. For smaller volume requirements, carbon monoxide can be supplied in high pressure steel cylinders with top pressures of (11–13.7 Mpa) and at purities ranging from 99.0% to 99.995%.

ECONOMIC ASPECTS

Carbon monoxide is manufactured as a syngas mixture or as purified gas by a number of chemical and industrial gas plants. A large majority of the carbon monoxide produced is used immediately downstream and at the plant site for chemical synthesis, or steel manufacturing. Carbon monoxide production has grown over the last few years and is expected to continue to grow.

Carbon monoxide pricing is dependent upon several factors: the price of by-product hydrogen, feedstock price, purity requirement, location, mode of supply, and volume. On-site carbon monoxide is the most economical supply method.

HEALTH AND SAFETY FACTORS

Hazards associated with the use of carbon monoxide derive primarily from its toxicity and its flammability.

Toxicity

Carbon monoxide is a chemical asphyxiant that acts toxically by combining with the hemoglobin of the red blood cells to form a stable compound called carbon monoxide–hemoglobin. This stable compound prevents the hemoglobin from taking up oxygen, thus depriving the body of the oxygen needed for metabolic respiration. The inhalation of a concentration as low as 0.04% causes headaches and discomfort within 2 to 3 hours. Inhalation of a 0.4% concentration in air is fatal in less than 1 hour. Carbon monoxide is odorless and colorless, which gives no warning of its presence, and inhalation of heavy concentration can cause sudden, unexpected collapse. The current eight-hour time-weighted average threshold limit value (TLV) adopted by the U.S. Occupational Safety and Health Administration is 35 ppm (or 40 mg/m^3) for exposure to carbon monoxide, and a ceiling limit of 200 ppm (229 mg/m^3).

According to the *Journal of the American Medical Association*, carbon monoxide is the leading cause of poisoning death in the United States. In concentrations of 12,800 parts per million (ppm) or 1.28 vol% unconsciousness is immediate with the danger of death in 1 to 3 minutes if not rescued. Domestic sources of CO are typically associated with home gas appliances (ovens, water heaters, clothes dryers), generators, furnace, fireplaces, charcoal grills, automobile exhaust fumes, power tools, etc. Only carbon monoxide detectors can detect lethal levels of CO in households. Industrial environments where carbon monoxide is used or stored should also be monitored for CO concentrations with CO detectors and alarms.

Flammability

Carbon monoxide is flammable in air over a wide range of concentration: a lower limit of 12.5% and an upper limit of 74% at 20°C and 101.3 kPa. In an industrial environment, special care should be taken to avoid storing carbon monoxide cylinders with cylinders containing oxygen or other highly oxidizing or flammable materials. It is recommended that carbon monoxide cylinders in use be grounded. Additionally, areas in which cylinders are in use must be free of all ignition sources and hot surfaces.

ENVIRONMENTAL CONCERNS

With the single exception of carbon dioxide, total yearly emissions of CO exceed all other atmospheric pollutants combined. Some of the potential sources of CO emission and exposure are foundries, petroleum refineries, kraft pulp mills, carbon black manufacturers, steel mills, formaldehyde manufacturers, coal combustion facilities, and fuel oil combustion operations. In the U.S., two-thirds of the carbon monoxide emissions come from transportation sources, with the largest contribution coming from highway motor vehicles. In urban areas, the motor vehicle contribution to carbon monoxide pollution exceeds 90%.

The Clean Air Act of 1990 gives state and local government primary responsibility for regulating pollution from power plants, factories, and other "stationary sources". The U.S. Environmental Protection Agency (EPA) has primary responsibility for "mobile sources" pollution control.

CO emitted from stationary sources, such as refineries, is under the EPA Clean Air Act 1990 regulation but enforced by state and local environmental agencies.

USES

Pure carbon monoxide is used in a number of chemical syntheses, as noted earlier. Applications of carbon monoxide include fuel gas alone, or in mixes as waste gas or producer gas; metallurgy as a reagent for manufacturing special steels, and as a reagent for reducing refractory oxides; as a reagent to make high-grade zinc white pigment for paints and varnishes; for electronics in dielectric etch recipes; and in infrared gas lasers used in solid state and molecular spectroscopy, nonlinear optics, and laser studies.

EPA 400-F-92-005; Automobiles and Carbon Monoxide, Jan. 1993.

Gas Encyclopedia, L'Air Liquide Division Scientifique, Elsevier, New York, 1976.

Handbook of Compressed Gases, 3rd ed., Compressed Gas Association, Inc., Van Nostrand Reinhold, New York, 1989.

M. M. T. Khan and A. E. Martell, *Homogeneous Catalysis by Metal Complexes*, Vol. 1, Academic Press, Inc., New York, 1974.

CHRISTINE GEORGE
Air Liquide America
Corporation

CARBOXYLIC ACIDS

Carboxylic acids from the smallest, formic, to the 22-carbon fatty acids, e.g., erucic, are economically important; several million metric tons are produced annually. The shorter-chain aliphatic acids are colorless liquids. Each has a characteristic odor ranging from sharp and penetrating (formic and acetic acids) or vinegary (dilute acetic) to the odors of rancid butter (butyric acid) and goat fat (the 6–10-carbon acids). At room temperature, the cis-unsaturated acids through C_{18} are liquids, and the saturated unbranched aliphatic acids from decanoic through the higher acids and trans-unsaturated acids are solids. The latter are higher melting because of their higher degree of linearity and greater degree of crystallinity, eg, elaidic acid.

Both odd and even numbered alkanoic acids of molecular formula $C_nH_{2n}O_2$ occur naturally. Until chromatographic techniques provided means to identify minor components in natural mixtures it was believed that only the even-numbered higher acids, most often the C_{18} acids, occurred naturally. Formic acid, acetic acid, propionic, and butyric acids are manufactured in large quantities from petrochemical feedstocks. The higher fatty acids are derived from animal fats, vegetable oils, or fish oils. Some higher saturated fatty acids with significant industrial applications are caprylic, pelargonic, capric, lauric, myristic, palmitic, stearic, and behenic acids (Table 1).

In the alkenoic series of molecular formula $C_nH_{2n-2}O_2$, acrylic, methacrylic, undecylenic, oleic, and erucic acids have important applications (Table 2). Acrylic and methacrylic acids have a petrochemical origin, and undecylenic, oleic, and erucic acids have natural origins.

The polyunsaturated aliphatic monocarboxylic acids having industrial significance include sorbic, linoleic, linolenic, eleostearic, and various polyunsaturated fish acids (Table 3). Of these, only sorbic acid is made synthetically. The other acids, except those from tall oil, occur naturally as glycerides and are used mostly in this form.

The shorter-chain alkynoic or acetylenic acids are common in laboratory organic syntheses, and several long-chain acids occur naturally (Table 4).

Many substituted, ie, branched, fatty acids, particularly methacrylic, 2-ethylhexanoic, and ricinoleic acids, are commercially significant. Several substituted fatty acids exist naturally (Table 5).

Some naturally occurring fatty acids have alicyclic substituents such as the cyclopentenyl-containing chaulmoogra acids, notable for their use in treating leprosy, and the cyclopropenyl or sterculic acids (Table 6).

The prostaglandins constitute another class of fatty acids with alicyclic structures. These are of great biological importance and are formed by *in vivo* oxidation of 20-carbon polyunsaturated fatty acids, particularly arachidonic acid.

Aromatic carboxylic acids are produced annually in amounts of several million metric tons. Several aromatic acids occur naturally, eg, benzoic acid, salicylic acid, cinnamic acid, and gallic acids, but those used in commerce are produced synthetically. These acids are generally crystalline solids with relatively high melting points, attributable to the rigid, planar, aromatic nucleus.

CHEMICAL PROPERTIES

The alkanoic acids, with the exception of formic acid, undergo typical reactions of the carboxyl group. Formic acid has reducing properties and does not form an acid chloride or an anhydride. The hydrocarbon chain of alkanoic acids undergoes the usual reactions of hydrocarbons except that the carboxyl group exerts considerable influence on the site and ease of reaction. The alkenoic acids in which the double bond is not conjugated with the carboxyl group show typical reactions of internal olefins. All three types of reactions are industrially important.

Reactions of the carboxyl group include salt and acid chloride formation, esterification, pyrolysis, reduction, and amide, nitrile, and amine formation.

Salt formation occurs when the carboxylic acid reacts with an alkaline substance.

MANUFACTURE

Carboxylic acids having 6–24 carbon atoms are commonly known as fatty acids. Shorter-chain acids, such as formic, acetic, and propionic acid, are not classified as fatty acids and are produced synthetically from petroleum sources. Fatty acids are produced primarily from natural fats and oils through a series of unit operations. Clay bleaching and acid washing are sometimes also included with the above operations in the manufacture of fatty acids for the removal of impurities prior to subsequent processing.

The predominant feedstocks for the manufacture of fatty acids are tallow and grease, coconut oil, palm oil, palm kernel oil, soybean oil, rapeseed oil, and cottonseed oil. Another large source of fatty acids comes from the distillation of crude tall oil obtained as a by-product from the Kraft pulping process.

Fatty Acids from Natural Fats and Oils

There are essentially four steps or unit operations in the manufacture of fatty acids from natural fats and oils: (1) batch alkaline hydrolysis or continuous high-pressure hydrolysis; (2) separation of the fatty acids, usually by a continuous solvent crystallization process or by the hydrophilization process; (3) hydrogenation, which converts unsaturated fatty acids to saturated fatty acids; and (4) distillation, which separates components by their boiling points or vapor pressures.

Synthetic Routes to Fatty Acids from Petroleum

The synthetic processes include catalytic oxidation for straight-chain paraffinic hydrocarbons, oxidation of straight-chain 1-olefins, carboxylation/oxidation of straight-chain 1-olefins, carboxylation of straight-chain 1-olefins, oxidation of straight-chain alcohols, and branched-chain carboxylic acids.

Table 1. Physical Properties of the Straight-Chain Alkanoic Acids, $C_nH_{2n}O_2$

n	Systematic name (common name)[a]	Mol wt	Mp, °C	Bp[b], °C	Density[c] d_4^{20}	Refractive index, n_D^d	ΔH°_f at 25°C, kJ/mol[e]	Specific heat, J/g[e]	Heat of fusion, kJ/mol[e]	Surface tension[f] mN/m (=dyn/cm)	Viscosity,[d] mPa·s (=cP)	ΔH_R^{25} kJ/mol[e]
1	methanoic (formic)	46.03	8.4	100.5	1.220	1.3714						
2	ethanoic (acetic)	60.05	16.6	118.1	1.049	1.3718						
3	propanoic (propionic)	74.08	−22	141.1	0.992	1.3874	−511.2 (l)	2.34 (l)		20.7	1.099	−1,536
4	butanoic (butyric)	88.11	−7.9	163.5	0.959	1.39906	−534.1 (l)	2.16 (l)		21.8	1.538	−2,194
5	pentanoic (valeric)	102.13	−34.5	187.0	0.942	1.4086	−559.2 (l)				2.30	−2,837.8
6	hexanoic ([caproic])	116.16	−3.4	205.8	0.929	1.4170	−584.0 (l)	2.23 (l)	15.1	23.4	3.23	−3,492.4
7	heptanoic ([enanthic])	130.19	−10.5	223.0	0.922	1.4230	−609.1 (l)		15.0		4.33	−4,146.9
8	octanoic ([caprylic])	144.21	16.7	239.7	0.910	1.4280	−635.2 (l)	2.62 (s)	21.4	23.7	5.74	−4,799.9
9	nonanoic (pelargonic)	158.24	12.5	255.6	0.907	1.4322	−658.5 (l)	2.91 (s)	20.3		8.08	−5,456.1
10	decanoic ([capric])	172.27	31.6	270.0	0.895[30]	1.4169[f]	−685.2 (l)		28.0	25.0	4.30	−6,108.7
11	undecanoic ([undecyclic])	186.30	29.3	284.0	0.9905[25]	1.4202[f]	−736.7 (s)		25.1		7.30	−6,762.3
12	dodecanoic (lauric)	200.32	44.2	298.9	0.883	1.4230[f]	−775.6 (s)	1.80 (s)	36.6	26.6	7.30	−7,413.7
13	tridecanoic ([tridecylic])	214.35	41.5	312.4	0.8458[80]	1.4252[f]						
14	tetradecanoic (myristic)	228.38	53.9	326.2	0.858[60]	1.4273[f]	−835.9 (s)	1.60 (s)	44.8	27.4	5.83[f]	−8,721.4
15	pentadecanoic ([pentadecyclic])	242.40	52.3	339.1	0.8423[80]	1.4292[f]						
16	hexadecanoic (palmitic)	256.43	63.1	351.5	0.8534[60]	1.4309[f]	−892.9 (s)	1.80 (s)	54.4	28.2	7.80[f]	−10,030.6
17	heptadecanoic (margaric)	270.46	61.3	363.8	0.853[60]	1.4324[f]						
18	octadecanoic (stearic)	284.48	69.6	376.1	0.847[f]	1.4337[f]	−949.4 (s)	1.67 (s)	63.2	28.9	9.87[f]	−11,342.4
19	nonadecanoic ([nonadecyclic])	298.51	68.6	299[100]	0.8771[25]	1.4512[25]	−1013.3 (s)		71.0			−12,646.2
20	eicosanoic (arachidic)	312.54	75.3	203[1]	0.8240[100]	1.4250[100]	−1063.7 (s)		78.7			−13,976
22	docosanoic (behenic)	340.59	79.9	306[20]	0.8221[100]	1.4270[100]						
24	tetracosanoic (lignoceric)	368.65	84.2		0.8207[100]	1.4287[100]						
26	hexacosanoic (cerotic)	396.70	87.7		0.8198[100]	1.4301[100]						
28	octacosanoic (montanic)	424.75										
30	triacontanoic (melissic)	452.81										
33	tritriacontanoic (psyllic)	494.89										
35	pentatriacontanoic (ceroplastic)	522.94										

[a] Brackets signify a trivial name no longer in use.
[b] At 101.3 kPa = 1 atm unless otherwise noted in kPa as a subscript.
[c] At 20°C unless otherwise noted by a superscript number (°C).
[d] At 20°C unless otherwise noted.
[e] To convert J to cal, divide by 4.184.
[f] At 70 C.
[g] Heat of combustion (liquid).
[h] At 50 C.
[i] To convert kPa to mm Hg, multiply by 7.5.

Table 2. Physical Properties of the Straight-Chain Alkenoic Acids, $C_n H_{(2n-2)}O_2$

n	Systematic name (common name)	Mol wt	Mp, °C	Bp, °C[a]	Density,[b] d_4^{20}	Refractive index,[b] n_D^{20}
3	propenoic (acrylic)	72.06	12.3	141.9	1.0621[16]	1.4224
4	trans-2-butenoic (crotonic)	86.09	72	189	1.018	1.4228[79.7]
4	cis-2-butenoic (isocrotonic)	86.09	14	171.9	1.0312[15]	1.4457
4	3-butenoic (vinylacetic)	86.09	−39	163	1.013[15]	1.4257[15]
5	2-pentenoic (β-ethylacrylic)	100.12				
5	3-pentenoic (β-pentenoic)	100.12				
5	4-pentenoic (allylacetic)	100.12	−22.5	188–189	0.9809	1.4281
6	2-hexenoic (isohydroascorbic)	114.14				
6	3-hexenoic (hydrosorbic)	114.14	12	208	0.9640[23]	1.4935
7	trans-2-heptenoic	128.17				
8	2-octenoic	142.20				
9	2-nonenoic	156.23				
10	trans-4-decenoic[c], cis-4-decenoic[c]	170.25				
10	9-decenoic (caproleic)	170.25				
11	10-undecenoic (undecylenic)	184.28	24.5	275	0.9075[25]	1.4464
12	trans-3-dodecenoic (linderic)	198.31				
13	tridecenoic	212.33				
14	cis-9-tetradecenoic (myristoleic)	226.36				
15	pentadecenoic	240.39				
16	cis-9-hexadecenoic (cis-9-palmitoleic)	254.41				
16	trans-9-hexadecenoic (trans-9-palmitoleic)	254.41				
17	9-heptadecenoic	268.44				
18	cis-6-octadecenoic (petroselinic)	282.47	30		0.8681[40]	1.4533[40]
18	trans-6-octadecenoic (petroselaidic)	282.47	54			
18	cis-9-octadecenoic (oleic)	282.47	13.6	234[d]	0.8905	1.4582[3]
18	trans-9-octadecenoic (elaidic)	282.47	43.7	234[d]	0.8568[70]	1.4405[70]
18	cis-11-octadecenoic	282.47	14.5			
18	trans-11-octadecenoic (vaccenic)	282.47	44		0.8563[70]	1.4406[70]
20	cis-5-eicosenoic	310.52				
20	cis-9-eicosenoic (godoleic)	310.52				
22	cis-11-docosenoic (cetoleic)	338.58				
22	cis-13-docosenoic (erucic)	338.58	34.7	281[e]	0.85321[70]	1.44438[70]
22	trans-13-docosenoic (brassidic)	338.58	61.9	265[d]	0.85002[70]	1.44349[70]
24	cis-15-tetracosenoic (selacholeic)	366.63				
26	cis-17-hexacosenoic (ximenic)	394.68				
30	cis-21-triacontenoic (lumequeic)	450.79				

[a]At 101.3 kPa = 1 atm unless otherwise noted.
[b]Superscript numbers indicate measurement at a temperature other than 20°C.
[c]The common name for both cis- and trans-4-decenoic is obtusilic.
[d]At 15 kPa = 113 mm Hg.
[e]At 30 kPa = 225 mm Hg.

ECONOMIC ASPECTS AND APPLICATIONS

Aliphatic carboxylic acids produced on a reasonably significant commercial scale range from acetic acid (two carbons or C2) through stearic acid (C18). Lesser amounts of commercially available shorter chain-length acids, such as formic (C1), and longer chain-length acids, such as erucic (unsaturated C22) and behenic (saturated C22), are also produced. As a general rule, all of the even chain-length, nonisomeric acids from C6 to C22 are produced from naturally occurring fats and oils. A significant proportion of the lower chain-length (C1–C6) and longer isomeric chain-length (C7–C10) acids are made synthetically.

Nonoleo-based Carboxylic Acids

Some of the more prominent carboxylic acids that are not fat- or oil-based include acetic, acrylic, and olefin-based propionic, butyric/isobutyric, 2-ethylhexanoic, heptanoic, pelargonic, neopentanoic, and neodecanoic.

Oleo-based Carboxylic Acids

Typically, fatty acids make up between 87 and 90% of the fat or oil from which they are made; the remaining 10–13% is glycerol. The most often used raw materials are coconut or palm kernel oil (lauric oils) for C8, C10, C12, and C14 acids; tallow, lard, and palm stearine for C16 and C18 acids; and soybean, sunflower, canola, and tall oil for

Table 3. Some Polyunsaturated Fatty Acids

Total number of carbon atoms	Systematic name (common name)	Mol wt	Mp, °C	Bp at kPa[a]	Refractive index, n_D^{20}
	Dienoic acids, $C_nH_{2n-4}O_2$				
5	2,4-pentadienoic (β-vinylacrylic)	98.10			
6	2,4-hexadienoic[b] (sorbic)	112.13	134.5		
10	*trans*-2,4-decadienoic	168.24			
12	*trans*-2,4-dodecadienoic	196.29			
18	*cis*-9,*cis*-12-octadecadienoic (linoleic)	280.45	−5	202 at 1.4	1.4699
18	*trans*-9,*trans*-12-octadecadi-enoic (linolelaidic)	280.45	28–29		
18	5,6-octadecadienoic (laballenic)	280.45			
22	5,13-docosadienoic	336.56			
	Trienoic acids, $C_nH_{2n-6}O_2$				
16	6,10,14-hexadecatrienoic (hiragonic)	250.38			
18	*cis*-9,*cis*-12,*cis*-15-octadeca-trienoic (linolenic)	278.44	−10–11.3	157 at 0.001	1.4800
18	*cis*-9,*trans*-11,*trans*-13-octa-decatrienoic (α-eleostearic)	278.44	48–49	235 at 12	1.5112
18	*trans*-9,*trans*-11,*trans*-13-octa-decatrienoic (β-eleostearic)	278.44	71.5		1.5002
18	*cis*-9,*cis*-11,*trans*-13-octade-catrienoic (punicic)	278.44			
18	*trans*-9,*trans*-12,*trans*-15-octadecatrienoic (linolenelaidic)	278.44			
	Tetraenoic acids, $C_nH_{2n-8}O_2$				
18	4,8,12,15 octadecatetraenoic (moroctic)	276.42			
18	*cis*-9,*trans*-11,*trans*-13,*cis*-15-octadecatetraenoic (α-parinaric)	276.42			
18	*trans*-9,*trans*-11,*trans*-13,*trans*-15-octadecate-traenoic (β-parinaric)	276.42			
20	5,8,11,14-eicosatetraenoic (arachidonic)	304.47			
	Pentaenoic acids, $C_nH_{2n-10}O_2$				
22	4,8,12,15,19-docosapentaenoic (clupanodonic)	330.51			

[a]To convert kPa to mm Hg, multiply by 7.5.
[b]$\Delta H_{298} = -393.5$ kJ/mol ($\frac{94.05}{4.184}$ kcal/mol); flash point (OC) = 127°C.

Table 4. The Acetylenic Fatty Acids

Total number of carbon atoms	Systematic name (common name)	Mol wt
3	propynoic (propiolic, propargylic)	70.05
4	2-butynoic (tetrolic)	84.07
5	4-pentynoic	98.10
6	5-hexynoic	112.13
7	6-heptynoic	126.16
8	7-octynoic	140.18
9	8-nonynoic	154.21
10	9-decynoic	168.24
11	10-undecynoic (dehydro-10-undecylenic)	182.26
18	6-octadecynoic (tariric)	280.45
18	9-octadecynoic (stearolic)	280.45
18	17-octadecene-9, 11-diynoic (isanic, erythrogenic)	274.40
18	*trans*-11-octadecene-9-ynoic (ximenynic)	278.44
22	13-docosynoic (behenolic)	336.56

whole cut unsaturated (lower melting point) acids. Fully hardened soya, canola, or edible tallow are usually used when high C18 food-grade stearic acid is needed, whereas edible tallow and tall oil are the primary raw materials for food-grade oleic. C22 acid is derived from rapeseed and/or marine oil (menhaden). Tall oil, a by-product of the kraft pulp and paper industry, is not a triglyceride and therefore does not contain glycerol. Castor oil is the primary source for ricinoleic acid or 12-hydroxystearic acid when hardened.

In 2001, production of 8-22 carbon acids in the United States, western Europe, and Japan was 2.5×10^6 t. North American production fell in 2001. New fatty acid plants were built in Southeast Asia. Major sources of raw materials, eg, coconut and palm, are found in this area of the world. Capacity of these plants was 1.5×10^6 t in 2001. A ca 9% growth in production and consumption in North America declined to 4% in 2001. A 1.6%/yr growth rate was expected through 2006.

Tall oil fatty acids are in decline because of reduced refinery output. Growth in western Europe was expected at the rate of 1.2% through 2006. Japan's consumption is expected to decline at the rate of 1%/yr.

Table 5. Some Substituted Acids

Total number of carbon atoms	Systematic name (common name)	Mol wt	Mp, °C	Bp, °C[a]	Density[b] d^{20}_4	Refractive index[b] n^{20}_D
4	2-methylpropenoic (methacrylic)	86.09	16	163	1.0153	1.4314
4	2-methylpropanoic (isobutyric)	88.10	−47	154.4	0.9504	1.3930
5	2-methyl-cis-2-butenoic (angelic)	100.12	45	185	0.9539^{76}	1.4434^{47}
5	2-methyl-trans-2-butenoic (tiglic)	100.12	65.5	198.5	0.9641^{76}	1.43297^{76}
5	3-methyl-2-butenoic (β,β-dimethyl acrylic)	100.12				
5	2-methylbutanoic	102.13				
5	3-methylbutanoic (isovaleric)	102.13	−37.6	176	$0.93319^{17.6}$	$1.40178^{22.4}$
5	2,2-dimethylpropanoic (pivalic)	102.13	35.5	163	0.905^{50}	
8	2-ethylhexanoic	144.21		220	0.9031^{25}	1.4255^{28}
14	3,11-dihydroxytetradecanoic (ipurolic)	260.37				
16	2,15,16-trihydroxyhexadecanoic (ustilic)	304.43				
16	9,10,16-trihydroxyhexadecanoic (aleuritic)	304.43				
16	16-hydroxy-7-hexadecenoic (ambrettolic)	270.41				
18	12-hydroxy-cis-9-octadecenoic (ricinoleic)	298.47	5.0,7.7,16.0	226^c	0.9496^{15}	1.4145^{15}
18	12-hydroxy-trans-9-octadecenoic (ricinelaidic)	298.47	52–53	240^c		
18	4-oxo-9,11,13-octadecatrienoic (licanic)	292.42				
18	9,10-dihydroxyoctadecanoic	316.48	90			
18	12-hydroxyoctadecanoic	300.48	79			
18	12-oxooctadecanoic	298.47	81.5			
18	18-hydroxy-9,11,13-octadecatrienoic (kamlolenic)	294.43	77–78			
18	12,13-epoxy-cis-9-octadecenoic (vernolic)	296.45				
18	8-hydroxy-trans-11-octadecene-9-ynoic (ximenynolic)	294.43				
18	8-hydroxy-17-octadecene-9,11-diynoic (isanolic)	290.40				
20	14-hydroxy-cis-11-eicosenoic (lesquerolic)	326.52				
	mixed isomers (isononanoic) acid	158	ca 70	232–246		
	2,2-dimethyloctanoic (neodecanoic) acid	172	<40	$147–150^d$		
	mixed isomers (isostearic) acid	284	ca 7	$192–204^e$		

[a]At 101.3 kPa = 1 atm unless otherwise noted.
[b]Superscript numbers indicate measurement at a temperature other than 20°C.
[c]At 1.3 kPa = 9.75 mm Hg.
[d]At 20 kPa.
[e]At 5 kPa.

Because they are made from renewable natural raw materials, oleo-based fatty acids are completely biodegradable and find widespread usage in a variety of applications and industries.

Table 6. Some Fatty Acids with Alicyclic Substituents

Total number of carbon atoms	Common name	n in (1) or (2)	Mol wt
	Cyclopentenyl compounds		
6	aleprolic	0	112.13
10	aleprestic	4	168.24
12	aleprylic	6	196.29
14	alepric	8	224.34
16	hydnocarpic	10	252.40
18	chaulmoogric	12	280.45
	Cyclopropenyl compounds		
18	malvalic (halphenic)	6	280.45
19	sterculic	7	294.48
19	lactobacillic[a]		296.49

[a]Saturated ring.

HEALTH AND SAFETY FACTORS

Carboxylic acid dust and vapors are generally destructive to tissues of the mucous membrane, eyes, and skin. The small molecules such as formic, acetic, propionic, butyric, and acrylic acids tend to be the most aggressive. Formic, acetic, propionic, acrylic, and methacrylic acids have time-weighted-average exposure limits of 20 ppm or lower. Acrylic acid showed an LD_{50} of 33.5 mg/kg from oral administration to rats.

The hazards of handling branched-chain acids are similar to those encountered with other aliphatic acids of the same molecular weight. Eye and skin contact as well as inhalation of vapors of the shorter-chain acids should be avoided.

ENVIRONMENTAL ASPECTS

Environmental regulation in the oleochemical industry addresses pollution of air, surface, and groundwater, along with land pollution and solid waste disposal. This is administered by the Environmental Protection Agency

(EPA) on the national level, an equivalent agency on the state level, and sometimes local agencies.

In-plant controls are perhaps the best approach to eliminating waste generation and pollution problems, and many times good payback exists on recovery of products lost because of poor process controls.

TRIALKYLACETIC ACIDS

Trialkylacetic acids are characterized by the following structure:

$$R-CH-COOH$$
$$|$$
$$NH_2$$

in which R, R′, and R″, are $C_xH_{2x}+1$ with $x \geq 1$. The lowest member of the series (R = R′ = R″ CH₃) is the C_5 acid, trimethylacetic acid or 2,2-dimethylpropanoic acid (also, neopentanoic acid, pivalic acid). For higher members in the series, the products are typically mixtures of isomers, resulting from the use of mixed isomer feedstocks and the chemical rearrangements that occur in the manufacturing process.

Trialkylacetic acids have been produced commercially since the early 1960s, in the United States by Exxon and in Europe by Shell, and have been marketed as neo acids (Exxon) or as versatic acids (Shell). The principal commercial products are the C_5 acid and the C_{10} acid (neodecanoic acid, or Versatic 10), although smaller quantities of other carbon numbers, such as C_6, C_7, and C_9, are also produced.

The trialkylacetic acids have a number of uses in areas such as polymers, pharmaceuticals, agricultural chemicals, cosmetics, and metal-working fluids. Commercially important derivatives of these acids include acid chlorides, peroxyesters, metal salts, vinyl esters, and glycidyl esters.

Trimethylacetic Acid

Physical Properties. 2,2-Dimethylpropionic acid, $(CH_3)_3$ CCOOH, also referred to as neopentanoic acid or pivalic acid, is a solid at room temperature with a pungent odor typical of many lower molecular weight carboxylic acids. It is commercially available at a purity greater than 99.5%. Neopentanoic acid is a single isomer with a high degree of symmetry and, thus, has a relatively high melting point (+34°C, compared to −34.5°C for n-pentanoic acid).

Chemical Properties. Neopentanoic acid undergoes reactions typical of carboxylic acids.

Acid Chloride Formation. Neopentanoic acid can be converted to neopentanoyl chloride by reaction with thionyl chloride, phosgene, phosphorus pentachloride, phosphorus trichloride, or by reaction with benzotrichloride in the presence of Friedel-Crafts catalysts. A laboratory procedure using tetramethyl-α-halogenoenamines at room temperature has also been reported.

Manufacture. Trialkylacetic acids are prepared using variants of the Koch reaction, a two-stage reaction for the preparation of carboxylic acids.

Shipment. Neopentanoic acid is shipped in heated tank cars, heated tank trucks, and drums.

Health and Safety Factors. Neopentanoic acid possesses low toxicity. The principal hazards associated with neopentanoic acid at ambient temperatures are from eye and skin irritation. At elevated temperatures, where concentrations of the vapor are significant, irritation of the respiratory tract can also occur. Contact with the material should be avoided.

Eye contact should be followed by flushing the eyes with large amounts of water. If irritation persists, medical attention should be obtained. Skin contact should be followed by flushing with water, using soap if available. Neopentanoic acid is combustible and will burn. Fire should be extinguished with foam, dry chemical, or water spray.

C_{10} Trialkylacetic Acids

Physical Properties. The C_{10} trialkylacetic acids, referred to as neodecanoic acid or as Versatic 10, are liquids at room temperature. Typical physical properties for commercially available material are given in Table 7.

Chemical Properties. Like neopentanoic acid, neodecanoic acid, $C_{10}H_{20}O_2$, undergoes reactions typical of carboxylic acids, including, metal salt formation.

Manufacture. The C_{10} trialkylacetic acids are prepared using the same process and catalysts as are used for the preparation of neopentanoic acid.

Shipment. The C_{10} acids are shipped in bulk sea vessels, tank cars, tank trucks, and drums.

Health and Safety Factors. The C_{10} trialkylacetic acids have toxicities similar to those for other neo acids: oral LD_{50} in rats is 2.0 g/kg, and dermal LD_{50} in rabbits is 3.16 g/kg.

Table 7. Physical Properties of Commercially Available Neodecanoic Acid

Property	Value
mp, °C	<−40
bp, °C	250−257
acid value, mg KOH/g	325
color	100 (Pt/Co)
specific gravity at 20/20°C	0.915
viscosity, mm²/s (=cSt)	
at 20°C	35.7
at 60°C	7
flash point, °C (Tag closed cup)	105
water, wt%	0.05
vapor pressure, kPa[a] at 60°C	0.012
solubility in water, g/100 mL H_2O at 25°C	0.017
heat of vaporization, kJ/kg[b], at the boiling point and 101.3 kPa[a]	249.5
ionization constant, $K_a \times 10^{-6}$ at 25°C	4.2

[a]To convert kPa to mm Hg, multiply by 7.5.
[b]To convert kJ to kcal, divide by 4.184.

Table 8. Physical Properties of Commercially Available Glycidyl Neodecanoates

Property	Cardura E10	GLYDEXX N-10
color (Pt/Co scale)	<60	45
density at 20°C, g/mL	0.958–0.968	
water content	<0.1% mass/mass	0.05 wt %
epoxy equivalent weight,[a] g	244–256	250
flash point, °C	126[b]	128[c]
residual epichlorohydrin	<10 mg/kg	<5 ppm
viscosity, mPa·s(=cP)		
25°C	7.13	
100°C	1.31	
150°C	0.72	
vapor pressure at 37.8°C, kPa[d]	0.9	
boiling range at 101.3 kPa,[d] °C	251–278	
freezing point, °C	<−60	
solubility in water at 20°C	0.01 % mass/mass	

[a]Grams of resin containing 1 g-equivalent of epoxide.
[b]PMCC = Pensky–Martin closed cup.
[c]Tagliabue closed cup.
[d]To convert kPa to mm Hg, multiply by 7.5.

Table 9. Physical Properties of VeoVa 10

Property	Value
color (Pt/Co)	<15
density at 20°C, g/mL	0.875–0.885
water content, % mass/mass	<0.1
acid value, mg KOH/g	<1.0
vinyl unsaturation, mol/kg	4.85–5.10
kinematic viscosity at 20°C, mm^{2}s(=cSt)	2.2
vapor pressure, kPa[a]	
at 30°C	<0.1
at 110°C	4.3
at 210°C	101
boiling range, °C at 13.3 kPa[a]	133–136
flash point,[b] °C	75
freezing point, °C	<−20
solubility in water at 20°C, % mass/mass	<0.1

[a]To convert kPa to mm Hg, multiply by 7.5.
[b]PMCC = Pensky–Martin closed cup closed cup.

Vinyl neodecanoate is shipped in bulk or in lined drums, stabilized with 5 ppm of the monomethyl ether of hydroquinone, MEHQ.

S. Abrahamsson, S. Stallberg-Stenhagen, and E. Stenhagen, in R. T. Holman, ed., *Progress in the Chemistry of Fats and Other Lipids*, Vol. 7, Pt. 1, Pergamon Press, Oxford, UK, 1964, pp. 1–164.

D. Chapman, *The Structure of Lipids*, John Wiley & Sons, Inc., New York, 1965, pp. 221–315.

R. W. Johnson and E. Fritz, eds., *Fatty Acids in Industry*, Marcel Dekker, Inc., New York, 1989.

W. J. McKillip, ed., *Advances in Urethane Science and Technology*, Vol. 3, Technomic Publishing Co., Inc., Westport, Conn., 1974, pp. 81–107.

The primary hazard associated with C$_{10}$ trialkylacetic acids is eye irritation. In contact with the eyes, the material is irritating and may injure eye tissue if not removed promptly. Any contact with the eyes should be immediately flushed with large amounts of water. Medical attention should also be obtained. For skin contact, flush with large quantities of water, using soap if available. To extinguish fires, use foam, dry chemical, or water spray.

Uses. The C$_{10}$ trialkylacetic acids are used in polymers, resins, and coatings, adhesion promoters, metal-working and hydraulic fluids, metal extraction, fuels and lubricants and electrial and electronic applications.

Glycidyl and Vinyl Esters. Glycidyl neodecanoate, sold commercially as GLYDEXX N-10 (Exxon) or as Cardura E10 (Shell), is prepared by the reaction of neodecanoic acid and epichlorohydrin under alkaline conditions, followed by purification. Physical properties of the commercially available material are given in Table 8. The material is a mobile liquid monomer with a mild odor and is used primarily in coatings.

Glycidyl neodecanoate is shipped in bulk or in drums and must be protected from contact with atmospheric water during storage.

Vinyl neodecanoate is prepared by the reaction of neodecanoic acid and acetylene in the presence of a catalyst such as zinc neodecanoate. Physical properties of the commercially available material, VeoVa 10 from Shell, are given in Table 9. The material is a mobile liquid with a typical mild ester odor used in a number of areas, primarily in coatings, but also in construction, adhesives, cosmetics, and a number of miscellaneous areas.

M. O. Bagby
United States Department of Agriculture
R. W. Johnson, Jr.
R. W. Daniels
Robert R. Contrell
Union Camp Corporation
E. T. Sauer
The Procter & Gamble Company
M. J. Keenan
M. A. Krevalis
Exxon Chemical Co.

CARDIOVASCULAR AGENTS

Over the last several decades, tremendous advances in basic and clinical research on cardiovascular disease have greatly improved the prevention and treatment of the nation's number-one killer of men and women of all races (it is estimated that ~40% of Americans (~60 million between the ages of 40–70) suffer from some degree of this disease. During the second half of the twentieth century, the problem of treating heart disease was at the forefront of the international medical communities' agenda. This was reflected in the World Health Organizations (WHO) 1967 classification of cardiovascular disease as the world's most serious epidemic. The problem of cardiovascular disease continues to be the leading cause of death in the United States and other industrialized countries as we progress into the new millennium. According to the American Heart Association, heart disease and stroke cost an estimated $329.2 billion in medical expenses and lost productivity in the United States in 2002– more than double the economic cost of cancer.

Cardiovascular disease encompasses a wide range of disorders, and the methods of its treatment are both vast and diverse. As such, this article covers a range of key therapeutic agents, including antiarrhythmic agents, antianginal agents, antilipemic agents, thrombolytic agents, agents used in the treatment of congestive heart failure, antiathersclerotic agents, and antihypertensive agents. The cardiac physiology, pathophysiology, and the causes of common cardiac diseases are reviewed before considering the drugs used in their treatment.

ANTIARRHYTHMIC AGENTS

Mechanisms of Cardiac Arrhythmias

The pumping action of the heart involves three principle electrical events: the generation, the conduction or propagation, and the fading away of the signal. When one or more of these events is disrupted, cardiac arrhythmias may arise.

In a healthy heart, cells located in the right atrium, referred to as the SA node or pacemaker cells, initiate a cardiac impulse. The spontaneous electrical depolarization of the sinoatrial (SA) pacemaker cells is independent of the nervous system; however, these cells are innervated by both sympathetic and parasympathetic fibers, which can cause increases or decreases in heart rate as a result of nervous system stimulation.

Disorders in the transmission of the electrical impulse can lead to conduction block and reentry phenomenon. Conduction block may be complete (no impulses pass through the block), partial (some impulses pass through the block), and bidirectional or unidirectional.

During another condition, known as heart block, the impulse signal from the SA node is not transmitted through either the atrioventricular (AV) node or lower electrical pathways properly. Heart block is classified by degree of severity: (1) first-degree heart block: all impulses moving through the AV node are conducted, but at a slower than normal rate; (2) second-degree heart block: some impulses fully transit the AV node, whereas others are blocked (as a result, the ventricles fail to beat at the proper moment); (3) third-degree heart block: no impulses reach the ventricles (automatic cells in the ventricles initiate impulses, but at a slower rate, and as a result the atria and ventricles beat at somewhat independent rates).

The most serious cause of life-threatening cardiac arrhythmias results from a condition known as reentry, which occurs when an impulse wave circles back through the heart, reenters previously excited tissue, and reactivates the cells. Under normal conditions, reentry does not occur, as cells become unable to accept an excitation impulse for a period of time that is sufficient for the original signal to abate. Hence, the cells will not contract again until a new impulse emerges from the SA node. However, there are certain conditions during which this does not happen and the impulse continues to circulate. The essential condition for reentry to occur involves the development of a cellular refractory period that is shorter than the conduction velocity. Consequently, any circumstance that either shortens the refractory period or lengthens the conduction time, can lead to reentry. Various types of alteration in automacity (enhanced and triggered automacity) can trigger cardiac arrhythmias.

Nearly all tachycardias, including fibrillation, are due to reentry. The length of the refractory period depends mainly on the rate of activation of the potassium current; the rate of conduction depends on the rate of activation of the calcium current in nodal tissue, and the sodium current in other myocytes. The channels controlling these currents are the targets for suppressing reentry. Many conditions can lead to reentry.

Types of Cardiac Arrhythmias

Arrhythmias can be divided into ventricular and supraventricular arrhythmias, and further defined by the pace of the heartbeats. Bradycardia indicates a very slow heart rate of <60 beats/min; tachycardia refers to a very fast heart rate of >100 beats/min. Fibrillation refers to fast, uncoordinated heartbeats.

The common forms of arrhythmias are grouped according to their origin in the heart. Supraventricular arrhythmias include (1) sinus arrhythmia (cyclic changes in heart rate during breathing); (2) sinus tachycardia (the SA node emits impulses faster than normal); (3) sick sinus syndrome (the SA node fires improperly, resulting in either slowed or increased heart rate); (4) premature supraventricular contractions (a premature impulse initiation in the atria causes the heart to beat prior to the time of the next normal heartbeat); (5) supraventricular tachycardia (early impulse generation in the atria speed up the heart rate); (6) atrial flutter (rapid firing of signals in the atria cause atrial myocardial cells to contract quickly, leading to a fast and steady heartbeat); (7) atrial fibrillation (electrical impulses in the atria are fired in a fast and uncontrolled manner, and arrive in the ventricles in an irregular fashion); and (8) Wolff-Parkinson-White syndrome (abnormal conduction paths between the atria and ventricles cause electrical signals to arrive in the ventricles too early, and subsequently reenter the atria).

Arrhythmias originating in the ventricles include (1) premature ventricular complexes (electrical signals from the ventricles cause an early heartbeat, after which the heart seems to pause before the next normal contraction of the ventricles occurs); (2) ventricular tachycardia (increased heart rate due to ectopic signals from the ventricles); and (3) ventricular fibrillation (electrical impulses in the ventricles are fired in a fast and uncontrolled manner, causing the heart to quiver).

Classification of Antiarrhythmic Drugs

No single classification system of antiarrhythmic agents has gained universal endorsement. At this time, the method proposed by Singh and Vaughan Williams continues to be the most widespread classification scheme. Since its initial conception, this classification method has undergone several modifications—calcium channel blockers have been added as a fourth class of compounds, and class I agents have been subdivided into three groups to account for their sodium channel blocking kinetics.

In recent years there have been many changes in the way the arrhythmia is treated. New technologies, including radiofrequency ablation and implantable devices for atrial and ventricular arrhythmias, have proven to be remarkably successful mechanical treatments. In addition, cardiac suppression trials (CAST) and numerous other studies have provided evidence indicating drugs that act mainly by blocking sodium ion channels—Class I agents under the Singh and Vaughan Williams system of classification—may have the potential to increase mortality in patients with structural heart disease. Since the CAST results were released, the use of Class I drugs has decreased, and attention has shifted to developing new Class III agents, which prolong the action potential and refractoriness by acting on potassium channels. Of the Class III antiarrhythmic agents, amiodarone has been studied extensively and has proven to be a highly effective drug for treating life-threatening arrhythmias.

In addition, new studies have indicated that combination therapies, eg, administration of amiodarone and Class II β-blockers, or concomitant treatment with implantable mechanical devices and drug therapies are effective avenues for treating arrhythmias.

Class I Antiarrhythmic Agents

Antiarrhythmic agents in this class bind to sodium channels and inhibit or block sodium conductance. This inhibition interferes with charge transfer across the cell membrane. Investigations into the effects of Class I antiarrhythmics on sodium channel activity have resulted in the division of this class into three separate subgroups referred to as IA, IB, and IC.

The basis for dividing the Class I drugs into subclasses resulted from measured differences in the quantitative rates of drug binding to, and dissociating from, sodium ion channels. Class IB drugs, which include lidocaine, tocainide, and mexiletine, rapidly dissociate from sodium channels, and consequently have the lowest potencies of the Class I drugs. Class IC drugs, which include encainide and lorcainide, are the most potent of the Class I antiarrhythmics; drugs in this class display a characteristically slow dissociation rate from sodium ion channels, causing a reduction in impulse conduction time. Agents in this class have been observed to have modest effects on repolarization. Drugs in Class IA—quinidine, procainamide, and disopyramide—have sodium

ion channel dissociation rates that are intermediate between Class IB and IC compounds.

The affinities of the Class I antiarrhythmic agents for sodium channels vary with the state of the channel or with the membrane potential. Sodium channels exist in at least three states: R = closed resting, or closed near the resting potential but able to be opened by stimulation and depolarization; A = open activated, allowing Na^+ ions to pass selectively through the membrane; and I = closed inactivated and unable to be opened. Under normal resting conditions, the sodium channels are predominantly in the resting or R state. When the membrane is depolarized, the sodium channels are active and conduct sodium ions. Next, the inward sodium current rapidly decays as the channels move to the inactivated (I) state. The return of the I state to the R state, referred to as channel reactivation, is voltage and time dependent. Class I antiarrhythmic drugs have a low affinity for R channels and a relatively high affinity for both the A or I channels.

Class IA Antiarrhythmic Agents

The class IA antiarrhythmic agents include quinidine, procainamide, and disopyramide.

Class IB Antiarrhythmic Agents

The class IB antiarrhythmic agents include lidocaine, tocainide, mexiletine, and phenytoin.

Class IC Antiarrhythmic Agents

Class IC antiarrhythmic agents include encainide, flecainide, lorcainide, propafenone, and moricizine.

Class II Antiarrhythmic Agents

The inhibitors in this class are all β-adrenergic antagonists that have been found to produce membrane-stabilizing or depressant effects on myocardial tissue. It has been hypothesized that the antiarrhythmic properties of these agents are mainly due to their inhibition of adrenergic stimulation of the heart by the endogenous catecholamines, epinephrine and norepinephrine. The principal electrophysiological effects of the β-blocking agents manifest as a reduction in the phase 4 slope potential of sinus or pacemaker cells, which decreases heart rate and slows tachycardias.

Propranolol is the prototype agent for this class of compounds. Due to the substitution pattern on its aromatic ring, it is not a selective β-adrenergic blocking agent. During propranolol-mediated β-receptor block, the chronotropic, ionotropic, and vasodilator responses to β-adrenergic stimulation are decreased. Propranolol exerts its antiarrhythmic effects in concentrations associated with β-adrenergic blockade. It has also been shown to possess membrane-stabilizing activity that is similar to quinidine. The β-adrenergic agents, including amiodarone are the only drugs proven to reduce death in patients with a history of myocardial information.

Nadolol and l-Sotalol are both nonspecific β-blockers, while para substitutions on the aromatic rings of atenolol, acetobutolol, esmolol, and metoprolol all confer β_1

antagonist selectivity. Each of these agents exerts electrophysiological effects that result in slowed heart rate, decreased AV nodal conduction, and increased AV nodal refractoriness.

Class III Antiarrhythmic Agents

The drugs in this class—amiodarone, bretylium, dofetilide, ibutilide, and (*d,l*) or racemic sotalol—all generate electrophysical changes in myocardial tissue by blocking ion channels. However, some are selective, while others are multichannel blockers. Importantly, all Class III drugs have one common effect—that of prolonging the action potential, which increases the effective refractory period without altering the depolarization or the resting membrane potential.

Racemic sotalol, dofetilide, and ibutilide are potassium channel blockers. Sotalol also possesses β-adrenergic blocking properties, while ibutilide is also a sodium channel blocker. The mechanisms of action of amiodarone and bretylium, which also prolong the action potential, remain unclear but both have sodium channel-blocking properties.

Of the compounds listed in this class, sotalol, dofetilide, and ibutilide are structurally similar.

Class IV Antiarrhythmic Agents

All of the calcium channel blockers in this class of agents—verapamil, diltiazem, and bepridil—also possess antianginal activity. With respect to cardiac arrhythmias, these agents affect calcium ion flux, which is required for the propagation of an electrical impulse through the AV node. By decreasing this influx, the calcium channel blockers slow conduction. This, in turn, slows the ventricular rate.

Miscellaneous Antiarrhythmic Agents

Two antiarrhythmic agents that do not fall within the Singh and Vaughan Williams classification are adenosine and digoxin, digitoxin.

Adenosine reduces SA node automaticity, slows conduction time through the AV node, and can interrupt reentry pathways. It is used to restore normal sinus rhythm in patients with paroxysmal supraventricular tachycardia, including Wolff-Parkinson-White syndrome. Digoxin and digitoxin are available both orally and through intravenous injection, and are used to treat and prevent sinus and supraventricular fibrillation, flutter, and tachycardia.

Trends in the Treatment of Arrhythmia

The treatment of antiarrhythmias has shifted away from Class I sodium channel blockers and now focuses on Class III drugs, which act by prolonging the action potential duration and the refractory period. Class III agents lack many of the negative side effects observed in other classes of antiarrhythmics, affect both atrial and ventricular tissue, and can be administered orally or intravenously. Members of this class, such as amiodarone (which has proven to be a clinically efficient therapeutic

for the treatment of a wide variety of arrhythmias) and racemic sotalol, have been the center of much attention in recent years and have led to the search for new Class III drugs with improved safety profiles. New and investigational Class III agents that are more selective for potassium channel subtypes include azimilide, dofetilide, dronedarone, ersentilide, ibutilide, tedisamil, and trecetilide.

Along with advances in the understanding and development of new therapeutic agents, the development of technological devices to treat arrhythmias has also evolved. One of the most important achievements has been the implantable cardioverter defibrillator (ICD). In the treatment of ventricular tachycardia and fibrillation no other therapy has been as effective in prolonging patient survival. ICD treatment is often used in combination with antiarrhythmic drug therapy. For frequent symptomatic episodes of ventricular tachycardia, administration of an adjuvant drug therapy is often required to provide maximum prevention and treatment of life-threatening arrhythmias. Combination therapies with ICD and both β-blockers and amiodarone have received the most attention.

Finally, new evidence suggests that combinations of therapeutics may be more effective at treating and controlling arrhythmias than using any single agent alone. Hence, the possibility of administering combination therapies will be an important aspect in the future development of therapeutic techniques for treating arrhythmias.

ANTIANGINAL AGENTS

Angina pectoris, the principal symptom of ischemic heart disease, is caused by an imbalance between myocardial oxygen demand and oxygen supply by coronary vessels. Such an imbalance could be the result of either increased myocardial oxygen demand due to exercise or decreased myocardial oxygen delivery or both. Angina pectoris is always associated with sudden, severe chest pain and discomfort. The location and character of the pain may vary but often radiates from the sternum to the left shoulder and over the flexor surface of the left arm to the tips of the medial fingers. However, some individuals do not experience pain with ischemia. Angina pectoris can be induced by exercise, anxiety, overeating, or stress and is often relieved quickly by rest. Other factors, such as decreased oxygen-carrying capacity of the blood or reduced aortic pressure, may be involved. The attack may be transient and damage to the ischemic myocardium may be minimal or it may result in an acute myocardial infarction (MI) and/or death. It is usually accompanied by ST segment changes in the ECG, depending on the condition. Angina occurs because the blood supply to the myocardium via coronary vessels is insufficient to meet the metabolic needs of the heart muscle for oxygen, either by a decrease in blood supply or an exceedingly large increase in oxygen requirements of the myocardium or both. For a drug to be efficacious in angina it should improve myocardial oxygen supply (increase blood flow) or reduce myocardial oxygen consumption or have both actions.

Nitrate Therapy of Angina

The various treatment modalities of different kinds of angina include (1) prevention of precipitating factors; (2) use of nitrates as vasodilators to treat acute symptoms; (3) utilization of prophylactic treatment using a choice of drugs among antianginal agents, calcium channel blockers, and β-blockers; (4) surgeries such as angioplasty, coronary stenting, and coronary artery bypass surgery; and (5) use of anticoagulants and antithromobolytic agents.

Some of the simple organic nitrates and nitrites find applications for both short- and long-term prophylactic treatment of angina pectoris, myocardial infarction, and hypertension. Most of these nitrates and nitrites are formulated by mixing inert suitable excipients such as lactose, dextrose, mannitol, alcohol, propylene glycol for safe handling, since some of these compounds are heat sensitive, very flammable, and powerful explosives, if used alone. The onset, duration of action and potency of organic nitrates could be attributed to structural differences. However, there is no relationship between the number of nitrate groups and activity (Table 1).

Vasodilating Agents

All vasodilators can be divided into three types depending on their pharmacological site of action. These include (1) cerebral; (2) coronary, and (3) peripheral vasodilators.

Amyl Nitrate. Amyl nitrate can be administered to patients with coronary artery disease by nasal inhalation for acute relief of angina pectoris. It has also been used to treat heart murmurs resulting from stenosis and aortic or mitral valve irregularities. Amyl nitrate acts within 30 s after administration and substantial hemodynamic effects such as increased heart rate and decreased diastolic pressure occurs within 30 s and duration of action persists ~3–5 min. However, this drug has a number of adverse side effects such as tachycardia and headache.

Glyceryl Trinitrate (GTN). Also called nitroglycerin, it is a powerful explosive. The undiluted drug occurs as a volatile, white-to-pale yellow, thick flammable liquid with a sweet burning taste, slightly soluble in water and soluble in alcohol. Glyceryl trinitrate is the only vasodilator drug known to stimulate the enhancement of coronary collateral circulation and capable of preventing myocardial infarction induced by coronary occlusion, and therefore is more widely used in preventing attacks of angina than in stopping them once they have instigated. Thus, nitroglycerin remains the drug of choice for treatment of angina pectoris. It has also been found useful for the treatment of congestive heart failure, myocardial infarction, peripheral vascular disease, such as Raynaud's disease, and mitral insufficiency, although the benefits of nitroglycerin in mitral insufficiency have been questioned. The principal side effects of nitroglycerin are headache, dizziness, nausea, vomiting, diarrhea, flushing, weakness, syncope, and tachycardia can result.

Pentaerythritol Tetanitrate (PETN). Since PETN is a powerful explosive, it is normally mixed and diluted with other inert materials for safe handling purposes and to prevent accidental explosions. PETN is mainly used in the prophylactic management of angina to reduce the severity and frequency of attacks.

Erythrityl Tetranitrate. This is a less potent antinaginal agent than nitroglycerin but has a more prolonged duration of action.

Isosorbide Dinitrate. Isosorbide dinitrate is routinely used for the treatment and relief of acute angina pectoris as well as in the short- and long-term prophylactic management of angina. It can also be used in combination with cardiac glycosides or diuretics for the possible treatment of congestive heart failure.

Isosorbide Mononitrate. Similar to isosorbide dinitrate, mononitrate is used for the acute relief of angina pectoris, for prophylactic management in situations likely to provoke angina attacks, and also for long-term management of angina pectoris.

Isoxsuprine Hydrochloride. This vasodilator causes vasodilation by direct relaxation of vascular smooth muscle cells. It acts by decreasing the peripheral resistance and at high doses is even known to reduce blood pressure. It also stimulates β-adrenergic receptors. It is used as an adjunct therapy in the management of peripheral vascular diseases such as Burger's disease, Raynaud's disease, arteriosclerosis obliterans, and for the relief of cerebrovascular insufficiency.

Nicorandil. This nicotinamide analogue is a balanced arterial and venous dilator and also offers cardioprotection. Nicorandil is used as an antianginal agent, known to improve the myocardial blood flow resulting in decreased systemic vascular resistance and blood pressure, pulmonary capillary wedge and left ventricular end-diastolic pressures. It is relatively well tolerated when used orally or intravenously in patients with stable angina in patients undergoing cardiopulmonary bypass surgery needs further evaluation.

Side Effects

The principal side effects of nitrates include dilation of cranial vessels causing headaches, which can limit the dose used. More serious side effects are tachycardia and hypotension resulting in corresponding increase in myocardial oxygen demand and decreased coronary perfusion, both of which have an adverse effect on myocardial oxygen balance. Another well-documented problem is the development of tolerance to nitrates.

Calcium Channel Blockers

Verapamil, the first calcium channel blocker (CCB), has been used since 1962 for its antiarrhythmic and coronary vasodilatory effects. The CCBs are widely used in the treatment of various types of angina, hypertension,

396   CARDIOVASCULAR AGENTS

Table 1. Antianginal Agents: Nitrates as Vasodilators

Chemical/ generic name	Molecular formula	Trade name	Uses	Side effects	Structure
amyl nitrate	$C_5H_{11}NO_2$	Inhalant	angina pectoris	tachycardia, CNS	
glyceryl trinitrate or nitroglycerin	$C_3H_5N_3O_9$	Nitrogard, Nitrolyn, Nitostat, Nitrol	angina pectoris, hypertension, acute MI	CNS	
pentaerythritol tetranitrate	$C_5H_8N_4O_{12}$		prophylactic anginal attacks	CNS	
isosorbide dinitrate	$C_6H_8N_2O_8$	Sorbitrate, Isordil, Isordil Titradose	angina pectoris, congestive heart failure, dysphasia	reflex tachycardia, CNS	
isosorbide mononitrate	$C_6H_9NO_6$	Monoket, Ismo, Imdur, Isotrate ER	angina pectoris, congestive heart failure	CNS, GI intolerance	
isoxsuprine	$C_{18}H_{23}NO_3$	Dilavase, Duvadilan, Isolait, Navilox, Suprilent, Vadosilan, Vasodilan, Vasoplex, Vasotran	peripheral vascular diseases, Burger's and Raynaud's diseases, arteriosclerosis obliterans	tachycardia, CNS	
nicorandil	$C_8H_9N_3O_4$		antianginal, hypotension		
erythrityl tetranitrate	$C_4H_6N_4O_{12}$	Tetranitrol, Tetranitrin, Cardilate, Cardiloid	coronary vasodilator		

certain arrhythmias, heart failure, acute myocardial infarction, cardioprotection, cerebral vasospasm, and cardiomyopathy.

Calcium plays a significant role in the excitation–contraction coupling processes of the heart and vascular smooth muscle cells as well as the conduction cells of the heart and failure to maintain intracellular calcium homeostatis results in cell death. The membranes of conduction cells contain a network of numerous inward channels that are selective for calcium, and activation of these channels leads to the plateau phase of the action potential of cardiac muscle cells.

Applications

Calcium channel blocking agents are the first drugs of choice for the management of Prinzmetal angina. Extended release or intermediate–long-acting calcium channel blocking agents may be useful in the management of hypertension in patients with diabetes mellitus, due to their fewer adverse side effects on glucose homeostasis, lipid, and renal function. Patients with impaired glucose metabolism receiving calcium channel blockers are however, at higher risks of nonfatal MI and other adverse cardiovascular events than those receiving ACE inhibitor or β-adrenergic agents.

Some of the new Ca^{2+} channel blockers in the treatment of hypertension have greater selectivity, since they can be used to treat hypertension in the presence of concomitant diseases, such as angina pectoris, hyperlipidemia, diabetes mellitus, or congestive heart failure. Reflex tachycardia and vasodilator headache are the major side effects that limit the use of these agents as antihypertensives.

The calcium channel blockers can be divided into four different classes of compounds based upon their pharmacophore and chemical structure. These include (1) arylalkylamines; (2) benzothiazepines; (3) 1–4 dihydropyridines; and (4) Mibefradil, which has been assigned its own class. These drugs have wide applications in cardiovascular therapy due to their effects, such as (a) arterial vasodilation resulting in reduced afterload; (b) slowing of impulse generation and conductance in nodal tissue; (c) reduction in cardiac work and sometimes myocardial contractility, i.e., negative inotropic effect so as to improve myocardial oxygen balance.

Arylalkylamines and Benzothiazepines. These drugs vary in their relative cardiovascular effects and clinical doses, but they have the most pronounced direct cardiac effects (eg, verapamil).

Mibefradil. This compound, assigned its own class, it is a T- and L-type CCB, primarily approved in 1997 by U.S. Food and Drug Adminstration (FDA) for management of hypertension and chronic stable angina. However, postmarketing surveillance discovered potential severe life-threatening drug–drug interactions between mibefradil and β-blockers, digoxin, verapamil, and diltiazem, especially in elderly patients, resulting in one death and three cases of cardiogenic shock with intensive support of heart rate and blood pressure. Therefore, the manufacturer voluntarily withdrew mibefradil from the U.S. market in 1998.

Dihydropyridine Derivatives. This important class of compounds is widely used as vasodilators. In general, 1,4-dihydropyridines demonstrate slight selectivity toward vascular versus myocardial cells, and therefore have greater vasodilatory effect than other calcium channel blockers. 1,4-Dyhydropyridines are also known to possess insignificant electrophysiological and negative inotropic effects compared to verapamil or diltiazem. The dihydropyridines have no significant direct effects on the heart, although they may cause reflex tachycardia. Most of the newer drugs have longer elimination half-lives but also higher rates of hepatic clearance and hence low bioavailability. Several metabolic pathways of DHP-type calcium channel blockers have been identified in humans.

Calcium antagonists are known to block calcium influx through the voltage-operated calcium channels into smooth muscle cells. Some of the compounds of 1,4-DHP category, such as Nifedipine, Nisoldipine, or Isradipine have been demonstrated to be useful in the management of coronary artery diseases. Nevertheless, these already available calcium antagonists have some major disadvantages: They are photosensitive and decompose rapidly, they are not soluble in water, and because of their depressive effects on myocardium they have negative inotropic effects. The CCBs account for almost $4 billion in sales and dihydropyridines like lercanidipine are the fastest growing class of CCB.

There are 13 derivatives of DHP calcium channel blockers currently licensed for the treatment of hypertension and widely used. Some of the most prescribed drugs include amlodipine, felodipine, isradipine, lacidipine, lercanidipine, nicardipine, nifedipine, and nisoldipine. Currently, thiazide diuretics or β-blockers are recommended as first line therapy for hypertension. Calcium channel blockers, ACE inhibitors or α-adrenergic blockers may be considered when the first-line therapy is not tolerated, contraindicated or ineffective.

ANTILIPEMIC AGENTS

Increased cholesterol levels, due to the consumption of a diet rich in saturated fat, stimulates the liver to produce cholesterol, $C_{27}H_{46}O$, a lipid needed by all cells for the synthesis of cell membranes and in some cells for the synthesis of other steroids. Cholesterol is the principal reversible determinant of risk of heart disease. Low density lipoproteins (LDLs, or "bad" cholesterol) transport cholesterol from the liver to other tissues, whereas high density lipoproteins (HDLs, or "good" cholesterol) transport cholesterol from tissues back to the liver to be metabolized. Triglycerides are transported from the liver to the tissues mainly as very low density lipoproteins (VLDLs). The VLDLs are the precursors of the LDLs. The LDLs are characterized by high levels of cholesterol, mainly in the form of highly insoluble cholesteryl esters. However, there exists a strong relationship between high LDL levels and coronary heart disease, and a negative correlation between HDL and heart disease. Total blood cholesterol is the most common measurement of blood cholesterol. In general, for people who have total cholesterol >200 mg/dL, heart attack risk is relatively low, unless a person has other risk factors. If the total cholesterol level is 240 mg/dL, the person has twice the risk of heart attack as people who have a cholesterol level of 200 mg/dL. If total cholesterol level is 240 or more, it is definitely high and the risk of heart attack and, indirectly, of stroke is greater. About 20% of the U.S. population has high blood cholesterol levels. The LDL cholesterol level also greatly affects risk of heart attack and, indirectly, of stroke. Some times the ratio of total cholesterol to HDL cholesterol is used as another measure. The goal is to keep the ratio below 5:1; the optimum ratio is 3.5:1. People with high triglycerides (>200 mg/dL) after have underlying diseases or genetic disorders. In such cases, the main treatment is to change the lifestyle by controlling weight and limiting the carbohydrate intake, since they raise triglycerides and lower HDL cholesterol levels.

During the last few years, there has been firm evidence that coronary artery disease (CAD) is a complex genetic disease involving a number of genes associated with lipoprotein abnormalities and genes influencing hypertension, diabetes, obesity, immune, and clotting systems play an important role in atherosclerotic cardiac disorders. Researchers have identified genes regulating LDL cholesterol, HDL cholesterol, and triglyceride levels based upon common *apo* E genetic variation. Many genes linked to CAD are involved in how the body removes LDL cholesterol from the bloodstream. If LDL is not properly removed, it accumulates in the arteries and can lead to CAD. The protein that removes LDL from the bloodstream is called the LDL receptor (LDLR). A mutation in this gene is responsible for familial hypercholesterolemia, or FH. People with FH have abnormally high blood levels of LDL.

As with LDLR, mutations in the *apo* E gene affect blood levels of LDL. Although, <30 mutant forms of *apo* E have been identified, people carrying the E4 version of the gene tend to have higher cholesterol levels than the general population, but levels in people with the E2 version are significantly lower. The *apo E* gene has also been implicated in Alzheimer's disease. Even though cardiovascular disease due to atherosclerosis remains the leading cause of death in the United States, most of the risk reduction strategies have traditionally focused on detection and treatment of the disease. However, some of the risk factors of cardiac diseases are reversible, and changes in life style could significantly contribute toward decreasing the mortality risk of CHD. One can reduce the risk of hypercholesterolemia by reducing the total amount of fat in diet, being physically active, since exercise can help to increase HDL, avoiding cigarette smoking and exposure to second-hand smoke, and by reducing sodium intake. In people whose cholesterol level does not respond to dietary intervention, and for those having genetic predisposition to high cholesterol levels, drug therapy may be necessary.

There are now several very effective medications available for treating elevated cholesterol levels and preventing heart attacks and death. These include hydroxymethylglutartaryl-coenzyme A (HMG-CoA) reductase inhibitors such as statins, namely, atorvastatin, cerevastatin, fluvastatin, lovastatin, pravastatin, simvastatin, and the most recently approved rosuvastatin (which lowers LDL cholesterol by 30–50% and increases HDL) and cholestyramine resin. Fibrates such as clofibrate, bezafibrate, micronized fenofibrate, and gemfibrozil also lower elevated levels of blood triglycerides and increase HDL.

HMG-CoA Reductase Inhibitors

In humans, biosynthesis of cholesterol from Acetyl CoA in the liver accounts for 60–70% of the total cholesterol pool. Statins are antilipemic agents that are structurally similar to HMG-CoA reductase, the enzyme that catalyzes the conversion of HMG-CoA to mevalonic acid, an early precursor of cholesterol. Statins produce selective and reversible competitive inhibition of HMG-CoA reductase by binding to two separate sites on the enzyme. All commercially available statins contain a nucleus that interacts with the coenzyme A recognition site of HMG-CoA reductase and a β,δ-dihydroxy acid side chain that competes with HMG-CoA for interaction with the enzyme. Statins introduced in the late 1980s are fast becoming the most widely prescribed drugs to lower cholesterol.

HMG-CoA reductase inhibitors (statins) are used as adjuncts to dietary therapy in the management of hypercholesterolemia to reduce the risks of acute coronary events such as CHD, atherosclerosis, MI, or angina. All statins are administered orally, once a daily, from 10–80 mg/day as per the individual requirements and response. At usual doses, statins are well tolerated and have very few adverse effects, the most common being GI disturbances, fatigue, localized pain, and headache. The most prescribed drugs currently on the markets are atorvastatin calcium (Lipitor), cerivastatin sodium (Baycol), fluvastatin sodium (Lescol), and lovastatin (Mevacor).

Simvastatin (Zocor), is the second most potent statin drug used to lower LDL cholesterol. Zocor has been found to be the most effective statin in raising HDL (good) cholesterol levels. It is the first statin used in a study indicating that statins can significantly reduce heart attacks and stroke in high risk patients regardless of cholesterol levels.

Pravastatin sodium (Pravachol) on HMG-CoA inhibitor, is also an efficient statin with a capability to prevent heart attacks and early mortality higher than any other statin. It does not cause drug–drug interactions.

Recently a new statin named Rosuvastatin (Crestor), superior to the most widely prescribed statins, including atorvastatin, has been approved by the FDA to lower LDL levels. It also increases the HDL significantly more than atorvastatin. Crestor is safe and well tolerated alone or in combination with fenofibrate, extended-release niacin and cholstyramine.

New ACAT Inhibitors

Although the statin class of compounds dominate the lucrative market of lipid-lowering drugs and are very successful in reducing LDL and total cholesterol, many patients still have higher than normal recommended levels of LDL and total cholesterol. However, in recent years increasing the levels of HDL and reducing levels of triglycerides is becoming the principal focus of antilipemic research. Since statins inhibit cholesterol biosynthesis in the liver, new agents that inhibit cholesterol absorption in the intestine would have a synergistic effect if used in combination with statins. Therefore, Acyl coA: cholesterol acyltransferase (ACAT), an enzyme responsible for cholesterol absorption in the intestines, is the target of choice for a new class of compounds that inhibit cholesterol absorption in the intestines.

Ezetimbie, the first member of a new class of cholesterol-lowering agents referred to as the cholesterol absorption inhibitors, was recently approved by the FDA for the reduction of cholesterol levels in patients with hyper-cholesterolemia. Ezetimibe is 18% more effective when used in combination with simvastatin at reducing LDL than simvastatin alone.

Bile Acid Sequestrants

The bile acid binding resins colestipol and cholestyramine are used as an adjunct therapy to decrease elevated serum and LDL–cholesterol levels in the management of type IIa and IIb hyperlipoproteinemia.

Fibric Acid Derivatives

Fibrates main action is to decrease serum triglycerides, LDL–cholesterol, serum VLDL, and to raise HDL–cholesterol.

The most well known compounds in this class are Clofibrate (Atromid-S), Bezafibrate (Bezalip), Fenofibrate (Tricor), Gemfibrozil (Lopid), and Niacin (Niacor)

ANTIHYPERTENSIVE AGENTS

The most common form of high blood pressure, essential hypertension, is one of the two major factors responsible for cardiovascular diseases, such as CHD, stroke, and CHF. Some 50 million Americans ages six and older have high blood pressure, and 31.6% do not know that they have it. Of all people with high blood pressure, 27.2% are on adequate medications, 26.2% are on inadequate medications, and 14.8% are not on any therapy. From 1999 to 2000, the death rate in United States due to high blood pressure increased 21.3%, while the actual number of deaths increased by 49.1%. The estimated annual cost of antihypertensive prescriptions in the United States is ~$15.5 billion.

Should nonpharmacological treatments, such as a change in diet and an increase in exercise fail to have an effect, drug therapy is the next option for controlling hypertension. The therapeutic regiment is tailored to the individual, and relevant factors affecting each patient are taken into consideration in choosing an antihypertensive drug. The first drug is usually an angiotensin converting enzyme (ACE) inhibitor, a β-adrenoceptor blocker, a calcium channel blocker, or a diuretic. The main goal of hypertension chemotherapy is to give the fewest number of drugs, using the smallest effective amounts having the lowest frequency of dosing and minimum side effects. An optimal drug regimen should reduce all risk factors of coronary heart diseases, reverse the hemodynamic abnormalities present by preserving cardiac output and tissue perfusion, and lower total peripheral resistance. The challenge is to choose the best antihypertensive drug for concomitant existing diseases while maintaining a good quality of life.

The drugs used as antihypertensive agents may be classified according to their mechanism of action as follows: (1) agents affecting the renin-angiotensin system, (2) agents affecting the adrenergic nervous system; (3) calcium channel blockers; (4) diuretics; (5) centrally acting agents; (6) vasodilators; and (7) potassium channel openers. To view structures of select agents used in the treatment of hypertension, the reader is referred to Table 2 and 3.

Table 2. Antihypertensive Agents

Chemical/generic name	Molecular formula	Trade name	Structure
ACE Inhibitors			
benazepril hydrochloride	$C_{24}H_{29}ClN_2O_5$	Lotensin, Lotrel, Lotensin HCT	
captopril	$C_9H_{15}NO_3S$	Capoten, Capozide	
enalapril maleate	$C_{20}H_{28}N_2O_5 \cdot C_4H_4O_4$	Vasotec, Vaseretic, Lexxel	

Table 2. (*Continued*)

Chemical/generic name	Molecular formula	Trade name	Structure
fosinopril sodium	$C_{30}H_{45}N \cdot NaO_7P$	Monopril	
lisinopril	$C_{21}H_{31}N_3O_5 \cdot 2\,H_2O$	Prinvil, Zestril, Prinzide, Zestoretic	
moexipril HCl	$C_{27}H_{34}N_2O_7 \cdot HCl$	Univasc, Uniretic	
perindopril erbumine	$C_{19}H_{32}N_2O_5 \cdot C_4H_{11}N$	Aceon	
quinapril	$C_{25}H_{30}N_2O_5 \cdot HCl$	Accupril	
ramipril	$C_{23}H_{32}N_2O_5$	Altace	
trandolapril	$C_{24}H_{34}N_2O_5$	Mavik, Tarka	

Table 2. (*Continued*)

Chemical/generic name	Molecular formula	Trade name	Structure
		Angiotensin II Receptor Blockers	
candesartan	$C_{24}H_{20}N_6O_3$	Atacand, Amias, Kenzen	
eprosartan	$C_{23}H_{24}N_2O_4S$	Teveten	
irbesartan	$C_{25}H_{28}N_6O$	Avapro, Avalide	
losartan	$C_{22}H_{22}ClKN_6O$	Cozaar, Hyzaar	
olmesartan	$C_{29}H_{30}N_6O_6$	Olmesatran, Medoxomil	
telmisartan	$C_{33}H_{30}N_4O_2$	Micardis, Pritor	

Table 2. (*Continued*)

Chemical/generic name	Molecular formula	Trade name	Structure
valsartan	$C_{24}H_{29}N_5O_3$	Diovan, Diovan HCT	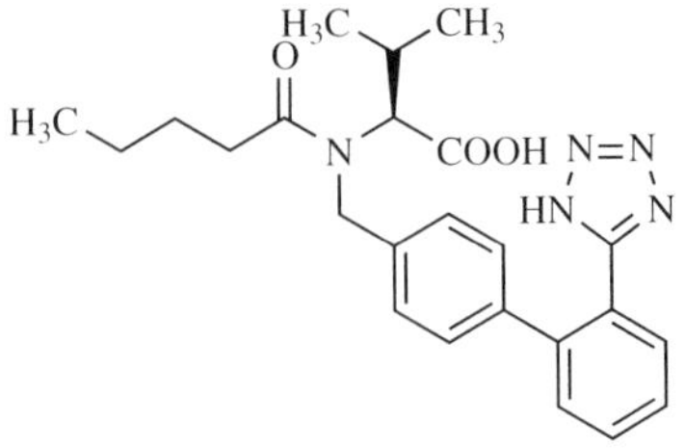

Aldosterone Receptor Antagonists

| spironolactone | $C_{24}H_{32}O_4S$ | Verospiron, Xenalon | 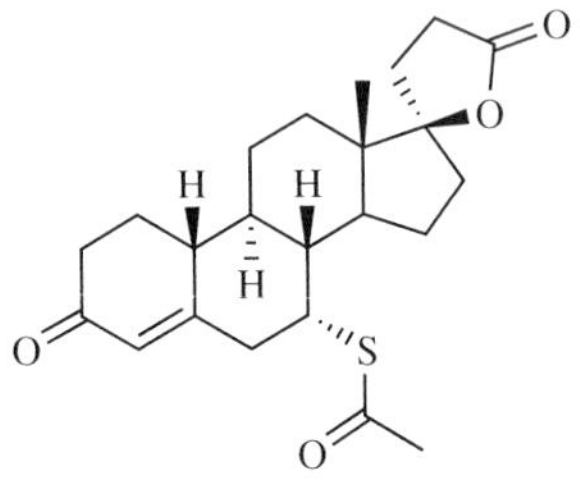 |

Vasopeptidase Inhibitors

| omapatrilat | $C_{19}H_{24}N_2O_4S_2$ | BMS-186716 | 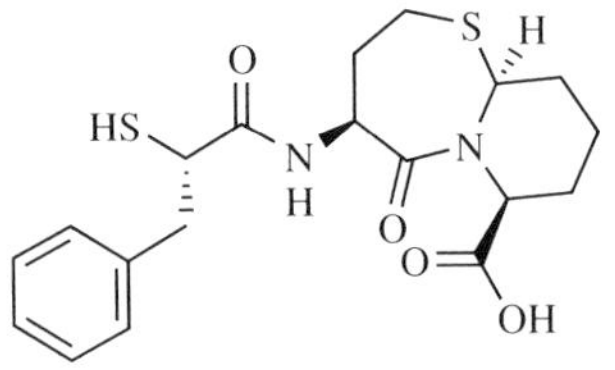 |

Neuronal Norepinephrine Depleting Agents

| reserpine | $C_{33}H_{40}N_2O_9$ | Rivasin, Sandril, Serfin, Serpasol | 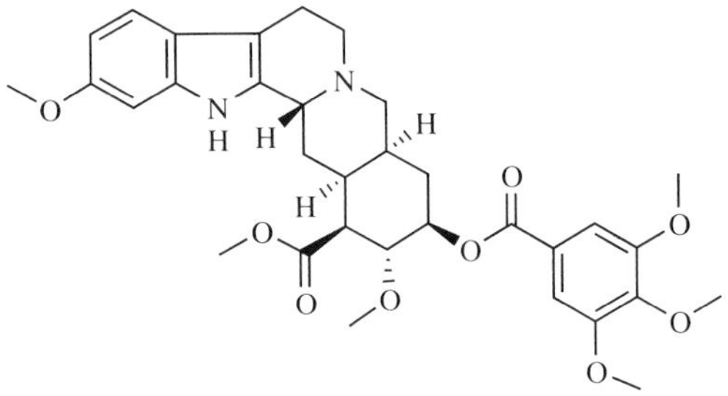 |

Adrenergic Neuronal Blockers

| guanethidine monosulfate | $(C_{10}H_{22}N_4)_2 \cdot H_2SO_4$ | Ismelin Sulfate | 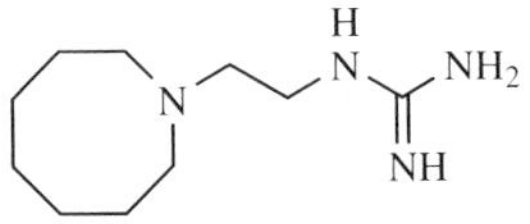 |

α-Adrenoceptor Blockers

| doxazosin mesylate | $C_{23}H_{25}N_5O_5 \cdot CH_3SO_3H$ | Cardura | 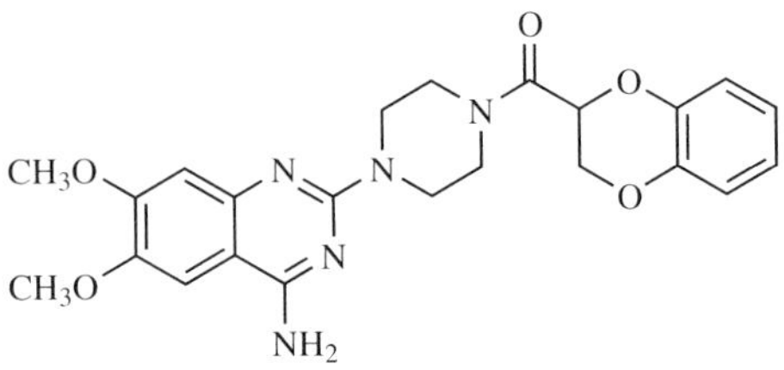 |

Table 2. (*Continued*)

Chemical/generic name	Molecular formula	Trade name	Structure
prazosin HCl	$C_{19}H_{21}N_5O_4 \cdot HCl$	Minipress, Minizide	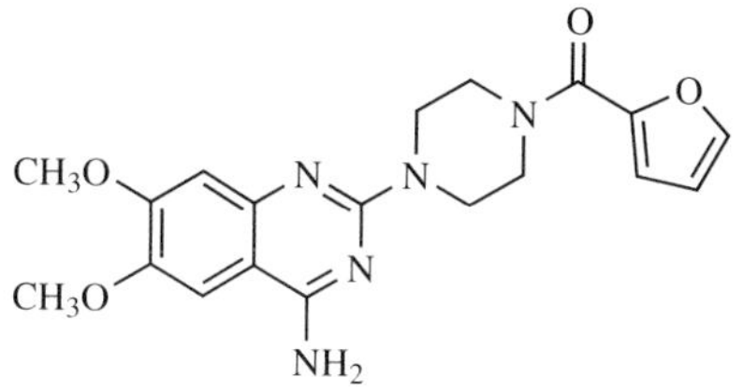
terazosin HCl	$C_{19}H_{25}N_5O_4 \cdot HCl$	Hytrin, Terazosin	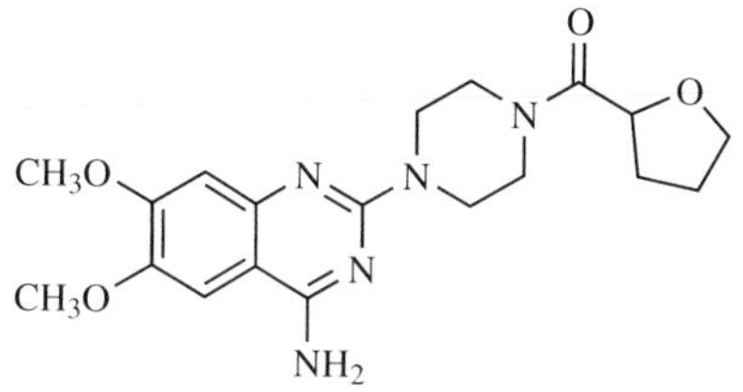

β-Adrenoceptor Blockers[*]

bisoprolol fumarate	$C_{18}H_{31}NO_4 \cdot 1/2\ C_4H_4O_4$	Zebeta, Ziac	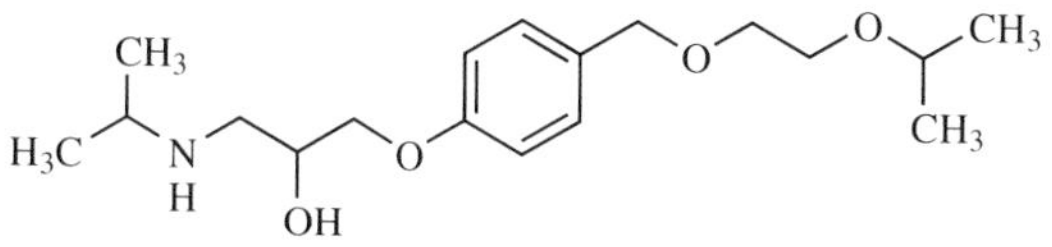
oxprenolol	$C_{15}H_{23}NO_3$	Coretal, Laracor, Paritane, Trasacor	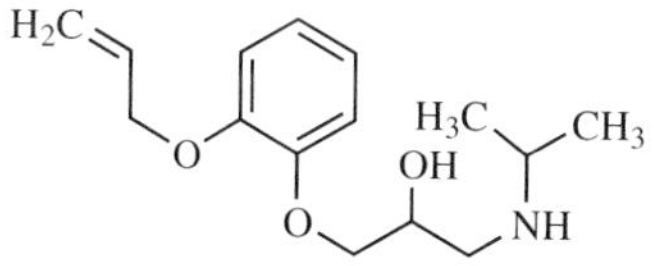
pindolol	$C_{14}H_{20}N_2O_2$	Visken	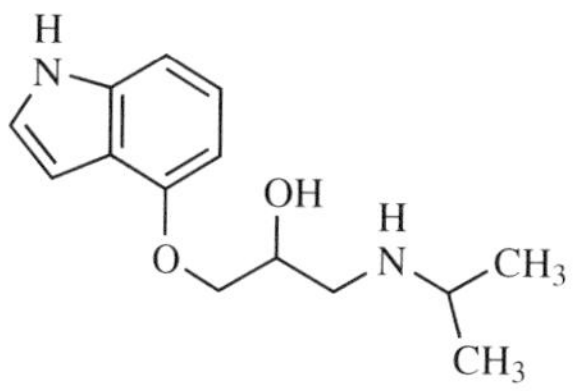

α- and β-Adrenoceptor Blockers

labetalol HCl	$C_{19}H_{24}N_2O_3 \cdot HCl$	Normodyne, Trandate	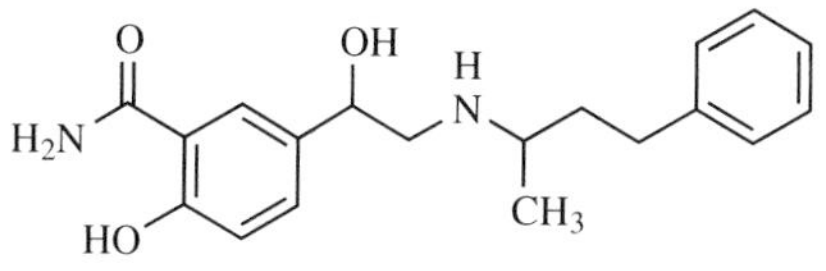

[*]Other β-blockers that are included as antihypertensive agents in the text are shown here, as they are also categorized as antiarrhythmics.

ANTITHROMBOLYTIC AGENTS

It has been well documented that the primary cause of acute myocardial infarction (AMI) is coronary arterial thrombosis. Thrombosis formation occurs at sites of ulcerated or fissured atheromatous plaques in the coronary circulation. Reperfusion of the coronaries to the ischemic–infarcted area within 6 h after the onset of AMI can salvage the myocardium and limit infarct size. Furthermore, the better the preservation of the ventricular function, the better the survival rate. Progressive irreversible myocardiac damage occurs 30 min after ischemia starts. Therefore, the faster the coronary arterial thrombus is lyzed, the less irreversible damage done on the myocardium. It is imperative that once an AMI is diagnosed, IV thrombolytic agent should be administered as quickly as possible.

Table 3. Antihypertensive Agents: Diuretics

Chemical/ generic name	Molecular formula	Trade name	Uses	Structure
		Thiazide Diuretics		
bendroflumethiazide	$C_{15}H_{14}F_3N_3O_4S_2$	Naturetin, Rauzide, Corzide	antihypertensive	
chlorothiazide Na	$C_7H_5ClN_3NaO_4S_2$	Diuril, Aldoclor	antihypertensive	
chlorthalidone	$C_{14}H_{11}ClN_2O_4S$	Thalitone, Combipress, Tenoretic	antihypertensive	
hydrochlorothiazide	$C_7H_8ClN_3O_4S_2$	Microzide, Esidrix, HydroDiural, Oretic, Aquazide-H etc.	antihypertensive	
hydroflumethiazide	$C_8H_8F_3N_3O_4S_2$	Diucardin, Saluron, Salutensin	antihypertensive	
Indapamide	$C_{16}H_{16}ClN_3O_3S$	Lozol	antihypertensive	
methylclothiazide		Enduron, Aquatensen, Dilutensen	antihypertensive	
metolazone	$C_{16}H_{16}ClN_3O_3S$	Mykrox, Zaroxolyn	antihypertensive	

Table 3. (*Continued*)

Chemical/ generic name	Molecular formula	Trade name	Uses	Structure
polythiazide	$C_{11}H_{13}ClF_3N_3O_4S_3$	Renese, Minizide	antihypertensive	
trichlormethiazide	$C_8H_8Cl_3N_3O_4S_2$	Aquazide, Naqua	antihypertensive	
bumetanide	$C_{17}H_{20}N_2O_5S$	Bumex	antihypertensive	
ethacrynic acid	$C_{13}H_{11}Cl_2NaO_4$	Edecrin	antihypertensive	
furosemide	$C_{12}H_{11}ClN_2O_5S$	Lasix	antihypertensive	
mannitol	$C_6H_{14}O_6$	Osmitrol	antihypertensive	
torsemide	$C_{16}H_{20}N_4O_3S$	Demadex	antihypertensive	
urea	CH_4N_2O	Ureaphil	antihypertensive	

Table 3. (*Continued*)

Chemical/generic name	Molecular formula	Trade name	Uses	Structure
Potassium Sparing Diuretics				
amiloride HCl	$C_6H_9Cl_2N_7O \cdot 2H_2O$	Midamor, Moduretic	antihypertensive	
spironolactone	$C_{24}H_{32}O_4S$	Aldactone, Aldactazide	antihypertensive	
triamterene	$C_{12}H_{11}N_7$	Dyrenium, Dyazide, Maxzide	antihypertensive	
Direct Vasodilators				
hydralazine	$C_8H_8N_4 \cdot HCl$	Apresoline, Alphapress		
diazoxide	$C_8H_7ClN_2O_2S$	Proglycem		
minoxidil	$C_9H_{15}N_5O$	Rogain		
nitroprusside		Nitropress		

THROMBOLYTIC AGENTS

Streptokinase, prourokinase, and acetylated plasminogen streptokinase activator complex (APSAC) agents are indirect plasminogen activators, t-PA and urokinase are direct plasminogen activators, and alteplase and tenecteplase are recombinant human t-PA. All these activators convert the formation of the active proteolytic enzyme plasmin from the proenzyme, plasminogen, by cleaving the Arg_{560}–Val_{561} bond. After intravenous administration, all five thrombolytic agents achieve approximately the same incidence of reperfusion rate (60–70%) if used optimally. When used at efficacious doses, each agent has similar degrees of bleeding complications. The least expensive thrombolytic agent is streptokinase. No single agent is superior when all factors are taken into consideration.

Whereas t-PA is clot selective, it has a very short half-life and simultaneous use of heparin is required. Bleeding is therefore as much of a problem with t-PA a less clot selective agent such as streptokinase. Streptokinase causes severe systemic fibrinogen degradation leading to extensive hypofibrinogenemia. Under this condition, reocclusion is less and the dissolution of the thrombus is more complete. This may explain why streptokinase has been shown, to be as efficacious as t-PA. The most recent comparison study reporting the equal efficacy of t-PA, APSAC, and streptokinase tends to confirm this point. Results of the largest clinical trials show that these agents reduce patient mortality from 16 to 30%.

Aspirin, a platelet aggregation inhibitor, has been demonstrated to decrease mortality of AMI by itself and it is further enhanced by the use of the thrombolytic agents if used concomitantly. The use of combinations of aspirin, heparin, vasodilator, and the thrombolytic agent decreases the incidence of rethrombosis.

NEW HUMAN T-PA THROMBOLYTIC AGENTS

Several new recombinant expressions are currently being explored toward improvement of thrombolytic agents. Some of the important methods include construction of mutants of PA, chimeric PA, conjugates of PA with monoclonal antibodies, and PA from either bacterial or animal origin. Some of these thrombolytic agents have shown promise in animal models of venous and arterial thrombosis and are being further investigated in clinical trials.

The introduction of new clot buster drugs or thrombolytics has revolutionized the treatment of heart attack or AMI, and current research is focused on identifying better thrombolytic and adjunctive agents or mechanical interventions such as the use of angioplasty or stents. A possible future approach that could universally be adopted consists of combination therapy and adjunctive/rescue percutaneous intervention at hospitals.

PHYSIOLOGY AND BIOCHEMISTRY OF CONGESTIVE HEART FAILURE

There can be a number of underlying causes of congestive heart failure (CHF). The most prevalent is the lack of oxygenated blood reaching the heart muscle itself because of coronary artery disease with myocardial infarction. Hypertension and valvular disease also contribute to CHF, but to a lesser extent.

Therapy of Congestive Heart Failure

Many of the drugs used to combat congestive heart failure are inotropic agents. Inotrope is a derivation of the Greek *ino* (fiber) and *tropikos* (changing or turning). A positive inotropic agent is therefore one that increases cardiac muscle contractility associated with CHF.

They included the cardiac glycosides, nondigitalis inotropic agents, catecholamines, agiotension-converting enzyme inhibitors , or α-blockers, phosphodiesterase inhibitors, and atrial natriuretic peptide.

FUTURE DIRECTIONS

The cardiovascular drug market is one of the larger pharmaceutical markets in the world, with global sales totaling >$50 billion/year and a number of drugs individually exceed $1 billion in annual sales. Even though copious drug classes are used to treat heart failure patients and other cardiovascular diseases, new cases of CHD are growing at >10%/year and the risk of death is also rising incessantly. Therefore, there is an unmet medical need for novel therapeutic agents.

There is increasing evidence of a relationship between apoptosis and pathophysiology of both ischemic and nonischemic cardiomyopathies. There has been a quest for a therapeutical agent that would delay the onset of apoptosis in the ischemic heart. In the future, several therapeutic interventions can be developed to prolong survival of smooth muscle and endothelial cells, and to enhance the vascular contractility, tone, and eventually delay the process of atherosclerosis.

Several new therapeutic approaches are under investigation for hypertension.

Clinical administration of drugs with negative inotropic activity is not desirable because of their cardiosuppressive effects, especially in patients with a tendency toward heart failure. Therefore, there has been a search for cardioprotective agents acting through entirely different mechanisms.

Additionally, there are a number of novel potential drug candidates undergoing various clinical studies. One of the most promising candidates is ranolazine.

A new class of hypolipidaemic drugs known as SCAP (escaping high cholesterol) ligands has been proposed. Since statins inhibit the rate-limiting enzyme in the cholesterol synthesis pathway, these new compounds act indirectly, by increasing the level of expression of the cell surface LDL receptor (LDLR), which removes cholesterol from circulation.

Finally, pharmacogenomics holds the promise that drugs might one day be tailormade for individual treatment and adapted to each person's own genetic makeup.

J. N. Delgado and W. A. Remers, eds., *Wilson and Gisvold's Textbook of Organic Medicinal Pharmaceutical Chemistry*, 9th ed., J. B. Lippincott, Philadelphia, 1991.

W. H. Frishman and S. Charlap, eds., *Medical Clinics of N.A.: Cardiovascular Pharmacotherapy III*, W. B. Saunders, Philadelphia, 1989.

A. Scriabine, ed., *Pharmacology of Antihypertensive Drugs*, Raven Press, New York, 1980.

R. E. Thomas, in M. E. Wolf, ed., *Burger's Medicinal Chemistry and Drug Discovery*, Wiley, New York, Vol. 2, 5th ed., 1996, pp. 153–261.

GAJANAN S. JOSHI
Allos Therapeutics, Inc.

JAMES C. BURNETT
Virginia Commonwealth University

CATALYSIS

Catalysis is the key to efficient chemical processing. Most industrial reactions and almost all biological reactions are catalytic. The value of the products made in the United States in processes that at some stage involve catalysis is approaching several trillion dollars annually, which is more than the gross national products of all but a few nations of the world. Products made with catalysis include food, clothing, drugs, plastics, detergents, and fuels. Catalysis is central to technologies for environmental protection by conversion of emissions.

A catalyst is a substance that increases the rate of approach to equilibrium of a chemical reaction without being substantially consumed itself. A catalyst changes the rate but not the equilibrium of the reaction.

It is well recognized that catalysts function by forming chemical bonds with one or more reactants, thereby opening up pathways to their conversion into products with regeneration of the catalyst. Catalysis is thus cyclic; reactants bond to one form of the catalyst, products are decoupled from another form, and the initial form is regenerated. The simplest imaginable catalytic cycle is therefore depicted as follows:

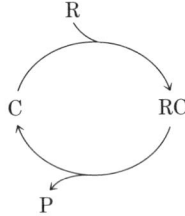

where R is the reactant, P the product, C the catalyst, and RC an intermediate complex. The intermediate complexes in catalysis are often highly reactive and not observable.

Ideally, the catalyst would cycle forever between C and RC without being consumed. But in reality there are competing reactions, and catalysts are converted into species that are no longer catalysts. In practice, catalysts must be regenerated and replaced. Catalyst manufacture is a large industry; catalysts worth some several billion dollars are sold annually in the United States.

Catalysts may be gases, liquids, or solids. Most catalysts used in technology are either liquids or surfaces of solids. Catalysis occurring in a single gas or liquid phase is referred to as homogeneous catalysis (or molecular catalysis) because of the uniformity of the phase in which it occurs. Catalysis occurring in a multiphase mixture such as a gas–solid mixture is referred to as heterogeneous catalysis; usually this is surface catalysis. Biological catalysts are proteins, ie, poly(amino) acids, called enzymes, and metalloenzymes, which are proteins incorporating inorganic components, eg, iron sulfide clusters.

The performance of a catalyst is measured largely by criteria of chemical kinetics, as a catalyst influences the rate and not the equilibrium of a reaction. The catalytic activity is a property of a catalyst that measures how fast a catalytic reaction takes place and may be defined as the rate of the catalytic reaction, a rate constant, or a conversion (or temperature required for a particular conversion) under specified conditions. The selectivity is a measure of the property of a catalyst to direct a reaction to particular products. Selectivity it is sometimes defined as a ratio of activities, such as the ratio of the rate of a desired reaction to the sum of the rates of all the reactions that deplete the reactants. Selectivity may also be represented simply as a product distribution. Because catalysts typically lose activity and/or selectivity during operation, they are also evaluated in terms of stability and lifetime. The stability of a catalyst is a measure of the rate of loss of activity or selectivity. In practical terms the stability might be measured as a rate of deactivation, such as the rate of change of the rate of the desired catalytic reaction or as the rate at which the temperature of the catalyst would have to be raised to compensate for the activity loss. Catalysts that have lost activity are often treated to bring back the activity, ie, reactivated; the regenerability is a measure, often not precisely defined, of how well the activity can be brought back. Technological catalysts are also evaluated in terms of cost.

The solids used as catalysts are typically robust porous materials with high internal surface areas, typically, hundreds of square meters per gram. Reaction occurs on the internal catalyst surface. The typical solid catalyst used in industry is a composite material with numerous components and a complex structure.

Catalytic processes are classified roughly according to the nature of the product and the industry of application, and to some degree separate literatures have developed. In chemicals manufacture, catalysis is used to make heavy chemicals, commodity chemicals, and fine chemicals. Catalysis is used extensively in the manufacture of pharmaceuticals. In fuels processing, catalysis is used in almost all the processes of petroleum refining and in coal conversion and related synthesis gas (CO and H_2) conversion. Most of the recent large-scale developments in industrial catalysis have been motivated by the need for environmental protection. Many processes for abatement of emissions are catalytic. Most of the applications of catalysis in biotechnology are fermentations, often carried out in stirred reactors with gases, liquids, and solids present; the catalysts are enzymes present in living organisms such as yeasts. There are applications of whole biological cells and of individual enzymes mounted on supports, ie, carriers, and used in fixed-bed reactors.

HOMOGENEOUS CATALYSIS

Characterization of Solution Processes

There are many important examples of catalysis in the liquid phase, but catalysis in the gas phase is unusual. From an engineering viewpoint, most of the liquid-phase processes have the following characteristics in common.

Pressure and Temperature. The pressure and temperature are relatively low, typically less than ~2 MPa (20 atm) and 150°C.

Corrosiveness. The catalyst solutions are corrosive, and the reactors, separation devices, etc., that come in contact with them must be made of expensive corrosion-resistant materials.

Separation Processes. Separation of the catalyst from the products is expensive; the process flow diagram and the processing cost are often dominated by the separations. The most common separation devices are distillation columns; extraction is also applied.

Gas Handling. The reactants are often gaseous under ambient conditions. To maximize the rate of the catalytic reaction, it is often necessary to minimize the resistance to gas–liquid mass transfer, and the gases are therefore introduced into the liquid containing the catalyst as swarms of bubbles into a well-stirred mixture or into devices such as packed columns that facilitate gas–liquid mixing and gas absorption.

Exothermicity. The catalytic reactions are often exothermic bond-forming reactions of small molecules that give larger molecules. Consequently, the reactors are designed for efficient heat removal. They may be jacketed or contain coils for heat-transfer media, or the heat of reaction may be used to vaporize the products and aid in the downstream separation by distillation.

There are also a number of generalizations about the chemistry of these processes. Often the reactants are small building blocks, many formed from organic raw materials, namely, petroleum, natural gas, and coal. The reactants include O_2, low molecular weight olefins, and synthesis gas (CO and H_2). Many reactions are catalyzed by acids and bases, usually in aqueous solution. Many reactions are catalyzed by transition-metal complexes, usually in nonaqueous organic solvents. The transition metal complex catalysts used in technology are often highly selective.

Influence of Mass Transport on Reaction Rates

When a relatively slow catalytic reaction takes place in a stirred solution, the reactants are supplied to the catalyst from the immediately neighboring solution so readily that virtually no concentration gradients exist. The intrinsic chemical kinetics determines the rate of the reaction. However, when the intrinsic rate of the reaction is relatively high and/or the transport of the reactant is relatively slow, as in a viscous polymer solution, the concentration gradients become significant and the transport of reactants to the catalyst cannot keep the catalyst supplied sufficiently for the rate of the reaction to be that corresponding to the intrinsic chemical kinetics.

Acid-Base Catalysis. Inexpensive mineral acids, eg, H_2SO_4, and bases, eg, KOH, in aqueous solution are widely applied as catalysts in industrial organic synthesis. Catalytic reactions include esterifications, hydrations, dehydrations, and condensations. Much of the technology is old and well established, and the chemistry is well understood. Reactions that are catalyzed by acids are also typically catalyzed by bases.

Metal Complex Catalysis. Most of the recent innovations in industrial homogeneous catalysis have resulted from discoveries of transition-metal complex catalysts. Thousands of transition-metal complexes (including organometallics, i.e., those with metal–carbon bonds), are known, and the rapid development of organotransition metal chemistry in recent decades has been motivated largely by the successes and opportunities in catalysis. The chemistry of metal complex catalysis is explained by the bonding and reactivity of organic groups (ligands) bonded to the metals. The important reactions in catalytic cycles are those of ligands bonded in the coordination sphere of the same metal atom. Bonding of ligands such as CO or olefin to a transition metal activates them and facilitates the catalysis.

One of the best understood catalytic cycles is that for olefin hydrogenation in the presence of phosphine complexes of rhodium, shown in Figure 1.

Phase-Transfer Catalysis

When two reactants in a catalytic process have such different solubility properties that they can hardly both be present in a single liquid phase, the reaction is confined to a liquid–liquid interface and is usually slow. However, the rate can be increased by orders of magnitude by application of a phase-transfer catalyst, and these are used on a large scale in industrial processing. Phase-transfer catalysts function by facilitating mass transport of reactants between the liquid phases. Often most of the reaction takes place close to the interface.

Industrial examples of phase-transfer catalysis are numerous and growing rapidly; they include polymerization, substitution, condensation, and oxidation reactions. The processing advantages, besides the acceleration of the reaction, include mild reaction conditions, relatively simple process flow diagrams, and flexibility in the choice of solvents.

HETEROGENEOUS CATALYSIS

Characterization of Surface Processes

Most of the largest scale catalytic processes take place with gaseous reactants in the presence of solid catalysts. From an engineering viewpoint, these processes offer the following advantages, in contrast to those involving liquid catalysts: (1) wide ranges of temperature and pressure are economically applied; (2) solid catalysts are only rarely corrosive; (3) the separation of gaseous or liquid products from solid catalysts is simple and costs little; (4) the mixing and mass transport in a fixed- or fluidized-bed reactor are facilitated by the solid catalyst particles through which the reactants and products flow; (5) strongly exothermic and strongly endothermic reactions are routinely carried out with solid catalysts.

Properties of Solid Catalysts

Most solid catalysts used on a large scale are porous inorganic materials. A number of these and the reactions they catalyze are summarized in Table 1. Catalysis takes place

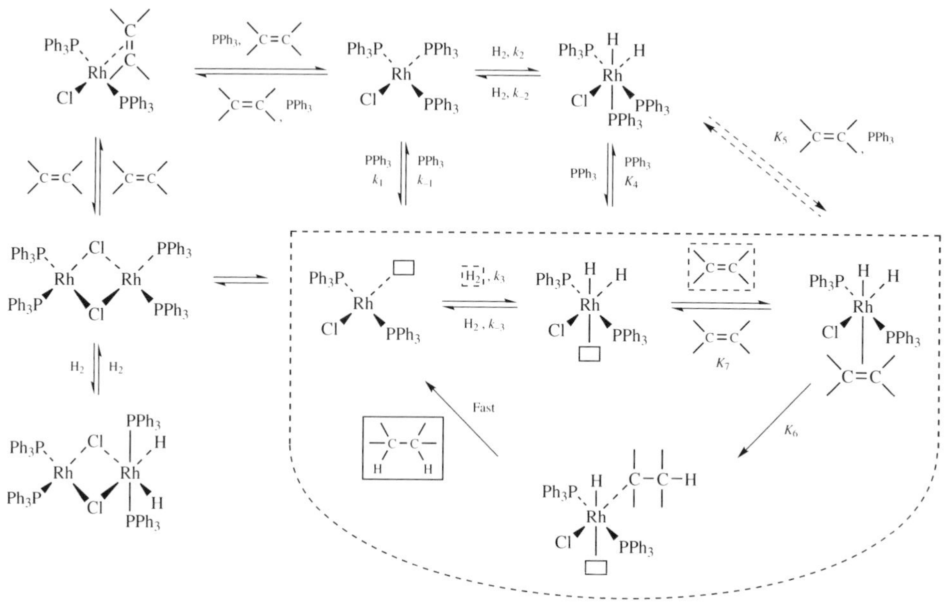

Figure 1. Catalytic cycle (within dashed lines) for the Wilkinson hydrogenation of olefin. Ph represents phenyl (C_6H_5). Values of rate constants and equilibrium constants are as follows: $k_1 = 0.68$ s^{-1}; $k_{-1} \geq 7 \times 10^4$ L/(mol·s); $K_1 \leq 10^{-5}$ mol/L; $k_{-1}/k_4 \cong 1$; $K_5 = 3 \times 10^{-4}$; $k_2 = 4.8$ L/(mol·s); $k_{-2} = 2.8 \times 10^{-4}$ s^{-1}; $K_2 = 1.7 \times 10^4$ L/mol; $k_3 \geq 7 \times 10^4$ L/(mol·s); $k_6 = 0.22$ s^{-1}.

Table 1. Some Large-Scale Industrial Processes Catalyzed by Surfaces of Inorganic Solids

Catalyst	Reaction
metals (eg, Ni, Pd, Pt, as powders or on supports) or metal oxides (eg, Cr_2O_3)	C=C bond hydrogenation (eg, olefin + H_2 → paraffin)
metals (eg, Cu, Ni, Pt)	C=O bond hydrogenation (eg, acetone + H_2 → 2-propanol)
metal (eg, Pd, Pt)	complete oxidation of hydrocarbons, oxidation of CO
Fe, Ru (supported and promoted with alkali metals)	$3\,H_2 + N_2 \rightarrow 2\,NH_3$
Ni	$CO + 3\,H_2 \rightarrow CH_4 + H_2O$ (methanation)
	$CH_4 + H_2O \rightarrow 3\,H_2 + CO$ (steam reforming)
Fe or Co (supported and promoted with alkali metals)	$CO + H_2 \rightarrow$ paraffins + olefins + $H_2O + CO_2$ + oxygen-containing organic compounds) (Fischer-Tropsch reaction)
Cu (supported on ZnO, with other components, eg, Al_2O_3)	$CO + 2\,H_2 \rightarrow CH_3OH$
Re + Pt (supported on η-Al_2O_3 and promoted with chloride)	paraffin dehydrogenation, isomerization and dehydrocyclization (eg, n-heptane → toluene + 4 H_2) (naphtha reforming)
solid acids (eg, SiO_2-Al_2O_3, zeolites)	paraffin cracking and isomerization; aromatic alkylation; polymerization of olefins
γ-Al_2O_3	alcohol → olefin + H_2O
Pd supported on zeolite	paraffin hydrocracking
metal-oxide-supported complexes of Cr, Ti, or Zr	olefin polymerization (eg, ethylene → polyethylene)
metal-oxide-supported complexes of W or Re	olefin metathesis (eg, 2-propylene → ethylene + butene)
V_2O_5 or Pt	$2\,SO_2 + O_2 \rightarrow 2\,SO_3$
Ag (on inert support, promoted by alkali metals)	ethylene + $\frac{1}{2}\,O_2$ → ethylene oxide (with $CO_2 + H_2O$)
V_2O_5 (on metal-oxide support)	naphthalene + $\frac{9}{2}\,O_2$ → phthalic anhydride + 2 CO_2 + 2 H_2O
	o-xylene + 3 O_2 → phthalic anhydride + 3 H_2O
bismuth molybdate, uranium antimonate, other mixed metal oxides	propylene + $\frac{1}{2}\,O_2$ → acrolein propylene + $\frac{3}{2}\,O_2$ + NH_3 → acrylonitrile + 3 H_2O
mixed oxides of Fe and Mo	$CH_3OH + O_2 \rightarrow$ formaldehyde (with $CO_2 + H_2O$)
Fe_3O_4 or metal sulfides	$H_2O + CO \rightarrow H_2 + CO_2$ (water gas shift reaction)
$\left\{ \begin{array}{l} Co - Mo/\gamma - Al_2O_3 \text{ (sulfided)} \\ Ni - Mo/\gamma - Al_2O_3 \text{ (sulfided)} \\ Ni - W/\gamma - Al_2O_3 \text{ (sulfided)} \end{array} \right\}$	olefin hydrogenation; aromatic hydrogenation; hydrodesulfurization; hydrodenitrogenation

as one or more of the reactants is chemisorbed (chemically adsorbed) on the surface and reacts there. The activity and selectivity of a catalyst depend strongly on the surface composition and structure.

Important physical properties of catalysts include the particle size and shape, surface area, pore volume, pore size distribution, and strength to resist crushing and abrasion.

Influence of Mass Transport on Catalyst Performance

Reactants must diffuse through the network of pores of a catalyst particle to reach the internal area, and the products must diffuse back. The optimum porosity of a catalyst particle is determined by tradeoffs: making the pores smaller increases the surface area and thereby increases the activity of the catalyst, but this gain is offset by the increased resistance to transport in the smaller pores; increasing the pore volume to create larger pores for faster transport is compensated by a loss of physical strength. If there is a significant resistance to transport of the reactant in the pores, a concentration gradient will exist at steady state, whereby the concentration of the reactant is a maximum at the particle periphery and a minimum at the particle center. The product concentration will be higher at the particle center than at the periphery. The concentration gradients provide the driving force for the transport.

Catalyst Components

Industrial catalysts are typically complex in composition and structure, consisting of catalytically active phases, supports, binders, and promoters.

Catalyst Treatments

Catalysts often require activation or regeneration and their disposal also requires special consideration.

Catalyst Preparation

Catalyst preparation is more an art than a science. Many reported catalyst preparations omit important details and are difficult to reproduce exactly, which has hindered the development of catalysis as a quantitative science. However, the art is developing into a science and there are now many examples of catalysts synthesized in various laboratories that have nearly the same physical and catalytic properties.

Supports are often prepared first and the catalyst and promoter components added later. Metal oxide supports are usually prepared by precipitation from aqueous solutions. Catalyst components are usually added in the form of precursor metal salts in aqueous solutions. In impregnation, the support may be dried, evacuated, and brought in contact with an excess of an impregnating solution containing metal salts. The processes are complex, possibly involving some dissolution of the support and reprecipitation of structures including mixed metal species.

Molecular Catalysis on Supports. The term molecular catalysis is commonly applied only to reactions in uniform

fluid phases, but it applies nearly as well to some reactions taking place on supports. Straightforward examples are reactions catalyzed by polymers functionalized with groups that closely resemble catalytic groups in solution. Industrial examples include reactions catalyzed by ion-exchange resins, usually sulfonated poly(styrene-divinylbenzene). This polymer is an industrial catalyst for synthesis of methyl *tert*-butyl ether (MTBE) from methanol and isobutylene and synthesis of bisphenol A from phenol and acetone, among others. The former application grew rapidly as MTBE became a component of high-octane-number gasoline.

Catalysis by Metals

Metals are among the most important and widely used industrial catalysts. They offer activities for a wide variety of reactions (Table 1). Atoms at the surfaces of bulk metals have reactivities and catalytic properties different from those of metals in metal complexes because they have different ligand surroundings. The surrounding bulk stabilizes surface metal atoms in a coordinatively unsaturated state that allows bonding of reactants. Thus metal surfaces offer an advantage over metal complexes, in which there is only restricted stabilization of coordinative unsaturation. Furthermore, metal surfaces provide catalytically active sites that are stable at high temperatures. For example, supported palladium catalysts (with gold) have replaced soluble palladium for vinyl acetate synthesis; the advantages of the solid include reduced corrosion and reduced formation of by-products.

Catalysis by Metal Oxides and Zeolites

Metal oxides are common catalyst supports and catalysts. Some metal oxides alone are industrial catalysts; an example is the γ-Al_2O_3 used for ethanol dehydration to give ethylene. But these simple oxides are the exception; mixed-metal oxides are more common.

Shape-Selective Catalysis. The zeolites are unique in their molecular-sieving character, which is a consequence of their narrow, uniform pores. The transport of molecules in such small pores is different from that in the larger pores of typical catalysts. When the pore diameters are large in comparison with the dimensions of the diffusing molecules, then molecular diffusion occurs. When the pores become smaller, the interactions of the molecules with the pore walls become dominant and Knudsen diffusion occurs. When the pores become so small that the molecules barely fit through them, configurational diffusion occurs.

Catalytic processes have been developed to take advantage of the unique transport and molecular sieving properties of zeolites. The zeolite that has found the most applications is the medium-pored HZSM-5. The term "shape-selective catalysis" is applied to describe the unique effects. There are different kinds of shape selectivity. Mass transport shape selectivity is a consequence of transport restrictions whereby some species diffuse more rapidly than others in the zeolite pores. In the simplest kind of shape-selective catalysis, small molecules in a mixture enter the pores and are catalytically converted, whereas large molecules pass through the reactor unconverted because they do not fit into the pores where the catalytic sites are located. Similarly, product molecules formed inside a zeolite may be so large that their transport out of the zeolite may be very slow, and they may be converted largely into other products that diffuse more rapidly into the product stream. A different kind of shape selectivity is called restricted transition state shape selectivity. It is not related to transport restrictions; rather, it is related to the size restriction of the catalyst pore that suppresses the formation of the transition state for a certain reaction, whereas it may not suppress the formation of a smaller transition state for another reaction. Mass transport selectivity is influenced by the particle size; restricted transition state shape selectivity is not.

Mass transport shape selectivity is illustrated by a process for disproportionation of toluene catalyzed by HZSM-5. The desired product is *p*-xylene; the other isomers are less valuable. The ortho and meta isomers are bulkier than the para isomer and diffuse less readily in the zeolite pores. This transport restriction favors their conversion to the desired product in the catalyst pores; the desired para isomer is formed in excess of the equilibrium concentration. Xylene isomerization is another reaction catalyzed by HZSM-5; the catalyst is preferred because of restricted transition state shape selectivity. An undesired side reaction, the xylene disproportionation to give toluene and trimethylbenzenes, is suppressed because it is bimolecular and the bulky transition state cannot readily form.

Mixed-Metal Oxides and Propylene Ammoxidation.

The best catalysts for partial oxidation are metal oxides, usually mixed-metal oxides. For example, phosphorus–vanadium oxides are used commercially for oxidation of *n*-butane to give maleic anhydride, and oxides of bismuth and molybdenum with other components are used commercially for oxidation of propylene to give acrolein or acrylonitrile.

Catalysis by Supported Metals

Metals used in industrial catalysis are often expensive and are predominantly used in a highly dispersed form. Metal species dispersed on supports may be as small as the mononuclear chromium and zirconium complexes used for olefin polymerization or may be clusters containing as few as $\sim$10 metal atoms, or larger particles that have three-dimensional structures and resemble small chunks of metal. The interactions between the metals and the support may be thought of as effects comparable to the ligand effects in molecular catalysis; the catalytic properties are sensitive to the structure and size of the metal cluster.

Catalysis by Metal Sulfides

Metal sulfides such as MoS_2, WS_2, and many others catalyze numerous reactions that are catalyzed by metals. The metal sulfides are typically several orders of magnitude less active than the metals, but they have the unique advantage of not being poisoned by sulfur compounds.

They are thus good catalysts for applications with sulfur-containing feeds, including many fossil fuels.

CATALYST DEVELOPMENT, TESTING, AND PRODUCTION

Catalysts are discovered to meet processing needs and opportunities, but the discovery of a catalytic application to take advantage of some newly discovered material almost never occurs. Catalyst development is largely a matter of trial and error testing. The methodology was defined by Mittasch and co-workers. in the development of the ammonia synthesis process. Catalyst developers benefit from an extensive and diverse literature and often can formulate good starting points in a search for candidate catalysts by learning what has been used successfully for similar reactions. Deeper insights, such as would arise from understanding of the mechanistic details of a catalytic cycle, are usually not attained; the exceptions to this rule largely pertain to molecular catalysis, usually reactions occurring in solution.

Catalyst testing and evaluation have been revolutionized by computers, automated and miniaturized test reactors, and analytical methods. Combinatorial methods are playing an increasing role in the preparation and testing with modern equipment. Researchers can systematically prepare and screen many catalysts in a short time and efficiently determine not only the initial catalytic activity and selectivity but also the stability and the appearance of trace products that may indicate some new catalytic properties worthy of further development.

Catalyst design is in a primitive stage. There are hardly any examples of true design of catalysts. However, development of improved catalysts has been guided successfully in instances when the central issues were the interplay of mass transport and reaction.

Almost all industrial catalysts are developed by researchers who are motivated to improve processes or create new ones. Catalysts are for the most part highly complex specialty chemicals, and catalyst manufacturers tend to be more efficient than others in producing them. Catalyst manufacturing is a competitive industry. Catalyst users often develop close relations with catalyst manufacturers, and the two may work together to develop and improve proprietary catalysts.

J. P. Collman, L. S. Hegedus, J. R. Norton, and R. G. Finke, *Principles and Applications of Organotransition Metal Chemistry*, 2nd ed., University Science Books, Mill Valley, Calif., 1987.

B. Cornils, W. A. Herrmann, R. Schlögl, and C. H. Wong, eds., *Catalysis from A to Z*, Wiley-VCH, Weinheim, Germany, 2000.

B. C. Gates, *Catalytic Chemistry*, John Wiley & Sons, Inc., New York, 1992.

C. M. Starks, C. L. Liotta, and M. Halpern, *Phase-Transfer Catalysis: Fundamentals, Applications, and Industrial Perspectives*, Chapman and Hall, New York, 1994.

BRUCE C. GATES
University of California, Davis

CATALYST DEACTIVATION AND REGENERATION

Catalyst deactivation, the loss over time of catalytic activity and/or selectivity, is a problem of great and continuing concern in the practice of industrial catalytic processes. Typically, the loss of activity in a well-controlled process occurs slowly. However, process upsets or poorly designed hardware can bring about catastrophic failure. While catalyst deactivation is inevitable for most processes, some of its immediate, drastic consequences may be avoided, postponed, or even reversed. Thus, deactivation issues (ie, extent, rate, and reactivation) greatly impact research, development, design, and operation of commercial processes. Accordingly, there is considerable motivation to understand and treat catalyst decay. This area of research provides a critical understanding that is the foundation for modeling deactivation processes, designing stable catalysts, and optimizing processes to prevent or slow catalyst deactivation.

MECHANISMS OF DEACTIVATION OF HETEROGENEOUS CATALYSTS

There are many paths for heterogeneous catalyst decay. As just one example, a catalyst solid may be poisoned by any one of a dozen contaminants present in the feed. Thus, the mechanisms of solid catalyst deactivation are many, grouped into six intrinsic mechanisms, (1) poisoning, (2) fouling, (3) thermal degradation, (4) vapor compound formation and/or leaching accompanied by transport from the catalyst surface or particle, (5) vapor–solid and/or solid–solid reactions, and (6) attrition/crushing. As mechanisms 1, 4, and 5 are chemical in nature while 2 and 5 are mechanical, the causes of deactivation are basically threefold: chemical, mechanical, and thermal. Each of the six basic mechanisms is defined briefly in Table 1. Mechanisms 4 and 5 are treated together, since 4 is a subset of 5.

Poisoning

Poisoning is the strong chemisorption of reactants, products, or impurities on sites otherwise available for catalysis. Thus, poisoning has an operational meaning; that is, whether a species acts as a poison depends upon its adsorption strength relative to the other species competing for catalytic sites. For example, oxygen can be a reactant in partial oxidation of ethylene to ethylene oxide on a silver catalyst and a poison in hydrogenation of ethylene on nickel. In addition to physically blocking adsorption sites, adsorbed poisons may induce changes in the electronic or geometric structure of the surface.

Catalyst poisons can be classified according to their chemical makeup, selectivity for active sites, and the types of reactions poisoned. Table 2 lists four groups of catalyst poisons classified according to chemical origin and their type of interaction with metals.

Organic bases (eg, amines) and ammonia are common poisons for acidic solids such as silica–aluminas and

Table 1. Mechanisms of Catalyst Deactivation

Mechanism	Type	Brief definition/description
poisoning	chemical	strong chemisorption of species on catalytic sites which block sites for catalytic reaction
fouling	mechanical	physical deposition of species from fluid phase onto the catalytic surface and in catalyst pores
thermal degradation	thermal	thermally induced loss of catalytic surface area, support area, and active phase-support reactions
vapor formation	chemical	reaction of gas with catalyst phase to produce volatile compounds
vapor–solid and solid–solid reactions	chemical	reaction of vapor, support, or promoter with catalytic phase to produce inactive phase
attrition/crushing	mechanical	loss of catalytic material due to abrasion loss of internal surface area due to mechanical-induced crushing of the catalyst particle

zeolites in cracking and hydrocracking reactions, while sulfur- and arsenic-containing compounds are typical poisons for metals in hydrogenation, dehydrogenation, and steam-reforming reactions. Metal compounds are poisons in automotive emissions control, catalytic cracking, and hydrotreating. Acetylene is a poison for ethylene oxidation, while asphaltenes are poisons in hydrotreating of petroleum residues.

"Selective" poisoning involves preferential adsorption of the poison on the most active sites at low concentrations. If sites of lesser activity are blocked initially, the

Table 2. Common Poisons Classified According to Chemical Structure

Chemical type	Examples	Type of interaction with metals
Groups VA and VIA	N, P, As, Sb, O, S, Se, Te	through s and p orbitals; shielded structures are less toxic
Group VII A	F, Cl, Br, I	through s and p orbitals; formation of volatile halides
toxic heavy metals and ions	As, Pb, Hg, Bi, Sn, Zn, Cd, Cu, Fe	occupy d orbitals; may form alloys
molecules that adsorb with multiple bonds	CO, NO, HCN, benzene, acetylene, other unsaturated hydrocarbons	chemisorption through multiple bonds and back bonding

poisoning is "antiselective." If the activity loss is proportional to the concentration of adsorbed poison, the poisoning is "nonselective."

Activity versus poison concentration patterns are based on the assumption of uniform poisoning of the catalyst surface and surface reaction rate controlling, i.e., negligible pore-diffusional resistance. These assumptions, however, are rarely.

Two important keys to reaching a deeper understanding of poisoning phenomena include (1) determining surface structures of poisons adsorbed on metal surfaces and (2) understanding how surface structure and hence adsorption stoichiometry change with increasing coverage of the poison.

The nature of reconstruction of a surface by a poison may depend on its pretreatment. For example, in a scanning tunneling microscopy (STM) study of room temperature H_2S adsorption on Ni, workers found that the S/Ni structure at saturation varied with the initial state of the surface, i.e., whether clean or oxygen covered (see Fig. 1). This study showed that no reconstruction occurs by direct exposure to H_2S at room temperature, rather only in the presence of O_2 or air. This emphasizes the complexities inherent in predicting the structure and stability of a given poison adsorbed on a given catalyst during a specified reaction as a function of different pretreatments or process disruptions, eg, exposure to air.

(a)	(b)
(c)	(d)

Figure 1. A series of in situ STM images recorded after exposure of Ni(110) to oxygen and then progressively higher exposures of H_2S: (**a**) $(2 \times 1)O$ overlayer; (**b**) white islands and black troughs with a $C(2 \times 2)S$ structure after exposure to 3 and 8 L of H_2S; (**c**) 25 L, islands transform to low-coordinated rows in the [001] direction; and (**d**) 50 L, stable, well-ordered $(4 \times 1)S$.

It is evident that structure and stoichiometry of sulfur adsorbed on nickel are complex functions of temperature, H_2S concentration, sulfur coverage, and pretreatment, phenomena that account at least in part for the complex nature of nickel poisoning by sulfur. Could one expect similar complexities in the poisoning of other metals? Probably, since poisoning of nickel is prototypical; i.e., similar principles operate and similar poisoning behaviors are observed in other poison/metal systems, although none have been studied to the same depth as sulfur/nickel.

There are a number of industrial processes in which one intentionally poisons the catalyst in order to improve its selectivity; for example, to minimize unwanted cracking reactions, to improve isomerization selectivity, and to minimize coking.

Fouling, Coking, and Carbon Deposition

Fouling is the physical (mechanical) deposition of species from the fluid phase onto the catalyst surface, which results in activity loss due to blockage of sites and/or pores. In its advanced stages it may result in disintegration of catalyst particles and plugging of the reactor voids. Important examples include mechanical deposits of carbon and coke in porous catalysts, although carbon- and coke-forming processes also involve chemisorption of different kinds of carbons or condensed hydrocarbons that may act as catalyst poisons. The definitions of carbon and coke are somewhat arbitrary and by convention related to their origin. Carbon is typically a product of CO disproportionation, while coke is produced by decomposition or condensation of hydrocarbons on catalyst surfaces and typically consists of polymerized heavy hydrocarbons. Coke forms may vary from high molecular weight hydrocarbons to primarily carbons such as graphite, depending upon the conditions under which the coke was formed and aged.

The chemical structures of cokes or carbons formed in catalytic processes vary with reaction type, catalyst type, and reaction conditions. Catalytic reactions accompanied by carbon or coke formation can be broadly classified as either coke-sensitive or coke-insensitive. In coke-sensitive reactions, unreactive coke is deposited on active sites, leading to activity decline, while in coke-insensitive reactions relatively reactive coke precursors formed on active sites are readily removed by hydrogen or other gasifying agents. The structure and location of a coke are more important than its quantity in affecting catalytic activity.

Not only the structure and location of coke vary but also its mechanism of formation varies with catalyst type, eg, whether it is a metal or metal oxide (or sulfide, sulfides being similar to oxides).

Carbon and Coke Formation on Supported Metal Catalysts.
Deactivation of supported metals by carbon or coke may occur chemically, owing to chemisorption or carbide formation, or physically and mechanically, owing to blocking of surface sites, metal crystallite encapsulation, plugging of pores, and destruction of catalyst pellets by carbon filaments. Blocking of catalytic sites by chemisorbed hydrocarbons, surface carbides, or relatively reactive films is generally reversible in hydrogen, steam, CO_2, or oxygen.

Coke Formation on Metal Oxide and Sulfide Catalysts.
Formation of coke on oxides and sulfides is principally a result of cracking reactions involving coke precursors (typically olefins or aromatics) catalyzed by acid sites. Dehydrogenation and cyclization reactions of carbocation intermediates formed on acid sites lead to aromatics, which react further to higher molecular weight polynuclear aromatics and condense as coke.

Coking reactions in processes involving heavy hydrocarbons are very complex; different kinds of coke may be formed and they may range in composition from CH to C and have a wide range of reactivities with oxygen and hydrogen, depending upon the time on stream and temperature to which they are exposed.

In addition to hydrocarbon structure and reaction conditions, extent and rate of coke formation are also a function of the acidity and pore structure of the catalyst. Generally, the rate and extent of coke formation increase with increasing acid strength and concentration. Coke yield decreases with decreasing pore size (for a fixed acid strength and concentration); this is especially true in zeolites where shape selectivity plays an important role in coke formation. However, in pores of molecular diameter, a relatively small quantity of coke can cause substantial loss of activity. Coke yield can vary considerably into the interior pores of a catalyst particle or along a catalyst bed.

The mechanisms by which coke deactivates oxide and sulfide catalysts are, as in the case of supported metals, both chemical and physical. However, some aspects of the chemistry are quite different. The principal chemical loss of activity in oxides and sulfides is due to the strong adsorption of coke molecules on acidic sites. But strong acid sites also play an important role in the formation of coke precursors, which subsequently undergo condensation reactions to produce large polynuclear aromatic molecules that physically coat catalytic surfaces. Physical loss of activity also occurs as coke accumulates, ultimately partially or completely blocking catalyst pores, as in supported metal catalysts.

Several studies have focused on coke formation during hydrocarbon reactions in zeolites, including (1) the detailed chemistry of coke precursors and coke molecules formed in zeolite pores and pore intersections (or supercages) and (2) the relative importance of adsorption on acid sites versus pore blockage. The principal conclusions from these studies are that (1) the formation of coke and the manner in which it deactivates a zeolite catalyst are shape-selective processes, (2) deactivation is mainly due to the formation and retention of heavy aromatic clusters in pores and pore intersections, and (3) while both acid-site poisoning and pore blockage participate in the deactivation, the former dominates at low coking rates, low coke coverages, and high temperatures, while the latter process dominates at high reaction rates, low temperatures, and high coke coverages. Thus, pore size and pore structure are probably more important than acid strength and density under typical commercial process conditions.

Thermal Degradation and Sintering

Thermally induced deactivation of catalysts results from (1) loss of the catalytic surface area due to crystallite growth of the catalytic phase, (2) loss of the support area due to support collapse and of the catalytic surface area due to pore collapse on crystallites of the active phase, and/or (3) chemical transformations of catalytic phases to noncatalytic phases. The first two processes, which are typically referred to as "sintering." Sintering processes which generally take place at high reaction temperatures (eg, >500°C), are generally accelerated by the presence of water vapor.

Three principal mechanisms of metal crystallite growth have been advanced: (1) crystallite migration, (2) atomic migration, and (3) (at very high temperatures) vapor transport. Crystallite migration involves the migration of entire crystallites over the support surface, followed by collision and coalescence. Atomic migration involves detachment of metal atoms or molecular metal clusters from crystallites, migration of these atoms over the support surface, and ultimately, capture by larger crystallites. The third mechanism, redispersion, the reverse of crystallite growth in the presence of O_2 and/or Cl_2, may involve (1) formation of volatile metal oxide or metal chloride complexes that attach to the support and are subsequently decomposed to small crystallites upon reduction and/or (2) formation of oxide particles or films that break into small crystallites during subsequent reduction.

In general, sintering processes are kinetically slow (at moderate reaction temperatures) and irreversible or difficult to reverse. Thus, sintering is more easily prevented than cured.

Factors Affecting Metal Particle Growth and Redispersion in Supported Metals. Temperature, atmosphere, metal type, metal dispersion, promoters/impurities and support surface area, texture, and porosity are the principal parameters affecting rates of sintering and redispersion. Sintering rates increase exponentially with temperature. Metals sinter relatively rapidly in oxygen and relatively slowly in hydrogen. Water vapor also increases the sintering rate of supported metals.

Promoters or impurities affect sintering and redispersion by either increasing (eg, chlorine and sulfur) or decreasing (eg, oxygen, calcium, cesium) metal atom mobility on the support; in the latter case this is due to their high resistance to dissociation and migration due to high melting points as well as their hindering dissociation and surface diffusion of other atoms.

Sintering studies of supported metals are generally of two types: (1) studies of commercially relevant supported metal catalysts and (2) studies of model metal–support systems. The former type provides useful rate data that can be used to predict sintering rates, the latter type insights into the mechanisms of metal particle migration and sintering, although the results cannot be quantitatively extrapolated to predict behavior of commercial catalysts.

Sintering of Catalyst Carriers. Single-phase oxide carriers sinter by one or more of the following processes: (1) surface diffusion, (2) solid-state diffusion, (3) evaporation/condensation of volatile atoms or molecules, (4) grain boundary diffusion, and (5) phase transformations. In oxidizing atmospheres, γ-alumina and silica are the most thermally stable carriers; in reducing atmospheres, carbons are the most thermally stable carriers. Additives and impurities affect the thermal properties of carriers by occupying defect sites or forming new phases.

Effects of Sintering on Catalyst Activity. Specific activity (based on catalytic surface area) can either increase or decrease with increasing metal crystallite size during sintering if the reaction is structure sensitive or can be independent of changes in metal crystallite size if the reaction is structure-insensitive. Thus, for a structure sensitive reaction, the impact of sintering may be either magnified or moderated, while for a structure insensitive-reaction sintering has in principle no effect on specific activity (per unit surface area).

Gas/Vapor–Solid and Solid-State Reactions

In addition to poisoning, there are a number of chemical routes leading to catalyst deactivation: (1) reactions of the vapor phase with the catalyst surface to produce (a) inactive bulk and surface phases (rather than strongly adsorbed species) or (b) volatile compounds that exit the catalyst and reactor in the vapor phase; (2) catalytic solid-support or catalytic solid-promoter reactions, and (3) solid-state transformations of the catalytic phases during reaction.

Gas/Vapor–Solid Reactions. Dispersed metals, metal oxides, metal sulfides, and metal carbides are typical catalytic phases, the surfaces of which are similar in composition to the bulk phases. For a given reaction, one of these catalyst types is generally substantially more active than the others, eg, only Fe and Ru metals are active for ammonia synthesis, while the oxides, sulfides, and carbides are inactive. If, therefore, one of these metal catalysts is oxidized, sulfided, or carbided, it will lose essentially all of its activity. While these chemical modifications are closely related to poisoning, the distinction here is that rather than losing activity owing to the presence of an adsorbed species, the loss of activity is due to the formation of a new phase altogether.

Metal loss through direct vaporization is generally an insignificant route to catalyst deactivation. By contrast, metal loss through formation of volatile compounds, eg, metal carbonyls, oxides, sulfides, and halides in CO, O_2, H_2S, and halogen-containing environments, can be significant over a wide range of conditions, including relatively mild conditions. While the chemical properties of volatile metal carbonyls, oxides, and halides are well known, there is surprisingly little information available on their rates of formation during catalytic reactions.

Some examples of catalyst deactivation due to volatile compound formation include (1) loss of the phosphorus promoter from the VPO catalyst used in the fluidized-bed production of maleic anhydride with an attendant loss of catalyst selectivity (2) vapor-phase loss of the potassium promoter from steam-reforming catalysts in the high temperature, steam-containing environment,

and (3) loss of Mo from a 12-Mo-V-heteropolyacid due to formation of a volatile Mo species during oxydehydrogenation of isobutyric acid to methacrylic acid.

The roles of kinetics and thermodynamics in deactivaton by formation of volatile metal compounds can be stated in general terms:

1. At low temperatures and partial pressures of the volatilization agent (VA), the overall rate of the process is limited by the rate of volatile compound formation.

2. At intermediate temperatures and partial pressures of the VA, the rate of formation of the volatile compound exceeds the rate of decomposition. Thus, the rate of vaporization is high, the vapor is stable, and metal loss is high.

3. At high temperatures and partial pressures of the VA, the rate of formation equals the rate of decomposition; ie, equilibrium is achieved. However, the volatile compound may be too unstable to form or may decompose before there is an opportunity to be transported from the system. From the previous work, it is also evident that besides temperature and gas phase composition, catalyst properties (crystallite size and support) can play an important role in determining the rate of metal loss.

Solid-State Reactions. Catalyst deactivation by solid-state diffusion and reaction appears to be an important mechanism for degradation of complex multicomponent catalysts in dehydrogenation, synthesis, partial oxidation, and total oxidation reactions. However, it is difficult in most of these reactions to know the extent to which the solid-state processes such as diffusion and solid-state reaction are affected by surface reactions. Recognizing this limitation, the focus here is on processes in which formation of a new bulk phase (and presumably the attendant surface phase) leads to substantially lower activity.

There are two basic principles underlying most solid-state reactions in working catalysts: (1) the active catalytic phase is generally a high-surface-area defect structure of high surface energy and as such acts as a precursor to more stable, but less active phases; and (2) the basic reaction processes may themselves trigger the solid-state conversion of the active phase to an inactive phase; for example, it may involve a redox process, part of which nucleates the inactive phase.

Mechanical Failure of Catalysts

Mechanical failure of catalysts is observed in several forms, including (1) crushing of granular, pellet, or monolithic catalyst forms due to a load; (2) attrition, the size reduction, and/or breakup of catalyst granules or pellets to produce fines, especially in fluid or slurry beds; and (3) erosion of catalyst particles or monolith coatings at high fluid velocities. Attrition is evident by a reduction in the particle size or a rounding or smoothing of the catalyst particles, easily observed under an optical or electron microscope. Washcoat loss is observed by scanning a cross-section of the honeycomb channel with either an optical or an electron microscope. Large increases in pressure drop in a catalytic process are often indicative of fouling, masking, or the fracturing and accumulation of attritted catalyst in the reactor bed.

Commercial catalysts are vulnerable to mechanical failure in large part because of the manner in which they are formed; that is, catalyst granules, spheres, extrudates, and pellets ranging in diameter from 50 μm to several millimeters are in general prepared by agglomeration of 0.02–2 μm aggregates of much smaller primary particles having diameters of 10–100 nm by means of precipitation or gel formation followed by spray drying, extrusion, or compaction. These agglomerates have in general considerably lower strengths than the primary particles and aggregates of particles from which they are formed.

Two principal mechanisms are involved in the mechanical failure of catalyst agglomerates: (1) fracture of agglomerates into smaller agglomerates of approximately $0.2d_0$–$0.8d_0$ and (2) erosion or abrasion of aggregates of primary particles having diameters ranging from 0.1 to 10 μm from the surface of the agglomerate. While erosion is caused by mechanical stresses, fracture may be due to mechanical, thermal, and/or chemical stresses. Mechanical stresses leading to fracture or erosion in fluidized or slurry beds may result from (1) collisions of particles with each other or with reactor walls or (2) shear forces created by turbulent eddies or collapsing bubbles (cavitation) at high fluid velocities. Thermal stresses occur as catalyst particles are heated and/or cooled rapidly; they are magnified by temperature gradients across particles and by differences in thermal expansion coefficients at the interface of two different materials. Chemical stresses occur as phases of different density are formed within a catalyst particle via chemical reaction.

Role of Properties of Ceramic Agglomerates in Determining Strength and Attrition Resistance.

The extent to which a fracture or erosion mechanism, participates in agglomerate size reduction depends upon (1) the magnitude of a stress, (2) the strength and fracture toughness of the agglomerate, (3) agglomerate size and surface area, and (4) crack size and radius. Erosion (abrasion) occurs when the stress exceeds the agglomerate strength. The erosion rate increases with decreasing agglomerate size.

Most heterogeneous catalysts are complex, multiphase materials that consist in large part of porous ceramic materials. When a tensile stress of a magnitude close to the yield point is applied, ceramics almost always undergo brittle fracture before plastic deformation can occur. Brittle fracture occurs through formation and propagation of cracks through the cross-section of a material in a direction perpendicular to the applied stress. Agglomerate fracture due to a tensile stress occurs by propagation of internal and surface flaws; these flaws created by external stresses or inherent defects are stress multipliers. Tensile stress multipliers may be microcracks, internal pores, and grain corners. The ability of a material to resist fracture is termed fracture toughness.

For ceramic materials in compression there is no stress amplification due to flaws or pores; thus ceramic materials (including catalytic materials) in compression are much stronger (approximately a factor of 10) than in tension.

Tensile Strengths and Attrition Resistance of Catalyst Supports and Catalysts.

The strengths of nonporous, annealed crystalline or polycrystalline materials do not necessarily apply to porous catalyst agglomerates even under compression; rather, agglomerate strength is dependent upon the strengths of chemical and physical bonds, including the cohesive energy between primary particles. Agglomerate strength would depend greatly on the preparation of the compact. Catalyst agglomerates are generally substantially weaker than polycrystalline ceramic materials prepared by high-temperature sintering such as alumina. Even subtle differences in preparation and pretreatment also affect agglomerate strength.

The mechanism by which attrition occurs (erosion or fracture) can vary with catalyst or support preparation, crush strength, reactor environment, and with the mechanical test method. In the presence of a large stress, weaker oxide materials are prone to failure by fracture, while stronger materials tend to erode. But fracture may be the preferred mechanism for strong TiO_2 agglomerates, while abrasion is favored for weaker agglomerates. Data used show that attrition mechanism and rate are independent of agglomerate strength but depend instead on the type of material. Which of the two attrition mechanisms predominates depends much more on material.

Implications of Mechanistic Knowledge of Attrition for Catalyst Design.

Several alternatives follow from this discussion for increasing attrition resistance: (1) increasing aggregate/agglomerate strength by means of advanced preparation methods and carefully controlled precipitation methods (2) adding binders to improve strength and toughness, (3) coating aggregates with a porous but very strong material such as ZrO_2 to significantly improve its attrition resistance, and (4) chemical or thermal tempering of agglomerates to introduce compressive stresses that increase strength and attrition resistance which of the two attrition mechanisms. Predominates depends much more on material composition and type than on agglomerate strength. However, irrespective of mechanism the rate of attrition is usually greater for the weaker material.

Summary of Deactivation Mechanisms for Solid Catalysts

The causes of solid (heterogeneous) catalyst deactivation are basically chemical, mechanical, and thermal. The mechanisms of heterogeneous catalyst deactivation can be classified into (1) chemical degradation, including volatilization and leaching; (2) fouling; (3) mechanical degradation; (4) poisoning; and (5) thermal degradation. Poisoning and thermal degradation are generally slow processes, while fouling and some forms of chemical and mechanical degradation can lead to rapid, catastrophic catalyst failure. Some forms of poisoning and many forms of fouling are reversible; hence, reversibly poisoned or fouled catalysts are relatively easily regenerated. On the other hand, chemical, mechanical, and thermal forms of catalyst degradation are rarely reversible.

It is often easier to prevent rather than cure catalyst deactivation. Many poisons and foulants can be removed from feeds using guard beds, scrubbers, and/or filters. Fouling, thermal degradation, and chemical degradation can be minimized through careful control of process conditions. Mechanical degradation can be minimized by careful choice of carrier materials, coatings, and/or catalyst particle forming methods.

While treating or preventing catalyst deactivation is facilitated by an understanding of the mechanisms, additional perspectives are provided by examining the route by which each of the mechanisms causes a loss of catalytic activity saved influences the reaction rate. Each of the deactivation mechanisms affects one or more of the factors comprising observed activity (see Table 3), but all of the mechanisms can effect a decrease in the number of catalytic sites.

HOMOGENEOUS CATALYSTS AND ENZYMES

Homogeneous Catalysts

Homogeneous catalysts may undergo degradation by routes similar to those of heterogeneous catalysts, eg, by chemical modification, poisoning, and thermal degradation. However, the specific details of these mechanistic routes are generally somewhat different, since the catalyst is a molecule rather than a solid; that is, an organometallic complex is quite different from a metal surface in terms of structure and scale.

Homogeneous catalysts are generally metal–ligand complexes. The metal center functions as the active site, while the ligands serve to influence site chemistry through electronic modifications of the metal that influence activity/selectivity and through geometric constraints that enhance selectivity. Hence activity and selectivity properties of homogeneous catalysts can be significantly influenced by processes that change the chemistry either of the metal center or the ligands or both.

Mechanisms (or causes) of homogeneous catalyst degradation can be classified as (1) metal deposition reactions, eg, decarbonylation of carbonyl complexes, loss

of protons from cationic species, or reductive elimination of C-, N-, or O-donor fragments; (2) decomposition of ligands attached to a catalytic complex; (3) reactions of metal–carbon and metal–hydride bonds with polar species (eg, water, oxygen, acids, alcohols, olefins, and halides); and (4) poisoning of active sites by impurities, reactants, or products or by dimerization of the catalyst.

Mechanisms 1 and 2 lead mainly to deactivation by either loss or modification of ligands, while mechanisms 3 and 4 cause deactivation largely by either modifying or poisoning the metal, although ligands are also clearly modified by type 3 mechanisms. Of the four mechanisms, deactivation by metal formation and deposition is the most common.

Mechanisms 1 and 4 are reversible to some extent. Mechanisms 2 and 3, involving breaking of active site bonds and formation of stable products, are largely irreversible. Products of ligand oxidation are generally more stable than the complexes from which they were formed.

Enzymes

Structural and Catalytic Properties of Enzymes. Enzymes are globular macromolecular polypeptide proteins (molecular weights of $10^4 - 10^6$) synthesized by living organisms. Each enzyme has a unique three-dimensional structure with a binding site or pocket that is chemically and geometrically compatible with a single reactant molecule (substrate) or group of chemically related reactants; in other words, enzymes have molecular-recognition capability. Enzymes are unique in their ability to catalyze biochemical reactions with high selectivity (essentially 100%) at extraordinarily high rates. These activities enable enzymes to be effective catalysts at extremely low concentrations.

The stereochemical specificity of enzymes is unmatched and absolute; ie, their sites can distinguish between optical and geometrical isomers, almost always catalyzing only the reaction of one isomer of an enantiomeric pair. Nevertheless, some enzymes catalyze reactions of chemically unrelated species; for example, nitrogenase reduces N_2 to NH_3 as well as hydrogenating acetylene to ethylene.

Table 3. How Deactivation Mechanisms Affect the Rate of a Catalyzed Reaction and the Rapidity and Reversibility of Deactivation Process

	Effects on reaction rate		Deactivation process		
Deactivation mechanism	Decrease in number of active sites	Decrease in intrinisic activity (k_{intr})	Decrease in effectiveness factor (η)	Fast or slow[a]	Reversible
chemical degradation	×	×	×[b,c]	varies	no
fouling	×		×	fast	yes
mechanical degradation	×			varies	no
poisoning	×	×		slow	usually
sintering	×	×[b,d]	×[b,e]	slow	sometimes
vaporization/leaching	×		×[b,f]	fast	sometimes

[a]Generally.
[b]In some cases.
[c]Chemical degradation can cause breakdown of support, pore plugging, and loss of porosity.
[d]If the reaction is structure-sensitive, sintering could either increase or decrease intrinsic activity.
[e]Sintering of the support may cause support collapse and loss of porosity; it may also increase average pore diameter.
[f]Leaching of aluminum or other cations from zeolites can cause buildup of aluminum or other oxides in zeolite pores.

There were an estimated 3000 known enzymes in 1996. An average cell contains some 3000 different enzymes, and as many as 25,000 different enzymes may exist.

While they are synthesized in vitro and are active only within a limited range of pH and temperature, enzymes otherwise have properties similar to synthetic homogeneous and polymer-supported catalysts. Moreover, they can be extracted from their biological source, purified, crystallized, and used in laboratory studies or industrial processes. Further, they can be attached to glass or ceramic supports and used as heterogeneous catalysts. Their application in industrial processes is rapidly increasing.

Distinctive catalytic characteristics of enzymes include (1) their flexible structure, which facilitates an "induced fit" of the substrate, the making and breaking of bonds, and the departure of products, and (2) their sensitivity to reaction *effectors* (inhibitors or activators), which function similarly to promoters of heterogeneous catalysts. Some enzymes require a *cofactor* that combines with the enzyme to form a catalytic site; metal ions are examples of cofactors. Enzymatic reactions may also require a *coenzyme* that reacts with the reactant to produce an enzyme-compatible substrate. Living organisms control and optimize biological processes using a variety of tools: (1) enzyme effectors, (2) regulation of enzyme growth or activation rates, (3) compartmentalization of enzymes within organs or organelles, and (4) destruction (editing) of undesired intermediates or products.

Deactivation of Enzymes Enzymes generally function only under mild conditions of temperature and pH observed in living organisms. Under typical commercial reaction conditions (40–60°C, 1 atm) enzymes otherwise stable in solution may lose activity rapidly as a result of only slight changes in their environment such as temperature, pressure, pH, and ionic strength that induce small free energy changes from native to denatured states. Moreover, their separation from the product is generally difficult and may cause further denaturation and loss of catalytic activity. The modest, largely reversible losses of activity resulting from small changes in reaction environment are largely due to modest changes in conformation of the active site. More severe changes in reaction conditions typically bring about the dissociation and unfolding of the quaternary and tertiary structures, respectively, into primary chains that subsequently order into fibrous protein bundles; in the process active sites are irreversibly destroyed. If further exposed to severe conditions of temperature and pH, the principal chain structure of the protein may undergo loss or modification of functional groups or amino acid residues.

The activity of a typical enzyme increases exponentially with temperature in accordance with the Arrhenius law up to about 50–60°C, passes through a maximum and then declines precipitously above about 60–70°C. Thus, catalyst life may be on the order of days to weeks at around 50°C; however, the deactivation rate is extremely high at only slightly higher temperatures. Nevertheless, a few enzymes are active and stable at temperatures exceeding 100°C; for example, α-amylase catalyzes starch liquifaction at 105–115°C. Because their deactivation rates are highly

temperature-dependent, enzymes are generally shipped and stored under refrigeration (0–4°C); at these low temperatures they are generally stable for months.

Causes of deactivation can be classified as chemical, mechanical, and thermal. However, for enzymes these causes are closely linked, since mechanically and thermally induced routes almost always effect chemical changes. Thermally induced chemical change (at elevated temperature) is the most likely scenario for enzyme deactivation.

Chemical deactivation mechanisms include (1) changes in stereo configuration by protons or hydroxyl ions at or near active sites, (2) structural modifications in aqueous or nonaqueous solvents, (3) poisoning of active sites by inhibitors, including "Trojan-horse inhibitors" that are activated by the target enzyme, (4) aggregation, (5) unfolding, (6) fragmentation due to solvolysis, hydrolysis in water, or self-hydrolysis (autolysis) of proteases, and (7) oxidation in air. Mechanisms 1–5 may be reversible, mechanisms 6 and 7 generally irreversible. Mechanical deactivation may be caused by hydrodynamic shear forces.

Thermal inactivation of enzymes, a well-studied phenomenon, may be either reversible or irreversible. Potentially reversible changes (due to small, short excursions in temperature near the characteristic unfolding temperature) include light aggregation, conformational changes, folding without further chemical change, disulfide exchange, and/or breaking of hydrogen bonds. Irreversible denaturation (due to prolonged, severe thermal treatment) may be caused by cleavage of disulfide bonds and/or cystinyl cross-links; unfolding followed by chemical change; chemical changes of the primary structure and/or active site; strong aggregation of inactive unfolded forms; and formation of rubbery, tough fibrous structures due to alignment and bundling of unfolded primary chains similar to that observed during the boiling of an egg. Chemical bonding of unfolded primary chains to form fibers is thermodynamically favorable because chemical bonding of hydrophobic functions exposed by unfolding lowers the entropy and hence free energy of the system.

Strategies to improve both chemical and thermal enzyme stability include (1) use of soluble additives, (2) immobilization, (3) protein engineering, and (4) chemical modification. Chemical modification and immobilization are probably the most successful, widely used methods. Enzyme stability can be greatly enhanced and recovery problems obviated by immobilizing (heterogenizing) enzymes.

PREVENTION OF CATALYST DECAY

General Principles

The catalyst inventory for a large plant may entail a capital investment of tens of millions of dollars. In such large-scale processes, the economic return on this investment may depend on the catalysts, remaining effective over a period of up to 3 to 5 years. This is particularly true of those processes involving irreversible or only partially reversible deactivation (eg, sulfur poisoning or sintering). In many processes more than one mechanism limits catalyst life. Moreover, there is a wide variation in catalyst lifetimes among different processes, ie, from 10^{-6} to

15 years. While there is clearly greater interest in extending catalyst lifetimes in processes where their life is short, great care must be exercised in protecting the catalyst in any process from process upsets that might reduce typical catalyst life from years to hours.

While complete elimination of catalyst deactivation is not possible. the rate of damage can be minimized in many cases through understanding of the mechanisms, thereby enabling control of the deactivation process.

Prevention of Chemical Degradation (by Vapor–Solid and Solid–Solid Reactions)

The most serious problems—oxidation of metal catalysts, overreduction of oxide catalysts, and reaction of the active catalytic phase with carrier or promoter—can be minimized or prevented by careful catalyst and process design. In Fischer–Tropsch synthesis, the oxidation of the active cobalt phase in supported cobalt catalysts to inactive oxides, aluminates, and silicates can be minimized by employing a two- or three-stage process in which product steam is moderated in the first stage by limiting conversion and in subsequent stages by interstage removal of water. It can also be moderated by addition of noble metal promoters that facilitate and maintain high reducibility of the cobalt and by coating the alumina or silica support with materials such as ZrO_2 that are less likely to react with cobalt to form inactive phases.

Prevention of Fouling by Coke and Carbon

Most of the general methods of preventing coke or carbon formation are based on one important fundamental principle, namely that *carbon or coke results from a balance between the reactions that produce atomic carbon or coke precursors and the reactions of these species with H_2, H_2O, or O_2 that remove them from the surface*. If the conditions favor formation over gasification, these species accumulate on the surface and react further to less active forms of carbon or coke, which either coat the surface with an inactive film or plug the pores, causing loss of catalyst effectiveness, pore plugging, or even destruction of the carrier matrix.

Methods to lower rates of formation of carbon or coke precursors relative to their rates of gasification vary with the mechanism of formation (ie, gas, surface, or bulk phase) and the nature of the active catalytic phase (eg, metal or oxide).

Coke deposition on oxide or sulfide catalysts occurs mainly on strongly acidic sites; accordingly, the rate of coking can be lowered by decreasing the acidity of the support.

As in the case of poisoning, there are certain reactor bed or catalyst geometries that minimize the effects of coking on the reaction. Choosing supports with relatively large pores minimizes pore plugging; choice of large-diameter, mechanically strong pellets avoids or delays reactor plugging. But in view of the rapidity at which coke and carbon can deposit on, plug, and even destroy catalyst particles, the importance of preventing the onset of such formation cannot be overemphasized.

Re-forming of naphtha provides an interesting case study of catalyst and process designs to avoid deactivation by coking. Naphtha reforming processes are designed for (*1*) high enough H_2 pressure to favor gasification of coke precursors while minimizing hydrocracking, (*2*) maintenance of Cl and S contents throughout the bed to ensure optimum acidity and coke levels, and (*3*) low enough overall pressure to thermodynamically and kinetically favor dehydrogenation and dehydrocylization. Accordingly, optimal process conditions are a compromise between case 1 and case 3. The above-mentioned improvements in catalyst technologies, especially resistance to coking, have enabled important process improvements such as optimal operation at lower pressure; thus, processes have evolved over the past two to three decades from conventional fixed-bed reactors at high pressure (35 bar) using nonregenerative Pt catalysts to low pressure (3.5 bar), slowly moving-bed, continuously regenerated units with highly selective Pt/Sn catalysts, resulting in substantial economic benefits.

Prevention of Poisoning

Since poisoning is generally due to strong adsorption of feed impurities and poisoned catalysts are generally difficult or impossible to regenerate, it is best prevented by removal of impurities from the feed to levels that will enable the catalyst to operate at its optimal lifetime. In cracking or hydrocracking reactions on oxide catalysts, it is important to remove strongly basic compounds such as ammonia, amines, and pyridines from the feed; ammonia in some feedstocks, for example, can be removed by aqueous scrubbing. The poisoning of catalysts by metal impurities can be moderated by selective poisoning of the unwanted metal. The poisoning of hydrotreating catalysts by nickel and vanadium metals can be minimized by (*1*) using a guard bed of inexpensive Mo catalyst or a graded catalyst bed with inexpensive, low-activity Mo at the top (bed entrance) and expensive, high-activity catalyst at the bottom and (*2*) depositing coke prior to the metals, since these metal, deposits can be physically removed from the catalyst during regeneration.

It may be possible to lower the rate of poisoning through careful choice of reaction conditions that lower the strength of poison adsorption or by choosing mass-transfer-limiting regimes that limit deposits to the outer shell of the catalyst pellet, while the main reaction occurs uninterrupted on the interior of the pellet. The manner in which the active catalytic material is deposited on a pellet can significantly influence the life of the catalyst.

Prevention of Sintering

Since most sintering processes are irreversible or are reversed only with great difficulty, it is important to choose reaction conditions and catalyst properties that avoid such problems. Metal growth is a highly activated process; thus by choosing reaction temperatures lower than 0.3–0.5 times the melting point of the metal, rates of metal sintering can be greatly minimized. The same principle holds true in avoiding recrystallization of metal oxides, sulfides, and supports. One approach to lowering reaction temperature is to maximize activity and surface area of the active catalytic phase.

Although temperature is the most important variable in the sintering process, differences in reaction atmosphere can also influence the rate of sintering. Water vapor in particular accelerates the crystallization and structural modification of oxide supports. Accordingly, it is vital to minimize the concentration of water vapor in high-temperature reactions on catalysts containing high surface area supports.

Besides lowering temperature and minimizing water vapor, it is possible to lower sintering rates through addition of thermal stabilizers to the catalyst.

Designing thermally stable catalysts is a particular challenge in high-temperature reactions such as automotive emissions control, ammonia oxidation, steam reforming, and catalytic combustion. The development of thermally stable automotive catalysts has received considerable attention, providing a wealth of scientific and technological information on catalyst design. The basic design principles are relatively simple: (1) utilize thermally and hydrothermally stable supports; (2) use PdO rather than Pt or Pt–Rh for high-temperature converters, since PdO is considerably more thermally stable in an oxidizing atmosphere because of its strong interaction with oxide supports; and (3) use multilayer strategies and/or diffusion barriers to prevent thermally induced solid-state reactions and to moderate the rate of highly exothermic CO and hydrocarbon oxidations.

Prevention of Mechanical Degradation

In terms of catalyst design it is important to (1) choose supports, support additives, and coatings that have high fracture toughness, (2) use preparation methods that favor strong bonding of primary particles and agglomerates in pellets and monolith coatings, (3) minimize (or rather optimize) porosity (thus maximizing density), and (4) use binders such as carbon to facilitate plastic deformation and thus protect against brittle fracture. Processes (and to some extent preparation procedures) should be designed to minimize (1) highly turbulent shear flows or cavitation that lead to fracture of particles or separation of coatings, (2) large thermal gradients or thermal cycling leading to thermal stresses, and (3) formation of chemical phases of substantially different densities or formation of carbon filaments leading to fracture of primary particles and agglomerates. Nevertheless, thermal or chemical tempering can be used in a controlled fashion to strengthen catalyst particles or agglomerates.

Recent scientific and patent studies focusing on the Fischer–Tropsch synthesis in slurry reactors indicate that (1) spray drying of particles improves their density and attrition resistance; (2) addition of silica and/or alumina into titania improves its attrition resistance, while addition of only 2000–3000 ppm of titania to γ-alumina improves alumina's attrition resistance; and (3) preformed alumina spheres promoted with La_2O_3 provide greater attrition resistance relative to silica. Increasing attrition resistance is apparently correlated with increasing density.

REGENERATION OF DEACTIVATED CATALYSTS

The loss of catalytic activity in most processes is inevitable. When the activity has declined to a critical level, a choice must be made among four alternatives: (1) restore the activity of the catalyst, (2) use it for another application, (3) reclaim and recycle the important and/or expensive catalytic components, or (4) discard the catalyst. The first alternative (regeneration and reuse) is almost always preferred; catalyst disposal is usually the last resort, especially in view of environmental considerations.

The ability to reactivate a catalyst depends upon the reversibility of the deactivation process. Carbon and coke formation is relatively easily reversed through gasification with hydrogen, water, or oxygen. Sintering is generally irreversible, although metal redispersion is possible under certain conditions in selected noble metal systems. Some poisons or foulants can be selectively removed by chemical washing, mechanical treatments, heat treatments, or oxidation; others cannot be removed without further deactivating or destroying the catalyst.

The decision to regenerate/recycle or discard the entire catalyst depends largely on the rate of deactivation. If deactivation is very rapid, as in the coking of cracking catalysts, repeated or continuous regeneration becomes an economic necessity. Precious metals are almost always reclaimed where regeneration is not possible. Disposal of catalysts containing nonnoble heavy metals is environmentally problematic and should be a last resort; if disposal is necessary, it must be done with great care, probably at great cost. Accordingly, a choice to discard depends upon a combination of economic and legal factors. Because of the scarcity of landfill space and an explosion of environmental legislation, there is a growing trend to regenerate or recycle spent catalysts. A sizable catalyst regeneration industry benefits petroleum refiners by helping to control catalyst costs and limiting liabilities.

The largest fraction of the enormous literature describes processes for regeneration of catalysts in three important petroleum refining processes: FCC, catalytic hydrocarbon reforming, and alkylation. However, a significant number of patents also claim methods for regenerating absorbents and catalysts used in aromatization, oligomerization, catalytic combustion, SCR of NO, hydrocracking, hydrotreating, halogenation, hydrogenation, isomerization, partial oxidation of hydrocarbons, carbonylations, hydroformylation, dehydrogenation, dewaxing, Fisher–Tropsch synthesis, steam reforming, and polymerization.

Regeneration of Catalyst Deactivated by Coke or Carbon

Carbonaceous deposits can be removed by gasification with O_2, H_2O, CO_2, and H_2. The temperature required to gasify these deposits at a reasonable rate varies with the type of gas, the structure and reactivity of the carbon or coke, and the activity of the catalyst. Rates of gasification of coke or carbon are greatly accelerated by the same metal or metal oxide catalysts upon which carbon or coke deposits.

Because catalyzed removal of carbon with oxygen is generally very rapid at moderate temperatures (eg, 400–600°C), industrial processes typically regenerate catalysts deactivated by carbon or coke in air. Air regeneration is used to remove coke from catalysts in catalytic cracking, hydrotreating processes, and catalytic reforming.

One of the key problems in air regeneration is avoiding hot spots or overtemperatures which could further deactivate the catalyst. The combustion process is typically controlled by initially feeding low concentrations of air and by increasing oxygen concentration with increasing carbon conversion; nitrogen gas can be used as a diluent in laboratory-scale tests, while steam is used as a diluent in full-scale plant operations.

Because coke burn-off is a rapid, exothermic process, the reaction rate is controlled to a large extent by film heat and mass transfer. Accordingly, burn-off occurs initially at the exterior surface and then progresses inward with the reaction occurring mainly in a shrinking shell consistent with a "shell-progressive" or "shrinking-core" model; Burn-off rates for coke deposited on SiO_2/Al_2O_3 catalysts has been found to be independent of initial coke level, coke type, and source of catalyst.

Regeneration of Poisoned Catalysts

Studies of regeneration of sulfur-poisoned Ni, Cu, Pt, and Mo with oxygen/air, steam, hydrogen, and inorganic oxidizing agents indicate that up to 80% removal of surface sulfur from Mg- and Ca-promoted Ni, steam reforming catalysts occurs at 700°C in steam.

Although this treatment is partially successful in the case of low-surface-area steam reforming catalysts, the high temperatures required for these reactions would cause sintering of most high-surface-area nickel catalysts.

Regeneration of sulfur-poisoned catalysts, particularly base metal catalysts, in air or oxygen has been largely unsuccessful. Nevertheless, sulfur can be removed as SO_2 at very low oxygen partial pressures, suggesting that regeneration is possible under carefully controlled oxygen or species such as CO_2 or NO that dissociate to oxygen. Regeneration of sulfur-poisoned noble metals in air is more easily accomplished than with steam, although it is frequently attended by sintering. Regeneration of sulfur-poisoned nickel catalysts using hydrogen is impractical because (1) adsorption of sulfur is reversible only at high temperatures at which sintering rates are also high, and (2) rates of removal of sulfur in H_2 as H_2S are slow even at high temperature.

Inorganic oxidizing agents such as $KMnO_4$ can be used to oxidize liquid phase or adsorbed sulfur to sulfites or sulfates. These electronically shielded structures are less toxic than the unshielded sulfides. This approach has somewhat limited application.

Redispersion of Sintered Catalysts

During catalytic reforming of hydrocarbons on platinum-containing catalysts, growth of 1-nm platinum metal clusters to 5–20-nm crystallites occurs. An important part of the catalyst regeneration procedure is the redispersion of the platinum phase by a high-temperature treatment in oxygen and chlorine, generally referred to as "oxychlorination."

Some guidelines and principles regarding the redispersion process follow:

1. In cases involving a high degree of Pt sintering or poisoning, special regeneration procedures may be required. If large crystallites have been formed, several successive oxychlorinations are performed.

2. Introducing oxygen into reactors in parallel rather than in series results in a significant decrease in regeneration time.

3. Introduction of hydrocarbons present in the reactor recycle after regeneration is said to stabilize the catalyst; solvents such as ammonium acetate, dilute nitric acid containing lead nitrate, EDTA and its diammonium salt are reported to dissolve out metal aggregates without leaching out the dispersed metal.

4. The procedures for redispersion of Pt/alumina are not necessarily applicable to Pt on other supports or to other metals. For example, Pt/silica is redispersed at lower temperature and higher Cl_2 concentration (150–200°C and 25% Cl_2). Pd/alumina can be redispersed in pure O_2 at 500°C. While Pt–Re/alumina is readily redispersed by oxychlorination at 500°C, Pt–Ir/alumina is not redispersed in the presence of O_2 unless the catalyst is pretreated with HCl.

An extensive scientific and patent literature of redispersion describes the use of chlorine, oxygen, nitric oxide, and hydrogen as agents for redispersion of sintered catalysts. Recent literature demonstrates the need for understanding the detailed surface chemistry in order to successfully develop and improve redispersion processes, especially in more complex catalyst systems such as alumina-supported bimetallics.

Redispersion of alumina-supported platinum and iridium crystallites is also possible in a chlorine-free oxygen atmosphere if chlorine is present on the catalyst. The extent of redispersion depends on the properties of the Pt/Al_2O_3 catalyst and temperature. The question whether redispersion of platinum occurs only in oxygen without chlorine present on the catalyst remains controversial.

Two models, "the thermodynamic redispersion model" and "the crystallite splitting model," have been advanced to explain the redispersion in oxygen. The "thermodynamic" redispersion model hypothesizes the formation of metal oxide molecules that detach from the crystallite, migrate to active sites on the support, and form surface complexes with the support. Upon subsequent reduction, the metal oxide complexes form monodisperse metal clusters. In the "crystallite splitting" model, exposure of a platinum crystallite to oxygen at 500°C leads to formation of a platinum oxide scale on the outer surface of the crystallite, which stresses and ultimately leads to splitting of the particle.

SUMMARY

1. The causes of deactivation are basically of three kinds: chemical, mechanical, and thermal. The five intrinsic mechanisms of catalyst decay, (a) poisoning,

(b) fouling, (c) thermal degradation, (d) chemical degradation, and (e) mechanical failure, vary in their reversibility and rates of occurrence. Poisoning and thermal degradation are generally slow, irreversible processes while fouling with coke and carbon is generally rapid and reversible by regeneration with O_2 or H_2.

2. Catalyst deactivation is more easily prevented than cured. Poisoning by impurities can be prevented through careful purification of reactants. Carbon deposition and coking can be prevented by minimizing the formation of carbon or coke precursors through gasification, careful design of catalysts and process conditions, and by controlling reaction rate regimes, eg, mass transfer regimes, to minimize effects of carbon and coke formation on activity. Sintering is best avoided by minimizing and controlling the temperature of reaction.

3. Prevention and monitoring are important engineering principles in "standard of care" practice. The prevention of catalyst decay is important in every aspect of a process including design, construction, operation, and regeneration. Careful monitoring of process variables is a necessity in understanding and preventing catalyst decay problems of either a slow or a catastrophic nature.

4. The optimization of a catalytic process considers optimum operation and regeneration policies subject to constraints of catalyst cost, operation cost, regeneration cost, and product value. The optimum operating policy maximizes the rate of formation of product during the operating period.

5. Catalyst deactivation kinetics for reactions involving relatively slow deactivation can be experimentally determined using a laboratory fixed-bed, mixed-fluid (CSTR) reactor. Reactors and processes involving a slowly deactivating catalyst can be designed using relatively simple numerical analysis of the design equations and a pseudo-steady-state approximation for the main reaction.

6. Modeling and experimental assessment of deactivation processes are useful in providing (a) accelerated simulations of industrial processes, (b) predictive insights into effects of changing process variables on activity, selectivity, and life, (c) estimates of kinetic parameters needed for design and modeling, (d) estimates of size and cost for scale-up of a process, and (e) a better understanding of the basic decay mechanisms. It is now possible to develop realistic mathematical models of most catalytic processes, which can be used in conjunction with short-term experimental tests to accurately predict catalyst life in a commercial unit. Proper application of this approach could save companies millions of dollars by alleviating the need for long-term deactivation tests and/or premature shutdown.

Trends

Research and development in catalyst deactivation have grown steadily in the past three decades.

Several other trends are evident that are likely to continue:

1. The increasing use of more sophisticated analytical tools to investigate the chemistry and mechanisms of deactivation. Surface science tools such as AES, quantitative HRTEM, XPS, and STM are now routinely applied to investigate deactivation mechanisms at very fundamental levels.
2. The increasing development of more sophisticated models of deactivation processes.

The combination of more sophisticated methods and models will hasten the practical application of models for predicting catalyst/process life.

Future Needs

Collection of Data. Few deactivation rate data are available for even the most important large-scale catalytic systems. Accordingly, there is a critical need for collection of such data at the laboratory, bench, and plant scale.

Data Analysis and Model Development. Much of the previously collected data were analyzed using outdated methods. There is much that could be learned by reanalyzing some of these data using new approaches such as GPLE and microkinetic modeling.

C. Bartholomew in I. I. Harvath, eds., *Encyclopedia of Catalysis*, Vol. 2, John Wiley & Sons, New York, 2003, pp. 182–315.

J. B. Butt and E. E. Petersen, *Activation, Deactivation, and Poisoning of Catalysts*, Academic Press, San Diego, 1988.

R. J. Farrauto and C. H. Bartholomew, *Fundamentals of Industrial Catalytic Processes*, Kluwer Academic Publishers, London, 1997.

B. E. Leach, ed., *Applied Industrial Catalysis*, Academic Press, New York, 1983.

CALVIN BARTHOLOMEW
Brigham Young University

CATALYSTS, SUPPORTED

The field of heterogeneous catalysis of organic reactions can be split into two distinct areas—heterogeneous catalysis of gas-phase reactions and of liquid-phase reactions. The former is a well-known, long established field, representing some of the largest chemical processes known—the cracking and other conversions of crude oil and gas, which lays the foundation for the remainder of organic chemistry. These processes are typically high temperature, high energy processes, operating on a megaton scale, and typically use zeolites as catalysts. This contribution, therefore, relates specifically to the use of heterogeneous catalysis to the synthesis of fine chemicals. The vast majority of these systems operate under moderate temperatures and in the liquid phase, although there are a few that take place in the gas phase.

PREPARATION METHODS FOR INORGANIC SUPPORTED CATALYSTS

There are several methods for the preparation of supported catalysts. One of the key concepts is the maximization of surface area; this is normally achieved by having a very porous support material, and the methods below aim to either generate materials with high surface areas, or to coat pre-formed high surface area supports with small amounts of catalytic species.

Given that the exposed surface area is where reaction takes place, it is clearly very important not to overload the material to the point where microcrystals are formed on the surface, although there may be instances where microcrystals are a necessary feature of the catalyst. The formation of the high surface area supports (which often may have catalytic activity themselves) is typically carried out by the condensation of inorganic precursors in the presence of a structure-directing agent, whose role is to control the formation of a porous or layered structure and thus maximise the surface area.

Mechanical and thermal properties are also very important features of supports. Mechanical attrition of particles in the aggressive environment of a large chemical reactor must be allowed for and may cause changes in performance on reuse (this is especially important when a solid is used in a high shear stirred tank reactor). Many commonly used support materials have more than adequate thermal stability to cope with the temperatures commonly encountered in liquid phase-organic reactions. Some materials are sensitive, however, notably those that are organic polymer based or have some surface-based organic functionalities.

Coating Preformed Supports

The simplest method, and often the most appropriate, is wet impregnation, where a solution of the material to be supported is dissolved in a solvent and mixed with the support. The incipient wetness technique involves filling the pages of the support with a solution of the reagent. Removal of the solvent then leads to the supported catalyst. The degree of dispersion depends on several factors such as the rate of removal of solvent—too rapid can lead to precipitation of clusters of catalyst, leading to reduced accessibility and possibly pore blockage. For many of the smaller pore size catalysts it may be sensible to allow a period of a few hours contact time to allow ingress of the solution into the pores (incipient wetness), but for many catalysts this is not always necessary. Loading is also a critical factor. Too high a loading can again lead to clustering (microcrystals) and reduced activity; too low can lead to the need for a larger quantity of catalyst than is optimal.

Choice of solvent is important, and is especially critical in strongly acidic systems such as supported $AlCl_3$, where the extreme sensitivity and reactivity of the catalyst limits the choice of solvent to simple aromatics such as ben-zene and chlorobenzene. Posttreatment of such catalysts is also vital, partly to remove loosely bound material that can desorb easily during reaction and contaminate the liquid phase, and partly to enhance the activity of the catalyst. Thorough washing with solvents is generally enough, and the problem can be mitigated by appropriate choice of loading. Increases in activity are often brought about by heating of the catalyst (usually to $100–150°C$), which serves to remove the last traces of solvent and water from the catalyst. More extreme heating usually is detrimental.

Other techniques have also been used, less frequently. These include grinding of two dry powders, or direct addition of both components to the reaction mix, where the supported catalyst forms *in situ*. Such approaches are occasionally of some value, but generally reproducibility is less good than with a carefully preformed system. Ion exchange can be an important method of preparation but is obviously limited to appropriate materials such as ion-exchange resins and solid acids that may be converted to sources of catalytically useful metal ions.

Formation of High Surface Area Supports/Catalysts

Catalysts can also be prepared by (co)-precipitation of oxide precursors, often brought about by a change in pH or by hydrolysis of alkoxides $M(OR)_n$. This route is typically carried out in water or aqueous alcohols, and can be used to prepare catalytic systems with one or more M atoms present.

An extension of this approach allows the incorporation of organic functionality directly into an inorganic framework. Catalysts based on the co-condensation of $Si(OR)_4$ and $R'Si(OR)_3$ can be achieved such that an organic–inorganic hybrid material is formed, where R' represents the active site of the catalyst, or a precursor thereto, or may be a group that modifies the polarity of the surface, improving adsorption–desorption phenomena and improving the activity of the catalyst. In these methods, it is important that the materials formed have a high surface area, in order to allow access to as much surface per unit mass catalyst as possible (in optimum cases surface areas of ~ 1000 m^2/g can be achieved) and it is also critical that the pores in the material be sufficiently large enough and open enough to allow free movement of the reactants and products. This is generally achieved by the use of templates, molecules that are added to the synthesis mixture and that encourage the formation of the solid around them, and that are subsequently removed. For purely inorganic materials, high temperature calcination steps are usually used to remove template, but for organic–inorganic hybrid materials, lower-temperature solvent extraction methods are used. These have the advantage of allowing recovery of the template and subsequent reuse, reducing the cost of the material.

While these template-assisted formations of high surface area oxides are now well established and lead to a range of materials with various pore systems, nontemplated methods are also known, which lead to structured and regular materials. Hydrotalcites are one such class of compounds, where a layered structure, reminiscent of the clays, is formed, again by alteration of pH.

Many of these methods lead directly to powdered solids that are directly usable in stirred tank reactors and that can be readily separated from reaction mixtures by centrifugation. However, some systems lend themselves to processing as films or monoliths, which may be more appropriate for use in continuous reactors by immobilisation onto, eg, a metal support or even, in some cases, the formation of a membrane.

SUPPORTED CATALYSTS AS SOLID ACIDS

Solid acids are generally categorized by their Brønsted and/or Lewis acidity, the strength and number of these sites, and the morphology of the support (eg, surface area, pore size). The synthesis of pure Brønsted and pure Lewis solid acids as well as the control of the acid strength are important objectives. Pure Lewis acidity is hard to achieve with hydroxylated support materials, especially as a result of polarization of those groups by supported Lewis acid centers.

Acids are the most widely used catalysts in organic chemistry, with applications covering all major sectors from petrochemicals to pharmaceuticals. The most commonly used include H_2SO_4, HF, $AlCl_3$, and BF_3 and are typically soluble in the organic reaction medium or remain as a separate liquid phase.

At the end of the reaction, such acids are normally destroyed in a water quench stage and require subsequent neutralization, thus consuming additional (alkaline) resources, and producing salt waste. These increasingly important environmental issues are incentives to utilize solid acids, which stay in a separate and easily recoverable phase from the organic components throughout the reaction. Acidic resins and clays have a proven track record in organic chemicals manufacturing but newer solid acids can substitute environmentally threatening soluble and liquid acids.

Supported Lewis Acids

The most widely used Lewis acid is aluminium chloride. Its immobilization onto typical support materials such as clays, aluminas, and silicas has been widely studied. This can be achieved from the vapor phase or from a suitable solvent (both aromatic hydrocarbons and chloroaliphatics have been suggested), but in all cases one can expect some combination of physisorbed and chemisorbed, and various levels and types of acidic sites on the surface species.

Supported Lewis acids, including supported aluminium chloride, has been described as good catalysts for the oligomerization of mixed refinery feedstocks leading to hydrocarbon resin products. This methodology can avoid the need for a water quench step (necessary in the familiar homogeneous Lewis acid continuous stirred tank reaction method) and enable catalyst reuse. These novel heterogeneous systems can also be expected to lead to new products with different and potentially useful properties such as softening point, molecular weights, and a degree of cross-linking.

Commercial Friedel-Crafts catalysts based on supported Lewis acids have been developed for use in typical batch reactions. These include acid clay supported zinc (II) salts for use in reactions including benzylations and iron salts for the more demanding benzoylations and sulfonylations.

The reactions are best carried out in the absence of solvent, adding to their environmental credentials. Acid clay supported zinc salts have also been reported as catalysts for aromatic brominations.

Supported Heteropoly Compounds

Heteropolyacids supported on silica show commercially viable activity and catalyst lifetime for the manufacture of ethyl acetate from ethylene and acetic acid. One of the major current commercial routes to ethyl acetate is based on $PdCl_2/CuCl_2$ catalysis (the Wacker process), but this has a number of drawbacks, including the use and subsequent disposal of high concentrations of HCl. A solid acid catalyst could be a viable option.

The hydrolysis of ethyl acetate can also be achieved using a supported heteropoly compound—$Cs_{2.5}H_{0.5}PW_{12}O_{40}$-silica. The catalyst is unchanged at the end of the reaction.

Supported Sulfur (VI) Acids

The widespread importance of sulfuric acid and other sulfur (VI) acids, notably methanesulfonic and triflic acids, has encouraged a substantial amount of research into heterogeneous analogues and not insignificant commercial application, notably via polymeric acidic resins.

Modern mesoporous silicas and other materials have been successfully functionalized by sulfonic acid and their activity proven in a number of important reactions, including esterifications, transeseterification and alkylations.

Bisphenol-A is manufactured on a commercial scale using sulfonic acid-modified polystyrenes. Acidic resins have been used for a range of other commercial scale chemical processes, including ester hydrolysis, olefin, hydration, alcohol dehydration, and the reaction of olefins with alcohols (notably the reaction of methanol with isobutylene to give MTBE. Commercially, MTBE is produced using a macroporous sulfonic acid resin catalyst in the liquid phase.

More specialty applications for sulfonic acid resins include the preparation of perfumery-grade methyl anthranilate from the reaction of methanol with anthranilic acid. Numerous other esterifications have also been show to be effectively catalyzed by these solid acids.

Polyfluorinated sulfonic acids are considerably stronger Brønsted acids than the conventional styrene-based materials ($\Delta H_0 \sim 10$) and are also more thermally robust. Successful applications of these strong solid acids include the bimolecular conversion of alcohols to ethers.

The alkylation of phenols such as the tertiary-butylation of p-cresol is a very important process in the manufacture of numerous chemical products, including fragrances, antioxidants, herbicides, and insecticides. Sulfonic acid resins have replaced homogeneous and liquid acids in many of these processes, especially where the alkylation is not too difficult.

For more challenging systems, the more acidic fluorinated analogues may be more suitable. Similarly, olefin isomerizations require the more acidic materials to achieve reasonable reaction rates.

Loss of activity on moving from a homogeneous acid to a heterogeneous analogues is a frequent cause for concern and has undoubtedly delayed the commercial uptake of many supported catalyst systems. One way around this problem is through the utilization of a moving bed of the liquid catalyst wherebye the catalyst slowly moves over the surface of a support while reactants are fed through the mixed-catalyst bed.

Acid Catalysis by Zeolites

The use of zeolites as acid catalysts is one of the most prominent areas, of gas-phase catalysis, predominantly in the field of catalytic cracking. In recent years, some applications have begun to emerge in the field of fine chemicals synthesis, mostly in the liquid phase, with one or two important exceptions.

The acid strength of a zeolite is a complex function of the aluminium content, with acid strength ranging from moderate (high number of Al sites, giving a large number of moderately acidic sites) to superacidic (low amount of Al, leading to a small number of extremely acidic centers).

The Beckmann rearrangement of cyclohexanone oxime has been the subject of a substantial amount of research using zeolitic catalysts. Selectivity is excellent and catalyst lifetime can be very long when a co-feed of carbon dioxide or methanol and water is used.

BASIC CATALYSTS

Solid base catalysts are less well developed than solid acids but nonetheless represent an important class of catalysts. They range from simple solids such as alumina, which has some basic activity, through more complex inorganic species such as zeolites, hydrotalcites, "blue alumina", and to the newer organic–inorganic hybrid materials based on silica/amine composites.

One generic difficulty inherent in the handling and use of basic catalysts (particularly with the most basic systems) is the presence of acidic impurities in the atmosphere. Both water, and especially CO_2, can poison basic catalysts rapidly and possibly irreversibly.

"Blue Alumina"

An interesting and very basic material that is finding industrial application is based on the reaction of alumina first with NaOH and then, in a high temperature step, with sodium metal, to produce the strongly basic "blue alumina". The basicity of this material is exceptionally high.

Superbases can be used to bring about the thermodynamic isomerization of alkenes, giving the most stable isomer.

The alkylation of alkyl aromatics at the benzylic position is an important class of reaction, which can often give different isomers when compared to acid catalyzed routes. Such superbase-catalyzed alkylations are used indust-

rially in the production of isobutylbenzene from toluene and propene.

Zeolites

While the majority of zeolites have pore sizes that are too small to allow all but the smallest molecules to enter and are consequently of little use in organic synthetic applications, a few examples exist where larger pore zeolites have been used as supports in base catalysis.

Hydrotalcites

Hydrotalcites are another group of catalysts that are proving efficient as medium-to-strong basic catalysts. As catalyst supports, they have been used in the synthesis of superbases, but are also interesting in their own right. Hydrotalcites are lamellar solids consisting of layers of mixed oxides, typically Mg and Al, but others are also known. The structure contains (partly) exchangeable anions such as hydroxide and carbonate that gives basicity to the structure and, combined with their lamellar structure, leads to them being referred to sometimes as anionic clays.

A range of C–C bond-forming reactions are known to be catalyzed by hydrotalcites and related materials. There also exist patents describing the use of hydrotalcites as catalysts for the formation of alkoxylated alcohols and related products, of use as nonionic surfactants, by oligomerization of alkylene oxides and to the preparation of crystalline stereoregular poly(propylene oxide) using a similar route.

Ni–Al hydrotalcites have been shown to have excellent properties for the industrially important reduction of nitriles to primary amines. The combination of basic character with reducing centers plays an important role in minimizing the production of secondary amines.

Metal Oxides

Metal oxides all have the potential for their surface oxygen atoms to act as basic sites. In practice, those with divalent cations are the most basic, and therefore the most studied. Tri- and tetravalent cations can give materials with basic sites, but these are often relatively mildly basic, and acidic sites can also be present. Such materials are very interesting for processes that require the presence of both types of sites.

Zirconium oxide possesses both mildly basic and mildly acidic sites, which makes it particularly suitable for certain reaction types.

KF-alumina

KF-alumina is the most studied of a group of MF-support catalysts, most of which are characterized by a strong interaction (or a much more fundamental reaction) between the two components. These catalysts are active in a wide range of organic reactions. Their basicity is such that many CH acids such as nitroalkanes and methyl ketones can undergo various C–C bond-forming reactions.

Organic–Inorganic Hybrid Catalysts

A range of mildly basic catalysts for fine chemicals synthesis exist where the basic centre is an amine or diamine group, attached to a silica matrix via a short hydrocarbon spacer. Two routes exist to these compounds, including grafting onto amorphous silica or the newer micelle templated silicas. The basicity of these materials is relatively mild, and thus these materials are being investigated for the condensation of C-acids such as diketones, dinitriles and ketoesters.

Ion-Exchange Resins

Ion-exchange resins, which are often used in acid catalysis, are much less commonly utilized in base catalyzed processes, due to their limited thermal stability.

SUPPORTED METAL COMPLEXES

The heterogenization of catalytically useful metal complexes is attractive both economically and in the context of green chemistry. In principle, this can be achieved via immobilization, typically on to a high surface area support material. However, limitations in the stabilities of supported metal complexes in liquid-phase reactions and uncertainties over possible parallel homogeneous catalysis due to leached metallic species makes this area controversial and still largely unproven.

J. H. Clark and D. J. Macquarrie, eds., *Handbook of Green Chemistry and Technology*, Blackwell Science, Abingdon, U. K., 2002.

G. Ertl, H. Knözinger, and J. Weitkamp, eds., *Handbook of Heterogeneous Catalysis*, Wiley-VCH, Weinheim, 1997.

R. A. Sheldon and H. van Bekkum, eds., *Fine Chemicals Through Heterogeneous Catalysis*, Wiley-VCH, Weinheim, 2001.

J. M. Thomas and W. J. Thomas, *Principles and Practice of Heterogeneous Catalysis*, VCH, Weinstein, 1997.

JAMES H. CLARK
DUNCAN J. MACQUARRIE
University of York

CELL CULTURE TECHNOLOGY

Cell culture processes, the *in vitro* growth of animal, insect, or plant cells on a large scale to manufacture biochemicals of commercial importance, have been used for some time for the manufacture of viral vaccines. Significant growth in this technology, primarily because of the advent of recombinant DNA methods for the production of therapeutic proteins and hybridoma technology for production of monoclonal antibodies occurred in the late 1980s and 1990s. The need for cell culture technology stems mainly from the fact that bacteria do not have the capability to perform many of the posttranslational modifications that most large proteins require for in vivo biological activity. These modifications include intracellular processing steps

Table 1. Examples of Therapeutic Products Manufactured by Cell Culture

Product	Company	Therapeutic use
OKT-3	Johnson & Johnson	prevention of transplant rejection
Zenapax	Hoffmann	prevention of transplant rejection
ReoPro	Johnson & Johnson	antiplatelet for high risk angioplasty
Rituxan	Idec Pharmaceuticals	non-Hodgkin's lymphoma
Remicade	Johnson & Johnson	rheumatoid arthritis
Herceptin	Genentech	breast cancer
Avakine	Johnson & Johnson	Crohn's disease
Synagis	Medimmune	RSV
Activase	Genentech	heart attacks, strokes
Epogen	Amgen	anemia in kidney dialysis patients
Aranesp	Amgen	anemia in kidney dialysis and cancer chemotherapy patients
Kogenate	Miles/Cutter	hemophelia
Pulmozyme	Genentech	cystic fibrosis
Avonex	Biogen	multiple sclerosis
Enbrel	Amgen	rheumatoid arthritis
Mylotarg	Celltech Chiroscience	acute myeloid leukemia
Zevalin	Idec Pharmaceuticals	non-Hodgkin's lymphoma
Xolair	Genentech	allergic asthma
Humira	Abbott	rheumatoid arthritis

such as protein folding, disulfide linkages, glycosylation, and carboxylation.

The major applications for cell culture technology are found in the production of monoclonal antibodies (MAbs), recombinant therapeutics and vaccines. Monoclonal antibodies were initially used in small quantities for diagnostic purposes. However, in the past few years several monoclonals have been developed for therapeutic purposes. Vaccines also continue to be produced by cell culture. Some examples of theropeutic products from cell culture are given in Table 1.

Finally, insect cell culture is being utilized increasingly for quickly producing research quantities of new proteins using the baculovirus expression system. The strong polyhedrin promoter and the insect's cells ability to perform many posttranslational modifications have made the system useful for the expression of mammalian proteins that cannot be produced in native form in *Escherichia coli*. Technology development for these products has centered around the differences in characteristics of mammalian versus microbial cells, notably the shear sensitivity and susceptibility to contamination of the mammalian lines.

Although the focus of this article is mainly on mammalian cells, the technologies described herein also apply in principle to insect and plant cells.

CHARACTERISTICS OF MAMMALIAN CELLS

Environmental Conditions

Mammalian cells *in vivo* are maintained in a carefully balanced homeostatic environment and thus have evolved to require fairly stringent environmental conditions. These cells differ significantly from bacterial cells in that they lack a rigid cell wall, and are hence much more shear sensitive. Many animal cells are also attachment dependent, needing a surface to grow on. Many of the cell culture technologies provide the low shear, high surface area environment needed for the mammalian cells. Another approach is to adapt cells to suspension culture and select cells that are less shear sensitive, permitting the use of fermentation technology for the culture of animal cells. The optimum environmental parameters depend on cell type and are specific to cell type.

Nutritional Requirements

The nutrient requirements of mammalian cells are many, varied, and complex. In addition to typical metabolic requirements such as sugars, amino acids, vitamins, and minerals, cells also need growth factors and other proteins. Some of the proteins are not consumed, but play a catalytic role in the cell growth process.

Cell growth kinetics of mammalian cells can be described by the typical lag, exponential growth, then stationary and death phases. The exponential phase may be described adequately by a Monod type of kinetic model when the growth rate is much larger than the death rate. At low growth rates, it is necessary to include cell death kinetics to account for the lower viability and to predict the cell viability. Toward the end of the exponential culture, cells are also subject to growth inhibition from metabolic by-products such as lactate and ammonia. Hence, for continuous processes, a comprehensive model should contain terms for cell growth, based on the limiting substrate concentration, cell death, and inhibition kinetics.

CELL CULTURE PROCESSES

A wide variety of mammalian cells are used in industrial practice and, to accommodate this diversity in cell lines, scale and products, several cell culture processes have evolved. Commonly used processes include batch (or fed batch) suspension culture, continuous perfusion culture, and microcarrier systems as well as a few other systems developed to meet specific needs. Figure 1 schematically illustrates the configuration of a few of these culture systems.

Other Systems

In the last 20 years, many different processes were developed that for one reason or another never found significant commercial applications. These include the fluidized-bed system and a ceramic matrix bioreactor. Other attempt have been made to scale-up existing T-flask processes linearly by increasing the available surface area in a compact space. A disposable bioreactor using wave induced agitation has also been described. These systems are commonly used in the laboratory environment for research purposes but have not been utilized for commercial production.

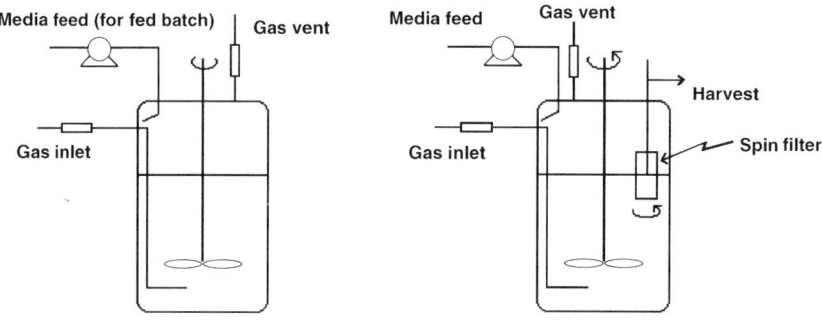

Batch/fed batch suspension culture reactor   Continuous perfusion culture with spin filter

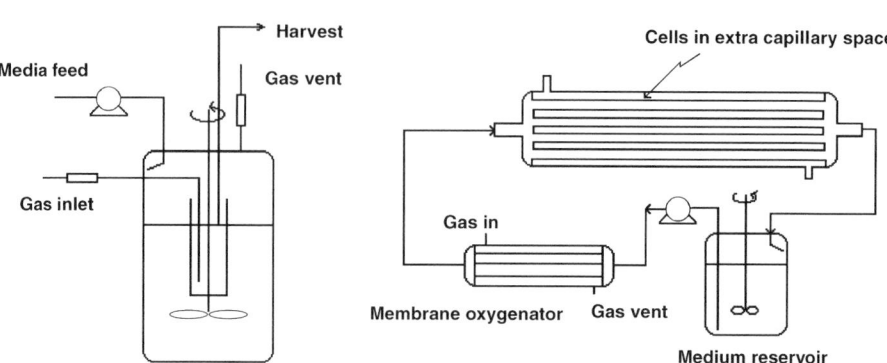

Figure 1. Commonly used cell culture processes.

Microcarrier perfusion culture with spin filter   Hollow fiber culture system with membrane oxygenator

ECONOMIC ASPECTS

The 2003 market for cell-culture-derived products was expected to exceed $15 billion/year. The market is expected to continue growing substantially throughout the next decade.

The relative economics of various cell culture processes depend heavily on the performance of the cell line in a system and on the cost of raw materials, particularly the medium. Often, for high-value products, the process that ensures the shortest time to market may be the process of choice because of other economic criteria. This is especially true for pharmaceuticals. Reliability concerns also often outweigh economic considerations in choosing a process for a high-value product.

Continuous processes have lower labor costs but higher failure risk. Batch processes can be started back up in a shorter period of time than can a complex continuous process. Batch processes are easier to take through the regulatory process than are continuous processes. Thus batch processes are often chosen for mammalian cell culture systems, even though continuous processes can offer significant cost advantages. Cell culture costs constitute only a small (10–30%) fraction of the overall cost of making a product.

REGULATIONS AND STANDARDS

Most of the products derived from cell culture technologies are for therapeutic or diagnostic applications and manufacture is regulated by the federal government through the Food and Drug Administration (FDA).

The biotechnology industry has no formal standards for equipment manufacture and quality control as of this writing. The American Society of Mechanical Engineers (ASME) has an active committee to develop standards for bioprocess equipment (Bioprocess Equipment Standards Committee; www.asme.org).

SAFETY CONSIDERATIONS

The fact that cell culture-derived products are often injected into humans as therapeutic agents makes it imperative that there be no component in the final product that can pose a potential health risk to the patient.

Some of the cells used in manufacturing are continuous or "immortal." Many of these have been shown to be tumorigenic in immunosuppressed animals. The cells also contain endogenous materials such as retroviruses and nucleic acids, both of which can induce tumorigenesis, and immunogenic foreign proteins. Serum used in media can also introduce adventitious agents such as viruses and mycoplasma into the product. Other process chemicals, including cleaning agents, are low-molecular-weight compounds that may be hazardous as well. Purification chemicals, such as monoclonals used for affinity purification, can be immunogenic to humans. Some of the potential health risks in mammalian cell culture processes and the methods used for risk reduction include:

cells	microfiltration
retroviruses	irradiation, sonication, heat, solvents, etc
nucleic acids	chromatography
cellular proteins	chromatography, ultrafiltration
bacterial contamination	microfiltration
process chemicals	diafiltration with appropriate buffers
serum proteins	affinity/ion-exchange chromatography

Most of these methods are commonly employed in the downstream processing of the desired cell culture technology product. Hence, most of the time it is only necessary to demonstrate that the designed process is reducing the putative risk factors to acceptable levels.

M. C. Flickinger and S. W. Drew, eds., *Encyclopedia of Bioprocess Technology: Fermentation, Biocatalysis, and Bioseparation*, John Wiley & Sons Inc., New York, 1999.

E. Lindner-Olsson, N. Chatzissavidou, and E. Lullau, eds., *Animal Cell Technology: From Target to Market*, Kluwer Academic Press, Dordrecht, The Netherlands 2001.

A. S. Lubiniecki, ed., *Large Scale Mammalian Cell Culture Technology*, Marcel Dekker, New York, 1990.

A. Prokop, R. K. Bajpai, and C. S. Ho, eds., *Recombinant DNA Technology and Applications*, McGraw Hill, New York, 1991.

Subhash B. Karkare
Scientia Consulting

CELLULOSE

Cellulose is the main molecule in cell walls of higher plants. It has even been reported in humans suffering from the rare disease of scleroderma. About 7.5×10^{10} t of cellulose grow and disappear each year, establishing it as the most abundant regenerated organic material on earth.

Natural cellulosic materials such as grass are eaten by grazing animals, and various species build nests or dens with wood. Cellulose in wood, in animal manure or in bagasse (the stalks of sugar cane after the juice has been pressed out) serves directly as fuel, while scientists are striving to develop efficient conversion of cellulose to alcohol and other fuels. After minimal processing, natural cellulosic materials are used as lumber, textiles, and cordage. After industrial treatment, with and without chemical derivatization, cellulose is made into diverse products, including paper, cellophane films, membranes, explosives, textiles (rayon and cellulose acetate), and dietary fiber. Besides its use in relatively simple products, cellulose fiber is being used to reinforce plastics in composites. This area can result in strong, lightweight, economical, biodegradable materials.

Cellulose is mostly insoluble in natural environments. Its fibers are relatively strong, with ramie and Fortisan both having specific breaking stress values of 0.59 Pa·mm³/g compared with steel wire, for example, with 0.26 Pa mm³/g. After factoring out their densities, these values can be converted to breaking stress values of 0.9

and 2.0 GPa, respectively. These attributes of strength and insolubility allow cellulose to stabilize the overall structure of plants. The stability of cellulose combined with lignin allows some trees to have very long lives. Some bristlecone pines (*Pinus longaeva (P. aristata)*) in cool, dry mountain regions of Colorado are more than 5,000 years old. On the other hand, cellulosic materials in damp, warm conditions are degraded naturally by enzymes collectively known as cellulases that are present in fungi, in bacteria that exist in soil, and in cattle rumen. Celluloses are also found in the guts of insects such as termites. Very strong acids can also degrade cellulose. The human digestive system has little effect on cellulose.

In industrial terminology, α-cellulose is mostly β-1,4-glucan, although some insoluble hemicelluloses may also be present. Hemicelluloses which occur along with cellulose in plant cell walls, are polysaccharides such as glucomannan and acetylated glucuronoxylan. Other hemicelluloses are natural derivatives of cellulose itself, with side chains of xylose, galactose, or fucose. Holocellulose is delignified cellulose that still contains the hemicellulose. The word "cellulose" means β-1,4-D-glucan, regardless of source.

SOURCES

Cellulose for industrial conversion comes from wood and scores of minor sources such as kenaf. Paper and rayon are now made mostly from wood pulp. Cotton rags were historically important for paper making, and cotton linters (short fibers not used to spin yarns) are now used in high quality writing and currency papers. The importance of cellulose recycling is increasing, especially for paper products. Some cellulose comes from the hairs (trichomes) on seeds, eg, cotton, kapok, and milkweed. Bast fibers are obtained from the stems of plants such as hemp, kenaf, ramie (a perennial Asian nettle), flax (linen), and jute. Besides the "soft" bast fibers, "hard" cellulose fibers are obtained from the leaves of plants such as agave (especially sisal), banana, and pineapple. In some cases, such as corn stover (the stalks and leaves of maize), the substantial amounts of cellulose present are interesting but not extensively exploited.

Celluloses from algae such as *Valonia ventricosa* are of considerable research interest because they occur in large, well-oriented crystallites. Ramie is also used for such experiments, because it represents textile celluloses better than algal or bacterial cellulose. Cellulose from the "tunic" of the tunicates has been processed to yield even larger crystals than those from algal celluloses, and these crystals have allowed the determination of the structure of one important form of native cellulose. Bacterial cellulose is of research interest because the synthesis of cellulose by an individual bacterium, *Acetobacter xylinium*, can be observed directly with a microscope. A single thread of bacterial cellulose can grow to a length of a meter, compared with a few centimeters for cotton fibers.

A commercial bacterial cellulose product (Cellulon) was introduced by Weyerhaeuser. The fiber is produced by an aerobic fermentation of glucose from corn syrup in an agitated fermentor. Because of its small particle diameter (10 μm), it has a surface area 300 times greater than normal wood cellulose and gives a smooth mouthfeel to formulations in which it is included. It has an unusual level of water binding and works with other viscosity builders to improve their effectiveness. It is anticipated that it will achieve GRAS (Generally Regarded As Safe) status and is neutral in sensory quality; microcrystalline cellulose has similar attributes. Other products made from bacterial cellulose include the oriental dessert, Nata di Coco, high quality loudspeaker cones, and Biofilm, a temporary skin substitute.

Recently, cellulose from sugar beet pulp and from citrus pulp has aroused interest for use as a fat substitute.

PREPARATION

Most large-scale manufacturers of cellulosic products other than textiles begin with cellulose that is in the form of pulp. Pulping partially separates cellulose from the lignin and hemicelluloses, leaving it in a fibrous form that is more susceptible to chemical treatment than the starting material. After pulping, the pulp is purified and otherwise treated to tailor it to the required specifications. Following drying, the pulp is shipped in large rolls.

STRUCTURE AND ITS RELATION TO CHEMICAL AND PHYSICAL PROPERTIES

Cellulose and starch can both be degraded to glucose, yet the properties and suitable uses of these two molecules are very different. The reason for the differences in the properties of starch and cellulose is largely the difference in the spatial arrangements of their molecules. The importance of structure was recognized at the same time as the discovery of cellulose. Some basic information from more than 70 years ago has been important in helping to understand many other polymers. Still, only at the time this is being written are the intra- and intermolecular arrangements of some of the important cellulose forms resolved well enough that most scientists can be comfortable with the results. Especially in the case of natural cellulose fibers, there are many different levels of structure. Ultimately, the properties depend on the structures of these levels.

Amorphous Cellulose

Some cellulose is amorphous, from either mechanical action (such as ball milling) or chemical treatment. Chemical and biochemical reactions of less crystalline cellulose are usually more rapid than those of highly crystalline materials. Most samples of cellulose have some amorphous character. Even cotton, the most crystalline of the commercial celluloses, has about 20% or more disordered material, including chain ends and crystallite surfaces.

Crystalline Cellulose

There are several different crystalline arrangements of cellulose, each giving a distinctive diffraction pattern. These polymorphs, or allomorphs, are denoted with Roman numerals I to IV, with some subclasses. Another form is called cellulose x. The particular crystalline form depends on source and treatment. In some cases, more than one form is present in a sample.

Cellulose I

Originally, most native cellulose was thought to have the same crystal structure, cellulose I. Then it became apparent from infrared spectroscopy and electron diffraction that cellulose I from the higher plants was somehow different from that of algal and bacterial cellulose. Now it is established that most native celluloses are mixtures of cellulose Iα, prevalent in algae and bacteria, and Iβ, prevalent in higher plants such as flax and cotton. After many years, the details of cellulose Iβ have been determined to a high level of reliability.

Cellulose II

Rayon, made from dissolved and regenerated wood and/or cotton, has the cellulose II structure, as does cellulose that has been treated with strong NaOH solutions. John Mercer found that mercerized cotton had improved luster and ability to take up dyes, and does not shrink. Treatments with alkali are still used to obtain these benefits, but in the textile industry more dilute solutions are used at higher temperatures. In the laboratory, concentrated, low temperature solutions are most effective at creating the cellulose II crystal structure. (Cold, highly concentrated NaOH can dissolve many cellulose samples.) Other systems, such as 65% nitric acid, also swell cellulose and can convert it to cellulose II.

Allomorphic form can be an important factor in biochemical reactions of cellulose as well as in some conventional chemical reactions. On the other hand, the improved receptivity of mercerized cotton to dyes may result more from the increased amorphous content rather than the packing arrangement of the chains and the hydrogen-bonding system of the new crystal structure.

Besides garments, rayon is used for tire cord and industrial belting. Typical rayon yarns have lower tensile strength than cellulose I yarns, but Fortisan, a heterogeneously saponified cellulose acetate, is a high strength rayon yarn, with its crystallites highly aligned along the fiber axis. The variables in the process of producing rayon fibers allow a large variety of performance characteristics.

Other Polymorphs

There are several other cellulose polymorphs. Cellulose exposed to amines or liquid ammonia forms complexes with the swelling agent. Upon removal of the swelling agent, cellulose III is produced. The actual form of III depends on whether the starting material is cellulose I or II, giving III$_I$ or III$_{II}$.

Similarly, cellulose IV$_I$ and IV$_{II}$, formed when cellulose is treated in glycerol at temperatures ca 260°C, can revert to the parent I and II structures and cellulose x results from strong hydrochloric or phosphoric acid treatments. No molecular arrangements have yet been proposed for cellulose x.

MICROCRYSTALLINE CELLULOSE

Pulverized forms of woodpulp have been widely used as fillers in some foods and pharmaceuticals. However, their utility is limited because the highly fibrous form results in poor mouthfeel. This problem can be overcome by reducing the woodpulp fibers to colloidal microcrystalline cellulose. It is made by reducing the particle size and molecular weight by hydrolysis with hydrochloric acid to the point of leveling off degree of polymerizaton. In aqueous suspensoids, these much finer particles have a smooth texture resembling uncolored butter and exhibit pseudoplastic properties, including stable viscosity, over a wide temperature range. It can therefore be used as a low-calorie substitute for fat. Microcrystalline celluloses are important for their heat stability; ability to thicken, with favorable mouthfeel; and flow control. They extend starches, form sugar gels, stabilize foams, and control formation of ice crystals. A few of the foods in which microcrystalline cellulose has been commercially successful are fillings, meringue (cold process), chocolate cake sauce (frozen), cookie fillings, whipped toppings, and imitation ice cream for use as a bakery filling. In the pharmaceutical industry, microcrystalline cellulose is used mostly for tableting.

In 2000, 55,000 t of microcrystalline cellulose were sold by the original vendor, FMC, and other companies also sell it. Its utility has led to development of other colloidal polymer microcrystals.

CHEMICAL REACTIVITY

Cellulose is chemically like other carbohydrate polymers that consist of pyranose rings bearing hydroxyl groups. These chains of glucose residues include a reducing end unit, a nonreducing end unit, and intermediate units. Most celluloses have a high degree of polymerization; the intermediate glucose residues determine the chemical and physical properties and the end units may be ignored. The glycosidic bonds in cellulose are strong and stable under a variety of reaction conditions. It is a generally insoluble, highly crystalline polymer.

CELLULOSE SOLVENTS

Solvents for cellulose are central to the rayon and cellophane industries as well as being necessary for many analyses. Despite the difficulty of dissolving cellulose in aqueous and organic liquids, several cellulose solvents have been been devised over the last 150 years. These solvents fall into several categories. The viscose process is the most important industrial method for dissolving cellulose. Because of the crystallite size, molecular weight and purity differences in cellulose from various sources, solvents that work well for some celluloses may not work for others. Cellulose subjected to high temperature and pressure during the steam explosion process can be dissolved in a strong base. It has been exceptionally difficult to find effective solvents that preserve the original molecular weights of high dp samples.

LIQUID CRYSTALS

Many cellulose derivatives form liquid crystalline phases, both in solution (lyotropic mesophases) and in the melt (thermotropic mesophases). A major reason for the interest in cellulosic liquid crystals is their role in the production of high strength, high modulus fibers.

T. Heinze and W. Glasser, eds., *Cellulose Derivatives: Modification, Characterization and Nanostructures* (ACS Symposium Series 688), American Chemical Society, Washington, D.C., 1998.

T. P. Nevell and S. H. Zeronian, eds., *Cellulose Chemistry and Its Applications*, Ellis Horwood, Chichester, U.K., 1985.

R. M. Rowell and H. P. Stout, in M. Lewin and E. M. Pearce, eds., *Handbook of Fiber Chemistry*, 2nd ed., Marcel Dekker, Inc., New York, 1998, p. 465.

C. Schuerch, ed., *Cellulose and Wood—Chemistry and Technology*, John Wiley & Sons, Inc., New York, 1989.

ALFRED D. FRENCH
NOELIE R. BERTONIERE
Southern Regional Research
    Center, USDA

R. MALCOLM BROWN
The University of Texas at Austin

HENRI CHANZY
CNRS-CERMAV

DEREK GRAY
Pulp and Paper Research Centre,
    McGill University

KAZUYUKI HATTORI
Kitami Institute of Technology

WOLFGANG GLASSER
Virginia Polytechnic Institute
    and State University

CELLULOSE ESTERS, INORGANIC ESTERS

Inorganic esters of cellulose have been known for well over 100 years and are still commercially viable products. Cellulose nitrate, commonly called nitrocellulose, is the ester with the most significant historical importance. It led to the birth of the plastics industry, the film and motion picture industry, and major improvements in explosives (eg, powderless ammunition) and coatings applications. Additionally, cellulose nitrate played a prominent role in numerous biochemical discoveries, including elucidation of the genetic code. Other valuable inorganic esters of cellulose include cellulose phosphate, cellulose sulfate, and cellulose sulfonates.

PREPARATION OF INORGANIC ESTERS OF CELLULOSE

Inorganic esters of cellulose include all esters where the atoms linked directly to the cellulosic oxygens are non-carbon. The following cellulose derivatives are of particular significance: cellulose nitrate, cellulose sulfate, cellulose sulfonate, and cellulose phosphate. A number of versions of cellulose nitrate are commercially available. Variations in nitrogen content and viscosity are the main differences between nitrocellulose product lines.

PROPERTIES OF INORGANIC ESTERS OF CELLULOSE

In general, inorganic esters of cellulose have significantly different solubility profiles than native cellulose. Cellulose has poor solubility in most organic or aqueous systems, while inorganic esters of cellulose are typically more soluble. Inorganic esters of cellulose that are soluble in organic or aqueous solvents are available. The solubility of inorganic esters of cellulose is dependent on the identity and degree of substitution of the substituents, the number of unreacted free hydroxyls, and the degree of polymerization of the cellulose backbone.

Solubility

Cellulose sulfate and cellulose phosphate can be either water-soluble or water-swellable depending on the conditions used during preparation. Cellulose nitrate and some cellulose sulfonates are generally more hydrophobic and are typically soluble in organic solvents. In general, the level of esterification of cellulose directly impacts the solubility of the product in various solvents and solvent blends.

Film Formation and Rheology

The ability to form films is the most important property of inorganic cellulose esters. When cellulose nitrate is dissolved in an organic solvent and then applied to a substrate (eg, wood, metal, glass) a clear film is formed on the substrate once the solvent evaporates. Coatings of this nature are referred to as lacquers. Coatings applications are the largest commercial applications for cellulose nitrate. Clear films of cellulose sulfate and cellulose phosphate have also been reported.

The rheological behavior of cellulose esters, both inorganic and organic, is an important property that contributes to the utility of these cellulosics, particularly in coatings applications. The semirigid cellulose nitrate displays a flow behavior similar to that of linear flexible polymers.

APPLICATIONS OF INORGANIC ESTERS OF CELLULOSE

Nitrocellulose is the most commercially important inorganic ester of cellulose. In most cases, nitrocellulose has been replaced over the years by less expensive or less flammable materials. Though considerable market share for nitrocellulose has been lost to alternate products, nitrocellulose remains a commercially viable product and should continue so for many years.

Coatings, Adhesives, and Inks

The largest market for cellulose nitrate is in the coatings industry. Relatively clear and workable (ie, sandable or removable) lacquers are generated when cellulose nitrate is used in a coatings application. Cellulose nitrate lacquers dry rapidly, have excellent flow properties, and the final coating has excellent appearance, with reasonable strength and durability. Cellulose nitrate has played such a significant role in lacquer coatings applications that cellulose nitrate-based lacquers are simply referred to as "lacquers" and it is assumed that a lacquer is cellulose nitrate-based unless stated otherwise. Cellulose nitrate has maintained much of its market share in applications where high-temperature curing is not acceptable, as for example in wood coatings, currently its primary use. Cellulose nitrate provides enhanced wood grain appearance, rapid drying, and easy damage repair to wood coatings. It has also been used as an additive to improve the performance, appearance, and drying time of coatings based on other resins, such as acrylics, vinyls, polyamides, epoxies, and polyesters. Cellulose nitrate aids in pigment dispersion and acts as a binder following solvent evaporation.

Cellulose nitrate-based lacquers have many advantages over other coatings. They are easily applied, have good adhesion, good solvent release properties, and excellent pigment dispersion. The problems with nitrocellulose lacquers include yellowing, flammability, and that they must be sprayed at low solids levels. Nitrocellulose-based lacquers have found use in cloth book bindings. Nitrocellulose is also used in adhesives and inks.

Explosives

Cellulose nitrate was initially developed for use in explosives and propellants; this remains one of the largest markets for it today.

Plastics

In 1847 it was discovered that cellulose nitrate could be dissolved in a mixture of ether and alcohol to produce a solution called Collodion, still in use today in pharmaceutical applications. Combination of nitrocellulose with castor oil and eventually camphor led to the production of Celluloid and ultimately the birth of the plastics industry.

Flame Retardants

Surface modification of cotton fibers generates flame-retardant fabrics by increasing the char content of the fiber, which in turn produces a lower percentage of flammable volatiles.

Separations Applications

Ion Exchange. Inorganic esters of cellulose have been used in ion exchange applications and as chromatographic adsorbents since the late 1950s. Cellulose sulfate and cellulose phosphate are excellent ion exchangers. Cellulose phosphate is the most widely used cellulose inorganic ester for ion exchange. The ion exchange nature of cellulose sulfate resulted in its use to concentrate viruses.

Molecular Weight Determination. The improved solubility of cellulose nitrate is advantageous over the relative insolubility of cellulose.

Chiral Chromatography. Cellulose trinitrate-impregnated silica beads have been used as a chiral stationary phase for the separation of several racemic aromatic compounds.

Membrane Applications

Cellulose esters are effective in membrane applications. Nitrocellulose membranes are used in numerous biochemical and diagnostic applications.

Synthetic Intermediates

Cellulose sulfonates have been evaluated as synthetic intermediates. Additionally, cellulose sulfonates can also function as a protecting group to prevent reaction of the C6 hydroxyl during esterification reactions. Though an effective synthetic tool, the expense of organic solvents utilized and the cost and corrosive nature of tosyl chloride have limited the use of this methodology on a commercial scale.

Medical Applications

Inorganic esters of cellulose have been evaluated in a number of medical–pharmaceutical applications: for treatment of hypercalciura, use as antiviral agents, reduction of cholesterol levels, orthopedic applications, and blood stabilization. Sodium cellulose phosphate is used in the treatment of adsorptive hypercalciura and nephrocalcinosis (both a common cause of kidney stones).

Cellulose sulfate displays heparinlike activity and to a lesser extent antiviral activity. High molecular weight cellulose sulfate lowers cholesterol levels in humans by inhibiting pancreatic cholesterol esterase.

Encapsulation of microorganisms, enzymes, plant cells, and animal cells (including mammalian cells) has been accomplished using a mixture of sodium cellulose sulfate with polydiallyldimethyl ammonium chloride. Encapsulation of hydridoma cells with cellulose sulfate has provided a means of subcutaneous delivery of monoclonal antibodies.

Cellulose phosphate has been evaluated in femoral implants to improve the mineralization of biomaterials used in orthopedic applications. Cellulose phosphate (with or without bound antimicrobial agents) and cellulose phosphate borate effectively stabilize blood for over four years.

Biochemical Applications

Cellulose phosphate paper, also called phosphocellulose paper, has been used for enzyme assay applications the ability of the material to act as an ion exchanger is the key property that leads to its use in this regard.

Cellulose nitrate filters and membranes have played key roles in numerous biochemistry and molecular biology discoveries and still play important roles in these laboratories today. For example, cellulose nitrate is an efficient binder of antibodies and can be used as a solid support for immunoaffinity chromatography.

Diagnostic Applications

The binding characteristics of cellulose nitrate membranes have led to their use in numerous biosensor applications, such as in an electrochemical microbial biosensor for ethanol and in glucose oxidase-based glucose sensors. Cellulose nitrate continues to play an important role in the development of new biosensors as both a solid support and a method for removal of unbound reagents.

Additionally, cellulose nitrate has been utilized in non-biological detectors, including an indoor radon gas detector. The cellulose nitrate portion of the film-based detector is the active portion for the detection of α-particles.

Inorganic esters of cellulose have a valuable place in history, from the birth of the plastics industry to elucidation of the genetic code, in a large number of applications: coatings, explosives, pharmaceuticals, membranes, and synthetic intermediates. Though inorganic esters of cellulose have been around for >100 years, these materials remain important tools for coatings scientists, explosives experts, and biochemists.

R. J. Brewer and R. T. Bogan, *Encyclopedia of Polymer Science and Engineering*, Vol. 2 Wiley, New York, 1998.

S. Subrahmanyam, K. Shanmugam, T. V. Subramanian, M. Murugesan, V. M. Madhar, and D. Jeyakumar, *Electoroanalysis* **13**, 11, 944 (2001).

R. Taylor, *Methods in Plant Electron Microscopy and Cytochemistry*, 101 (2000).

D. Voet and J. G. Voet, *Biochemistry*, 2nd ed., John Wiley & Sons, Inc., New York, 1995.

MICHAEL C. SHELTON
Eastman Chemical Company

CELLULOSE ESTERS, ORGANIC ESTERS

Cellulose esters of almost any organic acid can be prepared, but because of practical limitations esters of acids containing more than four carbon atoms have not achieved commercial significance.

Cellulose acetate is the most important organic ester because of its broad application in fibers and plastics; it is prepared in multi-ton quantities with degrees of substitution (DS) ranging from that of water-soluble monoacetates to those of fully substituted triacetate.

Although cellulose acetate remains the most widely used organic ester of cellulose, its usefulness is restricted by its moisture sensitivity, limited compatibility with other synthetic resins, and relatively higher processing temperature. Cellulose esters of higher aliphatic acids, C_3 and C_4, circumvent these shortcomings with varying degrees of success. They can be prepared relatively easily with procedures similar to those used for cellulose acetate.

Cellulose esters of aromatic acids, aliphatic acids containing more than four carbon atoms, and aliphatic diacids are difficult and expensive to prepare because of the poor reactivity of the corresponding anhydrides with cellulose; little commercial interest has been shown in

these esters. A notable exception, however, is the recent interest in the mixed esters of cellulose succinates, prepared by the sodium acetatecatalyzed reaction of cellulose with succinic anhydride. The additional expense incurred in manufacturing succinate esters is compensated by the improved film properties observed in waterborne coatings.

Mixed cellulose esters containing the dicarboxylate moiety, eg, cellulose acetate phthalate, have commercially useful properties such as alkaline solubility and excellent film-forming characteristics. These esters can be prepared by the reaction of hydrolyzed cellulose acetate with a dicarboxylic anhydride in a pyridine or, preferably, an acetic acid solvent with sodium acetate catalyst. Cellulose acetate phthalate for pharmaceutical and photographic uses is currently, produced commercially via the acetic acid–sodium acetate method.

PROPERTIES

The properties of cellulose esters are affected by the number of acyl groups per anhydroglucose unit, acyl chain length, and the degree of polymerization (DP) (molecular weight). The properties of some typical cellulose triesters are given in Table 1.

The common commercial products are the primary (triacetate) and the secondary (acetone-soluble, ca 39.5% acetyl, 2.45 DS) acetates; they are odorless, tasteless, and nontoxic. Their properties depend on the combined acetic acid content molecular weight. Solubility characteristics of cellulose acetates with various acetyl contents are given in Table 2.

In fibers, plastics, and films prepared from cellulose esters, mechanical properties such as tensile strength, impact strength, elongation, and flexural strength are greatly affected by the degree of polymerization and the degree of substitution. Mechanical properties significantly improve as the DP is increased from ca 100 to 250 repeat units.

Stabilization

After hydrolysis, precipitation, and thorough washing of the cellulose esters to remove residual acids, the esters must be stabilized against thermal degradation and color development, which may occur during processing, such as extrusion or injection molding. Thermal instability is caused by the presence of oxidizable substances and small amounts of free and combined sulfuric acid. The sulfuric acid combines with the cellulose almost quantitatively and most of it is removed during the latter stages of hydrolysis. The remaining sulfuric acid can be neutralized with alkali metal salts, such as sodium, calcium, or magnesium acetate, to improve ester stability. The combined sulfate ester may also be removed by treatment in boiling water or at steam temperatures in an autoclave. Treatment with aqueous potassium or calcium iodide reportedly stabilizes the cellulose acetate against thermal degradation.

Cellulose Acetate

Almost all cellulose acetate, with the exception of fibrous triacetate, is prepared by a solution process employing

Table 1. Properties of Cellulose Triesters[a]

| Cellulose ester | Shrinking point, °C | Mp,[c] °C | Water tolerance value[d] | Moisture regain[b], % | | | Density, g/mL | Tensile strength, Mpa[e] |
				50% rh	75% rh	95% rh		
cellulose[f]				10.8	15.5	30.5	1.52	
acetate		306	54.4	2.0	3.8	7.8	1.28	71.6
propionate	229	234	26.9	0.5	1.5	2.4	1.23	48.0
butyrate	178	183	16.1	0.2	0.7	1.0	1.17	30.4
valerate	119	112	10.2	0.2	0.3	0.6	1.13	18.6
caproate	84	94	5.88	0.1	0.2	0.4	1.10	13.7
heptylate[g]	82	88	3.39	0.1	0.2	0.4	1.07	10.8
caprate	82	86	1.14	0.1	0.1	0.2	1.05	8.8
caprate[h]	87	88		0.1	0.2	0.5	1.02	6.9
laurate	89	91		0.1	0.1	0.3	1.00	5.9
myristate	87	106		0.1	0.1	0.2	0.99	5.9
palmitate	90	106		0.1	0.1	0.2	0.99	4.9

[a]Courtesy of the American Chemical Society.
[b]At 25% rh moisture regain for cellulose is 5.4%; for the acetate, 0.6%; for the propionate and butyrate, 0.1%; all others are zero.
[c]Char point is 315°C or higher unless otherwise noted.
[d]Milliliters of water required to start precipitation of the ester from 125 mL of an acetone solution of 0.1% concentration.
[e]To convert MPa to psi, multiply by 145.
[f]Starting cellulose, prepared by deacetylation of commercial, medium viscosity cellulose acetate (40.4% acetyl content).
[g]Char point = 290°C.
[h]Char point = 301°C.

sulfuric acid as the catalyst with acetic anhydride in an acetic acid solvent. The acetylation reaction is heterogeneous and topochemical wherein successive layers of the cellulose fibers react and are solubilized in the medium, thus exposing new surfaces for reaction.

Recent Developments. A considerable amount of cellulose acetate is manufactured by the batch process. In order to reduce production costs, efforts have been made to develop a continuous process that includes continuous activation, acetylation, hydrolysis, and precipitation.

Demand for cellulose acetate flake in the United States was projected to decline slightly from 1988 to 1993. Cigarette-filter tow for export is the only market projected to grow. Cellulose acetate for textile fibers is expected to decline, as will flake demand for plastics, with the growth of photographic films somewhat offsetting declining markets in other plastics end uses.

HEALTH AND SAFETY FACTORS

The vapors of the organic solvents used in the preparation of cellulose ester solutions represent a potential fire, explosion, or health hazard.

Table 2. Solubility Characteristics of Cellulose Acetates

Acetyl, %	Soluble in	Insoluble in
43.0–44.8	dichloromethane	acetone
37–42	acetone	dichloromethane
24–32	2-methoxymethanol	acetone
15–20	water	2-methoxymethanol
≤13	none of the above	all of the above

Cellulose esters are considered nontoxic and may be used in food-contact applications. However, since cellulose esters normally are not used alone, formulators of coatings and films for use in food packaging should ensure that all ingredients in their formulations are cleared by the United States Food and Drug Administration for such use.

USES

The cellulose esters with the largest commercial consumption are cellulose acetate, including cellulose triacetate, cellulose acetate butyrate, and cellulose acetate propionate. Cellulose acetate is used in textile fibers, plastics, film, sheeting, and lacquers. The cellulose acetate used for photographic film base is almost exclusively triacetate; some triacetate is also used for textile fibers because of its crystalline and heat-setting characteristics.

Cellulose esters, especially acetate propionate and acetate butyrate mixed esters, have found limited use in a wide variety of specialty applications, such as in nonfogging optical sheeting, low profile additives to improve the surface characteristics of sheet-molding (SMC) compounds and bulk-molding (BMC) compounds, and controlled drug release via encapsulation.

G. D. Hiatt and W. J. Rebel, in N. M. Bikales and L. Segal, eds., *Cellulose and Cellulose Derivatives*, Part V of *High Polymers*, 2nd ed., Vol. 5, Wiley-Interscience, New York, p. 749.

C. J. Malm and co-workers, *Ind. Eng. Chem.* **5**, 81 (1966).

C. J. Malm and L. J. Tanghe, *Ind. Eng. Chem.* **47**, 995 (1955).

C. J. Malm, L. J. Tanghe, and J. L. Schmitt, *Ind. Eng. Chem.* **53**, 363 (1961).

Steven Gedon
Richard Fengl
Eastman Chemical Company

CELLULOSE ETHERS

GENERAL CONSIDERATIONS

Alkylation of cellulose yields a class of polymers generally termed cellulose ethers. Most of the commercially important ethers are water-soluble and are key adjuvants in many water-based formulations. The most important property these polymers provide to formulations is rheology control, ie, thickening and modulation of flow behavior. Other useful properties include water-binding (absorbency, retention), colloid and suspension stabilization, film formation, lubrication, and gelation. As a result of these properties, cellulose ethers have permeated a broad range of industries, including foods, coatings, oil recovery, cosmetics, personal care products, pharmaceuticals, adhesives, printing, ceramics, textiles, building materials, paper, and agriculture.

Cellulose ethers represent a mature industry, with annual sales of over one billion dollars and an annual growth rate averaging 2–3%. The highest-volume cellulose ethers, the industry workhorses, are sodium carboxymethylcellulose, hydroxyethylcellulose, and hydroxypropylmethylcellulose. Cellulose ethers as a class compete with a host of other materials, including natural gums, starches, proteins (qv), synthetic polymers, and even inorganic clays. They provide effective performance at reasonable cost and are derived from a renewable natural resource.

Cellulose ethers are manufactured by reaction of purified cellulose with alkylating reagents under heterogeneous conditions, usually in the presence of a base, typically sodium hydroxide, and an inert diluent. Cottonseed linter fiber and wood fiber are the principal sources for cellulose.

Many cellulose ethers contain mixed substituents (cellulose mixed ethers) in order to enhance or modify the properties of the monosubstituted derivative.

Health and Safety Factors

No adverse toxicological or environmental factors are reported for cellulose ethers in general. Some are even approved as direct food additives, including purified carboxymethylcellulose, methylcellulose, hydroxypropylmethylcellulose, and hydroxypropylcellulose.

The only known hazard associated with cellulose ethers is that they may form flammable dusts when finely divided and suspended in air, a hazard associated with most organic substances.

Table 1. Typical Properties of Purified CMC

Property	Value
Powder	
appearance	white to off-white
moisture, max %	8.0
charring temp, °C	252
molecular weight, M_w	$9.0\times10^4 - 7.0\times10^5$
Solution	
viscosity, Brookfield, 30 rpm, mPa·S(=cP)	
at 1% solids (high M_w)	~6000
at 4% solids (low M_w)	~50
sp gr, 2% at 25°C	1.0068
pH, 2%	7.5
surface tension, 1%, mN/m(=dyn/cm)	71

COMMERCIAL CELLULOSE ETHERS

Sodium Carboxymethylcellulose

Properties. Sodium carboxymethylcellulose (CMC), also known as cellulose gum, is an anionic, water-soluble cellulose ether, available in a wide range of substitution. The most widely used types are in the 0.7-1.2-DS range. DS is the degree of substitution. It refers to the number of hydroxyl groups substituted per anhydroglucose residue.

Some typical properties of commercial CMCs are given in Table 1.

Manufacture. Common to all manufacturing processes for CMC is the reaction of sodium chloroacetate with alkali cellulose complex.

Uses. CMC is an extremely versatile polymer, and it has a variety of applications. A sampling of significant applications is given in Table 2.

Hydroxyethylcellulose

Properties. Hydroxyethylcellulose (HEC), is a nonionic polymer. Low hydroxyethyl substitutions (MS=0.05-0.5) yield products that are soluble only in aqueous alkali. Higher substitutions (MS ≥1.5) produce water-soluble HEC. The bulk of commercial HEC falls into the latter category. Water-soluble HEC is widely used because of its broad compatibility with cations and the lack of a solution gel or precipitation point in water up to the boiling point. MS is the molar subsitution. It refers to the moles of reagent combined per anhydroglucose residue.

Some typical properties of HEC are given in Table 3.

Manufacture. Purified hydroxyethylcellulose is manufactured in diluent-mediated processes similar to those used to produce carboxymethylcellulose, except that ethylene oxide is used in place of MCA.

Uses. HEC is used as a thickener, protective colloid, binder, stabilizer, and suspending agent in a variety of industrial applications.

Table 2. Applications for CMC

Industry	Application	Function
foods	frozen desserts	inhibit ice crystal growth
	dessert toppings	thickener
	beverages, syrups	thickener, mouthfeel
	baked goods	water-binder, batter viscosifier
	pet food	water-binder, thickener, extrusion aid
pharmaceuticals	tablets	binder, granulation aid
	bulk laxatives	water-binder
	ointments, lotions	stabilizer, thickener, film-former
cosmetics	toothpaste	thickener, suspension aid
	denture adhesives	adhesion promoter
	gelled products	gellant, film-former
paper products	internal additive	binder, improve dry-strength
	coatings, sizes	water-binder, thickener
adhesives	wallpaper paste	adhesion promoter, water-binder
	corrugating	thickener, water-binder, suspension aid
	tobacco	binder, film-former
lithography	fountain, gumming	hydrophilic protective film
ceramics	glazes, slips	binder (promotes green strength)
	welding rods	binder, thickener, lubricant
detergents	laundry	soil antiredeposition aid
textiles	warp sizing	film-former, adhesion promoter
	printing paste, dye	thickener, water-binder

Table 3. Typical Properties of HEC

Property	Value
Powder	
appearance	white to light tan
ash content (as Na_2SO_4), %	5.5
molecular weight, M_w	$9 \times 10^4 - 1.3 \times 10^6$
Solution	
viscosity, Brookfield, 30 rpm, mPa·s(=cP)	
at 1% solids (high M_w)	5000
at 5% solids (low M_w)	75
sp gr, 2%, g/cm^3	1.0033
pH	7
surface tension, mN/m(=dyn/cm)	
MS 2.5 at 0.1%	66.8
at 0.001%	67.3
refractive index, 2%	1.336

Methylcellulose and its Mixed Ethers

Properties. Methylcellulose (MC) and its alkylene oxide derivatives hydroxypropylmethylcellulose (HPMC), hydroxyethylmethylcellulose (HEMC), and hydroxybutylmethylcellulose (HBMC) are nonionic, surface-active, water-soluble polymers. Each type of derivative is available in a range of methyl and hydroxyalkyl substitutions. The extent and uniformity of the methyl substitution and the specific type of hydroxyalkyl substituent affect the solubility, surface activity, thermal gelation, and other properties of the polymers in solution.

Typical properties of MC, HPMC, HEMC, and HBMC are given in Table 5.

Manufacture. Methylcellulose is manufactured by the reaction of alkali cellulose with methyl chloride.

Uses. There are numerous applications for methylcellulose and its derivatives. Some important ones are summarized in Table 6.

Ethylcellulose and Hydroxyethylethylcellulose

Properties. Ethyl cellulose (EC) is a nonionic, organosoluble, thermoplastic cellulose ether, having an ethyl DS in the range of ~2.2–2.7.

Mixed Ether Derivatives of HEC. Several chemical modifications of HEC are commercially available. The secondary substituent is generally of low DS (or MS), and its function is to impart a desirable property lacking in HEC. Derivatives include carboxymethylhydroxyethylcellulose (CMHEC) cationic hydroxyethylcelluloses, and hydrophobic hydroxyethylcelluloses (Table 4).

Table 4. Typical Properties of Mixed Ether Derivatives of HEC

Property	CMHEC	Cationic HEC	EHEC	HMHEC
Powder				
appearance	off-white	light yellow	off-white	off-white
ash content (as Na_2SO_4), %		3	3 (as NaCl)	10 max
Solution				
pH	6.5–10	7	6–7	6–8.5
flocculation temp in water, °C			~65	
surface tension, mN/m(=dyn/cm)			55	~62

Table 5. Typical Properties of Methylcellulose Ethers

Property	MC	HPMC	HEMC	HBMC
		Powder		
appearance	white	white	white	white
bulk density, g/cm^3		0.25–0.70		
ash content (as Na$_2$SO$_4$), %		2.5 max		
		Solution		
viscosity^a, mPa ċS(=cP)	10–15,000	5–70,000	100–70,000	
sp gr, 2% at 20°C		1.0032		
pH, 1%		5.5–9.5		
surface tension, 0.1%, mN/m(=dyn/cm)	47–53	44–56	46–53	49–55
interfacial tension^b, mN/m(=dyn/cm)	19–23	17–30	17–21	20–22

^{a} 2% Solution, Brookfield, 20 rpm.
^{b} Against paraffin oil.

Table 6. Applications for Methylcellulose and its Derivatives

Industry	Application	Function
construction	cements, mortars	thickener, water-binder, workability
foods	mayonnaise, dressing	stabilizer, emulsifier
	desserts	thickener
pharmaceuticals	tablets	binder, granulation aid
	formulations	stabilizer, emulsifier
adhesives	wallpaper paste	adhesive
ceramics	slip casts	binder (promotes green strength)
coatings	latex paints	thickener
	paint removers	thickener
cosmetics	creams, lotions	stabilizer, thickener

Table 8. Applications for EC and HEEC

Industry	Application	Function
coatings	lacquers, varnishes	protective film-former, additive to increase film toughness and durability, shorten drying time
printing	inks	film-former
adhesives	hot melts	additive to increase toughness

A large excess of sodium hydroxide and ethyl chloride and high reaction temperatures (up to 140°C) are needed to drive the reaction to the desired high DS values ($\geq$2.0).

Uses. A summary of the applications for ethylcellulose is given in Table 8.

Organo-soluble hydroxyethylethylcellulose (HEEC) is highly ethoxylated with small amounts of hydroxyethyl substitution. It is used in coating applications that require solubility in fast-drying aliphatic hydrocarbons. Table 7 gives typical properties for EC and HEEC.

Manufacture. Ethyl chloride undergoes reaction with alkali cellulose in high pressure nickel-clad autoclaves.

Table 7. Typical Properties of EC and HEEC

Property	EC	HEEC
	Powder	
appearance	white	white
bulk density, g/cm^3	0.3–0.35	0.3–0.35
softening point, °C	152–162	
	Film	
specific gravity	1.140	1.120
refractive index	1.470	1.47
tensile strength, MPa^a	46–72	34–41
elongation, %	7–30	6–10
flexibility^b	160–2000	500–900

^{a} To convert mPa to psi, multiply by 145.
^{b} MIT double folds.

Table 9. Typical Properties of HPC

Property	Value
	Powder
appearance	off-white
ash content (as Na$_2$SO$_4$), %	0.2–0.5
softening point, °C	100–150
molecular weight, M_w	8.0×10^4–11.15×10^6
	Solution
viscosity, Brookfield, 30 rpm, mPa ċs(=cP)	
at 1% (high M_w)	2500
at 10% (low M_w)	100
surface tension, 0.1%, mN/m(=dyn/cm)	43.6
interfacial tension^a, 0.1%, mN/m	12.5
	Film
tensile strength, MPa^b	14
elongation, %	50
flexibility^c (50 μm film)	10,000
refractive index	1.559

^{a} Against mineral oil.
^{b} To convert MPa to psi, multiply by 145.
^{c} MIT double folds.

Table 10. Applications for HPC

Industry	Application	Function
polymerization	PVC suspension polymerization	protective colloid
pharmaceutical	tablets	binder, film-former
coatings	paint remover	thickener
foods	whipped toppings	stabilizer
	processed foods	extrusion aid
ceramics	slip casts	binder (promotes green strength)

Hydroxypropylcellulose

Properties. Hydroxypropylcellulose (HPC) is a thermoplastic, nonionic cellulose ether that is soluble in water and in many organic solvents. HPC combines organic solvent solubility, thermoplasticity, and surface activity with the aqueous thickening and stabilizing properties characteristic of other water-soluble cellulosic polymers described herein.

Some typical properties of commercial HPC are given in Table 9.

Manufacture. HPC is manufactured by reaction of propylene oxide with alkali cellulose.

Uses. A summary of significant uses for HPC is given in Table 10.

N. M. Bikales and L. Segal, eds., *Cellulose and Cellulose Derivatives, High Polymers*, Vol. V, Wiley-Interscience, New York, 1971, Pts. IV–V.

R. L. Davidson, ed., *Handbook of Water-Soluble Gums and Resins*, McGraw-Hill, Book Co., Inc., New York, 1980.

E. K. Just and T. G. Majewicz, in J. I. Kroschwitz, ed., *Encyclopedia of Polymer Science and Engineering*, 2nd ed., Vol. **3**, John Wiley & Sons, Inc., New York, 1985, pp. 224–269.

E. Ott, M. Spurlin, and M. W. Graffin, ed., *Cellulose and Cellulose Derivatives, High Polymers*, Vol. V, Wiley-Interscience, New York, 1954–1955, Parts I–III.

THOMAS G. MAJEWICZ
THOMAS J. PODLAS
Aqualon Company

CEMENT

The term "cement" is used to designate many different kinds of substances that are used as binders or adhesives. The cement produced in the greatest volume and most widely used in concrete for construction is Portland cement. Masonry and oil well cements are produced for special purposes. Calcium aluminate cements are extensively used for refractory concretes. Such cements are distinctly different from epoxies and other polymerizable organic materials. Portland cement is a hydraulic cement; ie, it sets, hardens, and does not disintegrate in water. Hence, it is suitable for construction of underground, marine, and hydraulic structures, whereas gypsum plasters and lime mortars are not. Organic materials, such as latexes and water-soluble polymerizable monomers, are sometimes used as additives to impart special properties to concretes or mortars. The term "cements" as used here in is confined to inorganic hydraulic cements, principally Portland and related cements. The essential feature of these cements is their ability to form on hydration with water relatively insoluble bonded aggregations of considerable strength and dimensional stability.

CLINKER CHEMISTRY

Hydraulic cements are manufactured by processing and proportioning suitable raw materials, burning (or clinkering at a suitable temperature), and grinding the resulting hard nodules called clinker to the fineness required for an adequate rate of hardening by reaction with water.

The conventional cement chemists' notation uses abbreviations for the most common constituents: calcium oxide, $CaO, = C$; silicon dioxide, $SiO_2 = S$; aluminum oxide, $Al_2O_3 = A$; ferric oxide, $Fe_2O_3, = F$; magnesium oxide, MgO, $= M$; sulfur trioxide, $SO_3, = \bar{S}$; sodium oxide, $Na_2O, = N$; potassium oxide, $K_2O, = K$; carbon dioxide, CO_2, C; and water, $H_2O, = H$. Thus, tricalcium silicate, Ca_3SiO_5, is denoted by C_3S. Portland cement clinker is formed by the reactions of calcium oxide and acidic components to give C_3S, C_2S, C_3A, and a ferrite phase approximating C_4AF.

Clinker Formation

Portland cements are ordinarily manufactured from raw mixes, including components such as calcium carbonate, clay or shale, and sand. As the temperature of the materials increases during their passage through the kiln, the following reactions occur: evaporation of free water; release of combined water from the clay; decomposition of magnesium carbonate; decomposition of calcium carbonate (calcination); and combination of the lime and clay oxides.

Structure

The polarizing microscope has been used to determine the size and birefringence of alite crystals, and the size and color of belite to predict later age strength. The clinker phases are conveniently observed by examining polished sections selectively etched using special reagents.

Portland cement clinker structures vary considerably with composition, particle size of raw materials, and burning conditions, resulting in variations of clinker porosity, crystallite sizes and forms, and aggregations of crystallites.

Hydration Process

Portland cement is generally used at temperatures ordinarily encountered in construction; ie, from 5 to 40°C, temperature extremes have to be avoided. The exothermic heat of the hydration reactions can play an important part in maintaining adequate temperatures in cold environments and must be considered in massive concrete structures to prevent excessive temperature rise and cracking during subsequent cooling. Heat induced delayed expansion can also be controlled by keeping the concrete temperature <70°C.

The initial conditions for the hydration reactions are determined by the concentration of the cement particles (0.2–100 μm) in the mixing water (w/c = 0.3–0.7 on a wt% basis) and the fineness of the cement (300–600 m²/kg). Upon mixing with water, the suspension of particles, as shown in Figure 1, is such that these particles are surrounded by films of water. The anhydrous phases initially react by the formation of surface hydration products on each grain and by dissolution in the liquid phase. The solution quickly becomes saturated with calcium and sulfate ions, and the concentration of alkali cations increases rapidly. These reactions consume part of the anhydrous grains, but the reaction products tend to fill that space as well as some of the originally water-filled space. The hydration products at this stage are mostly colloidal (<0.1 μ), but some larger crystals of calcium aluminate hydrates, sulfoaluminate hydrate, and hydrogarnets form. As the reactions proceed, the coatings increase in thickness and eventually form

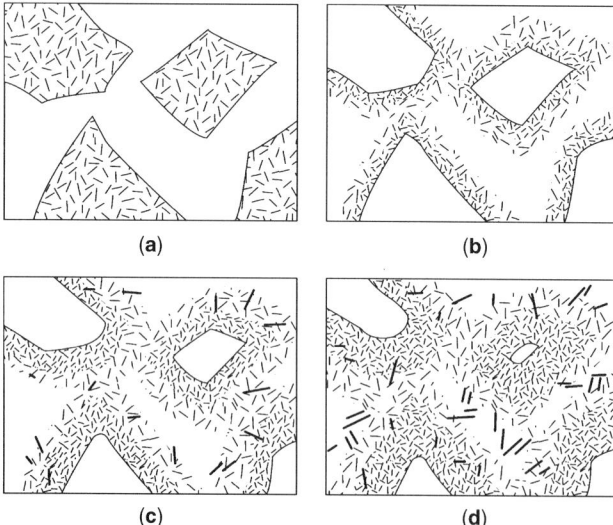

Figure 1. Four stages in the setting and hardening of Portland cement: simplified representation of the sequence of changes. (**a**) Dispersion of unreacted clinker grains in water. (**b**) After a few minutes; hydration products eat into and grow out from the surface of each grain. (**c**) After a few hours; the coatings of different clinker grains have begun to join up, the hydration products thus becoming continuous (setting). (**d**) After a few days; further development of hydration products has occurred (hardening). Courtesy of Academic Press Inc. (London).

bridges between the original grains. This is the stage of setting. Despite the low solubility and mobility of the silicate anions, growths of the silicate hydrates also form on the crystalline phases formed from the solution and become incorporated into the calcium hydroxide and other phases. Upon further hydration, the water-filled spaces become increasingly filled with reaction products to produce hardening and strength development.

MANUFACTURE

Portland cement manufacture consists of (1) quarrying and crushing the rock, (2) grinding the carefully proportioned materials to high fineness, (3) subjecting the raw mix to pyroprocessing in a rotary kiln, and (4) grinding the resulting clinker to a fine powder. A layout of a typical plant is shown in Figure 2.

Industrial by-products are becoming more widely used as raw materials for cement, eg, slags contain carbonate-free lime, as well as substantial levels of silica and alumina. Fly ash from utility boilers can often be a suitable feed component, because it is already finely dispersed and provides silica and alumina. Even vegetable wastes, such as rice hull ash, provide a source of silica. Probably 50% of all industrial byproducts are potential raw materials for Portland cement manufacture.

Clinker production requires large quantities of fuel. In the United States, coal and natural gas are the most widely used kiln fuels, but fuels derived from waste materials, eg, tires, solvents, etc, are increasing in importance. In addition to the kiln fuel, electrical energy is required to power the equipment, this energy, amounts to only about one-ninth that of the kiln fuel.

ENVIRONMENTAL ASPECTS

Cement plants in the United States are carefully monitored for compliance with Environmental Protection Agency (EPA) standards for emissions of particulates, SO_x, NO_x, and hydrocarbons. All plants incorporate particulate collection devices such as baghouses and electrostatic precipitators. The particulates removed from stack emissions are called cement kiln dust (CKD). CKD is characterized by low concentrations of metals which leach from it at levels far below regulatory limits.

SPECIFICATIONS AND TYPES

Portland cements are manufactured to comply with specifications established in each country. In the United States, several different specifications are used, including those of the American Society for Testing and Materials (ASTM) and American Association of State Highway and Transportation Officials (AASHTO).

In the United States, Portland cement is classified in five general types designated by ASTM Specification C150: Type I, when special properties are not required; Type II, for general use, especially when moderate

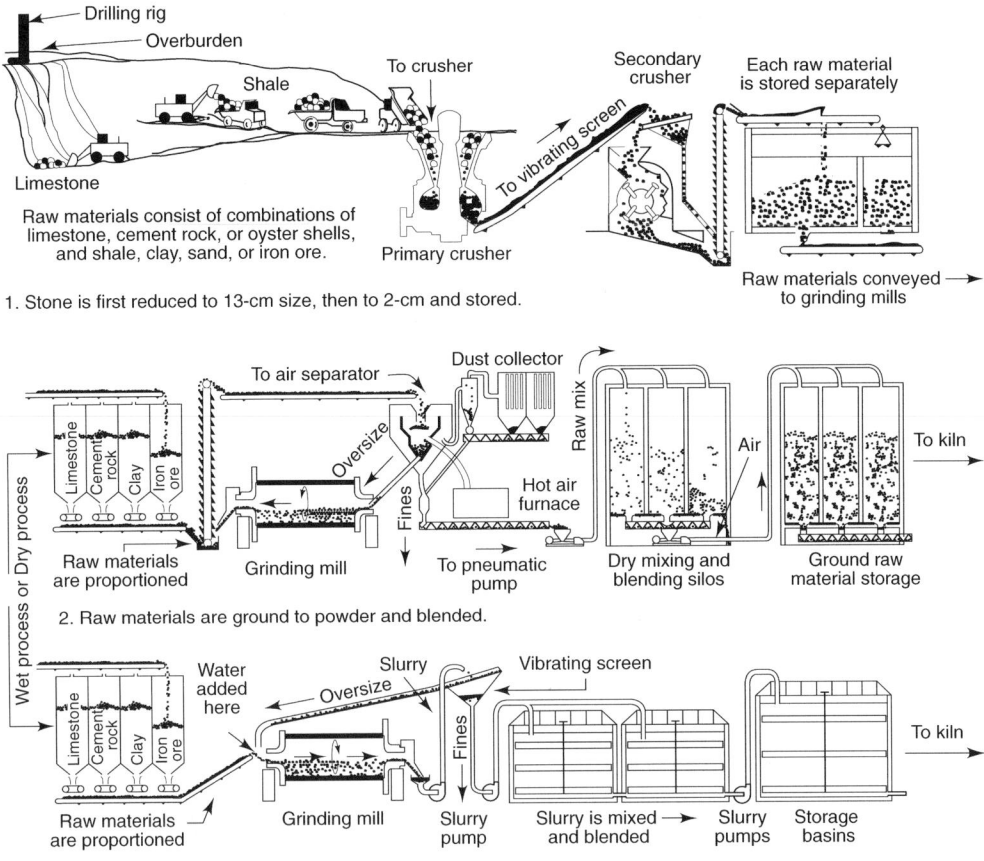

1. Stone is first reduced to 13-cm size, then to 2-cm and stored.

2. Raw materials are ground to powder and blended.

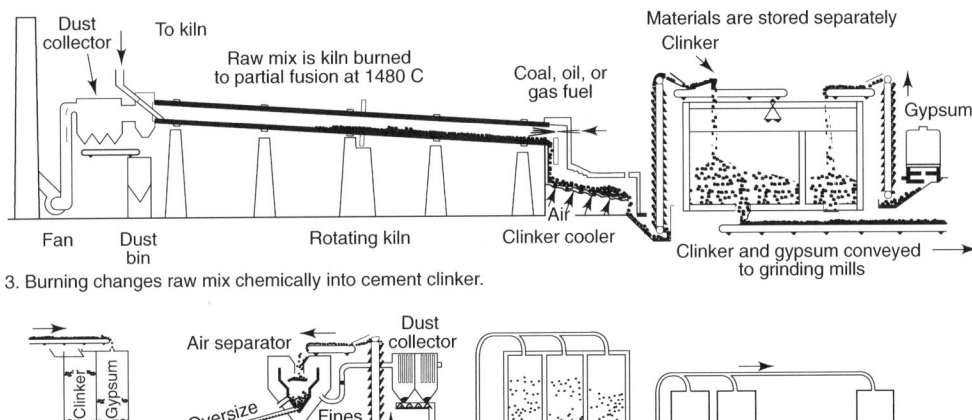

2. Raw materials are ground, mixed with water to form slurry, and blended.

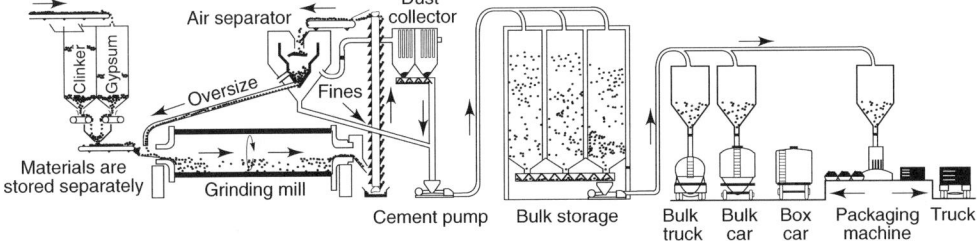

3. Burning changes raw mix chemically into cement clinker.

4. Clinker with gypsum is ground into Portland cement and shipped.

Figure 2. Steps in the manufacture of Portland cement. Courtesy of the Portland Cement Association.

sulfate resistance or moderate heat of hydration is desired; Type III, for high early strength; Type IV, for low heat of hydration; and Type V, for high sulfate resistance. Types I, II, and III may also be specified as air entraining. Chemical compositional, physical, and per-

formance test requirements are specified for each type; optional requirements for particular uses may also be specified. Portland cement can also be specified under ASTM C 1157 as general purpose cement, Type GU; high early strength cement, Type HE; moderate heat cement, Type MH; low heat cement, Type LH; moderate sulfate resistant cement, Type MS; and high sulfate resistant cement, Type HS of Portland and special cements. In Europe, cements are made to meet the requirements of EN 197.

USES

Hydraulic cements are intermediate products used for making concretes, mortars, grouts, and other composite materials. High early strength cements may be required for precast concrete products or in high rise building frames to permit rapid removal of forms and early load-carrying capacity. Cements of low heat of hydration may be required for use in massive structures, such as gravity dams, to prevent excessive temperature rise and thermal contraction and cracking during subsequent cooling. Concretes exposed to seawater or sulfate-containing ground waters require cements that are sulfate-resistant after hardening.

Air-entraining cements produce concretes that protect the concrete from frost damage. They are commonly used for concrete pavements subjected to wet and freezing conditions.

Low alkali cements may be used with certain concrete aggregates containing reactive silica to prevent deleterious expansions.

Expansive, or shrinkage-compensating cements, cause slight expansion of the concrete during hardening. The expansion has to be elastically restrained so that compressive stress develops in the concrete. Subsequent drying and shrinkage reduces the compressive stresses but does not result in tensile stresses large enough to cause cracks. Special highly expansive cements have been used for demolition purposes.

Finely ground cements, often called ultrafine cements, are used to grout soils with fine pore spaces, such as fine sand. These cements can be made with a wide combination of Portland cement, slag, or silica fume.

Regulated-set cement, called jet cement in Japan, is formulated to yield a controlled short setting time, <1 h, and very early strength. It is a modified cement that can be manufactured in a conventional Portland cement kiln. It incorporates set control and early strength development components.

Natural cements intermediate between Portland cements and hydraulic limes in hydraulic activity are no longer available in the United States.

Blended cements include Portland cement clinker is also interground with suitable other materials such as granulated blast-furnace slags and natural or artificial pozzolans. These substances also show hydraulic activity when used with cements, and the blended cements bear special designations such as *Portland blast-furnace slag cement* or *Portland–pozzolan cement*. Pozzolans are used in making concrete both as an interground or blended component of the cement or as a direct addition to the concrete mix. When the two materials are supplied as an intimate blend the mixture is referred to as Portland–pozzolan cement. Portland–pozzolan cements were developed originally to provide concretes of improved durability in marine, hydraulic, and underground environments; they may also prevent deleterious alkali–aggregate reactions. Blast-furnace slag cements may also reduce deleterious alkali–aggregate reactions and can be resistant to seawater if the slag and cement compositions are suitably restricted. Both cements hydrate and harden more slowly than Portland cement. This can be an advantage in mass concrete structures, where the lower rates of heat liberation may prevent excessive temperature rise, but when used at low temperatures the rate of hardening may be excessively slow. Portland blast-furnace slag cements may be used to advantage in steam-cured products that can have strengths as high as obtained with Portland cement. Interest in the use of blended cements is stimulated by energy conservation and solid waste utilization considerations.

Oil well cements are usually made from Portland cement clinker and may also be blended cements. The American Petroleum Institute Specification for Materials and Testing for Well Cements gives requirements for eight classes of oil well cements. Oil well cements are more coarsely ground than normal and contain special retarding admixtures.

Masonry cements are for use in mortars for masonry construction. They are formulated to yield easily workable mortars and contain special additives that reduce the loss of water from the mortar to the porous masonry units.

Calcium aluminate cement develops very high strengths at early ages. It attains nearly its maximum strength in one day, which is much higher than the strength developed by Portland cement in that time. At higher temperatures, however, the strength drops off rapidly. Heat is also evolved rapidly on hydration and results in high temperatures; long exposures under moist warm conditions can lead to failure. Resistance to corrosion in sea or sulfate waters, as well as to weak solutions of mineral acids, is outstanding. This cement is attacked rapidly, however, by alkali carbonates. An important use of high alumina cement is in refractory concrete for withstanding temperatures up to 1500°C. White calcium aluminate cements, with a fused aggregate of pure alumina, withstand temperatures up to 1800°C.

Trief cements, manufactured in Belgium, are produced as a wet slurry of finely ground slag. When activators such as Portland cement, lime, or sodium hydroxide are added in a concrete mixer, the slurry sets and hardens to produce concretes with good strength and durability.

Hydraulic limes may be used for mortar, stucco, or the scratch coat for plaster. They harden slowly under water, whereas high calcium limes, after slaking with water, harden in air to form the carbonate but not under water at ordinary temperatures. However, at elevated temperatures achieved with steam curing, lime–silica sand mixtures do react to produce durable products such as sand–lime bricks.

Specialty cements eg, are used for special architectural applications. White Portland cement with a very low iron oxide content can be produced. Colored cements are usually prepared by intergrinding 5–10% of pigment with white cement.

Numerous other specialty cements composed of various magnesium, barium, and strontium compounds as silicates, aluminates, and phosphates, as well as others, are also produced.

R. H. Bogue, *The Chemistry of Portland Cement*, 2nd ed., Rheinhold Publishing Corp., New York, 1955.

F. M. Lea, *The Chemistry of Cement and Concrete*, 3rd ed., Edward Arnold (Publishers) Ltd., London, 1971.

Minerals Yearbook: Cement, U.S. Bureau of Mines, Washington, D.C., 2001.

Iv. Odler, *Special Inorganic Cements*, E & FN Spon, New York, 2000.

STEVEN KOSMATKA
Portland Cement Association

CENTRIFUGAL SEPARATION

Centrifugal separation is a mechanical means of separating the components of a mixture of liquids that are immiscible or of liquids and insoluble solid particles. The material is accelerated in a centrifugal field that acts upon the mixture in the same manner as a gravitational field. The centrifugal field can, however, be varied by changes in rotational speed and equipment dimensions, whereas gravity is essentially constant. Commercial centrifugal equipment can reach an acceleration of 20,000 times gravity (20,000 G); laboratory equipment can reach up to 360,000 G. The ultracentrifuge and gas centrifuge represent special cases that establish separation gradients on a molecular scale. The usual gravitational operations, such as sedimentation or flotation of solids in liquids, drainage or squeezing of liquids from solid particles, and stratification of liquids according to density, are accomplished more effectively in a centrifugal field.

THEORY

Separation by Density Difference

A single solid particle or discrete liquid drop settling under the acceleration of gravity in a continuous liquid phase accelerates until a constant terminal velocity is reached. At this point the force resulting from gravitational acceleration and the opposing force resulting from frictional drag of the surrounding medium are equal in magnitude. The terminal velocity largely determines what is commonly known as the settling velocity of

the particle, or drop under free-fall, or unhindered conditions. For a small spherical particle, it is given by Stokes' law:

$$v_g = \frac{\Delta\delta d^2 g}{18\,\mu} \qquad (1)$$

where v_g = the settling velocity of a particle or drop in a gravitational field; $\Delta\delta = \delta_S - \delta_L$ = the difference between true mass density of the solid particle or liquid drop, and that of the surrounding liquid medium; d = the diameter of the solid particle or liquid drop; g = the acceleration of gravity; and μ = the absolute viscosity of the surrounding medium.

Stokes' law can be readily extended to a centrifugal field:

$$v_s = \frac{\Delta\delta d^2 \omega^2 r}{18\,\mu} = v_g \left(\frac{\omega^2 r}{g}\right) \qquad (2)$$

where v_s = the settling velocity of a particle or drop in a centrifugal field; ω = the angular velocity of the particle in the settling zone; and r = the radius at which settling velocity is determined.

Separation by Drainage

The theory covering drainage in a packed bed of particles is incomplete and requires more development for a centrifugal field. Liquid is held within the bed by various forces. Removal involves several flow mechanisms. In addition, the centrifugal acceleration changes with radius in the bed, causing changes in packing tendencies of particles and accelerating forces on the residual liquid.

LIQUID–LIQUID-PHASE BEHAVIOR

Liquid drops, suspended in a continuous liquid medium, separate according to the same laws as solid particles. After reaching a boundary, these drops coalesce to form a second continuous phase separated from the medium by an interface that may be well or ill defined. The discharge of these separated layers is controlled by the presence of dams in the flow paths of the phases. The relative radii of these dams determine the radius of the interface between the two separated layers.

CENTRIFUGE COMPONENTS

Power, Energy, and Drives

Centrifuges accomplish their function by subjecting fluids and solids to centrifugal fields produced by rotation. Electric motors are the drive device most frequently used; however, hydraulic motors, internal combustion engines, and steam or air turbines are also used. One power equation applies to all types of centrifuges and drive devices.

The total power, P_T, needed to run a centrifuge, ie, delivered by the drive device, is equal to sum of all losses:

$$P_T = P_P + P_S + P_F + P_W + P_{BD} + P_{CP} \qquad (3)$$

where P_P is the process power. Power for each liquid and the solid phase must be added to get P_P. P_s is the solids process power. The parameter P_F is the friction power, i.e., loss in bearings, seals, gears, belts, and fluid couplings. P_W is the windage power. Increased density owing to gas pressure increases the windage power, and this may be very significant for high-pressure applications. Windage power is a very important loss for large machines and must be determined. P_{CP} is the friction power consumed by the centripetal pump.

For scroll centrifuges having back-drives, P_{BD} is the back-drive power.

Materials of Construction and Operational Stress

Before a centrifugal separation device is chosen, the corrosive characteristics of the liquid and solids as well as the cleaning and sanitizing solutions must be determined. A wide variety of materials may be used.

Once the material choice based on corrosion is made, a careful analysis of the stresses produced by rotation for the particular type of centrifuge is required, so that for the given liquid and solids specific gravities a maximum operating speed can be determined.

The geometry of the bowl parts is important, and for intermittently discharging centrifuges, fatigue strength must be considered.

Equipment

Centrifugation equipment that separates by density difference is available in a variety of sizes and types and can be categorized by capacity range and the theoretical settling velocities of the particles normally handled. Centrifuges that separate by filtration produce drained solids and can be categorized by final moisture, drainage time, and physical characteristics of the system, such as particle size and liquid viscosity. For optimum results, a combination of several types of equipment may be used.

CENTRIFUGES

Sedimentation Equipment

Centrifugal sedimentation equipment is usually characterized by limiting flow rates and theoretical settling capabilities. Feed rates in industrial applications may be dictated by liquid handling capacities, separating capacities, or physical characteristics of the solids.

In general, solids-retaining batch and batch automatic machines are limited to low feed concentrations to minimize the time required to unload the solids. Continuous disk centrifuges can have higher feed concentration. The limit is the underflow concentration. Conveyor discharge centrifuges can handle high feed concentration and are limited only by the volume of solids displacement, or torque capacity.

Centrifugal Sedimentation Equipment

Commercial sedimentation centrifuges are characterized principally by how solids are discharged, and the general dryness of these solids. There are batch and automatic batch solid bowl machines which collect the solids at the bowl wall. Solids are removed very dry. Almost any solid is collectable, even those that are very soft and compressible.

Disk-type solid bowl machines are batch, batch automatic, and continuous. The solids are removed in many different ways but are usually wet. Scroll centrifuges discharge solids continuously and usually drier than disk and imperforate batch types. Generally, disk centrifuges are best for collecting fine particles at a high rate.

A bottle centrifuge is designed to handle small batches of material for laboratory separations, testing, and control. The basic structure is usually a motor-driven vertical spindle supporting various heads or rotors. A surrounding cover reduces windage, facilitates temperature control, and provides a safety shield. Accessories include a timer, tachometer, and manual or automatic braking. Bench-top bottle centrifuges operate at 500–5000 rpm, producing centrifugal fields up to 3000 G in the lower speed range, and operate up to 20,000 rpm with 34,000 G in the high-speed units. Larger models operate up to 6,000 rpm and develop 8,000 G, using special attachments that permit 40,000 G. These models may also be equipped with automatic temperature control down to $-10°C$ and other programmable controls to manage the cycle.

Specialty rotors permit ordinary bottle centrifuges to achieve some of the results previously considered possible only in ultracentrifuges. Preparation ultracentrifuges are suitable for a range of applications, such as processing quantities of subcellular particles, viruses, and proteins. Preparation ultracentrifuges range in operating speed from 20,000 rpm, generating $\sim$40,000 G, to 75,000 rpm and $\sim$500,000 G. The rotor is surrounded by a high-strength cylindrical casing and underdriven by an electric motor. To avoid overheating of the rotor by air friction at these speeds, the pressure in the casing is reduced to $\sim$0.13 Pa (1 mm Hg). Sensors monitor the temperature, and a cooling system controls the temperature in the range of -15 to $30°C$ within $\pm1°C$. Electronic controls maintain the rotor speed within a required narrow range and may be automatically programmed for sequential changes in speed, including control of the acceleration and deceleration.

The use of density gradients in centrifuge rotors greatly increases the sharpness of separations and the quantities of material that can be handled. In principle, the density gradient is established normal to the axis of rotation of the rotor and the highest density is located at the outer radius of the rotor. A natural gradient may be formed by introducing a homogeneous solution and centrifuging for long periods of time. Continuous or step gradients may also be formed by introducing successive layers of solution the composition of which varies continuously or stepwise from low to high density, where the latter displaces the former toward the center of the rotor.

Tubular centrifuges separate liquid–liquid mixtures or clarify liquid–solid mixtures having less than 1% solids content and fine particles. Liquid is discharged continuously, whereas solids are removed manually when sufficient bowl cake has accumulated. For industrial use, the cylindrical bowls are 100–180 mm in diameter with length/diameter ratios ranging from 4 to 8. Bowl speeds up to 17,000 rpm generate centrifugal accelerations up to 20,000 G at the bowl wall.

The tubular centrifuge was long used for the purification of contaminated lubricating oils because of the high centrifugal force developed and the simplicity of its operation.

Centrifuges that channel feed through a large number of conical disks to facilitate separation combine high flow rates with high theoretical capacity factors. For industrial units, flow rates up to 250 m³/h (1100 gpm) can be obtained on easy separations, and theoretical settling velocities may range from 8×10^{-6} to $\sim 5 \times 10^{-5}$ cm/s. Both liquid–liquid and liquid–solid separations are performed using feed solids concentration <15% and small particle sizes.

The outstanding feature of the disk bowl design is a stack of thin cones, commonly referred to as disks, which are separated by thin spacers. These are so arranged that the mixture to be clarified must pass through the disk stack before discharge. The resulting stratification of the liquid medium greatly reduces the sedimenting distance required before a particle reaches a solid surface and can be considered removed from the process stream. The angle of the cones to the axis of rotation is great enough to ensure that solid particles deposited on the surfaces slide, either individually, or as a concentrated phase according to the difference between their density and that of the medium.

Commonly used with disk centrifuges are centripetal pumps (Fig. 1) that discharge the clarified liquid phases under pressures up to 0.7 MPa (100 psi) at reduced aeration, and scoop the rotating liquid out by using a stationary impeller. Centripetal pumps are capable of discharging at rates to 250 m³/h (1100 gpm), and often eliminate the need for a tank and conventional pump.

Centrifugal Filtration Equipment

The important parameters of centrifugal filtration equipment are screen area, level of centrifugal acceleration in the final drainage zone, and cake thickness. The latter affects both residence time and volumetric throughput rate. The particle size of the solids and the kinematic viscosity of the mother liquor also strongly affect the final moisture content.

The simplest and most common form of centrifugal filter is a perforate-wall basket centrifuge, consisting of a cylindrical bowl having a diameter ranging from ~100–2400 mm and a diameter/height ratio ranging from 1 to 3. The wall is perforated with a large number of holes, more than adequate for the drainage of most liquid loads, and is lined with a filter medium. In the simplest case, the medium is a single layer of fabric or metal cloth or screen. In high-speed basket centrifuges, one or more backup screens of relatively large mesh support a finer mesh filter surface. The method of discharging accumulated solids distinguishes three types of basket centrifuge: those that are stopped for discharge, those that are decelerated to a very low speed for discharge, and those that discharge at full speed.

Another batch automatic horizontal perforated bowl centrifuge inverts the flexible filter to discharge the solids. Feed slurry may be deposited on the inside surface of a cloth filter, with the bucket end completely closed.

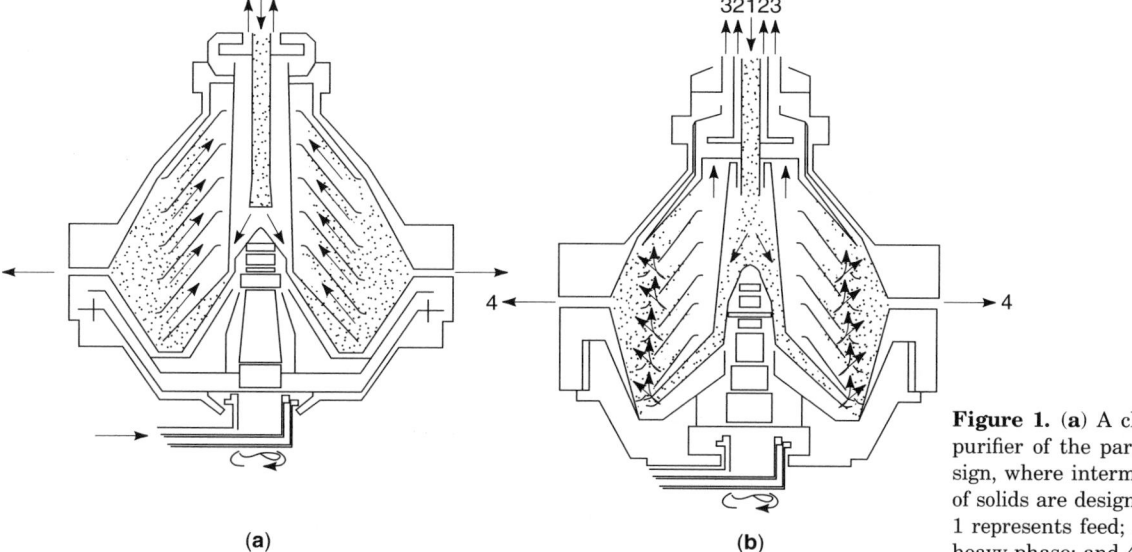

Figure 1. (**a**) A clarifier and (**b**) a purifier of the paring disk-type design, where intermittent discharges of solids are designated by →, and 1 represents feed; 2, light phase; 3, heavy phase; and 4, solids.

When the interior is full of dewatered solids, the bowl is decelerated to a slow speed and piston and closure plates move axially, inverting the filter cloth so that the solids reside on the outside diameter of the cloth. Very little residual material remains on the filter cloth surface for the next cycle.

Continuous filtering centrifuges are used for very fast draining that does not require extremely dry final products. Continuous centrifugal filters are equipped with either a cylindrical or a conical screen. Both types are made without a retaining lip on the solids discharge end of the bowl and employ various methods to move the solids through the bowl. The cylindrical screen centrifuge deposits solids at one end of the bowl in a layer 6–80 mm thick and pushes the annular ring of cake axially through the bowl by means of a reciprocating piston.

In conical screen centrifuges the angle of the bowl causes or assists the cake to move axially and redistributes it in an increasingly thin layer, which improves drainage characteristics. The feed slurry is deposited at the small end of the cone, where most drainage occurs. The drained solids are discharged from the large end, which has no retaining ring.

H. Axelsson "Cell Separation, Centrifugal" *Encyclopedia of Bioprocess Technology: Fermentation, Biocatalysis, and Bioseparation*, John Wiley & Sons, Inc., New York, 1999.

N. Corner-Walker, "The Dry Solids Decanter Centrifuge: Capacity Scaling", *Filtration-Separation*, May 2000, pp. 28–32.

F. W. Keith, Jr. and R. T. Moll, in R. A. Young and P. Cheremisinoff, eds., *Wastewater Physical Treatment Processes*, Ann Arbor Science Publishers, Ann Arbor, Mich., 1978.

E. S. Perry and C. F. van Oss, eds., *Progress in Separation and Purification*, Wiley-Interscience, New York, 1971.

Alan Letki
Alfa Laval Inc.
Nick Corner-Walker
Alfa Laval AB

CERAMIC–MATRIX COMPOSITES

Monolithic ceramics are brittle and are thus very sensitive to intrinsic flaws and damage produced by use. Failure of these materials occurs in a catastrophic manner and at low strain-to-failure ratios. However, the problem can be alleviated by reinforcing monolithic ceramics with a second phase which is itself capable of operating at high temperatures. Such systems are designated as ceramic matrix composites (CMC).

The reinforcing phases in ceramic matrix composites are usually also ceramic and have many possible morphologies: particulate, platelet, whisker, short-fiber, or continuous-fiber. Reinforcing entities are typically added to ceramic matrices to produce tough composites. In comparison, high strength reinforcements are added to polymer-based composites to increase strength and stiffness. To enhance toughness high strength reinforcements with high elastic modulus and weak interfaces with the matrix are required; to produce high strength and stiffness, strong interfaces along with high stress transfer are needed to allow efficient load transfer or shedding from the matrix to the reinforcement.

CERAMIC COMPOSITES SYSTEMS

With the appropriate choice of composite properties, such as reinforcement and matrix materials, reinforcing geometry and composite interface, an otherwise brittle mode of failure of a ceramic becomes more "ductile" and noncatastrophic in nature. Thus, the choice of the component materials is an important aspect of designing ceramic matrix composites. Two questions need to be addressed when making these choices. First, if a matrix crack encounters a potential bridging entity, will it deflect along the reinforcement/matrix interface or will fracture of the reinforcement occur? Second, if interface debonding occurs, will the interfacial sliding shear resistance, τ, be low enough to allow the bridge to slip in the matrix or will fracture of the bridging-reinforcement occur?

A partial answer to the first question has been provided by a theoretical treatment that examines the conditions under which a matrix crack will deflect along the interface between the matrix and the reinforcement. The calculations indicate that, for any elastic mismatch, interface failure will occur when the fracture resistance of the bridge is at least four times greater than that of the interface.

About the second question, concerning the relative strengths of the bridge and the interfacial sliding resistance, little is known a priori. The general recommendation is to have a high bridge strength and a low interfacial sliding shear resistance.

Various combinations of ceramic–matrix composites have been manufactured at the research level. Their properties are given in Table 1 for oxide-based matrices and in Table 2 for nonoxide matrices. Some commercial products are identified for information only. Such identification does not imply recommendation or endorsement by NIST, nor does it imply that the products are the best available for the purpose.

COMPOSITE REINFORCEMENTS

The structure of reinforcements can be either equiaxed or acicular. The nature of their placement within a composite, the composite architecture, is critical to the resultant composite properties. Possible architectures are summarized in Figure 1.

Equiaxed particles which are well dispersed in the ceramic matrix, tend to produce isotropic composite

Table 1. Oxide-Based Ceramic-Matrix Composites

Reinforcement

Type[a]	Amount, vol %	Strength,[b] MPa	Toughness,[c] Mpa√in.	Modulus GPa[d]	Density g/cm³ or % td[e]
Al_2O_3 matrix					
B_4C_p	50		4.5	380	3.28
SiC_w	20		2.5	400	
SiC_W/SiO_{2i}[f]	20		6.0	420	
SiC_w/Si_3N_{4w}	20	203	3.4		95% td
SiO_{2f}		6.3	28		
TiC_p	30		4.0	400	4.26
BN_p		24.6		490	91.1% td
Al_p	20		8.4		
$ZrO_2(t)_p$[g]		2000	5–8	333	4.54
Aluminosilicate glass matrix					
SiC_w			0.8	80	
$Al_{2}O_{3f}$		311	3.3		
Cordierite glass matrix					
SiC_f		128	1.6		2.44
Pyrex glass matrix					
Al_2O_{3f}		305	3.7		
SiC_p	30	171	1.79		
SiC_w	30	180	3.04		
SiC_p/SiC_w[h]		159	2.73		
Soda-lime silicate matrix					
SiC_w	20		0.7	72	
LASIII glass ceramic matrix					
SiC_w	35	327	5.1		
$3Al_2O_3 \cdot 2SiO_2$ matrix					
SiC_w	10	274	2.7	197	2.84
SiC_p		262	2.35	240	
$ZrO_2(t)_p$[g]		250	4.0	150	98.9% td
ZrO_2 matrix					
$ZrO_2(t)_P$[g]		400–600	10	200	6.08

[a]Subscripts denote reinforcement morphology; p = particulate, 1 = platelet, w = whisker, f = fiber, i = interlayer between reinforcement and matrix.
[b]Strength as measured in a four-point flexure test (modulus of rupture); to convert MPa to psi, multiply by 145.
[c]Fracture toughness; to convert MPa√m to psi√in., multiply by 910.048.
[d]To convert GPa to psi, multiply by 145,000.
[e]% td = percentage of theoretical density.
[f]20% SiC_w.
[g]Tetragonal.
[h]10% each.

behavior. The particles, either ceramic or metallic, may be single crystal or polycrystalline in nature.

Acicular reinforcements such as whiskers and platelets tend to produce rather more anisotropic composite properties. Whiskers and platelets are usually single crystals with aspect ratios up to 100 and with tensile strengths near their theoretical value. Composite processing can be tailored to produce either an aligned microstructure with the principal axis of all reinforcements lying in the same direction; a textured microstructure in which the principal axis is randomly arranged within a single plane; or an isotropic microstructure in which the reinforcements are randomly arranged in three dimensions. Aligned reinforcements produce a composite with highly unidirectional properties in the alignment direction, but with properties that are isotropic in the transverse direction. Such

Table 2. Nonoxide-Based Ceramic-Matrix Composites

| Reinforcement | | | | | |
Type[a]	Amount, vol %	Strength,[b] MPa	Toughness,[c] MPa$\sqrt{m}$	Modulus, GPa[d]	Density, g/cm^3
AlN matrix					
BN$_p$		65.5		480	
SiC matrix					
SiC$_w$			19.9	240	
TiB$_{2p}$	16		4.5	430	3.30
TiC$_p$	25		6.0	450	3.36
Si$_3$N$_4$ matrix					
SiC$_w$	10	620	7.8		
SiC$_w$	20	4.0	350		
SiC$_w$	10	436	5.7		
Si$_3$N$_{4p}$		680	7.6–8.6	160	
TiC$_p$		578	7.2	328	
TiC$_p$	30		4.5	350	3.7
TiN matrix					
Al$_2$O$_{3p}$/AlN		229	10.2		
WC matrix					
Co	20		16.9	442	

[a]Subscripts denote reinforcement morphology; p = particulate, 1 = platelet, w = whisker, f = fiber, i = interlayer between reinforcement and matrix.
[b]Strength as measured in a four-point flexure test; to convert MPa to psi, multiply by 145.
[c]Fracture toughness; to convert MPa $\sqrt{m}$ to psi $\sqrt{in}$., multiply by 910.048.
[d]To convert GPa to psi, multiply by 145,000.
[e]30% each.

microstructures are produced when whisker-reinforced composites are fabricated by extrusion or when platelet-reinforced composites are fabricated by tape-casting or hot-pressing techniques. Textured microstructures produce composites that have isotropic properties in the reinforcement plane. This tends to be the most common type of microstructure produced when a whisker-reinforced composite is fabricated by hot pressing or tape-casting techniques. Fabrication techniques required to produce a completely random microstructure with resulting isotropic properties are extremely difficult. Hence, most ceramic composites reinforced with acicular particles tend to have some form of texture and thus, some anisotropy of properties.

Fiber reinforcements can be amorphous, single crystal or polycrystalline in structure. They can be either short fibers producing similar composite architectures to those of whiskers or they can be continuous. Continuous fiber reinforced composites tend to have orthotropic properties. For unidirectional composites properties transverse to the fiber direction are significantly different from those parallel to the fiber direction.

The role of reinforcements in a ceramic-matrix composite system is to transfer stress from the matrix to the reinforcement, thereby shielding the crack tip from the applied load and providing an additional dissipative energy sink to resist crack propagation. This function is usually achieved via strong reinforcements with a weak interface between the reinforcement and the matrix. This combination allows ligament debonding and energy dissipation via frictional sliding of the reinforcement in the matrix. The matrix and reinforcement are usually chosen to allow weak interface debond stress. However, in practice, it is difficult to achieve this state because most ceramic systems react chemically. In fiber and whisker reinforced ceramic composite systems an interlayer coating of pyrolitic carbon is usually incorporated at the interface to facilitate easy debonding.

An alternative to the weak debond coatings is to create a mechanically weak debond interface. One that shows much promise is a porous coating of the matrix itself on the reinforcement. The coating is well-bonded to the reinforcement and the matrix but is mechanically much weaker than either because of its degree of porosity. A debond crack will thus run preferentially through the coating.

An alternative to the weak debond interface approach may lie in a ductile interface that is well-bonded to both the reinforcement and the matrix. Debonding of the interface then entails ductile yielding and shearing of the interface. Such a process potentially dissipates more energy than debonding and frictional interfacial sliding alone.

A related and important issue in choosing a reinforcing material is the chemical compatibility of the reinforcement with the matrix. The reinforcement must also have high strength that is retained to elevated temperatures. If the environment has access to the reinforcement, either at the surface or through matrix cracking, then the reinforcement must be sufficiently chemically inert in the service-environment. Table 3 presents the properties of a few of the currently available platelets,

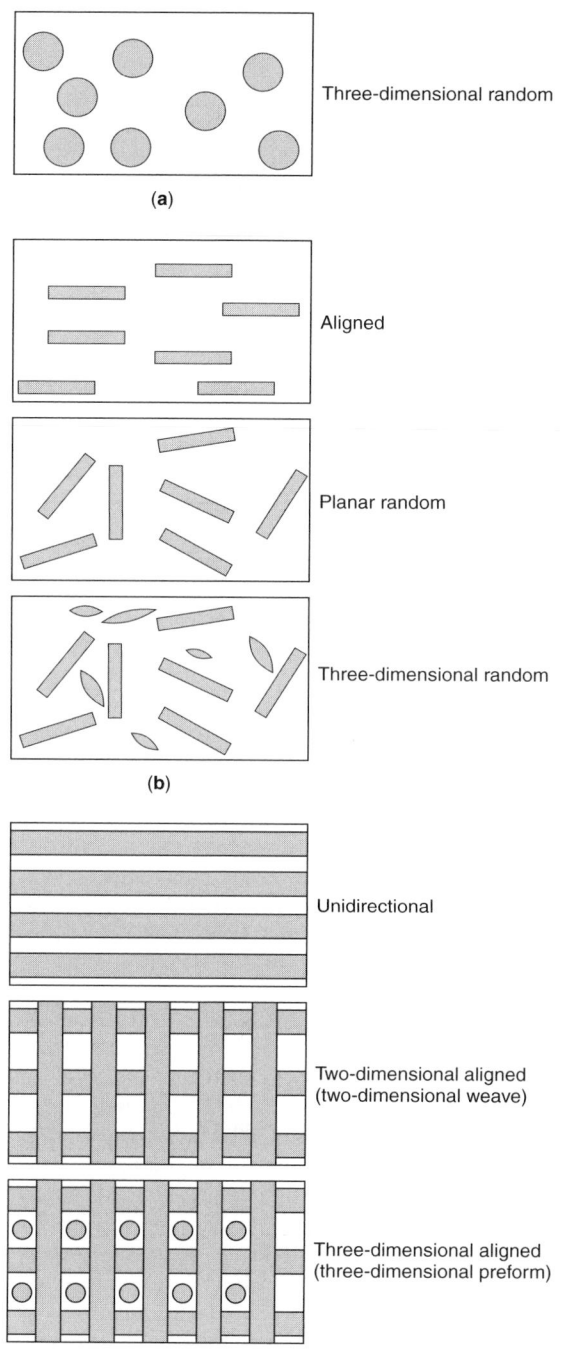

Figure 1. Reinforcement architectures for ceramic–matrix composites and corresponding composite properties, (**a**) Spherical particles; (**b**) platelets, whiskers, short fibers; and (**c**) continuous fibers.

whiskers, and fibers for use as reinforcements in ceramic composites.

CERAMIC MATRICES

Ceramic matrices are usually chosen on their merits as high temperature materials; reinforcements are added to improve their toughness, reliability, and damage tolerance. The matrix imparts protection to the reinforcements from chemical reaction with the high temperature environment. The principal concerns in choosing a matrix material are its high temperature properties, such as strength, oxidation resistance, and microstructural stability, and chemical compatibility with the reinforcement.

Another consideration is the difference in thermal expansion between the matrix and the reinforcement. Composites are usually manufactured at high temperatures. On cooling any mismatch in the thermal expansion between the reinforcement and the matrix results in residual mismatch stresses in the composite. These stresses can be either beneficial or detrimental: if they are tensile, they can aid debonding of the interface; if they are compressive, they can retard debonding, which can then lead to bridge failure.

Compressive interfacial stresses increase the interfacial shear resistance. Although usually detrimental to toughening, these stresses can enhance toughening if bridge pullout is the operative toughening process.

TOUGHENING PROCESSES

The toughness induced in ceramic matrices reinforced with the various types of reinforcements, that is, particles, platelets, whiskers, or fibers, derives from two phenomena: crack deflection and crack-tip shielding. These phenomena usually operate in synergism in composite systems to give the resultant toughness and noncatastrophic mode of failure.

Crack-Resistance Behavior

The goal of composite reinforcement is to produce tough, flaw-insensitive materials that fail in a "ductile" manner. Such materials are more damage tolerant than the monolithic ceramics because they can withstand larger cracks without fracture and the fracture strength may be independent of crack size within a certain flaw size range. This important property of flaw tolerance and stable crack growth results from a fracture resistance behavior known as $\mathcal{R}$-curve or T-curve behavior, in which the fracture resistance rises with crack extension. Fracture resistance can be formulated in terms of either stress intensity factor T or strain energy release rate $\mathcal{R}$ (or J). If stress intensity factor is used, then the ordinate is the square root of crack length and the plot is termed a T-curve. If, however, strain energy-release rate is used, $\mathcal{R}$; (or J_c) is plotted directly as a function of crack length and the curve is termed an $\mathcal{R}$-curve.

Crack Deflection Contribution to Toughening

Crack deflection is a phenomenon that leads both to toughening and to the formation of bridges that shield the crack tip from the applied stress. Little is known of

Table 3. Reinforcements for Ceramic-Matrix Composites

	Tensile strength, GPa^a	Modulus, GPa^a	Density, g/cm^3	Diameter, μm	Maximum use temperature, °C
Platelets					
Al_2O_3		400	3.96	5–15/1	2040
SiC	3	470	3.21	5–1500/1–15	1600
SiC	0.5	470	3.21	10-15	1600
whiskers					
Al_2O_3	20	450	3.96	4–7	2040
B_4	14	490	2.52		2050
SiC					
Silar SC	7	340–690	3.2	0.6	1760
VLS	8.3	580	3.2	4–7	1400
Tokamax		600	3.2	0.1–1.0	1400
SiC		600	3.2	0.5–1	1400
Si_3N_4					
SNWB	14	385	3.18	0.05–0.5	1900
Fibers					
Al_2O_3					
FP	1.38	380	3.90	21	1316
PRD166	2.07	380	4.20	21	1400
Sumitomo	1.45	190	3.9	17	1249
Safimax	2.0	300	3.30	3	1250
mullite					
Nextel312	3.12	1.55	150	2.70	1204
Nextel480	4.80	2.28	224	3.05	1200
SiC					
Nicalon	2.62	193	2.55	10	1204
SCS	2.80	280	3.05	6–10	1299
Sigma	3.45	410	3.40	100	1259
SiTiCO					
Tyrrano	2.76	193	2.5	10	1300
Si_3N_4					
TNSN	3.3	296	2.5	10	1204
SiO_2					
Astroquartz	3.45	69	2.2	9	993
Graphite					
T300R	2.76	2.76	1.8	10	1648
T40R	3.45	276	1.8	10	1648

^{a} To convert GPa to psi, multiply by 145,000.

the bridge formation process, but its effect, that is, crack-tip shielding, is considered in the following section.

The condition for propagation of a mode I edge crack, that is, a crack that is subjected to pure opening (tensile) stresses applied perpendicular to the crack plane, is given by:

$$K_a = Y\sigma_a\sqrt{c} = K_{IC} \qquad (1)$$

where Y is a dimensionless geometry term, σ_a is the applied stress, and c is the crack length. Once a crack is deflected from its original plane, further crack extension requires a higher driving force to accommodate

the mode II (shear) or mode III (tearing shear) contribution to the stress intensity factor on the new crack plane.

Crack-Tip Shielding

Crack-tip shielding has two origins: process-zone shielding and crack-wake bridging. Process-zone shielding derives from mechanisms occurring in a zone around the crack tip which extend to the crack wake as the crack advances, indirectly applying closure forces to the crack flanks. Crack-wake bridging derives from intact bridging elements in the wake of the crack, directly applying closure forces to the crack flanks.

MECHANICAL PERFORMANCE

Particle Reinforcement

Particle reinforcement is an excellent method for toughening brittle ceramic matrices. The toughness imparted to such composites is due to multiple toughening mechanisms including crack deflection, crack pinning, microcracking, residual stress, frictional bridging, particle pullout, and transformation toughening. The mechanisms important to any specific system depend on the physical properties of the particles: size; morphology; thermal expansion mismatch with the matrix; and strength, toughness, and ductility.

Brittle Particles. Reinforcement via small brittle particles exploits the toughening mechanisms of crack deflection, microcracking, crack pinning, and crack bowing. The toughening contribution from the mechanisms of crack bridging and frictional pullout may be significant if the reinforcing particles are of the order of the matrix grain size or larger. All of these mechanisms arise from, or are strongly enhanced by, thermal expansion mismatch stresses in the composite.

Ductile Particles. Ductile particle reinforced ceramic composites show promise as composite material for high strength–high toughness applications.

Ductile particles can act as bridging sites in the crack wake. Instead of fracturing in a brittle manner, they undergo plastic yielding as the crack opens up. A second toughening mechanism that operates simultaneously with crack bridging is the ductile yielding of particles in the crack-tip stress field within a process zone.

Transforming Particles. A special type of particulate-composite are those based on the tetragonal form of zirconia. Tetragonal zirconia has the ability to undergo a stress-induced martensitic phase-transformation from its tetragonal crystal form to a monoclinic form with an accompanying dilatation of 4% unconstrained. The toughening owing to the phase-transformation-induced dilatation in terms of a stress intensity factor approach has been calculated to be that shown in equation 2,

$$K_a = K_o + K_p = T = K_o + 0.3 E e^T V_f w^{1/2} \tag{2}$$

where E is the modulus of the matrix, e^T is the transformation strain per particle, V_f is the volume fraction of transforming particles, and w is the width of the process zone.

Whisker Reinforcement

Toughening for whisker-reinforced composites has been shown to arise from two separate mechanisms: frictional bridging of intact whiskers, and pullout of fractured whiskers, both of which are crack-wake phenomena.

Whisker reinforcement is a viable method of toughening composites. However, health considerations associated with the aspiration of fine, high-aspect-ratio whiskers raise serious concern about their widespread use.

Platelet Reinforcement

Ceramic composites reinforced with crystalline platelets show similar values of toughness as whisker-reinforced ceramic matrices. Platelets have the additional advantages of being at least one tenth the cost of whiskers, easier to process, and have higher thermal stability and none of the health hazards associated with the aspiration of whiskers. Toughness comes from a combination of crack deflection, frictional bridging and platelet pullout.

Fiber Reinforcement

The whiskers bridging mechanics apply also to short random fiber bridging mechanisms.

Composites reinforced by continuous fibers can fail in one of several possible modes depending on the interface properties and the fiber strength. When the distribution of fiber strengths is broad (as characterized by a low Weibull modulus) in the regime of low fiber strength/high shear resistance, fiber fracture in the crack wake occurs away from the crack mid-plane and the fibers pullout. The majority of toughening is a result of the frictional pullout mechanism, although, there may be some contribution from frictional bridging before fiber fracture occurs.

A transition to a different mode of composite failure, which is still within this low fiber strength/high shear resistance regime, occurs when the fiber strengths have a much tighter strength distribution as characterized by a high Weibull modulus. Initially the fiber strength is sufficient to allow the formation of a bridging zone in the crack wake before fiber fracture occurs at the point of highest stress in the fiber, that is, in the crack mid-plane. There is little to no fiber pullout contribution to toughening, and the contribution from fractional bridging predominates.

In the region of high fiber strength and low interfacial shear resistance, matrix cracks can propagate around the fibers, leaving them intact in the wake of the crack. The matrix can be completely cracked through with the fibers supporting all the load before fiber failure begins. In such a material the toughness is primarily due to bridging contribution, rather than fiber pullout.

CHEMICAL AND THERMAL STABILITY

Ceramic-matrix composites are a class of materials designed for structural applications at elevated temperature. Exposure at these temperatures will be for many thousands of hours. Therefore, the composite microstructure must be stable to both temperature and environment. Relatively few studies have been conducted on the high temperature mechanical properties and thermal and chemical stability of ceramic composite materials.

Reinforcement Integrity

Strength degradation with increasing temperature occurs to a much greater extent with ceramic reinforcements, particularly those of continuous fibers, than it does with monolithic materials. Reinforcements have high surface areas to volume so that they are more susceptible to strength degradation resulting from surface reactions with the atmosphere. These reactions can also decrease the toughness of the composite if crack-wake bridging and pullout are the predominant toughening mechanisms.

Composite Response

A majority of ceramic-matrix composites show strong trends in the manner in which the mechanical properties are affected by temperature. If the interface degrades, allowing strong bonding to occur between the reinforcement and the composite matrix, the toughness is considerably reduced. If the interface remains weak enough to allow debonding and pullout, composite strength and elastic modulus are reduced.

Studies on creep resistanceof particulate reinforced composites seem to indicate that such composites are less creep-resistant than are monolithic matrices. In contrast to the particulate-reinforced composites, all other reinforcement morphologies appear to provide enhanced creep resistance.

E. P. Butler, H. Cai, and E. R. Fuller, Jr., in *Engineering Ceramics Division, 16th Annual Conference on Composites & Advanced Ceramics*, ACerS, Cocoa Beach, Fla., 1992, pp. 475–482.

B. R. Lawn, *Fracture of Brittle Solids*, The Cambridge Press, Cambridge, 1992.

I. N. Sneddon and M. Lowengrub, *Crack Problems in the Classical Theory of Elasticity*, John Wiley & Sons, Inc., New York, 1969.

R. Warren, *Ceramic-Matrix Composites*, Blackie, Glasgow and London, 1992.

E. P. BUTLER
E. R. FULLER, JR.
National Institute of Standards
and Technology

CERAMICS AS ELECTRICAL MATERIALS

For most electroceramic applications the electrical conductivity, whether due to ionic, electronic, or mixed ionic–electronic conduction, is the dominant material property, that determines bulk insulation behavior. Even where the dominant material characteristic is, eg, magnetic, ferroelectric, piezoelectric, pyroelectric, electrooptic, or electrochemical, the underlying property of primary importance for device use of these materials is electrical conduction behavior. In oxide materials such as RuO_2 and $Bi_2Ru_2O_3$, metallic conduction does occur, making them ideal for use as components in thick-film pastes and in composite electrodes. Fast ion conduction in oxide materials such as $(Zr,Y)O_{2-}$ is made use of commercially in fuel cell applications. Superconductivity in ceramic oxides, based on the Y–Ba and Bi–Sr–Ca cuprate structures, is being exploited for use in microwave filters and for magnetic levitation. Other classes of ceramic materials that feature semiconducting properties are used in applications as varied as resistance heating elements, rectifiers, photocells, varistors, thermistors, and sensors. Ceramic materials serve equally important functions as electrical insulators.

ELECTRICAL CONDUCTION

The electrical conductivity ranges over many orders of magnitude, from 10^5 $(\Omega\text{-cm})^{-1}$ for conducting oxides such as rhenium(VI) oxide, ReO_3, and chromium(IV), CrO_2, to $10^{-14}(\Omega\text{-cm})^{-1}$ highly insulating materials such as steatite porcelains. Other compounds, such as TiO_2, may change conductivity by several orders of magnitude as a result of aliovalent doping, or of high temperature heat treatment in controlled pO_2 ambient.

The charge transport mechanisms for the electrical conduction modes in ceramic materials vary greatly, since the transport of current may be due to the motion of electrons, electron holes, or ions. Crystal structure may also significantly affect the mobility of the charged species.

IONIC CONDUCTION

For an ion to move through a crystalline lattice, there must be an equivalent vacancy or interstitial lattice site available, and it must also acquire sufficient thermal energy to surmount the free energy barrier, between the equivalent sites. Ionic conduction, which occurs through the transport of charge by mobile ions is, therefore, a diffusion activated process.

In crystalline ceramics the charge carriers are mainly mobile ions, with different mobility values for each specie. The mobility is related to the energetics of the site-to-site transport for each type ion, but this process can be enhanced significantly by forward biasing from an applied electric field. The effect of temperature, composition and structure on each of the terms in the general expression must be considered.

In polycrystalline materials, ion transport within the grain boundary must also be considered. For oxides with close-packed oxygen ions, the O^{2-} ion almost always diffuses much faster in the boundary region than in the

Table 1. Fast-Ion Conductors

Compound	Temperature, °C	Conducting ion	σ_{ion}, $(\Omega \cdot m)^{-1}$
$NaAl_{11}O_{17}$	300	Na	35
$Na_3OZr_2PSi_2O_{12}$	300	Na	20
$CeO_2 + 12$ mol% CaO	700	O	4
$ZrO_2 + 12$ mol% CaO	1000	O	0.8
$K_{1.4}Fe_{11}O_{17}$	300	K	2
$Li_2B_4O_7{}^a$	150	Li	10^{-4}
	400	Li	0.1
$Li_4B_7O_{12}Cl^a$	300	Li	0.2
crystal	300	Li	0.8

^{a} Glass.

bulk. This is due in part to second phases in the grain boundary region that are less closely packed, providing pathways for more rapid diffusion of ionic species. Thus the simplified picture of bulk ionic conduction becomes more complex when these additional effects are considered.

FAST-ION CONDUCTORS

Inorganic compounds which exhibit exceptionally high ionic conductivity ($t = 1$), described as fast-ion conductors, are of technological interest for a variety of applications, including use as solid electrolytes. As shown in Table 1, compounds in this category include (1) halides and chalcogenides of silver and copper, in which the metal atoms are disordered over a large number of interstitial sites; (2) oxides having the β-alumina, $NaAl_{11}O_{17}$, structure in which migration of the monovalent cation is aided by the presence of conduction channels, leading to high mobilities; and, (3) oxides of the fluorite, CaF_2, structure in which a large concentration of defects can be developed through incorporation of variable valence cations or by solid state substitution of cations of lower valence, eg, $CaO \cdot ZrO_2$ or $Y_2O_3 \cdot ZrO_2$.

CONDUCTION IN GLASSES

Electrical conduction in silicate glasses at ordinary temperatures can be attributed to the migration of univalent modifier ions such as Li^+, Na^+, K^+, H^+, and OH^-, under the influence of an applied field. At more elevated temperatures (~150°C), divalent ions, eg, Ca^{2+}, Mg^{2+}, and Pb^{2+} also contribute to conduction, although their mobility is generally low. Conduction in glass is an activated process and thus the number of ions contributing to conduction increases with both temperature and applied field.

There is no general consensus that universally explains all aspects of ionic transport in glasses. However, application of percolation theory within the well-known random-energy model leads to the most consistent explanation for both AC and DC conduction effects in glasses.

Explanations for the observed strong dependence of ionic conductivity on composition have mainly been based on changes in the activation energy for electrical conduction.

CERAMIC INSULATORS

Insulators are materials that offer effective resistance to current flow in an electric field, due to the very low concentration of mobile charge carriers. High resistivity values are the result of a large energy gap between a filled valence band and the next available energy level, where the promotion of an electron into a higher state is energetically unfavorable. The conductivity of such ceramics is, therefore significantly influenced by both ionic and electronic defects. In insulating oxides, ionic defects arise from the presence of impurities of different valence from the host cation.

Similarly, electronic conduction may arise in oxide materials from the natural loss of oxygen, which typically occur in oxides on heating to high temperatures. Electrons trapped at the vacancy can become partially or fully ionized, leading to weak n-type electronic conduction. Again, the conductivity is low.

Similar conduction mechanisms can be expected for most nitride materials. SiC, which has only a moderate band gap ($E_g = 2.8$–3.2 eV), can become semiconducting and has been developed for device use. SiC is also widely employed as heating elements for furnace applications. Electrically insulating SiC can also be fabricated using BeO dopant additions. This is an important material for laser heat sink applications because of its high thermal conducting and electrical insulating properties.

The primary function of insulation in electrical circuits is the physical separation of conductors and the regulation or prevention of current flow between them. Ceramic insulators are used in many demanding applications where high electrical resistance is a requirement, together with other important properties such as thermal conductivity, high operating temperatures, high dielectric strength, low dielectric loss, resistance to thermal shock, environmental resistance, thermal expansion, and long-life characteristics. Insulators of this type are known as linear dielectrics. The dielectric constant is a measure of the ability of the material to store charge relative to vacuum and is a characteristic material property.

Silicon dioxide (SiO_2) has the lowest dielectric loss properties of any inorganic material. It is commonly used in insulating fibers and in the development of electrical porcelains ($R_2O \cdot Al_2O_3 \cdot SiO_2$). These materials have high dielectric strengths with low loss and are therefore suitable for high voltage applications such as transmission line insulators, high voltage circuit breakers, and cutouts. Mullite, $3Al_2O_3 \cdot 2SiO_2$, MgO, and steatite, $MgO \cdot SiO_2$, are extensively used for high temperature electrical insulation and for high frequency insulation because of their low loss characteristics. For electrical insulating applications and heat sinks, Al_2O_3, AlN, SiC, and Si_3N_4 are the most commonly used materials. Both SiC and $Si_3 N_4$ are also industrially valuable as high temperature heat exchangers because of high thermal conductivity and electrical insulating behavior, high hardness, durability, excellent high temperature, corrosion, and thermal shock resistance. Films of these materials, including diamond, have been developed, where the properties obtained are similar to that of the bulk materials. The conduction processes in the films mainly result from impurity and electrode injection effects, which degrades the high resistivity and dielectric properties of the materials.

ELECTRONIC CONDUCTION

High electron and electron hole mobility in ceramic materials can contribute appreciably to electrical conductivity. In certain materials, metallic levels of conductivity can result, while in others the electronic contribution can be very small. In all cases, the total electrical conductivity for all modes of conduction (electronic and ionic) is given by the general equation

$$\sigma_i = \sigma(t_{ionic} + t_{electronic}) \quad (1)$$

where the total electronic conductivity is given as

$$\sigma - ne\mu_n + pe\mu_p \quad (2)$$

Where n and p denote the concentrations of electrons and holes, respectively, and μ_e and μ_h are the corresponding mobilities. The mobility of electrons and holes as charge carriers is generally much higher than for ionic carriers, because they are of lower mass and charge density and less confined to particular atomic sites. Scattering of electrons and holes occur by phonons and at point defects, dislocations, and grain boundaries. The conductivity is determined primarily by the concentration of electrons and holes and can be described by an energy band structure. Figure 1 schematically illustrates the band energy configurations corresponding to metal, intrinsic semiconductor, and insulator conduction. The highest energy band, which is completely filled at $T=0$ K, is called the valence band, and the next higher band, being empty at this temperature, is the conduction band. These energy levels are separated by an energy or band gap (E_g), which normally is not occupied by electrons. As shown in Figure 1, at $T=0$ K, metallic conduction can occur in ceramic materials where there are partially filled valence bands and a corresponding overlap with the unoccupied conduction band states. When a small energy band gap is present with no overlap, semiconduction results. If a large energy gap exists the material is insulating, because the energy gap is too great to thermally promote electrons into the conduction band.

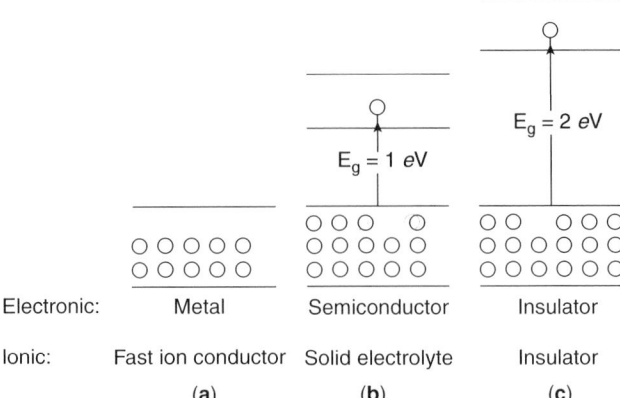

Figure 1. Schematic of band gap energy, E_g, for the three types of electronic and ionic conductors. For electronic conductors the comparison is made of the relative occupancy of valence and conduction bands. For ionic conductors, the bands correspond to the relative occupancy of ionic sublattices. For (**a**), $n = 10^{22}$ mL^{-1} for (**b**), $n = 10^{10}$ mL^{-1}; and for (**c**), $n = 1$ mL^{-1}.

ELECTRONIC CONDUCTING CERAMICS

Conducting Ceramics

Metals are not the only materials that exhibit high conductivity. Metallic conduction occurs also in transition-metal oxides. The high conductivity in these transition-metal oxides typically results from an overlap of unfilled d or f electron orbitals, forming energy bands with concentrations of quasifree electrons of the order of 10^{22}–10^{23}m^{-1}, equivalent to metallic conduction. The electronic conductivity in the perovskite and rutile oxide structures has been the most studied.

Very high conduction also occurs in many ceramic materials represented by the group formula MeC_x or MeN_x, where Me is a metal, C and N are carbon and nitrogen respectively, and x is the carbon or nitrogen/metal ratio. The high conductivity results from a redistribution of the orbitals and energy levels, resulting in band overlap at the Fermi energy level. Some of these materials, such as tantalum carbide and niobium carbide, can become superconducting near 10 K.

Semiconducting Ceramics

Defects can be created in many ceramic materials either by controlled doping, by heat-treated under conditions which create nonstoichiometric defects, or by uncontrolled impurity doping. Since mobility values for the defect transport mechanisms are often low and difficult to measure, the conductivity values that are reported for these materials are often at variance. This is in part due to the variable effects of impurities and thermal history which often overwhelm the expected dopant effects. Some impurity semiconductors are given in Table 2.

SPINELS FERRITES

The range of published resistivities for spinel and garnet ferrite materials is wide, from $\sim10^{-4}$ to $10^9\Omega$-cm at room

Table 2. Impurity Semiconductors

			n-Type		
Cds	$BaTiO_3$	Nb_2O_5	Fe_2O_3	WO_3	GeO_2
CdSe	$SrTiO_3$	Ta_2O_5	Tl_2O_3	TiO_2	MnO_2
ZnF_2	$PbCrO_4$	Fe_3O_4	In_2O_3	SnO_2	ZnO

			p-Type		
Se	CuI	Ag_2O	Hg_2O	NiO	PdO
Te	SnS	Cu_2O	MnO	FeO	CoO

			Amphoteric		
Si	Sn	PbSe	Ti_2S	Al_2O_3	Mn_3O_4
Ge	PbS	SiC	PbTe	Co_3O_4	UO_3

temperature. The low conductivity is typically associated with the simultaneous presence of Fe^{2+} and Fe^{3+} ions on equivalent lattice sites. In general, a condition for appreciable conductivity in the ferrite structure is the presence of ions having multiple valence states on like crystallographic sites.

There are three main commercial classes of ferrite spinels, namely nickel–zinc ferrite, $(NiZn)Fe_2O_4$; manganese–zinc ferrite, $(MnZn)Fe_2O_4$; and magnesium–manganese ferrites, $(MnMg)Fe_2O_4$. The electrical conduction is by small polaron mechanisms. Electrical resistivity primarily determines the utility of these materials in the high megahertz or microwave frequency ranges. Low frequency use requires a trade-off between high permeability and high resistivity. Nickel–zinc ferrites typically show an increase in permeability and a departure from stoichiometry in the iron-rich direction. Decreased resistivity results, therefore, when the formation of divalent iron becomes more probable, requiring close control of processing parameters is. Manganese–zinc and magnesium–manganese zinc ferrites are typically used in low-frequency devices such as pulse transformers and memory-core devices. For higher frequency use, the high DC resistivity needed for full magnetic penetration and low eddy current losses can be obtained with an iron deficient oxide powder.

SUPERCONDUCTIVITY

Superconductivity is partly typified by a perfect metallic conductor that has no resistance to current flow below a T_c. Besides the disappearance of electrical resistance, there is an expulsion of magnetic flux described as the Meissner effect.

The $YBa_2Cu_3O_{7-x}$ compound is currently the most intensively investigated high temperature oxide superconductor. It has an oxygen deficient, distorted orthorhombic, 1:1:3 (ABO_3) perovskite-type structure, with the *Pmmm* space group. The crystal structure shows the Y and Ba ions located in the center of the unit cell, Cu ions on the corners, and O ions on the edges. Based on neutron diffraction studies, the structure shows two important features for the Cu ions: (*1*) nonplanar CuO_2 planes extend in the crystallographic *ab* planes at $z = 0.36$ and -0.36; and (*2*) fencelike, square planar CuO_3 linear chains extending along the *b* axis at $z = 0$.

FERROELECTRICS

Barium titanate, $BaTiO_3$, has been a much studied ferroelectric, since it is the base material for large volume production of technologically important components such as disk and multilayer (MLC) capacitors; barrier layer (BL) and grain boundary barrier layer capacitors (GBBL); and, nonlinear PTCR. The ferroelectricity in $BaTiO_3$ and like (ABO_3) perovskite compounds arises from a dipolar shift in the B cations on cooling through the transition temperature. This results in a change in crystal symmetry from cubic to tetragonal and the appearance of spontaneous polarization charges, due to the development of a domain structure.

Lead-based materials such as PZT, PLZT, and PMN form a class of ceramics with important dielectric, relaxor, piezoelectric, or electrooptic properties, and are thus used for actuator and sensor devices. Common problems associated with their use are low dielectric breakdown, increased aging, and electrode injection, thereby decreasing the resistivity and degrading the dielectric properties.

VARISTORS

Varistors are primarily ZnO-based ceramic devices that exhibit high nonlinear current–voltage behavior, which is ideal for use in protecting electronic equipment against voltage surges. Varistors are processed so as to develop a microstructure consisting of conductive grains surrounded by a thin, resistive, Bi-rich second phase in the grain boundaries. At low voltages the grain boundary resistance is sufficiently high that little current leaks from the circuit. At the higher breakdown voltages a tunneling process occurs, which allows the overvoltage pulse to be rapidly gated away from the device circuitry. This is because the varistor resistance to ground decreases several orders of magnitude. Because the varistor action takes place across the ZnO grain boundaries, the tailoring of devices for specific breakdown voltages is done by fabricating the device with the appropriate number of grain boundaries in series between the electrodes.

PROCESSING EFFECTS

Electrical ceramic properties are ultimately structure dependent, especially in ceramic materials with variable valence cations. The processing of these materials therefore requires close control of composition and heat treatment conditions. Microstructure and grain size control are also crucial, since density changes and the presence of second phases can grossly affect properties.

Thermal history during processing can affect properties in important ways as well.

R. C. Buchanan, ed., *Ceramic Materials for Electronics*, 2nd ed., Marcel Dekker, Inc., New York, 1991.

Y. Chiang, D. Birnie, and W. D. Kingery, *Physical Ceramics*, John Wiley & Sons, Inc., New York, 1997.

D. C. Hill and H. L. Tuller, in R. C. Buchanan, ed., *Ceramic Materials for Electronics*, 2nd ed., Marcel Dekker, Inc., New York, 1991.

A. J. Moulson and J. M. Herbert, *Electroceramics*, Chapman and Hall, New York, 1990.

RELVA C. BUCHANAN
RODNEY D. ROSEMAN
University of Cincinnati

CERAMICS, MECHANICAL PROPERTIES

Structural ceramics are used in applications such as gas turbines, advanced heat engines, semiconductor processing equipment, armor, thermal barrier coatings, medical implants, as thin films for wear and electronic applications, heat exchangers, aerospace and weapons components, in high temperature solid oxide fuel cells, and as bearings components. Advantages of ceramics over metals include dimensional stability; low densities, which translate into weight savings and increased fuel efficiencies; high temperature capabilities; and corrosion resistance.

The overriding concern with regard to the mechanical performance of ceramics is their brittleness and hence sensitivity to flaws. There is usually little or no warning that failure is imminent because deformation strain prior to failure is usually <0.1%. As a result, a primary thrust of structural ceramics research has been the development of tougher and stronger ceramics. Ceramics are now routinely available that have toughness values of $7-10$ MPa·m$^{1/2}$ and strengths that exceed 1000 MPa (1.5×10^5 psi). These values compare to toughness values of $120-153$ MPa·m$^{1/2}$ and strengths of $1380-1790$ MPa for structural metals such as AF1410 high strength steel.

The mechanical properties of ceramics are sensitive to the starting materials, forming processes, heat treatment conditions, and surface preparation.

PROPERTIES AND BEHAVIOR

Elastic Behavior

Elastic deformation is defined as the reversible deformation that occurs when a load is applied. Most ceramics deform in a linear-elastic fashion; ie, the amount of reversible deformation is a linear function of the applied stress up to a certain stress level. If the applied stress is increased any further, the ceramic fractures catastrophi-

cally. This is in contrast to most metals, which initially respond elastically, then begin to deform plastic alloy. This plastic deformation allows stresses at stress concentrators to be dissipated rather than building to the point where bonds break irreversibly.

Strength

The elastic modulus describes how easily atoms in a solid can be moved together or apart for small deformations. The shape of the interatomic potential is such that beyond a certain spacing the atomic attraction is insufficient to hold the atoms together. If deformation of the material is continued, entire planes of atoms separate and the ceramic fractures.

The theoretical tensile strength of the material, σ_{theor}, has been approximated by

$$\sigma_{\text{theor}} = \left(\frac{E\gamma}{a}\right)^{1/2} \approx \frac{E}{10} \tag{1}$$

where $a =$ equilibrium spacing between planes of atoms, $\gamma =$ fracture surface energy, and $E =$ Young's modulus. If all bonds in a material were stressed equally up to the point of failure, the strength of a ceramic would be the theoretical strength. Large discrepancies between the theoretical and measured tensile strengths of ceramics result from the presence of imperfections. A typical ceramic such as alumina is as much as one hundred times weaker than the theoretical strength. These imperfections or flaws can raise the local stress to the point that bonds in the immediate vicinity of the flaw can fail a few at a time, as opposed to every atom in the plane failing simultaneously. The applied stress required to cause fracture, σ_{fracture}, can be written as

$$\sigma_{\text{fracture}} = \frac{K_C}{Y(c)^{1/2}} \tag{2}$$

The wide variety of flaw types and sizes in ceramics produces the large (typically $\pm 25\%$) variability in strength that has been one of the principal hurdles to the incorporation of ceramics in structural applications. This value compares unfavorably with the few percent for variability of the yield stress of a metal. The failure probability of a ceramic body at a given load depends on the probability of a flaw of a critical size being present in a location where it produces a stress concentration.

Many distribution functions can be applied to strength data of ceramics. The function that has been most widely applied is the Weibull function, which is based on the concept of failure at the weakest link in a body under simple tension. A normal distribution is inappropriate for ceramic strengths because extreme values of the flaw distribution, not the central tendency of the flaw distribution, determine the strength. One implication of Weibull statistics is that large bodies are weaker than small bodies because the number of flaws a body contains is proportional to its volume.

Ceramics are much stronger in compression than in tension and are frequently used in applications where they bear compressive loads. Under excessive compressive loads ceramics fail in a brittle manner just as they do in tension; however, the measured compressive stresses are typically eight times greater than the measured tensile stresses. In compression, the failure process can begin with microplastic deformation, not the growth of preexisting flaws. Measured compressive strengths often show an inverse relationship to the square root of the grain size.

Fracture Toughness

The fracture criterion is defined by a critical value of the crack tip stress intensity, known as the fracture toughness, K_C. Ceramics often fail in pure tension, designated the mode I stress intensity, and K_{IC} replaces K_C in equation 2. Thus $\sigma_{fracture}$, the applied tensile stress at which fracture occurs, is a function of the flaw size and K_{IC}. A crack propagates catastrophically when the stress intensity factor at the crack tip reaches the critical value required for bond breakage.

Ceramic toughening efforts have focused on the property of some ceramics to exhibit increased apparent fracture toughness as cracks grow. This increase is seen in terms of the far-field value of the stress intensity required to propagate the crack. An important consequence of this effect, which is commonly referred to as R- or T-curve behavior, is that the material has increased damage tolerance because there is a crack size regime in which the strength is independent of the crack size. The underlying basis of R-curve behavior is that the crack tip stress is redistributed, either to immediately adjoining material, as in the case of the process zone formed in transformation toughened materials, or to regions far removed from the crack tip, as in the case of fiber-reinforced ceramics. One reason for the interest in crack-bridging mechanisms is that these toughening mechanisms should operate over a broad range of temperatures. Characteristics of R-curve mechanisms are that they are activated only as the crack advances, the number of activated shielding elements increases as the crack grows, and the measured fracture toughness saturates when the generation rate of shielding elements equals the rate at which elements become inactive.

Transformation-toughened materials exhibit enhanced toughness because of a process zone at the crack tip that consumes energy that would otherwise be used in the creation of fracture surface. Various processes at the crack tip have been postulated to consume energy such as the phase transformation from tetragonal to monoclinic zirconia, microcracking, and deviation of the primary crack around the transformed particles. The zone can also be thought of as a region that partially shields the crack tip from the far-field stresses.

Significant toughening has been achieved by fabricating whisker- and fiber-reinforced ceramic composites, and metal–ceramic composites. Toughening primarily results from crack bridging and/or crack deflection, both of which reduce the crack-tip stress intensity.

Ferroelastic materials contain domains that can be switched by an applied stress in a manner analogous to magnetic domain switching in ferromagnetic materials. Many ceramics that exhibit ferroelasticity also exhibit ferromagnetism and/or ferroelectricity, such as $BaTiO_3$ and lead zirconate titanate (PZT). Toughening occurs because energy is absorbed in the switching of the domains.

Plasticity

Although even at elevated temperatures mechanical failure of ceramics is dominated by brittle fracture, plastic deformation mechanisms often precede brittle fracture. Plastic deformation is also important because of the role it plays in net shape forming operations such as extrusion, which require extremely high strain rates in a deformation regime known as superplasticity. In the lower range of elevated temperatures, plastic deformation of crystalline ceramics can occur by slip. In a slightly higher temperature range, plastic deformation can also occur by grain boundary sliding and softening of secondary phases such as glass.

Hardness

Hardness (H) is a measure of the resistance of a material to deformation, in particular the resistance to plastic deformation during surface penetration. Hardness is related to bond strength. Because covalent and multivalent ionic bonds are strong and highly directional in nature, slip is very difficult in these cases, and ceramics containing these bonds are generally the hardest materials. The ratio of the distance, a, between planes of atoms and the spacing, b, of the atoms in the plane also plays an important role in how easily planes of atoms slide past each other, and hence in the hardness.

Subcritical Crack Growth

At low and modest temperatures, under certain environmental conditions, the strength of many ceramics, especially glasses, can degrade with time under mechanical loading. This phenomenon, referred to as static fatigue or delayed failure, is the result of subcritical crack growth (SCG). A preexisting crack grows slowly at an applied stress intensity lower than that necessary; i.e., the critical value K_{IC}, to propagate a crack without environmental influences. SCG is pernicious because a flaw grows slowly at stresses far below the expected failure load until the flaw is large enough to satisfy the Griffith criterion, at which point failure occurs catastrophically. The mechanism of subcritical crack growth is the reaction of the corrosive medium with highly stressed bonds at the crack tip.

Impact and Erosion

Impact involves the rapid application of a load to a relatively small area. Much of the kinetic energy from the impacting object may be transformed into strain energy for crack propagation. If the impact is from a blunt indenter, a crack usually forms from excessive tensile stresses around the point of contact. If the impacting object still possesses a significant amount of kinetic energy, further damage may occur in the form of radial cracks and circumferential cracks. Sharp indenter impact can also produce cone cracks. If the impact load is relatively large and sustained, a macroscopic stress may be imposed on the target body, causing it to bend. This may lead to excessive tensile stresses on the opposite face of the body, crack initiation, and catastrophic failure. Failure can also occur if erosion reduces the cross section and load-bearing capacity of the component, causes a loss of dimensional tolerance, or causes the loss of a protective coating.

Tribological Behavior

Tribological performance of ceramic includes friction, adhesion, wear, and lubricated behavior of two solid materials in contact.

The coefficient of friction μ is the constant of proportionality between the normal force P between two materials in contact and the perpendicular force F required to move one of the materials relative to the other. Macroscopic friction occurs from the contact of asperities on opposing surfaces as they slide past each other. On the atomic level, friction occurs from the formation of bonds between adjacent atoms as they slide past one another.

Factors that affect μ include loading geometry, microstructure, crystal orientation, surface chemistry, environment, temperature, and the presence of lubricants.

Wear Ceramics generally exhibit excellent wear properties. Wear occurs by two mechanisms: adhesive wear and abrasive wear.

Generally, the harder the ceramic, the better its wear resistance; however, other properties such as fracture toughness may play a dominant role.

Thermal Stresses and Thermal Shock

Thermal stresses arise when a body is heated or cooled and constrained from expanding or contracting. Thermal stresses can lead to fracture and catastrophic failure when the magnitude of the thermal stress exceeds the strength of the ceramic. Factors that contribute to the generation of large thermal stresses and the failure of ceramics under these stresses are low thermal conductivities, which produce large temperature gradients, and the lack of a stress relief mechanism such as plastic deformation. Approaches used to minimize thermal stresses include matching the expansion of ceramics with the expansions of the materials to which they are joined, minimizing temperature gradients, minimizing cross-sectional thickness changes to ensure that the body heats or cools uniformly, keeping the body at its operating temperature, and heating and cooling slowly.

Cyclic Fatigue

Cyclic fatigue is the weakening and subsequent failure of a material during cyclic loading, often at stress levels significantly lower than those required to cause failure under static loading. Cyclic stresses can be produced by repeated heating and cooling, by vibrations, and in applications in which the component is repeatedly loaded and unloaded.

FRACTURE ANALYSIS

Fracture analysis, also known as fractography, plays an important role in understanding the relationships between the microstructure and mechanical properties, and the conditions that lead to failure. Systematic examination and interpretation of fracture markings and the crack path can often be used to reconstruct the sequence of events and stresses that led to failure. Fractography also plays an important role in the design and development of ceramic components, because it helps differentiate whether failure occurred because the material was weakened by the introduction of processing and handling flaws or because the applied stress exceeded the design stress.

R. W. Davidge, *Mechanical Behaviour of Ceramics*, Cambridge University Press, Oxford, U.K., 1986.

V. D. Fréchette, *Failure Analysis of Brittle Materials, Advances in Ceramics*, Vol. 28, The American Ceramic Society, Inc., Westerville, Ohio, 1990.

M. Srinivasan, in J. B. Wachtman, Jr., ed., *Treatise on Materials Science and Technology*, Vol. 29, Academic Press, Inc., New York, 1989, pp. 99–159.

J. Wolf, ed., *Aerospace Structural Metals Handbook*. Battelle Metals and Ceramics Information Center, Columbus, Ohio, 1988.

S. Jill Glass
Rajan Tandon
Sandia National Laboratories

CERAMICS, PROCESSING

In the manufacture of ceramic components, chemical composition and microstructure are specified to optimize the properties (eg, mechanical, electrical, and magnetic) of the finished product for a given application. Optimum properties are achieved by developing and refining processes to produce a target microstructure, and by controlling processes to minimize the concentration and scale of the defects in the finished part. The tolerance of the finished ceramic to defects determines the degree of control that must be exercised during processing. For example, advanced ceramics such as Si_3N_4 turbine blades require greater process control than traditional ceramic electrical porcelain insulators.

Ceramic component fabrication involves simultaneously optimizing multiple processes ranging from raw materials beneficiation to post-sinter machining

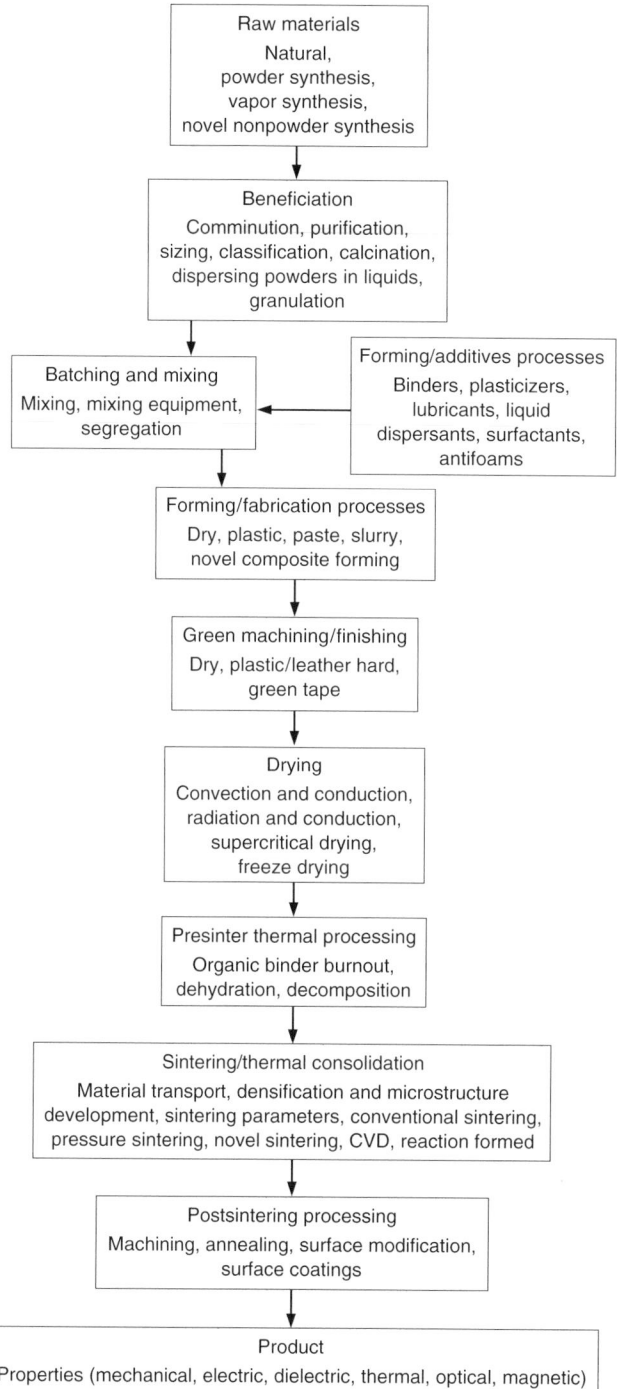

Figure 1. Flow diagram of the steps and processes involved in manufacturing a ceramic.

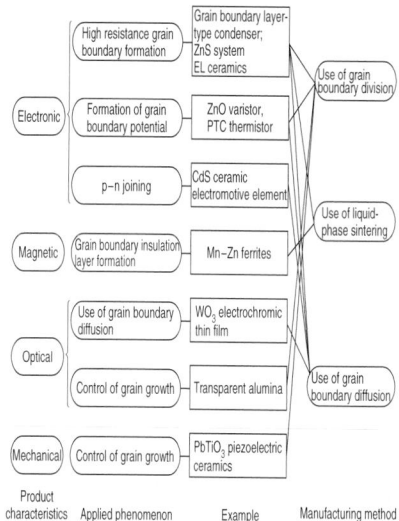

Figure 2. Process—microstructure—property relationships in advanced ceramics.

manufacturing processes and optimize properties, and to identify and correct process deficiencies when less than optimal properties are obtained. Examples of some process—microstructure—property relations in advanced ceramics are outlined in Figure 2.

Engineering Materials Handbook, Vol. 4, *Ceramics and Glasses*, ASM International, Materials Park Ohio, 1991.

N. Ichinose, *Introduction to Fine Ceramics, Applications in Engineering*, John Wiley & Sons, Inc., New York, 1987.

J. S. Reed, *Introduction to the Principles of Ceramic Processing*, John Wiley & Sons, Inc., New York, 1988.

D. W. Richerson, *Modern Ceramic Engineering*, 2nd ed., Marcel Dekker, Inc., New York, 1992.

KEVIN G. EWSUK
Sandia National Laboratories

CERIUM AND CERIUM COMPOUNDS

Cerium Ce, at no. 58, is the most abundant member of the series of elements known as lanthanides. Cerium ranks ca 25th in abundance in the earth's crust, and cerium, which occurs at 60 ppm crustal abundance, not lanthanum at 30 ppm, is the most abundant lanthanide.

RESOURCES

Whereas certain rocks of igneous origin formed by melting and recrystallization can include minerals enriched in the lanthanides, cerium is usually present as a trace element rather than as an essential component. Only a few minerals in which cerium is an essential structure-defining component occur in economically significant

(Fig. 1). To manufacture reliable ceramic components with reproducible properties, process—microstructure—property relationships must be understood and controlled during processing. These relationships can be determined by characterizing the ceramic raw materials, mixes, and the formed ceramic body intermittently during the various stages of processing and after final thermal consolidation. Established relationships between processing, microstructure, and properties can be applied to control deposits. Two minerals supply the world's cerium, bastnasite, $LnFCO_3$, and monazite, $(Ln,Th)PO_4$.

Bastnasite has been identified in various locations on several continents. The largest recognized deposit occurs mixed with monazite and iron ores in a complex mineralization at Baiyunebo in Inner Mongolia, China. The mineral is obtained as a by-product of the iron ore mining. The other commercially viable bastnasite source is the Mountain Pass, California, deposit where the average Ln oxide content of the ore is ca 9%. This U.S. deposit is the only resource in the world that is mined solely for its content of cerium and other lanthanides.

Several countries supply monazite concentrates for the world market. Extensive deposits along the coast of western Australia are worked for ilmenite and are the primary source of world monazite. Other regions of Australia, along with India and Brazil, also supply the mineral. Because monazite contains thorium, India and Brazil have embargoed its export for many years. In the United States, commerce in the mineral is regulated by the Nuclear Regulatory Commission.

World mine production of contained Ln oxide has been estimated at 119,000 t with ca 68% of this in China, mostly at Baiyunebo, ca 4.2% in the United States, and ca 22.6% in India.

PRODUCTION

The production of cerium derivatives begins with ore beneficiation and production of a mineral concentrate. Attack on that concentrate to create a suitable mixed lanthanide precursor for later separation processes follows. Then, depending on the relative market demand for different products, there is either direct production of a cerium-rich material or separation of the mixed lanthanide precursor into individual pure lanthanide compounds, including compounds of pure cerium, or both. The starting mineral determines how the suitable mixed lanthanide precursor is formed. In contrast, the separation technology, which involves liquid–liquid countercurrent extraction or solvent extraction (SX) for preparing the individual lathanides, is essentially independent of the starting mineral. Thus different feedstocks can ultimately be processed by the same separation routines and equipment.

CERIUM(IV) CHEMISTRY

The fluorite structure, which has a large crystal lattice energy, is adopted by CeO_2, preferentially stabilizing this oxide of the tetravalent cation rather than Ce_2O_3. Compounds of cerium(IV) other than the oxide, ceric fluoride, CeF_4, and related materials, although less stable, can be prepared.

The tetravalent ceric ion Ce^{4+}, is the only nontrivalent lanthanide ion, apart from Eu^{2+}, stable in aqueous solution. As a result of the higher cation charge and smaller ionic size, ceric salts are much more hydrolyzed in aqueous solution than those of the trivalent lanthanides. Ceric salt solutions are strongly acidic, basic salts tend to form readily, and there are no stable simple salts of weak acids.

The double salts, ceric ammonium nitrate, $(NH_4)_2[Ce(NO_3)_6]$, and ceric ammonium sulfate, $(NH_4)_2[Ce(SO_4)_3]$, are stable orange compounds prepared by dissolving freshly prepared hydrated oxide in excess of the appropriate acid and adding the correct amount of ammonium salt.

Cerium in the tetravalent state is a strong oxidizing agent and can be reduced by, eg, oxalic acid, halogen acids, ferrous salts, and hydrogen peroxide.

CERIUM(III) CHEMISTRY AND COMPOUNDS

Cerium is strongly electropositive, having a low ionization potential for the removal of the three most weakly bound electrons. The trivalent cerous ion Ce^{3+}, apart from its possible oxidation to Ce(IV), closely resembles, the other trivalent lanthanides in behavior.

Ce(III) forms a water-insoluble hydroxide, carbonate, oxalate, phosphate, and fluoride; sparingly soluble sulfate and acetate; and soluble nitrate and chloride (and bromide). In solution the salts are only slightly hydrolyzed. The carbonate is readily prepared and is a convenient precursor for the preparation of other derivatives. The sparingly soluble sulfate and acetate decrease in solubility with an increase in temperature. Calcination of most Ce(III) salts results in CeO_2.

Cerous salts in general are colorless because Ce^{3+} has no absorption bands in the visible. Trivalent cerium, however, is one of the few lanthanide ions in which parity-allowed transitions between $4f$ and $5d$ configurations can take place, and as a result Ce(III) compounds absorb in the ultraviolet region just outside the visible.

METAL

In bulk form cerium is a reactive metal. Pure metal is prepared by the calciothermic reduction of CeF_3.

$$2\,CeF_3 + 3\,Ca \longrightarrow 2\,Ce + 3\,CaF_2$$

On a fresh surface the metal has a steely lustre but rapidly tarnishes in air as a result of surface formation of oxide and carbonate species. For protection against oxidation the metal is usually stored in a light mineral oil. Cerium reacts steadily with water, readily dissolves in mineral acids, and is also attacked by alkali; it reacts with most nonmetals on heating.

Cerium metal has unique solid-state properties and is the only material known to have a solid–solid critical point. Three allotropes, α, β, γ, are stable at or close to ambient conditions and have complex structural interrelationships.

Mischmetal

Mischmetal contains, in metallic form, the mixed light lanthanides in the same or slightly modified ratio as occurs in the resource minerals. It is produced by the electrolysis of fused mixed lanthanide chloride prepared from

Table 1. Commercially Available Cerium-Containing Materials and Uses

Type	Material	Use	Cerium content
A	rare-earth chloride, mischmetal	FCC[a] catalysts iron metallurgy	principal component[b]
B	lanthanum concentrate, La–Ln chloride	FCC[a] catalysts	minor component[b]
C	cerium concentrate	glass polishing, glass decolorizing	dominant element[c]
D	oxide, nitrate, metal	autoemission catalysts, etc	> ca 90 wt%
E	oxide, salts	luminescence, catalysts	> ca 99 wt%

[a]FCC = fluid catalytic cracking.
[b]Of mixed-lanthanide composition.
[c]In oxide-type compound.

either bastnasite or monazite. Although the precise composition of the resulting metal depends on the composition of chloride used, the cerium content of most grades is always close to 50 wt%.

An alternative commercial form of a metallic mixed lanthanide-containing material is rare-earth silicide, produced in a submerged electric-arc furnace by the direct reduction of ore concentrate, bastnasite, iron ore, and quartz.

ECONOMIC ASPECTS

The worldwide production of lanthanides in 2002 was 85,500 t, measured as contained Ln oxide.

Forecast demand for rare earths is expected to grow at the rate of 4–9%/yr through 2007.

Cerium is used in several forms other than as the pure oxide. Only a small fraction of the 80,000 ton Ln total is produced as separated, relatively pure individual Ln derivatives, cerium included. The bulk of the material is consumed as concentrates, cerium included. The various cerium-containing derivatives available commercially are summarized in Table 1.

HEALTH AND SAFETY FACTORS

In general the lanthanides, including cerium, have a low toxicity rating (19), especially when they are present in material having low aqueous solubility.

USES

The technological applications of cerium may be categorized as occurring in metallurgy, glass and ceramics, catalysis, and chemicals, plus phosphors/luminescence.

J. B. Hedrick, "Rare Earths," *Mineral Commodity Summaries*, U.S. Geological Survey, Reston, Va., Jan. 2003.

G. W. Johnson and T. E. Sisneros, in G. McCarthy and co-eds., *The Rare Earths in Modern Science and Technology*, Vol. 3, Plenum Press, 1982, p. 525; K. Jyrkas and M. Leskela, *J. Less Common Metals* **126**, 291 (1986).

R. J. Lewis, Sr., *Sax's Dangerous Properties of Industrial Materials*, 10th ed., Vol. 2, John Wiley & Sons, Inc., New York, 2000.

W. H. Richardson, in K. B. Wiberg, ed., *Ceric Ion Oxidation of Organic Compounds*, Academic Press, New York, 1965.

BARRY T. KILBOURN
Molycorp Inc.

CESIUM AND CESIUM COMPOUNDS

Cesium, Cs, is a member of the Group 1 (IA) alkali metals. It resembles potassium and rubidium in the metallic state, and the chemistry of cesium is more like that of these two elements than like that of the lighter alkali metals.

Until the late 1970s, cesium continued to be little more than a research element. Much of its limited production was for research into thermionic power conversion, magnetohydrodynamics, and ion propulsion. Although the potential for these applications has not materialized, cesium chemical usage has increased significantly as catalysts in the chemical and petrochemical industries and in biotechnical engineering.

PHYSICAL PROPERTIES

Pure cesium is a silvery white, soft, ductile metal. Surface alteration by minute traces of oxygen results in the metal taking on a golden hue. Of the stable alkali metals, ie, excluding francium, cesium has the lowest boiling and melting points, the highest vapor pressure, the highest density, and the lowest ionization potential. These properties and the large radius of the monovalent cesium ion have important consequences directly related to applications.

CHEMICAL PROPERTIES

The ionization potential of the alkali metals decreases with increasing atomic number; consequently, cesium is generally far more reactive than the lower members of the alkali metal group. When cesium is exposed to air, the metal ignites spontaneously and burns vigorously, producing a reddish violet flame to form a mixture of cesium oxides. Similarly, cesium reacts vigorously with water to form cesium hydroxide, the strongest base known, as well as hydrogen; together with air and water a hydrogen explosion usually occurs as the burning cesium readily ignites the liberated hydrogen gas. Cesium, the most active of the alkali metals toward oxygen and the halogens, is the least reactive toward nitrogen, carbon, and hydrogen.

Cesium salts are, in general, chemically similar to other alkali metal salts. Cesium forms simple alkyl and aryl compounds that are similar to those of the other alkali metals. They are colorless, solid, amorphous, nonvolatile, and insoluble, except by decomposition, in most solvents except diethylzinc.

Cesium is the rarest of the naturally occurring alkali metals, ranking fortieth in elemental prevalence. Nevertheless, it is widely distributed in the earth's crust at very low concentrations. By far the most important commercial cesium source is pollucite, ideally $Cs_2O \cdot Al_2O_3 \cdot 4SiO_2$.

Economic concentrations of pollucite usually occur in highly zoned complex pegmatites, associated with lepidolite, petalite, and spodumene. The Bernic Lake orebody of Tantalum Mining Corp. of Canada Ltd. (Tanco) in southeastern Manitoba, Canada, is the world's largest cesium source, containing approximately two-thirds of the known ore. Other significant ore deposits are the Bikita pegmatite in Zimbabwe and in the Karibib desert of Namibia.

PROCESSING OF POLLUCITE

Pollucite preparation consists simply of mining the ore, crushing it to required size, followed in some instances by hand picking. No other concentration is required. Chemtall is the predominant processor worldwide, the only significant U.S. producer being Fluids (Woodland, Texas). Canada is the major source of cesium ores.

There are three basic methods of converting pollucite to cesium metal or compounds: direct reduction with metals; decomposition with bases; and acid digestion. In each case, grinding of the ore to 75 μm precedes conversion.

PRODUCTION OF CESIUM METAL

Cesium is produced by thermochemical methods, thermal decomposition, electrolytic reduction.

CESIUM ALLOYS

Eutectics melting at about -30, -47, and $-40°C$ are formed in the binary systems, cesium–sodium at about 9% sodium, cesium–potassium at ~25% potassium, and cesium–rubidium at about 14% rubidium. A ternary eutectic with a melting point of about $-72°C$ has the composition 73% cesium, 24% potassium, and 3% sodium. Cesium and lithium are essentially completely immiscible in all proportions.

Cesium does not alloy with or attack cobalt, iron, molybdenum, nickel, platinum, tantalum, or tungsten at temperatures up to 650°C.

CESIUM COMPOUNDS

Cesium compounds are manufactured and distributed by a comparatively large number of companies, considering the size of the total cesium market. Those companies that process pollucite produce their own range of products, some of which are then reprocessed and refined by other, smaller, specialty companies, many of which are located in the United States.

Compounds include carbonates cesium carbonate, Cs_2CO_3, mol wt 325.82; cesium hydrogen carbonate, $CsHCO_3$, mol wt 193.92; cesium chromate, Cs_2CrO_4; cesium halides cesium bromide, CsBr, mol wt 212.82; cesium chloride, CsCl, mol wt 168.36; cesium perchlorate, $CsClO_4$, mol wt 232.35; cesium fluoride, CsF, mol wt 151.90; cesium iodide, CsI, mol wt 259.81; cesium hydroxide, CsOH; cesium hydroxide monohydrate, $CsOH \cdot H_2O$, mol wt 167.93; cesium nitrate, $CsNO_3$, mol wt 194.91; cesium oxides cesium monoxide, Cs_2O, mol wt 281.81; the suboxides: cesium heptaoxide, CsO_7, tetracesium oxide, Cs_4O, heptacesium dioxide, Cs_7O_2, and tricesium oxide, Cs_3O; cesium peroxide, Cs_2O_2; and cesium superoxide, CsO_2, cesium permanganate, $CsMnO_4$, mol wt 251.84; cesium Sulfates cesium sulfate, Cs_2SO_4, mol wt 361.87; and cesium aluminum sulfate (cesium alum) $Cs_2SO_4 \cdot Al_2(SO_4)_3 \cdot 24H_2O$, mol wt 1136.39.

HANDLING, STORAGE, AND SHIPMENT

Because of the high reactivity of cesium metal, special precautions are required for its storage, transportation, and use. Small quantities are usually contained in evacuated glass ampuls, larger quantities in stainless steel containers that are themselves contained in an outer packing, ensuring that the metal is kept from moisture or air. Most cesium compounds are hygroscopic, especially the halides, and must therefore be stored dry. Other precautions that must be taken depend on the anion. Most products are sold in polyethylene bottles inside of clamping ring steel drums.

The toxicology, occupational health hazards, and transportation regulations of cesium compounds result from the anion rather than the cesium cation. Producers and distributors provide an MSDS as well as detailed shipping requirements for each product.

ECONOMIC ASPECTS

The cesium market is very small. Most current applications require relatively small quantities of cesium and hence annual world production of cesium and compounds is estimated to be on the order of 375 t of cesium chloride equivalents.

HEALTH AND SAFETY ASPECTS

The cesium ion is more toxic than the sodium ion but less toxic than the potassium, lithium, or rubidium ion. cesium has been studied as indicated medically in depressive disorders. Because of the small-scale production of cesium products, no significant environmental problems have been encountered.

USES

Cesium compounds are used in research and development, and commercially in electronic, photoelectric, and medical applications.

Electronic Applications

Electronic applications make up a significant sector of the cesium market. The main applications are in vacuum tubes, photoemissive devices, and scintillation counters.

Biotechnology and Medical Applications

Cesium chloride, and to a lesser extent the other halides, cesium trifluoracetate and cesium sulfate, are used in the purification of nucleic acids. ie, RNA and DNA, viruses, and other macromolecules. In medicine, cesium salts have been considered both as an antishock reagent following the administration of arsenical drugs, though a contraindication is the disturbance to heart rhythm, and for the treatment of epilepsy.

Chemical Applications

Cesium metal is used in carbon dioxide purification as an adsorbent of impurities; in ferrous and nonferrous metallurgy it can be used as a scavenger of gases and other impurities.

Catalyst doping is one of the principal commercial uses of cesium. The performance of many metal-ion catalysts can be enhanced by doping with cesium compounds. This is a result both of the low ionization potential of cesium and its ability to stabilize high oxidation states of transition-metal oxo anions.

A growing use of cesium compounds is in the field of organic synthesis, replacing sodium or potassium salts. Various cesium compounds are highly soluble in polar solvents. These compounds do not decompose as do many organic compounds, thus avoiding undesirable by-product formation. Additionally, the cesium component can be recovered and recycled.

Energy-Related Applications

Much research with regard to the use of cesium in energy related processes has resulted in little commercial application. The heightened awareness of the environmental degradation caused by fossil fuel power stations has resulted in increased research both into efficiency improvements for existing plants and into alternative power generation methods.

Cesium is ideally suited for use in magnetohydrodynamic (MHD) power generation. The metal can be used as the plasma seeding agent in closed-cycle MHD generators using high temperature nuclear reactors as the primary heat source.

One alternative method of generating electricity directly from a heat source is by the use of a cesium vapor thermionic convertor, which uses cesium to neutralize the space charge above a hot cathode that is emitting electrons toward a cooler anode.

Cesium may be useful in the fixation of radioactive waste in a cesium-based glass and in detoxification procedures for fugitive ^{137}Cs emissions, such as at Chernobyl, Ukraine. Methods for the removal of cesium from radioactive waste liquids have been patented.

CESIUM ISOTOPES

Naturally occurring cesium and cesium minerals consist of only the stable isotope ^{133}Cs. The radioactive cesium isotopes such as ^{137}Cs are generated in fuel rods in nuclear power plants.

M. Bick, *Cesium Chemicals from the World's Leading Producer*, Chemetall, Frankfurt, Germany.

F. M. Perel'man, *Cesium and Rubidium*, MacMillan, New York, 1953, pp. 5–11.

S. W. Pierce, in E. Bingham, B. Cohrssen, and C. H. Powell, eds., *Patty's Toxicology*, 5th ed., Vol. 3, John Wiley & Sons, Inc., New York, 2001, Chapt. 46.

R. G. Reese Jr., "Cesium", *Mineral Commodity Summaries*, U.S. Geological Survey, Jan. 2002.

WILLIAM FERGUSON
DENA GORRIE
Tanco Lac du Bonnet

CHELATING AGENTS

A chelating agent, or chelant, contains two or more electron donor atoms that can form coordinate bonds to a single metal atom. After the first such coordinate bond, each successive donor atom that binds creates a ring containing the metal atom. This cyclic structure is called a chelation complex or chelate, the name deriving from the Greek word *chela* for the great claw of the lobster.

Chelation is an equilibrium system involving the chelant, the metal, and the chelate. Equilibrium constants of chelation are usually orders of magnitude greater than are those involving the complexation of metal atoms by molecules having only one donor atom.

Chelating agents may be used to control metal ion concentrations. Chelation complexes usually have properties that are markedly different from both the free metal ion and the chelating agent. Consequently, chelating agents provide a means of manipulating metal ions through the reduction of undesirable effects by sequestration or through creating desirable effects such as in metal buffering, corrosion inhibition, solubilization, and cancer therapy.

Chelates and chelation reactions are abundant in nature, ranging from delicately balanced life processes depending on traces of metal ions to extremely stable metal chelates in crude petroleums. Examples of biochemical processes involving chelates include photosynthesis, oxygen transport by blood, certain enzyme reactions, ion transport through membranes, and muscle contraction. Technological applications include scale removal from steam boilers, water softening, ore leaching, textile processing, food preservation, treatment of lead poisoning, chemical analysis, tissue-specific medical procedures, and micronutrient fertilization of agricultural crops.

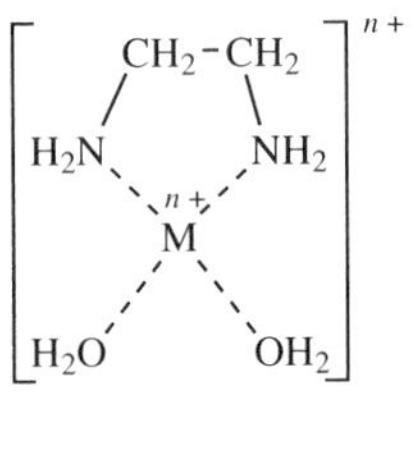

Figure 1. Types of chelates where (**1**) represents a tetracoordinate metal having the bidentate chelant ethylenediamine and monodentate water; (**2**), a hexacoordinate metal bound to two diethylenetriamines, tridentate chelants; (**3**), a hexacoordinate metal having triethylenetetramine, a tetradentate chelant, and monodentate water; and (**4**), a porphine chelate. The dashed lines indicate coordinate bonds.

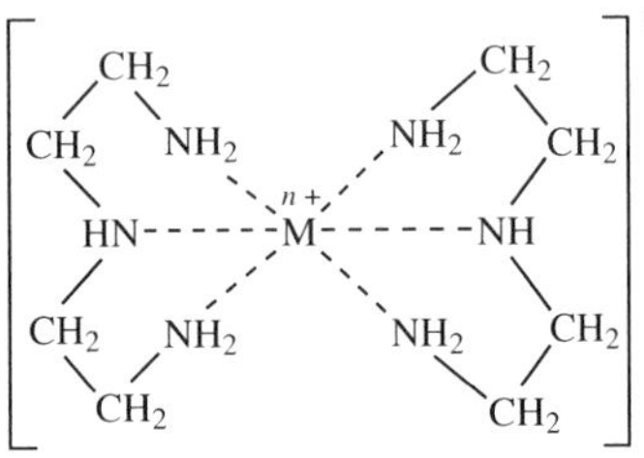

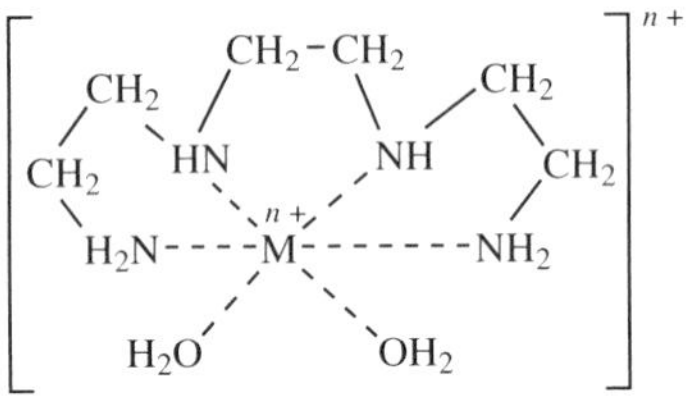

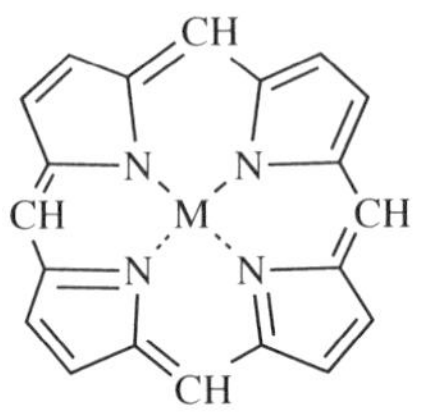

STRUCTURE AND TERMINOLOGY

The structural essentials of a chelate are coordinate bonds between a metal atom or a stable oxo cation, M, which serves as an electron acceptor, and two or more atoms in the molecule of the chelating agent, or ligand, L, which serve as the electron donors. A chelating agent may be bidentate, tridentate, tetradentate, and so on, according to whether it contains two, three, four, or more donor atoms capable of simultaneously complexing with the metal atom. Examples are shown in Figure 1.

COMPOUNDS HAVING CHELATING PROPERTIES

Compounds with chelating properties can be found in almost any class of structures containing two or more donor atoms spatially situated so that they can coordinate with the same metal atom. The chelate rings formed contain four or more members, but for the same donor atoms, the five- or six-membered chelate rings are usually the most stable and most useful.

Chelating agents may be either organic or inorganic compounds, but the number of inorganic agents is very small. The best known inorganic chelants are polyphosphates. The annual consumption of these compounds exceeds that of all the organic chelating agents combined. Although many hundreds of organic chelating agents are known, only a few members of a few classes of compounds find extensive industrial use. One important class of organic chelating agents is the group of phosphonic acids analogous to the amino- and hydroxycarboxylic acids. These phosphonate chelants possess many of the

complexing properties of the inorganic polyphosphates, particularly threshold-scale inhibition, effective at much less than stoichiometric ratios of chelant to metal ion, but unlike the polyphosphates, the phosphonates are stable in water at high temperature and pH.

NOMENCLATURE AND STRUCTURAL REPRESENTATION

Chelating Agents

Besides the conventional empirical and structural formulas, chelating compounds and chelates are often represented by type formulas; ie, formulas that show only generalized types of structural features.

For many macrocyclic ligands, simplified names are in common use. For example, crown ether nomenclature consists of four parts: (*1*) the number and type of fused rings on the polyether ring; (*2*) in square brackets the number of atoms in the polyether ring; (*3*) the word "crown"; and (*4*) the number of oxygen atoms in the macro ring. Ligand structures may be represented by any of the conventional means for depicting structure.

Chelates

Because of length and complexity, systematic names of chelates are little used except for special purposes, such as where unequivocal referencing is essential. Chelates are named in the literature in a variety of ways. The name of the ligand in a chelate is usually given a suffix -o or -ato if it is a negative group but remains unchanged if the ligand is electrically neutral. Prefixes indicate the number of bound ligand molecules. The central atom is given the name of the metal, or a derivative name having the suffix -ate; eg, cuprate and ferrate, if the complex is negatively charged. Oxidation states of the metals are indicated by roman numerals; e.g., iron(III), and ionic charges are shown as part of the name by Ewens-Bassett numbers; eg, (2+) or (1−). Chelates are often named merely as a complex; eg, cadmium complex with acetylacetone.

THE CHELATION REACTION

Chelate Formation Equilibria

In homogeneous solution the equilibrium constant for the formation of the chelate complex from the solvated metal ion and the ligand in its fully dissociated form is called the formation or stability constant. Whereas the ligand displaces solvent molecules coordinated to the metal, these solvent molecules do not generally enter into the equations. When more than one ligand molecule complexes with a metal atom, the reaction usually proceeds stepwise. For a metal having a coordination number of six and a bidentate chelating agent, the equations representing the equilibria are

$$M + L \rightleftharpoons ML \qquad K_1 = \frac{[ML]}{[M][L]} \qquad (1)$$

$$ML + L \rightleftharpoons ML_2 \qquad K_2 = \frac{[ML_2]}{[ML][L]} \qquad (2)$$

$$ML_2 + L \rightleftharpoons ML_3 \qquad K_3 = \frac{[ML_3]}{[ML_2][L]} \qquad (3)$$

$$overall: \quad M + 3L \rightleftharpoons ML_3 \qquad K = \frac{[ML_3]}{[M][L]^3} \qquad (4)$$

where the square brackets represent concentrations in units of molarity. The overall stability constant is the product of the step stability constants; ie, $K = K_1 K_2 K_3$, and is often designated by β. Experimentally determined equilibrium constants are usually calculated from concentrations rather than from the activities of the species involved.

Many experimental approaches have been applied to the determination of stability constants. Techniques include pH titrations, ion exchange, spectrophotometry, measurement of redox potentials, polarimetry, conductometric titrations, solubility determinations, and biological assay.

Displacement Equilibria

Species in solution are generally in formation—dissociation equilibrium, and displacement reactions of any given metal or ligand by another are possible. Thus,

$$ML + M' \rightleftharpoons M'L + M \qquad (5)$$

or

$$ML + L' \rightleftharpoons ML' + L \qquad (6)$$

If the stability constants for ML and M′L are K and K', respectively, then for the exchange shown in equation 5, the equilibrium constant is K_x.

$$K_x = \frac{[M][M'L]}{[ML][M']} = \frac{K'}{K} \qquad (7)$$

The extent of displacement depends on the relative stabilities of the complexes and the mass action effect of an excess of M′. For equivalent total amounts of M and M′, K_x must be on the order of 10^4 for 99% complete displacement to occur. Similar considerations apply for the displacement of L from ML by L′. The situation is quite analogous to the familiar competition of two bases for the hydrogen ion.

Metal exchange is the mechanism by which many foods, such as shortenings, shellfish, and dairy products, are stabilized against deleterious effects of trace metals by the addition of $Na_2CaEDTA$ ($\log K = 10.7$). Copper ($\log K = 18.8$) and iron ($\log K = 25.1$) displace calcium and become sequestered so that the remaining concentration of free iron or copper ions is too low for catalytic effects to occur at significant rates. Ligand exchange occurs when ascorbic acid bound to copper ($\log K = 1.57$) is displaced by EDTA, stabilizing this vitamin by disrupting an oxidation mechanism. Dyes and bleaches are similarly protected in the textile industry.

Rates of Reaction

The rates of formation and dissociation of displacement reactions are important in the practical applications of chelation. Complexation of many metal ions, particularly the divalent ones, is almost instantaneous, but reaction rates of many higher valence ions are slow enough to measure by ordinary kinetic techniques.

Factors Affecting Stability

Many factors influence the stability of chelates. Several of the stability factors common to all chelate systems are the size and number of rings, substituents on the rings, and the nature of the metal and donor atoms.

pH Effects

Being Lewis bases, the donor atoms of chelating agents react with Lewis acids such as metal and hydrogen ions. In the pH range of aqueous solutions, most of the well-known chelating agents exist as an equilibrium mixture of both protonated and unprotonated forms. Metal ions compete with hydrogen ions for the available donor atoms, and therefore simultaneous equilibria exist that are treated mathematically by the simultaneous equations for the formation constants of the chelates and the acid dissociation constants of the chelating agents.

Titration Behavior

Protonated chelating agents exhibit titration behavior typical of their respective acidic groups; eg, carboxyl phenolic hydroxyl, ammonium, or sulfhydryl moieties, if they are titrated with bases where the cations have a very weak or no tendency to form chelates. In the presence of a metal ion that coordinates with the donor atom of one of these acidic groups, hydrogen is displaced by the metal, and the acid strength of the group appears to be enhanced. The hydrogen ion concentration is then higher than in the absence of the metal. Strongly chelated metal ions can increase the acidity of an acidic group by several orders of magnitude.

Titration of the hydrogen ion liberated from a strong chelating agent is used to determine the concentration of metal ions in solution. The strength of chelation can also be determined from these data.

Metal Buffering

The equation for the formation constant of the reaction

$$M^{n+} + L^{m-} \rightleftharpoons^{K} ML^{n-m} \qquad (8)$$

can be rearranged to give

$$\frac{1}{[M^{n+}]} = \frac{K[L^{m-}]}{[ML^{n-m}]} \qquad (9)$$

from which, on taking logarithms and defining $-\log[M^{n+}]$,

$$pM = \log \frac{[L^{m-}]}{[ML^{n-m}]} + \log K \qquad (10)$$

The concentration of the metal ion can be controlled by adjusting the ratio of the concentrations of free ligand and metal chelate. If both species are present in appreciable amounts, moderate changes in either concentration have little effect on the ratio. The concentration of the metal ion can thus be buffered in a manner analogous to the buffering of pH by the presence of a weak acid and its anion

$$\mathrm{pH} = \log \frac{[A^-]}{[\mathrm{HA}]} + pK_a \qquad (11)$$

In the equation for pM, $\log K$ appears instead of pK because K is a formation constant, the reciprocal of the chelate dissociation constant, which is analogous to the acid dissociation constant K_a.

By choice of chelating agents, and thus $\log K$, pM can be regulated over a wide range. Two or more metals may be selectively buffered at different concentrations by a single chelating agent having different stability constants for the metals. Selective buffering of one metal to a low concentration in the presence of other metals is termed "masking". It is the ability to maintain a nearly constant concentration of metal ions at almost any level of concentration that is the basis of many of the commercial uses of chelating agents. The buffering capacity may be used to supply metal ions at a definite concentration as in electroplating and in nutrient media, or to remove or sequester metal ions in cleaning baths where the fresh stock entering the bath continually introduces additional amounts of metals.

Solubilization

The solubility product of a slightly soluble salt determines the concentration of metal ion that can be present in solution with the anion of that salt. For the salt MX the solubility product is

$$K_{sp} = [\mathrm{M}^{n+}][\mathrm{X}^{n-}] \qquad (12)$$

from which is obtained

$$\mathrm{pM} = \log[\mathrm{X}^{n-}] - \log K_{sp} \qquad (13)$$

The presence of a sufficiently strong chelating agent; i.e., one where K in equation 10 is large, keeps the concentration of free metal ion suppressed so that pM is larger than the saturation pM given by the solubility product relation (eq. 13) and no solid phase of MX can form even in the presence of relatively high anion concentrations. The metal is thus sequestered with respect to precipitation by the anion, such as in the prevention of the formation of insoluble soaps in hard water.

Deposits of an insoluble salt can be dissolved as a salt of the metal chelate.

$$\mathrm{MX\,(s)} + \mathrm{L}^{m-} \xrightleftharpoons{K_s} [\mathrm{ML}^{n-m}] + \mathrm{X}^{n-} \qquad (14)$$

In the presence of the chelating agent and the insoluble salt, MX, pM of the solution is subject to both the metal buffering and the solubility equilibria.

Electrochemical Potentials

The oxidation potential of a solution containing a metal in two of its valence states, M^{x+} and M^{x+n}, is given by

$$E = E^0 - \frac{RT}{nF} \ln \frac{[\mathrm{M}^{x+n}]}{[\mathrm{M}^{x+}]} \qquad (15)$$

In the presence of a chelating agent, the concentrations of the two forms of the metal are buffered according to the simultaneous equations

$$[\mathrm{M}^{x+n}] = \frac{[\mathrm{M}_{\mathrm{ox}}\mathrm{L}]}{K_{\mathrm{ox}}[\mathrm{L}]} \quad \mathrm{and} \quad [\mathrm{M}^{x+}] = \frac{[\mathrm{M}_{\mathrm{red}}\mathrm{L}]}{K_{\mathrm{red}}[\mathrm{L}]} \qquad (16)$$

where $\mathrm{M}_{\mathrm{ox}}\mathrm{L}$ and $\mathrm{M}_{\mathrm{red}}\mathrm{L}$ are the chelates of the oxidized and reduced forms of the ions and K_{ox} and K_{red} are the respective formation constants. Substituting these values in the potential equation gives

$$E = E^0 + \frac{RT}{nF} \ln \frac{K_{\mathrm{ox}}}{K_{\mathrm{red}}} - \frac{RT}{nF} \ln \frac{[\mathrm{M}_{\mathrm{ox}}\mathrm{L}]}{[\mathrm{M}_{\mathrm{red}}\mathrm{L}]} \qquad (17)$$

The first two terms of the right-hand side of the equation are sometimes combined and expressed as $E^{0\prime}$, which is called the standard oxidation potential for the chelate system.

ECONOMIC ASPECTS

Production and price estimates for the principal industrial chelating agents are given in Table 1.

ENVIRONMENTAL, HEALTH, AND SAFETY FACTORS

The primary industrial chelating agents are essentially environmentally benign and nontoxic under the conditions incidental to normal handling and use. With these, eye irritation is mainly a function of the acidity or alkalinity of the form of the product and its solubility. However, use of nitrilotrioracetic acid (NTA) is regulated in some jurisdictions. In medical uses the effects of the chelating agents on metal ion balances in the body tissues must be accommodated. Some of the commercial compounds used in smaller amounts as chelants, such as oxalic acid, are toxic, however. The hazards of using any chelant should be determined prior to use.

Solutions of iron chelates can be used to remove hydrogen sulfide and oxides of sulfur and nitrogen in industrial gas scrubbing processes before flue gases are released to the atmosphere.

There is increased interest in the development of chelating agents that have improved biodegradable characteristics.

APPLICATIONS

Concentration Control

Sequestration, solubilization, and buffering depend on the concentration control feature of chelation. Traces of metal

Table 1. United States Production of Industrial Chelating Agents

Agent	Production	Price	Producers[a]
EDTA	132[b]		Akzo, BASF, Dow
DTPA	46[b]		Akzo, BASF, Dow
HEDTA	13[b]		Akzo, BASF, Dow
NTA	49[b]		Akzo, BASF, Dow, Sol
EDTA(Na)$_4$ solution (38–39%)		0.39–0.42[c]	Akzo, BASF, Dow
DTPA(Na)$_5$ solution (40%)		0.50–0.51[c]	Akzo, BASF, Dow
HEDTA(Na)$_3$ solution (41%)		0.58[c]	Akzo, BASF, Dow
NTA(Na)$_3$ solution (40%)		0.31[c]	Akzo, BASF, Dow, Sol
gluconates	30[b]		Glu, PMP
gluconic acid solution (50%)		0.50[c]	Glu, PMP
glucoheptonates	21[b]		Cal, En
sodium glucoheptonate solution (50%)		0.20–0.21[c]	Cal, En
organophosphonates	55[b]		Cal, Rh, Sol
ATMP(H)$_6$ solution (50%)		0.80–0.88[c]	Cal, Rh
HEDP(H)$_4$ solution (60%)		1.21–1.23[c]	Cal, Rh
STPP	200[d]	0.46[e]	As, Pr, Rh
citric acid	460[f]	0.75[g]	C, ADM, TL

[a]AKZO = Akzo Nobel Chemicals, Inc.; BASF = BASF Corporation; DOW = The Dow Chemical Co.; Sol = Solutia Inc.; Glu = Glucona America, Inc.; PMP = PMP Fermentation Products, Inc.; Cal = Callaway Chemical Co.; En = EnCee Chemicals, Inc.; Rh = Rhodia, Inc.; As = Astaris, LLC.; Pr = Prayon Inc.; C = Cargill; ADM = Archer Daniels Midland Co.; TL = Tate & Lyle.
[b]Millions of pounds, 100% dry basis; estimated for 1998.
[c]Dollars per pound; estimated for 1990.
[d]Thousands of short tons; estimated for 2000.
[e]Dollars per pound; estimated for 2000.
[f]Millions of pounds, 100% dry basis; estimated for 1998.
[g]Dollars per pound; estimated for 1998.

ions are almost universally present in liquid systems, often arising from the materials of the handling equipment if not introduced by the process materials. Despite very low concentrations, some trace metals produce undesirable effects such as coloration or instability.

Special Properties of Chelates

Some of the principal applications of the preparative feature of chelation depend on the solubility properties, color, or catalytic effects of the chelates. Selection of a chelant to form a chelate that is soluble in the medium enables the solubilization of mineral deposits, pipe and boiler scale, films on surfaces, constituents of ores, and similar insoluble materials. Chelates having suitable solubilities can be designed to concentrate a metal into a particular phase by extraction from water into an organic solvent, by binding to a liquid but water-insoluble chelant, by precipitation of the chelate as a solid phase, or by ion exchange onto an insoluble, solid, chelating resin. Color and color fastness in some dyeing processes depend on the properties of chelates. Catalytic effects may result from a chelate, or chelation of the reactant may itself be part of the mechanism of catalysis by a metal. In biological systems the properties of some enzymes and vitamins (qv) involve chelation, and the activities of chlorophyll and hemoglobin are associated with their chelate structures.

In photography, specially designed chelates suppress or release metal ions to start or stop reactions at appropriate stages in processing sequences, sensitize or desensitize substances to radiation, function in optical and multiplex recording systems, and replace the less environmentally suitable ferricyanide for photographic bleaching. Chelates have been used in the preparation of superconducting compounds and as cross-linking agents in fracturing fluids and plugging gels for subterranean formations.

Medical Uses

A significant usage of chelation is in the reduction of metal ion concentrations to such a level that the properties may be considered to be negligible, as in the treatment of lead poisoning. However, the nuclear properties of metals may retain their full effect under these conditions, eg, in nuclear magnetic resonance or radiation imaging and in localizing radioactivity.

A. Catsch, A. E. Harmuth-Hoene, and D. P. Mellor, "The Chelation of Heavy Metals," in *International Encyclopedia of Pharmacology and Therapeutics*, Section 70, Pergamon Press, Oxford, UK, 1979.

S. Chaberek and A. E. Martell, *Organic Sequestering Agents*, John Wiley & Sons, Inc., New York, 1959, pp. 126–130.

Chemical Economics Handbook Marketing Research Reports, SRI International, Menlo Park, Calif.

G. Wilkinson, R. D. Gillard, and J. A. McCleverty, *Comprehensive Coordination Chemistry, The Synthesis, Reactions, Properties & Applications of Coordination Compounds*, Vols. 1–7, Pergamon Press, Oxford, New York, Beijing, Frankfurt, São Paulo, Sydney, Tokyo, Toronto, 1987.

WILLIAM L. HOWARD
Consultant
DAVID WILSON
The Dow Chemical Company

CHEMICAL METHODS IN ARCHAEOLOGY

Archaeology is the study of past human life, culture, and activities, which is shown by material evidence in the form of surviving artifacts, biological and organic remains, and a variety of other evidence recovered by archaeological excavation. Archaeology, tells about the past by studying materials that entered into the life of common people. Chemistry has developed methods to date archaeological material and has also allowed us to infer such historical items as trade routes, by studying ancient artifacts to shed light on the technology used to make them. The diet and customs of ancient peoples have also been discovered by applying chemical methods. Chemistry intervenes in the understanding of the mechanisms that cause archaeological material to degrade in order to set up procedures aimed at stabilizing decay and preventing further deterioration. It also tries to find the best way to restore ancient artifacts.

MATERIALS STUDY

When artifacts are investigated, some of the most important aims concern the study of the technology used to produce them, to reconstruct their distribution from the production areas, and to understand the use to which they were put in the past. Long-term storage often tends to obscure chemical information that contribute to understanding these goals.

Instrumental Methods for Chemical Analysis

Instrumental chemical methods used to study archaeological materials include elemental analytical techniques and molecular analytical techniques.

The composition of a great variety of solid archaeological materials, including stones, metals, glasses, ceramics, bones, paintings and other materials, is daily obtained in most archaeometry laboratories by making use of energy dispersive X-ray fluorescence (EDXRF) or wavelength dispersive XRF (WDXRF) equipment. Synchrotron radiation XRF (SRXRF) has been recently proposed as a technique that significantly improves the performance of the standard XRF. Particle-induced X-ray emission (PIXE) also has been widely used to study archaeological materials, by using multiproject facilities.

Trace element analyses contribute to chemically fingerprinting archaeological material. Optical emission spectroscopy (OES) has been used for analysis of elements present at concentration up to $\sim 0.001\%$ since the 1930s. It was largely replaced by atomic absorption spectroscopy (AAS) during the 1970s and by the inductively coupled plasma (ICP) techniques during the 1980s. Instrumental neutron activation analysis (INAA) has been the technique of choice in provenance investigation for a long time, largely due to its high sensitivity to many trace elements along multiple dimensions of element concentration. It also ensures good precision and accuracy of data compared to other techniques. However, ICP–mass spectrometry (ICP–MS), less costly equipment, today rivals INAA in provenance investigations.

Laser-induced breakdown spectroscopy (LIBS) has been proposed recently as a new microdestructive method for major, minor, and trace element analysis in the study of archaeological material.

Isotope analysis plays an important role in the study of archaeological material. The quantitative determination of the relative amount of isotopes of interest allows us to

hypothesize on the provenance of metal objects and also to support paleodietary research. Thermal ionization mass spectrometry (tims) has been used largely for the analysis of heavy metal isotopes for provenance purposes.

Molecular analytical techniques are used in archaeological science with a variety of purposes, but gas chromatography mass spectrometry (gc–ms) has become the workhorse of organic analysis in archaeology, thanks to its ability to separate and analyze mixtures of thermally stable volatile compounds or compounds that can be volatilized by the application of heat. Liquid chromatography mass spectrometry (lc–ms), more often in the form of high performance liquid chromatography mass spectrometry (hplc–ms), is used when thermally unstable or not volatile organic mixtures are going to be analyzed. Tandem mass spectrometry (ms^n) has also been shown to contribute to the study of organic remains.

Both Raman spectroscopy and Fourier transform infrared spectroscopy (ft–ir) have been used increasingly in archaeology to study a wide range of inorganic and organic archaeological materials.

Inorganic Archaeological Materials

Inorganic materials better survive the degradation processes that increase with time and thus have more easily been subjected to such investigations. The most studied such materials are stone, ceramics, glass, and metals.

Organic and Biomolecular Study of Archaeological Materials

This approach consists in identifying molecular markers capable of identifying unknown organic samples on the basis of their presence in contemporary natural substances. Lipids, in particular, have been shown to be particularly important as biomolecular markers.

Archaeological Lipids. The use of modern chromatographic techniques coupled with mass spectrometric analyzers has contributed to studies of artifact use patterns and food consumption through the identification of lipid residues. Lipids are extracted from the powdered original matrices by using organic solvents. They are properly derivatized and then analyzed by gc or gc/ms techniques or by gc–c–irms for isotope ratio studies. Evidence, for instance, of cholesterol, another lipid that persists in long-buried bones of humans and animals, can be used as a source of paelodietary information.

Proteins. Only under unusual burial environments have proteins survived microbial degradation, and in hard tissues such as tooth, bone, and shell they are prevalently protected. It should be expected that it will become possible to obtain protein residues from ancient ceramics that may have been in contact with protein-rich foodstuff for prolonged periods of time. However, protein extraction from mineral and ceramic surfaces is difficult, because most of the proposed methods disrupt the macromolecular structure of the protein residue.

Other Organic Residues. ft–ir in particular has provided chemical evidence of ancient food and beverages.

Ancient DNA. DNA entered the archaeological record from the second half of the 1980s. Before then it was not imagined that long-term preservation of DNA was possible. DNA is a record of ancestry. For this reason, ancient DNA can be used to determine kin relationship within a group of specimens. Moreover, ancient DNA can express some of the biological characteristics of an archaeological specimen. Biological sex or genetic diseases can be inferred by studying archaeological DNA. Studies carried out on DNA sequences older than 1 million years ago (antediluvian DNA) have concluded that such ancient sequences cannot be reproduced or derive from contaminations. A variety of studies on DNA sequences dated up to 100,000 years ago from extinct animals have revealed the phylogenetic relationships of extinct animals.

Amber Provenance. Understanding the provenance of amber made it possible to establish the earliest known trade routes that involved its transportation from northern to southern Europe ~5000 BC. In the 1960s, IR spectroscopy contributed greatly to this discovery by providing evidence of differences in composition between Baltic amber and Sicilian amber.

cp–mas–nmr has also been shown to be able to characterize both modern and fossil amber on a worldwide basis by distinguishing them in both their botanic as well as geographical differences.

Amino Acid Racemization Dating. Amino acids are the "building blocks" of proteins and up to several hundred of such building blocks can be contained in a protein. All of the amino acids that occur in proteins, except glycine, have at least an asymmetric carbon atom and thus can occur in two optical isomers called D and L. In life, the amino acids making up the proteins of higher eukaryotes consist solely of the L form. Metabolically active tissues contain specialized enzymes known as racemases that maintain a disequilibrium in our cells of only the L isomers. After death, the enzymes in living organisms cease their activity and amino acids undergo racemization, thus interconverting L isomers into D isomers at a time dependent rate.

DEGRADATION OF ARCHAEOLOGICAL MATERIALS

Degradation processes affect different materials to a different extent. Stone survives almost unaltered, while materials such as metal, glass, and certain organic material such as amber undergo some degradation but often survive in a recognizable form. Biological materials such as skin and hair survive only under exceptional condition. Biological hard tissues such as bone, tooth, and shell undergo complex degradation processes.

The overall degradation processes that act on organic remains after death are studied by *taphonomy*, from the Greek word ταφος (*taphos*, burial). Taphonomy studies all the natural and anthropogenic processes that affect the organism in its transferral from the living word (biosphere) to the sedimentary record (lithosphere). *Biostratinomy*, its first stage, includes all the interactions involved in the transferral of the living organism from the living world to the inorganic world, including burial. *Diagenesis* includes all the transformation occurring after burial. Diagenetic studies involve postdepositional changes that affect the structure of metal, glass or ceramic during burial. Great progress has been made in understanding taphonomic processes affecting bone, an important component of the archaeological record.

E. Ciliberto and G. Spoto, eds., *Modern Analytical Methods in Art and Archaeology*, John Wiley & Sons, Inc., New York, 2000.

Z. Goffer, *Archaeological Chemistry: A Sourcebook on Applications of Chemistry to Archaeology*, John Wiley & Sons, Inc., New York, 1980.

J. Henderson, *The Science and Archaeology of Materials*, Routledge, London, 2000.

J. B. Lambert, *Traces of the Past: Unraveling the Secrets of Archaeology Through Chemistry*, Addison-Wesley, Reading, Massachusetts, 1997.

GIUSEPPE SPOTO
Università di Catania

CHEMICAL PRODUCT DESIGN

Chemical product design is the process of choosing what product to make. It precedes chemical process design, which deals with how to make the product chosen. In the past, most of those involved in the chemical industry emphasized process design, because this enterprise was focused on perhaps 50 commodity products for which price was the key and efficient production the route to low prices. Now the goals of chemical product design have become much broader, to include also several thousand high value added chemicals, many of them pharmaceuticals.

What follows is a four-step template by which to decide which chemical products to make. The four steps are as follows: 1.) *Needs*: Decide what need the product will fill, 2.) *Ideas*: Generate chemical ideas to satisfy this need, 3.) *Selection*: Efficiently select the best ideas, and 4.) *Manufacture*: Design a process for making the product.

The first three steps of this template are unique to product design; the fourth step includes the more familiar process design.

NEEDS

Before considering how to define product needs, we can broadly identify two driving forces to product design: the pull of the market, typical for new consumer products, and technology push, an advance in technology, a new invention looking for an application.

The two driving forces of market pull and technology push result in rather different statements of product needs. In the case of a market pull product, we wish to define exactly what the market opportunity is. Product design then identifies an appropriate technology to exploit this opportunity. In the case of a technology push product, it is the other way around: The needs stage involves specifying the new technology and identifying areas of superiority to existing technology. The product design then consists of identifying markets where this new technology can be advantageously exploited.

Needs Identification

In defining product needs, we must make sure that the needs we define reflect the requirements of those who will ultimately use the product and not simply our own prejudices. The consensus among marketing organizations is that the best way to get this sort of information is by face-to-face interviews. One group of customers merits particular attention, the "lead users", who will benefit most by its improvement. In the case of a market pull product, the lead users are those who very much depend on existing and competing products. In the case of a technology push product, the lead users are those most responsible for the technological advance that has stimulated the product development. In addition to individual interviews, test panels and focus groups are sometimes used to help identify needs, particularly for consumer products.

Interpretation of Needs

The needs recorded from the information gathering will be a hodgepodge of conflicting and incomplete statements, of varying relevance and practicality. In editing them down it will be useful to rank them, for example, as essential, desirable, and useful. The essential needs are those without which the product cannot succeed. We will aim to achieve as many desirable needs as possible, particularly if a competitor's products fail to do so. We are unlikely to design explicitly for useful needs, although we will keep in mind that it will be a bonus if our product can fulfill these too.

The way in which needs are grouped and organized will depend on the product being considered. It will usually be an easy task if we aim to improve an existing product. The more innovative the proposal product, the harder it is to satisfactorily define the needs.

Quantification of Needs

The aim now is to convert the qualitative list of needs into specifications, including as much quantitative and chemical detail as possible. Doing so involves three steps: 1.) Write complete chemical reactions for any chemistry involved, 2.) Make mass and energy balances important to the product's use, 3.) Estimate the rates of any important changes that occur during the product's use.

Having produced a set of ideal product specifications in as much detail as possible, we should examine these carefully. If the only way of meeting customer requirements is to break a law of thermodynamics, we should stop product development now.

At the end of the Needs stage of product design, we have produced a ranked list of what our product needs to achieve and put this into quantitative and scientific terms as far as possible. We have also made a check that our aims are not wholly unreasonable and ideally we have a benchmark by which to judge the success of our product development. Up to this point, we have consciously avoided trying to think of solutions for our product needs. If we do already have an idea of the product's nature, we should try to keep it out of our considerations until the end of the Needs stage. Only now that we have well-established criteria for the success of any product should we begin to develop ideas for the product itself.

IDEAS

In principle, we need only one good product idea, the one that we will manufacture. In practice, product development requires up to 100 ideas in order to find one truly worth pursuing.

Idea Generation

To get our 100 or so product ideas, we will depend on people more than publications. The most important people are those on the team responsible for developing the specific product. In addition, we should pay special attention to customers who already are using existing, similar products. Some of these customers, the "lead users". Other human sources—consultants, private inventors, and the like—are often less useful. Patents and trade information from competitors are often more useful than archival literature. Other methods for ideas use forms of chemical synthesis.

Idea Sorting

We now have our 100 or so ideas, of widely varying quality. We must somehow sort through these ideas to locate quickly the best five or so for further developments.

First, without quantification, we should try to sort the ideas on completely qualitative grounds, reducing the number to perhaps twenty. To do so, we just make a list of all the ideas. We can then easily remove redundancy. Often this redundancy will occur because some ideas are simply special cases of others.

In addition, we want to drop ideas that are obvious folly. In doing so, we should be cautious, because some silly ideas may contain dreams. Sometimes, we can benefit from keeping a separate list of these flawed dreams. Normally, the efforts to remove redundancy and folly will still leave us up to 70 ideas.

To reduce the number of ideas further, we should try to organize them into categories, in a type of outline. How this should be done depends on the ideas generated. Once this outline is made, it may expose gaps. More often, we large groups of ideas will emerge as inconsistent with our organisation's objectives or strengths. These groups of ideas can be dropped. These last steps commonly cut the number of ideas to meet our target of twenty. However, many of our best ideas will often cluster under a single heading. Because we do not want to overspecialize too soon, we should choose at least one idea beyond this cluster for the next stage of product development.

Idea Screening

We must now reduce our 20 surviving ideas down to a still smaller number, normally five or fewer. We will still not have the resources to make more detailed calculations for the 20 survivors, so we need approximate but quantitative tools that let us continue the screening, but on a still more rational basis.

One commonly effective method for this screening is to choose five or fewer key attributes shared by most of the surviving ideas. Once these key attributes are chosen, we need to assign weighting factors to each. On the basis of these attributes and their weighting factors, we now score all of our ideas relative to a convenient benchmark. We find it convenient to assign the benchmark scores of 5, then choose scores from each new product between 1 (poor) and 10 (excellent). We then calculate an average weighted score for each product. The potential products with the highest scores are those to choose for further development.

SELECTION

With a handful of good ideas remaining, we next choose the best one to take forward for product development. This decision involves considerably more effort than we put into cutting down the number of ideas. As far as possible, we must now quantify how each idea will measure up to the criteria we set for a successful product at the Needs stage. This quantification will involve making estimates based on chemistry and engineering, and perhaps doing some simple experiments. At the same time, we still wish to develop our product as quickly as possible and do not want to put resources into exploring products we will end up rejecting. The key to success in the Selection stage is to make reliable choices with minimal effort.

Assessment

The first step in selecting between the remaining ideas is to estimate how each will perform relative to our earlier needs stage criteria. In order to do this, we must gather more information about each idea. This process will involve firming up exactly how each idea will work; we may need to do some simple experiments to achieve this, we will certainly need to explore the literature some more. As we generate more detailed information on each idea, the idea itself will change and become clearer. Thus there is an iteration between the Ideas and Selection stages—as we explore an idea in more detail, the idea evolves and new ideas may emerge.

Comparison

Having made an assessment of each idea against each criterion, we must now make an overall comparison among the ideas. In some cases, particularly where the identified needs are primarily technical, this will be easy once good estimates of performance are available. In other cases, subjective judgements will be necessary.

Getting Close to a Decision

At this point we should have a good indication of which idea looks the most promising in terms of fulfilling the needs defined in the first stage of product development. However, before proceeding, we need to pause to consider two important factors, intellectual property and risk.

The important point is to make sure that the ownership of the intellectual property is clear before large resources are invested. Often the profitability of a product (notably a pharmaceutical) depends on the exclusive license granted by patent protection. In such cases we must ensure there is at least a good prospect of obtaining such protection before proceeding. In all cases, we must at least ensure that our activities will not be restricted by any intellectual property held by others.

In assessing how well our product ideas measure up to the criteria set by the defined needs, we have largely ignored the issue of risk. However, the ideas we are choosing among may range from minor developments to an existing product to risky and untested new technology. We need to factor this into our decision. Risk may take three forms: the product may not work; the product may take a long time to develop; and external problems may occur because of local politics, fashion or a changing economic situation. The first of these, product function, should be unlikely; by this stage we should have eliminated product ideas that are likely to fail. The second, development time, is largely a technical issue that can often be translated into a financial risk—the longer a development programme is likely to be, the greater the uncertainty in cost and the larger the return must be. The third, external problems, is the hardest to estimate. In thinking about risk, one needs to consider both the probability of an event and the seriousness of its consequences.

Having identified the risks involved in our favored product, we have three possible responses. We might decide an idea, while attractive, is too risky to merit time and effort. Alternatively, we may decide simply to accept the risks and to proceed. This minimizes the time to product launch and is often the appropriate strategy if financial risks can be offset against the advantage of getting to market earlier. Proceeding in spite of risk is particularly apt where we hold a patent. Third, doing a little more research before committing resources to product manufacture will often result in a better product, but also in a longer (and more expensive) development period.

PRODUCT MANUFACTURE

By this stage, we have decided which chemical product we want to manufacture. We have identified a customer and that customer's product needs. We have generated ideas to fill that need and have selected the best idea. We are ready to decide how we will make the product.

Preparation

The very wide range of chemical products possible means that we will need to consider an equally wide range of manufacturing methods. To provide some initial guidance for manufacture, think of four types of products. The first type is devices, especially for medical applications, as the artificial kidney and the osmotic pump for drug delivery.

Another is commodity chemicals, made in amounts $>10^7$ kg/year^{-1}, which normally have molecular weights <100. As a result, their manufacture uses gaseous reagents, supported catalysts, and purification via distillation. This manufacture always has product cost as its primary focus.

The two other types of chemical products are much harder to describe. The more obvious type, 10,000 high value added chemicals, are exemplified by drugs. The last type of chemical product, which is the most heterogeneous, often has a useful microstructure. This large group of products includes most "specialty chemicals." Manufacturing ranges widely, and product properties are often a function of product history.

At the end of the Selection stage, we had sufficient information to convince ourselves that we had selected the best idea and that it had a high probability of success. Inevitably, there will have been gaps in our knowledge of the details of exactly how the product would work. In order to make the product, we not only need to be confident that the product can be made to work, but we must also establish in detail exactly how it is going to fulfil our aims. Before going further, we must fill in this missing information. This missing information may be obtained by literature searches, by consulting experts and by conducting experiments. It may be tedious, time consuming and expensive. That is why we put it off until after we had decided to proceed with product development. We do not wish to put this level of effort into more than one product idea or into an idea we later abandon if we can possibly avoid it.

Final Product Specification

Before designing the processing route, we need to produce a final specification for the product we have decided to make. We should by now have obtained all the information required to do this. For a typical chemical product, we need to define the physical structure, the chemical composition, the chemical reactions that occur during the product's operation and the thermodynamics of the product (including the microstructure). Which of these dominates depends on the type of product being made.

The key to a product is often how it responds to a change, which may involve dissolution (or precipitation) in a solvent, response to a temperature or other physical change, or a chemical reaction (pH change being the most common initiator). It is important to specify these chemical changes carefully as well as the nature of the product itself.

The Manufacturing Process

Finally, we must specify the process by which we will achieve the specifications we have just set. This is where chemical product design and traditional process design finally merge. The details of the process design will of course vary as widely as the nature of the products we have been discussing. For chemical products, function is key. For commodities, price is key. Because chemical products are defined by function, the design procedure

must start earlier: We cannot decide how to make something until we have decided what to make. A chemist or engineer involved in chemical product design must expect to participate in the identification of market needs, the generation of possible solutions and the selection of a product, in addition to the manufacturing decisions, which have been the traditional role of the engineer. This holistic approach to design is in sharp contrast to traditional process design, in which a specification is dictated to the chemist or engineer who then optimizes a process. Product design involves participation at a much earlier stage of product development, usually in a multidisciplinary team.

M. F. Ashby and K. Johnson, *Materials and Design. The Art and Science of Material Selection in Product Design*, Butterworth-Heinemann, Oxford, 2002.

J. M. Douglas, *Conceptual Design of Chemical Processes*, McGraw-Hill, New York, 1988.

S. A. Gregory, ed., *The Design Method*, Butterworths, London, 1966.

G. D. Moggridge
University of Cambridge
E. L. Cussler
University of Minnesota

CHEMICAL VAPOR DEPOSITION

Vapor deposition is a process that transfers gaseous molecules into solid-state materials (Fig. 1). From this process thin films can be grown on substrates and fine powders can be formed. There are two major types of vapor depositions, physical vapor deposition (PVD) and chemical vapor deposition (CVD). The difference is that in the PVD process a source is vaporized to deposit the film, usually the same as the source material. This process involves phase change but does not typically involve a chemical reaction. Major PVD techniques are vacuum evaporation, sputtering, ion plating, and molecular beam epitaxy (MBE). In the CVD process a molecular source is vaporized into a flow reactor. The precursor is activated by a form of energy and decomposed through steps of chemical reactions to deposit thin films. Using CVD technology it is

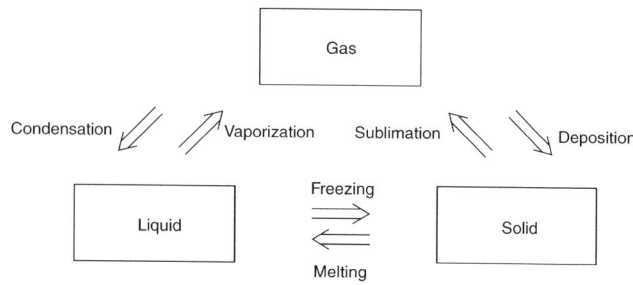

Figure 1. Definition of deposition.

possible to grow thin films of metals, semiconductors, and insulators to produce semiconductor devices and protective coatings on precision machine parts such as diamond films. Recently, CVD has been used to grow nanotubes, nanorods, and nanowires.

PRECURSORS AND DELIVERY SYSTEM

Traditionally, volatility has been the most important factor for selecting a CVD precursor. However, suitable simple molecules with high vapor pressure are relatively rare. Thus, only limited types of materials can be deposited by CVD.

Many new precursors have recently been synthesized. The volatility, the result of basic molecular structure and bonding, is a property difficult to predict and adjust. Other characteristics in choosing a suitable CVD precursor are its stability and reactivity, usually defined by the bond strength. An ideal precursor needs to be stable in storage and in delivery into the reactor. Also, it needs to be reactive enough to deposit thin films at low temperatures so that more substrates can be chosen. A compromise between stability and reactivity is frequently needed. Safety is another important issue. Many CVD precursors are toxic, poisonous, or flammable in the air. Thus, great caution should be paid in handling CVD precursors.

CVD REACTORS

CVD reactors are constructed from temperature and chemical resistant materials such as quartz and stainless steel. A source of energy is provided and controlled to supply thermal, plasma, or photo energy to the reactor.

Several types of reactors are the horizontal hot-wall reactor, the simplest type, so called because the furnace outside heats the tube wall, the vertical cold-wall reactor, so called because the reactor wall has a temperature lower than the central heating stage, the parallel plate plasma reactor, that has a radio frequency (rf) power supply, and the inductive tube remote plasma reactor, in which the plasma is generated by an induction coil outside the reactor.

Several types of reactors are the horizontal hot-wall reactor, the simplest type, so called because the furnace outside heats the tube wall, the vertical cold-wall reactor, so called because the reactor wall has a temperature lower than the central heating stage, the parallel plate plasma reactor that has a radio frequency (rf) power supply, and the inductive tube remote plasma reactor, in which the plasma is generated by an induction coil outside the reactor, away from the discharge zone.

SUBSTRATES

One advantage of CVD is that thin films can grow on complicated shapes, due to the fact that if the molecular diffusion rate is faster than the kinetic decomposition rate the precursor molecules can diffuse from the flow stream to the surface of the substrates uniformly.

FUNDAMENTAL PRINCIPLES

Thermodynamics

It is possible to establish theoretical feasibility for a CVD reaction by calculating the Gibbs energy of reaction and the equilibrium constant. In general, an ideal gas mixture in contact with pure condensed phases is assumed.

Kinetics

Growth of the thin film in a CVD process, after the precursor molecules are delivered into the reactor, is the result of several basic steps: 1.) diffusion of the precursor to the surface; 2.) gas-phase reactions of the precursor; 3.) adsorption–desorption of the precursor at the surface; 4.) reaction of the adsorbed precursor on the surface; 5.) incorporation of surface atoms into the bulk via surface diffusion; and 6.) desorption–adsorption and diffusion of the surface reaction by-products.

In a CVD system, the overall reaction rate is influenced by the rates of the precursor's feed, diffusion, are type of surface as well as gas-phase reactions.

USES

Just like PVD, CVD is an important and versatile technique to grow high quality, thin, solid films for many applications. In general, PVD is the preferred method of deposition in many industrial settings, because CVD is a more complicated process than PVD. However, the capability of conformal deposition of thin films on complex structures by CVD is its most important advantage. Thus, the CVD and PVD methods complement each other.

CVD is a bottom-up method of fabrication. It converts small molecules, or building blocks, into extended structures such as thin films. Recently, it has been found that in addition, many nano-sized materials in other shapes, such as particles, tubes, wires, and rods, can be grown effectively by CVD. Thus, CVD is expected to play an important role in the development of nanotechnology.

M. L. Hitchman and K. F. Jensen, eds., *Chemical Vapor Deposition: Principles and Applications*, Academic Press, London, 1993.

J. E. Mahan, *Physical Vapor Deposition of Thin Films*, Wiley-Interscience, New York, 2000.

C. E. Morosanu, *Thin Films by Chemical Vapour Deposition*, Elsevier, Amsterdam, The Netherlands, 1990.

H. O. Pierson, *Handbook of Chemical Vapor Deposition: Principles, Technology and Applications*, 2nd ed., Noyes Publications, Norwich, New York, 1999.

HSIN-TIEN CHIU
National Chiao Tung University

CHEMICAL WARFARE

Weapons of mass destruction have received increased attention recently. Terrorist groups, rogue states, and foreign governments possessing these weapons have raised major concerns to stability in different regions across the globe and about homeland security in the United States. Weapons of mass destruction are defined as nuclear, chemical, or biological weapons. While categorized together due to their potential destructive power, the nature of the threat of each of these types of weapons is very different, and the defensive measures taken in response to each vary greatly. This article focuses on the details of chemical warfare, primarily with U.S. capabilities. Due to the fact that the first step of defending against any threat is understanding the threat, a discussion of chemical agents follows. From there discussion turns to classes of chemicals that, among other uses, can be utilized to thwart attempts to detect the presence of chemical agents. Rapid and accurate determination of the presence of exceedingly small amounts of a chemical substance on a battlefield, an environment that is far from pristine, is difficult. However, certain chemicals can be used to further complicate detection by orders of magnitude. A short discussion describing contamination avoidance, protection, and decontamination materials is also included.

Chemicals used in war fit into five categories: toxic agents, riot-control agents, flame agents, incendiaries, and smokes and obscurants. Toxic chemical agents, used to achieve military objectives by producing casualties and discouraging enemy troops from approaching or remaining in certain areas of the battlefield (terrain denial), may be incapacitating or lethal. Army Field Manual 8-285 defines a chemical agent as "a chemical substance... intended for use in military operations to kill, seriously injure, or incapacitate humans (or animals) through its toxicological effects." Army and Department of Defense doctrine specifically excludes from this definition riot-control agents, which are considered to be legitimate law-enforcement tools. Also excluded are chemical herbicides and smoke and flame materials, which are considered legitimate military assets. Riot control agents are nonlethal tear agents most effective against unprotected personnel. Flame and incendiary agents can be used to harass and inflict casualties, and to destroy structures and material. Smokes and obscurants are employed for screening, signaling, and target marking.

Chemical warfare, a term used since 1917, is of vital interest not only to the world powers but also to many developing countries. The relative simplicity of production and ease with which existing pesticide or other chemical plants can be converted to a weapons facility make chemical weapons a continuing threat. Chemical weapons are not, as of this writing, banned. Whether employed or not, these materials exist as a potential weapon in any conflict. Toxic chemical agents may exist as chemical substances in a gaseous, liquid, or solid state intended to produce casualty effects ranging from harassment through degrees of incapacitation to death. A few such agents are true gases, but most are solids or liquids that are converted in use into a gaseous state or disseminated as aerosols. For contamination of terrain the agent can also be disseminated in bulk form, with or without additives to modify its physical properties.

LETHAL AGENTS AND INCAPACITANTS

Research on chemical agents after World War I led to the elimination of all but a handful of chemicals as being of practical battlefield significance. At the time of World War II, the only chemicals considered to be of practical significance included the mustard gases and phosgene.

Discovery of nerve agents in Germany led to the availability of a class of compounds at least one order of magnitude more lethal than previously known, where death might occur in a matter of minutes instead of hours.

Mustard and Related Vesicants

Mustard, bis(2-chloroethyl) sulfide (Chemical Agent Symbol HD), $Cl(CH_2)_2S(CH_2)_2Cl$, is a colorless, oily liquid when pure. Most samples have a characteristic garliclike odor. It is a potent alkylating agent that causes severe vesication of epidermal surface. It is favored for its ability to incapacitate opponents, lower their mobility, restrict the use of terrain, and reduce their ability to flight. At high-dose levels it exerts systemic cytotoxic effects, in particular involving the rapidly proliferative cells of the hematological system and intestinal mucosa. It is also genotoxic, mutagenic, and a dose-related carcinogen. Sulfur mustard is an oily liquid that becomes aerosolized when dispersed by spraying or by explosive blast from a shell or bomb. In temperate climates, it vaporizes slowly, posing particular risk in prolonged, closed-space, or below-grade exposures. At higher temperatures, vaporization increases markedly and contributes to more severe clinical effects. Because of its low volatility, mustard is persistent.

The procedure by which mustard is manufactured can be modified to yield either a mixture of mustard and Q (HQ) or a mixture of mustard and T (HT). These mixtures have several advantages over mustard alone, unless the agent is used only for vapor effects. HQ and HT are more toxic, more vesicant, more persistent, and have lower melting points than mustard alone.

Properties

The physical properties of the mustards are summarized in Table 1.

Table 1. Properties of Mustard Gases

Property	HD	Q	T	HN1	HN2	HN3
mol wt	159.08	219.08	263.25	170.08	156.07	204.54
bp, °C[a]	$80^{0.67}$		$120^{0.003}$	$85^{1.3}$	$87^{2.4}$	$144^{2.0}$
mp, °C	14.5	56	10	−34	−60	−4
density at 25°C, g/mL	1.2682		1.24	1.086	1.118	1.2347
volatility at 25°C, mg/m³	925	0.4	2.8	2s.29	3.581	0.120

[a]Pressure in kPa at which boiling point was determined is given as a superscript. To convert kPa to mm Hg, multiply by 7.5.

Physiological Effects

The sulfur and nitrogen mustards act first as cell irritants and finally as a cell poison on all tissue surfaces contacted. The first symptoms usually appear in 4–6 h. The higher the concentration, the shorter the interval of time between the exposure to the agent and the first symptoms. Local action of the mustards results in conjunctivitis (inflammation of the eyes); erythema (redness of the skin), which may be followed by blistering or ulceration; and an inflammatory reaction of the nose, throat, trachea, bronchi, and lung tissue. Injuries produced by mustard heal much more slowly and are much more liable to infection than burns of similar intensity produced by physical means or by other chemicals.

The rate of detoxification is slow, and the effects of even very small repeated exposures are cumulative or more than cumulative, owing to sensitization.

Uses

The nitrogen mustards are used clinically in the treatment of certain neoplasms. They have been used in treatment of Hodgkin's disease, lymphosarcoma, and leukemia.

Nerve Agents

Nerve agent refers to two groups of highly toxic chemical compounds that generally are organic esters of substituted phosphoric acid. The nerve agents inhibit cholinesterase enzymes and thus come within the category of anticholinesterase agents. The three most active G-agents are tabun, ethyl phosphorodimethylamidocyanidate (Chemical Agent Symbol GA), $(CH_3)_2NPOCNOC_2H_5$; sarin, isopropyl methylphosphonofluoridate (GB), CH_3PO-$FOCH(CH_3)_2$; and soman, pinacolyl methyl-phosphonofluoridate (GD), $CH_3POFOCHOCH_3C(CH_3)_3$.

The G-agent liquids under ordinary atmospheric conditions have sufficiently high volatility to permit dissemination in vapor form. They are generally colorless, odorless or nearly so, and are readily absorbable through not only the lungs and eyes but also the skin and intestinal tract without producing any irritation or other sensation on the part of the exposed individual. These agents are sufficiently potent so that even a brief exposure may be fatal. Death may occur in 1–10 min or be delayed for 1–2 h, depending on the concentration of the agent.

Another class of nerve agents, discovered after World War II, is the V-agents. These materials are generally colorless and odorless liquids that do not evaporate rapidly at normal temperatures. In liquid or aerosol form, V-agents affect the body in a manner similar to the G-agents.

Properties

Some physical properties of nerve agents are given in Table 2.

Physiological Effects

Inhalation of G-agent vapor at realizable field concentrations is immediately incapacitating. The symptoms in normal order of appearance are runny nose; tightness of chest; dimming of vision and pinpointing of the eye pupils (miosis); drooling and excessive sweating; nausea and vomiting; cramps and involuntary defecation or urination; twitching, jerking, and staggering; and headache, confusion, drowsiness, coma, and convulsion. These symptoms are followed by cessation of breathing and death. A person wearing skin covering and a gas mask is reasonably well protected.

Table 2. Properties of Nerve Agents

Property	GA	GB	GD	VX[a]
formula wt	162.13	140.10	182.18	267.38
bp, °C	246	147	167	298
mp, °C	−50	−56	unknown	below −51
density at 25°C, g/mL	1.073	1.0887	1.0222	1.0083
volatility at 25°C, mg/m³	610	21,900	3,060	10.5

[a] VX, $C_{11}H_{26}NO_2PS$, is a phosphonothioic acid ester.

Binary Munitions

Binary munitions contain two nonlethal components that are mixed during flight to form a nerve agent. Each component is manufactured separately, remaining in its own container until the munition components are assembled just prior to use. Mixing and subsequent agent formation occurs after firing or launch of the munition. In addition to greatly reduced storage and handling hazards, the binary components can be manufactured in ordinary chemical facilities, which need not be equipped with the stringent safety and environmental controls required for the older nerve agent munitions. The binary technology also overcomes the long-term storage problems associated with the unitary nerve agents.

Other Lethal Agents

A number of substances, many found in nature, known to be more toxic than nerve agents, from sources but none has been weaponized: shellfish poison, puffer fish poison, curare; "heart poisons" of the digitalis type, the sea cucumber, snake venom, and the protein ricin.

Incapacitants

An agent rendering an individual incapable of job performance may be classified as an incapacitating agent. Incapacitants are most suitable in limited warfare situations, e.g., when enemy troops are intermingled with a friendly population or in a city when the purpose is to capture the enemy without killing civilians. Incapacitating agents should produce no permanent aftereffects and allow for complete recovery. Agent BZ, 3-quinuclidinyl benzilate, $C_{21}H_{23}NO_3$, is a typical incapacitant. It disturbs the higher integrative functions of memory, problem solving, attention, and comprehension and has a gradual return to normalcy.

Use

Chemical agents can be adapted to a variety of munitions, including grenades, mines, artillery shells, bombs, bomblets, spray tanks, rockets, and missiles. Tactically, chemical agents have defensive and offensive capabilities in limited or general wars. Toxic chemical agents may be used alone or in conjunction with other types of weapons. Chemical weapons do not destroy matériel but allow physical preservation of industrial complexes and other facilities. Incapacitating agents may be used to preserve life and avoid permanent injury.

The use of chemical agents in battle imposes a significant burden on troops because of the cumbersome nature of the protective clothing and the attendant heat load in hot climate situations.

IRRITANTS

Riot control agent CS, a modern irritant compound, causes physiological effects that include extreme burning of the eyes, accompanied by a copious flow of tears; coughing; difficulty in breathing; chest tightness; involuntary closing of the eyes; stinging sensation of moist skin; runny nose; and dizziness or "swimming" of the head. Heavy concentrations also cause nausea and vomiting.

The effects of agent CS are immediate, even in extremely low concentrations.

A water-soluble white crystalline solid, CS is disseminated as a spray, as a cloud of dust or powder, or as an aerosol generated thermally from pyrotechnic compositions. The formulation designated CS1 is CS mixed with an antiagglomerant; when dusted on the ground, it may remain active for as long as five days. CS2 formulated from CS1 and a silicone water repellent may persist for as long as 45 days.

The principal uses of CS are in riot control and training; it has limited tactical use in defensive military modes.

FLAME

In the modern weapons arsenal flame agents are defined as various hydrocarbons, blends of hydrocarbons, and other readily flammable liquids, usually thickened with additives, that are easily ignited and can be projected to military targets. Although flame agents may be employed against buildings and other flammable targets, their primary role is against personnel in hardened structures or emplacements. In the United States, the principal application of flame agents is now in flame throwers and flame projectors, including flame rockets.

INCENDIARIES

Incendiary agents are designed for use in the planned destruction of buildings, property, and matériel by fire. Incendiaries burn with an intense, localized heat. They are very difficult to extinguish and are capable of setting fire to materials that normally do not ignite and burn readily. Although there are tactical applications for incendiary agents and munitions, they have played primarily a strategic role in modern warfare.

Incendiary Requirements

The mechanics of starting fires using incendiary agents involve a source of heat to act as a match to initiate combustion in a larger mass; combustible material to serve as kindling; and fuel. The match and the kindling are provided by the incendiary munition; the target is the fuel. All incendiary munitions, except for those containing materials that are spontaneously combustible, must have some sort of initiator such as a fuse or an ignition cup. The second element of the incendiary munition, the kindling, is the important factor, and both the amount and the nature of the combustible material in the munition have been the subject of much research and development.

Metal Incendiaries

Metal incendiaries include those of magnesium in various forms, and powdered or granular aluminum mixed with powdered iron(III) oxide. Magnesium is a soft metal that, when raised to its ignition temperature, burns vigorously in air. It is used in either solid or powdered form as an incendiary filling and in alloyed form as the casing for small incendiary bombs.

Oil and Metal Incendiary Mixtures

PT1 is a complex mixture composed of magnesium dust, magnesium oxide, and carbon, along with an adequate amount of petroleum and asphalt to form the paste.

SMOKES

Military smokes are aerosols of gaseous, liquid, or particulate matter that are tactically employed to defeat enemy surveillance, target acquisition, and weapons guidance devices.

Screening Smokes

Military smoke screens are produced by dispersing either finely divided solids or minute liquid droplets in air. To be useful, a smoke screen must be sufficiently opaque to provide the desired screening power and long-lasting enough to achieve effective military results.

Types of Screening Smokes

The generation of oil smoke is based on the production of minute oil droplets by purely physical means. The most desirable droplet size is 0.5–1.0 μm. The tiny droplets of oil scatter light rays and produce a smoke that appears to be white, and any individual droplet would be transparent under magnification. These droplets are produced as the vaporized oil passes through the nozzle of a generator and is subsequently cooled by the surrounding air.

Another type of smoke mixture, a volatile hygroscopic chloride for thermal generation, has the U.S. Army designation HC, type C. It is composed of ~6.7 wt% grained aluminum, 46.7 wt% zinc oxide ZnO, and 46.7 wt% hexachloroethane, C_2Cl_6.

A third screening smoke-type is white phosphorus (WP), -P_4 which reacts spontaneously with air and water vapor to produce a dense cloud of phosphorus pentoxide.

Signaling Smokes

Screening smokes can also be adapted for signaling purposes. However, a good signaling smoke must be clearly distinguished from the smoke incident to battle.

Volatilizing and condensing a mixture containing an organic dye produce colored signaling smokes. Of the dyes tested by the U.S. Army, the most satisfactory ones are the azo dyes, the azine dyes, the diphenylmethane dyes, and those of anthraquinone.

DEFENSE AGAINST TOXIC AGENTS

Defensive measures against toxic agents may be divided into four categories: agent detection and identification, individual and collective protection, decontamination, and medical defense. To these may be added a high degree of training in defensive measures and discipline in using them.

Detection

Two major classifications of detection exist. If the environment being probed for the existence of chemical agent concentrations is in the immediate vicinity, it is known as point detection. If the search for chemical agents is at some distance from the detector, it is referred to as remote or stand-off detection.

The United States has deployed a man-portable vapor alarm designed for advanced capability against standard blister and nerve agents. Another is a hand-held, real-time detector of nerve and mustard agents in vapor phase on personnel or equipment. And the Joint Chemical Agent Detector (JCAD) is a pocket-sized detector for use especially on aircraft and ships.

In the area of stand-off chemical detection, the Joint Service Lightweight Nuclear, Biological, Chemical Reconnaissance System is a vehicle-mounted system designed to sample, detect, and identify threats within a unit's area of responsibility.

Individual chemical sensors are much more effective if integrated into a network.

Individual and Collective Protection

The primary item of individual protection is the protective or gas mask.

The mask alone, however, does not provide protection from substances such as nerve and blister agents that penetrate through the skin. Thus, protection for the entire body should be provided. Airtight, impermeable clothing is available for personnel who must enter heavily contaminated areas. Such clothing is cumbersome and enervating because it retards release of body heat and moisture, and personnel efficiency is drastically lowered when it is worn for a long time.

Collective protection enclosures are required for groups of personnel. Such enclosures must be airtight to prevent inward seepage of contamination. They can be independent units or can be formed by adequately treating the interior walls of structures, tents, airplanes, or vehicles. A supply of uncontaminated air, provided by passing ambient air through high efficiency aerosol and carbon filters, must be provided.

Simplified collection protection equipment (SCPE) (U.S. designation M20) provides such protection using lightweight elements consisting of an inflatable enclosure, a hermetically sealed filter canister, motor blower, protective entrance, and a support kit.

Decontamination

If contaminated equipment or material does not have to be used immediately, natural aeration is an effective decontaminant procedure, as most chemical agents, including the blister and V-agents, are volatile to a certain degree.

If decontamination cannot be left to natural processes, chemical neutralizers or means of physical removal must be employed. In general, the neutralizers are of two types: chlorine-based oxidants or strong bases. Some neutralizers have been especially developed for the decontamination of chemical agents.

Medical Defense

The most important items of U.S. medical defense against organophosphorus nerve agents are atropine and pralidoxime chloride, (2-PAM), $C_7H_9ClN_2O_2$. These agents neutralize the effects of the anticholinesterase compounds and are capable of reactivating the inhibited enzymes.

Vesicant agents, such as mustard, require no special treatment once the burns have occurred. Copious washing is quite effective when used early for liquid contamination of the eyes, and soap and water removes the liquid agent from the skin. Burns resulting from mustard agent are treated like any other severe burn. The pulmonary injuries are treated symptomatically; antibiotics are used only if indicated for the control of infection.

J. E. Estes, *Remote Sensing Techniques for Environmental Analysis*, Hamilton Publishing Co., 1974.

J. H. Rothschild, *Tomorrow's Weapons*, McGraw-Hill Book Co., New York, 1964.

P. Taylor, ed. *Anticholinesterase Agents*, 8th ed, in A. G. Goodman, R. T. W, and A. S. Nies and co-eds., *The Pharmacological Basis of Therapeutics*, Pergamon Press, New York, 1990, pp. 131–140.

G. P. Wright, *Designing Water Pollution Detection Systems*, Ballinger Publishing Co., Cambridge, Mass., 1974.

Tom J. Evans
Cubic Defense Applications
Group

CHEMICALS FROM BRINE

Nearly every country in the world has a source of brine containing useful minerals. Many have underground ore bodies that may be turned into brine by solution mining. Oceans and seas of the world are the largest sources of brine. A second source of brine is found in terminal lakes. The Dead Sea in Israel and Jordan is an example of a large terminal lake with almost unlimited supplies of magnesium chloride, potassium chloride, and sodium chloride. More than two and one-half million tons of potassium chloride are extracted from the Dead Sea each year.

Great Salt Lake, Utah, is the largest terminal lake in the United States. From its brine, salt, elemental magnesium, magnesium chloride, sodium sulfate, and potassium sulfate are produced. A list of common minerals and their uses are listed in Figure 1.

A third source of brine is found underground. Underground brines are primarily the result of ancient terminal lakes that have dried up and left brine entrained in their salt beds.

A fourth source of brine is obtained through solution mining. Potash (KCl) is mined, in Moab, Utah, by solution mining. Much of the food-grade sodium chloride in the United States, Europe, and other parts of the world is solution mined.

The main metals in brines throughout the world are sodium, magnesium, calcium, and potassium. The nonmetals are chloride, sulfate and carbonate, with nitrate occurring in a few isolated areas. A major fraction of sodium nitrate and potassium nitrate comes from these isolated deposits.

All of these metallic and nonmetallic ions join together in a complicated array of salts and minerals called evaporites. Several evaporites usually crystallize simultaneously in a mixture, which often makes separation into pure chemicals difficult.

RECOVERY PROCESS

Solar Evaporation

Recovery of salts by solar evaporation is favored in hot, dry climates. Solar evaporation is also used in temperate zones where evaporation exceeds rainfall and in areas where seasons of hot and dry weather occur. Other factors affecting solar pond selection are wind, humidity, cloud cover, and land terrain.

Large solar salt operations can be found along the shores of Great Salt Lake and in the San Francisco Bay area. Salt production from solar ponds represents 14% of the total salt produced in the United States.

Seawater

Salt extraction from seawater is done in most countries having coastlines and weather conducive to evaporation. Seawater is evaporated in a series of concentration ponds until it is saturated with sodium chloride. At this point >90% of the water has been removed along with some impurities, $CaSO_4$ and $CaCO_3$, which crystallize at the bottom of the ponds. This brine, now saturated in NaCl,

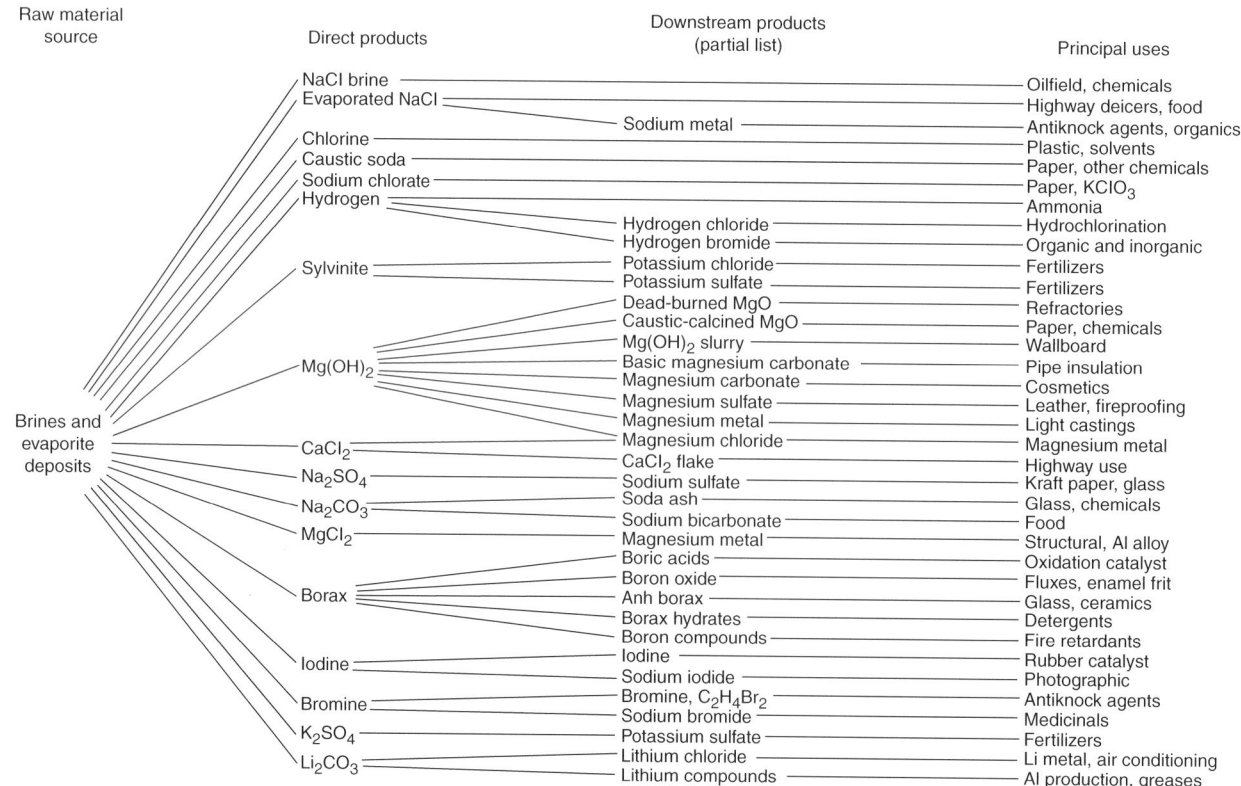

Figure 1. The brine chemical industry and some of its products.

is transferred to ponds, called crystallizers, where salt precipitates on the floor of the pond as more water evaporates.

The Great Salt Lake

All Great Salt Lake mineral extracting facilities have solar ponds as the first stage in processing minerals from brine. The first salt to saturate and crystallize is halite. This salt is successively followed by epsomite, schoenite, kainite, carnallite, and finally bischofite.

Solution Mining

Solution mining, also known as brining, is the recovery of sodium chloride (or any soluble salt) in an underground deposit by dissolving it *in situ* and forcing the resultant solution to the surface.

MINERALS FROM BRINE

A list of common minerals and their use is shown in Figure 1. This is only a small fraction of the tens of thousands of uses of these minerals extracted from brine.

BORON COMPOUNDS

Occurrence

Brine found in Searles Lake, California, is the only major brine source where boron from brine is produced commercially in the United States.

Boron is found in two underground ores, ulexite, and colemanite. Boron is found in over 200 minerals, but only four are of major importance: borax, kernite, colemanite, and ulixite.

Recovery Process

Boron is recovered from brine in one of three processes, liquid–liquid extraction, evaporator-crystallizers, borax recovers by refrigeration.

Economics and Uses

The principal producers in the United States are U.S. Borax and Chemical Corporation, North American Chemical, American Borate Corporation, and recently, Fort Cady. Their combined annual capacity in 2001 was reported to be 536,000 metric ton of equivalent B_2O_3. The United States was the largest producer of boron compounds in 2002.

BROMINE

Occurrence

Bromine is found in seawater and in underground brine deposits of marine origin. Bromine is also found in Dead Sea brine and is currently being produced there by the Dead Sea Works and the Jordan Bromine Co.

Recovery Process

Commercial processes depend on the oxidation of bromide to bromine.

Economics and Uses

The United States produces $\sim$225,000 tons of bromine per year at an estimated value of $200 million. All bromine is extracted from brine wells. In 2000 31% of bromine production was used for flame retardants; pharmaceuticals and agriculture, 18%; catalysts and additives, 21%; and performance chemicals, 30%.

CALCIUM CHLORIDE

Occurrence

Brines are the main commercial source of calcium chloride. Some brines of Michigan, Ohio, West Virginia, Utah, and California contain >4% calcium. Michigan is the leading state in natural calcium chloride production, with California a distant second.

Recovery Process

Because of its high solubility compared to that of other brine constituents, calcium chloride is the final constituent recovered in a multiproduct brine processing operation.

Economics and Uses

Most production of calcium chloride is from Michigan brines. The principle use of calcium chloride is to melt snow and ice from roads. It is also used in dust control, concrete setting control, waste water treatment, and various industrial uses.

IODINE

Occurrence

Iodine is widely distributed in the lithosphere at low concentrations ($\sim$0.3 ppm). It is present in sea water at a concentration of 0.05 ppm. Certain marine plants concentrate iodine to higher levels than occur in the sea brine and these plants have been used for their iodine content. A major source of iodine is caliche deposits of the Atacama Desert, Chile. In the United States, underground brine is the sole commercial source of iodine. Such brine can be found in the northern Oklahoma oil fields originating in the Mississippian geological system.

Recovery Process

By 2003, all of the iodine produced was being made from Oklahoma brines by a blowing out process.

Japan was the leading producer of iodine in the 1980s, producing nearly 7000 metric tons/year. Production in 2002 was 6,100 tons.

Economics and Uses

Most of the iodine used in the United States comes from Japan and Chile. The United States produces 7% of the world supply but consumes 28%. Iodine is produced in Woodward, and Vici, Oklahoma. These two locations produced an estimated 1,700 tons in 2002. Total world consumption is approximately 19,000 tons.

In 2001, estimated uses for iodine were sanitation 45%; animal feed, 27%; pharmaceuticals, 10%; catalysts, 8%; heat stabilizers, 5%; and other, 5%.

LITHIUM

Occurrence

Numerous brines contain lithium in minor concentrations. Commercially valuable natural brines are located at Silver Peak, Nevada (300 ppm), and at Searles Lake, California (50 ppm). Lithium brines with commercial potential are found in the Altiplano of Bolivia and Argentina, in salt beds of Chile, and in several salt beds in central and western China.

Recovery Process

Lithium is extracted from brine at Silver Peak Marsh, Nevada, and at the Salar de Atacama, Chile. The process consists of pumping shallow underground wells to solar ponds where brines are concentrated to 6000 ppm. Lithium is then removed by precipitation with soda ash to form lithium carbonate.

Economics and Uses

Lithium production at Searles Lake has been discontinued and the lithium concentration at Silver Peak is decreasing. The Salar De Atacama, Chile, operation is the largest lithium production in the world. Now nearly all lithium is produced from brine. The United States produces and consumes about one-half of all the world production.

MAGNESIUM COMPOUNDS

Occurrence

Magnesium hydroxide and magnesium chloride are two commercially important magnesium compounds recovered directly from natural brines. From these compounds many other compounds of magnesium are made, such as elemental magnesium and magnesia. Other important compounds containing magnesium are epsomite, schoenite, kainite, and carnallite. Major magnesium sources are seawater (1300 ppm Mg), Great Salt Lake (1.1%), underground brines near the surface east of Wendover, Utah (1%), subterranean brines in Michigan (0.7–2.5%), and brine from the Yates formation in the Midland Basin of West Texas (3%).

Recovery Process

Magnesium hydroxide can be recovered in relatively pure form either from brine or from an intermediate plant liquor by increasing alkalinity. Better recoveries can be made by replacing lime with dolomite that has been calcined. Recovery of magnesium chloride as a direct product is usually economically feasible only as a by-product.

Economics and Uses

Magnesium hydroxide and magnesium chloride are used as a basic feedstock to make elemental magnesium, refractories (MgO), and reactive chemicals.

POTASSIUM COMPOUNDS

Occurrence

There are two forms potassium compounds in brines of the world, muriate of potash (KCl) and sulfate of potash (K_2SO_4). The brine potash operations are located in Utah (Moab, Ogden, Wendover) and California (Searles Lake). Operations in Searles Lake have produced both muriate and sulfate of potash. The Ogden operation produces sulfate of potash. The others produce muriate.

Recovery Process

Moab Salt of Moab, Utah, produces KCl by solution mining. Production of KCl at the Wendover, Utah, operation employs a large 7,000-acre complex of solar ponds. Great Salt Lake Minerals Corporation near Ogden, Utah, produces potassium sulfate and several other products from Great Salt Lake brines.

Economics and Uses

Total world production of potassium products is 27,000,000 ton/year. More than 95% is used in fertilizer blends.

SODIUM CARBONATE

Occurrence

The brines of Searles Lake, California, are the sole brine source of sodium carbonate (soda ash) production in the United States. There is a large underground deposit of sodium carbonate brine in the Sua Pan area of Botswana, Africa.

Recovery Process

Presently, North American Chemicals Co. at Searles Lake is the only significant producer in the United States of sodium carbonate from brine. Production from brine represents ~10% of U.S. production.

SODIUM CHLORIDE

Occurrence

About one-half of all the sodium chloride produced in the world is from brine. Approximately 100 million tons a year are produced from brines of the ocean, terminal lakes, subterranean aquifers and solution mining. Sodium is found in large quantities in most areas of the world. Its quantity is so large that prices in some locations are only $25 a ton.

Recovery Process

There are two main processes. One is to flood solar ponds with brine and evaporate the water, leaving sodium chloride crystallized on the pond floor. The other is to artificially evaporate the brine in evaporative crystallizers. Industrial salt is made from solar ponds, where as food-grade salt made for human consumption is mostly made in evaporative crystallizers.

Economics and Uses

The United States is the largest producer of salt in the world. Salt continues to be one of the most heavily traded chemical ores in the world, representing >50% of all seaborne mineral trade. World consumption is >200 million tons/year.

SODIUM SULFATE

Occurrence

In the United States, natural sodium sulfate brines are found in California at Searles Lake, at the shallow Castile formation underlying Terry and Gains counties, Texas, and at the Great Salt Lake, Utah.

Other natural sodium sulfate brines of commercial importance are found in dry lake beds of southwestern Saskatchewan, Canada; Laguna Del Rey in Coahuila, Mexico; the Gulf of Kara-bogaz, Russia, and in western China.

Recovery Process

The process for making sodium sulfate is different at each facility where it is produced. One step common to all facilities is a cooling step to form Glauber's salt, followed by a purification and recrystallization step to form anhydrous sodium sulfate.

In Texas, brine is pumped from underground deposits. At Searles Lake, sodium sulfate is recovered as one of three coproducts in a series of complex operations where soda ash and borax are also recovered from the brine. In processing Great Salt Lake brine, Glauber's salt is crystallized in solar ponds by cooling during the winter.

Economics and Uses

About one-half of all sodium sulfate produced in the United States is from brine. In 2002, 500,000 tons was produced. All of the production is from North American Chemicals (Searles Lake) and Ozark-Mahoning at Brownfield and Seagraves, Texas.

W. J. Schlitt, ed., *Salts & Brines '85*, Port City Press, Baltimore, Md. 1985.

U.S. Geological Survey Minerals Yearbook, 2001, p. 64.6. www.minerals.usgs.gov.

J. Wallace Gwynn, ed., *Great Salt Lake*, Utah Geological Survey Special Publication, Department of Natural Resources, 2002, pp. 201–233.

F. Yaron, in Z. E. Jolles, ed., *Bromine and its Compounds*, Ernest Benn Ltd., London 1966, pp. 3–41.

DAVID BUTTS
DSB International Inc.

CHEMILUMINESCENCE, ANALYTICAL APPLICATIONS

Chemiluminescence (CL) has many advantageous features as a tool of detection in instrumental analysis, including sensitivity, selectivity, and simplicity. A CL reaction needs no excitation light source; thus, it is not accompanied by any scattering light. This results in a large signal-to-noise (S/N) ratio and, consequently, a substantial increase in the detector's sensitivity. To date, several kinds of CL reactions have been clarified on their reaction mechanisms. Generally, concentrations of the substrates or catalysts in each CL reaction can be determined by measuring the CL generated. Among the CL reactions, luminol (5-amino-2,3-dihydro-1,4-phthalazinedione) derivatives, acridine derivatives, aryloxalate derivatives, and ruthenium derivatives have frequently been utilized in analytical applications.

LUMINOL DERIVATIVES

Luminol as a CL Reagent

Luminol is generally used to determin compounds that catalyze the luminol CL reaction. Aliphatic alcohols, aldehydes, ethers, and sugars containing an oxygen atom in their molecules can be converted into H_2O_2 by oxygenation in photochemical reaction, using anthraquinone disulfonate as a sensitizer, which can then be detected by a CL reaction with Co(II) and luminol. The lower limit of detection is in the picogram level. After separation on a cation-exchange column, Co(II) in rice powder was determined by high-performance liquid chromatography (HPLC) in the same manner. The sensitivity is very high, with a detection limit of 0.5 ng/L. Although Cu(II) can catalyze the luminol–hydrogen peroxide CL, the coexisting proteins can quench the yield of emission. Based on this phenomenon, as low as 50 ng of ovalbumin has been determined by HPLC. Hydroperoxides, the primary lipid peroxidation products, were separated on a normal-phase HPLC, with selectively determined by reaction with a mixture of luminol and cytochrome c as hydroperoxide-specific postcolumn CL reagents. Phosphatidylcholine derivatives, which are important biological phospholipids, were separated on a normal-phase column and determined by the same CL reaction system. Phosphatidylcholine hydroperoxide in low density lipoprotein (LDL) in human plasma was determined by the same method. Quantification of phospholipid hydroperoxides in biological tissues is important in order to know the degree of peroxidation damage of membrane lipids. Determination of lipid peroxides in native LDL is clinically important, because these peroxides are considered to cause the pathogenesis of atherosclerosis. A rapid flow injection method was developed by utilizing the luminol–microperoxidase CL reaction for the determination of peroxides in LDL. Using the same CL system confirmed that the lipid peroxides in triacylglycerol, contained in butter or spreads, increased during storage. Cholesterol hydroperoxide in human red cell membrane was also determined, by a luminol–microperoxidase CL method. A unique CL method utilizing isoluminol and microperoxidase, developed for the measurement of lipid peroxidation in the cultivated medium, can be used to directly measure in vitro lipid peroxidation in cells. A highly sensitive method has been developed for determining proteins such as human serum albumin, human γ-IgG, and bovine serum albumin (BSA).

Luminol Derivatives

Hydrogen peroxide is an excellent oxidizing agent for luminol; the CL increases with the increase of H_2O_2 concentration. Therefore, H_2O_2 or substrates that produce it in an enzyme reaction can be determined by measuring the CL yielded, and enzyme activity can also be estimated. Glucose, cholesterol, and xanthines have been determined by measuring the H_2O_2 produced by enzyme reactions with glucose oxidase, cholesterol oxidase, and xanthine oxidase, respectively. Adenosine is converted to inosine by adenosine deaminase, which is further converted to hypoxanthine by nucleoside phosphorylase. The H_2O_2 formed by the reaction of hypoxanthine and xanthine oxidase can be determined by the luminol–peroxidase CL reaction.

The luminol CL reaction was also applied to immunoassays and DNA hybridization assays; by using microperoxidase and luminol as CL reagents, α-fetoprotein or ferritin as a tumor marker, thyroid-stimulating hormone (TSH), luteinizing hormone (LH), and follicle-stimulating hormone (FSH) were assayed. The HRP–luminol CL was utilized in hybridization assays of DNA on membrane, magnetic bead, polymer, etc.

Isoluminol Derivatives

Among newborn babies, women, and women at a late pregnancy stage, the concentrations and conjugated forms of bile acids in their urine are markedly different, and the metabolic reactions of fetus and newborn babies are clearly different. Because isoluminol has a higher luminescence efficiency then luminol, a series of isoluminol derivatives has been synthesized for the chemiluminescent immunoassay (CLIA). These CL reagents have been applied to the assays of biotin and thyroxine and steroid hormones (estradiol, estriol, progesterone, etc.). The CL reaction scheme is shown in Fig. 1.

Other Luminol Derivatives

4-Isothiocyanatophthalhydrazide and 6-isothiocyanobenzo[g]phthalazine-1,4-(2H,3H)-dione (IPO) have been used as highly sensitive CL labels of amino compounds. Twelve amino acids were sensitively determined by HPLC with 4-isothiocyanatophthalhydrazide and a detection limit of 10 fmol on column was achieved. The antidepressant maprotiline in plasma could be detected at a very low level of 0.1 ng/ml by HPLC with IPO. Primary amines such as n-hexylamine, n-butylamine, and n-octylamine were sensitively detected using IPO as a label over the range from 30 to 120 fmol at a S/N of 3. Secondary amines such as N-methyloctylamine, di-n-amylamine, di-n-hexylamine of at levels of 0.8–3 fmol (S/N = 3) were determined using the same approach.

The principle that analytes of interest can be determined after being converted to luminol derivatives was successfully introduced to the measurement of some biological components. For instance, 4,5-diaminophthalhydrazide

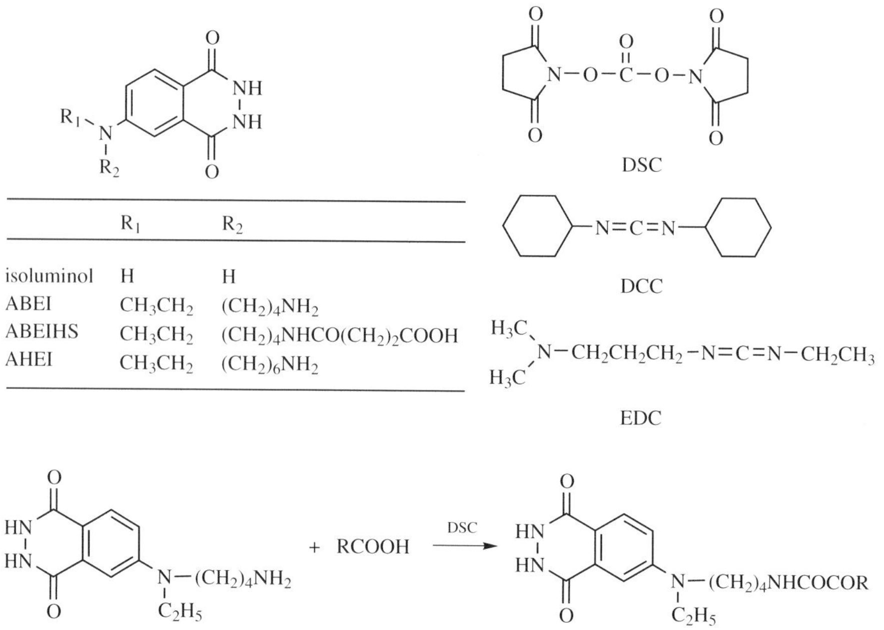

Figure 1. Structures of isoluminol derivatives, condensation agents, and derivatization reactions of ABEI with carboxylic acid.

	R_1	R_2
isoluminol	H	H
ABEI	CH_3CH_2	$(CH_2)_4NH_2$
ABEIHS	CH_3CH_2	$(CH_2)_4NHCO(CH_2)_2COOH$
AHEI	CH_3CH_2	$(CH_2)_6NH_2$

(DPH) gives CL compounds by reaction with α-keto acids, which are important intermediates in the biosynthesis of amino acids, carboxylic acids, and sugars.

Utilization of Enhancers for Luminol CL Reaction

In spite of efforts by analytical chemists to develop useful CL reagents, only a few new ones have been successfully developed. Thus, trials to enhance the efficiency of known CL reagents were introduced in CL analyses. In the luminol–peroxidase–H_2O_2 CL system, phenol compounds such as 4-iodophenol, aromatic compounds such as hydroxybenzothiazoles and dehydroluciferins, and phenylboronic acids have been found to serve as strong enhancers.

ACRIDINE DERIVATIVES

Lucigenin (N,N'-dimethyl-9,9'-biacridinium dinitrate), a representative chemiluminogenic acridine derivative, one of the oldest CL reagents, has been used for analytical purposes. Lucigenin reacts with H_2O_2 to yield an emission of light in a similar manner to that of luminol and can also produce emission with organic reductants such as reducing sugars. This finding is based on the reaction of lucigenin with the α-hydroxycarbonyl group of the organic reductants. Therefore, compounds bearing an α-hydroxycarbonyl group such as glyceraldehydes, cortisols, phenylacylalcohols, or phenylacylesters can be determined by lucigenin CL. The lucigenin CL was applied for the investigation of human seminal plasma, which has powerful antioxidant and immunosuppressive activities. A sensitive and simple CL assay for alkaline phosphatase (ALP) was developed by using dihydroxyacetone phosphate or its ketal as a substrate.

Acridinium esters or acid chlorides were developed as CL labeling reagents, to be applied to immunoassays and HPLC of trace amounts of biological components. Acridinium-labeled compounds have 100 times stronger CL intensity compared to luminol-labeled ones, and acridinium esters have the dominant feature that they do not loss the CL efficiency even after binding to an antigen or antibody. Acridinium esters, for instance, were applied to compensative DNA probe (cDNA) in hybridization assay (55), human gonadotropin and TSH (56), virus, and immunoassay of a cancer marker. By using a commercially available kit (TSA), as low as 0.01 mU/mL of TSH can be determined. This sensitivity makes it possible to clearly distinguish the TSH levels in normal human blood (0.53–3.05 μU) and Basedow's disease patients' blood (<0.10 μU). Also, 9-acridinecarbonylimidazole has been found useful as a reagent for measuring the activity of a number of enzymes that directly produce peroxide, including glucose oxidase, with the detection limits in the 1–10-amol range.

ARYLOXALATE DERIVATIVES

Aryloxalates and Fluorescent Compounds as Chemiluminescent Reagents

Chemiluminescents produced by the reaction of aryloxalate (or oxamides), H_2O_2, and fluorescent compounds are known as peroxyoxalate chemiluminescents (PO–CL). Many kinds of aryloxalates have been synthesized and evaluated as PO–CL reagents. Among these, bis(2,4,6-trichlorophenyl)oxalate (TCPO) is the most popular. The TCPO can be easily synthesized and is inexpensive. The solubility of this compound is low in organic solvents such as acetonitrile and methanol commonly used in HPLC, which may limit its use in some cases.

Determination of Fluorescent Compounds

Generally, highly fluorescent compounds, which should be suitable to PO–CL detection, are rare among biologically active components, environmental pollutants, etc. Therefore, derivatization of nonfluorescent or weakly fluorescent compounds to yield highly fluorescent ones is a commonly adopted practice in PO–CL. Many known fluorescence derivatization methods can be used to avoid this aim. A FIA method for determining perylene, anthracene, and pyrene was developed using TCPO and H_2O_2 in acetonitrile as CL reagents with the detection limits of 0.05, 65, and 75 pg (S/N = 5), respectively. Nitroarenes derived from diesel engine exhaust are known as carcinogens. In spite of being non-fluorescence, nitroarenes can be determined after reduction to highly fluorescent aminoarens. Air pollutants such as polycyclic aromatic hydrocarbons (PAHs) and their nitrated derivatives were determined over a period of 12 months by HPLC–PO–CL. The toxic equivalency factors adjusted concentration of total PAHs determined was 2.33 ng/m³ in Nagasaki city area. Fluorescent compounds such as 3,3'-diethylthiadicarbocyanine iodide (DTDCI) having their absorption wavelengths in the infrared (ir) region are preferable as fluorophores, because they can be excited with low excitation energy that generally the PO–CL reaction can produce.

A photographic method was devised for the detection of FL compounds and H_2O_2. Glucose in serum was semi-quantified by this method, and the CL immunoassay of the antihuman T-cell virus antibody was developed by a photographic method.

Determination of Amino Compounds

Dansyl chloride (Dns-Cl) was first used for FL labeling of amino acids in thin-layer chromatography (TLC). Dns-amino acids were quantified by HPCL–PO–CL, using TCPO as an oxalate reagent. Sixteen kinds of Dns-amino acids were separated within 30 min by a gradient elution, over a range from 2 to 5 fmol (S/N = 2). Dns derivatization was also applied to quantifiy abuse drugs (amphetamines) in urine samples. Methamphetamines in a single hair sample were detected at a very low level of 20 pg.

Mexiletine, an antiarrhythmic drug, was assayed in rat plasma using TDPO as an oxalate. A highly sensitive HPLC–PO–CL method for catecholamines, which act as neurotransmitters, was developed using fluorescemine as a label. Fluorescamine derivatives of histamine can be detected with FL 100 times more sensitively than CL. Fluorescamine has been utilized to quantify sulfamethazine (an antibacterial drug) in chicken eggs and serum. Methamphetamine and its related compounds were assayed by HPLC with TDPO and H_2O_2 as the postcolumn CL reagents. Dopamine and norepinephrine in urine were determined at sub–low femtomole levels with a sample size of only 20 μL. Fluvoxamine, an antidepressant, was also determined in the same manner. Ethylenediamine was applied for the online postcolumn derivatization approach of catecholamines, following PO–CL detection with TDPO and H_2O_2 as the postcolumn reagents.

Carbonyl Compounds

5-N,N'-Dimethylaminonaphthalene-1-sulfonohydrazide (Dns-H) has been used as a labeling reagent for carbonyl compounds. Carbonyl compounds in air, such as formaldehyde, acetaldehyde, and acetone, have been determined with HPLC–PO–CL after labeling with Dns-H. These compounds were derivatized by drawing an air sample through a small glass cartridge packed with porous glass particles impregnated with Dns-H; this method is very simple and can detect amounts at sub-ppb concentrations of carbonyl compounds.

Determination of Carboxylic Acids

Coumarin derivatives such as 4-(bromomethyl)-7-methoxycoumarin (Br-Mmc), 7-(diethylamino)cumarin-3-carbohydrazide (DCCH), and 7-(diethylamino)-3-[(4-iodoacetylamino)phenyl]-4-methylcumarine (DCIA) have been used for labeling carboxylic acids. These reagents were evaluated as PO–CL using a TCPO–H_2O_2 CL reagent.

Determination of Thiols, Alcohols, Phenols, and Others

Usable fluorescent derivatizing reagents for thiols and phenols are limited. N-[4-(6-Dimethylamino-2-benzofuranyl)phenyl]maleimide (DBPM), a maleimide-type fluorescent reagent, has been used for thiol compounds. The DBPM as well as other maleimide-type fluorescent labeling reagents, is natively nonfluorescent but gives strong fluorescent products upon reaction with thiol compounds.

Determination of Hydrogen Peroxide and Substrates That Produce Hydrogen Peroxide in Enzyme Reactions

Hydrogen peroxide and substrates from which H_2O_2 is produced in enzymatic oxidation can be sensitively determined by using an PO–CL system. For the determination of H_2O_2 by PO–CL, pyrimido[5,4-d]pyrimidines were synthesized and evaluated.

DIOXETANE DERIVATIVES

Adamantyl dioxetanes have been used as analytical reagents of CL. Owing to the very long lifetime of CL produced by adamantyl dioxetanes, they are recommended to photographic detection. Reverse transcriptase (RT) assays are very important for the detection of retroviruses, including human immunodeficiency virus (HIV) and for the development of new antiretroviral substances.

LOPHINE AND INDOLE DERIVATIVES

Lophine (2,4,5-triphenylimidazole) was the first synthetic CL compound. However, its analytical applications are relatively rare. Lophines show very strong FL and thus have been applied to FL rather than CL detection.

Indole derivatives have been much studied on their CL mechanisms, because their structures are well correlated with luciferin of *Cypridina hilgendorfi*, but studies of their analytical characteristics are few.

RUTHENIUM(II) COMPLEX

CL is produced by the oxidation of a ruthenium(II) complex and reducing compound on the surface of an electrode followed by the excitation of an oxidized ruthenium(III) complex with a radical derived from reducing compound. Therefore, reducing compounds can be determined after HPLC separation and following oxidation. Primary amines yield little CL from this system, but secondary and tertiary amines, especially cyclic aliphatic amines, can produce strong CL.

OTHER CL APPLICATIONS

In spite of its long history, gallic acid has only a few analytical applications. Recently, catechin and (−)-epigallocatechin 3-gallate, were found to yield CL by the reaction with acetaldehyde, horseradish peroxidase and H_2O_2. Pyrogallol is known as a potential CL compound. Analytical applications of such compounds are anticipated.

Adenine and guanidine compounds can produce emission of light by reaction with 9,10-phenanthrenequinone, which could be utilized for HPLC–CL of guanidino compounds. As well as adenine and its nucleosides and nucleotides, guanine and its nucleosides and nucleotides react with glyoxal derivatives to give CL compounds, which can intensely emit under alkaline conditions in aprotic solvents such as dimethylformamide (DMF).

Morphine, a narcotic analgesic, can yield CL by reaction with potassium permanganate in polyphosphoric acid solution (pH 1–2). Similarly, morphine and monoacetylmorphine been simultaneously determined. In a similar manner, ovalbumin-trypsin inhibitor and BSA-trypsin inhibitor were detected.

The CL of sulfur-containing compounds has also been reported. An FIA for thiopronine and its metabolite, 2-mercaptopropionic acid, was developed by utilizing the phenomenon that Ce(IV) CL can be enhanced by quinine. In the same manner, rhodamine B was used as an enhancer for determining preparations of captopril and hydrochlorothiazide.

CL is a versatile tool for sensitive and selective determination of trace amounts of organic and inorganic compounds, including biologically important components such as bioactive amines, nucleic acids, sugars, etc. Therefore, it is expected that CL will be utilized by such postgenome sciences as proteome, metabolome, etc.

A. M. Garcia Campana and W. R. G. Baeyens, eds., *Chemilumi-nescence in Analytical Chemistry*, Marcel Dekker, New York, 2001.

N. Kuroda and K. Nakashima, in T. Toyo'oka, ed., *Modern Deri-vatization Methods for Separation Sciences*, John Wiley & Sons, Inc., Chichester, 1999.

K. Nakashima, H. Yamasaki, R. Shimoda, N. Kuroda, and S. Akiyama, *Biomed. Chromatogr.* **11**, 63 (1997).

A. M. Osman, C. Laane, and R. Hilhorst, *Luminescence* **16**, 45 (2001).

KENICHIRO NAKASHIMA
Nagasaki University

CHEMOINFORMATICS

WHAT IS CHEMOINFORMATICS?

Different definitions of chemoinformatics have been given but, within the context of this article we will view it broadly as the management, analysis, and dissemination of data related to chemical compounds. Chemoinformatics results from the application of methods in information technologies to problems in chemistry.

During the last decade, chemoinformatics has become one of the essential tools in the early stages of pharmacological and agrochemical discovery. The reason for its importance is rooted in the emergence of high throughput screening and high throughput chemical synthesis as the dominant technologies for the discovery of starting points for chemical optimization.

Chemoinformatics work is and will continue to be multidisciplinary, because it acts at the interface between chemistry and informatics, as well as the multiple disciplines that use it. The main purpose of chemoinformatics is to provide tools for the efficient management of information, a critical step in any decision making process. Chemoinformatics transforms data into information and subsequently into knowledge, thus greatly facilitating all aspects of chemical research.

CHEMICAL INFORMATION STORAGE

Perhaps the most important task in the creation of a chemical database is the definition of the fields to be stored. The database design requires particular attention, because errors or lack of foresight when creating it are painful to correct as the systems are deployed throughout the organization. Yet, such foresight is challenging because chemical databases used in research are continuously evolving together with the data collected. Even at the earlier stages of the project, input from the end-users is a requisite for the design of any database.

The distinctive feature of a chemical database is that it allows the storage and retrieval of structural information as well as textual or numerical information on a chemical. All datatypes, including chemical, numerical, and textual information, can be combined when querying a chemical database. Simple queries may include combinations of datatypes such as 'Display all compounds with an imine functionality that cost less than a given amount and are currently in inventory.' Without the chemical structure component, the same searches could not be done within the framework of textual or numerical datatypes alone.

Representation of chemical structures in a computer searchable form requires the adoption of special formats. While multiple formats have been used over the years, the dominant chemoinformatics software provides a relationship between structure and either tables or lines that are intrinsically computer searchable. The two dominant file formats for structural representation are the SMILE strings and the MOL files, discussed below.

Line Notations

Alternatives to the valence representation of the molecular structure in the form of lines or strings have been pursued for decades, even before the use of computers in chemical information storage and management. Currently, SMILES (Simplified Molecular Input Line Systems) is one of the most popular notations in this class.

In a SMILES string, each atom is identified by its element symbol, as well as additional information that is placed into brackets, including chirality and net charge.

One of the limitations of the SMILE strings is that there could be more than one equivalent string for the same molecule, because they depend on the internal numbering system of the structure used. One given structure will not always yield the same representation, but it will depend on the algorithm used. A canonical representation is one where rules are defined to the extent that only one string is the correct representation for any chemical structure. "Unique SMILES" (USMILES) are such representations.

Table Representation

The most common alternative to the string or line notation is the use of connectivity or connection tables.

MOL (molecule) and SD (structure data) files contain connectivity tables. Both types of files have a "counts' line," after comments lines that specify the total number of atoms in the file, the number of bonds, atom lists, and information on chirality. In addition, since the format of the files has been mildly modified with time, the version of the file is included. The line is followed by an atom block that contains the atom symbol, charge stereochemistry, attached hydrogens for each atom, and a set of Cartesian coordinates for each atom. In two-dimensional (2D) representations, these coordinates can be used to plot a flat molecular structure. However, the coordinate fields can be used to store a three-dimensional (3D) representation of the molecule, in cases where a spatial arrangement of atoms is available. The ability to store (3D) information in some cases could be an advantage of table representations.

Other file formats centered on the connectivity tables are available. Reaction data (RD) files, are similar to the basic SD file but are able to contain structural data for the reactants and products of a reaction, as opposed to individual molecules.

CHEMICAL INFORMATION RETRIEVAL: DATA SEARCHING

The use of computer readable formats to store structural information is the key to generating software that will be capable of searching such data. The search and display of textual and numerical data can be done with Boolean operators (AND, OR, LESS THAN, GREATER OR EQUAL TO, etc). However, the unique feature of a chemical database is in the handling of structural information, and that is where we will focus our discussion.

Different types of searches on structural information can be carried out. Two-dimensional information is searched differently from 3D. Two-dimensional searches can be done with the purpose of (1) identifying an exact chemical structure; (2) identifying a molecule or molecules that contain a given structural feature, commonly referred to as a substructure search; and (3) to search for molecules that look like those in another used as a query, which is described as a similarity search. Three-dimensional searches can be carried out to identify molecules that have predetermined pharmacophoric features in a correct spatial arrangement, with or without explicit knowledge of the 3D structure of the target.

Construction of Three-Dimensional Chemical Databases

Structural information about the small molecule is necessary to build a 3D chemical database. The information can be obtained from experimental sources, such as the Cambridge Crystallographic Database (CSD), which contains crystal structure information for >230,000 organic and metallorganic compounds. Alternatively, computational techniques can be used to convert 2D representations of a molecule (a SMILES string or a MOL or SD file) to a 3D structure.

The conversion programs apply a knowledge base to construct the 3D structure. The pioneer program for the rapid conversion of 2D to 3D structures is CONCORD, which combines rules with energy estimation procedures in an attempt to produce the lowest energy 3D conformation for each structure.

CORINA, like CONCORD, is also a rule- and data-based algorithm, but with a superior ring handling technique, as well as organometallics. CORINA has been reported to have a higher conversion rate than CONCORD and other similar software for this purpose. A different approach to the problem is provided by Converter. It uses a distance geometry approach, coupled with upper and lower interatomic distance bounds, together with topological rules to generate the 3D structure. The procedure is somewhat slower than the purely rule-based algorithm, and may be less suitable for the conversion of large datasets or if structural information has to be generated as it will be used. That will be the case for software based online notations that do not store spatial information but generate it as required.

Diversity Searches

The need to carry out diversity searches emerged in the pharmaceutical and agrochemical industry because of the advances in automation in biological screening and high throughput chemical synthesis. Those technologies posed a new set of questions to be asked from a chemical database. The need to detect redundancies in a chemical library led to the concept of diversity analysis.

With few exceptions, diversity analysis of chemical collections evolved from the methodologies of structure-based drug design. Consequently, diversity analysis has been heavily dependent on the computation of physicochemical or structural properties. Chemical diversity is mostly associated with methods that allow the determination of how well libraries could represent portions of chemical property space. The scattering or clumping of representatives in a library can be surrogate indicators of the probability that that library would provide multiple, singular, or no hits for a set of targets. The challenges associated with the determination of chemical diversity are quite varied. The most significant issues are the selection of properties and the algorithms that are to be used to determine and select diverse compound sets.

Toxicology Prediction

The prediction of the toxicological character of a compound is of foremost importance throughout the chemical industry. A large number of approaches have been adopted and are employed to study the compound toxicity problem.

One of the most popular packages for that purposes is TOPKAT, which is a self-contained computational toxicology package that uses 2D descriptors and statistical models to generate reliable toxicological profiles of organic chemicals, one at a time.

An alternative to the statistical analysis of properties are knowledge-based systems that are computer programs to organize relevant experimental data to help a user make decisions about concrete issues. They require the use of a systematic database of information from which rules are derived, which allows the prediction of the property to be scrutinized. Examples are HazardExpert, which predicts different toxicity effects of compounds such as carcinogenic, mutagenic, teratogenic, membrane irritation, and neurotoxic effects, and DEREK, also a rule-based approach that can make predictions about a large set of toxicological properties including carcinogenicity, irritancy, lachrymation, neurotoxicity and thyroid toxicity, teratogenicity, respiratory and skin sensitization, and mutagenicity while the MULTICASE and CASE programs can automatically identify molecular substructures that have a high probability of being relevant or responsible for the observed biological activity of a learning set composed of a mix of active and inactive molecules of diverse composition.

CHEMICAL DATABASES

Chemical information itself is abundant, and the Chemical Abstract Service (CAS) has maintained the most comprehensive resource, in this field. SciFinder provides a desktop research tool that allows the exploration of research topics, with little training, containing information on >33 million substances. The STN service provides specialized information on >200 different subjects.

CrossFire Beilstein has extensive information on bioactivity and physical properties that makes it particularly useful when undertaking biological research. The database also provides information on the ecological fate of compounds.

Other databases are also worth noting because they contain information that is more specific. There are vast numbers of commercial databases with different focus. One of the most common is the ACD (Available Chemicals Directory), which contains information on price and availability on >300,000 compounds. This information includes not only a 2D representation of the molecule, but also 3D models that make it useable for pharmacophore searches or for docking purposes.

Chemical databases also deal with reactivity information. Multiple databases exist for this purpose: SpresiReact database is available from InfoChem GmbH and contains 2.5 million different reactions; RefLib is another broad collection of novel organic synthetic methodologies that covers functional group transformations, metal-mediated chemistry, and asymmetric syntheses, as well as reactions from Theilheimer's *Synthetic Methods of Organic Chemistry*; *Methods in Organic Synthesis* (MOS) is a selective current awareness database that focuses on important new methods in organic synthesis and comprises >3300 reactions per year, dating back to 1991; BioCatalysis is a selective, thematic database that focuses on chemical synthesis using biocatalysts, including pure enzymes, whole cells, catalytic antibodies, and enzyme analogues; the Failed Reactions database is a unique compilation of reactions with unexpected results, which may involve an unexpected product, an immediate further reaction, or simply no reaction; the SPORE and Solid-Phase Synthesis databases include data particular to solid-phase organic synthesis; the Protecting Groups database provides information on methods for protection, deprotection with the ability to search generically, based on functional groups, protected groups, tolerated groups, and reaction conditions. Other databases on chemical reactivity also exist

Apart from chemistry resources, there are a large number of content databases with information specific about different areas all linked by their chemical structure, which include material sciences, agrochemical, physicochemical, and biological activity.

DATA ANALYSIS AND PRESENTATION

Data mining is crucial when large amounts of data are generated. The idea is to integrate a number of visualization, statistical analysis tools through graphical user interfaces.

One of the most popular packages for this purpose is Spotfire, where chemical structure data can be combined with data from different sources, in order to provide insights into the property of interest, biological or physical. DIVA is a package with a similar purpose that was developed specifically for the pharmaceutical industry. DIVA allows users to retrieve and work with chemical structures, assay results, and other chemical and biological data in one

convenient spreadsheet. Powerful easy-to-use tools for data integration, visualization, analysis, and reporting save time and allow researchers to get more value from their data.

J. M. Barnard, in: P. von R. Schleyer, N. L. Allinger, T. Clark, J. Gasteiger, P. A. Kollman, H. F. Schaefer, and P. R. Shreiner, eds, *Encyclopedia of Computational Chemistry*, Wiley, Chichester, U.K. Vol 4, 1998, p. 2818.

E. M. Gordon and J. F. Kerwin, Jr., eds, *Combinatorial Chemistry and Molecular Diversity in Drug Discovery*, Wiley-Liss, 1998.

G. Klebe, in H. Kubinyi, ed, *3D QSAR in Drug Design: Theory Methods and Applications*, Escom, Leiden, 1993, p. 173.

HUGO O. VILLAR
Triad Therapeutics

CHEMOMETRICS

Chemometrics or more general multivariate regression methods are applied in many research fields from social science to measurement techniques. There are two competing and equivalent nomenclature systems encountered in the chemometrics literature. The first, derived from the statistical literature, describes instrumentation and data in terms of "ways" that an analysis is performed. Here a "way" is constituted by each independent and nontrivial factor that is manipulated with the data collection system. Multiway techniques (the section Multiway Analysis) have been investigated and applied to hyphenated measurement techniques. For example, with excitation–emission matrix fluorescence spectra, three-way data are formed by manipulating the excitation- way, emission-way, and the sample-way. Implicit in this definition is a fully blocked experimental design where the collected data forms a cube with no missing values. Equivalently, a second nomenclature is derived from the mathematical literature where data are often referred to in terms of "orders". In tensor notation a scalar is a zero-order tensor, a vector is first order, a matrix is second order, a cube is third order, etc. Hence, the collection of excitation–emission matrix fluorescence data would form a third-order tensor. However, it should be mentioned that the "way" and "order" based nomenclature are not directly interchangeable. By convention, "order" notation is based on the structure of the data collected from each sample. Analysis of collected excitation-emission fluorescence, forming a second-order tensor of data per sample, is referred to as second-order analysis compared to three-way analysis. In this work the "way" based notation will be adopted.

What is chemometrics? Simply put, chemometrics is the application of mathematical and statistical methods to the analysis of chemical data. However, it should be stressed that chemometrics is more than a subdiscipline of mathematics or statistics. The key to artfully practicing chemometrics is to extend the limitations of classical mathematics and statistics by understanding, and relying

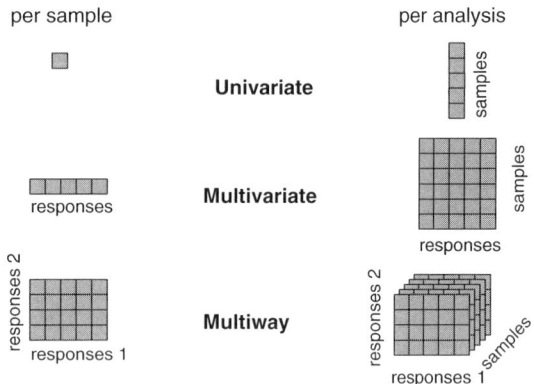

Figure 1. Matrix representation of data structure for three classes of data.

upon, the constraints that chemistry places on possible solutions to a statistically posed question. Chemometrics is a truly interdisciplinary science that does draw from mathematics, statistics, and information science; however, the tools from these disciplines cannot be directly applied without sound knowledge and understanding of the chemical system in question. Furthermore, the distinction between "statistical significance" and "practical significance" cannot be reliably made without an understanding of both statistics and chemistry.

From a chemometric standpoint, data and instrumentation can be classified based on the dimensionality of the data set obtained. Instrumentation can be designed to generate a single datum of information per sample analyzed, an ordered vector of data per sample analyzed, or a linked matrix of data per sample analyzed. In general, the higher the dimensionality or number of ways of the data set, the more powerful the instrument. And, consequently, more powerful data analysis methods can be applied to higher directional data set. The different ways of data are presented in Figure 1.

The most basic type of data is univariate data. However, the majority of literature in chemometrics addresses analysis of multivariate data.

BILINEAR CHEMOMETRIC METHODS

The reported successes of multivariate chemical analysis are based on three facts: (1) most, if not all chemical processes are multivariate in nature. Consequently, to be able to effectively perform in a multivariate world, multivariate data must be collected and analyzed. (2) Even if only a single piece of information is needed from a chemical system, it is very difficult to design a sensor that is fully selective to that property of interest. Therefore, to circumvent the lack of fully selective sensors, arrays of partially selective sensors can be constructed that rely on multivariate analysis methods to extract the information of interest. (3) There are inherent advantages associated with the redundancy of data when there are many more variables measured per sample than samples collected.

Classical Least Squares (CLS) versus Inverse Least Squares (ILS)

The physics oriented CLS approach considers the measured spectra as products of molar extinction coefficients $\mathbf{K}$ (unit spectra) and concentrations $\mathbf{C}_{cal}$. The spectral errors are contained in $\mathbf{E}_A$:

$$\mathbf{A}_{cal} = \mathbf{C}_{cal} \cdot \mathbf{K} + \mathbf{E}_A \qquad (1)$$

ILS uses a less intuitive calibration routine, which follows the introduction into multivariate least-squares fits as given in the section Multivariate Linear Regression:

$$\mathbf{C}_{cal} = \mathbf{A}_{cal} \cdot \mathbf{P}_{(N \times M)} + \mathbf{E}_C$$

Both methods have advantages and drawbacks: CLS minimizes spectral errors—ILS minimizes concentration errors. Usually, the spectroscopic data contain more noise then the calibration concentrations, which can be determined by very precise reference methods. Hence, the CLS calibration is the more appropriate one compared to the ILS calibration, since CLS calibration is based on the precisely known model. The ILS method, however, uses the less precise calibration based on noisier spectral data. Nonetheless, ILS is supposed to be the superior approach for practical applications since it only needs calibration concentrations of the analytes of interest. In order to make CLS a good predictor calibration concentrations of all analytes must be included that are expected during the measurement process. This restriction is severe, especially in process monitoring where usually a huge number of absorbers are involved.

Principal Component Analysis (PCA) and Principal Component Regression (PCR)

Factor analysis (FA) is employed to aid in visualization of sample (time) dependent trends and measurement (sensor) dependent trends in a multidimensional data space. In general, factor analysis does not give a physically meaningful model—only correlations among samples and measurements are determined. However, FA methods have been modified to apply constraints and assumptions based on previous knowledge of the chemical system being analyzed. These modified FA methods are useful for determining the underlying instrumental and/or sample (time) profiles of the chemical constituents of a process. Perhaps the most commonly applied method of FA is principal component analysis–regression (PCA–PCR)—an ILS based approach. The PCA method only extracts the principal component (PC) or loading vectors by means of which unknown measurement data will be represented. It takes a second step to relate such an abstract data representation to chemical properties, concentrations, for example.

The goal of PCA is to identify the major sources of correlated variance in a collection of data. Once these sources of variance have been identified, they can be exploited to aid in the visualization of the major trends throughout the data collection. The data collection can be reduced from a complicated multidimensional representation to a more easily visualized two- or three dimensional space that describes the majority of the variance (information) in the data collection.

There is a difference between factor analysis and calibration methods: Factor analysis extracts underlying factors or model by means of which the analyzed data can be described—calibration, however, extracts such factors and relates them to chemical or physical properties. A calibration enables a prediction of chemical or physical properties of unknown future samples, factor analysis analyzes the presented (calibration) data only.

Partial Least Squares (PLS)

Partial least-squares regression (PLS) is closely related to PCR and MLR. In fact, PLS can be viewed as a compromise midway between PCR (the section Principal Component Analysis and Principal Component Regression) and MLR.

PLS has two distinct advantages compared to PCR. First, PLS generally provides a more parsimonious model than PCR. The PCR calculates factors in decreasing order of $\mathbf{X}^{cal}$-variance described. Consequently, the first factors calculated, that have the least imbedded errors, are not necessarily most useful for calibration. On the other hand, the first few PLS factors are generally most correlated to concentration. As a result, PLS achieves comparable calibration accuracy with fewer loading vectors in the calibration model. This further results in improved calibration precision because the first factors are less prone to imbedded errors than are lower variance factors.

Second, the PLS algorithm is often faster to implement and optimize for a given application than is the PCR algorithm. The PLS calculates the factors one at a time. Hence, only the loading vectors needed for calibration are determined. The PCR, employing the singular value decomposition, calculates all possible loading vectors for $\mathbf{X}^{cal}$ prior to regression. For large data sets that require relatively few factors for calibration, PLS can be significantly faster than PCR.

Model Selection

The most difficult part of PCR and PLS is to determine the dimension R of the calibration model. In most applications the singular values drop to very small values belonging to nonsignificant PCs with a more or less pronounced step. Plotting these singular values can help to get a first guess at least. But this works only in limited cases and often a more sophisticated method is needed.

Usually cross-validation is applied that excludes one of the calibration samples for determining the number R of significant PCs. This is done by including iteratively more PCs and estimating the values of the response variables. If a sufficient quantity of calibration samples is available, the best method for selecting and validating a model is to divide the calibration set into three subsets. One set is employed to construct all of the models to be considered. The second set is employed to choose the best model in terms of accuracy and precision. The third set is employed to estimate the performance of the chosen model on future data. There are three statistics often employed

for comparing the performances of multivariate calibration models: root-mean-squared error of calibration (RMSEC), root-mean-squared error of cross-validation (RMSECV), and root-mean-squared error of prediction (RMSEP). All three method are based on the calculated root mean squared error (RMSE)

$$\text{RMSE} = \left(\sum_{k=1}^{K} (y_k^{cal} - \hat{y}_k^{cal})^2 / K \right)^{1/2} \quad (2)$$

where RMSEC, RMSECV, and RMSEP differ in the determination of $\hat{y}^{cal}$. The best estimate of future performance of a calibration model is the RMSEP. Estimates $\hat{y}^{cal}$ in the RMSEP are determined by applying the calibration model to a subset of data that was not employed in determining the model parameters. The RMSEP may be calculated for a 'validation set' in order to determine the optimal number of factors in a model or to a 'test set' in order to test the performance of the optimal model on future data. If an external subset of data is not available to optimize the calibration model, the RMSEP can be estimated by the RMSECV. The concentration estimates of equation 2 are determined in cross-validation. RMSEC is a measure of how well the calibration model fits the calibration set. This is potentially the least informative of the three statistics. The RMSEC is an extremely optimistic estimation of the model performance. In the limit, if every factor were included in the calibration model, the RMSEC would be zero.

Target Factor Analysis

The principal components (PCs) derived from PCA do not necessarily describe single, physically meaningful, effects. That is, while a set of data may consist of the NIR spectra of hydrocarbon mixtures, the PCs of the data set are not constrained to be NIR spectra of the constituent hydrocarbons. However, the multivariate space defined by the principal components is the same as the multivariate space defined by the pure (true) spectra of the chemical constituents of the data set plus any other forms of systematic variance. The difference is the basis used for representation: The PCs are rotated versions of the pure component spectra. Target factor analysis (TFA) is a method of testing whether the spectrum of a hypothesized chemical constituent, as defined by an assumed or recorded spectrum, lies in the PC space of the model.

Locally Weighted Regression

The global linear models calculated by PCR or PLS are not always the best strategy for calibration. Global models span the variance of all the samples in the calibration set. If the data are nonlinear, then the linear PCR and PLS methods do not efficiently model the data. One option is to use nonlinear calibration methods employing a global, nonlinear model. The second option is to employ linear calibration methods on small subsets of the data. The locally weighted regression (LWR) philosophy assumes that the data can be efficiently modeled over a short span with linear methods.

Nonlinear Methods

There are numerous nonlinear, multivariate calibration methods described in the chemometric literature. These methods can be divided into two classes. Alternating Conditional Expectations (ACE) and Projection Pursuit (PP) seek to transform the nonlinear data such that a linear calibration model is appropriate. Similarly, Global Linearizing Transformations (GLT) is employed to optimally linearize data prior to factor analysis by PCA. On the other hand, nonlinear-PLS (NPLS), Multivariate Adaptive Regression Splines (MARS), and Artificial Neural Networks (ANN) determine nonlinear global models that span the entire range of samples.

No single nonlinear method has demonstrated systematic superiority to the other methods. The safe conclusion is that calibration method superiority is application dependent. When the underlying type of nonlinearity implicit in the calibration method matches the latent nonlinearity in the data, the method will optimally model the data.

Nonlinear methods are much more prone to "overfitting" the calibration model than linear approaches. Overfitting occurs when the calibration begins to employ random variance (instrumental errors) for determining calibration parameters. The flexibility of the nonlinear models and the relatively large number of parameters that need to be estimated are the primary cause for this phenomenon. Consequently, the more complicated the model, the more prone the method is to overfitting (ie, ANN vs. PCR).

Multivariate Curve Resolution (MCR)

Where TFA (Target Factor Analysis) allows the analysis to test for the presence of a hypothesized constituent, TFA is limited in the ability to estimate the spectral profile of any constituents in the data set. This is due to the fact that TFA requires that the spectral profile of the target is available for target testing. If the profile is unavailable, TFA cannot be performed. On the other hand, multivariate curve resolution (MCR) methods allow for the estimation for both the hypothesized and unknown constituents in the data matrix, ie, spectral or chromatographic profile of the separated constituent as well as concentration profiles. Usually MCR techniques are applied in spectroscopy and chromatography.

Outlier Detection

Two important statistics for identifying outliers in the calibration set containing K samples are the "sample leverage" and the "studentized residuals". A plot of leverage versus studentized residuals makes a powerful tool for identifying outliers and assigning probable cause. The sample leverage is a measure of the influence, or weight, each sample has in determining the parameters of the calibration model. Samples near the center of the calibration set (average samples) will have a relatively low leverage compared to samples at the extreme edges of the experimental design and outliers.

MULTIWAY ANALYSIS

Introduction

There are six classes of three-way data and four of these classes can be appropriately modeled with the basic trilinear, or PARAFAC (PARAllel FACtor analysis) model.

In general, the number and form of factors are not constrained to be representative of any physical reality. With two-way factor analysis, PCA, this is often referred to as the rotational ambiguity of the factors; there is a continuum of factors that satisfy the PCA model and equivalently describe the data. This is different for three-way analysis. If the following four conditions are given, the factors $\hat{x}$, $\hat{y}$, and $\hat{z}$ of a chemical component are accurate and unique estimates of the true underling factors $\mathbf{x}$, $\mathbf{y}$, and $\mathbf{z}$ except for a scaling constant:

1. The true underlying factor in each of the three modes is independent from the state of the other two modes.
2. The true underlying factor in any of the three modes cannot be expressed by linear combinations of the true underlying factors of other components in the same mode.
3. Linear additivity of instrumental responses among the species present is given.
4. The proper number of factors N is chosen for the model.

Multiway Curve Resolution—PARAFAC/CANDECOMP

In PARAFAC and CANDECOMP, the two base algorithms are practically identical. The PARAFAC uses an alternating least squares (ALS) based algorithm for multivariate curve resolution applied to three-way data sets.

Two popular convergence criteria for the PARAFAC algorithm are based on changes in the residuals (unmodeled data) between successive iterations and changes in the predicted profiles between successive iterations. In the first case, the algorithm is terminated when the root average of the squared residuals between successive iterations agree to within an absolute or relative tolerance, say 10^{-6}. While such fit based stopping criteria are conceptually easy to visualize, a faster method for determining convergence relies on the correlation between the predicted X-, Y-, and Z-way profiles between successive iterations. When the product of the cosines between successive iterations in the X-, Y-, and Z-modes approach arbitrarily close to 1, say within 10^{-6}, the algorithm is terminated.

Tucker3 Models

The generalization of the PARAFAC model is the Tucker3 model. The PARAFAC model is intrinsically linear model and straightforward application thus assumes linear interactions and behavior of the samples. While many of the systems of interest to chemists contain nonlinearities that violate the assumptions of the models, the PARAFAC model forms an excellent starting point from which many subsidiary methods are constructed to incorporate non-linear behavior into calibration models constructed from three-way data collected with hyphenated methods. The trilinear model is actually a specific case of the Tucker3 model.

Solution Constraints

ALS algorithms are more flexible than rank annihilation based algorithms since constraints can be placed onto the solutions derived from ALS methods. The ALS algorithms implicitly constrain the estimated profiles to lie in the real space. Rank annihilation methods may fit factors with imaginary components to the data. Ideally, constraints are not needed for ALS to achieve accurate, meaningful concentration and spectral profile estimates. However, the presence of slight nonlinear interactions among the true underlying factors, of highly correlated factors, or of low signal to noise in the data will often result in profile estimates that are visually unsatisfying and large quantitative errors are derived from the model. These effects can often be minimized by employing constraints to the solutions that are based on a priori knowledge or assumptions of the data structure, eg, prior knowledge of sample concentrations or spectral profile characteristics.

Perhaps the most common constraint consciously placed on the PARAFAC or Tucker3 models is nonnegativity. A second constraint often applied in three-way calibration of chromatographic data is unimodality. This constraint exploits the knowledge that chromatographic profiles have exactly one maximum. The third common constraint is based on a priori knowledge of the three-way profiles. In this case, the known relative concentrations of the standards or the known spectral profiles of one or more components can be fixed as part of the solution. In the Tucker3 model, it is common to restrict some of the potential interactions between factors when they are known not to exist. Care must be employed when applying fixed values to the solutions, as the scaling of the factors must still be taken into account.

SELECTED TOPICS

Background Spectrum Correction

Background correction methods are often employed in spectroscopic applications to remove broad features from the data set. These features hinder calibration as a large source of variance compared to the analyte or as a seemingly random source of variance that consumes many factors in the model. Examples include fluorescence background in Raman spectroscopy and scattering backgrounds in near-ir reflectance spectroscopy.

Simple efforts at background correction include derivatives, polynomial curve fitting, and Fourier Transform (FT) filtering. Derivatives remove the portion of a background that can be modeled by a low order polynomial.

Multiplicative scatter correction (MSC) was developed to reduce the effect of scattered light on diffuse reflection and transmission NIR spectra. This method has also shown utility as a means of removing varying background spectra with nonscattering origins. Consequently, MSC sometimes appears as multiplicative signal correction.

Scattering theory states that scattering should have a multiplicative effect on reflection (and transmittance) spectra. That is, the observed spectra will contain a broad, changing background from differential scattering at each wavelength. Assuming a multiplicative model for the scattering, the scattering profile in a spectrum can be deduced from a plot of the spectrum of a standard scatterer versus a given spectrum at each wavelength. And ideal 'standard' would have no NIR absorbance (or transmittance) features; however, the mean spectrum from a collection of similar samples will suffice.

A different approach explicitly including drifts into the calibration was proposed to artificially extend the set of PCs with the so-called pseudo PCs. Advantage is taken from the fact that background drifts are usually very broad compared to the more localized absorption features. These pseudo-PCs have been defined to be polynomials up to a user selectable order, however, other linear independent functions could be used, too. It was shown that this combined set of PCs and pseudo PCs is able to determine considerably improved concentration results from highly drift affected spectra compared to a conventional PCR.

An alternative to pseudo-PCs utilizes a similar idea: Polynomials are fitted to the regular PCs and subtracted from them. In this case drift effects, which can be modeled by polynomials, are orthogonal to the PCs. This is due to the fact the "corrected" PCs, ie, original PCs minus fit polynomials, are the residuals of these fits. Hence, polynomial like drifts up to the considered order are orthogonal to the corrected PCs and cannot influence the concentration results. The pseudo-PCs method extracts additional information, ie, an estimate of the drift spectrum.

Instrument Standardization

One practical concern with multivariate calibration and prediction is the transport and stability of the calibration models. Ideally, a calibration model can be constructed in the laboratory on a bench-top instrument, then the model can be applied to many similar instruments in the field. Also, once a model is successfully transferred to the field, it will be robust to changes in instrumental sensitivity and alignment. Of course, the goal of a universal transferable and robust instrument–model has not been achieved. Seemingly identical spectrometers have slight wavelength resolution, and sensitivity differences that can prohibit reliable distribution of the calibration model among numerous instruments. Also, time-dependent instrumental drift eventually can render the calibration model obsolete for whichever instrument the model was constructed.

Individual calibration of each instrument is not an acceptable solution to the problem of model distribution. Calibration may be an expensive, time-consuming task when many calibration samples are needed, the calibration samples are not readily transportable, or the instrument is not easily accessible in the process stream. Concurrently, it is also unacceptable to repeat an entire calibration procedure whenever there are minor changes in the instrumental character.

Instrumental standardization strives to solve the problems derived from instrumental differences when constructing one calibration model for multiple instruments. The instrumental standardization philosophy is to construct the best model possible on one instrument then to build a second model that will transform the spectra from other instruments to appear as if they were recorded on the first instrument. Usually, this transfer function can be reliably calculated with less effort.

One standardization method popular in the literature is Piecewise Direct Standardization (PDS). A more useful method of standardization would not require transfer samples to be analyzed. There have been two approaches to this problem. When it can be safely assumed that the only spectral shifts (ie, wavelength or retention time) occur a PCA based method of standardization may be employed. A more general method based loosely on MSC has also demonstrated success when there are relatively minor performance differences between the original and second instruments.

Optical Computation

Most spectrometer concepts include moving parts like interferometers or scanning gratings. Such moving parts, however, limit the ruggedness of a field analyzer and the time resolution of the concentration runs. The strong point of such spectrometers combined with chemometric software packages is their versatility. For many applications this is not needed, though. In process analytics, eg, a measurement device is usually applied to one very specific task not needing versatility at all. Mechanical stability and good time resolution is of greater importance. In order to overcome both mentioned drawbacks it was proposed to design so-called multivariate optical elements (MOE). MOE are specially designed interference filter in a beam splitter arrangement.

Artificial Neural Networks Combined with Variable Selection

The measurement technique surface plasmon resonance (spr) is sensitive for analyzing refractive indexes of liquids or vapors. Since the matrix, water, eg, and a dissolved analyte have different refractive indices the refractive index of a sample is concentrations dependent. A change of the samples' refractive index is measured by a highly nonlinear wavelength shift of the plasmon absorbance. However, since only one property of a sample, ie, the refractive index, is measured, binary or ternary mixtures cannot be investigated without experimental adjustments. A polymer coating of the spr sensor head was proposed resulting in different, time-dependent enrichment or desorption processes depending on the molecule size. That means different analytes cause a time- and analyte-dependent change of the spr spectra. Usually, one wants a high time resolution, ie, a large number of variables, in order to capture fast and similar responses and not to lose information. However, there are several disadvantages of using a lot of information like hiding meaningful variables by irrelevant variables or overfitting. Furthermore, danger to change the correlation is increased with the number of variables and many

variables mean increased computation time for the neural net training.

Full-connected neural networks employing a large number of variables are prone to overfitting. Hence, so-called growing neural nets result in sparse, nonuniform structures optimized to a specific problem. However, the outcome of this procedure is still dependent on the way the calibration data are split into training and monitoring data. To overcome this ambiguity two strategies have been proposed: (1) To grow neural networks on a rather large number of different training–monitoring sets in parallel; (2) a certain training–monitoring set is defined and a small number of nets are grown with different initial weights. It was found that the procedure (2) resulted in better generalizations. Application of such grown networks to binary mixtures resulted in convincing concentration prediction.

As an alternative for variable selection, a genetic algorithm for selecting the optimum subset for neural networks based on the procedure (1) discussed above has been proposed. A genetic algorithm is applied to a rather large number of different training–monitoring sets in parallel resulting in a set of neural nets. Again, the variables are ranked in decreasing importance and added in the second step one after the other to a final network until the predictability is not improved anymore. The prediction of concentrations could be considerably improved using five selected variables compared to using all 50.

I. T. Jolliffe, *Principal Component Analysis*, 2nd ed., Springer-Verlag: New York, 2002.

C. L. Lawson and R. J. Hanson, *Solving Least Squares Problems*, Prentice-Hall, Englewood, Cliffs, New York, 1974.

E. Malinowski, *Factor Analysis in Chemistry*, 3rd ed., John Wiley & Sons, Inc., New York, 2002.

H. Martens and T. Næs, *Multivariate Calibration*, 2nd ed., John Wiley & Sons, Inc., New York, 1991.

FRANK VOGT
KARL BOOKSH
Arizona State University

CHIRAL SEPARATIONS

Chiral separations are concerned with separating molecules that can exist as nonsuperimposable mirror images. Examples of these types of molecules, called *enantiomers* or *optical isomers*, are illustrated in Figure 1. Although chirality is often associated with compounds containing a tetrahedral carbon with four different substituents, other atoms, such as phosphorus or sulfur, may also be chiral. In addition, molecules containing a center of asymmetry, such as hexahelicene, tetrasubstituted adamantanes, and substituted allenes or molecules with hindered rotation, such as some 2,2′ disubstituted binaphthyls, may also be chiral. Compounds exhibiting a center of asymmetry are called *atropisomers*. An extensive review of stereochemistry may be found under PHARMACEUTICALS, CHIRAL.

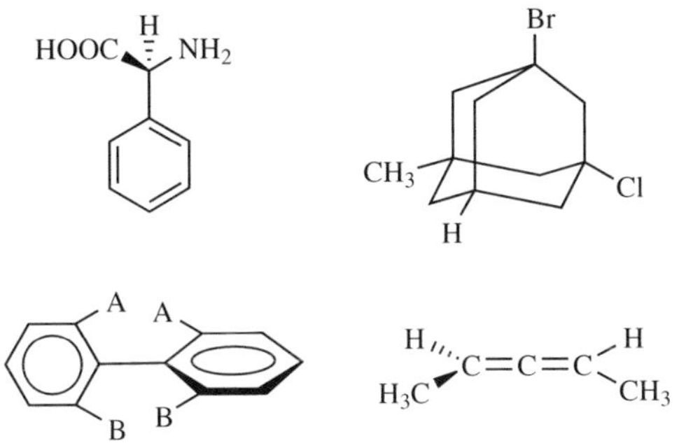

Figure 1. Examples of chiral molecules.

Although scientists have known since the time of Louis Pasteur that optical isomers can behave differently in a chiral environment (eg, in the presence of polarized light), it has only been since about 1980 that there has been a growing awareness of the implications arising from the fact that many drugs are chiral and that living systems constitute chiral environments. Hence, the optical isomers of chiral drugs may exhibit different bioactivities and/or biotoxicities.

In the case of enantiomerically pure chiral drugs, the possibility of racemization or inversion either *in vivo* or during storage cannot be ruled out. Ibuprofen is an example of a chiral drug which undergoes rapid inversion *in vivo*. In addition, there are several examples of achiral (or *prochiral*) drugs being biotransformed into chiral entities. In some cases, the enantiomeric ratios produced by laboratory animals may differ from that produced in humans. This raises the question of the suitability of laboratory animals as appropriate test models for a certain drug.

For those drugs that are administered as the racemate, each enantiomer needs to be monitored separately yet simultaneously, since metabolism, excretion or clearance may be radically different for the two enantiomers. Further complicating drug profiles for chiral drugs is that often the pharmacodynamics and pharmacokinetics of the racemic drug is not just the sum of the profiles of the individual enantiomers.

Although a great deal of the work currently being done in chiral separations is related to pharmaceuticals, the agricultural and the food and beverage industries are affected as well. For instance, several chiral pesticides are used commercially. It is possible that the enantiomers may differ in their persistence in the environment and their effectiveness against specific pests. In the food and beverage industry, many of the constituents that confer flavor or aroma in foods and beverages are chiral. For instance, the configuration of the 4-alkyl-substituted γ-lactones responsible for much of the flavor in fruits is almost exclusively R. Often, the two enantiomers have very different aromas or flavors. The presence of any of the "unnatural" enantiomer may confer an "off-flavor" to the substance and may be indicative of racemization under adverse storage conditions, adulteration, or formulation from nonnatural sources.

The growing awareness of the implications of chirality to the pharmaceutical industry has spurred tremendous effort toward stereoselective synthetic strategies and

the development of new chiral catalysts. However, the enantiomeric purity of these substances or their chiral precursors needs to be determined. Also, there are many chiral compounds for which no stereospecific synthetic pathways have been devised. Thus, there is a tremendous need not only for analytical scale (<5–10 mg), but bulk-scale chiral separations as well.

Whether analyzing drugs or synthetic precursors for enantiomeric purity, monitoring biological or environmental samples for chiral discrimination or trying to enantioresolve kilogram quantities of a racemic drug, there are a variety of reasons for performing chiral separations. The purpose of the separation dictates, to some extent, the method employed.

Traditionally, chiral separations have been considered among the most difficult of all separations.

A variety of strategies have been devised to obtain them. Although the focus of this article is on chromatographically based chiral separations, other methods include crystallization and stereospecific enzymatic-catalyzed synthesis or degradation. In crystallization methods, racemic chiral ions are typically resolved by the addition of an optically pure counterion, thus, forming diastereomeric complexes.

Enzymatically based methods depend upon the stereospecificity of an enzyme-catalyzed reaction, such as lipase-catalyzed esterification, to degrade enantioselectively the unwanted enantiomer or to produce the desired enantiomer. Because only one enantiomer undergoes the reaction, the subsequent separation is reduced to separating two different species. One disadvantage of enzymatically based methods is that only one enantiomer is obtained and there is usually no analogous method for producing the opposite enantiomer.

An alternative method of creating a chiral environmental is to derivatize a chiral analyte with an optically pure reagent, thus, producing diastereomers. The resultant diastereomers, containing more than one chiral center, have slightly different melting and boiling points and can often be separated using conventional methods. A number of chiral derivatizing agents, as well as the types of compounds for which they are useful, have been developed and are listed in Table 1. Limitations of this approach include lack of suitable functionality in the analyte that can be derivatized with an appropriate enantiomerically pure derivatizing agent, unavailability of a suitable derivatizing agent of sufficiently high or at least known optical purity, difficulty of removing the derivatizing group after the desired separation has been accomplished, enantiodiscrimination during derivatization, potential racemization either during derivatization or removal or the chiral derivatizing group (which is not always possible), and the additional validation required to confirm that the enantiomeric ratio of the final product corresponds to the original enantiomeric ratio.

USE OF CHIRAL ADDITIVES

Another method for creating a chiral environment is to add an optically pure chiral selector to a bulk liquid phase. Chiral additives have several advantages over

Table 1. Analyte Functional Groups and Chiral Derivatizing Reagents

Analyte functional group	Derivatizing agent	Product	Examples of derivatizing agents
carboxylic acid (acid or base catalyzed)	alcohol	ester	(−)-menthol
	amine	amide	1-phenylethylamine
			l-(l-naphthyl)ethylamine
amine (1°)	aldehyde	isoindole	o-phthaldialdehyde–2-mercaptoethanol
amine (1° and 2°)	anhydrides	amide	γ-butyloxycarbonyl-L-leucine anhydride
			O,O-dibenzoyltartaric anhydride
	acyl halides	amide	(R)-(−)-methylmandelic acid chloride
			α-methoxy-α-trifluoromethylphenylacetyl chloride
	isocyanates	urea	α-methylbenzyl isocyanate
			1-(1-naphthyl)ethyl isocyanate
	isothiocyanate	thiourea	2,3,4,6-tetra-O-acetyl-β-D-glucopyranosyl isothiocyanate
			α-methylbenzyl isothiocyanate
(1°, 2°; can N-dealkylate 3°)	chloroformates	carbamate	(−)-menthyl chloroformate
			(+)-1-(9-fluorenyl)ethylchloroformate
alcohols	acyl halides	ester	(−)-menthoxy acid chloride
			(S)-O-propionylmandelyl chloride
	anhydrides	ester	(S,S)-tartaric anhydride
	chloroformate	carbonate	(−)-menthyl chloroformate
	isocyanate	carbamate	α-methylbenzyl isocyanate

chiral stationary phases and continue to be the predominant mode for chiral separations by tlc and capillary electrophoresis (ce). First of all, the chiral selector added to a bulk liquid phase can be readily changed. The use of chiral additives allows chiral separations to be done using less expensive, conventional stationary phases. A wider variety of chiral selectors are available to be used as chiral additives than are available as chiral stationary phases thus, providing the analyst with considerable flexibility. Finally, the use of chiral additives may provide valuable insight into the chromatographic conditions and/or likelihood of success with a potential chiral stationary-phase chiral selector. This is particularly important for the development of new chiral stationary phases because of the difficulty and cost involved.

Chiral additives, however, do pose some unique problems. Many chiral agents are expensive or are not commercially available, and therefore, must be synthesized. The presence of the chiral additive in the bulk liquid phase may also interfere with detection or recovery of the analytes. Finally, the presence of enantiomeric impurity in the chiral additive may add analytical complications.

Thin-Layer Chromatography

Thin-layer chromatography (tlc) offers several advantages for chiral separations and in the development of new chiral stationary phases. Besides being inexpensive, tlc can be used to screen mobile-phase conditions rapidly (ie, organic modifier content, pH, etc), chiral selectors, and analytes. Several different analytes may be run simultaneously on the same plate. Usually, no preequilibration of the mobile phase and stationary phase is required. In addition, only small amounts of mobile phase, and therefore, chiral mobile-phase additive, are required. Another significant advantage is that the analyte can always be unambiguously found on the tlc plate.

Two mechanisms for chiral separations using chiral mobile-phase additives, analogous to models developed for ion-pair chromatography, have been proposed to explain the chiral selectivity obtained using chiral mobile-phase additives. In one model, the chiral mobile-phase additive and the analyte enantiomers form "diastereomeric complexes" in solution. As noted previously, diastereomers may have slightly different physical properties such as mobile phase solubilities or slightly different affinities for the stationary phase. Thus, the chiral separation can be achieved with conventional columns.

An alternative model has been proposed in which the chiral mobile-phase additive is thought to modify the conventional, achiral stationary phase in-situ, thus, dynamically generating a chiral stationary phase. In this case, the enantioseparation is governed by the differences in the association between the enantiomers and the chiral selector in the stationary phase.

Several different types of chiral additives have been used including (1R)-(−)-ammonium-10-camphorsulfonic acid, cyclodextrins, proteins, and various amino acid derivatives such as N-benzoxycarbonyl-glycyl-L-proline as well as macrocyclic antibiotics.

Chiral separation validation in tlc may be accomplished by recovering the individual analyte spots from the plate and subjecting them to some type of chiroptical spectroscopy such as circular dichroism or optical rotary dispersion. Alternatively, the plates may be analyzed using a scanning densitometer.

Capillary Electrophoresis

Capillary electrophoresis (ce) or capillary zone electrophoresis (cze), a relatively recent addition to the arsenal of analytical techniques, has also been demonstrated as a powerful chiral separation method. Its high resolution capability and lower sample loading relative to hplc makes it ideal for the separation of minute amounts of components in complex biological mixtures.

In a ce experiment, a thin capillary is filled with a run buffer and a voltage is applied across the capillary. The underlying impetus for separations in ce is, in general, derived from the fact that charged species migrate in response to an applied electric field proportionately to their charge and inversely proportionately to their size.

Chiral separations by ce have been performed almost exclusively using chiral additives to the run buffer. The advantages of this approach are identical to the advantages mentioned previously with regard to using chiral mobile-phase additives in tlc. Many of the chiral selectors used successfully as mobile-phase additives in tlc and as immobilized ligands in hplc have been used successfully in ce including proteins, native and functionalized cyclodextrins, various carbohydrates, assorted functionalized amino acids, chiralion pairing agents, and macrocyclic antibiotics.

Although chiral ce is most commonly performed using aqueous buffers, there has been some work using organic solvents such as methanol, formamide, *N*-methylformamide or *N,N*-dimethylformamide with chiral additives such as quinine or cyclodextrins. Nonaqueous ce requires that the background electrolyte be prepared using organic acids (eg, citric acid or acetic acid) and organic bases (eg, tetraalkylammonium halides or tris(hydroxymethyl)aminomethane).

CHIRAL STATIONARY PHASES

Most chiral chromatographic separations are accomplished using chromatographic stationary phases that incorporate a chiral selector. The chiral separation mechanisms are generally thought to involve the formation of transient diastereomeric complexes between the enantiomers and the stationary phase chiral ligand. Differences in the stabilities of these complexes account for the differences in the retention observed for the two enantiomers. Often, the use of a chiral stationary phase allows for the direct separation of the enantiomers without the need for derivatization. One advantage offered by the use of chiral stationary phases is that the chiral selector need not be enantiomerically pure, only enriched. In addition, for chiral stationary phases having a well understood chiral recognition mechanism, assignment of configuration (eg, *R* or *S*) may be possible even in the absence of optically pure standards. However, chiral stationary phases have some limitations. The specificity required for chiral discrimination limits the broad applicability of most chiral stationary phases; thus there is no "universal" chiral stationary phase. The cost of most chiral columns are typically much higher ($\sim 3 \times$) than for conventional columns. In contrast to conventional chromatographic columns, chiral stationary phases are generally not as robust, require more careful handling than conventional columns and usually, once column performance has begun to deteriorate, cannot be returned to their original performance levels. In many cases, chromatographic column choice or mobile phase optimization for chiral stationary phases is not as straightforward as with conventional stationary phases. For many of the chiral stationary phases, adequate chiral recognition models, used to guide selection of the appropriate column for a given separation, have yet to be developed. Column selection, therefore, is often reduced to identifying

structurally similar analytes for which chiral resolution methods have been reported in the scientific literature or chromatographic supply catalogues and adapting a reported method for the chiral pair to be resolved.

An additional complication, sometimes arising with the use of chiral stationary phases, may occur when the analytes either exist as *conformers* or can undergo inversion uring the chromatographic analysis.

Thin-Layer Chromatography

Chiral stationary phases in tlc have been primarily limited to phases based on normal or microcrystalline cellulose, triacetylcellulose sorbents or silica-based sorbents that have been chemically modified or physically coated to incorporate chiral selectors such as amino acids or macrocyclic antibiotics into the stationary phase. The cost of many chiral selectors, as well as the accessibility and success of chiral additives, may have inhibited widespread commercialization.

Of the silica-based materials, only the ligand-exchange phases are commercially available (Chiralplate, tlc plates are available through Alltech Associates, Inc.) Supelco, Inc., the Aldrich Chemical Company, and Bodman Industries are all based on ligand exchange. Typically in the case of the ligand-exchange type tlc plates, the ligand-exchange selector is comprised of an amino acid residue to which a long hydrocarbon chain has been attached (eg, (2*S*,4*R*, 2′*RS*)4-hydroxy-1-(2-hydroxydodecyl)proline). The hydrocarbon chain of the functionalized amino acid is either chemically bonded to the substrate or intercalates in between the chains of a reversed phase-stationary phase thus immobilizing the chiral selector. The bidentate amino acid chiral selector is thought to reside close to the surface of the stationary phase and participates as a ligand in the formation of a bi-ligand complex with a divalent metal ion (eg, Cu^{2+}) and the chiral bidentate analyte. Analytes enantioresolvable using ligand exchange are usually restricted to 1,2-diols, α-amino acids, α-amino alcohols, and α-hydroxyacids. Again, differences in the stabilities of the diastereomeric complexes thus formed give rise to the chiral separation.

High Performance Liquid Chromatography

The last decade has seen the commercialization of a large number of different types of chiral stationary phases including the cyclodextrin phases, the chirobiotic phases, the π-π interaction phases, the protein phases as well as the cellulosic and amylosic phases and chiral crown ether phases. Currently, there are over 50 different chiral columns that are commercially available for hplc. Table 2 briefly summarizes the types of columns available as well as typical applications and mobile-phase conditions. Each of these chiral stationary phases are very successful at separating large numbers of enantiomers, which in many cases, are unresolvable using any of the other chiral stationary phases. Unfortunately, despite the large number and variety of chiral stationary phases currently available, there remains a large number of enantiomeric compounds that are unresolvable by any of the existing chiral, stationary phases. In addition, incomplete

Table 2. Classes of Hplc Chiral Stationary Phases

Column chiral selector	Typical mobile phase conditions	Typical analyte features required
pirkle	nonpolar organic; 2-propanol-hexane	π-acid or π-basic moieties for charge transfer complex; hydrogen-bonding or dipole stacking capability near chiral center
protein	phosphate buffers	aromatic near chiral center; organic acids or bases; cationic drugs
cyclodextrin	aqueous buffers; polar organic	good "fit" between chiral cavity or chiral mouth of cyclodextrin and hydrophobic moiety; hydrogen-bonding capability near chiral center
ligand exchange	aqueous buffers	α-hydroxy or α-amino acids near chiral center; can do nonaromatic
chiral crown ether	0.01 N perchloric acid	primary amines near chiral center; can do nonaromatic
macrocyclic antibiotics	aqueous buffers, nonpolar and polar organic	amines, amides, acids, esters; aromatic; hydrophobic moiety
cellulosic and amylosic	nonpolar organic	aromatic

understanding of the chiral recognition mechanisms of many of these chiral stationary phases limits the realization of the full potential of the existing chiral stationary phases and hampers development of new chiral stationary phases.

LIGAND-EXCHANGE PHASES

Among the earliest reports of chiral separations by liquid chromatography were based on work done by Davankov using ligand exchange. These types of columns are available from Phenomenex, J. T. Baker, and Regis Technologies, Inc. Although almost any amino acid can form the basis for the chiral selector, proline and hydroxyproline exhibit the most widespread utility. Also, although other metals can be used, copper(II) is usually the metal of choice and is added to the aqueous buffer mobile phase.

The dependence of chiral recognition on the formation of the diastereomeric complex imposes constraints on the proximity of the metal binding sites, usually either an hydroxy or an amine α to a carboxylic acid, in the analyte. Principal advantages of this technique include the ability to: assign configuration in the absence of standards; enantioresolve nonaromatic analytes; use aqueous mobile phases; acquire a stationary phase with the opposite enantioselectivity; and predict the likelihood of successful chiral resolution for a given analyte based on a well-understood chiral recognition mechanism.

PIRKLE PHASES

Of all of the commercially available chiral stationary phases for liquid chromatography, the chiral recognition mechanism for the "Pirkle" phases are among the best understood. Chiral recognition on Pirkle phases is thought to depend upon complimentary interactions between the analyte and the selector. These interactions may be π-π, steric, hydrogen-bonding, or dipole—dipole interactions and contribute to the overall stability of the diastereomeric association complexes that form between the individual enantiomers and the chiral selector in the stationary phase.

Nonpolar organic mobile phases, such as hexane with ethanol or 2-propanol as typical polar modifiers, are most commonly used with these types of phases. Under these conditions, retention seems to follow normal phase-type behavior (eg, increased mobile phase polarity produces decreased retention). The normal mobile phase components only weakly interact with the stationary phase and are easily displaced by the chiral analytes thereby promoting enantiospecific interactions. Some of the Pirkle-types of phases have also been used, to a lesser extent, in the reversed phase mode.

Reciprocity, an important concept introduced by Pirkle, exploited the notion that analytes that were well resolved using a particular chiral selector would likely be good candidates for chiral selectors to enantioresolve analytes similar to the original chiral selector. This insight spawned a second generation of Pirkle phases based on *N*-(2-naphthyl)-α-amino acids. These phases were very successful at enantio-resolving analytes containing a 3,5-dinitrobenzoyl group, such as 3,5-dinitrophenyl carbamates, and ureas of chiral alcohols and amines. These columns are available through a variety of sources including Phenomenex, Regis Technologies, Inc., J. T. Baker, Inc., and Supelco, Inc.

The structure of the Whelk-O-1 phase, the most recent addition to this type of chiral stationary phase, is illustrated in Figure 2. The presence of both π-acid and π-base features, as well as the inherent rigidity of the chiral selector, confers greater versatility than any of the previous Pirkle-type phases, imposing fewer constraints on both analyte structural features required for successful enantioresolution and mobile phase conditions. Indeed, this chiral stationary phase has demonstrated considerable chiral selectivity for naproxen, warfarin, and its *p*-chloro analogue under nonaqueous reversed-phase conditions and reversed-phase conditions. An additional advantageous feature of this phase is its availability

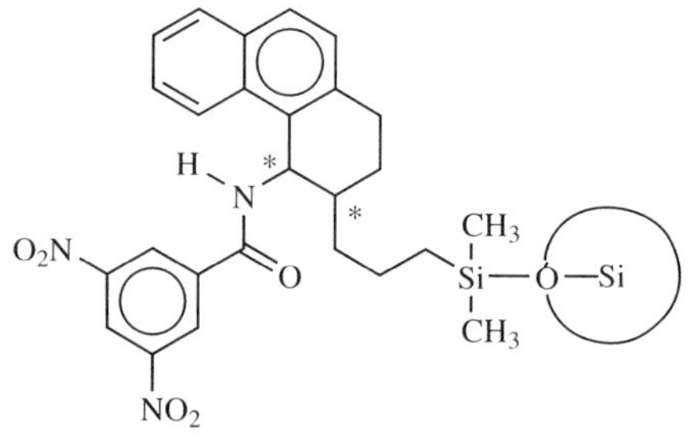

Figure 2. The structure of the chiral selector in the Whelk-O-1 chiral statiionary phase.

with either the (*R,R*) or (*S,S*) configuration, thus, permitting the enantiomeric elution order to be readily changed. The small size of the chiral selector also promotes fairly high bonded ligand densities in the stationary phase, which coupled with the high enantioselectivities often achieved with these phases, facilitates their use for preparative-scale separations.

CYCLODEXTRIN PHASES

Cyclodextrins are macrocyclic compounds comprised of D-glucose bonded through 1,4-α-linkages and produced enzymatically from starch. The greek letter which proceeds the name indicates the number of glucose units incorporated in the CD (eg, α = 6, β = 7, γ = 8, etc). Cyclodextrins are toroidal shaped molecules with a relatively hydrophobic internal cavity (Fig. 3).

Among the most successful of the liquid chromatographic reversed-phase chiral stationary phases have been the cyclodextrin-based phases, introduced by Armstrong and commercially available through Advanced Separation Technologies, Inc. or Alltech Associates. The most commonly used cyclodextrin in hplc is the β-cyclodextrin. In the bonded phases, the cyclodextrins are thought to be tethered to the silica substrate through one or two spacer ligands. The mechanism thought to be responsible for the chiral selectivity observed with these phases is based upon the formation of an inclusion complex between the hydrophobic moiety of the chiral analyte and the hydrophobic interior of the cyclodextrin cavity. Preferential complexation between one optical isomer and the cyclodextrin through stereospecific interactions with the secondary hydroxyls which line the

mouth of the cyclodextrin cavity results in the enantiomeric separation. Unlike the Pirkle-type phases, enantiospecific interactions between the analyte and the cyclodextrin are not the result of a single, well-defined association, but more of a statistical averaging of all the potential interactions with each interaction weighted by its energy or strength of interaction.

Vast amounts of empirical data suggest that chiral recognition on cyclodextrin phases in the reversed phase mode require the presence of an aromatic moiety that can fit into the cyclodextrin cavity, that there be hydrogen bonding groups in the molecule, and that the hydrophobic and hydrogen-bonding moieties should be in close proximity to the stereogenic center. Chiral recognition seems to be enhanced if the stereogenic center is positioned between two π-systems or incorporated in a ring.

Most of the chiral separations reported to date using the native cyclodextrin-based phases have been accomplished in the reversed-phase mode using aqueous buffers containing small amounts of organic modifiers. However, polar organic mobile phases have gained in popularity recently because of their ease of removal from the sample and reduced tendency to accelerate column degradation relative to the hydroorganic mobile phases. In these cases, because the more nonpolar component of the mobile phase is thought to occupy the cyclodextrin cavity, the analyte is thought to sit atop the mouth of the cyclodextrin much like a "lid".

Limitations with the chiral selectivity of the native cyclodextrins fostered the development of various functionalized cyclodextrin-based chiral stationary phases, including acetylated, sulfated, 2-hydroxypropyl, 3,5-dimethylphenyl-carbamoylated and 1-naphthylethylcaarbamoylated cyclodextrin. Each of the glucose residues contribute three

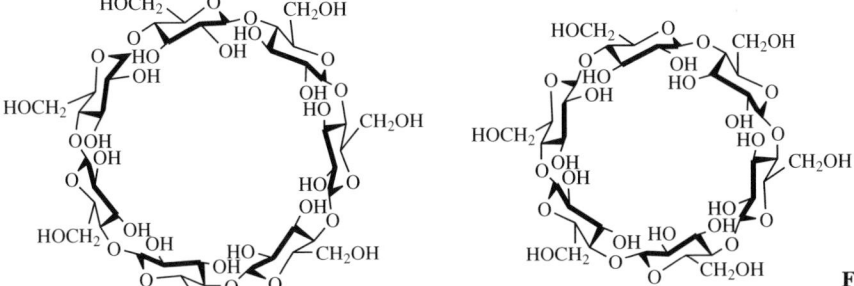

Figure 3. The structure of the three most common cyclodextrins.

hydroxyl groups to which a substituent may be appended; thus, each cyclodextrin contributes multiple sites for derivatization. The substituents of these functionalized cyclodextrins seem to play a variety of roles in enhancing chiral recognition.

CELLULOSIC AND AMYLOSIC PHASES

Cellulose and amylose are comprised of the same glucose subunits as the cyclodextrins. In the case of cellulose, the glucose units are attached through 1,4-β-linkages resulting in a linear polymer. In the case of amylose, the 1,4-α-linkages, as are found in the cyclodextrins, are thought to confer helicity to the polymeric chain.

Cellulosic phases as well as amylosic phases have been used extensively for enantiomeric separations recently. Most of the work in this area has been with various derivatives of the native carbohydrate. The enantioresolving abilities of the derivatized cellulosic and amylosic phases are reported to be very dependent upon the types of substituents on the aromatic moieties that are appended onto the native carbohydrate. Table 3 lists some of the cellulosic and amylosic derivatives that have been used.

PROTEIN-BASED PHASES

Proteins, amino acids bonded through peptide linkages to form macromolecular biopolymers, used as chiral stationary phases for hplc include bovine and human serum albumin, α_1-acid glycoprotein, ovomucoid, avidin, and cellobiohydrolase. The bovine serum albumin column is marketed under the name Resolvosil.

In most cases, the protein is immobilized onto γ-aminopropyl silica and covalently attached using a cross-linking reagent such as N, N'-carbonyldiimidazole. The tertiary structure or three dimensional organization of proteins are thought to be important for their activity and chiral recognition. Therefore, mobile phase conditions that cause protein "denaturation" or loss of tertiary structure must be avoided.

Typically, the mobile phases used with the protein-based chiral stationary phases consist of aqueous phosphate buffers. Often small amounts of organic modifiers, such as methanol, ethanol, propanol, or acetonitrile, are added to reduce hydrophobic interactions with the analyte and to improve enantioselectivity. In some cases, dra-

Table 3. Carbohydrate Derivatives Used as Hplc Chiral Stationary Phases

Cellulosic	Amylosic
triacetate	
tribenzoate	
tribenzylether	
tricinnamate	
triphenylcarbamate	triphenylcarbamate
tris-3,5-dichlorophenylcarbamate	
tris-3,5-dimethylphenylcarbamate	tris-3,5-dimethylphenylcarbamate
tris-1-phenylethylcarbamate	tris-1-phenylethylcarbamate

matic changes in chiral recognition occur when small amounts of organic modifiers, such as N,N-dimethyloctylamine or octanoic acid are added to the mobile phase. As in the case of the cyclodextrin and amylosic and cellulosic phases, the chiral recognition mechanism for these protein-based phases is not well understood. Optimization of chromatographic conditions and selection of analytes that can be successfully resolved on these phases is usually done empirically. In addition, the large molecular weight of these biopolymers dictates that the amount of chiral selector that can be immobilized on the column packing material is very small.

An interesting application of the protein-based phases is various protein binding and displacement experiments which can be done fairly routinely. For instance, differences in the enantioselectivity, toward a particular drug, of a column derived from human serum albumin and a column derived from some other animal serum albumin might be indicative that a particular species might not be a good animal model during drug development, thus, obviating the need for animal testing. Chiral separations on protein-based phases may also provide useful information on drug interactions.

CHIROBIOTIC PHASES

The chirobiotic chiral stationary phases are based on macrocyclic antibiotics such as vancomycin and teicoplanin. These chiral selectors, originally used as chiral additives in capillary zone electrophoresis, incorporate aromatic and carbohydrate, as well as peptide and ionizable moieties. The presence of aromatic groups, allowing for π-π interactions, and the macrocyclic rings, offering potential inclusion complexation, give these phases some of the advantages of the protein-based phases (eg, peptide and hydrogen bonding sites) and the carbohydrate-based phases but with greater sample capacity and greater mobile phase flexibility. Indeed, these phases seem to be truly "multimodal" in that they have demonstrated chiral selectivity in the normal, polar organic, and reversed-phase modes. In addition, the use of such well-defined chiral selectors facilitate method development and optimization.

CHIRAL CROWN ETHER PHASES

Chiral crown ethers based on 18-crown-6 (Fig. 4) can form inclusion complexes with ammonium ions and protonated primary amines. Immobilization of these chiral crown ethers on a chromatographic support provides a chiral stationary phase which can resolve most primary amino acids, amines and amino alcohols. However, the stereogenic center must be in fairly close proximity to the primary amine for successful chiral separation. Significantly, the chiral crown ether phase is unique in that it is one of the few liquid chromatographic chiral stationary phases that does not require the presence of an aromatic ring to achieve chiral separations.

Mobile phases used with this stationary phase are typically 0.01 N perchloric acid with small amounts of methanol or acetonitrile. One significant advantage of these

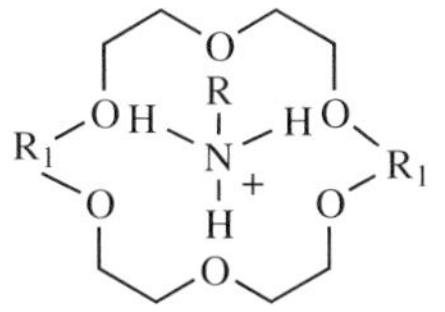

Figure 4. An inclusion complex formed between a protonated primay amine and a chiral crown ether.

phases is that both configurations of the chiral stationary phase are commercially available.

CHIRAL SYNTHETIC POLYMER PHASES

Chiral synthetic polymer phases can be classified into three types. In one type, a polymer matrix is formed in the presence of an optically pure compound to molecularly *imprint* the polymer matrix. Subsequent to the polymerization, the chiral template is removed, leaving the polymer matrix with chiral cavities. The selectivities achieved with these phases are generally excellent, thus, facilitating semipreparative separations. However, the applicability of these chiral stationary phases are generally limited to the analyte upon which the phase is based and a limited number of analogues. In addition, these types of phases generally exhibit poor efficiency in large part because the polymeric matrix contributes to nonsterespecific binding. Advantages of this approach include the ability to prepare reciprocal phases and the predictability of the enantiomeric elution order.

Another type of synthetic polymer-based chiral stationary phase is formed when chiral catalyst are used to initiate the polymerization.

A third type of synthetic polymer-based chiral stationary phase, developed by Blaschke, is produced when a chiral selector is either incorporated within the polymer network or attached as pendant groups onto the polymer matrix. Both are analogous to methods used to produce polymeric chiral stationary phases for gc.

In general, the synthetic polymeric phases seem to have polarities analogous to diol-type phases and a wide range of mobile phase conditions have been used including hexane, various alcohols, acetonitrile, tetrahydrofuran, dichloromethane and their mixtures, as well as aqueous buffers.

CHIRAL SEPARATION VALIDATION FOR HPLC

Chiral separations present special problems for validation. Typically, in the absence of spectroscopic confirmation (eg, mass spectral or infrared data), conventional separations are validated by analyzing "pure" samples under identical chromatographic conditions. Often, two or more chromatographic stationary phases, which are known to interact with the analyte through different retention mechanisms, are used. If the pure sample and the unknown have identical retention times under each set of conditions, the identity of the unknown is assumed to be the same as the pure sample. However, often the chiral separation that is obtained with one type of column may not be achievable

with any other type of chiral stationary phase. In addition, "pure" enantiomers are generally not available.

Most commonly, uv or uv–vis spectroscopy is used as the basis for detection in hplc. When using a chiral stationary phase, confirmation of a chiral separation may be obtained by either monitoring the column effluent at more than one wavelength or by running the sample more than once. Although not absolute proof of a chiral separation, this approach does provide strong supporting evidence.

As in tlc, another method to validate a chiral separation is to collect the individual peaks and subject them to some type of optical spectroscopy, such as, circular dichroism or optical rotary dispersion. Alternatively, a chiroptical spectroscopy can be used as the basis for detection on-line using commercially available optical rotary dispersion or circular dichroism-based detectors. Another method for validating chiral separations by lc is to couple the chromatographic system to a mass spectrometer.

CHIRAL STATIONARY PHASES FOR GAS CHROMATOGRAPHY

Gc chiral stationary phases can be broadly classified into three categories: diamide, cyclodextrin, and metal complex.

Diamide Chiral Separations

The first commercially available chiral column was the Chiralsil-val, which was introduced in 1976 for the separation of amino acid type compounds by gas chromatography. It is based on a polysiloxane polymer containing chiral side chains incorporating L-valine-*t*-butylamide. The polysiloxane backbone improved the thermal stability of these chiral stationary phases relative to the original coated columns and extended the operating temperatures up to 220°C. The column is effective for the separation of perfluoroac-ylated and esterified amino acids, amino alcohols, and some chiral sulfoxides. Another polysiloxane-based chiral stationary phase incorporating L-valine-(R)-α-phenyle-thylamide appended onto hydrolyzed XE-60 was found to be particularly successful at resolving perfluoroacetylated amino alcohol derivatives. Through judicious choice of derivatizing agent, chiral separations were obtained for a wider range of compounds, including amino alcohols, α-hydroxy acids, diols and ketones, than had previously been obtainable using these types of stationary phases.

Metal Complex

Complexation gas chromatography was first introduced by V. Schurig in 1980 and employs transition metals (eg, nickel, cobalt, manganese or rhodium) complexed with chiral terpenoid ketoenolate ligands such as 3-trifluoroacetyl-1R-camphorate (1), 1R-3-pentafluorobenzoylcamphorate or 3-heptafluorobutanoyl-(1R,2S)-pinanone-4-ate. This class of chiral columns is particularly adept at enantioresolving some olefins and oxygen-containing compounds such as ketones, ethers, alcohols, spiroacetals, oxiranes, and esters.

(**1**)

Cyclodextrins

As indicated previously, the native cyclodextrins, which are thermally stable, have been used extensively in liquid chromatographic chiral separations, but their utility in gc applications was hampered because their highly crystallinity and insolubility in most organic solvents made them difficult to formulate into a gc stationary phase. However, some functionalized cyclodextrins form viscous oils suitable for gc stationary-phase coatings and have been used either neat or diluted in a polysiloxane polymer as chiral stationary phases for gc.

CHIRAL SEPARATION VALIDATION FOR GAS CHROMATOGRAPHY

The special problems for validation presented by chiral separations can be even more burdensome for gc because most methods of detection (eg, flame ionization detection or electron capture detection) in gc destroy the sample. Even when nondestructive detection (eg, thermal conductivity) is used, individual peak collection is generally more difficult than in lc or tlc. Thus, off-line chiroptical analysis is not usually an option. Fortunately, gc can be readily coupled to a mass spectrometer and is routinely used to validate a chiral separation.

R. A. Kuzel, S. K. Bhasin, H. G. Oldham, L. A. Damani, J. Murphy, P. Camilleri, and A. J. Hutt, *Chirality* **6**, 607 (1994).

L. Pasteur, *Comptes Rendus de l'Academie des Sciences* **26**, 535 (1848).

S. A. Tobert, *Clin. Pharmacol. Ther.* **29**, 344 (1981).

W. J. Wechter, D. G. Loughhead, R. J. Reischer, G. J. Van Giessen, and D. G. Kaiser, *Biochem. Biophys. Res. Comm.* **61**, 833 (1974).

APRYLL M. STALCUP
University of Cincinnati

CHLORIC ACID AND CHLORATES

CHLORIC ACID

Chlorates are salts of chloric acid $HClO_3$.

Physical Properties

Aqueous chloric acid is a clear, colorless solution stable when cold up to ca 40 wt%. Upon heating, chlorine, Cl_2, and chlorine dioxide, ClO_2, may evolve. Concentration of chloric acid by evaporation may be carried to >40% un-

der reduced pressure. Decomposition at concentrations in excess of 40% is accompanied by evolution of chlorine and oxygen and the formation of perchloric acid, $HOCl_4$, in proportions approximating those shown in equation 1.

$$8\,HClO_3 \longrightarrow 4\,HClO_4 + 2\,H_2O + 3\,O_2 + 2\,Cl_2 \qquad (1)$$

Impurities such as chloride ion or other reducing agents generate chlorine dioxide when the chloric acid solution is heated. Transition-metal ions do not affect the stability of pure chloric acid at room temperature.

Chemical Properties

Chloric acid is a strong acid and an oxidizing agent. It reacts with metal oxides or hydroxides to form chlorate salts, and it is readily reduced to form chlorine dioxide.

$$ClO_3^- + 2\,H^+ + e^- \longrightarrow ClO_2 + H_2O \qquad (2)$$

Manufacture

Chloric acid is the precursor for generation of chlorine dioxide for pulp bleaching and other applications (see BLEACHING AGENTS), and is formed *in situ* by reaction of sodium chlorate, $NaClO_3$, and a strong acid. Any chloride present in the solution is oxidized to chlorine by the chloric acid.

Emerging technologies for the commercial manufacture of chloric acid fall into three categories: (*1*) generation of high purity chloric acid by thermal decomposition of pure solutions of hypochlorous acid, $HClO$. The chloric acid obtained is free of metal cations and chloride and sulfate anions. (*2*) Generation of chloric acid by passing a solution of sodium chlorate through a cation ion-exchange resin. The resulting $HClO_3$ solution contains some dissolved sodium chlorate as well as the impurities that were present in the initial sodium chlorate solution. (*3*) Hypochlorous acid can be oxidatively electrolyzed to chloric acid. Chloric acid prepared by oxidizing $HOCl$ is both metal- and chloride-ion free. It can be reduced to chlorine dioxide without the formation of solid by-products or chlorine. This reduction can be conducted electrochemically or chemically.

Shipment

Solutions of greater than 10 wt% chloric acid may be shipped using the label, "oxidizing substance, liquid, corrosive, n.o.s.," and using identification number UN3098, packing group II.

Health and Safety Factors

Chloric acid is a strong oxidizing agent and concentrated solutions ignite organic matter on contact. The acid must be stored apart from reducing agents and organic materials. Concentrated solutions are corrosive to the skin. It is a strong irritant by ingestion and inhalation.

Uses

Chloric acid is formed *in situ* by reaction of sodium chlorate and a strong acid during chlorine dioxide production.

Stoichiometric amounts of sodium salts are also formed as a by-product. The use of chlorine dioxide for pulp (qv) bleaching applications is growing and disposal of the by-product solids is a primary environmental concern. Use of chloric acid to generate chlorine dioxide can eliminate this problem. A process for bleaching pulp which employs chloric acid as the oxidizing agent, in the absence of a transition metal catalyst, has been reported.

Chloric acid also has found limited applications as a catalyst for the polymerization of acrylonitrile (qv), C_3H_3N, and in the oxidation of cyclohexanone, $C_6H_{10}O$ (see CYCLOHEXANOL AND CYCLOHEXANONE).

SODIUM AND POTASSIUM CHLORATE

Physical Properties

The physical properties of sodium chlorate and potassium chlorate, $KClO_3$, are summarized in Table 1.

Chemical Properties

On thermal decomposition, both sodium and potassium chlorate salts produce the corresponding perchlorate, salt, and oxygen. Mixtures of potassium chlorate and metal oxide catalysts, especially manganese dioxide, MnO_2, are employed as a laboratory source of oxygen. Mixtures of chlorates and organic materials have been employed as explosives. However, because of extreme shock sensitivity and unpredictability, such mixtures are not classed as permissible explosives in the United States. Chlorates also form flammable and explosive mixtures with phosphorus,

ammonium compounds, some metal compounds, and some metal salts. Chlorates in neutral and alkaline solutions at room temperature do not show oxidizing properties. Concentrated acidic solutions of chlorates are strong oxidants as a result of chloric acid formation and may also liberate chlorine dioxide gas.

Manufacture

Most chlorate is manufactured by the electrolysis of sodium chloride solution in electrochemical cells without diaphragms. Potassium chloride can be electrolyzed for the direct production of potassium chlorate, but because sodium chlorate is so much more soluble, the production of the sodium salt is generally preferred. Potassium chlorate may be obtained from the sodium chlorate by a metathesis reaction with potassium chloride.

The sodium chlorate manufacturing process can be divided into six steps: (1) brine treatment; (2) electrolysis; (3) crystallization and salt recovery; (4) chromium removal; (5) hydrogen purification and collection; and (6) electrical distribution.

The production of sodium chlorate is very energy intensive requiring between 4950–6050 kW·h of electricity per metric ton of sodium chlorate produced. Sodium chlorate manufacturing technology now incorporates a low chloride–chlorate solution manufacture coupled with a chromium removal system, or the use of a crystallizer to produce crystal chlorate as the final project.

Shipping

Sodium chlorate can be shipped either as solid crystals or preblended chlorate–chloride solution.

The crystalline sodium chlorate is usually dried in rotary driers to less than 0.2 wt% moisture content and is loaded into shipping containers or stored in moisture-free bins or silos prior to packaging. For conventional chlorine dioxide generators, sodium chlorate is shipped as a solution.

Health and Safety Factors

Sodium chlorate is harmful if swallowed, inhaled, or absorbed through the skin.

Sodium chlorate and potassium chlorate are human poisons by unspecified routes. They are moderately toxic by ingestion and intraperitoneal routes. They both damage red blood cells when ingested.

Chlorates are strong oxidizing agents. Dry materials, such as cloth, leather, or paper, contaminated with chlorate may be ignited easily by heat or friction. Chlorates should be stored separately from all flammable materials in a cool, dry, fireproof building.

Flammable resistant clothing such as Nomex should be worn when working with chlorates. Clothing splashed with chlorate solution should be removed before it dries. Shoes and gloves should be rubberized. Leather (qv) should not be worn. Goggles, face shields, and dust respirators should be worn when necessary to protect against dust, splashing, or spillage. Workers should bathe before leaving the working area.

Table 1. Physical Properties of Sodium and Potassium Chlorates

Properties	$NaClO_3$	$KClO_3$
molecular weight	106.44	122.55
crystal system	cubic	monoclinic
mp, °C	248–260	356–368
dec pt, °C	265	400
density, g/mL	2.487^a	2.338^b
affinity toward water	hygroscopic	nonhygroscopic
enthalpy of fusion, ΔH_{fus}, kJ/mol^c	21.3	
molar heat capacity, J/(mol·K)^c	$100^{b,d}$	99.8^b
standard enthalpy of formation, kJ/mol^c		
crystals	−365.8	−391
ideal solution of unit activity	−344.1	
standard entropy J/(mol·K)^c crystals	123.4	143
ideal solution of unit activity	22.3	
enthalpy of dissolution,^{e} kJ/mol^c	21.6	40.9

^{a}At 25°C.
^{b}At 20°C.
^{c}To convert J to cal, divide by 4.184.
^{d}From 298 to 533 K.
^{e}1 mol of chlorate per 200 mol H_2O at 25°C.

Only water is effective in the event of a spill or fire. Water cools and dilutes the sodium chlorate. Carbon dioxide, Halon, dry chemical or dry powder types of fire extinguishers are ineffective.

Uses

The primary (99%) use of sodium chlorate is in the production of chlorine dioxide for bleaching in the pulp (qv) and paper industry (qv).

The second most important use of sodium chlorate was as an intermediate in the production of other chlorates and of perchlorates.

The agricultural use of sodium chlorate is as a herbicide, as a defoliant for cotton (qv). Magnesium chlorate is used as a desiccant for soybeans to remove the leaves prior to mechanical picking (see DESICCANTS).

Sodium chlorate is used in uranium mixing. Minor uses of sodium chlorate include the preparation of certain dyes and the processing of textiles (qv) and furs.

Potassium chlorate is used mainly in the manufacture of matches (qv) and pharmaceutical preparation.

OTHER CHLORATES

Two other chlorates are barium chlorate monohydrate, which is used in pyrotechnics, and lithium chlorate, which has limited use in pyrotechnics.

J. A. Dean, ed., *Langes Handbook of Chemistry*, 12th ed., McGraw-Hill Book Co., New York, pp. 5–14.

W. Gerhartz, ed., *Ullmann's Encyclopedia of Industrial Chemistry*, 5th ed., Vol. A6, VCH Verlagsgesellscaft mbH, D-6940, Weiheim, Germany, 1985, pp. 483–503.

R. J. Lewis, Sr., ed., *Sax's Dangerous Properties of Industrial Materials*, 10th ed., John Wiley & Sons, Inc., New York, 2000.

D. M. Novak, B. V. Tilak, and B. E. Conway in *Modern Aspects of Electrochemistry*, Vol. 14, Plenum Press, New York, 1982.

SUDHIR K. MENDIRATTA
BUDD L. DUNCAN
Olin Corporation

CHLORINATED PARAFFINS

Chlorinated paraffins with the general molecular formula $C_xH_{(2x-y+2)}Cl_y$ have been manufactured on a commercial basis for over 50 years. The development of chlorinated paraffins into new and emerging technologies was constrained principally because of the limitations of grades based on paraffin wax and the lack of suitable alternative feedstocks to meet the demands of the new potential markets.

The availability of alpha olefins has enabled some manufacturers to offer a range of chlorinated alpha olefins alongside their existing range of chlorinated paraffins. Chlorinated alpha olefins are virtually indistinguishable from chlorinated paraffins but do offer the manufacturer a single-carbon number paraffinic feedstock and even greater flexibility in the product range.

CHEMICAL AND PHYSICAL PROPERTIES

By virtue of the nature of the paraffinic feedstocks readily available, commercial chlorinated paraffins are mixtures rather than single substances. The degree of chlorination is a matter of judgment by the manufacturers on the basis of their perception of market requirements.

The physical and chemical properties of chlorinated paraffins are determined by the carbon chain length of the paraffin and the chlorine content. This is most readily seen with respect to viscosity and volatility; increasing carbon chain length and increasing chlorine content lead to an increase in viscosity but a reduction in volatility.

Chlorinated paraffins vary in their physical form from free-flowing mobile liquids to highly viscous glassy materials. Chlorination of paraffin wax (C24–C30) to 70% chlorine and above yields the only solid grades. A key property associated with chlorinated paraffins, particularly the high chlorine grades, is nonflammability, which has led to their use as fire-retardant additives and plasticizers in a wide range of polymeric materials.

MANUFACTURE

Chlorinated paraffins are manufactured by passing pure chlorine gas into a liquid paraffin at a temperature between 80 and 100°C depending on the chain length of the paraffin feedstock. At these temperatures chlorination occurs exothermically and cooling is necessary to maintain the temperature at around 100°C. Catalysts are not usually necessary to initiate chlorination, but some manufacturers may assist the process with ultraviolet light. Failure to control the reactive exotherm during chlorination may lead to a colored and unstable product. The reaction is terminated by stopping the flow of chlorine when the desired degree of chlorination has been achieved. This is estimated by density, viscosity, or refractive index measurements. The reactor is then purged with air or nitrogen to remove excess chlorine and hydrochloric acid gas. Small quantities of a storage stabilizer, typically epoxidized vegetable oil, may be mixed in at this stage or later in a blending vessel.

SHIPMENT AND STORAGE

Liquid chlorinated paraffins are shipped in drums usually lacquer-lined mild steel or polyethylene and in road or rail barrels. Where appropriate larger quantities can be shipped by sea either in deck tanks of conventional cargo ships or in chemical parcel tankers for larger consignments.

The high viscosity of a number of grades generally precludes consideration of bulk supplies unless special transport, heating, pumping, and storage can be made.

The main points to be considered when designing a bulk storage installation are (*1*) the viscosity of all grades of chlorinated paraffin varies sharply with a change in temperature; (*2*) chlorinated paraffins should not be exposed to temperatures in excess of 40°C for prolonged periods of time; (*3*) chlorinated paraffin stability can be affected by contact with zinc and iron, therefore tanks of mild steel should generally be lined and galvanized steel avoided. Stainless steel, lacquer, or glass-lined mild steel tanks are recommended. The preferred linings are of the heat-cured phenol—formaldehyde type; and (*4*) chlorinated paraffins swell certain types of rubber and therefore rubber joints should be avoided. Polytetrafluorethylene joints are recommended.

As for storage tanks, stainless steel and lacquer-lined mild steel are suitable materials of construction for pipe lines. For pumps, valves, etc, various alloys are suitable, including phosphor bronze, gun metal, Monel, stainless steel, and certain nickel steel alloys.

HEALTH AND SAFETY FACTORS

The acute toxicity of chlorinated paraffins has been tested in a range of animals and was found to be very low. One comprehensive study demonstrated that the toxicity of chlorinated paraffins was related to carbon chain length and to a lesser degree chlorine content. The shorter chain-length chlorinated paraffins were more toxic than the longer chain chlorinated paraffins.

Because of the nature of some applications in which chlorinated paraffins are used, skin contact is inevitable and therefore an important potential route into the body. Skin absorption studies have shown that chlorinated paraffins are very poorly absorbed through the skin and should not cause significant systemic concentrations.

ENVIRONMENTAL ASPECTS

In general, chlorinated paraffins biodegrade; the rate is determined by chlorine content and carbon chain length. Microorganisms previously acclimatized to specific chlorinated paraffins show a greater ability to degrade the compounds than nonacclimatized organisms. Mammals and fish have been shown to metabolize chlorinated paraffins.

In the United States further information and advice is readily available from the Chlorinated Paraffin Industry Association (CPIA) based in Washington D.C.

APPLICATIONS

Chlorinated paraffins are versatile materials and are used in widely differing applications. As cost-effective plasticizers, they are employed in plastics particularly PVC, rubbers, surface coatings, adhesives, and sealants. Where required they impart the additional features of fire retardance, and chemical and water resistance. In conjunction with antimony trioxide, they constitute one of the most cost-effective fire-retardant systems for polymeric materials, textiles, surface coatings, and paper products. Chlorinated paraffins are also employed as components in fat liquors used in the leather industry, as extreme pressure additives in metal-working lubricants, and as solvents in carbonless copying paper.

H. J. Caesar and P. J. Davis "Flame Retardant Vinyl Compounds," *33rd Annual Technical Conference*, Atlanta, Ga., May 6, 1975.

"Chloroparaffins, Chemical Profile", *Chemical Market Reporter*, Aug. 26, 2002.

KELVIN L. HOUGHTON
ICI Chemicals and Polymers Ltd.

CHLORINE

Chlorine and caustic soda, manufactured by the electrolysis of aqueous sodium chloride (or brine), are among the top ten commodity chemicals, in terms of capacity, in the United States. When chlorine is produced by the electrolysis of salt, It is accompanied by 1.1 units of caustic soda. The combination of 1 ton of chlorine and 1.1 tons of caustic soda generated in this process is referred to as an electrochemical unit (ECU).

Some, chlorine processes do not produce caustic soda, such as the electrolysis of aqueous HCl, molten sodium or magnesium chloride, and potassium chloride solutions. Potassium hydroxide (KOH) is the coproduct in the last of these.

Figure 1 represents the various processes that produce chlorine. There are some processes that produce only caustic soda and no chlorine, such as the chemical caustic soda process that uses soda ash as the raw material. This process accounts for 1–2% of the total world capacity of caustic soda.

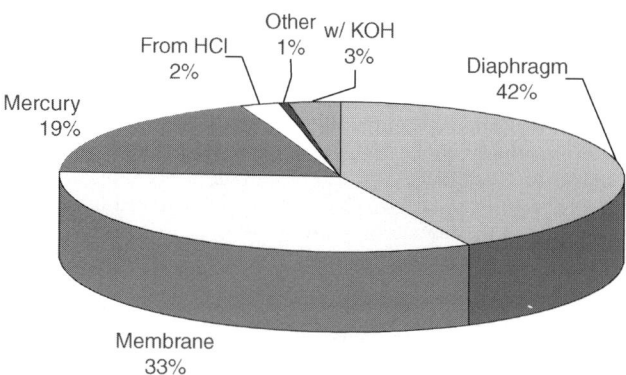

Figure 1. Technologies for manufacturing chlorine in 2000. Total capacity = 53×10^6 tpa.

504   CHLORINE

PHYSICAL PROPERTIES

Table 1 presents characteristic properties of chlorine.

Chlorine is somewhat soluble in water and in salt solutions, its solubility decreasing with increasing temperature or salt strength. It is partially hydrolyzed in aqueous solution:

$$Cl_2 + H_2O \rightleftharpoons HOCl + H^+ + Cl^- \qquad (1)$$

This hydrolysis increases the solubility by reducing the partial pressure corresponding to a given amount of dissolved chlorine. Taking the equilibrium of reaction 1 into account, chlorine's solubility in water at a given temperature can be expressed as a function of its partial pressure by

$$S = HP + (KHP)^{1/3} \qquad (2)$$

where

H = Henry's law solubility constant for molecular chlorine
K = equilibrium constant for reaction 1.

Chlorine forms a hydrate, containing 7–8 molecules of water per molecule of chlorine below 9.6°C at atmospheric pressure. These crystals are greenish-yellow in color. Chlorine hydrate forms greenish-yellow crystals below 9.6°C at atmospheric pressure. It does not exist above the quadruple point temperature of 28.7°C. It has a density of ~1.23 and therefore floats on liquid chlorine when deposited in a liquefaction system.

CHEMICAL PROPERTIES

Chlorine usually exhibits a valence −1 in compounds, but it also exists in the formal positive valence states of +1 (NaClO), +3 (NaClO$_2$), +5 (NaClO$_3$), and +7 (NaClO$_4$).

Molecular chlorine is a strong oxidant and a chlorinating agent, adding to double bonds in aliphatic compounds or undergoing substitution reactions with both aliphatics and aromatics. Chlorine is very reactive under specific conditions but is not flammable or explosive. Reactions with most elements are facile, but those with nitrogen, oxygen, and carbon are indirect. Chlorine reacts with ammonia to form the explosive compound NCl$_3$. It does not react with hydrogen at normal temperatures in the absence of light. However, at temperatures above 250°C, or in the presence of sunlight or artificial light of ~470-nm wavelength, H$_2$ and Cl$_2$ combine explosively to form HCl. Explosive limits of mixtures of pure gases are ~8% H$_2$ and ~12% Cl$_2$ (v/v). These limits depend on temperature and pressure, and they can be altered by adding inert gases such as nitrogen and carbon dioxide. Most plants control the process to keep the hydrogen concentration below 4%.

Dry chlorine reacts combustively with most metals at sufficiently elevated temperatures. Aluminum, arsenic, gold, mercury, selenium, tellurium, and tin react with dry chlorine in gaseous or liquid form at ordinary temperatures; carbon steel ignites at ~250°C, depending on physical shape; and titanium reacts violently with dry chlorine. Wet chlorine is very reactive and very corrosive because of the formation of hydrochloric and hypochlorous acids (see eq. 1). Metals stable to wet chlorine include platinum, silver, tantalum, and titanium. The apparent anomaly in the behavior of titanium, which is stable in wet chlorine but explosive with dry chlorine, is due to the formation of a protective layer of TiO$_2$ when the partial pressure of water vapor is sufficiently high. Tantalum is the metal most stable to both dry and wet chlorine.

Table 1. Physical Constants of Chlorine

Property	Value
atomic number	17
atomic weight (carbon-12 scale)	35.453
stable isotope abundance, atom %	
^{35}Cl	75.53
^{37}Cl	24.47
electronic configuration in ground state	[Ne]$3s^23p^5$
melting point, °C	−100.98
boiling point at 101.3 kPa(°C)	−33.97
gas density relative to air	2.48
critical density, kg/m^3	573
critical pressure, kPa	7977
critical volume, m^3/kg	0.001745
critical temperature, °C	143.75
critical compressibility	0.284777
gas density, kg/m^3 at 0°C and 101.3 kPa	3.213
gas viscosity, mPa.s at 20°C	0.0134
liquid viscosity, mPa.s at 20°C	0.346
gas thermal conductivity at 20°C, W/m K	0.00866
liquid thermal conductivity at 20°C, W/m K	0.120
latent heat of vaporization, kJ/kg	287.75
latent heat of fusion, kJ/kg	90.33
heat of dissociation, kJ/mol	2.3944
heat of hydration of Cl$^-$, kJ/mol	405.7
standard electrode potential, V	1.359
electron affinity, eV	3.77
ionization energies, eV	13.01, 23.80, 39.9, 53.3, 67.8, 96.6, and 114.2
specific heat at constant pressure	0.481
specific heat at constant volume	0.357
specific magnetic susceptibility at 20°C, m^3/kg	-7.4×10^{-9}
electrical conductivity of liquid at −70°C, (Ω cm)$^{-1}$	10^{-16}
dielectric constant at 0°C (wavelengths > 10 m)	1.97

Chlorine reacts with alkali and alkaline earth metal hydroxides to form bleaching agents such as NaOCl:

$$Cl_2 + 2\,NaOH \longrightarrow NaCl + NaOCl + H_2O \qquad (3)$$

Reaction of hypochlorite with ammonia produces hydrazine:

$$2\,NH_3 + NaOCl \longrightarrow N_2H_4 + NaCl + H_2O \qquad (4)$$

TiO_2 reacts with chlorine in the presence of carbon in the manufacture of $TiCl_4$, an intermediate in the production of titanium metal and pure TiO_2 pigment:

$$TiO_2 + 2\,Cl_2 + C(\text{or } 2C) \rightarrow TiCl_4 + CO_2(\text{or } 2\,CO) \qquad (5)$$

$SiCl_4$ is produced by a similar reaction with SiO_2.

Chlorine reacts with saturated hydrocarbons and certain of their derivatives by substitution of hydrogen to form chlorinated hydrocarbons and HCl. Thus, methanol and methane chlorinate to form methyl chloride, which can continue to react with chlorine to form methylene chloride, chloroform, and carbon tetrachloride. Reaction of chlorine with unsaturated hydrocarbons results in saturation of the double or triple bond.

MANUFACTURE AND PROCESSING

Anodes

Graphite was exclusively used as the anode for chlorine production for >60 years even though it exhibited high chlorine overpotential and dimensional instability caused by the electrochemical oxidation of carbon to CO_2, which led to an increased electrolyte ohmic drop, and hence to high energy consumption during use. In the late 1960s, H.B. Beer's invention of noble metal oxide coatings on titanium substrates as anodes revolutionized the chlor-alkali industry. The most widely used anodes were ruthenium oxide and titanium oxide-coated titanium, which operated at low chlorine overpotential with excellent dimensional stability. Presently, $RuO_2 + TiO_2$ coated titanium anodes, called Dimensionally Stable Anodes (DSA), are exclusively used as the anodes.

The DSA electrode is a titanium substrate coated with titanium and ruthenium oxide mixture containing >30 mol% Ruthenium oxide, the Ru loading ranging from 5 to 20 g/m^2. These coatings are formed on titanium mesh substrates by thermal decomposition. Ruthenium and titanium salts dissolved in organic solvent (e.g., butanol), are applied to the Ti mesh and fired at 500°C in air to obtain the mixed oxides.

DSA anodes exhibit long life, very low operating voltage, and high chlorine efficiency. They are tolerant to a wide range of operating conditions, although performance degradation can occur when the percentage of oxygen generated is high.

The mechanism of coating failure appears to depend on the type of cell in which the anode is operated. Life in diaphragm cells is at least 12 years; in mercury cells, it is considerably shorter, 3–4 years. The unavoidable occurrence of minor short circuits, through contact with the mercury cathode, causes gradual physical wear of the anode coating. In membrane cells, the anode life in about 5 years. The limiting factor in membrane and diaphragm cells, in the absence of impurity-related effects, is the loss of RuO_2 by dissolution that accompanies high oxygen production and cell shutdowns. Another explanation for the enhanced anode potentials is the formation (and buildup) of insulating TiO_2 layers between the coating and the substrate.

The anode suppliers recommend a feed brine concentration of 300 g/L and pH < 12, along with the following maximum limits of impurities in ionic form: Hg (0.04 ppm); Mn (0.01 ppm); heavy metals (0.3 ppm); total organic content (1 ppm); F (1 ppm); Ba (0.4 ppm).

Electrolytic Cell Operating Characteristics

Diaphragm cell technology is the source of the largest volume of chlorine, followed by the membrane cell process and then the mercury process.

The electrical energy consumption in diaphragm chlor-alkali electrolysis is ~10–20% higher than in membrane technology. In addition, diaphragm cells operate at a low caustic soda concentration of ~11wt%, and hence the energy requirements to achieve the commercial concentration of 50% wt of caustic soda are much higher compared to the membrane process. The overall plant efficiency, defined as the ratio of the heat of the reaction to the sum of the endothermic processes and the electrical energy in to the cell, with diaphragm chlor-alkali cells is ~23%, while the corresponding value for membrane cells is 35–40%.

Cell Technologies

As stated earlier, there are three primary electrolytic technologies based on mercury, diaphragm and membrane cells, that are used to produce chlorine and caustic.

Mercury Cells. The mercury cell process consists of two electrochemical cells: the electrolyzer to produce chlorine and sodium amalgam and the amalgam decomposer to decompose sodium amalgam to form caustic.

The mercury cell has a steel bottom with rubber coated steel sides, as well as end boxes for the brine and mercury feed and exit streams with a rubber or rubber-lined steel covers. Adjustable metal anodes hang from the top and mercury flows on the inclined bottom. The current flows from the steel bottom to the flowing mercury. Sodium chloride brine of 25.5% salt strength, fed from the inlet end box, is electrolyzed at the anode to produce chlorine gas, which leaves from the top of the cell.

Mercury cells operate efficiently because the hydrogen overpotential is high on mercury, favoring sodium amalgam formation over the hydrogen evolution reaction. The brine must be free of impurities such as vanadium or chromium that lower the hydrogen overvoltage, allowing hydrogen to form in dangerous quantities.

Recenter developments in mercury cells have been aimed toward lowering the energy consumption and minimizing mercury losses to the air, water, or the products of electrolysis.

Diaphragm Cells. In the diaphragm cell, brine flows continuously into the anolyte and subsequently through a diaphragm into the catholyte. The diaphragm separates the chlorine liberated at the anode from the sodium hydroxide and hydrogen produced at the cathode. Early cells used a horizontal asbestos sheet as diaphragms. Subsequently, three basic types of diaphragm cells have been developed: rectangular vertical electrode monopolar cells, cylindrical vertical electrode monopolar cells, and vertical electrode filter press bipolar cells.

Asbestos Diaphragms. The earliest diaphragms were made of asbestos paper sheets. Asbestos was selected because of its chemical and physical stability, its cost, and its abundance. Vacuum-deposited asbestos was the diaphragm of choice until 1971, when it was supplanted by the Modified Diaphragm (trademark of ELTECH Systems, Inc.), which, in its most common form contains fibrous polytetrafluoroethylene (PTFE) and a minimum of 75% asbestos. The polymer, following fusion, stabilizes the asbestos, lowers cell voltage, and allows the use of the expandable DSA anode, which further lowers the cell voltage.

The toxicological problems associated with asbestos have been widely publicized on its use in EPA ban, was overturned in 1991. In many parts of the world, this use is now being banned on tightly restricted.

Non-Asbestos Diaphragms. Polyramix fiber is a zirconia-PTFE deposited separator developed by ELTECH Systems in use in five cell rooms commercially. Three of these cell rooms have been converted to Polyramix (PMX) diaphragms, while two others are being converted. PMX separators offer longer diaphragm life and lower energy consumption. They can be reclaimed or cleaned with inhibited HCl to restore performance.

PPG Industries has also developed a non-asbestos diaphragm (Tephram). This technology uses vacuum depositing to produce a base diaphragm composed of PTFE fluoropolymer and inorganic particulate materials. This diaphragm has also exhibited greatly extended life in comparison to asbestos and equal or better voltage and current efficiency characteristics than the asbestos diaphragm.

Chlor-Alp has developed a two-layer diaphragm consisting of an activated nickel deposited sublayer with a synthetic separator deposited on top; their two cell rooms in France are being converted to that technology.

Electrolyzers. Various bipolar and monopolar diaphragm electrolyzers are in commercial use. These include the bipolar cells of Dow and the Glanor cells of PPG, and the monopolar cells of the Eltech/Uhde HU-type, which have not been licensed for use during the past 25 years. The sole supplier of diaphragm cell technology, at present, is ELTECH Systems, Inc., located in Chardon, Ohio.

ELTECH Monopolar Electrolyzers. ELTECH supplies monopolar diaphragm electrolyzers of two designs: the H-type and the MDC-type.

H-cells incorporate DSA anodes and operate at high current densities. The H-series employs cathode tubes having both ends open and extending across the cell, similar to the MDC cells. The MDC-series cathodes have been redesigned to replace the tube sheets and horizontal strap supports with copper corrugations that extend and are welded to the current-carrying side plates to which the intercell connectors are bolted.

New anode designs offer lower resistance and can be operated at zero gap with PMX diaphragms. The low voltage anode (LVA) features a solid conductor bar and 1-mm-thick expanders to reduce structural drop. The energy-saving anode features expanded mesh with micromesh continuously welded to the substrate, in addition to the LVA improvements.

The most effective improvements, however, involve rebuilding the cell to increase the electrode areas by reducing the cathode tube thickness and the spacing between the electrodes. Cell areas have been increased by 14–24%, depending on design.

The life of a PMX diaphragm is several times that of a polymer modified asbestos diaphragm. Furthermore, *in situ* diaphragm reclamation procedures have been developed that permit the life of the PMX diaphragm to be extended to match the life of the cathode internals. Titanium base covers, are replacing rubber covers to protect the base. Improved gaskets have been developed to extend the life of the modern diaphragm cell assembly to 5 years or more. Advanced diaphragm cell technology, incorporating these improved cell components, is claimed to offer electrical energy reduction of 10–15% over conventional diaphragm cell technology using expandable anodes and polymer modified asbestos diaphragms.

Ion-Exchange Membrane Cells. The key component of a membrane cell is the ion exchange membrane, which should be chemically stable, under the extremely aggressive conditions in an electrolysis cell offering excellent ion-exchange selectivity.

Principles. In a membrane cell, a cation-exchange membrane separates the anolyte and the catholyte. Ultrapure brine (containing <20 ppb of Ca^{2+} and Mg^{2+} ions combined) with a salt content of ~310 g/L NaCl is fed into the anolyte compartment, where chlorine gas is generated at the anode. The sodium ions, together with associated water molecules, migrate through the membrane into the catholyte. The membrane effectively prevents the migration of hydroxyl ions into the anolyte.

Unlike the separators used in the diagragm-cel process, the cation-exchange membrane mostly prevents the migration of chloride ions into the catholyte. Present-day membranes perform optimally at a caustic concentration of ~32–35%. The product caustic concentration is adjusted by adding demineralized water to the recirculating caustic.

Membranes. The membrane is the most critical component of the membrane-cell technology. It determines the current efficiency, cell voltage, and hence the energy consumption for the production of chlorine and caustic. An ideal ion-exchange membrane should exhibit:

- High selectivity for the transport of sodium or potassium ions,
- Negligible transport of chloride, hypochlorite, and chlorate ions,
- Zero back-migration of hydroxide ions,
- Low electrical resistance,
- Good mechanical strength and properties with long term stability.

An important characteristic of the membranes is their ability to achieve the desired transport of sodium ions while hindering the migration of the hydroxyl ions. Several theoretical descriptions have been proposed to understand the unique transport character of ions and water molecules across these polymeric membranes. Of these the most popular one is the cluster-network model (Figure 2).

A phenomenon of importance during electrolysis is the transport of water by electroosmotic mass transfer, driven by the electrical field between anode and cathode. While Na^+ ions migrate through the membrane in a hydrated state (3–5 molecules of water per ion), the OH^- ions cannot pass through the membrane in the opposite direction. As a result, there is a net transport of water to the catholyte. The amount of water carried over into the catholyte is inversely related to the salt concentration in the anolyte.

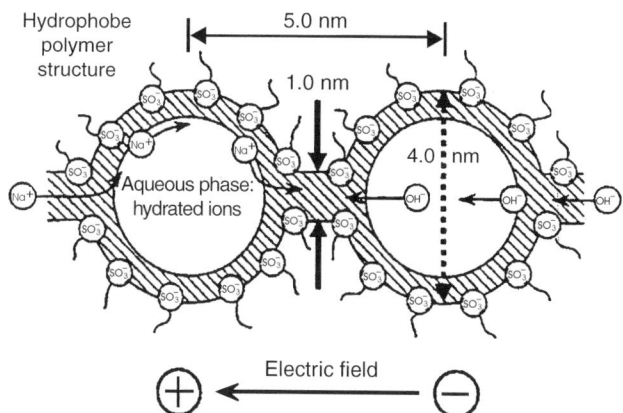

Figure 2. Principle of the cluster-network model from Bergner and co-workers.

Catalytic Cathodes. Although the first membrane electrolyzers were built with cathodes of low-carbon steel, a material used successfully in diaphragm cells, nickel is now preferred for its stability in concentrated caustic solutions.

Depending on the electrolyzer load, the hydrogen overpotential with carbon steel cathodes is in the range of 300–400 mV. Reductions of 200–280 mV are realizable, in principle, by using nickel cathodes with a catalytic coating. The various approaches include using materials that provide enhanced surface area and better electrocatalytic properties than steel. Composites generally chosen for coating are based on nickel or noble metals. Coatings commercially used in membrane cells include nickel–sulfur, nickel–aluminum, nickel–nickel oxide mixtures, and nickel containing platinum group metals. The technical problems confronting the use of catalytic cathodes in diaphragm cells have prevented commerialization. These probelms include selection of a coating technique for the complex cathode assembly that would not adversely influence the structural tolerances involved in the fabrication of cathodes and developing shutdown procedures that would eliminate hypochlorite as quickly as possible to preserve the catalytic activity of the coatings. Such problems do not exist in membrane cells because of the anion rejection properties of the membrane and effective anolyte flushing procedures during shutdowns.

Electrolyzers. Electrolyzers are classified as monopolar or bipolar, depending on the manner in which electrical connections are made between the electrolyzer elements. In the monopolar type, all anode and all cathode elements are arranged in parallel. Such an electrolyzer will operate at a high amperage and low voltage. While the amperage is set by the number of elements in an electrolyzer, the total voltage depends on how many electrolyzers are in an electric circuit. In a bipolar configuration, the cathode of one cell is connected to the anode of the next cell, and thus the cells are configured in series. The bipolar arrangement is advantageous for realizing low voltage drop between the cells. However, problems associated with current leakage and corrosion arise in bipolar operations, since the feed and discharge streams of electrolytes to and from the cells, having different electrical potentials, are hydraulically connected via common manifolds and collectors. These problems can be avoided by properly designing the electrolyzer (eg, by limiting the number of elements per electrolyzer) to limit the overall stack voltage to a safe value. The monopolar design, on the other hand, suffers from the voltage losses occurring in the interelectrolyzer connectors. This inevitable drawback can be minimized by a conservative design. While the bipolar systems allow shutdown for maintenance of a single electrolyzer unit, independently from the rest of the plant, monopolar electrolyzers have to be designed in such a way that an individual electrolyzer can be short circuited, enabling maintenance and membrane replacement without shutting down the entire circuit. It is also possible to combine these two configurations in a hybrid electrolyzer arrangement.

Commercial Electrolyzers. All membrane electrolyzers have common general design features like vertical membrane position, stacked elements and usage of similar materials of construction. Nevertheless, there are quite substantial differences in the cell design.

One of general design differences is the manner of achieving effective sealing of the electrolyzer. The most common principle is the filter press arrangement, where tightness is achieved by pressing together all elements of an electrolyzer from both ends. The relatively high sealing-forces are produced by means of hydraulic devices or tie rods. A different approach is the single element design, where each element is individually sealed, by a flanged connection between the cathode and the anode semishell. The preassembled elements are stacked to form an electrolyzer, but only moderate forces are applied to ensure electrical contact.

As membrane development allows increasing current densities, electrolyzer internals have to meet the related effects. Essential design targets are the minimization of structural voltage losses, homogenization of electrolyte concentration and temperature, smooth handling of the increased rate of evolution of gas.

The actual electrolyzer designs of the suppliers are listed below.

Asahi Kasei Acilyzer-ML Bipolar Electrolyzers. Asahi Kasei's electrolyzer (Acilyzer-ML) is of the bipolar type, composed of a series of cell frames.

Chlorine Engineers CME Monopolar Electrolyzers. Chlorine Engineers Corporation (CEC) (a subsidiary company of Mitsui and Company), produces the filter press type, monopolar membrane electrolyzer.

CEC BiTAC-800 Electrolyzer. Chlorine Engineers Corporation (CEC) adapted a bipolar electrolyzer first developed by Tosoh Corporation. The electrolyzer is of the filter press type and has an effective membrane area of ~ 3.3 m^2.

INEOS FM21-SP Monopolar Electrolyzer. INEOS' FM21-SP monopolar electrolyzer incorporates stamped electrodes.

ELTECH ExL Electrolyzers. ELTECH electrolyzers are modified versions of the earlier MGC electrolyzer (membrane gap cell). The electrodes, with areas of 1.5 m^2, can be assembled in monoplan on bipolar fashion.

Uhde Bipolar Electrolyzer. The characteristic feature of the Uhde membrane electrolyzer is its single element design. The cross-sectional area of an element in ~ 2.8 m^2.

Membrane Cells with Oxygen or Air Depolarized Cathodes. Substituting an oxygen reduction reaction for the hydrogen evolution reaction at the cathode in chlor-alkali electrolysis will reduce the total cell voltage by ~ 1 V (theoretically 1.23 V), thereby realizing a substantial saving of electrical energy.

The anode reaction is the same as in a conventional chlor-alkali cell, where the chloride ions are discharged to form chlorine gas product and the sodium ions migrate to the cathode compartment through the ion-exchange membrane. At the cathode, oxygen is reduced to the OH^- ions, which combine with the Na^+ ions to form sodium hydroxide. The advantages of using the oxygen reduction reaction include avoidance of costly downstream treatment of hydrogen and absence of the gas void fraction in the catholyte, reducing the ohmic drop in the cell. However, this scheme requires a high-performance air scrubbing system to remove all carbon dioxide from the air in order to prevent early failure from the accumulation of sodium carbonate.

The cathode side has several special requirements, the central one being the adjustment of the local differential pressure between the caustic and the oxygen compartment on the other side of the electrode. Due to the porous nature of the gas diffusion electrode, a pressure balance has to be established across the electrode in order to avoid flow from one side to the other. There is a certain range of differential pressure in which a cathode will operate property.

Chlorine Plant Auxiliaries

Although the auxiliary systems for each of the three electrolytic chlor-alkali processes differ somewhat in operation, the processes of brine purification and chlorine and hydrogen recovery are common to each.

Brine Preparation. There are several different major sources of the fundamental raw material, sodium chloride. It can be recovered from underground rock salt deposits. Methods of recovery include conventional mechanical mining, which has much in common with the mining of coal, and solution mining, in which water or weak brine is forced into a salt deposit to dissolve the material and carry it back to the surface. Salt is also obtained by evaporation of saline waters. Inland sources may be more concentrated and may contain other minerals of greater value. The Dead Sea and the Great Salt Lake are examples. The salt obtained from these waters is usually referred to as solar salt, since solar energy provides the driving force for evaporation.

However obtained, sodium chloride will have a characteristic set of impurities. The major impurity in nearly all salts is some form of calcium sulfate. While compositions from different sources vary widely, a few generalizations can be made. Solar salt, at least after the common operation of washing, is usually purer than rock salt. It is also more susceptible to caking and mechanical degradation. While rock salt contains more calcium sulfate, it contains in proportion less magnesium. The higher ratio of calcium to magnesium improves performance in the brine treatment process.

Brine Purification. Impurities in the brine can affect the performance of any type of cell. The most common cationic impurities, and those present in the greatest concentrations, are the hardness elements calcium and magnesium. In diaphragm and membrane cells, these can precipitate as the hydroxides when exposed to a sufficiently high pH, which occurs within the diaphragm or membrane as the ions approach the catholyte side. Membranes are especially sensitive to brine impurities, and the membrane-cell process requires a higher degree of brine quality than does either of the other technologies.

With a different operative mechanism in mercury cells, the same impurities do not cause permanent damage to the chlorine/caustic separator. Instead, traces of metals such as vanadium, molybdenum, chromium, iron, titanium, and tungsten must be avoided. These metals lower the hydrogen overvoltage at the mercury cathode and permit increased formation of hydrogen. This lowers the current efficiency and creates a hazard in the processing of chlorine.

Impurities that are soluble in the acidic anolyte can become insoluble once they have entered the membrane. In addition to the induced flow of cation impurities, neutral and anion impurities can enter by diffusion and by the considerable water flow from anolyte to catholyte. Internal precipitation of these impurities disrupts the structure of the membrane. This precipitation can increase the voltage drop through the membrane. If it occurs in the cathode-side layer, it will reduce the selectivity of the membrane and therefore the current efficiency of the cell. Even if the impurity is subsequently washed out, the void left behind still contributes to reduced current efficiency. There also are synergistic effects, such as the precipitation of complex compounds, usually involving calcium and magnesium in combination with silica and alumina.

The most important anionic impurity, because of its widespread distribution in salt, is sulfate.

Some brines also contain ammonium ions or organic nitrogen that can be converted to the explosive nitrogen trichloride. Ammonium ions in the brine are removed by treatment with chlorine or hypochlorite. Depending on its severity, this treatment produces NH_2Cl, which can be removed as a gas.

Chlorine Processing. The first step in the chlorine-handling process is the cooling of the gas, which is saturated with water at cell operating temperature. Most cells operate very close to atmospheric pressure, although membrane cells have the capability to run under positive pressures, up to several hundred kilopascals. Direct- and indirect-contact cooling is used. Direct cooling was more common in the past, when metals resistant to wet chlorine gas were not economically available. In this process, the gas from the cells is contacted with a countercurrent stream of brine or cooling water in a column. Indirect exchangers are single-pass devices with titanium tubes and carbon steel shells. Titanium plate exchangers are used in some smaller plants.

Cooling the chlorine incidentally removes most of the water by condensation. Lower temperatures promote the removal of water and reduce the consumption of drying acid in the next step of the process, which involves compression and liquefaction.

Hydrogen Processing. The hydrogen produced in all electrolytic chlor-alkali processes is relatively pure (>99.9%) and requires only cooling to remove water along with entrained salt and caustic. After compression to the proper level, the hydrogen can be used for organic hydrogenation, catalytic reductions, and ammonia or methanol synthesis. It can also be used to provide a reducing atmosphere in some applications or burned with chlorine to produce high quality HCl. The HCl is produced in a high temperature burner and then usually cooled and absorbed in chemically resistant graphite equipment to produce 30–38% acid.

Sodium Hydroxide Processing. Coproduct sodium hydroxide is usually sold and shipped as a 50% solution. A 73% solution and the anhydrous material are also marketed. The solutions produced by the various types of cell differ markedly, and the purity of the final product and the energy required to concentrate the cell product to 50% are major differentiators among the three technologies. Diaphragm cells produce a relatively weak liquor of ~10–12% caustic. Impurities that enter with the treated brine, such as sodium sulfate, and impurities generated in the cells, such as sodium chlorate, accompany the caustic. The unconverted NaCl, in concentrations of 13–16%, also is in the liquor. Evaporation of this mixture toward 50% caustic causes it to become supersaturated in NaCl, and most of the salt drops out of solution. This salt must be removed, usually by centrifuging, and returned to the brine plant for reuse. Cooling of the solution then rejects more NaCl, which is removed by filtration or centrifuging.

Nickel is the standard material of construction for caustic evaporators. Multiple-effect units are a universal choice in all but the smallest plants. Membrane-cell plants use double- or triple-effect evaporators. With the high evaporative load associated with the diaphragm-cell process, quadruple-effect systems often are justified. When liquor temperatures in the last effect are sufficiently low, stainless steel is sometimes used in place of nickel.

Other Chlorine Production Processes

Chlorine from Potassium Hydroxide Manufacture. Mercury and membrane cells can produce KOH instead of NaOH if KCl is used as the feedstock. Chlorine and hydrogen again are coproducts.

Chlorine from HCl. Reflecting the low cost, bulk chemical status of the two products, HCl, which otherwise is often made from chlorine, can sometimes be economically converted back to chlorine. There are two fundamentally different approaches, electrolysis and oxidation.

Electrolysis of HCl. Electrolytic decomposition of aqueous HCl to generate chlorine and hydrogen follows the overall reaction:

$$2\,HCl \longrightarrow H_2 + Cl_2 \qquad (6)$$

This process used in a number of plants, to produce about 10,000 tpa of chlorine.

Chemical Oxidation of HCl. Chlorine can also be produced from HCl by the following equilibrium reaction:

$$4\,HCl + O_2 \longleftrightarrow Cl_2 + 2\,H_2O \qquad (7)$$

Air or oxygen can be used as the oxidizing agent. This reaction is the basis for the Deacon process, which was the first continuous catalytic process carried out on large scale. It was the major source of chlorine before development of the electrolytic route.

The fundamental problem with the Deacon process is the inability to achieve very high degrees of conversion of the HCl.

Another variation on HCl oxidation is Mitsui Toatsu's MT Chlor process. This is essentially a Deacon reaction carried out in a fluidized bed, using a chromium oxide catalyst.

Chlorine from the Magnesium Process. Magnesium is produced by the fused salt electrolysis of $MgCl_2$.

Chlorine from the Titanium Process. Electrolysis of magnesium chloride is a step in the production of titanium. Titanium metal is produced by the reaction of titanium tetrachloride with magnesium. The magnesium chloride formed in this reaction is electrolyzed, as in the preparation of magnesium.

Energy Requirements

The minimum amount of electrical energy required for producing 1 ECU (electrochemical unit) is 5.75×10^6 Btu or 1686 kWh for 1 ton of chlorine and 1.1 tons of caustic soda. The corresponding value for 1 short ton of chlorine and 1.1 tons of sodium hydroxide is 5.23×10^6 Btu or 1534 kWh.

MATERIALS OF CONSTRUCTION

The choice of construction material for handling chlorine depends on equipment design and process conditions. Dry chlorine, with <50 ppm (w/w) of water, can be handled safely below 120°C in iron, steel, stainless steels, Monel, nickel, copper, brass, bronze, and lead. Silicones, titanium, and materials with high surface areas (eg, steel wool) should be avoided. Titanium ignites spontaneously at ordinary temperatures in dry chlorine; steel reacts at an accelerated rate at temperatures above 120°C, igniting near 250°C. The presence of rust or organic substances increases the risk of steel ignition because of the exothermic reactions of these materials with chlorine. Thorough cleaning, degreasing, and drying are essential before commissioning the steel equipment into chlorine service.

With chlorine gas, ordinary carbon steel is safe so long as the temperature is held safely below the ignition temperature. Good practice allows a wide margin to allow for hot spots and to extend the life of the equipment. In the chlorine production process, the only step in which dangerous temperatures normally may occur is compression. Limiting the compression ratio in each stage and cooling the gas before the compressor and between stages keeps the temperature low. In those applications of chlorine that require temperatures above 150°C, materials of construction other than carbon steel must be used.

The atmospheric pressure boiling point of chlorine (−34°C) demands the use of low temperature carbon steels. Because this temperature is possible whenever a higher pressure system is depressurized, most liquid chlorine systems are constructed of these materials. Liquid chlorine usually is stored in vessels made from unalloyed carbon steel or cast steel. Fine grain steel with limited tensile strength is used to facilitate proper welding.

A protective layer of ferric chloride on the metal prevents the reaction of dry chlorine with steel. This layer can be disrupted by excessive temperature (as above), the presence of moisture, or erosion.

Wet chlorine gas is handled primarily in fiberglass-reinforced plastics. Special constructions are necessary to achieve optimum performance. Rubber-lined steel is suitable for wet chlorine gas up to ∼100°C. At low temperature and pressure, PVC, CPVC, and reinforced polyester resins are also used. PTFE, PVDF (polyvinylidene fluoride), and FEP (fluorinated ethylene propylene) are resistant at higher temperatures, but suffer from permeability or poorer mechanical properties. Other materials stable in wet chlorine include graphite and glass. Among the metals, titanium is the usual choice for wet-chlorine service. As was the case with steel in dry chlorine, titanium depends on a protective layer (TiO_2) on the metal for its corrosion resistance. Maintenance of this layer requires a minimum water content in the gas. Tantalum is the most resistive metal over a wide range of conditions.

Gaskets in both dry gas and liquid chlorine have been made for many years from some form of compressed asbestos. It is an ideal material for the service, but restrictions on the use of asbestos have led to development of alternatives. For wet chlorine gas, rubber or synthetic elastomers are acceptable. PTFE within its serviceable temperature range is resistant to all forms of chlorine. Tantalum, Hastelloy C, PTFE, PVDF, Monel, and nickel are used in thin sections such as diaphragms, membranes, rupture disks, and bellows.

SHIPMENT AND STORAGE

There are two basically different approaches to storing chlorine. One is to store it under positive pressure; the other is to refrigerate it and store it essentially at atmospheric pressure. In a producing plant, the former usually is at liquefaction process pressure. The latter is mechanically more complex and requires a compression system for returning vaporized chlorine to the process. It is considered appropriate only for large systems in producers' plants. Pressurized storage tanks require relief and containment systems. It is common to use an empty expansion tank as a receptacle for escaping chlorine. This contains the material released during minor upsets and buffers the flow of gas to a scrubbing system when releases are too large to be completely contained.

The material of construction for atmospheric storage tanks is unalloyed low-temperature carbon steel. These tanks usually are approximately spherical and are contained within a shell. Overcoming differential thermal expansion is an important part of vessel design and fabrication. The annular space contains insulation and is monitored for chlorine leaking from the inner tank. While the temperature of the chlorine under pressurized storage may be above the minimum temperature for conventional steels (−29°C), such low temperatures are encountered whenever pressure is released from a tank. Normal practice therefore is to build these also from low temperature steel. Chlorine Institute Pamphlet 5 summarizes design principles and good practice.

Liquid chlorine has an unusually large coefficient of thermal expansion. The amount stored in a pressurized tank must be limited to allow for this expansion in case the temperature rises. A standard practice is to design for a filling density of 1.25 in order to give 5% freeboard at 50°C.

Chlorine loading and unloading are usually by one of two methods, pumping or pressurization with a dry gas. Pumps in large-scale storage systems usually are of the submerged type. This eliminates suction piping and the need for bottom or side connections to the storage tanks. Excess flow valves prevent release of chlorine in case of downstream maloperation or piping failure. Canned pumps are an alternative. Pressurization is mechanically simpler but creates the need for disposal of the padding gas, mixed with residual chlorine.

Chlorine is transported in cylinders and ISO (International Standards Organization) containers and by rail and road tankers. It is classified as a nonflammable compressed gas. U.S. Department of Transportation regulations call for a green label. Repackagers of chlorine supply it in small cylinders containing 45.4 or 68 kg. They also, along with some producers, supply ton lots in cylinders. These are pressurized and protected with fusible-plug relief devices. Quantities between 15 and 90 short tons are transported in tankers with covered manholes fitted with special angle valves. In the United States, shipping containers are fitted with special relief devices comprising a diaphragm-protected conventional relief valve mounted above a breaking-pin assembly.

USES

Phosgene is made by reacting carbon monoxide with chlorine in the presence of activated carbon. It is the starting material for the manufacture of polycarbonates, toluene diisocyanate (TDI), and methylene bisphenyl isoayanate (MDI). Polycarbonate (PC) resins are engineering thermoplastics that are produced by reacting phosgene and bisphenol A (BPA). The PC resins are typically characterized by high impact strength, dimensional stability, transparency, and excellent electrical properties. In addition to general purpose polycarbonate resins, a variety of specialty materials, blends, and alloys are also available. Polycarbonates are widely used, with applications in the automotive industry, glazing, electronics, computers and business machines, and for software, audio and video compact discs.

TDI is a major raw material for producing urethane foam, both flexible and rigid. Flexible urethane foams, by far the largest TDI end use, are used as bedding, furniture cushioning, carpet underlay, and packaging. MDI is used almost exclusively to make polyurethanes. MDI-based rigid urethane foams have the lowest thermal conductivity of any common insulation material. The largest insulation uses for these foams are in construction, refrigerators and freezers, and refrigerated rail cars and trucks. Rigid foams also have excellent buoyancy, and have been used in life-saving gear, swimming pools, sporting goods, and other flotation devices.

Most epoxy resins are made from BPA and epichlorohydrin (EPI). The chlorination of propylene produces allyl chloride which, when dehydrochlorinated with caustic soda, produces epi. Epoxy resins, used in making protective coatings, laminates, and fiber-reinforced composites, are the largest end use for EPI. The second largest end use is synthetic glycerin, produced by the hydrolysis of EPI with caustic soda. Glycerin is used in the pharmaceutical, tobacco, cosmetics, and food and beverage industries.

Chlorine and propylene are used in the chlorohydrin process to produce propylene oxide (PO). In the chlorohydrin process, propylene reacts with chlorine to make propylene chlorohydrin, which is then dehydrochlorinated with caustic soda to make propylene oxide. An alternative process competing with this route is the direct oxidation of propylene. PO is a highly reactive alkyl epoxide used principally as a chemical intermediate. The largest application for PO is as a raw material for polyether polyols. Polyether polyols are used in polyurethane applications including flexible and rigid foams, elastomers, coatings, adhesives, and reaction injection molding polymers. PO can also be reacted with water to form propylene glycol, a raw material for unsaturated polyester resins. A few applications of unsaturated polyester resins include tubs and showers, gasoline tanks, and boat hulls.

Chlor-alkali plants are built and operated to satisfy chlorine demand, not caustic demand. Chlorine demand is much more volatile than caustic demand and reacts more quickly to changes in economic conditions.

Compared to chlorine, the market for caustic soda adapts more slowly to changing economic conditions, because of the diversity of its end uses. Consumption of caustic soda in pulp mills is the largest single end use at 17%, which is about one-half of the size of the vinyl segment for chlorine consumption.

Alumina is the second largest end use for caustic soda at 8%, and demand for alumina is driven by aluminum metal. About one-third of aluminum is consumed by the transportation sector and ~15% by the construction sector, both of which are dependent on global economic health.

The organics category is a fairly large segment and is a collection of many organic chemicals, including PO, epi, TDI and MDI (isocyanates), and polycarbonates.

D. Bergner, M. Hartmann, and R. Staab, *Chem.-Ing.-Tech.* **66**(6), 783 (1994).

H. S. Burney, N. Furuya, F. Hine and K.-I. Ota, eds., *Chlor-Alkali and Chlorate Technology: R. B. Macmullin Memorial Symposium*, Proc. Vol. 99-21, The Electrochemical Society Inc., Pennington, N.J., 1999.

T. F. O'Brien, T. V. Bommaraju, and F. Hine, *Handbook of Chlor-Alkali Technology*, Springer, NY (2005).

P. Schmittinger, ed., *Chlorine—Principles and Industrial Practice*, Wiley-VCH, Weinheim, 2000.

J. S. Sconce, *Chlorine, Its Manufacture, Properties, and Uses*, Reinhold Publishing Corp., New York, 1962.

TILAK V. BOMMARAJU
Process Technology Optimization, Inc
BENNO LÜKE
GLEN DAMMANN,
Krupp Uhde GmbH
THOMAS F. O'BRIEN
Mudville 9, Inc.
MARY C. BLACKBURN
Chemical Market Associates, Inc.

CHLOROBENZENES

The chlorination of benzene can theoretically produce 12 different chlorobenzenes. With the exception of 1,3-dichlorobenzene, 1,3,5-trichlorobenzene, and 1,2,3,5-tetrachlorobenzene, all of the compounds are produced readily by chlorinating benzene in the presence of a Friedel-Crafts catalyst (see FRIEDEL-CRAFTS REACTIONS). The usual catalyst is ferric chloride either as such or generated *in situ* by exposing a large surface of iron to the liquid being chlorinated. With the exception of hexachlorobenzene, each compound can be further chlorinated; therefore, the finished product is always a mixture of chlorobenzenes. Refined products are obtained by distillation and crystallization.

With the discontinuation of some herbicides, eg, 2,4,5-trichlorophenol, based on the higher chlorinated benzenes, and DDT, based on monochlorobenzene, both for ecological reasons, the production of chlorinated benzenes has been reduced to just three with large-volume applications of (mono)chlorobenzene, *o*-dichlorobenzene, and *p*-dichlorobenzene. Monochlorobenzene remains a large-volume product, considerably larger than the other chlorobenzenes, in spite of the reduction demanded by the discontinuation of DDT.

PHYSICAL AND CHEMICAL PROPERTIES

The important physical properties of chlorobenzenes appear in Table 1. Only limited information is available for some chlorobenzenes:

Chlorobenzene	Mol wt	Melting point, °C	Normal boiling point, °C
1,2,3,5-tetrachloro	215.9	51	246
pentachloro	250.35	85	276

MANUFACTURE

The production of any chlorinated benzene is a multiple-product operation. All of the chlorobenzenes are now produced by chlorination of benzene in the liquid phase. Ferric chloride is the most common catalyst. Although precautions are taken to keep water out of the system, it is possible that the $FeCl_3 \cdot H_2O$ complex catalyst is present in most operations owing to traces of moisture in benzene entering the reactor. This $FeCl_3 \cdot H_2O$ complex is probably the most effective catalyst.

In the liquid-phase chlorination, 1,3-dichlorobenzene is found only in a small quantity, and 1,3,5-trichlorobenzene and 1,2,3,5-tetrachlorobenzene are undetectable.

Benzene chlorination reactors are subject to design and operating hazards. Because HCl is constantly present in most parts of the equipment, corrosion is always a potential problem. Chlorobenzene mixtures behave in distillation as ideal solutions. The dichlorobenzene isomers have very similar vapor pressures making separation by distillation difficult. Crystallization is generally used in combination with distillation to obtain the pure 1,2 and 1,4-dichlorobenzene isomers. HCl is a constant by-product in the manufacture of chlorobenzenes.

The HCl gas is absorbed in water to produce 30–40% HCl solution. Any plant at times produces unwanted isomers. This requires an incinerator, capable of burning chlorinated hydrocarbons to HCl, H_2O, and CO_2 equipped with an efficient absorber for HCl (see INCINERATORS). An alternative to burning is dechlorination using hydrogen over a suitable catalyst. Another alternative to burning is rearrangement of the undesired isomers.

STORAGE, SHIPMENT, AND HANDLING

Chlorobenzenes are stored in manufacturing plants in liquid form in steel containers. Mono-, 1,2-di- and 1,2,4-trichlorobenzenes are liquids at room temperature and are shipped in bulk in aluminum tank trucks and steel or stainless steel tank cars. In situations where chlorobenzenes are contained in aluminum at elevated temperatures, the product must be clean, ie, nonacidic and the moisture, low. The use of aluminum with a mixture of chlorobenzenes and strong oxidizers should be avoided. 1,4-Dichlorobenzene is shipped either in molten form in insulated steel tank cars with heater coils, or as flake or granular solid in suitably sealed containers, such as paper bags, fiber packs, or drums. Phenolic linings in all vessels offer protection over a wide range of conditions for all chlorobenzenes as well as the vessels themselves. For drums, the phenolic coating should be modified with epoxy for maximum impact

Table 1. Physical and Thermodynamic Properties of Chlorobenzenes

CAS Registry Number	Chloro-benzene	1,2-Dichloro-benzene	1,3-Dichloro-benzene	1,4-Dichloro-benzene	1,2,4-Trichloro-benzene	1,2,3,4-Tetrachloro-benzene	1,2,4,5-Tetrachloro-benzene	Hexachloro-benzene
mol wt	112.56	147.005	147.005	147.005	181.45	215.90	215.9	284.80
mp, °C	−45.34	−16.97	−24.76	53.04	17.15	46.0	139.5	228.7
bp at 101.3 kPa[a], °C	131.7	180.4	173.0	174.1	213.8	254.9	248.0	319.3
critical temperature, °C	359.2	417.2	415.3	407.5	453.3	450	489.8	551
critical pressure, kPa[a]	4519	4031	4864	4109	3718	3380	3380	2847
critical density, kg/L	0.3655	0.411	0.458	0.411	0.447	0.40	0.475	0.518
liquid density, kg/L	1.10118	1.3022	1.2828	1.2475	1.44829	1.70	1.833(s)	1.596
viscosity, mPa·s (= cP)	0.756	1.3018	1.0254			3.37		
heat capacity for liquid, J/g[b]	1.339	1.159		1.188	1.008	1.259	1.142	
heat of fusion, J/g[b]	90.33	86.11	85.98	123.8	85.78	64.52	112.2	89.62
heat of vaporization, J/g[b]	331.1	311.0	296.8	297.4	280.0	268.9	221.8	190.8
flash point,[c] °C	28	71		67	99		none	
standard entropy of formation, J/mol/K	197.5 (liq)			175.4 (cryst)				
thermal conductivity of liquid, W/(m·K)	0.127	0.121		0.105		0.108		
refractive index of liquid, n_D^{25}	1.5219	1.5492	1.54337	1.52849 (55°C)		1.56933		
dielectric constant of liquid	5.621	9.93	5.04	2.41		2.24		
surface tension, mN/m (= dyn/cm)	32.65	36.61	36.20	31.4		38.54		21.6
heat of combustion (25°C), kJ/mol	−3100	−2962	−2955	−2934				
coefficient of expansion of liquid, K^{-1}	95×10^{-5}	110×10^{-5}	111×10^{-5}	116×10^{-5}				
ignition temperature, °C	590	640	>500	>500	>500	>500	>500	

[a]To convert kPa to mm Hg, multiply by 7.5.
[b]To convert J to cal, divide by 4.184.
[c]ASTM method D56-70, closed cup.

513

resistance. 1,4-Dichlorobenzene has different labeling classifications depending on its intended use.

Chlorobenzenes are generally considered nonflammable materials with the exception of monochlorobenzene, which has a flash point of 34.5°C and is a flammable solvent based on DOT standards.

HEALTH AND SAFETY FACTORS

In general, all of the chlorobenzenes are less toxic than benzene. Liquid chlorobenzenes produce mild to moderate irritation upon skin contact. Continued contact may cause roughness or a mild burn. Solids cause only mild irritation. Absorption through the skin is slow. Consequently, with short-time exposure over a limited area, no significant quantities enter the body.

A few kidney and liver damage cases reported may have been caused by repeated exposure to some chlorobenzenes. Fires involving chlorobenzenes liberate HCl and possibly phosgene. When chlorobenzenes are involved in a fire, the proper protective equipment must be used for personnel involved in fighting the fire.

USES

Monochlorobenzene

The largest use of monochlorobenzene in the United States is in the production of nitrochlorobenzenes, both ortho and para, which are separated and used as intermediates for rubber chemicals, antioxidants (qv), dye and pigment intermediates, agriculture products, and pharmaceuticals. Other applications for monochlorobenzene include production of diphenylether, ortho- and para-phenylphenol, 4,4'-dichlorodiphenylsulfone, which is a primary raw material for the manufacture of polysulfones, diphenyldichlorosilane, which is an intermediate for specialty silicones, Grignard reagents, and in dinitrochlorobenzene and catalyst manufacture.

o-Dichlorobenzene

The principal use of o-dichlorobenzene in the United States is the manufacture of 3,4-dichloroaniline, a raw material used in the production of herbicides. A small amount of 3,4-dichloroaniline is used to produce 3,4,4'-trichlorocarbanilide (TCC) used as a bacteriostat in deodorant soaps.

p-Dichlorobenzene

p-Dichlorobenzene's largest and growing outlet is in the manufacture of poly(phenylene sulfide) resin (PPS). Other applications include room deodorant blocks and moth control. Small amounts of p-dichlorobenzene are used in the production of 1,2,4- trichlorobenzene, dyes, and insecticide intermediates.

m-Dichlorobenzene

Isolation of pure m-dichlorobenzene, produced at ~1% in the mixed dichlorobenzenes, is not economical. However, there is potential for m-dichlorobenzene in some new experimental agricultural chemicals.

Other Chlorobenzenes

1,2,4-Trichlorobenzene is used in the manufacture of the herbicide, Banvel. Trichlorobenzenes are used in some pesticides, as a dye carrier, in dielectric fluids, as an organic intermediate and a chemical manufacturing solvent, in lubricants, and as a heat-transfer medium.

J. A. Barter and R. S. Nair, *Review of the Scientific Evidence on the Human Carcinogenic Potential of Para-Dichlorobenzene*, Chlorobenzene Producers Association, Washington, D.C., 1990.

"Dichlorobenzene", *Chemical Products Synopsis*; A Reporting Service of Mannsville Chemical Products Corp., Adams, N.Y., Dec. 1995.

W. K. Johnson with A. Leder and Y. Sakuma, "CEH Product Review", *Chlorobenzenes Chemical Economics Handbook*, SRI International, Menlo Park, Calif., Oct 1989.

D R. Stull, E. F. Westrum, and G. C. Sinke, *The Chemical Thermodynamics of Organic Compounds*, John Wiley & Sons, Inc., New York, 1969.

RAMESH KRISHNAMURTI
Occidental Chemical Corporation

CHLOROCARBONS AND CHLOROHYDROCARBONS, SURVEY

Chlorinated hydrocarbons comprise a family of products widely used throughout the chemical and manufacturing industries. The most common of these products are the chlorinated derivatives of methane, ethane, propane, butane, and benzene. These compounds are used as chemical intermediates and as solvents. Solvent uses include a wide variety of applications, ranging from metal and fabric cleaning operations to reaction media for chemical synthesis. Due to environmental issues, the pattern of end uses for chlorocarbons and chlorohydrocarbons has evolved from primarily solvent applications to use as chemical intermediates for the production of other sustainable products.

PHYSICAL PROPERTIES

Physical properties of selected chlorocarbons and chlorohydrocarbons are listed in Table 1.

CHEMICAL PROPERTIES

Substitution Chlorination

The substitution of chlorine atoms for hydrogen atoms in organic molecules is the basis for many important chlorination processes. Substitution chlorination consists of a series of free-radical reactions, which can be further characterized into a series of steps—initiation, propagation, and termination. The chlorine radicals needed for this

Table 1. Physical Properties of Selected Chlorocarbons and Chlorohydrocarbons

Name	Molecular weight	Melting point (°C)	Boiling point (°C)	Density, 25°C (kg/m^3)	Refractive index, 20°C
methyl chloride	50.49	− 97.7	−24.09	911 ($p > 1$ atm)	1.3389
methylene chloride	84.93	− 97.2	40.0	1325	1.4242
chloroform	119.38	− 63.41	61.17	1479	1.4459
carbon tetrachloride	153.82	− 22.62	76.8	1594	1.4601
vinyl chloride	62.50	−153.84	−13.3	911 ($p > 1$ atm)	1.3700
vinylidene chloride	96.94	−122.56	31.6	1213	1.4249
trichloroethylene	131.39	− 84.7	87.21	1463	1.4773
perchloroethylene	165.83	− 22.3	121.3	1623	1.5059
ethyl chloride	64.50	−138.4	12.3	891	1.3676
ethylene dichloride	98.96	− 35.7	83.5	1256	1.4450
methyl chloroform	133.40	− 30.01	74.09	1338	1.4379
allyl chloride	76.52	−134.5	45.1	939	1.4157
chloroprene	88.54	−130	59.4	956	1.4583
monochlorobenzene	112.56	− 45.31	131.72	1107	1.5241
o-dichlorobenzene	147.00	− 17	180	1306	1.5515
p-dichlorobenzene	147.00	53.09	174	1241	1.5285

chemistry can be generated by thermal, photochemical, or chemical means.

Chlorine radicals obtained from the dissociation of chlorine molecules react with organic species to form hydrogen chloride and an organic radical. The organic radical can react with an undissociated chlorine molecule to produce the organic chloride and a new chlorine radical necessary to continue the reaction. This process may repeat many times before the reaction is finally terminated.

Chain terminations occur in a number of ways. One way is the collision between two chlorine atoms. Oxygen can act as a very strong inhibitor for the chlorine free-radical generation process. The process is inhibited by oxygen to the extent that only a few ppm of oxygen can drastically reduce the reaction rate.

The propagation reactions are highly exothermic and demand careful temperature control by cooling or dilution.

Addition Chlorination

Chlorination of olefins such as ethylene by the addition of chlorine is a commercially important process and can be carried out either as a catalytic vapor- or liquid-phase process. The reaction is influenced by light, the walls of the reactor vessel, and inhibitors such as oxygen. Addition chlorination may proceed by a radical-chain mechanism or an ionic addition mechanism. Ionic addition mechanisms can be maximized and accelerated by the use of a Lewis acid such as ferric chloride, aluminum chloride, antimony pentachloride, or cupric chloride.

Chlorination of olefins is an exothermic process.

Addition of Hydrogen Chloride

The addition of hydrogen chloride to alkenes in the absence of peroxides takes place by an electrophilic addition mechanism. The orientation is in accord with Markovnikov's rule in which the hydrogen atom adds to the side of the double bond that will result in the most stable carbonium ion. The addition occurs in two steps with formation of an intermediate carbonium ion.

Addition of the chloride ion completes the hydrochlorination mechanism.

Historically, metal chloride catalysts have been used in commercial processes for the hydrochlorination of olefinic derivatives. The hydrochlorination of olefins is a weakly exothermic reaction with heats of reaction ranging from 4 to 21 kJ/mol. The hydrochlorination of acetylene is more exothermic, at ~184 kJ/mol.

Elimination of Hydrogen Chloride

Dehydrochlorination of chlorinated hydrocarbons is useful as a means of producing unsaturated products. Dehydrochlorination can be accomplished by reaction with bases or Lewis acids, catalytic reactions, or by thermal noncatalytic chemistry.

Chlorinolysis

Chlorinolysis, also referred to as chlorolysis, is the process of inducing a rupture of carbon–carbon bonds through a combination of saturation or near saturation of the organic reactant with chlorine at elevated temperature. Reaction of C2 and C3 hydrocarbons with excess chlorine at high temperatures can cleave the C–C bonds of the hydrocarbon to give chlorinated derivatives of shorter chain length. Aromatic and aliphatic chlorohydrocarbons containing up to six carbons can be converted to carbon tetrachloride by chloronolysis, though pre-saturation of double bonds in the feedstock with chlorine is usually required in these cases.

Oxychlorination

The oxychlorination reaction consists of combining a source of chlorine atoms, such as hydrogen chloride or chlorine, oxygen (air), and a hydrocarbon or chlorohydrocarbon in the presence of a cupric chloride catalyst. This chemistry is of special importance in the manufacture of chlorinated hydrocarbons, since many of these process produce hydrogen chloride as a coproduct. Oxychlorination allows the hydrogen chloride to be captured for conversion to useful products. Oxychlorination is an exothermic reaction.

Chloro Dehydroxylation

Alcohols react with hydrogen halides to form alkyl halides and water. This chemistry can be carried out in the vapor phase by catalytic means or in the liquid phase by either catalytic or noncatalytic means.

Hydrolysis of Alkyl Halides

The reverse reaction of chloro-dehydroxylation of alcohols, or hydrolysis to replace a chlorine atom with an alcohol substituent group, can occur under appropriate conditions. This chemistry is used for the conversion of epichlorohydrin, which is derived from allyl chloride, into glycerol. This chemistry is an example of the introduction of chlorine into an organic molecule to allow the easy introduction of a different functionality, in this case an alcohol, at a later step in the synthesis.

Chlorination of Aromatics

Aromatic compounds may be chlorinated with diatomic chlorine in the presence of a catalyst such as iron, iron(III) chloride, or other Lewis acids. The halogenation reaction involves electrophilic displacement of the aromatic hydrogen by halogen. Introduction of a second chlorine atom into the monochloro aromatic structure leads to ortho and para substitution. The presence of a Lewis acid favors polarization of the chlorine molecule, thereby increasing its electrophilic character.

Other Chemistry of Chlorocarbons and Chlorohydrocarbons

Fluorination. The chlorine atoms in chlorocarbons can be replaced by fluorine through halogen exchange chemistry. A chlorine containing feedstock is typically reacted with hydrogen fluoride or other suitable inorganic fluoride. The selection of the halogenating agent is determined by the chemical structure of the chlorocarbon and the desired degree of fluorination of the product (see FLUORINATED ALIPHATIC COMPOUNDS).

Silation. Methyl chloride reacts with silicon in the presence of copper and promoters at high temperature to form methylchlorosilanes. Methylchlorosilanes are the building blocks of many silicone products (see SILICON COMPOUNDS, SILICONES).

Reaction with Aluminum. Many chlorinated hydrocarbons react readily with aluminum in the so-called bleeding reaction. A red aluminum chloride-chlorinated hydrocarbon complex is formed. Storage of uninhibited chlorinated solvents in aluminum vessels results in corrosion in a short period of time. For this reason, aluminum should not be used in the construction of storage vessels for chlorinated hydrocarbons.

Oxidation. All chlorinated hydrocarbons are susceptible to pyrolysis at high temperatures. This process liberates hydrogen chloride, water, and carbon dioxide.

Thermal oxidation processes have been used as means to destroy chlorinated by-products of chlorohydrocarbon manufacture that cannot otherwise be utilized. The Catalytic oxidation processes are also used for the safe disposal of waste chlorinated hydrocarbons.

Chlorinated hydrocarbons may be partially oxidized. This chemistry is used in the manufacture of dichloroacetyl chloride from vinylidene chloride. Partial oxidation can also cause undesired product degradation. Vinylidene chloride polymerizes to a solid very quickly if exposed to air without the presence of a proper inhibitor. Formation of a peroxide intermediate is an important step in the initiation of this process.

Hydrogenation. Chlorocarbons can be hydrogenated in the presence of suitable catalysts in order to replace chlorine atoms with hydrogen atoms. This chemistry can be employed to convert carbon tetrachloride to chloroform or perchloroethylene to trichloroethylene.

Polymerization. Olefinic chlorohydrocarbons react to form many useful polymers. The usual mechanism is a free-radical initiated chain reaction. The four principal steps of a free-radical chain reaction are initiation, propagation, chain transfer, and termination (see VINYL CHLORIDE POLYMERS; VINYLIDENE CHLORIDE MONOMER AND POLYMERS; POLYCHLOROPRENE).

Reimer-Timann Reaction. In the Reimer-Tiemann reaction, phenol is reacted with chloroform in the presence of a base to introduce an aldehyde group onto the aromatic ring.

MANUFACTURE

Chlorinated hydrocarbons are produced by a wide variety of chemical processes. An important aspect of the development of chlorinated hydrocarbon manufacture is the use of available by-products from the manufacture of other chlorinated products as raw materials. These by-products include products of over-chlorination as well as HCl that is coproduced as a part of many processes.

Chlorinated Methanes

Methyl chloride may be produced from either methanol or methane. The methanol-based route may be conducted in either the vapor or liquid phase. The vapor-phase process typically utilizes a reactor packed with catalyst. The catalyst is alumina or alumina based.

The liquid-phase methyl chloride process is the more widely used process. The liquid-phase process can be designed with multiple reactors to achieve both high methanol and high HCl conversion. In the liquid-phase process, methanol is contacted with HCl at $70-160°C$ and $200-1100$ kPa in a boiling bed reactor.

Methyl chloride is also produced by the thermal chlorination of methane in the gas phase at a temperature in the range of $490-530°C$. Methylene chloride, chloroform, carbon tetrachloride and coproduct HCl are produced by thermal chlorination of methane or methyl chloride.

Chlorinated Ethanes and Ethylenes

Ethyl Chloride. Ethyl chloride can be produced by the reaction of hydrogen chloride with ethanol, the chlorination of ethane, or the addition of hydrogen chloride to ethylene. The addition of HCl to ethylene is the most commonly practiced commercial process. Aluminum chloride is typically used as the catalyst.

EDC and Vinyl Chloride. Vinyl chloride is produced almost exclusively by the thermal dehychlorination of EDC, which is produced from ethylene. EDC may be produced by the direct chlorine addition to ethylene or by the oxychlorination of ethylene with oxygen and HCl.

Vinyl chloride is produced from EDC by a thermal dehydrochlorination process. Reactions are conducted in cracking furnaces, somewhat similar to the process for hydrocarbon cracking. Some vinyl chloride continues to be manufactured by the hydrochlorination of acetylene. This process has regained favor in areas where the abundance of coal allows production of acetylene at a cost competitive to ethylene.

1,1,1-Trichloroethane. The production of 1,1,1-trichloroethane has been sharply curtailed because of restrictions imposed by the Montreal Protocol.

Vinylidene Chloride. Vinylidene chloride can be manufactured from 1,1,1- or 1,1,2-trichloroethane. The 1,1,2-trichloroethane route is the more common. If 1,1,1-trichloroethane is used, it is dehydrochlorinated to vinylidene chloride by a thermal or catalytic process. 1,1,2-trichloroethehylene is converted to vinylidene chloride by reaction with caustic soda.

Trichloroethylene and Perchloroethylene. Trichloroethylene and perchloroethylene can be produced by many routes. An oxychlorination process has been used quite successfully to coproduce trichloroethylene and perchloroethylene. A wide range of chlorinated hydrocarbons may be used as feedstocks in this process. Ethylene, EDC, or the chlorinated by-products of EDC manufacture are the most common feedstocks. Either chlorine or HCl may be used as chlorinating agents in this process. The more heavily chlorinated organic species present in by-products of EDC manufacture can also serve as a chlorine source.

Trichloroethylene is also produced via a thermal chlorination process.

Like trichloroethylene, perchloroethylene can be produced in a thermal chlorination process. Historically, the thermal chlorination process for perchloroethylene has been used to coproduce carbon tetrachloride.

As with all thermal chlorination processes, the manufacture of perchloroethylene results in the coproduction of anhydrous HCl. A particular advantage of the thermal chlorination process for perchloroethylene is the wide range of feedstocks that can be used.

Chlorinated Propanes and Higher

Allyl Chloride. Allyl chloride is produced from chlorine and propylene in a thermal chlorination process, similar to that used for the production of chlorinated methanes, trichloroethylene, or perchloroethylene.

Chloroprene. Chloroprene is manufactured by a two-step process. The basic feedstocks are butadiene and chlorine. In the first step, butadiene is chlorinated to produce 3,4-dichloro-1-butene and other isomers. Historically, the chlorination was a vapor-phase process, but in recent years a liquid-phase chlorination process has also been developed.

Chlorinated Aromatics

Raschig developed the first commercial oxychlorination process to make chlorobenzene in 1928. The chlorobenzene product was then hydrolyzed to phenol.

SHIPMENT AND STORAGE

Because of the wide range in properties of the various chlorocarbons and chlorohydrocarbons, persons handling these substances should be trained to recognize and avoid the hazards of the specific compound they are using. The extreme difference in properties requires very different handling procedures.

Storage tanks should be of minimum carbon steel construction. Aluminum, zinc, and magnesium alloys should generally be avoided.

Storage tanks should be grounded and have provision for adequate pressure relief. Labeling should conform to local requirements to communicate the flammability, reactivity, and toxicity of the material being stored. Storage tanks should also have adequate spill protection. The spill protection system should be designed to prevent cross-contamination of incompatible materials. The vapor space of storage tanks used in chlorocarbon service should be oxygen-free and padded with nitrogen. The tank vent system should be designed to prevent cross-contamination with other chemicals stored at the same facility.

For chlorocarbons and chlorohydrocarbons the products that are normally liquids are shipped in drums, tank trucks, rail cars, barges, or ocean-going ships.

HEALTH AND SAFETY FACTORS

From an environmental standpoint, issues have been raised about the impact of these compounds both on the atmosphere and in groundwater.

Challenges continue to be waged about the potential impact that chlorinated hydrocarbon products have on human health. The earliest symptoms associated with these compounds were related to their impact on the central nervous system. The most significant findings in long-term studies are centered on effects on liver and kidney for trichloroethylene, perchloroethylene, and ethylene dichloride, while methylene chloride effects are mostly associated with the liver and the lungs. For vinyl chloride, the most significant toxic effect is carcinogenicity, specifically angiosarcoma of the liver, which occurs with extremely high exposure to vinyl chloride. Vinyl chloride is the only chlorinated hydrocarbon that has been classified as a known human carcinogen by authoritative bodies as a result of the findings of epidemiology studies.

USES

Because of their widely varying properties, chlorocarbons and chlorohydrocarbons are used in a broad range of applications. Many of these products have excellent solvent properties. For this reason, they continue to be in demand for metal cleaning and vapor degreasing operations and as reaction media for other chemical processes. They are also used as chemical intermediates for a variety of products, from polymers to silicones, fluorocarbons, and other speciality chemicals.

R. Bruckner, *Advanced Organic Chemistry:Reaction Mechanisms*, Harcourt Academic Press, New York, 2002, p. 69.

E. Linak and G. Toki, "C2 Chlorinated Solvents", *Chem. Econ. Handbook Product Review*, January, 2002.

J. March, *Advanced Organic Chemistry: Reactions, Mechanism and Structure*, 4th ed., John Wiley and Sons, Inc., New York, 1992, p. 544.

KENRIC A. MARSHALL
The Dow Chemical Company

CHLOROETHYLENES AND CHLOROETHANES

ETHYLENE DICHLORIDE

Introduction

1,2-Dichloroethane, ethylene chloride, ethylene dichloride, CH_2ClCH_2Cl, is a colorless, volatile liquid with a pleasant odor, stable at ordinary temperatures. It is miscible with other chlorinated solvents and soluble in common organic solvents as well as having high solvency for fats, greases, and waxes. It is most commonly used in the production of vinyl chloride monomer.

Physical and Chemical Properties

Pyrolysis. Pyrolysis of 1,2-dichloroethane in the temperature range of 340–515°C gives vinyl chloride, hydrogen chloride, and traces of acetylene and 2-chlorobutadiene. Reaction rate is accelerated by chlorine, bromine, bromotrichloromethane, carbon tetrachloride, and other free-radical generators.

Hydrolysis. Heating 1,2-dichloroethane with excess water at 60°C in a nitrogen atmosphere produces some hydrogen chloride. The rate of evolution is dependent on the temperature and volume of the aqueous phase. Hydrolysis at 160–175°C and 1.5 MPa (15 atm) in the presence of an acid catalyst gives ethylene glycol, which is also obtained in the presence of aqueous alkali at 140–250°C and up to 4.0 MPa (40 atm) pressure.

Oxidation. Atmospheric oxidation of 1,2-dichloroethane at room or reflux temperatures generates some hydrogen chloride and results in solvent discoloration. Addition of 0.1–0.2 wt.% of an amine, e.g., diisopropylamine, protects the 1,2-dichloroethane against oxidative breakdown.

Corrosion. Corrosion of aluminum, iron, and zinc by boiling 1,2-dichloroethane has been studied. Dry and refluxing 1,2-dichloroethane completely consumed a 2024 aluminum coupon in a 7-d study, whereas iron and zinc were barely attacked. Aluminum was attacked less than iron or zinc by refluxing with 1,2-dichloroethane containing 7% water.

Nucleophilic Substitution. The kinetics of the bimolecular nucleophilic substitution of the chlorine atoms in 1,2-dichloroethane with NaOH, $NaOC_6H_5$, $(CH_3)_3N$, pyridine, and CH_3COONa in aqueous solutions at 100–120°C has been studied. The reaction of sodium cyanide with 1,2-dichloroethane in methanol at 50°C to give 3-chloropropionitrile proceeds very slowly.

Manufacture

1,2-Dichloroethane is produced by the vapor- or liquid-phase chlorination of ethylene. Most liquid-phase processes use small amounts of ferric chloride as the catalyst. Other catalysts claimed in the patent literature include aluminum chloride, antimony pentachloride, and cupric chloride and an ammonium, alkali, or alkaline-earth tetrachloroferrate. The chlorination is carried out at 40–50°C with 5% air or other free-radical inhibitors added to prevent substitution chlorination of the product.

Health and Safety Factors

1,2-Dichloroethane at high vapor concentrations (above 200 ppm) can cause central nervous system depression and gastrointestinal upset characterized by mental confusion, dizziness, nausea, and vomiting. Liver, kidney, and adrenal injuries may occur at the higher vapor levels.

1,2-Dichloroethane is one of the more toxic chlorinated solvents by inhalation. Repeated skin contact can cause defatting of the skin, severe irritation, and moderate edema. Eye contact may have slight to severe effects.

Uses

Production of vinyl chloride monomer comprises 94% of EDC use. Three percent goes to the production of ethylene amines; 1% goes to the production of 1,1,1-trichloroethane; 1% goes to the production of vinylidene chloride; 1% goes to miscellaneous uses (includes production of trichloroethylene and perchloroethylene).

DICHLOROETHYLENE

Introduction

1,1-Dichloroethylene is more commonly known as vinylidene chloride and is covered in an article in the *Encyclopedia* by that title.

1,2-Dichloroethylene (1,2-dichloroethene) is also known as acetylene dichloride, dioform, α,β-dichloroethylene, and *sym*-dichloroethylene. It exists as a mixture of two geometric isomers: *trans*-1,2-dichloroethylene and *cis*-1,2-dichloroethylene.

Physical and Chemical Properties

1,2-Dichloroethylene consists of a mixture of the cis and trans isomers, as manufactured.

Manufacture

1,2-Dichloroethylene can be produced by direct chlorination of acetylene at 40°C. It is often produced as a by-product in the chlorination of chlorinated compounds and recycled as an intermediate for the synthesis of more useful chlorinated ethylenes. 1,2-Dichloroethylene can be formed by continuous oxychlorination of ethylene by use of a cupric chloride–potassium chloride catalyst, as the first step in the manufacture of vinyl chloride.

Storage and Handling

1,2-Dichloroethylene is usually shipped in 208-L (55 gal) and 112-L (30 gal) steel drums. Because of the corrosive products of decomposition, inhibitors are required for storage. The stabilized grades of the isomers can be used or stored in contact with most common construction materials, such as steel or black iron. Contact with copper or its alloys and with hot alkaline solutions should be avoided. However, the liquid, even hot, burns with a very cool flame which self-extinguishes unless the temperature is well above the flash point.

Health and Safety Factors

1,2-Dichloroethylene is toxic by inhalation and ingestion and can be absorbed by the skin. The odor does not provide adequate warning of dangerously high vapor concentrations. Thorough ventilation is essential whenever the solvent is used for both worker exposure and flammability concerns. Symptoms of exposure include narcosis, dizziness, and drowsiness. Currently no data are available on the chronic effects of exposure to low vapor concentrations over extended periods of time.

Uses

1,2-Dichloroethylene can be used as a low temperature extraction solvent for organic materials such as dyes, perfumes, lacquers, and thermoplastics. It is also used as a chemical intermediate in the synthesis of other chlorinated solvents and compounds.

TRICHLOROETHYLENE

Introduction

Trichloroethylene, trichloroethene, $CHCl=CCl_2$, commonly called "tri," is a colorless, sweet smelling (chloroformlike odor), volatile liquid and a powerful solvent for a large number of natural and synthetic substances. It is nonflammable under conditions of recommended use. In the absence of stabilizers, it is slowly decomposed (autoxidized) by air. The oxidation products are acidic and corrosive. Stabilizers are added to all commercial grades. Trichloroethylene is moderately toxic and has narcotic properties.

Physical and Chemical Properties

Trichloroethylene is immiscible with water but miscible with many organic liquids; it is a versatile solvent. It does not have a flash or fire point. However, it does exhibit a flammable range when high concentrations of vapor are mixed with air and exposed to high energy ignition sources.

The most important reactions of trichloroethylene are atmospheric oxidation and degradation by aluminum chloride. Atmospheric oxidation is catalyzed by free radicals and accelerated with heat and with light, especially ultraviolet.

Trichloroethylene is not readily hydrolyzed by water. Under pressure at 150°C, it gives glycolic acid, $CH_2OH-COOH$, with alkaline hydroxides.

In the presence of catalysts, trichloroethylene is readily chlorinated to pentachloro- and hexachloroethane. Fluorination with hydrogen fluoride in the presence of antimony trifluoride produces 2-chloro-1,1,1-trifluoroethane.

Manufacture

From Acetylene. Most trichloroethylene is made from ethylene, 1,2-dichloroethane, or ethylene dichloride. The acetylene-based process consists of two steps. First acetylene is chlorinated to 1,1,2,2-tetrachloroethane.

The product is then dehydrohalogenated to trichloroethylene at 96–100°C in aqueous bases such as $Ca(OH)_2$ or by thermal cracking, usually over a catalyst such as barium chloride on activated carbon or silica or aluminum gels at 300–500°C.

Chlorination of Ethylene. Dichloroethane, produced by chlorination of ethylene, can be further chlorinated to trichloroethylene and tetrachloroethylene.

Oxychlorination of Ethylene or Dichloroethane. Ethylene or dichloroethane can be chlorinated to a mixture of tetrachoroethylene and trichloroethylene in the presence of oxygen and catalysts. The reaction is carried out in a fluidized-bed reactor at 425°C and 138–207 kPa (20–30 psi). The most common catalysts are mixtures of potassium and cupric chlorides. Temperature control is critical. Reaction vessels must be constructed of corrosion-resistant alloys.

Shipping and Storage

Shipment of trichloroethylene is usually by truck or rail car and also in 208-liter (55-gallon) steel drums. Mild steel tanks, if appropriately equipped with vents and vent driers to prevent the accumulation of water, are adequate for storage. Precautions, such as diking, should be taken to provide for adequate spill containment at the storage tank. Seamless black iron pipes are suitable for transfer lines, gasketing should be of Teflon, Viton, or other solvent impermeable material. Aluminum should never be used as a construction material for any halogenated hydrocarbon. Glass containers, amber or green, are suitable for small quantities, such as in a laboratory.

Precautions in handling any waste products in conformance with federal, state, and local regulations should be included.

Although the flammability hazard is very low, ignition sources should not be present when trichloroethylene is used in highly confined or unventilated areas. Tanks in which flammable concentrations could develop should be grounded to prevent build-up of static electric charges. Under no circumstances should welding or cutting with a torch take place on any storage container or process equipment containing trichloroethylene.

Specifications and Standards

Commercial grades of trichloroethylene, formulated to meet use requirements, differ in the amount and type of added inhibitor.

Apart from added stabilizers, commercial grades of trichloroethylene should not contain more than the following amounts of impurities: water 100 ppm; acidity, ie, HCl, 5 ppm; insoluble residue, 10 ppm. Free chlorine should not be detectable. Test methods have been established by ASTM to determine the following characteristics of trichloroethylene: acid acceptance, acidity or alkalinity, color, corrosivity on metals, nonvolatile-matter content, pH of water extractions, relative evaporation rate, specific gravity, water content, water-soluble halide ion content, and halogen content.

Health and Safety Factors

Trichloroethylene is acutely toxic, primarily because of its anesthetic effect on the central nervous system. Exposure to high vapor concentrations is likely to cause headache, vertigo, tremors, nausea and vomiting, fatigue, intoxication, unconsciousness, and even death.

Exposure occurs almost exclusively by vapor inhalation, which is followed by rapid absorption into the bloodstream.

The distinctive odor of trichloroethylene may not necessarily provide adequate warning of exposure, because it quickly desensitizes olfactory responses. Ingestion of large amounts of trichloroethylene may cause liver damage, kidney malfunction, cardiac arrhythmia, and coma; vomiting should not be induced, but medical attention should be obtained immediately.

Protective gloves and aprons should be used to prevent skin contact, which may cause dermatitis. Eyes should be washed immediately after contact or splashing with trichloroethylene.

Uses

Approximately 67% of the trichloroethylene produced in the United States is consumed in the vapor decreasing of fabricated metal parts (see METAL SURFACE TREATMENTS); 30% is divided equally between exports and miscellaneous applications and 3% is used as a polymerization modifier.

A variety of miscellaneous applications include use of trichloroethylene as a component in adhesive and paint-stripping formulations, a low temperature heat-transfer medium, a nonflammable solvent carrier in industrial paint systems, and a solvent base for metal phosphatizing systems. Trichloroethylene is used in the textile industry as a carrier solvent for spotting fluids and as a solvent in waterless preparation dying and finishing operations. Cleaning or drying compositions as a replacement for CFCs has been described.

Trichloroethylene is widely used as a chain-transfer agent in the production of poly(vinyl chloride). There has been a significant increase in the use of trichloroethylene as a feedstock in the production of hydrofluorocarbons, the replacements for chlorofluorcarbons implicated in the depletion of stratospheric ozone.

TETRACHLOROETHYLENE

Introduction

Tetrachloroethylene, perchloroethylene, $CCl_2=CCl_2$, is commonly referred to as "perc" and sold under a variety of trade names. It is the most stable of the chlorinated ethylenes and ethanes, having no flash point and requiring only minor amounts of stabilizers.

Physical and Chemical Properties

Tetrachloroethylene dissolves a number of inorganic materials including sulfur, iodine, mercuric chloride, and appreciable amounts of aluminum chloride. Tetrachloroethylene dissolves numerous organic acids, including benzoic, phenylacetic, phenylpropionic, and salicylic acid, as well as a variety of other organic substances such as fats, oils, rubber, tars, and resins. It does not dissolve sugar, proteins, glycerol, or casein. It is miscible with chlorinated organic solvents and most other common solvents. Tetrachloroethylene forms approximately sixty binary azeotropic mixtures.

Stabilized tetrachloroethylene, as provided commercially, can be used in the presence of air, water, and light, in contact with common materials of construction, at temperatures up to about 140°C. It resists hydrolysis at temperatures up to 150°C. However, the unstabilized compound, in the presence of water for prolonged periods, slowly hydrolyzes to yield trichloroacetic acid and hydrochloric acid. In the absence of catalysts, air, or moisture, tetrachloroethylene is stable to about 500°C.

Manufacture

The following processes are commonly used today.

Chlorination of Ethylene Dichloride. Tetrachloroethylene and trichloroethylene can be produced by the noncatalytic chlorination of ethylene dichloride (EDC) or other two-carbon (C2) chlorinated hydrocarbons. This process is advantageous when there is a feedstock source of mixed C2 chlorinated hydrocarbons from other processes and an outlet for the by-product HCl stream.

Chlorination of C1–C3 Hydrocarbons or Partially Chlorinated Derivatives. Tetrachloroethylene and carbon tetrachloride are produced with or without a catalyst at high temperatures (550–700°C) from light hydrocarbon feedstocks or their partially chlorinated derivatives.

Oxychlorination of C2 Chlorinated Hydrocarbons. Tetrachloroethylene and trichloroethylene can be produced by reaction of EDC with chlorine or HCl and oxygen in the presence of a catalyst. When hydrochloric acid is used, additional oxygen is required. This process is advantageous in that no by-product HCl is produced, and it can be integrated with other processes as a net HCl consumer.

Shipping and Storage

Tetrachloroethylene is shipped by barge, tank car, tank truck, and 55-gallon (208-L) steel drums. It may be stored in mild steel tanks that are dry, free of rust, and equipped with a chemical (such as calcium chloride) vent dryer and controlled evaporation vent. Appropriate secondary containment including dikes and sealed surfaces should be provided in accordance with federal and local standards to prevent potential groundwater contamination in the event of a leak. Piping and centrifugal or positive displacement pumps should be constructed of ductile iron or carbon steel with gasket materials made of impregnated cellulose fiber, cork base materials, or Viton resin.

Health and Safety Factors

Overexposure to tetrachloroethylene by inhalation affects the central nervous system and the liver. Dizziness, headache, confusion, nausea, and eye and mucous tissue irritation occur during prolonged exposure to vapor concentrations of 200 ppm. These effects are intensified and include incoordination and drunkenness at concentrations in excess of 600 ppm. At concentrations in excess of 1000 ppm the anesthetic and respiratory depression effects can cause unconsciousness and death. Alcohol consumed before or after exposure may increase adverse effects.

Repeated exposure of skin to liquid tetrachloroethylene may defat the skin causing dermatitis. When frequent or prolonged contact is likely, gloves of Viton, nitrile rubber, or neoprene should be used, discarding them when they begin to deteriorate. Tetrachloroethylene can cause significant discomfort if splashed in the eyes. Although no serious injury results, it can cause transient, reversible corneal injury.

Ingestion of small amounts of tetrachloroethylene is not likely to cause permanent injury; however, ingestion of large amounts may result in serious injury or even death. If solvent is swallowed, consult a physician immediately. Do not induce vomiting. If solvent is aspirated it is rapidly absorbed through the lungs and may cause systemic effects and chemical pneumonia.

Exposure to tetrachloroethylene as a result of vapor inhalation is followed by absorption into the bloodstream.

Uses

Use of tetrachlorethylene as a feedstock for chlorofluorocarbon production accounts for 65% of demand. Approximately 15% is used in the dry-cleaning industry. Metal cleaning applications account for 10% of consumption and miscellaneous uses account for 10%. Miscellaneous uses include transformer insulating fluid, chemical maskant formulations, and in a process for desulfurizing coal.

"Ethylene Dichloride," *Chemical Economics Handbook*, 651.5000, Stanford Research Institute, Menlo Park, Calif., 2000.

D. D. Irish, in F. A. Patty, ed., *Industrial Hygiene and Toxicology*, 3rd Revised Ed., John Wiley & Sons, Inc., New York, 1963, pp. 3491–3497.

Specialty Chlorinated Solvents Product Stewardship Manual, 1991 ed, The Dow Chemical Company, Midland, Mich., form 100-6170-90HYC.

1990–1991 Threshold Limit Values for Chemical Substances and Physical Agents, American Conference of Governmental Industrial Hygienists, 1990.

GAYLE SNEDECOR
J. C. HICKMAN
The Dow Chemical Company
JAMES A. MERTENS
Dow Chemical USA

CHLOROFORM

Chloroform (trichloromethane, methenyl chloride), $CHCl_3$, at normal temperature and pressure is a heavy, clear, volatile liquid with a pleasant, ethereal, nonirritant odor. Although chloroform is nonflammable, its hot vapor in admixture with vaporized alcohol burns with a green tinged flame. Chloroform is miscible with the principal organic solvents and is slightly soluble in water. It is less stable in storage than either methyl or methylene chloride. Chloroform decomposes at ordinary temperatures in sunlight in the absence of air and in the dark in the presence of air. Phosgene is one of the oxidative decomposition products.

PHYSICAL AND CHEMICAL PROPERTIES

The physical properties of chloroform are listed in Table 1.

Chloroform dissolves alkaloids, cellulose acetate and benzoate, ethylcellulose, essential oils, fats, gutta-percha, halogens, methyl methacrylate, mineral oils, many resins, rubber, tars, vegetable oils, and a wide range of common organic compounds. Chloroform slowly decomposes on prolonged exposure to sunlight in the presence or absence of air and in the dark in the presence of air. The products of oxidative breakdown include phosgene, hydrogen chloride, chlorine, carbon dioxide, and water.

Chloroform resists thermal decomposition at temperatures up to about 290°C. Pyrolysis of chloroform

Table 1. Physical Properties of Chloroform

Property	Value
mol wt	119.38
refractive index at 20°C	1.4467
flash point, °C	none
mp, °C	−63.2
101 MPa[a]	−43.4
bp at 101 kPa[b], °C	61.3
sp gr at 0/4°C	1.52637
critical temperature, °C	263.4
critical pressure, MPa[a]	5.45
critical density, kg/m³	500
critical volume, m³/kg	0.002
thermal conductivity at 20°C, W/(m·K)	0.130
dielectric constant, 20°C	4.9
heat of combustion, MJ/(kg·mol)[c]	373
solubility of chloroform in water at 0°C, g/kg H_2O	10.62
viscosity, liq, at 0°C mPa·s (= cP)	0.700

[a]To convert MPa to atm, multiply by 9.87.
[b]To convert kPa to mm Hg, multiply by 7.5.
[c]To convert J to cal, divide by 4.184.

vapor occurs at temperatures above 450°C, producing tetrachloroethylene, hydrogen chloride, and a number of chlorohydrocarbons in minor amounts. Chloroform reacts with aniline and other aromatic and aliphatic primary amines in alcoholic alkaline solution to form isonitriles, ie, isocyanides, carbylamines.

MANUFACTURE

Chloroform can be manufactured from a number of starting materials. Methane, methyl chloride, or methylene chloride can be further chlorinated to chloroform, or carbon tetrachloride can be reduced, ie, hydrodechlorinated, to chloroform. Methane can be oxychlorinated with HCl and oxygen to form a mixture of chlorinated methanes. Many compounds containing either the acetyl (CH₃CO) or CH₃CH(OH) group yield chloroform on reaction with chlorine and alkali or hypochlorite. Methyl chloride chlorination is now the most common commercial method of producing chloroform. Many years ago chloroform was almost exclusively produced from acetone or ethyl alcohol by reaction with chlorine and alkali.

SPECIFICATIONS AND STANDARDS

Technical-grade chloroform generally contains one or more stabilizers, which vary according to specification requirements. The most common is 50 ppm 2-methyl-2-butene. Other stabilizers are industrial methylated spirit (0.2%), absolute alcohol (0.6–1%), thymol, t-butylphenol, or n-octylphenol (0.0005–0.01%).

HEALTH AND SAFETY FACTORS

The principal hazard in exposure to chloroform is damage to the liver and kidneys resulting from inhalation or ingestion. Inhalation of high concentrations may result in disturbances of equilibrium or loss of consciousness. Chloroform is mildly irritating to skin and mucous membranes upon contact, and to the alimentary tract upon ingestion. It is believed that medically significant quantities are not absorbed through intact skin.

Repeated or prolonged contact with the skin, especially under clothing, may result in local irritation and inflammation, and at elevated temperatures such as in the presence of an open flame, chloroform decomposes to form by-products, including phosgene, chlorine, and hydrogen chloride, all of which are severe irritants to the respiratory tract.

Ingestion of chloroform is followed immediately by a severe burning in the mouth and throat, pain in the chest and abdomen, and vomiting. Loss of consciousness and liver injury may follow depending on the amount swallowed.

The most serious hazard of repeated exposure to chloroform inhalation is injury to the liver and kidneys.

NIOSH recognizes chloroform as a carcinogen, and recommends a STEL of 2 ppm; ACGIH, TLV TWA is 10 ppm, with an A3 designation. OSHA has a ceiling limit of 50 ppm. WHO has a drinking water guideline of 30 μg/L. IARC classifies chloroform as a group 2B (possibly carcinogenic to humans, and EPA as a B2, probable human carcinogen.

Treatment of chloroform poisoning is symptomatic; no specific antidote is known. Adrenalin should not be given to a person suffering from chloroform poisoning.

USES

About 95% of the chloroform produced goes into the production of HCFC-22 (chlorodifluoromethane). Of this 95% about 70% is used as a refrigerant and about 30% is used as a starting material in the production of fluoropolymers, such as polytetrafluoroethylene (PTFE). Miscellaneous uses of the remaining 5% of the chloroform production include laboratory reagents and extractive solvents for pharmaceuticals.

The miscellaneous uses include extraction and purification of penicillin, alkaloids, vitamins, and flavors, and as an intermediate in the preparation of dyes and pesticides. A biologically active chloroform extract from mangrove plants has been described. Use as an antispasmodic, antiarrythmic agent, or anticholinergic agent are possible.

"Chloroform, Chemical Profile," *Chemical Market Reporter*, Oct. 16, 2000.

I. Mellan, *Source Book of Industrial Solvents*, Reinhold Publishing Corp., New York, 1957, p. 126.

J. B. Reid, in E. Bingham, B. Cohrssen, and C. H. Powell, eds., *Patty's Toxicology*, 5th ed., Vol. 5, John Wiley & Sons, Inc., New York, 2001.

MICHAEL T. HOLBROOK
Dow Chemical U.S.A.

CHLOROFORMATES AND CARBONATES

The reaction of phosgene (carbonic dichloride) with alcohols gives two classes of compounds, carbonic esters and carbonochloridic esters, commonly referred to as carbonates and chloroformates. The carbonic acid esters (carbonates), ROC(O) OR, are the diesters of carbonic acid. The carbonochloridic esters, also referred to as chloroformates or chlorocarbonates, ClC(O)OR, are esters of hypothetical chloroformic acid, ClCOOH.

CHLOROFORMATES

Physical Properties

In general, carbonochloridates or chloroformates are clear, colorless liquids with low freezing points and relatively high boiling points ($>100°C$). They are soluble in most organic solvents, but insoluble in water, although they do hydrolyze in water. The lower chloroformates, eg, methyl and ethyl chloroformate, hydrolyze rapidly in water at room temperature, whereas the higher chloroformates, eg, 2-ethylhexyl or aromatic chloroformates, hydrolyze slowly in water at room temperature. The physical properties of the most widely used chloroformate esters are given in Table 1.

Chemical Properties

Chloroformates are reactive intermediates that combine acid chloride and ester functions. They undergo many reactions similar to those of acid chlorides; however, the rates are usually slower. Those containing smaller organic (hydrocarbon) substituents react faster than those containing large organic (hydrocarbon) substituents. Reactions of chloroformates and other acid chlorides proceed faster and with better yields when means are employed to remove or capture HCl as it is formed. Classical acid scavengers include alkali hydroxides or tertiary amines, which act as stoichiometric acid acceptors rather than as true catalysts.

Stability. The ester moiety determines thermal stability generally in the following order of decreasing stability: aryl > primary alkyl > secondary alkyl > tertiary alkyl. In terms of mechanistic chemistry the chloroformates that produce stable carbonium ions on thermal decomposition, eg, benzyl, isopropyl, or tertiary butyl, are unstable and can cause increased pressure in closed containers.

Reactions with Oxygen Moieties. *Hydroxylic Compounds.* Chloroformates on reaction with water give the parent hydroxy compound, HCl, and CO_2 as well as the symmetrical carbonate formed by the reaction of the hydroxy compound with chloroformate.

Alkali Metal Hydroxides. Addition of base to aqueous chloroformates catalyzes hydrolysis to yield the parent hydroxy compound. However, the use of a stoichiometric amount of alkali metal hydroxides can lead to the symmetrical carbonate, especially from aryl chloroformates.

Aliphatic Alcohols and Thiols. Aliphatic alcohols on reaction with chloroformates give carbonates and hydrogen chloride. Frequently, the reaction proceeds at room temperature without a catalyst or hydrogen chloride acceptor. However, faster reactions and better yields are obtained in the presence of alkali metals or their hydroxides, or tertiary amines. Reactions of chloroformates with thiols yield monothiolocarbonates.

Heterocylic Alcohols. Their reactions with chloroformates lead to carbonates.

Table 1. Physical Properties of Selected Chloroformates

Chloroformate	Mol wt	Sp gr, d_4^{20}	Refractive index n^{20}_D	Flash point, °C TOC[a]	Flash point, °C TCC[b]	Viscosity, mPa·s (=cP), 20°C	Bp, °C at 2.67 kPa[c]	Bp, °C at 13.3 kPa[c]	Bp, °C at 101.3 kPac
methyl	94.5	1.250	1.3864	24.4	17.8				71
ethyl	108.53	1.138	1.3950	27.8	18.3				94
isopropyl	122.55	1.078	1.3974	27.8	23.3	0.65		47	105
n-propyl	122.55	1.091	1.4045	34.4		0.80	25.3	57.5	112.4
allyl	120.5	1.1394	1.4223	27.8	31.1	0.71	25	57	
n-butyl	136.58	1.0585	1.4106	52.2	46.0	0.888	44	77.6	
sec-butyl	136.58	1.0493	1.4560	35.6	38.0	0.897	36	69	
isobutyl	136.58	1.0477	1.4079	39.5	34.4	0.88	39	71	
2-ethylhexyl	192.7	0.9914	1.4307		86.0	1.774	98	137	
n-decyl	220.7	0.9732	1.4400	118.3	120.2	3.00	122	159	
phenyl	156.57	1.2475	1.5115		77.0	1.882	83.5	121	185
benzyl	170.6	1.2166	1.5175	80.0	107.9	2.57	103	123	152
ethylene bis	186.98	1.4704	1.4512	134.0	126.0	4.78	108	137	
diethylene glycol bis	231.0	1.388	1.4550	160.0	182.2	8.76	148	180	

[a]Tag open cup.
[b]Tag closed cup.
[c]To convert kPa to mm Hg, multiply by 7.50.

Phenols. Phenols are unreactive toward chloroformates at room temperature and at elevated temperatures the yields of carbonates are relatively poor (<10%) in the absence of catalysis or quantitative HCl scavengers. Many catalysts have been claimed in the patent literature leading to high yields of carbonates from phenol and chloroformates. Alternative systems include biphasic systems that employ alkali bases and phase transfer catalysts. The use of catalyst or an alkali base is even more essential in the reaction of phenols and aryl chloroformates. Among the catalysts claimed are amphoteric metals or their halides, magnesium halides, activated carbon, titanium oxide, magnesium or manganese, secondary or tertiary amines such as imidazole, pyridine, quinoline, picoline, heterocyclic basic compounds, and carbonamides, thiocarbonamides, phosphoroamides, and sulfonamides.

Carboxylic Acids. The reaction product of chloroformates and carboxylic acids is a mixed carboxylic–carbonic anhydride. The intermediate mixed anhydrides are very active acylating agents, but these agents may be isolated in cold temperatures for producing useful products. More often the anhydride is a transient intermediate that leads to the formation of a mixture of ester, carbonate and anhydride. The pathway is strongly dependent upon the anhydride itself and the choice of catalyst.

Pyrocarbonates or dicarbonates (anhydrides of carbonic acids) have been prepared from the alkali salt of the carbonate. Pyrocarbonates are useful as intermediates and protecting groups as well as effective radical scavengers, where they have established utility as polymer stabilizers and as preservatives in beverages such as wine and fruit juices.

Epoxides. Epoxy compounds react with chloroformates to yield β-chlorosubstituted carbonates. Ring opening is catalyzed with Lewis acids or sources of chloride anions.

Aldehydes. Aldehydes react with chloroformates in the presence of catalytic pyridine to yield 1-chloro carbonate esters. The reaction is highly dependant upon the stability of the chloroformate to the reaction conditions. The esters are useful intermediates for pharmaceuticals and insecticides.

Reactions with Nitrogen Compounds

Amines. Primary and secondary aliphatic amines yield carbamates in the presence of excess amine or other acid acceptors such as inorganic bases under conditions analogous to those used in the Schotten-Baumann reaction. Aromatic primary and secondary amines and heterocyclic amines react similarly, although slowly. Tertiary amines give crystalline quaternary ammonium compounds. The acyl ammonium salts provide activation of the carbonyl species and are potential precursors to secondary amines via dealkylation chemistry.

Amino Alcohols and Amino Phenols. Reaction with a chloroformate is much more rapid at the amino group than at the hydroxyl group.

Amino Acids. Chloroformates play a most important role for the protection of the amino group of amino acids (qv) during peptide synthesis. The protective carbamate formed by the reaction of benzyl chloroformate and amino acid can be cleaved by hydrogenolysis to free the amine after the carboxyl group has reacted further. The selectivity of the amino groups toward chloroformates results in amino-protected amino acids the other reactive groups remain unprotected. Methods for the preparation of protected amino acids generally involve a pH stat procedure.

Other Reactions of Chloroformates

Acylation. Aryl chloroformates are good acylating agents, reacting with aromatic hydrocarbons under Friedel-Crafts conditions to give the expected aryl esters of the aromatic (Ar) acid.

Dealkylation. Chloroformates such as vinyl chloroformates are used to dealkylate tertiary amines. Chloroformates are superior to the typical Von Braun reagent, cyanogen bromide, because of increased selectivity producing cleaner products and higher yields. Other chloroformates such as allyl, ethyl, methyl, phenyl, and trichloroethyl have also been used in dealkylation reactions. Although the dealkylation reaction using chloroformates is mostly carried out on tertiary amines, dealkylation of oxygen or sulfur centers, ie, ethers or thioethers, can also be achieved.

Miscellaneous Reactions. The reaction of chloroformates with hydrogen peroxide or metal peroxides results in the formation of peroxydicarbonates that are used as free-radical initiators of polymerization of vinyl chloride, ethylene, and other unsaturated monomers. The reaction of chloroformates with sodium xanthates results in the formation of alkyl xanthogen formates that are useful as flotation agents in extraction of metals from their ores. Methyleneaziridines undergo nucleophilic ring opening in the presence of alkyl chloroformates to generate enamide products. Alkyl chloroformates react with HCN in the presence of a tertiary amine, or with cyanide salts under phase-transfer conditions, to give cyanoformate esters. The reaction of chloroformates with aldehyde and ketone enolates has been shown as a selective route to enol carbonates.

Manufacture

Most alkyl chloroformates, especially those of low molecular weight alcohols, are prepared by the reaction of liquid anhydrous alcohols with molar excess of dry, chlorine-free phosgene at low temperature. Corrosion-resistant reactors, lines, pumps, and valves are required. Materials of construction include glass, porcelain, Hastelloy C, Teflon-lined steel, or chemically impregnated carbon on steel such as Karbate. Temperatures are kept at 0–10°C for the lower alcohols and may rise to 60°C for the higher aliphatic alcohols. Hydrogen chloride is evolved as the reaction proceeds and is then absorbed in a tower after recovering excess phosgene. The reactions are most often run in batch reactors, although some of the high volume chloroformates are produced in cascade-type continuous reactors in either cocurrent or countercurrent flow. The continuous reactors also ensure excess of phosgene at all times since both reactants are added simultaneously, thus minimizing dialkylcarbonate side products. Phenols react with phosgene with difficulty and usually require higher temperature and lead to a fair amount of diaryl carbonate as a side product. Many different catalysts have been used to reduce the reaction temperature and the carbonate side products.

Shipping and Storage

Chloroformates are shipped in nonreturnable 208-L (55-gal) polyethylene drums with carbon steel overpacks or high density polyethylene drums. For bulk shipments, insulated stainless steel tank containers and trucks provide secure protection. Bulk equipment is specially lined for protection from corrosion. Tank truck and rail car quantities are shipped using equipment dedicated for these types of products. Materials such as isopropyl chloroformate, benzyl chloroformate, and sec-butyl chloroformate that require refrigeration are precooled when shipped in bulk containers. Bulk shipments that are pre-cooled must proceed to the destination without layover. Drum shipments of IPCE, BCF, and SBCF must be shipped in refrigerated containers. Many of the chloroformates are only shipped in truckload shipments.

Chloroformates should be stored in a cool, dry atmosphere, preferably refrigerated, especially for prolonged storage. Drums must be stored out of direct sunlight. Chloroformate transfers to storage tanks or reactors should be made through a closed system, using stainless steel, nickel, glass or Hastelloy pumps, lines, and valves. Contact with iron oxides should be avoided.

Toxicity

Chloroformates, are pungent lachrymators and are irritating to the skin and mucous membranes producing severe burns and possible irreversible tissue damage.

Inhalation of vapors of lower chloroformates result in coughing, choking, and respiratory distress, and, with some chloroformates like methyl chloroformate, inhalation can be fatal as a result of the onset of pulmonary edema, which may not appear for several hours after exposure.

Uses

Derivatization of chloroformates with alcohols and amines is commonly practiced, and many industrial uses of the resultant carbamates and carbonates have been described. Chloroformates should be considered as intermediates for syntheses of pesticides, perfumes, drugs, foodstuffs, polymers, dyes, and other chemicals.

A significant use of chloroformates is for conversion to peroxydicarbonates, which serve as free-radical initiators for the polymerization of vinyl chloride, ethylene, and other unsaturated monomers.

Carbamates derived from chloroformates are used to manufacture pharmaceuticals, including tranquilizers, antihypotensives, and local anesthetics, pesticides, and insecticides (see CARBAMIC ACID).

Methyl chloroformate is the largest volume chloroformate used in the agricultural industry, primarily in the formation of carbamate functional groups.

Chloroformates derived from allyl alcohol and diethylene glycol are important starting materials for optical monomers, while the bis chloroformate of bisphenol A is the raw material used in the manufacture of polycarbonate plastics such as Lexan.

Another important use of chloroformates is the protection of amino and hydroxyl groups in the synthesis of complex organic compounds such as peptide-based pharmaceuticals.

Carbamate likages of poly(vinyl ether) carbamates, used as detergents additives in gasoline, have been derived from the suitable chloroformate. Ethyl chloroformate is used in the manufacture of ore flotation agents by reaction with various xanthates.

Cholesteric liquid crystal materials, useful as nondestructive indicators, are often derived from cholesterol chloroformate. Decarboxylation of bis(chlorofor-mates) to alkyl halides is used in the manufacture of the rubber component 1,6-dichlorohexane. Additionally, decarboxylation of alkoxyalkyl chloroformates provides alkyl chloride materials that are useful as surfactant intermediates.

By virtue of their exceptional reactivity, chloroformates are also valued as general purpose derivatizing agents for gas and liquid chromatographic analysis of molecules containing active functionality such as amines and carboxylic acids.

CARBONATES

Classically, the reaction of chloroformates and alcohols or phenols give carbonic diesters. Alternatively, higher diesters can be made from the lower ones by heat induced easter interchange. More recently, preparation of selected carbonic esters by nonphosgene routes, such as the metal catalyzed reaction of CO or NO_x with alcohols, or the catalytic reaction of CO_2 and oxiranes have been preferred. These oxidative carbonylation methods are more economic in many cases and naturally less hazardous than phosgene routes.

Properties

The lower alkyl carbonates are neutral, colorless liquids with a mild sweet odor. Aryl carbonates are normally crystalline compounds with relatively low melting points. Carbonic esters are soluble in polar organic solvents such as alcohols, esters, and ketones, but not soluble in water. An exception is lower molecular weight cyclic carbonates such as ethylene carbonate and propylene carbonate which readily dissolve in water. Several lower aliphatic carbonates form azeotropic mixtures with organic solvents.

Chemical Properties

The chemistry of carbonates is dominated by a reactivity similar to esters and a tendency to liberate CO_2. Carbonates undergo nucleophilic substitution reactions analogous to chloroformates except in this case, an ^-OR group (rather than chloride) is replaced by a more basic group. Normally these reactions are catalyzed by bases. Carbonates are sometimes preferred over chloroformates because formation of hydrogen chloride as a by-product is avoided, simplifying handling and in some cases eliminating impurities. However, the reactivity of carbonates toward nucleophiles is considerably less than chloroformates.

Reaction with Water. The alkyl carbonate esters, especially the lower ones, hydrolyze very slowly in water when compared to the carbonochloridic esters (chloroformates). Under alkaline conditions, the rates of hydrolysis are similar to those of the corresponding acetic acid esters. The net result is the formation of hydroxy compounds and CO_2.

Reaction with Alcohols and Thiols. The likeness of carbonates to esters is evident in their tendency to undergo transesterification with alcohol and thiols. Transesterification of both cyclic and acyclic carbonates is commonly practiced in industry. The process requires that the equilibrium be shifted in the desired direction.

Reaction with Phenols. Carbonates undergo carbonate interchange with aromatic hydroxy compounds. In cases involving aliphatic carbonates, the reaction is slow and thermodynamically unfavorable. The equilibrium heavily favors the aliphatic ester and high temperatures and very active catalysts are required to drive the reaction.

Reaction with Amines and Ammonia. Carbonates react with aromatic amines, aliphatic amines and ammonia to produce carbamates or ureas. For example, dimethyl carbonate reacts with ammonia in water to form methyl carbamate useful in coatings applications. The carbamate esters derived from carbonates can also be subjected to thermal conditions to eliminate alcohol thus leading to isocyanates.

Alkylation. Dialkyl carbonates react with a variety of functional groups to produce the alkylated derivatives. Alkylations have been noted on functionality ranging from thioorganics, phenols, anilines, amines, oximes, carboxylic acids, and silicon dioxide to C alkylation of CH_2 acidic species such as arylacetonitriles and alkyl aryl acetates.

Miscellaneous Acylation Reactions. Chloroalkyl alkyl carbonates are more activated acylating agents than their alkyl counterparts. An extreme example is bis trichloromethyl carbonate or triphosgene, which is often used as a solid source of phosgene.

It is important to note that triphosgene readily decomposes to phosgene at high temperatures, and in the presence of trace metals and small quantities of nucleophilic sources such as chloride ion. The less extreme acylating agent, 1-chloroalkyl alkyl carbonate, has been used as an entry to carbamates. Additionally, the 1-chloroalkyl carbonyl group is used as an acid labile, base stable alcohol protecting group. The reactivity of the 1-chloroalkyl carbonates also offers entry to fluoroformates that cannot be prepared by halogen exchange of their analogous chloroformate, such as the *tert*-butyl chloroformate. Thus, *tert*-butyl fluoroformate, which exhibits higher thermal stability than its chloroformate analogue, is used industrially for BOC protection of amino acids and peptides.

Manufacture

The most important and versatile method for producing carbonates is the phosgenation of hydroxy compounds. Manufacture is essentially the same method as chloroformates except that more alcohol is required in addition to longer reaction times and higher temperatures. The products are neutralized, washed, and distilled. The more acidic alcohols are less reactive, and in many cases organic base is included as catalyst. Corrosion-resistant equipment similar to that described for the manufacture of chloroformates is required.

Shipping and Storage

Dimethyl and diethyl carbonates are shipped in nonreturnable 208-L (55-gal) polyethylene drums with carbon steel overpack or high density polethylene drums. For bulk shipments, insulated stainless steel tank containers and trucks provide secure protection. Diethylene glycol bis(allyl) carbonate is shipped in drums as above. Diphenyl carbonate is delivered flaked in polyethylene sacks, or by tank car as a melt.

Health and Safety Factors

Unlike chloroformates, diethyl and dimethyl carbonates are only mildly irritating to the eyes, skin, and mucous membranes. Diethylene glycol bis(allyl carbonate) may be irritating to the skin, but it is not classified as a toxic substance; however, it is extremely irritating to the eyes.

Uses

The industrial utility of carbonates are widespread, ranging from pharmaceutical and cosmetic preparations to utility as specialty solvents and application in polymers. The major volume carbonates include diphenyl, ethylene, and propylene carbonates as well as the simple aliphatics: diethyl, dimethyl, and dipropyl. In many applications dimethyl carbonate is an effective carbonylating agent and is touted as a phosgene replacement. For instance, dimethyl carbonate and diphenyl carbonate are key components in the manufacture of polycarbonate resins such as Lexan[R]. Diethylene bis(allyl carbonate) is used as a monomer in many optical applications such as the manufacture of safety glasses and lightweight prescription lenses. Specialty carbonates derived from 1-haloalkyl chloroformates find important utility as pro drugs for antibiotics and other pharmaceitocals.

H.-J. Buysch, "Carbonic Esters", *Ullmann's Encyclopedia of Industrial Chemistry*, Wiley-VCH Verlag GmbH, Weinheim, Germany, 2002.

D. N. Kevill, in S. Patai, ed., *The Chemistry of Acyl Halides*, Wiley-Interscience, New York, 1972.

R. J. Lewis, Sr., ed., *Sax's Dangerous Properties of Industrial Materials*, 11th ed., Wiley, Hoboken, N.J., 2004; on-line.

J.-P. Senet, *The Recent Advance in Phosgene Chemistry* 1 & 2, L'Imprimerie GPA à Nanterre, Société Nationale des Poudres et Explosifs, Feb. 1999.

CHARLES B. KREUTZBERGER
PPG Industries

CHLOROTOLUENES, BENZYL CHLORIDE, BENZAL CHLORIDE, AND BENZOTRICHLORIDE

Nearly all of the benzyl chloride, benzal chloride, and benzotrichloride manufactured is converted to other chemical intermediates or products by reactions involving the chlorine substituents of the side chain. Each of the compounds has a single primary use that consumes the majority portion of the compound produced. Benzyl chloride is utilized in the manufacture of benzyl butyl phthalate, a vinyl resin plasticizer; benzal chloride is hydrolyzed to benzaldehyde; benzotrichloride is converted to benzoyl chloride. Benzyl chloride is also hydrolyzed to benzyl alcohol, which is used in the photographic industry, in perfumes (as esters), and in peptide synthesis by conversion to benzyl chloroformate (see BENZYL ALCOHOL AND β-PHENETHYL ALCOHOL; CARBONIC AND CARBONOCHLORIDIC ESTERS).

Several related compounds, primarily ring-chlorinated derivatives, are also commercially significant. In the case of p-chlorobenzotrichloride this can be further converted to p-chlorobenzotrifluoride by reaction with HF. p-Chlorobenzotrifluoride is an important intermediate in the manufacture of dinitroaniline herbicides (Trifluralin).

PHYSICAL PROPERTIES

Benzyl chloride [(chloromethyl)benzene, α-chlorotoluene], $C_6H_5CH_2Cl$, is a colorless liquid with a very pungent odor. Benzyl chloride vapors are irritating to the eyes and mucous membranes, and it is classified as a powerful lacrimator. The physical properties of pure benzyl chloride are given in Table 1. Benzyl chloride is insoluble in cold water, but decomposes slowly in hot water to benzyl alcohol. It is miscible in all proportions at room temperature with most organic solvents.

Benzal chloride (MW = 161.03) [(dichloromethyl)benzene, α,α-dichlorotoluene, benzylidene chloride], $C_6H_5CHCl_2$, is a colorless liquid with a pungent, aromatic odor. Benzal chloride is insoluble in water at room temperature but is miscible with most organic solvents.

Benzotrichloride (MW = 195.47) [(trichloromethyl)benzene, α,α,α-trichlorotoluene, phenylchloroform], $C_6H_5CCl_3$, is a colorless, oily liquid with a pungent odor. It is soluble in most organic solvents, but it reacts with water and alcohol.

CHEMICAL PROPERTIES

The reactions of benzyl chloride, benzal chloride, and benzotrichloride may be divided into two classes: (1) reactions taking place on the side chain containing the halogen; and (2) reactions taking place on the aromatic ring.

MANUFACTURE

Benzyl chloride is manufactured by the thermal or photochemical chlorination of toluene at 65–100°C. At lower temperatures the amount of ring-chlorinated by-products is increased. The chlorination is usually carried to no more than about 50% toluene conversion in order to minimize the amount of benzal chloride formed.

Benzal chloride can be manufactured in 70% yield by chlorination with 2.0–2.2 moles of chlorine per mole of toluene. The benzal chloride is purified by distillation.

Further chlorination at a temperature of 100–140°C with ultraviolet light yields benzotrichloride. The chlorination is normally carried to a benzotrichloride content of greater than 95% with a low benzal chloride content.

An understanding of competing reactions in the manufacturing process is important if by-products are to be minimized. Three competing reactions are possible under conditions of the reaction: free-radical substitution of the side chain of toluene, addition to the aromatic ring, and electrophilic substitution on the aromatic ring.

HANDLING AND SHIPMENT

As is the case during manufacture, contact with those metallic impurities that catalyze Friedel-Crafts condensation reactions must be avoided. The self-condensation reaction is exothermic and the reaction can accelerate producing a rapid buildup of hydrogen chloride pressure in closed systems.

Benzyl chloride is available in both anhydrous and stabilized forms. Both forms can be shipped in glass carboys, nickel and lined-steel drums, and nickel tank trucks and tank cars.

Benzyl chloride is classified by DOT as chemicals NO1BN, poisonous, corrosive and a hazardous substance (100 lbs = 45.45 kg). Benzal chloride is classified as poisonous and a hazardous substance (5000 lbs = 2270 kg). Benzotrichloride is classified under DOT regulation as a corrosive liquid NOS and a hazardous substance (10 lbs = 4.5 kg). The Freight Classification Chemical NOI applies. It is shipped in lacquer-lined steel drums and nickel-lined tank trailers. Benzal chloride is handled in a similar fashion.

HEALTH AND SAFETY FACTORS

Benzyl chloride is a severely irritating liquid and causes damage to the eyes, skin, and respiratory tract including pulmonary edema. Other possible effects of overexposure to benzyl chloride are CNS depression, liver, and heart damage.

Table 1. Physical Properties of Benzyl Chloride, Benzal Chloride, and Benzotrichloride

Property	Benzyl chloride	Benzal chloride	Benzo-trichloride
mol wt	126.58	161.03	195.48
freezing point, °C	−39.2	−16.4	−4.75
boiling point, °C	179.4	205.2	220.6
density, kg/m^3	1113.5_4^4	1256_4^{14}	1374_4^{20}
refractive index, n_D^t	1.5392^{20}	1.5502^{20}	1.55789^{20}
heat of combustion, kJ/mol^a	3708^b	3852^c	3684^c

[a] To convert J to cal, divide by 4.184.
[b] At constant volume.
[c] At constant pressure.

IARC states there is limited evidence that benzyl chloride is carcinogenic in experimental animals; epidemiological data were inadequate to evaluate carcinogenicity to humans.

Other toxicological effects that may be associated with exposure to benzyl chloride based on animal studies are skin sensitization and developmental embryo and/or fetal toxicity. A 1988 OSHA regulation has established a national occupational exposure limit for benzyl chloride of 5 mg/m^3 (1 ppm). Vapors of both benzal chloride and benzotrichloride are strongly irritating and lacrimatory.

For all three compounds, biological data relevant to the evaluation of carcinogenic risk to humans are summarized in the World Health Organization International Agency for Research on Cancer monograph which was updated in 1987.

USES

Nearly all uses and applications of benzyl chloride are related to reactions of the active halide substituent. More than two-thirds of benzyl chloride produced is used in the manufacture of benzyl butyl-phthalate, a plasticizer used extensively in vinyl flooring and other flexible poly(vinyl chloride) uses such as food packaging. Other significant uses are the manufacture of benzyl alcohol and of benzyl chloride-derived quaternary ammonium compounds, each of which consumes more than 10% of the benzyl chloride produced. Smaller volume uses include the manufacture of benzyl cyanide, benzyl esters such as benzyl acetate, butyrate, cinnamate, and salicylate, benzylamine, and benzyldimethylamine, and p-benzylphenol.

DERIVATIVES

Ring-Substituted Derivatives

The ring-chlorinated derivatives of benzyl chloride, benzal chloride, and benzotrichloride are produced by the direct side-chain chlorination of the corresponding chlorinated toluenes or by one of several indirect routes if the required chlorotoluene is not readily available.

Side-Chain Chlorinated Xylene Derivatives

Only a few of the nine side-chain chlorinated derivatives of each of the xylenes are available from direct chlorination. All three of the monochlorinated compounds, α-chloro-o-xylene [1-(chloromethyl)-2-methylbenzene, α-chloro-m-xylene (1-(chloromethyl)-3-methylbenzene, and α-chloro-p-xylene [1-(chloromethyl)-4-methylbenzene] are obtained in high yield from partial chlorination of the xylenes. 1,3-Bis(chloromethyl)benzene can be isolated in moderate yield from chlorination mixtures.

The fully side-chain chlorinated products, 1,3-bis(trichloromethyl)benzene and 1,4-bis(trichloromethyl)benzene, are manufactured by exhaustive chlorination of *meta* and *para* xylenes.

1-(Dichloromethyl)-2-(trichloromethyl)benzene, the end product of exhaustive side-chain chlorination of o-xylene

is an intermediate in the manufacture of phthalaldehydic acid.

F. M. Ashton and A. S. Crafts, *Mode of Action of Herbicides*, John Wiley & Sons, New York, 1973, pp. 10–24, 438–448.

Faith, Keyes, and Clark's Industrial Chemicals 4th ed., John Wiley & Sons, Inc., New York, 1975, pp. 145–148.

Handbook of Chemistry and Physics, 58th ed., CRC Press Inc., Cleveland, Ohio, 1977–1978, pp. C-522, 523, 527, 528, 738, D-198.

R. J. Lewis, Sr., ed., *Sax's Dangerous Properties of Industrial Materials*, 11th ed., Wiley, Hoboken, N. J., 2004, Online.

KARL W. SEPER
Occidental Chemical Corp.

CHLOROTOLUENES, RING

The ring-chlorinated derivatives of toluene form a group of stable, industrially important compounds. Many chlorotoluene isomers can be prepared by direct chlorination. Other chlorotoluenes are prepared by indirect routes involving the replacement of amino, hydroxyl, chlorosulfonyl, and nitro groups by chlorine and the use of substituents, such as nitro, amino, and sulfonic acid, to orient substitution followed by their removal from the ring.

Mono- and dichlorotoluenes are used chiefly as chemical intermediates in the manufacture of pesticides, dyestuffs, pharmaceuticals, and peroxides, and as solvents.

MONOCHLOROTOLUENES

Physical Properties

o-Chlorotoluene (1-chloro-2-methylbenzene, OCT) is a mobile, colorless liquid with a penetrating odor similar to chlorobenzene. It is miscible in all proportions with many organic liquids such as aliphatic and aromatic hydrocarbons, chlorinated solvents, lower alcohols, ketones, glacial acetic acid, and di-n-butylamine; it is insoluble in water, ethylene and diethylene glycols, and triethanolamine.

p-Chlorotoluene (1-chloro-4-methylbenzene, PCT) and m-chlorotoluene (1-chloro-3-methylbenzene, MCT) are mobile, colorless liquids with solvent properties similar to those of the ortho isomer.

Ortho and p-chlorotoluene form binary azeotropes with various organic compounds including alcohols, acids, and esters. They form stable ionic complexes with antimony pentachloride, as well as complexes with a number of organometallic derivatives, such as those of chromium, cobalt, iron, etc, many of which have synthetic utility.

Chemical Properties

The monochlorotoluenes are stable to the action of steam, alkalies, amines, and hydrochloric and phosphoric acids at moderate temperatures and pressures. Three classes of reactions, those involving the aromatic ring, the methyl group, and the chlorine substituent, are known for mono-chlorotoluenes.

Reactions of the Aromatic Ring. Ring chlorination of *o*-chlorotoluene yields a mixture of all four possible dichlorotoluenes, the 2,3-, 2,4-, 2,5-, and 2,6-isomers. The principal isomer, 2,5-dichlorotoluene, constitutes up to 60% of the product mixture. Similarly, nitration of *o*-chlorotoluene produces a mixture of the four corresponding nitrochlorotoluene isomers.

Reactions of the Methyl Group. Monochlorotoluenes are widely used to synthesize compounds derived from reactions of the methyl group. Chlorination under free-radical conditions leads successively to the chlorinated benzyl, benzal, and benzotrichloride derivatives (see CHLOROCARBONS AND CHLOROHYDROCARBONS—BENZYL CHLORIDE, BENZAL CHLORIDE, AND BENZOTRICHLORIDE). Oxidation to form chlorinated benzaldehydes and benzoic acids can be performed under both liquid- and vapor-phase conditions.

Halogen Reactions. Hydrolysis of chlorotoluenes to cresols has been effected by aqueous sodium hydroxide. Both displacement and benzyne formation are involved.

Preparation

Monochlorotoluenes have been prepared by chlorinating toluene with a wide variety of chlorinating agents, catalysts, and reaction conditions. The ratio of ortho and para isomers formed can vary over a wide range. Particular attention has been given to studies aimed at increasing the para isomer content owing to its greater commercial significance.

Chlorinations with Elemental Chlorine. Reaction of toluene with chlorine in the presence of certain Lewis acid catalysts including the chlorides of aluminum, tin, titanium, and zirconium give monochlorotoluene mixtures that contain more than 70% of the ortho isomer. A growing number of heterogeneous processes that employ zeolite-type catalysts for chlorination of aromatics have been discovered (see CATALYSIS). The majority of these catalysts are synthetic zeolites, specifically the L-type zeolites.

Noncatalytic ring chlorination of toluene in a variety of solvents has been reported.

Chlorination with Other Reagents. Chlorotoluenes can also be obtained in good yields by the reaction of toluene with stoichiometric proportions of certain Lewis acid chlorides such as iron(III) chloride, as the chlorinating agent. Generally, the product mixture contains *p*-chlorotoluene as the principal component. Toluene chlorination has also been effected with hydrogen chloride as the chlorinating agent.

Other methods for preparing *p*-chlorotoluene include α-elimination from an organotellurium(IV) halide, palladium-catalyzed decarbonylation of 4-methylbenzoyl chloride, and desulfonylation of *p*-toluenesulfonyl chloride catalyzed by chlorine, or chlorotris(triphenylphosphine)rhodium.

Pure monochlorotoluene isomers are prepared by diazotization of the corresponding toluidine isomers followed by reaction with copper(I) chloride (Sandmeyer reaction). This is the preferred method of obtaining *m*-chlorotoluene.

Both batch and continuous processes are suitable for commercial chlorination. The progress of the chlorination is conveniently followed by specific gravity measurements.

Handling and Shipment

Monochlorotoluenes are shipped in bulk in steel tank cars and tank trucks. Drum shipments are made using lined or unlined steel drums. Aluminum tanks can be used to store only acid-free material. The storage vessels are vented to a safe atmosphere and should be protected with suitable diking. Protection against static charge is essential when transferring material. Suitable ventilation should be provided and sources of ignition avoided as the vapor forms flammable mixtures with air.

Health and Safety Factors

Health and environmental issues related to chlorinated toluenes are included in the Hazardous Substances Data Bank.

HIGHER CHLOROTOLUENES

Dichlorotoluenes

There are six possible dichlorotoluene isomers, $C_7H_6Cl_2$, (mol wt 161.03) all of which are known.

2,4-Dichlorotoluene (2,4-dichloro-1-methylbenzene) constitutes 80–85% of the dichlorotoluene fraction obtained in the chlorination of PCT with antimony trichloride or zirconium tetrachloride catalysts. It is separated from 3,4-dichlorotoluene (1,2-dichloro-4-methylbenzene), the principal contaminant, by distillation.

Chlorination of OCT with chlorine at 90°C in the presence of L-type zeolites as catalyst reportedly gives a 56% yield of 2,5-dichlorotoluene. Pure 2,5-dichlorotoluene is also available from the Sandmeyer reaction on 2-amino-5-chlorotoluene.

2,3-Dichlorotoluene (1,2-dichloro-3-methylbenzene) is present in about 10% concentration in reaction mixtures resulting from chlorination of OCT.

2,6-Dichlorotoluene (1,3-dichloro-2-methylbenzene) is prepared from the Sandmeyer reaction on 2-amino-6-chlorotoluene. Other methods include ring chlorination of *p*-toluenesulfonyl chloride followed by desulfonylation, and chlorination and dealkylation of 4-*tert*-butyltoluene or 3,5-di-*tert*-butyltoluene.

Trichlorotoluenes

The chlorination of toluene and *o*- and *p*-chlorotoluenes produces a mixture of trichlorotoluenes, ($C_7H_5Cl_3$, (mol wt 195.48): the 2,3,6-isomer (1,2,4-trichloro-3-methylbenzene) and 2,4,5-trichlorotoluene (1,2,4-trichloro-5-methylbenzene) containing small amounts of 2,3,4-trichlorotoluene (1,2,3-trichloro-4-methylbenzene) and 2,4,6-trichlorotoluene (1,3,5-trichloro-2-methylbenzene). When toluene is chlorinated in the presence of iron(III) chloride catalyst, a mixture containing nearly equal amounts of 2,4,5- and 2,3,6-trichlorotoluenes is produced.

Tetra- and Pentachlorotoluenes

2,3,4,6-Tetrachlorotoluene, $C_7H_4Cl_4$ (mol wt 229.93) (1,2,3,5-tetrachloro-4-methylbenzene), is prepared from the Sandmeyer reaction on 3-amino-2,4,6-trichlorotoluene. 2,3,4,5-Tetrachlorotoluene (1,2,3,4-tetrachloro-5-methylbenzene) is the principal isomer in the further chlorination of 2,4,5-trichlorotoluene. Pentachlorotoluene (pentachloromethylbenzene), $C_7H_3Cl_5$ (mol wt 264.37), is formed in 90% yield by the ferric chloride-catalyzed chlorination of toluene in carbon tetrachloride or hexachlorobutadiene solution.

USES

Chlorotoluenes are used as intermediates in the pesticide, pharmaceutical, peroxide, dye, and other industries. Many side chain-chlorinated derivatives are converted to end products.

Mono and di-chlorotoluenes have been used as a solvent/desorbent to separate dichlorobenzenes using zeolites.

Chlorotoluene isomer mixtures, especially those containing a relatively high amount of o-chlorotoluene, are widely used as solvents in industry.

2,4-Dichlorotoluene is an intermediate for manufacture of herbicides. It has been used as a solvent in the preparation of trifluoromethylpyridine derivatives. 2,6-Dichlorotoluene is applied as a herbicide and dyestuff intermediate. 2,3,6-Trichlorotoluene is used as a herbicide intermediate. The other polychlorotoluenes have limited industrial application.

F. M. Ashton and A. S. Crafts, *Mode of Action of Herbicides*, John Wiley & Sons, Inc., New York, 1973, pp. 10–24, 438–448.

L. H. Horsley and co-workers, *Azeotropic Data III, Advances in Chemistry Series*, No. 116, American Chemical Society, Washington, D.C., 1973, pp. 197–198.

J. Timmerman, *Physico-Chemical Constants of Pure Organic Compounds*, Elsevier Science Publishing Co., Inc., New York, 1950, pp. 297–298.

PRAVIN KHANDARE
RON SPOHN
Occidental Chemical Corporation

CHOCOLATE AND COCOA

The name *Theobroma cacao*, food of the gods, indicating both the legendary origin and the nourishing qualities of chocolate, was bestowed upon the cacao tree by Linnaeus in 1720. All cocoa and chocolate products are derived from the cocoa bean, the seed of the fruit of this tree.

The terms cocoa and cacao often are used interchangeably in the literature. Both terms describe various products from harvest through processing. In this article, the term cocoa will be used to describe products in general and the term cacao will be reserved for botanical contexts.

STANDARDS FOR COCOA AND CHOCOLATE

In the United States, chocolate and cocoa are standardized by the U.S. Food and Drug Administration under the Federal Food, Drug, and Cosmetic Act. The current definitions and standards resulted from prolonged discussions between the U.S. chocolate industry and the Food and Drug Administration (FDA). These definitions and standards originally published in the *Federal Register* of December 6, 1944, have been revised only slightly.

The FDA announced in the *Federal Register* of January 25, 1989 a proposal to amend the U.S. chocolate and cocoa standards of identity. These amendments were in response to a citizen petition submitted by the Chocolate Manufacturers Association (CMA) and to better align U.S. standards with Codex. The new standards published as a final rule in the *Federal Register* of May 21, 1993, allow for the use of nutritive carbohydrate sweeteners, neutralizing agents, and emulsifiers; reduce slightly the minimum milkfat content and eliminate the nonfat milk solids/milkfat ratios in certain cocoa products including milk chocolate; update the language and format of the standards; and provide for optional ingredient labeling requirements. FDA has also received a proposal to establish a new standard of identity for white chocolate. Comments regarding the proposal amendments are under review by FDA, and a final ruling is expected to be issued in the near future.

White Chocolate

There is no standard of identity published for white chocolate. The presence of a standard would eliminate confusion over the content of products that are customarily referred to as white chocolate. A standard for white chocolate would also promote regulatory harmonization among nations that have already adopted a standard.

White chocolate has been defined by the European Economic Community (EEC) Directive 75/155/EEC as free of coloring matter and consisting of cocoa butter (not <20%); sucrose (not >55%); milk or solids obtained by partially or totally dehydrated whole milk, skimmed milk, or cream (not <14%); and butter or butter fat (not < 3.5%).

COCOA BEANS

The cocoa bean is the basic raw ingredient in the manufacture of all cocoa products. The beans are converted to chocolate liquor, the primary ingredient from which all chocolate and cocoa products are made. Figure 1 depicts the conversion of cocoa beans to chocolate liquor, and in turn to the chief chocolate and cocoa products manufactured in the United States, ie, cocoa powder, cocoa butter, and sweet and milk chocolate.

Significant amounts of cocoa beans are produced in ~30 different localities. These areas are confined to latitudes

20° north or south of the equator. Although cocoa trees thrive in this very hot climate, young trees require the shade of larger trees such as banana, coconut, and palm for protection.

Fermentation (Curing)

Prior to shipment from producing countries, most cocoa beans undergo a process known as curing, fermenting, or sweating. These terms are used rather loosely to describe a procedure in which seeds are removed from the pods, fermented, and dried. Some unfermented beans, particularly from Haiti and the Dominican Republic, are also used in the United States.

Fermentation plays a principal role in flavor development of beans by mechanisms that are not well understood.

Commercial Grades

Most cocoa beans imported into the United States are one of about a dozen commercial varieties that can be generally classified as Criollo or Forastero. Criollo beans have a light color, a mild, nutty flavor, and an odor somewhat like sour wine. Forastero beans have a strong, somewhat bitter flavor and various degrees of astringency. The Forastero varieties are more abundant and provide the basis for most chocolate and cocoa formulations.

Bean Specifications

Cocoa beans vary widely in quality, necessitating a system of inspection and grading to ensure uniformity. Producing countries have always inspected beans for proper curing and drying as well as for insect and mold damage. Recently, a procedure for grading beans has been established at an international level. It classifies beans into two principal categories according to the fraction of moldy, slaty, flat, germinated, and insect-damaged beans.

Blending

Most chocolate and cocoa products consist of blends of beans chosen for flavor and color characteristics. Cocoa beans may be blended before or after roasting, or nibs may be blended before grinding. In some cases, finished liquors are blended.

Production

Worldwide cocoa bean production has increased significantly over the past 10 years from ~2.4 million t in the 1989–1990 crop year to >3.0 million t in 2000.

Consumption

Worldwide cocoa bean consumption has increased over the past 10 years from ~2.3 million t in the 1989–1990 crop year to almost 2.6 million t in 1999–2000.

CHOCOLATE LIQUOR

Chocolate liquor is the solid or semisolid food prepared by finely grinding the kernel or nib of the cocoa bean. It is also commonly called chocolate, unsweetened chocolate, baking chocolate, or cooking chocolate. In Europe chocolate, liquor is often called chocolate mass or cocoa mass.

COCOA POWDER

Cocoa powder (cocoa) is prepared by pulverizing the remaining material after part of the fat (cocoa butter) is removed from chocolate liquor. The U.S. chocolate standards define three types of cocoas based on their fat content. These are breakfast, or high fat cocoa, containing not <22% fat; cocoa, or medium fat cocoa, containing <22% fat but >10%; and low fat cocoa, containing <10% fat.

Cocoa powder production today is an important part of the cocoa and chocolate industry because of increased consumption of chocolate-flavored products. Cocoa powder is the basic flavoring ingredient in most chocolate-flavored cookies, biscuits, syrups, cakes, and ice cream. It is also used extensively in the production of confectionery coatings for candy bars.

COCOA BUTTER

Cocoa butter is the common name given to the fat obtained by subjecting chocolate liquor to hydraulic pressure. It is the main carrier and suspending medium for cocoa particles in chocolate liquor and for sugar and other ingredients in sweet and milk chocolate.

The FDA has not legally defined cocoa butter, and no standard exists for this product under the U.S. Chocolate Standards. For the purpose of enforcement, the FDA defines cocoa butter as the edible fat obtained from cocoa beans either before or after roasting. Cocoa butter as defined in the *U.S. Pharmacopeia* is the fat obtained from the roasted seed of *Theobroma cacao Linne*.

Composition and Properties

Cocoa butter is a unique fat with specific melting characteristics. It is a solid at room temperature (20°C), starts to soften ~30°C, and melts completely just below body temperature. Its distinct melting characteristic makes cocoa butter the preferred fat for chocolate products.

Cocoa butter is composed mainly of glycerides of stearic, palmitic, and oleic fatty acids (see FATS AND FATTY OILS). The triglyceride structure of cocoa butter has been determined (12,13) and is as follows: tri-saturated, 3%; mono-unsaturated (oleo-distearin), 22%; oleo-palmitostearin, 57%; oleo-dipalmitin, 4%; di-unsaturated (stearo-diolein), 6%; palmito-diolein, 7%; tri-unsaturated, tri-olein, 1%.

Although there are actually six crystalline forms of cocoa butter, four basic forms are generally recognized as alpha, beta, beta prime, and gamma. Since cocoa butter is a natural fat, derived from different varieties of cocoa beans, no single set of specifications or chemical characteristics can apply.

Substitutes and Equivalents

In the past 25 years, many fats have been developed to replace part or all of the added cocoa butter in chocolate-flavored products. These fats fall into two basic categories commonly known as cocoa butter substitutes and cocoa butter equivalents.

Cocoa butter substitutes and equivalents differ greatly with respect to their method of manufacture, source of fats, and functionality; they are produced by several physical and chemical processes. Cocoa butter substitutes are produced from lauric acid fats such as coconut, palm, and palm kernel oils by fractionation and hydrogenation; from domestic fats such as soy, corn, and cotton seed oils by selective hydrogenation; or from palm kernel stearines by fractionation. Cocoa butter equivalents can be produced from palm kernel oil and other specialty fats such as shea and illipe by fractional crystallization; from glycerol and selected fatty acids by direct chemical synthesis; from edible beef tallow by acetone crystallization; or from domestic fats such as soy and cotton seed by enzymatic interesterification.

SWEET AND MILK CHOCOLATE

Most chocolate consumed in the United States is consumed in the form of milk chocolate and sweet chocolate. Sweet chocolate is chocolate liquor to which sugar and cocoa butter have been added. Milk chocolate contains these same ingredients and milk or milk solids.

U.S. definitions and standards for chocolate are quite specific. Sweet chocolate must contain at least 15% chocolate liquor by weight and must be sweetened with a nutritive carbohydrate sweetener. Semisweet chocolate or bittersweet chocolate, though often referred to as sweet chocolate, must contain a minimum of 35% chocolate liquor. These products, sweet chocolate and semisweet chocolate, or bittersweet chocolate, are often simply called chocolate or dark chocolate to distinguish them from milk chocolate.

Sweet chocolate can contain milk or milk solids (up to 12% max), nuts, coffee, honey, malt, salt, vanillin, and other spices and flavors as well as a number of specified emulsifiers. Many different kinds of chocolate can be produced by careful selection of bean blends, controlled roasting temperatures, and varying amounts of ingredients and flavors.

The most popular chocolate in the United States is milk chocolate. The U.S. Chocolate Standards state that milk chocolate shall contain no <3.39 wt% of milk fat and not <12 wt% of milk solids. Milk chocolate can contain spices, natural and artificial flavorings, ground whole nut meats, dried malted cereal extract and other seasonings that do not impart a flavor that imitates the flavor of chocolate, milk, or butter. In addition, chocolate liquor content must not be <10% by weight.

Production

The main difference in the production of sweet and milk chocolate is that in the production of milk chocolate, water must be removed from the milk. Many milk chocolate producers in the United States use spray-dried milk powder. Others condense fresh whole milk with sugar, and either dry it, producing milk crumb, or blend it with chocolate liquor and then dry it, producing milk chocolate crumb. These crumbs are mixed with additional chocolate liquor, sugar, and cocoa butter later in the process. Milk chocolates made from crumb typically have a more caramelized milk flavor than those made from spray- or drum-dried milk powder.

THEOBROMINE AND CAFFEINE

Chocolate and cocoa products, like coffee, tea, and cola beverages, contain alkaloids (qv). The predominant alkaloid in cocoa and chocolate products is theobromine, though caffeine, is also present in smaller amounts. Concentrations of both alkaloids vary depending on the origin of the beans. Published values for the theobromine and caffeine content of chocolate vary widely because of natural differences in cocoa beans and differences in analytical methodology.

NUTRITIONAL PROPERTIES OF CHOCOLATE PRODUCTS

Chocolate and cocoa products supply proteins, fats, carbohydrates, vitamins, and minerals. The Chocolate Manufacturers Association of the United States (Vienna, Virginia) completed a nutritional analysis from 1973 to 1976 of a wide variety of chocolate and cocoa products representative of those generally consumed in the United States.

Polyphenols

Chocolate and cocoa have been shown to be a rich source of antioxidant polyphenols. Flavonoids, a major subgroup, have been associated with the development of color and flavor. Flavonoids found in cocoa beans include epicatechin, catechin, and oligomers.

In fact, dark chocolate and cocoa were shown to contain more polyphenols on a dry defatted basis than 23 fruits, vegetables, and beverages. Data is growing indicating that polyphenols in the diet can help maintain cardiovascular health. Incorporating chocolate and cocoa in a diet that is rich in other food sources of antioxidants could reduce the risk of cardiovascular disease.

S. Coe, and M. Coe, *The True History of Chocolate*, Thames and Hudson Inc., New York, 1996, p. 36.

R. Heiss, *Twenty Years of Confectionery and Chocolate Progress*, AVI, Westport, Conn., 1970, p. 89.

New York Board of Trade http://www.nybot.com or http://www.csce.com.

Report of Codex Committee on Cocoa Products and Chocolate, Codex Alimentarious Commission, 18th Session, Fribourg, Switzerland, 2000.

B. L. ZOUMAS
C. D. AZZARA
J. BOUZAS
The Pennsylvania State
    University and
    Hershey Foods Corporation

CHROMATOGRAPHY

Chromatography is a technique used in many areas of science and engineering: petroleum chemistry, environmental studies, foods and flavorings, pharmaceuticals, forensics, and analysis of art objects (see BIOPOLYMERS, ANALYTICAL TECHNIQUES; FINE ART EXAMINATION AND CONSERVATION; FORENSIC CHEMISTRY), for separating and quantifying the constituents of a mixture. Since most chemical processes result in mixtures, separation techniques are essential for a successful characterization of chemical reactions. Most chemical laboratories employ one or more chromatographs for routine chemical analysis, and many processes involve the preparative use of chromatography for obtaining pure materials.

Chromatography relies on differential interactions of the components of a mixture with the phases of a chromatographic system to produce separation, eg, adsorption on the stationary phase. Thus, in addition to providing separation, the study of chromatographic parameters provides a means to determine fundamental quantities describing the interactions between the phases and the components, such as stability constants, vapor pressures, and other thermodynamic data.

A primary use of chromatography is the analysis of mixtures by passage through a column in which the differential interactions cause the components to pass through at different rates. The measurement of the rate of each material is a means of identification of the material, which is analytical chromatography.

PRINCIPLES

The principle of chromatographic separation is straightforward. A mixture is allowed to come into contact with two phases, a stationary phase and a mobile phase. The stationary phase is contained in a column or sheet through which the mobile phase moves in a controlled manner, carrying with it any material that may prefer to mix with it. Because of differences in the interactions of the mixture's constituents with stationary and mobile phases (the relative affinity of the constituents), the constituents are swept along with the mobile phase at different rates, so that they arrive at the end of the column at different times. This selective interaction is known as partitioning, and the different components are retained on the column for different times. To determine the retention time of substances on the column, a detector measures either the time required to travel to the end of the column or, as for thin-layer chromatography (tlc), the distance traveled in a fixed time. The detector may be as simple as the human nose or the human eye or as complex as a microsensor. A plot of detector response versus time of travel for a fixed distance is called a chromatogram. In preparative chromatography a device may be attached to the end of a column to collect the separated components of a mixture.

The nature of the stationary and mobile phases in a chromatographic experiment determines the efficacy of component separation in a particular mixture. A wide variety of stationary and mobile phases is used. The stationary phase may be a solid or a liquid supported on a solid. The mobile phase may be a gas, a liquid, or a material such as a supercritical fluid (see SUPERCRITICAL FLUIDS). One names a specific chromatographic technique by naming the mobile and the stationary phase, in that order. Thus, gas–liquid chromatography (glc) uses a gaseous mobile phase in contact with a film of liquid stationary phase.

Development of the Chromatogram

The term "development" describes the process of performing achromatographic separation. Because the processes that determine retention depend on the nature of the stationary and mobile phases, there are several ways in which separation may be made to occur, and several different ways for development of chromatograms.

Frontal chromatography (or frontal analysis) is a technique in which the sample is introduced onto a column continuously. In essence, the sample collected at the end of the column is free of materials that adsorb/absorb on the stationary phase.

In displacement chromatography a small sample on the column is displaced by a much more strongly held mobile phase. The sample is gradually pushed through the column as the mobile phase advances.

Both frontal and displacement chromatographies suffer a significant disadvantage in that once a column has been used, part of the sample remains on the column, so the column must be regenerated before reuse. In elution chromatography all of the sample material is usually removed from the column during the chromatographic process, allowing reuse of the column without regeneration. Most analytical applications of chromatography employ elution methods.

GAS CHROMATOGRAPHY

The most frequently used chromatographic technique is gc.

While solids and liquids are both used as stationary phases in gc, the most commonly used method is glc. Separation in a glc arises from differential partitioning of the sample's components between the stationary liquid phase bound on a porous solid, and the gas phase. In the other variant, gsc, preferential adsorption on the solid or, sometimes, exclusion of materials by size are the means of differentiating between components.

A second way of classifying gc separations is by use.

Theory

Most theoretical models of gas chromatographic processes are based on analogy to processes such as distillation or countercurrent extraction experiments. The separation process is viewed as a type of successive partitioning of the components of a mixture between the stationary and mobile phases similar to the partitioning that occurs in distillation columns. In gc, the equivalent measure of efficacy is the height equivalent to theoretical plates (HETP), which measures the ultimate ability of the column to separate like components. This quantity depends on many instrumental parameters such as wall or particle diameter, type of carrier gas, flow rate, liquid-phase thickness, etc.

Inlet Systems

The inlet (or injector) is the means by which the sample is introduced onto the gc column. Sample introduction requires one to create a representative aliquot of the sample at the beginning of the column without degradation or without discrimination among the components of the sample. Most inlets operate on the principle that a sample can be vaporized quickly, assuming it is not already a gaseous material, after being injected from a microliter syringe into a small, heated volume, usually $\sim$50°C hotter than the maximum temperature of the column during the experiment. Although other means are found in various applications, injection with a syringe is the most widely used technique. Inlets for syringe sampling are divided into two categories: packed and capillary columns.

LIQUID CHROMATOGRAPHY

Liquid chromatography (lc) refers to any chromatographic process in which the mobile phase is a liquid. Traditional column chromatography, tlc, pc, ce, cec, and high-performance liquid chromatography (hplc) are members of this class.

Liquid chromatography is complementary to gc because samples that cannot be easily handled in the gas phase, such as nonvolatile compounds or thermally unstable ones, eg many natural products, pharmaceuticals, and biomacromolecules, are separable by partitioning between a liquid mobile phase and a stationary phase, often at ambient temperature.

One advantage of lc is that the composition of the mobile phase, and perhaps of the stationary phase, can be varied during the experiment to provide a means of enhancing separation. In classical column chromatography the usual system consists of a polar adsorbent, or stationary phase, and a nonpolar mobile phase such as a hydrocarbon. In many instances, the polarities of the stationary and mobile phases are reversed for the separation, in which case the technique is known as reversed-phase liquid chromatography.

Paper chromatography and tlc are similar in the manner of development of the chromatogram. Paper and tlc may be further classified as one- or two-dimensional (1D or 2D), and as either analytical or preparative.

Affinity Chromatography

This technique, sometimes called bioselective adsorption, involves the use of a bioselective stationary phase placed in contact with the material to be purified, the ligate. Because of its rather selective interaction, sometimes called a lock-and-key mechanism, this method is more selective than other lc systems based on differential solubility.

Chiral Chromatography

Chiral chromatography is used for the analysis of enantiomers, and finds applications in the separations of

pharmaceuticals and biochemical compounds. There are several types of chiral stationary phases: those that use attractive interactions, metal ligands, inclusion complexes, and protein complexes.

Ion-Exchange Chromatography

In iec, the column contains a stationary phase having ionic groups such as a sulfonate or carboxylate. The charge of these groups is compensated by counterions such as sodium or potassium. The mobile phase is usually an ionic solution, eg, sodium chloride, having ions similar to the counterions.

Ion chromatography (ic), a variant of ion-exchange chromatography, is a technique in which a weak ion-exchange column is used for separation. After passing through the weak ion-exchange column, the eluent passes through a subsequent column called a stripper column, in which the stream, usually made acidic or basic in the ion-exchange column, is neutralized. Ion chromatography is a powerful technique for examining low concentrations of anions and cations. It has the advantage over selective ion-electrode analysis that it simultaneously gives information on many ions in a single experiment (see ELECTRO-ANALYTICAL TECHNIQUES).

Ion-pair chromatography (ipc), another variant of iec, is also sometimes called pic, soap chromatography, extraction chromatography, or chromatography with a liquid ion exchanger. In this technique the mobile phase consists of a solution of an aqueous buffer and an organic cosolvent containing an ion of charge opposite to the charge on the sample ion.

Size-Exclusion Chromatography

In sec or gpc, the material with which the column is packed has pores of a certain range of size. Molecules or solvent-molecule complexes too large to pass through these pores pass rapidly through the column, whereas molecules or complexes of suffiently small size are retained and are the last to exit the column. Molecules of intermediate size are partially retained and elute from the chromatographic column at intermediate times.

SUPERCRITICAL-FLUID CHROMATOGRAPHY

Supercritical-fluid chromatography is the link between gc and lc, because its mobile phase, a supercritical fluid, has physicochemical properties intermediate between a gas and a liquid (see SUPERCRITICAL FLUIDS). The physicochemical properties of the mobile phase are strong factors determining the selectivity, sensitivity toward a component, and efficiency of separation in the chromatographic process. Carbon dioxide is the mobile phase most often used in sfc.

This technique can be performed with either capillary or lc-like packed columns. Carbon dioxide is compatible with chromatographic hardware, is readily available, and is noncorrosive. The most important detector for sfc is the flame-ionization detector because the mobile phase does not give a significant background signal. More recent applications of sfc include separations in fields as diverse as natural products, drugs, foods, pesticides, herbicides, surfactants, and polymers. These are a direct result of the advantages that sfc has over other forms of chromatography because of low operating temperature, selective detection, and sensitivity to molecular weight.

J. S. Fritz and D. T. Gjerde, *Ion Chromatograph*, John Wiley & Sons, Inc., New York, 2000.

S. Mori and H. G. Barth, *Size Exclusion Chromatography*, Springer-Verlag, New York, 1999.

T. Provder, *Chromatography of Polymers*, American Chemical Society, Washington, D.C., 1999.

D. Rood, *A Practical Guide to Care, Maintenace, and Troubleshooting of Gas Chromatographic Systems*, John Wiley & Sons, Inc., New York, 1999.

CECIL DYBOWSKI
University of Delaware
MARY A. KAISER
Dupont Company

CHROMATOGRAPHY, AFFINITY

Affinity chromatography is a liquid chromatographic technique that uses a biologically related agent as the stationary phase. This makes use of the selective interactions that are common in biological systems, such as the binding of an enzyme with a substrate or the binding of an antibody with a foreign substance that has invaded the body. These interactions are used in affinity chromatography by immobilizing one of a pair of interacting molecules onto a solid support. This support is then placed into a column or onto a planar surface. The immobilized molecule is referred to as the *affinity ligand* and it represents the stationary phase of the chromatographic system.

In the most common scheme for performing affinity chromatography, a sample containing the compound of interest is first injected onto an affinity column in the presence of a mobile phase that has the right pH, ionic strength, and solvent composition for solute–ligand binding. This solvent, which represents the weak mobile phase of an affinity column, is referred to as the *application buffer*. As the sample passes through the column under these conditions, any compounds in the sample that are able to bind to the affinity ligand will be retained. However, due to the high selectivity of most such interactions, other substances in the sample tend to elute from the column as a nonretained peak.

After all nonretained or weakly retained substances have been washed from the column, the retained solutes are eluted by applying a solvent that displaces them from the column or promotes dissociation of the complex between each solute and affinity ligand. This second solvent, which acts as the strong mobile phase for the column, is usually referred to as the *elution buffer*. After

all of these substances have been removed from the system, the original application buffer is again applied and the affinity ligand is allowed to regenerate back to its original state prior to the application of another sample.

The wide range of ligands available for affinity chromatography makes this method a valuable tool for the purification and analysis of compounds present in complex samples.

PRINCIPLES OF AFFINITY CHROMATOGRAPHY

Theory of Affinity Chromatography

A number of factors affect how a compound is retained or eluted by an affinity column. These factors include the type of mobile phase that is being applied to the column, the strength of the solute–ligand interaction in this solvent, the amount of immobilized ligand present in the column, and the kinetics of solute–ligand association and dissociation. Furthermore, the type of support material that is used in the column can also play a role in determining the speed and efficiency of this chromatographic process.

General Types of Affinity Ligands

The most important item that determines the selectivity and retention of an affinity chromatographic system is the type of ligand used as the stationary phase. There are many biological agents and biological mimics that have been used for this purpose. However, all of these ligands can be placed into one of two categories: (1) high specificity ligands, and (2) general, or group-specific ligands.

The term *high specificity ligands* refers to compounds that bind only to one or a few closely related molecules, which is used in chromatographic systems when the goal is to analyze or purify a specific solute.

General, or group-specific, ligands are compounds that bind to a family or class of related molecules.

SPECIFIC TYPES OF AFFINITY CHROMATOGRAPHY

Bioaffinity Chromatography

Bioaffinity chromatography, or *biospecific adsorption*, is the oldest and most common type of affinity chromatography. This refers to affinity methods that use a biological molecule as the affinity ligand. This was the first type of affinity chromatography developed and represents the most diverse category of this technique.

Lectins represent a class of general ligands that are common in bioaffinity chromatography. The lectins are nonimmune system proteins that have the ability to recognize and bind certain types of carbohydrate residues.

Another useful class of bioaffinity ligands are bacterial cell wall proteins. These ligands have the ability to bind to the constant region of many types of immunoglobulins. This makes them useful in antibody purification.

Nucleic acids and polynucleotides can act as either general or specific ligands in bioaffinity chromatography. For instance, as high-specificity ligands they can be used to purify DNA/RNA-binding enzymes and proteins or to isolate nucleic acids that contain a sequence that is complementary to the ligand. As a group-specific ligand, an immobilized nucleic acid can be used to purify solutes that share a common nucleotide sequence.

Immunoaffinity Chromatography

The most common type of bioaffinity chromatography is that which uses an antibody or antibody-related agent as the affinity ligand. This set of methods is often referred to as *immunoaffinity chromatography (IAC)*. The high selectivity of antibody–antigen interactions and the ability to produce antibodies against a wide range of solutes has made immunoaffinity chromatography a popular tool for biological purification and analysis.

Dye-Ligand and Biomimetic Affinity Chromatography

Two other, related categories of affinity chromatography are the techniques of *dye-ligand affinity chromatography* and *biomimetic affinity chromatography*. In dye-ligand affinity chromatography, a synthetic substance like a triazine or triphenylmethane dye is used as the immobilized ligand.

Dye–ligand affinity chromatography is actually a subset of the more general technique known as biomimetic affinity chromatography. As the name of this latter method implies, it makes use of any ligand that acts as a mimic for a natural compound. This includes the use of synthetic dyes as ligands, as well as other types of agents. For instance, combinatorial chemistry and computer modeling have been used with peptide libraries to design biomimetic ligands for enzymes and other target compounds. Phage display libraries, aptamer libraries, and ribosome display have also been used for this purpose.

Immobilized Metal-Ion Affinity Chromatography

Another type of affinity chromatography that uses a ligand of nonbiological origin is *immobilized metal-ion affinity chromatography (IMAC)*. This method is also known as *metal chelate chromatography* or *metal ion interaction chromatography*. In this approach, the affinity ligand is a metal ion complexed with an immobilized chelating agent.

Boronate Affinity Chromatography

Boronic acid and its derivatives are another class of synthetic substances that have been used as affinity ligands. This makes use of the ability of such substances to form covalent bonds with compounds that contain cis-diol groups in their structure (Fig. 1).

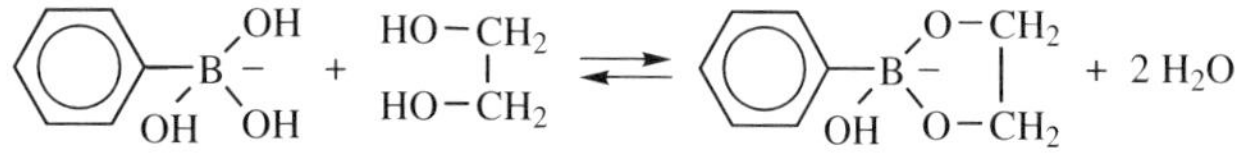

Figure 1. Reaction of boronate with a cis-diol, illustrating the mechanism of retention in boronate affinity chromatography.

Analytical Affinity Chromatography

Besides its use in separating molecules, affinity chromatography can also be employed as a tool for studying solute–ligand interactions. This particular application of affinity chromatography is called *analytical affinity chromatography* or *quantitative affinity chromatography*. Using this technique, information can be acquired regarding the stoichiometry, thermodynamics, and kinetics of biological interactions.

Two experimental formats that are used in this field are *zonal elution* and *frontal analysis*. Zonal elution involves the injection of a small amount of solute onto an affinity column in the presence of a mobile phase that contains a known concentration of competing agent. The equilibrium constants for binding of the ligand with the solute (and competing agent) can then be obtained by examining how the solute's retention changes with competing agent concentration.

Frontal analysis is performed by continuously applying a known concentration of solute to an affinity column at a fixed flow-rate. The moles of analyte required to reach the mean point of the resulting breakthrough curve is then measured and used to determine the equilibrium constant for solute-ligand binding. One advantage of this approach over zonal elution is that it simultaneously provides information on both the equilibrium constants and number of active sites involved in solute-ligand binding. The main disadvantage of this method is the need for a larger quantity of solute than is required by zonal elution.

Information on the kinetics of solute–ligand interactions can also be obtained using affinity chromatography. Recently, a new approach for such measurements has become possible through the availability of flow-through biosensors. *Surface plasmon resonance* is one detection scheme that has been used for this purpose. In these devices, an affinity ligand is immobilized or adsorbed at the sensor's surface, while the solute of interest is applied to the surface in a flow stream of the desired buffer. Changes in the optical properties of this surface are then monitored as the solute binds to the ligand.

Miscellaneous Methods

Other methods that are related to affinity chromatography include *hydrophobic interaction chromatography (HIC)* and *thiophilic adsorption*. Hydrophobic interaction chromatography is based on the interactions of proteins, peptides and nucleic acids with short nonpolar chains on a support.

Thiophilic adsorption is also known as *covalent chromatography* or *chemisorption chromatography*. This makes use of immobilized thiol groups for solute retention. Applications of this method include the analysis of sulfhydryl-containing peptides or proteins and mercurated polynucleotides.

D. S. Hage, in E. Katz, R. Eksteen, P. Shoenmakers, and N. Miller, eds., *Handbook of Liquid Chromatography*, Marcel Dekker, New York, 1998.

X-C. Liu and W. H. Scouten, in P. Bailon, G. K. Ehrlich, W.-J. Fung and W. Berthold, eds., *Affinity Chromatography*, Humana Press, Totowa, N. J., 2000, pp. 119–128.

B. Sellergren, *Molecularly Imprinted Polymers—Man-Made Mimics of Antibodies and Their Applications in Analytical Chemistry*, Elsevier, Amsterdam, The Netherlands, 2001.

D. J. Winzor and C. M. Jackson, in T. Kline, ed., *Handbook of Affinity Chromatography*, Marcel Dekker, New York, 1993.

DAVID S. HAGE
University of Nebraska

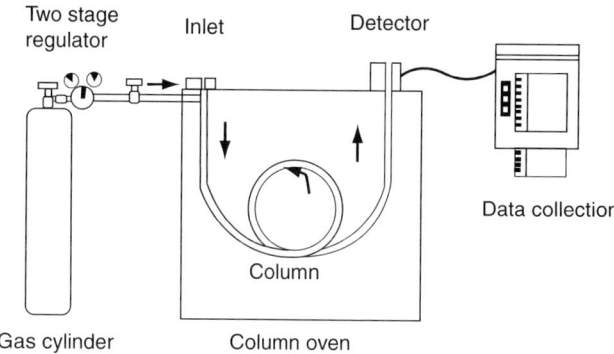

Figure 1. Schematic diagram of a gas chromatography. (Courtesy of Professor Harold McNair.)

CHROMATOGRAPHY, GAS

Gas chromatography (gc) is a physical method of separation in which compounds are separated using a moving gaseous phase (mobile phase) passing over or through a non-moving liquid or solid phase (stationary phase). Gc may be described as a form of column chromatography in that both the mobile and stationary phases are contained within a tube (column) and that the mobile phase is driven through the tube by a pressure drop between the two ends of the tube.

Gc is a high resolution, sensitive and relatively easy to use separation technique. Samples for gc must be volatile under conditions readily achieved in gc instruments, typically temperatures <350°C. They also are typically gases, solutes dissolved in an organic solvent, or sampled from head-space and must provide a signal from a Gc detector.

The data provided by an experiment in gc are called a chromatogram. There are a number of important pieces of information that are generated by analysis of every chromatogram. First, the retention time, indicated by the time elapsed from the point of injection to the maximum of a peak, is a physical property of the compound under the conditions of the experiment. Retention times, although not unique (many compounds may have the same retention time), are used for qualitative analysis by the matching of retention times of unknowns with those of known standards. The peak height, or peak area is related to the mass or concentration of the analyte present and is used for quantitative analysis. The gas hold-up time, defined as the retention time of a nonretained substance, is another important piece of information. The gas hold-up time is typically measured by injecting a small molecule gas, such as methane and recording the retention time.

In gc, column temperature is either maintained constant (isothermal gc) or the temperature is ramped (usually linearly) from a low value to a high value during the run (temperature programmed gc). Isothermal gc is much simpler, both instrumentally and conceptually, so it is often used in teaching and in process control environments, when method simplicity outweighs the need for high resolution. Due to the high thermal mass of the column and ovens used in packed column gc, isothermal conditions are often used in packed column methods. Temperature programming is most often employed with capillary columns, as they have low thermal mass and therefore, reach temperature equilibrium quickly. Also, temperature programming provides an excellent combination of improved resolution and analysis of compounds with a wide range of vapor pressures.

OVERVIEW OF INSTRUMENTATION

A schematic diagram of a modern instrument for gc is shown in Figure 1. A gc system consists of a carrier gas supply, pneumatics, and gas scrubbers, an instrument consisting of three separately controlled heated zones: inlet, column oven and detector, and a data collection and processing system. All of these can be microprocessor controlled and generally use solid-state pneumatics and controls. Modern gc, performed with capillary columns, requires that all ancillary equipment, such as gas supplies and equipment, syringes and devices for delivering samples and the samples be as free from contaminants as possible.

In capillary gc, helium is the most commonly employed carrier gas, with hydrogen used in cases where higher resolution is needed, or when the cost of helium is prohibitive. With packed columns, nitrogen is also used. The main requirements for the carrier gas are that it be of high purity and free of impurities such as water, hydrocarbons, and oxygen.

INLET SYSTEMS

The ability to transfer the analyte sample into a moving gas stream at elevated pressure, without causing the system to leak, is an important consideration in instrument design. Furthermore, the analyte must be transferred quantitatively, without losses or contamination and without decomposition. These requirements make the inlet and sampling system perhaps the most difficult part of the system to use and to understand for the average user. The common techniques for injecting samples into both packed and capillary gc include simple flash vaporizers and direct inlets used with packed columns and splitters, splitless techniques, on column and programmed temperature inlets for capillary columns. Also, there are myriad on-line sampling techniques for both liquid and nonliquid samples.

Direct Inlet

Packed column systems generally employ a simple inlet called a direct inlet or a flash vaporizer. This inlet is heated to enhance rapid vaporization of the injected sample and is pressurized to enhance rapid transfer of the sample to the stationary phase. The major advantage of this inlet is that a syringe needle will easily fit within a $\frac{1}{4}$- or $\frac{1}{8}$-in. outside-diameter packed column.

Split Inlet

In capillary gc, there are two fundamental problems with sample injection. First, the inside diameter of most capillary columns is too small to accommodate a typical syringe. Second, the small mass of stationary phase present in a capillary column can be very easily overloaded by a 1-μL liquid sample. Thus, a new inlet system, the inlet splitter, was conceived. In this inlet, the sample is injected by syringe into a pressurized, heated glass sleeve. The sample vaporizes and mixes homogeneously with the carrier gas stream. Finally, the mixture is passed to two possible exits: the capillary column and a larger diameter purge vent. The purge vent exit is controlled by a needle valve that controls the split ratio, which is the ratio of the amount of the vapor mixture transferred to the vent (waste) and the amount transferred to the column. Typical split ratios range from 10:1–100:1.

Split inlets are in very common use today. The main advantage of split injection lies in simplicity; there are three main variables: inlet pressure, inlet temperature, and split ratio. Split injection is also a very rapid technique, requiring only a few hundred milliseconds for the entire injection process to complete. This results in very sharp chromatographic peaks, necessary for high resolution separations and for good detector sensitivity. The main disadvantages of split injection are in the low final mass of sample reaching the column and in the potential for contamination or reaction of the sample components in the inlet. Gc analytical sensitivity and detection limits are relatively poor (concentration detection limits ~1 ppm) when split injection is used, since most of the injected sample is transferred to the split purge, rather than to the column. Split injection also suffers from the potential reaction of analytes with the inlet components themselves.

Splitless Inlets

Injection performed without splitting of the injected sample vapor, is termed "splitless" injection and has become the most commonly employed injection technique for trace analysis by gc over the past 30 years.

In fact, on most gcs, split and splitless inlets, use the same hardware, with the difference between the configurations being the position of an electronic solenoid valve on the purge vent line. A splitless injection is begun with the purge vent closed. The electronic pneumatic controller will maintain a constant head pressure, and therefore a constant column flow.

On-Column Inlet

On-column injection is receiving increased attention recently, as syringes and autoinjectors have been improved to accommodate the delicate handling required. In the on-column inlet the column extends all the way into the inlet and the syringe must be guided into the column by the inlet fittings. Also the low thermal mass allows the inlet to be temperature programmed along with the column, to ensure that the analytes elute. A syringe with a specially tapered needle is used to inject liquid sample directly into the column. During injection, the column oven and inlet are maintained below the boiling point of the sample solvent; the inlet temperature is usually increased to follow the column oven during a temperature-programmed analysis. The main advantage of on-column injection is that the entire injected sample reaches the column without the potential degradation that comes from the other hot injection techniques. The main disadvantage of on-column injection is also that the entire sample reaches the column, including any and all matrix and nonvolatile components that may be present. Column fouling and maintenance are often increased dramatically when using on-column injection techniques.

A modified on-column inlet has been used for large volume injections, sometimes of up to hundreds of microliters (remember that ~1 μL is typically injected), which allows an analogous increase in analytical sensitivity and lowering of detection limits.

Programmed Temperature Vaporization

Temperature programmed injection has also been used in combination with the classical split and splitless injections described above. The programmed temperature vaporization (PTV) inlet allows for the at-once injection of sample volumes up to 100 μL. The PTV inlet design is based on the classical splitless inlet, except that it has a low thermal mass to allow for rapid heating and cooling. There are a number of modes in which it can operate, including hot split and splitless, which are the same as the classical split and splitless techniques, and cold split and splitless, which involve a cool inlet during injection, which is temperature programmed to pass the injected material into the column.

COLUMNS AND LIQUID PHASES

In gc, the separation occurs in the column, in which the gaseous mobile phase passes over a solid or liquid stationary phase consisting of solid particles, or solid particles coated with a liquid, or consisting of a liquid or solid material coated onto the walls of a capillary tube. With packed columns, separation efficiency is relatively low, so there are a huge number of stationary phases available, to take advantage of the myriad available surface chemistries. Inherently, capillary columns have much higher separation power than packed columns, so there is less need for a wide variety of stationary phase chemistries.

Packed Column Instrumentation

A packed column typically consists of a $\frac{1}{4}$- or $\frac{1}{8}$-in. outside-diameter stainless steel or glass tube with length of 3–12 ft. The diameters of packed columns are generally determined by the availability of tubing and fittings from the suppliers of such equipment. The pressure drop that can be accommodated by a gas chromatograph limits the length of a packed column. Packed columns have a relatively large thermal mass, so temperature equilibration is a major factor in the development of methods. Commonly, to avoid this problem, packed column gcs are operated isothermally, so that temperature equilibrium and reproducibility is maintained.

Capillary Column Instrumentation

In contrast to packed columns, capillary columns, also called "wall coated open tubular" columns, are available in a huge variety of lengths and inside diameters, with relatively few (dozens, rather than hundreds) stationary phases available. Generally, capillary columns vary in length from 10 to 100 m, inside diameters of 0.10 to 0.53 mm, and liquid phase coating thickness 0.1 to 5.0 μm. Since they are open tubes, capillary columns do not share the pressure drop limitations with packed columns, allowing for very long lengths. However, the relatively small inside diameter places limitations on the carrier gas flow rate, injection system and sample capacity. Capillary column instrumentation is therefore more complex and expensive (a factor of 2–5) than packed column instrumentation, with the main differences being in the inlet systems, described earlier in this article. There are also minor differences in the operation of the common detectors flame ionization detector (FID), thermal conductivity detector (TCD), electron capture detector (ECD), mass selective (MS), between capillary and packed column instruments.

Stationary Phases

In both packed and capillary gc, the stationary phase may be either a liquid or solid. In a capillary column, the stationary phase is coated or chemically bound onto the capillary wall; in a packed column, the stationary phase consists of either solid particles or liquid coated solid particles.

DETECTORS

The purpose of the detector is to sense analytes as they elute from the column and record that information in the form of a chromatogram. The signals generated by the detector are received and recorded by a data collection device, such as a chart recorder, electronic integrator, computer data station, or central data collection system. The collected data is plotted as intensity versus time.

The ideal detector would be both universal, meaning that it is able to detect all compounds that elute from the column, and sensitive. However, in reality detectors are often either universal or selective. A selective detector is capable of only detecting certain types of compounds,

and this selectivity is often why the detector has a high sensitivity.

Flame Ionization Detector

The FID employs an ionization detection method invented specifically for gc. The main advantages are its simple design, affordability and reliability. Occasionally this detector is classed as a universal detector, but in fact the FID is only able to detect organic analytes, and will not detect compounds such as water, hydrogen, helium, nitrogen, carbon monoxide, and carbon dioxide.

Thermal Conductivity

The thermal conductivity detector is a universal detector and is frequently used with packed columns and for inorganic analytes that are not detected by FID. The TCD operates on the principle that a hot body (the filament) will lose heat at a rate that is proportional to the surrounding gas and this heat loss can be used to detect the elution of analytes from the column. Since any analyte, except for the carrier gas itself, that passes through the detector will change the rate of heat loss this detector is truly universal. An important advantage of the TCD is the ability to detect air, which is not retained by most gc columns.

Electron Capture

The ECD is very sensitive toward compounds that can capture electrons. Its primary use is in pesticide analysis.

The ECD is a quantitative detector, as the extent of electron capture is proportional to the analyte concentration. The ECDs are straightforward to use, but do require extra care in maintaining a clean system. If the detector is well maintained, conventional ECDs can easily detect picograms of analyte and micro ECD can detect as little as 4 fg of material. The detection limit is very dependent on the analyte's ability to capture electrons and thus the sensitivity of an ECD can drastically different from one analyte to another.

Mass Spectrometer

Perhaps the most useful detector for gc is mass spectrometry (ms). This detector provides both quantitative and qualitative analysis. State-of-the-art bench-top gc ms systems are capable of unit mass resolution.

The mass spectrometer ionizes the incoming sample and presents either the total ion chromatogram (TIC), or it will scan for only certain specified ions in selected ion monitoring (SIM) mode. The TIC of a sample contains all data necessary for compound identification and can be used to compare the mass spectrum of each individual peak in the chromatogram with reference spectra in a computer-based library. The SIM only monitors for one or a few ions in a sample, and so SIM data can be used to identify compounds with previously determined reference spectra.

Other Selective Detectors

Many more types of detectors exist. Two of these are, first, the nitrogen-phosphorous detector (NPD) is another ionization detector that was invented for use with gc. As the name suggests, this detector is selective for nitrogen- and phosphorous-containing compounds. The NPD is used in pharmaceutical labs for nitrogen-containing drugs, as it is the most sensitive detector available for nitrogen and phosphorous. Other applications of NPD are for the analysis of nitrogen- or phosphorous-containing pesticide residues, carcinogens, and amines.

The flame photometric detector (FPD) is primarily used for detection of sulfur and phosphorous. Applications for FPD include pesticide residue analysis, air pollution studies, and petroleum analysis. Infrared spectrophotometers have been successfully attached to a gc and used as a specific detector.

DATA COLLECTION AND HANDLING

The purpose of a chromatographic data system is to collect analogue data from an analytical instrument and convert it to digital data. This is accomplished by an analogue-to-digital converter (ADC). The important characteristics of ADCs are speed and accuracy of conversion. The three basic types of data collection currently used in gc are integrators, dedicated computer based instrument(s) data systems, and multiuser server networked systems.

Integrators and Recorders

In the 1980s, the primary data systems were digital electronic integrators, which combined strip chart recording with rudimentary computing capability.

Computer-Based Data Systems

Now stand-alone and networked computers, usually built around PC platforms, including servers, are also capable of instrument control, data collection, and archiving, generating detailed reports and documents, generating system suitability results and interfacing with laboratory information management systems (LIMS).

Regulatory Issues

In many industries, especially pharmaceutical, forensic, environmental and clinical analysis, data systems and instruments must be validated to assure that they are operating within established norms and procedures. Validation requirements may also go beyond instrumental and scientific concerns, to include data storage, retrieval, and security. The validation process is often lengthy and labor intensive.

MULTIDIMENSIONAL GC

In order to dramatically increase separation power, multidimensional gc, employing two columns, has been developed. In multidimensional gc, the column effluent from the first column, as it elutes, is transferred to a second column, typically with another stationary phase chemistry, for further separation. Multidimensional gc is most often employed in the petroleum industry, and sometimes for toxicology and environmental problems. There are two

common instrument configurations: traditional multidimensional gc, in which the effluents represented by single peaks from the first column are collected, trapped, then transferred to the second column, and comprehensive two dimensional gc, in which the column effluent is trapped continuously and transferred to the second column at regular intervals, using a trap combined with a switching valve.

Traditional Multidimensional Gc

In traditional multidimensional gc, column effluent representing one or more peaks in a separation is trapped and transferred to a second column, which is generally termed "heart-cutting" to represent the interesting portion of the chromatogram being further analyzed. Two-dimensional gc systems are often much more complex than traditional one-dimensional gcs, and are generally not in routine use.

Comprehensive Two-Dimensional Gc

In comprehensive two-dimensional gc, the effluent from a traditional column is continuously sampled into a short, narrow bore, thin film second column. The continuous use of the second dimension column generates tremendously high peak capacity. Notable applications of comprehensive two-dimensional gc include complex petroleum analysis and pesticides from biological samples.

FAST AND MICRO-Gc

Obtaining faster separations has been an interest of chromatographers since the pioneering work in the 1950s. The instrumental requirements for obtaining fast separations are more stringent than for traditional gc, as the columns are very short, the gas flows are high and require very precise control and the chromatographic peaks elute very quickly, requiring careful detector choices and optimization. Along with the drive toward faster separations in traditional bench-top systems, there has been a move toward smaller systems that are field portable. Systems for both applications became commercially available in the late 1990s.

SAMPLE PREPARATION

Almost all gc-based analytical methods in use today also involve some form of sample pretreatment prior to the injection and analysis. Some of the newer techniques for which texts are available, that are often employed online with analysis include pyrolysys, static and dynamic head-space, supercritical fluid extraction, solid-phase extraction, and solid-phase microextraction.

J. S. Fritz, *Analytical Solid Phase Extraction*, John Wiley & Sons, Inc., New York, 1999.

R. D. McDowall, in J. M. Miller and J. B. Crowther, eds., *Analytical Chemistry in a GMP Environment A Practical Guide*, John Wiley & Sons, Inc., New York, 2000, pp. 395–422.

H. M. McNair and J. M. Miller, *Basic Gas Chromatography*, John Wiley & Sons, Inc., New York, 1997.

N. H. Snow, in R. A. Myers, ed., *Encyclopedia of Analytical Chemistry*, John Wiley & Sons, Inc., Chichester, 2000, p. 10680.

Nicholas H. Snow
Seton Hall University

Gregory C. Slack
Clarkson University

CHROMATOGRAPHY, LIQUID

Liquid chromatography involves the separation of compounds by differential migration as a liquid mobile phase flows over a solid stationary phase. The mode of separation varies depending on the mobile and stationary phases (see Chromatography). In hplc, small stationary-phase particle sizes and highly controlled conditions are used to achieve high resolutions.

For analytical applications very small amounts of material are generally added to the column. Preparative hplc can also be used to isolate pure compounds from mixtures and this generally involves adding larger amounts of material to the column. For preparative separations columns of larger diameter (eg, 10 and 20 mm) are employed, and for industrial-scale separations even larger columns are available.

EQUIPMENT

A representative hplc instrument consists of a mobile-phase reservoir, a high pressure pump, an injection device, a separation column, a detector, and a data system. The equipment can be modular, with parts from different manufacturers connected together, or an integrated system from one manufacturer.

Computers are almost universally employed with modern equipment and can be used to control the pump, detector, and robotic sample preparation equipment as well as interpret the output from the detector. In large enterprises, eg, pharmaceutical companies, the systems are highly automated and are coupled together by Laboratory Information Management Systems (LIMS).

The mobile-phase reservoir can be as simple as a conical flask. Some provision should be made for degassing the mobile phase by filtration, sonication, application of a vacuum, sparging with helium, online membrane degassing, or some combination of these methods. A filter should be used to prevent particulates being drawn into the pump.

The pump should provide a constant flow of mobile phase at ~0.5–2 mL/min for analytical separations (with higher flow rates being used for preparative separations) at high pressure (up to 6000 psi). Ideally, pressure variations caused by the action of the pump should be as low as possible to minimize baseline noise.

The injection device can be a high pressure switching valve with a loop of narrow-bore tubing that can be wholly or partially filled with the sample.

The separation column, in which the actual separation of the sample into its components takes place, is typically a stainless steel tube 50–250 mm in length and 1–4.6 mm in internal diameter packed with small particles of modified or unmodified silica through which the liquid is pumped at a typical rate and pressure of 0.8–2 ml/min and 1000–3000 psi, respectively. The properties of the stationary and mobile phase (ie, polarity, size of particles, charge, etc) determine the mechanism of separation; thus, these components should be specifically selected to suit the nature of the sample. Other components that are useful include a pressure gauge, flow meter, and column heater.

It is good practice to employ a filter before the analytical column. Guard columns, packed with a small amount of material similar, or identical, to that in the analytical column, are frequently used.

SAMPLE PREPARATION

In some cases, sample preparation is as simple as filtering the sample before injection. However, in many cases complex sample preparation procedures are required to obtain reproducible results. All of these procedures can be automated using robotic equipment and this is desirable for long production runs. However, due to the effort involved in developing an automated procedure it is generally easier to use manual procedures when the number of samples to be processed is not large.

Liquid–Liquid Extraction

Liquid–liquid extraction generally involves the extraction of an aqueous phase (eg, urine, plasma, and serum) with an organic solvent. Separation, generally facilitated by the use of a centrifuge, results in an organic layer containing the drug and an aqueous layer containing most of the potentially interfering compounds. The extraction can be repeated several times.

More than one extraction can also be used to clean up the sample. Basification of the aqueous layer and extraction with an organic solvent will result in a cleaner sample.

The usual organic solvents can be used for extraction, eg, dichloromethane or heptane. Since most of the organic phase will eventually be vaporized toxic solvents such as benzene or chloroform should be avoided. Hexane is surprisingly toxic and should not be used. Also, appropriate safety precautions should be taken.

Precipitation

A variation of liquid–liquid extraction is to use a water-miscible organic solvent such as acetonitrile or methanol. Centrifuging the mixture causes the proteins to collect in the bottom of the tube and the aqueous/organic mixture may be removed, evaporated, and reconstituted.

Ultrafiltration

Plasma proteins may also be removed using a membrane filter having very small pores (eg, 0.2 µm) made of materials such as cellulose acetate, poly(tetrafluoroethylene)

(PTFE), or polysulfone. The resulting filtrate is clean enough to be injected directly on an hplc column. Low cost disposable equipment has been developed to make this procedure attractive.

Solid-Phase Extraction

Solid-phase extraction (SPE) is a technique whereby a crude chromatographic separation is used to effect an initial purification and produce a sample that is clean enough for injection onto the analytical hplc column. Low cost disposable SPE cartridges are generally used although they can also be constructed by the analyst, eg, in a disposable Pasteur pipet.

Generally, it is important not to let the SPE cartridge run dry between steps although some procedures call for all traces of liquid to be removed using a vacuum immediately before the final elution step. It is also important not to run liquid through the SPE cartridge at too fast a rate.

Solid-phase extraction can also be used to concentrate large amounts of relatively dilute solutions.

Column Switching

Column-switching techniques are closely related to solid-phase extraction and in some cases the distinction may become blurred. There are many variations but the simplest involves two columns, two pumps, and a switching valve.

Column-switching techniques can be used to accomplish achiral separation of a racemic mixture of a drug from contaminants on one column, and then chiral separation of the two enantiomers on another column.

Just as solid-phase extraction can be used to concentrate large amounts of relatively dilute solutions so can column-switching techniques.

Column-switching generally involves automation and complex plumbing and so is best suited to situations where many similar samples must be analyzed. Modern automation allows such systems to run with a minimum of attention once they have been set up.

Derivatization

In hplc, derivatization is generally used to improve the detectability of the compounds of interest although it may also be used to improve their chromatographic properties. The derivatization reaction usually involves a reaction that is simple and irreversible. Ideally, derivatization reactions should proceed rapidly, quantitatively, reproducibly, and irreversibly at room temperature but sometimes heating may be necessary.

Detectors

Refractive index and evaporative light-scattering detectors are so-called universal detectors, ie, they can detect all analytes, but they have some limitations and suffer from limited sensitivity. Refractive index detectors monitor the refractive index of the eluting mobile phase. In an evaporative light-scattering detector, the eluting mobile phase is nebulized and evaporated as it passes through a drift tube.

Uv Detectors

The most commonly used detectors detect eluting compounds by their uv absorbance. These detectors are sensitive, robust, and of relatively low cost. Although uv detectors are not universal detectors most compounds of interest can be detected. Other compounds that have essentially no uv chromophore, such as aliphatic alcohols, can be detected after derivatization.

In diode array detectors (DADs) [also known as photodiode array detectors (PDAs)], wavelengths between, say, 190 and 600–900 nm are continuously recorded.

Fluorescence Detectors

In a fluorescence detector, the mobile phase is illuminated with a beam of light and the light emitted by a fluorescent compound is picked up and quantitated by a detector placed at right angles to the light beam. Monochromators or filters may be used to set the excitation and emission wavelengths. Greater sensitivity may be obtained by using a laser as the excitation source but then the wavelength is not readily changed.

Electrochemical Detectors

Electrochemical detectors also offer great sensitivity and selectivity. However, only a limited range of compounds can be detected using this technique. Derivatization can be used to make the compounds of interest electroactive, ie, detectable using these detectors, but this is not commonly done. Generally, a working electrode is held at a fixed potential relative to a reference electrode and the current is monitored.

One problem with electrochemical detectors is the fouling of the electrode surface by reaction products and intermediates. To get around this problem, pulsed electrochemical detectors have been developed. A series of rapid positive and negative pulses are used to clean the electrode.

Mass Spectrometric Detectors

Historically, mass spectrometric (ms) detectors have been difficult to use and have lacked reliability. However, great advances have been made in recent years and LC–MS is now the method of choice for developmental pharmaceutical work, although production work and quality control generally use uv detectors. There are many different types of MS detectors and operating conditions tend to be instrument specific. Commonly used buffers, such as phosphate, that leave a residue on evaporation may not be used.

Other Detectors

Many other detectors have been described in the literature but they are not in widespread use. Conductivity detectors are useful for detecting inorganic ions in ion-exchange chromatography. Other detectors using principles such as infrared (ir), neclear magnetic resonance (nmr), radioactivity, polarimetry, and viscometry have specialized applications.

Postcolumn Reaction Detection

Postcolumn reaction detection involves the derivatization of compounds as they elute from the chromatographic column after separation. In many cases, the chemistry is the same as that used for precolumn derivatization.

Since chromatographic separation does not occur once the compound has eluted from the hplc column, postcolumn reaction detection can involve the use of reactions that do not necessarily lead to a single well-defined derivative.

Another technique that may lead to a mixture of products is postcolumn photochemical derivatization.

CHROMATOGRAPHIC SYSTEMS

Normal Phase

As originally developed, lc involved an unmodified polar solid stationary phase, and a nonpolar liquid mobile phase. Since this was the first type of lc to be developed it is referred to as "normal phase". As currently practiced normal-phase chromatography involving unmodified stationary phases uses silica particles of ~5–10 μm with a mobile phase consisting of organic solvents such as hexane, octane, dichloromethane, or methanol.

Currently normal-phase chromatography is not used extensively because of the expense of obtaining (and disposing of) high purity, possibly toxic organic solvents and the difficulty of maintaining the correct water level. However, normal-phase chromatography does have some advantages and in some cases separations may be obtained that are not feasible in any other way. Normal-phase chromatography is particularly useful for compounds that are unstable in aqueous solutions. It is also useful for preparative chromatography since the mobile phase is readily evaporated.

Reversed Phase

By far, the most common hplc technique is reversed-phase hplc. In reversed-phase hplc, stationary phases consisting of chemically modified silica are used with polar mobile phases.

Ion-Pair Chromatography

It is difficult to chromatograph ionic compounds by reversed-phase hplc because the polar ionic compounds prefer to stay in the polar mobile phase. However, addition of an ion-pair reagent such as a weak acid (eg, heptanesulfonic acid (as the sodium salt) or a weak base (eg, tetrabutylammonium phosphate)) leads to the formation of neutral ion-pair between the analyte and the reagent.

Ion-Exchange Chromatography

In ion-exchange chromatography, charged groups are covalently bound to the stationary phase. The charged ionic analyte is in competition with a counterion of the same charge in the mobile phase. Analytes that interact strongly with the stationary phase will be more retained (ie, elute later) than analytes that interact weakly.

Size-Exclusion Chromatography

Size-exclusion chromatography (sec) is also known at gel-permeation chromatography (gpc) and involves the separation of analytes by size.

Chiral Separations

The two main ways of effecting chiral separations ie, separating the enantiomers of optically active compounds, are derivatization and the use of columns containing chiral stationary phases. Chiral mobile phase additives have been used but such techniques are less common, perhaps because large quantities of expensive additives may be required. In contrast, chiral additives are commonly used in capillary electrophoresis where the quantity of "mobile phase" used in the course of a day is much less.

Derivatization is an indirect method of chiral analysis. The advantages are that conventional hplc equipment can be used. The disadvantages are increased sample preparation time and effort and the possibility that racemization or different reaction yields may occur during derivatization. This does not happen in every case but examples have been reported.

The use of a chiral column is a direct method of chiral analysis. A large variety of chiral columns are now available and many different types of compounds may be separated. In some cases, stationary phases that are chemically identical but of opposite chirality are available. Changing the chirality of the stationary-phase changes the order of elution of the analyte enantiomers. The order of elution should be arranged so that the trace enantiomer elutes first.

APPLICATIONS

Pharmaceutical Applications

Hplc remains the most important chromatographic method for pharmaceutical applications. Immunoassays are also widely employed but they are specific for a single drug. Microbiological methods are also used for antibiotics but have problems of specificity since active metabolites or other antibiotics will also give a response. Because of its specificity hplc remains the "gold standard" for drug analysis.

Pharmaceutical applications can be divided into two broad categories: analysis of the drug substance or dosage form and the analysis of drugs in bodily fluids (eg, blood or urine).

Nonpharmaceutical Clinical Applications

It is frequently desirable to measure compounds other than drugs in biological fluids. Generally, the procedures are similar to those used to measure drugs in biological fluids. In some cases, these compounds may indicate the presence of certain diseases. Hplc methods have been developed for many other clinically relevant analytes such as amino acids, lipids, or proteins.

Environmental Applications

Many methods have been published in the literature for the analysis of pollutants in the environment. Generally, the main problem is that of sensitivity and so solid-phase extraction is used extensively.

The opinions expressed in this article are those of the author and do not necessarily reflect the views or policies of the FDA.

V. R. Meyer, *Practical High-Performance Liquid Chromatography*, 3rd ed., John Wiley & Sons, Inc., Chichester, U.K., 1998.

P. C. Sadek, *Troubleshooting HPLC Systems: A Bench Manual*, John Wiley & Sons, Inc., New York, 2000.

J. Swadesh, ed., *HPLC: Practical and Industry Chromatography*, 2nd ed., CRC Press, Boca Raton, Fla., 2000.

Q. A. Xu and L. A. Trissel, *Stability-Indicating HPLC Methods for Drug Analysis*, American Pharmaceutical Association, Washington, D.C., 1999.

GEORGE LUNN
Center for Drug Evaluation and
Research, Food and Drug
Administration

CHROMIUM AND CHROMIUM ALLOYS

Chromium is one of the "newer" elements. It is used primarily in the metallurgical industry as an alloying element in steel. Chromium confers properties on the alloy that are not achievable with base metals alone. The most common use of chromium is with iron to make stainless steel, an iron–chromium alloy. Chromium confers oxidation resistance to stainless steel, making it "stainless." Stainless steel, in addition to being commonly found in home and commercial kitchens, is an important engineering alloy used throughout industry in machinery, containers, and pipes (see STEEL). Chromium is also used in chemicals for a variety of purposes. Chromite, the mineral from which chromium is extracted for use in the metallurgical and chemical industries, is used directly by the refractory industry to produce heat-, spalling-, corrosion-, and abrasion-resistant bricks for metallurgical and high-temperature industrial mineral processing applications.

Chromium has a wide range of uses in metals, chemicals, and refractories. Other applications are in alloy steel, plating of metals, pigments, leather processing, catalysts, surface treatments, and refractories.

The terms *chromium* and *chrome*, as used in the chemical industry, are synonymous. Similarly, the terms *dichromate* and *bichromate* are used interchangeably in the chemical industry.

OCCURRENCE

Only the mineral chromite occurs in large enough quantities to be a commercial source of chromium. Chromite can be found in many different rock types, but the host rocks for economically important chromite deposits are called *peridotite* and *norite*. These rocks occur primarily in two geologic settings: *layered intrusions*, which are large bodies of layered igneous rock that cooled very slowly in large underground chambers of molten rock; and *ophiolites*, which are large pieces of the oceanic crust and mantle that have been thrust over continental rocks by the same tectonic forces that cause continental drift. Because chromite deposits in layered intrusions tend to be tabular in form they are known as *stratiform deposits*, whereas those in ophiolites are typically podlike or irregular in form, are known as *podiform deposits* (see also MINERAL RECOVERY AND PROCESSING). Other sources of chromite are beach sands derived from chromite-containing rocks and laterites that are weathering products of peridotite, both minor sources.

Stratiform Deposits

Most of the world's chromite resources occur as stratiform deposits in layered intrusions. Clearly, chromite resources in layered intrusions are not evenly distributed worldwide, nor are they evenly distributed over geologic time.

Podiform Deposits

Although resources and reserves of podiform deposits are quite small compared to stratiform deposits, podiform deposits have been, and continue to be, important sources of chromite. As stated above, podiform deposits occur in *ophiolites*, which are pieces of the oceanic crust and mantle thrust up over continental rocks. Podiform deposits are found in many places in the world and throughout geologic time.

Podiform and stratiform deposits have different chemical characteristics, which has determined how they are used. Industry has classified chromite ore as high-chromium, high-iron, and high-aluminum.

Beach Sands

Beach sands that contain chromite exist as a result of erosion. The fact that chromite is ubiquitous in peridotite at low levels and peridotite can occur over large areas allows for the possibility of streams moving through peridotite to erode the rock and deposit chromite downstream. In addition, the fact that chromite is the most dense mineral in peridotite means that wave action will naturally concentrate the mineral in a beach environment.

Laterites

Laterite forms as the result of weathering of peridotite in a tropical or a forested, warm temperate climate. Laterite is a thick red soil derived from the rock below. It is red because of the high concentration of iron. The process of laterization leaches out most of the silicate minerals in the rock, leaving higher concentrations of elements that can fit in the structures of nonsilicate minerals. Thus laterites concentrate elements such as iron, nickel, cobalt, and chromium.

Chromite

The mineral chromite is jet black in color, has a submetallic luster, yields a brown streak, is generally opaque in thin section, and has no cleavage. It is a solid solution mineral of the spinel group, has cubic symmetry and a closely packed crystal lattice, hence the high density of the minerals of the spinel group. The six end-member compositions that combine to form chromite are hercynite ($FeAl_2O_4$), spinel ($MgAl_2O_4$), Fe–chromite ($FeCr_2O_4$), picrochromite ($MgCr_2O_4$), magnetite (Fe_3O_4), and magnesioferrite ($MgFe_2O_4$). Thus, the general formula is (Mg, Fe) (Cr, Al)$_2O_4$.

Terrestrial Chromium Abundance

Chromium is the 18th most abundant element in the earth's upper crust at 35 ppm. Chromium is most concentrated in rocks that constitute the upper mantle, from which crustal rocks are evolved.

PROPERTIES

The chemical symbol for chromium is Cr, and it has an atomic weight of 51.966 and atomic number of 24. Its melting point is 1907°C and its boiling point is 2671°C. At 20°C the specific gravity is 7.18–7.20 g/cm^3. Chromium is one of the so-called transition elements, meaning that it has valence electrons in two shells instead of one. Chromium is a steel gray metal, has cubic symmetry, and is very hard. It is soluble in H_2SO_4, HCl, HNO_3, and aqua regia. Chromium resists corrosion and oxidation. When used in steel at greater that 10 wt% it forms a stable oxide surface layer, which makes it particularly useful in making stainless steel and other specialty steels to ward off the corrosive effects of water. The ability of chromium to resist corrosion and accept a high polish has made it almost ubiquitous as a coating on household water faucets. See Table 1 for physical properties of chromium.

MINING AND PROCESSING

Both stratiform and podiform deposits are associated with ultramafic rocks even though the origins of these two types of chromite deposits differ. For stratiform deposits, the regular layering can be used to locate chromite deposits concealed by faulting or segmentation. Podiform deposits cannot be reliably inferred. So far, no consistently

reliable geophysical or geochemical exploration technique has been found for podiform deposits. Without chromite-specific physical indicators, the traditional methods of ore body location, outcrop analysis, trenching, and drilling remain the most reliable way to locate chromite deposits. Drilling and drifting are used to locate or extend underground deposits. When an ore body has been located, structural analysis may be used to locate deposit extensions if they exist.

Beneficiation

A wide variety of mining technology is applied to the surface and subsurface mining of chromite ore. Recovery includes surface and underground mining using unmechanized to mechanized methods.

Beneficiation to marketable chromite products varies from hand sorting to gravimetric and electromagnetic separation methods. The amount of beneficiation required and the techniques used depend on the ore source and end-use requirements. When the chromite is clean and mossive, only hand sorting of coarse material and gravity separation of fine material may be required. When the ore is lumpy and mixed with other minerals, heavy-media separation may be used. When the chromite mineral occurs in fine grains intermixed with other minerals, crushing may be used in conjunction with gravity separation and magnetic separation. Processing of chromite to produce chromium products for the refractory, chemical, and metallurgical markets includes crushing and grinding and size sorting by pneumatic and hydraulic methods, kiln roasting, and electric furnace smelting. Labeling of material as it moves from the earth to the consumer is not uniform. The terms *chromite* and *chromite ore* are used here to refer to material in the ground, run-of-mine ore (ie, material removed from the ground), or material supplied to the marketplace. For the purpose of trade, imports are called *chromite ore* and *concentrate made therefrom*. This description is frequently abbreviated to chromite ore and concentrate, chromite ore, or simply chromite. Some sources use chromite ore to refer to material in the ground and material removed from the ground before processing. The term *chromite products* is then used to refer to material supplied to the marketplace. Historically, mining operations supplied minimally processed material. Beneficiation and processing may be carried out at the mine site, at a plant that serves several mines in one geographic area, or at a plant associated with end users. This variety in processing further complicates labeling of material. Today, quality control leads consumers to seek chromite supplies that do not vary significantly in physical or chemical properties over time. As a result, chromite ore is typically beneficiated to produce a physically and chemically uniform product before it reaches the marketplace.

Mining methods are carefully chosen to meet the characteristics of a deposit, including the ore and its environment. Since both small and large, podiform and stratiform, high-grade and low-grade, subsurface and near surface, massive and disseminated chromite deposits are exploited, a variety of mining methods are used. Since,

Table 1. Physical Properties of Chromium

Property	Value
at no.	24
at wt	51.996
isotopes	
mass	50525354
relative abundance, %	4.3183.769.552.38
crystal structure	bcc
lattice parameter, a_o, nm	0.2888–0.2884
density at 20°C, g/mL	7.19
mp, °C	1875
bp, °C	2680
elastic modulus, GPa^a	250

^{a}To convert GPa to psi, multiply by 145,000.

typically, surface mining is less expensive than underground mining and ore bodies are found by their outcrops, surface mining at an outcrop precedes underground mining.

The purpose of beneficiation is to increase desirable ore attributes and decrease undesirable ones. For example, depending on end use, increasing chromic oxide content, chromium:iron ratio, or alumina content is desirable. Reducing silica or other minerals associated with chromite is desirable. Depending on end use, certain sizes may be selected or rejected. The techniques used to accomplish these tasks depend on the physical properties and sizes of the minerals present. Beneficiation does not change the chemical characteristics of the chromite mineral. However, since chromite ore is a mixture of minerals, the characteristics of the ore can be changed by altering its mineral mix. Beneficiation may also be selected to process tailings once sufficient quantities have been stockpiled and the technology of beneficiation and processing has been established.

Ferrochromium

Ferrochromium is produced from chromite ore by smelting a mixture of the ore, flux materials (eg, quartz, dolomite, limestone, and aluminosilicates), and a carbonaceous reductant (wood, coke, or charcoal) in an electric-arc furnace. If the ore is lumpy, it can be fed directly into the furnace. However, if the ore is not lumpy, it must be agglomerated before it is fed into the furnace. Efficient operations recover chromium lost to furnace fume by collecting and remelting the dust and recover chromium lost to slag by crushing and beneficiating the slag. The chromium content of the ferrochromium is determined by the chromium:iron ratio of the chromite ore.

The shift from high-chromium, low-carbon ferrochromium to low-chromium, high-carbon ferrochromium, commonly called *charge-grade ferrochromium*, permitted the use of low-chromium:iron ore for smelting to ferrochromium. Agglomeration technology has been developed to permit the use of fine chromite ore in the electric arc furnace. Both briquetting and pelletizing are practiced. Efficient production technology uses prereduced and preheated pelletized furnace feed.

Chromium Metal Production

Chromium metal is produced primarily through one of two production processes: electrodeposition process to produce electrolytic chromium metal and the reduction of chromic oxide with aluminum powder to produce aluminothermic chromium metal. The aluminothermic reduction process is more widely used and more easily installed or expanded. A wide variety of variations of reductants for the exothermic reduction process and of feed materials and electrolytes for the electrowinning process resulted in the current commercial production processes: aluminothermic reduction of chromic oxide to produce aluminothermic chromium metal and electrolytic deposition from a chromium–alum electrolyte made from high-carbon ferrochromium to produce electrolytic chromium metal.

MANUFACTURING AND PRODUCTION

Chromite ore mining and chromium material manufacturing is an international industry. The major industries associated with chromium are chemical, metallurgical, mining, and refractory. Mining is, of course, the first to process chromium in the form of chromite ore. The chemical industry processes chromite ore by kiln roasting to produce sodium dichromate initially and then other chromium chemicals. The metallurgical industry processes chromite ore mostly by electric-arc furnace smelting to produce ferrochromium. It also processes chromic oxide from the chemical industry and ferrochromium from the metallurgical industry into chromium metal. Ferrochromium and chromium metal are then incorporated into ferrous and nonferrous alloys. The refractory industry processes chromite ore into chromite-containing refractory materials. It also processes chromic oxide from the chemical industry into refractory materials (see also REFRACTORIES).

Chromium and Chromite

Chromite is used in the metallurgical, chemical, and refractory industries. In the metallurgical industry, chromite is processed into ferrochromium or chromium metal, then is used as an alloying metal to make a variety of ferrous and nonferrous alloys. The major end use is in stainless steel, a ferrous alloy made resistant to oxidation and corrosion by the addition of chromium. Chromite is used in the chemical industry to make sodium dichromate which is both a chemical industry product and an intermediate product used to make other chromium chemicals. Chromium chemicals find a wide variety of end uses, including pigments, and plating and surface finishing chemicals. Chromite is used in the refractory industry to produce refractory materials, including shapes, plastics, and foundry sands. These refractory materials are then used in the production of ferrous and nonferrous alloys, glass, and cement.

Chromite Consumption

Reported chromite consumption in the United States over the 5-year period from 1993 to 1997 averaged about 328,000 tons annually, indicating a decline from the 1970s, when annual production regularly exceeded 1 million tons annually. Virtually all of this chromite was imported. The chromite was used to make chromium ferroalloys and chemicals, and chromite refractory materials, including casting sand. The major reason for declining domestic chromite use is the shift from domestic to foreign ferrochromium supply as the source of chromium units for the metallurgical industry. Contributing to reduced chromite ore consumption is the decline in chromite- containing refractory use.

Metallurgical

The metallurgical industry consumed chromite ore to produce chromium ferroalloys and metal.

Refractory

Refractory materials resist degradation when exposed to heat. Chromite is a refractory material. Unlike the chemical and metallurgical industries, where chromite is processed to extract its chromium content, chromite is used chemically unmodified in the refractory industry. Chromic oxide, a chemical industry product, is also used to make refractories for the glass industry. Chromic oxide refractories are used in glass contact areas of glass melting furnaces to achieve long furnace life.

Refractories are broadly categorized according to their material composition into clay and nonclay refractories. The predominant nonclay refractory material is silica. Basic refractories are a type of nonclay refractory, so called because they behave chemically as bases. Basic refractories are made of chromite, dolomite, magnesite, or various combinations of magnesite and chromite. In the refractory industry, chromite-containing refractories are called chrome refractories. Chrome–magnesite refractories are those in which more chromite than magnesite is used. Magnesite–chrome refractories are those in which more magnesite than chromite is used. The terms chrome and magnesite are used in association with refractories to indicate that the refractory was made with chromite ore and magnesia.

Refractories are further categorized by the form in which they are supplied as shaped or unshaped. *Shaped refractories* are manufactured to fit together to form a desired geometric structure, like building blocks. *Unshaped refractories* include mortars (materials used to hold shaped refractories together), plastics (materials that may be formed into whatever shape is desired), and gunning (material that may be sprayed onto a surface). In the refractory industry, the term *monolithics* is commonly used to describe refractories that are not shaped.

The major end users for chromite refractories are in the cement, copper, glass, nickel, and steel industries. Basic refractories are used in copper and nickel furnaces. In the glass industry, chromite refractories are used in glass tank regenerators and chromic oxide refractories are used in melting furnaces for the production of reinforcing glass fibers and textiles. In the cement industry, chromite refractories are used primarily in the transition zones of cement kilns. Basic refractories are typically used in open-hearth and electric-arc steelmaking furnaces.

Chromite refractories were used heavily in steel production using the open-hearth furnace method. Contemporary steelmaking processes that use the basic oxygen furnace or the electric-arc furnace use much less chromite-containing refractories.

The general decline in refractory use results, at least in part, from the more cost-efficient use of refractories. Longer lasting refractories result in lower labor cost to change the refractories and higher production equipment availability because of less down time for relining. A specific reason for the decline in chromite-containing refractory use results from changes in steel industry production practice. The major end use for basic chrome refractories was in the production of steel in open-hearth furnaces. As steel production technology has shifted away from open-hearth furnace steel making, chromite refractory use has declined.

Foundry Sand

Foundry sand use of chromite is a modern application. Sand is used to contain molten metal in a desired shape until the metal has solidified. Sand used in the foundry industry is washed, graded, and dried. Since silica sand is common and inexpensive, it is the most commonly used mineral. However, when physical or chemical conditions dictate, other sands are chosen, such as zircon, olivine, or chromite. Chromite foundry sand is used in the ferrous and copper casting industries.

Casting sands are defined by function and by processing. Mold and core sands are designed for the exterior and interior of a casting, respectively. Facing sand is used on the surface of a core or mold. Flour or paint may be applied to the facing sand. As indicated by its name, flour is finer in size than sand. Before casting, sand is naturally or chemically bonded. There are a variety of methods for bonding sand before casting. Chromite sand is compatible with the commonly used methods. After casting, foundry sand is reclaimed.

Chromite sand is compatible with steel castings. It is typically used as facing sand in heavy section (>4 t) casting and enjoys a technical advantage over silica sand in casting austenitic manganese steel. Chromite sand does not react with the manganese in the steel. Chromite and zircon, each having a higher melting point than silica, are chosen when casting temperatures exceed those acceptable for silica sand. Chromite sand is also used in copper-base nonferrous casting.

Reclamation is an integral part of the foundry industry. It includes mechanical, pneumatic, wet, and thermal processes, and combinations thereof. Using these processes, as much as 90% of chemically bonded foundry sand (average over all minerals used) can be reclaimed. Chromite sand is adaptable to these processes. After casting, chromite sand, typically used as facing sand, becomes mixed with the bulk sand (silica). Since chromite sand has a size distribution similar to that of silica sand, mechanical separation is not applicable. Hydraulic spiral separation and magnetic separation are effective at separating chromite sand from silica and zircon sand. Silica and zircon sands are nonmagnetic. Some chromite sand was found to degrade during use. However, degraded sand tends to adhere to the castings, so it does not become part of the reclaimed sand. Reclaimed chromite sand was found to be interchangeable with new chromite sand. The actual amount of chromite sand reclaimed, like the amount used, is unknown.

SHIPMENT

Chromite ore is typically transported by trackless truck or conveyor belt from the mine face to storage or processing facilities on the mine site. From there, it is transported by truck from the mine site to the local railhead. It is then transported by rail to ports or to smelters. Smelters that do not have associated loading and unloading facilities for ships transport their product by rail to ports. Following transport by ship to consumer countries, chromium materials are typically hauled by barge, truck, or rail to end users who have no loading and unloading facilities for ships.

GRADES, SPECIFICATIONS, AND QUALITY CONTROL

Government and Industry Organization Specifications

U.S. industry sets chemical and physical specifications for chromium materials through the American Society for Testing and Materials (ASTM).

For the purpose of trade, the U.S. government has categorized chromium materials. The import category "chromite ore and concentrates made therefrom" is subdivided by chromic oxide content as follows: containing not more than 40% chromic oxide, containing more than 40% and less than 46% chromic oxide, and containing 46% or more chromic oxide. Producers of chromite ore and concentrate typically specify chromic oxide content; chromium:iron ratio; and iron, silica, alumina, magnesia, and phosphorous contents. They also specify the size of the ore or concentrate. Typically, chromic oxide content ranges from 36 to 56%; values in the 40–50% range are most common. Chromium:iron ratios typically range from about 1.5:1 to about 4.0:1, with typical values of about 1.5:1–3.0:1. In trade, the chromite ore is also called chromium ore, chromite, chrome ore, and chrome.

The import category "chromium ferroalloys" is subdivided into ferrochromium and ferrochromium–silicon. Ferrochromium–silicon, also called *ferrosilicon–chromium* and *chromium silicide*, is not further classified. Ferrochromium is classified by its carbon content as containing not more than 3% carbon, more than 3% but not more than 4% carbon, or more than 4% carbon. Producers of ferrochromium typically classify their material as low- or high-carbon or charge-grade ferrochromium. Charge-grade ferrochromium is also called *charge chrome*. Producers of chromium ferroalloys typically specify chromium, carbon, silicon, phosphorous, and sulfur contents and material size. Ferrochromium– silicon typically contains 24–40% chromium, 38–50% silicon, and 0.05–0.1% carbon. Ferrochromium typically contains 50–75% chromium and 0.05–8% carbon. Low-carbon ferrochromium typically contains 55–75% chromium and 0.02–0.1% carbon. High-carbon ferrochromium typically contains 60–70% chromium and 6–8% carbon. Charge-grade ferrochromium typically contains 50–55% chromium and 6–8% carbon.

By-Products and Coproducts

Chromite ore is a by-product only of platinum mining of the UG-2 layer of the Bushveld complex. No coproducts or by-products are associated with chromite mining operations. Here, by-product or coproduct is assumed to mean a mineral product that is different from the primary product and not different grades of the primary mineral product. A single mining operation is likely to produce more than one grade of its product. Grades of chromite products are distinguished by ore size and chemistry. Chromite recently became a by-product of platinum mining in South Africa.

ENVIRONMENTAL CONCERNS

In recognition of the development of environmental concerns about chromium worldwide and in response to a European Commission review of chromium occupational exposure limits, the International Chromium Development Association published industry guidelines on health, safety, and environment. The guidelines take account of extensive international changes and developments in legislation and regulation of chromium materials and is intended to help companies implement appropriate workplace practices and procedures for environmental protection.

Environmental concerns about chromium have resulted in a wide variety of studies to determine chemical characteristics, natural background levels, sources of environmental emission, movement of chromium in the environment, interaction of chromium with plants and animals, effect of chromium on plants and animals, measurement methods, and recovery technology. A broad review of many environmental factors and the role of chromium, among other metals, in the environment was published.

In the United States, the Environmental Protection Agency (EPA) regulates chromium releases into the environment. The Occupational Safety and Health Administration (OSHA) regulates workplace exposure.

Environmental Regulations

Chromium and chromium compounds are regulated by the EPA under the Clean Air Act (CAA), the Comprehensive Environmental Response, Compensation, and Liability Act of 1980 (also known as CERCLA or Superfund), National Primary Drinking Water Regulations (NPDWR), the Clean Water Act (CWA), and the Resource Conservation and Recovery Act (RCRA).

Effluent. Chromium in water effluents is manageable. The solubility of trivalent chromium compounds in neutral water usually results in a chromium concentration below that required by EPA for drinking water (0.1 ppm). Thus, when water is neutralized, chromium can be removed by filtration. If hexavalent chromium compounds are present, they must first be reduced to trivalent, a technically manageable operation.

Emissions. Congress enacted the Clean Air Act Amendments Law of 1990 (Public Law 101-549), completely revising the Air Toxics Program. Congress identified 189 hazardous air pollutants to be regulated. Chromium compounds—defined as any chemical substances that contain chromium as part of their structure—were included among those hazardous air pollutants. Under the revised Air Toxics Program, Congress instructed EPA to regulate hazardous air pollutants by regulating the source of those pollutants. EPA eliminated the use of chromium chemicals in comfort cooling towers and regulated chromium releases from the electroplating and anodizing industries.

Solid Waste. EPA regulates solid waste generated by the chemical industry in the production of sodium chromate and dichromate. Chromium-containing treated residues from roasting and/or leaching of chrome ore is regulated under Subtitle D of the Resource Conservation and Recovery Act. EPA found no significant danger associated with treated residue from roasting and/or leaching of chrome ore based on waste characteristics, management practices, and damage case investigations.

Resource Conservation and Recovery Act

The Resource Conservation and Recovery Act (RCRA) brought waste from the extraction, beneficiation, and processing (smelting and refining) of ores and minerals under the regulatory control of EPA. EPA listed emissions from the production of ferrochromium–silicon (RCRA waste number K090) and ferrochromium (RCRA waste number K091) as hazardous waste. EPA regulates treated residue from roasting and leaching of chromite ore under Section D of RCRA. An EPA study determined that treated residue from roasting and leaching of chromite ore does not pose an actual or potential danger to human health and the environment.

EPA regulates refractory material solid waste containing chromium. EPA determined that chromium-containing wastes exhibit toxicity. Therefore, they have established a policy that—if the extract from a representative waste sample contains chromium at a concentration greater than or equal to 5.0 mg/L (total chromium) as measured by a specified toxicity characteristics leaching procedure—it is hazardous. EPA promulgated a treatment standard for chromium-containing refractory brick wastes based on chemical stabilization. (Stabilization is a process that keeps a compound, mixture, or solution from changing its form or chemical nature.) EPA determined that some chromium-containing refractory brick wastes can be recycled as feedstock in the manufacture of refractory bricks or metal alloys.

EPA regulates the emission of chromium from toxic waste incinerators. Incineration is a desirable method of toxic waste disposal because organic waste is destroyed, leaving no future cost to society.

Chromium leaching behavior in soil derived from the kiln roasting and leaching of chromite ore was reported. It was found that (1) leaching was highly sensitive to pH and that the most chromium leached out at soil pH between 4 and 12 and (2) the presence of organic matter in the soil reduced the amount of chromium leached out.

Clean Air Act

In 1992, EPA identified chromium electroplaters and anodizers as an area source of hazardous air pollutants that warrant regulation under Section 112 of the Clean Air Act and described that source's adverse impact. The chromium electroplating industry includes hard chromium platers (usually a thick chromium coating on steel for wear resistance of hydraulic cylinders, zinc diecastings, plastic molds, and marine hardware), decorative chromium platers (usually over a nickel layer on aluminum, brass, plastic, or steel for wear and tarnish resistance of auto trim, tools, bicycles, and plumbing fixtures), and surface-treatment electroplaters or anodizers (usually a chromic acid process to produce a corrosion-resistant oxide surface on aluminum used for aircraft parts and architectural structures subject to high stress and corrosive conditions).

In 1994, EPA banned the use chromium chemicals for industrial process water-cooling towers for corrosion inhibition. It was reported that 90% of industrial cooling-tower operators had eliminated the use of chromium chemicals in anticipation of such an EPA ban.

Toxic Release Inventory

Under the Toxic Release Inventory program, EPA collected environmental release information since 1987 from manufacturing facilities that employ 10 or more persons and used a threshold amount of chromium contained in chromium compounds. Facilities report the amount of chromium released to the air, water, and earth environment; the amount of chromium recovered on site; and the amount transferred to offsite locations. The data are collectively referred to as the Toxic Release Inventory (TRI).

Water and Effluents

EPA promulgated its final rule on chromium contained in primary drinking water in 1991. EPA set the maximum contaminant level goal (MCLG) and the maximum contaminant level for chromium contained in primary drinking water at 0.1 mg/L. EPA identified the best available technologies to remove chromium(III) compounds to be coagulation with filtration, ion exchange, lime softening, and reverse osmosis. EPA identified the best available technologies to remove chromium(VI) compounds to be coagulation with filtration, ion exchange, and reverse osmosis. EPA concluded that chromium contained in drinking water should be minimized in recognition of its biological reactivity, including its potential for posing a carcinogenic hazard.

The EPA published a retrospective study on effluent guidelines, leather tanning, and pollution prevention. The report found that industry met the chromium limitations by modifying the tanning process to get more chromium out of the tanning wastewater and into the leather. Recycling was also done to meet guidelines.

RECYCLING

Chromium contained in stainless steel and other metal scrap is recycled. Both new and old scrap are collected by scrap processors and returned to stainless-steel manufacturers. Secondary production is calculated as chromium contained in reported stainless-steel scrap receipts.

Recycling (qv) is the only domestic supply source of chromium. Stainless-steel and superalloys are recycled, primarily for their nickel and chromium contents. As much as 50% of electric furnace stainless-steel production can result from recycled stainless-steel scrap. Advanced stainless-steel production technology such as continuous casting reduces prompt scrap generation and permits a higher product yield per unit of raw material feed.

Industry practice is to sort scrap for recycling. Chromium-containing stainless steel is collected, processed, and returned to stainless steel manufacturers for reuse. Processing may include changing the physical form of the scrap. Some materials require cleaning or sorting before they can be recycled. Some processors melt and combine several alloys to produce master alloy castings that meet stainless steel or other alloy manufacturers' chemical requirements. Superalloy (nickel- and cobalt-based alloys used in the aerospace industry) reuse is carried out by certified recycling companies in cooperation with alloy producers and product manufacturers. Superalloy scrap that cannot be reused is recycled in other alloys. Small quantities of chromium metal waste and scrap are also traded.

HEALTH AND SAFETY FACTORS

Chromium is a trace mineral required by the human body. As such, the National Research Council recommends a daily intake in the range of 50–200 µg. Chromium is a cofactor for insulin, a hormone that participates in carbohydrate and fat metabolism. A cofactor is a material that acts with the material. The dietary chemical form of chromium is as trivalent compounds. Because humans cannot convert trivalent chromium to hexavalent chromium, the carcinogenicity of hexavalent chromium compounds bears no relevance to the nutritional role of trivalent chromium.

The effect of an element on the human body depends on several factors. These factors include the chemical or class of chemical, the route of exposure, the quantity and duration of exposure, and characteristics of the exposed subject.

The chemical distinctions typically made with respect to chromium chemicals include whether the compound is synthetic or naturally occurring. Synthetic chromium compounds are typically classified by their oxidation state. Exposure to chromium compounds could typically occur through one or more of three routes: skin contact, ingestion, or inhalation. Chromium is one of those elements that is both essential to good health and detrimental to good health. The detrimental effects of chemical exposure are classified as acutely toxic when small amounts of the chemical cause significant damage in a short time, chronically toxic when exposure over a long time causes measurable damage, and carcinogenic when exposure can result in cancer.

USES

Chromium was first used in pigments and tanning compounds. Chromium plating, the electrodeposition of chromium from a solution of chromic acid, started in the early 1900s. A more recent use for chromium is in wood preservation. Chromium–copper–arsenate (CCA)

impregnated wood can be protected from weathering, insects, and rotting for 40 years. Chromium chemicals are also used to make biocides, catalysts, corrosion inhibitors, metal plating and finishing chemicals, refractories, and printing chemicals.

A chromium chemical end use with which many people are familiar is pigments. Chromium containing pigments are broadly classified as oxides or chromates. A rainbow of colors are produced by the pigment industry using a variety of mixed metal oxides with chromium. An important use of chromium pigments is in anticorrosion coatings. Chromium pigments that are used for corrosion control include lead, zinc, and strontium chromates. Chromate metal primers are used extensively by the federal government, in both civilian and military applications.

P. A. Lewis, ed., *Pigment Handbook*, Vol. I, *Properties and Economics*, 2nd ed., Wiley- Interscience, Inc., New York, 1988.

E. Merian, ed., *Metals and Their Compounds in the Environment. Occurrence, Analysis, and Biological Relevance*, VCH Publishers, Inc., New York, 1991.

National Research Council, *Recommended Dietary Allowances*, National Academy of Science, Washington, D.C., 1989, pp. 241–243.

T. P. Thayer and B. R. Lipin, A Geological Analysis of World Chromite Production to the Year 2000 A.D., *Proceedings of the Council of Economics*, 107th Annual Meeting, AIME, 1978, pp. 143–152.

JOHN F. PAPP
BRUCE R. LIPIN
U.S. Geological Survey

CHROMIUM COMPOUNDS

Kazakhstan and the Republic of South Africa account for more than half the world's chromite ore production. Almost all of the world's known reserves of chromium are located in the southeastern region of the continent of Africa. South Africa has 84% and Zimbabwe 11% of these reserves. The United States is completely dependent on imports for all of its chromium. The chromite's constitution varies with the source of the ore, and this variance can be important to processing. Typical ores are from 20 to 26 wt% Cr, from 10 to 25 wt% Fe, from 5 to 15 wt% Mg, from 2 to 10 wt% Al, and between 0.5 and 5 wt% Si. Other elements that may be present are Mn, Ca, Ti, Ni, and V. All of these elements are normally reported as oxides; iron is present as both Fe(II) and Fe(III).

PROPERTIES

Chromium compounds number in the thousands and display a wide variety of colors and forms. Examples of these compounds and the corresponding physical properties are given in Table 1.

Table 1. Physical Properties of Chromium Compounds

Compound	Density^a g/cm^3	Mp, °C	Bp, °C
chromium(0) hexacarbonyl	1.77_{18}	148.5	210^b
dibenzene chromium(0)	1.519	284–285	sub 150^c
bis(biphenyl) chromium(I) iodide	1.617_{16}	178	dec
chromium(II) acetate dihydrate	1.79		
chromium (II) chloride	2.88	815	1300
ammonium chromium(II) sulfate hexahydrate			
chromium(III) chloride	2.76_{15}	877	sub 947
chromium(III) acetylacetonate	1.34	216	340
potassium chromium(III) sulfate dodecahydrate	1.826_{25}	89^d	400^c
chromium(III) chloride hexahydrate	1.835_{25}	95	
chromium(III) oxide		90	
chromium(IV) oxide	5.22_{25}	2330	3000
chromium(IV) fluoride	4.98^f	dec	dec 300 to Cr_2O_3
barium chromate(V)	2.89	ca 277	ca 400
chromium(VI) oxide	2.7_{25}	197	dec
chromium(VI) dioxide dichloride	1.9145_{25}	−96.5	115.8
ammonium dichromate(VI)	2.155_{25}	dec 180	
potassium dichromate(VI)	2.676_{25}	398	dec 500
sodium dichromate(VI) dihydrate	2.348_{25}	356; 84.6^e	dec 400
potassium chromate(VI)	2.732_{18}	975	
sodium chromate(VI)	2.723_{25}	792	
potassium chlorochromate(VI)	2.497_{39}	dec	
silver chromate(VI)	5.625_{25}		
barium chromate(VI)	4.498_{25}	dec	
strontium chromate(VI)	3.895_{15}	dec	
lead chromate(VI)	6.12_{15}	844	

^{a}Measurement taken at temperature in °C noted in subscript.
^{b}Explodes.
^{c}In vacuum.
^{d}Incongruent.
^{e}Loses all water at temperature indicated.
^{f}Calculated value.

Chromium is able to use all of its $3d$ and $4s$ electrons to form chemical bonds. It can also display formal oxidation states ranging from Cr(−II) to Cr(VI). The most common and thus most important oxidation states are Cr(II), Cr(III), and Cr(VI). Although most commercial applications have centered around Cr(VI) compounds, environmental concerns and regulations in the early 1990s suggest that Cr(III) may become increasingly important, especially where the use of Cr(VI) demands reduction and incorporation as Cr(III) in the product.

Low Oxidation State Chromium Compounds

Cr(0) compounds are π-bonded complexes that require electron-rich donor species such as CO and C_6H_6 to stabilize the low oxidation state. A direct synthesis of $Cr(CO)_6$, from the metal and CO, is not possible. Normally, the preparation requires an anhydrous Cr(III) salt, a reducing agent, an arene compound, carbon monoxide that may or may not be under high pressure, and an inert atmosphere (see CARBONYLS).

Chromium(II) Compounds

The Cr(II) salts of nonoxidizing mineral acids are prepared by the dissolution of pure electrolytic chromium metal in a deoxygenated solution of the acid. It is also possible to prepare the simple hydrated salts by reduction of oxygen-free, aqueous Cr(III) solutions using Zn or Zn amalgam, or electrolytically. These methods yield a solution of the blue $Cr(H_2O)_6^{2+}$ cation. The isolated salts are hydrates that are isomorphous with Fe^{2+} and Mg^{2+} compounds. Examples are chromous sulfate heptahydrate, $CrSO_4 \cdot 7H_2O$, chromous chloride hexahydrate, $CrCl_2 \cdot 6H_2O$, and $(NH_4)_2Cr(SO_4)_2 \cdot 6H_2O$.

Chromium(III) Compounds

Chromium(III) is the most stable and most important oxidation state of the element. The $E°$ values (Table 2) show that both the oxidation of Cr(II) to Cr(III) and the reduction of Cr(VI) to Cr(III) are favored in acidic aqueous solutions. The preparation of trivalent chromium compounds from either state presents few difficulties and does not require special conditions. In basic solutions, the oxidation of Cr(II) to Cr(III) is still favored. However, the oxidation of Cr(III) to Cr(VI) by oxidants such as peroxides and hypohalites occurs with ease. The preparation of Cr(III) from Cr(VI) in basic solutions requires the use of powerful reducing agents such as hydrazine, hydrosulfite, and borohydrides, but Fe(II), thiosulfate, and sugars can be employed in acid solution. Cr(III) compounds having identical counterions but very different chemical and physical properties can be produced by controlling the conditions of synthesis.

Chromium(IV) and Chromium(V) Compounds

The formal oxidation states Cr(IV) and Cr(V) show some similarities. Both states are apparently intermediates in the reduction of Cr(VI) to Cr(III). Neither state exhibits a compound that has been isolated from aqueous media, and Cr(V) has only a transient existence in water. The majority of the stable compounds of both oxidation states contain either a halide, an oxide, or a mixture of these two. As of this writing, knowledge of the chemistry is limited.

Chromium(VI) Compounds

Virtually all Cr(VI) compounds contain a Cr–O unit. The chromium(VI) fluoride, CrF_6, is the only binary Cr^{6+} halide known and the sole exception. This fluoride, prepared by fluorinating Cr at high temperature and pressure, easily disproportionates to CrF_5 and F_2 at normal pressures, even at −100°C. The fluorination of chromium(VI) oxide or the reaction of KrF_2 and CrO_2F_2 in liquid HF produces chromium(VI) oxide tetrafluoride, $CrOF_4$. Only fluorine displays an oxyhalide having this formula.

The other Cr(VI) halides have the formula CrO_2X_2, where X = F, Cl, or Br. The mixed oxyhalides CrO_2ClY, where Y = F or Br, have been prepared but are not well characterized. The formula CrO_2X_2 also describes non-halide compounds, where X = ClO_4^-, NO_3^-, SO_3F^-, N_3^-, CH_3COO^-, etc. Compounds containing the theoretical cation CrO_2^{2+} are commonly named chromyl. All of the chromyl compounds are easily hydrolyzed to H_2CrO_4 and HX.

The primary Cr–O bonded species is chromium(VI) oxide, CrO_3, which is better known as chromic acid, the commercial and common name. This compound also has the aliases chromic trioxide and chromic acid anhydride and shows some similarity to SO_3.

MANUFACTURE

The primary industrial compounds of chromium made directly from chromite ore are sodium chromate, sodium dichromate, and chromic acid. Secondary chromium compounds produced in quantity include potassium dichromate, potassium chromate, and ammonium dichromate. The secondary trivalent compounds manufactured in quantity are chrome acetate, chrome nitrate, basic chrome chloride, basic chrome sulfate, and chrome oxide.

Sodium Chromate, Dichromate, and Chromic Acid

The basic chemistry used to process chromite ore has not changed since the early nineteenth century. However, modern technologies have added many refinements to the manufacturing techniques, and plants have been adapted to meet health, safety, and environmental regulations. A generalized block flow diagram for the modern chromite ore processing plant is given in Figure 1.

All stacks and vents attached to the process equipment must be protected to prevent environmental releases of hexavalent chromium. Electrostatic precipitators and baghouses are desirable on kiln and residue dryer stacks. Leaching operations should be hooded and stacks equipped with scrubbers (see AIR POLLUTION CONTROL METHODS). Recovered chromate values are returned to the leaching-water cycle.

Sodium chromate can be converted to the dichromate by a continuous process treating with sulfuric acid, carbon dioxide, or a combination of these two. Evaporation of the sodium dichromate liquor causes the precipitation of sodium sulfate and/or sodium bicarbonate, and these compounds are removed before the final sodium dichromate crystallization. The recovered sodium sulfate may be used for other purposes, and the sodium bicarbonate can replace some of the soda ash used for the roasting operation. The dichromate mother liquor may be returned to the evaporators, used to adjust the pH of the leach, or marketed, usually as 69% sodium dichromate solution.

Other Chromates and Dichromates

The wet operations employed in the modern manufacture of the chromates and dichromates are completely enclosed and all stacks and vents equipped with scrubbers and entrainment traps to prevent contamination of the plant and its environment. The continuous process equipment that is used greatly facilitates this task. The trapped material is recycled.

Potassium and ammonium dichromates are generally made from sodium dichromate by a crystallization process involving equivalent amounts of potassium chloride or ammonium sulfate. In each case the solubility relationships are favorable so that the desired dichromate can be separated on cooling, whereas the sodium chloride or sulfate crystallizes out on boiling. For certain uses, ammonium dichromate, which is low in alkali salts, is required.

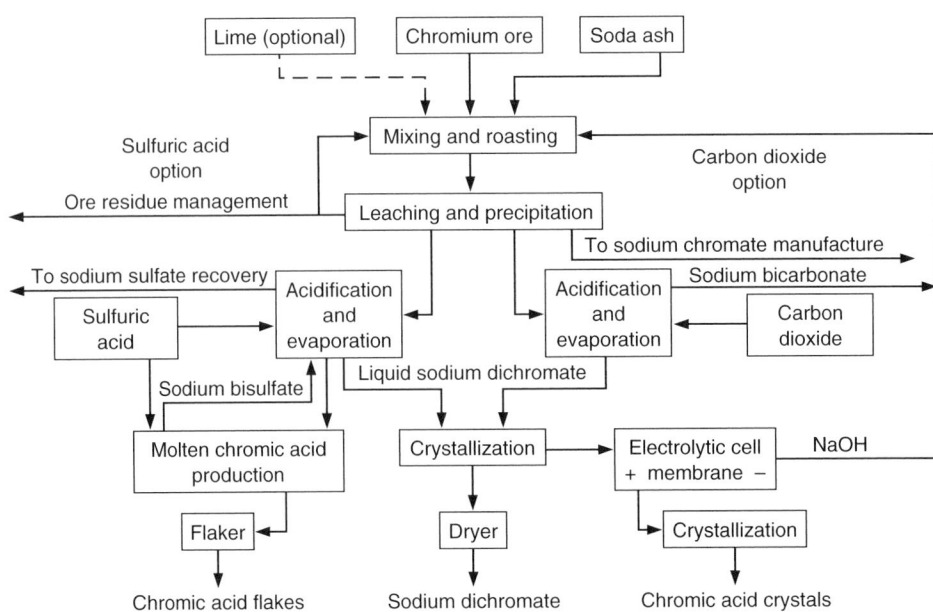

Figure 1. Flow diagram for the production of sodium chromate, sodium dichromate, and chromic acid flake and crystals.

This special salt may be prepared by the addition of ammonia to an aqueous solution of chromic acid. Ammonium dichromate must be dried with care, because decomposition starts at 185°C and becomes violent and self-sustaining at slightly higher temperatures.

Potassium chromate is prepared by the reaction of potassium dichromate and potassium hydroxide. Sulfates are the most difficult impurity to remove, because potassium sulfate and potassium chromate are isomorphic.

Water-Soluble Trivalent Chromium Compounds

Most water-soluble Cr(III) compounds are produced from the reduction of sodium dichromate or chromic acid solutions. This route is less expensive than dissolving pure chromium metal, it uses high quality raw materials that are readily available, and there is more processing flexibility. Finished products from this manufacturing method are marketed as crystals, powders, and liquid concentrates.

SPECIFICATIONS AND SHIPMENT

Chromates and dichromates are sold in both technical and reagent grades. Chlorides and sulfates are the principal impurities. Both manufacturers' and U.S. General Services Administration (GSA) specifications exist for the technical grades, and there are also producer specifications available for some trivalent chromium compounds.

Sodium dichromate, sodium chromate, and mixtures thereof are shipped as concentrated solutions in tank cars and trucks. The chloride and sulfate contents are usually somewhat higher than in the crystalline product. Sodium dichromate is customarily shipped at a concentration of 69% $Na_2Cr_2O_7 \cdot 2H_2O$, which is close to the eutectic composition freezing at −48.2°C.

Chromic acid is transported in steel drums and by rail in tank cars. Multiwall paper bags, fiber drums, as well as steel drums can be employed to ship the solid chromate salts, dichromate salts, and trivalent compounds. Trivalent chromium liquid concentrates are also available in polyethylene drums. The U.S. Department of Transportation (DOT) requires all packages having a capacity ≤416.4 L (110 gallons) to be marked with the proper shipping name and identification number of the chemical contained. The Occupational Safety and Health Administration (OSHA) requires all compounds containing chromium to be labeled as hazardous and all Cr(VI) compounds are required to contain an additional cancer hazard warning.

HEALTH AND SAFETY FACTORS

Chromium in its hexavalent oxidation state is identified by the United States Environmental Protection Agency as one of the seventeen high-priority toxic chemicals that can cause human health problems. Hexavalent chromium (chromium VI) is found in ammonium dichromate, chromic acid, sodium dichromate, sodium chromate, and potassium dichromate. Symptoms of acute dermal exposure to hexavalent chromium are irritated skin and mucous membranes. Ingestion can cause serious injury or death. Deep perforating nasal ulcers know as chrome holes can result form chronic inhalation. Chromium VI is mutagenic and carcinogenic in animals and is a Classification A human carcinogen.

All chromium compounds are considered hazardous substances under CERCLA (Comprehensive Response, Compensation and Liability Act.

Nutrition

Chromium, in the trivalent oxidation state, is recognized as an essential trace element for human nutrition, and the recommended daily intake is 50 to 200 micrograms. The transport of glucose via insulin's reaction with the cell membrane, a necessary mechanism of glucose metabolism, appears to be mediated by chromium. Increased coronary disease risk, glucose intolerance, elevated serum cholesterol and elevated insulin levels have been linked to chromium deficiency. Evidence is available that suggests dietary supplements of chromium(III) may improve glucose tolerance, and there is some indication that a correlation may exist between diabetes and chromium deficient diets.

ENVIRONMENTAL CONCERNS

All chromium-bearing waste materials in the United States are regulated by the EPA. Best practicable technology is required by the EPA to control chromium effluent in discharged industrial waters. Wastes are also subjected to designation as hazardous waste under the Resource Conservation and Recovery Act (RCRA). Chromium compounds are tracked by the EPA's Toxic Release Inventory.

Environmental regulations are expected to tighten with regard to air and water emissions and waste disposal.

Chromium containing solids from manufacturing and wastewater treatment sludges are classified as hazardous wastes and must be handled as such.

The EPA has established exposure levels for both Cr(III) and Cr(VI) for the general population. For exposures of short duration that constitute an insignificant fraction of the lifespan the acceptable intake subchronic (AIS) by ingestion is 979 mg/d for trivalent chromium and 1.75 mg/d for hexavalent chromium. There was insufficient data to calculate an AIS by inhalation for Cr(III), and the EPA believes this type of standard is inappropriate for hexavalent chromium. For lifetime exposures, an acceptable intake chronic (AIC) of 103 mg/d Cr(III) and 0.35 mg/d Cr(VI) is established for ingestion. The inhalation AIC is estimated to be 0.357 mg/d Cr(III). The EPA has calculated an inhalation cancer potency for Cr(VI) of 41 $[mg/(kg \cdot d)]^{-1}$ risk for a lifetime exposure to 1 μg/m^3 hexavalent chromium.

Where appropriate, the direct precipitation of hexavalent chromium with barium, and recovery of the Cr(VI) value can be employed. Another recycling (qv) option is ion exchange (qv), a technique that works for chromates and Cr^{3+}. Finally, recovery of the chromium as the metal or alloy is possible by a process similar to the manufacture of ferrochromium alloy and other metals.

USES

Metal Finishing and Corrosion Control

The exceptional corrosion protection provided by electroplated chromium and the protective film created by applying chromium surface conversion techniques to many active metals, has made chromium compounds valuable to the metal finishing industry. Cr(VI) compounds have dominated the formulas employed for electroplating (qv) and surface conversion, but the use of Cr(III) compounds is growing in both areas because of the health and safety problems associated with hexavalent chromium and the low toxicity of trivalent chromium (see CORROSION AND CORROSION INHIBITORS; METAL SURFACE TREATMENTS; METALLIC COATINGS).

Electroplating of Chromium. Until the middle to late 1970s, all of the commercially electroplated chromium was produced from plating baths prepared from chromic acid. Although these baths accounted for the majority of chromium electroplated products, decorative trivalent chromium baths are successfully operated in many installations.

Because the thickness of the plate deposited from trivalent baths is limited, these have only been employed for decorative applications. However, the bluish white deposit obtained from chromic acid baths can be closely matched by trivalent chromium baths.

Unlike most metals, chromium can be plated from solutions in which it is present as an anion in a high oxidation state. The deposition of chromium from chromic acid solutions also requires the presence of a catalyst anion, usually sulfate, although fluoride, fluosilicate, and mixtures of these two with sulfate have been extensively used. The amount of catalyst must be carefully regulated. Neither pure chromic acid or solutions containing excess catalysts produce a satisfactory plate. Even using carefully controlled temperature, current density, and bath composition, chromium plating is one of the most difficult electroplating operations. Throwing power and current efficiency are notably poor, making good racking procedures and good electrical practices essential.

In 1979, a viable theory to explain the mechanism of chromium electroplating from chromic acid baths was developed.

Decorative chromium plating, 0.2–0.5 μm deposit thickness, is widely used for automobile body parts, appliances, plumbing fixtures, and many other products. It is customarily applied over a nonferrous base in the plating of steel plates. To obtain the necessary corrosion resistance, the nature of the undercoat and the porosity and stresses of the chromium are all carefully controlled. Thus microcracked, microporous, crack-free, or conventional chromium may be plated over duplex and triplex nickel undercoats.

Functional or hard chromium plating is a successful way of protecting a variety of industrial devices from wear and friction. The most important examples are cylinder liners and piston rings for internal combustion engines. Functional chromium deposits must be applied to hard substrates, such as steel, and are applied in a wide variety of thicknesses ranging from 2.5 to 500 μm.

Black and colored plates can also be obtained from chromic acid baths. The plates are mostly oxides. Black chromium plating bath compositions are proprietary, but most do not contain sulfate. The deposit has been considered for use in solar panels because of its high absorptivity and low emissivity.

Chromium Surface Conversion. Converting the surface of an active metal by incorporating a barrier film of complex chromium compounds protects the metal from corrosion, provides an excellent base for subsequent painting, provides a chemical polish, and/or colors the metal. This conversion is normally accomplished by immersion, but spraying, swabbing, brushing, and electrolytic methods are also employed (see METAL SURFACE TREATMENTS). The metals that benefit from chromium surface conversion are aluminum, cadmium, copper, magnesium, silver, and zinc. Zinc is the largest consumer of chromium conversion baths, and more formulations are developed for zinc than for any other metal.

The compositions of the conversion baths are proprietary and vary greatly. They may contain either hexavalent or trivalent chromium, but baths containing both Cr(III) and Cr(VI) are rare. The mechanism of film formation for hexavalent baths has been studied, and it appears that the strength of the acid and its identity, as well as time and temperature, influences the film's thickness and its final properties, eg, color.

Clear-bright and blue-bright chromium conversion colors are thin films (qv) and may be obtained from both Cr(III) and Cr(VI) conversion baths. Iridescent yellows, browns, bronzes, olive drabs, and blacks are only obtained from hexavalent conversion baths, and the colors are listed in the order of increasing film thickness. Generally, the thicker the film, the better the corrosion protection (see FILM DEPOSITION TECHNIQUES).

Oxide films on aluminum are produced by anodizing in a chromic acid solution. These films are heavier than those produced by chemical conversion and thinner and more impervious than those produced by the more common sulfuric acid anodizing. They impart exceptional corrosion resistance and paint adherence to aluminum and were widely used on military aircraft assemblies during World War II. The films may be dyed. A typical anodizing bath contains 50 to 100 g/L CrO_3 and is operated at 35–40°C. The newer processes use about 20 volts dc and adjust the time to obtain the desired film thickness.

Dichromates and chromic acid are used as sealers or after-dips to improve the corrosion resistance of various coatings on metals. For example, phosphate coatings on galvanized iron or steel as well as sulfuric acid anodic coatings on aluminum can be sealed by hexavalent chromium baths.

Chromium compounds are used in etching and brightdipping of copper and its alloys.

Chromates are used to inhibit metal corrosion in recirculating water systems.

Steel immersed in dilute chromate solutions does not rust.

Pigments

Chromium pigments can be divided into chromate color pigments based on lead chromate, chromium oxide greens, and corrosion inhibiting pigments based on difficulty soluble chromate (see PIGMENTS, INORGANIC).

Chromate Pigments Based on Lead. Pigments based on lead can be further subdivided into primrose, lemon, and medium yellows, and chrome orange, molybdate orange, and normal lead silicochromates. Although earlier emphasis was on pure lead compounds, modern pigments contain additives to improve working properties, hue, light fastness, and crystal size and shape and to maintain metastable structures.

Chromium Oxide Greens. The chromium oxide green pigments comprise both the pure anhydrous oxide, Cr_2O_3, and hydrated oxide, or Guignet's green.

Chromic oxide green is the most stable green pigment known. It is used where chemical and heat resistance are required and is a valuable ceramic color (see COLORANTS FOR CERAMICS). It is used in coloring cement (qv) and granulated rock for asphalt (qv) roofing. An interesting application is in camouflage paints, as the infrared reflectance of chromic oxide resembles green foliage. A minor use is in the coloring of synthetic gem stones (see GEMS). Ruby, emerald, and the dichroic alexandrite all owe their color to chromic oxide.

Guignet's green, or hydrated chromic oxide green, is not a true hydrate, but a hydrous oxide, $Cr_2O_3 \cdot x\,H_2O$, in which x is about 2. Although Guignet's green is permanent, it does not withstand use in ceramics. It has poor tinting strength but is a very clean, transparent, bluish green. It is used in cosmetics (qv) and metallic automotive finishes (see COLORANTS FOR FOOD, DRUGS, COSMETICS, AND MEDICAL DEVICES).

Corrosion Inhibiting Pigments. Pigments inhibiting corrosion derive effectiveness from the low solubility of chromate. The principal pigment of this group is zinc chromate or zinc yellow. Others include zinc tetroxychromate, basic lead silicochromate, strontium chromate, and barium potassium chromate.

Zinc yellow is widely used fro corrosion inhibition on auto bodies, light metals, and steel, and in combination with red lead and ferric oxide for structural steel painting.

Zinc tetroxychromate, approximately $4\,ZnO \cdot ZnCrO_4 \cdot x\,H_2O$, has a somewhat lower chromate solubility than zinc yellow and has been used in wash primers.

Strontium chromate, $SrCrO_4$, is used increasingly despite its high cost. It works well on light metals, and is compatible with some latex emulsions where zinc compounds cause coagulation (see LATEX TECHNOLOGY). It is also an ingredient of some proprietary formulations for chrome plating.

Leather Tanning and Textiles

Although chromium(VI) compounds are the most important commercially, the bulk of the applications in the textile and tanning industries depend on the ability of Cr(III) to form stable complexes with proteins, cellulosic materials, dyestuffs, and various synthetic polymers.

Sodium dichromate and various chromic salts are employed in the textile industry. The former is used as an oxidant and as a source of chromium, for example, to dye wool and synthetics with mordant acid dyes, oxidize vat dyes and indigosol dyes on wool, aftertreat direct dyes and sulfur dyes on cotton to improve washfastness, and oxidize dyed wool. Premetallized dyes are also employed. Acid Black 63 (CI 12195) is a typical premetallized dye.

Another use of chromium compounds is in the production of water- and oil-resistant coatings on textiles, plastic, and fiber glass.

Wood Preservation

The use of chromium compounds in wood preservation is largely because of the excellent results achieved by chromated copper arsenate (CCA), available in three modifications under a variety of trade names. The treated wood (qv) is free from bleeding, has an attractive olive-green color, and is paintable. CCA is widely used, especially in treating utility poles, building lumber, and wood foundations.

Chromium compounds are also used in fire-retardant formulations where their function is to prevent leaching of the fire retardant from the wood and corrosion of the equipment employed.

Chromium compounds have a triple function in wood preservation. Most importantly, after impregnation of the wood the Cr(VI) compounds used in the formulations react with the wood extractives and the other preservative salts to produce relatively insoluble complexes from which preservative leaches only very slowly. Finally, although most of the chromium is reduced to chromium(III), there is probably some slight contribution of the chromium(VI) to the preservative value.

Miscellaneous Uses

A large number of chromium compounds have been sold in small quantities for a variety of uses, as follows: catalysts, photosensitive reactions, batteries, magnetic tapes, reagent-grade chemicals, and alloys.

F. A. Cotton and G. Wilkinson, *Advanced Inorganic Chemistry*, 5th ed., John Wiley & Sons, Inc., New York, 1988, pp. 679–697.

Health Assessment Document for Chromium, EPA-600/8-83-014F, United States Environmental Protection Agency (EPA), 1984.

J. F. Papp, "Chromium" *Minerals Yearbook*, U.S. Geological Survey, Reston, Va., 2001.

C. L. Rollinson, in J. C. Bailer, Jr., H. J. Emeléus, R. Nyholm, and A. F. Trotman-Dickenson, eds., *Comprehensive Inorganic Chemistry*, Vol. 3, Pergamon Press, Oxford, U.K., 1973, pp. 624–625.

BILLIE J. PAGE
GARY W. LOAR
McGean-Rohco, Inc.

CHROMOGENIC MATERIALS, ELECTROCHROMIC

A number of inorganic and organic materials exhibit redox states (reduced and/or oxidized forms) with distinct uv-visible (electronic) absorption bands. When electrochemical switching of these redox states is reversible and gives rise to different colors (ie, new or different visible region bands), the material is described as being electrochromic. The optical change is effected by a small electric current at low d-c potential. The potential is usually on the order of 1V, and the electrochromic material sometimes exhibits good open-circuit memory. Unlike the well-known electrolytic coloration in alkali halide crystals, the electrochromic optical density change is often appreciable at ordinary temperatures.

Electrochromic antiglare car rearview mirrors have already been commercialized. Other proposed applications of electrochromic materials including their use in controllable light, reflective or light-transmissive devices for optical information and storage, sunglasses, protective eyewear for the military, controllable aircraft canopies, glare-reduction systems for offices and "smart windows" for use in cars and in buildings.

The most exciting and attractive roles presently envisaged involve long-term display of information, such as at transport termini, reusable price labels, and advertising boards.

The major classes of electrochromic materials are metal oxides, Prussian blue systems, viologens, conducting polymers, transition metal and lanthanide coordination complexes metallopolymers and metal phthalocyanines. Although the latter two classes of metal coordination complexes might be considered as inorganic, they are included here because, mainly, the exhibited colors are a result of transitions that involve organic ligands.

Coloration occurs both cathodically and anodically, as well as in both organic and inorganic materials. Compounds of all types may be classified within one or the other of two general groups based on the nature of charge balancing. In one group, an electrolyte separates a cathode–anode pair, one or both of which may be chromogenically active. Typically the chromogenic material is a thin film on the cathode or anode. As charge neutrality must be preserved, and the electrochromic cathode or anode is a solid, insertion–extraction of ions, often H^+ or alkaline metal, accompanies reduction-oxidation within the electrode surface layer. Insertion/extraction in the cathode or anode is the distinguishing feature of the group. The second group is best described by referring to the viologens, a family of halides of quaternary ammonium bases derived from the 4,4'-bipyridinium structure. Viologens are recognized as the first important organic electrochromic materials. Some of these color deeply within solution by simple reduction; others are distinguished by their deep coloration when electrodeposited from solution onto a cathode. These colorations are characteristic of the noninsertion group, although incidental insertion may accompany electrodeposition.

Members of the ion-extraction group as inorganic or organic thin films, especially the former, have attracted the widest interest most recently. Tungsten trioxide was the earliest exploited inorganic compound, even before the mechanism of its electrochromic response was understood. It is still the best known of the important ion-insertion/extraction group.

OXIDATION–REDUCTION IN THE NONINSERTION/ EXTRACTION GROUP

The best known examples in this group are organic dyes, and the vehicle in which the oxidation–reduction takes place is in general, a liquid electrolyte. For displays, however, it is preferred that color is not developed within the liquid itself, but rather by electrodeposition. Otherwise there is drifting of the coloration and poor memory, which are especially troublesome for displaying information with high resolution.

Organic Compounds

Viologens. Viologens typically require a very low charge density of $2mc/cm^2$ to develop sufficient contrast for display applications. The best known viologen is 1,1'-diheptyl-4,4'-bipyridinium dibromide.

Use of Polymers Systems. A polymer electrochromic device has been made; however, the penalty for polymerization is a loss in device speed.

Chemical or electrochemical oxidation of many polyconjugated compounds such as pyrrole, thiophene, aniline, furan, carbazole, azulene and indole produces polymer films of polypyrrole, polythiophene, or polyaniline etc. doped with counter-ions. The doped polymers, which can adhere to the electrode surfaces are highly conducting, while the undoped, neutral forms are insulating. The oxidized (doped) and reduced (undoped) states of these polymers exhibit different colors, and electroactive conducting polymers are all potentially electrochromic as thin film.

Metallopolymers. Transition metal coordination complexes of organic ligands are potentially useful electrochromic materials because of their intense coloration and redox reactivity.

Chromophoric properties typically arise from low-enegy metal to ligand charge transfer (MLCT), intervalence CT, intraligand excitation, and related visible region electronic transitions. Because these transitions involve valence electrons, chromophoric characteristics are altered or eliminated upon oxidation or reduction of the complex. While these spectroscopic and redox properties alone would be sufficient for direct use of transition metal complexes in solution phase ECDS (electrochromic devices), polymers systems have also been investigated which have potential use in all-solid-state systems. Many schemes have been described for the preparation of thin-film "metallopolymers", including both the reductive and oxidative electropolymerization of suitable polypyridyl complexes. Spatial electrochromism has been demonstrated in metallopolymer films.

Metal Phthalocyanines. Phthalocyanines (Pc) are tetra-azatetrabenzo-derivatives of porphyrins with highly delocalized II-electron systems. Metallophthalocyanines are important industral pigments used primarily in inks and coloring plastics and metal surfaces. In addition to the familiar applications in the area of dyestuffs, the metallophthalocyanines have been intensely investigated in many fields including catalysis, liquid crystals, gas sensors, electronic conductivity, photosensitizers, nonlinear optics and electrochromism.

Inorganic Electrodeposition

From a comprehensive analysis of a variety of electrodepositable metals, the reversible cathodic electroplating of silver has been determined to be the best method. For the highest speeds and contrast intended at the time of analysis, several plating cells were required, back to back. Dependence on liquid-state electrolytes presents practical problems in cell assembly and scaling, so the solid-state is often prefered.

INSERTION/EXTRACTION COMPOUNDS

An important way to assess the many insertion/extraction films known is to compare spectral coloration efficiencies, for the visible region. Adherence to Lambert's law must either be assumed or tested to avoid pitfalls with thin films. The coloration efficiency is determined by spectroelectrochemistry, using a cell which employs a cathode–anode pair in a liquid electrolyte.

Cathodically Colored Inorganic Films

Dissolution of amorphous tungsten oxide films in sealed capsules has been reported to be 2.0–2.5 nm/d at $50°C$ in 10:1 glycerol/H_2SO_4. Dissolution is much easier in water–H_2SO_4, an even earlier electrolyte choice. With more recently and successfully developed sulfonic acid-functionalized polymer electrolytes, cell stability depends on minimizing the water content of the polymer. As with the film itself, however, some minimum water content is necessary in the electrolyte for achieving rapid electrochromic response. A balance must be struck for the water content of both film and electrolyte. One other critical issue with proton-based electrolytes, which applies when they are used with any ion-insertion/extraction film, is the likelihood for the water-containing electrolyte to contribute to electrochemical H_2 or O_2 gas evolution.

By way of contrast, using proton conductivity, at least two other workable solid-state designs have actually been demonstrated at scale. The function of neither one is dictated by the coloration efficiency of a counter electrode. These use sulfonic acid-functionalized polymer electrolytes and depend on unique counter electrodes. One has a high surface area carbon paper counter electrode for reversible proton storage in a display configuration. The other has a very fine, reversibly oxidized copper grid that permits vision through a large-area transparency.

Anodically Colored Inorganic Films

The important electrochromic films of this class include Prussian blue (PB) and the highly hydrated (h) oxides of iridium and nickel. Of lesser significance are the hydrous oxides of rhodium and cobalt. Like the cathodically coloring insertion films, the anodically coloring films depend, for useful darkening and bleaching rates, on having open porosity and hydration.

Doped/Undoped Organic Films

This class of electrochromic materials is probably the youngest and least thoroughly explored from a practical viewpoint. There has been more interest generally in the very high, metal-like conductivity of the oxidized state of some of its members than in the insertion/extraction electrochromism accompanying oxidation–reduction. Of interest have been applications for lightweight and moldable batteries and also for antistatic and electromagnetic shielding. On the other hand, there is not enough reported in the open literature to permit good comparisons of coloration efficiencies. Also, although the films themselves are solid state, almost all electrochromic work has been done with liquid electrolytes. This suggests that research and development are still at a fundamental stage.

SOL-GEL SYSTEMS

Gelling an electrolyte layer, which includes an electrochromic layer, has been suggested as a means of eliminating hydrostatic pressure concerns. However, not everything that is termed a gel is self-supporting, free-standing, and capable of eliminating hydrostatic pressure. The viscosity as a function of polymer content is highly dependent on crosslink density. If the crosslinking is provided by chemical bond formation the resulting gel can be termed a chemical gel by contrast to a physical gel.

APPLICATIONS

The increase in the interaction between humans and machine has made display devices indispensable for visual communication. Electrochromic display device (ECD) is one of the most powerful candidate for this purpose and has various merits such as multicolor, high contrast, optical memory and no visual dependence on viewing angle. A large number of electronic materials are avoidable from almost all branches of synthetic chemistry. The most important examples from major classes of electrochromic materials namely transitions metal oxides, Prussian blue, phthalocyanines, viologens, fullerenes, dyes and conducting polymers (including gels) are described.

N. Baba, M. Yamana, and H. Yamamoto, eds., *Electrochromic Display*, Sangyo Tosho Co. Ltd., Tokyo, 1991.

R. J. Mortimer, "Organic Electrochromic Materials", *Electrochimica Acta* **44**, 2971–2981 (1999).

N. M. Rowley and R. J. Mortimer, "New Electrochromic Materials", *Sciences Progress* **85**, 243–262 (2002).

P. R. Somami and S. Radhakrishman, "Electrochromic Materials and Devices : Present and Future," *Materials Chemistry and Physics* **77**, 117–133 (2002).

A. SAMAT
R. GUGLIELMETTI
Université de la Mediterannée

CHROMOGENIC MATERIALS, PHOTOCHROMIC

Photochromism can be simply defined as a light-induced reversible change of color. However the phenomenon is not limited to colored compounds. It applies to systems absorbing from the far uv to the ir and to very rapid or very slow reactions.

Photochromism concerns the reversible transformation of a chemical species (pigment) between two forms A and B having different absorption spectra and different molecular structure.

$$A \xrightleftharpoons[h\gamma_2 \text{ and (or) } \Delta]{h\gamma_1} B$$

The thermodynamically stable form A generally absorbs in the uv-region (200–400 nm) and is transformed by irradiation ($h\gamma_1$) into the form B which absorbs in the visible region (400–750 nm). The reverse reaction can occur thermally (Δ) or photochemically ($h\gamma_2$). Some systems can function both thermally and photochemically.

The most general phenomenon is referred to as positive photochromism when the colored form B has $\lambda_{max}(B) > \lambda_{max}(A)$. Photochromism is called negative or inverse when $\lambda max(A) > \lambda max(B)$.

The most important application of this phenomenon is in variable optical transmission materials such as lenses that darken in the sun and return to their initial transparency in diffuse light.

The photochromic phenomenon is generally observed in solution or in polymer matrix (viscous solution) and sometimes in the solid state depending on the system.

Fatigue (or photodegradation) is defined as a loss in photochromic behavior, as a result of the existence of side reactions that decrease the concentration of the active species (as A or B forms) or lead to the formation of by-products that inhibit the photochemical formation of B.

$$A \xrightleftharpoons[h\gamma_2 \text{ (or) } \Delta]{h\gamma_1} B \longrightarrow C$$
$$\text{degradation products}$$

These photochromic systems can be divided into two broad categories: organic and inorganic. The two types are quite different in their behavior and observable mechanisms and their characteristics are discussed giving different examples. It is noteworthy that the most important development of these organic systems has occurred during the twenty last years.

INORGANIC PHOTOCHROMIC SYSTEMS

Silver Halide-Containing Glasses

The most important examples of inorganic systems are those containing silver halide crystallites dispersed throughout a glass matrix. In general, these systems are characterized by broad absorption of visible light by the colored species and excellent resistance to fatigue.

The principle behind the generation of a photochromic glass with silver halide is the controlled formation of silver halide particles or crystallites suspended throughout the glass matrix. The formation of crystallites of the correct size and concentration is the key to a useful photochromic system. The general procedure involves the initial melting of a glass-forming mixture which is then cooled to a solid glass shape. Rapid cooling to room temperature results in a nonphotochromic glass. Holding the solid at a temperature in the range of 500–600°C for several minutes to hours causes the nucleation and growth of silver halide crystallites, the active photochromic species. Again, the size of the crystallites is important. With a size of less than 10 nm, significant darkening upon exposure to sunlight is not achieved. Above 20 nm, the scattering of visible light becomes a problem, leading to haziness. Also, with the larger particles the rate of thermal fading slows to an unacceptable rate.

Other Inorganic Metal Salt Systems

An effective silver-free photochromic system can be obtained by the dispersion of crystallites containing cadmium halide and copper halide throughout an inorganic glass matrix.

Another inorganic photochromic glass system was prepared by the addition of europium (II) or cerium(III) to a soda–silica glass with an approximate composition of $Na_2O-2.5\,SiO_2$. Photochromic silver–copper halide films were produced by vacuum evaporation and deposition of a mixture of the components onto a silicate glass substrate.

Simultaneous deposition of cadmium chloride and copper chloride by vacuum evaporation onto fused silica or optical glass resulted in photochromic thin films. Thin films of photochromic silver complex oxides were prepared by anodic oxidation of silver metal films.

Polyoxometalates

Photochromism of materials in which electron transfer and energy transfer within polyoxometalate and related metal oxide solids have been described.

Sols–Gels

The success of low temperature synthesizing sol-gel derived gel glasses doped with photoactive organics led to some new application opportunities in non-linear optics, the solid state tuneable laser, the visible display, and photochemical hole burning for example....

The main advantages of the use of hybrid organic–inorganic nanocomposites result from their high versatility in offering a wide range of possibilities to fabricate tailor-made materials in terms of their chemical and physical properties, and macroscopic shape molding. Such materials emerging in this field are known as sol-gel photonics. There were some striking examples of the use of room-temperature processed hybrids to design materials with emission, absorption, second order non-linear optical, and photochromic properties.

The interface in these nanocomposites is important and one key point of their synthesis is the control of this interface. These nanocomposites can be obtained by hydrolysis and condensation reactions of organically functionalized alkoxide precursors.

ORGANIC PHOTOCHROMIC SYSTEMS

The organic photochromic molecular systems are numerous and it is necessary to classify them in different families based upon the mechanism of the photochromic reaction. Their photochemical or thermal reverse reaction also links them to different types of applications. Other systems are bimolecular such as those involving photocycloaddition reactions.

Photochromism Based on Geometric Isomerism

The simplest example of a photochromic reaction involving a reversible cis-trans isomerization is azobenzene. This easy reaction produces a modest change in the absorption of visible light, largely because the visible absorption band of *cis*-azobenzene has a larger extinction coefficient than that of *trans*-azobenzene.

The cis-trans isomerization of stilbenes is another photochromic reaction of the same type. Although the absorption bands of the stilbene isomers occur at nearly identical wavelengths, the extinction coefficient of the lowest energy band of *cis*-stilbene is generally less important than that of *trans*-stilbene.

Photochromism Based on Tautomerism or Hydrogen Transfer

Several substituted anils (or imines or Schiff bases) of salicylaldehydes are photochromic, but only in the crystalline state.

Quinones are a class of organic photochromic compounds that have been known as photochromic substances quite recently as compared to other compounds. Their photochromism is explained by the reversible photoinduced para-ana-quinoïd transformation due to photochemical migration of different hydrogen, aryl, and acyl, groups.

Photochromism Based on Dissociation Processes

Both heterolytic and homolytic dissociation processes can result in the generation of photochromic systems.

Photochromism Based on Triplet Formation

Upon absorption of light, many polycyclic aromatic hydrocarbons and their heterocyclic analogues undergo transitions to their triplet state which has an absorption spectrum different from that of the ground state. In rigid glasses and some plastics, the triplet state which may absorb in the visible has a lifetime of up to 20 seconds (stabilization by the medium).

Photochromism Based on Redox Reaction

Although the exact mechanism of the reversible electron transfer is often not defined, several viologen salts (pyridinium ions) exhibit a photochromic response to uv irradiation in the crystalline state or in a polar polymer matrix.

Photochromism Based on Electrocyclic Reactions

The most common general class of photochromic systems involves electrocyclic reversible reactions. The most important applications are in the area of variable optical transmission materials, optical memories and switches.

Photochromism Based on Cycloaddition Reactions Involving a Bimolecular Mechanism

The photochemically reversible formation of endoperoxides is an example of this type of system involving $(4n + 2)$ electrons. The reaction is also accompanied by a rather drastic color change because of the disruption of the polycyclic chromophore during endoperoxide formation.

USES

The convenience of having lenses that darken automatically upon exposure of sunlight has proven appealing to spectacle wearers.

The desire for plastic lenses in the ophthalmic market has also accelerated the research on organic photochromic pigments that are more compatible with polymer matrices and more photoresponsive.

Other applications for photochromic light filters have been proposed including glazing applications for solar attenuation, variable transmission camera lenses, and shields for protection against the light flash from a nuclear explosion.

Besides the use of photochromic systems in light filters, their color development has also received considerable attention. For example, the introduction of photochromic components into product labels, tickets, credit cards, etc, aids in the verification of authenticity.

The color development of photochromic compounds can also be used as a diagnostic tool. The temperature dependence of the fading of 6-nitroindolinospiropyran served as the basis for a nondestructive inspection technique for honeycomb aerospace structures.

Photochromic compounds that can be thermally faded have also been used in engineering studies to visualize flows in dynamic fluid systems.

The erasable memory media developed so far have been inorganic materials which use the magneto-optic effect or phase change as the basis for optical recording, but the worldwide acceptance of CD-R (Compact Disk-Recordable) which uses organic dyes as the memory medium, has changed the situation and has given an impetus to find photochromic materials that change their refractive index by photoinduction.

Fulgides and diarylethenes are the main families involved in these applications. Nevertheless spiropyrans and spirooxazines could be used in association with polymers (PMMA or PVK) or liquid crystals.

The chirality is also used for chiroptical molecular switches in overcrowded alkenes, diarylethenes, binphthyl derivatives, fulgides, spiropyrans, photochromic polymers, polymer liquid crystals.

J. C. Crano, W. S. Kwak and C. N. Welch in C. B. Mc. Ardle, ed., *Applied Photochromic Polymer Systems*, Blackie, Glasgow and London, 1992.

H. Dürr and H. Bouas-Laurent, eds., *Photochromism: Molecules and Systems*, Elsevier, Amsterdam, The Netherlands, 1990.

M. Irie, in S. Kawata, M. Ohtsu and M. Irie, eds., *Nano-Optics*, Springer, Berlin, 2002, pp. 137–150.

V. Malatesta in J. C. Crano and R. J. Guglielmetti, eds., *Organic Photochromic and Thermochromic Compounds*, Vol. 2, Kluwer Academic/Plenum Publishers, New York, 1999.

A. Samat
R. Guglielmetti
Université de la Mediterannée

CHROMOGENIC MATERIALS, PIEZOCHROMIC

In its most general sense piezochromism is the change in color of a solid under compression. There are three aspects of the phenomenon. The first is, in a sense, trivial, but it is very general. The color of a solid results from the absorption of light in selected regions of the visible spectrum by excitation of an electron from the ground electronic state to a higher level. If the two electronic energy levels are perturbed differently by pressure, compression results in a color change. This is the basic definition of pressure tuning spectroscopy. Examples include, among others, increased splitting of the d orbitals of transition-metal ions in complexes with pressure, the shift to higher energy of a color center (a vacancy containing an electron) in an alkali halide or glass environment, and a change in the relative energy of bonding and antibonding orbitals as pressure increases.

A second aspect involves a discontinuous change of color when a crystalline solid undergoes a first-order phase transition from one crystal structure to another. The most obvious example is the change of the absorption edge. For example, CdS changes from yellow to deep red at 2.7 GPa (27 kbar) when the crystal structure changes from wurtzite to sodium chloride (face-centered cubic). CdSe, ZnS, ZnSe, and ZnTe undergo similar transitions with distinct color changes at pressures from 5–15 GPa (50–150 kbar). First-order phase transitions involving alterations in crystal structure only can change the electronic excitation energy associated with almost any kind of electronic process provided the two electronic states interact differently with the changing environment. However, for most molecular crystals at modest pressures, the coupling between the electronic states of the molecule and the lattice modes is rather small so these perturbations are usually not large, with few exceptions.

The phenomenon of most interest is a change in color of a solid as a result of a change in the molecular geometry of the molecules that make up the solid. The color change takes place because the change in geometry alters the relative energy of different electronic orbitals, and therefore the electronic absorption spectrum. Frequently it rearranges the order of these orbitals or provides new combinations of atomic orbitals because of symmetry changes. The rearrangements may be discontinuous at a given pressure, may occur over a modest range of pressures, or may occur gradually over the whole range of available pressure as for chemical equilibria in solution. A few examples, together with the principles or generalizations that arise from them, are discussed.

PRINCIPAL PIEZOCHROMIC SYSTEMS

Organic Molecules in Crystals and Polymer Films

The prototypes of piezochromic organic molecules are the salicylidene anils, eg, *N*-salicylidene-2-chloroaniline. At ambient pressure in the crystalline state they are either photochromic or thermochromic, but never both, depending on the side groups on the aromatic rings. When dissolved in a polymer film they are photochromic. The ground state has an OH group opposite a nitrogen on the adjacent ring, thus it is called the "enol" form. When heated, the thermochromic compounds exhibit an absorption in the visible spectrum which corresponds to a transfer of the H from O to N without other change of molecular geometry. This is the "*cis*-keto" form. The photochromic molecules, upon irradiation at low temperature, develop an absorption assigned to the "*trans*-keto" form.

Coordination Compounds and Metal Cluster Compounds

Compounds with metal–metal bonds stabilized by appropriate ligands constitute a second class of materials where a number of cases of piezochromic behavior have been observed. Compounds involving a Re–Re bond stabilized either by eight halides or by bridging (bidentate) ligands, such as the pivalate ion, are two well-established cases. Octahalodirhenates, Re$_2$X$_8^{2-}$ (X = Cl, Br, I), in crystals with a number of counterions, exhibit an absorption peak which corresponds to an excitation from a bonding to antibonding orbital with angular momentum two around the bond (δ orbital).

In the case of the bridged complexes, the process involves changing from a bidentate to a monodentate configuration. For these systems the mode of transformation is variable. In close-packed crystals the rearrangement is a first-order process, ie, it occurs discontinuously at a fixed pressure. For slightly less close-packed crystals the transformation occurs over some range of pressure, eg, 2–3 GPa (20–30 kbar). In the language of physics the process corresponds to a higher order phase transition.

Organometallic Complexes of Cu(II)

Complexes of Cu(II) occur in a wide variety of distorted geometries. The d^9 configuration is stabilized by distortions from a high to a slightly lower symmetry, ie, the Jahn-Teller effect. Cu(II) complexed to organic molecules such as ethylenediamine derivatives lie in a square planar configuration of nitrogens from the organic molecules.

Cu(II) complexed to four Cl$^-$ ions can adopt arrangements from tetragonal to square planar. The latter arrangement gives maximum Cu–Cl bonding; the former minimizes the Cl–Cl repulsion. In practice, complexes occur near both extremes and with almost all intermediate arrangements.

Piezochromic effects have been observed in a variety of other Cu(II) complexes. In some cases it can be shown that the structures of a series of related complexes follow a reaction pathway with the structure of one complex at, eg, 8 GPa (80 kbar) corresponding to that of a related complex at, for example, 2 GPa (20 kbar). The changes in color of the complex, of course, follow the same sequence.

H. G. Drickamer, in R. Pucci and J. Picatto, eds., *Molecular Systems under High Pressure*, North Holland Press, New York, 1991, p. 91 and references therein.

J. K. Grey and I. S. Butler, *Coordination. Chem. Rev.*, 219, 713 (2001).

A. SAMAT
R. GUGLIELMETTI
Université de la Mediterannée

CHROMOGENIC MATERIALS, THERMOCHROMIC

Thermochromism is the reversible change in the spectral properties of a substance that accompanies heating and cooling. Strictly speaking, the meaning of the word specifies a visible color change; however, thermochromism has come to also include some cases for which the spectral transition is either better observed outside of the visible region or not observed in the visible at all. Primarily, thermochromism occurs in solid or liquid phase, but it also describes a thermally dependent equilibrium between brown nitrogen dioxide, NO_2, and colorless dinitrogen tetroxide, N_2O_4, a rare example in the gas phase.

There are many materials, especially organic and metal-organic materials, which exhibit true thermochromism, with a variety of sometimes debatable structural transition mechanisms; it is difficult to summarize the whole with any continuity. For this reason, an effort is made to delineate the scope of the field by listing several thermochromic transitions (Table 1). Selected thermochromic material examples are accompanied in each instance by the corresponding transition stimulus for that case. Characteristically sharp transition temperatures, T_t, are indicated where appropriate. At the other extreme are examples of comparatively gradual transitions, associated for example with an equilibrium or a changing bandwidth. The sharpness of the transition is one aspect by which the several mechanisms could be classified. On the other hand, it is useful also to group materials into metal-complex, inorganic, and organic classes. In this way, the variety of thermochromic changes in each of the three material classes can easily be realized.

Thermochromic compounds such as Ag_2HgI_4 and Cu_2HgI_4 have long been known. These compounds color reversibly, exhibiting the discontinuous red shift of a charge transfer band edge during heating. As for VO_2, the characteristic hysteresis of reflectance in the visible suggests application for infrared image recording.

Table 1. Some Typical Thermochromic Compounds and Their Transitions

Thermochromic material	Thermochromic transition[a]	T_t^b °C
Co^{2+} solutions and glasses	equilibrium shift, two coordinations	
$[(C_2H_5)_2NH_2]_2CuCl_4$	square planar to tetrahedral	50
$[(CH_3)_2NH_2]_3CuCl_5$	variation in bandwidth	
$Cu_4I_4(Py)_4$	fluorescence variations	
$Al_{2-x}Cr_xO_3$(ruby)	lattice expansion/ contraction	
VO_2	monoclinic/tetragonal	68
Cu_2HgI_4	order/disorder	68
di-β-naphthospiropyran	close/open spiro ring	
poly(xylylviologendibromide)	hydration/dehydration	100
ETCD polydiacetylene[c]	side group rearrangement	~115

[a]When applicable, expressed as a change upon heating; various colors have been reported, often qualitatively.
[b]Transition temperatures for sharp transitions.
[c]Urethane-substituted polymer of where R = $(CH_2)_4OCONHCH_2CH_3$.

METAL COMPLEXES, SPECTRAL TRANSITIONS

Crystal field theory, which is simpler to present than the more comprehensive molecular orbital treatment, has been used to describe d–d orbital excitations of transition-metal ions and the effect on these excitations of ligand coordination geometry and field strength. Absorption bands in the visible region arise in energy states made nondegenerate by the crystal field. Color changes, such as induced by heating or cooling, are therefore a direct indication of change in the surrounding environment of the metal ion. Crystal field theory, even though it does not include charge-transfer processes, has proven to be qualitatively adequate for the $3d$ orbital transition metals because the $3d$ states are not well shielded from ligand field effects.

TRANSITIONS IN INORGANIC COMPOUNDS

There are not many oxides and sulfides that may be classified as truly thermochromic; again, however, compounds of transition metals dominate. Ruby exhibits a well-known, reversible, ligand-field thermochromism at different temperatures depending on the concentration of Cr^{3+} in the Al_2O_3 lattice. This is a manifestation of change in the ligand field strength as dependent on lattice expansion/contraction. A whole family of oxides is known to undergo reversible, nonmetal-metal, thermoresistive transitions with heating. The shifts in band structure have frequently been debated. Sometimes these transitions are associated with symmetry changes.

ORGANIC AND POLYMER COMPOUNDS

Simple organic molecules tend to be colorless with electronic transitions in ultraviolet light, whereas visible absorption, or color, is usually associated with electronic excitations in extended and conjugated structures. Color is influenced considerably by the extent of conjugation, as well as by the molecular environment imparted by substituents. So, thermochromism arises from critical, thermally induced changes in the existing structure.

Certain poly(di-n-alkylsilanes) and germanes, when in the solid state in particular, also exhibit large spectral changes that have been associated with side-chain influence on the backbone, but these (the polymer being saturated) occur at uv rather than visible wavelengths.

Spiropyranes and Spirooxazines

The thermochromism of spiropyranes discovered in 1926, has been extensively studied. Nearly every known compound of this class leads to deep color on melting (generally, red, purple, or blue). However heating solutions of spiropyrans also causes coloration.

The thermochromic mechanism in these classes has been assumed to involve a thermally sensitive equilibrium between the colorless spiroheterocyclic form (SP) and the quasi-planar open merocyanine-like structure (MC) obtained after the breaking of the C=O bond. For both these classes of spiroheterocyclic compounds, it seems certain that the most thermally stable photoinduced-colored form and the species formed thermally are spectroscopically and kinetically indistinguishable.

[2H]Chromenes (Benzo- or Naphthopyrans)

2-Phenylamino-2H-pyran and benzo- or heteroannellated 2H-pyrans have been reported to exhibit thermochromic behavior in solution.

Other Spiroheterocyclic Compounds

A series of novel photo- and thermochromic perimidine spirocyclohexadienones whose mechanism involved a ring-chain tautomerism with recently developed. Interesting zwitterionic spirocyclic compounds such as 2,4,6-trinitroaryl derivatives of o-hydroxyaldehydes and the corresponding imines show a negative thermochromism. Indeed in this case the colored spirocyclic form is more stable than the open colorless form even in an apolar solvent.

Schiff Bases and Related Nitrogen-containing Compounds

The Schiff bases of salicylaldehydes with arylamines, aminopyridines and aryl- or thienylalkylamines show thermochromism and photochromism in the solid state due to hydrogen transfer.

Thermochromism is restricted to planar molecules and is attributed to a shift of the tautomeric equilibrium toward the "NH" form absorbing at longer wavelengths. For a nonplanar molecule, much energy is required for hydrogen transfer in the ground state and the transfer occurs only in photochemically excited states.

Bianthrones and Other Overcrowded Ethenes

Bianthrones and related bianthrylidene systems undergo a reversible color change induced by light (photochromism), temperature (thermochromism) or pressure (piezochromism).

The thermochromism of bianthrone in solution has been shown to result from a thermal equilibrium between two distinct and interconvertible isomeric species A → B.

Besides bianthrone derivatives, other sterically overcrowded alkenes have generated considerable interest owing to their intriguing thermochromic (and photochromic) properties, such as dixanthenylidenes, bithioxanthylidenes, 9,9'-fluorenylidene anthrones, 9-diphenylmethyleneanthrones, xanthylideneanthrones, and [2-(thioxanthen-9-ylidene) indane-1,3-dione.

Miscellaneous Compounds

Some indano[1,2-b]aziridines are reported to exhibit thermo-(and photo-)chromism.

Thermochromism of hindered amino-substituted cyclohexadiones due to C−N bond cleavage has been described by Russian authors.

In the case of Mannich bases exhibiting thermo-(and photo-)chromism the proposed mechanism is quite similar.

USES

The optical properties of thermochromic coatings are strongly temperature dependent.

Films of vanadium oxide were produced by reactive RF magnetron sputtering. For the major technological applications, the undoped VO_2 has a high transition temperature. Addition of a dopant such as tungsten or molybdenum lowers the transition temperature.

Commercialy, only two types of thermochromic systems have been successfully applied to textiles: the liquid crystal type and the molecular rearrangement type.

Thermochromic coatings that reduce the transmission of solar energy as the temperature rises can prevent overheating and find application in the thermal control of building and housing sectors, satellites, and spatial equipment.

The scientific and technological challenges of large-area electrochromic and thermochromic devices are great, involving improvements such as low cost, high uniformity, high rate and large area thin-film deposition techniques (DC and RF sputtering, chemical vapor deposition [CVD, plasma-enhanced CVD and dip-coating]).

A. Mannschreck, K. Lorenz, and M. Schinabeck in J. C. Crano and R. J. Guglielmetti, eds., *Organic Photochromic and Thermochromic Compounds*, Vol. 2, Kluwer Academic/Plenum Publishers, New York, 1999, Chapt. 6.

A. Samat and V. Lokshin in J. C. Crano and R. J Guglielmetti, eds., *Organic Photochromic and Thermochromic Compounds*, Vol. 2, Kluwer Academic/Plenum Publishers, New York, 1999, Chapt. 10.

G. Shoham, S. Cohen, M. R. Suissa, and I. Agranat in J. J. Stezowski, J. L. Huang, and M. C. Shao, eds., *Molecular Structure, Chemical Reactivity and Biological Activity*, Oxford Univ. Press., Oxford, 1988, pp. 290–312.

K. Sone and Y. Fukuda, *Inorganic Thermochromism*, Vol. 10, Springer-Verlag, New York, 1987, pp. 2, 13.

R. C. Bertelson in G. H. Brown, ed., *Photochromism*, John Wiley & Sons, Inc., New York, 1971, pp. 45–431.

A. Samat
R. Guglielmetti
Université de la Mediterannée

CITRIC ACID

Citric acid (2-hydroxy-1,2,3-propanetricarboxylic acid), is a natural component and common metabolite of plants and animals. It is the most versatile and widely used organic acid in foods, beverages, and pharmaceuticals.

$$\begin{array}{c} CH_2-COOH \\ | \\ HO-C-COOH \\ | \\ CH_2-COOH \end{array}$$

Because of its functionality and environmental acceptability, citric acid and its salts (primarily sodium and potassium) are used in many industrial applications for chelation, buffering, pH adjustment, and derivatization. These uses include laundry detergents, shampoos, cosmetics, enhanced oil recovery, and chemical cleaning.

Citric acid specifications are defined in a number of compendia including *Food Chemicals Codex* (FCC), *United States Pharmacopoeia* (USP), *British Pharmacopoeia* (BP), *European Pharmacopoeia* (EP), and *Japanese Pharmacopoeia* (JP).

OCCURRENCE

Citric acid occurs widely in the plant and animal kingdoms. It is found most abundantly in the fruits of the citrus species, but is also present as the free acid or as a salt in the fruit, seeds, or juices of a wide variety of flowers and plants. The citrate ion occurs in all animal tissues and fluids. The total circulating citric acid in the serum of humans is approximately 1 mg/kg body weight. Normal daily excretion in human urine is 0.2–1.0 g.

Physiological Role of Citric Acid

Citric acid occurs in the terminal oxidative metabolic system of virtually all organisms. This oxidative metabolic system (Fig. 1), variously called the Krebs cycle (for its discoverer, H. A. Krebs), the tricarboxylic acid cycle, or the citric acid cycle, is a metabolic cycle involving the conversion of acetate derived from carbohydrates, fats, or proteins to carbon dioxide and water. This cycle releases energy necessary for an organism's growth, movement, luminescence, chemosynthesis, and reproduction. The cycle also provides the carbon-containing materials from which cells synthesize amino acids and fats. Many yeasts, molds, and bacteria conduct the citric acid cycle, and can be selected for their ability to maximize citric acid production in the process. This is the basis for the efficient commercial fermentation processes used today to produce citric acid.

PHYSICAL PROPERTIES

Citric acid, anhydrous, crystallizes from hot aqueous solutions as colorless translucent crystals or white crystalline powder. Its crystal form is monoclinic holohedra. Citric acid is deliquescent in moist air and is optically inactive. Some physical properties are given in Table 1.

Aqueous solutions of citric acid make excellent buffer systems when partially neutralized because citric acid is a weak acid and has three carboxyl groups, hence three pK_as. The buffer range for citrate solutions is pH 2.5 to 6.5. Buffer systems can be made using a solution of citric acid and sodium citrate or by neutralizing a solution of citric acid with a base such as sodium hydroxide.

Citric acid monohydrate has a molecular weight of 210.14 and crystallizes from cold aqueous solutions. When gently heated, the crystals lose their water of hydration at 70–75°C and melt in the range of 135–152°C. Citric acid monohydrate is available in limited commercial quantities since most applications now call for the anhydrous form.

CHEMICAL PROPERTIES

Citric acid undergoes most of the reactions typical of organic hydroxy polycarboxylates. Reactions such as esterification, salt formation and anhydride reactions can be easily perfomed. However, the tertiary hydroxyl group does not undergo all the common reactions. If a reaction requires more strenous conditions, dehydration to aconitic acid is promoted.

Decomposition

When heated above 175°C, citric acid decomposes to form aconitic acid, citraconic acid, itaconic acid, acetonedicarboxylic acid, carbon dioxide, and water.

Esterification

Citric acid is easily esterified with many alcohols under azeotropic conditions in the presence of a catalyst such as sulfuric acid, p-toluenesulfonic acid, or sulfonic acid-type ion-exchange resin. Alcohols boiling above 150°C esterify citric acid without a catalyst.

Oxidation

Citric acid is easily oxidized by a variety of oxidizing agents such as peroxides, hypochlorite, persulfate, permanganate, periodate, hypobromite, chromate, manganese dioxide, and nitric acid. The products of oxidation are usually acetone dicarboxylic acid, oxalic acid, carbon dioxide, and water, depending on the conditions used.

Reduction

The hydrogenation of citric acid yields 1,2,3-propanetricarboxylic acid. Hydrogenolysis yields carbon dioxide,

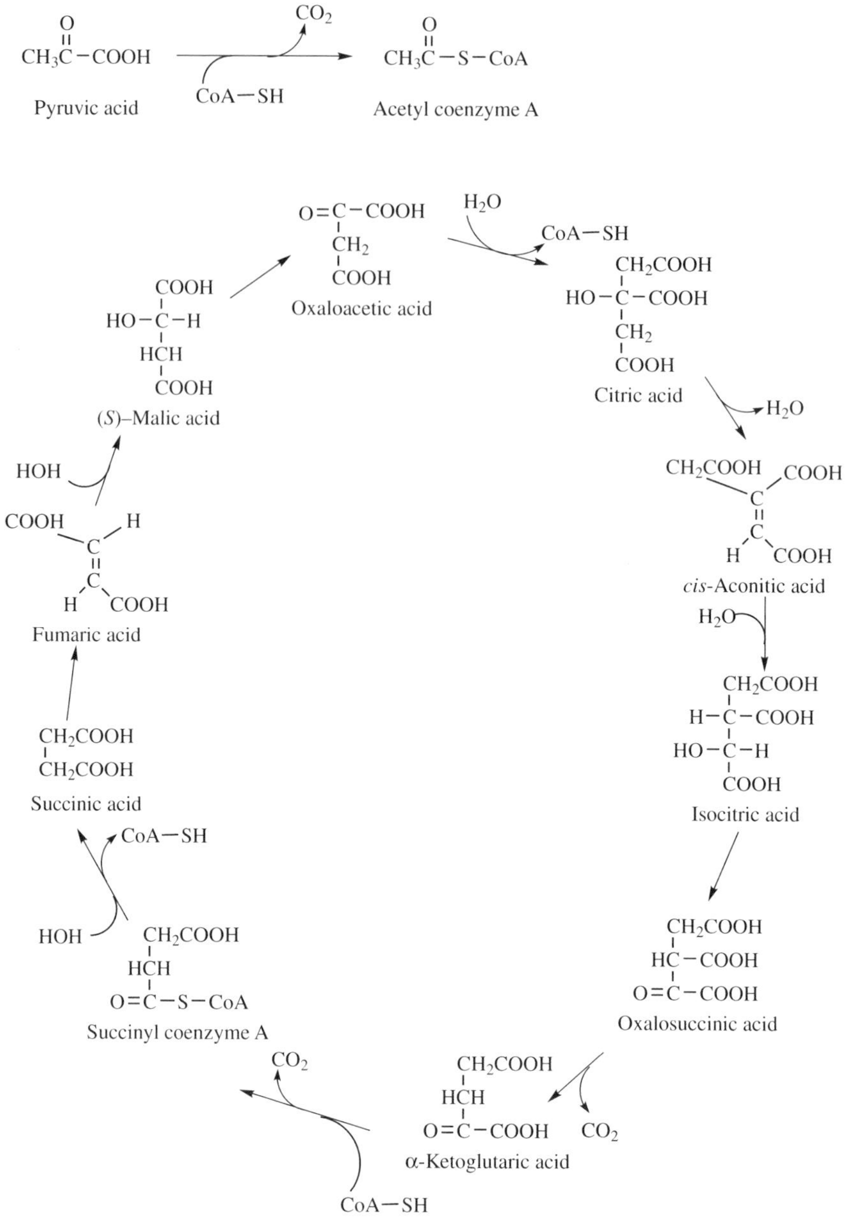

$$CH_3\overset{\overset{\text{O}}{\|}}{C}-COOH \xrightarrow[\text{CoA}-\text{SH}]{\text{CO}_2} CH_3\overset{\overset{\text{O}}{\|}}{C}-S-CoA$$

Pyruvic acid   Acetyl coenzyme A

Oxaloacetic acid

(S)–Malic acid

Fumaric acid

Succinic acid

Succinyl coenzyme A

α-Ketoglutaric acid

Citric acid

cis-Aconitic acid

Isocitric acid

Oxalosuccinic acid

Figure 1. Krebs (citric acid) cycle. Coenzyme A is represented CoA–SH. The cycle begins with the combination of acetyl coenzyme A and oxaloacetic acid to form citric acid.

water, methane, formic acid, acetic acid and a small amount of methyl succinic acid.

Salt Formation

Citric acid forms mono-, di-, and tribasic salts with many cations such as alkalies, ammonia, and amines. Salts may be prepared by direct neutralization of a solution of citric acid in water using the appropriate base, or by double decomposition using a citrate salt and a soluble metal salt.

Chelate Formation

Citric acid complexes with many multivalent metal ions to form chelates. This important chemical property makes citric acid and citrates useful in controlling metal contamination that can affect the color, stability, or appearance of a product or the efficiency of a process.

Table 1. Physical Properties of Citric Acid, Anhydrous

Property	Value
molecular formula	$C_6H_8O_7$
mol wt	192.13
gram equivalent weight	64.04
melting point, °C	153
thermal decomposition temp., °C	175
density, g/mL	1.665
heat of combustion,[a] MJ/mol[b]	1.96
heat of solution, J/g[b]	117

[a]At 25°C.
[b]To convert J to cal, divide by 4.184.

Corrosion

Aqueous solutions of citric acid are mildly corrosive toward carbon steels. At elevated temperatures, 304 stainless steel is corroded by citric acid, but 316 stainless steel is resistant to corrosion. Many aluminum, copper, and nickel alloys are mildly corroded by citric acid. In general, glass and plastics such as fiber glass reinforced polyester, polyethylene, polypropylene, poly(vinyl chloride), and cross-linked poly(vinyl chloride) are not corroded by citric acid.

MANUFACTURING AND PROCESSING

Fermentation

Several microorganisms (yeasts and molds) have been identified as citric acid accumulators. However, the most common microorganisms used for the commercial production for citric acid is *Aspergillus niger*. Surface fermentation is more labor intensive and is easier to maintain aseptically in industrial operations. Nowadays, the preferred method for large-volume industrial production is a submerged process known as deep tank fermentation.

Recovery

Citric acid fermentation broth is generally separated from the biomass using filtration or centrifugation. The citric acid is usually purified using either a lime-sulfuric acid method or a liquid extraction process. Choice between these two methods is dictated in part by the fermentation feedstock. Lime-sulfuric extraction is more traditional, so it is used in many of the older plants. Solvent extraction is currently viable where pure substrate is used for fermentation.

By-Products

The biomass from the fungal fermentation process is called mycellium and can be used as a supplement for animal feed since it contains digestible nutrients. The lime–sulfuric purification and recovery process results in large quantities of calcium sulfate cake.

Energy

In recent years the concern for energy conservation has resulted in many innovative process improvements to make the manufacture of citric acid more efficient.

Chemical Synthesis

None of the different synthetic routes have proven to be commercially feasible.

SHIPMENT AND STORAGE

Crystalline citric acid, anhydrous, can be stored in dry form without difficulty, although conditions of high humidity and elevated temperatures should be avoided to prevent caking. Storage should be in tight containers to prevent exposure to moist air. Several granulations are commercially available with the larger particle sizes having less tendency toward caking.

Although not as corrosive as the acid, the sodium and potassium salts of citric acid should be handled in the same type of equipment as the acid to avoid corrosion problems.

SPECIFICATIONS, STANDARDS, AND QUALITY CONTROL

Since citric acid is produced and sold throughout the world, it must meet the criteria of a variety of food and drug compendia.

HEALTH AND SAFETY FACTORS

Citric acid, as well as its common sodium and potassium salt forms, are Generally Recognized As Safe (GRAS) by the U.S. Food and Drug Administration as Multiple Purpose Food Substances. Citric acid is also approved by the Joint FAO/WHO Expert Committee on Food Additives for use in foods without limitation.

Tests have shown that citric acid is not corrosive to skin but is a skin and ocular irritant. For these reasons it is recommended that individuals use appropriate personal protection to cover the hands, skin, eyes, nose, and mouth when in direct contact with citric acid solutions or powders. This product is not hazardous under the criteria of the OSHA Hazard Communication Standard.

USES

Citric acid is utilized in a large variety of food and industrial applications because of its unique combination of properties. It is used as an acid to adjust pH, a buffer to control or maintain pH, in a wide range, a chelator to form stable complexes with multivalent metal ions, and a dispersing agent to stabilize emulsions and other multiphase systems (see DISPERSANTS). In addition, it has a pleasant, clean, tart taste making it in the acidulant of choice in food and beverage products applications. As well as food and beverages, citric acid has applications in the following areas.

Medical Uses

Citric acid and citrate salts are used to buffer a wide range of pharmaceuticals at their optimum pH for stability and effectiveness.

Agricultural Use

Citric acid and its ammonium salts are used to form soluble chelates of iron, copper, magnesium, manganese, and zinc micronutrients in liquid fertilizers. Citric acid and citrate salts are also used in animal feeds.

Industrial Uses

Laundry Detergents. Hard surface cleaners, reverse osmosis membrane cleaning, metal cleaning, petroleum, flue gas desulfurization, mineral and pigment slurries, electrodeposition of metals, concrete, mortar, and plaster, textiles, plastics, paper, tobacco, cosmetics and toiletries, and refractories and molds.

DERIVATIVES

Salts

The trisodium citrate salt is made by dissolving citric acid in water at a concentration of 50% w/w or higher. A 50% solution of sodium hydroxide is carefully added to pH 8.0–8.5. The reaction is exothermic and cooling is necessary to prevent boiling. The product crystallizes as the monohydrate. Ammonium salts of citric acid are made by adding either aqueous or anhydrous ammonia to citric acid dissolved in water.

Esters

The significant esters of citric acid are trimethyl citrate, triethyl citrate, tributyl citrate, and acetylated triethyl- and tributyl citrate. Many other esters are available but have not been used on a commercial scale. Citric acid esters are made under azeotropic conditions with a solvent, a catalyst, and the appropriate alcohol.

C. D. Barnett, *The Science & Art of Candy Manufacturing*, Harcourt Brace Jovanovich Publications, Duluth, Minn., 1978.
G. T. Blair and M. F. Zienty, *Citric Acid: Properties and Reactions*, Miles Laboratories Inc., 1979.
The Story of Soft Drinks, National Soft Drink Association, Washington, D.C., 1982.
R. C. Weast, *CRC Handbook of Chemistry and Physics*, 69th ed., CRC Press, Boca Raton, Fla., 1988, 1989, p. 163.

REBECCA LOPEZ-GARCIA
Tate & Lyle

CLAYS, SURVEY

The term "clay" is somewhat ambiguous unless specifically defined because it is used in three ways: (1) as a diverse group of fine-grained minerals, (2) as a rock term, and (3) as a particle size term. Clay is a natural earthy, fine-grained material comprised largely of a group of crystalline minerals known as the clay minerals. These minerals are hydrous silicates composed mainly of silica, alumina, and water. Several of the clay minerals also contain appreciable quantities of iron, alkalies, and alkaline earths. Many definitions include the statement that clay is plastic when wet, which is true because most clays do have this property but some clays are not plastic, eg, flint clays and halloysite. As a rock term many authors use the term clay for any fine-grained natural earthy argillaceous material that would include shale, argillite, and some argillaceous soils. As a particle

size term, clay is used for the category that includes the smallest particles. Even though there is no universally accepted definition of the term "clay", geologists, agronomists, ceramists, engineers, and others who use the term understand its meaning.

Because of the extremely fine particle size of clays and clay minerals, they require special techniques for identification. The optical microscope generally cannot resolve particles below 5 μ. The most useful instrument for identification and semiquantification of clay minerals is X-ray diffraction. Other methods of identification used are, differential thermal analysis, electron microscopy, and infrared (ir) spectroscopy.

Clay is an abundant natural raw material that has an amazing variety of uses and physical properties. Clays are among the leading industrial minerals in both tonnage produced and total value. Clays and clay minerals are important to industry, agriculture, geology, environmental applications, and construction. Clay minerals are important indicators in petroleum and metallic ore exploration and in reconstructing the geological history of deposits. However, clay minerals may be deleterious in aggregates and in oil reservoirs.

OCCURRENCE AND GEOLOGY OF MAJOR CLAY DEPOSITS

Some clay deposits are comprised of relatively pure concentrations of a particular clay mineral and others are mixtures of clay minerals. Kaolins, smectites, and palygorskite-sepiolite can occur in relatively pure concentrations whereas illite and chlorite usually occur in mixtures of clay minerals and non-clay minerals such as shales, which are the most common sedimentary rock.

Kaolins

Kaolins are hydrated aluminum silicates. Kaolin is a mineral group consisting of the minerals kaolinite, dickite, nacrite, and halloysite. Kaolinite is the most common of the kaolin group minerals.

Kaolin, in addition to being a group mineral name, is a rock term and is used for any rock that is comprised predominantly of one of the kaolin minerals. Kaolin deposits are classed as primary or secondary. Primary or residual kaolins are those that have formed by the alteration of aluminous crystalline rocks such as granite and remain in the place where they formed. Secondary kaolins are sedimentary rocks that were eroded, transported, and deposited as beds or lenses in association with other sedimentary rocks. The most common parent minerals from which kaolin minerals form are feldspars and muscovite.

There are special types of kaolinitic clays that are used primarily for ceramics. One of these types is ball clays. Ball clays are secondary and are characterized by the presence of organic matter, high plasticity, high dry and fired strength, long vitrification ranges, and light color when fired. Kaolinite is the principal mineral constituent of ball clay and typically comprises 70% or more of the minerals present. Other minerals commonly present are quartz, illite, smectite and feldspar, as well as lignitic material.

Smectites

Smectite is the name for a group of sodium, calcium, magnesium, iron, and lithium aluminum silcates. The group includes the specific clay minerals montmorillonite, saponite (magnesium smectite), nontronite (iron smectite), beidellite and hectorite (lithium smectite). Bentonite is the rock in which these smectite minerals are usually the dominant constituent.

Bentonites in which the smectite sodium montmorillonite is the major mineral component normally have a high swelling capacity, while bentonites in which calcium montmorillonite is the major mineral component commonly have a low swelling capacity.

Some less common bentonites are hectorite and saponite.

Palygorskite and Sepiolite

Hormite is a group name that has been used for the minerals palygorskite and sepiolite. Palygorskite and sepiolite are hydrated magnesium aluminum silicates that have an elongate shape. Because of their large surface area, they are sometimes referred to as sorptive clays called Fuller's earth. Fuller's earth is a term used for clays and other fine particle size earthy materials suitable for use as sorbent clays and bleaching earths.

Common Clay

Common clay includes a variety of clays and shales that are fine in particle size. The clay mineral composition of these common clays is mixed but usually illite and chorite are the most common clay minerals present. Kaolinite and smectites are usually present in smaller quantities. Illite is a hydrated potassium iron aluminum silicate and chlorite is a hydrated magnesium iron aluminum silicate. These common clays are used in many structural clay products such as brick, tile, pottery, stoneware, etc.

STRUCTURE AND COMPOSITION OF CLAY MINERALS

Clay minerals are phyllosilicates, which are sheet structures basically composed of silica tetrahedral layers and alumina octahedral layers. Each clay mineral has a different combination of these layers. Chemical substitutions of aluminum for silicon, iron and/or magnesium for aluminum, etc, generates positive or negative charges in the structure. The clay minerals are crystalline and their identification is determined by X-ray diffraction techniques. (1) Two layer types consist of sheet structures composed of one layer of silica tetrahedrons and one layer of alumina octahedron (termed "1:1 layer types"); (2) three-layer types consist of sheet structures composed of two layers of silica tetrahedrons and one central octahedral layer (termed "2:1 layer types"); (3) regular mixed-layer types, which are ordered stacking of alternate silica tetrahedral and alumina octahedrals layers, and (4) chain structure type, which are hornblend-like chains of silica tetrahedrons linked together by octahedral groups of oxygen and hydroxyls containing aluminum and magnesium ions.

The kaolin minerals are hydrous aluminum silicates with the composition of $2 H_2O \cdot Al_2O_3 \cdot 2 Si O_2$. Kaolinite is the most common of the kaolin minerals. The structure of kaolinite is a single silica tetrahedral sheet and a single alumina octahedral sheet combined to form the 1:1 kaolinite unit layer. These unit layers are stacked and are held together by hydrogen bonding. Variations in the orientation of the unit layers in stacking cause the differentiation between kaolinite, dickite, and nacrite. Halloysite is an elongate kaolin mineral that has a layer of water between the unit layers so the composition includes $4 H_2O$ instead of $2 H_2O$.

Smectite structures are comprised of two silica tetrahedral sheets with a central octahedral sheet. The structure has an unbalanced charge because of the substitution of aluminum for silicon in the tetrahedral sheet and iron and magnesium for aluminum in the octahedral sheet. In order to balance this negative charge, cations accompanied by water molecules enter between the 2:1 layers. Sodium montmorillonite is a smectite that has sodium ions and water molecules in the interlayer and calcium montmorillonite has calcium ions and water molecules in the interlayer. Nontronite has iron ions in the structure, saponite has magnesium ions in the structure, and hectorite has lithium and magnesium ions in the structure. Smectites expand when water and other polar molecules enter between the layers.

Illite is a clay mineral akin to mica. The basic structure is a 2:1 layer similar to smectite except that more aluminum ions replace silicon in the tetrahedral sheet, which results in a higher charge deficiency, balanced by potassium ions. These large diameter potassium ions act as a bridge between the 2:1 layers and so strongly bind them together that illite is nonexpandable. The composition of illite is a potassium aluminum silicate with the general structural formula of $(OH)_4 K_2 (Si_6 AL_2) Al_4 O_{20}$. In this structure, iron and magnesium can substitute for aluminum in the octahedral sheet.

Chlorite is a common clay mineral in shales and the structure consists of alternate silica tetrahedral layers and aluminum octahedral layers. The octahedral layer has considerable substitution of iron and magnesium for aluminum. However, the 2:1 mica sheet octahedral layer may have a different composition than the octahedral brucite layer between the 2:1 mica sheets.

Palygorskite and sepiolite have a chain-like structure, that consists of inverted ribbons of silica tetrahedral ribbons linked together by aluminum and magnesium octahedrals. The difference between palygorskite and sepiolite is that sepiolite has a higher content of magnesium and paylgorskite has more aluminum as well as a slightly different crystal structure.

Because most of these layer silicates are all made up of 1:1 and 2:1 layers, there are many possible unit layer mixtures of illite, chlorite, and smectite compositions. These mixed-layer clays are relatively common in occurrence. Illite/smectite, illite/chlorite, chlorite/smectite, and kaolinite/smectite have been described.

MINING AND PROCESSING

Clays of all types are mined principally by open pit and there are very few underground mines. An understanding of the geology and the origin of clays plays an important role in clay exploration. After a clay deposit is discovered, it must be evaluated to determine thickness, quality and quantity. This may be accomplished by testing either core or auger drilled clay. The spacing of the drill holes depends on the geologic and surface conditions associated with the particular clay deposit. A drilling pattern used to test a sedimentary clay deposit is very different from the pattern used to evaluate a hydrothermal clay deposit.

Testing the core or auger samples is an important second step in determining the quality of a particular clay deposit. The tests performed to evaluate a kaolin deposit are much different than those performed on a smectite or palygorskite clay deposit. The tests are related to the applications of the clay and to determine the type of processing that will be required to make the final product. General tests that are performed on clays are (1) mineralogy, (2) percent grit (plus 325 mesh or 44 micrometers), and (3) color. Special tests for kaolins are (1) particle size distribution, (2) brightness, (3) low and high shear viscosity, and (4) leach response to improve the brightness. Other special tests also may be performed such as magnetic separation, flotation, selective flocculation, and abrasion. Special tests for bentonites include ion exchange capacity, viscosity, swelling capacity in water, foundry tests, surface area, water and oil absorption, and bulk density. Special tests for paylgorskite and sepiolite include surface area, exchange capacity, viscosity, water and oil absorption, and bulk density. After drilling and testing a mining plan can be designed for obtaining the maximum quantity of clay.

Kaolin

Once drilling and testing have determined the quality and quantity of a deposit, the processing that is required to produce a saleable product is determined. Either a dry or wet process accomplishes beneficiation. The higher quality grades of kaolin that are used in the paper, paint, and plastics industries are prepared by wet processing because the product is more uniform, has better brightness and color, and is relatively free of impurities.

Dry Process. The dry process is simple and yields a lower cost and lower quality product than the wet process. In the dry process, the properties of the finished kaolin reflect the quality of the crude kaolin. At the processing it is shredded or crushed to about egg size. After crushing, it is dried, commonly with a rotary drier. After drying, the kaolin is pulverized in a roller mill, hammer mill, disk grinder, or some other grinding device. Commonly heat is applied during the grinding to further reduce the moisture content. The pulverized kaolin is then classified to separate the fine and coarse particles. The finished product can be loaded in bulk bags and shipped by railcar or truck.

Wet Process. The mined kaolin is either transported to the processing plant or fed into a stationary or mobile blunger. The blunger separates the kaolin into small particles, which are mixed with water and a dispersing chemical to form a clay–water slurry. This clay–water slurry is pumped from the blunger to rake classifiers or hydrocyclones and screens to remove the grit (material >325 mesh). The degritted slurry is collected into large storage tanks with agitators and is then pumped to the processing plant, which may be several miles away. The kaolin slurry is collected in large storage tanks at the plant before it is processed. The first step is to separate the kaolin particles into a coarse and fine fraction through continuous centrifuges. The degritted slurry is then passed through a high gradient magnetic separator prior to centrifugation to upgrade the crude clay by removing iron and titanium minerals. The high gradient magnetic separation (HGMS) process uses a canister filled with a fine stainless steel wool that removes iron, titanium, and some mica minerals from the kaolin slurry as it passes through the canister. Superconducting cryogenic magnets with field strengths up to 5 tesla are now being used in addition to the 2-tesla electromagnets. The superconducting magnets use very little power and are effective in removing ultrafine paramagnetic minerals. The coarse fraction can be used as a filler in paper, plastics, paint, and adhesives, as a casting clay for ceramics, or as a feed to delaminators. The fine particle size fraction is used in paper coatings, high gloss paints, inks, special ceramics, and rubber. Normally, the coarse kaolin takes one of two routes. (1) It can go directly to the leaching department, where it can be chemically treated to solubilize some of the iron if the brightness needs to be upgraded, or it can be flocculated so that the dewatering step is facilitated; or (2) it can go through magnetic separation and into the delaminators or directly to the delaminators and then to the magnetic separator or to the delaminators without any magnetic separation step.

Special Processes. Several special processes are used to produce unique and special quality grades of kaolin. One of these special processes is delamination, which is a process that takes a large kaolin stack and separates it into several thin, large-diameter plates. The process of delaminating involves attrition mills into which fine media, such as glass beads or nylon pellets, are placed along with the coarse kaolin stacks and intensely agitated. The brightness and whiteness of the delaminated kaolin is very good; the clean newly separated basal plane surfaces are white because they have been protected from ground water and iron staining.

Calcining is another special treatment that is used to produce special grades. One grade is thermally heated to a temperature just above the point where the structural hydroxyl groups are driven out as water vapor, which is between 650 and 700°C. A second grade is thermally heated to 1000–1050°C. By proper selection of the feed kaolin and careful control of the calcination and final processing, the abrasiveness of the calcined product can be reduced to acceptable levels. The brightness of this fully calcined, fine-particle kaolin is 92–95%, depending on the feed material.

Another special process is *surface treatment*. Generally, an ionic or a polar nonionic surfactant is used as a surface-treating agent.

Smectite

Virtually all bentonite, which is comprised mainly of a smectite mineral, is surface-mined.

Bentonite beneficiation and processing involves relatively simple milling techniques that includes crushing or shredding, drying, and grinding and screening to suitable sizes. The raw bentonite is passed through some sort of crushing or shredding device to break up the large chunks before drying. The bentonite properties can be seriously affected by overdrying, so the drying temperature must be carefully controlled. The dried bentonite is ground and sized in several ways. Granular bentonite is cracked by using roll crushers and screened to select the proper size range granules.

Palygorskite-Sepiolite

The mining of palygorskite clays is no different than mining bentonites. The processing is also similar. The clay is shredded or crushed, dried, screened, air-classified, and packaged. Some of this clay is thermally treated to harden the particles to prevent disintegration of the granules during bagging, transport, and handling prior to arrival to the customer. The temperature of this thermal treatment is <400°C so that no structural modification results, but all of the absorbed water is driven from the surface and interior of the particles.

ENVIRONMENTAL CONSIDERATIONS

Almost all clays are surface mined, so the industry is required to reclaim the disturbed land in most countries. Common practice is to open a cut and then deposit the overburden from the following panels or cuts into the mined out areas. The land is leveled or sloped to meet the governmental requirements and then planted with grasses or trees. In the processing plants the waste materials, which include chemicals and separated clay particles or other minerals, are collected in impounds. The clay and other particles are flocculated with alum or other chemical flocculants, and the clear water is released into streams after adjusting the pH to 6–8.

Air quality is maintained in the processing plants by using dust collectors on the dryers and by enclosing transfer points where dry clay may cause dust. In areas such as the bagging departments, the workers may be required to wear dust masks. None of the clays, kaolins, smectites, or palygorskite-sepiolite are health hazards.

PRODUCTION AND CONSUMPTION

Clays are used in construction, agriculture, process industries, and environmental applications. They are necessary ingredients in many products including kaolins in paper, paint, ceramics and rubber; bentonite in drilling fluids, foundries, and fluid barrier applications, palygorskite-sepiolite in special drilling fluids, carriers for agricultural chemicals, suspending agents in paints and pharmaceuticals, and common clays for making bricks and other structural clay products.

R. E. Grim, *Clay Mineralogy*, 2nd ed., McGraw-Hill Book Co., Inc., New York, 1968.

D. M. Moore and R. C. Reynolds, Jr., *X-Ray Diffraction and Identification of Clay Minerals*, Oxford University Press, 1989.

H. Murray, *Paper Coating Pigments*, TAPPI, Atlanta, Ga. 1984.

B. Velde, *Clay Minerals: A Physico-Chemical Explanation of Their Occurrence*, Elsevier, Amsterdam, The Netherlands, 1985.

H. H. MURRAY
Indiana University

CLAYS, USES

Clays are fine particle size materials comprised of clay minerals, which are basically hydrated aluminum silicates with associated alkali and alkaline earth elements. The clay mineral groups are kaolin, smectite, palygorskite-sepiolite, illite, chlorite, and mixed-layered clays. The properties of these clays are very different because of differences in their structure and composition. All are extremely fine and contain non-clay minerals.

The particle size, shape, and distribution are physical properties that are intimately related to the applications of the clay minerals. Other important properties are surface chemistry, area, and charge. These along with color and brightness affect many properties including low and high shear viscosity; absorption capacity and selectivity; plasticity; green, dry, and fired strength; casting rate; permeability; bond strength; and optical coating properties for paper and paint. In most every application the clays and clay minerals perform a function and are not just inert components in the system. Improved processing techniques, which have evolved over the past 40 years, have had a profound effect on the traditional and new applications.

KAOLINS

Kaolin is a group mineral name for kaolinite, dickite, nacrite, and halloysite. The most common mineral in the kaolin group is kaolinite. The uses are governed by several factors including the geological conditions under which the kaolin formed, the mineralogical composition of the kaolin deposit, and the physical and chemical properties of the kaolinite. Kaolin deposits can be sedimentary, residual or hydrothermal and in almost every instance the kaolin has different properties and thus must be fully tested and evaluated to determine its utilization. Sedimentary kaolins are called secondary deposits and residual or hydrothermal kaolins are called primary deposits.

Ceramic Products

A large proportion of the annual production of ball clay and fireclay and a large amount of kaolin are used in the manufacture of ceramic products. Ball clays, which are fine particle size kaolinitic clays, are used as a raw material in whiteware, sanitaryware and tile. Fireclays are used in the manufacture of refractories because of their high melting point and low shrinkage. Flint clays are very dense and brittle and are essentially pure, extremely fine-grained kaolinite.

Paper

Kaolin used in the paper industry has two main uses, as a filler where the kaolin is mixed with the pulp fibers and as a coating where the kaolin is mixed with water, adhesives, and various additives and coated onto the surface of the paper.

Kaolins used by the paper industry are of three types based upon the type of processing: air-floated, water washed, and calcined. Water washed kaolins are of higher value than air-floated because the more elaborate processing results in more uniform and higher quality products. Calcined kaolin, which is much less costly, can replace a majority of the titanium dioxide with little or no loss in brightness or opacity.

The properties of kaolin that make it useful in the paper industry are brightness, viscosity, and particle size and shape. Air-floated kaolins are at the lower end of the brightness range for fillers used by the paper industry. Improvement in paper sheet brightness and opacity results from the addition of kaolin filler clays. The best results are obtained using calcined clay followed by water washed kaolins and then air-floated kaolin.

Other properties of kaolin coating clays that are important to the paper coater are dispersion, opacity, gloss and smoothness, adhesive demand, film strength, and ink receptivity. In order to obtain the maximum efficiency of a coating clay, the individual clay particles must be completely dispersed. Kaolinite is easily dispersed because it is hydrophilic. Opacity is strongly influenced by particle packing that is dependent on particle size and shape and particle size distribution. Adhesive demand is related to surface area. In general, finer particle kaolins give higher gloss, opacity, and brightness.

Paint

Clays are widely used in both water and oil-based paints. Since they perform several important functions: They extend the much higher cost titanium dioxide opacifying pigment, control viscosity so as to prevent pigments from settling during storage, provide thixotropy so that the paint is easily applied yet does not sag or run after application, improve gloss, promote film strength, and aid in tint retention.

Plastics and Rubber

Kaolin is the only clay used in plastics and rubber as extenders and functional fillers. Various grades are produced specifically for use by the plastics and rubber industries: air-floated, water-washed, and calcined.

Properties that make kaolin useful in the plastics and rubber industries are color, particle size and shape, and viscosity. Clays used in the polymers industry are required to be white or nearly so except for applications in black compounds. In general, finer particle clays increase viscosity more than coarse particle clays. Depending on the polymer and its application, increased viscosity may be desirable or undesirable.

Halloysite

Halloysite is a member of the kaolin group and has a layer of water between the unit layers. It is used primarily in whiteware ceramics as an additive, and as a filler in paper and as a raw material to make synthetic zeolites.

Inks

Kaolins are a common ingredient in a large variety of printing inks.

Cracking Catalysts

Kaolins are used as a raw material to make zeolites and aluminum silicates for use as catalysts in the refining of petroleum.

Chemical Raw Materials

Air-floated kaolin is used as a chemical raw materials to make fiberglass, in some cement plants to whiten the color, to provide silica and alumina for the cement reactions, and increase cement strength, as well as raw material for producing zeolites.

Pencil Leads

Kaolin is used as an additive to graphite in making pencil leads.

Suspensions and Diluents

Kaolins are used as suspending agents in pharmaceuticals, cosmetics, enamels, and medicines.

Fertilizers, Dessicants, and Insecticides

Kaolin is used as a carrier for certain insecticides and fertilizers and is used as a dessicant to promote flow when mixed with highly deliquescent materials.

Foundry Binders

Some plastic, fine particle kaolins are used as a binder for sand to make high temperature resistant molds for special metals.

Roofing Granules and Polishing Compounds

Roofing granules are produced by calcining coarse granular particles of white kaolin.

SMECTITES

Smectite is the group name for a number of clay minerals including sodium montmorillonite, calcium montmorillonite, saponite, nontronite, and hectorite. The rocks in which the smectite minerals are dominant are called bentonites. Industrial quality bentonites are predominantly comprised of either sodium montmorillonite or calcium montmorillonite and to a much lesser extent, hectorite.

The largest uses of smectites by far are for drilling muds, foundry binders, iron ore pelletizing, cat litter, and sealants.

Smectites are very fine in particle size and the particles are extremely thin, which gives the material a high surface area. Smectites and particularly sodium montmorillonite have a high base exchange capacity on the order of 75–100 meq/100 g of clay. Sodium montmorillonite has a high swelling capacity of 10–15 times its dry volume.

Drilling Mud

The high swelling property of sodium montmorillonite makes it a necessary ingredient in freshwater drilling muds throughout the world.

Foundry Binders

The molding sands used in foundries are comprised of high silica sand and ~5–8% bentonite. The bentonite provides bonding strength and plasticity. A small amount of tempering water is added to the mixture to make it plastic.

Iron Ore Pelletizing

Sodium bentonites are used to pelletize iron ores.

Cat Litter

Both calcium and sodium bentonites are used in making cat litter.

Sealants

Sodium bentonites are used extensively for water impedance because of their high swelling capacity.

Slurry Trench Excavations

High swelling sodium bentonite is used in the slurry trench or diaphragm wall method of excavation in construction in areas of unconsolidated rocks and soils.

Absorbents

Calcium bentonites are excelent absorbent clays.

Suspension Aids

Sodium bentonite is used as suspension aids in cosmetics, medical formulations, pharmaceuticals, and for use in the distribution of suspension fertilizers.

Bleaching Clays

Calcium bentonites are treated with sulfuric and/or hydrochloric acid to remove ions from the surface and from the octahedral layer to increase the charge on the clay particle. These acid activated clays are widely used to decolorize mineral, vegetable, and animal oils.

Organoclays

Sodium montmorillonite and hectorite are processed so that the exchangeable sodium ions are replaced with alkylammonium cations to produce a hydrophobic surface.

Animal Feed Binders

Both sodium and calcium bentonites are used to bind animal feed into pellets.

Wine and Beer Clarification

Both calcium and sodium bentonites are used to remove colloidal impurities such as haze-forming compounds in wine and beer.

Nanoclays

A recent development using sodium montmorillonite is the separation of the unit layers into almost unit cell thickness (1 nm or 10 Å) for use in polymer compositions called nanocomposites.

PALYGORSKITE AND SEPIOLITE

Palygorskite and attapulgite are names for the same mineral. Also, attapulgite and calcium montmorillonites are classed as Fuller's earth. Palygorskite and sepiolite have a high surface area, a small octahedral layer charge, a fine particle size, and an elongate shape that gives palygorskite and sepiolite a high absorption capacity that makes these clays useful in many industrial applications.

Oil Well Drilling Fluids

Of prime importance among the characteristics of a clay for drilling mud is the ability of the clay to build a suitable viscosity at a relatively low solids level, and to maintain the desired viscosity throughout the drilling of the well. Sodium bentonite is widely used for this purpose, but can only be used with fresh water. Attapulgite does not depend on swelling for viscosity and is stable in the presence of these contaminants so it is the preferred clay when brines and other salt contaminants are encountered.

Liquid Suspension Fertilizers

Attapulgite is the obvious choice for this application because of its highly stable colloidal properties in high concentrations of salts.

Adhesives

Attapulgite develops viscosity under shear so the incorporation of attapulgite is an effective method for counteracting the loss of viscosity of the starch.

Colloidal and Suspension Applications

Attapulgite-based suspensions are very stable in the presence of salts and electrolytes.

Oil Refining

The mechanical and thermal stability and high surface area of granular attapulgite makes it useful as a percolation absorbent to remove high molecular weight compounds such as sulfonates, resins, and asphaltines in petroleum oils.

COMMON CLAYS

Common clays and shales are important raw materials for structural ceramic clay products in most every country in the world. The products include bricks, roof tiles, sewer pipe, conduit tile, structural tile, flue linings, and others. The properties that are important are plasticity, shrinkage, dry and fired strength, fired color, and vitrification range.

H. Murray, *Industrial Minerals and Rocks*, 6th ed., SME, Littleton, Cl., 1994.

S. Pickering and H. Murray, *Industrial Minerals and Rocks*, 6th ed., SME, Littleton, Cl., 1994.

D. G. Sekutowski, in J. D. Edenbaum, ed., *Plastics Additives and Modifiers Handbook*, Van Nostrand Reinhold, New York, 1992.

N. Trivedi and R. Hagemeyer, *Industrial Mineral and Rocks*, 6th ed., SME, Littleton, Colo., 1994.

HAYDN H. MURRAY
Indiana University

COAL

Coal is usually a dark black color, although geologically younger deposits of brown coal have a brownish red color (see LIGNITE AND BROWN COAL). The color, luster, texture, and fracture vary with rank, type, and grade. Coal is the result of combined biological, chemical, and physical degradation of accumulated plant matter over geological ages. The relative amounts of remaining plant parts leads to different types of coal, which are sometimes termed banded, splint, nonbanded (cannel and boghead); or hard or soft; or lignite, subbituminous, bituminous, or anthracite. In Europe, the banded and splint types are generally referred to as ulmic or humic coals. Still other terms refer to the origins of the plant parts through maceral names such as vitrinite, liptinite, and inertinite. The degree of conversion of plant matter or coalification is referred to as rank. Brown coal and lignite, subbituminous coal, bituminous coal, and anthracite make up the rank series with increasing carbon content. The impurities in these coals cause differences in grade.

Coal consists primarily of carbon, hydrogen, and oxygen, and contains lesser amounts of nitrogen and sulfur and varying amounts of moisture and mineral matter. The mode of formation of coal, the variation in plant composition, the microstructure, and the variety of mineral matter indicate that there is a mixture of materials in coal. The nature of the organic species present depends on the degree of biochemical change of the original plant material, on the historic pressures and temperatures after the initial biochemical degradation, and on the finely divided mineral matter deposited either at the same time as the plant material or later. The principal types of organic compounds have resulted from the formation and condensation of polynuclear and heterocyclic ring compounds containing carbon, hydrogen, nitrogen, oxygen, and sulfur. The fraction of carbon in aromatic ring structures increases with rank.

Nearly all coal is used in combustion and coking (see COAL CONVERSION PROCESSES). At least 80% is burned directly in boilers for generation of electricity (see MAGNETOHYDRODYNAMICS; POWER GENERATION) or steam for industrial purposes. Small amounts are used for transportation, space heating, firing of ceramic products, etc. The rest is essentially pyrolyzed to produce coke, coal gas, ammonia (qv), coal tar, and light oil products from which many chemicals are produced (see FEEDSTOCKS, COAL CHEMICALS). Combustible gases and chemical intermediates are also produced by the gasification of coal (see FUELS, SYNTHETIC), and differentcarbon (qv) products are produced by various heat treatments. A small amount of coal is used in miscellaneous applications such as fillers (qv), pigments (qv), foundry material, and water (qv) filtration (qv).

World reserves of bituminous coal and anthracite are $\sim 5.6 \times 10^{12}$ t of coal equivalent, ie, 29.3 GJ/t (12.6×10^3 Btu/lb), and subbituminous and lignite are 2.9×10^{12} t of coal equivalent. For economic and environmental reasons coal consumption has been cyclic.

ORIGIN OF COAL

Coal evolved from partially decomposed plants in a shallow-water environment. Various chemical and physical changes occurred in two distinct stages: one biochemical and the other physicochemical (geochemical). Because some parts of plant material are more resistant to biochemical degradation than others, optical variations in petrologically distinguishable coals resulted. The terms vitrain and clarain refer to bright coals; durain is a dull coal, and fusain is structured fossil charcoal. Exposure to pressure and heat during the geochemical stage caused the differences in degree of coalification or rank that are observable in the continuous series: peat, brown coal and lignite, subbituminous coal, bituminous coal, and anthracite.

Complete decay of plant material by oxidation and oxygen-based bacteria and fungi is prevented only in waterlogged environments such as swamps in regions where there is rapid and plentiful plant growth. A series of coal seams have been formed from peat swamps growing in an area that has undergone repeated subsidence followed by deposition of lacustrine or marine intrusion material. According to the autochthonous, *in situ*, theory of coal formation, peat beds and subsequently coal were formed from the accumulation of plants and plant debris in place. According to the allochthonous theory, the coal-producing peat bogs or swamps were formed from plant debris that had been transported, usually by streams or coastal currents, to the observed burial sites.

BIOCHEMICAL STAGE

The initial biochemical decomposition of plant matter depends on two factors: the ability of the different plant parts to resist attack and the existing conditions of the swamp water. Fungi and bacteria can cause complete decay of plant matter that is exposed to aerated water or to the atmosphere. The decay is less complete if the vegetation is immersed in water containing anaerobic bacteria.

Geochemical Stage

The conversion of peat to bituminous coal is the result of the cumulative effects of temperature and pressure over a long time. The sediment covering the peat provides the pressure and insulation so that the earth's internal heat can be applied to the conversion. The changes in plant matter are termed normal coalification.

The change in rank from bituminous coal to anthracite involves the application of significantly higher pressures, ie, as in mountain building activity, and temperatures, ie, as in volcanic activity. The more distant the coal from the disruption, the less proportionate the alteration. Tectonic plate movements involved in mountain building provide pressure for some changes to anthracite. As a general rule, the older the coal deposit, the more complete the coalification and the higher the rank of coal.

COAL PETROGRAPHY

The study of the origin, composition, and technological application of these materials is called coal petrology, whereas coal petrography involves the systematic quantification of the amounts and characteristics by microscopic study. The petrology of coal may involve either a macroscopic or microscopic scale.

On the macroscopic scale, two coal classifications have been used: humic or banded coals and sapropelic or nonbanded coals. Stratification in the banded coals, which result from plant parts, is quite obvious; the nonbanded coals, which derive from algal materials and spores, are much more uniform. The physical and chemical properties of the different layers in a piece of coal or a seam can vary significantly. Therefore the relative amounts of the layers are important in determining the overall characteristics of the mined product. Coal petrography has been widely applied in cokemaking and is important in coal liquefaction programs.

Macerals

Coal parts derived from different plant parts, are referred to as macerals. The maceral names end in "-inite". For example, vitrinite, derived from the woody tissues of plants, liptinite, derived from the waxy parts of spores and pollen, or algal remains, and inertinite, thought to be derived from oxidized material or fossilized charcoal remnants of early forest fires. A number of subdivisions of the maceral groups have been developed and documented by the International Commission on Coal Petrology.

The macerals in lower rank coals, eg, lignite and subbituminous coal, are more complex and have been given a special classification. The term huminite has been applied to the macerals derived from the humification of lignocellulosic tissues. Huminite is the precursor to the vitrinite observed in higher rank coals.

Vitrinite Reflectance

The amount of light reflected from a polished plane surface of a coal particle under specified illumination conditions increases with the aromaticity of the sample and the rank of the coal or maceral. Precise measurements of reflectance, expressed as a percentage, are used as an indication of coal rank.

Application of Coal Petrology and Petrography

Petrographic analysis is frequently carried out for economic evaluation or to obtain geologic information. Samples are usually lumps or more coarsely ground material that have been mounted in resins and polished. Maceral analysis involves the examination of a large number (usually >500) of particles during a traverse of a polished surface to identify the macerals at specified intervals. A volume percentage of each of the macerals present in a sample is calculated.

Seam correlations, measurements of rank and geologic history, interpretation of petroleum formation with coal deposits, prediction of coke properties, and detection of coal oxidation can be determined from petrographic analysis.

CLASSIFICATION SYSTEMS

Prior to the nineteenth century, coal was classified according to appearance, eg, bright coal, black coal, or brown coal. A number of classification systems have since been developed. These may be divided into two types, which are complementary: scientific and commercial. Both are used in research, whereas the commercial classification is essential industrially. In the scientific category, the Seyler chart has considerable value.

Systems include the Seyler classification (Fig. 1), the ASTM classification, National Coal Board classification for British coals, and International classification.

COMPOSITION AND STRUCTURE

The constitution of a coal involves both the elemental composition and the functional groups that are derived therefrom. The structure of the coal solid depends to a significant extent on the arrangement of the functional groups within the material.

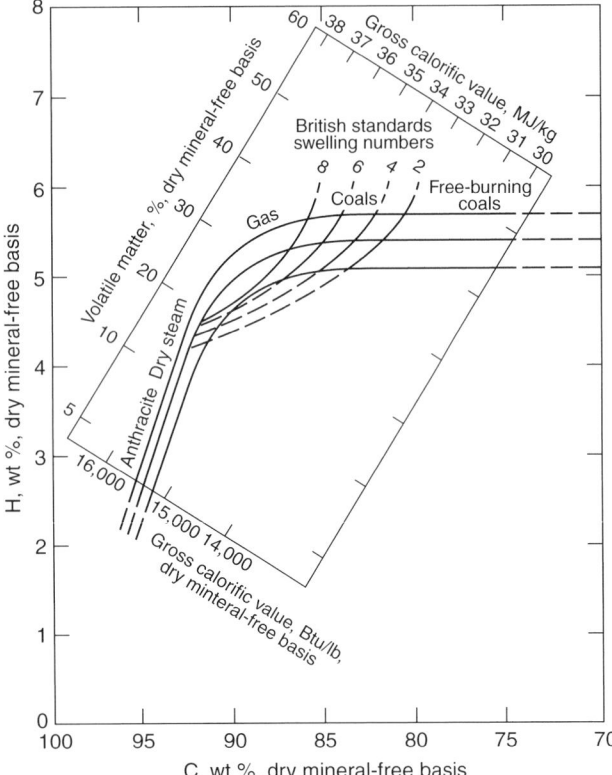

Figure 1. Simplified form of Seyler's coal classification chart. Note that ASTM uses the free-swelling index.

Composition

The functional groups within coal contain the elements C, H, O, N, or S. The significant oxygen-containing groups found in coals are carbonyl, hydroxyl, carboxylic acid, and methoxy. The nitrogen-containing groups include aromatic nitriles, pyridines, carbazoles, quinolines, and pyrroles. Sulfur is primarily found in thiols, dialkyl and aryl–alkyl thioethers, thiophene groups, and disulfides. Elemental sulfur is observed in oxidized coal. The relative and absolute amounts of the various groups vary with coal rank and maceral type. The principal oxygen-containing functional groups in vitrinites of mature coals are phenolic hydroxyl and conjugated carbonyls as in quinones.

Coal Structure

Conclusions regarding the chemical structure of the macromolecules within coal are generally based on experimental measurements and an understanding of structural organic chemistry.

Several requirements must be met in developing a structure. Not only must elementary analysis and other physical measurements be consistent, but limitations of structural organic chemistry and stereochemistry must also be satisfied. Mathematical expressions have been developed to test the consistency of any given set of parameters used to describe the molecular structure of coal and analyses of this type have been reported.

The macromolecules that make up the coal structure are held together by a variety of forces and bonds. The coal network model is one approach to describing the three-dimensional (3D) structure of the solid. Aromatic clusters are linked by a variety of connecting bonds, through oxygen, methylene or longer aliphatic groups, or disulfide bridges, and the proportions of the different functional groups change as the rank of the coal is progressively increased. Another type of linkage involves hydrogen bonds which, eg, hold hydroxy and ketogroups together in the solid.

Coal Constitution

Chemical composition studies indicate that brown coals have a relatively high oxygen content. About two-thirds of the oxygen is bonded carboxyl, acetylatable hydroxyl, and methoxy groups. Additionally, unlike in bituminous coals, some alcoholic hydroxyl groups are believed to exist.

The anthracites, which approach graphite in composition (see CARBON,GRAPHITE), are classified higher in rank, have less oxygen and hydrogen, and are less reactive than bituminous coals. Anthracites are also insoluble in organic solvents. These characteristics become more pronounced as rank increases within the anthracite group.

Mineral Matter in Coal

The mineral matter in coal results from several separate processes. Some comes from the material inherent in all living matter; some from the detrital minerals deposited during the time of peat formation; and a third type from secondary minerals that crystallized from water that has percolated through the coal seams. The various clay minerals are the most common detrital mineral (see CLAYS); however, other common ones include quartz, feldspar, garnet, apatite, zircon, muscovite, epidote, biotite, augite, kyanite, rutile, staurolite, topaz, and tourmaline. The secondary minerals are generally kaolinite, calcite, and pyrite.

PROPERTIES

Pieces of coal are mixtures of materials somewhat randomly distributed in differing amounts. The mineral matter can be readily distinguished from the organic, which is itself a mixture. Coal properties reflect the individual constituents and the relative proportions. By analogy to geologic formations, the macerals are the constituents that correspond to minerals that make up individual rocks. For coals, macerals, which tend to be consistent in their properties, represent particular classes of plant parts that have been transformed into coal. Most detailed chemical and physical studies of coal have been made on macerals or samples rich in a particular maceral, because maceral separation is time consuming.

The most predominant maceral group in U.S. coals is vitrinite. The other important maceral groups include inertinite, including fusinite, a dull fibrous material similar to charcoal, and the liptinite group, including sporinite, which is relatively fusible and volatile.

In the United States the commercial classification of coals is based on the fixed carbon (or volatile matter) content and the moist heating value. Table 1 indicates

Table 1. Composition of Humic Coals

| Type of coal | Composition, wt%[a] | | | | | | Calorific value, kJ/g[b] |
	C	H	O	N	Moisture as found	Volatile matter	
peat	45–60	3.5–6.8	20–45	0.75–3.0	70–90	45–75	17–22
brown coals and lignites	60–75	4.5–5.5	17–35	0.75–2.1	30–50	45–60	28–30
bituminous coals	75–92	4.0–5.6	3.0–20	0.75–2.0	1.0–20	11–50	29–37
anthracites	92–95	2.9–4.0	2.0–3.0	0.5–2.0	1.5–3.5	3.5–10	36–37

[a] Dry, mineral-matter-free basis except for moisture value.
[b] To convert kJ/g to Btu/lb, multiply by 430.2.

the usual range of composition of commercial coals of increasing rank.

Physical Methods of Examination

Physical methods used to examine coals can be divided into two classes that, in the one case, yield information of a structural nature such as the size of the aromatic nuclei, ie, methods such as X-ray diffraction, molar refraction, and calorific value as a function of composition; and in the other case indicate the fraction of carbon present in aromatic form, ie, methods such as ir and nuclear magnetic resonance (nmr) spectroscopies, and density as a function of composition.

Physical Properties

Most of the physical properties discussed herein depend on the direction of measurement as compared to the bedding plane of the coal. Additionally, these properties vary according to the history of the piece of coal. Properties also vary between pieces because of coal's brittle nature and the crack and pore structure.

The specific electrical conductivity of dry coals is very low, specific resistance 10^{10}–10^{14} Ωcm, although it increases with rank. Coal has semiconducting properties. The conductivity tends to increase exponentially with increasing temperatures.

The dielectric constant is also affected by structural changes on strong heating. Also the value is very rank dependent, exhibiting a minimum at ~88 wt% C and rising rapidly for carbon contents >90 wt%. Polar functional groups are primarily responsible for the dielectric of lower ranks. For higher ranks the dielectric constant arises from the increase in electrical conductivity.

Density values of coals differ considerably, even after correcting for the mineral matter, depending on the method of determination. The true density of coal matter is most accurately obtained from measuring the displacement of helium after the absorbed gases have been removed from the coal sample. Density values increase with carbon content or rank for vitrinites.

Thermal conductivity and thermal diffusivity are also dependent on pore and crack structure. Thermal conductivities for coals of different ranks at room temperature are in the range of 0.230.35 W/(m × K).

The specific heat of coal can be determined by direct measurement or from the ratio of the thermal conductivity and thermal diffusivity. The latter method gives values decreasing from 1.25 J/(g × K)[0.3 cal/(g × K)] at 20°C to 0.4 J/[g × K](0.1cal /(g × K)] at 800°C. The specific heat is affected by the oxidation of the coal.

Ultrafine Structure

Coal contains an extensive network of ultrafine capillaries that pass in all directions through any particle. The smallest and most extensive passages are caused by the voids from imperfect packing of the large organic molecules. Vapors pass through these passages during adsorption, chemical reaction, or thermal decomposition. The rates of these processes depend on the diameters of the capillaries and any restrictions in them. Most of the inherent moisture in the coal is contained in these capillaries. The porous structure of the coal and products derived from it have a significant effect on the absorptive properties of these materials.

Mechanical Properties

Mechanical properties are important for a number of steps in coal preparation from mining through handling, crushing, and grinding. The properties include elasticity and strength as measured by standard laboratory tests and empirical tests for grindability and friability, and indirect measurements based on particle size distributions.

Properties Involving Utilization

Coal rank is the most important single property for application of coal. Rank sets limits on many properties such as volatile matter, calorific value, and swelling and coking characteristics. Other properties of significance include grindability, ash content and composition, and sulfur content.

CHEMISTRY

Coal reactions, which on heating are important to the production of coke and synthetic fuels, are complicated by its structure.

Mature (>75 wt% C) coals are built of assemblages of polynuclear ring systems connected by a variety of functional groups and hydrogen-bonded cross-links. The ring systems themselves contain many functional groups. These polynuclear coal molecules differ one from another to some extent in the coal matter. For bituminous coal, a tarlike material occupies some of the interstices between the molecules. Generally, coal materials are nonvolatile except for some moisture, light hydrocarbons, and contained carbon dioxide. The volatile matter produced on carbonization reflects decomposition of parts of the molecule and the release of moisture. Rate of heating affects the volatile matter content such that faster rates give higher volatile matter yields.

Coal composition is denoted by rank. Carbon content increases and oxygen content decreases with increasing rank.

Partial oxidation as carried out in gasification produces carbon monoxide, hydrogen gas, carbon dioxide, and water vapor. Surface oxidation short of combustion, or using nitric acid or potassium permanganate solutions, produces regenerated humic acids similar to those extracted from peat or soil. Further oxidation produces aromatic acids and oxalic acid, but at least one-half of the carbon forms carbon dioxide.

Treatment with hydrogen at 400°C and 12.4 MPa (1800 psi) increases the coking power of some coal and produces a change that resembles an increase in rank. Treatment of coal with chlorine or bromine results in addition and substitution reactions. Hydrolysis using aqueous alkali has been found to remove ash material including pyrite.

The pyritic sulfur in coal can undergo reaction with sulfate solutions to release elemental sulfur (see SULFUR REMOVAL AND RECOVERY). Processes to reduce the sulfur content of coal have been sought.

Many of the products made by hydrogenation, oxidation, hydrolysis, or fluorination are of industrial importance. Concern about stable, low cost petroleum and natural gas supplies is increasing the interest in some of the coal products as upgraded fuels to meet air pollution control requirements as well as to take advantage of the greater ease of handling of the liquid or gaseous material and to utilize existing facilities such as pipelines (qv) and furnaces.

Reactions of Coal Ash

Mineral matter impurities have an important effect on the utilization of a coal. One of the constituents of greatest concern is pyrite because of the potential for sulfur oxide generation on combustion. Additionally, the mineral matter has a tendency to from stick deposits in a boiler.

Coal ash passes through many reactors without significant chemical change. Corrosion of boiler tubes appears to be initiated in some cases with the formation of a white layer of general composition $(Na,K)_3Al(SO_4)_3$.

Plasticity of Heated Coals

Coals having a certain range of composition associated with the bend in the Seyler diagram (Fig. 1) and having 88–90 wt% carbon soften to a liquid condition when heated. These materials are known as prime coking coals. The coal does not behave like a Newtonian fluid and only empirical measurements of plasticity can be made.

Pyrolysis of Coal

Most coals decompose below temperatures of ∼400°C, characteristic of the onset of plasticity. Moisture is released near 100°C, and traces of oil and gases appear between 100 and 400°C, depending on the coal rank. As the temperature is raised in an inert atmosphere at a rate of 1–2°C/min, the evolution of decomposition products reaches a maximum rate near 450°C, and most of the tar is produced in the range of 400–500°C. Gas evolution begins in the same range but most evolves >500°C. If the coal temperature in a single reactor exceeds 900°C, the tars can be cracked, the yields are reduced, and the products are more aromatic. Heating beyond 900°C results in minor additional weight losses but the solid matter changes its structure. The tests for volatile matter indicate loss in weight at a specified temperature in the range of 875–1050°C from a covered crucible. This weight loss represents the loss of volatile decomposition products rather than volatile components.

A predictive macromolecular network decomposition model for coal conversion based on results of analytical measurements has been developed called the functional group, depolymerization, vaporization, cross-linking (FG-DVC) model. Data are obtained on weight loss on heating (thermogravimetry, tg) and analysis of the evolved species by Fourier transform infrared (ftir) spectrometry. Separate experimental data on solvent swelling, solvent extraction, and Gieseler plastometry are also used in the model. Volatile matter yields decrease with increasing coal rank. An overall picture of the pyrolysis process is generally accepted but the detailed mechanism is controversial.

The mechanism of coal pyrolysis has been discussed in the literature. The early stages involve formation of a fluid through depolymerization and decomposition of coal organic matter containing hydrogen. Around 400–550°C aromatic and nonaromatic groups may condense after releasing hydroxyl groups. The highest yields of methane and hydrogen come from coals having 89–92 wt% C. Light hydrocarbons other than methane are released most readily <500°C; methane is released at 500°C. The highest rate for hydrogen occurs >700°C.

RESOURCES

World Reserves

Amounts of coal of some specified minimum deposit thickness and some specified maximum overburden thickness existing in the ground are termed resources. There is no economic consideration for resources, but reserves represent the portion of the resources that may be recovered economically using conventional mining equipment.

Comprehensive reviews of energy sources are published by the World Energy Conference, formerly the World Power Conference at 6-year intervals. The 1986 survey includes reserves and also gives total resources. In 1986, the total proven reserves of recoverable solid fuels were given as 6×10^{11} metric tons. One metric ton is defined as 29.2×10^3 MJ (27.7×10^6 Btu) to provide for the variation of calorific value in different coals. The total estimated additional reserves recoverable and total estimated additional amount in place are 2.2×10^{12} and 7.7×10^{12} metric tons, respectively. These figures are about double the

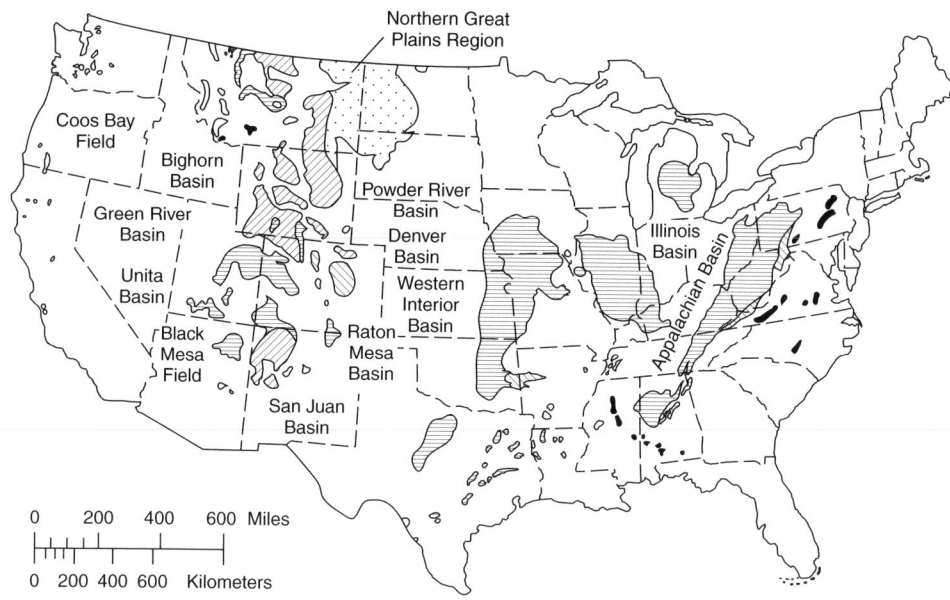

Figure 2. Coal fields of the conterminous United States where ■ represents anthracite and semianthracite; ▨, low volatile bituminous coal; ⩘, medium and high volatile bituminous coal; ◩, subbituminous coal; and ⊡, lignite.

1913 estimates, primarily because significantly increased reserves have been indicated for Russia.

The part of the resource that is economically recoverable varies by country. The estimates made in the survey show that the proven recoverable reserves would last ~1200 years at the 1988 annual rate of production and that the estimated additional amount in place represent almost 1700 years at 1988 annual consumption.

Coal is widely distributed and abundant in the United States as indicated in Figure 2.

Coal Production

World coal and lignite production rose to $\sim4.7 \times 10^9$t in 1988. Coal production in the United States has increased with fluctuations to about 1099×10^6t in 1999. The demand for energy is continually increasing and the highest energy consumption in the world occurs in the United States. Estimated coal consumption reduces the known recoverable reserves at ~1%/year. Whereas the use of bituminous coal is expected to continue to increase in terms of tonnage, the percentage of coal used in the United States has stabilized.

SAMPLE SOURCES

Basic coal research requires a variety of coal samples of different ranks that workers may access using a minimum of effort. Coal sample banks fill this need. Moreover, over the past decades it has become evident that the quality of samples degrades from atmospheric oxidation and the degradation has limited the ability of researchers to compare results. The U.S. Department of Energy Office of Basic Energy Sciences has sponsored the Argonne Premium Coal Sample Program to permit the acquisition of ton-sized samples of each of eight different coals representing a range of coal ranks, chemical composition, geography, and maceral content.

MINING AND PREPARATION

Mining

Coal is obtained by either surface mining of outcrops or seams near the surface or by underground mining depending on geological conditions, which may vary from thick, flat seams to thin, inclined seams that are folded and need special mining methods. Coal mining has changed from a labor intensive activity to one that has become highly mechanized.

Strip or open-pit mining involves removal of overburden from shallow seams, breaking of coal by blasting or mechanical means, and loading of the coal. The two methods of underground mining commonly used are room-and-pillar and longwall. In room-and-pillar mining the coal is removed from two sets of corridors that advance through the mine at right angles to each other. Regularly spaced pillars, constituting about one-half of the coal seam, are left behind to support the overhead layers in the mined areas. The pillars may later be removed, leading to probable subsidence of the surface. Longwall mining is used to permit recovery of as much of the coal as possible. Two parallel headings are made 100–200-m apart and at right angles to the main heading. The longwall between the two headings is then mined away from the main heading. The equipment provides a movable roof support system that advances as the coal is mined and allows the roof to collapse in a controlled manner behind it. This method also leads to subsidence of the overhead layers.

Preparation

Coal preparation is of significant importance to the coal industry and to consumers. Preparation normally involves some size reduction of the mined coal and the systematic removal of some ash-forming material and very fine coal.

In some areas, run-of-mine coal is separated into three products: a low gravity, premium-priced coal for metallurgical or other special use, a middlings product for possible boiler firing, and a high ash refuse. The complete preparation of coal usually requires several processes.

Jig washing is the most widely used of all cleaning methods. Froth flotation (qv) is the most important method for cleaning fine coal because very small particles cannot be separated by settling methods. Draining on screens removes substantial amounts of water from larger coal, but other dewatering (qv) methods are required for smaller sizes having larger surface areas. Vibrating screens and centrifuges are used for dewatering.

For utilities, two types of storage are used. A small amount of coal in storage meets daily needs and is continually turned over. This coal is loaded into storage bins or bunkers. However, long-term reserves are carefully piled and left undisturbed except as necessary to sustain production.

TRANSPORTATION

The usual means of transporting coal are railroad, barge, truck, conveyer belt from mine to plant, and slurry pipelines.

HEALTH AND SAFETY FACTORS

Coal mining has been a relatively dangerous occupation. In the 7 years after the passage of the Federal Coal Mine Health and Safety Act of 1969, the average fatality rate decreased to 0.58, and by 1989 the rate was 0.25.

The principal causes of fatalities are falling rock from mine roofs and faces, haulage, surface accidents, machinery, and explosions. For disabling injuries the primary causes are slips and falls, handling of materials, use of hand tools, lifting and pulling, falls of roof rock, and haulage and machinery.

Gases and Coal Dust Explosions

Gases can be hazardous in coal mines. Methane is of greatest concern, although other gases including carbon monoxide and hydrogen sulfide may be found in some mines. Methane must be detected and controlled because mixtures of air and 5–15% of methane are explosive.

Drainage

Some mines are located beneath subsurface streams, or the coal seams may be aquifers. These mines may become flooded if not continually pumped. Air or biological oxidation of pyrite leads to sulfate formation and dilute sulfuric acid in the mine drainage. Means of controlling this problem are under study.

Other Hazards

Rocks falling from the roofs of mines used to cause the largest number of accidents. Roof bolts are placed in holes drilled into the roofs of working areas to tie the layers of rock together and thus prevent rock falls. A disease called pneumoconiosis, also called black lung, results from breathing coal dust over prolonged periods of time. The coal particles coat the lungs and prevent proper breathing.

Regulations

The U.S. Bureau of Mines, Mining Enforcement and Safety Administration (MESA) studies hazards and advises on accident prevention. MESA also administers laws dealing with safety in mines. Individual states may also have departments of mines to administer state standards.

The Federal Coal Mine Health and Safety Act set standards for mine ventilation, roof support, coal dust concentrations levels, mine inspections, and equipment. As a part of this comprehensive act, miners must receive medical examinations at employer expense, and payments are made from the U.S. government to miners who cannot work because of black lung disease.

USES

Coal As Fuel

Coal is used as a fuel for electric power generation, industrial heating and steam generation, domestic heating, railroads, and coal processing. About 87% of the world's coal production is burned to produce heat and derived forms of energy. The balance is practically all processed thermally to make coke, fuel gas, and liquid by-products. Other uses of coke and fuel gas also contribute to coal consumption for heat.

Electric Power Generation. Coal is the primary fuel for thermal electric power generation. The reasons for increased coal demand include availability, relative stability of decreasing coal prices, and lack of problems with spent fuel disposal as experienced in nuclear power plants.

The overall efficiency of electric power plants consisting of coal-fired boilers and steam turbines has plateaued at ~39%. The addition of pollutant control equipment has increased the internal power use on the stations and lowered the effective efficiency of the plant. The increased efficiencies have been achieved through use of larger units (up to 1500 MW) and higher pressures to 24.1 MPa (3500 psi) and reheat, but concerns about reliability and ability to match power generation and demand have kept plant sizes below these values.

Almost all modern large coal-fired boilers for electric power generation use pulverized coal. One significant advantage of pulverized coal boilers is the ability to use any kind of coal, including run-of-mine or uncleaned coals. However, with the advent of continuous mining equipment, the ash content frequently is ~25%, and some preparation is frequently practiced.

Integration of coal gasification and a combination of a gas turbine for power generation and a waste heat boiler for power generation is termed integrated gasification combined cycle (IGCC). Efficiencies are currently ~42% and promise to be higher as gas turbine technology improves.

A primary concern in coal-fired power generation is the release of air pollutants.

Industrial Heating and Steam Generation. The principal industrial users of coal include the iron and steel industry and the food, chemicals, paper, engineering, bricks, and other clay products, and cement industries, and a group of miscellaneous consumers such as federal and local government installations, the armed services, and small industrial concerns. Most of the coal is burned directly for process heat, ie, for drying and firing kilns and furnaces, or indirectly for steam generation for process needs or for space heating, and for a small amount of electric power generation.

Coal Processing to Synthetic Fuels and Other Products

The primary approaches to coal processing or coal conversion are thermal decomposition, including pyrolysis or carbonization, gasification, and liquefaction by hydrogenation. The hydrogenation of coal is not currently practiced commercially.

In the United States, the Clean Coal Technology program was created to develop and demonstrate the technology needed to use coal in a more environmentally acceptable manner. Activities range from basic research and establishing integrated operation of new processes in pilot plants through demonstration with commercial-scale equipment.

High Temperature Carbonization. High temperatures and long processing times are used in carbonizing coking coals in coke ovens or gas retorts. Besides metallurgical or gas coke the products include fuel gas, crude tar, light oils (benzene, toluene and xylene, referred to as BTX, and solvent naphtha), and ammonia gas.

Low Temperature Carbonization. Lower temperature carbonization of lump coal at ~700°C, primarily used for production of solid smokeless fuel, gives a quantitatively and qualitatively different yield of solid, liquid, and gaseous products than does the high temperature processes. Although a number of low temperature processes have been studied, only a few have been used commercially. These have been limited in the types of coal that are acceptable, and the by-products are less valuable than those obtained from high temperature processing.

Gasification

Gasification of coal is used to provide gaseous fuels by surface and underground applications, liquid fuels by indirect liquefaction, ie, catalytic conversion of synthesis gas, and chemicals from conversion of synthesis gas. There are also applications in steelmaking.

Liquefaction

Liquefaction of coal to oil was first accomplished in 1914. Hydrogen was placed with a paste of coal, heavy oil, and a small amount of iron oxide catalyst at 450° and 20 MPa (200 atm) in stirred autoclaves. Since then the process has been through several improvements, but the most recent have not been commercialized. However, Processes for hydrogen gasification, hydrogen pyrolysis, or coking of coal usually produce liquid coproducts. Substitution of petroleum residuum for the coal-derived process oil has been used in studies of coal liquefaction and offers promise as a lower cost technology.

Bioprocessing and Biotreatment of Coal

The use of biotechnology to process coal to make gaseous and liquid fuels is an emerging field. Bacteria and enzymes have been studied to establish the technical feasibility of conversion. Efforts have focused on lower rank coals, lignite or brown coal and subbituminous coal, because of greater reactivity. The conversion processes frequently introduce chemically combined oxygen through hydrolysis or related reactions to make the solid soluble in the reaction mixture as an initial step. Further reaction involves biological degradation of the resulting material to form gases or liquids.

The large-scale processing of coal is expected to involve plants similar to sewage treatment facilities in the handling of liquid and solid materials. The reaction rates are substantially lower than those achieved in high temperature gasifiers and liquefaction reactors requiring much larger systems to achieve comparable coal throughput.

Biological processes are also being studied to investigate ability to remove sulfur species in order to remove potential contributors to acid rain. These species include benzothiophene-type materials, which are the most difficult to remove chemically, as well as pyritic material.

Other Uses

The quantity of coal used for purposes other than combustion or processing is quite small. Coal, especially anthracite, has established markets for use as purifying and filtering agents in either the natural form or converted to activated carbon.

Carbon black from oil is the main competition for the product from coal, which is used in filters. Carbon for electrodes is primarily made from petroleum coke, although pitch coke is used in Germany for this product. The pitch binder used for electrodes and other carbon products is almost always a selected coal tar pitch.

The preparation of pelletized iron ore represents a substantial market for coke and anthracite for sintering; Some minor uses of coal include the use of fly ash, cinders, or even coal as a building material; soil conditioners from coal by oxidizing it to humates, and a variety of carbon and graphite products for the electrical industry, and possibly the nuclear energy program. The growth of synthetic fuels from coal should also provide substantial quantities of by-products including elemental sulfur, fertilizer as ammonia or its salts, and a range of liquid products.

D. L. Crawford, ed., *Biotransformations of Low Rank Coals*, CRC Press Inc., Boca Raton, Fla., 1992.

W. Francis, *Coal*, 2nd ed., Edward Arnold & Co., London, 1961.

E. Stach and co-workers, *Stach's Textbook of Coal Petrology*, 3rd ed., Gebrüder Borntraeger, Berlin, Germany, 1982.

D. L. Wise, ed., *Bioprocessing and Biotreatment of Coal*, Marcel Dekker, New York, 1990, 744 pp.

KARL S. VORRES
Argonne National Laboratory

COAL GASIFICATION

Coal gasification is the process of reacting coal with oxygen, steam, and carbon dioxide to form a product gas containing hydrogen and carbon monoxide. Gasification is essentially incomplete combustion. The chemical and physical processes are quite similar; the main difference being the nature of the final products. From a processing point of view the main operating difference is that gasification consumes heat evolved during combustion. Under the reducing environment of gasification the sulfur in the coal is released as hydrogen sulfide rather than sulfur dioxide and the coal's nitrogen is converted mostly to ammonia rather than nitrogen oxides. These reduced forms of sulfur and nitrogen are easily isolated, captured, and utilized, and thus gasification is a clean coal technology with better environmental performance than coal combustion.

Depending on the type of gasifier and the operating conditions gasification can be used to produce a fuel gas suitable for any number of applications. A low heating value fuel gas is produced from an air blown gasifier for use as an industrial fuel and for power production. A medium heating value fuel gas is produced from enriched oxygen blown gasification for use as a synthesis gas in the production of chemicals such as ammonia, methanol, and transportation fuels. A high heating value gas can be produced from shifting the medium heating value product gas over catalysts to produce a substitute or synthetic natural gas (SNG).

COAL GASIFICATION CHEMISTRY

In a gasifier, coal undergoes a series of chemical and physical changes as shown in Figure 1.

Each of the steps is described in more detail below. As the coal is heated most of the moisture is driven out when the particle temperature is ~105°C. Drying is a rapid process and can be essentially complete when the temperature reaches 300°C depending on the type of coal and heating method used.

Devolatilization or pyrolysis accounts for a large percentage coal weight loss and occurs rapidly during the initial stages of coal heat up.

The volatile yield and composition depends on the heating rate and final temperature. At slow-heating rates (<1°C/s) the volatile yield is low due to repolymerization. Under rapid-heating rate (500–10^5°C/s) the volatile yield is 20–40% more than that at slow-heating rates. At any given temperature only a certain fraction of the volatiles is released. Significant devolatilization begins when the coal temperature is about 500°C. As the temperature is increased more volatiles are released. The maximum volatile yield occurs when the temperature is >900°C.

The fraction of the devolatilization gas that condenses at room temperature and pressure is called tar. It is a mixture of hydrocarbons with an average molecular weight ranging from 200–500 g/mol. The yield of tar depends on the coal rank; higher rank coals produce lesser amounts of tar. Higher gasifier temperature also reduces the amount of tar in the gasifier products because of increased cracking of tar into lighter gases. The amount of tar also decreases with increasing pressure and decreasing heating rates.

The solid product left over from devolatilization is char. The reactivity of char depends on properties of coal minerals, pyrolysis conditions, and gasification conditions. If the char porosity reaches a critical porosity (70–80%) the char will fragment into fine solids, which also increases the reactivity of char.

Char in an oxygen atmosphere undergoes combustion. In gasifiers partial combustion occurs in an oxygen-deficient, or reducing, atmosphere. Gasifiers use 30–50% of the oxygen theoretically required for complete combustion to carbon dioxide and water. Carbon monoxide and hydrogen are the principal products, and only a fraction of the carbon in the coal is oxidized completely to carbon dioxide.

High temperature favors endothermic reactions (increases the products on the right hand side). High temperature favors reactions in which there is a reduction in the number of moles, as in the methanation reaction (2 mol of hydrogen gives 1 mol of methane). Hydrogen and carbon monoxide production increases with decreasing oxygen in the feed, with decreasing pressure, and with increasing temperature. Hydrogen production increases

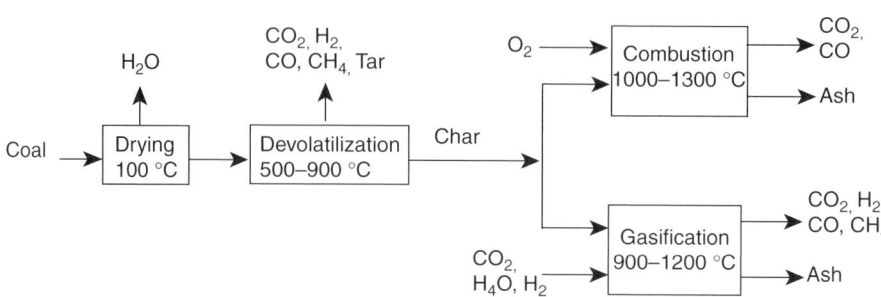

Figure 1. Chemical and physical changes of coal.

and carbon monoxide production decreases with increased steam rate. Methane production increases with decreasing temperature and increasing pressure. The product gas of air blown gasifiers is diluted by nitrogen. Upon heating bituminous coals become sticky and swell, which can cause problems in fixed-bed gasifiers. Such coals are more easily handled in fluidized and entrained bed gasifiers.

SYNGAS CHEMISTRY

Whereas near-term application of coal gasification is expected to be in the production of electricity through combined cycle power generation systems, longer term applications show considerable potential for producing chemicals from coal using syngas processing. Products include ammonia, methanol, synthetic natural gas, and conventional transportation fuels.

Ammonia

Ammonia is produced through the reaction of hydrogen and nitrogen. In a coal-to-ammonia facility, coal gasification produces the hydrogen and an air separation plant, which also provides oxygen for coal gasification, supplies the nitrogen.

The water gas shift reaction is used to increase the amount of hydrogen in the gas. For shifting coal-derived gas, conventional iron–chromium catalysts can be used. Because coal gas has a significantly higher concentration of carbon monoxide than is found in gas streams in conventional refineries, the catalyst must be able to withstand high thermal loads. However, potential catalyst poisons such as phenol and other hydrocarbons are not a concern in entrained-bed gasifiers.

Methanol

Methanol is produced by stoichiometric reaction of CO and H_2. The syngas produced by coal gasification contains insufficient hydrogen for complete conversion to methanol, and partial CO shifting is required to obtain the desired concentrations of H_2, CO, and CO_2.

Synthetic Natural Gas

Another potentially very large application of coal gasification is the production of SNG. The syngas produced from coal gasification is shifted to produce a H_2/CO ratio of approximately 3:1. The carbon dioxide produced during shifting is removed, and CO and H_2 react to produce methane (CH_4), or SNG, and water in a methanation reactor.

The tube wall reactor (TWR) system features the use of catalyst-coated tubes. Processes have also been developed for hydrogasification that maximize direct conversion of coal to methane. A good example is the HYGAS process, which involves the direct hydrogenation of coal in the presence of hydrogen and steam, under pressure, in two fluidized-bed stages.

A coal-to-SNG facility can be built at a coal mine-mouth location, taking advantage of low cost coal. SNG can then be pipelined to local distribution companies and distributed through the existing infrastructure. This approach is used in the Great Plains Coal Gasification Project in Beulah, North Dakota, which employs Lurgi gasifiers followed by shift and methanation steps. SNG has the advantage that it can directly displace natural gas to serve residential, industrial, and utility customers reliably.

Another technology that is being pursued for fuel utilization of coal is mild gasification. Similar to pyrolysis, mild gasification is performed at atmospheric pressure at temperatures <600°C. By drying and heating under controlled conditions, the coal is partially devolatilized and converted to gases and a solid residue. The gases can be used as fuel and partially condensed to produce a liquid fuel similar to residual fuel oil. The solid product is similar to low moisture, high heating value coal.

Conventional Transportation Fuels

Synthesis gas produced from coal gasification or from natural gas by partial oxidation or steam reforming can be converted into a variety of transportation fuels, such as gasoline, aviation turbine fuel, and diesel fuel.

The use of a fluidized-bed reactor is possible only when the reactants are essentially in the gaseous phase. Fluidized-beds are not suitable for middle distillate synthesis, where a heavy wax is formed. For gasoline synthesis processes such reactors are especially suitable when frequent or continuous regeneration of the catalyst is required. Slurry reactors and ebulliating-bed reactors comprising a three-phase system with very fine catalyst are, in principle, suitable for middle distillate and wax synthesis, but have not been applied on a commercial scale.

For the Fischer-Tropsch reaction in the first stage (heavy paraffin synthesis, or HPS) of the SMDS process, a tubular fixed-bed reactor has been chosen for its inherent simplicity in design and operation and also for its proven technology in other processes, such as methanol synthesis. The catalyst is located in the tubes, which are cooled by boiling water around them, and considerable heat can thus be removed by boiling heat transfer. The good stability of the SMDS catalyst makes it possible to use a fixed-bed reactor. In the next step, heavy paraffin cracking (HPC), the long-chain waxy paraffins are cracked to desired size under mild hydrocracking conditions using a commercial Shell catalyst. In the final step, by selection of the corresponding cut points, the product stream is split into fractions of the required specification. The products manufactured in the SMDS process are predominantly paraffinic and free of impurities such as nitrogen and sulfur.

COAL CHARACTERISTICS AFFECTING GASIFICATION

Developers of coal gasification technology have studied the impact of key coal properties on different parts of the gasification process. These tests have provided a good understanding of the influence of coal properties and have led to the development of process and equipment options.

Reactivity

Reactivity is used to describe the relative degree of ease with which a coal undergoes gasification reactions. The primary property affecting the ease of conversion is the coal rank, which in turn reflects its volatile matter content, oxygen content, level of maturity, extent of aromatic ring condensation, and porosity. The lower the rank the higher the volatile matter content and the more open the pore structure. Also, lower rank coals have more heteroatoms (oxygen, nitrogen, and sulfur) within the organic structure and the aromatic structures are poorly aligned. Such an amorphous and open structure contains more active sites making reaction with oxygen and steam easier. As the rank increases the carbon lattice becomes better aligned and the porosity reduces until, in anthracite coals the carbon structure becomes less reactive developing the flat basal structures found in graphite.

Other factors that have impact on reactivity are maceral distribution and the content of some catalytic mineral components. Vitrinites is the most common coal maceral derived from woody tissue. The properties and reactivity of vitrinites vary with the extent of geological maturation or coal rank. Fusinites originate from biodegraded or charred wood and are relatively unreactive C-rich macerals; while liptinites come from hydrogen-rich sources such as spores, leaf cuticles, and algal colonies. These liptinites are quite reactive.

Moisture and Oxygen Content

The moisture present in the coal is primarily a diluent. Although steam is used for gasification, there are several sources of water for this steam and only a small amount of the total water available is actually converted in the steam-carbon reaction. Steam is introduced with air and oxygen to moderate the temperature in the combustion zone of the gasifier. Some entrained gasifiers introduce coal into the reactor with the aid of water as a coal–water slurry. In addition, all coals have moisture content. Lower rank coals, which have been exposed to geological maturation over shorter periods of time and at lower temperatures, have more moisture than higher rank coals. Similarly, the higher oxygen content of lower rank coals reduces the heating value of these coals relative to higher rank coals. To offset the effectively higher oxidation state of the low rank coals a higher coal feed rate is required to obtain similar product gas quality.

Caking Properties

When bituminous coals are heated to 300–350°C the particles tend to swell and agglomerate producing a consolidated cake. The handling of this caking char and the heavy tars that accompany it has been critical to the development of gasification processes. The agglomerate that forms in a fixed or fluid bed disrupts gas flow patterns and lowers thermal efficiency.

In addition, to formation of cake, bituminous coals also produce high molecular weight tars. To handle the tars formed in coal process units, the downstream gas cleanup system must be engineered to avoid plugging and fouling of lines, heat exchangers, and filters.

Mineral Composition

The mineral content affects gasifier performance, especially for most slagging gasifiers, because minerals melt to form of slag and provide an insulating coverage on the wall of the gasifier, which reduces the heat transferred during the gasification reaction. Mineral content also influences the requirements of the slag tap and the slag handling system. A related parameter is the slagging efficiency, which is the percentage of mineral solids recovered as slag out of the bottom of the gasifier relative to the total mineral solids produced by the process.

HISTORY

Early Coal Use and Gasification

Coal has been used for centuries (see Table 1).

Table 1. Significant Events Related to Coal

Year AD	Event
589	First recorded use of coal in China
852	Coal first mentioned in the "Saxon Chronicle" of the Abbey of Petersborough
1180	Coal systematically mined in England
1250	Coal recognized as a commercial commodity
1316	Royal Proclamation forbidding use of coal in London due to its "noisome smell"
1609	Van Helmont identifies gas production from coal combustion.
1659	Shirley investigated "natural gas" released from a well in Lancashire England
1675	Coal was distilled for the production of tar
1780	Fontana proposes making "blue water gas" by passing steam over incandescent carbon.
1792	Murdoch lights his Scotch home with gas from coal heated in an iron retort
1803	Huge gas powered lamp installed on 40 ft. high tower on Main Street, Richmond, Virginia
1812	London Streets illuminated by the London and Westminster Gas Light and Coke Co.
1859	Drake drills first oil well near Titusville, Pennsylvania
1872	Lowe invents carbureted gasifier
1880	Development of modern day fixed bed coke ovens
1920	Fischer and Tropsch develop catalysts to convert coal synthesis gas to liquids.
1926	Rheinbraun develops fluid bed gasifier
1936	Development of the modern day entrained bed Koppers-Totzek gasifier
1950	Production of gasoline and diesel fuel using Lurgi gasifier in South Africa
1970	Clean Air Act
1973	Arab oil embargo
1983	Syngas production for chemical production using Texaco gasifier at Tennessee Eastman
1984	Production of synthetic natural gas at Dakota Gasification Plant using Lurgi gasifiers
1996	Clean coal demonstration power production plants using Texaco and E-Gas gasifiers
2001	Power shortages in California

3

3

To understand the development of coal conversion technologies, such as coal gasification, one must understand the factors that drove its development. Coal was the energy source that fueled the Industrial Revolution. Coal has a higher heating value than wood making it less bulky and easier to store and transport to the marketplace.

Coke Manufacturing Processes

These processes were developed to convert the softer bituminous coals into a strong hard coke ideally suited for iron making. It was found that when bituminous coals are heated slowly in the absence of air and then cooled the solid coke is both hard and inherently strong. The *coking process* requires heating to temperatures approaching 1000°C. This process heat is provided either by burning a portion of the coal directly and allowing the hot products of combustion to pass through a bed of coal, or by indirectly heating the exterior walls of a vessel containing the coal. The indirect method of heating a coke oven is still the most commonly used coking process because it maximizes the formation and quality of the coke product.

Manufacture of "Blue Water Gas" and "Town Gas"

The "blue water gas" produced from *cyclic gas generators* were not strictly suitable for street or household lighting. The first coal gasifiers built for this purpose undoubtedly suffered several setbacks. Although the heating value was high enough to sustain combustion, the blue flame was not sufficiently bright to illuminate a street or room in a house. Early developers learned that the coal gas needed to contain components called "illuminates" to provide a bright luminous yellow flame. These components consisted of higher hydrocarbons with hydrocarbon chain lengths of 2 or more. However, a second problem was observed if too many or the wrong types of higher hydrocarbons were added; condensable species, tars and naphthalene caused fouling and plugging in the piping and transfer lines. In addition, noxious and poisonous gases were also produced.

The solution to produce a brighter yellow flame was to simply add "illuminates" as the product gases left the gasifier. In the *carbureted gasifier process* developed in 1872, both an "oil gas" and "blue water gas" were produced. The "oil gas" contained condensable coal tars or cheap petroleum distillate oil and was sprayed onto a hot brick matrix or checker work. After the "blue water gas" was formed it was passed over this checker work and mixed with the volatile products of "oil gas" decomposition.

As is still true today, the coal gas in those early gasifiers was produced in several different reactors depending on the desired products and end use. Coal technologies besides combustion included: *pyrolysis, coking, cyclic gas generators*, and *gas producers*. All of these processes are heated with insufficient air to convert the coal to the final products of combustion. The products of all of these processes are a solid fuel, condensables and tars, and flammable gases.

The development of *gas producers* did not, however, displace the *cyclic gas generators*. By firing air and steam simultaneously the product gas in these producers now contained the unreactive nitrogen that is present in the air stream fed to the gasifier. Air is 79% N_2 and only 21% O_2 by volume. This added nitrogen is merely a diluent and plays no role in the combustion or gasification reactions. In addition, this diluent must be heated both during the gasification process and later as the fuel is burned. As a result the producer gas has only about one-half of the heating value compared to "town gas".

By 1930, there were over 11,000 coal gasifiers operating in the United States and in the early 1930s over 11 million metric tons of coal were gasified annually. Three modern coal gas producer types are shown schematically in Figure 2, called Fixed Bed, Fluid Bed, and Entrained-Flow gasifiers. Each of these gasifiers has advantages, disadvantages, and potential for process improvement.

Gasification for Liquid and Chemical Feedstock

In the 1930s oil and natural gas began to effectively displace manufactured gas from pipeline distribution networks. However, coal gasification remained a strong industry as a result of the emerging new transportation and chemical industries.

During the first quarter of the twentieth century coal was king, providing: heat for homes and industry, fuel gas for town lighting, process gas for industry, coke for the iron industry, and fuel for steam locomotives and shipping. By 1900, electric power was beginning to compete with the coal gas industry for public lighting. The electric power industry was also coal based, although large hydro-electric power plants from dams would be built between 1900 and 1930.

The first liquid fuels developed from coal used a liquid extraction process that heated pulverized coal with oil under high pressures and with hydrogen. This is called direct hydrogenation or direct liquefaction of coal.

The capacity for oil and gasoline manufacture from direct liquefaction plants quadrupled during the war years in Germany. Elsewhere, however, oil resources were sufficient to meet demands and coal was not extensively used to produce liquids. After the war the technology used in the German synthetic fuels industry was studied, but considered uneconomical given the more competitive costs of natural gas and oil production.

GASIFIER TYPES

Fixed-Bed Gasification

The early gasification processes were developed using a countercurrent, *fixed bed gasifier*. Within the bed the fuel is not actually fixed but in fact moves, by gravity flow, as the combusted ash is withdrawn from the gasifier. Typically, the air and steam are introduced at the bottom and travel upward through the coal bed. The coal is fed onto the top of the bed and travels downward countercurrent to the flow of gases. The gas outlet and coal feed inlets fix the upper level of the bed, while the bottom of the bed is most commonly fixed by the presence of a rotating grate.

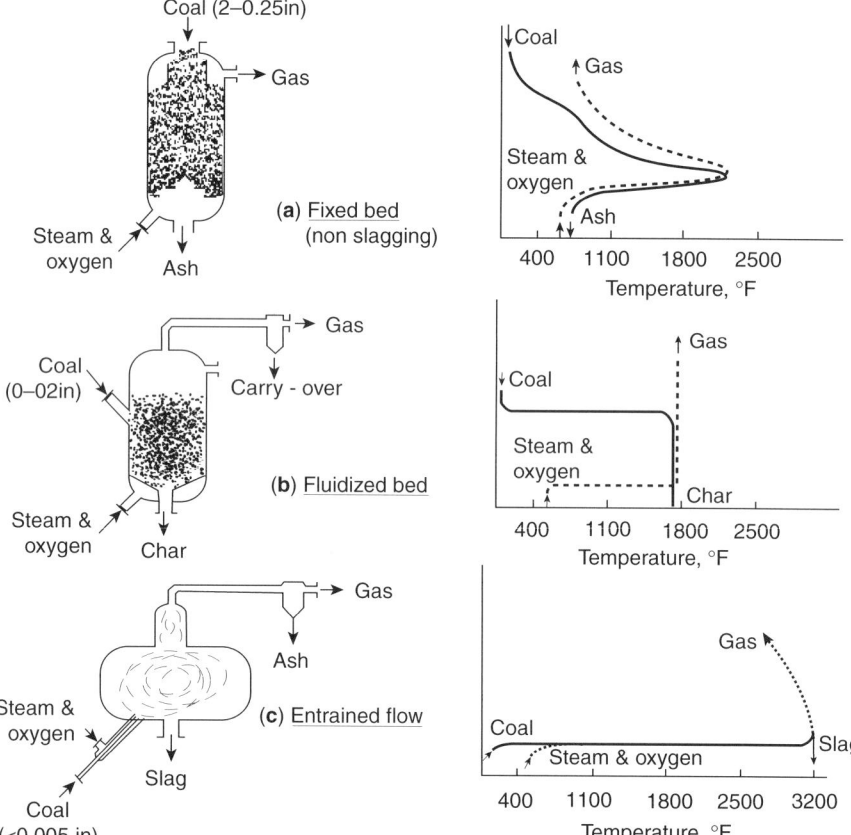

Figure 2. Types of gasifiers: (**a**) moving-bed (dry ash), (**b**) fluidized bed, and (**c**) entrained-flow.

The *fixed bed* was long considered to be the most efficient method of converting carbonaceous fuels to fuel gas. The fuel bed is generally divided into different temperature zones corresponding to the following:

1. The topmost layer where coal is dried and preheated and volatile hydrocarbons are released.
2. The reduction, or gasification, zone where the hot char reacts with steam and carbon dioxide to produce hydrogen and carbon monoxide.
3. The oxidation zone where the residual carbon reacts with oxygen producing heat for gasification reactions.
4. The ash cooling and air preheat zone at the bottom of the gasifier.

Fixed beds have several advantages. The flow of the hot gases up from the combustion zone preheats the coal leading to maximum heat economy. High carbon conversion is assured by plug flow of solids through the gasification and combustion zones and the relatively long residence times of the fuel in the vessel. The product gas exits at relatively cool temperatures and without contamination of solids.

Fixed-Bed gasifiers have been developed to handle a variety of solid fuels. The factors important in affecting a fuels performance in a fixed-bed gasifier are particle size and size distribution, tendency for coal to melt and form an agglomerated mass, the temperature at which the ash melts and fuses, and the reactivity of the coal.

The most suitable coals are uniformly sized crushed particles without tendency to agglomerate and with a minimum of fines but having reasonable mechanical strength. With such fuels the flow of gases through the bed is distributed uniformly through the bed resulting in uniform temperature distribution and stable reaction zones. Reactivity of the coal or coke feedstock affects the size of the reaction zones. Less reactive coals need larger bed depths, higher reaction temperatures, and longer residence times to achieve complete carbon conversion.

The disadvantage of the *fixed-bed gasifier* is the inability to process caking coals.

In fixed-bed gasifiers the devolatilization products exit the gasifier with the syngas, because of low temperatures and lack of oxygen in the devolatilization zone. This causes increased amount of methane in the product gas, which increases the heating value of the gas. But the low temperature and the countercurrent operation also allows the tar to escape, which is detrimental to the down stream equipment.

Fluid-Bed Gasifiers

Fluidized-bed gasifiers provide better mixing and uniform temperatures that allow oxygen to react with the devolatilization products. These products also undergo thermal cracking, primarily on hot char surfaces, reacting with steam and H_2. In dry fluidized-bed gasifiers, temperatures have to be maintained below the ash melting point, which leads to incomplete carbon conversion for unreactive coals. Agglomerating ash gasifiers operate at higher temperatures, near the ash softening point, which provides improved carbon conversion.

The primary advantage of the fluid-bed gasifier is the flexibility to use caking coals as well as low quality coals of high ash content. In addition, a fluid-bed gasifier is able to operate over a wide range of operating loads or outputs without significant drop in process efficiency. This fluid-bed process has a large inventory of hot solids that stabilizes the temperature and eliminates the potential of oxygen breaking through and burning with the fuel gases in the event of an inadvertent loss of coal feed. Fluid beds also have high heat transfer rates and good solids and gas mixing.

In addition, fluid-bed gasifiers can include inexpensive disposable sorbents, such as limestone, to absorb sulfur, reducing air emissions. The temperature regime is ideal for capture of hydrogen sulfide using limestone or dolomite.

One drawback, actually a result of good mixing in the fluid-bed gasifier, is the high temperature of the fuel gas at the exit of the reactor.

Entrained Flow Gasifiers

Like fluidized beds, entrained coal gasifiers were developed to improve the gas production rate and operate with a wider range of fuel feedstocks. In an entrained gasifier the coal is introduced into air or oxygen in a dilute stream and heated to high temperatures, 1300–1475°C, over only a very short period, 2–3 s. Entrained-bed slagging gasifiers provide uniform high temperatures, resulting in complete conversion of all coals to hydrogen, carbon monoxide, and carbon dioxide, and producing no tars, oils, or phenols. As a result the throughput and capacity of the entrained reactor is the highest of all gasifiers. Coal friability does not affect operations since the coal must be pulverized for the entrained flow gasifiers. Likewise, coal swelling and agglomeration do not influence gasification performance since the particles are separated from each other in the flowing gas stream. The product stream contains no tars and very little methane because the heavy volatiles are rapidly released and cracked at the high temperature and within the short time available in the reactor.

GASIFIER PERFORMANCE

Operating Parameters. The primary gasifier operating parameters are coal composition, coal throughput, oxygen/coal ratio and steam/oxygen ratio. The amount of oxygen and steam fed to the gasifier depends on the coal composition. In general, low rank coals are very reactive and require less oxygen and little to no steam, whereas high rank coals are relatively unreactive, requiring more oxygen and a moderate amount of steam. Steam provides an alternative source of oxygen for the gasification reaction and helps to moderate the gasification temperature. Gasifier performance is evaluated in terms of syngas production and composition, carbon conversion, and cold gas efficiency.

Cold Gas Efficiency (CGE). Cold gas efficiency, a key measure of the efficiency of coal gasification, represents the chemical energy in the syngas relative to the chemical energy in the incoming coal. Cold gas efficiency on a sweet gas basis is calculated as the percentage of the heating value in coal that is converted to clean product syngas after removal of H_2S and COS.

Carbon Conversion. Carbon conversion on a once-through basis is a function of the coal composition and is strongly influenced by the oxygen/coal ratio.

Gas Composition and Heating Value. The primary gas components of syngas are CO and H_2, ranging from 59 to 67% and from 25 to 31%, respectively. Generally the gas composition is constant within a fairly narrow band for all coals including petroleum coke. The product syngas, also called medium-Btu gas (MBG), makes an excellent fuel for commercial gas turbines.

Heat Balance. Mass and heat balances are calculated around the gasification block, which includes the gasifier, quench, syngas cooler, and solids removal systems.

GASIFICATION SYSTEMS

System Configurations

The flexibility of gasification technology allows it to be integrated into a variety of system configurations to produce electrical power, thermal energy, fuels, or chemicals. The heart of the system is the gasifier. It converts a carbonaceous feedstock (such as coal) in the presence of steam and oxygen (or air) at high temperatures and moderate pressure, into synthesis gas, a mixture of carbon monoxide and hydrogen (with some carbon dioxide and methane).

The other significant mode that this technology can be configured into is coproduction. This term refers to the coproduction of power, fuels, or chemicals. Products can be produced by either processing the feedstock prior to gasification to remove valuable components or by converting the feedstock into synthesis gas and later into products.

Attributes of Gasification Technology

Gasification has many positive attributes that make it a desirable technology for the production of power, fuels, and/or chemicals. Some of those attributes that have helped to stimulate the current market and provide for a promising future are as follows:

Fuel Flexibility. In general, gasification has the ability to utilize all carbon-containing feedstocks. In addition to primary fuels such as coal, gasification can process hazardous wastes, municipal solid waste, sewage sludge, biomass, etc, after proper preparation to produce clean synthesis gas for further processing. Because of its ability to use low-cost feedstocks, gasification is the technology of choice for many industrial applications such as the gasification of petroleum coke in refineries.

Product Flexibility. Gasification is the only technology that offers both upstream (feedstock flexibility) and downstream (product flexibility) advantages. Integrated gasification combined cycle, and gasification processes in general, is the only advanced power generation technology capable of coproducing a wide variety of commodity and premium products (eg, methanol, higher alcohols, diesel fuel, naphtha, waxes, hydrogen...) in addition to electricity, to meet future market requirements.

Cleanup. Because gasification operates at high pressure with a reducing atmosphere, the products from the gasifier are more amenable to cleaning to reduce ultimate emissions of sulfur and nitrogen oxides as well as other pollutants than those from combustion processes.

By-Product Utilization. Unlike that from combustion processes, the by-product ash and slag from the gasification technologies have also been shown to be nonhazardous. The material can be readily used for landfill without added disposal cost or can be used in construction materials or further processed to produce value-added products, leading to a zero discharge plant. Sulfur can also be readily removed and converted into elemental sulfur or sulfuric acid as a saleable product.

Efficiency. Compared to combustion systems, gasification is the most efficient and environmentally friendly technology for producing low cost electricity from solid feedstocks, and IGCC can be made to approach the efficiency and environmental friendliness of natural gas combined cycle plants. Further increases in efficiency can be achieved through integration with fuel cells and other advanced technologies. In addition, the gasification process can be readily adapted with advanced technologies for the concentration of CO_2 with minimized impact on cost and thermal efficiency.

System Flexibility. Gasification technology can be configured into a wide variety of systems to maximize efficiency, achieve fuel/product flexibility, or emphasize environmental performance.

ENVIRONMENTAL PERFORMANCE

One advantage of modern IGCC systems is excellent environmental performance. Not only are regulatory standards met, but also emissions and effluents are well below acceptable levels.

Acid Rain Emissions

Integrated gasification combined cycle represents a superior technology for controlling SO_2 and NO_x emissions. Emissions are much lower than those from traditional coal combustion technologies. During gasification, the sulfur in the coal is converted to reduced sulfur compounds, primarily H_2S and a small amount of carbonyl sulfide, COS. Because the sulfur is gasified to H_2S and COS in a high pressure concentrated stream, rather than fully combusted to SO_2 in a dilute-phase flue gas stream, the sulfur content of the coal gas can be reduced to an extremely low level using well-established acid gas treating technology.

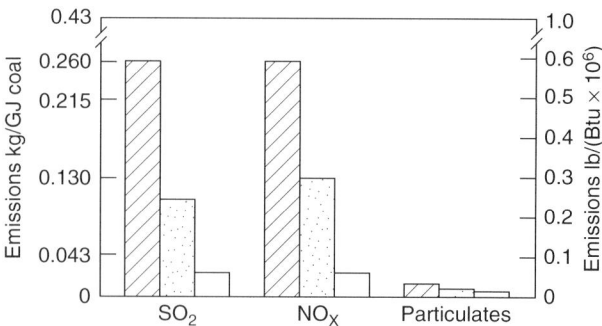

Figure 3. Environmental emissions, where ▨ represents new source performance standards (NSPS) requirements; ▢ represents a pulverized coal (PC) plant; and □ represents SCGP-1.

The sulfur is recovered from the gasification plant as salable, elemental sulfur. A small quantity of sulfur can also be captured in the slag as sulfates.

Figure 3 compare emissions from a coal gasification plant with a modern pulverized coal (PC) power plant. New technologies are being developed for removing sulfur and other contaminants at high temperature. One hot-gas cleanup process uses metal oxide sorbents to remove $H_2S + COS$ from raw gas at high ($>500°C$) temperature and system pressure.

During coal gasification the nitrogen content of coal is converted to molecular nitrogen, N_2, ammonia, NH_3, and a small amount of hydrogen cyanide, HCN. In moving-bed gasifiers, some of the nitrogen also goes into tars and oils. The NH_3 and HCN can also be removed from the coal gas using conventional (cold) gas treating processes. Other techniques are being investigated in hot-gas cleanup technologies. After removal of HCN and NH_3, combustion of the coal gas in the gas turbine produces no fuel-based NO_x. Only a small amount of thermal NO_x is formed and this can be controlled to low levels through turbine combustor design and, if necessary, steam or nitrogen addition. Based on tests using SCGP-type coal gas fired in a full-scale GE-frame 7F combustor, a NO_x concentration of no >10 ppm in the gas turbine flue gas is attainable. See Figure 1 for a comparison of NO_x emissions from a PC plant equipped with low NO_x burners.

Criteria Air Pollutants

Moving-bed gasifiers produce tars, oils, phenols, and heavy hydrocarbons, the concentrations in the gas product are controlled by quenching and water scrubbing. Fluidized-bed gasifiers produce significantly lower amounts of these compounds because of higher operating temperatures. Entrained-flow gasifiers operate at even higher temperatures, typically in excess of 1650°C. SCGP-1 experience has confirmed that carbon conversions of >99.5% are easily attainable for any coal and that essentially no organic compounds heavier than methane are produced. Emissions of volatile organic compounds (VOC) from a IGCC plant are expected to be ~300 times lower than those from a similarly sized coal-fired steam plant equipped with low NO_x burners and an FGD unit.

Hazardous Air Pollutants

The total emissions of hazardous air pollutants from an IGCC plant having wet cleanup are expected to be at least an order of magnitude lower than those achievable from a modern coal-fired steam plant. Metals removal in hot-gas cleanup systems is still under development.

Water Consumption and Effluent Characterization

Another advantage of IGCC power generation is derived from lower water requirements. Because more than one half of the power generated in a IGCC plant comes from the gas turbine, the water requirement is only 70–80% of that required for a coal-fired power plant, where all of the power is generated from steam turbines.

Whereas moving-bed gasifiers require complex water-treatment systems to address tars, phenols, and metals, this complexity is mostly alleviated for fluidized-bed gasifiers and is eliminated for entrained-flow gasifiers.

Solid By-Products

Coal gasification power generation systems do not produce any scrubber sludge, a significant advantage over both direct coal combustion processes that use limestone-stack gas scrubbers and fluidized-bed combustion processes that use solid absorbents for sulfur capture. In coal gasification, the sulfur in the coal is recovered as bright yellow elemental sulfur for which there are several commercial applications, the largest being in the phosphate fertilizer industry. The ash in the coal is converted to slag, fly slag, or fly ash. Environmental characterization of SCGP-1 slag and fly slag was performed for several coals using the extraction procedure (EP) toxicity tests and the toxicity characteristic leaching procedure test (TCLP), confirming that toxic trace metal concentrations in the leachate were well below Resource Conservation and Recovery Act (RCRA) requirements.

As part of a solids utilization program at SCGP-1, gasifier slag has been used as a principal component in concrete mixtures (Slagcrete) to make roads, pads, and storage bins. Other applications of gasifier slag and fly slag that are expected to be promising are in asphalt aggregate, Portland cement kiln feed, and lightweight aggregate (see CEMENT). Compressive strength and dynamic creep tests have shown that both slag and fly slag have excellent construction properties.

CO₂ Emissions and Global Warming

The high coal-to-busbar efficiency of an IGCC system provides a significant advantage in responding to CO_2 emissions and thus to global warming concerns. High efficiency translates to lower coal consumption and lower CO_2 production per unit of electricity generated.

EEI, "Fuel Diversity — Key to Affordable and Reliable Electricity", Edison Electric Institute, April 2001.

D. Gray and G. Tomlinson, "Co-production of Electricity and High Quality Transportation Fuels in One Facility", *Proceedings Gasification 4: The Future, Noordwuk*, The Netherlands, April 11–13, 2000.

NCC, "Increasing Electricity Availability From Coal-Fired Generation in the Near-Term", The National Coal Council, May 2001.

R. F. Probstein and R. E. Hicks, *Synthetic Fuels*, McGraw-Hill Book Co., New York, 1979, pp. 1–256.

LAWRENCE J. SHADLE
DAVID A. BERRY
MADHAVA SYAMLAL
U.S. Department of Energy
The National Energy Technology
Laboratory

COAL LIQUEFACTION

Liquefaction is the generic term for converting coal to fuels and chemicals (see also FUELS, SYNTHETIC, GASEOUS FUELS; FUELS, SYNTHETIC, LIQUID FUELS). Coal (qv) has been described variously, depending on the context, as "nature's dump" and "nature's storehouse." The reason is that while the primary constituents of coal are carbon and hydrogen, one also finds oxygen, sulfur, and nitrogen (generally classified as "heteroatoms"). Lesser amounts of many other elements can be detected (as inorganic oxides, or "mineral matter") as well. All methods for converting coal to fuels, and most methods of converting coal to chemicals, require both an increase in the hydrogen/carbon (H/C) ratio (from ~0.2 to ~2, order of magnitude, both molar) and removal of sulfur, nitrogen, and the other elements.

Coal can be converted to liquid and gaseous fuels and chemicals by two different processing routes, normally termed "direct" and "indirect." Direct liquefaction processes result in primary products (liquids or solids) of molecular weight greater than, or of the order of magnitude of, the fuels and chemicals desired. Catalysts may be used. Secondary processing is usually required to form fuels and chemicals. Some direct liquefaction schemes also involve chemical pretreatment of the coal. Other schemes involve a second feed source, generally heavy fractions of petroleum (coal–oil coprocessing), sometimes recyclable wastes (coal–waste coprocessing). In indirect liquefaction (IL) processes, on the other hand, the first step is always gasification of coal to synthesis gas ("syngas," $CO + H_2$), and this is followed by additional steps in which the syngas is catalytically recombined to form hydrocarbons and/or oxygenates.

In the 1990s, the U.S. Department of Energy (USDOE) considered catalytic two-stage liquefaction and coal/oil and coal–waste coprocessing as the two major elements of its direct coal liquefaction (DCL) program. Major elements of the indirect coal liquefaction program were advanced Fischer-Tropsch technology for transportation fuels and processes for oxygenated fuel additives and high value chemicals. At the turn of the century, USDOE's Vision 21 Concept had as a goal the development of a suite of "modules" that can be interconnected to design a plant that takes advantage of local resources and supply local needs. The object is for the plant to be able to use one or more fuel types (coal, natural gas, biomass, petroleum coke from oil refineries, waste from municipalities) and, coupled with carbon sequestration techniques, to produce multiple products (one or more of electricity, heat, fuels, chemicals, hydrogen) at high efficiencies with no emission of greenhouse gases.

DIRECT COAL LIQUEFACTION

From the above, there are two chemical concerns in direct coal liquefaction (DCL)—introduction of hydrogen to the parent coal molecule, and removal of heteroatoms (nitrogen, sulfur, and oxygen) and mineral matter. A third concern is the transport of solid and slurry-phase material. Though a physical rather than a chemical problem, it has ramifications in the economics of commercial–scale plant design.

Hydrogenation (hydroprocessing, hydrorefining, solvent refining, or hydroliquefaction) and pyrolysis are the two means used for DCL. In hydrogenation, the organic components of coal are dissolved under a moderate-to-high hydrogen pressure using a solvent, generally a coal-derived heavy aromatic material. Here the primary reactions are a combination of homogeneous thermal cracking (ie, free-radical generation) and heterogeneous hydrogenation (involving hydroaromatics in the slurry vehicle and/or the coal itself as hydrogen-transfer agents). Rapid and efficient capping of the primary free radicals generated by heating is thought to be necessary in order to prevent retrogressive reactions leading to formation of solid char. Other theories of coal liquefaction suggest that hydrogen can engender reactions involving scission of strong bonds in the coal macromolecule, and hence can act as an active bond-cleaving agent rather than simply a passive radical quencher. Typically, the object of hydrogenation is to maximize the yield of distillate fractions that can subsequently be converted into fuels.

Pyrolysis normally involves heating in an inert or reducing atmosphere and produces char and oil, and often a low-BTU gas. The relative proportion of char to the other products can be quite high, hence the rationale for liquefaction by pyrolysis is often not production of coal-derived distillate materials but rather the solid. Hydropyrolysis (heating in the presence of hydrogen) and/or pyrolysis under conditions of rapid heating can, however, generate yields of distillate products significantly in excess of the volatile matter content of the starting coal.

Hydrogenation

Examples of hydrogenation processes include the following:

Solvent-Refined Coal Process. There are two solvent-refined coal (SRC) processing schemes: SRC-I for production of low ash solid boiler fuels and SRC-II for distillates, eg, "syncrude."

The SRC-I technology was tested at a large-scale (50 t/day) demonstration plant. Total solids [SRC plus two-stage liquefaction (TSL) solids] were reduced to ∼27%, resulting in an increase in distillate materials. An overall thermal efficiency (defined loosely as the energy capable of being generated by combustion of product from a unit weight of feed per energy generated by combustion of a

Table 1. Comparison between Products of the SRC-I and SRC-II Processes[a]

Process	SRC-I	SRC-II
Product yield, wt %		
C_1–C_4	10.5	16.1
total oil	25.9	38.9
SRC solids	42.7	21.0
insoluble organic matter	4.1	5.1
H_2[b]	−2.4	−5.6

[a]High-volatile Kentucky bituminous coal.
[b]The negative sign indicates that hydrogen is being consumed.

unit weight of feed) of this proposed facility was calculated to be 70%.

The SRC-II process, was developed in order to minimize the production of solids from the SRC-I coal-processing scheme. The principal variations were incorporation of a recycle loop for the heavy ends of the primary liquefaction process and imposition of more severe conditions during hydrogenation. It was quickly realized that minerals concentrated in this recycle stream served as heterogeneous hydrogenation catalysts that aid in the production of distillate. In particular, pyrrhotites, Fe_xS_y (where x and y are ∼1), nonstoichiometric iron sulfides produced by reduction of iron pyrite (FeS_2), were identified as being especially important. Pyrite was subsequently added for cases where the inherent pyrite content of the coal was low. A yield comparison between the products of the SRC-I and SRC-II processes is given in Table 1.

Changing the process configuration to SRC-II was successful in producing ∼50% additional oil. However, a large increase in light hydrocarbon gas make accompanied this increase, with an attendant reduction in hydrogen utilization efficiency. Problems persisted using many coals, particularly subbituminous coals.

Exxon Donor Solvent Process. The principal difference between this technology and SRC was the incorporation of a fixed-bed catalytic hydrogenating unit for the recycle solvent stream. This additional unit was required to keep the hydrogen donating/shuttling capacity of the recycle solvent oil at an acceptably high value. The use of bottoms slurry recycle to increase the solvent "make" fraction by taking advantage of the catalytic properties of minerals was also investigated, and improved yields in the bottoms recycle mode were generally reported. Recycle of this fraction was also reported to improve operability of the process dramatically, especially using low-rank coals where viscosity of the bottoms stream was a significant problem. The primary liquefaction part of the reaction system operated at temperatures of 425–480°C and pressures of 10–14 MPa (1450–2030 psi), using mean residence times in the range of 15 min–2 h, depending on coal reactivity and process configuration.

H-Coal. A significantly different scheme for DCL, developed by Hydrocarbon Research Inc. (HRI; now Hydrocarbon Technologies Inc., HTI, a wholly owned subsidiary of Headwaters Inc), was based on research and development on the H-Oil ebullated-bed catalytic reactor for hydrotreating and hydrocracking heavy oil. The heart of this process is the reactor, where coal, catalyst, solvent, and hydrogen are all present in the same vessel. The reactor is maintained in a "bubbling" or ebullated, ie, well-mixed, state by internal agitation coupled with the action of the gas bubbling through the fluid. The process consists of slurry preparation followed by catalytic hydrogenation/hydrocracking at 450°C and 15 MPa (2200 psi) in the ebullated bed reactor.

A principal focus of this project was research and development for catalysts that were tolerant of the coal-derived mineral matter in the reactor. Typical early catalysts showed rapid deactivation because of coking and loss of surface area, presumably from pore–mouth blockage by coke and metals laydown. Although one of the primary advantages of the H-Coal processing scheme was the ability to add and withdraw catalyst continuously from the reactor in order to maintain a stable level of activity, catalyst replacement and consumption rates were unacceptably high under these conditions.

Wilsonville Coal-Liquefaction Facility. The plant began operation in 1974 in the SRC-I mode, but evolved to a two-stage operation utilizing two ebullating bed catalytic reactors. Initial efforts in TSL focused on catalytic upgrading of the thermal products, or Nonintegrated Two-Stage Liquefaction (NTSL). This configuration, termed nonintegrated because the coal-derived resid hydrocracking step did not interact with the primary thermal part of the plant, was excessively inefficient because of high hydrogen consumptions associated with the thermal part of the operation.

In Integrated Two-Stage Liquefaction (ITSL), a short contact-time thermal reactor was close coupled to an ebullated-bed catalytic reactor and process solvent was generated by distillation of the hydrocracked products. The thermal resid produced in the ITSL at short contact times was more reactive toward expanded-bed hydrocracking, thus permitting operation of the ebullated-bed reactor at lower severity and minimizing gas make.

Coal throughput, ie, space velocity per unit reactor volume, was substantially improved in going to the ITSL mode. The higher reactivity of the coal-derived resid permitted operation of the hydrocracker at lower temperature; this would be expected to reduce the rate of coke lay-down on catalyst, and to improve hydrogen utilization efficiency by minimizing formation of light hydrocarbon gases (higher distillate selectivity). A 35% increased yield of C_4^+ distillate was obtained.

Also explored were Reconfigured Integrated Two-Stage Liquefaction (RITSL), where solvent deashing was practiced after the hydrocracking step, and Close-Coupled Integrated Two-Stage Liquefaction (CC–ITSL), where the two reactors (thermal/catalytic) were linked directly without any intervening processing steps. Incremental improvements in distillate yield and selectivity were realized by changing the process configuration, but at the expense of increased hydrogen consumption.

From 1985 to 1992, process development at Wilsonville focused on development of a Catalytic/Catalytic Two-Stage Liquefaction (CTSL) scheme utilizing ebulating-bed catalytic reactors in both stages. Initial work indicated that distillate yields as high as 78% could be obtained by operating the first stage at low severity (399°C) and by using a large-pore bimodal NiMo catalyst having a mean pore diameter in the 11.5–12.5-nm range. These data show the significant improvement in distillate production that can be achieved by use of catalyst in both stages, but (again) at the cost of increasing levels of hydrogen consumption.

Kohleoel-Integrated Gross Oil Refining (IGOR+). In this process conditions in the primary reactor are maintained at 30 MPa and 470°C, and an iron oxide catalyst is used. Products are separated hot, and the vapor products are hydrotreated at 30 MPa and 350–420°C. The liquid from the first separator is recycled as part of the solvent. Liquid from the second separator is distilled at atmospheric pressure to yield a light oil and a medium oil.

Bench-Scale Research on Catalysts for DCL. Bench-scale test results are generally looked upon with skepticism because it is not clear how they relate to processes going on in industrial-scale reactors. However, a comparison of various types of bench-scale reactors and found that results similar to those from a large-scale ebullated-bed reactor could be obtained from a microautoclave reactor shaken at 400 cpm and containing a steel ball for efficient mixing.

Pyrolysis and Hydropyrolysis

The second category of DCL aimed at producing distillate materials from coal is pyrolysis and hydropyrolysis. Pyrolysis, sometimes called destructive distillation, essentially involves heating the coal in an inert atmosphere, followed by recovery of coal-derived tars and distillates in the off-gas stream. Pyrolysis carried out in a hydrogen atmosphere is termed hydropyrolysis; pyrolysis at extremely rapid heating rates is termed flash pyrolysis.

Pyrolysis. Large-scale research and development on coal pyrolysis was carried out on the Char Oil Energy Development (COED) process. This scheme involved temperature-staged pyrolysis in a dryer–separator and three interacting fluidized beds, and was tested in a 36-t/day process demonstration unit during the early 1970s. Pyrolysis temperatures ranged from 191 to 871°C in the COED process, and the long residence times associated with the fluid beds mandated low yields of liquid products. The yield structure is heavily weighted toward production of char and gas. Production of coal-derived liquids ranged from 0.04 to 0.21 m^3/t of coal as compared to 0.61–0.79 m^3/t for direct hydrogenation. Further, the liquids produced were high in heteroatoms (especially nitrogen) and required extensive hydrotreating before use as a synthetic crude oil.

Process development on fluidized-bed pyrolysis was also carried out by the Consolidation Coal Co., culminating in operation of a 32-t/day pilot plant. The resulting CONSOL pyrolysis process incorporated a novel stirred carbonizer as the pyrolysis reactor, which made operation of the system feasible even when using strongly agglomerating Eastern-U.S. bituminous coals. This allowed the process to bypass the normal preoxidation step that is often used with caking coals, and resulted in a nearly 50% increase in tar yield. Use of a sweep gas to remove volatiles rapidly from the pyrolysis reactor gave overall tar yields of nearly 25% for a coal that had tar yields of only 15% as measured by the Fischer assay, a standardized test to measure the amount of liquids produced by pyrolysis.

Hydropyrolysis. Process development of the use of hydrogen as a radical quenching agent for the primary pyrolysis was conducted. This process was carried out in a fluidized-bed reactor at pressures of 3.7–6.9 MPa (540–1000 psi), and a temperature of 566°C. The reactor was designed to minimize vapor residence time in order to prevent cracking of coal volatiles, thus maximizing yield of tars. Average residence times for gas and solids were quoted as 25 s and 5–10 min. A typical yield structure for hydropyrolysis of a subbituminous coal at 6.9 MPa (1000 psi) total pressure was char 38.4, oil 29.0, water 19.2, and gas 16.2, on a wt% maf coal basis. Tar yields of ~0.32 m^3/t were quoted. Because the scheme used hydrogen, the liquids generally exhibited lower heteroatom contents than conventional tars derived from coal pyrolysis in an inert atmosphere.

Flash Pyrolysis. The rapid, ie, flash, pyrolysis process was designed to heat coal at rates in excess of 5000 °C/s. Rapid heating, and hence high tar yields, could be obtained with this system. However, rapid quenching of reaction products proved to be a significant problem, especially as the process was scaled up from the laboratory.

Coprocessing

The main difference between coprocessing and hydroliquefaction is that the solvent is not simply a recycled stream from the process but is a separate feed stream, either a resid fraction (or other fraction, typically heavy) of petroleum or a waste (such as postconsumer plastic material, tire rubber or even municipal waste). The motivations for the additional feed are to reduce the severity of the liquefaction conditions compared to coal-alone hydroliquefaction, to recycle to extinction the heavy fraction, to take advantage of synergies of operation, and to take advantage of the favorable economics and/or politics in eliminating an unneeded stream.

Coal–Oil Coprocessing. Chevron, CANMET, and Ohio-Ontario Clean Fuels are among the organizations that developed strategies and technologies for coal–oil coprocessing. The two-stage coprocessing scheme of HRI (now HTI/Headwaters) was used on coal ranks from lignite through high volatile bituminous and with a variety of resids. As an example of the synergistic benefits of coprocessing, resid-based organometallic Ni and V compounds (that would serve as poisons for downstream processing if present in the liquid product) were found to be included in the solid (ash) phase contributed by the coal, and thereby removable before downstream processing.

Coal–Waste Coprocessing. The use of catalysts, iron-based and others, on coal-waste coprocessing has been quantified in tests at the bench scale and larger. Bench-scale tests have been carried out on a standard commingled waste plastic developed by the American Plastics Council, as well as on pure low density polyethylene, high density polyethylene, polypropylene and poly- (vinyl chloride); these were used alone, with coal, and with coal and resid. Sawdust and farm manure have also been used in bench-scale coprocessing. A two-stage process was suggested for coprocessing waste rubber (from tires) with coal—the tire would be liquefied noncatalytically at relatively low severity conditions to obtain a tire oil and (marketable) carbon black, and the tire oil would be combined with coal containing *in situ* ferric sulfide-based catalyst at higher temperatures and pressures.

HTI used a proof-of-concept bench-scale unit to study the effect of adding waste plastics to either coal/resid feedstock, or resid alone. After the second-stage reactor, the product was flashed and the light ends hydrotreated to yield a naphtha-like fraction. The addition of the waste plastics was found to increase the yield of distillate and to decrease the consumption of hydrogen, regardless of whether coal/resid or resid alone was used as the feedstock.

INDIRECT LIQUEFACTION

The second category of coal liquefaction involves those processes that first generate synthesis gas (syngas), a mixture of CO and H$_2$, by steam gasification of coal, $C(s) + H_2O \longrightarrow CO + H_2$, followed by production of solid, liquid, and gaseous hydrocarbons and oxygenates via catalytic reduction of CO in subsequent stages of the process. Whereas coal is usually the preferred feedstock, other carbon-containing materials such as coke, biomass, or natural gas can also be used.

In the general process, syngas from the gasifier is first cleaned to remove gasifier tars, hydrogen sulfide and organic sulfur. The composition of the gas is then adjusted in a catalytic shift converter to increase the H$_2$/CO ratio via the water–gas shift reaction, $CO + H_2O \rightleftharpoons CO_2 + H_2$. This clean and shifted gas is finally converted to hydrocarbons and/or other products in a series of catalytic reactors. The synthesis reaction is usually carried out using two or three reactors in series because of the highly exothermic nature of the overall reaction.

Production of Hydrocarbon Fuels

By convention, only the production of hydrocarbons is termed Fischer-Tropsch (FT) synthesis. Hydrocarbons are typically used as fuels or fuel enhancers, generally diesel fuel.

In processes that operated at relatively low pressures, in the range of 100–200 kPa (1–2 atm), catalysts were primarily cobalt based. However, catalyst lifetimes were short and deactivation was difficult to reverse. At the other extreme, high pressure synthesis has been carried out at pressures in the range 5–100 MPa (50–1000 atm) and temperatures of 100–400°C. Supported ruthenium catalysts are used, and the products are typically straight-chain paraffin waxes.

The greatest successes, including the processes used by the South African Coal Oil and Gas Corporation Ltd. (SASOL), have occured at medium pressures, typically in the range 0.5–5 MPa (5–50 atm). Cobalt catalysts, similar to those used for the low pressure synthesis, were typically used at temperatures of 170–200°C. Iron catalysts, usually promoted, have also been used in the SASOL process, but at temperatures of 220–340°C. The primary differences between low and medium pressure synthesis are increased catalyst life for the medium pressure process, more diesel fuel, and a slightly higher hydrocarbon yield.

Production of Alcohols and Other Oxygenates

Methanol is used as a fuel in its own right, as an octane extender for gasoline, and as a feed stock for the production of polymers and other chemicals. The original process for obtaining methanol from syngas operated at high temperatures and pressures (350–450°C, 25–35MPa) using a zinc oxide/chromium catalyst. Current ICI plants operate at low temperatures and pressures (220–280°C, 5–10 MPa) using a $Cu/ZnO/Al_2O_3$ catalyst in a multiquench reactor. Lurgi plants operate under similar conditions except in a multitubular reactor. In both cases, naphtha or natural gas is preferred to coal as a feed material for the syngas.

Higher molecular weight alcohols ("higher alcohols," HAs) are preferred as fuel additives because of their lower vapor pressure.

Production of Other Chemicals

Many of the chemicals attributed to the indirect liquefaction of coal are formed from methanol. However, it is worth noting the hydroformylation (oxo) reaction. Here aldehydes are produced by reacting olefins with CO using complexes of groups 8–10 (Group VIII) metals such as Co or Rh as a homogeneous catalyst. Hydroformylation is the fourth-largest use of synthesis gas, after the production of hydrogen, methanol synthesis and FT synthesis. As an example of hydroformylation, propylene can be converted to n-butyraldehyde:

$$CH_3CH{=}CH_2 + CO + H_2 \longrightarrow$$
$$CH_3CH_2CH_2CHO + CH_3CH(CH_3)CHO$$

used in the synthesis of 2-ethyl-hexanol, a plasticizer.

Developments in Indirect Liquefaction

Much of the research and process development on indirect liquefaction after the 1990s has been aimed at matching the synthesis conditions with modern, efficient coal gasifiers. Whereas the newer gasifiers are considerably more efficient, the gas produced has a much lower H_2/CO ratio. The slurry reactor has been shown to be capable of using this type of feedstock, under the right conditions.

Optimization of the performance of the slurry-bed reactor requires work on improved catalysts and on the separation of catalyst and wax in the product stream.

In bench-scale tests, molybdenum- or nickel-based catalysts have been used for the production of high molecular weight alcohols. A promising development is the introduction of a high boiling inert solvent, such as tetraglyme, in concurrent flow with the syngas in a fixed-bed reactor. The solvent absorbs methanol as it is produced, and shifts the chemical equilibrium to the "right" so that more methanol is produced.

N. Berkowitz, *An Introduction to Coal Technology*, 2nd ed., Academic Press, New York, 1994.

J. C. Hoogendoorn, in *Clean Fuels from Coal, Institute of Gas Technology Symposium Series*, IGT, Chicago, Ill., 1973, p. 353.

C. C. Kang and E. S. Johanson, in R. T. Ellington, ed., *Liquid Fuels from Coal*, Academic Press, New York, 1977.

G. W. Parshall and S. D. Ittel, *Homogeneous Catalysis*, John Wiley & Sons, New York, 2nd ed., 1992.

DADY B. DADYBURJOR
West Virginia University
ZHENYU LIU
Institute of Coal Chemistry

COATING PROCESSES

Coating process technology is in widespread use because there are few single materials that are suitable for the intended final use without treating the surface to meet all the functional needs and requirements of the product. The modification is accomplished by applying a coating or series of coatings to the material—the substrate—to improve its performance and make it more suitable for use or give it different characteristics. The coating process is defined here as replacing the air at a substrate with a new material—the coating.

Typical coatings are the paints and the diverse surface coatings used to protect houses, bridges, appliances, and automobiles. These coatings protect the surface from corrosion and degradation, and may provide other functional advantages such as making the materials waterproof or flameproof and improving the appearance. Adhesives are applied to paper or plastic to produce labels and tapes for a variety of uses. A thin layer of adhesive is coated onto paper to produce self-sticking note pads. Glass windows are coated with a variety of materials to make them stronger and to control light penetration into the structure. High-energy lithium batteries contain coated structures. Plastic food wrap has layers to reduce oxygen penetration and retain moisture. Packaging materials for electronic products are coated with antistatic compounds to protect sensitive components. Other important coated products are photographic films for medical, industrial, graphic arts, and consumer use; optical and magnetic media for audio and visual use data storage; printing plates; and glossy paper for magazines.

COATING MACHINES

The basic steps in continuously producing a coated structure are (1) preparing the coating solution or dispersion; (2) unwinding the roll of substrate; (3) transporting it through the coater; (4) applying the coating from a solvent, or as a liquid to be cross-linked, or from the vapor; (5) drying or solidifying the coating; (6) winding the final coated roll; and (7) converting the product to the final size and shape needed. Other operations that are often used are surface treatment of the substrate to improve adhesion, cleaning of the substrate prior to coating to reduce contamination, and lamination, where a protective cover sheet is added to the coating structure.

Depending on the substrate, the can be web coaters, sheet coaters, and coaters for nonflat applications. Web coaters, the most prevalent, coat onto continuous webs of material. Magnetic tapes, window films, wallpaper; barrier coatings for plastic films, and many printed goods are all produced using this process.

COATING PROCESSES

The widely used commercial coating methods are reverse roll, wire-wound or Mayer rod, direct and offset gravure, slot die, blade, hot melt, curtain, knife over roll, extrusion, air knife, spray, rotary screen, multilayer slide, coextrusion, meniscus, comma and microgravure coaters (based on analysis of methods reported by coaters in various trade sources). The choice of method depends on the nature of the support to be coated, the rheology of the coating fluid, the solvent, the wet-coating weight or coverage desired, the needed coating uniformity, the desired coating width and speed; the number of layers to be coated simultaneously,

cost and environmental considerations, and whether the coating is to be continuous or intermittent.

The first step in the selection process is to establish the requirements for the product to be coated and then matching then with the capabilities of the process and evaluating the best methods experimentally to determine the one to use. Some of the basic characteristics of the principal coating processes are listed in Table 1.

The processes are grouped according to the principle used to control the coverage or coating weight of the coating and its resulting uniformity, sometimes called self-metered, premetered, and hybrid. Self-metered processes are those in which the coverage is a function of the liquid properties and the system geometry, the web speed, the roll speeds, and any doctoring device. Premetered processes deliver a set flow rate of solution per unit width to the applicator and all the material is transferred to the web. If a smooth coating is obtained then the coverage is fixed. Hybrid processes use features of both self- and premetered coating to achieve coating weight control, as gravure.

In most coating operations a single layer is coated. When more than one layer must be applied one can make multiple passes or use tandem coaters (where the next layer is applied at another coating station immediately following the dryer section for the previous layer) or a multilayer coating station can be used. Slot, extrusion, slide, and curtain coaters are used to apply multiple layers simultaneously. Slide and curtain coaters can apply an unlimited number of layers simultaneously, whereas slot coaters are limited by the complexity of the die internals and extrusion coaters by the ability of the combining adapter, ahead of the extrusion die, to handle many layers.

The precision or uniformity of the coating is very important for some products such as photographic or magnetic coatings. Some processes are better suited for precise control of coverage. When properly designed,

Table 1. Summary of Coating Methods[a]

Process	Viscosity, mPa·s (= cP)	No. of layers	Wet thickness, μm	Coating accuracy, %	Max. speed, m/min
Self-metered					
rod	20–1000	1	5–50	10	250
dip	20–1000	1	5–100	10	150
forward roll	20–1000	1	10–200	8	150
reverse roll	100–50,000	1	5–400	5	400
air knife	5–500	1	2–40	5	500
knife over roll	100–50,000	1	25–750	10	150
blade	500–40,000	1	25–750	7	1500
Premetered					
slot	5–20,000	1–3	15–250	2	500
extrusion	50,000–5,000,000	1–3	15–750	5	700
slide	5–500	1–18	15–250	2	300
curtain	5–500	1–18	2–500	2	300
Hybrid					
gravure, direct	1–5000	1	1–25	2	700
gravure, offset	100–50,000	1	5–400	5	300
microgravure	1–4000	1	1–40	2	100

[a]Values given are only guidelines.

slot, slide, curtain, gravure, and reverse-roll coaters are able to maintain coverage uniformity to within 2%.

Web coating can be used for intermittent coatings, such as in the printing process and to form coated batteries, as well as for the more common continuous coatings like photographic films. In general, there is an ideal coater arrangement for any given product. However, most coating machines produce many different products and coating thickness and the machine is therefore usually a compromise made for the several applications.

Limits of Coatability

In any coating process there is a maximum coating speed above which coating does not occur. At higher speeds air is entrained, resulting in many bubbles in the coating, or in ribs and finally rivulets, or in wet and dry patches. Above the critical speed the minimum thickness depends only on the gap. Above some higher speed a coating cannot be made. Lower viscosity liquids can be coated faster and thinner. Polymer solutions can be coated at higher speeds than Newtonian liquids.

DISCRETE SURFACE-COATING METHODS

A variety of coating techniques are available to coat surfaces that are planar and have irregular surfaces.

Spray Coating

Coatings may be applied by spraying the coating material onto the object to be coated, which may be irregularly shaped with compound curves and with sharp edges.

Dip Coating

The dip-coating technique used for webs can also be used to coat discrete surfaces such as toys and automotive parts. The item to be coated is suspended from a conveyor and dipped into the coating solution, removed, drained, dried, or cured in an oven.

Spin Coating

Spin coating is used to produce a thin uniform coating on discrete supports. In this process the coating fluid, usually a colloidal suspension, is placed on a horizontal substrate which rests on a rotating platform. The speed of the platform is increased to the desired level, which can be as high as 10,000 rpm. Centrifugal forces drive much of the coating off the support, leaving a thin, uniform film behind. In addition, the coating is drying during the process and as a result the viscosity increases, resistance to flow occurs, and a level thin coating is left.

Vacuum Deposition Techniques

Thin coatings are applied to a variety of substrates for use on semiconductors, ceramics, and electrooptical devices, using a wide variety of vacuum deposition techniques. Vacuum deposition is a rapidly advancing area of coating technology.

PATCH COATING

It is sometimes necessary to coat patches of material on a web, such as coating the anode and cathode in batteries and in fuel cell membranes. Gravure coating is well suited for this purpose because the desired pattern can be etched into the gravure cylinder. Slot-coating techniques are also used.

COATING PROCESS MECHANISMS

One of the principal advances in the coating process area in the 1980s and 1990s was the development of techniques to understand and define basic coatings mechanisms. This led to improved quality and a wider range of utility for most coating techniques. This has involved the computer modeling of the coating process and the development of visualization techniques to actually see the flows in the coating process. The flow patterns predicted by the computer models can be verified by the visualization techniques.

DRYING AND SOLIDIFICATION

The drying process after the application of the coating is as important as the coating process itself. The properties of the coating are not complete until solidification has occurred. The coated film or web is transported through the dryer, where the properties of the coating can either be enhanced or deteriorated by the drying process. Drying of coatings involves the removal of the inert inactive solvent used to suspend, dissolve, or disperse the active ingredients of the coating, which include polymeric binder, pigments, dyes, slip agents, hardener, coating aids, etc. Coating solvents range from the easy-to-handle water to flammable and toxic organic materials. Drying must occur without adversely affecting the coating formulation while maintaining the desired physical uniformity of the coating.

While drying is a physical process involving only solvent removal, solidification can occur by cross-linking liquid monomer or liquid low molecular weight polymer. This process can be accelerated by catalysts or can be accomplished by an electron-beam or uv radiation. This cross-linking process is called *curing*. Material coated from solution often also undergoes curing to improve the physical properties of the dried coating. Thus both curing and drying may occur in the dryer.

Air Impingement Dryers

Air impingement dryers, the most widely used for drying coated webs, basically consist of a heat source and heat exchangers (unless hot flue gas from combustion of natural gas is used), fans to move the air, ducts and nozzles or air delivery devices positioned close to the web, and solvent removal ducts.

In these single-sided dryers, the air impinges only on the coated side, heating and drying from that side only.

The two-sided or floater dryer is now most often used. In this configuration the roll transport system in the dryer is replaced with air nozzles on the back side of the web so that the air transports and supports the web as well as heating and drying it from both sides.

Contact or Conduction Dryers

Coatings on webs, as well as sheets of newly formed paper, can be dried by direct contact with the surface of a hot drum.

Radiation Drying

Infrared or microwave radiation can supply concentrated energy to the web to evaporate the solvent, but air must still be used to carry away the solvent. These techniques provide a high heat input over short distances.

Pollution Control

The solvent removed during drying is frequently a pollutant, so the exhaust air must be treated to ensure that it meets government standards before being discharged to the atmosphere. The two basic approaches to treating the air are to recover the solvent for reuse and to convert it by burning to compounds which can safely be discharged.

Modeling Convection Drying

Models of the drying process have been developed to estimate whether a particular coating can dry under the conditions of an available dryer. These models can be run on personal computers. To model convection drying in the constant rate period, both the heat transfer to the coated web and the mass transfer from the coating must be considered.

Commercial Availability

All of the many types of coaters and dryers discussed herein are commercially available from many different vendors. These vendors usually have pilot facilities so that new coating and drying techniques can be easily tested. Contract coating companies, specializing only in coating, also exist.

E. D. Cohen and E. B. Gutoff, eds., *Modern Coating and Drying Technology*, Wiley-VCH, New York, 1992.

E. D. Cohen and E. B. Gutoff, *Coating and Drying Defects: Troubleshooting Operating Problems*, John Wiley & Sons, Inc., New York, 1995.

R. H. Perry and C. H. Chilton, eds., *Chemical Engineers' Handbook*, 5th ed., McGraw-Hill, New York, 1973.

R. J. Stokes and D. Fennell Evans, *Fundamentals of Interfacial Engineering*, Wiley-VCH, New York, 1997.

EDWARD D. COHEN
Technical Consultant
EDGAR B. GUTOFF
Consulting Chemical Engineer

COATING PROCESSES, POWDER

Powder coating is a process for applying coatings on a substrate using heat-fusible powders. Materials used in the process are referred to as coating powders, finely divided particles of organic polymer, either thermoplastic or thermosetting, which usually contain pigments, fillers, and other additives. After application to the substrate, the individual powder particles are melted in an oven and coalesce to form a continuous film having decorative and protective properties associated with conventional organic coatings.

In the fluidized-bed coating process, the coating powder is placed in a container having a porous plate as its base. Air is passed through the plate, causing the powder to expand in volume and fluidize. In this state, the powder possesses some of the characteristics of a fluid. The part to be coated, which is usually metallic, is heated in an oven to a temperature above the melting point of the powder and dipped into the fluidized bed where the particles melt on the surface of the hot metal to form a continuous film or coating. By using this process, it is possible to apply coatings ranging in thickness from $\sim$250–2500 μm (10–100 mils). It is difficult to obtain coatings thinner than $\sim$250 μm and, therefore, fluidized-bed applied coatings are generally referred to as thick-film coatings, differentiating them from most conventional paintlike thin-film coatings applied from solution or as a powder at thicknesses of 20–100 μm (0.8–4 mils).

In the electrostatic spray process, the coating powder is dispersed in an air stream and passed through a corona discharge field where the particles acquire an electrostatic charge. The charged particles are attracted to and deposited on the grounded object to be coated. The object, usually metallic and at room temperature, is then placed in an oven where the powder melts and forms a coating. By using this process, it is possible to apply thin-film coatings comparable in thickness to conventional paint coatings, i.e., 20–75 μm. A hybrid process based on a combination of high voltage electrostatic charging and fluidized-bed application techniques (electrostatic fluidized bed) has evolved as well as triboelectric spray application methods.

Compared to other coating methods, powder technology offers a number of significant advantages. These coatings are essentially 100% nonvolatile; ie, no solvents or other pollutants are given off during application or curing. They are ready to use; ie, no thinning or dilution is required. Additionally, they are easily applied by unskilled operators and automatic systems because they do not run, drip, or sag, as do liquid (paint) coatings. The reject rate is low and the finish is tougher and more abrasion resistant than that of most conventional paints. Thicker films provide electrical insulation, corrosion protection, and other functional properties. Powder coatings cover sharp edges for better corrosion protection.

Coating powders are frequently separated into decorative and functional grades. Decorative grades are generally finer in particle size and color and appearance are important.

Coating powders are based on both thermoplastic and thermosetting resins. For use as a powder coating, a resin should possess: low melt viscosity, which affords a smooth continuous film; good adhesion to the substrate; good physical properties when properly cured; eg, high toughness and impact resistance; light color, which permits pigmentation in white and pastel shades; good heat and chemical resistance; and good weathering characteristics.

The volume of thermosetting powders sold exceeds that of thermoplastics by a wide margin. Thermoplastic resins are almost synonymous with fluidized-bed applied thick-film functional coatings and find use in coating wire, fencing, and corrosion resistant applications, whereas thermosetting powders are used almost exclusively in electrostatic spray processes and applied as thin-film decorative and corrosion resistant coatings.

THERMOPLASTIC COATING POWDERS

Thermoplastic resins used in coating powders must melt and flow at the application temperatures without significant degradation. The principal polymer types are based on plasticized poly(vinyl chloride) (PVC), polyamides, and other specialty thermoplastics.

THERMOSETTING COATING POWDERS

Thermosetting coating powders, with minor exceptions, are based on resins that cure by addition reactions. Thermosetting resins are more versatile than thermoplastic resins in the formulation of coating powders in that: many types are available varying in melt viscosity, functional groups, and degree of functionality; numerous cross-linking agents are available, thus the properties of the applied film can be readily modified; the resin/curing agent system possess a low melt viscosity allowing application of thinner, smoother films, and necessary level of pigments and fillers required to achieve opacity in the thin films can be incorporated without unduly affecting flow; gloss, textures, and special effects can be produced by modifying the curing mechanism or through the use of additives; and manufacturing costs are lower because compounding is carried out at lower temperatures and the resins are friable and can be ground to a fine powder without using cryogenic techniques.

Formulation

In addition to the binder resin(s), curing agents, which can range from crystalline solids to polymers, flow agents, additives, pigments, and fillers are utilized in coating powders. Care must be exercised when using powders with differing flow additives or significantly different binder resins since cross-contamination can occur resulting in loss of gloss, surface imperfections, and loss of smoothness.

Fillers such as calcium carbonate, blanc fixe and barium sulfate and wollastonite are used in coating powders to modify gloss, hardness, permeability, and other coated film characteristics and to reduce costs. Clays and talcs are seldom used, except in textured coatings, because of their high binder demand and adverse affect on flow and smoothness. Silicas are usually avoided due to their abrasiveness during extrusion and grinding, with the exception of colloidal silica used as a postextrusion additive.

Matting or flattening agents are employed to control gloss, which is dependent on microscopic surface smoothness. There are many methods of gloss control available, covering the full range of gloss in both interior and exterior durable systems.

Special Finishes

Clear coatings are formulated using curing agents and flow additives, which have a high degree of compatibility with the resin. Conventional uv and hindered amine light stabilizers can be added to improve exterior durability. Metallic finishes can also be prepared.

Thermosetting coating powders based on epoxy resins $C_{15}H_{16}O_2 \cdot (C_3H_5ClO)$, have been used longer than any other resin system.

Many other types of curing agents have been evaluated in formulating epoxy-based coating powders but have found use in only specialized applications.

MANUFACTURE

The vast majority of thermosetting coating powders are prepared by melt mixing, but some thermoplastic powders are produced by this method as well as the dry blend process, as shown in Figure 1.

APPLICATION METHODS

Applications include fluidized-bed coating, electrostatic fluidized-bed coating, electrostatic spray coating, and hot flocking.

ECONOMIC ASPECTS

Powder coatings represent only ~4–5% of the industrial paint market in North America. Globally, the penetration of powder coatings in the industrial paints sector is 6%, with Europe leading at ~9%. By 2010, global penetration of powder coatings is expected to reach the 10% level.

The vast majority of acrylic powders in both Europe and North America find use in the automotive industry, as clear coats for wheels and bodies and as pigmented coatings in exterior trim, "blackout" coatings, and primer surfacers.

The automotive market is the largest and fastest-growing segment of the powder coating market in North America accounting for over 16% market share, followed by appliance coatings at 15%, architectural at 3%, lawn and garden at 7%, and general metal finishing at 58%.

ENVIRONMENTAL AND ENERGY CONSIDERATIONS

In addition to the environmental advantages, the low volatile emissions of powder coatings during the baking operation have economic and energy saving advantages. Fewer air changes per hour in the baking oven are required for powder coatings than for solvent-based coatings, which saves fuel. Further, in the coating operation almost all powder is recovered and reused, resulting in higher

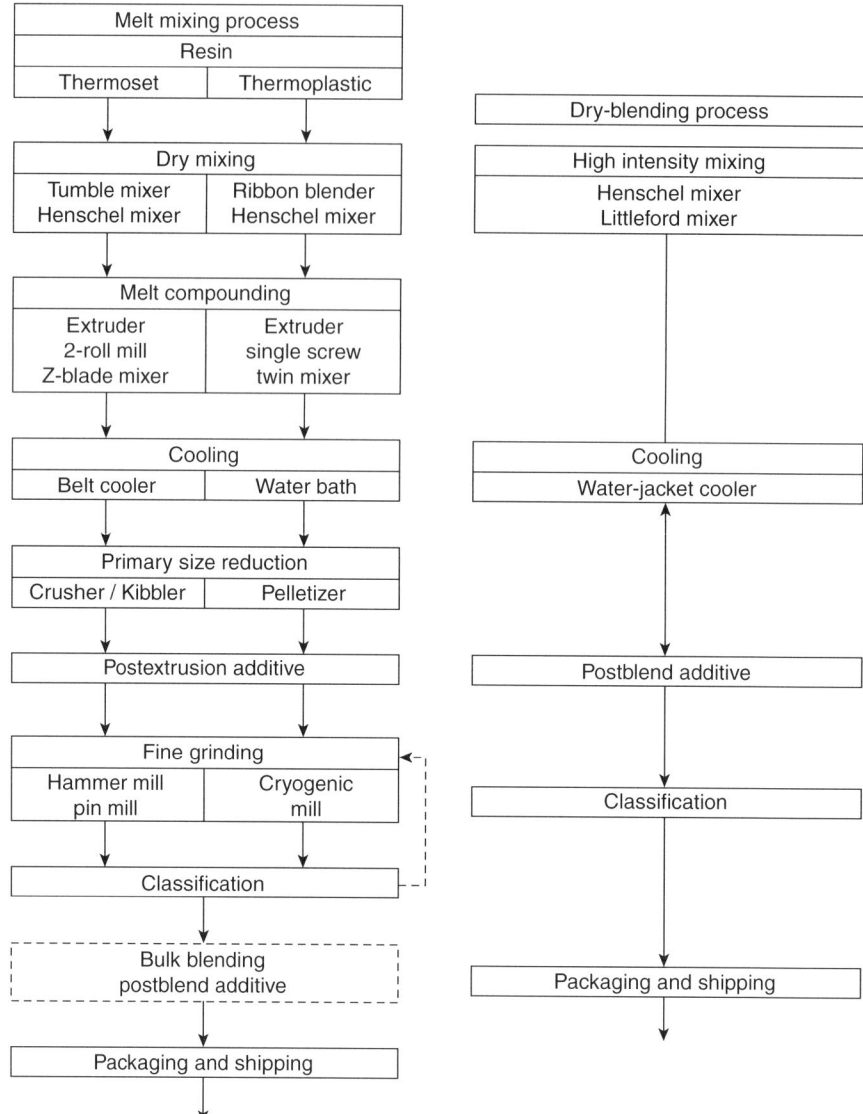

Figure 1. Flow diagram for coating powder manufacture.

material utilization, and waste minimization. The air used in the coating booths during application is filtered and returned to the workplace atmosphere, reducing heating and cooling demands. Additionally, because of the need for more sophisticated devices to control emission of volatile organic compounds (VOCs) in liquid systems, the capital investment to install a new powder coating line is becoming increasingly more economically favorable. The savings in material and energy costs of powder coating systems has been documented in a number of studies.

The only components in a coating powder that might cause the waste to be classified as hazardous are certain heavy-metal pigments sometimes used as colorants.

HEALTH AND SAFETY FACTORS

The most significant hazard in the manufacture and application of coating powders is the potential of a dust explosion. The severity of a dust explosion is related to the material involved, its particle size, and its concentration in air at time of ignition. The lower explosive limit

(LEL) is the lowest concentration of a material dispersed in air that explodes in a confined space when ignited. In powder coating installations, the design of the spray booth and duct work, if any, should be such that the powder concentration in air is always kept below the LEL, employing a wide margin of safety. The use of flame detection systems in all automatic powder coating installations is required.

The health hazards and risk associated with the use of powder coatings must also be considered. Practical methods to reduce employee exposure to powder, such as the use of long-sleeved shirts, and gloves to prevent skin contact, should be observed. Furthermore, exposure can be minimized by good maintenance procedures to monitor and confirm that the spray booth and dust collection systems are operating as designed. Ovens should be properly vented and operated under negative pressure so that any volatiles released during curing, eg, caprolactam, do not enter the workplace atmosphere.

In general, the raw materials used in the manufacture of powder coatings are relatively low in degree of hazard. None of the epoxy, polyester, or acrylic resins normally used in the manufacture of thermoset powder coatings are defined as hazardous materials by the OSHA Hazard Communication Standard. Most pigments and fillers used in powder coatings generally have no hazards other than those associated with particulates.

Although coating powders do not appear to pose significant hazards to personnel working with them, worker exposure should nevertheless be minimized, primarily through environmental controls.

D. A. Bate, *The Science of Powder Coatings*, Vol. 1, *Chemistry Formulation and Application*, 1990, Vol. 2, *Applications* 1994. Sita Technology, London.

D. Howell, *The Technology, Formulation and Application of Powder Coatings*, Sita Technology Ltd., London, Vol. 1, 2000.

N. P. Liberto, ed., *Powder Coating—The Complete Finishers Handbook*, 2nd ed., The Powder Coating Institute, Alexandria, Va., 1999.

T. Misev, *Powder Coatings—Chemistry and Technology*, John Wiley & Sons, Inc., New York, 1991.

DOUGLAS S. RICHART
D.S. Richart Associates

COATING PROCESSES, SPRAY

A coating may be applied to articles, ie, workpieces, by spraying. This application method is especially attractive when the articles have been previously assembled and have irregularly shaped and curved surfaces. The material applied is frequently a paint (qv), ie, a combination of resin, solvent, diluent, additives, and pigment. The material can also be a hot thermoplastic, an oil, or a polymer dissolved in a solvent. Many types of spray equipment are available. Methods can be used in combinations, and most of the techniques can be used for simple one-applicator manual systems or in highly complex computer-controlled automatic systems having hundreds of applicators. In an automatic installation, the applicators can be mounted on fixed stands, reciprocating or rotating machines, or even robots.

ATOMIZATION

Airless Atomization

In airless or pressure-atomizing systems, the coating is atomized by forcing the coating (or the liquid) through a small-diameter nozzle under high pressure. The fluid pressure is typically between 5 and 35 MPa (700–5000 psi); fluid flow rates are between 150–1500 cm^3/min. In most commercial applications, a pump designed for the type of material sprayed is used to develop the high pressure.

A variation of airless atomization is called air-assisted airless. A small amount of compressed air at 35–170 kPa (5–25 psi) is introduced adjacent to the airless nozzle and impinges upon the thin sheet of fluid as it exits from the nozzle. This air aggravates the turbulence in the fluid and results in improved atomization at lower fluid pressures.

Often, material that cannot be properly atomized using straight airless atomization can be using the air-assisted airless method. In some cases, the introduction of the air allows some control of the fan size.

Air Atomization

In an air atomizer, an external source of compressed air, usually supplied at pressures of 70–700 kPa (10–100 psi), is used to atomize the liquid. Air atomization is perhaps the most versatile of all the atomization methods. It is used with liquids of low to medium viscosity, and flow rates of 50–1000 cm^3/min are common. Medium-to-fine particle sizes are produced, and the resulting surface finish is very good. It is sometimes difficult to penetrate small recessed areas, however, because the atomization air forms a barrier in the recess that the coating particles must then penetrate. When higher air pressures are used, air atomization produces considerable misting and overspray, which can be a disadvantage under some conditions. Air-atomizing devices can be of internal- or external-mix design.

Electrostatic Atomization

The atomization of the coating material by electrostatic forces occurs when an electrical charge is placed on a filament or thin sheet of coating, and the mutual repulsion of the charges tears the coating material apart. For this process to produce acceptable atomization, the physical properties of the coating material must be within a relatively narrow range. The material is charged by an external source of high voltage, either prior to or as it is forced to flow over a knife edge, through a thin slot, or orifice, or it is discharged from the edge of a slowly rotating disk or bell (cup). The thin sheet or small diameter stringers or cusps of coating material are torn apart by the mutual repulsion of the charges on the material.

Rotary Atomization

In rotary atomization, a bell (cup) or disk rotates at a speed of 10,000–40,000 rpm. In contrast to electrostatic atomization, mechanical forces dominate. The coating material is introduced near the center of the rotating device, and centrifugal force distributes it to the edge, where the material has an angular velocity close to that of the rotating member. As the coating material leaves the surface, its main velocity component is tangential, and it is spun off in the form of a thin sheet or small cusps. The material is then atomized by turbulent or aerodynamic disintegration, depending on exact conditions. Rotary atomization produces the most uniform atomization of any of the aforementioned techniques, and produces the smallest maximum particle size. It is almost always used with electrostatics to aid deposition, and at lower rotational speeds the electrostatics assist the atomization.

Recent developments in rotary atomization include the use of semiconductive composites for the rotary cup, permitting the construction of a unit that does not produce an ignition spark when brought close to a grounded workpiece yet has the transfer efficiencies associated with a rotary atomizer.

ELECTROSTATIC SPRAYING

Use of electrostatic spraying or electrostatic deposition increases the efficiency of material transfer to the workpiece (see Coating processes, powder technology). The cost of solvents and coating materials and the emphasis on reducing emissions to the atmosphere have both increased dramatically since the late 1970s. These factors have effected an emphasis on increased transfer efficiency, ie, the fraction of the material removed from the coating bucket that is placed on the workpiece. The transfer efficiency is affected by the painting technique, workpiece geometry, the coating material, how the workpiece is presented to the atomizer, the ambient air movement, and other variables.

Electrostatic forces can be very effective in increasing the transfer efficiency. An electrical charge, usually negative, is placed on the coating material before atomization or as the coating particles are being formed. This is accomplished either by direct charging, where the coating material comes in contact with a conductor at high voltage, or by an indirect method, where the air in the vicinity of the coating particles is ionized and these ions then attach themselves to the coating particles.

An electrostatic force is exerted on each coating particle equal to the product of the charge it carries and the field gradient. The trajectory of the particle is determined by all the forces exerted on the particle. These forces include momentum, drag, gravity, and electrostatics. The field lines influencing the coating particles are very similar in arrangement to the alignment of iron particles when placed between two magnets. Using this method, coating particles that would normally pass alongside the workpiece are attracted to it, and it is possible to coat part or all of the back side of the workpiece.

All of the atomization techniques that produce a spray can be used with electrostatic spraying.

J. M. Lipscomb, *Surface Coating '83*, Chemical Coaters Association, *Finishing '83* Conference Proceedings, Association for Finishing Processes of SME, pp. 10–1,10-10.

F. A. Robinson, Jr., G. Pickering, and J. Scharfenberger, *Paint Con. '84*, Hitchcock Publishing Co., Carol Stream, Ill., 1984.

J.Schrantz and J. M. Bailey, *Indust. Finish.* (June 1989).

K. J. COELING
Nordson Corporation

COATINGS

Coatings are ubiquitous in an industrialized society. Coatings are used for decorative, protective, and functional treatments of many kinds of surfaces. The low gloss paint on the ceiling of a room is used for decoration, but also diffuses light. Exterior automobile coatings fulfill both decorative and protective functions. Still others provide friction control on boat decks. Some coatings control the fouling of ship bottoms, others protect food and beverages in cans. Other coatings reduce growth of barnacles on ship bottoms, protect optical fibers against abrasion, etc.

Each year tens of thousands of coating types are manufactured. In general, these are composed of one or more resins, a mixture of solvents (except in powder coatings), commonly one or more pigments, and frequently one or more additives. Coatings can be classified into thermoplastic and thermosetting coatings. Thermoplastic coatings contain at least one polymer having a sufficiently high molecular weight to provide the required mechanical strength properties without further polymerization. Thermosetting coatings contain lower molecular weight polymers that are further polymerized after application in order to achieve the desired properties.

FILM FORMATION

Coatings are manufactured and applied as liquids and are converted to solid films after application to the substrate. In the case of powder coatings, the solid powder is converted after application to a liquid, which in turn forms a solid film. The polymer systems used in coatings are amorphous materials and therefore the term "solid" does not have an absolute meaning, especially in thermoplastic systems such as lacquers, most plastisols, and most latex-based coatings. A useful definition of a solid film is that it does not flow significantly under the pressures to which it is subjected during testing or use. Thus a film can be defined as a solid under a set of conditions by stating the minimum viscosity at which flow is not observable in a specified time interval. For example, it is reported that a film is dry-to-touch if the viscosity is greater than $\sim 10^6$ mPa$\cdot$s($=$cP). However, if the definition of dry is that the film resists blocking, ie, sticking together, when two coated surfaces are put against each other for 2 s under a mass per unit area of 1.4 kg/cm^3 (20 psi), the viscosity of the film has to be $>10^{10}$ mPa$\cdot$s($=$cP).

For practical coatings, it is not sufficient just to form a film; the film must also have a minimum level of strength, depending on product use. Film strength depends on many variables, but one critical factor is molecular weight. This weight varies according to the chemical composition of the polymer and the mechanical properties required for a particular application.

Solvent Evaporation from Solutions of Thermoplastic Polymers

A solution of a copolymer of vinyl chloride (chloroethene), vinyl acetate (acetic acid ethenyl ester), and a hydroxy-functional vinyl monomer having a number average molecular weight (M_n) of 23,000 and a T_g of 79°C, gives coatings having good mechanical properties without crosslinking. A simple coating having only the resin and 2-butanone (methyl ethyl ketone, MEK) (C_4H_8O) as the sole solvent would give a polymer concentration of ~ 19 wt% solids or ~ 12 vol% in order to have a viscosity of ~ 100 mPa$\cdot$s($=$cP) for spray application. Because of the relatively high vapor pressure under application conditions, MEK evaporates rapidly and a substantial fraction of the solvent evaporates in the time interval between the coating leaving the orifice of the spray gun and its arrival on the surface being coated. As the solvent evaporates, the viscosity increases and the coating reaches the dry-to-touch stage very rapidly after application and does not block under the conditions discussed. However, if the film is formed at 25°C, the dry film contains several percent retained solvent.

In the first stages of solvent evaporation from such a film, the rate of evaporation depends on the vapor pressure at the temperatures encountered during the evaporation, the ratio of surface area to volume of the film, and the rate of air flow over the surface and is essentially independent of the presence of polymer. However, as the solvent evaporates the T_g increases, free volume decreases, and the rate of loss of solvent from the film become at some point dependent not on how fast the solvent evaporates but on how rapidly the solvent molecules can diffuse to the surface of the film. In this diffusion-control stage, solvent molecules must jump from free-volume hole to free-volume hole to reach the surface where evaporation can occur. As solvent loss continues, T_g increases and free volume decreases. When the T_g of the remaining polymer solution approaches the temperature at which the film is being formed, the rate of solvent loss becomes very slow.

The rate of solvent diffusion through the film depends not only on the temperature and the T_g of the film but also on the solvent structure and solvent–polymer interactions.

Film thickness is a factor in solvent loss and film formation. In the first stage of solvent evaporation, the rate of solvent loss depends on the first power of film thickness. However, in the second stage, when the solvent loss is diffusion rate controlled, it depends on the square of the film thickness.

Thermoplastic polymer-based coatings have low solids contents because the relatively high molecular weight requires large amounts of solvent to reduce the viscosity to that required for application. Air pollution regulations limiting the emission of volatile organic compounds (VOCs) and the increasing cost of solvents has led increasingly to replacement of such coatings with types that require less solvent for application.

Film Formation from Solutions of Thermosetting Resins

Substantially less solvent is required in formulating a coating from a low molecular weight resin that can be further polymerized to a higher molecular weight after application to the substrate and evaporation of the solvent. Theoretically, difunctional reactants could be used. However, this is not feasible for coatings where the close control of stoichiometric ratio and purity required to achieve a desired molecular weight reproducibly with difunctional reactants is impractical. Therefore, the average functionality must be >2 in order to ensure that the molecular weight of the final cured film is high enough for good properties. Not only should the average functionality be >2, it is usually preferable for the number of monofunctional molecules to be at a minimum because these terminate polymerization. If any of the resin molecules have no functional groups, they cannot react and remain in the film as a plasticizer. The reactions are commonly called cross-linking reactions. A cross-linked film not only has very high molecular weight, it is also insoluble in solvents. For many applications, this solvent resistance is an advantage of thermosetting coatings over thermoplastic coatings.

The mechanical properties of the cross-linked film depend on many factors; two of the most important are the lengths of the segments between cross-links and the T_g of the cross-linked resin. Segment length depends on the average equivalent weight and the average functionality of the components and the fraction of cross-linking sites actually reacted.

Film Formation by Coalescence of Polymer Particles

Latex paints have low solvent emissions as well as many other advantages. A latex is a stabilized dispersion of high molecular weight polymer particles in water. Because the latex polymer is not in solution, the rate of water loss by evaporation is almost independent of concentration until near the end of the evaporation process. When a dry film is prepared from a latex, the forces that stabilize the dispersion of latex particles must be overcome and the particles must coalesce into a continuous film. As the water evaporates, the particles come closer and closer together. As they approach each other, they can be thought of as forming the walls of capillary tubes in which surface tension leads to a force striving to collapse the tube. The smaller the diameter of the tube, the greater the force. When the particles get close enough together so that the force pushing them together exceeds the repulsive forces holding them apart, coalescence is possible. A surface tension driving force also promotes coalescence because of the decrease in surface area when the particles coalesce to form a film. Both factors have been shown to be important in film formation from latexes. Coalescence, however, also requires that the polymer molecules in the particles be free to intermingle with those from adjoining particles. This movement can occur only if there are a sufficient number and size of free-volume holes in the polymer particles into which the polymer molecules from other particles can move. In other words, the T_g of the latex particles must be lower than the temperature at which film formation is being attempted.

The rate of coalescence is controlled by the free-volume availability, which in turn depends mainly on $(T - T_g)$. The viscosity of the coalesced film is also dependent on free volume.

Powder coatings form films by coalescence. Because the powder must not sinter during storage, the free volume at storage temperature must be sufficiently low to avoid coalescence at this stage.

FLOW

Rheological properties; ie, flow and deformation, of coatings have significant impacts on application and performance properties. The application and film formation of liquid coatings require control of the flow properties at all stages. The mechanical properties of the applied coating films are controlled by the viscoelastic responses of the films to stress and strain.

Viscosity of Resin Solutions

The viscosity of coatings must be adjusted to the application method to be used. It is usually between 50 and 1000 mPa·s(= cP) at the shear rate involved in the application method used. The viscosity of the coating is controlled by the viscosity of the resin solution, which is in turn controlled mainly by the free volume. The factors controlling free volume are temperature, resin structure, solvent structure, concentration, and solvent–resin interactions.

Viscosity of Systems with Dispersed Phases

A large proportion of coatings are pigmented and therefore have dispersed phases. In latex paints, both the pigments and the principal polymer are in dispersed phases. The viscosity of a coating having dispersed phases is a function of the volume concentration of the dispersed phase and can be expressed mathematically by the Mooney equation.

$$\ln \eta = \ln \eta_e + \frac{K_E V_i}{1 - V_i/\phi}$$

where η_e is the viscosity of the external phase, K_E is a shape constant (2.5 for spheres), V_i is the volume fraction of internal phase, and ϕ is the packing factor, ie, the volume fraction of internal phase when the V_i is at the maximum close-packed state possible for the system. The Mooney equation assumes rigid particles having no particle–particle interaction. It fits pigment dispersions and latexes that exhibit Newtonian flow.

MECHANICAL PROPERTIES

Coating films should withstand use without damage. The coating on the outside of an automobile should not break when hit by a piece of flying gravel. The coating on aluminum siding must be flexible enough for fabrication of the siding and resist scratching during installation on a house. In ideal elastic deformation, a material elongates under a tensile stress in direct proportion to the stress applied. When the stress is released, the material returns to its original dimensions essentially instantaneously. An ideal viscous material elongates when a stress is applied in direct proportion to the stress but does not return to its original dimensions when the stress is released. Almost all coating films are viscoelastic—they exhibit intermediate behavior.

Many coated products are subjected to mechanical forces either to make a product, as in forming bottle caps or metal siding, or in use, as when a piece of gravel strikes the surface of a car with sufficient force to deform the steel substrate. To avoid film cracking, the elongation-at-break must be greater than the extension of the film.

When a cross-linked film on a metal substrate is deformed by fabrication, it is held in the deformed state by the metal substrate. As a result, there is a stress within the film acting to pull the film off the substrate. Stress within films can also arise during the last stages of solvent loss and/or cross-linking of films. It is common for coatings

to become less flexible as time goes on. Particularly in air dry coatings, loss of the last of the solvent may be slow. If the cross-linking reaction was not complete, the reaction may continue, decreasing flexibility. Another possible factor with baked films is densification. If a coating is heated above its T_g and then cooled rapidly, the density is commonly found to be lower than if the sample had been cooled slowly.

Abrasion is the wearing away of a surface, marring is a disturbance of a surface that alters its appearance. A study of the mechanical properties of a series of floor coatings with known wear life concluded that work-to-break values best represented the relative wear lives. Studies of automobile clear coats have shown that wear resistance increases as energy-to-break of films increases. The coefficient of friction of a coating can also affect abrasion resistance.

Marring is a near-surface phenomenon; even scratches <0.5 µm deep can degrade appearance. Marring is a major problem with automobile clear topcoats. In going through automatic car washes, the surfaces of some clear coats are visibly marred and lose gloss. Plastic deformation and fracture lead to marring. In general, MF cross-linked acrylic clear coats are more resistant to marring than isocyanate cross-linked coatings, but MF cross-linked coatings have poorer environmental etch resistance. Coatings can be made hard enough that the marring object does not penetrate into the surface, or they can be made elastic enough to bounce back after the marring stress is removed.

Many empirical tests are used to test coatings. In most cases, they are more appropriate for quality control than performance prediction.

EXTERIOR DURABILITY

For many coatings, an important performance requirement is exterior durability. There are many potential modes of failure when coatings are exposed outdoors. Commonly, the first indication of failure is reduction in gloss, resulting from surface embrittlement and erosion leading to the development of roughness and cracks in the surface of the coating. In some cases, the next step is "chalking," the erosion of resin from the surface of the coating leaving loose pigment particles on the surface. Various kinds of chemical attack, such as those resulting from acid rain and bird droppings, can result in film degradation and discoloration. Although many failure mechanisms are involved, the two most common modes are hydrolysis and photochemical oxidation by free-radical chain reactions. In general, resins that have backbone linkages that cannot hydrolyze provide better exterior durability than systems having, eg, ester groups in the backbone.

Susceptibility to free-radical induced photoxidation varies with structure of the resins and pigments and, in some cases, with the interactions between pigment and resin. In general terms, resistance of resins to photochemical failure is related to the ease of abstraction of hydrogens from the resin molecules by free radicals. The greatest resistance is shown by fluorinated resins and silicone resins, especially methyl-substituted silicone resins. The greatest sensitivity to degradation is shown by resins having methylene groups between two double bonds; methylene groups adjacent to amine nitrogens, ether oxygens, or double bonds; and methine groups.

Pigment selection can also be critical in formulating for high exterior durability. Some pigments are more susceptible to color change on exposure than others. Some pigments act as photosensitizers to accelerate degradation of resins in the presence of uv and water.

The exterior durability of relatively stable coatings can be enhanced by use of additives. Ultraviolet absorbers reduce the absorption of uv by the resins, and hence decrease the rate of photodegradation. Further improvements can be gained by also adding free-radical trap antioxidants such as hindered phenols and especially hindered amine light stabilizers (HALS).

ADHESION

In most cases, it is desirable to have a coating that is difficult to remove from the substrate to which it has been applied. An important factor controlling this property is the adhesion between the substrate and the coating. The difficulty in removing a coating also can be affected by how difficult it is to penetrate through the coating and how much force is required to push the coating out of the way as the coating is being removed from the substrate as well as the actual force holding the coating onto the substrate. Furthermore, the difficulty of removing the coating can be strongly affected by the roughness of the substrate. If the substrate has undercut areas that are filled with cured coating, a mechanical component makes removal of the coating more difficult. This is analogous to holding two dovetailed pieces of wood together.

Surface roughness affects the interfacial area between the coating and the substrate. The force required to remove a coating is related to the geometric surface area, whereas the forces holding the coating onto the substrate are related to the actual interfacial contact area. Thus the difficulty of removing a coating can be increased by increasing the surface roughness. However, greater surface roughness is only of advantage if the coating penetrates completely into all irregularities, pores, and crevices of the surface. Failure to penetrate completely can lead to less coating-to-substrate interface contact than the corresponding geometric area and leave voids between the coating and the substrate, which can cause problems.

Adhesion to Metals

For interaction between coating and substrate to occur, it is necessary for the coating to wet the substrate. Somewhat oversimplified, the surface tension of the coating must be lower than the surface tension of the substrate. In the case of metal substrates, clean metal surfaces have very high surface tensions and any coating wets a clean metal substrate.

Penetration of the vehicle of the coating as completely as possible into all surface pores and crevices is critical to achieving good adhesion. The critical viscosity is that of the continuous phase because many of the crevices in the surface of metal are small compared to the size of pigment particles. Because penetration takes time, the initial viscosity of the external phase should be low and the viscosity should be kept as low as possible for as long as possible. Slow-evaporating solvents are best for coatings that are to be applied directly on metal.

Adhesion is strongly affected by the interaction between coating and substrate. On a clean steel substrate, hydrogen bond or weak acid–weak base interactions between the surface layer of hydrated iron oxide that is present on any clean steel surface and polar groups on the resin of the coating provide such interaction.

Fracture mechanics affect adhesion. Fractures can result from imperfections in a coating film that act to concentrate stresses. In some cases, stress concentration results in the propagation of a crack through the film, leading to cohesive failure with less total stress application. Propagating cracks can proceed to the coating–substrate interface, then the coating may peel off the interface, which may require much less force than a normal force pull would require.

Adhesion of coatings is also affected by the development of stresses as a result of shrinkage during drying of the film.

The formation of covalent bonds between resin molecules in a coating and the surface of the substrate can enhance adhesion.

Adhesion to Plastics and Coatings

Wetting can be a serious problem for adhesion of coatings to plastics. Some plastic substrates have such low surface tension that it may be difficult to formulate coatings having a sufficiently low surface tension to wet the substrate. Polyolefin plastics, in particular, are difficult to wet. Frequently, the surface of the polyolefin plastic must be oxidized to increase surface tension and provide groups to interact with polar groups on the coating resin. Adhesion to plastics can be enhanced if resin molecules from the coating can penetrate into the surface layers of the plastic.

The same considerations apply to intercoat adhesion to other coatings as to plastics.

Testing for Adhesion

Because of the wide range of exposures to stresses in actual use, there is no really satisfactory laboratory test for the adhesion of a coating film to the substrate during use. A useful guide for an experienced coatings formulator is the use of a penknife to see how difficult it is to remove the coating and to observe its mode of failure. Many tests for adhesion have been devised. From the standpoint of obtaining a measurement related to the work required to separate the coating from the substrate, the direct pull test is probably the most widely used. The accuracy of the test is subject to considerable doubt and the precision is not very good. Even for experienced personnel, reproducibility variations of 15% or more must be expected. A compressive shear delamination test has proven to be useful in studying the adhesion and cohesion properties of clear coat–base coat–primer coatings on TPO. The most common specification test, cross-hatch adhesion, is of little value beyond separating systems having very poor adhesion from others.

CORROSION CONTROL

An important function of many coatings is to protect metals, especially steel, against corrosion. Corrosion protection is required to protect steel against corrosion with intact coating films. Another objective is to protect the steel against corrosion even when the film has been ruptured.

Protection by Intact Films

In the case of intact films, the key factors responsible for corrosion protection are adhesion of the coating to the steel in the presence of water, oxygen and water permeability, and the resistance of the resins in the film to saponification. If the resins in the coating are adsorbed to cover the steel surface completely and if the adsorbed groups cannot be displaced by water, oxygen and water permeating through the film cannot contact the steel and corrosion does not occur. Only a monolayer is required to protect against corrosion, if the monolayer stays in place over the period of exposure.

Protection by Nonintact Films

It is also possible to achieve corrosion control by coatings after a film has been ruptured. Because the coatings used to achieve this control generally give poorer protection when their films are not ruptured, such systems should be used only when film rupture must be anticipated or when complete coverage of the steel interface cannot be achieved. There are two techniques used on a large scale: primers containing corrosion inhibiting pigments and zinc-rich primers.

RESINS FOR COATINGS

Latexes

Latexes are aqueous dispersions of solid polymer particles, generally made by emulsion polymerization. Latex paints are sometimes called emulsion paints but that terminology should be avoided since there are paints made with emulsions of resin solutions.

Latexes are the principal vehicle of a large fraction of architectural coatings and a small but rapidly growing fraction of industrial and special-purpose coatings. The largest volume of latexes are polymers of acrylic esters, of which almost all will form films at ambient temperatures.

Acrylic latexes are widely used in exterior latex coatings and in higher-performance interior coatings. Vinyl acetate based latexes are widely used in interior coatings such as flat wall paint because of their lower cost. Latex coatings form films by coalescence. It has been common to reduce film formation temperature by including a coalescing solvent in a formula. That is a solvent that dissolves in the latex particles reducing their T_g that slowly diffuses and evaporates out of the film after application.

There is increasing use of thermosetting latexes; i.e., latexes having functional groups that can be cross-linked after application. Such latexes are prepared with lower T_g to permit lower temperature film formation and then cross-linked after application to increase their modulus. Some are used in two package coatings; eg, hydroxy-functional latexes with urea–formaldehyde or melamine–formaldehyde resins.

Thermosetting latexes stable enough at ambient temperature for use in architectural coatings are more difficult to make. Several workers have prepared latexes with allylic substituents that cure by autoxidation.

Amino Resins

Melamine–formaldehyde (MF) resins are the most widely used cross-linking agents for baking enamels. They are made by reacting melamine (1,3,5-triazine-2,4,6-triamine) and formaldehyde, followed by etherification of the methylol groups using an alcohol. Two classes of MF resins are used. Class I resins are made using excess formaldehyde and a high fraction of the methylol groups are etherified with alcohol.

Class II MF resins are made using a lower ratio of formaldehyde to melamine and a significant fraction of the nitrogens have one alkoxymethyl group and a hydrogen.

Urethane Systems

Isocyanates react with a wide variety of functional groups to give cross-links. The most widely used coreactants are hydroxy-functional polyester and acrylic resins. The isocyanate group reacts with the hydroxyl group to generate a urethane cross-link. The reaction proceeds relatively rapidly at ambient or modestly elevated temperatures. In addition to the low curing temperatures, significant advantages of urethane coatings are generally excellent abrasion and impact resistance combined with solvent resistance.

Epoxy Resins

Epoxy resins are used to cross-link other resins with amine, hydroxyl, carboxylic acid, and anhydride groups. The epoxy group, properly called an oxirane, is a cyclic three-membered ether group. By far the most widely used epoxy resins are bisphenol A (BPA) [4,4-(1-methylethylidene)bisphenol], epoxy resins.

An important use for epoxy resins is as a component in two-package primers for steel. One package contains the epoxy resin and the other a polyfunctional amine. In coatings, generally low molecular weight polyamines are not useful because the equivalent weight is so low that the ratio of the two components would be very high, increasing the probability of mixing ratio errors; furthermore, low molecular weight amines tend to have greater toxic hazards. Amine-terminated polyamides, sometimes called amido-amines but frequently just polyamides, are wodely used. Amide groups do not react readily with epoxy groups.

Epoxy resins also are used to cross-link phenolic resins. Such coatings are used in interior can linings. The hydrolytic stability and adhesion of the coatings are critical. Adhesion is further improved by the incorporation of a small amount of epoxy phosphate in the coatings.

Epoxy resins are widely used in powder coatings. Probably the largest volume usage is of BPA epoxy resins cross-linked with dicyanodiamide (cyanoguanidine). Because BPA epoxy resins are easily photoxidized, they are not useful in coatings requiring exterior exposure. Triglycidylisocyanurate (TGIC) [1,3,5-tris(oxiranylmethyl)-1,3,5-triazine-2,4,6(1H,3H,5H)-trione] has been used in carboxylic acid terminated polyesters in powder coatings that require exterior durability.

Epoxy resins are raw materials to make epoxy esters by reacting BPA epoxy resins and drying oil fatty acids; each epoxy ring can potentially react with two fatty acid molecules and hydroxyl groups on the backbone of the epoxy resin can also esterify.

Acrylic Resins

Acrylic resins are the largest volume class of coatings resins. Thermosetting acrylic resins are copolymers of acrylic or methacrylic esters and a hydroxy-functional acrylic ester. Other monomers such as styrene (ethenylbenzene), vinyl acetate, and others may be included in the copolymer. There is increasing emphasis on high solids and water-reducible types. The main advantages of acrylic coatings involve the high degree of resistance to thermal and photoxidation and to hydrolysis, giving coatings that have superior color retention, resistance to embrittlement, and exterior durability.

Hydroxy-functional thermosetting acrylics are widely used in baking enamels for automobile and appliance topcoats, exterior can coatings, and coil coating.

Polyester Resins

The term "polyester" is used in the coatings field almost entirely for low molecular weight hydroxy, or sometimes carboxylic acid, terminated oil-free polyesters. Polyesters have been a class of replacements for alkyd resins in MF cross-linked baking enamels. Hydroxy-terminated polyesters are also used with polyfunctional isocyanates in making air-dry and force-dry coatings as well as with blocked isocyanates in coatings for higher baking temperature and in powder coatings. Carboxylic acid-terminated polyesters are used predominantly with epoxy cross-linkers in powder coatings. When adhesion directly to metal is required, polyesters are generally preferred over acrylic resins. When highest exterior durability is needed over primers, acrylic resins are generally preferred over polyesters because of the greater hydrolytic stability of acrylic resin coatings.

Alkyd Resins

Although no longer the principal class of resins in coatings, alkyds are still important and a wide range of types of alkyds are manufactured. Whereas some nonoxidizing alkyds are used as plasticizers in lacquers and cross-linked with MF resins in baking enamels, the majority are oxidizing alkyds for use in coatings for air-dry and force-dry applications. The principal advantages of alkyds are their low cost and relatively foolproof application characteristics, resulting from low surface tensions. The principal shortcomings of these resins are embrittlement and discoloration upon aging and relatively poor hydrolytic stability.

Other Coatings Resins

A wide variety of other resin types are used in coatings. Phenolic resins; ie, resins based on reaction of phenols and formaldehyde, have been used in coatings for many years. Silicone resins provide coatings having outstanding heat resistance and exterior durability. Polyfunctional 2-hydroxyalkylamides serve as cross-linkers for carboxylic acid-terminated polyester or acrylic resins. A range of acetoacetylated resins has been introduced.

VOLATILE COMPONENTS

In most coatings, solvents are used to adjust the viscosity to the level required for the application process and to provide for proper flow after application. Most methods of application require coating viscosities of 50–1000 mPa·s(= cP). Many factors must be considered in the selection of the solvent or, more commonly, solvent mixtures. Except for water-borne systems, solvents are usually chosen that dissolve the resins in the coating formulation. Solubility parameters have been recommended as a tool for selecting solvents that can dissolve the resins. However, the concept of solubility parameters for the prediction of polymer solubility is an oversimplification, and the old principle that like dissolves like is the most useful selection criterion. The problem of solvent selection is most difficult for high molecular weight polymers such as thermoplastic acrylics and nitrocellulose in lacquers. As molecular weight decreases, the range of solvents in which resins are soluble broadens. Even though solubility parameters are inadequate for predicting all solubilities, they can be very useful in performing computer calculations to determine possible solvent mixtures as replacements for a solvent mixture that is known to be satisfactory for a formulation.

An important characteristic of solvents is rate of evaporation. Rates of solvent loss are controlled by vapor pressure and temperature, partial pressure of the solvent over the surface, and thus the air-flow rate over the surface, and the ratio of surface area to volume. Tables of relative evaporation rates, in which n-butyl acetate is the standard, are widely used in selecting solvents.

Final adjustment of solvent selection must be done under actual field conditions.

Toxic hazards, environmental considerations, flammability, odor, surface tension, and viscosity also affect solvent selection.

PIGMENTS

Pigments in coatings provide opacity and color. Pigment content governs the gloss of the final films and can have important effects on mechanical properties. Some pigments inhibit corrosion. Pigmentation affects the viscosity, and hence the application properties of coatings. Pigment manufacturing processes are designed to afford the particle size and particle size distribution that provide the best compromise of properties for that pigment. In the process of drying the pigment, the particles generally aggregate. The coatings manufacturer must disperse these dry pigment aggregates in such a way as to achieve a stable dispersion where most, if not all, of the pigment is present as individual particles.

Pigments can be divided into four broad classes: white, color, inert, and functional pigments. The ideal white pigment, when dispersed in the coating, would absorb no visible light and would efficiently scatter light entering the film.

Color pigments selectively absorb some wavelengths of light more strongly than others. A wide variety of color pigments are used in coatings.

Inert pigments, also called extenders and fillers, do not exhibit significant absorption or scattering of light when incorporated into coatings. In most cases, inert pigments are used to occupy volume in the coating composition. A wide variety of clays, silica, and carbonates are used as inert pigments. The most widely used functional pigments are the corrosion-control pigments. These pigments inhibit the corrosion of steel.

Pigment Dispersion

The dispersion of pigments involves wetting, separation, and stabilization. Wetting; ie, displacement of air and water from the surface of the pigment by the vehicle, requires that the surface tension of the dispersion medium be lower than that of the pigment surface.

Pigment aggregates are separated into individual particles by a variety of dispersion equipment, which transmit shear stress of sufficient magnitude to break up the aggregates.

Pigment dispersions are stabilized by charge repulsion and entropic repulsion. Although both types of stabilization force may be present in most cases, for pigment dispersions in solvent-borne coatings entropic repulsion is usually the most important mechanism for stabilization.

Dispersion of pigments for latex paints is the principal application of aqueous dispersions. Commonly, three surfactants are used in preparing the dispersion of the white and inert pigments: potassium tripolyphosphate, an anionic surfactant, and a nonionic surfactant. The final colored latex paint formulations, which are very complex, commonly contain seven pigments, some having high surface energies (inorganic pigments) and some having low surface energies (organic pigments). The latex polymer is present as a dispersion that must be stabilized against coalescence and flocculation, and latexes themselves commonly contain two or more surfactants and a water-soluble polymer.

Pigment Volume Relationships

Pigmentation can have profound effects on the properties of coating films, depending on the level of pigmentation of these films. Variations in these effects are best interpreted in terms of volume relationships rather than weight relationships. Pigment volume concentration (PVC) is defined as the volume of pigment in a dry film divided by the total volume of the dry film, commonly expressed as a percentage.

As the volume of pigment in a series of formulations is increased, properties change, and at some PVC there is a fairly drastic change in a series of properties. This PVC is defined as the critical pigment volume concentration

(CPVC) for that system. The CPVC is the maximum PVC that can be present in a dry film of that system, having sufficient solvent-free resin to adsorb on all the pigment surfaces and fill all the interstices between the pigment particles. In other words, when the PVC is above the CPVC, there are voids in the film. Gloss decreases as PVC increases. Hiding generally increases as PVC increases, but above CPVC there is a rapid increase in the rate of increase of hiding because of the presence of voids. Tinting strength of white paints increases rapidly above CPVC. Tensile strength of films increases with increasing PVC, passing through a maximum at CPVC. Stain resistance is poorer and the ease of removing stains becomes more difficult for coatings above CPVC. Blistering of films on wood is less likely to occur above CPVC. The intercoat adhesion to a primer is improved when the primer has a PVC > CPVC.

Color Matching

Most pigmented coatings must be color matched to a standard. Poor color matching is a common source of customer complaints. Problems and costs can be minimized if established specifications for the initial color match and for judging the acceptability of production batches are made. Acceptance of a color recommendation made by the coatings supplier effectively eliminates the time and cost involved in an initial color match and ensures selection of a pigment combination appropriate to the coating use.

PRODUCT COATINGS

About 30% of the total volume of coatings produced in the United States in 1999 (1.45×10^9 L) were applied in factories to a very large variety of products ranging from automobiles to toys. They are often called original equipment market (OEM) coatings.

Coatings for Metal

A large fraction of the product coatings are applied to metal. The essential first step in metal coating is the preparation of the metal. Oil and related contaminants must be removed by detergent or solvent washing. Solvent degreasing is the most effective. In detergent washing, the last trace of detergent must be rinsed off before drying preparatory to painting. For best adhesion and corrosion resistance, the surface of the metal should be treated. For steel, phosphate conversion coating treatments are used. Aluminum, for applications when the product will not be exposed to salt, needs no treatment. Surface contaminats can result in crawling and/or blistering of the coating.

Primers. If reasonably high performance is required in the end product and unless cost is of paramount importance, a minimum of two coats, usually a primer and a topcoat, should be applied to metal. For highest performance, primer vehicles should provide good wet adhesion, be saponification resistant, and have low viscosity to permit penetration of the vehicle into microsurface irregularities in the substrate. Color, color retention, exterior durability, and other such properties are generally not important in primers. Resin systems such as those including BPA epoxy resins that provide superior wet adhesion can thus be used in spite of their poor exterior durability.

Electrodeposition Primers. Primers for automobiles and a significant part of primers for household appliances are applied by electrodeposition. Almost all electrodeposition primers are now cationic. The compositions are proprietary, but the vehicles in some primers are epoxy/amine resins neutralized with volatile organic acids such as lactic acid; an alcohol-blocked isocyanate is used as a cross-linking agent. The pigments are dispersed in the resin system and the coating is reduced with water.

Topcoats. The selection of a topcoat depends on cost, method of application, product use and performance requirements, among other factors. As a result of increasingly stringent air quality standards and increased solvent costs, approaches to reduction of solvent emissions are being sought.

High Solids Coatings. There is no agreement on a definition of high solids coatings. For any particular use, it means higher solids than previously used in that application. In automotive acrylic metallic coatings, high solids usually is taken to be ~45 vol%. In the case of clear coats, ~70 vol% can be achieved with reasonable film properties.

Whereas the main driving force behind the development of higher and higher solids coatings has been the reduction of VOC emissions, solvent cost is also a factor. A further advantage is that the same dry film thickness can be applied in less time. High solids coatings are made using lower molecular weight resins having fewer average functional groups per molecule as compared to conventional coatings. As a result, more complete reaction of the functional groups is necessary to achieve good film properties.

Waterborne Coatings. Two classes of waterborne systems are used: water-reducible and latex systems. Water-reducible systems are used on a larger scale, but the consumption of latex product coatings is increasing.

Because the molecular weight and average functionality of the resins used in water-reducible coatings are comparable to those of conventional solution thermosetting coatings, film properties obtained after curing are fully equivalent.

Coil Coatings. An important segment of the product coatings market is sold for application to coiled metal, both steel and aluminum. In this process, the metal is first coated and then fabricated into the final product rather than fabricating first into product. The method offers considerable economic advantages because coatings are applied by direct or reverse roll-coating at high (up to 400 m/min) speeds to wide (up to 3 m) coils in a continuous process. The metal is cleaned, conversion coated, and coated with primer (if desired) and topcoat in an in-line operation. Coatings can be applied to both sides of the metal during the same run through the coil coating line. The labor cost of coating application is much less than for application to a previously fabricated product. There is essentially 100%

effective utilization of coating, and the loss by overspray involved in coating fabricated products is avoided.

A wide range of resin compositions is used in coil coatings, depending on the product performance requirements and cost limitations. The lowest cost coatings are generally alkyd coatings. They are appropriate for indoor applications where color requirements are not stringent. Alkyd–MF coatings are sometimes used outdoors when long-term durability is not needed. Greater durability and better color retention on overbaking are obtained using polyester–MF.

For greater exterior durability, silicone-modified polyesters or silicone-modified acrylic resins are used.

In the can industry, large-volume three-piece cans, used for packing many fruits and vegetables, are made from coil-coated stock. Oil-modified phenolic resins are used for coating tinplated steel coils on the side that becomes the interior of the can. Coil stock for making ends for two-piece beverage cans is coated with epoxy–MF, epoxy–phenolic, or a cationic uv cure epoxy coating.

Wood Products Coating

Furniture is an important class of wood products that is industrially coated. The appearance standards are set by the fine furniture industry where the flat areas are composed of plywood having high quality top veneer; the legs and rails are solid wood; and there are frequently carved wood decorative additions. The finishing process for fine furniture is long and requires significant artistic skill. If the final overall color of the furniture is lighter than the color of any part of the wood, the first step is bleaching, using a solution of hydrogen peroxide in methanol. The bleached wood, or unbleached for darker color finishes, is given a wash coat of size to stiffen the fibrils so they can be cleanly removed by sanding. The wood is then coated with stain, a solution of acid dyes in methanol, to give a desired overall color tone to the piece of furniture. A wash coat, generally a low solids vinyl chloride copolymer lacquer, is applied over the stain, partially to seal the stain in place but also to prepare the surface for the next operation, filling. The filler, a dispersion of pigments, usually in linseed oil with mineral spirits solvent, is sprayed over the whole piece of furniture. It is then wiped off, leaving filler only in the pores of the wood. The colors of fillers are commonly dark brown; the filler serves to emphasize the grain pattern in the hardwood veneer. Next, shading stains are selectively sprayed on the wood to give different colors to various sections. It is common to distress the surface to resemble antique furniture. Then a sanding sealer is applied and finally a topcoat is applied and polished smooth.

The primary binder used for many years in the sanding sealer and topcoat is nitrocellulose. Topcoats generally include fine particle-size silicon dioxide to reduce the gloss. In making a low gloss topcoat, it is essential to retain the transparency.

Considerable efforts have been invested in developing waterborne coatings for wood furniture. The two approaches are water-reducible acrylic resins with UF cross-linkers and hydroxy-functional vinyl acetate copolymer latexes with UF cross-linking.

Uv-cure coatings are widely used in European furniture manufacture but have found more limited applications in the United States.

The other principal component of industrial wood finishing is the panel industry. The highest cost segment of this industry, fine hardwood veneer paneling for executive office walls, is comparable to the fine furniture industry where similar nitrocellulose lacquer finishing systems are used. However, the bulk of the industry requires less expensive finishing operations. Some plywood is stained followed by roll coating with a low gloss nitrocellulose lacquer topcoat. It is common to print grain patterns from woods such as walnut onto inexpensive, relatively featureless veneers such as luan before applying a lacquer topcoat. Lacquer topcoats are being replaced by alkyd–UF coatings.

ARCHITECTURAL COATINGS

About 45% of the volume of all coatings produced in the United States in 1999 were architectural coatings (2.16×10^9 L). These coatings are designed to be applied to residences and offices, and for other light-duty building purposes. In contrast to most product coatings, they are designed to be applied in the field, in some cases by contractors but in large measure by do-it-yourself consumers. A wide range of products is involved.

Flat Wall Paint

The largest volume of architectural coatings is flat wall paint. In the United States, almost all flat wall paint is latex-based. Latex paints have the advantages of low odor, fast drying, easy clean-up when wet, durability of color and film properties, and lower VOC emissions as compared to oil- or alkyd-based paints. They are manufactured primarily as white base paints, which are tinted by the retailer to the color selected by the customer from large collections of color chips. Two or three base whites are made, one for pastel shades, one for deep shades, and sometimes one for medium color shades. This method of marketing has the advantage of being able to supply a wide choice of colors while carrying a relatively low inventory. It is critical to maintain the same level of hiding when formulating new white base paints.

Exterior House Paints

Latex paints dominate the exterior house paint market in the United States because of superior exterior durability of latex paints (resistance to chalking, checking, and cracking) compared to oil or alkyd paints. Another advantage of latex paints used on wood surfaces is that because of their high moisture vapor permeability they are much less likely to blister than oil or alkyd paints.

Another application where latex paints show outstanding performance is over masonry such as stucco or cinder block construction. This performance results from saponification resistance in the presence of the alkali from the cement.

There are limitations to the applicability of exterior latex house paints providing a continuing market for oil or alkyd exterior house paints. Because film formation from latex paints occurs by coalescence, there is a temperature limit below which the paint should not be applied. In the United States, most latex paints are formulated for application at temperatures $>5-7°C$. If painting must be done when the temperature is lower, oil or alkyd paint is preferable.

Another limitation is that latex paints do not give good adhesion when applied over a chalky surface such as weathered oil or alkyd paint.

Gloss Enamels

Professional painters particularly favor the continued use of alkyd gloss paints. The need for reduction of VOC emission levels, especially in California, has led to efforts to increase the solids content of alkyd paints or overcome the disadvantages of latex gloss paints. It is not possible to make latex enamels that have as high a gloss as solution-based coatings.

An important limitation of gloss latex paints is not gloss, but rather the greater difficulty of getting adequate hiding from one coat.

The largest factor affecting hiding by gloss latex paints is probably poor leveling.

The rheological properties of gloss latex paints influence leveling and hiding. Latex paints have exhibited a much higher degree of shear thinning than alkyd gloss paints, leading to paints having viscosity that is too low at high shear rate and a subsequent applied film thickness that is too thin. Latex paints formulated with associative thickeners have increased high shear viscosity allowing the application of thicker wet films. The thickeners afford reduced low shear rate viscosity and a slower rate of recovery of the low shear rate viscosity that improves leveling at the same time.

Another shortcoming of latex paints, which is particularly evident in gloss paints, is the time required to develop full film properties. Latex paints dry to touch and even to handling much more rapidly than do alkyd paints, but the latex requires a much longer time to reach the full dry properties.

Another important potential problem for use of gloss latex paints is adhesion to an old gloss paint surface when water is applied to the new dry paint film.

SPECIAL PURPOSE COATINGS

Special purpose coatings include those coatings that do not fit under the definition of architectural or product coatings. About 25% of the volume (1.22×10^9 L) of U.S. coatings production in 1999 falls into this category.

They include heavy-duty maintenance coatings are applied to bridges, off-shore drilling rigs, chemical or petroleum refinery tanks, and similar structures plus automotive refinish paints.

ECONOMIC ASPECTS

The value of coatings has been growing, but the volume of coatings has been growing slowly or declining. Coating volume is also affected by the need to reduce Voc and by other technologies that can substantially reduce the need for coatings. For example, in many cases coatings are not required on molded plastics products that have replaced coated metal products. High-pressure laminates are increasingly used as furniture tops. In some cases, the replacements have been from one kind of coating to another.

The effect of environmental regulation and the increasing recognition of immediate or potential toxicity hazards of coating components has led to a technical revolution in the coatings field and will continue as strictness of environmental regulations increases. In general, requirements in different countries are quite different and there is generally a need for relatively close contact between consumer and supplier so that the U.S. industry faces little competition from imported coatings and, conversely, exports play a minor role in the field.

J. V. Koleske, *Paint and Coatings Testing Manual*, 14th ed., ASTM, Philadelphia, Pa., 1995.

T. A. Misev, *Powder Coatings Chemistry and Technology*, John Wiley & Sons, Inc., New York, 1991, p. 346.

P. A. Reynolds, in A. Marrion, ed., *The Chemistry and Physics of Coatings*, Royal Society of Chemistry, London, 1994.

Z. W. Wicks, Jr., F. N. Jones, and S. P. Pappas, *Organic Coatings: Science and Technology*, 2nd ed., John Wiley & Sons, Inc., New York, 1999.

ZENO W. WICKS, JR.
Consultant

COATINGS, ANTIFOULINGS

When living organisms attach and grow on the underwater surfaces of ships there is either a loss of speed or an increase in the propulsive energy required to counteract the speed loss which presents an enormous economic problem. There are also other submerged surfaces on which fouling can create problems. Offshore oil platforms (which are designed to stay for long periods of time in the world's oceans) can become more susceptible to damage from the added weight that fouling contributes to the structure and from the increased resistance to tidal and water flow. Fouling growth in conduits for conveying cooling water to power stations can lead to serious and costly downtime for cleaning. On yachts and pleasure craft, fouling is not only unsightly but also reduces maneuverability and speed.

One of the most important fouling organisms is the barnacle. As barnacles grow they exert pressure on the surface to which they are attached and can penetrate and undermine protective coatings. There are also marine organisms that can bore into underwater structures of wood, such as the "shipworm" teredo, which bores its way through the wood in which it lives.

MARINE BIOFOULING

Many organisms can contribute to marine fouling communities; from microscopic bacteria and diatoms, through shelled invertebrates such as barnacles and tubeworms, to kelps >10 m long. Microfouling includes microbial organisms such as bacteria, fungi and microalgae (notably diatoms), and their secretions. Microfouling organisms are able to form tenacious films of exuded extracellular polymeric materials, which chelate inorganic ions. A wide range of factors affects the fouling rate and composition of microfouling communities, including water chemistry, water temperature, pressure, shear stress, and substratum composition and structure.

CURRENT ANTIFOULING TECHNOLOGIES

There are currently only two principal ways that marine fouling is controlled on underwater hulls. The first of these is based on the historical method of dispersing a biocide in a binder system that is then released slowly from the coating surface once it is immersed in seawater. The second type does not use biocides but relies on the surface being "nonstick." This is referred to as "foul release" technology.

Biocidal Antifoulings

There are two key factors in the development of a successful biocidal antifouling: the toxicity of the biocide, or biocide combinations; and the delivery mechanism of the biocide(s) to the marine environment.

Biocide Toxicity. Copper compounds were the first biocides used in the large scale industrial production of antifouling paints and are still the most common biocides employed in antifoulings. The most commonly used copper compounds are cuprous oxide (Cu_2O), which is red; cuprous thiocyanate (CuSCN), a pale cream compound used for making brightly colored antifoulings; and metallic copper, either in the traditional sheet form or as a powder.

Copper by itself is, however, limited in its effectiveness. It tends to work well as a biocide against animal (shell) fouling, but algal (weed) fouling is more resistant to copper, and therefore antifouling chemists have spent much time and effort searching for additional biocides that can be added to the copper to boost performance. These are referred to as boosting biocides. A key characteristic of these is that they should have very low seawater solubility (ideally <10 ppm) so that they are not released too quickly from the antifouling paint film. Since the discovery of organotin compounds in the 1960s there have been very few other boosting biocides developed. Not only has it proven to be very difficult to improve on the efficiency and effectiveness of the organotin boosters, but also the relatively small size of the overall antifouling market (~50 million liters of paint worldwide, annually) makes it difficult to justify the very high cost of developing and registering new biocides.

Some commonly used booster biocides currently in use are 2-methylthio-4-*tert*-butylamino-6-cycloproylamino-*s*-triazine (eg, Irgarol 1051); dichlorophenyl dimethyl urea (eg, Diuron); zinc hydroxypyridmethione (eg, zinc omadine); copper hydroxypyridmethione (eg, copper omadine); 4,5-Dichloro-2*N*-octyl-4-isothiazol-3-one (e.g., SeaNine 211); *N*-Dimethyl-*N*-phenyl-*N*-fluorodicholoromethylthiosulphamide (eg, Prevetol A4); tolylfluanid (eg, Preventol A5S); and zinc ethylene-1,2-bisdithiocarbamate (Zineb). Only those of these that show rapid degradation in seawater and in sediments will be likely to survive the close regulatory scrutiny to which they are being increasingly subjected.

Research and development in antifouling coatings is focused on maximizing the efficiency of the few biocides that are available. This involves studies to get the maximum toxic effect from the minimum quantities, and on the best mechanism to control the release of the biocides, to maximize the lifetimes.

Biocide Release Mechanisms. The standard method for measuring biocide release from antifoulings is the leaching rate, the amount of biocide released from a given surface area in a given time, expressed as $\mu g/cm^2/day$. The release mechanism itself depends on the technology of the coating system employed. There are three main technologies: Rosin based, self-polishing copolymers (SPCs), and hybrid SPC/rosin systems.

Foul Release Antifoulings

From an environmental perspective, the most desirable approach to fouling control is one that does not rely on the release of biocides to achieve its effect. A plethora of ideas for how this can be achieved have been proposed, and numerous patents have appeared, but only the foul release or low-adherence systems have been commercialised successfully. These operate by a "nonstick" principle, having surface characteristics that minimize the adhesion strength of fouling organisms. Any fouling that does settle is removed by hydrodynamic forces on the hull as it moves through water.

Despite their beneficial environmental profile, the market penetration of foul release coatings has been limited. In part, this is due to the increased initial cost of installation, but it is also due to the fact that the majority of the world's fleet of crude oil tankers and bulk carriers do not operate at high enough speeds or activity for foul release coatings to perform at their best. As the technology matures it is very likely that products will emerge that both work at lower ship speeds and are less expensive to install.

USES

The use of "natural product" antifoulants has received wide attention, along with attempts to mimic the surfaces of marine animals such as seals, dolphins, and whales. The successful commercial development of any such novel approach will require knowledge of the biological, chemical, and physical processes involved as well as an understanding of the operational requirements of the end-users.

In the meantime, copper-based systems are certain to dominate the antifouling coatings market for the foreseeable future, but it is anticipated that foul release systems will become increasingly important as the environmental pressure on the use of biocides increases.

R. F. Brady, J. R. Griffith, K. S. Love, and D. E. Field, "Non-toxic Alternatives to Antifouling Coatings", *Polymers in a Marine Environment*, The Institute of Marine Engineers, Paper 26, 1989.

D. J. Crisp and G. Walker, "Marine Organisms and Adhesion", *Polymers in a Marine Environment*, The Institute of Marine Engineers, Paper 34, 1985.

A. Milne, M. E. Callow, and R. Pitchers, in Evans and Hoagland, eds., *Algal Biofouling*, Elsevier, 1986.

S. Seagrave, *Lords of the Rim*, Corgi Books, London, 1996.

COLIN ANDERSON
International Coatings, Ltd.

COATINGS FOR CORROSION CONTROL, ORGANIC

Corrosion is a process by which materials, especially metals, are worn away by electrochemical and chemical actions. Metals have anodic and cathodic areas and in the presence of oxygen, water, and a conducting medium, corrosion results. The metal is oxidized to form metal ions at the anode and oxygen is reduced at the cathodes to form hydroxy ions. Since conductivity of the water is an important factor in the rate of corrosion, salts, such as sodium chloride, tend to increase the rate of corrosion. The principal metals that corrode are steel and to a lesser extent aluminum. Cold-rolled steel has more internal stresses than hot-rolled steel and is generally more susceptible to corrosion. Aluminum is higher in the electromotive series than iron and is more easily oxidized, yet aluminum generally corrodes more slowly than steel.

Three strategies are employed to control electrochemical corrosion by coatings: covering the metal with a barrier coat to prevent water and oxygen from contacting the surface; suppressing the anodic reaction; and suppressing the cathodic reaction.

In many cases, a metal object can be completely covered by intact coating films, but where the metal surface cannot be completely covered or where an intact film is damaged during use, but even with nonintact films it is still essential to control corrosion.

CORROSION PROTECTION BY INTACT COATINGS

Coatings can be effective barriers to protect steel when the coating can be applied to cover essentially all of the substrate surface and the film remains intact in service. However, when there will not be complete coverage of the substrate or the film may be ruptured in service, alternative strategies using coatings that can suppress electrochemical reactions involved in corrosion may be preferable.

The key to maintaining corrosion protection by a coating is sufficient adhesion to resist displacement forces. If the coating covers the entire surface of the steel on a microscopic as well as macroscopic scale, and if perfect wet adhesion could be achieved at all areas of the interface, the coating would indefinitely protect steel against corrosion. It is difficult to achieve both of these requirements in applying coatings, however, so a high level of wet adhesion is important but is not the only factor affecting corrosion protection by coatings. Low water and oxygen permeability also help increase corrosion protection. In any case, if wet adhesion is poor, corrosion protection is also poor. However, if the adhesion is fairly good, a low rate of water and oxygen permeation may delay loss of adhesion long enough so that there is adequate corrosion protection for many practical conditions.

Primers made with saponification-resistant vehicles give better corrosion protection than ones made with vehicles that saponify readily. When water and oxygen permeate through a film and water displaces some of the adsorbed groups of the coating resin from the surface of the steel, corrosion starts. Hydroxide ions are generated at the cathodic areas. Hydroxide ions catalyze hydrolysis (saponification) of such groups as esters. If the backbone of the vehicle resin is connected by ester groups, hydrolysis results in polymer network degradation, leading to poorer wet adhesion and, ultimately, catastrophic failure.

Adhesion for Corrosion Protection

Good dry adhesion is required for corrosion protection, if there is no coating left on the substrate, it cannot protect the steel. It has not been so obvious, however, that good wet adhesion is required, so that the adsorbed layer of the coating will not desorb when water permeates through the film.

The first step to obtain good wet adhesion is to clean the steel surface, removing any oils and salts. Zinc and iron phosphate conversion coatings have been the standard treatment for many years. Before drying, the treated metal is rinsed with a chromic acid solution. Waste disposal problems are severe, especially considering this wash step.

For many applications, aluminum does not have to be treated for corrosion control due to the coherent aluminum oxide surface of the aluminum. But if exposure to salt is to be expected, the surface must be treated before applying a coating. Chromate surface treatments have been the industry standard, but with the concern about carcinogenicity of chromium(VI), proprietary chromium-free treatments have been developed.

Clean steel is treated with an aqueous solution of bis(-triethoxysilyl)- ethane (BTSE). The BTSE reacts with water and hydroxyl groups on the steel to give a water-resistant anchor on the steel. After drying, the treated metal can be coated and baked.

After cleaning and treating, the surface should not be touched and should be coated as soon as possible. Finger-

prints leave oil and salt on the surface. After exposure to high humidity, fine blisters can form, disclosing the identity of the miscreant by the fingerprints.

It is also critical to achieve as nearly complete penetration into the micropores and irregularities in the surface of the steel as possible. If any steel is left uncoated, when water and oxygen reach the surface, corrosion will start, generating an osmotic cell that can lead to blistering. The viscosity of the external phase of the coating should be as low as possible and remain low long enough to permit complete penetration. It is desirable to use slow evaporating solvents, slow cross-linking coatings, and, when possible, baking primers. Macromolecules may be large compared to the size of small crevices, so lower molecular weight components may give better protection.

Wet adhesion requires that the coating not only be adsorbed strongly on the surface of the steel, but also that it not be desorbed by water that permeates through the coating. Wet adhesion is enhanced by having several adsorbing groups scattered along the resin chain, with parts of the resin backbone being flexible enough to permit relatively easy orientation and other parts rigid enough to assure that there are loops and tails sticking up from the surface for interaction with the rest of the coating. Another reason baking primers commonly provide superior corrosion protection is that at the higher temperature, there may be greater opportunity for orientation of resin molecules at the steel interface. Amine groups are particularly effective polar substituents for promoting wet adhesion.

Saponification resistance is another important factor. Corrosion generates hydroxide ions at the cathode, raising pH levels as high as 14. Ester groups in the backbone of a binder can be saponified, degrading the polymer near the interface and reducing wet adhesion. Epoxy-phenolic primers are an example of high bake primers that are completely resistant to hydrolysis.

Water-soluble components that stay in primer films should be avoided, because they can lead to blister formation. Passivating pigments cannot function unless they are somewhat soluble in water; their presence in coating films can, therefore, lead to blistering. Hydrophilic solvents, which become immiscible in the drying film as other solvents evaporate, can be retained as a separate phase and lead to blister formation.

Factors Affecting Oxygen and Water Permeability

Many factors affect permeability of coating films to water and oxygen. Water and oxygen can permeate, to at least some extent, through any amorphous polymer film, even though the film has no imperfections such as cracks or pores. Small molecules travel through the film by jumping from free volume hole to free volume hole. Free volume increases as temperature increases above glass-transition temperature T_g. Therefore, normally, one wants to design coatings with a T_g above the temperature at which corrosion protection is desired.

Permeability is also affected by the solubility of oxygen and water in a film. The variation in oxygen solubility is probably small, but variation in water solubility can be large. Salt groups on a polymer lead to high solubility of water in films. This makes it difficult to formulate high-performance, air-dry, water-reducible coatings that are solubilized in water by amine salts of carboxylic acids.

Pigmentation can have significant effects on water and oxygen permeability. Oxygen and water molecules cannot pass through pigment particles; therefore, permeability decreases as pigment volume concentration (PVC) increases. Pigments should be used that are as free as possible of water-soluble impurities and use of hydrophilic pigment dispersants should be avoided, or at least minimized.

Pigments with platelet shaped particles can reduce permeability rates as much as fivefold when they are aligned parallel to the coating surface. A factor favoring alignment is shrinkage during solvent evaporation. Since oxygen and water vapor cannot pass through the pigment particles, the presence of aligned platelets can reduce the rate of vapor permeation through a film. If platelets are not aligned, permeability may be increased, especially if the film thickness is small relative to the size of the platelets.

A Monte Carlo simulation model of the effect of several variables on diffusion through pigmented coatings indicates that finely dispersed, lamellar pigment particles at a concentration near, but below CPVC, give the best barrier performance.

There are advantages to applying multiple layers of coatings. The primer can be designed so that it has excellent penetration into the substrate surface, has excellent wet adhesion, and saponification resistance without particular concern about other properties. The topcoat(s) can provide for minimum permeability and other required properties. The primer film does not need to be thick, as long as the topcoat is providing barrier properties; the lower limit is probably controlled by the need to assure coverage of the entire surface. Another advantage of applying multiple coats is the decrease in probability that any area of the substrate will escape having coating applied.

The corrosion protection afforded by intact films would be expected to be essentially independent of film thickness. However, since film thickness affects the mechanical performance of films, there may be an optimal film thickness for the maintenance of an intact film. In air-dry, heavy-duty maintenance coatings, there is generally a film thickness, dependent on the coating that provides a more than proportional increase in corrosion protection relative to thinner films. Commonly, this film thickness is as much as 400 μm or more. The use of barrier platelet pigments permits a reduction in the required film thickness without loss of protection. The platelets may minimize the probability of defects propagating through the film to the substrate. Such defects are less likely to occur in baked films.

CORROSION PROTECTION BY NONINTACT FILMS

Even with coatings designed to minimize the probability of mechanical failure, in many end uses there will be breaks in the films during their service life. When it is not possible to have full coverage of all the steel surface, it is generally desirable to design coatings to suppress electrochemical reactions, rather than primarily for their barrier properties.

Minimizing Growth of Imperfections

If there are gouges through the film down to bare metal, water and oxygen reach the metal and corrosion starts. If the wet adhesion of the primer to the metal is not adequate, water creeps under the coating and the coating comes loose. Poor hydrolytic stability can be expected to exacerbate the situation. This mode of failure, cathodic delamination, requires wet adhesion and saponification resistance. Blisters are also likely to develop under a film near the location of a gouge.

When wet adhesion varies on a local scale, filiform corrosion can occur, characterized by development of thin threads of corrosion wandering randomly under the film but never crossing another track. Formation of these threads often starts from the edge of a scratch. At the head of the thread, oxygen permeates through the film and cathodic delamination occurs. The head grows following the directions of poorest wet adhesion. Behind the head, oxygen is consumed by oxidation of ferrous ions and ferric hydroxide precipitates, passivating the area, explaining why threads never cross. Since the ion concentration decreases, osmotic pressure drops and the thread collapses, leaving a discernible rust track.

Passivating Primers

A passivator suppresses corrosion above some critical concentration, but may accelerate corrosion at lower concentrations by cathodic depolarization. The mechanism of passivation has not been fully elucidated.

Passivating pigments promote formation of a barrier layer over anodic areas, passivating the surface. To be effective, such pigments must have some minimum solubility. However, if the solubility is too high, the pigment would leach out of the coating film too rapidly, limiting the time that it is available to inhibit corrosion. For the pigment to be effective, the binder must permit diffusion of water to dissolve the pigment. Therefore, the use of passivating pigments may lead to blistering after exposure to humid conditions. Such pigments are most useful in applications in which the need to protect the steel substrate after film rupture has occurred outweighs the desirability of minimizing the probability of blistering. They are also useful when it is not possible to remove all surface contamination or when it is not possible to achieve complete coverage of the steel by the coating.

A fairly new approach to passivation is applying a film of electrically conductive polymer to a steel surface to protect it from corrosion.

Many organic compounds are corrosion inhibitors for steel. Most are polar substances that tend to adsorb on high energy surfaces. Amines are particularly widely used. Clean steel wrapped in paper impregnated with a volatile amine or the amine salt of a weak acid is protected against corrosion. The reason for their effectiveness is not clear.

To protect uncoated steel and aluminum components during shipment and storage, strippable coatings can be used. Typically, the coating is easily removed and can be recycled.

Red lead pigment, Pb_3O_4 containing 2–15% PbO, has long been used as a passivating pigment, as have chromate pigments.

"Zinc chromates" have been widely used as passivating pigments, but this terminology is poor, since zinc chromate itself is too insoluble and could promote corrosion rather than passivate.

Zinc chromates—and, presumably, other soluble chromates—are carcinogenic to humans, so must be handled with appropriate caution. In some countries, their use has been prohibited; prohibition worldwide is probable in the future. Substantial efforts have been undertaken to develop less hazardous passivating pigments.

Basic zinc and zinc–calcium molybdates are said to act as passivating agents in the presence of oxygen, apparently leading to precipitation of a ferric molybdic oxide barrier layer on the anodic areas. Barium metaborate, the salt of a strong base and a weak acid, may act by increasing the pH, thus lowering the critical concentration of oxygen required for passivation. Zinc phosphate, $Zn_3(PO_4)_2 \cdot 2 H_2O$, has been used in corrosion protective primers and may act by forming barrier precipitates on the anodic areas. Calcium and barium phosphosilicates and borosilicates are being used increasingly. Calcium tripolyphosphate has also been recommended.

These pigments are all inorganic pigments; a wider range of potential oxidizing agents would seem to be available if organic pigments were used. An example of a commercially available organic pigment is the zinc salt of 5-nitroisophthalic acid, said to be as effective as zinc yellow at lower pigment levels.

Cathodic Protection by Zinc-Rich Primers

Zinc-rich primers are another approach to protecting steel with nonintact coatings. They are designed to provide the protection given by galvanized steel, but to be applied to a steel structure after fabrication. The primers contain high levels of powdered zinc, >80% by weight usualy.

Vehicles for zinc-rich primers must be saponification resistant. Two classes of binders are used: organic resins and inorganic resins. Alkyds are not appropriate resins for this application, since they are readily saponified. Urethane coatings have adequate saponification resistance and have the advantage over epoxy/amine coatings of being one-package coatings. Water-borne urethane coatings are also available.

The most widely used vehicles are tetraethyl orthosilicate and oligomers derived from it by controlled partial hydrolysis with a small amount of water. Ethyl or isopropyl alcohol is used as the principal solvent, since an alcohol helps maintain package stability. After application, the alcohol evaporates and water from the air completes

the hydrolysis of the oligomer to yield a film of polysilicic acid partially converted to zinc salts.

Properly formulated and applied, zinc-rich primers are very effective in protecting steel against corrosion. Their useful lifetime is not completely limited by the amount of zinc present, as one might first assume. Initially, the amount of free zinc decreases from the electrochemical reaction; later, loss of zinc metal becomes slow, but the primer continues to protect the steel.

Zinc is expensive, especially on a volume basis. Early attempts to replace even 10% of the zinc with low-cost inert pigment caused a serious decrease in performance, presumably due to decrease in metal to metal contact. A relatively conductive inert pigment, iron phosphide (Fe_2P), has shown promise.

Zinc-rich powder coatings cannot be applied with the PVC above CPVC. Lower zinc content powder coatings have been used, but corrosion protection is inferior to liquid applied coatings because of low penetration of water through the coating and inferior contact between zinc particles and between them and the steel substrate. Addition of sufficient carbon black increases electrical conductivity between zinc particles and with the substrate, providing good performance.

Zinc-rich primers are frequently topcoated to minimize corrosion of the zinc, protect against physical damage, and improve appearance.

TYPES OF COATINGS

Primers for Baked Coatings

In most cases, baking primers are designed to be barrier coatings to minimize the risk of blistering. Primers formulated with epoxy and phenolic resins provide excellent corrosion over steel; they show good wet adhesion and are not subject to hydrolysis or saponification. Epoxy ester based primers cure at a somewhat lower temperature and approach epoxy-phenolic coatings in performance. Polyurethane primers can provide good properties. To minimize cost they are prepared with aromatic polyisocyanates, but even then their cost tends to be higher than epoxy/amine primers. Alkyd resins are used only when performance requirements are limited. They have the advantage of low cost and ease of application but are subject to saponification.

Coil coating is sometimes an exception to the general statement that barrier coatings are used for corrosion protection with baked coatings. When coil-coated steel is fabricated there are cut edges exposed. Corrosion can start at these cut edges. Epoxy ester primers with a chromate passivating pigment have been used.

Electrodeposition Coatings

Cationic electrodeposition coatings are used as primers on all automobiles and trucks as well as many other steel substrates.

The coatings must be designed so that all components are in the aggregate particles so that all components in composition are attracted to the electrode at the same rate. The electrolysis of water leads to the release of the neutralizing acid at the anode, which must be removed. The rate of deposition and deposition in restricted areas is affected by the conductivity of the aqueous phase. As the coated object on the conveyor moves out of the tank it must be rinsed so that the system is being continually diluted.

The rate of deposition increases with increasing voltage. As film thickness increases, the rate of deposition decreases and finally stops as the electrical resistance increases. If the voltage is too high, hydrogen gas evolved below the surface of the coating erupts through the coating. If the conductivity of the aqueous phase is too high, the rate of deposition will decrease. The balance of conductivity and voltage is critical in controlling "throw power," that is, the distance of deposition into a closed-end narrow hollow in the object.

The corrosion resistance of automobiles coated by cationic E-coating is excellent. This results because of the strong interaction between the amine groups on the coating and the steel and its phosphate coating leads to excellent wet adhesion. Also, the film completely covers the surface and penetrates into all the micropores on the surface by the driving force of the electrocoating and the high temperature baking. The urethane cross-linked epoxy resin is almost impervious to saponification.

The film thickness applied is uniform, which means that any irregularities in the surface of the metal are copied in the surface of the primer coat. Also, since the pigmentation must be lower than usual in primers, adhesion to the surface of topcoats can be difficult to achieve. In general, a coat of a primer-surfacer is applied over the primer before topcoating. Another approach is to apply two E-coats, the first designed to provide the excellent wet adhesion to the steel, the second to provide a level surface to which topcoats will adhere.

Maintenance Paints

Maintenance paints are coatings applied to field installations such as highway bridges, factories, and tank farms. Composition of the coating is a main variable, but surface preparation and application procedures are also critical to performance. The most commonly used method of cleaning has been sandblasting, but, conventional sand blasting is being rapidly replaced by less hazardous modifications like abrasive blasting with materials like steel grit or water-soluble abrasives like sodium bicarbonate or salt to replace the sand that can cause silicosis. Ultra-high-pressure hydroblasting at pressures >175 mPa (25,000 psi) is very effective in removing oil and surface contaminants such as salt.

At least two coats, a primer and a topcoat, are applied; in some cases a primer, intermediate coat, and topcoat are required. The primer provides the principal protection against corrosion but the other coats also help reduce water and oxygen permeability as well as protecting the primer and providing the desired surface properties. Three types of primers are used depending on the applications: barrier primers, zinc-rich primers, and primers containing passivating pigments.

Automotive Refinishing

When major damage has been experienced on a component such as a door, it is common to use a replacement that has been E-coated by the car manufacturer. When bare metal has to be refinished, wash primers are most often used. The binders of wash primers are poly(vinyl butyral) and a phenolic resin, generally pigmented with zinc tetroxychromate. Acid is added just before application and a thin wash coat is applied to the bare metal. Wash primers give excellent adhesion for topcoats and provide excellent corrosion resistance but have very high VOC emissions.

Aircraft Coatings

Primers for exterior surfaces are 2K BPA epoxy/amino-amide-based coatings with a passivating pigment. Strontium chromate has been the pigment of choice. Interior components such as baggage areas are coated with a novolak epoxy amine coating pigmented with strontium chromate and generally not topcoated. Increasingly, water-borne epoxy primers are being used.

Another approach has been the use of electrically conductive polymer coatings on aluminum. Aluminum coated with poly(3-octyl)pyrrole and topcoated with a urethane topcoat has shown promising results in preliminary evaluations.

EVALUATION AND TESTING

There is no laboratory test available that can be used to predict corrosion protection performance of a new coating system.

Use testing is the only reliable test of a coating system—ie, to apply it and then observe its condition over years of actual use. Simulated tests are the next most reliable for predicting performance. One common approach is to expose laboratory-prepared panels on test fences on beaches in the South.

The lack of laboratory test methods that reliably predict performance puts a premium on collection of databases permitting analysis of interactions between actual performance and application and formulation variables. In time, it may be possible to predict performance better from a knowledge of the underlying theories than from laboratory tests.

R. A. Dickie and F. L. Floyd, eds., *Polymeric Materials for Corrosion Control*, ACS Symp. Ser. No. 322, American Chemical Society, Washington, D.C., 1986.

W. Funke, in H. Leidheiser, Jr., ed., *Corrosion Control by Coatings*, Science Press, Princeton, N.J., 1979, pp. 35–45.

C. C. Munger, *Corrosion Prevention by Protective Coatings*, National Association of Corrosion Engineers, Houston, Tex., 1997.

B. M. Perfetti, *Metal Surface Characteristics Affecting Organic Coatings*, Federation of Societies for Coatings Technology, Blue Bell, Pa., 1994.

ZENO W. WICKS, JR.
Consultant

COATINGS, MARINE

The selection and application of marine coatings has become a highly specialized discipline in which governmental regulations are a dominant influence. Many paints and painting procedures that were used in the past are no longer permitted or are extremely costly to use.

Corrosion Control Plan

A corrosion control plan for each ship or structure, which is designed to control deterioration in the most economical and practical manner and to include all appropriate mechanisms for corrosion control, must be developed before construction begins. For steel in the marine environment, the chief methods available are protective coatings and cathodic protection.

Protective Coatings

Each coating in the protective coating system is designed to perform a specific function and to be compatible with the total system. The effectiveness of a coating is directly related to its ability to maintain adhesion to the substrate, its integrity, and its thickness. Areas that cannot be easily or safely repaired, especially those which require drydocking for repair, need to be given the best available coating.

Fouling organisms attach themselves to the underwater portions of ships and have a severe impact on operating costs. Because fouling is controlled best by use of antifouling paints, it is important that these paints be compatible with the system used for corrosion control and become a part of the total corrosion control strategy.

ENVIRONMENTAL CONCERNS

Local environmental regulations have significantly affected the production, transportation, use, and disposal of coatings.

Volatile Organic Compounds

As coatings dry, solvents are released into the atmosphere, where they undergo chemical reactions in sunlight and produce photochemical smog and other air pollutants (see Air POLLUTION). As a general rule, the volatile organic compound (VOC) content of marine coatings is restricted to 340 g/L. In the locations where ozone (qv) levels do not conform to the levels established by the Environmental Protection Agency, regulations require an inventory of all coatings and thinners from the time they are purchased until they are used.

Heavy-Metal Pigments

Lead and chromate pigments, used for many years as corrosion inhibitors in metal primers and topcoats for marine coating systems, have been linked to adverse health and environmental effects. Because of these concerns, lead- and chromate-containing pigments are not used in marine coatings.

Organotins

In the mid-1970s compounds based on derivatives of triphenyl- or tributyltin (see Tin COMPOUNDS) known generically as organotins, were found to be much more effective than cuprous oxide paints in controlling fouling, and numerous products were introduced. These fell into two classes. Free-association coatings containes a tributyltin salt, eg, acetate, chloride, fluoride, or oxide, physically mixed into the coating. Copolymer coatings contain organotin which is covalently bound to the resin of the coating and is not released until a tin–oxygen bond hydrolyzes in seawater. This controlled hydrolysis produces a low and steady leach rate of organotin and creates hydrophilic sites on the binder resin. This layer of resin subsequently washes away and exposes a new layer of bound organotin.

However, there is now considerable evidence that sufficiently high concentrations of organotins kill many species of marine life and affect the growth and reproduction of others. Thus many nations restrict organotin coatings to vessels greater than 25 meters in length. In the United States, laws prohibit the retail sale of copolymer paints containing greater than 7.5% (dry weight) of tin, and of free-association paints containing greater than 2.5% (dry weight) of tin, but do not restrict the size of the ship to which the paints may be applied.

Abrasive Blast Cleaning

Removal of paint by abrasive blasting may lead to adverse health effects for workers who breathe dust formed during the operation. Regulations restrict blasting operations to such procedures as blasting within enclosures, using approved mineral abrasives, using a spray of water to reduce dust, and blasting with alternative materials such as ice, plastic beads, or solid carbon dioxide.

Debris from the removal of paint may contain lead, chromium, or other heavy metals. Collection of such debris is required to prevent release of these metals into the environment and to avoid exposure and contamination of workers.

Reactive Coatings

Coatings that cure by chemical reaction of two component parts are the most widely used in marine applications and protection of workers from the reactive ingredients is required.

SURFACE PREPARATION FOR MARINE PAINTING

Surface preparation, always important in obtaining optimal coatings performance, is critical for marine coatings (see METAL SURFACE TREATMENTS). Surface preparation usually comprises about half of the total coating costs, and if inadequate may be responsible for early coating failure. Proper surface preparation includes cleaning to remove contaminants (eg. abrasive blasting) and roughening the surface to facilitate adhesion.

TYPES OF COATINGS

Coatings ingredients fall into four principal classes. Resins form a continuous solid film after curing, bind all ingredients within the film, and provide adhesion to the substrate. The properties of a coating are determined principally by the resins it contains. Pigments are metals or nearly insoluble salts that impart opacity, color, and chemical activity to the coating. Solvents are used primarily to facilitate manufacture and application, but are lost by evaporation after application and are not a permanent part of the coating. Additives used in small (1–50 ppt) amounts give the coating such desirable additional properties as ease of manufacture, stability in shipment and storage, ease of application, or increased performance of the dried film.

These ingredients may be formulated to give coatings that protect against corrosion in different ways. Barrier coatings physically separate oxygen, water, ions, and other corrosive agents from the steel surface. Inhibitive coatings prevent corrosion by absorbing or neutralizing corrosive agents, or by slowly releasing protective ions. Sacrificial coatings contain a metal (usually zinc) that is oxidized more rapidly than steel, thereby providing protection for the substrate by electrochemical action. Conversion coatings chemically oxidize the surface of the substrate to a depth of 7–10 μm, producing a passive layer which resists corrosion better than the metal itself. Modern marine coatings fall into eight generic categories: epoxies, urethanes, alkyds, inorganic silicate coatings, vinyls, chlorinated rubber, coal tar, and powder coatings.

APPLICATION METHODS

The application of marine coatings is a critical factor in achieving maximum performance. Protective clothing and breathing equipment should be worn during application. Because of large surface areas, ships are usually spray painted (see COATING PROCESSES, SPRAY COATINGS). Three techniques are widely used: air, airless, and electrostatic spraying. Manual painting occurs mostly during touch-up or repair, and is best suited for piping, railings, and other hard to spray places.

Transfer Efficiency

Many components of ships and marine structures are now coated in the shop under controlled conditions to reduce the amount of solvents released into the atmosphere, improve the quality of work, and reduce cost. Regulations designed to limit the release of volatile organic compounds into the air confine methods of shop application to those having transfer efficiencies of 65%. Transfer efficiency is defined as the percent of the mass or volume of solid coating that is actually deposited on the item being coated.

SELECTION OF COATINGS

Underwater Hull

Hull coatings consist of two layers of an anticorrosive coating topped with one layer of an antifouling coating. The coating system must resist marine fouling, severe corrosion, the cavitation action of high speed propellers, and the high current densities near the anodes of the cathodic protection system. Epoxy and coal-tar epoxy systems are commonly used as anticorrosive coatings.

Anticorrosive systems require an antifouling topcoat. Arsenic, cadmium, and mercury are proscribed and organotins are severely restricted. Cuprous oxide has always been and remains the most widely used toxicant.

The system of hull coatings, including antifouling paint, must be compatible with the cathodic protection system. Thus the coating system must have good dielectric properties to minimize cathodic protection current requirements and must be resistant to the alkalinity produced by the electric current. The cathodic protection system should prevent corrosion undercutting of coatings that become damaged, and the current density should be able to be increased easily to meet the increased electrical current needed as the coating deteriorates.

Boottop and Freeboard Areas

The boottop suffers mechanical damage from tugs, piers, and ice, and experiences intermittent wet and dry periods with nearly constant exposure to sunlight. Thus coatings for this area require resistance to sunlight and mechanical damage, good adhesion, and flexibility. Frequently the hull coating system is used in the boottop area, and one or two extra topcoats are applied for added strength. In the freeboard areas, commercial ships use organic zinc-rich primers extensively and usually topcoat with them a two- or three-coat epoxy system.

Weather Decks

Coatings for decks must be resistant to abrasion by pedestrians and small vehicles, and must be slip-resistant. Inorganic zinc primers overcoated with epoxy coatings for additional corrosion protection perform well on steel decks, or a multiple-coat all-epoxy system can be used. Nonskid coatings are used on aircraft carrier landing and hangar decks and in passageways of all ships to maintain traction during wet and slippery conditions. The coatings contain epoxy resins and a coarse grit and are applied using a roller over epoxy primers to produce a textured finish. Nonskid coatings are 6 to 10 mm thick when dry.

Superstructure

Deck hardware and machinery, masts, and booms are coated with an inorganic zinc primer, an intermediate coat of epoxy, and a finish coat of aliphatic polyurethane or silicone-alkyd enamel. Powder coatings are used effectively on antennas and other equipment on the superstructure. This equipment, as well as exhaust stacks, steam riser valves and piping, and other hot surfaces, can also be coated using thermal-sprayed aluminum.

Tanks

Coatings for liquid cargo tanks are selected according to the materials that the tanks are to contain. Tank coatings protect the cargo from contamination and must be compatible with the material carried. Epoxy systems are most frequently selected because they perform well with both aqueous and organic products.

Machinery Spaces, Bilges, and Holds

Machinery spaces and bilges are so inaccessible that surface preparation is a significant problem and damage to machinery that cannot be removed must be avoided. Chemical cleaning by aqueous citric acid solutions, followed by degreasing using a nonflammable solvent, is widely used. Surfaces are best protected using a two- or three-coat epoxy-polyamide system having a total thickness of 250–300 μm. Alkyd enamels perform well in dry machinery spaces. Holds for carrying cargo may be painted with either of these systems, but the epoxy system is preferred for chemical and abrasion resistance.

Living Areas

Living areas are generally painted with nonflaming coatings, or with intumescent coatings which foam when heated and produce a thick char that lessens damage to the substrate. Epoxy systems are generally used in damp areas such as galleys, washrooms, and showers where moisture causes enamels to deteriorate.

H. R. Bleile and S. D. Rogers, *Marine Coatings*, Federation of Societies for Coatings Technology, Blue Bell, Pa., 1989.

R. F. Brady, Jr., "Marine Applications," in J. I. Kroschwitz, ed., *Encyclopedia of Polymer Science and Engineering*, Vol. 9, John Wiley & Sons, Inc., New York, 1988, pp. 295–300.

Journal of Coatings Technology, published monthly by the Federation of Societies for Coatings Technology, Blue Bell, Pa.

Journal of Protective Coatings and Linings, published monthly by the Steel Structures Painting Council, Pittsburgh, Pa.

ROBERT F. BRADY JR.
U.S. Naval Research Laboratory

RICHARD W. DRISKO
U.S. Naval Civil Engineering Laboratory

COBALT AND COBALT ALLOYS

Cobalt, a transition-series metallic element having atomic number 27, is similar to silver in appearance.

Cobalt and cobalt compounds have expanded from use as colorants in glasses and ground coat frits for pottery to drying agents in paints and lacquers animal and human nutrients, electroplating materials, high-emperature alloys hardfacing alloys, high-speed tools, magnetic alloys, alloys used for prosthetics, and uses in radiology. Cobalt is also used as a catalyst for hydrocarbon refining from crude oil for the synthesis of heating fuels.

OCCURRENCE

Cobalt is the thirtieth most abundant element on earth and comprises approximately 0.0025% of the earth's crust. It occurs in mineral form as arsenides, sulfides, and oxides; trace amounts are also found in other minerals of nickel and iron as substitute ions. Cobalt minerals are commonly associated with ores of nickel, iron, silver, bismuth, copper, manganese, antimony, and zinc.

The cobalt resources of the United States are estimated to be about 1.3 million tons. Most of these resources are in Minnesota, but other important occurrences are in Alaska, California Missouri, Montana, and Idaho. Although large, most U.S. resources are in subeconomic concentrations. The identified world cobalt resources are about 11 million tons. The vast majority of these resources are in nickel-bearing laterite deposits, with most of the rest occurring in nickel-copper deposits hosted in mafic and ultramafic rocks in Australia, Canada, and Russia, and in the sedimentary copper deposits of Congo (Kinshasa) and Zambia. In addition, millions of tons of hypothetical and speculative cobalt resources exist in manganese nodules on the ocean floor.

PROPERTIES

The electronic structure of cobalt is [Ar] $3d^7 4s^2$. At room temperature the crystalline structure of the α (or ε) form, is close-packed hexagonal (cph) and lattice parameters are $a = 0.2501$ nm and $c = 0.4066$ nm. Cobalt is magnetic up to $1123°C$ and at room temperature the magnetic moment is parallel to the c-direction. Physical properties are listed in Table 1.

Metallic cobalt dissolves readily in dilute H_2SO_4, HCl, or HNO_3 to form cobaltous salts. Cobalt cannot be classified as an oxidation-resistant metal. Scaling and oxidation rates of unalloyed cobalt in air are 25 times those of nickel.

ECONOMIC ASPECTS

In 2000, approximately 45% of U.S. cobalt use was in superalloys, which are used primarily in aircraft gas turbine engines; 9% was in cemented carbides for cutting and wear-resistant applications, another 9% in magnetic alloys, and the remaining 37% in various chemical uses. The total estimated value of cobalt consumed in 2000 was 300×10^6.

During much of its history, the price of cobalt metal was set primarily by the producers. Prices from 1990 on reflect changes in supply/demand, political unrest in leading producer countries, selling from stockpiles, delayed purchases, etc.

HEALTH AND SAFETY FACTORS

Low levels of cobalt, as part of the vitamin B_{12} complex, are necessary to maintain good health, and the Food and Drug Administration has recognized a number of cobalt compounds as safe for use in materials that come in contact with food. Cobalt also stimulates the production

Table 1. Properties of Cobalt

Property	Value		
at wt	58.93		
transformation temperature, °C	417		
heat of transformation, J/g[a]	251		
mp, °C	1493		
latent heat of fusion, ΔH_{fus}, J/g[a]	259.4		
bp, °C	3100		
latent heat of vaporization, ΔH_{vap}, J/g[a]	6276		
specific heat, J/(g · °C)[a]			
15–100°C	0.442		
molten metal	0.560		
coefficient of thermal expansion, °C^{-1}			
cph at RT	12.5		
fcc at 417°C	14.2		
thermal conductivity at RT, W/(m · K)	69.16		
thermal neutron absorption, Bohr atom	34.8		
resistivity, at 20°C[b], 10^{-8} Ω · m	6.24		
Curie temperature, °C	1121		
saturation induction, $4\pi I_S$, T[c]	1.870		
permeability, μ			
initial	68		
max	245		
residual induction, T[c]	0.490		
coercive force, A/m	708		
Young's modulus, GPa[d]	211		
Poisson's ratio	0.32		
hardness,[f] diamond pyramid, of % Co	99.9	99.98[e]	
at 20°C	225	253	
at 300°C	141	145	
at 600°C	62	43	
at 900°C	22	17	
strength of 99.9% cobalt, MPa[g]	as cast	annealed	sintered
tensile	237	588	679
tensile yield	138	193	302
compressive	841	808	
compressive yield	291	387	

[a]To convert J to cal, divide by 4.184.
[b]Conductivity = 27.6% of International Annealed Copper Standard.
[c]To convert T to gauss, multiply by 10^4.
[d]To convert GPa to psi, multiply by 145,000.
[e]Zone refined.
[f]Vickers.
[g]To convert MPa to psi, multiply by 145.

of red blood cells and, accordingly, has been widely used as a treatment for anemia, particularly in pregnant women. For these reasons, cobalt is unlikely to produce adverse health effects in the general population at levels typically found in the environment. General population exposure to cobalt is very low.

USES

The largest consumption of cobalt is in metallic form in magnetic alloys, cutting and wear-resistant alloys, and superalloys. Alloys in this last group are used for components requiring high strength as well as corrosion and oxidation resistance, usually at high temperatures.

The second largest use of cobalt is in the form of salts which have the largest application as raw material for electroplating baths and as highly effective driers for lacquers, enamels, and varnishes. Addition of cobalt salts to paint greatly increases the rate at which paint hardens.

One of its most widespread uses is in sterilization facilities. Food irradiation is an application of ^{60}Co with great potential, but growth in this area is slow because of the public perception of radiation. Cobalt is an essential ingredient in animal nutrition.

COBALT ALLOYS

Mechanical properties depend on the alloying elements. Addition of carbon to the cobalt base metal is the most effective choice. The carbon forms various carbide phases with the cobalt and the other alloying elements. The presence of carbide particles is controlled in part by such alloying elements such as chromium, nickel, titanium, manganese, tungsten, and molybdenum that are added during melting. The distribution of the carbide particles is controlled by heat treatment of the solidified alloy.

Cobalt alloys are strengthened by solid-solution hardening and by the solid-state precipitation of various carbides and other intermetallic compounds. Minor phase compounds, when precipitated at grain boundaries, tend to prevent slippage at those boundaries, thereby increasing creep strength at high temperatures. Aging and service under stress at elevated temperature induce some of the carbides to precipitate at slip planes and at stacking faults, thus providing barriers to slip. If carbides are allowed to precipitate to the point of becoming continuous along the grain boundaries, they often initiate fracture.

Cobalt-Base Alloys

As a group, the cobalt-base alloys may generally be described as wear resistant, corrosion resistant, and heat resistant; ie, strong even at high temperatures. Many of the alloy properties arise from the crystallographic nature of cobalt, in particular its response to stress; the solid-solution-strengthening effects of chromium, tungsten, and molybdenum; the formation of metal carbides; and the corrosion resistance imparted by chromium. Generally, the softer and tougher compositions are used for high temperature applications such as gas-turbine vanes and buckets. The harder grades are used for resistance to wear.

Cobalt-Base Wear-Resistant Alloys

The main differences in the Stellite alloy grades from the 1990s on versus those of the 1930s are carbon and tungsten contents, and hence the amount and type of carbide formation in the microstructure during solidification. Carbon content influences hardness, ductility, and resistance to abrasive wear. Tungsten also plays an important role in these properties.

Types of Wear

There are several distinct types of wear that can be divided into three main categories: abrasive wear, sliding wear, and erosive wear. The type of wear encountered in a particular application is an important factor influencing the selection of a wear-resistant material.

Alloy Compositions and Product Forms

Stellite alloys 1, 6, and 12, derivatives of the original cobalt–chromium–tungsten alloys, are characterized by their carbon and tungsten contents. Stellite alloy 1 is the hardest, most abrasion resistant, and least ductile.

Stellite alloy 21 differs from the first three alloys in that molybdenum rather than tungsten is used to strengthen the solid solution. Stellite alloy 21 also contains considerable less carbon. Each of the first four alloys is generally used in the form of castings and weld overlays. Haynes alloy 6B differs in that it is a wrought product available in plate, sheet, and bar form.

The Tribaloy alloy T-800 is from an alloy family developed by DuPont in the early 1970s, in the search for resistance to abrasion and corrosion. Excessive amounts of molybdenum and silicon were alloyed to induce the formation during solidification of hard and corrosion-resistant intermetallic compounds, known as Laves phase.

The physical and mechanical properties of the six commonly used cobalt wear alloys are presented in Table 2.

Cobalt-Base High-Temperature Alloys

For many years, the predominant user of high temperature alloys was the gas-turbine industry. In the case of aircraft gas-turbine power plants, the chief material requirements were elevated-temperature strength, resistance to thermal fatigue, and oxidation resistance. For land-base gas turbines, which typically burn lower-grade fuels and operate at lower temperatures, sulfidation resistance was the primary concern. The use of high-temperature alloys has become more diversified as more efficiency is sought from the burning of fossil fuels and waste and as new chemical processing techniques are developed.

Cobalt-base alloys are not as widely used as nickel and nickel–iron alloys in high-temperature applications. Nevertheless, cobalt-base high-temperature alloys play an important role because of excellent resistance to sulfidation and strength at temperatures exceeding those at which the γ'- and γ''-precipitates in the nickel and nickel–iron alloys dissolve. Cobalt is also used as an alloying element in many nickel-base high-temperature alloys.

Alloy Compositions and Product Forms

Stellite 21, an early type of cobalt-base high temperature alloy, is used primarily for wear resistance. The use of tungsten rather than molybdenum, moderate nickel contents, lower carbon contents, and rare-earth additions typify cobalt-base high-temperature alloys from the 1990s on.

Cobalt-Base Corrosion-Resistant Alloys

To satisfy the industrial need for alloys that exhibit resistance to aqueous corrosion yet share the attributes of cobalt as an alloy base; ie, resistance to various forms of wear and high strength over a wide range of temperatures, several low carbon, wrought cobalt–nickel–chromium–molybdenum alloys are produced. In addition, the cobalt–chromium–molybdenum alloy Vitallium is widely used for prosthetic devices and implants, owing to excellent compatibility with body fluids and tissues.

Economic Aspects

With cobalt historically being approximately twice the cost of nickel, cobalt-base alloys for both high-temperature and corrosion service tend to be much more expensive than competitive alloys. In some cases of severe service their performance increase is, however, commensurate with the cost increase and they are a cost-effective choice. For hardfacing or wear applications, cobalt alloys typically compete with iron-base alloys and are at a significant cost disadvantage.

Table 2. Mechanical and Physical Properties of Cobalt-Base Wear-Resistant Alloys

Property	Alloy					
	1	6	12	21	6B	T-800
hardness, Rockwell	55	40	48	32	37^a	58
yield strength, MPa[b]		541	649	494	619^a	
ultimate tensile strength, MPa[b]	618	896	834	694	998^a	
elongation, %	<1	1	<1	9	11	
thermal expansion coeff, μm/(m·°C)						
20–100°C	10.5	11.4	11.5	11.0	13.9^c	
20–500°C	12.5	14.2	13.3	13.1	15.0^c	12.6
20–1000°C	14.8		15.6		17.4^c	15.1
thermal conductivity, W/(m·K)					14.8	14.3
specific gravity	8.69	8.46	8.56	8.34	8.39	8.64
electrical resistivity, μΩ·m	0.94	0.84	0.88		0.91	
melting range, °C						
solidus	1255	1285	1280	1186	1265	1288
liquidus	1290	1395	1315	1383	1354	1352

[a]3.2 mm (1/8 in.) thick sheet.
[b]To convert MPa to psi, multiply by 145.
[c]Starting temperature of 0°C.

ASM Metals Handbook, Vol. 1, 10th ed., ASM International, *Metals* Park, Ohio, 1990.

C. S. Hurlbut Jr., *Dana's Manual of Mineralogy*, 17th ed., John Wiley & Sons, Inc., New York, 1966.

K. B. Shedd, "Cobalt," *Mineral Commodity Summaries*, U.S. Geological Survey, Reston, Va., Jan. 2001.

What is the Health Benefit of Radiation?, Canadian Nuclear Association fact sheet, Jan. 1998.

F. Galen Hodge
Haynes International, Inc.

Larry Dominey
OM Group, Inc., (retired)

COBALT COMPOUNDS

Cobalt forms numerous compounds and complexes of industrial importance.

Cobalt exists in the +2 or +3 valence states for the majority of its compounds and complexes. A multitude of complexes of the cobalt(III) ion exist, but few stable simple salts are known. Cobalt(II) forms numerous simple compounds and complexes, most of which are octahedral or tetrahedral in nature; cobalt(II) forms more tetrahedral complexes than other transition-metal ions. Because of the small stability difference between octahedral and tetrahedral complexes of cobalt(II), both can be found in equilibrium for a number of complexes. Typically, octahedral cobalt(II) salts and complexes are pink to brownish red; most of the tetrahedral Co(II) species are blue.

PREPARATION AND PROPERTIES

Cobalt(II) Salts

Cobalt(II) acetate tetrahydrate $Co(C_2H_3 O_2)_2 \cdot 4H_2O$, occurs as pink, deliquescent, monoclinic crystals. It can be prepared by reaction of cobalt carbonate or hydroxide and solutions of acetic acid, by reflux of acetic acid solutions in the presence of cobalt(II) oxide, or by oxygenation of hot acetic acid solutions over cobalt metal. It is used as a bleaching agent and drier in inks and varnishes, and in pigments catalysis agriculture, and the anodizing industries.

The mauve colored cobalt(II) carbonate of commerce is a basic material of indeterminate stoichiometry, $(CoCO_3)_x \cdot (CO(OH)_2)_y \cdot zH_2O$, that contains 45–47% cobalt. It is prepared by adding a hot solution of cobalt salts to a hot sodium carbonate or sodium bicarbonate solution.

Cobalt(II) acetylacetonate, cobalt(II) ethylhexanoate cobalt(II) oleate, cobalt(II) linoleate, cobalt(II) formate, and cobalt(II) resinate can be produced by metathesis reaction of cobalt salt solutions and the sodium salt of the organic acid, by oxidation of cobalt metal in the presence of the acid, and by neutralization of the acid using cobalt carbonate or cobalt hydroxide.

Cobalt(II) chloride hexahydrate $CoCl_2 \cdot 6H_2O$, is a deep red monoclinic crystalline material that deliquesces. It is prepared by reaction of hydrochloric acid with the metal, simple oxide, mixed valence oxides, carbonate, or hydroxide.

Cobalt(II) hydroxide, $Co(OH)_2$, is a pink, rhombic crystalline material containing about 61% cobalt. It is insoluble in water, but dissolves in acids and ammonium salt solutions. The material is prepared by mixing a cobalt salt solution and a sodium hydroxide solution. The hydroxide is a common starting material for the preparation of cobalt compounds. It is also used in paints and lithographic printing inks and as a catalyst.

Cobalt(II) nitrate hexahydrate, $Co(NO_3)_2 \cdot 6H_2O$, is a dark reddish to reddish brown, monoclinic crystalline material containing about 20% cobalt. Cobalt nitrate can be prepared by dissolution of the simple oxide or carbonate in nitric acid, but more often it is produced by direct oxidation of the metal with nitric acid. The nitrate is used in electronics as an additive in nickel–cadmium batteries, in ceramics, and in the production of vitamin B_{12}.

Cobalt(II) oxalate CoC_2O_4, is a pink to white crystalline material that absorbs moisture to form the dihydrate. It precipitates as the tetrahydrate on reaction of cobalt salt solutions and oxalic acid or alkaline oxalates. It is used in the production of cobalt powders for metallurgy and catalysis, and is a stabilizer for hydrogen cyanide.

Cobalt(II) phosphate octahydrate $Co_3(PO_4)_2 \cdot 8H_2O$, is a red to purple amorphous powder. The product is obtained by reaction of an alkaline phosphate and solutions of cobalt salts. The phosphate is used in glazes, enamels, pigments and plastic resins, and in certain steel phosphating operations.

Cobalt(II) sulfamate, $Co(NH_2SO_3)_2$, is generally produced and sold as a solution containing about 10% cobalt. The product is formed by reaction of sulfamic acid and cobalt(II) carbonate or cobalt(II) hydroxide, or by the aeration of sulfamic acid slurries over cobalt metal. Cobalt(II) sulfamate is used in the electroplating industry and in the manufacture of precision molds for record and compact discs.

Cobalt(II) sulfate heptahydrate, $CoSO_4 \cdot 7H_2O$, is a reddish pink monoclinic crystalline material that effloresces in dry air to form the hexahydrate. Cobalt(II) sulfate can be prepared by solution of cobalt(II) carbonate, cobalt(II) hydroxide, or cobalt(II) oxide in sulfuric acid. Cobalt sulfate heptahydrate and cobalt(II) sulfate monohydrate are the most economical sources of cobalt ion and are used in feed supplements as well as in the electroplating industry, in storage batteries, in porcelain pigments, glazes, and as a drier for inks.

Cobalt Oxides

Cobalt(II) oxide, CoO, is an olive green, cubic crystalline material. The simple oxide is most often produced by oxidation of the metal at temperatures above 900°C. It is used in glass decorating and coloring and is a precursor for the production of cobalt chemicals.

Cobalt(II) dicobalt(III) tetroxide, Co_3O_4, is a black cubic crystalline material containing about 72% cobalt. It is prepared by oxidation of cobalt metal at temperatures below 900°C or by pyrolysis in air of cobalt salts, usually the nitrate or chloride. It is used in enamels, semiconductors, and grinding wheels. Both oxides adsorb molecular oxygen at room temperatures.

Cobalt Carbonyls

Dicobalt octacarbonyl, $Co_2(CO)_8$, is an orange-red solid that decomposes in air. It is prepared by heating cobalt metal to 300°C under 20–30,000 kPa (3–4000 psi) of carbon monoxide, by reduction of cobalt(II) carbonate with hydrogen under pressure at high temperatures, or by heating a mixture of cobalt(II) acetate and cyclohexane to 160°C in the presence of carbon monoxide and hydrogen at 30,000 kPa (4000 psi) pressure.

HEALTH AND SAFETY FACTORS

Cobalt is one of twenty-seven known elements essential to humans. It is an integral part of the cyanocobalamin molecule; ie, vitamin B_{12}, the only documented biochemically active cobalt component in humans.

Cobalt compounds can be classified as relatively nontoxic. There have been few health problems associated with workplace exposure to cobalt. The primary workplace problems from cobalt exposure are fibrosis, also known as hard metal disease, asthma, and dermatitis. Finely powdered cobalt can cause silicosis. There is little evidence to suggest that cobalt is a carcinogen in animals and no epidemiological evidence of carcinogenesis in humans.

USES

Cobalt in Catalysis

About 80% of cobalt catalysts are employed in three areas: (1) hydrotreating/desulfurization in combination with molybdenum for the oil and gas industry; (2) homogeneous catalysts used in the production of terphthalic acid or dimethylterphthalate; and (3) the high-pressure oxo process for the production of aldehydes and alcohols.

Cobalt in Driers for Paints, Inks, and Varnishes

The cobalt soaps; eg, the oleate, naphthenate, resinate, linoleate, ethylhexanoate, synthetic tertiary neodecanoate, and tall oils, are used to accelerate the natural drying process of unsaturated oils such as linseed oil and soybean oil.

Cobalt as a Colorant in Ceramics, Glasses, and Paints

Cobalt(II) ion displays a variety of colors in solid form or solution ranging from pinks and reds to blues or greens. It has been used for hundreds of years to impart color to glasses and ceramics or as a pigment in paints and inks.

Miscellaneous Uses

Adhesives in the Tire Industry. Cobalt salts are used to improve the adhesion of rubber to steel.

Adhesion of Enamel to Steel. Cobalt compounds are used both to color and to enhance adhesion of enamels to steel.

Agriculture and Nutrition. Cobalt salts, soluble in water or stomach acid, are added to soils and animal feeds to correct cobalt deficiencies. In soil application the cobalt is readily assimilated into the plants and subsequently made available to the animals. Plants do not seem to be affected by the cobalt uptake from the soil. Cobalt salts are also added to salt blocks or pellets.

Electroplating. Cobalt is plated from chloride, sulfate, fluoborate, sulfamate, and mixed anionic baths. Cobalt alloyed with nickel, tungsten, iron, molybdenum, chromium, zinc, and precious metals is plated from mixed metal baths. A cobalt phosphorus alloy is commonly plated from electroless baths. Cobalt tungsten and cobalt molybdenum alloys are produced for their excellent high-temperature hardness. Magnetic recording materials are produced by electroplating cobalt from sulfamate baths and phosphorus-containing baths or by electroless plating of cobalt from baths containing sodium hypophosphite as the reducing agent. Cobalt is added to nickel electroplating baths to enhance hardness and brightness or for the production of record and compact discs.

Electronic Devices. Small quantities of cobalt compounds are used in the production of electronic devices such as thermistors, varistors, piezoelectrics, and solar collectors. Cobalt salts are useful indicators for humidity. The blue anhydrous form becomes pink (hydrated) on exposure to high humidity. Cobalt pyridine thiocyanate is a useful temperature indicating salt. A conductive paste for painting on ceramics and glass is composed of cobalt oxide.

Batteries and Fuel Cells. Cobalt salts are used as activators for catalysts, fuel cells, and batteries. Thermal decomposition of cobalt oxalate is used in the production of cobalt powder. Cobalt compounds have been used as selective absorbers for oxygen, in electrostatographic toners, as fluoridating agents, and in molecular sieves.

Cure Accelerator. Cobalt ethylhexanoate and cobalt naphthenate are used as accelerators with methyl ethyl ketone peroxide for room-temperature curing of polyester resins.

R. Brugger, *Nickel Plating: A Comprehensive Review of Theory, Practice, Properties, and Applications Including Cobalt Plating*, Draper, Teddington, UK, 1970.

J. D. Donaldson, S. J. Clark, and S. M. Grimes, *Cobalt in Chemicals, The Monograph Series*, Cobalt Development Institute, London, 1986; Guilford, Surrey, 2003.

K. B. Shedd, "Cobalt", *Mineral Commodity Summaries*, U.S. Geological Survey, Reston, Va., Jan. 2003.

R. S. Young, *Cobalt in Biology and Biochemistry*, Academic Press, Inc., London, 1979.

H. Wayne Richardson
CP Chemicals, Inc.
Dayal T. Meshri
Advance Research Chemicals, Inc.

COFFEE

Commercial coffees are grown in tropical and subtropical climates at altitudes up to $\sim$1,800 m; the best grades are grown at high elevations. Most individual coffees from different producing areas possess characteristic flavors. Commercial roasters obtain preferred flavors by blending or mixing the varieties before or after roasting. Colombian, and washed Central American or East African coffees, are generally characterized as mild, winey-acid, and aromatic; Brazilian coffees as heavy body, moderately acid, and aromatic; and African and Asian robusta coffees as heavy body, neutral to slightly earthy, slightly acid, and slightly aromatic. Premium coffee blends contain higher percentages of Colombian, East African, and Central American coffees.

GREEN COFFEE PROCESSING

The coffee plant is a relatively small tree or shrub belonging to the family Rubiaceae. It is often controlled to heights between 3 and 5 m. *Coffea arabica* (milds) accounts for 70–80% of world production and *Coffea Canephora* (robustas) for most of the remaining world's production. *Coffea liberica* and >20 others comprise the remaining species.

Green coffee processing is effected by one of three methods: the wet method (washed or semiwashed, which appears to be growing in popularity) and the natural or dry process method.

In the wet (fully washed) method (practiced in Colombia and many Central American origins), the harvested ripe coffee cherry is passed through a tank for washing that removes stones and other foreign matter. The coffee is then passed through a depulper, removing the outer covering or skin and most of the pulp. Some pulp mucilage remains on the parchment layer covering the bean. This remaining pulp is removed by utilizing a fermentation process. Depending on the amount of mucilage remaining on the bean, fermentation may last from 12 h to several days. Because excessive fermentation may cause development of undesirable characters in the flavor and odor of the beans, enzymes may be added to speed the process. The fermentation breaks down the remaining pulp so it can then be easily washed away.

The wet (semiwashed) method is similar. Here, a partial or reduced fermentation after hulling may be used. The coffee is then washed to remove the remaining pulp (other variations include aggressively mechanically washing the beans to remove the pulp without fermentation).

From both wet methods, the resultant beans (covered with parchment) will be dried to uniform moisture using natural or mechanical methods before going through the milling process. The mechanical methods have gained in popularity despite higher costs. They are faster and not dependent on weather conditions. A potential negative of this method is the use of wood fuel for the dryers that may impart smoke/ash flavors to the coffee.

After drying, the coffee is further mechanically processed to remove the parchment and silverskin before grading. The wet process generally produces coffee of more uniform quality and flavor.

The dry method is favored by Brazil and other countries where available water is limited and a relatively uniform flowering allows strip picking with a predominance of ripe coffee. This method involves setting the harvested

coffee cherries on a patio to dry naturally (mechanical methods can be used but are much more costly). The coffee is periodically turned to ensure uniform drying and to help prevent the occurrence of mold. Once proper drying has occurred (7–21 days based upon conditions) the resultant coffee may then be stored or hulled to remove the husk, parchment, and silverskin.

After being processed by any of the above methods, the coffee may need to be further dried to bring it to the desired 11–13% moisture range.

Coffee prepared by either method is then machine graded using oscillating screens, density separating tables, and airveyors to separate the coffee into varying sizes. The beans are then further processed to remove foreign matter, and any undesirable and damaged beans. This is accomplished by hand picking, machine separating, or color sorting or any combination of the aforementioned.

Unroasted green coffee, once processed and properly stored, can be expected to maintain its quality for 6–12 months.

Coffee is valued and offered for sale based on its grade as determined by the number of imperfections (beans that are black, broken, insect damaged or otherwise unsound, and foreign material) contained in a 300-g sample. Coffee processors and roasters will also grade the coffee based on final moisture and bean color as well as the cup quality of the prepared beverage.

Consumers more conscious of health and also those interested in conservation have driven niche markets for coffee that has been organically grown, farms that are sustainable, and fair-traded coffee.

Organic coffee is coffee that has been certified by an acceptable agency as being grown under conditions considered to be organic in nature and have not used synthetic fertilizers and pesticides.

Sustainable coffee farms employ techniques beneficial to the environment. Effective land management, the use of fertilizers and pesticides that are safe for animal and bird populations, and following clean water practices ensures farms remain sustainable for future generations.

Fair-traded coffee is coffee that has a guaranteed minimum price, provides credit to producers, and establishes long-term relationships directly with cooperatives. Fair-traded coffee encourages organic and sustainable farming by providing financial incentives to the farmers and growers through the elimination of the intermediaries (middlemen).

COFFEE CHEMISTRY

Chemical Composition of Green Coffee

The composition of the green coffee bean is complex; carbohydrate polymers and protein make up ~60% of the bean; the rest consists of low molecular weight compounds of various types, including lipids, acids and sugars. The composition can vary according to species, variety, growing environment, postharvest handling (including wet and dry processing), and storage time, temperature, and humidity. Table 1 summarizes the analyses of robusta and arabica green beans.

Table 1. Typical Analyses of Green Coffee (% Dry Weight Basis)

Constituent	Type[a]	
	Robusta	Arabica
lipids	11.5 (9–13)	16 (15–18)
ash	5.0 (4.5–5.3)	4.7 (4.2–5.2)
caffeine	2.3 (1.8–2.8)	1.3 (1.2–1.5)
chlorogenic acids	10.5 (9–11)	9 (8–10)
other acids	3.0 (2.7–3.3)	3.5 (3.2–3.8)
trigonelline	1.0 (0.7–1.2)	1.3 (1.1–1.5)
protein	12 (11–14)	11 (10–13)
free amino acids	0.2	0.2
sucrose	4 (2–5)	7 (5–8.5)
other sugars	0.5 (0.2–1.0)	0.5 (0.1–0.8)
polymeric carbohydrate		
mannan	22	22
arabinogalactan	17	14
cellulose	8	8
others		
other compounds	2.0	2.0

[a]Robustas generally have lower levels of lipid, trigonelline, sucrose and phytic acid but higher levels of caffeine chlorogenic acids (mainly 5-caffeoylquinic acid and arabinogalactan than arabicas.

Table 2. Approximate Analyses of Roasted, Brewed, and Instant Coffee (% Dry Weight Basis)

Constituent	Roasted	Brewed	Instant
lipids	17 (16–20)	0.2[a]	0.1
ash	4.5 (4–5)	15	8
caffeine	1.2 (1.0–1.6)	5	2.5 (2–3)
chlorogenic acids	3.5 (1.5–5)	15	5 (3–9)
other acids	3 (2.5–3.5)	10	5.5 (4–8)
trigonelline	0.8 (0.5–1.0)	2 (1–2.5)	1.5 (1–2)
protein	8–10	5	3 (2–5)
sucrose	0.1 (0–0.3)	0.2 (0–0.5)	0.1 (0–0.3)
reducing sugars	0.1 (0–0.3)	0.2 (0–0.5)	2 (1–5)
polymeric carbohydrate[b]			
mannan	22	1	15 (10–18)
arabinogalactan	13	2	17 (15–20)
cellulose	8	0	0
other compounds[c]	18	45	40

[a]Maximum level in brew (filtered), espresso levels higher, even higher levels in "Scandinavian" brews.
[b]Much of polymer is converted in instant coffee processing to lower molecular weight (oligosaccharides) material.
[c]Most of "other" material is in the form of "browning products", such as the melanoidins.

Roast Coffee Composition

The pleasant taste and aroma characteristics as well as the brown color of roast coffee are developed during the roasting process. The chemical and physical changes associated with this process are very complex. In the first stage of roasting, loss of free water (~12% in the green bean) occurs. In the second stage, chemical dehydration, fragmentation, recombination, and depolymerization reactions occur. Many of these reactions are of the Maillard type and lead to the formation of lower molecular weight compounds such as carbon dioxide, free water, and those associated with flavor and aroma as well as higher molecular weight compounds termed melanoidins that are responsible for the brown color. Most of the lipid, caffeine, inorganic salts, and polymeric carbohydrate survive the roasting process.

Table 2 indicates some of the chemical changes that occur in arabican green coffee as a result of processing.

Chemistry of Brewed Coffee

The chemistry of brewed coffee is dependent on the extraction of water-soluble and hydrophobic aromatic components from the coffee cells and lipid phase, respectively. Factors that affect extraction and flavor quality of brewed coffee are degree of roast; blend composition; grinding technique; particle size and density; water quality; water to coffee ratio; and brewing technique or device, such as drip filter, percolator, or espresso, which defines the water temperature, steam pressure, brewing time, water recycle, etc. Extraction yields of home brewing range from ~9 to 28% and typically ~23% dry basis roasted and ground (R&G). The trend in the United States, with some notable exceptions, has been toward weaker brew strengths, with typical recipes of ~5 g of R&G coffee per 6 oz cup (brew solids concentration ~0.7%), compared to ~10 g a generation ago (brew solids ~1.2%). In comparison, espresso typically uses ~8–12 g of coffee per 2 oz of beverage (brew solids ranging from ~3 to 5%).

Instant Coffee

The chemistry of instant or soluble coffee is dependent on the R&G blend and processing conditions. This is indicated in Table 2 by the wide range of constituents. In addition to the atmospherically extractable solids found in brewed coffee, commercial percolation generates water-soluble carbohydrate by hydrolysis that contributes to the yield.

ROASTED AND GROUND COFFEE PROCESSING AND PACKAGING

The main processing steps in the manufacture of roast and ground coffee products are blending, roasting, grinding, and packaging.

Roasting Technology

Commercial roasting is generally by hot combustion gases in either rotating cylinders or fluidized-bed systems. Though steam pressure, infrared (ir) and microwave roasting systems are found in the patent literature, none are believed to be used to a significant extent.

Grinding

Grinding of the roasted coffee beans is tailored to the intended method of beverage preparation. Average particle size distributions range from very fine (500 µm or less) to very coarse (1100 µm).

Packaging

Most roasted and ground coffee sold directly to consumers in the United States is vacuum packed with

0.33–0.37 kg in 10.3-cm (4 1/16 in.) diameter metal cans or with three times as much in 15.7-cm (6 3/16 in.) diameter metal cans. Whether packing 0.33 or 0.37 kg of coffee, the 10.3-cm diameter cans generally have a fill volume of ~1000 cm^3. Vacuum-packed coffee retains a high-quality rating for at least 1 year. Coffee vacuum-packed in flexible, bag-in-box packages has gained wide acceptance in Europe and is the prevalent format for prepackaged ground coffee.

INSTANT COFFEE PROCESSING AND PACKAGING

Instant coffee is the dried water-extract of ground, roasted coffee.

Green beans for instant coffee are blended, roasted, and ground similarly to those for roasted and ground products. A concentrated coffee extract is normally produced by pumping hot water through the coffee in a series of cylindrical percolator columns. The extracts are further concentrated prior to a spray- or freeze-drying step, and the final powder is packaged in glass or other suitable material. Some soluble coffees, both spray- and freeze-dried, are manufactured in producing countries for export.

DECAFFEINATED COFFEE PROCESSING

Decaffeinated coffee products represented 9% of the coffee consumed in 2001 in the United States.

Until the 1980s, synthetic organic solvents commonly were used in the United States to extract the caffeine, either by direct contact as above or by an indirect secondary water-based system. In each case, steaming or stripping was used to remove residual solvent from the beans and the beans were dried to their original moisture content (10–12%) prior to roasting.

In the 1980s, manufacturers' commercialized processes that utilized either naturally occurring solvents or solvents derived from natural substances to position their products as naturally decaffeinated. The three most common systems use carbon dioxide under supercritical conditions, oil extracted from roasted coffee, or ethyl acetate, an edible ester naturally present in coffee.

In all the above-mentioned processes of coffee decaffeination, changes occur that affect the roast flavor development.

To make an instant decaffeinated coffee product, the decaffeinated roast and ground coffee is extracted in a manner similar to nondecaffeinated coffee. Alternatively, the caffeine from the extract of untreated roasted coffee is removed by using the solvents described previously.

ECONOMIC IMPORTANCE

Coffee ranks second only to petroleum as the world's largest traded commodity. The total world production of green coffee in the 1999–2000 growing season was 113.5 million bags; exportable production was 88.2 million bags with an export value of $7.5 billion.

COFFEE BIOTECHNOLOGY

There has recently been considerable interest in applying biotechnology to coffee, through both conventional breeding using tissue culture techniques and molecular markers and through the application of genetic modification (GM) technology.

Advances in Coffee Breeding and Culture Methods

Application of modern plant-breeding methods to coffee improvement, including production of F$_1$ hybrids, has led to improvements in crop performance. Breeding targets remain a combination of agronomic factors such as disease resistance and yield, and quality; eg, cup quality and caffeine content. Disease resistance targets include coffee leaf rust, coffee berry disease, and nematodes.

Application of Molecular Markers to Coffee Breeding

Like other tree-based crops, coffee's long generation time makes it a good candidate for application of marker-assisted selection in breeding. Programs seeking to identify genes controlling traits such as host resistances to diseases and pests, caffeine content and cup quality, and to develop molecular markers for these traits are under way.

Advances and Applications of Coffee Molecular Biology

Coffee transformation using Agrobacterium sp. vectors has been reviewed recently. Arabica coffee has proved to be more amenable to transformation than robusta coffee. Rapid progress has been made in isolating and identifying genes from coffee.

Despite the technical progress that has been made, there are currently no genetically modified (GM) coffee varieties in the marketplace. Until there are clear consumer benefits, this is unlikely to change.

Biotechnology has been applied to roast and ground and soluble coffee processing. The flavor profile of roast and ground coffee has been manipulated by treating green or partially roasted coffee with a cocktail of hydrolytic enzymes. The process is claimed to generate a novel and pleasant flavor profile.

COFFEE REGULATIONS AND STANDARDS

Soluble and roast and ground coffee is covered by a range of international and national legislation (eg, EEC, United States, Germany) and voluntary Codes of Practice (eg, National Coffee Association). The standards and regulations provide definitions for green and processed coffee and potential health risks (eg, ochratoxin in coffee, decaffeination of coffee). Additionally, the International Standards Organization (ISO) maintains standards for coffee and coffee-based products.

COFFEE SUBSTITUTES

Coffee substitutes, which include roasted chicory, chick peas, cereal, fruit, and vegetable products, have been used in all coffee-consuming countries. It is not unusual for consumers in some of the coffee-producing countries to blend coffee with noncoffee materials.

R. J. Clarke and R. Macrae, eds., *Coffee*, Vol. 1, *Chemistry*, 1985; Vol. 2, *Technology*, 1987; Vol. 3, *Physiology*, 1988; Vol. 4, *Agronomy*, 1988; Vol. 5, *Related Beverages*, 1987; Vol. 6, *Commercial and Technico-Legal Aspects*, 1988; Elsevier Applied Science Publishers, Ltd., Barking, U. K.

R. J. Clarke and O. G. Vitzthum, eds., *Coffee, Recent Developments*, Blackwell Science, London, U.K., 2001.

M. Sivetz and N. Desrosier, *Coffee Technology*, AVI Publishing Co., Inc., Westport, Conn., 1979.

GERALD S. WASSERMAN
ALLAN BRADBURY
THEODORE CRUZ
SIMON PENSON
Kraft Foods Corporation

COLLOIDS

Matter of colloidal size, just above atomic dimensions, exhibits physicochemical properties that differ from those of the constituent atoms or molecules yet are also different from macroscopic material. The atoms and molecules of classical chemistry are extremely small, usually having molar masses <1000 g/mol and measurable by freezing point depression. Macroscopic particles fall into the realm of classical physics and can be understood in terms of physical mechanics. Residing between these extremes is the colloidal-size range of particles whose small sizes and high surface-area-to-volume ratios make the properties of their surfaces very important and lead to some unique physical properties. Their solutions may have undetectable freezing point depressions, and their dispersions, even if very dilute, may sediment out very slowly, and not be well described by Stokes' law. Whereas the particles of classical chemistry may have one or a few electrical charges, colloidal particles may carry thousands of charges. With such strong electrical forces, complete dissociation is the rule rather than the exception. In addition, the electric fields can strongly influence the actions of neighbouring particles. In industrial practice it is very common to encounter problems associated with colloidal-sized particles, droplets, or bubbles.

A variety of types of colloidal dispersions occur, as illustrated in Table 1. In practice, many colloidal dispersions are more complex and are characterized by the nature of the continuous phase and a primary dispersed phase.

One reason for the importance of colloidal systems is that they appear in a wide variety of practical disciplines, products, and processes. The colloidal involvement in a process may be desirable, as in the stabilizing of emulsions in mayonnaise preparation, or undesirable, as in the tendency of very finely divided and highly charged particles to resist settling and filtration in water treatment plants.

Table 1. Types of Colloidal Dispersion

Dispersed phase	Dispersion medium	Name	Examples
liquid	gas	liquid aerosol	fog, mist
solid	gas	solid aerosol	smoke, dust
gas	liquid	foam	soap suds
liquid	liquid	emulsion	milk, mayonnaise
solid	liquid	sol, suspension	ink, paint, gel
gas	solid	solid foam	polystyrene foam, pumice stone
liquid	solid	solid emulsion	opal, pearl
solid	solid	solid suspension	alloy, ruby-stained glass

The problems associated with colloids are usually interdisciplinary in nature and a broad scientific base is required to understand them completely.

DISPERSED SPECIES CHARACTERIZATION AND SEDIMENTATION

The characterization of colloids depends on the purposes for which the information is sought. Among the properties to be considered are the nature and/or distributions of purity, crystallinity, defects, size, shape, surface area, pores, adsorbed surface films, internal and surface stresses, stability, and state of agglomeration.

INTERFACIAL ENERGETICS

In colloidal dispersions, a thin intermediate region or boundary, known as the interface, lies between the dispersed and dispersing phases. Each interface has a certain free energy per unit area that has a great influence on the stability and structure of the colloidal dispersion, and in practical areas such as mineral flotation, detergency, and waterproofing.

ELECTROKINETICS

Charged Interfaces

Most substances acquire a surface electric charge when brought into contact with a polar medium such as water. The origin of the charge can be ionization, ion adsorption, ion dissolution, or ion diffusion.

Surface charge influences the distribution of nearby ions in a polar medium: ions of opposite charge (counterions) are attracted to the surface while those of like charge (co-ions) are repelled. Together with mixing caused by thermal motion a diffuse electric double layer is formed. The electric double layer (EDL) can be viewed as being composed of two layers: an inner layer that may include adsorbed ions and a diffuse layer where ions are distributed according to the influence of electrical forces and thermal motion.

Electrokinetic Phenomena

Electrokinetic motion occurs when the mobile part of the electric double layer is sheared away from the inner layer (charged surface). There are several types of electrokinetic measurements; electrophoresis, electroosmosis, streaming potential, sedimentation potential, and two electroacoustical methods. Of these the first finds the most use in industrial practice. The electroacoustical methods involve detection of the sound waves generated when dispersed species are made to move by an imposed alternating electric field, or vice versa. In all of the electrokinetic measurements either liquid is made to move across a solid surface or vice versa. Thus the results can be interpreted only in terms of charge density (σ) or potential (zeta potential, ζ) at the plane of shear.

COLLOID STABILITY

A consequence of the small size and large surface area in colloids is that quite stable dispersions of these species can be made. That is, suspended particles may not settle out rapidly and droplets in an emulsion or bubbles in a foam may not coalesce quickly. Charged species, when sedimenting, present a challenge to Stokes' law because the smaller counterions sediment at a slower rate than do the larger colloidal particles. This creates an electrical potential that tends to speed up the counterions and slow down the particles. At high enough electrolyte concentrations the electric potentials are quickly dissipated and this effect vanishes.

Although some lyophobic colloidal dispersions can be stable enough to persist for days, months, or even years, they are not thermodynamically stable. Rather, they possess some degree of kinetic stability, and one must consider the degree of change and the timescale in the definition of stability. Having distinguished coalescence and aggregation as processes in which particles, droplets, or bubbles are brought together with and without large changes in surface area respectively, it is clear that there can be different kinds of kinetic stability. Finally, stability depends upon how the particles interact when this happens, since encounters between particles in a dispersion can occur frequently due to diffusion (as in Brownian motion), sedimentation, or stirring.

Electrostatic and Dispersion Forces

Several repulsive and attractive forces operate between colloidal species and determine their stability. In the simplest example of colloid stability, particles would be stabilized entirely by the repulsive forces created when two charged surfaces approach each other and their electric double layers overlap. The overlap causes a Coulombic repulsive force that will act in opposition to any attempt to decrease the separation distance. Another important repulsive force, called the Born repulsion, causes a strong repulsion at very small separation distances where the atomic electron clouds overlap. There also exist dispersion, or London-van der Waals, forces that molecules exert toward each other.

DLVO Theory

The quantitative theory for the stability of lyophobic colloids referred to as DLVO theory can be used to calculate the energy changes that take place when two particles approach each other by estimating the potential energies of attraction and repulsion versus interparticle distance. These are then added together to yield the total interaction energy V_T. The theory has been developed for two special cases, the interaction between parallel plates of infinite area and thickness, and the interaction between two spheres. The classic DLVO models are for flat planes and spheres, but more complex shapes arise in practice.

Steric and Hydrodynamic Effects

Additional influences on dispersion stability beyond those accounted for by DLVO theory, like steric, surface hydration, and hydrodynamic effects, have received considerable attention over the past several decades. Generally, the stability of a dispersion can be enhanced (protection) or reduced (sensitisation) by the addition of material that adsorbs onto particle surfaces. The protective agents can act in several ways. They can increase double-layer repulsion if they have ionisable groups. The adsorbed layers can lower the effective Hamaker constant. An adsorbed film may necessitate desorption before particles can approach closely enough for van der Waals forces to cause attraction.

KINETIC PROPERTIES

In principle, any distinctive physical property of the colloidal system in question can be used, at least empirically, to monitor changes in the dispersed state. The more complex a system is, the less likely it is that a single property uniquely and completely describes changes in the colloidal state. Aggregation and/or coalescence of colloidal material can be monitored by a wide variety of techniques, including light scattering, neutron scattering, microscopy, rheology, conductivity, filtration, sedimentation, and electrokinetics.

APPLICATIONS

General Uses

Although colloids may be undesirable components in industrial systems, particularly as waste or by-products and, in nature, in the forms of fog and mist, they are desirable in many technologically important processes such as mineral beneficiation and the preparation of ceramics, polymers, composite materials, paper, foods, textiles, photographic materials, drugs, cosmetics, and detergents.

Colloidal Solids

Some uses of solid colloids include reinforcement aids in metals, ceramics, and plastics; as adhesion promoters in paints and thermoplastics; as nucleating agents in cloud seeding; as activated powder catalysts; as thickening agents in gels and slurries; and as abrasives in toothpastes. Alumina and thoria are used to reinforce aluminum and nickel, respectively, and zirconia and silicon carbide to reinforce a variety of ceramics. Stable but ordered suspensions can be regarded as precursory systems for ordered, prefired, ceramic components; outlets for such systems include various processing techniques such as slip, tape, freeze, pressure, centrifugal, and ultrasonic casting, as well as isostatic, and hot pressing. Asbestos, crystalline silicas, and organic solids are added to concrete to improve its strength by providing an interlocking particulate structure within the concrete matrix; asbestos, various oxides, and carbon black are added to reinforce polymers by inducing a stiffened or high yield matrix.

Particle suspensions have long been of great practical interest because of their widespread occurrence in everyday life. Some important kinds of familiar suspensions include those occurring in foods (batters, puddings, sauces), pharmaceuticals (cough syrups, laxative suspensions), household products (inks, paints, "liquid" waxes) and the environment (suspended lake and river sediments, sewage). In the petroleum industry alone, suspensions may be encountered throughout each of the stages of petroleum recovery and processing.

Other applications of colloidal solids include the preparation of rigid, elastic and thixotropic gels, aerogel-based thermal insulators, and surface coatings. Commercial uses of silica gel and sol–gel processing often focus on rigid gels having 20–30 vol% SiO_2. Elastic gels are commonly associated with cellophane, rubber, cellulosic fibers, leather, and certain soaps. Many thixotropic gels and surface coatings contain colloidal solids; eg, clays, alumina, ferric oxide, titania, silica, and zinc oxide. Consumer and industrial pastes—putty, dough, lubricating grease, toothpaste, and paint are some examples.

Colloidal Liquids

These fluids are commonly used in emulsions by many industries. Permanent and transient antifoams consisting of an organic material; eg, polyglycol, oils, fatty materials, or silicone oil dispersed in water, is one application important to a variety of products and processes: foods, cosmetics, pharmaceuticals, pulp and paper, water treatment, and minerals beneficiation. Other emulsion products include foods, insecticides and herbicides, polishes, drugs, biological systems, asphalt paving emulsions, personal care creams and lotions, paints, lacquers, varnishes, and electrically and thermally insulating materials.

Other common applications of colloidal liquids include liquid aerosols, such as those occurring in the environment (fog, mist, cloud, smog), agriculture (crop sprays), manufacturing (paint sprays), medicine (nasal sprays), and personal care (hair spray).

Colloidal Gases

Fluid foams are commonplace in foods, shaving cream, fire-fighting foam, mineral flotation, and detergents. Thus, in view of the fact that the concentration of bubbles greatly affects the properties of foams, the production, dispersion, and maintenance of colloidal gas bubbles are basic to foams and related materials. Often, natural and synthetic soaps and surfactants are used to make fluid foams containing colloidal gas bubbles. These agents reduce the interfacial tension and, perhaps, increase the viscosity at the gas–liquid interface, making the foam stable. Also, some soluble proteins that denature upon adsorption or with agitation of the liquid phase can stabilize foams by forming insoluble, rigid layers at the gas–liquid interface.

A class of enhanced oil recovery processes involves injecting a gas in the form of a foam.

Microfoams (also termed colloidal gas aphrons) comprise a dispersion of aggregates of very small foam bubbles in aqueous solution. Some interesting potential applications have been reported: soil remediation and reservoir oil recovery.

Many agents will act to reduce the foam stability of a system (termed foam breakers or defoamers) while others can prevent foam formation in the first place (foam preventatives, foam inhibitors).

Other common applications of colloidal gases include solid foams, such as those occurring in the areas of food (leavened breads), geology (pumice stone, zeolites), manufacturing (polystyrene foam, polyurethane foam), and personal care (synthetic sponges). Solid foams such as polyurethane foam contain dispersed gas bubbles that are often produced via viscoelastic polymer melts within which gas (eg, carbon dioxide) bubbles are nucleated.

HAZARDS OF COLLOIDAL SYSTEMS

The occurrence of some materials in the form of a colloidal dispersion can introduce or enhance safety hazards. Considering that the dispersion of a material down to colloidal size results in a high specific surface area and colloidal chemical reactivity may differ considerably from that of the identical macroscopic material with less surface area. This is particularly important if the colloidal surface is easily and rapidly oxidized. Dust explosions and spontaneous combustion are potential dangers whenever certain materials exist as finely divided dry matter exposed to oxidizing environments. Dispersions of charged colloidal particles in nonaqueous media occur throughout the petroleum industry. The flow of petroleum fluids in tanks or pipelines, combined with the low conductivity of the petroleum fluids themselves, can allow the build-up of large potential gradients and a separation of charges. Sufficient charging for there to be an electrostatic discharge can cause an explosion.

Health problems can be caused by solids and liquids suspended in air or water. Specific potential hazards have been associated with a diverse spectrum of colloidal materials, including chemicals, coal, minerals, metals, pharmaceuticals, plastics, and wood pulp. The effects of the colloidal solids and liquids that comprise smog are widespread and well known. Liquid droplets may also constitute a hazard; eg, smog can contain sulfuric acid aerosols. Elements such as lead, zinc, and vanadium that are released into the atmosphere as vapors can subsequently condense or be removed as solid particulates by rain. Similarly, exposure to airborne pollutants found indoors and in confined spaces, many of which are particulates or microbes of colloidal size, can lead to complex physiological responses.

There is a need to understand the environmental properties and risks associated with any large-volume chemicals. The mass of surfactants that could ultimately be released into the environment, for example, is significant.

In addition to products, many industries produce waste containing significant amounts of suspended matter for which treatment incurs significant technological challenges and costs. Large fractions of readily hydrolyzable metals exist as adsorbed species on suspended (colloidal) solids in fresh and marine water systems and can also be anticipated to exist in industrial wastewater.

EMERGING AREAS IN COLLOID SCIENCE

The full potential to control colloids is not presently realized. There are several types of mixed colloids that are only poorly understood. For example, the properties of colloids in which more than one type of colloidal particle is dispersed may be dominated by the behavior of the minor dispersed phase component, and the nature and properties of colloids within colloids, such as suspended solids in the dispersed phase of an emulsion, or emulsified oil within the aqueous lamellae of a foam, are only beginning to be understood.

P. Becher, *Dictionary of Colloid and Surface Science*; Dekker, New York, 1990.

P. Hiemenz and R. Rajagopalan, *Principles of Colloid and Surface Chemistry*, 3rd. ed., Manel Dekker, New York, 1997.

L. L. Schramm, *The Language of Colloid and Interface Science*, American Chemical Society, Washington, and Oxford University Press, New York, 1993.

G. B. Sukhorukov, in D. Möbius and R. Miller, eds., *Novel Methods to Study Interfacial Layers*, Elsevier, New York, 2001, pp. 383–414.

LAURIER L. SCHRAMM
Saskatchewan Research Council

COLOR

COLOR FUNDAMENTALS

An immediate complexity is illustrated in the two early and apparently incompatible theories of color vision. Trichromatic theory, first proposed in 1801 by Thomas Young and later refined by Hermann von Helmholtz, postulated three types of color receptors in the eye. This explained many phenomena, such as various forms of color blindness, and was confirmed in 1964, when three types of blue-, green-, and red-sensitive cones were reported to be present in the retina. Yet Ewald Hering's 1878 opponent theory, which used three pairs of opposites, light–dark, red–green, and blue–yellow, also offered much insight, including the explanation of contrast and after-image effects and the absence of some color combinations such as reddish greens and bluish yellows. In the modern zone theories, it is now recognized that the data from three trichromatic detectors in the eye are processed on their way to the brain into opponent signals, thus removing the apparent inconsistencies.

In color technology and measurement, both types of approaches are used. Color printing, eg, generally employs three colors (usually plus black), and the ever useful CIE system was based on experiments in which colors were matched by mixtures of three primary color light beams, blue, green, and red. Yet transmitted television signals are based on the opponent system, with one intensity and two color-balance signals, as are the modern representations of color, such as the CIELAB and related color spaces based on red–green and yellow–blue opponent axes.

Light and Color

Visible light is that part of the electromagnetic spectrum, shown in Figure 1, with wavelengths between the violet limit of about 400 nm and the red limit at about 700 nm. Depending on the observer, light intensity, etc., typical values for the spectral colors are blue, 450 nm; green, 500–550; yellow, 580; orange, 600; and red, 650. For example, there is no "brown light"; are orange surface viewed in the aperture mode against a brighter surround is preceived as brown.

The appearance of color clearly depends significantly on the exact viewing circumstances. Normally one thinks of viewing a colored object under some type of illumination, the object mode (Table 1). When viewing a light source there is the illuminant or luminous mode. Viewing through a hole in a screen is the aperture mode, and so on. Color perception differs significantly in these modes.

Interactions of Matter with Light

In the most generalized interaction of light with matter the many phenomena of Figure 2 are possible.

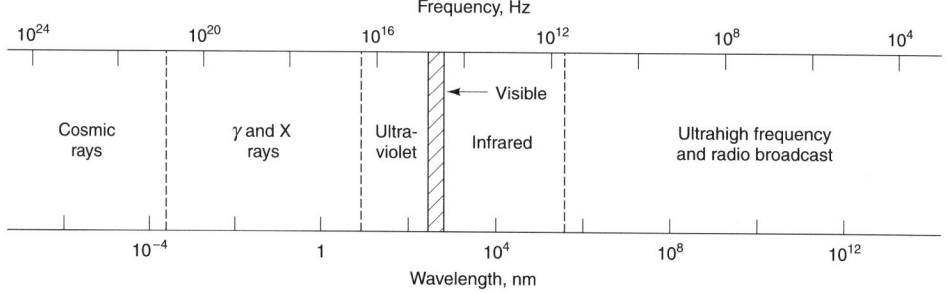

Figure 1. The electromagnetic spectrum.

Table 1. Object Mode Perceptions

Object	Dominant perception	Secondary attributes[*]
opaque metal, polished	specular reflection	reflectivity, gloss, hue
opaque metal, matte	diffuse reflection	hue, saturation, brightness, gloss
opaque nonmetal, glossy	diffuse and specular reflections	hue, saturation, gloss, brightness
opaque nonmetal, matte	diffuse reflection	hue, saturation, brightness
translucent nonmetal	diffuse transmission	translucency, hue, saturation
transparent nonmetal	transmission	hue, saturation, clarity

[*]In approximate sequence of importance.

COLOR VISION

The Eye

Light passes through the cornea, the transparent outer layer of the eye, through the lens and the aqueous and vitreous humors, and is focused onto the retina. The iris, forming the pupil, acts as a variable aperture to control the amount of light that enters the eye, varying from $\sim f/2.5$ to $f/13$ with a 30:1 light intensity ratio. The two humors serve merely as neutral transmission media and to keep the eyeball distended. The retina is a layer ~0.1 mm thick that contains the light-sensitive rods and cones. Only the rods function in low levels of illumination of about <1 lux, providing an achromatic, noncolor image. Each set of cones is sensitive to a wide range of wavelengths with extensive overlap. Appropriate designations are short-, medium-, and long-wavelength sensitive cones.

The trichromatic theory, subsequently confirmed by the existence of the three sets of cones, must be combined with the opponent theory, which is involved in the signal sent along the retinal pathway. A third approach, the appearance theory or the retinex theory, must be added to explain color constancy and other effects. As one

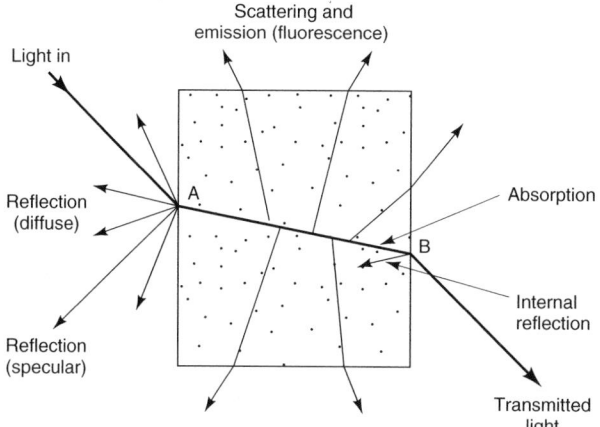

Figure 2. The adventures of a beam of light passing through a block of partly transparent substance.

example of this last, consider an area in a multicolored object such as a Mondrian painting perceived as red when illuminated with white light. If the illumination is changed so that energy reflected by this same area is greater at shorter wavelengths than the energy reflected at longer wavelengths, this area is still perceived as red within the overall visual context. If all other colors are now covered so that only our "red" area is visible, corresponding to the aperture mode, then this area is perceived as green. Clearly, all three approaches must be melded to give a full description of color perception, a process that is not yet fully understood.

Color Vision Defects

Anomalous color vision is present, eg, if one of the three sets of cones is inoperative (dichromacy) or defective (anomalous trichromacy). There are other color vision defects.

COLOR ORDER SYSTEMS

Many one-, two-, and three-dimensional systems have been developed over the years to order colors in a systematic way and provide specimen colors for visual comparison. Coordination has now been achieved with computer programs between essentially all of these systems, including the CIE systems described below, and conversions can easily be made between them.

The Munsell System

The best known and most widely used color order system is the Munsell system, developed by the artist A. H. Munsell in 1905 and modified over the years. This is a three-dimensional space. With interpolations some 100,000 colors can be distinguished within the Munsell system by visual comparison under carefully standardized viewing conditions.

Other Color Order Systems

The Natural Color System, NCS, developed in Sweden is an outgrowth of the Hesselgren Color Atlas and uses the opponent color approach. There are also the Colorcurve, the Coloroid, and the German DIN systems, among others. The Optical Society of America has published the OSA Uniform Color Scale System with 558 equally spaced color chips. For many purposes a much simpler set of 267 color regions as provided by the ISCC-NBS Centroid System, a joint project of the Inter-Society Color Council and the U.S. National Bureau of Standards, now NIST, is convenient.

BASIC COLORIMETRY

The International Commission on Illumination (abbreviated CIE from the French name) over the years has recommended a series of methods and standards in the field of color.

Consider light of a certain spectral energy distribution falling on an object with a given spectral reflectance and perceived by an eye with its own spectral response. To obtain the perceived color stimulus, it is necessary to multiply these factors together. Standards are clearly required for both the observer and the illuminant.

The CIE Standard Observer

The CIE standard observer is a set of curves giving the tristimulus responses of an imaginary observer representing an average population for three primary colors arbitrarily chosen for convenience. The 1931 CIE standard observer was determined for $2°$ foveal vision, while the later 1964 CIE supplementary standard observer applies to a $10°$ vision; a subscript 10 is usually used for the latter. The standard observers were defined in such a way that of the three primary responses the value of $\bar{y}$ corresponds to the spectral photopic luminous efficiency, i.e., to the perceived overall lightness of an object.

Chromaticity Diagrams

The CIE 1931 chromaticity diagram uses the chromaticity coordinates

$$x = X/(X+Y+Z) \qquad y = Y/(X+Y+Z)$$
$$\text{and} \quad z = Z/(X+Y+Z)$$

It is not actually necessary to specify z, since $x+y+z=1$. Here Newton's pure spectral colors in fully saturated form the horsehoeshaped curve with wavelengths from 400–nm violet to 700–nm red.

One of several defects of the chromaticity diagram is that the minimum distinguishable colors are not equally spaced; ie, that equal changes in x, y, and Y do not correspond to equally perceived color differences.

Standard Illuminants

Clearly, standardized light sources are desirable for color matching, particularly in view of the phenomenon of illuminant metamerism described below. Over the years, CIE has defined several standard illuminants, some of which can be closely approximated by practical sources. In 1931 there was Source A, defined as a tungsten filament incandescent lamp at a color temperature of 2854 K. Sources B and C used filtering of source A to simulate noon sunlight and north sky daylight, respectively. Subsequently a series of D illuminants was established to better represent natural daylight. Of these the most important is Illuminant D_{65}.

Gloss and Opacity

Attributes such as gloss, transparency, translucency, opacity, haze, and luster may apply to some materials (Table 1), and these are relevant in that they may influence the judgment of color differences. As one example, gloss can produce veiling reflections that change the apparent contrast. When present, these attributes can be measured using specialized approaches.

Light Mixing

Light or additive mixing applies to light beams, such as stage lighting or on a television screen. White results from mixtures of two complementary colors or when any suitable set of three-color beams of the appropriate intensity are mixed.

Colorant Mixing

A colorant, whether a dye dissolved in or absorbed by a medium or pigment particles dispersed in a paint, produces color by absorbing and/or scattering part of the transmitted or reflected light. If only absorption is present, the Beer-Lambert law applies:

$$A(\lambda) = \log 1/T(\lambda) = \sum_i a_i(\lambda)bc_i$$

where A is the absorbance, T is the transmittance, $a_i(\lambda)$ is the absorptivity or specific absorbance of absorber i at wavelength λ, b is the length of the absorbing path, and c_i the concentration of the absorber. When colorants are mixed, they function by each independently absorbing light and the subtractive mixing rules merely specify this additivity, a mixing of the three primary colorants to produce black.

When both absorption and scattering are present, the Beer-Lambert law must be replaced by the Kubelka-Munk equation employing the absorption and scattering coefficients K and S, respectively. This gives the reflectivity R_∞

$$(1-R_\infty)^2/2R_\infty = K/S = \sum_i c_i K_i(\lambda) \Big/ \sum_i c_i S_i(\lambda)$$

where c_i is the concentration and $K_i(\lambda)$ and $S_i(\lambda)$ are the specific absorbance and scattering parameters, respectively, of absorber and scatterer i at wavelength λ.

The color of an opaque paint depends both on the size of the pigment particles and on refractive index considerations.

The ready availability of computers has led to the detailed analysis of the colorant formulation problems faced every day by the textile, coatings, ceramics, polymer, and related industries. The resulting computer match prediction has produced improved color matching and reductions in the amounts of colorants required to achieve a specific result with accompanying reductions of cost.

Metamerism

There are several types of metamerism, the phenomenon in which two objects perceived as having a perfect color match under one set of conditions are found to differ in color under other conditions. Most common is illuminant metamerism, which occurs when a change in illuminant is the cause. This originates from the situation that a given visually identified hue can be caused by many different stimuli. Another, related cause would be the presence of an ultraviolet component in one of the sources (usually actual daylight) causing a fluorescent object to emit light in addition to that reflected compared to a visible-light-only source such as an incandescent lamp. The standard illuminant D_{65} specifies significant intensity at wavelengths less than 400 nm in the uv.

Observer metamerism derives from the significant differences in spectral response found among persons even with normal color vision. Finally, there can also be a change in perceived color with a change in viewing angle, as with some metallic paints and with interference-based color-changing inks.

THE MEASUREMENT OF COLOR DIFFERENCES

Two systems of colorimetry were eatablished by CIE designated the CIELUV and CIELAB color spaces, and have come into widespread use.

The 1976 CIELUV Color Space

The CIELUV space, properly designated CIE $L^*u^*v^*$, uses a white object or light source designated by the subscript n as the reference standard and employs the transformations

$$L^* = (116Y/Y_n)^{1/3} - 16; \quad Y/Y_n > 0.008856$$
$$L^* = 903.3Y/Y_n; \quad Y/Y_n \leq 0.008856$$
$$u^* = 13L^*\left(u' - u'_n\right)$$
$$v^* = 13L^*\left(v' - v'_n\right)$$

where L^* is the perceived lightness.

The CIELUV space preserves a property of the CIE 1931 chromaticity space which is important in the field of color reproduction; eg, in the television industry. This is the characteristic that the chromaticities of additive mixtures of color stimuli lie on the straight line connecting the chromaticities of the component stimuli; this is true of the 1976 metric chromaticity diagram but not of the CIELAB space that follows.

The 1976 CIELAB Color Space

Defined at the same time as the CIELUV space, the CIELAB space, properly designated CIE $L^*a^*b^*$, is a nonlinear transformation of the 1931 CIE x, y space. It also uses the metric lightness coordinate L^*, together with

$$a^* = 50\left[(X/X_n)^{1/3} - (Y/Y_n)^{1/3}\right]$$
$$b^* = 200\left[(Y/Y_n)^{1/3} - (Z/Z_n)^{1/3}\right]$$

These equations apply for X/X_n, Y/Y_n, and Z/Z_n all >0.008856. For $X/X_n \leq 0.008856$, the term $(X/X_n)^{1/3}$ is replaced by $[7.787(X/X_n) + 16/116]$ and similarly for Y and for Z in these two equations.

This transformation results in a three-dimensional space that follows the opponent color system with $+a^*$ as red, $-a^*$ as green, $+b^*$ as yellow, and $-b^*$ as blue.

CIELAB is closely related to the older Adams-Nickerson and other spaces of the L,a,b type, which it replaced.

The 1976 CIE Metric Color Spaces

Both the CIELUV and CIELAB spaces can have their Cartesian coordinates converted to cylindrical coordinates, called metric or hue-angle coordinates, with L^* unchanged.

Hunter L,a,b and Other Color Spaces

This was the earliest practical opponent-based system that is still occasionallyt used. In this system, for illuminant C and the $2°$ standard observer

$$L = 10Y^{1/2} \qquad \text{(lightness coordinate)}$$
$$a = 17.5(1.02X - Y)/Y^{1/2} \qquad \text{(red-green coordinate)}$$
$$b = 7.0(Y - 0.847Z)/Y^{1/2} \qquad \text{(yellow-blue coordinate)}$$

There are other equations for other illuminants and other observers and various modifications for special conditions.

Color Difference Assessment

Early color difference scales include those of Judd-Hunter, Macadam, Adams-Nickerson, ANLAB, and ANLAB40. All of these have limitations in some way or another.

COLOR MEASURING INSTRUMENTS

There has been a tremendous change in the last two decades as computers have taken over the tedious calculations involved in color measurement. Indeed, microprocessors either are built into or are connected to all modern instruments, so that the operator may merely need to specify, for example, x, y, Y or L^*, a^*, b^* or L^*, C^*, h, either for the $2°$ or the $10°$ observer, and for a specific standard illuminant, to obtain the desired color coordinates or color differences, all of which can be stored for later reference or computation. The use of high-intensity filtered xenon flash lamps and array detectors combined with computers has resulted in almost instantaneous measurement in most instances.

Instruments

Spectrophotometers, the most sophisticated color measuring instruments, provide the most detailed and accurate information. They may provide continuous spectral data of reflectance and transmittance against wavelength or use up to 20-nm wavelength steps, with high precision in reflectance (down to 0.01%) and tristimulus values (down to 0.01). There may be dual beams with one for the sample and a second as reference for the most stable and precise operation. Many spectrophotometers use an integrating sphere, although some use other geometries or permit alternative ones.

High-end spectrophotometers may use xenon flash lamps and large diode detector arrays, while lower end units may use light-emitting diodes and small arrays. Slightly less sophisticated are spectrocolorimeters that determine spectral response curves for the computation of colorimetric values, but from which the spectral curve itself is not available.

Colorimeters, also known as tristimulus colorimeters, are instruments that do not measure spectral data but typically use four broadband filters to approximate the $\bar{y}$, $\bar{z}$ and the two peaks of the $\bar{x}$ color-matching functions of the standard observer curves.

THE FIFTEEN CAUSES OF COLOR

No less than 15 distinct chemical and physical mechanisms explain the various causes of color, ordered into five groups as in Table 2. In the first group, covered by

Table 2. Fifteen Causes of Color

Cause	Examples
Vibrations and simple excitations	
incandescence	flames, lamps, carbon arc, limelight
gas excitations	vapor lamps, flame tests, lightning, auroras, some lasers
vibrations and rotations	water, ice, iodine, bromine, chlorine, blue gas flame
Transitions involving ligand field effects	
transition-metal compounds	turquoise, chrome green, rhodonite, azurite, copper patina
transition-metal impurities	ruby, emerald, aquamarine, red iron ore, some fluorescence and lasers
Transitions between molecular orbitals	
organic compounds	Most dyes, most biological colorations, some fluorescence and lasers
charge transfer	blue sapphire, magnetite, lapis lazuli, ultramarine, chrome yellow, Prussian blue
Transitions involving energy bands	
metals	copper, silver, gold, iron, brass, pyrite, ruby glass, polychromatic glass, photochromic glass
pure semiconductors	silicon, galena, cinnabar, vermillion, cadmium yellow and orange, diamond
doped semiconductors	blue and yellow diamond, light-emitting diodes, some lasers and phosphors
color centers	amethyst, smoky quartz, desert amethyst glass, some fluorescence and lasers
Geometrical and physical optics	
dispersive refraction	prism spectrum, rainbow, halos, sun dogs, green flash, fire in gemstones
scattering	blue sky, moon, eyes, skin, butterflies, bird feathers, red sunset, Raman scattering
interference	oil slick on water, soap bubbles, coating on camera lenses, some biological colors
diffraction	diffraction gratings, opal, aureole, glory, some biological colors, most liquid crystals

quantum theory, there are incandescence, simple electronic excitations, and vibrational and rotational excitations. Most chemical compounds contain only paired electrons that require very high energies to become unpaired and form excited energy levels; this requires uv, hence there is no visible absorption and no color. Absorption color can, however, be derived from the easier excitation of lower energy unpaired electrons in transition-metal compounds and impurities, covered by ligand-field theory in the second group. Absorptions from paired electrons can be shifted into the visible by increasing the size of the region over which the electrons are localized, as in organic compounds, covered by molecular orbital theory in the third group; this also explains various forms of charge transfer. In the fourth group, there is color in metals and alloys as well as in semiconductors such as yellow cadmium sulfide, both pure and doped, covered by band theory; this also covers color centers. In the final group there are four color-causing mechanisms explained by geometrical and physical optics.

R. S. Berns, *Billmeyer and Saltzman's Principles of Color Technology*, 3rd ed., John Wiley & Sons, Inc., New York, 2000.

P. K. Kaiser and R. M. Boynton, *Human Color Vision*, 2nd ed., Optical Society of America, Washington, D.C., 1996.

K. Nassau, "Colour" in, *Encyclopedia Britannica*, 15th ed., Macropedia Vol. 4, pp. 595–604, 1988 on.

K. Nassau, *The Physics and Chemistry of Color*, 2nd ed., John Wiley & Sons, Inc., New York, 2001.

KURT NASSAU
Consultant

COLORANTS FOR CERAMICS

Any product that depends on aesthetics for consideration for purchase and use will be improved by the use of color. Hence, many ceramic products, such as tile, sanitary ware, porcelain enameled appliances, tableware, and some structural clay products and glasses, contain colorants.

For both economic and technical reasons, the most effective way to impart color to a ceramic product is to apply a ceramic coating that contains the colorant. The most common coatings, glazes, and porcelain enamels are vitreous in nature. Hence, most applications for ceramic colorants involve the coloring of a vitreous material.

There are a number of ways to obtain color in a ceramic material. First, certain transition-metal ions can be melted into a glass or dispersed in a ceramic body when it is made. Although suitable for bulk ceramics, this method is rarely used in coatings because adequate tinting strength and purity of color cannot be obtained this way.

A second method to obtain color is to induce the precipitation of a colored crystal in a transparent matrix. Certain materials dissolve to some extent in a vitreous material at high temperatures, but when the temperature is reduced, the solubility is also reduced and precipitation occurs. This method is used to disperse nonoxide precipitates of gold, copper, or cadmium sulfoselenide in bulk glass. In coatings it is used for opacification, the production of an opaque white color. Normally, some or all of the opacifier added to the coating slip dissolves during the firing process and recrystallizes upon cooling. For oxide colors other than white, however, this method lacks the necessary control for reproducible results and is seldom used.

The third method to obtain color in a vitreous matrix is to disperse in that matrix an insoluble crystal or crystals that are colored. The color of the crystal is then imparted to the transparent matrix. This method is the one most commonly used to introduce color to vitreous coatings.

CERAMIC PIGMENTS

The principal method of coloration of ceramic coatings is dispersal of a ceramic pigment in a vitreous matrix. To be suitable as a ceramic pigment, a material must possess a number of properties that fall in two categories: strength of pigmentation and stability. Another desirable property for a ceramic color is a high refractive index.

Manufacturing

Although a number of different pigment systems exist, most are prepared by similar manufacturing methods. The first step in pigment manufacture is close control over the selection of raw materials. Most of these raw materials are metallic oxides or salts of the desired metals.

The raw materials are weighed and then thoroughly blended. The reaction forming the pigment crystal occurs in a high-temperature calcining operation.

Following calcination, the product may require milling to reduce particle size to that necessary for use.

It has been found that there are several advantages to modification of pigments by addition to the calcined product of a small quantity of a dispersing agent.

Pigment Systems

Most of the crystals used for ceramic pigments are complex oxides, owing to the great stability of oxides in molten silicate glasses. The one significant exception to the use of oxides is the family of cadmium sulfoselenide red pigments. This family is used because the red and orange colors obtained cannot be obtained in oxide systems; thus, it is necessary to sustain the difficulties of a nonoxide system.

There are no oxides that can be used to give a true red pigment that is stable to the firing of ceramic coatings. Hence, orange, red, and dark red colors are obtained by the use of cadmium sulfoselenide pigments. The cadmium sulfoselenides are a group of pigments based on solid solutions of cadmium selenide, cadmium sulfide, or zinc sulfide.

Although red is not available in oxide systems, pink and purple shades are obtained in several ways. One such system is the chrome alumina pinks. Chrome alumina pinks are combinations of ZnO, Al_2O_3, and Cr_2O_3. Depending on the concentration of zinc, the crystal structure may be either spinel or corundum.

The most stable pink pigment is the iron-doped zircon system. This pigment is made by calcining a mixture of ZrO_2, SiO_2, and iron oxide at a stoichiometry to produce zircon in shades from pink to coral.

The chrome–tin system is the only family to produce purple and maroon shades, as well as pinks. The system can be defined as pigments that are produced by the calcination of mixtures of small amounts of chromium oxide with substantial amounts of tin oxide. In addition, most formulations contain substantial amounts of silica and calcium oxide.

The most important brown pigments used in ceramic coatings are the zinc iron chromite spinels. This pigment system produces a wide palette of tan and brown shades.

There are several systems for preparing a yellow ceramic pigment. Moreover, there are valid technical and economic reasons for the use of a particular yellow pigment in a given application. The pigments of greatest tinting strength, the lead antimonate yellows, cadmium sulfide, and the chrome titania maples, do not have adequate resistance to molten ceramic coatings. Thus, other systems must be used if the firing temperature is $>1000°C$.

Zirconia vanadia yellows are prepared by calcining ZrO_2 with small amounts of V_2O_5. They are economical pigments for use with a broad range of coatings firing at temperatures $>1000°C$. They are brighter and stronger in glazes low in PbO and B_2O_3.

Tin vanadium yellows are prepared by introducing small amounts of vanadium oxide into the cassiterite structure of SnO_2. Tin vanadium yellows develop a strong color in all ceramic coating compositions.

The praseodymium zircon pigments are formed by calcination of ~5% of praseodymium oxide with a stoichiometric zircon mixture of ZrO_2 and SiO_2 to yield a bright yellow pigment.

Just as there are several alternative yellow pigments, there are several alternative green pigments. Formerly, the chromium ion was the basis for green pigments.

Higher quality results are obtained if chromium oxide is used as a constituent in a calcined pigment. One such system is the cobalt zinc alumina chromite used to produce blue-green pigments.

Victoria green is prepared by calcining silica and a dichromate with calcium carbonate to form the garnet $3CaO-Cr_2O_3-3SiO_2$. This pigment gives a transparent bright green color.

Because of all the difficulties with the use of chromium-containing pigments and because there is a definite limitation on the brilliance of green pigments made with chromium, many green ceramic glazes are now made with zircon pigments.

The use of copper is of little interest to most industrial manufacturers, but the colors obtained from them are of great interest to artists because of the many subtle shades that can be obtained.

The traditional way to obtain blue in a ceramic coating is with cobalt, which has been used as a solution color since antiquity. Today, cobalt may react with Al_2O_3 to produce the spinel $CoAl_2O_4$ or with silica to produce the olivine Co_2SiO_4. The silicate involves higher concentrations of cobalt, with only modestly stronger color.

In glazes, the cobalt pigments have been largely replaced by pigments based on vanadium-doped zircon. These pigments are turquoise in shade and are less intense than the cobalt pigments. Therefore, they are not applicable when the greatest tinting strength is required or when a purple shade is called for. Where they are applicable they give vastly improved color stability.

Black ceramic pigments are formed by calcination of several oxides to form the spinel structure. The prototype black is a cobalt iron chromite.

It is easiest to obtain a uniform gray color when a calcined pigment is used that is based on zirconia or zircon as a carrier for various ingredients of blacks such as Co, Ni, Fe, and Cr oxides. This pigment is called cobalt nickel gray periclase.

USE OF PIGMENTS IN COATINGS

There are several additional factors that must be considered in selecting pigments for a specific coating application. These factors include processing stability requirements, pigment uniformity and reproducibility, particle size distribution, dispersibility, and compatibility of all materials to be used.

ECONOMIC ASPECTS

Owing to the limited market and the variety and complexity of the products, ceramic pigments are manufactured by specialist firms, not by the users. The principal producers are Ceramic Color and Chemical Corp., Englehard Corp., Ferro Corp., General Color and Chemical Corp., Mason Color and Chemical Corp., and Pemco Corp. Estimated annual production is ~2500–3000 metric tons. This figure does not include some of the same and similar products manufactured for use in nonceramic applications. The costs of ceramic pigments range from $10 to $60/kg or higher, depending on the elemental composition and the required processing. The most expensive pigments are those containing gold and the cadmium sulfoselenides.

HEALTH AND SAFETY FACTORS

Properly handled, ceramic colorants should not cause unacceptable problems of health and safety. Preventive measures to avoid inhalation of fine particulate matter should invariably be used. Care should be taken to avoid ingestion of pigments by thorough washing before eating or smoking. Particular care should be taken in handling cadmium sulfoselenide pigments and lead antimonate pigments, which are highly toxic if ingested or inhaled.

When these pigments are used with lead-containing glazes, care should be exercised to use lead-safe glaze materials.

R. A. Eppler, and D. R. Eppler, *Glazes and Glass Coatings*, American Ceramic Society, Westerville, Ohio, 2000.

W. D. Kingery, H. K. Bowen, and D. R. Uhlmann, *Introduction to Ceramics*, John Wiley & Sons, Inc., New York, 1976, pp. 677–689.

A. Paul, *Chemistry of Glasses*, Chapman & Hall, Ltd., London, 1982, pp. 233–251.

W. Vogel (trans. N. Kriedl), *Chemistry of Glass*, American Ceramic Society, Columbus, Ohio, 1985, pp. 163–177.

Richard A. Eppler
Eppler Associates

COLORANTS FOR PLASTICS

Incorporating colorants into plastics requires that the colorants meet "zero or no reaction chemistry." If any chemical reaction takes place any time in a products fabrication or lifetime, a color change is most likely to take place and the product will have failed to meet its technical and commercial design objectives.

Historically, colorants, have been divided into three distinct groups: organic pigments, inorganic pigments, and soluble dyes. Soluble dyes go into solution in the plastic system, are at the molecular level, and have no particulate identity. Unfortunately, this distinction is not as clear-cut as one would like. Inorganic pigments are thought to be inherently insoluble. Indeed, most organic pigments are essentially insoluble. However, there are situations where under very specific circumstances an organic pigment may become soluble or partially soluble, causing product failure. To perform properly, soluble dyes must be totally soluble in the plastic system to be effective and meet their design criteria.

An organic pigment is made up of carbon, hydrogen, oxygen, and sulfer and nitrogen in combinations of atoms bonded together. Sulfur, chlorine, bromine, fluorine, calcium, strontium, barium, and manganese may also be present. The molecules are crystalline in nature and selectively absorb light to provide color. The organic pigment must remain in particulate form to perform its color functionality.

An inorganic pigment is made up of a combination of metalloid or metallic elements combined with oxygen, sulfur, and/or selenium. The molecules will selectively absorb light to provide color. The inorganic pigment must remain in its original particulate form to perform its color functionality.

Soluble dyes dissolve in the plastic system, losing any crystalline characteristics, and operate at the molecular level. Since soluble dyes properly dissolved in a system have no particulate characteristics they impart color only through selective absorption. The soluble dye must be 100% in solution to perform as designed.

Pigments can vary from almost transparent (slightly hazy to translucent) to totally opaque, while soluble dyes, used properly, will impart complete transparency.

Pigments and soluble dyes predominantly leave their manufacturing sites as powders. There may be a number of intermediate steps, however, to improve value before the colorant reaches a company for fabrication into a useful consumer product.

DISPERSIONS

Colorant dispersions bring value-added effects to the plastics processes, avoiding many of the problems associated with dry colors. Dispersions for plastics fall into two categories: solid pellets of various sizes and shapes, and liquids or pastes.

Liquid color concentrates play the same role as dry concentrates, repeat the carrier is a liquid rather than a solid polymer. The dispersion can be very intense and superior to dry color concentrates in terms of reducing aggregates and agglomerates.

COLORANT PROPERTIES

A number of general property issues should be considered every time an individual colorant is added to a system: dispersibility, lightfastness, migration, chemical resistance, blooming, or sublimation, toxicity, solvent resistance, environmental effects, compatibility, batch uniformity, particle integrity, particle size, heat stability, weatherability, bleeding or migration, electrical enviroment, fire resistance,

availability, economics–cost, color values, and ease of cleanup.

ORGANIC PIGMENTS

There are hundreds of organic pigment chemical families in commerce. Figure 1 presents a schematic of the basic issues involved.

INORGANIC PIGMENTS

These pigments are unique in that they all contain a metal in their composition. Inorganic pigments are essentially insoluble; therefore, migration and bleeding are nonexistent. Another characteristic is that their lightfastness, weatherability, chemical resistance, heat stability, and opacity usually come as a package: If a particular inorganic pigment exhibits excellent heat stability, the chances are good that its other properties will probably be good to excellent as well. Figure 2 is an initial guide to the usage of inorganic pigments.

SOLUBLE DYES

There are many chemical types of soluble dyes to choose from, similar to organic and inorganic pigments (see Figure 3). A full range of colors is available in each chemical type that figuratively covers the entire visual color spectrum. Therefore, soluble dye selection is critically and fundamentally based upon what chemical type performs the best in a system and the target color desired. Pigments have some translucency or opacity, i.e., pigments scatter some light, soluble dyes do not.

Widely used solvent dye chemical types include anthrapyridone, anthraquinone, azo, nigrosine, perinone, pyrazolone, quinoline, quinophthalone, and xanthene.

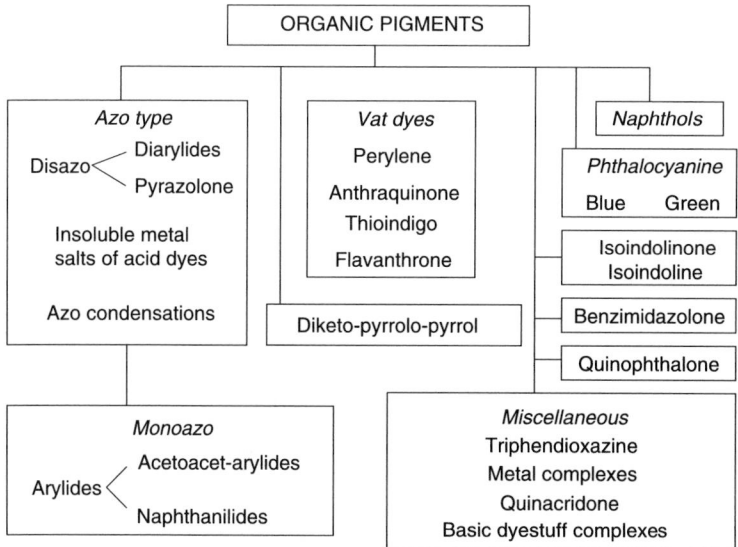

Figure 1. Organic pigments.

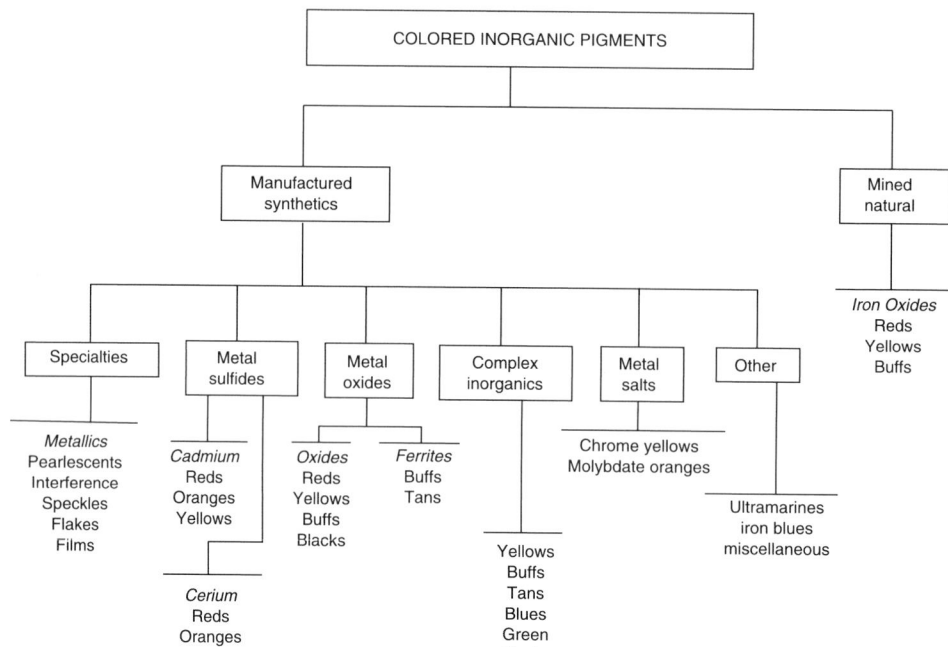

Figure 2. Colored inorganic pigments.

EFFECT COLORANTS

Effect colorants cover a wide gamut of optical effects. They include metallic, pearlescent, fluorescent, phosphorescent, speckles, marble, granite, and others. There is such a wide variety in this group it is impossible to characterize them as done with the organic, inorganic pigment, and soluble dye groups.

SPECIAL OR UNUSUAL EFFECT COLORANTS

These groups of colorants are difficult to classify, some being pigments or distinctly different materials and others soluble dyes. Every colorant brings an extraordinary visual effect not found in typical colorants. These colorants take many forms: die cut or small strips of cellophane or mylar films, special light interference colorants, chemiluminescent colorants, bioluminescence, thermochromic colorants found in digital thermometer tapes, photochromic colorants change hue when subjected to particular kinds of light, electroluminescent colorants that emit visible light when an electrical charge or voltage is applied, and others such as piezo-luminescent and hydrochromatic materials that can produce strange color effects that find some use in commerce.

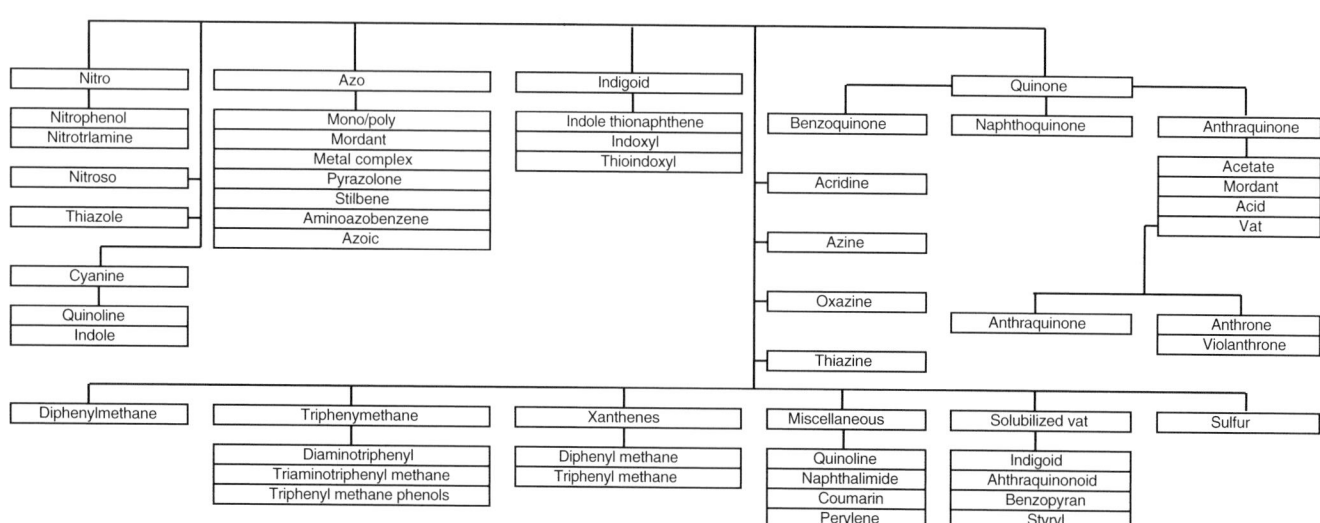

Figure 3. Soluble dyes.

G. Buxbaum, *Industrial Inorganic Pigments*, VCH Verlagsgesellschaft mbH, D-69451 Weinheim, Federal Republic of Germany, 1993.

Colour Index, and its Additions and Amendments, 3rd ed., Society of Dyers and Colourists, London, and American Association of Textile Chemists and Colorists, Durham, N.C.

W. Herbst and K. Hunger, *Industrial Organic Pigments*, VCH Verlagsgesellschaft mbH, D-6940 Weinheim, Federal Republic of Germany, 1993.

T. C. Patton, *Pigment Handbook, Volume I*, John Wiley & Sons, Inc., New York, 1973.

ROBERT A. CHARVAT
Charvat and Associates, Inc.

COMBINATORIAL CHEMISTRY

In the late 1990s Symyx technologies developed and marketed "combinatorial methods," or high throughput experimentation (HTE), for advanced materials, and created a new paradigm in materials research for the chemical process industry (CPI) and advanced materials producers. This discontinuity, or step-change, reflected an earlier response by the pharmaceutical industry toward significant market demands for new products—reduced product innovation cycle time, increased return on research and development (R&D) investment, and industry consolidation. The methodology known as combinatorial chemistry is now relatively ubiquitous in companies conducting research in advanced materials, catalysts, and polymers.

Today many chemical and advanced materials companies have implemented some form of HTE in their discovery research phases. Although HTE for materials often has little similarity with methods developed in the drug discovery arena, researchers have effectively leveraged many methods and tools from pharmaceutical applications. This is evident in the area of industrial catalysis, where HTE utilized in the discovery and process development phases have cut concept-to-launch cycle times in half.

THE NEW PARADIGM FOR MATERIALS RESEARCH

Applications

The transfer of technology from drug discovery has resulted in the development of HTE for inorganic materials, fine and specialty chemicals, and advanced materials. Because HTE techniques are especially suited to complex mixtures containing many different components and processing conditions, this methodology lends itself to the discovery of new, higher performance materials that contain multicomponent formulations; eg, polymer blends in engineering plastics. In addition, HTE permits the screening of compositions that would not otherwise be attempted. Broad economic benefits are envisioned from the downstream impact of these methodologies, as indicated in Table 1.

High throughput experimentation utilizes synthesis and analysis procedures wherein "libraries" of many tens, hundreds, or thousands of discrete samples are fabricated, processed, and characterized in parallel in hours and days, rather than months and years, at a fraction of the cost of traditional serial approaches. An alternative methodology referred to as compositional spread fabricates continuous multidimensional gradients of different material; eg, by codeposition of two or more components at different concentrations onto a substrate surface. New materials with unique chemical, optical, and electronic properties have been discovered, using a variety of parallel screening approaches, for new products.

Methodology

The conversion of the traditional discovery process to high throughput experimentation was enabled by the convergence of computer software running on more cost-effective computers; robotics; MEMS (microelectromechanical systems) technologies; and sensors. In effect, research was transformed from the traditional "serial"

Table 1. Application Areas Open to High Throughput Experimentation

Application areas	Technical challenges	Affected products	Economic benefits
• industrial chemicals and monomers	• faster catalyst screening • capability of screening • extremely diverse combinations of catalyst ingredients	• industrial chemicals • engineering thermoplastics • other plastics	• lower cost • lower energy usage • new products based on newly affordable raw materials
• polymers – catalysts – polymer blends – surface modifiers	• faster screening • process optimization	• engineering plastics • thermoplastics	• new products • new markets such as automotive glazing • reduced domestic energy consumption
• ceramics	• thermal barrier coating • optimization • electronic properties • higher strength	• aircraft engines • advanced power machines • conductors • semiconductors • dielectrics • machine tools	• higher engine temperatures • increased service life, and reduction of downtime • higher component densities and speeds • machining of new alloys, increased productivity

processes into parallel "factory" processes. Traditional, serial discovery processes rely on the preparation and characterization of individual samples from bench scale (milligrams or grams) to pilot scale (grams to kilograms). HTE methods can more rapidly sample the same preparation and characterization space *in parallel*, using automated laboratory instrumentation at the microgram-to-milligram scale. Eventually, HTE will be applied to product development and customer service when it becomes more automated, simpler, and faster.

The materials and chemicals industries can leverage, to some extent, the tools developed for drug discovery, ie, from being focused on solution-state synthesis to solid-state materials fabrication. The processing of materials typically involves energetic reaction environments, with temperature requirements of hundreds or thousands of degrees Celsius (°C), and pressures of hundreds or thousands of kilopascals (kP). Microscale solid-state samples may also be subject to significant influence from the substrate onto which they are deposited—interfacial effects such as diffusion can produce phenomena that are not reproducible in bulk samples of manufacturing scale. Thus, advanced materials suffer from "scalability", or differences in properties observed in microscale vs. lab-, pilot-, or commercial scale. Finally, solid-state compositions may develop into different (kinetically controlled) metastable structures, depending on the processing and testing conditions.

A general process flow for materials discovery is shown in Figure 1. Target definition utilizes expert opinion, hypothesis, market need, and knowledge based on computational chemistry to develop an experimental target. Library design reduces the number of samples in defined sample spaces within the experimental universe or to direct screening to other spaces within the universe. The design of the sample library requires rational chemical synthesis or process information to reduce the number of samples and experiments without increasing the probability for endless searches, false positives, or false negatives. Library fabrication involves the automated deposition or processing of an *n*-dimensional matrix of physical samples. Library characterization in a parallel or massively parallel mode involves the use of robotics and sensors to rapidly and automatically analyze the library of targets for desired properties. Data collection and analysis uses database and artificial intelligence tools—"informatics," software that collects and stores raw data and converts it into information in such a way that it is easily interpreted by researchers. An informatics engine sits at the front end of the HTE installation to provide hardware control and collect analytical data from instrumentation; it stores and manages large databases and provides a suitable human-computer interface for visualizing the data to yield knowledge.

Commercial Environment

Materials manufacturers are utilizing several business scenarios to obtain HTE capabilities: (*1*) by developing internal capabilities; (*2*) by contracting with service providers having a core competency in high throughput discovery methods; (*3*) by developing an independent consortium or alliance partnership with individual tools providers; and (*4*) by various combinations of these scenarios. These events signal a clear trend that the materials industry is beginning to outsource (subcontract) its front-end discovery efforts to smaller external entities. There are currently seven companies known to be either performing front-end R&D on a contract basis or developing integrated systems for large materials manufacturers utilizing HTE R&D.

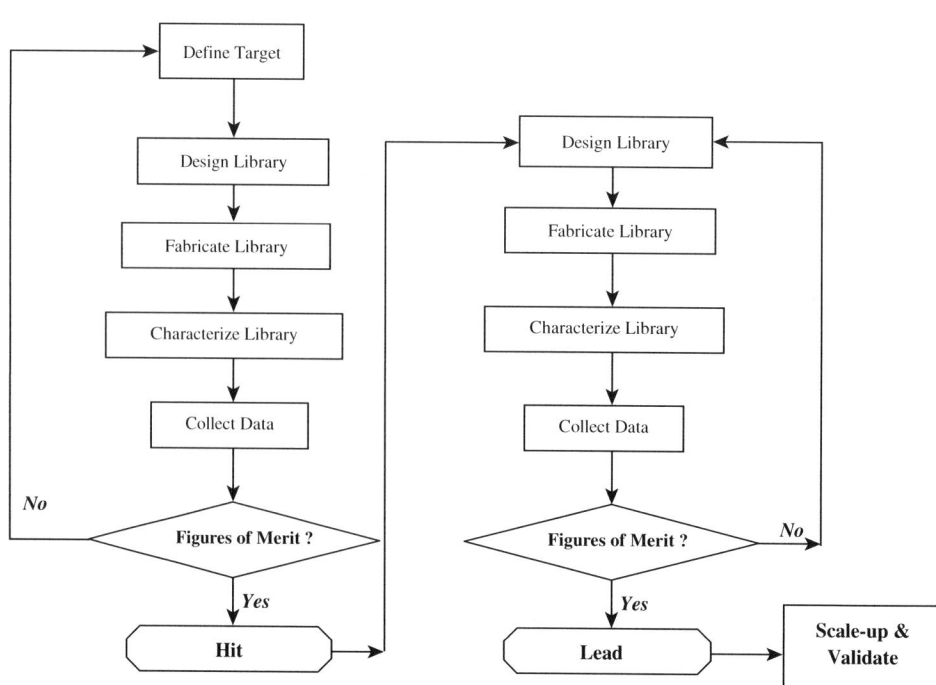

Figure 1. The key steps of a typical high throughput RD&E process.

HIGH THROUGHPUT EXPERIMENTATION FOR INDUSTRIAL CATALYSIS

Economic Benefits of HTE Methods in the Chemicals Industry

The penetration of HTE methods in industry can be measured by the amount of research money expended on this technology. Modest growth might also be expected because of the lower revenues from materials-based industries, where profits and revenues are significantly lower, and development times are shorter, compared to the pharmaceutical industry.

Another measure of the growth of HTE methods is the commercial impact of the materials discovered using this technology. For the end-user, there will be nothing to distinguish a material discovered using HTE from a material discovered using "traditional" technology. Over time it is likely that materials discovery using HTE itself will become the "traditional" research methodology.

With rising economic demands for higher efficiency and productivity in research and development, HTE methods are increasingly being implemented to bring more catalysts per unit time to the marketplace. High throughput automated synthesis and advanced screening technologies are now being applied to the discovery of more efficient homogeneous as well as heterogeneous catalysts and materials. The HTE process allows the exploration of large and diverse compositional and parameter spaces by establishing an integrated workflow of rapid parallel or combinatorial synthesis of large numbers of catalytic materials, subsequent high throughput assaying of these compounds, and large-scale data analysis. The number of experiments that can be screened has risen by orders of magnitude, resulting in a much higher probability of discovering new catalysts or materials.

Demands for more effective catalysts in the polymer, specialty chemical, and environmental markets are driving growth by 5%/year, and catalyst suppliers are reacting to lowered sales in the commodity markets through consolidations and strategic alliances with process developers, suppliers, and customers. High throughput experimentation is speeding up the discovery process, with the expectation that new catalysts and corresponding intellectual property will be produced faster and more efficiently.

The entry costs for obtaining HTE capabilities can be very high—in the neighborhood of $10–20 million for very high throughput systems. Many companies have centralized HTE facilities that were developed internally to service their business units. Firms of all sizes are exploiting alliances with external contract research firms as a route to obtaining competitive advantages.

The period 2000–2002 showed significant advances in the development of HTE approaches to the discovery and optimization of catalysts. The highlights include the development of modular approaches to the synthesis of libraries of organometallic complexes and catalysts, novel indirect screening methods, and automated syntheses and screening of high-density solid-state libraries.

Methodologies

The application of high throughput discovery and process development for chemicals and materials will drive the enabling integration of a hardware- and software-based infrastructure toward specific product applications. The long-term vision shared by many researchers is to have high throughput research become part of expanded enterprise widesystems that include tools for hardware interfaces, technology assessment/decision, and logistics (Fig. 2). Because HTE is currently highly capital intensive, with start-up costs in the range of $8–20 million, discontinuous innovation in generic or modular hardware and software technologies will be necessary to drive down costs and facilitate its implementation in the industrial sectors that have lower returns on R&D investment such as exist in the CPI and materials sectors.

The methodologies inherent in high throughput experimentation for catalysts can be categorized according to their design of the previously mentioned experimental

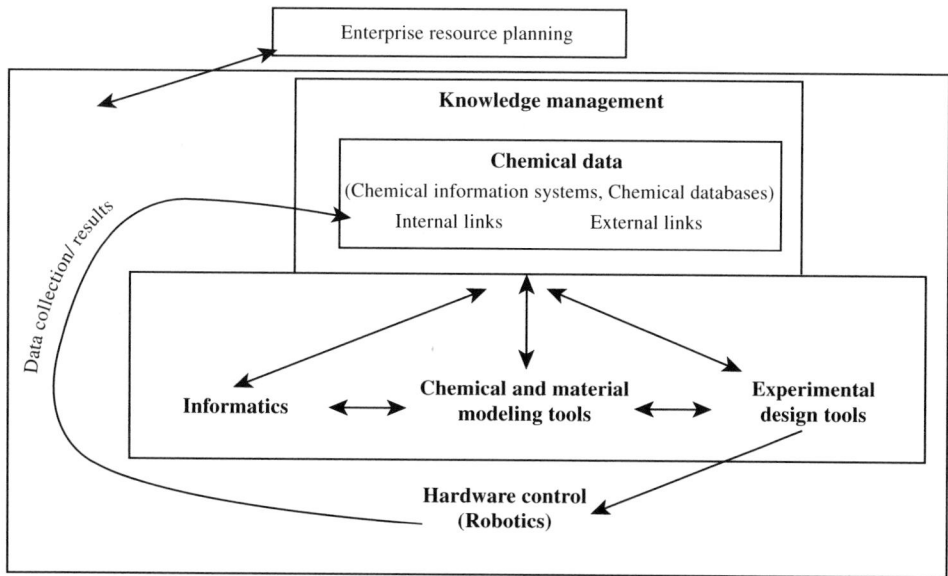

Figure 2. The integration of software systems.

space (library design), data capture and information retrieval (informatics), library synthesis or fabrication, and library characterization.

POLYMERIC MATERIALS

As applied to polymeric materials research, combinatorial or HTE methodologies allow efficient characterization of polymer properties and optimization of polymer processing parameters in addition to synthesis of new polymers and accelerated development of new materials. This approach has led to scientific discoveries related to polymer material properties. Consequently, a new feature for polymer materials is that combinatorial and high throughput methods are rapidly extending knowledge discovery into fundamental polymer science as well as industrially important application areas like organic light-emitting diodes and coatings.

Application Areas and Needs

The complex phenomena and large variable spaces present in multicomponent, multiphase, interfacial regions in polymers strain the capabilities of conventional one-sample for one-measurement polymer characterization hence the need to develop high throughput techniques for efficient synthesis and characterization of complex polymeric systems. For polymers, the advantages of high throughput methods include efficient characterization of novel regimes of thermodynamic and kinetic behavior, rapid testing and identification of structure–property relationships, testing hypotheses for accelerated development of functional materials, and reduced experimental variance (many measurements at the same environmental conditions). High throughput screening has been successfully used for measurements of phase behavior in polymer blends, block copolymer ordering, polymer dewetting, polymer crystallization and organic light-emitting devices.

Methodologies and Techniques

There has been an increasing effort to apply HTE to materials science, but the widespread adaptation of HTE to polymers research has been hindered by a lack of techniques for preparing polymer libraries with systematically varied composition (Φ) thickness (h), and processing conditions; eg, temperature (T). In addition, much of polymer characterization instrumentation is not suitable for high throughput screening. For this reason, a number of research groups have begun developing library preparation and screening approaches suited for polymer characterization.

HIGH THROUGHPUT SCREENING: INORGANIC MATERIALS

The use of complex inorganic materials is increasing in advanced electronic, optoelectronic, magnetic, and structural applications. These materials are typically composed of three or more elements and may have a crystalline structure, in which atoms are in a periodic arrangement, or an amorphous structure, in which atoms have no long-range order. Inorganic materials are used in a variety of forms, including thin films and coatings, bulk masses, and powders. The fabrication processes typically involve a number of variables, such as temperature, pressure, and atmosphere. It is necessary to search large composition and process spaces to optimize the properties and structure of an inorganic material for a specific application or to discover a new material. High throughput experimentation (HTE) methods are ideally suited to such searches of multiparameter space.

Applications

The use of HTE methods to increase the efficiency of materials discovery was first reported in 1970 and then not reported again until 1995, when it was demonstrated that known high temperature superconducting compounds could be readily identified using HTE methods. Subsequently, combinatorial studies have been performed on materials for electronic, magnetic, photonic, and structural applications.

Tools and Methodologies

A number of novel library fabrication and characterization techniques have been developed specifically for HTE studies of inorganic materials. Some tools used in conventional studies have been adapted or modified to increase fabrication or measurement throughput.

Library Fabrication. The primary combinatorial variable in most inorganic libraries is chemical composition; additional processing parameters such as temperature, pressure, and atmosphere may also be varied during library fabrication. Various techniques have been used to synthesize thin film, bulk, and powder libraries.

High Throughput Characterization. Ideally, characterization methods in HTE involve the measurement of a small area or volume of material at a very high measurement rate, or throughput. The measurements may be made sequentially (in serial) or simultaneously (in parallel). Imaging techniques are inherently high throughput, since data are collected in parallel from a large area, resulting in a map of information. For any measurement technique, rapid data analysis methods must be developed to extract the required information from the results. In order to compare the performance of the elements in a given library, it is desirable to formulate a figure of merit that includes the critical properties. Composition and structure of inorganic materials, which often have a strong effect on the properties, must also be determined in library characterization.

Several methods have been used for high throughput electrical measurements. The four-point probe contact technique illustrated in Figure 3 is a common method for measuring electrical resistance, particularly for low-resistance samples.

High throughput photoluminescence and birefringence measurements have been used to evaluate the optical properties of combinatorial library samples. Photoluminescence

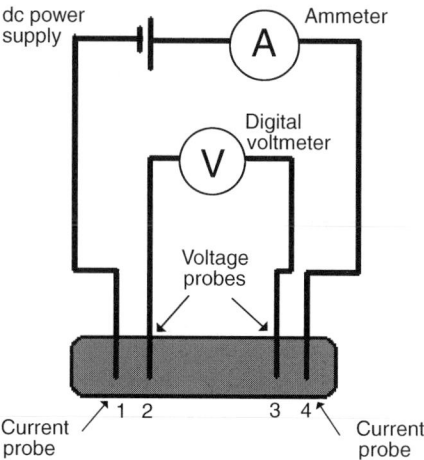

Figure 3. Conventional four-point contact probe method for electrical resistance measurements. Current is applied across two electrodes 1,4 and voltage is measured across the other two electrodes, 2,3.

is an optical phenomenon resulting from the excitation of an electron to a higher electronic state by the absorption of a photon from a high energy light source, typically a laser. The emitted light or luminescence has a specific wavelength of a specific color, such as red, green, or blue. Combinatorial phosphor libraries have been screened for luminescence efficiency and color by a photoluminescence parallel imaging technique. In this method, the entire library is illuminated with an ultraviolet lamp, yielding a color intensity map of visible emission that can be quantitatively analyzed using materials with known efficiency as calibration standards. This technique measures the absorption, transmission, or reflectance of light from a material as a function of wavelength.

Electrooptic measurement methods probe the effect of an electric field on the optical properties of a material. A high throughput birefringence technique was developed to measure the electrooptic coefficient, defined as the change in refractive index with applied electric field.

Table 2. Technology Challenges

Software	
technology	*challenge*
library design	
statistics modeling literature/patent databases	• development of higher order designs
	• (review by NSF is pending)
	• query languages; visualization; integration
informatics	
QSPR (structure-property predictions)	• property prediction, large-scale correlations, integration into experimental design tools
database query engines	• new languages, genetic programs
	• interoperability, enterprise-wide integration
Hardware	
technology	*challenge*
screening	
thermal properties optical characterization	• electrical and thermal conductivity
	• fluorescence, luminescent properties, X-ray diffraction
mechanical properties electrical properties chemical properties	• modulus, tensile strength, impact resistance
	• capacitance, conductance
	• molecular weight
	• polymer architecture/morphology
	• catalyst turnover, selectivity, conversion
processing	
control of physical environment	• control of temperature and pressure over library array with control over individual sample sites or wells
	• sample size—control of interfacial diffusion, mass transport properties, etc
deposition	
thermally driven (e-beam, laser, etc)	• delivery of finite samples or composition spreads of consistent, known composition at microscopic sizes: reproducibility
laser ablation	
microjet	
vacuum deposition	
robotic pipetting of nanoaliquots	• reduction of cross-talk of sample properties through diffusion, etc. across substrate surface
micro-reactor technologies	
systems integration	• development of standards for modular component interconnections
	• process control devices tuned to MRT
	• modeling for fluid flow, heat and mass transfer
reactor design	• substrates: glass, polymer, metal
manufacturing processes	• surface micromachining using wet or dry chemical etching, laser ablation, mechanical micromilling, LIGA processes

X-ray diffraction is commonly used to study the structure of inorganic materials. In this technique, the angle of the sample (theta) and the angle of the detector (two-theta) are scanned and the intensity at each value of two-theta is measured. The resulting data are then used to identify the crystalline phases present in the sample. X-ray fluorescence measurements are used to determine the chemical composition of multicomponent materials.

A wide variety of methods have been used to fabricate thin-film, bulk, and powder libraries of varying composition. High throughput characterization tools to measure electrical, optical, and structural properties have been developed for inorganic libraries. These fabrication and analysis tools will facilitate the discovery and optimization of multicomponent inorganic materials for advanced applications.

THE PATH FORWARD

The future of high throughput experimentation will require that the basic underlying software technology be capable of defining profitable experimental spaces, visualizing complex data relationships; and of correlating high throughput experimentation target materials with properties to permit database queries from a broad spectrum of data mining engines and the development of structure–property relationships. This requires interfacing with data visualization tools at the back end and statistical experimental design engines on the front end while remaining compliant with enterprisewide systems for knowledge management and maintaining control of experimental hardware (see Table 2).

Success using HTE in the CPI and materials sectors has already resulted in the reduction of idea-to-commercialization cycle times to three to five years from seven to ten years. HTE has already reached commercial validation.

F. P. Boer, *The Valuation of Technology: Business and Financial Issues in R&D*, John Wiley & Sons, Inc., New York, 1999.

E. G. Derouane, *Combinatorial Catalysis and High Throughput Catalyst Design and Testing*, Kluwer Academic Publishers, Inc., Boston, 2000.

V. V. Guliants, *Current Developments in Combinatorial Heterogeneous Catalysis*, Elsevier Science, New York, 2001.

J. C. Meredith, A. Karim, and E. J. Amis, in R. Malhotra ed., *ACS Symposium Series: Combinatorial Approaches to Materials Development*, American Chemical Society, Washington, D.C., 2001.

N. K. Terrett, *Combinatorial Chemistry*, Oxford, 1998.

JOHN D. HEWES
Honeywell International, Inc.

DEBRA KAISER
ALAMGIR KARIM
ERIC AMIS
National Institute of Standards
and Technology

COMBUSTION SCIENCE AND TECHNOLOGY

Fuel combustion is a complex process, the understanding of which involves knowledge of chemistry (structural features of the fuel), thermodynamics (feasibility and energetics of the reactions), mass transfer (diffusion of fuel and oxidant molecules), reaction kinetics (rate of reaction), and fluid dynamics of the process. Therefore, the design of combustion systems involves utilizing information and data generated in a range of disciplines. Often, the design of practical combustion systems is based on experience rather than on fundamental mechanistic understanding. However, for certain fuels such as methane, the combustion mechanism is better understood than for other more complex fuels such as coal. To accommodate the variety of approaches used to solve practical problems, this article is divided into two subsections: combustion science and combustion technology.

COMBUSTION SCIENCE

Higher Heating Value

The higher heating value of a fuel is the heat of combustion at constant pressure and temperature (usually ambient) determined by a calorimetric measurement in which the water formed by combustion is completely condensed.

FAR or AFR

The composition of a mixture of fuel and air or oxidant is often specified according to the Fuel to Air Ratio (FAR), and can be expressed on a mass, molar, or volume basis.

Flammability Limits

Any given mixture of fuel and oxidant is flammable (explosive) within two limits referred to as the upper (rich) and lower (lean) limits of flammability (Fig. 1).

Flash Point

The flash point is the lowest temperature at which vapor, given off from a liquid, is in sufficient quantity to enable ignition to take place.

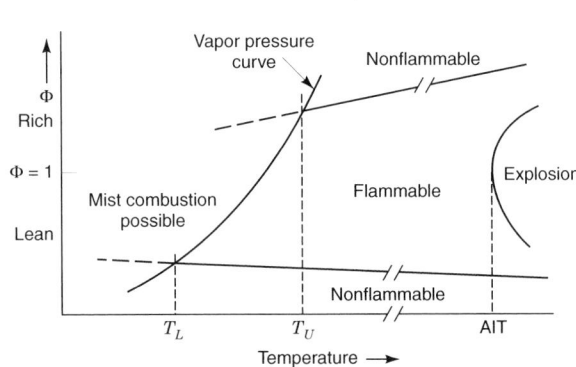

Figure 1. Effect of temperature on limits of flammability of a pure liquid fuel in air, where T_L = lean (or lower) flash point; T_U = rich (or upper) flash point; and AIT = autoignition temperature.

Ignition

To understand the phenomenon of ignition it is necessary to consider the following concepts: ignition source, gas temperature, flame volume, and presence of quench wall surfaces. In general, there are two main methods of igniting a flammable mixture. In the self-ignition method, the mixture is heated slowly so that the vapor released as the temperature is raised ignites spontaneously at a particular temperature. In the forced ignition method, a small quantity of combustible mixture is heated by an external source and the heat released during the combustion of this portion results in propagation of a flame. The external ignition source can be an electric spark, pilot flame, shock wave, etc. For ignition to take place the following conditions should be satisfied: (1) the amount of energy supplied by the ignition source should be large enough to overcome the activation energy barrier; (2) the energy released in the gas volume should exceed the minimum critical energy for ignition; and (3) the duration of the spark or other ignition source should be long enough to initiate flame propagation, but not too long to affect the rate of propagation. Ignition models fall into two categories: the thermal model explains the ignition as resulting from supplying the mixture with the amount of heat sufficient to initiate reaction. In the chemical diffusion model the main role of the ignition source is attributed to the formation of a large number of free radicals in the preheat zone, where their diffusion to the surrounding region initiates the combustion process. The thermal model is applied more widely in the literature and shows better agreement with experimental data.

Cool Flames

Under particular conditions of pressure and temperature, incomplete combustion can result in the formation of intermediate products such as CO. As a result of this incomplete combustion, flames can be less exothermic than normal and are referred to as cool flames. An increase in the pressure or temperature of the mixture outside the cool flame can produce normal spontaneous ignition.

Flame Temperature

The adiabatic flame temperature, or theoretical flame temperature, is the maximum temperature attained by the products when the reaction goes to completion and the heat liberated during the reaction is used to raise the temperature of the products. Flame temperatures, as a function of the equivalence ratio, are usually calculated from thermodynamic data when a fuel is burned adiabatically with air.

Flame Types and Their Characteristics

There are two main types of flames: diffusion and premixed. In diffusion flames, the fuel and oxidant are separately introduced and the rate of the overall process is determined by the mixing rate. Examples of diffusion flames include the flames associated with candles, matches, gaseous fuel jets, oil sprays, and large fires, whether accidental or otherwise. In premixed flames, fuel and oxidant are mixed thoroughly prior to combustion. A fundamental understanding of both flame types and their structure involves the determination of the dimensions of the various zones in the flame and the temperature, velocity, and species concentrations throughout the system.

The structure of a one-dimensional premixed flame is well understood. By coupling the rate of heat release from chemical reaction and the rate of heat transfer by conduction with the flow of the unburned mixture, an observer moving with the wave would see a steady laminar flow of unburned gas at a uniform velocity, S_u, into the stationary wave or flame. Hence S_u is defined as the burning velocity of the mixture based on the conditions of the unburned gas. The thickness of the preheating zone and the equivalent reaction zone is found to be inversely proportional to the burning velocity. By considering the heat release from the chemical reaction, it is possible to calculate the thickness of the effective reaction zone.

There are a number of sources of instability in premixed combustion systems. Laminar flame instabilities are dominated by diffusional effects that can only be of importance in flows with a low turbulence intensity, where molecular transport is of the same order of magnitude as turbulent transport. Flame instabilities do not appear to be capable of generating turbulence. They result in the growth of certain disturbances, leading to orderly three-dimensional structures which, though complex, are steady.

Combustion processes and flow phenomena are closely connected and the fluid mechanics of a burning mixture play an important role in forming the structure of the flame. Laminar combusting flows can occur only at low Reynolds numbers, defined as

$$Re = \frac{\rho u d}{\mu} \qquad (1)$$

where, ρ = density, kg/m^3; u = velocity, m/s; d = diameter, m; and μ = kinematic viscosity, kg/ms. When $Re > Re_{cr}$ the laminar structure of a flow becomes unstable and when the Reynolds number exceeds the critical value by an order of magnitude, the structure of the flow changes. Along with this change of structure from an orderly state to a more chaotic state, the following parameters begin to fluctuate randomly: velocity, pressure, temperature, density, and species concentrations. Overall, with increasing Reynolds number, laminar flow becomes unstable as a result of these fluctuations and breaks down into turbulent flow. Laminar and turbulent flames differ greatly in appearance. For example, while the combustion zone of a laminar Bunsen flame is a smooth, delineated and thin surface, the analogous turbulent combustion region is blurred and thick.

Turbulent flame speed, unlike laminar flame speed, is dependent on the flow field and on both the mean and turbulence characteristics of the flow, which can in turn depend on the experimental configuration.

In high speed dusted premixed flows, where flames are stabilized in the recirculation zones, the turbulent flame speed grows without apparent limit, in approximate proportion to the speed of the unburned gas flow.

In the reaction zone, an increase in the intensity of the turbulence is related to the turbulent flame speed.

The balanced equation for turbulent kinetic energy in a reacting turbulent flow contains the terms that represent production as a result of mean flow shear, which can be influenced by combustion, and the terms that represent mean flow dilations, which can remove turbulent energy as a result of combustion. Some of the discrepancies between turbulent flame propagation speeds might be explained in terms of the balance between these competing effects.

To analyze premixed turbulent flames theoretically, two processes should be considered: (1) the effects of combustion on the turbulence, and (2) the effects of turbulence on the average chemical reaction rates.

A unified statistical model for premixed turbulent combustion and its subsequent application to predict the speed of propagation and the structure of plane turbulent combustion waves is available.

Many different configurations of diffusion flames exist in practice. Laminar jets of fuel and oxidant are the simplest and most wellunderstood diffusion flames.

The discussion of laminar diffusion flame theory addresses both the gaseous diffusion flames and the single-drop evaporation and combustion, as there are some similarities between gaseous and liquid diffusion flame theories. A frequently used model of diffusion flames has been developed, and despite some of the restrictive assumptions of the model, it gives a good description of diffusion flame behavior.

The Displacement Distance theory suggests that since the structure of the flame is only quantitatively correct, the flame height can be obtained through the use of the displacement length or "displacement distance" (eq. 2), where h = flame height, m; V = volumetric flow rate, m^3/s; and D = diffusion coefficient.

$$h = \frac{V}{2\pi D} \qquad (2)$$

As the velocity of the fuel jet increases in the laminar-to-turbulent transition region, an instability develops at the top of the flame and spreads down to its base. This is caused by the shear forces at the boundaries of the fuel jet. The flame length in the transition region is usually calculated by means of empirical formulas of the form (eq. 3): where l = length of the flame, m; r = radius of the fuel jet, m; v = fuel flow velocity, m/s; and C$_1$ and C$_2$ are empirical constants.

$$l = \frac{r}{C_1 - \frac{c_2}{v}} \qquad (3)$$

Laminar diffusion flames become turbulent with increasing Reynolds number. Some of the parameters that are affected by turbulence include flame speed, minimum ignition energy, flame stabilization, and rates of pollutant formation. Changes in flame structure are believed to be controlled entirely by fluid mechanics and physical transport processes.

The various studies attempting to increase our understanding of turbulent flows comprise five classes: moment methods disregarding probability density functions, approximation of probability density functions using moments, calculation of evolution of probablity density functions, perturbation methods beginning with known structures, and methods identifying coherent structures.

Fundamentals of Heterogeneous Combustion

The discussion of combustion fundamentals so far has focused on homogeneous systems. Heterogeneous combustion is the terminology often used to refer to the combustion of liquids and solids. From a technological viewpoint, combustion of liquid hydrocarbons, mainly in sprays, and coal combustion are of greatest interest.

Most theories of droplet combustion assume a spherical, symmetrical droplet surrounded by a spherical flame, for which the radii of the droplet and the flame are denoted by r_d and r_p, respectively. The flame is supported by the fuel diffusing from the droplet surface and the oxidant from the outside. The heat produced in the combustion zone ensures evaporation of the droplet and consequently the fuel supply. Other assumptions that further restrict the model include (1) the rate of chemical reaction is much higher than the rate of diffusion and hence the reaction is completed in a flame front of infinitesimal thickness; (2) the droplet is made up of pure liquid fuel; (3) the composition of the ambient atmosphere far away from the droplet is constant and does not depend on the combustion process; (4) combustion occurs under steadystate conditions; (5) the surface temperature of the droplet is close or equal to the boiling point of the liquid; and (6) the effects of radiation, thermodiffusion, and radial pressure changes are negligible.

In order to obtain an expression for the burning rate of the droplet, the following parameters are needed: physical constants such as the specific heat, and the thermal conductivity of the droplet, the radius of the flame, and the temperature of the flame. To determine these quantities, heat conduction, diffusion, and the kinetics of the chemical processes associated with droplet combustion need to be analyzed. This is achieved mathematically by solving the equations of mass continuity, mass continuity for components, and the energy equation. The solving of these equations can be facilitated if the following simplifying assumptions are made: the flame surrounding the droplet is a diffusion flame and, by definition, is formed where the fuel and oxidant meet in stoichiometric proportions; the temperature of this flame is very close to the adiabatic flame temperature; and the heat required for evaporation of the droplet and the heat loss to the surroundings through the burned gas are small and can therefore be neglected. These equations are usually solved in spherical coordinates for a one-dimensional case. However, since the flame is relatively thick, and the droplet is relatively small, the one-dimensional model of the process may not be a particularly accurate representation. Nevertheless, the values obtained for burning rates provide useful information.

The amount of data available on droplet combustion is extensive. However, the results can be easily summarized, because the burning rate constants for the majority of fuels of practical interest fall within the narrow range of 7 to 11×10^{-3} cm^2/s. An increase in oxygen concentration results in an increase in the burning rate constant. If the burning takes place in pure oxygen, the values for burning rate are increased by a factor of about 2.0, compared to when the burning takes place in air.

The convective gas flow around a burning particle affects its burning rate. It has been postulated that in the absence of convection, the burning rate is independent of pressure. Forced convection, on the other hand, is believed to increase the burning rate.

During the final stages of the combustion of a droplet, coke remains, and although it represents a relatively small percentage of the mass of the original oil droplet, the time taken for the heterogeneous reaction between the oxygen-depleted combustion air and the coke particle is generally the slowest of all the combustion steps.

The reaction between a porous solid, such as a coke sphere, and a gas, such as oxygen, occurs in the following stages: (1) the main reactant species diffuse thoroughly through the boundary layer toward the solid surface and the products of reaction diffuse away from the surface; (2) diffusion and simultaneous chemical reaction take place within the pores of the solid proceeding from the external surface toward the interior, and gaseous products diffuse in the opposite direction; and (3) at the participating surfaces, the reacting gas chemisorbs, some intermediate species are formed, then the final products of the reaction desorb from the surface. Thus the observed reaction rate is a function of the individual resistances—boundary layer diffusion, pore diffusion, and chemical kinetics, and the rate controlling process is the slowest of them or a combination of these processes. Even though a number of gas reactions may take place at the surface of a burning carbonaceous solid, the reaction forming CO is most often assumed, $C + 1/2\ O_2 \rightarrow CO$. Coke combustion is treated mathematically like char combustion.

In a practical combustion chamber, the droplets tend to burn in the form of sprays, hence it is important to understand the fundamentals of sprays. In the most simple case, a fuel spray suspended in air will support a stable propagating laminar flame in a manner similar to a homogeneous gaseous mixture. In this case, however, two different flame fronts are observed. If the spray is made up of very small droplets they vaporize before the flame reaches them and a continuous flame front is formed. If the droplets are larger, the flame reaches them before the evaporation is complete, and if the amount of fuel vapor is insufficient for the formation of a continuous flame front, the droplets burn in the form of isolated spherical regions. Flames of this type are referred to as heterogeneous laminar flames. Experimental determination of burning rates and flammability limits for heterogeneous laminar flames is difficult because of the motion of droplets caused by gravity and their evaporation before the arrival of the flame front.

In modern liquid-fuel combustion equipment the fuel is usually injected into a high velocity turbulent gas flow. Consequently, the complex turbulent flow and spray structure make the analysis of heterogeneous flows difficult and a detailed analysis requires the use of numerical methods.

The combustion of a coal particle occurs in two stages: (1) devolatilization during the initial stages of heating with accompanying physical and chemical changes, and (2) the subsequent combustion of the residual char. The burning rate or reactivity of the residual char in the second stage is strongly dependent on the process conditions of the first stage.

The ignition mechanism is rather complex and is not well understood in terms of actually defining the ignition temperature and reaction mechanisms. The ignition temperature is known, however, not to be a unique property of the coal and depends on a balance between heat generated and heat dissipated to the surroundings around the coal particle. Measuring the ignition characteristics is complicated by the fact that they are strongly dependent on the physical arrangement of the particles, eg, single particle, clouds of coal dust, or coal piles. Reported ignition temperatures range from 303 to 373 K in the case of spontaneous ignition of coal piles at ambient temperature to 1073–1173 K in the case of single coal particle ignition. Characteristics such as coal type, particle size and distribution of mineral matter, and experimental conditions such as gas temperature, heating rate, oxygen, and coal dust concentration are some of the important factors that influence values obtained for ignition temperatures.

A variety of techniques has been used to determine ignition temperatures: fixed beds, the crossing point method, the critical air blast method, photographic techniques, entrained flow reactors, electric spark ignition, luminous glow observations, plug flow reactors, shock tubes, and thermogravimetric analysis. The techniques mostly used are constant temperature methods, in which coal particles are introduced into a preheated furnace maintained at a fixed temperature.

The structure of residual char particles after devolatilization depends on the nature of the coal and the pyrolysis conditions such as heating rate, peak temperature, soak time at the peak temperature, gaseous environment, and the pressure of the system.

The rate limiting step in the combustion of char is either the chemical kinetics (adsorption of oxygen, reaction, and desorption of products) or diffusion of oxygen (bulk and pore diffusion). Variations in the reaction rate with temperature for gas-carbon reactions have been grouped into three main regions or zones depending on the rate limiting resistance.

COMBUSTION TECHNOLOGY

Technology addresses the more applied, practical aspects of combustion, with an emphasis on the combustion of gaseous, liquid, and solid fuels for the purpose of power production. In an ideal fuel burning system (1) there should be no excess oxygen or products of incomplete combustion, (2) the combustion reaction should be initiated by the input of auxiliary ignition energy at a low rate,

(3) the reaction rate between oxygen and fuel should be fast enough to allow rapid rates of heat release and it should also be compatible with acceptable nitrogen and sulfur oxide formation rates, (4) the solid impurities introduced with the fuel should be handled and disposed of effectively, (5) the temperature and the weight of the products of combustion should be distributed uniformly in relation to the parallel circuits of the heat absorbing surfaces, (6) a wide and stable firing range should be available, (7) fast response to changes in firing rate should be easily accommodated, and (8) equipment availability should be high and maintenance costs low.

Combustion of Gaseous Fuels

In any gas burner some mechanism or device (flame holder or pilot) must be provided to stabilize the flame against the flow of the unburned mixture. This device should fix the position of the flame at the burner port. Although gas burners vary greatly in form and complexity, the distribution mechanisms in most cases are fundamentally the same. By keeping the linear velocity of a small fraction of the mixture flow equal to or less than the burning velocity, a steady flame is formed. From this pilot flame, the main flame spreads to consume the main gas flow at a much higher velocity. The area of the steady flame is related to the volumetric flow rate of the mixture by equation 4:

$$V_{mix} = A_f S_u \qquad (4)$$

where $\dot{V}_{mix}$ = volumetric flow rate, m^3/s; A_f = area of the steady flame, m^2; and S_u = burning velocity, m/s.

The volumetric flow rate of the mixture is, in turn, proportional to the rate of heat input (eq. 5):

$$\dot{V}_{mix} \cdot (HHV) = \dot{Q} \qquad (5)$$

where $\dot{V}$mix = volumetric flow rate, m^3 s; HHV = higher heating value of the fuel, J/kg; and Q = rate of heat input, J/s.

Most of the commercial gas-air premixed burners are basically laminar-flow Bunsen burners and operate at atmospheric pressure.

Atmospheric pressure industrial burners are made for a heat release capacity of up to 50 kJ/s (12 kcal/s), and despite the varied designs, their principle of stabilization is basically the same as that of the Bunsen burner.

Gas burners that operate at high pressures are usually designed for high mixture velocities and heating intensities and therefore stabilization against blowoff must be enhanced. This can be achieved by a number of methods such as surrounding the main port with a number of pilot ports or using a porous diaphragm screen.

It is often desired to substitute directly a more readily available fuel for the gas for which a premixed burner or torch and its associated feed system were designed. Satisfactory behavior with respect to flashback, blowoff, and heating capability, or the local enthalpy flux to the work, generally requires reproduction as nearly as possible of the maximum temperature and velocity of the burned gas, and of the shape or height of the flame cone. Often this must be done precisely, and with no changes in orifices or adjustments in the feed system.

Turbulence in the flow of a premixture flattens the velocity profile and increases the effective burning velocity of the mixture; eg, at a pipe-Reynolds number of 40,000 the turbulent burning velocity is several times the laminar burning velocity and it can be perhaps fifty times larger at very high Reynolds numbers.

Combustion of Liquid Fuels

There are several important liquid fuels, ranging from volatile fuels for internal combustion engines to heavy hydrocarbon fractions, sold commercially as fuel oils. The technology for the combustion of liquid fuels for spark-ignition and compression-ignition internal combustion engines is not described here. The emphasis here is primarily on the combustion of fuel oils for domestic and industrial applications.

In general, the combustion of a liquid fuel takes place in a series of stages: atomization, vaporization, mixing of the vapor with air, and ignition and maintenance of combustion (flame stabilization). Recent advances have shown the atomization step to be one of the most important stages of liquid fuels combustion. The main purpose of atomization is to increase the surface area to volume ratio of the liquid. This is achieved by producing a fine spray. The finer the atomization spray the greater the subsequent benefits are in terms of mixing, evaporation, and ignition. The function of an atomizer is twofold: atomizing the liquid and matching the momentum of the issuing jet with the aerodynamic flows in the furnace.

Atomizers for large boiler burners are usually of the swirl pressure jet or internally mixed twin-fluid types, producing hollow conical sprays. Less common are the externally mixed twin-fluid types.

Combustion of fuel oil takes place through a series of steps, namely vaporization, devolatilization, ignition, and dissociation, which finally lead to attaining the flame temperature.

The study of the combustion of sprays of liquid fuels can be divided into two primary areas for research purposes: single-droplet combustion mechanisms and the interaction between different droplets in the spray during combustion with regard to droplet size and distribution in space. The wide variety of atomization methods used and the interaction of various physical parameters has made it difficult to give general expressions for the prediction of droplet size and distribution in sprays. The main fuel parameters affecting the quality of a spray are surface tension, viscosity, and density, with fuel viscosity being by far the most influential parameter.

Combustion of Solid Fuels

Solid fuels are burned in a variety of systems, some of which are similar to those fired by liquid fuels. In this article the most commonly burned solid fuel, coal, is discussed. The main coal combustion technologies are fixed-bed, eg, stokers, for the largest particles; pulverized-coal for the smallest particles; and fluidized-bed for medium size particles (see COAL).

Fixed-Bed Technology

Fixed-bed firing of coal by means of stokers consists of a solid bed of large (2–3-cm) coal particles on grates with combustion air passing through the grates and ash removal from the end of the grate.

Pulverized-Coal Firing

This is the most common technology used for coal combustion in utility applications because of the flexibility to use a range of coal types in a range of furnace sizes. Nevertheless, the selection of crushing, combustion, and gas-cleanup equipment remains coal-dependent.

Prior to being fed to a pulverized fuel burner, coal is ground to a size generally specified such that at least 70% passes a 200 mesh screen (75 μm) and less than 2% is retained on a 50 mesh screen (300 μm). The top size is determined by the classifying component of the crushing mill, oversize material being retained for further grinding.

Suspensions of pulverized coal or coal dust in air can be explosive; hence, it is essential to have adequate guidelines and procedures to ensure safe and stable operation during pulverized-coal (PC) firing.

As for oil and gas, the burner is the principal device required to successfully fire pulverized coal. The two primary types of pulverized-coal burners are circular concentric and vertical jet-nozzle array burners.

The self-igniting characteristics of pulverized coal vary from one coal to another, but for most coals it is possible to maintain ignition without auxiliary fuel when firing above the capacity of the boiler. The igniters may have to be activated in the following cases: (1) when firing pulverized coal with volatile matter less than about 25%, (2) when firing excessively wet coal, and (3) when feeding coal sporadically into the pulverizer.

Compared to natural gas and oil, complete combustion of coal requires higher levels of excess air, about 15% as measured at the furnace outlet at high loads, and this also serves to avoid slagging and fouling of the heat absorption equipment.

As pulverized-coal combustion potentially has a significant impact on the environment, the 1980s saw the employment of techniques such as coal washing and beneficiation to reduce the emissions of fly-ash, SO_x, and water-soluble metallic oxides.

The environmental impact associated with pulverized-coal firing has given rise to efforts to develop other combustion technologies such as fluidized beds or the use of coal-water slurry fuels (CWSF), which can be burned as substitutes for certain liquid fuels. CWSFs were developed as alternatives to more expensive and increasingly scarce conventional hydrocarbon fuels. The main challenge in the utilization of CWSFs is obtaining stable mixtures that can be successfully atomized and burned.

Fluidized-Bed Technology

In fluidized-bed combustion of coal, air is fed into the bed at a sufficiently high velocity to levitate the particles. This velocity is referred to as the minimum fluidizing velocity, u_{mf}. At this velocity, the volume occupied by the bed increases abruptly and the bed exhibits some of the characteristics of a fluid. The two predominant designs of fluidized beds are bubbling and recirculating, with most theories of fluidization being based on the simpler bubbling bed concept.

Fluidized combustion of coal entails the burning of coal particles in a hot fluidized bed of noncombustible particles, usually a mixture of ash and limestone. Once the coal is fed into the bed it is rapidly dispersed throughout the bed as it burns. The bed temperature is controlled by means of heat exchanger tubes. Elutriation is responsible for the removal of the smallest solid particles and the larger solid particles are removed through bed drain pipes. To increase combustion efficiency the particles elutriated from the bed are collected in a cyclone and are either reinjected into the main bed or burned in a separate bed operated at lower fluidizing velocity and higher temperature.

Fluidized beds are ideal for the combustion of high sulfur coals since the sulfur dioxide produced by combustion reacts with the introduced calcined limestone to produce calcium sulfate. The chemistry involved can be simplified and reduced to two steps, calcination and sulfation.

$$Calcination \quad CaCO_3 \rightarrow CO_2 + CaO$$
$$Sulfation \quad SO_2 + CaO + 1/2O_2 \rightarrow CaSO_4$$

The main steps associated with coal combustion (heating, devolatilization, volatiles combustion, and char burnout), occur sequentially to some extent; however, there is always some overlap between the stages.

Design Considerations in Fossil Fuel Combustion Systems

One of the most important considerations in the design of a combustion chamber for a boiler is the fuel that is to be burned in the chamber (see FURNACES, FUEL-FIRED). Although all fuels burn and release heat during combustion, the rate at which a fuel burns and releases heat, and the impurities associated with the fuel have to be considered.

Furnaces for Oil and Natural Gas Firing

Natural gas furnaces are relatively small in size because of the ease of mixing the fuel and the air; hence, there is a relatively rapid combustion of gas. Oil also burns rapidly with a luminous flame. To prevent excessive metal wall temperatures resulting from high radiation rates, oil-fired furnaces are designed slightly larger in size than gas-fired units in order to reduce the heat absorption rates.

Furnaces for Pulverized Coal Firing

The main differences between boilers fired with coal and those fired with oil or natural gas result form the presence of mineral matter in coals. The volume of the coal-fired furnace is higher because of the longer residence time required for the complete combustion of coal particles,

Table 1. Comparison of Design Parameters for Fossil Fuel Boilers

Parameter	Gas	Oil	Coal Grate	Coal Fluid bed	Coal Pulverized coal
heat rate, mW (t)	0.03–3000	0.03–3000	0.3–30	up to 30	30–3000
volumetric combustion intensity, kW/m^3	250–450	250–450	250–750^a	up to 2000^a (based on bed volume)	150–250
area combustion intensity, kW/m^2	280–500	280–500	2000	3000	7500
fuel firing density					
kg/m^{3}h			30–100	$\approx$250	15–30
kg/m^{2}h	6–11	6–11	40–250	up to 500	up to 1000
practical combustion temperature, °C	1000–1600	1100–1700	1200–1300	850–950	1600–1700
combustion time, s	10×10^{-1}	$20\ 25 \times 10^{-1}$	up to 5000	100–500	$\approx$1–2
particle heating rate, °C/s			>1	10^3–10^4	10^4–10^5

^{a}Based on the total combustion volume which includes space between the bed and the convective tubes.

the requirement of a controlled combustion rate to reduce NO$_x$ formation, the provision for a larger heat-transfer surface area resulting form decreased heat-transfer rates because of ash deposits on the surfaces, and increased spacing of heat-transfer tubes to reduce flue gas velocities and thereby erosion of heat-transfer surfaces. Even when firing coal, depending on the reactivity of the coal (rank), the size of the combustion chamber required can vary. Table 1 provides some design parameters for fossil fuel burners.

Environmental Considerations

Atmospheric pollutants released by combustion of fossil fuels fall into two main categories: those emitted directly into the atmosphere as a result of combustion and the secondary pollutants that arise from the chemical and photochemical reactions of the primary pollutants (see AIR POLLUTION).

The main combustion pollutants are nitrogen oxides, sulfur oxides, carbon monoxide, unburned hydrocarbons, and soot. Combustion pollutants can be reduced by three main methods depending on the location of their application: before, after, or during the combustion. Techniques employed before and after combustion deal with the fuel or the burned gases. A third alternative is to modify the combustion process in order to minimize the emissions.

Diffusion Flame Chemistry

Since most combustion systems employ mixing-controlled diffusion flames, which are characterized by very high pollutant emissions, it is imperative to look into the chemistry occurring in diffusion flames. In a typical diffusion flame the mixture composition in the reaction zone is close to the stoichiometric proportion and the temperature is at a maximum resulting from the large volume of this zone, thus NO$_x$ production is favored. If, however, the surrounding gas cools the combustion products rapidly, further reactions of CO and NO are eliminated. This fixes the concentrations of these pollutants at unfavorable levels. Furthermore, the fuel diffuses into the combustion zone through the burned gases and thus is heated in the

absence of oxygen. This creates ideal conditions for the formation of soot and the reduction of the CO$_2$ produced in the combustion zone to CO. Additionally, diffusion flames have low combustion intensity and efficiency and hence release large amounts of unburned hydrocarbon emissions. In general, despite the fact that the structure of the diffusion flame is more complex and difficult to analyze, the same basic description of soot formation and oxidation should apply to diffusion flames as for premixed flames.

Emissions Control

More advanced techniques for emissions control include electrical or plasma jet augmentation of flames based on radical production. Because in two-phase, heterogeneous combustion the flames are always diffusion flames on the microscale, ie, the individual droplets or particles burn as diffusion flames, and because at the characteristic times for evaporation, decomposition and burning of individual particles can be comparable to the characteristic times for mixing and pollutant formation, prevaporization, or gasification of the fuel can reduce pollutant emissions. For this reason catalytic systems for liquid-fuel decomposition and coal gasification are being considered seriously as alternatives to conventional combustion technology.

J. M. Beer and N. A. Chigier, *Combustion Aerodynamics,* Applied Science, London; John Wiley & Sons, Inc., New York, 1972.

N. Chigier, "Energy," *Combustion and Environment,* McGraw-Hill, Inc., New York, 1981.

R. H. Essenhigh, in M. A. Elliot, ed., Fundamentals of Coal Combustion, in *Chemistry of Coal Utilization,* 2nd Suppl. Vol., John Wiley & Sons, Inc., New York, 1981.

S. C. Stultz and J. B. Kitto, eds., *Steam, Its Generation and Use,* 40th ed., Babcock and Wilcox Co., Barberton, Ohio, 1992.

REZA SHARIFI
SARMA V. PISUPATI
ALAN W. SCARONI
Pennsylvania State University

COMPOSITE MATERIALS

Composite materials are multiphase materials obtained by artificial combination of different materials so as to attain properties that the individual components by themselves cannot attain. The broad concept of improved performance includes increased strength or reinforcement of one material by the addition of another material, as well as increased toughness, a decreased coefficient of thermal expansion, and increased thermal or electrical conductivity. An example of a composite material is a lightweight structural material (as used for aircraft) that is obtained by embedding continuous carbon fibers in one or more orientations in a polymer matrix. The fibers provide the strength and stiffness, while the polymer serves as the binder. Another example is highway concrete, a structural composite obtained by combining cement sand, gravel, and optionally other ingredients known as admixtures.

In the early 1960s there was a significant increase in interest in the science of composite materials when very strong and stiff, but brittle, ceramic, boron, and carbon fibers became available. These new fibers were at least as strong as the glass fibers of the same diameter but ~5–10 times stiffer. The term *advanced composites* has been coined to cover reinforced plastics containing continuous, strong, stiff fibers such as carbon (graphite), boron, aramid, or glass. Advanced composites have been developed primarily for the aerospace industry, in which the demand for strong, stiff lightweight structures overcame the prohibitive costs of early composite materials systems. Currently, advanced composite materials, mainly carbon fiber reinforced epoxy composites, are used widely in military aircraft and increasingly in civil aircraft. The largest-volume usage of these materials has, however, been in the recreational area in applications such as tennis rackets and golf club shafts.

COMPOSITE CLASSIFICATION ACCORDING TO THE MATRIX MATERIAL

Composites can be classified, according to their matrix material, as fibrous, laminated, and particulate.

Fibrous Composites

These composites consist of fibers in a matrix. The fibers may be short or discontinuous and randomly arranged, continuous filaments arranged parallel to each other, woven rovings (collections of bundles of continuous filaments), or braided sheaths.

Laminates

Two or more layers of material bonded together form a laminated composite, eg, automobile windshields (laminated glass) and bimetal thermostats.

Plastic laminated sheets produced in 1913 led to the formation of the Formica Products Company and the commercial introduction, in 1931, of decorative laminates consisting of a urea–formaldehyde surface on an unrefined (kraft) paper core impregnated with phenolic resin and compressed and heated between polished steel platens. Since 1937, the surface layer of most decorative laminates has been fabricated with melamine–formaldehyde, which can be prepared with mineral fillers, thus offering improved heat and moisture resistance and allowing a wide range of decorative effects.

Plywood is a laminate consisting of thin sheets of wood arranged with the grain in alternate sheets at right angles.

Fiber composite laminates often consist of unidirectional (parallel) continuous fibers in a polymer matrix, with the individual layers, plies, or laminae stacked with selected fiber angles so as to produce specific laminate stiffness and strength values. Laminates are also made by stacking layers of mats or woven fabric, which have many possible weave configurations.

Particulate Composites

These composites encompass a wide range of materials. As the word *particulate* suggests, the reinforcing phase is often spherical or at least has dimensions of similar order in all directions. Examples are concrete, filled polymers, solid rocket propellants, and metal and ceramic particles in metal matrices.

APPLICATIONS

Structural materials are predominantly metal and polymer-based materials, although they also include carbon- and ceramic-based materials, which are valuable for high-temperature structures. Among the metal-based structural materials, steel and aluminum alloys are dominant. Among the cement-based structural materials, concrete is dominant. Among the polymer-based structural materials, fiber-reinforced polymers are dominant, due to their combination of high strength and low density.

The dominant material for electrical connections is solder (eg, Sn–Pb alloy). Polymer–matrix composites in paste form and containing electrically conducting fillers are being developed to replace solder, because thermal fatigue can lead to failure of the solder joint.

Heat sinks, materials with high thermal conductivity, are used to dissipate heat from electronics. Materials exhibiting both a high thermal conductivity and a low coefficient of thermal expansion (CTE) are needed for heat sinks. Copper has high thermal conductivity, but its CTE is too high, so it is reinforced with continuous carbon fibers, molybdenum particles, or other fillers of a low CTE.

REINFORCEMENTS

The choice of reinforcement for a particular engineering application is likely to depend on a large number of parameters, including strength, stiffness, environmental stability, long-term characteristics, and cost. At present, carbon, glass, and aramid fibers account for >95% of the industrial market, with increasing use in areas such as

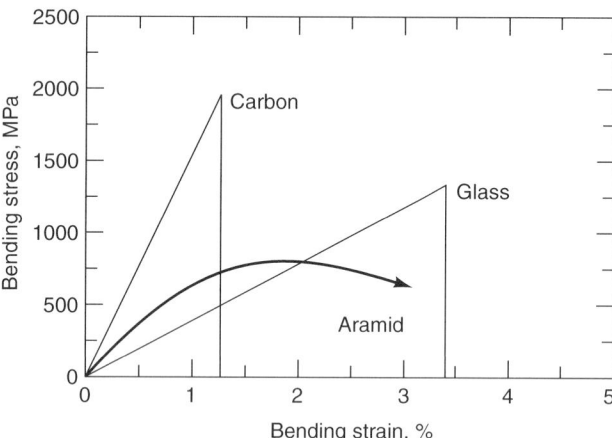

Figure 1. Bending stress versus bending strain for typical carbon, glass, and aramid fibers. To convert MPa to psi, multiply by 145.

the aerospace, automotive, construction, biomedical, and sport sectors. No single fiber type can be said to be truly superior to another; each has its own merits as well as shortcomings. The stress–strain responses obtained by bending typical carbon, glass, and aramid fibers are summarized in Figure 1. Carbon fibers generally offer the highest strengths and stiffnesses but can be brittle, failing at relatively low applied strains. Glass fibers offer intermediate strengths and higher failure strains but exhibit lower moduli. Aramid fibers such as Kevlar have lower strengths but are capable of absorbing considerable energy without fracture. Rather than using strength and stiffness as the performance parameters for material selection, specific properties obtained by normalizing the property with respect to the density (ρ) of the fiber are often quoted. For tensile or compressive members, the specific strength (σ/ρ) yields the largest load-carrying capacity for a given mass and the specific stiffness (E/ρ) produces the smallest deflection for a given mass. A convenient way of comparing fibers is to plot its specific strength against its specific modulus.

Glass Fibers

Glass fibers represent the most frequently used reinforcement in modern polymer composites. This popularity results from their relatively low cost and high tensile strength. In bulk form, glass is brittle, having relatively low strength. However, when extruded and drawn into fine fiber the strength of glass increases enormously, by as much as two orders of magnitude. Glass-fiber-reinforced composites are currently used in a wide variety of applications, including boat construction and the automotive and aerospace industries. Many types of glass fiber are available, each having specific properties and characteristics. The most commonly used fibers, E-glass, contain ~14% Al_2O_3, 18% CaO, 5% MgO, 8% B_2O_3, and 1% Na_2O + K_2O as well as SiO_2 (silica).

Glass fibers are produced by dry mixing the individual components and then heating them to form a melt. The temperature of the melt depends on the glass composition but is typically ~1250°C. It is common practice to apply a size, a surface coating, to protect the fibers during handling and prevent damage during any subsequent processing stages, such as weaving.

Carbon Fibers

The current technology for manufacturing carbon fibers is based on the thermal decomposition of organic precursor materials such as polyacrylonitrile (PAN) and pitch. Generally, the stages in the production of carbon fibers from the various organic precursors can be identified as spinning, stabilization, carbonization, and graphitization.

Surface treatment of fibers is an important stage in the manufacturing process. The primary aim of such treatment is to improve the adhesion between the fiber and the matrix and to improve handleability. The types of treatment are oxidative gas or dry oxidation, wet oxidation, and aqueous electrolytic or anodic oxidation.

Carbon-fiber-reinforced plastic composites are currently used in primary and secondary aircraft structures, helicopter rotor blades, sporting goods such as fishing rods and tennis rackets, and certain biomedical applications.

Aramid Fibers

Aramid fibers are formed by mixing a polymer, poly (p-phenyleneterephthalamide), with a strong acidic solution and extruding the mixture through spinnerets at a temperature between 50 and 100°C. The fibers are cooled and then washed thoroughly before being dried on bobbins. The properties of the aramid fibers can be modified by using solvent additives or by using a postspinning heat treatment. Aramid fibers with relatively low moduli, such as Kevlar 29, find use in energy-absorbing applications such as bullet-resistant and other protective clothing, helmets, and ropes. The higher modulus counterpart, Kevlar 49, is used in high-performance engineering applications such as the manufacture of load-bearing components for the aerospace industry, where it is used in filament-wound rocket motor cases.

Other Fibers

Currently, boron fibers are commonly used in metal matrices such as aluminum and magnesium. Boron-reinforced plastics, however, still find use as repair patches for damaged aircraft structures as well as in certain sporting goods, such as fishing rods and golf club shafts.

Polyethylene fibers are attracting considerable interest, owing to their high specific strength and stiffness as well as their excellent energy-absorbing capability. Two methods are used to manufacture polyethylene fibers. The first involves the extrusion and drawing of a

Table 1. Mechanical Properties and Relative Costs of Thermosetting and Thermoplastic Composites

Matrix material[a]	Young's modulus GPa[b]	Tensile strength, MPa[c]	Heat distortion temp, °C
Thermosets			
polyester	3.6	60	95
vinyl ester	3.4	83	110
epoxy	3.0	85	110
phenolic	3.0	50	120
Thermoplastics			
PES	2.8	84[d]	203
PEI	3.0	105[d]	200
PEEK	3.7	92[d]	140

[a] PES = polyethersulfone,
PEI = polyetherimide, PEEK = polyetheretherketone.
[b] To convert GPa to psi, multiply by 145,000.
[c] To convert MPa to psi, multiply by 145.
[d] Yield stress.

medium-molecular-weight polyethylene. The second method involves dissolving the polymer in a solvent and spinning at a temperature of ~140°C.

Silicon carbide fibers exhibit high-temperature stability and therefore find use as reinforcments in certain metal matrix composites. Silicon carbide fibers can be made in a number of ways, eg, by vapor deposition on carbon fibers.

MATRIX MATERIALS

The mechanical properties of composites based on the fibers discussed depend not only on the characteristics of the fibers but also on those of the matrix itself as well as on the fiber-matrix interface.

Polymer Matrices

The matrix in a polymer composite serves both to maintain the position and orientation of the fibers and to protect them from potentially degrading environments. Polymer matrices may be thermosets or thermoplastics. Thermosetting polymers are rigid, cross-linked materials that degrade rather than melt at high temperatures; thermoplastics are linear or branched molecules that soften upon heating. A comparison of the mechanical properties and relative cost of various thermosetting and thermoplastic materials is given in Table 1. Thermoset-based composites are somewhat less expensive than thermoplastic-based composites but have lower heat distortion temperatures and poorer toughness when tested in an interlaminar mode. The majority of present day composite components are still based largely on thermosetting matrices, such as unsaturated polyesters, epoxies, and phenolic resins.

Other Matrix Materials

Advanced materials, eg, structural components, in aerospace vehicles also employ ceramics and metals as composite matrices.

FABRICATION

A large number of methods are presently available for manufacturing long-fiber composites. The cost of finished components depends not only on the price of the raw materials but also on labor costs and energy requirements. Many composite components are manufactured by hand, involving relatively long manufacturing cycles. Large engineering components, such as boat hulls, are frequently manufactured using hand lay-up or spray-up techniques. The whole cycle may last several months, involving a large work force. Considerable effort is being made to find cheaper, more efficient ways to manufacture composite parts and structures. Some of the more commonly employed techniques used to manufacture such parts are fabrication methods such as hand and spray lay-up, filament winding, autoclave molding, compression molding, pultrusion, resin transfer molding, and composite interface engineering.

THEORIES OF REINFORCEMENT

The advantages of composites include improved stiffness and strength of the composite material system compared with the base line or unreinforced material. In composites, enhanced toughness can be achieved by increasing the energy required to initiate and propagate a crack through the brittle matrix. Increased work of fracture has been obtained in composites through debonding at the fiber–matrix interface; frictional interaction between the fiber and the matrix as fibers bridging a matrix crack are pulled out of the matrix as the crack extends and opens; deformation of the fibers; and fiber fracture. In the case of brittle matrix composites, only small quantities of reinforcing material are needed to achieve the desired performance. Fiber volume fraction is a quantitative measure of degree of reinforcement of the matrix material in a fiber-reinforced composite.

Strength predictions of composites are in general quite complex and somewhat limited. This is particularly true of compressive and shear strengths, which are needed, together with the tensile strengths, in composite failure prediction.

The tensile strength of a unidirectional lamina loaded in the fiber direction can be estimated from the properties of the fiber and matrix for a special set of circumstances. If all of the fibers have the same tensile strength σ_f and the composite is linear elastic until failure of the fibers, then the strength of the composite is given by

$$\sigma_c = \sigma_f v_f + \epsilon_{mf} E_m v_m$$

where ϵ_{mf} is the strain in the matrix when the fibers fail. For carbon fibers in epoxy resin the tensile strength of the composite is predicted to be approximately proportional to the fiber volume fraction. The assumption of a constant failure stress of the fibers is unrealistic. This strength prediction relates to a fiber-dominated model. However, at low values of fiber volume fraction the fiber-dominated model is invalid. For very low values of fiber volume fraction, the composite tensile strength is given approximately by the matrix-dominated model: $\sigma_c = \sigma_m(1 - v_f)$. The stress–strain relationship is used in conjunction with the rules for determining the stress and strain components with respect to some angle θ relative to the fiber direction to obtain the stress–strain relationship for a lamina loaded under plane strain conditions where the fibers are at an angle θ to the loading axis. Classical laminated-plate theory is used to determine the stiffness of laminated composites. Details of the Kirchoff-Love hypothesis on which the theory is based can be found in standard texts.

The strength of laminates is usually predicted from a combination of laminated plate theory and a failure criterion for the individual lamina. A general treatment of composite failure criteria is beyond the scope of the present discussion. Broadly, however, composite failure criteria are of two types: noninteractive, such as maximum stress or maximum strain, in which the lamina is taken to fail when a critical value of stress or strain is reached parallel or transverse to the fibers in tension, compression, or shear; or interactive, such as the Tsai-Hill or Tsai-Wu type, in which failure is taken to be when some combination of stresses occurs.

ECONOMIC ASPECTS

In the form of fiber-reinforced unidirectional and multidirectional composites very high values of strength and stiffness can be achieved with fiber volume fractions of ~60%. Excellent fatigue properties can also be obtained. Transverse impact damage tolerance, which was an early limiting factor in carbon fiber composites, has been improved greatly through the development of high strain to failure fibers and tough matrix and interleaf materials. However, these improvements in properties have been associated with significant increases in material costs. There have been many applications in which cost has been a secondary factor to performance, such as in the military aerospace fields. However, cost has become the critical issue in the continued development and application of advanced composite materials. Composite components can be fabricated through various routes, depending on the quality of the end product, and economies in the finished cost can be made by reducing the number of parts and attachments and assembly operations. In comparing costs it is important to include the life-cycle costs. In the oil and gas industry significant reductions in the life-cycle costs can be achieved by replacing steel with fiber-reinforced plastics. Another very large market with enormous potential for the applications of medium technology composites is the automotive industry.

F. S. Galasso, *Advanced Fibers and Composites*, Gordon and Breach Publishers, London, 1989.

M. Lewin and J. Preston, eds., *Handbook of Fiber Science and Technology*, Vol. 3, *High Technology Fibers*, Part A, Marcel Dekker, Inc., New York, 1985.

M. M. Schwartz, *Composite Materials Handbook*, McGraw-Hill, Inc., New York, 1984, Chapt. 7, Section 7.7.

C. S. Smith, *Design of Marine Structures in Composite Materials*, Elsevier Applied Science Publishers, London, 1990.

DEBORAH D. L. CHUNG
University at Buffalo,
State University of New York
Buffalo, N.Y.

COMPUTER-AIDED CHEMICAL ENGINEERING

Computer aids are now used at every stage from conception through design to operation. Virtually all chemical engineering is now "computer-aided chemical engineering." At the conceptual stage, software is used to plan and analyze laboratory experiments. At the design stage, processes (developed manually or by computer synthesis) are simulated in detail to ensure safe and economic operation.

A wide range of computer tools may be applied to operating plant. Computers may be applied online or offline.

In practice, an engineer may have up to 200 different computer programs that can be applied to aid the efficient design and operation of chemical processes. The breath of a typical range of computer software employed by a large manufacturing company is given in Table 1.

DEVELOPING ENGINEERING SOFTWARE

In general, engineering software should meet defined engineering goals, be based on sound principles of chemistry and physics, be testable and maintainable, be tested to ensure that it correctly codes the models on which it is based, and take account of the finite precision of computer arithmetic to give numerically accurate results.

Program Design

The following guidelines are broadly based on European Space Agency (ESA) software engineering standards. Figure 1 presents an outline of the software design process.

The initial stage is to specify exactly what the user wants, the User Requirements. Where the software

Table 1. Examples of Computer-Aided Engineering Software

Data Correlation and Prediction of Physical and Chemical Properties

Fitting experimental data for physical properties. Predicting physical properties

Fitting experimental reaction equilibrium and rate data

Prediction of flammability, toxicity, and other data important for safety computation

Prediction of ozone depletion, greenhouse effect, and other data for environmental studies

Prediction of equipment failure and repair rates

Unit Operations

Heat exchanger process and mechanical design (includes multi-pass, multifluid, tube, plate). Fired heaters

Evaporator design

Design of pressure vessels according to various national and international standards

Simulation and design of distillation columns, batch, and continuous distillation

Absorption and stripping column simulation and design

Design of column internals, packed columns, valve tray, sieve tray, and bubble-cap

Reboiler and condenser design

Compressor and expander design and simulation

Liquid–liquid extractor simulation and design

Modeling Rankine cycle systems

Gas and steam turbine modeling

Analysis and performance of agitated vessels

Pressure drop calculations for liquids and for gases in isothermal and adiabatic flow

Two-phase pressure drop calculations in pipes and conduits

Two-phase choked flow

Instability in two-phase flow

Surge analysis

Estimation of tube vibration

Tubular and fluidized bed reactor simulation and design

Fluidized bed drier modeling

Driers, indirectly heated, directly heated. Spray driers

Restrictor orifice design

Combustion calculations

Non-Newtonian flow and heat transfer calculations

Furnace design and radiant heat transfer

3-D fluid flow, heat, and mass transfer through Computational Fluid Dynamics

Environmental Calculations

Dispersion of gases, aerosols and smokes from stacks, ruptures, and fires

Dry and rain-enhanced deposition of aerosols and smokes from plumes

Leaching from landfill sites and dispersion of leakages through groundwater

Modeling river networks for accumulation of pollutants

Concentration of pollutants in land and marine life (vegetable, animal and microbial)

Integrated effect of releases to air, water, and land

Safety Studies

Hazard analysis, fire and explosion, toxic chemical release

Bursting disk and pressure relief valve computations

Adiabatic and isothermal relief in piping networks

Design and simulation of flare release systems

Process Availability and Reliability

Plant availability computed from equipment failure and maintenance statistics

Availability with stand-by equipment

Process Simulation

Simulation and design of steady-state processes

Data reconciliation (estimation of statistically most likely performance from measurements)

Simulation and design of unsteady processes: Control, start-up, shut-down, upset conditions

Design of batch and semi batch processes; dedicated, multi purpose and multi-product plant

Discrete dynamic simulation of batch and semi-batch processes

Simulation of linked distillation columns

Simulation of heat-exchanger networks

Simulation of evaporator trains

Site simulation: energy use, utilities requirements, major intermediates

Economic evaluation of plants and sites

Optimization of design and operating conditions

On-line Computation

On-line optimization to compensate for performance, market, and raw material changes

On-line data reconciliation

On-line regulation control to ensure stable performance in the face of disturbances

On-line fault diagnosis

Condition monitoring (prediction of equipment deterioration from noise etc. measurements)

Optimization of start-up, shut-down and load change trajectories

On-line scheduling, eg, of batch-process operation

Aids to the Design Process

Intelligent piping and instrumentation diagram systems

Design rationale systems (knowledge-based design)

Interchange software based on Process industries

STEP standards

Process Synthesis

Minimal energy or utility requirements for a process

Minimal energy or utility requirements for a site

Optimal heat-exchanger network design

Optimal separator network design

Optimal process design by process synthesis

---

automates a procedure well known to the intended end-users, these requirements should be drawn up in consultation with the end-users. Where the software can give functionality beyond the experience of current users, the input of innovators in the field is required. The rationale (reasoning) behind each requirement should be recorded to ensure that the program is written efficiently. This requirement is easier to meet with simpler, faster, and more easily tested software.

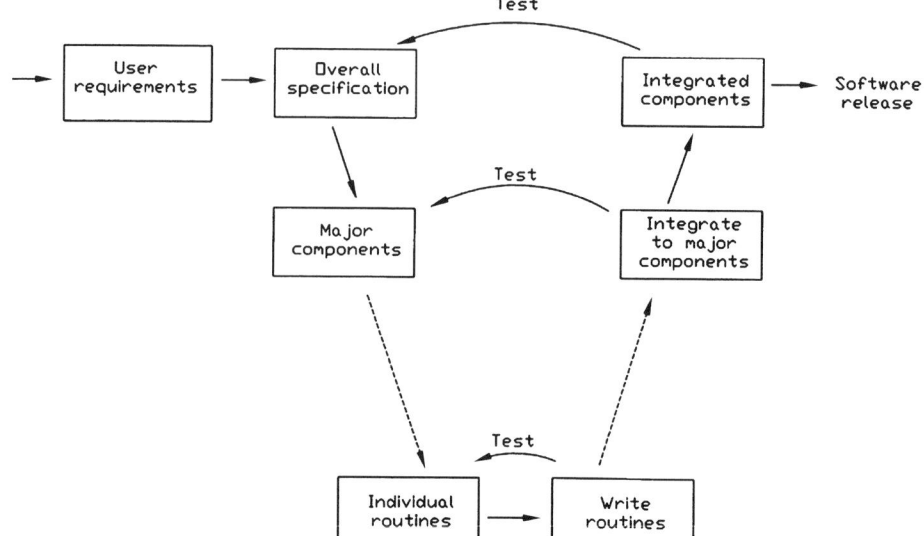

Figure 1. Software design process.

The User Requirements cover:

- The functionality, that is the chemical engineering problems to be solved.
- The user interface, ie, how the user will interact with the program (the appearance of any windows, the use of buttons and keys etc).
- The environment in which the program will be used. Will it be stand-alone, accessed over the web or used in some other way, eg, as an additional function to be attached to a spreadsheet?
- The programs to which it must be interfaced. Does it import data from another program and/or export results to a further program?
- The services that it must use. For example, must it employ a defined physical properties package?

The User Requirements prioritize the required functionality. *Essential* requirements must be met in the first release; without this functionality, the software does not meet its basic goals. *Desirable* requirements would add value to the program, but could be deferred or omitted.

The User Requirements should translate directly into tests that will verify that (at least for the specific values tested) the software meets the requirements. These tests form part of a validation plan for the software.

The overall software package is divided into its major components: subroutines, classes (objects), components dynamically linked by middleware, or separate programs that will be invoked as needed. The interfaces between these components must then be defined so that they can work together. At this stage, separate acceptance tests for each major software component are defined. Then major software components are further subdivided and acceptance tests for each subcomponent defined. When the program is written, each component is verified against the pre-defined tests before it is integrated into the final program. Similarly, subcomponents are tested both before and after integration. The test procedure is an integral part of the program design.

Before programming starts, a "build" sequence is defined. The build methodology provides early delivery of a simple program offering a subset of the requirements. The program is enhanced in each subsequent build as more functionality is added.

The build approach has benefits both in meeting delivery dates and in improved program quality. Delivery dates are improved because any delay in the first build signals problems that can be identified and corrected early.

The build discipline extends beyond the initial release into the support and development phase. User experience provides suggestions for enhancements and uncovers bugs that escaped pre-release testing. In response to this experience, the User Requirements document is updated, the changes prioritized, and new releases planned each with its own test and build schedule. Each build and release should be archived so as to remain accessible over extended periods. Versions of major components should similarly be separately archived. Dynamically linked components may follow separate, unsynchronized, build and release patterns. An effective software design and validation procedure lasts throughout the life of the software.

Programming Languages and Modeling Systems

Chemical engineering software may be written in a conventional high-level language, an artificial intelligence (AI) language, a general-purpose or specific equation-based modeling system, or using spreadsheet tools. Many engineers construct simulations using modeling tools rather than writing special-purpose programs. Simulations developed using these tools must be designed and tested as for conventional programs. There is the same scope for logical and numerical errors.

Among, the most-used language for engineering are general-purpose languages, from ada to BASIC to C/C++ to Delphi, etc., languages designed for artificial intelligence like LISP and Prolog; equation-based modeling systems from equation manipulation systems such as Mathematica and MathCad through general tools for solving and optimizing problems defined as equations to tools specifically for the chemical engineer, to spreadsheet tools such as Excel and Lotus 1-2-3.

Programming

Programs should be written to minimize programming errors, with dimensional consistency, limiting of side effects, limited run time, limited use of computer memory, minimal arithmetic failures, a division into component parts, checking of data, little or no handling uncertainty, and a minimizing of error messages.

Numerical Analysis

Four areas that can give rise to numerical problems are expressions that evaluate to 0/0, convergence of simple iterations, solution of equations and matrix inversion, and numerical solution of differential equations.

Arranging Expressions for Computation

The general principles to be applied in arranging equations for computation are as follows: (1). The equations should be as nearly linear as possible over as wide a range as possible; (2). The equations should be computable for all values of the right-hand side variables; and (3). Explicitly computable expressions should be preferred to expressions that need to be solved iteratively.

Linearity is required because expressions frequently form part of larger iterative schemes. Such iterations are nearly always solved by local linearization (eg, Newton–Raphson iteration). The size of the region within which rapid convergence is achieved is determined by the size of the region within which the relevant equations are (nearly) linear. These considerations apply equally to conventional (assignment-type) programs and to equation-based modeling systems.

If the right-hand side (or equation) is not computable, no iterative scheme can make sensible progress to a converged solution.

Explicit expressions are preferred both because they are faster and because the risk of an iteration failing to converge is eliminated.

In order to put the equations into the best form for computation, it may be necessary to derive them from first principles rather than employ a conventional textbook formula.

USING ENGINEERING SOFTWARE

The moral and professional responsibility for engineering decisions rests with the engineer making the decisions rather than the authors of any software employed.

Usually, the legal responsibility also rests with the engineer making the decisions.

Engineers using computer programs written by others must thoroughly understand the application to which the program is applied. They should provide a decision audit trail so that all recommendations can be checked. The audit trail should include the following:

1. A clear statement of the problem being tackled.
2. A statement of the assumptions made and their justification.
3. A review of the software applicable to the problem.
4. Identification of the chemical species that might arise.
5. Review of the data sources and the range of temperature, pressure and composition over which they are valid.
6. Review of the models employed by the software, their validity and applicability.
7. Estimation of the errors that might be introduced through the data or the models.
8. Sensitivity tests to assess the consequences of possible data or model errors.
9. The alternative solutions that have been generated.
10. A critical assessment of the performance and risks of the alternative solutions.
11. The recommended solution.

The audit trail should also include all error messages generated by the software used and a critical assessment of the implications of the messages.

Use of Spreadsheets

Quality assurance of results obtained from spreadsheets requires special attention. In particular, there is the potential for the user to inadvertently change formulas in cells. Strict guidelines on spreadsheet use should be issued. Caution is required for safety-critical applications. The decision audit trail should include a full copy of the spreadsheet program, not just its results.

Neural Networks

Neural networks contain a large number of fitted parameters. Statistically, the larger the number of parameters, the less significance any such fit has. In many cases, the statistical significance of neural net predictions is minimal. Consequently, the Net may be an effective method of interpolating the conditions in which it was trained but it may not be a good basis for predicting behavior outside the training region. Applications in control may be valid because there is a constant stream of data enabling retraining to be undertaken and the data are likely to span the conditions being predicted. In other applications, however, engineers should use neutral networks with caution.

Integrating Multiple Uncertainties

The sensitivity of performance to assumptions can be assessed. The resulting uncertainties should include both data uncertainties and model uncertainties. Where specific interactions are important, users may explore perturbing several uncertain parameters in the same simulation. However, the combinatorial problem of exploring all possible combinations of uncertainties makes it impracticable to explore more than a few multiple uncertainty cases.

The uncertain distribution functions for most parameters found in engineering studies are, to sufficient accuracy, symmetrical. However, there is obvious asymmetry in parameters such as mass transfer coefficients. These may have large uncertainties, but they clearly cannot be negative. In most engineering problems, estimates of standard deviation are little more than guesses.

An approach to integrating uncertainties can be built into software. However, more often it will be used as part of the quality assurance tests employed by the end user. For example, if an engineer records sensitivity results in a spreadsheet, equations can be used to generate additional columns giving the integrated effect of uncertainty.

CURRENT ADVANCES

There are number of areas of computer-aided chemical engineering that are gaining in importance as the pressure for a more environmentally friendly chemical industry grows.

Computer-Aided Molecule Design (CAMD)

CAMD has been developing rapidly during the last 10 to 15 years. Techniques such as the group contribution method enable the properties of molecules to be predicted before they have been synthesised. CAMD exploits these abilities to design molecules that have the desired properties. Initial applications have been in the development of selective solvents, particularly nonhalogenated solvents with reduced toxicity and reduced ozone depletion potential. Potential applications include the development of economic, effective, safer, and less polluting chemical products.

Computer-Aided Process Synthesis

Process synthesis has been studied for over 30 years, but until recently applications have been limited to energy reduction studies. Process synthesis differs from process optimization in the variables selected for optimization. Process optimization adjusts only continuously variable parameters such as lengths, temperatures, and pressures. Process synthesis optimizes also discrete variables. Recent advances in computer hardware and software promise a much wider range of application.

The benefit of computer-aided process synthesis is that very large numbers (eg, millions) of process alternatives

can be implicitly evaluated. The evaluation identifies processes that are economic and have reduced environmental emissions. It is impracticable to evaluate such a large range of alternatives by hand. Regulatory authorities increasingly demand that processes be evaluated to ensure that there are not competitive alternatives with lower environmental impact. Process synthesis enables these demands to be met in a rigorous manner.

Flexible Design

Parameters such as physical size are expensive to change after a process plant is built. However, flow rates, temperatures, and pressures can be changed. Thus, if the desired purity cannot be achieved at the nominal throughput, it may be possible to achieve it at a reduced throughput. Flexible design recognizes that operation can be optimized after production starts and that retrofit modification is possible where initial production targets cannot be met. It understans that it may be uneconomic to design to ensure that, even under the worst combination of uncertain outcomes, target production rate is met. Market size is often one of the most uncertain parameters on which a chemical engineering design is based. As the market builds up, there may be the opportunity to debottleneck a plant that initially underperforms. Flexible design explores the trade-off between applying excessive design margins and running a risk that a target production rate cannot be met. Flexible design requires optimization under uncertainty.

There is active research in automated methods for optimal design under uncertainty. The technology is, however, not yet available for use outside the relevant research schools.

The use of computer aids does not reduce the responsibility of engineers. Indeed, with fewer simplifying assumptions, there is a greater requirement that the engineer have a fundamental understanding of the technology. Furthermore, more detailed models demand more extensive data that also need to be obtained and critically assessed. The speed and accuracy of the computations makes it practicable to design better processes. Processes can be more thoroughly evaluated and their risks and opportunities more thoroughly assessed. Recent developments promise better, cleaner, safer processes.

R. Best, G. Goltz, J. Hulbert, A. Lodge, F. A. Perris, and M. Woodman, *The Use of Computers by Chemical Engineers.* IChemE (CAPE Subject Group), Rugby, UK, 1999.

G. Booch, I. Jacobson, and J. Rumbaugh, *Unified Modeling Language User Guide*, Addison Wesley Longman Publishing Co., Reading, Mass, 1998.

C. Mazza, J. Fairclough, B. Melton, D. DePablo, A. Scheffer, R. Stevens, M. Jones, and G. Alvin, *Software Engineering Guides*, Pearson Education, Harlow, UK, 1995.

W. E. Perry, *Effective Methods for Software Testing*, Wiley, New York, 1995.

W. R. JOHNS
Chemcept Limited

CONDUCTING POLYMERS

The discovery that polyacetylene could be prepared with high electronic conductivity initiated a major research effort on organic conducting polymers was recognized by the awarding of the Nobel Prize for Chemistry in 2000 to Alan Heeger, Alan MacDiarmid, and Hideki Shirakawa. A variety of organic conducting polymer materials has now been developed for applications ranging from electromagnetic shielding and "plastic electronics" to light-emitting devices and corrosion-inhibiting paints.

Considered here are intrinsically conductive polymers—excluding polymer composite materials containing metal particles or carbon black, where the electrical conductivity is the result of percolation of conducting filler particles in an insulating matrix or tunneling between the particles.

The key feature of the electrically conductive organic polymers is the presence in them of conjugated bonds with π-electrons delocalized along the polymer chains. In the undoped form, the polymers are either insulating or semiconducting with a large band gap. The polymers are converted to electrically conductive, or doped, forms via oxidation or reduction reactions that form delocalized charge carriers. Charge balance is accomplished by the incorporation of an oppositely charged counterion into the polymer matrix. The conductivity is electronic in nature and no concurrent ion motion occurs in the solid state. The redox doping processes are reversible and can be accomplished electrochemically. During electrochemical switching, ions do move into and out of the polymers as charge-balancing species for the charge carriers on the polymer backbone. Most applications of conducting polymers utilize their electronic (electrically conducting and optical) properties, but some (eg, battery or sensor electrodes) involve their ionic properties.

SYNTHESIS OF ELECTRICALLY CONDUCTIVE POLYMERS

A number of synthetic routes have been developed for the preparation of conjugated polymers. The diversity has been driven by the desire to examine many different types of conjugated polymers and attempts to improve material properties. Material property enhancement has centered on the synthesis of polymers that are processible in various forms. Five primary classes of conjugated polymers have been shown to exhibit high levels of electrical conductivity in the doped state. These include polyacetylenes, polyarylenes and polyheterocycles, poly(arylene vinylenes), and polyanilines. In addition, a number of multicomponent materials, usually polymer blends and composites, have been prepared in which at least one of the components is a conducting polymer.

Polyacetylenes

Even with improvement in the properties of polyacetylenes prepared from acetylene, the materials remained intractable. To avoid this problem, soluble precursor polymer methods for the production of polyacetylene have been developed. The most highly studied system utilizing this method is the Durham technique.

Polyphenylenes

Poly(p-phenylene) (PPP), synthesized using direct polymerization methods, yields oligomers of a largely intractable material. Although it is generally difficult to use intractable polymers in practical applications, sintering techniques are available that may make this polymer technologically useful.

Polypyrroles

Heterocyclic monomers such as pyrrole and thiophene form fully conjugated polymers with the potential for doped conductivity when polymerization occurs in the 2,5 positions. The heterocycle monomers can be polymerized by an oxidative coupling mechanism, which can be initiated by either chemical or electrochemical means.

The electrochemical polymerization of pyrrole is generally believed to follow a radical step-growth mechanism.

Polythiophenes

In contrast to the intractability of many polypyrroles, a substantial number of substituted polythiophenes have been found to be processible both from solution and in the melt. The most studied of these systems are the poly(3-alkylthiophenes) (P3ATs).

Poly(arylene vinylenes)

The use of the soluble precursor route has been successful in the case of poly(arylene vinylenes), both those containing benzenoid and heteroaromatic species as the aryl groups. The simplest member of this family is poly(p-phenylene vinylene) (PPV).

Polyanilines

Polyaniline (PAni) is commonly prepared by polymerization of aniline using $(NH_4)_2S_2O_8$ in HCl. As prepared, it has a structure known as emeraldine hydrochloride. In this form, PAni is highly conductive but completely insoluble. When emeraldine hydrochloride is deprotonated with NH_4OH, the highly soluble emeraldine base is produced. As was the case for polythiophenes, substitution along the PAni backbone has been utilized as a means of improving processibility. Many derivatives are known for PAni because of the possibility of substitution of the monomers at either the main-chain nitrogen atoms, or on the aromatic ring. Substituents studied have included alkyl, aryl, sulfonyl, and amono groups.

Polypyridine and Ladder-Structured Polyquinoxalines

Unlike most other heterocyclic monomers for conducting polymers, pyridine cannot be polymerized directly by oxidation. However, polypyridine, poly(2,5-pyridinediyl), can be prepared by chemical or electrochemical reductive debromination of 2,5-dibromopyridine. Polypyridine is somewhat soluble in polar solvents such as formic acid and is found to be relatively electron accepting; hence it can be doped n-type much more easily than p-type and has been found to have a doped conductivity in the n-type form of ~ 0.1 S/cm. A 95% regioregular polymer has been synthesized, that displays significantly higher conductivities.

Ladder polymers are those having double strands rather than single linkages between the monomer units; conjugated ladder polymers are of considerable interest through being expected to have small energy gaps. Some of them, like the polypyridines, also have potential commercial interest as possible electron-transporting layers in light-emitting devices.

Liquid Crystalline Conducting Polymers (LCCPs)

Since the electronic properties of conjugated polymers are sensitive to the extent of π-orbital delocalization, and hence to chain alignment and planarity, there has been considerable interest in the effects of solvation (in solution) and of drawing (in the solid state). An alternative method of modifying the backbone conformation is to use the self-organizing properties of liquid crystals (LCs) to control it, either by blending or by chemical functionalization of the conducting polymer with mesogenic groups. This offers potential benefits from switchable electronic properties or from the use of electric or magnetic fields as an aid to orienting conjugated polymers during processing.

Conducting Polymer Blends, Composites, and Colloids

Incorporation of conducting polymers into multicomponent systems allows the preparation of materials that are electroactive and also possess specific properties contributed by the other components. Dispersion of a conducting polymer into an insulating matrix can be accomplished as either a miscible or phase-separated blend, a heterogeneous composite, or a colloidally dispersed latex. When the conductor is present in sufficiently high composition, electron transport is possible.

There are several approaches to the preparation of multicomponent materials; and the method utilized depends largely on the nature of the conductor used. In the case of polyacetylene blends, *in situ* polymerization of acetylene gas into a polymeric matrix has been a successful technique.

Because of the aqueous solubility of polyelectrolyte precursor polymers, another method of polymer blend formation is possible. The precursor polymer is codissolved with a water-soluble matrix polymer, and films of the blend are cast. With heating, the fully conjugated conducting polymer is generated to form the composite film. This technique has been used for poly(arylene vinylenes) with a variety of water-soluble matrix polymers, including polyacrylamide, poly(ethylene oxide), polyvinylpyrrolidinone, methylcellulose, and hydroxypropylcellulose. These blends generally exhibit phase-separated morphologies.

The true thermoplastic nature of poly(3-alkylthiophenes) and of some polyanilines, ie, solubility and fusi-

bility, allows the use of compounding methods commonly used in the plastics industry for the preparation of composites of these polymers. The polymers can be codissolved with a matrix polymer, then processed from organic solution. Again the resulting blends are phase separated, but if the composition of conducting polymer is high enough, the conducting component forms the matrix with the insulating polymer dispersed within it and high conductivity is possible.

Electrochemical polymerization of heterocycles is useful in the preparation of conducting composite materials. Conducting polymer composites have also been formed by coelectrodeposition of matrix polymer during electrochemical polymerization. The preparation of molecular composites by electropolymerization of heterocycles in solution with polyelectrolytes is an extremely versatile technique, and many polyelectrolyte systems have been studied.

DOPING AND OPTICAL PROPERTIES

In order to induce high electrical conductivity in organic conjugated polymers, charge carriers must be introduced. These are created by removing electrons from, or adding electrons to, the delocalized π-electron network of the polymer, creating a conducting unit that is now a polymeric ion rather than a neutral species. The charges introduced are compensated by ions from the reaction medium. This process is called doping by analogy to the changes that occur in inorganic semiconductors upon addition of small quantities of electronic defects. However, it proceeds through a different mechanism and is more precisely termed a redox reaction. The doping level, or ratio of charge carriers per polymer repeat unit, which is generally between 0.2 and 0.4 in most polyarylenes, can be determined by measuring the content of charge-balancing counterions. The ability to control the electrical properties of conducting polymers over wide ranges, by adjusting the redox doping level, has created interest in these materials for a number of emerging applications.

Charge Carriers in Conducting Polymers

The mechanism for the conductivity increase resulting from doping in inorganic semiconductors involves the formation of unfilled electronic bands. Electrons are removed from the top of the valence band during oxidation, called p-type doping, or added to the bottom of the conduction band during reduction, termed n-type doping. Extension of this argument to the case of conjugated organic polymers was found to be inaccurate, as the conductivity in many conducting polymers was found to be associated with spinless charge carriers. *In situ* electron spin (epr)/electrochemistry techniques have shown that the conducting entity in polyacetylene, polypyrrole, polythiophene, and poly(p-phenylene) can be spinless, although evidence exists for mixed-valence charge carriers as well.

The conductivity increase following doping in conjugated polymers is explained in terms of local lattice distortions and localized electronic states. In this case the valence band remains full and the conduction band empty so that there is no appearance of metallic character. When the polymer chain is redox doped, a lattice distortion results and the equilibrium geometry for the doped state is different than the ground-state geometry.

Doping Processes

Redox doping can be accomplished through both chemical and electrochemical means. Vapor-phase doping of neutral polyacetylene has been carried out using AsF_5 and I_2 as oxidants. These conjugated polymers can be chemically and electrochemically reduced and reoxidized in a reversible manner. In all cases, the charges on the polymer backbone must be compensated by ions from the reaction medium, which are then incorporated into the polymer lattice. The rate of the doping process is dependent on the mobility of these charge compensating ions into and out of the polymer matrix.

Electrogenerated conducting polymer films incorporate ions from the electrolyte medium for charge compensation. Electrochemical cycling in an electrolyte solution results in sequential doping and undoping of the polymer film.

Optical Properties

The energy difference between the valence band and the conduction band in conjugated polymers, referred to as the band gap can be determined using optical spectroscopy. In the neutral (or insulating) form, conducting polymers exhibit a single electronic absorption in either the visible or ultraviolet (uv) region attributed to the electronic transition between the HOMO and LUMO levels.

In the doped form, transitions between the band edges and newly formed intragap electronic states are observed in the optical spectra of conducting polymers. When polarons are present as charge carriers, an additional transition is apparent that corresponds to the electronic transitions between the two gap states. Since the intragap electronic states are taken from the band edges, the band gap increases with increasing doping levels. Also, since a bipolaron creates a larger lattice distortion than a polaron, the gap states are further away from the band edges in the bipolaron model.

The changes in the optical absorption spectra of conducting polymers can be monitored using optoelectrochemical techniques.

ELECTRICAL CONDUCTION PROPERTIES

The electrical conductivity, σ, of a material, which is the inverse of its specific resistivity, ρ, is a measure of a material's ability to transport electrical charge.

Quite generally, the magnitude of the conductivity of organic conducting polymers depends not only on the doping level but also on the degree of disorder and structural aspects of the polymer. Sample morphology plays a key role in limiting or assisting in charge transport. Important factors are the ratio of crystalline-to-amorphous material content and whether the polymer has a bulk dense morphology or a highly open fibrillar morphology.

No reproducible observation of superconductivity has so far been made in organic polymers. Superconductivity does, however, occur in the inorganic polymer poly(sulfur nitride) at temperatures below about 0.3 K.

Metallic Polymers

Room-temperature conductivity of several of the common organic conducting polymers reaches remarkably high values, considering that the density of carriers is much less than that in conventional metals. The highest conductivities are for polyacetylene stretched by a factor of ~6 to improve alignment of the polymer chains—the conductivity perpendicular to the stretch direction is much lower.

Apart from the high conductivity values, the metallic character of the highly conducting polymers is such that their conductivity remains nonzero as the temperature tends to absolute zero. In contrast, semiconductors show zero conductivity at the absolute zero of temperature, owing to the gap between valence and conduction bands or localization of states.

In spite of this metallic character, the temperature dependence of the conductivity shows a predominantly nonmetallic sign, ie, the conductivity increases as temperature increases. A similar pattern of temperature dependence is seen for polyaniline, while for polypyrrole, polythiophene, and PPV the nonmetallic sign for temperature persists even up to room temperature.

Semiconducting Polymers

As the doping level of conducting polymers is reduced, the conductivity changes from mixed metallic–nonmetallic to disordered semiconductor behavior. For a wide variety of semiconducting polymers, the shape of the temperature dependence agrees with that for variable-range hopping conduction (ie, tunneling between localized electronic states assisted by thermally excited lattice vibrations).

Dispersions of metallic conducting polymer particles in an insulating polymer matrix show a similar temperature-dependent conductivity, which is consistent with charging-limited tunneling between mesoscopic metallic islands. Owing to the similarity of the predicted conductivity for this mechanism and variable-range hopping, it is difficult to identify the precise nature of the tunneling conduction mechanism in semiconducting polymers.

STABILITY

Although polyacetylene has served as an excellent prototype for understanding the chemistry and physics of electrical conductivity in organic polymers, its instability in both the neutral and doped forms precludes any useful application. In contrast to polyacetylene, both polyaniline and polypyrrole are significantly more stable as electrical conductors. When addressing polymer stability it is necessary to know the environmental conditions to which it will be exposed; these conditions can vary quite widely.

APPLICATIONS

The novel and varied electrical, electrochemical and chemical properties of conducting polymers are leading to their development for a large variety of applications. In many cases the polymers have unique advantages over other materials; eg, low density, mechanical flexibility, tunable optical properties, and their ability to be functionalized for different purposes. There are of course disadvantages: eg, the poor stability of polyacetylene severely limits its usefulness for applications.

The range and versatility of conducting polymers in technology can be seen from an abbreviated list of their uses.

Electromagnetic Shielding and Charge Dissipation

One of the first applications was the general area of electromagnetic shielding.

Polyaniline has been used as a charge dissipation layer for electron-beam lithography as a removable SEM discharge layer as a conducting electrode for electrolytic metallization of copper on through-holes in circuit boards and to provide a solderable finish in printed circuit board technology. The processability of some of the conducting polymers has enabled them to be prepared in forms applicable to typical industrial processes for applications.

Corrosion-Inhibiting Paints

An application of PAni is its use to inhibit corrosion. Studies have shown a reduction of corrosion of steel plates exposed to brine and acid environments when coated with a thin layer of PAni. The mechanism of this effect could be that the PAni leads to the formation of a thin, uniform layer of unhydrated iron oxide that acts as a barrier to the rusting process. Polyaniline-based anticorrosion paints have been developed and are in commercial use.

Charge Storage Batteries

The reversible redox chemistry of conductive polymers has allowed the materials to be used as electrodes in rechargeable storage batteries. Initially, it was hoped that the low density of polymer electrodes would yield lightweight high-energy-density batteries. This has not been realized because the polymers have low charge densities relative to metal electrodes.

Actuators

Conducting polymers can be used in electromechanical applications, since they show a volume change of up to 10% on oxidation or reduction as ions and solvent move into the polymeric structure. Bilayer constructions can act as artificial muscles, and the compatibility of conducting polymers with the human body makes implants feasible. Polypyrrole–gold bilayer microactuators that bend out of the plane of the wafer can be used to position other microcomponents.

Chemical and Biochemical Sensors

The sensitivity of the electrical properties of conductive polymers to chemical stimuli has made them useful in a number of sensing applications. One early program carried out at Allied Signal proposed the use of conductive polymers in remotely readable indicators. Systems designed to detect time–temperature, temperature limit, radiation dosage, mechanical abuse, and chemical exposure were developed.

Conductive polymers can change their electrical conductivities by many orders of magnitude when exposed to specific gaseous reagents. The reactivity of polypyrrole to NH_3, NO_2, and H_2S was tested to examine this phenomenon. These sensors proved to be quite sensitive with large resistance changes occurring with exposure to an atmosphere containing 0.01–0.1% of the reactive gas. In all three cases, the reactions were shown to be reversible as removal of the gas from the atmosphere led to recovery of the original conductivity. "Electronic noses" to detect odors at low concentrations have been marketed.

Entrapment of biochemically reactive molecules into conductive polymer substrates can be used to develop electrochemical biosensors. This has proven especially useful for the incorporation of enzymes that retain their specific chemical reactivity. Electropolymerization of pyrrole in an aqueous solution containing glucose oxidase (GO) leads to a polypyrrole in which the GO enzyme is codeposited with the polymer. These polymer-entrapped GO electrodes have been used as glucose sensors. A direct relationship is seen between the electrode response and the glucose concentration in the solution that was analyzed.

Gas Separation Membranes

Because of the fixed ionic sites along the polymer backbones, conducting polymers are useful as gas separation membranes. The ability to switch the polymers between charge states (doped and undoped) suggests that external control of gas permeability and selectivity may be possible.

Electrochromic Applications

During electrochemical switching, distinct changes occur in the optical properties (color and extinction coefficient) of conducting polymer films. This has led many investigators to examine the feasibility for use in electrically switchable, electrochromic, or "smart windows" with polymer layers that darken on application of a voltage. As exemplified by polyisothianaphthene, thin polymer films are relatively transparent when held in the conducting state and opaque in the insulating state.

Polymer OLEDs and Diode Lasers

One of the most exciting developments in applications of conducting polymers is the use of their electroluminescent properties in organic light-emitting diodes (OLEDs).

OLEDs also include devices based on organic molecules rather than polymers, so OLEDs based on conducting polymers are identified as "polymer OLEDs" or "PLEDs".

Polymer OLEDs have the advantage of flexibility, coverage of a wide area, low power consumption, high intensity at low voltages, fast response times, thinness, and wide viewing angles. The thin film of light-emitting polymer is typically sandwiched between a transparent electrode and a metal electrode.

In 2002, Philips claimed the first high-volume production polymer LED product (the battery charging indicator for an electric razor), based on technology developed by CDT.

Photovoltaic Cells

The inverse effect to electroluminescence, ie, the generation of electricity from incident light (the photovoltaic effect), also operates in conducting polymers, which can therefore be used to make photovoltaic cells. Quantum efficiencies of up to 29% with overall power conversion of approximately 2% (for a simulated solar spectrum) have been obtained using polythiophene and a PPV derivative. Hybrid photovoltaic cells incorporating fullerenes and conducting polymers are under development, and solar cells using inorganic CdSe nanorods in semiconducting organic polymers could make solar power more affordable.

Plastic Electronics

Electrolytic capacitors using polypyrrole as a solid-state electrolyte can achieve very high values of capacitance, producing a very thin dielectric layer without short circuits.

Cheap disposable all-polymer integrated circuits (ICs) being developed by are likely to find application in smart labels to replace bar codes in supermarkets, eg, that allow automatic pricing without the need to unload goods from the trolley.

H. Shirakawa, *Handbook of Conducting Polymers*, 2nd ed., Marcel Dekker, New York, 1998.

M. P. Stevens, *Polymer Chemistry: An Introduction*, 2nd ed., Oxford University Press, Oxford, 1990, p. 91.

W. M. Wright and G. W. Woodham, in J. M. Margolis, ed., *Conductive Polymers and Plastics*, Chapman and Hall, New York, 1989.

PETER J. S. FOOT
Materials Research Group,
Kingston University

ALAN B. KAISER
MacDiarmid Institute for
Advanced Materials and
Nanotechnology and Victoria
University of Wellington

CONTROLLED RELEASE TECHNOLOGY, AGRICULTURAL

PESTICIDE-CONTROLLED RELEASE FORMULATIONS

Controlled release formulations (CRF) aim to make available pesticides at rates appropriate for efficient control of pests under field conditions. These formulations are combinations of the pesticidal active agent with inert materials that protect, and release, the active agent according to the pest control needs. In conventional formulations, complete availability of the active agent is usually considered to be immediate or rapid following deployment.

Controlled release formulations can be used with a wide range of pesticides, including inorganic substances, conventional low molecular weight organic substances, high molecular weight substances such as peptides or proteins, microbials, and semiochemicals that modify pest behavior. Applications may be found in agriculture, veterinary, and public health sectors and may be aimed at controlling a variety of pest organisms such as insects, mites, rodents, nematodes, weeds, and microorganims as well improving crop production with plant growth regulators. There exist a variety of release types, such as extended, slow, fast, delayed, programmed, sustained, pulsed, etc.

As for all pesticide formulations, CRF need to be applied, or placed, in the field appropriately for targeting the pests. The greatest advances in CRF in agriculture have been found with sprayable, and to a lesser extent, granular methods. More specialized methods, eg, based on pheromones or baits, have been commercially possible using larger devices.

Controlled release technology aims to manipulate the bioavailability of the pesticide in the local environment following application. This approach has many benefits compared to conventional formulations that include increased safety to the environment, workers, and consumers. Having lower concentrations of released pesticide in the environment leads to reduced losses, such as leaching, evaporation, degradation, and binding. Reduced losses may mean better pest control, less nontarget impacts, reduced crop phytotoxicity, and safer formulations.

Among the numerous benefits controlled release formulations are protection of active ingredients from environmental degradation; manipulation of bioavailability and persistence; reduction of toxicity and operator hazards; reduction of phytotoxicity to seeds and crops; improved selectivity between target and nontarget organisms and usage in integrated pest management (IPM); reduction in repellency (also reduction in odors); it allows coformulation, especially of incompatible pesticides (eg, of chemical and microbial pesticides); permits elimination of solvents; improves formulation of actives with phase changes near ambient temperatures; improves handling qualities of formulations and ease of cleaning sprayers; and furnishes possibly reduced application rates.

PRINCIPLES OF CRF FOR USE IN THE ENVIRONMENT

Pesticide Delivery

As with all treatments based on biologically active molecules, targeting is fundamental to CRF's success. If the substances do not reach the pest, no control will be achieved. Sometimes the pesticide moves toward the target, and sometimes the target moves toward the pesticide. Thus the efficiency of this delivery process can be defined as the ratio of the amount of pesticide reaching the pest divided by the amount applied. This delivery process is the sum of the placement methods, i.e., spraying, granules, bait, etc., and the subsequent movement of the pesticide combined with the movement and growth of the pest itself. Sometimes, the pesticide is activated following application, in which it is chemically transformed into a more pesticidal substance. Transfer of the pesticide occurs through contact and also in mobile phases such as air and water.

Losses

During delivery, the pesticide is quickly lost by a multitude of processes, including spray drift, evaporation, leaching, run-off, sorption, dispersal, and dilution below active concentrations. Slower loss mechanisms include degradation of the pesticide caused by light (photodegradation), by biological processes (especially by microorganisms in soil) and chemical processes (such as hydrolysis and oxidation).

Controlled Delivery and Half-life

As pesticide loss is concentration dependent, reducing environmental concentrations will reduce losses. If the concentration at the target pest could be kept at the minimum (or just above) for effective pest control by continuous supplementing for that portion lost or dissipated, overall losses could be minimized. Keeping this minimum for the duration of control needed would represent the ideal approach with the highest possible level of efficiency of delivery. To maintain the concentration at the target, pesticide needs to be supplied at the same rate at which it is dissipated.

Controlled release formulations, combined with other aspects of pesticide application, thus offer the feasibility of improving pesticide delivery, reducing losses and benefitting the environment. Compounds of short persistence are half-life (such as insect pheromones) may be used effectively in place of long persistent compounds when appropriately formulated. Loss kinetics for any individual pesticide vary depending on the environmental location considered; eg, loss by evaporation or photodegradation at the surface of plants may be much more rapid than published data giving half-life values for soils.

TYPES OF FORMULATION: CHEMICAL, PHYSICAL, BIOLOGICAL

For environmental application of pesticide delivery, controlled release formulations have traditionally been divided into chemical and physical types. More recently,

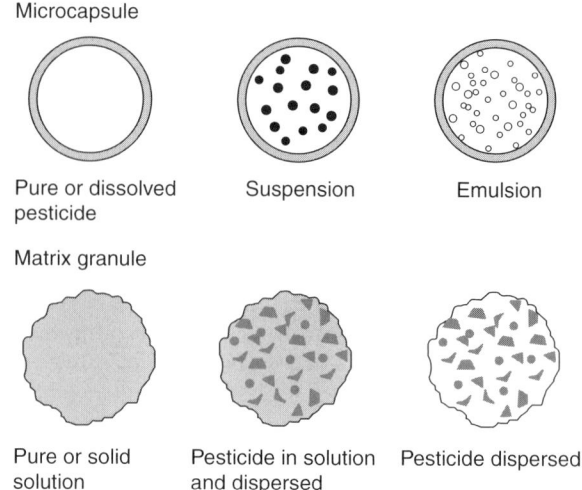

Figure 1. Various configurations of capsule and matrix formulations.

a third approach has appeared, biological, partly in response to delivery requirements for genetically engineered pesticides. The chemical types of controlled release formulations described to date can be categorized as backbone linking, side-chain bonding, matrix degradation, and carrier molecules such as cyclodextrins. The physical types of CRFs are reservoir, with or without a membrane; and monolith or matrix (films, paint, sheets, slabs, pellets, strips, granules, microparticles, powders, and microspheres). Biological CRFs use living or dead, cells (microorganisms) as delivery mechanisms.

All formulation types have been prepared and tested, but not all have reached commercial practice. The basic configurations of CRF are given in Figure 1.

Kinetics and Characteristics

There is a great deal of variation in the release kinetics of pesticides from the various formulation types described above. In order to discern the kinetics, the rate-controlling step needs to be identified, under controlled conditions, in the laboratory, but it does not necessarily follow that this will be true in the environment where the formulation is to be used. Ideally, release kinetics ought to be determined under field conditions, but this often presents insurmountable problems and instead the biological response to the released pesticide is observed over time to validate the performance of any formulation.

Mechanisms of Release

The main release mechanisms are constant release (zero order), in which transport through the polymer membrane (or matrix) occurs by a dissolution–diffusion process where the active ingredient first dissolves in the polymer and then diffuses across the polymer to the external surface where the concentration is lower.

The rate remains constant as long as the internal and external concentrations of the pesticide and the concentration gradient are constant.

Another is release by erosion, where the rate is independent of the concentration of pesticide remaining in the device. Erosion can occur by dissolution of surface polymer or by degradation of the matrix.

In practice an release from reservoir microcapsules small particles become depleted before large particles and the overall release from a population of particles will decrease with time. This has been shown in laboratory and field tests.

In nonsurface erodable matrix systems, diffusion of the active ingredient occurs from the interior of the particle to the surface. This gives rise to a declining rate of release. The pesticide may be dissolved or dispersed in the polymer; for dissolved pesticide the second phase of release is by first-order kinetics.

Swellable Matrices

In matrix systems, where water uptake or swelling can occur, such as may be possible in moist soil or water, the rate- controlling step may be solid-state diffusion or relaxation of the polymer by incoming water or a combination. Thus, the time exponent of the equation characterizing the release rate may vary for swelling, according to the nature of the matrix and the pesticide. Generally, the higher the water solubility of the pesticide, the faster will be its rate of release. Less polar molecules with high partition coefficients tend to transport slower. In the case of irregular particles, such as granules, and where polydispersity exists, the overall time exponent will typically be less than the corresponding value for microspheres.

Finally, in situations where a chemical reaction liberates the active species or where boundary conditions are rate limiting, the rate of release depends on the concentration in the solid phase. Where more than one mechanism (including diffusion) operates, complex release patterns occur.

DESIGN AND PREPARATION OF CONTROLLED-RELEASE FORMULATIONS

Chemical Methods

Chemical methods involve the formation of a chemical bond with the pesticide and another molecule; this bond is then broken in the field to allow the release of the pesticide. The bond energy relates to the ease of breaking and thus the rate of release of the pesticide. Where the structure of the pesticide permits, it can be homopolymerized through a condensation reaction and the pesticide forms the backbone of a resulting high molecular weight polymer, which is in effect a polymeric propesticide. In the environment, this polymer depolymerizes to release the original pesticide, usually from each end of the chain (unzipping). Often breakdown of such homopolymers is slow and there is need for copolymerization with another appropriately functional monomer. Pesticides capable of homo- or copolymerization are few and include those containing functional groups such as amino, hydroxyl, and carboxyl groups.

A second approach is where the pesticide is attached to a side chain of a high molecular weight polymer or macromolecule. This polymer may be either preformed, and the pesticide is then bound to appropriate side-chain functional groups, or the pesticide is first attached to a polymerizable monomer that is subsequently polymerized to yield the pesticide bound polymer. The third approach is where the active is trapped in a network of a cross-linked polymer. Chemical breakdown of this polymer then allows the release of the pesticide.

The first two mechanisms usually involve covalent bonding of the pesticide and the formation of a molecular species different to the original structure. This approach implies considerable additional registration costs that outweigh the putative formulation benefits. Thus, true chemical approaches are often proscribed in favor of physical methods.

Physical Methods

Physical methods are divided into two general approaches. The pesticide is entrapped within a physical structure either at a molecular or microdomain level or the pesticide in the form of a reservoir is enclosed within a polymeric envelope. In the first, the pesticide is mixed with the polymer or other material to form a monolithic structure or matrix. Release is normally by means of diffusion through the matrix or dissolution and erosion of the matrix. In the second approach, structures are based upon a reservoir of the pesticide enclosed by the polymer. The shapes of these devices are varied and include spherical and laminar or layered structures with the reservoir bounded by permeable membranes. These membranes provide a permeable barrier that controls the release rate. These "physical" methods provide the most important technologies for CRF of pesticides.

Reservoir-based Formulations with Membrane

In this method, a reservoir or depot of the pesticide is bounded by a polymeric membrane, which protects it from the environment and provides a mechanism for its release.

The method most suitable for use for pesticides is microencapsulation, where particle sizes are of the order of 10–100 μm and can be delivered by standard agricultural spraying. Microencapsulation is the placing of a layer on the surface of a single *liquid* droplet. Coating is covering a single *solid* particle, whereas a matrix particle contains the solid or liquid active agent dispersed throughout a binding material. These matrix particles should be considered as matrices.

Microencapsulation

Microencapsulation has now been commercially practiced for >40 years. Production of microcapsules is based on three main methods. The oldest, phase separation or coacervation, uses emulsification to produce core droplets containing the pesticide dispersed in an immiscible phase in which the wall material is dissolved, but then precipitates around the core droplets. Interfacial encapsulation is done by emulsifying or dispersing the pesticide solution in a continuous phase and a polymerization reaction takes place at the interface. Finally, in the physical methods the wall material is spread around the pesticide-containing core to make the microcapsule.

Coating of Solid Particles

Various methods may be used to cover a solid particle with a polymeric wall. Pan coating is well established, in which core particles are tumbled in a rotating drum while the coating solution is sprayed slowly; warm air circulates to remove the solvent. For small particles, spray drying in which the core material and the coating solution is atomized and the droplets dried rapidly in hot air gives poorer quality of encapsulation. There are numerous other methods for encapsulation encluding fluidized bed techniques.

Laminate Formulations

The laminate system comprises a reservoir layer of pesticide-containing polymer enclosed between two other plastic layers. The two outer layers of this multilaminate structure protect and release the active ingredient by diffusion driven by the concentration gradient. Often one of the layers is impermeable and functions as a support for adhesion to suitable surfaces. At the surface, the pesticide is continually removed by evaporation, degradation, leaching, or by mechanical contact by humans, insects, moisture, wind, dust, or other agents.

An important application is for release of insect pheromones and attractants for insect control. A combination of the insecticide propoxur with the cockroach attractant periplanone-B provides 1 month of control in a laminate bait strip. Delivery of volatile compounds to the atmosphere surrounding crops is a crucial part of the mating disruption technique for many insect pests. The number and disposition of devices releasing the volatile pheromones depends on the pest, crop, and other environmental factors. Laminates can thus be dispersed in the crop as individual large devices (adhesive strips) or as small flakes or confetti applied by aircraft (applied with adhesive to ensure retention toward the top of the crop canopy) according to the control requirements. Among a number of crop pests, an important use has been for pink bollworm in cotton. The use of a plastic film for controlling release of pheromones can take the form of the laminate, or film enclosing a reservoir of the active agent on a porous substrate, or even in the form of polyethylene bags, vials, tubes, and caps.

Reservoir-based Formulations Without Membrane

Reservoir systems that lack a bounding membrane to protect and regulate the release are usually designed for liquid actives that are volatile. The liquid is held in place through capillary forces and is released in the vapor phase. The rate of evaporation is regulated by diffusion of the vapor through the static air phase above the liquid surface. The simplest example is the hollow fiber, which is a fine polymeric capillary closed at one end and filled (or partially filled) with the liquid active. This diffuses through the air column to the opening from where it disperses. This method has been mostly developed to deliver many of the volatile sex pheromones for insect pest control to maintain a minimum concentration in the air surrounding the crop to be protected. Operating by a similar process are porous and foam polymers.

Matrix Formulations

These formulations consist of a uniform continuous phase with the pesticide dissolved or dispersed throughout. Their preparation is generally easier, requiring less process control, but can exhibit a rich variety of release types according to the material and structure of the matrix. An almost endless selection of materials is available for the matrix. Generally, a range of additives such as plasticizers, light protectants, pigments, antioxidants, processing aids, etc, are usually included. The products can be produced in a number of forms or shapes, especially sheets, ropes, extruded cylinders, slabs, and granules. As release is inversely related to device size, the production of simple powders involves difficulty in uniformity as well as minimal reduction in release kinetics compared to conventional formulations. Many of the large formulation types are, or have been, popular for aquatic applications, such as for insect and mollusc vectors of human disease causing organisms, with few applications in agriculture.

Pesticide-Containing Films

In the agricultural field, the use of plastic mulch and plastic films for plant growing has become widespread, as it advances and enhances cropping when temperatures are low. It also encourages pests, particularly weeds and disease-causing agents. The use of pesticides in these conditions can cause problems, not least a result of the need for reapplication after the film has been laid. Incorporation of the pesticide into the film obviates both of these problems. Using coating processes, films may be applied to seeds that provides a very effective means of controlled delivery of pesticides to the seed region and to the emerging seedling.

Matrix Particles

Small particles based on a matrix can range in size from powders (microparticles) to granules (fine to macrogranules) to pellets. Microspheres can be considered as the matrix equivalent to microcapsules. Most controlled release granules are matrix based, although some have a solid core or reservoir of pesticide with a coating.

Microparticles

Size matters; release rates depend on surface area, and thus larger particles release for longer and are able to manipulate the external availability of the pesticide. Small microparticles are therefore limited in their scope for controlling release but can be used in traditional spraying of dispersions onto soils and crops as well as for seed dressing.

Granules

Granular controlled-release formulations can be achieved by coating as well as from matrices. Although controlled release granules have not been as popular as microcapsules, there have been significant developments for applications to soil, especially where extended control is required, as with shoot borers in sugar cane and forestry, termites and weevils in forestry and ornamentals, and borers and nematodes in a number of crops.

BIOLOGICAL METHODS

Finally, the use of living cells (eg, yeast) as encapsulating materials has been under investigation for many years. The problems associated with the encapsulation of pesticides within preformed cells have been overcome by using proteinaceous pesticides such as the toxin from *Bacillus thuringiensis* (Bt). The genes for the production of the toxin have been introduced into the soil bacterium *Pseudomonas fluorescens*. The toxin is then expressed, seen as a crystalline inclusion. Following production by fermentation, the cells are killed and fixed to provide the capsule formulation that has been registered for use on brassicas.

A. F. Kydonieus, ed., *Controlled Release Technologies: Methods, Theory and Applications*, CRS Press, Boca Raton, Fla., 1980.

B. A. Leonhardt, in R. M. Wilkins, ed., *Controlled Delivery of Crop-Protection Agents*, Taylor and Francis, London, 1990.

H. B. Scher, ed., *Controlled-release Delivery Systems for Pesticides*, Marcel Dekker, New York, 1999.

R. M. Wilkins, ed., *Controlled Delivery of Crop Protection Agents*, Taylor and Francis, London, 1990.

Richard M. Wilkins
Newcastle University

COORDINATION COMPOUNDS

A coordination compound typically consists of a metal atom or ion (Lewis acid) surrounded by a number of electron-pair donors (Lewis bases) called ligands. Coordination compounds, also known as metal complexes, are pervasive throughout chemistry, biochemistry, and chemical technology. The metallic elements, which constitute 80% of the Periodic Table, exhibit predominantly coordination chemistry. Whereas the ligands may have charges equal in magnitude and opposite in sign to the charge of the metal ion, in which case a neutral coordination compound or inner complex results, often the metal plus ligands result in a charged entity or complex ion. In these situations, the coordination compound may include neutral ligands or counterions, either simple or complex. Generally, coordination compounds have properties which are unique relative to both the ligands and the metal ion itself.

Coordination compounds are used as catalysts in nature, ie, metal enzymes, and in industry in situations where the metal ion or ligand would not work alone. The ligands often modify the properties of the metals or metal ions and vice versa. For example, the metal deactivators in gasoline modify the chemistry of dissolved copper to the extent that the copper does not promote gum formation. Some ligands react with different metal ions

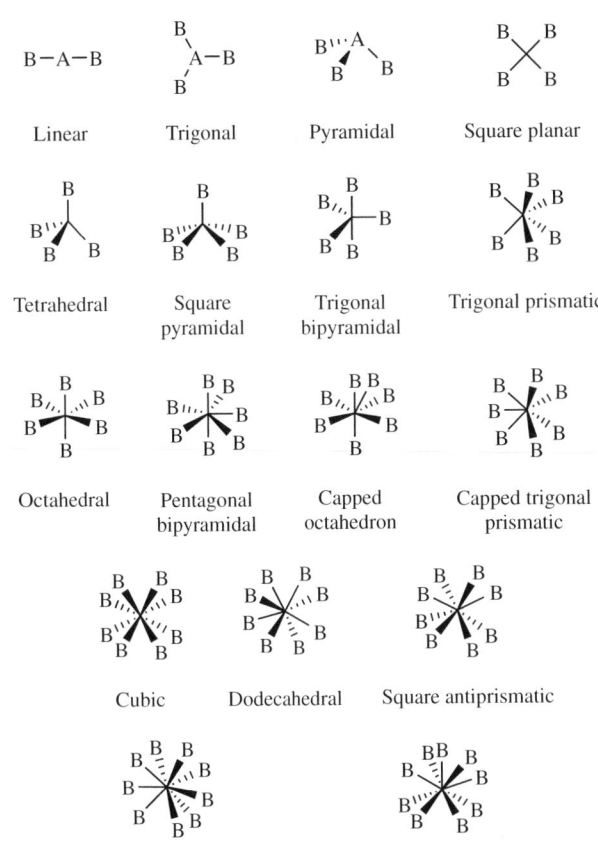

Figure 1. Common geometries for metal coordination numbers from two through nine, where A represents the metal and B a ligand donor atom. For higher coordination numbers, A is in the center of the structure. The principal axis orientation is vertical.

such that selective analysis of metallic elements is possible through coordination followed by solvent extraction, spectroscopy, gravimetry, electrochemistry, etc. Furthermore, complexes of copper and zinc in water allow brass to be electroplated. Conversely, metals are sometimes used to modify the properties of ligands. For example, azo dyes are metalated to give more permanence and/or alter color tones; the coordination of zinc to bactericides modifies the properties of the bactericides. The modification includes template and neighboring group effects, promotion of nucleophilic substitution, enhanced ligand acidity, and strain modification. Common geometries for coordination numbers from two through nine are shown in Figure 1. International Union of Pure and Applied Chemistry (IUPAC) and American Chemical Society (ACS) polyhedral symbols for the geometries are provided in Table 1.

CLASSIFICATION OF LIGANDS

The ligands can be classified as one of four types.

Unidentate ligands coordinate to the metal ion via one donor atom featuring at least one electron pair. Typical unidentate ligands act as terminal or as bridging ligands.

Table 2. Polyhedral Symbols for Common Coordination Geometries[a]

Coordination geometry	Coordination number	Polyhedral symbols	
		IUPAC	ACS
linear	2	L-2	L-2
triangular plane	3	TP-3	TP-3
tetrahedron	4	T-4	T-4
square plane	4	SP-4	SP-4
trigonal bipyramid	5	TBPY-5	TB-5
square or tetragonal pyramid	5	SPY-5	SP-5
octahedron	6	OC-6	OC-6
trigonal prism	6	TPR-6	TP-6
pentagonal bipyramid	7	PBPY-7	PB-7
octahedron, face capped	7	OCF-7	OCF-7
trigonal prism, square face monocapped	7	TPRS-7	TPS-7
cubic	8	CU-8	CU-8
dodecahedral	8	DD-8	DD-8
square antiprism	8	SAPR-8	SA-8
trigonal prism, square face tricapped	9	TPRS-9	TPS-9

[a]See Figure 1.

In the latter case, they usually possess more than one lone electron pair, so that each pair can coordinate to a different metal ion.

Chelating ligands possess more than one donor atom and can coordinate a metal ion by two or more donor atoms. Usually, the more stable five- and six-membered rings are formed on coordination. Coordination compounds with chelating ligands are usually more stable than unidentate ligands if the same donor atoms are concerned. Typical chelating ligands are bidentate ethylene diamine, polydentate polyethyleneglycol dimethyl ethers, as well as the hexadentate ethylenediaminetetraacetate, known as $EDTA^{4-}$ (**1**).

(1)

The latter plays a predominant role for the efficient complexation of metal ions with an octahedral coordination sphere and is used to prevent precipitation of metal salts, ie, in water treatment (bathroom cleaners, eg) for complexing calcium and magnesium ions, to prevent blood clots, to remove heavy metals from the body on poisoning, to solubilize iron in plant fertilizers, or to remove the iron taste from mayonnaise.

Macrocyclic ligands are chelating ligands that are closed to a ring. Compounds in which the metal ion is coordinated by a macrocycle are more stable than compounds in which the metal ion is coordinated by the analogue open chelating ligand. The gain in energy is mainly entropic and due to the preorganization of the ligand. A typical example for a macrocyclic ligand is the porphyrin (**2**), whose derivative forms the heme with iron in human blood, and cyclic polyethers such as dibenzo-[18]-crown-6 (**3**), which are used as extraction agents for metal ions and in biomimetics as compound mimicking ion transport through a cell membrane. The size of the macrocyclic ligand can be used as a selective tool in extraction and separation processes in industry.

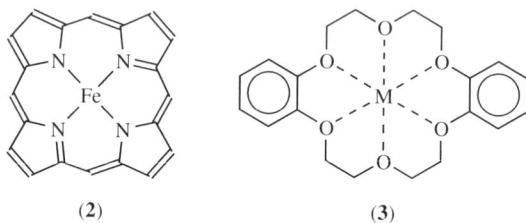

(2)   **(3)**

Cryptates are macropolycyclic ligands, and their size plays an important role as to which metal ions they are capable of complexing. The coordination compounds formed are usually more stable than the corresponding macrocyclic ones. This phenomenon is called the cryptate effect. A typical example is the [2,2,2]-cryptate (**4**), the numbers in brackets indicating the number of oxygen atoms present in the chain connecting two bridgehead nitrogen atoms.

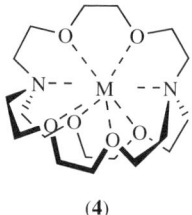

(4)

COORDINATION THEORIES AND BONDING

The quest for a comprehensible theory of coordination chemistry has given rise to the use of valence-bond, crystal-field, ligand-field, and molecular orbital theories. Ligand-field theory incorporates covalency with the electrostatic crystal field. The symmetry-induced separation of energy levels that is perceived from the chemical and physical properties of coordination compounds is inherent in all four theories, but symmetry effects are more apparent in the latter three. Molecular orbital treatments of complexes range from the very simple semiempirical, such as Hückel, angular overlap, etc, through so-called *ab initio* methods.

PROPERTIES

Stability

The thermodynamic stability of coordination compounds in solution has been extensively studied. The equilibrium constants may be reported as stability or formation constants: $M + n\,L \rightleftharpoons ML_n$. This compound, ML_n, has a cumulative stability constant β_n related to the activities a of the species by $\beta_n = a_{ML_n}/(a_M a_L^n)$ and the stepwise constant $K_n = a_{ML_n}/(a_{ML_{n-1}} a_L)$. Alternatively, instability or dissociation constants are sometimes used to describe compounds, and caution is necessary when comparing values from different sources.

Steric Selectivity

In addition to the normal regularities that can be rationalized by electronic considerations, steric factors are important in coordination chemistry.

Coordination stereochemistry (including various forms of isomerization) is an area of significant research interest. This aspect of coordination is important for stereospecific catalytical applications.

Reactivity

Coordination species are often categorized in terms of the rate at which they undergo substitution reactions. Complexes that react with other ligands to give equilibrium conditions almost as fast as the reagents can be mixed by conventional techniques are termed labile. Included are most of the complexes of the alkali metals, the alkaline earths, the aluminum family, the lanthanides, the actinides, and some of the transition-metal complexes. On the other hand, numerous transition-metal complexes that kinetically resist substitution reactions are termed inert. These terms refer to substitution reactivity and not to thermodynamic properties.

Oxidation–Reduction. Redox or oxidation reduction reactions are often governed by the hard–soft base rule. For example, a metal in a low oxidation state (relatively soft) can be oxidized more easily if surrounded by hard ligands or a hard solvent. Metals tend toward hard-acid behavior on oxidation. Redox rates are often limited by substitution rates of the reactant so that direct electron transfer can occur.

Photochemistry. Substitution rates of many complexes are enhanced by irradiation of the low energy $d–d$ transitions, such as $t_{2g} \rightarrow e_g$ in octahedral coordination compounds. Quantum yields, Φ, defined as the ratios of moles of product formed (or reactant depleted) to the moles of photons absorbed, vary from very good, e.g., chromium(III) ~ 0.5, to poor, eg, cobalt(III) (0.01, for ligand substitution. The substituted ligand is normally the stronger of the two on the axis with the weakest net pair of ligands as determined by spectrochemical relationships, ie, $CN^- > NO_2 > NH_3 > H_2O$, $F^- > Cl^-$. Exceptions do occur. Photochemical ligand dissociation is useful in the synthesis of multinuclear metal complexes such as diiron nonacarbonyl from iron pentacarbonyl.

Magnetism. Many coordination compounds are paramagnetic; ie, they have unpaired electrons. Combining several metal ions in one coordination compound can lead to a coupling of the individual magnetic moments. Their interaction depends on the distance between the magnetic ions and on the orientation of the magnetic moments. The magnetic moments can become aligned parallel to each other via exchange coupling below the Curie temperature, in which case one obtains ferromagnetism, or antiparalleism to each other, leading to antiferromagnetism. Nonparallel alignment leads to ferrimagnetism.

APPLICATIONS

Coordination compounds are used in bactericides and fungicides for catalysis, in coordination polymers, dyes and pigments, photography, electroplating, fuel additives, medical applications, chemosensors, electronic and magnetic materials mimetics of biological systems, and liquid crystals.

B. F. G. Johnson and M. Schroeder, eds., *Advances in Coordination Chemistry*, Royal Chemical Society, UK, 1996.

J.-M. Lehn, *Supramolecular Chemistry: Concepts and Perspectives*, VCH, Germany, 1995.

G. Wilkinson, R. D. Gillard, and J. A. McCleverty, eds., *Comprehensive Coordination Chemistry*, Pergamon Press, Oxford, UK, 1987, 7 Vols., Vol. 1, theory and background; Vol. 2, ligands; and Vol. 6, applications.

A. F. Williams, C. Floriani, and A. E. Merbach, *Perspectives in Coordination Chemistry*, VCH, Weinheim, Germany, 1992.

KATHARINA M. FROMM
University of Fribourg

COPOLYMERS

Over the last 50 years, synthetic polymers have replaced traditional materials such as metals, ceramics, wood, and natural fibers in a large number of applications, including automotive, construction, appliances, and clothing. In addition, complete new markets have been opened. This revolution has been possible because polymers have many interesting features, including a high strength/weight ratio, chemical inertness, and easy processability. Another important feature of the synthetic polymers is that their properties can be tuned to match the needs of a given application. This tuning process is often done by preparing new materials through copolymerization.

COPOLYMER STRUCTURES

Copolymers are macromolecules formed by polymerization of two or more different monomers (comonomers). Homopolymers are formed by polymerization of a single class of monomers, which are molecules of low molecular mass.

Copolymers differ in the sequence arrangements of the monomer species in the copolymer chain. In terms of monomer sequence distribution, different classes of copolymers can be distinguished (Table 1).

Table 1. Classes of Copolymers in Terms of Monomer Sequence Distribution

Type of copolymer	Structure	Examples		
		Comonomers/reactants	Polymerization method	Name
statistical (random)	...ABBABABAAABBABAAB...	methyl methacrylate, butyl acrylate	free-radical polymerization	poly(methyl methacrylate-*stat*-butyl acrylate)
alternating	...ABABABABABAB...	styrene, maleic anhydride	free-radical polymerization	poly(styrene-*alt*-maleic anhydride)
periodic	...ABBABBABB... or $-(ABB)_n$	formaldehyde, ethylene oxide	free-radical polymerization	poly(formaldehyde-*per*-ethylene oxide-*per*-ethylene oxide)
gradient	...AAAABAAABAABBABBBBABBBB...	styrene, butyl acrylate	controlled free-radical polymerization	polystyrene-*co*-butylacrylate
block	AAAAAABBBBBBBBAAAAA	styrene, butadiene	ionic polymerization	polystyrene-*block*-polybutadiene-*block*-polystyrene
graft	AAAAAAAAAA B C B C C B	styrene/acrylonitrile, polybutadiene	free-radical polymerization	polybutadiene-*graft*-poly(styrene-*stat*-acrylonitrile)

Statistical copolymers are copolymers in which the sequential distribution of the monomeric units obeys known statistical laws. Within this category are copolymers named *random copolymers* because the probability of finding a given monomeric unit at any given site of the chain is independent of the nature of the adjacent units. Nevertheless, random copolymer is often used in a broader sense as here, to refer to copolymers in which the comonomer units are evenly distributed along the polymer chain.

An alternating copolymer is a copolymer comprising two species of monomeric units distributed in alternating sequence. A gradient copolymer is formed by polymer chains whose composition changes gradually along the chain.

Block copolymers are polymers having a linear arrangement of blocks of different monomer composition. In other words, a block copolymer is a combination of two or more polymers joined end-on-end. Any of these polymers or blocks is comprised by monomeric units that should have at least one constitutional unit absent in the other blocks. The blocks forming the block copolymer can be different homopolymers, a combination of homopolymers and copolymers, or copolymers of different chemical composition in adjacent blocks.

A graft copolymer is a polymer comprising one or more blocks connected to the backbone as side chains, having constitutional or configurational features that make them different from the main chain.

COPOLYMERIZATION REACTIONS

Chain-growth polymerization involves polymer chain growth by reaction of an active polymer chain with single monomer molecules. In step-growth polymerization, polymer growth involves reactions between macromolecules. In addition, nonpolymeric by-products may be formed in both types of polymerizations. Condensative chain polymerization is very rare.

Chain-Growth Copolymerization

In chain-growth copolymerization, monomers can only join active chains. The activity of the chain is generated by either an initiator or a catalyst.

Step-Growth Copolymerization

Monomer molecules consisting of at least two functional groups would undergo step-growth polymerization. These functional groups would be capable of reacting with each other, eg, a −COOH group would react with −OH and −NH$_2$ groups. The two reacting functional groups could be on the same monomer molecule, type A−B (eg, an aminoacid) or on two separate molecules, types A−A and B−B (eg, a diacid and a diol).

Step-growth polymerization proceeds by reaction of the functional groups of the reactants in a stepwise manner: monomers react to form dimer; monomer and dimer react to form trimer; dimer and trimer form pentamer. A number of different chemical reactions may be used in step-growth polymerization, including esterification, amidation, the formation of urethanes, aromatic substitution, and carbonate bond formation.

FREE-RADICAL COPOLYMERIZATION

Free-radical copolymerization is attractive because of the huge number of monomers that can be copolymerized, the different media that can be used (both organic and aqueous), and the relative robustness of this technique to impurities. In classical free-radical copolymerization, only a few polymer chains are growing at the same time and the time spent in building a chain is very short (typically 0.5–10 s). This is convenient to produce random and alternating copolymers, but well-defined complex copolymer topologies (eg, gradient and block copolymers in Table 1) are not accessible through classical free-radical polymerization. In order to produce a well-defined block copolymer, all chains should start at the same time and should grow for some time in the presence of monomer 1 until the first block is formed. Then monomer 1 should be removed and a different monomer added to produce the second block. In order to conduct such a process successfully, termination should be avoided during polymerization.

In recent years, the development of controlled radical polymerization (CRP) methods has made it possible to conduct a free-radical polymerization minimizing the extent of termination. This allows preparing almost any kind of copolymer microstructure by means of a free-radical mechanism. All CRP methods have in common that a rapid dynamic equilibrium is established between a tiny amount of growing free radicals and a large majority of dormant polymer chains. In these processes, each growing chain stays for a long time in the dormant state, then is activated and adds a few monomer units before becoming dormant again. As the activation–polymerization–deactivation process is a random process, the molecular weight distribution of the growing chains becomes narrower as they grow longer. The composition of the polymer chain can be easily modified by controlling the monomer composition in the reactor. Termination between active radicals is minimized by simply maintaining its concentration at a low value. The CRP methods differ in the way in which these dormant species are formed. The most efficient CRP methods are stable free-radical polymerization, best represented by nitroxide mediated polymerization (NMP); atom-transfer radical polymerization (ATRP), and reversible addition-fragmentation chain transfer (RAFT).

The key feature of CRP processes is that molecular termination is minimized by maintaining the concentration of active radicals at a low value by establishing a rapid equilibrium between this small fraction of active radicals and a large majority of dormant polymer chains. The extent of bimolecular termination can be lowered by reducing the concentration of active chains, but at the expense of a longer process time. Therefore, there is a practical lower limit in the concentration of active chains.

Controlled radical polymerizations can be carried out in bulk, solution, and suspension polymerization. Mass transfer limitations of the radical trapping agent make difficult the implementation in conventional emulsion polymerization. This limitation has been overcome working in miniemulsion polymerization.

The chemical composition distribution of the chains formed by CRP offers considerably from that of the polymer chains produced in the classical free- radical polymerization. In a batch-controlled process in which

monomers with different reactivity ratios are polymerized, all polymer chains have a very similar chemical composition distribution (CCD) with a composition gradient along the chain. In the corresponding classical free-radical polymerization, the CCD of the chains formed at the beginning of the polymerization may differ greatly from that of the chains formed at the end of the process, and the composition along each chain is constant.

The concepts of instantaneous and cumulative copolymer composition in CRP are different from those in classical free-radical polymerization. In CRP the instantaneous copolymer composition, refers to the composition of the part of the chains that is being formed in a given moment. In classical free-radical polymerization, the instantaneous copolymer composition is the composition of the chains being formed in a given moment. On the other hand, the cumulative copolymer composition in a CRP represents the average composition of the fraction of each chain formed up to the time considered, whereas in classical free-radical copolymerization it is the average composition of all chains formed up to this time.

ANIONIC COPOLYMERIZATION

Anionic polymerization is not spontaneous and requires the presence of initiators that provide the initiator anions. Anions can attack only monomers whose electrons can be moved in such a way that a monomer anion results. Therefore, anionically polymerizable monomers should contain electron-accepting groups.

In purified systems, most macroanions grow until all of the monomer present in the reactor is polymerized. Such polymerization is called living polymerization, because if additional monomer is added into the reactor the polymer chains undergo further growing. A characteristic of living polymerization is that, provided that initiation is quick enough, all polymer chains grow to a similar extent, yielding very narrow molecular weight distributions. In addition, block copolymers can be produced by adding a second monomer once the first one has completely reacted. Triblock and multiblock copolymers can be prepared by subsequent additions of different monomers. Also, graft, star, and hyperbranched polymers can be obtained by means of this technique, simply by using suitable initiation systems.

CATIONIC POLYMERIZATION

Cationic polymerization, like anionic, ha sits initiators formed from carbocation salts, Brønsted acids, or Lewis acids, which react with a monomer to give monomer cations that upon addition of more monomer become macrocations. Both monomer cations and macrocations are fairly reactive and may react with the counterions instead of with the monomer. Therefore, the counterions should not be too nucleophilic. Relatively few cationic polymerizations are performed industrially, because macrocations are highly reactive and prone to suffer termination and chain-transfer reactions.

RING-OPENING COPOLYMERIZATION

Ring-opening copolymerization (ROP) consists in opening a cyclic compound to produce a linear polymer.

Ring-opening polymerization can proceed via (a) electrophilic propagating species, mainly cationic polymerization; (b) nucleophilic propagating species, mainly anionic polymerization; and (c) Zwitterion intermediates, in which the propagating chain and the monomer bear both cationic and anionic species. This type of polymerization is always catalyzed by an initiator at very low temperatures to avoid depropagation.

Copolymerization is predominantly employed to produce block and also graft copolymers. In these cases, the concepts discussed for copolymerization via cationic and anionic polymerization are also applicable.

CATALYTIC POLYMERIZATION

Free-radical, anionic, and cationic polymerizations proceed by addition of monomer units to the active end of the growing polymer chain that in the course of polymerization separates from the bound initiator fragment. Catalytic polymerizations proceed instead by mechanism in which the monomer units are inserted between the catalytic site and the growing polymer chain. Over 40% of yearly polymer production is obtained by catalytic polymerization.

Ziegler-Natta Catalysts

A typical commercial Ziegler-Natta catalyst for ethylene polymerization is produced by reacting $TiCl_4$ with a finely divided $MgCl_2$ stabilized with an electron donor and activating the resulting system with trialkylaluminum compounds. In propylene polymerization, the addition of another electron, external donor is usually needed to achieve high stereospecificity.

There is some debate about the mechanisms involved in the growth of polymer chains on the active center. The most widely accepted mechanism is one proposed in which propagation occurs at the transition-metal–alkyl bond.

Ziegler-Natta catalysts present multiplicity of active centers. Each center may have a different activity, and hence different reactivity ratios. Therefore, the average copolymer composition of chains of different lengths is different, namely, there is a correlation between chain molecular weight and chain composition. In addition, the catalytic activity changes with time. Consequently, the reaction scheme and the copolymerization equation may require modification to account for these features of the Ziegler-Natta catalysts. Reactivity ratios depend on operating conditions.

Transition-Metal Catalysts

The Phillips catalyst is the most common example of these catalysts. It encompasses two families of supported chromium catalysts organochromium compounds and chromium oxide. Suitable supports include silica, aluminophosphates, and silica–titania. Polymerization of ethylene on these catalysts yields a linear polymer (HDPE). Copolymerization of ethylene and α-olefins yields linear polymers with short branches. The crystallinity of the copolymer, and consequently its density, decreases as the copolymer content in α-olefins increases. Phillips catalysts are able to produce a wide range of polyethylenes, from HDPE to LLDPE.

Metallocene Catalysts

Metallocenes are single-center catalysts that represent the most versatile way of synthesizing almost any kind of stereoregular polymer. Polymers with entirely new properties may be produced by varying the components (type of metal and its substituents, type and length of the bridge, and substituents of the rings) of the metallocene molecule. Because metallocenes are single active center catalysts and hence the composition of all then copolymer chains is statistically the same, independent of chain length, this opens the possibility of a much tighter control of copolymer composition. In addition, metallocene catalysts exhibit higher reactivity for α-olefines than do conventional Ziegler-Natta catalysts the molecular weight distribution of the polymer produced with metallocene catalysts is also narrower than that obtained with Ziegler-Natta catalysts.

RANDOM COPOLYMERS

Random copolymers are by far the largest class of copolymers made and used today. They are mainly produced by catalytic polymerization and by classical free-radical polymerization. Some copolymers are also produced by ionic polymerization.

Catalytic Random Copolymers

Polyolefins produced by catalytic polymerization account for more than one-third of the yearly production of synthetic polymers. Polyolefins are made from a surprisingly short list of monomers, mainly ethylene and propylene, but also α-olefins (1-butene, 1-hexane, 1-octane), isobutylene and a few other monomers. These monomers are produced in large-scale petrochemical units and hence are available in large quantities at low cost. Polyolefins are extremely versatile materials with properties ranging from elastomers to thermoplastics to high-strength fibers. One of the key characteristics of the polyolefins is crystallinity, which depends on the regularity of the chemical structure of the polymer chains. Highly regular polymers such as linear polyethylene and isotactic polypropylene are crystalline polymers, whereas some copolymers of ethylene–propylene and ethylene–α-olefin are amorphous materials.

The main olefin copolymers are LLDPE, some grades of HDPE, ethylene–propylene elastomers, propylene copolymers, and high-impact polypropylene (hiPP). LLDPE is mainly used in film applications, with smaller markets in injection molding and wire-cable. Metallocene-based grades (mLLDPE) are also available in the market. mLLDPE provides a higher shock resistance (downgauged films), lower content of extractables (better organoleptic properties), and better optical properties (clarity and gloss). However, the processability of mLLDPE is in general worse than that of the conventional LLDPE.

Ethylene/propylene elastomers (EPM) are copolymers of ethylene and propylene with intermediate levels of each comonomer. These materials are completely amorphous and rapidly recover than shape after removal of a strain of at least 50%. Products called EPDM (ethylene, propylene, diene monomer) are meal by using both Ziegler-Natta and metallocene catalysts. EPDMs are used in construction (roof sheeting, insulation sponge, seals, hoses-tubes, reservoir linings), automotive (sealing systems, hoses), impact modification of plastics, and wire and cable.

In propylene–ethylene copolymers with a low ethylene content, the insertion of ethylene in the propylene chain reduces the regularity of the chain leading to a semicrystalline polymer. The properties of these materials, called propylene random copolymers, depend not only on the total ethylene content, but also on the monomer sequence distribution. The main uses are cast film and rigid and flexible packing.

High-impact polypropylene (hiPP) is a multiphase material in which an ethylene–propylene elastomeric phase is finely dispersed in an isotactic polypropylene matrix. In principle, hiPP can be produced by blending iPP and the elastomeric copolymer, but it is simpler and more versatile in terms of product properties achievable to produce the heterophase material during polymerization.

The main applications of hiPP are automotive (interior and exterior trims, bumpers), rigid packaging, consumer goods, pails, and corrugated pipes. Grades containing ~70 wt% of amorphous, propylene-rich ethylene–propylene copolymers are supersoft polypropylene alloys suited for roofing and geomembranes.

Random Free-Radical Copolymers

Styrene–butadiene rubber (SBR) is the largest volume synthetic rubber. It is produced in emulsion polymerization by free-radical polymerization and in solution by means of an anionic polymerization mechanisms. There are two broad SBR types, the so-called "cold" grades produced at low polymerization temperatures (~5°C) and the "hot" grades produced at ~50°C.

The main use of cold SBR is for tires. Other applications include conveyor belts, rubber articles, and footwear. Hot polymerized SBR is used in sporting goods, shoe soles, and adhesives.

Styrene–acrylonitrile copolymers (SANs) are manufactured by emulsion, suspension, and bulk free-radical polymerization. The properties of the SAN copolymers strongly depend their on copolymer composition. SAN is used in applications that require better chemical resistance, toughness, and heat distortion temperature such as shower doors, cosmetic bottles, and toys. Most SAN is consumed in the production of acrylonitrile–butadiene-styrene (ABS resins) copolymers. Other applications are appliances, housewares, automotive, and packaging. Grades containing a high amount of acrylonitrile (60–80%) are used as barrier plastics.

Ethylene–vinyl acetate (EVA) copolymers are produced to practically cover the whole range of comonomer ratios from 98/2 (ethylene–vinyl acetate) to 4/96. Copolymers containing 3–25 wt% of vinyl acetate are used in film applications (blow and cast films, shrink films, stretch film, disposable surgical gloves). Grades with higher vinyl acetate content are used as adhesives, diesel fuel additives, and asphalt modifiers.

Vinyl acetate–ethylene dispersions are used in paint and coatings, adhesives, paper coating, and binders for nonwoven materials such as cleaning cloths and sanitary products, moist wipes, bed covers, winter clothing, and tablewear. Redispersable powders based on vinyl acetate–ethylene dispersions are used in construction (ceramic tile adhesives, mortars, plasters) and as powder paints. Redispersable powders for nonwoven textiles are also available. Copolymers of vinyl acetate and branched vinyl esters are used for architectural paints. The most distinguishing feature imparted by this branched vinyl ester to the polymer is its resistance to hydrolysis.

Dispersions of vinyl chloride and ethylene are used for paints and flame retardant systems. Redispersable powders based on vinyl chloride–ethylene copolymers are also available.

Acrylic copolymers are used for high-temperature automotive coatings. Coating quality requires the use of solvent-borne systems, but the use of solvents is limited by environmental regulations. Therefore, there is strong pressure to increase the solids content. In order to increase the solids content maintaining the viscosity in manageable values, low molecular weight polymers are used.

Acrylic dispersions are extensively used for coatings textiles and nonwoven fabrics, adhesives, floor care, caulks, and sealants.

Fluorinated polymers give excellent performance under highly demanding conditions requiring resistance to high temperature, chemical inertness, and low surface tension. Fluorinated homopolymers are often crystalline. Copolymerization induces disorder of the polymer chain and consequently reduces crystallinity. Depending on the copolymer composition, fluorinated copolymers range from thermoplastic to elastomers.

Fluoroelastomers are designed to maintain rubberlike elasticity under extremely severe conditions (high temperatures and in contact with chemicals).

Thermoplastic fluoropolymers are used in applications that require chemical inertness, excellent dielectric properties, nonaging characteristics, antistick properties, low coefficient of friction, and performance at extreme temperatures. These applications include electrical applications (wire-cable, molded electrical parts), lined pipes and fittings, heat exchangers, and conveyor belts.

The main uses of the fluorinated elastomers are O-rings and gaskets, shaft and oil seals, diaphgrams, hoses, and profiles.

Random Copolymers Produced by Ionic Polymerization

Butyl rubber is a copolymer of isobutylene and isoprene (1–3%) formed by cationic polymerization. Butyl rubber cures well, although more slowly than polyisoprene and polybutadiene. This causes problems when butyl rubber is blended with these polymers. Halogenation of some of the isoprene groups greatly increases the cure rate of butyl rubber. The main application of polyisobutylene copolymers is for the innerliners of tubeless tires and inner tubes of tires. Other uses are adhesives, binders, pipe wrap, caulking, and sealing compounds.

SBR can be also be produced by a solution process using organolithium initiators. The main application is for tires. SBR gives low rolling resistance (less fuel consumption), improved wet grip, and good wear properties.

ALTERNATING COPOLYMERS

A limited number of alternating copolymers are commercially available. Alternating styrene–maleic anhydride copolymers (SMAs) are produced by free-radical copolymerization. Low molecular weight SMA with a high maleic anhydride content (25–50 wt%) is alkali soluble and used in paper and textile sizing, floor polishes, printing ink, pigment dispersants, and coatings. High molecular weight SMA with low maleic anhydride content (<25 wt%) is used for molding and extrusion applications.

Tetrafluoroethylene and ethylene are copolymerized in emulsion (using fluorinated surfactants) or in a nonaqueous media (using a fluorinated solvent) to yield an alternating copolymer that has high tensile strength, moderate stiffness, outstanding impact strength, a low dielectric constant, excellent resistivity, good thermal stability, and excellent chemical resistance. The main uses of this copolymer are power and automotive wiring, injection-molded electrical components, pump impellers, and molding articles.

BLOCK COPOLYMERS

Synthetic Methods

Block copolymers can be produced by means of several polymerization mechanisms, including anionic polymerization, ring-opening polymerization, step-growth polymerization, catalytic polymerization, controlled free-radical polymerization, and combinations of some of these methods.

Properties and Commercial Products

Block copolymers present unique structure–property relationships that are useful for a variety of applications, including thermoplastic elastomers (TPEs), elastomeric fibers, toughened thermoplastic resins, adhesives, compatibilizers for polymer blends, membranes, and surfactants. Table 2 presents some commercially available block copolymers.

Block copolymers, particularly styrenic triblock copolymers, are used as adhesives and sealants. Block copolymers are efficient compatibilizing agents for polymer

Table 2. Commercially Available Block Copolymers

Hard segments	Soft segments	Type	Trade name	Applications
styrene	butadiene or isoprene	A–B–A	Kraton D (Kraton polymers)	footwear, bitumen/asphalt modification, adhesives, sealants, household appliances, toys, tubing
styrene	ethylene–butylene isobutylene	A–B–A	Kraton G (Kraton polymers)	bitumen/asphalt modification, sealants, high performance adhesives, automotive, sports, medical equipment
styrene (high styrene content)	butadiene	triblock	Finaclear (ATOFINA)	health products, packaging
styrene	butadiene	linear/star	Finaprene (ATOFINA)	footwear, compounds, bitumen modification, adhesives, plastic modification
polyamide	polyether	multiblock	Pebax (ATOFINA)	footwear, sporting goods, protective films, waterproof breathable materials
polyester	polyether	multiblock	Hytrel (DuPont)	automotive, fluid power, sporting goods, furniture and off-road transportation. thin flexible membranes, tubing, hose jackets, wire and cable electrical connectors
polyamide	polyether	multiblock	Vestamid (Degussa-Hüls)	fuel lines, air brake tubing, hydraulic tubes, catheters, cable and wire, plastic-rubber components for the automotive industry, sport shoes
polyurethane	polyester/polyether/ polycarbonate	multiblock	Estane/Estagrip/ Estaloc (Noveon)	automotive, power and handtools, wire and cable, footwear, consumer goods, health care
polystyrene/ poly(methyl methacrylate)	polybutadiene	A–B–C	(SBM) ATOFINA	nanostructurated thermoplastic and thermoset materials, compatibilization of minerals and carbon black with polymers
polyester	polyester	multiblock	Arnitel V (DSM)	automotive, tubing, cable insulation, injective molding, films
poly(ethylene oxide) (hydrophilic block)	poly(propylene oxide) (hydrophobic block)	diblock	Synperonic (Uniquema)	surfactants for resin emulsification and emulsion polymerization

blends. Block copolymers are used in the surface modification of fillers and as coatings for metal and glass surfaces. Amphiphilic block copolymers consist of hydrophilic polyoxyethylene segments and various hydrophobic parts (polypropylene oxide, polystyrene, etc) are efficient surfactants. New nanostructured materials can be prepared by assembly of block copolymers.

GRAFT COPOLYMERS

High impact polystyrene (HIPS) and ABS resins are important graft copolymers produced by polymerizing vinyl monomers in the presence of polybutadiene through a free-radical mechanism. Grafting occurs by participation of the double bonds of the polybutadiene in the propagation reaction and by chain transfer to the polybutadiene, followed by addition of monomer to the resulting allylic radical.

ABS copolymers are produced in bulk polymerization by means of a process almost identical to that of the HIPS but using styrene and acrylonitrile as polymerizing monomers. ABS polymers can also be produced in emulsion polymerization by polymerizing styrene and acrylonitrile on a polybutadiene seed. Commercial ABS is often a blend of this reactor produced ABS with SAN.

Graft copolymers have been used for improving processability, compatibility, dyeability, and water repellency. Because of their inherent surface activity, other uses for these copolymers are coatings, adhesives, fibers, films, and moldings. Recently, applications of graft copolymers for 100% solids in situ curing resins (by using macromonomers that cure under electron beam or ultra violet, uv, radiation) have been reported. The low cost and nonpolluting nature of the resulting coatings and the possibilities to improve film properties such as adhesion, tensile strength, and flexibility make them very attractive products.

Also, hydrogels were recently formed by grafting hydrophobic polymers in hydrophilic backbones. These have been employed for biomedical applications such as controlled release of drugs, enzyme immobilization, and contact lenses.

STAR AND HYPERBRANCHED COPOLYMERS

Star copolymers, from a wide range of monomers can be prepared by controlled radical polymerization. Hyperbranched polymers are commonly produced by step-growth polycondensation of AB_x monomers, self-condensing vinyl polymerization of AB* monomers, and multi-branching ring-opening polymerization of AB_x monomers.

Hyperbranched copolymers can, in general, be considered a subclass of dendritic polymers with the advantage that their production is less costly and more suitable for mass production. Hyperbranched polymers, in comparison to their counterpart linear polymers, offer better solubility in organic solvents, lower viscosities because of their spherical shape, and the choice of controlling the T_g by chemical modification of the end functional groups. Hyperbranched polymers cannot engage in chain entanglements, and hence their use in conventional structural applications is futile. Hyperbranched polymers with

acrylate, vinyl ester, alkyl ether, epoxy, and OH functions are used as cross-linkers in coatings and thermosets. Hyperbranched polymers provide an exceptional film hardness that allows the use of low viscous–high solids–low molecular weight resins without compromising the coatings performance and also allows the use of entirely aliphatic monomers, which in turn result in excellent weatherability. Hyperbranched polymers are used as melt modifiers, to increase the toughness of glass and carbon-reinforced composites, and have been used as dye carriers, for nonlinear optical materials, in molecular imprinting, and for the synthesis of nanoporous polymers with low dielectric constant.

CHARACTERIZATION OF COPOLYMERS

Polymers are complex materials whose complete characterization requires the determination of a number of characteristics, which in turn determine the application performance. Polymer characteristics can be classified by their molecular structure, molecular size, molecular organization, and mechanical, electrical, and physical properties.

FUTURE TRENDS

Projections of future trends in polymers are highly speculative. For high-tonnage copolymers, olefin-based copolymers will likely remain and even further strengthen their position in the market by entering other fields by outperforming other plastic materials or replacing more expensive and/or problematic conventional materials. The potential of the polyolefins is based on the easy availability of cheap monomers, the economics of large-scale production, and the fact that they are both environmentally benign and extremely versatile in properties and applications. Developments in catalysts and polymerization processes have expanded the properties envelope, making it possible for the polyolefins share of the global plastics market to have grown from 35 to 62% in the last 25 years. This process will be reinforced with the development of new metallocene catalysts, which may be accelerated by the extensive application of high throughput screening techniques. Combinatorial materials research will have an important impact in the development of new polymeric materials. However, the development of new materials produced by new monomers is not expected unless these materials adapt to the existing production technologies and/or facilities. The development of water-resistant catalysts will bring new products to the market.

Metallocene catalysts will also affect nonolefinic polymers because the highly ordered polymers obtained, have properties not attained by traditional products.

Nanotechnology will likely have a critical influence on polymeric materials. Polymer nanocomposites will further expand the properties envelope bringing dramatic improvements in stiffness and gas-barrier properties.

In the specialty market, commercialization of controlled radical polymerization processes will bring a whole portfolio of new polymer materials. Block, graft, and hyperbranched copolymers of well-defined topology will have opportunities in markets such as coatings, adhesives, elastomers, sealants, lubricants, imaging materials, powder binders, dispersants, personal care products, detergents, photopatternable materials, and biological sensors.

Polymer producers will continue to suffer strong competition and increasing social pressure to achieve sustainable growth. Therefore, polymerization processes will be run more efficiently to achieve a consistent production of high-performance polymers under safe and environmentally friendly conditions. This will require the development of robust, accurate online sensors for polymerization monitoring and efficient on-line optimization and control strategies.

J. Brandrup, E. M. Immergut, and E. A. Grulke, eds., *Polymer Handbook* 4th ed., Wiley-Interscience, New York, 1998.

H. G. Elias, *An Introduction to Polymer Science*, VCH, Weinheim, Germany, 1997.

J. I. Kroschwitz, ed., *Polymers: Polymer Characterization and Analysis*, Wiley- Interscience, New York, 1990.

R. G. Nelb and A. T. Chen, in G. Holden, N. R. Legge, R. P. Quirk, and H. E. Schroeder, eds., *Thermoplastic Elastomers-A Comprehensive Review*, Hanser & Hanser/Gardner, Munich, Germany, 1996, Chapt. 9.

D. I. Lee, in J. M. Asua, ed., *Polymerization Dispersions, Principles and Applications*, Kluwer, Dordrecht, The Netherlands, 1997.

José R. Leiza
José M. Asua
The University of the Basque Country

COPPER

Copper, Cu, was one of the first metals to be used by early humans because, like gold and silver, it occurs in nature in the metallic state.

Today, the large porphyry copper ore bodies in Chile and Peru have made South America the world's leading copper-producing region. The United States ranks second and comprises the only region in which production and consumption are in balance. Central Africa, the Congo and Zambia, and the CIS follow the United States in production.

OCCURRENCE

Copper is one of the most abundant of the metallic elements in the earth's crust. Only a small fraction of the copper in the uppermost mile of the continental crust is relatively concentrated. At the current global mine production rate, this fraction represents a million years' supply of copper theoretically available in the mineable portion of the earth's crust. There are also appreciable quantities of copper in deep-sea clays, in manganese-based nodules scattered on the ocean floor, and in effluents from undersea fumaroles. None of these undersea sources have been economically exploited.

Copper ore minerals are classified as primary, secondary, oxidized, and native copper.

Secondary minerals formed when copper sulfides exposed at the surface became weathered, leached (dissolved) by groundwater, and precipitated near the water table. Copper, like gold and silver, can also occur as a primary mineral, in its metallic (native) form. A good example of the latter is the lava-associated deposits of the Keweenaw Peninsula in Upper Michigan, which formed a significant portion of global copper production in the nineteenth Century.

The five classifications of economically viable copper ore include porphyry deposits and vein replacement deposits; strata-bound deposits in sedimentary rocks; massive sulfide deposits in volcanic rocks; magmatic segregates associated with nickel in mafic intrusives; and native copper. A sixth type of deposit, oxide minerals, is now recognized as a result of the development of leaching and solution purification technologies.

Almost two-thirds of the world's copper resources are porphyry deposits, a copper deposit that is hydrothermal in origin and in which a large portion of the copper minerals is uniformly distributed as small particles in fractures and small veins. Porphyry deposits usually contain $\leq 1\%$ copper. The most extensive of these deposits are located in western Canada, the southwestern United States, Mexico, and the Andes Mountains of western South America. In addition to the porphyries, there are large bedded copper deposits in Germany, Poland, the CIS, Australia and central Africa. Most near-surface copper sulfide deposits have an oxidized cap of secondary mineralization. Since these oxide minerals are not treatable in a smelter, such deposits were previously not regarded as ores. Modern leaching technology now makes their recovery possible, and these minerals, together with oxidized portions of previously mined material, currently account for about 15–20% of primary-copper production.

PROPERTIES

Copper, with symbol Cu and atomic number 29, is one of the "noble" metals, and like gold and silver, it is a member of subgroup IB in the periodic chart of the elements. Its atomic electronic structure is described by the notation $1s^2 2s^2 p^6 3s^2 3p^6 3d^{10} 4s^1$, which depicts an argon core plus a filled $3d$ orbital and a single $4s$ electron. Copper owes its unique properties to this structure. For example, the filled $3d$ states limit compressibility and, consequently, scattering of conductance electrons due to thermal lattice vibrations. Lattice incompressibility and the loosely bound $4s$ and $3d$ electrons give copper its high electrical and thermal conductivity.

A compilation of copper's physical properties can be found in Table 1.

From both technical and commercial standpoints, copper's most important physical property is its high electrical conductivity, highest among the "engineering" metals and second only to silver.

Copper's high thermal conductivity, is exploited in applications ranging from ordinary heat exchangers to

Table 1. Physical Properties of Pure Copper

Property	Value
atomic weight	63.546
atomic volume, cm^3/mol	7.11
mass numbers, stable isotopes	63 (69.1%), 65 (30.9%)
oxidation states	1, 2, 3
standard electrode potential, V	$Cu/Cu^+ = 0.520$
	$Cu^+/Cu^{2+} = 0.337$
electrochemical equivalent, mg/C	0.3294 for Cu^{2+}
	0.6588 for Cu^+
electrolytic solution potential, V(SCE)	0.158 $(Cu^{2+}+e^- = Cu^+)$
	0.3402 $(Cu^{2+}+2e^- = Cu)$
	0.522 $(Cu^+e^- = Cu)$
density, g/m^3	8.95285 (pure, single crystal)
	8.94 (nominal)
metallic (Goldschmidt) radius, nm	0.1276 (12-fold coordination)
ionic radius, M^+, nm	0.096
covalent radius, nm	0.138
crystal structure	fcc, $A1$, $Fm3m$: $cF4$
lattice parameter	0.0361509 ± 0.000004 nm (25°C)
electronegativity	2.43
ionization energy, kJ/mol	
first	745
second	1950
ionization potential, eV	7.724 Cu(I)
	20.29 Cu(II)
	36.83 Cu(III)
Hall effect[a]	
Hall voltage, V at 0.30–0.8116 T	-5.24×10^{-4}
Hall coefficient, mV/(mA)(T)	-5.5
heat of atomization, kJ/mol	339
thermal conductivity, W/(m)(K)	394[b], 398
electrical resistivity at 20°C, $n\Omega \cdot m$	16.70
temperature coefficient of electrical resistivity, 0–100°C	0.0068
melting point	1358.03 K (1084.88°C)
	1356 K (1083°C)
heat of fusion, kJ/kg	205, 204.9, 206.8
	212[b]
boiling point	2868 K (2,595°C), 2,840 (2,567°C)
	2595 K (2,868°C)
heat of vaporization, kJ/kg	7369[b]
	4729, 4726, 4793
specific heat, kJ/(kg)(K)	0.255 (100 K)
	0.384 (293 K)
	0.386 (293 K)
	0.494 (2000 K)
coefficient of expansion, linear, μm/m	16.5
coefficient of expansion, volumetric, $10^{-6}\,K^{-1}$	49.5
tensile strength, MPa	230 (annealed)
	209 (annealed)
	344 (cold drawn)
elastic modulus, GPA	125 (tension, annealed)
	102–120 (tension, hard-drawn)
	128 (tension, cold drawn)
	46.4 (shear, annealed)
	140 (bulk)
magnetic susceptibility, 291 K, mks	-0.086×10^{-6}
	-1.08×10^{-6}
emissivity	0.03 (unoxidized metal, 100°C)
	0.8 (heavily oxidized surface)
spectral reflection coefficient, incandescent light	0.63
nominal spectral emittance, $\lambda = 655$ nm, 800°C	0.15
absorptivity, solar radiation	0.25

Table 1. (*Continued*)

Property	Value
viscosity, mPa·s (cP)	3.36 (1085°C)
	3.33 (1100°C)
	3.41 (1145°C)
	3.22 (1150°C)
	3.12 (1200°C)
surface tension, mN/m (dyn/cm)	1300 (99.99999% Cu, 1084°C, vacuum)[a]
	1341 (99.999% Cu, N_2, 1150°C)
	1104 (1145°C)
	(see Ref. 15 for additional data)
coefficient of friction	4.0 (Cu on Cu) in H_2 or N_2
	1.6 (Cu on Cu) in air or O_2
	1.4 (clean)
	0.8 (in paraffin oil)
velocity of sound, m/s	4759 (longitudinal bulk waves)
	3813 (irrotational rod waves)
	2325 (shear waves)
	2171 (Rayleigh waves)

superconducting cables used in magnetic resonance imaging (MRI) equipment, although copper itself is not a superconductor. Copper's relatively high coefficient of thermal expansion is one reason why bonds between copper and silicon semiconductor chips always include a transitional layer, typically of a metal, usually nickel, which has an intermediate expansion coefficient.

Copper is diamagnetic, although a number of ferromagnetic copper alloys are known. Ferromagnetism in such alloys arises not from copper itself but from the alloys' crystal structure or from precipitated ferromagnetic phases. Copper's low magnetic permeability avoids energy losses in electromagnetic devices such as motors and generators. Special heat treatments are employed to avoid traces of ferromagnetism in copper alloys in critical applications such as minesweepers.

Uses of pure copper based on its mechanical properties alone are quite limited, and most applications in which high strength, ductility or other such properties are required—usually often in combination with other attributes—are served by copper alloys, of which several hundred are now produced worldwide. In the late twentieth century, copper alloy development centered on improvements in machining characteristics (lead-free brasses), corrosion resistance (marine and plumbing alloys), elevated-temperature resistance (electrical connector- and leadframe alloys), and, as in ancient times, castability.

The relative instability of copper's oxides is responsible for the occurrence of native, that is, metallic, copper in nature. Cu(I) forms compounds with the anions of both strong and weak acids. Many of these compounds are stable and insoluble in water.

After its electrical and thermal conductivity, copper's third most important property is its intrinsically high corrosion resistance. To a degree, copper exhibits the thermodynamic stability of silver and gold. Native copper nuggets found in Michigan, for example, are estimated to have remained stable in subterranean brines for as long as 10^9 years. Numerous copper and bronze archaeologic artifacts, some of them retrieved from the sea, have sur-

vived for centuries. Although copper is thermodynamically stable in a few environments, its corrosion resistance is more often based on kinetics, that is, the low rates of speed at which corrosion reactions proceed. Thus, in many potentially corrosive media (including, importantly, air, steam, and potable and marine waters) the metal tends to form adherent and protective surface films that, once established, effectively inhibit further attack unless the films are damaged or altered by changes in the environment. Copper's long and successful history of use for plumbing tube rests on this phenomenon, as does the formation of protective—and attractive—patina on copper roofs and bronze statuary. Copper roofing in certain rural atmospheres corrodes less than 0.4 mm in 200 years. The copper skin of the Statue of Liberty has lost only 0.1 mm to corrosion in more than 100 years, despite exposure to a marine/industrial atmosphere. Pure copper resists aerated alkaline solutions except in the presence of ammonia. Copper does not displace hydrogen from acid but dissolves readily in oxidizing acids such as nitric acid or sulfuric acid solutions containing an oxidizer like ferric sulfate. Copper is also attacked by soft, low pH waters, ammonia solutions, amines, cyanide solutions, nitrates and nitrites, oxidizing heavy-metal salts, and certain sulfides. Aqueous solutions containing ammonia, amines, cyanide, nitrates, and nitrites attack copper and can, under certain conditions, generate stress corrosion cracking. Although copper is biostatic, it can exhibit microbially induced corrosion (MIC). Chlorides, including seawater, do not seriously attack copper and its alloys. This has led to many applications for which, for example, stainless steels are less well suited. On the other hand, titanium alloys have begun to replace copper alloys in naval and seawater-cooled power plant condensers.

SOURCES AND SUPPLIES

Copper enters into trade primarily in the form of concentrates, blister, anodes, refined copper ingots and cathodes, and copper semis. The former are provided by copper producers (miners, smelters, and refiners) the latter by fabricators.

RECOVERY AND PROCESSING

Most copper is produced using a combination of mining, concentrating, smelting, and refining. Worldwide about 19% and in the United States about 45% of refined copper is produced through the leaching of mine and concentrator wastes, depleted tailings, and oxide ores with sulfuric acid derived from the smelting process's off-gases. The copper-containing leach solutions are purified and concentrated by solvent extraction (SX), and copper is recovered by electrowinning (EW). In recent years, *bioleaching*, the use of microorganisms to convert copper sulfides to acid-leachable oxides and thus, avoid expensive crushing and smelting, has gained considerable papularity. Whether from the traditional or SXEW paths, the final product is a high purity copper cathode.

Copper ore is obtained from either open-pit or underground mines; the open-pit source is predominant today. In some instances a combination of open-pit and underground mining is used once a depth is reached in which open-pit mining is uneconomical.

The recovery technique called concentrating consists of crushing and grinding to liberate the copper mineralization from the host rock, followed by *flotation*, in which copper mineralization and waste materials, or gangue are separated. Milling refers to the crushing and grinding steps.

Copper concentrate usually contains a mixture of copper and iron sulfides together with small amounts of gangue minerals. Smelting comprises the two operations needed to extract copper from concentrate using heat, flux, and oxidation.

In a converter or *converting* furnace, matte from the smelting furnace is further oxidized to form copper and slag. The final product is *blister copper* containing >98% Cu plus some sulfur, oxygen, and other impurities. Several continuous (ie, combined smelting–converting) technologies were introduced in the late 1990s.

The environmentally driven need to convert smelter gases to acid has been the most important force driving the development of new smelting technologies. Today, those gases are normally captured and sent to a sulfuric acid plant, where they are catalytically converted to sulfur trioxide, and contacted with water to form sulfuric acid. The most modern smelters produce gases containing ≥35% sulfur dioxide. Simple scrubbers, the devices first used to remove sulfur dioxide from offgases, are now employed primarily to treat gases with very low sulfur dioxide concentrations, such as those from dryers, anode furnaces, and building ventilation. Lime is the most commonly used absorbent; the resulting product is calcium sulfate.

Fire refining adjusts the sulfur and oxygen levels in blister copper, removing impurities as slag or volatile products.

Copper intended for electrical uses then requires further refining by electrolysis in order to raise its purity and electrical conductivity to the degree needed for electrical products. Total impurity content in the highest-grade copper is restricted to fewer than a few parts per million.

Copper ores have been leached in the United States since the 1920s, but hydrometallurgical processing did not become a major part of the copper industry until the availability of large amounts of sulfuric acid from smelters. Hydrometallurgy was further enhanced by adoption of a solvent extraction and solution purification process later developed. Combination of the solvent extraction process with electrowinning led to the development of the (solvent extraction/electrowinning (SXEW)) process. Previous leaching processes used cementation, the replacement of copper in solution for iron by contacting acidic copper-rich solutions with scrap iron. The product, known as cement copper, was fed to the smelter. The SXEW process enables the direct recovery of high purity copper in cathode form.

An increasing portion of U.S. copper production is being obtained by the leaching of waste rock and tailings. Hydrometallurgy is also gaining importance in other parts of the world. By 2005, leaching and SXEW accounted for an estimated 20% of the newly mined copper in the world.

FABRICATION

Copper and copper alloy shapes, known in the trade as semifabricates, or more commonly, semis, are made from refined copper or copper and copper alloy scrap, with or without the addition of alloying ingredients. Fabricator's facilities are classified as wire mills, brass mills, foundries, and powder plants. Wire mills and foundries produce wire products and castings, respectively. The general term brass mills encompasses a variety of plants that produce, respectively, extruded rods and shapes, tube and fittings, and sheet and strip. Only some of these products are made from brass.

High purity copper wire for electrical uses is by far the principal product manufactured by wire mills. The principal products of brass mills are not limited to brass alone but include copper itself and many alloys. Most mills now use continuous casting machines to produce the intermediate products of slabs called cakes and cylinders called logs.

Rod, bar, and other shapes are normally made by extrusion, a process frequently likened to squeezing toothpaste from a tube. Plate, sheet, strip, and foil are made by rolling continuously cast slabs of copper or copper alloys.

Plumbing tube and commercial tube are made from refined (cathode) copper and/or high-grade copper scrap that has been fire refined and deoxidized with phosphorous.

Tubular products are produced by extrusion and drawing, by seam welding sheet that has been roll-formed into a tube, by continuous casting a shell and rolling the shell into a tube, and by the centrifugal casting process. Welded tube is made by drawing copper strips through a conical die such that the strip assumes a tubular shape.

Centrifugal and, occasionally, continuous-casting processes are used to produce thick-walled pipe. Many copper and copper alloy products can be produced by forging or hot stamping, as the process is called in the United Kingdom.

Many copper alloys, particularly those that share copper's crystal structure, are also ductile and easy to cold form into useful products.

Copper alloys are also known for their excellent castability: Statues, bells, and works or art are still cast in bronze.

Machined products constitute an important copper market.

High machinability in free-cutting brass (60–63% copper) and most other alloys derives from the addition of a few percent of lead. And, finally, products made from pressed and sintered metal powders constitute a small (1% of total consumption) but interesting market for copper.

ECONOMIC ASPECTS

The basis for copper pricing is simply the state of the world economy, which establishes the level of demand, counterbalanced against the installed capacity to produce and the unit costs of production, which establish the level of supply. Copper is an energy-intensive material; hence, the cost of energy-producing commodities, coal, oil and natural gas, have a direct effect on copper supply.

ENVIRONMENTAL CONCERNS

Copper production, and to some extent copper use and disposal practices, are regulated in the United States and elsewhere by both state and federal regulations. The concerns regarding copper and the byproducts of its production in the natural environment can be divided into four categories: natural flux to the atmosphere and the oceans; effluents from production; effluents from uses; and lifecycle environmental impact.

Copper is relatively abundant in the earth's crust. It is estimated that as much copper is dispersed into the natural environment by natural forces as is by anthropogenic point sources—specifically, a total of ~20,000–51,000 tons/year. During the smelting processes, sulfides are oxidized to form gaseous sulfur dioxide. Copper mining operations have always been faced with a large solid-waste disposal problem. And water effluents include process water and wastewater related to specific mining, milling, and smelter operations, plus storm water runoff that may come in contact with a facility's operations.

On the other hand, a large portion of all of the copper in use is ultimately recycled for other uses although there is a finite amount of attrition into the environment. Copper contamination in the soil is not a major issue.

The major concern over contamination due to copper uses is introduction of the metal to bodies of freshwater and saltwater where aquatic life may be adversely affected. Because copper is an essential element, adverse effects appear only at copper intake levels higher than those required for metabolism. Toxicity levels vary widely among species.

Road traffic–based sources are reportedly the main contributors of copper to the environment at large and are major contributors to aquatic systems as well. Copper-based chemicals used as antifoulants, algicides, and similar products are the principal contributors to the aquatic environment and a major contributor to the total environment. These are followed in magnitude by landfills and agricultural sources. Plumbing systems, roofing and facades contribute only minor amounts of copper to the environment. There is now growing interest in quantifying the environmental impact of materials by performing "cradle to grave" analyses of the energy consumed in the materials' production and use, and by the materials' impact on the environment at each stage of their lifecycle.

RECYCLING AND WASTE DISPOSAL

Because of its high value, copper is among the most thoroughly recycled metals. In North America, copper scrap is graded, according to standard definitions established by the Institute of Scrap Recycling Industries (ISRI). A large and sophisticated industry gathers, grades, remelts, and returns copper-base metals to fabricators for reuse in commercial products. Recycled copper is known as *secondary copper* or simply *scrap*, of which there are three categories:

- *Prompt, home,* or *direct scrap*, which originates in smelting and/or fabrication operations. This scrap is recycled within the plant.
- *New scrap*, which is metal returned to the fabricator by manufacturers as leftover material. Machine shop turnings (chips) are one example.
- *Old scrap*, which includes all worn out, discarded, or salvaged products, such as telephone cable and plumbing tube from demolished buildings.

HEALTH AND SAFETY FACTORS

Copper is an essential trace element required for human metabolism. The total amount of copper in humans, based on individual tissue analyses, is estimated at 50–120 mg for adults and 14 mg for a full-term newborn infant. The concentration of copper in adults and infants is 0.7–1.71 milligrams per kilograms (mg/kg) of body weight and 4 mg/kg of body weight, respectively. The higher concentration of copper in newborn infants compensates for the lack of copper in mothers' milk. The liver, brain, and heart have the largest concentrations of copper of all the organs in the body. All body tissues require copper in their metabolism, but some have greater metabolic needs than others. Copper is normally obtained from food and to a lesser extent, water intake; however, it is also provided in some vitamin and mineral supplements and is added to baby formulae.

Copper deficiency is rare in Western countries, but copper deficiency is more of a potential problem than copper toxicity in humans. Acute environmentally related copper toxicosis is not a major problem in humans. Since copper is an essential nutrient, humans and other forms of life requiring copper in their metabolism have a variety of mechanisms that function to prevent overdose. Tolerance mechanisms (eg, homeostasis) make copper generally innocuous.

Copper has been used in medicine for thousands of years. Modern research has shown certain copper compounds to be valuable in the treatment of inflammatory diseases, in some forms of cancer, and as an anticonvulsant agent. Likewise, copper plumbing tube has been found to be effective in restricting the growth of bacteria, including *Legionella pneumophila*, the cause of Legionnaires' disease. Copper–silver ionizers have been found effective in controlling *L. pneumophila* in hospital drinking water. Such systems have been widely used in the disinfecting of swimming pools. Further, research strongly suggests that the pathogen *Escherichia coli* O157 is killed within several hours' exposure to copper surfaces. The bacteriostatic properties of copper-bearing surfaces such as ordinary brass doorknobs and pushplates have also been credited with reducing infection in hospitals and other public buildings.

With the exception of vineyard sprayer's disease, no significant chronic effects of copper have been reported as a result of occupational exposure.

USES

Electrical and communications wire and cable together with plumbing products account for 85% of refined copper consumption and 63% of total consumption, including scrap. Electrical products are primarily manufactured from refined copper, whose degree of purity ensures high electrical conductivity. About 65% of the copper in plumbing products is derived from copper scrap. Most other applications, including most copper alloys, except those used in coinage, are also based mainly on scrap as a raw material. Electrical uses constitute the largest single market for refined copper. In rapidly emerging economies where power and communications infrastructures are being installed, the fraction of refined copper consumed as wire and cable can be as high as 80%.

Copper's use in homes and in commercial and industrial buildings is by far the metal's largest market, amounting to approximately 41% of total consumption. Electrical products, plumbing goods, and roofing sheet, respectively, make up the bulk of copper's building-construction applications.

Plumbing tube, faucets, valves, and fittings for both potable water and heating use constitute copper's second largest use, accounting for approximately 14% of all copper consumed. Roofing sheet and related products such as gutters and downspouts are important (and highly visible) uses for copper. Much of the copper used in such applications is refined from scrap. Copper, in the form of brass and bronze alloys, is widely used in builders' hardware, such as lock sets, doorknobs, hinges, and push- and kickplates.

Copper's use in electrical and electronic products constitutes about 26% of the total market for copper in the United States. Power utilities contribute to about 9% of total consumption.

Copper magnet wire—that used in electric motors, transformers, and similar devices—continues to be an important copper product. Approximately 4.5% of the copper and copper alloy products are used in the electronics industry in the United States. Electrical contacts, connectors, and switch components are important outlets for copper and copper alloy strip. Pure copper foil, produced either by rolling, or more often by electrodeposition, is used in many printed circuits. Copper interconnects for integrated circuits promise to increase processing speed several fold over traditional circuitry, which rely on aluminum.

Transportation uses constitute about 12% of the market for copper and copper alloys in the United States. Automobile wiring harnesses represent the largest portion, about 8%, of this copper. Thin-gauge copper and brass strip for automotive radiators, once an important market, declined steadily after the introduction of aluminum radiators in the 1970s. Copper–brass radiators for new cars are preferred in tropical countries where use of corrosion-inhibiting antifreeze compounds is uncommon.

Copper alloys are widely used in marine applications ranging from propellers to seawater piping, pumps, heat exchangers, condensers, and other equipment. Alloys most commonly used include copper–nickels (condenser tubes), leaded brasses (tubesheets), aluminum bronzes (pipe, pump, and valve components), nickel–aluminum bronzes (propellers), aluminum brasses (tube), manganese bronzes (pump and valve components), and silicon bronzes (pump and valve components). Copper –nickel sheathing applid to the legs of offshore platforms prevents buildup of algae. Ship hull sheathing, historically important, has in modern times been limited to small craft. Copper, in the form of cuprous oxide and other specialty compounds is widely used in antifouling hull paints. Copper hull paints are environmentally superior to organotin coatings, which once threatened copper in this market but are now widely banned.

Use of copper for in-plant equipment, nonelectrical instruments, and off-highway vehicles is its fourth largest in the United States, representing approximately 11% of total consumption.

Approximately 9% of the total copper and copper alloys used in the United States are found in consumer and general products. Copper used in home appliances constitutes about 3% of this volume. Also included in this category are copper chemicals that, as a product form, represent about 1% of the copper market. Copper's unique chemical, toxicologic, and nutritional properties enable it to serve a number of niche markets in both the agricultural and industrial sectors. Agriculturally, copper compounds are used as fungicides (55% of consumption) livestock feeds (8%), crop nutrients (6%), and other uses. Among industrial uses, wood preservative (12%), antifouling paints (7%), mining and metallurgy (4%) chemicals and petroleum processing (4%), and textile and leather treatment (2%) are major uses. Copper sulfate, copper oxides and oxychlorides, and copper hydroxide are the dominant chemical product forms.

R. F. Mikesell, *The World Copper Industry: Structure and Economic Analysis*, Johns Hopkins Press, Baltimore, Md., 1979.

R. F. Prain, *Copper: The Anatomy of an Industry*, Mining Journal Books, London 1975.

S. D. Strauss, *Trouble in the Third Kingdom*, Mining Journal Books, London, 1986.

U.S. Bureau of Mines, annual, *Minerals Yearbook*, Washington, D.C.

K. J. A. KUNDIG
Metallurgical Consultant

W. H. DRESHER
WHD Consulting

COPPER ALLOYS, WROUGHT

Among the metals of commercial importance, copper and its alloys are surpassed only by iron and aluminum in worldwide consumption. Typically, copper is alloyed with other elements to provide a broad range of mechanical, physical, and chemical properties that account for its widespread use. The principal characteristics of copper alloys are their moderate to high electrical and thermal conductivities combined with good corrosion resistance, good strength, good formability, unique decorative appearance, and moderate cost. Most copper alloys are readily hot and cold formed, joined (soldered, brazed, and welded), and plated.

Chief consumers of copper and copper alloys are the building construction industry for electrical wire, tubing, builder's hardware, plumbing, and sheathing; electrical and electronic products for motors, connectors, printed circuit copper foil, and lead frames; and the transportation sector for radiators and wiring harnesses. Other industries include ordnance, power utilities, coinage, decorative hardware, musical instruments and flatware.

ALLOY DESIGNATIONS

Elements typically added to copper are zinc, tin, nickel, iron, aluminum, silicon, silver, chromium, titanium, and beryllium.

Copper and its alloys are classified in the United States by composition according to the Unified Numbering System (UNS) for metals and alloys. The designations of wrought copper alloys are given in Table 1. Designations that start with numeral 8 or 9 are reserved for cast alloys.

Most wrought alloys are provided in conditions that have been strengthened by various amounts of cold work or heat treatment. Cold-worked tempers are the result of cold rolling or drawing by prescribed amounts of plastic deformation from the annealed condition. Alloys that respond to strengthening by heat treatment are referred to as precipitation or age hardenable. Cold-worked conditions can also be thermally treated at relatively low temperatures (stress relief annealed) to produce a slight decrease in strength to benefit other properties, such as corrosion resistance, formability, and stress relaxation.

Temper

The system for designating material condition, whether the product form is strip, rod, or wire, is defined in

Table 1. UNS Designation for Copper and Copper Alloy Families

Alloy group	UNS designation	Principal alloy elements
coppers[a]	C10100–C15999	Ag, As, Mg, P, Zr
high coppers[b]	C16000–C19999	Cd, Be, Cr, Fe, Ni, P, Mg, Co
brasses	C20500–C28580	Zn
leaded brasses	C31200–C38590	Zn—Pb
tin brasses	C40400–C486	Sn, Zn
phosphor bronzes	C50100–C52400	Sn—P
leaded bronzes	C53200–C54800	Sn—P—Pb
phosphorus–silver	C55180–C55284	P, Ag—P
aluminum bronze	C60600–C64400	Al, Fe, Ni, Co, Si
silicon bronze	C64700–C66100	Si, Sn
modified brass	C662–C69950	Zn, Al, Si, Mn
cupronickels	C70100–C72950	Ni, Fe, Sn
nickel silvers	C73150–C77600	Ni—Zn
leaded nickel silvers	C78200–C79900	Ni—Zn—Pb

[a]Contains a minimum of 99.3 wt% copper.
[b]Contains a minimum of 96 wt% copper.

ASTM Recommended Practice B601. The ASTM system uses an alpha-numeric code for each of the standard temper designations. This system replaces the historical terminology of one-half hard, hard, spring, etc.

PRODUCT FORMS AND PROCESSING

The output from brass mills in the United States is split nearly equally between copper and the alloys of copper. Copper and dilute copper alloy wrought products are made using electrically refined copper so as to maintain low impurity content. Copper alloys are commonly made from either refined copper plus elemental additions or from recycled alloy scrap. Copper alloys can be readily manufactured from remelted scrap while maintaining low levels of impurities. A greater proportion of the copper alloys used as engineering materials are made from recycled materials than are most other commercial materials.

Wrought alloy product forms are varied and include plate, sheet, strip and foil, round and special cross-section bars, rod, and wire.

ALLOYING FOR STRENGTHENING

Elements added to pure copper can remain in solid solution in the copper or can form second displaced, phases separate from the copper, which forms the bulk of the alloy.

Solid Solution Alloys

Copper dissolves other elements to varying degrees to produce a single-phase alloy that is strengthened relative to unalloyed copper. The contribution to strengthening from an element depends on the amount in solution and by its particular physical characteristics such as atom size and valency. Tin, silicon, and aluminum show the highest strengthening efficiency of the common solute additives, whereas nickel and zinc are the least efficient.

Dispersed Phase Alloys

The presence of finely dispersed second-phase particles in copper alloys contributes to strength, by refining the grain size and increasing the amount of hardening due to cold working.

Precipitation (Age) Hardening Alloys

Only a few copper alloy systems are capable of responding to precipitation or age hardening. Those that do have the constitutional characteristics of being single-phase (solid solution) at elevated temperatures and are able to develop into two or more phases at lower temperatures that are capable of resisting plastic deformation. The copper alloy systems of commercial importance are based on individual additions of Be, Cr, Ti, Zr, or Ni + X, where X = Al, Sn, Si, or P.

Special Addition Alloys

The most notable of the special additives to copper alloys are those added to enhance machinability. Lead, tellurium, selenium, and sulfur are within this group of additives. Because of increasing concern over lead toxicity, interest has centered on use of bismuth, which is nearly as effective as lead for improving machinability. The alloys that contain such additives are limited because of the difficulty they cause to hot and cold working.

PROPERTIES

Strength

An increase in strength and an accompanying decrease in electrical conductivity derives from the alloying of copper. This trend is clearly apparent among solid solution strengthened alloys, namely those that contain zinc, tin, nickel, and zinc plus nickel as their principal alloying constituents. Notable exceptions to this trend are precipitation-hardening alloys where the precipitating phases remove elements from the solid solution leaving a leaner, higher conductivity matrix; eg, C1751, C182, and C7025. For the latter group, strength–conductivity combinations not possible with solid solution alloys are achieved.

Formability

Copper and most of the wrought alloys are readily formed by bending, drawing, upset forging, stamping, and coining. The maximum formability condition is the fully soft or annealed condition. When additional strength or hardness is desired in the final part, the forming step is done starting with a cold-worked temper or the part is not annealed between forming steps. Cold-forming operations always work-harden alloys, and in many cases sufficient strength for the application is produced.

A variety of tests have been established to indicate whether a specific forming operation can be safely accomplished without failure. Three of the most commonly used tests are bend testing around a mandrel, limiting draw ratio testing, and bulge height formability testing. Temper, strip thickness, grain size, crystallographic texture, and alloy composition are important variables in these tests. An alloy in the annealed temper that can be formed into a particular shape may not be able to be similarly formed starting from a work-hardened temper.

Softening Resistance

The ability of being readily annealed or softened in a controlled manner to restore ductility is beneficial in mill processing, but resistance to softening of the wrought product is often preferred during fabrication and subsequent service. Joining operations such as welding, brazing, and soldering are prime examples where softening resistance is essential. Most alloying elements and impurities increase softening resistance. The amount by which softening resistance is increased is specific to the element being added and also depends on whether the element remains in solid solution or forms a second-phase particle.

Stress Relaxation Resistance

The amount of stress relaxation depends on the alloy, its temper, the temperature of exposure, and the duration of the exposure. Resistance to stress relaxation of copper is improved by alloying with solid solution elements, as well as by dispersion and precipitation strengthening. Changing temper to higher strength for a given alloy results in some loss in relaxation resistance. Relief annealing where yield strength is decreased slightly while causing little or no change in tensile strength is used to improve relaxation resistance. Relief annealing where yield strength is decreased slightly while causing little or no change in tensile strength is used to improve relaxation resistance.

The highest stress relaxation resistance, at both high strength and moderate conductivity, is available from precipitation hardened alloys.

Fatigue Resistance

Imposed cyclic stressing of metals may result in localized cracking that leads to fracture. The pattern of stressing can be in reversed bending, reversed torsion, and tension–compression, or half cycles of these such as bending in only one direction. The number of cycles of stressing that can be endured without fracture depends on the magnitude of the peak applied stress, the pattern of stressing, and the alloy's mechanical properties.

Fatigue strengths for several copper alloys are listed in Table 2. Generally, fatigue strength increases with the tensile strength of the material. The rule of thumb is that the fatigue strength of a copper alloy is about one-third of its tensile strength.

Corrosion Resistance

Copper and selected copper alloys perform admirably in many hostile environments. Copper alloys with the appropriate corrosion resistance characteristics are recommended for atmospheric exposure (architectural and

Table 2. Fatigue Strengths of Copper Alloys[a]

Alloy	Average 0.2% yield strength, MPa[b]	Fatigue strength 10^8 cycles, MPa[b]
C172[c]	760	275
C194	480	150
C260	615	185
C510	690	235
C762	725	205

[a]All are H08, spring temper unless indicated.
[b]To convert MPa to psi, multiply by 145.
[c]TM02 (1/2 HM).

builder's hardware, cartridge cases), for use in fresh water supply (plumbing lines and fittings), in marine applications (desalination equipment and biofouling avoidance), for industrial and chemical plant equipment (heat exchangers such as condensers and radiators), and for electrical/electronic applications (connectors and semiconductor package lead frames).

Atmospheric exposure, fresh and salt waters, and many types of soil can cause uniform corrosion of copper alloys. The relative ranking of alloys for resistance to general corrosion depends strongly on their environment and is relatively independent of temper. Atmospheric corrosion, the least damaging of the various forms of corrosion, is generally predictable from weight loss data obtained from exposure to various environments.

Hydrogen Embrittlement

Copper alloys that contain cuprous oxide in their microstructures, as in C110, are potentially susceptible to embrittlement when heated in hydrogen-containing gases. Accordingly, susceptible alloys are annealed during processing in nitrogen or very low hydrogen-containing gas, at low temperatures, and short times, to avoid embrittlement.

Solderability

Most copper alloys have good solderability (tinability), meaning that they are wet by molten tin, tin–lead, and modifications of these to produce a continuous coating that has few to no pinhole-sized nonwet areas. This characteristic of copper and its alloys accounts for the significant use of tin and solders (both lead containing and lead-free) to provide corrosion resistance and in joining.

Brazeability

Brazing is, by definition, elevated temperature soldering, that is, soldering above the arbitrarily defined temperature of 425°C. Copper and its alloys are readily, and often are, brazed to take advantage of the stronger and more stable brazed joint compared to the soldered joint. In addition, it is easy to match the properties of the filler metal to the copper alloy to be brazed because many of the brazing alloys are themselves copper-base alloys.

Filler alloys for brazing of copper alloys are usually copper base. These include copper–zinc alloys (RBCuZn), copper–phosphorus alloys (BCuP-1), and copper–silver (Zn, Cd, or Li) alloys (BAg).

Weldability

There are four primary concerns for the successful welding of copper. The first is the high thermal conductivity of copper and its resulting ability to conduct heat away from the weld zone. The second concern is the chemical, and in some cases toxic, properties of the typical elements alloyed with copper. Adequate ventilation must be provided for the lead vapor, the zinc fumes, or the toxic compounds of Be, As, or Sb that can be emitted from the copper alloys containing these elements. The third concern is the ready solubility of oxygen in copper at elevated temperatures and the subsequent precipitation of cuprous oxide particles at grain boundaries during solidification and cooling, which if uncontrolled cause reduced strength and ductility in the weld zone. The fourth concern is that alloys containing Pb, Te, and S (eg, the free-machining alloys) are prone to hot shortness (hot cracking) during cooling.

Arc welding has long been used to join copper alloys. The gas tungsten-arc welding (GTAW) and the inert gas metal arc welding (GMAW) methods are the preferred arc welding methods for copper alloys.

Resistance welding has been successfully applied to copper alloys in all of its various spot, seam, or butt joining modes. Because the process depends on ohmic (I^2R) heating at the interface to be joined, the ability to resistance weld is inversely related to the electrical conductivity of the alloys being welded.

Induction welding also depends on I^2R heating. In this case, the electrical current is induced in the workpiece via an imposed high-frequency magnetic field designed to concentrate the induced eddy currents near the surface at the edges of the workpiece to be joined. This method has been effectively used to make the longitudinal seam in the high-speed manufacture of welded tube from copper alloy strip.

Machinability

Copper and its alloys can be machined with differing degrees of ease. Special additives such as lead, bismuth, tellurium, selenium, and sulfur, are added, to enhance machinability, although other properties, such as formability and tensile ductility, normally suffer. These particular alloying elements form second-phase particles that promote chip fracture and the development of lubricating films at the tool-to-chip interface. Smaller, easier to handle chips and lower cutting forces leading to longer tool life result from use of these additives. Notable uses for special alloys having high machinability are rod, for high production rate screw machine items such as fasteners, connectors and plumbing components, and strip for keys.

ALLOY SPECIFIC PROPERTIES

Coppers

The coppers represent a series of alloys ranging from the commercially pure copper, C101, to the dispersion hardened alloy C157. The difference within this series is the specification of small additions of phosphorus, arsenic,

Table 3. Physical Properties of Copper[a]

Parameter	Value
mp, °C	1083
density, g/mL	8.94
electrical conductivity, % IACS min	101
electrical resistivity, $n\,\Omega\cdot m$	17.1
thermal conductivity, W/(m · K)	391
coefficient of thermal expansion from 20–300°C, μm/(m · K)	17.7
specific heat, J/(kg · K)[b]	385
elastic modulus, GPa[c]	
tension	115
shear	44

[a]All properties at 20°C unless otherwise noted.
[b]To convert J to cal, divide by 4.184.
[c]To convert GPa to psi, multiply by 145,000.

cadmium, tellurium, sulfur, magnesium, zirconium, as well as oxygen. To be classified as one of the coppers, the alloy must contain at least 99.3% copper. The physical properties of pure copper are given in Table 3.

The mechanical properties of coppers having UNS designations between C10100 and C13000 are listed in Table 4. The coppers include high purity copper (C101, C102), electrolytic tough pitch (C110), phosphorus deoxidized (C120, C122), and silver-bearing copper (C115, C129, etc). The mechanical properties of these alloys are essentially the same. Other coppers, C142 through C157, offer higher strength at usually lower conductivity.

Excellent resistance to saltwater corrosion and biofouling are notable attributes of copper and its dilute alloys. High resistance to atmospheric corrosion and stress corrosion cracking, combined with high conductivity, favor use in electrical–electronic applications.

Copper–Zinc Brasses

Copper–zinc alloys have been the most widely used copper alloys since the 1990s.

Brass alloys fall within the designations C205 to C280 and cover the entire solid solution range of up to 35 wt% zinc in the Cu–Zn alloy system. Zinc, traditionally less expensive than copper, does not too greatly impair conductivity and ductility, as it solution hardens copper.

Table 4. Tensile Properties of Copper (C10100–C13000)

Temper	Tensile strength, MPa[b]	Yield strength,[a] MPa[b]	Elongation, %	Hardness, HRF[c]
OS 025[d]	235	76	45	45
H01	260	205	25	70
H02	290	250	14	84
H04	345	310	6	90
H08	380	345	4	94
H10	395	365	4	95

[a]0.5% offset.
[b]To convert MPa to psi, multiply by 145.
[c]Hardness is on the Rockwell-F scale.
[d]Annealed to average grain size of 0.025 mm.

Brass alloys are highly formable, either hot or cold, and provide moderate strength and conductivity. Moreover, the alloys have a pleasing yellow "brass" color at zinc levels >20 wt%. The material is amenable to polishing, buffing, plating, and soldering. By far the best known and most used composition is the 30 wt% zinc alloy Cartridge Brass. Door knobs and bullet cartridges are the best known applications, illustrating the material's excellent formability and general utility.

Brasses are susceptible to dezincification in aqueous solutions when they contain >15wt% zinc. Stress corrosion cracking susceptibility is also significant >15 wt% zinc. Over the years, other elements have been added to the Cu–Zn base alloys to improve corrosion resistance. For example, a small addition of arsenic, antimony, phosphorus, or tin helps prevent dezincification to make brasses more useful in tubing applications.

Tin Brasses

The tin brass series of alloys consists of various copper–(2.5–35 wt%) zinc alloys to which up to ~4 wt% tin has been added. These are solid solution alloys that have their own classification as the C40000 series of alloys. Tin provides better general corrosion resistance and strength without greatly reducing electrical conductivity. As with all the brasses, these alloys are strengthened by cold work and are available in a wide range of tempers. These alloys offer the combined strength, formability, and corrosion resistance required by their principal applications, namely, fuse clips, weather stripping, electrical connectors, heat exchanger tubing, and ferrules. Several tin brasses have lead additions to enhance machinability.

Tin Bronzes

Tin bronzes may be the most familiar of copper alloys, with roots going back into ancient times. Whereas bronze is still used for statuary, these alloys are found in many modern applications, such as electrical connectors, bearings, bellows, and diaphragms. The wrought tin bronzes are also called phosphor bronzes, because 0.03–0.35 wt% phosphorus is commonly added for deoxidation and improved melt fluidity during casting.

The several wrought alloys of commercial importance span the range of 1.0–10 wt% tin and are mostly used in work-hardened tempers. These alloys are single phase and offer excellent cold working and forming characteristics. Strength, corrosion resistance, and stress relaxation resistance increase with tin content. Unfortunately, conductivity decreases, cost increases (tin has been historically more costly than copper), and the capability of being hot processed is impaired as the tin level is increased.

Aluminum Bronzes

Aluminum bronze alloys comprise a series of alloys (C606 to C644) based on the copper–aluminum (2–15 wt%) binary system, to which iron, nickel, and/or manganese are added to increase strength.

Aluminum bronze alloys are used for their combined good strength, wear, and corrosion-resistance properties where high electrical and thermal conductivity are not required. Corrosion resistance results from the formation of an adherent aluminum oxide that protects the surface from further oxidation. Mechanical damage to the surface is readily healed by the redevelopment of this oxide. The aluminum bronzes are resistant to nonoxidizing mineral acids such as sulfuric or hydrochloric acids but are not resistant to oxidizing acids such as nitric acid. However, these alloys must be properly heat treated to be resistant to dealloying and general corrosion.

The aluminum oxide surface layer that provides wear and corrosion resistance is not without drawbacks. This adherent film is difficult to remove during industrial cleaning of the alloys. Furthermore, excessive tool wear in stamping and shearing equipment is caused by the presence of this film. Soldering and brazing are also made difficult by this oxide film.

Two single-phase, binary alloys are used commercially: C606, containing 5 wt% Al, and C610, 8 wt% Al. Both alloys have a golden color and are used in rod or wire applications, such as for bolts, pump parts, and shafts.

Silicon Bronzes

Silicon bronzes have long been available for use in electrical connectors, heat exchanger tubes, and marine and pole line hardware because of their high solution-hardened strength and resistance to general and stress corrosion. As a group, these alloys also have excellent hot and cold formability. Unlike the aluminum bronzes, the silicon bronzes have moderately good soldering and brazing characteristics. Their compositions are limited to below 4.0 wt% silicon, because above this level an extremely brittle phase (kappa) is developed that prevents cold processing. Electrical conductivities of silicon bronzes are low.

Modified Copper–Zinc Alloys

The series of copper–zinc base alloys identified as C664 to C698 have been modified by additions of manganese (manganese brasses and the manganese bronzes), aluminum, silicon, nickel, and cobalt. Each of the modifying additions provides some property improvement to the already workable, formable, and inexpensive Cu—Zn brass base alloy. Aluminum and silicon additions improve strength and corrosion resistance. Nickel and cobalt form aluminide precipitates for dispersion strengthening and grain size control. The high zinc-containing alloys are formulated to facilitate hot processing by transforming to a highly formable (beta) phase at elevated temperature. C674 and C694 are commonly used in rod and wire forms.

Copper–Nickels

The copper–nickel alloy system is essentially single phase across its entire range. Alloys made from this system are easily fabricated by casting, forming, and welding. They are noted for excellent tarnishing and corrosion resistance. Commercial copper alloys extend from 5 to 40 wt% nickel.

Iron is added in small (usually 0.4–2.3 wt%) amounts to increase strength. More importantly, iron additions also enhance corrosion resistance, especially when precautions are taken to retain the iron in solution. A small (up to 1.0 wt%) amount of manganese is usually added to react with and remove both sulfur and oxygen from the melt. These copper alloys are most commonly applied where corrosion resistance is paramount, as in condenser tube or heat exchangers.

Nickel–Silvers

Nickel–silver alloys, once called German silver, are a series of solid solution Cu–Ni–Zn alloys the compositions of which encompass the ranges of 3–30 wt% Zn and 4–25 wt% Ni. This family of alloys falls within the UNS designation numbers C731 to C770. Leaded nickel–silvers that contain from 1.0 to 3.5 wt% lead for improved machining characteristics are designated as C782 to C799.

Nickel–silver alloys are not readily hot worked but have excellent cold-fabricating characteristics. Because of the high nickel and zinc contents, these alloys exhibit good resistance to corrosion, good strength, and usually adequate formability but low electrical conductivity. Wire and strip are the dominant forms used for hardware, rivets, nameplates, hollowware, and optical parts.

Precipitation-Hardening Alloys

Copper alloys that can be precipitation hardened to high strength are limited in number. In addition to the metallurgical requirement that the solubility of the added element(s) decreases with decreasing temperature, the precipitated phase that forms during aging must be distributed finely and have characteristics that act to resist plastic deformation.

Commercial precipitation-hardening copper alloys are based on beryllium, chromium, titanium, and nickel, the latter in combination with silicon or tin. The principal attributes of these alloys are high strength in association with adequate formability. Electrical conductivity varies according to alloy and ranges from ~20–80% IACS.

ECONOMIC ASPECTS

One of the factors affecting the relative pricing of commercial copper alloys is the metal value. Other factors affecting costs include the metal value of the constituent alloy elements, order quantities, supplier, and specified dimensional tolerances. Oxygen-free copper, C102, is sold as a premium product. Brass, eg, C260, is less costly than copper, because zinc is historically lower priced than copper. Tin-containing phosphor bronzes, such as C510, and nickel containing alloys such as the 700-series are more costly than brass and copper because of higher tin and nickel metal cost.

APPLICATIONS

Alloy selection for a particular application generally involves consideration of a combination of physical and mechanical properties as well as cost. Heat exchangers, including automobile radiators and domestic heating systems, are prime examples. Alloys used for such applications require not only good corrosion resistance and reasonable thermal conductivity but also the capabilities of being formed into a variety of shapes and joined easily by soldering or brazing. C110 and C122 coppers are therefore widely used in such applications.

Electrical interconnections that range from home wall receptacles to miniature connectors used in electronic products all require good formability and other properties specific to their application. Automotive connectors are frequently situated in the harsh environment of the engine compartment, where resistance to stress relaxation is needed to provide stable contact force for reliable performance. Alloys like the beryllium coppers and C7025 are supplanting C260 brass and the phosphor bronzes in such elevated temperature applications.

Strength, thermal conductivity, and formability account for the selection of particular alloys for use as lead frames in plastic-encapsulated devices. Leads having high strength are needed to ensure against damage to the electronic package in the handling before and during assembly to circuit boards.

Other applications that take advantage of the high electrical and thermal conductivity of copper and its alloys include superconducting wires and RF/EMI shielding. Controlled electrical resistance is required for applications such as resistance heating, as in heated car seats.

An additional virtue of many copper alloys is their capacity for deep drawing, such as in cartridge cases and flexible bellows. C260 and C510 are well known for their drawability. Be–Cu that is drawn and then aged is used for bellows that have excellent fatigue life.

Finally, coinage and architectural uses take advantage of the variety of distinctive colors inherent to copper and its alloys, from red, through gold-yellow and silver-white, as well as tarnish and corrosion resistance. Among the alloys used are C713, a white-appearing alloy used in U.S. coinage; gold-appearing aluminum–bronze (C6155) and brass (C260); and red-brown copper–tin bronze.

The ability to be plated or coated with a wide range of materials, including tin, solder, chromium, nickel, gold, and palladium, is an important attribute of copper alloys which contributes to their selection for many applications. Automotive connectors may be tin or solder coated, electronics connectors nickel and gold plated, with palladium.

Annual Book of ASTM Standards, Vol. 02. 01, *Copper and Copper Alloys*, American Society for Testing and Materials, Philadelphia, Pa., 1992.

Copper and Copper-Base Powder Alloys, Metal Powder Industries Federation, Princeton, N.J., 1976.

A. Fox and E. O. Fuchs, *J. Testing Eval.* **6**, 211 (1978).

Source Book on Copper and Copper Alloys, ASM International, Materials Park, Ohio, 1979.

RONALD N. CARON
PETER W. ROBINSON
Olin Corporation

COPPER COMPOUNDS

Copper compounds, which represent only a small percentage of all copper production, play key roles in both industry and the biosphere. Copper, mol wt $= 63.546$, $[Ar]3d^{10}4s^1$, is a member of the first transition series and much of its chemistry is associated with the copper(II) ion, $[Ar]3d^9$. Copper forms compounds of commercial interest in the $+1$ and $+2$ oxidation states.

PROPERTIES AND MANUFACTURE OF COMMERCIALLY IMPORTANT COMPOUNDS

Copper(II) Carbonate Hydroxide

Basic copper carbonate, also named copper(II) carbonate hydroxide, occurs in nature as the green monoclinic mineral malachite. The approximate stoichiometry is $CuCO_3 \cdot Cu(OH)_2$. There are two grades available commercially, the light and the dense. The light grade is produced by adding a copper salt solution to a concentrated solution of sodium carbonate, usually at 45–65°C. The blue, voluminous azurite, $C_2H_2Cu_3O_8$, forms initially and converts to the green malachite within two hours. The dense product can be produced by boiling an ammoniacal solution to copper(II) carbonate or by addition of a copper salt solution to sodium bicarbonate at 45–65°C. A dense product can also be produced by simultaneous addition of copper(II) salt solutions and soda ash solutions at controlled pH. Pure $CuCO_3$ has not been isolated.

Basic copper carbonate is essentially insoluble in water but dissolves in aqueous ammonia or alkali metal cyanide solutions. It dissolves readily in mineral acids and warm acetic acid to form the corresponding salt solution.

Copper Chloride

Copper(I) chloride, CuCl, is a colorless or gray cubic crystal that occurs in nature as the mineral nantokite. The commercial product is white to gray to brown to green and of variable purity. Copper(I) chloride is usually produced at 450–900°C by direct combination of copper metal and chloride gas to yield a molten product. The molten product is variously cast, prilled, flaked, or ground, depending on final use.

Copper(I) chloride is insoluble to slightly soluble in water. Solubility values between 0.001 and 0.1 g/L have been reported. CuCl is insoluble in dilute sulfuric and nitric acids but forms solutions of complex compounds with hydrochloric acid, ammonia, and alkali halide. Copper(I) chloride is fairly stable in air at relative humidities of less than 50% but quickly decomposes in the presence of air and moisture.

Cupric chloride or copper(II) chloride, $CuCl_2$, is usually prepared by dehydration of the dihydrate at 120°C. The anhydrous product is a deliquescent, monoclinic yellow crystal that forms the blue-green orthorhombic, bipyramidal dihydrate in moist air. Both products are available commercially.

Copper(II) oxychloride, $Cu_2Cl(OH)_3$, is found in nature as the green hexagonal paratacamite or rhombic atacamite. It is usually precipitated by air oxidation of a concentrated sodium chloride solution of copper(I) chloride.

Copper(II) Fluorides

Copper(II) forms several stable fluorides; eg, cupric fluoride, CuF_2, copper(II) fluoride dihydrate, $CuF_2 \cdot 2\,H_2O$, and copper hydroxyfluoride, $CuOHF$, all of which are interconvertible.

Copper Hydroxide

Copper(II) hydroxide, $Cu(OH)_2$, produced by reaction of a copper salt solution and sodium hydroxide, is a blue, gelatinous, voluminous precipitate of limited stability. Usually ammonia or phosphates are incorporated into the hydroxide to produce a color-stable product.

Copper hydroxide is almost insoluble in water (3 μg/L) but readily dissolves in mineral acids and ammonia forming salt solutions or copper amine complexes.

Copper Nitrates

The most common commercial forms of copper nitrate are the trihydrate and solutions containing about 14% copper. Copper nitrate can be prepared by dissolution of the carbonate, hydroxide, or oxides in nitric acid. The trihydrate is very soluble in water and ethanol.

Copper Oxides

Copper(I) oxide is a cubic or octahedral naturally occurring mineral known as cuprite. It is red or reddish brown in color. Commercially prepared copper(I) oxides vary in color from yellow to orange to red to purple as particle size increases. Usually copper(I) oxide is prepared by pyrometallurgical methods.

Copper(I) oxide is stable in dry air but reacts with oxygen to form copper(II) oxide in moist air. Cu_2O is insoluble in water but dissolves in ammonia or hydrochloric acid. The product disproportionates to copper metal and copper(II) in dilute sulfuric or nitric acid.

Copper(II) oxide, CuO, is found in nature as the black triclinic tenorite or the cubic or tetrahedral paramelaconite. The black product of commerce is most often prepared by evaporation of $Cu(NH_3)_4CO_3$ solutions or by precipitation of copper(II) oxide from hot ammonia solutions by addition of sodium hydroxide.

Copper(II) oxide is insoluble in water but readily dissolves in mineral acid or in hot formic or acetic acids. CuO slowly dissolves in ammonia solution, but alkaline ammonium carbonate solubilizes it quickly.

Copper(II) Sulfates

Copper(II) sulfate pentahydrate, $CuSO_4 \cdot 5\,H_2O$, occurs in nature as the blue triclinic crystalline mineral chalcanthite. It is the most common commercial compound of copper.

Copper(II) sulfate can be prepared by dissolution of oxides, carbonates, or hydroxides in sulfuric acid solutions.

Copper(II) sulfate monohydrate, $CuSO_4 \cdot H_2O$, which is almost white in color, is hygroscopic and packaging must contain moisture barriers. This product is produced by dehydration of the pentahydrate at 120–150°C.

Anhydrous copper(II) sulfate is a gray to white rhombic crystal that occurs in nature as the mineral hydrocyanite. $CuSO_4$ is hygroscopic. It is produced by careful dehydration of the pentahydrate at 250°C.

The basic copper(II) sulfate that is available commercially is known as the tribasic copper sulfate, $CuSO_4 \cdot 3\,Cu(OH)_2$, which occurs as the green monoclinic mineral brochantite. This material is essentially insoluble in water but dissolves readily in cold dilute mineral acids, warm acetic acid, and ammonia solutions. Tribasic copper sulfate is usually prepared by reaction of sodium carbonate and copper sulfate.

ECONOMIC ASPECTS

Copper sulfate, the most important compound of copper, is the starting material for the production of many other copper salts. However, 75% of production is used in agriculture, principally as a fungicide.

Today there are more than 100 manufacturers and the world's consumption is 200×10^3 t/yr.

HEALTH AND SAFETY FACTORS

Copper is one of the twenty-seven elements known to be essential to humans. The daily recommended requirement for humans is 2.5–5.0 mg. Copper is probably second only to iron as an oxidation catalyst and oxygen carrier in humans. It is present in many proteins, such as hemocyanin, galactose oxidase, ceruloplasmin, dopamine β-hydroxylase, monoamine oxidase, superoxide dismutase, and phenolase. Copper aids in photosynthesis and other oxidative processes in plants.

Copper is toxic in exceedingly low concentrations to most fungi, algae, and certain bacteria and can be lethal to higher life forms in relatively high doses. The acute oral toxicity in humans, LD_{LO}, is 100 mg/kg, but recovery from ingestion of 600 mg/kg has occurred. The symptoms of copper poisoning are nausea, vomiting, cramps, gastric disturbances, apathy, anemia, convulsions, coma, and death.

Inhalation of dusts can cause metal fume fever, and ulceration or perforation of the nasal septum. The workplace standard ACGIH (TLV) for copper dusts or mist is $1\,mg/m^3$ and $0.2\,mg/m^3$ for copper fume. OSHA PEL TWA for dust, mist is $1\,mg/m^3$ and for fumes and respirable particles is $0.1\,mg/m^3$.

USES

The majority of copper compounds are used as fungicides, nutritionals, and algicides. Of the ~70,000 t/yr as copper in compounds used in agriculture, almost 75% is used in the control of fungi. Copper is an effective broad-spectrum fungicide, although its action is more prophylactic in nature.

Copper is one of seven micronutrients identified as essential to the proper growth of plants.

Copper compounds are also used as feed additives in Europe and the United States, primarily for chickens and swine, to increase the rate of gain and feed efficiencies of the animals.

Copper sulfate provides an effective and ecologically sound method to control algal blooms. Other copper-containing algicides for use in domestic applications, such as swimming pools, are usually chelated to prevent hydrolysis and precipitation of the copper.

Copper is also one of the primary ingredients used in wood preservation. In combination with chromium and arsenic or zinc or borates it has largely replaced penta-chlorophenol and creosote. As an antifouling pigment, copper(I) oxide has found renewed use because of regulatory problems of organotin compounds. Copper is much more amenable to the environment than the tin or lead alternatives. Copper phthalocyanines are excellent color-stable blue and green pigments for paints. As an oxidation catalyst, copper with chrome is used extensively. Mixed oxides of chromium and copper show promise as a replacement for platinum group metals in emission control devices. Copper used with zinc or with zinc and chromium is employed as a low temperature shift catalyst for the synthesis of fuels and methanol from coal gases. Electroplating uses copper sulfate baths extensively, and electroless copper baths are used to produce circuit boards. Copper oxide is used in air-bag technology, and oxides and sulfides are used as solar collectors. Copper compounds are also used as fuel additives to minimize sulfide and carbon monoxide emissions.

R. J. Lewis, Sr., *Sax's Dangerous Properties of Industrial Materials*, 10th ed., Vol. 2, John Wiley & Sons, Inc., New York, 2000.

J. O. Nriagu, ed., *Copper in the Environment*, Part 1, John Wiley & Sons, Inc., New York, 1979.

C. P. Poole, Jr., T. Datta, and H. A. Farach, with M. M. Rigney and C. R. Sanders, *Copper Oxide Superconductors*, John Wiley & Sons, Inc., New York, 1988.

"Uses of Copper Compounds," Copper Development Asssociation, Inc., www.copper. org/compounds, accessed May 2003.

H. WAYNE RICHARDSON
CP Chemicals, Inc.

DAYAL T. MESHRI
Advance Research Chemicals, Inc.

COPYRIGHTS

Copyright has been grouped with other forms of legal protection under the general term "intellectual property." It is a means of protecting that particular form of creativity that has been variously referred to as originality of authorship, expression of ideas, or writings of an author. It is distinct from other forms of intellectual property that do not protect original expression of authorship; eg, patents, which protect novel inventions or discoveries; trademarks, which protect terms and symbols identifying the source or origin of goods and services; and trade secrets, which protect confidential, proprietary information.

COPYRIGHTABILITY

Under United States law, a work is either protected (copyrighted), or unprotected and free for all to use (in the public domain).

The Copyright Act specifies that copyright extends to "original works of authorship fixed in any tangible medium of expression, now known or later developed, from which they can be perceived, reproduced or otherwise communicated, either directly or with the aid of a machine or device." Many of the requirements for copyrightability may be gleaned from this provision.

The work must be an "original work of authorship." Thus, unlike patent rights, originality, and not novelty, constitutes the touchstone of protection. The work must also be the product of an author. Finally, the work must be fixed in a tangible medium of expression. To a very limited extent, there are some works that are not so fixed, such as purely improvised and unrecorded pieces of music or choreography, extemporaneous speeches, or live, unrecorded and ephemeral broadcasts. Unfixed works are protected, but by state common law copyright, not the federal statute.

The Subject Matter of Copyright

The copyright law specifies, by way of example, the types of works that are covered: literary works, musical works (including lyrics), dramatic works (including accompanying music), pantomimes and choreographic works; pictorial, graphic, and sculptural works; motion pictures and other audiovisual works; sound recordings; and architectural works. This list is nonexhaustive. It matters not whether the work is "good" or "bad" art, or even obscene. Such considerations are not relevant. The law also specifies that works of the United States government are not subject to copyright protection, which is to say that works created by government employees are common property and so are not copyrightable, although the government may own copyrights.

The Idea/Expression Dichotomy

Copyright protects the expression of ideas, but not ideas themselves. Thus, copyright will not protect any procedure, process, system, method of operation, concept, principle, or discovery, regardless of the form in which it is described, explained, illustrated, or embodied. Thus, copyright does not extend to titles, phrases, or it forms, nor does extend to facts or news—only to the particular form of expression of those facts. Similarly, copyright does not protect research: It protects creativity, not effort. Where the idea and its expression are not distinguishable, as in simple sweepstakes rules, copyright will not protect the work, for that would protect not only the expression but the idea itself.

Utilitarian Works

Utilitarian works may be copyrightable, but only to the extent that they contain copyrightable subject matter. The copyrightable subject matter must be physically or conceptually separable from the purely utilitarian object.

Compilations

Copyright extends not only to works that can exist on their owns but also to compilations of such works or even of public domain material. The Copyright law imposes a three-step test for such copyrightable compilations. They must first constitute the collection and assembling of preexisting data or materials. Second, those materials must be selected, coordinated, or arranged in a particular fashion. And, third, that selection, coordination, or arrangement must itself possess sufficient originality and creativity to constitute an original work of authorship.

COPYRIGHT OWNERSHIP

Copyright is a property right. Although it differs from most other forms of property in that it is intangible, it nevertheless has the essential elements of property and is governed by the principles of property ownership.

The intangible nature of copyright requires a distinction between the intangible property of the copyright (called a "work") and the material object in which the copyrighted work is embodied (termed a "copy" or "phonorecord," terms that include such diverse media as paper-and-ink, computer disks, and audiotapes). Ownership of the copyrighted work does not constitute ownership of the material object in which it is embodied, and vice versa. Copyright ownership vests initially in the author or authors of the work.

Joint Authorship

When more than one author has created a work, the work is said to be a joint work. Under the law, such a joint work is one prepared by two or more authors with the intention that their contributions be merged into inseparable or interdependent parts of a unitary whole.

Joint Ownership

Joint ownership of copyright occurs when there is joint authorship, but it may also occur in other ways—eg, by transfer of a copyright to two or more individuals, such as when an author bequeaths his/her copyright to two children.

Works Made for Hire

In many instances a person has created a work at the behest of another. In such circumstances, the creator does not own the copyright. As an easy example, consider a company that manufactures an appliance and has one of its employees write an instruction manual for the appliance. Logically, the company, and not the employee, should own the copyright.

Such situations are governed by the work-made-for-hire doctrine of the copyright law. Under this law, copyright ownership vests initially in the author of a work. In cases of works made for hire, the law specifies that the employer or other person for whom the work is prepared is deemed to be the author. Thus, the appliance company would be deemed to be the author, and hence the initial copyright owner, of the copyrighted instruction manual.

TRANSFERS AND LICENSES OF COPYRIGHT

Like other forms of property, copyright may be freely transferred. There are, however, certain special rules for the transfer of copyrights, and certain aspects of the law concerning transfer of property are of special importance to copyright.

COPYRIGHT FORMALITIES

Changes to the copyright law, starting with the 1976 Copyright Act and continuing with the Berne Convention Implementation Act of 1988, have radically changed United States copyright law regarding copyright formalities, many formalities that previously were of paramount importance have now been eased or entirely eliminated.

Copyright Notice

In the past, the law contained an absolute requirement that each copy of a published work bear a proper copyright notice. An amendment to the law abolished the notice requirement for all works first published on or after March 1, 1989. For such works, no copyright notice is required. Notice, however, is still required on all copies of works publicly distributed before that date. And notice still widely used even when it is not required, so as to inform the world of the copyright status of the work.

Copyright Deposit

The law requires that copies of every published work be submitted to the United States Copyright Office, which is a branch of the Library of Congress.

Copyright Registration

The term "copyrighting a work" is usually inexactly used. Although it refers to registering the work with the Copyright Office, copyright registration is not required for copyright protection. Federal copyright protection exists from the moment a work is created, that is, fixed in a tangible medium of expression, even if it is never registered. Although copyright protection is not dependent on registration, registration does have important advantages in litigation.

Copyright Duration

Two different regimes of copyright duration apply in the United States: one for works first created, published, or registered for copyright on or after January 1, 1978 (new law works) and one for works published or registered before that date (old law works). In all cases, copyright terms run through December 31 of their anniversary year. A 20-year extension of all existing copyrights became effective October 27, 1998.

New Law Works. For new law works, the basic copyright term is the life of the author and 70 years after the author's death. In the case of joint authors, the life in question is that of the longest surviving joint author.

For works where the duration of the author's life is not known—anonymous and pseudonymous works, and works made for hire—the term is 95 years from publication or 120 years from creation, whichever expires first.

Old Law Works. Protection for pre-1978 registered or published works endures under a system of dual terms. There is an initial term of 28 years from the earlier of publication or registration, followed by a renewal term of an additional 67 years, for a total of 95 years of protection. For works first published or copyrighted before 1964, renewal required registration in the Copyright Office in the last year of the initial term. If renewal was not made, the work fell into the public domain. For works first published or copyrighted from 1964 to 1977, renewal is automatic, but, in the last year of that initial term an application for renewal of copyright may be filed in the Copyright Office, which will provide certain benefits to the renewal claimant.

COPYRIGHT RIGHTS

The Copyright Act grants copyright owners six exclusive rights. These rights include not only the right to do the specified actions but also to authorize them: the right to reproduce in copies, the right to prepare derivative works, the right of public distribution, the right of public performance, the right of public display, and the performance right in sound recordings.

MORAL RIGHTS

In addition to copyright rights, the copyright law was amended effective June 1, 1991, to grant very limited additional rights to authors of certain types of works, even if they have parted with copyright ownership. These "moral rights" are applicable only to works of visual art that exist in single copies or multiples of up to 200. Even within this limited category of works, there are many exceptions—eg, moral rights do not apply to works made for hire. The moral rights are those of attribution (the right to have the author's name attached to or deleted from the work) and integrity (the right to prevent mutilation or distortion of the work which would prejudice the author's honor or reputation).

LIMITATIONS AND EXEMPTIONS

Fair Use

Certain uses of copyrighted works that would otherwise be infringements are excused from liability because they are "fair." The law gives examples in uses for purposes such as criticism, comment, news reporting, teaching, scholarship, and research.

First Sale Doctrine

Although the copyright owner has the exclusive right to distribute copies to the public, the bona fide possessor of a particular copy may, in most circumstances, further dispose of that copy without the copyright owner's consent. Thus, eg, the purchaser of a book may freely resell the copy she purchased (hence, used book stores do not violate copyright law). The Copyright Act has been amended to prohibit the rental of sound recordings or computer software, even though that rental would have been permitted by the first sale doctrine.

INFRINGEMENT

Anyone who violates the exclusive rights of a copyright owner is liable for infringement, in a lawsuit brought in federal court. There is a three-year statute of limitations on copyright infringement actions.

The Test for Infringement

It is rare that actual evidence of copying exists. Proof of copying is usually circumstantial and is shown by a two-part test. First, the alleged infringer must be shown to have had access to the copyrighted work. Second, the two works must appear to their hypothetical intended audience to be substantially similar.

Remedies

A copyright owner successfully proving infringement has three types of remedies available: recovery of monetary damages, injunctive relief, and recovery of costs, including attorneys' fees.

INTERNATIONAL COPYRIGHT

Because copyright easily transcends national boundaries, several international copyright conventions have been developed to protect copyrights internationally. The best known and most widely effective conventions are the Berne Convention for the Protection of Literacy and Artistic Works and the Universal Copyright Convention; the United States is a signatory to both. More recently, two treaties dealing with the use of copyrighted works in the digital environment have been created, the WIPO Copyright Treaty and the WIPO Performances and Phonograms Treaty. In varying degrees, the treaties specify minimum standards which each member country's copyright law must meet. Even with adoption of those minimum standards, national copyright laws vary significantly from country to country.

International copyright treaties generally follow the principle of national treatment. Each member country treats nationals of other countries at least as well as it treats its own nationals.

U.S. Const., Art. I, Sec. 8, Cl. 8.

I. Fred Koenigsberg
White & Case LLP

CORROSION AND CORROSION CONTROL

Corrosion is the natural degradation of materials in the environment through electrochemical or chemical reaction. Traditionally, the definition of corrosion refers to the degradation of metals and has not included the degradation of nonmetals such as wood (rotting) or plastics (swelling or crazing), but increasingly, natural degradation of any engineering material is being regarded as corrosion. The vast majority of the technologically significant corrosion involves the deterioration of metallic materials. Only the corrosion of metallic materials is discussed here.

ELECTROCHEMICAL NATURE OF CORROSION

Ores are mined and then refined in an energy-intensive process to produce pure metals, which in turn are combined to make alloys. Corrosion occurs because of the tendency of these refined materials to return to a more thermodynamically stable state. The key reaction in corrosion is the oxidation or anodic dissolution of the metal to produce metal ions and electrons: $M \xrightarrow{k_1} M^{n+} + n\,e^-$. The ions, M^{n+}, formed by this reaction at a rate, k_1, may be carried into a bulk solution in contact with the metal or may form insoluble salts or oxides. In order for this anodic reaction to proceed, a second reaction that uses the electrons produced, ie, a reduction reaction, must take place. This second, cathodic, reaction occurs at the same rate because the electrons produced by the anodic reaction must be consumed by the cathodic reaction to maintain electroneutrality. Therefore, $I_c = I_a$, where I_c and I_a are the cathodic and anodic currents, respectively. In most cases, the cathodic reaction, is hydrogen evolution or oxygen reduction.

The four elements necessary for corrosion are an aggressive environment, an anodic and a cathodic reaction, and an electron-conducting path between the anode and the cathode. Other factors such as mechanical stress also play a role. The thermodynamic and kinetic aspects of corrosion determine, respectively, if corrosion can occur and the rate at which it does so.

MANIFESTATIONS OF CORROSION

The most common form of corrosion is uniform corrosion, in which the entire metal surface degrades at a near uniform rate. Often the surface is covered by the corrosion products. The rusting of iron in a humid atmosphere or the tarnishing of copper or silver alloys in sulfur-containing environments are examples. High temperature, or dry, oxidation is also usually uniform in character. Uniform corrosion, the most visible form of corrosion, is the least insidious because the weight lost by metal dissolution can be monitored and predicted.

An especially insidious type of corrosion is localized corrosion, which occurs at distinct sites on the surface of a metal, while the remainder of the metal is either not attacked or is attacked much more slowly. Localized corrosion is usually seen on metals that are passivated, ie, protected from corrosion by oxide films, and occurs as a result of the breakdown of the oxide film. Generally the oxide film breakdown requires the presence of an aggressive anion, the most common of which is chloride. Localized corrosion can cause considerable damage to a metal structure without the metal's exhibiting any appreciable loss in weight. Localized corrosion occurs on a number of technologically important materials such as stainless steels, nickel-base alloys, aluminum, titanium, and copper.

Two types of localized corrosion are pitting and crevice corrosion. Pitting corrosion occurs on exposed metal surfaces, whereas crevice corrosion occurs within occluded areas on the surfaces of metals, such as the areas under rivets or gaskets, or beneath silt or dirt deposits. Crevice corrosion is usually associated with stagnant conditions within the crevices. A common example of pitting corrosion is evident on household storm window frames made from aluminum alloys.

Another type of corrosion is dealloying, also called parting or selective leaching. Dealloying is the preferential removal of one of the alloying elements from an alloy, resulting in the enrichment of the other alloying element(s). Common examples are the loss of zinc from brasses (dezincification) and the loss of iron from cast irons (graphitization).

Corrosion may also appear in the form of intergranular attack; ie, preferential attack of the boundaries between the crystals (grains) in metals and alloys. Intergranular attack generally occurs because the grain boundary and the grain have different corrosion tendencies; ie, different potentials. Intergranular corrosion often leads to a loss in strength or ductility of the metal.

Corrosion also occurs as a result of the conjoint action of physical processes and chemical or electrochemical reactions. The specific manifestation of corrosion is determined by the physical processes involved. Environmentally induced cracking (EIC) is the failure of a metal in a corrosive environment and under a mechanical stress. The observed cracking and subsequent failure would not occur from either the mechanical stress or the corrosive environment alone. Specific chemical agents cause particular metals to undergo EIC, and mechanical failure occurs below the normal strength (yield stress) of the metal.

When a stress is cyclic rather than constant, the failure is termed corrosion fatigue. Fretting corrosion results from the relative motion of two bodies in contact, one or both being a metal. The motion is small, such as a vibration. Erosion corrosion results from the action of a high-velocity fluid impinging on a metal surface. Metals and alloys can also experience cracking in liquid metal environments. This form of corrosion is referred to as liquid metal cracking (LMC).

Galvanic corrosion occurs as a result of the electrical contact of different metals in an aggressive environment. The driving force is the electrode potential difference between the two metals. One metal acts principally as a cathode, the other as the anode. Galvanic corrosion can result from the presence of a second phase in a metal.

Microbiologically influenced corrosion results from the interaction of microorganisms and a metal. The action of microorganisms is at least one of the reasons why natural seawater is more corrosive than either artificial seawater or sodium chloride solutions. Microorganisms attach to the surfaces of metals and can, for example, act as diffusion barriers; produce metabolites that enhance or initiate corrosion; act as sinks or sources for species involved in cathodic reactions, such as oxygen and hydrogen; increase the pH at the surface as a result of photosynthesis; or decrease the pH by production of acid metabolites.

ELECTROCHEMICAL EQUILIBRIUM DIAGRAMS

The thermodynamic data pertinent to the corrosion of metals in aqueous media have been systematically assembled in a form that has become known as Pourbaix diagrams. The data include the potential and pH dependence of metal, metal oxide, and metal hydroxide reactions and, in some cases, complex ions. The potential and pH dependence of the hydrogen and oxygen reactions are also supplied, because these are the common corrosion cathodic reactions.

KINETICS OF ELECTROCHEMICAL REACTIONS

Even in uniform corrosion, a corroding metal surface has numerous local anodes and cathodes. The sites of these local reactions may be fixed by microstructural features or may change as corrosion proceeds. The oxidation reaction at anodic sites on the metal surface can be represented as in the equation 1 above under "Electrochemical Nature of Corrosion". A corresponding reduction reaction must be occurring at cathodic sites. The potentials of these two reactions would be moved toward each other, away from the respective equilibrium potentials, and the metal surface would assume an overall uniform potential.

Galvanic corrosion can be used to a corrosion advantage. If, eg, a metal such as zinc or magnesium, having a low corrosion potential in most environments, is coupled with a steel component, the zinc or magnesium pulls the potential of the steel down, causing the steel to corrode less. When a sacrificial metal or alloy, called a sacrificial anode, is attached to a structure having a higher corrosion potential to intentionally pull the potential of the higher potential metal down and thus decrease the corrosion rate, it is called cathodic protection. This method of corrosion mitigation is common for underground pipelines, residential hot water heaters, and hulls and tanks of ships. The lowering of potential can also be achieved by the application of external electrical current; ie, impressed current cathodic protection.

The lower potential metal in a galvanic couple does not always have its corrosion rate accelerated. For metals that form a passive film, coupling with another metal of higher potential can cause the film-forming metal's potential to shift from a value at which it corrodes to one at which it passivates and therefore corrodes less. When this is done intentionally the procedure is referred to as anodic protection; ie, achieving protection by intentionally shifting the potential in the positive direction. Anodic protection is generally achieved by adding oxidizers to the electrolyte or by an external electrical circuit.

ENVIRONMENTAL EFFECTS

The environment plays several roles in corrosion. It acts to complete the electrical circuit; ie, it supplies the ionic conduction path; provides reactants for the cathodic process; removes soluble reaction products from the metal surface; and/or destabilizes or breaks down protective reaction products such as oxide films that are formed on the metal. Some important environmental factors include the oxygen concentration; the pH of the electrolyte; the temperature; and the concentration of anions.

Reduction of oxygen is one of the predominant cathodic reactions contributing to corrosion.

Very often the environment is reflected in the composition of corrosion products; eg, the composition of the green patina formed on copper roofs over a period of years. The determination of the chemical composition of this green patina was one of the first systematic corrosion studies ever made. The composition varied considerably, depending on the location of the structure, as shown in Table 1.

Table 1. Composition of Green Patina on Copper from Different Locations

Location of structure	Age of structure, year	Composition of green patina, %			
		$CuCO_3$	$CuCl_2$	$Cu(OH)_2$	$CuSO_4$
urban	30	14.6		9.6	49.8
rural	300	1.4		58.5	25.6
marine	13	12.8	26.7	52.5	2.5
urban–marine	38		4.6	61.5	29.7

ALLOY COMPOSITION AND METALLURGICAL FACTORS

A primary factor in determining the corrosion behavior of metals and alloys is their chemical composition. A good example of how corrosion resistance can be successfully changed by altering the composition can be seen in the alloying of steels. The corrosion resistance of these alloys is based on the protective nature of the surface film, which in turn is based on the physical and chemical properties of the oxide film. There are, however, other metallurgical factors, such as crystallography, grain size and shape, grain heterogeneity, second phases, impurity inclusions, and residual stress, that influence corrosion. The technologically important structural materials are polycrystalline aggregates. Each individual crystal is referred to as a grain. Grain orientation can affect corrosion resistance, as evidenced by metallographic etching rates and pitting behavior. Grain shape may likewise vary greatly, depending on the alloy and processing history.

Stainless Steels

The most common and serious metallurgical factor affecting the corrosion resistance of stainless steels is termed sensitization. This condition is caused by the precipitation of chromium-rich carbides at the grain boundaries, giving rise to chromium depleted grain boundary areas. These areas are anodic to the grain interior and tend to dissolve, thereby causing intergranular corrosion. There are several measures available to mitigate sensitization. Low carbon grades such as AISI 304L and 316L are available that have much less tendency toward sensitization. Also, alloying additions of titanium or niobium and tantalum can be used to tie up the carbon.

Copper Alloys

Brasses are susceptible to dealloying in the form of dezincification; ie, the preferential loss of zinc from the alloy. Brasses having zinc concentrations of 15% or greater are prone to dezincification, and dezincification is generally more severe in brasses that have two metallurgical phases.

Conditions that favor dezincification include stagnant solutions, especially acidic ones, high temperatures, and porous scale formation. Additions of small amounts of arsenic, antimony, or phosphorus can increase the resistance to dezincification.

Aluminum Alloys

Copper, silicon, magnesium, zinc, and manganese are some common alloying additions to aluminum. Most of the alloying elements are added to aluminum to produce alloys having improved mechanical properties. However, the strengthening phases that result from the alloying can disrupt the passive oxide layer on aluminum and lead to localized corrosion. Also, the second-phase constituents can produce local galvanic cells.

ENVIRONMENTALLY INDUCED CRACKING

Environmentally induced cracking (EIC) is a brittle fracture process caused by the conjoint action of a mechanical stress and a corrosive environment. There are several different types of EIC: stress–corrosion cracking (SCC), corrosion–fatigue cracking (CFC), and hydrogen-induced cracking (HIC).

The cracking in EIC can proceed either intergranularly; i.e., between the grains, or transgranularly; ie, across the grains.

Copper Alloys

Copper alloys under an applied or residual stress are susceptible to SCC in environments containing ammonia or ammonium compounds. Trace quantities of nitrogen oxides may also cause SCC. The nitrates settle as dust on the brass parts. The susceptibility to SCC can be minimized by (1) proper alloy selection; (2) thermal stress relief; (3) avoiding contact with ammonia and ammonia compounds; and (4) using an inhibitor.

Aluminum Alloys

Both the 2000 and the 7000 series aluminum alloys have experienced significant SCC in service. The alloys are most vulnerable to SCC if stressed parallel to the short transverse grain direction; ie, parallel to the thinnest dimension of the grain and most resistant if stressed only parallel to the longest grain dimension. In practice, it is the short transverse-direction stresses that cause SCC problems. Hence, prudent practice is to avoid designs in which high sustained stresses are imposed across the short transverse direction.

In addition to the possibility of selecting alloys having minimum SCC susceptibility while retaining other properties as needed, there are other steps possible to reduce the SCC probability: (1) avoid designs that permit water to accumulate; (2) avoid conditions in which salts, especially chlorides, can concentrate; and (3) where available and otherwise acceptable, use a clad alloy.

High-Strength Steels

Steels that owe their strength to heat treatment, whether martensitic, precipitation hardened, or maraging, and whether stainless or not, are susceptible to SCC in aqueous environments, including water vapor. The primary factor in determining the degree of SCC susceptibility of a given steel is its strength.

Stainless Steels

Austenitic stainless steels undergo SCC when stressed in hot aqueous environments containing chloride ion. The oxygen level of the environment can be important, probably through its effect in establishing the electrode potential of the steel. The standard methods for avoiding chloride SCC in austenitic stainless steels include avoidance of fabrication stresses; minimizing chloride ion level; and minimizing oxygen concentration in the environment.

INHIBITORS

Corrosion inhibitors are substances that slow or prevent corrosion when added to an environment in which a metal usually corrodes. Corrosion inhibitors are usually added to a system in small amounts, either continuously or intermittently. The effectiveness of corrosion inhibitors is partly dependent on the metals or alloys to be protected as well as the severity of the environment.

Inhibitors act and are classified in a variety of ways. Types of inhibitors include (1) anodic, (2) cathodic, (3) organic, (4) precipitation, and (5) vapor-phase inhibitors.

COATINGS FOR CORROSION PREVENTION

Coatings are applied to metal substrates to prevent corrosion. Generally, the coating protects the metal by imposing a physical barrier between the metal substrate and the environment. However, the coating can also act to provide cathodic protection or by serving as holding reservoirs for inhibitors. Coatings may be divided into organic, inorganic, and metallic coatings.

L. S. V. Delinder, *Corrosion Basics: An Introduction*, National Association of Corrosion Engineers, Houston, Tex., 1984.

N. E. Hamner in L. S. V. Delinder, ed., *Corrosion Basics: An Introduction*, NACE, Houston, Tex., 1984.

D. A. Jones, *Principles and Prevention of Corrosion*, Macmillan Publishing, New York, 1992.

S. R. Taylor, H. S. Isaacs, and E. W. Brooman, *Environmentally Acceptable Inhibitors and Coatings*, The Electrochemical Society, Inc., Pennington, 1997.

PAUL NATISHAN
Naval Research Laboratory
PATRICK MORAN
U.S. Naval Academy

COSMETICS

Cosmetics are products created by the cosmetic industry and marketed directly to consumers. The cosmetic industry is dominated by manufacturers of finished products but also includes manufacturers who sell products to distributors as well as suppliers of raw and packaging materials. Cosmetics represent a large group of consumer products designed to improve the health, cleanliness, and physical appearance of the human exterior and to protect a body part against damage from the environment. Cosmetics are promoted to the public and are available without prescription.

Cosmetics, regardless of form, can be grouped by product use into the following seven categories: (1) skin care and maintenance, including products that soften (emollients and lubricants), hydrate (moisturizers), tone (astringents), protect (sunscreens), etc, and repair (antichapping, antiwrinkling, antiacne agents); (2) cleansing, including soap, bath preparations, shampoos, and dentifrices (qv); (3) odor improvement by use of fragrance, deodorants, and antiperspirants; (4) hair removal, aided by shaving preparations, and depilatories; (5) hair care and maintenance, including waving, straightening, antidandruff, styling and setting, conditioning, and coloring products (see HAIR PREPARATIONS) (6) care and maintenance of mucous membranes by use of mouthwashes, intimate care products, and lip antichapping products; and (7) decorative cosmetics, used to beautify eyes, lips, skin, and nails.

REGULATION OF THE COSMETIC INDUSTRY

In the United States, the 1938 revision of the Federal Food and Drug Act regulates cosmetic products and identifies these materials as:

(1) articles intended to be rubbed, poured, sprinkled, or sprayed on, introduced into, or otherwise applied to the human body or any part thereof for cleansing, beautifying, promoting attractiveness, or altering the appearance, and (2) articles intended for use as a component of any such articles, except that such term shall not include soap.

This definition establishes the legal difference between a drug and a cosmetic. It is clearly the purpose of, or the claims for, the product, not necessarily its performance, that legally classifies it as a drug or a cosmetic in the United States. For example, a skin-care product intended to beautify by removing wrinkles may be viewed as a cosmetic because it alters the appearance and a drug because it affects a body structure.

The FDA is responsible for enforcing the 1939 act as well as the Fair Packaging and Labeling Act. In light of the difficulty of differentiating between cosmetics and drugs, the FDA has in recent years implemented its regulatory power by concluding that certain topically applied products should be identified as OTC drugs. As a group, these OTC drugs were originally considered cosmetics and remain among the products distributed by cosmetic companies. They include acne, antidandruff, antiperspirant, astringent, oral-care, skin-protectant, and sunscreen products.

The use or presence of poisonous or deleterious substances in cosmetics and drugs is prohibited. The presence of such materials makes the product "adulterated" or "misbranded" and in violation of good manufacturing practices (GMP), which are applicable to drugs and, with minor changes, to cosmetics.

In contrast to prescription drugs, OTC drugs and cosmetics are not subject to preclearance in the United States. However, the rules covering OTC drugs preclude introduction of untested drugs or new combinations. A "new chemical entity" that appears suitable for OTC drug use requires work-up via the new drug application

(NDA) process. In contrast, the use of ingredients in cosmetics is essentially unrestricted and may include less well known substances.

Color Additives

The FDA has created a unique classification and strict limitations on color additives (see also COLORANTS FOR FOODS, DRUGS, COSMETICS, AND MEDICAL DEVICES). Certified color additives are synthetic organic dyes that are described in an approved color additive petition. Each manufactured lot of a certified dye must be analyzed and certified by the FDA prior to usage.

Hair colorants, the fourth class of color additives, may be used only to color scalp hair and may not be used in the area of the eye. Use of these colorants is exempt, that is, coal-tar hair dyes may be sold with cautionary labeling, directions for preliminary (patch) testing, and restrictions against use in or near the eye.

Under the Fair Packaging and Labeling Act, the FDA has instituted regulations for identifying components of cosmetics on product labels. To avoid confusion, the CTFA has established standardized names for about 6000 cosmetic ingredients.

European Regulations

Regulations for cosmetics differ from country to country but, in general, are similar to or patterned after U.S. regulation.

Japanese Regulation

Japanese regulations of cosmetics are similar to those already discussed. The safety and quality of cosmetic products are regulated under the Pharmaceutical Affairs Law with detailed requirements for approval and licensing of manufacturing and import, for labeling and advertising standards, and for reporting safety data to the Ministry of Health and Welfare.

PRODUCT REQUIREMENTS

Safety

Cosmetic products must meet acceptable standards of safety during use, must be produced under sanitary conditions, and must exhibit stability during storage, shipment, and use.

For many years the safety of cosmetic ingredients has been established using a variety of animal safety tests. More recently, animal welfare organizations have urged that this type of safety testing be abandoned. Despite widespread use of cosmetics without professional supervision, the incidence of injury from cosmetic products is rare.

In vitro safety testing technology is becoming more common. Validation of these methods is based on comparisons with early animal safety data. Whether these *in vitro* tests can ensure the safety of all products that reach the consumer cannot be predicted. As of this writing, the principle that *in vitro* testing may be substituted for *in vivo* testing for complete safety substantiation has not been accepted by regulatory agencies. In the United States, the CTFA

created the Cosmetic Ingredient Review (CIR) for the purpose of evaluating existing *in vitro* and *in vivo* data and reviewing the safety of the ingredients used in cosmetics.

In addition to the CIR process the cosmetic industry has instituted a second, important self-regulatory procedure: the voluntary reporting of adverse reactions, which is intended to provide data on the type and incidence of adverse reactions noted by consumers or by their medical advisors.

Safety testing of a finished cosmetic product should be sufficient to ensure that the product does not cause irritation when used in accordance with directions, neither elicits sensitization nor includes a sensitizer, and does not cause photoallergic responses.

Production Facilities

The manufacture of acceptable cosmetic products requires not only safe ingredients but also facilities that maintain high standards of quality and cleanliness. Most countries have established regulations intended to assure that no substandard product or batch is distributed to consumers. Good Manufacturing Practices (GMP) represent workable standards that cover every aspect of drug manufacture, from building construction to distribution of finished products. GMPs in the United States that have been established for drug manufacture are commonly used in cosmetic production.

Contamination

Manufacturers of cosmetics must be careful to guard against chemical and microbial contamination. Compendial specifications and publications by the CTFA and other professional societies form the basis of most intracompany raw material specifications. Moreover, all packaging components must meet not only physical and design specifications but also such chemical requirements as extractables and absence of dust and similar contaminants.

Stability

An additional mandatory requirement for cosmetic products is chemical and physical stability. Interactions between ingredients that lead to new chemical entities or decomposition products are unacceptable. In the absence of an expiration date, a cosmetic product or an OTC drug should be stable for 60 months at ambient temperature.

Performance

Consumer acceptance is a criterion on which cosmetic marketers cannot compromise. Performance is tested by *in vitro* techniques during formulation, but the ultimate test of a product's performance requires in-use experience with consumers and critical assessment by trained observers.

INGREDIENTS

Manufacturers of cosmetics employ a surprisingly large number of raw materials. Some of these ingredients are active constituents that have purported beneficial effects on the skin, hair, or nails, for example, acting as moisturizers

Table 1. Cosmetic Functions and Representative Ingredients

Function	Ingredient[a]	Molecular formula
Biologically active agents		
antiacne	salicylic acid	$C_7H_6O_3$
anticaries	monosodium fluorophosphate	Na_2HPO_3F
antidandruff	zinc pyrithione	$C_{10}H_8N_2O_2S_2Zn$
antimicrobial	benzalkonium chloride	
antiperspirant	aluminum chlorohydrate	$Al_2ClH_5O_5$
biocides	triclosan	$C_{12}H_7Cl_3O_2$
sunscreen	octyl methoxycinnamate	$C_{18}H_{26}O_3$
skin protectant	dimethicone	$(C_2H_6OSi)_nC_4H_{12}Si(C_2H_6OSi)_n$
external analgesic	methyl salicylate	$C_8H_8O_3$
Nonbiologically active agents		
abrasive		
skin	oatmeal	
teeth	dicalcium phosphate	$Ca_2(HPO_4)_2$
antifoam	simethicone	
antioxidant	ascorbic acid	$C_6H_8O_6$
antistatic agent	dimethylditallow alkylammonium chlorides	
binder	hydroxypropylcellulose	
chelator	hydroxyethyl ethylenediamine triacetic acid (HEDTA)	$C_{10}H_{18}N_2O_7$
colorant		
pigment	ultramarine	$Na_7Al_6Si6O_{24}S_2$
dye	FD&C Red No. 4	$C_{18}H_{16}N_2O_7S_2\cdot2$ Na
emulsion stabilizer	xanthan gum	
film former	PVP	$(C_6H_9NO)_x$
hair colorant	p-phenylenediamine	$C_6H_8N_2$
hair conditioner	sodium lauroamphoacetate	$Na_2C_{18}H_{35}N_2O_3\cdot HO$
humectant	glycerol	$C_3H_8O_3$
deodorant		
mouth	zinc chloride	$ZnCl_2$
external	cetylpyridinium chloride	$C_{21}H_{38}ClN$
preservative	propylparaben	$C_{10}H_{12}O_3$
emollient	octyl stearate	$C_{26}H_{52}O_2$
skin-conditioning agent		
general	pyrrolidinone carboxylic acid (PCA)	$C_5H_7NO_3$
occlusive	petrolatum	C_nH_{2n+2}
film forming	hyaluronic acid	
solvent	ethanol	C_2H_6O
cleansing agent	sodium lauryl sulfate	$C_{12}H_{25}NaO_4S$
emulsifying agent	polysorbate 65	
foam booster	cocamide DEA	
suspending agent	sodium lignosulfonate	
hydrotrope	sodium toluenesulfonate	
viscosity-controlling agent		
decrease	propylene glycol	$C_3H_8O_2$
increase	hydroxypropylmethyl cellulose	

[a]CTFA adopted names are used; this notation is used for cosmetic labeling.

or conditioners. These substances are generally used in limited quantities. Other ingredients are used to formulate or create the vehicle. These are bulk chemicals used in comparatively large amounts. The resulting combination of various substances affects the nature (viscosity, oiliness, etc) of the finished cosmetic. As a rule numerous combinations and permutations are tested to optimize textural characteristics and to match these to consumers' preferences. Finally, cosmetics may include substances added primarily to appeal to consumers. These ingredients need not contribute appreciably to product performance.

About 6000 different cosmetic ingredients have been identified. These can be divided into smaller groups according to chemical similarity or functionality. Table 1 represents a breakdown by functionality on the skin or in the product.

Antioxidants

The operant mechanisms of antioxidants are interference with radical propagation reactions, reaction with oxygen, or reduction of active oxygen species. Antioxidants are intended to protect the product but not the skin against oxidative damage resulting from ultraviolet radiation or singlet oxygen formation.

Preservatives

Several microorganisms can survive and propagate on unpreserved cosmetic products. Preservatives are routinely added to all preparations that can support microbial growth.

Lipids

Natural and synthetic lipids are used in almost all cosmetic products. Lipids serve as emollients or occlusive agents, lubricants, binders for creating compressed powders, adhesives to hold makeup in place, and hardeners in such products as lipsticks. In addition, lipids are used as gloss-imparting agents in hair-care products. The primary requirements for lipids in cosmetics are absence of excessive greasiness and ease of spreading on skin.

Solvents

Solvents can be added to cosmetics to help dissolve components used in cosmetic preparations. Water is the most common solvent and is the continuous phase in most suspensions and water/oil (w/o) emulsions. Organic solvents are required in the preparation of colognes, hair fixatives, and nail lacquers. Selected solvents are used to remove soil, sebum, and makeup from skin.

Surfactants

Substances commonly classified as surfactants (qv) or surface active agents are required in a wide variety of cosmetics. Prolonged contact with anionic surfactants can cause some swelling of the skin. Although this is a temporary phenomenon, skin in this swollen condition allows permeation of externally applied substances. Nonionic surfactants as a group are generally believed to be mild even under exaggerated conditions. The more hydrophobic nonionics, those that are water dispersible (not water-soluble), can enhance transdermal passage. Amphoteric surfactants as a group exhibit a favorable safety profile. Finally, cationic surfactants are commonly rated as more irritating than the anionics, but the evidence for generalized conclusions is insufficient.

Colorants

Color (qv) is used in cosmetic products for several reasons: the addition of color to a product makes it more attractive and enhances consumer acceptance; tinting helps hide discoloration resulting from use of a particular ingredient or from age; and finally, decorative cosmetics owe their existence to color.

The importance of coal-tar colorants cannot be overemphasized. The cosmetic industry, in cooperation with the FDA, has spent a great deal of time and money in efforts to establish the safety of these dyes (see COLORANTS FOR FOOD, DRUGS, COSMETICS, AND MEDICAL DEVICES). Contamination, especially by heavy metals, and other impurities arising from the synthesis of permitted dyes are strictly controlled. Despite this effort, the number of usable organic dyes and of pigments derived from them has been drastically curtailed by regulatory action.

In addition to the U.S. certified coal-tar colorants, some noncertified naturally occurring plant and animal colorants, such as alkanet, annatto, carotene, $C_{40}H_{56}$, chlorophyll, cochineal, saffron, and henna, can be used in cosmetics. In the United States, however, natural food colors, such as beet extract or powder, turmeric, and saffron, are not allowed as cosmetic colorants.

The terms FD&C, D&C, and External D&C (Ext. D&C), which are part of the name of colorants, reflect the FDA's colorant certification. FD&C dyes may be used for foods, drugs, and cosmetics; D&C dyes are allowed in drugs and cosmetics; and Ext. D&C dyes are permitted only in topical products.

In addition to various white pigments, other inorganic colorants are used in a number of cosmetic products. These usually exhibit excellent lightfastness and are completely insoluble in solvents and water.

For many years nacreous pigments were limited to guanine (from fish scales) and bismuth oxychloride. Mica, gold, copper, and silver, in flake form, can also provide some interesting glossy effects in products and on the face. An entirely new set of colored, iridescent, inorganic pigments, which may be described as mixtures of mica and titanium dioxide (sometimes with iron oxides), has been created by coating mica flakes with titanium dioxide.

SKIN PREPARATION PRODUCTS

Products for use on the skin are designed to improve skin quality, to maintain (or restore) skin's youthful appearance, and to aid in alleviating the symptoms of minor diseases of the skin. Many of these products are subject to different regulations in different countries. Skin products are generally formulated for a specific consumer purpose.

Skin-Care Products

Preparations are generally classified by site of application and purpose. The smoothing or emollient properties of creams and lotions are critical for making emulsions the preferred vehicles for facial skin moisturizers, skin protectants, and rejuvenating products. On the body, emollients provide smoothness and tend to reduce the sensation of tightness commonly associated with dryness and loss of lipids from the skin. Although a wide variety of plant and animal extracts have been claimed to impart skin benefits, valid scientific evidence for efficacy has been provided only rarely.

Emulsion components enter the stratum corneum and other epidermal layers at different rates. Most of the water evaporates, and a residue of emulsifiers, lipids, and other nonvolatile constituents remains on the skin. Some of these materials and other product ingredients may permeate the skin; others remain on the surface. If the blend of nonvolatiles materially reduces the evaporative loss of water from the skin, known as the transepidermal water loss (TEWL), the film is identified as occlusive.

The ability to moisturize the stratum corneum has also been claimed for the presence of certain hydrophilic polymers, for example, guar hydroxypropyl trimonium chloride, on the skin. By far the most popular way to moisturize skin is with humectants. It is claimed that humectants attract water from the environment and thereby provide moisture to the skin.

Antiacne Preparations

Antiacne products are designed to alleviate the unsightly appearance and underlying cause of juvenile acne.

As of 1991 in the United States, OTC antiacne preparations may contain only a few active drugs, for example, sulfur, resorcinol acetate, resorcinol, salicylic acid, and some combinations. OTC antiacne constituents may be included in a variety of conventional cosmetic preparations, which then become OTC drugs.

Sunscreens

The use of uv light absorbing substances is accepted worldwide as a means of protecting skin and body against damage and trauma from uv radiation. These colorless organic substances are raised to higher energy levels upon absorption of uv light, but little is known about mechanisms for the disposal of this energy. These substances can be classified by the wavelengths at which absorbance is maximal. Absorption throughout the incident uv range (285 to about 400 nm) affords the best protection against erythema.

It is also possible to deflect uv radiation by physically blocking the radiation using an opaque makeup product. A low particle-size titanium dioxide can reflect uv light without the undesirable whitening effect on the skin that often results from products containing, for example, zinc oxide or regular grades of titanium dioxide.

A list of uv absorbing substances found useful in protective sunscreen products is provided in Table 2. Some information on the levels permitted in products in both the United States and the EEC is included. Descriptions and specifications of sunscreens have been published.

In principle, emulsified sunscreen products are similar to emollient skin-care products in which some of the emollient lipids are replaced by uv absorbers.

Facial Makeup

This classification applies to all products intended to impart a satinlike tinted finish to facial skin and includes liquid makeups, tinted loose or compressed powders, rouges, and blushers.

In modern liquid makeups and rouges, the required pigments are extended and ground in a blend of suitable cosmetic lipids. This magma is then emulsified, commonly as o/w, in a water base. Soaps, monostearates, conditioning lipids, and viscosity-increasing clays are primary components. Nonionic emulsifiers can replace part or all of the soap.

Tinted dry powders form the second type of facial makeup. Commonly, the blended solids are compressed into compacts.

Table 2. Cosmetic Uv Absorbers

Ingredient	Quantity approved, % U.S.	EEC
uvA Absorbers		
benzophenone-8	3	
menthyl anthranilate	3.5–5	
benzophenone-4	5–10	5
benzophenone-3	2–6	10
uvB Absorbers		
p-aminobenzoic acid (PABA)	5–15	5
pentyl dimethyl PABA	1–5	5
cinoxate	1–3	5
DEA p-methoxycinnamate	8–10	8
digalloyl trioleate	2–5	4
ethyl dihydroxypropyl PABA	1–5	5
octocrylene	7–10	
octyl methoxycinnamate	2–7.5	10
octyl salicylate	3–5	5
glyceryl PABA	2–3	5
homosalate	4–15	10
lawsone (0.25%) plus dihydroxyacetone (33%)		
octyl dimethyl PABA	1.4–8	8
2-phenylbenzimidazole-5-sulfonic acid	1–4	8
TEA salicylate	5–12	2
sulfomethyl benzylidene bornanone	10	
urocanic acid (and esters)		2
Physical barriers		
red petrolatum	30–100	
titanium dioxide	2–25	

Skin Coloring and Bleaching Preparations

Products designed to simulate a tan, to lighten skin color in general, or to decolorize small hyperpigmented areas such as age spots either impart to or remove color from the skin. Skin stains are intended to create the appearance of tanned skin without exposure to the sun. The most widely used ingredient is dihydroxyacetone (2–5% at pH 4 to 6) which reacts with protein amino groups in the stratum corneum to produce yellowish brown Maillard products. Lawsone and juglone are known to stain skin directly. Stimulation of melanin formation is another approach to artificial tanning. Commercialization, which is limited, depends primarily on topical application of products containing tyrosine or a tyrosine precursor.

The number of cosmetically acceptable bleaching ingredients is very small, and products for this purpose are considered drugs in the United States. The most popular ingredient is hydroquinone at 1–5%; the addition of uv light absorbers and antioxidants reportedly helps to reduce color recurrence.

ASTRINGENTS

Astringents are designed to dry the skin, denature skin proteins, and tighten or reduce the size of pore openings on the skin surface. These products can have antimicrobial effects and are frequently buffered to lower the pH of skin. They are perfumed, hydro-alcoholic solutions of weak acids, such as tannic acid or potassium alum, and various plant extracts, such as birch leaf extract. In the United States, some astringents, depending on product claims, are considered OTC drugs. Only three ingredients, aluminum acetate, aluminum sulfate, and hamamelis, are considered safe and effective.

Antiperspirants and Deodorants

There are many forms of antiper-spirants and deodorants: liquids, powders, creams, and sticks. Deodorants do not interfere with the delivery of eccrine or apocrine secretions to the skin surface but control odor by reodorization or antibacterial action. Deodorant products, regardless of form, are antimicrobial fragrance products. An important antimicrobial or cosmetic biocide used in many products is triclosan. Other active agents include zinc phenol-sulfonate, p-chloro-m-xylenol, and cetrimonium bromide. There have been claims that ion-exchange polymers and complexing agents provide protection against unpleasant body odors. In addition, delayed-release, that is, liposomal or encapsulated, substances of diverse activity have been employed.

The mechanism of antiperspirant action has not been fully established but probably is associated with blockage of ducts leading to the surface by protein denaturation by aluminum salts. The FDA has mandated that an antiperspirant product must reduce perspiration by at least 20% and has provided some guidelines for testing finished products.

CLEANSING PREPARATIONS

Cleansing preparations are products, based on surfactants or abrasives, that are designed to remove unwanted oil and debris from skin, hair, and the oral cavity. Soaps (qv) are the best-known cleansers but are not considered cosmetic products unless they are formulated with agents that prevent skin damage or contain antimicrobial agents.

Hair Cleansers

Except for a few specialty preparations, hair cleansers, or shampoos, are based on aqueous surfactants. The most popular surfactants in shampoos are alkyl sulfates and alkylether sulfates, commonly used at about 10–15% active. These ingredients by themselves do not provide the dense, copious foam desired by consumers, and additives are required, especially for use on oily hair or scalps. The foam boosters usually found in finished shampoos are fatty acid alkanolamides, fatty alcohols, and amine oxides.

Excessive degreasing by shampoos can be overcome by treatment with an after-shampoo (cream) rinse or a hair dressing. A wide variety of hair-conditioning additives

have been recommended and tested. Only a few have gained wide acceptance, for example: dialkyl (C_{12}–C_{18}) dimethylammonium chloride, hydrolyzed collagen polymeric quaternary derivatives, potassium cocoyl hydrolyzed collagen, sodium cocoamphoacetate, sodium lauroyl glutamate, and stearamidopropyl betaine.

Dandruff, a benign scaling skin disease of the scalp, is commonly viewed as a hair problem. The etiology and therapy of dandruff are similar to those of seborrheic dermatitis. Antidandruff shampoos are formulated using antimicrobial or desquamating agents to reduce the lipophilic yeasts (qv) widely believed to be the cause of scalp flaking. In the United States, shampoos for which antidandruff claims are made are OTC drugs. The choice of active agents is limited to coal tar, zinc pyrithione, salicylic acid, selenium sulfide, and sulfur, which can be added to shampoos or other scalp preparations.

Oral Cleansing Products

Toothpastes and mouthwashes are considered cosmetic oral cleansers as long as claims about them are restricted to cleaning or deodorization. Because deodorization may depend on reduction of microbiota in the mouth, several antimicrobial agents, either quaternaries, such as benzethonium chloride, or phenolics, such as triclosan, are permitted. Products that include anticaries or antigingivitis agents or claim to provide such treatment are considered drugs.

Mouthwashes are hydro-alcoholic preparations in which flavorants, essential oils (see OILS, ESSENTIAL), and other agents are combined to provide long-term breath deodorization. Palatability can be improved by including a polyhydric alcohol such as glycerin or sorbitol (see ALCOHOLS, POLYHYDRIC).

Dentifrices (qv), or toothpastes, depend on abrasives to clean and polish teeth. The principal ingredients in toothpastes or powders are 20–50% polishing agents, such as calcium carbonate, di-or tricalcium phosphate, insoluble sodium metaphosphate, silica, and alumina; 0.5–1.0% detergents, for example, soap or anionic surface-active agents; 0.3–10% binders (gums); 20–60% humectants, such as glycerol, propylene glycol, and sorbitol; sweeteners (saccharin, sorbitol); preservatives, such as benzoic acid or p-hydroxybenzoates; flavors, for example, essential oils; and water. The most widely used surfactant is sodium lauryl sulfate, which is available with high purity. It produces the desired foam during brushing, acts as a cleansing agent, and has some bactericidal activity.

SHAVING PRODUCTS

Cosmetic shaving products are preparations for use before, during or after shaving.

Preshaves

Preshave products are used primarily for dry (electric) shaving. Solid preshaves are usually compressed-powder sticks based on lubricating solids, such as talc or zinc or glyceryl stearate. Liquid preshaves are intended to remove perspiration residues and tighten and lubricate the skin. The alcohol content is relatively high (50–80%) to accelerate drying. The remaining ingredients may be polymeric lubricants, such as 1–2% polyvinylpyrrolidinone (PVP), emollients, such as 1–5% diisopropyl adipate, and up to about 5% propylene glycol.

Shaving Creams

Shaving creams and soaps are available as solids, that is, bars; creams, generally in tubes; or aerosols.

The principal ingredients of shaving creams and aerosols are liquid soaps, usually a blend of potassium, amine, and sodium salts of fatty acids, formulated to create a foam with the desired consistency and rinsing qualities. The soap blend may include synthetic surfactants, skin-conditioning agents, and other components.

The objectives of shaving creams include protecting the face from cuts by cushioning the razor. Beard-softening qualities are attributable almost exclusively to hair hydration, which also depends on pH.

After-Shaves

After-shave preparations serve the same function as and are formulated similarly to skin astringents.

NAIL-CARE PRODUCTS

Over the years the cosmetic industry has created a wide variety of products for nail care. Some of these, such as cuticle removers and nail hardeners, are functional; others, such as nail lacquers, lacquer removers, and nail elongators, are decorative.

Functional Nail-Care Products

Cuticle removers are solutions of dilute alkalies that facilitate removal, or at least softening, of the cuticle.

Nail hardeners have been based on various protein cross-linking agents. Only formaldehyde is widely used commercially. Contact with skin and inhalation must be avoided to preclude sensitization and other adverse reactions.

Decorative Nail-Care Products

Nail lacquers, or nail polishes, consist of resin, plasticizer, pigments, and solvents. The most commonly used resin is nitrocellulose. Ethyl acetate, butyl acetate, and toluene are typical solvents. Toluenesulfonamide–formaldehyde resin and similar polymers are the resins of choice as secondary film formers for optimal nail adhesion.

Camphor, dibutyl phthalate, and other lipidic solvents are common plasticizers. Nail lacquers require the presence of a suspending agent because pigments have a tendency to settle. Most tinted lacquers contain a suitable flocculating agent, such as stearalkonium hectorite.

The blend of pigments used to create a particular shade must conform to regulations covering pigments and dyes in cosmetics.

Nail lacquer removers are simply acetone or blends of solvents similar to those used in nail lacquers. It is commonly accepted that solvents have a drying effect on nails, and nail lacquer removers are often fortified with various lipids such as castor oil or cetyl palmitate.

Nail elongators are products intended to lengthen nails. These have become extremely popular. Nail elongation is achieved by adhering a piece of nonwoven nylon fabric (referred to as nail wrap) to the nail with a colorless lacquer. This process may be repeated until the desired nail thickness has been reached. After shaping, the artificial nail is further decorated.

HAIR PRODUCTS

Hair Conditioners

Hair conditioners are designed to repair chemical and environmental damage, replace natural lipids removed by shampooing, and facilitate managing and styling hair. The classical hair-conditioning products were based on lipids, which were deposited on hair either directly, with oils or pomades, or from emulsions.

An entirely different, and in the 1990s more popular, type of hair conditioning is achieved by treating hair with substantive quaternary compounds or polymers. Quaternaries are sorbed by hair, retained despite rinsing with water, and removed only by shampooing. The most widely used quaternary is stearalkonium chloride, which has been used at 3–5% concentration in cream rinses for many years.

A third type of hair-conditioning product relies on the use of proteins, amino acids (qv), botanicals, and amphoterics. Many ingredients have been identified as hair conditioners. Some of them are claimed to be substantive to the hair, whereas others are claimed to penetrate into the hair and repair previously incurred damage. Some of these hair-conditioning substances have been incorporated into newer delivery systems, such as mousses.

Hair Fixatives

These products are designed to assist in hair styling and in maintaining the style for a period of time. In contrast with hair dressings, hair fixatives do not leave an oily residue on the hair but tend to coat the hair with film-forming residues after drying. As in the case of hair dressings, style-holding qualities depend primarily on fiber–fiber adhesion and to a minor extent on fiber coating. The products may be conveniently divided into two groups: those that are applied to damp or wet hair, hair–setting products, and those that are applied to hair after styling, hair sprays.

Wave-setting products can be applied to wet hair and should not interfere with or delay drying. Wave sets can be formulated with water-soluble polymeric substances or with polymers that show solubility only in hydro-alcoholic media. Some of the preferred hair-fixative polymers are combined with lubricants or emollients and other excipients. The viscosity of these products can vary from that of a water-thin fluid to a rather firm gel.

Hair-spray products containing little or no water are preferred, because the presence of significant levels of water tends to soften a preexisting style. Environmental regulations today preclude the use of propellant solvents. Thus higher levels of alcohols are now used and the propellants of choice are low concentrations of hydrocarbons.

Hair Colorants

Hair colorants are commonly divided into temporary, semi-permanent, and permanent types. Decolorizing (bleaching) represents a fourth type of hair coloring.

Hair bleaching removes the pigment melanin from the hair shaft by oxidative destruction. Alkaline hydrogen peroxide is the agent of choice. Thickening is required in order to retain the blended oxidizing mixture on the hair. Surfactants are required to assure that every hair fiber is thoroughly wet by the blended moisture. Bleaching by hydrogen peroxide is enhanced by the presence of a peroxydisulfate, such as potassium persulfate. Bleaching damages the hair by converting some cystine to cysteic acid. In addition, the high pH induces swelling and cuticular damage. These adverse effects are counteracted by conditioning after treatment or by including some protectants in the hair-bleach product.

Temporary hair colorants are removed from the hair by a single shampoo. Temporary hair colorants usually employ certified dyes that have little affinity for hair. They are incorporated into aqueous solutions, shampoos, or hair-setting products.

Semipermanent hair colorants employ dyes that are absorbed directly by the hair. These dyes add color to the preexisting (natural) hair color and are useful primarily for blending in gray fibers. These dyes may fade significantly owing to exposure to sunlight and also are gradually removed by shampooing. Typically, temporary hair colorings are distributed as pourable lotions. Formulations may include alkanolamides, polymeric substances, fatty alcohols, thickeners, and conditioners commonly employed in all hair cosmetics. The CTFA lists temporary hair dyes with other substances as "Color Additives—Hair Colorants".

Permanent hair colorants, frequently identified as oxidation dyes, show much greater resistance to fading and shampoo loss than do semipermanent hair colorants. As a rule, these dyes remain on the hair; it is common practice to dye only that portion of the hair shaft that has emerged from the scalp since the last application. In permanent dyeing, a lotion containing developers and couplers is blended with hydrogen peroxide and then applied to the hair. The objective is uniform penetration of the various components into the hair, oxidation of the developer to a reactive intermediate, and formation of a colored dye stuff with the coupler. The dyes are synthesized within the fiber and migrate outward only slowly because of their size. In addition, the reactions occur in alkaline media, and some of the peroxide bleaches the hair. Thus it is possible to generate colored hair lighter than the original shade.

Hair-Waving and Straightening Products

The development of hair-waving and hair-straightening products requires a careful balance between product performance and hair damage. The hair-waving process essentially depends on converting some cystine cross-links in keratin to cysteine residues, which are reoxidized after the configuration of the hair has been changed. Sometimes hair straightening can also be achieved by a similar, relatively innocuous chemical change in the hair. As a rule, however, much more chemical destruction is required to achieve rapid and permanent straightening than to achieve permanent waving.

Permanent waving depends on the metathesis of a mercaptan and the cystine in hair while the hair is held in a curly pattern on a suitable device (curler). The most commonly used mercaptan is thioglycolic acid, although some other nonvolatile mercaptans can be employed. The active species is the mercaptide anion. Thus, adjustment to a pH between about 8.8 and 9.5 using amines or especially ammonia is required. A typical hair-waving product may consist of a 0.5–0.75 N solution of thioglycolic acid adjusted to a pH of about 9.1 with ammonia. The product generally includes a nonionic surfactant, to ensure thorough wetting of the wound hair tress, and a fragrance. Opaque lotion products can be created by adding the actives to mineral oil or other lipid-containing emulsions. The thioglycolate lotion is allowed to remain on the hair for about 10–30 min; then the hair is rinsed with water. Next, an oxidizing solution consisting of a dilute (1.5–3%) acidified hydrogen peroxide solution or of a potassium or sodium bromate is applied to the hair. This so-called neutralizing solution oxidizes the cysteine residues to cystine (without bleaching) in a new configuration within about 5–10 min. The neutralizer may contain a variety of hair conditioners and is removed from the hair by thorough rinsing with water after unwinding.

Hair straightening is more difficult than hair waving. Kinky hair has a tight crimp that cannot be straightened by winding over a rod or curler. Two processes for straightening exist. One, based on thioglycolates, effects the same chemical change as that occurring during permanent waving. The other, more aggressive process is based on (1–8%) sodium hydroxide (or guanidine). In order to hold the hair straight, hair-straightening products are viscous. The chemical reactions with sodium hydroxide involve formation of cysteine and dehydroalanine residues in the hair with some loss of sulfur. The cysteine and dehydroalanine can subsequently react to form the thioether, lan-thionine, which helps repair the mechanical strength of the fiber to some extent. Similar chemical reactions occur when steam is allowed to interact with hair, such as during hot pressing, which was an earlier technique for straightening hair.

Conditioners, lipids, acid rinses, and related cosmetics have been developed to minimize hair damage from these rather destructive processes.

Hair Removers

Hair removers are designed to remove hair from the skin surface without cutting in order to avoid undesirable stubble. Cosmetic products have been developed for chemical destruction of hair, that is, depilation, and for facilitating mechanical hair removal, that is, epilation.

Depilatories epitomize the chemical destruction of hair and allow hair removal by scraping with a blunt instrument or by rubbing with terry cloth. Chemical depilatories are based on 5–6% calcium thioglycolate in a cream base (to avoid runoff) at a pH of about 12. The pH is maintained with calcium or strontium hydroxide. This type of treatment does not destroy the dermal papilla, and the hair grows back.

Epilation is required for permanent hair removal. The most effective epilation process is electrolysis or a similar procedure. Epilation can also be achieved by pulling the fibers out of the skin. For this purpose, wax mixtures (rosin and beeswax) are blended with lipids, for example, oleyl oleate, which melt at a suitable temperature (about 50–55°C).

DECORATIVE COSMETICS

Decorative cosmetics are products intended to enhance appearance by adding color or by hiding or deemphasizing physical defects. In Western cultures, most decorative cosmetics are for use on the face. Products in this category are various types of powders, facial makeups, and lip- and eye-coloring products.

CTFA International Cosmetic Ingredient Dictionary, 5th ed., CTFA, Washington, D.C., 1993.

CTFA International Cosmetic Ingredient Handbook, 2nd ed., CTFA, Washington, D.C., 1992.

E. Jungermann, ed., *Cosmetic Science and Technology Series*, Marcel Dekker, New York, continuing series.

MARTIN M. RIEGER
M & A Rieger Associates

COTTON

Cotton is both a fiber (lint) and food (cottonseed) crop. For each 45.36 kg (100 lb) of fiber produced, the plant also produces about 68.04 kg (150 lb) of cottonseed. Cotton fiber (see FIBERS, VEGETABLE) is the most important natural vegetable textile fiber used in spinning to produce apparel, home furnishings and industrial poducts. In 2004, worldwide ~42% of the textile fiber consumed was cotton. In its marketed form, raw cotton consists of masses of fibers packaged in bales of ~85–230 kg (187–507 lb). A single kilogram (2.2 lb) of cotton may contain 200 million or more individual fibers.

Cottonseed, can be fed as whole seed (16% oil, ~45% protein) to dairy cattle or crushed at a cottonseed oil mill to obtain oil, hulls, meal, linters, and manufacturing loss. The oil is used for human consumption; the hulls and meal are sources of vegetable protein feed for animals; and the linters are used as a chemical cellulose source and in batting for upholstered furniture and mattresses as well as for high quality paper.

The origin, development, biology/breeding, production, morphology, chemistry, physics, and utilization of cotton have been discussed in many publications. Cotton fibers are seed hairs from plants of the *Malvaceae* family, the tribe *Gossypieae*, and the genus *Gossypium*. It is a warm-weather shrub or tree that grows naturally as a perennial but for commercial purposes is grown as an annual. The principal domesticated species of cotton of commercial importance are *hirsutum, barbadense, arboreum*, and *herbaceum*. Many different varieties of these species have been developed through conventional breeding to produce cotton plants with improved agronomic properties and cotton fibers with improved length, strength, and uniformity. In addition to conventional breeding methods, genetic engineering is being used to produce transgenic cottons with insect resistance and herbicide tolerance.

At present, cotton is grown in eivronments that range from arid to tropical, with long to very short growing seasons. About 80 countries in the world grow cotton-about 59 countries grow at least 5000 ha. About 40–50% of the 26 million metric tons (115 million 480 lb U.S. bales) of cotton production in the world in the 2005/06 crop year was biotech cotton. Less than 0.1% (~25,000 MT; 110,000 bales) was produced by certified organic agriculture practices. Cotton typically requires a growing season of at least 160 days when minimum temperatures are >15°C (60°F). Under normal climatic conditions, cotton seeds germinate and seedlings emerge in 7–10 days after planting. Flower buds (known as squares) appear 35–45 days later followed by open white (Upland cotton) or creamy to dark-yellow (Pima cotton) flowers 21– 25 days later. One day after the flower opens the cotton boll begins to grow rapidly, if the flower has been fertilized. Mature bolls open 40–80 days after flowering, depending on variety and environmental conditions. Within the boll are three to five divisions called locks or locules, each of which normally has seven to nine seeds that are covered with both lint and linters. The linters form a short, shrubby undergrowth beneath the lint hairs on the seed. At least 13,000–21,000 fibers are attached to each seed and there are close to 500,000 fibers in each boll.

Each cotton fiber is a single cell that originates in the epidermis of the seed coat at about the time the flower opens; it first emerges on the broad, or chalazal, end of the seed and progress by degrees to the sharp, or micropylar, end. As the boll matures, the fiber grows until it attains its maximum length, which averages ~2500 times its diameter.

The seed hairs of cultivated cottons are divided into two groups (lint and fuzz fibers or linters) that differ in length, width, pigmentation, and strength of adherence to the seed. The growth of linters is much the same as that of lint, but elongation is initiated about 4 days after flowering. For each 45.36 kg (100 lbs) of lint fiber produced, about 2.7 kg (6 lbs) of linters are produced. They are usually ~1.3 cm (0.5 in.) long compared with the 2.5 cm (1 in.) average length of lint fibers and are twice as thick, or ~32 μm. Their color is usually greenish brown to gray. After lint fibers have been ginned off the seed, the linters remain. Removal of linters is usually done at the cottonseed oil mill and requires a machine similar to that used at the saw cotton gin to remove the fiber from the seed.

COTTON FIBER BIOSYNTHESIS

During the cell elongation stage of fiber development, a primary cell wall envelopes the growing fiber. The principal components of fiber primary cell walls are pectins, hemicelluloses, cellulose, and proteins. In higher plants, pectins and hemicelluloses are produced in Golgi bodies and are deposited in the wall by fusion of Golgi-derived vesicles with the cell membrane. Cell wall proteins are synthesized in association with the endoplasmic reticulum and may be glycosylated in the Golgi. In contrast, the enzyme complex responsible for cellulose biosynthesis is associated with the cell membrane in structures known as rosettes.

Cellulose biosynthesis has been extremely difficult to characterize biochemically. At maturity, cotton fibers are nearly pure cellulose and should be a rich source of the enzyme cellulose synthase. Unfortunately, it has been difficult to separate a β-1,4-glucan (cellulose) producing activity from a large background of β-1,3-glucan (callose) synthesis. With the advent of molecular genetic approaches to study genes expressed during cotton fiber development, a break-through has been achieved. By determining the sequence of many messenger RNA (m RNA) molecules produced by immature cotton fibers, two gene transcripts with regions similar to those found in bacterial cellulose synthases were discovered: CesA1 and Ces A2, which are produced concomitantly with the initiation of secondary cell wall biosynthesis in fiber. A third CesA gene from cotton has also been described and is expressed both during the cell elongation and secondary wall thickening stages, but it alone will not produce cellulose. In addition, a membrane-associated cellulase gene has also been implicated in cellulose biosynthesis by induced mutations in *Arabidopsis*. It seems paradoxical that cellulase, an enzyme capable of degrading cellulose, is involved in cellulose biosynthesis.

PRODUCTION

Field Preparation

Field preparation practices very considerably in the various cotton-growing regions of the world and reflect the varied environments and production systems encountered in these regions.

Planting

Less than 5% of the cottonseed produced is used for planting seed.

Irrigation

Approximately 35% of the U.S.A. cotton acreage receives supplemental irrigation and about 55% of world cotton production comes from irrigated land.

Fertilization

Cotton normally is grown under intense production systems; many fields are planted in cotton year after year. Therefore, an efficient fertilization program must be based on results of soil and tissue tests and the yield desired from the crop. Supplying nutrients according to crop demands has replaced traditional methods, as soil and tissue testing have become widespread. Unnecessary and undesirable applications are, therefore, avoided, reducing the risk of off-site discharge of nitrates.

Crop Protection

Cotton can be affected by insects, weeds, diseases, nematodes, and mycotoxins. About 90% of the U. S. cotton uses Integrated Pest Management (IPM) practices. This approach optimizes the total pest management system by utilizing all available tools, including rotation, crop residue destruction, maximum crop competitiveness, earliness, pest scouting, action thresholds, releases of beneficial insects, sterile insect releases and selective crop protection chemistry. New plant protection options including new chemical, biological, and transgenic technologies coupled with good IPM schemes are helping to reduce use of broad spectrum pesticides favored in the past.

HARVESTING

Except for the cotton gin, the introduction of the mechanical harvester has probably had a greater effect on cotton production than any other single event.

When the cotton boll reaches full maturity, it begins to lose moisture and opens. As the boll opens, the drying fiber fluffs or expands outward. After the seed cotton (lint and linters on the seed) has dropped to a moisture content of ~12% it is ready for harvest. If the cotton is to be mechanically harvested, the plant is usually treated with a harvest-aid chemical (ie, a defoliant or desiccant). Once the plant is ready, the cotton is either handpicked (about 75% of the world) or mechanically harvested with a spindle picker or cotton stripper.

Ginning

After harvesting the seed cotton is transported to the gin to finish the harvesting process by separating the cotton fiber/lint from the seed. Because the gin capacity is usually not sufficient to keep up with the harvesters, the harvested cotton is often stored in a compacted module that allows it to be ginned at a later date.

CLASSIFICATION/MEASUREMENT OF FIBER QUALITY

Classification is a standardized set of procedures for measuring the quality/physical attributes of raw cotton fiber that affect the quality of finished products and/or manufacturing efficiency.

Classing U.S. Uplant Cotton

In the United States, the quality of cotton is described (classed) in terms of color, leaf, extraneous matter, fiber length, length uniformity, strength, and micronaire according to the Official Cotton Standards ("universal standards"). Research to rapidly measure other important fiber characteristics, such as maturity, stickiness, and short fiber content, continues. Measurements for fiber length, length uniformity, fiber strength, micronaire (fineness), color grade, and trash are performed by precise high volume instruments (commonly referred to as "HVI" classification). There are 25 official color grades (15 physical standards and 10 descriptive) for American Upland cotton, plus 5 categories of below-grade. Micronaire reading is determined by an airflow measurement. Classification for leaf grade, preparation, and extraneous matter are still based on subjective (classer) determinations performed by visual observation.

Classing U.S. Long Staple (Pima)

Pima (ELS cotton) and Upland cotton-grade standards differ. The most significant difference is that the American Upland grade/fiber quality is determined by instrument measurement and the American Pima grade/fiber quality by expert trained appraisal classers. Pima is naturally of a deeper yellow color than Upland cotton. The leaf content of Pima standards do not match Upland standards. Pima cotton is ginned on roller gins and has a more stringy and lumpy appearance. Pima cotton grades range from a grade 1 (highest) to 10 (lowest).

PHYSICAL PROPERTIES

Fiber length is universally accepted as the most important fiber property, because it greatly affects processing efficiency and yarn quality. The recognized reference machine method for fiber length information is the Suter–Webb Comb Sorter. Variations of length are unique to specific varieties of cotton and range from <2.5 cm (1 in.) for short-staple Upland varieties to 2.6–2.8 cm (1.02–1.10 in.) for medium-staple Uplands, to >2.85 cm (1.12 in.) for long-staple varieties (Pima, Egyptian, and Sea Island).

Next to length, *fiber strength* is the most important physical property that relates to fiber and yarn quality.

The recognized reference method for fiber strength is based on measurements made on bundles of parallel fibers. Fiber strength is expressed as breaking stress or force to break per linear density of the bundle. These units are newtons [or gram force (gf)] per linear density (tex), where 1.0 tex $= 1$ g/1000 m. Variations in fiber strength are also unique to specific varieties of cotton and range from 0.176 to 0.216 N/tex (18 to 22 gf/tex) for short-staple Upland varieties to 0.235–0.275 N/tex (24–28 gf/tex) for some medium-staple Uplands to 0.314–0.373 N/tex (32–38 gf/tex) for long-staple varieties (Pima, Egyptian, and Sea Island).

Another important characteristic property of cotton is its *fineness*, or *linear density*, or weight per unit length. The normal units for cotton fineness are millitex (the units of tex are g/km). Fineness is directly related to the amount of cellulose in the fiber, which is a function of the fiber wall area, excluding the hollow center (lumen), and the fiber length. Recent developments in techniques for preparing excellent thin cross-sections of cotton coupled with advances in computerized microscopic image analysis allow for rapid and accurate measurements of fiber wall area and perimeter.

In addition to fiber length, strength, and fineness, two other properties that have significant bearing on fiber and yarn properties are *color* and *trash* measurements.

TEXTILE PROCESSING

Yarn Manufacturing

Cotton is received by the textile mill in the form of highly compressed bales ($\sim$450 kg/m^3), weighing $\sim$227 kg (480 lb). Although the seed and a large portion of the plant trash are removed at the gin, baled cotton still contains various forms of trash, including stem, leaf, and seed coat fragments that must be removed in the manufacturing process. In yarn manufacturing, the cotton goes through the processes of opening, blending, carding to form sliver, which is spun into yarn.

Fabric Manufacturing (Weaving and Knitting)

Yarns manufactured in the spinning process are used to make woven or knitted fabrics. Weaving and knitting are the two pimary textile processes for manufacturing fabrics. Most woven and knitted cotton fabrics are produced from single yarns. However, for the manufacture of industrial fabrics such as canvas, it is necessary to combine, or ply twist, several strands of single yarns together to obtain increased strength and resilience. Sewing thread and cordage are also produced from multiple plies of single yarns twisted together.

Nonwoven Manufacturing

Cotton staple is readily processed to form carded, air laid, or carded/crossed-lapped webs that can be bonded by various techniques to form useful nonwoven materials. Many times a combination of these processes is used to produce hybrid structures and other products.

CHEMICAL COMPOSITION AND MORPHOLOGY

The cotton fiber is a single biological cell, 15–24 μm in width and 12–60 mm (4.7–23.6 in.) long. It has a central canal, or lumen, down its length except at the tip. It is tapered for a short length at the tip, and along its entire length the dried fiber is twisted frequently and the direction of twist reverses occasionally. These twists (referred to as convolutions) are important in spinning because they contribute to the natural interlocking of fibers in a yarn.

Raw cotton fiber after ginning and mechanical cleaning is essentially 95% cellulose. The noncellulose materials, consisting mostly of waxes, pectinaceous substances, and nitrogenous matter (mainly protein), are located to a large extent in the primary wall, with small amounts in the lumen.

Of the noncellulose constitutents, nitrogen-containing compounds (mostly protein) normally occur in the largest amounts, almost entirely in the lumen, and are most likely protoplasmic residue left behind after the gradual drying up of the living cell. Most of the pectin in the cotton fiber is in the primary wall. Removal of the pectic substances is accomplished by scouring, which does not change the properties of the cotton greatly.

Chemical modification of the cotton fiber must be achieved within the physical framework of this rather complicated architecture. Uniformity of reaction and distribution of reaction products are inevitably influenced by rates of diffusion, swelling, and shrinking of the whole fiber, and by distension or contraction of the fiber's individual structural elements during finishing processes.

STRUCTURE AND REACTIVITY

Chemical Structure

The raw cotton fiber produced in the bolls of the cotton plant is composed almost entirely of the polysaccharide cellulose, a $1 \longrightarrow 4$ linked polymer of β-D-glycopyranose.

Molecular and Supramolecular Physical Structure

The chains of cellulose molecules associate with each other by forming intermolecular hydrogen bonds and hydrophobic bonds. They coalesce to form microfibrils also called crystallites. In cotton, the microfibrils can organize into macrofibrils 60–300 nm wide. The macrofibrils are organized into fibers. Cotton fibers have a complex, reversing, helical arrangement of macrofibrils. There are several different forms or polymorphs [cellulose I to IV and X with recent subclasses Iα and Iβ], depending on the source and treatment. There are both different unit cells and different packing arrangements in the unit cell. Native cotton is cellulose I.

Pore Structure and Affinity for Water

The cotton fiber is a porous, hydrophilic material that accounts for the comfort of cotton clothing. Moisture is retained tenaciously in cotton. The moisture absorbed from the atmosphere and held under ambient conditions is expressed either as moisture content (amount of moisture as the percentage over the oven-dried weight) or more commonly as moisture regain (amount of moisture as a percentage of the oven-dry sample). Under ordinary atmospheric conditions, moisture regain is 7–11%.

Pores accessible to water molecules are not necessarily accessible to chemical agents. Many uses of cotton, eg, easy care fabric, depend on chemical modification to impart the desired properties. Knowledge of accessibility to dyes and other chemical agents of various sizes under water-swollen conditions is required for better control of the various chemical treatments applied to cotton textiles.

Availability of Hydroxyl Groups

The regular occurrences of intermolecular and intramolecular hydrogen bonds in the crystalline regions of cotton cellulose render the involved hydroxyl groups unavailable to chemical agents under mild reaction conditions. Chemical agents that have access to the interior pores of the cotton fiber thus find potential reactive sites unavailable for reaction. The order of decreasing availability of hydroxyl groups in cotton is 2-OH > 6-OH ≫ 3-OH.

REACTIONS FOR PRACTICAL OBJECTIVES

Chemical modification has assisted in building cotton's position in the market place despite the advent of synthetic fibers.

Mercerization

One of the earliest known modifications of cotton that had commercial potential was mercerization. Traditionally, the process employed a cold concentrated sodium hydroxide (caustic soda) treatment of yarn or woven fabric followed by washing and a mild acetic acid neutralization. Maintaining the fabric under tension during the entire procedure was integral to achieving the desired properties. The resultant mercerized cotton has improved luster and dyeability and strength. If the cotton is allowed to shrink freely during contact with mercerizing caustic, slack mercerization takes place; this technique produces a product with greatly increased stretch (stretch cotton) that has found application in both medical and apparel fields.

Etherification

The accessible, available hydroxyl groups on the 2, 3, and 6 positions of the anhydroglucose residue are quite reactive and provide sites for much of the current modification of cotton cellulose to impart special or value-added properties. The two most common classes into which modifications fall, include etherification and esterification of the cotton cellulose hydroxyls as well as addition reactions with certain unsaturated compounds to produce cellulose ethers (see CELLULOSE ETHERS). One large class of cellulose-reactive dyes ("fiber-reactive" dyes) in commercial use attaches to the cellulose through an alkali-catalyzed etherification by nucleophilic attack of the chlorotriazine moiety of the dyestuff.

Cross-Linking. By far, the most important commercial modifications of cotton cellulose are those that occur through etherification. For example, commercial modification of cotton to impart durable-press, smooth drying, or shrinkage resistance properties involves cross-linking adjacent cellulose chains through amidomethyl ether linkages.

Resiliency. Base-catalyzed reactions of cotton cellulose with either monoepoxides or diepoxides to form cellulose ethers also result in fabrics with increased resiliency. Monoepoxides, believed to result only in cellulose hydroxyalkyl ethers or linear graft polymers, produce marked improvement in resiliency under wet conditions, but little improvement under ambient conditions.

Other Cellulose Ethers. Other cotton cellulose ethers include carboxymethyl, carboxyethyl, hydroxyethyl, carbamoylethyl, cyanoethyl, sulfoethyl, and aminoethyl (aminized cotton) products. Most, with the exception of cyanoethylated and aminized cotton, are of interest in applications requiring solubility or swellability in water or alkali.

Flame Resistance. The chemical treatment of cotton with fire retardants to make it flame resistant is discussed elsewhere (see FLAME RETARDANTS FOR TEXTILES). Although certain cellulose esters, such as the ammonium salt of phosphorylated cotton and cellulose phosphate, are flame resistant, the attachment of most currently used durable polymeric flame retardants for cotton is through ether linkage to the cellulose or the use of an ammonia precondensate, which forms an insoluble polymer in the fiber. Nondurable fire retardants based on applications of boric acid or methyl borate are used in treatment of cotton batting for upholstery, bedding, and automotive cushions.

Water Repellency. The development of water-repellent cellulose ethers has been reviewed elsewhere (see WATERPROOFING). A typical example of a commercial etherification for waterproofing cotton is with stearamidomethylpyridinium chloride:

$$C_{17}H_{36}\overset{\overset{O}{\|}}{C}-NCH_2-\overset{+}{N}\langle\bigcirc\rangle \ + \ cell-OH \ \longrightarrow$$

$$Cl^-$$

$$C_{17}H_{36}\overset{\overset{O}{\|}}{C}-NCH_2O-cell \ + \ HCl \ + \ N\langle\bigcirc\rangle$$

Cyanoethylation. One of the earliest examples of etherification of cellulose by an unsaturated compound through vinyl addition is the cyanoethylation of cotton. Cyanoethylation can impart a wide variety of properties to the cotton

fabric, such as rot resistance, heat and acid resistance, and receptivity to acid and acetate dyes.

Irradiation

The effects of high energy radiation (eg, gamma radiation) on cotton properties have also been investigated. Depolymerization of cellulose occurs with increasing energy absorption; carbonyl formation, carboxyl formation, and chain cleavage occur in the ratio of 20:1:1. With these chemical changes, there is a corresponding increase in solubility in water and alkali and a decrease in fiber strength. The induction of cellulose free radicals by near uv irradiation forms the basis for photofinishing with vinyl monomers to produce graft polymers on the cotton. Another useful reaction of cotton cellulose occurs in an ionized atmosphere, which is essentially a surface reaction. Glow discharge treatment of cotton yarn in air increases water absorbency and strength, and surface-dependent properties of cotton fabric are drastically changed by exposure to low temperature–low pressure plasma generated by radio-frequency radiation.

INSOLUBILIZATION

Dyes

With the exception of fiber-reactive dyes discussed under Etherification earlier, other cotton dyes, ie, vat and sulfur, are insolubilized within the fiber after an oxidization step.

Antimicrobial

Insolubilization and five other methods for imparting antimicrobial properties to cotton. These methods can all be classified under one or more of the chemical reactions of cotton cited earlier; they include fiber reactions to form metastable bonds, grafting through thermosetting agents, formation of coordination compounds, ion-exchange methods, polymer formation with possible grafting, and a regeneration process. Also a commercialized process for antibacterial cotton fabrics uses insoluble peroxide complexes of zirconyl acetate.

Temperature Adaptable Fabrics

Altering the chemical composition of the fabrics such that large amounts of heat are absorbed and released in repeatable cycles of controllable temperature ranges produces fabrics that are described as temperature adaptable. The process insolubilizes polyethylene glycols by cross-linking with methylolamides on the cotton fabric.

Esterification

There are both inorganic and organic esters of cellulose. Of the three most common inorganic esters, cellulose nitrate, phosphate, and sulfate, only cellulose sulfate is soluble in water.

Acetylation of cotton to an acetyl content slightly >21% produces a material with greatly increased resistance to fungal and microbiological degradation, in addition to tolerance of high temperatures not exhibited by native cotton.

In the 1960's, esterification of cotton cellulose with polycarboxylic acids to produce smooth-drying fabrics was investigated. In the late 1980's, better catalyst systems were discovered for the ester cross-linking of cellulose; inorganic salts of phosphorus-containing acids were found to give ester cross-links that are durable to multiple home launderings. Because of the improved catalysis, certain tricarboxylic and tetracarboxylic acids have shown promise for commercialization.

ENZYMATIC MODIFICATION

The industrial use of enzymes in the textile industry has increased substantially in recent years. Lipases, proteases, and cellulases are being used. Lipases and proteases are used to assist in cleaning textiles. Treatments involving cellulases, which hydrolyze the cellulose polymer, are relatively new and are of particular importance. Cellulases obtained from both bacterial and fungal sources are being used to give fabrics a soft hand, to give cellulosic fabric surfaces a smooth and clear appearance by removing fabric fuzz fibers (biopolishing), and to provide a stone-washed appearance to denim (biostoning). Cellulases are also being added to detergents to maintain the color appearance of cotton cellulose fabrics by removing fabric fuzz fibers and pills that form on wear and laundering. One of the main reasons for using enzymes instead of other chemicals as finishing agents for cotton cellulose is that they are environmentally safer.

HEALTH AND SAFETY FACTORS

Respiratory Disease

Byssinosis is an occupational lung disease that can affect a small number of textile workers after repeated inhalation of the dust generated during the processing of cotton and some other vegetable fibers (eg, flax and soft hemp). Byssinosis may cause progressive and disabling airway narrowing. Appropriate engineering controls in cotton textile processing areas or washing cotton essentially can eliminate incidence of workers' reaction to cotton dust. The U.S. Occupational Safety and Health Administration (OSHA) issued revised standards for occupational exposure to cotton dust in 1985.

Formaldehyde

Formaldehyde is a component of resins used to impart durable-press and other properties to cotton fabrics. It is classified as a "probable human carcinogen", because it has been shown to be an animal carcinogen and there is limited evidence to indicate that it is a carcinogen in humans. Exposure to formaldehyde from cotton textiles is controlled by the chemical technology for low emitting formaldehyde resin technology and nonformaldehyde finishes discussed earlier and by increased ventilation in the workplace. OSHA has issued standards for control of occupational exposure to formaldehyde.

A. S. Basra, ed., *Cotton Fibers Developmental Biology, Quality Improvement and Textile Processing*, Food Products Press, Binghamton, N.Y., 1999.

The Classification of Cotton, Agriculture Handbook 566, U.S. Department of Agriculture, Cotton Program, Agricultural Marketing Service, Washington, D.C., revised April 2001.

G. P. Fitt (Chair), P. J. Wakelyn (Vice Chair) and co-workers *Report of the Second Expert Panel on Biotechnology of Cotton*. International Cotton Advisory Committee (ICAC), Washington, D.C., Nov. 2004.

P. J. Wakelyn and co-workers, *Cotton Fiber Chemistry and Technology*, Series: International Fiber Science and Technology. CRC Press (Taylor & Francis Group), Boca Raton, Fla., 2006.

Phillip J. Wakelyn
The National Cotton Council of
America

CRYOGENIC TECHNOLOGY

In present day usage, the word "cryogenics" refers to "all phenomena, processes, techniques, or apparatus occurring or using temperatures <120 K" (B.A. Hands, Cryogenic Engineering).

Cryogenic technology has contributed greatly to scientific research and is widely used in many industrial applications. The ability to condense a gas such that it can be stored and shipped as a cryogenic liquid rather than as a pressurized gas has found several applications. Natural gas is stored and transported as liquified natural gas (LNG) to many countries in the world. Liquid hydrogen is used as fuel for space vehicles. Oxygen, nitrogen, and argon are shipped as liquids by truck and rail car to the point of end use.

Another major use of cryogenic technology has been to produce low cost, high purity gases through fractional condensation and distillation. Cryogenic air separation is used for the production of pure oxygen, nitrogen, and argon. These gases are used in primary metals manufacturing (eg, steel), chemical manufacturing, glass manufacturing, electronic industries, partial oxidation and coal gasification processes, enhanced oil recovery and many other applications. Cryogenic methods are also used for the purification of hydrogen, helium, and carbon monoxide. Hydrogen and carbon monoxide gases are used in chemical manufacturing and some metal industries. Helium is used in welding, medicine, gas chromatography, and diving.

Cryogenic processes that provide low temperatures to refrigerate other materials or to alter their properties have been used in many applications. Liquid nitrogen is used for freezing food such as hamburgers and shrimp. Rubber tires and scrap metal from old cars are reclaimed using cryogenic cooling techniques to make them brittle for easier fracturing and component separation. Biological materials such as bone marrow, blood, animal semen, tissue cultures, tumor cells and skin are preserved by cryogenic freezing and storage.

Magnetic resonance imaging (mri) uses cryogenics to cool high conductivity magnets for nonintrusive body diagnostics. Low temperature infrared (ir) detectors are utilized in astronomical telescopes. Cryogenic refrigerators have been applied industrially for cryopumping to yield high pumping speeds and ultrahigh vacuum. With the advent of high temperature superconductivity, it is anticipated that applications of superconductivity at near liquid nitrogen temperature will have great potential for electric power transmission, magnetic transportation systems, and magnets for energy generation in fusion processes.

REFRIGERATION METHODS

Refrigeration for cryogenic applications is produced by absorbing or extracting heat at low temperature levels and rejecting it to the atmosphere at higher temperatures. Three general methods for producing cryogenic refrigeration in large scale commercial applications are (1) the liquid vaporization cycle; (2) the Joule-Thomson (J-T) expansion cycle; and (3) the engine expansion cycle. The first two are similar in that they both utilize irreversible isenthalpic expansion of a fluid, usually through a valve. Expansion in an engine approaches reversible isentropic expansion with the performance of work.

Liquid Vaporization Cycle

In this process, a refrigerant fluid with the desired low temperature boiling point is compressed to a pressure at which it can be condensed with ambient air, cooling water, or another refrigerant fluid with a higher boiling temperature.

Joule-Thomson Expansion Cycle

In this process, a refrigerant fluid is compressed and precooled below its inversion temperature, ie, the temperature below which a pressure reduction results in a temperature decrease. The cold refrigerant fluid is isenthalpically (J-T) expanded to a lower pressure to produce the required low temperature.

Engine Expansion Cycle

In this process, refrigeration is supplied by expanding a pressurized stream through a work-producing device (an engine). Whereas expansion though a J-T device does not cause the enthalpy of the stream to change, expansion though an engine causes the enthalpy of the stream to be reduced. This reduction in enthalpy produces the refrigeration for the process.

Other Refrigeration Methods

Cryocoolers provide low temperature refrigeration on a smaller scale by a variety of thermodynamic cycles.

Thermoacoustic and pulse tube refrigeration both use a closed refrigerant inventory (typically helium) to produce refrigeration by a thermodynamic cycle. The Gifford-McMahon cryocooler consists of displacer, regenerator, compressor, and intake/exhaust valves that can be staged to reach the required cryogenic temperatures. Magnetic refrigeration uses the magnetocaloric effect to produce cooling. When ferromagnetic or paramagnetic materials are put into a magnetic field, their temperature rises: when removed from the field, their temperature falls. The amount of temperature change depends on the material, its temperature, and the strength of the magnetic field.

APPLICATIONS

Air Separation

A considerable number of cryogenic concepts were developed near the end of the nineteenth and beginning of the twentieth century to liquefy and separate air into its major constituents (composition of air: 78% nitrogen, 21% oxygen, and 1% argon by volume). By 1905, Carl von Linde's double distillation column process to produce practically pure oxygen and nitrogen was already known. Argon was produced industrially by 1912. Trace quantities of other rare gases such as neon, krypton, and xenon are present in air and can be recovered by proper modifications to a cryogenic double distillation column air separation plant.

In the field of semiconductor device manufacturing, on-going trends toward device miniaturization and the need for high production yield are leading to gas specifications with ultrahigh purities. This finding requires that the concentration of all impurities in each of the nitrogen, oxygen, and argon products be <10 parts per billion (ppb).

Liquid Nitrogen, Oxygen, and Argon

Large quantities of liquid nitrogen are typically produced by liquefying gaseous nitrogen from an air separation plant.

Liquid argon can be produced directly from a crude argon condenser by increasing the flow of air through the expander. However, refrigeration supply by expanding air into the low pressure column is not economical for production of large quantities of liquid oxygen.

Liquefied Natural Gas

Liquefied natural gas plants can be categorized as "peakshaving" or "base-load". Peakshaving LNG plants are built at the consumer end of natural gas pipelines to accumulate LNG in storage tanks for later vaporization and send out into the local grid during periods of peak demand. Base-load LNG plants provide a steady "base" supply of natural gas to utility companies, generally by transportation of LNG by ship from one country to another.

Base-load LNG plants have predominantly used variations of the mixed refrigerant cycle, although some have utilized the cascade cycle. The mixed refrigerant cycle can be precooled by a high stage cycle that may be a propane refrigerator, a second separate mixed refrigerant cycle, or even an ammonia absorption system. Base-load plants generally recover propane and heavier components from the natural gas for internal use in charging the refrigeration systems, or for external use as LPG, butane, or natural gasoline.

Hydrogen Purification

Cryogenic separation is used extensively for recovery and purification of hydrogen from refinery and petrochemical plant off-gases. The most common applications are in recovery of hydrogen from catalytic reformer gas, ethylene plant off-gas, hydrodealkylation (HDA) recycle gas, hydrodesulfurization (HDS) off-gas and methanol/ammonia synthesis purge gases. Hydrogen can also be coproduced when carbon monoxide is recovered by cryogenic separation of a steam-methane reformer off-gas.

Hydrogen Purification with Light Olefins and Liquefied Petroleum Gas (LPG) Recovery

The relatively simple cryogenic purification process for hydrogen recovery can easily be adapted to recover a crude light olefin or propane and heavier hydrocarbon (LPG) stream. The pretreated feed gas is cooled to an intermediate temperature, in the range of 240–200 K for propylene/LPG recovery or 180–150 K for ethylene recovery. The uncondensed vapor is then further cooled to condense the remaining methane and residual heavy hydrocarbons.

Hydrogen Liquefaction

Hydrogen can be produced from caustic/chlorine electrolytic cells, by decomposition of ammonia or methanol, by steam-methane reforming, or by partial oxidation of hydrocarbons. Hydrogen recovered by these methods must be further purified prior to liquefaction. This is generally achieved by utilizing pressure swing adsorption methods whereby impurities are adsorbed on a solid adsorbent.

Light Olefins and LPG Recovery

As described above, light olefins and LPG can be recovered by making proper modifications to a hydrogen recovery process. Dephlegmators (fractionating condensers) can also be used for light olefin and LPG recovery. Dephlegmators are usually specially designed brazed aluminum plate-fin heat exchangers in which liquid condensed from an upward cooling vapor stream flows counter-currently within the heat exchanger to act as a reflux stream.

Nitrogen Rejection and Helium Recovery

Cryogenic distillation has been used extensively in the processing of natural gas for nitrogen removal and for helium recovery. Three basic processes have been used for nitrogen rejection from natural gas; the single-column heat-pumped process, the double-column process, and the dual column cycle. Earlier processes utilized multistage flash columns for helium recovery from natural gas.

Helium Purification and Liquefaction

Helium is commercially recovered from natural gas. If helium is present in a natural gas it usually occurs in concentrations between 0.2 and 2%, although it has been found in concentrations up to 8%.

Helium is normally concentrated in stages, first from the initial field concentration to crude, then to pure helium for liquefaction or sale as pure gas. As shown in the previous section, crude helium is easily recovered from a plant rejecting nitrogen from natural gas. Final upgrading of helium from crude to pure is typically done by Pressure Swing Adsorption (PSA) in combination with cryogenic partial condensation.

EQUIPMENT

Machinery

Compressor selection for a cryogenic process plant depends on the fluid, the volumetric flowrate, the pressures involved, the compressor efficiency, and the cost of energy and capital. Centrifugal compressors are lower in installed cost than reciprocating machines, and are preferred if the volumetric flows and pressures of the process allow them to be applied. For large volumetric flows, axial compressors are used. At very high pressures with small volumetric flows, reciprocating machines are required. Sometimes a combination of more than one kind of compression stage is used to yield the most cost effective and efficient system.

Expanders provide refrigeration by extracting work from a fluid, thereby reducing its enthalpy. Gas expanders can be either reciprocating or centrifugal, and can be loaded (braked) by electric generators, gas blowers or compressors, oil-film "cups", or oil pump brakes. The extracted work can be usefully recovered in an electric generator or compressor. Centrifugal expanders are lower in first cost and in maintenance cost. Reciprocating expanders may be required for low volumetric flows and/or high expansion pressure ratios of low molecular weight gases. Bearings for gas expanders can be oil or gas bearing type (static or dynamic). Reverse-running liquid pumps have been used as liquid expanders. Isentropic efficiencies for gas expanders can be as high as 85–90% for machines with high discharge volumetric flow.

Heat Exchangers

The two most prominent types of heat exchangers used in cryogenic service are the coil wound, tube-in-shell exchanger and the brazed aluminum plate and fin (core) exchanger.

Distillation Columns

In a cryogenic air separation plant, distillation accounts for the major fraction of the total energy consumption.

The low relative volatility characteristic of many cryogenic separations requires the use of many stages. Columns operating at low pressures are consequently designed for a low pressure drop per theoretical stage of separation.

In some cryogenic hydrocarbon separation plants, such as a nitrogen rejection unit, both flowrate and feed composition may change over the life of the plant. This requires the use of valve or bubble cap trays with high turndown capability for distillation.

Insulation

Cryogenic insulation should economically reduce heat leak into the system so that its impact on the overall refrigeration requirement is minimized. Insulation can be categorized as unevacuated bulk type (eg, purged rockwool or perlite), rigid foam (eg, foam glass or urethane), vacuum-jacketed (VJ), evacuated powder (eg, perlite), and multilayer insulation (MLI) (e.g., evacuated aluminized mylar).

SAFETY

The possibility of an uncontrolled release of a cryogenic fluid such as liquid oxygen, methane, or hydrogen from storage and during handling must be carefully considered during the design of a cryogenic facility. The level of risk may be reduced by providing dikes for secondary containment of liquid spills. Procedures to protect personnel from cryogenic burns and asphyxia and to protect the nearby equipment from embrittlement failure must be carefully considered and followed. When trace impurities in the feed streams can lead to the combination of an oxidant with a flammable cryogen (eg, solid oxygen in liquid hydrogen) or a combustible with an oxidant (eg, acetylene in liquid oxygen), special precautions must be taken to eliminate them. Many materials react with pure oxygen, so care must be taken in the selection of any materials that may be in contact with oxygen and in the cleaning of oxygen systems prior to use. Potential ignition sources must be minimized, particularly in oxygen compression and in systems for handling oxygen at elevated pressures.

R. Agrawal and D. M. Herron, "Air Liquefaction: Distillation", in *Encyclopedia of Separation Science*, Academic Press, London, 2000.

B. A. Hands, ed., *Cryogenic Engineering*, Academic Press, New York, 1986.

S. W. Van Sciver, *Helium Cryogenics*, Plenum Press, New York, 1986.

J. G. Weisend, *Handbook of Cryogenic Engineering*, Taylor & Francis, Philadelphia, Pa., 1998, Chapt. 7.

RAKESH AGRAWAL
D. MICHAEL HERRON
HOWARD C. ROWLES
GLENN E. KINARD
Air Products and Chemicals, Inc.

CRYSTAL ENGINEERING

Crystal engineering (CE hereafter) is the bottom-up construction of functional materials from the assembly of molecular or ionic components. CE applies the concepts of supramolecular chemistry to the solid state. In the supramolecular approach to crystalline solids, the crystals are seen as networks of interactions. These interactions can be covalent bonds between atoms (eg, diamond, silica, and graphite) as well as coordination bonds between ligands and metal centers, Coulombic attractions and repulsions between ions, and noncovalent bonds between neutral molecules (van der Waals, hydrogen bonds, etc) or—of course—any combination of these linkages. The difference in bonding types offers a practical way to subdivide CE target materials as a function of the energy involved in *local* bond breaking–bond forming processes. These bonding interactions follow an approximate ranking in energy: from the high enthalpies involved in breaking and forming of covalent bonds between atoms to the tiny energies involved in the van der Waals interactions between neutral atoms in neutral molecules.

HISTORICAL BACKGROUND

The qualifier *engineering* associated to crystals was first employed by G. Schmidt and collaborators at the Weitzmann Institute in the early 1970s to describe the photodimerization reaction of cinnamic acid and derivatives in the solid state.

Modern crystal engineering draws its strength from the synergistic interaction between design and synthesis of supermolecules on the one hand, and design and synthesis of crystalline materials with desired solid-state properties, on the other hand. In a way, the definition of supramolecular chemistry put forward by J.-M. Lehn in his Nobel lecture (*chemistry beyond the molecule bearing on the organized entities of higher complexity that result from the association of two or more chemical species held together by intermolecular forces*) seems to encompass crystal engineering. What is a (molecular) crystal if not an "organized entity of higher complexity held together by intermolecular forces"? Rather than thinking of a crystal as a "molecular container", ie, a box, in which molecules and ions with identical characteristics and properties can be conveniently packed, synthetic chemists have begun to think in "supramolecular" terms. The *collective properties* of the aggregate depend on the choice of intermolecular and interionic interactions between components and on the convolution of the properties of the building blocks with the periodicity of the crystal.

THE RANKING IN ENERGY AND THE SYNTHETIC STRATEGIES

A topological distinction needs to be made between molecular crystal engineering, where the building blocks are clearly recognizable molecular or ionic species, and coordination and covalent crystal engineering, which often utilize building blocks that do not exist as separate entities. Coordination crystal engineering, in particular, can be seen as periodic coordination chemistry, as the ligand–metal bonding capacity is projected in two (2D) or three dimensions (3D) to form extended networks (coordination polymers) by using polydentate ligands.

A second broad difference arises from the energies involved in the construction of the different types of crystalline materials. The construction of covalent networks (e.g., synthetic zeolites, or intercalates) usually requires larger energies than those required to prepare coordination networks or to assemble molecular crystals. Next, we will use the term *intermolecular* as a synonym of *noncovalent*, with this encompassing all types of secondary interionic or intermolecular interactions that do not imply two-electron σ bonds (eg, Coulombic interactions, hydrogen bonds, van der Waals interactions, and their combination). *Making crystals* on purpose requires an appreciation of the different energetic factors involving molecular crystals, which are held together by van der Waals interactions of the order of very few kilojoules per mole (kJ/mol), and those involving covalent bonds, which require hundreds of kilojoules per mole of energy to be broken and formed.

CRYSTAL ENGINEERING AND POLYMORPHISM

Crystal polymorphism, ie, the existence of more than one packing arrangement for the same molecular or ionic substance(s), could be a major drawback for the purposed bottom-up construction of functional solids. However, although the discovery of polymorphs of molecular crystals or of their diverse solvate forms (*pseudo*-polymorphs) is often serendipitous, crystal polymorphism can, to some extent, be controlled. Polymorphic and pseudo-polymorphic modifications of the same substance can also be obtained by thermal and mechanical treatment and by solvation and desolvation. An important discrimination is between polymorphs that interconvert via a solid–solid phase transition (enantiotropic systems) and those that melt before interconversion takes place (monotropic systems).

Conformational polymorphism occurs when a molecule possesses internal degrees of freedom, which allow the existence of different low-energy conformations, as in organic species, or different relative disposition of ligands, in a metal–organic species. Conformational polymorphism is a common characteristic of coordination and organometallic species, because of the often delocalized nature of the ligand–metal interactions and the consequent high conformational freedom. *Concomitant polymorphs* are those obtained from the same crystallization process. Pseudo-polymorphism refers to cases in which a given substance is known to crystallize with different amounts or types of solvent molecules. Even though polymorphic modifications contain exactly the same substance, they usually differ in chemical and physical properties such as density, diffraction pattern, solid-state spectroscopy, melting point, stability, reactivity, and mechanical properties.

AN OVERVIEW OF CRYSTAL ENGINEERING STRATEGIES

For the sake of this discussion, we have decided to describe different CE strategies on the basis of the energetics of the interactions involved, namely, very weak noncovalent interactions (van der Waals interactions), hydrogen bonds between neutral molecules and ions, coordination bonds, etc. It should be clear, however, that all intermediate situations are possible. Irrespective of the nature of the *principal* interaction, it should be kept in mind that every crystal represents a compromise between several, often nonconverging factors, such as the optimization of intramolecular interactions versus intermolecular interactions, together with that of less directional interactions, such as those of van der Waals nature, the electrostatic terms arising from dipoles, etc, or other interactions. Moreover, formation and rupture of weak or very weak interactions between component subunits in noncovalent syntheses imply small ΔH values. Cooperativity is thus required to overcome unfavorable entropy terms in order for the supramolecular aggregation process to become thermodynamically spontaneous.

CRYSTAL ENGINEERING BASED ON VAN DER WAALS INTERACTIONS

In solids made of discrete molecules without strong dipolar moments (often oversimplified as "van der Waals solids"), the attractive forces acting between *molecules*, regarded as ensembles of atoms, fall off very rapidly with the distance. Repulsions are effective at very short distances and much dependent on the nature of the peripheral atoms, which determine the electrostatic potential hypersurface surrounding the molecule. In this way, the bulk of the molecule provides attraction, while surface atoms determine recognition, optimum relative orientation, and interlocking of molecules in the solid state. In general, a given supramolecular arrangement in the solid state can be seen as the result of the minimization of short-range repulsions, rather than the optimization of attractions. It is therefore important, when considering a molecular crystal, to focus on the relationship between molecular shape and nature of the peripheral atoms.

In the absence of directing interactions, resulting, eg, from the presence of strong dipoles or hydrogen bonding donor–acceptor groups (see below), the recognition process will be controlled by the outer shape of the molecule and by the nature of the peripheral atoms. The formation of a stable dimolecular aggregate—as the initial step of a crystallization process—whether formed by the same molecule, ie, AA, or by two different molecules/ions, ie, AB, or $A^{+/-}B^{-/+}$, will depend primarily on the complementarity of shape.

An example of CE based only on van der Waals interactions is the preparation of one-dimensional (1D) van der Waals networks via calix[4]arene derivatives, bearing two receptor cavities arranged in a divergent fashion, and neutral molecules employed as linear connectors. The resulting 1D network, or *koilate*, is obtained by translation of the assembling core defined by the inclusion connector into the cavity of the receptor. Recognition, self-assembly, and cohesion of the solid-state networks are all based on van der Waals interactions.

CRYSTAL ENGINEERING BASED ON HYDROGEN BONDS

The hydrogen bond is the interaction of choice in molecular crystal engineering because it combines strength and directionality. Strength is synonym of cohesion and stability, while directionality implies topological control and selectivity, which guarantee reproducibility to the supramolecular assembly process. A directional, ie, selective, intermolecular interaction possesses specific topological properties and its performance within different structural environments can be predictable.

For most purposes, the hydrogen bond can be described as a stable interaction of essentially electrostatic nature between an X–H donor and a Y acceptor, being X and Y electronegative atoms or electron rich groups. The hydrogen-bonding interaction is generally stronger than the strongest van der Waals interaction. H···Y and X···Y separations shorter than van der Waals contact distances and X–H···Y angles that tend to linearity are considered diagnostic of the presence of strong hydrogen bonds. The same topological rules are followed by hydrogen-bonding interactions between ions, even though the energetic scale is different.

In general, strong donor–acceptor groups such as –COOH and –OH systems, as well as primary –$CONH_2$ and secondary –CONHR amido groups, form essentially the same type of hydrogen-bonding interactions whether as part of organic molecules or of metal coordinated ligands. This is not surprising, as hydrogen bonds formed by such strong donor and acceptor groups are at least one order of magnitude stronger than most noncovalent interactions and are most often already present in solution.

For the purposes of CE, the utilization of a single very strong interaction, such as the O–H···$O^{(-)}$, is not necessarily the best or only way to provide cohesion. The "Gulliver effect" can also be exploited: The collective strength of weaker bonds may be equivalent, in terms of cohesion, to the strength of a single bond, although the directionality component may be lost or greatly diminished.

Hydrogen Bonding and Crystal Engineering Involving Neutral Molecules

The usefulness of hydrogen-bonding interactions in CE is a direct function of the strength and predictability of the interaction. for this reason CE applications based on weak hydrogen bonds are less frequent. Weak hydrogen bonds, such as C–H···O, C–H···N, or C–H···π, are more important as ancillary interactions, whose optimization often determines the fine-tuning of the crystal packing, while molecular recognition and self-assembly are controlled by the stronger and more directional interactions. Nonetheless, optimization of weaker interactions may have dramatic consequences on the molecular arrangement in the solid state.

Crystal Engineering with Hydrogen Bonds Between Ions

A practical instrument in devising new solids is provided by the combined use of ionic charges (viz, Coulombic interactions) and hydrogen-bonding interactions. Since the hydrogen bond has a fundamentally electrostatic nature, the presence of ionic charges on the building blocks can be exploited to strengthen the interaction. Charge assistance to hydrogen bond is the enhancement of donor and acceptor systems polarity by utilizing cationic donors and anionic acceptors instead of neutral systems, ie, $X–H^{(+)}···Y^{(-)}$ rather than X–H···Y. The favorable location of ionic charges enhances both proton acidity and acceptor basicity in the solid state. Hydrogen-bonding interactions between ions optimally convolute the strength of the Coulombic field generated by the ions with the high level of directionality afforded by the X–H···Y interaction.

There are essentially two distinct strategies that utilize acid–base reactions to construct crystals via charge-assisted hydrogen bonds between ions: (1) Formation of hetero-ionic interactions, and (2) formation of homo-ionic interactions.

Crystal Engineering Involving Metal-Containing Species

In recent years the utilization of metal-containing compounds in CE application has represented the turning point of the discipline, because of the wide variety of combination of spin, charge, oxidation state, topology, let alone the specific chemical reactivity afforded by coordination compounds. The role of metal atoms in CE is both electronic and structural. Distinct functions of metal atoms can be identified: (1) A topological function, (2) an electronic function, (3) a (tunable) electrostatic function, (4) direct participation of metal atoms in intermolecular bonds: and (5) a templating function.

Crystal Engineering with Coordination Networks

Nowadays, coordination network engineering takes the "lion's share" of the scientific endeavors in the field of CE. The basic idea is that of utilizing the coordination bonding capacity of transition-metal atoms to build supramolecular arrangements in 3D; the result is the convolution of coordination chemistry with crystal periodicity, ie, periodic coordination chemistry.

The phenomena of self-entanglement and interpenetration represent the major obstacles to the preparation of crystalline materials with large and accessible empty space.

The most popular ligands are bidentate bipyridyl-type ligands, because of their well-known capacity for strongly binding late transition metals, leading to robust superstructures.

An important area of coordination networks is that constituted by porphyrin and metalloporphyrin systems. These supramolecular assemblies not only afford alternative ways for the construction of crystalline materials with large channels and cavities that mimic inorganic zeolites but are also investigated as model systems of light-harvesting and as molecular receptors and sieves.

Crystal Engineering with Coordination Networks and Hydrogen Bonds

Metal–ligand coordination and hydrogen-bonding interactions, although very different in electronic nature, possess high directionality features. Since directionality is essential for a controlled assembly of the components, the topological properties of both types of interactions can be exploited simultaneously in the construction of molecular solids with predefined architectures. The simplest approach utilizes ligands that, besides being able to coordinate to the metal centers, can also establish intermolecular hydrogen-bonding interactions. The simultaneous utilization of coordination bonds and hydrogen bonds affords an intermediate strategy whereby coordination complexes are linked via intermolecular hydrogen bonds. Whether these interactions are between neutral molecules or between charged species will, of course, depend on the electronic nature (oxidation state) of the metal center and on the formal charge carried by the ligands. This strategy allows combining the chemical and physical properties of coordination compounds with the features of typical organic solids. Since many coordination complexes are ions, the counterions often play a fundamental role in determining the topology of the superstructures that can be constructed.

SOLID-STATE REACTIVITY

Besides applications in molecular materials chemistry, however, CE encompasses some traditional branches of solid-state sciences, eg, CE initiated from an investigation of solid-state reactions, and indeed the understanding of the way molecules self-recognize and self-aggregate in the solid is the first step to devise novel solid-state processes. Importantly, most reactions occurring between solids or involving solids are solvent-free. Because of the strive for environmentally benign reaction conditions, the use of solvent-free conditions is attracting a wide interest. Another goal of great interest is the exploitation of solid–gas reactions as alternative routes for the construction of molecular traps, sieves, and sensors. The investigation of the reactivity of molecular crystals lies close to the origins of crystal engineering. The idea is that of organizing molecules in the solid state using the principles of molecular recognition and self-assembly.

Diverse applications of host–guest chemistry in a variety of crystalline organic inclusion compounds have been described. For example, inclusion compounds, in which chiral crystal structures are obtained from racemic or achiral molecules, have been investigated, with application of such compounds to the synthesis of species in which the crystal structure chirality is imprinted upon the achiral molecular components. When achiral molecules cannot be arranged in a chiral form in the crystal, they can be arranged in a chiral form in inclusion complex crystals with a chiral host compound. Reaction of the inclusion complex in the solid state has been shown to give the optically active compound.

Another important application of CE is in the investigation of reactions between engineered molecular solids and molecules in the vapor phase.

Uptake and release of solvent molecules (solvation, hydration) can often be paralleled to solid–gas reactions, whereby the reactants are, respectively, the molecules in the crystalline solid and in the gas phase, and the product is the solvated crystal. Clearly, the same reasoning applies to the reverse process, ie, generation of a new crystalline form by means of gas release. In gas–solid reactions, gases are reacted directly with crystals or amorphous phases to give solid products, often in quantitative yields.

Another relevant example of the use of crystalline coordination compounds to sense and trap molecules is that self-assembled organoplatinum(II) complexes, containing N,C,N terdentate coordinating anion "pincers", reversibly and quantitatively bind gaseous SO_2 in the solid state by Pt–S bond formation and cleavage, giving five-coordinate adducts. The five-coordinate adduct is also crystalline and the reverse reaction, namely, the release of SO_2, does not destroy the crystalline ordering. The Pt-complex can thus be seen as a crystalline supermolecule able to switch "on" and "off" as a direct response to gas uptake and release.

It can be argued that the *reaction* of a molecular solid, whether formed of organic, organometallic molecules or coordination compounds, with a vapor is conceptually related to the *supramolecular reaction* of a crystalline material with a volatile solvent to form a new crystalline solid. Indeed, the two processes, solid–gas reaction and solid–gas solvation, differ only in the energetic ranking of the interactions that are broken or formed through the processes. In solvation–desolvation processes, one is dealing mainly with noncovalent van der Waals or hydrogen-bonding interactions, while in chemical reactions covalent bonds are broken or formed.

One may purposefully plan to assemble molecules that are capable of absorbing molecules from the gas phase and, possibly, to react with them. Reaction implies *sensing* and could be exploited to detect molecules, if there is a measurable response from the solid state. If the reaction is quantitative and reversible, the same processes can be used to trap gases and deliver them where appropriate. The control on solid-state reactions, that can be used to trap environmentally dangerous or poisonous molecules, is an attractive goal for solid-state chemistry and crystal engineering.

D. Braga, F. Grepioni, and A. G. Orpen, eds., *Crystal Engineering: from Molecules and Crystals to Materials*, Kluwer Academic Publishers, Dordrecht, 1999.

D. Braga, F. Grepioni, and G. R. Desiraju, *Chem. Rev.* **98**, 1375 (1998); I. Haiduc and F. T. Edelmann, eds., *Supramolecular Organometallic Chemistry*, Wiley-VCH, Weinheim, 1999.

Dario Braga
Università degli Studi di Bologna
Fabrizia Grepioni
Università di Sassari

CRYSTALLIZATION

Crystallization is one of the oldest unit operations in the portfolio of industrial and/or laboratory separations. Almost all separation techniques involve formation of a second phase from a feed, and processing conditions must be selected that allow relatively easy segregation of the two or more resulting phases. This is also a requirement for crystallization, and there are a variety of other properties of the solid product that must be considered in the design and operation of a crystallizer. Interactions among process, function, product, and phenomena important in crystallization are illustrated in Figure 1.

There are several possible functions that can be achieved by crystallization: separation, concentration, purification, solidification, and analysis.

Products

In all of the instances in which crystallization is used to carry out a specific function, product requirements are a central component in determining the ultimate success of the process. These result as a consequence of how the product is to be used and the processing steps between crystallization and recovery of the final product. Key determinants of product quality are the size distribution (including crystal mean size and variance), the morphology (including habit or shape and form), and purity. Of these, only the last is important with other separation processes.

Process

In each of the systems discussed above there is a need to form crystals, to cause the crystals to grow, and to separate the crystals from residual liquid. There are various ways to accomplish these objectives leading to a multitude of processes, batch or continuous, that are designed to meet requirements of product yield, purity, and, uniquely, CSD.

Phenomena

The critical phenomena in crystallization are, as shown in Figure 1, generation of supersaturation, nucleation and growth kinetics, interfacial phenomena, breakage, and agglomeration.

SOLID–LIQUID EQUILIBRIA AND MASS AND ENERGY BALANCES

Solubility

Solid–liquid equilibrium, or the solubility of a chemical compound in a solvent, refers to the amount of solute that can be dissolved at constant temperature, pressure, and system composition; in other words, the maximum concentration of the solute in the solvent at static conditions. In a system consisting of a solute and a solvent, specifying system temperature and pressure fixes all other intensive variables. In particular, the composition of each of the two phases is fixed, and solubility diagrams of the type shown for a hypothetical mixture of R and S in Figure 2 can be constructed. Such a system is said to form an eutectic, ie, there is a condition at which both R and S crystallize into a solid phase at a fixed ratio that is identical to their ratio in solution. Consequently, there is no change in the composition of residual liquor as a result of crystallization.

Several features of the hypothetical system in Figure 2 can be used to illustrate proper selection of crystallizer operating conditions and limitations placed on the operation by system properties. Suppose a saturated solution at temperature T_1 is fed to a crystallizer operating at temperature T_2. Because the feed is saturated, the weight fraction of S in the feed is given as shown in Figure 2.

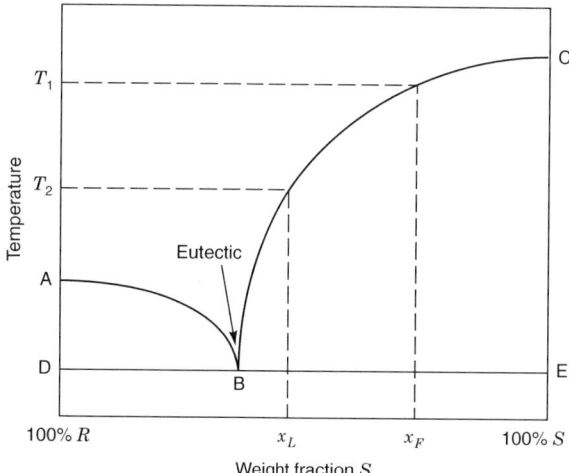

Figure 2. Solubility diagram for a hypothetical system. The curves AB and BC represent solution compositions that are in equilibrium with solids whose compositions are given by the lines AD and CE. If AD and CE are vertical along the respective axes, the crystals are pure R and S, respectively. Crystallization from any solution whose composition is to the left of the vertical line through point B produces crystals of pure R, whereas solutions to the right of the line produce crystals of pure S. A solution whose composition falls on the line through B produces a solid mixture that has a composition identical to the liquid solution.

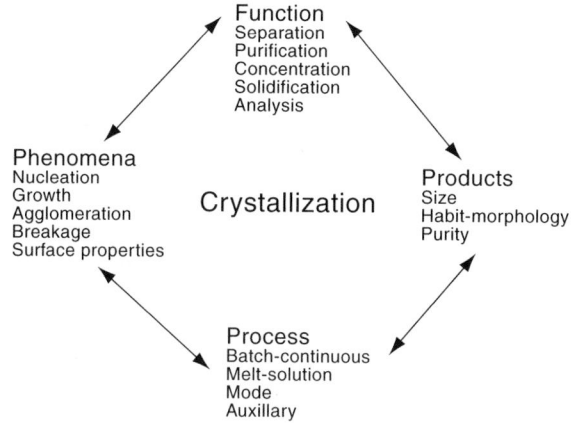

Figure 1. Crystallization.

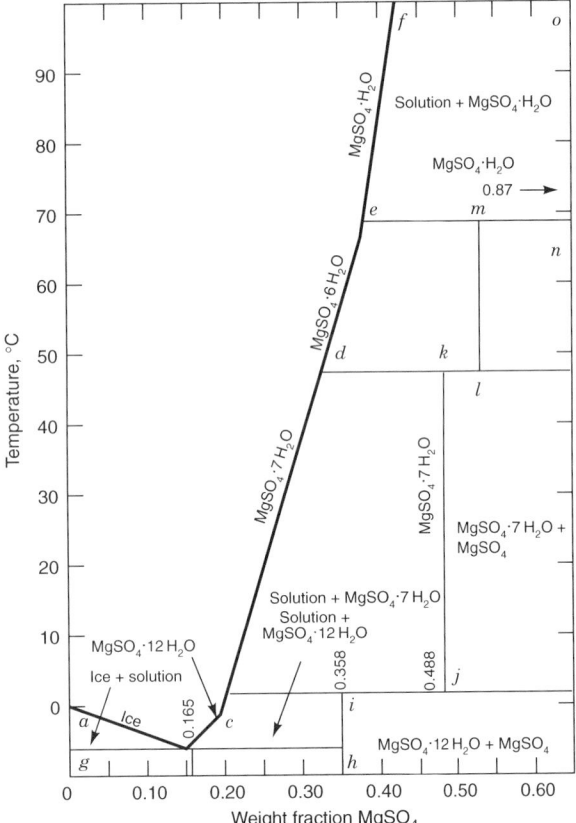

Figure 3. Solubility diagram for magnesium sulfate in water.

The maximum crystal production rate $P_{\max}$ from such a process depends on the value of T_2 and is given by

$$P_{\max} = F_{\chi F} - L_{\chi L} \qquad (1)$$

where F is the feed rate to the crystallizer, and L is the solution flow rate leaving the crystallizer. No other stream is fed to or removed from the crystallizer. Note that the lower limit on T_2 is given by the eutectic point B.

Figure 3 presents the equilibrium behavior of magnesium sulfate in water, and it is illustrative of systems that form hydrated salts. Equilibrium solution concentrations are plotted as curves ab, bc, cd, de, and ef; the solid phases that are in equilibrium with these solutions have compositions given by the lines ag, hi, jk, lm, and no, respectively. Ice is the solid phase whose composition is given by ag, and crystals containing differing ratios of water of hydration to magnesium sulfate constitute the solids represented by the other lines. Specifically, the line no represents magnesium sulfate monohydrate ($MgSO_4 \cdot H_2O$), which has one water molecule per molecule of magnesium sulfate, whereas the lines ml, kj, and ih represent the hexahydrate, heptahydrate, and dodecahydrate forms, respectively. The weight fraction of $MgSO_4$ in each of the crystal forms is shown in Figure 3, and as with all crystalline materials having water of hydration, the solute balance of equation 1 must be modified to read

$$\chi_c P_{\max} = F_{\chi F} - L_{\chi L} \qquad (2)$$

where x_c is the mass fraction of solute in the crystal, eg, x_c is 0.488 when the crystalline substance is magnesium sulfate heptahydrate. Differences in the forms of magnesium sulfate crystals affect the dependence of solubility on temperature, which is reflected by the slopes of the solution composition curves.

Supersaturation

The thermodynamic driving force for both crystal nucleation and growth is the key variable in setting the mechanisms and rates by which these processes occur. Supersaturation is defined rigorously as the deviation from thermodynamic equilibrium, which is the difference between the chemical potential of the solute in solution μ and the chemical potential of the solution in equilibrium with the solid phase μ^*. Less abstract definitions involving measurable system properties such as temperature, concentration, or mass or mole fraction also have been used to express supersaturation.

Mass and Energy Balances

The formulation of mass and energy balances follows procedures outlined in many basic texts. The use of solubilities to calculate crystal production rates from a cooling crystallizer is given by equations 1 and 2. Subsequent to determining the yield, the rate at which heat must be removed from such a crystallizer can be calculated from an energy balance:

$$F\hat{H}_F = P\hat{H}_C + L\hat{H}_L + Q \qquad (3)$$

where F, P, and L are feed rate, crystal production rate, and mother liquor flow rate, respectively; $\hat{H}$ is the specific enthalpy of the stream corresponding to the subscript; and Q is the required rate of heat transfer. As F, P, and L are known or can be calculated from a simple mass balance, determination of Q requires methods of estimating specific enthalpies.

If appropriate enthalpy data are unavailable, estimates (with such assumptions as constant heat capacities in the temperature range considered) can be obtained by first defining reference states for both solute and solvent. Often the most convenient reference states are crystalline solute and pure solvent at an arbitrarily chosen reference temperature. The reference temperature selected usually corresponds to that at which the heat of crystallization $\Delta\hat{H}_c$ of the solute is known. The heat of crystallization is approximately equal to the negative of the heat of solution.

The mass balance on a crystallizer is related to the growth kinetics occurring within the unit. This may be simplified by considering systems in which crystal growth kinetics are sufficiently fast to deplete essentially all of the supersaturation provided by the crystallizer. Under such conditions (referred to in the crystallization literature as class II or fast-growth behavior), the solute concentration in the mother liquor can be assigned a value corresponding to saturation. Alternatively, should supersaturation in the mother liquor be so great as to affect the solute balance, the operation is said to follow class I or slow-growth behavior. An expression coupling the rate of growth to a solute balance must be used to describe such a system.

CRYSTALLIZATION KINETICS

Along with operating variables of the crystallizer, nucleation and growth kinetics determine such crystal characteristics as size distribution, purity, and shape or habit.

Nucleation

Crystal nucleation is the formation of an ordered solid phase from a liquid or amorphous phase. Nucleation sets the character of the crystallization process, and it is, therefore, the most critical component in relating crystallizer design and operation to CSD.

Mechanisms. Classical nucleation theory is based on homogeneous and heterogeneous mechanisms, both of which involve the formation of crystals through a process of combining the constituent units (atoms, ions, or molecules) that form a crystal sequentially. Heterogeneous and homogeneous mechanisms are referred to as primary nucleation because existing crystals play no role in the nucleation.

Both homogeneous and heterogeneous mechanisms require relatively high supersaturation, and exhibit a high order dependence on supersaturation. These factors often lead to production of excessive fines in systems where primary nucleation mechanisms are important. The classical theoretical treatment of primary nucleation results for spherically shaped nuclei in the expression

$$B^0 = A \exp\left(-\frac{16\pi\sigma^3 v^2}{3k^3 T^3 [\ln(s+1)]^2}\right) \tag{4}$$

where B^0 is the nucleation rate at zero size, k is the Boltzmann constant, σ is surface energy per unit area, v is molar volume, and A is a constant. This equation can be simplified by recognizing that $\ln(s+1)$ approaches s as s approaches 0. So for small supersaturations,

$$B^0 = A \exp\left(-\frac{16\pi\sigma^3 v^2}{3k^3 T^3 s^2}\right) \tag{5}$$

The most important variables affecting nucleation rate are shown by equations 4 and 5 to be interfacial energy, temperature, and supersaturation.

Secondary nucleation is crystal formation through a mechanism involving the solute crystals; crystals of the solute must be present for secondary nucleation to occur. Several features of secondary nucleation make it more important than primary nucleation in industrial crystallizers. First, continuous crystallizers and seeded batch crystallizers have crystals in the magma that can participate in secondary nucleation mechanisms. Second, the requirements for the mechanisms of secondary nucleation to be operative are fulfilled easily in most industrial crystallizers. Finally, low supersaturation can support secondary but not primary nucleation, and most crystallizers are operated in a low supersaturation regime that improves yield and enhances product purity and crystal morphology.

Secondary nucleation can occur as the result of several mechanisms that have been identified in selected systems and include. *initial breeding, contact nucleation, and shear breeding.*

Process Variables Affecting Contact Nucleation. Pioneering studies elucidated many factors affecting contact nucleation. The number of crystals produced by a controlled impact of an object with a seed crystal depends on energy of impact, supersaturation at impact, supersaturation at which crystals mature, hardness of the impacting object, area of impact, angle of impact, and system temperature.

Crystal Growth

In most of the literature and in textbooks the model of two resistance's determining growth kinetics is promoted: (*1*) those associated with integration or incorporation of the crystalline unit (eg, solute molecules) into the crystal surface (lattice), and (*2*) molecular or bulk transport of the unit from the surrounding solution to the crystal face. However, it has been proven that the heat resistance is also of equal importance. This is especially important for melts and can in most cases be reduced to the a. m. two resistance's in the case of solution crystallization. The primary concern here is with surface incorporation.

Growth Models. Numerous models have been proposed to describe surface reaction kinetics, including those that assume crystals grow by layers and others that consider growth to occur by the movement of a continuous step. Each model results in a specific relationship between growth rate and supersaturation, but none can be used for a priori predictions of growth kinetics. Insights regarding the roles of certain process variables can be obtained, however, and with additional research, predictive capabilities may be achieved.

Models used to describe the growth of crystals by layers call for a two-step process: (*1*) formation of a two-dimensional (2D) nucleus on the surface and (*2*) spreading of the solute from the 2D nucleus across the surface. The relative rates at which these two steps occur give rise to the mononuclear 2D nucleation theory and the polynuclear 2D nucleation theory. In the mononuclear 2D nucleation theory, the surface nucleation step occurs at a finite rate, whereas the spreading across the surface is assumed to occur at an infinite rate. The reverse is true for the polynuclear 2D nucleation theory.

The screw dislocation theory, often referred to as the BCF theory (after its formulators Burton, Cabrera, Frank), shows that the dependence of growth rate on supersaturation can vary from a quadratic relationship at low supersaturation to a linear relationship at high supersaturation.

All the models described above indicate the importance of system temperature on growth rate. Dependencies of growth kinetics on temperature are often expressed in terms of an Arrhenius expression:

$$k_G = k_G^0 \exp\left(-\frac{\Delta E_G}{RT}\right) \qquad (6)$$

where k_G is a growth rate coefficient of the type required in equation 23, K_G^0 is a constant, and ΔE_G is an activation energy.

Effects of Impurities and Solvent. The presence of impurities usually decreases the growth rates of crystalline materials, and problems associated with the production of crystals smaller than desired are commonly attributed to contamination of feed solutions. Strict protocols should be followed in operating units upstream from a crystallizer to minimize the possibility of such occurrences. Equally important is monitoring the composition of recycle streams to prevent possible accumulation of impurities. Furthermore, crystallization kinetics used in scaleup should be obtained from experiments on solutions as similar as possible to those expected in the full-scale process.

The effects of a solvent on growth rates have been attributed to two phenomena: one has to do with the effects of solvent on mass transfer of the solute through changes in viscosity, density, and diffusivity; the second is concerned with the structure of the interface between crystal and solvent.

Crystal Growth in Mixed Crystallizers. Multicrystal magma studies usually involve examination of the rate of change of a characteristic crystal dimension or the rate of increase in the mass of crystals. The characteristic dimension depends on the method used in the determination of size; eg, the second-largest dimension is measured by sieve analyses, whereas electronic-zone-sensing instruments provide estimates of an equivalent spherical diameter, and laser-light-scattering gives a dimension close to the largest dimension of a particle, if it is randomly oriented relative to the laser beam path.

Anomalous growth means that growth rates of crystals in a magma are not identical or that the growth rate of an individual crystal or mass of crystals is not constant. Two theories have been used to explain growth rate anomalies: size-dependent growth and growth rate dispersion. Both alter the form of the population density function obtained from perfectly mixed continuous crystallizers; unfortunately, such behavior cannot be used to distinguish between size-dependent growth and growth rate dispersion, as both have the same qualitative effects on population density.

Size-Dependent Crystal Growth

A number of empirical expressions correlate the apparent effect of crystal size on growth rate. The most commonly used correlation uses three empirical parameters to correlate growth rate with crystal size:

$$G = G^0(1 + \gamma L)^b \qquad b < 1 \qquad (7)$$

where G^0, γ, and b are determined from experimental data. There have been attempts to relate the kinetic parameter b to crystallizer variables. The only success in this regard showed a qualitative dependence on crystallizer volume. The conclusion is that there is no size-dependent growth, however, the use of such a model leads in most cases to sufficiently precize results to work with for industrial use.

Growth Rate Dispersion

This phenomenon is the exhibition of different growth rates by crystals in a magma, even though they may have the same size and are exposed to identical conditions. It is now generally accepted that many observations originally attributed to size-dependent growth were due to growth rate dispersion. Such erroneous interpretations were the result of similarities in the effects of the two types of behavior on CSD.

Two distinctly different mechanisms leading to growth rate dispersion have experimental support. The first assumes that all crystals have the same time-averaged growth rate, but the growth rates of individual crystals fluctuate about some mean value. The second assumes that crystals are formed with a characteristic distribution of growth rates, but individual crystals retain a constant growth rate throughout their residence in a crystallizer.

Although evidence exists for both mechanisms of growth rate dispersion, separate mathematical models were developed for incorporating the two mechanisms into descriptions of crystal populations: random growth rate fluctuations and growth rate distributions. Both mechanisms can be included in a population balance to show the relative effects of the two mechanisms on crystal size distributions from batch and continuous crystallizers.

CRYSTAL CHARACTERISTICS

The morphology (including crystal shape or habit), size distribution, agglomeration, and purity of crystalline materials can determine the success in fulfilling the function of a crystallization operation.

Morphology

If atoms, molecules, or ions of a unit cell are treated as points, the lattice structure of the entire crystal can be shown to be a multiplication in three dimensions of the unit cell. Only 14 possible lattices (called Bravais lattices) can be drawn in three dimensions. These can be classified into seven groups based on their elements of symmetry. Moreover, examination of the elements of symmetry (about a point, a line, or a plane) for a crystal shows that there are 32 different combinations (classes) that can be grouped into seven systems. The correspondence of these seven systems to the seven lattice groups is shown in Table 1.

The general shape of a crystal is referred to as its habit. The appearance of the crystalline product and its processing characteristics (such as washing and filtration) are affected by crystal habit. Relative growth rates of the

Table 1. The 14 Bravais Lattices

Type of symmetry	Lattice	Crystal system
cubic	cube	regular
	body-centered cube	
	face-centered cube	
tetragonal	square prism	tetragonal
	body-centered square prism	
orthorhombic	rectangular prism	orthorhombic
	body-centered rectangular prism	
	rhombic prism	
	body-centered rhombic prism	
monoclinic	monoclinic parallelepiped	monoclinic
	clinorhombic	
triclinic	triclinic parallelepiped	triclinic
rhomboidal	rhombohedron	triclinic
hexagonal	hexagonal prism	hexagonal

faces of a crystal determine its shape. Faster growing faces become smaller than slower growing faces and, in the extreme case, may disappear from the crystal altogether. Growth rates depend on the presence of impurities, rates of cooling, temperature, solvent, mixing, and supersaturation. Furthermore, the importance of each of these factors may vary from one crystal face to another.

Polymorphism is a condition in which chemically identical substances may crystallize into different forms. Transitions from one polymorphic form to another may be accompanied by changes in process conditions (temperature, pessure, shear or solution composition), transitions from one polymorphic form to another and lead to formation of a solid product with unacceptable properties (eg, melting point or dissolution rate).

Agglomeration

Many of the analyses of industrial crystallizers require that the particle recovered from the crystallizer consist of a single crystal. Many of the properties of the crystal are affected deleteriously by agglomeration. Purity, eg, typically is diminished when agglomeration occurs. Countering the negative aspects of agglomeration is recognition that in many systems the single crystals produced by normal crystal growth would be too small to be separable using conventional solid–liquid separation equipment. In such instances, there would be no recoverable product without agglomeration.

Purity

Although crystallization has been employed extensively as a separation process, purification techniques using crystallization have become of growing importance. Mechanisms by which impurities can be incorporated into crystalline products include adsorption of impurities on crystal surfaces, solvent entrapment in cracks, crevices and agglomerates, and inclusion of pockets of liquid. It has been noted that the key to producing high purity

crystals was to maintain the supersaturation at a low level so that large crystals were obtained. Others have found that reducing the size of ammonium perchlorate crystals resulted in a substantial decrease in moisture due to inclusion.

Crystal Size Distributions

Particulate matter produced by crystallization has a size distribution that varies in a definite way over a specific size range. A CSD is most commonly expressed as a population (number) distribution relating the number of crystals at each size to a size or as a mass (weight) distribution expressing how mass is distributed over the size range. The two distributions are related and affect many aspects of crystal processing and properties, including appearance, solid–separation, purity, reactions, dissolution, and other properties involving surface area.

POPULATION BALANCES AND CRYSTAL SIZE DISTRIBUTIONS

Population balances and crystallization kinetics may be used to relate process variables to the CSD produced by the crystallizer. Such balances are coupled to the more familiar balances on mass and energy. It is assumed that the population distribution is a continuous function and that crystal size, surface area, and volume can be described by a characteristic dimension L. Area and volume shape factors are assumed to be constant, which is to say that the morphology of the crystal does not change with size.

Determination of Crystallization Kinetics

From a series of runs at different operating conditions, a correlation of nucleation and growth kinetics with appropriate process variables can be obtained; the resulting correlation can then be used to guide either crystallizer scaleup or the development of an operating strategy for an existing crystallizer. The variables affecting nucleation and growth kinetics include temperature, supersaturation, magma density, and external stimuli, such as agitation or circulation rate of the magma.

Mass Balance Constraints

The following mass balance on solute can be constructed from the schematic diagram of a continuous crystallizer:

$$Q_i c_i = Q_0 c_0 + Q_0 M_T \tag{8}$$

c_0 is determined by system kinetics and constrained by a solid–liquid equilibrium (solubility) relationship, which gives the equilibrium concentration c^* at the system conditions. The system (solute–solvent and crystallizer) is characterized by the magnitude of the supersaturation $(c_0 - c^*)$ remaining in the solution exiting the crystallizer. If the mass balance is closed by substituting c^* for c_0 in equation 8, the system is said to be a fast-growth or class II system. If the mass balance is not closed, significant supersaturation remains in the solution, the system is said to be a slow-growth or class I system.

CSD Characteristics for MSMPR Crystallizers

The perfectly mixed crystallizer described in the preceding discussion is highly constrained and the functional form of CSD produced by such systems is fixed. Such distributions have the following characteristics:

1. Moments of the distribution can be calculated for MSMPR crystallizers by the simple expression

$$m_j = j \ln^0 (G\tau)^{j+1} \qquad (9)$$

2. The dominant crystal size L_D is given by $L_D = 3 \, G\tau$. This quantity is also the ratio m_3/m_2, which is often given the symbol $\bar{L}\bar{L}_{3,2}$.

3. The spread of the mass density function about the dominent gives a cv of 50% for an MSMPR crystallizer.

4. The magma density M_T (mass of crystals per unit volume of slurry or liquor) may be obtained from the third moment of the population density function and is given by

$$M_T = 6\rho k_v n^0 (G\tau)^4 \qquad (10)$$

5. Kinetic parameters for nucleation and growth rate can be used to predict the CSD for a given set of crystallizer operating conditions. Variation in one of the kinetic parameters without changing the other is not possible. Accordingly, the relationship between these parameters determines the ability to alter the characteristic properties (such as dominant size) of the distribution obtained from an MSMPR crystallizer.

Preferential Removal of Crystals

Crystal size distributions produced in a perfectly mixed continuous crystallizer are highly constrained; the form of the CSD in such systems is determined entirely by the residence time distribution of a perfectly mixed crystallizer. Greater flexibility can be obtained through introduction of selective removal devices that alter the residence time distribution of materials flowing from the crystallizer. The role of classified removal is best described in terms of idealized models of clear-liquor advance, classified-fines removal, and classified-product removal.

Clear-liquor advance is simply the removal of mother liquor from the crystallizer without simultaneous removal of crystals. The primary objective of *fines removal* is preferential withdrawal from the crystallizer of crystals whose size is below some specified value. Such crystals may be redissolved and the resulting solution returned to the crystallizer. *Classified-product removal* is carried out to remove preferentially those crystals whose size is larger than some specified value.

Batch Crystallization

Crystal size distributions obtained from batch crystallizers are affected by the mode used to generate

supersaturation and the rate at which supersaturation is generated.

CRYSTALLIZERS AND CRYSTALLIZATION OPERATIONS

Crystallization equipment can vary in sophistication from a simple stirred tank to a complicated multiphase column, and the operation can range from allowing a vat of liquor to cool through exchanging heat with the surroundings to the complex control required of batch cyclic operations. In principle, the objectives of these systems are all the same: to produce a pure product at a high yield in an acceptable retention time with, in many cases, a desired CSD. However, the characteristics of the crystallizing system and desired properties of the product often dictate that a specific crystallizer be used in a particular operating mode.

Crystallization from Solution

Crystallization techniques are related to the methods used to induce a driving force for solids formation and to the medium from which crystals are obtained. Approaches include cooling crystallizers, evaporative crystallizers, evaporative-cooling crystallizers, salting-out or antisolvent crystallization, reactive crystallization, and supercritical fluid solvents.

Crystallizers

The basic requirements of a system involving crystallization from solution are as follows: (*1*) a means of generating supersaturation in a fashion commensurate with the requirements of producing a satisfactory CSD, (*2*) a vessel to provide sufficient residence time for crystals to grow to a desired size, and (*3*) mixing to provide a uniform environment for crystal growth. There are numerous manufacturers of crystallization equipment; in addition, many chemical companies design their own crystallizers based on expertise developed within their organizations. Crystallizers include the forced-circulation crystallizer, the feed in cooling crystallizers, the draft-tube-baffle (DTB) crystallizer, and the Oslo-type unit.

Melt Crystallization

The use of a solvent can be avoided in some systems. In such cases, the system operates with heat as a separating agent, as do several processes involving crystallization from solution, but formation of crystalline material is from a melt of the crystallizing species rather than a solution.

For the following reasons, melt crystallization holds great promise in situations in which it can substitute for crystallization from solution: (*1*) Without the need to recover and maintain the purity of a solvent, processing costs are reduced substantially. (*2*) Because there is no contaminated solvent to handle, melt crystallization may be more environmentally benign. (*3*) Energy costs found in evaporative crystallization obviously would be reduced if it is possible to produce a desired solid without the need to evaporate solvent (*4*). Melt crystallization can yield high selctivities (5). Melt crystallization may be a reasonable alternative to other separation and purification processes, because the heat of vaporization of most volatile organic materials is between two and five times their heat of fusion (in case of water, seven) and the temperature level is much lower than in atmospheric evaporive processes. An analysis of the energy requirements in melt processes has shown that such processes can compete with other thermal separation techniques only if the plant is well designed and the process precisely controlled.

S. Henning, S. Niehörster, and J. Ulrich, in A. S. Myerson, D. A. Green, and P. Meenan, eds., *Crystal Growth of Organic Materials*, American Chemical Society, Symposium Series, Washington, D. C., 1996, pp. 163–171.

S. Henning and J. Ulrich, in J. Ulrich, ed., *Proceedings of the 4th International Workshop on Crystal Growth of Organic Materials*, Shaker Verlag, Aachen, Germany, 1997, pp. 269–276.

D. Kashehiev, *Nucleation: Basic Theory with Application*, Butterworth-Heinemann, Oxford, 2000.

A. S. Myerson, *Molecular Modeling Applications in Crystalization*, Cambridge University Press, Cambridge, U.K., 1999.

JOACHIM ULRICH
Martin Luther University

CUMENE

Cumene (1-methylethylbenzene, 2-phenylpropane, isopropylbenzene), C_9H_{12}, is an aromatic compound. It is a clear liquid at ambient conditions. High purity cumene is manufactured from propylene and benzene. It is used primarily for the manufacturing of phenol and its coproduct acetone, which are two important petrochemicals with many uses in the chemical and polymer industries.

PROPERTIES

Physical, chemical, and thermodynamic properties of cumene are listed in Table 1.

MANUFACTURE

The three main manufacturing processes are the solid phosphoric acid (SPA) process, the aluminum chloride process, and the Mobil/Badger Zeolite-based cumene process. Most existing aluminum chloride cumene plants can be converted to the Mobil/Badger process with a capacity expansion of 30 to >100% with minimal capital investment.

Table 1. Some Properties of Cumene

Property	Value
freezing point, °C	−96.03
boiling point, °C	152.39
density, g/cm³	
refractive index, n_D^{20}	1.4915
thermal conductivity at 25°C, W/(m · K)	0.124
viscosity, mPa · s (= cP)	
surface tension at 20°C, mN/m (= dyn/cm)	28.2
vapor pressure, kPa[a]	
35°C	1
180°C	196
flash point, °C	33
relative molar mass	120.2
critical temperature, °C	358.0°C
critical pressure, kPa[a]	3220
critical density, g/cm³	0.280
heat of vaporization at bp, J/g[c]	312
heat of formation (liquid) at 25°C, J/mol[c]	−44,150
heat capacity (liquid) at 25°C, J/(mol · K)[c]	197
odor threshold, ppmv	1.2
threshold limit value, ppmv	50

[a]Calculated from the equation: $\ln P = A - B/(t + C)$; where t = temp, °C; P = vapor pressure, kPa[b]; A = 13.99; B = 3400; and C = 207.78.
[b]To converted kPa to mm Hg, multiply by 7.5, to atm, divide by 101.3.
[c]To convert from J to cal, divide by 4.184.

HEALTH AND SAFETY FACTORS

Cumene is a significant fire hazard when exposed to flame or sparks and is in the class of liquids that can be ignited under almost all normal temperature conditions. In fighting cumene fires, use water, dry chemical, foam, or carbon dioxide. Use water spray to keep fire-exposed containers cool. Water may be ineffective in controlling or extinguishing cumene fires.

Cumene is a primary skin and eye irritant. The recommended threshold limit value (TLV) is 50 ppm (245 mg/m³), which is an 8-h time-weighted average for exposure to cumene. The permissible exposure limit for cumene, which is an 8-h time-weighted average given by the Occupational Safety and Health Administration (OSHA), is also 50 ppm (245 mg/m³), with a skin designation.

USES

Virtually all the cumene produced is used as feedstock for the production of phenol (qv) and its coproduct acetone (qv). Phenol, in its various purity grades, is used for bisphenol-A employed in making epoxy resins (qv) and polycarbonates (qv) for phenolic resins to bond construction materials and for caprolactam (qv), the starting material for nylon-6. Minor amounts are used for alkylphenols (qv) and others. Cumene in minor amounts is used as a thinner for paints, enamels, and lacquers. It is also used to produce acetophenone, dicumylperoxide, and DIPB and is a good solvent for fats and resins. As such, cumene has been suggested as a replacement for benzene in many of its industrial applications.

E. Camara, "Cumene" in *Chemical Economics Handbook*, Stanford Research Institute International, Menlo Park, Calif., March 1999.

Encyclopedia of Occupational Health and Safety, Vols. I and II, Geneva, Switzerland, 1983.

Hazardous Substances Databank (HSDB), a database of the National Library of Medicine's TOXNET system (http://toxnet.nlm.nih.gov), April 10, 2001.

K. Verschueren, *Handbook of Environmental Data on Organic Chemicals*, 3rd ed., New York, N.Y., Van Nostrand Reinhold Co., 1996.

S. Y. HWANG
S. S. CHEN
Washington Group International

CYANAMIDES

It has been suggested that under primordial conditions, cyanamide could have acted as the original peptide-forming and phosphorylating reagent at the beginning of life on earth. Structural formulas of cyanamide, CH_2N_2 (**1**), and its dimer $C_2H_4N_4$ (**2**), and trimer $C_3H_6N_6$ (**3**), are given as follows:

CYANAMIDE

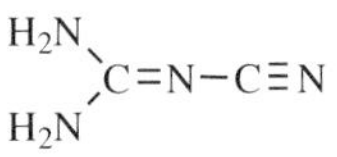

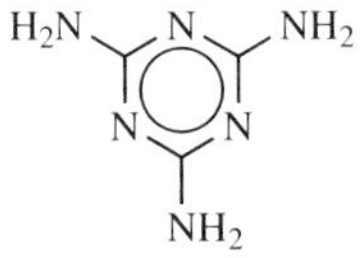

Properties

Cyanamide, also called carbamodiimide or carbamic acid nitrile, crystallizes from a variety of solvents as somewhat unstable, colorless, orthorhombic, deliquescent crystals. The properties of cyanamide are listed in Table 1.

Cyanamide is a weak acid with a very high solubility in water. It is completely soluble at 43°C, and has a minimum solubility (eutectic) at −15°C. It is highly soluble in polar organic solvents, such as the lower alcohols, esters, and ketones, and less soluble in nonpolar solvents.

Table 1. Properties of Cyanamide

Property	Value
molecular weight	42.04
mp, °C	46
bp, 101 kPa,[a] °C	dec
specific heat at 0–39°C, J/(g K)[c]	2.288
heat of formation at 25°C, kJ/mol[c]	58.77
heat of solution[d] at 15°C, kJ/mol[c]	−15.05
heat of combustion at 25°C, kJ/mol[c]	−737.9

[a]To convert kPa to mm Hg, or Pa to μm Hg, multiply by 7.5.
[b]Calculated.
[c]To convert J to cal, divide by 4.184.
[d]In 1000 parts H_2O.

Reactions

Reactions of cyanamide are either additions to the nitrile group or substitutions at the amino group. Both are involved in the dimerization to dicyandiamide.

Manufacture

Calcium cyanamide can be manufactured by either the Frank–Caro batch oven process or continuous processes such a rotary furnace.

Shipment and Handling

In general, cyanamide should be added to a reaction mixture at such a rate that it is used up as it is added, otherwise a high concentration of cyanamide results which could react violently.

Health and Safety Factors

Manufacture of cyanamide and calcium cyanamide does not present any serious health hazard. Ingestion of alcoholic beverages by workmen within several hours of leaving work sometimes results in a vasomotor reaction known as cyanamide flush.

Commercial grades of calcium cyanamide contain lime and are moderate skin irritants where contact is repeated or prolonged.

Contact or ingestion of cyanamide must be avoided, and precautions taken to prevent inhalation of dust or spray mist. The compound is, therefore, considered to be moderately toxic both by ingestion in single doses and by single-skin applications.

Uses

The most important use of calcium cyanamide is as fertilizer, but it is also effective as a herbicide and defoliant. It was used as a starting material for ammonia until it was displaced by the Haber process.

DICYANDIAMIDE

Properties

Dicyandiamide (**2**) (cyanoguanidine) is the dimer of cyanamide and crystallizes in colorless monoclinic prisms. It is amphoteric, and generally soluble in polar solvents and insoluble in nonpolar solvents. Its properties are listed in Table 2.

Table 2. Properties of Dicyandiamide

Property	Value
mol wt	84.08
mp, °C	208
bp, °C	dec
heat of formation at 25°C, kJ/mol[a]	24.9
heat of combustion at 25°C, kJ/mol[a]	−1382
heat of solution at 15°C, kJ/mol[a]	−24.1

[a]To convert J to cal, divide by 4.184.

Reactions

The reactions of dicyandiamide resemble those of cyanamide. However, cyclizations take place easily and the nitrile group is less reactive.

Manufacture

Dicyandiamide is manufactured by dimerization of cyanamide in aqueous solution.

Uses

Dicyandiamide is used as a raw material for the manufacture of several chemicals, such as guanamines, biguanide and guanidine salts, and various resins. Since 1975, it has also been used in the manufacture of potassium or sodium dicyanamide which are used as insecticides and in chemotherapy. Melamine has extensive applications in the resin and plastic industry; guanamines are used as copolymers (qv) in many resin compositions. Guanidine phosphate is employed as a fire retardant in applications where water solubility is not a drawback.

MELAMINE

Properties

The outstanding characteristic of melamine, usually a white crystalline material, is its insolubility in most organic solvents. This property is also evident in melamine resins after they are cured. On the other hand, melamine is appreciably soluble in water, its solubility increasing with increased temperature.

Manufacture

Dicyandiamide is converted into melamine by heating.

Toxicity

Extensive toxicity investigations performed with melamine in experimental animals suggest that the compound may have a low order of biological activity.

Human subjects were given patch tests with melamine. No evidence of either primary irritation or sensitization was found. Such results suggest that melamine crystal may be handled in ordinary industrial use without special hygienic precautions.

Uses

Most of the melamine produced is used in the form of melamine–formaldehyde resins (see AMINO RESINS AND PLASTICS). Other applications include the use of melamine pyrophosphate in fire retardant textile finishes, chlorinated melamine as a bactericide, and melamine as a tarnish inhibitor in detergent compositions, in papermaking, and manufacture of adhesives.

Cyanamide, Technical bulletin, American Cyanamid Co., Wayne, N.J., 1966.

J. W. Lyons, *The Chemistry and Use of Fire Retardants*, John Wiley & Sons, Inc., New York, 1970, p. 136.

WILLIAM CAMERON
William Cameron Consulting

CYANIDES

HYDROGEN CYANIDE

Hydrogen cyanide (hydrocyanic acid, prussic acid, formonitrile), HCN, is a colorless, poisonous, low viscosity liquid having an odor characteristic of bitter almonds. The compound has been known and used as a poison for decades. Today, hydrogen cyanide is used in the manufacture of many important chemicals.

It is theorized that hydrogen cyanide played the key role in the origin of plant and animal life on earth via formation of amino acids. Hydrogen cyanide is present in the normal human being's blood. People who smoke or who consume vegetables having relatively high cyanide content have slightly higher blood concentrations.

Hydrogen cyanide is a basic chemical building block for such chemical products as adiponitrile to produce nylon, methyl methacrylate to produce clear acrylic plastics (see ACRYLIC ESTER POLYMERS), sodium cyanide for recovery of gold (see GOLD AND GOLD COMPOUNDS), triazines for agricultural herbicides, methionine for animal food supplement (see FEEDS AND FEED ADDITIVES), chelating agents for water treatment, and many more.

Properties

The physical properties of hydrogen cyanide are listed in Table 1.

Hydrogen cyanide is a weak acid. Its structure is that of a linear, triply bonded molecule, $HC{\equiv}N$.

Hydrogen cyanide, as the nitrile of formic acid, CH_2O_2, undergoes many of the typical nitrile reactions.

Hydrogen cyanide can be oxidized by air at 300–650°C over silver or gold catalyst to give yields of up to 64% cyanic acid, HOCN, and 26% cyanogen, $(CN)_2$. Reaction with chlorine in the liquid phase gives cyanogen chloride, CClN, which is the basic route to triazines of which melamine, $C_3H_6N_6$, is an important derivative (see CYANAMIDES; UREA). Bromine reacts similarly, but the reaction with iodine is incomplete.

Hydrogen cyanide adds to an olefinic double bond most readily when an adjacent activating group is present in the molecule, eg, carbonyl or cyano groups.

Hydrogen cyanide adds across the carbonyl group of aldehydes and ketones and Ketones and opens the oxirane ring of epoxides, both under mildly basic conditions. Several of these cyanohydrins are commercially important.

Hydrogen cyanide reacts with formaldehyde and aniline to form *N*-phenylglycinonitrile, and with formaldehyde alone to form glycolonitrile. Hydrogen cyanide reacts with NaOH, KOH, and $Ca(OH)_2$ to form the

Table 1. Physical Properties of Hydrogen Cyanide

Property	Value
molecular weight	27.03
melting point, °C	−13.24
boiling point, °C	25.70
density, g/mL	
0°C	0.7150
20°C	0.6884
specific gravity of aqueous solutions[a]	
10.04% HCN	0.9838
60.23% HCN	0.829
vapor pressure, kPa[b]	
−29.5°C	6.697
27.2°C	107.6
vapor specific gravity, at 31°C[c]	0.947
surface tension at 20°C, mN/m (= dyn/m)	19.68
liquid viscosity at 20.2°C, mPa · s(= cP)	0.2014
specific heat, J/mol[d]	
−33.1°C, liquid	58.36
27°C, gas	36.03
heat of fusion at −14°C, kJ/mol[d]	7.1×10^3
heat of formation, ΔH_f, kJ/mol[d]	
gas at 25°C	−130.5
liquid at 25°C	−105.4
heat of combustion, net, kJ/mol[d]	642
heat of vaporization, kJ/mol[d]	25.2
heat of polymerization, kJ/mol[d]	42.7
flash point, closed cup, °C	−17.8
explosive limits in air at 100 kPa[b] and 20°C, vol %	6–41
autoignition temperature, °C	538

[a]Measured at 18°C, compared to water at 18°C.
[b]To convert kPa to mm Hg, multiply by 7.5.
[c]Air = 1.
[d]To convert J to cal, divide by 4.184.

corresponding cyanides. Amines can be derived from olefins and hydrogen cyanide via the Ritter reaction.

Cyanohydrins (qv) are formed by the reaction of glucose and similar compounds with hydrogen cyanide. The corresponding aminonitrile from methyl isobutyl ketone can be formed with ammonia and hydrogen cyanide.

Dimethylformamide can be produced from the reaction of hydrogen cyanide and methanol. Adenine can be prepared from hydrogen cyanide in liquid ammonia. Thioformamide can be produced from hydrogen cyanide and hydrogen sulfide.

Under certain conditions hydrogen cyanide can polymerize to black solid compounds, eg, hydrogen cyanide homopolymer and hydrogen cyanide tetramer, $C_4H_4N_4$.

Although hydrogen cyanide is a weak acid and is normally not corrosive, it has a corrosive effect under two special conditions: (1) water solutions of hydrogen cyanide cause transcrystalline stress cracking of carbon steels under stress even at room temperature and in dilute solution and (2) water solutions of hydrogen cyanide containing sulfuric acid as a stabilizer severely corrode steel (qv) above 40°C and stainless steels above 80°C.

Manufacture and Processing

Hydrogen cyanide has been manufactured from sodium cyanide and mineral acid, and from formamide by catalytic dehydration. As of this writing, primarily because of high raw material costs, only one manufacturer uses the formamide route and one plans to use the sodium cyanide route for small quanities.

Two synthesis processes account for most of the hydrogen cyanide produced. The dominant commercial process for direct production of hydrogen cyanide is based on classic technology involving the reaction of ammonia, methane (natural gas), and air over a platinum catalyst; it is called the Andrussow process. The second process involves the reaction of ammonia and methane and is called the Blausäure-Methan-Ammoniak (BMA) process; it was developed by Degussa in Germany. Hydrogen cyanide is also obtained as a by-product in the manufacture of acrylonitrile (qv) by the ammoxidation of propylene (Sohio process).

The Shawinigan process uses a unique reactor system. The heart of the process is the fluohmic furnace, a fluidized bed of carbon heated to 1350–1650°C by passing an electric current between carbon electrodes immersed in the bed. Feed gas is ammonia and a hydrocarbon, preferably propane. High yield and high concentration of hydrogen cyanide in the off gas are achieved. This process is presently practiced in Spain, Australia, and South Africa.

Health and Safety Factors

The cyanides are true noncumulative protoplasmic poisons, ie, they can be detoxified readily. Cyanide combines with those enzymes at the blood tissue interfaces that regulate oxygen transfer to the cellular tissues. Unless the cyanide is removed, death results through insufficient oxygen in the cells. The warning signs of cyanide poisoning include dizziness, numbness, headache, rapid pulse, nausea, reddened skin, and bloodshot eyes. More prolonged exposure can cause vomiting and labored breathing followed by unconsciousness; cessation of breathing; rapid, weak heart beat; and death. Severe exposure by inhalation can cause immediate unconsciousness. Hydrogen cyanide can enter the body by inhalation, oral ingestion, or skin absorption.

First Aid and Medical Treatment. Action should be fast and efficient. With the protection of a gas mask remove or drag the victim to fresh air. Remove contaminated clothing and rinse contaminated body areas. Keep victim warm. If the victim is conscious and speaking, no treatment is necessary. If the victim is unconscious but breathing, break an ampul of amyl nitrite in a cloth and hold it under the victim's nose for 15 s. Repeat five or six times. Use a fresh ampul every 3 min. Continue until the victim regains consciousness. If the patient is not breathing, apply artificial respiration; this can best be done using an oxygen resuscitator. The amyl nitrite antidote should also be administered during resuscitation. Mouth-to-mouth resuscitation is the next-best method followed by the Holger-Mielsen arm-lift method. Notify a physician immediately.

Disposal. Small quantities of concentrated hydrogen cyanide can be burned in a hood in an open vessel. Large-scale burning in outdoor pans can be performed, but special safety precautions must be employed. A cyanide solution can be decontaminated by making the solution strongly basic (pH 12) with caustic and pouring it into ferrous sulfate solution. The resulting ferrocyanide is relatively nontoxic. Cyanide solution can be converted to less toxic cyanate by treatment with chlorine, sodium or calcium hypochlorite, or ozone at pH 9 to 11. A solution of 10% hypochlorite maximum should be used. The final solution should be checked for absence of free cyanide. The hypochlorite or $Cl_2 + NaOH$ method is by far the most widely used commercially.

Environmental. The toxicity of cyanide in the aquatic environment or natural waters is a result of free cyanide, ie, as HCN and CN^-. Much work has been done to establish stream and effluent limits for cyanide to avoid harmful effects on aquatic life, as fish are extremely sensitive to very low concentrations.

Another important environmental issue is the fate of cyanide. Hydrogen cyanide, if spilled, evaporates quite readily butis not accumulating in the atmosphere. That which does not evaproate is soon decomposed or rendered nonhazardous by complexing with iron in the soil, biological oxidation, or polymerization.

General Safety Aspects. Laboratory work with hydrogen cyanide should be carried out only in a well-ventilated fume hood. Special safety equipment such as air masks, face masks, plastic aprons, and rubber gloves should be used. A chemical-proof suit should be available for emergency. Where hydrogen cyanide is handled inside a building, suitable ventilation must be provided. The people involved should be thoroughly trained in first aid. The most important rule when working with hydrogen cyanide is never to work alone. A second person must be in view at all times about 9 to 10 m away, must be equipped to make a rescue, and must be trained in first aid for hydrogen cyanide exposure.

Besides toxicity, hydrogen cyanide presents other hazards. Hydrogen cyanide undergoes an exothermic polymerization at conditions of pH 5 to 11. This polymerization can become explosively violent, especially if confined. The reaction is between hydrogen cyanide and cyanide ions, so the presence of water and heat contribute to the onset of this polymerization. Stored hydrogen cyanide should contain less than 1 wt % water, should be kept cool, and should be inhibited with sulfuric, phosphoric, or acetic acid. Manufacturers recommend a maximum of 90-day storage even for inhibited hydrogen cyanide.

Explosively violent hydrolysis can occur if an excess of a strong acid (H_2SO_4, HNO_3, or HCl) is added to hydrogen cyanide. Because of its low boiling point, hydrogen cyanide can be a fire and explosion hazard.

SODIUM CYANIDE

Sodium cyanide, NaCN, is a white cubic crystalline solid commonly called white cyanide.

Sodium cyanide is made by the niutralization or wet process in which liquid hydrogen cyanide and sodium

Table 2. Physical Properties of Sodium Cyanide

Property	Value
molecular weight	49.015
melting point, °C	562
boiling point, °C	1530
density of 30% NaCN solutions, at 25°C, g/mL	1.150
heat capacity,^{a} 25–72°C, J/(g·K)^b	1.40
heat of fusion, J/g^b	179
heat of formation, ΔH_f°, NaCN(c), J/mol^b	-89.9×10^3
heat of solution, ΔH_{soln}, J/mol^c	-1548

^{a}The heat capacity of sodium cyanide has been measured between 100 and 345 K (48).
^{b}To convert J to cal, divide by 4.184.
^{c}In 200 mol H_2O.

hydroxide solution reat and water is evaproate. The resulting crystals are briquetted or made into granular form. The principal applications of sodium cyanide are gold and silver extraction, electroplating, synthesis of iron blues, and synthesis of a large number of chemicals.

Properties

The physical properties of sodium cyanide are listed in Table 2.

Sodium cyanide is soluble in liquid ammonia. At 15°C, 100 g anhydrous methanol dissolves 6.44 g anhydrous sodium cyanide; at 67.4°C, it dissolves 4.10 g. Sodium cyanide is slightly soluble in formamide, ethanol, methanol, SO_2, furfural, and dimethylformamide.

Sodium chloride and sodium cyanide are isomorphous and form an uninterrupted series of mixed crystals.

When heated in a dry CO_2 atmosphere, sodium cyanide fuses without much decomposition. A brown-black color appears when water vapor and CO_2 are present at temperatures of 100°C below the fusion point. This color is presumably from the hydrogen cyanide polymer.

In the presence of a trace of iron or nickel oxide, rapid oxidation occurs when cyanide is heated in air, first to cyanate and then to carbonate. Case hardening of steels using a sodiu cyanide molten bath depends on these reactions, wherein the actiove carbon and nitrogen are absorbed into the steel surface; hencfe the names craburizing and nitriding.

When sodium cyanide and sodium hydroxide are heated in the absence of water and oxygen above 500°C, sodium carbonate, sodium cyanamide, sodium oxide, and hydrogen are produced. In the presence of small amounts of water at 500°C decomposition occurs with the formation of ammonia and sodium formate, and the latter is converted into sodium carbonate and hydrogen by the caustic soda. In the presence of excess oxygen, sodium carbonate, nitrogen, and water are produced.

Molten sodium cyanide reacts with strong oxidizing agents such as nitrates and chlorates with explosive violence. In aqueous solution, sodium cyanide is oxidized to sodium cyanate by oxidizing agents such as potassium permanganate or hypochlorous acid. The reaction with chlorine in alkaline solution is the basis for the treatment of industrial cyanide waste liquors.

Table 3. Physical Properties of Potassium Cyanide

Property	Value
molecular weight	65.11
melting point, °C	634
density, g/ml	
cubic	1.55
orthorhombic at −60°C	1.62
specific heat, 25–72°C, J/g[a]	1.01
heat of fusion, J/mol[a]	14.7×10^3
heat of formation, ΔH_f°, J/mol[a]	113×10^3
heat of solution, ΔH_{soln}, J/mol[a]	−12550
solubility at 25°C, g/100 g H_2O	71.6

[a]To convert J to cal, divide by 4.184.

Sodium cyanide, when fused with sulfur or a polysulfide, is converted into sodium thiocyanate; this compound is also formed when a solution of sodium cyanide is boiled with sulfur or a polysulfide. A solution of sodium cyanide shaken with freshly precipitated ferrous hydroxide is converted to a ferrocyanide.

Aqueous solutions of sodium cyanide are slightly hydrolyzed at room temperature. At temperatures above 50°C, irreversible hydrolysis to formate and ammonia becomes important.

Hydrogen cyanide is a weak acid and can readily be displaced from a solution of sodium cyanide by weak mineral acids or by reaction with carbon dioxide, eg, from the atmosphere; however, the latter takes places at a slow rate.

In the presence of oxygen, aqueous sodium cyanide dissolves most metals in the finely divided state, with the exception of lead and platinum. This is the basis of the MacArthur process for the extraction of gold and silver from their ores.

Economic Aspects

Sodium cyanide is sold as granular or powder, pillow-shaped briquettes of 15-g and 30-g sizes, tablets of 30 g, and 30% aqueous solution. Sodium cyanide is packed in mild steel or fiber drums and in 1.4 t Flo-bins. Dry sodium cyanide is also shipped in wet-flo tank cars and trucks of up to 32 t net. At destination, water is circulated through the wet-flo car or trailer to dissolve the dry sodium cyanide at delivery. This type of shipment reduces freight costs and reduces environmental risks compared with 30% aqueous solution shipment. Safety regulations are imposed by the various shipping lines and by the countries in which cyanide is transported.

Health and Safety Factors

Handling, storage, and the use of the alkali metal cyanides must be carried out by trained people. Most serious injuries and fatalities have been caused by inadvertently mixing these cyanides with acids, thereby releasing hydrogen cyanide. The present threshold limit value for 8-h exposure to cyanide dust is 5 mg/m^3 calculated as CN. Cyanide salts also must be protected from large concentrations of carbon dioxide to avoid hydrogen cyanide liberation. Carbon dioxide fire extinguishers should not

be used. Cyanide salts as solids or solutions must be stored in tightly closed containers that must be protected from corrosion or damage.

Rubber gloves should be worn when handling dry salts. In addition, the following protective items should be used with solution or dusty salts: protective sleeves, aprons, shoes, boots or overshoes made of rubber, chemical safety goggles, full-face shield, and filter-type respirator (where dust is present). Cyanide spills should be flushed to a contained area where treatment to destroy the cyanide can be carried out. In the event that cyanide salts or solutions contact the eyes, they should be flushed for 15 min with a copious, gentle flow of water followed by immediate medical attention. Eating, smoking, and chewing should be forbidden in areas where cyanide salts are handled. Employees should be required to wash carefully after working with cyanide salts and before eating, smoking, or chewing.

POTASSIUM CYANIDE

Potassium cyanide, KCN, a white crystalline, deliquescent solid, was initially used as a flux, and later for electroplating. With the decline in the use of alkali cyanides for plating the demand for potassium cyanide continues to decline.

Commercial potassium cyanide made by the neutralization or wet process contains 99% KCN; the principal impurities are potassium carbonate, formate, and hydroxide. To prepare 99.5+%KCN, high quality hydrogen cyanide and KOH must be used.

Properties

The physical properties of potassium cyanide are given in Table 3. Unlike sodium cyanide, potassium cyanide does not form a dihydrate.

The solubility of potassium cyanide in nonaqueous solvents is as follows: in anhydrous liquid ammonia, 4.55 g/100 g NH$_3$ at −33.9°C; 4.91 g/100 g methanol at 19.5°C; 0.57 g/100 g ethanol at 19.5°C; 146 g/L solution in formamide at 25°C; 41 g/100 g hydroxylamine at 17.5°C; 24.24 g/100 g glycerol of specific gravity 1.2561 at 15.5°C; 0.73 g/L solution in phosphorus oxychloride at 20°C; 0.017 g/100 g liquid sulfur dioxide at 0°C; and 0.22 g/100 g dimethylformamide at 25°C.

At room temperature, potassium cyanide has fcc crystal structure.

Potassium cyanide is readily oxidized to potassium cyanate by heating in the presence of oxygen or easily reduced oxides, such as those of lead or tin or manganese dioxide, and in aqueous solution by reaction with hypochlorites or hydrogen peroxide.

Dry potassium cyanide in sealed containers is stable for many years. An aqueous solution of potassium cyanide is slowly converted to ammonia and potassium formate.

Many reactions can be carried out between potassium cyanide and organic compounds with the alkalinity of the KCN acting as a catalyst; these reactions are analogous to reactions of sodium cyanide. The reactions of potassium cyanide with sulfur and sulfur compounds are also analogous to those of sodium cyanide. Potassium cyanide is reduced to potassium metal and carbon by heating it out of contact with air in the presence of powdered magnesium. Beryllium, calcium, boron, and aluminum act in a similar manner. Malonic acid is made from monochloroacetic acid by reaction with potassium cyanide followed by hydrolysis. The acid and the intermediate cyanoacetic acid are used for the synthesis of polymethine dyes, synthetic caffeine, and for the manufacture of diethyl malonate, which is used in the synthesis of barbiturates. Most metals dissolve in aqueous potassium cyanide solutions in the presence of oxygen to form complex cyanides.

OTHER CYANIDES

Lithium, Rubidium, and Cesium Cyanides

Lithium cyanide, rubidium cyanide, and cesium cyanide are white or colorless salts, isomorphous with potassium cyanide. In physical and chemical properties these cyanides closely resemble sodium and potassium cyanide. As of this writing these cyanides have no industrial uses.

All of these alkali metal cyanides may be prepared by passing hydrogen cyanide into an aqueous solution of the hydroxide or by precipitating a solution of barium cyanide with lithium, rubidium, or cesium sulfate. A product with fewer contaminants may be obtained by the reaction of the base in absolute alcohol or dry ether with anhydrous hydrogen cyanide. In another method of preparation, a suspension of rubidium, cesium, or lithium metals in anhydrous benzene is treated with anhydrous hydrogen cyanide, and the benzene subsequently removed by evaporation under reduced pressure.

These cyanides are all soluble in water.

Lithium cyanide melts at 160°C. In the fused state the specific gravity at 18°C is 1.075. It is highly hygroscopic. Rubidium cyanide is not hygroscopic and is insoluble in alcohol or ether. Cesium cyanide is highly hygroscopic.

Ammonium Cyanide

Ammonium cyanide, NH_4CN, a colorless crystalline solid, is relatively unstable, and decomposes into ammonia and hydrogen cyanide at 36°C. Ammonium cyanide reacts with ketones (qv) to yield aminonitriles. Reaction of ammonium cyanide with glyoxal produces glycine. Because of its unstable nature, ammonium cyanide is not shipped or sold commercially.

Ammonium cyanide may be prepared in solution by passing hydrogen cyanide into aqueous ammonia at low temperatures. It may also be prepared from barium cyanide and ammonium sulfate, or calcium cyanide with ammonium carbonate.

Calcium Cyanide

Crude calcium cyanide, about 48 to 50 eq % sodium cyanide, is the only commercially important alkaline-earth metal cyanide, and output tonnage has been greatly reduced. This product, commonly called black cyanide, is marketed in flake form as a powder or as cast blocks un-

der the trademarks Aero and Cyanogas of the American Cyanamid Company.

Properties. Because of decomposition, the melting point of calcium cyanide can only be estimated by extrapolation to be 640°C.

Calcium cyanide diammoniate, $Ca(CN)_2 \cdot 2NH_3$, is formed in liquid ammonia by reaction of calcium hydroxide or nitrate with ammonium cyanide. Deammoniation under heat and high vacuum yields calcium cyanide, a white powder, which is readily hydrolyzed to hydrogen cyanide.

Aqueous solutions of calcium cyanide prepared even at low temperature turn yellow or brown owing to the formation of HCN polymer. Calcium cyanide hydrolyzes readily.

Ferrocyanides are produced by reaction of ferrous salts. With sulfur in aqueous medium, calcium cyanide forms calcium thiocyanate.

Manufacture. Calcium cyanide is made commercially from lime, CaO, coke, and nitrogen. The reactions are carried out in an electric furnace.

Safety Precautions. Precautions similar to those used for sodium cyanide should be used for black cyanide.

Uses. The extraction or cyanidation of precious metal ores was the first, and is still the largest, use for black cyanide.

R. J. Cicerone and R. Zellner, *J. Geophys. Res.* **88**(C15), 10689 (1983).

D. Hasenberg, *HCN Synthesis on Polycrystalline Platinum and Rhodium*, dissertation, University of Minnesota, 1984.

J. L. Huiatt and co-workers, eds., *Proceedings of a Workshop—Cyanide From Mineral Processing*, University of Utah, Salt Lake City, 1982.

Sodium Cyanide Material Safety Data Sheet, E.I. du Pont de Nemours & Co., Inc., Wilmington, Del., 1990.

LAWRENCE D. PESCE
E.I. du Pont de Nemours & Co., Inc.

CYANURIC AND ISOCYANURIC ACIDS

Cyanuric acid is a white crystalline solid, which achieved commercial importance primarily via its derivatives. The chemistry of cyanuric acid is diversified because of multiple reaction sites leading to a number of useful products, e.g., chloroisocynaurates and organoisocyanurates. In addition, the thermal dissociation product of cyanuric acid, isocyanic acid, is very useful in organic synthesis and in other applications, eg, diesel engine pollution control.

STRUCTURE

Cyanuric acid is a heterocyclic compound containing a 1,3,5 triazine ring that exists in two isomeric foms, ie, enol and keto tautomers, which can be considered as cyclic trimers of cyanic (HOCN) and isocyanic acid (HNCO), respectively.

enol
cyanuric acid

keto
isocyanuric acid

Infrared (ir), Raman, and ultraviolet (uv) spectroscopic, and X-ray crystallographic data suggests that the keto (iso) form prevails in the solid state and in neutral or acidic solutions. In alkaline solution, the anion formed is that of the enol tautomer. Through common usage, both forms are collectively called cyanuric acid (CA).

PROPERTIES

Acidic Dissociation Constants

Cyanuric acid is a titrable weak acid: $pK_{a1} = 6.88$, $pK_{a2} = 11.40$, $pK_{a3} = 13.5$.

Solubility

Solubility of CA in water is only 0.2% at 25°C but increases to 2.6% at 90°C and 10.0% at 150°C. In aqueous alkali (eg, NH_4OH, NaOH, KOH) solubility increases due to salt formation. Cyanuric acid is only slightly soluble (~0.1%) at room temperature in common organic solvents such as acetone, benzene, diethyl ether, ethanol, and hexane. Solubility is significant in basic nitrogen compounds (e.g., dimethylformamide 7.2%) or unusual solvents such as DMSO (17.4%). By contrast with CA, its chlorinated derivatives di- and trichloroisocyanuric acids (DCCA and TCCA) are readily soluble in polar organic solvents (eg, ketones, nitriles, and esters).

Hydrates

In aqueous media CA forms a dihydrate, which crystallizes as colorless monoclinic prisms that effloresce in dry air. The transition temperature from dihydrate to anhydrous CA is ~57°C. The densities of anhydrous CA and CA dihydrate at 25°C are 1.75 and 1.66 g/mL, respectively.

Sublimation/Dissociation

Cyanuric acid, $H_3(NCO)_3$, is an odorless, white, infusible, crystalline solid that does not melt up to 330°C. At higher temperatures it sublimes and dissociates to isocyanic acid (HNCO).

The possible formation of dicyanic acid from HNCO has been studied in the gas, liquid, and solid phases by NMR and MS.

Vapor Pressure

The vapor pressure of CA at 167–200°C is given by log P (kPa) = 5552/T + 11.54 (T in kelvin). Over the 295–360°C range the vapor pressure is given by: log P (kPa) = 6740/ T + 13.38.

Thermodynamic Data

Thermodynamic values for CA are: ΔH_f° −690.8, −660.2 kJ mol^{-1}, ΔH_{comb} 918.4 kJ mol^{-1}, $\Delta H_{subl}^{25°C}$ 133.6 kJ mol^{-1} $C_p^{25°C}$ 132.9 kJ mol^{-1} K^{-1}; and ΔH_{neut} (NaOH) 1st H 28.2, 2nd H 17.2 kJ mol^{-1}.

CHEMISTRY

Cyanuric acid is a cyclic triimide, and undergoes reactions at N, O, or C, eg, salt formation, hydrolytic and oxidative ring cleavage, C-halogenation, N-halogenation, and alkylation. Reaction at nitrogen produces isocyanurates $R_3(NCO)_3$, whereas reaction at oxygen forms cyanurates $(RO)_3(NC)_3$. Mixed derivatives are possible, as in the case of the sodium salt of dichloroisocyanuric acid. Virtually all of the organo derivatives of CA are produced by reactions characteristic of a cyclic imide, wherein isocyanurate nitrogen (frequently as the anion) nucleophilically attacks a positively polarized carbon of the second reactant.

Although numerous mono-, di-, and trisubstituted organic derivatives of cyanuric and isocyanuric acids appear in the literature, many are not accessible via cyanuric acid. Cyanuric chloride 2,4,6-trichloro-s-triazine, is generally employed as the intermediate to most cyanurates. Trisubstituted isocyanurates can also be produced by trimerization of either aliphatic or aromatic isocyanates with appropriate catalysts (see ISOCYANATES, ORGANIC). Alkylation of CA generally produces trisubstituted isocyanurates even when a deliberate attempt is made to produce mono- or disubstituted derivatives. There are exceptions, as in the production of mono(2-aminoethyl) isocyanurate in nearly quantitative yield by reaction of CA and aziridine in DMF.

Hydrolysis

Compared to its precursor (HNCO), CA is unusually stable to hydrolysis.

Oxidation

Although the triazine ring of cyanuric acid is stable to oxidizers such as peroxydisulfate, it can be cleaved by alkaline hypochlorite. Chloroisocyanurates are similarly decomposed by alkaline hypochlorite.

Although CA is stable to ozone, it can be oxidized by ozone and uv-ozone under hydrothermal-supercritical conditions and by O_3–H_2O_2. In addition, CA can be oxidized by sulfate ion radicals generated by laser photolysis.

Salt Formation

Although much weaker than the parent compound isocyanic acid ($pK_a = 3.7$), CA is sufficiently acidic to form salts.

Chloroisocyanurates

The N-chloro derivatives are the most important commercial products derived from CA. Trichloroisocyanuric acid (TCCA) is produced by chlorination of trisodium cyanurate. Another commercial process is based on reaction of HOCl and chlorine with monosodium cyanurate. TCCA also can be formed by reaction of preformed or in situ generated HOCl with CA slurry or by reaction of dichlorine monoxide with finely powdered CA. By contrast, reaction of melamine with excess HOCl produces hexachloromelamine which is realtively stable to HOCl.

Chloroisocyanurate Double Salts

Two double salts of potassium dichloroisocyanurate KCl_2Cy (PDCC) and TCCA are known, ie, $Cl_3Cy \cdot KCl_2Cy$ $Cl_3Cy \cdot 4\,KCl_2Cy$ (where Cy is the tri-isocyanurate anion). The latter has been produced commercially. A number of mixed metal dichloroisocyanurate double salts have been prepared. Other double salts also are possible, including those containng DCCA as well as bromoisocyanurates.

Bromoisocyanurates

Bromo and bromochloro derivatives of CA include $HBrClCy$, HBr_2Cy, $NaBrClCy$, $NaBr_2Cy$, KBr_2Cy, $BrCl_2Cy$, Br_2ClCy, and Br_3Cy.

Organometallic Derivatives

Organogermanium, organosilicon, and organotin substituted isocyanurates have been prepared by reaction of CA with the appropriate reagent containing a reactive chlorine or hydroxyl.

Triazine Chemistry via Cyanuric Chloride

Conversion of CA into cyanuric chloride $(ClCN)_3$ by PCl_5 is an example of reaction at carbon.

PREPARATION

A convenient laboratory synthesis of high purity CA is hydrolysis of cyanuric chloride. On a commercial scale, CA is produced by pyrolysis of urea.

Crude CA Purification

The crude product containing aminotriazines can be purified by digestion with acids (eg, hydrochloric, nitric, or sulfuric); this hydrolyzes the acyclic impurities to carbon dioxide and ammonia and the aminotriazines to CA and ammonia. Other options for the purification of CA include dissolution in hot water, aqueous ammonia, aqueous formaldehyde, or hot dimethylformamide followed by filtration to remove most of the impurities.

By-product Reduction

By-product formation can be reduced by use of a stripping gas or vacuum to facilitate removal of ammonia; however, sublimation of urea becomes excessive if the pressure is too low. Addition of ammonium salts (eg, F^-, Br^-, Cl^-, NO_3^-, or SO_4^{2-}), acids, or pyrolysis of preformed urea salts, eg, urea hydrochloride, sulfate, or nitrate significantly reduces aminotriazine formation.

Pyrolysis in Organic Solvents

Aminotriazine formation can be reduced to an acceptable level ($<1\%$), thus eliminating the need for acid digestion, by pyrolysis of urea in certain high boiling solvents. Desirable solvents are good solvents for urea, poor solvents for CA, high boiling, and stable to pyrolysis intermediates, ammonia, oxygen, and heat. Although no perfect solvent has been identified, some solvents, eg, dinitriles, pyrrolidinones, and sulfones largely meet these requirements.

MANUFACTURE

The majority of the cyanuric acid produced commercially is made via pyrolysis of urea in directly or indirectly fired stainless steel rotary kilns. Small amounts of CA are produced by pyrolysis of urea in stirred batch or continuous reactors, over molten tin, or in sulfolane. Beside continuous horizontal kilns, numerous other methods for dry pyrolysis of urea have been described, eg, use of stirred batch or continuous reactors, ribbon mixers, ball mills, etc, heated metal surfaces such as moving belts, screws, rotating drums, etc, molten tin or its alloys, dielectric and microwave heating, and gas and vibrational fluidized beds (with performed urea cyanurate). All of these modifications yield impure CA.

HEALTH AND SAFETY FACTORS

Acute toxicities (LD_{50} g/kg, rats) for CA and chloroisocyanurates are CA >5.0, SDCC 1.67, PDCC 1.22, and TCCA 0.75. Toxicological studies on CA and its chlorinated derivatives show that the compounds are safe for use in swimming pool and spa/hot tub disinfection, sanitizing, and bleaching applications when handled and used as directed. Most uses of chloroisocyanurates are regulated by the EPA under FIFRA.

Cyanuric acid is stable and relatively inert. Chloroisocyanurates also are stable when dry and uncontaminated and kept away from fire or a source of high heat.

USES

Cyanuric Acid

Cyanuric acid is widely used to stabilize av Cl_2 (from chlorine, hypochlorites, or chloroisocyanurates) against decomposition by sunlight (ie, uv light) in swimming pools; it is used on a small scale for reducing nitrogen oxides (NO_x) in stationary diesel engine exhaust gases. It can also be used for NO_x reduction in coal, oil, or gas fired boilers. Cyanuric acid in the form of the adduct melamine cyanurate is useful as a flame retardant additive in various plastics, and in combination with other additives can be used as a flame retardant in resins. It also is useful as a solid lubricant and in preparation of abrasive slurries.

Chloroisocyanurates

Most of the CA produced commercially is chlorinated to produce TCCA, anhydrous SDCC and its dihydrate (SDCC $\cdot$ 2H). They have become standard ingredients in formulations for scouring powders, household bleaches, institutional and industrial cleansers, automatic dishwasher compounds, and general sanitizers, and most importantly, in swimming pool and spa/hot tub disinfection.

Organo(iso)cyanurates

Tris(2-hydroxyethyl)isocyanurate (THEIC) is used as an additive in the production of high performance polyester magnet–wire enamels and in electrical varnishes.

Homopolymers of triallyl cyanurate (TAC) and triallyl isocyanurate (TAIC) are brittle and find little application. However, TAC and TAIC can be used as crosslinking agents and as curing agents.

Tris(2,3-epoxypropyl)isocyanurate (ie, triglycidyl isocyanurate) is used as a cross-linking agent for carboxylated polyesters and as curing agent for weather-resistant powder coatings (qv).

Chlorinated Isocyanurates, Chemical Economics Handbook, SRI International, Menlo Park, CA, Feb. 2001.

Pool/Spa Operators Handbook, National Swimming Pool Foundation, San Antonio, Tex., 1988.

C. E. Schildknecht, in J. I. Kroschwitz, ed., *Encyclopedia of Polymer Science and Engineering*, Vol. 4, John Wiley & Sons, Inc., New York, 1986, pp. 802–811.

E. M. Smolin and L. Rapoport, in A. Weissberger, ed., *The Chemistry of Heterocyclic Compounds*, Vol. 13, Wiley-Interscience, New York, 1967, pp. 17–146.

JOHN A. WOJTOWICZ
Olin Corporation, Retired

CYCLOPENTADIENE AND DICYCLOPENTADIENE

Cyclopentadiene (CPD), C_5H_6, (**1**) and its more stable dimer, dicyclopentadiene (DCPD), $C_{10}H_{12}$, (**2**) are the major constituents of hydrocarbon resins, cyclic olefin polymers, and a host of specialty chemicals:

(**1**)   (**2**)

They can be transformed into many chemical intermediates used in the production of pharmaceuticals, pesticides, perfumes, flame retardants, and antioxidants. Because of their wide industrial uses, their chemistry has been extensively investigated and documented. In additional to the classical organic reactions, CPD forms organic metallic complexes, ferrocene, with transition metals. Some of these complexes have been established as excellent olefin polymerization catalysts.

Table 1. Compressive Strength of Dental Plasters, Investments

Materials	Compressive strength, MPa[a,b,]	ISO standard no.
Plasters		
impression	4.0^c–8.0^d	6873
model	9.0^c	6873
dental stone	20^c	6873
high-strength dental stone	35^c	6873
Investments		
inlay and crown[e]	2.3 (1.7)	7490
partial denture[e]	5.8 (3.3)	7490
phosphate-bonded[f]	3.0	7490
ethyl silicate-bonded[f]	1.5	proposed
Dies		
gypsum refractory	13.0	proposed
phosphate refractory	13.0	proposed

[a] To convert MPA to psi, multiply by 145.
[b] Value given is at room temperature. Value in parenthesis is at casting temperature.
[c] Value given is minimum value.
[d] Value given is minimum value.
[e] Low temperature casting investments.
[f] High temperature casting investments.

PHYSICAL PROPERTIES

The physical properties of CPD and DCPD are given in Table 1. DCPD, $3a,4,7,7a$-tetrahydo-4,7-methano-1H-indene, can exist in two stereoisomers, the endo and exo forms. Because commercially available DCPD is mostly the endo isomer, the properties in Table 1 are pertinent to those of the endo isomer.

DCPD decomposes rapidly at its normal boiling point to two molecules of cyclopentadiene. Purification of DCPD by distillation must be conducted under vacuum conditions. The dimer is the form in which CPD is sold commercially.

CHEMICAL REACTIONS

Cyclopentadiene is very reactive. In addition to the vast number of reactions one expects from a conjugated double bond structures which include Diels–Alder addition, hydrogenation, halogen addition, etc, the highly acidic methylene group promotes a number of condensation reactions. On the other hand, DCPD behaves like nonconjugated dienes, with the exception that the double bond in the bicycloheptene ring is more reactive than that in the five-member ring because of the bond angle strain in the bicycloheptene ring.

Diels–Alder Addition

In Diels–Alder addition, the conjugated double bonds of CPD react with the π bond of a dienophile, a compound containing ethylenic or acetylenic unsaturation. Both CPD and DCPD can be used in the reactions. The reaction needs to be carried out at temperatures above 175°C in order for the DCPD to dissociate rapidly to the CPD monomer. The rate of the reaction is strongly affected by other substituents of the dienophile. Diels–Alder addition is stereochemically specific.

Polymerization

Cyclopentadiene dimerizes spontaneously and exothermally at ambient temperature to endo DCPD via the Diels–Alder addition mechanism, in which one of the CPD molecule acts as the dienophile. At temperatures above 100°C, CPD polymerizes thermally to trimers, tetramers, and higher oligomers. Since either one of the nonconjugated double bonds in an oligomer can participate in the Diels-Alder addition to the CPD to form a higher oligomer, the higher oligomers can have several structural isomers. In contrast to DCPD, higher thermal oligomers are crystalline compounds with little odor. In addition to thermal polymerization, CPD can be polymerized rapidly at low temperatures with the aid of Lewis acid catalysts.

Condensation Involving the Methylene Group

Because of the resonance stabilization of the π-electron system, the cyclopentadienyl anion forms readily from CPD after the dissociation of the acidic methylene proton. Under alkaline conditions, CPD undergoes condensation reaction with carbonyl compounds, such as ketones and aldehydes, yielding a family of highly colored fulvene derivatives.

Hydrogenation

Cyclopentadiene can be hydrogenated stepwise through cyclopentene to cyclopentane in the presence of hydrogen and noblemetal catalysts such as palladium. Similarly DCPD is hydrogenated first to the dihydro derivative and then to the tetrahydro DCPD. Oligomers of CPD can also be hydrogenated.

Hydrogenation of CPD and DCPD with noble metal catalysts tends to produce a mixture of dihydro and tetrahydro derivatives.

Oxidation

Cyclopentadiene reacts spontaneously with oxygen to form a brown, gummy substance that contains a substantial amount of peroxides. Dicyclopentadiene reacts with air slowly, yielding a gummy deposit.

Halogenation

Halogens and hydrogen halides react readily with the conjugated double bonds of CPD, producing a series of halogenated compounds ranging from the monohalocyclopentene to tetrahalocyclopentane.

Alkylation

Cyclopentadiene can be multiply alkylated in high yields using alkyl halides, oxo alcohols, and Guerbet alcohols.

SOURCE AND PRODUCTION

Steam crackers for the production of ethylene are the primary source of cyclopentadiene and dicyclopentadiene, although a small amount is still recovered from coal tar distillation. The amount of CPD produced depends on the feedstock to the cracker. The yield from a naphtha cracker is 6 to 8 times of that from a gas cracker, which uses ethane and propane feed mixture.

There are two ways by which DCPD can be obtained from the pyrolysis gasoline, depending on whether the pyrolysis gasoline is processed immediately after leaving the debutanizer. When the pyrolysis gasoline is processed immediately, the amount of DCPD present is small.

If the pyrolysis gasoline is placed in storage before it is processed or if the process is supplemented with purchased pyrolysis gasoline, a significant amount of the CPD in the pyrolysis gasoline is already in the form of DCPD, since cyclopentadiene dimerizes to DCPD at a rate of 9 mol%/h at 35°C.

The first process requires less demanding fractionation but it has the disadvantage of leaving in the DCPD all the oligomers produced during the heat-soak step. The second process scheme also has the advantage of flexibility in adjusting the DCPD product to a desirable purity by simply varying the fractionation conditions in the last column.

STORAGE AND HANDLING

Because CPD monomer dimerizes spontaneously at room temperature and a large quantity of heat (75 kJ/mol or 18 kcal/mol) is released in the process, material containing a substantial amount of CPD should not be stored in any sealed container without the provision of removing the heat. Commercial quantity of CPD is usually produced, stored, and shipped in the form of the stable dimer, DCPD. Because of the high freezing point of DCPD, the material is transported in tank cars, rail cars, or barges equipped with heating elements. The heating of the DCPD must be regulated. Otherwise, excess temperature will lead to the formation of the undesirable oligomers. For commercial applications, which require the cyclopentadiene monomer, the CPD is obtained from DCPD by thermal cracking.

HEALTH AND SAFETY FACTORS

DCPD is a toxic substance. Toxicdogical studies indicate that DCPD has no deleterious effects on the blood and blood-forming organs. Toxicological effects are similar to terpenes rather than to benzene. Chronic exposure causes damages to the liver, kidneys and lungs.

USES OF CYCLOPENTADIENE AND DICYCLOPENTADIENE

There are two general categories of industrial end uses of cyclopentadiene and dicyclopentadiene: (1) commodity resins and polymers, which include hydrocarbon resins, unsaturated polyester resins, and ethylene propylene diene rubbers (EPDM); (2) specialty polymers and fine chemicals, which include cyclic olefin copolymers, flame retardants, agrochemicals, specialty norbornenes, flavor and fragrance intermediates.

Specialty Uses and Applications

Dicyclopentadiene alcohol can be used as a component of unsaturated polyester resins, perfumes, and plasticizers. Some dicyclopentadiene esters are also used as perfume components.

Cyclopentadiene oligomers are employed as coatings for controlling release rates of fertilizers. Thermal addition of sulfur to a mixture of DCPD and CPD oligomers has led to a number of beneficial applications such as waste water oil adsorbent powdery foam, plasticized backing for carpets and artificial turfs, and in modified sulfur cements for encapsulating low-level radioactive wastes.

Cyclopentadiene itself has been used as a feedstock for carbon fiber manufacture.

Adamantane, tricyclo[3.3.1.1]decane, is the base for drugs that control German measles and influenza.

Other specialty applications of DCPD and CPD derivatives includes dicyclopentadiene dioxide, dicyclopentadiene diepoxide and dicyclopentadiene dicarboxylic acid in surface coatings, and the tetrahydrogenated DCPD and CPD as high energy fuels for racing cars, missiles, and jet S.

A. S. Onishchenko, *Diene Synthesis*, Old Bourne Press, London, 1964, pp. 274–320.

Process Evaluation/Research Planning Dicyclopentadiene and Derivatives 97/98S7 Chem Systems Inc., Tarrytown, New York, 1998.

T. T. Peter Cheung
Phillips Petroleum Company

D

DEFOAMERS

The control or elimination of the foam that occurs in many industrial processes is a vital factor in their efficient operation. Additives for this pupose are the largest single category of process aids used in the chemical industry. They are used in low concentration to achieve this effect and are known variously as defoamers, antifoaming agents, foam inhibitors, foam supressants, air release agents, and foam control agents. Defoaming implies breaking a preexisting foam whereas antifoaming or foam inhibition indicates prevention of the formation of that foam. Such distinctions call for different product features. A defoamer is expected to exhibit rapid knockdown of a foam, whereas longevity of action might be the key requirement in many antifoam applications. Despite these varying performance features, many applications require both preventive and control functions, and in practice the same types of materials are used both for antifoaming and defoaming.

Many industries rely on the efficient and economical use of defoamers both as a process aid in product manufacture and to increase the quality of the finished product in its subsequent application. The most obvious use of defoamers as process aids is to increase holding capacity of vessels and improve efficiency of distillation or evaporation equipment. They are also used to improve filtration, dewatering, washing, and drainage of suspensions, mixtures, or slurries. Examples of industrial operations that benefit in these ways from the use of defoamers include oil well pumping; gas scrubbing at petrochemical plants; polymer and chemical synthesis and processing, particularly in monomer stripping; textile dyeing and finishing; leather processing; paint and adhesive manufacture; phosphoric acid production; control of wastewater and sewage; food preparation, notably the refining of sugar; the brewing of beer; and penicillin production by fermentation. Among the finished products that are improved in quality or efficacy by the proper inclusion of defoamers are lubricants, particularly cooling lubricants in metal working; diesel fuel, hydraulic and heat-transfer fluids; paints and other coatings; adhesives; inks; detergents; and antiflatulence tablets.

Most modern defoamers are complex, formulated specialty chemicals, whose composition is often proprietary. In addition to control of foam and associated features such as rate of foam knockdown and the persistence of the effects, other frequently needed application requirements of these specialty materials include adequate shelf life, absence of adverse effects on and by the products being treated, ease of handling, lack of toxicity to manufacturing personnel and users, environmental acceptability, and cost-effectiveness. Defoamers range from relatively inexpensive mineral oils to costly fluorinated polymers but it is not the cost per kilogram of defoamer that matters, but rather the cost per unit produced using this processing aid. Another factor that strongly influences the choice of a specific defoamer is its ancillary surface properties such as wetting, dispersion, and leveling.

DEFOAMER COMPONENTS

Active Ingredients

These are the components of the formulation that do the actual foam control work.

Liquid-Phase Components. The four most common liquid-phase components found in defoamers are hydrocarbons, polyethers, silicones, and fluorocarbons.

Solid-Phase Components. Dispersed solids are vital ingredients in commercial antifoam formulations. Much of the current theory on antifoaming mechanism ascribes the active defoaming action to this dispersed solid phase with the liquid phase primarily a carrier fluid, active only in the sense that it must be surface-active in order to carry the solid particles into the foam films and cause destabilization. It is only when compounded with hydrophobic silica to give the so-called silicone antifoam compounds that highly effective aqueous defoamers result. The three main solid-phase component classes are hydrocarbons, silicones, and fluorocarbons.

In most cases, these active defoaming components are insoluble in the defoamer formulation as well as in the foaming media, but there are cases that function by the cloud-point mechanism. These products are soluble at low temperature and precipitate when the temperature is raised. When precipitated, these defoamer-surfactants function as defoamers; when dissolved, they may act as foam stabilizers.

Ancillary Agents

Surface-Active Materials. The active defoamer components are necessarily surface active materials, but this ancillary category covers the surfactants that are often incorporated in the formulation for other effects such as emulsification or to enhance dispersion. Emulsifiers are essential in the common oil-in-water emulsion systems but they are also required where mixtures of active liquid components are used. Examples of emulsifying agents used in defoamer compositions are fatty acid esters and metallic soaps of fatty acids; fatty alcohols and sulfonates, sulfates and sulfosuccinates; sorbitan esters; ethoxylated products such as ethoxylated octyl or nonylphenols; and silicone-polyether copolymers.

729

Carriers. The function of the carrier is to provide an easily handleable, readily dispersible system for delivering the active defoamer components to the foaming system and also to tie the complex defoamer formulation together, ie, coupling agents, compatibilizers, or solubilizers. Sometimes the carrier is used simply as an extender to lower the cost of the final product. Many of the low viscosity organic solvents that are used also exhibit some antifoaming properties in cases where they are both insoluble and of lower surface tension than the medium to which they are applied. Any of the usual paraffinic, naphthenic, aromatic, chlorinated, or oxygenated organic solvents can be used, but aliphatic hydrocarbons are the most common. Water is often used as a carrier fluid. In these cases, the defoamer product is typically an oil-in-water emulsion. With growing concern over unrecovered solvents, this has become a preferred type of defoamer formulation. Such products usually require preservatives to prevent bacterial spoilage in storage.

Sometimes the defoamer is required in a solid form; eg, to be suitable for incorporation into a low-sudsing detergent powder or agricultural chemical composition. Water soluble inorganic sorbent carriers such as sodium sulfate, sodium carbonate, or sodium tripolyphosphate, are used as well as organic polymers such as methylcellulose.

DEFOAMING THEORY

Foams are thermodynamically unstable. To understand how defoamers operate, the various mechanisms that enable foams to persist must first be examined. There are four main explanations for foam stability: (1) surface elasticity; (2) viscous drainage retardation effects; (3) reduced gas diffusion between bubbles; and (4) other thin-film stabilization effects from the interaction of the opposite surfaces of the films.

The stability of a single foam film can be explained by the Gibbs elasticity E, which results from the reduction in equilibrium surface concentration of adsorbed surfactant molecules when the film is extended. This extension produces an increase in equilibrium surface tension that acts as a restoring force.

A stable foam possesses both a high surface dilatational viscosity and elasticity. In principle, defoamers should reduce these properties. Ideally a spread duplex film, one thick enough to have two definite surfaces enclosing a bulk phase, should eliminate dilatational effects because the surface tension of an insoluble, one-component layer does not depend on its thickness.

Both high bulk and surface shear viscosity delay film thinning and stretching deformations that precede bubble bursting. The development of ordered structures in the surface region can also have a stabilizing effect. Liquid crystalline phases in foam films enhance stability. In water–surfactant–fatty alcohol systems the alcohol components may serve as a foam stabilizer or a foam breaker depending on concentration. On the other hand, too rigid a surface will also be prone to rupture.

In addition to having a lower surface energy than the foaming medium, defoamers must be insoluble in that medium, but also readily dispersible in it. There are five basic processes involved in the rupture of foam films by defoamers: entering, spreading, bridging, dewetting, and rupture.

HEALTH AND SAFETY FACTORS

Defoamers are usually added at low bulk concentrations ranging from a few to 1000 ppm of the foaming medium. Often the health risk posed by such additives is negligible compared to that of the material being defoamed. Such is the case in the defoaming of asphalt (qv) and phosphoric acid. Sometimes a specific defoamer type/foaming medium combination presents a particular problem, so the supplier should always be involved in defoamer selection. Health and safety concerns arise primarily in applications in the food and drug industries. U.S. government regulations governing the use of additives such as defoamers in food and drugs are listed in the *Code of Federal Regulations*.

APPLICATIONS

Defoamers are used in adhesives and sealants, chemical processing, cleaning compounds, the construction industry, fermentation processes, fertilizers, food and beverages preparation, leather processing, metal working, oil recovery and petrochemical operations, coatings, polymer production, pulp and paper manufacturing, textile and processing, and wastewater treatment.

H. Ferch and W. Leonhardt, in P. R. Garrett, ed., *Defoaming: Theory and Industrial Applications, Surfactant Sci. Ser.* **45**, Marcel Dekker, New York, 1993, pp. 221–268.

R. Hofer and co-workers, in B. Elvers, J. F. Rounsaville, and G. Schulz, eds., *Ullman's Encyclopedia of Industrial Chemistry*, Vol. A11, 5th ed., VCH Publishers, New York, 1988, pp. 465–490.

D. T. Wasan, K. Koczo, and A. D. Nikolov, in L. L. Schramm, ed., *Foams: Fundamentals and Applications in the Petroleum Industry, Adv. Chem. Ser.* **242**, 1994, pp. 47–114.

D. T. Wasan, and S. P. Christiano, in K. S. Birdi, ed., *Handbook of Surface and Colloid Chemistry*, CRC Press, Boca Raton, Fla., 1997, p. 179.

MICHAEL J. OWEN
Dow Corning

DENDRIMERS

Prior to 1984, only three synthetic polymer architectures were known: linear, cross-linked, and branched-type configurations. The "dendritic state" is a new, fourth class of polymer architecture, consisting of five subclasses: random hyperbranched polymers, dendrigrafts, dendrons,

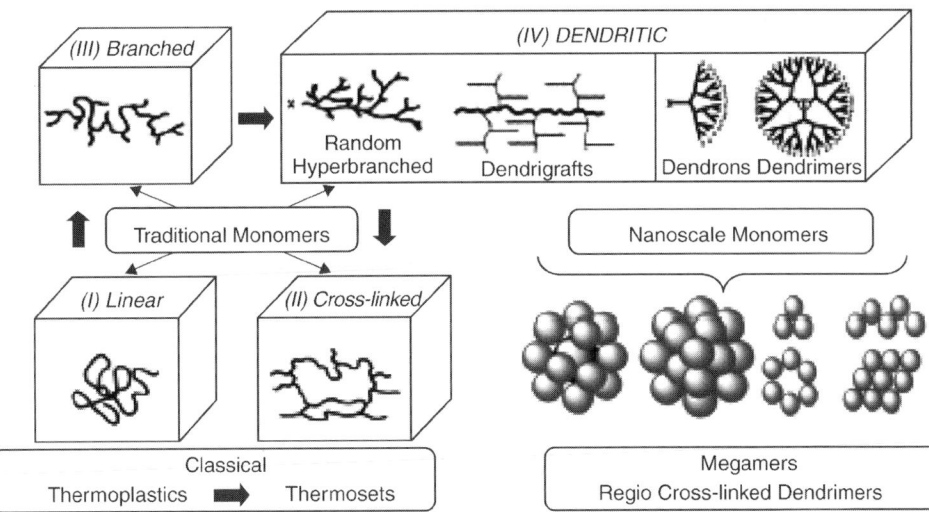

Figure 1. Schematic representation of classical and dendritic polymer structures. (D. A. Tomalia, with permission, Elsevier)

dendrimers, and tecto(dendrimers) or megamers (Fig. 1). The monodisperse nature of dendrimers makes them important building blocks for applications in nanomedicine and nanotechnology. These structures differ from traditional polymers in critical nanoscale parameters, such as size, shape, and presentation of chemical functionality. Their architecture can be precisely controlled. The core determines size, shape, directionality, and multiplicity of surface functionality. Within the interior, the length and amplification of branches define the volume and type of containment space enclosed by the terminal groups, offering a variety of possible guest–host interactions. Finally, the surface consists of reactive or passive terminal groups. These may serve as polyvalent nanoscaffolding upon which new generations of dendrimers can be covalently attached for further growth. Alternatively, the surface groups may function as control gates for the entry and departure of guest molecules from the interior. The core, interior, and surface determine all the properties of dendrimers. With the exception of biological polymers, or perhaps fullerenes, no other covalent structure offers such "bottom up" control. Precise dendritic synthesis strategies have been reported, leading to >100 dendrimer compositional families possessing >1000 different surface/interior chemistries.

DENDRIMER SYNTHESIS

The three traditional macromolecular architectural classes generate rather polydisperse products of different molecular weights. In contrast, the synthesis of dendrimers offers an opportunity to generate monodisperse macromolecular architectures similar to those observed in biological systems. Two major strategies have evolved for dendrimer synthesis. In the first, the *divergent method*, growth of a dendron originates from a core site. This approach involves assembling monomeric modules in a radial, branch-upon-branch motif according to certain dendritic rules and principles. The second method follows a *convergent growth process* from what will become the

dendrimer surface inward to a reactive focal point, leading to the formation of a single reactive dendron. To obtain a dendrimer structure, several dendrons are reacted with a multifunctional core to yield such a product. Recently, two new breakthrough approaches in dendrimer synthesis have been reported. The first, coined "lego" chemistry, utilizes highly functionalized cores and branched monomers to create phosphorus dendrimers. Several variations of the general synthetic scheme have been developed, allowing, eg, multiplication of the number of terminal surface groups from 48 to 250 in one step. These dendrimers require just one step per generation performed in a minimum volume of solvent, allow easy purification, and produce environmentally benign by-products, such as water and nitrogen. The second approach is based on "click" chemistry, ie, the near-perfect reliability of the Cu(I)-catalyzed synthesis of 1,2,3-triazoles from azides and alkynes to produce dendrimers with various surface groups in high purity and excellent yield.

DENDRIMERS AS NANOSCALE CONTAINERS

The core and interior shells of dendrimers represent well-defined nanoenvironments which, in the case of higher-generation dendrimers, are protected from the outside by the dendrimer surface (nanoscale containers). These three domains can be tailored for a specific purpose, ie, to function as a dendritic sensor, drug carrier, or as a drug. The high density of exo-presented surface functionalities makes the dendritic surface well suited as a nanoscaffold, where the close proximity of functional groups is important (polyvalency), or for receptor-mediated targeting purposes. The interior is well suited for host–guest interaction and encapsulation of guest molecules. Transmission electron micrographs (TEM) of sodium carboxylate poly(amidoamine) (PAMAM) dendrimers have revealed topologies reminiscent of regular classical micelles. The micelle-mimetic behavior of dendrimers was also observed in recent molecular dynamic studies. Depending on the conditions of the bulk solution, ie, its polarity, ionic strength, and pH, dendrimers adopt conformations of different shape and density. For example, poly(propylene imine) (PPI) and PAMAM poly(amidoamine) dendrimers with primary amines as surface groups exhibit extended conformations upon lowering the pH.

Dendrimers specifically tailored to bind hydrophobic guests to the core, coined "dendrophanes," have been shown to be excellent carriers of steroids. Very recently, an impressive approach has been presented, using tandem mass spectrometry to investigate the dynamic behavior of host–guest dendrimer complexes, which offers the potential to provide better insights into these constructs.

MEDICAL APPLICATIONS

Gene Transfection

Dendrimers and dendrons are actively under investigation for the delivery of DNA. Numerous reports have been published describing the use of amino-terminated PAMAM or PPI dendrimers as nonviral gene-transfer agents, enhancing the transfection of DNA by endocytosis and, ultimately, into the cell nucleus. The influence of dendrimer generation and core structure on gene transfection efficiency has been studied, as well as the nature of the dendrimer–DNA complexes. Dendrimers of high structural flexibility and partially degraded high-generation dendrimers (ie, hyperbranched architectures) appear to be better suited for certain gene delivery applications than intact high-generation symmetrical dendrimers. Furthermore, it has been found that maximum transfection efficiency is obtained with a net positive charge on the complexes (ie, an excess of primary amines over DNA phosphates). Very recently, dendrons with spermine surface and PAMAM dendron–poly(L-lysine) copolymers have been produced and enhanced transfection efficiency has been reported. Several recent studies have combined dendrimers or dendritic structures with amino acids and peptides to improve the delivery ability of amino acid based devices or to create more biocompatible molecules.

Drug Delivery

In addition to DNA, dendrimers have been successfully utilized to carry a variety of small molecule pharmaceuticals. Encapsulation of the well-known anticancer drug cisplatin within PAMAM dendrimers gave conjugates that exhibited slower release, higher accumulation in solid tumors, and lower toxicity compared to free cisplatin. Similarly, the encapsulation of silver salts within PAMAM dendrimers produced conjugates exhibiting slow silver release rates and antimicrobial activity against various Gram positive bacteria. The PAMAM dendrimers carrying a biocompatible hydroxyl-surface were able to encapsulate small acidic molecules, such as benzoic acid and 2,6-dibromo-4-nitrophenol, but did not form a complex with the nonacidic drug tioconazole. In another study, two polyester-based dendrimers genera-

tion 4 were synthesized, one carrying a hydroxyl surface, the other a tri(ethylene glycol) monomethyl ether surface. These dendrimers were compared to a three-arm poly(ethylene oxide) star polymer, carrying generation 2 polyester dendrons at the surface. The star polymer gave the most promising results regarding low cytotoxicity and long systemic circulatory half-life (72 h). When the anticancer drug doxorubicin was covalently bound to this carrier via an acid-labile hydrazone linkage, the cytotoxicity of the doxorubicin was reduced 80–98%, and the drug was successfully taken up by several cancer cell lines. And when the nonsteroidal antiinflammatory drug (NSAID) ibuprofen was used as a model compound to study its complexation and encapsulation into generations 3 and 4 PAMAM dendrimers and a hyperbranched polyester having ~128 surface OH groups, it was found that up to 78 ibuprofen molecules were complexed by the PAMAM dendrimers through electrostatic interactions between the dendrimer amines and the carboxyl group of the drug. The drug was successfully transported into lung epithelial carcinoma cells by the dendrimers.

Two recent studies employed dendrimers as building blocks in the construction of new drug delivery devices. In one, the PAMAM–PSS capsules are expected to allow the selective encapsulation of the drug into the capsule core and the dendrimers localized within the shell of the capsule, thus providing a dual release system of either two different drugs (ie, a drug cocktail) or of one drug released following two different time protocols (ie, fast and sustained release).

IMAGING AGENTS

Dendrimers are under investigation as imaging agents for magnetic resonance imaging (mri), in vivo oxygen imaging, and computed tomography (ct) imaging. In vivo oxygen imaging offers the potential for diagnosing complications from diabetes and peripheral vascular diseases, as well as the detection of tumors and the design of their therapeutic treatment.

Cell Targeting

The surface of dendrimers provides an excellent platform for the attachment of cell-specific ligands solubility modifiers, stealth molecules reducing the interaction with macromolecules from the body defense system, and imaging tags. The ability to attach any or all of these molecules in a well-defined, controllable manner onto a robust dendritic surface clearly differentiates dendrimers from other carriers such as micelles, liposomes, emulsion droplets, and engineered particles.

Nanodrugs

Dendrimers have been studied extensively as antitumor, antiviral, and antibacterial drugs. As antitumor drugs, dendrimers are being investigated for their use in photodynamic therapy (PDT). In this application, they are constructed around a light-harvesting core, such as a porphyrin. To reduce the toxicity under nonirradiative conditions (dark toxicity), these dendrimers are then encapsulated into micelles, for example poly (ethylene glycol)-*b*-poly(aspartic acid) micelles that are stable under certain physiological conditions but disintegrate in a acidic environment of an intracellular endosome.

Poly(lysine) dendrimers, modified with sulfonated naphthyl groups, have been found useful as antiviral drugs against the *herpes simplex* virus. Earlier studies found that those dendrimers also exhibited antiviral activity against HIV.

The general mode of action of antibacterial dendrimers is to adhere to and damage the anionic bacterial membrane, causing bacterial lysis. PPI dendrimers with tertiary alkyl ammonium groups attached to their surface have been shown to be potent antibacterial biocides against Gram positive and Gram negative bacteria. Poly(lysine) dendrimers with mannosyl surface groups are effective inhibitors of the adhesion of *Escherichia coli* to horse blood cells in a haemagglutination assay, making these structures promising antibacterial agents, while chitosan–dendrimer hybrids have been found useful as antibacterial agents, carriers in drug delivery systems, and other biomedical applications.

Biocompatibility Studies of Dendrimers

Dendrimers in medical applications have to exhibit low toxicity and be nonimmunogenic in order to be widely useable. To date, the cytotoxicity of dendrimers has been primarily studied *in vitro*. Those studies have shown that dendrimers are bicompatible as many accepted formulation ingredients. Only a few systematic studies on the *in vivo* toxicity of dendrimers have been reported so far.

Technical Applications

Technical Applications of dendrimers include luminescence, quantum dot stabilization, catalytic applications and dendrimers as chelators. Luminescence is an important phenomenon for practical applications such as lasers, displays and sensors. Coupling luminescence with dendrimer chemistry can lead to systems capable of (i) light harvesting, (ii) changing of light color, (iii) sensing with signal amplification, (iv) quenching and sensitization processes, and (v) shielding effects. Luminescent metal complexes of dendrimers have extensively been studied. Quantum dots are semiconductor nanocrystals that fluoresce when excited by a light source, emitting bright colors that can identify and track properties and processes in various applications. By varying the size of the crystals one can cause a rainbow of colors. However, these nanoparticles have a strong tendency to aggregate, forming particle clusters with different size-dependent properties. Close packing of dendrimers around a nanoparticle would create a dense shell at the metal interface, which may be more efficient in preventing cluster formation and passivating the surface than traditional ligand shells formed from flexible, linear polymers.

Noble metal nanoparticles are attractive catalysts because of their size effects. However, having atoms with very active surfaces often leads to aggregation of the nanoparticles and decrease in catalytic activity and selectivity. Recently it was reported that dendrimer-encapsulated noble metal clusters exhibited highly catalytic activity, and the dendrimers acted as both templates and porous nanoreactors. The possibility of attaching various functional groups such as carboxyl, hydroxyl, and so forth to their surface make dendrimers attractive as high capacity chelating agents for metal ions not only in catalytic applications but also as metal ion scavengers for waste remediation. Removal of copper from contaminated sandy soil by use of only 0.1% (w/w) dendrimers of various generations and surface groups resulted in as much as 90% of copper removal. Dendrimers carrying chelating groups for lanthanides and actinides on their surfaces were able to reduce the presence of europium and americium by factors 50-400 compared to particles without dendrimer structure.

J.M.J. Fréchet and D. A. Tomalia, *Dendrimers and Other Dendritic Polymers*, John Wiley & Sons, Inc., Chichester, 2001.

G. R. Newkome, C. N. Moorefield, and F. Vögtle, *Dendrimers and Dendrons, Concepts, Synthesis and Applications*, Wiley-VCH, Weinheim, 2001.

S.; Svenson, D.A. Tomalia, "Dendrimers in Biomedical Applications - Reflections on the Field", *Adv. Drug Deliv. Rev.* **2005**, *57*, 2106–2129.

SÖNKE SVENSON
Dendritic NanoTechnologies,
Inc.

DENDRIMERS, LUMINESCENT

Dendrimers are well defined, tree-like macromolecules, with a high degree of order and the possibility to contain selected chemical units in predetermined sites of their structure. Dendrimers are currently attracting the interest of a great number of scientists because of their unusual chemical and physical properties and the wide range of potential applications (Newkome and co-workers).

Luminescence can be defined as the emission of light [intended in the broader sense of ultraviolet (uv), visible (vis), or near-infrared (ir) radiation] by electronic excited states of atoms or molecules. Luminescence is an important phenomenon from a basic viewpoint (eg, for monitoring excited-state behavior) as well as for practical applications (lasers, displays, sensors, etc).

Luminescent units can be incorporated in different regions of a dendritic structure and can also be noncovalently hosted in the cavities of a dendrimer or associated at a dendrimer surface as schematically shown in Figure 1.

Coupling luminescence with dendrimer chemistry can lead to systems capable of performing very interesting

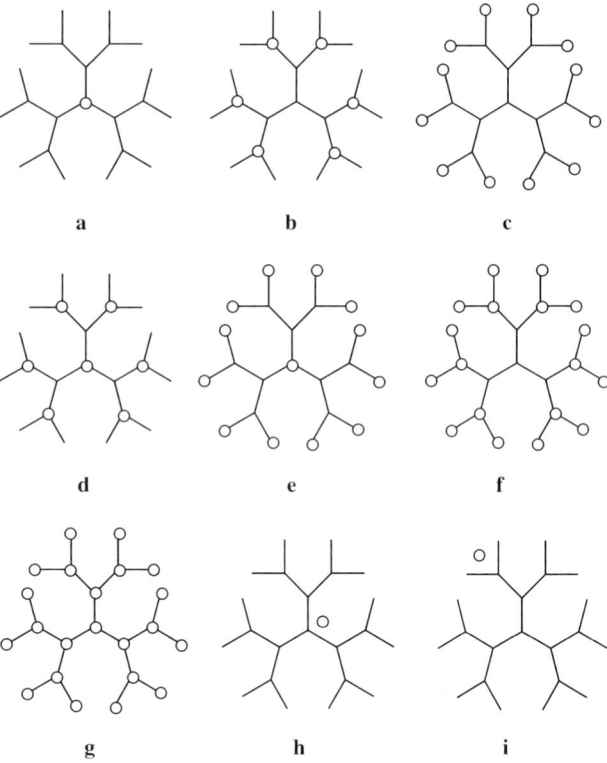

Figure 1. Schematic illustration of the possible location of photo-active units, represented by circles, covalently linked (types **a–g**) or associated (types **h–i**) to a dendrimer.

functions such as (a) light harvesting, (b) changing the "color" of light, (c) sensing with signal amplification, (d) quenching and sensitization processes, (e) shielding effects. Luminescence can also be used for elucidating dendritic structures and superstructures, and for investigating dendrimer rotation in solution.

LUMINESCENCE

It is worthwhile recalling a few elemental principles of molecular luminescence. A necessary, but not sufficient, condition to obtain light emission from a molecule is the population of electronic excited states. When this result is achieved by light absorption, the light emission that follows is called luminescence (strictly speaking, *photo*luminescence). When the population of excited states (and the consequent light emission) is obtained by suitable chemical or electrochemical reactions, the terms *chemi*luminescence and *electro*chemiluminescence are used. In most cases, the ground state of a molecule is a singlet state (S_0), and the excited states are either singlets (S_1, S_2, etc) or triplets (T_1, T_2, etc).

In principle, transitions between states having the same spin value are allowed, whereas those between states of different spin are forbidden. Therefore, the absorption bands observed in the uv-vis spectrum of molecules usually correspond to $S_0 \rightarrow S_n$ transitions. Excited states are unstable species that undergo fast deactivation by intrinsic (first-order kinetics) photo-

physical processes. When a molecule is excited to upper singlet excited states, it usually undergoes a fast and 100% efficient radiationless deactivation (internal conversion, ic) to the lowest excited singlet, S_1. Such an excited state undergoes deactivation via three competing photophysical processes: (1) nonradiative decay to the ground state (internal conversion, rate constant k_{ic}); (2) radiative decay to the ground state (luminescence, that in the specific case of a spin-allowed process is called *fluorescence*, rate constant k_{fl}); (3) conversion to the lowest triplet state T_1 (intersystem crossing, rate constant k_{isc}). In its turn, T_1 can undergo deactivation to the ground state S_0 via a nonradiative process (intersystem crossing, rate constant k'_{isc}) or a radiative process (luminescence, which in the specific case of a spin-forbidden transition is called *phosphorescence*, rate constant k_{ph}). Thermally activated back-conversion $T_1 \rightarrow S_1$ and triplet–triplet annihilation with formation of S_1 and S_0 can also occur, giving rise to delayed fluorescence. When the molecule contains heavy atoms, the formally forbidden (and, therefore, slow) intersystem crossing and phosphorescence processes become faster.

Because of their proximity, the various functional groups of a dendrimer may easily interact with one another. Interaction can also occur between dendrimer units and molecules hosted in the dendritic cavities or associated to the dendrimer surface.

In most cases, quenching of an excited state (eg, by energy transfer) takes place by a weak interaction. It may happen, however, that the excited state and the quencher undergo a strong interaction yielding a new chemical species, which is called an excimer (from <u>exci</u>ted <u>di</u>mer) or exciplex (from <u>exci</u>ted complex) depending on whether the two interacting units have the same or different chemical nature. It is important to notice that excimer and exciplex formation is a reversible process and that both excimers and exciplexes sometimes (but not always!) can give luminescence. Compared with the "monomer" emission, the emission of an excimer or exciplex is always displaced to lower energy (longer wavelengths) and usually corresponds to a broad and rather weak band.

Excimers are usually obtained when an excited state of an aromatic molecule interacts with the ground state of a molecule of the same type. Exciplexes are obtained when an electron donor (acceptor) excited state interacts with an electron acceptor (donor) ground-state molecule.

DENDRIMERS WITH A LUMINESCENT CORE

Luminescent metal complexes, particularly porphyrins, have extensively been used as cores to obtain large dendrimers. Even when the dendrons do not exhibit relevant light absorption and light emission properties, the luminescent properties of the dendrimer may differ from those of the naked core since the appended dendrons can shield the luminescent excited state of the core from quenching by dioxygen or other impurities.

DENDRIMERS WITH LUMINESCENT UNITS IN THE PERIPHERY

The dendrimers of the poly(propylene amine) family do not exhibit any absorption band in the visible and near-uv spectral region and do not show any luminescence. These dendrimers, however, can be easily functionalized in the periphery with luminescent units (Vögtle and co-workers). Compound **1** represents the fourth generation dendrimer 4D of the nD family where the generation number n goes from 1 to 5. These dendrimers contain 2^{n+1} (i.e., 64 for 5D) dansyl functions in the periphery and $(2^{n+1}-2)$ (ie, 62 for 5D) tertiary amine units in the interior. The dansyl units behave independently from one another so that the dendrimers display light absorption and emission properties characteristic of dansyl, ie, intense absorption bands in the near-uv spectral region $\lambda_{max} = 252$ and 339 nm; $\partial_{max} \approx 12{,}000$ and 3,900 L mol^{-1} cm^{-1}, respectively, for each dansyl unit) and a strong fluorescence band in the visible region ($\lambda_{max} = 500$ nm; $\Phi_{em} = 0.46$, $\tau = 16$ ns). Poly(propylene amine) dendrimers functiona-

lized with dansyl units at the periphery can coordinate metal ions by the aliphatic amine units contained in the interior of the dendrimer and can thus be used as fluorescent sensors for metal ions. The advantage of a dendrimer for this kind of application is related to the fact that a single analyte can interact with a great number of fluorescent units, which results in signal amplification.

DENDRIMERS WHOSE LUMINESCENCE IS GOVERNED BY ENERGY-TRANSFER PROCESSES

A dendrimer constituted by a $[Ru(bpy)_3]^{2+}$ - type core (bpy = 2,2' bipyridine) carries branches containing the 1,3-dimethoxybenzene- and 2-naphthyl type units (Pleovets and co-workers).

In the $[Ru(bpy)_3]^{2+}$ metal complex, the lowest excited state is a metal-to-ligand charge-transfer triplet, ^{3}MLCT, which, because of the presence of the heavy metal atom, is populated with unitary efficiency from the upper lying excited state. The ^{3}MLCT excited state

(1)

emits a relatively strong phosphorescence ($\lambda_{max} = 610$ nm). In fact, all the three types of chromophoric groups present in the dendrimer, namely, $[Ru(bpy)_3]^{2+}$, dimethoxybenzene, and naphthalene, are potentially luminescent species. In the dendrimer, however, the fluorescence of the dimethoxybenzene- and naphthyl-type units is almost completely quenched in acetonitrile solution, with concomitant sensitization of the $[Ru(bpy)_3]^{2+}$ phosphorescence.

Much attention has been focused on light harvesting antenna dendrimers constituted by a porphyrin core and properly chosen dendrons. Morphology-dependent antenna properties have been reported for a series of dendrimers consisting of a free-base porphyrin core bearing different numbers (from 1 to 4) of poly(benzylether) dendrons at the meso position of the central porphyrin. In dichloromethane solutions, excitation of the chromophoric groups of the dendrons causes singlet–singlet energy-transfer processes that lead to the excitation of the porphyrin core. It was also postulated that cooperativity between dendrons, which decreases with increasing conformational mobility, is necessary for efficient energy transfer. Such a behavior would mimic that of natural photosynthetic systems, where energy migration within "wheels" of chromophoric groups results in an efficient energy transfer to the reaction center. More recently, the morphology effect was investigated by using much larger porphyrin dendrimers consisting of a free-base porphyrin core with appended up to four dendrons, each containing seven zinc porphyrin units.

Development in synthetic procedure has recently brought to the construction of dendrimers containing, in selected region of their architecture, three different types of properly chosen chromophoric units, among which energy-transfer processes can take place.

Other interesting examples of dendrimer with a cascade energy transfer among three different types of chromophoric units are (1) a shape-persistent dendrimer with a polyphenylene backbone, a terrylene tetracarboxdiimide core, perylene and naphthalene dicarboxmonoimide chromophores in the scaffold and at the periphery, respectively (Weil and co-workers); and (2) the dendrimer containing dansyl, dimethoxybenzene, and naphthalene units discussed below (Hahn and co-workers), and (3) a dendrimer made of Os(II)- and Ru(II)-based chromophores and pyrene units (McClenaghan and co-workers).

LUMINESCENT DENDRIMERS HOSTING LUMINESCENT GUESTS

An important property of dendrimers is the presence of internal cavities where ions or neutral molecules can be hosted.

Energy transfer from the numerous chromophoric units of a suitable dendrimer to a luminescent guest may be exploited to construct systems for changing the color of the incident light and for light harvesting. An advantage shown by such host–guest systems compared with dendrimers with a luminescent core is that the wavelength of the sensitized emission can be tuned by changing the luminescent guest hosted in the same dendrimer.

U. Hahn and co-workers, *Angew. Chem. Int. Ed.* **41**, 3595 (2002).

N. D. McClenaghan, F. Loiseau, F. Puntoriero, S. Serroni, and S. Campagna, *Chem. Commun.* 2634 (2001).

G. R. Newkome, C. N. Moorefield, and F. Vögtle, *Dendrimers and Dendrons, Concepts, Syntheses, Applications*, Wiley-VCH, Weinheim, 2001.

M. Pleovets, F. Vögtle, L. De Cola, and V. Balzani, *New J. Chem.* **23**, 63 (1999).

F. Vögtle and co-workers, *J. Am. Chem. Soc.* **122**, 10398 (2000).

T. Weil, E. Reuther, and K. Müllen, *Angew. Chem. Int. Ed.* **41**, 1900 (2002).

Vincenzo Balzani
Paola Ceroni
Mauro Maestri
Università di Bologna

DENTAL MATERIALS

DENTAL CERAMICS

Ceramic materials are well suited for biomedical applications requiring tissue compatibility, compressive strength, durability, radiopacity, and inertness toward sterilization chemicals and conditions (see Ceramics). In addition, glass-matrix ceramics, containing appropriate crystalline fillers and colorants, can richly mimic many of the aesthetic and optical properties of natural teeth (see Colorants for ceramics). Limitations to the use of ceramics include design restrictions based upon the sometimes low and variable tensile strengths of ceramic products; susceptibility to stress corrosion, eg, crack extension in the presence of water or other low molecular weight species; and shrinkage during processing. Ceramics have also been associated with high levels of abrasive wear to opposing tooth structure.

The principal use of ceramics in dentistry is for the aesthetic restoration of missing teeth or tooth structure, ie, crowns and bridges (prostheses), primarily as glassy ceramic coatings fused to an underlying cast metal framework. Additional aesthetic use is made of ceramics not supported by metal substructures. These uses include single-tooth crowns, ie, restoring all the external surfaces of a tooth; ceramic fillings, eg, inlays and onlays; veneers, ie, replacement of only the visible surface of anterior teeth; denture teeth; and orthodontic brackets. Some semiaesthetic generally more highly crystalline and polycrystalline ceramics are utilized as core materials in place of cast metal substructures and are generally veneered with more aesthetic ceramics.

Ceramic prostheses are generally fabricated from one or two ceramics that derive from three basic categories of material; beneficiated feldspathic minerals, glass–ceramics, and polycrystalline ceramics. Dental fillings and simple prostheses are often fabricated from only one type of ceramic. Prostheses requiring more structural performance are fabricated from two materials, involving an aesthetic outer coating over a stronger ceramic or metal core.

Feldspathic Matrix Ceramics

The majority of aesthetic ceramics are based on mined feldspar minerals such as potassium feldspar, eg, orthoclase and microcline, $K_2O \cdot Al_2O_3 \cdot 6SiO_2$; sodium feldspar, eg, albite, $Na_2O \cdot Al_2O_3 \cdot 6SiO_2$; and nepheline syenite, which contains ~50% albite, 25% nepheline, $Na_2O \cdot Al_2O_3 \cdot 2SiO_2$, and 25% microcline. These feldspars are processed to remove iron-bearing impurities, melted with various additives, quenched in water, and ground to yield powders (frits) of alkali-modified aluminosilicate glasses containing little or no crystalline filler (see also GLASS).

Natural tooth colors are developed in these formulations by the addition of color frits, ie, vitreous powders to which colored and/or opaquing pigments (qv) have been added. Because aluminosilicate glasses are very aggressive solvents at elevated temperatures, pigments must have inertness toward dissolution as well as color stability at elevated temperatures. Crystalline minerals in the spinel family have been found to be successful pigments.

Additional raw materials, such as kaolin clay and quartz, may be used as processing aids for factory-based processing, eg, of denture teeth.

Most ceramics designed to be fused to metals are derived from either high potassium feldspars, or from compositions to which potassium oxide, K_2O, has been added to above ~11 mol%. This places these ceramics into a phase region from which feldspars melt incongruently to form leucite, $K_2O \cdot Al_2O_3 \cdot 4SiO_2$, plus liquid.

Dental Glass–Ceramics

Glass–ceramics (qv) are highly crystalline ceramics having some residual glass matrix prepared by the controlled crystallization of glasses. Crystalline phase(s) are formed from elements within the parent glasses by subjecting them to carefully regulated thermal treatments, called ceramming. Generally, ceramming treatments involve separate control, at two different temperatures, over both the nucleation and growth stages of crystallization within the glass.

Glass-ceramics have been developed for restorative dentistry based on four different crystalline phases, ie,

tetrasilicic fluoromica, $KMg_{2.5}Si_4O_{10}F_2$, β-quartz crystals, leucite, and lithium disilicate. Fluoromica develop with a plate-like morphology having a relatively low energy cleavage path along their basal plane. These tendencies towards basal-plane cleavage and the interlocking of crystals at high volume fraction provide an effective strengthening mechanism. Fluoromica crystals and the residual glass matrix have relatively similar indices of refraction.

Polycrystalline Dental Ceramics

Pure polycrystalline ceramics were not widely used in dentistry until computer-based systems became available enabling the "net shape" fabrication of prostheses, compensating for the shrinkage inherent during the sintering of such structural ceramics. Such ceramics include alumina and transformation toughened zirconia, in which the tetragonal zirconia (high temperature) phase has been stabilized at room temperature with the addition of yttria or ceria (2–3%). Dental prostheses are custom devices, each one unique, and are required to fit prepared tooth structure to within certain tolerances (eg, 20–30 μm over 10 mm to 30 mm). Three-dimensional (3D) data sets captured from replicas of patients by tactile or optical scanning are used to make oversized greenware of well-controlled ceramic powders having predictable shrinkage (~30 vol%).

DENTAL CEMENTS

Dental cements are composites, ie, they have a continuous or matrix phase linked to a discontinuous or reinforcing phase by an interphase. Based on their chemistry, dental cements can be divided into acid–base cements, acrylic or resin cements, and resin-modified acid–base cements, ie, hybrid cement composites. Acid–base cements are either aqueous-based cements or nonaqueous-based cements, although small amounts of water or protic agents are essential to the acid–base setting mechanism of the latter. Acrylic cements are resin-based composites modified for cement applications. The monomer(s) used in these cements can be simply methyl methacrylate or various types of multifunctional monomers, eg, BIS–GMA, triethylene glycol dimethacrylate. The setting mechanism is by free-radical addition polymerization, either chemical, photochemical, or a combination of the two. The hybrid cement composites include a vinyl resin system as well as the usual acid–base components of the cement. They have a dual, ie, ionic and free radical, setting mechanism.

Although used in relatively small quantities, dental cements are essential in a number of dental applications as temporary, intermediate and, in some cases, more permanent restoratives; cavity liners and bases; luting agents to bond preformed restorations and orthodontic devices; pulp capping agents and endodontic sealers;

components in periodontal dressings; and impression pastes. Because this wide range of dental uses makes it virtually impossible for one type of cement to have all the necessary properties demanded, tailored dental cements exist to meet certain specific objectives. Generally, cements are sought that excel in strength, oral environmental resistance, durability, adhesiveness, and biocompatibility.

DENTAL PLASTERS

Gypsum is widely distributed naturally as calcium sulfate dihydrate, $CaSO_4 \cdot 2H_2O$. Plaster is the rehydrated calcined gypsum. The ADA classifies five types of dental plaster according to the physical properties: type I, impression plaster; type II, model plaster; type III, dental stone; type IV, high strength dental stone; and type V, high strength, high expansion dental stone. These different types are the result of various calcining methods.

Although plaster has been a very successful and serviceable material, it is seriously lacking in hardness, edge strength, chip resistance, abrasion resistance, and strength to fulfill many needs of dentistry. Some of these requirements have been partially filled by the development of the types III and IV plasters. Table 1 lists the compression strength of dental plasters.

DENTAL INVESTMENTS

Dental investments are comprised of refractory materials capable of withstanding elevated temperatures.

Table 1. Compressive Strength of Dental Plasters, Investments

Materials	Compressive strength, MPa[a,b]	ISO standard no.
Plasters		
impression	$4.0^c - 8.0^d$	6873
model	9.0^c	6873
dental stone	20^c	6873
high-strength dental stone	35^c	6873
Investments		
inlay and crown[e]	2.3 (1.7)	7490
partial denture[e]	5.8 (3.3)	7490
phosphate-bonded[f]	3.0	7490
ethyl silicate-bonded[f]	1.5	proposed
Dies		
gypsum refractory	13.0	proposed
phosphate refractory	13.0	proposed

[a] To convert MPA to psi, multiply by 145.
[b] Value given is at room temperature. Value in parenthesis is at casting temperature.
[c] Value given is minimum value.
[d] Value given is minimum value.
[e] Low temperature casting investments.
[f] High temperature casting investments.

They are used as casting investments and model investments.

Casting Investments

Casting investments are used to form molds into which molten metal may be cast. The cavity for receiving the metal is formed by the lost wax process. The composition of investments used for alloys cast from low (1100°C) temperatures are different from those used for alloys cast from higher (1300°C) temperatures.

A casting investment must provide sufficient expansion to compensate for shrinkage (up to 0.4%) of the wax (or plastic) pattern during its fabrication, and shrinkage (1.2–2.0%) of the cast alloys resulting from solidification and cooling. The higher the solidification temperature of an alloy the greater is its casting shrinkage.

Investments for Low-Temperature (T ≤ 1100°C) Casting Alloys

Low temperature casting alloys are usually comprised of gold, silver, and copper and are used for inlays, crowns, and fixed partial dentures. At the temperature tested, the compressive strengths specified for inlay or partial denture investments are adequate to prevent fracture during handling and in casting. The strength may vary at casting temperature. The strengths of gypsum-bonded investments at high temperatures vary with the additive used to reduce shrinkage during heating. The type of binder also influences the effect of the temperature on strength. Table 1 presents the mechanical property limits defined by International Organization for Standardization (ISO) standards for casting investment for low temperature dental casting alloys.

These investments consist of a binder of calcined gypsum, ie, calcium sulfate hemihydrate, $CaSO_4 \cdot \frac{1}{2}H_2O$; to which is added silica and modifying agents such as finely ground gypsum, soluble sulfates, and chlorides, ie, accelerators of setting; and low solubility salts, ie, retarders of setting. The refractory base consists of the silica in some crystalline form; quartz and crystobalite are the main constituents.

Investments for High-Temperature Casting Alloys

These investments are used for alloys that require high (1300°C) casting temperatures. The alloys vary greatly in nobility. Some are rich in gold and/or palladium, and others are predominately cobalt, nickel, and chromium. Some of these alloys are used for crowns and fixed partial dentures with porcelain veneers fused to the metal; others are used for removable partial denture frameworks. The noble-based alloys for fixed partial dentures require casting temperatures in the neighborhood of 1300°C, and phosphate bonded investments are used. The cobalt, nickel, and chromium alloys require casting temperatures of ~1400–1500°C, and investments that use ethyl silicate or stabilized silica solutions are used.

Model Investments

Model investments are materials used for noncasting operations in the fabrication of dental protheses. They differ from casting investments in various ways depending on the prosthetic device being constructed. For low temperature operations, such as soldering, gypsum is used; phosphate-bonded materials are employed for higher solder temperatures or for the fabrication of porcelain veneers.

Investment Casting Rings and Liners

Casting rings and ring liners are needed prior to investing a wax pattern. Casting rings are cylindrical tubes that mold and retain the casting investment. Ring liners are paper or fabric sheets adapted to the inside of the rings to allow the investment to expand within the ring. Three types of castings rings commonly used are metal, split, and disposable rings. Liners are either cellulose, which readily absorbs water, or ceramic, which does not absorb water or investment liquid.

DENTAL WAXES

Pattern Waxes

The pattern waxes, ie, inlay casting, base-plate, and sheet and shape waxes, are used to construct the prototype or pattern from which a finished dental restoration is produced.

Inlay Casting Waxes

The three types of inlay casting waxes, ie, types A, B, and C, are used to produce wax patterns for the lost wax casting process in the production of cast gold inlays, crowns, and bridges. Some inlay wax is also used to produce patterns for acrylic restorations.

Type A waxes are hard waxes used when an extra hard wax is preferred by the dentist in making patterns in the mouth. Type B waxes are medium waxes used by most dentists to make patterns in the mouth. Type C waxes are soft waxes used for making patterns outside the mouth.

The exact formulations for inlay casting waxes are considered trade secrets, and little has been published on the subject.

Inlay waxes are generally dark blue, green, black, or purple; however, inlay waxes for acrylic work are uncolored and are a natural ivory. The waxes are generally produced in stick form, ~75 mm long and 6.5 mm across; the cross section may be round or hexagonal.

Base-Plate Waxes

Base-plate waxes are used to substitute for, or be used in conjunction with, a base plate to form a pattern for the production of complete or partial dentures and certain orthodontic appliances to be molded of acrylic resin, modified vinyl resins, or other denture-base polymers.

Base-plate waxes are formulated for specific uses or working conditions into types I, II, and III. Consequently, the flow requirements differ. Type I waxes are soft waxes for building contours and veneers, type II waxes are medium waxes used for pattern production in the mouth in temperate weather, and type III waxes are hard waxes used for production in the mouth in hot weather.

Base-plate wax compositions are generally regarded as trade secrets.

Sheet and Shape Waxes

Sheet and shape waxes are used to produce patterns from which complete or partial dentures are cast of gold or base metal alloys. They are used to fabricate the restoration prototype directly upon a refractory investment cast.

The flow of sheet and shape wax is much higher than that of inlay wax and base-plate wax, reflecting increased pliability, ductility, and plasticity.

The compositions of sheet and shape waxes are also trade secrets. However, they are blends of various proportions of paraffin, microcrystalline waxes, carnauba wax, ceresin, beeswax, gum dammar, mastic gum, and possibly other resins. Sheet waxes are marketed in square sheets ~80 × 90 mm. Various thicknesses are available from 32 gauge (0.5 mm) to 14 gauge (1.63 mm).

Shape wax is similar in composition and properties to the nontacky sheet waxes. It is processed into preformed shapes or definite shape and gauge to facilitate the waxing-up of partial-denture patterns. The shapes and sizes conform to those most needed for lingual bars, palatal bars, clasps, saddle construction, retainers, etc.

Impression Waxes

Impression waxes include those waxes used to obtain a negative cast of the mouth structure (impression waxes), waxes used to establish tooth articulation (bite-registration waxes), and waxes used to detect tooth interference and high spots or improper fit of denture bases (disclosing waxes). They must be plastic and moldable at mouth temperatures, and chill to a still nonplastic mass upon cooling within a few degrees below mouth temperature.

Processing Waxes

The extensive amount of handwork and craftsmanship necessary in the fabrication of most dental restorations and appliances has created a need for several types of wax compositions. These are known as boxing, sticky, utility, or study waxes.

Wax Manufacture

Wax compositions for dental usage are usually compounded in simple melting and blending equipment.

Wax compositions are processed into rods, sheets, cakes, and special forms by a variety of processes. Casting, roll forming, and extrusion, by mechanical or hydraulic piston-type extruders or by continuous screw extrusion, are operations in practice. Stamping, roll cutting, or molding are also employed.

DENTAL ALLOYS AND METALS

The chemical and physical properties of some metals and alloys, such as hardness, strength, stiffness, toughness, corrosion resistance, and biocompatibility have provided materials capable of withstanding the most severe demands of restorative dentistry, namely, the harsh corrosive environment of the mouth and high stresses on the small cross-sectional areas of teeth.

Amalgams

Dental amalgam is a novel alloy. Within a few seconds of the start of trituration, the alloy and mercury must amalgamate to a smooth plastic mass. Within 3–5 min it must set to a carvable mass and remain so for 15 min. Within 2 h it must develop sufficient strength, hardness, and toughness to resist mild biting and chewing forces. It should not tarnish or corrode in the mouth, nor react to produce toxic or soluble salts, and must maintain its color.

The mercury content in dental amalgam is generally restricted to a mass fraction of ~50% or less. Excessive mercury in the final restoration results in a low strength amalgam with poor creep resistance, excessive setting expansion, and poor corrosion resistance. Properly proportioned, mixed, and processed amalgams will have compressive strengths in the range of 210–345 MPa (30,000–50,000 psi) in 24 h.

Composition

The composition of powdered alloys used in preparing dental amalgams usually includes mass fractions of 66.7–75.5% silver; 25.3–27.0% tin; 0.0–6.0% copper; and 0–1.9% zinc. These are commonly referred to as conventional alloys, and the amalgams made from them are conventional amalgams. This composition range was in use for almost a century until two Canadian researchers designed an alloy consisting of a mixture of two powders; one having the customary flakes of the aforementioned composition range and the other, spherical shapes of a silver–copper eutectic, ie, 72% Ag, 28% Cu. The ratio of the customary powder to the spherical one was ~2:1 by mass, giving an approximate composition (mass fractions) of 70% silver, 18% tin, 11% copper, and 1% zinc. Amalgams made with this powder mix have high compressive strengths and low creep, have better clinical performance than conventional amalgams, and are sometimes called dispersed amalgams.

Gallium-Based Alloys

Gallium-based alloys has been introduced commercially as substitutes for dental amalgam, but problems with high levels of corrosion and latent expansion have prevented their continued use as a direct filling material.

Gold and Gold Alloys

Gold foil, crystal powder, a gold–calcium alloy, and combinations thereof in the noncohesive states are used in dentistry as direct-filling materials; an adsorbed layer of ammonia renders the gold noncohesive for packaging, and heating returns it to a cohesive state for use. Gold and gold alloys serve the needs of dentistry better than any other metals or alloy systems. Gold alloys have a broad range of working characteristics and physical properties, coupled with excellent resistance to tarnish and corrosion in the mouth.

Gold Casting and Wrought Alloys

Gold alloys useful in dentistry may contain gold, silver, platinum, palladium, iridium, indium, copper, nickel, tin, iron, and zinc. Other metals occasionally are found in minor amounts. The effect of each of the constituents is empirical, but some observations have been made.

Gold Alloys, Cast Types. Four types of gold alloys have been recognized for cast dental restorations. They provide desired material for specific uses. The appropriate specifications for these alloys is ANSI/ADA specification No. 5.

Type I, soft alloys (20–22-kt golds), are used for inlays of simpler nonstress-bearing types.

Type II, medium-hard alloys, are harder, stronger, and have lower elongation than type I alloys. They are used for moderate stress application, eg, three-quarter crowns, abutments, pontics, full crowns, and saddles.

Type III, hard alloys, are the hardest, strongest, and least ductile of the inlay casting alloys. Their use is indicated for restorations required to resist large forces such as three-quarter crowns, abutments, pontics, supports for appliances, and precision-fitting inlays.

Type IV, extra hard (partial denture) alloys, are indicated where high strength, hardness, and stiffness are required. These partial-denture alloys are used for cast removable partial dentures, precision-cast fixed bridges, certain three-quarter crowns, saddles, bars, arches, and clasps.

Gold Alloys, Wrought Type. Two types of wrought gold alloys were formerly recognized by the ADA specification No. 7 for the fabrication of orthodontic and prosthetic dental appliances. Alloys of this type are seldom used in the United States; they have been replaced by stainless steels and nickel–titanium alloys.

Platinum and Platinum Alloys

Platinum has excellent resistance to strong acids and, at elevated temperatures, to oxidation. Under reducing conditions at high temperatures it must be protected from low fusing elements or their oxides. Easily reduced metals at high temperatures may form low fusing alloys with platinum.

Platinum, a bluish-white metal, is soft, tough, ductile, and malleable. It has a melting point of 1773–C. The coefficient of thermal expansion is $9 \times 10^{-6}/°C$.

Platinum has many uses in dentistry. Pure platinum foil serves as the matrix in the construction of fused-porcelain restorations. Platinum foil may be laminated with gold foil for cold-welded foil restorations. Platinum wire has found use as retention posts and pins in crown and bridge restorations. Heating elements and thermocouples in high fusing porcelain furnaces are usually made of platinum or its alloys (see PLATINUM-GROUP METALS, COMPOUNDS).

Palladium and Palladium Alloys

Palladium is not used in the pure state in dentistry. However, it is a useful component of many gold casting alloys.

Alloys based on Ag–Pd have been used for a number of years and are available from most gold alloy manufacturers. The palladium content is a mass fraction of 22–50%; silver content is from 35 to 66%.

Base-Metal Alloys

Base-metal casting alloys are inferior to gold-based casting alloys in some dental applications but are superior in others. Base-metal alloys and low noble alloys have made inroads into dentistry at the expense of the noble metal alloys. Base-metal alloys containing sufficient chromium to make them passive have essentially displaced gold-based alloys in the casting of skeletons for partial dentures. Some inlays, crowns, or bridges are made of non-noble alloys. Wrought stainless steels of the 18Cr–8Ni type are used infrequently as denture bases but have supplanted gold alloys in orthodontic appliances for preventing and correcting malocclusion and associated dental and facial disharmonies. Chromium-containing alloys and titanium-based alloys are the only base metal alloys that can be used with dental techniques and have almost no tarnish or corrosion in the mouth.

Cobalt–Chromium Alloys

Co–Cr and Ni–Cr alloys are used predominately for the casting of removable partial dentures; fixed partial dentures (bridges), and crowns. Because of high hardness, corrosion resistance, and wear resistance, cobalt–chromium alloys are used for bite adjustments and as serrated inserts in plastic teeth used in full dentures. These alloys are well tolerated by the body and also are used for dental implants and orthopedic implant alloys.

Nickel–Chromium Alloys

Because gold is so expensive, there was an intensive development in the 1970s and 1980s of alloys based mostly on nickel–chromium with as many as eight modifying elements. Nickel–chromium alloys cannot be cast as easily or as accurately as gold-based alloys, nor can they be fabricated as easily by soldering. The Ni–Cr alloys primarily contain nickel with a mass fraction of 12% or more of chromium. They are used mainly for the casting of fixed partial dentures (bridges), crowns, or inlays, either with or without aesthetic porcelain or plastic veneers. However, some of them are used for removable partial dentures.

Titanium-Based Casting and Wrought Alloys

Titanium-based alloys offer an attractive alternative to gold alloys and to the base-metal alloys that contain nickel or chromium.

Many of the technical problems of fabrication that formerly inhibited the use of titanium alloys in dental castings have been effectively solved, and titanium castings may now be obtained for virtually any type of dental appliance at prices that are increasingly competitive.

The property most frequently cited in connection with the use of Ti dental or medical appliances is titanium's unique biocompatibility. This helps practitioners avoid occasional allergic reactions that occur with nickel or chromium alloys, and removes concerns about the toxic or carcinogenic potential of appliances that contain nickel, chromium, or beryllium. Wrought alloys of titanium are used for orthodontic wires because of their unique elastic properties, and are used to fabricate most dental implants.

Properties

The casting shrinkage of titanium alloys is comparable, but not identical, to that of chromium-containing alloys, and greater than that of gold alloys. The ability of commercially pure titanium to be cast in relatively thin sections, together with its low density, permits light-weight appliances in bulky restorations.

The Vickers hardness of Ti alloys is strongly dependent on the presence of elements such as oxygen, nitrogen, and carbon that dissolve in hot or molten titanium through contact with residual air molecules or with the surface of refractory molds. Under optimum conditions, castings of commercially pure titanium show yield strengths of 400–500 MPa and ultimate strengths of 490–650 MPa with elongations of 20–30%. Yield strengths and tensile strengths are increased by the presence of dissolved elements.

The stiffness of pure titanium can be increased slightly by alloying; alloys such as Ti-6Al-4V may be specified for partial dentures requiring additional rigidity. Titanium appliances do not tarnish or corrode in the mouth, have no metallic taste, and are easy to clean. The relatively low thermal conductivity of titanium (relatively close to that of tooth enamel) gives these appliances a seemingly natural feel in the mouth and minimizes thermal sensitivity.

Composition

Acceptable composition limits for titanium dental castings have not yet been established, but current practice favors the use of commercially pure or unalloyed titanium.

Oxygen, nitrogen, and carbon promote brittle behavior in all titanium alloys, and it is important to control the cumulative effect of these three elements during casting operations.

Stainless Steels (Iron-Based Alloys)

Today, stainless steels find their primary use in wrought form for temporary applications such as orthodontic wires, brackets, and temporary crowns. The temporary crowns are obtained in preformed sizes/shapes and then are trimmed by the dentist with shears to fit over prepared teeth that are awaiting the fabrication of permanent cast crowns.

Orthodontic wires and brackets owe their strength to work hardening during their formation, contrary to the heat treatment process used for wrought gold alloy wires. The 18–8 stainless steels are the most commonly used alloys; their strength, ductility, and elastic modulus are generally higher than the wrought gold alloys.

Aluminum

Aluminum and aluminum alloys play a limited role in dental applications. Aluminum is used to make lightweight impression trays, articulator facebows, and radiograph alignment holders. Soft aluminum is used as preformed temporary crowns where shaping and bending can be easily done at chairside.

Electrodeposited Metals

Electrodeposited, electroplated, or electroformed metals are used to produce an accurate metal-clad die, or cast, from a compound polysulfide rubber, or silicone rubber impression (see ELECTROPLATING). Copper and silver are used to give accurate dies and casts with an improved working surface having strength, abrasion resistance, and chip resistance.

Solders and Fluxes

Dental solders, like all dental alloys, must be biologically tolerated in the oral environment. They are specifically designed or employed for the purpose of fusing two pieces of dental alloy through the use of intermediate low temperature filler metal.

The soldering process involves the use of solders and fluxes. Solders are materials that fuse and join two dental alloys together. Fluxes coat the surfaces of alloys to be joined in order to produce clean metal surfaces at the temperatures of joining. This enables the molten solder to fuse with those surfaces.

Nonconventional Solder Systems

Nonconventional solder systems are developed for use with newer alloys, especialy base metal alloys. They are few in number and will probably remain the exception rather than the rule.

POLYMERIC DENTAL MATERIALS

Acrylic Resins

Because of the unique combination of properties, eg, aesthetics and ease of fabrication, acrylic resins based on methyl methacrylate and its polymer and/or copolymers have received the most attention since their introduction in 1937. However, deficiencies include excessive polymerization shrinkage and poor abrasion resistance. Polymers used in dental application should have minimal dimensional changes and stress development during and subsequent to polymerization; excellent chemical, physical, and color stability; processability; and biocompatibility and the ability to blend with contiguous tissues.

Denture Bases

Dentures require accurate fit, reasonable chewing efficiency, and lifelike appearance. The chewing efficiency of artificial dentures is one-sixth that of natural dentition. Acrylic resins are generally used as powder–liquid formulations for denture base, bone cement, and related applications. Polymerization is achieved by using thermal initiators; photochemically using photoactive initiators and/or sensitizers and either ultraviolet (uv) or visible light irradiation; and at ambient temperatures using chemical initiator–activator systems.

Properties of Denture Base Materials. Physical properties of acrylic denture base materials are given in Table 2. Mechanical properties of denture bases can vary considerably, and depend on composition, mode of polymerization, and degree of interaction with the oral environment, especially water sorption.

Special-Purpose Resins, Repair Resins. Fractured acrylic dentures can be repaired with materials similar in composition to cold-cured denture resins. These materials generally cure more rapidly because of the relative simple manipulations involved. The process is quick and

Table 2. Physical Properties of Denture-Base Resins

Property	Poly(methyl methacrylate)	Vinyl–acrylics
tensile strength, MPa[a]	48–62	51
compressive strength, MPa[a]	75	61–75
elongation, %	1–2	7–10
elastic modulus, MPa[a]	3.8×10^3	2.3×10^3
impact strength, N · m	1050	3150
transverse strength, MPa[a]	41–55	41–55
flexural strength, MPa[a]	83–117	69–110
Knoop hardness	16–22	14–20
thermal coefficient expansion, °C^{-1}	81×10^{-6}	71×10^{-6}
heat-distortion temp, °C	160–195	130–170
polymerization shrinkage, %	6	6
24-h water sorption	0.3–0.4	0.07–0.04

[a]To convert MPa to psi, multiply by 145.

there is little dimensional change, but the strength of the repaired denture may be only one-half that of the original appliance.

Denture Reliners

A denture can be readapted to the changing contours of soft tissue by relining it with rigid or resilient materials.

Resilient Liners. Resilient liners reduce the impact of the hard denture bases on soft oral tissues. They are designed to absorb some of the energy produced by masticatory forces that would otherwise be transmitted through the denture to the soft basal tissue. The liners should adhere to but not impair the denture base. Other critical properties include total recovery from deformation, retention of mechanical properties, good wettability, minimal absorption of fluids, nonsupport of bacterial or fungal growth, and ease of cleaning. At present, no material fulfills all of these requirements.

Resilient liner materials are generally supplied as powders, liquids, or ready-to-use sheets. The commercial materials currently available are plasticized acrylics or silicones (see Silicon compounds, silicones).

Tissue Conditioners

Tissue conditioners are gels designed to alleviate the discomfort from soft-tissue injury, eg, extractions. Under a load, they exhibit viscous flow, forming a soft cushion between the hard denture and the oral tissues. The polymer in tissue conditioners is often the same as that used for resilient liners. The liquid is a plasticizer containing an alcohol of low volatility.

Crown and Bridge Resins

These materials, based on methyl methacrylate, higher molecular weight methacrylates, and the epimine resin system, are used as interim tooth coverage during fabrication of permanent prostheses. They are used to maintain the correct biting relationship, stop teeth drifting, and protect the prepared tooth against pulpal irritation and fracture. The epimine resin system requires cationic polymerization rather than free radical polymerization common to acrylic materials.

Resins are also used for permanent tooth-colored veneers on fixed prostheses, ie, crown and bridges. Compositions for this application include acrylics, vinyl–acrylics, and dimethacrylates, as well as silica- or quartz-microfilled composites.

Plastic Teeth

Plastic teeth are manufactured by an injection- or transfer-molding process. In addition to acrylics, vinyl resins, polycarbonates, and polysulfones have been suggested for molded teeth. Newer compositions contain very finely dispersed spheres such as pyrogenic silica as reinforcing fillers, ie, microfillers; a urethane dimethacrylate resin;

nonfilled highly cross-linked copolymers with interpenetrating networks; and a layered tooth with an exterior containing a BIS–GMA-type resin and silica filler.

Mouth Protectors

The widespread use of protective mouth guards in contact sports has greatly reduced the incidence of orofacial injuries. Guards are produced from ready-made stock or are custom-fabricated. Natural rubber, poly(vinyl chloride), poly(vinyl acetate-co-ethylene), polyurethane, and silicone elastomers are the materials of choice.

Maxillofacial Prosthetic Materials

Extraoral or external maxillofacial prosthetics (EMFP) is the science of using polymeric biomaterials for the restoration of missing and/or defective facial tissues. The synthesis of materials that can be easily fabricated into lifelike facial devices has long challenged researchers. Ideally, maxillofacial polymeric materials should mimic as closely as possible the mechanical, physical, chemical, and aesthetic properties of the natural tissues they replace.

Maxillofacial polymers include the chlorinated polyethylenes, poly (ether-urethanes), polysiloxanes (see Elastomers), and conventional acrylic polymers. These are all deficient in a number of critical performance and processing characteristics. It is generally agreed that there is a need for improved maxillofacial polymers that can be conveniently fabricated into a variety of prostheses.

Elastomer Impression Materials

Dentistry requires impression materials that are easily handled and accurately register or reproduce the dimensions, surface details, and interrelationship of hard and soft oral tissues. Flexible, elastomeric materials are especially needed to register intraoral tooth structures that have undercuts. The flexibility of these elastomers allows their facile removal from undercut areas while their elasticity restores them to their original shape and size.

Impression materials based on natural polymers, eg, reversible hydrocolloids (agar), irreversible hydrocolloids (alginates), combinations of agar/alginate, and, to a lesser extent, zinc oxide–eugenol cements, are still employed. However, the growing use of elastic impression materials by the dental profession has spurred a continuous search for better elastomers. The primary emphasis has been in the development of nonaqueous elastomeric impression materials.

Nonaqueous Impression Materials

Polysulfide Impression Materials. Significant improvements in strength, toughness, and especially dimensional stability of the set polysulfide elastomers over the aqueous elastic impression materials made these materials popular.

The materials are available in three basic grades, ie, light-bodied or syringe, regular, and heavy-bodied types. The products are supplied as a two-part paste system, usually packaged in collapsible tubes.

The polysulfide impression materials can be formulated to have a wide range of physical and chemical characteristics by modifying the base (polysulfide portion), and/or the initiator system. Further changes may be obtained by varying the proportion of the base to the catalyst in the final mix. Variations in strength, toughness, and elasticity can also be achieved.

The second catalyst paste of the two-paste product is a curing agent. A wide variety of materials convert the liquid polysulfide polymers to elastomeric products. Alkali salts, sulfur, metallic oxides, metallic peroxides, organic peroxides, and many metal–organic salts, ie, paint driers, are all potential curing agents.

Only three types of systems have found application in dentistry. Lead peroxide is the curing agent most frequently used for the polysulfide polymers that serve as dental impression materials. Lead peroxide converts the liquid polymer to elastic solid within a time short enough for oral applications.

Condensation Silicones. Odor, color, and stickiness of the polysulfide rubbers have deterred universal acceptance. The development of room temperature vulcanizing (RTV) silicone rubbers, or siloxanes, has made another acceptable elastomeric system available to dentistry.

The silicone bases have a mild, pleasant odor, and are generally white, although some manufacturers add a pink pigment. The materials are nontacky and can be wiped away from instruments, hands, etc, at any stage of the mix or set. The silicones offer a selection of setting rates and mix viscosities. Regardless of base viscosity, the set rubbers are about the same hardness, ie, a Shore A Durometer value of $\sim$55.

Both the silicone base and the catalyst compositions are moisture sensitive and subject to deterioration when exposed to the atmosphere.

Addition Silicones. Perhaps the most important development in the area of elastic impression materials has been the addition siloxane (or silicone) system. Several reviews have been published on the materials.

Addition siloxanes are the kind most frequently used in the dental clinic and are supplied as a two-paste system. One paste contains a low molecular weight silicone with terminal vinyl groups and reinforcing filler, and the other consists of a hydrogen-terminated siloxane oligomer, filler, and chloroplatinic acid catalyst. On mixing, the catalyzed the vinyl-hydrogen addition reaction results in a cross-linked elastomer. Hydroxyl-containing silicones evolve hydrogen in the setting reaction. Some formulations contain palladium to absorb this gas. Addition silicones have the best elastic properties and lowest dimensional change on setting of all elastomeric impression materials.

Polyether Impression Materials. A polyether-base polymer elastomeric impression material was introduced in 1964. This material is related to the epimine resin and cured by the reaction between aziridine (cyclic imine) rings, which are at the ends of branched polyether molecules. The main chain is probably an ethylene oxide–tetrahydrofuran copolymer. Setting is by cross-linking brought about by an aromatic sulfonate ester catalyst that produces cross-linking by cationic polymerization via the opening of the cyclic imine end groups. These polyethers are supplied as two pastes.

A novel impression material that has a polyether backbone with urethane–acrylic end groups and that cures free radically by visible light irradiation has been developed.

Restoratives

Polymer resins were introduced as tooth restorative materials in the early 1940s. These materials can be classified as unfilled tooth restorative resins, composite or filled restorative resins, and pit and fissure sealants.

Resin Composite Restorative Materials

Improvements in the properties of resin-based restoratives have been brought about primarily by the addition of reinforcing silane-treated silica fillers to unfilled resins. These modifications have made resin composites the most popular aesthetic filling material used today. Techniques are now available that permit resin composite inlays to be fabricated by a laboratory technician, and then adhesively cemented into the tooth cavity by the dentist.

The addition reaction product, Bis–GMA, of bisphenol, A and glycidyl methacrylate can be a compromise between epoxy and methacrylate monomers. Thus, Bis–GMA polymerizes through the free-radical induced covalent bonding of methacrylate groups rather than by ionic opening of the oxirane (epoxide) ring of epoxy resins. Mineral fillers, coated with a silane coupling agent, which bond the powdered silica-type fillers chemically to the resin matrix, are incorporated into Bis–GMA monomer diluted with other methacrylate monomers to make it less viscous. A second base monomer commonly used to make composites is UDMA, a urethane dimethacrylate derived from the reaction of 2-hydroxyethyl methacrylate and aliphatic diisocyanate.

Composite resins can be cured using a variety of methods. Intraoral curing can be done by chemical means, where amine–peroxide initiators are blended by mixing two components to start the free-radical reaction. By contrast, single paste systems can be formulated with photoinitiators, eg, camphorquinone- tertiary amines, and then cured by exposure to visible light irradiation (400–500 nm) spectrum. In contrast to chemically activated composites, single paste, light cured composites have virtually an unlimited amount of working time. Single pastes activated by uv light were previously used in some early materials but are no longer employed for direct restorations. Laboratory curing of indirect restorations can be done by the above methods as well as the additional application of heat and pressure.

Thus composites can be dispensed as paste systems in one or two parts.

Pit and Fissure Sealants

The BIS–GMA or urethane dimethacrylate portion of the composite restorative material has been further diluted with methyl methacrylate monomer or other low viscosity monomers and used to seal developmental pits and fissures on natural tooth enamel.

Adhesives

The fundamentals of adhesion have an important role in modern restorative dentistry (see ADHESIVES). The retention of restorative materials to tooth structure, in addition to holding the restoration to the teeth, seals the interface between the tooth and restorative material. Maintenance of this seal reduces pulpal irritation and the potential for marginal discoloration and recurrent decay. The application of high performance adhesives reduces the need for mechanically retentive cavity designs, thereby minimizing removal of healthy tooth structure. Adhesion also plays an important role in holding prosthetic materials to one another. Polymer adhesive materials can be classified as composite or filled resin cements, porcelain or ceramic coupling agents, metal coupling agents, or enamel and dentin adhesives.

ABRASIVES

Dental abrasives range in fineness from those that do not damage tooth structure to those that cut tooth enamel. Abrasive particles should be irregular and jagged so that they always present a sharp edge, and should be harder than the material abraded. Another property of an abrasive is its impact strength, ie, if the particle shatters on impact it is ineffective; if it never fractures, the edge becomes dull. Other desirable characteristics include the ability to resist wear and solvation.

Dental abrasives can be classified either according to their use or according to the degree of their ability to abrade. They include aluminum oxide and silicates.

THERAPEUTIC DENTAL MATERIALS

Fluorides

Most worldwide reductions in dental decay can be ascribed to fluoride incorporation into drinking water, dentifrices, and mouth rinses. Numerous mechanisms have been described by which fluoride exerts a beneficial effect. Fluoride either reacts with tooth enamel to reduce its susceptibility to dissolution in bacterial acids or interferes with the production of acid by bacterial within dental plaque. The multiple modes of action with fluoride may account for its remarkable effectiveness at concentrations far below those necessary with most therapeutic materials. Fluoride release from restorative dental materials follows the same basic pattern. Fluoride is released in an initial short burst after placement of the material, and decreases rapidly to a low level of constant release. The constant low level release has been postulated to provide tooth protection by incorporation into tooth mineral.

Composite Resins

Many composite restorative resins have incorporated fluoride into the filler particles.

Glass-Ionomers

Glass–ionomers show fluoride release at levels that are usually higher than those found in composite materials.

Other Fluoride Containing Materials

Many forms of fluoride are used strictly as preventive or therapeutic materials. Varnishes containing sodium fluoride are available to place in a cavity prior to filling. These varnishes can also be applied as protective agents on tooth crown and root surfaces. Cavity rinses containing stannous fluoride and/or acidulated phosphate fluoride are applied topically to a cavity just before filling. The largest groups of therapeutic fluorides are the topical formulations used for treatment of both primary and permanent teeth. These materials are topically applied to tooth surfaces as gels or rinses. Prescription fluoride supplements are supplied as rinses, drops, and tablets containing sodium fluoride or acidulated phosphate fluoride in varying doses. These are used as both topical treatments and dietary supplements. Fluoride is also available at varying concentrations in many over-the-counter mouth rinses, toothpastes, and chewable vitamins. The most common and widespread form of fluoride is that contained in drinking water supplies.

Chlorhexidine Gluconate

Chlorhexidine gluconate (1,1'-hexamethylene bis[5-(p-chlorophenyl) biguanide] di-D-gluconate) is used as an antimicrobial against both aerobic and anaerobic bacteria in the oral cavity. It is used as a therapeutic supplement in the treatment of gingivitis, periodontal disease, and dental caries. A mouth rinse form is available as a 0.12 wt% aq solution.

CALCIUM PHOSPHATE MATERIALS

Hydroxyapatite

The mineral of teeth and bone comprises impure forms of hydroxyapatite (HA), $Ca_{10}(PO_4)_6(OH)_2$. Because of its excellent biocompatibility, synthetic HA, in ceramic forms, has been used clinically for filling bony defects since the mid-1970s. Dense and porous types of ceramic HA are available. Dense HA ceramics are made by compressing calcium phosphate solids from a very fine powder form, known as the green state, into a pellet that is then subjected to a heat treatment that causes the powder particles to fuse together by means of solid-state diffusion. Porous HA ceramics are derived from certain species of coral in the genus *Protes* that have regularly patterned 200-μm diameter pores.

Other Ceramic Calcium Phosphate Materials

Other ceramic calcium phosphate materials for repairing bony defect include β-tricalcium phosphate (β-TCP),

β-Ca$_3$(PO$_4$), and biphasic calcium phosphate (BCP) ceramics that consist of both β-TCP and HA.

Calcium Phosphate Cements

Self-setting calcium phosphate cements have been a subject of considerable interest in recent years. Materials that are totally biocompatible and also harden like cement at the site of application are highly desirable in a wide range of biomedical applications. Data in the literature show that cementation can occur in mixtures containing a variety of calcium phosphate compounds. The products formed in these systems included a dibasic calcium phosphate known as dicalcium phosphate dihydrate (DCPD), CaHPO$_4 \cdot 2$ H$_2$O, octacalcium phosphate (OCP), Ca$_8$H$_2$ (PO$_4$)$_6 \cdot 10$ H$_2$O, and HA.

DENTAL IMPLANTS

The use of dental implants is increasing due to improved success rates and is considered a predictable procedure. Success criteria for dental implants include: implant immobility, absence of a radiolucency between bone and implant, no progressive bone loss, and no other adverse symptoms such as pain and infection. A number of present day implants have means for mechanical fixation such as grooved and porous surfaces. Currently, dental implants can be expected to function for 12 years or more.

Regulation

Dental implants are regulated by the FDA.

Requirements

Requirements for dental implant materials are similar to those for orthopedic uses. The first requirement is that the material used in the implant must be biocompatible and not cause any adverse reaction in the body. The material must be able to withstand the environment of the body, and not degrade and be unable to perform the intended function.

Materials

Currently, titanium, either commercially pure or a Ti-6 Al-4 V alloy, is the most widely used metal for implants. The cobalt–chromium–molybdenum was used for many years and still is used for casting individual implants and for some of the blade implants. Ceramic materials including hydroxyapatite, high density alumina, and single-crystal sapphire (alpha alumina) also have been studied and/or used as dental implants.

Hydroxyapaite, the mineral constituent of bone, is applied to the surfaces of many dental implants for the purpose of increasing initial bone growth. However, titanium and its alloy, Ti-6Al-4V, are biocompatible and have osseointegrated as successfully as dental implants without the hydroxyapatite coating.

Surface preparation of the dental implant prior to implantation will have an effect on corrosion behavior, initial metal ion release, and interface tissue response. The titanium and titanium alloy dental implants presently in use have many configurations designed to facilitate bone ingrowth including cylinders with holes, screw threaded surfaces, porous surfaces, and other types of roughened surfaces. Methods used to produce porous surfaces include arc plasma spraying of metal (titanium) onto the implant and the sintering of spheres to the implant surface.

Types of Dental Implants

Indications for a specific type of implant are based primarily on the amount of bone available to support the implant. Also to be considered is the implant proven most successful. Three types of implants are discussed here.

Endosseous

The implants that are anchored in the bone are commonly referred to as being osseointegrated, and are the most successful implants. Endosseous implant types are root form (cylinder, screw, cone; bladeform, and ramus frame).

Subperiosteal

The subperiosteal implants are placed on the residual bony ridge and are not osseointegrated. This implant is most commonly used in the mandible but sometimes is used in the maxilla.

Transosteal and Staple Implants

In this implant the posts are intended to go through the mandible and are intended to support a denture. The mandibular bone staple plate has replaced the transosteal pin.

Certain commercial materials and equipment may be identified in this article for adequate definition of subject matter. In no instance does such identification imply recommendation or endorsement by the National Institute of Standards and Technology, or that the material or equipment is necessarily the best available for the purpose.

A. C. Fraker and co-workers, in M. J. Fagan, Jr., and co-workers, eds., *Implant Prosthodontics: Surgical and Prosthetic Techniques for Dental Implants*, Yearbook Medical Publishers, Littleton, Mass., 1991, pp. 293–304.

L. Linden, in J. C. Salmone, ed., *Polymeric Materials Encyclopedia*, 1996, p. 1854.

R. W. Phillips, *Skinner's Science of Dental Materials*, 7th ed., W. B. Saunders Co., Philadelphia, 1973, Chaps. 31–33.

M. E. Ring, *Dentistry, an Illustrated History*, Abrams, New York, 1985, pp. 160–81, 193–211.

JOHN A. TESK
JOSEPH M. ANTONUCCI
CLIFTON M. CAREY
FRED C. EICHMILLER
J. ROBERT KELLY
NELSON W. RUPP
RICHARD W. WATERSTRAT
ANNA C. FRAKER
LAWRENCE C. CHOW
LAURIE A. GEORGE
GARY E. SCHUMACHER
JEFFREY W. STANSBURY
EDWARD E. PARRY
National Institute of Standards and Technology (NIST)

DESICCANTS

Some materials have sufficient capacity for water or efficiency for drying as well as appropriate physical and chemical properties that they are classed as drying agents, or desiccants. These substances are widely used for removing water from gases, liquids, and solids. Desiccants may be liquids or solids. They may be used repetitively by regenerating the desiccant after use to return it to its active state, or they may be used only once. If the desiccant is used only once, it may last the life of the article being dried or may be discarded when spent. Drying agents are used either in a static (batchwise) or dynamic (continuous or semicontinuous) mode. Their use may be further classified as open system, if fluid flows through the system, or closed system if it does not. Examples of the industrial uses of desiccants, designated as dynamic or static and open- or closed-system applications, are given in Table 1. The list is not all inclusive and ignores various laboratory uses.

Desiccants have varied fundamental characteristics in terms of water capacity and the rate of water sorption. The degree of water removal achieved, or efficiency, is usually given in terms of the water content remaining in the substance that has been dried. This water content can be expressed in several ways, such as humidity ratio, relative humidity (at atmospheric pressure only), relative saturation (at elevated pressure or in liquids), dew point (used at any temperature), ice or frost point (used below 0°C), or parts per million (ppm) by weight or volume. The effectiveness of any drying agent can be measured in terms of its water capacity. In static applications, this capacity is usually the true equilibrium capacity. In dynamic systems, the rate of water removal must be taken into account. Usually, to allow for mass-transfer zones, an additional amount of drying agent is used. In these instances the dynamic capacity, also termed breakthrough capacity, falls short of the true equilibrium capacity.

MECHANISM

The drying mechanisms of desiccants may be classified as follows: Class 1: chemical reaction, which forms either a new compound or a hydrate; Class 2: physical absorption with constant relative humidity or vapor pressure (solid + water + saturated solution); Class 3: physical absorption with variable relative humidity or vapor pressure (solid or liquid + water + diluted solution); and Class 4: physical adsorption.

These mechanisms are characterized by the relative magnitudes of the heats of reaction, solution, or adsorption (see ADSORPTION, GAS SEPARATION). All useful drying mechanisms are exothermic.

Adsorption (qv) is a phenomenon in which molecules in a fluid phase spontaneously concentrate on a solid surface without any chemical change. The adsorbed molecules are bound to the surface by interactions varying from weak to relatively strong between the solid and gas, similar to condensation (van der Waals) forces. Because adsorption is a surface phenomenon, all practical adsorbents possess large surface areas relative to their mass.

Table 1. Applications of Desiccants

Industry	Application	Classification
compressed air	prevent freeze-up and corrosion in air-actuated components	dynamic, open
air separation	prevent ice formation in heat exchangers before cryogenic distillation	dynamic, open
natural gas	prevent corrosion and hydrate formation in pipelines, remove water before cryogenic hydrocarbon recovery, dry liquefied petroleum gas (LPG) to prevent freeze-ups during vaporization	dynamic, open
petrochemical	remove moisture before low temperature fractionation, remove moisture during the rejuvenation or burnoff of spent catalysts, prevent side reactions during catalytic refining	dynamic, open
chemical	remove water that is a diluent or contaminant of some finished product, remove water prior to or during polymerization reactions, prevent caking and corrosion	static or dynamic, closed or open
storage and shipping	prevent food deterioration and corrosion of equipment by relative humidity control	static or dynamic, closed or open
moisture vapor control	lower the dew point in sealed spaces where condensation could occur	static, closed
vapor compression refrigeration	remove moisture from circulating refrigerants	dynamic, closed
space cooling	dry air to permit cooling by evaporation of water	dynamic, open
dehumidification	dry ambient air for air conditioning or for storage, manufacture, or drying of moisture-sensitive parts or materials	dynamic, open
corrosion control	reduce the dew point in automobile exhaust systems during cool down to reduce internal cold condensate corrosion	static, semiclosed
absorption refrigeration	cyclic absorption and stripping of water with liquid desiccants to produce chilled water for air conditioning[a] and process cooling	dynamic, open

[a]See AIR CONDITIONING.

Desiccants can lose water capacity and drying efficiency by taking up moisture during storage. They should therefore be analyzed before use. If necessary, the materials should be reactivated (regenerated) before putting them in service.

WATER ANALYSIS

Hand in hand with desiccants goes the analysis for low levels of water (moisture). There are a number of proven methods, ranging from color change indicators to electronic hygrometers to the wet chemical and Coulometric Karl Fischer titration methods. The method chosen depends on the phase being analyzed, the precision needed, and convenience.

Color change indicators are chemicals often impregnated into a paper substrate or a solid desiccant itself. The indicator typically changes color at a fixed value of the relative saturation of the fluid in contact with it. However, color change indicators are relatively imprecise.

Hygrometers are instruments capable of continuous on-line measurement of water in gases. Some hygrometers also work in liquids and saturated vapors. Hygrometers are particularly useful for continuous analysis in a flowing stream, such as dried air in a heating, ventilating, and air conditioning (HVAC) system. Some can be placed directly in the flowing stream while others must be installed outside the stream and exposed to a continuous sample of the fluid.

For analysis of water in a solid or a static liquid or gas sample, the standard technique is the Karl Fischer method.

COMPATIBILITY

Desiccants must be chemically compatible with the material being dried. Ideally, the desiccant and the material should not react because such a reaction may produce harmful or undesirable by-products.

STATIC DRYING

Many liquids are dried batchwise rather than continuously. The drying agent is added to the liquid and sufficient time is allowed to dry the product. The liquid is then separated from the drying agent by filtration, decantation, or distillation. Drying agents employing Class 1 or 2 mechanisms are generally used for these applications.

Desiccants Used in Static Drying

The most commonly used desiccants are discussed in this section are barium oxide, calcium chloride, calcium oxide, calcium sulfate, lithium chloride, perchlorates, and sodium and potassium hydroxides. Activated alumina, silica gel, and molecular sieves, which are discussed later under dynamic, solid drying agents, are also widely used in static or batch-drying situations.

Capacity and Efficiency

The higher capacity desiccants go through the various hydrate levels to yield fairly broad ranges of constant relative humidities as moisture is picked up. However, these compounds do not produce very low relative humidities or dew points. The best performance, or lowest dew point, occurs with excess drying agent.

From a study of the efficiency of some 25 desiccants for drying several families of laboratory solvents and reagents it was concluded that molecular sieves are the desiccants of choice in most cases.

Closed-System Drying

Equilibrium capacity is the principal consideration in the design of closed nonregenerative, relatively static drying systems. The total amount of moisture to be removed must first be calculated from the volume of the system and the initial water concentration. Depending on the final moisture content desired, a drying agent can be selected based on its compatibility with the material to be dried and its ability to produce the final dew point. The equilibrium capacity of the drying agent must be determined at the system temperature and the final water concentration.

Although equilibrium capacity is the prime concern, few of these closed systems are truly static. Even closed systems have dynamic features or other nonsteady-state aspects, such as temperature fluctuations, moisture ingression, and drydown rates.

Closed systems are usually nonregenerative: The desiccant charge is designed for the life of the system or is replaceable.

Because the system likely is nonisothermal, the analysis of a closed-desiccant system requires knowledge of the temperature of the desiccant as well as the dew point (ice point) or water concentration (partial pressure) specification. Indeed, the whole system may undergo periodic temperature transients that may complicate the analysis.

Another aspect to consider in the design of closed-drying systems is the drydown time. The drydown time is the period required for the system to dry down from its initial water concentration (or partial pressure) to a concentration that approaches equilibrium with the desiccant. During this time, the system is not fully protected from the negative effects of the moisture that the desiccant is designed to remove. In such a system, the instantaneous drying rate is proportional to the water content at any time.

If required, the drydown can be hastened by increasing desiccant mass, particle surface area, or mass-transfer coefficient. The mass-transfer coefficient can be altered to some extent by the design of the desiccant container.

An interesting and novel use of a solid desiccant, the reduction of cold condensate corrosion in automotive exhaust systems, illustrates a hybrid closed–open system.

DYNAMIC DRYING

Continuous drying is employed when drying a volume of gas or liquid in a batchwise fashion is not practical. When a solid, dynamic desiccant is used, the fluid stream is passed over a fixed bed of the drying agent, which must have the physical properties to allow the fluid to pass readily through. When liquid desiccants are used, the drying is usually achieved by countercurrent contact of the gas (flowing up) against the liquid (flowing down). The desiccants are usually regenerable.

Glycols and sulfuric acid are the principal examples of liquid desiccants.

Solid Desiccants

The solid desiccants used in dynamic applications fall into a class called adsorbents. Because they are used in large packed beds through which the gas or liquid to be treated is passed, the adsorbents are formed into solid shapes that allow them to withstand the static (fluid plus solid head) and dynamic (pressure drop) forces imposed on them. The most common shapes are granules, extruded pellets, and beads (spheres). Solid desiccants include activated alumina, silica gel, and molecular sieves.

DESIGN OF DYNAMIC ADSORPTION DRYING SYSTEMS

Adsorbent drying systems are typically operated in a regenerative mode with an adsorption half-cycle to remove water from the process stream and a desorption half-cycle to remove water from the adsorbent and to prepare it for another adsorption half-cycle. Usually, two beds are employed to allow for continuous processing. In most cases, some residual water remains on the adsorbent after the desorption half-cycle because complete removal is not economically practical. At cyclic steady state the amounts of water removed during the adsorption and desorption half-cycles is the same. Divided by the adsorbent weight, this amount is termed the differential loading, which is the working capacity available for dehydration.

The two most common types of drying systems operate on either a pressure-swing cycle or a thermal-swing cycle to take advantage of the difference in water loading on the desiccant with changes in pressure or temperature.

A growing application is the use of a rotating desiccant wheel for drying air in dehumidification systems. The wheel may consist of a packed bed or a rolled corrugated paper (or polymer) monolith with desiccant particles adhered to the surface.

The desiccant wheel can be used in two ways—for dehumidification and for energy exchange.

In designing a gas drying system, the engineer must often estimate the water content of the gas to be dried. Although this estimate will sometimes be sufficient, it may well fail at higher pressures.

Adsorption Plots

Isotherm plots are the most common method of presenting adsorption data. An isotherm is a curve of constant temperature: The adsorbed water content of the adsorbent is plotted against the water partial pressure in equilibrium with the adsorbent. An isostere plot shows curves of constant adsorbed water content: The vapor pressure in equilibrium with the adsorbent is plotted against temperature.

Mass Transfer and Useful Capacity

The term useful capacity, also referred to earlier as breakthrough capacity, differs from the equilibrium capacity. The useful capacity is a measure of the total moisture taken up by a packed bed of adsorbent at the point where moisture begins to appear in the effluent. Thus the drying process cycle must be stopped before the adsorbent is fully saturated. The portion of the bed that is not saturated to an equilibrium level is called the mass-transfer zone.

The parameters affecting the size and shape of a mass-transfer zone are adsorbent type, adsorption isotherm shape, flow rate, packed-bed depth, adsorbent particle size, physical properties of the carrier fluid, temperature, pressure, and the concentration of water in the carrier fluid.

ECONOMIC ASPECTS

The largest U.S. manufacturer of molecular sieves for adsorbent and desiccant use is UOP, which has a production capacity of ~30 million kg/year. W.R. Grace and Zeochem have ~5 and 9 million kg/year capacity, respectively. W.R. Grace is the largest producer of silica gel desiccants. Activated alumina for use as adsorbent and desiccant is produced by UOP (formerly LaRoche Chemicals), Aluminum Company of America, and Porocel.

The largest users of molecular sieve desiccants are the natural gas processing, insulating glass, and refrigeration industries. Silica gel dominates the packaging industry, where the material is used to protect electronic equipment and pharmaceuticals, eg, from moisture. Silica gel is also used in dehumidification of buildings. The principal uses of activated alumina are in refrigeration, where its primary function is to adsorb organic acids rather than water, air drying, and alkylation feed stream drying in oil refineries.

EMERGING APPLICATIONS

Energy Storage

Reactivating a desiccant stores the reactivation energy in the dehydrated desiccant. This energy-storage feature is useful if the energy source is intermittent or seasonal, such as solar energy, or interruptable. Research suggests that this energy-storage feature is especially useful if the desiccant is used to dry air for agricultural applications.

Desiccant Cooling

Desiccant cooling is a continuing area of research. In these systems, a desiccant is used to produce an extremely dry air stream. The dry air is then cooled in a heat exchanger and humidified with a water spray. The evaporation of the water absorbs heat from the air and produces a cold, almost saturated air stream for air conditioning. The desiccant is thermally regenerated with exhaust air, which is heated with solar energy, natural gas, or waste heat from power generation. Both liquid and solid desiccant systems have been studied.

R. S. Boynton, *Chemistry and Technology of Lime and Limestone*, John Wiley & Sons, Inc., New York, 1966.

A. W. Czanderna, *Polymers as Advanced Materials for Desiccant Applications: 1987*, Solar Energy Research Institute, Golden, Colo., 1988.

L. Harriman, ed., *Desiccant Cooling and Dehumidification*, American Society of Heating, Refrigerating, and Air-Conditioning Engineers, Atlanta, Ga., 1992.

I. Mellan, *Polyhydric Alcohols*, Spartan Books, Washington, D.C., 1962.

ALAN P. COHEN
UOP

DESIGN OF EXPERIMENTS

The main reason for designing an experiment statistically is to obtain unambiguous answers to questions of major interest at a minimum cost. The need to quantify relationships, learn about interactions among variables, and to measure experimental error are a few of the added reasons for designing an experiment statistically.

Many chemists and engineers think of experimental design mainly in terms of standard plans for assigning treatments to experimental units, such as the Latin square, factorial, fractional factorial, and central composite (or Box) designs. These designs are described in books, such as those summarized in the general references of this article, and catalogued in various reports and articles. Additionally, numerous commercial software packages are available for generating such experimental designs, as well as to aid the experimenter in analyzing the resulting data. Important as such formal plans are, the final selection of test points represents only the proverbial tip of the iceberg, the culmination of a careful planning process.

Statistically planned experiments are characterized by (*1*) the proper consideration of extraneous variables; (*2*) the fact that primary variables are changed together, rather than one at a time, in order to obtain information about the magnitude and nature of interactions and to gain improved precision in estimating the effects of these variables; and (*3*) built-in procedures for measuring the various sources of random variation and for obtaining a valid measure of experimental error against which one can assess the impact of the primary variables and their interactions.

A well-planned experiment is often tailor-made to meet specific objectives and to satisfy practical constraints. The final plan may or may not involve a standard textbook design. If possible, a statistician knowledgable in the design of experiments should be called in early, made a full-fledged team member, and be fully apprised of the objectives of the test program and of the practical considerations and constraints. He or she may contribute significantly merely by asking probing questions. However, designing an experiment is often an iterative process requiring rework as new information and preliminary data become available. With a full understanding of the problem, the statistician is in an improved position to respond rapidly if last-minute changes are required, to help experimenters gain understanding in a sequential manner, and to provide meaningful analyses of the experimental results, including statements about the statistical precision of any estimates.

PURPOSE AND SCOPE OF THE EXPERIMENT

Designing an experiment is like designing a product. Every product serves a purpose; so should every experiment. This purpose must be clearly defined at the outset. An understanding of this purpose is important in developing an appropriate experimental plan.

In addition to defining the purpose of a program, one must decide on its scope. An experiment is generally a vehicle for drawing inferences about the real world, as expressed by some, usually quantitative, response (or performance) variable. Since it is highly risky to draw inferences about situations beyond the scope of the experiment, care must be exercised to make this scope sufficiently broad. In any case, one need keep in mind that the statistical inferences one can draw from the experiment apply only to the range of conditions under which the experiment was conducted.

EXPERIMENTAL VARIABLES

An important part of planning an experimental program is the identification of the controllable or "independent" variables (also known as factors) that affect the response and deciding what to do about them. Controllable or independent variables in a statistical experiment can be dealt with in four different ways, as described next. The assignment of a particular variable to a category often involves a trade-off among information, cost, and time.

Primary Variables

The most obvious variables are those whose effects on the mean or the variability of the response variable(s) are to be evaluated directly; these are the variables that, most likely, created the need for the investigation in the first place. Such variables may be quantitative, such as catalyst concentration, temperature, or pressure, or they may be qualitative, such as method of preparation, catalyst type, or batch of material.

Quantitative controllable variables are frequently related to the response variable by some assumed statistical relationship or model. The minimum number of conditions or levels per variable is determined by the form of the assumed model. However, it is recommended that additional points, above the minimum, be included so as to allow assessment of the adequacy of the assumed model.

Qualitative variables can be broken down into two categories. The first consists of those variables whose specific effects on the mean response are to be compared directly; e.g., comparison of the impact on average performance of two proposed preparation methods or of three catalyst types. The required number of conditions for such variables is generally evident from the context of the experiment. Such variables are sometimes referred to as fixed effects or Type I variables.

The second type of qualitative variables are those whose individual contributions to variability or noise in the responses are to be evaluated. The specific conditions of such variables are generally randomly determined. Material batch is a typical example. The batches used in the experiment are selected randomly (or as close to randomly as is practically feasible) from a large population of batches. It is desirable to have a reasonably large sample (eg, five or more batches) in order to obtain an adequate degree of precision in estimating the variability in response attributable to such variables. These variables are generally referred to as random effects or Type II variables. Differentiation between fixed and random effect variables is an important consideration both in the design of the experiment, and in the analysis of the resulting data.

When there are two or more variables, they might interact with one another, ie, the effect of one variable upon the response depends on the value of the other variable.

An important purpose of a designed experiment is to obtain information about interactions among the primary variables which is accomplished by varying factors simultaneously rather than one at a time.

Background Variables and Blocking

In addition to the primary controllable variables there are variables, such as day-to-day differences, that cannot, and perhaps should not, be held constant in the experiment. It is crucial that such background variables are separable from (in technical terms, not "confounded" with) the primary variables in the experiment. Such background variables are often introduced into the experiment in the form of experimental blocks.

An experimental block represents a relatively homogeneous set of conditions within which different conditions of the primary variables are compared.

A main reason for running an experiment in blocks is to ensure that the effect of a background variable does not contaminate evaluation of the effects of the primary variables. Blocking, moreover, removes the effect of the blocked variables from the experimental error also, thus allowing more precise estimation of the experimental error and, as a result, more precise estimates of the effects of the primary variables. Finally, in many situations, the effect of the blocking variables on the response can also be readily evaluated, an important added bonus for blocking.

Uncontrolled Variables and Randomization

A number of further variables, such as ambient conditions (temperature, pressure, etc), can be identified but not controlled, or are only hazily identified or not identified at all but affect the results of the experiment. To ensure that such uncontrolled variables do not bias the results, randomization is introduced in various ways into the experiment to the extent that this is practical.

Randomization means that the sequence of preparing experimental units, assigning treatments, running tests, taking measurements, etc., is randomly determined, based, e.g., on numbers selected from a random number table. The total effect of the uncontrolled variables is thus lumped together as unaccounted variability and part of the experimental error.

Variables Held Constant

Finally, some variables should be held constant in the experiment. Holding a variable constant limits the size and complexity of the experiment but, as previously noted, can also limit the scope of the resulting inferences.

EXPERIMENTAL ENVIRONMENT AND CONSTRAINTS

The operational conditions under which the experiment is to be conducted and the manner in which each of the factors is to be varied must be clearly spelled out.

In many programs, variables are introduced at different operational levels, resulting in so-called split-plot experimentation. A clear statement is required of the performance characteristics or dependent variables to be evaluated as the experimental response. The various ways of obtaining repeat results in the experiment need to be specified and initial estimates of repeatability obtained. Clear procedures for recording all pertinent data from the experiment must be developed and documented.

STAGEWISE EXPERIMENTATION

Contrary to popular belief, a statistically planned experiment does not require all testing to be conducted at one time. Instead, the design of experiments is a catalyst for the general scientific learning process. Thus, much experimentation should be sequential. This permits changes to be made in later tests based on early results and allows preliminary findings to be reported.

Whether or not to conduct an experiment in stages depends on the program objectives and the specific situation; a stagewise approach is recommended when units are made in groups or one at a time and a rapid feedback of results is possible. Running the experiment in stages is also attractive in searching for an optimum response, because it might permit moves closer to the optimum from stage to stage. On the other hand, a single-stage experiment may be desirable if there are large start-up costs at each stage or if there is a long waiting time between the start of the experiment and the measurement of the results.

If the experiment is conducted in stages, precautions must be taken to ensure that possible differences between the stages do not invalidate the results. Appropriate procedures to compare the stages must be included, both in the test plan and in the statistical analysis.

OTHER CONSIDERATIONS

Many other questions must be considered in planning the experiment: (1) What is the most meaningful way to express the controllable or independent variables? (2) What is a proper experimental range for the selected quantitative controllable variables? (3) What is a reasonable statistical model, or equation form, to approximate the relationship, sometimes referred to as the response surface, between the independent variables and each response variable? (4) What is the desired degree of precision of the statistical estimates and conclusions based upon the analyses of the experimental results? (5) Are there any benchmarks of performance? (6) What statistical techniques are required for the analysis of the resulting data, and can these tools be rapidly brought to bear after the experiment has been conducted?

FORMAL EXPERIMENTAL PLANS

The test plan is developed to best meet the goals of the program. This might involve one of the standard plans developed by statisticians and practitioners, such as a blocking design, a full or fractional factorial design, a central composite (or Box) design, or a Box-Behnken design. Another type of experimental design, used principally for response surface exploration, and favored by some statisticians, is a computer-generated optimal design. Mixture designs arise when the variables under consideration are the ingredients of a product that must add to 100%. Also, Genichi Taguchi, an eminent Japanese engineering professor, has stressed the role of experimental design in identifying the process conditions that show the greatest consistency, or are most robust, in the face of variability in manufacturing conditions and customer use. A series of articles that describe, illustrate, and critique Taguchi methods is provided in the August 1988 issue of the Journal Quality Reliability Engineering International (see references). As already suggested, combinations of plans in a stagewise approach are often used—eg, a factorial experiment conducted in blocks or a central composite design using a fractional factorial base.

COMBINATORIAL CHEMISTRY

Technology now allows us to search vast experimental regions to identify materials with desired properties.

This is the realm of "combinatorial chemistry", also known as high throughput screening (or "finding a needle in a haystack" experimentation). In fact, combinatorial chemistry "is now practical because of the convergence of low-cost computer systems, reliable robotic systems, sophisticated molecular modeling, statistical experimental strategies, and database software tools" Cawse book.

Early applications of combinatorial chemistry were targeted at identifying potentially promising drugs by pharmaceutical companies. However, in recent years, similar approaches are being used by chemical and other companies for new product discovery and development. Frequently, the goal is to establish a library of performance responses for different material combinations.

Combinatorial chemistry for materials improvement has been made possible by the ability to make up and assess, possibly, hundreds of vials a day. However, the "experimental space" that one wishes to explore may consist of millions of possible combinations of conditions.

General design of experiment concepts have much applicability for combinatorial chemistry. In an initial stage, one might consider the entire response surface to identify potential regions that warrant further exploration. In subsequent testing, one would home in on such promising regions with various stages of further testing.

STATISTICAL TOOLS FOR THE ANALYSIS OF DESIGNED EXPERIMENTS

The analysis of most designed experiments involves a combination of three major types of tools: the analysis of variance, regression analysis, and graphical analysis.

The analysis of variance is, probably, the most frequently used tool for analyzing the results of a designed experiment. It is a formal statistical method that breaks down the total observed variability in a response variable into separate parts attributable to each of the sources of variation, based on an assumed statistical model.

Regression analysis allows one to fit a relationship between a series of experimental variables (x's) and a response variable (y). This is especially relevant for data from response surface designs. In fact, regression analysis has been referred to as "the workhorse of statistical data analysis."

We strongly encourage graphical analyses to supplement the more formal statistical analyses, and, on occasion, to take their place. Also, half-normal probability plots provide information similar to a formal analysis of variance concerning the relative importance of the individual sources of variation, but in graphical form.

Finally, we note that improved methods for the analysis of designed experiments, frequently capitalizing on increased computations capabilities, continue to be developed.

MULTIPLE RESPONSE VARIABLES

Often one is concerned with two or more performance variables. Multiple response variables complicate the analyses as well as the stagewise development of the experimental plan, especially if one is seeking an optimum experimental region (versus just developing a response library) Myers and Montgomery (2002).

COMMERCIAL SOFTWARE FOR THE DESIGN AND ANALYSIS OF EXPERIMENTS

Numerous computer software packages have been developed to help generate experimental designs. These are especially useful in performing the mechanical tasks associated with the design and analysis of experiments.

FURTHER READING

The classic book by Box, Hunter and Hunter (2005) is, perhaps, the one that has become best known among practitioners, and provides a good starting point for those with limited knowledge of statistics. An elementary overview is provided in the 175-page introductory book by Del Vecchio (1997). Schmidt and Launsby (2000) give a basic-to-intermediate discussion. The 684-page volume by Montgomery (2005) is an in-depth, but still applications-oriented, treatment. Daniel (1976) is especially strong on fractional factorial designs, and Myers and Montgomery (2002) specialize in response surface experiments.

G. E. P. Box, W. G. Hunter, and J. S. Hunter, *Statistics for Experimenters: An Introduction to Design, Data Analysis, and Model Building*, John Wiley & Sons, Inc., New York, 1978, 2nd ed. 2005 (1st ed., 1978).

J. N. Cawse, ed., *Experimental Design for Combinatorial and High Throughput Materials Development*, John Wiley & Sons, Inc., New York, 2002.

C. Daniel, *Applications of Statistics to Industrial Experimentation*, John Wiley & Sons, Inc., New York, 1976.

R. J. Del Vecchio, *Understanding Design of Experiments*, Hanser/Gardner Publications, Inc., Cincinnati, 1997.

D. C. Montgomery, *Design and Analysis of Experiments*, 6th ed., John Wiley & Sons, Inc., New York, 2005 (1st ed., 1976).

R. H. Myers and D. C. Montgomery, *Response Surface Methodology: Process and Product Optimization Using Designed Experiments*, 2nd ed., John Wiley & Sons, 2002 (1st ed., 1995).

S. R. Schmidt and R. G. Launsby, *Understanding Industrial Designed Experiments*, 4th ed., Air Academy Press, Colorado Springs, Co. 2000 (1st ed., 1988).

GERALD J. HAHN
ANGELA N. PATTERSON
General Electric

DETERGENCY AND DETERGENTS

The term detergency is limited to systems in which a liquid bath is present and is the main cleaning component of the system. The cleaning is enhanced primarily by the presence in the bath of a special solute, the surfactant, which alters interfacial effects at the various phase boundaries within the system.

In the cleaning or washing process in a typical detersive system the soiled substrate is immersed in or brought into contact with a large excess of the bath liquor. Enough bath is used to provide a thick layer over the whole surface of the substrate. During this stage, air is displaced from soil and substrate surfaces, ie, they are wetted by the bath. The system is subjected to mechanical agitation, either rubbing or shaking, which provides the necessary shearing action to separate the soil from substrate and disperse it in the bath. Agitation also promotes mass-transfer in the system, just as in a heterogeneous chemical reaction. The bath carrying the removed soil is drained, wiped, squeezed, or otherwise removed from the substrate. The substrate is rinsed free of the remaining soiled bath. This rinsing step determines the final cleanliness of the substrate. The cleaned substrate is dried or otherwise finished.

A meaningful discussion of detergency requires a definition of clean. In the physiochemical sense, a surface is clean if it contains no molecular species other than those in the interior of the two adjoining phases. It is difficult to achieve such a state even under the most exacting laboratory conditions. Practically, a surface is clean if it has been brought to a desired state with regard to foreign matter present upon it, as judged by agreed upon criteria. Most standards for cleanness involve a visual or optical judgment for the presence of foreign matter.

COMPONENTS OF DETERSIVE SYSTEMS

Substrates

Solid objects to be cleaned vary widely in chemical composition and surface configuration. With few exceptions, however, they can be divided into fabrics and fibrous materials, and hard surfaces.

Soils

Soils vary greatly. They may be a single solid or liquid phase but usually are two or more phases, intimately and randomly mixed and irregularly disposed over the substrate. In a large number of important detersive systems, the nature of soil and the quantity present are well known.

As a result of many painstaking investigations, the soils on apparel encountered in laundering have been shown to be complex mixtures containing both oily and finely divided solid material. The oily material consists largely of fatty acids and polar fatty material but a considerable proportion of neutral nonpolar oil is also present. The solid components vary widely with the locale in which samples are taken, and resemble local street dust in composition.

Particle size is one of the most important factors in determining the ease with which solid soil can be removed from a substrate. Particles of >5 μm dia are generally easily removed. Particles of <10-μm dia cannot be removed by ordinary detersive processes once they are attached to a typical textile fabric. Such particles are responsible for the gradual irreversible graying of white goods with continued wear and laundering. Particles in this size range tend to form clumps and clusters before they reach the fiber surface. These clusters behave like individual large particles. Particles or clusters in the range of 100 nm dia resist removal by simple agitation in liquids that are not surface active, but these particles are removable by normal detersive processes. This is the size range of greatest interest.

Soil may include material that is soluble in the bath, such as encrusted sugar residues and molecularly dispersed material such as fruit juice stains. Removal of these soils is an important aspect of cleaning but is not generally considered in discussions of detergency.

Baths

The baths discussed here are aqueous solutions. Some nonaqueous systems, however, are true detersive systems. Modern dry cleaning baths, for example, contain solutes that are surface active in the conventional hydrocarbon or chlorinated hydrocarbon medium and aid soil removal. The physical chemistry of such systems differs considerably from that of aqueous systems. Among bath components the solute that is effective in cleaning, usually a mixture of several components, is called the detergent. The term detergent is also used frequently in the restricted sense of a surfactant of high detersive power. In many hard-surface systems, however, nonsurfactants such as alkaline silicates and phosphates exert a true detersive effect. They are, in fact, the principal detergents in these systems even in the complete absence of any surfactants. In the cleaning of fabric systems, the most important detersive component in the bath is the surfactant. Nonsurfactant components that augment the cleaning effect of surfactants are called builders. Many materials that act as builders in fabric systems, eg, phosphates and silicates, are the primary detergents in hard-surface systems, although their primary contribution to the cleaning process may differ in the two cases.

FORMULATION

Detergents are formulated to clean a defined set of soiled substrates under an expected range of washing conditions. Some detergents, the familiar bar or toilet soap, for example, consist essentially of only one component. There are few systems, however, in which a suitably formulated detergent consisting of several components does not outperform the best single-component system. Although detergents for hand dishwashing rarely contain builders, those currently used in the U.S. contain at least three surfactants, and may contain up to six. Ingredients of laundering detergent formulations for fabrics may be divided into the following groups: surfactants, including soap and various others; the inorganic salts, acids, and bases, including builders, and other compounds that do not contribute to detergency but provide other functions, such as regulating density and assuring crispness of powdered formulations; organic additives that enhance detergency, foaming power, emulsifying power, or soil-suspending effect of the composition; and special purpose additives, such as bleaching agents, fluorescent whitening agents, antimicrobial agents, blueing agents, or starch, which provide desirable performance functions but have no direct effect on soil removal (see also INDUSTRIAL ANTIMICROBIAL AGENTS).

Fabric detergent formulations for special applications, such as the various specific operations within the textile mill, are frequently much simpler. They tend to contain little if any builder or special-purpose additive. The indispensable ingredient in fabric detergency is the organic surfactant. Formulations for hard surface detergency such as those used in automatic machine dishwashing, are simpler than fabric-washing compositions. An organic surfactant is frequently not needed and inorganic salts are the detersive ingredients.

Surfactants

The most important components of detersive systems are, of course, the surfactants described elsewhere in the *Encyclopedia*.

Builders

Builders are substances that augment the detersive effects of surfactants. Most important is the ability to remove hardness ions from the wash liquor (ie, soften the water) and thus to prevent them from interacting with the surfactant. They include phosphates, sodium carbonate, silicates, zeolites, clays, nitrilotriacetic acid, alkalies, and neutral soluble salts.

Organic Additives

Certain nonsurfactant organic additives improve cleaning performance and exhibit other desirable properties.

Antiredeposition agents contribute to the appearance of washed fabrics. Sodium carboxymethylcellulose, NaCMC is the most widely used, and on cotton fabrics, the most effective. With the advent of synthetic fabrics, other cellulose derivatives, eg, methylcellulose, hydroxybutylecllulose, hydroxypropyl- and mixed methyl and hydroxybutycellulose ethers have been shown to be more effective than NaCMC (see CELLULOSE ETHERS).

Fluorescent whitening agents (qv) absorb ultraviolet radiation and subsequently emit some of the radiation energy in the blue part of visible spectrum. As a result, they confer enhanced whiteness to the appearance of washed articles. For synthetic fabrics such as polyester, it has proved to be more effective to prebrighten the fabric by incorporating the fluorescent whitening agent in the spin-melt during manufacture rather than depend on adsorption from the detergent bath. As a result, the usage of fluorescent whitening agents in formulated laundry products has decreased in recent years.

Blueing agents, which are dyes, provide another approach to maintaining fabric whiteness by a mechanism in which a yellow cast of washed fabrics is covered by the blue dye. Since this approach reduces reflectance, it is less desirable than the use of fluorescent whitening agents that increase reflectance.

Proteolytic enzymes in particular are widely usfed in premium products. They degrade proteinaceous stains and aid the cleaning performance of other formulation ingredients. Amylases and lipases have been used in a few U.S. detergents, the former to remove starches and the latteer, fatty esters and triglycerides.

One U.S. detergent manufacturer has introduced detergents with the bleach activator sodium nonanoyloxybenzene sulfonate. This activator forms the surface active species pernonanoic acid that does provide a bleach benefit under U.S. conditions. In automatic-dishwashing formulations, bleaching agents are needed to remove food stains from dishware and break down proteinaceous soil. Chlorine is the most cost-effective agent available for this purpose and is present in all U.S. products as chlorinated isocyanurate

Foam regulators such as amine oxides, alkanolamides, and betaines are present in products where high foam value is functionally or esthetically desirable, mainly hand-dishwashing liquids and shampoos. In automatic dishwashing products, on the other hand, copious foam volumes interfere with the efficiency of the mechanical rotors during operation. In this type of product, a foam depressant is often present.

Organic sequestering agents serve the same purpose as the sequestering phosphates, ie, to remove interfering metal ions from the detergent bath. They would appear to also provide some benefits through ionic strength and soil suspending effects. They are used where the less expensive phosphates are, for one reason or another, not applicable.

FACTORS INFLUENCING DETERGENCY

Detergency is mainly affected by the concentration and structure of surfactant, hardness and builders present, and the nature of the soil and substrate. Other important factors include wash temperature; length of time of washing process; mechanical action; relative amounts of soil, substrate, and bath, generally expressed as the bath ratio, ie, the ratio of the bath weight to substrate weight; and rinse conditions.

MECHANISMS

Even the simplest detersive system is surprisingly complex and heterogeneous. It can nevertheless be conceptually resolved into simpler systems that are amenable to theoretical treatment and understanding. These simpler systems are represented by models for substrate-solid soil and substrate-liquid soil. In practice, many soil systems include solid–liquid mixtures. However, removal of these systems can generally be analyzed in terms of the two simpler model systems. Although these two systems

differ markedly in behavior and structure, and require separate treatment, there are certain overriding principles that apply to both.

The first principle is that soil systems can be regarded and treated as classical systems of colloid and surface chemistry. A second principle applying to these model systems is derived from their colloidal nature. With the usual thermodynamic parameters fixed, the systems come to a steady state in which they are either agglomerated or dispersed. No dynamic equilibrium exists between dispersed and agglomerated states. In the solid-soil systems, the particles (provided they are monodisperse, ie, all of the same size and shape) either adhere to the substrate or separate from it. In the liquid-soil systems, the soil assumes a definite contact angle with the substrate, which may be anywhere from $0°$ (complete coverage of the substrate) to $180°$ (complete detachment). The governing thermodynamic parameters include pressure, temperature, concentration of dissolved components, and electrical conditions.

In applying this concept, the factor of particle size must be continuously borne in mind. A heterodisperse system can reach a steady state wherein the smaller particles are agglomerated and the larger particles are dispersed, giving the apparent effect of an equilibrium. In ideal monodisperse systems under steady conditions, however, no such effects are noted.

Purely mechanical disturbances (which are not usually considered thermodynamic variables) may influence the state of aggregation of a colloidal system; for example, floc size in carbon and iron oxide suspensions varies with the degree of agitation being imposed on the system.

A final consideration in resolving practical detersive systems into their simpler components relates to soil removal versus redeposition. Superficially, it would appear that the redeposition phenomenon contradicts the all-or-nothing concept that the system must exist in either the agglomerated or the dispersed state. Keeping in mind both the composite nature and the kinetics of a practical system, it is readily shown that no such contradiction exists. The soil particle that redeposits is in a different state from what it was during its initial removal from the substrate. Thus, the initial group of agglomerated systems, which composes the soil–substrate–bath complex before soil removal, is quite different from the agglomerated system composed of substrate and redeposited soil.

SOLID-SOIL DETERGENCY

Adsorption

Many studies have been made of the adsorption of soaps and synthetic surfactants on fibers in an attempt to relate detergency behavior to adsorption effects. Relatively fewer studies have been made of the adsorption of surfactants by soils. Plots of the adsorption of sodium soaps by a series of carbon blacks and charcoals show that the fatty acid and the alkali are adsorbed independently, within limits, although the presence of excess alkali reduces the sorption of total fatty acids. No straightforward relationship was noted between detergency and adsorption.

In a study of the adsorption of soap and several synthetic surfactants on a variety of textile fibers, it was found that cotton and nylon adsorbed less surfactant than wool under comparable conditions. Among the various surfactants, the cationic types were adsorbed to the greatest extent, whereas nonionic types were adsorbed least. The adsorption of nonionic surfactants decreased with increasing length of the polyoxyethylene chain. When soaps were adsorbed, the fatty acid and the alkali behaved more or less independently just as they did when adsorbed on carbon.

Adsorption of bath components is a necessary and possibly the most important and fundamental detergency effect. Adsorption (qv) is the mechanism whereby the interfacial free energy values between the bath and the solid components (solid soil and substrate) of the system are lowered, thereby increasing the tendency of the bath to separate the solid components from one another. Furthermore, the solid components acquire electrical charges that tend to keep them sperated, or acquire a layer of strongly solvated radicals that have the same effect.

Mass Transfer near the Substrate Surface

Mechanical action has a great effect on soil removal, probably by influencing mass transfer, ie, the diffusion of soluble material away from the immersed fibers. Mechanical action tends to maintain a high concentration gradient near the fiber, and the resulting increased diffusion causes stronger diffusion currents to flow. These diffusion currents are presumably responsible for carrying away the soil particles that have already been detached or loosened from the fiber surface by physico-chemical action.

Colloidal Stabilization

Surfactant adsorption reduces soil–substrate interactions and facilitates soil removal. For a better understanding of these interactions, a consideration of colloidal forces is required.

The model solid-soil detersive system is advantageously treated as a sol-agglomerate colloid system or, in more general terms, a lyophobic colloid. In the typical lyophobic colloid, consisting of a single disperse phase in an aqueous suspending medium, only one type of liquid–solid interface and one type of solid–solid interface is present. The simplest detersive system, however, has an added degree of complexity in the presence of two types of liquid-solid interface: soil-bath and substrate–bath. Also present are two effective types of solid–solid interface: soil–substrate and soil–soil. The soil-soil interface relates to flocculation or dispersion of soil particles remote from the substrate, and is not of primary concern in the present discussion. The soil–soil interface is, of course, important in practical detergency since soil aggregates can be regarded as large single particles.

There are two general theories of the stability of lyophobic colloids, or, more precisely, two general mechanisms controlling the dispersion and flocculation of these colloids. Both theories regard adsorption of dissolved species as a key process in stabilization. However, one theory is based on a consideration of ionic forces near the interface, whereas the other is based on steric forces. The two theories complement each other and are in no sense contradictory. In some systems, one mechanism may be predominant, and in others both mechanisms may operate simultaneously. The fundamental kinetic considerations common to both theories are based on Smoluchowski's classical theory of the coagulation of colloids.

OILY-SOIL DETERGENCY

Roll-up

The principal means by which oily soil is removed is probably roll-up. The applicable theory is simply the theory of wetting.

Solubilization

The role of micellar solubilization (as the term is used in the physical chemistry of surfactants) in oily-soil removal has been debated for many years. The amount of oily soil that could be present in a normal wash load could not all be removed and held in micellar solution by anionic surfactants. On the other hand, nonionic surfactants could do so, because of their greater solubilizing ability. High solubilizing power is definitely linked with good detergency. Thus, a very direct relationship between the solubilizing power of a surfactant for the test dyestuff Orange OT and its ability to remove polar solid from steel surfaces was established.

Phase Changes at the Soil-Bath Interface

Closely related to solubilization is a phenomenon that involves polar organic soils and surfactant solutions. If a complete phase diagram is plotted for a ternary system containing sodium dodecyl sulfate (or glycerol oleate), and water, several important and unusual features are noted. A large area represents a liquid phase consisting of a microemulsion, where the dispersed particles are so small that the system is isotropic, like the familiar soluble oils. Also, over another large area, a liquid crystalline phase is formed, containing all three components. This liquid crystalline phase flows like a liquid, at least in one directions. Flow perpendicular to the oriented planes is accomplished by folding the planes cylindrically, but the physical flow is still of the purely viscous type, with no yield point evident. These two phases, particularly the liquid–crystal phase, play an important part in detergency. Furthermore, liquid-crystal formation lowers interfacial tension. Although this phenomenon was demonstrated in tertiary oil recovery, the principles could also apply to oily-soil detergency.

When the polar organic component is a solid at ordinary temperatures, the addition of detergent and water markedly lowers the melting point; more specifically, as the temperature is raised, a point is reached where surfactant and water penetrate the solid. Thus, the ternary liquid–crystal phase might form spontaneously at room temperature by mixing the components, or, more precisely, an aqueous detergent solution can literally melt and liquefy a relatively large proportion of solid polar fatty matter (see LIQUID CRYSTALLINE MATERIALS).

In a detersive system containing a dilute surfactant solution and a substrate bearing a solid polar soil, the first effect is adsorption of surfactant at the soil–bath interface. This adsorption is equivalent to the formation of a thin layer of relatively concentrated surfactant solution at the interface, which is continuously renewable and can penetrate the soil phase. Osmotic flow of water and the extrusion of myelin forms follows the penetration, with ultimate formation of an equilibrium phase. This equilibrium phase may be microemulsion rather than liquid crystalline, but in any event it is fluid and flushable from the substrate surface. This phase change effect explains the detersive behavior of sucrose fatty esters in admixture with alkylarenesulfonates.

MEASUREMENT OF DETERGENCY

The measurement of detergency can be approached from two different points of view. The theoretical approach is concerned with the relative quantity of soil bound to the substrate before and after washing. In this case, measurement is a necessary analytical procedure in the study of the detergency mechanism. The second approach emphasizes the development of reproducible laboratory methods that predict the results of practical cleaning operations.

The measurement of detergency in the laboratory requires the following components: a means of measuring or estimating the amount of soil on the substrate or the degree of cleanness both before and after washing; satisfactory substrates and soiling composition; a means of applying soil to substrate in a realistic manner; and a realistic and reproducible cleaning device. These fundamental requirements apply regardless of the particular type of substrate that is being cleaned.

FABRIC DETERGENCY

Laundering

Reflectance is the most commonly used measurement for the whiteness of fabrics, although the transmittance of light by fabric specimens can also be used. The most commonly used instrument for reflectance measurement is the Gardner colorimeter, although the Zeiss Elrepho is also used. For general detergency, the grayness of the fabric is measured. Color effects can also be measured, and fabric yellowing is especially important. It is masked by fluorescent whitening agents (FWA). Special filters are available to eliminate this effect, and whitening caused by soil removal can be distinguished from that of FWA deposition.

Textile Mill Operations

Detergency is important in textile finishing because small quantities of foreign matter on the goods can interfere seriously with dyeing and other finishing treatments. Furthermore, the goods are expected to be uniformly and thoroughly clean when sold. Many detergency tests in this area are of the semipractical type, ie, test swatches are analyzed for soil content. This analysis generally consists of gravimetric determination of the soil content either directly from the fabric weight, or by extraction (see TEXTILES).

HARD SURFACE DETERGENCY

Despite the variety of hard-surface objects that are purposefully cleaned at regular intervals, detergency has been studied quantitatively in relatively few cases only. The small-scale user normally judges washing results as satisfactory or unsatisfactory. If satisfactory results are obtained with the amount of detergent and the degree of mechanical action employed, the user is not interested in minor qualitative differences. In those areas where specifications are important and where differences among detergents or mechanical washing equipment are readily perceivable, quantitative methods for measuring detergency have been developed.

In specific cases of metal cleaning where small amounts of residual soil must be detected and are difficult to measure by conventional means, radiotracer methods have been employed. Interest in these techniques has been stimulated by the development of methods for decontaminating hard surfaces subjected to atomic fallout.

Quantitative measurements have been obtained for ceramics and glass, metals, and organic surfaces such as painted and plastic tile.

Glassware and Dishwashing

Dishes are washed either by hand or in an automatic dishwashing machine. Hand-dishwashing detergents are generally high foaming compositions containing organic surfactants as the main ingredient. The consumer judges efficiency not only by the cleanness of plates but also by foam persisting throughout the operation. Evaluation of hand-dishwashing products by manufacturers simulates this procedure. The number of plates that can be washed clean, judged visually without or with a color or fluorescence indicator, and the number of plates necessary to kill the foam in the dishpan is taken as the measure of detersive efficiency.

Metal Cleaning

The purpose of cleaning steel is to remove dirt and leave the article in a state where it can be delivered for use without further finishing (see METAL SURFACE TREATMENTS). The surface must therefore be covered with a tenacious corrosion-resistant coating as it emerges from the cleaning bath. Many emulsion cleaners remove lubricants and other unwanted dirt while depositing an anticorrosive coating on the metal. The primary test for efficacy in this situation is a corrosive test of the cleaned article.

Organic Surfaces

Tests for detergency on organic surfaces such as painted walls and plastic tile generally include a rubbing or sponging step corresponding to the manner in which such surfaces are cleaned in practice.

DETERGENT MANUFACTURE

Liquid Products

The manufacture of liquid detergent products is generally a straightforward process requiring batch equipment with provisions for metered addition of individual ingredients, agitation, and if needed, heating and cooling. Capital cost can vary depending on the degree of automation.

Spray-Dried Products

The manufacture of powdered product is more complicated. High-pressure spray-drying of an aqueous slurry has replaced the earlier process in which a solidified cake of the product had to be broken up mechanically. Spray-drying equipment requires a relatively high capital outlay.

The first step in preparing spray-dried products involves producing a slurry of liquid and soil ingredients. Two processes are available, a batch process or a continuous process.

Acids such as fatty acids and alkylbenzenesulfonic acids are neutralized with NaOH during slurry preparation to form soap and sodium alkylbenzenesulfonate, respectively.

After preparation, the slurry is transported to an aging vessel. During residence time of 20–30 min, the neutralization process, hydration of sodium tripolyphosphate and structural changes in the slurry are completed to provide a homogeneous composition. By means of a high pressure pump, the slurry is conveyed to the spray-drying tower under a pressure of ca 10 MPa (100 atm).

The slurry, at 80–100°C, is forced through nozzles of 2.5–3.5 mm dia arranged on a nozzle ring. In the tower, the slurry encounters hot air that has entered the tower at 250–350°C. Upon exit, the powder is conditioned during passage via a belt conveyer and an airlift to the packaging machinery.

Because of stringent air pollution rules the exit gases are wet-scrubbed in brine because high NaCl concentrations reduce foam formation. NaOH scrubs out SO_2 from sulfur-containing fuel. The scrubbing solution, saturated with detergent fines, is recycled to the water tank for slurry preparation.

Although spray-drying accommodates relatively high content of surfactants, certain types, such as the alkanolamides and some nonionic surfactants are best added to the product after spray-drying.

Dry-Blended Product

In addition to lower capital outlay, dry-blending requires considerably less processing energy. Final product density, which is usually near unity, depends on the density of the starting materials and the nature of equipment used to blend these materials. Modern mixing and blending equipment, if properly controlled, can give product density and particle sizes comparable to spray-dried products.

Agglomerated Products

The process of agglomeration is intermediate between spray-drying and dry-blending. In agglomeration, a liquid is sprayed onto a continuously agitated powder. Equipment designs include stationary mixers, rotating mixers with spray nozzles, and rotating blenders with a liquid dispersion bar, either twin shell or continuous zigzag. Automatic dishwashing detergents are manufactured mostly by agglomeration.

HEALTH AND SAFETY FACTORS

As a class, surfactants and detergent products are among the most widely used chemical compositions. Almost everyone is exposed to these products on a daily basis in situations that range from ingestion of food-grade emulsifiers to intimate contact of skin and eyes with personal-care and laundry products. Safety is therefore a matter of great importance. Ranges of surfactant LD_{50} values are shown in Table 1.

Under conditions of normal use, detergent products are not hazardous to users. Nonetheless, surfactants possess some toxicity, and they are mild irritants.

The manufacture of surfactants and of detergent products is regulated by OSHA. Dust concentration in detergent plants as well as factory noise levels are the primary areas of relevance, since the individual components in these products are essentially nonhazardous. Of more immediate concern to the detergent industry is the Federal Hazardous Substances Act (FHSA) of 1960 and the Consumer Products Safety Act (CPSA) of 1972. The FHSA defines specific labeling requirements, such as "Danger" for extremely flammable, corrosive, or highly toxic substances, and "Warning" or "Caution" for less hazardous materials.

ENVIRONMENTAL CONSIDERATIONS

The introduction of surfactant products into the environment, after use by consumers or as part of waste disposed

Table 1. Rat Oral LD_{50} Values of Surfactant

Type of compounds	Oral LD_{50}, mg/kg
alkylbenzenesulfonates	700–2,480
alcohol ethoxylates	1,600 to greater than 25,000
sulfated alcohol ethoxylates	7,000 to greater than 50,000
alcohol sulfates	5,000–15,000

during manufacture, is regulated by the Clean Water Act, and Clean Air Act, and the Resource Conservation and Recovery Act. In this respect, surfactants are subject to the same regulations as chemicals in general. There are, however, two areas of specific relevance to surfactants and detergent products, ie, biodegradability and eutrophication.

Extensive investigations led to the conclusion that a branched hydrophobe impedes the rate and extent of degradation of surfactants by microorganisms. The most immediately apparent remedy, therefore, was to replace the propylene tetramer in ABS with a straight hydrocarbon chain giving straight-chain alkylbenzenesulfonate, so-called linear alkanesulfonate (LAS). At the same time, commercialization of the Ziegler process for the oligomerization of ethylene provided another route to straight-chain hydrophobes that could easily be converted to detergent alcohols and straight-chain nonionic surfactants. By 1965, the U.S. detergent industry had completed a voluntary switch from hard to soft surfactants at a cost that has been estimated at ca $150×106. In addition to ABS, other surfactants based on propylene oligomers, such as alkylphenol derivatives, have largely disappeared from U.S. consumer laundry products.

Even though the biodegradability problem has been solved for alll practical purposes, the subject continues to receive considerable attention. The biodegradation of LAS has been studied intensively, and several mechanistic pathways have been identified such as β- and ω−oxidation as well as reductive and oxidative desulfonation. Investigation of the biodegradation of LAS, alcohol ethoxylates, and alkylphenol ethoxylates in the laboratory and under sewage plant operating conditions showed that LAS and straight-chain alcohol ethoxylates and their sulfates degrade to CO_2 and H_2O.

Eutrophication

This term, which denotes excessive nutrition or overfertilization, has been applied to the contribution excesssive amounts of phosphorus may make to the growth of algae under certain contain conditions. Phosphours in water supply originates from run-off of agricultural fertilizers, human excrement, and sodium tripolyphosphate present in detergent formulations. It has been extimated that 25–30% of the phosphorus in waste water comes from laundry detergents, and that detergents contribute about 3% of the phosphorus annually entering U.S. surface waters. In the United States, and later in Western Europe, detergent phosphates were singled out in the early 1960s as the cause of eutrophication, and their removal from consumer laundry formulations was proposed as a feasible approach to improvement of environmental water quality. Many states and a number of local jurisdications have banned detergent products containing phosphate.

The efforts of the detergent industry toward solution of its part of the eutrophication problem are, at this point, less complete than its response to the biodegradability problem. Soda ash, Na_2CO_3, sodium silicate, and, to a lesser extent, sodium citrate formed the basis of the early formulations marketed inthe areas where phosphates were baneed. technically, these substances are considerably less effetive than sodium tripolyphosphats. As a precipitant builder, soda ash can lead toundesirable deposits of calcium carbonate on textiles and on washing machines.

A. W. Adamson, *Physical Chemistry of Surfaces*, 5th ed., John Wiley & Sons, Inc. New York, 1990.

McCutcheon's Emulsifiers & Detergents, North Am. & International Ed., Glen Rock, N.J., 1991.

M. J. Rosen, *Surfactants and Interfacial Phenomena*, 2th ed., Wiley-Interscience, New York, 1989.

A. M. Schwartz, in E. Matijevic, ed., *Surface and Colloid Science*, Vol. 5, John Wiley & Sons, Inc., New York, 1972.

JESSE L. LYNN, JR.
Lever Brothers Company

DEUTERIUM AND TRITIUM

DEUTERIUM

Deuterium (symbol ^2H or D) occurs in nature in all hydrogen-containing compounds to the extent of about 0.0145 atom%.

Molecular deuterium, D_2, was first isolated in relatively pure form in 1931, and nearly pure D_2O was prepared shortly thereafter by electrolysis. Subsequently applications of deuterium as a tracer for the path of hydrogen in biological systems were developed and became widely used. In physical organic chemistry, the differences in rates of reaction between corresponding ^1H and D compounds became an important tool for the elucidation of organic reaction mechanisms. The significance of deuterium isotope effects in a biological context attracted attention very early. The discovery in 1959 that it was possible to grow fully deuterated organisms opened new areas of isotope chemistry and biology for exploration.

The recognition in 1940 that deuterium as heavy water has nuclear properties that make it a highly desirable moderator and coolant for nuclear reactors fueled by uranium of natural isotopic composition stimulated the development of industrial processes for the manufacture of heavy water.

Physical Properties

As in the case of hydrogen and tritium, deuterium exhibits nuclear spin isomerism. However, the spin of the deuteron is 1 instead of ½ as in the case of hydrogen and tritium. As a consequence, and in contrast to hydrogen, the ortho form of deuterium is more stable than the para form at low temperatures, and at normal temperatures the ratio of ortho- to para-deuterium is 2:1 in contrast to the 3:1 ratio for hydrogen. The physical and thermodynamic properties of

Table 1. Properties of Liquid H$_2$ and D$_2$ at 20.4 K

Property	H$_2$[a]	D$_2$[b]
Equilibrium states		
density, g/L	70	169
viscosity, mPa · s($=$ cP)	1.4×10^{-2}	4.0×10^{-2}
surface tension, mN/m($=$ dyn/cm)	2.17	3.72
thermal conductivity, W/(cm · K)	11.6	12.64
dielectric constant	1.226	1.275
Normal states		
molar volume at 20 K, mL	28.3	23.5
heat of vaporization, J/mol[c]	904	1226
heat of fusion, J/mol[c]	117	197

[a]The equilibrium state of hydrogen has 0.21% *o*-H$_2$O; the normal state has 75% *o*-H$_2$O.
[b]The equilibrium state of deuterium has 97.8% *o*-D$_2$; the normal state has 67% *o*-D$_2$.
[c]To convert J to cal, divide by 4.184.

Table 2. Physical Properties of Light and Heavy Water

Property	H$_2$O	D$_2$O
molecular weight	18.015	20.028
melting point, T_m, °C	0.00	3.81
normal boiling point, T_b, °C	100.00	101.42
critical constants		
temperature, °C	374.1	371.1
pressure, MPa[a]	22.12	21.88
volume, cm^3/mol	55.3	55.0
density at 25°C, g/cm^3	0.99701	1.1044
vapor pressure, liquid at 25°C, kPa[b]	3.166	2.734
coefficients of thermal expansion,°C^{-1}		
solid at T_m	1.39×10^{-4}	1.39×10^{-4}
liquid at T_m	-5.9×10^{-5}	-3.2×10^{-5}
compressibility at 20°C, Pa^{-1b}	4.45	4.59
length of the hydrogen bond, nm	0.2765	0.2760
dielectric constant at 25°C	78.304	77.937
refractive index, n^{20}_D	1.3330	1.3283
viscosity at 25°C, mPa · s($=$ cP)	0.8903	1.107
surface tension at 25°C, mN/m ($=$ dyn/cm)	71.97	71.93
ion product constant at 25°C	1.01×10^{-14}	1.11×10^{-15}
heat of ion product at 25°C, kJ/mol[c]	56.27	60.33

[a]To convert MPa to atm, multiply by 10.1.
[b]To convert kPa to mm Hg, multiply by 7.5.
[c]To convert Joules to calories, divide by 4.184.

elemental hydrogen and deuterium and of their respective oxides illustrate the effect of isotopic mass differences.

Properties of Light and Heavy Hydrogen. The equilibrium state for these substances is the low temperature ortho–para composition existing at 20.39 K, the normal boiling point of normal hydrogen. The normal state is the high (above 200 K) temperature ortho–para composition, which remains essentially constant.

Thermodynamic data on H$_2$, the mixed hydrogen–deuterium molecule, HD, and D$_2$, including values for entropy, enthalpy, free energy, and specific heat have been tabulated. Some physical properties of liquid H$_2$ and D$_2$ at 20.4 K are presented in Table 1.

Properties of Light and Heavy Water. Selected physical properties of light and heavy water are listed in Table 2. At room temperature both light and heavy water appear to be highly structured, ie, extensively hydrogen bonded. Heavy water is the more structured.

Deuterium as Neutron Moderator. Deuterium has very desirable properties as a moderator for neutrons. As illustrated in Table 3 heavy water has a much greater moderating ratio than any of the other materials commonly used as moderators.

Kinetic Isotope Effects. The principal difference in chemical behavior between H and D derives from the generally greater stability of chemical bonds formed by D. The most important factor contributing to the difference in bond energy and the kinetic isotope effect is the lower (5.021–5.275 kJ/mol (1.2–1.5 kcal/mol)) zero-point vibrational energy for D bonds.

Biological Effects of Deuterium. Replacement of more than one third of the hydrogen by deuterium in the body fluids of mammals or two thirds of the hydrogen in higher green plants has catastrophic consequences for the organisms. Extensive replacement of H by D in living organisms

is, however, not invariably fatal to the organisms. Numerous green and blue-green algae have been grown in which >99.5% of the hydrogen has been replaced by D. Numerous varieties of bacteria, molds, fungi, and even a protozoan have been successfully grown in fully deuterated form. These organisms of unnatural isotopic composition and the deuterated compounds that can be extracted from them have found uses in many areas of biological research.

Kinetic isotope effects are an important factor in the biology of deuterium. Isotopic fractionation of hydrogen and deuterium in plants occurs in photosynthesis. The lighter isotope ^{1}H is preferentially incorporated from water into carbohydrates and lipids formed by photosynthesis. Hydrogen isotopic fractionation has thus become a valuable tool in the elucidation of plant biosynthetic pathways.

Fully deuterated griseofulvin and benzylpenicillin have antifungal and antibiotic potencies at least as

Table 3. Properties of Neutron Moderators

Moderator	Slowing-down power $\xi \times \Sigma_s$, cm^{-1}	Macroscopic absorption cross section Σ_a, cm^{-1}	Moderating ratio, $\xi \times \Sigma_a/\Sigma_a$
water	1.28	2.2×10^{-2}	58
heavy water	0.18	8.5×10^{-6}	21,000
helium	10^{-5}	2.2×10^{-7}	45
beryllium	0.16	1.2×10^{-3}	130
graphite	0.065	3.3×10^{-4}	200

great as their ordinary hydrogen analogues, and are probably metabolized *in vivo* more slowly because of the enhanced stability of C—D relative to C—H bonds. Synthetic deuterated drugs have also been considered as therapeutic agents. Considering the well-documented toxic effects of deuterium in mammals, long-term administration of D_2O to humans is not likely to become a routine procedure. However, D_2O confers significant protection to yeast against hydrostatic pressure damage, and pretreatment with D_2O has been claimed to protect cultured cells against x-ray damage.

Production of Heavy Water

Because of the low natural abundance of deuterium, very large amounts of starting material, which is water, must be processed to produce relatively small amounts of highly enriched deuterium. No water or other hydrogen compound has been found either in nature or as a by-product of an industrial operation that is significantly enriched in deuterium. The cost of subsequent enrichment to 99% is negligible compared to the costs incurred in the initial enrichment from natural abundance to 1%. For small-scale preparations, a highly efficient but very expensive process such as electrolysis can be used. For large-scale use, however, a high enrichment factor per stage is of only secondary importance to the overall costs of operation both in power and in capital investment. The isotope separation methods that have attracted the greatest interest include chemical exchange between water and hydrogen sulfide, hydrogen and water, and hydrogen and ammonia; distillation of water or hydrogen; and electrolysis of water in combination with other procedures.

Uses

The only large-scale use of deuterium in industry is as a moderator, in the form of D_2O, for nuclear reactors.

TRITIUM

Tritium, the name given to the hydrogen isotope of mass 3, has symbol ^3H or more commonly T. Its isotopic mass is 3.0160497. Moletecular tritium, T_2, is analogous to the other hydrogen isotopes. The tritium nucleus is energetically unstable and decays radioactively by the emission of a low-energy β particle. The half-life is relatively short (~12 yr), and therefore tritium occurs in nature only in equilibrium with amounts produced by cosmic rays or man-made nuclear devices.

Physical Properties

Tritium is the subject of various reviews, and a book provides a comprehensive survey of the preparation, properties, and uses of tritium compounds. Selected physical properties for molecular tritium, T_2, are given in Table 4. All components appear miscible in both liquid and solid phases from 17 to 22 K. The T—T bond energy has been estimated at 4.5881 eV. The entropy of T_2 at 298.15 K is 164.8562 kJ/mol (39.4016 kcal/mol), the specific heat

Table 4. Physical Properties of Molecular Tritium

Property	Value
melting point, at 21.6 kPa^a, K	20.62
boiling point, K	25.04
critical temperature, K	40.44
critical pressure, MPa^b	1.850
critical volume, cm^3/mol	57.1^c
heat of sublimation, J/mol^d	1640
heat of vaporization, J/mol	1390
entropy of vaporization, J/(mol · K)^d	54.0
molar density of liquid, mol/L	
20.62 K^a	45.35
29 K	39.66

^{a}Value represents the triple point (162 mm Hg).
^{b}To convert MPa to psi, multiply by 145.
^{c}Value is calculated.
^{d}To convert J to cal, divide by 4.184.

is 29.1997 J/(mol · °C) (6.9789 cal/(mol · °C)), and the Gibbs free energy is 135.9083 kJ/mol (32.4829 kcal/mol).

Ortho-Para Tritium. As in the case of molecular hydrogen, molecular tritium exhibits nuclear spin isomerism. The spin of the tritium nucleus is ½, the same as that for the hydrogen nucleus, and therefore H_2 and T_2 obey the same nuclear isomeric statistics. Below 5 K, molecular tritium is 100% para at equilibrium. At high (100°C) temperatures the equilibrium concentration is 25% para and 75% ortho.

Properties of T_2O. Some important physical properties of T_2O are listed in Table 5. Tritium oxide can be prepared by catalytic oxidation of T_2 or by reduction of copper oxide using tritium gas. T_2O, even of low (2–19% T) isotopic abundance, undergoes radiation decomposition to form HT and O_2.

Nuclear Properties

Radioactivity. Tritium decays by β emission, $^3T \longrightarrow ^3He + \beta^-$.

Nuclear Fusion Reactions. Tritium reacts with deuterium or protons (at sufficiently high temperatures) to undergo nuclear fusion. Nuclear fusion using tritium can be initiated and sustained at the lowest temperature (at least in principle) of any nuclear fusion reaction known. Tritium thus becomes the key element both for controlled thermonuclear energy sources and in the uncontrolled release of thermonuclear energy in the hydrogen bomb.

Table 5. Physical Properties of T_2O

Property	Value	Reference
mol wt	22.032	
temperature of maximum density, °C	13.4	
boiling point, °C	101.51	121
density at 25°C, g/mL	1.2138	122
ionization constant at 25°C	ca 6×10^{-16}	124

Nuclear Magnetic Resonance. All three hydrogen isotopes have nuclear spins, $I \neq 0$, and consequently can all be used in nmr spectroscopy. Tritium is an even more favorable nucleus for nmr than is ^{1}H, which is by far the most widely used nucleus in nmr spectroscopy.

Chemical Properties

Most of the chemical properties of tritium are common to those of the other hydrogen isotopes. However, notable deviations in chemical behavior result from isotope effects and from enhanced reaction kinetics induced by the β-emission in tritium systems. Isotope exchange between tritium and other hydrogen isotopes is an interesting manifestation of the special chemical properties of tritium.

Production

Nuclear Reactions. The primary reaction for the production of tritium is

$$^6\text{Li} + n^1 \longrightarrow {}^3\text{T} + {}^4\text{He} + 4.78 \text{ MeV}.$$

Production in Target Elements. Tritium is produced on a large scale by neutron irradiation of ^{6}Li.

Production in Heavy Water Moderator. A small quantity of tritium is produced through neutron capture by deuterium in the heavy water used as moderator in the reactors.

Production in Fission of Heavy Elements. Tritium is produced as a minor product of nuclear fission.

Production-Scale Processing. The tritium produced by neutron irradiation of ^{6}Li must be recovered and purified after target elements are discharged from nuclear reactors. In the recovery process the gaseous constituents of the target are evolved, and the hydrogen isotopes separated from other components of the gas mixture.

Isotopic Concentration. A number of techniques have been reported for concentrating tritium from naturally occurring sources, eg, low (20–25 K) temperature distillation, concentration by gas chromatography, separation by a cryogenic thermal diffusion column, by diffusion through a palladium–silver–nickel membrane, and by chromatography on coated molecular sieves.

Health and Safety Factors

Because tritium decays with emission of low-energy radiation ($E_{av} = 5.7$ keV), it does not constitute an external radiation hazard. However, tritium presents a serious hazard through ingestion and subsequent exposure of vital body tissue to internal radiation. A widely used instrument for air monitoring is a type of ionization chamber called a Kanné chamber. Uptake of tritium by personnel is most effectively monitored by urinalyses normally made by liquid scintillation counting on a routine or special basis. Environmental monitoring includes surveillance for tritium content of samples of air, rainwater, river water, and milk.

The radiological hazard of tritium to operating personnel and the general population is controlled by limiting the rates of exposure and release of material. Personnel are protected in working with tritium primarily by containment of all active material. Metal hydride technology has been developed to store, purify, pump, and compress hydrogen isotopes. Conversion to or extraction from metal triteride would offer flexibility and size advantages compared to conventional processing methods that use gas tanks and mechanical compressors, and should considerably reduce the risk of tritium gas leaks (see HYDRIDES).

Personnel who must work in areas in which tritium contamination exceeds permitted levels are safeguarded by protective clothing, such as ventilated plastic suits.

Uses

The development of a tritium fuel cycle for fusion reactors is likely to be the focus of tritium chemical research. Tritium is widely used as a tracer in molecular biology (see RADIOACTIVE TRACERS).

J. Bigeleisen in P. A. Rock, ed., *Isotopes and Chemical Principles*, American Chemical Society Symposium Series No. 11, 1975, pp. 1-43.

E. A. Evans, *Tritium and Its Compounds*, 2nd ed., John Wiley & Sons, Inc., New York, 1974.

G. M. Murphy, ed., *Production of Heavy Water*, McGraw-Hill Book Co, Inc., New York, 1955.

H. C. Urey, *Science* **78**, 566 (1933).

JOSEPH J. KATZ
Argonne National Laboratory

DIAMINES AND HIGHER AMINES, ALIPHATIC

The aliphatic diamine and polyamine family encompasses a wide range of multifunctional, multireactive compounds. This family includes ethylenediamine (EDA) and its homologues, the polyethylene polyamines (commonly referred to as ethyleneamines), the diaminopropanes and several specific alkanediamines, and analogous polyamines. The molecular structures of these compounds may be linear, branched or cyclic, or combinations of these.

The ethyleneamines have found the broadest commercial application and are the primary focus of this article. The lower molecular weight ethylenediamines, ie, EDA, diethylenetriamine (DETA), piperazine (PIP), and N-(2-amino-ethyl)-piperazine (AEP), are available commercially as industrially pure products. The tetramine (TETA), pentamine (TEPA), hexamine (PEHA), and higher polyamine products are commercially available as boiling point fractions consisting of natural mixtures of linear, branched, and cyclic compounds. Their compositions are largely determined by the chemical processes used in their production. The individual components in these higher ethyleneamines are generally not available in industrial quantities.

Table 1. Properties of Commercial Diamines and Higher Amines

Commercial name	Molecular weight	Freezing point, °C	Bp[a], °C	ΔH_{vap}[a]kJ/mol	Refractive index, n_D^{20}	Viscosity at 20°C, mPa·s (=cP)
ethylenediamine	60.1	10.8	117.0	40.7	1.4565	1.8
diethylenetriamine	103.2	−39	206.9	54.0	1.4859	7.2
triethylenetetramine	146.2[b]	−35	277.4	56.4	1.4986	26
tetraethylenepentamine	189.3[b]	−40	315		1.5067	76
pentaethylenehexamine[c]	232.4[b]	−30	180−280[d]			100−300
aminoethylpiperazine	129.2	−17	221	41.2	1.5003	15
piperazine	86.1	109.6	144.1			
1,2-propylenediamine	74.1	−27	120−123	38.2	1.4455	1.6
1,3-diaminopropane	74.1	−12	137−140	46.4[e]	1.4555	2.0
iminobispropylamine	131.2	−16	110−120	76.2[e]	1.4791	9.6
N-(2-aminoethyl)-1,3-propylenediamine	117.2		80			
N,N'-bis-(3-aminopropyl)-ethylenediamine	174.3		170			
dimethylaminopropylamine	102.2	−56	134.9	35.6	1.4350	1.1
menthanediamine	170.3	−45	107−126[f]	1.479	17.5[g]	
triethylenediamine	112.2	158	174	61.9[h]		
hexamethylenediamine[i]	116.2	41	204.0	51.0		

[a]At 101.3 kPa = 1 atm unless otherwise noted. [b]Linear component. Commercial product consists of a mixture of linear, branched, and cyclic structures with the same number of nitrogen atoms. [c]Commercial higher polyamine products contain up to about 40% PEHA. [d]At 0.67 kPa, 10−60% distills in this range. [e]At 93.3°C. [f]At 1.3 kPa. [g]At 25°C. [h]Heat of sublimation, below 78°C. [i]For manufacture of HMDA in preparation of nylon-6,6, see POLYAMIDES.

The predominant commercial diaminopropanes are 1,2-propylenediamine (1,2-PDA), 1,3-diaminopropane (1,3-PDA), iminobispropylamine (IBPA), and dimethylaminopropylamine (DMAPA). Other commercially important products include other higher alkylenediamines, such as hexamethylenediamine (HMDA); certain cyclic amines, such as triethylenediamine (TEDA); and various alkyl- and hydroxyalkyl-derivatives.

PHYSICAL PROPERTIES

Physical properties of some commercially available polyamines appear in Table 1. Generally, they are slightly to moderately viscous, water-soluble liquids with mild to strong ammoniacal odors. Although completely soluble in water initially, hydrates may form with time, particularly with the heavy ethyleneamines (TETA, TEPA, PEHA, and higher polyamines), to the point that gels may form or the total solution may solidify under ambient conditions. The amines are also completely miscible with low mol wt alcohols, acetone, benzene, toluene and ethyl ether, but only slightly soluble in heptane. Piperazine, the lowest mol wt cyclic diamine, freezes above room temperature. As such, it is available commercially as either the anhydrous solid or an aqueous solution.

CHEMICAL PROPERTIES

The aliphatic alkyleneamines are strong bases exhibiting behavior typical of simple aliphatic amines. Additionally, dependent on the location of the primary or secondary amino groups in the alkyleneamines, ring formation

with various reactants can occur. This same feature allows for metal ion complexation or chelation. The alkyleneamines are somewhat weaker bases than aliphatic amines and much stronger bases than ammonia as the pK_b values indicate.

Inorganic Acids

Alkyleneamines react vigorously with commonly available inorganic acids forming crystalline, water soluble salts. The free alkyleneamines can be regenerated by reaction of their salts with aqueous caustic.

Alkylene Oxides and Aziridines

Alkyleneamines react readily with epoxides, such as ethylene oxide (EO) or propylene oxide (PO), to form mixtures of hydroxyalkyl derivatives. Aziridines react with alkyleneamines in an analogous fashion to epoxides.

Aliphatic Alcohols and Alkylene Glycols

Simple aliphatic alcohols, such as methanol, can be used to alkylate alkyleneamines.

Organic Halides

Alkyl halides and aryl halides, activated by electron withdrawing groups (such as NO_2) in the ortho or para positions, react with alkyleneamines to form mono- or disubstituted derivatives.

Aldehydes

Alkyleneamines react exothermically with aliphatic aldehydes. The products depend on stoichiometry, reaction conditions, and structure of the alkyleneamine.

Organic Acids and Their Derivatives (Anhydrides, Nitriles, Ureas)

Alkyleneamines react with acids, esters, acid anhydrides or acyl halides to form amidoamines and polyamides. Various diamides of EDA are prepared from the appropriate methyl ester or acid at moderate temperatures.

Sulfur Compounds

EDA reacts readily with two moles of CS_2 in aqueous sodium hydroxide to form the bis sodium dithiocarbamate.

Olefins

Olefins can be aminomethylated with carbon monoxide (CO) and amines in the presence of rhodium-based catalysts.

Environmentally Available Reactants

Under normal conditions ethyleneamines are considered to be thermally stable molecules. However, they are sufficiently reactive that upon exposure to adventitious water, carbon dioxide, nitrogen oxides, and oxygen, trace levels of by-products can form and increased color usually results.

MANUFACTURE

Ethyleneamine Processes

Present industrial processes are based on ethylene and ammonia. The seventy year old ethylene dichloride (EDC) process is still the most widely practiced industrial route for producing ethyleneamines.

Alternative processes for the manufacture of ethyleneamines have been actively sought since the late 1960s. The catalytic reductive amination of monoethanolamine (MEA), which was the first such process to appear, produces the lower mol wt ethyleneamines (EDA, DETA, PIP, and AEP) and coproduct aminoethylethanolamine (AEEA), and hydroxyethylpiperazine (HEP).

The condensation of MEA with EDA over heterogeneous catalysts to form primarily DETA represents the newest commercial technology for making ethyleneamines.

A process for the production of ethylenimine, a suspect carcinogen, by the vapor phase dehydration of monoethanolamine has been developed.

Diaminopropane Processes

1,2-Propylenediamine can be produced by the reductive amination of propylene oxide, 1,2-propylene glycol, or monoisopropanolamine. 1,3-Propanediol can be used to make 1,3-diaminopropane. Various propaneamines are produced by reducing the appropriate acrylonitrile–amine adducts. Polypropaneamines can be obtained by the oligomerization of 1,3-diaminopropane.

ECONOMIC ASPECTS

Worldwide growth for most ethyleneamines is expected to parallel GDP. Some regional and certain applications demands will show somewhat higher growth rates.

Of the worldwide ethyleneamines capacity, well over 50% is EDC-based; the balance is monoethanolamine-derived.

STORAGE AND HANDLING

By virtue of their unique combination of reactivity and basicity, the polyamines react with, or catalyze the reaction of, many chemicals, sometimes rapidly and usually exothermically. Some reactions may produce derivatives that are explosives (eg, ethylenedinitramine). The amines can catalyze a runaway reaction with other compounds (eg, maleic anhydride, ethylene oxide, acrolein, and acrylates), sometimes resulting in an explosion.

As commercially pure materials, the ethyleneamines exhibit good temperature stability, but at elevated temperatures noticeable product breakdown may result in the formation of ammonia and lower and higher mol wt species.

Like many other combustible liquids, self-heating of ethyleneamines may occur by slow oxidation in absorbent or high-surface-area media, eg, dumped filter cake, thermal insulation, spill absorbents, and metal wire mesh (such as that used in vapor mist eliminators). In some cases, this may lead to spontaneous combustion; either smoldering or a flame may be observed. These media should be washed with water to remove the ethyleneamines, or thoroughly wet prior to disposal in accordance with local and Federal regulations.

Since ethyleneamines react with many other chemicals, dedicated processing equipment is usually desirable. Amines slowly absorb water, carbon dioxide, nitrogen oxides, and oxygen from the atmosphere, which may result in the formation of low concentrations of by-products and generally increased color. Storage under an inert atmosphere minimizes this sort of degradation.

Galvanized steel, copper and copper-bearing alloys are unacceptable for all ethyleneamine service. A 300 series stainless steels or aluminum are recommended for the storage of the lighter amines (Particularly EDA or DETA) to maintain quality. Carbon steel generally can be used for storage of the heavier ethyleneamines without noticeable impact on product quality if the storage temperature is modest ($<60°C$), nitrogen blankets are maintained to exclude air, and the material is anhydrous. A 300 series stainless steel is often specified for heating coils, transfer lines and small agitated tanks, because carbon steel can suffer enhanced corrosion due to the erosion of the passive film by the product velocity. Similar logic suggests cast 316 stainless steel for pumps and valves in ethyleneamine service.

Baked phenolic-lined carbon steel is acceptable for storage of many pure ethyleneamines, except EDA. Gaskets utilized in ethyleneamine service generally are made of Grafoil flexible graphite or polytetrafluoroethylene (TFE).

Most common thermal insulating materials are acceptable for ethylene-amine service. However, porous insulation may introduce the hazard of spontaneous combustion if saturated with ethyleneamines from a leak or external spill.

Certain ethyleneamines require storage above ambient temperature to keep them above their freezing points (EDA and PIP) or to lower the viscosity (the heavy amines). As a result, the vapors "breathing" from the storage tank can contain significant concentrations of the product. Water scrubbers may be used to capture these vapors.

Solid ethyleneamine carbamates, formed by the reaction of the amines with carbon dioxide, can foul tank vents and pressure relief devices. Vent fouling can be minimized by using a nitrogen blanket that prevents atmospheric CO_2 from being drawn in, or by steam-tracing the vents ($>160°C$) to decompose the carbamates.

Although the ethyleneamines are water soluble, solid amine hydrates may form at certain concentrations that may plug processing equipment, vent lines, and safety devices. Hydrate formation usually can be avoided by insulating and heat tracing equipment to maintain a temperature of at least $50°C$. Water cleanup of ethyleneamine equipment can result in hydrate formation even in areas where routine processing is nonaqueous. Use of warm water can reduce the extent of the problem.

HEALTH AND SAFETY FACTORS

Ethyleneamine vapors are painful and irritating to the eyes, nose, throat, and respiratory system. Extremely high vapor concentration may cause lung damage. Prolonged or repeated inhalation may lead to kidney, liver, and respiratory system injury. Contact with the liquids will severely damage the eyes and may cause serious burns to the skin. When swallowed, the concentrated liquid materials may produce considerable local injury. Both vapors and liquid can cause sensitization in some individuals, resulting in contact dermatitis and/or the development of an asthmatic respiratory response. This may occur in certain susceptible individuals following exposure to extremely low concentrations of ethyleneamines, even below the irritation threshold.

The ACGIH has adopted TLVs of 10 ppm (25 mg/m^3) and 1 ppm (4.2 mg/m^3) for EDA and DETA, respectively. Strict precautions should be observed to prevent direct contact with all ethyleneamines, including eye, skin, and respiratory protection. If contact is made, medical treatment should be obtained immediately, in addition to flushing and washing with copious amounts of water. Vomiting is not to be induced following ingestion.

APPLICATIONS

Polyalkylene polyamines find use in a wide variety of applications by virtue of their unique combination of reactivity, basicity, and surface activity. With a few significant exceptions, they are used predominantly as intermediates in the production of functional products.

Fungicides

The ethylenebisdithiocarbamates (EBDCs) are a class of broad-spectrum, preventive, contact fungicides first used in the early 1940s. They have found application on many fruits, vegetables, potatoes, and grains for prevention of mildew, scab, rust, and blight. The EBDCs are prepared by reaction of EDA with carbon disulfide in the presence of sodium or ammonium hydroxide initially, then with zinc and/or manganese salts, as appropriate. A continuous process has recently been reported.

Other materials based on EDA have also been suggested as fungicides. The most important of the imidazoline type (162) is 2-heptadecyl-2-imidazoline, prepared from EDA and stearic acid.

Lubricant and Fuel Additives

The preparation of ashless dispersants for motor oil and other lubricants, and of certain detergents for motor fuels, is one of the largest applications for polyamines. The most widely used derivatives are the mono- and bis-polyisobutenylsuccinimides.

In the fuel additives area, EDA and DETA, as well as N-(2-hydroxyethyl) ethylenediamine, have found significant commercial application as dispersant detergent additives for gasoline after reacting with chlorinated polybutylenes. Improved antirust properties are reported when these compounds are neutralized with carboxylic acids. Numerous similar products made by alkylating or acylating EDA or DETA have also been suggested as fuel detergent and deposit control additives.

Epoxy Curing Agents

A variety of polyamines and their derivatives that contain primary and secondary amine functionality are used as epoxy resin hardeners in various functional coatings, adhesives, castings, laminates, and grouts.

Polyamide Resins

Another class of polyamide resins, in addition to the liquid resins used as epoxy hardeners, are the thermoplastic type, prepared generally by the condensation reaction of polyamines with polybasic fatty acids. These resins find use in certain hot-melt adhesives, coatings, and inks.

Paper Pulping, Resins and Additives

Considerable interest has been generated in the sulfur-free delignification of wood chips with EDA–soda liquors since the late 1970s, with more recent interest in EDA–sulfide pulping. Another significant end-use for polyamines is in preparation of paper wet-strength resins.

The most widely used paper wet-strength resins are the modified polyamide variety commonly prepared from DETA (sometimes TETA), adipic acid, and epichlorohydrin. Waterproofing and sizing agents are also made with polyamines.

Chelates and Chelating Agents

Poly(aminoacetic acid)s and their salts derived from polyamines are used in a variety of systems where metal ions, commonly iron, need to be inactivated, buffered, concentrated, or transported.

Other developments in chelating resins include fibers made from poly(ethylene glycol) and poly(vinyl alcohol) to which EDA was attached with epichlorohydrin; and a styrene–divinylbenzene resin with pendant EDTA or DETPA groups.

Fabric Softeners, Surfactants and Bleach Activators

Mono- and bisamidoamines and their imidazoline counterparts are formed by the condensation reaction of one or two moles of a monobasic fatty acid (typically stearic or oleic) or their methyl esters with one mole of a polyamine. Imidazoline formation requires that the ethyleneamine have at least one segment in which a secondary amine group lies adjacent to a primary amine group. These amidoamines and imidazolines form the basis for a wide range of fabric softeners, surfactants, and emulsifiers. Commonly used amines are DETA, TETA, and DMAPA, although most of the polyethylene and polypropane polyamines can be used.

The most common alkyleneamine-based fabric softeners use bisamidoamines made from DETA, which are either cyclized by further dehydration to the corresponding imidazoline or lightly alkoxylated to convert the central secondary amine group to a tertiary amine.

Many of the surfactants made from ethyleneamines contain the imidazoline structure or are prepared through an imidazoline intermediate. Several cleaning formulations for specific uses contain unreacted polyamines.

Petroleum Production and Refining

Specific polyamine derivatives are used in the petroleum production and refining industries as corrosion inhibitors, demulsifiers, neutralizers, and additives for certain operations.

Asphalt Additives and Emulsifiers

Mono- and bisamidoamines, imidazolines and their mixtures are commonly used in formulating antistrip additives used to promote adhesion between the asphalt and mineral aggregate in bituminous mixtures for road paving, surfacing, and patching. Similar polyamine derivatives are used, generally as their HCl or acetic acid salts, to make stable asphalt-in-water emulsions.

J. A. Deyrup and A. Hassner, eds. *Chemistry of Heterocyclic Compounds*, Vol. 42 of *Small Ring Heterocycles. Pt. 1: Aziridines*, John Wiley & Sons, Inc., New York, 1983, pp. 1–214.

L. S. Goodman and A. Gilman, eds., *The Pharmacological Basis of Therapeutics*, 5th ed., MacMillan Publishing Co., New York, 1975, 1027–1028.

SRIVASAN SRIDHAR
RICHARD G. CARTER
The Dow Chemical Company

DIAMOND, NATURAL

Naturally occurring diamond is a relatively rare polymorph of carbon characterized by a three-dimensional (3D) arrangement of tetrahedrally coordinated carbon atoms. On the Earth's surface, diamonds occur in several major kinds of deposits: primary and secondary (both alluvial and littoral). In primary deposits, they are enclosed in host rocks of kimberlite or lamproite that form "pipes" that are downward-tapering, cone-shaped structures of igneous (volcanic) origin. Subsequent erosion of these pipes and fluvial transport of the diamonds lead to the formation of alluvial deposits in river beds and river terraces. The final resting place of diamonds is in littoral or ocean floor deposits. Separation of diamonds from their enclosing or associated rocks include crushing, screening and sieving, use of grease belts, suspension in heavy or dense media, and sorters that use the luminescence of diamonds when they are exposed to X-rays. Diamond crystals are sorted into many categories depending on size, color, clarity, and shape for valuation purposes. The classification of diamonds into four types (Ia, Ib, IIa, IIb) is based upon the presence or absence of nitrogen and boron distributed in the lattice.

OCCURRENCE AND EXPLORATION

Exploration of alluvial deposits can be done by panning of stream sediments, or by drilling, pitting, and trenching of streambed and terrace deposits, in conjunction with a search for the heavy mineral assemblages that accompany diamond. Recovery of diamonds from the ocean floor requires the use of sophisticated underwater equipment to collect the diamonds and return them to a processing ship. Alluvial deposits account for ~20% by weight (but 41% by value) of the world's annual production of diamonds.

Upstream exploration has sometimes led to the discovery of the primary sources of alluvial diamonds, namely, kimberlite or lamproite pipes that occur in cratons. These pipes are the principal sources of natural diamonds.

Both the origins of kimberlite (or lamproite) and the enclosed diamonds are still a source of controversy and continued geological study. The growth history of each diamond crystal is complex, often involving intermittent periods of growth and dissolution as revealed by studies of the internal structure. There may also have been a problem in maintaining a supply of carbon to the growing crystal over long periods of time, and dissolution of material during the strenuous trip upward from the stability region of diamond at depth likely contributed to decreasing the size of some crystals. Overpressure is known to increase the nucleation frequency in synthesized diamond, and this might translate to many small crystals at great depths.

Diamonds, usually in the size of microcrystals <0.5 mm in diameter, have also been found in non-kimberlitic and non-lamproitic host rocks. Diamonds also occur in meteorites as well as meteorite impact structures (astroblemes), probably as a result of high pressures produced dynamically by impact. The shock or explosive mode of synthesis is a viable process for producing fine diamond powders of both the cubic and hexagonal (lonsdaleite) polymorphs naturally or otherwise. Some diamonds in space appear to have formed in supernovas by processes more closely related to the low pressure chemical vapor deposition processes.

RECOVERY

Alluvial diamonds are recovered from gravel-rich layers of sediments deposited by streams in terraces or along channels. Often the removal of the contents from depressions, potholes, or "pockets" in the streambed where fluvial concentrations are very great has produced spectacularly large amounts of diamonds. High quality diamonds (95% gemstones) are recovered from ocean deposits in the littoral zone off the western coast of southern Africa.

The original mining of the kimberlite pipes begins at the surface in the softer weathered "yellow ground", but most of the subsequent recovery is by removal from depth of the harder, unweathered "blue" kimberlite that is accomplished by open-pit and usually followed by underground mining methods. Recovery of diamond from its gangue is carried out by a variety of techniques depending on the size of the deposit and physical properties of the host rock. Nonstick diamonds can be recovered by electrostatic methods. The method most often used to concentrate heavy minerals (including diamonds) from the crushed rock is by suspension in a heavy or a dense medium [hence, HMS (heavy media separation) and DMS (dense media separation) plants at the major diamond mines]. The more modern method for the separation of diamonds from the heavy mineral concentrate is by an X-ray sorter that uses the emission of light by luminescence of diamond when exposed to X-rays.

PROPERTIES

The lattice constant of the unit cell of the face-centered cubic (fcc) form of diamond is 0.3567 nm with a density of 3.52 g/cm^3. For the hexagonal form lonsdaleite, $a = 0.252$ nm and $c = 0.412$ nm with the same density as the cubic form. The growth morphology of natural diamond is typically octahedral, but dodecahedral crystals are common, perhaps because of subsequent solution after growth.

Besides the single crystals usually implied in a discussion of diamond, it also occurs in nature in the form of polycrystalline aggregates. Diamond is nominally pure carbon with a $^{12}C/^{13}C$ ratio of ~99:1. The classification of diamond into four principal types is based on the presence of nitrogen (type Ia and Ib) and boron (type IIb), and the effects of these impurities on the infrared (ir) and ultraviolet (uv) transmission. Diamond is thermodynamically unstable at ambient conditions, but unless heated to 650°C in air (oxidation to CO_2) or 1700°C in vacumm or an inert atmshpher (graphitization), it wil remain as diamond indefinitely, also it is chemically inert to inorganic acids, but can be etched in oxidizing molten salts such as KNO_3 at 600°C. Carbon as diamond or graphite is soluble in several metals, particularly Fe, Ni, Co, and other Groups 8–10 (VIII) elements.

Diamond is the hardest material known because of its combination of a three-dimensional (3 D) arrangement of tetrahedrally coordinated C–C bonds with a bond distance of 0.154 nm. Although hard, diamond is also very brittle and cleaves readily on the (111) plane and also on other planes under certain conditions.

The thermal conductivity value of 2000 W/(m·K) at room temperature for type IIa natural diamond is about five times that of Cu, and recent data on 99.9% isotopically pure ^{12}C type IIa synthesized crystals are in the range of 3300–3500 W/(m·K). More impure forms of diamond (type Ia) have lower thermal conductivities [600–1000 W/(m·K)]. The averaged value of the coefficient of linear thermal expansion of diamond over the range 20–100°C is 1.34×10^{-6}cm/cm/°C and 3.14×10^{-6} from 20 to 800°C. At room temperature, the values for silica glass and diamond are 0.5×10^{-6} and 0.8×10^{-6}, respectively.

The high refractive index (2.42 at 589.3 nm) and dispersion (0.044) are the basis for the "brilliance" (white light return) and "fire" (flashes of colored light return) of a properly cut diamond gemstone.

APPLICATIONS

The selected applications of diamond are based on its high hardness, wear resistance, ability to self-sharpen as it cleaves, thermal conductivity, and chemical inertness. Loose abrasive grain is used in lapping and polishing operations, eg, on the scaife used for facetting gem crystals and polishing rock materials for monuments and buildings. Other uses for single diamonds are as phonograph needles, bearings, surgical knives, and wire dies. Abrasive grains are also used in grinding, cutting, drilling, and sawing.

Besides the gem qualities dependent on optical properties, diamond is very useful as a light-transmitting window for lasers, and for simple windows for monitoring chemical processes in corrosive and otherwise hostile environments. A significant impact on research at high pressure has come about with the use of gem quality diamonds as Bridgman-type anvils in a small compact high-pressure device.

L. L. Copeland and co-workers, *The Diamond Dictionary*, Gemological Institute of America, Los Angeles, 1960.

E. Hahn, *Diamond*, Doubleday & Co., Garden City, N.Y., 1957.

R. Maillard ed., *Diamonds: Myth, Magic, Reality*, Crown Publishers, New York, 1980.

A. F. Williams, *The Genesis of Diamond*, Ernst Benn Ltd., London, 1932.

A. N. Wilson, *Diamonds from Birth to Eternity*, Gemological Institute of America, Santa Monica, Calif., 1982.

JAMES E. SHIGLEY
Gemological Institute of America

DIAMOND, SYNTHETIC

REPRODUCIBLE LABORATORY DIAMOND SYNTHESIS

In 1955, a team of research workers at General Electric developed the necessary high pressure equipment and discovered solvent–catalytic processes by which ordinary forms of carbon could be changed into diamond.

CATALYZED SYNTHESIS

In this process, a mixture of carbon (eg, graphite) and catalyst metal is heated high enough to be melted, while the system is at a pressure high enough for diamond to be stable. Graphite is then dissolved by the metal and diamond is produced from it. Effective catalysts are Cr, Mn, Fe, Co, Ni, Ru, Rh, Pd, Os, Ir, Pt, and Ta, and their alloys and compounds.

Apparatus

Many kinds of apparatus have been devised for simultaneously producing the high pressures and temperatures necessary for diamond synthesis. An early successful design is the belt apparatus. In this apparatus, two opposed, conical punches, made of cemented tungsten carbide and carried in strong steel binding rings, are driven into the ends of a short, tapered chamber that is also made of cemented tungsten carbide supported by strong steel rings. A compressible gasket, constructed in a sandwich-fashion of stone, usually pyrophyllite, and steel cones, seals the annular gap between punch and chamber, distributes stress, provides lateral support for the punch, and permits axial movement of the punches to compress the chamber and contents. The reaction zone, usually a cylinder, is buried in pyrophyllite stone in the chamber. The pyrophyllite, a good thermal and electrical insulator, is easily machined and transmits pressure fairly well. The reaction zone is heated electrically with a heavy current.

A belt apparatus is capable of holding pressures of 7 GPa (70 kbar) and temperatures of up to 3300 K for periods of hours. The maximum steady-state temperatures are limited by melting of the refractory near the reaction zone.

Figure 1 shows an arrangement of carbon and catalyst metal. As the sample is heated at high pressure, the metal next to the graphite usually melts and diamond begins to form there. An exceedingly thin film of molten metal (at most a few thousandths of a cm thick) separates the newly formed diamond from the unchanged graphite. This film advances like a wave through the mass of graphite and transforms it to diamond.

Crystal Morphology

Size, shape, color, and impurities are dependent on the conditions of synthesis. Lower temperatures favor dark-colored, less pure crystals; high temperatures promote paler, purer crystals. Low pressures (5 GPa) and tem-

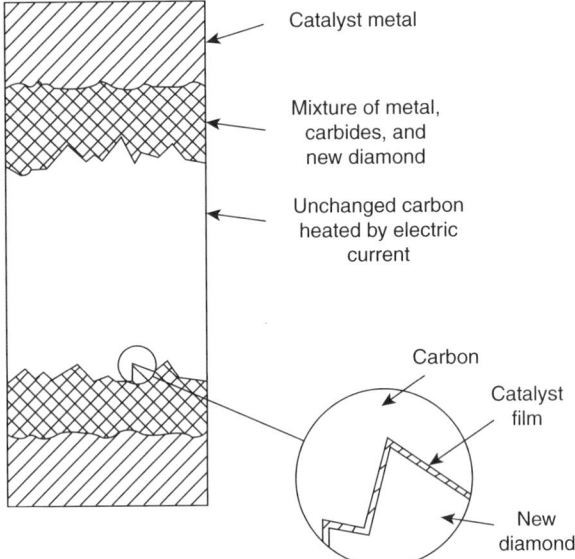

Figure 1. Diamond synthesis cell.

peratures produce octahedral faces. Nucleation and growth rates increase rapidly as the process pressure is raised above the diamond–graphite equilibrium pressure.

Crystal Growth

If diamond seed crystals are placed in the active diamond growing zone of a typical graphite–catalyst metal apparatus, new diamond usually forms on the seed crystals. Excellent growth can be obtained if pressure and composition are held relatively constant while the change of composition with temperature is employed as a driving force.

DIRECT GRAPHITE-TO-DIAMOND PROCESS

In this process, diamond forms from graphite without a catalyst. The refractory nature of carbon demands a fairly high temperature (2500–3000 K) for sufficient atomic mobility for the transformation, and the high temperature in turn demands a high pressure (above 12 GPa; 120 kbar) for diamond stability. The combination of high temperature and pressure may be achieved statically or dynamically. During the course of experimentation on this process a new form of diamond with a hexagonal (wurtzitic) structure was discovered.

Shock Synthesis

When graphite is strongly compressed and heated by the shock produced by an explosive charge, some (up to 10%) diamond may form. These crystallite diamonds are small (on the order of 1 μm) and appear as a black powder. The peak pressures and temperatures, which are maintained for a few microseconds, are estimated to be about

30 GPa (300 kbar) and 1000 K. It is believed that the diamonds found in certain meteorites were produced by similar shock compression processes that occurred upon impact.

The annual production of diamond by this process is only a small fraction of total industrial diamond consumption.

Static Pressure Synthesis

Diamond can form directly from graphite at pressures of about 13 GPa (130 kbar) and higher at temperatures of about 3300–4300 K. No catalyst is needed. The transformation is carried out in a static high pressure apparatus in which the sample is heated by the discharge current from a capacitor. Diamond forms in a few milliseconds and is recovered in the form of polycrystalline lumps.

Crystal Structure. Diamonds prepared by the direct conversion of well-crystallized graphite at pressures of about 13 GPa (130 kbar) show certain unusual reflections in the x-ray diffraction patterns. They could be explained by assuming a hexagonal diamond structure (related to wurtzite) with $a = 0.252$ and $c = 0.412$, nm, space group P_{63}/mmc – D_{6h}^4 with four atoms per unit cell. The calculated density would be 3.51 g/cm^3, the same as for ordinary cubic diamond, and the distances between nearest neighbor carbon atoms would be the same in both hexagonal and cubic diamond, 0.154 nm.

METASTABLE VAPOR-PHASE DEPOSITION

Metastable growth of diamond takes place from gases rich in carbon and hydrogen at low pressures where diamond would appear to be thermodynamically unstable.

In a typical use of this method, a mixture of hydrogen and methane is fed into a reaction chamber at a pressure of about 1.33 kPa (10 torr). The substrate upon which diamond forms is at about 950°C and lies about 1 cm away from a tungsten wire at 2200°C. Small diamond crystals, 1 mm or so in size, nucleate and grow profusely on the substrate at a rate around 0.01 mm/h to form a dark, rough polycrystalline layer with exposed octahedral or cubic faces, depending on the substrate temperature. This method has been actively studied since 1993.

THE SYNTHETIC DIAMOND INDUSTRY

Soon after the first successful diamond synthesis by the solvent–catalyst process, a pilot plant for producing synthetic diamond was established, the efficiency of the operation was increased, production costs declined, and product performance was improved while the uses of diamond were extended. Today (1993) the price of synthesized diamond is competitive with that of natural diamonds.

The bulk of synthetic industrial diamond production consists of the smaller crystal sizes up to 0.7-mm particle size (25 mesh). This size range has wide utility in industry, and a significant fraction of the world's need for diamond abrasive grit is now met by synthetic production yielding thousands of kilograms per year.

Semiconducting Diamonds

Semiconducting diamonds are prepared by adding small amounts of boron, beryllium, or aluminum to the growing mixture, or by diffusing boron into the crystals at high pressures and temperatures.

Sintered Diamond Masses

Some natural diamonds known as carbonado or ballas occur as tough, polycrystalline masses (see CARBON, DIAMOND, NATURAL). The production of synthetic sintered diamond masses of comparably excellent mechanical properties has only been achieved recently. The essential feature is the presence of direct diamond-to-diamond bonding without dependence on any intermediate bonding material between the diamond grains, since no extraneous bonding material can match the stiffness, thermal conductivity and hardness of diamond.

C_{60} Conversion

Buckminsterfullerene can be crushed to diamond by high pressure applied at room temperature. The process is highly efficient and fast at room temperature, suggesting industrial potential.

ECONOMIC ASPECTS

About 90% of industrial diamond is synthesized at high pressures because its price is relatively low, and it can be tailor-made for efficiency in each application. They are made in countries. The market is very competitive, and manufacturers are reluctant to disclose detailed sales information.

F. P. Bundy, H. M. Strong, and R. H. Wentorf, Jr., in P. L. Walker and P. A. Thrower, eds., *Chemistry and Physics of Carbon*, Marcel Dekker, Inc., New York, 1973, pp. 213–263.

F. P. Bundy and J. S. Kasper, *J. Chem. Phys.* **46**, 3437 (1967).

P. S. DeCarli and J. C. Jamieson, *Science* **133**, 1821 (1961).

R. H. Wentorf, R. C. DeVries, and F. P. Bundy, *Science* **208**, 873–888 (1980).

R. H. WENTORF, JR.
General Electric Company

DICHLORINE MONOXIDE, HYPOCHLOROUS ACID, AND HYPOCHLORITES

Chlorine has formal positive oxidation numbers (states) in oxychlorine compounds since its appreciable electronegativity (2.83) on the Allred-Rochow scale is exceeded by that of oxygen. All positive chlorine compounds are strong

Table 1. Chlorine Oxides^a and Oxo Acids

Formula	Oxidation state	Stability
		Oxides
Cl_2O	+1	anhydride of HOCl; yellowish brown gas at 25°C; explodes when heated or sparked
Cl_2O_2	+2	$t_{1/2}$ of gas = 4.8 day at −78°C
Cl_2O_3	+3	anhydride of $HClO_2$; explodes below 0°C
ClO_2	+4	odd electron molecule; yellow gas; explodes at >6.7 kPa^b
Cl_2O_4	+1,+7	pale yellow liquid; decomposes to Cl_2, O_2, and Cl_2O_6
Cl_2O_6	+6	red liquid; gas decomposes to Cl_2O_4+ O_2
Cl_2O_7	+7	anhydride of $HClO_4$; colorless liquid; can be distilled under reduced pressure
		Acids
HOCl^c	+1	very weak acid, pK_a 7.54; cannot be concentrated
$HClO_2$	+3	decomposes rapidly at 25°C; pK_a ~2.0; cannot be concentrated
$HClO_3$	+5	decomposes slowly at ~40% and 25°C; cannot be concentrated
$HClO_4$	+7	can be concentrated

^{a}Anhydride of chloric acid, Cl_2O_5, is unknown. Oxides with even number of oxygen atoms are mixed anhydrides. Other chlorine oxides such as the radicals ClO, ClO_3, and ClO_4 are known. Chlorine monoxide, ClO plays a key role in depletion of the ozone layer.
^{b}To convert kPa to mm Hg, multiply by 7.5.
^{c}The isomeric HClO is an unstable compound formed in the earth's ozone layer.

oxidants because the transfer of electrons to the orbitals of electronegative chlorine is favored in reactions with compounds of less electronegative elements. Chlorine oxides and oxo-acids exhibit the lack of stability expected of compounds having bonds between two strongly electronegative elements. The decomposition reactions of these compounds are, therefore, always energetic and violent in many cases. The chemical properties of chlorine oxides and oxo acids display a trend toward greater thermodynamic and kinetic stability with increasing oxidation state. It is possible to isolate pure perchloric acid and its anhydride, Cl_2O_7, whereas pure hypochlorous, chlorous, and chloric acids have not been obtained. The reduction potentials of the oxo acids exhibit a similar trend in that the strongest oxidants have chlorine in its lower states of oxidation. Compounds of chlorine having intermediate oxidation states exhibit a strong tendency to disproportionate.

The oxo anions of chlorine are weaker oxidants than the corresponding acids. Because they are also more stable, it is not too difficult to isolate certain salts of those acids that can be obtained only in aqueous solution. Hypochlorites and chlorites are hydrolyzed in aqueous solution since HOCl and $HClO_2$ have acid dissociation constants of ~10^{-8} and 10^{-2}, respectively;

however, aqueous chloric and perchloric acids are fully ionized.

The chlorine oxides are anhydrides or mixed anhydrides of chlorine oxo acids; oxides with an odd number of oxygens are simple anhydrides, whereas those with an even number are mixed anhydrides.

Chlorine in dichlorine monoxide, hypochlorous acid, and ionic hypochlorites is in the +1 formal oxidation state. Other compounds where univalent chlorine is bonded to oxygen are the alkyl, aryl, and acyl hypochlorites and other unipositive chlorine compounds such as $ClOSF_5$, $ClOSO_2F$, ClONO, $ClONO_2$, and $ClOClO_3$. The latter compounds are chlorine derivatives of the corresponding acids and can also be considered as mixed anhydrides of HOCl and the corresponding acid. The polar bonding in these compounds imparts a partial positive charge on the first chlorine atom. Along with HCl, $ClONO_2$ forms a temporary chlorine reservoir in the upper stratosphere that reduces depletion of the ozone layer. The compound HOCl is also an important intermediate in the chemistry of the stratosphere.

DICHLORINE MONOXIDE

Dichlorine monoxide is the anhydride of hypochlorous acid; the two compounds are interconvertible in the gas and aqueous phases via the equilibrium: $Cl_2O + H_2O \rightleftharpoons 2HOCl$. Like other chlorine oxides, Cl_2O has an endothermic heat of formation and is thus thermodynamically unstable with respect to decomposition into chlorine and oxygen. Dichlorine monoxide typifies chlorine oxides in being highly reactive and potentially explosive. Nevertheless, it can be handled safely with proper precautions.

Properties

Gas–Liquid–Solid. Physical properties of gaseous, liquid, and solid dichlorine monoxide are summarized in Table 2.

Chemistry

Explosivity. Explosion of gaseous dichlorine monoxide can be initiated by spark or heat. The liquid is shock sensitive but less so than ClO_2. The minimum explosive

Table 2. Gas–Liquid–Solid Properties of Dichlorine Monoxide

color (gas, liquid, solid)	yellow-brown, red-brown, cherry-red
density (solid)	2.089 g/cm^3 at 90 K
freezing point	−120.6°C
boiling point	2.0°C
vapor pressure (173–288 K)	$\log P$ (kPa) = 6.995 − 1373T^{-1}
solubility (g/100 g H_2O) at −9.4°C	143.6

concentration of gaseous Cl_2O in oxygen at 23°C and 101 kPa (1 atm) in faint daylight is 23.5 mol%. Explosions are mild in the 25–30% range but become progressively more violent at higher concentrations. The effect of various diluent gases on the explosion limit at various total pressures has been determined in the complete absence of light. The extrapolated explosion limit in O_2 is ~33 mol%. The threshold for spark initiated decomposition of pure Cl_2O is 0.53 kPa (4.0 Torr).

Thermal and Photolytic Decomposition. Dichlorine monoxide decomposes thermally and photochemically into Cl_2 and O_2. Gaseous Cl_2O decomposes thermally in 12–24 h at 60–100°C, but at 150°C the reaction is complete in only a few minutes; above 110°C the reactions terminate in explosion. The decomposition is preceded by an induction period inversely proportional to the starting Cl_2O concentration. Several mechanisms, some involving chains, have been proposed for this heterogeneous reaction. The photolytic decomposition, initiated by cleavage into ClO, O, and Cl free radicals, is sensitized by Cl_2. Significant concentrations of ClO_2 are generated from Cl_2O by controlled thermal decomposition or irradiation of CCl_4 solutions. Photolysis of matrix-isolated Cl_2O yields ClClO, $(ClO)_2$, and ClO.

Inorganic Reactions. Dichlorine monoxide reacts with a variety of inorganic substances. The reaction of Cl_2O with various free radicals (F, Cl, Br, O, N, and OH) has been studied kinetically. The low temperature, condensed phase reaction with N_2O_5 is a convenient route to $ClNO_3$. In solution or in the gas phase both $ClNO_2$ and $ClNO_3$ are formed. Chlorine dioxide is an intermediate in certain reactions of Cl_2O. The transformation of metal halides into oxy halides by Cl_2O apparently involves hypochlorite intermediates. Dichlorine monoxide can be used for generation of singlet oxygen in the chemical oxygen iodine laser (COIL).

Dichlorine monoxide reacts explosively with ammonia forming nitrogen, chlorine, and water. It also reacts with $Ca(OH)_2$ and moist CaO forming calcium hypochlorite. Dry hypochlorites are oxidized to chlorates by Cl_2O.

Dichlorine monoxide exists in very low equilibrium concentrations in dilute HOCl solutions, nevertheless, it is a kinetically significant reactant.

Organic Reactions. Dichlorine monoxide reacts primarily as a chlorinating agent toward organic matter. By contrast with chlorine, which forms HCl as a by-product, dichlorine monoxide forms water.

Methane is quantitatively converted to CCl_4 by reaction with dichlorine monoxide at 350°C. Lower temperatures give a mixture of chlorinated methanes. The gas-phase reaction of propane with Cl_2O at 100°C gave 1- and 2-chloropropane (7:1 molar ratio) and HOCl. The liquid-phase reaction with saturated compounds produces 2 mol of chlorinated products per mol substrate. The photoinduced reaction proceeds via a free-radical chain mechanism, giving a mixture of chlorination products. In contrast, the dark reaction favors tertiary chlorination.

Dichlorine monoxide is a powerful and selective reagent for either ring or side-chain chlorination of deactivated aromatic substrates providing excellent yields under mild conditions where conventional reagents fail or require harsh conditions.

Dichlorine monoxide reacts with finely divided cyanuric acid in a fluidized bed forming dichloro- and trichloroisocyanuric acids and with monosodium cyanurate monohydrate yielding sodium dichloroisocyanurate monohydrate. Ammelide and ammeline also are converted to trichloroisocyanuric acid by dichlorine monoxide; the exocyclic amino groups being converted to NCl_3.

Uses

Dichlorine monoxide has been used as an intermediate in the manufacture of calcium hypochlorite and in sterilization for space applications. Its use in the preparation of chlorinated solvents and chloroisocyanurates has been described. Dichlorine monoxide has been shown to be effective in bleaching of pulp (qv) and textiles. It also can be used as an etchant in semiconductor manufacture.

HYPOCHLOROUS ACID

Hypochlorous acid is a highly reactive and relatively unstable compound known both in solution and in the gas phase. In solution it is the most stable and the strongest of the hypohalous acids and is one of the most powerful oxidants among the chlorine oxy acids. Although it is not an item of commerce, it is an important intermediate in the manufacture of various inorganic and organic chemicals and also performs an important role as a disinfectant and oxidant in water treatment.

Properties

Dissociation. Hypochlorous acid is a very weak acid (an order of magnitude weaker than carbonic acid).

The temperature (kelvin) dependence is given by $pK_a = 0.0253T + 3000T^{-1} - 10.0686$. The calculated value of pK_a at 25°C is 7.54. The relative concentrations of HOCl and ClO^- are a function of pH.

Phase Equilibria. The phase diagram of aqueous HOCl-Cl_2O shows a eutectic at −40°C (11.7 mol% Cl_2O). Below the eutectic concentration the solid phase in equilibrium with the liquid phase is ice and at higher concentrations the solid phase is $HOCl \cdot 2H_2O$. Liquid Cl_2O and aqueous HOCl are only partially miscible.

Vapor–Liquid Equilibria. Hypochlorous acid and dichlorine monoxide coexist in the liquid and vapor phase. Dilute HOCl solutions are colorless; at higher concentrations the color ranges from yellow to yellow-orange due to small equilibrium amounts of Cl_2O. Vapor pressure measurements of aqueous HOCl solutions show that HOCl is the main chlorine species in the vapor phase over ≤5% solutions, whereas at higher concentrations, Cl_2O becomes dominant.

Chemistry

Stratospheric. The primary source of HOCl in the stratosphere is the gas-phase reaction:

$$HO_2 + ClO \longrightarrow HOCl + O_2 \qquad k^{25^\circ C} = 3.0 \times 10^9 \ M^{-1}s^{-1}$$

HOCl is rapidly photolyzed by solar uv radiation producing HO and Cl radicals. Although these radicals can decompose HOCl, the rates are much slower than HOCl photolysis. However, these radicals can decompose ozone. The estimated photolysis lifetime of ~2 h (at 30 km) suggests that HOCl may be unimportant as a reservoir for stratospheric chlorine. HOCl is also an intermediate in activation of chlorine reservoirs via its heterogeneous formation and decomposition in aerosols present in stratospheric clouds.

Photochemical Decomposition. Under laboratory conditions, gaseous HOCl is decomposed by uv light into Cl_2, O_2, HCl, and H_2O and is initiated by cleavage into HO + Cl radicals. Photolysis of aqueous HOCl is also initiated by formation HO and Cl radicals that undergo a series of further reactions producing hydrochloric and chloric acids and oxygen. The half-life of aqueous chlorine ($HOCl/ClO^-$) depends primarily on the wavelength of light and pH. Organic matter in water can consume HO and Cl radicals formed by HOCl photolysis, eg, ethanol gave acetaldehyde and acetic acid, whereas n-butanol and benzoic acid yielded numerous chlorinated products.

Thermal Decomposition. Dilute, Cl^--free, aqueous hypochlorous acid solutions are quite stable if pure, especially if kept cool and in the dark. The presence of sodium chloride greatly increases the decomposition rate.

Inorganic Reactions. Hypochlorous acid oxidizes numerous inorganic substrates.

Hypochlorous acid has important uses in synthesis. Hypochlorous acid, preformed or formed *in situ*, is an intermediate in formation of hypochlorites. Organic solutions of HOCl have been used to prepare $Ca(OCl)_2$, by reaction with CaO or $Ca(OH)_2$, and hydrazine by reaction with NH_3.

Available chlorine as HOCl in waste streams can be destroyed by reaction with hydrogen peroxide, which occurs via the postulated intermediate formation of peroxyhypochlorous acid.

Organic Reactions. Hypochlorous acid undergoes a variety of reactions with organic substances including oxidation, addition, C- and N-chlorination, and ester formation. In many of its reactions, HOCl is generated *in situ* via chlorine hydrolysis:

Hypochlorous acid oxidizes many organic ions such as formate, oxalate, cyanide, cyanate, and is often involved in oxidation of unwanted or potentially toxic substances. Secondary alcohols are oxidized at room temperature to ketones in high yields by HOCl generated *in situ* from aqueous NaOCl and acetic acid. Selective oxidation in the presence of a primary alcohol is possible.

The industrial scale reaction of propylene with aqueous chlorine forms primarily the α-propylene chlorohydrin isomer.

Reaction of HOCl, formed from calcium hypochlorite and CO_2, with highly substituted alkenes in CH_2Cl_2 is a convenient route to allylic chlorides. Ketones are chlorinated to α-chloroketones by reaction with HOCl.

Chlorination of amino acids forms unstable N-chloro compounds. By contrast, stable N-chloro compounds are formed by reaction of hypochlorous acid and appropriate N−H compounds. For example, chloroisocyanurates can be prepared from isocyanuric acid or monosodium cyanurate and preformed HOCl.

Isocyanates are prepared via N-chlorination of amides with HOCl followed by phase-transfer catalyzed Hoffman rearrangement.

Insoluble resins that slowly release HOCl in water can be prepared by reaction of HOCl with appropriate N-H containing polymers.

Under the proper conditions, alcohols can be esterified with HOCl forming isolable alkyl hypochlorites: ROH + $HOCl \rightarrow ROCl + H_2O$.

Uses

Hypochlorous acid, preformed or generated *in situ* from chlorine and water, is an intermediate in the manufacture of chlorohydrins (qv) from olefins (en route to epoxides). It also is used in the production of hypochlorites and chloroisocyanurates from cyanuric acid. HOCl is used in water treatment for disinfection and oxidation of contaminants such as ammonia. It is the active species that kills bacteria and other microorganisms in municipal water treatment or in swimming pool sanitation when Cl_2, hypochlorites, or chloroisocyanurates are used. Hypochlorous acid can be used for oxidation of waste streams containing formaldehyde.

METAL HYPOCHLORITES

Hypochlorites, either as solutions or solids, are much more stable than hypochlorous acid. One of the novel uses of hypochlorites was for disinfection of Apollo Eleven on its return from the moon.

The only known stable solid neutral hypochlorites are those of lithium, calcium, strontium, and barium. Calcium also forms two stable basic hypochlorites (calcium hydroxide hypochlorites): $Ca(OCl)_2 \cdot 0.5\,Ca(OH)_2$ and $Ca(OCl)_2 \cdot 2\,Ca(OH)_2$. Sodium hypochlorite does not have good stability and is potentially explosive. Potassium hypochlorite exists only in solution. Attempts to isolate the solid have resulted in decomposition. Neutral magnesium hypochlorite has not been isolated, but two stable basic hypochlorites have been prepared. Impure silver and basic zinc hypochlorite compositions have been prepared.

Properties

The solubilities of Li, Na, and Ca hypochlorites in H_2O at 25°C are 40, 45, and 21%, respectively.

Solution Chemistry

Photochemical Decomposition. The ultimate products of the photolysis of alkaline sodium hypochlorite by uv radiation are chlorate and chloride ions and oxygen. The primary processes: $ClO + h\nu \rightarrow Cl^- + O(^3P)$, $ClO + h\nu \rightarrow Cl^- + O(^1D)$, and $ClO^- + h\nu \rightarrow Cl + O^-$ generate transient radicals that react further producing the observed products: Cl^-, ClO_2^-, ClO_3^-, and O_2. The quantum yield (ϕ) and product selectivity vary with the wavelength of the radiation. Product selectivities (based on oxygen) at 254 nm are O_2 47% and ClO_3^- 53%.

Thermal Decomposition. Although hypochlorite solutions are much more stable than HOCl, they are subject to decomposition, which is influenced by concentration, ionic strength, pH, temperature, and impurities. Decomposition of sodium hypochlorite produces oxygen and chloride and chlorate ions.

Inorganic Reactions. Hypochlorites yield HOCl when treated with excess CO_2 or stoichiometric amounts of mineral acids and are converted to Cl_2 when excess HCl is used. Hypochlorite is easily reduced to chloride ion by various reducing agents. It is also reduced by hydrogen peroxide. The oxidation of various inorganic anions by hypochlorite has been studied kinetically.

Organic Reactions. Hypochlorite ion acts as a chlorinating and oxidizing agent toward organic compounds and has numerous interesting and useful synthetic applications.

Solid-State Chemistry

Thermal decomposition of calcium hypochlorite produces $CaCl_2$, O_2, and small amounts of $Ca(ClO_3)_2$. By contrast, sodium hypochlorite produces primarily sodium chlorate.

Commercial Hypochlorite Solutions

Commercial strength liquid bleach used by industries, laundries, and in swimming pool sanitation, contains 12–15% av Cl_2 and is sold in 3.8- and 7.6-L polyethylene bottles and 23–57-L carboys, 205-L drums, and tank trucks of ~3-kL capacity and greater. Household bleach contains about 5% av Cl_2 and is sold in 1–5.7-L polyethylene containers. Shipping is limited within a short radius of the plant because of transportation costs.

Uses. Some uses of commercial hypochlorite solutions are household bleach (5.25% NaOCl): residential laundry bleach and sanitizer; institutional and restaurant sanitizer; residential pool and spa sanitizer. Industrial bleach (12.5% NaOCl): municipal and industrial water treatment (potable water, wastewater, cooling towers, etc); commercial and municipal swimming pool sanitizer; commercial laundry bleach; liquid dishwasher detergents; textile bleaching; chemicals (organic synthesis, chlorinated trisodium phosphate, hydrazine, etc); food processing, fungicide in oil drilling, sweetening agent in oil refineries, etc.

Commercial Solid Hypochlorites

Sodium Hypochlorite-Trisodium Phosphate (TSP). This product can be obtained by crystallization from a liquor having the proper Na_2O/P_2O_5 ratio containing an excess of NaOCl.

The principal uses are in scouring cleansers and acid metal cleaners for dairy equipment. Use in dishwasher detergents has been supplanted by chlorinated isocyanurates, which are more cost-effective, more stable in hot water, and possess water softening properties.

Lithium Hypochlorite. The commercial product contains a significant amount of diluent by-products such as salt and sodium and potassium sulfates that reduce dustiness, increase bulk density, reduce reactivity, and improve storage stability.

Applications. Lithium hypochlorite has limited use in swimming pool and spa sanitation and dry laundry bleaches.

Calcium Hypochlorite. High assay calcium hypochlorite was first commercialized in the United States in 1928.

Calcium hypochlorite is used for disinfection in swimming pools and drinking water supplies. It is also used for treatment of industrial cooling water for slime control of bacterial, algal, and fungal origin, and for disinfection, odor control, and BOD reduction in sewage and wastewater effluents. Calcium hypochlorite is employed as a sanitizer in households, schools, hospitals, and public buildings. It is also used for microbial control in restaurants and other public eating places. Calcium hypochlorite is used for bacterial control, odor control, and general sanitation in dairies, wineries, breweries, canneries, food processing plants, and beverage bottling plants. It is employed as a mildewcide in and around the home, boats, campers, and trailers.

Calcium hypochlorite is heated to ~90°C during drying and thus is stable at normal temperatures encountered in transportation, storage, and use. It should be stored in a cool, dry, and well-ventilated area. Since calcium hypochlorite can react vigoruously and sometimes explosively with certain organic and inorganic materials, they should be kept away from it during shipment, storage, use, or disposal.

Hemibasic Calcium Hypochlorite. This basic hypochlorite produced by chlorination of hydrated lime. The actual product contains a higher lime content because of the presence of some dibasic compound. Because of the simplicity of the manufacturing process it is cheaper than neutral calcium hypochlorite, and because of its higher av Cl_2 and better stability, it is a superior alternative to bleaching powder.

Hemibasic calcium hypochlorite can be used in the textile and paper making industries, industrial sewage treatment, household sanitation, potable water treatment, general sanitation, etc.

Bleaching Powder

Bleaching powder is made by chlorination of powdered calcium hydroxide.

Bleaching powder can be employed for general sanitation and may also be used to disinfect seawater, reservoirs, and drainage ditches where the volume of insolubles is not important. It can be used as a decontaminating agent for areas sprayed with chemical warfare agents such as mustard gas.

Organic and Nonmetal Hypochlorites

Alkyl Hypochlorites. Alkyl hypochlorites, esters of hypochlorous acid, are nonpolar, volatile liquids with irritating odors and are extremely lachrimatory. Physical property data are limited. Primary and secondary hypochlorites are very unstable but tertiary hypochlorites exhibit good stability. *tert*-Butyl hypochlorite, can be prepared by chlorination of an alkaline solution of the alcohol followed by phase separtion of the oily product.

tert-Butyl hypochlorite has been found useful in upgrading vegetable oils, in the preparation of α-substituted acrylic acid esters, and esters of isoprene halohydrins.

Fluoroalkyl Hypochlorites. Numerous fluorinated and perfluorinated alkyl hypochlorites have been synthesized and characterized. Fluoroalkyl hypochlorites are extremely susceptible to hydrolysis but are much more thermally stable than the corresponding parent compounds and can be prepared by reaction of ClF with the appropriate carbonyl compound or alcohol. The fluoroalkyl hypochlorites readily react with CO and SO_2 to form the corresponding chloroformates and chlorosulfates.

Acyl Hypochlorites. Although no acyl hypochlorites [RCO_2Cl] have been isolated in pure form, they have been characterized in solution and employed as reactants via *in situ* generation from Cl_2O or HOCl and carboxylic acids, or from Cl_2 and silver salts of carboxylic acids. Perfluoroacyl hypochlorites have also been prepared. Although they are thermally unstable and explosive, CF_3CO_2Cl and $C_3F_7CO_2Cl$ are easily handled and are well characterized.

Other Hypochlorites. Inorganic nonmetal unipositive oxychlorine compounds include $ClONO_2$, $ClOClO_3$, and the hypochlorites derived from monobasic fluorine-containing oxyacids of the group VIA elements S, Se, and Te; eg, $HOSO_2F$, $HOSO_2CF_3$, $HOSF_5$, $HOSeF_5$, and $HOTeF_5$. In addition, two members of a new class of unipositive chlorine compounds derived from the hydroperoxides CF_3OOH and SF_5OOH have been prepared.

A Practical Guide to Chlorine Bleach Making, Technical and Engineering Service Bulletin 72–19, Allied Chem. Corp., Morristown, N.J., 1974; Soda Bleach Solutions, Diamond Alkali Co., 1952.

Sodium Hypochlorite Chemical Profile, *Chemical Marketing Reporter*, Mar. 24, 2003; Calcium Hypochlorite Chemical Profile, *Chemical Marketing Reporter*, Oct. 31, 2002.

JOHN A. WOJTOWICZ
Olin Corp., Retired

DIMENSIONAL ANALYSIS

Dimensional analysis is a technique that treats the general forms of equations governing natural phenomena. It provides procedures of judicious grouping of variables associated with a physical phenomenon to form dimensionless products of these variables; therefore, without destroying the generality of the relationship, the equation describing the physical phenomenon may be more easily determined experimentally. It guides the experimenter in the selection of experiments capable of yielding significant information and in the avoidance of redundant experiments, and makes possible the use of scale models for experiments (see also DESIGN OF EXPERIMENTS). The method is particularly valuable when the problems involve a large number of variables. On such occasions, dimensional analysis may reveal that, whatever the form of the inaccessible final solution, certain features of it are obligatory. The technique has been utilized effectively in engineering modeling.

UNITS AND DIMENSIONS

The concepts used to describe natural phenomena are based on the precise measurement of quantities. The quantitative measure of anything is a number that is found by comparing one magnitude with another of the same type. It is necessary to specify the magnitude of the quantity used in making the comparison if the number is to be meaningful. The statement that "the length of a car is 6 meters" implies that a length has been chosen, namely, 1 m, and that the ratio of the length of the car to the chosen length is 6. The chosen magnitudes, such as the meter, are called *units* of measurement.

Classical physics is built on the foundation of the laws of motion. It was felt at the time that the entire subject could be based on the laws of classical mechanics, and further work would undoubtedly make electromagnetism another branch of mechanics. Under these circumstances, it was natural to regard length *l*, mass *m*, and time *t* as the fundamental, primary, or reference dimensions. However, such designations lead to dimensional ambiguity in that two distinct concepts may possess the same dimensions.

Over the years, the number of reference dimensions in physics has evolved from the original three, to four, to five, and then gradually downward to an absolutely necessary one, and then upward again through an understanding that, though only one is absolutely necessary, a considerable convenience can stem from using three, to four, or five reference dimensions depending on the problem at hand. There is nothing sacrosanct about the number of reference dimensions, and dimensional analysis is merely a tool that may be manipulated at will. This principle of free choice of the reference dimensions has been widely accepted, although one still finds references to true dimensions. Thus, an important step in dimensional analysis is the selection of reference dimensions in such a way that the others, called the secondary or derived dimensions, can be expressed in terms of them. The relation between reference and derived dimensions is generally established either through the fundamental law or equation governing the phenomenon or through definitions.

DIMENSIONAL MATRIX AND DIMENSIONLESS PRODUCTS

An appropriate set of independent reference dimensions may be chosen so that the dimensions of each of the variables involved in a physical phenomenon can be expressed in terms of these reference dimensions. In order to utilize the algebraic approach to dimensional analysis, it is convenient to display the dimensions of the variables by a matrix. The matrix is referred to as the dimensional matrix of the variables and is denoted by the symbol $\boldsymbol{D}$. Each column of $\boldsymbol{D}$ represents a variable under consideration, and each row of $\boldsymbol{D}$ represents a reference dimension. The ith row and jth column element of $\boldsymbol{D}$ denotes the exponent of the reference dimension corresponding to the ith row of $\boldsymbol{D}$ in the dimensional formula of the variable corresponding to the jth column. As an illustration, consider Newton's law of motion, which relates force F, mass M, and acceleration A by (eq. 1):

$$F = \text{constant} \times MA \qquad (1)$$

If length l, mass m, and time t are chosen as the reference dimensions, the dimensional formulas for the variables F, M, and A are as follows:

Variables	Dimensional formulas
F	$m^1 l^1 t^{-2}$
M	$m^1 l^0 t^0$
A	$m^0 l^1 t^{-2}$

The dimensional matrix associated with Newton's law of motion is obtained as (eq. 2)

$$D = \begin{array}{c} \\ m \\ l \\ t \end{array} \begin{array}{ccc} F & M & A \\ \left[\begin{array}{ccc} 1 & 1 & 0 \\ 1 & 0 & 1 \\ -2 & 0 & -2 \end{array}\right] \end{array} \qquad (2)$$

The validity of the method of dimensional analysis is based on the premise that any equation that correctly describes a physical phenomenon must be dimensionally homogeneous. An equation is said to be dimensionally homogeneous if each term has the same exponents of dimensions. Such an equation is of course independent of the systems of units employed provided the units are compatible with the dimensional system of the equation. It is convenient to represent the exponents of dimensions of a variable by a column vector called dimensional vector represented by the column corresponding to the variable in the dimensional matrix. In equation 2, the dimensional vector of force F is $[1,1,-2]'$ where the prime denotes the matrix transpose.

Suppose that there are n variables $Q_1, Q_2, \ldots, Q_n$ that are involved in a physical phenomenon whose dimensional vectors are $\boldsymbol{D}_1, \boldsymbol{D}_2, \ldots, \boldsymbol{D}_n$, respectively. This phenomenon can generally be expressed by (eq. 3):

$$f(Q_1, Q_2, \ldots, Q_n) = 0 \qquad (3)$$

When such a function is established or assumed, it will still exist even after the variables are intermultiplied in any manner whatsoever. This means that each variable in the equation can be combined with other variables of the equation to form dimensionless products whose dimensional vectors are the zero vector. Equation 3 can then be transformed into the nondimensional form as (eq. 4):

$$f(\pi_1, \pi_2, \ldots, \pi_n) = 0 \qquad (4)$$

where the dimensionless products π_i $(i = 1, 2, \ldots, n)$ can generally be expressed as the power products of the form (eq. 5):

$$\pi_i = Q_1^{x_{1i}} Q_2^{x_{2i}} \ldots Q_n^{x_{ni}} \qquad (5)$$

Let $R_1, R_2, \ldots, R_m$ be a set of chosen reference dimensions. Then the dimensional formulas for the variables Q_i are given by (eq. 6):

$$R_1^{d_{1i}} R_2^{d_{2i}} \ldots R_m^{d_{mi}} \qquad (6)$$

where the exponents of dimensions are represented by the dimensional vectors as (eq. 7):

$$\boldsymbol{D}_i' = [d_{1i}\, d_{2i} \ldots d_{mi}] \qquad i = 1, 2, \ldots, n \qquad (7)$$

By using eq. 6, the dimensional formulas for π_i of equation 5 can be written to give (eq. 8):

$$\left[R_1^{d_{11}} R_2^{d_{21}} \ldots R_m^{d_{m1}}\right]^{x_{1i}} \left[R_1^{d_{12}} R_2^{d_{22}} \ldots R_m^{d_{m2}}\right]^{x_{2i}} \ldots \left[R_1^{d_{1n}} R_2^{d_{2n}} \ldots R_m^{d_{mn}}\right]^{x_{ni}} \qquad (8)$$

Since π_i are dimensionless products having dimensional vectors equal to the zero vector, the exponents of the R_j $(j = 1, 2, \ldots, m)$ must add up to zero, giving (eq. 9):

$$\begin{aligned} d_{11}x_{1i} + d_{12}x_{2i} + \cdots + d_{1n}x_{ni} &= 0 \\ d_{21}x_{1i} + d_{22}x_{2i} + \cdots + d_{2n}x_{ni} &= 0 \\ &\vdots \\ d_{m1}x_{1i} + d_{m2}x_{2i} + \cdots + d_{mn}x_{ni} &= 0 \end{aligned} \qquad (9)$$

In terms of the dimensional vectors of equation 8, equation 10 can be written as (eqs. 10–12):

$$[\boldsymbol{D}_1 \boldsymbol{D}_2 \cdots \boldsymbol{D}_n]X_i = 0, \qquad i = 1, 2, \ldots, n \qquad (10)$$

where

$$X_i' = [x_{1i}\, x_{2i} \ldots x_{ni}] \qquad (11)$$

or more compactly

$$DX = 0 \qquad (12)$$

where $X = X_i$, $i = 1, 2, \ldots, n$. Thus, the product of a set of variables is dimensionless if, and only if, the exponents of these variables are a solution of the homogeneous linear algebraic equation. A vector X is said to be a B-vector of D if it is a solution of equation 13. The corresponding dimensionless product associated with the variables of a B-vector is called a B-number. Frequently, the term pi number is also used by many authors because it was first introduced by Buckingham in 1914, who used the symbol π for a dimensionless product or group. In fact, the term pi was even attached to his contributions to dimensional analysis, and is known as Buckingham's pi theorem. But this usage is deprecated because of possible confusion with the universal constant of $\pi = 3.14159$. Therefore, the choice of his initial B is preferred to that of the term π.

SYSTEMATIC CALCULATION OF A COMPLETE B-MATRIX

Once the dimensional matrix has been set up and the number of products in a complete set of B-numbers is determined, a complete set of B-vectors must be computed.

Theorem 1

The transpose C'_{23} of C_{23} is a complete B-matrix of equation 13.

It is advantageous if the dependent variables or the variables that can be regulated each occur in only one dimensionless product, so that a functional relationship among these dimensionless products may be most easily determined. In other words, it is sometimes desirable to have certain specified variables, each of which occurs in one and only one of the B-vectors. Theorem 2 gives a necessary and sufficient condition for the existence of such a complete B-matrix. This result can be used to enumerate such a B-matrix without the necessity of exhausting all possibilities by linear combinations.

Theorem 2

Let A_1 be a given complete B-matrix associated with a set of variables. Then there exists a complete B-matrix A_2 of these variables such that certain specified variables each occur in only one of the B-vectors of A_2 if, and only if, the rows corresponding to these specified variables in A_1 are linearly independent.

OPTIMIZATION OF THE COMPLETE B-MATRICES

With the exception that certain variables may each be required to occur in only one dimensionless product, the selection of a complete B-matrix is totally arbitrary. In order to simplify the formulas associated with a complete B-matrix and to provide a procedure for establishing an explicit set of B-numbers, it is necessary to impose additional constraints in the selection of these B-vectors in forming a complete B-matrix. To avoid the fractional exponents of the formulas, the elements of the matrices are restricted to integers. In addition, the following criteria are proposed for the optimization of the B-matrices: (1) maximize the number of zeros in a complete B-matrix, and (2) minimize the sum of the absolute values of all the integers of a complete B-matrix. These criteria are chosen so that the formulas associated with a physical phenomenon are in their simplest form, otherwise they are completely arbitrary. Evidently the order of the two-optimization criteria is important. For the purpose of this article the sequence consisting of criterion 1 followed by criterion 2 is assumed.

Theorem 3

The associated B-vectors of a set of submatrices of a complete B-matrix are linearly independent if and only if the associated vectors of these submatrices are linearly independent.

CONCLUSION

Dimensional analysis provides a means of judicious grouping of variables associated with a physical phenomenon to form dimensionless products of these variables without destroying the generality of the relationship. Consequently, the equation describing the physical phenomenon may be more easily determined experimentally. In addition, it guides the experimenter in the selection of experiments capable of yielding significant information and in the avoidance of redundant experiments, and makes possible the use of scale models for experiments. The technique is particularly useful when the problems involve a large number of variables.

R. Bellman, *Introduction to Matrix Analysis*, McGraw-Hill Book Co., New York, 1960.

J. F. Douglas, *An Introduction to Dimensional Analysis for Engineers*, Sir Isaac Pitman & Sons, London, 1969.

S. J. Kline, *Similitude and Approximation Theory*, McGraw-Hill Book Co., New York, 1965.

L. I. Sedov, *Similarity and Dimensional Methods in Mechanics*, Academic Press, Inc., New York, 1959.

WAI-KAI CHEN
University of Illinois at Chicago

DISINFECTION

Agents that served as disinfectants and antiseptics were known and utilized by ancient peoples. Soldiers disinfected equipment and clothing with fire or boiling water, houses were fumigated with burning sulfur, drinking water was purified by storing it in silver or copper vessels, and food was preserved by drying, salting, acidifying, and treating with spices. These methods worked, but it was not known why they worked.

The Environmental Protection Agency (EPA) and the Centers for Disease Control (CDC) reported in the early

1990s that diseases caused by viruses and parasites are on the increase. The cause may be the drinking water supply, because these organisms are not always destroyed by the water-treatment processes. Wider, intelligent use of disinfectants and antiseptics can greatly aid in removing microbes to limit the chance of infection. Disinfectants find additional use in preventing spoilage of products such as food, pharmaceuticals, cosmetics, paints, wood, cloth, and even in helping to keep office buildings from becoming uninhabitable.

DEFINITIONS

Disinfectant. A disinfectant is a chemical or physical agent that frees from infection and kills bacteria, fungi, viruses, and protozoa, but may not kill or inactivate bacterial spores, and is used only on inanimate objects, not on or in living tissue. A bactericide, fungicide, virucide, etc, is a disinfectant intended to kill the organisms indicated in the term. A germicide claims to kill pathogenic microorganisms, or germs.

Antiseptic. An antiseptic is a chemical substance that prevents or inhibits the action or growth of microorganisms but may not necessarily kill them, and is used topically on living tissue. The distinction between a disinfectant and an antiseptic is that the former is expected to kill all vegetative cells and is used only on inanimate objects, whereas the latter may not kill all cells and is used on the body.

Antiseptics are used in the home for simple cuts and wounds, and in the hospital for treating patient's skin and surgeon's hands prior to operative procedures. Soap, mouthwash, lotions, ointments, nose drops, suppositories, and vaginal creams that contact the skin and mucous membranes are often treated with germ-killing antiseptics.

DISINFECTION METHODS, MEANS, AND TECHNOLOGIES

Disinfection refers to the selective destruction of disease-causing organisms. All the organisms are not destroyed during the process. This differentiates disinfection from sterilization, which is the destruction of all organisms. In the field of wastewater treatment, the three categories of human enteric organisms of the greatest consequence in producing disease are bacteria, viruses, and amoebic cysts.

Description of Disinfection Methods and Means

An ideal disinfectant would have to possess a wide range of characteristics. It is also important that the disinfectant be safe to handle and apply and that its strength or concentration in treated waters be measurable. Disinfection is most commonly accomplished by the use of chemical agents, physical agents, mechanical means, and radiation.

Chemical Agents. Chemical agents that have been used as disinfectants include (*1*) chlorine and its compounds, (*2*) bromine, (*3*) iodine, (*4*) ozone, (*5*) phenol and phenolic compounds, (*6*) alcohols, (*7*) heavy metals and related compounds, (*8*) dyes, (*9*) soaps and synthetic detergents, (*10*) quaternary ammonium compounds, (*11*) hydrogen peroxide, (*12*) various alkalis and acids (eg, peracetic acid), and (*13*) antimicrobial nanoemulsions.

Of these, the most common disinfectants are the oxidizing chemicals. Chlorine is the one most universally used. Chlorine dioxide is another bactericide, equal to or greater than chlorine in disinfecting power. Ozone is a highly effective disinfectant, and its use is increasing even though it leaves no residual. Bromine and iodine have also been used for wastewater disinfection. Highly acidic or alkaline water can also be used to destroy pathogenic bacteria because water with pH >11 (basic) or <3 (acidic) is relatively toxic to most bacteria.

Physical Agents. Physical disinfectants that can be used are heat and light. Sunlight is also a good disinfectant. In particular, uv radiation can be used. Special lamps that emit uv rays have been used successfully to sterilize small quantities of water. The efficiency of the process depends on the penetration of the rays into water. The contact geometry between the uv-light source and the water is extremely important because suspended matter, dissolved organic molecules, and water itself, as well as the microorganisms, will absorb the radiation.

Mechanical Means. Bacteria and other organisms are also removed by mechanical means during wastewater treatment.

Radiation. The major types of radiation are uv, electromagnetic, acoustic, and particle. Gamma rays are emitted from radioisotopes such as Co 60. Because of their penetration power, uv, gamma rays, and high energy electron-beam devices have been used to disinfect (sterilize) water, wastewater or sludge, meat, poultry, fish and other foods, and air borne microorganisms or bioaerosols.

Mechanisms of Disinfectants

Four mechanisms that have been proposed to explain the action of disinfectants are (*1*) damage to the cell wall, (*2*) alteration of cell permeability, (*3*) alteration of the colloidal nature of the protoplasm, and (*4*) inhibition of enzyme activity.

Analysis of Factors Influencing the Action of Disinfectants

In applying the disinfection agents or means that have been described, the following factors must be considered: (*1*) contact time, (*2*) concentration and type of chemical agent, (*3*) intensity and nature of physical agent, (*4*) temperature, (*5*) number of organisms, (*6*) types of organisms, and (*7*) nature of suspending liquid.

DISINFECTION BY CHLORINATION

Disinfection has received increased attention over the past several years from regulatory agencies through the establishment and enforcement of rigid bacteriological effluent standards. In upgrading existing wastewater treatment facilities, the need for improved disinfection as well as the elimination of odor problems are frequently encountered. Adequate and reliable disinfection is essential in ensuring that wastewater treatment plants are both environmentally safe and aesthetically acceptable to the public.

Chlorine is the most widely used disinfectant in water and wastewater treatment. It is used to destroy pathogens, control nuisance microorganisms, and for oxidation. As an oxidant, chlorine is used in iron and manganese removal, for destruction of taste and odor compounds, and in the elimination of nitrogen as ammonia. It is, however, a highly toxic substance and recently concerns have been raised over handling practices and possible residual effects of chlorination. Recent shortages and price escalation of liquid chlorine have also emphasized the need to consider other alternative methods of disinfection.

Background and Properties of Chlorine

Chlorine (Cl_2) is a greenish-yellow colored gas having a specific gravity of 2.48 as compared to air under standard conditions of temperature and pressure. In nature, it is found in the combined state only, usually with sodium as common salt ($NaCl$), carnallite ($KMgCl_3 \cdot 6 H_2O$), and sylvite (KCl).

Chlorine is a member of the halogen (salt-forming) group of elements and is derived from chlorides by the action of oxidizing agents and, most frequently, by electrolysis. As a gas, it combines directly with nearly all elements.

In addition to being the most widely used disinfectant for water treatment, chlorine is extensively used in a variety of products, including paper products, dyestuffs, textiles, petroleum products, pharmaceuticals, antiseptics, insecticides, foodstuffs, solvents, paints, and other consumer products. Most chlorine produced is used in the manufacture of chlorinated compounds for sanitation, pulp bleaching, disinfectants, and textile processing. It is also used in the manufacture of chlorates, chloroform, and carbon tetrachloride and in the extraction of bromine.

As a liquid, chlorine is amber colored and is 1.44 times heavier than water. In solid form, it exists as rhombic crystals.

Chlorine gas is a highly toxic substance, capable of causing death or permanent injury due to prolonged exposures via inhalation. It is extremely irritating to the mucous membranes of the eyes and the respiratory tract. It will combine with moisture to liberate nascent oxygen to form hydrochloric acid. If both these substances are present in sufficient quantity, they can cause inflammation of the tissues with which they come in contact.

Chlorine gas has an odor detectable at a concentration as low as 3.55 ppm. Irritation of the throat occurs at 15 ppm. A concentration of 50 ppm is considered dangerous for even short exposures. Chlorine can also cause fires or explosions upon contact with various materials. It emits highly toxic fumes when heated and reacts with water or steam to generate toxic and corrosive hydrogen chloride fumes.

FUNDAMENTALS OF CHLORINE CHEMISTRY

This section describes the chemistry of chlorine gas molecules and their reactions when dissolved in aqueous solutions. The purpose of this presentation is to show all of the fundamental reactions so that the practical application of chlorine to potable water, industrial process water, and wastewater can be better understood and analyzed.

Our knowledge of the fundamental chemistry of chlorination has been enlarged considerably in the past 50 years, which has contributed to significant advancement in the field. However, the more we learn, the more we realize how fortuitous it is that chlorine, applied in its simplest form (Cl_2), can be a very potent disinfectant. The phenomenon of chemical simplicity must surely be an important contributing factor to its germicidal efficiency. As a disinfectant, it is without equal, despite its shortcomings.

Hydrolysis of Chlorine Gas

When chlorine gas is dissolved in water, it hydrolyzes rapidly according to the following equation:

$$Cl_2 + H_2O \longrightarrow HOCl + H^+ + Cl^- \qquad (1)$$

The rapidity of this reaction has been studied by many investigators. Complete hydrolysis occurs in a few tenths of a second at 64°F; and at 32°F, only a few seconds are needed. This unusually rapid rate of reaction is best explained if the mechanism is a reaction of the chlorine molecule with the hydroxyl ion rather than with the water molecule.

Chemistry of Hypochlorous Acid

Effect of pH. The next most important reaction in the chlorination of an aqueous solution is the formation of hypochlorous acid. This specie of chlorine is the most germicidal of all chlorine compounds with the possible exception of chlorine dioxide.

Hypochlorous acid is a "weak" acid, which means that it tends to undergo partial dissociation as follows:

$$HOCl \longrightarrow H^+ + OCl^- \qquad (2)$$

to produce a hydrogen ion and a hypochlorite ion. In waters of pH between 6.5 and 8.5 the reaction is incomplete, and both species are present to some degree.

The percent distribution of the OCl^- ion (hypochlorite ion) and undissociated hypochlorous acid can be calculated for various pH values as follows:

DECHLORINATION WITH SULFUR DIOXIDE

The practice of dechlorination has seen dramatic growth in the past decade. Dechlorination removes all or part of the chlorine residual and halogenated organics remaining after chlorination, and reduces or eliminates toxicity harmful to aquatic life in receiving waters.

Sulfur dioxide gas successively removes free chlorine, monochloramine, dichloramine, nitrogen trichloride, and poly-n-chlor compounds. For the overall reaction between sulfur dioxide and chlorine, the stoichiometric weight ratio of sulfur dioxide to chlorine is 0.9:1. In practice, it has been found that ~1.0 mg/L of sulfur dioxide will be required for the dechlorination of 1.0 mg/L of chlorine residue (expressed as Cl_2). Because the reactions of sulfur dioxide with chlorine and chloramines are nearly instantaneous, contact time is not usually a factor and contact chambers are not used; however, rapid and positive mixing at the point of application is an absolute requirement.

The ratio of free chlorine to the total combined chlorine residual before dechlorination determines whether the dechlorination process is partial or proceeds to completion. If the ratio is <85%, it can be assumed that significant organic nitrogen is present and that it will interfere with the free residual chlorine process.

There is no question that dechlorination removes most of the total residual chlorine from disinfected wastewaters. Consequently, it reduces the toxicity of disinfected wastewater effluent to aquatic wildlife. In most situations, sulfur dioxide dechlorination is a very reliable unit process in wastewater treatment, provided that the precision of the combined chlorine residual monitoring service is adequate.

CHLORINE DIOXIDE

Chlorine dioxide is a highly selective oxidant that is more similar to ozone than it is to chlorine. It is unstable as a compressed gas, and must be generated at the point of use. It cannot be stored in steel containers like chlorine. New technology, which reacts to chlorine gas with specially processed solid sodium chlorite, has substantially resolved all of the problems historically associated with chlorine dioxide generation, making chlorine dioxide more of a wastewater treatment candidate, especially for tertiary treatment of water intended for reuse. Chlorine dioxide does not combine with the nitrogen as ammonia normally present. Therefore, in a nitrogen-laden wastewater it is reputed to have a disinfection efficiency for both bacterial and viral destruction comparable to that of free chlorine.

Disinfection with Chlorine Dioxide

Chlorine dioxide is another bacteriocide, equal to or greater than chlorine in disinfecting power. Chlorine dioxide has proven to be more effective in achieving inactivation of viruses than chlorine.

Chlorine Dioxide Generation

Chlorine dioxide is an unstable and explosive gas, and for this reason it must be generated on site. Generation of chlorine dioxide involves reacting sodium chlorite ($NaClO_2$) with chlorine to produce gaseous chlorine dioxide according to the following reaction:

$$2\,NaClO_2 + Cl_2 \longrightarrow 2\,ClO_2 + 2\,NaCl \qquad (3)$$

Summary

Advantages of Chlorine Dioxide

- Chlorine dioxide is an effective, fast-acting, broad-spectrum bactericide.
- It is superior as a viricide to chlorine, which makes it a promising candidate for water reuse disinfection.
- It kills chlorine-resistant pathogens—eg, encysted parasites *Giardia* and *Cryptosporidium*.
- It does not react with nitrogen as ammonia or primary amines.
- It does not react with oxidizabale organic material to form trihalomethane (THM).
- It destroys THM precursors and enhances coagulation.
- It is excellent for the destruction of phenols, which cause taste and odor problems in potable water supplies.
- It has a long track record in the removal of iron and manganese. It is superior to chlorine, particularly when the iron and manganese occur in complexed compounds.

Disadvantages of Chlorine Dioxide

- The cost of chlorine dioxide, several times more expensive than chlorine, may make its use prohibitive for certain applications, especially in economically deprived (eg, Third World) regions where even chlorination is not readily affordable.
- Chlorine dioxide cannot be transported as a compressed gas; it must be generated on-site.
- The chlorine dioxide prepared by some processes may contain significant amounts of free chlorine, which could defeat the objective of using ClO_2 to avoid the formation of THMs.

OZONE

Ozone is an unstable gas that must be produced at the point of use. It is made commercially by the reaction of an oxygen-containing gas (air or pure oxygen) in an electric discharge. It is a powerful oxidant and has been used since the early 1990s for odor and color removal as well as disinfection of potable-water supplies in Western Europe and Canada. It appears that ozone in combination with either chlorine or chlorine dioxide could solve the disinfection problem of both bacterial and viral contamination in tertiary wastewater effluents. This is particularly significant where there is consideration of wastewater reuse.

Disinfection with Ozone

Today nearly 1000 ozone disinfection installations exist (primarily in Europe), almost entirely for treating water supplies. A common use for ozone at these installations is to control taste-, odor-, and color-producing agents. Although historically used primarily for the disinfection of water, recent advances in ozone generation and solution technology have made the use of ozone economically more competitive for wastewater disinfection. Ozone can also be used in wastewater treatment for odor control and in advanced wastewater treatment for the removal of soluble refractory organics, in lieu of the carbon-adsorption process.

Ozone Generation

Because ozone is chemically unstable, it decomposes to oxygen very rapidly after generation, and thus must be generated on-site. The most efficient method of producing ozone today is by electrical discharge. Ozone is generated either from air or pure oxygen when a high voltage is applied across the gap of narrowly spaced electrodes. The high energy corona created by this arrangement dissociates one oxygen molecule, which re-forms with two other oxygen molecules to create two ozone molecules. The gas stream generated by this process from air will contain ~0.5–3% ozone by weight and from pure oxygen about twice that amount, or 1–6% ozone.

Ozone Chemistry

Some of the chemical properties displayed by ozone may be described by its decomposition reactions. The free radicals HO_2 and HO have great oxidizing powers and are probably the active form in the disinfection process. These free radicals also possess the oxidizing power to react with other impurities in aqueous solutions.

Effectiveness of Ozone

Ozone is an extremely reactive oxidant, and it is generally believed that bacterial kill through ozonation occurs directly because of cell wall disintegration (cell lysis). Ozone is also a very effective virucide and is generally believed to be more effective than chlorine. Ozonation does not produce dissolved solids and is not affected by the ammonium ion or pH influent to the process. For these reasons, ozonation is considered a viable alternative to either chlorination or hypochlorination, especially where dechlorination may be required.

Other Benefits of Using Ozone

An additional benefit associated with the use of ozone for disinfection is that the dissolved oxygen concentrations of the effluent will be elevated to near saturation levels as ozone rapidly decomposes after application to oxygen. Further, because ozone decomposes rapidly, no chemical residual that may require removal, as is the case with chlorine residuals, persists in the treated effluent.

BROMINE, BROMINE CHLORIDE, AND IODINE

Bromine (Br₂)

All bromine species used in water and wastewater treatment revert to bromides after being consumed in the oxidation process. Therefore, the use of bromine as an alternative or a supplement to chlorination is not considered

practical or acceptable from an environmental point of view.

Oxidation of bromide to bromine can be accomplished either chemically or electrochemically. Chemical oxidation can be affected by either chlorine compounds or oxygen-containing compounds such as manganese dioxide, bromate, or chlorate.

The extraction of bromine from bromide compounds requires four steps: (1) oxidation of bromide to elemental bromine (Br_2); (2) separation of the bromine from solution; (3) condensation and isolation of the bromine vapor; and (4) purification.

Physical and Chemical Properties

Bromine (Br_2; atomic number, 35; molecular weight, 159.83; specific gravity, 3.12) weighs 26.0 lb/gal, and has a boiling point of 58.78°C. Of the metals used to handle bromine, lead is the most versatile. Bromine is a dark brownish red, heavy, mobile liquid. It gives off, even at ordinary temperatures, a heavy, brownish red vapor with a sharp, penetrating, suffocating odor.

Bromine is three times as soluble as chlorine (ie, 3.13 g/100 mL water at 30°C), which is an important characteristic when one is considering the physical aspects of applying bromine to a process stream.

Chemistry of Bromine in Water and Wastewater

Bromine is unique in being the only nonmetallic element that is liquid at ordinary temperatures. It reacts with ammonia compounds in solution to form bromamines and displays the breakpoint phenomenon similarly to chlorine.

Reactions with Chlorine

The reactions with chlorine and the bromide ion are the only ones of particular interest. It is important to realize that free chlorine (HOCl) has the ability to oxidize bromide ions to form hypobromous acid (HOBr) in the pH range 7–9. However, chloramines residuals will not oxidize the bromide ion at pH 4, but free chlorine will.

Other applications of this phenomenon are described below.

Use of Bromine in Water Processes

Potable Water. The use of free bromine (Br_2) in potable water is probably nonexistent.

Wastewater. The only use of bromine in wastewater or water reuse situations in the United States or Canada that would be acceptable would be in a wastewater discharging into seawater.

Cooling Water. One test, inconclusive at this time has been run to see it using bromine in condenser cooling water treatment was variable. The system adopted by some of the steam generating plants amounted to pumping a bromide salt solution into the chlorine solution discharge of existing conventional chlorination equipment. The chlorine oxidizes the bromide ion in the salt solution to free bromine, which goes into solution as hypobromous acid, the active ingredient.

Swimming Pools. The most widespread use of bromine as a disinfectant is the use of liquid bromine in the elemental form as a substitute for chlorine in swimming pools.

BROMINE CHLORIDE

Physical and Chemical Properties

Bromine chloride, BrCl, is classified as an interhalogen compound because it is formed from two different halogens. These compounds resemble the halogens themselves in their physical and chemical properties except where differences in electronegativity are noted. Bromine chloride at equilibrium is a fuming dark-red liquid <5°C. It can be withdrawn as a liquid from storage vessels equipped with dip tubes under its own pressure (30 psig 25°C). Liquid BrCl can be vaporized and metered as a vapor in equipment similar to that used for chlorine.

Bromine chloride is an extremely corrosive compound in the presence of low concentrations of moisture. Although it is less corrosive than bromine, great care must be exercised in the selection of materials for metering equipment. Like chlorine and bromine, it may be stored in steel containers—assuming that the BrCl or Br_2 is packaged in an environment when the air is dry.

When in contact with skin and other tissues, liquid BrCl, like Br_2, causes severe burns.

Bromine chloride exists in equilibrium with bromine and chlorine in both gas and liquid phases.

The density of BrCl is 2.34 g/cm^3 at 20°C, and the solubility of BrCl in water is 8.5 g/100 cm^3 at 20°C.

Chemistry of Bromine Chloride in Water

Bromine chloride vapor appears to hydrolyze exclusively to hypobromous acid and hydrochloric acid, whereas bromine vapor (or liquid) hydrolyzes to hypobromous acid and hydrogen bromide. The formation of HBr represents a significant loss in the disinfecting potential of the expensive bromine molecules.

IODINE

Physical and Chemical Characteristics

Iodine, is a nonmetallic element with an atomic weight of 126.92—the heaviest of the halogen group. It is the only halogen that is solid at room temperature, and it can change spontaneously into the vapor state without first passing through a liquid phase.

Occurrence

Iodine is always found combined, as in the iodides. It is prepared from kelp and from crude Chile saltpeter. This saltpeter ($NaNO_3$) contains ~0.2% sodium iodate ($NaIO_3$).

Uses

Iodine and its compounds are used as catalysts in the chemical industry (production of synthetic rubber), and in food products, pharmaceutical preparations, stabilizers (as nylon precursors), antiseptics, medicine (treatment of cretinism and goiter), inks and colorants, and industrial and household disinfectants.

Iodination of water supplies has been limited largely to emergency treatment by the military. The use of iodine as a disinfectant for water has been recognized for a long time, but has never generated enough interest to displace the popular use of chlorine. As compared to chlorine, the very high cost of iodine and its possible physiologic effects are the main reasons for its limited use.

PERACETIC ACID IN WASTEWATER TREATMENT

Commercially available peracetic acid, also known as ethaneperoxoic acid or peroxyacetic acid, is available in a quaternary equilibrium mixture containing acetic acid, hydrogen peroxide, peracetic acid, and water. Although peracetic acid is known to be a potent antimicrobial agent, quantitative information on the activity of peracetic acid against typical wastewater microorganisms had been scarce until Solvay (Warrington, Chesire, England) conducted a series of laboratory and field tests to study its effectiveness in disinfecting wastewaters and effluents. He concluded that the potential of using peracetic acid for wastewater disinfection application is significant, and cannot be overlooked.

ANTIMICROBIAL NANOEMULSION TECHNOLOGY

The NanoBio nanoemulsions destroy microbes effectively without toxicity or harmful residual effects. The nanoemulsions also can be formulated to kill only one or two classes of microbes. Due to a large part to the low toxicity profile, the nanoemulsions are a platform technology for any number of topical, oral, vaginal, cutaneous, preservative, decontamination, veterinary, and agricultual antimicrobial applications.

In case of bioterrorism attacks, the nanoemulsions can potentially destroy pathogens such as anthrax, ebola, and many others. Since it is nontoxic and noncorrosive, it can be used to decontaminate personnel, equipment, terrain, structure, and water.

THERMAL DISINFECTION

The application of heat is the oldest disinfection technique. Variations in chemical impurities in water do not interfere with the disinfecting efficiency of heat as they do with chlorine and other chemical disinfectants. In addition to bacterial pathogens, heat readily destroys a variety of microorganisms that are highly resistant to chemical disinfectants. These organisms include amoebic cysts, worms, and viruses.

The Inate Heat Resistance of the Strain

The more heat resistant a strain, the more likely it is to have a pronounced activation requirement.

The Temperature of the Treatment

The lower the treatment temperature, the more pronounced the initial lag is likely to be because of a relatively slow activation rate. Conversely, as treatment temperatures increase, the activation time reduces, the hump becomes larger, and the lag time shorter (eventually approaching extinction because of the inability to observe the effect over very short increments of time). This may partially explain some of the nonlinearity that exists at the initial portion of survival curves.

Thermal Resistances

Heat resistance varies among microbial species, proteins, and bacterial spores. There are a number of bacteria spore species that are highly resistant to most chemical, physical killing agents, as well as heat. There is a significant difference between the thermal resistances to moist and dry heat. A spore species resistant to one form of heat is not necessarily resistant to the other.

Bacterial death rates are higher in acid or alkaline media than in neutral suspensions. Also, a higher recovery of survivors occurs in the neutral pH zone. Citrate, phthalate, or ammonium buffers reduce thermoresistance of spores compared to those in a phosphate buffer. Spores are more readily inactivated at a low pH, since pH can influence the type of ions that absorb onto the spore surface (this absorption alters the heat stability).

Applications of Heat Treatment

The terms sterilization and disinfection should not be used interchangeably. It is only when heat is used to affect microorganisms that both have the same meaning. Probably the oldest disinfection application is that of boiling water. The application of thermal treatment for disinfection is not limited to water treatment. There are numerous applications in the canning and food industries as well. There are three specific applications in which thermal treatment is widely used. These are in the processing of dry goods, pasteurization of water, and the thermal conditioning of sludges.

DISINFECTION WITH SOUND

Sound waves are an alteration in pressure, stress, particle displacement, particle velocity, or a combination of the above that is propagated in an elastic medium. Sound waves thus require a medium for transmission (ie, they cannot be transmitted in a vacuum). The sound spectrum covers all possible frequencies.

In water purification, ultrasonic waves have been used to effect disintegration by cavitation and mixing of organic materials. The waves themselves have no germicidal effect. However, when used with other treatment methods they can provide the necessary mixing and agitation for purification.

Ultrasonic Generators and Bubble Dynamics

Among the possible applications that can be derived from ultrasonics are

- Faster rate of coagulation of suspended particles in water and sewage.
- Sterilization of water and sewage effluents.
- Lower treatment costs for sewage and industrial wastes.

Application to Microorganism Destruction

Induced cavitation through ultrasonics can be used to destroy bacteria, provided the cavitation implosion occurs in the immediate vicinity of the bacteria cell. The effective range of a shock wave generated by the implosion is $\sim 1-2\,\mu$.

To assure an adequate bacteria kill rate, a cavitation nucleus must exist in the molecular structure of the water surrounding the bacterial cell. The weaker this nucleus becomes, the greater the probability that cavitation will occur; thus, the greater the probability that a bacteria cell will be destroyed.

The sonoration causing cavitation enhances inactivation by reducing the high surface tension caused by organic material. Although the sonication process reduces the time needed for a complete inactivation, by itself it would not have the desired effect. It is desirable to combine sonication and ozonation. This combination results in a synergistic effect that inactivates microorganisms.

Miscellaneous Wastewater Applications

Since ultrasonics can be used to facilitate coagulation, it makes a perfect method for hardness removal. Ultrasound can be used to precipitate calcium oxide (CaO) and magnesium oxide (MgO), which are major constituents contributing to hard water.

Ultrasound can be used in conjunction with ozonation. Ultrasound will break particle aggregates into discrete particles. This results in a much greater surface area exposed for ozone to oxidize contaminants. In terms of design requirements for ozonation vessels, smaller volume requirements are needed to achieve the same degree of germicidal destruction than a system without ultrasonics.

Ultrasound can be effectively used in the bacteriological and viricidal fields. It can also reduce turbidity if used with a CaSO$_4$ solution. Ultrasound is an effective method for wastewater disinfection. It provides residual disinfection much as chlorine does. However, ultrasound does have disadvantages. Production of bubbles or cavities is needed for this technique to be a powerful disinfection.

ULTRAVIOLET DISINFECTION OF WASTEWATER

What Is uv Light and the Mechanism of Germicidal Action?

Effective disinfection in air, on surfaces and in water has been accomplished by exposure to the direct rays of the

Table 1. Major Parameters Affecting the Uv Disinfection of Wastewater

Parameters	Acceptable values
percent transmittance (T) or absorbance	35–65
total suspended solids (TSS), mg/L	5–10
particle size, μm	10–40
flow rate/condition or hydraulics	ideal plug flow
iron, mg/L	<0.3
hardness, mg/L	<300

sun. Sunlight is an important factor in the self-purification of water in streams and in impounding reservoirs.

The term "ultraviolet light" or simply "ultraviolet (uv)" is applied to electromagnetic radiation emitted from the region of the spectrum lying beyond the visible light and before X-rays. The uv light causes molecular rearrangements in the genetic material of microorganisms and this prevents them from reproducing.

Effects of Wastewater Quality Parameters on Uv Efficiency

The efficiency of a uv disinfection system strongly depends on effluent quality that acts to decrease the uv intensity in wastewater. Table 1 shows the major parameters that must be taken into consideration when a uv disinfection system is being designed for wastewater.

Summary

The germicidal efficiency of an uv disinfection system strongly depends on effluent quality—high TSS concentration in effluent will reduce the uv transmittance and as a result higher coliform count. If dissolved iron concentration in effluent is >0.3 ppm, iron will absorb uv light, as well as will precipitate out on the sleeve and absorb uv light before it transmit to the wastewater that would limit the germicidal efficiency. However, studies reported that dissolved aluminum salts have no effect on uv transmittance and germicidal efficiency. Waters that contain hardness >300 mg/L will have an adverse effect on the overall uv germicidal efficiency. Higher uv dose (product of light intensity and exposure time) will increase higher germicidal efficiency.

ULTRAVIOLET GERMICIDAL IRRADIATION IN INACTIVATING AIRBORNE MICROORGANISMS

Three factors are important in the efficacy of upper room ultraviolet germicidal irradiation (UVGI): the upper room disinfection rate, air volume ratio for the irradiated upper room and the nonirradiated lower room, and the air mixing rate between the upper and lower room. The disinfection rate in the upper room depends on uv dose and uv susceptibility of the microorganism. The uv dose is the product of uv irradiance and exposure time. Susceptibility of microorganisms to uv depends on the complexity of the microorganism's structure, its repairability, and its general sensitivity. As the uv dose becomes higher and microbial susceptibility to UVGI increases, the efficacy of upper room UVGI increases. The desirable scenario is

to maintain the maximum amount of uv irradiance in the upper part of the room while minimizing people's exposure to UVC in the lower part of the room.

The mixing rate between air in the upper and the lower part of the room is also important in the efficacy of upper room UVGI. This air mixing occurs mainly by convection caused by a vertical temperature gradient in nonmechanically ventilated rooms.

Upper room UVGI can significantly reduce the concentration of airborne microorganisms (S. marcescens and BCG) aerosolized in a saliva stimulant in a typical mechanically ventilated hospital isolation room. Because upper room UVGI varied significantly with environmental factors such as temperature, air mixing, and air exchange rate, it is important to optimize environmental conditions to produce the best and most persistent effect in reducing the risk of TB in high risk settings.

ELECTROMAGNETIC RADIATION TECHNIQUES

Electromagnetic radiation is the propagation of energy through space by means of electric and magnetic fields that vary in time. Electromagnetic radiation can be specified by frequency, vacuum wavelength, or photon energy.

Effects of Ionizing Radiation on Microorganisms

- Radappertization—the commercial sterilization of materials.
- Radurization—the reduction of organisms, mainly vegetative cells, to a very low level and subsequent storage of the foodstuff at temperatures above freezing to ensure longer shelf life. Normally, a dose of <10^6 rad is used. An equivalent term is radiopasteurization.
- Radicidation—the removal of pathogens or organisms significant in public health (eg, Salmonella, from a food or foodstuff) by substerilizing doses of radiation.

Ionizing energy affects microorganisms both directly and indirectly. Theory describing direct affects is referred to as the target theory. Here, the organism is visualized as a target that is hit directly by an ionizing particle or ray. Mathematical analysis shows that, in general, when targets (in this case microorganisms) are destroyed by direct hits, this involves a probability concentration depending solely on the number of particles or rays (ie, the dose) and the number of targets (ie, number of microorganisms).

Characterization of Dose Requirements

The dosage of ionizing radiation required for a specific effect depends on a number of parameters. These parameters are the species of microorganism, population size, medium properties, temperature, the gas phase, water activity, and the presence of sensitizing compounds. Among these, microorganism species plays the dominant role.

Summary

Although the main effect of ionizing energy on microorganisms is direct, much of the lethal effect is also due to indirect action of the radiation mediated through free radicals and activated molecule formation. As a result, the nature of the medium or menstruum in which the organisms are suspended, gaseous environment, water activity, temperature of the substrate, etc—in fact all factors that influence indirect effects—are important in determining the degree of effect on the microorganism and, thus, on the sterilizing doses of radiation needed.

In studies directed at determining required sterilizing doses for a given process, materials should be inoculated with a large enough number of the most radioresistant species one would expect to encounter. Also, the material should be packed under conditions that are commercially encountered.

OTHER STERILIZATION METHODS

Ethylene Oxide

Ethylene oxide (EtO) is, for many products, an effective sterilization agent. It is especially useful for custom procedure kits containing unit dose drugs in hermetically sealed packages and for products that discolor, distort, or otherwise degrade when processed with radiation. The EtO sterilization process normally requires a product conditioning step in which product is placed in a highly humidified area for a specified period of time. This process humidifies the product, allowing the sterilizing agent, ethylene oxide gas, to penetrate more effectively.

Gamma Rays (Co-60)

Gamma irradiation involves exposure of products in their final shipping containers to a radioactive isotope known as Co-60. Cobalt-60 is processed almost exclusively worldwide by the Canadian government and provided to gamma irradiation plants by Nordion International, who processes and packages the isotope in cobalt "pencils." The pencils are shipped, under extremely tight control and under "hazardous materials" shipping regulations, to gamma plants as needed.

The cobalt pencils are housed in "source racks" within the gamma facility. The racks are placed in the gamma cell, and products in their final shipping configurations travel through the cell on a conveyor system. This process takes several hours to complete. Cobalt-60 decays with time, and appropriate cycle timer setting adjustments must be made at the gamma plant to account for the current cobalt inventory quantity. When the cobalt source is sufficiently decayed, the pencils must be replaced. Upon completion of this "resourcing," dose mapping must be performed on all products which will be irradiated in the cell.

The effect of gamma irradiation to the product microbial population (bioburden) is the same as the effect of electron beam processing. The fundamental difference between gamma and e-beam sterilization is the manner in which the radiation energy is delivered to the material being irradiated.

Other Sterilization Methods

Several other sterilization methods, such as gas plasma, steam, and others are available and/or under development.

Advantages and Benefits of Electron Beam

Electron Beam Compared with Ethylene Oxide

- **Significantly faster turnaround time**. Electron beam requires no prehumidification, no poststerilization aeration, and no sterility testing after the product is processed.
- **Reduced inventory carrying costs**. Faster turnaround time requires less product to be held in inventory.
- **Product consistency**. Electron beam processing variables consists of only "dose delivered;" other processing variables inherent to EtO, (eg, gas concentration, vacuum, pressure, etc) are not variables in electron beam processing.
- **Lower cost of sterilization**. Electron beam processing is extremely cost competitive; furthermore, biological indicator testing is not necessary in electron beam processing, thus eliminating an additional testing charge.
- **Safety**. Electron beam processing does not utilize potentially dangerous chemicals or gases, nor is there the potential that chemicals or gases might remain on or in the product.
- **Cost savings in product packaging**. Electron beam does not require the special "breathable" packaging necessary for EtO to allow for noxious gas aeration; nonpermeable plastic packaging is less expensive and easily penetrated with electron beam.

M. W. First, E. A. Nardell, W. Chaisson, and R. Riley. *Guidelines for the Application of Upper-Room Ultraviolet Germicidal Irradiation for Preventing Transmission of Airborne Contagion Part I: and part II.*

R. J. Lewis, Sr., *Sax's Dangerous Properties of Industrial Materials*, Wiley, Hoboken, N. J., Online 2005.

E. A. Murano, ed., *Food Irradiation: A Sourcebook*, Iowa State University Press, Ames, Iowa, 1995.

G. C. White, *Handbook of Chlorination and Alternative Disinfectants*, 4th ed., John Wiley & Sons, Inc., New York, 1999.

Tapas K. Das
Washington State Department
of Ecology

DISPERSANTS

Dispersants are compounds that are used to maintain particles suspended in a liquid medium, or assist in dispersing them. They are usually polymeric or oligomeric. Dispersants have many applications. They are used to disperse pigments in paints and inks, in water treatment, cement, builder assists, etc.

FUNCTION OF DISPERSANTS

Suspended particles attract each other due to van der Waals attractive forces. This causes them to form clusters of particles, called flocculates. These flocculates often detract from the properties of the system. Flocculation of pigments reduces gloss and color strength in paint systems. Flocculates of fouling or scale in cooling or heating water systems may precipitate and cause clogging or reduced heat conductivity.

Dispersants provide repulsion forces to counteract the particle–particle attraction, reducing or eliminating the flocculation or coagulation. Moreover, the viscosity of dispersions is reduced and a higher particle loading can be obtained. They adsorb onto particle surfaces, thus building a shell around them, which forms a barrier around the particle that can stabilize the particles against flocculation in two ways: sterically and electrostatically. Very often, a combination of steric and electrostatic stabilization (usually called electrosteric stabilization) is used.

Electrostatic Stabilization

Particles suspended in water usually have some surface charge. This charge of course is pH dependent. At low pH values, the particle or ionic groups on the surface will be protonated, or OH^- ions will desorb, and the particle will be more positively or less negatively charged. Conversely, at high pH values the charge will become more negative or less positive. The overall stability of a particle dispersion depends on the sum of the attractive and repulsive forces as a function of the distance separating the particles. The net energy between two particles at a given distance is the sum of the repulsive and attractive energies.

Steric Stabilization

When a polymer is adsorbed to a particle, it usually does not lie flat on the surface, but (in a good solvent) parts of the chains will protrude into the solvent or water. This solvated polymer builds a shell around the particle. When two particles approach each other their adsorbed polymer layers begin to interpenetrate.

In the overlapping part, the polymer concentration is higher and the degree of solvation is lower than in the rest of the shells. Moreover, due to the higher concentration, the polymer chains have a volume restriction, which causes a loss of entropy. As a result, the particles must separate to allow the chains more freedom of movement, while the solvent moves in to resolvate the polymer layer. As with electrostatic repulsion, an energy barrier is created. A common approximation used is that the strength of the energy barrier rises steeply at slightly less than the adsorbed layer thickness.

Some of the practical differences between sterically and electrostatically stabilized dispersions may be summarized as follows:

Steric stabilization	Electrostatic stabilization
insensitive to electrolyte	coagulation occurs with increased electrolyte concentration
effective in aqueous and nonaqueous media	more effective in aqueous media
effective at high and low concentrations	more effective at low concentrations
often reversible flocculation	coagulation often irreversible
good freeze–thaw stability	freezing often induces irreversible coagulation

STRUCTURE OF DISPERSANTS

Classical dispersants are simple linear or branched, random polymers. Trains of segments of these polymers adsorb to the particle surface, whereas other segments build loops or tails. These tails and loops can provide steric stabilization. In addition, in aqueous systems the polymer may have charged groups, both in the trains and in the tails and loops, providing electrosteric stabilization.

Anchoring Groups

Adsorption is usually due to anchoring groups in the dispersants, which have specific interactions with the particle surface, eg, by hydrogen bonds, acid–base interactions, complex-, or ligand formation, overlapping orbitals, etc.

The specific character of these interactions makes it difficult to find universal anchoring groups, ie, anchoring groups that adsorb on every surface, which is particularly important in the paint industry, where a broad variety of pigments are used. A method for the selection of anchoring groups has been described in the literature. In this method, the adsorption of simple model compounds with anchoring groups is measured and from this process the free energy of adsorption ΔG is calculated. The parametric ΔG is a measure for the strength of anchoring. It allows the rapid screening of anchoring groups for a particle surface, without the need of preparing and testing a whole range of dispersants.

Alternatively, anchoring can take place through nonspecific interactions. Many dispersants have segments that are insoluble in the solvent (or water) in which they are used. In this case, dispersants do not adsorb due to their affinity for the particles, but to their insoluble segment precipitating on the surface. Such dispersants are much more universal, ie, applicable to a broader range of particles, but their interaction with the particle is much weaker.

Soluble Tails

The soluble parts of a polymeric dispersant must extend into the solution, in order to contribute to the stabilization of the dispersion of particles. Whether they do so or lie flat or coiled on the particle surface depends on whether the continuous phase solvent is a good solvent (polymer–solvent interactions energetically favored) or a poor solvent (polymer–polymer and solvent–solvent interactions favored).

On the other hand, if the polymer–solvent interaction is so strong that it is greater than the anchoring energy, dispersant may be adsorbed only weakly.

COMPARISONS WITH OTHER MATERIALS

Surfactants vs Dispersants

Surface-active agents or surfactants (qv) are occasionally used as dispersants, but are mainly used in other applications. Surfactants are used in a whole range of two-phase interface systems: water-in-oil and oil-in-water emulsions, change the surface tension at the air–water interface, improve wetting at the solid-in-water interface. Dispersants are only used to stabilize particles either in water or in solvents. As indicated, surfactants are sometimes used for this purpose either because the dispersions are not stable enough or the surfactants are too expensive. Consequently, the terms dispersant and surfactant are frequently confused. However, there are several important differences between the two classes of materials. Surfactants tend to orient at the air–water interface, oil–water interface, or sometimes at a liquid–solid interface depending on the length of the hydrophobic portion and the nature of the hydrophilic part (anionic, cationic, or nonionic). In contrast, dispersants, which tend to be larger polymeric molecules, are defined more by their use, which is to disperse a solid in a liquid. Surfactants adsorb at surfaces, preferring to be out of the water phase. Dispersants mostly adsorb by means of chemisorption or electron transfer, using specific anchoring groups.

Chelants and Precipitation Inhibitors vs Dispersants

Dispersants can inhibit crystal growth, but chelants, can be more effective under certain circumstances. Chelants can prevent scale by forming stoichiomctric ring structures with polyvalent cations (such as calcium) to prevent interaction with anions (such as carbonate). Chelants interact stoichiometrically with polyvalent cations, in preference to adsorbing on surfaces.

Flocculants vs Dispersants

In direct contrast to dispersants, flocculants or coagulants are used to aggregate fine particles or liquid droplets in aqueous media to improve separation of the two components. Flocculants function by charge neutralization, double-layer compression, particle bridging, or by forming large nets that engulf masses of particles (sweep flocculation). Dispersants normally function by charge repulsion, steric repulsion, or both. Some polymeric flocculants can be chemically similar to dispersants, differing only in molecular weight and dosage level used.

In summary, dispersants are effective for particle dispersion and crystal-growth inhibition, but do not normally have surface-active properties such as oil emulsification. Chelants and antiprecipitants frequently inhibit crystal growth better than dispersants, but are ineffective for particle dispersion. Flocculants are effective for aggregating particles, the opposite function of a dispersant.

USES

Recirculating Cooling Water

Water used to cool plant processes and buildings contains contaminants that can accelerate corrosion of metal surfaces and leave scale or particle deposits on pipes and heat-transfer surfaces. To prevent corrosion and scale, cooling water treatment formulations contain corrosion inhibitors, biocides, phosphonates, and dispersants. Dispersants aid in the prevention of inorganic fouling (silt, iron oxide), scaling (calcium carbonate, calcium sulfate), and corrosion.

Boiler Water

Dispersants are used to prevent scale buildup on the boiler tubes and drums of boilers that operate at pressures $\lesssim 10.3$ MPa (1500 psig). Boiler water treatments using dispersants generally fall into three categories: precipitation programs, where water hardness ions are preferentially precipitated as calcium carbonate or hydroxyapatite (calcium phosphate hydroxide) and are dispersed in the bulk water rather than at the metal surface; dispersant and chelant combinations, where the dispersant is used to control precipitates formed from excess hardness leakage from water pretreatment; and dispersant only or dispersant plus sequestrant, where dispersants bind hardness ions to prevent precipitation and scaling with polyvalent anions.

Geothermal Fluids

Geothermal fluids are used to provide energy for power generation and home heating. Dispersants are used to change the surface of the precipitates formed so that they no longer adhere to the other surfaces.

Seawater Distillation

The principal thermal processes used to recover drinking water from seawater include multistage flash distillation, multieffect distillation, and vapor compression distillation. Dispersants such as poly(maleic acid) inhibit scale formation, or at least modify it to form an easily removed powder, thus maintaining cleaner, more efficient heat-transfer surfaces.

Reverse Osmosis

In contrast to distillation, reverse osmosis (qv) (RO) uses hydraulic pressure as its energy source to purify water. In RO, a fraction of the water content of seawater or brackish water is driven under pressure through a semipermeable membrane. Dispersants are used to minimize fouling and decrease the frequency of flushing or cleaning. Dispersants are also used in proprietary membrane cleaning agents.

Sugar Processing

Dispersants are used in the production of cane and beet sugar to increase the time between evaporator clean outs. Only certain dispersants, conforming to food additive regulations, can be used, since a small amount of the dispersant may be adsorbed on the sugar crystals.

Oilfield

Scales can plug a producing well, requiring expensive remediation or even requiring a new well to be drilled. Scale also forms on topside equipment and piping, which is usually less difficult to handle. Polymeric dispersants and organic phosphonates are most often used to prevent oilfield scaling by delaying precipitation and preventing scale adherence on pipes and surface equipment.

Drilling Muds

Aqueous drilling muds normally consist of bentonite clay, weighting agents such as barite, dispersants such as lignite, lignosulfonate, and various polymers, and fluid loss agents. A dispersant neutralizes the positive edge charges of bentonite particles, allowing them to lay flat against the sides of the drilled hole to minimize water intrusion into the formation. Weighting agents, for increasing the density of the mud, must also be dispersed. Seawater muds (offshore drilling) and gypsum muds require specialty dispersants that have higher divalent cation tolerance than dispersants used in freshwater muds. Chrome lignosulfonate has been the preferred dispersant for these wells, but is now being replaced by polymeric acrylic dispersants because of the harmful effects of chromium on the environment.

Cement

Although water is needed in the hydration reactions of cement (qv), excess water added for workability of the concrete and mortar creates voids that decrease strength and increase water permeability. Dispersants are used as plasticizers in cements to cut water demand by up to 40% and decrease void volume.

Paints and Inks

Pigments must be dispersed in the liquid medium and stabilized at their primary particle size to provide maximum hiding and film properties.

A special requirement in the stabilization of pigments in paints arises from the fact that nearly always a combination of different pigments is used, to obtain the required color. Very often, differently colored paints or color concentrates are mixed at the point of sale or at the point of application. Since not every pigment can be stabilized with the same dispersant, in many cases different dispersants have to be used for different pigments. Although much progress has been made in the prediction of color properties of mixtures, experimental verification of the final color is still required.

Mineral Processing

Dispersants are used as mineral processing aids for grinding and improving slurry stability. Key mineral processes using dispersants include calcium carbonate and kaolin manufacture, and gold beneficiation.

Caulks, Sealants, and Roof Coatings

These have similar technology to paints, except that they are formulated at higher solids to produce a thicker coating. As with paints, two types of dispersant are used; a primary dispersant such as KTPP (potassium tripolyphosphate) to disperse pigment agglomerates, and a secondary dispersant, such as sodium polymethacrylate, to provide storage stability of the formulation. Caulks, sealants, and roof coatings differ from paints in that they must minimize water leakage (permeance), but still allow water-vapor transport. Although water permeability of these films lessens with time due to leaching of water-sensitive materials such as dispersants, a variety of less sensitive dispersants have been developed to overcome this problem. Less water-sensitive dispersants include zinc sodium hexametaphosphate (primary dispersant), and hydrophobic polymeric dispersants (secondary dispersant).

Agricultural Uses

Dispersants are used to formulate pesticides into aqueous dispersions (flowables), wettable powders, water-dispersible granules, and dry flowables.

Detergents and Cleaners

Dispersants function as builder assists in cleaning formulations to increase particulate soil removal, prevent redeposition of soils to maintain whiteness or eliminate residues (spots) on hard surfaces, prevent precipitation of inorganic salts (carbonates, phosphates, and silicates), increase water-wettability of soiled surfaces, and promote physical stability of slurried formulations. They improve spray drying of powders by improving slurry (crutcher) homogeneity, increasing solids of crutcher mix (time and energy savings), reducing dusty fines, and increasing bead strength. Dispersants increase the rate of solution for powdered detergents and buffer the water to maintain optimum cleaning pH.

ENVIRONMENTAL CONSIDERATIONS

Most reviews on biodegradable polymers suggest that, with the exception of poly(vinyl alcohol) and poly(ethylene glycol)s, most synthetic organic dispersants are recalcitrant in the environment. More recently developed dispersants displaying biodegradability are polymers containing ester linkages and ether linkages. There is currently a great deal of research activity to develop dispersants that are both effective and biodegradable.

R. J. Hunter, *Foundations of Colloid Science*, Vol. I, Oxford Science Publications, New York, 1988.

E. A. Johnson and E. J. Schaller, *Additives for Coatings*, Center for Professional Advancement, E. Brunswick, N. J., 1991.

Th. F. Tadros, *The Effect of Polymers on Dispersion Properties*, Academic Press, London, 1982.

D. Napper, *Polymeric Stabilization of Colloidal Dispersions*, Academic Press, London, 1983.

HENK J. W. VAN DEN HAAK
Akzo Nobel

DISPERSIONS

A dispersion is a mixture of particles suspended in a liquid. Examples are blood, ceramics, concrete, grease, inks, paints, paper coatings, pesticides, and photographic "emulsions." The particles are called the discontinuous or internal or dispersed phase. The liquid phase is called the continuous or external phase, serum, or medium. When the particles have at least one dimension less than a micron (1000 nm), they are called colloidal particles and the mixture is called a colloidal dispersion. The lower particle size limit is 1 nm (a billionth of a meter, a thousandth of a micron). Dispersions finer than this are considered molecular solutions.

The descriptor, dispersion, is sometimes used broadly to include mixtures of liquids in liquids (emulsions), gases in liquids (foams), powders in gases (dusts), and liquids in gases (aerosols).

A dispersion of submicron particles has a large liquid/solid interface. The properties of this interface are different than those of either bulk phase. Some components of the liquid may be preferentially adsorbed at the interface. These are properly called surface-active solutes but more commonly called surfactants. If these surface-active solutes are ions, then the particle acquires a net electric charge. If surface-active solutes are polymers, then a barrier to close approach is created on the particle surfaces. If an ionic component of the solid is preferentially dissolved into the liquid, then the particles also acquire an electric charge.

The smaller the particles, the more they respond to thermal fluctuations of the liquid. The randomness of thermal fluctuations creates random motion of particles called Brownian motion. Brownian motion leads to diffusion of the particles, which in turn leads to particle–particle collisions. If the particles bounce apart after they collide, the average particle size remains the same, and the dispersion is called stable. If the particles stick when they collide, the average particle size grows with time, and the dispersion is called unstable.

A concentrated dispersion is sometimes called a slurry or a slip. Dispersions of film-forming polymer particles are called latexes. Natural rubber is a latex. Dispersibility is the ease with which a dry powder may be dispersed in a liquid.

Unfortunately, the term "dispersion" is also used in colloid science to refer to a type of interparticle attractive force. This attraction is a consequence of coordinated electronic fluctuations in particles close to each other and is also referred to as London, van der Waals, Hamaker, or Lifshitz attraction after key theoretical contributors.

A word of warning: Names can be misleading. A grave error is to name a component by its putative role, wetting agent, dispersant, film former, etc. These functions are interrelated and many components participate in multiple roles.

CLASSIFICATION

Dispersions can be classified by the degree of dispersion, the state of the dispersed state, the dispersion medium, the interaction between the dispersed phase and the medium, and the interaction between the particles.

The degree of dispersion is generally the average particle size. If all the particles are the same size, the dispersion is called monodisperse. If the particles have a variety of sizes, the dispersion is called polydisperse. If the least dimensions of the particles are $>1 \mu$ (1000 nm), the dispersion is called coarse. If at least one dimension of the particles is $<1 \mu$, the dispersion is called submicron. If at least one dimension is on the order of a few nanometers, the dispersion is called nanoparticle.

The growth of particle size as particles collide and stick is called flocculation or sometimes agglomeration, and the cluster is called a floc or agglomerate. Generally flocs can be redispersed by strong agitation. However, when particles bind tightly and cannot be redispersed, the growth is called aggregation and the cluster is called an aggregate. The distinction between agglomeration and aggregation is the strength of the interparticle forces in the cluster. Aggregates are redispersed by attrition. Coagulation is the term used for the merging of emulsion drops upon contact to form a larger drop.

Dispersions are also classified by the nature of the medium; these are aqueous and nonaqueous dispersions. The former are dispersions in water, the later are dispersions in organic liquids. This distinction is made because in aqueous dispersions the role of ions is significant and in most organic liquids they are less so. Dispersions are also called polar or nonpolar based on the dielectric constant of the medium, again reflecting the role of ions.

When particle–medium interactions are weak, the dispersion is called lyophobic. When particle–medium interactions are strong, the dispersion is called lyophilic (from the Greek lyos—to loosen, phobos—a fear, and philia—an attraction). These terms were originally used to distinguish dispersions that were sensitive to salt, lyophobic, from those that were not, lyophilic. Sensitivity to salts is also called electrocratic. Dispersions of silver halides or carbon blacks in water are lyophobic. Dispersions of albumin or gum Arabic in water are lyophilic.

The interaction between particles is a combination of attractive and repulsive forces. The repulsive forces are either electrostatic, the repulsion of like-charged particles or steric, the repulsion of interpenetrating polymer layers. The attractive and repulsive forces are all dependent on interparticle distance so that the total force varies with distance. If the particles are held apart by electric forces, the dispersion is called electrostatically stabilized.

If the particles are held apart by polymer coatings, the dispersion is called sterically stabilized. Sometimes dispersions are stabilized by a combination of electric charges and steric barriers. These are called electrosterically stabilized.

Lyophobic dispersions often flocculate with the addition of salt or by waiting a period of time. These dispersions are kinetically stabilized and the flocculation is irreversible. Lyophilic dispersions often flocculate with a change in temperature or solvent composition, but when the temperature or solvent composition is restored, they redisperse. These dispersions are called reversible or thermodynamically stabilized.

METHODS TO PRODUCE DISPERSIONS

Producing a satisfactory dispersion is not easy. Specialized equipment and chemicals are used, and an understanding of some fundamental principles is necessary for optimum results. To fracture particles under a micron requires applying significant forces to small areas and so appropriate equipment is needed. Newly created interfaces have high energies and will recombine unless the adsorption of dispersant is quick enough.

Dry Grinding and Powder Flow

Dry grinding, sometimes called micronizing, is first used to reduce particles to under a few millimeters or so. The most common equipment is the air-impact pulverizer in which particles are entrained in air streams that are directed toward walls or each other. This grinding is usually combined with a cyclone separator to recycle the large particles. Dusts of micron-sized particles can be an inhalation health hazard and an explosion and fire threat.

Wetting and Millbases

The next dispersion step is wetting of the powder with liquid. The role of the adsorbed wetting agent is to reduce the contact angle as close to zero as possible while keeping the surface tension of the liquid as high as possible. Wetting agents are low molecular-weight surface-active solutes, hence they adsorb on particle surfaces quickly.

Subsequent processing in high energy mills is more efficient if the dispersion is concentrated. Therefore most manufacturers wet pigment with a minimum of liquid to form a high volume-loaded dispersion called a millbase. This millbase is fed to the high energy mixers for subsequent particle-size reduction.

High Speed Stirrers

The most obvious method to produce a dispersion is to stir the powder and the liquid together. Experience has shown that even the best high speed stirrers do not usually reduce the particle size below a few microns. The reason for this is that the breakup of agglomerates requires a significant shear rate (velocity gradient) on the size scale of the agglomerates. The smaller the particles, the higher the necessary shear rate. The shear rates in high speed stirrers are the ratios of their blade velocities (limited by cavitation) to the distance over which the velocities drop (approximately tens of centimeters). These are relatively low shear rates.

Much higher shear rates are attained by surrounding a spinning, slotted rotor with a shield. It is common practice to use a blade stirrer to mix dry powder with the liquid and then a rotor-stator disperser to reduce agglomerate size to the micron range.

High Shear Mills

The key idea to produce finer and finer dispersions is to attain higher and higher shear rates. Liquid velocity is limited by cavitation. Therefore the design criterion is to force the liquid to flow through narrower and narrower gaps.

The device used is called a colloid mill, which is constructed with a spinning rotor on a movable shaft inside a chamber. To attain higher shear rates, the rotor is moved closer to the walls.

The homogenizer obtains even higher shear rates. The deeper the plunger is in the channel, the faster the fluid flows and the higher the shear rate. The advantage of the homogenizer over the colloid mill is that there are no fast moving parts in the milling zone. Homogenizers have nearly completely replaced colloid mills.

Media Mills

Media mills stir shot or rods through the dispersion to break aggregates by impact. The simplest of the media mills is the ball mill. It may be as simple as a glass jar partially filled with steel shot and placed on a roller. A pilot and manufacturing media mill is steel and is driven by a variable speed motor.

The milling comes from the mass of balls cascading through the dispersion. If the rotation is too fast, the media just carry over and do not cascade. If the rotation is too slow, the media just slide across the bottom and do not cascade. The proper speed is selected by listening to the mill. When the media are cascading properly, they can be heard. The proper rate of rotation varies with the viscosity of the dispersion. Since the viscosity of the dispersion varies with particle size and temperature, the rotation speed needs to be adjusted during the milling run, which also means that multiple samples cannot be put on a laboratory roller mill with the hope that all samples will get equivalent milling.

Mills for High Viscosities

The high speed stirrers, high shear mills, and impact mills only work with low viscosity dispersions. Some dispersions, such as rubber and plastic compounds, are too viscous to be made in that equipment. For these applications heavy-duty mixers, such as three-roll mills, double-blade mixers (like dough mixers), kneaders, and screw mixers of various designs, are used. Although the dispersions produced are adequate, none of these pieces of equipment produces a fine, submicron dispersion. That requires high shear rates unobtainable when the viscosity is too high.

Special Methods

Many dispersions, especially latex or polymeric dispersions are prepared synthetically. That is, the particles are formed *in situ* rather than by attrition. The particles may be formed by controlled nucleation and growth or the particles may be formed by dispersion or emulsion polymerization.

CHEMICAL PROCESSING AIDS

Chemical processing aids are used to improve the wetting of the dry powder by a liquid, to stabilize newly created surfaces during milling, to stabilize particles against flocculation on long time scales, to induce flocculation when needed, to improve dispersability in polymers, and to modify the flow of dispersions. Each of these processes requires solutes to move to the particle/liquid interface, so these aids are called surface-active solutes or, more commonly, surfactants.

The first criterion for a chemical processing aid is that it accumulates at the particle/liquid interface and does this even in the presence of other materials. This process is called preferential adsorption. A solute can only be preferentially adsorbed when it displaces other molecules from the interface, including the solvent. In this sense, preferential adsorption is a kind of insolubility.

Work is required to create new particle/liquid interfaces during milling. If nothing changes after new surfaces are created, they recombine and nothing is gained. Surface-active solutes are adsorbed at the particle/liquid interface. If the adsorption of surface-active solute is slight, the particles recombine slowly. If the adsorption is substantial, the particles may not recombine for a very long time, if at all.

Solutes can be graded in terms of how hydrophobic they are and how hydrophilic they are. Those that stabilize water-in-oil emulsions (lipophilic solutes) are assigned low numbers. These materials also stabilize dispersions in organic solvents, polymers, and plastics. Those that stabilize oil-in-water emulsions (hydrophilic solutes) are assigned high numbers. These materials also stabilize water dispersions. The scale is called the HLB scale for hydrophilic–lipophilic balance.

Flocculants are also surface-active solutes, but they are designed to bind particles by bridging them. At excessive concentrations, flocculant molecules cover all bare surfaces, particle bridging no longer occurs, and flocs do not form.

Donor–Acceptor Interactions

Adsorption of solutes from solution by a particle surface is a complex process. Energy is required to displace solvent from both the surface-active solute and the particle surface. Energy is gained when the surface-active solute is adsorbed and the released solvent molecules mix. Adsorption is spontaneous only if the free energy decreases. If the particle is strongly wetted by the solvent or the solute is very soluble, adsorption is minimal.

Understanding and predicting which molecules are adsorbed requires considering the balance of acid–base interactions between all three components: the surface, the solute, and the solvent.

Creation of Surface Charge

Some surface-active solutes are also ionic: these are called ionic. The surface-active solutes whose anions are adsorbed are called anionic surfactants. Anionic surfactants create negatively charged particles. The surface-active solutes whose cations adsorb are called cationic surfactants. Cationic surfactants create positively charged particles. Those surface-active solutes that do not dissociate into ions are called nonionic. Nonionic surfactants do not confer charge to particles.

Creation of Steric Barriers

When molecules are adsorbed by a solid surface, they coat the surface. Small, uncharged molecules on the particle surface have little effect on whether particles stick during collisions because the strength of the dispersion force attraction is strong at short separations. However, if the molecule is large, ie, polymeric, then the surface coating may be thick enough to keep particles far enough apart that dispersion forces do not hold them together. This produces a stable dispersion. The adsorbed polymer film is called a steric barrier and the resulting stabilization is called steric stabilization.

The essential requirement for a good steric stabilizer is that it forms a thick adsorbed film. Homopolymers are not good steric stabilizers since, if they are adsorbed, they will adsorb flat against the surface. The exceptions are some naturally occurring polymers whose molecular weights are so high that the time required for them to uncoil and lay flat is long.

Copolymers are better steric stabilizers, but the most efficient steric barriers are obtained by grafting the polymer to the particle surface.

Selection of Dispersion Aids

The general principles are understood: The surface-active solute needs to adsorb to the particle surface. It must have a stronger affinity for the particle than the solvent. If the surface-active solute is ionic, then enough must adsorb to produce a high enough electric charge to keep the particles apart. If the surface-active solute is nonionic, then it must have a high enough molecular weight to form a thick enough surface coating to keep particles apart.

A good start at selecting dispersing aids is from the experience of colleagues who have prepared similar formulations. By examining what dispersants have worked before or what ones are described in patent literature, clues are found about classes of materials to try. The important information is not the trade name but rather the chemical composition. Another source of information is McCutcheon's *Emulsifiers* & Detergents, an annual trade publication.

The Daniel flow point test is commonly used to pick efficient dispersants. The dispersant requiring the least addition to make a free-flowing paste is considered best.

Flocculants

Flocculation is the formation of clusters of particles called flocs. Flocs usually settle quickly and are easy to filter. Flocs can be redispersed by relatively weak mechanical forces. Aggregation is the formation of clusters of particles that cannot be redispersed except with strong mechanical forces.

One method to induce flocculation is to remove the stabilizing mechanism, not necessarily by removing the dispersant. If the particles are stabilized by electric charge, then neutralizing the charge by adsorbing ions of the opposite charge flocculates the dispersion. A dispersion may be flocculated by adding enough salt to shield the electric fields around charged particles. This shielding allows particles to approach each other closely enough to be trapped by dispersion forces.

A key factor in the use of flocculants is control of the dosage. The optimum dosage is lower than the dosages used for dispersants. Polymer flocculants work when they are able to connect bare surfaces on two particles. If their concentration is too high, they cover the particle surfaces and no bridges are formed.

A general guide in the evaluation of both flocculants and dispersants is to test over a wide range of concentrations, especially for polymers. At too low a concentration, many dispersants can be adsorbed on two particles simultaneously and act as flocculants. Similarly, at too high a concentration, many flocculants cover all the particle surfaces and act as dispersants.

Selective Flocculation

Selective flocculation is used to separate selected particles from mixed-particle dispersions. The original use was to separate valuable minerals, such as iron, copper and tungsten, from the gangue with which they are mixed. More recent uses are for the separation of low and high density cholesterol and in the separation of treated paper from pulp.

Flushing

Flushing is the direct transfer of dispersed particles from one liquid to another. Flushing is commonly used to transfer pigment particles from the water dispersion in which they are synthesized to an oil phase. Flushing is often done at high pigment loadings and is surprisingly sudden.

High volume loaded pigment dispersions in resins are called pigment chips. They are manufactured by dispersion into molten resin at high shear, often on two- or three-roll mills. Pigments provided as concentrated dispersions in low viscosity liquids are called master batches.

DISPERSED PARTICLES

Particle Motion

The pressure of a gas against the walls of its container is a clear indication of the motion of the gas molecules; the higher the temperature, the faster the motion. Liquid molecules are also in constant random motion, but this is not so obvious and it took the molecular thermodynamic theory of Boltzmann to explain it.

Optical Properties

Light is a traveling electric and magnetic wave. As light passes through a particle, the electric field moves the electrons in the particle. The electric field is sinusoidal so that the electrons accelerate back and forth. This oscillation of electrons produces light. The light scattered by particles is the light generated by the oscillating electrons. It is not light reflected off a particle. Most particles have little magnetic susceptibility so the interaction with the magnetic field of the light is ignored. The scattered light that is seen by eye or is measured by a detector comes from the summation of the electric fields emanating from all the illuminated electrons.

Particle Size

The most common measurement made on a dispersion is the average particle size. The most certain technique is microscopy, if applicable. No other technique approaches this level of information. The limitation of microscopy is its lower size limit.

Sizing by Flow

Velocities are not easy to measure directly. The usual procedure is to shake the dispersion, pour it into a glass cylinder, and let it settle. The concentration of the dispersion is measured at a known distance from the top of the liquid as a function of time.

Sizing by Single Particle Detection

A simple particle-sizing technique is to pass a dilute suspension through a small sensing volume so that only one particle is in the sensing zone at a time. A light is mounted opposite to an optical detector and the cross-sectional area of each particle is measured as it passes through the sensing volume. This technique is called photozone detection.

Sizing by Light Scattering

For particles less than $\sim$100-nm Rayleigh theory is used. Fraunhofer diffraction is used for particles greater than $\sim$1 μm. Rates of flocculation are easily measured by Fraunhofer diffraction.

Quasielastic light scattering (QELS) is used for particles less that $\sim$1 μm. QELS is so accurate for monodisperse spheres that it is used to measure the size of the latex standards that are used to calibrate all other techniques. As long as the scattered light intensity is strong enough, QELS can be used down to the nanoparticle range.

Sizing by Acoustic Scattering and Absorption

Light is a traveling transverse wave (the oscillations are perpendicular to the direction of propagation). Sound is a traveling longitudinal wave (the oscillations are parallel to the direction of propagation.) Light scattering depends on the electric and magnetic properties of matter. Acoustic scattering depends on the mechanical and thermal properties of matter. The similarity is that both are the scattering of waves. If the scattering of light gives particle size information so does acoustic scattering, although they depend on different material properties. The advantage of light scattering is that it depends practically only on one material property, the ratio of the refractive index of the particle to that of the medium. One advantage of acoustic scattering is that data can be taken as a function of wavelength. This means that, for any particle size, the wavelength of sound can be adjusted to bring the scattering data into a range appropriate to theory. A second advantage is particle sizing can be done on concentrated dispersions so dilution is usually not necessary. A third advantage is that the dispersions and even the containers can be opaque.

Particle Charge

Particles dispersed in water acquire charge chiefly by the preferential adsorption of surface-active ions, usually anions, or by dissociation of surface ionogenic groups.

Measurements of surface potentials in nonaqueous dispersions are no different that in aqueous dispersions. Care needs to be taken to avoid space charge, a charge accumulation in some part of the apparatus or sample, because space charge changes the electric field.

FLOW OF DISPERSIONS

Quantitative information on how a dispersion flows under various conditions is useful in optimizing manufacturing steps such as coating, draining, and pumping. Rheology (from the Greek rheos, a stream) is the study of flow. A rheological test is often used as a quality control or quality assurance test in manufacturing. The effectiveness of many products depends on proper flow, eg, cosmetics, paints, and lubricants. Rheology is the tool of choice for an analysis of the interactions of components of a dispersion.

Effect of Particle Interactions

Dispersions are mixtures of phases and it is rare that their internal structures remain the same no matter what the flow. The most common observation is a decreasing viscosity with increasing shear stress or shear rate. Such a flow is called shear thinning. This decrease in viscosity implies that internal structure is being destroyed. The usual explanation is that flocs are being broken apart. If the structure has some minimum shear stress before it starts to flow, it is said to have a yield point and the flow is called plastic.

A fourth type of rheogram is the increase in structure with flow called dilatant flow.

Effects of Polymers

Most commercially important dispersions are pigment dispersions in polymer solutions. The polymers usually have two roles, first to aid in the stabilization of the dispersion as steric surface-active solutes and second to be film formers when the formulation is dried. One polymer is often not optimum for both roles so complex formulations are developed.

The solubility of polymers is sensitive to other components, especially other polymers. The general rule is that two polymers are not both soluble. Even small molecules change the solubility of polymers since they act like a second solvent.

At low polymer concentrations, particle–particle interactions can dominate the flow. If polymers are adsorbed, they can stabilize dispersions, hence lower viscosity. If the polymers adsorb simultaneously onto two particles, they are flocculants and raise viscosity.

At high polymer concentrations, the polymer solution rheology can dominate the flow.

Time-Dependent Flow

Shear thinning and dilatancy arise from structural changes in the dispersion. The loss of structure from shear thinning rebuilds as particles reflocculate, but the recovery may be slow. Built-up structure from dilatancy takes time to dissipate. A complete picture of the flow of dispersions accounts for the time structures take to form and to relax. These times can be important practically.

Some materials do not flow steadily after stress is applied, they merely stretch and stop. This phenomenon is called elastic flow.

However, the structures in dispersions are brittle, that is, they break with only a small strain, so the experiment is difficult. A useful technique generates a small-amplitude torsional shear strain at one point in a dispersion and measures the speed of propagation of the shear wave through the dispersion.

STABILITY OF DISPERSIONS

Dispersed particles collide with each other frequently because of Brownian motion. If the particles stick when they collide, the dispersion is called unstable and the process is called flocculation. If the particles separate after a collision then the composition of the dispersion remains constant and the dispersion is called stable.

MODEL DISPERSIONS

Dispersions in which all the particles are the same size are called monodisperse. Monodisperse dispersions are important for testing theories but also for their interesting optical and electronic properties. Charged, monodisperse dispersions, carefully prepared at low ionic strength and high volume loadings, settle into well-ordered structures.

A. W. Adamson and A. P. Gast, *The Physics and Chemistry of Surfaces*, 6th ed., John Wiley & Sons, Inc., New York, 1997.

I. D. Morrison and S. Ross, *Colloidal Dispersions: Suspensions, Emulsions, and Foams*, John Wiley & Sons, Inc., New York, 2002.

P. C. Hiemenz and R. Rajagopalan, *Principles of Colloid and Surface Chemistry*, 3rd ed., Marcel Dekker, New York, 1997.

D. Myers, *Surfaces, Interfaces, and Colloids: Principles and Applications*, 3rd ed., John Wiley & Sons, Inc., Hoboken, N.J., 2006.

R. M. Pashley and M. E. Karaman, *Applied Colloid and Surface Chemistry*, John Wiley & Sons, Inc., Hoboken, N.J., 2004.

IAN MORRISON
Cabot Corporation

DISTILLATION

Distillation is a method of separation that is based on the difference in composition between a liquid mixture and the vapor formed from it. This composition difference arises from the dissimilar effective vapor pressures, or volatilities, of the components of the liquid mixture. Distillation as normally practiced involves condensation of the vaporized material, usually in multiple vaporization/ condensation operations, and thus differs from evaporation (qv), which is usually applied to separation of a liquid from a solid but which can be applied to simple liquid concentration operations.

Distillation is the most widely used industrial method of separating liquid mixtures and is at the heart of the separation processes in many chemical and petroleum plants (see SEPARATION PROCESS SYNTHESIS). The most elementary form of the method is simple distillation, in which the liquid is brought to boiling and the vapor formed is separated and condensed to form a product. If the process is continuous with respect to feed and product flows, it is called flash distillation. If the feed mixture is available as an isolated batch of material, the process is a form of batch distillation and the compositions of the collected vapor and residual liquid are thus time dependent. The term fractional distillation, which may be contracted to fractionation, was originally applied to the collection of separate fractions of condensed vapor, each fraction being segregated. In modern practice the term is applied to distillation processes in general, where an effort is made to separate an original mixture into several components by means of distillation. When the vapors are enriched by contact with counterflowing liquid reflux, the process is often called rectification. When fractional distillation is accomplished with a continuous feed of material and continuous removal of product fractions, the process is called continuous distillation. When steam (qv) is added to the vapors to reduce the partial pressures of the components to be separated, the term steam distillation is used.

Most distillations conducted commercially operate continuously, with a more volatile fraction recovered as distillate and a less volatile fraction recovered as bottoms or residue. If a portion of the distillate is condensed and returned to the process to enrich the vapors, the liquid is called reflux. The apparatus in which the enrichment occurs is usually a vertical, cylindrical vessel called a still or distillation column. This apparatus normally contains internal devices for effecting vapor–liquid contact; the devices may be categorized as plates or packings.

VAPOR–LIQUID EQUILIBRIA

The equilibrium distributions of mixture component compositions in the vapor and liquid phases must be different if separation is to be made by distillation. It is important, therefore, that these distributions be known. The compositions at thermodynamic equilibrium are termed vapor–liquid equilibria (VLE) and may be correlated or predicted with the aid of thermodynamic relationships. The driving force for any distillation is a favorable vapor–liquid equilibrium, which provides the needed composition differences. Reliable VLE are essential for distillation column design and for most other operations involving liquid–vapor phase contacting.

Thermodynamic Relationships

A closed container with vapor and liquid phases at thermodynamic equilibrium may be depicted as in Figure 1, where at least two mixture components are present in each phase. The components distribute themselves between the phases according to their relative volatilities. A distribution ratio for mixture component i may be defined using mole fractions:

$$K_i = y_i^{*}/x_i \qquad (1)$$

where the asterisk is used to denote an equilibrium condition. This K term, known as the vapor–liquid equilibrium ratio, or often the K value, is widely used, especially in the

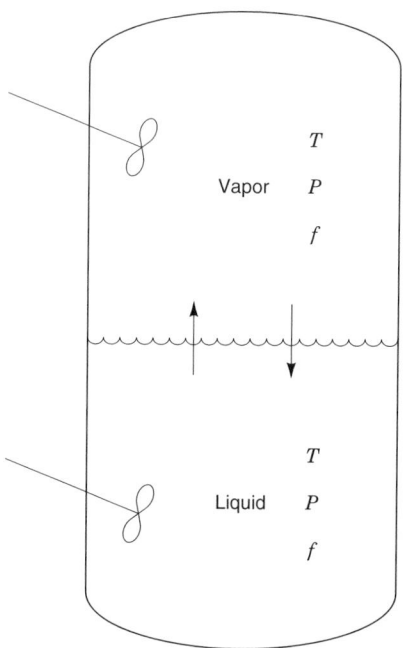

Figure 1. Equilibrium between vapor and liquid. The conditions for equilibrium are $T^{V} = T^{L}$ and $P^{V} = P^{L}$. For a given T and P, phase fugacities are equal, ie, $f^{V} = f^{L}$ and $f_i^{V} = f_i^{L}$.

petroleum (qv) and petrochemical industries. For any two mixture components i and j, their relative volatility, often called the alpha value, is defined as

$$\alpha_{ij} = \frac{K_i}{K_j} = \frac{y_i x_j}{x_i y_j} = \frac{y_i(1-x_i)}{x_i(1-y_i)} \quad (2)$$

The relative volatility, α, is a direct measure of the ease of separation by distillation. If $\alpha = 1$, then component separation is impossible, because the liquid- and vapor-phase compositions are identical. Separation by distillation becomes easier as the value of the relative volatility becomes increasingly greater than unity.

When both phases form ideal thermodynamic solutions, ie, no heat of mixing, no volume change on mixing, etc, Raoult's law applies:

$$p_i^V = x_i P_i^0 \quad (3)$$

where P_i^0 is the vapor pressure of i at the equilibrium temperature. Combining this expression with Dalton's law of partial pressures, K values and relative volatilities may be obtained:

$$K_i = P_i^0/P \quad (4)$$

$$\alpha_{ij} = P_i^0/P_j^0 \quad (5)$$

The development and thermodynamic significance of activity coefficients is discussed in most chemical engineering thermodynamics texts. The liquid-phase coefficients are strong functions of liquid composition and temperature and, to a lesser degree, of pressure. A system with positive deviation, ie, the two components having activity coefficients greater than one such that the logarithm of the coefficient is positive, is shown in Figure 2**a**; a system with negative deviation, the coefficients are less than unity and logarithms negative, is shown in Figure 2**b**. In a few cases one component of a binary mixture has a positive deviation and the other a negative deviation. Most commonly, however, both coefficients have positive deviations.

Terminal activity coefficients, γ_i^∞, are noted in Figure 2. These are often called *infinite dilution coefficients* and for some systems are given in Table 1.

If the molecular species in the liquid tend to form complexes, the system will have negative deviations and activity coefficients less than unity, eg, the system chloroform–ethyl acetate. In azeotropic and extractive distillation (see DISTILLATION, AZEOTROPIC AND EXTRACTIVE) and in liquid–liquid extraction, nonideal liquid behavior is used to enhance component separation (see EXTRACTION, LIQUID–LIQUID).

Azeotropic Systems

An azeotropic mixture is one that vaporizes without any change in composition. In homogeneous azeotropic systems, Positive activity coefficients tend to produce minimum boiling azeotropes, and negative coefficients tend to produce maximum boiling azeotropes.

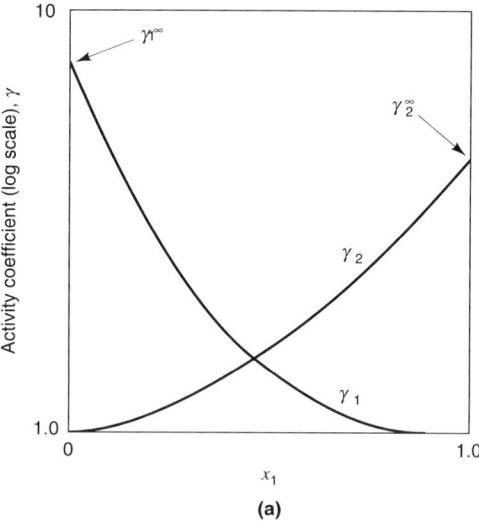

(a)

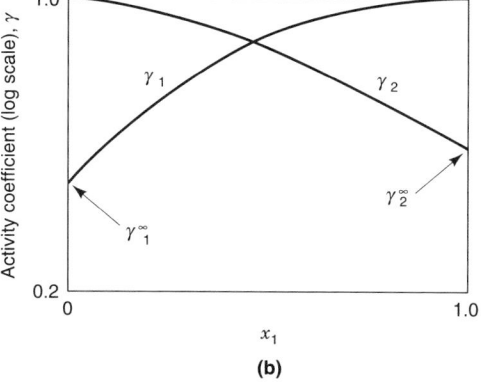

(b)

Figure 2. Binary activity coefficients for two component systems having (**a**) positive and (**b**) negative deviations from Raoult's law. Conditions are either constant pressure or constant temperature and terminal coefficients, γ_i^∞, are noted.

Heterogeneous azeotropes are formed when the positive activity coefficients are sufficiently large to produce two liquid phases that exist at the boiling point, and a constant boiling mixture that is formed at some composition, generally within the liquid immiscibility composition range.

Table 1. Terminal Activity Coefficients at Atmospheric Pressure[a]

Component 1	Component 2	γ_1^∞	γ_2^∞
chloroform	ethyl acetate	0.3	0.3
chloroform	benzene	0.9	0.7
n-hexane	n-heptane	1.0	1.0
ethyl acetate	ethanol	2.5	2.5
ethanol	toluene	6.0	6.0
benzene	methanol	9.0	9.0
ethanol	isooctane	11.0	8.0
methyl acetate	water	20.0	7.0
ethyl acetate	water	100.0	15.0
hexane	water	>100.0	>100.0

[a]Values are approximate.

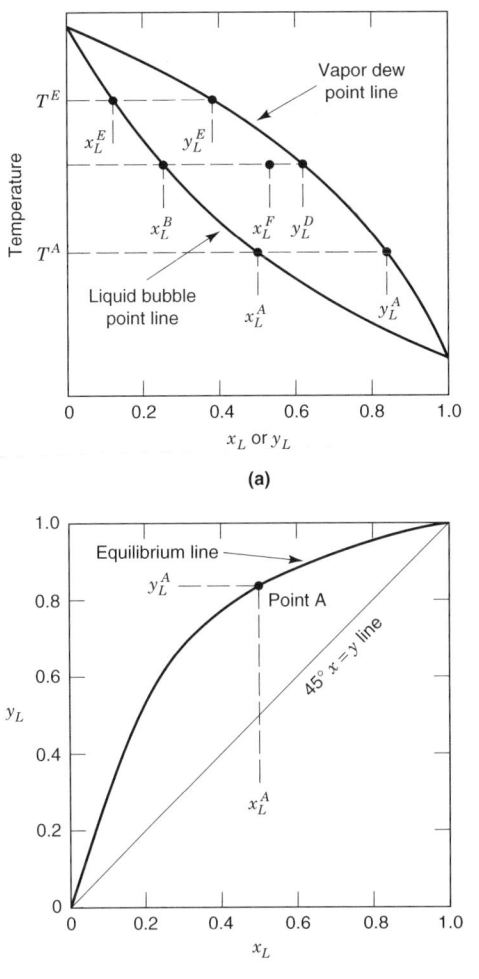

(a)

(b)

Figure 3. Isobaric VLE diagrams: (**a**) dew and bubble point; (**b**) vapor–liquid (y–x) equilibrium.

DISTILLATION PROCESSES

For ease of presentation and understanding, the initial discussion of distillation processes involves binary systems. Examining the binary boiling point (Fig. 3**a**) and phase (Fig. 3**b**) diagrams, the enrichment from liquid composition x_L to vapor composition y_L represents a theoretical step, or equilibrium stage.

Simple Distillations

Simple distillations utilize a single equilibrium stage to obtain separation. Simple distillation, also called differential distillation, may be either batch or continuous, and may be represented on boiling point or phase diagrams. In Figure 3**a**, if the batch distillation begins with a liquid of composition x_L^A the initial distillate vapor composition is y_L^A. As the distillate is removed, the remaining liquid becomes less rich in the low boiler, L, and the boiling liquid composition moves to the left along the bubble point line. If the distillation is contin-

ued until the liquid has a composition of x_L^E, the last vapor distillate has a composition of y_L^E. Simple batch distillation is not widely used in industry, except for the processing of high-valued chemicals in small production quantities, or for distillations requiring regular sanitization.

Simple continuous distillation, also called flash distillation, has a continuous feed to a single equilibrium stage; the liquid and vapor leaving the stage are considered to be in phase equilibrium. On the boiling point diagram (Fig. 3a), the feed is represented by x_L^F, the bottoms liquid by x_L^B, and the equilibrium vapor distillate by y_L^D. The overall mass balance is $F = D + B$ and the component L balance is $x_L^F F = y_L^D D + x_L^B B$. Flash distillations are widely used where a crude separation is adequate.

Multiple Equilibrium Staging

The component separation in simple distillation is limited to the composition difference between liquid and vapor in phase equilibrium. To overcome this limitation, multiple equilibrium staging is used to increase the component separation. Figure 4 schematically represents a continuous distillation that employs multiple equilibrium stages stacked one upon another. The feed, F, enters the column at equilibrium stage f. The heat q^s required for vaporization is added at the base of the column in a reboiler or calandria. The vapors V^T from the top of the column flow to a condenser from which heat q^c is removed. The

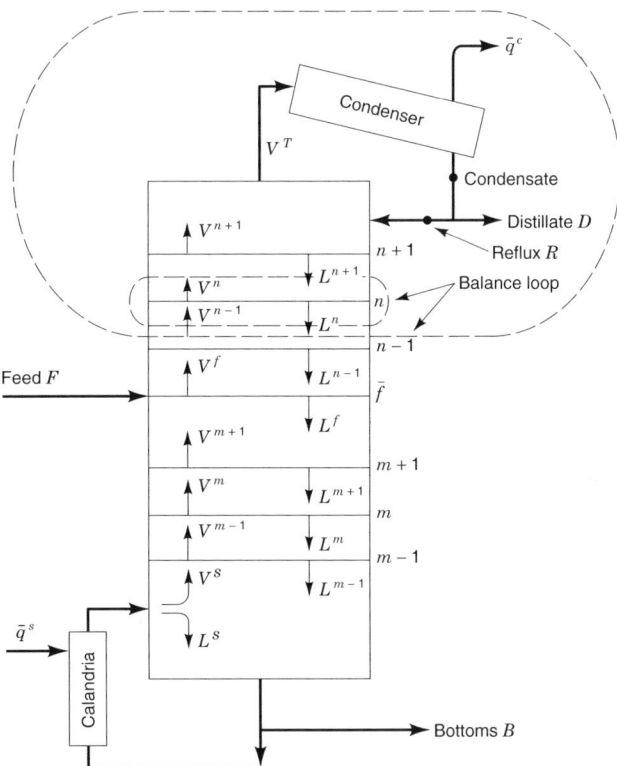

Figure 4. Distillation column with stacked multiple equilibrium stages. Terms are defined in text.

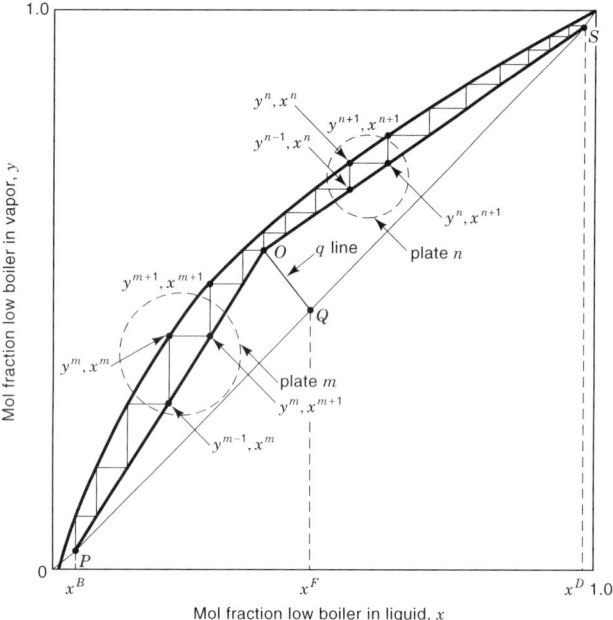

Figure 5. McCabe–Thiele diagram. Terms are defined in text.

liquid condensate from the condenser is divided into two streams: the first, a distillate D, which is the overhead product (sometimes called heads or make), is withdrawn from the system, and the second, a reflux R, which is returned to the top of the column. A bottoms stream B is withdrawn from the reboiler. The overall separation is represented by feed F separating into a distillate D and a bottoms B.

The graphical McCabe–Thiele design method facilitates a visualization of distillation principles while providing a solution to the material balance and equilibrium relationships. Here, the subscripts L and H are not used, and x and y refer to the lower boiler, ie, more volatile component, in the binary system. A McCabe–Thiele diagram is given in Figure 5 where P, Q, and S are the x^B, x^F, and x^D compositions on the $y = x$, $45°$ construction line, respectively. Line OP is the stripping operating line and line OS is the rectifying operating line.

The McCabe–Thiele method employs the simplifying assumption that the molal overflows in the stripping and the rectification sections are constant. The method is based on the simplifying assumption that the molal overflow is constant in both the rectifying and stripping sections. For many problems this assumption is not valid and more precise calculations are necessary. For the more general case, detailed enthalpy balances are made around individual stages or groups of stages. Standard distillation texts discuss the internal enthalpy calculations by algebraic balances or by graphical procedures; eg, the stage-to-stage mass and enthalpy balances with equilibrium calculations and also by means of the graphical Ponchon–Savarit procedure. Hand algebraic and graphical methods requiring internal enthalpy calculations have been largely superseded by simulations performed on modern computing

devices, including personal computers (see COMPUTER TECHNOLOGY).

There are infinite combinations of reflux ratios and numbers of theoretical stages for any given distillation separation. The larger the reflux ratio, the fewer the theoretical stages required. For any distillation system with its given feed and its required distillate and bottoms compositions, there are two constraints within which the variables of reflux ratio and number of theoretical stages must lie: the minimum number of theoretical stages and the minimum reflux ratio. The minimum reflux ratio occurs when the reflux ratio is reduced so that the upper and lower operating lines and the q line are coincident at a single point on the equilibrium line. When this condition exists, an infinite number of theoretical stages would be required to make the separation. The minimum number of theoretical stages occurs when the system is at total reflux: no feed, distillate, or bottoms. In this case, the two operating lines are coincident with the digonal line.

Simple analytical methods are available for determining minimum stages and minimum reflux ratio. Although developed for binary mixtures, they can often be applied to multicomponent mixtures if the two key components are used. These are the components between which the specification separation must be made; frequently the heavy key is the component with a maximum allowable composition in the distillate and the light key is the component with a maximum allowable specification in the bottoms. On this basis, minimum stages may be calculated by means of the Fenske relationship:

$$N_{\min} = \frac{\ln[(y_i/y_j)D(x_j/x_i)_B]}{\ln \alpha_{ij,\mathrm{avg}}} \qquad (6)$$

where i and j are the light and heavy components of a binary mixture, or the light key and heavy key in a multicomponent mixture.

For minimum reflux ratio, the following equations may be used:

$$\sum_i \frac{\alpha_i x_{if}}{\alpha_i - \phi} = 1 - q \qquad (7)$$

$$\sum_i \frac{\alpha_i (x_{id})}{\alpha_i - \phi} = R_{\min} + 1 \qquad (8)$$

where the value of q is determined as in the McCabe–Thiele procedure. Equation 7 is solved for root ϕ, the value of which must lie between 1.0 and the light key volatility. The root value so determined is then used in equation 8 to obtain the value of $R_{\min}$.

Both of these limits, the minimum number of stages and the minimum reflux ratio, are impractical for useful operation, but they are valuable guidelines within which the practical distillation must lie. A representative plot of the number of theoretical stages vs reflux ratio for some distillation separation is shown in Figure 6. Both minimum limits may be calculated for any distillation, thereby bracketing the practical design. Actual operating

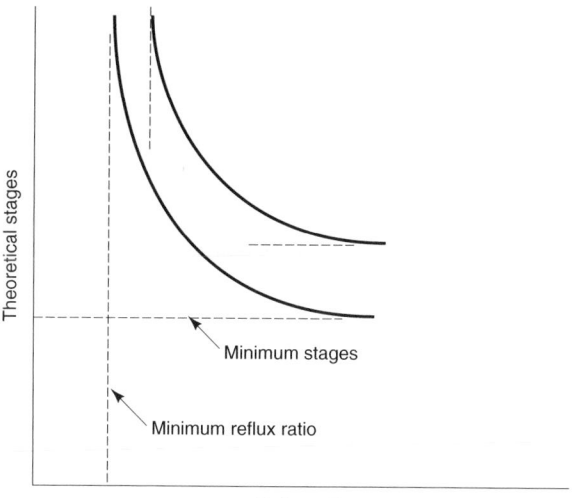

Figure 6. Representative plot of theoretical stages vs reflux ratio for a given separation. Each curve is the locus of points for a given separation. Note the limiting conditions of minimum reflux and minimum stages.

reflux ratios for most commercial columns are in the range of 1.1 to 1.5 times the minimum reflux ratio.

The operating, fixed, and total costs of a distillation system are functions of the relation of operating reflux ratio to minimum reflux ratio. Figure 7 shows a typical plot of costs; as the operating to minimum reflux ratio increases, the operating cost (principally energy cost for the boil-up) increases almost linearly. Similarly, the fixed costs at first decrease from the infinite number of stages, pass through a minimum, and then increase again as the diameter of column increases with increased

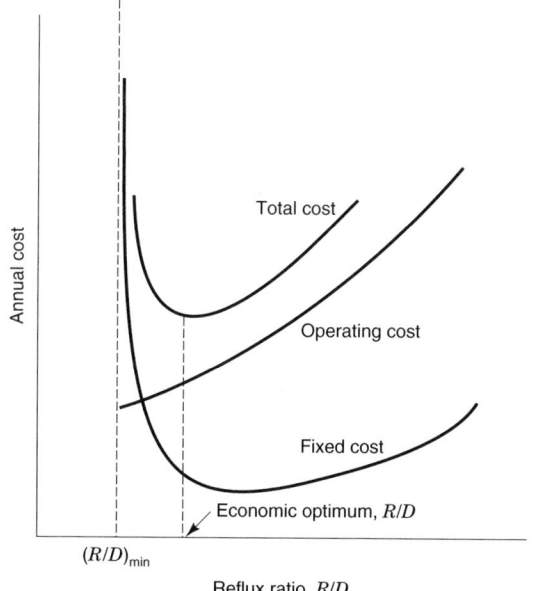

Figure 7. Fixed, operating, and total costs of a typical distillation, as a function of reflux ratio.

reflux ratio. These costs for typical distillations have been calculated; the ratio of the economic optimum reflux to the minimum reflux is often 1.2 or less.

The calculations that determine the reflux and stage requirements are more difficult to make for multicomponent systems than for binary systems. When the concentration of a component in the distillate and in the bottoms is specified for the overall solution of a binary distillation, the component balance around the column also is completely specified. In the multicomponent case, only a single high-boiling key component can be specified in the distillate and a single low-boiling key component in the bottoms; the split of other components can be determined only by detailed calculations. These require a series of trial and error computations to obtain the solution at any given reflux ratio and number of stages. As the number of components and number of stages become large, the mathematical problem becomes formidable. Two approaches may be followed: use of approximate, ie, shortcut, methods, or use of a suitable computer program.

Multiple Products

If each component of a multicomponent distillation is to be essentially pure when recovered, the number of columns required for the distillation system is N^*-1, where N^* is the number of components. Thus, in a five-component system, recovery of all five components as essentially pure products requires four separate columns. However, those four columns can be arranged in 14 different ways.

The number of columns in a multicomponent train can be reduced from the N^*-1 relationship if side-stream draw-offs are used for some of the component cuts. The feasibility of multicomponent separation by such draw-offs depends on side-stream purity requirements, feed compositions, and equilibrium relationships.

DISTILLATION COLUMNS

Distillation columns are vertical, cylindrical vessels containing devices that provide intimate contacting of the rising vapor with the descending liquid. This contacting provides the opportunity for the two streams to achieve some approach to thermodynamic equilibrium. Depending on the type of internal devices used, the contacting may occur in discrete steps, called plates or trays, or in a continuous differential manner on the surface of a packing material. The fundamental requirement of the column is to provide efficient and economic contacting at a required mass-transfer rate. Individual column requirements vary from high vacuum to high pressure, from low to high liquid rates, from clean to dirty systems, and so on. As a result, a large variety of internal devices has been developed to fill these needs.

Packed vs Plate Columns

Relative to plate towers, packed towers are more useful for multipurpose distillations, usually in small (under 0.5 m) towers or for the following specific applications:

severe corrosion environment where some corrosion-resistant materials, such as plastics, ceramics, and certain metallics, can easily be fabricated into packing but may be difficult to fabricate into plates; vacuum operation where a low pressure drop per theoretical plate is a critical requirement; high (e.g., above 49,000 kg/(h m²) [~10,000 lb/(h ft²)] liquid rates; foaming systems; or debottlenecking plate towers having plate spacings that are relatively close, under 0.3 m.

Plate columns have the advantage of lower fabrication cost, less dependence on good liquid and gas distribution, and protection against vapor bypassing the liquid in critical zones, eg, regions of extremely low impurities. Further, methods for the design on plate columns are somewhat more reliable than those for many of the packings, especially those packings of a proprietary nature.

There are notable cases where plate columns have been converted to packed columns to gain advantage of the low pressure drop exacted from the vapor stream. More recently the packings have been largely of the structured type. Illustrative of this is the trend toward the use of structured packing in ethylbenzene–styrene fractionators, some of which have diameters of 10 m or higher.

MOLECULAR DISTILLATION

Molecular distillation occurs where the vapor path is unobstructed and the condenser is separated from the evaporator by a distance less than the mean-free path of the evaporating molecules. This specialized branch of distillation is carried out at extremely low pressures ranging from 13–130 mPa (0.1–1.0 µm Hg) (see VACUUM TECHNOLOGY). Molecular distillation is confined to applications where it is necessary to minimize component degradation by distilling at the lowest possible temperatures. Commercial usage includes the distillation of vitamins (qv) and fatty acid dimers (see DIMER ACIDS).

Distillation as a Separation Method

Distillation is the most important industrial method of separation and purification of liquid components. Liquid separation methods in less common use include liquid–liquid extraction (see EXTRACTION, LIQUID–LIQUID), membrane diffusion (see DIALYSIS; MEMBRANE TECHNOLOGY), ion exchange (qv), andadsorption (qv). However, distillation does not require a mass-separating agent such as a solvent, adsorbent, or membrane, and distillation utilizes energy in a convenient heating medium (often steam). Also, a wealth of experience with design and operations makes distillation column performance prediction more reliable than equivalent predictions for other methods. At times distillation also competes indirectly with methods involving solid–liquid separations such as crystallization (qv). An extensive discussion of the selection of alternative separation methods is available (see SEPARATION SYSTEMS SYNTHESIS).

The suitability and economics of a distillation separation depend on such factors as favorable vapor–liquid equilibria, feed composition, number of components to be separated, product purity requirements, the absolute pressure of the distillation, heat sensitivity, corrosivity, and continuous vs batch requirements. Distillation is somewhat energy inefficient because in the usual case heat added at the base of the column is largely rejected overhead to an ambient sink. However, the source of energy for distillations is often low-pressure steam, which characteristically is in long supply and thus relatively inexpensive. Also, schemes have been devised for lowering the energy requirements of distillation and are described in many publications.

COLUMN CONTROL

Distillation columns are controlled by hand or automatically. The parameters that must be controlled are (1) the overall mass balance, (2) the overall enthalpy balance, and (3) the column operating pressure. Modern control systems are designed to control both the static and dynamic column and system variables.

J. R. Fair, in R. W. Rousseau, ed., *Handbook of Separation Process Technology*, John Wiley, New York, 1987, Chapt. 5.

J. R. Fair, in Y. A. Liu, H. A. McGee, and W. R. Epperly, eds., *Recent Developments in Chemical Process and Plant Design*, John Wiley & Sons, Inc., New York, 1987, Chapt. 3.

H. Z. Kister, *Distillation—Design*, McGraw-Hill, Inc., New York, 1992.

P. A. Schweitzer, ed., *Handbook of Separation Techniques for Chemical Engineers*, 3rd ed., McGraw-Hill Book Co., Inc., New York, 1997.

JAMES R. FAIR
The University of Texas at Austin

DISTILLATION, AZEOTROPIC, AND EXTRACTIVE

Distillation (qv) is the most widely used separation technique in the chemical and petroleum industries. Not all liquid mixtures are amenable to ordinary fractional distillation, however. Close-boiling and low relative volatility mixtures are difficult and often uneconomical to distill, and azeotropic mixtures are impossible to separate by ordinary distillation. Yet such mixtures are quite common and many industrial processes depend on efficient methods for their separation (see also SEPARATIONS PROCESS SYNTHESIS).

Vapor–liquid phase equilibrium (VLE) in a c-component mixture can be represented by

$$y_i \phi_i^V P = x_i \gamma_i P_i^{\text{sat}} \phi_i^{\text{sat}} \exp \int_{P_i^{\text{sat}}}^{P} \frac{V_i^L dP}{RT} \quad \text{for} \quad i = 1, \ldots, c$$

where y_i is the mole fraction of component i in the vapor phase; ϕ_i^V is the vapor-phase fugacity coefficient of component i; P is the total system pressure; x_i is the mole fraction of component i in the liquid phase; γ_i is the liquid-phase activity coefficient of component i; P_i^{sat} is the vapor pressure of component i; ϕ_i^{sat} is the saturation fugacity coefficient of component i; and the exponential term is the Poynting correction factor (Poy_i). At low to moderate pressures, the value of the term $\phi_i^{\text{sat}}/\phi_i*Poy_i$ is typically close enough to unity that it can be ignored and equation 1 reduces to

$$y_i P = x_i \gamma_i P_i^{\text{sat}} \qquad \text{for} \quad i = 1, \ldots, c \qquad (2)$$

For simplicity, this is the VLE equation that will be used throughout the rest of this article.

The activity coefficient (γ_i) is a measure of the liquid-phase nonideality of a mixture and its value varies with both temperature and composition. When $\gamma_i = 1$ the liquid phase is said to form an ideal solution and equation 2 reduces to Raoult's law. Nonideal mixtures ($\gamma_i \neq 1$) can exhibit either positive ($\gamma_i > 1$) or negative ($\gamma_i < 1$) deviations from Raoult's law. Positive deviations are more common and occur when the molecules of the different compounds in the solution are dissimilar and have no preferential interactions between them. Negative deviations occur when there are preferential attractive forces (hydrogen bonds, etc.) between the molecules of the different species that do not occur in the absence of the other species. If these deviations are large enough, the pressure-composition (P-x,y) and temperature-composition (T-x,y) phase diagrams exhibit a minimum or maximum point. At these minima and maxima the liquid phase and its equilibrium vapor phase have the same composition, ie,

$$y_i = x_i \qquad \text{for} \quad i = 1, \ldots, c \qquad (3)$$

the mixture boils at constant temperature, and the dew-point (vapor) and bubble-point (liquid) curves are tangent with zero slope. These are the defining conditions for a homogeneous azeotrope where a single liquid phase is in equilibrium with a vapor phase.

Partially miscible mixtures are more nonideal than completely miscible mixtures and thus are more likely to form azeotropes. Usually, these will be heterogeneous azeotropes where two (or more) liquid phases are in equilibrium with a vapor phase, the mixture boils at constant temperature and at constant composition, but the composition of the equilibrium phases are all different. The *overall* liquid-phase composition is, however, identical to the vapor-phase composition. Since positive deviations from Raoult's law are necessary for liquid phase immiscibility, only minimum boiling heterogeneous azeotropes can exist. When the immiscibility exists only over a limited composition range, the azeotrope can lie outside of the two-liquid-phase region, giving rise to a homogeneous azeotrope.

Separation by distillation depends on the vapor and liquid phases having different compositions when a liquid (vapor) mixture is partially vaporized (condensed). The vapor phase becomes enriched in the more volatile components and depleted in the less volatile components while the opposite occurs in the equilibrium liquid phase. By successively repeating these partial vaporizations and condensations in a countercurrent cascade, it is often possible to achieve the desired degree of separation. A common measure of the degree of enrichment or the ease of separation is the relative volatility defined as:

$$\alpha_{ij} = \frac{y_i x_j}{x_i y_j} = \frac{\gamma_i P_i^{\text{sat}}}{\gamma_j P_j^{\text{sat}}} \qquad (4)$$

Most methods for distilling azeotropic and low relative volatility mixtures rely on the addition of specially chosen chemicals or "mass separating agents" to facilitate the separation. These separating agents can be divided into distinct classes that define the principal distillation techniques used to separate such mixtures. The five categories of distillation-based methods for separating low relative volatility and/or azeotropic mixtures are (1) extractive distillation and homogeneous azeotropic distillation where the liquid separating agent is completely miscible. In extractive distillation, the separating agents, variously known as solvents, extractive agents, extractants, and sometimes entrainers, are high boiling, non-azeotrope-forming, liquids that alter the relative volatility of the mixture to be separated. (2) Heterogeneous azeotropic distillation or, more commonly, azeotropic distillation, where the liquid separating agent, called the entrainer, forms one or more azeotropes with the other components in the mixture and causes two liquid phases to exist over a broad range of compositions. This immiscibility is the key to making the distillation sequence work. (3) Salt-effect distillation, which is a variation of extractive distillation, where the separating agent is an ionic salt. The salt dissociates in the liquid mixture and alters the relative volatilities to make the separation possible. (4) Pressure-swing or pressure-sensitive distillation where a series of distillation columns operating at different pressures is used to separate mixtures containing azeotropes whose composition changes appreciably over a moderate pressure range. (5) Reactive distillation where the separating agent reacts preferentially and reversibly with one of the components in the mixture. The reaction product is then distilled from the nonreacting components and the reaction is reversed to recover the initial component.

Distillation can also be combined with other separation methods such as liquid–liquid extraction (see EXTRACTION, LIQUID–LIQUID), absorption, adsorption/molecular sieves (qv), melt crystallization (qv), or membrane-permeation-based methods like pervaporation to separate azeotropic mixtures. Ordinary fractional distillation is used for the separation over the composition range where the relative volatility is appreciable and the separation is easy. One of the other separation techniques, whose efficacy does not depend on relative volatility, is then used to achieve the separation over the "pinched" (low relative volatility) or azeotropic composition range where distillation would be difficult or impossible.

RESIDUE CURVE MAPS

The most basic form of distillation, called simple distillation, is a process in which a multicomponent liquid mixture is slowly boiled in an open pot and the vapors are continuously removed as they form. At any instant in time the vapor is in equilibrium with the liquid remaining in the still. Because the vapor is always richer in the more volatile components than the liquid, the liquid composition changes continuously with time, becoming more and more concentrated in the less volatile species. A simple distillation residue curve is a graph showing how the composition of the liquid residue in the pot changes over time. A residue curve map is a collection of residue curves originating from different initial compositions. Residue curve maps contain the same information as phase diagrams, but present it in a way that is more useful for understanding how to create a distillation sequence to separate a mixture. Simple distillation residue curves represent the liquid composition profiles of packed distillation columns operating at total (infinite) reflux.

Residue curves can only originate from, terminate at, or be deflected by the pure components and azeotropes in a mixture. Pure components and azeotropes that residue curves move away from are called unstable nodes (UN), those where residue curves terminate are called stable nodes (SN), and those that deflect residue curves are called saddles (S).

The simplest residue curve map for a ternary mixture is shown in Fig. 1. All ternary nonazeotropic mixtures, including ideal and constant relative volatility mixtures, are qualitatively represented by this map. Residue curves point in the direction of increasing temperature and must always move in such a way that the mixture boiling temperature continuously increases along every curve. (Distillation lines are typically given the opposite orientation. That is, the arrows on a distillation line point in the direction of decreasing boiling temperature. Consequently, the discussion in the text has to be reversed if

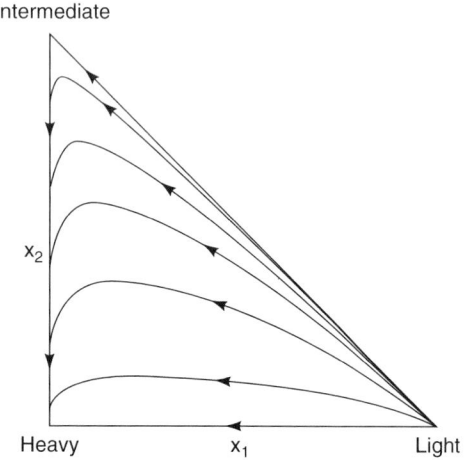

Figure 1. Residue curve map for a ternary nonazeotropic mixture.

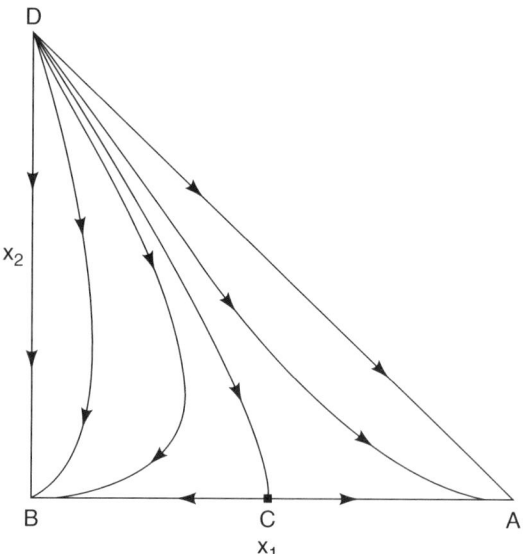

Figure 2. Residue curve map for a ternary mixture with a distillation boundary running from pure component D to the binary azeotrope C.

distillation lines are used). From this property and the direction of the arrows in Fig. 1, the light component is an unstable node; the intermediate component, which deflects the residue curves, is a saddle; and the heavy component is a stable node.

Many different residue curve maps are possible when azeotropes are present. For example, there are six possible residue curve maps for ternary mixtures containing a single azeotrope. These six maps differ by the binary pair forming the azeotrope and by whether the azeotrope is minimum or maximum boiling. Figure 2 represents the case where the intermediate and heaviest components (A and B) form a minimum boiling binary azeotrope. Pure component D is an unstable node, pure components A and B are stable nodes, the minimum boiling binary azeotrope C is a saddle, and the boiling point order from low to high is D→C→A or B. The residue curve connecting component D to the azeotrope C has the special property that it divides the composition triangle into two separate distillation regions. Any initial mixture charged to the still-pot of the simple distillation apparatus with a composition lying to the left of the curve D–C will result in the last drop of liquid being pure B; whereas any initial mixture with a composition lying to the right of the curve D–C will yield pure A as the last drop of liquid. Residue curves like D–C that divide the composition space into different distillation regions are called simple distillation boundaries, infinite reflux boundaries, separatrices, or, most commonly, distillation boundaries. The residue curve map for a binary mixture is the one-dimensional line connecting the two pure components and an azeotropic point is a boundary in the composition space that limits how far a distillation can proceed.

Residue curve maps would be of limited usefulness if they could only be generated experimentally. Fortunately that is not the case. The simple distillation process can be described by the set of equations:

$$\frac{dx_i}{d\xi} = x_i - y_i \qquad \text{for} \quad i = 1, \dots, c \qquad (3)$$

where x_i and y_i are the liquid and vapor mole fractions, respectively, of component i, ξ is a nonlinear time scale that is related to the fraction of liquid remaining in the pot at any instant in time, and c is the number of components in the mixture. Note that the pure components and azeotropes in the mixture correspond to the steady-states or singular points of equation 3 (where the derivative goes to zero) and they are all either nodes or saddles.

Even though the simple distillation process is not a practical method for separating mixtures, simple distillation residue curve maps are extremely useful. They can be used to test the consistency of experimental azeotropic data; to predict the sometimes nonintuitive order and content of the cuts from batch distillations of azeotropic mixtures; to determine whether a given mixture is separable by distillation; to help identify feasible separating agents; to predict the attainable product compositions; and to synthesize the corresponding distillation sequences. By helping to identify the limits of distillation, residue curve maps can also be useful in synthesizing separation sequences where distillation is combined with other methods.

Residue curve maps exist for mixtures having any number of components, though they cannot be visualized when there are more than four components.

HOMOGENEOUS AZEOTROPIC DISTILLATION

The most general definition of homogeneous azeotropic distillation is the distillation of any mixture containing one or more azeotropes into the desired pure component or azeotropic products without exploiting any liquid-phase immiscibilities that might be present. Thus, homogeneous azeotropic distillation includes the distillation of azeotropic mixtures in which the desired separation can be achieved without the addition of a separating agent, azeotropic mixtures to which a separating agent that may or may not form new azeotropes is added to facilitate the separation, and non-azeotropic mixtures to which an azeotrope-forming separating agent is added to more easily distill one or more of the compounds away from the others in the original mixture.

As mentioned above, simple distillation boundaries are the same as the distillation boundaries of a distillation column operated at total reflux. While the two are not equivalent at finite reflux ratios, the simple distillation boundaries remain a very good approximation to the finite-reflux distillation boundaries, except in regions where the boundaries are extremely curved. Although the composition profile of a distillation column operating at finite reflux can cross a curved simple distillation (infinite reflux) boundary to a limited degree, the resulting distillation sequences typically require large reflux (reboil) ratios and/or large recycle flowrates and, thus, are not normally

economical. Mixtures such as nitric acid–water–sulfuric acid that have extremely curved boundaries can be exceptions. A good rule-of-thumb, therefore, is that the compositions of the distillate and bottom streams from a distillation column should lie in the same simple distillation region in order for the separation to be feasible. Crossing of simple distillation boundaries should only be attempted as a last resort and only with caution, because the feasibility of such column designs can be sensitive to small inaccuracies in the VLE model and, once built, the column may be sensitive to disturbances.

An overall material balance for a continuously operated distillation column requires that the feed, distillate, and bottoms compositions lie on a straight line in the composition space (the composition triangle for ternary mixtures). Thus, a feasible distillation sequence for separating a homogeneous azeotropic mixture can be identified by determining whether or not the desired products lie in the same distillation region and then can be synthesized by superimposing material balance lines onto the distillation residue curve map. When setting the compositions of the distillate and bottom streams leaving each column in the sequence: (1) the distillate composition must have a lower boiling temperature than the bottoms composition, however the component with the lowest (highest) boiling point is not necessarily removed as the distillate (bottoms); (2) pure components and azeotropes that are nodes on the residue curve map are easier to obtain as pure products than saddles; and (3) a double-feed column is almost always required in order for a saddle on the residue curve map to be the product from a distillation column, eg, extractive distillations.

The overwhelming majority of all ternary mixtures that can potentially exist are represented by only 125 different residue curve maps. (The rare mixtures in which a pair of compounds form two binary azeotropes are exceptions). For each type of separation objective, these 125 maps can be subdivided into those that can potentially meet the objective (ie, the residue curve maps where the desired pure component and/or azeotropic products lie in the same distillation region) and those that cannot. Knowing the structure of the residue curve map for a mixture is thus sufficient to determine if a given separation objective is feasible. (There is, however, no guarantee that a feasible separation will be economical).

The seven most favorable residue curve maps for the common task of distilling a mixture containing a binary minimum-boiling azeotrope into its two constituent pure components by adding a suitable separating agent are shown in Fig. 3. (Favorable residue curve maps for separating binary maximum-boiling azeotropes can also be identified). Other maps that exploit distillation boundary curvature, namely those with a low boiling separating agent that forms a highly curved boundary ending at the azeotrope, have the potential to meet the stated objective. However, as described earlier, distillation sequences based on such maps have a number of potentially serious drawbacks. Thus, for initial screening purposes, only compounds that yield one of these seven maps need to be considered as possible

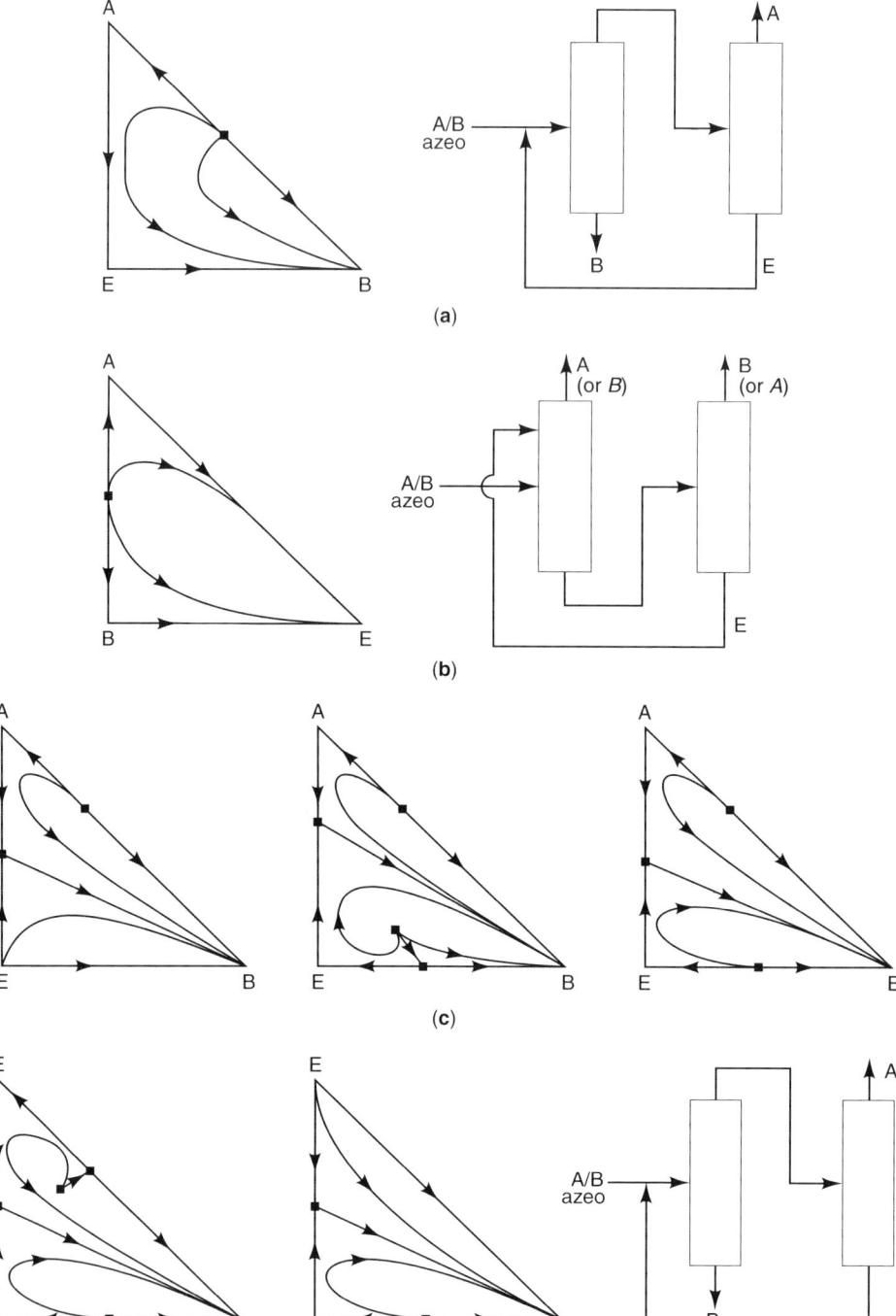

Figure 3. The seven most favorable residue curve maps and corresponding column sequences for the homogeneous azeotropic distillation of a minimum-boiling binary azeotrope using a mass separating agent E to recover the two constituent pure components A and B. The symbol ■ represents an azeotrope. (**a**) Case I, where the separating agent is intermediate boiling and does not introduce any new azeotropes: (**b**) Case II, extractive distillation with a heavy solvent that does not introduce any new azeotropes. In some cases, B will come off the top of the first column instead of A; (**c**) Case III, where the separating agent is intermediate boiling and forms a maximum boiling azeotrope with the lighter of the two pure components (ie, A). The agent may or may not form a minimum boiling azeotrope with B, and there may or may not be a minimum boiling ternary azeotrope lean in A; and (**d**) the same column configuration as case III, but the separating agent is lower boiling.

separating agents for separating binary minimum-boiling azeotropes. Five of the candidate maps require the separating agent to form a maximum-boiling azeotrope with one of the other components. Since maximum-boiling azeotropes are much less common than minimum-boiling azeotropes, these five maps will also be relatively rare.

For other separation objectives or when homogeneous azeotropic distillation is combined with other non-

distillation separation methods, additional residue curve maps can lead to successful separations and need to be considered.

When neither or only one of the components needs to be recovered as a pure product, separating agents that form azeotropes and introduce distillation boundaries that divide the pure components into different distillation regions can sometimes be used without resorting to a nondistillation step.

EXTRACTIVE DISTILLATION

Extractive distillation is defined as distillation in the presence of a miscible, higher boiling, liquid separating agent, commonly called the solvent, which does not form azeotropes with the other components in the mixture. Extractive distillation is widely used in the chemical and petrochemical industries for separating azeotropic, close-boiling, and other low relative volatility mixtures, including those whose phase diagram shows the presence of a severely "pinched" (low relative volatility) region over a limited composition range (ie, a tangent pinch).

Extractive distillation works because the solvent is specially chosen to interact differently with the components of the original mixture, thereby altering their relative volatilities. Because these interactions occur predominantly in the liquid phase, the solvent is continuously added near the top of the extractive distillation column so that it is present at an appreciable concentration in the liquid phase on all of the trays below.

Extractive distillations can be divided into three general categories: (1) the separation of minimum-boiling azeotropes, (2) the separation of maximum- boiling azeotropes, and (3) the separation of low relative volatility nonazeotropic mixtures. For separating binary mixtures into pure components, each category is represented by a single, unique residue curve map.

Minimum Boiling Azeotropes

Every extractive distillation that separates a binary minimum boiling azeotropic mixture into its pure components has a residue curve map and column sequence like the one shown in Fig. 3b.

Extractive distillations exhibit unexpected behavior at large reflux ratios. In ordinary distillations, the number of stages required to achieve a given separation decreases monotonically as the reflux ratio increases. Equivalently, in an operating distillation column where the number of stages is fixed, the distillate purity monotonically increases as the reflux ratio increases. Consequently, it is very common to increase the reflux when higher purity products are desired. In extractive distillations, however, the behavior is more complex.

Maximum Boiling Azeotropes

Maximum boiling azeotropes are far less common than minimum boiling azeotropes and so are successful extractive distillations of maximum boiling azeotropes using high boiling solvents. Adding a high boiling solvent to a mixture containing a maximum boiling azeotrope causes a distillation boundary running from the maximum boiling azeotrope to the heavy solvent, which divides the desired pure components into different distillation regions. The only way a high boiling solvent can yield an economically viable means for separating a maximum boiling azeotrope is if the resulting distillation boundary is extremely curved.

Nonazeotropic Mixtures

Extractive distillation is widely used in the petrochemical industry to separate close-boiling but nonazeotropic mixtures. Common examples include the purification of aromatics, the separation of olefins and paraffins, diolefins and olefins, paraffins and naphtalenes, styrenes and aromatics, and a number of other hydrocarbon mixtures.

The extractive distillation of any close-boiling or other low relative volatility, but non-azeotropic, mixture using a high boiling solvent is represented by the residue curve map shown in Figure 1. Although this map is different than the one for the extractive distillation of minimum boiling azeotropes, the distillation sequence is identical (see Fig. 3b) and the process works for the same reason. The solvent alters the relative volatilities of the components to be separated via liquid-phase interactions. Depending on the nature of these interactions, either the low boiling or the intermediate boiling pure component will be distilled overhead in the extractive column.

Because there is no azeotrope, these mixtures could in theory be separated without adding a solvent, though it would be a difficult and expensive separation. Consequently, there is no minimum feed ratio (minimum solvent flow). Curves relating the number of stages in the extractive column to the feed ratio end at a large but finite number of stages when the solvent flow rate goes to zero. By using the heuristics provided earlier, optimization of extractive distillation sequences for separating non-azeotropic mixtures usually reduces to a one-variable optimization of the feed ratio. These extractive distillations can also exhibit inverse response at sufficiently large reflux ratios, due to dilution of the solvent concentration in the column.

Solvent Selection

The most important step in developing a successful (ie, economical) extractive distillation sequence is identifying an effective solvent for the separation. The solvent selection procedure is typically carried out in several stages. In the first step, simple qualitative methods are used to identify general classes of compounds or functional groups that may make effective solvents for a given separation. In the next step, individual compounds are evaluated, typically by experimental methods, and then ranked to identify the most promising candidates. In the final step, detailed vapor–liquid equilibrium measurements are made for several of the top candidates, the separation is simulated using chemical process simulation software and/or tested in a lab-scale or mini-works column, and then the final solvent selection is made. The final selection process should include an economic comparison of the optimal separation sequence with each of the final solvent candidates, because differences in the solvent recovery costs can influence the decision.

As described earlier, extractive distillation works because the presence of the solvent alters the relative volatilities of the components being separated.

Homologues, polarity, and hydrogen-bonding tendencies can be used for a crude screening of candidate solvents by qualitatively predicting the types of deviations from ideality that can be expected. When the compounds being separated are chemically dissimilar, suitable solvents may be found among the higher boiling homologues of either compound, though homologues of the less volatile compound are generally favored, as they enhance the natural volatility difference. The molecular similarity of two members of the same chemical family causes them to form relatively ideal solutions, while the molecular dissimilarity of the homologue and the other compound tend to cause positive deviations.

Mixtures of liquids with similar polarity typically form nearly ideal solutions while liquids with different polarities exhibit positive deviations from ideality. The degree of deviation from ideality is roughly proportional to the difference in polarity.

Carlson and Stewart's guidelines for choosing potential extractive distillation solvents based on polarity arguments are given below. Be aware, however, that exceptions do occur. (1) If the mixture to be separated is highly polar (like acetic acid and water), choose a polar solvent or a nonpolar solvent (like a hydrocarbon) that is more soluble with the higher boiling component. (2) For nonpolar mixtures (like cyclohexane–benzene) choose a polar solvent (like aniline) that is more soluble with the higher boiling component. (3) For a mixture where the more volatile component is polar and the less volatile component is nonpolar, choose a nonpolar or a polar solvent. Be aware that polar solvents may reverse the relative volatility. (4) For mixtures where the more volatile component is nonpolar and the other component is polar, choose a polar solvent. (5) For mixtures of compounds that are moderately polar, choose a polar solvent or a nonpolar solvent that is more soluble with the higher boiling component. For cases where either a polar or a nonpolar solvent is recommended, polar compounds are often preferred because they typically cause the solution to be more nonideal.

Strong deviations from ideality are often associated with hydrogen bonding between molecules and many successful extractive distillation solvents are compounds capable of forming strong hydrogen bonds such as phenols, aromatic amines like aniline, higher alcohols, glycols, etc.

Once several potentially promising families of compounds have been identified using the above qualitative methods, individual solvent candidates need to be identified and ranked. As a starting point, elminate all compounds with boiling points lower than the compounds to be separated. This ensures that, throughout the column, the solvent will be predominantly present in the liquid phase where it can alter the activity coefficients of the mixture being separated. The likelihood that a potential solvent will form azeotropes with any of the mixture components can be minimized by selecting compounds that boil 30–50°C or more above the mixture. On the other hand, the solvent should not boil so high that excessive temperatures or high vacuum are required in the solvent recovery column.

Group contribution methods such as modified UNIFAC (Dortmund) can be used to predict activity coefficients in mixtures and thus provide a nonexperimental method for *estimating* solvent selectivities. Bear in mind, however, that all group contribution methods have weaknesses and they sometimes give, not only quantitatively inaccurate, but also qualitatively incorrect results.

The infinite-dilution selectivity provides a convenient quantitative method for comparing the effectiveness of different solvents.

Gas–liquid chromatography (glc) is one of the most common experimental methods used for screening potential extractive distillation solvents and/or for determining infinite-dilution activity coefficients.

Potential extractive distillation solvents can also be evaluated by experimentally measuring the selectivity or relative volatility of a fixed composition mixture of the components to be separated (often 50% each) in the presence of a constant amount of each candidate solvent. Solvent-to-feed ratios of 1:1 to 3:1 are typically used. The preferred experimental apparatus is a modified Othmer still operated to yield one theoretical stage of separation. As before, the objective is to find the candidate solvents that cause the largest increase in relative volatility (i.e., the highest selectivities).

While good selectivity at reasonable solvent concentrations is essential, it is not the only important property when selecting an extractive distillation solvent. A solvent's *capacity* (ie, its ability to solubilize the components in the mixture being separated) is also important.

In addition to having a high selectivity at reasonable concentrations and a high capacity, desirable extractive distillation solvents should: (1) not form azeotropes with the components in the mixture to be separated—this can almost be guaranteed by picking solvents with boiling points 30–50°C or more above the other compounds, (2) be easily separated from the other components at reasonable temperatures and pressures, (3) be thermally stable at the temperatures encountered in the distillation columns, (4) be nonreactive with the other compounds, (5) be nontoxic, (6) inexpensive, (7) readily available (perhaps a compound already in the process), (8) noncorrosive to commonly used materials of construction, and (9) have a low latent heat of vaporization. In practice, some compromise in solvent properties is almost always required.

SALT-EFFECT DISTILLATION

Salt-effect distillation (also called salt extractive distillation or salt rectification) is another form of extractive distillation, where, rather than using high boiling liquids to alter the relative volatility of the mixture being separated, the separating agent is a soluble, nonvolatile ionic salt or mixture of salts. Because the added salt is completely nonvolatile, no stages are needed above the salt feed point to keep it out of the overhead product.

The principal advantage of an ionic salt over a liquid separating agent is that the salt ions typically cause much larger changes in relative volatility than do liquid separating agents. Consequently, much less salt is

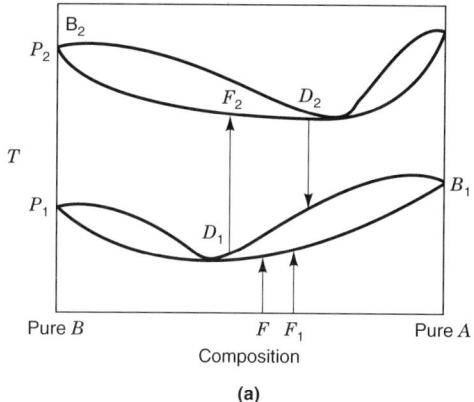

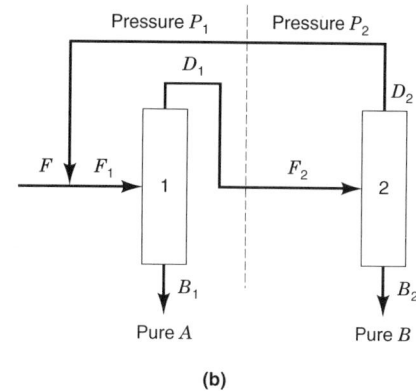

Figure 4. Pressure-swing distillation of a minimum boiling binary azeotrope. (**a**) Temperature–composition phase diagram showing the effect of pressure on the azeotropic composition; (**b**) column sequence.

typically needed—sometimes as little as a few percent as compared with 50–90% of the liquid phase, as can be the case with some liquid separating agents. This, in turn, leads to shorter, smaller diameter columns and/or higher purity products, and noticeably smaller energy requirements.

Salt-effect distillation also has several significant disadvantages compared with traditional extractive distillation. Foremost is the relatively limited number of systems for which there is a known salt that is both soluble in the components being separated and effective at altering their relative volatility.

PRESSURE-SWING DISTILLATION

It is well known that varying the system pressure can affect the composition of an azeotrope. When the change in azeotropic composition is appreciable over a moderate pressure range, this effect can be exploited to separate both minimum- and maximum-boiling binary azeotropes using a two-column sequence with the columns operated at different pressures. This process, called pressure-swing or pressure-sensitive distillation, is illustrated by Fig. 4.

Systems amenable to pressure-swing distillation include: tetrahydrofuran (THF) and water, acetonitrile and water, a variety of alcohol–ketone azeotropes, including the methanol–acetone azeotrope, which is known to disappear at both low and high pressure, methanol and methyl ethyl ketone, the alcohol–acetate azeotropes arising from transesterification reactions, ethanol and water, and the maximum-boiling hydrogen chloride–water azeotrope. Only a fraction of the known azeotropes, however, are sufficiently pressure-sensitive for pressure-swing distillation to be economical.

Pressure-swing distillation can also be applied to multicomponent mixtures containing distillation boundaries as long as at least one end of the boundary terminates at an azeotrope whose composition changes with pressure.

HETEROGENEOUS AZEOTROPIC DISTILLATION

Heterogeneous azeotropic distillation, or simply azeotropic distillation, is widely used for separating nonideal mixtures. The technique uses minimum boiling azeotropes and liquid–liquid immiscibilities in combination to overcome the effect of other azeotropes or tangent pinches in the mixture that would otherwise prevent the desired separation. The azeotropes and liquid heterogeneities that are used to make the desired separation feasible may either be induced by the addition of a separating agent, usually called the entrainer, or they may be intrinsically present, in which case the mixture is sometimes called self-entrained. The most common case is the former; it includes such classic separations as ethanol dehydration using either benzene, cyclohexane, toluene, heptane, ethyl ether, etc, as the entrainer, and acetic acid recovery from water using either 1-propyl acetate, isopropyl acetate, 1-butyl acetate, or isobutyl acetate as the entrainer. In ethanol dehydration the entrainer is used to break the homogeneous minimum boiling azeotrope between ethanol and water; in the acetic acid recovery process the entrainer is used to overcome the tangent pinch between acetic acid and water.

Phase Diagrams

For binary mixtures, it is well known that when a liquid–liquid envelope merges with a minimum boiling vapor–liquid phase envelope the resulting azeotropic phase diagram has the form shown in Fig. 5.

The properties of ternary heterogeneous vapor–liquid–liquid equilibrium (VLLE) phase diagrams are important for understanding azeotropic distillation.

The typical phase equilibrium problem encountered in distillation is to calculate the boiling temperature and the vapor composition in equilibrium with a liquid phase of specified composition at a given pressure.

Residue Curve Maps

Residue curve maps are useful for representing the infinite reflux behavior of continuous distillation columns and for getting quick estimates of the feasibility of carrying out a desired separation.

Using the new generation of VLLE computation techniques, it is possible to calculate residue curve maps for heterogeneous liquid systems.

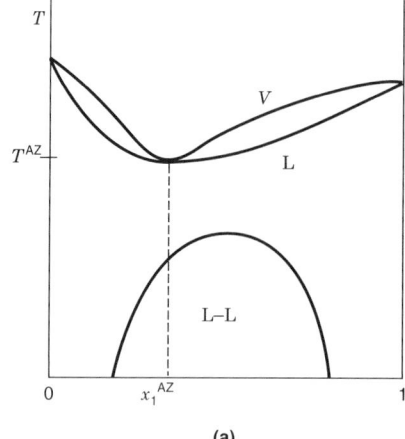
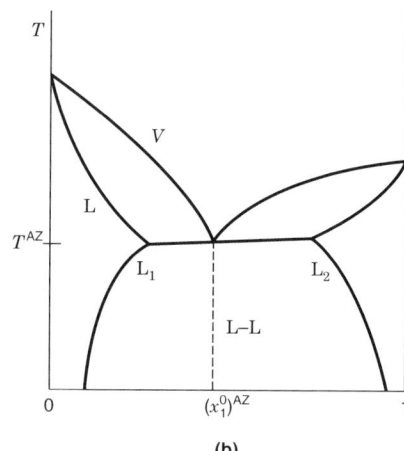

Figure 5. Schematic isobaric phase diagrams for binary azeotropic mixtures (AZ). (**a**) Homogeneous azeotrope; (**b**) heterogeneous azeotrope.

Binary or ternary heterogeneous azeotropes are restricted to being either unstable nodes or saddles in the residue curve map.

Column Sequences

The analysis of residue curve maps and distillation boundaries for homogeneous azeotropic mixtures provides a simple and useful technique for distinguishing between feasible and infeasible sequences of distillation columns, and many of the same insights apply to heterogeneous mixtures. Residue curves cross continuously through the liquid boiling envelope from one side to the other. Thus a simple distillation boundary inside the heterogeneous region does not stop abruptly or exhibit a discontinuity at the liquid boiling envelope but passes continuously through it, becoming a homogeneous distillation boundary thereafter.

Binary Mixtures. A binary mixture containing a homogeneous azeotrope can be distilled up to, but not beyond, the azeotropic composition. If the binary azeotrope is heterogeneous, however, the situation is more favorable and a simple sequence of columns is capable of isolating each pure component.

If the process feed does not lie in the liquid–liquid region it can be made to do so by deliberately feeding one of the pure components to the decanter, as required. This may only be necessary during start-up or for control purposes because the recycled azeotrope has the beneficial effect of dragging the decanter composition further into the liquid–liquid region.

Ternary Mixtures. When the binary mixture containing the minimum boiling azeotrope is completely homogeneous, i.e., the liquid is homogeneous for all compositions, a third component, called the entrainer, is added which induces a liquid–liquid phase separation over a limited portion of the ternary composition diagram. Many options for sequencing ternary heterogeneous azeotropic distillation systems exist. These sequences generally consist of two, three, or four columns using various techniques for

handling the entrainer recycle stream. The feasibility of such sequences rests on the use of liquid–liquid phase splits to provide each column with a feed composition in a different distillation region. In this regard, the sequences for ternary mixtures resemble the sequences for binary mixtures. In all cases the heart of the process is the azeotropic column and its decanter.

Kubierschky Three-Column Sequence. If only simple columns are used, ie, no side-streams, side-rectifiers/strippers etc, then the separation sequence consists of an azeotropic column (and its decanter) together with an entrainer recovery column to recycle the entrainer, and a preconcentrator column to bring the feed to the azeotropic column up to the composition of the binary azeotrope.

Other Sequences. The Kubierschky sequence is not the only way to perform the separation. Alternatives include: (1) If the process feed already has a composition at or near the composition of the binary azeotrope then the preconcentrator is not needed. (2) The distillate stream from the entrainer recovery column can be recycled directly to the decanter. (3) Use of the Kubierschky two-column sequence, in which the preconcentration and entrainer recovery steps are combined into a single column. (4) Use of the Steffen three-column sequence, the basic layout of which is the same as the Kubierschky three-column sequence. The essential new feature is to replace the single decanter in the Kubierschky sequence by multiple decanters, adding fresh or recycled water to each. (5) Use of the Ricard-Allenet four-column sequence. The first two columns in this sequence are the same as the first two columns in the Kubierschky three-column sequence. The third column in the Ricard-Allenet sequence takes the water-rich phase from the decanter and splits it into a distillate stream containing the ternary azeotrope, and by material balance, a bottoms stream consisting of water and ethanol. (6) Use of the Ricard-Allenet three-column sequence in which the fourth column is eliminated and the bottom stream from the third column (containing only ethanol and water) is recycled to the preconcentrator column.

In summary, for systems of the ethanol–water–benzene type, the three most attractive sequences for carrying out azeotropic distillation are the Kubierschky three-column sequence, the Kubierschky two-column sequence, and the Ricard-Allenet three-column sequence. For each of these there is the added possibility of replacing the decanter with a liquid–liquid extraction step.

Other Classes of Entrainers. Not all heterogeneous azeotropic mixtures are of the ethanol–water–benzene type. The number of azeotropes in the mixture may vary from system to system as may their character, ie, maximum or minimum boiling, hetereogeneous or homogeneous. In addition, the size and shape of the liquid–liquid region varies greatly from system to system. Any entrainer that does not divide the components to be separated into different distillation regions is normally a feasible entrainer, and remains feasible even if it induces a region of partial miscibility.

Wentworth Process for Ethanol–Water Separation. In the Wentworth process ethyl ether is used as the entrainer for ethanol–water separation, producing a residue curve map. Ethyl ether and water form a minimum boiling heterogeneous azeotrope at $34.15°C$ containing 98.75 wt% ether and 1.25 wt% water at atmospheric pressure. There is no azeotrope between ethyl ether and ethanol and no ternary azeotrope.

Rodebush Sequence for Ethanol–Water Separation. In the Rodebush sequence, water is continuously added to the decanter in order to shift the overall composition into the two-liquid phase region.

Self-Entrained Systems. It is quite common to find that the feed mixture to an azeotropic distillation column is not a binary azeotropic mixture but a ternary mixture that already contains a third component that can act as an entrainer. Such systems are called self-entrained. The presence of the third component may be unavoidable on account of the reaction chemistry (eg, acetic acid–water–vinyl acetate; ethanol–water–diethoxy methane), or may be present on purpose due to the choice of process technology.

M. F. Doherty and M. F. Malone, *Conceptual Design of Distillation Systems*, McGraw-Hill, New York, 2001.

E. J. Hoffman, *Azeotropic and Extractive Distillation*, Interscience Publishers, New York, 1964.

J. D. Seader, J. J. Siirola, and S. D. Barnicki, in R. H. Perry, D. W. Green and J. O. Maloney eds., *Perry's Chemical Engineer's Handbook*, 7th ed., McGraw-Hill, New York, 1997, Section 13, pp. 54–81.

MICHAEL F. DOHERTY
University of California, Santa Barbara

JEFFREY P. KNAPP
E. I. du Pont de Nemours & Co.

DRILLING FLUIDS

A range of specialty chemicals is used by the petroleum industry for oil and gas exploration, drilling and production, and for enhanced oil recovery. Broadly speaking, drilling fluids are used to lubricate the drill bit, to control formation pressure, and to remove formation cuttings. Workover and completion fluids are utilized when operating in producing formations. Chemicals are used to cement steel pipes or casing to the sides of the bore hole and to encourage the flow of crude oil to the well. Chemicals find applications at all stages, from oil production at the well bore to delivery of crude oil to the refinery. During Enhanced Oil Recovery (EOR), chemicals are used for the various techniques that renew the oil flow from fields that have ceased to produce by conventional methods.

Drilling fluids or muds are considered an essential component of the rotary drilling process used in drilling for oil and gas on land and in offshore environments. This fluid performs a variety of functions that influence the drilling rate, cost, efficiency, and safety of the operation. Some of the most important functions of drilling fluids include (*1*) transporting the drilling cuttings to the surface, (*2*) balancing the subsurface and formation pressures to prevent of well blowouts, and (*3*) cooling, lubricating, and supporting part of the weight of the drill and drill pipe. During the drilling of a well, the drilling fluid mud is pumped from the mud tanks down the hollow drill pipe and through nozzles in the drill bit. The flowing mud sweeps the crushed rock cuttings from under the bit and carries them back up the annular space between the drill pipe and the bore hole or casing to the surface. After reaching the surface, the drilling fluid is passed through a series of vibrating screens, settling tanks or pits, hydrocyclones and centrifuges to remove the cuttings brought up. It is then treated with additives to obtain a set of desired physical and chemical properties. Once treated, the fluid is pumped back into the well and the cycle is repeated.

Drilling fluids generally are composed of liquids; eg, water, petroleum oils, and other organic liquids; dissolved inorganic and organic additives; and suspended, finely divided solids of various types. The chemistry of the liquid phase and the level of the suspended solids together determine the treatment strategy and the efficiency of the mud-handling equipment. This chemistry also influences the type and amount of materials needed to maintain or change the cuttings' density, viscosity, and other properties. Drilling fluids are dynamic systems. These muds constantly change in response to changing conditions while the well is being drilled and downhole temperatures and pressures increase. Occasionally, the downhole environment requires replacement of one fluid with another of a different type.

Drilling fluid costs range from several thousand to several million dollars per well and depend on the nature of the well being drilled. The length of the drilling time may vary from a few days to more than a year. On the average, ~6–8% of the total drilling cost arises directly from the drilling fluid and additives. Additional fluid and total well costs may arise from improperly formulated or treated fluids that can prolong the drilling time.

Drilling fluid materials are also used as completion and workover fluids. When the drilling reaches the oil-producing zone, the subsequent operation is referred to as completion. These fluids are formulated to enhance production rates of crude oil and minimize permeability damage to the formation while drilling through a production zone. Workover fluids are used on existing oil and gas wells with the aim of improving or maintaining current production levels and during remedial work on older wells. Some of the workover operations in oil well drilling can include sand control measures, casing repair, reperforation, and repair or replacement of subsurface equipment. With the increasingly difficult conditions encountered during the drilling of very deep wells, special water- or oil-based fluids are being developed to maximize hydrocarbon recovery. Workover fluids also find use in many abandoned wells that are being reworked and converted back to producing wells.

CLASSIFICATION OF DRILLING FLUIDS

Drilling fluids are classified as to the nature of the continuous phase: gas, water, oil, or synthetic. Within each classification are divisions based on composition or chemistry of the fluid or the dispersed phase.

Gas-based Muds

Gas-based drilling fluids are used mostly for hard-rock drilling. These fluids range from compressed dry air or natural gas to water-based mist and stable foams. Foam is considered gas based, since the gas comprises the bulk of the fluid volume. Bottomhole pressures imposed by a gas-based fluid are low, and therefore formation strengths must be relatively high, having little or no influx or formation fluid. Air, gas, or mist drilling requires a high annular velocity to remove drill cuttings. On the other hand, large cuttings can be removed at low annular velocities when using stable foams.

Chemical additives for gas-based drilling fluids are generally biodegradable mixtures of surfactants, certain polymers, and occasionally salts such as sodium or potassium chloride. No additives are used in dry air or gas drilling operations. Gas-based fluids are not recirculated, and materials are added continuously. As the fluid exits the well, air or water vapor escapes to the atmosphere, gas and oil are burned, and water and formation solids are collected into a pit for later disposal. Stable foams must be destabilized to separate the air from the liquid phase for disposal.

Water-based Muds

The vast majority of all drilling fluids are water-based systems. The types depend on the composition of the water phase (pH, ionic content, etc), viscosity builders (clays or polymers), and rheological control agents (deflocculants or dispersants).

Freshwater fluids can range from clear water having no additives to high density muds containing clays, barite, and various organic additives. Onshore wells typically use freshwater muds, as do some offshore wells where highly weighted muds are needed. Freshwater is ideal for formulating stable drilling fluids, as many mud additives are most effective in a system of low ionic strength. Inorganic or organic additives control the rheological behavior of the clays, particularly at elevated temperatures. An organic polymer may be used for filtration control.

Many offshore wells are drilled using a seawater system, because of its ready availability. Seawater muds generally are formulated and maintained in the same way that freshwater muds are used. However, because of the presence of dissolved salts in seawater, more additives are needed to achieve the desired flow and filtration properties.

The high salinity of saltwater muds may require different clays and organic additives than those used in fresh- or seawater muds. Saltwater clays and organic polymers contribute to viscosity. Filtration properties are adjusted using starch or cellulosic polymers. Alternatively, clays used primarily in fresh and seawater muds can be prehydrated in freshwater and then added to a salt mud for viscosity and some filtration control.

Fresh- or seawater muds may be treated with gypsum or lime to alleviate drilling problems that may arise from drilling water-sensitive shale or clay-bearing formations. Potassium-treated systems combine one or more polymers and a potassium ion source, primarily potassium chloride, in order to prevent problems associated with drilling certain water-sensitive shales. Potassium muds have been applied in most active drilling regions around the world. Environmental regulations in the United States have limited the use of potassium muds in offshore drilling owing to the apparent toxicity of high potassium levels in that process.

Freshwater, clay, and polymers for viscosity enhancement and filtration control make up low solid/nondispersed muds. Low-solids muds are maintained using minimal amounts of clay and require removal of all but modest quantities of drill solids. These are called nondispersed systems, because no additives are used to further disperse or deflocculate the viscosity-building clays. Most water-based muds are considered dispersed, because deflocculating additives are used to control the flow properties. Nondispersed muds can be weighted to high densities, but are used primarily in the unweighted state. The main advantage of these systems is the high drilling rate that can be achieved because of the lower colloidal solids content. These are normally applied in hard formations where increasing the penetration rate can reduce drilling costs significantly and the tendency for solids buildup is minimal.

Oil-based Muds

Oil-based drilling fluids have diesel or mineral oil as a continuous phase with both internal water and solid phases. Fluids having no or very low water content are usually called oil-based muds or all-oil muds; fluids having higher water content are called invert oil–emulsion muds, or simply inverts. Most oil muds maintain a fixed oil–water ratio, depending on the desired properties. Oil muds are employed for high-angle wells where good lubricity is required, for high-temperature wells where water-based systems might be thermally unstable, for drilling water-sensitive shale formations, or where corrosive gases such as hydrogen sulfide and carbon dioxide could be encountered. Environmental restrictions and cost often limit use, although the higher drilling rates achievable using oil muds and polycrystalline diamond compact (PDC) bits can often offset high fluid and disposal costs.

Oil-based muds have diesel or mineral oil as the continuous phase and are formulated using no internal aqueous phase or have only minimal water content. Organophilic clay or colloidal asphalt are used to control viscosity and filtration rates. The internal water phase, either added as part of the formulation or incorporated while drilling, is stabilized using emulsifiers. The desired density is reached by adding a powdered, high specific gravity solid. A wetting agent ensures the oil wetting of added or drilled formation solids.

Oil muds are usually characterized by very low filtration rates. Relaxing the normally tight filtration control specifications on oil muds can result in higher drilling rates without loss of emulsion or mud stability. These relaxed fluid-loss muds are also termed low colloid oil muds or relaxed filtration oil muds because their higher filtration rates are achieved by omitting some of the colloidal solids from the formulation and reducing the concentration of emulsifiers and surfactants in the fluid. In most other respects these fluids are similar to a conventional invert oil–emulsion mud and can easily be converted to an invert. Use of relaxed filtration oil muds has dropped considerably owing to improvements in bit design, where allow high drilling rates using the more conventional low filtration-rate oil muds.

Synthetic-based Models.. The synthetic-based drilling fluids are a relatively new class of drilling muds that are particularly useful for deepwater and deviated hole drilling. These muds were developed to provide an environmentally superior alternative to oil-based drilling muds as well as offering an alternative to the high costs associated with the disposal of drill cuttings generated when diesel or mineral oil-based muds are used. The synthetic muds have a continuous phase that consists of synthetic organic liquid. The synthetic-based liquids, because of their similarity with the oil-based muds, are often called pseudo-oil muds outside the United States. While the technical and economic benefits of using such fluids in the deepwater Gulf of Mexico are recognized, the environmental impacts of the discharged drill cuttings produced are not well understood. The high cost of synthetic-based fluids can be offset by their enhanced drilling performance and the on-site discharge of synthetic-based fluids when this is permitted. Current concerns with toxicity, biodegradation rates, environmental impacts, and costs have virtually eliminated the use of synthetic-based fluids except for olefins and esters in offshore drilling.

PROPERTIES

The type of drilling fluid required for a particular well is determined by the geological formation at the site.

The functions of the drilling fluids include removal of formation cuttings from beneath the bit and transporting these cuttings to the surface, prevention of gases in the formation from escaping through the bore holes, cooling and lubrication of the bit and drill string at points of contact with the cased or uncased bore hole, prevention of an influx of formation fluids into the wellbore, sealing of exposed permeable formations, maintenance of the stability of exposed formations, helping to suspend the weight of the drill string and casing, and control of pressure.

The fluid should not damage productive formations, it should not be corrosive to the equipment, and be cost effective. The fluid must be safe for handling and be compatible with the environment or be disposable in an environmentally sound manner. How well the drilling fluid fulfills these functions is determined primarily by the response of the well. The whole drilling operation suffers if the fluid is not adequate. In extreme cases, drilling may be stopped or a hole may have to be redrilled.

Density

The density of the drilling fluid is adjusted using powdered high density solids or dissolved salts to provide a hydrostatic pressure against exposed formations in excess of the pressure of the formation fluids. In addition, the hydrostatic pressure of the mud/clear brine column prevents collapse of weak formations into the bore hole.

Flow Properties

Fluid viscosity and annular flow velocity must be high enough to remove cuttings generated by the drill bit and other formation material that may fall into the wellbore. These solids are carried up the annulus to the surface, where they are separated with varying degrees of efficiency and then disposed of. In order to accomplish this, low viscosity drilling fluids are circulated at high flow rates or high viscosity fluids at low flow rates. The varying demands on flow properties are best met by fluids exhibiting non-Newtonian rheological characteristics. Drilling fluids are normally shear thinning, having apparent viscosity decreasing with increasing shear rate.

The gel strength (thixotropy) is a measure of the capability of a drilling fluid to hold particles in suspension after the mud circulation ceases. It results from the electrical charges on the individual clay platelets. When the mud pump is shut off and flow ceases, the attraction between the clay particles causes the platelets to bond to each (referred to as flocculation). This edge-to-face flocculation results in an open card-house structure capable of suspending cuttings and sand and gravel particles. The ability to keep cuttings in suspension prevents sandlocking (sticking) of the tools in the bore hole while drill rods are added to the string and minimizes sediment collecting in the bottom of the hole after reaming and before going back in the hole with a sampler. A drawback to this property is that cuttings do not readily settle out of the drilling mud in the mud pit and may be recirculated, thus resulting in grinding of particles by the drill bit, increased mud density, increased mud pump wear, and a lower penetration rate.

Filtration Properties

Drilling fluids have a natural tendency to flow into permeable formations, because the bore hole pressure is generally higher than that in the formation. To prevent excessive leak-off, a thin, low permeability filter cake is formed using additives. Filtration occurs under both dynamic (during circulation) and static (no circulation) conditions. The filtration rate is adjusted using colloidal solids and organic polymers to reduce loss of filtrate to the formation and prevent buildup of a thick filter cake, which would restrict the wellbore.

Large solid particles of various sizes and shapes may be added to control circulation loss where natural or induced fractures, highly porous formations, or vugular zone; ie, formations containing small cavities (vugs) larger than the matrix grain size, are encountered. Particle sizes from those large enough to bridge the opening down to those fine enough to seal small spaces between the larger bridging particles may be required to prevent drilling fluid losses into these zones.

Water Chemistry

Water is present in all but purely gaseous or oil drilling fluids, both of which comprise only a small percentage of drilling fluid applications. The water may be present as fine droplets in a mist, emulsified in an organic continuous phase, or as is most common, comprise the continuous phase of the drilling fluid. Water added to drilling fluids may be freshwater, seawater, or saturated salt solutions.

DRILLING FLUID MATERIALS

The pressure exerted by a column of drilling fluid in a well balances formation pressures to prevent in uncontrolled influx of formation fluids that could result in a blowout. The mud density must be controlled accurately by suitable weighting materials that do not adversely affect the other properties. Most important is the specific gravity of the weighting agent as well as its insolubility in water and its chemical inertness. The weighting material should be ground to the preferred particle-size distribution and be relatively nonabrasive.

Barite, predominantly composed of $BaSO_4$, meets the overall requirements for weighting material better than other materials and is used for increasing the density of drilling fluids throughout the world. It is virtually insoluble in water and does not react with other mud constituents. Drilling technology favors systems having the lowest possible solids content, because of its resulting higher penetration rate, easier control of mud properties, and fewer problems experienced during drilling. Alternative weighting materials such as hemotite having specific gravities higher than that of barite offer this advantage.

Calcite and siderite are used occasionally because of their solubility in hydrochloric acid, which offers a method of removing mud filter cake deposited on productive formations. Calcite and siderite are used most frequently in workover or completion fluids when a nondamaging fluid is required; ie, one that can either be removed by acidizing.

Solid salt, ground and packaged in several particle size grades, can be used in saturated salt brines to increase the fluid density. However, sized salt is also a weighting agent, having specific use in sensitive reservoirs. In such cases, an oversaturated solution is used to form a filter cake that can easily be removed with freshwater flushing.

The use of solids-free fluids or clear brine fluids (CBF) is occasionally resorted to for achieving high drilling rates, and also in workover and completion operations.

The chemical and mechanical dispersion of drilled solids tends to increase the percentage of small-sized solids in a mud as drilling progresses. Incorporation of a limited amount of drilled solids is an economical way of increasing the density of low-density muds, but it also reduces penetration rates; hence, drilled solids are usually kept to a minimum.

Viscosity Buildup

The drilling fluid removes cuttings from the wellbore as drilling progresses. This process is governed by the angle of the hole and the velocity at which fluid travels up the annulus, as well as by the fluid viscosity or flow properties and fluid density. The cuttings-removal efficiency usually increases with increasing viscosity and density, although at high wellbore angles a less viscous fluid may be desirable if high flow rates can be achieved. The viscosity depends on the concentration, quality, and state of dispersion of suspended colloidal solids.

Water-based muds have three basic components: water, reactive solids, and inert solids. The water forming the continuous phase may be freshwater, seawater, or saltwater. The reactive solids are composed of commercial clays, incorporated hydratable clays and shales from drilled formations, and polymeric materials, which may be suspended or dissolved. Solids, such as barite and hematite, are chemically inactive in most mud systems. Oil and synthetic muds contain, in addition, an organic liquid as the continuous phase plus water as the discontinuous phase.

The most important commercial clays used for increasing the viscosity of drilling fluids are bentonite, attapulgite, and sepiolite. For oil-based and synthetic-based muds, organophilic clays are used, to produce viscosity and help suspend weighting materials.

When the water phase of the drilling fluid contains substantial amounts of electrolyte, saltwater clays such as attapulgite and sepiolite are added to raise viscosity. Attapulgite is used solely for its suspending qualities.

Sepiolite is more stable at higher temperatures than attapulgite and therefore is used in geothermal drilling fluids. Oil-dispersable or organophilic clay provides viscosity and suspending qualities in oil-based muds. Oil-dispersable clays can suspend solids in oil without requiring additional soaps and emulsifying agents.

A wide variety of organic polymers serve a number of useful purposes in drilling fluids, the most important of which are to increase viscosity and control filtration rates. These polymers are either natural polysaccharides like starch, guar gum, xanthan gum, and other biopolymers or derivatives of natural polymers, such as cellulose, lignosulfonate, and lignite and synthetic polymers. The most commonly used polymeric viscosity builders are the cellulosics, xanthan gum, and polyacrylamides.

Sodium carboxymethyl cellulose (CMC) and hydroxyethyl cellulose (HEC) are the cellulosics most widely used in drilling fluids. The effectiveness of sodium carboxymethyl cellulose, an anionic polymer, as a viscosity builder decreases with increasing electrolyte concentration. This polymer can be coprecipitated with calcium and magnesium by raising the pH of the mud. The polyanionic cellulose (PAC), which has a higher degree of substitution than CMC, was introduced to overcome some of these limitations. The primary application of both CMC and PAC is in the control of filtration rates.

Hydroxyethyl cellulose (HEC), a nonionic thickening agent, is used in drilling muds, but more commonly in completion fluids, where its acid-degradable nature is advantageous.

Xanthan gum a viscosity builder and suspending agent that can be used in almost any type of water. Xanthan gum is widely used for drilling, workover, and completion fluids. Two other biopolymers, succinoglucan gum and welan gum, are also finding some use in drilling fluids.

Guar gum produces viscous solutions in fresh or saltwater at certain concentrations and is used in solids-free and low-solids muds but degrades rapidly above 80°C, limiting its use to shallow wells.

High molecular weight polyacrylamides are used as viscosity builders in freshwater muds or as bentonite extenders.

Occasionally, polymers are used to increase the viscosity of oil-based and synthetic-based muds. Such polymers are usually used in conjunction with an organophilic clay.

Viscosity Reduction

Proper control of viscosity and gel strengths is essential for efficient cleaning of the bore hole, suspension of weight material and cuttings when circulation is interrupted, and to minimize circulating pressure losses and swab/surge pressures owing to axial movement of the drill string. Reduced viscosity can be achieved by thinning or deflocculating clay–water suspensions. Thinning is measured as a reduction of plastic viscosity, yield point, or gel strength, or a combination of these properties. Viscosity is reduced by decreasing the solids content and the number of particles per unit volume or by neutralizing the attractive forces between particles. Typical mud-thinning chemicals are polyanionic materials that are adsorbed on positive edge sites of the clay particles, thereby reducing the attractive forces between the particles without affecting clay hydration.

Thinners or deflocculants for clay–water muds include polyphosphates, tannins, lignites, lignosulfonates, and low molecular weight polyacrylates and their derivatives. These materials also can remove chemical contaminants by precipitation or chelation. Sodium polyphosphates are effective deflocculants for clays in freshwater and were among the first thinners used in drilling fluids.

Quebracho plant extract is an acidic mud thinner that performs best at high pH. It is an excellent thinner for lime-treated and cement-contaminated muds.

Lignite, generally leonardite, and lignite derivatives are applied in water-based muds as thinners and filtration control agents. Natural lignite is not a good thinner for low pH water-based muds at moderate temperatures but can be an excellent thinner or mud conditioner at high pH and high temperatures where the humic acids are solubilized. It has better temperature stability than most plant tannins or lignins. Lignosulfonate thinners are among the most versatile and important chemicals used in water-based drilling fluids, and are the largest-volume additives used in this category worldwide. The chrome-based lignosulfonates have been banned in many European countries due to their toxicity.

Chrome and ferrochrome lignosulfonates are effective deflocculants in most water-based muds over a wide range of salinity, hardness, and pH (8.5–12.5). Lignosulfonates at high concentrations provide some filtration control and inhibit disintegration and dispersion of shale cuttings. Low molecular weight (1000–5000) polyacrylates and copolymers of acrylic acid and AMPS are used as dispersants for weighted water-base muds. These materials are particularly useful where high temperatures are encountered or in muds, which derive most of their viscosity from fine drill solids, and polymers such as xanthan gum and polyacrylamide. Another high temperature polymer, a sulfonated styrene maleic–anhydride copolymer, is provided in powdered form. All of these materials are used in relatively low concentrations in the mud.

Filtration control is particularly important in permeable formations where the mud hydrostatic pressure exceeds the formation pressure. Proper filtration control reduces drill-string sticking and drag, and rotary torque, as well as minimizing damage to protective formations; in some formations it improves bore hole stability. Several types of materials available for water-based muds and application varies include clays, organic polymers, and lignite derivatives. The bentonite present in the system often acts as the primary filtration control agent. It not only develops viscosity, but also lowers the filtration rate, particularly in freshwater muds. The ability of bentonite clay to control filtration is attributed to its flat, platelike particle shape, the capacity to disperse and hydrate, its ability to form a compressible filter cake, and the colloidal to near-colloidal particle size.

Although a combination of bentonite clay and an organic thinner provides filtration control in many water-based muds, additional control generally is needed. Filtration additives for both fresh- and saltwater muds are usually organic polymers and lignites.

Numerous modifications and derivatives of drilling fluid starch have been made for application in drilling and workover fluids. Most modified starches are cross-linked to some degree to improve thermal stability. Carboxymethyl and hydroxypropyl starches are finding increasing application in drilling fluids. There are some applications involving cationic starches as well, although toxicity must be considered for offshore use.

Lignite products, mined, ground, and possibly treated with sodium or potassium hydroxide, are economical filtration control additives for some water-based muds, in addition to improving flow properties. A sulfonated lignite complexed with a sulfonated phenolic resin is an effective high-temperature, filtration control additive, for both fresh- and seawater muds that contain high concentrations of soluble calcium, and it does not affect viscosity. A high molecular weight polyanionic lignin has also found application for high temperature muds with a high electrolyte content. These and similar products can provide filtration control for high density, high solids muds above 120°C. A number of synthetic polymers having the ability to control filtration rates at high temperature and in the presence of calcium and magnesium have also been developed.

Filtration control in oil- and synthetic-based fluids is achieved by the emulsified aqueous phase, by the emulsifier package, and by additions of powdered solid materials. The powdered solids used for this purpose consist of asphalt, gilsonite, and amine-treated lignite. The choice of additive depends on the nature of the base fluid, the emulsifier package, and most importantly the downhole temperature. Styrene–butadiene copolymers in aqueous dispersions have also found application as filtration control agents in oil-base muds.

Alkalinity Control

Water-base drilling fluids are generally maintained at an alkaline pH. Most mud additives require a basic environment to function properly, and corrosion is reduced at an elevated pH. The primary additive for pH control is sodium hydroxide. The second most common alkalinity control agent is lime, used in brine systems containing substantial quantities of soluble calcium and in high pH lime muds. Potassium hydroxide is occasionally used for alkalinity control, particularly for some polymer and lime muds where a low sodium level is desired. A fourth alkalinity control additive is magnesium oxide, which is used in clay-free polymer-based fluids. Magnesium oxide provides an alkaline environment and, as it is only slightly soluble, also has a buffering effect. It enhances the thermal stability of polymer solutions by preventing a pH decrease to neutral or slightly acidic conditions at elevated temperatures. It is mainly applied in completion or workover operations where clay-free acid-soluble fluids are desired.

Removal of Contaminants

A drilling fluid contaminant is any material or condition encountered during drilling that adversely affects the performance of the fluid. Elevated temperatures and drill solids are encountered in every drilling operation. In most wells these are handled easily, but in some, one or both can seriously reduce efficiency. Temperature problems normally are treated by using viscosity or filtration control additives, materials having better thermal stability, or possibly by replacement of the mud system with an oil or synthetic mud. Drill solids are removed mechanically by various combinations of screens, hydrocyclones, and centrifuges, or chemically by flocculants. Dilution or replacement of part or all of the mud system may reduce drill solids to tolerable levels.

Various inorganic chemicals remove soluble contaminants such as salt encountered during drilling. The adverse effects of salt, primarily clay flocculation, can be overcome by a deflocculant such as a lignosulfonate or sulfomethylated tannin. Calcium is removed using a phosphate, sodium carbonate, sodium bicarbonate, and occasionally oxalic acid.

Gypsum and lime are used to control bicarbonate/carbonate ions that can cause mud gelation and rheological problems, particularly in high density muds in deep, hot wells.

Stabilization of Water-sensitive Formations

Many subsurface formations encountered during drilling are water-sensitive shales containing various amounts of clay minerals. A variety of methods have been devised to stabilize shales. The most successful uses an oil or synthetic mud that avoids direct contact between the shale and the emulsified water. High initial cost and environmental restrictions prevent the use of oil and synthetic muds in many cases where shale problems are expected. It is necessary then to treat a water-base mud to minimize the destabilizing effect of the drilling fluid.

Sodium chloride has long been used as a shale stabilizer because of its low cost, wide availability, and presence in many subsurface formations. The inhibitive nature of salt muds increases as the salt content increases from seawater to saturated sodium chloride. This material has been used more for minimizing washouts in salt zones than for stabilizing shales. High salt levels have found application in deep-water drilling.

Calcium sources, such as gypsum and lime, promote cation exchange from sodium clay to a less-swelling calcium clay. Calcium concentrations are normally low and osmotic swelling is reduced only if other salts are present. Calcium chloride has been used infrequently for this purpose, but systems are available that allow high calcium chloride levels to be carried in the mud system.

A variety of shale-protective muds are available that contain high levels of potassium ions. Potassium chloride is generally preferred as a potassium source because of its low cost and availability.

Ammonium chloride, ammonium sulfate, and diammonium phosphate have also been used for shale stabilization. Ammonium ions have essentially the same effect on shales as potassium ions, but use of ammonium salts is often objectionable because of the alkaline nature of the mud.

A number of nonionic and anionic polymers are employed in water-based muds to stabilize shales. These may be added to a freshwater mud or to a system containing one of the salts mentioned. Typically, shale-stabilization polymers include modified starches; cellulosic polymers such as CMC and HEC; gums such as guar, xanthan, and flax meal; and high molecular weight polyacrylamides of varying degrees of hydrolysis.

A number of cationic muds have been developed and used. Some of these additives may require a salt such as sodium or potassium chloride for best results.

A number of glycol and glycerol-base additives are being used to formulate shale protective muds, usually in conjunction with a salt and/or a polymer. A low molecular weight poly(amino acid) has also been touted for its shale-stabilizing properties.

Solid materials, such as gilsonite and asphalt, and partially soluble sulfonated asphalt may also be added to plug small fractures in exposed shale surfaces and thereby limit water entry into the formation. The asphalts are oxidized or treated to impart partial solubility. These materials may be softened by the downhole temperature, causing them to deform and squeeze into small openings exposed to the bore hole.

Surfactants

Depending on the type of fluid, a surfactant may be added to emulsify oil in water (o/w) or water in a nonaqueous liquid (w/o), to water-wet mud solids or to maintain the solids in a nonwater-wet state, to defoam muds, or to act as a foaming agent.

Lignites and lignosulfonates can act as o/w emulsifiers but generally are added for other purposes. Various anionic surfactants, including alkylarylsulfonates and alkylaryl sulfates and poly(ethylene oxide) derivatives of fatty acids, esters, and others, are used. Very little oil is added to water-based muds in use offshore, for environmental reasons.

Emulsifiers are incorporated in oil and synthetic mud formulations to maintain a stable emulsion of the internal brine phase. These materials include calcium and magnesium soaps of fatty acids and polyamines and amides and their mixtures.

Solids in oil and synthetic muds must be kept wet with the nonaqueous phase to prevent coagulation and settling and mud instability. Oil-wetting agents are normally incorporated in the basic mud package. These materials are typically amines or quaternary ammonium salts having hydrocarbon chains of 10 or more carbon atoms.

Defoamers such as aluminum stearate are frequently needed for salty muds, although in very small quantities. Foaming agents maintain stable drilling foams in areas where minimal bottomhole pressures are required. The foaming agent must be chosen to handle a variety of possible contaminants (salt, crude oils, solids) and downhole temperatures.

Lost Circulation Control

To function properly, a drilling fluid must be circulated through the well and back to the surface. Occasionally, highly permeable or cavernous formations and fractured zones, both natural and induced by the mud pressure, are encountered and circulation is partially or completely lost. Loss of drilling fluid, owing to openings in the formation, can result in loss of hydrostatic pressure at the bottom of the hole and allow an influx of formation fluids and possibly loss of well control. A wide variety of materials can be added to the drilling fluid to seal off the lost circulation zones.

Lost circulation materials (LCMs) are flake, fiber, or granular-shaped particles. Materials of different shapes and sizes are often blended into the mud at the well site. Some common flake-shaped LCMs consist of shredded cellophane and paper, mica, rice hulls, cottonseed hulls, or laminated plastic. These materials lie flat across the opening to be sealed or are wedged into an opening such as a fracture.

Granular LCMs, generally much stronger than the other types, include ground rubber, nylon, plastics, limestone, gilsonite, asphalt, and ground nut shells. Granular-shaped particles enter the opening, bridge it, and form a tight seal against further mud losses. Particle size and distribution are important for this mechanism to be effective.

Removal of Solids

Solids incorporated in the mud during drilling generally are separated mechanically, reduced by dilution, or removed chemically by flocculation. Polymers used for flocculating drill solids generally are high molecular weight polyacrylamides of varying degrees of hydrolysis. The polymer may be cationic, nonionic, or anionic, depending on the chemistry of the drilling fluid and the nature of the solids. At higher concentrations, some polymers act as protective colloids that stabilize and enhance the viscosity of bentonite suspensions and protect water-sensitive shales.

Lubricants and Spotting Fluids

To overcome the many difficulties of pipe torquing, sticking, and even twisting, off drilling fluids are treated with a variety of mud lubricants available from various suppliers. They are mostly general purpose, low toxicity, nonfluorescent types that are blends of several anionic or nonionic surfactants and products such as glycols and glycerols, fatty acid esters, synthetic hydrocarbons, and vegetable oil derivatives. Extreme pressure lubricants containing sulfurized or sulfonated derivatives of natural fatty acid products or petroleum-base hydrocarbons can be quite toxic to marine life and are rarely used, for environmental reasons.

COMPLETION AND WORKOVER FLUIDS

Completion and workover fluids are used during the completion phase of the well or when performing a workover during the life of the well. Completion fluids should (1) cause minimal formation damage, especially to reservoir permeability; (2) maintain long-term stability at downhole temperatures; (3) be resistant to contaminants; and (4) protect down-hole equipment against corrosion. These fluid types can be divided into solids-free and solids-laden. The solids-free completion/workover fluids are generally brines ranging in composition from potassium chloride to zinc bromide. The solids-laden completion and workover fluids are generally composed of a brine as a base fluid and solids, in the form of water or acid-soluble particles, for density and fluid-loss control as well as polymers for suspension-carrying capacity and filtrate control. In general, the solids-laden fluids cover the same density range as the solids-free fluids. Solids-free fluids are the predominant completion/workover fluids in use. Use of these fluids is dictated in many completion operations, especially in gravel pack or frac pack type completions. Use of a solids-laden fluid in such operations can lead to failure. Solids-free fluids cause a minimum amount of formation damage and allow tools to be run into and pulled out of the well with ease. Solids can penetrate into the permeable media of the formation and become lodged, thus causing reduced permeability and ultimately lower production. Solids-laden fluids also have typically increased viscosities, which can inhibit tools being run into and pulled out from the well. In order to keep solids-free fluids free of solids, filtration, either cartridge or diatomaceous earth, is generally used while these fluids are being employed. If there are no milling or drilling operations to be carried out, then a solids-free fluid can be used. The particular solids-free fluid/soluble salt to be employed also depends on a number of factors.

One inherent problem with solids-free brines is that since they have no solids and in general have low viscosities, there is nothing to prevent the liquid from leaking off into the formation. Therefore, an important part of planning in the use of the solids-free completion and workover fluids is fluid loss control pills. These pills are generally a blend of the brine being used and polymers for imparting viscosity.

ENVIRONMENTAL ASPECTS

At present, the major challenge in formulating drilling fluids is the need to satisfy the increasingly demanding conditions of high temperature and pressure that are found in some deep wells and horizontal wells while avoiding harm to the environment. In addition, the health of rig workers has become an important concern in the development and use of new products.

NEW DIRECTIONS IN DRILLING FLUIDS

In the last decade, a number of new formulations have been developed for use in the various stages of oil and gas well drilling. These have involved water-based drilling fluids, organic-based drilling fluids, viscosifiers, densifiers.

N. J. Adams and T. Charrier, *Drilling Engineering, A Complete Well Planning Approach*, PennWell Books, Tulsa, Okla., 1985.

A. T. Bourgoyne, K. K. Millheim, M. E. Chenevert, and F. S. Young, *Applied Drilling Engineering*, Society of Petroleum Engineers, Richardson, Tex., 1991.

G. R. Gray and H. C. H. Darley, *Composition and Properties of Oil Well Drilling Fluids*, 5th ed., Gulf Publishing Co., Houston, Tex., 1988.

P. L. Moore, *Drilling Practices Manual*, PennWell Books, 2nd ed., Tulsa, Okla., 1985.

SHMUEL D. UKELES
BARUCH GRINBAUM
IMI (TAMI) Institute for
Research and Development,
DSBG (Israel)

DRUG DELIVERY SYSTEMS

For many decades, pharmaceuticals have primarily consisted of simple, fast-acting chemical compounds that are dispensed orally for the treatment of an acute disease or a chronic illness and have been mostly facilitated by drugs in various pharmaceutical dosage forms, including tablets, capsules, pills, suppositories, creams, ointments, liquids, aerosols, and injections. Even today these conventional dosage forms are the primary mode of drug administration for prescription and over-the-counter drug products. In the last decade, several technical advancements have resulted in the development of new technologies capable of controlling the administration of a drug at a targeted site in the body in an optimal concentration-versus-time profile. The term "drug delivery" covers a very broad range of techniques used to get therapeutic agents into the human body. These techniques are capable of controlling the rate of drug delivery, sustaining the duration of therapeutic activity, and/or targeting the delivery of a drug to tissue.

The rapid advancement of biomedical research has led to many creative applications for biocompatible polymers. As modern medicine discerns more mechanisms of both physiology and pathophysiology, the approach to healing is to mimic or, if possible, to re-create the physiology of healthy functioning. Thus, the area of responsive drug delivery has evolved. Also called "smart" polymers, for drug delivery, the developments fall in to two categories: externally regulated or pulsatile systems (also known as "open loop" systems) and self-regulated ("closed loop") system.

PHYSIOLOGICAL ROUTES FOR DRUG DELIVERY

Design of a drug delivery device is dictated by the properties of the physiological barrier, the effective plasma levels, and the total dosage.

Oral

The oral route for drug delivery includes the gastrointestinal (GI) tract and the oral cavity, including the buccal mucosa. The buccal mucosa is considered separately because of differences in the approach to drug delivery via this route.

The primary function of the GI tract is the digestion and absorption of food. Thus, drugs entering the GI tract are exposed to a wide range of pH values, from 1–2 in the stomach to 5.0–6.5 in the small intestine, as well as high levels of various enzymes involved in the digestion of proteins, fats, and carbohydrates.

The transit of a dosage form through the GI tract can have a profound influence on its performance. Total GI transit time is between 24 and 48 h on average.

Absorption of drugs across the wall of the GI tract is primarily the result of passive diffusion. Absorption is believed to take place by partitioning of the drug from the aqueous GI environment into the lipoidal membrane, diffusion through the membrane, and partitioning into the blood and body fluids.

Drugs, such as opiates, may undergo metabolism both in the intestinal wall and in the liver (first-pass metabolism). The metabolism may be extensive and considerably reduce the amount of drug reaching the systemic circulation. Alternatively, the metabolite may be metabolically active and contribute significantly to the action of the parent drug. Some compounds undergo enterohepatic circulation in which they are secreted into the GI tract in the bile and are subsequently reabsorbed. Enterohepatic circulation prolongs the half-life of a drug.

For compounds absorbed from a small part of the intestine, the amount of drug absorbed can be increased by extending the residence time of the dosage form in the GI tract. The two basic approaches used are gastric flotation or retention devices, and bioadhesive delivery systems.

Drug absorption from the colon has become the subject of much attention. The development of dosage forms that released drug for 16 to 24 h depends on the drug being absorbed from the colon, because the bulk of the delivery period may be spent there. Only compounds that exhibit good colonic absorption are suitable for extended delivery dosage forms; e.g., metoprolol. Protein and peptide drugs are more readily available since the advent of recombinant DNA technology, and the oral delivery of these compounds has become the holy grail of drug delivery. However, proteins present several challenges because of their size, hydrophilicity, and susceptibility to hydrolysis and degradation by proteases. Compared to the upper GI tract, proteolytic activity is lower in the colon and the residence time is longer, which has led to interest in the development of dosage forms targeted to the colon. However, it appears unlikely that significant absorption of proteins occurs from the colon in the absence of either protease inhibitors, absorption enhancers, or both. Targeting drugs to the colon has followed two basic approaches, ie, delayed release and exploitation of the colonic flora. Delayed release generally relies upon enteric coating to ensure safe passage through the stomach, and a delay of 4 to 6 h before drug release.

Rectal

The rectal route for drug delivery is an extremely unpopular one in the United States but may present advantages in certain situations. Enemas containing either steroids or 5-acetylsalicylic acid for the treatment of proctitis, ie, inflammation of the rectum, offer good therapy in inflammatory bowel disease. The rectal route may be used when gastric stasis or vomiting is present, making the oral route of drug delivery untenable; eg, ergotamine for the treatment of migraine. The vascular drainage of the rectum may partially avoid first-pass metabolism, which offers definite advantages for drugs undergoing extensive metabolism.

Transdermal

The skin offers a formidable barrier to the entry of foreign compounds, including drugs, into the body, both in terms of a physical barrier and an immunological one. The principal barrier to drug diffusion lies in the outer few layers of the epidermis, the stratum corneum, which is 10–20 μm thick in humans and consists of sheets of kerati-nized epithelial cells joined by tight junctions. The remainder of the epidermis, which is about 100 μm thick in humans, consists of living cells that are metabolically active. A drug applied to the skin must therefore diffuse through the epidermis to reach the blood capillaries in the dermis for distribution to the systemic circulation. Blood supply to the skin can vary tremendously, from 200 to 4,000 mL/(m^2·min) as a result of its role in the control of body temperature. Drug delivery by the transdermal route avoids presystemic metabolism in the gastrointestinal tract or first-pass metabolism in the liver. The permeability of skin is low, which limits the usefulness of this route to highly permeable, potent compounds. Permeability varies somewhat with regions of the body. The greatest permeability is in the scrotum.

The use of absorption enhancers for transdermal delivery may be necessary as a result of the low permeability of a drug through skin. Ethanol has been the only enhancer in use in a commercially available system, and the flux of estradiol and nitroglycerin is linearly correlated with the flux of ethanol. Other absorption enhancers such as 1-dodecylazacycloheptan-2-one (Laurocapram), terpenes, oleic acid, pyrrolidones, n-alkanols, and alkyl esters are candidates.

Dermal irritation and sensitization are issues specific to the transdermal route of drug delivery and can result in the cessation of therapy.

Transdermal drug delivery is associated with a relatively long time lag before the onset of efficacy, and removal of the system is followed by a correspondingly extended fall in plasma concentration, which probably results from formation of a drug depot in the skin that dissipates slowly. The time lag is approximately 3 to 5 h for many drugs that have low binding in the skin, but it may be considerably longer. In contrast, plasma drug levels may be obtained between 2 and 5 min by the oral, buccal, or nasal routes.

Despite the limitations imposed by the physiology of the skin, several marketed controlled release transdermal drug delivery systems are available in the United States; for example, nitroglycerin for angina, estradiol for the relief of postmenopausal symptoms and osteoporosis, clonidine for the treatment of hypertension, fentanyl as an analgesic, and nicotine as an aid to smoking cessation. These systems are designed to deliver drug for periods of one to seven days.

Buccal

Buccal mucosa has a high blood flow of 20–30 mL/min for each 100 g of tissue and has good lymphatic drainage. Vascular drainage is directly into the systemic circulation, and thus first-pass metabolism is avoided. The buccal mucosa is readily accessible to the patient for self-administration of drugs, as well as rapid removal of the dosage form should it be necessary.

The buccal route may prove useful for peptide or protein delivery because of the absence of protease activity in the saliva. However, the epithelium is relatively tight, based on its electrophysiological properties. Absorption of proteins and peptides is generally low and somewhat erratic. The judicious use of absorption enhancers may be necessary and can be accomplished in a very controlled manner in this area.

Commercially available buccal or sublingual dosage forms include nitroglycerin for angina, buprenorphine for pain relief, ergotamine for the treatment of migraine, methyltestosterone for hypogonadism, captopril for hypertensive emergencies, and nifedipine for hypertensive emergencies and acute angina. Nicotine gum is available as a smoking cessation aid. Absorption is predominantly from the oral cavity, with a minor contribution from intestinal absorption of a swallowed drug. These dosage forms are essentially tablets that dissolve rapidly over a few minutes. An alternative approach is the use of a bioadhesive, polymeric system that would provide sustained drug delivery over an extended period of time. The use of a backing material that is impermeable to the drug and saliva directs the drug toward the mucosa and prevents drug loss because of swallowing. The feasibility of this approach has been demonstrated in clinical trials.

Nasal

The nose has good vascular drainage and an estimated blood supply of 40 mL/min for each 100 g of tissue. The nasal cavity is obviously accessible, absorption is very rapid, and first-pass metabolism in the liver is avoided. A potential disadvantage is the rapid mucociliary clearance rate for removal of trapped particles from the nose. The estimated turnover rate is 15 minutes. Both the common cold and conditions such as allergic rhinitis can affect clearance as well as the extent of absorption. The nasal route is used primarily for topical delivery of drugs, generally in aerosol form, for the treatment of allergic rhinitis and cold/flu symptoms. This route may also have utility for rapid delivery of proteins or peptides; ie, compounds that may require pulsatile rather than sustained delivery.

The permeability of the nasal mucosa is similar to that of the ileum and is therefore a leaky epithelium. The structural requirements for drug absorption from the nasal cavity have been analyzed. Examination of data for 24 compounds has shown that the nasal route is suitable for the efficient, rapid delivery of many drugs having mol wts <1,000. Mean bioavailability is 70% without the use of adjuvants. Another approach is to prolong the residence time by using bioadhesive agents such as methylcellulose, carboxymethylcellulose, hydroxypropyl cellulose, and polyacrylic acid, or bioadhesive microspheres that also protect proteins from degradation.

The effects of drugs and adjuvants must be assessed, both in short-term administration and during chronic treatment. Local effects include changes in mucociliary clearance, cell damage, and irritation. Chronic erosion of the mucous membrane may lead to inflammation, hyperplasia, metaplasia, and deterioration of normal nasal function.

Pulmonary

Drug delivery to the lung is primarily for local therapy, but pulmonary delivery may offer opportunities for systemic delivery of compounds, including vaccines.

Ocular

Drug delivery to the eye presents several challenges, based on anatomy and physiology. Drugs have to cross two lipid layers and an aqueous layer to enter the eye, and compounds such as acetazolamide that are readily absorbed elsewhere cannot effectively cross the corneal barrier. The epithelium is rate limiting for most drugs; the aqueous region; ie, the stroma, is rate limiting for very lipophilic drugs.

The eye is highly innervated, and patient comfort is of paramount importance in order to achieve good compliance. The eye is designed to keep the surface free of foreign bodies by blinking, tear production, and rapid drainage into the nasolacrimal duct. Two approaches used to increase the residence time of drugs in the eye, and consequently the amount of drug absorbed, are increasing the viscosity of the solution and the use of an implant, such as Ocusert, or hydrogel contact lenses loaded with a drug. Polymers that undergo a phase change from a liquid to a gel in response to temperature, pH, or ionic strength also show promise in this field.

Vaginal

The vaginal mucosa consists of stratified squamous epithelium, thrown into numerous transverse folds or rugae. The area is well supplied with both blood and lymphatic drainage. This route may offer opportunities for systemic delivery for the treatment of diseases, such as osteoporosis, in which the patient population is predominantly female.

NEED FOR CONTROLLED DRUG RELEASE SYSTEMS

Controlled drug release formulations (CDRFs) offer several advantages over conventional dosage forms. Some of the salient features of controlled release formulations are as follows: (1) the drug is released in a controlled fashion that is most suitable for the application. The control could be in terms of onset of release (delayed vs. immediate), duration of release, and release profile itself; (2) the frequency of doses can be reduced, thereby enhancing patient compliance; (3) the drug can be released in a targeted region, by tailoring the formulation to release the drug in that particular environment or by a timed release of the drug, (4) by targeting the drug to the desired site, systemic exposure of the drug can be reduced, decreasing systemic side effects, (5) the drug can be protected from the physiological environment for a longer time, extending the effective residence time of the drug.

However, controlled release products do not always provide positive effects for every type of formulation design. Negative effects outweigh benefits in (1) dose dumping; (2) less accurate dose adjustment; (3) increased potential for first-pass metabolism; (4) dependence on residence time in the gastrointestinal (GI) tract; and (5) delayed onset.

The limitations of CDRFs making some drugs unsuitable for formulations are as follows: (1) a risk of drug accumulation in the body if the administered drug has a long half-life, causing the drug to be eliminated slower than it is absorbed; (2) some drugs have a narrow therapeutic index and thus need to be administered repeatedly to maintain the serum drug level within a narrow range; (3) if the GI tract limits the absorption rate of the drug, the effectiveness of the CDRF is limited for oral controlled release; (4) if a drug undergoes extensive first-pass clearance, its controlled release formulation may suffer from lower bioavailability; and (5) The cost of the CDRF may be substantially higher than that of the conventional form.

Especially from the point of view of cost, improving the safety and efficacy of the new products alone has not been enough to justify introducing new CDRF products.

DESIGN OF CONTROLLED RELEASE SYSTEMS

A controlled release system comprises a drug and the material in which the drug is loaded. This system must be biocompatible and friendly with the body.

Before designing a controlled drug release system, one has to select the route of drug delivery, considering the physical and chemical properties of the drug, doses, route of administration, type of drug delivery system desired,the desired therapeutic effect, physiologic release of the drug from delivery system, the bioavailability of the drug at the absorption site, and the pharmacodynamics of the drugs. To control drug release, one can employ a variety of approaches, such as dissolution, diffusion, swelling, osmotic pressure, complexation, ion exchange, and magnetic field.

Physicochemical Properties of Drugs

Physicochemical properties of the drug affect the drug release performance of a controlled drug release system in the body. These properties, which include aqueous solubility, drug stability, molecular size, partition coefficient, and protein binding, may prohibit/restrict placement of drug in controlled release, restrict the route of drug administration, and significantly restrict the drug release performance.

Biological Properties of Drugs

At the time of designing a system, a comprehensive picture of drug deposition must be very clear, based on a complete examination of pharmacological action of the drug in *in vivo* experiments. The pharmacological action of a drug can be correlated better with the concentration–time course of the drug (or its active metabolite) in the blood or some other biophase than with the absolute dose administered and involves pharmacokinetics and pharmacodynamics of a drug in the body. Pharmacokinetics facilitates predictions of time course of drug concentrations and drug action in the body. Pharmcodynamics offer as a quantitative assessment of the time course of the drug effect on the body after administration, by any route.

The pharmacodynamics of a drug has a significant impact on the design and development of sustained release products. Each drug is characterized by its own pharmacokinetic–pharmacodynamic (PK-PD) profile (as a part of the drug-approval process) on the basis of the physicochemical properties, conformation, and other structural attributes that govern the transport within the body and across various barriers.

Factors That May Make a Drug Unsuitable for CDRF Use

Some drugs are not fit for controlled release because of the nature of drug action, physical limitations (large dose, duration of drug release), and alternative administration; ie, oral daily doses vs. monthly implant, etc. Drugs that are given at acute situations are usually not useful for controlled extended release; for example, tissue plasminogen activator (TPA), given during a heart attack to dissolve blood clotting and allow blood flow. Any delay in medication may result in death. Drugs that have a beneficial effect in the body at specific times during the day should be given only at that time, not delivered in large doses of controlled release formulations, which may result in dose dumping.

CLASSIFICATION OF CDRFs

Drug delivery systems have been classified on the basis of route administration; for example, parenteral, eternal, respiratory, transdermal, and miscellaneous flow. Controlled release systems (see Figure 1) are based on their release mechanisms, which may be erosion, diffusion, or chemically controlled. Thus, these are classified under the heading of a various categories of drug delivery. For example, under eternal drug delivery systems, release of a drug can be controlled by various mechanisms like diffusion, osmosis, or a chemically controlled mechanism. Broadly, these devices are of two types: reservoir devices and matrix devices. The former involve the encapsulation of a drug within the polymeric shell, the latter a system in which a drug is well dispersed throughout within the polymer matrix.

Diffusion Controlled

Two types of diffusion controlled systems have been used, including reservoir systems (drug coated by a polymer membrane) and matrix systems (drug dispersed in a polymer matrix). In reservoir systems, the drug is encapsulated by a polymeric membrane through which the drug is released by diffusion. In monolithic systems, the drug is dissolved or dispersed homogeneously throughout the water-insoluble polymer matrix, which may be microporous or nonporous.

Dissolution Controlled

Dissolution controlled systems can also be classified as reservoir and matrix devices. Polymers used for these devices are generally water-soluble, but water-insoluble polymers can also be used. In reservoir devices, drug particles are coated with water-soluble polymeric membranes. The solubility kinetics of the membrane depends on the thickness of the membrane and type of the polymer used. Thus, drug release can be achieved and controlled

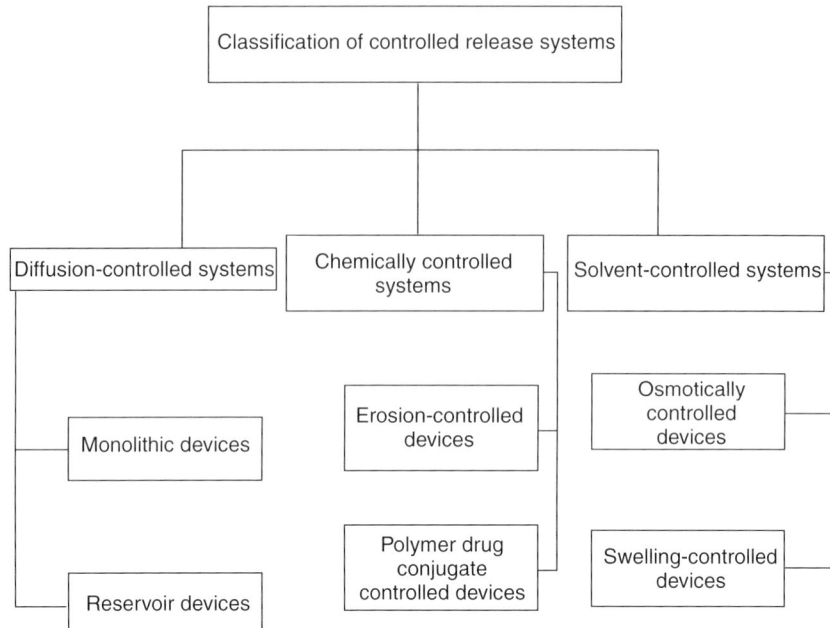

Figure 1. Classification of controlled release systems.

by preparing devices with alternating layers of drug and polymeric coats or by preparing a mixture of particles which have different coating characteristics. Matrix-dissolution devices are generally prepared by compressing a powder mix of a drug and a water-soluble or water-swellable polymer. The use of drugs with higher solubility leads to a slight acceleration of release.

Degradation/Erosion-based Systems

Degradable polymers are preferred over nondegradable for parenteral drug delivery applications. Degradation of the polymer eliminates the need for surgery to recover the spent polymer after the entire drug is released. It also reduces issues related to the long-term safety of the polymer. For biodegradable polymers, release of the drug is often intricately tied up with the polymer degradation profile.

Osmotic Delivery Systems

Osmosis-controlled devices comprises a core reservoir of drugs, with or without osmotically active salt, coated with a semipermeable membrane. The presence of salt or drug molecules creates an osmotic pressure gradient across the membrane. This plus the diffusion of water into the device gradually forces the drug molecules out through an orifice made in the device. For a durable device, the mechanical strength of semipermeable membrane should be strong enough to resist the stress building inside the device. The drug release rate from the osmotic devices, which is directly dependent on the rate of external water diffusion, can be controlled by the type, thickness, and area of the semipermeable membrane.

Ion-Exchange Systems

Polyelectrolytes have been used as a cross-linker to form water-insoluble ion-exchange resins. The drug is bound to the ionic groups by salt formation during absorption and released after being replaced by appropriately charged ions in the surrounding media. For cationic drug delivery, poly(styrene sulfonic acid) and poly(acrylic acid) can be used as anionic ion-exchange resin where sulfonic and carboxylic groups make the complexes with cationic drugs and hydrogen ions and/or other cation such as sodium or potassium ions activate the release of cationic drugs by replacing them from the drug–resin complex. On the other hand, cationic ion-exchange resins like poly(dimethylamino ethyl methacrylate) have been used for the delivery of an anionic drug. Sometimes the ion-exchange resins are coated with a polymer film such as acrylic acid and methacrylate copolymer or ethylcellulose to regulate the swelling of the resin and further control the drug release.

Polymeric Prodrugs

Many water-soluble polymers possess functional groups to which drug molecules can be covalently attached, and thus these polymers that have no therapeutic effect serve as drug carriers. The drug molecules are gradually released from the polymer by hydrolytic or enzymatic cleavage.

Magnetically Stimulated Systems

The two principal parameters controlling the release rates in these systems are the magnetic field's characteristics and the mechanical properties of the polymer matrix. The mechanical properties of the polymeric matrix also affect the extent of magnetic enhancement.

Photostimulated Systems

Photoresponsive gels reversibly change their physical or chemical properties upon photoradiation. A phase transition in polymer gels induced by visible light, where the transition mechanism is due to the direct heating of the network polymer by light, has been reported.

Ultrasonically Stimulated Systems

Researchers over the last decade have proposed that cavitation and acoustic streaming are responsible for the augmented degradation and release of biodegradable polymers. Others have speculated that the ultrasound caused increased temperature in their delivery system, which may facilitate diffusion.

REPRESENTATIVE APPLICATIONS

Controlled Release of Peptides and Proteins

Sustained release applications are especially useful for proteins/peptides because of their short half-lives. In 2000 a human growth hormone sustained release formulation became the first polymer-based sustained release formulation of a therapeutic protein to receive marketing approval from the U.S. Food and Drug Administration. This commercialized formulation encapsulates a zinc-complexed form of recombinant human growth hormone in a PLGA matrix. Other therapeutic proteins being tested for encapsulation in polymer matrices include bone morphogenic protein, erythropoietin, and nerve growth factor.

Controlled Release of Antirestenotic Agents from Stent Coatings

Stents are tiny wire scaffoldlike devices, which have become the most successful, and widely used innovation in the interventional cardiology of the last decade. More than 500,000 Americans are treated for restenosis artery scar tissue annually. Stents coated with a biocompatible polymer, encapsulating an antirestenotic agent, have proven to be a successful therapy to reduce or eliminate restenosis.

Polymeric Systems for the Treatment of Cancer

By providing sustained release at the desired site, high doses of toxic drugs can be delivered to the site without introducing the drug into systemic circulation. This results in a major advantage for the administration of chemotherapeutic drugs wherein the drugs can be encapsulated into polymer matrices and administered directly into the tumor area.

The biggest success of polymer-controlled release systems in cancer therapy, however, has been in the area of prostate cancer treatment. This unique therapy uses a hormone-suppressant rather than a chemotherapeutic agent to minimize cancer cell growth in the prostate.

Sustained Release of Drugs for CNS-Related Disorders

Another area of application for polymer-based controlled release technologies is for the sustained delivery of drugs for central nervous system (CNS)–related disorders. CNS-related drugs include a wide range of therapeutics, including pain management agents, drugs to prevent substance abuse, as well as drugs for conditions such as schizophrenia, Parkinson's and Alzheimer's disease. Drugs such as lidocaine and bupivicaine have been studied for sustained local anesthesia at the site of surgery. For this application, the anesthetic agent is encapsulated in a biodegradable polymer and injected in the proximity of the site of pain. The drug is released facilitating high concentrations at the local site, without reaching the threshold levels of systemic toxicity. In other cases such as schizophrenia, Parkinson's and Alzheimer's diseases, sustained release of the medication may reduce the chance of missed doses and aid in an effective dose regimen.

Gene Therapy

In gentherapy a gene is delivered to cells, allowing them to produce their own therapeutic proteins. Because of their highly evolved and specialized components, viral systems are by far the most effective means of DNA delivery, achieving high efficiencies (>90%) for both delivery and expression. The most promising nonviral gene delivery system thus far is the DNA vaccine application.

Nonviral vectors can be divided into physical methods, taking plasmids and forcing them into cells through such means as electroporation or particle bombardment, and chemical methods, using lipids, polymers, or proteins that will complex with DNA, condensing it into particles and directing it to the cells.

CONTROLLED RELEASE SYSTEMS ON THE MARKET

Controlled release system have gained acceptability among the various drug delivery technologies because of patient compliances, the safety of the drug, and the minimum side effects. Total sales of drug delivery products were expected to more than double from 2000, to $104 billion in 2005. Among them, more than a 20% share is for controlled drug release technologies.

Y. W. Chien, *Novel Drug Delivery Systems: Fundamentals, Developmental Concepts and Biomedical Assessments*, Marcel Dekker, New York, 1982.

L. S. Olanoff and R. E. Gibson, *Controlled-Release Technology, Pharmaceutical Applications*, American Chemical Society, Washington, D.C., 1987, 301–309.

B. D. Rattner, A. S. Hoffman, F. J. Schoen, and J. E. Lemons, eds., *Biomaterials Science: An Introduction to Materials in Medicine*. Academic Press, New York, 1996.

R. Siegel, in K. Park, ed., *Controlled Release: Challenges and Strategies*, American Chemical Society, Washington, D.C., 1997.

JOSEPH KOST
SMADAR A. LAPIDOT
Ben-Gurion University
    of the Negev
NEERAJ KUMAR
University of Tennessee Health
    Science Center
MAHESH CHAUBAL
Drugdel.com
A. J. DOMB
The Hebrew University of
    Jerusalem
RAVI KUMAR
N. V. MAJETI
University of Saarland

DRYING

Thermal drying converts a solid, semisolid, or liquid feedstock into a solid product by evaporation of the liquid into the vapor phase via application of heat. In freeze drying, drying occurs by sublimation of the solid phase directly into the vapor phase. A thermally induced phase change and the production of a solid phase as an end product are essential features of thermal drying (often termed dehydration). It is an essential operation in chemical, agricultural, biotechnology, food, polymers, ceramics, pharmaceuticals, pulp and paper, textiles, and the mineral processing as well as wood-processing industries. Over 500 types of dryers have been reported in the literature; some 100 distinct types are commercially available. It is the most energy-intensive unit operation, competing with distillation as one of the major energy-consuming industrial operations.

Gas drying is the separation of condensable vapors from noncondensable gases by cooling, adsorption or absorption. Evaporation differs from drying in that the feed and the product are both pumpable fluids.

Reasons for drying include user convenience, shipping cost reduction, product stabilization, removal of noxious or toxic volatiles, waste recycling, and disposal. Environmental factors, such as emission control and energy efficiency, increasingly influence equipment choices. Drying operations involving toxic, noxious, or flammable vapors employ gas-tight equipment combined with recirculating inert gas systems having integral dust collectors, vapor condensers, and gas reheaters.

Drying is an applied science; ie, drying theory is based on the laws of physics, physical chemistry, and the principles underlying the transfer processes of chemical and mechanical engineering: heat, mass and momentum transfer, vaporization, sublimation, crystallization, fluid mechanics, mixing, and material handling. Drying is one of several unit operations involving simultaneous heat and mass transfer. However, drying is complicated by the presence of solids that interfere with heat, liquid, and vapor flow and retard the transfer processes, at least during the final drying stages or when a solids phase is continuous.

Because all drying operations involve processing of solids, equipment material handling capability is of primary importance. In fact, most industrial dryers are derived from material handling equipment designed to accommodate specific forms of solids. If possible, liquid separation from solids as liquid, by dewatering in a mechanical separation operation, should precede drying.

Considering mode of heat transfer as the criterion for the classification, the heat transfer mechanisms used in drying are (1) convection from a hot gas that contacts the material, used in direct-heat or convection dryers; (2) conduction from a hot surface that contacts the material, used in indirect-heat or contact dryers; (3) radiation from a hot gas or hot surface that contacts or is within sight of the material, used in radiant-heat dryers; and (4) dielectric and microwave heating in high-frequency electric fields that generate heat inside the wet material by molecular friction, used in dielectric, or radio frequency, and microwave dryers. In the last group, high

internal vapor pressures develop and the temperature inside the material may be higher than at the surface.

Some of the important factors that govern the selection of industrial dryers are (1) personnel and environmental safety; (2) product moisture and quality attainment; (3) material handling capability; (4) versatility for accommodating process upsets; (5) heat- and mass-transfer efficiency; and (6) capital, labor, and energy costs.

Costs are determined by energy, labor, capacity, and equipment materials of construction. Continuous dryers are less expensive than batch dryers, and drying costs rise significantly if plant size is <500 t/year. Vacuum batch dryers are four times as expensive as atmospheric pressure batch dryers, and freeze dryers are five times as costly as vacuum batch dryers. Once-through air dryers are half as costly as recirculating inert-gas dryers. Per unit of liquid vaporization, freeze and microwave dryers are the most expensive. The cost difference between direct- and indirect-heat dryers is minimal because of the former's large dust recovery requirement. Drying costs for particulate solids at rates of $1 \times 10^3 - 50 \times 10^3$ t/year are about the same for rotary, fluid bed, and pneumatic conveyor dryers, although few applications are equally suitable for all three.

PSYCHROMETRY

Before drying can begin, a wet material must be heated to such a temperature that the vapor pressure of the contained liquid exceeds the partial pressure of vapor already present in the surrounding atmosphere. The effect of a dryer's atmospheric vapor content and temperature on performance can be studied by construction of a psychrometric chart for the particular gas and vapor. Figure 1 is a standard chart for water vapor in air.

The wet bulb or saturation temperature curve indicates the maximum weight of vapor that can be carried by a unit weight of dry gas. For any temperature on the abscissa, saturation humidity is found by reading up to the saturation temperature curve, then across to the ordinate, kg/kg dry air.

DRYING MECHANISMS

Drying Periods

The goal of most drying operations is not only to separate a volatile liquid but also to produce a dry solid of a desirable size, shape, porosity, density, texture, color, or flavor. An understanding of liquid and vapor mass-transfer mechanisms is essential for quality control. Mass-transfer mechanisms are best understood by measuring drying behavior under controlled conditions in a prototypic, pilot-plant dryer. No two materials behave alike, and a change in material handling method or any operating variable, such as temperature or gas humidity, also affects mass transfer.

Figure 2**a** shows drying time profiles for one material dried under three conditions. Corresponding rate profiles are in Figure 2**b**. Three products having uniquely

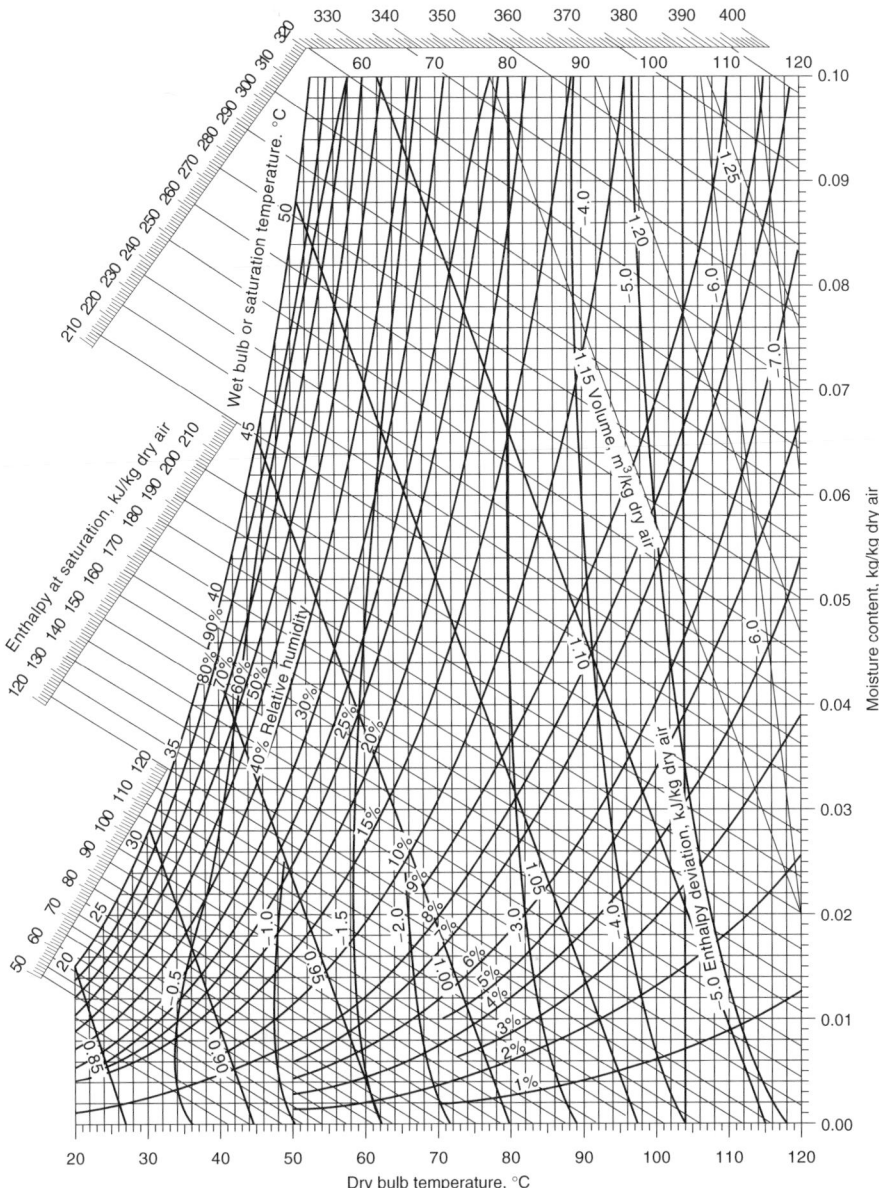

Figure 1. Carrier psychrometric chart for air and water vapor at 101.325-kPa (1-atm) total pressure. Courtesy of Carrier Corp.

different characteristics were produced by three different kinds of agitation. Other controllable drying conditions were constant. These profiles show that during drying several distinct periods may occur, which depend on how the material is handled. These are (*1*) an induction period during which wet material is heated to drying temperature; (*2*) a constant-rate drying period indicated by the horizontal portions of the profiles in Figure 2**b**; (*3*) a period of decreasing rate shown by the sloping portions of the two rate profiles during which the drying rate appears proportional to moisture content; and (*4*) a period of decreasing rate shown by the curved portions of the two rate profiles during which the drying rate is evidently a more complex function of moisture content than just being a simple proportionality.

The moisture content at the end of the constant-rate drying is the critical moisture content. The drying periods that follow are falling rate periods. The curved portion of profile B in Figure 2**b** is a second falling rate period;

moisture content at the second break is the second critical moisture content. Profile C shows that drying may occur almost entirely in a falling rate period; a slight change in specified product moisture content can have a significant effect on drying time.

Constant Rate Drying

During constant rate drying, vaporization occurs from a liquid surface of constant composition and vapor pressure. Material structure has no influence, except moisture movement from within the material must be fast enough to maintain the wet surface. The vaporization rate is controlled by the heat-transfer rate to the surface. The mass-transfer rate adjusts to the heat-transfer rate and the wet surface reaches a steady-state temperature. The drying rate remains constant, therefore, as long as external conditions are constant. If heat is supplied solely by convection, the steady-

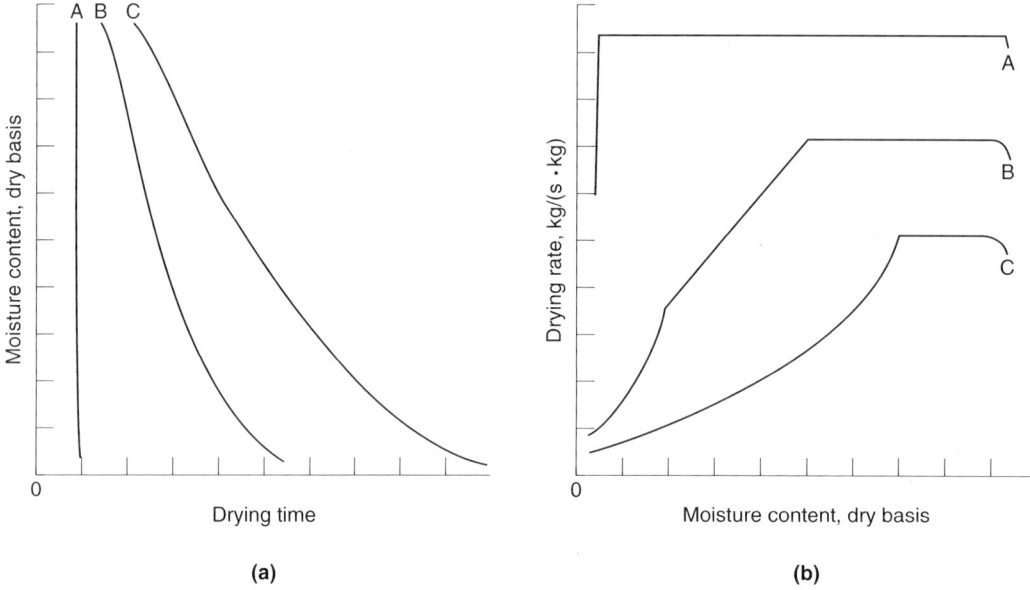

Figure 2. (**a**) Profiles of moisture content versus drying time; and (**b**) drying rate versus moisture content for a slightly soluble, water-wet organic powder centrifuge cake at 4.0-kPa (0.58 psi) absolute pressure using 120°C indirect heat. Profile A was produced in a continuous, high-speed agitator dryer provided with scrapers to maintain a clean heating surface. Drying time was 45 s and at an almost entirely constant rate because of high solids surface exposure to the heating surface; the product particle size was 100% <150 μm. Profile B was produced in a paddle-agitated batch dryer also having scrapers. Drying time was 70 min, including periods of constant rate, capillary, and diffusion drying. Because of the much slower agitator, the product was a porous, 100–500-μm powder having some dust. Profile C was produced in a double-cone batch dryer using some dry recycle. Drying time was 120 min, and was almost entirely liquid and vapor diffusion-controlled because turning of the double-cone pelletized the wet material early in the cycle; the product was composed of rather dense (200–800 μm) spheres, having negligible dust.

state temperature is the gas wet-bulb temperature. When conduction and radiation contribute; eg, the material contacts and/or receives radiation from a warm surface, a liquid surface temperature between the wet-bulb temperature and the liquid's boiling point is obtained. In indirect-heat and radiant-heat dryers, where conduction and radiation predominate, surface liquid may boil regardless of ambient humidity and temperature. During constant rate drying, material temperature is controlled more easily in a direct-heat dryer than in an indirect- heat dryer, because in the former the material temperature does not exceed the gas wet-bulb temperature as long as all surfaces are wet. For convection, all principles relating to simultaneous heat and mass transfer between gases and liquids apply.

Contact Drying

Contact drying occurs when wet material contacts a warm surface in an indirect-heat dryer. A sphere resting on a flat heated surface is a simple model. The heat-transfer mechanisms across the gap between the surface and the sphere are conduction and radiation.

Critical Moisture Content

Critical moisture content, is the average material moisture content at the end of constant rate drying. Critical moisture content cannot be determined except by a prototypic drying test.

Particle size distribution determines surface/mass ratios and the distance internal moisture must travel to reach the

surface. Large pieces thus have higher critical moisture contents than fine particles of the same material dried under the same conditions. Pneumatic-conveyor flash dryers work because very fine particles are produced during initial dispersion and these have low critical moisture contents.

Case hardening refers to a circumstance in which a mass of nonporous, soluble, or colloidal material is dried at such a high rate during initial constant rate drying that the surface overheats and shrinks. Because liquid diffusivity decreases with moisture content, the barrier formed by the overdried surface prevents moisture flow from the interior of the mass to the surface. Case hardening of nonporous materials can be minimized by initially maintaining a high relative humidity environment and consequently a high surface equilibrium moisture content until internal moisture has time to escape.

Equilibrium Moisture Content

Equilibrium moisture content is the steady-state equilibrium reached by the gain or loss of moisture when material is exposed to an environment of specific temperature and humidity for a sufficient time. The equilibrium state is independent of drying method or rate. It is a material property. Only hygroscopic materials have equilibrium moisture contents. Clean beach sand is nonhygroscopic and has an equilibrium moisture content of 0. The same rules apply to organic vapors. Hygroscopic material retains a constant fraction of moisture under specific ambient humidity and temperature conditions. At constant temperature, if ambient humidity increases or decreases, an increase or decrease in moisture content follows, which is called equilibrium moisture, because it is held in vapor pressure equilibrium with the partial pressure of vapor in the atmosphere. The reason it is retained even when the atmosphere is quite dry is that the retention mechanism reduces effective liquid vapor pressure. It is bound moisture because it is bound to material in solution or by adsorption, and bound moisture behaves as if the atmosphere were saturated even when it is not, relative to the unbound liquid's normal vapor pressure. Chemically combined liquid may behave like bound moisture, depending on the nature of the chemical bond. Because equilibrium is influenced by partial vapor pressure in the atmosphere and the effective vapor pressure of the bound liquid, temperature and humidity are both important. For many materials in the 15–50°C temperature range, equilibrium moisture content can be plotted versus relative humidity as an essentially straight line.

Falling Rate Drying

Heat transfer is limited by material conductivity, but the drying rate usually is controlled by internal liquid and vapor mass transfer. The principal mass-transfer mechanisms are (1) liquid diffusion in continuous, homogeneous materials; (2) vapor diffusion in porous and granular materials; (3) capillarity in porous and fine granular materials; (4) gravity flow in granular materials; (5) flow caused by shrinkage-induced pressure gradients; and (6) pressure flow of liquid and vapor when porous material is heated on one side, but vapor must escape from the other.

Usually, only one mass transfer mechanism predominates at any given time during drying, although several may occur together. In most materials, the mechanisms of internal liquid and vapor flow during falling rate drying are complex. Simultaneous heat transfer is a factor, and falling rate drying rarely can be described with mathematical precision. Computer models for some materials are published, but most employ data from actual drying tests. In the absence of tests, the falling rate drying periods usually are studied on the assumption that internal mass transfer is controlled either by diffusion or capillarity depending on whether the material is porous or nonporous, soluble or not.

Drying Profiles

An application of diffusion principles to falling rate drying is exemplified in Figure 3. Single drops of whole milk were dried by suspension in a warm air stream. Because of the rapid formation of surface films, drying was mostly by vapor diffusion.

DRYERS

Industrial dryers may be broadly classified by heat-transfer method as being either direct or indirect heat. Dryers evolved from material handling equipment and thus most types of industrial dryers are specially suited for certain forms of material. Dryers are also classified as being batch or continuous.

The material suitability of industrial dryers may be summarized as follows.

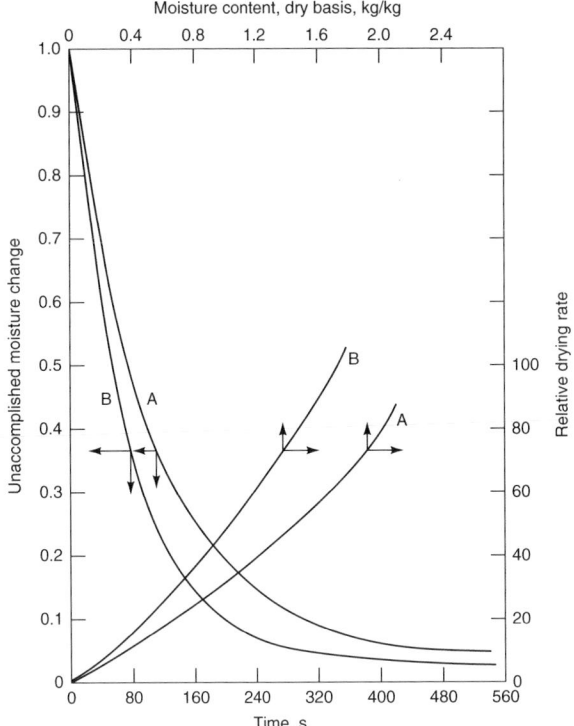

Figure 3. Drying profiles for single drops of whole milk at 94°C and 0.6-m/s relative air flow. The initial diameter of drop A = 1900 μm, initial moisture content = 2.6 kg/kg, dry basis; drop B = 1470-μm initial diameter and 2.4-kg/kg moisture.

Dryer	Material form
spray dryer	pumpable, heat-sensitive pastes, slurries, and solutions; all pumpables at high capacities
indirect-heat drum dryer	pumpable, heat-insensitive pastes, slurries, and solutions
pneumatic conveyor dryer	materials instantly dispersible into discrete particles in the drying gas
fluid-bed dryer	fluidizable particulate materials
spouted-bed dryer	particulate materials too coarse or uniform in size to fluidize adequately
hopper dryer	preheated coarse or uniform materials and beds pervious to gas throughflow
direct-heat rotary dryer	particulate materials too coarse, sticky, or unpredictable to be fluidized or spouted
indirect-heat rotary dryer	fine, dusty materials
double-cone (vacuum) dryer	particulate materials that do not stick together, ball, or pelletize during drying
agitator (vacuum) dryer	particulate materials that may stick together, ball, or pelletize until almost dry
through-circulation or band dryer	materials that can be formed into static beds pervious to gas throughflow
continuous conveyors	continuous webs, paper, fabric, film, and fiber tow
batch, cabinet, and tray dryers	small lots, batch identification, and single-product plants <500 t/year

Direct Heat Dryers

In direct-heat dryers, steam-heated, extended-surface coils are used for gas heating up to ~200°C. Electric and hot oil or vapor heaters are added for higher temperatures. Diluted combustion products are used for all temperatures. An increasingly popular technique for producing inert gas is to recycle the dryer exit gas and vapor as secondary dilution gas for incoming combustion products. Thereby the oxygen level in the dryer gas stream is reduced to safe levels for organic materials. This is usually less than 10% oxygen, but always material dependent. These are called self-inerting heaters.

Indirect-Heat Dryers

In indirect-heat dryers, heat is transferred mostly by conduction, but heat transfer by radiation is significant when conducting surface temperatures exceed 150°C.

Radiant-Heat Dryers

Heat transfer by radiation occurs in all dryers to some degree and is controlled by the temperature and emissivity of the source and the temperature and absorptivity of the receiver. For drying, sources may consist of a number of incandescent lamps, reflector-mounted quartz tubes, electrically heated ceramic surfaces, and ceramic-enclosed gas burners. Usual source temperatures are 800–2,500 K. Radiant energy does not penetrate most material surfaces. Heat penetration below the surface is dependent on material conductivity. Radiant heaters are most suitable for the drying of thin films; eg, paint films.

Dielectric and Microwave Dryers

Dielectric, also called radio frequency, dryers operate in the frequency range of 1–100 MHz. Microwave dryers in the United States operate at 915 and 2450 MHz.

Microwave applicators are single, like a microwave oven, or multimode cavities in which material is placed or through which it is conveyed, or rectangular waveguides, which in effect surround material as it is conveyed. Rapid reversal of electrode polarity generates heat in the material. In a mechanism called dipole rotation, dipoles, which normally are in random orientation, become ordered in the electrical field. As the field dies, they return to random orientation; as the field reverses, they again become ordered, but in the opposite direction. Electrical energy is converted to potential energy, to random kinetic energy, and to heat. In ionic conduction, ions are accelerated by the electrical field. They collide with nonionized molecules in random billiard ball fashion. Electrical energy is converted to kinetic energy and to heat. These are two primary mechanisms of energy conversion.

Industrial applications of dielectric and microwave energy for drying are many; however, response to high-frequency electromagnetic radiation depends on a material's dielectric constant and dissipation factor, the product of which is its loss factor. A material having a loss factor >0.05 is a potential drying candidate.

The cost of microwave equipment per kilowatt output is about twice that of dielectric. For irregular shapes, microwaves are preferable, because to avoid hot spots during heating, dielectric electrodes are needed that conform to the material shape. Industrial dielectric dryers are employed for lumber drying, plywood bonding and drying, furniture parts drying, textile skeins and package drying, paper moisture leveling, tire cord drying, and many food products. Dielectric heating frequently is combined with radiant heat and hot air for print and coating drying. Microwave dryers are employed for drying cloth, lumber, and foods. Microwaves are used as an energy source in vacuum and freeze dryers.

Dielectric and microwave heating are generally more costly than alternative methods. Thus many applications involve material preheating and second-stage drying where energy demand is low and cycle times can be reduced significantly. Dielectric and microwave heating are chosen mostly when other methods will not work or are impractical.

Superheated Steam Dryers

Superheated steam drying (SSD) has emerged as a viable drying technology with major potential only in recent years. Essentially SSD involves use of superheated steam in a direct convective dryer of any design. One of the obvious advantages of SSD is that the dryer exhaust is also superheated steam, although of a lower specific enthalpy. In air drying, the latent heat of water vapor in the exhaust is difficult and expensive to recover. On the other hand, if it is pure steam the exhaust heat can be recovered more cost effectively by partial recycling, compression, or condensation.

Among the current commercial applications of steam drying are drying of coal, pulp, biomass, peat, waste sludges, wood, and some meat products. There is potential for application of SSD for numerous products.

C. G. J. Baker, ed., *Industrial Drying of Foods*, Blackie Academic & Professional, London, 1997.

S. Devahastin, ed., *Mujumdar's Practical Guide to Industrial Drying*, Exergex, Brossard, Canada, 2000.

S. J. Kowalski, *Thermodynamics of Drying Processes*, Springer, London, 2003.

A. S. Mujumdar, ed., *Drying Technology in Agriculture and Food Sciences*, Science Publishers, Enfield, N.H, 2001.

PAUL Y. MCCORMICK
Drying Unincorporated
ARUN S. MUJUMDAR
National University of
Singapore

DRYING OILS

Drying oils are liquid vegetable or fish oils that react with oxygen to form solid films. They have been used since prehistoric times; in the nineteenth and early twentieth centuries, binders of most paints and printing inks were drying oils. Their use has decreased; however, they still have applications. Most importantly, they are raw materials for other binders such as alkyd resins, epoxy esters, and uralkyds. These resins can be considered to be synthetic drying oils.

COMPOSITION OF NATURAL OILS

Naturally occurring oils are triglycerides. The reactivity of drying oils with oxygen results from the presence of diallylic groups (ie, two double bonds separated by methylene groups, $-CH=CHCH_2CH=CH-$) or conjugated double bonds. Typical fatty acid contents of some oils are given in Table 1.

AUTOXIDATION AND CROSS-LINKING

Drying, semidrying, and nondrying oils are often defined based on their iodine value, that is, the number of grams of iodine required to saturate the double bonds of 100 g of an oil. Although iodine values can serve as satisfactory quality control specifications, they are not useful and can be misleading in defining a drying oil or for predicting reactivity.

Nonconjugated oils that have a drying index >70 are drying oils. The drying index is calculated as follows: Drying index=(% linoleic acid)+2(% linolenic acid).

Table 1. Typical Fatty Acid Compositions of Selected Oils

Oil	Saturated[a]	Oleic	Linoleic	Linolenic	Other
linseed	10	22	16	52	
safflower	11	13	75	1	
soybean	15	25	51	9	
sunflower, mn	13	26	61	trace	
sunflower, tax.	11	51	38	trace	
tung	5	8	4	3	80[b]
tall oil fatty acids[c]	8	46	41[d]	3	2[e]
tall oil fatty acids[f]	2.5	30	45	1	14[g]
castor	3	7	3		87[h]

[a] Saturated fatty acids are mainly mixtures of stearic (C_{18}) and palmitic (C_{16}) acids.
[b] α-Eleostearic acid.
[c] North American origin.
[d] Linoleic plus geometric and conjugated isomers.
[e] Rosin.
[f] European origin.
[g] Pinolenic acid.
[h] Ricinoleic acid.

If we use this formula with the data from Table 1, the drying index of linseed oil is 120, it is a drying oil; the drying index of soybean oil is 69, it is a semidrying oil. The reactivity of nonconjugated drying oils is related to the average number of diallylic groups per molecule. The methylene groups are activated by their allylic relationship to two double bonds and are much more reactive than methylene groups allylic to only one double bond. The reactions taking place during drying are complex with many side effects.

The rates at which uncatalyzed nonconjugated drying oils dry are slow. Metal salts (driers) catalyze drying. The most widely used driers are oil-soluble cobalt, manganese, lead, zirconium, and calcium salts of octanoic or naphthenic acids. Salts of other metals, including rare earths, are also used. Cobalt and manganese salts, so-called top driers or surface driers, primarily catalyze drying at the film surface. Lead and zirconium salts catalyze drying throughout the film and are called through driers. Mixtures of lead with cobalt and/or manganese are particularly effective but, as a result of toxicity control regulations, lead driers can no longer be used in consumer paints sold in interstate commerce in the United States. Combinations of cobalt and/or manganese with zirconium, frequently with calcium, are commonly used. The amounts of driers needed are system specific. Their use should be kept to the minimum possible level, since they not only catalyze drying, but also catalyze the reactions that cause postdrying embrittlement, discoloration, and cleavage.

Oils containing conjugated double bonds, such as tung oil, dry more rapidly than nonconjugated drying oil. Free radical polymerization of the conjugated diene systems can lead to chain-growth polymerization, rather than just a combination of free radicals to form cross-links. In general, the water and alkali resistance of films derived from conjugated oils are superior, presumably because more of the cross-links are stable carbon–carbon bonds. However, since the α-eleostearic acid in tung oil has three double bonds, discoloration on baking and aging is severe.

SYNTHETIC AND MODIFIED DRYING OILS

Heat-Bodied Oils, Blown Oils, and Dimer Acids

Both nonconjugated and conjugated drying oils can be thermally polymerized by heating under an inert atmosphere to form bodied oils that have higher viscosities and are used in oil paints to improve application and performance characteristics.

The viscosity of drying oils can also be increased by passing air through oil at relatively moderate temperatures, 140–150°C, to produce blown oils. Presumably, reactions similar to those involved in cross-linking cause autoxidative oligomerization of the oil. The oligomerization increases functionality, and hence accelerates the drying of linseed oil and makes a drying oil from soybean oil.

Polyunsaturated acids dimerize or oligomerize by heat treatment; the reactions are acid catalyzed. The products obtained are called dimer acids, used to make polyesters and aminoamides.

Varnishes

The drying rate of drying oils can be increased by dissolving a solid resin in the oil and diluting with a hydrocarbon solvent. Such a solution is called a varnish. The solid resin serves to increase the T_g of the solvent-free film so that film hardness is achieved more rapidly.

In varnish manufacture, the drying oil (usually linseed oil, tung oil, or mixtures of the two) and the resin are cooked together to high temperature to obtain a homogeneous solution of the proper viscosity. The varnish is then thinned with hydrocarbon solvents to application viscosity. Varnishes were widely used in the nineteenth and early twentieth centuries but have since been almost completely replaced by a variety of other products, especially alkyds, epoxy esters, and uralkyds. The term "varnish" has come to be used more generally for transparent coatings, even though few of them today are varnishes in the original meaning of the word.

Synthetic Conjugated Oils

Tung oil dries rapidly but is expensive, and its films discolor rapidly, due to the presence of three double bonds. These shortcomings led to efforts to synthesize conjugated oils, especially those containing esters of fatty acids with two conjugated double bonds.

Esters of Higher Functionality Polyols

When oil-derived fatty acids are reacted with polyols with more than three hydroxyl groups per molecule, the number of cross-linking sites per molecule increases relative to the corresponding natural triglyceride oil. Alkyds, epoxy esters, and uralkyds made with fatty acids from such oils as soybean and linseed oils can be considered higher functionality synthetic drying oils.

Maleated Oils

Both conjugated and nonconjugated oils react with maleic anhydride to form adducts. The products of these reactions, termed maleated oils, or sometimes, maleinized oils, react with polyols to give moderate molecular weight derivatives that dry faster than the unmodified drying oils. Such products have not found significant commercial use, but the process is used to make water-reducible alkyds and epoxy esters.

Vinylated Oils

Both conjugated and nonconjugated drying oils react in the presence of a free-radical initiator, usually benzoyl peroxide, with such unsaturated monomers as styrene, vinyltoluene, and acrylic esters. High degrees of chain transfer result in the formation of a variety of products, for example, a low molecular weight homopoly-mer of the monomer, short-chain graft copolymers, and dimerized drying oil molecules. The reaction of drying oils with such monomers is no longer commercially important, but the same principle is used to make modified alkyds.

USES

The use of drying oils has decreased substantially since the early part of the twentieth century. Some formerly large uses have been completely replaced with other materials: Oil cloth and linoleum have been completely converted to vinyl resins or other polymers. Drying oil vehicles for putty have been virtually completely replaced by acrylic latexes. Use of oils in paints and coatings, printing inks, and artists' colors have been substantially reduced. The largest use for drying oils now is in the manufacture of alkyd resins, epoxy esters, uralkyds, and modified drying oils.

In the paint and coatings industry, relatively small amounts of drying oils are used in the manufacture of penetrating stains. These stains are made with an oil such as linseed oil. They have been replaced to a large extent with stains formulated with latexes but oil stains give greater penetration and more clarity. Tung oil is used as a penetrating finish for wood furniture by the do-it-yourself trade. Some red-lead-in-oil corrosion protective primers are still used in applications where there is no risk of lead poisoning. Linseed oil and modified linseed oils have been used as concrete coatings to protect against spalling. Very little drying oil is used in wall paints or exterior house paints now, replaced with alkyd resins or latexes.

Some drying oils are used in printing inks, particularly in sheet-fed lithographic printing, primarily after being converted to alkyd resins. Drying oils and modified drying oils are so-called renewable resources. Polyesters prepared with dimer acids are used in making vehicles for coil coating with exceptional flexibility and formability.

Linseed oil has been used as a binder for replacement brake linings and as a core oil to bond sand for metal-casting molds.

F. L. Fox, *Oils for Organic Coatings*, Federation of Societies for Coatings Technology, Blue Bell, Pa., 1965.

A. E. Rheineck and R. O. Austin, in R. R. Myers and J. S. Long, eds., *Treatise on Coatings*, Vol. I, No. 2, Marcel Dekker, New York, 1968, pp. 181–248.

ZENO W. WICKS, JR.
Consultant

DYEING

The global consumption of textiles as the twenty-first century began was estimated at ~50 million tons per year, with polyester and cotton accounting for ~75% of the total. Fiber use has generally increased at a rate faster than the growth of population. Dye is applied to fibers at an average rate of 1–2%; ie, annual production of synthetic dyes is ~750,000 tons.

Dyeing occurs when a soluble colorant is adsorbed at the surface of a fibrous substrate, and then diffuses into the substrate. Colorants are typically divided into dyes pigments, based chiefly on solubility: A pigment relies on insolubility in the medium in which it is dispersed, while a dye requires some degree of solubility that will allow it to diffuse into the polymeric matrix of a textile fiber. In order for a colored substance to be regarded as a useful dyestuff, factors beyond solubility are required. A dyestuff must be substantive for a textile and thus be preferentially taken up by the fiber, usually from an aqueous solution. The uptake should be high enough to be economic, and the rate at which the dyeing occurs should be controllable to give a uniform, level result. The dyed textile should have satisfactory fastness for the intended end use. The process of dyeing therefore combines chemistry, application technology, economics, and customer needs.

The *Colour Index* assigns CI generic names to commercial dyes. The main CI dye application types are acid, mordant, direct, reactive, sulfur, vat, azoic, disperse, and basic. The colors are restricted to yellow, orange, red, violet, blue, green, brown, black and (for pigments) white.

THE DYEING PROCESS

Fibers

For dyeing to take place, the polymeric substrate must be accessible to dye molecules. Fibers may be natural or manufactured, hydrophilic or hydrophobic, nonionic or ionic. Natural fibers vary with (sub-)species of plant or animal and with the growing conditions experienced as the fiber forms. Synthetic fibers are generally more consistent, but are available in a myriad of variations. Most manufactured fibers are drawn after extrusion to obtain better mechanical properties by the development of increased orientation and crystallinity of the polymer chains, but subtle variations soon in the rate or extent of dye uptake, to the frustration of a dyer trying to achieve a level dyeing.

The general chemical rule of thumb that "like dissolves like" corresponds in dyeing to "like dyes like": Hydrophilic substrates are dyed with hydrophilic dyes, hydrophobic dyes dye hydrophobic fibers.

Hydrophilic Fibers

Cellulose and protein substrates (cotton, rayon, lyocell, linen, silk and wool, etc.) are hydrophilic and absorb water readily. The absorption of water by wool is less immediate because of hydrophobic surface scales, but in a dyebath the fiber is swollen with water. The model for the uptake of dyes by these fibers is thus one of water-filled pores through which soluble dye diffuses.

Hydrophobic Fibers

Hydrophobic fibers absorb comparatively little water, are typically thermoplastic, and undergo a glassy–rubbery transition at a characteristic temperature, the so-called glass-transition temperature T_g. Above this temperature the polymer chain segments are mobile, and at any given time there is a free volume within the polymer matrix. The fiber is thus better regarded as a system of continuously changing regions of "free volume" through which dye can diffuse.

Zeta Potential. When a textile is immersed in water and relative motion exists between the two a negative charge called the electrokinetic or zeta potential is developed on its surface. Negatively charged dyes therefore are coulombically repelled, and uptake is reduced unless electrolytes are present to diffuse this charge. Similarly, cationic dyes are ionically attracted, and initial uptake (strike) may be rapid and unlevel unless precautions are taken. Since dyeing is a penetration of dye within the fiber structure, this surface charge is not useful in providing an attractive force between dyes and fibers. The zeta potential of a fiber is, however, useful for "exhausting" cationic materials onto the surface of a fiber. Thus cationically charged pigment and/or binder materials can be attracted to a fiber surface, and fixed in a subsequent curing process. This is the basis of the oxymoronically titled process of "pigment dyeing". The attraction of a cationic material to this fiber surface charge is also commonly used in household laundry when a fabric softener is applied in a rinse cycle.

Water: The Medium for Dyeing

Water is a polar solvent that interacts with both dyes and fibers. Dyes are often ionic, and the colored ion is solvated by water. Nonionic dyes typically have polar groups that interact with water, and provide limited but essential solubility. Other parts of a dye molecule may be hydrophobics and have positive or negative effects on the local water structure. Overall, the transition from a dye dissolved in water to one bound to a substrate is accompanied by changes in the binding of water with both components, with concomitant effects on the energy changes that drive the process.

Ideally, dyeing would take place from a pure medium, but in practice, it is uneconomic to purify large volumes of water when experience has shown that certain levels of impurity can be tolerated. There is on one specificaton forsatisfactory dyehouse water. Generally of concern, however, are alkalinity, hardness, and heavy metals.

Hardness, heavy metals and to some extent pH can be dealt with by the use of ion exchange resins. Where these impurities are present, plant water can be routinely treated. Water from municipal sources is often treated with low levels of chlorine but even low levels of chlorine can affect the shade of some dyes, and a prudent dyer would add an antichlor to a dyebath if chlorine is present.

Solvent Dyeing. The use of media other than water for dyeing has not materialized into practical acceptance, chiefly on cost and environmental grounds.

Attractive Forces Between Dyes and Fibers

Assuming that a fiber is accessible, the forces of attraction between a dye and a fiber are those that apply in any organic chemical system. While covalent bonds do ultimately contribute to the linking of reactive dyes and fibers, these occur only after secondary forces have brought the dye and fiber together. For most dye–fiber systems, only noncovalent bonding operates.

For dye-fiber bonding to take place, such bonding has to replace the bonding of both the dye and the fiber with water: In the case of the dye, the interaction with water represents its solubility. For dyeing to take place to an economical extent the *net* energy (enthalpy and entropy) change must drive the process forward, and any positive contribution from the formation of dye-fiber bonding must outweigh any negative contribution from the loss of bonding with water.

Aggregation of Dye. While simple ionic dye molecules dissolve readily, larger and more complex dye molecules tend to aggregate in solution as ionic repulsion of groups in one part of the molecule are counterbalanced by the attractive intermolecular forces that occur between other parts. Dyes can thus mimic surfactants in combining hydrophilic and hydrophobic properties in the same molecule, and dye aggregation shares some characteristics of micelle formation in surfactants. Practical problems arise when aggregates break down over a limited temperature range, leading to a rapid availability of dye for uptake by the fiber and possibly unlevel absorption.

Once the dye is sorbed into the fiber (or even concurrent with its sorption) a dye may undergo a chemical reaction with functional groups within the fiber to form covalent bonds. This is the characteristic of reactive dyes. The covalent bonding of the dye and fiber subsequent to sorption provide the dyeing with good wet fastness properties.

By definition, dyes are soluble entities during the dyeing process. However, dyes may, after sorption into the fiber, undergo a chemical change to become insoluble, or revert to insolubility. This is the case with azoic/ingrain dyes and vat or sulfur dyes, respectively.

Processes for Dyeing

For the most part, dyeing is accomplished in batch processes. When large amounts of a single color (shade) are required, however, continuous processing is more efficient. In this process uncolored material is fed into the system and colored material withdrawn continuously.

Printing

When textile substrates are printed with colored designs, dyes or pigments can be used. Currently, the market is approximately evenly divided between the two. Most printing is a continuous process. When dyes are used, the dye is applied to the fiber in a paste thickened to control its rheology and thus its transfer onto the fabric and the sharpness of the printed design obtained. The paste may also contain the chemicals required for fixation.

GENERAL DYEING PRACTICE

Dyers' Controls

A practical dyeing process requires stirring to achieve initial level sorption and for subsequent migration, and to make diffusion through the fiber the rate-determining step.

The liquor ratio may be dependent on the dye machine and amount of substrate to be dyed and so not be fully under the dyer's control. As the liquor ratio increases, the dyebath is more dilute and the dye has a longer and slower diffusion path through it. Because a higher liquor ratio means that less dye will exhaust, but more energy used, for efficiency the trend is to lower liquor ratios.

Since dyeing is an exothermic process, an increase in temperature will displace equilibrium in favor of the dyebath. In other words, equilibrium exhaustion is lower at higher temperatures. More importantly, a higher temperature of dyeing increases the rate of dyeing and the rate at which equilibrium is attained.

Auxiliaries are chemical additions to bath to aid the dyeing process. They may interact with the fiber or with the dye. They may increase or decrease the rate and extent of dyeing.

The Aims of a Dyer

The aims of a dyeing can be summarized as achieving the correct shade, a level dyeing, in which the dye is distributed evenly throughout the substrate, fastness, the resistance a dye on a substrate has to removal or destruction, no damage to the substrate, and efficiency.

DYEING OF CELLULOSIC FIBERS

Preparation for Dyeing

Cotton may be made suitable for dyeing in a variety of forms, such as raw stock, yarn, or piece goods. Cotton fibers are coated with natural waxes and pectins, and are contaminated with motes. Raw stock is normally dyed without thorough dewaxing, since the natural waxes aid in subsequent spinning operations. Surfactants are employed to aid the penetration of dyestuffs through the protective waxes. Flaws in the dye levelness are overcome by the mixing that takes place in subsequent carding.

Careful preparation of cotton piece goods is essential to achieve suitable dye penetration, fastness, and general appearance. Fabric construction dictates whether the fabrics will be processed in rope or open-width forms. Heavy piece goods, and those which are subject to rubs and crease marks, are handled in open width.

Before dyeing light or bright shades, the goods should be bleached with hydrogen peroxide and caustic soda. This operation also helps in the removal of trace impurities that remain after scouring.

Mercerizing is accomplished by passing the cotton fabric under tension through 15–30% caustic soda, and then rinsing out (and recycling) the alkali. While knitted fabrics do not contain size, they are typically contaminated with knitting oils. Viscose rayon, because of its low wet strength, must be processed under minimum tension at all stages of preparation. Yarns contain few impurities and require only light scouring.

The Dyeing Process

When cellulose fiber is immersed in water it develops a negative charge, the zeta potential. In order for dyes to show good affinity on cellulose the dyes must be soluble, and have planar, aromatic structures. Solubility is generally achieved via sulfonic acid or other negatively charged groups and the dyes show long-range, Coulombic (ionic) repulsion, but very strong short-range van der Waals forces of attraction. Thus there is a potential barrier that a dyestuff molecule has to overcome.

The use of electrolyte is critical in the dyeing of cellulose. Sodium chloride or sodium sulfate are most often used, although any electrolyte is effective.

Direct Dyes

The simplest way of coloring cellulosic fibers is with direct dyes. Direct dyes provide reasonably bright shades that cover most of the color gamut. The addition of salt is used to allow dyestuff to be absorbed on the fiber. This is done carefully to ensure that level dyeing is achieved, especially during the early stages of dyeing.

Leveling Power. Direct dyes are classified according to their leveling characteristics. Class A direct dyes migrate well and have high leveling power; ie, they have low affinity and high diffusion. Class B direct dyes level less well than Class A, and exhaustion must be controlled by careful salt addition.

Class C direct dyes have high affinity where, resulting from the complexity of the molecules, the nonionic forces of attraction dominate and can overcome the ionic repulsion of the surface. They are thus dyes of poor leveling power which exhaust well without added salt and the rate of exhaustion is controlled by temperature.

Wetfastness. Class A direct dyes offer the most trouble-free process for dyeing cellulose. However, they do not always provide sufficient wetfastness.

The Class B and C dyes show better resistance to desorption, ie, they show higher wetfastness, but they do not

overcome it fully, and even the Class C direct dyes show inadequate wetfastness and poor staining of adjacents in more severe fastness tests as a result of the reversible nature of the dyeing process.

The general utility of direct dyes and their relatively simple application processes have made aftertreatments to increase their fastness attractive propositions. A wide variety of treatments was developed for dyes with appropriate functional groups, but with the rise in popularity of reactive dyes, such treatments have declined in use. However, the application after dyeing of a cationic surface-active agent to form a sparingly soluble complex with the dye will improve wetfastness slightly and is still of some use.

Fiber-Reactive Dyes

Reactive dyes form the most recently introduced class of dye. Their good wetfastness and extensive shade range have made them the dominant dye type for dyeing cellulose fibers.

A reactive dye contains a chemical group capable of reaction with cellulose to form a covalent bond. It is absorbed onto cellulose in the same manner as a direct dye, with the aid of an electrolyte. When alkali is added to the dyebath, ionization of cellulose and the reaction between dye and fiber is initiated. As the reaction proceeds, dye is removed from the equilibrium between active dye in the dyebath and fiber phases, which is reestablished by the absorption of more dye. The addition of alkali produces both secondary exhaustion and reaction (fixation).

At the end of the dyeing process very little, if any, active dye remains: There is fixed dye on the fiber, and hydrolyzed dye in both the dyebath and fiber. Any hydrolyzed dye that remains absorbed by the fiber will tend to desorb (like a direct dye) in subsequent laundering, and negate the benefit of the covalently bonded (and thus fast) dye. Even very small amounts of residual hydrolyzed dye (1–2% of the original amount applied) give heavy stains on adjacent fabrics used in fastness tests. The same forces that promoted initial sorption make the removal more than a simple rinsing process. Removal is a relatively difficult procedure that is a critical part of the total dyeing process. The final steps in a fiber-reactive dyeing are thus a series of rinses that remove first electrolyte and alkali, and then encourage the desorption of hydrolyzed dye. Once a dye is fixed it cannot migrate and therefore dyeings achieved using fiber-reactive dyes must be level before they are fixed.

The Ideal Fiber-Reactive Dye Profile. Figure 1 shows the general profile for the application of a reactive dye. In addition to showing the rate profile of fixation between dye and fiber, three other practical parameters (A–C) are noted.

The overall objective is to make the fixation (C) as high as possible for economic reasons. The closer the fixation (C) is to the total dye on the fiber (B) then the smaller the concentration of hydrolyzed material [dye-OH] and

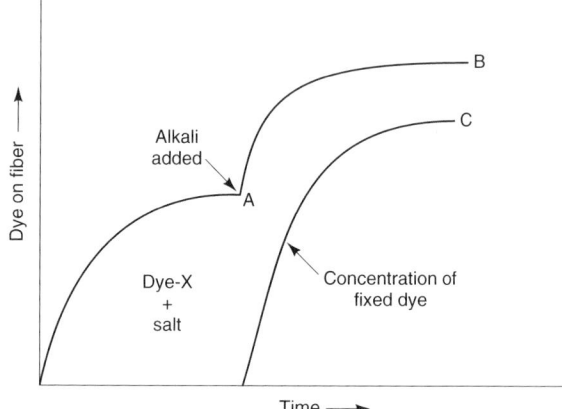

Figure 1. Amounts and forms of fiber-reactive dye on the fiber as a function of time for a low affinity dye, where X represents the leaving group. Point A represents the amount of dye exhausted in neutral conditions; B is the total amount of dye exhausted at the end of the dyeing process, ie, [dye-OH] + [dye-X] + [dye-O-cell]; and C is the amount of dye fixed [dye-O-cell].

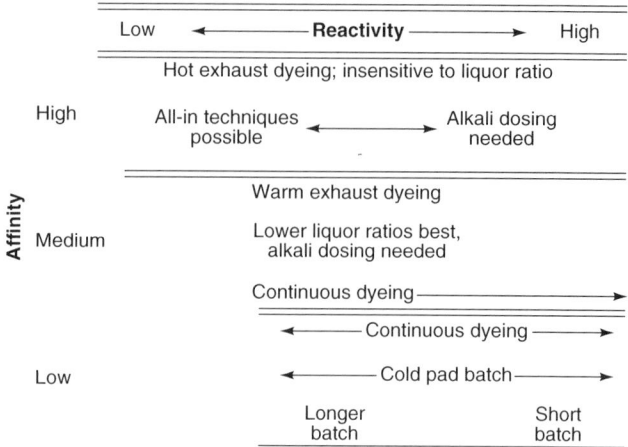

Figure 2. Summary of dyeing techniques related to dye reactivity and affinity characteristics.

the less that needs to be removed in the washing-off process after dyeing.

Application Methods. There are many detailed application methods used for applying reactive dyes, all described in detail. Examples of the main methods include cold exhaust dyeing, warm, hot exhaust dyeing, migration exhaust technique, the all-in method, continuous dyeing and cold pad-batch dyeing.

Chemical Types. A wide range of reactive groups has been investigated, with 20–30 used commercially and >200 patented. Because these reactive groups differ chemically the activation of the reactive systems is different, as are the rates of reaction with cellulose, from one reactive system to another. This rate of reaction with cellulose, or reactivity, dictates the temperature and pH needed for dyeing.

The most important reactive groups are those based on halotriazine or halopyrimidine systems, where an activated halogen substituent undergoes a nucleophilic substitution reaction with ionized cellulose, or dyes based on sulfatoethylsulfonyl groups.

Bifunctional fiber-reactive dyes making use of two different types of reactive group in each molecule have also been developed.

Correlation of Application, Affinity, and Reactivity. Figure 2 correlates fiber-reactive dye application suitability to reactivity and affinity.

Vat Dyes

Vat dyes are solubilized by reduction with agents such as sodium dithionite under alkaline conditions. Reconversion to the original quinone structure of the insoluble pigment is achieved by oxidation, in conjunction with removal or neutralization of the alkali. The dyes are applied by either exhaust or continuous dyeing techniques. In both cases, the process is comprised of five stages: preparation of the dispersion, reduction, dye exhaustion, oxidation, and soaping.

Uses. The main characteristic of vat dyes is their excellent fastness to light, water, and other agents, eg, chlorine. Vat dyes are therefore widely used in outlets demanding high lightfastness such as outerwear, furnishings, drapes, etc; high wetfastness and fastness to repeated washing such as workwear; high chlorine fastness such as institutional articles (eg, towels); or where general excellent fastness is required, as in the case of sewing threads, where it is impossible to know the use of final garment they will be used to construct. Vat dyes can achieve a wide range of shades, but there are few bright reds and yellows, and they are particularly strong in the blue, green, brown, and black regions.

Indigo.. The principal use for indigo is in denim, where indigo is dyed continuously on cotton warp yarns which are subsequently woven with white cotton weft yarns to give the typical denim look. Special effects such as "stone wash," "ice wash," etc, are obtained by chemical treatments of the dyed fabric, eg, with sodium hypochlorite or potassium permanganate and pumice. The "washdown" behavior, where the shade reduces during each wash, is achieved by deliberately ring dyeing the yarn by altering the conditions of reduction and oxidation. The more penetration of indigo into the yarn the less wash-down.

Sulfur Dyes

Sulfur dyes behave in an analogous manner to vat dyes and are applied in a reduced form that is later reoxidized into an insoluble dye on the fiber. The principal differences between sulfur and vat dyes derive from the greater stability and lower affinity of the reduced form of the sulfur dyes.

Sulfur dyes, despite their tintorial weakness, are cheap. They penetrate heavy weight and tightly woven materials well, and it is comparatively easy to produce level dyeings. They are used on towels, corduroy, heavy twills for work wear, etc. They produce only dull shades in a limited range of colors and have a notable sensitivity to hypochlorite bleach. If stored in humid conditions, sulfur-dyed cellulose can weaken drastically on storage.

Azoic Dyes

In applying azoic dyes to cellulose, the dyer becomes the dye maker, and the (azo) dye is synthesized within the fiber. Azo dyes derive from the reaction between a diazonium salt and a coupler.

The dyes are especially strong in the orange/red/bordeaux shade area, and since they are insoluble, tend to have good wet-fastness. Light fastness is also good in dark shades and thus they are a traditional complement to vat dyes that are weak in the bright red shades. Like sulfur dyes, they are good for penetrating heavy materials like toweling. They are economical, but they have tended to be replaced by reactive dyes.

DYEING OF WOOL

Preparation for Dyeing

Raw wool must be cleaned before it can be efficiently carded, combed, otherwise processed, or dyed.

Dyeing Mechanism. Wool is a complex protein polymer based on some 20 amino acids. The polymer chains are cross-linked by disulfide groups of cystine. The presence of aspartic acid, glutamic acid, arginine, and lysine leads to the presence of amino and carboxylic acid terminated side chains, as well as at the chain ends.

Acid Dyes

Classes. Three classes of acid dyes are usually recognized in wool dyeing: acid leveling, acid milling, and super milling or neutral dyeing. Acid leveling dyes are molecular dispersions at low temperatures (true solutions) and are simple molecules. Acid milling dyes are colloidal dispersions at low temperatures and true solutions at high temperatures. Super milling dyes are colloidal dispersions at both low and high temperatures and are complex molecules that often contain hydrophobic alkyl chains that increase attraction through van der Waals forces, charge transfer, and hydrophobic interaction, and thus enhance fastness.

Controlling Dyeing Behavior. As with all other dyes, the dyeing process concentrates on obtaining level dyeings within an economic time period, and once again slower dyeing means better control and level dyeing is enhanced. When dyeing wool with acid dyes four factors control the dyeing behavior: the pH of the dyebath, presence and concentration of electrolyte and/or leveling agents, temperature of the dyebath, and choice of dyestuff class.

Practical Processes. With acid leveling dyes no real problems exist, because the dyes show good migration. Electrolyte is added from the beginning, and rather like Class A direct dyes, level dyeing is achieved by prolonging the times at the boil.

The other extreme is found with super milling dyes, when at the start ammonium acetate, sulfate, or an organic ester is present without any electrolyte. Dyeing is carried out more slowly, taking some 60 min to reach the boil, and often the dye is applied with a leveling agent.

Acid milling dyes are intermediate in behavior, being applied with acetic or formic acid in the presence of sodium sulfate. A disadvantage of acid leveling dyes is that their wetfastness depends on the formation and maintenance of a salt linkage between the charged wool and dye.

Metal Complex Dyes

With the monosulfonated dyes, the mode of attraction is by the lone pair of electrons on the nitrogen groups in wool acting as a ligand and coordinating with the metal in the dye. This provides both excellent fastness and poor level dyeing properties and migration.

Mordant Dyes

Many natural dyes were mordant dyes. The range of shades was extended by using different metal salts as mordants. As synthetic analogs were introduced, a wide range of colors could be produced by varying the dye molecule. Mordant dyeing in general is usually reserved for standard repeating shades, especially black, navy, and bordeaux.

Fiber-Reactive Dyes for Wool

Fiber-reactive dyes are by no means as popular for dyeing wool as they are for cotton but do provide the means of obtaining shades that are both bright and fast: Other fast dyeings on wool (from 1:2 metal complex dyes or chrome dyes) are dull. Such bright and fast shades are required less than on cotton because wool is rarely subjected to more than mild laundering conditions.

Silk

Because it is also a protein, silk can be dyed in the same way as wool, and acid dyes are typically used. Basic dyes have also been used satisfactorily, but only succeed because much silk is only dry cleaned. The main difference between wool and silk is in the preparation of the fiber for dyeing. Silk in its raw state is coated with sericin. It is necessary to remove this gum in order to develop the silk luster and dyeability. Synthetic detergent systems such as higher alcohol sulfates and soda ash and boric acid have replaced soap to a large extent for degumming.

DYEING OF SYNTHETIC POLYAMIDES

Nylon is synthetic, it has defined chemical structure depending on the manufacturing process, and it is hydrophobic. As manufactured fibers, polyamides (nylons) generally require much less preparation than natural fibers that are contaminated with impurities that reflect their agricultural origin. They are usually given an alkaline scour and, since they are thermoplastic, given a heat setting treatment to maintain dimensional stability in the subsequent dyeing. When bleaching is required, sodium chlorite is preferred.

Nylon is prone to physical and chemical variations that affect the dyeing of the fiber. Physical (morphological) differences arise from the "heat history" of the polymer and occur in fiber spinning, yarn texturizing and fabric heat setting. These affect the application of all dye types, with the differences generally being greater with dyes of larger molecular size. Chemical differences involve changes to the amine groups present and arise from oxidation, in bleaching, for example. These variations only affect dyes that interact specifically with amine groups, most notably acid dyes. Physical and chemical variations will affect the rate and extent of dye uptake and a nylon fabric will tend to dye unlevel if it contains fibers with these variations.

Acid Dyes

Many of the same acid dyes that are applied to wool are also used on nylon, and since nylon is more widely used than wool, nylon now represents the major outlet for the use of these dyes. The same range of types exist, ranging from the easier-to-level, lower fastness to the higher fastness dyes prone to unlevelness. A commonly used way of dividing acid dyes on nylon is into Groups 1, 2, and 3, roughly corresponding to the Acid Leveling, Acid Milling, and Supermilling dyes for wool. Group 1 includes dyes with little affinity at neutral or acidic pH but that exhaust under strongly acidic conditions. Group 2 is the largest group of dyes, which exhaust onto nylon in the pH range 4.0–5.0. Group 3 dyes have a high affinity for nylon under neutral or weakly acidic pH. Only dyes within one group should be used together. In contrast to wool, where the tendency is to dye at a constant pH, the use of a sliding pH to control exhaustion while achieving levelness is common in nylon dyeing.

Tanning Agents. The wetfastness of acid dyes can be increased by aftertreatment. The original method was to apply tannic acid and tartar emetic (potassium antimony tartrate) in a sequential process under slightly acid conditions. Today synthetic anionic tanning agents (syntans) are used.

Metal Complex Dyes. The 1:1 metal–dye complexes cannot be used on nylon as they are on wool because the low pH values needed cause fiber degradation: However, some of these dyes can be used satisfactorily at higher pH values. The 1:2 metal–dye complexes have excellent lightfastness in pale shades and provide the main way to achieve dark dull shades, especially black. These macromolecules are difficult to apply level in pale shades and are sensitive to both chemical and physical variations. In their application they are treated as the Group 3 acid dyes.

Other Soluble Hydrophilic Dyes

Some direct dyes have profiles on nylon very similar to Group 3 dyes and therefore to supplement the range of shades available they are sometimes applied with those dyes.

Disperse Dyes

The insoluble, hydrophobic disperse dyes readily dye nylon, and because their mode of attraction is completely nonionic they are insensitive to chemical variations. Small molecular-sized disperse dyes ($\sim$mw 400) of the type originally developed for acetate show very high rates of diffusion and excellent migration properties and are insensitive to physical variations in the nylon. As the molecular size of disperse dyes increases they show increasing fastness, but at the same time, greater sensitivity to physical variation.

Although disperse dyes are readily absorbed at temperatures up to the boil, they are also readily desorbed.

The main use for disperse dyes is where excellent coverage of fibers likely to have physical and chemical variations is needed and where wetfastness is not critical. The small molecular weight dyes are therefore widely used for pale shades on continuous filament yarns used in hosiery. There is also some use made in exhaust dyeing of carpets made from continuous bulk filament nylon to give good coverage.

Dyeing is relatively simple. The disperse dye is added to a dyebath containing a nonionic dispersing agent, sodium hexametaphosphate, and sometimes acetic acid is added to give pH 5.5 to prevent decomposition of some disperse dyes. Dyeing is carried out by bringing the dyebath to the boil and continuing until exhaustion is completed.

Carpet Coloration

Some 50% of all nylon is in the form of carpets, almost exlusively colored with acid dyes, and $\sim$50% of the carpet manufacturing industry is located in the United States. The acid dyes from Group 1 are those most widely used because they exhibit the rapid diffusion needed to penetrate the bulky yarns used in carpets, especially bulk continuous filament yarn used in tufted constructions, with high exhaustion. Their wetfastness properties are generally adequate for most outlets. Where high wetfastness is needed; eg, in hotel lobbies and bars where liquid spillages are likely, the higher fastness acid dyes (Groups 2 and 3) and even metal complex dyes are used.

DYEING OF ACRYLIC FIBERS

In order to make fibers of commercial interest, acrylonitrile is copolymerized with other monomers such as methacrylic acid, methyl methacrylate, vinyl acetate, etc., to improve mechanical, structural, and dyeing properties. Fibers based on at least 85% of acrylonitrile monomer are termed acrylic fibers; those containing between 35–85% acrylonitrile monomer, modacrylic fibers. Acrylic and modacrylic are in general dyed similarly, although the type and number of dye sites generated by the fiber manufacturing process have an influence.

Like nylon, acrylic can be dyed with disperse dyes. While the fastness is a little better, the saturation values can be low. Disperse dyes are therefore only used for pale shades where excellent levelness is needed.

Preparation for Dyeing

Fabrics are scoured with a synthetic detergent at 45–65°C and are rinsed before further processing to remove tints, size, wax, grease, spinning oils, or other impurities that were applied or picked up during manufacturing. Bleaching, when required, is usually accomplished by means of a sodium chlorite bleach, a selected optical brightener, or a suitable combination of the two. Acrylic-blend fabrics may require other bleaching agents if chlorine-sensitive fibers are present. Most acrylic fibers require a presetting in open-width in boiling water to avoid dimensional stability problems during subsequent wet-processing steps.

Level Dyeing

It is exceptionally difficult to obtain level dyeings on acrylic, and temperature and pH control depend on fiber type and are not always adequate. Sodium sulfate is usually added, to slow the rapid strike of dye at the surface.

Cationic products are a popular method for the control of leveling. If amounts of colored modified basic dye and colorless modified basic dye equal to the saturation value of the fiber are uniformly dissolved in the dyebath then level dyeing behavior is promoted.

DYEING OF POLYESTER

Polyester fibers are based on poly(ethylene terephthalate) (PET); some modified versions are formed by copolymerization; eg, basic dyeable polyester. The modified forms dye in analogous manner to other fibers of similar charge.

Preparation for Dyeing

A hot alkaline scour with a synthetic surfactant and with 1% soda ash or caustic soda is used to remove size, lubricants, and oils. Sodium hypochlorite is sometimes included in the alkaline scouring bath when bleaching is required. After bleaching, the polyester fabric is given a bisulfite rinse and, when required, a further scouring in a formulated oxalic acid bath to remove rust stains and mill dirt which is resistant to alkaline scouring.

Disperse Dyes

There is a general correlation between heat fastness, the propensity to desorb under conditions of dry heat onto a white piece of polyester, and the dyeing properties of disperse dyes. Low energy dyes are not usually used in thermofixation, as their low heat fastness at the thermofixation temperatures used (200–210°C) results in them subliming from the hot fabric.

Medium energy dyes are based on larger-sized molecules than the low energy dyes. They have slower rates of dyeing, better heat fastness, and generally higher wet-fastness. Few of them are suitable for carrier dyeing. Their main application methods are exhaust dyeing at temperatures of 125–135°C, and at this temperature their medium molecular size allows dyeing to take place rapidly: 30 min or less.

High energy dyes are based on larger molecules with polar groups that further reduce their volatility. Thus they have excellent heat fastness resulting from extremely low rates of sublimation. Their main use is in dyeing fabrics that are to be given a subsequent high temperature heat treatment; eg, permanent pleating finish, or sewing threads whose future use and treatments are unknown.

Dyeing Processes

Polyester yarns and fabrics are usually dyed by exhaust techniques; continuous dyeing is largely confined to blends with cellulose. The dyeing process is relatively simple. The dyebath is set with a disperse dye and an additional dispersing agent. A leveling agent, usually a nonionic or anionic surface-active agent (or a blend of these) may also be added. The temperature is slowly raised up to the dyeing temperature (125–135°C) and kept there to complete exhaustion and promote migration. Once exhaustion and diffusion are complete, the bath is cooled.

During the cooling process, the solubility of any disperse dye remaining in the dyebath decreases rapidly and it can precipitate onto the surface of the polyester fibers. If it is not removed, the resulting dyeing will exhibit both poor fastness to rubbing and poor wetfastness. This precipitated dye is removed by a "reduction clearing" using hot (70°C) caustic soda and sodium dithionite (hydrosulfite), optionally in the presence of a detergent. This process, based on strong reducing agents, can be avoided by the use of disperse dyes that are removed by aqueous alkali alone.

Subsequent Migration. Although it is possible to obtain excellent fastness properties by either reduction clearing of traditional dyes or alkali clearing of novel alkali-sensitive dyes, this fastness can be short lived. In any subsequent heat treatment of the polyester such as heat-setting to stabilize the fiber or fabric, the polyester is again taken above its glass-transition temperature and dyestuff molecules again have mobility within the fiber. Additionally, nonionic surfactants (leveling agents, etc) that are insufficiently rinsed out (especially when the rinsing is carried out above the cloud point) or finishing agents such as a softener or antistat can dissolve dye from the fiber over time and make it available to stain other materials.

DYEING OF CELLULOSE ESTERS

Secondary cellulose acetate was developed as a textile fiber (acetate silk) in the 1920s and cellulose triacetate became viable in the 1950s. Triacetate is becoming obsolete and reference to "acetate" implies the acetone-soluble secondary acetate.

Acetate fibers are now dyed almost exclusively with low energy disperse dyes:

Cellulose Diacetate

When preparing cellulose diacetate for dyeing, strong alkalies that would saponify the fiber surface must be avoided in scouring. Large quantities of acetate filament are found in satin, taffeta, and tricot fabrics; these are usually dyed open-width on a jig owing to their inclination to crease or crack easily.

Cellulose Triacetate

Cellulose acetate having 92% or more of the hydroxyl groups acetylated is referred to as triacetate. This fiber is characteristically more resistant to alkali than the usual acetate and may be scoured, generally, in open width, with aqueous solutions of a synthetic surfactant and soda ash. Triacetate is a more hydrophobic fiber than secondary acetate, and dyes diffuse more slowly through it. The rate of diffusion of the disperse dye into the fiber is increased by increasing the dyeing temperature to 110–120°C or using a dye accelerant or carrier.

DYEING OF FIBER BLENDS

Fiber blends combine the advantageous properties of two or more fibers into one fabric. They are available as blends of natural fibers, synthetic fibers, or natural fibers blended with synthetic. The most important blend is that of polyester with cotton. The differences in dyeability between the many fibers on the market open a wide field of multicolored yarns and fabrics to the stylist. Whichever way the fibers are arranged, the dyeing on each fiber can be aimed at the same shade (union dyeing), a similar shade in different depths (tone-on-tone), or a different shade (cross-dyeing). Sometimes efforts are made to leave one fiber undyed (reserving), although a complete reserving of a fiber is not possible in all cases.

Depending on the fibers involved and the effect required, as well as the dyes used and the fastness requirements, it may be possible to dye the two fibers simultaneously in the same bath (one-bath, one-stage). The same bath may be used sequentially (one-bath, two-stage), or it may be necessary to conduct two separate processes in separate baths (two bath, two stage). For the last two, the order in which the fibers are dyed can vary. Two-bath, two-stage processes are time-consuming and inefficient of water and energy but may be needed to achieve the shade and the required levels of fastness. Conditions can be chosen to give the maximum dyeing efficiency for each dye–fiber combination.

Cellulosic Fiber Blends

Cellulosic–Polyester Fibers. One of the most important fiber blends on the market is the mix of cotton–polyester in roughly equal amounts, usually from 35 to 65% of one or other of the fibers. Many apparel knitgoods consist of this fiber blend as do sheeting, shirting, and work-cloth fabrics. High wet modulus (modal) viscose fibers are sometimes used instead of cotton. Although the knitgoods are dyed in exhaust dyeing procedures, most of the woven fabrics are dyed according to one of the continuous dyeing processes. The choice of dyes and hence dyeing method is determined by the fastness properties required.

Cellulosic–Acrylic Fibers. Commonly, this blend is used in knitgoods including socks and hosiery, and woven fabrics for slacks, drapery, and upholstery. Since anionic direct dyes are used for the cellulosic fiber and cationic dyes for the acrylics, a one-bath dyeing process is suitable only for light-to-medium shades. Auxiliaries are needed to prevent precipitation of any dye–dye complexes.

In two-bath processes, either the cotton or the acrylic can be dyed first. If the cotton is dyed first, cationic dye can form a salt linkage with the sulfonic acid groups on the anionic dye on cotton leading to poor fastness. Heavy shades are best dyed by first dyeing the acrylic and then dyeing the cotton under neutral or slightly alkaline conditions. In order to prevent desorption of the cationic dye the dyeing temperature for the cotton dyeing must be below the glass-transition temperature for the acrylic of 80°C.

Cotton–acrylic fiber blends are also used for high quality upholstery pile fabrics. Besides the one-bath exhaust dyeing procedure involving a very high ratio of liquor to fabric, a continuous pad–steam process is used to dye these fabrics.

Cellulosic Fiber–Nylon Blends. These blends are used in fabrics for apparel, corduroy, socks, and swimwear. If wet-fastness requirements are relatively low, the nylon portion can be dyed with disperse dyes and the cellulosic fiber with direct dyes and a one-bath procedure can be employed. For better wetfastness, the nylon portion is dyed with level dyeing acid colors together with the direct dyes in one bath at 95°C using a reserving agent (typically a phenolsulfonic acid condensation product) to prevent the direct dyes from dyeing the nylon. An aftertreatment with a cationic fixative improves the wetfastness properties of the direct portion of the dyeing. For swimwear, the cotton portion is dyed with fiber-reactive dyes. After rinsing hot and cold and soaping at the boil, selected acid and/or acid milling colors are applied to the nylon portion with a phosphate buffer system. An aftertreatment with a syntan results in best wetfastness properties.

Wool Blends

Wool–Cellulosic Fibers.. One of the oldest fiber blends in the textile market is the combination of wool and cotton or wool and viscose. Selected direct dyes, which dyed both fibers from a neutral bath in a uniform shade, or a combination of neutral-dyeing acid dyes with direct dyes, were used to dye this fiber blend. In the one-bath process, selected direct and acid dyes are applied at pH 4.5–5.0 at 98–100°C. A reserving agent is used (as in cellulose–nylon blends), to prevent the direct dyes from dyeing the wool under acid conditions.

If optimum wetfastness properties are required, fiber-reactive dyes can be applied to both fibers by use of a two-bath process.

Wool–Nylon. Nylon has been blended with wool in order to give additional strength to the yarn or fabric. It is used mainly in the woolen industry for coats and jackets and, to a lesser extent, for socks and carpet yarns. Both fibers are dyed with the same products; however, the fibers have different affinity to them. Generally, level dyeing acid dyes are applied.

Wool–nylon upholstery fabrics and carpet yarns require higher light- and wetfastness properties. Neutral premetallized dyes are used in these cases. However, they have a much higher affinity to the nylon than the wool. Therefore, stronger retarding agents have to be employed; eg, phenolsulfonic acid condensation products.

Wool–Acrylic Fibers. This blend is being used for industrial and hand-knitting yarns. Special precautions are necessary, since the two fibers are colored with dyes of opposite ionic type. Usually, level dyeing acid dyes are used for the wool portion in combination with the cationic dyes for acrylic fiber.

Wool–Polyester Fibers. The 45–55 wool–polyester blend is the most common fiber combination in the worsted industry. Disperse dyes for polyester and acid or neutral premetallized dyes for wool are employed in a one-bath process.

Blends of Synthetic Fibers

Polyester Fiber Blends. Disperse dyeable and cationic dyeable polyester fibers are frequently combined in apparel fabrics for styling purposes. Whereas the disperse dyes dye both fibers, but in different depths, selected cationic dyes reserve the disperse dyeable fiber completely, resulting in color/white effects.

Polyester Fiber–Nylon Blends. This fiber blend is used in apparel fabrics as well as in carpets. Disperse dyes dye both fibers; however, they possess only marginal fastness properties on nylon. Therefore it is important to select disperse dyes that dye nylon least under the given circumstances. The nylon is dyed with acid dyes, selected according to the fastness requirements.

Polyester Fiber–Acrylic Fiber Blends. This fiber blend is dyed in a similar fashion to that of the blends of the different polyester fibers. The selection of cationic dyes is substantially larger for the acrylic blend.

Nylon Blends. Differential dyeing nylon types and cationic dyeable nylon blends are used primarily in the carpet industry. The selection of cationic dyes for nylon is rather limited; most products have very poor fastness to light. These blends are dyed in a one-bath procedure at 95–100°C. Selected acid dyes are used for differential dyeing. Disperse dyes will dye all different types in the same depth.

Elastomeric Fibers. Elastomeric (spandex, elastane) fibers are mostly polyurethanes. They are combined with nonelastic fibers to produce fabrics with controlled elasticity. Processing chemicals and conditions must be carefully selected to protect all fibers present in the blend.

Polyurethanes are dyeable with many acid dyes and disperse dyes, and in blends with nylon both components are colored by the acid dyes used. A post-dyeing application of fixatives is helpful.

Specialty Uses

In addition to polyester–cellulosic and cellulosic fabrics, pigments may also be applied to 100% synthetic fibers of special construction for unique uses. Examples are 100% filament polyester for draperies and glass fabrics for bedspread, curtain, and upholstery uses.

Solvent Dyeing

Solvent dyeing generally refers to dyeing in nonaqueous media. The technique has not achieved importance.

DYEING MACHINERY

Certain aspects of batch dyeing equipment are common to whatever type of material is being processed. Stainless steel is the universal material of choice for dyeing machinery. Heat is supplied to dyeing machines in the form of steam: a boiler is an essential item of equipment in a dyehouse. Modern machines are equipped with heat exchangers: These are essential for pressurized machines. The same exchangers can also provide cooling. The major control is that of temperature: The installation of thermometers was followed by remote recording, and later by preprogrammed heating and cooling cycles, first by mechanical and now by electronic means. Systems for similarly controlling pH have been developed but have been far less widely adopted. Pressures, water volume, flow rates, and fabric speed and tension can all be measured and controlled. There are many examples of completely automated computer-controlled exhaust dyehouses that control loading, unloading, and drying of the material being dyed.

In most situations, no-add dyeing is not (always) practicable, and the shade of the dyeing must be checked before the process is ended and the material removed. Machines are thus designed with some facility for sampling the dyeing. It is not always easy to provide an easily retrievable sample that has undergone the same conditions as the bulk of the material, and levelness cannot be checked this way.

Dyeing Fiber and Tow

Fiber dyeing is usually accomplished by pumping dye liquor through a stationary mass of fibers prepacked into some form of perforated basket, which is then loaded into the kettle or kier containing the dyebath. Several fiber baskets per machine are used, so that the unloading and loading of fiber can be accomplished at the same time as dyeing is taking place.

Sliver and top are loose, untwisted ropes of staple fibers: tow is the equivalent version of filament fibers. These are often dyed in continuous form, but batch methods are available, particularly for top.

Dyeing Yarn

Yarn is dyed in one of two forms: skein or package. Today the dyebath is moved through yarn that is held in place. Skeins (or hanks) are hung from poles. Skein dyeing involves winding the skeins from cones, and backwinding onto cones after dyeing. Together with the loading and unloading onto the poles, the overall process tends to be labor intensive.

In package dyeing, yarn is wound onto perforated metal or plastic formers in the form of conical or parallel sided cylinders, called cones or cheeses. The whole carrier, with many spindles carrying packages, is placed in the kettle, where dye liquor is pumped through the spindles and through the yarn.

Skeins and packages are bulky forms of textile materials and drying them can be slow. Cold air is used first. Final drying, by blowing warm air through a package or suspending the skein or package in an oven, may take several hours.

Dyeing Fabric

Probably the majority of textile material is dyed in the form of fabric. Four different machine types have been developed for the batch dyeing of such piece goods beam dyeing, jig dyeing, which or beck dyeing, and jet dyeing.

Dyeing Garments

Several types of garment-dyeing machines were developed many years ago. Garments do not lend themselves to close and orderly packing in a dyeing process and garments must be constructed with dyeing in mind, with the materials used for sewing thread, zipper facings, buttons, and other trim chosen carefully.

Among the older machines are paddle machines in which a bath is circulated by a paddle. The bath may be an annular oval, with the axle holding the paddle blades just above the liquor surface. As the paddle rotates it moves liquor and goods around the annulus. Newer machines tend to be based on front-loading commercial washers/drycleaning machines.

Continuous and Semicontinuous Dyeing

Continuous dyeing separates the individual steps of a batch process into different machine elements of a dyeing range. As the name implies, uncolored material and dye are fed in, and colored material emerges, continuously. The multiplicity of components in the range means that at the end of dyeing one color, several different pieces of machinery must be cleaned before a new color can be dyed. Continuous dyeing is thus used in cases where a large volume of the same color is to be dyed. Recent trends have tended to reduce the yardage per shade and thus the importance of continuous dyeing.

Continuous dyeing can be applied to fiber, yarn, and fabric. Continuous dyeing is most suited to fabric, and most woven goods, other than stretch fabrics, can be continuously dyed. Continuous yarn dyeing is most commonly used in the application of indigo to warp yarns to be woven into denim.

Fully continuous dyeing requires continuous fixation. This may simply be via heated air which is common for the thermosol dyeing of disperse dyes, the curing of binders in pigment/binder coloration, and can also be used for fixing reactive dyes on cellulose. Most often it is accomplished with saturated steam that provides the moisture required to dissolve the dye, swell a hydrophilic fiber, and allow diffusion.

In the continuous dyeing of blend fabrics, dyes must be fixed on both fibers. For polyester/cotton blends, this usually means two fixation steps with chemical application in between.

After fixation, the fabric must be rinsed and finally dried, in a tenter frame, or over heated cylinders.

TEXTILE PRINTING

In printing on textiles, a localized dyeing process takes place, in which the usual chemical and physical interactions of dyeing apply. The major differences lie in the need to provide a clearly defined design, often with sharp edges, and at the same time achieve levelness within areas of color. Printing, like continuous dyeing, generally consists of application, fixation, and washing off steps.

Most paper printing is carried out using the four process colors, yellow, magenta, cyan and black. In comparison, most textile printing does not allow for reliable mixing on the substrate, and each color in the design must be separately premixed and applied: in paper printing these would be referred to as "spot colors." Designs typically contain from 1 to 12 colors, although higher number color prints are possible. Taking a design on paper or in an electronic file, and separating the colors prior to printing is an important part of the process.

The colorant (dye or pigment) is applied as a dispersion or solution from a vehicle called print paste or printing ink. The paste must have the appropriate flow properties (rheology) to give good transfer from the print machine onto the fabric, good levelness in large printed areas, but a sharp mark for the details of the design. It should also hold the colorant in place until fixation occurs.

After application and any intermediate drying comes fixation. The printed textile material is exposed to heat, heat and steam, or chemical solutions to allow fixation to take place.

In contrast to pigment prints, dye prints must be afterscoured. In this step, the prints are rinsed in a detergent solution to remove auxiliary chemicals, thickening agents, and portions of unfixed dyes remaining on the surface of the printed fibers.

Colorants for Textile Printing.

Pigments.. Pigment-printed textiles represent the highest percentage of all printed textiles, accounting for between 40 and 50% of all cellulose and >90% of polyester–cotton blend prints.

Disperse Dyes. Disperse dyes are used in powder or paste form, or ready-to-prepare aqueous dispersions for incorporation into a thickener solution.

Acid and Premetallized Dyes. Acid dyes are most widely used in printing polyamide: woven or knitted fabrics for apparel, swimwear, drapery, and upholstery materials. Dyes are usually of the higher fastness types: acid leveling dyes are rarely used in printing.

Fiber-Reactive Dyes. This class represents the main dye group for cellulosic fibers; ie, cotton and rayon.

Vat Dyes. Vat dyes yield prints on cellulosic fibers with excellent fastness properties. They are used to print furnishings, drapes, and camouflage where their infrared reflectance resembles natural terrain and foliage.

Azoic Dyes. These are used to produce cost-effective heavy yellow, orange, red, maroon, navy blue, brown, and black shades, and are printed alongside other dye classes to extend the coloristic possibilities for the designer.

Basic (Cationic) Dyes. The use of basic dyes is confined mainly to acrylic fibers and as complementary dyes for basic-dyeable polyester fibers. They can be applied to silk and acetate as a means of achieving bright prints, but have very poor wetfastness on those fibers.

Other Dyes.. A few selected direct dyes are used to complement the acid dyes in printing of polyamide. Printing of cellulosic fibers with direct dyes is of little importance. Sulfur dyes are rarely used in printing. The calico printing of the nineteenth century, heavily based on the use of mordant dyes, and a few synthetic mordant dyes survived for special shades until the 1970s but have little importance today. Ingrain colorants that produce phthalocyanine dyes in situ on the fiber after application of precursors have become obsolete; likewise, aniline black (an oxidation dye) survived in printing long after its use in dyeing had all but disappeared but is obsolete today.

Styles of Printing

The different styles of printing include direct printing, overprinting, discharge printing, and resist printing.

Printing Machinery

Screen Printing. This printing process essentially consists of the transfer of print paste through a mesh to the substrate to be printed. The early use of silk is reflected in the term "silk screen printing." The pattern or design is produced blocking those parts that should not transfer the print paste with an impervious material.

Roller Printing. Roller printing produces designs with half-tones, fine lines and details very well but this advantage is now challenged by fine-mesh screens. The machine is massive, running such a machine is highly skilled, and there are high capital costs associated with it, so the roller machine is rapidly becoming obsolete.

Other Printing Methods

Heat transfer printing employs the intermediate step of printing dye onto a temporary substrate, usually paper. The dye is transferred to the textile while printed paper and textile are in close contact. The method has achieved commercial success only with disperse dyes. Short runs of textile can be printed as needed from paper held in stock. Ink-jet printing was developed for printing on paper and is a viable method only when controlled by computer. Much attention in recent years has focused on the application of inkjet printing on textiles. The chief limitation of the method has been its slow speed: some two orders of magnitude slower than conventional textile printing. In xerography research sought to eliminate water from textile coloration and finishing by achieving printed designs on textile substrates, but limitations in the production of toners and the variable nature of substrates prevented this technique from achieving commercial viability.

PAPER COLORING

Colorants for Paper

While many textile dyes are supplied as powders, the continuous application of dyes to paper makes liquid forms preferred. Direct dyes represent the majority of the dyes used, and with basic dyes make up 90% or more of the market. Acid dyes are also used. Paper dyeings with pigments have outstanding fastness properties, but poor affinity, low tinctorial strength, and two-sidedness problems limit their usefulness.

Acid Dyestuffs. These are used for paper that does not require wetfastness, such as construction grades, and are most suitable for calendar staining or surface coloring because of their solubility and brightness of shade. With increased environmental concern their use has declined, since they readily stain the backwater.

Direct Dyestuffs. Direct dyestuffs generally have a high affinity for cellulose and are therefore the most useful dyestuff type for unsized or neutral pH dyeings. Their bonding ability to nonligneous pulps and excellent fastness properties to light and bleeding make them useful for all fine papers. The shades of direct dyestuffs are not as bright as those of acid or basic dyestuffs, and in blended furnishes (bleached–ligneous pulps) mottling or graniting may occur.

Basic Dyestuffs. Basic dyestuffs are usually used for dyeing unbleached pulp in mechanical pulp such as wrapping paper, kraft paper, box board, news, and other inexpensive packaging papers. Their strong and brilliant shades also make them suitable for calendar staining and surface coloring where lightfastness is not critical.

Dyeing Processes for Paper

Paper may be colored by dyeing the fibers in a water suspension by batch or continuous methods. The classic process is by batch dyeing in the beater, pulper, or stock chest. Continuous dyeing of the fibers in a water suspension is more appropriate to modern paper machine processes with high production speeds in modern mills. Solutions of dyestuffs can be metered into the high density or low density pulp suspensions in continuous operation.

Paper may also be colored by surface application of dyestuff solutions after the paper has been formed and dried or partially dried by utilizing size-press addition, calendar staining, or coating operations on the paper machine. In addition, paper may be colored in off-machine processes by dip dyeings or absorption of dyestuff solution and subsequent drying, such as for decorative crepe papers.

DYEING OF LEATHER

Leather is a less homogeneous product than textiles or paper. Leather has, compared to textile substrates, a more three-dimensional character with a thickness that prevents full penetration of most colorants: the extent of penetration is thus a variable in practical dyeing. The two sides (grain and flesh) of a leather also differ in their uptake of dye. The application of colorants can therefore vary with the needs of the final product: A one-sided coloration emphasizing the grain side, a more fully penetrated coloration for clothing or upholstery, or coloration for suede products.

Tanning

Vegetable tannage is employed to a fairly large degree for specific types of leather. Syntans and aldehydes; eg, glutaraldehyde, are applied combined with chrome or vegetable agents, or as a retan or second tannage over either the chrome or vegetable agents.

Colorants for Leather

The main classes of dyes employed in the coloring of leather are the acid, direct, and basic types. Acid and direct dyes vary considerably in their penetrating properties on chrome leather. Although it is not possible to make a distinct division, the direct dyes as a class are more surface dyeing than the acid dyes. As with textiles, in leather dyeing it is usually necessary to employ two or more dyestuffs to produce a given shade, and the components of the mixture should be compatible; this is an important factor in the adjustment of shades.

FASTNESS TESTS FOR TEXTILES

Dyed textiles are evaluated with regard to their fastness (resistance to changes in color or propensity to stain other materials) when challenged by destructive agents, cleaning treatments or general use. Tests are designed to standardize or mimic these conditions in a laboratory setting.

The satisfactory performance of a textile item derives from more than the durability of its coloration, and a myriad of other properties can be tested: the tests to be used for a particular item, and the level of performance required on those tests comprise the specification that an item has to meet.

The principal active bodies in the field of colorfastness testing have been national bodies such as the American Association of Textile Chemists and Colorists (AATCC). Increasingly, tests are becoming standardized across international borders, and the European Community and the International Organisation for Standardization (ISO) have been active in promoting international versions of tests.

Colorfastness refers to the resistance of the textile coloration to the different agencies; eg, light or chemicals, to which these materials may be exposed during manufacture or subsequent use. A lack of fastness may be due to destruction of the colorant. This is exemplified by the effects of sunlight or bleach, and only the change of color of the substrate is of interest. More often, the color is removed and is available to stain uncolored adjacent materials, in which cases both the change of color of the substrate and the extent of staining of a white material are measured. In some cases, only the staining of a white material is of concern (eg, rubbing).

Fastness to Light

While most fastness properties are assessed on a 1–5 scale, exposure to light is assessed by comparing the change in color of the test specimen with that of blue wool references and is given the rating of the reference that fades most like the test specimen.

Fabrics are subjected to a wide range of light conditions: real life light exposure varies with season, latitude, altitude, cloud cover, and may or may not be behind window glass. For controlled and reproducible exposures, some artificial light source operating under known conditions of temperature and humidity is required. Some dyes change color on exposure to light but revert to the original color on storage, and tests for this photochromism have been developed.

Fastness to Cleaning and Refurbishing Processes

Consumers subject textile materials to a wide range of washing conditions, and the range of conditions in tests reflects this. Heat exposure, as a result of ironing or other hot conditions, can also cause color change, and fastness tests to simulate these have been developed. A similar test simulates steam pleating. In colorfastness to dry cleaning, a specimen of the textile is placed in a cotton fabric bag together with stainless steel disks and agitated in perchloroethylene. The effect on the shade is assessed using the gray scale.

Fastness to Other Agencies

Textile materials are subjected to a variety of aqueous media that can cause loss of color or staining. Several of these are mimicked by wetting the test material and an adjacent material with the medium of interest, placing them between glass or plastic plates, and exposing them to standard conditions (often body temperature) for several hours.

Colorfastness to acids and alkalis is determined by exposing specimens to various acid, of specified concentrations. Colorfastness to atmospheric contaminants and to rubbing and abrasion are similarly tested.

SAFE HANDLING OF DYES

The Ecological and Toxicological Association of Dyes and Organic Pigments Manufacturers (ETAD), an international body of all primary manufacturers based in Europe but also with standing committees in the United States, Brazil, and Japan, issues clear guidelines for the safe handling of dyes. In December 1991, the United States Operating Committee of ETAD joined with the United States Environmental Protection Agency in publishing a pollution prevention guidance manual for the dye manufacturing industry. Most countries have health and safety legislation that covers the way in which industrial chemicals are handled, and dyes and dyeing auxiliaries are covered under these regulations.

R. S. Berns, *Principles of Color Technology*, 3rd ed., John Wiley & Sons, Inc., New York, 2000.

A. D. Broadbent, *Basic Principles of Textile Coloration*, Society of Dyers and Colourists, Bradford, U.K., 2001.

K. Nassau, *The Physics and Chemistry of Color*, 2nd ed., Wiley-Interscience, New York, 2001.

J. Shore, ed., *Colorants and Auxiliaries, Organic Chemistry and Application Processes*, 2nd ed. Vol. 1, *Colorants*, Society of Dyers and Colourists, Bradford, U.K., 2002.

MARTIN BIDE
University of Rhode Island

DYES AND DYE INTERMEDIATES

CLASSIFICATION SYSTEMS FOR DYES

Dyes may be classified according to chemical structure or by their usage or application method. The former approach is adopted by practicing dye chemists who use terms such as azo dyes, anthraquinone dyes, and phthalocyanine dyes. The latter approach is used predominantly by the dye user, the dye technologist, who speaks of reactive dyes for cotton and disperse dyes for polyester. Very often, both terminologies are used, for example, an azo disperse dye for polyester and a phthalocyanine reactive dye for cotton.

CLASSIFICATION OF DYES BY USE OR APPLICATION METHOD

The classification of dyes according to their usage is summarized in Table 1, which is arranged according to the *Colour Index* (CI) application classification. It shows the

Table 1. Usage Classification of Dyes[a]

Class	Principal substrates
acid	nylon, wool, silk, paper, inks, and leather
azoic components and compositions	cotton, rayon, cellulose acetate, and polyester
basic	paper, polyacrylonitrile-modified nylon, polyester, and inks
direct	cotton, rayon, paper, leather, and nylon
disperse	polyester, polyamide, acetate, acrylic, and plastics
fluorescent brighteners[b]	soaps and detergents, all fibers, oils, paints, and plastics[c]
food, drug, and cosmetic[d]	foods, drugs, and cosmetics
mordant[e]	wool, leather, and anodized aluminum
natural[f]	food
oxidation bases	hair, fur, and cotton
pigments[g]	paints, inks, plastics, and textiles
reactive[h]	cotton, wool, silk, and nylon
solvent	plastics, gasoline, varnish, lacquer, stains, inks, fats, oils, and waxes
sulfur	cotton and rayon
vat	cotton, rayon, and wool

[a] *Encyclopedia* articles on specific chemical types of dyes are AZINE DYES; AZO DYES; CYANINE DYES; DYES, ANTHRAQUINONE; PHTHALOCYANINE COMPOUNDS; POLYMETHINE DYES; STILBENE DYES; SULFUR DYES; TRIPHENYLMETHANE AND RELATED DYES; XANTHENE DYES.
[b] See FLUORESCENT WHITENING AGENTS.
[c] See COLORANTS FOR PLASTICS.
[d] See COLORANTS FOR FOOD, DRUGS, COSMETICS, AND MEDICAL DEVICES.
[e] See DYES, APPLICATION AND EVALUATION.
[f] See DYES, NATURAL.
[g] See PAINT; PIGMENTS; INKS.
[h] See DYES, REACTIVE.

principal substrates, the methods of application, and the representative chemical types for each application class.

Although not shown in Table 1, dyes are also being used in high technology applications, such as in the medical, electronics, and especially the reprographics industries.

NOMENCLATURE OF DYES

Dyes are named either by their commercial trade name or by their *Colour Index* (CI) name. In the *Colour Index* these are cross-referenced.

The commercial names of dyes are usually made up of three parts. The first is a trademark used by the particular manufacturer to designate both the manufacturer and the class of dye, the second is the color, and the third is a series of letters and numbers used as a code by the manufacturer to define more precisely the hue, and also to indicate important properties the dye possesses.

CLASSIFICATION OF DYES BY CHEMICAL STRUCTURE

The two overriding trends in dyestuffs research for many years have been improved cost-effectiveness and increased technical excellence. Improved cost-effectiveness usually means replacing tinctorially weak dyes such as anthraquinone, the second largest class after the azo dyes, with tinctorially stronger dyes such as heterocyclic azos, triphendioxazines, and benzodifuranones. This theme will be pursued throughout this section discussing dyes by chemical structure.

Azo Dyes

These dyes are by far the most important class, accounting for over 50% of all commercial dyes, and having been studied more than any other class (see AZO DYES). Azo dyes contain at least one azo group ($-N=N-$) but can contain two (disazo), three (trisazo), or, more rarely, four or more (polyazo) azo groups. The azo group is attached to two radicals of which at least one, but, more usually, both are aromatic.

In monoazo dyes, the most important type, the A radical often contains electron-accepting groups, and the E radical contains electron-donating groups, particularly hydroxy and amino groups.

Almost without exception, azo dyes are made by diazotization of a primary aromatic amine followed by coupling of the resultant diazonium salt with an electron-rich nucleophile.

In theory, azo dyes can undergo tautomerism: azo/hydrazone for hydroxyazo dyes; azo/imino for aminoazo dyes, and azonium/ammonium for protonated azo dyes. A more detailed account of azo dye tautomerism can be found elsewhere.

The three metals of importance in azo dyes are copper, chromium, and cobalt. The most important copper dyes are the 1:1 copper(II): azo dye complexes of formula; they have a planar structure.

In contrast, chromium(III) and cobalt(III) form 2:1 dye:metal complexes that have nonplanar structures. Geometrical isomerism exists.

Premetallized dyes are now used widely in various outlets to improve the properties of the dye, particularly lightfastness. However, this is at the expense of brightness, because metallized azo dyes are duller than nonmetallized dyes.

Carbocyclic azo dyes are the backbone of most commercial dye ranges. Based totally on benzene and naphthalene derivatives, they provide yellow, red, blue, and green colors for all the major substrates such as polyester, cellulose, nylon, polyacrylonitrile, and leather.

The carbocyclic azo dye class provides dyes having high cost-effectiveness combined with good all-around fastness properties. However, they lack brightness, and consequently, they cannot compete with anthraquinone dyes for brightness. This shortcoming of carbocyclic azo dyes is overcome by heterocyclic azo dyes.

One long-term aim of dyestuffs research has been to combine the brightness and high fastness properties of anthraquinone dyes with the strength and economy of azo dyes. This aim is now being realized with heterocyclic azo dyes, which fall into two main groups: those derived from heterocyclic coupling components, and those derived from heterocyclic diazo components.

All the heterocyclic coupling components that provide commercially important azo dyes contain only nitrogen as the hetero atom. They are indoles, pyrazolones, and especially pyridones; they provide yellow to orange dyes for various substrates.

In contrast to the heterocyclic coupling components, virtually all the heterocyclic diazo components that provide commercially important azo dyes contain sulfur, either alone or in combination with nitrogen (the one notable exception is the triazole system). These S or S/N heterocyclic azo dyes provide bright, strong shades that range from red through blue to green, and therefore complement the yellow-orange colors of the nitrogen heterocyclic azo dyes in providing a complete coverage of the entire shade gamut.

Anthraquinone Dyes

Anthraquinone dyes are based on 9,10-anthraquinone, which is essentially colorless. To produce commercially useful dyes, powerful electron-donor groups such as amino or hydroxy are introduced into one or more of the four alpha positions (1,4,5, and 8). The most common substitution patterns are 1,4-, 1,2,4-, and 1,4,5,8-. To maximize the properties, primary and secondary amino groups (not tertiary) and hydroxy groups are employed.

Anthraquinone dyes are prepared by the stepwise introduction of substituents on to the performed anthraquinone skeleton or ring closure of appropriately substituted precursors.

The principal advantages of anthraquinone dyes are brightness and good fastness properties, but they are both expensive and tinctorially weak. However, they are still used extensively, particularly in the red and blue shade areas, because other dyes cannot provide the combination of properties offered by anthraquinone dyes, albeit at a price.

Benzodifuranone Dyes

BDFs are unusual in that they span the whole color spectrum from yellow through red to blue, depending on the electron-donating power of the R group on the phenyl ring of the aryl acetic acid, ie, $Ar==C_6H_4R$ ($R=H$, yellow-orange; $R=$alkoxy, red; $R=$amino, blue). The first commercial BDF, Dispersol Red C-BN, a red disperse dye for polyester, is already making a tremendous impact. Its brightness even surpasses that of the anthraquinone reds, while its high tinctorial strength (ca 3–4 times that of anthraquinones) makes it cost-effective.

Polycyclic Aromatic Carbonyl Dyes

Structurally, these dyes contain one or more carbonyl groups linked by a quinonoid system. They tend to be relatively large molecules built up from smaller units, typically anthraquinones. Since they are applied to the substrate (usually cellulose) by a vatting process, the polycyclic aromatic carbonyl dyes are often called the anthraquinonoid vat dyes.

Although the colors of the polycyclic aromatic carbonyl dyes cover the entire shade gamut, only the blue dyes and the tertiary shade dyes, namely, browns, greens, and blacks, are important commercially. As a class, the polycyclic aromatic carbonyl dyes exhibit the highest order of lightfastness and wetfastness.

Indigoid Dyes

Like the anthraquinone, benzodifuranone, and polycyclic aromatic carbonyl dyes, the indigoid dyes also contain carbonyl groups. They are also vat dyes.

Indigoid dyes represent one of the oldest known classes of dyes. Although many indigoid dyes have been synthesized, only indigo itself is of any importance today. Indigo is the blue used almost exclusively for dyeing denim jeans and jackets and is held in high esteem because it fades in tone to give progressively paler blue shades.

Polymethine and Related Dyes

Cyanine dyes are the best known polymethine dyes. Nowadays, their commercial use is limited to sensitizing dyes for silver halide photography. However, derivatives of cyanine dyes provide important dyes for polyacrylonitrile. They include azacarbocyanines, hemicyanines, and diazahemicyanines.

Styryl Dyes

The styryl dyes are uncharged molecules containing a styryl group $C_6H_5-CH=C$ usually in conjugation with an N,N-dialkylaminoaryl group. Styryl dyes were once a fairly important group of yellow dyes for a variety of substrates. They are synthesized by condensation of an active methylene compound, especially malononitrile with a carbonyl group, especially an aldehyde. As such, styryl dyes have small molecular structures and are ideal for dyeing densely packed hydrophobic substrates such as polyester.

Yellow styryl dyes have now been largely superseded by superior dyes such as azopyridones, but there has been a resurgence of interest in red and blue styryl dyes. The addition of a third cyano group to produce a tricyanovinyl group causes a large bathochromic shift: the resulting dyes are bright red rather than the greenish yellow color of the dicyanovinyl dyes. These tricyanovinyl dyes have been patented by Mitsubishi for the transfer printing of polyester substrates. Two synthetic routes to the dyes are shown: one is by the replacement of a cyano group in tetracyanoethylene, and the second is by oxidative cyanation of a dicyanovinyl dye with cyanide. The use of such toxic reagents could hinder the commercialization of the tricyanovinyl dyes (see CYANOCARBONS).

Di- and Triaryl Carbonium and Related Dyes

As a class, these dyes are bright and strong, but are generally deficient in lightfastness. Consequently, they are used in outlets where brightness and cost-effectiveness,

rather than permanence, are paramount, for example, the coloration of paper. Many dyes of this class, especially derivatives of pyronines (xanthenes), are among the most fluorescent dyes known.

Resurgence of interest in triphendioxazine dyes arose through successful modification of the intrinsically strong and bright triphendioxazine chromogen to produce blue reactive dyes for cotton. These blue reactive dyes combine the advantages of azo dyes and anthraquinone dyes. Thus they are bright, strong dyes with good fastness properties.

Phthalocyanines

Apart from the recent discoveries of benzodifuranone dyes and diketopyrrolopyrrole pigments, phthalocyanine is the only novel chromogen of commercial importance discovered since the nineteenth century.

Phthalocyanines are analogues of the natural pigments chlorophyll and heme. However, unlike these natural pigments, which have extremely poor stability, phthalocyanines are probably the most stable of all the colorants in use today. Substituents can extend the absorption to longer wavelengths, into the near infrared, but not to shorter wavelengths, and so their hues are restricted to blue and green.

Of all the metal complexes evaluated, copper phthalocyanines give the best combination of color and properties and consequently the majority of phthalocyanine dyes are based on copper phthalocyanine.

Besides being extremely stable, phthalocyanines are bright and tinctorially strong ($\varepsilon_{max} \sim 100,000$); this renders them cost-effective. Consequently, phthalocyanines are used extensively in printing inks and paints.

Quinophthalones

Like the hydroxy azo dyes, quinophthalone dyes can, in theory, exhibit tautomerism. The dyes are synthesized by the condensation of quinaldine derivatives with phthalic anhydride. Quinophthalones provide important yellow dyes for the coloration of plastics and for the coloration of polyester.

Sulfur Dyes

These dyes are synthesized by heating aromatic amines, phenols, or nitro compounds with sulfur or, more usually, alkali polysulfides. Sulfur dyes are used for dyeing cellulosic fibers. They are insoluble in water and are reduced to the water-soluble leuco form for application to the substrate by using sodium sulfide solution. The sulfur dye proper is then formed within the fiber pores by atmospheric oxidation. Sulfur dyes constitute an important class of dye for producing cost-effective tertiary shades, especially black, on cellulosic fibers.

Nitro and Nitroso Dyes

These dyes are now of only minor commercial importance, but are of interest for their small molecular structures. The most important nitro dyes are the nitrodiphenylamines. Their small molecules are ideal for penetrating dense fibers such as polyester, and are therefore used as disperse dyes for polyester. All the important dyes are yellow. Although the dyes are not terribly strong ($_{max} \sim$ 20,000), they are cost-effective because of their easy synthesis from inexpensive intermediates.

Nitroso dyes are metal-complex derivatives of *o*-nitrosophenols or naphthols. Tautomerism is possible in the metal-free precursor between the nitrosohydroxy tautomer and the quinoneoxime tautomer. The only nitroso dyes important commercially are the iron complexes of sulfonated 1-nitroso-2-naphthol. These inexpensive colorants are used mainly for coloring paper.

Miscellaneous Dyes

Other classes of dyes that still have some importance are the stilbene dyes and the formazan dyes. Stilbene dyes are in most cases mixtures of dyes of indeterminate constitution that are formed from the condensation of sulfonated nitroaromatic compounds in aqueous caustic alkali either alone or with other aromatic compounds, typically arylamines. The sulfonated nitrostilbene is the most important nitroaromatic, and the aminoazobenzenes are the most important arylamines.

Formazan dyes bear a formal resemblance to azo dyes, since they contain an azo group. The most important formazan dyes are the metal complexes, particularly copper complexes, of tetradentate formazans. They are used as reactive dyes for cotton.

DYE INTERMEDIATES

The precursors of dyes are called dye intermediates. They are obtained from simple raw materials, such as benzene and naphthalene, by a variety of chemical reactions. Usually, the raw materials are cyclic aromatic compounds, but acyclic precursors are used to synthesize heterocyclic intermediates. The intermediates are derived from two principal sources, coal tar and petroleum (qv).

Intermediates Classification

Intermediates may be conveniently divided into primary intermediates (primaries) and dye intermediates. Large amounts of inorganic materials are consumed in both intermediates and dyes manufacture.

Inorganic materials include acids (sulfuric, nitric, hydrochloric, and phosphoric), bases (caustic soda, caustic potash, soda ash, sodium carbonate, ammonia, and lime), salts (sodium chloride, sodium nitrite, and sodium sulfide) and other substances such as chlorine, bromine, phosphorus chlorides, and sulfur chlorides. The important point is that there is a significant usage of at least one inorganic material in all processes, and the overall tonnage used by, and therefore the cost to, the dye industry is high.

Primary intermediates are characterized by one or more of the following descriptions, which associate them with raw materials rather than with intermediates.

1. Manufactured in a dedicated plant.
2. At least 1000 t/yr capacity from a single plant.

3. Manufacturing process and/or operation is continuous or semicontinuous.
4. A primary intermediate has established usage in basic industries such as rubber, polymers, or agrochemicals in addition to dyes.

All the significant primaries, about 30 different products, are derived from benzene, toluene, or naphthalene. The primaries are listed here with a reference to the *Encyclopedia* article that covers them in detail.

The following amines are covered under the title AMINES, AROMATIC: aniline, *p*-nitroanilineaniline, *o*-toluidine, *p*-toluidine, dimethylaniline, *m*-phenylenediamine, and *p*-phenylenediamine. The article NITROBENZENE AND NITROTOLUENES covers the primaries: nitrobenzene, *p*-chloronitro benzene, *o*-chloronitrotoluene, and *p*-nitrotoluene.

Some primaries have articles devoted to them and their derivatives, ie, BENZOIC ACID, PHENOL, SALICYCLIC ACID, and phthalic anhydride as one derivative of PHTHALIC ACIDS. The primary β-naphthol is discussed in NAPHTHALENE DERIVATIVES.

Dye intermediates are defined as those precursors to colorants that are manufactured within the dyes industry, and they are nearly always colorless. Colored precursors are conveniently termed color bases. As distinct from primaries they are only rarely manufactured in single-product units because of the comparatively low tonnages required. Fluorescent brightening agents (FBAs) are neither intermediates nor true colorants.

There are at least 3000 different intermediates in current manufacture (over half that number are specifically mentioned in the *Colour Index*), and in addition there is a comparatively small number of products manufactured by individual companies for their own specialties.

Intermediates vary in complexity, usually related to the number of chemical and operational stages in their manufacture, and therefore cost. Prices may be classed as cheap (less than $1500/t, as with primaries), average ($1500 to $5000/t) or expensive (more than $5000/t).

THE CHEMISTRY OF DYE INTERMEDIATES

The chemistry of dye intermediates may be conveniently divided into the chemistry of carbocycles, such as benzene and naphthalene, and the chemistry of heterocycles, such as pyridones and thiophenes.

Chemistry of Aromatic Carbocycles

Benzene and naphthalene are by far the most important aromatic carbocycles used in the dyes industry. The hundreds of benzene and naphthalene intermediates used can be prepared from these parent compounds by the sequential introduction of a variety of substituents eg, NO_2, NR^1R^2, Cl, SO_3H, etc. Introduction of these groups are known as unit processes. The substituents are introduced into the aromatic ring by either electrophilic or nucleophilic substitution. In general, aromatic rings, because of their inherently high electron density, are much more susceptible to electrophilic attack than to nucleophilic

Table 2. Unit Processes in Dyes Manufacture

Process	Primaries[a]	Intermediates (common usage)	Colorants (common usage)
nitration	6	✓	
reduction	8	✓	
sulfonation	4	✓[b]	✓
oxidation	5	✓	
fusion/hydroxylation	3	✓	
animation	3	✓[c]	
alkylation	2	✓	✓
halogenation	2	✓	✓
hydrolysis	2	✓	
condensation	1	✓	✓
alkoxylation	1	✓	
esterification	1	✓	
carboxylation	1	✓	
acylation	1	✓	✓
phosgenation	1	✓	✓
diazotization	1	✓	✓
coupling (azo)	1	✓	✓

[a] Number of occurrences within 30 identified product manufactures.
[b] Includes chlorosulfonation.
[c] Includes the Bucherer reaction.

attack. Nucleophilic attack only occurs under forcing conditions unless the aromatic ring already contains a powerful electron-withdrawing group, eg, NO_2. In this case, nucleophilic attack is greatly facilitated because of the reduced electron density at the ring carbon atoms.

Unit Processes

The unit processes encountered in intermediate and dye chemistry are summarized in Table 2.

Chemistry of Aromatic Heterocycles

In contrast to the benzenoid intermediates, it is unusual to find a heterocyclic intermediate that is synthesized via the parent heterocycle. They are synthesized from acyclic precursors.

The most important heterocycles are those with five- or six-membered rings; these rings may be fused to other rings, especially a benzene ring. Nitrogen, sulfur, and to a lesser extent oxygen, are the most frequently encountered heteroatoms. They are often considered in two groups: those containing only nitrogen, such as pyrazolones, indoles, pyridones, and triazoles which, except for triazoles, are used as coupling components in azo dyes, and those containing sulfur (and also optionally nitrogen), such as thiazoles, thiophenes, and isothiazoles, that are used as diazo components in azo dyes. Triazines are treated separately since they are used as the reactive system in many reactive dyes.

EQUIPMENT AND MANUFACTURE

The basic types of dye (and intermediate) manufacture are shown in Figure 1. There are usually several reaction steps or unit processes.

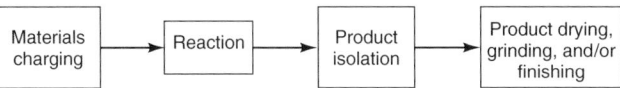

Fig. 1. Operation sequence in dye and intermediate manufacture.

The reactions for the production of intermediates and dyes are carried out in bomb-shaped reaction vessels made from cast iron, stainless steel, or steel lined with rubber, glass (enamel), brick, or carbon blocks. Wooden vats are also still used in some countries, eg, India. These vessels have capacities of 2–40 m^3 (ca 500–10,000 gal) and are equipped with mechanical agitators, thermometers or temperature recorders, condensers, pH-probes, etc, depending on the nature of the operation. Jackets or coils are used for heating and cooling by circulating through them high boiling fluids (eg, hot oil, or Dowtherm), steam, or hot water to raise the temperature, and air, cold water, or chilled brine to lower it. Unjacketed vessels are often used for aqueous reactions, where heating is affected by direct introduction of steam, and cooling by addition of ice or by heat exchangers. The reaction vessels normally span two or more floors in a plant to facilitate ease of operation (see REACTOR TECHNOLOGY).

Products are transferred from one piece of equipment to another by gravity flow, pumping, or by blowing with air or inert gas. Solid products are separated from liquids in centrifuges, on filter boxes, on continuous belt filters, and perhaps most frequently, in various designs of plate-and-frame or recessed plate filter presses. The presses are dressed with cloths of cotton, Dynel, polypropylene, etc. Some provide separate channels for efficient washing, others have membranes for increasing the solids content of the presscake by pneumatic or hydraulic squeezing.

The plates and frames are made of wood, cast iron, or now usually hard rubber, polyethylene, and polyester.

When possible, the intermediates are taken for the subsequent manufacture of other intermediates or dyes without drying because of savings in energy costs and handling losses. There are, however, many cases where products, usually in the form of pastes discharged from a filter, must be dried. Where drying is required, air or vacuum ovens (in which the product is spread on trays), rotary dryers, spray dryers, or less frequently, drum dryers (flakers) are used. Spray dryers have become increasingly important.

The final stage in dye manufacture is grinding or milling. Dry grinding is usually carried out in impact mills (Atritor, KEK, or ST); considerable amounts of dust are generated, and well-established methods are available to control this problem. Dry grinding is an inevitable consequence of oven drying, but more modern methods of drying, especially continuous drying, allow the production of materials that do not require a final comminution stage. Wet milling has become increasingly important for pigments and disperse dyes.

In the past the successful operation of batch processes depended mainly on the skill and accumulated experience of the operator. This operating experience was difficult to codify in a form that enabled full use to be made of it in developing new designs. The gradual evolution of better instrumentation, followed by the installation of sequence control systems, has enabled much more process data to be recorded, permitting maintenance of process variations within the minimum possible limits.

Full computerization of multiproduct batch plants is much more difficult than with single-product continuous units because the control parameters vary fundamentally with respect to time. The first computerized azo and intermediates plants were brought on stream by ICI Organics Division (now Zeneca Specialties) in the early 1970s, and have now been followed by many others. The additional cost (ca 10%) of computerization has been estimated to give a saving of 30 to 45% in labor costs. However, highly trained process operators and instrument engineers are required.

HEALTH AND SAFETY FACTORS

Toxicology and Registration

The toxic nature of some dyes and intermediates has long been recognized. Acute, or short-term, effects are generally well known. They are controlled by keeping the concentration of the chemicals in the workplace atmosphere below prescribed limits and avoiding physical contact with the material. Chronic effects, on the other hand, frequently do not become apparent until after many years of exposure.

The positive links between benzidine derivatives and 2-naphthylamine with bladder cancer prompted the introduction of stringent government regulations to minimize such occurrences in the future. Currently, the three principal regulatory agencies worldwide are European Core Inventory (ECOIN) and European Inventory of Existing Commercial Substances (EINECS) in Europe, Toxic Substances Control Act (TOSCA) in the United States, and Ministry of Technology and Industry (MITI) in Japan. Each of these has its own set of data and testing protocols for registration of a new chemical substance.

Environmental Concerns

Dyes, because they are intensely colored, present special problems in effluent discharge; even a very small amount is noticeable. However, the effect is more aesthetically displeasing rather than hazardous. Of greater concern is the discharge of toxic heavy metals such as mercury and chromium.

Effluents from both dye works and dyehouses are treated both before leaving the plant, eg, neutralization of acidic and alkaline liquors and heavy metal removal, and in municipal sewage works. Various treatments are used.

Biological treatment is the most common and most widespread technique used in effluent treatment, having been employed for over 140 years. There are two types of treatment, aerobic and anaerobic.

Removal of color by adsorption using activated carbon is also employed. Activated carbon is very good at removing low levels of soluble chemicals, including dyes. Its main drawback is its limited capacity. Consequently, activated carbon is best for removing color from dilute effluent (see CARBON–ACTIVATED CARBON).

Chemical treatment of the effluent with a flocculating agent is the most robust and generally most efficient way to remove color.

Chemical oxidation is a more recent method of effluent treatment, especially chemical effluent. This procedure uses strong oxidizing agents like ozone, hydrogen peroxide, chlorine, and potassium permanganate in order to force degradation of even some of the more resilient organic molecules. As of this writing (ca 1993), these treatments remain very expensive and are of limited size, thought they may have some promise in the future.

Additional strategies being implemented to minimize dye and related chemical effluent include designing more environmentally friendly chemicals, more efficient (higher yielding) manufacturing processes, and more effective dyes, eg, reactive dyes having higher fixation.

Colour Index, 3rd ed., The Society of Dyers and Colorists, Bradford, U.K., 1971.

P. F. Gordon and P. Gregory, *Organic Chemistry in Color*, Springer-Verlag, Berlin, 1983.

K. Venkataraman, *The Chemistry of Synthetic Dyes*, Vols. I–VIII, Academic Press, Inc., New York, 1952–1974.

H. Zollinger, *Color Chemistry: Synthesis, Properties and Applications of Organic Dyes and Pigments*, 2nd ed., VCH, 1991.

PETER GREGORY
Zeneca Specialties

DYES, ANTHRAQUINONE

The synthesis of an anthraquinone dye generally involves a large number of steps. Highly toxic metals such as mercury or chromium(VI) are sometimes required. Some processes need to employ a large amount of organic solvent, and others involve a great quantity of waste acids. With the increasing demand for environmental protection, the regulation of pollutant effluents has become more stringent year after year, which has caused a sharp increase in the costs for wastewater treatment. This situation has led to intensive improvement of conventional methods and the development of new synthetic routes as well.

Efforts have also been made to overcome complicated processes. Methods to reduce the number of steps or to use new starting materials have been studied extensively.

Because of their small extinction coefficients, anthraquinone dyes have less tinctorial strength than azo dyes; that is the intrinsic disadvantage of anthraquinones. This fact and the complexity of preparation have made their production costs higher than those of azo dyes. However, the anthraquinone dyes have excellent properties that are not attainable by azo dyes, such as brilliancy of color, fastness, and excellent dyeing properties (leveling and dye bath stability). Thus the anthraquinone dyes have been widely used in the areas where these properties are required. Cotton or polyester–cotton blend fibers for military wear and working wear that require extreme fastness are dyed mainly with anthraquinone vat dyes. Most polyester fabrics for automobile seats are dyed with anthraquinone disperse dyes, since the requirement for lightfastness is extremely high and, simultaneously, bright shades are needed.

World dye manufacturers have already begun to develop new types of dyes that can replace the anthraquinones technically and economically. Some successful examples can be found in azo disperse red and blue dyes. In the reactive dye area intensive studies have continued to develop triphenodioxazine compounds, to replace anthraquinone blues.

METHOD OF SYNTHESIS

Anthraquinone dyes are derived from several key compounds called dye intermediates. The methods for preparing these key intermediates can be divided into two types: (1) introduction of substituent(s) onto the anthraquinone nucleus, and (2) synthesis of an anthraquinone nucleus having the desired substituents, starting from benzene or naphthalene derivatives (nucleus synthesis). The principal reactions are nitration and sulfonation, which are very important in preparing α-substituted anthraquinones by electrophilic substitution. Nucleus synthesis is important for the production of β-substituted anthraquinones such as 2-methylanthraquinone and 2-chloroanthraquinone. Friedel-Crafts acylation using aluminum chloride is applied for this purpose. Synthesis of quinizarin (1,4-dihydroxyanthraquinone) is also important.

COLOR AND STRUCTURE

The uv–vis spectrum of anthraquinone shows an absorption maximum at 323 nm ($\varepsilon = 4{,}500$) due to a π-π^* transition and very weak absorption in the visible range, 405 nm ($\varepsilon = 60$) due to a n-π^* transition. Thus anthraquinone is almost colorless. Introduction of electron-donating substituents causes a bathochromic shift. This is due to the charge-transfer band from the lone pair of amino or hydroxyl groups to the oxygen atom of the carbonyl group. By increasing the electron-donating ability of substituents, the bathochromic shifts are enhanced (Table 1). In the case of the same substituent, the bathochromic shift is larger when the substituent is in the 1 position rather than in the 2 position. The introduction of an electron-withdrawing group has little effect on the absorption maximum of the spectrum.

The absorption maximum of a disubstituted anthraquinone greatly depends on the substituents and their positions. The 1,4-disubstituted compound shows a remarkable bathochromic shift. Larger bathochromic shifts are observed with increasing electron-withdrawing ability of β-substituents.

1,4,5,8-Tetrasubstituted anthraquinones give a slightly reddish blue tint to greenish blue color, depending on the substituents and their positions.

In addition to the color and the tinctorial strength, which are very important factors for the molecular design

Table 1. Spectral Data for Some Monosubstituted Anthraquinones[a] in Methanol

Substituent	1-position		2-position	
	λ_{max}, nm	ε	λ_{max}, nm	ε
Electron-donating groups				
OCH₃	378	5200	363	3950
OH	402	5500	368	3900
NHCOCH₃	400	5600	367	4200
NH₂	475	6300	440	4500
NHCH₃	503	7100	462	5700
N(CH₃)₂	503	4900	472	5900
Electron-withdrawing groups				
NO₂	325	4300	323	5200
Cl	333	5000	325	3900

[a]Unsubstituted anthraquinone $\lambda_{max} = 323$ nm; $\varepsilon = 4{,}500$.

of anthraquinone dyes, affinity for fibers, various kinds of fastness (light, wet, sublimation, nitrogen oxides (NO_x) gas, washing, etc.), and application properties (sensitivity for dyeing temperature, pH, etc) must be considered thoroughly as well.

KEY INTERMEDIATES

1-Aminoanthraquinone and Related Compounds

1-Aminoanthraquinone is the most important intermediate for manufacturing acid, reactive, disperse, and vat dyes. It has been manufactured from anthraquinone-1-sulfonic acid by ammonolysis of the sulfo group with aqueous ammonia in the presence of an oxidizing agent such as nitrobenzene-3-sulfonic acid. In this process the starting material can be obtained only by mercury-catalyzed sulfonation of anthraquinone with oleum. For improved ecology, the alternative route based on 1-nitroanthraquinone was established. 1-Nitroanthraquinone is prepared from anthraquinone by nitration in sulfuric acid or organic solvent. 1-Aminoanthraquinone can be prepared from 1-nitroanthraquinone by reduction with sodium sulfide, sodium hydrogen sulfide in water, in organic solvent, with hydrazine hydrate, or by catalytic hydrogenation. Highly purified product is manufactured by continuous vacuum distillation.

1-Amino-4-bromoanthraquinone-2-sulfonic acid (bromamine acid) is the most important intermediate for manufacturing reactive and acid dyes. Bromamine acid is manufactured from 1-aminoanthraquinone-2-sulfonic acid by bromination in aqueous medium or in concentrated sulfuric acid.

1-Amino-2-bromo-4-hydroxyanthraquinone (bromo pink) is one of the most important intermediates for manufacturing red disperse dyes. It is prepared by dibrominating 1-aminoanthraquinone in concentrated sulfuric acid and subsequent hydrolysis in the presence of boric acid.

1-Amino-2-chloro-4-hydroxyanthraquinone (chloro pink) is another important intermediate in red disperse dye manufacture. 1-Amino-2-chloro-4-hydroxyanthraquinone is prepared via a route from chlorobenzene and phthalic anhydride as the raw materials.

1,4-Dihydroxyanthraquinone

This anthraquinone, also known as quinizarin, is of great importance in manufacturing disperse, acid, and vat dyes. It is manufactured by condensation of phthalic anhydride with 4-chlorophenol in oleum in the presence of boric acid or boron trifluoride.

1,4-Diaminoanthraquinone and Related Compounds

Leuco-1,4-diaminoanthraquinone (leucamine), an important precursor for 1,4 diaminoanthraquinone, is prepared by heating 1,4-dihydroxyanthraquinone with sodium dithionite in aqueous ammonia under pressure. 1,4-Diaminoanthraquinone is an important intermediate for vat dyes and disperse dyes and is prepared by oxidizing leuco-1,4-diaminoanthraquinone with nitrobenzene in the presence of piperidine.

1,4-Diaminoanthraquinone-2-sulfonic acid is a possible precursor of 1,4-diamino-2,3-dicyanoanthraquinone. It is prepared from 1-amino-4-bromoanthraquinone-2-sulfonic acid by reaction with liquid ammonia in the presence of copper catalyst.

Instead of liquid ammonia, aqueous ammonia is also used together with a polar aprotic solvent such as formamide. It is also prepared by sulfonating 1,4-diaminoanthraquinone with chlorosulfonic acid, sulfuric acid, or oleum.

1,4-Diamino-2,3-dichloroanthraquinone (CI Disperse Violet 28) is an important compound as an intermediate for CI Disperse Blue 60 and CI Disperse Violet 26. It is prepared by chlorination of leuco-1,4-diaminoanthraquinone with chlorine gas or sulfuryl chloride in an inert organic solvent such as nitrobenzene.

1,4-Diamino-2,3-dicyanoanthraquinone is the key intermediate for manufacturing CI Disperse Blue 60. 1,4-Diamino-2,3-dicyanoanthraquinone is manufactured by reaction of 1,4-diaminoanthraquinone-2,3-disulfonic acid with alkali metal cyanide.

1,4-Diaminoanthraquinone-2,3-dicarboxyimide is the intermediate for CI Disperse Blue 60, in which the imide H is replaced by the R group $-CH_2CH_2CH_2OCH_3$. Compound is prepared by hydrolysis of 1,4-diamino-2,3-dicyanoanthraquinone in concentrated sulfuric acid.

Anthraquinone-1-sulfonic acid and Its Derivatives

Anthraquinone-1-sulfonic acid has become less competitive than 1-nitroanthraquinone as the intermediate for 1-aminoanthraquinone. However, it still has great importance as an intermediate for manufacturing vat dyes via 1-chloroanthraquinone.

Anthraquinone-1-sulfonic acid is prepared from anthraquinone by sulfonation with 20% oleum in the presence of mercury catalyst, a Hg(II) salt such as $HgSO_4$ or HgO, at 120°C.

1-Chloroanthraquinone is an intermediate for manufacturing vat dyes such as CI Vat Brown 1. 1-Chloroanthraquinone is prepared by chlorination of anthra quinone-1-sulfonic acid with sodium chlorate in hydrochloric acid at elevated temperature.

1-Methylaminoanthraquinone, is an important intermediate for manufacturing solvent dyes and acid dyes, is prepared from anthraquinone-1-sulfonic acid by replacing the SO_3H group with methylamine.

Anthraquinone-α,α′-disulfonic acids and Related Compounds

Anthraquinone-α,α′-disulfonic acids and their derivatives are important intermediates for manufacturing disperse blue dyes (via 1,5-, or 1,8-dihydroxyanthraquinone or 1,5-dichloroanthraquinone) and vat dyes (via 1,5-dichloroanthraquinone).

Anthraquinone-1,5-disulfonic acid, and anthraquinone-1,8-disulfonic acid are produced from anthraquinone by disulfonation in oleum.

1,5-Dichloroanthraquinone is an important intermediate for vat dyes and disperse blue dyes. It is prepared by the reaction of anthraquinone-1,5-disulfonic acid with $NaClO_3$ in hot hydrochloric acid solution.

1,5-Dihydroxyanthraquinone (anthrarufin) is an important intermediate for manufacturing disperse blue dyes', eg, CI Disperse Blue 73, and is prepared from anthraquinone-1,5-disulfonic acid by heating with an aqueous suspension of calcium oxide and magnesium chloride under pressure at 200–250°C.

α,α′-Dinitroanthraquinones and Related Compounds

1,5- and 1,8-Dinitroanthraquinone are the key intermediates for manufacturing disperse blue dyes via dinitrodihydroxyanthraquinone and vat dyes via diaminoanthraquinones. 1,5-Dinitroanthraquinone and 1,8-dinitroanthraquinone are prepared by nitration of anthraquinone with nitric acid in sulfuric acid. α,β′-Dinitroanthraquinones are also formed in the reaction.

1,5-Diaminoanthraquinone is prepared from 1,5-dinitroanthraquinone by ammonolysis in organic solvents, in aqueous ammonia, by catalytic hydrogenation in an organic solvent, or by reduction with sodium sulfide. It is also prepared from anthraquinone 1,5-disulfonic acid by ammonolysis. 1,5-Diaminoanthraquinone is an important intermediate for manufacturing vat dyes.

1,5-Dihydroxy-4,8-dinitroanthraquinone is an important dye precursor for CI Disperse Blue 56 and is prepared from 1,5-diphenoxyanthraquinone by hexanitration in sulfuric acid and subsequent hydrolysis with aqueous alkali. 1,5-Dinitro-4,8-dihydroxyanthraquinone is also prepared from 1,5-dimethoxyanthr aquinone.

2-Methylanthraquinone and Related Compounds

2-Methylanthraquinone and its derivatives are important as intermediates for manufacturing various kinds of vat dyes and brilliant blue (turqoise blue) disperse dyes. 2-Methylanthraquinone is prepared from phthalic anhydride and toluene via a benzoylbenzoic acid.

2-Chloroanthraquinone and Its Derivatives

2-Chloroanthraquinone and its derivatives are the most important intermediates for vat dyes and high performance organic pigments.

2-Chloroanthraquinone is prepared by Friedel-Crafts reaction of chlorobenzene and phthalic anhydride in the presence of aluminum chloride, followed by ring closure in concentrated sulfuric acid. 2-Amino-3-hydroxyanthraquinone is prepared by heating 5-benzoylbenzoxazolone-2'-carboxylic acid in sulfuric acid. This compound is an intermediate for CI Vat Red 10.

Benzanthrone and Related Compounds

Benzanthrone is prepared by the reaction of anthraquinone with glycerol, sulfuric acid, and a reducing agent such as iron. Benzanthrone is an important intermediate for manufacturing vat dyes.

N-Methylanthrapyridone and Its Derivatives

6-Bromo-3-methylanthrapyridone is an important intermediate for manufacturing dyes soluble in organic solvents. These solvent dyes are prepared by replacing the bromine atom with various kinds of aromatic amines 6-Bromo-3-methylanthrapyridone is prepared from 1-methylamino-4-bromoanthraquinone by acetylation with acetic anhydride followed by ring closure in alkali. The starting material of this route is anthraquinone-1-sulfonic acid.

REACTIVE DYES

Most of the anthraquinone reactive dyes are derived from bromamine acid. These dyes give a bright blue shade and excellent lightfastness. A great number of reactive groups have been proposed; typical examples include sulfatoethylsulfone, dichlorotriazine, monochlorotriazine, monofluorotriazine, and other heterocyclic groups.

DISPERSE DYES

Disperse dyes are water-insoluble, aqueous dispersed materials that are used for dyeing hydrophobic synthetic fibers, including polyester, acetate, and polyamide.

By introducing amino, hydroxy, or methyl groups onto the anthraquinone moiety as the principal auxochromes, dyes that have yellow through greenish blue shades are obtained. Among these dyes many that have brilliant red, violet, blue, and greenish blue shades have great industrial importance in view of their affinity for polyester or cellulose acetate fibers and lightfastness and sublimation resistance. On the contrary, yellow or orange dyes are not satisfactory because of their rather simple molecular structure. Therefore, these shades are obtained from other chromophores.

On the basis of the kind and the position of their substituents and their color range, the anthraquinoid disperse dyes may be classified as follows:

Color range	Chemical description
red	1-amino-4-hydroxyanthraquinones
blue, greenish blue	1,4,5,8-substituted anthraquinones
greenish blue	1,4-diaminoanthraquinone-2,3-dicarboximides
violet, blue	1,4-diaminoanthraquinone derivatives
violet, blue	N-substituted 1-amino-4-hydroxyanthraquinones

ACID DYES

Acid dyes are used for dyeing wool, synthetic polyamides, and silk in aqueous media. Anthraquinone acid dyes give brilliant reds, violets, blues, and greens and exhibit excellent lightfastness. Because of their relatively high cost, they are used to dye high grade textiles in pale and moderate shades. Various kinds of anthraquinone acid dyes have been developed so far, mainly by IG-Farbenindustrie in Germany applying chemical reactions that were studied in developing vat dyes. However, the number of commercial products has declined because of poor properties or unavailable raw materials. Anthraquinone acid dyes may be classified into two groups: bromamine acid derivatives and quinizarin derivatives.

VAT DYES

Anthraquinone vat dyes have been used to dye cotton and other cellulose fibers for many decades. Despite their high cost, relatively muted colors, and difficulty in application, anthraquinone vat dyes still form one of the most important dye classes of synthetic dyes because of their all-around superior fastness.

Anthraquinone vat dyes are water-insoluble dyes. They are converted to leuco compounds (anthrahydroquinones) by reducing agents such as sodium hydrosulfite in alkaline conditions. These water-soluble leuco compounds have an affinity to cellulose fibers and penetrate them. After reoxidation by means of air or other oxidizing agents, the dye becomes water-insoluble again and fixes firmly on the fiber.

The anthraquinone vat dyes can be classified into several groups on the basis of their chemical structures: (1) benzanthrone dyes, (2) indanthrones, (3) anthrimides, (4) anthrimidocarbazoles, (5) acylaminoanthraquinones, (6) anthraquinoneazoles, (7) anthraquinone acridones, (8) anthrapyrimidines, and (9) highly condensed ring systems. Recently, research and development efforts have focused on improved manufacturing of traditional vat dyes.

MORDANT DYES

Mordant dyes have hydroxy groups in their molecular structure that are capable of forming complexes with metals. Although a variety of metals such as iron, copper, aluminum, and cobalt have been used, chromium is most preferable as a mordant. Alizarin or CI Mordant Red 11 (CI 58000), the principal component of the natural dye obtained from madder root, is the most typical mordant dye. Many mordant dyes have given way to the vat or the azoic dyes, which are applied by much simpler dyeing procedures.

Acid–mordant dyes have characteristics similar to those of acid dyes which have a relatively low molecular weight, anionic substituents, and an affinity to polyamide fibers and mordant dyes. In general, brilliant shades cannot be obtained by acid–mordant dyes because they are used as their chromium mordant by treatment with dichromate in the course of the dyeing procedure. However, because of their excellent fastness for light and wet treatment, they are predominantly used to dye wool in heavy shades (navy blue, brown, and black).

FUNCTIONAL DYES

The investigation of new dyes has always been focused on the development of fast, brilliant, inexpensive, and easy applicable dyes. Because great emphasis has been placed especially on fastness, the dyes with poorer fastness have been ignored in the past. However, in recent years new needs for dyes that change color in response to low energy stimuli, including light, electricity, or heat, have arisen in the electronics industry. This application includes information recording, information display, and energy conversion. The term "functional dye" has been applied to dyes that are used in advanced technologies based on optoelectronics since 1981, when *The Chemistry of Functional Dyes* was published in Japan.

In order to develop dyes for these fields, characteristics of known dyes have been re-examined and some anthraquinone dyes been found usable. One example is in thermal-transfer recording, where the sublimation properties of disperse dyes are applied. Anthraquinone compounds have also been found to be useful dichroic dyes for guest-host liquid crystal displays when the substituents are properly selected to have high-order parameters. These dichroic dyes can be used for polarizer films of LCD systems as well. Anthraquinone derivatives that absorb in the near-infrared region have also been discovered, which may be applicable in semiconductor laser recording.

HEALTH AND SAFETY FACTORS

In general, anthraquinone dyes and their intermediates have not been reported as strongly toxic substances, but for many compounds safety data have not been evaluated. 1-Nitroanthraquinone, 1-chloroanthraquinone, and benzanthrone are reported to cause mild skin irritation in a test with rabbits, 500 mg/24 h. Some eye irritation data have been reported.

There are some tumorigenic data for anthraquinone dyes and intermediates which have been evaluated thoroughly. Data for 2-aminoanthraquinone and 2-methyl-1-nitroanthraquinone are available. 2-Aminoanthraquinone has been assessed by the United Nations International Agency for Research on Cancer (IARC) from studies on animals and is judged to fall into the *Animal: Limited Evidence* group. 2-Aminoanthraquinone has been evaluated by EPA (Genetic Toxicology program) and a positive carcinogenic effect for the rat and mouse is designated. 2-Methyl-1-nitroanthraquinone has been assessed by the IARC and judged as belonging in the *Animal: Sufficient Evidence* group. 2-Methyl-1-nitroanthraquinone has been

evaluated by the National Cancer Institute (NCI) and clear evidence of carcinogenicity for the rat and mouse is demonstrated.

Most anthraquinone dyes and their intermediates are handled in a powder form. Their dust poses the threat of contact to eyes and skin or contamination of surroundings. Attention must be paid to avoid these hazards. Special attention should be paid to avoid contact with compounds that are recognized to have probable carcinogenicity.

In the case of handling in relatively small quantities, ie, for laboratory use, normal personal equipment, ie, dust masks, safety glasses, and gloves, and hoods with local exhaust ventilation should be used. In plant operations, special technical handling measures should be taken, because the possibility of contact is extremely high, especially when charging the raw materials or isolating or packaging the intermediates or final products.

S. Abeta and K. Imada, *Kaisetsu Senryo Kagaku (Comprehensive Dyestuff Chemistry)*, Shikisensha, Osaka, 1989.

G. Booth, *The Manufacture of Organic Colorants and Intermediates*, Society of Dyers and Colourists, Bradford, UK, 1988.

G. Hallas in J. Shore, ed., *Colorants and Auxiliaries*, Society of Dyers and Colourists, Bradford, UK, 1990, 230–267.

H. Zollinger, *Color Chemistry*, 2nd rev. ed., VCH Verlagsgesellshaft mbH, Weinheim, Germany, 1991.

MAKOTO HATTORI
Sumitomo Chemical Company
Revised by Staff

DYES, AZO

The term "azo dyes" is applied to synthetic organic colorants that are characterized by the presence of the chromophoric azo group (−NN−). This divalent group is attached to sp^2 hybridized carbon atoms: on one side, to an aromatic or heterocyclic nucleus; on the other, it may be linked to an unsaturated molecule of the carbocyclic, heterocyclic, or aliphatic type. No natural dyes contain this chromophore. Commercially, the azo dyes are the largest and most versatile class of organic dyestuffs. There are more than 10,000 *Colour Index* (CI) generic names assigned to commercial colorants, approximately 4,500 are in use, and over 50% of these belong to the azo class. Synthetic dyes are derived in whole or in part from cyclic intermediates. Approximately two-thirds of the dyes consumed in the United States are used by the textile industry to dye natural and synthetic fiber or fabrics, about one-sixth is used for coloring paper, and the rest is used chiefly in the production of organic pigments and in the dyeing of leather and plastic. Dyes are sold as pastes, powders, and liquids; concentrations vary from 6 to 100%. The concentration, form, and purity of a dye is determined largely by the use for which it is intended.

CLASSIFICATION AND DESIGNATIONS

The most authoritative compilation covering the constitution, properties, preparations, manufacturers, and other coloring data is the publication *Colour Index*, which is edited jointly by the Society of Dyers and Colourists and the American Association of Textile Chemists and Colorists (AATCC).

THE CHEMISTRY OF SYNTHESIS

Peter Griess synthesized the first azo dye soon after his discovery of the diazotization reaction in 1858. The two reactions that form the basis for azo dye chemistry are diazotization (eq. 1) and coupling (eq. 2).

$$ArNH_2 + 2\,HX + NaNO_2 \longrightarrow ArN_2^+X^- + NaX + 2\,H_2O \quad (1)$$

$$\text{where } X = Cl^-,\ Br^-,\ NO_3^-,\ HSO_4^-,\ BF_4^-,\ \text{etc.}$$

$$ArN_2^+X^- + RH \longrightarrow ArN{=}N{-}R + HX \quad (2)$$

where R represents an alkyl or aryl radical whose conjugate acid RH is capable of coupling.

Azo Coupling

The coupling reaction between an aromatic diazo compound and a coupling component is the single most important synthetic route to azo dyes. Of the total dyes manufactured, about 60% are produced by this reaction. Other methods include oxidative coupling, reaction of arylhydrazine with quinones, and oxidation of aromatic amines. These methods, however, have limited industrial applications.

All coupling components used to prepare azo dyes have the common feature of an active hydrogen atom bound to a carbon atom. Compounds of the following types can be used as azo coupling components: (*1*) aromatic hydroxy compounds such as phenols and naphthols; (*2*) aromatic amines; (*3*) compounds that possess enolizable ketone groups of aliphatic character.

The broad principles governing the activity of coupling components may be summarized as follows:

1. Diazo coupling follows the rules of orientation of substituents in aromatic systems in accordance with the mechanism of electrophilic aromatic substitution and the concept of resonance.
2. Generally, phenols (as the phenolate anion) couple more readily than amines, and members of the naphthalene series more readily than the members of the benzene series.
3. Electron-attracting substituents in the coupling components such as halogen, nitro, sulfo, carboxyl, and carbonyl, are deactivating and tend to retard coupling.
4. A lower alkyl or alkoxy group substituted in the ortho or meta position to an amino group may promote coupling. Good couplers are obtained from dimethy-

laniline when lower alkyl, lower alkoxy, or both groups are present in the 2 and 5 position.

5. It is possible for diazo compounds to attack both the ortho and para position of hydroxyl and amino coupling components when these positions are not already occupied.

Technologically, the most important examples of such couplers are 1-naphthylamine, 1-naphthol, and sulfonic acid derivatives of 1-naphthol. Of great importance in the dyestuff industry are derivatives of 1-naphthol-3-sulfonic acid, such as H-acid (8-amino-1-naphthol-3,6-disulfonic acid, J-acid (6-amino-1-naphthol-3-sulfonic acid (**1**), and gamma acid (7-amino-1-naphthol-3-sulfonic acid (**2**).

R = H, R′ = NH₂
R = NH₂, R′ = H

The azo coupling reaction proceeds by the electrophilic aromatic substitution mechanism.

CLASSIFICATION

In addition to classification according to the number of azo groups, further subdivision is achieved, first according to whether the compound is water-soluble and, secondly, according to the types of components used. Another system of classification is based on dyeing classes. All colorants are divided to indicate the chief method of application or to indicate principal use. Azo dyes are found in the acid, basic (cationic), direct, disperse, mordant, and reactive dyeing classes.

In the disazo group of azo dyes, primary and secondary types are distinguished. The former covers compounds made from two molecules of a diazo derivative and one molecule of a bifunctional coupling component. In both cases, the monofunctional reagent may consist of two molecules of one compound or one molecule of each of two substances used stepwise, the first alternative yielding symmetrical products.

Secondary Disazo Dyes

There 250 or so dyes of in this group are made by diazotizing an aminoazo compound the amino group of which derives from the original coupling component, and coupling it to a suitable intermediate.

Miscellaneous Disazo Dyes

Another group of disazo dyes is prepared by condensation of two identical or different aminoazo compounds commonly with phosgene, cyanuric chloride, or fumaryl dichloride, the fragments of which act as blocking groups between chromophores.

Trisazo and Polyazo Dyes

These are mostly direct dyes, the hues of which are predominantly brown, black, or dark blue or green. Some are leather dyes. Benzidine, which used to be an important bisdiazo component, has been replaced by 4,4'-diaminobenzanilide 4,4'-diaminodiphenylamine-2-sulfonic acid, etc. Benzidine dyes are almost never produced any longer because of their carcinogenicity.

DYEING CLASSES: STRUCTURE, APPLICATION, USES

Of all classes of dyestuffs, azo dyes have attained the widest range of usage, because their variations in chemical structure are readily synthesized and their methods of application are generally not complex. There are azo dyes for dyeing all natural substrates, such as cotton, paper, silk, leather, and wool; and there are azo dyes for synthetics, such as polyamides, polyesters, acrylics, polyolefins, viscose rayon, and cellulose acetate; for the coloring of paints, varnishes, plastics, printing inks, rubber, foods, drugs, and cosmetics; for staining polished and absorbed surfaces; and for use in diazo printing and color photography. The shades of azo dyes cover the whole spectrum.

Acid Dyes

Commercial acid dyes contain one or more sulfonate groups, thereby providing solubility in aqueous media. These dyes are applied in the presence of organic or mineral acids (pH 2–6). Such acids protonate any available cationic sites on the fiber, thereby making possible bonding between the fiber and the anionic dye molecule.

There are three general classifications of acid dyes, depending on their method of application: acid dyes that dye directly from the dyebath, mordant dyes that are capable of forming metallic lakes on the fiber when aftertreated with metallic salts, and premetallized dyes. Yellow and orange azo acid dyes are listed in Table 1. Other azo acid dyes are listed in Table 2.

Table 1. Yellow and Orange Commercial Azo Acid Dyes

CI name	CI number	Chemical type
Acid Yellow 34	18890	monoazo
Acid Yellow 36	13065	monoazo
Acid Yellow 49	18640	monoazo
Acid Yellow 59	18690	monoazo, metallized
Acid Yellow 65	14170	monoazo
Acid Yellow 99	13900	monoazo, metallized
Acid Yellow 135	14255	monoazo
Acid Yellow 151	13906	monoazo, metallized
Acid Yellow 200	18930	monoazo
Acid Orange 7	15510	monoazo
Acid Orange 10	16230	monoazo
Acid Orange 24	20170	disazo
Acid Orange 60	18732	monoazo, metallized
Acid Orange 156	26501	disazo

Table 2. Other Commercial Azo Acid Dyes

CI name	CI number	Chemical type
Acid Red 4	14710	monoazo
Acid Red 14	14720	monoazo
Acid Red 18	16255	monoazo
Acid Red 73	27290	disazo
Acid Red 85	22245	disazo
Acid Red 88	15620	monoazo
Acid Red 114	23635	disazo
Acid Red 137	17755	monoazo
Acid Red 151	26900	disazo
Acid Red 186	18810	monoazo, metallized
Acid Red 266	17101	monoazo
Acid Violet 3	16580	monoazo
Acid Violet 7	18055	monoazo
Acid Violet 12	18075	monoazo
Acid Blue 92	13390	monoazo
Acid Blue 113	26360	disazo
Acid Blue 118	26410	disazo
Acid Green 20	20495	disazo
Acid Brown 14	20195	disazo
Acid Black 1	20470	disazo
Acid Black 52	15711	monoazo, metallized
Acid Black 60	18165	monoazo, metallized
Acid Black 63	12195	monoazo, metallized

Metal Complexes of Azo Dyes

Metal complexes of certain o,o'-dihydroxyazo, o-carboxy-o'-hydroxyazo, o'-amino-o'hydroxyazo, arylazosalicyclic acid, and formazan compounds are used as dyes for wool, nylon, and cotton with generally much improved washfastness and lightfastness properties when compared to their respective unmetallized precursors. Dyes that are chelated with the metal on the substrate during the dyeing process are termed metallizable or mordant dyes. Conversely, dyes that have been metallized by the dye manufacturer prior to use by the dyer are classified as premetallized dyes. The two important types of premetallized dyes are the 1:1 and 2:1 complexes; eg, complexes with 1:1 and 2:1 ligand-to-metal ratios, respectively.

Chromium is the principal metal used with mordant dyes for wool, whereas both chromium and cobalt are used extensively in premetallized types for wool and nylon. Copper(II) is employed almost exclusively as the chelating metal ion in both metallizable and premetallized direct dyes for cotton.

Direct Dyes

Direct dyes are defined as anionic dyes substantive to cellulosic fibers (cotton, viscose, etc), when applied from an aqueous bath containing an electrolyte.

Direct dyes are one of the most versatile classes of dyestuff. In worldwide usage for cellulosic textiles, direct dyes are the second largest class of dyestuff. The important direct yellows and oranges of revealed chemical composition are listed in Table 3.

Direct Oranges. All principal commercially produced direct oranges are of disazo or stilbene chemical composition (Table 3).

Table 3. Yellow and Orange Shade Commercial Direct Dyes

CI name	CI number	Chemical type
Direct Yellow 4	24890	disazo
Direct Yellow 6	40001–40006	stilbene
Direct Yellow 11	40000	stilbene
Direct Yellow 28	19555	thiazole
Direct Yellow 34	29060	disazo
Direct Yellow 44	29000	disazo
Direct Yellow 106	40300	stilbene
Direct Yellow 118	29042	disazo
Direct Orange 6	23375	disazo
Direct Orange 8	22130	disazo
Direct Orange 15	40002–40003	stilbene
Direct Orange 26	29150	disazo
Direct Orange 34	40215–40220	stilbene
Direct Orange 39	40215	stilbene
Direct Orange 72	29058	disazo
Direct Orange 102	29156	disazo

Direct Reds. The principal commercially produced direct reds, with revealed chemistry, are of disazo or polyazo composition.

Direct Blues. Direct Blue 86, a phthalocyanine direct dye, represents a small but important segment of the direct dye structure groups. The dyes are brilliant greenish blue or turquoise shades. Table 4 shows some direct blues.

Direct Violets, Greens, Browns, and Blacks. Direct violets and greens are small-volume products.

Two important browns, other than benzidine derivatives, are of azo chemical composition: Direct Brown 30 and Direct Brown 44.

Table 5 lists some direct blacks.

Azoic or Naphthol Dyes. Azoic dyes (known also as ice colors and ingrain colors) are water-insoluble azo pigments, free from solubilizing groups, formed on the fiber by reaction of a diazo component with a coupling component, a so-called Naphthol AS compound, such as an arylide of 3-hydroxy-2-naphthoic acid. The discovery that 3-hydroxy-2-naphthoic acid arylides have greater substantivity for cotton than 2-naphthol tremendously increased the range of bright and fast shades and led to

Table 4. Blue Shade Commercial Direct Dyes

CI name	CI number	Chemical type
Direct Blue 15	24400	disazo
Direct Blue 22	24280	disazo
Direct Blue 25	23790	disazo
Direct Blue 75	34220	trisazo
Direct Blue 76	24410	disazo
Direct Blue 80	24315	disazo
Direct Blue 86	74180	phthalocyanine
Direct Blue 98	23155	disazo
Direct Blue 108	51320	oxazine
Direct Blue 218	24401	disazo metallized

Table 5. Black Shade Commercial Direct Dyes

CI name	CI number	Chemical type
Direct Black 19	35255	polyazo
Direct Black 22	35435	polyazo
Direct Black 150	32010	trisazo
Direct Black 166	30026	trisazo

the introduction of an extensive line of Naphthol AS derivatives, fast color salts, rapid fast dyes, and the rapidogen dyes from diazoamino compounds.

Disperse Azo Dyes

Generally speaking, disperse dyes for acetate possess only moderate fastness to gas, light, sublimation (heat), and washing. There were no suitable blue azo disperse dyes available in the early years and, as a consequence, the aminoanthraquinone types gained prominence in this shade range. The latter, however, were subject to gas fading, ie, on exposure to oxides of nitrogen and ozone acetate dyed blue became pink. This deficiency led to the development of many antigas fading additives. Selected commercially important disperse yellow and orange dyes are listed in Table 6.

The disperse reds are second only to the blues as the most important disperse color manufactured. All commercial disperse reds are monoazo dyes. Table 7 lists the commercial disperse dyes of chief importance.

Blues are the single most important color in the disperse class, both in terms of amounts produced and dollar sales volume. The only disperse brown of commercial importance is Disperse Brown 1.

Dispersion Technology.. Manufacturing procedures for producing dye dispersions are generally not disclosed. The principal dispersants in use include long-chain alkyl

Table 6. Yellow and Orange Shade Commercial Disperse Dyes

CI name	CI number
Disperse Yellow 3	11855
Disperse Yellow 4	12770
Disperse Yellow 5	12790
Disperse Yellow 7	26090
Disperse Yellow 8	12690
Disperse Yellow 10	12795
Disperse Yellow 23	26070
Disperse Yellow 60	12712
Disperse Orange 1	11080
Disperse Orange 3	11005
Disperse Orange 5	11100
Disperse Orange 13	26080
Disperse Orange 25	11227
Disperse Orange 29	26077
Disperse Orange 30	11119
Disperse Orange 56	12650
Disperse Orange 62	11239
Disperse Orange 138	11145

Table 7. Commercial Disperse Dyes

CI name	CI number
Disperse Red 1	11100
Disperse Red 5	11215
Disperse Red 7	11150
Disperse Red 13	11115
Disperse Red 17	11210
Disperse Red 19	11130
Disperse Red 31	11250
Disperse Red 32	11190
Disperse Red 58	11135
Disperse Red 65	11228
Disperse Red 72	11114
Disperse Red 73	11116
Disperse Red 90	11117
Disperse Violet 24	11200
Disperse Violet 33	11218
Disperse Blue 11	11260
Disperse Blue 79	11345
Disperse Blue 165	11077
Disperse Blue 183	11078
Disperse Brown 1	11152
Disperse Black 1	11365

sulfates, alkaryl sulfonates, fatty amine–ethylene oxide condensates, fatty alcohol–ethylene oxide condensates, naphthalene–formaldehyde–sulfuric acid condensates, and the lignin sulfonic acids.

All dispersions are thermodynamically unstable, since the interfacial area and hence the surface energy tend to decrease; ie, agglomeration occurs. The primary function of dispersing agents is to stabilize dispersions.

Application Techniques, Structural Variations, and Fastness Properties.. Since 1950 there has been a steady development of new disperse dyes to meet the demands imposed by the changing application methods and to provide the much needed improvement in fastness properties. Six different methods of applying disperse dyes have been developed since the introduction of polyester fibers: (1) dyeing at the boil in the presence of a carrier, for delicate fabrics, polyester–wool blends, etc; (2) dyeing at 120–135°C in pressurized vessels, for better exhaustion and often improved fastness to light, rubbing, and perspiration; (3) thermofixation techniques at 190–220°C, for continuous processing of certain types of fabrics; (4) transfer printing, generally at 210°C for 30 seconds, an important development; (5) solvent dyeing methods that are available but not popular; and (6) printing and continuous dyeing processes developed for polyester–cotton blends using specialist dyes and application techniques.

Disperse dyes are classified as high energy or low energy types. The use of higher dyeing temperatures for polyester fibers compared with those used for cellulose acetate has made possible the use of dyes of higher molecular weight, the so-called high energy dyes.

A use for disperse dyes that has undergone rapid growth since 1970 is in inks for the heat-transfer printing of polyester, especially double-knit fabrics. This simple method consists of printing the desired design on paper and then transferring the design from the paper to the fabric with heat. In the heat-transfer process the dye volatilizes, is adsorbed onto the fiber surface, and then diffuses into the fiber. Generally, low and medium energy types are used for this purpose.

Generally, the dye structures have been modified to achieve these desirable properties, by either increasing the molecular weight or introducing more polar groups or both, and by decreasing further their slight solubility in water.

Heterocyclic Disperse Dyes. Diazotizable aminoheterocyclic compounds are also used in the production of disperse dyes.

Oil-Soluble Azo Dyes

The oil soluble, water-insoluble, azo dyes dissolve in oils, fats, waxes, etc. Generally, yellow, orange, red, and brown oil colors are azo structures, and greens, blues, and violets are primarily anthraquinones. Blacks are usually nigrosines and indulines of the azine type. Substitution by chloro, nitro, and similar groups increases the molecular weight and improves sublimation fastness but lowers the oil solubility of this group of dyes.

Spirit-Soluble Azo Dyes

Spirit-soluble azo dyes dissolve in polar solvents, such as alcohol and acetone, and find application in the coloring of lacquers, plastics, printing inks, and ball-point pen inks. Of the two principal types of azo structures used, the most important are the insoluble salts of azo dyes containing sulfo groups and relatively complex organic amines. Mono- and dicyclohexylamine, isoamylamine, and the arylguanidines often serve as the amine, and the anionic component is chosen from the class of acid dyes for wool.

The second type is comprised of 2:1 metal complexes of o,o′-dihydroxy azo dyes, which generally do not contain sulfo or other strongly hydrated groups as found in the premetallized 2:1 complexes for wool. Thus their solubility in esters, ketones, and alcohols is relatively increased.

Basic (Cationic) Azo Dyes

Basic dyes of the azo class are the simplest and oldest known synthetic dyes. Current cationic dyes are used for modified acrylics, modified nylons, modified polyesters, leather, unbleached papers, and inks. An important application is for conversion into pigments. Principal chemical classes include azo, anthraquinone, triarylmethane, methine, thiazine, oxazine, etc. The dyes are applied in acidic dyebaths to fibers made of negatively charged polymer molecules.

Cationic azo dyes carry a positive charge in the chromophore portion of the molecule. The salt-forming counterion is usually a chloride or acetate. CI basic dyes are ammonium, sulfonium, or oxonium salts. Commercial basic azo dyes for which chemical structures are revealed by U.S. producers are listed in Table 8.

Table 8. Commercial Basic Azo Dyes

CI name	CI number
Basic Yellow 15	11087
Basic Yellow 24	11480
Basic Yellow 25	11450
Basic Yellow 57	12719
Basic Orange 1	11320
Basic Orange 2	11270
Basic Red 18	11085
Basic Red 22	11055
Basic Red 24	11088
Basic Red 29	11460
Basic Red 39	11465
Basic Red 76	12245
Basic Blue 41	11105
Basic Blue 54	11052
Basic Blue 65	11076
Basic Blue 66	11075
Basic Blue 67	11185
Basic Brown 1	21000
Basic Brown 2	21030
Basic Brown 4	21010
Basic Brown 16	12250
Basic Brown 17	12251
Basic Black 2	11825

Table 9. Safety Profiles for Some Azo Dyes

Dye	Safety profile
Acid Blue 41	eye irritant
Acid Blue 62	moderately toxic by intraperitoneal route, mutation data reported
Acid Blue 129	skin and eye irritant
Acid Blue 185	moderately toxic by intraperitoneal route, low toxicity by ingestion
Acid Green 40	mildly toxic by ingestion, eye irritant
Acid Red 98	mutation data reported
Acid Violet 7	low toxicity by ingestion
Acid Yellow 3	low toxicity by ingestion
Acid Yellow 7	poison by intravenous route
Basic Orange 1	mutation data reported
Basic Orange 21	moderately toxic by ingestion
Basic Red 13	moderately toxic by ingestion
Basic Red 29	mutation data reported
Direct Blue 2	low toxicity by ingestion, eye irritant mutation data reported
Direct Blue 218	moderately toxic by ingestion, low toxicity by skin contact, mutation data reported
Direct Brown 31	mutation data reported
Direct Red 39	mutation data reported
Direct Red 81	moderately toxic by interperitoneal route, mutation data reported
Direct Green 1	mutation data reported
Disperseb Blue 27	mutation data reported
Disperse Red 29	mutation data reported
Disperse Yellow 7	moderately toxic by unspecified route

Azo Pigments

Organic pigments are an important class of organic colorants. Expanding areas of usage include the mass coloration of synthetic fibers and textile printing in the textile field, and in the nontextile area, plastics. A pigment is insoluble in the medium in which it is used. The physical properties of pigments are of great significance, since the coloring process does not involve solution of the colorant. Azo pigments can be grouped as metal toners, metal chelates, and metal-free azo pigments.

Metal toners usually contain one sulfonic acid group and often a carboxylic acid group. Metal chelation is also a means of insolubilizing organic molecules.

In the metal-free class of azo pigments it is remarkable that the simplest derivatives of 2-naphthol such as Hansa Red B (CI Pigment Red 3; CI 12120, Toluidine Red)Toluidine Red [2425-85-6] and Para Red, both known since 1905, are still of importance.

Because these pigments are organic in nature, they tend to bleed in resins and solvents. Increasing the molecular weight often reduces this tendency.

ECONOMIC ASPECTS

Consumption of dyes is governed by the demand for textiles, leather, and colored paper.

Environmental laws covering air, water, and waste emissions are the largest concerns for the dye industry. The U.S. Toxic Substances Control Act has made development of new dyes costly. Western Europe and Japan have similar problems. Pollution abatement costs are rising in these countries and is a growing concern in other countries such as China as well.

The European Community has added two new azo dyes (o-anisidine and 4-aminobenzene) to the banned list. Products must be free of these amines to be exported to the EU. All products made of textiles and leather (if they come in direct contact with human skin and oral cavity) are covered in this directive.

HEALTH AND SAFETY FACTORS

Safety profiles for some representative dyes are listed in Table 9. All of the dyes listed here emit toxic vapors when heated to decomposition.

"Dyes," *Chemical Economics Handbook*, Stanford Research Institute, Menlo Park, Calif., Aug. 2000.

G. Hallas, in J. Griffiths, ed., *Developments in the Chemistry and Technology of Organic Dyes*, Blackwell, Scientific Publishing, Oxford, UK, 1984.

G. D. Parfitt, *Dispersions of Powders in Liquids*, Elsevier Publishing Co., Inc., New York, 1969, p. 259.

K. Venkataraman, *The Chemistry of Synthetic Dyes*, Vols. 1 and 2, Academic Press, New York, 1952.

RASIK J. CHUDGAR
BASF Corporation
JOHN OAKES
John Oakes Associates

DYES, ENVIRONMENTAL CHEMISTRY

EFFLUENT TREATMENT METHODS

Methods of effluent treatment for dyes may be classified broadly into three main categories: physical, chemical, and biological.

Physical	Chemical	Biological
adsorption	neutralization	stabilization ponds
sedimentation	reduction	aerated lagoons
flotation	oxidation	trickling filters
flocculation	electrolysis	activated sludge
coagulation	ion exchange	anaerobic digestion
foam fractionation	wet-air oxidation	bioaugmentation
polymer flocculation		
reverse osmosis/ ultrafiltration		
ionization radiation		
incineration		

There are four stages: preliminary, primary, secondary, and tertiary treatment processes, which differ mainly by the number of operations performed on the waste steams.

Preliminary treatment processes of dye waste include equalization, neutralization, and possibly disinfection. Primary stages are mainly physical and include screening, sedimentation, flotation, and flocculation. The objective is to remove debris, undissolved chemicals, and particulate matter. Secondary stages are used to reduce the organic load, which essentially is a combination of physical/chemical separation and biological oxidation. Tertiary stages are important because they serve as a polishing of effluent treatment. These methods are adsorption, ion exchange, chemical oxidation, hyperfiltration (reverse osmosis), electrochemical, etc.

FATE OF DYES

The chemical reaction of Methyl Violet (CI 42535) and Indigo Sulfonate (CI 73015) with ozone leads to the formation of colorless carboxyl-containing products. The interaction between the large-volume reactive dye Reactive Blue 19, and the strong oxidant potassium peroxydisulfate via the process known as direct chemical oxidation has also been examined. This process led to the formation of products arising from replacement of the amino ($-NH_2$) and anilino groups by $-OH$ groups and an unusual oxidative coupling.

Bioaugmentation and bioremediation treatments using white rot fungi can be used to degrade azo, anthraquinone, and vat dyes. This is of interest because of the potential for the release of toxic metal ions such as Cu^{2+} as a product of biodegradation.

In a very detailed report on the fate of the high-volume commercial dye disperse blue 79:1 in anaerobic environments the expected products arising from hydrosulfite-

induced reductive-cleavage of the azo bond and nitro group reduction were observed.

The fate of several anthraquinone disperse dyes in anoxic sediments has been studied. The identified reaction products arise from O-dealkylation and an unusual and significantly slower deamination process.

ANALYTICAL METHODS

The key to success in studies pertaining to the fate of dyes has been the development of suitable methods for characterizing products produced from the chemical and biological degradation methods. In this regard the uniqueness of dyes, among the many classes of organic compounds, has necessitated the development of specific analytical methods. This was especially important in the case of sulfonated structures, which are involatile and often insoluble in traditional nuclear magnetic resonance (nmr) solvents.

To facilitate studies in this area, environmental samples of organic colorants have been analyzed by using various extraction, spectrophotometric, and chromatographic methods. The chromatographic methods include paper, thin-layer, gas and high performance liquid chromatography (hplc). In addition, capillary electrophoresis and mass spectrometry (direct probe, fast-atom bombardment, electrospray, particle beam, field desorption, and laser desorption) were employed. Dye classes used in this comprehensive study included acid, basic, disperse, and vat dyes.

An analysis of basic, disperse, and solvent dyes in wastewater using mass spectrometry has also been undertaken. In this regard, suitable methods for determining the structures of dyes present in textile effluents have been reported.

POLLUTION PREVENTION

In view of the tough regulations facing companies that manufacture and use inorganic and organic colorants, it quickly became evident that the best way to address these safeguards responsibly would involve adopting pollution prevention measure involving waste minimization and/or source reduction. This approach is the most effective mechanism because it provides a reduction in waste management and treatment costs, a lowering of raw material costs, enhanced public standing in the community, added protection for the health and safety of workers and consumers, and substantial reductions in the environmental management needs connected with manufacturing operations.

Process Optimization

In some cases, especially in dye manufacturing, components that must be removed from wastewater arise from reactions that do not go to completion. In the case of azo dyes, this leaves residual couplers, diazo components, or by-products in the liquors produced in filtration. When disperse dyes based on heteroaromatic amines are used, the presence of excess coupler provides a solvent for the dye produced, enhancing the amount of unprecipitated dye in the filtrates and giving deeply colored wastewater.

Effective measures in this area have involved developing ways to push reaction yields as close to the theoretical level (100%) as possible. In this regard, it has been possible to synthesize azo dyes and pigments such as disperse red 167:1, disperse red 177, pigment red 3, pigment red 48, Pigment Red 21, pigment yellow 14, and pigment yellow 1 in yields of 97–100%. To achieve these yields, efficient diazotization and azo coupling steps were combined with neutralization and heat stabilization.

Process optimization in the textile industry has included the use of initiatives such as product substitution, process analysis, effluent reduction, chemical use reduction, and process modification, to reduce costs associated with effluent wastewater treatment and disposal. Similarly, a reduction in water usage was achieved by implementing a closed-loop system that involved the reuse of vacuum pump cooling water rather than discharging it as wastewater after a single use.

Heavy Metals

Metals such as copper, chromium, mercury, nickel, and zinc are priority pollutants. A number of compounds containing these metals have long been used as catalysts and complexing agents in the synthesis of dyes and key intermediates.

A number of studies have been conducted on the removal of heavy metals from processing waters. The methods used involve coagulation, polymer adsorption, ultrafiltration, carbon adsorption, electrochemistry, incineration and land disposal, powdered activated carbon treatment, neutralization, and bisulfite-catalyzed borohydride reduction. In addition, the removal of heavy metals by methods involving ion exchange, sulfide precipitation, chelation with trimercaptotriazine, mercaptobenzothiazole, diethyldithiocarbonate, carbonotrithioic acid, adsorptive filtration, and carbon adsorption followed by stripping the metals with an acid regenerant and recovering them by electrolysis have been explored.

An emerging effective and inexpensive method for removing metals such as cobalt, chromium, and copper is phytoremediation. This natural process has been carried out by using vegetation for an in situ treatment of contaminated soils and sediments.

Reactive and acid dyes complexed with chromium, copper, and nickel have been decolorized by ozonation followed by treatment with chelating resins. This led to a 77–86% removal of the heavy metals.

Studies on the fate of copper in textile dyehouse wastewater have indicated that the use of biological solids in the wastewater afforded efficient sorption of copper containing direct and reactive phthalocyanine dyes. Therefore, little of the copper bound in dyes of these types would be discharged in the wastewater in soluble form. Also, sequestrants in wastewater would compete with biological solids for free and bound copper ions, decreasing the tendency of copper to attach to the solid phase during wastewater treatment.

Questions about the toxicity of textile mill discharges to plant life in receiving waters led to an evaluation of the toxicity of 46 dyes to fresh-water green algae. All except two of the dyes were anionic colorants, many of which were reactive and/or metal complexes. Only the two cationic dyes exhibited toxicity.

Metal-Complexed Dyes

One of the most serious environmental problems in the dye, textile, and leather industries is associated with the manufacture and use of metallized azo dyes that are complexed with chromium or cobalt to obtain desirable fastness properties. This issue arises because the superior lightfastness required for applications involving polyamide fibers can be achieved only by using metallized azo dyes in which the azo linkage is protected from light-induced degradation.

Although metallized dyes can be removed from wastewater using various chemical, physical, and biological methods, such treatments can be expensive and may result in sludges that must be disposed of by incineration or land filling. In view of an emphasis on pollution prevention instead of waste treatment, the merits of substituting iron (Fe) for chromium (Cr) and cobalt (Co) in commercially important acid dyes has come to the forefront. Although Fe is innocuous, until relatively recently little had been published about the utility of Fe-complexed dyes as environmentally friendly alternatives to widely used Cr and Co complexed acid dyes.

Dye Manufacturing

At least one major dyestuff manufacturer has developed a method for eliminating mercury as a catalyst in the sulfonation step employed in the synthesis of key intermediates for anthraquinone dyes. Similarly, one of the most important pollution prevention measures to date has been termination of the use of toxic and carcinogenic compounds such as 2-naphthylamine and benzidine that were historically associated with the production of certain azo dyes and pigments. In the case of benzidine, the termination of its use has curtailed the manufacture and sale of 67 dyes.

No doubt the most stringent pollution prevention measure is one imposed by Germany, where Consumer Goods Ordinance has barred the distribution of fabric containing azo dyes capable of forming an established carcinogenic aromatic amine.

The presence of certain benzidine congeners on this list will impact the manufacture and use of such important dyes as direct blue 218 and direct blue 281, as well as large-volume diarylide yellow pigments.

Pigment Replacements

In pigments, naphthol AS pigments such as pigment red 31 and 112, high-performance polycyclic pigments such as perylenes and quinacridones serve as nontoxic alternatives to certain barium lakes. Similarly, substitution of environmentally safe synthetic organic pigments such as pigment red 48:2 and pigment red 49:2 for toxic inorganic pigments containing cadmium, lead, nickel, or copper is also deemed prudent.

Aromatic Amines

Various research groups have conducted studies aimed at the development of nonmutagenic benzidine analogues for use in the synthesis of organic dyes and pigments. It has been shown that the mutagenicity of benzidine could be removed by incorporating bulky alkyl or alkoxy groups *ortho* to the amino groups. This work led to the design of nonmutagenic azo dyes for inkjet application that had good wetfastness but lightfastness that requires improvement.

Recovery and Reuse

Some reactive dyes are used to provide dyeings on cotton that possess high wetfastness. Unfortunately, dye-fiber bond formation is always accompanied by alkaline hydrolysis of the reactive group in the dyes employed, so that the resultant hydrolyzed dye can no longer react with cotton. It is this color that must be removed in the waste treatment process. An alternative to the destructive decolorization of dyebaths remaining following the application of reactive dyes to cotton has been explored.

Salt (NaCl) is essential in the isolation of water-soluble acid, direct, and reactive dyes from their aqueous reaction mass during the manufacturing process. The most effective and efficient methods for removing salt are reverse osmosis, ultrafiltration, and hyperfiltration. This produces clean wastewater for use in other plant processes. As an indirect salt removal method, reactive dyes requiring less salt in the dye application process have been developed.

Energy Savings

An interesting approach in this area has involved the preparation of dye filter cakes having a higher percent of solids (less water). This has been accomplished by modifying the crystal form or habit (size), resulting in less energy (steam or electricity) required to dry dye cakes prior to grinding and making powders. Also, in the case of disperse dyes, having a higher percent solids has eliminated the need for drying prior to standardization into disperse dye pastes or spray-dried granules.

Similarly, it has been shown that disperse dyes such as disperse blue 79 and disperse brown 2 can be prepared at high percent solids in the filter cake. Such products are more stable and have better dispersion properties following the addition of a nonionic surfactant such as polyoxyethylene-9-octadecenoic acid ester. As a source reduction measure, reactive dyes that require less salt in the dye application process have been developed. Such dyes have also been referred to as "environmentally friendly" reactive dyes because they leave behind less color in the final dyebaths. In this regard, dyes having higher affinity for cotton than the traditional low molecu-

lar weight reactive dyes are employed. While these dyes give wastewater containing less salt and dye requiring removal, they can produce hydrolyzed forms with enhanced affinity for cotton, which leads to lengthy wash off times and increased water consumption.

Dyeing Medium

The conventional method for dyeing polyethylene terephthalate (polyester) fibers involves the use of dyebaths containing water, disperse dye, dispersing agent, and other chemical auxiliaries needed to enhance the efficiency of the dyeing process. After the dyeing step, dyeing auxiliaries remain in the wastewater, adding to the cost and complexity of treating the effluent stream. Since the cost of wastewater treatment and the value of water as a raw material are important considerations in dye application, there has been renewed interest in exploring approaches to the coloration of textiles that involve alternatives to water as the dye transport medium.

CORPORATE PROGRAMS

EPA Initiatives

The EPA's P2 Recognition Project is looking for safer substitutes for chemical and biological products currently in commerce that are either less toxic or are derived from substances that have lower toxicity. Other considerations include pollution prevention, source reduction, or recycling processes that reduce exposures or releases, environmentally beneficial uses of the product, and conservation of energy and water during its manufacture, processing, or use.

ISO 14000

In the future, a significant factor in the implementation of pollution prevention and waste minimization/source reduction measures in the dye, textile, and pigment industries worldwide will be the International Organization for Standardization (ISO) and its ISO 14000 standards. More than 60 countries have agreed to participate in the voluntary standards of this organization, which cover environmental auditing and labeling, environmental aspects of product standards, life cycle assessment, and emphasize a strong commitment to pollution prevention.

Eco-Efficiency

The corporate program known as eco-efficiency has been established to achieve a balance between environmental and economic considerations, to manufacture cost-effective products with a minimum amount of raw materials and energy, and to minimize emissions. This initiative has also been used as a strategic instrument for determining which product lines and processes are appropriate for future investment from an environmental and economical perspective.

N. P. Cheremisinoff, *Handbook of Pollution Prevention Practices*, Marcel Dekker, Inc., New York, 2001.

D. A. Hammer, *Constructed Wetlands for Wastewater Treatment*, Lewis Publishers Inc., Chelsea, Mich., 1989.

R. A. Meyers, ed., *Encyclopedia of Environmental Analysis and Remediation*, John Wiley & Sons, Inc., New York, 1998.

A. Reife and H. S. Freeman, eds., *Environmental Chemistry of Dyes and Pigments*, John Wiley & Sons, Inc., New York, 1996.

HAROLD S. FREEMAN
North Carolina State University
ABRAHAM REIFE
Environmental Consultant

DYES, REACTIVE

The concept of producing a dye-fiber covalent bond to achieve dyeings of very high wetfastness has been around for a century. Most of the earlier studies were directed at producing a covalently bonded color on cellulosic fiber substrates, but significant early developments also took place in the field of covalently attaching dye chromophores to wool. The first attempts to improve the wetfastness of cellulosic fibers by covalently bonding chromophores pretreated cotton with benzoyl chloride, nitrated the benzoyl ester, reduced the nitro group to an amine, and diazotized and coupled it to 2-naphthol, producing an orange coloration resistant to severe washing.

Then, in 1948, when dyes patented that gave wool dyeings of excellent wetfastness, the patentees evidently did not realize that these were in fact reactive dyes. However that may be, by the late 1950s the patenting of new reactive dye systems had become intensive as major dye manufacturing companies rushed to get into this promising area. The major drive was to develop molecules to dye cellulose to the widest possible shade gamut, giving dyeings of good washfastness; thus, wool, where much of the reactive dye innovation started, was temporarily side-lined. There is no doubt that reactive dyes have been a major success for cellulose fibers, in 2000 some 130,000 tons of reactive dye were sold for cotton and regenerated cellulosic fiber dyeing and printing, representing nearly 40% of the cellulose fiber dye market.

EARLY REACTIVE DYES FOR WOOL

Only when developments took place that allowed covalent fixation of dyes on the major textile fiber, cotton, did a major marketing effort commence to sell the reactive dye concept, probably because wool had only a small share of the twxtile fiber market. Also, the early reactive dyes for wool gave considerable problems of unlevel dyeing, especially in piece dyeing and yarn hank dyeing, so that their usage was mainly for loose stock and top dyeing. Many of the unlevelness problems associated with the reactive dyeing of wool were alle-

viated by dyeing with the novel 1966 product Lanasol dyes. By happy coincidence the launch of the Lanasol dyes coincided with the marketing of truly machine washable wool, produced by the continuous chlorine–Hercosetttreatment of wool tops. Dyeings produced on the chlorine–Hercoset-treated substrate with acid dyes, acid milling dyes, and premetallized dyes did not show adequate washfastness properties, but dyeings produced with reactive dyes gave outstanding washing performance.

DYEING SYNTHETIC POLYAMIDE FIBERS WITH REACTIVE DYES

Attempts to dye nylon fibers with reactive dyes have been frustrated by the paucity of nucleophilic sites available for reaction. This factor alone has meant that achieving build-up of sulfonated dyes even in moderate depths of shade is impossible, since every dye molecule covalently fixed means that, depending on the chromophoric component, one, two, or even three strongly anionic sulfonate groups become fixed at the same time, resulting in a build-up of negative charge on the fiber that acts as a resist to further anionic dye uptake.

MODERN REACTIVE DYES

The early reactive dyes were applied to cellulosic fibers by a variety of processes, including long liquor (so-called "exhaustion" dyeing), pad-batch, pad-steam, pad-bake, print-steam, and print-bake. Currently, padding and printing processes account for ~30% of the market, the residual, most popular application method being "exhaustion" dyeing. The reason for the popularity of this method lies in the short-run, high-fashion nature of the textile industry which also requires that coloration be delayed to the piece goods or even garment stage. Dye is lost to dye-house effluent for a number of reasons, reactive dye hydrolysis during application being one of the most important.

In medium shades, up to 30% of the dye applied ends up in the effluent, whereas in full depths up to 50% of the dye may be wasted. Given that reactive dyes are highly water soluble and thus difficult to remove from dye-house effluent at the water treatment works, color in rivers is a problem associated with the practice of cellulose fiber dyeing using reactive dyes.

The major use of reactive dyes is in long-liquor dyeing processes. For this process to operate well three factors apply, but only in the case of cellulosic fibers:

- Dye substantivity must be as high as possible during the so-called neutral exhaustion phase (typically 30 min at the required dyeing temperature).
- Following addition of alkali (usually sodium carbonate), dyeing is continued for a further 30 min to bring about covalent bonding with the fiber.
- At the end of dyeing, repeated rinsing in cold water and then in boiling water, until no more color is

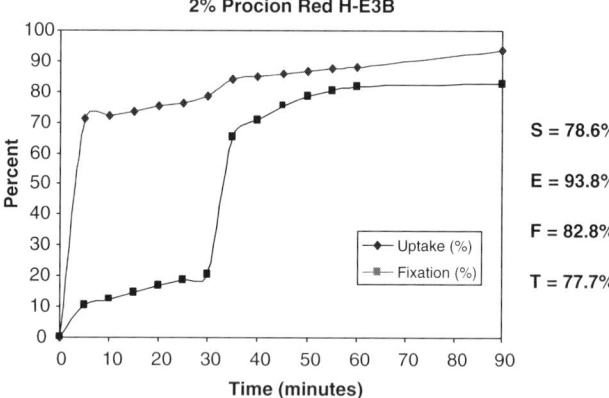

Figure 1. Substantivity(S), exhaustion(E), and fixation(F) values versus time(T) (C I Reactive Blue 19).

removed, is carried out to remove all noncovalently bound dye.

A typical exhaustion–fixation curve for a sulphatoethylsulfonyl dye, C.I. Reactive Blue 19 (Remazol Brilliant Blue R), is reproduced in Figure 1.

Unlevelness during reactive dyeing can be a problem; therefore, machines should be selected that give good mechanical interchange between the dye-liquor and the goods being dyed.

Polyfunctional Reactive Dyes for Cellulosic Fibers

The concept of polyfunctionality in reactive dyes was around from the start, insofar as the dichloro-s-triazinyl dyes contained two reactive sites, but the likelihood of actually achieving two dye-fiber bonds was low, due to their ease of hydrolysis. The first efficient bifunctional dye, Remazol Black B (C I Reactive Black 5), marketed in 1957, is now by far the biggest selling dye in the world for producing navy blue and black shades.

In cold pad-batch applications this dye is capable of giving total fixation values >90% in moderate depths of shade, but in long-liquor applications total fixation is only of the order of 70%, because of its relatively moderate substantivity.

In the mid-1990s Ciba launched Cibacron LS bifunctional dyes based on bis(monofluoro)-triazines; these dyes were such a molecular size as to be highly substantive to cellulose and thus could be dyed in the presence of reduced salt, hence the suffix LS (Low Salt).

In 1980, Sumitomo launched the first complete range of hetero-bifunctional reactive dyes that contained both a monochloro-triazine and a sulfatoethylsulfone group in the same dye molecule. As well as being of high combinability these dyes showed repeatable fixation in the temperature range 60–80°C; thus, if there was temperature variation front to back in a dyeing machine, shade reproducibility would not be compromised.

Trifunctional dyes capable of fixation efficiencies from long-liquor dyeing processes approaching 90% in medium depths of shade have been developed. Everlight (Everzol ED) has a trifunctional range and it is believed that some of the Procion HE-XL+ dyes are tris(monochloro)-triazines.

Tetrafunctional reactive dyes have not yet been prepared by replacement of the residual monochloro-triazine in the above scheme with a further mole of bis(chloroethylsulfone)-ethylamine; the problem lies in the conditions required. Less reactive vinylsulfone blocking groups may lead to a resolution of this problem.

Tetrafunctional reactive dyes have been prepared from the reaction of dichloro-triazine dyes (or difluoro-monochloro-pyrimidine dyes) with cysteamine and their subsequent reaction with cyanuric chloride.

When applied, in medium shade depths, by long-liquor processes at 50°C in the presence of 40 g/dm^3 sodium sulfate and then fixed with sodium carbonate additions, these dyes are claimed to give total fixation efficiencies >95%; in addition, its large hydrophobic side chain gives the dyes very good combinability properties in tertiary shades.

Neutral Fixing Reactive Dyes for Cellulosic Fibers

The chemistry behind this range of dyes was based on an early observation that the reactivity of monochloro-s-triazine dyes could be greatly enhanced by dyeing in the presence of tertiary amines; thus, alkaline fixation at 50°C rather than 80°C was possible in long-liquor dyeing processes. Subsequently, catalytic quantities of tertiary amines such as diazabicyclo-octane (DABCO) were promoted as dyebath additives to convert hot dyeing systems to warm dyeing systems; a good example was Cibacron Catalyst CC1 from Ciba.

Acid Fixing Reactive Dyes for Cellulosic Fibers

In order to improve the efficiency of cotton–polyester blend coloration, with mixtures of reactive and disperse dyes, mildly acidic fixation conditions would be desirable, as many disperse dyes are alkali sensitive.

In the ICI Procion Resin process, dichloro-s-triazine dyes are mixed with a cotton durable press cross-linking agent such as N,N′-dimethylol-dihydroxy-ethylene urea (DMDHEU), and magnesium chloride. When the mix is padded on to cotton fabric, dried, and cured at 170–180°C for 30 s, excellent fixation values are obtained. The procedure still finds practical use in the printing of cotton–polyester sheeting materials. The main problem on 100% cotton fabrics is loss of strength due to the acidic conditions employed.

Improved Dyeing of Cotton Cellulose by Fiber Pretreatment

The fixation of reactive dyes on substrates containing amino groups is much higher than on cellulosic substrates. Thus wool, which contains amino terminal groups on individual amino acid side chains, can be dyed with reactive dyes from baths set at pH 5–7, to give very high fixation efficiencies. In some cases, total overall color yields (fixation T) of 95–100% are recorded, even in moderate depths of shades. These excellent results on wool are achieved without the need to add electrolyte. It is therefore not surprising that a large amount of work has been done to prepare modified cellulosic substrates containing amino residues in order to improve dyeability.

At one time, it was thought that a cheap convenient way of modifying cotton to make it readily reactive dyeable was to pretreat it, from a long liquor, with a reactive cationic polymer under alkaline conditions. Such treatments imparted a very high neutral substantivity for reactive dyes, in the absence of electrolyte, and gave dyeings of good wetfastness. However, such approaches have not met with commercial success, as the treated fabric dyed to duller shades than those produced by the conventional salt–alkali process and a significant drop in lightfastness of 1–2 points was noted. Bright, light-fast dyeings could, however, be obtained by using a variety of methods to incorporate amino residues on cellulosic substrates.

The ease of neutral dyeing of this substrate with reactive dyes led to typical dyeing conditions of pH 7, with no salt, the bath being raised to the boil over 30 min and dyeing continued at the boil for a further 60 min. Dyeing, eg, with a 2% o.m.f. shade of C1 Reactive Red 5 (dichloro-s-triazine dye) gave exhaustion values of 85%. Even more interestingly, 99% of the absorbed dye was apparently covalently fixed; thus, soaping off the dyeing gave rise to hardly any color removal. Color yields, compared to the conventional salt–alkali method on untreated cotton, were doubled.

Due to perceived health and safety problems in handling the epoxide form, recent efforts to modify cotton in this manner have concentrated on a pad-batch procedure using a mixture of the chlorohydrin analogue, 3-chloro-2-hydroxy-N,N,N-trimethyl propanaminium chloride and sodium hydroxide. This product converts the above agent, during the batching procedure, to the corresponding epoxide that then fixes covalently to the fiber. In the United States, cotton modified by this method can be dyed with selected reactive dyes in the absence of salt. Dyebath exhaustion is 99–100%, even in full depths of shade, and fixation values also equal or approach 100%.

Reactive Fibers

Existing reactive dye systems depend on two basic elements: an electron deficient "reactive" carbon center in the dye and an electron rich "nucleophilic" center in the fiber. This concept may be reversed by producing dyes containing pendant nucleophiles. Aminoalkyl dyes are conveniently prepared from the reaction of aliphatic diamines with existing reactive dyes.

Reactive cellulosic fibres may be prepared by the reaction of cotton with N-methylol-acrylamide and reaction of 2,4-dichloro-6-(2-pyridino-ethylamino)-s-triazine (DCPEAT) with cotton. This reactive substrate shows excellent dye-fiber fixation when dyed with amino-alkyl dyes. Good substantivity and fixation are achieved at the higher levels of DCPEAT application slightly better

results are achieved using the bifunctional aminoethyl dye derived from C1 Reactive Red 120.

Covalent Fixation of Dyes on Fibers Using Cross-Linking Systems

The earliest commercial success in this area was achieved by a system that contained an unreacted double bond capable of undergoing further reaction with either the nucleophilic sulfonamide residue or with the nucleophilic cellulosate anion. This system was highly novel, but the dyes were withdrawn from the marketplace, because of its restriction to padding and printing processes, due to the nonsubstantive character of the cross-linker and possible toxicity problems. The clear advantage of such a system was that the dyes were not susceptible to hydrolysis and thus, in theory, very high levels of fixation should be achievable, provided sufficient cross-linker was present.

If such systems are to be further developed for covalently fixing dyes on cellulosic materials, experience appears to indicate that fiber-substantive cross-linking agents should be developed along with suitably nucleophilic dyes.

Modern Reactive Dye Systems for Wool Fibers

In the light of increasing environmental concerns with heavy metals, it is desirable to use reactive dyes to match deep shades of black and navy blue in order to offer the dyer a real alternative to chrome dyes. In this context, dye-manufacturers have increased their efforts to offer wool dyers ranges of reactive dyes that are attractively priced.

Black and navy blue shades are often based on the popular and widely available dye CI Reactive Black 5. This dye is a bifunctional sulphatoethylsulfone, which activates to vinylsulfone on boiling at pH 5.5.

DEVELOPMENTS WITH REACTIVE DYES AND ANTISETTING AGENTS

Of particular interest in the chemistry of reactive dyes for wool is the ability of certain reactive dyes to interfere with setting. Permanent setting is a contributory factor to loss in wool fiber strength during dyeing, is the main cause of increased hygral expansion of wool fabrics following piece dyeing, is the reason for surface marks such as "crowsfeet" in piece dyeing, and is the source of reduced bulk or yarn leanness following package dyeing. Apart from these wholly negative effects, permanent setting can sometimes be seen as beneficial. Such instances include the setting of wool yarn in hank form when using hank dyeing machines, which gives extra bulk and resilience to yarns and explains why this dyeing route remains popular for the production of wool carpets.

Reactive dyes actively prevent damage in wool dyeing, especially those dyes that contain activated carbon–carbon double bonds and that thus react with fiber nucleophiles via a Michael addition mechanism (these dyes include acrylamido dyes and vinylsulfone dyes). The magnitude of this effect increases with the increasing amounts of reactive dye applied being optimum at ~3% dye o.m.f.

Reactive dyes based on reactive halogenated heterocycles, which react with wool fiber nucleophiles by a nucleophilic substitution reaction, are less effective in controlling wool damage in dyeing than are the activated carbon–carbon double-bond type of reactive dye. Successful control of damage and set go hand in hand, and it is thus necessary to look carefully at the reactivity–stability of the reactive dye–cysteinyl residue covalent bond.

A. J. Farnworth and J. Delminico, *Permanent Setting of Wool*, Merrow Publishing, 1971.

D. M. Lewis, *Wool Dyeing*, Society of Dyers and Col., Bradford, 1992.

D. M. Lewis and D. G. Marfell, *Nylon Dyeing in The Dyeing of Synthetic Fibers*, Society of Dyers & Col., Bradford, 2003.

D. M. LEWIS
University of Leeds

DYES, SENSITIZING

Spectral sensitizing dyes extend the wavelengths of light to which semiconductors or chemical reactions can respond. The silver halides in photography require spectral sensitizers to respond to blue, green, red, and infrared wavelengths, because the silver halide intrinsic sensitivity reaches only partially into the blue. Colorless organic semiconductors in electrophotography require dyes to sensitize the electrophotographic response to red and infrared wavelengths. Infrared dyes in general serve both biological and data-storage commercial markets. Biological photochemical or photothermal effects require dyes that match the infrared transmission wavelengths of body tissue. Data-storage sensitizing dyes are matched to output wavelengths of inexpensive (infrared) solid-state lasers and infrared-sensitized color films have been historically important in environmental survey photography.

Spectral sensitizing dyes are considered "functional" dyes to distinguish their photosensitizing actions from the light-absorbing (filtering) action of conventional colorants. The absorption of light by a functional dye causes some additional function(s) to occur, such as a chemical reaction, donation (or acceptance) of an electron by the dye, or a change in the electronic state of a nearby molecule. Large commercial enterprises like the photographic imaging industry have created substantial dye libraries. Manufacturers and suppliers have prepared and tested a large number of dyes in research quantities to meet the continuing demands of color films, graphic arts, color printing, medical imaging, astronomy, and scientific photography.

SENSITIZATION WAVELENGTH AND EFFICIENCY

The availability of dye-sensitized imaging materials for the blue, green, red, and near-infrared segments of the electromagnetic spectrum has created vast imaging opportunities over the past eight or nine decades. Recent estimates by photographic industry leaders suggest that some 90-billion images are captured on film worldwide each year.

The demand for spectral sensitizers extends well beyond photography, into many systems that have somewhat fewer and quite different limitations. Dye-sensitized organic photoconductors, coated on flexible supporting belts, have been important for over two decades in electrophotography, following the discovery of an effective dye-aggregate-polymer system for this application. Photodynamic therapy research continues to develop biologically suitable dyes, which especially absorb the wavelengths of light most readily transmitted by human tissue. Optical data storage and optoelectronics are additional areas where light-absorbing dye layers for writable optical disks and voltage-sensitive dyes for telecommunications continue to be of interest.

STRUCTURAL CLASSES OF SPECTRAL SENSITIZERS

Several broad classes of spectral sensitizing dye structures are useful to distinguish for chemists who design these dyes and for databases containing patents. Such classes include cyanine dyes, merocyanine dyes, oxonol dyes, polymethine dyes, azine dyes, azo-methine dyes, arylmethane dyes, quinone dyes, porphin dyes, and pyrylium dyes, xanthene dyes.

Syntheses of polymethine dyes like the cyanines, merocyanines, or oxonols have some advantageous features. A single heterocycle can provide a spectral series of dyes, prepared by varying the length of the connecting methine chain from one to three to five or more methine carbons.

SPECTRAL SENSITIZATION FOR COLOR AND BLACK-AND-WHITE IMAGING

Silver Halides

The high-volume imaging industry worldwide requires efficient and high-productivity imaging materials to meet consumer needs for images. Silver halide films capture a large proportion of these images. The color films, which dominate this market, have undergone several decades of improvement: highly accurate color reproduction, high information content, high photographic speed, and less dependence on the source of illumination.

The image content captured in the exposure and development (or readout) steps can go through digital or optical pathways to give a final image for display. High productivity pathways like optical printing from a film image are capable of several thousand prints per hour and use the environmentally improved silver chloride color papers, which have rapid, low-solution-volume processing. Optical printing from a digital image also benefits from this high productivity.

Recently, synthetic methods have focused less on formation of the dye chromophore and more on the synthetic challenges that exist to provide substituents to improve nonchromophoric properties such as aggregation and solubility.

Also of current interest are analytical techniques that give reliable electrochemical potentials or help visualize the locations and sizes of dye aggregates on the surfaces of silver halide microcrystals.

Photographic products can be subjected to wide variations in temperature, humidity, and time delays after exposure until processing. Consequently, high priorities are given to spectral sensitizers that do not degrade either the film's physical state or its image quality, as well as provide efficient spectral sensitization at desired wavelengths. Although there are many available dyes and pigments, commercial silver halide films, papers, and plates are sensitized efficiently by just a few types of chromophores. Color films, color infrared films, and color paper now employ combinations of spectral sensitizers having polymethine structures in which the heterocyclic groups can be benzothiazole, benzoxazole, or benzimidazole. Effective commercial sensitizers can also have two different heterocyclic groups at the end of the polymethine chain. Black-and-white films and papers for camera, medical, or laser-scanner uses employ cyanine and merocyanine dyes related to structure.

Dye Design for Silver Halides

The suitability of cyanine and merocyanine dyes as sensitizers rests largely in their being able to simultaneously optimize three properties (electrochemical potentials, J-aggregation, and solubility) by choosing among various substituent and heteroatom combinations. Silver halides in color films and papers must also exhibit narrow (color-selective) absorption bands. Substantial dye efficiency challenges remain.

Co-Aggregation, Solubility, and Substituent Effects

Practical color films use more than one dye to achieve the correct spectral response for color reproduction. Newer films have evolved to a shorter wavelength sensitization peak for the red-sensitive layers.

Other Dye Technology Trends for Silver Halides

Spectral sensitizers are expected to be chemically inert, nondiffusible between layers in unexposed films, highly efficient photoparticipants in forming latent images, and rapidly decolorizable (or soluble) with few other effects during photographic development, (develop, bleach, fix steps). To the extent that these desirable properties are not equally achieved by each and every dye structure, there are opportunities for selective tailoring of dyes to achieve properties other than spectral sensitization at a desired wavelength. Dyes are known, for example, to alter the course of silver halide precipitation. Other specific dyes can serve as antifoggants.

Since many photographic processing laboratories can process color negative film, black-and-white films and portrait proofing papers now employ the same developer (chromogenic development), with broad spectral sensitization and multiple color couplers in a single layer. Some current graphic arts color-separation films are designed to use one film for cyan, magenta, yellow, or black. In these materials, a typical color film having blue-, green-, and red-sensitive layers is modified to include a fourth emulsion (infrared sensitive) in all the layers, and this infrared emulsion receives the exposure if a black dot image is desired.

Dipolar Conjugated Dyes: Merocyanines

The merocyanines were often used for graphic arts and other black-and-white applications where broad spectral sensitization bands were acceptable. Because of their dipolar tendency, the merocyanines show significant solvent sensitivity of their absorption wavelength (solvatochromic behavior).

Color Filter Arrays, Antihalation Dyes, InkJet Inks

Color filter array, antihalation, and visual image inks are not strictly spectral sensitization uses that employ "functional" dyes. Although commercially adequate digital systems exist, the dye-technology segment of digital color photography is in its early stages.

Hard Copy Systems

Electrophotographic copy processes using organic photoconductors may be spectrally sensitized by many dye classes, in sharp contrast to the limited choices for silver halides. Most current photoreceptors are organic dyes or metal complex dyes, the useful physical states being dye aggregates, pigment dispersions, or vacuum-deposited pigments.

Thermal Sensitization

Photothermal actions of dyes are important in sensitizing optical disk data storage layers, medical photothermal effects with lasers, and spectrally sensitized superconductors.

Biological stains allow wavelength-selective destruction of dye-stained tumors by transfer of absorbed light energy to heat the nearby tumor cells. Photothermal dye sensitization of temperature-sensitive materials like superconductors makes oxide superconductors spectrally sensitive.

SPECTRAL SENSITIZATION IN PHOTOCHEMICAL TECHNOLOGY

Functional dyes of many structural types are important photochemical sensitizers for oxidation, photopolymerization, (polymer) degradation, isomerization, and photodynamic therapy. Often, dyes from several structural classes can fulfill a similar technological need.

USES AND SUPPLIERS

Sensitizing dyes are used primarily for specialty purposes: photography, electrophotography, lasers, optical disks, and medicine. Because of this, their manufacture is limited to significantly smaller quantities than for fabric dyes or other widely used coloring agents. The photographic, laser, and medical uses place high demands on the degree of purity required, and the reproducibility of synthetic methods and purification steps is very important. Suppliers of sensitizing dyes including cyanines and merocyanines include manufacturers who supply other specialty chemicals: Sigma-Aldrich Chemical Company (Milwaukee, WI), Japanese Research Institute for Photosensitizing Dyes (Okayama, Japan), Molecular Probes (Eugene, OR), NK Dyes (Japan), Clariant Corporation (Coventry, RI), Allied-Signal Specialty Chemicals (Morristown, NJ), H. W. Sands Corporation (Jupiter, FL), and Crompton Corporation (Groton, CT). More importantly, these firms provide sources of generally useful reagents that, in two or three synthetic steps, lead to many of the commonly used sensitizers.

SENSITIZING DYE TOXICOLOGY

Most significantly, large classes of dyes or pigments have some members that function as sensitizers. Toxicological data are often included in surveys of dyes, reviews of toxic substance identification programs, data coordination by manufacturers' associations, and in material safety data sheets provided by manufacturers of dyes.

F. J. Green, *The Sigma-Aldrich Handbook of Stains, Dyes, and Indicators*, Aldrich Chemical Company, Milwaukee, Wis., 1990.

A. T. Peters and H. S. Freeman, eds., *Colour Chemistry (The Design and Synthesis of Organic Dyes and Pigments)*, Elsevier Applied Science, New York, 1991.

P. N. Prasad and D. J. Williams, *Introduction to Nonlinear Optical Effects in Molecules and Polymers*, Wiley-Interscience, New York, 1991.

C. W. Thomas and C. N. Proudfoot, *Handbook of Photographic Science and Engineering*, Society for Imaging Science and Technology, Springfield, Va., 1997.

DAVID M. STURMER
Eastman Kodak Company
(retired)

E

ECONOMIC EVALUATION

Economic evaluation is a quantitative estimate of the expected profitability of a venture scenario, often in comparison with other choices, including the competition. It is part of the analysis that supports the decisionmaking task in engineering and management. The four essentials of an economic study are problem definition, cost estimation, revenue estimation, and profitability analysis, as well as a characterization of the uncertainty and risk.

COST ESTIMATION

The three cost estimation categories that are important for economic analysis of chemical process facilities are equipment cost, capital investment cost, and product cost.

Equipment Costs

Equipment cost includes the purchased cost of process and materials-handling equipment, storage facilities, waste treatment equipment, structures, and site service facilities. Installation costs such as insulation, piping, painting and finishing, foundations, equipment setting, process structures, instrumentation, and electrical service connections are estimated or factored separately. Actual quoted prices from suppliers provide the best data, but these are not usually available when estimates are made. Consequently, equipment cost estimates tend to be based on personal files, internal company data, or published correlations.

Capital Investment Cost

The capital investment involved in a proposed project is important because it represents the money that must be raised to get the project started, is used in profitability forecasts, and is reflected in the estimated manufacturing cost of a product. Capital investment is classified here for discussion as fixed capital, working capital, and land cost. Fixed capital can be classified as direct plant, indirect plant, and nonplant costs. Working capital is the money required for the day-to-day operation of the venture over and above the fixed investment. Land cost, while part of the direct plant cost, is placed in a separate capital category because it is not depreciable. Land cost is also site specific and highly variable.

Product Cost

An estimate of total product cost is an important part of economic evaluation and management planning from R&D phases through the entire operating life cycle. The total product cost can be viewed as the sum of the manufacturing cost and the general expense.

The manufacturing cost consists of direct, indirect, distribution, and fixed costs. Direct costs are raw materials, operating labor, production supervision, utilities, supplies, repair, and maintenance. Typical indirect costs include payroll overhead, quality control, storage, royalties, and plant overheads such as safety, protection, personnel, services, yard, waste, and environmental control, and other plant categories. The principal distribution costs are packaging and shipping. Fixed costs, which are insensitive to production level, include depreciation, property taxes, rents, insurance, and, in some cases, interest expense.

General expense consists of corporate services such as administration, sales and shipping departments, marketing, financial, technical service, research and development, engineering, legal, accounting, purchasing, public relations, human resources, and communications.

"Joint product costing", when more than one product is manufactured by a single process, is a common situation, which can be treated on a weight or a volume basis. If one product is largely a by-product, it might not be assigned any manufacturing cost and be accounted for in terms of a by-product credit based on its sales volume. The suitable allocation of costs should reflect such factors as production level, changing markets, and raw materials. Another problem is the allocation of costs when a raw material for one process operation is produced internally by another process operation of the same organization.

Product cost can be computed on an annual or daily basis but is frequently reported on a product unit basis ($/product unit). An estimate of the cost for various levels of production is often needed. This can be done by separating the manufacturing cost into fixed and variable components, with the variable components being those that vary with production level. On occasion, manufacturing costs are computed on an incremental or marginal basis instead of the usual allocated basis.

Unit cost data should be carefully assessed to ensure that process type, size, and raw materials are similar to the proposed venture. Operating cost data sometimes are reported for separate categories such as operating labor, maintenance labor, supervision, and utilities.

A more detailed product cost estimation method is to relate manufacturing cost items to a few calculated items, such as raw materials, labor, and utilities by means of simple factors. Internal accounting groups often develop factors for use with this popular method.

REVENUE ESTIMATION

Revenues are money inflows from venture activities. Product sales, based on the unit product price ($/unit) and the yearly sales volume (units/year), are the only types of revenue considered here. The price and sales volume are related to each other and typically contribute most of the uncertainty in economic forecasts of chemical processes.

Product Price

If a need is not met by any other available product, then the price can be set as high as the need can support. However, a very high price tends to limit market growth and encourage the introduction of competitive or substitute products. The preferred strategy is to establish a moderately high initial price, but to plan on future price reductions to help expand the market and meet any competitive pressures.

If a need is met by a variety of products, then the price should reflect both the price of these competitive products and any unique features of the product being priced.

It is desirable that the price be high enough to generate an adequate rate of return, which is reflected in various cost or margin-based pricing policies. The percentage markup is defined as 100 (selling price-cost)/selling price. In situations where a dominant position exists, such as a patented pharmaceutical without significant competition, the price is market-driven and may not be directly related to cost.

Sales Volume

The quantity of annual sales is often called the sales volume, even though the units may be mass quantities instead of volume units. The estimation of the annual sales volume over the expected life of a production facility is extremely difficult in most cases. It requires an estimate of the total market, production capabilities, costs of competition, expected market share, selling strategy, and future economic conditions.

The expected annual sales volume is important not only for estimating sales revenue, but also for the selection of plant capacity, process type, and planned expansions.

PROFITABILITY ANALYSIS

Profitability analysis involves the generation of criteria that characterize the expected financial yield of a proposed investment, especially in comparison with other choices, as well as an assessment of the uncertainty and risk involved. The emphasis herein is on the quantitative profitability analysis of proposed multiyear investments.

If an investor buys a computer in the morning and sells it in the afternoon for more, a return on investment is realized. The return, in dollars, is the reward for the effort and risk involved in undertaking the venture. The desirability is influenced by how an observer assesses the many factors involved and can even be based on irrational decision criteria. There is no single correct answer, only future events determine the wisdom of the selection.

Multiyear Venture Analysis

The multiyear case introduces a new complication—money flows at different times cannot be compared directly, because there is a time cost associated with capital. The factors that must therefore be considered are net present value. (NPV), in which corresponds to a discounted net return, discounted total capital (DTC), which provides a translation of the investment flows to the same common time point used in the NPV calculation, and net return rate (NRR): NPV divided by DTC, for the return rate (%) on the investment over the life of the venture. If this, in turn, is divided by the venture life, the result is an *effective annual return on investment* called the Net Return Rate (%/year).

The actual profitability analysis of a multiyear venture should be based on the ventures "cash flow". The money flows are assumed to be end-of-year flows for simplification. Investment and other outflows are negative; sales revenues and other inflows are positive.

The annual cash flow of a venture, which is where its profitability analysis starts, is the actual net cash generated by the project in a given year. For any project year it is the sum of four items: (+ after-tax income + tax basis depreciation + end-of-life items − yearly investment). Other cash flow definitions, such as the EBITDA (earnings before interest, taxes, depreciation, amortization, and other items), are used by entertainment and Internet companies to emphasize annual growth and enhance annual reports. The prediction of profitability is based on expected cash flows instead of earnings.

BREAK-EVEN CHARTS

A break-even chart is a visual tool for analyzing operating profitability at various levels of production. In this type of diagram, annual expenses are separated into fixed, variable, and semivariable categories.

Break-even charts are used with production models to predict optimum production levels, break-even points, and shutdown conditions under various scenarios. These models tend to involve a reasonable amount of approximation.

INFLATIONARY EFFECTS

Inflation can have a significant effect on the profitability of a venture. However, the U.S. federal tax laws do not allow for indexing the inflationary effects of depreciation schedules, salvage values, replacement costs, or taxable income. Inflation rates can vary unpredictably with time and can differ for certain revenues or expenditures.

Prevailing interest rates and capital cost rates tend to reflect an estimate of future inflation and contain a component that can be attributed loosely to inflationary expectations. Different discount/encount rates could be used for various time intervals in a cash flow analysis, although future rate values are rarely known with any accuracy.

UNCERTAINTY AND RISK

Most economic analysis is based on a "best-estimate" scenario in which the most probable estimates of all inputs are used. However, uncertainty in these estimates can introduce uncertainty into the profitability parameters. Various approaches have been developed to treat uncertainty and to characterize the associated risk: extreme-case analysis, sensitivity analysis, and statistical methods.

In order to treat probability, statistical methods can be employed to characterize probability distributions. Because most distributions in profitability analysis are not accurately known, the common assumption is that normal distributions are adequate. The normal distribution of a quantity can be characterized by two parameters: the expected value and the variance. These usually have to be estimated from meager data.

Decision trees, Monte Carlo simulations, and other approaches have led to the development of quantitative risk assessment methods in economic analysis. However, the historical database is limited and there are no sound methods for projecting the historical database to future scenarios. In spite of these difficulties, statistical approaches have found increasing use in recent years.

J. R. Couper, O. T. Beasley, and W. R. Penney, *The Chemical Process Industries Infrastructure: Function and Economics*, Marcel Dekker, New York, 2000.

M. Gerrard, *Capital Cost Estimating*, 4th ed., Institution of Chemical Engineers, Rugby, UK, 2000.

R. C. Higgins, *Analysis for Financial Management*, 2nd ed., Irwin-McGraw-Hill, Homewood, Ill., 2000.

J. A. White, K. E. Case, D. B. Pratt, and M. H. Agee, *Principles of Engineering Economic Analysis*, 4th ed., John Wiley & Sons, Inc., New York, 1998.

THOMAS J. WARD
Clarkson University

ELASTOMERS, SYNTHETIC

The purpose of this article is to provide a brief overview of the materials designated synthetic elastomers and the elastomeric or rubbery state. Subsequent entries describe the individual classes of elastomers in detail. Table 1 provides a fundamental description of the principal classes of synthetic elastomers.

DEFINITION OF ELASTOMERS

The term elastomer is the modern word to describe a material that exhibits rubbery properties, ie, that can recover most of its original dimensions after extension or compression. Ever since the pioneering work of Staudinger in the early 1900s, it has been accepted that such rubbery behavior results from the fact that the material is composed of cross-linked, long-chain, flexible polymer molecules. When such a material is extended or stretched, the individual long-chain molecules are partially uncoiled, but will retract or coil up again when the force is removed because of the kinetic energy of the segments of the polymer chain. The flexibility of such polymer-chain molecules is actually the result of the ability of the atoms comprising the chain to rotate around single bonds. Theories of rubberlike elasticity are well-developed.

The properties of elastomeric materials are also greatly influenced by the presence of strong interchain, ie, intermolecular, forces which can result in the formation of crystalline domains. Irregular polymer chains will not crystallize, have weak interchain interaction and have the properties of an amorphous material.

A most interesting class of materials is comprised of elastomers with chain regularity that undergo a temporary crystallization when stretched to a high extension, thus virtually becoming fibers, but that retract to their original dimension when the force is removed. Such strain-crystallizing rubbers can thus demonstrate unusually high tensile strength, but revert to the amorphous state when the force is relaxed because of relatively weak interchain forces.

EFFECT OF TEMPERATURE ON POLYMER PROPERTIES

There are two principal forces that govern the ability of a polymer to crystallize: the interchain attractive forces, which are a function of the chain structure, and the countervailing kinetic energy of the chain segments, which is a function of the temperature. The fact that polymers consist of long-chain molecules also introduces a third parameter, ie, the imposition of a mechanical force, eg, stretching, which can also enhance interchain orientation and favor crystallization.

In addition to the phenomena of crystallization and melting, which both represent a change of state in the material, there is a third transition which plays a strong role in the behavior of polymers, although it is by no means absent in the behavior of simple liquids. This is the glass-transition temperature, T_g.

Polymers fall into two classes, those that are capable of crystallization and those that are not. A noncrystalline (amorphous) polymer is considered a liquid (although a highly viscous one), which becomes a glass at reduced temperatures. Thus, atactic polystyrene is always amorphous, but it is in a "glassy" state at room temperature, since its T_g is about 105°C. Above its T_g it becomes rubbery although its chemical stability in air at that temperature is so poor as to render it useless. On the other hand, polyethylene and isotactic polypropylene are crystalline, to a greater or lesser extent, at room temperature, and hence do not exhibit rubbery behavior even though above T_g.

The class of rubbers that show the ability to crystallize when stretched represent a special class of rubbers.

COMPOUNDING AND VULCANIZATION OF ELASTOMERS

In order to "cure" or "vulcanize" an elastomer, ie, cross-link the macro-molecular chains (Fig. 1), chemical reactants are mixed or compounded with the rubber, depending on its nature. The mixing process depends on the type of elastomer: a high viscosity type, eg, natural rubber, requires powerful mixers (such as the Banbury type or rubber mills), while the more liquid polymers can be handled by ordinary rotary mixers, etc (see RUBBER COMPOUNDING).

Table 1. Elastomers^a and Their Characteristics

Name	Chemical name	Repeat unit structure	Vulcanizing agent	Stretching crystallization	Gum tensile strength
natural rubber	*cis*-1,4-polyisoprene (>99%)	^{b}	sulfur	good	good
styrene–butadiene rubber	poly(butadiene-*co*-styrene)	$-\!\!(CH_2-CH=CH-CH_2)_m(CH_2-CH)_n$ $\quad C_6H_5$	sulfur	poor	poor
butadiene rubber	polybutadiene (>97% *cis*-1,4)	$-\!\!(CH_2-CH=CH-CH_2)\!-$	sulfur	poor to fair	poor to fair
isoprene rubber	*cis*-1,4-polyisoprne (>97%)	$\quad CH_3$ $-\!\!(CH_2-C=CH-CH_2)\!-$	sulfur	good	good
EP(D)M	poly(ethylene-*co*-propylene-*co*-diene)	$-\!\!(CH_2-CH_2)_m(CH-CH_2)_n(\quad CH=CH_2\quad)_o$ $\qquad CH_3$	sulfur	poor	poor
butyl rubber	poly(isobutyene-*co*-isoprene)	$CH_3 \qquad CH_3$ $-\!\!(CH_2-C)_{50}(CH_2-C=CH-CH_2)\!-$ CH_3	sulfur	good	good
nitrile rubber	poly(butadiene-*co*-acrylonitrile)	$-\!\!(CH_2-CH=CH-CH_2)_m(CH_2-CH)_n$ $\qquad\qquad CN$	sulfur	poor	poor
chloroprene rubber	polychloroprene (mainly trans)	$-\!\!(CH_2-C=CH-CH_2)\!-$ $\qquad Cl$	MgO or ZnO	good	good
silicones	polydialkylsiloxane (mainly poly-dimethyl siloxane)	$\quad R$ $-\!\!(Si-O)\!-$ $\quad R$	peroxides	poor	poor
fluorocarbon elastomers	poly(vinylidene fluoride-*co*-hexafluoropropene)	$-\!\!(CH_2-CF_2)_x(CF_2-CF)_y$ $\qquad\qquad CF_3$	diamines	poor	poor
polysulfide rubber	poly(alkylene sulfide)	$-\!\!(CH_2-CH_2-S_{2-4})\!-$	metal oxides	fair	poor
polyurethanes	polyurethanes	$HO-\!\!(R-OCONHR'NHCOO)_x R-OH$	diisocyanate	fair	good

^{a}Not inclusive; see also Acrylic elastomers, Phosphazenes, Chlorosulfonated polyethylene, Ethylene–acrylic elastomers, and Polyethers under the title Elastomers, synthetic.
^{b}See Isoprene.

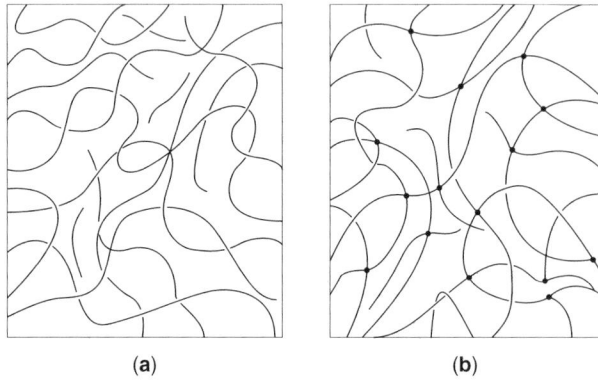

Figure 1. Vulcanization of rubber macromolecules: (**a**), before cross-linking; (**b**), after cross-linking.

Typical tire rubbers, eg, natural, SBR, or polybutadiene, being unsaturated hydrocarbons, are subjected to sulfur vulcanization, and this process requires certain ingredients in the rubber compound, besides the sulfur, eg, accelerator, zinc oxide, and stearic acid. Accelerators are catalysts that accelerate the cross-linking reaction so that reaction time drops from many hours to perhaps 20–30 min at about 130°C. There are a large number of such accelerators, mainly organic compounds, but the most popular are of the thiol or disulfide type. Zinc oxide is required to activate the accelerator by forming zinc salts. Stearic acid, or another fatty acid, helps to solubilize the zinc compounds.

In addition to the ingredients that play a role in the actual vulcanization process, there are other components that make up a typical rubber compound (see Rubber chemicals). Softeners and extenders, generally inexpensive petroleum oils, help in the mastication and mixing of the compound. Antioxidants (qv) are necessary because the unsaturated rubbers can degrade rapidly unless protected from atmospheric oxygen. They are generally organic compounds of the amine or phenol type (see also ANTIOZONANTS). Reinforcing fillers, eg, carbon black or silica, can help enormously in strengthening the rubber against rupture or abrasion. Nonreinforcing fillers, eg, clay or chalk, are used only as extenders and stiffeners to reduce cost.

Styrene–Butadiene Rubber (SBR)

This is the most important synthetic rubber and represents more than half of all synthetic rubber production (see STYRENE–BUTADIENE RUBBER). It is a copolymer of 1,3-butadiene, $CH_2=CH-CH=CH_2$, and styrene, $C_6H_5CH= CH_2$, and is a descendant of the original Buna S first produced in Germany during the 1930s. The polymerization is carried out in an emulsion system where a mixture of the two monomers is mixed with a soap solution containing free-radical initiators. The final product is an emulsion of the copolymer, ie, a fluid latex (see LATEX TECHNOLOGY).

Polybutadiene

The homopolymer polybutadiene (PB) is next in importance to SBR, as shown by its world production. It is prepared by polymerization of butadiene in solution, using organometallic initiators, either of the Ziegler-Natta type or lithium compounds. It is, therefore, a result of the discovery of stereospecific polymerization during the 1950s and 1960s (see ELASTOMERS, SYNTHETIC–POLYBUTADIENE).

Polyisoprene (Synthetic)

Polyisoprene has four possible chain unit geometric isomers: cis- and trans-1,4-polyisoprene, 1,2-vinyl, and 3,4-vinyl.

$$\sim CH_2-\underset{\underset{CH_3}{|}}{C}=CH-CH_2\sim$$

$$\sim CH_2-\underset{\underset{CH=CH_2}{|}}{\overset{\overset{CH_3}{|}}{C}}\sim$$

$$\underset{CH_3-C=CH_2}{\overset{\sim CH-CH_2\sim}{|}}$$

As in the case of polybutadiene, the Ziegler-Natta type of initiators are used to produce a polymer of high cis-1,4 structure (~98–99%) (natural rubber is 100% cis-1,4).

AGE-RESISTANT ELASTOMERS

Ethylene–Propylene (Diene) Rubber

The age-resistant elastomers are based on polymer chains having low unsaturation not in the backbone. This is sufficient for sulfur vulcanization but low enough to reduce oxidative degradation and ozone attack does not cause chain cleavage and subsequent cracking. Because of its irregular chain structure, EPDM is amorphous at <60 wt% ethylene and shows no crystallization on stretching. Hence, as is the case with all amorphous elastomer, it exhibits poor strength and requires carbon black reinforcement. The presence of unsaturation in the side chain makes it possible to use sulfur vulcanization, but a higher proportion of accelerators must be used because of the low unsaturation. Its excellent aging and low temperature properties make it ideal for use as a sheet rubber for roofing applications, where it has shown a rapid growth.

Butyl Rubber

Butyl rubber is a copolymer of isobutylene and isoprene, with just enough of the latter to provide cross-linking sites for sulfur vulcanization. The polymerization system is of the cationic type, using coinitiators such as $AlCl_3$ and water at very low temperatures ($-100°C$) and leading to an almost instantaneous polymerization.

SOLVENT-RESISTANT ELASTOMERS

Nitrile Rubber (NBR)

This is the most hydrocarbon oil swelling resistant of the synthetic elastomers. This elastomer is prepared by emulsion polymerization, similar to that used for SBR, but generally carried out to high conversion. As for SBR, the chain irregularity leads to a noncrystallizing rubber, so that this polymer requires carbon black reinforcement for strength.

Acrylic Elastomers

These materials are based principally on an acrylate chain structure, as follows:

$$\sim CH_2-\underset{\underset{\underset{OR}{|}}{\underset{C=O}{|}}}{CH}\sim$$

where R is generally an alkyl group. Because of the absence of unsaturation, vulcanization is carried out by amine compounds instead of sulfur, but this absence of unsaturation also confers good aging properties. This rubber also shows good resistance to hydrocarbon solvents because of the polar acrylic ester groups present. Practically all commercial acrylic elastomers are produced by free-radical polymerization. Of the four processes available, ie, bulk, solution, suspension, and emulsion, only aqueous suspension and emulsion polymerization are used to produce the acrylic elastomers present in the market (see ELASTOMERS, SYNTHETIC–ACRYLIC ELASTOMERS; ETHYLENE–ACRYLIC ELASTOMERS).

Chloroprene Rubber

Polychloroprene can be represented by the formula:

$$\sim CH_2-\underset{\underset{Cl}{|}}{C}=CH-CH_2\sim$$

The polymer is prepared by emulsion polymerization of chloroprene. Polychloroprene has a variety of uses, both in latex and dry rubber form. In addition to its excellent solvent resistance, it is also much resistant to oxidation or ozone attack than natural rubber. It is also more resistant to chemicals and has the additional property of flame resistance from the chlorine atoms.

TEMPERATURE-RESISTANT ELASTOMERS

Silicone Rubber

These polymers are based on chains of Si–O rather than carbon atoms, and owe their temperature properties to their unique structure. The most common types of silicone rubbers are specifically and almost exclusively the polysiloxanes, eg, poly-dimethylsiloxane. The Si–O–Si bonds can rotate much more freely than the C–C bond, or even the C–O bond, so the silicone chain is much more flexible and less affected by temperature (see SILICON COMPOUNDS, SILICONES). The polymer chain is formed by a ring-opening reaction caused by the action of alkalies on the monomer, a cyclic siloxane.

Fluorocarbon Elastomers

These elastomers are the most resistant elastomers to heat, chemicals, and solvents known, but they are also the most expensive. The most common types are copolymers of vinylidene fluoride and hexafluoropropene. Emulsion polymerization is used, but the latex is too unstable for use and all the latex is coagulated to dry rubber.

LIQUID RUBBER TECHNOLOGY

An entirely new concept was introduced into rubber technology with the idea of "castable" elastomers, ie, the use of liquid, low molecular-weight polymers that could be linked together (chain-extended) and cross-linked into rubbery networks. This was an appealing idea because it avoided the use of heavy machinery to masticate and mix a high viscosity rubber prior to molding and vulcanization. In this development three types of polymers have played a dominant role, ie, polyurethanes, polysulfides, and thermoplastic elastomers.

J. P. Queslel and J. E. Mark, in J. I. Kroschwitz, ed., *Encyclopedia of Polymer Science and Engineering*, 2nd ed., Vol. 5, John Wiley & Sons, Inc., New York, 1986, pp. 365–408.

MAURICE MORTON
The University of Akron

ELECTROANALYTICAL TECHNIQUES

Electroanalysis employs electrochemical cells, the most common of which are batteries, consisting of two electrodes—an anode and a cathode. The principles that govern batteries and electroanalytically useful cells are the same. Electroanalytical cells typically have at least two electrodes: a working electrode and a reference electrode. The working electrode may serve as either an anode or a cathode, depending on the applied voltage, or as a source or sink for ion exchange, as in the case of ion-selective electrodes. The reference electrode invariably serves as a source or sink for ions at its interface with the solution, and as a source or sink for electrons at its interface with an external circuit. The solution, or electrolyte in batteries, has a very high ionic strength in order to carry enormous amounts of current. In electroanalytical cells, the solution is generally the sample and must typically contain supporting electrolyte that supports much lower current densities. Supporting electrolytes may also be used to define ionic strength for potentiometric measurements.

Cells useful for electroanalysis typically consist of two or more electrodes dipping into the solution to be analyzed. Sample solutions can range from water to blood, but can also include virtually all organic solvents. Analyte concentrations are usually in the picomolar to millimolar range. Numerous choices exist for working electrodes, ranging from electron exchangers to ion exchangers, including metals of many kinds (Pt, Au, Ag, and Hg), semiconductors (Si, Ge, and TiO_2) and plastics, ion exchangers, and sequesterers in poly(vinyl chloride) (PVC). Choices for reference electrodes are more limited. The problems involving reference electrodes are often more profound, and availability of the right reference electrode may ultimately dictate the feasibility of an assay. They are inherently unstable and may drift, leak, become foul or plugged, and frequently need to be replaced.

Electrochemical measurements may be either active or passive, depending on whether or not a signal, typically a current or a voltage, must be actively applied to the sample in order to evoke an analytically useful response. Electroanalytical techniques have also been divided into two broad categories, static and dynamic, depending on whether or not current flows in the external circuit. In the static case, the system is assumed to be at equilibrium. The term "dynamic" indicates that the system has been disturbed and is not at equilibrium when the measurement is made. These definitions are often inappropriate because active measurements can be made that hardly disturb the system at all and passive measurements can be made on systems that are far from equilibrium. The terms "static" and "dynamic" also imply some sort of artificial time constraints on the measurement. Furthermore, "active" and "passive" are terms that nonelectrochemists seem to understand more readily than "static" and "dynamic".

ACTIVE TECHNIQUES

Active techniques are classified by the method of collection and display of data. If a voltage is applied to the cell and the resultant current measured and displayed as a function of time, the technique is called chronoamperometry. If a current is applied to the cell and the resultant voltage is measured and displayed as a function of time, the technique is called chronopotentiometry. Similarly, if current is measured and displayed as a function of applied potential, the technique is called voltammetry. Subcategorizing depends on such things as the geometry and size of the working electrode, its physical treatment, usually its rotation, and the functionality of the applied waveform. Therefore there are rotating disk voltammetry, cyclic voltammetry, pulse voltammetry, etc. Even postcollection treatment of the data may result in a renaming of the technique. Integration of the current, either during analogue data acquisition or digitally from computer stored currents, results in chronocoulometry.

PASSIVE TECHNIQUES

Perhaps the most precise, reliable, accurate, convenient, selective, inexpensive, and commercially successful electroanalytical techniques are the passive techniques, which include only potentiometry and the use of ion-selective electrodes, coated wire electrodes, and ion-selective field effect transistors (ISFETs), either directly or in potentiometric titrations. Whereas these techniques receive only cursory or no treatment in electrochemistry textbooks, the subject is regularly reviewed and treated. There is a journal, *Ion-Selective Electrode Reviews*, devoted solely to the use of ion-selective electrodes.

According to the definition, a passive technique is one for which no applied signal is required to measure a response that is analytically useful. Only the potential (the equilibrium potential) corresponding to zero current is measured. Because no current flows, the auxiliary electrode is no longer needed. The two-electrode system, where the working electrode may or not be an ion-selective electrode, suffices.

The equilibrium potential may be forced to change by the addition of a titrant, as in potentiometric titrations employing nonspecific working electrodes. But the equilibrium potential can also be analytically useful in its own right if the response of the electrode is highly selective to the analyte. Chemically modified electrodes are being researched that may provide this high selectivity; eg, for use in active techniques such as the glucose sensor. However, ion-selective electrodes, of which the pH glass electrode is probably the most common example, dominate applications involving high selectivity even though their response mechanisms have nothing to do with oxidation and reduction. Potentiometry employing ion-selective electrodes without the addition of titrants is termed "direct potentiometry."

STATIC AND DYNAMIC MEASUREMENTS

The definition herein of static electroanalytical measurements implies a time independent response, regardless of whether or not that response is generated by an applied signal. If an applied signal is needed, the method is active; if not, it is passive. These terms should not be confused with other definitions, where static more or less equates to passive and dynamic to active.

ECONOMIC ASPECTS

Companies that sell and service electroanalytical instrumentation are few in number and small in size or are parts of much larger companies. Bioanalytical Systems, Brinkman, Princeton Applied Research, Radiometer, and Solartron Analytical are among the companies that supply electrochemical instrumentation.

Electroanalytical chemistry does not yet generate the kind of revenue that battery technology and electrowinning do. Electroanalysis, even at the most sensitive limits, can be performed using simple instruments that cost between $100 and $1,000. However, assays generally require highly skilled technicians and are thus correspondingly expensive even if the equipment used is not, and electrochemical instrumentation, fully supported with warranties, can cost $10,000 or more. The future of electroanalytical chemistry probably lies in two areas: processes control and biosensors. If it is ever to be a competitive industry, rather than a support technique, the latter of these opportunities will most probably be the greater source of future revenue. In the area of consumer products, amperometric glucose sensors hold high potential. Industrially, process monitors for the manufacture of consumer chemicals are under development. However, replacement of defective reference electrodes, which in a laboratory environment may be trivial, may be prohibitively difficult *in vivo* or in an industrial process environment. In any event, the reference electrode contributes heavily to the economics of electroanalytical chemistry.

A. Evans, in A. M. James, ed., *Potentiometry and Ion-Selective Electrodes*, John Wiley & Sons, Inc., New York, 1987.

J. Janata, *Principles of Chemical Sensors*, Plenum Press, New York, 1989, Chapt. 4.

B. W. Rossiter and J. F. Hamilton, eds., *Physical Methods of Chemistry*, Vol. II, John Wiley & Sons, Inc., New York, 1986.

J. Wang, *Electroanalytical Techniques in Clinical Chemistry and Laboratory Medicine*, VCH, New York, 1988.

JAMES R. SANDIFER
Eastman Kodak Company
(Retired)

ELECTROCHEMICAL MACHINING

High-strength heat-resistant alloys can be difficult to machine by conventional techniques. Electrochemical machining (ECM) offers an alternative. Hard metals can be shaped electrolytically, and the rate of machining does not depend on their hardness. The tool electrode used in the process does not wear, and therefore soft metals can be used as tools to form shapes on harder workpieces, unlike conventional practice. The process is used to smooth surfaces, to drill holes, and form complex shapes. Its combination with other techniques yields fresh applications in diverse industries. Its effect on the environment lies with gas generation and waste metal disposal. Recent advances lie in computer-aided tool design, and the use of pulsed power, which has led to greater accuracy for ECM-produced components.

PRINCIPLES OF ELECTROCHEMICAL MACHINING

Electrochemical machining (ECM) relies on the laws of electrolysis. An anodic (positively polarized) iron metal is plated a distance away from a cathode metal (of negative polarity), both of which are placed in aqueous electrolyte solution. When a voltage is applied between the two electrodes, metal is dissolved electrochemically from the anode metal at a rate that is proportional to the product of the chemical equivalent (atomic weight divided by the valency of dissolution) of the latter metal and the current. The outcome of these reactions is that the metal ions precipitate out as iron hydroxide. At the cathode, only gas, normally hydrogen, is generated.

The rate of evolution of the gas generated at the cathode can also be determined from Faraday's law. The process therefore offers an alternative way of removing material from a workpiece, especially when the latter is made of an alloy, the hardness of which makes it difficult to machine by conventional methods. Thus, if a nickel anode were used it would dissolve, yielding nickel hydroxide and hydrogen gas. If brass, steel, or copper is used as the cathode, the reaction still generates hydrogen gas. The hardness or other mechanical properties of the anode do not affect the rate of metal removal from that electrode.

In the ECM process, a cathode tool is produced from a soft metal, such as brass or copper, to a shape that is approximately the image of that required on the anode workpiece, which usually would be a tough alloy metal. An aqueous solution of electrolyte, often sodium chloride or nitrate, is then pumped between the two electrodes. When a voltage of ~10 V is applied between them, the interelectrode gap tends to an equilibrium width as the tool is moved mechanically toward the workpiece. A shape complementary to that of the tool is thus reproduced on the workpiece.

The main advantages of ECM are that there is no tool wear; the rate of machining is not affected by the hardness of the workpiece, and complicated shapes can be machined with a single tool-electrode, which can be of a softer metal than the workpiece.

RATES OF METAL REMOVAL

One can calculate the rates at which metals can be electrochemically machined using Faraday's law:

$$m = \frac{A}{Z} \cdot \frac{It}{F}$$

where m is the mass of metal electrochemically machined by a current I (A) passed for a time t(s). The quantity A/ZF is called the electrochemical equivalent of the anode metal; F is Faraday's constant (96,500°C).

Some metals are more likely than others to machine at the Faraday rates of dissolution. In addition to current, factors such as electrolyte type and the rate of electrolyte flow influence the rate of machining. With the growing industrial use of titanium alloys and of ECM to shape them, much effort is being expended to find appropriate electrolyte solutions and process conditions which yield better efficiencies.

If the rates of electrolyte flow are too low, the current efficiency of even the most readily electrochemically machined metal is reduced; with insufficient flow the products of machining cannot be so readily flushed from the machining gap. The accumulation of debris within the gap impedes further dissolution of metal. The buildup of cathodically generated gas can lead to short-circuiting between tool and workpiece, causing termination of machining, with both electrodes becoming damaged. When complex shapes have to be produced, the design of tooling incorporating the right kind of flow ports for the electrolyte becomes a considerable problem.

AQUEOUS ELECTROLYTE SOLUTIONS

The current efficiency depends greatly on the choice of the electrolyte solution. In ECM, the main electrolytes used are aqueous solutions of sodium chloride and nitrate, and occasionally acid electrolytes.

The electrolyte solution warms during ECM, owing to electrical heating caused by the passage of current. The machining action is often easier to control if the electrolyte is maintained at a higher temperature from the outset.

Although sodium chloride electrolyte generally has a higher efficiency than sodium nitrate over a wide range of current densities, the latter is often preferred in practice, as it permits closer dimensional accuracy of ECM, particularly in hole drilling. Other electrolytes include 5% hydrochloric acid solution, useful in fine-hole drilling, since it dissolves the metal hydroxides as they are produced. Sodium chlorate solution has also been investigated, as has sodium silicate solution, an electrolyte known to produce sparking conditions in hole-drilling applications, leading to an electrodischarge erosion type of action.

SURFACE FINISH

The choice of electrolyte will also influence the quality of surface finish obtained in ECM. Depending on the metal, some electrolytes leave an etched finish. Sodium chloride tends to produce an etched, matte finish with steels and nickel alloys.

The production of an electrochemically polished surface is usually associated with the random removal of atoms from the anode workpiece, whose surface has become covered with an oxide film. This is governed by the metal–electrolyte combination used. The mechanisms controlling high current density electropolishing in ECM are still not completely understood.

Occasionally, metals that have undergone ECM have a pitted surface, while the remaining area is polished or matte. Pitting normally stems from gas evolution at the anode; the gas bubbles rupture the oxide film.

Process variables also affect surface finish. For example, as the current density is raised the finish generally becomes smoother on the workpiece surface. A similar effect is achieved when the electrolyte velocity is increased.

METAL SHAPING BY ECM

ECM can be used to shape metals in three main ways: deburring, hole drilling, and three-dimensional shaping. In deburring, or smoothing, of surfaces, a plane-faced cathode tool is placed opposite a workpiece that has an irregular surface. The current densities at the peaks of the surface irregularities are higher than those in the valleys. The former are therefore, removed preferentially and the workpiece becomes smooth, admittedly at the expense of stock metal (which is still machined from the valleys of the irregularities, albeit at a lower rate).

Hole drilling is another principal way of using ECM. The cathode-tool is usually made in the form of a tubular electrode. Electrolyte is pumped down the central bore of the tool, across the main machining gap, and out between the sidegap that forms between the wall of the tool and the hole. Reversal of the electrolyte flow can often produce considerable improvement in machining accuracy.

The main machining action is carried out in the gap formed between the leading edge of the drill tool and the base of the hole in the workpiece. ECM also proceeds laterally between the side walls of the tool and component, where the current density is lower than at the leading edge of the advancing tool. Since the lateral gap width becomes progressively larger than that at the leading edge, the side-ECM rate is lower. The overall effect of the side-ECM is to increase the diameter of the hole produced.

The third main application is three-dimensional ECM contour shaping. The effective use of this technique requires that a constant equilibrium gap be maintained between the two electrodes by use of a constant rate of mechanical feed of one electrode toward the other. To achieve the required dimensional accuracies, electrolytes such as sodium nitrate are commonly employed.

Electrolyte flow plays a significant role in contour shaping. The electrolyte usually has to be maintained at high pressures to flush away the products of machining before they can interfere with the machining action. Thus, careful design of tooling is necessary to provide the right entry and exit ports.

INDUSTRIAL APPLICATIONS OF ECM

Typical applications of ECM from industry are in the aircraft engine industry, to manufacture of turbine blades, in the medical industry, in the manufacture of artificial hip implants; in the car industry, to remove pre-

cisely the right amount metal in combustion chambers so that the volumes of each chamber are similar, in using seawater, which even with its specific conductivity of about one-fifth of that of standard ECM electrolytes has proved to be a useful working fluid for electrolytic grinding (ECG); in the manufacturing industry, for rubber mold production, and in the electronics industry, for the etching of microparts.

HYBRID ECM PROCESSES

The combination of ECM with other processes is occasionally used to provide advantage not possible with the single techniques, as in electrochemical grinding, electrolytic in-process dressing (ELID), Hybrid ECM–with Lasers, Ultrasonics, and Electrolyte Jets and EDM/ECM.

ENVIRONMENTAL EFFECT OF ECM

ECM can affect the environmental through generation of waste substances during electrolysis and subsequently their disposal. A consequence of the electrochemical process is evolution of cathodic (and anodic) gaseous by-products, eg, hydrogen gas. Anodic metal dissolution gives rise to solid, eg, hydroxides and solid–liquid mixture, acids, nitrates, oils, or heavy metals. The subsequent processing of waste electrolyte, and disposal of its ECM machining products, or "sludge" is carried out with electrolyte settling tanks, centrifuges, and filter presses. The ECM of alloys containing chromium can present considerable environmental implications.

THE FUTURE OF ECM

High rate anodic electrochemical dissolution is a practical method of smoothing and shaping hard metals by employing of simple aqueous electrolyte solutions without wear of the cathodic tool. ECM can offer substantial advantages in a wide range of cavity-sinking and shaped-hole production operations.

Control of the ECM process is improving all the time, with more sophisticated servo-systems, and better insulating coatings. However, there is still a need for basic information on electrode phenomena at both high current densities and electrolyte flow rates. Tool design continues to be of paramount importance in any ECM operation.

The advent of new technology for controlling the ECM process and the development of new and improved metal alloys, which are difficult to machine by conventional means, will assure the future of electrochemical machining.

A. E. DeBarr and D. A. Oliver, eds., *Electrochemical Machining*, McDonald & Co., London, 1975.

V. K. Jain, *Advanced Machining Processes*, Allied Publishers PVT Limited, New Delhi, 2002.

J. A. McGeough, *Principles of Electrochemical Machining*, Chapman and Hall, London, 1974.

J. A. McGeough, ed., *Micro-machining of Engineering Materials*, Marcel Dekker, New York, 2001.

JOSEPH A. MCGEOUGH
University of Edinburgh

ELECTROCHEMICAL PROCESSING

Electrochemical systems convert chemical and electrical energy through charge–transfer reactions. These reactions occur at the interface between two phases. Consequently, an electrochemical cell contains multiple phases, and surface phenomena are important. Electrochemical processes are sometimes divided into two categories: electrolytic, where energy is supplied to the system, eg, the electrolysis of water and the production of aluminum; and galvanic, where electrical energy is obtained from the system, eg, batteries and fuel cells.

The industrial economy depends heavily on electrochemical processes. Electrochemical systems have inherent advantages such as ambient temperature operation, easily controlled reaction rates, and minimal environmental impact. Electrosynthesis is used in a number of commercial processes. Batteries and fuel cells, used for the interconversion and storage of energy, are not limited by the Carnot efficiency of thermal devices. Corrosion, another electrochemical process, is estimated to cost hundreds of millions of dollars annually in the United States alone (see CORROSION AND CORROSION CONTROL). Electrochemical systems can be described using the fundamental principles of thermodynamics, kinetics, and transport phenomena.

THERMODYNAMICS OF ELECTROCHEMICAL CELLS

Consider the cell

$$
\begin{array}{c|c|c|c|c|c|c}
\alpha & \beta & \gamma & & \in & \beta' & \alpha' \\
Pt(s) & Fe(s) & NaOH & membrane & NaCl & TiO_2(s)\cdot & Pt(s) \\
& H_2 & in\ H_2O & & in\ H_2O & Cl_2 &
\end{array}
$$

for which the electrode reactions are

$$2\,Cl^- \rightarrow Cl_2 + 2e^- \qquad (1)$$

$$2\,H_2O + 2e^- \rightarrow H_2 + 2\,OH^- \qquad (2)$$

The electrode where oxidation occurs is the anode; the electrode where reduction occurs is called the cathode. Electrons released at the anode travel through an external circuit and react at the cathode. The vertical lines denote phase separation; the squiggly lines separate a junction region. Although adjacent phases are in equilibrium, not all species are present in every phase. The membrane provides an ionic path for sodium ions that are transported from the anode to the cathode, but also separates the chlorine and hydrogen gases. Within the

junction region, ie, the membrane, transport processes occur (see also ALKALI AND CHLORINE PRODUCTS; MEMBRANE TECHNOLOGY).

Determining the cell potential requires knowledge of the thermodynamic and transport properties of the system. The analysis of the thermodynamics of electrochemical systems is analogous to that of neutral systems. For ionic species, however, the electrochemical potential replaces the chemical potential.

KINETICS AND INTERFACIAL PHENOMENA

The rate of an electrochemical process can be limited by kinetics and mass transfer. Before considering electrode kinetics, however, an examination of the nature of the interface between the electrode and the electrolyte, where electron-transfer reactions occur, is in order.

Because some substances may preferentially adsorb onto the surface of the electrode, the composition near the interface differs from that in the bulk solution. If the cell current is zero, there is no potential drop from ohmic resistance in the electrolyte or the electrodes. The question from where this potential arises can be answered by considering the interface.

At the interface between phases, there is a region called the electrical double layer where potential variation occurs. Figure 1 shows this region. Although the electrode and electrolyte are overall electrically neutral, the metal electrode may have a net charge near the surface. The solution layer closest to the electrode contains specifically adsorbed ions and is called the inner Helmholtz plane (IHP). Ions that are hydrated, generally the cations, can approach the metal surface only to a finite distance, and comprise the outer Helmholtz plane (OHP). The nonspecifically adsorbed ions are distributed by thermal agitation; this region is called the diffuse double layer and lies just outside the OHP.

Even in the absence of Faradaic current, ie, in the case of an ideally polarizable electrode, changing the potential of the electrode causes a transient current to flow, charging the double layer. The metal may have an excess charge near its surface to balance the charge of the specifically adsorbed ions. These two planes of charge separated by a small distance are analogous to a capacitor. Thus the electrode is analogous to a double-layer capacitance in parallel with a kinetic resistance.

In electrode kinetics a relationship is sought between the current density and the composition of the electrolyte, surface overpotential, and the electrode material. This microscopic description of the double layer indicates how structure and chemistry affect the rate of charge-transfer reactions. Generally in electrode kinetics the double layer is regarded as part of the interface, and a macroscopic relationship is sought. For the general reaction

$$O + ne^- \rightleftharpoons R \qquad (3)$$

the cathodic and anodic reaction rates can be written as

$$r_a = k_a c_R \exp\left[(1-\beta)\frac{nF}{RT}V\right] \qquad (4)$$

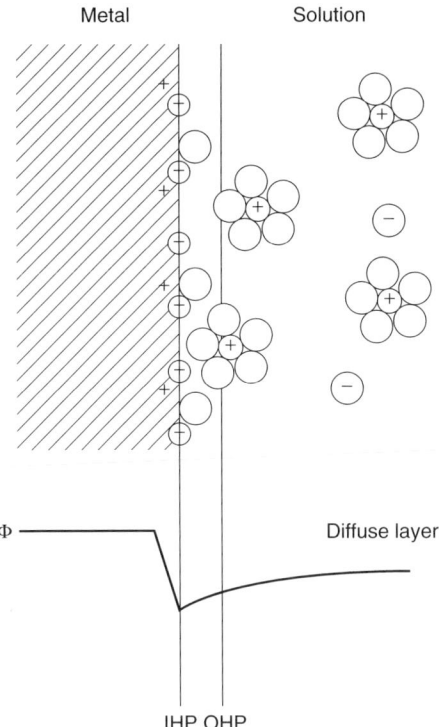

Metal   Solution

Φ ——————————   Diffuse layer

IHP OHP

Figure 1. The structure of the electrical double layer where ◯ represents the solvent; ⊖, specifically adsorbed anions; ⊖, anions; and ⊕, cations. The inner Helmholtz plane (IHP) is the center of specifically adsorbed ions. The outer Helmholtz plane (OHP) is the closest point of approach for solvated cations or molecules. Φ, the corresponding electric potential across the double layer, is also shown.

and

$$r_c = k_c c_o \exp\left[-\beta \frac{nF}{RT} V\right] \qquad (5)$$

β is a symmetry factor equal to the fraction of the potential that promotes the cathodic reaction. The reaction rate and current are related through Faraday's law:

$$\frac{i}{nF} = r_a - r_c \qquad (6)$$

These two reactions may be generalized to the Butler-Volmer equation:

$$i = i_o \left\{ \exp\left(\frac{\alpha_a F \eta_s}{RT}\right) - \exp\left(-\frac{\alpha_c F \eta_s}{RT}\right) \right\} \qquad (7)$$

The exchange current density, i_o, depends on temperature, the composition of the electrolyte adjacent to the electrode, and the electrode material. The exchange current density is a measure of the kinetic resistance. High values of i_o correspond to fast or reversible kinetics. The three parameters, α_a, α_c, i_o, are determined experimentally. The surface overpotential, η_s, is the difference in potential of the metal and the potential of an electrode

of the same kind in the electrolyte measured adjacent to the electrode, ie, just outside the double layer, but passing no current. The surface overpotential appears in the exponential terms for both the anodic and cathodic reactions, and can be considered the driving force for the electrochemical reaction.

TRANSPORT PROCESSES

In addition to electrode kinetics, the rate of an electrochemical reaction can be limited by the rate of mass transfer of reactants to and from the electrode surface. In dilute solutions, four principal equations are used. The flux of species i is

$$\mathbf{N}_i = -z_i u_i F c_i \nabla \Phi - D_i \nabla c_i + c_i \upsilon \qquad (8)$$

These three terms represent contributions to the flux from migration, diffusion, and convection, respectively. The bulk fluid velocity is determined from the equations of motion. Equation 8, with the convection term neglected, is frequently referred to as the Nernst-Planck equation. In systems containing charged species, ions experience a force from the electric field. This effect is called migration. The charge number of the ion is z_i, F is Faraday's constant, u_i is the ionic mobility, and Φ is the electric potential. The ionic mobility and the diffusion coefficient are related:

$$D_i = RTu_i \qquad (9)$$

This relation, discovered by Nernst and Einstein, applies in the limit of infinite dilution.

A material balance on an element of the fluid gives

$$\frac{c_i}{t} = -\nabla \cdot \mathbf{N}_i + R_i \qquad (10)$$

where R_i is the homogeneous reaction rate. Except near the diffuse double layer, to a good approximation the solution is electrically neutral

$$\sum_i z_i c_i = 0 \qquad (11)$$

The current density is given by

$$\mathbf{i} = F \sum_i z_i \mathbf{N}_i \qquad (12)$$

Equations 8, 10–12 using the appropriate boundary conditions, can be solved to give current and potential distributions, and concentration profiles. Electrode kinetics would enter as part of the boundary conditions. The solution of these equations is not easy and often involves detailed numerical work.

A. J. Bard and L. R. Faulkner, *Electrochemical Methods*, John Wiley & Sons, Inc., New York, 1980.

J. Newman, *Electrochemical Systems,* Prentice-Hall, Inc., Englewood Cliffs, N.J., 1991.

G. Prentice, *Electrochemical Engineering Principles,* Prentice-Hall, Inc., Englewood Cliffs, N.J., 1991.

K. J. Vetter, *Electrochemical Kinetics,* Academic Press, Inc., New York, 1967.

THOMAS F. FULLER
JOHN NEWMAN
University of California,
Berkeley

ELECTROCHEMICAL PROCESSING, INORGANIC

The electrochemical production of inorganic chemicals and metals in the United States consumes about 5% of all the electricity generated annually, and about 16% of the electric power consumed by industry. This includes the production of such commodity chemicals as sodium hydroxide, NaOH, and chlorine, Cl_2.

HARDWARE FOR ELECTROCHEMICAL PROCESSING

Power supplies and electrolytic cells are the distinguishing features of electrochemical processes. Nearly all electric power is generated and transported as high voltage multiple-phase alternating current. Industrial electrochemical processes require direct current; thus transformers are needed to decrease voltage, and rectifiers are required to convert the alternating current to direct current. Rectifiers may be rated for voltages up to 400 V or more and for any amperage up to hundreds of thousands of amperes. Rectifier efficiencies generally increase as voltages increase. Very high amperages are achieved by connecting rectifier units in parallel. State-of-the-art rectifiers use thyristers which are semiconductor devices that conduct only when a triggering potential is applied (see SEMICONDUCTORS). Thyristers can be made to conduct for half of an alternating current cycle and not conduct the other half-cycle, rectifying alternating current to direct current.

Electrochemical processes require feedstock preparation for the electrolytic cells. Additionally, the electrolysis product usually requires further processing. This often involves additional equipment, as is demonstrated by the flow diagram shown in Figure 1 for a membrane chlor-alkali cell process (see ALKALI AND CHLORINE PRODUCTS). Only the electrolytic cells and components are discussed herein.

Design possibilities for electrolytic cells are numerous, and the design chosen for a particular electrochemical process depends on factors such as the need to separate anode and cathode reactants or products, the concentrations of feedstocks, desired subsequent chemical reactions of electrolysis products, transport of electroactive species to electrode surfaces, and electrode materials and shapes. Cells may be arranged in series and/or

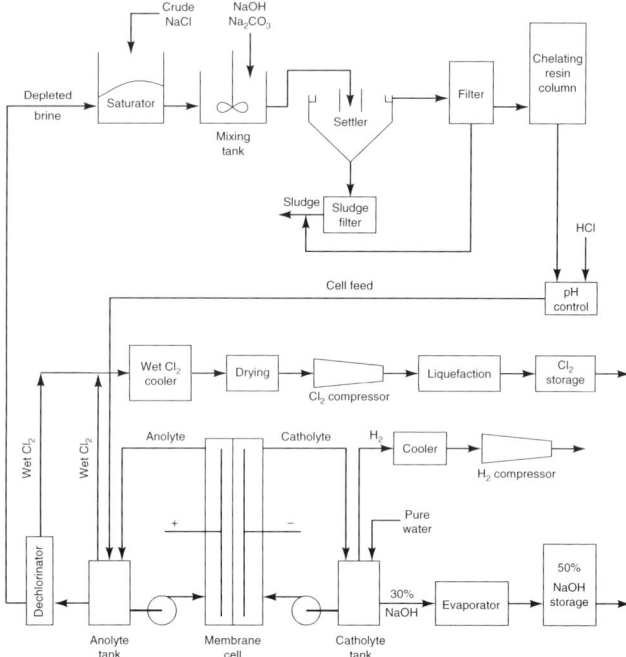

Figure 1. Flow diagram for chlor-alkali production by a membrane cell process.

parallel circuits. Some cell design possibilities for electrolytic cells are

Cell type	Process
one-compartment cells	
open-top inert tank	
monopolar electrodes	MnO_2
bipolar electrodes	chlorate
open-top cathodic tank	chlorate and perchlorate
enclosed horizontal	chlorine–caustic and
liquid metal cathode	aluminum
two-compartment cells	
diaphragm cells	
monopolar electrodes	chlorine–caustic and Mn metal
bipolar electrodes	chlorine–caustic
membrane cells	
monopolar electrodes	chlorine–caustic
bipolar electrodes	chlorine–caustic
pressurized cells	
filter press	
bipolar electrodes	hydrogen–oxygen
rotating cylindrical cathode	continuous production of metal
particle-bed electrodes	removal of low concentration of metals from waste streams

Electrode materials and shapes may have a profound effect on cell designs. Anode materials encountered in electrochemical processes are

Anode material	Process
carbon	fluorine
	aluminum
graphite	chlorine–caustic
	chlorate
DSA	chlorine–caustic
	chlorate
Pt	perchlorates
	persulfates
lead dioxide, PbO_2, on graphite	chlorate
or titanium	perchlorate
lead alloy (Pb, 1% Ag)	zinc electrowinning
	manganese electrowinning

INDUSTRIAL PROCESS CONDITIONS

Electrolysis of Chloride Solutions

Chloride may be oxidized electrochemically to chlorine or hypochlorite, chlorate, and perchlorate.

The electrolysis of NaCl brine for the production of chlorine and caustic soda is one of the oldest and certainly one of the most important industrial electrochemical processes. The overall reaction is

$$2\,NaCl + 2\,H_2O \xrightarrow[\text{energy}]{\text{electrical}} 2\,NaOH + Cl_2 + H_2 \qquad (1)$$

There are three main technologies available for carrying out this process: diaphragm cells, mercury cells, and membrane cells. Membrane cells are the most recent development, and are generally chosen for new production capacity.

Chlorates

Sodium chlorate is produced by the electrolysis of sodium chloride at pH 6.5–7.5 in a one-compartment cell. DSA anodes and steel cathodes are generally used in chlorate cells. The electrolysis products, hypochlorous acid and hypochlorite ions, react chemically to produce chlorate.

Sodium dichromate, $Na_2Cr_2O_7$, is added to chlorate cell electrolytes. The chromate minimizes the reduction of hypochlorite and chlorate at the cathode. Chromate also minimizes the corrosion of cathodes when or where cathodic current densities are inadequate to cathodically protect the steel cathodes from the very corrosive electrolyte.

Nearly 95% of the sodium chlorate produced in North America is used to produce chlorine dioxide, ClO_2, for pulp (qv) bleaching. Minor amounts are used to produce other chemicals such as $KClO_3$, $NaClO_2$, $NaClO_4$, etc, to recover uranium, U, and for agricultural uses as a defoliant or herbicide.

Perchlorates

Concentrated solutions of sodium chlorate are electrolyzed to produce sodium perchlorate, $NaClO_4$. Most of the $NaClO_4$ produced is converted to ammonium perchlorate, NH_4ClO_4.

$$NaClO_4 + HCl + NH_3 \rightarrow NH_4ClO_4 + NaCl \qquad (2)$$

The products of equation 2 are separated by controlled crystallizations to produce high purity crystalline anhydrous ammonium perchlorate and sodium chloride. The main use for ammonium perchlorate is as an oxidizer in the propellant of rockets and missiles (see EXPLOSIVES AND PROPELLANTS).

Manganese Dioxide

High performance alkaline batteries and some lithium batteries require pure electrolytic manganese dioxide (EMD), MnO_2. The production of EMD involves the following unit operations: first, high quality manganese ores are reduced. Reduced ore is then leached using acidic electrolyte from the cells and makeup sulfuric acid, H_2SO_4. Crude manganese sulfate, $MnSO_4$, solution is treated to remove impurities. Purified manganese sulfate solution is fed into an electrolytic cell where MnO_2 is deposited on the anode. Deposited MnO_2 is periodically harvested, milled, neutralized, and packaged for market.

Water Electrolysis

The electrolysis or water for hydrogen and oxygen production is economically attractive only in those areas where electric power is available at very low cost. Hydrogen is usually the primary product (see HYDROGEN ENERGY); oxygen (qv) is a co-product. Hydrogen in large quantities is produced by steam (qv) reforming methane, ie, natural gas (see GAS, NATURAL). Hydrogen produced in chlor-alkali and chlorate cells is available in some areas, and in some cases is compressed and sold in cylinders and as liquid hydrogen. Oxygen is commonly produced by the separation of air (see CRYOGENICS).

Cells for the electrolysis of water are available from several sources. Cells are usually of a filter-free design incorporating bipolar electrodes, porous diaphragms or ion-exchange membranes, alkaline electrolyte, KOH, and catalyzed electrodes.

Recent developments in water electrolysis cells include the use of ion-exchange membranes, optimization of pressure and temperature, and suggestions for better electrolysis cell designs. Interest in water electrolysis is associated with load leveling and space exploration. Electric power available during periods of low usage, ie, off-peak periods, can be used to electrolyze water. The hydrogen and oxygen produced can be stored and used in fuel cells to produce electric power when peak power demands occur.

Heavy water, D_2O, was produced by a combination of electrolysis and catalytic exchange reactions. Some nuclear reactors (qv) require heavy water as a moderator of neutrons.

Cold fusion has been reported to result from electrolyzing heavy water using palladium, Pd, cathodes. Experimental verification of the significant excess heat output and various nuclear products are still under active investigation.

Fluorine

Fluorine is the most reactive product of all electrochemical processes. The principal use of fluorine continues to be the production of UF_6 from UF_4.

$$UF_4 + F_2 \rightarrow UF_6$$

The electrolyte used in fluorine cells is KF–HF in a ratio that minimizes melting point, HF vapor pressure, and corrosion of materials. Fluorine has been compressed, liquified, and shipped. However, most fluorine is produced and used on-site. Fluorine production in the United States is based on electrolytic cells developed in the 1940s. Modern type "E" cells are rated for 6 kA.

Permanganate

Potassium permanganate, $KMnO_4$, is produced in commercial quantities. It may be prepared from manganese dioxide or directly from manganese metal, ferromanganese, or other manganese alloys. The Carus Chemical Company produces potassium permanganate from manganese dioxide in the United States.

Hydrogen Peroxide

Peroxydisulfuric acid, $H_2S_2O_8$, is one of the strongest oxidizing agents known. It and other peroxydisulfates are produced electrochemically. The production of peroxydisulfates was once important for the manufacture of hydrogen peroxide (qv), H_2O_2. The development of the autoxidation of alkyl anthraquinones led to a rapid increase in the production of H_2O_2 but a sharp decline in the importance of the electrolytic process. No H_2O_2 is produced by the electrolytic peroxydisulfate process. The last plant using this process closed in 1983.

Rapid growth in demand for H_2O_2 in the late 1980s created widespread interest in new processes for H_2O_2 production. New process development efforts were generally focused either on direct combination of hydrogen and oxygen to produce H_2O_2 or the electrolytic reduction of oxygen to produce H_2O_2. HD Tech Inc., a joint venture of Dow Chemical Canada Inc. and Huron Technologies Inc., has developed and commercialized an electrolytic reduction of oxygen process.

ELECTROWINNING OF METALS

The metals that are produced by electrolysis are included in Table 1. Fused salt processes are used when the reactivity of the metal does not allow electrowinning from aqueous solutions. Manganese is the most reactive metal that is produced by electrolysis of an aqueous solution.

Electrowinning from Aqueous Solutions

The aqueous processes for electrowinning of metals from ores have the following common unit operations or steps: (1) the metal in the ore is converted to an acid-soluble form and this may be an oxidizing roast or a reduction; (2) ores from step 1 are leached, usually in sulfuric acid; (3) metal solutions from step 2 are purified and in some cases concentrated; (4) purified metal solutions are electrolyzed in cells where the metal is deposited on the cathode; and (5) acid is produced at the anode and recycled to the leaching step 2. Some acid values are lost, usually in the purification step, 3. Makeup acid is added in the leaching step, 2. In most cases the metal solution from leaching step 2 contains impurities, other metals. Many of these metals have the characteristic of low hydrogen over-voltage. Codeposition of the impurity metals causes contamination of the desired product and decreases

Table 1. Production of Metals by Electrolysis of Aqueous Solutions

Metal	Anode	Diaphragm	Cathode	Cell feed[a], g/L	Electrolyte[a], g/L	Temperature, °C	Cell voltage, V	Cathode current density, A/m^2	Energy requirement, kW·h/kg	Current efficiency, %
Cd	Pb–Ag	no	Al	90–200 Cd, 20–40 Zn	10–20 Cd, 20–40 Zn, 60–140 H$^+$ [b]	30–35	2.5–2.7	80	1.5	90
Cr	Pb–Ag	yes	316 stainless steel			53	4.2	700	18	45
Cu	Pb–Sb–Ag	no	Cu	20–70 Cu, 20–70 H$^+$		30–35	2.0–2.2	130	2.2	80–90
Mn	Pb–Ag	yes	stainless steel or Ti	30–40 Mn, 125–150 NH$_4^+$, 0.1 SO$_2$ + glue	[c]	35	5.1	400–600	8–9	60–68
Ni	Pb	yes	Ni 99.9%		[d]	52	3.4	180		91–96
Zn	Pb–Ag	no	Al	100–200 Zn	100–200 H$^+$, 20–40 Zn	35	3.2–3.6	350–1000	3.3	90

[a] As sulfates unless otherwise noted.
[b] Anolyte in g/L: 13 Cr(VI), 2 Cr(III), 24 NH$_3$, 280 H$_2$SO$_4$. Catholyte in g/L: 11.5 Cr(III), 12.5 Cr(II), 84 NH$_3$; pH = 2.1–2.4.
[c] Anolyte in g/L: 10–20 Mn, 25–40 H$_2$SO$_4$, 125–150 NH$_4^+$. Catholyte in g/L: pH 6–7.2.
[d] Catholyte in g/L: 70 Ni(II) + H$_3$BO$_3$, Na$_2$SO$_4$; pH = 3.0–3.5. Anolyte in g/L: 40 H$_2$SO$_4$.

current efficiencies. The removal of impurities before electrolysis is very important. This is especially true in the case of the more reactive metals such as zinc, Zn, and manganese, Mn. These metals have deposition potentials close to the hydrogen evolution potential. The current efficiency of manganese electrowinning is about 60 to 68%. The principal inefficiency is hydrogen evolution.

The electrowinning of metals from aqueous solutions is generally carried out in tank cells. Developments in the electrowinning of metals form aqueous solutions have been directed toward improved anodes, improved additives, higher current densities, the use of ion-exchange membranes, better electrolyte quality control, and computer modeling of the processes.

Electrowinning from Fused Salts

Aluminum, Al, is produced world-wide by the Bayer-Hall-Heroult process. This process involves the electrolysis of alumina, Al_2O_3, dissolved in molten cryolite, Na_3AlF_6.

Sodium is produced by the electrolysis of a fused salt mixture of calcium chloride, $CaCl_2$, and NaCl in a Downs cell. An improved cell for the production of sodium was patented. This cell utilizes an electrode separator, solid electrolyte tubes that are permeable to the flow of sodium ions but impermeable to fluids and the flow of other ions. Production of sodium from mixtures of NaCl and aluminum chloride, $AlCl_3$, has been described.

Several processes for lithium, Li, metal production have been developed. The Downs cell with LiCl–KCl electrolyte produces lithium in much the same manner as sodium is produced. Lithium metal or lithium–aluminum alloy can be produced from a mixture of fused chloride salts. Granular Li metal has been produced electrochemically from lithium salts in organic solvents.

There are three electrolytic processes for magnesium, Mg, production: the Dow process, a process developed by I.G. Farbenindustrie in Germany, and an Alcan process. All processes involve the electrolysis of magnesium chloride, $MgCl_2$.

Research and development efforts have been directed toward improved cell designs, theoretical electrochemical studies of magnesium cells, and improved cathode conditions. A stacked-type bipolar electrode cell has been operated on a lab scale. Electrochemical studies of the mechanism of magnesium ion reduction have determined that it is a two-electron reversible process that is mass-transfer controlled.

Beryllium Be, metal is produced by electrolysis of $KCl–NaCl–BeCl_2$ melts.

ELECTROCHEMICAL WASTE TREATMENT

In many instances the metal processing industry produces aqueous effluents containing dissolved metals. Many of these metals are toxic and controlled by the U.S. Environmental Protection Agency (EPA) regulations. Most water treatment plants do not remove toxic metals, or they concentrate the toxic metals in sludges that are classified as hazardous waste. Concentrations of the metals in waste-

water are generally very low. The flow rates may be relatively high. Electrodes with large areas operating at low current densities are desirable for this application. Various porous electrodes, particulate-bed electrodes, fluidized-bed electrodes, and roll cells have been developed for metals recovery from dilute wastewater streams. Electrochemical processing has advantages over other chemical processes because the electrochemical process usually requires no addition of materials. Electrochemical processes for wastewater treatment often provide recovery of metal resources that partially offset processing costs. Silver has been recovered electrolytically from spent photographic liquors for many years. Silver recovery was motivated by economics. More recently, heavy-metal recovery has been motivated by environmental regulations.

In addition to metal recovery, which generally involves cathodic reduction, some waste may be treated by anodic oxidation processes. Many organic contaminants in wastewater can be oxidized by electrochemical treatments. High over-voltage anodes have been developed to improve the efficiency of these oxidations. Electrochemical oxidation of halogenated hydrocarbons has been achieved using barium peroxide, BaO_2, in aqueous NaCl solutions containing cationic surfactants.

Electrolytically generated hypochlorite may by used for the oxidative destruction of cyanides (qv) or the sterilization of domestic wastes. Several on-site systems for swimming pool sterilization and municipal waste treatment works have been developed. On-site production and immediate use of chlorine is considered safer than the transportation of chlorine.

Other electrochemical processes such as electrodialysis, electroflotation, and electrodecantation are also used in waste treatment.

SAFETY AND ENVIRONMENTAL CONSIDERATIONS

The electrochemical process industries are confronted with a wide range of hazards. These include electrical hazards, various explosion hazards, and the hazards associated with exposure to reactive chemicals.

J. O'M. Bockris and co-workers, eds., *Comprehensive Treatise of Electrochemistry*, Vol. 2, Plenum Press, New York, 1981, pp. 3–9.

A. T. Kuhn, *Industrial Electrochemical Processes*, Elsevier Publishing Co., Amsterdam, The Netherlands, 1971, Chapt. 3.

N. M. Prout and J. S. Moorhouse, eds., *Modern Chlor-Alkali Technology*, Vol. 4, Society of Chemical Industry, Elsevier Applied Science Publishers, Ltd., London, 1990, pp. 40–42, 97, and 178.

"Report of the Electrolytic Industries," published annually in Aug./Sept. by The Electrochemical Society, Inc.

J. B. Talbot and S. D. Fritts, *J. Electrochem. Soc.* **139**, 2981–3018 (1992).

MORRIS P. GROTHEER
Kerr-McGee Corporation

ELECTROCHEMICAL PROCESSING, ORGANIC

An electroorganic reaction is the chemical transformation of an organic substance by the action of an electric current. It takes place at an electrode and is therefore a heterogeneous process. Like other processes taking place at surfaces, its selectivity is influenced by the nature of the electrode material, mass transport, mixing rates, and the nature and composition of the surface and bulk regimes. A significant theme in electrochemical engineering in recent years has been process design to minimize environmental impact or to replace environmentally undesirable chemical processes with more benign electrochemical ones.

In an electrochemical cell electrons are given up to some component of the medium (reduction) at the cathode and taken up from another component (oxidation) at the anode, both processes driven by a power source. Of necessity, therefore, reduction and oxidation take place simultaneously. Most organic electrode reactions involve initial one-electron reduction to a radical anion or one-electron oxidation to a radical cation, although there are processes that can involve direct addition or loss of two electrons. Not surprisingly, the nature of the reactions which take place is a function of the substance and electrolysis conditions such as concentration, pH, and the nature of the electrode material, particularly where starting materials, intermediates, or products are adsorbed on the electrode surface.

Electrochemical processes must compete with standard reduction or oxidation reactions using reformer hydrogen, atmospheric oxygen, or chemical reagents. An electrochemical process can be a viable alternative to such chemical redox reactions if it involves a cheaper feedstock, reduction of the number of steps, or elimination of a difficult, hazardous, or environmentally adverse reaction step, avoidance of waste disposal, toxic materials, permits recycling of reagents, or allows for the possibility of generating useful products simultaneously at both anode and cathode. For an electroorganic process to be profitable, desirable features are high product yield and selectivity, current efficiency >50%, electrolysis energy consumption less than about 8 kWh/kg of product, lifetime >1000 hours, ease of isolation of product and ease of recycling of the electrolyte, product stream having >10% concentration of product, and of course production of a useful material.

Most successful applications have been in the fine chemical area. Numerous applications, some of which have been commercialized, have been developed in medium-to-small-scale processing. Especially attractive on this scale is production of special intermediates in pharmaceutical synthesis or high added-value chemical synthesis. In these processes multistep reactions are common, and an electrochemical reaction step can aid in process simplification.

CELL DESIGN

Materials

Cell Construction and Composition. Because most electroorganic syntheses are carried out at or near room temperature, a wide range of plastics can be used in cell construction. A number of readily available polymers such as polyethylene, poly(vinylidene fluoride (PVDF), polypropylene, and polytetrafluoroethylene (PTFE; Teflon) are used for cell frames or vessels, while elastomers such as neoprene, Viton, and ethylene–propylene–diene monomer (EPDM) and PTFE are used as gaskets to seal cell parts.

Electrodes. At least three factors have to be considered: (1) cost: the lowest cost form of the electrode material that produces the desired electrode reaction; (2) stability: the electrode material must be stable over long times—slow dissolution or physical breakdown of the electrode may be unnoticeable in the laboratory but become a significant problem over months of use; (3) conductivity: if the electrode material is not a good conductor, energy losses during electrolysis will add to the costs of operation. The most widely used cathode materials are C, Fe, Ni, Al, Hg, Pb, Pb, Zn, Cu, Sn, and Cd.

Steel and nickel are generally used in making anodes for oxygen evolution in alkaline electrolytes. Graphite or nickel is frequently used as anodes for anodic halogenation of organics. Platinum cannot be used as an anode in the presence of halide ions because of its dissolution to form tetrahaloaluminates. A major development in anode technology was introduction of the so-called dimensionally stable anode (DSA) in the chlor-alkali process for chlorine and sodium hydroxide production, which represents the largest use of electricity for electrochemical processing worldwide.

The best way to view the electrode is as a catalyst in the process. Its costs and effectiveness should be assessed on the amount of electrode required per kg of product produced. Electrode surface contamination can be circumvented by (1) periodic mechanical and/or chemical cleaning; (2) periodic current reversal; (3) stringent raw material specifications with respect to problem impurities; (4) purging and processing the electrolyte recycle stream; or (5) periodic replacement of the electrodes.

Electrical conductivity is important and the cell conditions should be arranged to maximize it.

Diaphragms. In many electroorganic systems, the use of a diaphragm to separate anolyte and catholyte is necessary to minimize the possibility of substances being formed at the anode migrating to the cathode and being reduced, or vice versa. In the development of a commercially feasible process, the selection of a diaphragm for large-scale cells often presents a serious obstacle. The preferred electrolysis diaphragm would have low cost, low electrical resistance, high resistance to mass transfer between anolyte and catholyte while allowing movement of ions to carry the current, long operating life, good dimensional stability, and be resistant to plugging and fouling. The likelihood of finding all of these properties in a single material is small. Diaphragm materials are available in two types: permeable and permselective. The permeable material is a porous matrix offering a limited barrier to all species in an electrolyte and having little variation in transport numbers among different species.

Permselective materials, or ion-exchange membranes, allow the passage of ions of one charge, either negative or positive, with a high degree of exclusion of ions of the opposite charge.

Electrolyte. The ideal electrolyte, ie, the conductive solvent of the cell, for organic synthesis would have high solubility for the organic component(s), possess good conductivity, have low cost, be easily recovered and purified, and be noncorrosive. Quaternary ammonium salts provide many of the above criteria in aqueous systems. A concise compilation of solvents and salts used in electroorganic chemistry is available.

Transport Phenomena

Electrochemical reactions are heterogeneous and are governed by various transport phenomena, which are important features in the design of a commercial electroorganic cell system. As for other heterogeneous reactions, the electrochemical reaction is affected by heat and mass transport. The electrochemical reaction, however, is unique in that it also has charge-transport characteristics. More comprehensive works on transport phenomena are available.

Mass Transport. Probably the most investigated physical phenomenon in an electrode process is mass transfer in the form of a limiting current; ie, one that is controlled by the rate at which the electroactive arrives at the electrode surface, not by the applied electrode potential.

Methods proposed for improving mass-transfer rates in large-scale cells are (1) rotation of cylindrical or disk electrodes, including wiping of the surface; (2) use of turbine or propeller agitators; (3) fluidized beds of electrode particles; (4) fluidized beds of nonconducting particles; (5) vibration of the electrode; (6) gas sparging; and (7) external pumping of electrolyte in open channels or channels having turbulence promoters.

Charge Transport. Side reactions can occur if the current distribution and therefore electrode potential along an electrode is not uniform. The side reactions can take the form of unwanted by-product formation or localized corrosion of the electrode. The problem is addressed by the analysis of charge transport in cell design. The path of current flow in a cell is dependent on cell geometry, activation overpotential, concentration overpotential, and conductivity of the electrolyte and electrodes. Three types of current distribution can be described when these factors are analyzed, a nontrivial exercise even for simple geometries. The three factors are primary current distribution, secondary current distribution, and tertiary current distribution.

Heat Transfer. Heat removal for a commercial electrolysis system is generally carried out by internal or external evaporative cooling, circulation of electrolyte through external heat exchangers, or internal cooling with coils, jackets, or tubes that may also act as electrodes.

Reaction Engineering

Electrochemical reaction engineering considers the performance of the overall cell design in carrying out a reaction. The joining of electrode kinetics with the physical environment of the reaction provides a description of the reaction system. Both the electrode configuration and the reactant flow patterns are taken into account. More in-depth treatments of this topic are available.

Electrochemical Cells as Reactors. The electrochemical reactor readily parallels the chemical reactor, of which there are two types used as ideal models. The first is the plug flow reactor (PFR), where reactants and products move through the reactor in a pluglike manner. No mixing occurs in the PFR, which presumes that concentration changes occur only in the direction of electrolyte flow. The second model is referred to as a back-mix or stirred tank reactor (STR). The STR assumes perfect mixing and uniform composition in all zones of the reactor. An E is inserted for the electrochemical version of these abbreviations to produce PFER and STER. Idealized models are not attained in practice: mixing occurs in the PFER, whereas imperfect mixing occurs in the STER. Thus reactors are often described as having either PFER or STER features.

Scale-up of Electrochemical Reactors. The intermediate scale of the pilot plant is frequently used in the scale-up of an electrochemical reactor or process to full scale. Dimensional analysis has been used in chemical engineering scale-up to simplify and generalize a multivariant system and may be applied to electrochemical systems, but has shown limitations. It is best used in conjunction with mathematical models. Scale-up often involves seeking a few critical parameters. For electrochemical cells, these parameters are generally current distribution and cell resistance. The characteristics of electrolytic process scale-up have been described.

Cell Geometries

Uniform electrode potential, short interelectrode gaps, and good mixing and mass transport benefit many electrochemical reactions. These are best achieved with narrow spaced rectangular plates with turbulent flow electrolyte. Several types of cell designs are available.

COMMERCIALLY AVAILABLE CELLS

The lack of commercially available electrochemical cells was formerly a significant obstacle to electrode process development. This has changed because a number of companies now fabricate a range of cells for general use in electroorganic synthesis. These are parallel plate cells that may be divided or undivided. All use electrolyte recirculation for convection, and some include turbulence promoters to enhance mass transport. They follow the plate and frame design with external, or internal electrolyte manifolding. A variety of electrode materials is offered; eg, pure metals, PbO_2, DSA, graphite, and alloys. The cells are constructed of polypropylene, polyvinylidenedifluoride, or Teflon, and a range of elastomers. Peripheral components such as pumps, piping, etc., are also made available from cell manufacturers. Manufacturers of commercial cells include ElectroCell AB (Karlskoga, Sweden), Electrochemical Technology Business (a branch of Ineos Chlor Ltd, previously part of ICI, Runcorn, Cheshire, UK), and USFilter Electrocatalytic (Union, N.J.).

PRODUCT RECOVERY

Comparison of the electrochemical cell to a chemical reactor shows the electrochemical cell to have two general features that affect product recovery. Cell product is usually liquid, can be aqueous, and is likely to contain a salt; ie, a supporting electrolyte added to improve conductivity. In addition, there is a second product from the counter electrode, even if this is only a gas. Supporting electrolyte conservation and purity are requirements. Because product separation from the starting material may be difficult, carrying out the reaction to completion is desirable; cells would be run batch or plug flow. The water balance over the whole flow sheet needs to be considered, especially for divided cells where membranes transport a number of moles of water per Faraday. At the inception of a proposed electroorganic process, feasibility and cost of product recovery and refining should be included in the evaluation to determine true viability. Thus, early cell work needs to be carried out with the preferred electrolyte/solvent and degree of conversion. The economic aspects of product recovery strategies have been discussed. Some process flow sheets are also available.

COMMERCIAL ELECTROORGANIC PROCESSES

Commodity-Scale Processes

Adiponitrile. The most significant commercial electroorganic synthesis process is Monsanto's electrohydrodimerization (EHD) of acrylonitrile to adiponitrile. The importance of adiponitrile is as a precursor to hexamethylenediamine, which is used in the manufacture of nylon-6,6. The cost of manufacturing nylon-6,6 is critically dependent on the cost of the intermediates used, which has maintained the pressure to produce improvements in the EHD process:

Main cathode reaction:

$$2\ Cl^- \rightarrow Cl_2 + 2e^-$$

$$Cl_2 + H_2O \rightarrow HOCl + H^+ + Cl^-$$

$$2\ HOCl + 2\ CH_2{=}CHCH_3 \longrightarrow \underset{HOCH_2\overset{Cl}{\underset{|}{C}}HCH_3}{} + \underset{ClCH_2\overset{OH}{\underset{|}{C}}HCH_3}{}$$

Anodic water decomposition:

$$2\ e^- + 2\ H_2O \rightarrow 2\ OH^- + H_2$$

$$HOCH_2\overset{Cl}{\underset{|}{C}}HCH_3 + ClCH_2\overset{OH}{\underset{|}{C}}HCH_3 + OH^- \longrightarrow 2 \underset{O}{\overset{CH_3}{\triangleleft}} + 2\ Cl^- + H_2O$$

Two principal cathode by-products:

Overall $$CH_2{=}CHCH_3 + H_2O \longrightarrow \underset{O}{\overset{CH_3}{\triangleleft}} + H_2$$

Propionitrile, C_3H_5N, and an acrylonitrile trimer, 1,3,6-tricyanohexane are also produced at the cathode as by-products.

Smaller-Scale Processes

Although the Monsanto process is the only really large-scale operation, more than 50 small-scale processes appear to be commercial worldwide and about an equal number have been shown to be feasible in the pilot plant.

Fluorination. Perfluorinated organic compounds are important industrial surfactants and textile-treating agents.

The discovery that a number of organic compounds can be fluorinated by electrolysis in a solution of anhydrous hydrogen fluoride (HF) was made around 1940. In the Simons process, fluorination of the dissolved organics takes place at a nickel anode, generally without generation of free fluorine; hydrogen is evolved at the iron cathode of a diaphragmless cell. Alkali fluorides may be added for improved electrolyte conductivity. Electrochemical fluorinations of a wide range of compounds, including hydrocarbons, alcohols, ketones, carboxylic acids, and amines, have been studied.

Sebacic Acid. Sebacic acid, $C_{10}H_{18}O_4$, is an important intermediate in the manufacture of polyamide resins. It has an estimated demand worldwide of approximately 50,000 t/yr. The alkaline hydrolysis of castor oil, which historically has shown wide fluctuations in price, is the conventional method of preparation. Because of these price fluctuations, there has been considerable interest in an electrochemical route to sebacic acid based on adipic acid as the starting material. The electrochemical step involves the Kolbé-type or Brown-Walker reaction where anodic coupling of the monomethyl ester of adipic acid forms dimethyl sebacate.

BASF, Asahi Chemical Industry, and a CIS group have carried out pilot-plant studies of the electrochemical route to sebacic acid. Yields of dimethyl sebacate reported for the BASF and CIS processes are in the range of 80–85%. Asahi claims product yields as high as 92% at current efficiencies in the range of 85–90%.

Maltol. Otsuka Chemical Co. in Japan has operated several electroorganic processes on a small commercial scale. It has used plate and frame and annular cells at currents in the range of 4500–6000 A. The process for the synthesis of maltol, a food additive and flavor enhancer, starts from furfural. The electrochemical step is the oxidation of α-methylfurfural to give a cyclic acetal. The remaining reaction sequence is acid-catalyzed ring expansion, epoxidation with hydrogen peroxide, and then acid-catalyzed rearrangement to yield maltol.

Paired Synthesis of Phthalide and 4-t-Butylbenzaldehyde.

This BASF process is interesting synthetically, economically, and environmentally. Every electrochemical cell requires both an anode and a cathode. In almost all processes the product of interest is formed at one of these, and the product at the other electrode is of little interest and may even be deleterious, requiring separation of the two electrodes by a diaphragm. If one could design a process where both products are desirable materials, then the electrical power would in effect be used twice, saving overall operational costs as well as eliminating the need to dispose of the undesired counter electrode product. There are stringent requirements for such a process to be feasible commercially: the two substances must be produced under the same conditions, the two half reactions must not interfere with each other, and the overall cell reaction must be balanced, including consumption of the same amount of current at each electrode, and the two products must be in approximately equal market demand. The chlor-alkali process, in which an aqueous sodium chloride solution is electrolyzed to produce sodium hydroxide at the cathode and chlorine or hypochlorite is formed at the anode, is a good example of a commercial paired inorganic synthesis. A number of paired syntheses have been reported on the laboratory scale, including even a few which produce the same product at both electrodes, thus doubling the current efficiency.

A. J. Fry, *Synthetic Organic Electrochemistry*, 2nd ed., John Wiley & Sons, Inc., New York, 1989.

A. J. Fry and Y. Matsumura, eds., *New Directions in Organic Electrochemistry*, The Electrochemical Society, Pennington, N.J., 2000.

F. Goodridge and K. Scott, *Electrochemical Process Engineering*, Plenum Press, New York, 1995.

C. A. Sequeira, ed., *Environmentally Oriented Electrochemistry*, Elsevier, Amsterdam, 1994.

ALBERT J. FRY
Wesleyan University

ELECTROLESS DEPOSITION

"Electroless processes" is used to describe methods of depositing metals by means of chemical reduction. These systems are "autocatalytic" or self-catalyzing from their ability to plate onto their own deposits to build metallization thickness. This mechanism is a result of the chemical reduction based upon the reducing agent used in the process solution. Electroless plating dates to 1835 with the reduction of silver salts by reducing aldehydes. Progress remained slow until 1944, when modern electroless plating began with the rediscovery that hypophosphite could deposit nickel.

The growth of electroless plating is traceable to (1) the discovery that some alloys produced by electroless deposition, notably nickel phosphorus, have unique properties; (2) the growth of the electronics industry, especially the development of printed circuits; and (3) the large-scale introduction of plastics and other types of substrates benefiting from electroless coatings to meet many types of engineering requirements.

THEORY OF ELECTROLESS PLATING

The theory and practice of electroless plating parallels that of electrolytic plating. The actual metal reduction and film development occurs at the interface of the solution and the item being plated in both electrolytic and electroless processes. The main difference is that the electrons in electroless plating are supplied by a chemical reducing agent present in solution. This means that electroless solutions are not inherently thermodynamically stable, because the reducing agent and the metal ions are always present and ready to react.

Electroless solutions contain a metal salt, a reducing agent, a pH adjuster, a buffer, one or more complexing agents and one or more additives to control solution stability, film properties, deposition rate, etc.

Of the large number of potential reducing agents, the principal commercial materials are formaldehyde for copper and silver, hypophosphite for nickel and palladium, and organoborane compounds for gold, nickel, palladium, silver, and copper. Borohydride and hydrazine may be also used for nickel and formate for palladium. Most reducers also contribute to the composition of the deposit to some degree by introducing phosphorus, boron, or carbon into the corresponding film. Likewise the stabilizing materials used in a process typically codeposit with the metal into the deposit.

Most reducing agents are too slow, giving insufficient plating rates, or too fast, resulting in bulk decomposition. Each combination of metal and reducing agent requires a specific pH range and bath formulation. The metal ion and reducer are consumed in the reaction and so must be replenished at periodic intervals. Any components lost through spillage or drag-out also must be replenished along with adequate additional complexer to compensate for the build-up of salts and reaction by-products. Stabilizers are used to prevent spontaneous decomposition and are typically used at a few to a few tens part per million (ppm). They may affect deposition rate, deposit color or reflectivity, ductility, hardness and internal stress. Other additives may be introduced specifically to improve the smoothness or reflectivity of the coating.

The ideal electroless solution deposits the coating only on an immersed article, never on the sides of the tank or as a fine powder. State-of-the-art compositions can approach this ideal but the plating reaction and side reactions never allow coating with a very high efficiency. The primary industrial metals that are plated include nickel, copper, gold, palladium, and silver.

Electrolytic plating depends on current flow from the anode to the substrate surface to supply the electrons.

Plating rates are controlled by the current density at the metal-solution interface. This results in a coating that will have great variations in thickness and coverage. Electroless plating has the ability to uniformly coat articles, especially in recesses or blind holes, since the reducing potential is equal all over the part surface. Film thicknesses range from <0.1 mm, where only conductivity or reflectivity is desired to 1 mm or more for functional applications.

Electroless plating rates are affected by the rate of reduction of the dissolved reducing agent and the dissolved metal ion which diffuse to the catalytic surface of the object being plated. When an initial continuous metal film is deposited, the whole surface is at one potential determined by the mixed potential. The current density is the same everywhere on the surface as long as flow and diffusion are unrestricted so the metal deposited is uniform in thickness over the whole surface. However, maximum plating rates are lower for electroless plating than those possible for electrolytic plating. Extremely thin films of electroless coatings are not uniform because the initial deposition is confined to discrete nucleation sites that grow and coalesce into a film.

ELECTROLESS TECHNOLOGIES COMMERCIALLY AVAILABLE

As electrolytic plating is often employed as an ideal means to provide a thin surface coating which has some property (or properties) superior to that of the substrate, electroless deposited coatings offer a multitude of advantages that make them the preferred finish for a wide range of applications.

Nickel Phosphorus Alloys

Commercial nickel phosphorus alloy systems exhibit many desirable engineering characteristics. Their deposits have a low coefficient of friction and are antigalling. These systems offer the ability to provide metallization over nonconductors and are applied for improving the ease of soldering. They have a good "as plated" deposit hardness compared to other types of plated coatings, and the hardness can be further increased by subsequent heat treatment. By their nature as cathodic coatings, all electroless nickel alloys exhibit good corrosion protection, especially over steel and aluminum substrates. Because these deposits provide a uniformity of coverage and thickness over all surfaces, irregular components and shapes can be protected from corrosion in areas where typical electroplated deposits would be thin or absent because of current density distribution limitations. Electroless processes provide the ability for small parts to be metallized in mass quantities, utilizing a basket or barrel. Large articles the size of automobiles can also be processed with electroless technology.

For many applications, softer parts or components that have poor abrasion resistance can be given a hard, wear-resistant electroless nickel deposit. In some cases, EN-plated components can be substituted for expensive, more difficult to machine materials such as stainless steel. In other applications, nonconductors can be plated with electroless nickel and copper to provide EMF shielding or a base metallization for subsequent decorative copper, nickel, and chromium electrolytic plating.

Knowing how the phosphorus content relates to specific deposit properties is the single most important aspect of understanding electroless nickel systems. High phosphorus deposits provide the greatest corrosion protection and corrosion resistance in the widest variety and types of environments to which they are exposed. Medium phosphorus deposits are typically used and suited for many types of applications. Medium-low phosphorus alloy deposits are characterized by high hardness and excellent wear resistance both in the as plated and as heat-treated condition. They offer engineers new opportunities for utilization, while the metal finishers who use these technologies recognize improved operation, deposition rates and throughput, and as a result, they continue to become more versatile.

Low phosphorus alloy deposits are primarily utilized in specialized wear or electronic applications. They are ideal for soldering or brazing and can be welded. The deposits are compressively stressed, have good elongation (ductility), and provide excellent hardness and wear resistance, as plated or heat treated. In alkaline corrosive environments they outform high phosphorus deposits.

Nickel–Boron Deposits

The investigation of alternative reducing agents to sodium hypophosphite led to the use of borohydrides and amine boranes to provide alloys of nickel–boron. Nickel–boron alloys are deposited from formulations utilizing dimethylamine borane (DMAB) as the reducing agent and usually contain from 0.2 to 3.0% by weight boron in the deposits. But the lower boron alloys are especially suited for use in the electronics industry because of their electrical and physical qualities. Boron alloys in general are sometimes used in industrial wear applications because of their high hardness levels. Applications for 2 to 5% boron alloy include wear applications, aircraft engines, landing gear components, valves, pumps, resistance to erosion and galling.

The lower concentration boron deposits (<1% by weight) provide high electrical conductivity, exhibit low contact resistance and show good "as plated" hardness and wear resistance. They are easily soldered or brazed and maintain their ability to remain solderable and wire bondable for extended periods. EN boron deposits have found many applications to reduce or eliminate the usage of gold in some electronics industry applications with ultrasonic methods and provide excellent wetting of brazing alloys. Higher concentration boron containing alloys (>3% by weight) are very hard and wear resistant coatings and tend to further retard the formation of oxides on their surfaces. Nickel-boron deposits also provide an excellent strike coating for aluminum or plastics.

In high-circuit-density, multilayer ceramic (MCL) modules, which are characterized by narrow line width and spacing, a Ni-B deposit enhances the circuit characteristics while avoiding the potential for bridging between circuit patterns. In hybrid circuitry designs, Ni-B deposits assure continuous and dense conducting patterns and pads while improving solderability, brazeability and wire bonding.

Molybdenum–manganese and other metallizing substrate materials are subject to moisture where corrosion and loss of adhesion can occur. An electroless layer will protect these substrates from corrosive attack.

Electroless Copper

Electroless copper metallization is an important technology throughout the electronics and plastics industries. The chemistries typically use formaldehyde as a reducing agent. The processes have been used for metallization of holes in two-sided and multilayer printed circuit or printed wiring boards in order to provide copper on nonconductive plastics surfaces exposed after drilling through-holes (PTH) for maintaining internal electrical paths. Printed wiring boards (PWB) are manufactured from an epoxy glass laminate that is clad with an electrolytic copper foil. Other PTH purposes include holding a component lead wire and to interconnect circuitry or printed wires. Electroless copper deposits are ideally suited for providing a uniform thickness when electrically discontinuous surfaces or very high aspect ratio holes require plating. In another application, fabrication of multichip modules requires an electroless copper to define and build up conductive traces on the board surfaces.

Electroless copper technologies will continue to be favored for many applications as electronics components continue to shrink with a corresponding increasing complexity requirement of circuit boards.

Electroless copper deposits provide an initial strike layer for plating onto plastics (POP) for decorative purposes. For POP applications, electroless nickel or electroless copper metallization may be used to produce an electrically conductive deposit. Automotive industry studies have shown that utilizing an electroless copper layer for exterior applications produces better durability in various humid and corrosive environments.

Attenuation of electromagnetic signals generated either internally or externally to a packaged component requires electromagnetic interference (EMI) shielding for the component. EMI is electrical noise generated by electronic equipment that interferes with the operation of other electronic equipment. Plastic enclosures do not reduce EMI signals appreciably, so electroless copper and electroless nickel deposits have been used to address these requirements.

Electroless copper deposits have recently been utilized to encapsulate particles or oxide layers with a pure copper deposit. After encapsulation, the coated particles can be mixed with resins to be formed into components to provide electromagnetic shielding.

Electroless Silver

Electroless silver deposits provide electrical conductivity or as a strike undercoating for plating, as to a substrate prior to electroless gold plating.

Electroless Gold

Electroless gold deposits are applied mostly in the electronics industry. The metallization of components for printed circuit boards is a primary use of this technology, on flex circuits, ceramics, three-dimensional circuits, and metals.

Composite Coatings

The excellent wear resistance of EN deposits can be further improved by co-deposition of particulate matter called composite coatings. These systems represent specialty types of deposits for specific applications and enhanced wear and abrasion resistance. Composites consist of particles dispersed throughout the electroless deposit. Composite coatings have improved frictional and release properties and overall improved wear resistance in many applications.

Ternary Alloy Coatings

Also called poly alloys, these are electroless nickel deposits, which contain more than two elements. Each ternary alloy is designed to maximize a given deposit property such as corrosion resistance, high hardness, high temperature resistance, electrical properties, and ferromagnetism or brightness. Today the more common systems are based upon Ni-P-Cu, Ni-P-Sn, and Ni-B-Tl alloy deposits.

Encapsulation of Particles

Today, iron powders, ceramic powders, diamond powders, and other types of particles are being plated with nickel phosphorus and copper deposits. Often, these particles are plated so they can be molded or pressed with other polymers into complexly shaped grinding wheels and dicing blades or to provide a more corrosion resistant part when pressed into parts utilizing powder metal metallurgy. The more common reason to plate particles is to enhance their ability to bond to other materials. Many types of particles can be plated up to 60% by weight of electroless nickel phosphorus or copper deposits.

INDUSTRIES AND APPLICATIONS FOR ELECTROLESS DEPOSITION

Among the many industry applications for Ni-P deposits are in the petroleum and chemical industries, medical, dental, and pharmaceutical industries, aerospace industry, mining industry, electronics industry, for salvage and repair applications, automotive industry, for molds and dies, and in the printing and textile industries.

KEY PROPERTIES

A clear understanding of the interrelationships concerning the properties for any of the electroless nickel-phosphorus and nickel-boron coatings is important for the control of the overall performance characteristics of both the process and the resulting coating. Mechanical properties of electroless nickel–phosphorus alloys are dependent upon the amount of phosphorus in the deposit, other electroless systems are similar. For Ni-P alloy deposits, the phosphorus content of the alloy is the most significant parameter to control. And changing one property will often change a different property. Then again, sometimes a different property is evaluated for the actual one desired. In some wear applications, hardness tests are used as a substitute test method. Typically, deposit hardness has some relationship to wear resistance, and deposit hardness testing may be considered a more practical test.

Corrosion Protection

Electroless nickel, as a cathodic coatings, protects substrates by a mechanism of encapsulation from the environment. Once this barrier is penetrated, the protective value of the deposit is lost with the resulting development of corrosion to the substrate. As a contrast, the mechanism for anodic coatings, such as zinc plating over steel, provides protection to the substrate by sacrificially corroding relative to the substrate.

Aluminum substrates are especially susceptible to galvanic attack due to their high electro-potential difference. Corrosion of the aluminum will occur if the substrate is exposed to corrosive elements through pores of the nickel deposit. For this reason, especially high phosphorus deposits provide good protection to aluminum substrates.

The two inherent properties of nickel phosphorus deposits are their low porosity (corrosion protection) and resistance to chemical attack (corrosion resistance). High phosphorus deposits have less porosity compared to lower ranges of phosphorus. Despite this, all ranges of nickel phosphorus deposits provide some level of corrosion protection to substrates, given that their porosity is minimal.

Corrosion Resistance

Electroless nickel deposits provide corrosion protection from their cathodic nature, but their degree of corrosion resistance is determined by many factors, including the environment. The exposure temperature, concentration of media, whether wet or dry, whether an oxidizing or reducing environment affects the performance. In severe corrosion service environments, duplex or double-layer EN coatings are used to further enhance the corrosion resistance.

Not all high phosphorus EN deposits perform alike in all corrosion environments. For example, the codeposition of impurities or tramp constituents that may be present in an electroless nickel bath are even more important to its corrosion resistance than its phosphorus content.

The potential for corrosion is very high in petroleum production facilities as a result of several types of corrosive media present, with complications from the effects of temperature, pressure, and velocity. The most common corrosive materials are saltwater, carbon dioxide, and hydrogen sulfide. The presence of oxygen and other acid mediums will further exacerbate the conditions. High-phosphorus deposits have been used extensively in oil environments to retard corrosion.

Wear Resistance

Their resistance to friction and wear is often cited as one of the primary reasons to use electroless coatings in many applications. For nickel phosphorus alloys, wear resistance is primarily a function of the deposit hardness, which is related to phosphorus alloy content. However, deposit hardness and wear resistance are not always synonymous, and some data show that a softer coating can provide longer wear life. With metal-to-metal contact, the condition and hardness of the contacting surface must be taken into account, especially when the deposit is heat treated.

It has been pointed out by many that a persistent problem in engineering wear testing procedures involves the correlation of lab bench testing to actual wear behavior in field service conditions. In many actual field wear situations the precise loads, temperatures, contact geometries and corrosive environments are difficult to know or predict. In many cases, more than one mode of wear may be operating and the contact conditions could change with time. However, the ultimate success or failure of any EN coating is best determined by testing the deposit in the specific application. The wear test data presented here and in other publications provide indications of the characteristics of EN plated deposits ideally in a comparative environment. There are many combinations and conditions that can be tested. Without direct correlations to a specific test, it may be difficult to accurately compare wear and friction data from different sources.

There are four wear methods utilized to comparison test EN coatings, the most common being the Taber abrasion wear tester (abrasion wear), the Falex wear tester (adhesive wear, pin and V block), the Alpha LFW-1 ring/block, and the cross-cylinder test for determining of the adhesive wear and coefficients of friction.

SURFACE PREPARATION FOR ELECTROLESS DEPOSITION

For successful application of electroless deposition, the process cycle has a significant impact on the final quality of the deposit. Matching the substrate to the proper cycle ensures success with any electroless application. Inadequate surface preparation can result in lack of adhesion, deposit roughness, excessive coating porosity, and premature failure of the coating. The methods used to prepare surfaces for electroless deposits are similar to those used prior to conventional electroplating.

ENVIRONMENTAL CONCERNS

Waste Management

The solution life of most electrolytic plating baths is indefinite if the proper care is taken to maintain the chemistry balance and limit the introduction of contamination. By contrast, electroless plating baths have a short life, are not purified in most cases, and must be discarded after plating as little as 24 g/L and up to 60 g/L of metal. As a result, facilities with high production will generate high volumes of waste solution.

Among more common methods of waste management for electroless solutions include taking the plate out of the metal, hauling away the spent solutions to recycle for reclaiming the metal, autocatalytic precipitation, ion exchange, and combinations of these as well as other methods.

There are fundamental approaches to extend the useful life of electroless nickel baths. Since the formation of by-products is the major reason a bath has a limited life, limiting their formation or removing them adds to the useful life of the bath.

Additional strategies for the management of electroless waste today include reformulation to make solutions easier for treatment. Some of the modifications to avoid waste issues include elimination of strong chelating agents like EDTA, avoiding the use of ammonium hydroxide for the manufacture of components, and control of process pH regulation. Low-metal EN formulations operating at 3 to 4 g/L of nickel versus the conventional 6 g/L will also help minimize waste.

Legislative Restrictions

The newer regulations of concern today are the *End of Life Vehicle* (ELV) ELV Directive 2000/53/EC Annex II June 2002 and Waste Electrical and Electronic Equipment (WEEE) WEEE Directive 2002/96/EC initiatives. The ELV and WEEE directives seek to eliminate materials, chemistries, or deposits that do not allow easy recycling in the automotive and electronics industries.

G. O. Mallory and J. B. Hajdu, eds., *Electroless Plating Fundamentals & Applications*, American Electroplaters Society, 1990.

T. B. Massalski, ed., *Binary Phase Diagrams*, American Society for Metals, Metals Park, Ohio, 1986.

R. Suchentrunk, ed., *Metallizing of Plastics-A Handbook of Theory and Practice*, ASM International, Finishing Publications Ltd., 1993.

J. L. N. Violette, D. R. J. White, M. F. Violette, *Electromagnetic Compatibility Handbook*, Van Nostrand Reinhold, New York, 1987.

Brad Durkin
Carl Steinecker
MacDermid, Inc.

ELECTRONIC MATERIALS

Electronic materials are those exquisitely pure crystal structures which form the basis for the information technology of the 1990s. The chemistry and chemical engineering required for making electronic materials, ie, these very specific inorganic chemical bonding structures, have come about by techniques employing chemical vapor deposition to effect material growth and through an increased control over surface and interfacial chemistry.

SEMICONDUCTOR ENERGY LEVELS

Semiconductor materials are rather unique and exceptional substances. The entire semiconductor crystal is one giant covalent molecule. In semiconductors, the electron wave functions are delocalized, in principle, over an entire macroscopic crystal. Because of the size of these wave functions, no single atom can have much effect on the electron energies, ie, the electronic excitations in semiconductors are delocalized.

Good semiconductors are drawn from the central columns, Groups 13, 14, and 15 (III, IV, and V), of the Periodic Table, where the atoms tend to be nonpolar. For this reason, and because of the giant size of the wave functions, the electron–atom interaction is very weak. The electrons move as if in free space, colliding with the atomic lattice rather infrequently.

In a semiconductor the available energy levels are the valence and conduction bands, which are generally filled and empty, bonding and antibonding, respectively, separated by a forbidden gap. The electrons in the conduction and valence bands act as two separate subsystems. Not only do the electrons ignore the crystallographic lattice of atoms, the electrons in one band tend to ignore those in the other subsystem. This property is unique to electronic materials.

Electrons excited into the conduction band tend to stay in the conduction band, returning only slowly to the valence band. The corresponding missing electrons in the valence band are called holes. Holes tend to remain in the valence band. The conduction band electrons can establish an equilibrium at a defined chemical potential, and electrons in the valence band can have an equilibrium at a second, different chemical potential. Chemical potential can be regarded as a sort of available voltage from that subsystem. Instead of having one single chemical potential, ie, a Fermi level, of all the electrons in the material, the possibility exists for two separate quasi-Fermi levels in the same crystal.

The possibility of two separate electronic equilibria, ie, the establishment of two quasi-Fermi levels or two different chemical potentials, requires a very slow decay of electrons from the conduction band back into the valence band. The very weak electron–atom coupling, resulting from the large delocalized wave functions in nonpolar materials, permits the slow decay. Electron-hole recombination requires getting rid of the electronic band gap energy, which is usually around one volt, and dumping it off as heat of atomic motion. In relation to characteristic energies of atomic motion, one volt is a huge amount of energy to dissipate in a single step. The weak electron–atom coupling makes nonradiative decay an extremely unlikely event and the low probability for semiconductor materials to dissipate electronic energy as heat is probably their most unique property. By contrast, in organic molecules, decay by nonradiative recombination is sufficiently likely to occur that it is given the name internal conversion.

In fact, nonradiative recombination does occur in semiconductors, but primarily as a result of chemical defects that introduce new energy levels into the forbidden gap.

These defect levels act as stepping stones, permitting conduction electrons to cascade down to the valence band in two smaller steps rather than one improbable leap.

Thus a principal goal of semiconductor materials science has been to create chemically perfect semiconductor structures. Any defects that disturb the perfect valence bonding structure, allowing energy levels in the forbidden gap, must be eliminated as far as possible. Even the utmost extrinsic chemical purity is insufficient, however, because intrinsic defects such as broken bonds, self-interstitials, and vacancies are also proscribed. In particular, unsaturated chemical bonds on the material surface, or in the bulk, contribute nonbonding orbitals having unwanted energy levels in the forbidden gap. The rigid, tetrahedrally coordinated semiconductor crystal structures of silicon, Si; germanium, Ge; and gallium arsenide, GaAs, have a tendency to reject both extrinsic and intrinsic defects, contributing to their technological success.

SEMICONDUCTOR SURFACES

Semiconductor surfaces are the most likely location for intrinsic defects such as dangling or weak bonds to occur. The bulk chemical defect densities that can be tolerated in solid-state electronics range from 1 in 10^6 to 1 in 10^{11}, depending on the specific application. Corresponding surface defect densities that can be tolerated range form 1 in 10^4 to 1 in 10^7, ie, nearly all of the semiconductor surface atoms must be cleanly saturated with strong covalent bonds, because defects introduce energy levels into the forbidden gap. These requirements for semiconductor surfaces and interfaces give rise to a chemical figure of merit, ie, equivalent to a surface chemical reaction having a 99.99% to 99.99999% yield.

THE ROLE OF SILICON

The Si–SiO₂ Interface

Beginning in the mid-1950s, thermal oxidation of silicon at high temperatures in oxygen was begun, coating the silicon with a thin layer of silicon dioxide, SiO_2, glass. The thermal oxidation recipe was gradually perfected and by the late 1960s the figure of merit for the Si–SiO₂ interface had been improved to 1 defective chemical bond in 10^6. The Si–SiO₂ interface is an amorphous/crystalline heterojunction. The interfacial bonds can be 99.9999% saturated. Thus, in short order the microprocessor (1969), the memory chip, and the pocket calculator, were developed.

Purification of Silicon

Chemical purity plays an equally important role in the bulk of materials as on the surface. To approach the goal of absolute structural perfection and chemical purity, semiconductor Si is purified by the distillation of trichlorosilane, $SiHCl_3$, followed by chemical vapor deposition (CVD) of bulk polycrystalline silicon (at 1100°C), $SiHCl_3 + H_2 \rightarrow Si(s) + 3HCl(g)$. Purified polycrystalline CVD silicon from this reaction is then melted and a single-crystal boule weighing as much or more than 50 kg,

and having a diameter up to 20 cm, is pulled from the melt by Czochralski growth. Metallurgical-grade silicon is not sufficiently pure for applications in electronics (see SILICON AND SILICON ALLOYS).

III–V SEMICONDUCTORS

For optoelectronics the binary compound semiconductors drawn from Groups 13 and 15 (III and V) of the Periodic Table are essential. These often have direct rather than indirect band gaps, which means that, unlike Si and Ge, the lowest lying absorption levels interact strongly with light. The basic devices of optical communications, light-emitting diodes (LEDs) (see LIGHT GENERATION-LIGHT-EMITTING DIODES) and semiconductor lasers, are made of these III–V semiconductors. Aluminum arsenide, AlAs, GaAs, and the alloys of these compounds have historically been the most important III–V material system. The reason once again derives from the need to control interfacial chemical bonding structures.

The double heterostructure, invented in the early 1960s, is essentially a crystalline sandwich: the bread is made of AlAs and the filling of GaAs. Because the band gaps of AlAs and GaAs are 2.2 eV and 1.4 eV, respectively, the GaAs wave functions are sandwiched in by the 2.2 eV potential barriers. Although the electrons and holes are prevented from seeing any external surface, they do see the AlAs–GaAs interface. However, the cubic unit cell dimensions of GaAs and AlAs are 0.56533 nm and 0.56605 nm, respectively. Thus the mismatch at the interface is less than 0.1%, meaning that the crystal structures can match up nearly perfectly, leaving only a few unsaturated chemical bonds at the interface. Generally, the interfacial bonds in the $Ga_{1-x}Al_xAs$ system are 99.999% saturated. Although this is not as good as the best Si–SiO₂ interfaces, it is excellent nonetheless. The growth of successive atomic layers of semiconductor material, in perfect registry with atoms in the underlying crystal, is called epitaxy. The perfect atomic registry between layers of differing composition is called heteroepitaxy.

Physics and chemistry researchers approach III–V synthesis and epitaxial growth differently. The physics approach, known as molecular beam epitaxy (MBE), is essentially the evaporation of the elements. The chemistry approach, organometallic chemical vapor deposition (OMCVD) is exemplified by the typical chemical reaction: (at 580°C) $Ga(CH_3)_3 + AsH_3 \rightarrow GaAs(s) + 3CH_4(g)$. Thin-film epitaxy by OMCVD is generally more flexible, faster, lower in cost, and more suited for industrial production than MBE. An OMCVD system usually consists of two principal components, a gas manifold for blending the gas composition, and a graphite substrate holder which is usually inductively heated.

F. L. Carter, ed., *Molecular Electronic Devices*, Marcel Dekker, Inc., New York, 1982.

A. S. Grove, *Physics and Technology of Semiconductor Devices*, John Wiley & Sons, Inc., New York, 1967.

M. A. Herman and H. Sitter, *Molecular Beam Epitaxy: Fundamentals and Current Status*, Springer-Verlag, Berlin, 1989.

G. B. Stringfellow, *OMVPE: Theory and Practice*, Academic Press, Boston, Mass., 1989.

ELI YABLONOVITCH
University of California
at Los Angeles

ELECTROPHORESIS

The term "electrophoresis" refers to the movement of a solid particle through a stationary fluid under the influence of an electric field. The study of electrophoresis has included the movement of large molecules, colloids, fibers, clay particles, latex spheres, and basically anything that can be said to be distinct from the fluid in which the substance is suspended. This wide range in particle size to which electrophoresis has been applied has required that the theory describing electrophoresis be very general.

The fundamental principle behind electrophoresis is the existence of charge separation between any surface and the fluid in contact with it. The surface carries an immobilized charge, and the electrolyte fluid in contact with the charged surface balances the electric charge with an increased density of ions of the opposite charge. An applied electric field can act on the resulting charge densities, causing the particle to move, the fluid around the particle to move, or both. An applied electric field also generates heat, through resistive heating, and gases, through electrolysis reactions.

There are three distinct modes of electrophoresis: zone electrophoresis, isoelectric focusing, and isotachophoresis. These three methods may be used alone or in combination.

Distinction is also made among electrophoretic techniques in terms of the type of matrix employed for analysis. Matrices include polymer gels such as agarose and polyacrylamide, paper, capillaries, and flowing buffers. Each matrix is used for different types of mixtures and has its own unique advantages.

There are a variety of techniques for detecting separated sample compounds using chemical stains, photographic media, and immunochemistry. Each detection technique gives different information about the identity, quantity, and physical properties of the molecules in the mixture. Detection is often the focus of electrophoresis and usually yields basic information about the mixture being studied.

PRINCIPLES

Electrophoresis uses the force of an applied electric field to move molecules or particles, often through a polymer matrix. The electric field acts on the intrinsic charge of a substance. The force on each substance is proportional to the substance's charge or surface potential. The resulting force on the substance gives a distinct velocity for the substance that is proportional to the substance's surface potential. If two different substances have two different velocities, an electric field applied for a fixed amount of time results in different locations on the matrix for these substances.

The application of an electric field to a gel matrix or capillary tube results in heating in the media and gas generation at the electrodes. Thus special attention in the design and use of electrophoretic equipment is required.

MODES OF ELECTROPHORETIC SEPARATIONS

Zone electrophoresis is by far the most commonly practiced analytical technique, and there are several different such methods and techniques available for different separation goals. Isoelectric focusing is also useful for separation, and possibly more useful for determining charge characteristics of sample proteins. Isotachophoresis takes advantage of the continuity of current across a medium to segregate a sample into contiguous zones of high purity. It is not useful as an analytical technique is applied as a potential preparative method.

Zone Electrophoresis

In zone electrophoresis, multicomponent samples are applied to an electrophoretic medium, most commonly a gel, an electric field is applied, and after a predetermined length of time or after a certain level of power, current, or voltage has been applied the electrophoretic medium is inspected for resolution of the sample components. Bands that migrate the same distance in different sample mixtures typically represent the same substance. Standards are usually run concurrently with samples that are to be compared. In zone electrophoresis, proteins or sample components are completely separated into discrete zones as they migrate through the media onto which they are applied.

The use of standards with samples makes zone electrophoresis particularly useful as an analytical tool. However, when samples cannot be analyzed on the same gel, differences in the experimental conditions from experiment to experiment make direct comparison more difficult.

Isoelectric Focusing

Isoelectric focusing (IEF) is a technique used for protein separation, by driving proteins to a pH where they have no mobility. Resolution depends on the slope of a pH gradient that can be achieved in a gel.

Isoelectric focusing is an electrophoretic technique in which amphoteric samples are separated according to their isoelectric points along a continuous pH gradient. IEF analyses are carried out in various matrices: in acrylamide, agarose, and capillaries. The agarose or acrylamide gels that are used must be prepared with carrier ampholytes bracketing a specific pH range. After some time, the ampholytes separate and there is a pH gradient which covers the range of all the ampholytes' isoelectric points. Initial research was primarily directed toward evaluating the properties of synthetic ampholytes in solution. Later work refined the technique to apply gel matrices and provided the basis for IEF methodologies.

Because protein samples are actually ampholytes, when samples are loaded onto the gel and a current is applied the compounds migrate through the gel until they come to their isoelectric point where they reach a steady state. This technique measures an intrinsic physicochemical parameter of the protein, the pI, and therefore does not depend on the mode of sample application. In IEF, the highest sample load of any electrophoretic technique may be used; however, the sample load affects the final position of a component band if the load is extremely high; ie, high enough to titrate the gradient ampholytes or distort the local electric field.

Isoelectric focusing takes a long time (from ~3 to 30 h) to complete because sample compounds move more and more slowly as they approach the pH in the gel that corresponds to their isoelectric points. Because the gradient ampholytes and the samples stop where they have no mobility, the resistivity of the system increases dramatically toward the end of the experiment, and the current decreases dramatically. For this reason, isoelectric focusing is usually run with constant voltage, which can lead to an overheating of the system.

Another form of IEF is a method called direct tissue isoelectric focusing (DTIF), where isoelectric focusing in agarose is used to evaluate tissues. The tissue to be analyzed is placed directly onto the gel. Using the tissue itself and not tissue extracts has advanced the study of proteins that are difficult to extract from tissue or are damaged by the extraction procedure. DTIF is an important advancement in the area of sample handling and application where direct application of a solid to a gel matrix may actually enhance resolution.

Isotachophoresis

Isotachophoresis takes advantage of the fact that electroneutrality must be maintained in an electrophoretic system in order to support an electric field. If a current passes through a medium, that current must be constant from one electrode to the other, regardless of the local ion concentration or mobility; ie, dilute ions must move faster to keep up with a zone of more concentrated ions. Electric fields compensate for this because the electric fields strength does not have to be constant along the length of the medium. The electric fields strength is lowest where the ions are most concentrated and most mobile. Isotachophoresis takes advantage of this phenomenon by lining up the ions of interest, from fastest (most mobile) to slowest. This is a highly specialized technique that requires detailed knowledge of the properties of the sample to be separated and is generally not applicable to analytical separations of unknown constituents.

This separation technique has been employed primarily for preparative types of separations because detailed knowledge of the properties of the sample is required. Also, because this separation results in discrete zones of sample ions that are virtually pure, it makes sense to use this technique when the sample size is large. This technique is ineffective when the levels of impurities are small with respect to the target compound; small amounts of sample ions do not form zones well and tend to mix with the target compound.

ELECTROPHORETIC MATERIALS AND MATRICES

Various support media may be employed in electrophoretic techniques. Separation on agarose, acrylamide, and paper is influenced not only by electrophoretic mobility but also by sieving of the samples through the polymer mesh. The finer the weave of the selected matrix, the slower a molecule travels. Therefore, the molecular weight or molecular length, as well as charge, can influence the rate of migration.

In addition to polymeric support media, capillaries and flowing buffers have been used as support media for electrophoresis. Although these are not used as frequently, there are definite advantages for certain types of samples and applications.

The use of agarose, produced from the processing of red seaweed, as an electrophoretic method is widespread.

Polyacrylamide gel electrophoresis is one of the most commonly used electrophoretic methods. Analytical uses of this technique center around protein characterization; eg, purity, size, or molecular weight, and composition of a protein. Polyacrylamide gels can be used in both reduced and nonreduced systems as well as in combination with discontinuous and IEF systems.

Both the ease of use of this method for characterization of proteins and nucleic acids, and the ability to analyze many samples simultaneously for comparative purposes, have led to the prevalence of this technique. The drawbacks of a polyacrylamide matrix is that acrylamide is a neurotoxin, the reagents must be combined extremely carefully, and the gels are not as pliable as are most agarose gels.

Paper as an electrophoretic matrix was employed in some of the first electrophoretic techniques developed to separate compounds. Paper is easier than a gel matrix because the paper matrix requires no preparation. Besides being easy to obtain, paper is a good medium because it does not contain many of the charges that interfere with the separation of different compounds.

In paper electrophoresis, the sample is placed directly onto chromatographic or filter paper and then exposed to a buffer solution at each end and an electric field is applied. As in most electrophoretic techniques, charged dyes are combined with samples and standards to see the progress of the electrophoresis. The movement of samples on paper is best when the current flow is parallel to the fiber axis in the paper. The paper has high resistance, so voltages are typically much higher than in agarose and polyacrylamide matrices. Like agarose and polyacrylamide matrices, paper is combined with other analytical tools to enhance separation and identification of sample components. Paper electrophoresis has been combined with chromatography.

The difference between paper electrophoresis and paper chromatography is that electrophoresis separates by charge, whereas chromatography separates by polarity. This combined technique was used to evaluate polymorphisms of the hemoglobin molecule, ie, normal A-type versus the sickle cell S-type. It has been called peptide fingerprinting. This method was later modified to further evaluate peptides in a technique known as peptide mapping.

Some advantages of paper in electrophoresis are that paper is readily available, easy to handle, and new methodologies can be developed rapidly. The disadvantages of paper electrophoresis are that the porosity of paper cannot be controlled, the technique is not very sensitive, and it is not easily reproducible.

Free-flow electrophoresis is the most common technique for scaling up electrophoresis for commercial application. In this technique, sample compounds are injected into a curtain of buffer that flows between two flat plates, with electrodes parallel to the flow at each end. The electric field is then applied perpendicularly to the flow direction, so that as compounds flow down between the electrodes they separate horizontally and exit the flow field at different locations. The main challenge for this technique is stabilizing the flow to both heating and electroosmotic forces. Sometimes this is done by dividing the flow curtain into cells using semipermeable membranes, which allow proteins and other sample compounds to migrate from chamber to chamber but restrict flow. Another method is to apply a very low electric field so that little heating and electroosmosis occur, but then to recycle the material through coolers. The material is then sent back through the separating cell.

DETECTION TECHNIQUES

Most sample components analyzed with electrophoretic techniques are invisible to the naked eye. Thus, methods have been developed to visualize and quantify separated compounds. These techniques most commonly involve chemically fixing and then staining the compounds in the gel. Other detection techniques can sometimes yield more information, such as detection using antibodies to specific compounds, which gives positive identification of a sample component either by immunoelectrophoretic or blotting techniques, or enhanced detection by combining two different electrophoresis methods in two-dimensional (2D) electrophoretic techniques. Some techniques successfully employed are chemical staining, immunoelectrophoretic techniques, two-dimensional (2D) electrophoresis, and various blotting techniques.

Blotting techniques may be used in a variety and combination of electrophoretic systems that makes their use widespread and convenient when protein concentrations are minimal and agarose or polyacrylamide is the matrix choice.

B. A. Baldo and E. R. Tovey, *Protein Blotting: Methodology, Research, and Diagnostic Applications*, Karger, Switzerland, 1989.

R. F. Boyer, *Modern Experimental Biochemistry*, The Benjamin/Cummings Publishing Co., Inc., 1986.

N. Catsimpoolas, *Methods of Protein Separation*, Plenum Press, New York, 1975.

M. J. Dunn, ed., *Gel Electrophoresis of Proteins*, Wright, Bristol, U.K., 1986.

J. F. Robyt and B. J. White, *Biochemical Techniques: Theory and Practice*, Waveland Press, Inc., Prospect Heights, Ill., 1987.

SCOTT RUDGE
FeRx Incorporated

ELECTROPLATING

The term "electroplating" means just what it sounds, coating an object with a thin layer of metal by use of electricity. The metals most often used are gold, silver, chromium, copper, nickel, tin, and zinc, but many others are used also. The object to be plated is usually made of different metal, but can be the same metal or a nonmetal, such as a plastic grille for an automobile. The three basic reasons for surface finishing/electroplating are: (1) to improve appearance; (2) to slow or prevent corrosion (rust); and (3) to increase strength and resistance to wear (in the case of "engineering" finishes).

Progress in electroplating, particularly on larger-dimension objects, is also linked to improvements in materials of construction, power supplies, and other plating equipment, purer industrial chemicals and anodes, and improved analytical test and control methods. The quality of electroplating is dependent on the basis metal surface. Cleaner, less porous castings and better casting alloys, and improved steel and steel finishes have helped significantly. Recent great strides toward a full understanding of the detailed atomic level processes responsible for electroplating make it possible to apply the technics to ever more and more complex systems (eg, nanotechnology).

MATERIALS

In general, any electrically conductive surface can be electroplated. Specific techniques may be required to make a given surface electrically conductive. Many techniques may be used to metallize nonconductive surfaces. These can range from coating with metallic-loaded paints or reduced-silver spray to autocatalytic processes on tin–palladium activated surfaces or vapor-deposited metals. Preparation steps must be optimized and closely controlled for each substrate being electroplated.

Although metals and alloy substrates account for a good portion of the volume in electroplating, there is a large and rapidly growing amount of plastic and other surfaces being plated, both for decorative trim as well as for many modern electronic applications. On a far smaller scale, other materials that are plated include wood, plaster, fibers, cloth materials, and plant and animal tissue, such as leaves, leather, paper, and seashells.

The metals commonly though not exclusively electroplated on a commercial scale from specially formulated aqueous solutions include cadmium, chromium, cobalt, copper, gold, indium, iron, lead, nickel, platinum-group metals, silver, tin, and zinc. Although it is possible to electroplate some metals, such as aluminum, from nonaqueous solutions as well as some from molten salt baths, these processes have achieved only limited commercial use, due to the stringent conditions (such as high temperature) required for successful results and severe environmental restrictions.

In addition to the metals listed above, many alloys are commercially electroplated: brass, bronze, many gold alloys, lead–tin, nickel–iron, nickel–cobalt, nickel–phosphorus, tin–nickel, tin–zinc, zinc–nickel, zinc–cobalt, and zinc–iron. Electroplated alloys in somewhat lesser use include lead–indium, nickel–manganese, nickel–tungsten, palladium alloys, silver alloys, and zinc–manganese. Ternary and many other alloys can feasibly be electroplated and have of late found commercial applications in the area of lead-free solders.

Still another type of electrodeposit in commercial use is the composite form, in which insoluble materials are codeposited along with the electrodeposited metal or alloy to produce particular desirable properties. Polytetrafluoroethylene (PTFE) particles are codeposited with nickel to improve lubricity. Thus, eg, silicon carbide and other hard particles, including diamond, are codeposited with nickel to improve wear properties or to make cutting and grinding tools.

ECONOMIC ASPECTS

Electroplating is done less and less in job shops, where a customer's work is plated, and more and more in captive (in-house) shops. A reduction, particularly in the number of smaller job shops, is the consequence of a number of factors, chief among them the problems in meeting the waste regulations imposed on plating-shop effluents.

USES

Electroplated objects and materials are required for a number of specific properties or functions. There is in this, of course, some overlap; eg, a decorative use certainly requires some degree of corrosion resistance. Various usages and the principal plating metals employed are as listed in Table 1. There are also smaller

Table 1. Uses and Typical Plating Metals

Property/Function/ Application	Typical plating metals
decoration	chromium, copper, nickel, brass, bronze, gold, silver, platinum-group, zinc
corrosion resistance	nickel, chromium, electroless nickel, zinc, cadmium, copper and copper alloys, gold
wear, lubricity, hardness	chromium, electroless nickel, bronze, nickel, cadmium, metal composites
bearings	copper and bronze, silver and silver alloys, lead–tin
joining, soldering, brazing, electrical contact resistance, conductivity	nickel, electroless nickel, electroless copper, copper, cadmium, gold, silver, lead–tin, gold–tin, tin, cobalt
barrier coatings, antidiffusion, heat-treat, stop-off	nickel, cobalt, iron, copper, bronze, tin–nickel
electromagnetic shielding	copper, electroless copper, nickel or electroless nickel, zinc
paint/lacquer base, rubber bonding	zinc, tin, chromium, brass
manufacturing; electroforming	copper, nickel
manufacturing; electronic circuitry	electroless copper, copper, electroless nickel, nickel electroless gold, gold
dimensional buildup, salvage of worn parts	chromium, nickel, electroless nickel, iron

amounts of other metals and alloys used for specific applications.

Functional Plating

Plating for functional properties has grown to surpass decorative applications. The introduction of electroplating techniques in the ever-burgeoning electronic industry has a great deal to do with this. Nickel coating used as diffusion barriers beneath precious-metal deposits in electronic applications, nickel electroplating of strip steel and hardware in the production of batteries, and nickel-electroplated steel coin blanks are but a few examples of functional applications that have grown in importance. In general, platings for functional properties are extremely varied and are present in almost every industry. Work is plated in a variety of shapes and forms, but the simpler shapes are less troublesome in achieving adequate plate thickness distribution.

Plating Methods

Strips. Materials such as strip steel are plated on machines where coils of steel are unrolled on a continuous, high-production basis, fed through a sequence of preparation steps, and into the plating tank(s) on a series of rolls and rollers. Short plating times, high current densities, and relatively thin deposits are the rule.

Wires. Wire is moved at high speeds through a plating solution, causing rapid wire motion relative to the solution, not only parallel to the pull direction but also perpendicular to it as a result of wire vibration through the plating cell. Wire is plated commercially with several metals. Among these are copper and copper alloys, zinc, iron and iron allyoys, nickel and nickel alloys, gold, and silver.

Rack Platers. Stampings, moldings, and castings are usually mounted onto specially designed plating racks to allow rigid and proper positioning of the part.

Barrel Plating. When the parts to be plated are small enough, bulk plating methods may be used. Although some work may be electroplated in dipping baskets, plating barrels are more effective. Plating barrels come in many shapes, sizes, and styles. All are partially or entirely immersed and capable of rotating while going through the plating process. They are shaped to tumble small parts so that these are continually mixed, and are perforated to allow for the passage of current and an exchange of solutions. Plating barrels are made of inert plastic materials with means for electrical contact to the parts, often with flexible probes called danglers.

Brush (Decorative and Engineering) Plating. When parts are larger than the smaller areas of the part that require plating or when the part is in a fixed position away from a plating shop, brush plating may be employed. Other terms used for this method include selective, contact, swab, and out-of-tank plating. Specially designed plating tools, which are essentially shaped-anode materials covered with some absorbent material saturated with a plating (electrolyte) solution, are used.

Automation. Electric devices and in particular computers have revolutionized the converstion of sensor signals to a useful form for monitoring and control. Further, plating information can be gathered and analyzed quickly to ensure that quality is maintained at all times. The design and engineering of individual sensor devices has improved greatly in recent years. Many incorporate a microprocessor for signal conditioning, and some now can also carry out control of certain plating conditions directly without relying on a PLC (programmable logic controller) or computer. Automation of electroplating utilizing monitoring and control has resulted in lower manufacturing cost while improving quality and the reproducibility of processing.

FUNDAMENTALS OF ELECTROPLATING

The deposition of a metallic coating onto an object is achieved by putting a negative charge on the object to be coated and immersing it into a solution that contains a salt of the metal to be deposited (in other words, the object to be plated is made the cathode of an electrolytic cell). The metallic ions of the salt carry a positive charge and are thus attracted to the object. When they reach the negatively charged object that is to be electroplated, it provides electrons to reduce the positively charged ions to a metallic form.

Electroplating is performed in a liquid solution called an electrolyte, otherwise referred to as the plating bath; a specially designed chemical solution that contains the desired metal (such as gold, copper, or nickel) dissolved in a form of submicroscopic metallic particles (positively charged ions). In addition, various substances (additives) are introduced in the bath to obtain smooth and bright deposits. The object to be plated is submerged into the plating bath; usually at the center, where it acts as a negatively charged cathode. The positively charged anode(s) completes the electric circuit. A power source in the form of a battery or rectifier (which converts AC electricity to regulated low voltage DC current) provides the necessary current. This type of circuit arrangement directs electrons (negative charge carriers) into a path from the power supply (rectifier) to the cathode (the object to be plated). Now, in the bath the electric current is carried largely by the positively charged ions from the anode(s) toward the negatively charged cathode. This movement makes the metal ions in the bath migrate toward extra electrons at or near the cathode's surface layer. Finally, by way of electrolysis the metal ions are removed from the solution and deposited on the surface of the object as a thin layer. It is this process to which we refer as "electrodeposition."

The thickness of the electroplated layer on the substrate is determined by the time duration of the plating and the amount of current applied. According to Faraday's law, the overall amount of chemical change produced by any given quantity of electricity can be exactly accounted for. Thus, the current efficiency is the ratio between the actual amount of metal deposited to that expected theoretically

from Faraday's law. The ratio of the weight of metal actually deposited to the weight that would have resulted if all the current had been used for depositing, called the cathode efficiency, should be kept as close to 100% as possible.

Thickness of Deposit

The thickness of the deposit may be predicted by taking into account the cathode efficiency of the particular plating solution, the current density, and the duration time of the plating. Corrosion resistance, porosity, wear, appearance, and a number of other properties are all proportional to plate thickness. Minimum or maximum plate thicknesses should be specified in different applications.

Throwing Power. Plate thickness distribution does not always follow the primary current distribution. Throwing power, a term used to describe the relative plate thickness distribution, was originally defined as the improvement in percentage of the metal distribution ratio above the primary current distribution ratio. Plating solutions vary widely in ability to exhibit good throwing power. As current is increased in a given plating process, a point is reached where the metal ion being deposited is not replaced in the solution film nearest the surface fast enough and a concentration polarization occurs, shifting some of the current to the unpolarized, lower current density areas. The effect is to reduce the plating rate in these higher current areas. Examples of plating solutions having good throwing power include cyanide plating baths such as copper, zinc, cadmium, silver, and gold, and noncyanide alkaline zinc baths. Examples of poorer throwing power baths are acid baths such as copper, nickel, zinc, and hexavalent chromium.

Covering Power. Covering power refers to the lowest current density area where plating appears on a part. There are qualitative and semiquantitative tests for covering power, but no standard method is described by ASTM. One method, used in the past, simulated plating a tubular part by rolling a metal foil into a tubular shape, plating the rolled foil, unrolling, and measuring the distance that the plating covered on the inside of the foil. With the application of electroplating techniques in the rather sophisticated areas of electronics, angle cathodes with different angles and various side lengths, sloted cathodes, and slit cells are normally used to determine covering power.

Current Density. As the current density is increased on an object to be plated, a point where the deposit becomes rough, coarse grained, and takes on what is termed a burned and generally unacceptable appearance is eventually reached. The range from the minimum covering power to just below this burn-producing current is the practical usable-practical current density range. In electroplating, this range is not the same for all metal-plating baths and is influenced by metal ion concentration, chemical compositions, temperature, agitation, and anode-to-cathode spacing, as well as by the shape of the plated object and how it is positioned in the bath. In production plants, it is usually preferred to plate at the highest possible current density to shorten production time. This is not always done, however, especially when using lower throwing power plating baths, because increasing current decreases throwing power even further. For parts that require more even deposit thicknesses, lower current densities and longer plating times are to be used.

The Plating Tank

The materials that make up the plating tank as well as the ones for the auxiliary equipment required should be totally inert to the plating solution, or at the least, lined with inert material to protect the tank. In addition, it is advisable to have a nonconductive inner surface of the tank to provide electrical as well as chemical resistance. Unlined steel tanks are used, in case of alkaline plating solutions, in a number of plating outfits. However, under certain adverse conditions the steel may become susceptible to corrosion from the electrolyte. Raising free cyanide content in some electrolytes may serve to protect the steel. A recent requirement in many states is a containment surrounding the plating tank area to prevent accidental chemical spills from entering the environment. In acid plating solutions, other materials are used, depending on the chemical composition of the plating bath. Titanium and various stainless-steel alloys, PTFE, Karbate, Hastalloys, zirconium alloys, and others are among the choices.

Tank Sizes. Commercial electroplating lines come in practically all sizes, from liter-sized tanks in benchtop wire plating lines to ones holding 100,000 L and more. Tank width should allow for adequate anode-to-cathode spacing. In strip and wire lines, where the work is of a simple shape, anode-to-cathode spacing may be on the order of several millimeters to keep voltage requirements low. For plating work on racks, consideration should be given to the size and shape of the work to be plated. More complicated parts having deeply recessed areas require greater anode-to-cathode spacing to achieve uniform coverage. In any case, the plating tanks have to be of such size that the parts being plated are positioned 10–25 cm (4–10 in.) from the sides and the bottom of the tank as well as from the surface.

Filtration. Continuously filtering plating solutions are, becoming rather common, except for hexavalent chromium plating solutions, and even for those filtering is recommended. Baths are filtered remove any fine particulate matter that could codeposit with the electroplate and cause roughness. Filter media have to be chosen carefully to avoid introducing any soluble impurities and to avoid dissolving any of the filter media itself.

Temperature Control. Temperature control is important to all electroplating baths where good quality is required, because many properties are affected with only a few degrees of temperature change. Solution temperatures can be monitored with a thermocouple, thermistor, RTD (resistance temperature detector), or silicon integrated circuit device.

Anodes. There are two types of anodes: soluble (made of the metal being deposited) and insoluble (made of inert material). Soluble anodes are designed to dissolve efficiently with current flow and, preferably, not to dissolve when the system is idle. A plating solution having the anode efficiency close to the cathode efficiency provides a balanced process that has fewer control problems and is less costly. If the anode efficiency is much greater than the cathode efficiency, and there are only small solution losses, the dissolved metal concentration rises until at some time the bath has to be diluted back or the excess metal has to be reduced by some other means. If, the anode efficiency is less than the cathode efficiency, the dissolved metal decreases, pH decreases, and eventually metal salt additions and other solution corrections are required. Based on the cost of metal, it is usually considerably more economical to plate from the anode rather than add metal salt.

Faraday's law applies to the anode as well as to the cathode. Much like the cathode, the anode efficiency varies with the current density. As the current on the anode is increased, the anode efficiency decreases, slightly at first, until it reaches a point at which the anode metal cannot dissolve fast enough through the anodic surface-related film. The first stage of dissolution for the soluble anode is the oxidation of the metal followed by dissolution of the oxide. When the oxide dissolution rate is less than the oxidation rate, polarization of the anode takes place. The oxide film builds up in sufficient thickness to form an insulating coating, and the current decreases rapidly. The thick anode films may dislodge at the reduced current and remain as particulate matter for a while until they slowly dissolve. As particulate matter, the dislodged anode films form a significant source of roughness because they can codeposit with the metal on the work before redissolving. While the anode is polarized, oxygen and other gases can be given off from the anode. Solutions containing chlorides or bromides can emit chlorine or bromine, which are hazardous. Anode current densities should be maintained well below the point at which polarization occurs. This can be accomplished by increasing the anode area or lowering the current.

Electrolyte agitation around the anodes and some bath chemical variations may increase the anode current density range before polarization sets in. Some platers use anode bags to prevent particulate anode and anode film material from reaching the work. However, the solution flow through the bag material may be inadequate and add to the polarization problem.

Insoluble anodes are used exclusively in some plating baths such as chromium plating solutions that utilize lead–tin, lead–antimony, or lead anodes. Gold and other precious metal plating processes use stainless-steel anodes, keeping precious inventory costs down.

The consequence of insoluble anodes usage is that the pH of the plating solution decreases, along with the metal ion concentration. In some plating baths, therefore, a portion of the anodes is replaced with insoluble anodes in order to prevent metal ion buildup or to reduce metal ion concentration.

Insoluble anodes can cause undesired side effects. In alkaline cyanide solutions, the generation, and build-up of carbonates is accelerated significantly, together with a reduction in alkalinity. In acid solutions the pH decreases, which requires frequent adjustments. In sulfamate nickel plating solutions, insoluble anodes, and even slightly passive soluble anodes, partially oxidize the sulfamate ion to form sulfur-bearing compounds that change the character and performance of the deposit.

Power Supplies

A variety of power supplies are now available to the plater. Continuous controls, constant current, and constant voltage are commonplace, preprogrammed automatic current manipulation is also getting more use. The newer forms of current manipulation are called pulse cycles because the cycle is typically on the order of milliseconds. There is now increased interest in pulse deposition techniques.

The d-c output from a rectifier can have a portion of modulation from the a-c input, called ripple. Modern rectifiers usually have ~5% or less ripple at full load. Ripple can have some bad effects, especially in chromium plating, where ripple can cause dullness and poor coverage. In all cases, even when one uses an old-fashioned d-c power supply, it needs a means of control, stepless preferred, and an appropriate ammeter and voltmeter. Ampere-hour recorders are convenient for making additions and keeping records.

PREPARATION FOR PLATING

Inadequate precleaning of the substrate to be plated can result in poor adhesion, pitting, roughness, lower corrosion resistance, smears, and stains. Since electroplating takes place at the single molecular surface layer of an object, it is mandatory that the substrate surface be absolutely clean and receptive to the plating. In the effort to get the substrate into this condition, several separate steps may be required, and it is in these cleaning steps that most of the problems associated with plating arise.

Cleaning Methods

No simple, universal cleaning cycle exists for electroplating. The key is in knowing which combinations of metals and processes are effective in specific situations.

Mechanical Cleaning. employs some form of abrasion, such as dry blasting with sand, aluminum oxide, silicon carbide grits, glass beads, or similar abrasives, wet blasting, using vapor honing, wet tumbling with abrasive media, and vibratory finishing. Alternatively, a strong pickling acid treatment, must be used.

Solvent cleaning is still possible within the current environmental and safety/health regulations. Three solutions without an ozone depletion capability are methylene chloride, perchloroethylene, and trichloroethylene.

Aqueous cleaners normally contain surfactants to remove intrinsic oil film and extrinsic soils. Semiaqueous cleaners are an alternative to solvent cleaning.

A common visual test to determine the cleanliness of the work is to observe how the surface rinses; if the surfaces drain evenly when rinsed using clean water, holding a sheet of water, it is said to be water-break-free and clean. If rinse water beads up in droplets on the part surface or does not hold a sheet of water as it drains, it is said to water-break and is in need of additional cleaning. However, if this test is run on surfaces having surfactant-solubilized oil films, the water-break test is not definitive. Thus, the part has to be thoroughly rinsed, acid dipped in clean acid, and thoroughly rinsed again before using the water-break test. Neither does the water-break test reliably detect inorganic films. Thus, the water-break test indicates when a part is dirty, but not necessarily when it is clean.

Electrocleaning. This type of cleaning, which is comprised of the application of a low voltage DC current through a conductive alkaline cleaning solution using the work itself as one electrode and usually steel suspended from busbars on either side of the work as the other, is known as electrocleaning. The equipment and process has the characteristic elements of the electroplating process.

Cleaning Cycles

Depending on the nature of the object to be plated, the cleaning cycle usually starts with a soak cleaner followed by electrocleaning. If the work is heavily soiled, a precleaning step may be required. Buffing compound residues fall into this class. Solvents, diphase mixes, power washers, and other presoaks are used. If the work is received in a fairly clean state, the high surfactant soak cleaner steps is omitted from the cleaning cycle. The electrocleaner formulation has less tendency toward leaving substantive films and is preferred.

Soak-and-electrocleaner combined products are promoted for use when the normal soak cleaner is too much and a normal electrocleaner is too little. These combination cleaners are often nothing more than electrocleaners with additional surfactants. The undesired results often are a hard-to-rinse electrocleaner or a low capacity soak.

Rinsing

Postdeposit rinsing is extremely important to the overall integrity of the plating operation. Work transferred from a processing solution carries a film on its surface from that solution. This amount of drag-out varies with the shape and surface area of the part, along with the nature of the solution. Drag-out can reach considerable quantities; eg, from 20 to 80 mL/m^2.

To reduce costs of the more expensive plating solutions and to decrease the amount of hazardous or regulated material in the waste stream, recovery and reuse of the drag-out is a common practice. This is done simply by closing off the water flow to the first rinse tank following the plating process tank, and periodically returning the accumulated solution to the process tank.

Recycling drag-out losses can have the unwanted effect of concentrating impurities in the plating tank. Thus, it is considered a good practice to purify the drag-out before returning it to the process tank. Filtration through activated carbon is helpful on most nonchromium solutions; cation exchange has been useful with chromic acid drag-out solutions, before it is evaporated.

Rinsing requires "good" water. The water should not be too cold for vigorously agitated, and time must be provided.

Dual- or triple-rinse tanks follow process solutions. Additionally, many plating lines utilize counterflow or cascade rinsing with the multiple tanks plumbed so that freshwater flows from one tank to the next counter to the direction of the work. This method of conserving water is a strongly recommended practice for plating lines.

Spray rinsing can be efficient because of impingement; misting at the exit of a processing tank can reduce drag-out. Barrel plating lines need much more rinsing time and special consideration because of the relatively high drag-out rates.

Acid Dips and Pickles

Acids should be used to remove inorganic soils. An important distinction is to be made between acid dips and acid pickles. Where there is no rust or scales, a dilute acid dip is sufficient for activation and as a rinse aid. Caustic residues from cleaners are notoriously difficult to remove with water rinsing alone. By contrast, strong acid pickles should be used to remove rust and scale.

The two kinds of acids most in use for preplate processes are hydrochloric acid, HCl, used mainly for rust and scale removal, and sulfuric acid, H_2SO_4.

Electropickling. Electrolytic pickling is used in some instances; steel can be treated anodically in strong sulfuric acid, 70%.

PROPERTIES, SPECIFICATIONS, AND TEST METHODS

Standard test methods are required to determine the relevant parameters of electroplated materials. Documents on plating specifications for many phases of the plating process are published by a number of organizations; eg, the federal government, the military, ASTM, ISO, SAE, etc. An excellent cross-index of these is available.

ENVIRONMENTAL ASPECTS OF ELECTROPLATING

No single topic is as important to the plating industry as waste management, which has become the key to economic survival. Waste management includes recycling, waste reduction, and waste treatment and disposal. Metal finishers (electroplaters) are regulated under a federal Environmental Protection Agency (EPA) program, the National Pollutant Discharger Elimination System (NPDES), often administered by the states.

The Resource Conservation and Recovery Act (RCRA) is the most comprehensive pollution control law for metal finishers. One requirement assigns cradle-to-grave responsibility for waste to the generato; ie, the plater. Unless the waste is made into another form or recycled as another product, the original generator is still responsible for future treatment or liability costs. Other requirements of the RCRA include a generator's license, proper waste containers, labels, and manifest shipping documents, licensed haulers, record maintenance, and an annual waste minimization report.

For the foreseeable future, the plating industry has to expect tighter restrictions and more regulation of recovery and recycling operations. Solid wastes, the sludge from waste treatment processes, have to pass stringent leaching tests to be allowed in landfills, and costs for the disposal of solid wastes are increasing dramatically. Some earlier legislation (eg, the Clean Air Act of 1990) is concerned with air pollutants. This act restricted chromium in air exhausts from chromium-plating and chromic acid anodizing plants to $0.01-0.03$ mg/m^3. Also, the Pollution Prevention Act of 1990 stressed reduction of chemicals that could enter the waste stream before treatment.

PLATING BATH FORMULATIONS

The specific formulations of plating baths are flexible in some systems while very sensitive to variation in others. Many of the more recent changes have resulted from waste treatment and safety requirements. Besides the ability to deposit a coating having acceptable appearance and physical properties, the desired properties of a plating bath would include high metal solubility, good electrical conductivity, good current efficiencies for anode and cathode, noncorrosivity to substrates, nonfuming, stable, low hazard containing, low anode dissolution during downtime, good throwing power, good covering power, a wide current density plating range, ease of waste treatment, and economy of use. Few formulas have all these attributes. Only a few plating solutions are, commercially used without special additives, but chemical costs often constitute only a relatively low percentage of the total cost of electroplating. Additives are used to brighten, reduce pitting, or otherwise modify the character of the deposit or performance of the solution. Preferred formulations are normally specified by the suppliers of the proprietary additives.

Plating Bath Purification

Once a plating bath is in use, purification is, required periodically to maintain the plating solutions' useful service. Alkaline zinc plating solutions are sensitive to a few milligrams per liter of heavy-metal contamination, which may be precipitated using sodium sulfide and filtered out. Nickel plating solutions may contain excess iron as well as unknown organic contaminants. Iron is removed by peroxide oxidation, precipitation at a pH of ~5, then filtered out. The more complex, less water-soluble organic contaminants, along with some trace metals, are removed with activated carbon treatments in separate treatment tanks.

Another, common purification treatment used on both new and used plating solutions is dummying. Heavy-metal impurities are removed by electrolyzing, usually at moderate current densities, using large disposable steel cathodes. Good agitation and lower (acid) pH speed the process.

Testing

Analysis and testing are required whenever a new plating solution is made up, and thereafter at periodic intervals. Frequency of analyses and testings should be adjusted so

Table 2. Cadmium-Plating Solutions[a]

Parameter	Cyanide	Acid sulfate	Fluoborate
Cd metal, g/L	20–30	15–30	75–150
NaCN, g/L	90–150		
Na_2CO_3, g/L	30–60		
H_2SO_4, g/L		45–90	
NH_4BF_4, g/L[b]			60–120
NaOH, g/L	10–20		
pH	12.5–13.5	very acid	2.5–3.0
anodes	Cd and steel	Cd and graphite	Cd
temperature, °C	15–40	18–23 max	10–40
current density, A/m^2	50–900	300–800[c]	100–600

[a] Addition agents for all three plating solutions are as required.
[b] Some baths contain 20–30 g/L $NaBF_4$ and 20–30 g/L H_3BO_3 instead of NH_4BF_4.
[c] Current densities from 100–200 A/m^2 are used for barrel plating.

that corrections can be held to 10% or less from established concentrations.

INDIVIDUAL PLATING BATHS

Cadmium and Cadmium Compounds

The suggested makeup of cadmium plating baths is shown in Table 2. Whereas cadmium provides better corrosion resistance to steel and other substrates than zinc when exposed to marine environments, zinc is better in industrial, sulfur-bearing atmospheres. The resistance of cadmium is improved with chromate conversion coatings; bright, yellow iridescent, olive drab, and black finishes are used. Cadmium is readily solderable, provides lubricity on threaded fasteners, and has been used as a high-temperature protective coating in the aircraft industry when diffused into nickel plating. Olive-drab chromated cadmium plate is used on nickel-plated aluminum electrical connectors to extend salt spray resistance. For diffusion processes, for low hydrogen embrittlement, and for plating directly on cast iron, cadmium–cyanide plating solutions are used without brighteners; in other applications, brighteners produce attractive lustrous deposits.

The future of cadmium plating is in some doubt because of cadmium's toxicity. Cyanide-free cadmium plating systems, have experienced some limited growth. Acid cadmium, based on cadmium sulfate compositions, is replacing some cyanide baths in the United States. Fluoborate cadmium is reported in use in the United Kingdom,

especially in barrel plating. Cadmium plating is covered by ASTM, U.S. government, and ISO specifications.

Chromium

Chromium plating can be separated into two areas: hard chromium, also called functional, industrial, or engineering chromium; and decorative chromium. The plating bath compositions may be the same for both. In most cases, the differentiating factor is plate thickness. Decorative chromium is usually less than ~1 μm; hard chromium can be from ~1–500 μm or more.

Formulations and conditions for operating hard chromium plating solutions are shown in Table 3.

Chromic-acid-based plating solutions differ from the other common metal-plating baths in that chromium solutions have poor current efficiencies, poor covering power, and poor throwing power, as well as a need to use much higher current densities to plate. Modifications developed in attempts to improve these weaknesses incorporate a second catalyst and are called cocatalyzed baths. The second catalyst is often fluoride or, more commonly, the fluosilicate ion.

Metallic Impurities. Chromic acid plating solutions are affected by metallic impurities that are usually accompanied by chromium reduction, increasing electrical resistance. Iron is a common impurity that increases rapidly, say, when the chromium plating tank is used as the reverse current chromium etch tank. A separate chromic acid etch tank should be used to extend the life of the plating solution. Limits of the trivalent chromium that can be tolerated vary from 10–60 g/L. Treatment using an electrolytic porous pot is used sometimes to purify chromium solutions.

Additional purification processes include cation-exchange treatments of diluted solutions, followed by evaporation back to working bath strength; and electrodialysis directly in the working solution.

Cobalt

Cobalt-plating is of limited use because most high volume applications can be satisfied using nickel at considerably less cost. Cobalt has been used to coat tungsten carbide components that needed to be brazed and exposed to high temperatures in use; and as a barrier coating to inhibit the migration of copper into gold. Cobalt is sputtered onto computer memory disks for its magnetic properties. Cobalt is used in nickel electroforming solutions to increase strength and hardness. Cobalt plating solutions are so similar to

Table 3. Chromium-Plating Solutions[a]

Parameter	Conventional baths		Cocatalyzed baths	
	Range	Typical	Range	Typical
CrO_3, g/L	150–400	240–260	150–400	150–180
SO_4^{2-}, g/L	1.25–5	2.4–2.6	0.6–1.3	0.9–1.0
SiF_6^{2-}, g/L			0.3–0.8	0.5–0.6
CrO_3/SO_4^{2-} ratio	80–120:1	90–110:1	150–200:1	170–180:1
anodes	Pb-7% Sn or Pb-6% Sb		Pb-7% Sn	
current density, A/m^2	400–4000	1000–3000	400–4500	1000–3500

[a] All decorative chromium baths are run at 38–43°C, hard chromium baths at 40–60°C.

Table 4. Copper Cyanide Solution Formulations

Constituent	Strike, g/L		Rochelle, g/L		High efficiency, g/L	
	Typical	Limits	Typical	Limits	Typical	Limits
CuCN	22	15–30	26	19–45	75	49–127
NaCN	33	23–48	35	26–68	102	62–154
or						
KCN	43	31–64			136	76–178
Na_2CO_3	15	0–15	30	15–16		
NaOH					15	22–37
or						
KOH					15	31–52
Rochelle salt ($KNaC_4H_4O_6 \cdot 4H_2O$	15	15	45	30–60		
By analysis						
copper	16	11–21	18	13–32	55	34–89
free Cyanide	9	6–11	6	4–9	19	10–20

nickel plating solutions in composition that in most cases the equivalent cobalt salts can be substituted for nickel salts in published nickel bath formulations.

Copper

Copper plating is used as a final finish in some applications and is widely employed as an undercoat for other deposits. Of the several types of copper baths, the two most popular in the United States are acid sulfate and cyanide baths.

Cyanide copper baths are, a target for replacement, and alkaline, noncyanide baths are becoming increasingly popular as replacements for cyanide copper because of environmental issues. Care is needed, in choosing a noncyanide copper systems to avoid trading one waste treatment problem for another. Copper cyanide baths are shown in Table 4. Cyanide solutions are known for their cleaning ability, which is enhanced using low efficient, high gasing cyanide copper strike baths.

Brass, the first alloy to be electroplated, remains a popular decorative finish. Plated as an undercoating for nickel–chrome, brass provides good corrosion protection compared to copper, especially on aluminum and zinc substrates. There has been considerable use of brass plating on steel wire and other shapes because of improved rubber bonding characteristics. Several brass copper–zinc alloys sare commercially stated; the more common are yellow brass (~30% zinc), red brasses (10–15% zinc), and white brass (~72% zinc).

Electroplated bronze electrodeposits are used for decorative applications as a final finish, with and without subsequent coloring or antiquing treatments. A clear lacquer helps preserve the deposit. On a limited basis, bronze has also been used as an undercoating for nickel–chromium finishers on steel, zinc, and aluminum. In some functional applications, plated bronze is used as a stop-off material in the selective nitriding of steel parts. This system also finds frequent use as a nongalling bearing material.

Gold

Gold is used both for decorative and industrial or engineering purposes. The important properties of industrial gold are corrosion resistance, low electrical contact resistance, and good solderability.

Gold is usually plated over electro- or electroless nickel-plated substrates. A small amount of gold is used in electroforming with substantial thicknesses. Aerospace applications for both thin and thick deposits are among the many newer uses for gold.

There are literally hundreds of gold-plating bath formulations, most of which are proprietary. Gold flash, gold strikes, soft golds, hard golds, bright golds, and golds of all purities and colors account for the wide selection. Formulations can be classified as alkaline, neutral, or acidic.

Indium

Indium diffuses readily into copper, silver, and lead, increasing hardness and imparting good antifriction properties. Indium plating is used for bearing materials and in electronics for contact pads for flip chip bonding. Indium may be plated from cyanide, fluoborate, and sulfate baths, but the standard bath recommended is based on sulfamate.

Iron and Iron Alloys

Practically all iron is plated from acidic solutions of iron(II) ferrous salts, but its presence in the baths lowers the cathode efficiency for depositing the metal and may cause the deposit to be brittle, stressed, and pitted. Copper-alloy soldering iron tips are plated with a thick (175–200 μm) iron layer to prevent the solder from dissolving the copper. Iron has also been used to plate stereotype printing plants, and in plating of aluminum pistons and cylinders in automotive engine blocks. Iron has been plated from chloride, sulfate, mixed chloride–sulfate, fluoborate, and sulfamate solutions; chloride is the most common.

Lead and Lead (Tin) Alloys

Lead is plated from fluoborate solutions. Lead–tin alloys of 3–15% tin are called terne; for hot-dip coatings, terneplate can be 20% tin. Terneplate has been used to protect steel in gasoline tanks. Solder deposits are also plated. Lead–tin baths are similar to lead baths. Stannous fluoborate is added to supply the tin. Lead is under strict regulation in the environment and waste streams. In the future the use of lead plating will be reduced to a minimum.

Nickel

Nickel electroplating is a commercially important and versatile surface-finishing process. The applications of nickel electroplating fall into three main categories: decorative, functional, and electroforming. A great variety of plating baths have been formulated, but most nickel plating is done in either Watts or sulfamate baths. Typical bath compositions and conditions are shown in Table 5.

Both Watts and sulfamate baths are used for most applications. The difference in the deposits is in the much lower internal stress obtained, without additives, from the sulfamate solution.

Bright Nickel. Electroplated from pure nickel salts, nickel is a dull gray, satinlike deposit that has to be buffed to obtain a bright finish. Brighteners should be used to obtain bright deposits directly from the bath. The additives currently used fall into two classes, variously labeled primary and secondary, first class and second class, carrier and brightener. The last is more commonly used in plating plants.

The additives that go into the fairly commonplace nickels bath are almost always proprietary; carrier portions may include the sodium salts of benzene disulfonic acid,

naphthalene 1,5-disulfonic acid, saccharin, and allyl sulfonic acid. Generally, these compounds are characterized by a $=C-SO_2$ group. Brightener leveler portions include a large number of possible materials. A few are based on small amounts of zinc, cadmium, lead, selenium, or tellurium, etc., and many are based on organic compounds containing unsaturated groups such as $C=O$, $C=C$, $C\equiv C$, $C=N$, $C\equiv N$. Nickel brighteners and the theories of the brightening processes are well covered in patents and other literature.

Dual Nickel. Sulfur-free systems are referred to as semibright nickel. The more well known of these are based on coumarin. In the past, these soft, hazy deposits were buffed bright and a layer of bright nickel was deposited for appearance. It was later observed that better corrosion protection was obtained using this dual nickel system, not all of which could be attributed to the buffing. This improvement is ascribed to the sulfurfree nickel being much more noble than the more anodic top layer. Improvements in sulfurfree additives have resulted in much brighter deposits.

Special Nickel Plating Baths. Nickel for nonreflective, decorative, or solar absorptive uses is plated from nickel solutions containing ammonium, zinc, and thiocyanate salts. Nickel composites containing hard particles for wear, or abrasive particles for cutting and grinding tools, are plated from conventional Watts or sulfamate solutions to which the particles are added.

Table 5. Nickel Electroplating Solutions

	Electrolyte composition[a] (g/L)		
	Watts nickel	Nickel sulfamate	Basic semibright bath[b,c]
nickel sulfate, $NiSO_4 \cdot 6H_2O$	225–400		300
nickel sulfamate, $Ni(SO_3NH_2)_2$		300–450	
nickel chloride, $NiCl_2 \cdot 6H_2O$	30–60	0–30	35
boric acid, H_3BO_3	30–45	30–45	45
Operating conditions			
temperature, °C	44–66	32–60	54
agitation	air or mechanical	air or mechanical	air or mechanical
cathode current density (A/dm^2)	3–11	0.5–30	3–10
anodes	nickel	nickel	nickel
pH	2–4.5	3.5–5.0	3.5–4.5
Mechanical properties			
tensile strength, MPa	345–485	415–610	
elongation, %	10–30	5–30	8–20
Vickers hardness, 100-g load	130–200	170–230	300–400
internal stress, MPa	125–185 (tensile)	0–55 (tensile)	35–150 (tensile)

[a] antipitting agents formulated for nickel plating are added to control pitting.
[b] Organic additives available from plating supply houses are required for semibright nickel plating.
[c] Typical properties of *full-bright* nickel deposits are as follows: Elongation percentage—2–5; Vickers hardness, 100-g load—600–800; internal stress, MPa—12–25 compressive.

Nickel Alloys

Nickel alloys containing 10–35% iron are useful in decorative applications as a cost-saving substitute for all-nickel deposits. One Watts-based variation contains about 45 g/L nickel, 3 g/L iron, and boric acid and additives such as citrates and gluconates. Hydroxy carboxylic acids act as stabilizers for the ferrous iron.

Nickel–cobalt alloys are used for engineering properties, especially in electroforming. Alloys over the entire range (0–100% cobalt) have been obtained from various baths containing simple salt solutions. Nickel–cobalt plating solutions are similar to those used for nickel deposits with cobalt salts substituted for some of the nickel salts. Sulfate, chloride, and sulfamate salts of cobalt are used. Cobalt adds strength and hardness to nickel deposits.

Interest in electrodeposited nickel–phosphorus came with the realization of the benefits of the electroless nickel–plated alloy. The properties of the alloys appear to be the same from either process and are related to the phosphorus content. Deposits using 2–15% phosphorus and higher can be electroplated; lower phosphorus deposits are dull while higher phosphorus deposits are bright. Low (2%) phosphorus deposits can be obtained simply using phosphorus acid additions to a Watts bath and operating the bath at 0–2 pH with a temperature of 75–95°C and current of 500–4000 A/m².

Nickel–Tungsten alloys may be deposited from acid Watts baths that contain sodium tungstate-producing deposits with up to ∼5% tungsten. Alkaline baths with ammonium salts in the nickel solution, along with Rochelle salts, citrates, or glycolates, can deposit 10–20% tungsten. The properties of the alloy are being considered as a possible substitute for decorative chromium. Nickel–tungsten alloys having codeposited silicon carbide particles are considered as a possible replacement for hard chromium plating.

Platinum-Group Metals

Rhodium has useful properties for engineering applications: good corrosion resistance, stable electrical contact resistance, wear resistance, heat resistance, and reflectivity. Platinum plating has found application in the production of platinized titanium, niobium, or tantalum anodes used as insoluble anodes in many other plating solutions.

Palladium is used in telephone equipment and in electronics applications as a substitute for gold in specific areas. Palladium is usually plated from ammoniacal and acid baths, available along with chelated variations as proprietary processes.

Ruthenium, the least expensive of the platinum group, is the second best electrical conductor after silver, has the hardest deposit, and has a high melting point.

Silver

Cyanide baths are the standard for silver plating. Only minor modifications and improvements in brighteners have occurred since silver electroplating was first patented

Table 6. Typical Silver-Plating Bath Composition (ln g/L¹)ᵃ

	Bright silver (conventional)	High speed (thick deposit)
silver (metal)	20–45	35–120
silver cyanide	31–55	45–150
potassium cyanide (total)	50–80	70–230
potassium cyanide (free)	35–50	45–160
potassium carbonate	15–90	15–90
potassium nitrate		40–60
potassium hydroxide		4–30
current density, A/dm²	0.5–1.5	0.5–10.0
temperature, °C	20–28	35–50

ᵃ Note: Brighteners are added as required.

in 1840. Typical plating baths are shown in Table 6. Silver strikes are necessary to obtain adhesion on most metals. A large proportion of silver plating is used for decorative purposes, such as tableware, jewelry, art work, musical instruments and the like, but a growing amount is found in industrial applications. These uses include bearings, reflective surfaces, contact areas on busbar and other electrical and electronic contacts, solderable surfaces, in thermocompression bonding, selective plating on semiconductor components to replace gold on many devices, and other applications where its high thermal and good electrical conductivity (best among metals) are useful.

Tin and Tin Alloys

Tin may be plated from fluoborate, sulfuric acid/sulfate, or alkaline stannate, baths. The deposits are matte when plated without brighteners. Brighteners for use in sulfate baths have been vastly improved, and very bright, decorative deposits can be obtained directly from the bath.

A major application of tin-lead alloys is in the production of electronic components. There coatings are a prerequisite for reliable soldering when making interconnections on PCBs. They are plated from fluoborate baths and more recently methanesulfonate solutions. Solder, 60% tin plate, is the most often used one on contacts and as a solderable etch resist on printed circuit board (PCB)s. Higher (96–98% tin) alloys are used in semiconductor and other electronic applications.

Tin–nickel alloy deposits having 65% tin exhibit good resistance to chemical attack, staining, and atmospheric corrosion, especially when plated copper or bronze undercoats are used. This alloy has a low coefficient of friction. Deposits are solderable, hard (650–710 HV₅₀), act as etch resists, and find use in printed circuit boards, electronics in general, watch parts, and as a substitute for chromium in some applications. The most common plating bath of tin–nickel alloys uses fluoride to complex the tin.

Although alloys of all concentrations are possible, 80% tin-20% zinc gives the best combination of properties. This alloy has a low coefficient of friction, low electrical contact resistance, is solderable, slightly anodic to steel, and does not form voluminous corrosion products. In addition, the tin–zinc alloy has good paint adhesion qualities, good ductility, and is easily spotwelded. A typical bath is based on stannate and cyanide.

Zinc and Zinc Alloys

Zinc is plated on pipe, conduit, strip, sheet, and wire, as well as sheet metal stampings, all sorts of hardware, and fasteners. Zinc plating is used on electronic housings and cabinetry to reduce high frequency radiation interference.

Plating solutions for continuous strip, called electrogalvanizing, sheet, and wire are usually simple zinc sulfate solutions, although chloride and mixed variations have found some use.

Chloride Baths. Chloride baths'zincs are brighter, can plate directly on cast iron, and operate at high efficiency but have poor throwing power, require acid-proof equipment, and as such are more corrosive to surrounding steel materials. A typical chloride zinc may contain ~35-g/L zinc, 135-g/L Cl⁻, 30-g/L boric acid, proprietary brighteners and wetting agents.

Zinc Baths. Zincate are simple, economical, noncorrosive to steel, and have superior throwing power. Some recent systems are showing better tolerance to operating variables, with less tendency toward flaking or delayed blistering, and good brightness. Zincate baths, however, require good cleaning, frequent analysis, close chemical control, and have low cathode efficiency.

Cyanide Baths. Cyanide baths produce zinc deposits that have excellent covering power and throwing power, good cleaning ability, deposit relatively pure zinc, are capable of thick deposits, and require little control. Zinc deposits from cyanide baths are purer than from the other baths but still contain traces of the metal contaminants, brightener components, sulfur, hydrogen, and other gases. Pure, sulfur-free zinc is resistant to hydrochloric acid.

Zinc Alloys. Recent decades have seen considerable worldwide activity in the area of plating zinc alloys. This interest results from efforts to improve the corrosion resistance of automobiles, automotive components, and components in the aerospace arena, without using the rather toxic cadmium. Three zinc alloys dominate the practical interest: zinc–nickel, zinc–iron, and zinc–cobalt. Alloys are generally 6–12% nickel.

Zinc–nickel usually shows better results than zinc–cobalt in salt-spray tests.

Alloys of zinc–cobalt usually contain 0.3–0.8% cobalt. Alloys with higher cobalt content, 4–8%, have shown better salt spray resistance, but the commonly plated alloy is still 0.3–0.8%.

Posttreatment. Many posttreatments are used over plated metals, but chromate conversion coatings remain one of the most popular. Chromates are used to improve corrosion resistance, provide good paint and adhesive base properties, and produce brighter or colored finishes.

ELECTROFORMING AND ELECTROFABRICATION

Electroforming is the production or reproduction of articles by electrodeposition upon a mandrel or mold that is subsequently separated from the deposit. The separated electrodeposit becomes the manufactured article. Of all the metals, copper and nickel are most widely used in electroforming. Mandrels are of two types: permanent or expendable. Permanent mandrels are treated in a variety of ways to passivate the surface so that the deposit has very little or no adhesion to the mandrel, and separation is easily accomplished without damaging the mandrel or the deposit. Expendable mandrels are used where the shape of the electroform would prohibit removal of the mandrel without damage. Low-melting alloys, metals that can be chemically dissolved without attack on the electroform, and plastics that can be dissolved in solvents are typical examples.

L. J. Durney, ed., *Electroplating Engineering Handbook*, 4th ed., Van Nostrand Reinhold, New York, 1984.

W. Goldie, *Metallic Coating of Plastics*, Vols. 1 and 2, Electrochemical Publications Ltd., Ayr., Scotland, 1968–1969.

M. Paunovic and M. Schlesinger, *Fundamentals of Electrochemical Deposition*, John Wiley & Sons, Inc., New York, 1998.

M. Schlesinger, in M. Schlesinger and M. Paunovic, eds., *Modern Electroplating*, 4th ed., John Wiley & Sons, Inc., New York, Chapter 5 and references cited therein, 2000.

MORDECHAY SCHLESINGER
University of Windsor

EMBEDDING

Advances in electronic technology have had great technological and economic impact on the electronic industry throughout the world. The rapid growth of the number of components per chip, the rapid decrease of device dimension, and the steady increase in integrated circuit (IC) chip size have imposed stringent requirements, not only on IC physical design and fabrication, but also in electronic packaging and embedding. Electronic embedding is one of the most common processes used to encapsulate and protect these electronic components. With advances in very large-scale integration (VLSI) technology and multichip module packaging, embedding of this high density package in electronics has become a challenge. Various polymeric encapsulants are used for embedding these types of electronic components.

MATERIALS FOR ELECTRONIC EMBEDDING

Silicones

Polydimethylsiloxanes, polydiphenylsiloxanes, and polymethylphenylsiloxanes are generally called silicones. With a repeating unit of alternating silicon–oxygen, the siloxane chemical backbone structure, silicone possesses excellent thermal stability and flexibility that are superior to most other materials. Polydimethylsiloxane provides a very low glass-transition temperature T_g material but is suitable for use at temperatures up to 200 °C. The basis of commercial production of silicones is that chlorosilanes are readily hydrolyzed to give disilanols which are unstable and condense to form siloxane oligomers and polymers. Depending on the reaction conditions, a mixture of linear polymers and cyclic oligomers is produced. The cyclic components can be ring-opened by either an acid or base to become linear polymers, and it is these linear polymers that are of commercial importance. The linear polymers are typically liquids of low viscosity and, as such, are not suited for use as encapsulants. These must be cross-linked (or vulcanized) in order to increase the molecular weight sufficiently to achieve useful properties. For electronic applications, only the high-purity room temperature vulcanized (RTV) condensation cure silicone, which uses an alkoxide-cure system with noncorrosive alcohol by-products, and platinum-catalyzed addition heat-cure (hydrosilation) silicone systems are suitable for device encapsulation.

Epoxies

Their unique chemical and physical properties such as excellent chemical and corrosion resistances, electrical and physical properties, excellent adhesion, thermal insulation, low shrinkage, and reasonable material cost have made epoxy resins very attractive in electronic applications. The commercial preparation of epoxies is based on bisphenol A, which upon reaction with epichlorohydrin produces diglycidyl ethers. The reactant ratio (bisphenol A:epichlorohydrin) determines the final viscosity of the epoxies. In addition to the bisphenol A resins, the novolak resins with multifunctional groups which lead to higher cross-link density and better thermal and chemical resistance have gained increasing acceptance in electronic applications.

Polyurethanes

Recent work has focused on the use of intermediates, which are low molecular weight polyethers with reactive functional groups such as hydroxyl or isocyanate groups able to further cross-link, chain extend, or branch with other chain extenders to become higher molecular weight polyurethanes. Diamine and diol are chain extended with the prepolymer, either polyester or polyether, to form polyurethanes with urea or urethane linkages, respectively. The morphology of polyurethane is well characterized. Hard and soft segments from diisocyanates and polyols, respectively, are the key to the excellent physical properties of this material.

High performance polyurethane elastomers are used in conformal coating, potting, and in reactive injection molding (or reaction impingement molding) of IC devices. Furthermore, rigid polyurethane foams, most often in free-foam densities of 128–288 kg/m^3 [8–18 lbs/ft^3 (pcf)], are useful for embedding complex electronic systems.

Polyesters

Polyester is used in embedding resins for electronic components because of its low cost compared to silicones and epoxies. Polyesters are condensation products of dicarboxylic acids and dihydroxy alcohols; the reaction provides a wide range of viscosities for polyesters.

$$HOOC-R-COOH + HO-R'-OH \xrightarrow{H_2O} \left(O-\overset{O}{\underset{\|}{C}}-R-\overset{O}{\underset{\|}{C}}-O-R' \right)_n$$

There are electronic-grade polyesters with relatively low viscosities that are suitable for embedding of coils, as in transformers. Furthermore, free-radical-cured polyesters are popularly used for display castings.

Polysulfides

Polysulfides are organic polymers that contain sulfur in disulfide linkages (S–S), mercaptans (S–H), or thiol groups (S–R). Low molecular weight polysulfides (3,000–4,000) may oxidatively react with free-radical reactants such as lead, tin, cobalt octoate, p-quinone, dicumene hydroperoxide, lead, and manganese dioxides to yield rubbery flexible polysulfides with good water, gas, and moisture sealer properties. Most polysulfides have excellent adhesion to coated substrates and resistance to oxidation, solvents, and ozone. However, the electrical properties of this material are marginal as an insulator, and its dielectric constant is relatively high ($\epsilon_1 \leq 7$). It is a low cost embedding material for transformers and connector sealing, but not too common for microelectronic embedding.

Advanced Thermoplastics Materials

Thermoplastics and linear plastics of finite molecular weight that can be fabricated into very complex structures by hot melt or injection molding are different from the thermoset materials that require cross-linking to build up infinite molecular weight to form network (cross-link) structures. Advances in thermoplastic engineering materials include amorphous thermoplastics, crystalline thermoplastics, liquid crystal thermoplastics, and fluorinated thermoplastics.

EMBEDDING MATERIALS PROPERTIES

The ability of a given material to perform as an electronic embedding encapsulant depends largely on its properties. Ultrapure chemical properties with a low level of mobile ions such as sodium, potassium, and chloride are essential. Furthermore, the material's electrical, mechanical, and rheological properties are critical.

MATERIAL PROCESSES

Material processes consist of cavity-filling and saturation coating. The cavity-filling process involves molding, potting, and coating.

MATERIAL CURING

In order to optimize each embedding material property, complete cure of the material is essential. Various analytical methods are used to determine the complete cure of each material. Differential scanning calorimetry, Fourier transform-infrared (ftir), and microdielectrometry provide quantitative curing processing of each material.

NEW TECHNOLOGY ALUMINUM-FILLED COMPOSITES

A novel aluminum-filled composite for embedded passive applications has recently been developed. Passive components are devices that do not generate voltage or current, but their electrical characteristics are essential to form a complete circuit together with power sources and active components like ICs in an electronic system. Discrete passive components occupy more and more substrate space as the active components continue to miniaturize. Replacing discrete passive components with integral or embedded passives, in which passive components are integrated inside the substrate, has become an urgent target for next generation electronics.

Embedded passives offer many advantages. Miniaturization and lower board cost are anticipated since the board area previously occupied by discrete passives is now reduced. Also, with embedded passives, internal parasitic resistance and inductance associated with surface-mounted discrete components can be suppressed, due to the elimination of solder joints. Embedded passives can also improve reliability. Embedded passives not only provide increased silicon packaging efficiency and reduced assembly cost but also improved system performance.

Embedded capacitors are especially interesting because of their wide range of applications. Currently surface-mounted discrete capacitors are used in all levels of electronic packing, including for signal decoupling, noise suppression, filtering, tuning, bypassing, termination, and frequency determination. Embedded capacitors are particularly favored by decoupling application, because of their ability to reduce parasitic properties to an extent beyond the limit of discrete capacitors.

By using self-passivation aluminum as a filler, the thickness of the passivation oxide layer formed outside the aluminum sphere determines the ultimate electrical properties of the corresponding composites. With the proper insulating oxide layer, the high loading level of aluminum can be used while the composite materials remains insulated. For composites containing 80 wt% 3.0 μm aluminum a dielectric constant of 109 and a low dissipation factor of about 0.02 (at 10 kHz) have been achieved. Compared with aluminum oxide-filled composites, interfacial polarization is found to be the major mechanism underlying the high dielectric constant, which is supported by bulk resistivity measurement, TGA analysis, and high-resolution TEM observation. Aluminum composites have good reliability and good adhesion toward a copper-laminated substrate. Bimodal aluminum-filled composites have also been systematically studied. A rheology study shows that the minimum shear viscosity occurs at a weight ratio of 76/24 for a system with 10.0 μm plus 3.0 μm aluminum, at a weight ratio of 80/20 for a system with 10.0 μm plus 100 nm aluminum, and at a weight ratio of 79/21 for a system with 3.0 μm plus 100 nm aluminum. Such weight ratios can provide the highest filler loading for the specific systems. Using the optimized filler weight ratio from rheology study, the highest dielectric constant obtained at 10 kHz is 88 for a system filled with 10 μm plus 3.0 μm aluminum, 136 for a system filled with 10 μm plus 100 nm aluminum, and 160 for a system filled with 3.0 μm plus 100 nm aluminum.

L. T. Manzione, *Plastic Packaging of Microelectronic Devices*, Van Nostrand Reinhold, New York, 1990.

R. R. Tummala and E. J. Rymaszewski, eds., *Microelectronic Packaging Handbook*, Van Nostrand Reinhold Co., Inc., New York, 1989.

C. P. Wong, *Application of Polymers in Encapsulation of Electronic Parts, Advances of Polymer Science*, Vol. 84, Springer-Verlag, Berlin, 1988, pp. 63–83.

C. P. Wong, ed., *Polymers for Electric and Photonic Applications*, Academic Press, San Diego, Calif., 1993.

C. P. WONG
JIANWEN XU
Georgia Institute of
Technology

EMISSION CONTROL, AUTOMOTIVE

The over 210 million U.S. automobiles and trucks on the road produce exhaust that consists primarily of carbon dioxide, water, unburned hydrocarbons, carbon monoxide, oxides of nitrogen (NO_x), remaining oxygen, and nitrogen. The latter three atmospheric pollutants have been regulated since the 1970s by the U.S. government and, more stringently, by the state of California. Automobile companies have developed fuel metering and exhaust systems using the catalytic converter to meet emission regulations. Carbon dioxide emissions are indirectly controlled by corporate average fuel economy (CAFÉ) standards for passenger cars and small trucks.

By 2004, the Tier 2 exhaust emission standards mandated by the Clean Air Act Amendments of 1990 required automobiles and small trucks to reduce emissions of hydrocarbons by 98%, carbon monoxide by over 90%, and oxides of nitrogen by 98% as compared to 1970 pre-control emissions.

The key components of spark-ignited engine emission control systems are the three-way catalytic converter and closed loop oxygen sensor controlled fuel metering system. Exhaust gas recirculation is used to reduce formation of engine NO_x and provides antiknock values. Emission control is achieved without negatively affecting fuel economy or performance.

Very few light-duty diesel engines are now in service, and their emissions control technology has not been selected. Light-duty and medium-duty diesel engine powered vehicles have to meet the same Tier 2 standards as gasoline-fueled engines. Heavy-duty diesel engine trucks must meet very stringent standards starting with the 2007 model year.

EMISSION REGULATION AND TESTING

United States federal regulations require automobile manufacturers to obtain a certificate of conformity indicating compliance with all applicable emission standards over a vehicle's useful life period. The Tier 2 useful life for car and light-duty vehicles is 160,000 km or 10 years, or 11 years and 193,000 km for heavier light-duty vehicles.

Clean Air Act Amendments

The Clean Air Act Amendments of 1970 set strict emission standards requiring 90% reduction of emissions of hydrocarbons, carbon monoxide, and oxides of nitrogen. The Clean Air Act Amendments of 1990 required more stringent standards that resulted in Tier 1 and Tier 2 standards—the later being initiated in 2004 through 2007 for light-duty vehicles and fully implemented for heavy light-duty vehicles in 2009. The Tier 2 standards essentially required all passenger cars and light-duty trucks (LDT) including sports utility vehicles (SUV) up to 3,856 kg (8,500 lb) gross vehicle weight (GVW) to meet the same level standards as passenger cars for 193,000 kilometers or 10 years with a 25, 50, 75, and 100% phase-in schedule over four years. Heavier trucks and passenger-carrying vehicles up to 4,536 kg (10,000 lb) GVW have to meet the same standards for 193,000 km or 11 years with a 50, 100% phase-in schedule starting in 2008. Tier 2 also set fuel sulfur content in gasoline to an average of 30-ppm with a cap at 80-ppm. For a complete list of federal emission standards, go to www.epa.gov/otaq/standards.htm (the Tier 2 program); for the state of California LEV-II emission standards see www.arb.ca.gov. Highly corrosive leaded gasoline has been completely banned in the United States since 1996. The U.S. Tier 2 and California LEV-II program will dominate advances in emission control in this next decade.

Test Procedures

Originally, vehicle manufacturers had to actually run test track vehicles for 80,000 km as durability vehicles. Periodic exhaust emissions tests using the U.S. Federal Test Procedure (FTP) determined emissions at several mileages between 6,400 km and 80,000 km and a deterioration factor (DF) was computed. A simplified modern verification has been proposed for new emission durability procedures using a new standard road cycle (SRC) for whole vehicle durability or an alternative method of new standard engine bench cycle (SBC) for accelerated aging of the catalytic converter and oxygen sensor and other exhaust system components. The objective is to use durability procedures that are at least as severe as actual vehicle on-road use. Auto manufacturers are permitted to develop their own SRC and SBC programs and have them approved. Tier 2 emission regulations require field surveys by automobile manufacturers to assure that vehicles in actual use are in compliance with the emission standards.

The Federal Test Procedure (FTP) specifies that a test vehicle be fueled with commercial unleaded gasoline and stored in a temperature between 20 and 29°C for at least 12 hours immediately prior to the emission test. The vehicle is placed on a chassis dynamometer calibrated for the vehicle's weight and road load. The vehicle is started cold and driven for 41 minutes on a prescribed cycle of accelerations, cruises, decelerations, idle periods, a 10-minute shutdown (called the hot soak), and a hot start and period of rerun that matches the driving conditions of the first 505 seconds of the cold start. The total distance traveled is 17.8 km (11.115 miles). Exhaust emissions are sampled, analyzed, and computed with a constant volume sampling system (CVS). All engine exhaust gases from test start to test end are fed into the CVS system, where it is diluted with a large amount of air drawn in by the CVS constant-flow pump. Since the CVS flow rate is constant throughout the test, the air drawn in modulates inverse to vehicle exhaust flow. Total CVS system flow is recorded for each test period. Diluted exhaust samples are continuously collected from the well-mixed diluted exhaust CVS stream for each portion of the entire test.

From analyzing the accumulation of emissions from a vehicle under test using the FTP it is seen that HC accumulation is high during the first $\sim$50 seconds from engine start, before the catalyst has started to function. Very little accumulation of HC occurs after the engine and catalyst are up to operating temperature. Also, NO_x emissions occur during cold and hot starts and are coincident with high-speed modes. Clearly, Tier 2 engines will have to be designed to heat up the catalytic unit in about 10 to 15 seconds. The, Tier 2–compliant engine design factors being examined to produce hotter exhaust gas are (1) reduced exhaust system thermal mass; (2) lean air/fuel cold start; (3) retarded ignition timing; and (4) in some cases, variable valve timing. The catalytic converter is located close to the exhaust manifold (close coupled) to make it heat up faster. New catalytic units have been developed with an increased geometric surface area per unit of volume, a reduced thermal mass, and improved thermal/physical properties of the catalyst layer. Clearer fuels with very low sulfur content are also needed.

EXHAUST GAS COMPOSITION

The exhaust composition from gasoline/air combustion is dependent on many factors. Total combustion in the internal combustion engine is not possible even when excess oxygen is present. The quality and formation of the air/fuel mixture as well as the design of the combustion chamber influence the combustion process, as do engine power and ignition timing. However, in modern engines, for emission control the main factor affecting exhaust gas composition is the air/fuel mixture or ratio, for a standard gasoline fuel the hydrogen to carbon ratio (H/C) is 1.86. The exhaust gas composition will change substantially as the H/C ratio changes.

Sources of exhaust unburned hydrocarbons are crevices in the combustion chamber, such as gaps between the piston and cylinder wall, where the combustion flame cannot burn. Unburned hydrocarbon composition is dictated by fuel composition. Carbon monoxide results from areas of insufficient oxygen. Oxides of nitrogen are produced in high-temperature flame zones during combustion by reaction of nitrogen molecules and oxygen atoms thermally produced from oxygen and oxygen-containing species.

Hydrocarbons and carbon monoxide emission can be minimized by lean air/fuel mixtures, but lean air/fuel mixtures maximize NO_x emissions. Very lean mixtures (>20 air/fuel) result in reduced CO and NO_x but encounter increased HC due to unstable combustion at the lean limit. Improvements in lean-burn engines have extended the lean limit. Rich mixtures, containing excess fuel and insufficient air, produce high HC, CO, and H_2 emissions in the exhaust. Very rich mixtures have been used in some engine applications at high engine speeds and loads where the cooling effect of gasoline as it is vaporized in the cylinder is used to limit high cylinder temperature. Rich engine calibrations result in toxic CO exhaust concentrations of 4 to 5% or more.

The best engine power is achieved at a slightly rich setting and the best fuel economy is achieved slightly lean, whereas the stoichiometric air/fuel ratio is the optimal point for the best power, fuel economy, and emissions.

Over 150 hydrogen and carbon species are present in the exhaust mix of a gasoline fueled engine, including methane, various paraffins, olefins, aldehydes, aromatics, and polycyclic hydrocarbons as well as unburned gasoline. Competitive catalytic oxidative reactions favor oxidation of the latter four classes before paraffins and methane. Loss of catalytic performance is higher for the methane and paraffin hydrocarbons than for more reactive species.

Sulfur dioxide is present from the combustion of sulfur contained in gasoline. The average U.S. gasoline sulfur content graduallay changed from 2004 to 2006 to an average of 30-ppm with 80-ppm maximum cap.

Exhaust gas also contains small amounts of hydrogen cyanide and ammonia, depending on the air/fuel ratio.

EMISSION CONTROL SYSTEMS

Figure 1 is a schematic of the major components of the three-way-conversion (TWC) Closed-loop Fuel Metering System. The catalytic converter is located at the engine exhaust manifold (in a close coupled position for Tier 2 compliant engines). An O_2 sensor is located ahead of the catalyst unit inlet, in the converter inlet shell or the exhaust manifold. A second O_2 sensor is located at the catalytic converter outlet. The inlet O_2 sensor generates an operating control signal to the electronic control unit (ECU). The outlet O_2 sensor provides a means for trimming the control point to correct for system aging changes over the life of the engine. Both the inlet and outlet O_2 sensors also perform a required emission control quality on-board diagnostic (OBD) function. The operating O_2 sensor is exposed to raw exhaust emitted from the engine and generates an electrical signal indicative of the concentration of oxygen in the exhaust. The signal is sent to the ECU. The ECU also obtains an inlet airflow measurement signal from an engine inlet combustion air measurement meter. Inlet air flow varies in response to the throttle position and engine load (manifold vacuum). In addition, the ECU (1) obtains information from the throttle position switch and engine temperature sensor; (2) has a stored library of engine speed and load maps; (3) calculates the stoichiometric amount of fuel to be injected into the measured amount of inlet air; and (4) activates the fuel injector opening time to deliver the proper amount of fuel to a location just ahead of the inlet valve. After combustion in the cylinder, the exhaust gas is measured by the O_2 sensor ($\sim$3 millisecond response time), producing a millivolt signal indicating rich (insufficient oxygen) or lean (excess oxygen) combustion products. The ECU responds to deviations from the stoichiometric control point by correcting the quantity of fuel injected in the

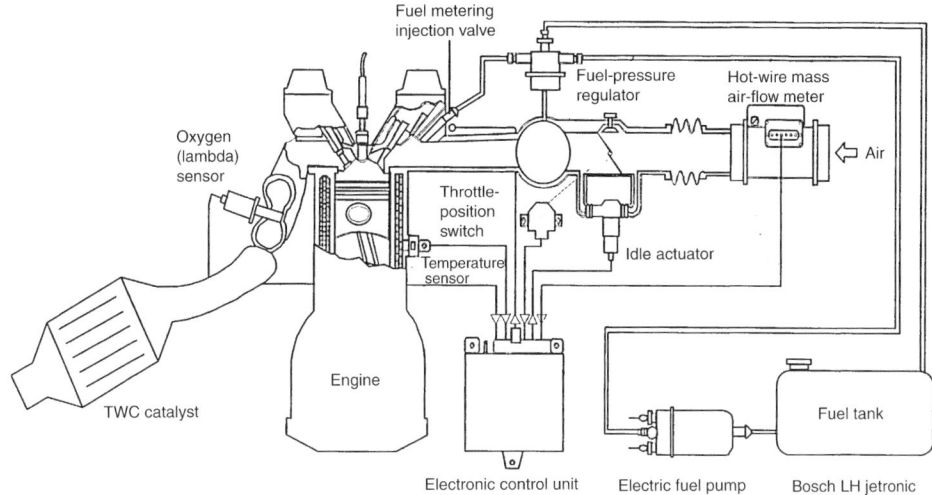

Figure 1. Basic component of the TWC closed-loop fuel metering system. Courtesy of Robert Bosch.

904   EMISSION CONTROL, AUTOMOTIVE

next fueling sequence. Thus the air/fuel ratio is constantly being adjusted slightly rich and slightly lean of the stoichiometric mixture. The three-way conversion catalyst receives exhaust gas that reflects this constant change back and forth in an intake air/fuel mixture that is stoichiometrically centered but because of the nature of the control system deviates widely in periodic fashion. The TWC is designed to operate under these conditions to convert NO_x by reduction and HC and CO by oxidation by over 95%.

THREE-WAY CATALYTIC CONVERTER: PHYSICAL DESCRIPTION

Catalytic Converter

The modern catalytic converter consists of three main components: the catalytic unit, a metal shell, and resilient mounting media. The converter shell assembly surrounds and encloses the fragile ceramic or delicate metallic catalytic unit so that exhaust gas can easily flow into the inlet through the catalytic channels and out of the outlet. The steel shell serves to protect and insulate the catalytic unit from the elements and harsh road physical environments. For the ceramic catalytic unit version, a compliant resilient layer is positioned between the outside of the catalytic unit and the inside of the converter shell grips the catalytic unit with sufficient force to prevent movement of the catalytic unit within the canister under all conditions of vibration and road shock forces and to compensate for thermal expansion differences of the catalytic unit and shell materials.

The monolithic catalytic converter now universally used by automobile manufacturers is smaller, lighter and can be mounted in any orientation with a low pressure drop. Originally it was thought that the catalyst would have to be replaced sometime during engine life, but experience has shown that a monolithic honeycomb catalytic converter with remain active and fully functional for the life of the vehicle.

Catalytic Unit

Automotive catalysts for gasoline engines are known as three-way conversion (TWC) catalysts because they destroy all three regulated pollutants: HC, CO, and NO_x. As clean vehicle technology spread throughout the world after the 1977 introduction of TWC catalysts they became the dominant system for gasoline spark-ignited engine automobiles and trucks. The active component of a catalytic unit is a thin activated catalyst layer that is applied uniformly through all flow channels of the monolithic substrate. The catalytic unit is designed to provide sufficient geometric surface so that all exhaust gas pollutants contact hot catalyst surfaces as they pass through its channels. In order to rapidly heat to operating temperature upon engine start, the catalytic unit has a low thermal mass and good heat-exchange and mass-transfer properties to extract heat and efficiently exchange pollutants to the catalyst surfaces and the resulting products back to the exhaust gas stream.

All catalytic chemical reactions occur on catalyst sites within the activated catalyst layer, which layer contains all the catalytically active ingredients consisting of precious metals, base metal promoter additives, stabilized alumina support particles, and oxygen storage components. The activated catalyst layer of a TWC is applied uniformly to all channel surfaces of the high geometric surface area substrate (ceramic or metallic). The activated catalyst layer consists of finely divided high BET surface area particles that are porous and thermally resistant.

The activated catalyst layer must adhere tenaciously to the substrate surface under a variety of extreme conditions, including rapid thermal changes, high flow, and moisture condensation, evaporation, or freezing. It must have an open porous structure to permit easy gas passage into the catalyst layer and back into the main exhaust stream. It must be able to maintain this porous structure up to exhaust temperature above 900°C.

Precious metals are deposited (impregnated) on and within the small particles throughout the TWC-activated layer. Rhodium plays an important role in the reduction of NO_x, and is Combined with platinum and/or palladium for oxidation of HC and CO. Palladium-only TWC catalysts have also been developed. Only a small amount of these expensive materials is used. The metals are dispersed on the high surface area as precious metal salt solutions, then reduced to small metal crystals. Catalytic reactions occur on the precious metal surfaces. A typical Palladium catalyst with a 20% dispersion could contain 3.0×10^{21} catalyst sites within one liter of catalyst. The small precious metal crystals can agglomerate through in-use thermal processes into a much larger diameter structure with consequent loss of catalyst site surface area. The small precious metal crystals can exist as metal crystallites or as metal oxides, both of which can be catalytic.

Heat and Mass Transfer

The heat and mass exchange processes of exhaust gas to catalytic surfaces takes place under hot pulsating exhaust flow conditions. Turbulent flow conditions (Reynolds numbers above 2000) exist in the exhaust manifold and inlet cone of the catalytic converter in response to the exhaust stroke of each cylinder (about 6 to 25 times per second) times the number of cylinders under fluctuating gas pressure.

Heat and mass transfer are driven by temperature and concentration differences, respectively, between catalyst surfaces and the bulk gas. Effective mass transfer of reactants and products commence at operating temperatures over 250°C as catalyst reactions take place with the catalytic layer.

Depending on the location of the catalytic unit in the exhaust system, total thermal mass before and including the catalytic unit, and exhaust flow and temperature, it can take from 10 to 120 seconds for the catalyst unit to reach ignition temperature of approximately 250–300°C where heat exchange processes dominate. At temperatures above the ignition point, catalyst function increases directly with temperature increase as catalyst activity increases sharply, involving both heat exchange and reactant/product mass transfer processes. Exothermic reactions contribute to temperature rise within the catalyst layer. Some catalysts reach a point where the sharp increase in catalytic activity with inlet temperature abruptly takes on an undesirable mild positive slope. Then a point is reached at which catalytic performance improves only slightly in response to further increases in temperature, where reactant mass transfer processes dominate.

THREE-WAY CATALYTIC CONVERTER: CHEMICAL REACTIONS AND SURFACE CHEMISTRY

Automobile exhaust catalysts are perfect examples of materials that accelerate chemical reactions toward thermodynamic equilibrium but are not consumed by the reactions. Reactions are completed on the catalyst surface and the products leave. Thus the auto exhaust three-way catalyst performs its functions over and over again for the life of the vehicle. The catalyst permits reactions to occur at considerably lower temperatures. For instance, CO reacts with oxygen in bulk gas phase above $700°C$ at a substantial rate. An automobile exhaust catalyst enables the CO oxidation reaction to occur at a temperature of about $250°C$ and at a much faster rate and in a smaller reactor volume. This is also the case for the oxidation of hydrocarbons and the reduction of NO_x.

The reduction of NO_x in an automotive three-way catalyst involves a catalyst property called selectivity. Selectivity occurs when several reactions are thermodynamically possible but one reaction proceeds at a faster rate than another. The catalyst chooses the route with lowest activation energy.

Two classes of metals have been examined for potential use as catalytic materials for automobile exhaust control: some of the transitional base metal series (cobalt, copper, chromium, nickel, manganese, and vanadium) and the precious metal series consisting of platinum, Pt; palladium, Pd; rhodium Rh; iridium Ir; and ruthenium Ru. The precious metals possess much higher specific catalytic activity than do the base metals.

The principal chemical reactions of a TWC catalyst are as follows:

Oxidation reactions:

$$H_2 + 1/2\,O_2 \longrightarrow H_2O$$

$$CO + 1/2\,O_2 \longrightarrow CO_2$$

$$C_mH_n + (m + n/4)\,O_2 \longrightarrow m\,CO_2 + n/2\,H_2O$$

TWC reaction at stoichiometric A/F mixtures:

$$CO + NO \longrightarrow 1/2\,N_2 + CO_2$$

$$C_mH_n + 2(m + n/4)\,NO \longrightarrow (m+n/4)\,N_2 + n/2\,H_2O + m\,CO_2$$

$$H_2 + NO \longrightarrow 1/2\,N_2 + H_2O$$

$$CO + H_2O \longrightarrow CO_2 + H_2 \quad \text{water gas shift}$$

$$C_mH_n + m\,H_2O + m\,CO + \left(m + \frac{n}{2}\right)H_2 \quad \text{steam reforming}$$

Other NO-reduction reactions:

$$2\,NO + 5\,CO + 3\,H_2O \longrightarrow 2\,NH_3 + 5\,CO_2$$

$$2\,NO + 5\,H_2 \longrightarrow 2\,NH_3 + 2\,H_2O$$

$$NO + \text{hydrocarbons} \longrightarrow N_2 + H_2O + CO_2 + CO + NH_3$$

$$\left.\begin{array}{l} 2\,NO + CO \longrightarrow N_2O + CO_2 \\ 2\,NO + H_2 \longrightarrow N_2O + H_2O \end{array}\right\} \begin{array}{l} \text{at } 200°C \text{ (below exhaust} \\ \text{gas temperature)} \end{array}$$

Fuel sulfur reactions:

$$S + O_2 \longrightarrow SO_2$$

$$SO_2 + 1/2\,O_2 \longrightarrow_{Pt} SO_3$$

$$3\,SO_3 + Al_2O_3 \underset{}{\overset{>600\,C}{\rightleftharpoons}} Al_2(SO_4)_3$$

$$Al_2(SO_4)_3 + 12\,H_2 \longrightarrow_{PM\ catalyst} 3\,H_2S + Al_2O_3 + 9\,H_2O$$

$$SO_2 + 3\,H_2 \longrightarrow_{PM\ catalyst} H_2S + 2\,H_2O$$

Where PM is precious metal.

FACTORS RELATED TO LONG-TERM DURABILITY

Catalyst Durability

The actual in-use overall durability of the three-way catalyst has proven to be very good even under extreme operating conditions. Automobile catalysts last for the life of the vehicle and still function well at the time the vehicle is scrapped. However, there is potential for decline in total catalytic performance from exposure to very high temperatures, accumulation of catalyst poisons, or loss of the active layer. Catalyst poisons and inhibitors can come from the fuel, the lubricating oil, from engine wear and corrosion products, and from air ingestion of various materials.

OTHER EMISSION CONTROL SYSTEMS

NO_x control within the engine can be achieved by recycling a fraction of the exhaust gases into the engine air inlet stream. The resultant mixture upon combustion results in lower peak combustion temperature and lowers the amount of thermal NO_x produced within the combustion flame. The degree of NO_x suppression is dependent on the amount of exhaust gas recirculation (EGR).

Gasoline fumes from the on-board fuel tank, from fuel left in the fuel delivery system, and from running losses are regulated by U.S. EPA standards. The vehicle evaporative control system typically consists of a charcoal canister that connects to the intake manifold. At engine shutoff a valve permits hydrocarbon fumes to be absorbed and stored by a carbon canister. When the engine is operating, the stored hydrocarbons are purged into the inlet manifold, where they enter the combustion process.

Exhaust gases enter the crankcase by what is known as blow-by gases that escape the combustion chamber past the piston rings or inlet and outlet valve stem seals. The exhaust gases thereby enter the atmosphere via the crankcase breather tube. As the engine wears, blow-by increases. A positive crankcase ventilation (PCV) valve is required by regulation (with specified test procedure) to assure that these blow-by gases are fed into the inlet manifold so that they pass through the engine and exhaust emission control system.

EMISSIONS CONTROL FOR ALTERNATIVE FUELS

The National Energy Policy Act of 1992 requires that nonpetroleum-based transportation fuels steadily replace petroleum fuels in the United States. Such fuels include natural gas, liquid petroleum gas (LPG), methanol, ethanol and hydrogen.

The preferred emission control TWC catalyst for natural gas fueled engines is based on Pd and Rh rather than Pt and Rh. Liquid Petroleum Gas (LPG) is a transportation fuel used for special vehicle and equipment applications. Emission control systems for LPG and typically use Pt-based (Pt-Rh) TWC catalysts which exhibit lower lightoff temperature for propane (C_3H_8), the major HC species in LPG. Emission control systems developed for methanol and ethanol transportation fuels normally contain 15% gasoline and are known respectively as M85 and E85 alcohol fuels.

DIESEL ENGINE EMISSION CONTROL

Light-duty, high speed, diesel compression ignition (CI) engine powered vehicles must meet the same emission standards as gasoline spark ignited (SI) engines. The most difficult standards for diesel engines are the particulate matter (PM) and NO_x standards whereas HC and CO are easily removed with diesel oxidation catalysts.

Emission Control Technology

A light-duty vehicle emission control system conform to Tier 2 emission light-duty vehicle emission standards is currently under development. The two major challenges are: control of NO_x and control of particle emissions.

R. B. Bird and co-workers, *Transport Phenomena*, John Wiley & Sons, Inc., New York, 1965.

Code of Federal Regulations, CFR 40 Part 86, 86.1863–07 Appendix V to Part 86—Standard Road Cycle (SEC) and Appendix VII—Standard Bench Cycle (SBC). FR, Vol. 67, No 64, April 2, 2004, pp. 17564–17568.

C. N. Satterfield, *Heterogeous Catalysis in Practice*, 1st ed., McGraw-Hill Book Co., Inc., New York, 1980.

Tier2/LEV Emission Control Technologies For Light-Duty Gasoline Vehicles, Manufacturers of Emission Controls Association, 1660 L St., NW, Washington DC 20035 www.meca.org. August 2003.

JOHN J. MOONEY
Environmental and Energy
Technology and Policy
Institute

EMISSION CONTROL, INDUSTRIAL

There are 10 potential sources of industrial exhaust pollutants that may be generated in a production facility: (*1*) unreacted raw materials; (*2*) impurities in the reactants; (*3*) undesirable by-products; (*4*) spent auxiliary materials such as catalysts, oils, solvents, etc; (*5*) off-spec product as; (*6*) maintenance; ie, wastes and materials; (*7*) exhausts generated during start-up and shutdown; (*8*) exhausts generated from process upsets and spills; (*9*) exhausts generated from product and waste handling, sampling, storage, and treatment; and (*10*) fugitive sources.

Exhaust streams generally fall into two general categories, intrinsic and extrinsic. Intrinsic wastes represent impurities present in the reactants, by-products, coproducts, and residues as well as spent materials used as part of the process. These materials must be removed from the system if the process is to continue to operate safely. Extrinsic wastes are generated during operation of the unit but are more functional in nature. These are generic to the process industries overall and not necessarily inherent to a specific process. Waste generation may occur as a result of unit upsets, selection of auxiliary equipment, fugitive leaks, process shutdown, sample collection and handling, solvent selection, or waste-handling practices.

CONTROL STRATEGY EVALUATION

There are two broad strategies for reducing volatile organic compound (VOC) emissions from a production facility: (*1*) altering its design, operation, maintenance, or manufacturing strategy so as to reduce the quantity or toxicity of air emissions produced or (*2*) installing aftertreatment controls to destroy the pollutants in the generated air emission stream. Whether the exhaust stream contains a specific hazardous air pollutant, a VOC, a nitrogen oxide, or carbon monoxide, the best way to control the pollutant is to prevent its formation in the first place. Many technologies are being developed that seek to minimize the generation of undesirable by-products by modifying specific process materials or operating conditions. Whereas process economics or product quality may restrict the general applicability of these approaches, an increased understanding of the mechanisms and conditions by which a pollutant is created is leading to significant breakthroughs in burner design and operation (for nitrogen oxide control), equipment design, maintenance, and operation for fugitive and vent VOC emission control, and product and waste storage and handling design and operation (for VOC emission control). One source category of increasing importance is that associated with fugitive emissions from equipment leaks.

FUGITIVE EMISSIONS FROM EQUIPMENT LEAKS

Emissions can occur from process equipment whenever there is contact between the process chemicals and the atmosphere. Such fugitive emissions are typically due to leakage from general plant equipment such as pumps, valves, pressure relief valves, flanges, agitators, and compressors.

Because these emissions generally occur randomly and may vary in intensity over time, they are difficult to predict, much less manage. Since a chemical process unit may contain anywhere from 5,000 to 100,000 components, measurements of equipment leak emissions actually represent only a snapshot of the leaking process. Predicting whether any one component will leak is difficult, but the statistical predictability over large numbers of components in any one facility produces a fairly consistent picture of the distribution of leak rates. The distribution of loss rates varies predictably, with only a few components leaking significantly, many leaking notably, and most others leaking a little.

Considering that these losses generally represent valuable raw materials or final products lost, controlling such emissions can represent a substantial increase in performance to the operating facility.

Rates of product loss from individual components at facilities depend on factors such as equipment design and standard of installation; age of components; access to component by operators; time since last major maintenance turnaround; specific inspection and maintenance procedures; external environmental variability (temperature, pressure) and stress; specific process temperatures and pressures; nature of chemicals in process unit; and rates and variability of production operation.

The approaches used to reduce equipment leak emissions fall into two categories: (1) modifying or replacing existing equipment with a more effective design and (2) implementing a leak detection and repair (LDAR) program. The focus of such a program will vary depending on component type, the service (gas, light liquid or heavy liquid) and previous leak history.

Pumps

Chemicals transferred by pumps can leak at the point of contact between the moving shaft and the stationary casing. Equipment modifications for pumps include (1) routing leaking vapors to a closed-vent system, (2) installing a dual mechanical seal containing a barrier fluid, or (3) replacing the existing pump with a sealless type. The control efficiency of sealless pumps and a dual mechanical seal is essentially 100% if both the inner and outer seal do not fail simultaneously.

Valves

Most common valve designs contain a valve stem to restrict or limit fluid flow. Emissions from valves occur at the stem or gland area of the valve body. Emissions from process valves can be eliminated when the valve stem can be isolated from the process fluid using sealless valves. Two types of sealless valves, diaphragm valves and sealed bellows, have been developed. The control efficiency of both types is essentially 100%.

Compressors

Compressors to transport gases through a process unit are typically driven with rotating or reciprocating shafts. The sealing mechanisms for compressors are similar to those for pumps. The dual mechanical seal system has an emissions control efficiency of 100%.

Pressure Relief Devices

Pressure relief safety devices are commonly used in chemical facilities to prevent operating pressures from exceeding the maximum allowable working pressures of the process equipment. Equipment leaks from pressure relief devices occur when material escapes from the pressure relief device during normal operation, either because the device does not properly reseat after a release, from general deterioration of the seal, or when the process pressure is operating too close to the set pressure of the pressure relief valve (PRV).

There are two primary equipment modifications that can be used for controlling equipment leaks from pressure relief devices: (1) a closed-vent system, or (2) use of a rupture disk in conjunction with the PRV. The control efficiency of a rupture disk/PRV combination is essentially 100% when operated and maintained properly. Rupture disks cannot be used on processes where solids or polymers might collect around the disk.

Flanges and Connectors

Flanges are bolted, gasket-sealed connectors that are normally used for pipes with diameters of 2.0 in. or greater. The primary causes of flange leakage are poor installation, aging, deterioration of the sealant, and thermal stress. Flanges can also leak if improper gasket material is chosen. Threaded fittings (connectors) are normally used to connect piping and equipment having diameters of 2.0 in. or less. Seals for threaded fittings are made by coating the male threads with a sealant before joining it to the female piece. Leakage can occur as the result of poor assembly or sealant application or from thermal stress.

Where connectors are not required for safety, maintenance, process modification, or periodic equipment removal, emissions can be eliminated by removing the connector and welding the pieces together.

Leak Detection and Repair Program

The LDAR program is structured to monitor equipment on a regular schedule and repair it when identified as leaking. An LDAR program is more effective on valves and pumps than for flanges and connectors, because these sources have a greater leak potential relative to their fewer numbers.

In a typical LDAR program, a portable screening device is used to identify (monitor) pieces of equipment that have the potential to emit amounts of material at levels that warrant reduction of the emissions through simple repair techniques. These programs are best applied to equipment that can be repaired online, resulting in immediate emissions reduction. The control efficiency of an LDAR program is dependent on three factors: (1) the leak definition, (2) the monitoring frequency of the LDAR program, and (3) the final leak frequency after the LDAR program is implemented.

The monitoring frequency may vary from as short as a week to several years for low leaking components such as flanges. Leak frequency is the percentage of components found to be leakers, which will decrease substantially as defective or worn-out components are identified and replaced.

The most widely used approach to control exhaust emissions is the application of add-on control devices. For organic vapors, these devices can be one of two types, combustion or capture. Applicable combustion devices include thermal incinerators, catalytic oxidization devices; flares; or boilers/process heaters. Primary applicable capture devices include condensers, adsorbers, and absorbers. A comparison of the primary control alternatives is shown in Table 1.

The most desirable of the control alternatives is capture of the emitted materials followed by recycle back into the process. However, the removal efficiencies of the capture techniques generally depend strongly on the physical and chemical characteristics of the exhaust gas and the pollutants considered. Combustion devices are the more commonly applied control devices, because these are capable of a high level of removal efficiencies.

Capture devices, discussed extensively elsewhere, include oxidation devices that are either thermal units that use heat alone or catalytic units in which the exhaust gas is passed over a catalyst, usually at an elevated temperature. The latter speed oxidation and are able to operate at temperatures well below those of thermal systems. One developing technology that is a combination of the capture and oxidaton approaches is known as biofiltration.

BIOFILTRATION SYSTEMS

Biofiltration is a new technology being increasingly considered for use in certain chemical process industries. A biofiltration system uses microbes to remove and consume pollutants from contaminated exhaust air. Such systems include bioreactors, biotrickling filters, bioscrubbers, biomembranes, and bioadsorption systems. These systems vary in the design of their beds and the amount and type of auxiliary equipment.

Almost any substance will, with the help of microbes, decompose, given the proper environment. This is especially true for organic compounds, but certain microbes can also consume inorganic compounds such as hydrogen sulfide and nitrogen oxides. The most common usage is in addressing odor nuisance where it is not necessary to meet high control efficiencies.

Table 1. Emission Control Technologies

Technology	Reduction effectiveness	Recovery	Waste generation	Advantages	Disadvantages
activated carbon adsorption	90–98%	chemical recovery possible with regeneration	spent carbon or regenerant	good for wide variety of VOCs	carbon replacement, regeneration costs, potential for bed fires
adsorption in wet scrubbers	75–90%+	chemical recovery possible through decanting/distillation	spent solvent or regenerant	simple operation	not efficient at low concentration
vapor condensation	50–80%	chemical recovery possible through decanting/treatment	liquid wastes, needs off-gas treatment	simple operation, effective for high VOC concentration	low removals applicability limits to some VOCs, high power costs
thermal oxidation	99%	heat recovery	NO_x generation, CO_2 generation	handles any VOC concentration	high operating costs, capital costs, temperatures, and maintenance
catalytic oxidation	95–99%	heat recovery	spent catalyst regeneration acids and alkalines	simple systems, lower T than thermal economical operation	fouling of catalysts, temperature limits
biofiltration	65–99%	none	some contaminated water and inert solids	low installation and operating cost, minimal waste and air pollutant generation, particularly suited for odor reduction	large area required for installation, low removals for heavy organics, some pretreatment required, potential for upsets with variable exhaust stream.

An efficient biofiltration system should use only small amounts of electrical power and some micronutrients in order to operate, and will generate water and CO_2 as by-products. Material that is indigestible becomes residue. However, because microbes are sensitive to environmental conditions, a biofiltration system is suitable only for selected applications.

Variables that effect the operation and efficiency of a biofiltration system include temperature, pH, moisture, pollutant mix, pollutant concentration, macronutrient feeding, residence time, compacted-bed media, and gas channeling.

The advantages of biofiltraion are that its installation and operation costs are usually low; secondary waste generation, pollutants, and other greenhouse gases are minimal; natural-bed media used in biofilters usually need to be replaced only every 2–5 years; and it has high destruction efficiencies for problematic odorous compounds such as aldehydes, organic acids, nitrous oxide, sulfur dioxide, and hydrogen sulfide.

Biofiltration's disadvantages include the large land requirement for traditional design; it has no continuous internal liquid flow in which to adjust bed pH or to add nutrients; excessive conditions of the exhaust stream may require pretreatment to remove water, solids, or alter temperature; bed replacement can take 2–6 weeks, depending on bed size; information on recovery of the system after an upset remains limited; it has low destruction efficiencies for some organic compounds; and exhaust organic concentrations in excess of 500 ppmv may overtax the system.

OXIDIZATION DEVICES

Thermal Oxidation

Thermal oxidation is one of the best known methods for industrial waste gas disposal. Unlike the capture methods, thermal oxidation destroys the objectionable combustible compounds in the waste gas rather than collecting them. There is no solvent or adsorbent of which to dispose or regenerate. On the other hand, there is no product to recover. A primary advantage of thermal oxidation is that virtually any gaseous organic stream can be safely and cleanly incinerated, provided that proper engineering design is used.

A thermal oxidizer is a chemical reactor in which the reaction is activated by heat and characterized by a specific rate of reactant consumption. There are at least two chemical reactants, an oxidizing agent and a reducing agent.

Thermal oxidization devices are widely used and generally provide a high degree of assurance that the process oxidizes the material in the exhaust gas. The high-temperature operation causes other problems, however, especially compared to alternatives such as catalytic oxidation. Thermal oxidation devices often incur higher fuel costs because of the higher temperatures necessary, and require exotic high temperature materials. In addition, equipment durability is reduced by the extent of thermal cycling. Further, a thermal oxidizer may produce nitrogen oxides (NO_x) and sometimes undesirable by-products such as dioxins from chlorinated materials.

Catalytic Oxidization

A principal technology for control of exhaust gas pollutants is the catalyzed conversion of these substances into innocuous chemical species, such as water and carbon dioxide. This typically thermally activated process, commonly called catalytic oxidation, is a proven method for reducing VOC concentrations to mandated levels. Catalytic oxidation is also used for treatment of industrial exhausts containing halogenated compounds.

As an exhaust control technology, catalytic oxidation enjoys significant advantages over thermal oxidation. The former often occurs at temperatures that are less than half those required for the latter, consequently saving fuel and maintenance costs. Lower temperatures allow use of exhaust stream heat exchangers of a low-grade stainless steel rather than the expensive high-temperature alloy steels. Furthermore, the lower temperatures tend to avoid the emissions problems arising from the thermal oxidation processes.

Critical factors that need to be considered when selecting an oxidation system include (1) waste stream heating value and explosive properties; (2) waste gas components that might affect catalyst performance; (3) the type of fuel available and optimum energy use; and (4) space and weight limitations on the control technology. Catalysts are favored for their small, light systems.

There are situations where thermal oxidation may be preferred over catalytic oxidation: for exhaust streams that contain significant amounts of catalyst poisons and/or fouling agents, thermal oxidation may be the only technically feasible control; where extremely high VOC destruction efficiencies of difficult to control VOC species are required, thermal oxidation may attain higher performance; and for relatively rich VOC waste gas streams, the gas stream's explosive properties and potential for catalyst overheating may require the addition of dilution air to the waste gas stream.

EXHAUST CONTROL TECHNOLOGIES

In addition to VOCs, specific industrial exhaust control technologies are available for nitrogen oxides, NO_x, carbon monoxide, CO, halogenated hydrocarbon, and sulfur and sulfur oxides, SO_x.

Nitrogen Oxides

In many cases, NO_x has a significant role in the formation of tropospheric ozone, a heavily regulated atmospheric pollutant. The production of nitrogen oxides can be controlled to some degree by reducing formation in the combustion system. Combustion control technologies include operational modifications, such as low excess air, biased firing, and burners-out-of-service, which can achieve 20–30% NO_x reduction; and equipment modifications such as low NO_x burners, overfire air, and reburning, which can achieve 40–60% reduction.

When NO_x destruction efficiencies approaching 90% are required, some form of postcombustion technology applied downstream of the combustion zone is needed to reduce the NO_x formed during the combustion process. Three postcombustion NO_x control technologies are utilized: selective catalytic reduction (SCR); nonselective catalytic reduction (NSCR); and selective noncatalytic reduction (SNCR).

Carbon Monoxide

The carbon monoxide emitted by gas turbine power plants, reciprocating engines, and coal-fired boilers and heaters can be controlled by a precious-metal oxidation catalyst on a ceramic or metal honeycomb. The catalyst promotes reaction of the gas with oxygen to form CO_2 at efficiencies that can exceed 95%. Carbon monoxide oxidation catalyst technology is broadening to applications requiring better catalyst durability, such as the combustion of heavy oil, coal, municipal solid waste, and wood. Research is under way to help cope with particulates and contaminants, such as flyash and lubricating oil, in gases generated by these fuels.

Halogenated Hydrocarbons

The first step in any control strategy for halogenated hydrocarbons is recovery and recycling. However, even upon full implementation of economic recovery steps, significant halocarbon emissions can remain. In other cases, halogenated hydrocarbons are present as impurities in exhaust streams. Impurity sources are often intermittent and dispersed.

The principal advantage of a catalytic oxidation system for halogenated hydrocarbons is in operating cost savings. Catalytically stabilized combusters improve the incineration conditions but still must employ very high temperatures as compared to VOC combusters. Alternatively, the low temperature catalytic oxidation process is typically designed for a maximum adiabatic temperature rise of only 200°C. This would correspond to only ~1,500 ppm of an organic compound in the exhaust stream. But, with the lower heat of combustion, up to 40,000 ppm of carbon tetrachloride could be treated in the same temperature rise, or with less dilution air.

Groundwater contaminated with chlorinated hydrocarbons is being remediated by a conventional air stripper or a rotary stripper, producing an air stream containing the halogenated hydrocarbon vapors and saturated with water vapor, which is then passed through a catalyst bed.

At least two catalytic processes have been used to purify halogenated streams. Both utilize fluidized beds of probably nonnoble metal catalyst particles.

USES

Catalytic oxidation of exhaust streams is increasingly used in industries involved in the following:

Surface coatings: aerospace, automobile, auto refinishing, can coating, coil coating, fabric coating, large appliances, marine vessels, metal furniture, paper coating, plastic parts coating, wire coating and enameling, wood furniture.

Printing inks: flexographic, lithographic, rotogravure, screen printing.

Solvent usage: adhesives, disk manufacture, dry cleaning, fiber glass manufacture, food tobacco manufacture, metal cleaning, pharmaceutical, photo finishing labs, semiconductor manufacture.

Chemical and petroleum processes: cumene manufacture, ethylene oxide manufacture, acrylonitrile manufacture, caprolactam manufacture, maleic anhydride manufacture, monomer venting, phthalic anhydride manufacture, paint and ink manufacture, petroleum product refining, petroleum marketing, resin manufacture, textile processing.

Industrial/commercial processes: aircraft manufacture, asceptic packaging, asphalt blowing, automotive parts manufacture, breweries/wineries, carbon fiber manufacture, catalyst regeneration, coffee roasting, commercial charbroiling, electronics manufacture, film coating, filter paper processing, food deep frying, gas purification, glove manufacture, hospital sterilizers, peanut and coffee roasting, plywood manufacture, rubber processing, spray painting, tire manufacture, wood treating.

Engines: diesel engines, lean burn internal combustion, natural gas compressors, oil field steam generation, rich burn internal combustion, gas turbine power generation.

Cross media transfer: air stripping, groundwater cleanup, soil remediation (landfills), hazardous waste treatment, odor removal from sewage gases.

The most important factors affecting performance are operating temperature, surface velocity, contaminant concentration and composition, catalyst properties, and the presence or absence of poisons or inhibitors.

Air Stripping of Groundwater

Treatment of exhaust streams from the air stripping of contaminated groundwater is a particular challenge, because the emissions from air stripping units may consist of a complex mixture of both fuel and solvent fractions.

Printing and Graphic Arts

In the graphic arts industry, the catalyst in the oxidizer needs to be monitored regularly, because it is susceptible to contamination by phosphorus from fountain solutions, silica from silicone gloss enhancer sprays, and chlorides from chlorinated solvents or blanket wash solutions. Phosphorus and silica accumulate most rapidly on the leading edge of the catalyst bed, deactivating the catalyst by masking the precious metals. In a fluidized-bed configuration, the catalyst surface is continually renewed by abrasion and the problem of masking the catalyst surface with silicones is avoided.

Chemical Processing

The control of exhaust from production of pure terephthalic acid (PTA) has been a challenge. Vent gases from the process of producing high-grade polyester contain such by-products as methyl acetate, organic acids, and often methyl bromide. The presence of methyl bromide limited the use of fixed-bed catalytic oxidation as a control technology using precious-metal catalysts. Thus, base metal catalysts in fluidized-bed reactors have been the primary catalytic technology of choice. In this application, the continuous abrasion of the outer layer of the catalyst particles exposes a fresh surface of unpoisoned material to the reactants, allowing the catalyst to treat the exhaust stream effectively.

Coatings Industries

Surface coating processes produce similar air pollution problems in a number of different industries.

An internal coating is necessary to protect the purity and flavor of can contents for beverages or any edible product that might react with the container metal. Catalytic oxidation systems are used by the principal can manufacturers to treat coatings exhaust streams. Coil coating is the prefinishing of many sheet metal items with protective and decorative coatings that are applied by roll coating on one or both sides of a fast-moving metal strip. Coil coaters operate equipment continuously and, in most cases, operate catalytic fume abaters 6,000–7,000 h/year. Under these conditions the anticipated catalyst life is years, with an annual aqueous solution cleaning. However, the catalyst may last no more than two years if frequent maintenance is needed, such as in-place air lancing every 60–90 days to remove noncombustible particulates.

Catalysis is utilized in the majority of new paper fuel oil and air filter cure ovens as part of an oven recirculation/burner system designed to keep the oven interior free of condensed resins and provide an exhaust without opacity or odor. The application of catalytic fume control to the exhaust of paper-impregnation dryers permits a net fuel saving by oxidation of easy-to-burn methyl or isopropyl alcohol, or both, at adequate concentrations to achieve a 110–220°C exotherm.

ENVIRONMENTAL TECHNOLOGY VERIFICATION

In 1995, the U.S. Environmental Protection Agency (EPA) established the Environmental Technology Verification (ETV) program to evaluate the truth of a commercially ready environmental technology's performance under specific, predetermined criteria and adequate quality assurance procedures. As of 2003, this program had five operational centers to evaluate the technologies being reviewed, including greenhouse gas technology, air pollution control technology, and pollution prevention processes.

R. L. Berglund, "Maintenance Operations and Pollution Prevention" H. F. Freeman, ed., in *Industrial Pollution Prevention Handbook*, McGraw-Hill, Inc., New York, 1995.

T. A. Melnick, R. Sommerlad, R. Payne, and T. Sommer, "Low NO$_x$ Burner Modifications for Cost Effective NO$_x$ Control", Forum 2000 on Cutting NO$_x$ Emissions, Institute of Clean Air Companies, Arlington, Va., March 2000.

R. G. Silver, J. E. Sawyer, and J. C. Summers, eds., "The 1990 Clean Air Act and Catalytic Emission Control Technology for Stationary Sources," *Catalytic Control of Air Pollution: Mobile and Stationary Sources*, ACS Symposium Series 495, 1992.

United States Environmental Protection Agency, Using Bioreactors to Control Air Pollution, EPA-456/R-03-003, September 2003.

RONALD L. BERGLUND
Terracon

EMULSIONS

Emulsions are mixtures of immiscible liquids wherein one liquid is finely dispersed and suspended in the other. A (macro)emulsion has a broad distribution of droplet diameters in the 0.05–100-μm range and is turbid and minimally stable. A microemulsion has structural diameters in the 10–50-nm range and is clear, isotropic, and thermodynamically stable. Much theoretical and experimental work has been done on microemulsions, but their practical application has been somewhat limited compared to macroemulsions. The term "miniemulsion" has also appeared, to describe a subset of macroemulsions in which the particle diameter range of 50–400 nm is achieved by the use of mixed emulsifiers comprised of a surfactant and water-insoluble long-chain alcohol costabilizer such as cetyl alcohol. In this discussion, "emulsion" will refer to macroemulsions and "oil" and "water" to the two, immiscible phases.

Emulsifiers aid in the formation and stabilization of emulsions. Emulsifiers are amphiphilic molecules, containing both hydrophilic and lipophilic groups, which provide the molecule with some affinity for both the disperse phase and the continuous phase. Emulsion stabilizers are polymeric molecules of higher molecular weight which form a protective steric layer around the dispersed droplets and also have some affinity for both phases.

FORMATION AND STABILIZATION

There is a very large increase in interfacial surface area when an emulsion is formed from layers of two immiscible liquids. To accomplish this, work must be done, some of which remains in the system as potential energy and some of which is dissipated as heat. Emulsions are inherently thermodynamically unstable. Upon formation, the system naturally tries to minimize its energy by minimizing the interfacial surface area, which results in spherical droplets. In most cases, when mechanical energy is used to mix two pure immiscible liquids, the droplets have such high interfacial tension that the system reverts quickly to two layers, the lowest energy configuration. The addition of an emulsifier tends to lower the interfacial tension, thus allowing emulsions to form more easily. Emulsifiers also stabilize the droplets enough to have a finite, useful lifetime by providing a protective barrier against coalescence (ie, recombination).

Because emulsions are metastable, it is a question of when, not if, they will lose stability. Therefore, the preparation of emulsions is the challenge of formulating a system that is at least stable enough to meet the demands of the particular application for the required timescale. In practice, some emulsions can be stable for years, whereas others exist for a very short time. For example, a simple salad dressing of vinegar, vegetable oil, and spices often barely exists as an emulsion for long enough, after shaking, to pour it onto a salad. In contrast, mayonnaise (an oil/water, or o/w emulsion) is stable for months. The stability and physical characteristics of emulsions depend on many variables, including the nature of the liquids, temperature, additive types and amounts, order of addition of additives, and the type and amount of mechanical or chemical energy used to form the emulsion.

Emulsifiers

Emulsifiers are the key to producing stable emulsions. Emulsifiers have two main functions: lowering interfacial tension, which makes droplet formation less energy intensive, and providing colloidal stability to the droplet by forming an electrically charged layer at its interface with the continuous phase. There are three main types of stabilizers for emulsions: monomeric surfactants, polymers, and particles. To attain stability, the emulsifiers need to cover the surface of the droplets completely and remain firmly adsorbed. The amount of emulsifier needed for a given emulsion is highly dependent on the droplet size (ie, surface area), the emulsifier efficiency, and the composition of the two phases.

Formation

The most common method of preparing emulsions employs mechanical shear to achieve droplet breakup (Figure 1). When the emulsion components are subjected to high mechanical shear and turbulent flow, it is the size distribution of the micro-eddies in the flow pattern that influences the emulsion droplet size distribution. The process of emulsification is extremely complex from a modeling perspective due to the hydrodynamics and the multitude of interacting variables.

Stabilization

In order for an emulsion to be stable, the droplets need to exist as discrete entities uniformly dispersed in the contin-

uous phase. If the droplets are not sufficiently stabilized, they will coagulate when they collide during mixing or through Brownian motion. This process, called coalescence, leads to gross separation of the two liquid phases or "breaking" of the emulsion. Simple mixing usually cannot reestablish the emulsion. It is also possible for droplets to maintain their integrity but rise to the top due to gravity if the droplets are less dense than the continuous phase. This process, known as creaming, can usually be reversed, at least temporarily, by simple mixing.

Raising the viscosity of the continuous phase will slow the creaming process, as will reducing the droplet size or adjusting the density of the droplet to match closely the density of the external phase. Droplets may also retain their identity, but form multidroplet aggregates. This is known as flocculation and can usually be reversed, at least temporarily, by simple mixing. Flocculated droplets are not actually in physical contact with each other but are held in close proximity in a shallow potential energy minimum.

Emulsion droplets can possess kinetic energy from several sources, including mechanical mixing, gravity, and thermal (Brownian) motion. When two droplets collide, they will coalesce unless there is a sufficiently high potential energy barrier to keep them apart. This is the key to emulsion stability. When no emulsifier is present, the main force acting between the two droplets in close proximity is van der Waals attraction, which is basically the universal phenomenon that "all matter attracts matter". The attractive force pulls the droplets together into the primary energy minimum resulting in coalescence.

In addition to electrostatic stabilization, droplets can also be stabilized sterically by using nonionic oligomeric surfactants or polymers at the interface to form a physical rather than a charge repulsive barrier.

If both electrostatic and steric stabilizers are used, a compound curve results. Electrosteric stabilization is often the most robust type, because it takes advantage of both mechanisms in case the emulsion is subjected to multiple instability stresses such as high electrolyte and temperature fluctuations.

Emulsions are stabilized by establishing electrostatic and/or steric barriers at the droplet surface. Both types of barriers are effective for o/w emulsions, but only steric barriers are effective in w/o emulsions due to the low dielectric constant of the continuous phase. The structure of the interfacial region is quite complex.

Emulsions are never perfectly unimodal, so there are both larger and smaller droplets in a given sample. There is a driving force for the migration of the internal phase from the smaller to the larger droplets. The pressure (or chemical potential) of molecules inside a droplet is inversely proportional to the radius of the droplet.

Figure 1. When a droplet undergoes shear, it distorts and breaks into smaller droplets. Very small droplets are formed by tip-streaming, leading to an overall broad size distribution.

The result is a coarsening of the emulsion as it minimizes interfacial surface area. It is surprising that this can happen even when the solubility of the internal phase is as low as a few ppm in the external phase. The growth of the large droplets at the expense of the small droplets can lead to changes in rheology, efficacy, and physical appearance (eg, creaming).

Biodegradation

Emulsions often contain additives that are subject to biodegradation. The destruction of biodegradable stabilizers and rheology modifiers will lead to degradation of performance properties at best and complete destabilization of the emulsion at worst. Sometimes emulsifiers with biocidal properties such as those based on quaternary ammonium salts can be used in the formulations. If that is not possible, a small amount of biocide may be required to preserve the integrity of the emulsion. When oxidative degradation is an issue, antioxidants will also have to be part of the formulation.

Microemulsion Stability

Microemulsions are thermodynamically stable isotropic solutions. Microemulsions differ from macroemulsions in that their oil–surfactant components can exist as cylindrical or bicontinuous structures in addition to spheroids. The stability and structure of nonionic microemulsions is governed by temperature. Microemulsions may also lose stability when diluted.

Destabilization

In certain instances, destabilizing an emulsion (ie, demulsification) is desirable. For example, in tertiary oil recovery the w/o emulsion needs to be broken to recover the crude oil. Also, unwanted emulsions may form during a process and will need to be eliminated. Purely mechanical means of destabilization include centrifugation, filtration, freezing, and application of mechanical shear. Addition of a surfactant that will disrupt the balance of the system and the interfacial structure can also lead to destabilization.

PREPARATION

The formulator must know the purpose of the emulsion formulation and how it will be used. Some important considerations are stability requirements (ie, shelf life), cost, rheology, ease of preparation, environmental conditions for use and storage (eg, temperature extremes), regulatory restrictions, and safety.

Formulation

Bancroft's rule has been known for almost a century: Surfactants that partition preferentially into the water phase tend to form o/w emulsions and surfactants that partition preferentially into the oil phase tend to form w/o emulsions. In addition, experience has shown that two or more emulsifiers are usually needed to attain the most robust stability and this often includes both ionic

and nonionic emulsifiers. The most useful system for choosing emulsifiers is the HLB system where HLB stands for hydrophile–lipophile balance and is a way to match the characteristics of the liquid to be emulsified with the most effective emulsifiers. Unfortunately, just getting an HLB match is often not sufficient to produce an acceptable emulsion formulation. The chemical compositions of the emulsifiers and the liquid to be emulsified also need to be matched to gain the compatibility needed for stability. One problem with the HLB system is that it is temperature sensitive, so the formulator needs to know the stable temperature range of the emulsion.

Equipment and Methods

Mechanical energy is needed to turn the components of a formulation into an actual emulsion. The process is called emulsification; the energy requirements can cover a wide range, depending on the application. The most common mechanical emulsification equipment includes colloid mills, high pressure homogenizers, and jet impact devices. Colloid mills are rotor-stator devices that rely on very high shear rates (speeds up to 20,000 rpm) to accomplish droplet breakup.

Traditional colloid mills utilize a cone-shaped rotor, whereas more recent designs can have concentric cylinders with complicated tooling to aid in droplet breakup and turbulent flow (eg, Ultra-Turrax). In high pressure homogenizers, the emulsion components are forced through a small orifice at between 1000 and 5000 psi. Other emulsification methods include using ultrasonics, electrostatics, electrocapillarity, and aerosols. Sometimes chemical energy is sufficient to produce an emulsion, but use of mechanical energy is still the usual method for achieving emulsification.

HEALTH AND SAFETY AND ENVIRONMENTAL ISSUES

Emulsions can help reduce VOCs by replacing organic solvents with water. A good example of this is the use of latex paints in place of solvent-based paints for many coatings applications, from house paint to automotive coatings. Latex-based adhesives is another example. As technologists continue to improve the chemistry, more emulsion-based systems will be introduced, thus improving the quality of our work and living environment. A serious environmental issue is the cleanup of oil spills. Oil tends to form relatively stable w/o emulsions with sea water. Cleanup and demulsification of this viscous, persistent pollutant is a challenge because emulsion compositions vary with type of crude oil and the conditions of formation. Demulsifiers in the HLB range of 8–11 have been found to have some utility in this application.

USES

Emulsions are encountered in a very wide range of applications in our daily lives. Some emulsions such as milk and latex from the rubber tree have been around for millions of years, whereas many emulsions were developed during the technology revolution of the nineteenth and twentieth centuries.

Food

The most familiar food emulsions are dairy products. These include the o/w emulsions milk and cream and the w/o emulsions butter and margarine. Mayonnaise is an o/w emulsion prepared in a colloid mill wherein the oil is stabilized by egg yolk and mustard. The egg yolk provides the emulsifiers lecithin and cholesterol and the mustard serves as a particle-type stabilizer. Creamy salad dressings are w/o emulsions thickened with xanthan gum. Typical thickeners found in food emulsions are gums such as xanthans, carageenans, pectinates, and alginates. The type of stabilizers that are most often used include proteins, lecithin, cellulosics, starches, and gelatin. Some food products like baked goods start as emulsions but end up as a more complex system.

Personal Care and Household Products

Cold cream, vanishing cream, hand lotion, and deodorant cream are among the emulsions found in the cosmetic and personal care product line. Some of the advantages of emulsions include efficient cleansing action, ease of application, and the ability to apply both water- and oil-soluble ingredients at the same time. Cosmetic emulsions tend to have many ingredients such as fragrances, humectants, pigments, moisturizers, thickeners, preservatives, and pH adjusters, in addition to the usual oils, waxes, emulsifiers, and water. Emulsions can also be found in the area of household cleaning and maintenance products. Many furniture, floor, and automobile polishes are oil or wax emulsions in water.

Emulsion (Latex) Polymers

Latex polymers are used as binders in latex paint, in adhesives, and in paper coatings.

Pharmaceuticals

Emulsions are being used increasingly as delivery systems for pharmaceutical treatments. The emulsion allows intravenous administration of hydrophobic active ingredients, which are then absorbed more readily due to the small droplet size. Slow release can also be built into the emulsion formulation to prolong the effect of the drug or to reduce its toxicity. Oil and water soluble components can be mixed and delivered in the same formulation.

Agricultural Chemicals

A common type of agricultural formulation is known as emulsifiable concentrates (ECs), nonaqueous formulations comprised of a pesticide active ingredient, a solvent, and an emulsifier mixture. They are not emulsions as supplied, but easily and quickly form emulsions when added to water with a minimum amount of agitation. The ECs are sold to both farmers and homeowners as a convenient way to deliver pesticide to the target crop.

Oil Recovery

When oil is recovered from sedimentary formations by conventional means, more than one-half of it can be left behind in the rock. This oil is very difficult to remove because it is coating the rock surfaces and not free-flowing. Surfactant-based systems that have been developed to enhance the recovery of the trapped oil are pumped underground, to form microemulsions, bicontinuous structures, and possibly very fine macroemulsions, with the oil. Once the emulsified oil is removed from the ground, the emulsion is broken and the oil recovered.

Asphalt Emulsions

Asphaltic bitumen emulsions are used for a number of applications where water repellency is important as in road surfacing, where the relatively low viscosity of the ~50% internal phase emulsion allows for easy application of the very viscous asphalt. Final properties depend on the emulsion breaking so that a durable highway surface is formed. Other uses for asphalt emulsions include roof coatings, insulating coatings, and as a water resistant treatment for paper and fabric.

FUTURE TRENDS

Emulsion formulation and application is still a very active technology. Work will no doubt continue on lowering VOCs, thus reducing our dependence on organic solvents. Efforts are being made to develop "smart" emulsions that could react or adjust to their environment. Applications could be in materials science, biomedical engineering, or drug delivery systems. Self-assembly of colloidal systems is also an area of intense activity.

P. Becher, ed., *Encyclopedia of Emulsion Technology*, Vols. 1–4, Marcel Dekker, New York, 1983–1996.

D. McClements, *Food Emulsions: Principles, Practice, and Techniques*, CRC Press, 1998.

D. Meyers, *Surfaces, Interfaces, and Colloids*, VCH Publishers, New York, 1991, p. 222.

C. Solans, R. Pons, and H. Kunieda, in C. Solans and H. Kunieda, eds., *Industrial Applications of Microemulsions*, Marcel Dekker, New York, 1997, p. 1.

EDWARD KOSTANSEK
Rohm and Haas Co.

ENERGY MANAGEMENT

The chemical industry is inherently energy intensive, and in the U.S. energy costs for the industry are equivalent to ~5% of the value of shipments, although this percentage is typically higher for bulk chemicals. Effective energy management is therefore an important factor in ensuring profitability.

Energy management has many dimensions. Its overall goal is to provide, at minimum cost, the heat and power needed to operate a chemical facility. However, in pursuing that goal the energy manager needs to understand the basic physics and chemistry, as well as the commercial aspects, of providing energy. Reliability, environmental impacts, and selection of utility systems and process equipment are also important issues. Finally, energy management often includes organizing and implementing energy efficiency programs, which are usually focused on improving the operation of existing facilities.

ENERGY AND THE CHEMICAL INDUSTRY

The chemical industry uses energy to supply heat and power for plant operations. In addition, hydrocarbons are used as a raw material for the production of petrochemicals, plastics, and synthetic fibers, and these are customarily reported in energy-equivalent terms.

Excluding hydrocarbon feedstocks, most of the energy use in chemical processing falls into one of two categories: (1) thermal energy (heating and cooling), which is used primarily to drive reaction and separation processes; and (2) mechanical energy (often derived from electrical power), which is used primarily to move materials, most commonly using pumps and compressors.

Most energy is supplied to processing facilities either as imported fuel or electricity. Steam is often used as a medium for transporting energy, and in many cases steam turbines or gas turbines drive individual pumps or compressors, rather than using electricity. Electric power is also often generated within the chemical plant or by a third-party cogenerating facility, to reduce dependence on imported electric power.

By-product Energy

In a number of commercially important chemical processes, by-product energy from feedstock oxidation dominates purchased fuel and electricity.

Fuels

Based on 1998 data, the industry is the largest single consumer of natural gas (>26% of the domestic manufacturing total) and uses virtually all the liquefied petroleum gas (LPG) consumed in U.S. manufacturing. Nearly all LPG and about one-fourth of natural gas are used as feedstocks. Other energy sources include by-products produced onsite (eg, off-gases from the acrylonitrile process) hot water, and purchased steam.

Electricity

Electricity, including the losses associated with production, represents 24% of the total energy used by the chemical industry. Increases in electrical costs have provided the driving force for increased cogeneration, ie, the combined production of heat and power.

Energy Efficiency Improvements

Two major forces are largely responsible for the improvement in energy efficiency: technological progress and cost optimization. Technological progress is a long-term trend that yields improved designs that reduce the inherent demand for energy in chemical processes. Cost optimization is a short-term trend that responds to price swings.

Whereas energy conservation is an important component of cost reduction for the chemical industry, conservation is rarely the only driving force for technological change. Much of the increased energy efficiency comes as a by-product of changes made for other reasons, such as higher quality, increased product yield, lower pollution, increased safety, and lower capital.

Energy and the Environment

The impact of energy usage on gaseous emissions is an important environmental issue, and regulatory action has required emission reductions in NO_x and SO_2. The best way to reduce NO_x emissions is by improving energy efficiency and thus eliminating the need to fire fuel. Where fired heating cannot be eliminated, generation of NO_x can be reduced by lowering the temperature of combustion and limiting excess oxygen. Technologies such as low NO_x burners (and now "Ultra-low NO_x" burners) and flue gas recirculation are typically used. Alternatively, NO_x emissions can be reduced after the combustion process, using either selective catalytic reduction (SCR) or selective noncatalytic reduction (SNCR).

The most desirable way to reduce SO_2 emissions is also by eliminating the need to fire fuel. Where this is not possible, SO_2 can be controlled either by changing to a low sulfur fuel or by flue gas scrubbing.

The Kyoto Accord of 1997 calls on nations to reduce their overall emissions of greenhouse gases to at least 5% below 1990 levels in the period 2008–2012. Regulations at the national, state, and local levels also require industry to reduce emissions. Energy conservation directly reduces hydrocarbon combustion. The elimination of fugitive hydrocarbon emissions as a result of improved maintenance procedures is also a tangible step that the industry is taking.

ENERGY TECHNOLOGY

Energy management requires the merging of such technologies as thermodynamics, process synthesis, heat transfer, combustion chemistry, and mechanical engineering.

Thermodynamics

The first law of thermodynamics, which states that energy can neither be created nor destroyed, dictates that the total energy entering an industrial plant equals the total of all of the energy that exits it. Feedstock, fuel, and electricity count equally, and a plant should always be able to close its energy balance to within 10%. If the energy balance does not close, there probably is a big opportunity for saving.

The second law of thermodynamics focuses on the quality, or value, of energy. The measure of quality is the fraction of a given quantity of energy that can be converted to work. What is valued is the ability to do work.

Unlike the conservation guaranteed by the first law, the second law states that every operation involves some loss of work potential. The second law is a very powerful tool for process analysis, because it tells what is theoretically possible and pinpoints the quantitative loss in work potential at different points in a process. Typically, the biggest loss that occurs in chemical processes is in the combustion step. The second law can also suggest appropriate corrective action. For example, in combustion, preheating the air or firing at high pressure in a gas turbine, as is sometimes done for an ethylene cracking furnace, improves energy efficiency by reducing the lost work of combustion.

There has been a historic bias in the chemical industry to think of energy use in terms of fuel and steam systems. A more fundamental understanding of the factors that affect energy efficiency and cost is to minimize the loss of work potential or excess use of work, which is much harder to spot but often is larger and more easily corrected.

Steam Systems and Power Recovery

Steam is the most common medium for distributing energy within chemical complexes. Steam at high pressure can be let down through a turbine for power. The shaft work developed by the turbine is sometimes referred to as by-product power, the process as cogeneration.

By-product power from a steam turbine typically takes only 40% as much incremental energy to produce as on-purpose firing for power only. In some cases there are cost savings if steam turbines can be directly coupled to power consumers, such as pumps or compressors. However, newer plants typically utilize back-pressure turbines only in applications where efficiencies >70% can be attained. This is often accomplished by installing large (>1000 kW) turbines driving electric generators and distributing the electricity, rather than using steam turbines to drive individual items of equipment. Small steam turbines are used only where they are necessary for the safe shutdown of the unit. Multistage turbines are used even on these smaller loads, to improve efficiency. Many large plants also have some condensing turbines to handle process and seasonal swings and provide some flexibility to the steam blance.

However, the scope for generating by-product power is limited by the demand for process heating, and for large process operations the demand for power is usually far greater than the simple steam cycle can produce. Many steam system design decisions fall back on the question of how to raise the ratio of by-product power to process heat. One simple approach, which allows a modest increase in power generation, is to limit the turbines that are used to extract power to large sizes, where high efficiency can be obtained. Another way to raise the power/heat ratio is by raising the pressure of the steam system.

The combined cycle first fires fuel into a gas turbine and greatly increases the power extracted per unit of steam produced. The big advantage of the gas turbine in cogeneration is that it permits a much higher ratio of power to heat. This ratio, which is routinely >0.8, gets bigger as the unit size of the turbine increases. The ratio of power output to heat input is larger for aero-derivative systems, which are basically jet engines exhausting into power recovery turbines.

Gas turbine cogeneration is typically higher cost in capital for power delivered than steam turbine systems, but has inherently low comparative costs for heat delivered. Rising natural gas prices have driven the need for higher efficiencies in both power generation and heat recovery from gas turbine exhausts.

Most gas turbine applications in the chemical industry are tied to the steam cycle, but gas turbines can be integrated anywhere in the process where there is a large requirement for fired fuel. The combined cycle is also applicable to dedicated power production.

Steam turbines offer by far the largest opportunity for power recovery from pressure letdown in chemical plants. However, there are opportunities to recover power using expanders on various process vapor streams, such as tail-gas in nitric acid plants and on catalytic crackers.

Energy Balances and Heat Recovery

An energy balance is a summary of all of the energy sources and all of the energy sinks for a unit operation, a process unit, or an entire manufacturing plant. Table 1 gives an energy balance for a simple propane-fired product dryer. The energy balance is the basic tool for analyzing an operation for energy conservation opportunities such as operational changes, system reconfiguration, and equipment alterations. Development of an energy balance is therefore an essential step in any meaningful energy efficiency program.

Table 1. Product Dryer Analysis Heat Balance

Material	Mass, kg/h	Energy, MJ/h[a]
Inputs		
fuel, C_3H_8	130	6,553
air		
combustion	6,817	106
secondary	14,846	232
in-leakage	4,289	67
water with product	1,731	354
dry product solids	4,478	249
Totals in	*32,291*	*7,561*
Outputs		
water vapor		
from product	1,445	3,808
from combustion of H_2	212	560
air and combustion products	25,870	1,933
dry product solids	4,478	458
water with product out	286	146
heat losses		656
Totals out	*32,291*	*7,561*

[a]To convert J/h to Btu/h, multiply by 0.95×10^{-3}.

The goal of heat recovery is to ensure that energy does the maximum useful work as it cascades to ambient. In most chemical process plants, the steam system is the integrating energy system. Recovering waste heat by generating steam makes the heat usable in any part of the plant served by the steam system.

Heat exchange is commonly used to cool the product of a thermal process by preheating the feed to that process, thus providing a natural stabilizing, feed-forward type of process integration. Product-to-feed interchange is common on reactors as well as distillation trains.

Flue gas to air exchange is extremely important because of the large loss associated with the combustion of unpreheated air. This exchange process has generated unique types of hardware such as the Ljungstrom or rotary wheel regenerator; the brick checkerwork regenerators used in metallurgical furnaces; hot oil, or hot water belts (also called "liquid runarounds") and heat pipes. Liquid runaround systems make it practical to use a finned surface on both gas exchange surfaces. These are particularly useful for retrofits because of the ability to move the heat to physically separated units.

Heat pumps use a compressor to boost the temperature level of rejected heat. This can be accomplished in several different ways. Heat pumps can be a very effective means of recovering heat and making it reusable, especially in small plants where there are few opportunities for conventional heat interchange.

DESIGN OF UTILITY SYSTEMS

Steam

The steam system serves as the integrating energy system in most chemical process plants. Steam holds this unique position because it is an excellent heat-transfer medium over a wide range of temperatures. Water gives high heat-transfer coefficients whether in liquid phase, boiling, or in condensation. In addition, water is safe, nonpolluting, inexpensive and, if proper water treatment is maintained, noncorrosive to carbon steel.

The steam balance is usually the most important plant-wide energy balance. A complete balance should show each service requirement, including the use of steam as a working fluid to develop power.

In a process plant, various condensate drainage devices can be used to recover and return condensate. These devices include steam traps, combination pump/traps, level pots and condensate pumps. Proper selection and installation of these products is required to ensure the system operates with maximum energy efficiency and operating productivity.

Electrical

A plant's electrical system consists of the utility company's entry substation, any in-plant generating equipment, primary distribution feeders, secondary substations and transformers, final distribution cables, and various items of switch-gear, protective relays, redundant systems, motor starters, motors, lighting control panels, and capacitors to adjust the power factor.

By far the largest single opportunity for improving overall energy efficiency in the chemical industry is cogeneration. There are also opportunities to reduce electricity costs associated with facilities improvements.

Other Energy Systems

Chemical plants usually require cooling water, compressed air, and fuel distribution systems. Refrigeration, pressurized hot water, or specialized heat-transfer fluids such as Therminol liquid or condensing vapor are sometimes also needed. Each of these systems serves the process, reliability being the most important characteristic.

KEY PROCESS EQUIPMENT ITEMS

Virtually all chemical processing is energy driven, but in separations such as distillation, drying, and evaporation this is particularly clear. All three of these processes are simple thermal operations that involve separation through vaporization and only a minor change in the chemical energy of the products. Energy, this time in the form of mechanical work, is also central to compression and pumping—operations that physically move materials. Boilers and furnaces are the main unit operations for obtaining heat from fuel, and heat exchangers transfer heat between process streams, so they are also considered key process equipment items for energy management.

A major concern is to design and operate these equipment items, together with their ancillaries, at the highest possible energy efficiency. However, in almost all cases there is a balance between capital and energy costs, and typically one is traded against the other to achieve the lowest overall cost.

Distillation

The optimum reflux rate for a distillation column depends on the value of energy but is generally between 1.05 and 1.25 times the minimum reflux rate that could be used with infinite trays. At this level, excess reflux is a secondary contributor to column inefficiency. However, when designing to this tolerance, correct vapor-liquid equilibrium data and adequate controls are essential.

Energy savings from improved control are surprisingly high, because advanced control schemes, based on process computers and on-line analyzers, permit a reduction in the margin of safety that the operators use to handle changes in feed conditions. One key element is the use of feed-forward capability, which automatically handles changes in feed flow and composition.

The real work used in a distillation column varies with the temperature difference between the heating medium and the cooling medium.

An important factor that sets the temperature differential across a distillation column is the pressure drop in the column and its auxiliaries. One way to reduce this is by using special structured packings, which give an extremely low (10% of an equivalent column with trays) pressure drop. This energy benefit can show up in an overhead temperature high enough to permit generation of by-product steam. It can also show up in a variety of other ways including lower bottoms temperature, yielding less fouling and product degradation to by-products, as in the styrene–ethylbenzene separation.

One way to reduce energy use in distillation is by means of double-effect distillation, which uses the overhead vapor from one column as the heat source for another column such that the second column's reboiler becomes the first column's condenser. This reduces the energy requirement by roughly one-half, because external heat is supplied to only one of the units.

Drying

The typical dryer mass and energy balance shown earlier shows that the heat loss is 10% of the fuel input. Improving insulation is one of the simplest ways to reduce energy input. Another simple way to reduce energy input is improving the dewatering of the feed.

Some of the other energy conservation approaches applicable to dryers are heat interchange between the stack vapor and the incoming dryer air; recovering sensible heat from the product; use of waste heat from another operation for air preheat; and using less, but hotter, drying air. This last is limited to non-heat-sensitive materials.

Evaporation

A single-effect evaporator produces slightly less than a kilogram of water vapor per kilogram of steam. By using the vapor produced by the first-effect as the heat source for a second-effect evaporator, steam use can be essentially halved. The performance can be improved almost in proportion to the number of effects employed.

In some cases reverse osmosis can be a viable alternative to process evaporation. It can be used in desalination applications, and is particularly attractive where the inlet stream is >99% water.

Compressors

Compression equipment accounts for a large fraction of power use as well as a large fraction of installed capital in many chemical plants. Usually, the energy bill for a compressor is large enough to warrant a very visible monitor of the driver, such as a control room electric meter.

Pumps

Energy use for pumps can best be controlled by designing pipes, fittings and control systems for low pressure drop, using high efficiency pumps and motors, and ensuring the pumps themselves are not over-sized for the service. In some case, especially where flow rates are changeable, variable frequency drives can be effective in reducing power consumption for both pumps and compressors.

Vacuum Systems

Vacuum systems in chemical facilities are often overspecified. It is therefore often possible to achieve significant energy savings with little or no adverse impact on process

operations by derating the vacuum system. This has the added benefit of reducing the amount of effluent that has to be treated.

Boilers and Process Furnaces

Boilers and process-fired heaters are the entry point for the energy released from burning fuel. Fuel combustion is irreversible, and fired heaters are typically the principal loss point for work potential. Air preheat cuts energy losses by reducing fuel firing and increasing the flame burst temperature.

A more obvious energy loss is the heat to the stack flue gases. The sensible heat losses can be minimized by reduced total air flow; ie, low excess air operation. Lowering the discharge temperature via increased heat recovery in economizers or air preheaters also minimizes flue gas losses.

Heat Recovery Equipment

Factors that limit heat recovery applications are corrosion, fouling, safety, and cost of heat-exchange surface. Most heat interchange utilizes shell and tube-type units because of the rugged construction, ease of mechanical cleaning, and ease of fabrication in a variety of materials. However, there is a rich assortment of other heat exchangers. Examples found in chemical plants in special applications include plate heat exchangers, brazed-fin aluminum cores, and spiral plate construction.

Insulation

A surprisingly important capital element of energy management is insulation. On large projects the capital cost of insulation is in the same range as that for heat exchangers or distillation towers and trays.

Insulation provides other functions in addition to energy conservation. A key role is safety. It protects personnel from burns and minimizes hot surfaces that could ignite inflammables. It also protects equipment, piping, and contents in event of fire.

ENERGY EFFICIENCY PROGRAMS AND ACTIVITIES

The most obvious reason for pursuing energy efficiency is cost reduction. A second reason is good stewardship of resources, which is closely linked to sustainable development, waste minimization and pollution prevention. Environmental standards have risen and continue to rise; it is no longer socially, politically, or legally acceptable for companies to be seen as polluters, and this includes the pollution associated with inefficient energy use.

In general, there are three main dimensions to energy efficiency activities in the industry; (1) to operate existing facilities optimally and efficiently through applications of best practices; (2) to identify economic investment opportunities for step-change improvements; and (3) to implement strong management systems to sustain progress and drive continuous improvement.

Typical activities in the first area (optimal operation of existing facilities) include minimizing the cost of electric supply by exploiting preferential contract terms, improving steam system maintenance to minimize leaks and losses, steam balance optimization, proper maintenance of compressed air systems, optimizing cleaning cycles for heat exchangers, and tuning fired heaters.

In the second area (identifying economic investment opportunities), approaches include process reviews to identify inefficiencies and find ways of correcting them, pinch analysis (which provides a systematic way of identifying heat integration opportunities), equipment changes to eliminate inefficiencies in the steam system, and employee contests.

The most tangible part of the third area (management systems) consists of computer-based systems to monitor plant performance. However, this area also includes management practices such as developing corporate energy policies, increasing employee awareness of energy issues, securing, training and retaining suitable human resources, and integrating energy management into overall business management.

Chemicals Industry Analysis Brief, U.S. Department of Energy, Energy Information Administration. Accessed 12 April 2004 at: http://www.eia.doe.gov/emeu/mecs/iab98/chemicals/.

V. Ganapathy, *Waste Heat Boiler Deskbook*, Prentice Hall, Inc., Englewood Cliffs, N.J., 1991.

W. F. Kenney, *Energy Conservation in the Process Industries*, Academic Press, Inc., New York, 1984.

Kyoto Protocol To The United Nations Framework Convention On Climate Change, Kyoto, 1–10 December, 1997. Accessed 03/19/2004 at: http://www.carleton.ca/~tpatters/teaching/climate-change/kyoto/kyoto1.html.

ALAN ROSSITER
Rossiter & Associates

ENGINEERING THERMOPLASTICS

The general definition today describes engineering plastics as high performance materials that provide a combination of high ratings for mechanical, thermal, electrical, and chemical properties. For present purposes, then, (1) thermoplastics generally are considered to be produced on an industrial scale; (2) with some exceptions, their predominant application is as solid parts or films, not fibers or cellular materials; and (3) sophisticated derivations of such commodities as reinforced polypropylene, ultra-high-molecular-weight polyethylene, etc., widely used in engineering applications are excluded. Occasionally copolymers, blends, and reinforced polymers are included.

In Table 1, the qualitative dependence of some properties of polymeric materials as a function of their morphological state is reported. Such properties are determined directly or indirectly by the different response of chains to solicitations (chemical, thermal, and so on) when they are in either an ordered arrangement or a random one. Totally crystalline (100%) polymers are impossible to obtain, due to the unavoidable presence of chain folds; further, the

Table 1. Relationships Between Polymer Properties and Morphology

Property	Crystalline	Amorphous
light transmission	high	none to low
solvent resistance	high	low
lubricity	high	low
dimensional stability	high	low
mold shrinkage	high	low
resistance to dynamic fatigue	high	low
facility to form high strength fibers	high	none
thermal expansion coefficient	high	low
melting temperature	sharp	absent
dependence of properties on temperature	high	low

crystallinity degree can change under the effect of thermal, mechanical, or chemical operations.

PROPERTIES OF THERMOPLASTICS

Because some processing is often necessary to prepare testing specimens, intrinsic properties can be difficult to measure. Herein mainly intrinsic and processing properties are considered, divided into four conventional groups: physical, electrical, thermal, and mechanical.

Physical Properties

Physical properties include density, properties connected to the combustion tendency (flammability and oxygen index), optical properties (refractive index and yellow index), and the ability to absorb water. Polyolefins, composed of C and H only, have densities in the range 0.85–1; organic polymers containing heteroatoms rarely have densities >2. Crystalline phases are generally more dense than the amorphous phases. Only a few high-performance polymers like the polyetherimides have been classified as inherently nonflammable. Most of the engineering plastics here considered are opaque and/or inherently colored. Some polymers like polyamides absorb water from air humidity and hold water molecules rather firmly by hydrogen bonding. Absorbed water causes a slow variation of properties like electrical characteristics, mechanical strength, and dimensions. Polymers or specific grades insensitive to water must therefore be employed in moist environments.

Electrical Properties

Electrical properties include the dielectric constant, dielectric strength, dissipation factor, and volume resistivity. All of them depend on temperature and water absorption.

Thermal Properties

Thermal properties include some transitions like melting temperature (T_m) and glass-transition temperature (T_g), the heat deflection temperature (HDT), specific heat capacity, thermal conductivity, coefficient of thermal expansion, and upper working temperature.

Mechanical Properties

Mechanical properties include tensile properties (modulus and strength), flexural properties (modulus and strength), compressive strength, elongation at break, impact resistance, hardness, and the friction coefficient. Other relevant properties are creep and fatigue, but it is difficult to find comparative data among materials.

Rheological Properties

Rheological properties, describing the deformation of materials under stress and concerning their flow properties, must be considered in all processing techniques for the fabrication of plastics articles. In order to give operators necessary rheological information, melt viscosity vs. shear plots are commonly included in data sheets provided by plastics producers.

Chemical Resistance

Chemical resistance is less rigidly defined than the properties discussed previously. Measurement methods include immersion in selected vapors or liquids of a test specimen, then determining the variation of mechanical properties after and before treatment. Optical properties are also considered, particularly in the case of transparent materials. The test results are generally indicated as excellent, good, fair or poor, or other arbitrary scale units. Chemical agents are chosen in order to simulate possible real situations: strong and weak acids, alkalies, saline solutions, hydrocarbons (aliphatic or aromatics), oils and greases, alcohols, aldehydes, ketones, etc. Engineering plastics are generally difficult to dissolve in most solvents. In strict correlation to chemical resistance is weathering resistance, where a combination of a particular environment, temperature, time, and ultraviolet (uv) irradiation is considered, as also are cyclic experiments.

PROCESSING OF THERMOPLASTICS

Processing of thermoplastic materials can be classified in four main categories: extrusion, post-die processing, forming, and injection molding. Extruders are very often used at the end of the polymerization reactor, to obtain polymer pellets by chopping an extruded strand. Extruders are also currently used to mix in the proper additives for the polymer, to obtain mixed-polymer blends intimately, to devolatilize the material from the monomers or solvent residues, and in some special cases as a chemical reactor. Post-die processing includes a number of operations carried out at the exit of the extruder die in a free-surface way. Forming processes use a mold to confer the final form to the article. Blow molding is widely used in the manufacture of bottles or other containers for liquids, widely using engineering polymers like PET and PC. Injection molding is the most commonly used processing technique for engineering thermoplastics. Typically, the polymer pellets are melted and the melt fill a mold under appropriate pressure. Very complex article shapes can be obtained by this technique. The main problem in injection molding is shrinkage, caused by the volume changes during transition from the melt to the solid.

HYDROCARBON MATERIALS

Cyclic Olefin Polymers

Cyclic olefin polymers (COP) and copolymers (COC) exhibit unique optical properties, accompanied by excellent moisture resistance, high T_g, and toughness.

Syndiotactic Polystyrene

The outstanding properties of syndiotactic polystyrene (sPS) have allowed its a fast penetration of the engineering polymers market. It is a semicrystalline polymer having a T_g similar to that of atactic polystyrene. However, whereas atactic polystyrene is amorphous, syndiotactic polystyrene has a T_m of ~270°C; the crystallization rate is comparable to that of PET. Due to its crystallinity, sPS is highly resistant to acids, alkali, and most organic solvents except chlorinated and aromatic hydrocarbons that cause swelling.

The material is conveniently processed by using conventional transformation methods, including extrusion (also into films), injection molding, and thermoforming. The aptitude of sPS for being very easily injection molded, combined with its material properties and extremely low affinity to moisture, make sPS a serious competitor of polyamides and thermoplastic polyesters (PET, PBT) in automotive (under-the-hood) and electric applications. Typical applications are coolant circuit components, impellers for water pumps, lighting systems, electrical interconnects, printed circuit board connectors, and coil bobbins. sPS films may find application in capacitors. Medical equipment, eg, sterilization trays and surgical and dental devices, are other end uses for which sPS is gaining increasing attention.

OXYGEN-CONTAINING POLYMERS

Acetal Resins

Acetal resin is the common name for homopolymers of formaldehyde. This material is also referred to as polyoxymethylene, or copolymers of formaldehyde and oxiranes.

Acetal resins are translucent, white, highly crystalline materials. The homopolymer and the copolymer have a melting temperature of 175 and 165°C, respectively. They are soluble in phenols and insoluble in hydrocarbons, alcohols, esters, and ketones. Acid and alkali solutions cause degradation. Acetal resins provide a high modulus of elasticity combined with high strength, stiffness, resistance to abrasion, and good wear properties; dimensional stability is also excellent because moisture absorption is very low. Their high surface gloss gives acetals a pleasant appearance. The copolymer exhibits slightly inferior mechanical properties compared to the homopolymer, but improved thermo-oxidative stability. High flammability, low uv light stability and low impact resistance are weak points of this material. Its dielectric properties are comparable to those of other heteroatom-containing engineering plastics.

Applications range from gears, bearings, and cranks in the automotive field to engineering components, appliance casings, cassette cartridges, container valves, and pump components. PTFE-filled grades are used for moving parts, where low friction is important.

Polyesters

Thermoplastic polyesters are polymers (or copolymers) obtained from aromatic diacids and aliphatic diols. Terephthalic acid is the most common aromatic diacid monomer; several different diols are used as comonomers. The most commercially important polyesters are PET, PEN, PBT, poly(1,3-propylene terephthalate) (PPT), and the copolymer of terephthalic acid, ethylene glycol and 1,4-cyclohexanedimethanol (PETG). Other important thermoplastic polyesters are poly(1,4-cyclohexylenedimethylene terephthalate) and poly(1,4-cyclohexylenedimethylene terephthalate-co-isophthalate).

Poly(ethylene terephthalate)(PET) was the first polyester commercially developed, in the 1940s. PBT was introduced some years after PET, as an engineering plastic for its better processability and toughness. PEN was studied and commercialized starting in 1994 as a packaging material for films and liquid containers, in applications where better performances than PET are required. PPT, the latest polyester of the family, is currently under development.

PET is a stiff, hard, dimensionally stable material with excellent wear resistance and a low friction coefficient. It absorbs low amounts of water and its resistance to solvents is good, except to alkali, especially at high temperatures. Its barrier properties toward gases and vapors are remarkably good.

PET 1,4-Phenylene unit

Applications of PET are primarily as fiber and packaging material (bottles and food films). Other end uses include ovenable food trays, electrical components, fuel pump components, fuel system rotor and connectors, carousels, test equipment manifolds, and industrial tire cords. PET and its blends for technical applications are available in a wide range of grades (filled, flame retarded, nucleated, etc).

Poly(ethylene 2,6-naphthalenedicarboxylate) (PEN) is a rather new material introduced to overcome some limitations of PET. The synthesis of PEN on an industrial scale has become feasible after the discovery of convenient routes for the preparation of 2,6-dimethylnaphthalene as

an intermediate for 2,6-dimethyl naphthalenedicarboxylate (DMNDC). PEN is much stiffer than PET and its transition temperatures are higher: the T_g is 120°C and the T_m is 265°C. Gas permeability is greatly reduced as compared to PET (almost one order of magnitude) and the elastic modulus is improved, whereas its tensile strength is similar. PEN is also a much more efficient uv barrier than PET.

PEN films are used in electronic and electric fields. Liquid containers made of PEN can be used for oxygen-sensitive products, like beer. Refillable bottles can also be conveniently manufactured with PEN or PEN copolymers or blends. The market growth of PEN partially continues to be limited by the availability and cost of the monomer (DMNDC).

Poly(1,4-butylene terephthalate) (PBT) is a semicrystalline polyester having a T_g of 52°C and a T_m of 223°C. Due to its very high crystallization rate it is not generally possible to obtain PBT in the amorphous state. For the same reason, PBT can be injection molded without the need of nucleating agents.

PBT has good impact resistance and toughness combined with good chemical resistance toward water, steam, chlorine, and chemicals. Its coefficient of friction and wear are low. The strength and stiffness are slightly inferior to those of PET. The shrinkage is anisotropic, so manufacturing tolerances cannot be too tight.

Typical end uses are automotive parts—including underhood—(housings and connectors, distributor caps, air-blower deflectors), power tool casings, food and fuel pump components, valves and valve bodies, gears, and cams. Blends with PET and PC have been developed for such uses as in the electric industry (bobbins, switch components, connectors) and for medical and food applications. PBT/PC blends have outstanding resilience and are used for auto bumpers and other automotive applications.

Poly(trimethylene terephthalate)(PTT), like PET and PBT, is based on terephthalic acid as the diacid monomer component, but the diol monomer is 1,3-propanediol, which has a number of carbon atoms intermediate between ethylene glycol and 1,4-butanediol. Because of this molecular structure, PTT's properties are between those of the two other polyesters. PTT competes with PBT in engineering polymer applications but finds its main use in fibers.

Poly(ethylene-co-1,4-cyclohexylenedimethylene terephthalate) (PETG) is commercially available as a clear amorphous resin with a T_g of 82°C having good toughness (up to −40°C), a fair resistance to organic solvents and chemicals, and excellent gas barrier properties. Extrusion blow molding is the most convenient technique for using it to manufacture liquid containers. Sheets, which exhibit an impact strength 10 times that of acrylics, can also be obtained by extrusion. The high transparency and gloss of PETG provides an excellent alternative to polycarbonate at lower cost. Other advantages are good blendability and printability, superior in some cases to acrylics. Applications are in store fixtures (shelf dividers, racks), material handling and storage, food packaging, medical lab equipment, interior sign and price markers, graphic devices, and industrial components.

Polyarylates

The ardel polyarylate resins(PARs) based on isophthalic/terephthalic acids are amorphous, transparent, amber materials whose properties are dependent on the diacids ratio (generally not far from 1:1). They are tough resins, with excellent flexural recovery and high rigidity, exhibiting a mechanical and electrical behavior similar to polycarbonate but better heat resistance. A characteristic of polyarylate is its inherent uv stability. Polyarylates have fair heat stability but low hydrolytic stability; for this reason, they must be carefully dried before processing. Their solvent resistance is fair. Processing techniques for PARs are the same as those used for polycarbonate. The material does not exhibit outstanding moldability, but it can be easily shaped into rods, sheets, films, and slabs. PARs can be blended with other resins, such as polyamides, to increase their solvent resistance, or poly(phenylene sulfide) to improve their dielectric properties and hydrolytic stability.

Typical applications are as automotive trim and headlight housings, exterior lighting, solar energy components, semiconductor components, appliance parts, snap lock connectors, and fire helmets.

Liquid-Crystal Polyesters

Liquid-crystal polymers (LCPs) are rodlike macromolecules that are oriented in the melt state and organized into a paracrystalline state as a consequence of their structure. Such orientation is further promoted by shear. On cooling, the strong orientation is preserved and the resulting material exhibits an exceptionally high stiffness and strength.

The mechanical properties of LCPs are strongly dependent on the orientation and, thus, on the processing parameters adopted during molding. The tensile strength and elastic modulus are extremely high in the flow direction. Mechanical properties are retained after exposure to radiation and weathering. Flame resistance is inherent, dielectric strength is good, and dimensional stability is excellent. The resistance to aggressive solvents—acids, bases, polar organic solvents, chlorinated solvents, and aromatic hydrocarbons—is very high. Properties are significantly lost only after prolonged contact with concentrated sulfuric acid, boiling alkali, or high-temperature steam.

Processability is easy, due to the very low melt viscosity, but high temperatures must be applied. Thin walls and complex paths are easily filled; low warpage allows manufacturing of high-precision parts, where close tolerances are required. Typical applications are as electric and electronic components such as connectors and medical devices. Aerospace applications can be conceived.

Poly(phenylene ether)(PPE)

The most common and commercially available poly(phenylene ether) is poly(2,6-dimethyl-1,4-phenylene ether) (trade name PPO, General Electric), obtained through oxidative coupling polymerization of 2,6-dimethylphenol in the presence of a catalyst.

PPE–PS blends (trade name Noryl, General Electric) present an excellent combination of properties. Blends with polystyrene homopolymer are somewhat brittle; to improve the toughness, blends with high impact polystyrene have been developed. The Noryl EF alloy family is the most recent development in this class of materials.

Noryl resins are cream or pale gray colored, opaque materials exhibiting good flexural fatigue properties, humidity absorption as low as 0.07%, and an excellent melt-flow behavior that allows a wide processing temperature range. Dimensional stability at elevated temperatures and dielectric properties at any frequency are also outstanding. The resistance to aromatic and chlorinated solvent is poor and limits some applications. Filled, pigmented, and flame-retarded grades containing phosphorus-based halogen-free additives are available.

Noryl blends are suited to injection molding processes on conventional machines equipped with general-purpose screws. The high melt strength allows fabrication of cellular products through foam molding with gas-injection or chemical blowing agents. Blow molding is possible due to the melt strength to produce large parts. Noryl sheets can be easily extruded and thermoformed.

Applications include appliances' internal components, structural components in office products, housings (personal computers and printers), automotive coil forms, instrument parts and wheel covers, and electrical switch boxes and connectors. Noryl EF beads can be used in such car interior applications, as door paddings, instrument panels, knee bolsters, and pillar trims.

Another important family of PPO blends is Noryl GTX, PPO–polyamide blends, used in the automotive market (mainly bodypanels and under-the-hood).

Polycarbonates

Polycarbonates (PCs) are obtained industrially by interfacial polycondensation of phosgene and aromatic diols.

The most widespread polycarbonate is bisphenol A polycarbonate (BPA–PC), due to the large commercial availability of bisphenol A; in some applications where higher heat distorsion, improved optical properties or flame resistance, greater weatherability are required, other aromatic polycarbonates based on different diphenols or containing diacid comonomers (isophthalate or terephthalate) have been proposed.

BPA–PC is a colorless crystalline polymer, but due to its extremely low crystallization rate the commercial pelletized product is amorphous. It has excellent moldability, structural strength, and toughness, even at-low temperatures (down to $-40°C$) and high transparency. Other outstanding properties are its broad temperature resistance and the ease of colorability and compounding with inorganic fillers and other polymers (eg, ABS). Properties are reported in Table 2. Its drawbacks are poor chemical resistance to alkali, hydrocarbons, ketones, and halogens, sensitivity to environmental stress cracking, and poor wear and fatigue behavior.

Polycarbonates find many applications in substitution of glass, steel, wood, and other materials. Some examples

Table 2. Properties of PC Resins

Property	Neat	20% GF
density, g/mL	1.2	1.34
flammability	V0–V2	V1/V0
oxygen index (LOI), %	25–27	25–36
refractive index	1.585	
water absorption, 24 h, 23°C, %	0.1	0.1
dielectric constant, 1 MHz	2.9	3.2
dielectric strength, kV/mm	15–67	35
dissipation factor, 1 kHz	0.01	0.007
volume resistivity, 23°C, dry, $\Omega \cdot$cm	10^{14}–10^{16}	$>10^{15}$
HDT at 0.45 MPa, °C	140	147
specific heat, J/(kg·K)	1.2	1.13
thermal conductivity, 23°C, W/(m·K)	0.19–0.22	
thermal expansion coefficient, K^{-1}	$66–70\times10^{-6}$	$30–40\times10^{-6}$
upper working temperature, °C	115–130	115–165
elastic modulus, GPa[a]	2.3–2.4	6
tensile strength, MPa[b]	55–75	90
compressive strength, MPa[b]	>80	
elongation at break, %	100–150	3
notched Izod, 3.2 mm, J/m[c]	600–850	100
hardness (Rockwell)	M70	M91
abrasion (10^3 cycles), mg	10–15	

[a] To convert GPa to psi, multiply by 145,000.
[b] To convert MPa to psi, multiply by 145.
[c] To convert J/m to lbf·ft/in., divide by 53.38.

of applications are in architectural glazing (eg, for roofings), for helmets and safety shields manufacture, lighting devices, housings, kitchenware (microwaveable), compact disks, and sterilizable medical devices (syringes, catheters, etc). PC blends are used for computer housings and panels (PC–ABS), in the automotive industry (PC–polyesters and PC–ASA), and in construction (PC–ASA). PC–polyetherimide is especially used where higher heat resistance or low flammability are required (automotive reflectors, aircraft interiors).

Glass-filled PC, which exhibits improved stiffness and creep, is used for electrical enclosure manufacture. Carbon-filled PC is used for conductive film manufacture.

Aliphatic Polyketones

Aliphatic polyketones (PKs) contain ketone groups along a main aliphatic chain. As engineering polymers, only perfectly alternating olefin/CO copolymers can be considered; polymers containing <50% CO groups are photodegradable and consequently not suitable for engineering applications.

The properties of aliphatic polyketones derive from their tight packing, favored by hydrogen bonds between chains. They are quite rigid and exhibit excellent elastic recovery, good sliding and abrasion properties, and impact resistance at low temperature. The peculiar molecular and crystalline structures confer to such materials a high hydrolytic stability, good chemical resistance against dissolution, degradation and swelling, and excellent gas permeation resistance.

Aliphatic polyketones have found application mainly in the automotive (snap-on assemblies, spring elements, fuel tanks and lines, grabs, clamps, and clips), electrical and electronics (connectors and switch gears), and medical fields (disposable breathing systems, test tube racks). Glass-fiber reinforced and lubricated grades have been offered.

Poly(ether ketones)

Poly(ether ketones) (PEKs) include a variety of aromatic high performance polymers characterized by the presence of ether bridges and ketone groups in the main chain, linking together arylene groups. Currently, the only product manufactured worldwide is Victrex PEEK. PEEK has found application in the transport, teletronics, and aerospace sectors, with the fabrication of injection molded engineering components and circuit boards. Recently, PEEK materials have also found some space in medical technologies.

Acrylic Resins

Poly(methyl methacrylate) (PMMA) is the only industrial product based on acrylic resins that has found rigid applications.

PMMA exhibits excellent optical properties—clarity, and transparency—and is used widely as a glass substitute in many sectors. Because of its high refractivity index, sophisticated applications in advanced sectors (optical lenses, prisms, fibers, etc) are also possible. The high surface gloss and good scratch resistance of PMMA favored its penetration in the sanitary wares field. By the addition of up to 60% of an inorganic filler like aluminum hydroxide, synthetic marble like resin products can be prepared.

SULFUR-CONTAINING POLYMERS

Poly(phenylene sulfide)

Poly(phenylene sulfide) (PPS) is the most significant high performance polymer in terms of tonnage but is commercialized only as a filled or reinforced compound. PPS is mainly used for injection molding. Due to its excellent melt viscosity, monofilaments, multifilaments, fabrics, and fibers are also produced for technical uses. PPS is applied for precision mechanical parts, electrical components, and components for use in aggressive environments. Among the manufactured parts it is used for are engine components, valves, high-pressure nozzles, fuel manifolds, exterior light reflectors, coil bobbins, and flow-measuring units for hot and cold water.

Polysulfones

Polysulfones (PSUA) are a group of aromatic resins containing sulfone and ether linkages. The commercially relevant structures also include PES and PAS.

Applications are mainly in electrical and electronics components, medical equipment requiring repeated sterilization procedures, household equipment, automotive applications, and autos under-the-hood. A very interesting application is in synthetic membranes, where polysulfones found market space for blood dialysis, water treatments, and food preparation, due to their superior separating capacity. Injection molding and extrusion grades are available for all the polysulfone families. PSU and PES are available also as glass-reinforced grades (up to 30% by weight).

NITROGEN-CONTAINING POLYMERS

Plastics based on styrene as a monomer include standard polystyrene (homopolymer), expanded polystyrene, high impact polystyrene, SAN and ABS. The first three hydrocarbon materials are considered commodity plastics, the last two as engineering plastics because of their better performance and use in more demanding applications.

Styrene-Acrylonitrile Copolymer (SAN)

The chemical structure of SAN is represented below. SAN was initially put on the market in the mid-1930s but became attractive for end users only after the development of the modern injection molding machine in the 1950s. The most widespread process for SAN manufacture is bulk continuous polymerization.

SAN

SAN is an amorphous, transparent resin having a T_g ~105°C. Its water absorption is ~0.3%. The polymer is not colored, but because the monomer units are randomly distributed along the chain, the occurrence of consecutive acrylonitrile monomer units causes a pronounced yellowing after prolonged exposure at high temperatures. Its outstanding properties are the high rigidity and hardness (up to 80°C), and good resistance to instantaneous and sustained loads and to cyclic temperatures. Chemical resistance is good, particularly against stress cracking. Heat resistance and dimensional stability are excellent, but fatigue behavior is not satisfactory. Mechanical properties are affected by composition and, to a minor extent, by molecular weight. Tensile strength, flexural strength, and impact strength reach a maximum around the azeotropic composition. SAN has good resistance to apolar solvents (aliphatic hydrocarbons, greases, oils, petrol, fats), but lower resistance to polar solvents than polystyrene; the nitrile groups are also sensitive to acids and alkali. Several grades of SAN are available on the market: general purpose, easy-flow grade for thin walls and super clarity grade with high gloss.

SAN is used to manufacture housings for electronic and electric devices (computers, batteries), casings of domestic appliances (refrigerators), chair shells, domestic tools,

instrument lenses, medical syringes, toys, etc. Glass-filled grades are used in autos and appliances. Packaging materials (food, pharmaceuticals, cosmetics) are another important application. Special grades of SAN with improved weatherability, uv stability, and vapor barrier properties have also been developed.

Butadiene–Acrylonitrile–Styrene Polymer

Due to the inadequacy of SAN materials for several applications, a range of materials referred to as ABS polymers came out in the early 1950s.

The methods applied currently for ABS preparation are based on the radical grafting of styrene and acrylonitrile monomers onto butadiene rubber. The butadiene content of ABS usually ranges from 10 to 25% by weight (for some special grades up to 45%), and the styrene/acrylonitrile ratio can be controlled as in the case of SAN.

ABS is a translucent-to-opaque resin exhibiting a high gloss. Careful control of the rubber particles' morphology even allows attainment of transparent materials. The presence of rubber gives ABS good toughness and high ductility compared to that if SAN, even at low temperatures. Its rigidity and hardness are comparable to that of SAN. Chemical resistance (water, salts, alkalies, acids do not affect the properties), impact and abrasion resistance, dimensional stability, and surface hardness are also good; the polymer is very sensitive to uv radiation.

All the common molding technologies can be applied to ABS, which is available in different neat, filled, and fire retarded grades; also, several blends (in particular that with polycarbonate) have been commercially developed. Electroplating is used with some special grades for finishing.

ABS is used mainly in the automotive industry, electrical and electronic equipment, and in home appliances. Prolonged exposure to sunlight causes discoloration, but outdoor applications are wide nonetheless. Some examples of ABS parts are vehicle components (instrument panels and other interior components), casings and cabinets, communication equipments, pipes and pipe fittings, shower trays, and baths.

Styrene–Maleic Anhydride Copolymer (SMA)

Styrene–maleic anhydride copolymer is an alternate copolymer obtained by radical bulk or solution polymerization of the two monomers. SMA is located here as a styrene copolymer of minor importance. The high content of maleic anhydride make the polymer outstandingly resistant to aliphatic and aromatic hydrocarbons. Polar organic solvents and alkalies attack the material. The polymer is transparent and has a T_g 120°C. As for SAN, the applications of SMA are in the automotive industry, casings, and appliances.

Polyamide (Aliphatic) Resins

The name "nylon" was never registered as a trademark but is now universally used instead of the chemical name "polyamide."

Polyamides contain the amide –CONH– group as a common feature. From a commercial point of view, PA6,6 and PA6 are the most important materials.

Generally, aliphatic polyamides are white, semicrystalline materials. Their properties vary greatly with the degree of crystallinity, which in turn is strongly dependent on processing conditions. It is easy to obtain moldings of PA6 having 60% or, on the contrary, 10% crystallinity. Amorphous, transparent materials can also be obtained, by suppression of the crystallization process. They exhibit high strength and stiffness, lubricity, resistance to abrasion, and chemical resistance. Because of the presence of amide groups, water absorption is relatively high. Its occurrence lowers the mechanical and electrical properties. The properties of PA6,6 and PA6 are rather equivalent. PA6 is slightly easier to process because of its higher melt viscosity and shows higher impact strength but lower tensile strength. Its water absorption is the highest of all the polyamides. PA11 and PA12 have similar properties (PA12 slightly inferior). They absorb less water with respect to PA6,6 and PA6. Consequently, dimensional stability and electrical properties are improved. PA4,6 is colored (from yellowish to red). It maintains properties at high temperatures.

Extrusion and blow molding are common processing techniques for polyamides. Films find large application in packaging foodstuffs and pharmaceutical goods. Glass-fiber-filled grades are widely fabricated, obtaining materials able to replace metal parts for their rigidity, creep resistance, low coefficient of friction, and high HDT. Toughened polyamides are offered for high impact resistance.

PAs are applied in mechanical engineering, mainly in the sectors of transportation and electrical/electronic appliances. Gears, bearings, valve seats, nuts, bolts, rivets, wheels, and casings are typically fabricated by injection molding, taking advantage of several properties like low viscosity at processing temperatures.

Semiaromatic polyamides are also present in the market, in relatively low amounts.

Aromatic Polyamides

Aromatic polyamides (aramids, ArPA) are a family of high molecular weight wholly aromatic polyamides, including poly(p-phenylene terephthalamide) and poly(m-phenylene isophthalamide). The chemical structure of these polymers is shown below.

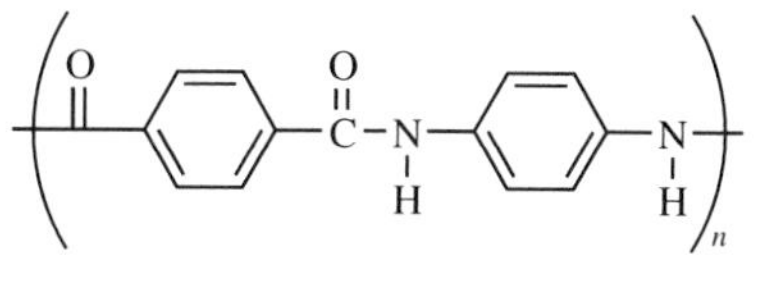

poly(p-phenylene terephthalamide)

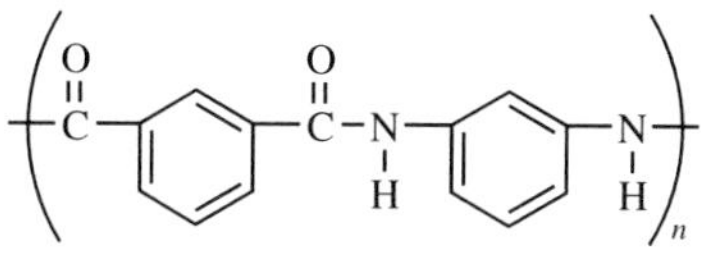

poly(m-phenylene isophthalamide)

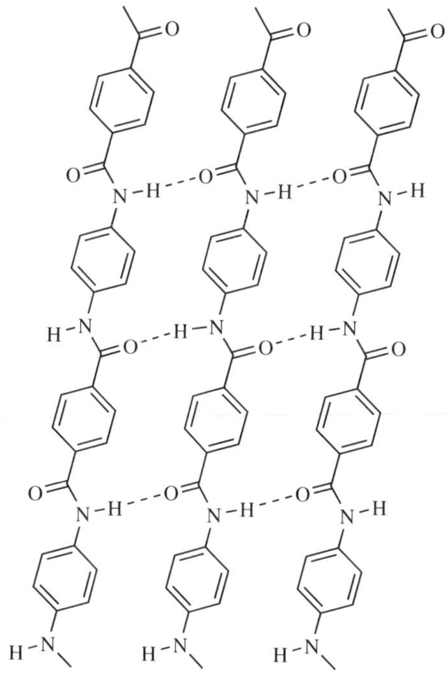

Figure 1. Interchain hydrogen bonding in aromatic polyamides.

Aramids are crystalline, infusible, very insoluble materials. They are commercially available only in the form of fibers, generally spun from pure sulfuric acid solutions using in air-gap technique. Aramid solutions are generally fluid and opaque because they form lyotropic liquid-crystalline phases. Poly(p-phenylene terephthalamide) chains exist in the all trans conformation and reach a high degree of orientation with strong interchain hydrogen bonding (Fig. 1), that imparts to the fiber an exceptionally high modulus: Kevlar is about five times stronger than steel by weight.

Mechanical properties are remarkably better along the fiber longitudinal direction than in the axial direction, Kevlar is elastic in tension and ductile in compression, elongation at break and thermal shrinkage are low.

One of the most important applications of Kevlar is in ballistics and defense sectors, for body armors, anti-mine boots, and helmets to break the material, a fracture must traverse over different planes, which makes for difficult penetration by projectiles or cutting objects. It is also widely used in composites for aircraft structural body parts and cabin panels. In addition, it is used for rope reinforcement, reinforcement for automotive and industrial hoses, and belts.

Nomex fibers (poly(m-phenylene isophthalamide) can be converted to continuous-filament yarns, staples, spun yarn, flocs, paper, needle felt and fabric suited for protective devices (racing drivers' suits, firefighters suits, flight suits, space suits, fire blankets), filtration media (hot gas filtration, laundry filters), electrical paper and pressboard (transformers, microwave ovens, electric motors), and mechanical parts of various kinds (engine nacelles, helicopter blades, aircraft parts).

Polyimides

Polyimides (PIs) contain cyclic imide groups in their molecular structure, the branched nature of which facilitates the build-up of multiring structures. Due to their importance in advanced fields like electronics and aerospace, polyimides, continue to be the subject of extended research.

Polyimides are generally colored and exhibit excellent thermal properties, high radiation resistance, low flammability, high wear resistance, and low creep.

Due to their high price, applications of polyimides are limited to very demanding cases, eg, aerospace. Films are used for capacitors, insulation, printed circuit boards, engine components, and bearings. Fillers are added to enhance electrical or thermal properties.

Polyamideimides

Polyamideimides (PAIs) are a class of amorphous thermoplastic polyimides. Commercialized in the early 1970s by Amoco.

PAI resins have an amber to gray color. Due to their molecular rigidity and hydrogen bonding, they present several interesting properties, in particular high tensile strength and stiffness, excellent dimensional stability, and creep resistance (among the best for thermoplastics), are good solvent (aliphatic, aromatic, chlorinated, fluorinated hydrocarbons), and have good chemical resistance (dilute acids, aldehydes, ketones, ethers, esters). Only their resistance to alkalis is poor. Electrical insulation properties as well as flame resistance are also very good.

PAIs possess high melt viscosity, so special screws are necessary for injection molding. Compression moldable powder is also available for fabricating small, simple parts. PAIs are used in demanding applications like connectors, gears, relays, ball bearings, engine components (piston skirts and rings, valve stems), marine winches, and switches.

Polyphthalamides (PPAs)

PPAs are a class of semiaromatic polyamides introduced in 1991 with a chemical nature that has not been fully disclosed but is believed to be the result of polycondensation of hexamethylenediamine and a mixture of terephthalic and isophthalic acids.

PPAs' interesting properties with respect to other high performance-polymers are its thermal resistance, toughness, chemical resistance, ease of processing (mainly by injection molding), and dimensional stability. Water absorption is very much decreased compared to aliphatic polyamides. PPA has found application in mechanical engineering, for the fabrication of fasteners, pump housings, motor end frames, connectors, engine parts, electric motor brush holders, pump and fan impellers, cooling circuits, etc.

Polyetherimide

Polyetherimides are thermoplastic polymers containing imidic moieties, which impart the outstanding thermal, mechanical, and electrical properties to the material, and ether groups, which facilitate processability.

The main applications of PEI are in automotive components (under-the-hood, electrical, heat-exchange components), electrical and electronics (circuit boards, switches, connectors), and aerospace (lighting, seating, wiring, microwave oven trays). Other fields of application are as medical reusable tools, requiring repeated sterilization procedures; in high-frequency printed wiring boards, where the low dissipation factor at high frequency plays a key role; and as fibers for dry filtration, aircraft fabrics, and protective clothing.

FLUORINE-CONTAINING POLYMERS

The interest in fluoropolymers like PTFE as engineering polymers resides in their excellent chemical inertness, heat resistance, electrical insulation properties, and low friction coefficient.

PTFE is a white-to-translucent polymer that is highly crystalline and exhibits unique chemical inertness, chemical resistance, and electrical insulation properties, as well as a very low coefficient of friction.

Fluoropolymers are widely used for fabricating bearings, seals, gaskets, packings, valve, and pump parts. They are also used for wire insulation, vessel lining, and laminates for printed circuitry. Because large objects made with PTFE cannot be fabricated, metal objects are conveniently coated by a PTFE layer.

INTERPOLYMER COMPETITION

To select the right polymeric material for a specific application requires consulting four main groups of technical considerations: mechanical, electrical, environmental, and appearance. In addition, cost and specifications are two other elements of importance. Environmental considerations include the operating temperature, chemical environment, weathering exposure, and humidity. Appearance includes style, shape, color, transparency, and surface finish. Appearance can vary under service conditions. In most of applications of engineering thermoplastics, the following characteristics and properties are considered: price, mechanical properties, thermal properties, electrical properties, and chemical resistance.

D. P. Bashford, *Thermoplastics Directory and Databook*, Chapman and Hall, New York, 1997.

C. P. MacDermott and A. V. Shenoy, *Selecting Thermoplastics for Engineering Applications*, 2nd ed., Marcel Dekker, Inc., New York, 1997.

D. V. Rosato, *Rosato's Plastics Encyclopedia and Dictionary*, Hanser Publishing, Munich, 1993.

J. Scheirs, *Fluoropolymers: High Performance Polymers for Diverse Applications*, John Wiley & Sons, Inc., New York, 1997.

FABIO GARBASSI
RICCARDO PO
Eni Chem Research Center

ENVIRONMENTAL IMPACT ASSESSMENT

The field of environmental impact assessment (EIA) is becomingly increasingly important to the chemical industry and to those using chemical technologies. EIA requirements are sometimes applied directly to some chemical industries or to projects with certain chemical technologies, such as hazardous waste treatment. Although requirements vary greatly, chemical industries can be subject to EIA requirements because the proponent is public, the proposed action occurs on public lands, the proposed action is publicly funded or funded by a development bank, or a specified government approval requirement "triggers" EIA requirements. A given chemical industry can also be affected if the proposed action is likely to adversely affect environmentally significant areas or species.

GENERAL

EIA is customarily viewed as beginning with the introduction of the National Environmental Policy Act (NEPA) in the United States in 1969. Since that time EIA has expanded rapidly, particularly over the past two decades, to include social impact assessment (SIA). By 1996 it had spread to over 100 countries on six continents. It is widely applied at different government levels and by international aid agencies to both public and private undertakings.

The initial emphasis of EIA was on large capital projects, but more stress is now being placed on assessing the effects of policies, plans, programs, legislative proposals, technologies, products, and trade agreements. Over time, EIA methods have become more sophisticated, forging with related fields that include planning, risk assessment, and environmental management. Increased attention has been devoted to transboundary effects, EIA applications in developing countries, EIA legislation by developing countries, institutional capacity building, and such global issues as climate change, biodiversity and sustainability.

EIA has sought to incorporate from early on environmental information and interpretations into planning and decision-making procedures and documents. The regulatory regime associated with EIA is expected to alter organizational values, attitudes and behavior, and to contribute to more open, systematic, accountable, and effective organizations and decision-making. Increasingly, EIA is viewed as one instrument among many for achieving broader environmental objectives such as sustainability.

There are numerous impact assessment types that fall under the EIA umbrella. Examples are given in Table 1.

EIA INSTITUTIONAL ARRANGEMENTS

EIA institutional arrangements begin with legislation, regulations, guidelines, and case law. Increasingly, EIA requirements from senior government levels can be accessed at government Web sites. EIA legislation and

Table 1. Impact Assessment Types

Impact assessment	What is assessed?
ecological	potential ecosystem impacts
social (SIA)	consequences on people and on how people and communities interact with their surroundings
economic	impacts on how people make a living, on material well-being, and on economic activities
strategic environmental (SEA)	environmental impacts of a policy, plan or program and its alternatives, generally within policy sectors
cumulative effects (CEA)	impacts of an action when combined with other past, present and reasonably foreseeable future human activities
technology (TA)	effects on society from new or modified technology
human health impact (HIA)	human health impacts of a proposed action
sustainability appraisal or SA	extent to which action contributes to or undermines ecological and societal sustainability
life cycle (LCA)	environmental effects of products, processes, systems and services during their life cycles
integrated environmental (IEA)	the ecological, economic, social and institutional effects of societal activities and government policy, across policy sectors

regulations identify the purpose and objectives of the governmental legislation, and define the environment and environmental effects. In most jurisdictions social and economic effects are either not considered or only indirect social and economic effects are addressed.

EIA requirements generally specify what "triggers" the process action type, proponent type, environment type, or a combination, and may indicate what does not trigger EIA requirements. The procedures for making discretionary judgments regarding the application of EIA requirements are usually outlined. EIA requirements detail the content requirements for different EIA document types, eg, overview documents, detailed analyses, and SEAs. They identify procedural stages, specify decision points, describe the roles of the major parties in the process, outline agency review procedures and detail document circulation, review, and approval procedures. They indicate other government requirements that may be integrated into EIA requirements. They include provisions for public access to project-related information and documents. They commonly outline decision-making criteria, include timing limits, describe documentation requirements, specify agency and public notification and involvement procedures, indicate links to other government levels, include procedures for addressing transboundary effects, provide for postapproval follow-up, and cross-reference other policies and requirements.

EIA guidelines generally address procedures for preparing various EIA document types, for interagency coordination, and for public and other government notification and involvement. More specific advice is sometimes provided for various EIA activities such as scoping, alternatives analysis, cumulative effects assessment (CEA), project characteristics descriptions, significance interpretation, for various impact types (which can include climate change, ecological, heritage, health, for certain project types; eg, mining, pipelines), for public policy links (eg, environmental justice), the precautionary principle, biodiversity, and to facilitate good practice with effective public participation.

EIA systems tend to start with the questions of what should trigger the application of EIA requirements and which set of requirements should be applied. Such "screening" questions can focus on various actions (what), on proponents (who) or on environments (where). Each screening decision involves significance judgments, whether it is important enough to institute EIA requirements, or to warrant EIA requirement "A" or "B". Most EIA systems involve action, proponent and environment combinations. How these elements are combined depends on whether the role of EIA requirements is primarily seen as building environmental considerations into proponent or action-related decision-making or protecting and enhancing the environment.

EIA regulatory systems, to varying degrees, include requirements and provide guidance concerning individual EIA activities. EIA requirements tend to identify objectives, specify minimum requirements, and include general performance standards or criteria. EIA guidelines tend to offer good-practice guidance. A balance is generally sought between ensuring a consistent level of adequate practice and not stifling innovation.

Integration and coordination are central attributes of EIA regulatory systems. EIA requirements and guidelines tend to refer briefly to interconnections among EIA process activities. They generally include extensive horizontal coordination procedures pertaining to, for example, links to related laws, regulations and permits, connections to related policies, programs and plans, interactions with related projects and activities, and interrelationships with the actions of other government departments and agencies. EIA regulatory systems also refer to vertical coordination mechanisms.

Extraterritorial connections are sometimes cross-referenced in EIA requirements. Such links include, for example, impacts on neighboring countries, impacts on the global commons such as oceans, impacts from development aid and the application of international environmental standards and protocols.

Most jurisdictions provide for the periodic review and reform of EIA requirements. Sometimes one EIA regulatory system is replaced with a fundamentally different system, or supplementary EIA requirements are introduced, as new SEA requirements in Europe.

The nature of EIA regulatory reform varies among jurisdictions. The general thrust in the United States and Canada, at the federal level, has been toward higher quality data, documents and analyses, clarified requirements, improved guidance, more efficient and effective interagency and intergovernmental procedures, better communi-cations and coordination, and enhanced public access to information and involvement.

EIA PROCESSES AND METHODS

Figure 1 presents an example of the EIA process, starting at the screening stage to focus on what matters environmently. In practice there are multiple EIA process design and management choices. EIA process activities are highly interactive and often occur in different forms at different stages in the EIA process.

The EIA process can be reduced to a few basic activities and events or be extremely complex. EIA activities can be subdivided, combined, or rearranged. There are numerous possibilities available for feed forward and feedback loops. Some interconnections are more important than others. The process can be linked (in different ways) to proposal planning, decision-making, related environmental decisions, related fields and related activities.

Choices are available regarding the treatment of inputs and outputs. Example inputs include EIA requirements, public and agency concerns and preferences, roles and responsibilities, environmental substance, knowledge, values and experience, and methods. Example outputs include documents, decisions, and environmental changes. The EIA process can vary depending on the effect types being assessed, the proposal type, the setting type, and the proponent type. Adjustments are always necessary.

General EIA methods vary in their characteristics, strengths, limitations, and suitability for different EIA activities, settings, and proposal types. Many EIA methods are derived from other fields of practice. There are task and participant-oriented EIA methods, often the two are merged.

Numerous methods are adapted and applied in EIA practice by natural scientists involved in geology and soils, climatology, hydrology, terrestrial, freshwater and coastal ecology, and by social scientists involved in sociology, economics, anthropology, psychology, archaeology, and applied scientists. Technical personnel apply risk assessment, noise analysis, air quality analysis, groundwater analysis, water quality analysis, visual and landscape analysis, and transportation analysis methods in EIA. Procedural specialists use project management, public participation, alternative dispute resolution, legal requirements and procedures, and document preparation methods in EIA. Synthesis skills, methods, and procedures are critical to coordinate and integrate individual specialty analyses and to prepare focused and understandable synthesis and summary documents.

Chemical industry proposals and proposals involving chemical technologies, subject to EIA requirements, can involve many of these specialties. Public and government agencies are often particularly concerned with human health and ecological risks from chemical or chemical waste emissions or effluents. Addressing the risks and uncertainties associated with acute and chronic air emissions and with potential surface and groundwater contamination are often especially important. Comparisons against regulatory standards often only partially address the pathways between chemicals and receptors.

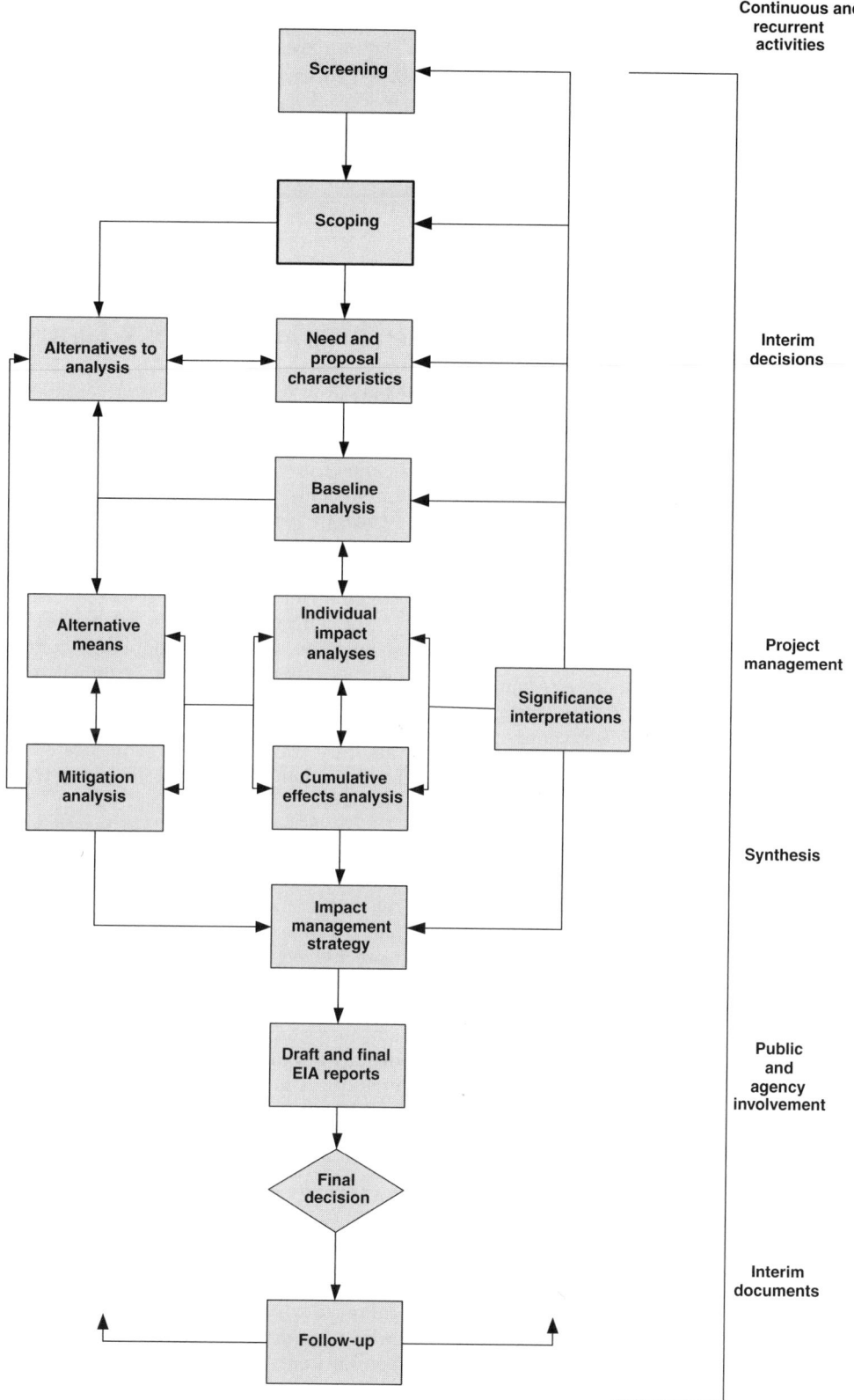

Figure 1. Example of the EIA Process.

Frequently, quantitative risk assessment, the systematic consideration of uncertainties, risk and uncertainty management, and an open and collaborative approach to perceived risks also are necessary.

EMERGING PRIORITIES

Private corporations and public agencies are increasingly using environmental management systems (EMS) to proactively and independently demonstrate their performance in addressing environmental and social concerns. An EMS can frame a proposed project EIA by providing environmental policies, planning and management review procedures, checking and corrective action protocols, and implementation and operational procedures. Having an EMS ensures that an environmental management structure with clearly defined environmental policies, responsibilities, training procedures, controls, communications, and monitoring and auditing procedures is in place.

EMS can provide an environmental and impact prediction baseline, a focused means of effectively anticipating, avoiding, and minimizing potential adverse environmental impacts, a structure for instituting mitigation and monitoring measures, and a departure point for community liaison and agency review. Life cycle assessment (LCA), an EMS tool, can help systematically and comprehensively assess a proposed action and its alternatives over the life of an activity.

The increased stress on EIA quality, effectiveness, and good practice demonstrates that simply meeting current EIA requirements is necessary but not sufficient. Environmental professionals also need to consider evolving standards of good regulatory and applied practice. By doing so they will satisfy professional ethical standards and anticipate emerging general, and likely future, project-specific EIA requirements and guidelines.

EIA commentators and practitioners commonly acknowledge that physical/biological and social/economic impacts are highly interrelated, that cumulative effects cannot be adequately addressed if social and economic effects are not fully considered, and that public concerns often focus on potential social and economic impacts. EIA requirements, to this point, have largely lagged behind in adequately addressing social and economic concerns.

Increasingly, EIA processes are becoming more collaborative, to involve more than periodic, tightly circumscribed, public involvement. They now include two-way information exchanges, incorporate more continuous forms of public participation, and provide for and facilitate dialogue, mutual education, negotiations, and joint and creative planning, management, and collaboration. Facilitators, mediators and other third parties often aid the procedure. Sometimes measures, such as participant funding, facilitate the involvement of traditionally under-represented parties.

R. E. Bass, A. I. Herson and K. M. Bogden, *The NEPA Book: A Step-by-Step Guide on How to Comply With The National Environmental Policy Act*, Solano Press Books, Point Arena, Calif., 2001.

H. Becker and F. Vanclay, *The International Handbook of Social Impact Assessment: Conceptual and Methodological Advances*, Edward Elgar, Northhampton, Mass., 2003.

D. P. Lawrence, *Environmental Impact Assessment*, John Wiley & Sons, Inc., Hoboken, N.J., 2003.

C. Wood, *Environmental Impact Assessment: A Comparative Review*, 2nd ed., Longman Group, United Kingdom, 2002.

DAVID P. LAWRENCE
Lawrence Environmental

ENZYME APPLICATIONS, INDUSTRIAL

Enzymes are designed to accelerate reactions taking place in the cell and its immediate surroundings, and are essential for the development and maintenance of life. These biomolecules are proteins composed of up to 20 different amino acids. By varying the structure and, for some enzymes, by incorporating cofactors such as metal ions and organic molecules like flavins and porphyrin a surprising diversity of activities is displayed among the several thousand enzymes known.

The effective catalytic properties of enzymes have opened the way for introducing them into a number of products and processes. Applied enzymology forms an important branch of industrial biotechnology. In all areas of application, enzymes are used to effectively facilitate transformations that are not technically feasible by other means or to replace traditional chemical processes, as enzyme-based operations inherently offer cleaner processes with less waste and impact on the environment.

The present industrial applications of enzymes are the result of rapid development seen primarily over the past four decades thanks, first of all, to the evolution of modern biotechnology. The latest developments in this technology introducing protein engineering and directed evolution have further revolutionized the development of industrial enzymes and made it possible to tailor-make enzymes to display new activities and adapt to new process conditions enabling a further expansion of their use. The result is a highly diversified industry that is still growing in terms both of size and complexity.

CATALYTIC ACTIVITY

Enzymatic Catalysis in General

Enzymes are catalysts; ie, they increase the rate of chemical reactions without undergoing permanent change themselves by doing so and without affecting the reaction equilibrium.

The characteristics of enzymes are their catalytic efficiency and their specificity. Enzymes increase reaction rates by factors of often 10^6 or more compared to the uncatalyzed reaction. Many enzymes are also highly specific, and consequently a vast number exist. For example, most hydrolytic enzymes acting on carbohydrates are so specific that even the slightest change in the stereochemical configuration of the substrate is sufficient to make the enzyme unable to catalyze hydrolysis.

Another characteristic of enzymes is their frequent need for cofactors or prosthetic groups. A cofactor is a non-protein compound that combines with the otherwise inactive enzyme to give the active enzyme.

According to common theory, enzymes accelerate reactions by stabilizing the transition state, the highest energy species on the reaction pathway, thereby decreasing the activation barrier. In other words, the combination of enzyme and substrate creates a new reaction pathway whose transition-stae energy is lower than it would be if the reaction were taking place without the participation of the enzyme. Compounds that mimic the structure of the transition state often bind strongly to the active site of the enzyme and may be potent inhibitors of the enzyme-catalyzed reaction.

Enzyme Kinetics

A simple enzyme-catalyzed reaction with a single substrate can often be described as

$$\mathrm{E} + \mathrm{S} \underset{k_{-1}}{\overset{k_1}{\rightleftharpoons}} \mathrm{ES} \xrightarrow{k_{cat}} \mathrm{E} + \mathrm{P}$$

The enzyme, E, and the substrate, S, initially combine to form an enzyme–substrate complex, ES. In the second step the catalytic process occurs, whereby the enzyme is released again and the product or products, here denoted P, are formed. This step is controlled by a first-order rate constant k_{cat}, called the catalytic constant or the turnover number. When deriving kinetic expressions, it is generally assumed that the concentration of enzyme is negligible compared to the concentration of substrate. Furthermore, it is assumed that what is being measured is the initial rate V of formation of products; ie, the rate of formation of the first few percent of the products. Under these conditions the products have not accumulated, the substrates have not been depleted, and the reaction rate is generally constant over time. It is often found experimentally that V is directly proportional to the concentration of enzyme $[E]_0$ and varies with the substrate concentration $[S]$. At low $[S]$, V is proportional to $[S]$. At higher $[S]$, however, this relation begins to break down, and at sufficiently high $[S]$, V tends toward a limiting value V_{max}. The Michaelis-Menten equation expresses this relation quantitatively.

Enzyme Inhibition

Enzyme inhibitors are reagents that bind to the enzyme and cause a decrease in the reaction rate. Irreversible inhibitors bind to the enzyme or change its molecular structure by an irreversible reaction and consequently then effect cannot be reverted by such processes as dilution or dialysis.

Reversible inhibition is characterized by equilibrium between enzyme (E), inhibitor (I) and the enzyme–inhibitor complex: $E + I \rightleftharpoons EI$. Many reversible inhibitors are substrate analogues and bear a close relationship to the normal substrate or to the transition state. When the inhibitor and the substrate compete for the same site on the enzyme, the inhibition is called competitive.

Effect of Temperature and pH

The temperature dependence of enzymes often follows the rule that a $10°C$ increase in temperature approximately doubles the activity. However, this is true only as long as the enzyme is not deactivated by the thermal denaturation characteristic for enzymes and other proteins. Most enzymes have temperature optima between 40 and $60°C$. However, enzymes with optima near $100°C$ and with optima at low temperatures exist as well.

The pH dependency of an enzyme-catalyzed reaction also typically exhibits an optimum. Currently applied detergent proteases have a broad pH optimum in the alkaline range. Other enzymes have a narrow pH optimum. The nature of the pH profile often gives clues to the mechanism of the enzyme-catalyzed reaction. Generally speaking, the pH optima for different enzymes span a wide range on the pH scale. Temperature profiles may vary with varying pH, and pH profiles may vary with varying temperature. Both types of profile may also vary with other reaction conditions such as the identity of the substrate, substrate concentration and buffer composition.

Enzyme Assays

An enzyme assay determines the amount of active enzyme present in a sample. Typically, enzymes are present at very low levels in application situations, so it is not practical to determine enzyme content on a stoichiometric basis. Enzyme activity is usually determined from a rate assay and expressed in activity units. As mentioned, a change in temperature, pH, and/or substrate concentration affects the reaction rate. These parameters must therefore be carefully controlled in order to achieve reproducible results.

Spectrophotometry, a simple and reliable technique, is often used in rate assays. Potentiometry is another useful method for determining enzyme activity in cases where the reaction liberates or consumes protons.

ENZYME NOMENCLATURE

Enzymes are classified on the basis of the reactions they catalyze. Six main reaction types are recognized, leading to six main classes of enzymes (Table 1).

Table 1. The Six Main Classes of Enzymes

Number	Class	Type of reaction catalyzed
1	oxidoreductases	transfer of electrons
2	transferases	group-transfer reactions
3	hydrolases	hydrolysis, i.e. cleavage of substrate with the addition of water
4	lyases	addition of groups to double bonds or elimination of groups to leave double bonds or rings
5	isomerases	transfer of groups within substrate molecules to yield isomers
6	ligases	formation of C–C, C–S, C–O, and C–N bonds by condensation reactions coupled to ATP cleavage

Each class is divided into subclasses and subsubclasses according to details of the reaction catalyzed. These categories are then characterized by the nature of the substrate (eg, 3.1.1 comprises carboxylic ester hydrolases). To the three digits are added serial numbers, and the resulting four digits are prefixed by "EC"; thus one arrives at classification numbers such as EC 3.1.1.3 for esterases acting on triacylglycerols (triglycerides). A systematic name is given to reflect the described classification, in this case "triacylglycerol acylhydrolase". If the systematic name proves cumbersome, the enzyme list often includes "other names", in this case "lipase".

PRODUCTION OF INDUSTRIAL ENZYMES

Today almost all industrial enzymes are produced by microorganisms by submerged fermentation. A tiny fraction are still made by surface culture fermentation. A handful of enzymes are extracted from animal or plant sources.

Enzymes are usually sensitive to harsh physical and chemical conditions, and care must be taken during the whole production process to avoid inactivation of the enzyme. This demands careful selection of production processes and conditions for each individual enzyme. Different formulation methods are subsequently applied to assure the stability and activity of the enzymes during storage and application.

Recovery

The principal purpose of recovery is to remove nonproteinaceous material from the enzyme preparation. Most industrial enzymes are produced as extracellular enzymes; ie, released into the medium by the microorganism with only a few other components. The first recovery step is often the removal of whole cells and other particulate matter by centrifugation or filtration. In the case of centrifugation, different versions of centrifuges or decanters are used. Filtration can either be done by pressure, vacuum, or cross-flow filtration. In the rare cases of cell-bound enzymes, the harvested cells can be used as is or disrupted by physical (eg, bead mills, high-pressure homogenizers), and/or chemical (eg, solvent, detergent, lysozyme, or other lytic enzyme) techniques. Enzymes can be extracted from disrupted microbial cells, and ground animal (trypsin) or plant (papain) material by dilute salt solutions or aqueous two-phase systems.

Ultrafiltration is increasingly used to remove water, salts, and other low molecular-weight impurities after the first recovery step. Water may be added to wash out impurities; ie, diafiltration.

Purification. Enzyme purity, expressed in terms of the percent active enzyme protein of total protein, is primarily achieved by the strain selection and fermentation method. In some cases, however, removal of nonactive protein by purification is necessary. A key purification method is crystallization, which is controlled by the combination of pH, salt, concentration, etc. The method has to be carefully developed in order to obtain the desired purity, yield, and acceptable processing times. Other techniques described to purify proteins are often too costly for industrial enzymes

and are used only in situations where the application demands a high purity. These include chromatography, aqueous two-phase extraction, and ion exchange.

Formulation

When an enzyme is recovered from the fermentation broth, it is usually present in an aqueous solution or processed to a dried state. Both types of preparation have to be formulated into a final product to comply with requirements appropriate to their final application.

A very important requirement is related to the storage of enzymes from the time of manufacture to the time of application; ie, the catalytic activity of the enzyme must be maintained after prolonged storage at relevant temperatures and humidity. In the case of a liquid formulation, other stability issues like microbial and physical stability are important as well. As industrially recovered enzymes in aqueous solution are potentially excellent growth media for microorganisms, it is usually necessary to prevent microbial growth; also it is usually required that the enzyme remain in solution to avoid inhomogeneous dosage. Some applications have special requirements. These can be of a technical nature such as having no precipitate and specifications regarding color and odor, or the product should not give rise to any unwanted effects in the application when eg, mixed into a liquid detergent. Side activities are often also a problem to address; eg, transferase activity must be absent in saccharifying enzymes like amyloglucosidase, and protease must be absent in cell wall degrading enzymes for the upgrading of vegetable proteins. Other requirements can be dust level, particle size, and flow ability in solid formulations. Finally, certain requirements derive from approval considerations; eg, only food-grade ingredients, absence of certain microorganisms, and kosher/halal restrictions on enzymes for food applications.

Any formulation is a compromise between the previously mentioned requirements and the necessary cost associated with formulation of the final industrial enzyme product. For example, the fermentation broth may contain enzyme-stabilizing substances, but the application of the enzyme or precipitation problems in the formulation may demand a high degree of purification that eliminates the stabilizers.

Immobilization

Cost reduction is the primary argument for using immobilized enzymes, especially when comparing this method with soluble enzyme or nonenzymatic methods; nevertheless, satisfactory technical solutions can be found among the latter two alternatives. However, enzyme production methods are constantly being improved and costs reduced accordingly. Costs in this context also include process costs; eg, equipment, energy, product recovery, and enzyme inactivation. Enzyme stability factors; eg, temperature, pH, proteases, oxidation, and solvents/organics, are also important but are not often regarded as a cost issue because the desired stability is not always found with soluble enzymes. Continuous processes involving immobilized enzymes enable large substrate volumes to be handled by comparatively small reactors and allow the reuse of enzymes. Further, the advantage associated with a continuous process is achieved; eg, better process control, optimizing product yield, and quality. A significant advantage of immobilized enzymes is the total absence of catalytic activity in the product.

Immobilization Methods. Because enzymes can be intracellularly associated with cell membranes, whole microbial cells, viable or nonviable, can be used to exploit the activity of one or more types of enzyme and cofactor regeneration; eg, alcohol production from sugar with yeast cells. Viable cells may be further stabilized by entrapment in aqueous gel beads or attached to the surface of spherical particles. Otherwise cells are usually homogenized and cross-linked with glutaraldehyde to form an insoluble yet penetrable matrix.

Extracellular microbial enzymes can be immobilized in the form of proteins purified to varying degrees. Other immobilization methods are based on chemical and physical binding to solid supports; eg, polysaccharides, polymers, glass, and other chemically and physically stable materials, which are usually modified with functional groups such as amine, carboxy, epoxy, phenyl, or alkyl to enable covalent coupling to amino acid side chains on the enzyme surface.

Membrane reactors, where the enzyme is adsorbed or kept in solution on one side of an ultrafiltration membrane, provide a form of immobilized enzyme and the possibility of product separation.

Microemulsions or reverse micelles are composed of enzyme-containing, surfactant-stabilized aqueous microdroplets in a continuous organic phase. Such systems may be considered as a kind of immobilization in enzymatic synthesis reactions.

The choice of a suitable immobilization method for a given enzyme and application is based on a number of considerations, including previous experience, new experiments, enzyme cost and productivity, process demands, chemical and physical stability of the support, approval and safety issues regarding support, and chemicals used. Enzyme characteristics that greatly influence the approach include intra- or extracellular location; size; surface properties, eg, charge/pI, lysine content, polarity, and carbohydrate; and active site; eg, amino acids or cofactors. The size, charge, and polarity of the substrate should also be considered.

Industrial-Scale Applications of Immobilized Products. When immobilized glucose isomerase, based on cross-linking with glutaraldehyde of cells, was introduced in the early 1970s, it was believed that other industrial applications of immobilized enzymes would soon be found; however, this has turned out not to be true. One of the reasons is the high and expensive demands to the purity and solubility of the substrates. In the case of glucose isomerase, the cost of purifying the syrup before entering it into the enzyme reactor is substantially higher than the enzyme cost. Further, molecular biology techniques have improved dramatically in efficiency over the last decades resulting in significant reduction of enzyme production costs, omitting the need for immobilization. So instead of initiating an era of immobilized products, immobilized glucose isomerase today stands as a special case of efficient use of an intracellular enzyme. A few other immobilized enzymes have been commercialized, but so far not to the same scale as glucose isomerase. They include aminoacylases for production of amino acids, penicillin acylase for production of 6-aminopenicillanic acid, and lipases for both interesterification of fats and oil as well as for synthesis of esters. They have demonstrated the feasibility of using immobilized enzyme technology in nonaqueous or low water systems.

Solid Formulations—Granulation

Although the trend is to market industrial enzymes as liquid products, solid enzymes still account for a significant part of the total volume of industrial enzymes, especially in the segment of enzymes for solid detergents, animal feed, textile, and flour improvement. Granulation is the generic term for a particle-size enlargement process. A granulate is preferred to powders, especially in order to secure a dust-free product.

Several different methods have been used for the granulation of enzymes. In general, the development of granulation methods focuses on parameters such as the cost of the process and additives as well as the solubility of the granulate. The mechanism of granulation is very complex and difficult to model, making it very empirically based.

To reduce the enzyme dust level in detergent factories to an absolute minimum, the majority of detergent enzyme granulates are coated with a layer of inert material, typically an organic polymer and/or salt. This coating layer can also be used for coloring purposes.

Another important segment for solid formulated enzymes is the feed industry. The same granulation methods are used as for detergent enzymes, except in this case stability of the enzyme during high temperature pellet formation of the final feed is a very critical parameter. A good pelleting stability can be achieved by applying to the granulate an extra coating of vegetable oil or other materials. In the case of enzymes for food application, some of the additives needed for the production of a rigid granulate may not be accepted. Fortunately, for such applications the handling of the enzyme is more gentle, and the requirement for physical stability less. For these enzymes, a fluid-bed granulation performed as an agglomeration of powder with a liquid binder, integrated spray drying and fluidized-bed granulation, or a simple fluidized-bed granulation of the enzyme onto inert carrier particles of a selected size often gives the desired quality of product. These enzyme preparations may also be coated.

INDUSTRIAL APPLICATIONS OF ENZYMES

Detergent Enzymes

The term "detergents" here means cleaning products in a broad sense: not only products for household laundering, including soaking and topical spot removal, but also automatic dishwashing detergents (ADDs) and products for a wide range of industrial and institutional (I&I) cleaning functions.

The penetration of enzymes into laundry detergent products is close to 95% in Europe and Japan and 75% in the United States and is increasing rapidly in other regions, including Latin America and China. Thus, enzymes have become an important ingredient, along with surfactants, builders, and bleaching systems.

An important task for any detergent is to remove visible soilings. Stains with good water solubility are easily removed during the washing process. All other stains are partially removed by the surfactant/builder system of a detergent, although the result is often unsatisfactory, depending on the conditions. In many cases, a suitable detergent enzyme may help. Contrary to the purely physical action of the surfactant system, enzymes work by degrading the dirt into smaller and more soluble fragments. However, total removal of a stain requires the joint effects of the enzyme, surfactant system, and mechanical agitation. In cases of "anchored dirt" in laundry, an enzyme may assist in removing the dirt even though it does not attack it directly.

A still more sophisticated effect of the degradation of such invisible films of sticky materials is the prevention of redeposition of dirt and colorants. Some items in a wash load may release particulate soils or colorants (whether originating in food stains or excess dye in dyed fabrics) to the wash liquor, and these may then bind to the film-covered areas. Inclusion of enzymes that keep away the sticky residues prevents this kind of dingy build-up. This may be termed an *antiredeposition effect*.

Not all enzymes with a potential for stain degradation and/or removal are suitable for inclusion in detergent products. A detergent enzyme must have good activity at the pH of detergent solutions (between 7 and 11) and at the relevant wash temperatures (20–60°C), and must be compatible with detergent components (eg, surfactants, builders, bleaches, and other enzymes) during storage as well as during the wash process. In particular, such an enzyme must be resistant toward protease degradation under these conditions. With enzymes like proteases and lipases, for which the average load of dirty laundry contains a multitude of different substrates, broad substrate specificity is demanded.

A given enzyme may be assayed analytically by its action on soluble or insoluble substrates under chemical and physical conditions different from those encountered in a real-life wash. Such experiments may give information on the enzyme's relative activity as a function of pH and temperature or its compatibility with other soluble substances, etc. The analytical data thus obtained, however, do not necessarily correlate with the wash performance of the enzyme, which must be evaluated in more realistic wash trials.

Detergent enzyme performance is often reported in the form of what are called dose-response curves. The performance increases dramatically at the beginning but reaches a maximum level at higher enzyme concentrations. The extent to which the enzyme is able to remove stains from the fabric depends on the detergent system, temperature, pH, washing time, wash load, etc. Enzyme wash performance in particular varies between liquid and powder detergents and with the composition of the soiling.

Enzyme Types. *Proteases.* A protease is an enzyme that hydrolyzes proteins into smaller fragments, ie, peptides or even amino acids. A detergent protease must degrade the insoluble protein in stains into fragments that can be removed or dissolved by the detergent. All currently used detergent proteases belong to the class of serine proteases and are based on enzymes naturally produced by *Bacillus* strains. Some are highly alkaline, with a maximum activity in the high pH range (~9.5–11); others are low alkaline, work best at pH <9.5. They all have a molecular weight between 20,000 and 30,000.

Protease performance is thus strongly influenced by the detergent's pH and also by ionic strength; the two parameters are not independent.

In general, surfactants influence both enzyme performance and stability in the wash solution, and this is particularly true for proteases. On the one hand, surfactants are often needed for a protease to exert its cleaning effect to a measurable degree. On the other hand, surfactants may be aggressive toward the enzyme molecules. All in all, however, the stability of proteases in a washing context is not problematic in practice.

Chlorine bleach (sodium hypochlorite, NaOCl) is not incorporated into laundry detergents but is used separately in some parts of the world as an additive. Added in this way, it quickly oxidizes most enzymes, and certainly proteases, resulting in a complete loss of activity. In fact, such inactivation may happen already at the much lower levels of hypochlorite encountered in areas with tap water chlorination. Addition of low levels of a hydrogen peroxide source to the detergent may counteract this by reducing hypochlorite to chloride.

Amylases. Commercial laundry amylases such as the amylase from *B. amyloliquefaciens* and the heat-stable amylase from *B. licheniformis* are α-amylases and are characterized by attacking the amylose part of starch in an endo fashion, randomly cleaving internal glycosidic α-1,4-bonds to yield shorter, water-soluble dextrins. The α-amylases have become a common ingredient in both powder and liquid formulations in many countries. They boost overall detergent performance at lower wash temperatures and with milder detergent chemical systems. They improve the removal of stains containing starch, such as a wide range of food stains, and contribute to general whiteness maintenance by preventing redeposition. A noticeable amylase effect is obtained in the main wash as well as with prespotting.

Examples of artificially soiled test pieces used to test the performance of amylases include cocoa/milk/sugar, cocoa/sugar/potato starch, cocoa/milk/sugar/potato starch, and starch/carbon black, all on cotton or polyester/cotton.

Bacterial α-amylases used in laundry detergents are fully compatible with detergent proteases; ie, the two enzymes work together in the wash process. In fact, on combination soilings like the cocoa test materials mentioned, a synergy may be observed between the amylase and the protease action. Also, during storage in both powder and liquid detergents the amylases are very stable in the presence of proteases.

The main problems with amylases are their dependency on calcium ions for structural stability and their sensitivity toward bleaching agents.

Lipases. The first detergent on the market to contain a lipase was a compact powdered detergent launched in 1988. Since then, the development of lipases for detergents has continued.

Animal and vegetable fats consist mainly of triglycerides or tri-*O*-acylglycerols (TAGs), which are natural lipase substrates. The enzymes catalyze hydrolysis of the ester bonds in TAGs to give a mix of free fatty acids, diacylglycerols, monoacylglycerols, and glycerol. The more degraded the fatty stain becomes, the easier it is to remove from the fabric due to its increased hydrophilicity. Because of the presence of free fatty acids in the mix of hydrolysis products, pH strongly influences the removal of decomposed stains. Thus, for optimum removal, a pH value >8 is required. Conversely, the liberation of acid makes it possible to monitor the lipolytic process at constant pH by using the technique of pH-stat titrations.

The lipase itself needs to be well suited for approaching the hydrophobic substrate that often is a solid at the wash temperature and needs to have good activity when located at the interface. Many lipases in their resting conformation in an aqueous environment have a lid or flap that covers the active site and that needs to be opened by appropriate interaction of the enzyme molecule with the surface.

Effective decomposition of fatty soils is now possible in a single wash cycle and at low temperature, thanks to the application of lipases. Detailed electron microscopy and radiotracer studies have shown that lipases may even contribute to the removal of fatty residues in the interior parts of cotton fibers where the surfactants that remove the fat from fiber surfaces apparently do not work effectively. Also, the antiredeposition effects described above; such effects can be very pronounced with lipases and hydrophobic colorants that get absorbed into fatty residues.

Cellulases. The term "cellulases" covers a broad class of enzymes produced by microorganisms (both fungi and bacteria) with the purpose of degrading cellulose to low-molecular saccharides. Typically, cellulases are secreted by the organisms in question as complexes comprising several specialist hydrolases that each contribute in their own way to the overall degradation process by either attacking internal positions of the polymer chain or cutting off terminal oligosachharides.

A special feature of many cellulases is their cellulose-binding module (CBM), a part of the enzyme molecule that has the function of binding to the initially insoluble substrate. The rest of the molecule, including the active site region, is tethered to the substrate surface via a flexible linker region.

Cellulases may be assayed in the laboratory and classified according to their effects across a series of cellulosic substrates. A favorite water-soluble substrate is CMC (carboxymethylcellulose). Insoluble ones include cross-linked and dyed, and acid-swollen preparations of real cellulose.

Taken as a class, cellulases provide a range of benefits when washing cotton and cotton blends, especially over several wash cycles. These include color care (retainment of bright colors; rejuvenation of colors that have become dull through wear); fabric care (providing softness); cleaning (removal of soilings, including particulate matter as well as other materials); and antiredeposition (preventing the redeposition of soils from the wash liquor).

Mannanases. A great number of household products, foodstuffs, and cosmetics contain natural gums; complex carbohydrates that play a role as formulation aids to provide the right texture or consistency in products. Guar gum, a polysaccharide, is one such example. Mannanases, enzymes that degrade mannans, are further examples of glycoside hydrolases, in addition to amylases and cellulases. The inclusion of a mannanase in a detergent may provide both stain-removal and antiredeposition effects just as with proteases, amylases, and lipases.

Oxidoreductases (redox enzymes). It has long been thought that one could construct a detergent bleaching system using one or more of the hydrogen peroxide or dioxygen-based types of enzymes. The mere generation of hydrogen peroxide using an oxidase and a suitable substrate has never materialized in detergents, probably because it is difficult to compete against the traditional hydrogen peroxide sources (sodium perborates and sodium percarbonate), both cost-wise and in terms of space taken up in the detergent formulation. Also, the substrate itself and its oxidation product would have to be in all ways compatible with the wash process. An alcohol oxidase using methanol as substrate and producing formaldehyde would hardly be acceptable, even though the substrate would be very weight-effective.

Peroxidase systems should be useful for bleaching fugitive dyes in solution. Some textile dyes are substrates themselves for peroxidases, and by introducing a mediator, an auxiliary compound that becomes oxidized by a peroxidase (or laccase) and then itself oxidizes dye molecules, a broad range of dyes can be bleached in this way. However, no such system has not implemented in commercial detergents.

Stain bleaching is a complicated task for any bleaching system because there needs to be a good effect on the unwanted stains combined with a low risk of bleaching textile dyes on garments. Peroxycarboxylic acids satisfy these criteria and are tough to compete with. Although many patent applications have proposed the use of peroxidase and laccase systems for this purpose, no such system has been commercialized. A laccase system that bleaches dyes located in the textile fibers has been put to good use in the denim industry. Enzymatic systems for producing the technically attractive peroxycarboxylic acids have been proposed, but there are good arguments why hydrolytic systems have not succeeded in producing these acids.

Automatic Dishwashing. There are many differences between laundering and automatic dishwashing (ADW). The hard surfaces present in the latter process differ from textiles because they are impermeable to soils; therefore, cleaning fluids have better access to the soils. ADW detergent compositions are quite different, with only very little surfactant, present primarily to prevent excessive foaming.

The automatic dishwashing detergents (ADD) enzymes in current use are heat-stable proteases and α-amylases. The actions of the enzymes in ADDs are similar to those in laundering. Starch soils are considered the most stubborn kind of soil on kitchenware. This applies to freshly formed deposits as well as to the starchy film that tends to build up on plates, leaving them with a dull appearance. Several α-amylases exist with a high temperature optimum, ~70°C, which are efficient at removing starch film even under the harsh conditions in a dishwasher.

There are also commercial proteases with a high temperature optimum (~60°C) that can remove most protein soils in a dishwasher if the latter are not totally denatured. Some protein residues present particular challenges, most notably eggs that contain, particularly in the egg white, inhibitors that reduce the performance of many serine proteases.

Some soils, such as food residues cross-linked by heat, are difficult to remove and are note obvious substrates at all for enzymes.

Industrial and Institutional (I&I) cleaning. The application of enzymes has grown substantially within the I&I sector. The primary field here is the use of detergent enzymes for laundry purposes, but a broad range of other applications has been investigated; I&I dishwashing, membrane cleaning, drain and bowl cleaning, cleaning of septic tanks and sewage plants, hard surface cleaning of walls or machinery parts where a cleaning-in-place procedure can be used, and cleaning of apparatus parts like endoscopes and electrodes. The choice of the relevant enzyme type is directly related to the composition of the soils, waste, or deposit that has to be dissolved and removed. Thus, as for household laundering, proteases, lipases, and various types of glycoside hydrolases can be considered. Proteases are suitable for use on fabrics heavily soiled with blood and/or meat residues; eg, from hospitals and the food industry, in particular slaughterhouses. Fatty stains are removed efficiently from restaurant tablecloths and napkins by the addition of a lipase to the detergent. Residues of starchy foods such as mashed potatoes, spaghetti, hot oatmeal, and chocolate are cleaned with the use of amylases.

An interesting situation occurs with burnt milk residues on heat exchangers and ultrafiltration modules in the dairy industry. Here, by applying a combination of a protease and a lipase, the milk-based substrate itself upon hydrolysis provides both emulsifiers and foaming agents for the cleaning process, in the form of peptides and fatty acids.

I&I cleaning procedures, compared with household laundering, are characterized by huge variations in the composition of the soils, the types of surface to which they adhere, the cleaning time available, etc. The optimum choice of enzyme type and dosage level normally has to be established through cooperation between the customer, the manufacturer of the detergent, and the enzyme producer.

Enzymes for Grain and Biomass Processing

Enzymes are used by grain-based industries in the manufacturing of a variety of products. The primary substrates are starch-containing grains, such as corn, wheat, and tapioca, that are converted into value-added food and industrial ingredients. In the brewing industry, enzymes are added to supplement the natural enzymes in malt to provide a number of benefits. Furthermore, enzymes are used to generate fermentable sugars in the production of ethanol from various starch-containing crops. Currently, intense efforts are being devoted to the development of enzyme-based processes for the production of ethanol based on lignocellulosic materials to provide a bio-fuel competitive with fossil fuels.

By controlling the enzymatic reactions, valuable syrups and modified starches with different compositions and physical properties can be obtained. These tailor-made glucose syrups are used in a wide variety of foodstuffs such as soft drinks, confectionery, meat products, baked products, ice cream, sauces, baby food, and canned fruit. Depending on the degree of degradation or modification desired, the processes of liquefaction, saccharification, and glucose isomerization may be applied.

Enzymes for Ethanol Production

The main process stages in alcohol production from starch-containing crops comprise: (*1*) dry milling; (*2*) gelatinization; (*3*) dextrinization; (*4*) saccharification; (*5*) fermentation; (*6*) distillation; and (*7*) drying of the stillage.

Ethanol for Fuel. Over the last decade, there has been an increasing interest in fuel alcohol as a result of increased environmental concern, higher crude oil prices and, more acutely, by the ban in certain regions of the gasoline additive MTBE (*tert*-butyl methyl ether), that can be interchanged directly with ethanol. Therefore, intense efforts are currently being undertaken to develop improved enzymes that can enable utilization of cheaper and currently not fully utilized substrates such as lignocellulose to make bio-ethanol more competitive with fossil fuels. The cost of enzymes required to turn lignocellulose into a suitable fermentation feedstock is a major issue. Current developments focus both on the development of enzymes with increased activity and stability as well as on their efficient production. Huge governmental programs have been launched in the United States by the Department of Energy to support these developments.

Ethanol Production Using Lignocellulose-Based Raw Materials. Degradation of biomass using cellulases has been a major research area for >30 years; however, it has not yet been economically feasible, partly due to the complex substrates, cost of enzymes, and overall lack of efficiency. To improve process economy, the major focus areas at present are to reduce the cost of industrial cellulases significantly and to improve the activity and thermostability of the enzymes.

Food Applications

A number of features make enzymes ideal catalysts for the food industry. They are all natural, efficient, and specific; they work under mild conditions; they have a high degree of purity; and they are available as standardized preparations. Because enzymatic reactions can be conducted at moderate temperatures and pH values, simple equipment can be used, and few by-products are formed. Furthermore, enzymatic reactions are easily controlled and can be stopped when the desired degree of conversion is reached.

Dairy. Milk is processed into a variety of products. Lactase (β-galactosidase) is used to manufacture milk products with a reduced content of lactose by hydrolyzing it to glucose and galactose. Many people are lactose intolerant and do not have sufficient lactase to digest lactose. By using lactase, lactose can be broken down and a whole range of lactose-free milk products made. Manufacturers of milk-based drinking products, ice cream, yogurt, and frozen desserts use lactase to improve digestibility, sweetness, scoop, and texture of the products.

Another important application of enzyme technology used in the dairy industry is the modification of proteins with proteases to reduce the allergenicity of cow milk for infant formula products. A smaller enzyme application within the dairy industry is the hydrolysis of milk with lipases for the development of flavors in specialty cheeses.

Baking. Wheat flour contains enzymes, the most important being amylases. However, the quantities of these enzymes are not always ideal for baking purposes, and supplementary enzymes are often added. Traditional applications of enzymes are for improvement of the dough properties during processing (especially industrial processing of bread), loaf volume, crumb structure, and shelf-life. The enzyme products used are either microgranulates that are easy to handle and freely mixed with flour or liquid formulated enzymes. Another important application of the enzyme within baking is the use of a bacterial maltogenic amylase preparation for antistaling of bread, which increases the shelf life from days to weeks by maintaining the softness and elasticity of the bread. A less important application is the use of neutral bacterial endoprotease used to weaken the gluten in wheat flour, if necessary, or to provide the plastic properties required in a dough used for biscuits.

Protein Modification. Treatment of vegetable or animal protein with proteases is a way to obtain controlled hydrolysis and thus change the flavor, functional, and nutritional properties of food proteins. Different protein raw materials are used with different purposes in mind. Extraction processes with enhanced yields include soy milk, scrap meat recovery, bone cleaning, gelatin, fish/meat stick-water, and rendering of fat. Processes for producing protein hydrolysates used as ingredients in processed foods include soluble soy protein, soluble wheat gluten, foaming wheat gluten, blood cell hydrolyzate, whey protein hydrolyzates, casein hydrolyzates, soluble meat and fish proteins, and gelatin hydrolyzates.

Extraction Processes. Many ingredients used by the food and brewing industries are produced by extraction from plant matter: proteins, starch, sugar, fruit juice, oil, flavor, color, coffee, and tea. Enzyme preparations capable of attacking plant cell walls contain different enzyme activities, eg, pectinase, hemicellulase, and cellulase. Conventional enzyme products within these groups are, however, unable to degrade completely the rhamnogalacturonan backbone of the pectic substances.

Pectinases have been used in fruit juice processing since the 1930s. The enzymes are used to improve the yield of juice, liquefy the entire fruit for maximal utilization of the raw material, improve color and aroma, clarify juice, and break down all insoluble carbohydrates like pectins, hemicellulose, and starch.

Oil from rape seed, coconut, corn germ, sunflower seed, palm kernel, and olives is traditionally produced by a combined process using pressing followed by extraction with organic solvents. Cell-wall degrading enzymes may be used to extract vegetable oil in an aqueous process.

Proteases are widely applied in extraction processes of heparin and chondroitin sulfates and for making yeast extracts.

Animal Feed Application

Enzymes are today widely applied in the feed industry for enhancing the digestibility and thus nutritional value of feed, especially in monogastrics like pigs and chickens, which, unlike ruminants, are unable to utilize fully components in plant-based feed stock. Enzymes like β-glucanases and xylanases or mixtures of various carbohydrases are well known in the industry for increasing the digestibility of feed with a high content of cellulose and hemicellulose. Benefits include increased daily weight gain, increased feed conversion ratio, reduced mortality, and a reduced amount of sticky droppings in chicken farming. In short, enzymes provide increased output at a given feed cost.

During the last 10 years, a significant enzyme application within the feed industry has developed to allow monogastrics to utilize the phosphorus bound in phytic acid in cereal-based feed. Phytase is applied to liberate inorganic phosphorus from phytic acid. Phytic acid contains ~85–90% of the total phosphorus in the plant and is largely inaccessible to monogastrics. Two particular factors have contributed to this development; the ban of bone meal as a source for cheap inorganic phosphorus due to the bovine spongiform encephalopathy (BSE) and the increasing focus on the phosphorus and nitrogen outlet from intensive animal production in many coutries. The addition of phytase results in a significant reduction in the phosphorus outlet from monogastrics. Furthermore, the use of recombinant DNA technology allows for highly cost-efficient production of the phytase, which to some degree closes the gap between utilizing phytase and adding inorganic phosphorus.

Enzymes can be added to the feed together either with the premix, or the granulated enzyme products may be mixed with feed components and subjected to pelleting. Alternatively, liquid enzyme can be applied directly on the final feed.

Enzymes for the Textile Industry

The textile industry was one of the first industries to use enzymes. Crude amylases were introduced as early as the turn of the twentieth century to "desize" (remove) starch from woven fabric, overcoming the fiber-damaging effects of conventional acid-based processes. Since then, enzymes have been introduced in a number of steps in the manufacturing of textiles to provide a variety of benefits.

Enzymes in Pulp and Paper Production

Enzyme-modified starch has been used for adhesives to strengthen paper bases and for surface coating. Developments since the late 1980s of further uses of enzymes in papermaking include pitch control and bleach boosting.

Other Industrial Enzyme Applications

Leather Processing. The main benefits of using enzymes during the different stages of leather manufacturing are reduced process time, increased opening up of fibrous structure, a cleaner surface, increased softness, improved area yield, and a reduced need for chemicals.

Modification of Fats and Oils. A new process for immobilizing lipases based on granulation of silica has dramatically lowered process cost, and processes based on this new material are now used for the production of commodity fats and oils with no content of trans-fatty acids. Another recently introduced process is the removal of phospholipids in vegetable oils (degumming), using a highly selective microbial phospholipase. This enzymatic application has enabled savings of energy as well as water.

Personal Care. Safe applications of enzymes have been developed in several areas. Proteases are used to clean dentures and contact lenses effectively under very mild conditions. Lipases have found some use in contact lens cleaning. The residual hydrogen peroxide used for disinfections can be neutralized using a catalase. Finally, enzymes such as glucoamylase and glucose oxidase are used in certain toothpastes to provide more effective cleaning and an antimicrobial effect.

Oil Drilling. In underground oil and gas drilling, different types of drilling mud are used to cool the drilling head, to transport stone and grit up to the surface, and for controlling the pressure underground. Biopolymers in the mud glue particles together during the drilling process to make a plastic-like coating, which acts as a filter. Enzymatic clean-up processes have emerged as an attractive alternative to strong acid based on enzymes such as cellulases, mannanases, and amylases that effectively degrade the biopolymers used.

Organic Synthesis. Enzymes offer many potential advantages when used as catalysts for chemical synthesis. Their unique properties are an often outstanding chemo-, regio-, and, in particular, stereoselectivity. Furthermore, enzymes are highly efficient catalysts working under very mild conditions. Enzymes do, however, also have some drawbacks that may limit their potential use, such as the ability to accept only a limited number of substrates and a moderate operational stability. Ways of overcoming most of these potential limitations exist and they pose in most cases more of a perceived than a real problem. Well over 100 different enzyme-based processes have now been implemented on an industrial scale.

ENVIRONMENTAL AND SAFETY ASPECTS

The industrial use of microbial enzymes is an important contribution to the development of industrial or white biotechnology. Enzymes have a positive impact on the environment because they replace conventional chemical-based technologies and conventional energy-intensive manufacturing processes, originate from natural biological systems, are totally biodegradable, and leave no harmful residues.

The safety and environmental impact of the production of industrial enzymes can be evaluated at three different levels: the potential risk if the microorganisms, their products, or both are released into the environment; the possible health hazards to staff working with the microorganisms, their products, or both; and safety when products are used by the consumer.

Enzymes are totally biodegradable, and their release into the environment does not cause problems. The release of the production organism itself is controlled by two categories of safety measures that are complementary. The first is physical containment in a fermentor system and recovery plant with a high standard of hygiene. The second is biological containment. Being specially bred, either by traditional techniques or by modern genetic engineering techniques, to produce one specific substance, the production organisms are adapted to grow optimally only under the defined conditions during fermentation. The growth of strains of production organisms in nature is handicapped in comparison with microorganisms already existing in the environment. Their chances of survival in the environment are extremely limited.

Like other proteins, enzymes are potential allergens. In addition, proteases may act as skin and eye irritants. However, during the production and handling of industrial enzymes, the occupational health risks entailed by these properties can be avoided by protective measures and by the form in which the enzyme preparations are supplied. In order to avoid dust generation, enzymes are supplied as liquids, granulates, or immobilized preparations along with safe handling guidance.

To guarantee that enzymes can be used safely by the consumer, microbial enzymes are obtained from nonpathogenic and nontoxinogenic microorganisms grown on raw materials that do not contain compounds hazardous to health.

Genetically engineered microorganisms can be used under the same conditions of containment, and the same safety rules apply as for equivalent, naturally occurring microorganisms. Provided an enzyme is produced by a harmless host, the contained use of recombinant microorganisms does not warrant any special provisions concerning production conditions, worker protection, environmental assessment, field monitoring, or product approval.

Regulatory Aspects

For the use of genetically modified microorganisms in containment, harmonized EU (European Union) legislation applies, and similar regulation is in place in other parts of the world. For the enzyme products, specific legislation is quite limited. Most national authorities have preferred to use or adapt existing legislation on chemicals, food, and feed additives. In the EU, however, specific enzyme guidelines have been prepared for the evaluation of food and feed enzymes, and product approval applies in Denmark and France. For the adaptation of the food additive regulations to fit the processing aids applications of enzymes, guidance has been available in the recommendations of the Joint FAO/WHO Expert Committee on Food Additives (JECFA), and the Food Chemicals Codex (FCC). The enzyme manufacturers' associations, the Association of Manufacturers and Formulators of Enzyme Products (AMFEP) in Europe and the Enzyme Technical Association (ETA) in the United States, work nationally as well as internationally for a harmonization of regulations. The Codex Committee on Food Additives and Contaminants (CCFAC) plays an important role in this work.

Product specifications for microbial food enzymes have been established by the JECFA and the FCC. They limit or prescribe the absence of certain ubiquitous contaminants such as lead, coliforms, *E. coli*, and *Salmonella*. Furthermore, they prescribe the absence of antibacterial activity and, for fungal enzymes only, mycotoxins.

Enzymes are used in feed primarily as active additives. The JECFA/FCC purity specifications for food-grade enzyme have been adopted. Harmonized EU legislation and guidelines apply for the assessment and approval of enzymes used as feed additives.

All enzymes are classified as potential respiratory sensitizers, due to their protein nature. When an enzyme is used for a technical application, its regulatory status is generally determined by its compliance with chemical substance inventories. In some cases, they are exempted as naturally occurring substances. Enzyme manufacturers have developed nondusting product formulations and safe handling guidelines that effectively prevent human exposure, sensitization, and allergies.

ECONOMIC ASPECTS

Worldwide consumption of industrial enzymes amounted to ~$2.15 billion in 2003; about one-third was accounted for by the U.S. market.

The growth in volume of the enzyme business from 1990 to 2003 is estimated to have been 5–10%/year. The technical industry segment, covering a broad range of industries, with the detergent, starch, textile, fuel alcohol, leather, and pulp and paper industries as the most important, accounts for ~60% of total enzyme sales. The detergent industry is still the most important segment accounting for ~35% of total enzymes sales. A few industries, such as the fuel alcohol industry, are, growing at double-digit growth rates. The food and feed segments are both expected to grow at a rate of 10–15%.

W. Aehle, ed., *Enzymes in Industry*, 2nd ed., Wiley-VCH Verlag, Weinheim, 2004.

J. E. Bailey and D. F. Ollis, *Biochemical Engineering Fundamentals*, 2nd ed., McGraw-Hill Book Co., Inc., New York, 1986, Chapts. 3–4.

T. Palmer, *Understanding Enzymes*, 1st ed., Ellis Horwood, New York, 1991.

R. Whitehurst and B. A. Law, eds., *Enzymes in Food Technology*, Sheffield Food Technology, Vol. 8, 2001.

OLE KIRK
TURE DAMHUS
TORBEN VEDEL BORCHERT
CLAUS CRONE FUGLSANG
HANS SEJR OLSEN
TOMAS TAGE HANSEN
HENRIK LUND
HANS ERIK SCHIFF
LONE KIERSTEIN NIELSEN
Novozymes A/S

ENZYME INHIBITORS

Any substance that reduces the velocity of an enzyme-catalyzed reaction can be considered to be an inhibitor. In everyday life, enzyme inhibitors masquerade as drugs, antibiotics, preservatives, poisons, and toxins.

Reversible enzyme inhibition (noncovalent binding) is divided into two types: complete (linear), in which the plots of reciprocal velocity versus inhibitor concentrations give a straight line; and partial (hyperbolic), in which this plot is a hyperbola. With complete enzyme inhibition, the velocity tends to zero when the concentration of the inhibitor increases; with partial inhibition, the enzyme is converted into a modified, but still functional, enzyme–substrate–inhibitor (ESI) complex.

There are three types of inhibition: competitive, noncompetitive, and uncompetitive. The first deals with a substance that combines with free enzymes in a manner that prevents substrate binding. The second has no effect on substrate binding, as the substrate and inhibitor bind reversibly, randomly, and independently at different sites. The third is illustrated by a compound that binds to the enzyme–substrate complex to yield an inactive ESI complex.

Irreversible enzyme inhibition (covalent binding) creates a covalent bond between the inhibitor and the enzyme and may be divided into two phases: The inhibitors first bind to the enzyme in a noncovalent fashion and then undergo subsequent covalent bond formation. There are two fundamental classes: suicide inhibition and mechanism-based inhibitors.

When selective irreversible inhibitors are used to label active site residues of an enzyme to aid in their identification, they are called affinity labels. A suicide inhibitor, on the other hand, is an affinity label that is unreactive until it is acted upon by the enzyme, at which point it binds irreversibly. The fundamental action of a mechanism-based inhibitor depends on the enzymes' catalytic mechanism as the inhibitor, a substrate analogue, which irreversibly modifies the enzyme at a particular step in the catalytic cycle. A knowledge of the catalytic mechanism of the enzyme with its normal substrate, and the introduction of an appropriate latent functional group into the substrate is important, as a fairly unreactive species is converted into a highly reactive one only during the specific catalytic step. Furthermore, the inhibitor must fulfill the binding specificity requirements for the ligand recognition site.

INHIBITOR DESIGN

De Novo

The de novo design of inhibitors requires the three-dimensional (3D) structure of the target enzyme or of a model constructed from related enzymes; or the biological activities and structures of related inhibitors for the particular enzyme; or the pharmacore, which consists of the chemical groups of a ligand and their relative orientations that are important for binding. A compound is then designed that will bind to a molecular site on the enzyme in such a way as to alter its behavior. This compound is next chemically synthesized and tested. A structure–activity relationship is needed to determine the properties of a molecule based on its structure, followed by building a structure based on the desired properties.

Computer Assisted

In most current applications, attempts are made to find an inhibitor (drug) that will interact favorably with an enzyme that represents the target site. Binding of the inhibitor to the enzyme may include hydrophobic, electrostatic, or hydrogen-bonding interactions, and solvation energies that optimize the fit of an inhibitor to an enzyme. Many computer aided inhibitor design systems choose to predict the properties of either the inhibitors (inhibitor based) that operate on the enzyme or the enzyme itself (enzyme based), but not usually both.

The former approach is applicable when the structure of the enzyme site is unknown but a series of compounds has been identified that exerts the activity of interest. One should have structurally similar compounds with high activity, with no activity, and with a range of intermediate activities. In recognition site mapping, an attempt is made to identify a pharmacophore, a template derived from the structures of these compounds, represented as a collection of functional groups in 3D space that is complementary to the geometry of the enzyme.

In applying this approach, conformational analysis will be required. One strategy is to find the lowest energy conformers of the most rigid compounds and superimpose them. Conformational searching on the more flexible compounds is then done while applying distance constraints derived from the structures of the more rigid compounds. Ultimately, all of the structures are superimposed to generate the pharmacophore that is then used as a template to develop new compounds with functional groups in the desired positions. In applying this strategy, one assumes that it is the minimum energy conformers that will bind most favorably in the receptor site though there is no *a priori* reason to exclude higher energy conformers as the source of activity.

The second half of the challenge in developing a new inhibitor molecule is to take the important properties and determine the structure that will have them. Computational and rule-based equations suffer from drawbacks arising from the combinatorial complexity of the search space, design knowledge acquisition difficulties, nonlinear structure–property correlations, and problems incorporating higher-level biological knowledge.

The enzyme-based approach applies when a reliable model of the enzyme is available, as from X-ray diffraction, nuclear magnetic resonance (nmr), or homology modeling. With the availability of the enzyme, the problem is to design inhibitors that will interact favorably at the site. Enzyme-based inhibitor design incorporates a number of molecular modeling techniques, a few of which follow.

QSAR. Enzyme inhibitors may exert their biological effects by participating in a series of events that include transport to the enzyme's active region, binding with the enzyme, and subsequent metabolism to an inactive species. Since the interaction mechanisms between the inhibitor and the putative enzyme are unknown in most cases inferences are made, to explain these interactions, from molecular properties and descriptors for known molecules. Once the relationship is defined, it can be used to aid in the prediction of new or unknown molecules. Since the purpose of a QSAR is to highlight relationships between activity and structural features, it is necessary to find one or more structural features that relate these molecules to their associated activity. Additionally, it would be necessary to find a parameter that works consistently for all molecules.

There are several potential classes of parameters used in QSAR studies. Electronic effects such as electron-donating and electron-withdrawing tendencies, partial atomic charges and electrostatic field densities are defined by Hammett sigma values, resonance parameters (R values), inductive parameters (F values), and Taft substituent values (*, Es). Steric effects such as molecular volume and surface area are represented by values calculated for molar refractivity and the Taft steric parameter. Enthalpic effects are calculated using partition coefficients (log P) or the hydrophobic parameter, which is derived from this partition coefficient. In addition, an assortment of structural indices are used to describe the presence of specific functional groups at positions within the molecule.

CoMFA. This analysis utilizes partial least squares (PLS) and cross-validation to develop inhibitor models for activity predictions. The approach used requires that alignment rules for the series of inhibitor molecules be defined that overlap the putative pharmacophore for each molecule. Once aligned, each molecule is fixed into a 3D grid and the electrostatic (estat) and steric (ster) components of the molecular mechanics force field, arising from interaction with a probe atom, are calculated at intersecting lattice points within this grid.

PLS essentially relies closely on the fact that the correlations among parts of a molecule are similar so that the real dimensionality is smaller than the number of grid points. Since these coefficients are position dependent, substituent patterns for the series are elucidated that define regions of steric bulk and electrostatic charge associated with increased or decreased activity. The number of components needed for the best model and the validity as a predictive tool are assessed using cross-validation.

QSAR with CoMFA. This technique provides tools to build statistical and graphical models of activity from molecular structure, then uses these models to make accurate predictions for the activity of untested compounds. It organizes structures and their associated data into molecular spreadsheets, calculates molecular descriptors, structural, conformational, geometric, electronic, thermodynamic, hydrophobic, and molar refractivity, highest occupied molecular orbital (HOMO) or lowest unoccupied molecular orbital (LUMO) values, and specialised 2D fingerprints (HQSAR), and also performs sophisticated statistical analyses that reveal patterns in structure–activity data.

CoMSIA (Comparative Molecular Shape Indexes Analysis). This method is similar to CoMFA but uses a Gaussian function rather than Coulombic and Lennard-Jones potentials to assess steric, electrostatic, hydrophobic, and hydrogen-bond donor–acceptor fields. If the correct conformation of a molecule is not known, multiple conformers can be stored in a molecular spreadsheet to allow alternative conformers to be considered in a CoMFA or CoMSIA analysis. The results of CoMFA or CoMSIA analyses are displayed as color-coded contours around molecules, allowing visual identification of regions responsible for favorable or unfavorable interactions with the receptor.

HQSAR. This method does not require exact 3D information for the inhibitors but is reflected by a molecular fingerprint that encodes the frequency of occurrence of various molecular fragment types. The fragment size controls both the minimum and maximum length of the fragments to be included in the fingerprint. Molecular holograms are produced by generating all the linear and branched fragments such as atoms, bonds, the number

of hydrogen atoms, and chirality parameters. HQSAR identifies the patterns of substructural fragments related to activity in sets of bioactive molecules by identifying color-coded molecular fragments that have a positive or negative impact on activity.

EDDFA. This method is an improved CoMFA that utilizes eight molecular property fields instead of two. EDDFA uses steric, electrostatic, electronic, kinetic energy densities, fukui function, laplacian, local average ionization potential, and binary node potential fields. Each property field is rapidly generated using properties of the transferable atom equivalent (TAE) electron density distribution. Since inhibitor molecules would interact with enzymes via noncovalent interactions involving electron densities, descriptors using this principle would show high correlation to biological activity. A fine-grained version of the cross-validated guided region section (R2-GRS) routine of Cho and Tropsha is used to identify important regions of space surrounding each molecule in the dataset. Field values in the selected regions are used as descriptors in a PLS regression analysis.

Pharmacophore Perception. *Apex-3D.* This detailed description of the fundamental architecture for computer assisted molecular modeling is an automated pharmacophore identification system that can identify possible pharmacophores from a set of biologically active molecules using statistical techniques and 3D pattern matching algorithms.

Pharmacophores are defined by different chemical centers (atom-centered functional groups) and the distances between these centers. They include aromatic ring centers, electron donor ability hydrogen-bonding sites, lipophilic regions, and having a partial atomic charge. The information for each molecule is stored in a knowledge base in the form of rules that can be used to predict the activity of novel structures. Depending on the type of biological activity available, it is possible to identify pharmacophores for different binding orientations, enzymes, or agonist versus antagonist activity by building a knowledge base.

Once the knowledge base has been constructed, it can predict biological activity for inhibitors not included in the training set. The pharmacophores defined can be used to build 3D QSAR models by correlating indexes calculated for biophore sites, secondary sites, or whole molecule properties.

HipHop. This method searches for types of chemical functional groups within potential inhibitor molecules and chooses conformations and alignments that overlap the groups in space and ignores any intervening backbone atoms in making these comparisons. The result is a series of hypotheses and alignments of inhibitor molecules for possible pharmacophores that ultimately can be used to generate an enzyme–inhibitor model. The advantage of HipHop is that the use of chemical properties in generating alignments is often more realistic in reproducing the details of the molecular recognition mechanism as compared to atom–atom matching procedures. HipHop and atom–atom matching are complementary techniques;

both would be used in suggesting pharmacophores for future study.

Catalyst. This method generates structure–activity hypotheses from a set of molecules of various activities. Once molecular connectivity and activity values are specified for all molecules. Catalyst creates a set of generalized chemical functions at specified relative positions. Up to 10 hypotheses are produced and ranked by estimated statistical significance then examined graphically, "fitted" to new molecules, or fed directly to a flexible 3D database search. In the first step, a set of representative conformers is found that covers the low energy conformational space of each molecule. The second step locates a list of candidate hypotheses common among active and rare among inactive compounds. The theory of minimum complexity estimation indicates that a predictive hypothesis will be the least number of descriptors required as well as a minimum of errors in the activities.

MOLCAD. This method creates graphical images of an enzyme's active region to illustrate the properties of inhibitors that would be necessary for molecular recognition. van der Waals and solvent-accessible enzyme surfaces can be shown, as well as broad range of properties such as lipophilic potential, electrostatic potential, hydrogen-bonding ability, local curvature, and distance. MOLCAD reveals the underlying secondary and tertiary structure, and maps onto these fundamental physical properties such as residue lipophilicity, flexibility based on atomic temperature factors, and packing density. It can characterize the size, shape, and physical properties of intramolecular cavities and channels and examines the specificity of enzyme–inhibitor interactions.

SiteID. The recent explosion in the ability to predict and/or solve macromolecule structures has led to increased interest in methods for modeling the interaction of inhibitors with enzymes. The first hurdle that must be surmounted, however, is to locate the active site or pocket in which the inhibitor binds. SiteID provides analysis and visualization tools leading to the identification of potential binding sites within or at the surface of enzyme targets. SiteID can automatically create a color-mapped database of solvent exposure, hydrogen-bonding character, hydrophobicity, and local surface curvature.

Modeller. Knowledge of the 3D structure of an enzyme is a prerequisite for the rational design of site-directed mutations in the enzyme and can be of great importance for the design of inhibitors. Structural information often greatly enhances understanding of how enzymes function, and model-building on the basis of the known three dimensional structure of a homologous protein is at present the only reliable method to obtain structural information. Modeller is a computer package used for homology or comparative modeling of protein 3D structures.

FlexX. This method is a fast algorithm for flexibly docking inhibitors, using incremental construction to actually build the inhibitors within the binding site of the enzyme. FlexX incorporates enzyme–inhibitor interaction scores, fragmentation of the inhibitor along natural dividing points, inhibitor core placement in the active site, and reconstruction of the complete inhibitor from the fragments.

Dock. The more recent versions of DOCK allow score parameters to be based on force fields, which include both van der Waals and electrostatic interactions. These results illustrate the potential for programs to search objectively for inhibitors that are complementary to enzyme active sites, thereby assisting researchers in identifying potential drugs that may be considerably different from existing ones. Once potential inhibitors have been identified by such methods, other molecular modeling techniques like geometry optimization may be used to "relax" the structures and identify low energy orientations of inhibitors bound to enzyme active site. Molecular dynamics may also assist in exploring the energy landscape, and free energy simulations can be used to compute the relative binding free energies of a series of putative drugs. Many enzymes are membrane bound, making it extremely difficult to determine their 3D structure by nmr or X-ray crystallography. Furthermore, complications may arise since enzymes may change shape as they bind, a process called "induced fit". Existing methods for constructing predictive models are unable to model steric interactions accurately, particularly when these interactions involve large regions of the molecular surface. Likewise, QSAR techniques are accurate only on a small scale, determining properties of specific regions but failing to produce an accurate global description of the molecule. Pharmacophore models attempt to combine some of the advantages of QSAR techniques with the idea of identifying substituents, and advances in 3D QSAR have led to superior characterization of molecules and better calculation of their properties.

LUDI

This method fits inhibitor molecules into the active site of an enzyme by matching complementary polar and hydrophobic groups and uses an empirical scoring function.

Affinity

This method automatically docks inhibitors to enzymes identifying low energy orientations of the inhibitor within the active site and using force-field-based methods to automatically find the best binding mode. This energy-driven method is especially useful in structure-based inhibitor design where the experimentally determined structure of an enzyme–inhibitor complex is often unavailable.

Texture Mapping

Texture mapping is a graphic design by which a 2D surface, or texture map, is wrapped around as 3D object. It is a technique that applies a computer-derived image of an enzymes active region (texture space) by individual elements called texels.

Evolutionary Computing

Currently, most major pharmaceutical companies use rational inhibitor design and evolutionary techniques such as genetic algorithms or genetic function algorithm (GFA) as part of the inhibitor discovery process. SYBYL offers genetic algorithm-based conformational search tools for exploring 3D shapes that inhibitors attain.

Genetic software techniques automatically design inhibitor molecules under the control of a fitness function that must be capable of determining which of two arbitrary molecules is better for a specific task. A population of random molecules is first generated and these are then evolved toward greater fitness by randomly combining parts of the better existing molecules to create new molecules that eventually replace some of the less-fit molecules in the population. A unique genetic crossover operator such as sets of atoms and connector bonds are represented by genetic graphs and can evolve any possible molecule given an appropriate fitness function. Inhibitors are generally small molecules and it is known that they fit precisely into enzyme active sites to block normal molecular processes that may occur in a living system. Furthermore, the inhibitor molecules must survive within such living system. One approach to inhibitor design is to find molecules that are similar to good drugs that have fewer negative side effects and consequently a candidate replacement drug is sufficiently similar to have the same beneficial effect but is different enough to avoid the side effects. Genetic software techniques used for enzyme–inhibitor design, describes the obvious parts of mapping standard genetic algorithm techniques to inhibitor design and the nonobvious portions: the crossover algorithm and fitness function. These use two parameters: the digestion rate that breaks bonds, and the dominance rate that controls how many parts of each parent appear in the descendants. Inevitably this algorithm produces fragments rather than completely connected molecules.

Enzyme–inhibitor design may be viewed as searching the space of all possible molecules for inhibitors with particular properties. The key point in deciding whether or not to use genetic algorithms for a particular problem centers around the question; what is the space to be searched? If that space contains a structure that can be exploited by special-purpose search techniques, the use of genetic algorithms (GAs) is generally computationally less efficient. If the space to be searched is relatively unstructured, and if an effective GA representation of that space can be developed, then GAs provide a surprisingly powerful search technique. It is reasonable to presume that searching the space of molecules using genetic graphs will be profitable in a number of domains.

FUTURE PROSPECTS

Molecular Nanotechnology

With particular reference to enzyme-inhibition, molecular nanotechnology can be defined as the 3D positional control of molecular structure to create the enzyme inhibitors to molecular precision. The availability of molecular nanotechnology permits dramatic progress in human medical services and inhibitor design. More than just an extension of "molecular medicine," nanomedicine is the preservation and improvement of human health using molecular tools and molecular knowledge. It has extraordinary, far-reaching implications for the medical profession, for the definition of disease, for the diagnosis and treatment of medical conditions, and ultimately for the improvement and extension of natural human biological structure and function.

Cures for the major life-threatening diseases could be in sight within the next few years if inhibitors against the enzymatic processes involved in the molecular basis of the diseases could be manufactured using revolutionary nanotechnology. Computer-assisted nanoinhibitor design could be at the cutting edge of the technology and, in conjunction with nanobots, be delivered at relatively high but safe concentrations to any biologically active region. In this way they can "close down" any external pathogen or virus, act as biosensors in the detection of a threat by biological warfare or prevent tumor growth.

Molecular Engineering

Though there are differences of opinion when referring to molecular engineering and the creation of microassemblers, the development of the nanoinhibitor as a microchip and the role of the computer as a delivery vehicle cannot be too far away. Once molecular engineered machines (nanobots) are the order of the day, one can exploit this to make absolute copies of themselves, thereby creating a second level of mass production limited only by the materials and information therein.

H.-J. Böhm and G. Schneider, eds., *Protein-Ligand Interactions: From Molecular Recognition to Drug Design*, 2003.

H. Kubinyi, in R. Mannhold, ed., *Methods and Principles in Medicinal Chemistry*, VCH, Weinheim, 1993.

C. Silippo and A. Vittoria, eds., *Rational Approaches to the Design of Bioactive Compounds*, Pharmacochemistry Library Vol. 16, Elsevier, Amsterdam, The Netherlands, 1991.

C. G. Whiteley, *Cell Biochem. Biophys.* **33**(3), 217 (2000).

CHRIS G. WHITELEY
Rhodes University

EPOXY RESINS

Epoxy resins are an important class of polymeric materials, characterized by the presence of more than one three-membered ring known as the epoxy, epoxide, oxirane, or ethoxyline group:

By strict definition, epoxy resins refer only to uncrosslinked monomers or oligomers containing epoxy groups. However, in practice, the term "epoxy resins" is loosely used to include cured epoxy systems.

Epoxies are one of the most versatile classes of polymers with such diverse applications as metal can coatings, automotive primers, printed circuit boards, semiconductor encapsulants, adhesives, and aerospace composites. Most cured epoxy resins provide amorphous thermosets with excellent mechanical strength and toughness; outstanding chemical, moisture, and corrosion resistance; good thermal, adhesive, and electrical properties; no volatiles emission and low shrinkage upon cure; and dimensional stability—a unique combination of properties generally not found in any other plastic material. These superior performance characteristics, coupled with outstanding formulating versatility and reasonable costs, have gained epoxy resins wide acceptance as the materials of choice for a multitude of bonding, structural, and protective coatings applications.

Commercial epoxy resins contain aliphatic, cycloaliphatic, or aromatic backbones and are available in a wide range of molecular weights from several hundreds to tens of thousands. The most widely used epoxies are the glycidyl ether derivatives of bisphenol A (> 75 % of resin sales volume). The capability of the highly strained epoxy ring to react with a wide variety of curing agents under diverse conditions and temperatures imparts additional versatility to the epoxies. The major industrial utility of epoxy resins is in thermosetting applications. Treatment with curing agents gives insoluble and intractable thermoset polymers. In order to facilitate processing and to modify cured resin properties, other constituents may be included in the compositions: fillers, solvents, diluents, plasticizers, catalysts, accelerators, and tougheners.

Epoxy resins were first offered commercially in the late 1940s and are now used in a number of industries, often in demanding applications where their performance attributes are needed and their modestly high prices are justified. However, aromatic epoxies find only limited uses in exterior applications, because of their poor ultraviolet (uv) light resistance. Highly cross-linked epoxy thermosets sometimes suffer from brittleness and are often modified with tougheners for improved impact resistance.

The largest use of epoxy resins is in protective coatings (> 50 %), with the remainder being in structural applications such as printed circuit board (PCB) laminates, semiconductor encapsulants, and structural composites; tooling, molding, and casting; flooring; and adhesives. New, growing applications include lithographic inks and photoresists for the electronics industry.

CLASSES OF EPOXY RESINS AND MANUFACTURING PROCESSES

Most commercially important epoxy resins are prepared by the coupling reaction of compounds containing at least two active hydrogen atoms with epichlorohydrin followed by dehydrohalogenation:

$$R-H + CH_2CHCH_2Cl \longrightarrow R-CH_2CHCH_2Cl$$
$$\underset{O}{\diagdown\diagup} \qquad\qquad\qquad \underset{OH}{|}$$

$$R-CH_2CHCH_2Cl \xrightarrow{-HCl} R-CH_2CHCH_2$$
$$\underset{OH}{|} \qquad\qquad \underset{O}{\diagdown\diagup}$$

These include polyphenolic compounds, mono and diamines, amino phenols, heterocyclic imides and amides, aliphatic diols and polyols, and dimeric fatty acids. Epoxy resins derived from epichlorohydrin are termed "glycidyl-based resins".

Alternatively, epoxy resins based on epoxidized aliphatic or cycloaliphatic dienes are produced by direct epoxidation of olefins by peracids:

$$RCH=CHR' + R''COOOH \longrightarrow RCH-CHR' + R''COOH$$
$$\overset{O}{\diagup\diagdown}$$

Approximately 75 % of the epoxy resins currently used worldwide are derived from DGEBA, diglycidyl ether of bisphenol A.

LIQUID EPOXY RESINS (DGEBA)

The most important intermediate in epoxy resin technology is the reaction product of epichlorohydrin and bisphenol A, often referred to as liquid epoxy resin (LER), which can be described as crude DGEBA where the degree of polymerization, n, is very low ($n \cong 0.2$).

Pure DGEBA is a crystalline solid (mp 43 °C) with an epoxide equivalent weight (EEW) of 170. EEW, the weight of resin required to obtain one equivalent of an epoxy functional group, is widely used to calculate reactant stoichiometric ratios for reacting or curing epoxy resins.

The outstanding performance characteristics of the resins are conveyed by the bisphenol A moiety (toughness, rigidity, and elevated temperature performance), the ether linkages (chemical resistance), and the hydroxyl and epoxy groups (adhesive properties and formulation latitude; reactivity with a wide variety of chemical curing agents).

LERs are used in coatings, flooring, and composites formulations where their low viscosity facilitates processing. A large majority of LERs are used as starting materials to produce higher molecular weight (MW) solid epoxy resins (SERs) and brominated epoxy resins, and to convert to epoxy derivatives such as epoxy vinyl esters, epoxy acrylates, etc. Some of LERs' outstanding properties are their superior electrical properties, chemical resistance, heat resistance, and adhesion. Cured LERs give tight cross-linked networks having good strength and hardness but have limited flexibility and toughness. The manufacturing processes for LERs can be divided into two broad categories according to the type of catalyst used to couple epichlorohydrin and bisphenol A. One is the caustic coupling process, the other the phase-transfer catalyst process.

Pure DGEBA is a solid melting at 43 °C. The unmodified commercial liquid resins are supercooled liquids with the potential for crystallization, depending on purity and storage conditions. This causes handling problems, particularly for ambient cure applications. Addition of certain reactive diluents and fillers can either accelerate or retard crystallization. Crystallization-resistant, modified resins are available. A crystallized resin can be restored to its liquid form by warming.

SOLID EPOXY RESINS BASED ON DGEBA

High molecular weight (MW) SERs based on DGEBA are characterized by a repeat unit containing a secondary

$$CH_2-CHCH_2 \left(O-C_6H_4-\underset{\underset{CH_3}{|}}{\overset{\overset{CH_3}{|}}{C}}-C_6H_4-O-CH_2CHCH_2 \right)_n O-C_6H_4-\underset{\underset{CH_3}{|}}{\overset{\overset{CH_3}{|}}{C}}-C_6H_4-O-CH_2CH-CH_2$$

hydroxyl group with degrees of polymerization, ie, n values ranging from 2 to about 35 in commercial resins; two terminal epoxy groups are theoretically present.

The epoxy industry has adopted a common nomenclature to describe the SERs. They are called type "1," "2" up to type "10" resins, which correspond to the increased values of n, the degree of polymerization, EEW, MW, and viscosity. A comparison of some key properties of LERs and SERs is shown in Table 1.

SERs based on DGEBA are widely used in the coatings industry. The longer backbones give more distance between cross-links when cross-linked through the terminal epoxy groups, resulting in improved flexibility and toughness. Furthermore, the resins can also be cured through the multiple hydroxyl groups along the backbones using cross-linkers such as phenol–formaldehyde resoles or isocyanates to create different network structures and performance.

SERs are prepared by two processes: the taffy process and the advancement or fusion process. The first is directly from epichlorohydrin, bisphenol A, and a stoichiometric amount of NaOH. This process is very similar to the caustic coupling process used to prepare liquid epoxy resins. Lower epichlorohydrin to bisphenol A ratios are used to promote formation of high MW resins. The term taffy is derived from the appearance of the advanced epoxy resin prior to its separation from water and precipitated salts.

An alternative method is the chain-extension reaction of liquid epoxy resin (crude DGEBA) with bisphenol A, often referred to as the advancement or fusion process, which requires an advancement catalyst.

In a typical advancement process, bisphenol A and a liquid DGEBA resin (175–185 EEW) are heated to ca 150–190 °C in the presence of a catalyst and reacted (ie, advanced) to form a high MW resin. The oligomerization is exothermic and proceeds rapidly to near completion. The exotherm temperatures are dependent upon the targeted EEW and the reaction mass. In the cases of higher MW resins such as type 7 and higher, exotherm temperatures of > 200 °C are routinely encountered. Advancement

reaction catalysts facilitate the rapid preparation of medium and high MW linear resins and control the prominent side reactions inherent in epoxy resin preparations. Nuclear Magnetic Resonace (nmr) spectroscopy can be used to determine the extent of branching. Conventional advancement catalysts include such basic inorganic reagents, as NaOH, KOH, Na_2CO_3, or LiOH, and amines and quaternary ammonium salts.

Branched epoxies are prepared by advancing LER with bisphenol A in the presence of epoxy novolac resins. Such compositions exhibit enhanced thermal and solvent resistance.

SERs are available commercially in solid form or in solution. MW distributions of SERs have been examined by means of theoretical models and compared with experimental results. The major differences between taffy-processed resins and advancement-processed resins are in the higher α-glycol content and the repeating units of oligomers. Resin viscosity and softening points are also lower with taffy resins. In addition, certain formulations based on taffy resins exhibit different behavior in pigment loading, formulation rheology, reactivity, and mechanical properties compared to those based on advancement resins.

SER Continuous Advancement Process

One of the major deficiencies of the traditional batch advancement process is its long reaction time, resulting in EEW and viscosity drift, variable product quality, and gel formation. Shell has patented several versions of the continuous resin advancement process using modified reactor designs and Dow-holds patents covering the uses of reactive extrusion (REX) produce SERs and other epoxy thermoplastic resins. The latter process is claimed to be very efficient and particularly suitable for the production of high molecular weight SERs, phenoxy resins, and epoxy thermoplastic resins. Compared to the traditional taffy processes used to produce phenoxy resins, the chemistry is salt-free. The resins made via the REX process are fully converted in a matter of minutes, significantly reducing manufacturing costs.

Table 1. DGEBA-Based Epoxy Resins

Resin type	n value[a]	EEW	Mettler softening point, °C	Molecular weight (M_w)[b]	Viscosity at 25 °C, mPa·s (= cP)
Low viscosity LER	< 0.1	172–176		~350	4,000–6,000
Medium viscosity LER	~0.1	176–185		~370	7,000–10,000
Standard grade LER	~0.2	185–195		~380	11,000–16,000
Type 1 SER	~2	450–560	70–85	~1,500	160–250[c]
Type 4 SER	~5	800–950	95–110	~3,000	450–600[c]
Type 7 SER	~15	1,600–2,500	120–140	~10,000	1,500–3,000[c]
Type 9 SER	~25	2,500–4,000	145–160	~15,000	3,500–10,000[c]
Type 10 SER	~35	4,000–6,000	150–180	~20,000	10,000–40,000[c]
Phenoxy resin	~100	> 20,000	> 200	>40,000	

[a] n value is the number-average degree of polymerization which approximates the repeating units and the hydroxyl functionality of the resin.
[b] Molecular weight is weight average (M_w) measured by gel-permeation chromatography (GPC) using polystyrene standard.
[c] Viscosity of SERs is determined by kinematic method using 40% solids in diethylene glycol monobutyl ether solution.

Phenoxy Resins

Phenoxy resins, thermoplastic polymers derived from bisphenol A and epichlorohydrin, are offered as solids, solutions, and waterborne dispersions. The majority of phenoxy resins are used as thermoplastics, but some are used as additives in thermoset formulations. Their high MW provides improved flexibility and abrasion resistance. Their primary uses are in automotive zinc-rich primers, metal can/drum coatings, magnet wire enamels, and magnetic tape coatings. However, the zinc-rich primers are being phased out in favor of galvanized steel by the automotive industry. Smaller volumes of phenoxy resins are used as flexibility or rheology modifiers in composites and electrical laminate applications, and as composite honeycomb impregnating resins. A new, emerging application is fiber sizing, which utilizes waterborne phenoxies. Literature references indicate their potential uses as compatiblizers for thermoplastic resins such as polyesters, nylons, and polycarbonates because of their high hydroxyl contents.

Epoxy-Based Thermoplastics

The thermoplastic resins based on epoxy monomers; like polyhydroxy amino ether, trade named BLOX, is produced by the reaction of DGEBA with monoethanol amine using the reactive extrusion process. The high cohesive energy density of the resin gives it excellent gas-barrier properties against oxygen and carbon dioxide. It also possesses excellent adhesion to many substrates, optical clarity, excellent melt strength, and good mechanical properties. The product has been evaluated as a barrier resin for beer and beverage plastic bottles, as thermoplastic powder coatings, and as a toughener for starch-based foam.

HALOGENATED EPOXY RESINS

A number of halogenated epoxy resins have been developed and commercialized to meet specific application requirements. Brominated epoxies evaluated for flame retardancy properties were commercialized in the late 1960s.

Brominated Bisphenol A–Based Epoxy Resins

Many applications of epoxy resins require the system to be ignition-resistant. A common method of imparting this ignition resistance is the incorporation of tetrabromobisphenol A (TBBA), 2,2-bis(3,5-dibromophenyl)propane, or the diglycidyl ether of TBBA, 2,2-bis[3,5-dibromo-4-(2,3-epoxypropoxy)phenyl]propane, into the resin formulation. The diglycidyl ether of TBBA (ca 50 wt% Br) is used for critical electrical/electronic encapsulation where high flame retardancy is required. Brominated epoxies are also used to produce epoxy vinyl esters for structural applications. Very high MW versions of brominated epoxies are used as flame-retardant additives to engineering thermoplastics used in computer housings.

Fluorinated Epoxy Resins

Fluorinated epoxy resins are highly resistant to chemical and physical abuse and should prove useful in high-performance applications, including specialty coatings and composites, where their high cost may be offset by their special properties and long service life.

MULTIFUNCTIONAL EPOXY RESINS

The multifunctionality of the epoxy resins provides higher cross-linking density, leading to improved thermal and chemical resistance properties over bisphenol A epoxies.

Epoxy Novolac Resins

Epoxy novolacs are multifunctional epoxies based on phenolic formaldehyde novolacs. Both epoxy phenol novolac resins (EPN) and epoxy cresol novolac resins (ECN) have attained commercial importance. The improved thermal stability of EPN-based thermosets is useful in elevated temperature services, such as aerospace composites. Filament-wound pipe and storage tanks, liners for pumps and other chemical process equipment, and corrosion-resistant coatings are typical applications that take advantage of the chemical resistant properties of EPN resins. However, the high cross-link density of EPN-based thermosets can result in increased brittleness and reduced toughness.

Other Polynuclear Phenol Glycidyl Ether–Derived Resins

In addition to the epoxy novolacs, there are other epoxy resins derived from phenol–aldehyde condensation products. New applications that require increased performance from the epoxy resin, particularly in the electronics, aerospace, and military industries, have made these types of resins more attractive despite their relatively high cost.

The semisolid resins are used in advanced composites and adhesives where toughness, hot-wet strength, and resistance to high-temperature oxidation are required. Their purity, formulated stability, fast reactivity, and retention of electrical properties over a broad temperature range make the solid resins suitable for use in the semiconductor molding powders industry. The trisphenol-based epoxies command significantly high prices that limit their uses.

Aromatic Glycidyl Amine Resins

Among the multifunctional epoxy resins containing an aromatic amine backbone, only a few have attained commercial significance. Their higher costs limit their use to critical applications where their costs are justified. Glycidyl amines contain internal tertiary amines in the resin backbone, hence their high reactivity. Epoxy resins with such built-in curing catalysts are less thermally stable than nitrogen-free multifunctional epoxy resins.

SPECIALTY EPOXY RESINS

Crystalline Epoxy Resins Development

A number of new epoxy resins used in epoxy molding compounds (EMCs) have been developed in response to the increased performance requirements of the semiconductor industry. Most notable are the 2002 commercialization of crystalline epoxies based on biphenol.

The very low viscosity of these crystalline, solid epoxies when molten allows very high filler loading (up to 90 wt%) for molding compounds. This reduces the coefficient of thermal expansion (CET) and helps manage thermal shock and moisture and crack resistance of molding compounds used in new, demanding semiconductor manufacturing processes such as Surface Mount Technology (SMT). Cured thermosets derived from these crystalline resins do not retain crystallinity.

Weatherable Epoxy Resins

One of the major deficiencies of the aromatic epoxies is their poor weatherability, attributable to the aromatic ether segment of the backbone, which is highly susceptible to photoinitiated free-radical degradation. Numerous efforts have been devoted to remedy this issue, resulting in a number of new weatherable epoxy products. Their commercial success has been limited, primarily because of higher resin costs and the fact that end users can topcoat epoxy primers with weatherable coatings based on other chemistries such as polyesters, polyurethanes, or acrylics. Some epoxy products like hydrogenated DGEBA, when formulated with appropriate reactants can provide certain outdoor weatherability.

Elastomer-Modified Epoxies

Epoxy thermosets derive their thermal, chemical, and mechanical properties from their highly cross-linked networks. Consequently, toughness deficiency is an issue in certain applications. To improve the impact resistance and toughness of epoxy systems, elastomers such as BF Goodrich's CTBN rubbers (carboxyl-terminated butadiene nitrile) are often used as additives or prereacted with epoxy resins. This process has resulted in products with improved toughness, peel adhesion, and low temperature flexibility over unmodified epoxies. Primary applications are adhesives for aerospace and automotive and as additives to epoxy vinyl esters for structural composites.

MONOFUNCTIONAL GLYCIDYL ETHERS AND ALIPHATIC GLYCIDYL ETHERS

A number of low MW monofunctional, difunctional, and multifunctional epoxies are used as reactive diluents, viscosity reducers, flexiblizers, and adhesion promoters. Recent trends toward lower VOCs, higher solids and 100% solids epoxy formulations have resulted in increased utilization of these products. However, the uses of reactive diluents, especially at high levels, often result in decreased chemical resistance and thermal and mechanical properties of the cured epoxies.

CYCLOALIPHATIC EPOXY RESINS AND EPOXIDIZED VEGETABLE OILS

A combination of aliphatic backbone, high oxirane content, and no halogens gives resins with low viscosity, weather-

ability, low dielectric constant, and high cured T_g. This class of epoxy is popular for diverse end uses, including auto topcoats, weatherable high-voltage insulators, uv coatings, acid scavengers, and encapsulants for both electronics and optoelectronics. The largest end uses of cycloaliphatic epoxies in order of volume are electrical, electronic components encapsulation, and radiation-curable inks and coatings. A potentially large volume application is uv-curable metal can coatings for beer can exteriors and ends. Other uses include acid scavengers for vinyl-based transformer fluids and lubricating oils; filament winding for aerial booms and antennas; and as a viscosity modifier for bisphenol A LERs in tooling compounds. An epoxy silicone containing cycloaliphatic epoxy end groups and a silicone backbone is used as radiation-curable release coatings for pressure-sensitive products.

EPOXY ESTERS AND DERIVATIVES

Epoxy Esters

The esterification of epoxy resins with commercial fatty acids is a well-known process that has been employed for industrial coatings for many years. The carboxylic acids are esterified with the terminal epoxy groups or the pendant hydroxyls on the polymer chain.

A wide variety of saturated and unsaturated fatty acids are utilized to confer properties useful in air-dried, protective, and decorative coatings. Typical fatty acids include tall oil fatty acids, linseed oil fatty acid, soya oil fatty acid, and castor oil fatty acid. A medium molecular weight SER, a so-called 4-type, is commonly used. Catalysts such as alkaline metal salts (Na_2CO_3) or ammonium salts are essential to prevent chain branching and gelation caused by etherification of the epoxy groups.

Metallic driers are incorporated in unsaturated ester solutions to promote cure via air-drying. Chemical resistance is generally lower than that of unmodified epoxy resins cured at ambient temperatures with amine hardeners. Epoxy esters are also used to produce anodic electrodeposition (AED) coatings by further reaction with maleic anhydride followed by neutralization with amines to produce water-dispersable coatings. Epoxy esters were widely used as automotive primer-surfacer and metal can ends coatings for many years but are now being replaced by other technologies. Their high viscosity limited their uses in low solids, solvent-borne coatings. Waterborne epoxy esters are now available and are used in flexographic inks for milk cartons.

Glycidyl Esters

Glycidyl esters are prepared by the reaction of carboxylic acids with epichlorohydrin followed by dehydrochlorination with caustic.

A commercially important glycidyl ester is glycidyl methacrylate (GMA), a dual functionality monomer containing both a terminal epoxy and an acrylic C=C bond. The dual functionality of GMA brings together the desirable properties of epoxies and acrylics; eg, the weatherability of acrylics and the chemical resistance of epoxies, in one product. GMA is useful as a comonomer in the synthesis of epoxy-containing polymers via free-radical polymerization. The resultant epoxy-containing polymers can be further cross-linked. GMA-containing polymers are also used as compatiblizers for engineering thermoplastics, in adhesives and latexes, and as rubber and asphalt modifiers.

Epoxy Acrylates

Epoxy resins are reacted with acrylic acid to form epoxy acrylate oligomers, curable via free-radical polymerization of the acrylate C=C bonds initiated by light. This is a fast-growing market segment for epoxy resins because of the environmental benefits of the uv cure technology: low to zero VOC, and low energy requirements. Major applications include coatings for overprint varnishes, wood substrates, and plastics. Radiation-cured epoxy acrylates are also growing in importance in inks, adhesives, and photoresists applications.

Epoxy Vinyl Esters

A major derivative of epoxy resins, the epoxy vinyl ester resin, is a high performance resin used in glass-reinforced structural composites, particularly for its outstanding chemical resistance and mechanical properties.

The vinyl ester functionality of the epoxy vinyl esters provides outstanding hydrolysis and chemical resistance properties, in addition to the inherent thermal resistance and toughness properties of the epoxy backbone. These attributes have made epoxy vinyl esters a material of choice in demanding structural composite applications such as corrosive chemicals storage tanks, pipes, and ancillary equipment for chemical processing. Other applications include automotive valve covers and oil pans, boats, and pultruded construction parts. Significant efforts have been devoted to improve toughness and to reduce levels of styrene in epoxy vinyl ester formulations because of environmental concerns.

Epoxy Phosphate Esters

Epoxy phosphate esters are reaction products of epoxy resins with phosphoric acid. Epoxy phosphate esters can be made to disperse in water to produce waterborne coatings. They are used primarily as modifiers to improve the adhesion property of nonepoxy binders in both solvent-borne and waterborne systems for container and coil coatings.

CHARACTERIZATION OF UNCURED EPOXIES

Epoxy resins often contain isomers, oligomers, and other minor constituents. The proper stoichiometric amount of cross-linker(s) needs to be calculated, and a successful thermoset formulation must also have the proper reactivity, flow, and performance. Consequently, other epoxy resin properties are required.

Liquid epoxy resins are mainly characterized by epoxy content, viscosity, color, density, hydrolyzable chloride, and volatile content. Less often analyzed are α-glycol content, total chloride content, ionic chloride, and sodium.

Solid epoxy resins are characterized by epoxy content, solution viscosity, melting point, color, and volatile content. Less often quoted are phenolic hydroxyl content, hydrolyzable chloride, ionic chloride, sodium, and esterification equivalent.

Gel-permeation chromatography (gpc), high-performance liquid chromatography (hplc), and other analytical procedures such as nuclear magnetic resonance (nmr) and infrared spectroscopy (ir) are performed to determine the MW, MW distribution, oligomer composition, functional groups, and impurities.

Resin components such as α-glycol content and chloride types and levels are known to influence formulation reactivity and rheology, depending on their interactions with the system composition such as basic catalysts (tertiary amines) and/or amine curing agents. Knowing the types and levels of chlorides guides formulators in the adjustment of their formulations for proper reactivity and flow.

For instand, the viscosity of epoxy resins is an important characteristic affecting handling, processing, and application of the formulations. Viscosities of liquid resins are typically determined with a Cannon–Fenske capillary viscometer at 25 °C or a Brookfield viscometer. The viscosity depends on the temperature.

The "ball and ring" and Durran's methods traditionally measure the softening point of SERs, which is important in applications such as powder coatings.

CURING OF EPOXY RESINS

With the exception of the very high MW phenoxy resins and epoxy-based thermoplastic resins, almost all epoxy resins are converted into solid, infusible, and insoluble three-dimensional thermoset networks for their uses by curing with cross-linkers. Optimum performance properties are obtained by cross-linking the right epoxy resins with the proper cross-linkers, often called hardeners or curing agents. Selecting the proper curing agent is dependent on the requirements of the application process techniques, pot life, cure conditions, and ultimate physical properties. Besides affecting viscosity and reactivity of the formulation, curing agents determine both the types of chemical bonds formed and the degree of cross-linking that will occur. These in turn affect the chemical resistance, electrical and mechanical properties, and heat resistance of the cured thermosets.

Epoxy resins contain two chemically reactive functional groups: epoxy and hydroxy. Low MW epoxy resins such as LERs are considered difunctional epoxy monomers or prepolymers and are mostly cured via the epoxy group. However, as the MW of SERs increases, the epoxy content decreases, whereas the hydroxyl content increases. High molecular weight SERs can cross-link via reactions with both the epoxy and hydroxyl functionalities, depending on the choice of curing agents and curing conditions. Reaction of the epoxy groups involves opening of the oxirane ring and formation of longer, linear C–O bonds. This feature accounts for the low shrinkage and good dimensional stability of cured epoxies. The polycondensation curing via hydroxyl groups is often accompanied by generation of volatile by-products, such as water or alcohol, requiring heat for proper cure and removal of volatiles.

It is the unique ability of the strained epoxy ring to react with a wide variety of reactants under many diverse conditions that gives epoxies their versatility. Detailed discussions on the probable electronic configurations, molecular orbitals, bond angles, and reactivity of the epoxy ring are available.

Compared to noncyclic and other cyclic ethers, the epoxy ring is abnormally reactive. The highly strained bond angles, along with the polarization of the C–C and C–O bonds, may account for the high reactivity of the epoxide. The electron-deficient carbon can undergo nucleophilic reactions, whereas the electron-rich oxygen can react with electrophiles. It is customary in the epoxy industry to refer to these reactions in terms of anionic and cationic mechanisms.

Curing agents are either catalytic or coreactive. A catalytic curing agent functions as an initiator for epoxy resin homopolymerization or as an accelerator for other curing agents, whereas a coreactive curing agent acts as a comonomer in the polymerization process. The majority of epoxy curing occurs by nucleophilic mechanisms. The most important groups of coreactive curing agents are those with active hydrogen atoms, eg, primary and secondary amines, phenols, thiols, and carboxylic acids (and their anhydride derivatives). Lewis acids, eg, boron trihalides; and Lewis bases; eg, tertiary amines, initiate catalytic cures.

The functional groups surrounding the epoxide resin also affect the curing process. In general, aromatic and brominated aromatic epoxy resins react quite readily with nucleophilic reagents, whereas aliphatic and cycloaliphatic epoxies react sluggishly toward nucleophiles.

Clearly the epoxy structure dramatically influences the cure response of the epoxy as a function of pH. Cycloaliphatic epoxies are fast-reacting under low pH conditions. Aromatic glycidyl ethers are faster under high pH conditions. These results generally agree with "practical" cures: Aromatic epoxies are easily cured with amines and amidoamines. Cycloaliphatics are cured with acids and superacids. The behavior of the aliphatic epoxies is more complex but on balance is similar to that of cycloaliphatics. The most commonly used curing agents are amines, followed by carboxylic-functional polyesters and anhydrides.

COREACTIVE CURING AGENTS

Commercially, epoxy resins are cured predominantly with coreactive curing agents. The following are the important classes of epoxy coreactive curing agents.

The amine functional curing agents include primary and secondary amines, the aliphatic amines, ketimines, Mannich base adducts (the reaction product of an amine with phenol and formaldehyde), polyetheramines, cycloaliphatic amines, aromatic amines, arylyl amines, polyamides, amidoamines, and dicyandiamide.

The class of carboxylic functional polyester and anhydride curing agents includes carboxylic functional polyesters and acid anhydrides.

Phenolic-terminated curing agents contain phenolic hydroxyls capable of reacting with the epoxy groups.

They include phenol-, cresol-, and bisphenol A terminated epoxy resin hardener. More recent additions include bisphenol A based novolacs.

Melamine–formaldehyde, urea–formaldehyde, and phenol–formaldehyde resins react with hydroxyl groups of high MW epoxy resins to afford cross-linked networks. There are two types of phenol–formaldehyde condensation polymers: resoles phenol–formaldehyde polymers prepared from the base-catalyzed condensation of phenol and excess formaldehyde and novolacs. The melamine- and urea–formaldehyde resins are also called amino resins.

These formaldehyde-based resins are widely used to cure high MW solid epoxy resins at elevated temperatures (up to 200 °C) for metal can, drum, and coil coatings applications. The resultant coatings have excellent chemical resistance, good mechanical properties, and no effects on taste. The vast majority of the food and beverage cans produced in the world today are coated internally with epoxy–formaldehyde resin coatings. The phenol–formaldehyde resoles are also used with epoxies in coatings for high-temperature service pipes and to protect against hot, corrosive liquids.

The mercaptan group of curing agents includes polysulfide and polymercaptan compounds that contain terminal thiols.

The tertiary amine accelerated polymercaptan/epoxy systems are used in high lap-shear adhesion applications such as concrete patch repair adhesives. One disadvantage of polymercatans is their strong odor. Aliphatic amine/polysulfide co-curing agent systems are widely used as building adhesives for their excellent adhesion to glass and concrete. Both systems lose some flexibility on aging.

Cyclic amidine curing agents are typically used in epoxy powder coating formulations and in decorative epoxy–polyester hybrid powder coatings to produce a matte surface for furniture and appliance finishes.

Isocyanate curing agents react with epoxy resins via the epoxy group to produce an oxazolidone structure or with a hydroxyl group to yield a urethane linkage. The urethane linkage provides improved flexibility, impact, and abrasion resistance. The oxazolidone products have been successfully commercialized in high temperature resistance coating and composite applications. Blocked isocyanates are used as cross-linkers for epoxy in PPG's cathodic electrodeposition (CED) coatings. Isocyanates are also used to cure epoxies in some powder coatings, but their toxicity has limited their use.

Cyanate ester curing agents can be used to cure epoxy resins to produce highly cross-linked thermosets with a high modulus and excellent thermal, electrical, and chemical resistance properties. They are used in high-performance electrical laminate and composite applications but the high costs of cyanate esters limit their uses.

CATALYTIC CURING

The catalytic curing agents are a group of compounds that promote epoxy reactions without being consumed in the process. In some of the epoxy literature, catalysts are referred to as "accelerators."

Lewis Bases

Lewis bases contain an unshared pair of electrons in an outer orbit and seek reaction with areas of low electron density. They can function as nucleophilic catalytic curing agents for epoxy homopolymerization; as co-curing agents for primary amines, polyamides, and amidoamines; and as catalysts for anhydrides. Tertiary amines and imidazoles are the most commonly used nucleophilic catalysts. Several different mechanisms are possible.

Lewis Acids

Lewis acids; eg, boron trihalides, contain an empty outer orbit and therefore seek reaction with areas of high electron density. Boron trifluoride, BF_3, a corrosive gas, reacts easily with epoxy resins, causing gelation within a few minutes. Complexation of boron trihalides with amines enhances the curing action. Reasonable pot lives using these complexes can be achieved because elevated temperatures are required for cure. Reactivity is controlled by the choices of the halide and the amine. The amine choice also affects other properties such as solubility in resin, and moisture sensitivity. Boron trifluoride monoethylamine ($BF_3 \cdot NH_2 C_2H_5$), a crystalline material commonly used as a catalyst, cures epoxy resins at $80-100\,°C$. A chloride version is also commercially available. Other Lewis acids used in epoxy curing include stannic chloride and tin octanate.

Photoinitiated Cationic Curing

Photoinitiated cationic curing of epoxy resins is a rapidly growing method for the application of coatings from solvent-free or high solids systems. This technology allows the formulation of epoxy coatings and adhesives with essentially "infinite" shelf life but almost "instantaneous" cure rates. Cycloaliphatic epoxies are widely cured using photoinitiated cationic initiators.

Photoinitiators used for epoxy curing include aryldiazonium salts ($ArN_2^+X^-$), diaryliodonium salts ($Ar_2I^+X^-$), and onium salts of Group VIa elements, especially salts of positively charged sulfur ($Ar_3S^+X^-$). The anions must be of low nucleophilicity, such as tetrafluoroborates or hexafluorophosphates, to promote polymer chain growth rather than chain termination. Upon uv irradiation, photoinitiators yield a "super" acid, which polymerizes the epoxy resins by a conventional electrophilic mechanism.

Significant interest in thermal cationic care of epoxies, especially cycle aliphatic epoxies, has developed.

FORMULATION DEVELOPMENT WITH EPOXY RESINS

The most important step in using epoxy resins is to develop the appropriate epoxy formulation, since most are used as precursors to a three-dimensional cross-linked network. With the exception of the very high MW phenoxy resins and the epoxy-based thermoplastics, epoxy resin is rarely used by itself. It is usually formulated with modifiers such as fillers and used in composite structures with glass fiber or metal substrates (coatings).

The development of an epoxy formulation containing a high number of components can be very resource-and time-consuming. Techniques such as design of experiments (DOE) are useful tools to facilitate the formulation development process and to obtain optimum performance. Future developments should include application of high throughput techniques to epoxy formulation development and optimization.

Selection of Epoxy Resins

Successful performance of epoxy-based systems depends on the proper selection and formulation of components. The components that have the most significant influences are the epoxy resins and the curing agents. The numerous choices of epoxy resins and curing agents present a wide variety of structure and functionality. Figure 1 shows the general attributes of common types of epoxy resins.

EPOXY CURING PROCESS

The epoxy curing process is an important factor affecting cured epoxy performance. Consequently, it is imperative to understand the curing process and its kinetics to design the proper cure schedule to obtain optimum network structure and performance.

The curing of a thermoset epoxy resin can be expressed in terms of a time–temperature-transformation (TTT) diagram (Figure 2). In the TTT diagram, the time to gelation and vitrification is plotted as a function of isothermal cure temperature. Important features are the gel point and the onset of vitrification. The gel point is defined as

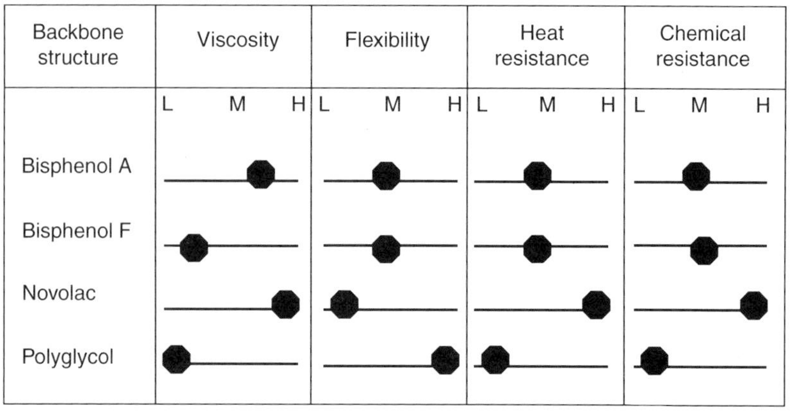

Backbone structure	Viscosity			Flexibility			Heat resistance			Chemical resistance		
	L	M	H	L	M	H	L	M	H	L	M	H
Bisphenol A		●			●			●			●	
Bisphenol F	●				●			●			●	
Novolac			●	●					●			●
Polyglycol	●					●	●				●	

Figure 1. Comparison of relative properties of common epoxy resins. L, low; M, medium; H, high.

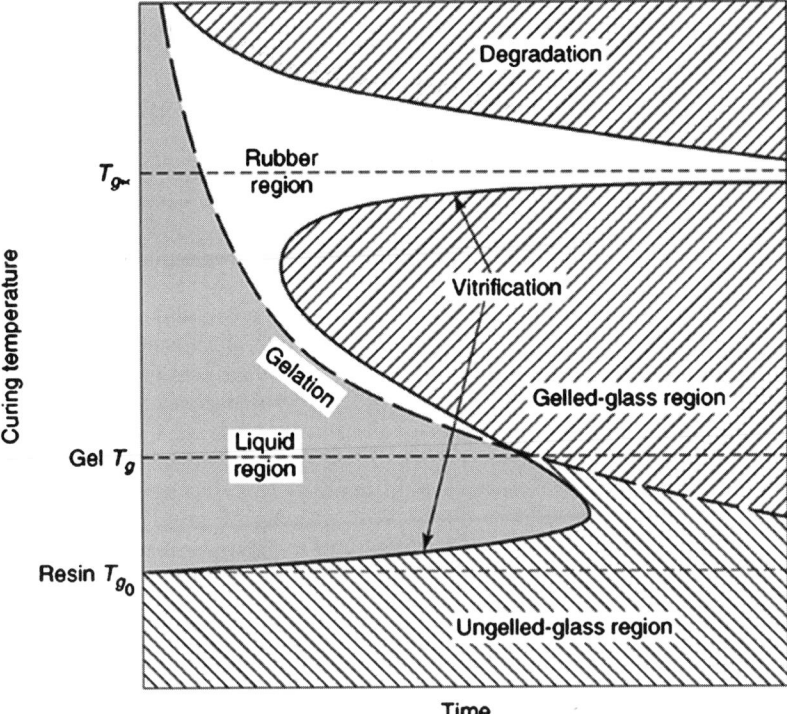

Figure 2. Time–temperature-transformation (TTT) diagram.

the onset of the formation of insoluble, cross-linked polymer (gel fraction) in the reaction mixture. However, a portion of the sample may still be soluble (sol fraction). The onset of vitrification is when the glass-transition temperature (T_g) of the curing sample approaches the curing temperature, T_c. Ideally, a useful structural thermoset would cure until all monomers are built into the network, resulting in no soluble fraction.

The S-shaped vitrification curve and the gelation curve divide the time–temperature plot into four distinct states of the thermosetting-cure process: liquid, gelled rubber, ungelled glass, and gelled glass. T_{g0} is the glass-transition temperature of the unreacted resin mixture; $T_{g\infty}$ the glass-transition temperature of the fully cured resin; and gel T_g the point where the vitrification and gellation curves intersect.

In the early stages of cure prior to gelation or vitrification, the epoxy curing reactions are kinetically controlled. When vitrification occurs the reaction is diffusion controlled, and the reaction rate is orders of magnitude below that in the liquid region. With further cross-linking of the glass, the reaction rate continues to decrease and is eventually quenched. In the region between gelation and vitrification (rubber region) the reaction can range from kinetic to diffusion control. This competition causes the minimum in vitrification temperature seen in the TTT diagram between gel T_g and $T_{g\infty}$. As the cure temperature is raised the reaction rate increases and the time to vitrification decreases until the decrease in diffusion begins to overcome the increased kinetic reaction rate. Eventually, slower diffusion in the rubbery region decreases the overall reaction rate and thus the increase in time to vitrify. Below $T_{g\infty}$, the reaction does not go to completion. As curing proceeds, the viscosity of the system increases as a result of increasing molecular weight, and the reaction becomes diffusion-controlled and eventually is quenched

as the material vitrifies. After quenching, the cure conversion can be increased by raising the temperature. This is often practiced as postcure for certain epoxy systems to achieve maximum cure and performance. Post-cure is effective only at temperatures higher than $T_{g\infty}$ but one must be careful about potential overcuring.

The TTT diagram is useful in understanding the cure kinetics, conversion, gelation, and vitrification of the curing thermoset. Gelation and vitrification times can be determined from the intersections of the storage and loss moduli and the maxima in the loss modulus of an isothermal dynamic mechanical spectrum, respectively. Recently, techniques have been developed using rheological and dynamic mechanical analysis instruments to determine the gel point and vitrification. Understanding the gelation and vitrification characteristics of an epoxy/curing agent system is critical in developing the proper cure schedule/process to achieve optimum performance.

Characterization of Epoxy Curing and Cured Networks

Cured thermoset polymers are more difficult to analyze than thermoplastics since they are insoluble and generally intractable. Their properties are influenced by factors at the molecular level, such as the backbone structures of epoxy resin and curing agent; the nature of the covalent bond developed between the epoxy resin and the curing agent during cross-linking; and the density and extent of cross-linking.

Epoxy resin formulators are concerned with formulation reactivity and flow during application. Reactivity tests or gel time tests are used to determine the proper reactivity of the formulations. Formulators have also developed flow tests to check for the formulation rheology profile. The coatings industry widely uses MEK (methyl ethyl ketone) double rubs as an indication of cure. While the test does give a relative indication of cure for a certain system, caution must be exercised when comparing different systems, which may have very different inherent resistance against MEK. In general, these end-use tests do not provide insights on the structure–property relationship of the system.

The epoxy curing process can be monitored by a number of different techniques: analysis of the disappearance and/or formation of functional groups; indirect estimation of cure conversion; and measurements of changes in thermal, physical, and mechanical properties of the system.

FORMULATION MODIFIERS

The processing behavior (mainly viscosity and substrate wetting) and other properties of an epoxy system can be modified by diluents, fillers, toughening agents, thixotropic agents, etc. Most commercial epoxy resin systems contain modifying agents.

Diluents

Diluents affect the properties of the cured resin system and, in particular, lower the viscosity in order to improve handling and wetting. They are often used in the range of 2–20 wt% based on the epoxy resin. Diluents can be classified into reactive and nonreactive types. The reactive diluents are products with low viscosity (1–500 cP at 25 °C) used to lower the viscosity of standard epoxy formulations. Lower viscosity allows higher filler loading, lower costs, and/or improved processability.

Solvents and plasticizers are nonreactive diluents. The most common nonreactive diluents are nonyl phenol, furfuryl alcohol, benzyl alcohol, and dibutyl phthalate. These materials have the advantage of being able to add to the amine side of the system to better balance mix ratios. Nonyl phenol and furfuryl alcohol also improve wet-out and accelerate cure slightly. They are also capable of reacting with the epoxy group under high temperature cure conditions. Benzyl alcohol is a popular diluent used with amine-cured systems to reduce viscosity and increase cure speed.

Thixotropic Agents

Thixotropy is the tendency of certain colloidal gels to flow when subjected to shear, then return to a gel when at rest. A thixotropic gel can be produced through the addition of either high surface area fillers such as colloidal silicas and bentonite clays or of chemical additives. Thixotropy is desirable in applications such as encapsulation where the coating is applied by dipping. The resin will wet out and coat the object being dipped, but will not run off when the object is removed from the dipping bath.

Fillers

Fillers are incorporated in epoxy formulations to enhance or obtain specific desired properties in a system. Fillers can also reduce the cost of epoxy formulations. Inert commercial fillers can be organic or inorganic, and spheroidal, granular, fibrous, or lamellar in shape. For certain applications, fillers can have significant effects on thermoset morphology, adhesion, and resulting performance.

Other properties that can be affected with the proper choice of fillers for a specific application include compressive strength, adhesion, arc and tracking resistance, density, and self-lubricating properties.

Epoxy Nanocomposites

Significant recent developments in polymer property enhancement involve polymer nanocomposites. This is a special class of fillers (mostly clay derivatives) in which the nanoscale, highly oriented particles are formed in the polymer matrix through monomer intercalation and particle aggregate exfoliation. The objective is to combine the performance attributes of both hard inorganic and plastic materials. Significant recent efforts in developing epoxy nanocomposites in the past decade have seen improvements in electrical and mechanical properties, chemical resistance, high temperature performance, and flame retardancy.

The emerging field of nanotechnology has produced new materials such as carbon nanotubes, which are filaments of carbon with atomic dimensions. However, cost remains a barrier for commercialization.

Toughening Agents and Flexiblizers

Some cross-linked, unmodified epoxy systems exhibit brittleness, poor flexibility, and low impact strength and fracture resistance. Modifiers can be used to remedy these shortcomings. However, there will usually be some sacrifices of properties. In general, there are two approaches used to modify epoxies to improve these features. Flexibilization can enhance elongation of the system but is often accompanied by a reduction of glass-transition temperature, yield stress, and elastic modulus.

Toughening refers to the ability to increase resistance to failure under mechanical stress. Toughening approaches for epoxies include the dispersion of preformed elastomer particles into the epoxy matrix and reaction-induced phase separation of elastomers or thermoplastic particles during cure.

Through the proper selection of resin, curing agent, and modifiers, the cured epoxy resin system can be tailored to specific performance characteristics. The choice depends on cost, processing, and performance requirements. Cure is possible at ambient and elevated temperatures. Cured epoxies exhibit good combinations of outstanding properties and versatility at moderate cost: excellent adhesion to a variety of substrates; outstanding chemical and corrosion resistance; excellent electrical insulation; high tensile, flexural, and compressive strengths; good thermal stability; relatively low moisture absorption; and low shrinkage upon cure. Consequently, epoxies are used in diverse applications.

COATINGS APPLICATIONS

Commercial uses of epoxy resins can be generally divided into two major categories: protective coatings

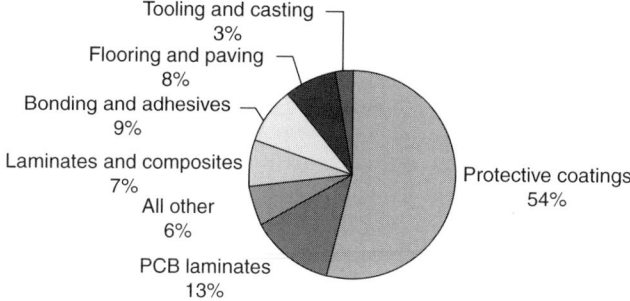

Figure 3. End-use markets of epoxy resins (U.S. data, 2000).

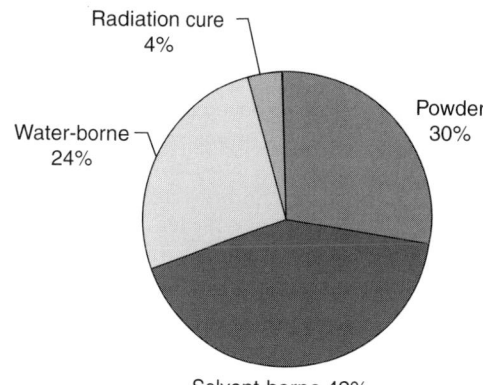

Figure 4. Global epoxy coating application technologies.

and structural applications. U.S. consumption of epoxy resins is given in Figure 3. The largest single use is in coatings (> 50 %). Among the structural composite applications, electrical laminates contribute the largest epoxy consumption. While the overall epoxy markets have continued to grow at a steady pace over the past two decades, more rapid growth has occurred in powder coatings, electrical laminates, electronic encapsulants, adhesives, and radiation-curable epoxies.

The majority of epoxy coatings are based on DGEBA or modifications of DGEBA. As a class, epoxy coatings exhibit superior adhesion (both to substrates and to other coatings), chemical and corrosion resistance, and toughness. However, epoxy coatings have been employed mainly as primers or undercoats, because of their tendency to yellow and chalk on exposure to sunlight.

Epoxy-based coatings are the preferred and dominant choices for cathodic electrodeposition of automotive primers, marine and industrial maintenance coatings, and metal container interior coatings. Use of epoxy flooring for institutions and industrial buildings has been growing at a steady rate.

Solvents are commonly used to facilitate dissolution of resins, cross-linkers, and other components, and for ease of handling and application. Although most of the epoxy coatings sold in the 1970s were solvent-borne types, they made up only 40 % of epoxy coating consumption in 2001. Economic and ecological pressures to lower the volatile organic content (VOC) of solvent-borne coatings have stimulated the development of high solids, solvent-free systems (powder and liquid), and waterborne and radiation-curable epoxy coatings technologies. These environmentally friendly coating technologies have experienced rapid growth in the past decade. Epoxy powder coatings have been growing at rates exceeding those of other coating technologies as new applications such as automotive primer-surfacer and low-temperature-cure coatings for heat-sensitive substrates are developed. Radiation-curable liquid coatings based on epoxy acrylates and cycloaliphatic epoxies have also been growing significantly over the last decade. The current distribution of coating technologies is summarized in Figure 4.

Coatings Application Technologies

Low Solids Solvent-Borne Coatings. The traditional low solids coatings contain less than 60 % solids by volume (typically 40 %). Their advantages include established application equipment and experience, fast drying and cure at ambient temperatures, and excellent film formation at extremely fast cure conditions. However, because of stricter VOC regulations, solvent-based coatings have been losing market share steadily to more environmentally friendly technologies.

High Solids Solvent-Borne Coatings. High solids coatings contain 60–85 % by volume of solids. They are mostly based on standard LERs or low molecular weight SERs modified by reactive diluents, low viscosity multifunctional aliphatic epoxies, or bisphenol F epoxy resins. High film build is one key advantage of high solids coatings.

Solvent-Free Coatings (100 % Solids). Ecological concerns have led to increasing uses of these materials. Low viscosity LERs based on bisphenol A and bisphenol F epoxies are often used in combination with reactive diluents. The advantages include high buildup in a single application, minimization of surface defects owing to the absence of solvents, excellent heat and chemical resistance, and lower overall application costs. Disadvantages include high viscosity, difficulties to apply and produce thin films, poor impact resistance and flexibility, short pot life, and increased sensitivity to humidity. Weatherable cycloaliphatic epoxies can be used to formulate solvent-free thermally curable coatings because of their low viscosities.

Waterborne Coatings. In the switch from solvent-borne to waterborne systems, epoxies are successfully bridging the gap, largely by adaptation of conventional resins. Waterborne coatings accounted for almost 25 % of epoxy coating consumption in 2001. In addition to the waterborne epoxy dispersions which are typically supplied by epoxy resin producers, significant advances in waterborne coatings have been made by coatings producers.

Recent developments include the elimination of co-solvents in some epoxy dispersions. Custom synthesized acrylic latexes have shown promise when thermally cured with cycloaliphatic epoxies. While global overall volume is still relatively modest, it is expected that future growth rate for this segment will be much higher than that of standard epoxy resins, particularly in Europe, where environmental pressures are stronger.

Powder Coatings. Epoxy-based powder coatings exhibit useful properties such as excellent adhesion, abrasion resistance, hardness, and corrosion and chemical resistance. The application possibilities are diverse, including metal finishing, appliances, structural rebars, pipes, machinery and equipment, furnitures, and automotive coatings. Together, these applications accounted for 30% of epoxy coatings and 17% of epoxy resin consumption globally in 2001. This is a high growth segment of epoxy coatings, mainly because of the 100% solids feature, improved coverage, and recyclability of overspray materials. Pipeline projects, important because of worldwide energy problems, are significant consumers of epoxy powder coatings. The value of improved service life is being increasingly accepted even at the somewhat higher material cost of epoxy systems.

EPOXY COATINGS MARKETS

The marine and industrial protective coating is the largest market for epoxy coatings, followed by powder coatings, automotive, and container coatings. Epoxy coatings are known to have poor weatherability and often chalk when exposed to sunlight for long periods of time. Today, most industrial structures are only coated with epoxy coatings that can last up to 10–15 years. When appearance is critical, epoxy primers are often top-coated with aliphatic isocyanate-based polyurethane coatings.

Other new technology developments in this market segment include surface-tolerant epoxy coatings for aged or marginally prepared surfaces, interval-free epoxy coatings to extend coating service life, mineral spirit-soluble epoxy coatings for shipping container repairs, and styrene-free coatings to replace foul-smelling and regulated organic solvents such as toluene and xylene in coatings for new shipping containers.

Metal container and coil coatings represent a major outlet for epoxy resins, considering that there are more than 100 billion beverage cans and 30 billion food cans produced annually in the United States. Globally, the metal can market is estimated at over 300 billion cans. While the majority of metal containers coated with epoxy coatings are aluminum and steel food and beverage cans, coatings for drums, pails, and aerosol spray cans are included in this market segment. Coil coating is a highly efficient, automated coating process used to produce precoated metal coils, which are subsequently stamped and fabricated to parts. The majority of epoxy coil coatings are used to produce metal can ends and can bodies, with smaller amounts going to building products, appliance panels, transportation, and metal furniture applications.

Coil coatings have also been gaining in the appliance market. More OEMs have turned to precoated metal coils as an efficient manufacturing alternative to produce appliance panels, eliminating the need for postformed coating processes. PVC organosol (copolymers of vinyl chloride and vinyl acetate) coatings for coil-coated can ends and bodies have been under environmental pressures, and epoxy has been gaining as PVC coatings are replaced.

Automotive coatings are another major application for epoxy resins. The excellent adhesion and corrosion resistance properties of epoxies make them the overwhelming choice for automotive primers. One new, growing application is the use of epoxy–polyester or acrylic–GMA powders in primer-surfacer coatings. In addition, glycidyl methacrylate (GMA) is used as a comonomer in etch-resistant liquid top coats containing acrylic acid/anhydride and in GMA-acrylic powder coatings for clear coats and automotive parts. Epoxy powder coatings for automobiles are expected to grow significantly in the near future.

Inks and Resists

Inks and resists comprise a relatively small but high value, growing market for epoxies and epoxy derivatives. Epoxies are often used with other resins such as polyester acrylates and urethane acrylates in these formulations. The largest applications are lithographic and flexographic inks, followed by electronic inks and resists.

STRUCTURAL APPLICATIONS

Next to coatings, structural applications account for the second largest share of epoxy resin consumption (~40%). Epoxy resins in structural applications can be divided into three major areas: fiber-reinforced composites and electrical laminates; casting, encapsulation, and tooling; and adhesives. Within this segment, the largest applications are electrical laminates for PCBs and composites made of epoxy and epoxy vinyl ester for structural applications.

Structural Composites

Epoxy resins and epoxy vinyl ester resins are well suited as fiber-reinforcing materials because they exhibit excellent adhesion to reinforcement, cure with low shrinkage, provide good dimensional stability, and possess good mechanical, electrical, thermal, chemical, fatigue, and moisture-resistance properties. Epoxy composites are formed by aligning strong, continuous fibers in an epoxy resin-curing agent matrix. Processes currently used to fabricate epoxy composites include hand lay-up, spray-up, compression molding, vacuum bag compression molding, filament winding, resin transfer molding reaction, injection molding, and pultrusion.

The most important market for epoxy composites is for corrosion-resistant equipment where epoxy vinyl esters is the dominant material of choice. Other smaller markets are automotive, aerospace, sports/recreation, construction, and marine uses. Because of their higher costs, epoxy and epoxy vinyl esters composites found applications where their higher mechanical strength and chemical and corrosion resistance properties are advantageous.

Epoxy Composites. Composites made with glass fibers usually have a bisphenol A based epoxy resin–diamine matrix and are used in a variety of applications including automotive leaf springs and drive shafts, where mechanical strength is a key requirement. A large and important application is for filament-wound glass-reinforced pipes used in oil fields, chemical plants, for water distribution, and as electrical conduits. Low viscosity liquid systems having good mechanical properties when cured are preferred. Similar systems are used for filament-winding pressure bottles and rocket motor casings. Other applications that use fiber-reinforced epoxy composites include sporting equipment, such as tennis racquet frames, fishing rods, and golf clubs, as well as industrial equipment. The wind energy field is emerging as a potential high growth area for epoxy composites, particularly in Europe.

In the aerospace industry, the use of graphite fiber-reinforced composites has been growing because of their high strength-to-weight ratios. Some newer commercial airliners now contain up to 10% by weight of composite materials.

Epoxy Vinyl Ester Composites.
Epoxy vinyl ester composites are widely used to produce chemically resistant glass-reinforced pipes, stacks, and tanks by contact molding and filament-winding processes. Epoxy vinyl ester resins provide outstanding chemical resistance against aggressive chemicals such as aqueous acids and bases and are materials of choice for demanding applications in petrochemical plants, oil refineries, and paper mills. Epoxy vinyl ester composites are also used in demanding automotive applications such as engine and oil pan covers, where high-temperature performance is required. Exterior panels and truck boxes are also growth automotive applications for vinyl esters. However, in less demanding automotive applications, cheaper thermoplastics and thermosets such as unsaturated polyesters or furan resins are often used. In general, epoxy vinyl ester is considered to be a premium polyester resin with higher temperature and corrosion resistance properties at higher costs. It is used where the cheaper unsaturated polyesters cannot meet performance requirements. For the same reason, epoxy vinyl ester has not grown significantly in less demanding civil engineering applications. Other uses of epoxy vinyl ester composites include boat hulls, swimming pools, saunas, and hot tubs.

Mineral-Filled Composites.
Epoxy mineral-filled composites are widely used to manufacture laboratory equipment such as laboratory bench tops, sinks, hoods, and other laboratory accessories. The excellent chemical and thermal resistance properties of epoxy thermosets make them ideal choices for this application.

Civil Engineering, Flooring, and Construction

Civil engineering is another large application for epoxies, accounting for up to 13% of total global epoxy consumption. This application includes flooring, decorative aggregate, paving, and construction. Key attributes of epoxies such as ease of installation, fast ambient cure, good adhesion to many substrates, excellent chemical resistance, low shrinkage, good mechanical strength, and durability make them suitable for this market.

Epoxy resins are used for both functional and decorative purposes in monolithic flooring and in factory-produced building panel applications. Products include floor paints, self-leveling floors, trowelable floors, and pebble-finished floors. Epoxy floorings provide wear-resistant and chemical-resistant surfaces for dairies and food processing and chemical plants where acids normally attack concrete. Epoxies are also used in flooring for walk-in freezers, coolers, kitchens, and restaurants because of their good thermal properties, slip resistance, and ease of cleanup. In commercial building applications, such as offices and lobbies, terrazzolike surfaces can be applied in thin layers. Continuous seamless epoxy floors are competitive with ceramic tiles. Semiconductive epoxy/carbon black floorings are used in electronics manufacturing plants because of their ability to dissipate electrical charges. Decorative slip-resistant coatings are available for outdoor stair treads, balconies, patios, walkways, and swimming-pool decks. Epoxy aggregates containing up to 90% of stones or minerals are used for decorative walls, floors, and decks.

Epoxy systems for roads, tunnels, and bridges are effective barriers to moisture, chemicals, oils, and grease. They are used in new construction as well as in repair and maintenance applications. Typical formulations consist of liquid epoxy resins extended with coal tar and diethylenetriamine curing agent. Epoxy resins are widely used in bridge expansion joints and to repair concrete cracks in adhesive and grouting (injectable mortar) systems. Epoxy pavings are used to cover concrete bridge decks and parking structures. Formulations of epoxy resins and polysulfide polymers in conjunction with polyamine curing agents are used for bonding concrete to concrete.

Recent developments in the construction and civil engineering industry include the development of "intelligent concrete" with self-healing capability in Japan. Some of the systems are based on epoxy resins encapsulated in concrete which, when triggered by cracks, open and cure to repair the concrete.

Electrical Laminates

Printed wiring boards (PWB) or printed circuit boards (PCB) are used in all types of electronic equipment. In noncritical applications such as inexpensive consumer electronics these components are made from paper-reinforced phenolic, melamine, or polyester resins. For more critical applications such as high-end consumer electronics, computers, complex telecommunication equipment, etc, where higher performance materials are required, epoxy-resin-based glass fiber laminates fulfill the requirements at reasonable cost. This application constitutes the single largest volume of epoxies used in structural composites.

In recent years, environmental concerns over toxic smoke generation during fire and end-of-life incineration of electronic equipment containing brominated products, particularly in Europe and Japan, have driven development efforts on halogen-free resins. This has resulted in a number of alternative products such as phosphorous additives and phosphor-containing epoxies.

Other Electrical and Electronic Applications

Casting, Potting, and Encapsulation. Since the mid-1950s, electrical-equipment manufacturers have taken advantage of the good electrical properties of epoxy and the design freedom afforded by casting techniques to produce switchgear components, transformers, insulators, high voltage cable accessories, and similar devices.

In casting, a resin-curing agent system is charged into a specially designed mold containing the electrical component to be insulated. After cure, the insulated part retains the shape of the mold. In encapsulation, a mounted electronic component such as a transistor or semiconductor in a mold is encased in an epoxy resin based system. Coil windings, laminates, lead wires, etc, are impregnated with the epoxy system. Potting is the same procedure as encapsulation except that the mold is a part of the finished unit. When a component is simply dropped into a resin-curing agent system and cured without a mold, the process is referred to as dipping. It provides little or no impregnation and is used mainly for protective coatings.

Transfer Molding. Epoxy molding compounds (EMC) are solid mixtures of epoxy resin, curing agent(s) and catalyst, mold-release compounds, fillers, and other additives. An important application of epoxy molding compounds is encapsulation of electronic components such as semiconductor chips, passive devices, and integrated circuits by transfer molding.

Adhesives

Epoxy-based adhesives provide powerful bonds between similar and dissimilar materials such as metals, glass, ceramics, wood, cloth, and many types of plastics. In addition, epoxies offer low shrinkage, low creep, high performance over a wide range of usage temperatures, and no by-products (such as water) release during cure.

The broad range of epoxy resins and curing agents on the market allows a wide selection of system components to satisfy a particular application. Although the majority of epoxy adhesives are two-pack systems, heat activated one-pack adhesives are also available. Low molecular weight DGEBA liquid resins are the most commonly used.

Tooling

Tools made with epoxy are used for producing prototypes, master models, molds and other parts for aerospace, automotive, foundry, boat building, and various industrial molded items. Epoxy tools are less expensive than metal ones and can be modified quickly and cheaply. Epoxy resins are preferred over unsaturated polyesters and other free-radical cured resins because of their lower shrinkage, greater interlaminar bond strength, and superior dimensional stability.

HEALTH AND SAFETY FACTORS

There have been many investigations of the toxicity of various classes of epoxy-containing materials (glycidyloxy compounds). The use and interpretation of the vast amount of data available has been obscured by two factors: (1) proper identification of the epoxy systems in question and (2) lack of meaningful classification of the epoxy materials. In general, the toxicity of many of the glycidyloxy derivatives is low, but the diversity of compounds found within this group does not permit broad generalizations for the class.

All suppliers provide material safety data sheets (MSDS), which contain the most recent toxicity data. These are the best sources of information and should be consulted before handling the materials.

B. Ellis, ed., *Chemistry and Technology of Epoxy Resins*, 1st ed., Blackie Academic & Professional, Glasgow, U.K., 1993.

H. Lee and K. Neville, *Handbook of Epoxy Resins*, McGraw-Hill, Inc., New York, 1967. Reprinted 1982.

C. A. May and Y. Tanaka, eds., *Epoxy Resins Chemistry and Technology*, 2nd ed., Marcel Dekker, Inc., New York, 1988.

P. K. T. Ording, *Waterborne and Solvent Based Epoxies and Their End User Applications*, John Wiley & Sons, Inc., New York, 1996, pp. 57, 100.

B. Sedlacek and J. Kahovec, eds., *Crosslinked Epoxies*, Walter de Gruyter, Berlin, 1987.

Ha. Q. Pham
Maurice J. Marks
Dow Chemical

ESTERIFICATION

This article describes methods for the production of carboxylic esters:

$$\underset{\quad}{R-\overset{\displaystyle O}{\overset{\|}{C}}-OR'}$$

For the properties of these compounds, see Esters, organic. For esters of inorganic acids, see the articles on nitric acid, phosphoric acids, sulfuric acid, etc.

Esters are most commonly prepared by the reaction of a carboxylic acid and an alcohol with the elimination of water. Esters are also formed by a number of other reactions utilizing acid anhydrides, acid chlorides, amides, nitriles, unsaturated hydrocarbons, ethers, aldehydes, ketones, alcohols, and esters (via ester interchange).

On the basis of bulk production, poly(ethylene terephthalate) manufacture is the most important ester producing process. This polymer is produced by either the direct esterification of terephthalic acid and ethylene glycol, or by the transesterification of dimethyl terephthalate with ethylene glycol. Dimethyl terephthalate is produced by the direct esterification of terephthalic acid and methanol.

Other large-volume esters are vinyl acetate (VAM), methyl methacrylate (MMA), and dioctyl phthalate (DOP). VAM is produced for the most part by the vapor-phase oxidative acetoxylation of ethylene. MMA and

DOP are produced by direct esterification techniques involving methacrylic acid and phthalic anhydride, respectively.

The acetates of most alcohols are also commercially available and have diverse uses. Because of their high solvent power, ethyl, isopropyl, butyl, isobutyl, amyl, and isoamyl acetates are used in cellulose nitrate and other lacquer-type coatings (see CELLULOSE ESTERS). Butyl and hexyl acetates are excellent solvents for polyurethane coating systems (see COATINGS; URETHANE POLYMERS). Ethyl, isobutyl, amyl, and isoamyl acetates are frequently used as components in flavoring (see FLAVORS AND SPICES), and isopropyl, benzyl, octyl, geranyl, linalyl, and methyl acetates are important additives in perfumes (qv).

Effect of Structure

The rate at which different alcohols and acids are esterified as well as the extent of the equilibrium reaction are dependent on the structure of the molecule and types of functional substituents of the alcohols and acids.

In making acetate esters, the primary alcohols are esterified most rapidly and completely, ie, methanol gives the highest yield and the most rapid reaction. Under the same conditions, the secondary alcohols react much more slowly and afford lower conversions to ester products; however, wide variations are observed among the different members of this series. The tertiary alcohols react slowly and the conversions are generally low (1–10% conversion at equilibrium).

The introduction of a nitrile group on an aliphatic acid has a pronounced inhibiting effect on the rate of esterification.

Substitutions that displace electrons toward the carboxyl group of aromatic acids diminish the rate of the reaction.

Kinetic Considerations

Extensive kinetic and mechanistic studies have been made on the esterification of carboxylic acids since Berthelot and Saint-Gilles first studied the esterification of acetic acid. A number of mechanisms for acid- and base-catalyzed esterification have been proposed. One possible mechanism for the bimolecular acid-catalyzed ester hydrolysis and esterification is shown below.

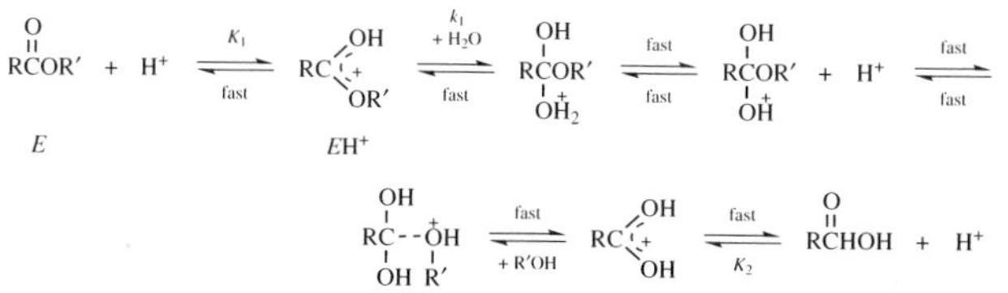

This mechanism leads to the rate equation for hydrolysis and to an analogous expression for the esterification:

$$-\frac{d[E]}{dt} = \frac{k_1 K_1 [E][H_2O][H^+]}{1+\alpha} - \frac{k_2 K_2 [A][R'OH][H^+]}{1+1/\alpha}$$

In this expression, α depends on those rate coefficients in the above mechanism whose values are assumed to be high. Other mechanisms for the acid hydrolysis and esterification differ mainly with respect to the number of participating water molecules and possible intermediates.

Applications of kinetic principles to industrial reactions are often useful. Initial kinetic studies of the esterification reaction are usually conducted on a small scale in a well stirred batch reactor. In many cases, results from batch studies can be used in the evaluation of the esterification reaction in a continuous operating configuration.

Equilibrium Constants

The reaction between an organic acid and an alcohol to produce an ester and water is expressed as:

This was first demonstrated in 1862 by Berthelot and Saint-Gilles, who found that when equivalent quantities of ethyl alcohol and acetic acid were allowed to react, the esterification stopped when two-thirds of the acid had reacted. Similarly, when equal molar proportions of ethyl acetate and water were heated together, hydrolysis of the ester stopped when about one-third of the ester was hydrolyzed. By varying the molar ratios of alcohol to acid, yields of ester >66% were obtained by displacement of the equilibrium. The results of these tests were in accordance with the mass action law shown. $K = [ester][water]/[acid][alcohol]$. However, in many cases the equilibrium constant is affected by the proportion of reactants. The temperature as well as the presence of salts may also affect the value of the equilibrium constant.

Completion of Esterification

Because the esterification of an alcohol and an organic acid involves a reversible equilibrium, these reactions usually do not go to completion. Conversions approaching 100% can often be achieved by removing one of the products formed, either the ester or the water, provided the esterification reaction is equilibrium limited and not rate limited. A variety of distillation methods can be applied to afford ester and water product removal from the esterification reaction (see DISTILLATION). Other methods such as reactive extraction and reverse osmosis can be used to remove the esterification products to maximize the reaction conversion. In general, esterifications are divided into three broad classes, depending on the volatility of the esters: (1) Esters of high volatility, such as methyl formate, methyl acetate, and ethyl formate, have lower boiling points than those of the corresponding alcohols, and therefore can be readily removed from the reaction mixture by distillation. (2) Esters of medium volatility are capable of removing the water formed by distillation. Examples are propyl, butyl, and amyl formates, ethyl, propyl, butyl, and amyl acetates, and the methyl and ethyl esters of propionic, butyric, and valeric acids. (3) Esters of low volatility are accessible via several types of esterification.

Use of Azeotropes to Remove Water. With the aliphatic alcohols and esters of medium volatility, a variety of azeotropes is encountered on distillation. Removal of these azeotropes from the esterification reaction mixture drives the equilibrium in favor of the ester product.

Use of Desiccants and Chemical Means to Remove Water. Another means to remove the water of esterification is calcium carbide supported in a thimble of a continuous extractor through which the condensed vapor from the esterification mixture is percolated (see CARBIDES). A column of activated bauxite (Florite) mounted over the reaction vessel has been used to remove the water of reaction from the vapor by adsorption.

Catalysts

The choice of the proper catalyst for an esterification reaction is dependent on several factors. The most common catalysts used are strong mineral acids such as sulfuric and hydrochloric acids. Lewis acids such as boron trifluoride, tin and zinc salts, aluminum halides, and organo–titanates have been used. Cation-exchange resins and zeolites are often employed also.

Acid-Regenerated Cation Exchangers. The use of acid-regenerated cation resin exchangers (see ION EXCHANGE) as catalysts for effecting esterification offers distinct advantages over conventional methods. Several types of cation-exchange resins can be used as solid catalysts for esterification. In general, the strongly acidic sulfonated resins comprised of copolymers of styrene, ethylvinylbenzene, and divinylbenzene are used most widely. With the continued improvement of ion-exchange resins, such as the macroporous sulfonated resins, esterification has become one of the most fertile areas for use of these solid catalysts.

Despite the higher cost compared with ordinary catalysts such as sulfuric or hydrochloric acid, the cation exchangers present several features that make their use economical. The ability to use these agents in a fixed-bed reactor operation makes them attractive for a continuous process. Cation-exchange catalysts can be used also in continuous stirred tank reaction (CSTR) operation.

BATCH ESTERIFICATION

Batch esterification is used to produce ethyl acetate (Fig. 1) and n-butyl acetate.

CONTINUOUS ESTERIFICATION

The law of mass action, the laws of kinetics, and the laws of distillation all operate simultaneously in a process of this type. Esterification can occur only when the concentrations of the acid and alcohol are in excess of equilibrium values; otherwise, hydrolysis must occur. The equations governing the rate of the reaction and the variation of the rate constant (as a function of such variables as temperature, catalyst strength, and proportion of

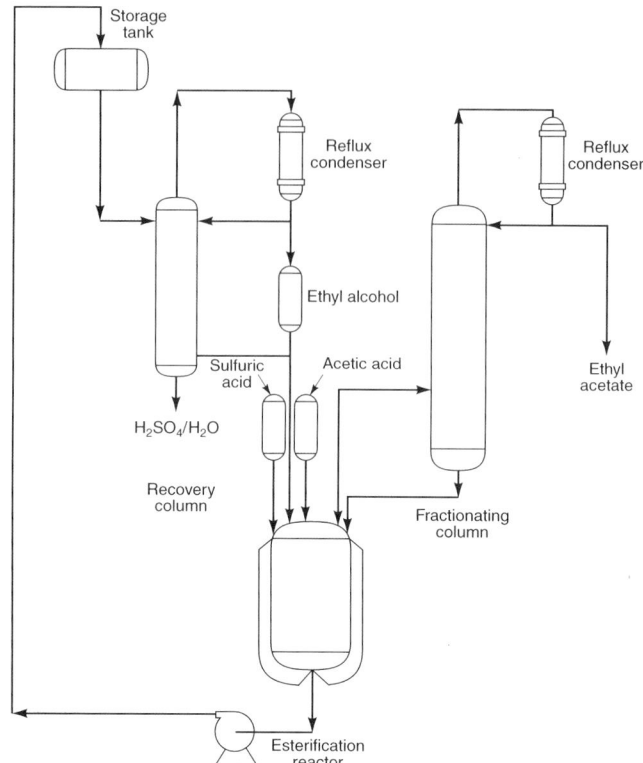

Figure 1. Batch ethyl acetate process.

reactants) describe the kinetics of the liquid-phase reaction. The usual distillation laws must be modified, since most esterifications are somewhat exothermic and reaction is occurring on each plate. Since these kinetic considerations are superimposed on distillation operations, each plate must be treated separately by successive calculations after the extent of conversion has been determined (see DISTILLATION).

Continuous esterification of acetic acid in an excess of n-butyl alcohol with sulfuric acid catalyst using a four-plate single bubblecap column with reboiler has been studied. The rate constant and the theoretical extent of reaction were calculated for each plate, based on plate composition and on the total incoming material to the plate. Good agreement with the analytical data was obtained.

A continuous distillation process has been studied for the production of high boiling esters from intermediate boiling polyhydric alcohols and low boiling monocarboxylic aliphatic or aromatic acids. The water of reaction and some of the organic acid were continuously removed from the base of the column.

Continuous esterification is used to produce methyl acetate (Fig. 2).

VAPOR-PHASE ESTERIFICATION

Catalytic esterification of alcohols and acids in the vapor phase has received attention because the conversions obtained are generally higher than in the corresponding liquid-phase reactions.

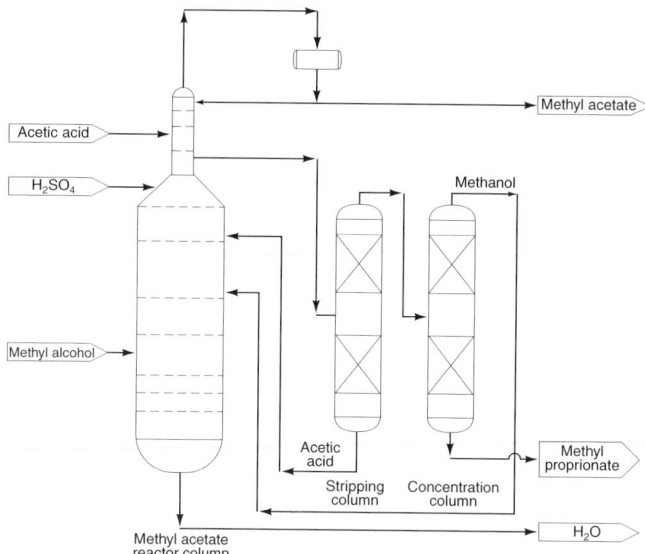

Figure 2. Continuous methyl acetate process.

Physicochemical Considerations

The determination of the equilibrium constant K_G for the reaction $C_2H_5OH + CH_3COOH \rightarrow C_2H_5OOCH_3 + H_2O$ has been the subject of a number of investigations over the temperature range of 40–300°C. The values of the equilibrium constant range from 6–559 with 71–95% ester as the equilibrium concentration from an equimolar mixture of ethyl alcohol and acetic acid, depending on the technique used. A study of the reaction mechanism indicates that adsorption of acetic acid is the rate-controlling step; the molecularly adsorbed acetic acid then reacts with alcohol in the vapor phase.

Ethyl Acetate

Catalysts proposed for the vapor-phase production of ethyl acetate include silica gel, zirconium dioxide, activated charcoal, and potassium hydrogen sulfate. More recently, phosphoric-acid-treated coal and calcium phosphate catalysts have been described.

ESTERIFICATION OF OTHER COMPOUNDS

Acid Anhydrides

Acid anhydrides react with alcohols to form esters (in high yields in many cases) with a carboxylic acid formed as by-product:

$$\underset{\overset{\|}{RC}-O-\overset{\|}{CR}}{\overset{O\quad\ O}{}} + R'OH \longrightarrow \underset{\overset{\|}{RC}-OR'}{\overset{O}{}} + \underset{\overset{\|}{RC}-OH}{\overset{O}{}}$$

However, this method is applied only when esterification cannot be effected by the usual acid–alcohol reaction because of the higher cost of the anhydrides. The production of cellulose acetate (see FIBERS–CELLULOSE ESTERS), phenyl acetate (used in acetaminophen production), and aspirin (acetylsalicylic acid) (see SALICYLIC ACID AND RELATED COMPOUNDS) are examples of the large-scale use of acetic anhydride.

Formic anhydride is not stable. However, formate esters of alcohols and phenolics can be prepared using formic–acetic anhydride. Dibasic acid anhydrides such as phthalic anhydride and maleic anhydride readily react with alcohols to form the monoalkyl ester. Ketene, like acid anhydrides, reacts with alcohols to form (acetate) esters.

Acid Chlorides

Acid chlorides react with alcohols to form esters.

Amides

Alcoholysis of amides provides another method for synthesizing esters.

Other methods of converting amides to esters have been described. Alkyl halides can be treated with amides to give esters. Also, esters can be synthesized from N-alkyl-N-nitrosoamides, which are derived from the corresponding amides.

Nitriles

Alcoholysis of nitriles offers a convenient way to produce esters without isolating the acid. Catalysts such as hydrogen chloride, hydrogen bromide, and sulfuric acid have been employed. One of the most important applications of this process is that of methyl methacrylate manufacture.

Unsaturated Hydrocarbons

Olefins from ethylene through octene have been converted into esters via acid-catalyzed nucleophilic addition.

Most of the vinyl acetate produced in the United States is made by the vapor-phase ethylene process.

Ethers

In the presence of anhydrous agents such as ferric chloride, hydrogen bromide, and acid chlorides, ethers react to form esters. Esters can also be prepared from ethers by an oxidative process.

Unsaturated esters can be prepared from the corresponding acetylenic ethers with yields in most cases of >50%. β-Hydroxyethyl esters can be prepared from carboxylic acids and ethylene oxide.

Aldehydes and Ketones

Esters are obtained readily by condensation of aldehydes in the presence of alcoholate catalysts such as aluminum ethylate, $Al(OC_2H_5)_3$, by the Tishchenko reaction.

Alcohols

The direct synthesis of esters by dehydrogenation or oxidative hydrogenation of alcohols offers a simple method for the preparation of certain types of esters, such as ethyl acetate.

TECHNICAL PREPARATION OF ESTERS

Esterification is generally carried out by refluxing the reaction mixture until the carboxylic acid has reacted with the alcohol and the water has been split off. The water of the ester is removed from the equilibrium by distillation. The choice of the esterification process to obtain a maximum yield is dependent on many factors, ie, no single process has universal applicability.

Methyl Esters

Methyl esters are obtained in good yield using methylene dichloride or ethylene dichloride as solvent.

Medium Boiling Esters

Esterification of ethyl and propyl alcohols, ethylene glycol, and glycerol with various acids, eg, chloro- or bromoacetic, or pyruvic, by the use of a third component such as benzene, toluene, hexane, cyclohexane, or carbon tetrachloride to remove the water produced is quite common.

High Boiling Esters

The following procedure can be used for making diethyl phthalate and other high boiling esters. Phthalic anhydride (1 equiv) and 2.5 equivalents of ethanol are refluxed for 2 h in the presence of 1% of concentrated H_2SO_4.

Difficulty Esterifiable Acids

The sterically hindered acids, such as 2,6-disubstituted benzoic acids, cannot usually be esterified by conventional means. Several esters of sterically hindered acids such as 2,4,6-triisopropylbenzoic acid have been prepared by dissolving 2 g of the acid in 14–20 mL of 100% H_2SO_4. After standing a few minutes at room temperature, when presumably the acylium cation is formed, the solution is poured into an excess of cold absolute methanol. Most of the alcohol is removed under reduced pressure, about 50 mL of water is added, and the distillation is continued under reduced pressure to remove the remainder of the methanol. The organic matter is extracted with ether and treated with sodium carbonate solution. The ester is then distilled. Yields of esters made in this manner are 57–81%.

ESTER INTERCHANGE

Ester interchange (transesterification) is a reaction between an ester and another compound, characterized by an exchange of alkoxy groups or of acyl groups, and resulting in the formation of a different ester. The process of transesterification is accelerated in the presence of a small amount of an acid or a base.

Three types of transesterification are known: (1) exchange of alcohol groups, commonly known as alcoholysis, (2) exchange of acid groups, acidolysis, and (3) ester–ester interchange. These reactions are reversible and ordinarily do not involve large energy changes.

Applications

Transesterifications via alcoholysis play a significant role in industry as well as in laboratory and in analytical chemistry. The reaction can be used to reduce the boiling point of esters by exchanging a long-chain alcohol group with a short one, eg, methanol, in the analysis of fats, oils, and waxes.

An industrial example of acidolysis is the reaction of poly(vinyl acetate) with butyric acid to form poly(vinyl butyrate). Often a butyric acid–methanol mixture is used and methyl acetate is obtained as a coproduct.

R. C. Larock, *Comprehensive Organic Transformations*, VCH Publishers, Inc., New York, 1989.

K. S. Markley, in K. S. Markley, ed., *Fatty Acids*, part 2, Wiley-Interscience, New York, 1961.

S. Patai, *The Chemistry of Carboxylic Acids and Esters*, Wiley-Interscience, New York, 1969.

E. E. Reid, in P. Grotggins, *Unit Processes in Organic Synthesis*, 5th ed., McGraw-Hill Book Co., Inc., New York, 1958.

Mohammad Aslam
G. Paull Torrence
Edward G. Zey
Hoechst Celanese Corporation

ESTERS, ORGANIC

Organic carboxylic esters are compounds that are formed by a condensation reaction between a molecule of carboxylic acid and a molecule of alcohol or phenol with elimination of water as depicted in the following equation. $R^1COOH + R^2OH \rightleftharpoons R^1COOR^2 + H_2O$, where R^1 and R^2 are the same or different hydrocarbon radicals, including unsaturated ones.

NOMENCLATURE

The names of esters consist of two words that reflect their formation from an alcohol and a carboxylic acid. According to the IUPAC rule, the alkyl or aryl group of the alcohol is cited first, followed by the carboxylate group of the acid with the ending "-ate" replacing the "-ic" of the acid. For example, $CH_3CH_2COOCH_3$, the methyl ester of propanoic acid, is called methyl propanoate (or methyl propionate, if the trivial name propionic acid, is for the carboxylic acid).

PHYSICAL PROPERTIES

The physical properties of organic esters vary according to the molecular weight of each component. Lower molecular weight esters are colorless, mobile, highly volatile liquids that usually have pleasant odors. As the molecular weight increases, volatility decreases and the consistency becomes waxy, then solid, and eventually even brittle, often with formation of lustrous crystals. The melting

Table 1. Physical Properties of Some Common Esters

Ester	Mol wt	n^{20}_D	d^{20}_{20}	Bp, °C[a]	Freezing point, °C	Flash point, °C[b]
methyl formate	60.05	1.344	0.0975	32	−99.8	−9
ethyl formate	74.08	1.3598	0.9236	54.3	−80	−20
butyl formate	102.13	1.3889	0.8885[c]	106	−91.9	−18
methyl acetate	74.08	1.3594	0.933	57	−98.1	−10
ethyl acetate	88.1	1.3723	0.0902	77.1	−83.6	−4
vinyl acetate	86.1	1.3959	0.932	72.2	−93.2	−8
propyl acetate	102.13	1.3844	0.887	101.6	−92.5	13
isopropyl acetate	102.13	1.3773	0.872	90	−73.4	2
butyl acetate	116.16	1.3951	0.882	126	−73.5	22
isobutyl acetate	116.16	1.3902	0.871	117.2	−98.6	18
sec-butyl acetate	116.16	1.3877	0.8758[d]	112		31.1[e]
tert-butyl acetate	116.16	1.3855	0.8665[c]	97		
pentyl acetate	130.18	1.4023	0.876	149.3	−70.8	25
isoamyl acetate	130.18	1.4000	0.872	142	−78	25
sec-hexyl acetate	144.22	1.4014[f]	0.8651[g]	157	0	
2-ethylhexyl acetate	172.26	1.4204	0.873	199.3	−93	71
ethylene glycol diacetate	146.14	1.415	1.128	191	−31	88
2-methoxyethyl acetate	118.13	1.4019	1.0067	145	−65.1	44
2-ethoxyethyl acetate	132.16	1.4058	0.975	156.4	−61.7	47
2-butoxyethyl acetate	160.12	1.42	0.943	187.8	−32	81
2-(2-ethoxyethoxy)ethyl acetate	176.21	1.423	1.011	217.4	−25	107
2-(2-butoxyethoxy) ethyl acetate	204.27	1.4265	0.981	247	−32.2	110
benzyl acetate	150.18	1.5232	1.055	215.5	−51.5	90
glyceryl triacetate	218.23	1.4296	1.161	258	−78	138
ethyl 3-ethoxypropionate	146.19		0.95	165–172	−50	58
glyceryl tripropionate	260.3	1.4318	1.100[h]	176	−58	167[e]
methyl acrylate	86.09	1.4040	0.953	80.5	<−75	−3
ethyl acrylate	100.11	1.4068	0.923	99.8	<−72	10
butyl acrylate	128.17	1.4185	0.898	69	−64.6	29
2-ethylhexyl acrylate	184.28		0.887	130[i]	−90	82[e]
methyl methacrylate	100.12	1.4119	0.944	100	−48	10[e]
methyl butyrate	102.13	1.3878	0.898	102.3	−84.8	14
ethyl butyrate	116.16	1.4000	0.878	121.6	−100.8	24
butyl butyrate	144.22	1.4075	0.871	166.6	−91.5	53
methyl isobutyrate	102.13	1.3840	0.891	92.6	−84.7	
ethyl isobutyrate	116.16	1.3870	0.869	110	−88	<21
isobutyl isobutyrate	144.22	1.3999	0.875	148.7	−80.7	38
methyl stearate	298.5	1.457	0.836	215	40	153
ethyl stearate	312.52	1.429	1.057	213–215	33.7	
butyl stearate	340.58		0.855	343	27.5	160
dodecyl stearate	440.8	1.433			28	
hexadecyl stearate	496.91	1.441			57	
dimethyl maleate	144.13	1.4409	1.152	204		91
dimethyl oxalate	111.09	1.4096	1.148	185	−41	76
dimethyl adipate	174.2	1.4283	1.0600	115	10.3	
diethyl adipate	202.25	1.4372	1.008	245	−19.8	
di(2-ethylhexyl) adipate	370.58	1.4472	0.927	214	−60	206
methyl benzoate	136.15	1.517	1.094	199.5	−12.5	83
ethyl benzoate	150.18	1.505	1.051	212.9	−34.2	88
methyl salicylate	152.15	1.536	1.184	223.3	−8.6	96
ethyl salicylate	166.18	1.522	1.137	231.5	1.3	107
dimethyl phthalate	194.19	1.515	1.190	282	−2	146
diethyl phthalate	222.24	1.499	1.118	295	−33	161
dibutyl phthalate	278.35	1.4911	1.0465	340	−35	157
di(2-ethylhexyl) phthalate	390.56	1.486	0.9861	231[j]	−50	218.3
dimethyl isophthalate	194.19	1.5168	1.194[c]	124	67	138
dimethyl terephthalate	194.19			288	140	153
methyl anthranilate	151.17	1.584	1.168	132	24	>100

Table 1. (*Continued*)

Ester	Mol wt	n^{20}_D	d^{20}_{20}	Bp, °C[a]	Freezing point, °C	Flash point, °C[b]
benzyl cinnamate	238.29		1.109[g]	244[j]	39	110
dimethyl carbonate	90.08	1.3682	1.0694[c]	90	3	19[e]
diethyl carbonate	118.13	1.3854	0.9752[c]	127	−43	25

[a]At 101.3 kPa (760 mm Hg) unless otherwise stated.
[b]Closed cup determination unless otherwise stated.
[c]d^{20}_2.
[d]d^{16}_4.
[e]Open cup determination.
[f]n^{25}_D.
[g]d^{15}_{15}.
[h]d^{20}_{18}.
[i]At 6.7 kPa (50 mm Hg).
[j]At 0.67 kPa (5 mm Hg).

point of an ester is generally lower than that of the corresponding carboxylic acid. However, the boiling point depends on the chain length of the alcohol component and eventually exceeds that of the acid. Lower molecular weight esters are relatively stable when dry and can be distilled without decomposition. Organic esters are generally insoluble in water but soluble in various organic solvents. Lower esters are themselves good solvents for many organic compounds. The physical properties of commercially important aliphatic and aromatic organic esters are listed in Table 1.

CHEMICAL PROPERTIES

The reactions of esters have been extensively reviewed. The chemical properties of esters may differ considerably, as they are composed of a large number and variety of acid and alcohol moieties. Only typical reactions relevant to the majority of esters are described below.

Hydrolysis

Esters are cleaved (hydrolyzed) into their parent acid and alcohol in the presence of water. This hydrolysis is catalyzed by acids or bases. The mechanistic aspects of ester hydrolysis have received considerable attention and have been reviewed.

Enzymatic Hydrolysis. Enzymatic hydrolysis has received enormous attention. The enzymes generally employed are lipases from microorganisms, plants, or mammalian liver. The great advantage of the enzymatic process is its high chemo- and stereoselectivity.

Transesterification

When esters are heated with alcohols, acids, or other esters in the presence of a catalyst, the alcohol or acid groups are exchanged. The reaction is accelerated by the presence of a small amount of acid or alkali. Three types of transesterification are known: (*1*) exchange of alcohol groups (alcoholysis); (*2*) exchange of acid groups (acidolysis), and (*3*) ester–ester interchange.

Ammonolysis and Aminolysis

Esters and ammonia react to form amides and alcohols. If primary or secondary amines are used, N-substituted amides are formed. This reaction is called aminolysis.

Reduction

Esters can be reduced to alcohols by catalytic hydrogenation, using molecular hydrogen, or by chemical reduction.

Reaction of Enolate Anions

In the presence of certain bases; eg, sodium alkoxide, an ester having hydrogen on the α-carbon atom undergoes a wide variety of characteristic enolate reactions. Mechanistically, the base removes a proton from the α-carbon, giving an enolate that can then react with an electrophile.

Grignard and Related Reactions

Esters react with alkyl magnesium halides in a two-stage process to give alcohols. The reaction involves nucleophilic substitution of R^3 for OR^2 and addition of R^3MgX to the carbonyl group. With 1,4-dimagnesium compounds, esters are converted to cyclopentanols. Lactones react with Grignard reagents and give diols as products. Many other organometallic compounds also react with carbonyl groups.

α-Halo esters react with aldehydes or ketones in the presence of zinc to form β-hydroxy esters. This is known as the Reformatsky reaction.

Preparation of Acyloins

When aliphatic esters are allowed to react with metallic sodium, potassium, or sodium–potassium alloy in inert solvents, acyloins (α-hydroxyketones) are formed.

Pyrolysis

The pyrolysis of simple esters of the formula R^1COOCR_2-CHR_2 to form the free acid and an alkene is a general reaction that is used for producing olefins.

Carbonylation Reaction

The carbonylation of methyl acetate is an important industrial reaction for producing acetic anhydride.

Substitution, Alkylation, and Rearrangement

The reaction of alkaline phenoxides with alkyl (*S*)-2-(chloro)- or (*S*)-2-(mesyloxy)propionate gives optically active

(R)-2-aryloxyalkanoic acid esters in good chemical and optical yields (>97% ee). The reaction is utilized in the synthesis of several phenoxy herbicides.

Optically active 2-arylalkanoic acid esters have been prepared by Friedel-Crafts alkylation of arenes with optically active esters, such as methyl (S)-2-(chlorosulfonoxy)- or (S)-2-(mesyloxy)propionate, in the presence of aluminum chloride.

The Fries rearrangement of phenol esters gives a mixture of 2- and 4-acylphenols. Similarly, enol esters undergo rearrangement to give the corresponding 1,3-diketones.

STABILITY AND STORAGE

All organic esters are unstable in the presence of an acid or base and nucleophiles such as water or alcohols. However, if stored anhydrous, they are stable. Storage vessels can be constructed of steel, aluminum, or other metallic materials, but plastic storage tanks are unsuitable because the highly lipophilic esters can sometimes permeate into the container boundary and soften or even dissolve it.

The properties of flash point, autoignition temperature, and flammable limit should be considered when an ester is to be handled in any fashion.

HEALTH AND SAFETY FACTORS

Toxicity

The degree of toxicity of organic esters covers a wide range. These toxicities are usually described in terms of threshold limit values (TLVs), or permissible exposure limits (PELs). Both the PEL and the TLV describe the average concentration over an 8 h period to which a worker may be exposed without adverse effects. The lethal dosages for 50% of the exposed animals, LD_{50}s, are also used as an indicator of the relative toxicity. The LD_{50}s of organic esters for small mammals range between 0.4 and 16 g/kg. The TLVs of organic esters range between 5 and 400 ppm.

When ingested or absorbed, organic esters are likely to be hydrolyzed to the corresponding alcohols and carboxylic acids. Therefore, the toxicities of the hydrolysis products should also be considered. Some organic esters are highly volatile and can act as an asphyxiant or narcotic. Also, skin absorption and inhalation are among the hazards associated with esters that are volatile or have good solvent action. Because of the high solubility of fats and oils in organic esters, prolonged or repeated exposure to skin can cause drying and irritation.

Acetates generally do not cause any physiological effects unless high exposure occurs, since they are usually converted into or occur naturally as metabolites. However, large enough exposure to acetate esters can cause narcotic effects.

Propionates and higher aliphatic esters generally become less toxic as the size of the alkyl carboxylate increases.

The acrylate esters are more physiologically hazardous than their saturated homologues. They are usually lachrymators and irritants, and their toxicities decrease with increasing molecular weights.

Among adipates, oxalates, malonates, and succinates, the adipates are the least toxic.

Benzoate esters, like most organic esters, are not very toxic. They are not absorbed through the skin as rapidly as alkyl esters but are more potent physiologically. They are also moderate skin irritants.

The phthalate esters are one of the most widely used classes of organic esters, and fortunately they exhibit low toxicity.

More information on the toxicities of a range of organic esters is available in the literature.

Exposure Limits

The Occupational Safety and Health Act (OSHA) lists a multitude of acetates, phthalates, formates, and acrylates along with the corresponding permissible exposure limits and threshold limit values. If there is potential for exposure to an organic ester for which PEL or TLV data have been identified, an exposure limit lower than that listed is usually selected for working in that environment.

REGULATION AND WASTE

Waste from production of organic esters is usually not a problem, since the method of synthesis often involves a carboxylic acid condensation with an alcohol and the only by-product is water. Any organic remnants lost to the process water can usually be biologically degraded. The biochemical oxygen demand (BOD) or chemical oxygen demand (COD) should be measured if biological treatment is used on the process waste from ester production. Organic ester vapor emitted in processing usually can be burned. Extensive federal environmental regulations exist that govern organic esters as well as many other substances. These regulations must always be consulted for complete information before using large amounts of organic esters. State and local regulations must also be met, which in some cases are more stringent than federal regulations.

USES

Organic esters find use in wide range of fields encompassing solvents, extractants, diluents, plasticizing agents, lubricants, perfumes, medicinals, herbicides and pesticides, and photoresists.

S. Patai, ed., *The Chemistry of Carboxylic Acids and Esters*, Wiley-Interscience Inc., New York, 1969.

M. B. Smith and J. March, *Advanced Organic Chemistry: Reactions, Mechanisms, and Structure*, 5th ed., John Wiley & Sons, Ltd., New York, 2000.

I. O. Sutherland, in D. Barton and W. D. Ollis, eds., *Comprehensive Organic Chemistry*, Vol. 2, Pergamon Press, Inc., Elmsford, New York, 1979, pp. 869–956.

B. M. Trost and I. Fleming, eds., *Comprehensive Organic Synthesis: Selectivity, Strategy and Efficiency in Modern Organic Chemistry*, Vol. 1–9, Pergamon Press, Inc., Elmsford, New York, 1992.

RAJ SAKAMURI
Clariant Corporation

ETHANOL

Ethanol or ethyl alcohol, CH_3CH_2OH, is one of the most versatile oxygen-containing organic chemicals because of its unique combination of properties as a solvent, a germicide, a beverage, an antifreeze, a fuel, a depressant, and especially as a chemical intermediate for other organic chemicals. The use of fermentation-derived ethanol as an automotive fuel additive to enhance octane and reduce emissions has seen explosive growth. Ethanol is a replacement for methyl *tert*-butyl ether (MTBE) as an oxygenate for fuels. MTBE has been deemed an environmental risk because of its seepage into groundwater. The high cost of gasoline containing ethanol, however, is an issue.

PHYSICAL PROPERTIES

A summary of the physical properties of ethyl alcohol is presented in Table 1. Detailed information on the vapor pressure, density, and viscosity of ethanol can be obtained from many references.

CHEMICAL PROPERTIES

The chemistry of ethyl alcohol is largely that of the hydroxyl group, namely reactions of dehydration, dehydrogenation, oxidation, and esterification.

Table 1. Physical Properties of Ethanol

Property	Value
freezing point, $^\circ$C	-114.1
normal boiling point, $^\circ$C	78.32
critical temperature, $^\circ$C	243.1
critical pressure, kPa^a	6383.48
critical volume, L/mol	0.167
critical compressibility factor, z, in $PV = znRT$	0.248
density, d_4^{20}, g/mL	0.7893
refractive index, n_D^{20}	1.36143
$\Delta n_D/\Delta t$, 20–30°C, per $^\circ$C	0.000404
surface tension, at 25°C, mN/m (=dyn/cm)	23.1
viscosity, at 20°C, mPa$\cdot$s(=cP)	1.17
solubility in water, at 20°C	miscible
heat of vaporization, at normal boiling point, J/g^b	839.31
heat of combustion, at 25°C, J/g^b	29676.69
heat of fusion, J/g^b	104.6
flammable limits in air, vol %	
lower	4.3
upper	19.0
autoignition temperature, $^\circ$C	423.0
flash point, closed-cup, $^\circ$C	14
specific heat, at 20°C, J/(g$\cdot^\circ$C)^b	2.42
thermal conductivity, at 20°C, W/(m$\cdot$K)	0.170
dipole moment, liq at 25°C, C$\cdot$m^c	5.67×10^{-30}
magnetic susceptibility at 20°C	0.734×10^{-6}
dielectric constant at 20°C	25.7

^{a}To convert kPa to atm, divide by 101.3.
^{b}To convert J to cal, divide by 4.184.
^{c}To convert C$\cdot$m to debye, divide by 3.336×10^{-30} (esu = D $\times 10^{-18}$).

MANUFACTURE

Industrial ethyl alcohol can be produced synthetically from ethylene, as a by-product of certain industrial operations, or by the fermentation of sugar, starch, or cellulose, by 2001 fermentation processes accounted for of total production in the U.S., western Europe, and Japan.

There are two main processes for the synthesis of ethyl alcohol from ethylene. The indirect hydration process is variously called the strong sulfuric acid–ethylene process, the ethyl sulfate process, the esterification–hydrolysis process, or the sulfation–hydrolysis process. The other synthesis process, the catalytic vapor-phase hydration of ethylene, is designed to eliminate the use of sulfuric acid and, since the early 1970s, has completely supplanted the old sulfuric acid process in the United States. Other synthetic methods have been investigated but have not become commercial.

RECOVERY AND PURIFICATION

Various distillation and equipment modifications are used to ensure a pure water azeotrope of ethanol (95% by weight ethanol).

Purification schemes have generally emphasized the following techniques: (1) extractive distillation using water reflux to distill a large share of the impurities and concentrate the crude alcohol–water mixture; (2) efficient fractionation to produce approximately 190-proof alcohol; (3) hydrogenation to convert aldehyde impurities to alcohols, together with the use of chemicals such as inorganic bases and sodium sulfite; and (4) ion-exchange resins or azeotropic distillation to dehydrate 190-proof to 200-proof or absolute alcohol.

SHIPMENT

Commercial ethyl alcohol is shipped in railroad tank cars, tank trucks, 208-L (55-gal) and 19-L (5-gal) drums, and in smaller glass or metal containers having capacities of 0.473 L (one pint), 0.946 L (one quart), 3.785 L (one U.S. gal), or 4.545 L (one Imperial gal). The 208-L drums may be of the unlined iron type. If a guarantee of more meticulous quality is desired, the drums may be lined with phenolic resin. All containers, of course, must comply with the specifications of the U.S. Department of Transportation. Both 190-proof and 200-proof ethyl alcohol are considered red label (flammable) materials by the DOT, as both have flash points below 37.8°C by the Tag closed-cup method.

ECONOMIC ASPECTS

In Brazil and the U.S., fermentation ethanol is used primarily in motor fuels. Small amounts of ethanol are used in motor fuels in Western Europe, however, more than two-thirds of the industrial market is supplied by fermentation alcohol. In Japan, both synthetic and fermentation sources are supplied. Ethanol is not used in motor fuels in Japan.

Demand for ethanol for fuel consumption was expected to amount to 10×10^6 kiloliters by 2005. Much uncertainty exists in this market because it is highly sensitive to political influences.

HEALTH AND SAFETY FACTORS

Ethyl alcohol is a flammable liquid requiring a red label by the DOT and Coast Guard shipping classifications; its flash point is 14°C (Tag, closed cup). Vapor concentrations between 3.3 and 19.0% by volume in air are explosive. Liquid ethyl alcohol can react vigorously with oxidizing materials. Ethyl alcohol has found wide application in industry, and experience shows that it is not a serious industrial poison. If proper ventilation of the work environment is maintained, there is little likelihood that inhalation of the vapor will be hazardous.

The threshold limit (TLV) value for ethyl alcohol vapor in air has been set at 1,000 ppm for an 8-h time-weighted exposure by the ACGIH; the OSHA PEL is TWA 1,000 ppm. The minimum identifiable odor of ethyl alcohol has been reported as 350 ppm. Exposure to concentrations of 5,000–10,000 ppm result in irritation of the eyes and mucous membranes of the upper respiratory tract and, if continued for an hour or more, may result in stupor or drowsiness.

Ethyl alcohol is oxidized completely to carbon dioxide and water in the body; thus it is not a cumulative poison. Alcohol poisoning and alcohol intoxiction are almost invariably the result of using alcohol as a beverage, rather than inhalation as a vapor.

USES

As of 2003, the uses for industrial ethanol were as solvents, 60% (toiletries and cosmetics, 33%; coatings and inks, 30% detergents and household cleaners, 15%; processing solvents, 10%; external pharmaceuticals, 7%, miscellaneous, 5%) and chemical intermediates, 40% (ethyl acrylate, 27%; distilled vinegar, 25%; ethylamines, 13%; ethyl acetate, 10%; glycol ethers, 8%; miscellaneous 17%). Uses for fermentation ethanol were as fuels, 92%; industrial solvents and chemicals, 4%; beverages, 4%.

As an intermediate for other chemical derivatives, ethanol has been displaced by lower cost alternatives.

Denatured Ethanol

For hundreds of years alcoholic beverages have been taxed all over the world to generate government revenue. When ethanol emerged as a key industrial raw material, the alcohol tax was recognized as a burden to many essential manufacturing industries. To lift this burden, the Tax-Free Industrial and Denatured Alcohol Act of 1906 was passed in the United States. The U.S. Treasury, Bureau of Alcohol, Tobacco, and Firearms (BATF), now oversees the production, procurement, and use of ethanol in the United States.

The concern of the government is to prevent tax-free industrial ethanol from finding its way into beverages. To achieve this end, the regulations call for a combination of financial and administrative controls (bonds, permits, and scrupulous record keeping) and chemical controls (denaturants that make the ethanol unpalatable). Regulations establish four distinct classifications of industrial ethanol. The classifications with the most stringent financial and administrative controls call for little or no chemical denaturants. The classifications that call for the most effective chemical denaturants require the least financial and administrative controls. For a list of denaturants currently authorized, see the appropriate code of Federol Regulations.

Chemicals Derived From Ethanol

Acetaldehyde. By 1977, the ethanol route to acetaldehyde had largely been phased out in the United States as ethylene and ethane became the preferred feedstocks for acetaldehyde production. Acetaldehyde usage itself has also changed; two primary derivatives of acetaldehyde, acetic acid, and butanol, are now produced from feedstocks other than acetaldehyde. Acetaldehyde is still produced from ethanol in India.

Ethylene. Where ethylene is in short supply and fermentation ethanol is made economically feasible, as in India and Brazil, ethylene is manufactured by the vapor-phase dehydration of ethanol.

Glycol Ethers. The addition of one mole of ethylene oxide to ethanol gives ethylene glycol monoethyl ether. Addition of two moles of oxide gives the monoethyl ether of diethylene glycol. The oxide–alcohol route is the only commercially important route to glycol ethers now in use. Anhydrous alcohols must be used; otherwise the water present forms contaminating glycols.

Vinegar. Dilute solutions of alcohol as fermented worts are oxidized by air at 30–40°C in the presence of various organisms such as *Mycoderma aceti, B. aceti, and B. xylinus*, to produce dilute acetic acid as vinegar. Vinegar based on synthetic ethanol has a fully acceptable aroma and taste and has gained a healthy share of the market for vinegar used in such products as pickles, ketchup, and mustard. However, vinegar based on fermentation ethanol is gaining back some of this market because of the "all natural ingredients" trend in advertising.

Ethylamines. Mono-, di-, and triethylamines, produced by catalytic reaction of ethanol with ammonia, are a significant outlet for ethanol.

Ethyl Acrylate. The esterification of acrylic acid is a primary use for ethanol. Acrylic acid can also react with either ethylene or ethyl esters of sulfuric acid. These processes have supplanted the condensation reaction of ethanol, carbon monoxide, and acetylene as the principal method of generating ethyl acrylate.

Ethyl Ether. Most ethyl ether is obtained as a by-product of ethanol synthesis via the direct hydration of ethylene.

Ethyl Vinyl Ether. The addition of ethanol to acetylene gives ethyl vinyl ether.

Ethyl *tert*-Butyl Ether. Ethanol can react with isobutylene to form ETBE much the same way as methanol is now processed into MTBE, methyl *tert*-butyl ether.

Ethyl Acetate. The esterification of ethanol by acetic acid was studied in detail over a century ago, and considerable literature exists on determinations of the equilibrium constant for the reaction. The usual catalyst for the production of ethyl acetate is sulfuric acid, but other catalysts have been used, including cation-exchange resins, α-fluoronitrites, titanium chelates, and quinones and their partly reduced products. Ethyl acetate is made industrially by both batch and continuous processes.

Other Derivatives and Reactions. The vapor-phase condensation of ethanol to give acetone has been well documented, however, acetone is usually obtained as a by-product from the cumene process, by the direct oxidation of propylene, or from 2-propanol.

The Guerbet reaction involving condensation of ethanol in the presence of sodium ethoxide, catalyzed by potassium hydroxide and boric anhydride or alkaline phosphates, gives *n*-butanol.

"Ethanol, Chemical Profile," *Chemical Market Reporter* **263**(31), 27 (Jan. 20, 2003).

V. C. Mehlenbocher, in J. Mitchell, Jr., ed., *Organic Analysis*, Vol. 1, Interscience Publishers, Inc., New York, 1953, pp. 1–65.

A National Plan for Energy Research, Development, and Demonstration: Creating Energy Choices for the Future, Vol. 1 ERDA, 1976.

JOHN E. LOGSDON
Union Carbide Corporation
updates by staff

ETHERS

Ethers are compounds of the general formula Ar−O−Ar′, Ar−O−R, and R−O−R′ where Ar is an aryl group and R is an alkyl group.

PHYSICAL PROPERTIES

In general, ethers are neutral, pleasant-smelling compounds that have little or no solubility in water but are easily soluble in organic liquids. Their boiling points approximate those of hydrocarbons having comparable molecular weights and geometries. Table 1 gives the basic physical properties for many of the ethers. More detailed physical properties for the ethers most commonly used as solvents are listed in Table 2.

CHEMICAL PROPERTIES

Most ethers, particularly the dialkyl ethers, are comparatively unreactive compounds because the carbon-oxygen bond is not readily cleaved. For this reason, ethers are fre-

quently employed as inert solvents in organic synthesis. However, within the ether family the cyclic ethers are more reactive, the most reactive being the olefin oxides or epoxides. Epoxides are generally used as intermediates for producing other small molecules or polymers. Ethers do react with exceptionally powerful basic reagents, particularly certain alkali metal alkyls, to give cleavage products. Ethers react with less powerful bases to give the same cleavage products, but only under the forcing conditions of high temperature and pressure. The ether linkage can also be cleaved by strong acids, generally at high temperatures.

Most ethers are potentially hazardous chemicals because, in the presence of atmospheric oxygen, a radical-chain process can occur, resulting in the formation of peroxides that are unstable, explosion-prone compounds.

Ethers are weakly basic and are converted to unstable oxonium salts by strong acids such as sulfuric acid, perchloric acid, and hydrobromic acid; relatively stable complexes are formed between ethers and Lewis acids such as boron trifluoride, aluminum chloride, and Grignard reagents. They also undergo Friedel-Crafts alkylation and acylation.

FUEL PROPERTIES

In addition to MTBE, two other ethers commonly used as fuel additives are *tert*-amyl methyl ether (TAME) and ethyl *tert*-butyl ether (ETBE). There are a number of properties that are important in gasoline blending.

MANUFACTURE

The most versatile method of preparing ethers is the Williamson ether synthesis, particularly in the preparation of unsymmetrical alkyl ethers.

Alkyl tertiary alkyl ethers can be prepared by the addition of an alcohol or phenol to a tertiary olefin under acid catalysis (reycler reaction); sulfuric acid, phosphoric acid, hydrochloric acid, and boron trifluoride have all been used as catalysts. Commercially, sulfonic acid ion-exchange resins are used in fixed-bed reactors to make these tertiary alkyl ethers.

Commercially Important Ethers

Ethyl Ether. Much of the diethyl ether manufactured is obtained as a by-product when ethanol is produced by the vapor-phase hydration of ethylene over a supported phosphoric acid catalyst. Such a process has the flexibility to adjust to some extent the relative amounts of ethanol and diethyl ether produced in order to meet existing market demands. Diethyl ether can be prepared directly to greater than 95% yield by the vapor-phase dehydration of ethanol in a fixed-bed reactor using an alumina catalyst.

Methyl *tert*-Butyl Ether. MTBE is easily made by the selective reaction of isobutylene and methanol over an acidic ion-exchange resin catalyst, in the liquid phase and at temperatures below 100°C. To be economically competitive, MTBE's use as an octane enhancer in gasoline has

Table 1. Physical Properties of Some Representative Ethers

Systematic name	Common name	Bp,[a] °C	d_4^{20b}	n_D^{20}
		Saturated		
symmetrical				
methyl ether		−23.7	1.617[c]	
2-methoxyethyl ether		162	0.9451	1.4097
ethyl ether		34.5	0.7146	1.3527
1-chloroethyl ether		116–117	1.1060[25]	1.4185[25]
n-propyl ether		90.5	0.7360	1.3809
isopropyl ether		68.5	0.7257	1.3682
n-butyl ether		142.0	0.7704	1.3981
sec-butyl ether		122.0	0.7590[25]	1.3931
isobutyl ether		123.0	0.7612[15]	
tert-butyl ether		108.0	0.7622[15]	1.3946
n-amyl ether		188.0	0.7849	1.4119
isoamyl ether		173.0	0.7777	1.4085
sec-amyl ether		161.0	0.7830	1.4058
n-hexyl ether		223[102]	0.7936	1.4204
n-heptyl ether		258.5[102]	0.8008	1.4275
n-octyl ether		286–287	0.8063	1.4327
unsymmetrical				
methyl *n*-propyl ether		38.9	0.738	1.3579
methyl isopropyl ether		32.5[104]	0.7237[15]	1.3576
methyl *n*-butyl ether		70.5	0.7443	1.3736
methyl isobutyl ether		105–106[96]	0.7549	1.3852[25]
methyl *tert*-butyl ether	MTBE	55.1	0.7406	1.3690
methyl *tert*-amyl ether	TAME	85–86	0.770	1.3896
ethyl isopropyl ether		53.5	0.7211	1.3624
ethyl *n*-butyl ether		91.5	0.7490	1.3818
ethyl *tert*-butyl ether	ETBE	72–73	0.742	1.3756
ethyl *n*-amyl ether		118.0	0.7622	1.3927
ethyl *tert*-amyl ether		101	0.7657	1.3912
isopropyl *tert*-butyl ether		87.6	0.7365[25]	1.3799
2-ethoxyethanol	cellosolve	135	0.931[20]	1.406[25]
2-(2-ethoxy) ethoxyethanol ether	carbitol	196	0.9855[25]	1.4273
		Unsaturated		
vinyl ether		28–31	0.767[25]	
vinyl methyl ether		5–6	0.7511	
vinyl ethyl ether		35.0	0.7533	1.3739[25]
vinyl *n*-butyl ether		93.5	0.7735[25]	1.3997[25]
allyl ether		95.0	0.8053	1.4163
bis(2-methallyl) ether		105.4	0.8627	1.4206
allyl ethyl ether		67.6	0.765	1.3881
allyl glycidyl ether		75.0[6.7]		1.4310[30]
ethynyl ethyl ether		49.0	0.8001	1.3796
ethynyl butyl ether		104.0	0.8078[25]	1.4033[25]
		Cyclic		
ethylene oxide	oxirane	13.5[99.4]	0.8824[10]	1.3597[7]
1,2-propylene oxide	methyloxirane	34.3	0.8590[10]	1.3670
1,3-propylene oxide	oxetane	47.8	0.8930[25]	1.3961
tetrahydrofuran	oxolane	64.5	0.8892	1.4050
furan	oxole	31.36	0.9514	1.4214[2]
tetrahydropyran	oxane	88	0.8810	1.4200
1,4-dioxane		101[100]	1.0337	1.4224
		Aromatic		
methyl phenyl ether		153.8	0.9954	1.5179
4-methoxytoluene		176.5	0.9689	1.5124
ethyl phenyl ether		172	0.9792	1.5076
1-methoxy-4-*trans*-propenylbenzene	*trans*-anethole	253	0.9882	1.5615
		21[d]		
1-methoxy-4-allylbenzenephenyl	estragole	215	0.9645	1.5230
		258	1.0863	1.5780
		28[d]		

Table 1. (*Continued*)

Systematic name	Common name	Bp,[a] °C	d_4^{20b}	n_D^{20}
2-methoxyphenol	guaiacol	205.5	1.1287	1.5385
1,2-dimethoxy-benzene	veratrole	206.7	1.084	1.5385
		$22\text{--}23^d$		
1,4-dimethoxy-benzene		212.6	1.0526_{55}^{55}	
2-methoxy-4-allylphenol	4-allylguaiacol	255	1.0664	1.5410
1,2-dimethoxy-4-allylbenzene	4-allylveratrole	248	1.055	1.532
1-allyl-3,4-methylenedioxybenzene	safrole	234.5	1.0950	1.5383
1-propenyl-3,4-methylene- dioxybenzene	isosafrole	252	1.1224	1.5782
2-methoxy-4-*cis*-propenylphenol	*cis*-isoeugenol	$134_{1.7}$	1.0851	1.5700
2-methoxy-4-*trans*-propenylphenol	*trans*-isoeugenol	$141_{1.6}$	1.0852	1.5782
1-benzyloxy-2-methoxy-4-*trans*-propenylbenzene	benzyl isoeugenol	58^d		
butyrated hydroxyanisole (BHA),		$264\text{--}270_{98}$		
a mixture of: 2-*tert*-butyl-4-methoxyphenol		$48\text{--}55^d$		
and 3-*tert*-butyl-4-methoxyphenol				

[a] At 101.3 kPa unless otherwise noted by subscript in kPa; to convert kPa to mm Hg, multiply by 7.5.
[b] Unless otherwise noted as superscript, °C.
[c] Specific gravity (air).
[d] Melting point.

been dependent on low-cost isobutylene. There are a number of possible isobutylene sources for making MTBE.

Tetrahydrofuran. Almost all the THF in the United States is currently produced by the acid-catalyzed dehydration of 1,4-butanediol.

Other Ethers

***n*-Butyl Ether.** *n*-Butyl ether is prepared by dehydration of *n*-butyl alcohol by sulfuric acid or by catalytic dehydration over ferric chloride, copper sulfate, silica, or alumina at high temperatures.

Isopropyl Ether. Isopropyl ether is manufactured by the dehydration of isopropyl alcohol with sulfuric acid. It is obtained in large quantities as a by-product in the manufacture of isopropyl alcohol from propylene by the sulfuric acid process, very similar to the production of ethyl ether from ethylene. Isopropyl ether is of moderate importance as an industrial solvent, since its boiling point lies between that of ethyl ether and acetone.

Tertiary-Amyl Methyl Ether

Like MTBE, TAME is produced by the simple reaction of methanol and isoamylenes (2-methyl-1-butene and 2-methyl-2-butene).

Table 2. Typical Physical and Chemical Properties of Commonly Used Ethers

Property	Ethyl	MTBE	THF	Isopropyl	*n*-Butyl
chemical formula	$C_4H_{10}O$	$C_5H_{12}O$	C_4H_8O	$C_6H_{14}O$	$C_8H_{18}O$
molecular weight	74.12	88.15	72.10	102.17	139.22
boiling point at 101.3 kPa,[a] °C	34.5	55.0	66.0	68.4	142.0
vapor pressure at 20°C, kPa[a]	56	27	17	16	0.67
evaporation rate[b]	11.8	8.5	5.7	8.0	0.66
viscosity at 20°C, mPa · s(= cP)	0.23	0.35	0.48	0.38	0.65
surface tension in air, 25°C, mN/m (= dyn/cm)	17.0	18.3	26.4	32.0	22.9
dipole moment, 25–50°C, C · m[c]	4.3×10^{-30}	4.7×10^{-30}	5.3×10^{-30}	4.7×10^{-30}	4×10^{-30}
Hilderbrand solubility parameter	7.4	7.4	9.1	7.1	7.7
water solubility, wt%, 20°C					
ether in water	6.9	4.8	inf.	1.07	0.03
water in ether	1.3	1.4	inf.	0.53	0.19
flash point	−40	−30	−17	−13	25
autoignition temp, °C	160	426	321	440	185
flammability limits in air, vol%					
lower	1.9	1.6	1.8	1.0	0.9
higher	48.0	8.4	11.8	21.0	8.5
reactivity with hydroxyl radical[d]	13.6	3.2	17.8		27.8

[a] To convert kPa to mm Hg, multiply by 7.5.
[b] *n*-Butylacetate = 1.
[c] To convert CC · m to debyes, multiply by 3×10^{29}.
[d] k_{OH} (298 K), $10^{12} \cdot cm^3 \cdot mol^{-1}sec^{-1}$

Ethyl *tertiary*-Butyl Ether

Similar to methanol in the MTBE reaction, ethanol can react with isobutylene to produce ETBE.

Vinyl Ether. Vinyl ether is manufactured by the pyrolytic dehydrochlorination of 1,1'-dichloroethyl ether.

2-Methoxyphenol. This ether is prepared by methylating 1,2-dihydroxybenzene.

Butylated Hydroxyanisole. 2- and 3-*tert*-Butyl-4-methoxyphenol (butylated hydroxyanisole (BHA)) is prepared from 4-methoxyphenol and *tert*-butyl alcohol over silica or alumina at 150°C or from hydroquinone and *tert*-butyl alcohol or isobutene, using an acid catalyst and then methylating.

SHIPMENT

Handling

The handling of ethyl ether is hazardous because of its highly flammable properties. Not only is it highly volatile, but it is also has a low autoignition temperature and, as a nonconductor, can generate static electrical charges that may result in ignition or vapor explosion. The area in which ethyl ether is handled should be considered a Class 3 hazardous location as defined by the National Electrical Code. All tools used in making connections or repairs should be of the nonsparking type. All possible care should be taken in loading and unloading tank cars.

Special containers have been developed for anesthetic either to prevent deterioration before use. Their effectiveness as stabilizers usually depends on the presence of a lower oxide of a metal having more than one oxidation state. Thus the sides and the bottoms of tin-plate containers are electroplated with copper, which contains a small amount of cuprious oxide. Stannous oxide is also used in the linings for the containers. Instead of using special containers, iron wire or certain other metals and alloys or organic compounds have been added to other to stablize it.

Peroxide Formation

Except for the methyl *tert*-alkyl ethers, most ethers tend to absorb and react with oxygen from the air to form unstable peroxides that may detonate with extreme violence when concentrated by evaporation or distillation, when combined with other compounds that give a detonable mixture, or when disturbed by heat, shock, or friction.

HEALTH AND SAFETY FACTORS

Although ethers are not particularly hazardous, their use involves risks of fire, toxic effects, and several unexpected reactions. Since almost all ethers burn in air, an assessment of their potential hazards depends on flash points and ignition temperatures. If an ethyl ether fire occurs, carbon dioxide, carbon tetrachloride, and dry chemical fire extinguishers meeting National Fire Prevention Association Code 1 and 2 requirements may be used successfully. Water may also be effectively applied.

USES

Alkyl ethers are used for organic reactions and extractions, as plasticizers, as vehicles for other products, as anaesthetics, and octane (and oxygen) enhancers in gasoline. Most ethers have very low solubility in water but dissolve most organic compounds and therefore have found wide application in paint and varnish removers; as high boiling solvents for gums, resins, and waxes; in lubricating oils; as an extraction solvent in the fragrance industry; and as an inert reaction medium in the pharmaceutical industry. The vapors of certain ethers are toxic to insects and are useful as agricultural insecticides and industrial fumigants. Except for those used in gasoline, the lower mol wt technical-grade ethers are the least expensive and most commonly used in industry.

Aryl ethers have distinctive, pleasant odors and flavors which make them valuable to the perfume and flavor industries. Because of their heat stability, they are useful as heat-transfer fluids. Other aryl ethers are useful as food preservatives and antioxidants.

E. W. Flick, *Industrial Solvents Handbook*, 3rd ed., Noyes Data Corp., Park Ridge, N.J., 1985.

Material Safety Data Sheet E 2340, Ethyl Ether, Mallinckrodt Baker Inc., May 17, 2001.

"MTBE Chemical Profile," *Chemical Market Reporter* **263**(1), 27 (Jan. 6, 2003).

S. Patai, ed., *The Chemistry of the Ether Linkage*, Wiley-Interscience, New York, 1967.

LAWRENCE KARAS
W. J. PIEL
ARCO Chemical Company

ETHYL CHLORIDE

Ethyl chloride (chloroethane), C_2H_5Cl, is a colorless, mobile liquid of bp 12.4°C that has a nonirritating ethereal odor and a pleasant taste. It is flammable and burns with a green-edged flame, producing hydrogen chloride fumes, carbon dioxide, and water. Ethyl chloride has primarily been used in the manufacture of tetraethyl lead (TEL), an antiknock additive in engine fuel, but it also serves as an ethylating agent, solvent, refrigerant, and local and general anaesthetic. It is less toxic than the chloromethanes. The manufacture of automotive engines that could use TEL was gradually phased out until the end of 1978, after which no new car manufactured for sale in the United States was allowed to use TEL as a fuel. Because of this phasing out of leaded fuels, production of ethyl chloride has reduced steadily since 1979 and imports of ethyl chloride have been essentially zero since 1983. Ethyl chloride demand is expected to continue to diminish.

PHYSICAL AND CHEMICAL PROPERTIES

Ethyl chloride dissolves many organic substances, such as fats, oils, resins, and waxes, and is also a solvent for sulfur and phosphorus. It is miscible with methyl and ethyl alcohols, diethyl ether, ethyl acetate, methylene chloride, chloroform, carbon tetrachloride, and benzene. Butane, ethyl nitrite, and 2-methylbutane each have been reported to form a binary azeotrope with ethyl chloride, but the accuracy of these data are uncertain.

Ethyl chloride can be dehydrochlorinated to ethylene using alcoholic potash. Condensation of alcohol with ethyl chloride in this reaction also produces some diethyl ether. Ethyl chloride yields butane, ethylene, water, and a solid of unknown composition when heated with metallic magnesium for about six hours in a sealed tube. Ethyl chloride yields ethyl alcohol, acetaldehyde, and some ethylene in the presence of steam with various catalysts; eg, titanium dioxide and barium chloride.

When ethyl chloride is chlorinated under light, both ethylidene and ethylene chlorides are formed. Reaction of ethyl chloride with an alcoholic solution of ammonia yields ethylamine, diethylamine, triethylamine, and tetraethylammonium chloride. In the presence of Friedel-Crafts catalysts, gaseous ethyl chloride reacts with benzene at about 25°C to give ethylbenzene, three diethylbenzenes, and other more complex compounds.

Good technical-grade ethyl chloride should not contain more than the following quantities of the indicated impurities: water, 15 ppm; acid (as HCl), 120 ppm; residue on evaporation at 110°C, 50 ppm. Ethyl chloride does not require added stabilizers.

MANUFACTURE

Three industrial processes have been used for the production of ethyl chloride: hydrochlorination of ethylene, reaction of hydrochloric acid with ethanol, and chlorination of ethane. Hydrochlorination of ethylene is used to manufacture most of the ethyl chloride produced in the United States. Because of its prohibitive cost, the ethanol route to ethyl chloride has not been used commercially in the United States since about 1972. Thermal chlorination of ethane has the disadvantage of producing undesired by-products and has not been used commercially since about 1975.

SHIPMENT AND HANDLING

Ethyl chloride is a very dangerous fire hazard when exposed to heat or flame and can react vigorously with oxidizing materials. There is a severe explosion hazard when exposed to flame. It reacts with water or steam to produce toxic and corrosive fumes. It is incompatible with potassium. To fight fire, use carbon dioxide. When heated to decomposition, it emits toxic fumes of phosgene and Cl.

Ethyl chloride is handled and transported in pressure containers under conditions similar to those applied to methyl chloride. In the presence of moisture, ethyl chloride can be moderately corrosive. Carbon steel is used predominantly for storage vessels, and prolonged contact with copper should be avoided.

ECONOMIC ASPECTS

The only important demand for ethyl chloride, other than its use in TEL manufacture, arises from the ethylcellulose industry, in, eg, cellulose ethers.

HEALTH AND SAFETY FACTORS

Ethyl chloride is readily absorbed into the body through the mucous membranes, lungs, and skin. Although it is rapidly excreted by the lungs, its high solubility in blood prolongs total elimination from the body. Ethyl chloride is apparently not metabolized to any significant degree. Recovery of consciousness after exposure to ethyl chloride often entails an unpleasant hangover period. Experiments with animals provide evidence of kidney irritation and promotion of fat accumulation in the kidneys, cardiac muscle, and liver. Concentrations of 15–30 vol% in air are quickly fatal to animals; a concentration of 2% causes some unsteadiness; exposure to 1% concentration has no observable effect. The Environmental Protection Agency stated in 1987 that ethyl chloride is one of the least toxic of the chloroethanes.

USES

Ethyl chloride is used in the production of ethyl cellulose, used as a solvent, refrigerant, and topical anaesthetic, in the manufacture of dyes, chemicals, and pharmaceuticals, and as a medication to alleviate pain associated with insect burns and stings.

Ethylcellulose produced by the reaction of ethyl chloride with alkali cellulose, is used mainly in the plastics and lacquer industries.

Ethyl chloride is used to some extent as an ethylating agent in the synthesis of dyestuffs and fine chemicals. Ethyl chloride is used as a solvent in the polymerization of olefins using Friedel-Crafts catalysts, and as a polymerization activator to produce polyquinoline from quinoline at high temperature (121–160°C).

Ethyl chloride can also be used as a feedstock to produce 1,1,1-trichloroethane by thermal chlorination at temperatures of 375–475°C or by a fluidized-bed reactor at similar temperatures.

Other minor uses of ethyl chloride include use as blowing agents for thermoplastic foam and styrene polymer foam, the manufacture of polymeric ketones used as lube oil detergents, the manufacture of acetaldehyde, as an aerosol propellant, as a refrigerant (R-160), in the preparation of acid dyes, and as a local or general anaesthetic.

Health Effects Assessment for Ethyl Chloride, Report EPA/600/8-88/036, United States Environmental Protection Agency, Environmental Criteria Assessment Office, Cincinnati, Ohio, 1987.

L. H. Horsley and co-workers, "Azeotropic Data," in *Advances in Chemistry Series*, Vols. 1 and 2, Nos. 6 and 35, American Chemical Society, Washington, D.C., 1952 (Vol. 1) and 1962 (Vol. 2).

R. J. Lewis, Sr., *Sax's Dangerous Properties of Industrial Materials*, 10th ed., John Wiley & Sons, Inc., New York, 2000.

MATT C. MILLER
Dow Chemical U.S.A.

ETHYLENE

Ethylene (ethene), $H_2C=CH_2$, is the largest-volume building block for many petrochemicals. This olefin is used to produce many end products such as plastics, resins, fibers, etc. Ethylene is produced mainly from petroleum-based feedstocks by thermal cracking, although alternative methods are also gaining importance.

PHYSICAL PROPERTIES

Ethylene is the lightest olefin. It is a colorless, flammable gas with a slightly sweet odor. Physical and thermodynamic properties are given in many references and are briefly summarized in Table 1.

CHEMICAL PROPERTIES

Structure

Ethylene is a planar molecule with a carbon–carbon bond distance of 0.134 nm, which is shorter than the C–C bond length of 0.153 nm found in ethane.

Reactivity

Ethylene is a very reactive intermediate and hence is involved in many chemical reactions. Ethylene reacts with electrophilic reagents like strong acids (H^+), halogens, and oxidizing agents but not with nucleophilic reagents such as Grignard reagents and bases. Some of the reactions have commercial significance but others have only academic interest.

The principal reactions with commercial significance include polymerization, oxidation, and addition, including halogenation, alkylation, oligomerization, hydration, and hydroformylation.

Polymerization

Very high purity ethylene (>99.9% plus) is polymerized under specific conditions of temperature and pressure in the presence of an initiator or catalyst.

Oxidation

Ethylene oxide is produced by oxidizing ethylene.

$$CH_2{=}CH_2 + 0.5\,O_2 \longrightarrow \underset{O}{CH_2{-}CH_2}$$

Addition

Addition reactions of ethylene have considerable importance and lead to the production of ethylene dichloride,

Table 1. Physical Properties of Ethylene

Property	Value
mol wt	28.0536
triple point	
temperature, °C	−169.164
pressure, kPa[a]	0.12252
latent heat of fusion, kJ/mol[b]	3.353
normal freezing point	
temperature, °C	−169.15
latent latent heat of fusion, kJ/mol[b]	3.353
normal boiling point	
temperature °C	−103.71
latent heat of vaporization, kJ/mol[b]	13.548
density of liquid	
mol/L	20.27
d_4^{-104}	0.566
specific heat of liquid, J/(mol · K)[b]	67.4
viscosity of the liquid, mPa · s (= cP)	0.161
surface tension of the liquid, mN/m (= dyn/cm)	16.4
specific heat of ideal gas at 25°C, J/(mol · K)[b]	42.84
critical point	
temperature, °C	9.194
pressure, kPa[a]	5040.8
density, mol/L	7.635
compressibility factor	0.2812
gross heat of combustion of gas at 25°C, MJ/mol[b]	1.411
limits of flammability at atmospheric pressure and 25°C	
lower limit in air, mol%	2.7
upper limit in air, mol%	36.0
autoignition temperature in air at atmospheric pressure, °C	490.0
Pitzer's accentric factor	0.278
standard enthalpy of formation at 25°C, kJ/mol[b]	52.3
standard Gibbs energy of formation at 25°C for ideal gas at atmospheric pressure, kJ/mol[b]	68.26
solubility in water at 0°C and 101 kPa[a], mL/mL H_2O	0.226
speed of sound at 0°C and 409.681 kPa[a], m/s	224.979
standard entropy of formation, J/(mol · K)[b]	219.28
standard heat capacity, J/(mol · K)[b]	42.86

[a] To convert kPa to mm Hg, multiply by 7.5.
[b] To convert J to cal, divide by 4.184.

ethylene dibromide, and ethyl chloride by halogenation–hydrohalogenation; ethylbenzene, ethyltoluene, and aluminum alkyls by alkylation; α-olefins by oligomerization; ethanol by hydration; and propionaldehyde by hydroformylation.

BIOLOGICAL PROPERTIES

Ethylene is slightly more potent as an anaesthetic than nitrous oxide; the smell of ethylene causes choking.

MANUFACTURE BY THERMAL CRACKING

Although ethylene is produced by various methods, as follows, only a few are commercially proven: thermal cracking of hydrocarbons, catalytic pyrolysis, membrane dehydrogenation of ethane, oxydehydrogenation of ethane, oxidative coupling of methane, methanol to ethylene, dehydration of ethanol, ethylene from coal, disproportionation of propylene, and ethylene as a by-product.

Thermal cracking of hydrocarbons is the principal route for the industrial production of ethylene. The chemistry and engineering of thermal cracking has been reviewed. In thermal cracking, valuable by-products including propylene, butadiene, and benzene are also produced. Commercially less valuable methane and fuel oil are also produced in significant proportions. An important parameter in the design of commercial thermal cracking furnaces is the selectivity to produce the desired products.

Mechanism

The thermal cracking of hydrocarbons proceeds via a free-radical mechanism.

Conversion

The terms "severity" or "conversion" are used to measure the extent of cracking. Conversion can easily be measured for a single component (C) feed and is defined as follows, where the quantities are measured in weight units: conversion $= \frac{C_{in} - C_{out}}{C_{in}}$ When a mixture is cracked, one or more components in the feed may also be formed as products.

Instead of conversion, some producers prefer to use other identifications of severity, including coil outlet temperature, propylene to methane ratio, propylene to ethylene ratio, or cracking severity index. Of course, all these definitions are somewhat dependent on feed properties, and most also depend on the operating conditions.

Industrial Furnaces

Thermal cracking of hydrocarbons is accomplished in tubular reactors commonly known as cracking furnaces, crackers, cracking heaters, etc.

Environmental. Stringent environmental laws require that nitrogen oxides (NO_x) and sulfur oxides emission from furnaces be drastically reduced. In many parts of the world, regulations require that NO_x be reduced to 70 vol ppm or lower on a wet basis. Conventional burners usually produce 100 to 120 vol ppm of NO_x. Many vendors (McGill, John Zink, Callidus, Selas, and North American) are supplying low NO_x burners.

Product Distribution. In addition to ethylene, many by-products are also formed. The product distribution is strongly influenced by residence time, hydrocarbon partial pressure, steam-to-oil ratio, and coil outlet pressure.

Kinetic Models Used for Designs. Numerous free-radical reactions occur during cracking; therefore, many simplified models have been used.

Many researchers have correlated the overall decomposition as an nth order reaction, with most paraffins following the first order and most olefins following a higher order. In general, isoparaffin rate constants are lower than normal paraffin rate constants. To predict the product distribution, yields are often correlated as a function of conversion or other severity parameters.

Instead of radical reactions, models based on molecular reactions have been proposed for the cracking of simple alkanes and liquid feeds like naphtha and gas oil. However, the validity of these models is limited and cannot be extrapolated outside the range with confidence.

With the introduction of Gear's algorithm for integration of stiff differential equations, the complete set of continuity equations describing the evolution of radical and molecular species can be solved even with a personal computer. Many models incorporating radical reactions have been published.

Run Length. Coke is produced as a side product that deposits on the radiant tube walls. This limits the heat transfer to the tubes and increases the pressure drop across the coil. The coke deposition not only limits the heat transfer but also reduces the olefin selectivity. Periodically, the heater has to be shut down and cleaned. Typical run lengths are 40 to 100 days between decokings. Prediction of run length of a commercial furnace is still an art, and various mechanisms are postulated in the literature.

Recovery and Purification

The pyrolysis gas leaves the transferline exchanger at 300 to 400°C for gaseous and light naphtha feeds, and at 550 to 650°C for heavy liquid feeds. In order to minimize any further cracking for liquid feeds, the temperature must be quickly reduced. This is achieved by spraying quench oil directly into the effluent. For naphtha-based plants, quenching is performed before primary fractionation. In gas oil plants, quenching is done immediately after the transferline exchanger, resulting in two-phase flow in the transferline.

For all feeds the effluent is separated into desired products by compression in conjunction with condensation and fractionation at gradually lower temperatures.

Energy Efficiency Improvement

Modern plants are more energy efficient than those designed in the 1980s. Reduction in energy consumption was made possible by improvements in cracking coil technology and in recovery section design. Some improvements may not only reduce energy but can also increase the capacity of an existing plant. They include quench oil viscosity control, feed saturation, predemethanization, demethanizer overhead expander and multifeed fractionation, tower internals and equipment modification, and dephlegmaters.

OTHER ROUTES TO ETHYLENE PRODUCTION

In addition to conventional thermal cracking in tubular furnaces, other thermal methods and catalytic methods to produce ethylene have been developed. None of these are as yet commercialized. Other routes include the advanced cracking reactor, the adiabatic cracking reactor, fluidized-bed cracking, catalytic pyrolysis, the membrane reactor, dehydrogenation, oxydehydrogenation, oxidative coupling of methane, methanol to ethylene, ethanol to ethylene, ethylene from coal, propylene disproportionation, and use of ethylene as a by-product.

ADVANCED COMPUTER CONTROL SYSTEMS AND TRAINING SIMULATORS

An ethylene plant contains more than 300 equipment items. Traditionally, operators were trained at the site alongside experienced co-workers. With the advent of modern computers, the plant operation can be simulated on a real-time basis and the results displayed on monitors. Computers are used in a modern plant to control the entire operation; eg, they are used to control the heaters and the recovery section.

SHIPMENT AND STORAGE

In the United States, the Gulf Coast produces and consumes the majority of the U.S. ethylene production. The plants are located along the Texas southeast coast extending into Louisiana. The plants are served by a system of pipelines connecting the production and consuming plants.

HEALTH AND SAFETY FACTORS

Although ethylene is a colorless gas with a mild odor that is not irritating to the eyes or respiratory system, it is a hydrocarbon and therefore a flammable gas. All vessels must be designed for handling the liquids and gases during operation at the temperatures and pressures that exist, and safety and depressuring valves must be provided to relieve excessive pressure. Releasing of hydrocarbons into the air in large amounts must be avoided because of health and fire hazards. If hydrocarbons must be released into the air, it is done under a blanket of steam. To protect the plant and personnel in case of fire, a complete fire-fighting system is provided with tanks grouped to minimize fire and provided with foam makers and deluge systems. Reviews at various stages of a project assure that safety is given constant attention in the plant design. An ethylene plant produces liquid, gaseous, and solid wastes that must be disposed of in an environmentally safe manner.

USES

Almost all ethylene produced is consumed as a feedstock for manufacturing other petrochemicals. Only a very small amount has been used in the agricultural industry for ripening fruits.

Although some ethylene is shipped across the oceans in large quantities, the preference is to ship first-generation products such as polyethylene, ethylbenzene, etc.

K. M. de Reuck and co-workers, eds., *Ethylene (Ethene)-International Thermodynamic Tables of the Fluid State-10*, Blackwell Scientific Publishers, Oxford, UK, 1988.

L. Kane, ed., *Advanced Process Control Handbook, Hydrocarbon Process.*, Sept. 1991.

L. Kniel, O. Winter, and K. Stork, *Ethylene-Keystone to the Petrochemical Industry*, Marcel Dekker, Inc., New York, 1980.

H. Ulrich, *Raw Materials for Industrial Polymers*, Hanser Publishers, Munich, 1988.

K. M. SUNDARAM
M. M. SHREEHAN
E. F. OLSZEWSKI
ABB Lummus Global

ETHYLENE OXIDE

Although early manufacture of ethylene oxide was accomplished by the chlorohydrin process the direct oxidation process has been used almost exclusively since 1940. The primary use for ethylene oxide is in the manufacture of derivatives such as ethylene glycol, surfactants, and ethanolamines.

PHYSICAL PROPERTIES

Ethylene oxide (C_2H_4O) is a colorless gas that condenses at low temperatures into a mobile liquid. It is miscible in all proportions with water, alcohol, ether, and most organic solvents. Its vapors are flammable and explosive. The physical properties of ethylene oxide are summarized in Table 1.

Clathrate Formation

Ethylene oxide forms a stable clathrate with water. The maximum observed melting point is 11.1°C.

Table 1. Some Physical Constants of Ethylene Oxide

Property	Value
molecular weight	44.05
bp, °C	
at 101.3 kPa[a]	10.4
Δbp/pressure at 100 kPa, K/kPa[b]	0.25
coefficient of cubical expansion at 20°C, per °C	0.00158
critical pressure, MPa[c]	7.19
critical temperature, °C	195.8
dielectric constant at 0°C	14.50
dipole moment, C·m[d]	6.30×10^{-30}
explosive limits in air, %	
upper	100
lower	3
flash point, Tag open cup, °C	$\leftarrow$ 18
freezing point, °C	−111.7
heat of combustion at 25°C, kJ/mol[e]	−1218
heat of fusion, kJ/mol[e]	5.17
heat of solution in pure water at 25°C and constant pressure, kJ/mol[e]	−6.3
heat of reaction with water at 25°C, kJ/mol[e]	−87.9
ionization potential, J[e]	
experimental	$1.73 - 1.80 \times 10^{-18}$
calculated	1.65×10^{-18}
refractive index, n_D^7	1.3597

[a] To convert kPa to mm Hg, multiply by 7.5.
[b] To convert K/kPa to K/mm Hg, divide by 7.5.
[c] To convert MPa to atm, multiply by 9.87.
[d] To convert C·m to debyes, multiply by 3.0×10^{29}.
[e] To convert kJ to kcal, divide by 4.184.

CHEMICAL PROPERTIES

Ethylene oxide is a highly reactive compound used industrially as an intermediate for many chemical products. The three-membered ring is opened in most of its reactions. These reactions are very exothermic because of the tremendous ring strain in ethylene oxide, which has been calculated.

Polymerization

The reaction of ethylene oxide with a nucleophile introduces the hydroxyethyl group:

$$ROH \ + \ \triangle\!\!\!\!O \longrightarrow ROCH_2CH_2OH$$

The product of this reaction can also react with ethylene oxide; if this process is repeated many times, a polymer is formed.

Low molecular weight polymers of ethylene oxide, poly(ethylene glycol), are formed by allowing ethylene oxide to react with water or alcohols under the proper conditions. The average molecular weight can be varied from 200 to 14,000.

Crown Ethers

Ethylene oxide forms cyclic oligomers (crown ethers) in the presence of fluorinated Lewis acids such as boron trifluoride, phosphorus pentafluoride, or antimony pentafluoride. Hydrogen fluoride is the preferred catalyst.

Other Chemical Reactions

Hydration is slow at ambient temperatures and neutral conditions but is much faster with either acid or base catalysis. By-products, namely diethylene and triethylene glycol, are also formed in this reaction.

Reactions with alcohols parallel those of ethylene oxide with water. The primary products are monoethers of ethylene glycol; secondary products are monoethers of poly(ethylene glycol). Most are appreciably water-soluble.

The carboxyl group of an organic acid reacts with ethylene oxide to give the corresponding ethylene glycol monoester. Ethylene glycol diesters may be obtained directly by the reaction of ethylene oxide with the acid anhydride.

Ethylene oxide reacts with ammonia to form a mixture of mono-, di-, and triethanolamines.

Complex nitrogen compounds are formed from the reaction of alkylamines with ethylene oxide. Primary and secondary aromatic amines react with ethylene oxide to give the corresponding arylaminoethanols.

Ethylene oxide reacts with hydrogen sulfide to yield 2-mercaptoethanol and thiodiglycol (bis-2-hydroxyethyl sulfide). The reaction of ethylene oxide with long-chain alkyl mercaptans yields polyoxyethylene mercaptans, some of which are nonionic surfactants.

Ethylene oxide reacts with Grignard reagents, RMgX, to yield the corresponding two-carbon homologue, RCH_2CH_2OH.

Ethylene oxide reacts with acetyl chloride at slightly elevated temperatures in the presence of hydrogen chloride to give the acetate of ethylene chlorohydrin.

Compounds containing active $-CH_2-$ or $-CH-$ groups, such as malonic and monosubstituted malonic esters, ethyl cyanoacetate, and β-keto esters, react with ethylene oxide under basic conditions.

The 2-hydroxyethyl aryl ethers are prepared from the reaction of ethylene oxide with phenols at elevated temperatures and pressures. 2-Phenoxyethyl alcohol is a perfume fixative. The water-soluble alkylphenol ethers of the higher poly(ethylene glycol)s are important surface-active agents.

Ethylene oxide reacts readily with hydrogen cyanide in the presence of alkaline catalysts, such as diethylamine, to give ethylene cyanohydrin.

Autodecomposition of ethylene oxide vapor occurs at ~500°C at 101.3 kPa (1 atm) to give methane, carbon monoxide, hydrogen, and ethane. Isomerization of ethylene oxide to acetaldehyde occurs at elevated temperatures in the presence of catalysts such as activated alumina, phosphoric acid, and metallic phosphates.

MANUFACTURE

The direct oxidation technology, as the name implies, utilizes the catalytic oxidation of ethylene with oxygen over a silver-based catalyst to yield ethylene oxide. The process can be divided into two categories, depending on the source of the oxidizing agent: the air-based process and the oxygen-based process.

Several companies have developed technologies for direct oxidation plants. All the ethylene oxide plants that were built during the late 1990s were oxygen-based processes, and a number of existing ethylene oxide plants were converted from the air to the oxygen-based process during the same period.

Process Technology Considerations

Innumerable complex and interacting factors ultimately determine the success or failure of a given ethylene oxide process. Those aspects of process technology that are common to both the air- and oxygen-based systems are reviewed below, along with some of the primary differences.

Of all the factors that influence the utility of the direct oxidation process for ethylene oxide, the catalyst used is of the greatest importance. There are four basic components in commercial ethylene oxide catalysts: the active catalyst metal; the bulk support; catalyst promoters that increase selectivity and/or activity and improve catalyst life; and inhibitors or anticatalysts that suppress the formation of carbon dioxide and water without appreciably reducing the rate of formation of ethylene oxide.

Silver-containing catalysts are used exclusively in all commercial ethylene oxide units, although the catalyst composition may vary considerably.

The chemical and physical properties of the support strongly dictate the performance of the finished catalyst. Although nonsupported silver catalysts have been advocated in some patents, it is unlikely for them to be used commercially, since pure silver tends to sinter at reaction temperatures, with a resultant activity loss. For commercial operation, the preferred supports are alundum (a-alumina) and silicon carbide. Other supports are glass wool, quartz, carborundum, and ion-exchange zeolites.

Silver alone on a support does not give rise to a good catalyst. However, addition of minor amounts of promoter enhance the activity and the selectivity of the catalyst, and improve its long-term stability. The most commonly used promoters are alkaline-earth metals, such as calcium or barium, and alkali metals such as cesium, rubidium, or potassium.

Many organic compounds, especially the halides, are very effective for suppressing the undesirable oxidation of ethylene to carbon dioxide and water, although not significantly altering the main reaction to ethylene oxide. These compounds, referred to as catalyst inhibitors, can be used either in the vapor phase during the process operation or incorporated into the catalyst manufacturing step.

Temperature is used to control two related aspects of the reaction: heat removal from the reactor bed and catalyst operating temperature.

Space velocity has a strong effect on the process economics. It establishes the reactor size and pressure drop, affecting compression costs. The optimum space velocity is a function of energy costs, reaction rate, and selectivity.

Operating pressure has only a marginal effect on the economics of the ethylene oxide process. High pressure increases production due to higher gas density, increases heat transfer, increases ethylene oxide and carbon dioxide recovery in the absorbers, and lowers compression costs. Also, since the total number of moles decreases in the formation of ethylene oxide from ethylene and oxygen, high pressure is consistent with high conversion. However, high pressures reduce the flammable limit of the process gas as well as increase equipment costs. Typical commercial pressures are 1–2 MPa (10–20 atm).

Air process technology uses nitrogen as the diluent gas. The choice of diluent for the oxygen process is based on the thermal properties of the gas. The small process purge makes it economically possible for the process to operate under a wide variety of ballast gases. Several gases have been proposed in the patent literature, including methane and ethane.

The oxygen process has four main raw materials: ethylene, oxygen, organic chloride inhibitor, and cycle diluent. The purity requirements are established to protect the catalyst from damage due to poisons or thermal runaway, and to prevent the accumulation of undesirable components in the recycle gas. The latter can lead to increased cycle purging and consequently higher ethylene losses.

Typical ethylene specifications call for a minimum of 99.85 mol% ethylene. The primary impurities are usually ethane and methane. Impurities that strongly affect catalyst performance and reactor stability include acetylene, propylene, hydrogen, and sulfur. Acetylene causes catalyst coking at very low concentrations.

Oxygen specifications can vary, depending on the economics of the process. The dominant impurity in oxygen is argon. As the argon concentration in the reactor feed increases, the flammable limit of the gas decreases.

Organic chloride and cycle diluent specifications are less critical, since the flows are significantly less. The organic chloride specifications must prevent gross contamination as well as the potential of solids that would lead to plugging. The cycle diluent must also be free of gross contamination as well as significant catalyst poisons such as sulfur.

The air process has similar purity requirements to the oxygen process. The ethane content of ethylene is no longer a concern, due to the high cycle purge flow rate. Air purification schemes have been used to remove potential catalyst poisons or other unwanted impurities in the feed.

An economic recovery scheme for a gas stream that contains less than 3 mol% ethylene oxide (EO) must be designed. It is necessary to achieve nearly complete removal, since any ethylene oxide recycled to the reactor would be combusted or poison the carbon dioxide removal solution. Commercial designs use a water absorber followed by vacuum or low-pressure stripping of EO to minimize oxide hydrolysis. Several patents have proposed improvements to the basic recovery scheme. Other references describe how to improve the scrubbing efficiency of water or propose alternative solvents.

The main impurities in ethylene oxide are water, carbon dioxide, and both acetaldehyde and formaldehyde. Water and carbon dioxide are removed by distillation in columns containing only rectifying or stripping sections. Aldehydes are separated from ethylene oxide in large distillation columns.

Process Safety Considerations. Unit optimization studies combined with dynamic simulations of the process may identify operating conditions that are unsafe regarding fire safety, equipment damage potential, and operating sensitivity.

The safe operating ranges of the unit are dependent on all of the process parameters: temperature, pressure, residence time, gas composition, unit dynamic responses, instrumentation system, and the presence of ignition sources. The ethylene oxide reaction cycle operates close to the flammable limit. Higher oxygen concentrations yield higher activity and efficiency but more closely approach the flammable limit. One of the more sensitive areas of the unit's design is thus oxygen mixing. Another sensitive area is the final refining of high-purity ethylene oxide. As with any reactive chemical, it is prudent to use the lowest temperature heat source practical when refining ethylene oxide. Proper operation of the reboiler is critical. The presence of specific forms of iron oxide in contact with ethylene oxide vapor can lead to highly exothermic reactions that can initiate the explosive decomposition of ethylene oxide. Proper selection of insulation is critical. The selection of the safe operating conditions and design of effective process safety systems is a complex task that requires extensive laboratory testing to determine the effect of the various process parameters on explosibility as well as proven commercial experience.

Environmental Considerations. A detailed study of the environmental considerations in the manufacture of ethylene oxide by the direct oxidation of processes is available. The primary air emissions from the formation of ethylene oxide by direct oxidation are ethylene, ethylene oxide, carbon dioxide, and ethane. Traces of NO_x and SO_x from pollution control and process machinery operations have also been reported.

Liquid emissions from ethylene oxide units originate in the recovery section. The water of reaction from complete combustion of ethylene must be purged from the oxide absorber water cycle. This stream contains glycol, organic salts, aldehydes, and ethylene oxide. The location of the purge stream is selected to minimize ethylene oxide and glycol emissions. This stream is readily biodegradable. Several technologies have been proposed to reduce the amount of organics in the waste, either by distillation or the use of membranes to recover the contained glycol.

Air vs Oxygen Process Differences and Economics. The relative economics of the air vs. oxygen process have been reported. Two process characteristics dictate the difference in the capital costs for the two processes. The air process requires additional investment for the purge reactors and their associated absorbers, and for energy recovery from the vent gas. However, this is offset by the need for an oxygen production facility and a carbon dioxide removal system for an oxygen-based unit. In a comparison of necessary investments for medium to large capacity units ($>20,000$ t/yr), oxygen-based plants have a lower capital cost even if the air-separation facility is included. However, for small- to medium-scale plants, the air process investment is smaller than that required for the oxygen process and the air-separation unit, unless the oxygen is purchased from a large air-separation unit serving many customers.

Other Processes

Other processes for producing ethylene oxide include the chlorohydrin process, the arsenic-catalyzed liquid-phase process, the thallium-catalyzed epoxidation process, the Lummus hypochlorite process, liquid-phase epoxidation with hydroperoxides, the electrochemical process, the unsteady-state direct oxidation Process, the fluid-bed direct oxidation process, and biological routes.

SHIPMENT AND STORAGE

Small shipments of ethylene oxide are made in either compressed gas cylinders up to ~ 0.1 m^3 (30 gal) or in 1A1 steel drums (61 gal). Very large shipments >40 m^3 (10,000–25,000 gal) are made in insulated, type 105J100W or other DOT-approved tank cars.

Storage

Carbon steel and stainless steel should be used for all equipment in ethylene oxide service.

Ethylene oxide storage tanks are pressurized with inert gas to keep the vapor space in a nonexplosive region and prevent the potential for decomposition of the ethylene oxide vapor.

ECONOMIC ASPECTS

In 2002, world consumption of ethylene oxide was 14.7×10^6 t, 73% of which was used to make ethylene glycols. The second largest use of ethylene oxide is in surface active agents. Ethylene oxide is used in the production of nonionic alkylphenols ethoxylates and detergent ethoxylates.

HEALTH AND SAFETY FACTORS

Toxicology

An excellent review of the toxicity and health assessment of ethylene oxide has been compiled. Ethylene oxide (EO) can be relatively toxic as both a liquid and a gas. Inhalation of ethylene oxide in high concentrations may be fatal.

Inhalation exposure to high concentrations of ethylene oxide has been reported to result in respiratory system irritation and edema. There is some evidence that occupational exposure to high levels of ethylene oxide can result in cataracts. Neurological effects have also been reported in association with recurrent human and animal inhalation exposures to ethylene oxide. Again, depending on the degree of exposure, headache, nausea, vomiting, diarrhea, dizziness, loss of coordination, convulsion, or coma may occur. The onset of illness is rapid in severe exposures but may be delayed after moderate exposure. In the reports of human peripheral neurotoxic effects or central nervous system toxicity, most cases have shown a marked improvement on removal from further exposure.

Ethylene oxide has been shown to produce mutagenic and cytogenic effects in a variety of test systems.

When the data as a whole are reviewed for studies on humans exposed to ethylene oxide, no conclusion can be made that there is an increase in mortality associated with those exposed to ethylene oxide.

Developmental toxicity inhalation tests have been conducted using rats or rabbits. No teratogenic effects were observed. The only developmental effects noted were decreased fetal weights and delayed ossification.

Dermal exposure information has been collected from case reports of industrial accidents. Concentrated ethylene oxide evaporates rapidly from the skin and produces a freezing effect, often compared to frostbite, leaving burns ranging from first-to third-degree severity.

OSHA PEL; TWA $= 1$ ppm and listed as a cancer hazard, ACGIH (TLV) $= 1$ ppm and is listed as a suspected human carcinogen.

Explosibility and Fire Control

As in the case of many other reactive chemicals, the fire and explosion hazards of ethylene oxide are system dependent. Each system should be evaluated for its particular hazards, including start-up, shut-down, and failure modes. Storage of more than a threshold quantity of 5,000 lb ($\sim 2,300$ kg) of the material makes ethylene oxide subject to the provisions of OSHA 29 CFR 1910 for "Highly Hazardous Chemicals."

Liquid Hazards

Pure liquid ethylene oxide will deflagrate, given sufficient initiating energy either at or below the surface, and a propagating flame may be produced. Liquid mists of ethylene oxide will decompose explosively in the same manner as the vapor.

Liquid ethylene oxide under adiabatic conditions requires about 200°C before a self-heating rate of 0.02°C/min is observed. However, in the presence of contaminants such as acids and bases, or reactants possessing a labile hydrogen atom, the self-heating temperature can be much lower.

Ethylene oxide is an electrically conductive liquid that does not accumulate static electricity in grounded equipment. Static electricity can, however, accumulate in liquid mist produced by splashing and spraying. Although the vapor alone has a large minimum ignition energy, mixtures with oxidants such as air can be very sensitive to ignition.

Vapor Hazards

Ethylene oxide vapor can decompose explosively and propagate a decomposition flame when its pressure is greater than about 300 mm Hg (depending on temperature). To prevent decomposition, dilution using an inert gas (N_2 or CO_2) or an extinguishant such as a halocarbon have been used. The amount of dilution required depends on temperature, pressure, and the anticipated ignition source.

Hazards of Mixtures with Air

Pools of liquid ethylene oxide will continue to burn until diluted with at least 22 parts of water by volume. This must be increased to about 100 parts of water if the vapor is confined, such as in a sewer.

Mixtures of ethylene oxide with air are far easier to ignite and burn much faster than the pure vapor.

When ethylene oxide vapor or liquid leaks into porous refractory insulation such as mineral wool or calcium silicate, it reacts with water contained in the insulation forming low molecular weight poly(ethylene glycol)s. Whereas ethylene oxide is volatile, the glycols can accumulate and self-heat in the presence of air. The event may lead to a fire in the insulation that might overheat small-diameter lines, causing internal decomposition. To prevent this, cellular glass insulation may be used, since it is nonporous and cannot accumulate glycols.

Catalysts such as iron oxides cause isomerization of the ethylene oxide to acetaldehyde with the evolution of heat. The acetaldehyde has a much lower autoignition temperature in air than does ethylene oxide, and the two effects may lead to hot-spot ignition.

USES

Ethylene oxide is an excellent fumigant and sterilizing agent. It is used as an antimicrobial pesticide to fumigate spices and to sterilize medical devices, such as sutures, bandages, catheters, endoscopes, and cardiac pacemakers. Ethylene oxide has been used as a fumigant for corrosion-promoting microbes, particularly in fire protection sprinkler systems. Over half the medical devices made in the United States are sterilized using ethylene oxide; every hospital performing surgery has at least one ethylene oxide sterilizer. Ethylene oxide gas sterilants permit convenient sterilization of delicate instruments and supplies made of almost any material. Ethylene oxide readily penetrates deep pores and narrow crevices, and passes through wrappings of most polymers, paper, and cloth. The ethylene oxide sterilization process requires relatively low temperatures and pressures and does not damage the materials or packaging being sterilized. Finally, ethylene oxide has been studied for use as a rocket fuel and as a component in munitions. It has also been reported to be used as a fuel in FAE (fuel air explosive) bombs.

DERIVATIVES

Derivatives include ethylene glycol, di-, tri-, and tetraethylene glycols, nonionic surface-active agents, ethanolamines, glycol ethers, poly(ethylene glycol)s, poly (ethylene oxide) and other derivatives.

Ethylene Oxide, Brochure F-ICD23, Union Carbide Corp., Danbury, Conn., 1993.

J. Farakawa and T. Saegusa, *Polymerization of Aldehydes and Oxides*, Wiley-Interscience, New York, 1963.

M. S. Malinovskii, *Epoxides and Their Derivatives*, Daniel Davey & Co., Inc., New York, 1965 (especially for references from Eastern European countries).

B. J. Zwolinski and R. Wilhoit, in D. E. Gray, ed., *American Institute of Physics Handbook*, 3rd ed., McGraw-Hill Book Co., Inc., New York, 1972, pp. 316–342.

J. P. Dever
K. F. George
W. C. Hoffman
H. Soo
Union Carbide Corporation

ETHYLENE OXIDE POLYMERS

Poly(ethylene oxide) (PEO) is a water-soluble, thermoplastic polymer produced by the heterogeneous polymerization of ethylene oxide. The white, free-flowing resins are characterized by the following structural formula: $-(CH_2CH_2O)_n-$.

The resins are available in a broad range of molecular weight grades, from as low as 100,000 to over 7×10^6. Although most commonly known as poly(ethylene oxide) resins, they are occasionally referred to as poly(ethylene glycol) or polyoxyethylene resins.

PHYSICAL PROPERTIES

Crystallinity

At molecular weights of 1×10^5–1×10^7, poly(ethylene oxide) forms a highly ordered structure. The polymer chain contains seven structural units per fiber identity period (1.93 nm). The diffraction pattern of the monoclinic unit cell of poly(ethylene oxide) contains four molecular chains, in which $a = 0.796$ nm, $b = 1.311$ nm, $c = 1.939$ nm, and b $= 124°48'$. Infrared studies show that the oxygen atoms of the crystalline polymer are in the gauche configuration. The high molecular weight poly(ethylene oxide) resins are of the spherulitic structure. Proper annealing of a melt-cast film produces a distinct lamellar structure. The molecular conformation of poly(ethylene oxide), as determined by the use of X-ray diffraction, ir, and Raman spectroscopic methods, is shown in Figure 1.

Density

Although the polymer unit cell dimensions imply a calculated density of 1.33 g/cm^3 at 20°C, and extrapolation of melt density data indicate a density of 1.13 g/cm^3 at 20°C for the amorphous phase, the density actually measured is 1.15–1.26 g/cm^3, which indicates the presence of numerous voids in the structure.

Glass-Transition Temperature

The glass-transition temperature, T_g, of poly(ethylene oxide) has been measured over the molecular weight range of 10^2–10^7. The highest percentage of crystalline character develops at a molecular weight of 6,000, and it is at that point that T_g is the highest.

Solubility

Poly(ethylene oxide) is completely soluble in water at room temperature. However, at elevated temperatures (>98°C) the solubility decreases. It is also soluble in several organic solvents, particularly chlorinated hydrocarbons. Aromatic hydrocarbons are better solvents for poly(ethylene oxide) at elevated temperatures.

Aqueous poly(ethylene oxide) solutions of higher molecular weight (ca. 10^6) become stringy at polymer concentrations less than 1 wt%. At concentrations of 20 wt.%, solutions become nontacky elastic gels; above this concentration, the solutions appear to be hard, tough, water-plasticized polymers.

Thermoplasticity

High molecular weight poly(ethylene oxide) can be molded, extruded, or calendered by means of conventional

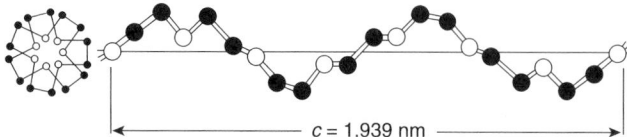

Figure 1. Molecular conformation of poly(ethylene oxide).

Table 1. Typical Physical Properties of Poly(ethylene oxide) Film

Property	Value
specific gravity	1.2
tensile strength, MP^a	
machine direction	16
transverse direction	13
secant modulus, MP^a	
machine direction	290
transverse direction	480
elongation, %	
machine direction	550
transverse direction	650
tear strength, kN/m	
machine direction	100
transverse direction	240
dart impact at 50% failure, kN/m^b	80
release time in water, s	15
O$_2$ transmission, μmol/(m · s · Gpa)^c	85.8
melting point, °C	67
heat-sealing temperture, °C	71–107
cold-crack resistance, °C	−46

^{a}To convert MPa to psi, multiply by 145.
^{b}To convert kN/m to lbf/in., multiply by 57.14.
^{c}To convert μmol/(m · s · Gpa) to cm^3 · mil/(in.^{2} · d · atm), multiply by 5.

thermoplastic processing equipment. Films of poly(ethylene oxide) can be produced by the blown-film extrusion process and, in addition to complete water solubility, have the typical physical properties shown in Table 1. Films of poly(ethylene oxide) tend to orient under stress, resulting in high strength in the draw direction.

At 100–150°C above the melting point, the melt viscosities of these polymers may exceed 10 kPa · s (10^5 P). These high melt viscosities indicate an extremely high molecular weight. Melt viscosities are relatively unaffected by temperature changes but are directly proportional to the molecular weight of the polymer. Thus, polymers with molecular weights of $(1–3) \times 10^5$ are usually used for applications involving thermoplastic forming processes.

Polymer Blends

The miscibility of poly(ethylene oxide) with a number of other polymers has been studied; eg, with poly(methyl methacrylate), poly(vinyl acetate), polyvinylpyrrolidinone, nylon, poly(vinyl alcohol), phenoxy resins, cellulose, cellulose ethers, poly(vinyl chloride), poly(lactic acid), polyhydroxybutyrate, poly(acrylic acid), polypropylene, polyethylene, and poly(styrene-co-maleic anhydride). The crystallization behavior of representative PEO blends have been studied using time-resolved wide- and small-angle X-ray scattering.

CHEMICAL PROPERTIES

Association Complexes

The unshared electron pairs of the ether oxygens, which give the polymer its strong hydrogen bonding affinity, can also take part in association reactions with a variety of monomeric and polymeric electron acceptors. These include poly(acrylic acid), poly(methacrylic acid), copolymers of maleic and acrylic acids, tannic acid, and naphtholic and phenolic compounds, as well as urea and thiourea.

When equal amounts of solutions of poly(ethylene oxide) and poly(acrylic acid) are mixed, a precipitate, which appears to be an association product of the two polymers, forms immediately. This association reaction is influenced by hydrogen-ion concentration. The highest yield of insoluble complex usually occurs at an equimolar ratio of ether and carboxyl groups.

Poly(ethylene oxide) associates in solution with certain electrolytes. Complexes with electrolytes, in particular lithium salts, have received widespread attention on account of the potential for using these materials in a polymeric battery.

Oxidation

Because of the presence of weak C—O bonds in the backbone, high molecular weight polymers of ethylene oxide are susceptible to oxidative degradation in bulk, during thermoplastic processing, or in solution.

Several stabilizers are useful in minimizing oxidative degradation during thermoplastic processing or in the bulk solid. Aqueous solutions can be stabilized against viscosity loss by addition of 5–10 wt% anhydrous isopropyl alcohol, ethanol, ethylene glycol, or propylene glycol. The manganous ion (Mn^{2+}) also is an effective stabilizer.

MANUFACTURE AND PROCESSING

Heterogeneous Catalytic Polymerization

The polymerization of ethylene oxide to produce high molecular weight polymer involves a heterogeneous reaction with propagation at the catalyst surface. The polymerization can involve anionic or cationic reactions of ethylene oxide that generally produce lower molecular weight products. The mechanism for production of extremely high molecular weight polymers is thought to involve a coordinate anionic reaction where ethylene oxide is coordinated with a metal atom of the catalyst and is then attacked by an anion. Catalysts capable of polymerizing ethylene oxide to high molecular weight polymers include many metal compounds; eg, alkaline-earth carbonates and oxides, alkyl zinc compounds, alkyl aluminum compounds and alkoxides, and hydrates of ferric chloride, bromide, and acetate.

Polymer Suspensions

Poly(ethylene oxide) resins are commercially available as fine granular solids. However, the polymer can be dispersed in a nonsolvent to provide better metering into various systems. Production processes involve the use of high-shear mixers to disperse the solids in a nonsolvent vehicle.

Thermoplastic Processing

Poly(ethylene oxide) resins can be thermoplastically formed into solid products; eg, films, tapes, plugs, retainers, and fillers. Through the use of plasticizers, PEO can be extruded, molded, and calendered on conventional thermoplastic processing equipment. Sheets and films of this resin are heat sealable.

Irradiation and Cross-linking

Exposure of poly(ethylene oxide) to ionizable radiation (gamma irradiation, electron beam, or ultraviolet light) can result in molecular weight breakdown or cross-linking, depending on the environmental conditions. If oxygen is present, hydroperoxides are formed and chain scission leads to an overall decrease in molecular weight. However, in the absence of oxygen, cross-linking becomes the preferred reaction. The resulting polymer network exhibits hydrogel properties of high water capacity.

SPECIFICATIONS, STANDARDS, AND QUALITY CONTROL

The primary quality control measure for PEO resins is the concentrated aqueous solution viscosity, which is related to molecular weight. Additional product specifications frequently include moisture content, particle-size distribution, and the residual catalyst by-product level.

HEALTH AND SAFETY FACTORS

Poly(ethylene oxide) resins are safely used in numerous pharmaceutical and personal care applications. Poly(ethylene oxide) resins show a low order toxicity in animal studies by all routes of exposure. Because of their high molecular weight, they are poorly adsorbed from the gastrointestinal tract and are completely and rapidly eliminated. The resins are not skin irritants or sensitizers, nor do they cause eye irritation.

Considerable interest has been shown in poly(ethylene oxide) for diverse applications in food, drug, and cosmetic products. Such uses fall within the scope of the Federal Food, Drug, and Cosmetic Act. The U.S. Food and Drug Administration (FDA) has recognized and approved the use of poly(ethylene oxide) for specific food and food packaging uses. USP/NF grades of Polyox water-soluble resins that meet all requirements of the U.S. Pharmacopeia/National Formulary (USP/NF) are available for pharmaceutical applications.

USES

The significant use properties of poly(ethylene oxide) are its complete water solubility, low toxicity, unique solution rheology, complexation with organic acids, low ash content, and thermoplasticity.

Pharmaceutical and Biomedical Applications

On account of its low toxicity and unique properties, poly(ethylene oxide) is utilized in a variety of pharmaceutical and biomedical applications.

Denture Adhesives. Fast hydration and gel-forming properties are ideally mated to produce a thick, cushioning fluid between the dentures and gums. The biologically inert nature of poly(ethylene oxide) helps reduce unpleasant odors and taste in this type of personal care product.

Mucoadhesives. Poly(ethylene oxide) has good adhesive properties to mucosal surfaces because of its high molecular weight, linear molecules, and fast hydration properties. PEO's mucoadhesive properties have been utilized in the design of buccal-sustained drug delivery systems and occular delivery systems.

Ophthalmic Solutions. The viscoelastic properties of poly(ethylene oxide) produce unique benefits for vitreous fluid substitution for ophthalmic surgery. Solutions of high molecular weight poly(ethylene oxide) have been used as vehicles for therapeutics for the eye and as a contact lens fluid for hard or gel-type lenses. A treated lens appears to have a high viscosity layer at the low shear rates that occur on the inside surface. This provides a thick, comfortable cushioning layer. At the high shear rates caused by the blinking eyelid, the apparent viscosity is much lower. This allows the lid to move smoothly and effortlessly over the outside surface of the lens. Unlike the cellulosics, PEO does not support bacterial growth.

Wound Dressings. Cross-linked poly(ethylene oxide) solutions form hydrogels containing about 90–97% of water. These hydrogels are clear, transparent, permeable to gases, and absorb 5–100 times their weight in water. Such characteristics make these hydrogels interesting materials for wound dressings compared to other occlusive dressings.

Oral Drug Release. The dissolution rate of tableted poly(ethylene oxide) depends on the molecular weight and particle-size distribution. High molecular weight resin provides an excellent tablet binder for sustained drug release from matrix tablets. The good flow properties and compressibility of PEO powder can be advantageously exploited in preparing tablets by direct compression. The high swelling capacity of high molecular weight poly(ethylene oxide) tablets when exposed to intestinal fluids has been successfully used in osmotic delivery systems for water-insoluble drugs. A zero-order drug release has been reported from films produced from PEO and polycaprolactone. The thermoplastic properties of PEO lend themselves to the use of extrusion methods in the preparation of oral delivery systems.

Biomaterials with Low Thrombogenicity. Poly(ethylene oxide) exhibits extraordinary inertness toward most proteins and biological macromolecules. The polymer is therefore used in bulk and surface modification of biomaterials to develop antithrombogenic surfaces for blood-contacting materials. Such modified surfaces result in reduced concentrations of cell adhesion and protein adsorption when compared to the nonmodified surfaces.

Lubricious Coatings for Biomaterials. Coatings of poly(ethylene oxide) when dry are tactile. If brought into contact with water, the poly(ethylene oxide) hydrates rapidly and forms a lubricious coating. This type of technology is of great interest for biomedical devices introduced into the human body, such as catheters and endotracheal tubes, and for sutures.

Industrial Applications

Poly(ethylene oxide)s have numerous industrial uses.

Flocculation. Poly(ethylene oxide)s of molecular weights greater than 4 million have been used as specialty flocculants. Some of the end uses for PEO as a flocculant are as a fines retention aid in the paper industry, a low pH flocculant of silica in beryllium, uranium, and copper mines that use acid leaching, and as a dewatering aid in industrial waste treatment. In the newsprint industry, PEO is widely used as a retention aid and pitch control agent. In the mining industry, PEO is used to flocculate siliceous substrates at pH <2 during the acid leaching operation.

Drag Reduction. The addition of 0.03% of high molecular weight PEO (greater than 4 million) to aqueous solutions has resulted in a 100% increase in the flow rate at fixed pump pressures. Drag reduction properties have been demonstrated by trials using fire hoses, which show that water travels 50% farther because of the addition of small quantities of PEO.

Binders in Ceramics, Powder Metallurgy, and Water-based Coatings for Fluorescent Lamps. When PEO is used as a binder in aqueous suspensions, it is possible to remove PEO completely in less than 5 min. by baking at temperatures of 400°C. This property has been successfully commercialized in several ceramic applications, in powder metallurgy, and in water-based coatings of fluorescent lamps.

Personal Care. The addition of PEO provides a silky feel to solid and liquid products. This unique lubricious property has been successfully exploited in formulation of razor strips and in shampoos, detergents, and other personal care applications.

Adhesives. High concentration (>10%) solutions of poly(ethylene oxide) exhibit wet tack properties that are used in several adhesive applications. The tackiness disappears when the polymer dries. This property can be successfully utilized in applications that require adhesion only in moist conditions.

Acid Cleaners. The addition of PEO can significantly increase the viscosity of acid solutions. Highly viscous acid solutions are used in cleaning formulations for glass, ceramic, and metal surfaces.

Drift and Mist Control. The pseudoplastic properties of PEO solutions reduce mist formation during spraying of aqueous solutions that contain PEO. This property is used in metal-working fluids to lower worker exposure to mists from the cutting and grinding aids.

Construction. Research studies and the patent literature suggest that PEO can be used as a pumping aid to concrete, where its lubricity allows concrete to be pumped to longer distances. In addition, PEO is also used to disperse the water more uniformly in the concrete mixture.

Batteries. It is believed that the 7C2 helical structure of PEO allows an ideal structure for ion transport and leads to its effective use as a battery. The crystallinity of PEO at room temperature has limited the use of this technology to batteries that are used at temperatures higher than 65°C, the melting point of PEO.

Other Applications. PEO has also been used as an antistat additive, a water-soluble packaging material for seeds and fertilizers, a rheology modifier in aqueous flexographic printing inks, and for the production of nanofibers containing multiwalled carbon nanotubes.

F. E. Bailey Jr. and J. V. Koleske, *Poly(Ethylene Oxide)*, Academic Press, Inc., New York, 1976, p. 105.

F. M. Gray, *Polymer Electrolytes, RSC Material Science Monograph*, The Royal Society of Chemistry, 1997, p. 20.

B. Scrosati, ed., *Second International Symposium on Polymer Electrolytes*, Elsevier Science Publishing Co., Inc., New York, 1990.

F. W. Stone and J. J. Stratta, in N. Bikales, ed., *Encyclopedia of Polymer Science and Technology*, Vol. 6, Wiley-Interscience, New York, 1967, pp. 103–145.

DARLENE M. BACK
ROBERT L. SCHMITT
The Dow Chemical Company

ETHYLENE–ACRYLIC ELASTOMERS

Ethylene–acrylic elastomers are best known for their excellent heat and oil resistance, but they also possess a good balance of compression set resistance, flex resistance, physical strength, low-temperature flexibility, and weathering resistance. Special compounded attributes include uniquely temperature-stable vibrational dampening properties and the ability to produce flame-resistant compounds with combustion products having an exceptionally low order of smoke density, toxicity, and corrosiveness. Because of this balance of properties, ethylene–acrylic elastomers have found ready acceptance in many high-performance applications, especially in the automotive market.

POLYMER PROPERTIES

Polymer Composition

Ethylene–acrylic elastomer terpolymers are manufactured by the addition copolymerization of ethylene and methyl acrylate in the presence of a small amount of an alkenoic acid to provide sites for cross-linking with diamines.

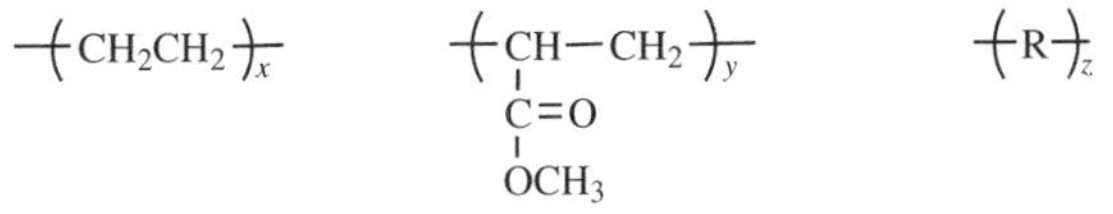

| Ethylene | Methyl acrylate | Cure-site monomer |

The polymerization process yields a random, amorphous terpolymer or a random, amorphous dipolymer (no cure-site monomer). The polymer backbone is fully saturated, making it highly resistant to ozone attack even in the absence of antiozonant additives. The fluid resistance and low-temperature properties of ethylene–acrylic elastomers are largely a function of the methyl acrylate-to-ethylene ratio.

Commercial Forms

Four different base polymers of Vamac ethylene–acrylic elastomers are commercially available (Table 1).

PROCESSING

Mixing

Ethylene–acrylic elastomers are processed in either an internal mixer with an upside down process for large-scale production or a rubber mill for smaller scales.

Extrusion and Calendering

Most compounds of ethylene–acrylic elastomers have low nerve and yield smooth extrusions or calendered sheets.

Molding

Parts can be produced from ethylene–acrylic elastomers using compression, transfer, or injection-molding techniques.

Postcuring

Whenever production techniques or economics permit, it is recommended that compounds based on terpolymer grades be postcured for optimum properties. Relatively short press cures can be continued with an oven cure in order to develop full physical properties and maximum resistance to compression set.

Table 1. Vamac Ethylene–Acrylic Elastomer Polymers

Commercial designation	Monomers[a]	Methyl acrylate level	Type of cure system
Vamac G	E/MA/CS	average	amine
Vamac HVG	E/MA/CS	average	amine
Vamac GLS	E/MA/CS	high	amine
Vamac DP	E/MA	average	peroxide

[a] E is ethylene; MA, methyl acrylate; and CS, proprietary cure-site monomer.

Adhesion

Commercially available one- or two-coat adhesive systems produce cohesive rubber failure in bonds between ethylene–acrylic elastomer and metal. Adhesion to nylon, polyester, or aramid fiber cord or fabric is greatest when the cord or fabric have been treated with carboxylated nitrile rubber latex.

ECONOMIC ASPECTS

Several new products are under development, including a more flex-resistant polymer and a lower T_g polymer. These new products, when commercialized, will expand the serviceability of the ethylene–acrylic elastomers and improve their processability.

Uses

The favorable balance of properties of ethylene–acrylic elastomers has gained commercial acceptance for these elastomers in a number of demanding applications, especially in the automotive industry and in wire and cable jacketing.

Industrial applications include pipe seals, hydraulic system seals, dampers for machinery and high speed printers, and motor lead wire insulation. The fact that the polymer contains no halogens along with certain unique compounding techniques for flame resistance prompts the selection of ethylene–acrylic as jacketing material on certain transportation/military electrical cables and in floor tiles.

H. J. Barager, K. Kammerer, E. McBride, L. C. Muschiatti, and Y. T. Wu, *Increased Cure Rates of Vamac* [R] *(AEM) Dipolymers and Terpolymers using Peroxides*, American Chemical Society Rubber Division, Cincinnati, Ohio, Oct. 2000.

T. M. Dobel, *New Development in Ethylene/Acrylic Elastomers*, Paper No. 28, American Chemical Society Rubber Division, Detroit, Mich., Oct. 1991.

R. E. Vaiden, *Elastomeric Materials for Engine and Transmission Gaskets*, Paper No. 920132, Society of Automotive Engineers, Detroit, Mich., Feb. 1992.

C. Williams, *Vamac Ethylene/Acrylic Elastomer, A Survey of Properties, Compounding and Processing*, Bulletin H-34753, DuPont Polymers, Stow, Ohio, Jan. 1992.

Yun-Tai Wu
Edward McBride
E.I. du Pont de Nemours
& Company, Inc.

ETHYLENE–PROPYLENE POLYMERS

Copolymers of ethylene and propylene (EPM) and terpolymers of ethylene, propylene, and a diene (EPDM) as manufactured today are rubbers that can be vulcanized radically by means of peroxides. A small amount of built-in third nonconjugated diene monomer in EPDM permits conventional vulcanization with sulfur. Among the variety of synthetic rubbers, EPM and EPDM are particularly known for their excellent ozone resistance in comparison with natural rubber (*cis*-1,4-polyisoprene) and its synthetic counterparts IR (isoprene rubber), SBR (styrene–butadiene rubber), and BR (butadiene rubber). Further, EPDM– rubber can be extended with fillers and plasticizers to a very high level in comparison with other elastomers and still give good processability and properties. This leads to an attractive price/performance ratio for these polymers.

Even though EPM– and EPDM–rubbers have been commercially available for >40 years, the technology concerning these products, both their production and their applications, is still very much under development.

POLYMER PROPERTIES

The properties of EPM–copolymers are dependent on a number of structural parameters of the copolymer chains: the relative content of comonomer units in the copolymer chain, the way the comonomers are distributed in the chain (more or less randomly), the variation in the comonomer composition of different chains, the average molecular weight, and the molecular weight distribution. In the case of EPDM–terpolymers, there are additional structural features to be considered: the amount and type of unsaturation introduced by the third monomer, the way the third monomer is distributed (more of less randomly) along the chain, and long-chain branching. These structural parameters can be regulated via the operating conditions during polymerization and the chemical composition of the catalyst.

Although the rubbery properties of ethylene–propylene copolymers are exhibited over a broad range of compositions, weight percentages of commercial products generally range from 45:55 to 80:20 ethylene/propylene. On the high propylene side, the polymer fails on thermal and ozone stability, because of the lower oxidative stability of the propylene units relative to ethylene units; on the high ethylene side, the polymer is too highly crystalline and loses its rubbery character. Depending on the catalyst system and polymerization conditions used, the ethylene units may tend to group together to form blocky or sequential structures. This tendency is the more pronounced, the higher the ethylene/propylene ratio. Crystallinity renders the EP(D)M rubber a certain green strength:tensile strength in the unvulcanized state, making the polymer easier to handle and to store. Moreover, it adds to the strength of EP(D)M vulcanizates in those cases, where carbon blacks cannot be used as reinforcing fillers and only less reinforcing light colored fillers can be applied. On the other side, these blocky structures or crystallinity have a detrimental effect on the rubbery properties of the polymer, particularly at subambient temperatures, and they enhance the thermoplastic nature of the polymer. In addition to the ethylene/propylene ratio, the average molecular weight of the rubber is controlled by polymerization variables.

The structure of EPM shows it to be a saturated synthetic rubber. There are no double bonds in the polymer

Table 1. Properties of Raw Ethylene–Propylene–Diene Co- and Terpolymers

Property	Value
specific gravity	0.86–0.88
appearance	glassy-white
ethylene/propylene ratio by wt	
amorphous types	45/55
crystalline or sequential types	80/20
onset of crystallinity,°C	
amorphous types °C[a]	Below −50
crystalline types	below ~30
glass-transition temperature, °C[a]	−45 to −60
heatcapacity, kJ/(kg · K)	2.18
thermal conductivity, W/(m · K)	0.335
thermal diffusivity, m/s	1.9×10^{-5}
thermal coefficient of linear expansion per°C	1.8×10^{-4}
Mooney viscosity, ML (1 + +4) 125°C[b]	10–90

[a] Dependent on third monomer content.
[b] Oil-extended grades, when viscosity >100 for the raw polymer.

chain as there are in the case of natural rubber and most of the common commercial synthetic rubbers. The main-chain unsaturation in these latter materials introduces points of weakness. When exposed to the degrading influences of light, heat, oxygen, and ozone, the unsaturated rubbers tend to degrade through mechanisms of chain scission and cross-linking at the points of carbon–carbon unsaturation. Since EPM does not contain any carbon–carbon unsaturation, it demonstrates an inherently higher resistance to degradation by heat, light, oxygen, and, in particular ozone.

As a saturated elastomer, EPM cannot be cured or cross-linked using conventional chemical accelerators and sulfur but can be vulcanized into useful rubber products using peroxides.

EPDM is a more commercially attractive product that retains the outstanding performance features of heat, oxygen, and ozone resistance. It includes some carbon–carbon unsaturation—pendent to the main chain—from a small amount of an appropriate third nonconjugated diene monomer to accommodate it to conventional sulfur vulcanization chemistry.

As EPM, EPDM shows outstanding resistance to heat, light, oxygen, and ozone because one double bond is lost when the diene enters the polymer and the remaining double bond is not in the polymer backbone but is external to it. Properties of typical EPDM rubbers are shown in Table 1.

MANUFACTURE

The two principal raw materials for EPM and EPDM, ethylene and propylene, both of which are gases, are available in abundance at high purity. Propylene is commonly stored and transported as a liquid under pressure. Although ethylene can also be handled as a liquid, it is generally transported in pipelines as a gas. A third monomer, DCPD, is also available in large quantities. ENB is produced in a two-step process: a Diels-Alder reaction of cyclopentadiene (in equilibrium with DCPD) and buta-

diene; the resulting product VNB is rearranged to ENB via proprietary processes.

EPM and EPDM rubbers are produced in continuous processes. All EPDM manufacturing processes are highly proprietary and differ greatly between various suppliers.

COMPOUNDING

EPM/EPDM grades have to be compounded with reinforcing fillers if high levels of mechanical properties are required. Carbon blacks are usually used as fillers. The semireinforcing types, such as FEF (fast extrusion furnace) and SRF (semireinforcing furnace) give the best performance. To lower the cost and improve the processibility of light-colored compounds or to lower the cost of black compounds, calcined clay or fine-particle-size calcium carbonates are used. The most widely used plasticizers are paraffinic oils. For applications that specify high use temperatures, or for peroxide cures, paraffinic oils of low volatility are recommended.

Although EPM can be cross-linked only with peroxides, peroxide, or sulfur plus accelerators, other vulcanization systems like resins can be used for EPDM. The choice of chemicals used in an EPDM vulcanizate depends on many factors such as the mixing equipment, mechanical properties, cost, safety, and compatibility.

PROCESSING

EPM/EPDM compounds are almost exclusively mixed in internal mixers. In the latter case, the cycles and dump temperatures are about the same as would be used for SBR. The speed of carbon black dispersion is greatly dependent on molecular weight, and long-chain branching parameters within practical mixing cycles.

In general, EPM/EPDM compounds can be extruded easily on all commercial rubber extruders. For extrusion purposes, both long-chain branching and broad molecular weight distribution again have a positive effect. The higher shear thinning character of these polymers permits higher extrusion speeds, while the presence of branched molecules or molecules with very high molecular weight increase resistance to collapse via chain entanglements. EPDM compounds can be calendered both as unsupported sheeting and onto a cloth substrate.

EPM/EPDM compounds are cured on all of the common rubber-factory equipment: compression molding, transfer molding, steam cure, hot-air cure, and injection molding are all practical. UHF heating is applicable only for black-filled compounds. In the case of peroxide curing, open-air vulcanization in the presence of free oxygen in UHF and hot air is problematic as in all other rubbers.

PROPERTIES OF EPM AND EPDM VULCANIZATES

Mechanical properties depend considerably on the structural characteristics of the EPM/EPDM and the type and amount of fillers in the compound. A wide range of hardnesses can be obtained with EPM/EPDM vulcanizates. The elastic properties are by far superior to those of many other synthetic rubber vulcanizates, particularly of butyl rubber, but they do not reach the level obtained with NR or SBR vulcanizates. The resistance to compression set is surprisingly good, in particular for EPDM with a high ENB content and cross-link density or when cured with peroxide. The resistance to heat and aging of optimized EPM/EPDM vulcanizates is better than that of SBR and NR. Particularly noteworthy is the ozone resistance of EPM/EPDM vulcanizates. EPM/EPDM vulcanizates have an excellent resistance to chemicals, such as dilute acids, alkalis, alcohol, etc, in contrast to their resistance to aliphatic, aromatic, or chlorinated hydrocarbons.

The electrical-insulating and dielectric properties of the pure EPM/EPDM are excellent, but in compounds they are strongly dependent on the proper choice of fillers and their dispersion. The electrical properties of vulcanizates are also good at high temperatures and after heat aging. Because EPM/EPDM vulcanizates absorb little moisture, their good electrical properties suffer only minimally when they are submerged in water.

HEALTH AND SAFETY FACTORS

EP(D)M is not classified as a hazardous material. It is not considered carcinogenic according to Occupational Safety and Healthy Administration (OSHA) Hazard Communications Standard and IARC Monographs. In handling EPM/EPDM, normal industrial hygienic procedures should be followed. It is advisable to minimize skin contact. The use of EPM/EPDM is permitted for food contact under the conditions given in the respective FDA paragraphs.

USES

Part of the growth of EPM and EPDM comes from their replacement of the general purpose rubbers such as NR and SBR, by virtue of their better ozone and thermal resistance. The main uses of EPM and EPDM are in automotive applications such as weatherstrip profiles, radiator hoses, and seals; in building and construction as sealing profiles, roofing foil, and seals; in cable and wire as cable insulation and jacketing; and in appliances as a wide variety of mostly molded articles.

Another application for EPDM is in blends with general purpose rubbers. Ozone resistance is thus provided, with the host rubber comprising the principal portion of the blend. This technique has been applied in enhancing the ozone and weathering resistance of tire sidewalls and cover strips. There are no all-EPDM tires currently being produced. Economic factors favor the use of natural and general purpose synthetic rubbers in tires.

Considerable amounts of EPM and EPDM are also used in blends with thermoplastics; eg, as an impact modifier in quantities up to ~25% wt/wt for polyamides, polystyrenes, and particularly polypropylene. The latter products are used in many exterior automotive applications such as bumpers and body panels. In blends with polypropylene, wherein the EPDM component may be

increased to become the larger portion, a thermoplastic elastomer is obtained, provided the EPDM phase is vulcanized during the mixing with polypropylene to suppress the flow of the EPDM phase and give the end product sufficient set.

The addition of relatively small quantities of EPM to lubricating oil raises the viscosity of the solution to an overall less temperature-dependent level. Due to the thinning character of its molecular chains with decreasing temperature, the natural increase of oil viscosity with decreasing temperature is thereby largely compensated. The leading polymer for this application is EPM, because of its excellent heat and shear stability under the operating conditions of automobile engines.

J. Boor, *Ziegler Natta Catalysts and Polymerizations*, Academic Press, New York, 1979, p. 563.
G. Natta, A. Valvasori, and G. Satori, in J. P. Kennedy and E. G. M. Törnquist, eds., *Polymer Chemistry of Synthetic Elastomers*, John Wiley & Sons, Inc., New York, 1969, p. 687.
G. Stella and G. Wouters, *Int. Rubber Conf. 2000, Helsinki, Finland*, June 12–15, 2000.

Jacobus W. M. Noordermeer
DSM Elastomers

EXPLOSIVES AND PROPELLANTS

Propellants and explosives are both classed as combustible materials. They contain both oxidizer and fuel in their compositions. During combustion, propellant and explosive substances will liberate a large amount of gas at high temperatures and will self-sustain combustion without the presence of oxygen in the surrounding atmosphere. The combustion process of propellant and explosive substances can be defined as a self-sustaining exothermic rapid-oxidizing reaction. The chemical composition of propellants and explosives is essentially the same; consequently, some propellants can be used as explosives and some explosives used as propellants. Propellants in general generate combustion gases by the deflagration process. The combustion process of propellants is usually subsonic, whereas explosives generate combustion gases by deflagration or detonation. The combustion process of explosives during detonation is supersonic.

THE COMBUSTION PROCESS

Deflagrating explosives burn faster and more violently than ordinary combustible materials. They burn with a flame or sparks, or a hissing or crackling noise. Propellants are classified as deflagrating explosives. A small amount of a deflagrating explosive in an unconfined condition will ignite when subjected to a flame, spark, shock, friction, or high temperatures. On initiation, local finite "hot spots" are developed, either through friction between the solid particulates, by the compression of voids or bubbles in the liquid component, or by plastic flow of the material. These hot spots produce heat and volatile intermediates which then undergo highly exothermic reactions in the gaseous phase, producing more than enough energy and heat to initiate the decomposition and volatilization of newly exposed surfaces. Therefore, deflagration is a self-propagating process.

If a deflagrating explosive is initiated in a confined state (completely enclosed in a casing) the gaseous products will not be able to escape. Hence the pressure will increase with a consequent rapid increase in the rate of deflagration. When the rate of deflagration reaches a velocity of 1,000–1,800 m s^{-1} it becomes classed as a low order detonation. At a velocity of 5,000 m s^{-1} the detonation becomes classed as a high order detonation. Therefore, a given explosive may behave as a deflagrating explosive when unconfined and as a detonating explosive when confined and suitably initiated.

The burning of a deflagrating explosive is therefore a surface phenomenon similar to other combustible materials except that it does not need a supply of oxygen to sustain the burning. The propagation of an explosion reaction through a deflagrating explosive is based on thermal reactions that are relatively slow. The explosive material surrounding the initial exploding site is warmed above its decomposition temperature, resulting in small explosions. The transfer of energy through the deflagrating explosive, by thermal means through a temperature difference, depends very much on external conditions such as ambient pressure. The speed of the explosion process in deflagrating explosives is always subsonic; that is, less than the speed of sound traveling through an explosive material.

Detonating explosives on initiation decompose via the passage of a shockwave rather than thermal processes. The velocity of the shockwave in solid or liquid explosives is between 1,500 and 9,000 m s^{-1}, an order of magnitude higher than that for the deflagration process. The rate at which the material decomposes is governed by the speed at which the material will transmit the shockwave, not by the rate of heat transfer as in deflagrating explosives. The speed of the explosion process in detonating explosives is always supersonic. Detonation can be achieved either by burning to detonation or by an initial shock.

When an explosive substance is confined in a tube and ignited at one end, burning to detonation may take place. The gas generated from the chemical decomposition of the explosive becomes trapped in the tube, resulting in an increase in pressure at the burning surface; this in turn raises the linear burning rate. If the linear burning rate is raised high enough by pressure pulses generated at the burning surface, detonation will take place when it exceeds the velocity of sound. An explosive which burns to detonation will show an appreciable delay between the initiation of burning and the onset of detonation. This delay will vary according to the nature of the explosive composition, its particle size, density and conditions of confinement.

Some explosive substances can also be detonated when subjected to a high-velocity shockwave. The shockwave forces the particles to compress; this gives rise to adiabatic heating, which raises the temperature to above the decomposition temperature of the explosive material. At these high temperatures the explosive crystals undergo an exothermic chemical decomposition, which accelerates the shockwave. Detonation will take place only if the velocity of the shockwave in the explosive composition exceeds the velocity of sound. Although initiation to detonation does not take place instantaneously, the delay is negligible.

Explosives can therefore be classified by the ease with which they can be ignited and then exploded. Primary explosives are readily ignited or detonated by a small mechanical or electrical stimulus. Secondary explosives are not so easily initiated; they require a high-velocity shockwave generally produced from the detonation of a primary explosive. Propellants are generally initiated by a flame, and do not detonate, only deflagrate.

EXPLOSIVES

The majority of substances classed as explosives generally contain oxygen, nitrogen, and fuels such as carbon and hydrogen. The oxygen is generally attached to nitrogen (ie, NO, NO$_2$, and NO$_3$). Azides, such as lead azide (PbN$_6$), and nitrogen compounds, such as nitrogen triiodide (NI$_3$) and azoimide (NH$_3$NI$_3$), are the exceptions because they contain no oxygen. In the event of a chemical reaction, the nitrogen and oxygen molecules separate and then unite with the fuel components (carbon and hydrogen).

During this reaction large quantities of energy are liberated, which is generally accompanied by the evolution of hot gases. The heat given out during the reaction is the difference between the heat required to break up the explosive molecule into its elements and the heat released on recombination of these elements to form CO$_2$, H$_2$O, N$_2$, etc.

Chemical explosives can be divided into two groups, depending on their chemical nature: those that are classed as substances which are explosive, and those which are explosive mixtures such as black powder. An alternative way of classifying explosives is by their performance and uses. Using this classification, explosives can be divided into three classes; (1) primary explosives, (2) secondary explosives, and (3) propellants, as shown in Figure 1.

Primary explosives have a high degree of sensitivity to initiation through shock, friction, electric spark, or high temperatures. They explode whether confined or unconfined. They differ considerably in their sensitivity to heat and in the amount of heat they produce on detonation. Primary explosives differ from secondary explosives in that they undergo a very rapid transition from burning to detonation and have the ability to transmit the detonation shockwave to less sensitive explosives. On detonation the molecules in the explosive disassociate and produce a tremendous amount of heat and/or shock. This will in turn initiate a second, more stable explosive. For these reasons, they are used in initiating devices. The heat and shock on detonation can vary but are comparable to that for secondary explosives. Their detonation velocities are in the range of 3,500–5,500 m s^{-1}.

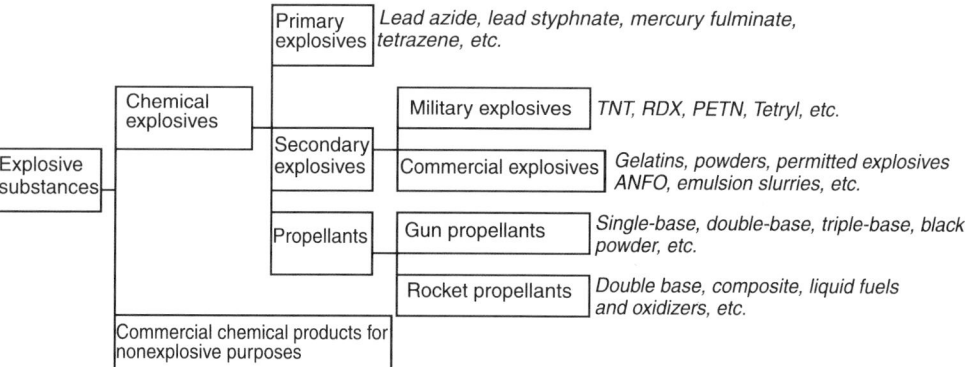

Figure 1. The classes of explosives. Pyrotechnic compositions can also be classed as chemical explosives. Reproduced by permission of the Royal Society of Chemistry.

Secondary explosives are generally more powerful than primary explosives and cannot be detonated readily by heat or shock. Secondary explosives are less sensitive than primary explosives and can be initiated to detonation only by the shock produced by the explosion of a primary explosive. On initiation, the secondary explosive compositions disassociate almost instantaneously into other, more stable components.

The detonation of a secondary explosive is so fast that a shockwave is generated. This shockwave acts on its surroundings with a great shattering effect (brisance) before the pressure of the expanding gaseous products can take effect. Some secondary explosives can be very stable-Rifle bullets can be fired through them or they can be set on fire without detonating. Secondary explosives that detonate at very high velocities exert a much greater force during their detonation than the explosive materials used to initiate them. The values of their detonation velocities are in the range of 5,500–9,000 m·s^{-1}, considerably higher than primary explosives.

PROPELLANTS

A propellant is an explosive material that undergoes rapid, predictable combustion (without detonation), resulting in a large volume of hot gas. This gas can be used to propel a projectile or in gas generators. In order to produce gas quickly a propellant, like a high explosive, must carry its own oxygen, together with suitable quantities of fuel elements such as carbon, hydrogen, etc. A homogeneous propellant is where the fuel and oxidizer are in the same molecule; eg, nitrocellulose, whereas a heterogeneous propellant has the fuel and oxidizer in separate compounds. Gun propellants are traditionally known to be homogeneous, whereas rocket propellants are heterogeneous.

Gun Propellants

Gun propellants have traditionally been fabricated from nitrocellulose-based materials. These fibrous materials have good mechanical properties and can be fabricated in granular or stick form (known as grains) to give a constant burning surface without detonation. The size of the propellant grains will depend on the size of the gun. Larger guns require larger grains, which take more time to burn. The shape of the propellant grain is also very important. Grains with large surface areas will burn at a faster rate than those with low surface areas.

Apart from the size and shape of the propellant grains, their composition also plays an important role. There exist three basic types of solid gun propellant: single-base, double-base, and triple-base. Other types of gun propellants that are less common are high energy, liquid, and composite gun propellants.

Rocket Propellants

Rocket propellants are very similar to gun propellants in that they are designed to burn uniformly and smoothly without detonation. Gun propellants, however, burn more rapidly, due to the higher operating pressures in the gun barrel. Rockets propellants are required to burn at a chamber pressure of ~7 MPa compared to ~400 MPa for a gun propellant. A rocket propellant must also burn for a longer time, to provide a sustained impulse.

Solid rocket propellants are manufactured in the form of geometrical shapes known as grains. For short-range missiles the grains are larger and fewer in number than in a gun cartridge and are designed to burn over their entire surface to give a high mass burning rate. For larger and longer distance missiles the rocket motor will contain only one or two large grains.

Double-base rocket propellants are homogeneous and contain nitrocellulose plasticized with nitroglycerine. Composite rocket propellants are two-phase mixtures comprising a crystalline oxidizer in a polymeric fuel/ binder matrix. The oxidizer is a finely dispersed powder of ammonium perchlorate suspended in a fuel. The fuel is a plasticized polymeric material that may have rubbery or plastic properties. Liquid rocket propellants are subdivided into monopropellants and bipropellants. Monopropellants are liquids that burn in the absence of external oxygen. They have comparatively low energy and specific impulse and are used in small missiles that require low thrust. Bipropellants consist of two components, a fuel and an oxidizer, that are stored in separate tanks and injected into a combustion chamber where they come into contact and ignite. Some bipropellants that are gaseous at room temperature need to be stored and used at low temperatures so that they are in the liquid state, as the bipropellant hydrogen and oxygen. These types of bipropellants have very high specific impulses and are used on the most demanding missions such as satellite launch vehicles.

Gas-Generating Propellants

Solid propellants can be used in systems where large quantities of gas are required in a very short period of time. Such systems include airbags for cars and ejector seats in aircraft. The advantage of using propellant compositions is the speed at which the gas is generated.

Some of the more common explosive materials are mercury fulminate, lead azide, lead styphnate, silver azide, tetrazene, nitroglycerine, nitrocellulose, picric acid, tetryl, TNT, nitroguanidine, PETN, RDX, HMX, TATB, HNS, NTO, and TNAZ.

There are many other ingredients added to explosive compositions that which in themselves are not explosive but can enhance the power of explosives, reduce the sensitivity, and aid in processing. Aluminum powder is frequently added to explosive and propellant compositions to improve their efficiency. Ammonium nitrate is used extensively in commercial explosives and propellants. Small glass or plastic spheres containing oxygen can be added to emulsion slurries to increase its sensitivity to detonation. Polymeric materials can be added to secondary explosives to produce polymer-bonded explosives (PBXs). The polymers are generally used in conjunction with compatible plasticizers to produce insensitive PBXs. The polymers and plasticizers can be in the nitrated form which will increase the power of the explosive. These nitrated forms are known as energetic polymers and energetic plasticizers. Phlegmatizers are added to explosives to aid in processing and reduce impact and friction sensitivity of highly sensitive explosives. Phlegmatizers can be waxes that lubricate the explosive crystals and act as a binder.

RECENT DEVELOPMENTS

The research into energetic molecules producing a large amount of gas per unit mass led to molecular structures having a high hydrogen to carbon ratio, as HNF (hydrazinium nitroformate) and ADN (ammonium dinitramide). ADN is a dense nonchlorine containing a powerful oxidizer and is an interesting candidate for replacing ammonium perchlorate as an oxidizer for composite propellants. ADN is less sensitive to impact than RDX and HMX but more sensitive to friction and electrostatic spark. HNIW, commonly called CL20, belongs to the family of high energy dense caged nitramines. Octanitrocubane (ONC) and heptanitrocubane (HpNC) were successfully synthesized in 1997 and 2000, respectively. The basic structure of ONC is a cubane molecule where all the hydrogens have been replaced by nitro groups. HpNC is denser than ONC and is predicted to be a more powerful, shock-insensitive explosive.

Recent developments of novel explosive materials have concentrated on reducing the sensitivity of the explosive materials to accidental initiation by shock, impact, and thermal effects. The explosive materials with this reduced sensitivity are called insensitive munitions, IM. Although these explosive materials are insensitive to accidental initiation they still perform very well when suitably initiated.

Historically, waste explosive compositions (including propellants) have been disposed of by dumping the waste composition in the sea, or by burning or detonating the composition in an open bonfire. In 1994 the United Nations banned the dumping of explosive waste into the sea, and due to an increase in environmental awareness, burning the explosive waste in an open bonfire will soon be banned. Methods are currently being developed to remove the waste explosive compositions safely from casings using high-pressure water jets. The recovered material then has to be disposed of; one method is to reformulate the material into a commercial explosive. In the future, when formulating a new explosive composition, scientists must not only consider its overall performance but must make sure that it falls into the "insensitive munitions" category that can easily be disposed of or recycled in an environmentally friendly manner.

H. Ellern, *Military and Civilian Pyrotechnics*, Chemical Publishing Company Inc., New York, 1968.

B. T. Fedoroff and O. E. Sheffield, *Encyclopedia of Explosives and Related Items*, Vol. 3, Picatinny Arsenal, Dover, N. J., 1966; Vol. 4, 1969; Vol. 5, 1972; Vol. 6, 1974, Vol. 7, 1975.

S. M. Kaye, *Encyclopedia of Explosives and Related Items*, Vol. 8, U.S. Army Armament Research and Development Command, Dover, N. J., 1978; Vol. 9, 1980; Vol. 10, 1983.

T. Shimizu, *Fireworks, The Art, Science and Technique*, Maruzen Co Ltd, Tokyo, Japan, 1981.

JACQUELINE AKHAVAN
Royal Military College
of Science,Cranfield
University

EXTRACTION, LIQUID–LIQUID

The physical process of liquid–liquid extraction separates a dissolved component from its solvent by transfer to a second solvent, immiscible with the first but having a higher affinity for the transferred component. The latter is sometimes called the consolute component. Liquid–liquid extraction can purify a consolute component with respect to dissolved components which are not soluble in the second solvent, and often the extract solution contains a higher concentration of the consolute component than the initial solution. In the process of fractional extraction, two or more consolute components can be extracted and also separated if these have different distribution ratios between the two solvents.

The principle of liquid–liquid extraction, and some of the special terminology, are illustrated in Figure 1 which

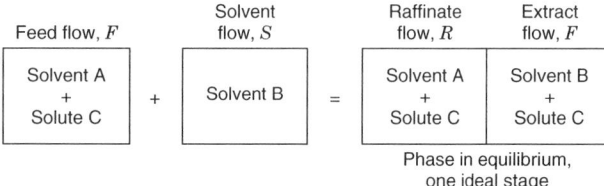

Figure 1. Single contacting stage.

shows a single contacting stage. If equilibrium is fully established after contact, the stage is defined as an ideal or theoretical stage. The two resulting liquid phases are the raffinate from which most of solute C has been removed, and the extract, consisting mainly of solvent B and C.

In the simplest case, the feed solution consists of a solvent A containing a consolute component C, which is brought into contact with a second solvent B. For efficient contact there must be a large interfacial area across which component C can transfer until equilibrium is reached or closely approached. On the laboratory scale this can be achieved in a few minutes simply by hand agitation of the two liquid phases in a stoppered flask or separatory funnel. Under continuous flow conditions it is usually necessary to use mechanical agitation to promote coalescence of the phases. After sufficient time and agitation, the system approaches equilibrium which can be expressed in terms of the extraction factor ϵ for component C:

$$\epsilon = \frac{\text{quantity of C in B-rich phase}}{\text{quantity of C in A-rich phase}} = m\,\frac{\text{B}}{\text{A}} \qquad (1)$$

where B and A refer to the quantities of the two solvents and m is the distribution coefficient.

The component C in the separated extract from the stage contact shown in Figure 1 may be separated from the solvent B by distillation (qv), evaporation (qv), or other means, allowing solvent B to be reused for further extraction. Alternatively, the extract can be subjected to back-extraction (stripping) with solvent A under different conditions, eg, a different temperature; again, the stripped solvent B can be reused for further extraction. Solvent recovery is an important factor in the economics of industrial extraction processes.

Whereas Figure 1 assumes a physical extraction based on different solubilities as expressed by the distribution coefficient, many extractions depend on chemical changes. In such cases the component C in the feed solvent may not itself have any solubility in the extracting solvent B, but can be made to react with an extractant to produce a compound or species which is soluble in B. Many metals can be extracted from aqueous solutions of their salts into organic carrier solvents by using organic extractants which can form organometallic compounds or complexes. Stripping of the metals from the organic to an aqueous phase can be effected by changing a chemical condition such as pH.

PRINCIPLES

Physical Equilibria and Solvent Selection

In order for two separate liquid phases to exist in equilibrium, there must be a considerable degree of thermodynamically nonideal behavior. If the Gibbs free energy, G, of a mixture of two solutions exceeds the energies of the initial solutions, mixing does not occur and the system remains in two phases. For the binary system containing only components A and B, the condition for the formation of two phases is

$$\frac{d^2G}{dx_A^2} > \qquad (2)$$

The selection of solvents for a given separation depends largely 0 on equilibrium considerations. Other important factors include cost, ease of solvent recovery by distillation (qv) or other means, safety and environmental impact, and physical properties which must permit easy phase dispersion and separation. Solvent selection is therefore a broad-based exercise which is hard to quantify. However a useful quantitative approach has been proposed for comparing simplified equilibrium estimations on the basis of regular solution theory.

Chemical Equilibria

In many cases, mass transfer between two liquid phases is accompanied by a chemical change. The transferring species can dissociate or polymerize depending on the nature of the solvent, or a reaction may occur between the transferring species and an extractant present in one phase.

In addition to the liquid–liquid reaction processes, there are many cases in both analytical and industrial chemistry where the main objective of separation is achieved by extraction using a chemical extractant. The technique of dissociation extraction is very valuable for separating mixtures of weakly acidic or basic organic compounds.

In hydrometallurgical separations, a metal ion in aqueous solution can be selectively converted to an organometallic compound or complex which is soluble in an organic carrier solvent.

Chelating extractants owe effectiveness to the attraction of adjacent groups on the molecule for the metal. Anionic extractants are commonly based on high molecular weight amines. Solvating extractants contain one or more electron donor atoms, usually oxygen, which can supplant or partially supplant the water which is attached to the metal ions.

Interfacial Mass-Transfer Coefficients

Whereas equilibrium relationships are important in determining the ultimate degree of extraction attainable, in practice the rate of extraction is of equal importance. Equilibrium is approached asymptotically with increasing contact time in a batch extraction. In continuous extractors the approach to equilibrium is determined primarily by the residence time, defined as the volume of the phase contact region divided by the volume flow rate of the phases.

The rate of mass transfer depends on the interfacial contact area and on the rate of mass transfer per unit interfacial area, ie, the mass flux. The mass flux very close to the liquid–liquid interface is determined by molecular diffusion in accordance with Fick's first law:

$$N = -D \frac{\partial c}{\partial z} \qquad (3)$$

where N refers to the flux in the z direction, c is the concentration of the consolute component, and D is its molecular diffusivity in the solvent.

Mass-Transfer Coefficients with Chemical Reaction

Chemical reaction can occur in any of the five regions shown in Figure 2, ie, the bulk of each phase, the film in each phase adjacent to the interface, and at the interface itself. Irreversible homogeneous reaction between the consolute component C and a reactant D in phase B can be described as

$$C + zD \rightarrow products \qquad (4)$$

The equations of combined diffusion and reaction, and their solutions, are analogous to those for gas absorption (qv). It has been shown how the concentration profiles and rate-controlling steps change as the rate constant increases. When the reaction is very slow and the B-rich phase is essentially saturated with C, the mass-transfer rate is governed by the kinetics within the bulk of the B-rich phase. This is defined as regime 1. For a slow reaction defined as regime 2, the consolute component C is

almost entirely depleted in the bulk of the B-rich phase and the mass transfer of C between the phases controls the rate of the reaction. For a very fast reaction the depletion of C affects the concentration profile in the diffusion film. The steepening of the concentration profile for regime 3 leads to an enhancement in the film mass-transfer coefficient in the B-rich phase. Finally, the case of an instantaneous reaction (regime 4) leads to the formation of a thin reaction zone to which components C and D diffuse in stoichiometric amounts.

Interfacial Contact Area and Approach to Equilibrium

Experimental extraction cells such as the original Lewis stirred cell are often operated with a flat liquid–liquid interface the area of which can easily be measured. In the single-drop apparatus, a regular sequence of drops of known diameter is released through the continuous phase. These units are useful for the direct calculation of the mass flux N and hence the mass-transfer coefficient for a given system. In industrial equipment, however, it is usually necessary to create a dispersion of drops in order to achieve a large specific interfacial area, a, defined as the interfacial contact area per unit volume of two-phase dispersion. Thus the mass-transfer rate obtainable per unit volume is given as

$$(N \cdot a) = K_A a (c_A - c_A^*) \qquad (5)$$

Calculation of Equilibrium Stages

Multistage contacting can be arranged in a concurrent, crosscurrent, or countercurrent manner. The sequence of stages is sometimes referred to as a cascade, referring to the early use of gravity overflow from stage to stage. The countercurrent arrangement represents the best compromise between the objectives of high extract concentration and a high degree of extraction of the solute, for a given solvent-to-feed ratio. For the case of a partially miscible ternary system, the number of ideal stages in a countercurrent cascade can be estimated graphically on a triangular diagram, using the Hunter-Nash method.

Fractional Extraction

Fractional extraction is the separation of two or more consolute components by solvent extraction. Single-solvent fractional extraction has been known for many years, but the range of solvents available is limited because of the requirement that the solvents must be sparingly miscible with each of the feed components.

Dual solvent fractional extraction makes use of the selectivity of two solvents (A and D) with respect to consolute components.

Differential Contacting

Although the equilibrium stage concept has proved extremely useful in describing the performance of mixer-settlers and plate columns having discrete stages, it is not appropriate for spray towers, packed columns, etc, in which no discrete stages can be identified. In such differential types of contactors, equilibrium between

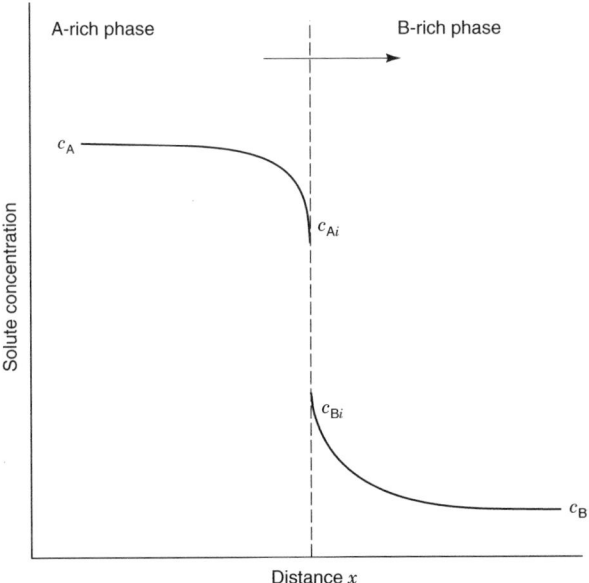

Figure 2. Concentration profiles near an interface where the arrow represents the direction of mass transfer, c_A = concentration of C in A-rich phase, c_B = concentration of C in B-rich phase, and the subscript i denotes the interface.

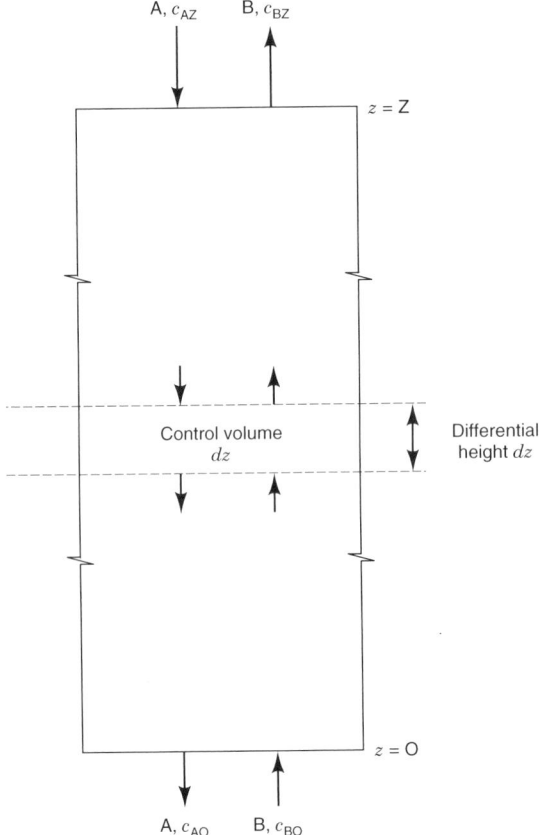

Figure 3. Mass transfer in a differential contactor. Terms are defined in the text.

phases is never reached and therefore the mass-transfer rate is important in the design procedure.

A differential countercurrent contactor operating with a dilute solution of the consolute component C and immiscible components A and B is shown in Figure 3. Under these conditions, the superficial velocities of the A-rich and B-rich streams can be assumed not to vary significantly with position in the contactor, and are taken to be U_A and U_B, respectively. The concentration of C in the A-rich stream is c_A and that in the B-rich stream is c_B.

A steady-state material balance can be carried out on a small section of length dz and volume dz (on the basis of unit cross-sectional area) in the contactor:

$$U_B dc_B = U_A dc_A = K_A a(c_A - c_A)dz \qquad (6)$$

Axial Dispersion

Elementary texts assume that all the fluid in each phase has the same resident time in a countercurrent extractor. In practice, the two phases rarely move countercurrently in plug flow because of axial mixing which arises from the action of turbulent eddies, circulation currents, or the effects of drop wakes. The effect is to flatten the axial concentration profiles within each phase. Axial mixing can lead to a reduction in the effective driving force for mass transfer which in turn reduces the NTU below that expected for the plug flow case. An important feature of

the profile is the discontinuity or "jump" in concentration which occurs at entry to the contactor when the liquid in the feed line enters the mixed region of the column.

Two alternative approaches are used in axial mixing calculations. For differential contactors, the axial on model is used:

$$N = -E \frac{\partial c}{\partial z} \qquad (7)$$

For contactors in which discrete well-mixed compartments can be identified, for example sieve-plate columns, axial mixing effects are incorporated into the stagewise model by means of the backflow ratio α which is defined as the fraction of the net interstage flow of one phase which is considered to flow in the reverse direction. For a contactor in which there are many compartments, the axial dispersion coefficient and the backflow ratio, α, are interrelated as follows:

$$E = \frac{UH}{\ln((1 + \alpha)/\alpha)} \qquad (8)$$

where H is the height of one compartment and U is the superficial velocity. The detailed calculations of concentration profiles and mass-transfer rates with axial mixing require the solution of a fourth-order differential equation (dispersion model) or the equivalent difference equation (backflow model) along with appropriate boundary conditions.

Drop Diameter

In extraction equipment, drops are initially formed at distributor nozzles; in some types of plate column the drops are repeatedly formed at the perforations on each plate. Under such conditions, the diameter is determined primarily by the balance between interfacial forces and buoyancy forces at the orifice or perforation.

In many types of contactors, such as stirred tanks, rotary agitated columns, and pulsed columns, mechanical energy is applied externally in order to reduce the drop size and thereby increase the rate of mass transfer.

Holdup and Flooding

The volume fraction of the dispersed phase, commonly known as the holdup h, can be adjusted in a batch extractor by means of the relative volumes of each liquid phase added. However, in a countercurrent column contactor, the holdup of the dispersed phase is considerably less than this, because the dispersed drops travel quite fast through the continuous phase and therefore have a relatively short residence time in the equipment. The holdup is related to the superficial velocities U of each phase, defined as the flow rate per unit cross section of the contactor, and to a slip velocity U_s:

$$U_s = U_d/h + U_c/(1 - h) \qquad (9)$$

As the throughput in a contactor represented by the superficial velocities U_c and U_d is increased, the holdup

h increases in a nonlinear fashion. A flooding point is reached at which the countercurrent flow of the two liquid phases cannot be maintained. The flow rates at which flooding occurs depend on system properties, in particular density difference and interfacial tension, and on the equipment design and the amount of agitation supplied.

The nonuniformity of drop dispersions can often be important in extraction. This nonuniformity can lead to axial variation of holdup in a column even though the flow rates and other conditions are held constant.

Coalescence and Phase Separation

Coalescence between adjacent drops and between drops and contactor internals is important for two reasons. It usually plays a part, in combination with breakup, in determining the equilibrium drop size in a dispersion, and it can therefore affect holdup and flooding in a countercurrent extraction column. Secondly, it is an essential step in the disengagement of the phases and the control of entrainment after extraction has been completed.

Membrane Extraction

An extraction technique which uses a thin liquid membrane or film has been introduced. The principal advantages of liquid-membrane extraction are that the inventory of solvent and extractant is extremely small and the specific interfacial area can be increased without the problems which accompany fine drop dispersions.

Supercritical Extraction

The use of a supercritical fluid such as carbon dioxide as extractant is growing in industrial importance, particularly in the food-related industries. The advantages of supercritical fluids (qv) as extractants include favorable solubility and transport properties, and the ability to complete an extraction rapidly at moderate temperature.

Two-Phase Aqueous Extraction

Liquid–liquid extraction usually involves an aqueous phase and an organic phase, but systems having two or more aqueous phases can also be formed from solutions of mutually incompatible polymers such as poly(ethylene glycol) (PEG) or dextran.

Because of the growth in biotechnology, two-phase aqueous extraction is becoming more important industrially. Two-phase aqueous systems have low interfacial tension, low interphase density difference, and high viscosity in comparison with most aqueous–organic systems. Although interfacial contact is very efficient, the separation of the phases after contact can be slow, requiring centrifugation. The performance of a spray column for two-phase aqueous extraction has also been reported.

EQUIPMENT AND PROCESSING

Laboratory Extractors, Pilot-Scale Testing, and Scale-Up

Several laboratory units are useful in analysis, process control, and process studies. The AKUFVE contactor incorporates a separate mixer and centrifugal separator. It is an efficient instrument for rapid and accurate measurement of partition coefficients, as well as for obtaining reaction kinetic data. Miniature mixer–settler assemblies set up as continuous, bench-scale, multistage, countercurrent, liquid–liquid contactors are particularly useful for the preliminary laboratory work associated with flow-sheet development and optimization because these give a known number of theoretical stages.

Because the factors relating to mass transfer and fluid dynamics of the systems in an extractor are extremely complex, particularly for mixed solvents and feedstocks of commercial interest, pilot-scale testing remains an almost inevitable preliminary to a full-scale contactor design. These tests provide (1) total throughput and agitation speed; (2) HETS or HTU; (3) stage efficiency; (4) hydrodynamic conditions, such as droplet dispersion, phase separation, flooding, emulsion layer formation, etc; (5) selection of direction of mass transfer; (6) solvent-to-feed ratio; (7) material of construction and its wetting characteristics; and (8) confirmation of the desired separation in cases where equilibrium data are not available.

For design of a large-scale commercial extractor, the pilot-scale extractor should be of the same type as that to be used on the large scale. Reliable scaleup for industrial-scale extractors still depends on correlations based on extensive performance data collected from both pilot-scale and large-scale extractors covering a wide range of liquid systems. Only limited data for a few types of large commercial extractors are available in the literature.

Commercial Extractors

Extractors can be classified according to the methods applied for interdispersing the phases and producing the countercurrent flow pattern. Figure 4 summarizes the classification of the principal types of commercial extractors.

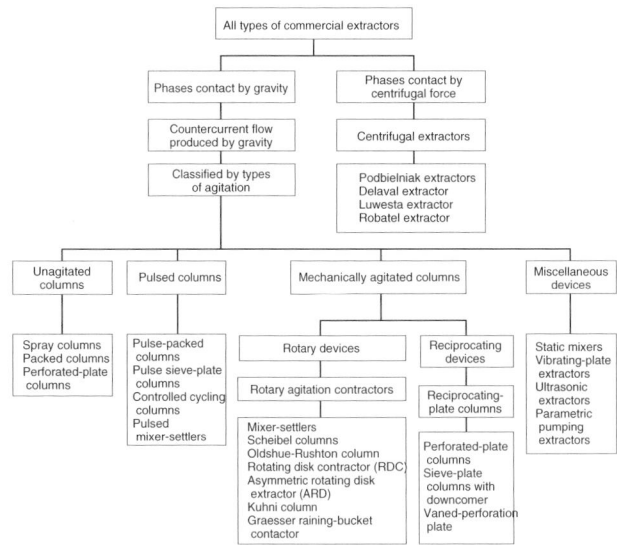

Figure 4. Classification of commercial extractors.

ORGANIC PROCESSES

Petroleum and Petrochemical Processes

The first large-scale application of extraction was the removal of aromatics from kerosene to improve its burning properties. Solvent extraction is also extensively used to meet the growing demand for the high purity aromatics such as benzene, toluene, and xylene (BTX) as feedstocks for the petrochemical industry. Additionally, the separation of aromatics from aliphatics is one of the largest applications of solvent extraction.

Pharmaceutical Processes

The pharmaceutical industry is a principal user of extraction because many pharmaceutical intermediates and products are heat-sensitive and cannot be processed by methods such as distillation. A useful broad review can be found in the literature. Extraction is used in the production of antibiotics, vitamins, sulfa drugs, methaqualone, phenobarbital, antihistamines, cortisone, estrogens and other hormones (qv), and reserpine and alkaloids (qv).

Food Processing

Food processing (qv) makes use of solvent extraction in several ways. Industrial refining of fats and oils using propane is known as the Solexol process. Solvent extraction is used in many protein refining processes, for example the extraction of fish protein from ground fish using i-propyl alcohol. Recovery of lactic acid by an extractive fermentation has recently been reported. The applications of extraction in the food industry have been reviewed.

Other Organic Processes

Solvent extraction has found application in the coal-tar industry for many years, as for example in the recovery of phenols from coal-tar distillates by washing with caustic soda solution. Solvent extraction of fatty and resimic acid from tall oil has been reported. Dissociation extraction is used to separate m-cresol from p-cresol and 2,4-xylenol from 2,5-xylenol. Solvent extraction can play a role in the direct manufacture of chemicals from coal, treatment of industrial effluents, biopolymer extraction, and difficult separations.

INORGANIC PROCESSES

Nuclear Fuel Reprocessing

Spent fuel from a nuclear reactor contains ^{238}U, ^{235}U, ^{239}Pu, ^{232}Th, and many other radioactive isotopes (fission products). Reprocessing involves the treatment of the spent fuel to separate plutonium and unconsumed uranium from other isotopes so that these can be recycled or safely stored (see NUCLEAR REACTORS–NUCLEAR FUEL RESERVES).

Copper

The recovery of copper, Cu, from ore leach liquors as a stage in the hydrometallurgical route to the pure metal is one of the largest applications of liquid–liquid extraction.

Nickel and Cobalt

Often present with copper in sulfuric acid leach liquors are nickel and cobalt. In the case of chloride leach liquors, separation of cobalt from nickel is inherently simpler because cobalt, unlike nickel, has a strong tendency to form anionic chloro-complexes. Thus cobalt can be separated by amine extractants, provided the chloride content of the aqueous phase is carefully controlled. A successful example of this approach is the Falconbridge process developed in Norway.

Extraction of Nonmetallic Inorganic Compounds

Phosphoric acid is usually formed from phosphate rock by treatment with sulfuric acid, which forms sparingly soluble calcium sulfate from which the phosphoric acid is readily separated. However, in special circumstances it may be necessary to use hydrochloric acid.

J. C. Godfrey and M. J. Slater, eds., *Liquid–Liquid Extraction Equipment*, John Wiley & Sons, Ltd., Chichester, U.K., 1994.

T. C. Lo, M. H. I. Baird, and C. Hanson, eds., *Handbook of Solvent Extraction*, Wiley-Interscience, New York, 1983.

J. D. Thornton, ed., *The Science and Practice of Liquid–Liquid Extraction*, Oxford University Press, Oxford, U.K., 1992.

R. E. Treybal, *Liquid Extraction*, 2nd ed., McGraw-Hill, New York, 1963.

TEH C. LO
T. C. Lo & Associates
MALCOLM H. I. BAIRD
McMaster University

F

FATS AND FATTY OILS

Fats and oils are composed primarily of triacylglycerols (**1**), esters of glycerol and fatty acids. However, some oils such as sperm whale jojoba, and orange roughy are largely composed of wax esters (**2**). Waxes are esters of fatty acids with long-chain aliphatic alcohols, sterols, tocopherols, or similar materials.

(1)   (2)

Fatty acids derived from animal and vegetable sources generally contain an even number of carbon atoms, since they are biochemically derived by condensation of two carbon units through acetyl or malonyl coenzyme A. However, odd-numbered and branched fatty acid chains are observed in small concentrations in natural triacylglycerols, particularly ruminant animal fats through propionyl and methylmalonyl coenzyme A, respectively. The glycerol backbone is derived by biospecific reduction of dihydroxyacetone. Chain lengths of even-numbered fatty acids range from 4 to 22.

Structure (**1**) shows the stereochemistry of the triacylglyceride molecule. Positions are numbered by the stereochemical numbering (sn) system. In chemical processes the 1 and 3 positions are not distinguishable. However, for biological systems, the enantiomeric (*R* or *S*) form is important.

Fatty acids may be saturated, monounsaturated, or polyunsaturated, according to the number of double bonds in the alkyl chain. Naturally occurring double bonds are almost exclusively cis (*Z*) in configuration. The most common fatty acids in animal and vegetable fats and oils are dodecanoic (lauric, 12:0), hexadecanoic (palmitic, 16:0), octadecanoic (stearic, 18:0), 9-*cis*-octadecenoic (oleic, 18:1), 9-*cis*, 12-*cis*-octadecadienoic acid (linoleic, 18:2), and 9-*cis*, 12-*cis*,15-*cis*-octadecatrienoic acid (linolenica, 18:3).

Fats and oils are distinguished by their physical state; fats are solid at ambient temperature, whereas oils are liquid. Some edible triacylglycerols, such as butter, lard, vegetable oils, shortenings, and margarines, have substantial quantities of both liquid and solid components at ambient temperature. Commercial products may be derived from animal carcasses by rendering, or vegetable sources by pressing or solvent extraction.

COMPOSITION

Natural fats and oils are composed principally of triacycerols, but other components may be present in minor quantities. These components may have important effects on the nature and quality of the oil or fat.

Free Fatty Acids and Partial Glycerides

After harvest, many crude oil crops contain lipase enzymes that cleave triacylglycerols into fatty acids and partial glycerides. Elevated free fatty acid concentrations are undesirable because they cause high losses during further processing of the oil. Diacylglycerols and monoacylglycerols are formed by hydrolysis.

Free fatty acids are removed by alkali refining, physical refining, or deodorization. Mono- and diacylglycerols are not removed by alkali refining or bleaching and may have an adverse effect on the quality of the oil.

Phospholipids

Glycerides esterified by fatty acids, at the 1,2-positions, and a phosphoric acid, a residue at the 3-position, constitute the class called phospholipids. The identity of the moiety (other than glycerol) esterified to the phosphoric group determines the specific phospholipid compound. The three most common phospholipids in commercial oils are phosphatidylcholine or lecithin, phosphatidylethanolamine or cephalin, and phosphatidylinositol. These materials are important constituents of plant and animal membranes. The phospholipid content of oils varies widely. Most oils contain 0.1–0.5%. Some phospholipids, such as dipalmitoylphosphatidylcholine (R = R′ = palmitic; R″ = choline), form bilayer structures known as vesicles or liposomes. The bilayer structure can microencapsulate solutes and transport them through systems where they would normally be degraded. This property allows their use in drug delivery systems.

Sterols

Sterols are tetracyclic compounds derived biologically from terpenes. They are fat-soluble and therefore are found in small quantities in fats and oils. Cholesterol is a common constituent in animal fats such as lard, tallow, and butterfat.

Tocopherols and Tocotrienols

Algae and plants used as sources of edible oils contain tocopherols, phenolic materials that function as antioxidants. Mammals do not synthesize these compounds, and residues present in their bodies are present because of ingestion. Tocopherols are designated as being vitamin E active. Tocotrienols differ from tocopherols by the presence of three isolated double bonds in the branched alkyl side chain.

Several other naturally occurring antioxidants have been identified in oils. Sesamol occurs as sesamoline, a glycoside, in sesame seed oil. Ferulic acid is found esterified to cycloartenol in rice bran oil and to β-sitosterol in corn oil. Although it does not occur in oils, rosemary extract has also been found to contain powerful phenolic antioxidants.

Carotenoids and Other Pigments

Carotenoids contain conjugated double bonds, a strong chromophore that produces red and yellow coloration in vegetable oils. Carotenoids are tetraterpene hydrocarbons formed by the condensation of eight isoprene units. Another class of compounds, the xanthophylls, is produced by hydroxylation of the carotenoid skeleton. β-Carotene is the best known component of the carotenoids because it is the precursor for vitamin A.

Green coloration, present in many vegetable oils, poses a particular problem in oil extracted from immature or damaged soybeans. Chlorophyll is the compound responsible for this defect.

PROCESSING OF FATS AND OILS

Fats and oils are derived from animals, plants, or fish by rendering (animal tissues), pressing, or solvent extraction. Crude oils from these processes are often of insufficient quality to be used directly, particularly for edible products. Impurities such as pigments, phosphatides, volatile odorous compounds, and certain metals must be removed by further processing, eg, by degumming and dewaxing, refining, bleaching, hydrogenation, randomization/interesterification, physical fractionation, and deodorization.

SOURCES OF FATS AND OILS

Fats and oils may be synthesized in enantiomerically pure forms in the laboratory or derived from vegetable sources (mainly from nuts, beans, and seeds), animal depot fats, fish, or marine mammals. Oils obtained from other sources differ markedly in their fatty acid distribution. One variation in composition is the chain length of the fatty acid. Another variation of the fatty acid is the degree of unsaturation.

Castor oil contains a predominance of ricinoleic acid, which has an unusual structure inasmuch as a double bond is present in the 9 position while a hydroxyl group occurs in the 12 position. The unusual structure of ricinoleic acid affects the solubility and physical properties of castor oil.

Solid fats may show drastically different melting behaviors.

Biotechnology is rapidly revolutionizing the edible oils industry. Crop breeding, genetic engineering, tissue culture, and mutation selection are avenues being pursued to deliver desirable fatty acid compositions into agronomically favored plants. Oils from microbial sources may offer unique fatty acid compositions. Canola, high oleic sunflower, and high oleic canola oils are recent successes in harnessing the biosynthetic factories.

Separation of a fat or oil from its source material can be accomplished by several different methods.

Selection of an extraction process is based on: (1) obtaining oil substantially undamaged and relatively free of undesirable impurities, (2) achieving the highest practical yield, and (3) obtaining the maximum economic return on the oil and coproducts. The processes involved include rendering, mechanical pressing, and solvent extraction.

PHYSICAL PROPERTIES

The physical properties of fats and oils have been reviewed.

Crystallization and Melting Behavior

Pure compounds usually display sharp melting points and impure compounds show broad melting behavior. However, even pure triglycerides show complex melting behavior because of their tendency to pack in several different crystal lattice forms (polymorphism). Triacylglycerols having three identical fatty acids pack into three distinct polymorphs: (1) β, the most stable form shows a triclinic subcell; (2) β', a less stable crystal that suggests orthorhombic packing; and (3) α, a loosely packed triacylglycerol that packs hexagonally. Rapid cooling of the triacylglycerol leads initially to the α form, followed by slow reorganization to β' and β forms. Mixed glycerides, with more than one type of saturated fatty acid, pack with defects in the structure, and then chains appear to tilt to correct for these defects. Glycerides with unsaturated fatty acids must pack to accommodate the bend in the alkyl chain caused by the cis double bond. Perhaps the most widely studied fat, cocoa butter, may show as many as seven distinct polymorphic forms.

Some general trends in specific heats have been suggested: (1) for solid fats, there is little variation in specific heat for saturated fats and their fatty acids as chain length varies. Specific heat varies directly with the degree of unsaturation. The specific heat of a solid is less than that of the liquid at the same temperature; (2) the specific heat of liquid fatty acids and glycerides increases with increasing chain length but decreases with increasing unsaturation. For both liquids and solids, specific heat increases with increasing temperature; and (3) mixed-acid glycerides have lower specific heats than their corresponding simple glycerides.

Viscosity

Fats, oils, fatty acids, and other fatty acid derivatives show relatively high viscosities compared to other liquids because of the intermolecular interaction of long alkyl chains. Some general trends are that longer chain lengths and lower unsaturation produce higher viscosities. Fatty acids are more viscous than esters because of their greater tendency to form hydrogen bonds. Castor oil is in a class by itself because of its side chain hydroxyl group that can form hydrogen bonds. Derivatives of castor oil are consequently useful as specialty lubricants. Fats and oils behave as Newtonian liquids except at very high shear rates, where degradation may begin to occur.

Surface and Interfacial Tension

Commercial oils tend to have lower surface and interfacial tensions because of the presence in them of polar surface-active components such as monoacylglycerols, phospholipids, and soaps. Purification of oils on a Florisil column can be used to obtain higher and more consistent values. Monoacylglycerols and phospholipids can reduce the interfacial tension between an oil and water. Emulsions may be formed that are relatively stable. Food products such as mayonnaise, margarine, and non-separating salad dressings are commercial examples of stable emulsions.

Density

The density of liquid oils at 15°C does not vary markedly with changes in composition. Values generally range from 0.912 to 0.964 g/mL. Density increases with decreasing molecular weight and increasing unsaturation.

Smoke, Flash, and Fire Points

These thermal properties may be determined under standard test conditions. These values are profoundly affected by minor constituents in the oil, such as fatty acids, mono- and diacylglycerols, and residual solvents. These factors are of commercial importance where fats or oils are used at high temperatures as in lubricants or edible frying fats.

Refractive Index

The refractive index of a fat or oil increases with its molecular weight and unsaturation.

Absorption Spectra

The infrared (IR) spectra of fats and oils are similar regardless of their composition. The principal absorption seen is the carbonyl stretching peak, which is virtually identical for all triacylglycerol oils. The most common application of IR spectroscopy is the determination of trans fatty acids occurring in a partially hydrogenated fat.

Solubility Properties

Fats and oils are characterized by virtually complete lack of miscibility with water. However, they are miscible in all proportions with many nonpolar organic solvents.

CHEMICAL PROPERTIES

Most triacylglycerol fats and oils have only two reactive functional groups: the ester linkage joining the fatty acid to the glycerol backbone and double bonds in the alkyl side chain. There is a free hydroxyl group in the side chain of ricinoleic acid found in castor oil and a carbonyl group in the licanic acid side chain of oiticica oil. The double bond influences the reactivity of the adjacent allylic carbon atom, particularly when multiple double bonds are present.

Chemical reactions can cause serious quality problems for oils. Hydrolytic and oxidative rancidity can cause oils to become unacceptable to consumers for edible or other uses.

ANALYTICAL METHODS

Specifications

The quality of individual crude oils is specified by trading rules established by organizations such as the National Soybean Processors Association, National Renderers Association, or National Institute of Oilseed Processors. Standardized tests are defined by the American Oil Chemists Society (AOCS), the Association of Official Analytical Chemists (AOAC), and the American Society for Testing Materials (ASTM). Crude oils must contain only minimal amounts of foreign material, protein, volatile or toxic solvents, pesticides, heat-transfer media, moisture, and foreign adulterating fats. They must also not be abused or mishandled, which causes them to become oxidized. Crude oils must not show excessive loss on refining, which adds to costs and waste disposal problems. Oil processors must also meet specifications for their customers, which include measures of oil quality, such as free fatty acid level, color, and oxidation. Other specifications may relate to functionality in the customer's product, such as melting range or fatty acid composition.

USES

Fats and oils are used in food components and cooking oils, soaps and detergents, drying oils, and the manufacture of fatty acids and derivatives.

G. D. Nelson, ed., *Health Effects of Dietary Fatty Acids*, American Oil Chemists' Society, Champaign, Ill., 1991, pp. 50–135.

R. D. O'Brien, *Fats and Oils: Formulating and Processing for Applications*, 2nd ed., CRC Press, Boca Raton, Fla., 2004.

R. D. O'Brien, W. E. Farr, and P. J. Wan, eds., *Introduction to Fats and Oils Technology*, 2nd ed., AOCS Press, Champaign, Ill., 2000.

F. Shahidi and A. E. Bailey, *Bailey's, Industrial Oil and Fats Products*, 6th ed., John Wiley & Sons, Inc., New York, 2004.

G. L. Hasenhuettl
Consultant

FEEDS AND FEED ADDITIVES, NONRUMINANT

FEED INGREDIENTS

Both swine and poultry diets are comprised primarily of grains such as corn and grain sorghum, with occasional use of wheat, barley, and other small grains. Soybean meal is the primary source of protein in these diets, but animal by-products such as meat and bone meal, poultry by-product meal, feather meal, and fish meal contribute significant amounts of protein and provide some of the minerals required for growth, maintenance, reproduction, and lactation. Many human and industrial by-products are also used in swine and poultry feeds; eg, dried bakery products, produced from leftover bread and other bakery waste; inedible fats from the processing of vegetable oils for human consumption; large amounts of fats and oils from the restaurant and fast food trade that are produced as by-products of cooking food; and numerous other products that would otherwise go unused.

Because of the simplicity of swine and poultry feeds, most feed manufacturers add vitamins and trace minerals to ensure an adequate supply of essential nutrients. Amino acids such as methionine lysine, threonine and tryptophan, produced by chemical synthesis or by fermentation are used to fortify swine and poultry diets. The use of these supplements to provide the essential amino acids permits diets with lower total crude protein content.

Virtually all broiler and turkey diets, and much of the swine feeds, are pelleted prior to feeding. Pelleted feeds are consumed in greater quantity than are feeds in a mash or meal form and generally result in more rapid weight gain and better feed conversion. Proper pelleting of feed also aids in reducing the potential of salmonellas or other bacterial contamination of feeds. Pelleting is accomplished by forcing the feed through a die having many small holes. Pelleting is improved by steaming the feed to gelatinize the starch provided by the grains, by adding low (<2%) levels of fat to the feed, or by addition of various types of pellet binders. There are several types of pellet binders, including bentonite clays, lignosulfonates, and grain starch products, that result in firmer, more durable pellets able to withstand the rigors of mechanical feed handling systems.

NUTRIENT REQUIREMENTS OF SWINE AND POULTRY

Numerous researchers at state and governmental research institutes have defined the requirements of swine and poultry for virtually all known nutrients. In addition, the nutrient composition of common ingredients has been determined. Through the use of a mathematical technique known as linear programming, aided by the use of high-speed computers, poultry and swine nutritionists are able to formulate nutritionally balanced diets for all species of animals. When ingredient prices change as a result of supply and demand, diet composition can be changed almost instantly.

There is no best feed composition, because animals thrive on diets composed of many different types of ingredients. Swine and poultry generally adapt readily and rapidly to changes in ingredient composition, as long as the diets provide adequate levels of essential nutrients.

Poultry can be grown on many diverse types of diets. Because of the high percentage of chickens grown under an integrated system, it is sometimes difficult to purchase small quantities of high-quality feeds. Persons who sometimes utilize chickens for laboratory or research animals may require information regarding formulas that can be mixed from readily available ingredients.

FEED ADDITIVES FOR NONRUMINANTS

Feed additives are common to swine and poultry feeds for a number of purposes. Antibiotics used to promote growth and improve feed utilization, are poorly digestible and function primarily by controlling the bacterial flora of the intestinal tract. Antibiotics fed to animals for this purpose generally are not used for human antibiotic therapy. Other antibiotics, used for disease therapy, are generally injected or are absorbed into the body tissues, where they are effective against the disease-causing organisms. Other feed additives include antioxidants to protect feeds against oxidative rancidity, mold inhibitors to prevent development of potentially toxic mold products, anticoccidial compounds that protect chickens against this severe parasitic disease, anthelmintics, and other types of growth promoters.

Official information concerning FDA approval of antibiotics and other drugs is available in the *Code of Federal Regulations*. This document is revised at least once per year and updated in individual issues of the *Federal Register*. An effective and less expensive way to attain information regarding the status of approved feed additives for animal feeds is to subscribe to the *Feed Additive Compendium*, published yearly and updated monthly.

There are a large number of feed additive products classified as generally recognized as safe (GRAS). Some restrictions are placed on the quantity of some products, such as selenium and ethoxyquin. GRAS products include a wide range of materials, ranging from ammoniated cottonseed meal to xanthan gum.

The GRAS listing does not include widely used historical products such as grains, sugar, salt, etc. In general, feed ingredients listed in the *American Association of Feed Control Officials* (AAFCO) official publication are considered in the GRAS category.

A number of products designated GRAS are being scrutinized by the FDA because of advertisements and claims made by producers or manufacturers of these products. Statements that indicate that feeding such products improve animal performance may require substantive data to support such claims in the future.

American Association of Feed Control Officials (AAFCO) Directory, AAFCO, College Station, Tex.

Feedstuffs Reference Issue, Miller Publishing Co., Minnetonka, Minn., 1991.

Feed Situation and Outlook Report, U.S. Department of Agriculture, Washington, D.C., 1991.

PARK W. WALDROUP
University of Arkansas

FEEDS AND FEED ADDITIVES, PET FOODS

All pet foods sold in the United States are subject to scrutiny by both competitors and feed control officials, including the Food and Drug Administration (FDA), Association of American Feed Control Officials (AAFCO), U.S. Department of Agriculture (USDA), Federal Trade Commission (FTC), American Animal Hospital Association (AAHA), American Veterinary Medical Association (AVMA), and Pet Food Institute (PFI). A European group organized to assure fair trade and free circulation of products through Europe (FEDIAF) also monitors every aspect of U.S. pet foods and follows American trends. More is known about the nutrition of dogs and cats than is known about the nutrition of humans.

Pet foods are different from other animal feeds. Most pet foods are processed in highly sophisticated plants using equipment, sanitation, and quality control exceeding standards observed in many plants producing human-grade foods. Pet foods may be stored for up to a year following manufacture before being consumed. This possible delay in the consumption of pet foods requires more careful ingredient selection, preservation of freshness with antioxidants (qv), packaging and processing to avoid insect infestations and rancidity, and careful storage. Pet foods may contain expensive ingredients to provide desirable promotional and marketing copy.

Pets are fed a wide range of commercial foods, which vary on a dry basis from 15 to 60% protein and 5 to 50% fat. Some pet foods are expensive, nutrient-rich foods that contain twice as much nutrition density as needed. Although small quantities of a high calorie food may be consumed, approximately equivalent nutrition may be obtained by pets consuming larger quantities of a less concentrated food.

TYPES OF COMMERCIAL PET FOODS

Pet foods are produced in canned, semimoist, and dry forms. Canned pet foods contain approximately 78 to 82% water and have a strong appeal to both pets and owners. Semimoist foods have moisture contents of 25 to 50%. Dry-type foods contain 10 to 12% moisture and supply about 90% of the nutrition consumed by dogs and 72% of the nutrition eaten by cats.

Therapeutic foods have been developed to meet the needs of pets that have nephritic failure, allergies, thyroid problems, geriatric difficulties, and obesity. Most of these therapeutic diets are dispensed by veterinarians, though some are available in pet food outlets and human-food stores stocking pet foods. Treats are usually snacks that may be nutritionally complete or may provide a tasty morsel as a reward. The number of treat products has escalated rapidly.

Canned and Semimoist Foods

Canned and dry foods are nutritionally comparable on a moisture-free basis. Some canned foods are basically dry foods to which gravy, moisture, and flavor enhancers have been added. Almost all animals tend to prefer moist foods to dry, and canned foods are desirable for geriatric dogs and cats, particularly those having gum and dental deterioration. Canned foods can be gulped by dogs and consumed quickly by cats.

Dry Foods

Dry foods are concentrated sources of nutrition and provide the most economical nutritional value because water in canned foods is expensive. Dry foods tend to scrape the teeth as pets eat, minimizing tartar deposition. When dry food is moistened prior to being consumed, tartar accumulates in a manner comparable to deposits observed with canned foods. Approximately 95 to 98% of dry-type cat and dog foods are made by the extrusion process; the remainder is made by pelleting or baking.

PET FOOD FORMULATION

Weights of adult cats in normal physical condition vary from 2 to 6 kg, which is contrasted with the 1 to 100 kg encountered in adult dogs of different breeds. Dogs have proportionately longer digestive tracts and can digest foods more efficiently than cats. This difference in digestibility helps account for the requirement by cats for higher protein diets.

Animal food ingredients are selected to provide desirable contributions of nutrient availability, digestibility, droppings condition, palatability, processing characteristics, ethical desirability, and economics. Modern commercial pet foods contain about 50 nutrient and nonnutrient additives. Each nutrient is supplied at a near-optimum bioavailable level.

Ingredients used in pet foods are usually high in nutritional quality but generally not desirable as human foods primarily because they do not conform to human taste or processing expectations. By-products such as rendered proteins and fat converted into pet foods may have a derivation unappealing to humans, yet after processing may actually be more free of microorganisms and toxins than foods consumed by humans.

Nutritive Ingredients

Nutrients include amino acids (qv), fats, carbohydrates, fibers, minerals, and vitamins (qv). Some ingredients, such as niacin, supply only niacin, whereas salt provides both essential sodium and chlorine. Meat and bone meal may contain all of the nutrients, but not the correct quantities and ratios needed by dogs and cats, and the minimum required level of some nutrients for some species may be toxic to others.

Proteins. Huge amounts of concentrated proteins, available as oilseed plant by-products from the brewing, distilling, starch, and oil industries, when properly supplemented provide excellent sources of amino acids for pets. The world production of oilseed meals is estimated to be 109×10^6 t. Horses, sheep, cattle, swine, and poultry also use oilseeds efficiently and provide intense competition for the use of these plant proteins. Plant proteins are heated during processing to inactivate enzymes that could otherwise be detrimental.

Soybean products that have been processed to remove a portion or all of the carbohydrates and minerals are used to make textured vegetable proteins which can be formed into various shapes and textures (see SOYBEANS AND OTHER OILSEEDS). Many canned dog foods utilize the textured vegetable protein chunks with added juices, flavor enhancers, vitamins, and minerals to produce canned dog foods that have the appearance of meat chunks. Similarly, those proteins can be combined with uncolored ingredients to imitate marbling and form pet foods with chunk-meat appearance. This processing is commonly used in semimoist pet foods.

Plant proteins from single sources, such as soybean meal, may be abundant in specific amino acids that are deficient in some cereal grains. Thus a combination of soybean meal and corn with their amino acid symbiosis may provide an excellent amino acid profile for dogs. Plant protein mixtures alone do not meet the amino acid needs for cats, because taurine is not generally present in plant proteins.

In the United States, more than 16.3×10^6 kg of human-inedible raw materials are available each year, and the rendering industry is a valuable asset in diverting these into valuable ingredients for use primarily in animal foods. For example, fish meal production worldwide in 1986 was estimated at 6.23×10^6 t, which with the 125×10^6 t of meat and bone meal plus 6.67×10^6 t of feather meal and poultry by product meal is the primary source of animal proteins used by the pet food industry.

Milk and egg products are highly desired in pet foods since they supply the highest quality amino acid profiles, with nearly 100% digestibility.

Meat derived from crippled, old, discarded, injured animals, and those that have recently died (designated as 4-D beef), as well as USDA rejected meats, are used in canned pet foods. Fresh meats of human-grade are produced in excess of human needs and are used including wing-tips, gizzards, livers, necks, backs, and meat still attached to bones.

Fats and Oils. Fats and oils from rendering animal and fish of fal and vegetable oilseeds provide nutritional by-products used as a source of energy, unsaturated fatty acids, and palatability enhancement. Fats influence the texture in finished pet foods. The use and price of the various melting point fats is determined by the type and appearance of the desired finished food appearance.

Large quantities of fat are used from the fast food industry; these fats may have dissolved plastics from restaurant wrappers which can restrict spray nozzle orifices as the fats cool during spraying on pet foods (see FATS AND FATTY OILS).

Carbohydrates and Plant Products. The world supply of excess grains provides desirable sources of carbohydrates (qv) and fibers (qv) for animals, including pets. Most grains are relatively low in proteins and, unless processed for starch or alcohol, are generally ground whole and used in animal feeds. Thus the contribution of the accompanying protein, vitamins, minerals, and fibers can be accounted for advantageously during pet food formulation. Corn, wheat, and rice are the most desirable common grains and are used extensively in pet foods.

Fibers and Fiber Sources. Fibers are present in varying amounts in food ingredients and are also added separately (see DIETARY FIBER). Some fibers, including beet pulp, apple pomace, citrus pulp, wheat bran, corn bran, and celluloses are added to improve droppings (feces) form by providing a matrix that absorbs water. Some calorie-controlled foods include fibers, such as peanut hulls, to provide gastrointestinal bulk and reduce food intake.

Nonnutrient Additives

Nonnutritional dietary additives provide antioxidants to preserve freshness, flavor enhancers to stimulate food selection, color to meet the owner's expectations, pellet binders to minimize fine particles, mycostats to minimize mold growth, and ingredient-flow enhancers.

Cat-Specific Additives

Cats are more sensitive to some nutritional deviations than are dogs. A dietary deficiency of arginine is more severe in cats than in dogs. This difference is associated with the higher dietary protein in cat foods.

Taurine. Taurine is a sulfonic amino acid derived from methionine and cystine and functions in many biological systems. Although taurine is plentiful in most mammalian tissues as a free acid, the cat's synthesis of taurine is insufficient to meet its biological needs.

Because heat processing during canning inactivates considerably more taurine or forms an inhibition against taurine uptake by the feline, less taurine is required in dry foods than in canned foods. The Feline Nutrition Expert Subcommittee of the Association of American Feed Control Officials (AAFCO), in the nutrient profiles for complete and balanced cat foods, suggests 0.1% taurine in extruded food and 0.2% in canned foods as a result of the extra loss of taurine during canning processing.

Feather meal, first hydrolyzed and then oxidized, produces cysteic acid an excellent precursor for taurine in cats. Hydrolyzed feather meal may supplement the taurine provided by other dietary animal proteins and help replace part or all of the synthetic taurine in cat food formulations with considerable cost savings.

Phosphoric Acid. To provide safe levels of dietary magnesium and also prevent feline urinary syndrome (FUS), ingredients such as phosphoric acid, which acidulates the urine, are added at carefully controlled levels to produce an acidic urine of approximately pH 6.5.

Other Additives. Cats cannot convert tryptophan to niacin, or protene to vitamin A in sufficient amounts to meet their needs. These deviations, as compared with other animals, need not produce problems because added dietary sources of niacin and vitamin A provide the needs of cats.

AAFCO Nutrient Profiles

Pet food products provide package claims of complete and balanced for specific physiological states" to provide the pet owner with confidence and to assure that pets receive nutritionally desirable foods. Before the promulgation and acceptance of the Association of American Feed Control Official (AAFCO) Nutrient Profiles, a number of references were used for complete and balanced recommendations. The Canine Nutrition Expert (CNE) subcommittee was formed to establish new profiles for complete and balanced dog foods. The AAFCO–CNE nutrient profiles are considered the AAFCO-recognized authority on canine nutrition. The Feline Nutrition Expert (FNE) subcommittee was appointed following the development of AAFCO–CNE recommendations to compile profiles for complete and balanced cat foods. These AAFCO–FNE nutrient protocols for cats were published, and include protocols for adequate testing of cat food products, which are monitored by AAFCO.

ECONOMIC ASPECTS

The annual production of pet foods is approximately 6.35×10^6 t, valued at \$8.6 billion. It has been estimated that there are as many as 15,000 different brand labels and package sizes of pet foods, marketed by 3,000 manufacturers. Conservative estimates are closer to 5,000 brands and sizes, with 1,800–1,900 registered in the state of Texans.

National Research Council, *Nutrient Requirements of Cats*, National Academy Press, Washington, D.C., 1986.

National Research Council, *Nutrient Requirements of Dogs*, National Academy Press, Washington, D.C., 1974 and 1985.

Official Publication of Association of American Feed Control Officials, Association of American Feed Control Officials, Georgia Department of Agriculture, Capitol Square, Atlanta, Ga.

JAMES CORBIN
University of Illinois
at Urbana-Champaign

FEEDS AND FEED ADDITIVES, RUMINANT

Ruminants that consume plant materials grown on land unsuitable for crop farming need not compete with humans and nonruminant livestock for feed resources. At least one-third of the world's land area is more suitable for grazing than for cultivation. The typical high-fiber forage produced on such land is practically indigestible by nonruminants and its thus best utilized by ruminant animals. Ruminant animals, whose ruminal microflora ferment and digest cellulose, the predominant component of fiber and the most abundantly produced carbohydrate, utilize much of the plant energy produced on this land. Anatomical differences between monogastric and ruminant animals allow the ruminant to be more efficient in digesting cellulose but generally less efficient in gaining weight and in converting feed to

gain because of energetic losses resulting from the fermentation process.

FEEDS

Forages/Roughages

Approximately 75–80% of the feed fed to ruminants during their lifetime production cycle is forage/roughage material. Roughages are made up predominantly of the stem or stalk portion of plants and usually include the seeds and leaves of such plants. Such feeds, typically, are higher in crude fiber, >50% neutral detergent fiber; low in starch, <4%; and moderately low in crude protein, <20%. Roughages not only are a source of nutrients to the ruminant but also help maintain normal rumen function. Roughages play a role in such practices as weaning the young ruminants from milk on to solid feed; preventing metabolic diseases; ie, bovine ketosis and ovine pregnancy toxemia; and, maintaining proper fat level (~3.5% in milk produced by Holstein dairy cows). In the young ruminant weaning process, forage in the diet helps establish a normal gastrointestinal tract microbial population. Ruminants consume forages either by grazing or by being fed harvested material.

The moisture content at which a plant is harvested usually determines its storage method. Low (15–25%) moisture forages often are stored in some type of baled (and tied) form and no attempt is made to exclude oxygen from it. Various types of oxygen limitation are utilized for storing higher (40–75%) moisture forages; ie, stave silos, oxygen-limiting silos, concrete bunker silos, in-ground pit silos, and large plastic bags.

Several different sources of low-moisture forages; eg, prairie hay, alfalfa, bromegrass, orchard grass, and blends of hay, are grown specifically for the purpose of harvesting. Crop residues such as corn stalks, soybean stubble, or small grain straws are also available.

High-Energy Feeds

Concentrated sources of energy are fed to ruminants to allow young animals to grow more rapidly and efficiently. Feedstuffs of this nature generally are high in readily fermentable carbohydrates; ie, they are high starch-containing feedstuffs. Feedstuffs containing high amounts of starch often are from the seeds of plants such as corn, grain sorghum, oats, and barley.

By-products of agricultural commodities are used as readily-fermentable energy-containing feedstuffs. Molasses, a by-product of the sugar-refining industry, is an excellent source of carbohydrates for ruminant feeding. The addition of molasses to rations increases feed acceptability, reduces dustiness of the mixture, and improves feed pelleting. Molasses from citrus and wood processing are also utilized as ruminant feeds. Other useful energy-containing by-products include wheat bran, wheat middlings and shorts, dried citrus pulp, dried beet pulp, dried bakery waste, hominy, potato meal, whey, corn gluten feed, and rice bran. Since tremendous amounts of agricultural by-products and residues exist, a great quantity of ruminant feedstuffs is available, which is not utilizable for human food or monogastric animal feeding. Such by-products often are used as a feed nutrient source for ruminant feeding because of availability and low cost. Some of these may present problems such as being lower in energy content or possibly less acceptable by animals than is true for corn.

Feed processing methods influence the availability of energy from feedstuffs, probably by influencing the sites of digestion and absorption in the ruminant animal. Fermentation products produced in the rumen and absorbed through the ruminal wall do not contain as much energy as carbohydrates absorbed through the small intestine. Methods of processing include grinding, rolling, cracking, extruding, steam flaking, roasting, heating, wetting, and gelatinization.

Various sources of lipids have been incorporated into ruminant rations to increase energy density. This practice is followed in high-producing dairy cattle. Addition of fat will reduce the dustiness of rations, increase ease of pelleting, and increase feed acceptability. The predominant source of lipid utilized in ruminant rations is of animal origin. Animal fat typically is higher in saturation and often is referred to as grease. Various sources of vegetable lipids are also available.

Lipids present in the ration may become rancid fairly quickly. When included at levels >4–6%, lipids may decrease acceptability, increase handling problems, result in poorer pelleting quality, cause diarrhea, reduce feed intake, and decrease fiber digestibility in the rumen. To alleviate the fiber digestibility problem, calcium soaps or prilled fatty acids have been developed to escape ruminal fermentation. Such fatty acids then are available for absorption from the small intestine.

SUPPLEMENTS

Protein

Although most feedstuffs contain protein, supplemental protein or nitrogen often is required to meet animal physiological requirements. Practical situations in which supplemental protein is required include the feeding of growing/immature animals, lactating females, females in the last trimester of pregnancy, and those grazing non-leguminous forage, such as on range land pasture. Soybean meal is the protein supplement utilized most frequently in the United States. Oilseed meals used in lesser quantities include cottonseed meal (contains gossypol that restricts its use for poultry and swine), canola meal (derived from rapeseed), and linseed meal (derived from flax seed). Additional meals derived from the oil-extraction process include peanut, sunflower, safflower, sesame, coconut, and palm kernel.

Raw soybeans may be used as a source of supplemental protein, but because of urease content it cannot be used in rations containing urea. Also, raw soybeans contain a trypsin-inhibitor that prevents their use in poultry and swine diets. Dry beans; ie, beans normally harvested in the green, immature state, such as fava, lupins, field peas, lentils, and other grain legumes, contain sources of supplemental protein.

Various milling, distilling, and brewing by-products are available as supplemental protein sources. Legume forages, such as alfalfa or clover, are considered high quality, available sources of protein for ruminant animals. The most widely used animal proteins include hydrolyzed poultry feather meal, blood meal, fish meal, and meat-and-bone scraps from nonruminant animals. Meat-and-bone-scraps from ruminant animals may not be fed to ruminants because of possible transmission of BSE, or "Mad Cow Disease."

Supplemental energy may be needed when the majority of an animal's diet is from bulky feedstuffs such as poorer quality roughages. In the case of cows grazing rangeland, protein intake may not be adequate; furthermore, the ruminal microflora may lack a fermentable source of carbohydrates. In such case, a highly fermentable starch source such as corn or sorghum grain may need to be supplied.

Minerals

The most universally deficient mineral element(s) in ruminant diets is salt (sodium chloride), because very little is found in such diets. However, salt is so economical and can be supplied so readily that it should not be overlooked. Calcium and phosphorus usually are border-line deficient under most ruminant feeding conditions. A condition known as grass tetany is associated with a magnesium deficiency often occurring in cattle on pasture during cooler seasons of the year. It can be supplied from either magnesium oxide or magnesium sulfate. Potassium usually is not deficient. Sulfur seldom is deficient in the diet of ruminants. Cobalt, copper, molybdenum, iodine, iron, manganese, nickel, selenium, and zinc sometimes need to be provided to ruminant animals.

Vitamins

The B-vitamins and vitamin K, $C_{31}H_{46}O_2$, are synthesized by the ruminal microorganisms and their supplementation usually is not necessary. However, there are times when B-vitamin supplementation may be essential. Polioencephalomalacia, sometimes called circling disease in cattle, is a nervous disorder alleviated by intravenous injection of thiamine hydrochloride, $C_{12}H_{18}Cl_2N_4OS$, vitamin B_1. Niacin supplementation has been shown to alleviate subclinical ketosis, partially increase milk production, and increase average daily weight gain under some conditions. Vitamins A, D, and E are required by ruminants and therefore supplementation is sometimes needed.

PERFORMANCE MODIFIERS

Several feed additives and implants are available for use with ruminants. Use of hormones results in significant improvements in rate of gain, feed intake, and efficiency of feed conversion.

Ionophores are additives that alter rumen fermentation and change the relative proportions of fermentative products produced by the bacteria; ie, acetate production decreases and propionate production increases. Ionophores accomplish this by altering the proportions of various ruminal bacteria present. The FDA has approved two ionophores for use for non-lactating ruminants: monensin, $C_{36}H_{62}O_{11}$, and lasalocid, $C_{34}H_{54}O_8$. Use of effective ionophores usually results in improved efficiency of feed utilization. Other potential benefits include decreased incidence of lactic acidosis, control of coccidiosis, control of feedlot bloat, and reduction in the number of face fly and horn fly larvae in feces. Direct-fed microbials are feed additives composed of microbes and/or ingredients to stimulate microbial growth, which allegedly results in more favorable microbial population.

A problem common to animals consuming a high energy diet or lush, immature legume vegetation is increased susceptibility to bloat, a condition where gas either is formed too rapidly or else the animal is not able to release gas sufficiently rapidly. Antifoaming agents available to prevent this condition include silicones, detergents, vegetable oils, animal fats, animal mucins, and liquid paraffins.

Buffers are used to stabilize ruminal pH at 6.0–6.8. Available buffers include sodium bicarbonate, calcium carbonate (limestone), and bentonite.

Many ruminal bacteria require one or more branched-chain volatile fatty acids (VFA) for proper growth. Supplementation with these VFAs has been researched, but the results have not been clear-cut.

Bovine somatotropin (BST) is a naturally occurring protein hormone produced by the pituitary gland of cattle and is a major regulator of growth and milk production. It is produced in commercial quantity using recombinant DNA technology. Increases in milk production from the use of varying levels of BST (5–50 mg/cow/day) range from 3–6 kg/cow/day. Persistency of lactation is improved. Use of supplemental BST has increased milk production in all breeds of cattle.

The rumen is not functional at birth and ingested milk is shunted to the abomasum. Within one to two weeks after birth the neonate will start to consume solid feed if offered. A calf or lamb that is nursing tends to nibble its mother's feed. An alternative method of raising the neonate is to remove it from its dam at a very early age, <1 week. Such a neonate must be supplied with complete supplementation provided by a milk replacer. Sources of milk replacer protein traditionally have been skim milk but may also include soybean protein, fish protein concentrates, field proteins, pea protein concentrates, and yeast proteins.

Approximately eight weeks after birth, the ruminant has developed a fully functional rumen, capable of extensive fermentation of feed nutrients.

Several sources of energy feeds can be utilized by the neonate. Lipids generally are ~90% digestible, and lipid sources include milk fat, tallow, and corn oil. Carbohydrates are another source of available energy. Lactose, glucose, and galactose are utilized efficiently, whereas starch, maltose, sucrose, and fructose are not. Hydrolyzed starch has been used successfully to replace a portion of the energy in diets fed to young ruminants. Protein sources given young ruminants just beginning to consume solid food should contain high quality (good balance of amino acids)

from plant sources, or else a combination of such with milk by-products.

Creep feeding often is used in the production of beef cattle. This practice involves offering feedstuffs to the young that are not accessible to the dam. Since the dam's milk is rich in protein and minerals, the greatest supplemental need is for energy. Young calves relish such high-energy grains as shelled corn and other similar grains. If it is possible, allowing young calves access to pasture where the dams are not admitted is another form of creep feeding.

P. R. Cheeke, *Applied Animal Nutrition: Feeds and Feeding*, MacMillan Publishing Co., New York, 1991.

D. C. Church, ed., *The Ruminant Animal*, Prentice-Hall, Inc., Englewood Cliffs, N. J., 1988.

M. E. Heath and C. J. Kaiser, in M. E. Heath, R. F. Barnes, and D. S. Metcalfe, eds., *Forages: The Science of Grassland Agriculture*, 4th ed., The Iowa State University Press, Ames, Towa, 1985, p. 3.

T. W. Perry, A. E. Cullison, and R. S. Lowrey, *Feeds and Feeding*, 6th ed., Prentice-Hall, Upper Saddle River, N. J., 2003.

TILDEN WAYNE Perry
Purdue University

FERMENTATION

Fermentation is most commonly defined as the anaerobic evolution of carbon dioxide from microorganisms (such as yeast or bacteria) growing on energy-rich nutrients (such as sugar, malted grain, or fruit) to produce ethanol and/or organic acids (such as beer, wine, or vinegar). The term is derived from the Latin verb *fervere*, which means "to boil" and describes the bubbling gas emissions commonly observed during the process. More generally, fermentation is defined as the overall activity of microorganism cultivation in a container. Although often used interchangeably, the word "fermentor" refers to the vessel and the word "fermenter" refers to the organism itself.

Anaerobic processes now are in the minority with aerobic processes much more commonplace. Although a large number of processes have been researched and developed to various extents, fewer have been commercialized. Culture types utilized in fermentation include single-cell bacteria, filamentous bacteria (actinomycetes), yeast, fungi, as well as mammalian, insect, and plant cells. The key components of a fermentation process include the (1) development of medium and conditions best suited to organism growth and production; (2) preparation of sterilized medium, fermentors, and related equipment; (3) propagation of active culture as seed (inoculum) for the production fermentor; (4) production fermentor growth and product accumulation; (5) product separation and isolation; and (6) treatment and waste disposal. The term "biotechnology" was introduced originally around World War I (WWI) to describe the transformation of raw materials to fermentation products by living organisms. Subsequently, it has been more closely associated with genetic engineering products of the late twentieth century and the process of redesigning what Nature has provided to improve its benefits. The term "biochemical processing" or "bioprocessing" broadly includes fermentation (both aerobic and anaerobic) and animal cell cultivation as well as whole cell and enzymatic biotransformations.

Teams of microbiologists, biochemists, chemical engineers work on the development of a fermentation process. Best results are obtained using cross-disciplinary and integrated approaches to problem solving. In addition to the microbial reaction of interest, the nature of the microenvironment around the cell, and the microbial reaction to this environment must be considered. The total process impact of any proposed solution should be considered before changes are implemented.

Many fermentation processes are conducted with cells circulating in liquid medium (submerged culture); several processes utilize cells immobilized within or upon either an inert support (immobilized culture) or a nutritive substrate (surface or solid state culture).

TYPES OF PRODUCTS

Primary Metabolites: Organic Acids, Solvents, Amino Acids, Polyols

Primary metabolites are intermediates of pathways directly involved in growth processes. They are small molecules, typically under 1500 Da in molecular weight. Beer, wine, and other distilled beverages are produced through finely tuned fermentation processes employing highly developed strains. Nonalcoholic beer can be manufactured via arrested fermentation, in which the fermentation is halted before significant alcohol is produced. Examples of higher volume organic acids and salts include citric acid (a food and beverage acidulant), acetic acid (vinegar), lactic acid (for food preservation such as pickling), propionic acid (food preservation), butyric acid (a dietary supplement), succinic acid (a flavoring agent; alternative fermentation process to petrochemical route developed using agricultural waste carbon sources and consuming or fixing carbon dioxide), itaconic acid (an intermediate in polymeric resins and fibers), and sodium gluconate (an industrial cleaner). Key solvents examples include ethanol (fermented from cornstarch instead of petrochemically derived), 2,3-butanediol (a precursor to 1,3-butadiene for rubber manufacture), acetone, and isopropyl alcohol. Glycerol (used for sealing compounds, antifreeze, and personal care products) is produced by altering the yeast fermentation process to favor its accumulation. Major amino acids, commonly produced using *Corynebacterium glutamicum*, are L-glutamic acid (monosodium glutamate, MSG, for flavor enhancement), L-lysine (key supplement in feed grains for livestock and poultry), L-phenylalanine (one component of the artificial dipeptide sweetener, aspartame), D,L- methionine, and L-aspartic acid. D-Sorbitol (a low calorie food sweetener and precursor for vitamin C synthesis) is the most common polyol produced via fermentation. Additional pri-

mary metabolite classes include vitamins [vitamin B_{12} and riboflavin (vitamin B_2)] and nucleotides (5'-inosinic and 5'-guanylic acids) that are used to enhance meat flavor. Finally, water-soluble viscous polysaccharides (such as xanthan and gellan gums used for thickening of foods) and biodegradable polymers (such as polyhydroxyalkanoate and polylactic acid), are synthesized in fermentations possessing complex broth rheology.

Secondary Metabolites: Antibiotics and Other Natural Products

Secondary metabolites also are known as idiolites (peculiar metabolites). Most have molecular weights <1500 Da and have unusual extended ring structures. They are not involved in growth processes, are strain specific, and are produced as a mixture of a chemical family with slight differences in side chains (structural analogs). Sometimes the natural compound has been modified via chemical synthesis or biotransformation (semisynthetic compound) to improve potency and selectivity. It often has been debated whether secondary metabolites have survival functions in nature or sometimes serve no role whatsoever.

Most commonly, the term secondary metabolites is meant to indicate antibiotics or antiinfectives. These compounds are produced by microorganisms that kill other microorganisms at low concentrations. About two-thirds of known antibiotics are made by actinomyces. Other types of nonantibiotic secondary metabolites include cholesterol-lowering drugs (such as lovastatin), immunosuppressants (such as cyclosporin), enzyme inhibitors (such as acarbose acid and clavulanic acid), herbicides (such as bialaphos), and anticancer–antitumor compounds (such as taxol and daunorubicin).

Biomass

The production of biomass has been undertaken to manufacture cell mass for the food industry. Food manufacturing applications include starter cultures such as Bakers' yeast for cooking and cheese cultures. Animal feed applications include Brewers' yeast, a by-product of the brewing industry, obtained via broth flocculation. In addition, lactic acid bacteria are used to inoculate silage, corn and hay crops harvested for animal feed and then fermented to lower pH, thus inhibiting the growth of microorganisms and preserving nutritional value. Cell mass from one fermentation even has been used as a nutrient source for another fermentation. Single-cell protein (SCP) has been developed as an animal feed and potential human food source.

In the field of bioremediation, cell mass is used to degrade toxic chemicals by bioconverting substrates into less harmful compounds (algae and bacteria for activated sludge), consumption of residual oil (*Pseudomonas putida*), destruction of microbial and synthetic polyesters by fungi, and extraction of metals from low grade ores (acid or base produced by *Thiobacillus* strains releases soluble metal). In addition, the biodegradation of dairy waste high in fat has been accomplished by the combination of lipases that digest the fat and bacteria which use the fatty acids liberated for growth.

Enzymes (and Whole-Cell Biocatalysts)

Bioconversions or biotransformations are fermentation processes performed by organisms in which the products and substrates are similar. They utilize enzymes (protein catalysts) either as whole cells, isolated enzymes, or enzyme preparations (crude, pure liquid, or solid). Enzymes used as whole cells must have a cell wall that is permeable to the substrate of interest. Each of these sources of enzymes can be used in either free or immobilized forms or in either aqueous or solvent solutions. The nature of the enzyme source depends on the application as well as influences process cost and product impurities from side reactions. Batch-to-batch and vendor-to-vendor specific activity (activity per unit weight of enzyme preparation) variations can be difficult to minimize. Enzyme selection is affected by process economics, specifically product recovery, enzyme recycle/reuse, coenzyme requirements, side reaction tolerance, and activity stability.

There are >2000 known microbial and mammalian enzymes, excluding catalytic antibodies (Abzymes), which have been created for specific substrate specificity. They catalyze a broad spectrum of reactions and current areas of application are varied. One-third of all enzymatic reactions require cofactors such as adenosine triphosphate (ATP) or reduced nicotinamide adenine dinucleotide phosphate (NAD(P)H), which then need to be regenerated.

The current areas of enzyme application are varied, with the major enzymes used industrially being proteases (added to detergents for protein stain removal) and amylases (for starch breakdown to dextrins, corn syrup extraction from wet grain, and bread shelf-life improvement). In addition, glucose is converted to high fructose corn syrup (HFCS) by immobilized glucose isomerase, juice and wine clarification is done by pectinase, progesterone is converted to cortisone by *Rhizopus nigricans*, and ethanol is converted to vinegar by *Acetobacter* and *Gluconobacter* cultures. Microbial rennins (also known as chymosin, a protease) have been developed for cheese manufacture to replace animal-sourced rennin. Finally, due to the European ban on animal proteins in animal feeds, demand for plant-protein hydrolyzing enzymes such as phytase has increased.

In the preparation of chiral synthons for specialty and fine chemicals, several examples of enzymes use are evident. Enzymes have been used in the preparation of protein pharmaceuticals such as the modification of porcine insulin to human insulin and in gene splicing for the construction of plasmids in biotechnology.

Biologicals

Biologicals are large molecular weight (often >5 kDa) therapeutic proteins or vaccines with complex structures that are produced by living systems but are not metabolites. Therapeutic proteins can be constitutive (naturally occurring in the host cell) or recombinant (not normally synthesized by the host cell). Most commonly, the recombinant protein production route in genetically engineered cells has been selected since it results in higher product concentrations and permits alternate forms of the natural protein to be expressed. Therapeutic proteins include growth hormone, human insulin, interferons, erythropoietin, factor VIII, colony stimulating factors, interleukins, glucocerebrosidase, and humanized–human monoclonal antibodies [eg, for rheumatoid arthritis and pediatric respiratory syncytial virus (RSV) infection]. Proteins also are produced for diagnostic assays.

Vaccines can be of three major types: live attenuated viral (measles, mumps, rubella, oral polio, chicken pox), inactivated attenuated viral (hepatitis A), cellular–subcellular component (polysaccharide conjugate vaccines), or recombinant epitope (hepatitis B). Viral vaccines are grown in mammalian cell culture using anchorage-dependent primary cell lines (cells that require surface attachment to grow and that eventually reach senescence after a set number of generations). Attenuated or weakened viruses (viruses cultured for generations to select for strains weakened in their ability to cause disease) are used to inoculate these cultures. Cells then are lysed, viruses or antigens purified to varying degrees, and finally inactivated as necessary. In contrast, only the genetic information necessary to create the viral epitope (the subunit of the virus which causes an immune response) is expressed in a recombinant epitope vaccine.

ADVANTAGES OVER ORGANIC SYNTHESIS

Fermentation-derived compounds can have greater specificity for the desired target since only the desired enantiomer is produced that avoids yield losses incurred by discarding the wrong enantiomer. A single step can replace several steps of a chemical synthesis, reducing the number of overall reaction steps. Biochemical conversion is able to be achieved at positions difficult to alter chemically due to lack of activation. In addition, natural carbon sources, such as corn starch, already have an oxygen on each carbon atom and thus avoid having to oxidize petrochemical feedstocks. Mutation and genetic engineering techniques can enhance the versatility of biochemical routes, increasing the variety of reactions performed.

Adverse and extreme chemical synthesis conditions are reduced substantially since biochemical reactions are performed at moderate conditions of pH, pressure, and temperature. Biochemical routes utilize cheaper and more environmentally friendly raw materials with the toxic and hazardous reagents often necessary in chemical synthesis less common.

For fine chemical synthesis, the main drawback of biochemical processes when directly compared to synthetic processes is their dilute nature, with typical product concentrations less than the peak values of 10–15 wt% observed for high volume primary metabolite organic acid and alcohol fermentations. A second drawback can be the efficiency and speed of the discovery of the biochemical route itself. Although in general there is a high ratio of biocatalyst surface area to liquid volume which facilitates rapid uptake of nutrients in biochemical systems, limitations in liquid mass transfer, mixing and possibly heat transfer at lower cultivation temperatures can challenge scale-up performance.

ECONOMIC ASPECTS

Fermentation products span the gamut from high value, low volume products to low value, high volume products. Products that have a high market value are normally required in smaller quantities. Costs generally are inversely proportional to annual tonnage demand since higher demands force greater production efficiency for both the strain (maximal specific productivity) and the process (maximal volumetric productivity).

To maintain filled pipelines, larger fermentation companies have expanded from small molecules and in some cases vaccines into therapeutic proteins (biotechnology products) either by acquisition/alliances or by developing in-house discovery and production capabilities.

MICROBIOLOGICAL AND FERMENTATION ASPECTS

The overall goal is to overproduce the product of interest and minimize impurities due to differing minor by-products. In the case of secondary metabolites, the challenge can be reducing structurally similar analogs with small differences in side groups such as a hydroxyl group instead of a methyl group at a specific site.

Strain

Prompt identification of the production strain is key. Extensive search efforts for new compounds from biological sources are executed with a variety of therapeutic targets. It is critical to design the screening assay carefully since hits can be low—on the order of 0.01%. High throughput screening in miniature microtiter plates using automated robotic systems efficiently identifies cultures producing desired compounds either as secondary metabolites, enzymatic transformations or recombinant protein clones. To minimize false negative assays, it is often necessary to ensure that any synthetic enzymes needed are induced by adding appropriate precursors, there is no catabolite repression, and the product is not degraded further after formation.

Microbial isolates from nature often are a heterogeneous population of clones with abilities to produce notably different titers of secondary metabolites. The highest producing colony must be isolated to assure consistent cultivation and to use for further strain development work. Mutation and strain improvement efforts are undertaken with multiple goals. It is important to select and prioritize the objective carefully since strain development is costly and time consuming despite the fact that microtiter plates have sped up the process considerably.

A mutation permanently alters one or more nucleotides at a specific DNA site either by substitution, deletion or rearrangement. Mutation is conducted by exposing cells to chemical mutagens or ultraviolet (uv) radiation (wavelengths ~260 nm).

Mutations for secondary metabolites cultures are performed to overproduce the product of interest often by severalfold over wildtype levels. Alternatively, the goal can be to produce the product more efficiently by reducing carbon utilization, lowering analog levels, blocking a metabolic pathway leading to impurities or product degradation, or creating analogs (new metabolites) that might have enhanced therapeutic value. In some cases, mutations are conducted to elucidate the secondary metabolite synthetic pathway. Mutations for primary metabolites are conducted to alter feedback resistance (such as regulation, inhibition and/or repression), which increases product concentration. Mutation also can be done to produce an auxotroph that requires specific nutrient(s) for cell growth or an idiotroph that requires specific nutrient(s) for secondary metabolite production.

Genetic engineering is being used in addition to mutation for secondary metabolites. Advances in the genetics of *Streptomyces* and other fungi are improving the toolbox contents available. Genetic engineering has been used to introduce a macrolactone, normally synthesized by an actinomycetes organism, into a *Streptomyces* host and then to alter the associated genes to produce novel analogs.

For recombinant proteins, the selection of the host cell fundamentally influences major aspects of the process and the nature of the product, and the efficiency of the expression system determines the size of the facility (capital investment and operating costs). Although early cell line engineering can minimize future expenditures, it also can delay development and clinical material manufacture. Typical host cells include single cell bacteria (*Escherichia coli, Bacillus*), filamentous bacteria (*Streptomyces*), yeast (*Saccharomyces cerevisiae*), fungal (*Aspergillus*), animal (insect—*Spodoptera*, mammalian—Chinese Hamster Ovary (CHO) cells, hybridomas) and plant—*Nicotiana tabacum* (tobacco) cell lines.

Medium Development and Feeding

By necessity strain development proceeds in parallel with medium development and often is the key to realizing the full potential of a new mutant. In the most significant well-published example, it was process, media, and strain development together that improved penicillin G titers to 15,000-fold greater than the wild-type original process. Economics and raw material consistency influence the selection of medium components.

Media can contain complex and/or, chemically defined ingredients. Complex components are "natural" ingredients, not well characterized with typical compositions available as rough percentages of carbohydrate, protein, and lipid. Their composition can vary according to the manufacturer and the growth–harvest conditions, especially for ingredients of agricultural origin. In some cases, complex ingredients are by-products from food processing. Chemically defined components are synthetic, well-characterized, pure chemicals. Although defined medium has been used successfully for a variety of processes, process sensitivity to any medium variations can be heightened. A hybrid approach is the use of semi-defined medium which has mostly defined ingredients with one (or possibly two) complex ingredients. Regardless of the types of ingredients selected (complex or defined), adequate raw material specifications are necessary and sometimes incorporate small scale use tests.

Kinetics of Growth/Product Formation

Growth and production kinetics vary considerably depending on the specific cultivation conditions and the organism being cultivated. Figure 1 illustrates an example of the kinetics of growth, substrate utilization, product and by-product appearance as well as other fermentation parameters.

Secondary metabolite cultivations can be divided into the tropophase or growth phase and idiophase or production phase with morphological differentiation often occurring during the idiophase. When these phases are distinct, the product is nongrowth associated and when, in some cases the tropophase and idiophase overlap, the product is growth associated or mixed. Secondary metabolites are produced at low growth rates and often need an induction phase resulting from a nutrient deficiency (such as carbon, nitrogen, phosphate, or sulfur) during which there are specific enzymatic changes in the cell. The culture itself is resistant to the substances it secrets primarily because antibiotic production is delayed until the rapid growth period finishes.

EQUIPMENT

The design of fermentation equipment must consider the culture type and its characteristics as well as the potential process requirements. It affects culture growth, morphology, and production. Although the requirements for inoculum or seed fermentors are typically simpler than those of production fermentors, often the need for interchangeable operation necessitates the design of the more complex capabilities. Fermentation equipment used for microbial cultivation has been distinguished from bioreactor equipment used for animal cell cultivation (or mixed use) by some practitioners, with the term bioreactor implying more broad based cell host application. A typical fermenter/bioreactor is shown in Figure 2.

Three fermentation scales generally are required: the bench or laboratory scale for process development; the pilot scale for clinical material manufacture, initial scale up, and subsequent process development; and the production scale for commercial manufacture.

INOCULUM DEVELOPMENT AND SCALE-UP

Inoculum Development

Inoculum development passes through several stages, starting from a frozen vial containing a few milliliters and moving to the production fermentation volume containing tens of thousands of liters. Actively growing

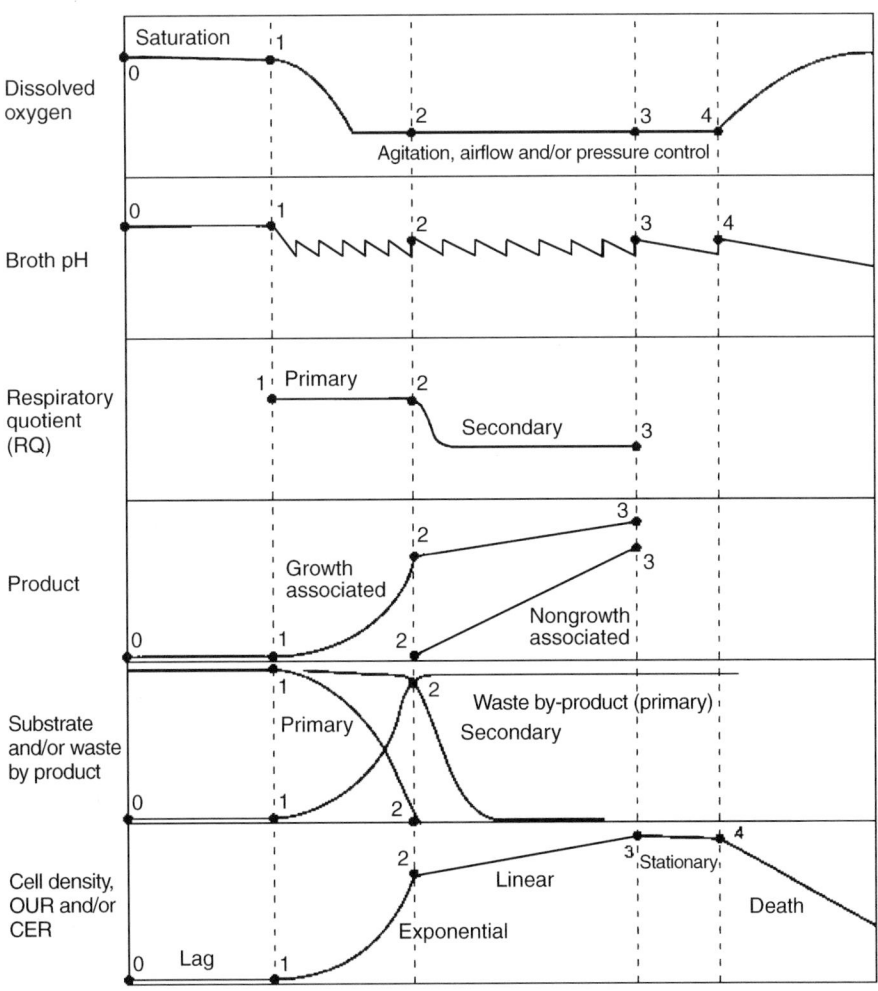

Figure 1. Typical fermentation kinetic profiles with emphasis on secondary metabolite production.

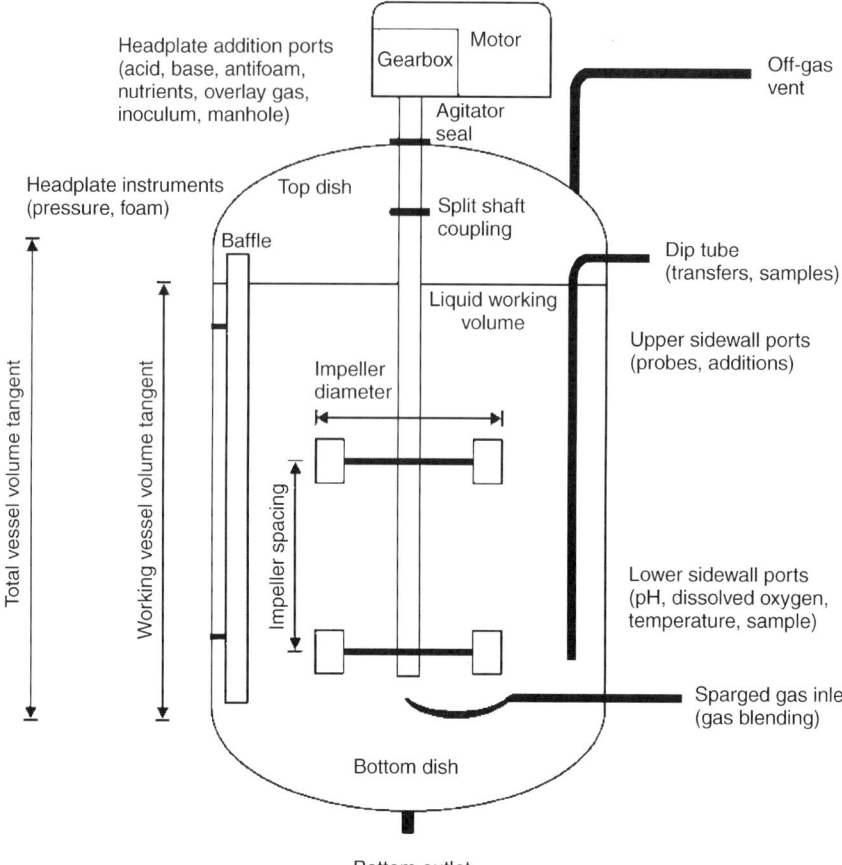

Headplate addition ports
(acid, base, antifoam,
nutrients, overlay gas,
inoculum, manhole)

Gearbox Motor

Agitator
seal

Off-gas
vent

Headplate instruments
(pressure, foam)

Top dish

Split shaft
coupling

Baffle

Dip tube
(transfers, samples)

Liquid working
volume

Upper sidewall ports
(probes, additions)

Impeller
diameter

Total vessel volume tangent

Working vessel volume tangent

Impeller spacing

Lower sidewall ports
(pH, dissolved oxygen,
temperature, sample)

Sparged gas inlet
(gas blending)

Bottom dish

Bottom outlet

Figure 2. Characteristics of a typical fermentation vessel.

cultures in mid-exponential growth are preferred for transfer to the next stage that serves to minimize the lag phase in the subsequent stage before growth is noticeable. Inoculum quality (eg, viability, morphology) can impact production performance although this impact can be difficult to quantify directly. Although the elapsed cultivation time is a common indicator of when the culture should be transferred to the next stage, it is not always a reliable one. Quantitative measures of transfer times can be based on cell mass, glucose depletion, oxygen uptake rate, or other on-line or off-line measurements.

Since the microorganism can be the most valuable part of the entire process, the culture storage method must assure longevity, specifically high cell viability.

Transfer times, cell concentrations, and inoculum volume percentages depend on the seed stage itself, the culture growth requirements and/or the facility staffing constraints. Generally, 1–4 days per stage is optimal with a minimum of 20 h desirable so that adequate time exists to obtain initial culture purity results prior to transfer. If a seed stage shorter than 20 h exists then it may be possible to alter the pre- and poststages to omit it from the process. Seed morphology must be consistent as it can influence production culture morphology and productivity.

Scale-Up

For scale-up to production fermentors, there are many process dependent factors to consider. Key variables

must be selected to be maintained constant upon scale-up since if all variables were held constant the production fermentor design would be overconstrained.

Oxygen transfer can change radically upon scale-up since surface aeration becomes less of an influence as the surface area at broth level to broth volume ratio decreases and mass transfer rates requirements increase. Superficial velocity (gas flowrate divided by fermenter cross-sectional area) increases upon scale-up for the same air flowrate to vessel volume ratio (VVM), which can increase foaming.

The design and operation of the agitator is critical to providing adequate mixing to disperse additions and minimize gradients (such as pH, dissolved oxygen, carbon dioxide, nutrient/substrate, and temperature). Mixing tends to become less uniform and takes longer with increases in scale.

ISOLATION

Product concentrations in the broth can be as low as 0.1 wt% for some processes (such as therapeutic proteins and vitamins), but can be as high as 5–15 wt% for organic acids and solvents (such as lactic acid and ethanol). Postfermentation the first isolation step is the cell harvest, a liquid–solid separation. Commonly, unit operations of centrifugation and filtration (with flocculation used to a lesser extent) are used to separate the cells from the broth. The nature of harvested fermentation broth (eg, level of antifoam added, residual quantities of medium components) influences the harvest conditions.

For primary or secondary metabolites, if the product is excreted, the broth can be extracted either before or after cell separation, the extract concentrated, then a crude crystallization performed. Since some hydrophobic secondary metabolites are attached to the outer cell wall, acid, or base adjustment may be used to release the product prior to extraction if necessary. For excreted protein products, cells are separated from the product-containing broth and discarded. Ultrafiltration can be used to concentrate the protein of interest and discard lower molecular weight contaminants.

For protein products that accumulate intracellularly, cells must be separated then lysed by a homogenizer (or by chemical means) in such a fashion as to release the product without becoming overly disintegrated. This step then is followed by centrifugation or microfiltration to remove cell debris. Ammonium sulfate or polyethylene glycol precipitation followed by low pressure column chromatography (such as ion exchange, affinity, and/or size exclusion) then might be used to eliminate contaminants. Any inclusion bodies present can be easily separated from lysed cells by ultracentrifugation, dissolved in a strong denaturing solution, and then renatured (refolded into their tertiary form) usually through dilution by adding high volumes of buffer.

UTILITIES

Product Contact

Product contact utilities are those that directly mix with the process fluid, environment, or equipment. Since they are utilized continuously, back up units usually are installed to provide redundancy for planned and unplanned maintenance outages.

Required water quality is influenced by both process and regulatory factors. Process (city) water often is used for medium make up for larger scale lower cost products if trace element composition is not a medium issue and/or has been quantified. Although least expensive, it can be subject to seasonal variations, the magnitude of which depend on the municipal water company practices and the water source (well, reservoir, surface run-off). Deionized or USP (*United States Pharmacopeia*) water is more expensive to use at the large scale but assures a consistently low amount of trace elements (so that known amounts can be added back to the medium). It can be useful for process development and for fermentation products requiring higher water quality. The most expensive but highest quality water is low bioburden and low endotoxin water-for-injection (WFI), which is made by condensation of clean steam. WFI is most suitable for animal cells which can be sensitive to endotoxin content but is used for many other microbially sourced biological products as well.

Compressed oil-free air is required for sparging and surface aeration at a supply pressure of at least 30 psig

or equivalent to/slightly higher than the product contact steam pressure. Enormous quantities of air are needed for commercial scale aeration creating capital and operating expenses due to the required compressor size, redundancy, and energy consumption.

HVAC-supplied air contacts equipment during open transfers (when the product is exposed directly to the environment) and surrounds equipment during closed transfers. In some facilities, 100% once through air (no air recirculation) is implemented with heat recovery either avoided or implemented using heat exchangers to avoid any direct contact of inlet and outlet steams. Separate air handling systems may be used for areas needing additional segregation such as inoculum preparation, live virus, or β-lactam antibiotic suites. Typically, 10–40 room air changes per hour are utilized depending on level of cleanliness anticipated to attain the appropriate air quality required for the processing step being executed in the area. Tight humidity limits may be implemented depending on the hygroscopic nature of raw materials/product and the level of cleanliness required.

Inlet and outlet terminal HEPA filtration, room air changes and laminar air flow all are used to attain viable and nonviable particulate reduction. It is usually desirable to minimize open transfers in the developed process and to ensure that the fermentor is a closed system with no leaks and sterilization of all in-process connections.

Room air supply rates and returns are balanced so that air pressurization forms containment barriers. Pressurization, typically ~0.03 in water or higher, is used to minimize cross-contamination among different process suites, between clean and less-clean areas, and between live culture/virus and inactivated culture/virus areas.

Nonproduct Contact

Although nonproduct contact utilities often are assumed to be less critical, their reliability and the presence of back up equipment can dramatically influence facility operation. The nonproduct contact utilities of cooling water, steam, and power are major cost factors for antibiotics and industrial enzymes.

For fermentation processes requiring a high efficiency of biological waste destruction, all product contact streams must be treated using a HTST biowaste sterilization system prior to disposal. Depending on the facility design and operation, the flows to the biowaste system may become limiting during certain peak periods.

REGULATORY ASPECTS

Product Quality

Product quality is determined by purity as well as by titer and product characteristics. In addition to the desired compound, similar molecules may be produced during the cultivation. Structural analogues, differing by as little as a single methyl or hydroxyl group, are found in secondary metabolite fermentations; heterogeneity in glycosylation (glycoforms) or minor differences in amino acid sequence can emerge during protein production. Glycosylation differences affect parameters such as protein solubility, stability, and biological clearance.

Culture purity, contamination or the presence of foreign cells is influenced mainly by equipment design (eg, size of pockets or deadlegs, crevices), standard operating procedures, training, preventative maintenance (eg, instrument calibration, gasket/o-ring replacement) and validation. Some fermentation processes are self-protected and enjoy repeatedly low levels of contamination.

Safety Considerations

Fermentation containment levels, established by the National Institute of Health for the United States, contemplate the potential health risk to workers based on prior experiences with the microorganism and any known pathology. Increasing levels of containment are devised to reduce the likelihood of the release of live organisms from aerosol generation in off-gas or during operations such as sampling and waste disposal.

Environmental

Fermentation medium components rarely are toxic to the environment and usually have few environmentally adverse consequences and processing conditions than synthetic alternatives. However, fermented broths can have appreciable aquatic toxicity, both to organisms in municipal sewage treatment facilities and to freshwater–saltwater species in local waterways.

Fermentor off-gas can have strong odors during media SIP, waste broth heat inactivation and during the cultivation itself. Process parameters as well as vent line "stack" locations can greatly impact the degree of objectionable odor. Alternatively, off-gas can be treated through scrubbing or other means.

Anonymous, *The Pharmaceutical Century: Ten Decades of Drug Discovery*, ACS Publications Supplement, American Chemical Society, 2000.

A. Demain and J. E. Davies, eds., *Manual of Industrial Microbiology and Biotechnology*, 2nd ed., American Society for Microbiology, Washington, D.C., 1999.

D. G. Springham, ed., *Biotechnology: The Science and the Business*, Harwood, Amsterdam, The Netherlands, 1999.

V. A. Vinci and S. R. Parekh, *Handbook of Industrial Cell Culture: Mammalian, Microbial and Plant Cells*, Humana Press, Totawa, N.J., 2000.

M. C. Flickinger and S. W. Drew, *Encyclopedia of Bioprocess Technology: Fermenation, Biocatalysis and Bioseparation*, Vols. 1–5, John Wiley & Sons, Inc., New York, 1999.

BETH JUNKER
Merck Research Laboratories

FERRITES

The term ferrite is commonly used generically to describe a class of magnetic oxide compounds that contain iron oxide as a principal component. In metallurgy (qv), however, the term ferrite is often used as a metallographic indication of the α-iron crystalline phase. Some representatives of the ferrite family have long been known as magnetic minerals.

Ferrites can be classified according to crystal structure, ie, cubic vs hexagonal, or magnetic behavior, ie, soft vs hard ferrites. A systematic classification as well as some applications are given in Table 1 and Figure 1.

COMMON PROPERTIES OF SPINEL FERRITES AND M-TYPE FERRITES

The commercial sintered spinel and M-type ferrites have a porosity of 2–15 vol% and a wide range in grain sizes (1–40 μm). In addition, these materials usually contain up to about 1 wt% of a second phase, eg, $CaO + SiO_2$ on grain boundaries, originating from impurities or sinter aids.

Ferrites are oxides and thus rather inert with respect to water, bases, and organic solvents. However, they may be attacked by acids having sufficiently high strength (pH <2).

Being ceramic materials, ferrites are also resistant to high temperatures, at least up to the sintering temperature (1200–1400°C). However, noticeable reduction may take place at temperatures >1100°C and an oxygen partial pressure.

Ceramic ferrites cannot explode or release poisonous gases, and generally do not contain toxic elements. However, permanent magnets based on Sr-ferrite contain strontium, Sr, which is in principle toxic. In dense (porosity <10%) materials the Sr is firmly bound; however, in porous (porosity >10%) materials the second phase may dissolve partially in water or acids giving rise to release of Sr. Even in the latter case the effect is limited.

CRYSTAL CHEMISTRY AND PHYSICAL PROPERTIES

The magnetic properties of ferrites result from the electronic configuration and mutual interactions of the ions present. Thus investigation of the crystal structure is fundamental to the understanding of these materials. Although the specific structures of spinel ferrites and M-type ferrites differ, both classes can be considered to be composed of two basic sublattices: An anionic lattice having relatively large anions in a closest packing and a cationic lattice containing the smaller cations, which fill interstitial sites.

Spinel Ferrites

In spinel ferrites having the composition AB_2O_4, where A and B are metals, cubic close-packed (CCP) oxygen ions leave two kinds of interstitial sites for the cations: tetrahedral or A-sites, surrounded by four oxygen ions; and octahedral or B-sites, surrounded by six oxygen ions. A wide variety of transition-metal cations can be fit into these interstitial sites. The most important family of spinels is $Me^{2+}Fe_2^{3+}O_4$, where Me = Mg, Mn, Fe, Co, Ni, Zn, Cu, etc, either singly or in combination. But similar ferrites having less than two Fe-ions per formula unit are

Table 1. Systematic Classification of Ferrites

Crystal chemistry[a]	Formula[b]	Magnetic nature			Appearance			Application
		Soft	Intermediate	Hard	Polycrystalline	Single crystalline	Bonded powder	
Cubic								
spinel	$MeFe_2O_4$	X			X	X		recording heads
		X			X			core material for various inductors, transformers, and TV deflection units
	$CoFe_2O_4$, γ-Fe_2O_3	X					X	recording tape
garnet	RFe_5O_{12}	X				X		microwave, magnetooptics, bubble-information storage
perovskite	$R'FeO_3$					X		electroceramic devices
orthoferrite[c]	$RFeO_3$	X				X		bubble-information storage
Hexagonal								
magnetoplumbite	$R'Fe_{12}O_{19}$			X[d]	X		X	permanent magnets
	$R'Fe_{12-x}Me'O_{19}$			X[e]			X	tape for perpendicular recording
W (= MS)	$R'Me_2Fe_{16}O_{27}$			X[d]	X		X	permanent magnets[f]
	$R'Co_2Fe_{16}O_{27}$		X[e]			X		microwaves, shielding[f]
X (= M$_2$S)	$R'MeFe_{28}O_{46}$			X[d]	X	.	X	permanent magnets[f]
	$R'CoFe_{28}O_{46}$		X[e]			X		microwaves, shielding[f]
Y (= ST)	$R'_2Me_2Fe_{12}O_{22}$		X[e]		X	X		microwaves, shielding
Z (= MST)	$R'_3Me_2Fe_{24}O_{41}$			X[d]				permanent magnets[f]
	$R'_3Co_2Fe_{24}O_{41}$		X[e]		X			microwaves, shielding

[a] $M = BaFe_{12}O_{19}$; $S = Me_2Fe_4O_8$; $T = Ba_4Fe_8O_7$. See Figure 1.
[b] $Me = Fe^{2+}$, Ni^{2+}, Mn^{2+}, Co^{2+}, Zn^{2+}, etc; $Me' = Mn^{3+}$, $(Ti^{4+} + \{Co^{2+}\})$, etc; $R = Y$, Nb, etc; $R' = Ba$, Sr, Pb, Ca.
[c] These materials are orthorhombic.
[d] Preferred axis (uniaxial anisotropy).
[e] Preferred plane (planar anisotropy).
[f] Potential application.

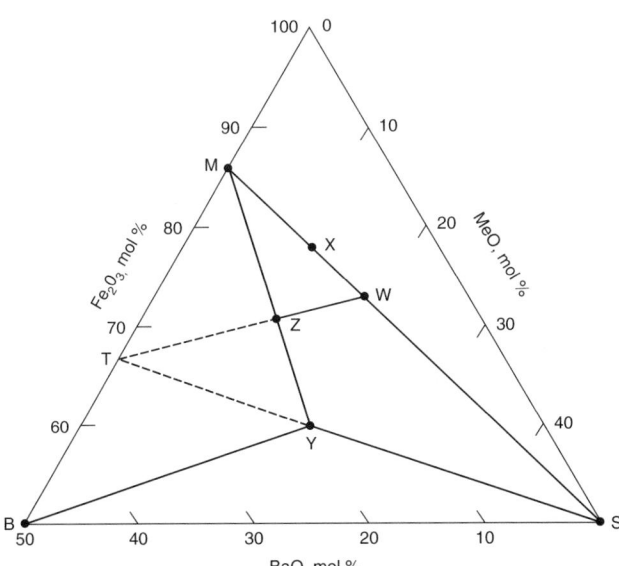

Figure 1. Composition diagram for hexagonal ferrites where $B = BaFe_2O_4$, $M = BaFe_{12}O_{19}$, $S = MeFe_2O_4$, $T = BaFe_4O_7$, $W = BaMe_2Fe_{16}O_{27}$, $X = BaMeFe_{14}O_{23}$, $Y = BaMeFe_6O_{11}$, $Z = Ba_3Me_2Fe_{24}O_{41}$, and Me is as defined in Table 1.

also of industrial significance because of the high electrical resistivity.

In the cases of existing unpaired d electrons, transition-metal ions possess a net magnetic moment. In a spinel, these magnetic moments interact through the anions (super exchange), resulting in a situation where the moments of both A- and B-site ions are aligned, ie, A—A and B—B parallel, but A—B antiparallel, a ferrimagnetic ordering. The net magnetic moment per unit formula can be calculated from the distribution of the cations over these sites.

It is possible to systematically alter the net magnetic moment of ferrites by chemical substitutions. A very important industrial application is the increase of the magnetic moment in mixed MnZn-ferrites and NiZn-ferrites.

The direction of the alignment of magnetic moments within a magnetic domain is related to the axes of the crystal lattice by crystalline electric fields and spin-orbit interaction of transition-metal d ions.

The net macroscopic B and the resulting μ_i result from two types of magnetization processes. First, there is a contribution from the rotation of the magnetization inside each individual magnetic domain, from the preferred direction toward the direction of the applied magnetic field until the sum of the magnetostatic energy (minimal if B lies along H) and the anisotropy energy has reached its minimum value. Second, domain walls move. Domains having favorable magnetization directions with respect to the applied magnetic field grow at the expense of others, thus further minimizing the total magnetostatic energy.

Stresses, which can, eg, be introduced during cooling after ceramic sintering, during machining of sintered products, or simply when product parts are clamped together before use, lead to anisotropy by the magnetostriction effect. Magnetocrystalline anisotropy, magnetostriction, and magnetic permeability depend markedly on chemical composition. These dependencies have been extensively investigated within the ternary diagrams $MnFe_2O_4$–$ZnZnFe_2O_4$–Fe_3O_4 and $NiFe_2O_4$–$ZnFe_2O_4$–Fe_3O_4.

Properties can also be manipulated by adding specific dopants: Co^{2+} ions are, eg, introduced for extra anisotropy compensation; and Ti^{+4} ions are substituted to form pairs with Fe^{2+} ions and thus to reduce electron hopping. Extensive investigations have been carried out involving the addition of dopants such as CaO (typically 0.1 mol%) and SiO_2 (0.01 mol%) in order to provide ceramic grains having electrically insulating grain boundaries, thus markedly increasing the effective resistivity of the ferrite product.

At high frequencies, ferrites exhibit energy losses resulting from various physical mechanisms at different frequencies and appearing as heat dissipation. Hysteresis losses arise from irreversible domain wall jumps. During each cycle of the H- and B-fields one hysteresis loop is completed and the loss per cycle is proportional to the area of the loop. A way to reduce hysteresis losses, ie, prevent domain wall jumps, is to reduce the number of inhomogeneities able to pin domain walls, eg, pores and impurities, and to reduce magnetocrystalline and stress anisotropy. Another method is to deliberately pin the walls, for instance by addition of Co^{2+} and Ti^{4+} ions or by using ceramic microstructures having small grains. A second important loss contribution comes from eddy currents, induced by alternating magnetic fluxes. A third main loss contribution is from magnetic resonances.

M-type Ferrites

The magnetism of $BaFe_{12}O_{19}$ comes from the Fe^{3+} ions, each carrying a magnetic moment of $5\mu_B$. These are aligned by either parallel or antiparallel ferromagnetic interaction. Ions of the same crystallographic position are aligned parallel, constituting a magnetic sublattice. The interaction between neighboring ions of different sublattices is a result of superexchange by oxygen. It is the magnetic structure in terms of sublattices and their mutual orientation that governs magnetic behavior, which in turn is described in terms of intrinsic and material properties.

The intrinsic magnetic properties may be subdivided into primary and secondary. The primary properties are directly related to the magnetic structure; the secondary properties, derived from the primary ones, govern the actual magnetic behavior.

Substitution may affect the intrinsic properties. Ba can be fully replaced by Sr or Pb and partly by Ca (<40 mol%).

CaM, stabilized with 0.03 mol % La_2O_3, is also possible. The intrinsic properties of these M-ferrites vary somewhat and other factors such as sintering behavior and price of raw materials often dictate the commercial viability. Large-scale production is concentrated on BaM and SrM. High performance magnets are based on SrM, and cheap magnets are usually based on BaM.

Substitution for Fe^{3+} has a drastic effect on intrinsic magnetic properties. Partial substitution by Al^{3+} or Cr^{3+} decreases J_s without affecting K_1 seriously, resulting in larger H_A and H_c values. Substitution by Ti^{4+} and Co^{2+} causes a considerable decrease in K_1; the uniaxial anisotropy ($K_1 > 0$) may even change into planar anisotropy ($K_1 < 0$). Intermediate magnetic structures are also possible.

PROCESSING

Commercial ferrites are produced by a ceramic process involving powder preparation, shaping, firing, and finishing (see CERAMICS). The powder preparation is usually the classical one involving the mixing of powder raw materials, prefiring (or calcination), milling, and granulating. The raw materials are oxides or carbonates, the main component always iron oxide. The morphology and purity of the raw materials are an important factors with respect to the processing and final quality. Mixing can be done in different ways, depending on the nature and quality of raw materials and of the final product. During prefiring the different compounds react in the solid state to form the final compound and intermediate compounds, loosing the volatile substances such as CO_2. Mostly this process is accompanied by homogenization on a local scale. In addition, there is some densification. To limit the effect of densification and to facilitate the handling of the material, it is often granulated before prefiring. To enable shaping and sintering, the prefired material has to be milled down to micrometer-sized particles. The last milling step is generally wet milling to prevent agglomeration effects. During or after milling, binders and lubricants are usually added to facilitate granulating and pressing.

In some cases, it may be advantageous to deviate from the classical technology. For example, in wet-chemical preparation better chemical and morphological control may be achieved by starting from salt solutions.

Shaping is often done by dry pressing, which in fact is a simple and effective method to make the variety of shapes needed for electronic applications.

During firing, formation of the proper compound is completed and densification occurs from ~50 to ~90% solid by volume, implying a linear shrinkage of 10–25%.

Spinel Ferrites

Prefiring is usually carried out in an air atmosphere in a continuous rotary kiln. In such a kiln the material is transported through a heat zone typically of 900–1100°C, in a rotating tube inclined at a small angle, which transports the powder downward along its length by a tumbling action. The angle is predesigned for a proper heating time and an economical throughput. When the mixture of raw materials enters the heat zone, carbonates and higher oxides decompose and a sequence of solid-state reactions occurs, starting with the formation of Zn-ferrite and ending with the partial formation of the desired MnZn-ferrite or NiZn-ferrite. Usually, the aim is not a 100% spinel structure after prefiring. A 50–80% one usually suffices because the remaining conversions take place during the final sintering step after the forming step. Too high prefiring temperatures would result in considerable shrinkage in this stage, which makes the ferrite hard and thus difficult to mill. The prefired powder is characterized by X-ray diffraction, by the BET specific surface or the Fisher number, and sometimes by the inductance of a coil wound on a toroid pressed from the prefired powder. The prefired and subsequently milled powder has to be such that it results in a predictable and very constant shrinkage of pressed products during final sintering, in order to satisfy tight demands normally imposed on final product dimensions or to be able to realize these dimensions by grinding.

M-Type Ferrites

There are a variety of processing routes for the four main classes of M-ferrite magnets: (1) sintered, (2) plastic bonded, (3) anisotropic, and (4) isotropic. As raw materials, dry powders of strontium carbonate, $SrCO_3$, Fe_2O_3, and additives such as silica, SiO_2, and boric trioxide, B_2O_3, are employed.

The most important microstructural demands are high aligning degree, high density, and small grain size with minimum platelet shape. These impose demands on the sintering process and the preceding operations. During sintering considerable densification must be realized without allowing significant grain growth. These more or less contradictory demands are fairly well realized by using sinter additives such as SiO_2 or B_2O_3 and SiO_2. SiO_2 has been investigated extensively. Important demands for the pressed products are high and homogeneous pressed density and high degree of aligning. The powder particles in the milling slurry have to be single-crystalline and free-movable in view of the aligning process, and sufficiently small in view of the sintering process. To enable a good milling performance, the grain size in the prefired granule must not be too large. In addition, the conversion to SrM must be sufficient, otherwise a high B_r is no longer attainable. Calcination with sufficient conversion and controlled grain size can be realized in two ways (1) nonstoichiometric (Sr-excess) basic composition with SiO_2 and B_2O_3 addition, and (2) stoichiometric basic composition. In the former case, there is some liquid second phase during calcination that promotes the conversion and, hence, compensates for poor mixing.

USES

The number of applications of spinel ferrites is very large and growing. In radio, television, and measuring equipment ferrite cores are extensively applied as inductors in LC-filters. In telecommunication ferrites serve the same purpose, but have clearly higher demands on quality factor Q and temperature-plus-time stability of the inductor.

As electronic equipment is increasingly used electromagnetic compatibility (EMC) has become an issue of fast-growing importance (see ELECTRONIC MATERIALS). Emission of unwanted signals has to be limited, as does the sensitivity of equipment to incoming interferences. These signals are subjected to international and national regulations. Ferrites are being increasingly used to limit unwanted signals and to make eloctronic equipment sensitive to incoming interference, because ferrites can supply electromagnetic interference (EMI) suppression as inductor cores in low pass LC-filters, as well as serve as selective impedance inductors in series with circuit load impedances, without the use of capacitors.

High permeability ferrites are used in closed cores, such as ring cores, or composed cores having carefully polished contact surfaces.

Recent digital transmission systems, denoted as DSL systems, make use of wide-band transformers with more severe requirements on low signal distortion. For this purpose new ferrites have been developed with an improved lineairy of the BH loop resulting in a lower total harmonic distortion.

Another area having a wide range of applications for modern ferrites is power conversion such as switched mode power supplies (SMPS) and LCD backlight inverters.

M-Type Ferrites

A large variety of M-ferrite magnets and applications is available. Bonding of the microscopic crystallites to solid bodies is performed either by sintering or by plastic bonding. The latter produces a plasto-ferrite. In both cases, the crystallites may be either randomly oriented (isotropic) or aligned with the c axis in one direction (anisotropic). Depending on the binder used, plasto-ferrites may be flexible or rigid. Special advantages of sintered materials are high magnetic quality, close dimensional control when machined, and relatively high mechanical strength. When these properties are not of prime importance, plasto-ferrites may be preferred because of lower price or special mechanical properties and shaping possibilities. Anisotropic materials are applied when high magnetic quality is essential. The less expensive isotropic material is preferred when low magnetic quality is acceptable or when the properties have to be isotropic.

The anisotropic sintered form is by far the most important one.

Products, whether or not anisotropic, may have different magnetization modes. That is, magnetization may be from one side to the opposite one, or along one side (lateral magnetization). The resulting polar surfaces may contain regions having opposite polarity (poles), separated by a neutral zone. Different pole numbers and configurations are possible. Combinations of these possibilities give rise to a variety of magnetization modes. All products may be anisotropic and products of the same appearance may have a different aligning mode. When a complex magnetization mode is desired, isotropic materials are preferred. Lateral magnetization, always in multipole, is only applied to isotropic materials.

As compared to the classical alnico-magnets, M-ferrite magnets have some distinct advantages. These are relatively high H_c, high resistivity, low price, low density, high chemical resistance, and the suitability of being applied as (flexible) plastoferrite. The relatively high H_c and, in particular, the low price have contributed to economic success.

The most important disadvantages are moderate B_r and $(BH)_{max}$, relatively high temperature coefficients αB_r and αH_c, and poor mechanical properties (low strength, brittleness).

G. Cryssis, *High Frequency Switching Power Supplies*, McGraw-Hill Book Co., Inc., New York, 1984.

A. Goldman, *Modern Ferrite Technology*, Van Nostrand Reinhold Co., Inc., New York, 1990.

S. Mulder, *Application Note on the Design of Low Profile High Frequency Transformers: A New Tool in SMPS*, Philips Components, Eindhoven, The Netherlands, 1990.

E. J. Pateer, *Soft Ferrite Selection Guide*, Ferroxcube, Eindhoven, The Netherlands, 2003.

FRANS KOOLS
Technology University
P. J. VAN DER VALK
Ferroxcube

FERROELECTRICS

Polarization which can be induced in nonconducting materials by means of an externally applied electric field $\overline{E}$ is one of the most important parameters in the theory of insulators, which are called dielectrics when their polarizability is under consideration. Experimental investigations have shown that these materials can be divided into linear and nonlinear dielectrics in accordance with their behavior in a realizable range of the electric field. The electric polarization $\overline{P}$ of linear dielectrics depends linearly on the electric field $\overline{E}$, whereas that of nonlinear dielectrics is a nonlinear function of the electric field. The most important materials among nonlinear dielectrics are ferroelectrics which can exhibit a spontaneous polarization $\overline{P}_{\overline{S}}$ in the absence of an external electric field and which can split into spontaneously polarized regions known as domains. It is evident that in the ferroelectric the domain states differ in orientation of spontaneous electric polarization, which are in equilibrium thermodynamically, and that the ferroelectric character is established when one domain state can be transformed to another by a suitably directed external electric field. It is the reorientability of the domain state polarizations that distinguishes ferroelectrics as a subgroup of materials from the 10-polar-point symmetry group of pyroelectric crystals.

PROPERTIES

It is this high intrinsic dielectric susceptibility response that is the phenomenon most used in the practical application of polycrystalline ceramic ferroelectrics. Ferroelectric ceramics having relative permittivities $\epsilon_{ij}/\epsilon_0 = K_{ij}$ ranging up to 10,000, where ϵ_0 is the dielectric permittivity of vacuum, are widely used in many types of capacitors including the multilayer variety (see ADVANCED CERAMICS; CERAMICS; CERAMICS AS ELECTRICAL MATERIALS).

The piezoelectric and electrostrictive voltage coefficients, d_{ijm} and M_{mnij}, of the ferroelectrics are very large because of the large polarizability. Thus a second principal application of the ferroelectrics uses this high electromechanical coupling for efficient transduction between electrical and mechanical signals in sonic and ultrasonic transducers and filter applications.

Both the spontaneous polarization $\overline{P}_{\overline{s}}$ and the remanent polarization $\overline{P}_R$ are strong functions of temperature, particularly near the transition temperature T_c in ferroelectrics; $\Delta \overline{P}_R = \pi \Delta T$, where π is the pyroelectric coefficient. Many ferroelectrics have large pyroelectric coefficients and can be used in thermometry and in bolometry sensing devices of infrared radiation (see INFRARED TECHNOLOGY AND RAMAN SPECTROSCOPY; SENSORS).

Many ferroelectrics are high band gap insulating crystals and have good transparency in both the visible and near-ir spectral regions. In the single-domain state, many ferroelectric crystals also exhibit high optical nonlinearity and this, coupled with the large standing optical anisotropies (birefringences) that are often available, makes the ferroelectrics interesting candidates for phase-matched optical second harmonic generation (SHG).

One area of application utilizes the interaction between the dielectric polarization and the electrical transport processes in ferroelectrics. In single dielectric crystals the effects of the domain polarizations on the drift and retrapping of photogenerated carriers give most interesting photoferroelectric effects. Of more immediate applicability, however, are the large effects of the dielectric changes at the ferroelectric phase transition on the potential barriers at grain boundaries in suitably prepared semiconducting ceramic ferroelectrics. These barium titanate-based compositions show strong positive temperature coefficients of resistivity (PTC effects) and are finding widespread use in temperature and current control for domestic, industrial, and automotive applications.

MATERIALS

Oxygen Octahedra

An important group of ferroelectrics is that known as the perovskites. The perfect perovskite structure is a simple cubic, having the general formula ABO_3, where A is a monovalent or divalent materials such as Na, K, Rb, Ca, Sr, Ba, or Pb, and B is a tetra- or pentavalent cation such as Ti, Sn, Zr, Nb, Ta, or W.

$PbZrO_3$–$PbTiO_3$-Based Materials. Since the mid-1950s, solid solutions of $PbZrO_3$–$PbTiO_3$ (PZT) ceramics having the perovskite structure have gained rising interest because of the superior piezoelectric properties.

PREPARATION OF FERROELECTRIC MATERIALS

Ceramics

The properties of ferroelectrics, basically determined by composition, are also affected by the microstructure of the densified body which depends on the fabrication method and condition. The ferroelectric ceramic process is comprised of the following steps: (1) selection of raw oxide materials, (2) preparation of a powder composition, (3) shaping, (4) densification, and (5) finishing.

Powder Preparation. *Mixing.* The most widely used mixing method is wet ball milling, which is a slow process, but it can be left unattended for the whole procedure.

Calcination. Calcination involves a low (<1000 °C) temperature solid-state chemical reaction of the raw materials to form the desired final composition and structure such as perovskite for $BaTiO_3$ and PZT.

Shaping. The calcined powders must be milled and a binder (usually organic materials) added if necessary for the forming procedure.

Densification. Sintering, hot-pressing, or hot-isostatic-pressing methods may be used to densify the shaped green ferroelectric ceramics to ~95–100% of theoretical value.

Finishing. The densified ferroelectric ceramic bodies usually require machining and metallizing for dimension and surface roughness control and electrical contact.

Thin-Film Ferroelectrics

The trends in integrated circuits (qv) and packaging technologies toward miniaturization have stimulated the development of ferroelectric thin films (see PACKAGING, ELECTRONIC MATERIALS; THIN FILMS). Advances in thin-film growth processes offer the opportunity to utilize the material properties of ferroelectrics such as pyroelectricity, piezoelectricity, and electrooptic activity for useful device applications. The primary impetus of the activity in ferroelectric thin-film research is the large demand for the development of nonvolatile memory devices, also called FERRAMS (ferroelectric random access memories).

Several techniques have been investigated for the preparation of ferroelectric thin films. The thin-film growth processes involving low energy bombardment include magnetron sputtering, ion beam sputtering, excimer laser ablation, electron cyclotron resonance (ECR) plasma-assisted growth, and plasma-enhanced chemical vapor deposition (PECVD) (see PLASMA TECHNOLOGY). Other methods are sol-gel, metal organic decomposition (MOD), thermal and e-beam evaporation, flash evaporation, chemical vapor deposition (CVD), metal organic chemical vapor deposition (MOCVD), and molecular beam epitaxy (MBE).

Ferroelectric Ceramic–Polymer Composites

The development of active ceramic–polymer composites was undertaken for underwater hydrophones having hydrostatic piezoelectric coefficients larger than those of the commonly used lead zirconate titanate (PZT) ceramics. It has been demonstrated that certain composite hydrophone materials are two to three orders of magnitude more sensitive than PZT ceramics while satisfying such other requirements as pressure dependency of sensitivity. The idea of composite ferroelectrics has been extended to other applications such as ultrasonic transducers for acoustic imaging, thermistors having both negative and positive temperature coefficients of resistance and active sound absorbers.

APPLICATIONS

Multilayer Capacitors

Multilayer capacitors (MLC), at greater than 30 billion units per year, outnumber any other ferroelectric device in production.

Piezoelectric and Electrostrictive Device Applications

Devices made from ferroelectric materials utilizing their piezoelectric or electrostrictive properties range from gas igniter to ultrasonic cleaners (or welders).

Composite Devices

Composites made of active-phase PZT and polymer-matrix phase are used for the hydrophone and medical imaging devices (see COMPOSITE MATERIALS–POLYMER-MATRIX; IMAGING TECHNOLOGY).

Relaxor Ferroelectrics

The general characteristics distinguishing relaxor ferroelectrics, eg, the $PbMg_{1/3}Nb_{2/3}O_3$ family, from normal ferroelectrics such as $BaTiO_3$, are summarized in Table 1.

Table 1. Properties of Relaxor and Normal Ferroelectrics

Property	Normal ferroelectrics	Relaxor ferroelectrics
permittivity temperature dependence $\epsilon = \epsilon(T)$	sharp first- or second-order transition above	broad–diffuse phase transition about Curie maxima Curie temperature
permittivity temperature and frequency dependence $\epsilon = \epsilon(T, \omega)$	weak frequency dependence	strong frequency dependence
remanent polarization	strong remanent polarization	weak remanent polarization
scattering of light	strong anisotropy (birefringent)	very weak anisotropy (pseudocubic)
diffraction of x-rays	line splitting owing to spontaneous deformation from paraelectric to ferroelectric phase	no x-ray splitting giving a pseudocubic structure

Relaxor ferroelectrics have been extensively investigated since the late 1970s because of the ability to generate large electrically induced strains, minimal hysteresis of the strain–electric field response, and minimal thermal strain. These materials also show promise in capacitor applications because of large dielectric permittivities. The field-induced piezoelectric and elastic properties of relaxor ferroelectrics have been investigated for transducer applications, including three-dimensional medical ultrasonic imaging devices.

Polymer Ferroelectrics

Polymer ferroelectrics are used as audio frequency transducers, such as microphones, headphones, and loudspeaker tweeters having excellent frequency response and low distortion because of the low density, lightweight transducer film; ultrasonic transducers for underwater applications, such as hydrophones, and for medical imaging applications; electromechanical transducers for computer and telephone keypads and a variety of other contactless switching applications; and pyroelectric detectors for infrared imaging and intruder detection.

L. E. Cross, *Ferroelectrics* **76**, 241–267 (1987).

B. Jaffe, W. R. Cooke, Jr., and H. Jaffe, *Piezoelectric Ceramics,* Academic Press, New York, 1971.

A. J. Moulson and J. M. Herbert, *Electroceramics,* Chapman and Hall, London, 1990.

R. E. Newnham, D. P. Skinner, and L. E. Cross, *Mater. Res. Bull.* **13**, 525 and 599 (1978).

SEI-JOO JANG
The Pennsylvania State
University

FERTILIZERS

Fertilizers are added to soils to supplement the supply of inorganic nutrients required for plant growth in amounts necessary to eliminate the deficiencies that limit profitable crop and livestock production. The three principal nutrients required for plant growth are nitrogen, phosphorus—as orthophosphate—and potassium.

The chemistry of nitrogen in soils is complex, involving cycling through ammoniacal, nitrate, and organic forms. The total nitrogen content of many soils is of the order of 2 to 4 tons/hectare, but this is almost entirely bound to organic matter and mineral soil material; at any one time only a few kg/hectare are present in available forms as nitrates or exchangeable ammonium. Except in the case of leguminous plants, which have the capacity for the "biological fixation" of atmospheric nitrogen, this amount of available nitrogen is insufficient to meet the needs of high yielding crops, although continually replenished by "mineralization" of the organic forms.

Soil phosphorus exists in a number of forms, including organic compounds, precipitated minerals, and adsorbed

forms, with a very small amount present in the soil solution. Only a small fraction of the total is readily available for uptake by plants. The phosphorus itself is always present as orthophosphate (PO_4) minerals or organically bound orthophosphate.

The primary sources of potassium in soils are the potassium bearing minerals in the parent rock and the clay minerals present in the soil. As these weather, the potassium is released and retained as exchangeable potassium associated with the negatively charged clay minerals and organic matter. This exchangeable potassium is readily available for plant uptake but may not equal the amount required by the crop over the whole growing season, so supplemental additions from fertilizers are needed.

Calcium and sulfur are also essential plant nutrients present in fertilizer materials but are not regularly included in listings of composition and are not used to calculate application rates. Calcium is present in superphosphate fertilizers as calcium phosphate or calcium sulfate; it may also be added to soils in limestone, which is not considered as a fertilizer but a soil additive used to correct soil acidity. Where ordinary superphosphates are used, the sulfur needs of crops are usually more than adequately met by the sulfate which these contain. In areas where sulfur deficiencies do occur, these may be overcome by supplemental additions of gypsum, which is generally regarded as a soil amendment rather than a fertilizer material.

NITROGEN FERTILIZERS

Nitrogen Fertilizer Materials

The nitrogen compounds present in fertilizers are listed in Table 1.

PHOSPHATE FERTILIZERS

Source Materials

Mineral deposits of phosphate rock are the primary source material for phosphate fertilizers. The principal phosphate mineral is apatite or fluorapatite, $Ca_{10}(PO_4)_6(OH,F)_2$. Secondary minerals include silica, silicates, and carbonates, usually as calcite or dolomite ($CaCO_3$); these must be removed by beneficiation treatment before the rock is processed.

Table 1. Nitrogen Compounds in Fertilizers, Formulas, and % Nitrogen Content

Compound	Formula	Nitrogen Content, %
anhydrous ammonia	NH_3	82
aqua ammonia	NH_4OH	varied
ammonium nitrate	NH_4NO_3	35.0
ammonium sulfate	$(NH_4)_2SO_4$	21.2
ammonium phosphates	$(NH4)H_2PO_4$	16.1
	$(NH4)_2HPO_4$	21.2
calcium nitrate	$Ca(NO_3)_2$	17.0
urea	$(NH_2)_2CO$	46.6

Table 2. Compounds Present in Fertilizer Materials, with Representative P₂O₅ Content and Water-Soluble Fraction

Fertilizer material	% P_2O_5	% P_2O_5 Water-soluble	Compounds present
superphosphates			
ordinary	21	85	$Ca(H_2PO_4)_2$, $CaSO_4$
triple	45	87	$Ca(H_2PO_4)_2$
ammoniated superphosphates			
ordinary	14	35	$(NH_4)H_2PO_4$, $CaHPO_4$, reprecip. apatite, $CaSO_4$
triple	48	50	$(NH_4)H_2PO_4$, $CaHPO_4$, reprecip. apatite,
ammonium phosphates			
monoammonium	52	100	$(NH_4)H_2PO_4$
diammonium	46	100	$(NH_4)_2HPO_4$
polyphosphate	43	100	$(NH_4)_2HPO_4$, $(NH_4)_3HP_2O_7$ and other polyphosphates
urea-ammonium phosphate	28	100	$CO(NH_2)_2$, $(NH_4)H_2PO_4$, $(NH_4)_2HPO_4$
nitric phosphate	9	40	$CaHPO_4$, $(NH_4)H_2PO_4$, $Ca(NO_3)_2$, reprecip. apatite,
dicalcium phosphate	48	3	$CaHPO_4$

Source: International Fertilizer Industry Association, 2000.

Phosphate Fertilizer Materials

The principal phosphate compounds present in fertilizers are listed in Table 2.

Agronomic Effectiveness of Phosphate Fertilizers

The effectiveness of phosphate fertilizers depends on a number of factors, including their chemical composition, their physical state, the way in which they are used, and the crop and farming systems in which they are employed. Two chemical indices have been found to be broadly useful in relating their relative agronomic effectiveness to their chemical composition. These are the size of the water-soluble phosphate fraction and the overall solubility in neutral ammonium citrate solution. The latter includes the water-soluble fraction but also serves as an additional index of the value of the water-insoluble fraction.

POTASSIUM FERTILIZERS

Potassium fertilizers are produced by the refinement of mineral deposits of soluble potassium salts found in salt lakes or underground deposits. The principal ore is sylvinite, a mixture of crystals of sylvite (potassium chloride) and halite (sodium chloride). The potassium salt is separated by flotation or fractional crystallization to give a product containing about 50% K or 60% K_2O. The mineral langbeinite ($K_2SO_4 \cdot 2MgSO_4$) is also used to produce potassium sulfate by treatment with KCl and removal of the magnesium chloride by fractional crystallization. The fertilizer grade of K_2SO_4 contains about 41% K or 49% K_2O.

USE

Fertilizer applications include a very wide variety of usage patterns and materials with a considerable range of compositions. Fertilizers are applied as fluids, dry bulk, or bagged granulated materials. Mixed formulations containing two or three nutrients are used as dry materials made by the blending of granular materials in bulk or in fluid forms made by blending fertilizer solutions.

Application Techniques

Fertilization practices depend upon a number of factors including crops grown, soil types, and weather conditions in a region or area. The timing of applications in relation to rainfall events and the rate of crop growth is always of major importance. Practices vary greatly between regions with different rainfall patterns, with major differences between temperate and tropical regions.

There are several principal techniques for fertilizer application in the field. These include: 1) surface broadcasting without incorporation into the soil, 2) surface application with immediate incorporation and placement in the soil by an attachment to the plow or cultivator, and 3) the injection of liquids and gaseous formulations by pumping through cultivator knives. Applications are also made in irrigation water for crops grown with such water supplies.

In some cases the use of high analysis fertilizer materials may cause problems if the risks associated with their physical or chemical properties are not understood. Where surface applications of formulations containing urea or some ammonium salts are not plowed or cultivated into the soil and the fertilizer is not rapidly washed into the soil by rain, they may lose a significant fraction of their nitrogen to the atmosphere; the risk of such losses is particularly high on calcareous or more alkaline soils.

ENVIRONMENTAL IMPACTS

The function of fertilizer applications is to raise the level of soil fertility and biological activity in the soils and farming areas where they are used by increasing the amounts of available plant nutrients, in particular nitrogen and phosphorus. Adverse environmental impacts may occur where the amounts of these nutrients are increased to levels at which losses of nutrients carried in drainage water or air from the fertilized land cause undesirable changes in water quality and the ecological environment in wetlands, streams, rivers, and lakes that receive surface drainage and groundwater.

It should, however, be noted that, while fertilizer applications represent major inputs to agricultural systems, they are not the only sources of nutrients contributing to the decay of water quality.

S. E. Allen, in R. D. Hanck, ed., *Nitrogen in Crop Production*, American Society of Agronomy, Madison, Wisc., 1984, Chap. 13.

L. G. Bundy and J. J. Meisinger, in R. W. Weaver, ed., *Methods of Soil Analysis, Part II: Microbiological and Biochemical Properties*, Soil Science Society of America, Madison, Wisc., 1994.

V. Smil, "Fritz Haber, Carl Bosch and the Transformation of World Food Production", *Enriching the Earth*, MIT Press, Cambridge, Mass., 2001.

R. D. Young and C. H. Davies, in M. Stelly, ed., *The Role of Phosphorus in Agriculture*, American Society of Agronomy, Madison, Wisc., 1980, Chap. 7.

ALAN W. TAYLOR
USDA-ARS

FIBER OPTICS

Optical communication was long considered as a possibility for high speed data transmission because light at terahertz frequency is capable of enormous bandwidth. The power of lightwave communication could not be tapped, however, until the obstacle of a suitably transparent transmission medium was overcome.

The development of digital encoding solved the problem of signal degradation noise. Using this method, the amplitude of an analogue waveform is sampled at frequent intervals and coded into a binary (0s and 1s) sequence of digits. A voice waveform requires the transmission of 64,000 bits/s, which are digitally transmitted by a series of positive and negative electrical pulses. The optical equivalent uses pulses of light transmitted over glass fiber. Unlike the electrical signal, laser light may be modulated at a frequency high enough (10s of gigahertz) for any conceivable communication need.

Figure 1 gives a comparison of analogue and digital transmission schemes. The incoming signal (Fig. 1a) can be transformed directly into an intensity variation of the light beam (Fig. 1b). A photodetector at the receiver converts this varying intensity into an electrical signal which is then amplified to reproduce the original waveform. Such a signal becomes increasingly degraded and distorted during transmission and amplifica-

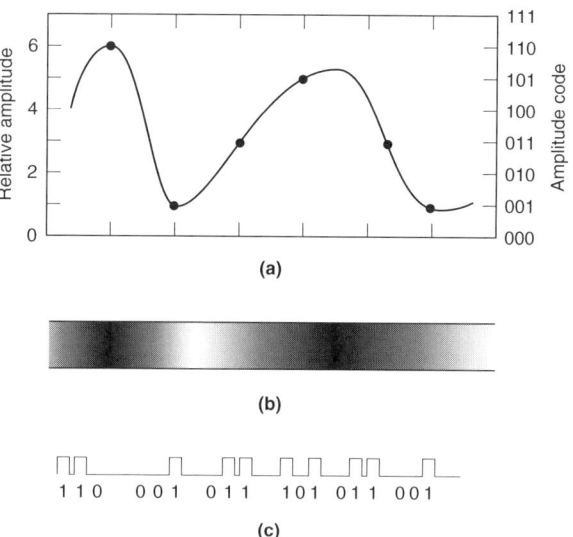

Figure 1. (a) A transmission signal and its (b) analogue and (c) digital encoding.

tion. Improved fidelity is provided by digital encoding (Fig. 1c). In the digital scheme the signal is encoded by flashes of light at regularly timed intervals. The sampling rate must be twice that of the highest frequency component for accurate representation of the waveform. A voice signal having a maximum frequency of 4000 Hz must be sampled at a rate of 8000/s. The binary coding of 0s and 1s corresponds to the absence and presence of light. Representation of the height of a voice waveform requires eight bits (a bit is a 0 or a 1). Therefore, to sample a voice wave for one second the digital system requires 64,000 bits (8,000 samples × 8 bits/sample). Although the intensity of the light signal diminishes over distance, as long as it remains above the threshold of the detector the signal can be cleanly regenerated because a pulse is either present or absent. In this manner, noise is eliminated.

PRINCIPLES OF LIGHT GUIDANCE

Light guidance is governed by the structure of the lightguide itself. The refractive index, n, of a material is defined as the ratio of the speed of light in a perfect vacuum to the speed of light through that material. This property is a function of both the composition of the material and the wavelength of the transmitted light. The higher the refractive index of a material, the more light is retarded, or slowed, in passing through it. At the interface of two materials of different refractive indexes, light is refracted, or bent toward the higher index medium by an angle the sine of which is proportional to the relative indexes of the two media. This property is known as Snell's law. Light that travels in a medium of higher refractive index at an angle less than the critical angle, $_c$, is totally reflected. This property of total internal reflection provides a means to transmit light over long distances without radiative losses.

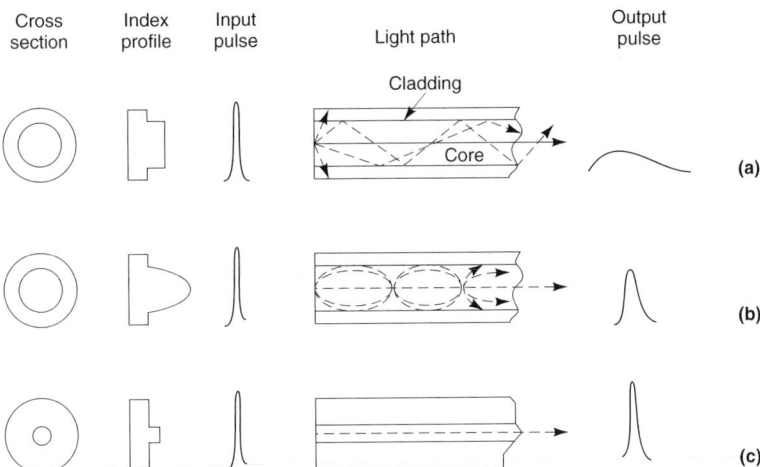

Cross section | Index profile | Input pulse | Light path | Output pulse

Cladding

Core

(a)

(b)

Figure 2. Types of optical fiber: (**a**) multimode stepped index, (**b**) multimode graded index, and (**c**) single-mode stepped index.

(c)

The ability of a waveguide to collect light is determined by the numerical aperture (NA) which defines the maximum angle at which light entering the fiber can be guided. (NA $= \sin\theta_c = \left(n_1^2 - n_2^2\right)^{1/2} - n_1(2\Delta)^{1/2}$) where $\Delta = (n_1 - n_2)/n_1$ and typically $\Delta \ll 1$; n_1 is the refractive index of the core; and n_2 is the index of the cladding. Lightguide structures are shown in Figure 2. Whereas Figures 2a and 2b are multimode structures having relatively higher Δs and core diameters on the order of 50 μm, Figure 2c is a single-mode fiber of lower refractive index and a core diameter $< 10\,\mu\,\mu$m. The number of modes that can propagate in a fiber is lgoverned by Maxwell's equations for electromagnetic fields, and is related to a dimensionless quantity V called the normalized frequency: $V = \left(2\pi a/\lambda\right)\text{NA} \simeq (2\pi a n_1/\lambda)\,(2\Delta)^{1/2}$, where λ is the wavelength of light in vacuum and a is the radius of the fiber core.

Attenuation

The exceptional transparency, or low attenuation, of silica-based glass fibers has made them the predominant choice for optical transmission because of the low level of absorption and scattering of light as it traverses the material. Together these comprise optical attenuation, or loss, measured in dB where loss (dB) $= 10\log(I_0/I)$, where I_0 is the input intensity and I is the output intensity. Values for loss are typically given per kilometer of fiber.

Additional optical attenuation may result from large-scale imperfections or defects in the glass structure as well as waveguide imperfections formed during processing of the glass. Fluctuations longer than the wavelength of light, such as diameter variations, may cause Mie scattering. Even in a nearly perfect glass, absorptions from low level cation impurities are detrimental. Similarly, point defects in the anion network can play a role in controlling loss; suboxides of germanium can be formed at high temperature in an oxygen-deficient reaction and result in coloration, especially after exposure to radiation. If these extrinsic losses are avoided, as is typical of current production, then fiber attenuation is dominated by Rayleigh scattering and decreases as λ^{-4} until the multiphonon edge is intersected.

Dispersion

The effects of dispersion on the ultimate system performance are as important as the attenuation. Dispersion arises from the variation of the velocity of light with the wavelength of the light. Two types of dispersion which occur are intermodal, found only in multimode fibers, and chromatic, which is important for single-mode performance.

Waveguide dispersion depends on how much of the power travels in the cladding of the fiber relative to the amount traveling in the core. Proper design of the fiber's core diameter and refractive index profile may be fine-tuned to completely cancel the material dispersion at a given wavelength, or to flatten the total dispersion over a range of wavelengths.

OPTICAL FIBER FABRICATION

Viable glass fibers for optical communication are made from glass of an extremely high purity as well as a precise refractive index structure.

Double Crucible

The earliest attempts at producing high purity glass employed the double crucible technique. In spite of the elegance of this technique, contamination occurred during processing, leading to impurity levels in the glass on the order of parts per million rather than the ppb levels necessary.

High silica glasses having lower losses at wavelengths from the visible to the infrared became available through technology using vapor-phase techniques where silicon tetrachloride, $SiCl_4$, is the precursor to silicon oxide, SiO_2, of near-intrinsic purity. Other compositions could be produced as well using other chloride vapors such as phosphorus oxychloride, $POCl_3$, and germanium tetrachloride, $GeCl_4$, to dope the silica and provide changes in the refractive index of the glass. Two methods evolved. Inside vapor-phase deposition followed from chemical vapor deposition (CVD) processes used in the electronics industry (see ELECTRONIC MATERIALS; INTEGRATED CIRCUITS), and outside vapor-phase techniques such as OVPO and VAD.

Inside Processes

Inside processes such as modified chemical vapor deposition (MCVD) followed CVD techniques. In these the concentration of the reactants was kept low to inhibit homogeneous gas-phase reaction in favor of heterogeneous wall reactions which produce a vitreous (amorphous), particle-free deposit on the tube wall. Deposition was continued until a sufficient thickness of glass was produced; then the tube was collapsed to form a solid rod, or preform. This preform was then drawn to a relatively low loss fiber. Deposition rates by this method were impractically low, and attempts to increase them invariably led to the formation of particles and, subsequently, bubbles, resulting in excess scattering loss. The solution was to radically increase the reactant concentration and flow rates to more than 10 times those used for CVD. This led to the formation of SiO_2 and germanium oxide, GeO_2, particles in the gas stream that were then deposited on the tube wall interior and fused by a traversing external torch. Multiple layers were deposited to achieve the desired refractive index profile before the tube was collapsed to a preform.

One of the areas critical to the MCVD process was understanding the chemistry of the oxidation reactions. It was necessary to control the incorporation of GeO_2 while minimizing OH^- formation. Additionally, understanding the mechanism of particle formation and deposition was critical to further scale-up of the process.

Thermophoresis is a process by which particles in a temperature gradient travel toward the cooler region when bombarded by more energetic particles on the hot side. In the MCVD process the reactants enter the tube, are reacted in the hot zone of the torch, deposit thermophoretically downstream of the torch, and are subsequently sintered to a clear glass as the torch passes over the deposited particulate layer. Once the desired structure has been deposited, the direction of the torch is reversed and the tube is collapsed to form a solid preform.

The chemistry of the MCVD process has been studied using a variety of techniques. Infrared (ir) spectroscopy was used to investigate the oxidation reaction of $SiCl_4$ and $GeCl_4$ (see INFRARED TECHNOLOGY; RAMAN SPECTROSCOPY). It was concluded from these results that at temperatures lower than 1600 K the degree of reaction of the $SiCl_4$, $GeCl_4$, and $POCl_3$ is controlled by reaction kinetics, whereas at higher temperatures the reaction is dominated by thermodynamic equilibria. Rate studies have shown that the residence times typically experienced in the hot zone are sufficient to reach equilibrium above 1700 K.

Incorporation of OH is another critical aspect of the oxidation chemistry. Reduction to the ppb level is necessary for the manufacture of low loss optical fiber. During the deposition phase of MCVD the chlorine level is between 3 and 10% owing to the oxidation of the chloride reactants. This level of chlorine leads to a reduction in the OH incorporation by a factor of about 4000. During the collapse phase of the process the chlorine level is significantly reduced and OH incorporation can be high as a result of the presence of hydrogen from either the oxy-hydrogen torch or impurities in the starting gases.

The process by which the glass consolidates was found to be viscous sintering controlled by the viscosity of the glass at the consolidation temperature. The driving force is the reduction of surface energy via a decrease in the surface area.

Another process closely related to MCVD is plasma chemical vapor deposition (PCVD) (see PLASMA TECHNOLOGY). This inside process uses the same precursor chemicals as MCVD to form a glass coating inside a tube which is collapsed to form a solid glass preform. The reaction within the tube, however, is quite different. Here the reaction is initiated by a nonisothermal microwave plasma which traverses the inside of the tube. The plasma requires a pressure of a few hundred Pascals and is generated by a microwave cavity which operates at 2.45 GHz. The glass is deposited not as a soot, but as a thin glass layer with efficiencies approaching 100%. In addition to the high efficiency, complex waveguide structures may be formed because many thin layers are produced by rapidly traversing the plasma. This method provides a smoother refractive index variation, which is especially advantageous for multimode fiber preforms.

Outside Processes

The outside vapor deposition (OVD) process developed by Corning Glass Works. Soot is deposited layer by layer on a rotating mandrel at a temperature such that the soot particles are partially sintered. The precursor chemicals are the same as those used in the MCVD process but are oxidized by a gas–oxygen torch by similar chemical reactions.

The vertical axial deposition (VAD) process was developed by a consortium of Japanese cable manufacturers and Nippon Telephone and Telegraph (NTT). This process also forms a cylindrical soot form. However, deposition is achieved end-on without use of a mandrel and subsequent formation of a central hole. Both the core and cladding are deposited simultaneously using more than one torch. Whereas the OVD, PCVD, and MCVD processes build a refractive index profile layer by layer, the VAD process uses gaseous constituents in the flame to control the shape and temperature distribution across the face of the growing soot boule.

Although the control in VAD is more difficult, this process has an advantage over OVD, especially for multimode fiber. The thermal mismatch between core and cladding materials caused by the heavy doping necessary to achieve the desired refractive index profile causes cracking in the OVD preforms at the inner surface as the glass cools through the glass transition, T_g. The VAD preforms withstand the stress because they possess no central hole to result in tensile stress. The primary obstacle for VAD was the creation of an optimized refractive index profile to minimize intermodal dispersion. Then it was discovered that the composition could be graded by control of the boule's surface-temperature distribution. This temperature distribution depends on the shape of the boule's growing face. Through understanding the relationship between the temperature and germania concentration, fiber properties equivalent to those formed by OVD and MCVD were achieved.

FIBER DRAWING AND STRENGTH

Preforms manufactured by MCVD, PCVD, OVD, and VAD all must be drawn into fiber in a similar manner. Standard fibers are drawn to 125 µm in diameter from preforms on the order of 2 to 7.5 cm diameter. Fibers are drawn by holding the preform vertically and lowering it into a furnace. The preform is heated to a temperature at which the glass softens (2200°C) until a gob of glass stretches from the tip of the preform and drops under the force of gravity. A neck-down region is formed at this point, providing the transition between preform and fiber. Fiber is drawn by means of a capstan system, and its diameter is controlled by a diameter monitor that adjusts the draw speed at a fixed furnace temperature. The result is long lengths of uniform fiber. To preserve the intrinsic strength of the pristine glass surface, a polymer coating must be applied before the fiber is contacted by the capstan.

Overcladding

Fiber manufacturing and drawing technology have advanced to the point that the optical losses are limited almost entirely by the intrinsic loss of the glass. Initially all of the fiber manufacturing processes (MCVD, PCVD, OVD, and VAD) produced preforms yielding on the order of 10 km of fiber. This situation changed as single-mode fiber usage grew. The proportion of core glass to the total amount of glass in single-mode fiber is much lower than in multimode fiber. Single-mode core diameter is ca 8 µm, multimode core diameter is 50 or 62.5 µm. This led to a desire for a method of manufacturing larger preforms by shrinking a second silica tube over the preform or depositing a thick soot layer on the preform to provide additional cladding. This procedure, known as overcladding, increased the length of fiber drawn from a single VAD preform to more than 100 km, for example, and significantly reduced the cost of producing fiber.

Sol–Gel Processing

Fibers can be designed so that light only travels in the inner 30–40 µm of the fiber, which accounts for only about 5% of the fiber mass. Thus, using a core rod, the remaining 95% could be manufactured from less expensive, lower purity materials typically obtained by sol–gel processing (see Sol–gel technology).

Defects

The ever-increasing demand for high data rate systems is forcing the search for an even greater understanding of those defects which produce attenuation of only hundredths of a dB/km. Profile control to produce zero dispersion at operating wavelengths is necessary, and environmental effects such as radiation and hydrogen exposure

must be minimized. In addition, for reliability concerns, higher strength must be achieved with a narrow distribution, necessitating the understanding of flaw distributions and growth of flaws (fatigue).

OPTICAL AMPLIFIERS

Throughout the first two decades of their existence optical fibers served a passive role, ie, in the transmission of encoded light signals. In the late 1980s erbium-doped fiber amplifiers (EDFAs) were introduced, making it possible to amplify a 1.55 µm optical signal without first converting it to an electronic signal. The amplifier consists of a section (tens of meters) of single-mode optical fiber having about 100 ppm of erbium incorporated into the core. This fiber section becomes an amplifier when a continuous source of pump light, usually 0.98 or 1.48 µm, wavelength is propagating through the fiber. As the optical signal, usually 1.53 to 1.6 µm, travels through the length of fiber containing excited erbium ions, amplification occurs by the stimulated emission of photons from the excited state. Noise in the form of broad-band spontaneous emission accompanies this process; however, the signal-to-noise ratio is kept to an acceptable level even when cascading many of these devices for system applications.

A number of means have been developed to produce erbium-doped optical fibers. The compatibility of aluminum and erbium extends to the vapor phase where complex aluminum—erbium chlorides exist at vapor-pressure orders of magnitude higher than $ErCl_3$. The passage of aluminum chloride, Al_2Cl_6, vapor over a heated erbium oxide, Er_2O_3, or erbium chloride, $ErCl_3$, source permit doping in MCVD reactions. In addition, erbium chelates and other organic precursors can be introduced into VAD or OVD flows to produce doped soot. Finally, doping with solutions containing rare-earth ions or sol–gel doping of soot or MCVD substrate tubes prior to the final collapse to a solid preform also provide the means for controlled introduction of the ions.

EDFAs are being introduced into long-distance, particularly undersea, systems which operate at wavelengths near 1.6 µm. In addition to providing an inexpensive means of amplification EDFA also make it possible to amplify numerous wavelengths near 1.6 µm, whereas semiconductor amplifiers suffer crosstalk when amplifying more than one wavelength (see Semiconductors).

FIBER OPTIC SENSORS FOR SMART STRUCTURES

Man-made structures can be made "smart" by duplicating the essential elements of the system that consists of embedded sensors, data links, a programmed data processor, and actuators. Fiber-optic sensors offer embedded sensor capability and natural connections to fiber-optic data links that can be used in a wide variety of composite materials to act as the structure's "nervous system." These sensors have a series of important advantages over conventional electronic sensors. (1) They are small and often made in overall diameters of 125 microns or less that results in a hair thin sensor that can be embedded in many types of composite structures without changing mechanical properties. (2) Fiber-optic sensors can be made environmentally rugged and can withstand the temperatures and pressures in manufacturing composite structures. (3) The glass fibers are passive dielectric devices that enable their usage in organic composite materials like carbon epoxy and thermoplastics where electrical discharge hazards such as lightning on aircraft and spacecraft require eliminating conductive paths. The passive nature of fiber-optic sensors also allows embedding these sensors successfully into metal structures. (4) Many fiber-optic sensors can be made that have a high degree of immunity to electromagnetic interference and eliminate the need for costly and bulky shielding. (5) Fiber-optic sensors may be multiplexed, so that many sensors lie along a single fiber line. (6) Fiber sensors are naturally compatible with fiber-optic data links that have the bandwidth necessary to support large number of sensors. (7) There is a high degree of synergy between fiber-optic sensors and the telecommunication and optoelectronics industry that leads to continuously improve components and lower costs.

B. Culshaw and J. Dakin, *Optical Fiber Sensors: Systems and Applications*, Vol. 2. Artech House, Norwood, Mass., 1989.

J. Dakin and B. Culshaw, eds., *Optical Fiber Sensors: Principles and Components*, Vol. 1, Artech House, Boston, 1988.

E. Udd, ed., *Fiber Optic Sensors: An Introduction for Engineers and Scientists*, John Wiley & Sons, Inc., New York, 1991.

SANDRA KOSINSKI
AT&T Bell Laboratories
JOHN B. MACCHESNEY
AT&T Bell Laboratories
ERIC UDD
Blue Road Research

FIBERS

This overview of fibers and fiber products introduces the underlying concepts that govern the properties, manufacture, and utilization of these materials. For thousands of years in human history through the last century, fibers have been mainly used in constructing woven, nonwoven, knitted textiles for clothing, home and interior furnishing, and in papermaking. With the advent of polymer chemistry, fiber spinning technologies, product conversion processing, and the ever increasing demands for better quality goods, the field of fiber materials has evolved with new chemistry, processing technologies, and new product development strategies.

CLASSIFICATION

A fiber is generally defined as a flexible, macroscopically homogeneous body having a high length/width ratio and

a small cross-section. Fibers can be incorporated as whole or parts of materials and products of various forms and performance properties for wide-ranging applications. The unique combination of fibers characteritics determine not only how fibers can be put together and made into products, but also the performance of these products. For thousands of years, textile and paper have been the main products of fibers and the textile and paper industries are the prime converters of fibers. Paper is made from aqueous suspension of short celulose fibers held together by hydrogen bonding after water is removed. Textile fibers are traditionally classified according to their origin as follows:

Naturally occurring fibers
Plant: based on cellulose, eg, cotton, flax, hemp, jute, and ramie.
Animal: based on proteins, eg, silk, wool, mohair, vicuna, other animal hairs.
Mineral: eg, asbestos.

Regenerated fibers
Based on natural cellulose and proteins.
Rayon: regenerated cellulose.
Acetate: partially acetylated cellulose derivative.
Triacetate: fully acetylated cellulose derivative.
Azlon: regenerated protein.

Synthetic fibers
Based on synthesized organic polymers.
Acrylic: polyacrylonitrile (also modacrylic).
Aramid: aromatic polyamides.
Nylon: aliphatic polyamides.
Olefin: polyolefins.
Polyester: polyesters of an aromatic dicarboxylic acid and a dihydric alcohol.
Spandex: segmented polyurethane.
Vinyon: poly(vinyl chloride).
Vinal (or vinylon): poly(vinyl alcohol).
Specialty fibers: poly(phenylene sulfide) and polyetheretherketone.

Inorganic fibers
Carbon/graphite: derived from polyacrylonitrile, rayon, or pitch.
Glass, metallic, ceramic.

Natural fibers are those derived directly from the plant, animal, and mineral kingdoms. A significant segment of the world's agricultural activity is concerned with the growth and harvesting of natural fibers. With the exception of silk, which is extruded by the silkworm as continuous filaments, natural fibers are of finite length and are known as staple fibers. Natural fibers usually require certain levels of separation and cleaning before being used in the manufacturing of textiles and other fibrous products. The factors that affect the utility and quality of natural fibers for textile purposes are fineness, length, color, and spinnability.

Manufactured fibers, the other main category of textile fibers, are generated from natural polymers, synthesized polymers, and inorganic substances. Glass fiber is the main inorganic fiber in common use, although other ceramic and metallic fibers have been developed for niche specialized applications. Regenerated and synthetic fibers, are generated from organic polymers of natural origins or organic syntheses via spinning processes. The three principal fiber spinning methods are melt spinning, dry spinning, and wet spinning, although there are many variations and combinations of these basic processes.

Synthetic fibers are based on synthesized polymers and are important products of the worldwide petroleum chemical industry. The most widely used synthetic fibers are based on polyesters, polyamides, polyolefins, acrylics, polyurethanes, and polyaramids. A relatively small number of polymer types can produce synthetic fibers with wide ranges in fiber properties and characteristics because of the enormous versatility of fiber manufacturing processes.

FIBER PROPERTIES

The most significant fibers properties are the geometric, physical, and chemical categories.

Geometric Properties

These properties pertain primarily to staple fibers and include various aspects of fiber dimensions and form such as length, width, cross-sectional area, cross-sectional shape, and crimp. They are of particular importance in processing and for product quality.

Fiber cross-sectional area or fineness also affects textile processing efficiency and the quality of the end product. To produce fine fabrics requires fine yarns. Yarn spinning is only possible with a certain minimum number of fibers. With a minimum number of fibers, the finer the fibers, the finer the yarns. However, fibers with too small a cross-section cannot be processed efficiently. The number of fibers in a cross-section, of yarn of a given size determines yarn properties. The smaller the fiber size or cross-section, the more fibers can be packed in the yarn. The yarns would be more compact and denser in appearance. Other factors being equal, yarn strength increases as the number of fibers in the yarn cross-section increases.

Crimp is a form factor that describes the waviness of a fiber or its longitudinal shape.

Physical Properties

Thermal and light properties are among the primary physical properties of fibers. For textile applications, fibers should be optically opaque so their refractive indexes need to be significantly different from those of their most common environments, namely, air and water. Luster and color are two optical properties that relate to a fiber's aesthetic quality and consumer acceptance. Textile fibers are always used in aggregates of a large number of single fibers that are caused to interact with each other through their surfaces. Electrical properties also affect fiber utility.

Chemical Properties

The ability of fibers to interact with water is among the most significant chemical property. The ability of fibers to absorb moisture or to be wetted by water ranges from hydrophilic, to those that are essentially hydrophobic. All natural and regenerated fibers and some synthesized fibers are hygroscopic, ie, they are capable of absorbing moisture from the atmosphere. This water vapor absorption property is directly related to the chemical structure of the polymers and the molecular arrangement of the polymers within the fibers.

Swelling is a chemical property closely related to a fiber's sorption characteristics. It may be defined as the reversible dimensional changes that occur when fibers undergo an absorption process.

Fibers must be resistant to the effects of acids, alkalies, reducing agents, and oxidizing agents, as well as to electromagnetic and particulate irradiation.

FIBER MORPHOLOGY

Fiber structure is described at three levels of molecular organization, each relating to certain aspects of fiber behavior and properties. First is the organochemical structure, which defines the chemical composition and molecular structure of the repeating unit in the base polymer, and also the nature of the polymeric link. This primary level of molecular structure is directly related to chemical properties, dyeability, moisture sorption, and wetting characteristics, and indirectly related to all physical properties. The chemical structure also determines the magnitude of intermolecular forces, which is important in terms of overall molecular arrangement in the fiber and many physical properties. The macromolecular level of structure describes the chain length, chain-length distribution, chain stiffness, molecular size, and molecular shape. The supramolecular organization is the arrangement of the polymer chains in three-dimensional (3D) space. The influenced by the organization of polymeric chains into crystalline and noncrystalline domains, and the disposition of these domains with respect to each other. In the case of natural fibers, a further level of structural organization related to the natural growth and development of the fiber must be considered.

PRODUCTION OF THE MAJOR COMMODITY FIBERS

Cotton

The most essential cotton fiber qualities related to dry processing (yarn spinning, weaving, and knitting) are length, strength, and fineness. For wet processing such as scouring, dyeing, finishing, fiber structure related to maturity or the level of development, plays a major role.

Wood

Wood fibers are much shorter and contain less cellulose than cotton. Wood fibers are derived from trees and categorized by the tree types. Fibers from most softwood, ie,

Gyomnosperns or evergreens, are average 3–3.6 mm in length and 25–60 mm wide. Fibers from certain southern pines are longer, up to 7 mm in length. Hardwood fibers from *Angiosperms* or deciduous trees are in shorter lengths of 0.1–1.5 mm. Both contain cellulose, hemicellulose, xylans, lignin, extractives, and ash. Softwood fibers contain slightly higher cellulose (45–50%), hemicellulose (20–25%) and xylans (25–35%) than hardwood (40–50% cellulose, 2–5% hemicellulose, 18–25% lignin).

Wool

Wool is the hair (fleece) of various breeds of sheep (*Ovis areis*). However, the term wool has been used more generally, sometime to describe the hair of all animals, including goat (cashmere and angora "mohair"), camel, alpaca, rabbit, yak, and vicuna. Depending on the breeds, environment, individual animals and body location on each animal, wool fibers vary significantly in their physical and performance properties. Wool fibers have complex chemical compositions and morphological structures; both have been well documented.

Silk

Silk fibers are spun from the glands of adult silkworms *Bombyx mori*. Each of the two glands of a worm produces a single strand of silk and are bound together by sericin. The silk filaments wrap around to form the protective cocoon for it chrysalis, the intermediate stage between larva and moth. Silk is the thinnest and longest natural fibers, this unique combination of characteristics putting it in a class of fibers all by itself. The fineness of the silk fibers yields soft hand. Silk's triangular cross-section and smooth surfaces offer excellent light reflection and a lustrous appearance.

Manufactured Fibers

Manufactured fibers, with the exception of glass, asbestos, ceramic, and the specialty metallic fibers, are organic polymers. The major *regenerated* fibers are cellulosics, ie, rayon, lyocell, and cellulose acetates. *Synthetic* fibers are spun from fiber-forming polymers. Fiber-forming polymers are linear polymers that can be oriented and partially crystallized under the spinning conditions. Fiber spinning is the engineering process by which polymer solutions or melts are mechanically forced or extruded through spinnerets followed by drawing to finer, orientated and stronger fibers.

To a large extent, the structural characteristics of polymers that allow them to crystallize under appropriate conditions have been summarized as follows. *Regularity*, the polymer chains must be uniform in chemical composition and stereochemical form. *Shape and interaction*, the shape of the polymer chains must allow close contact or fit to permit effective and strong intermolecular interaction. This is generally achieved by linear macromolecules with no bulky side groups or with side groups that are regularly spaced along the backbone chain. *Repeat length*, the ease of crystallization decreases with increasing length of the polymer repeating unit. *Chain directionality*,

since certain polymer chains have a directionality, the mode of chain packing in a crystallite can take either parallel or antiparallel configurations. *Single-chain*, *conformation*, crystallization is favored if the chain conformation is compatible with its form in the crystallite. *Chain stiffness*, an optimal stiffness of the polymer is necessary.

USES

Fibers are used in the manufacture of a wide range of products that can generically be referred to as fibrous materials. Since the properties of fibrous materials depend both on the properties of the fibers themselves and on the spatial arrangement of the fibers in the assembly, a given type of fiber may be used in many different end products. Similarly, a given end product can be produced from different fiber types.

Most manufacture fibers, ie, regenerated and synthetic, are obtained in continuous form or filaments. Schematic representations of typical yarns are given in Figure 1.

Yarns are used principally in the formation of textile fabrics either by weaving or knitting processes.

Knitted fabrics are produced from one set of yarns by looping and interlocking processes to form a planar structure.

The term nonwoven simply suggests a textile material that has been produced by means other than weaving, but these materials really represent a rather unique class of fibrous structure. The type of nonwoven material that is produced depends largely on the fiber type used and on the method of manufacture. Generally, air-laid nonwovens are less dense and compact and tend to be softer, more deformable, and somewhat weaker. Wet-laid or paper-like nonwovens are more dense, but brittle. However, nonwovens by either process can be made into a wide variety of products with a broad range of physical properties.

NEW FIBERS PROPERTIES

New fibers and new properties can be created by either chemical or physical means. Changing fiber geometries, specifically reducing fiber size or fineness is among the simplest fiber modification approaches to create significant changes to fiber properties and product performance. Current melt and solution spinning technologies are capable of producing fibers with diameters typically in the 20 μm range and higher at relatively high speeds and with good uniformity. Through the development of new fiber spinning technologies, fibers <1 denier have been generated from polyester, nylon, rayon and most recently acrylic. *Micro-denier* fibers, also called microfibers, are finer than any naturally occurring fibers.

Submicrometer size fibers have also been generated by electrospinning technologies. Electrospinning presents many advantages to fiber formation. It is versatile for forming fine fibers from a wide range of polymers, some of which cannot be converted into fibers via conventional

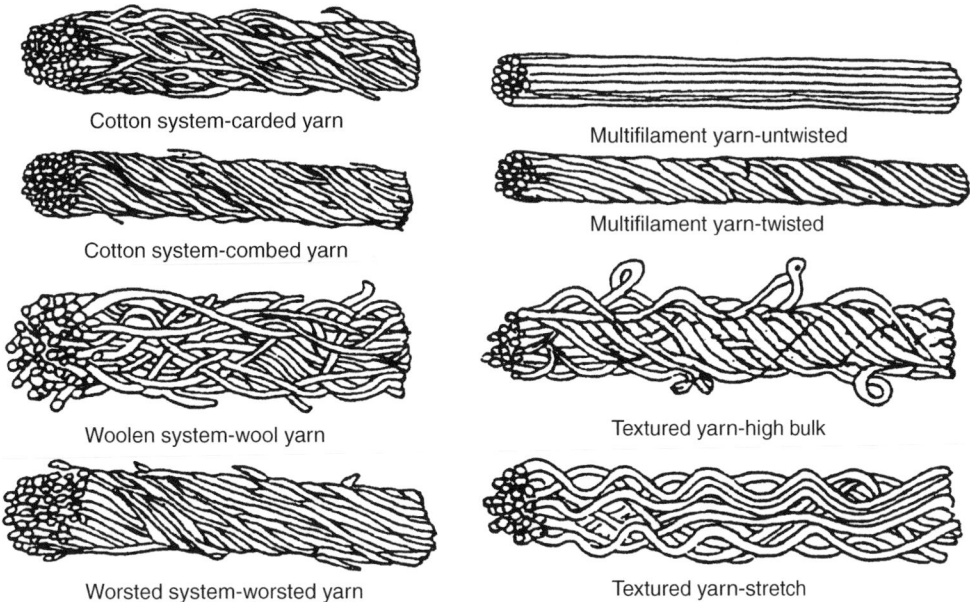

Cotton system-carded yarn

Cotton system-combed yarn

Woolen system-wool yarn

Worsted system-worsted yarn

Multifilament yarn-untwisted

Multifilament yarn-twisted

Textured yarn-high bulk

Textured yarn-stretch

Figure 1. Schematic description of spun and multifilament yarns.

fiber spinning processes. It requires very small quantities of polymer, as low as submilligram levels, making it possible to transform research-quantity polymer to fibers.

C. J. Biermann, *Handbook of Pulping and Papermaking*, 2nd ed., Academic Press, New York, 1996.

A. R. Horrocks and S. C. Anand, *Handbook of Technical Textiles*, The Textile Institute, Woodhead Publishing Ltd. and CRC Press, New York, 2000.

J. Lunenschloss and W. Albrecht, *Nonwoven Bonded Fabrics*, Halstead Press, a division of John Wiley & Sons, Inc., New York, 1985.

T. L. Vigo, *Textile Processing and Properties, Preparation, Dyeing, Finishing and Performance*, Textile Science and Technology 11, Elsevier, New York, 1999.

You-Lo Hsieh
University of California, Davis

FIBERS, ACRYLIC

In the 1970s, there was rapid growth of acrylic fiber production capacity in Japan, eastern Europe, and developing countries. By 1981, an estimated overcapacity of ~21% had developed. The 1990s saw significant shrinkage of acrylic production in the United States as DuPont and Mann Industries (formerly Badische) exited the business. Significant change has continued into the new century. Explosive industry growth has taken place in the Far East, particularly China, where plants based on DuPont and Sterling (Cytec) processes have proliferated. Modacrylics have all but disappeared from the marketplace, as demand for flame retardant textiles have been met by treated cotton or other synthetics at lower cost. A few new

markets have emerged, such as carbon fiber precursors and asbestos replacement fibers, but the volume is small compared to that of the markets lost. For acrylic producers, profit will continue to be sparse.

PHYSICAL PROPERTIES

Acrylic fibers are sold mainly as staple and tow. Staple lengths may vary from 25 to 150 mm, depending on the end use. Fiber fineness may vary from 1.0 to 22 dtex (0.9 to 20 dpf); 2.2 dtex (2.0 dpf) and 1.3 dtex (1.2 dpf) are the most common forms.

The fiber cross-section under microscopic examination is generally one of three shapes—round (wet-spun, slow coagulation), bean (wet spun fast coagulation), or dog-bone-shaped (dry spun). It is also possible to produce acrylics with special shapes such as ribbon or mushroom by use of shaped or bicomponent spinnerettes. The physical properties of these fibers are compared with those of natural fibers and other synthetic fibers in Table 1.

CHEMICAL PROPERTIES

Among the outstanding properties of acrylic fibers is their very strong resistance to sunlight. Acrylic fibers are also resistant to all biological and most chemical agents. In resistance of fibers to oxidizing agents, Orlon acrylic was compared to cotton and acetate yarns. Acrylic fibers discolor and decompose rather than melt when heated.

Flammability

Most apparel uses either do not have any flammability standard, or only a modest one which serves to eliminate "torch" fabrics. More rigorous standards are applied for

Table 1. Physical Properties of Staple Fibers

Property	Acrylic	Modacrylic	Nylon-6,6	Polyester	Polyolefin	Cotton	Wool
sp gr	1.14–1.19	1.28–1.37	1.14	1.38	0.90–1.0	1.54	1.28–1.32
tenacity, N/tex[a]							
dry	0.09–0.33	0.13–0.25	0.26–0.64	0.31–0.53	0.31–0.40	0.18–0.44	0.09–0.15
wet	0.14–0.24	0.11–0.23	0.22–0.54	0.31–0.53	0.31–0.40	0.21–0.53	0.07–0.14
loop/knot tenacity	0.09–0.3	0.11–0.19	0.33–0.52	0.11–0.50	0.27–0.35		
breaking elongation, %							
dry	35–55	45–60	16–75	18–60	30–150	<10	25–35
wet	40–60	45–65	18–78	18–60	30–150	25–50	
average modulus, N/tex[a] dry	0.44–0.62	0.34	0.88–0.40	0.62–2.75	1.8–2.65		
elastic recovery, %							
2% stretch	99	99–100	99	67–86	96	74	99
10% stretch		95		57–74			
20% stretch							65
electrical resistance	high	high	very high	high	high	low	low
static buildup	moderate	moderate	very high	high	high	low	low
flammability	moderate	low	self-extinguishing	moderate	moderate	spontaneous ignition at 360°C	self-extinguishing
limiting oxygen index	0.18	0.27	0.20	0.21		0.18	0.25
char/melt	melts	melts	melts, drips	melts, drips	melts	chars	chars
resistance to sunlight	excellent	excellent	poor; must be stabilized	good	poor; must be stabilized	fair; degrades	fair; degrades
resistance to chemical attack	excellent	excellent	good	good	excellent	attacked by acids	attacked by alkalies, oxidizing, and reducing agents
abrasion resistance	moderate	moderate	very good	very good	excellent	good	moderate
index of birefringence	0.1		0.6	0.16			0.01
moisture regain, 65% rh, 21°C, %	1.5–2.5	1.5–3.5	4–5	0.1–0.2	0	7–8	13–15

[a]To convert cN/tex to gf/den, multiple by 0.113.

end uses such as carpet, children's sleepwear, drapery, and bedding. Fibers for these applications must be self-extinguishing after removal from the ignition source. Cotton, rayon, and acrylics burn with the formation of a char. The char acts as a wick that feeds additional fuel to the flame. Nylon and polyester meet some flammability tests by melting away from the ignition source. Modacrylics self-extinguish by generation of chlorine radicals which interfere with the flame-propogation mechanism. This is generally achieved by incorporating vinylidene chloride or vinyl chloride comonomers. Blends of a char-forming fiber with a meltable one require incorporation of an active fire-retardant to meet any stringent flammability test.

FIBER CHARACTERIZATION

In addition to characterizing the properties introduced by the choice of comonomers and the polymerization process itself, further characterization is required to describe the properties imparted by spinning and subsequent down-stream processing. These properties relate to the order and microstructure of the fibers, and the resultant performance characteristics, such as crimp retention, abrasion resistance and mechanical properties.

ACRYLONITRILE POLYMERIZATION

Virtually all acrylic fibers are made from acrylonitrile combined with at least one other monomer. The comonomers most commonly used are neutral comonomers, such as methyl acrylate and vinyl acetate to increase the solubility of the polymer in spinning solvents, modify the fiber morphology, and improve the rate of diffusion of dyes into the fiber. Sulfonated monomers, such as sodium p-(vinylbenzene) sulfonate (SSS), sodium methallyl sulfonate (SMAS), and sodium p-(sulfophenyl) methallyl ether (SPME) are used to provide additional dye sites or to provide a hydrophilic component in water-reversible-crimp bicomponent fibers. Halogenated monomers, usually vinylidene chloride or vinyl chloride, impart flame resistance to fibers used in the home furnishings, awning, and sleepwear markets.

Polymerization Methods

Aqueous dispersion polymerization is the most common commercial method; solution polymerization, where the spin solvent serves as the polymerization medium, is the other commercial process. Emulsion polymerization is used for certain modacrylic compositions.

Aqueous Dispersion Polymerization. By far, the most widely used method of polymerization in the acrylic fibers industry is aqueous dispersion (also called suspension). When inorganic compounds such as persulfates, chlorates, or hydrogen peroxide are used as radical generators, the initiation and primary radical growth steps occur mainly in the aqueous phase. Chain growth is limited in the aqueous phase, however, because the monomer concentration is normally low and the polymer is insoluble in water. Nucleation occurs when aqueous chains aggregate or collapse after reaching a threshold molecular weight.

Solution Polymerization. Solution polymerization is used by a few producers in the acrylic fiber industry. The reaction is carried out in a homogeneous medium by using a solvent for the polymer. The homogeneous solution polymerization of acrylonitrile follows the conventional kinetic scheme developed for vinyl monomers.

Of the two common comonomers incorporated in textile-grade acrylics, methyl acrylate is the least active in chain transfer whereas vinyl acetate is as active in chain transfer as DMF. Vinyl acetate is also known to participate in the chain transfer-to-polymer reaction. This occurs primarily at high conversion, where the concentration of polymer is high and monomer is scarce.

The advantage of solution polymerization is that the polymer solution can be converted directly to spin dope by removing the unreacted monomer. Incorporation of nonvolatile monomers, such as the sulfonated monomers can be a problem. The sulfonated monomers must be converted to a soluble form such as the amine salt. Nonvolatile monomers are difficult to recover or purge from the reaction medium.

Bulk Polymerization. The idea of bulk polymerization is attractive, since the polymer would not require water removal and the process would not have the low propagation rates and high-chain transfer rates of solution processes. But bulk polymerization of acrylonitrile is complex. Even after many investigations into the kinetics of the polymerization, it is still not completely understood. The complexity arises because the polymer precipitates from the reaction mixture barely swollen by its monomer. The heterogeneity leads to kinetics that deviate from normal.

Emulsion Polymerization. The use of emulsion polymerization in the acrylic fiber industry is limited to the manufacture of modacrylic compositions. The ability of emulsion polymerization to segregate radicals from one another is of great importance commercially. The effect is to minimize the rate of radical recombination, allowing high rates of polymerization to be achieved along with high molecular weight. This is important in modacrylic polymerizations where chain-transfer constants of the halogen monomers are high.

Copolymerization

Homogeneous Copolymerization. Virtually all acrylic fibers are made from acrylonitrile copolymers containing one or more additional monomers that modify the properties of the fiber. When carried out in a homogeneous solution, the copolymerization of acrylonitrile follows the normal kinetic rate laws of copolymerization.

Copolymer composition can be predicted for copolymerizations with two or more components, such as those employing acrylonitrile plus a neutral monomer and an ionic dye receptor.

Heterogeneous Copolymerization

When copolymer is prepared in a homogeneous solution, kinetic expressions can be used to predict copolymer composition. Bulk and dispersion polymerization are somewhat different since the reaction medium is heterogeneous and polymerization occurs simultaneously in separate loci. In bulk polymerization, eg, the monomer-swollen polymer particles support polymerization within the particle core as well as on the particle surface. In aqueous dispersion or emulsion polymerization, the monomer is actually dispersed in two or three distinct phases: a continuous aqueous phase, a monomer droplet phase, and a phase consisting of polymer particles swollen at the surface with monomer. Where polymerization occurs predominantly in the organic phases these relatively water-soluble monomers may incorporate into the copolymer at lower levels than expected.

In a CSTR dispersion process, the percentage of the less reactive monomer increases until steady state is reached.

SOLUTION SPINNING

Dry Spinning

This was the process first employed commercially by DuPont in 1950. For acrylic fibers, the only dry spinning solvent used commercially is DMF. The DMF spin dope coming from the dope preparation unit is filtered and then heated to approximately $140°C$. It is pumped through spinnerets of up to 2800 holes placed at the top of a solvent removal tower. Consequently, the fiber from the bottom of the tower contains 10-25% solvent. In discontinuous processes, the fiber exiting the tower is wet with water and combined with the product from other threadlines into a rope; the rope is plaited into a can. The residual DMF is removed in a second step by passing the rope via roll sets through a series of hot water baths. A more modern process, introduced by Bayer, washes the fiber by sprays while passing on a belt. The as-spun fiber has little orientation, so it is stretched 3–6X either before or concurrent with the washing step. The fiber is crimped to improve bulk and textile processing, then dried by heated air on a moving belt. During drying the fiber structure collapses to the same density as solid polymer and the length decreases as the structure relaxes. A "finish" comprising an antistatic agent and a lubricant are applied by spray or kiss rolls and the product is either cut to staple or packaged directly as tow.

Wet Spinning

Wet spinning differs from dry spinning primarily in the way solvent is removed from the extruded filaments. Instead of evaporating the solvent in a drying tower, the fiber is spun into a liquid bath containing a solvent–nonsolvent mixture called the coagulant. The solvent is the same as the dope solvent and the non solvent is usually water.

Air Gap Spinning

This process, also termed dry-jet wet spinning, is used to provide filament yarn either for textile use or as a carbon fiber precursor. It is suitable for producing the small bundles required for these end uses because the filament has been drawn before it enters the bath, so drag forces are less likely to cause breakage; thus much higher line speeds can be achieved. In theory any acrylic solvent can be used in air gap spinning.

Solvent Recovery

Efficient use of solvent and water are key elements in an economic process. With most spinning processes practiced on a large scale, <1% solvent is not recycled, based on fiber produced.

The main means of solvent recycling is distillation, either atmospheric or vacuum. With the organic solvents, the water is distilled, perhaps in several steps, then the higher boiling solvent is distilled, leaving behind dissolved salts and low molecular weight polymer.

Melt Spinning

Compared to most other synthetic fibers, acrylics have always had the disadvantage of extra process steps and cost incurred because they could not be melt spun. Several approaches to eliminating this limitation have been proposed.

A true melt spinning process has been developed by a group at Standard Oil. Their approach was to make a polymer containing substantial comonomer content by a process which minimized "blocking" of AN groups. The resultant polymer was melt processible without degradation. Possible limitations of this approach are that the high comonomer content leads to high relaxation shrinkage, lower softening and sticking temperatures.

Modifications of Properties

Reduced Pilling. Staple fabrics, in general, develop small balls of fiber or pills on the fabric surface as a result of abrasive action on the fabric surface. However, the pills build up more on acrylic fabrics than on comparable woolens. Pilling can be reduced by increasing the likelihood that the pills will break or wear off. Thus the most effective approaches include reducing fiber strength, incorporating defects in the fiber, increasing fiber brittleness, and reducing shear strength.

Improved Abrasion Resistance. Abrasion resistance is generally improved by reducing the microvoid size and increasing the initial fiber density. Abrasion resistant fibers have been produced by incorporating hydrophilic comonomers or comonomers with small molar volumes.

COMMERCIAL PRODUCTS

The majority of acrylic fiber production is 1.0–5.6 dtex (0.9–5 den) staple and tow furnished, undyed, in either bright or semidull ($\sim$0.5% TiO_2) luster. The principal markets are in apparel and home furnishings. Within the apparel sector these fibers are used in sweaters and in single jersey, double-knit, and warp-knit fabrics for a variety of knitted outerwear garments such as dresses, suits, and children's wear. Other markets for acrylics in the knit goods area are hand-knitting yarns, deep-pile fabrics, circular knits, fleece fabrics and half-hose. Acrylics also find uses in broadwoven fabric categories such as blankets, drapery and upholstery. Minor tufted end uses include area rugs and carpets.

Acrylic Tow

A significant proportion of acrylic fiber, perhaps 25% in the United States and >50% in Europe, is sold as tow for conversion to yarn through stretch-breaking using the Superba, Seydel or similar equipment. The larger the tow package, the higher the tow customer's productivity.

Acrylic Filament Yarns

Continuous filament acrylic yarns face stiff competition from nylon and polyester. Since they are more costly, acrylics have penetrated only those markets where they have a clear advantage in a critical property.

Producer Dyed Fiber

The largest volume "specialty" product offered by acrylic producers is producer-dyed fiber (PDF). Producer dyeing decreases systems cost by elimination of a process step for the customer, but it complicates the inventory. The dye is applied using a device which promotes rapid penetration of the fiber mass, as acrylics have a high strike rate and very poor leveling qualities. The process is automated to maintain constant color (shade and depth) by real-time color analysis and correction. Since the final color is influenced by relaxation and crimp, further monitoring and testing may be required. Alternately, the dyes are premixed and a mixed stream is injected into the dyeing device.

Pigmented Fiber

Pigmented acrylic and a small amount of modacrylic are used in outdoor applications where outstanding light stability provides a competitive advantage. Pigmentation provides more stable coloration than dyeing through the lifetime of the fabric. End uses include awnings, tents and lawn furniture. Modacrylic is used where local codes require a flame-retardant fabric. The technology involves mixing of the pigments with the spin dope prior to extrusion. The same feedback mechanism of color control described for PDF may be used with pigmented fiber.

Fibers with High Bulk and Pile Properties

High-bulk acrylic fibers are commonly made by blending high shrinkage and low shrinkage staple fibers or by blending relaxed and unrelaxed sliver from tow. The two staple products are made by variation in the fiber stabilization process. When the resulting yarn is allowed to relax, the high-shrink component causes the low shrinkage (relaxed) fiber to buckle and add bulk to the yarns.

Another method of producing high bulk yarns is the use of bicomponent fibers.

Flame-Resistant Fibers

Acrylics have relatively low flame resistance, comparable to cotton and regenerated cellulose fibers. Additional flame resistance is required for certain end uses, such as children's sleepwear, blankets, carpets, outdoor awnings, and drapery fabrics. The only feasible route is copolymerization of acrylonitrile with halogen-containing monomers such as vinyl chloride, vinyl bromide, or vinylidene chloride. Modacrylics were developed for uses where a high resistance to burning is required. In such fibers, the level of halogen-containing units was up to 60%.

High Strength Fibers by Conventional Solution Spinning

As a reinforcing material for ambient-cured cement building products, acrylics offer three key properties: high elastic modulus, good adhesion, and good alkali resistance. The high modulus requires an unusually high stretch orientation. This can be accomplished by stretching the fiber 8 to 14X above its glass-transition temperature, T_g. Normally, this is done in boiling water or steam to give moduli of 8.8 to 13 N/tex (100–150 gf/den). Alternatively, the stretch orientation can be achieved by a combination of wet stretch at 100°C and plastic stretch on hot rolls or in a heat-transfer fluid such as glycerol. This technique is reported to give moduli as high as 17.6 N/tex (200 gf/den).

Carbon Fiber

Carbon fibers are valued for their unique combination of extremely high modulus and strength and low specific gravity. Precursors for carbon fiber can be pitch, rayon or acrylic fiber. Rayon offers a very low yield of carbon fiber and is no longer used as a precursor. Pitch is useful for generating carbon fiber of exceptionally high modulus, but the predominant precursor is acrylic.

Other Specialty Fibers

Other specialty fibers include microdenier, antimicrobial, fibrillated fibers and conductive fibers.

F. W. Billmeyer, Jr., *Textbook of Polymer Science*, 3rd ed., John Wiley & Sons, Inc., New York, 1984, Chapt. 8.

Fibre Consumption in the Main End-uses, International Rayon & Synthetic Fibers Committee Tables, 2000.

Manufactured Fiber Handbook, Fiber Economics Bureau, SRI International, 2000.

J. C. Masson, ed., *Acrylic Fiber Technology and Applications*, Marcel Dekker, New York, 1995.

GARY J. CAPONE
Solutia Inc.

JAMES C. MASSON
JCM Consulting

FIBERS, OLEFIN

Olefin fibers, also called polyolefin fibers, are defined as manufactured fibers in which the fiber-forming substance is a synthetic polymer of at least 85 wt% ethylene, propylene, or other olefin units. Several olefin polymers are capable of forming fibers, but only polypropylene (PP) and, to a much lesser extent, polyethylene (PE) are of practical importance. Olefin polymers are hydrophobic and resistant to most solvents. These properties impart resistance to staining, but cause the polymers to be essentially undyeable in an unmodified form.

The growth of polyolefin fibers continues. Advances in olefin polymerization provide a wide range of polymer properties to the fiber producer. Inroads into new markets are being made through improvements in stabilization, and new and improved methods of extrusion and production, including multicomponent extrusion and spunbonded and meltblown nonwovens.

PROPERTIES

Physical Properties

Table 1 shows that olefin fibers differ from other synthetic fibers in two important respects: (1) olefin fibers have very low moisture absorption and thus excellent stain resistance and almost equal wet and dry properties, and (2) the low density of olefin fibers allows a much lighter weight product at a specified size or coverage. Thus one kilogram of polypropylene fiber can produce a fabric, carpet, etc, with much more fiber per unit area than a kilogram of most other fibers.

Tensile Strength

Tensile properties of all polymers are a function of molecular weight, morphology, and testing conditions. Lower temperature and higher strain rate result in higher breaking stresses at lower elongations, consistent with the general viscoelastic behavior of polymeric materials. Similar effects are observed on other fiber tensile properties, such as tenacity or stress at break, energy to rupture, and extension at break. Under the same spinning, proces-

sing, and testing conditions, higher molecular weight results in higher tensile strength. The effect of molecular weight distribution on tensile properties is complex because of the interaction with spinning conditions. In general, narrower molecular weight distributions result in higher breaking tenacity and lower elongation. The variation of tenacity and elongation with draw ratio for a given spun yarn correlates well with amorphous orientation. However, when different spun yarns are compared, neither average nor amorphous orientation completely explains these variations.

Creep, Stress Relaxation, Elastic Recovery

Olefin fibers exhibit creep, or time-dependent deformation under load, and undergo stress relaxation, or the spontaneous relief of internal stress. High molecular weight and high orientation reduce creep.

Chemical Properties

The hydrocarbon nature of olefin fibers, lacking any polarity, imparts high hydrophobicity and consequently resistance to soiling or staining by polar materials, a property important in carpet and upholstery applications. Unlike the condensation polymer fibers, such as polyester and nylon, olefin fibers are resistant to acids and bases. At room temperature, polyolefins are resistant to most organic solvents, except for some swelling in chlorinated hydrocarbon solvents. At higher temperatures, polyolefins dissolve in aromatic or chlorinated aromatic solvents, and show some solubility in high boiling hydrocarbon solvents. At high temperatures, polyolefins are degraded by strong oxidizing acids.

Thermal and Oxidative Stability

In general, polyolefins undergo thermal transitions at much lower temperatures than condensation polymers, thus the thermal and oxidative stability of polyolefin fibers are comparatively poor. Preferred stabilizers are highly substituted phenols such as Cyanox 1790 and Irganox 1010, or phosphites such as Ultranox 626 and Irgafos 168 (see ANTIOXIDANTS, POLYMERS; HEAT STABILIZERS).

Table 1. Physical Properties of Commercial Fibers

Polymer	Standard tenacity, GPa[a]	Breaking elongation, %	Modulus, GPa[a]	Density, kg/m³	Moisture regain[b]
olefin	0.16–0.44	20–200	0.24–3.22	910	0.01
polyester	0.37–0.73	13–40	2.1–3.7	1380	0.4
carbon	3.1	1	227	1730	
nylon	0.23–0.60	25–65	0.5–2.4	1130	4–5
rayon	0.25–0.42	8–30	0.8–5.3	1500	11–13
acetate	0.14–0.16	25–45	0.41–0.64	1320	6
acrylic	0.22–0.27	35–55	0.51–1.02	1160	1.5
glass	4.6	5.3–5.7	89	2490	
aramid	2.8	2.5–4.0	113	1440	4.5–7
fluorocarbon	0.18–0.74	5–140	0.18–1.48	2100	
polybenzimidazole	0.33–0.38	25–30	1.14–1.52	1430	15

[a]To convert GPa to psi, multiply by 145,000.
[b]At 21°C and 65% rh.

Ultraviolet Degradation

Polyolefins are subject to light-induced degradation; polyethylene is more resistant than polypropylene. Because polyolefins readily form hydroperoxides, the more effective light stabilizers are radical scavengers. Hindered amine light stabilizers (HALS) are favored, especially high molecular weight and polymeric amines that have lower mobility and less tendency to migrate to the surface of the fiber. This migration is commonly called bloom.

Flammability

Most polyolefins can be made fire retardant using a stabilizer, usually a bromine-containing organic compound, and a synergist such as antimony oxide. However, the required loadings are usually too high for fibers to be spun. Fire-retardant polypropylene fibers exhibit reduced light and thermal resistance. Commercial fire-retardant polyolefin fibers have just recently been introduced, but as expected the fibers have limited light stability and poor luster. Where applications require fire retardancy it is usually conferred by fabric finishes or incorporation of fire retardants in a latex, such as in latex-bonded nonwovens and latex-coated wovens.

Dyeing Properties

Because of their nonionic chemical nature, olefin fibers are difficult to dye. A broad variety of polymeric dyesites have been blended with polypropylene; nitrogen-containing copolymers are the most favored. In apparel applications where dyeing is important, dyeable blends are expensive and create problems in spinning fine denier fibers. Hence, olefin fibers are usually colored by pigment blending during manufacture, called solution dying in the trade.

MANUFACTURE AND PROCESSING

Olefin fibers are manufactured commercially by melt spinning, similar to the methods employed for polyester and polyamide fibers.

Slit-Film Fiber

A substantial volume of olefin fiber is produced by slit-film or film-to-fiber technology. For producing filaments with high linear density, above 0.7 tex (6.6 den), the production economics are more favorable than monofilament spinning. The fibers are used primarily for carpet backing and rope or cordage applications.

Several more recent variations of the film-to-fiber approach result in direct conversion of film to fabric. The film may be embossed in a controlled pattern and subsequently drawn uniaxially or biaxially to produce a variety of nonwoven products. Addition of chemical blowing agents to the film causes fibrillation upon extrusion. Nonwovens can be formed directly from blown film using a unique radial die and control of the biaxial draw ratio (see NONWOVEN FABRICS, STAPLE FIBERS).

Bicomponent Fibers

Polypropylene fibers have made substantial inroads into nonwoven markets because they are easily thermal bonded. Further enhancement in thermal bonding is obtained using bicomponent fibers. In these fibers, two incompatible polymers, such as polypropylene and polyethylene, polyester and polyethylene, or polyester and polypropylene, are spun together to give a fiber with a side-by-side or core–sheath arrangement of the two materials. The lower melting polymer can melt and form adhesive bonds to other fibers; the higher melting component causes the fiber to retain some of its textile characteristics. Bicomponent fibers have also provided a route to self-texturing (self-crimping) fibers.

Meltblown, Spunbond, and Spurted Fibers

A variety of directly formed nonwovens exhibiting excellent filtration characteristics are made by meltblown processes, producing very fine, submicrometer filaments. A stream of high velocity hot air is directed on the molten polymer filaments as they are extruded from a spinneret. This air attenuates, entangles, and transports the fiber to a collection device. Because the fiber cannot be separated and wound for subsequent processing, a nonwoven web is directly formed.

In the spunbond process, the fiber is spun similarly to conventional melt spinning, but the fibers are attenuated by air drag applied at a distance from the spinneret. This allows a reasonably high level of filament orientation to be developed. The fibers are directly deposited onto a moving conveyor belt as a web of continuous randomly oriented filaments.

Pulp-like olefin fibers are produced by a high pressure spurting process developed by Hercules Inc. and Solvay, Inc. Polypropylene or polyethylene is dissolved in volatile solvents at high temperature and pressure. After the solution is released, the solvent is volatilized, and the polymer expands into a highly fluffed, pulp-like product. Additives are included to modify the surface characteristics of the pulp. Uses include felted fabrics, substitution in whole or in part for wood pulp in papermaking, and replacement of asbestos in reinforcing applications.

High Strength Fibers

The properties of commercial olefin fibers are far inferior to those theoretically attainable. A number of methods, including superdrawing, high pressure extrusion, spinning of liquid crystalline polymers or solutions, gel spinning, and hot drawing, but these methods are tedious and uneconomical for olefin fibers. A high modulus commercial polyethylene fiber with properties approaching those of aramid and graphite fibers is prepared by gel spinning.

Hard-Elastic Fibers

Hard-elastic fibers are prepared by annealing a moderately oriented spun yarn at high temperature under tension. They are prepared from a variety of olefin polymers, acetal copolymers, and polypivalolactone.

APPLICATIONS

Olefin fibers are used for a variety of purposes from home furnishings to industrial applications. These include carpets, upholstery, drapery, rope, geotextiles, and both disposable and nondisposable nonwovens. Olefin fiber use in apparel has been restricted by low melting temperatures, which make machine drying and ironing of polyethylene and polypropylene fabrics difficult or impossible.

Polypropylene fibers are used in every aspect of carpet construction from face fiber to primary and secondary backings. Polypropylene's advantages over jute as carpet backing are dimensional stability and minimal moisture absorption. Drawbacks include difficulty in dyeing and higher cost. Bulked-continuous-filament (BCF) carpet yarns provide face fiber with improved crimp and elasticity; BCF carpet yarns are especially important in contract carpets.

Olefin fiber is an important material for nonwovens. Disposable nonwoven applications include hygienic coverstock, sanitary wipes, and medical roll goods.

A special use for meltblown olefin fiber is in filtration media such as surgical masks and industrial filters.

Other applications, including rope, cordage, outdoor furniture webbing, bags, and synthetic turf, make up the remaining segments of the olefin fiber market. Spunbond polyethylene is used in packaging applications requiring high strength and low weight. Specialty olefin fibers are employed in asphalt and concrete reinforcement. Hollow fibers have been tested in several filtration applications. Ultrafine fibers are used in synthetic leather, silk-like fabrics, and special filters. These fibers are also used in sports outerwear, where the tight weaves produce fabrics that are windproof and waterproof, but still are able to pass vapors from perspiration and, thus, keep the wearer cool and dry.

D. J. Carlsson, A. Garton, and D. M. Wiles, in G. Scott, ed., *Developments in Polymer Stabilisation*, Applied Science Publishers, London, 1979, p. 219.

F. Hadjuk, T. Sasano, and S. Schlag, "Nonwoven Fabrics," *Chemical Economics Handbook*, Menlo Park, Calif., Jan. 2003.

J. Zhou and J. E. Spruiell, in *Nonwovens—An Advanced Tutorial*, TAPPI Press, Atlanta, Ga., 1989.

C. J. WUST, JR.
Fibervisions

FIBERS, REGENERATED CELLULOSE

Cellulose is the natural polymer that makes up the living cells of all vegetation. It is the material at the center of the carbon cycle, and the most abundant and renewable biopolymer on the planet. Regenerated cellulose fiber producers have converted it from the fine short fibers that come from trees into the fine long fibers used in textiles and nonwovens for over a century. Regenerated cellulosics nevertheless remain unique among the mass-produced fibers because they are the only ones to use the natural polymer *directly*.

Fibers manufactured from cellulose are either *derivative* or *regenerated*. A *derivative fiber* is one formed when a chemical derivative of a natural polymer, eg, cellulose, is prepared, dissolved, and extruded as a continuous filament, and the chemical nature of the derivative is retained after the fiber formation process (see CELLULOSE ESTERS, INORGANIC). A *regenerated fiber* is one formed when a natural polymer or its chemical derivative is dissolved and extruded and the chemical nature of the natural polymer is either retained or regenerated after the fiber formation process. The difficulties of making solutions of natural cellulose from which fibers can be spun has led to most fabricated cellulosic fibers being regenerated from more readily soluble derivatives of cellulose.

THE VISCOSE PROCESS

The main raw material required for the production of viscose is cellulose, a natural polymer of D-glucose (Fig. 1). The repeating monomer unit is a pair of anhydroglucose units (AGU).

Cellulose is the most abundant polymer, an estimated 10^{11} t being produced annually by natural processes. Supplies for the rayon industry can be obtained from many sources; but in practice, the wood-pulping processes used to supply the needs of the paper and board industries have been adapted to make the necessary specially pure grade. The trees used to make dissolving pulp are fast-growing hard or soft woods farmed specifically for their high quality pulping.

The final properties of the rayon fiber and the efficiency of its manufacturing process depend crucially on the purity of the pulp used. High tenacity fibers need high purity pulps, and this means pulping to get up to 96% of the most desirable form of cellulose (known as alpha cellulose), and removing most of the unwanted hemicellulose and lignin. Cellulose molecular weight must be tightly controlled, and levels of foreign matter such as resin, knots, shives, and silica must be very low. Of the two main pulping processes, the sulfite route produces higher yields of lower alpha, more reactive pulp, suitable for regular staple fiber. Prehydrolyzed kraft pulps are preferred for high strength industrial yarn or modal fiber production. The highest purity pulps, up to 99% alpha cellulose, are obtained from cotton fiber. These are no longer used in viscose production but are now the main raw material for cuprammonium rayon.

The flow diagram for the viscose process is given in Figure 2. The sequence of reactions necessary to convert

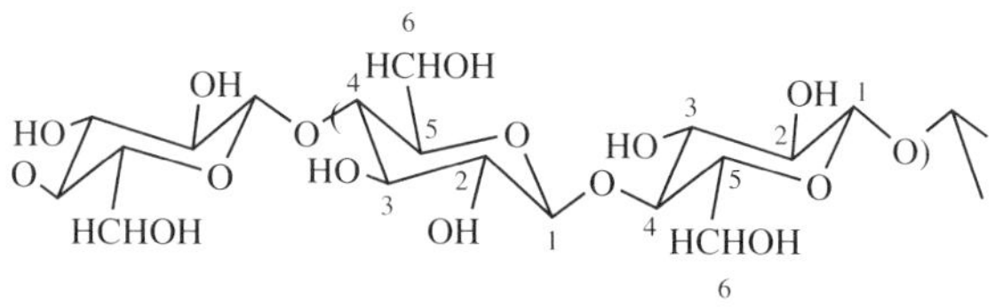

Figure 1. Anhydroglucose units with 1–4 beta linkages as in cellulose.

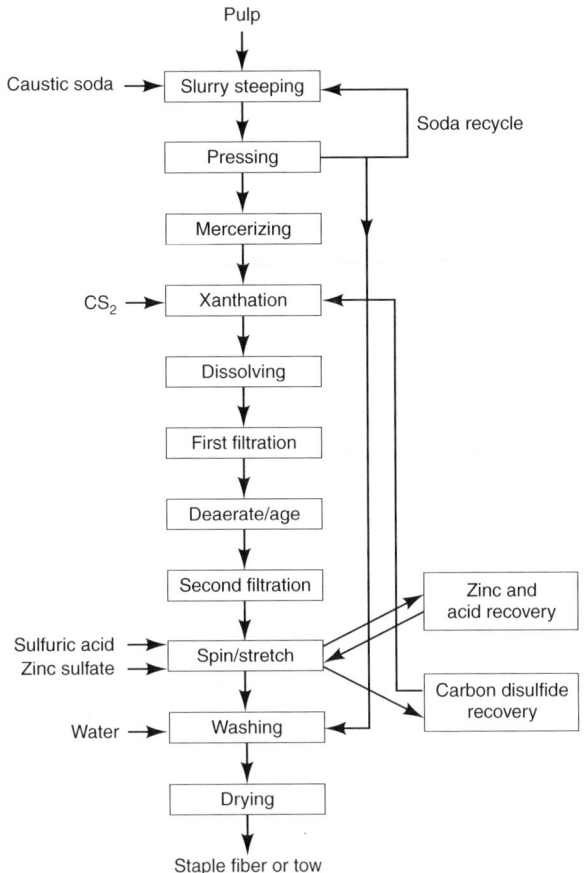

Pulp

Caustic soda → Slurry steeping

Pressing   Soda recycle

Mercerizing

CS₂ → Xanthation

Dissolving

First filtration

Deaerate/age

Second filtration   Zinc and acid recovery

Sulfuric acid → Spin/stretch
Zinc sulfate   Carbon disulfide recovery

Water → Washing

Drying

Staple fiber or tow

Figure 2. The viscose process.

cellulose into its xanthate and dissolve it in soda used to be performed batchwise. Fully continuous processes, or mixtures of batch and continuous process stages, are more appropriate for high volume regular viscose staple production.

Modified Viscose Processes

The need for ever stronger yarns resulted in the first important theme of modified rayon development and culminated—technically if not commercially—in the 0.88 N/tex (10 gf/den) high wet modulus industrial yarn process.

Tire Yarns. A method to increase the strength of viscose yarn from the 0.2 N/tex (2.2 gf/den) standard to levels needed in tires was first patented by Courtaulds in 1935. The next significant strength improvement followed the 1950 Du Pont discovery of monoamine and quaternary ammonium modifiers, which, when added to the viscose, prolonged the life of the zinc cellulose xanthate gel, and enabled even higher stretch levels to be used. Modifiers have proliferated since they were first patented and the list now includes many poly(alkylene oxide) derivatives, polyhydroxypolyamines, and dithiocarbamates.

Fully modified yarns had smooth, all-skin cross sections, a structure made up of numerous small crystallites of cellulose, and filament strengths around 0.4 N/tex (4.5 gf/den). They were generally known as the Super

tire yarns. Improved Super yarns [0.44–0.53 N/tex (5–6 gf/den)] were made by mixing modifiers, and one of the best combinations was found to be dimethylamine with poly(oxyethylene glycol) of about 1500 mol wt. Ethoxylated fatty acid amines have now largely replaced dimethylamine because they are easier to handle and cost less.

The strongest fibers were made using formaldehyde additions to the spin bath while using a mixed modifier system or using highly xanthated viscoses (50% + CS₂). Unfortunately, problems associated with formaldehyde side reactions made the processes more expensive than first thought, and the inevitable brittleness that results whenever regenerated cellulose is highly oriented restricted the fibers to nontextile markets. The commercial operations were closed down in the late 1960s.

The formaldehyde approach is still used by Futamura Chemical (Japan). They make spun-laid viscose nonwovens where the hydroxymethylcellulose xanthate derivative formed from formaldehyde in the spin bath allows the fibers to bond after laying.

High Tenacity Staple Fibers. When stronger staple fibers became marketable, the tire yarn processes were adapted to suit the high productivity staple fiber processes. Improved staple fibers use a variant of the mixed modifier approach to reach 0.26 N/tex (3 gf/den).

The full potential of the mixed modifier tire yarn approach is, however, achieved in the modal or HWM (high wet modulus) staple processes using special viscose-making and -spinning systems. Lenzing still makes Modal staple in Austria. These fibers are most popular in ladies' apparel for their soft handle and easy dyeing to give rich coloration. They are made in finenesses down to 1 dtex (0.9 den) for fine yarns and hydroentangled nonwoven production.

Polynosic Rayons. Another strand of development began in 1952 when Tachikawa patented a method for making strong, high modulus fibers that needed neither zinc nor modifier. The process depends on the fact that minimally aged alk-cell can, after xanthation with an excess of carbon disulfide, be dissolved at a low cellulose concentration to give very viscous viscoses containing high molecular weight cellulose. These viscoses have sufficient structure to be spun into cold, very dilute spin baths containing no zinc and low levels of sodium sulfate. The resulting gel filaments can be stretched up to 300% to give strong, round-section, highly ordered fibers, which can then be regenerated.

The polynosic rayons enjoyed a revival of fortunes in the 1990s when the new lyocell fiber produced by Courtaulds (Tencel)—a solvent spun fiber with polynosic properties—became popular in ladies' apparel and denim.

Bulky Rayons. Permanent chemical crimp can be obtained by creating an asymmetric arrangement of the skin and the core parts of the fiber cross section. Skin cellulose is more highly ordered than core cellulose and shrinks more on drying. If, during filament formation in the spin bath, the skin can be forced to burst open to expose fresh viscose to the acid, a fiber with differing shrinkage potential from side to side is made, and crimp should be obtained. Crimp is most important in rayon used for hygienic absorbent products.

Process conditions that favor chemical crimp formation are similar to those used for improved tenacity staple (zinc/modifier route). However, spin-bath temperature should be as high as possible (ca 60°C) and the spin-bath acidity as low as possible (ca 7%).

Cross-sectional modifications of a more extreme nature than skin-bursting, which nevertheless do not form crimp, have grown in importance since the early 1980s. These yield a permanent bulk increase, which can be translated into bulky fabrics without the need for special care.

Inflation had long been known as an intermittent problem of textile yarn manufacture. It was caused by the skin forming too quickly to allow the escape of gases liberated by the regeneration reactions, or as a result of air in the viscose at the jet.

All these early inflation processes were difficult to control, and after World War II they were neglected until the 1960s. However, their development led to an increased understanding of the inflation process and the identification of conditions that could yield a continuously hollow staple fiber in large-scale production, such as the development of solid Y- and X-shaped multilimbed fibers, which performed like SI fiber but had much lower levels of water imbibition than the inflated version. Their shape and relative stiffness enabled them to absorb more fluid between, as opposed to inside, the fibers. They were therefore as absorbent in use as the inflated versions, but did not require the extra process chemicals, and were easier to wash and dry in production and use. They are the most important bulky rayons now in production.

Alloy Rayons. It is possible to produce a wide variety of effects by adding materials to the viscose dope. The resulting fibers become mixtures or alloys of cellulose and the other material. The two most important types of alloy arise when superabsorbent or flame-retardant fibers are made.

American Enka and Avtex both produced superabsorbent alloy rayons by adding sodium polyacrylate, or copolymers of acrylic and methacrylic acids, or sodium carboxymethylcellulose to the viscose. They performed particularly well in tampons, the only real market that developed, declined after the Toxic Shock Syndrome outbreak in the early 1980s. Other polymers that have formed the basis of absorbent alloys are starch, sodium alginate, poly(ethylene oxide), poly(vinyl pyrrolidinone), and sodium poly(acrylamido-2-methyl-2-propane sulfonic acid).

Flame retardancy can be obtained by adding flame-retardant chemicals to make up about 20% of the fiber weight. Propoxyphosphazine (Ethyl Corp.) retardants were later used in Avtex's PFR fiber, and a bis(5,5-dimethyl-2-thiono-1,3,2-dioxaphosphorinanyl) oxide powder (Sandoz) was the basis of later European FR fiber developments. Alloys with inorganic salts such as silicates or aluminates are possible, the salts being converted to fibrous polyacids when the cellulose is burnt off.

This latter approach is the basis of the Visil flame-retardant fiber introduced by Kemira Oy Saeteri.

CUPRAMMONIUM RAYON

Asahi Chemical Industries (ACI, Japan) is now the leading producer of cuprammonium rayon. Its continuing success with a process that has suffered intense competition from the cheaper viscose and synthetic fibers owes much to their developments of high speed spinning technology and of efficient copper recovery systems.

DIRECT DISSOLUTION PROCESSES

Regenerated fibers produced via the direct dissolution of cellulose in organic solvents are generically known as lyocell fibers.

The production routes already described cope with the difficulties of making a good solution by going through an easy-to-dissolve cellulose derivative, eg, xanthate or a cellulose complex, such as cuprammonium. The ideal process, one that could dissolve the cellulose directly from ground wood, is still some way off, but since the early 1980s significant progress has been made.

The efforts to dissolve cellulose directly as a *base* using phosphoric, sulfuric, and nitric protonic acids, or zinc chloride, thiocyanates, iodides, and bromides as Lewis acids are recorded. With regard to cellulose acting as an *acid*, sodium zincate, hydrazine, and sodium hydroxide are listed as inorganic solvents, and quaternary ammonium hydroxides, amines, dimethylamine–DMSO mixtures, and amine oxides as organic bases. However, despite early promise, the problems of developing fiber production routes using these systems have, with the exception of the amine oxide route, proved insurmountable.

The Courtaulds Lyocell Process

The increasing costs of reducing the environmental impact of the viscose process coupled with the increasing likelihood that the newer cellulose solvents would be capable of yielding a commercially viable fiber process led Courtaulds Research to embark on a systematic search for a new cellulosic fiber process in the late 1970s.

By 1980, N-methyl morpholine-n-oxide (NMMO) was shown to be the best solvent, provided well-known difficulties associated with its thermal stability could be avoided by appropriate chemical engineering.

Filaments obtained from the first single-hole extrusion experiments had promising properties, so Courtaulds committed the resources in 1982 to build a small pilot plant to test the feasibility of overcoming the solvent handling and recovery problems that had prevented earlier commercial exploitation. Scale-up to 1000 kg/week pilot line was possible in 1984, and in 1988 a 25,000 kg/week semicommercial line was commissioned to allow a thorough test of the engineering and end-use development aspects.

Dry-jet wet-spinning of lyocell fiber has been described as being similar to the extrusion of lyotropic rigid-rods, giving rise to a highly microfibrillar structure that is consistent across the entire cross section of the fiber—ie, differing from the skin–core structure observed in viscose rayon. However, microfocus small-angle X-ray scattering has recently shown a definite skin–core structure, with the thin skin having better-oriented voids than the core. This observation is more in keeping with the macro observations of fibrillation in paper-making—the surface appearing to fibrillate easily—whereas breaking up the core region of the fiber proves very difficult.

The new fiber thus had physical properties (Table 1) sufficiently different from regular rayon to allow an initial market development strategy that did not erode the position of the traditional viscose fiber. The unique strength, texture, and coloration potential of the fiber enabled it to command premium prices in upmarket mens' and ladies' outerwear.

OTHER SOVENT ROUTES

Work on other routes to cellulosic fibers continues, some driven by a desire to identify an environmentally benign route to cellulosic fibers that can utilize the large capital investment in the viscose process equipment and hence cost less than a completely new fiber process such as lyocell.

The Finnish viscose producer Kemira Oy Saeteri collaborated with Neste Oy on the development of a carbamate derivative route. This system is based on work that showed that the reaction between cellulose and urea gives a derivative easily dissolved in dilute sodium hydroxide:

$$\text{Cell—OH} + \text{NH}_2\text{—CO—NH}_2 \longrightarrow$$
$$\text{Cell—O—CO—NH}_2 + \text{NH}_3$$

Neste patented an industrial route to a cellulose carbamate pulp that was stable enough to be shipped into rayon plants for dissolution as if it were xanthate. The carbamate solution could be spun into sulfuric acid or sodium carbonate solutions, to give that when completely regenerated had similar properties to viscose rayon's. When incompletely regenerated they were sufficiently self-bonding for use in paper-making. The process was said to be cheaper than the viscose route and to have a lower environmental impact. It has not been commercialized, so no confirmation of its potential is yet available.

Asahi has been applying the steam explosion treatment to dissolving pulp to make it dissolve directly in sodium hydroxide, and they claim a solution of 5% of steam-exploded cellulose in 9.1% NaOH at 4°C being spun into 20% H_2SO_4 at 5°C. The apparently poor fiber properties [best results being 0.16 N/tex (1.8 gf/den) tenacity dry, with 7.3% extension] probably arise because the fibers were syringe-extruded at 8.3 tex/filament (75 den/fil). Asahi feels that this could be the ultimate process for large-scale production of regenerated cellulose fibers.

Patents published in February 1996 describe anisotropic cellulose solutions at concentrations up to 40% in a "solvent containing from 72 to 79 wt% of phosphorous pentoxide." Claiming a highly economical route for the production of tire yarns, the process described therein appears to be simple and elegant.

FIBER PROPERTIES

The bulk properties of regenerated cellulose are the properties of Cellulose II, which is created from Cellulose I by alkaline expansion of the crystal structure (see CELLULOSE). The key textile fiber properties for the most important current varieties of regenerated cellulose are shown in Table 1. Fiber densities vary between 1.53 and 1.50.

A discussion of the fiber properties is complicated by the versatility of cellulose and its conversion routes. Many of the properties can be varied over wide ranges, depending on the objectives of the producer.

Thermal Properties

Fibers are not thermoplastic and are stable at temperatures below 150°C, with the possible exception of slight yellowing. They begin to lose strength gradually above 170°C, and decompose more rapidly above 300°C. They ignite at 420°C and have a heat of combustion of 14,732 J/g (3.5 kcal/g).

Moisture Regain

The fibers are all highly hydrophilic, with moisture regains at 65% rh, ranging from 11 for the polynosics to 13 for regular rayon.

Table 1. Properties of Selected Commercial Rayon Fibers

Property	Cuprammonium	Regular rayon	Improved rayon	Modal	Polynosic	Y-shaped rayon[a]	Solvent spun rayon[b]
dry tenacity, cN/tex[c]	15–20	20–24	24–30	34–36	40–65	18–22	40–44
extensibility at break (dry), %	7–23	20–25	20–25	13–15	8–12	17–22	14–16
wet tenacity, cN/tex[c]	9–12	10–15	12–16	19–21	30–40	9–12	34–38
extensibility at break (wet), %	16–43	25–30	25–35	13–15	10–15	23–30	16–18
water imbibition, %	100	90–100	90–100	75–80	55–70	100–110	65–70
cellulose DP[d]	450–550	250–350	250–350	300–500	550–700	250–350	550–600
initial wet modulus[e]	30–50	40–50	40–50	100–120	140–180	35–45	250–270

[a] The Y-shaped rayon data are based on Courtaulds Galaxy fiber.
[b] The solvent-spun rayon data are based on Courtaulds Tencel fiber.
[c] To convert cN/tex to gf/den, divide by 8.82.
[d] DP = degree of polymerization.
[e] The load required to extend the wet fiber by 5% × 20.

Chemical Properties

The fibers degrade hydrolytically when contacted with hot dilute or cold concentrated mineral acids. Alkalies cause swelling (maximum with 9% NaOH at 25°C) and ultimately disintegration. They are unaffected by most common organic solvents and dry-cleaning agents. They are degraded by strong bleaches such as hypochlorite or peroxide.

Optical Properties

The fibers are birefringent.

Electrical Properties

The electrical properties of the fiber vary with moisture content. The specific resistance of the fibers is around 3×10^6 W/cm at 75% rh and 30°C compared with 1×10^{18} W/cm for pure dry cellulose. The dielectric constant (100 kHz) is 5.3 at 65% rh and 3.5 at 0% rh. The zeta potential in water is -25 mV.

ENVIRONMENTAL ISSUES

Viscose Process

As noted in the introduction, rayon is unique among manufactured fibers because it is the only one to use a natural polymer (cellulose) directly. Polyesters, nylons, polyolefins, and acrylics all come indirectly from vegetation; they come from the polymerization of monomers obtained from reserves of fossil fuels, which in turn were formed by the incomplete biodegradation of vegetation that grew millions of years ago. The extraction of these nonrenewable reserves and the resulting return to the atmosphere of the carbon dioxide from which they were made is one of the most important environmental issues of current times. Cellulosic fibers therefore have much to recommend them provided that the processes used to make them have minimal environmental impact.

Liquid Effluents. Recycling of acid, soda, and zinc have long been necessary economically, and the acid–soda reaction product, sodium sulfate, is extracted and sold into other sectors of the chemical industry.

Gaseous Effluents. Twenty percent of the carbon disulfide used in xanthation is converted into hydrogen sulfide (or equivalents) by the regeneration reactions. Ninety to 95% of this hydrogen sulfide is recoverable by scrubbers that yield sodium hydrogen sulfide for the tanning or pulp industries, or for conversion back to sulfur. Up to 60% of the carbon disulfide is recyclable by condensation from rich streams, but costly carbon-bed absorption from lean streams is necessary to recover the remaining $20 + \%$. The technology to deal with this is becoming available, but there remains the danger that cost increases resulting from the necessary investments will make the fibers unattractively expensive compared with synthetics based on cheap nonrenewable fossil fuels.

Energy Use. Energy consumption in the xanthate process compares favorably with the synthetics. The methodology of assessing the energy usage of products and processes is currently the subject of much debate, and a standardized approach has yet to emerge. Not surprisingly, most of the published work on fibers was carried out during the last energy crises in the 1973–1981 period.

Fiber Disposal. Cellulosic fibers, like the vegetation from which they arise, can become food for microorganisms and higher life forms; ie, they biodegrade.

It is also possible to liberate and use some of the free solar energy that powered the manufacture of sugars and cellulose during photosynthesis. This can be achieved by burning or by anaerobic digestion. If future landfills are lined and operated with moisture addition and leachate recycling, then energy generation and the return of landfill sites to normal use can be accelerated.

FUTURE CONSIDERATIONS

For 60 years, cotton production has grown almost entirely because of increased fiber yield per hectare. Land area under cotton cultivation has been constant for that period, and pressure for the same top quality agricultural land will increase because of the need to feed increased population. Better irrigation, higher pesticide use, higher chemical fertilizer use, and genetic improvements achieved the cotton yield increase. Howerver, in recent years cotton yield per acre has leveled off.

It is taken for granted that the implied need for an additional 50 million tons of synthetics could be readily provided. The predicted 20 million tons shortfall in absorbent or "comfort" fibers is likely to be filled by a combination of a variety of technologies: (1) development of more comfortable blends containing progressively less cellulosic fiber; (2) some increased yield of cotton per hectare from genetic engineering advances yet to be achieved; (3) some increased use of prime agricultural land for cotton growing; (4) development of low cost viscose and lyocell fibers; (5) development of soft flax fibers encouraged by agricultural subsidies and improved fiber-separation methods; (6) development of truly hydrophilic synthetics; (7) the direct use of wood fiber in textiles.

In order for rayon to benefit from the opportunity offered by this increasing demand for fibers, the price differential between cellulosics and polyester must be reduced.

W. Albrecht and co-workers, in *Man Made Fiber Yearbook*, CTI, Maryland, 1991, pp. 26–44.

D. C. Coleman, *Courtaulds: An Economic and Social History*, Vols. 2 and 3, Clarendon Press, New York, 1980.

J. Dyer and G. C. Daul, *The Handbook of Fiber Science and Technology*, Vol. 4: *Fiber Chemistry*, Marcel Dekker, Inc., New York, 1985, Chap. 11.

C. R. Woodings, *Regenerated Cellulose Fibers*, CRC Press, 2001, (ISBN 0-8493-1147-0), and Woodhead Publishing Ltd., 2001 (ISBN 1-85573-459-1).

CALVIN WOODINGS
Calvin Woodings Consulting Ltd.

FIBERS, VEGETABLE

Vegetable fibers, as the name implies, are derived from plants. The principal chemical component in plants is cellulose, and therefore they are also referred to as cellulosic fibers. The fibers are usually bound by a natural phenolic polymer, lignin, which also is frequently present in the cell wall of the fiber; thus vegetable fibers are also often referred to as lignocellulosic fibers, except for cotton, which does not contain lignin.

Vegetable fibers are classified according to their source in plants as follows: (1) the bast or stem fibers, which form the fibrous bundles in the inner bark (phloem or bast) of the plant stems, are often referred to as soft fibers for textile use; (2) the leaf fibers, which run lengthwise through the leaves of monocotyledonous plants, are also referred to as hard fibers; and (3) the seed-hair fibers, the source of cotton, which is the most important vegetable fiber. There are >250,000 species of higher plants; however, only a very limited number of species have been exploited

for commercial uses (<0.1%). The commercially important fibers are given in Table 1.

World markets for vegetable fibers have been steadily declining in recent years, mainly as a result of substitution with synthetic materials. Jute has traditionally been one of the principal bast fibers (tonnage basis) sold on the world market; however, the precipitous decline in jute exports by India (Fig. 2) indicate the decreasing market demand for this fiber that has been vitally important to the economies of India (West Bengal), Bangladesh, and Pakistan.

GENERAL PROPERTIES

Chemical Composition

Chemically, cotton is the purest, containing >90% cellulose with little or no lignin. The other fibers contain 40–75% cellulose, depending on processing. Boiled and bleached flax and degummed ramie may contain >95% cellulose. Kenaf and jute contain higher contents of lignin, which contributes

Table 1. Vegetable Fibers of Commercial Interest

Commercial name	Source	Botanical name of plant	Growing area
		Bast or soft fibers	
China jute	Abutilon	*Abutilon theophrasti*	China
flax		*Linum usitatissimum*	north and south temperate zones
hemp		*Cannabis sativa*	all temperate zones
jute		*Corchorus capsularis; C. olitorius*	India
kenaf		*Hibiscus cannabinus*	India, Iran, CIS, South America
ramie		*Boehmeria nivea*	China, Japan, United States
roselle		*Hibiscus sabdariffa*	Brazil, Indonesia (Java)
sunn		*Crotalaria juncea*	India
urena	cadillo	*Urena lobata*	Zaire, Brazil
		Leaf or hard fibers	
abaca		*Musa textilis*	Borneo, Philippines, Sumatra
cantala	Manila maguey	*Agave cantala*	Philippines, Indonesia
caroa		*Neoglaziovia variegata*	Brazil
henequen		*Agave fourcroydes*	Australia, Cuba, Mexico
istle		*Agave* (various species)	Mexico
mauritius		*Furcraea gigantea*	Brazil, Mauritius, Venezuela, tropics
phormium		*Phormium tenax*	Argentina, Chile, New Zealand
pineapple	piña	*Ananas comasus*	Hawaii, Philippines, Indonesia, India, West Indies
sansevieria	bowstring hemp	*Sansevieria* (entire genus)	Africa, Asia, South America
sisal		*Agave sisalana*	Haiti, Java, Mexico, South Africa, Brazil
		Seed-hair fibers	
coir	coconut husk fiber	*Cocos nucifera*	tropics, India, Mexico
cotton		*Gossypium* sp.	United States, Asia, Africa
kapok		*Ceiba pentandra*	tropics
milkweed floss		*Chorisia* sp.	North America
		Other fibers	
broom root	roots	*Muhlenbergia macroura*	Mexico
broom corn	flower head	*Sorghum vulgare technicum*	United States
crin vegetal	palm leaf segments	*Chamaerops humilis*	North Africa
palmyra palm	palm leaf stem	*Brossus flabellifera*	India
pissava	palm leaf base fibers	*Attalea funifera*	Brazil
raffia	palm leaf segments	*Raphia raffia*	East Africa

to their stiffness. Although the cellulose contents are fairly uniform, the other components, eg, hemicelluloses, pectins, extractives, and lignin vary widely without obvious pattern. These differences may characterize specific fibers.

Fiber Dimensions

Except for the seed-hair fibers, the vegetable fibers of bast or leaf origins are multicelled and are used as strands. In contrast to the bast fibers, leaf fibers are not readily broken down into their ultimate cells. The ultimate cells are composites of microfibrils, which, in turn, are comprised of groups of parallel cellulose chains.

Physical Properties

Bast and leaf fibers are stronger (higher tensile strength and modulus of elasticity) but lower in elongation (extensibility) than cotton. Vegetable fibers are stiffer but less tough than synthetic fibers. Kapok and coir are relatively low in strength; kapok is known for its buoyancy.

Among the bast textile fibers, the density is close to 1.5 g/cm^3, or that of cellulose itself, and they are denser than polyester, as shown in Table 5. Moisture regain (absorbency) is highest in jute at 14%, whereas that of polyester is <1%. The bast fibers are typically low in elongation and recovery from stretch. Ramie fiber has a particularly high fiber length/width ratio.

The microfibrils in vegetable fibers are spiral and parallel to one another in the cell wall. The spiral angles in flax, hemp, ramie, and other bast fibers are lower than cotton, which accounts for the low extensibility of bast fibers.

PROCESSING AND FIBER CHARACTERISTICS

Bast Fibers

Bast fibers occur in the phloem or bark of certain plants. The bast fibers are in the form of bundles or strands that act as reinforcing elements and help the plant to remain erect. The plants are harvested and the strands of bast fibers are released from the rest of the tissue by retting, common for isolation of most bast fibers. The retted material is then further processed by breaking, scutching, and hackling.

Leaf (Hard) Fibers

Hard or cordage fibers are found in the fibrovascular systems of the leaves of perennial, monocotyledonous plants growing in Central America, East Africa, Indonesia, Mexico, and the Philippines. They are generally of the Agave and Musa genera. The leaf elements are harvested by cutting at the base with a sickle-like tool, and bundled for processing by hand or by machine decortication. In the latter case, the leaves are crushed, scraped, and washed. The fibers are generally coarser than the bast fibers and are graded for export according to national rules for fineness, luster, cleanliness, color, and strength.

Seed- and Fruit-Hair Fibers

The seeds and fruits of plants are often attached to hairs or fibers or encased in a husk that may be fibrous. These fibers are cellulosic based and of commercial importance, especially cotton, the most important natural textile fiber.

Palm and Other Fibers

These palm (palm family, Arecaceae) fibers are obtained from the palm leaf base of *Attalea funifera* growing in Brazil and the palm leaf segments of *Chamaerops humilis* growing in North Africa. The former are used for cordage and brushes, the latter for stuffing.

Broom root fiber is obtained from the root of the bunch grass, *Muhlenbergia macroura* (Poaceae), found in Mexico, where it grows 1–3 m high. The long fibers are bleached in fumes of burning sulfur before grading according to length. They are used in stiff scrubbing and scraping brushes and whisk brooms.

Broom corn is the fiber obtained from the flower head of another grass, *Sorgum vulgare technicum*, grown in the United States. The fibers are less stiff than those of the broom root and are used in brooms.

USES

Vegetable fibers have application in a broad range of fibrous products, including textiles and woven goods, cordage and twines, stuffing and upholstery materials, brushes, paper and new biobased composites. The uses for each of the specific fibers have been discussed in the designated sections. The traditional uses for the vegetable fibers have been eroded by substitution with synthetics on the world market. The declining uses include cordage, mats, filling material, brushes, etc. However, the unique properties of the bast fibers have allowed continued use in such specialty papers as bank notes, some writing papers, and cigarette papers.

Work at the USDA and Kenaf International (Texas) during the past decade demonstrated the potential of both growing and processing kenaf fibers for newsprint and other paper products in the United States. Another promising use for vegetable fibers is in the new biobased composites that are now marketed in various parts of the industrialized world. The vegetable fibers are mixed with thermoplastic or thermosetting resin matrices and either extrusion or compression molded into a variety of useful shapes. Such products are already utilized in the automotive industry for automobile interior door and head liners and as trunk liners.

D. M. Colling and J. E. Grayson, *Identification of Vegetable Fibres*, Chapman and Hall Ltd., London, 1982.

R. M. Rowell and R. A. Young, eds., *Paper and Composites from Agro-Based Resources*, CRC/Lewis Pub., Boca Raton, Fla., 1997.

R. A. Young and M. Akhtar, eds., *Environmentally Friendly Technologies for the Pulp and Paper Industry*, John Wiley & Sons, Inc., New York, 1998.

RAYMOND A. YOUNG
University of Wisconsin-Madison

FILLERS

By definition, fillers are used to extend a material and to reduce its cost. However, despite this fact that few inexpensive fillers, such as walnut shells, fly ash, wood flour, and wood cellulose, are still being used purely for filling purposes, nearly all fillers employed provide more than space filling. Considering their relative higher stiffness to the material matrix, they will always modify the mechanical properties of the final filled products, or composites. Fillers can constitute either a major or a minor part of a composite. The structure of filler particles ranges from precise geometrical forms, such as spheres, hexagonal plates, or short fibers, to irregular masses. Fillers are generally used for nondecorative purposes in contrast to pigments, although they may incidentally impart color or opacity to a material.

Fillers can be classified according to their source, function, composition, and/or morphology. No single classification scheme is entirely adequate due to the overlap and ambiguity of these categories. Extensive usage of particulate fillers in many commerical polymers is for the enhancement in stiffness, strength, dimensional stability, toughness, heat distortion temperature, damping, impermeability, and cost reduction, although not all of these desirable features are found in any single filled polymer. The properties of particulate-filled polymers are determined by the properties of the components, by the shape of the filler phase, by the morphology of the system, and by the polymer-filler interfacial interactions.

PHYSICAL PROPERTIES OF FILLERS

The overall value of a filler is a complex function of intrinsic material characteristics, such as average particle size, particle shape, intrinsic strength, and chemical composition; of process-dependent factors, such as particle-size distribution, surface chemistry, particle agglomeration, and bulk density; and of cost. Abrasion and hardness properties are also important for their impact on the wear and maintenance of processing and molding equipment.

Particle Morphology, Shape, Size, and Distribution

Filler particles come in a variety of shapes and sizes. In general, for most polymer applications, the filler size required is <40μ. Finer particles, <3μ, provide stronger enhancements in properties. Nanoparticles, with dimensions ranging up to 100 nm, deliver the strongest enhancements when they all are properly dispersed.

The shape of an individual particle has great impact on the flexural modulus, permeability, and flow behavior of a filled polymer. Almost all fillers do not exist as the discrete individual particles of their primary structure. They form aggregates, ie, secondary structure, which can agglomerate into tertiary structures in the material to be filled. An aggregate is a collection of primary particles that are chemically bonded together. In any commercial filler grade, there exists a collection of multiple shapes and sizes. The particle size and shape distributions of fillers can be best measured by direct microscopic inspection together with image processing although there are other methods to determine particle sizes and shapes.

Intrinsic Strength, Hardness, and Abrasivity

The intrinsic strengths and moduli of some crystalline fillers can be calculated along the crystallographic axes using molecular simulation. Particles with sharp edges or rod shapes are more abrasive than those of smooth and round particles, and large particles are more abrasive that smaller particles of the same shape. Additionally, the coefficient of friction, surface treatment, surface energy, and purity of a filler all affect its abrasivity.

Surface Area, Chemistry, Wetting, and Coupling

Available surface areas of fillers include surfaces of filler aggregates and agglomerates and surfaces in their pores, crevices, and cracks. The chemical compatibility between the filler surface and a polymer to be filled is critically important in both the wetting and dispersion of this filler by the polymer and the final physical performance of the resulting filled polymer. Organosilane wetting agents are also called silane coupling agents and, in general, consist of a trialkoxy group and a functional group having a $(RO)_3-Si-R^*$ structure.

Loading and Density

The amount of filler in a filled polymer is termed the loading and is always expressed quantitatively although the quantitative measures vary from industry to industry.

The optimal loading of fillers in a polymer is a balance between physical property enhancement and trade-off, and processing and material cost over the filler loading range.

The average mass per unit volume of the individual particle is the true density or specific gravity of the filler. Densities of finely divided, porous, and irregular fillers are typically measured by a gas pycnometer that ensures all pores and crevices of filler agglomerates are penetrated. Bulk density is used in weighting fillers during filler purchasing, shipping, and storage.

FILLED POLYMERS

It is probably true that all polymers contain some form of additive, ranging from small fractions of catalyst residue to large-scale incorporation of fillers. In paper, fillers, such as clay, kaolin, talc, calcium carbonate, or silica, and pigments, such as titanium dioxide, are typically added in a pulp slurry prior to its deposition onto the wire for paper making. Similarly, these types of fillers along with pigments are incorporated into paint formulations prior to paint application. All fillers are employed for such reasons as processing improvements, mechanical reinforcements, thermal stability, optical properties, permeability reduction, or cost saving.

Rheology and Processability

The immediate effect of a filler is to increase viscosity, interfere with the polymer flow pattern in a given process, produce thixotropy, and to give rise to machine wear. The relevant properties of the filler are concentration, size, aspect ratio, stiffness, strength, and specific interaction between filler and the polymer matrix. Within these are the special cases of easily deformable fillers that could easily be broken down or shaped during flow.

With an increase in filler concentration that exceeds the filler percolation threshold, a loose filler network is formed that would show elastic behavior, and in particular, a yield stress or thixotropy. This non-Newtonian behavior is not limited to filled polymers with high filler concentrations. Many polymer melts are viscoelastic and non-Newtonian. In these filled polymer melts, flow behavior is governed by viscoelastic properties of polymer compounded by the diverse properties of densely packed fillers and cannot be modeled and predicted theoretically. However, fine and particulate fillers are known to suppress elasticity of viscoleastic gum rubbers and render them better processability, such as less die swell, less shrinkage, and less melt fracture.

Mixing and Dispersion

In preparation of filled polymers, the process of uniformly distributing fillers without forming any filler composition gradients is most important. The process of mixing consists of three basic elements: incorporation, distribution, and dispersion. These elements occur simultaneously throughout the mixing cycle.

Mechanical Properties

In addition to the hydrodynamic effects of particulate fillers on polymer flow behavior, an enhanced stiffening effect is observed in filled polymers. For soft polymers, such as elastomers, with diluted filler loading, effects of filler on modulus are proportional to that on viscosity. However, this viscosity to modulus relationship only holds when the polymer is incompressible, such as elastomers with Poisson's ratio of 0.5, and when the rigidity of the filler is very much greater than that of the polymer.

Thermal Properties

Since polymers generally have a much larger thermal expansion coefficient than most rigid fillers, there is a significant mismatch in thermal expansion in a filled polymer. This mismatch could lead to generation of thermal stresses around filler particles during fabrication and, most severely, induce microcracks at the filler interface that could lead to premature failure of the filled polymer. As for the thermal expansion coefficient of a filled polymer, it generally falls below the value calculated from the simple rule of mixtures but follows the Kerner equation for nearly spherical particles.

Optical Properties

The use of fillers has indirect effect, or direct in the case of pigments, on the optical properties of polymers. The light transmissivity in a filled polymer is controlled primarily by light scattering which, in turn, depends on the differences in refractive indexes between the polymer and filler.

A filler with a refractive index near to that of the polymer, such as silica in polyester, can provide a translucent filled polymer. If the filler is optically anisotropic, such as calcite and talc, the corresponding filled polymers may appear in color under the polarized light or by optical interference. When the filler particle size is smaller than the wavelength of the light, $<0.4 \mu$, the filled polymer becomes transparent, which is one reason for the interest in using nanofillers.

Permeability

High aspect-ratio plate-like fillers can drastically lower the diffusivity or permeability of gases in solid polymers. Without considering the possible changes in the local permeability values due to molecular-level transformation by the presence of silicates, this permeability reduction simply arises from the increase in diffusion path lengths.

FILLER TYPES

Mineral Fillers

Mineral fillers are naturally occurring materials that are mined and are ground to a specified particle size. Grinding may be done dry using mechanical mills. For a finer product the ore is ground wet.

Nanoclays

Nanocomposites are materials that contain nanofillers, or fillers of nanometer dimensions. The successful synthesis of nylon–clay nanocomposites ushered in nylon nanocomposites that could attain high modulus, heat distortion temperature, dimensional stability, impermeability, and strength with only a few percent modified clay nanofillers.

Synthetic Fillers

Synthetic fillers are generally manufactured by precipitation of soluble materials under carefully controlled conditions to provide tailored properties. They may be found as colloidal particles that may be spherical, ellipsoid, rod, or tube shaped, as aggregates which are covalently bonded groupings of individual particles, or as agglomerates that are loosely held associations of aggregates physically interacting. Reinforcement properties are a function of the colloidal particle size and shape, the aggregate dimensions and morphology, and the ability of agglomerates to break down during mixing. Additionally, the composition and surface chemistry of the filler plays a significant role.

Nanoscale Oxides and Metals

Using vapor-phase plasma-based techniques, precipitation sol–gel reaction or simple grinding, nanoscale metal oxides are being produced in competition with existing fumed oxide products.

Carbon Fillers

The application of carbon black in rubber compounds is over a hundred years old. Unlike the well-known crystalline forms of carbon, such as diamond and graphite, carbon black is amorphous and is a manufactured product. Carbon blacks are prepared by incomplete combustion of hydrocarbons or by thermal cracking.

Carbon Nanotubes

Carbon nanotubes are graphene cylinders end capped with pentagonal rings. Single-walled nanotubes have fewer defects and much improved performance compared to multiwalled tubes. The use of single-walled tubes in flat panel displays, conductive plastics, and high performance fibers are recent application developments.

HEALTH AND SAFETY FACTORS

The principal hazard involved in the handling and use of many fillers is inhalation of airborne particles (dusts) in the respirable size range, ie, 10 µm and below. Filler dusts may be classified as nuisance particulates, fibrogens, and carcinogens.

The American Conference of Governmental and Industrial Hygienists (AC-GIH) establishes TLVs for the airborne concentration of many fillers in workroom air. In addition, concern for the toxicity of many metals and their compounds is limiting the use of many fillers, eg, Pb, Co, Cr, and Ba compounds, and possibly the use of certain organometallic surface coatings. Suppliers have information on proper usage and handling of their products. The use of NIOSH–OSHA-approved dust masks or respirators is required when dust concentrations exceed permissable exposure limits.

S. Laube, S. Monthey, and M.-J. Wang, in J. Dick, ed., "Compounding with Carbon Black and Oil," *Rubber Compounding and Testing for Performance*, Hanser Publishers, Munich, 2000.

R. G. Larson, *The Structure and Rheology of Complex Fluids*, Oxford University Press, New York, 1999.

J. C. Russ, *The Image Processing Handbook*, 2nd ed., CRC Press, Boca Raton, Fla., 1995.

D. Sekutowski, in J. Ededbaum, ed., "Fillers, Extenders, and Reinforcing Agents," in *Plastics Additives and Modifiers Handbook*, revised edition, Chapman & Hall, London, 1996.

ANDY H. TSOU
WALTER H. WADDELL
ExxonMobil Chemical Company

FILTRATION

Filtration is a fundamental unit operation in the chemical process industry. In the simplest sense, filters separate a suspension into its component solid and liquid phases. The end result of the filtration may be one of several possible aims:

- Recovery of the liquid phase and discarding of the solid phase.
- Recovery of the solid phase and discarding of the liquid phase.
- Recovery of both the solids and liquid phases.
- Recovery of neither phase, as, eg, when water is cleaned prior to discharge to control water pollution.

The end result is sometimes as important as the material properties in selecting the best filter for the separation.

There are many different types of filtration equipment. In general the filter consists of a *filter housing* that holds a *filter medium* through which the fluid must pass. The fluid discharge from the medium is called *filtrate*. The operation may be performed on incompressible liquids or compressible gases. The physical mechanisms controlling the filtration can vary with the degree of fluid compressibility.

Types of Liquid Filtration

There are two major types of liquid filtration: surface and depth filtration. In surface filtration, particles are captured on the surface of the filter medium, such as shown in Figure 1. The particles are captured on the surface due to a straining effect, where the particles are too large to fit through the pore openings, or the concentration of particles is large enough that they bridge over the pores.

In depth filtration, the particles penetrate into the medium and are captured within its depth, as shown in Figure 2. Typical depth media have pores that are much larger than the particle sizes it is meant to remove. Ideally the particles penetrate the medium and are captured by attachment to the filter medium fibers or granules due to various interception mechanisms.

Driving Forces in Filtration Operations

While pressure is the most common mechanism for driving the filtration operation, there are a number of

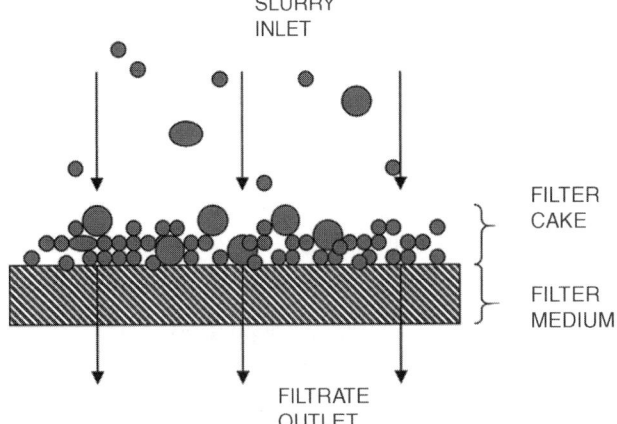

Figure 1. Surface filtration. The particles in the slurry are stopped at the inlet surface of the filter medium. As the surface becomes covered with particles, the particles themselves stop the approaching particles. The layers of particles build up to form the filter cake.

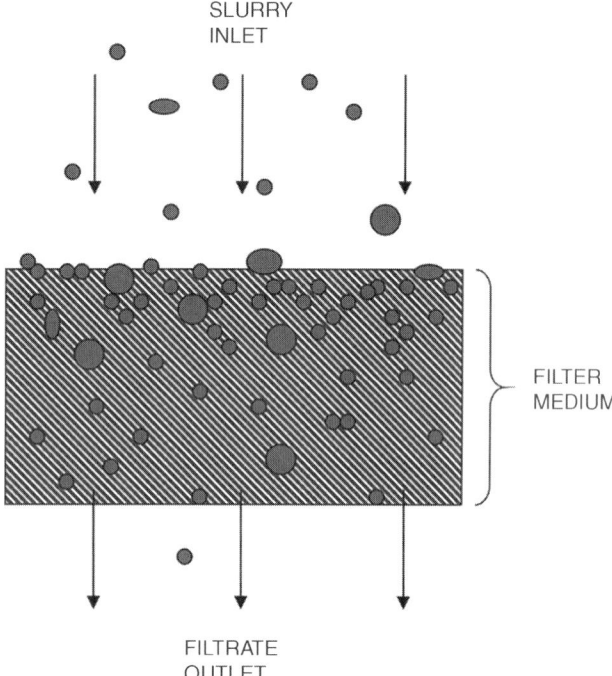

SLURRY
INLET

FILTER
MEDIUM

FILTRATE
OUTLET

Figure 2. Depth filtration. Particles in the slurry penetrate into the depth of the filter medium and are captured within the medium.

externally applied forces that can be used: gravity, vacuum, pressure, and centrifugal.

In addition to the above forces, there are secondary mechanisms that may also be applied directly cause the separation or indirectly to enhance one of the other methods: electric, magnetic, and acoustic.

FILTER MEDIA

Filter media control the filtration process. For a particular slurry, the filter media determine whether the filtration will produce a cake, capture particles within the depth of the media, whether the particles in the slurry will clog the media, or whether the media will be ineffective in particle removal.

Filter media come in a wide range of materials and forms. Some of the major types are listed in Table 1. A successful filter medium often requires a combination of various properties.

Most of the pertinent properties of filter media can be reasoned out with common sense. When the particles have sizes close to the size of the pore openings the filter medium can clog. Clogging can cause excessive pressure drop to maintain a fixed flow rate. Conversely, if the pressure drop is maintained constant, clogging of the medium will cause the flow rate to rapidly decrease, thus shortening the useful life of the filter medium. To remedy a clogging problem one can go with a more open medium and capture the particles in the depth, or with a more closed medium and capture particles on the surface. If the particles are rigid the latter is the method typically used. Alternatives to remedy clogging problems are to use filter-aids as a body feed or a precoat on the filter medium.

Table 1. Summary of Filter Media Types and Typical Size Range of Smallest Particles Captured

Type	Examples	Effective smallest particle retained, μ
solid fabrics	flat wedge-wire screens	100
	wire bound tubes	100
	stacks of rings	5
rigid porous media	ceramics and stoneware	1
	carbon	?
	plastics	?
	sintered metals	3
cartridges	sheet fabrics	3
	yarn wound	2
	bonded beds	2
metal sheets	perforated	100
	woven wire	5
plastic sheets	woven monofilaments	
	fibrillated film	
	porous sheets	
	membranes	0.1
woven fabrics	textile fabrics	10
nonwoven media	felts and needle felts	10
	paper (cellulose or glass fiber)	2
	bonded media	10
	meltblown	<2
	nanofiber sheets	<1
loose media	fibers, powders	<1

Soft deformable gel-like particles can be problematic. Soft particles can deform and spread to cover the pore openings at the surface of the medium and cause blinding of the medium. Normally, gel-like particles can only be effectively filtered at low applied pressure drops with large surface area filter media.

Particle Shapes and Particle Sizes

Here, the term "particle" represents a small body (of sand, salt, soil, silt, etc). Individual solid particles are characterized by their density, size (and size distribution), and shape. Particles of homogeneous solids have the same density as the bulk material. Particles obtained by breaking up a composite solid, such as an ore, may have varying densities. The size and shape for regular particles such as spheres and cubes are easily specified. The size and shapes of irregular particles are not so clear.

Irregular particles of even the same material have a wide range in sizes and shapes. The most common method of characterizing size is sieving. Particles are characterized by the size of the screen opening through which the particles can pass. Separating particles by size, known as *classifying*, provides a size distribution of particles. For sieving the particles are weighed to determine the mass fraction that pass through one screen but not through the next smaller screen. Sieving provides a well-defined method for sizing the particles but it may not be the most convenient method in terms of speed, accuracy, or convenience. Screening also has a lower practical limit of ~20 μ. Many methods have been developed using laser light diffraction, light absorption, electrical resistance, stokes settling velocity, and others.

Particle size and size distribution are important to filter performance. A filter cake of smaller particles usually requires a higher pressure for fluid flow of a given rate than a cake of larger particles. Particles of sizes similar to the pores of a filter medium tend to clog the medium, also causing an increased pressure drop. A slurry with a wide particle size distribution may be problematic to filter because of the difficulty to find a medium that does not clog. For slurries with small particles and wide size distributions pretreatment techniques such as flocculation or filter aid body feed may be necessary.

VOLUME FRACTION VERSUS MASS FRACTION

Some measurement techniques use gravimetric measurements that result in data in terms of mass fractions. The void volume fraction (porosity) depends on the sphericity (shape) of the particles, their size distribution, and their geometric packing arrangement. The densities of the individual particles have no effect on their packing arrangement. Fluid flow through a filter cake depends on the characteristics and dimension of the pores and properties of the liquid and not on the density of the particles. In this context, volume fraction is more useful than mass fraction information.

FILTER PERFORMANCE

Similar to other unit operations in chemical engineering, filtration is never 100% complete. The separation of solids is usually measured by the fractional recovery of how much of the incoming solids are collected by the filter. The separation of the liquid is measured by how much liquid remains with the filter cake, ie, moisture content.

One method that can account for both the conditions of the cake and the filtrate, or more generally to any separation process with one inlet and two outlet streams, is the method of entropy generation.

With depth filter media, the number of particles that pass through the filter depends on the particle size. Two common methods are available to characterize such filter performance. The first method is the beta ratio and the second method is the quality factor.

The beta ratio is the ratio of number concentrations particle of a particular diameter and larger in the feed stream to the filtrate stream, hence the beta ratio is normally greater than unity.

The beta ratio and efficiency only consider the capture of particles. The quality factor considers the capture of particles as well as pressure drop. Ideally, the pressure drop through a depth filter is proportional to the thickness of the filter. Conversely, the log of the particle penetration (ratio of the number of particles passing through the filter to the number challenging the filter) is proportional to the thickness of the filter. The quality factor provides a means for comparing performance between different types and thickness of filter media.

CAKE FILTRATION THEORY

Cake filtration is an example of flow through a porous medium. To increase the filtration rate one naturally will try increasing the pressure drop. While this works in many cases, there are situations in which increasing the applied pressure drop is not advisable: in cases of severe clogging of the filter medium and increasing the applied pressure for filtering gel-like particles.

The location of the largest pressure gradient usually controls the filtration process. In a clogging cake, the largest pressure gradient normally occurs in the filter medium. In this case, to improve the filtration rate requires either selecting a different filter medium that does not clog or changing the cake properties, such as with a body feed, so that the cake particles do not penetrate and clog the medium. To better understand and predict the performance of filter cakes several theories are briefly discussed in the following sections.

Darcy's Law for Flow Through Porous Media

This relation has limited *a priori* predictability, but it does serve as a useful reference for comparing the rate of flow through porous materials such as filter cakes and filter media.

In its customary present-day form, the macroscopic Darcy's equation is

$$\frac{Q}{A} = \frac{k}{\mu} \frac{\Delta P}{L} \qquad (1)$$

where ΔP is the pressure drop across a porous material of thickness L and cross-sectional area A. The fluid has viscosity μ and flows at rate Q through the material. The permeability, k, is defined by equation (1) to relate the pressure drop to the flow rate.

Ergun Equation to Predict Permeabillity

Coulson and co-workers have shown that for regular-shaped particles, such as spheres, the permeability also depends on surface area of the particles and the void fraction. When the flow rate is high enough, the permeability also becomes a function of the fluid velocity.

One popular correlation that accounts for all of these factors that control the permeability is the Ergun equation. In its derivation, the particles are assumed to be spherical and the surface area effect is accounted for by relating specific surface area to the particle diameter. The void fraction of the porous material is explicitly introduced into the correlation in the derivation. The Ergun equation applies to porous material with void fractions <0.9. The Ergun equation is modified to irregularly shaped particles through the sphericity parameter, Φ_s. The Ergun equation is commonly written as,

$$\frac{\Delta P}{L} = \frac{180 V_o \mu}{\Phi_s^2 d_p^2} \frac{(1-\varepsilon)^2}{\varepsilon^3} + \frac{1.80 \rho V_o^2}{\Phi d_p} \frac{(1-\varepsilon)}{\varepsilon^3} \qquad (2)$$

where ε is the void fraction (porosity).

The Ergun equation applies for porosities <0.9. Above 0.9, the Langmuir model is recommended. The Ergun equation also does not work well for fibers and plate-like objects (low sphericities).

Advanced Cake Filtration Modeling

Advanced cake filtration models normally account for other mechanisms that affect the cake formation, such as sedimentation or geometric affects. Advanced models also allow for modeling of non-Newtonian behavior such as power-law fluids or yield-stress fluids.

Multiphase continuum theory obtained from volume averaging of the single-phase continuum equations provides a rigorous and complete set of equations to account for mass, momentum, energy, and chemical species for modeling flows through porous media. The continuum equations are easily applied to more complex processes, such as cake filtration combined with expression, sedimentation, or chemical reactions through appropriate use of boundary conditions and constitutive functions.

The disadvantage to the volume averaged approach is it requires effort to learn the notation and to understand how to apply the concepts at different scales of phenomena. The volume averaging theory is less intuitive than the *ad hoc* approach.

Continuum modeling of cake filtration, whether from volume averaging theory or from an *ad hoc* approach, requires appropriate constitutive relations. Constitutive theory has been applied to the multiphase continuum equations to generalize the constitutive equations. Other constitutive relations can be determined using approaches applied in other disciplines such as soil mechanics to account for phenomena such as creep and swelling.

Another advanced approach to model cake filtration is by a pore-structure model. The pore-structure model has the advantage that the single-phase rheological properties of the fluid are applied to the flow through the channels as defined by the structure of the solid phase. Pore structure models circumvent the need for constitutive equations at the mixture scale by modeling the system at the single-phase scale. Disadvantages to this approach are the pore structure may not represent the particular medium of interest and they tend to be computationally intensive and only a limited number of pore structures have been investigated.

DEPTH FILTRATION THEORY

Depth filtration occurs when the particles in the slurry are much smaller than the pore openings in the filter medium. The particles penetrate into the filter medium and are captured on the surfaces of the pore walls. Depth filtration is typically applied to polish fluids such as beverages.

Single Collector Efficiency

The filter coefficient may be modeled as the combination of many single collectors per unit volume. The single collectors may be individual granules or fibers.

Mechanisms of Capture

There are a number of mechanisms that contribute to the capture of particles by a collector. Most of the mechanism models assume that if a particle comes into physical contact with a collector surface that the particle is captured. Particle bounce and particle release are possible but are often ignored.

The five primary mechanisms of single collector capture of a particle are (1) direct interception, (2) inertial impaction, (3) Brownian diffusion, (4) sedimentation, and (5) electrostatic forces.

The efficiency of each mechanism may be predicted with appropriate correlations.

Most Penetrating Particle Size

Of interest to many applications is the most penetrating particle size. The most penetrating particle size is the size particle for which the overall net filter efficiency is lowest.

As particle sizes increase, the single collector efficiencies tend to decrease. However, as the particle size increases a straining mechanism also contributes to the net filter efficiency.

Advanced Modeling of Depth Filtration

As in cake filtration, depth filtration may be modeled using the multiphase continuum theory derived from volume averaging of the single phase equations. In this approach, the captured particles may be treated as a separate phase (hence a three phase system: the liquid as one phase, the particles carried in the liquid and captured in the medium as a second phase, and the medium itself as the third phase). Or, the particles can be modeled analogously as a chemical species.

STAGES OF SOLID–LIQUID SEPARATIONS

Solid–liquid separations often involve a wide variation in material properties and operating conditions. One item of equipment or process technique often is not sufficient to achieve the desired separation.

The stages are defined based on the intent of the operation and not necessarily on the type of equipment involved. A summary of the four stages is as follows:

Pretreatment

Pretreatment is normally used with slurries that are difficult to separate. The intent of the pretreatment is to change the slurry properties, by changing the properties of the liquid or the solid particles, to improve the separation by a subsequent process. Pretreatment may be separated into physical or chemical treatments. For physical treatments, the simplest approach may be to change the viscosity. Another pretreatment for the liquid is degassing.

Physical pretreatment of the solid particles include crystallization to increase the particle sizes, freezing or other phase changes to increase size or improve shape;

and the application of filter aids. Filter aids may be used to precoat the filter medium (in essence, change the surface of the filter medium to prevent clogging), or by adding a body feed. The filter aid material particles are usually larger than the particles in the slurry and serve to capture or trap the smaller particles, yet the filter aid material is highly porous and allows the fluid to pass through.

Chemical pretreatment of the slurry is mostly concerned with causing the particles to agglomerate to form larger particles. There are two primary ways that this can be accomplished. The first is by means of coagulation in which very fine particles of colloidal size adhere directly to each other as a consequence of Brownian motion.

The second method is by flocculation. Flocculation is the formation of more open aggregates that form by high molecular weight polymers acting as bridges between particles.

Solids Concentration

Concentrated slurries are more likely to form filter cakes, by particle bridging across pore spaces, and thus reduce filter medium clogging. A number of methods may be used for concentrating solids. Some of the more common are cross-flow filters, hydrocyclones, and gravity settlers. More recently, other field-assisted techniques have been developed that may be useful for thickening.

Solids Separation

Solids separations involve filters, which are classified as cake filters or depth filters. There are many types and variations on each of these categories.

Posttreatment

Posttreatment is used to improve on the quality of the final product in the filtration. If the product is a low turbidity liquid, then posttreatment may require polishing the filtrate to remove particles to below a specified standard. Posttreatment of filter cakes commonly includes deliquoring, washing, and drying. Deliquoring physically reduces the liquid content, such as by expression. Washing is done to remove soluble components from the cake. Drying is done to further reduce the moisture content of the filter cake.

FILTER CYCLES

Cake filtration by its nature is a batch operation. There are a few methods, such as rotary vacuum filters, that can be designed for continuous throughput.

Continuous Filter Cycles

A horizontal rotating tilting pan filter operates quasi-continuously by sequentially filtering, deliquoring, washing, drying, and removing the filter cake from the pan in a series of stages. The pans are lined on the bottom with the filter media. The slurry is loaded into the first pan and a vacuum is pulled on the bottom of the pan to form the cake. In the subsequent stages, the cake is processed until it is finally discharged by tilting the pan for the cake to fall out.

Batch Filter Cycles

A cycle for a batch filter includes a combination of steps, such as cake formation, deliquoring, washing, expression, drying, discharge, and cleaning.

SOLID–LIQUID SEPARATION EQUIPMENT SELECTION

Solid–liquid separations (SLS) are often viewed as a necessary evil that must be tolerated to produce a product, clarify liquids, dispose of wastes, recover raw materials, decontaminate process fluids, etc. Because of these multiple objectives, a plant engineer is faced with a large array of SLS equipment from which to choose. Compounding the selection problem, various industries have adopted certain specialized equipment. Furthermore, standards vary from industry to industry. Vendors may intensify the selection problem by publishing ambiguous or misleading information. A number of equipment selection guides have been published to help a plant engineer through the maze to select SLS equipment.

FILTER MEDIA SELECTION

Selection

Media selection can be one of the most important items in the SLS selection process, considering that the medium controls the initial filtration process. Specific media such as general nonwovens, filter press cloths, ceramics, flash-spin Tyvek®, and wire mesh are examples of the wide array of available media.

Types of Media

There are many types of media. In addition to the ones already mentioned, there are also deep-bed grains, sintered metals (powder, mesh, and fibers), membranes, filter papers, needle felts, a whole array of nonwovens, perforated and wedge-wire metals, as well as filter aids of diatomaceous earth (DE), perlite, cellulose fibers, coal fines, and calcined rice hull ash. Many of these woven–nonwoven, sintered, or paper filter media can be fabricated into filter bags, socks, and cartridges.

Testing

There are hundreds of test methods for evaluation and characterizing filter media, but the most prevalent today are air permeability coupled with porometry (ie, pore size distribution) with automated porometers.

VACUUM FILTERS

In vacuum filters, the driving force for filtration results from a suction on the filtrate side of the medium. Although the theoretical pressure drop available for vacuum filtration is 1 bar in practice it is often limited to 0.7–0.8 bar (532–608 mm Hg).

Vacuum filters are available in a variety of types, and are usually classified as either batch operated or continous. An important distinguishing feature is the position of the filtration area with respect to gravity, ie, horizontal or nonhorizontal filtering surface. Vacuum filters generally produce lower cake solids than pressure filters.

A number of vacuum filter types use a horizontal filtering surface with the cake forming on top. This arrangement offers a number of advantages: gravity settling can take place before the vacuum is applied, and in many cases may prevent excessive blinding of the cloth due to action of a precoat formed by the coarser particles; if heavy or coarse materials settle out from the feed they do so onto the filter surface, and can be filtered; and fine particle penetration through the medium can be tolerated because the initial filtrate can be recycled back onto the belt. Top-feed filters are ideal for cake washing, cake dewatering, and other process operations such as leaching. Horizontal filter surfaces also allow a high degree of control over cake formation.

There are, however, two significant drawbacks to horizontal filters, ie, such filters usually require large floor areas, their capital cost is high, and they are somewhat difficult to enclose for hazardous materials.

Nutsche Filter

The nutsche filterb is simply an industrial-scale equivalent of the laboratory Buechner funnel. Nutsche filters consist of cylindrical tanks divided into two compartments of roughly the same size by a horizontal medium supported by a filter plate.

Nutsche filters are simple in design, but laborious in cake discharge. They are prone to high amounts of wear due to the digging out operation. They are completely open and quite unsuited for dealing with inflammable or toxic materials.

Enclosed Agitated Vacuum Filters

These filters, often called mechanized nutsches, are circular vessels provided with a cover through which passes a shaft carrying a stirrer. The stirrer can sweep the whole area of the filter cake and can be lowered or raised vertically as required. The pressure version of the enclosed agitated filter is known as the Rosenmund filter; it uses a screw conveyor to convey the cake to a central cake discharge hole.

Vacuum Leaf Filter

The vacuum leaf, or Moore, filter consists of a number of rectangular leaves manifolded together in parallel and connected to a vacuum or compressed air supply by means of a flexible hose. Each leaf is composed of a light pervious metal backing, usually of coarse wire grid or expanded metal set in a light metal frame and covered on each surface with filter cloth or woven wire cloth. The leaves, which are carried by an overhead crane during the filtration sequence, are dipped successively into a feed slurry tank, where the filtration takes place; a holding tank, where washing occurs; and a cake-receiving container, where cake discharge is performed by back-blowing with compressed air. An alternative arrangement is to move the tanks rather than the leaf assembly.

Simple design, general flexibility, and good separation of the mother liquor and the wash are important advantages of vacuum leaf filters. On the other hand, they are labor-intensive, require substantial floor space, and introduce the danger of the cake or leafs falling off during transport. Vacuum leaf filters are particularly useful when washing is important, but washing is difficult with very fine solids, such as occur in titanium dioxide manufacture.

Tipping Pan Filter

This is a nutsche filter with a small filtrate chamber, in the form of a pan built so that it can be tipped upside down to discharge the cake. A separate vessel is used to receive the filtrate; this allows segregation of the mother and wash liquor if necessary.

A variation on this type of filter is the double tipping pan filter, which is a semicontinuous type consisting of two rectangular pans fitted with a filter cloth and pivoted about a horizontal axis.

In general, pan filters are selected for freely filtering solids and thick filter cakes.

Horizontal Rotating Pan Filters

These filters represent a further development of the tipping pan filter for continuous operation. They consist of a circular pan rotating around the central filter valve. The pan is divided into wedge-shaped sections covered with the filter medium. Vacuum is applied from below. Each section is provided with a drainage pipe that connects to a rotary filter valve of the same type as in drum filters. This allows each section, as it rotates, to go through a series of operations such as filtration, dewatering, cake washing, and discharge. Two basic designs exist, depending on the method of solids discharge: The horizontal pan filter with scroll discharge and the horizontal rotary tilting pan filters.

An important variation of the horizontal pan filter is based on replacing the rigid outer wall necessary for containing the feed and the cake on the rotating table by an endless rubber belt.

Horizontal Belt Vacuum Filters

This type of filter is another development of the pan filter idea. A row of vacuum pans arranged along the path of an endless horizontal belt was the original patented design. This has been superseded by the horizontal belt vacuum filter, which resembles a belt conveyor in appearance. The top strand of the endless belt is used for filtration, cake washing, and drying, whereas the bottom return is used for tracking and washing of the cloth. There is appreciable flexibility in the relative areas allocated to filtration, washing, and drying. Hooded enclosures are available wherever necessary. Modular construction of many designs allows field assembly as well as future expansion if process requirements change.

Horizontal belt filters are well suited to either fast or slowly draining solids, especially where washing requirements are critical. Multistage countercurrent washing can be effectively carried out due to the sharp separation of filtrates available. Horizontal belt vacuum filters are classified according to the method employed to support the filter medium.

One common design is typified by a rubber belt mounted in tension. The belt is grooved to provide drainage toward its center. Covered with cloth, the belt has raised edges to contain the feed slurry, and is dragged over stationary vacuum boxes located at the belt center. Wear caused by friction between the belt and the vacuum chamber is reduced by using replaceable, secondary wear belts made of a suitable materials such as (PTFE), etc.

Another type of horizontal belt vacuum filter uses reciprocating vacuum trays mounted under a continuously traveling filter cloth, and the indexing cloth machines are a further development. Some horizontal belt vacuum filter designs incorporate a final compression stage for maximum mechanical dewatering.

Rotary Vacuum Drum Filters

This is the most popular vacuum filter. There are many versions available and they all incorporate a drum that rotates slowly, ~1–10 min/revolution, about its horizontal axis and is partially submerged in a slurry reservoir. The perforated surface of the drum is divided into a number of longitudinal sections of ~20 mm in thickness. Each section is an individual vacuum chamber, connected through piping to a central outlet valve at one end of the drum. The drum surface is covered with a cloth filter medium and the filtration takes place as each section is submerged in the feed slurry. A rake-type slowly moving agitator is used to keep the solids in suspension in the slurry reservoir without disturbing the cake formation. The agitator usually has a variable speed drive.

Rotary Vacuum Disk Filters

An alternative to the drum filter is the disk filter, which uses a number of disks mounted vertically on a horizontal shaft and suspended in a slurry reservoir. This arrangement provides a greater surface area for a given floor space, by as much as a factor of 4, but cake washing is more difficult and cloth washing virtually impossible.

BATCH PRESSURE FILTERS

Excluding variable chamber presses, pressure filters may be grouped into two categories: plate-and-frame filter presses, and pressure vessels containing filter elements. The latter group also includes cartridge filters. Pressure vessel filters (leaf-type) handle incompressible or slightly compressible cakes. Filter presses handle both compressible and incompressible cakes, especially with the flexibility potential of membranes. Cylindrical element filters, are used for clarification applications, using membrane socks (or tubes), precoat and often body-feed, resulting in cakes that are slightly compressible. Cartridge filters are for clarification only, with little if any cake formed.

Plate-and-Frame Filter Presses

In the conventional plate-and-frame press, a sequence of perforated square, or rectangular, plates alternating with hollow frames is mounted on suitable supports and pressed together with hydraulic or screw-driven rams. The plates are covered with a filter cloth that also forms the sealing gasket. The slurry is pumped into the frames and the filtrate is drained from the plates.

The drainage surfaces are usually made in the form of raised cylinders, square-shaped pyramids, or parallel grooves in materials such as stainless steel, cast iron, rubber of coated metal, polypropylene, rubber, or wood. Designs are available with every conceivable combination of inlet and outlet location, ie, top feed, center feed, bottom feed, corner feed, bottom external feed, and side feed, with a similar profusion of possible positions of discharge points.

Both flush plates and recessed plates can be specified. Recessed plates obviate the need for the frames but are tougher on filter cloths due to the strain around the edges. These presses are more suitable for automation because of the difficulty of the automatic removal of residual cake from the frames in a plate-and-frame press.

Pressure Vessel Filters

The several designs of pressure vessel filters all consist of pressure vessels housing a multitude of leaves or other elements that form the filtration surface and that are mounted either horizontally or vertically. With horizontal leaves most suitable where thorough washing is required, there is no danger of the cake falling off the cloth; with vertical elements, a pressure drop must be maintained across the element to retain the cake. The disadvantage of horizontal leaf types is that one-half the filtration area is lost because the underside of the leaf is not used for filtration because of the danger of the cake falling off. Discharge of the cake also may be more difficult in this case.

Cylindrical Element Filters. These filters, often referred to as candle filters, have cylindrical elements or sleeves mounted vertically and suspended from a header sheet, which divides the filter vessel into two separate compartments. The filtration takes place on the outside of the sleeves, but in some designs, filtration takes place on the inside generally used for semidry cake discharge. The inlet is usually in the bottom section of the vessel and the filtrate outlet in the top section above the header (or tube) sheet.

The advantage of candle filters is that as the cake grows on the outside of the tubular elements the filtration area increases and the thickness of a given volume of cake is therefore less than it would be on a flat element.

A new type of candle filter (Amafilter's Cricketfilter, Holland) has been introduced that has flattened candles that reportedly permit more elements per unit volume and allow for better backwashing or backpulsing.

An even more unique candle filter (DrM's Fundabae, Switzerland) uses multiple small tubes for enhanced cake removal from their usual woven cloth socks. Both of these designs can provide for dry cake discharge.

Vertical Vessel, Vertical Leaf Filters. These are the cheapest of the pressure leaf filters and have the lowest volume/area ratio. Their filtration areas are limited to <80 m². Large bottom outlets, fitted with rapid-opening doors, are used for dry cake discharge, and smaller openings are used for slurry discharge. As with all vertical leaf filters, these are not suited for cake washing.

Horizontal Vessel, Vertical Leaf Filters. In a cylindrical vessel with a horizontal axis, the vertical leaves can be arranged either laterally or longitudinally. The latter, less common, arrangement may be designed as the vertical vessel, vertical leaf filters but mounted horizontally. Its design is suitable for smaller duties and the leaves can be withdrawn individually through the opening end of the vessel.

Horizontal Vessel, Horizontal Leaf Filters. These filters consist of a horizontal cylindrical vessel with an opening at one end. A stack of rectangular horizontal trays is mounted inside the vessel; the trays can usually be withdrawn for cake discharge, either individually or in the whole assembly. The latter case requires a suitable carriage. One alternative design allows the tray assembly to be rotated through 90° so that the cake can fall off into the bottom part, designed in the shape of a hopper and fitted with a screw conveyor.

Cartridge Filters

Cartridge filters use easily replaceable, tubular cartridges made of paper, sintered metal, woven cloth, needle felts, activated carbon, or various membranes of pore size down to ~0.1 µm. Cartridge filtration is limited to liquid polishing or clarification, ie, removing very small amounts of solids, in order to keep the frequency of cartridge replacements down.

Cartridge filters are either depth or surface type, according to where most of the solids separate; the precise demarcation line is difficult to assess. The most common depth cartridge is the yarn-wound type that has a yarn wound around a center core in such a way that the openings closet to the core are smaller than those on the outside. The yarn may be made of any fibrous material, ranging from cotton or glass fiber to the many synthetic fibers such as polypropylene, polyester, nylon, or Teflon.

Mechanical Batch Compression Filters

In mechanical compression filtration the liquid is expelled from the slurry by reduction of the volume of the retaining chamber.

Membrane Plate Presses. Membrane presses are closely related to conventional plate and frame presses. They consist of a recessed plate press in which the plates are covered with an inflatable diaphragm that has a drainage pattern molded into its outside surface. The filter cloth is placed over the diaphragm. One advantage is in the reduction of the washing time and washwater requirements. Another advantage of the membrane plate is its flexibility to cake thickness, ie, thinner cakes can be easily handled with increased cake dryness.

Cylindrical Presses. Another group of filters that utilize the variable chamber principle are those with a cylindrical filter surface. There are two designs in this category, both of which originate from the United Kingdom.

The VC filter consists of two concentric hollow cylinders mounted horizontally on a central shaft. The inner cylinder is perforated and carries the filter cloth, the outer cylinder is lined on the inside with an inflatable diaphragm. The slurry enters into the annulus between the cylinders and conventional pressure filtration takes place, with the cake forming on the outer surface of the inner cylinder. The filtration can be stopped at any cake thickness or resistance, as required by the economics of the process, and hydraulic pressure is then applied to the diaphragm that compresses the cake.

CONTINUOUS PRESSURE FILTERS

A continuous pressure filter may be defined as a filter that operates at pressure drops >1 bar and does not require interruption of its operation to discharge the cake; the cake discharge itself, however, does not have to be continuous. There is little or no downtime involved, and the dry solids rates can sometimes be as high as 1750 kg/m²h with continuous pressure filters.

Most continuous pressure filters available have their roots in vacuum filtration technology. A rotary drum or rotary disk vacuum filter can be adapted to pressure by enclosing it in a pressure cover; however, the disadvantages of this measure are evident. The enclosure is a pressure vessel which is heavy and expensive, the progress of filtration cannot be watched, and the removal of the cake from the vessel is difficult. Other complications of this method are caused by the necessity of arranging for two or more differential pressures between the inside and outside of the filter, which requires a troublesome system of pressure regulating valves.

Despite the disadvantages, the advantages of high throughputs and low moisture contents in the filtration cakes have justified the vigorous development of continuous pressure filters.

Horizontal or vertical vessel filters, especially those with vertical rotating elements, have undergone rapid development with the aim of making truly continuous pressure filters, particularly but not exclusively for the filtration of fine coal. There are basically three categories of continuous pressure filters available, ie, disk filters, drum filters, and belt filters including both hydraulic and compression varieties.

The advantages of continuous pressure filtration are clear and indisputable, particularly with slow-settling slurries and fairly incompressible cakes.

Disk Filters

The McGaskell and Gaudfrin Disk Filters. One of the earliest machines in this category, the McGaskell rotary pressure filter, is essentially a disk-type filter enclosed in a pressure vessel. The rotating disks are each composed of several wedge-shaped elements connected to a rotary filter valve at the end of the shaft, similar to the vacuum rotary disk filters.

The Gaudfrin disk filter is also similar in design to a vacuum disk filter, but it is enclosed in a pressure vessel with a removable lid. The disks are 2.6 m in diameter, composed of 16 sectors. The cake discharge is by air blowback, assisted by scrapers if necessary, into a chute where it may be either reslurried and pumped out of the vessel or, for pasty materials, pumped away with a monopump without reslurrying.

The KDF Filter. The KDF filter (Amafilter, Holland) is based on the same principle as disk filters. It was developed for the treatment of mineral raw materials, like coal flotation concentrates or cement slurries, and can produce a filter cake of low moisture content at very high capacities, up to 1750 kg/m^{2}h. The pressure gradient is produced by pressurized air above the slurry level that provides the necessary driving force for the filtration and also is used for displacement dewatering of the cake.

The KHD Pressure Filter. Another development of the disk filter has been reported (KHD Humboldt Wedag AG, Germany). A somewhat different system, probably a predecessor, was patented. The newer version has significant advantages over the patented version.

Drum Filters

The rotary drum filter, also borrowed from vacuum filtration, makes relatively poor use of the space available in the pressure vessel, and the filtration areas and capacities of such filters cannot possibly match those of the disk pressure filters. In spite of this disadvantage, however, the pressure drum filter has been extensively developed. Two drum filter types are the TDF drum filter, and the BHS-Fest Filter.

Horizontal Belt Pressure Filters

Horizontal belt filters have a great advantage in cake washing application due to their horizontal filtration surface. In the context of pressure filtration and the requirements of good dewatering, however, they have a significant disadvantage because the cake is not very homogeneous; gravity settling on the belt and the inevitable problems of distribution of the feed suspension over the belt width cause particle stratification and nonhomogeneous cakes.

Continuous Compression Filters

The variable chamber principle applied to batch filtration, as described before, can also be used continuously in belt presses and screw presses.

Thickening Pressure Filters

The most important disadvantage of conventional cake filtration is the declining rate due to the increased pressure drop caused by the growth of the cake on the filter medium. A high flow rate of liquid through the medium can be maintained if little or no cake is allowed to form on the medium. This leads to thickening of the slurry on the upstream part of the medium; filters based on this principle are sometimes called filter thickeners.

The methods of limiting cake growth are classified into five groups, ie, removal of cake by mass forces (gravity or centrifugal), or by electrophoretic forces tangential to or away from the filter medium; mechanical removal of the cake by brushes, liquid jets, or scrapers; dislodging of the cake by intermittent reverse flow; prevention of cake deposition by vibration; and cross-flow filtration by moving the slurry tengentially to the filter medium so that the cake is continuously sheared off. The extent of the commercial exploitation of these principles in the available equipment varies, but cross-flow filtration is exploited most often.

CENTRIFUGAL FILTERS

The driving force for filtration in centrifugal filters is centrifugal forces acting on the fluid. Such filters essentially consist of a rotating basket equipped with a filter medium. Similar to other filters, centrifugal filtration does not require a density difference between the solids and the suspending liquid. If such density difference exists sedimentation takes place in the liquid head above the cake. This may lead to particle size stratification in the cake, with coarser particles being closer to the filter medium and acting as a precoat for the fines to follow.

In centrifuges, in addition to the pressure due to the centrifugal head due to the layer of the liquid on top of the cake, the liquid flowing through the cake is also subjected to centrifugal forces that tend to pull it out of the cake. This makes filtering centrifuges excellent for dewatering applications. From the fundamental point of view, there are two important consequences of these additional dewatering forces. First, Darcy's law and all of the theory based on it is incomplete because it does not take into account the effect of mass forces. Second, pressures below atmospheric can occur in the cake in the same way as in gravity fed deep bed filters. The conventional filtration theory has been modified to make it applicable to centrifugal filters.

Due to good performance and high cost, centrifuges are often referred to as the ultimate in SLS. They have parts rotating at high speeds and require high engineering standards of manufacture, high maintenance costs, and special foundations or suspensions to absorb vibrations. Another feature distinguishing the filtering centrifuges from other cake filters is that the particle size range they are applicable to is generally coarser, from 10 μm to 10 mm. In particular, cake filters that move the cake across the filter medium are restricted to using metal screens, which by their very nature are coarse. No cloth can withstand the abrasion due to the cake forced on the cloth and pushed over its surface. Only the fixed-bed, batch-operated centrifuges can use cloth as the filtration medium and be used, therefore, with fine suspensions.

Fixed-Bed Centrifuges

The simplest of the fixed-bed centrifuges is the perforated basket centrifuge that has a vertical axis, a closed bottom, and a lip or overflow dam at the top end. In the industrial versions, the basket housing is often supported by a three-point suspension called the three-column centrifuge.

Moving Bed Centrifuges. The continuously fed, moving bed machines are available with conical or cylindrical screens. The conical screen centrifuges have a conical basket rotating either on a vertical or horizontal axis, with the feed suspension fed into the narrow end of the cone. If the cone angle is sufficiently large for the cake to overcome its friction on the screen, the centrifuge is self-discharging. Such machines cannot handle dilute slurries and require high feed concentrations of the order of 50% by mass, eg, as occurs in coal dewatering. Different products, however, require different cone angles. Unnecessarily large angles shorten the residence time of the solids on the screen surface and thus lead to poor dewatering. The movement of the solids along the screen may be assisted by vibrating the basket either in the axial direction or, in a tumbler centrifuge, in a tumbling action.

The pusher-type centrifuge has a cylindrical basket with its axis horizontal. The feed is introduced through a distribution cone at the closed end of the basket and the cake is pushed along the basket by means of a reciprocating piston that rotates with the basket. The screen is made of self-cleaning trapezoidal bars of at least 50 μm in spacing, so that only particles larger than ∼50 μm can be efficiently separated by this machine.

Scale-Up

The scale-up of filtration centrifuges is usually done on an area basis, based on small-scale tests. A test procedure has been described with a specially designed filter beaker to measure the intrinsic permeability of the cake. The best test is, of course, with a small-scale model, using the actual suspension. The scale-up is most reliable if the basket diameter does not increase by a factor of >2.5 from the small scale. Newer more modern bucket-type centrifuges have found widespread use for filtering centrifuge scale-up.

Newer designs are also now available (FIMA and Heinkel) that combine centrifugal dewatering with hyperbaric or steam-assisted drying in order to achieve even lower moisture contents.

T. Allen, *Particle Size Measurement*, Vol. 1. Powder Sampling and Particle Size Measurement, 5th ed., Chapman & Hall, London, 1997.

R. B. Bird, W. E. Stewart, and E. N. Lightfoot, *Transport Phenomena*, 2nd ed., John Wiley & Sons, Inc., New York, 2002.

S. Middleman, *An Introduction to Fluid Dynamics: Principles of Analysis and Design*, John Wiley & Sons, Inc., New York, 1998.

D. B. Purchas, *Handbook of Filter Media*, Elsevier Science, Ltd., Oxford, U.K., 1998.

GEORGE G. CHASE
The University of Akron
ERNEST MAYER
E. I. du Pont Nemours & Company

FINE ART EXAMINATION AND CONSERVATION

A small number of specialist scientists, called conservation scientists, work with museum curators and art conservators in examining, from a technical point of view, works of art. Chemical and materials analysis is directly relevant to the treatment and preservation of fine art since its conservation is strictly dependent on the materials that were used in its creation. The knowledge of the materials and methods used by artists allows us to gain an insight into the original appearance and artists' intention. The results of the conservation scientists' research are used to specify appropriate environments for storage, exhibition, and to help choose appropriate cleaning of the works. The work of conservation scientists is important also for understanding the process of creation of a work. This knowledge can help identify forgeries, fakes, and imitations. It is also useful to art historians in addressing issues related to attribution. Studying the materials and methods used by artists provides a way to understand the technology of the times in which they created their work.

TECHNICAL INVESTIGATION OF WORKS OF ART

The range of science and analytical techniques involved in fine art examination and conservation is applied to works of art made in every medium: paint, paper, photography, stones, glass, metals, plastic, etc. The works themselves may be classed as fine art, decorative art, cultural property or ethnographic material. The work of the conservation scientist depends also on information from the fields of art history, conservation studies, and the history of technology.

Methods for Investigation of Fine Art

Instrumental analysis of every type has been applied to studying fine art. Limitations often relate less to the desire to apply a technology to the investigation of art than to the hurdles of expense, moving art to the instrumentation, and other practical difficulties.

Nondestructive analysis in this field implies that no samples are removed from the object. There was a restricted range of techniques that were employed for nondestructive analysis of fine art until recently, when the development and refinement of many spectroscopic and imaging methods has changed this. These include, examination of paintings in ultraviolet(uv), visible(vis), and ir radiation. Despite remarkable refinements in non-invasive analysis, there are many times when samples are necessary to analyze the materials used in fine art and to understand the structure of the work.

Nondestructive Analysis. *Stereomicroscopy and Scanning.* In low power stereomicroscopy magnifications of ×2.5–×350 typically are used. Microscopy of paintings can reveal surface defects in paint films, small cracks in surfaces, and through them the underlying layers of paint. A system developed at INOA is capable of scanning the surface of a work of art with submillimeter precision generating a virtual three-dimensional(3D) object with the surface roughness, cracks, and holes all thoroughly documented.

Colorimetry. Measurement of color is beginning to find application in the field to identify pigments on the surface of a work of art and for matching the color of losses. Advances in fiber optics and detector sensitivity have allowed the technique to be used to measure the color of very small areas of paint and thus to determine the pigments present on the surface of a work of art.

Infrared Reflectography (IRR). This technique is used to examine the artists' drawings below the surface of paintings (underdrawings). Infrared radiation can penetrate the layers of paint due to the transmissivity of many pigments in the ir region. This is exploited to show the black underdrawings on white primings used for painting. The use of focal plane array detectors provides a digital output that is more easily handled and provides more resolution than the television output of a vidicon detector.

X-Radiography. X-Radiography is an important tool for nondestructive analysis of works of art. The X-ray tube voltage and current chosen depend on the nature of the object being examined. Low voltage is appropriate for studying low density works such as paintings, ceramics, and small sculptures made from organic materials such as wax. High voltage, "hot," units are used for examining large bronzes.

Neutron Activation Autoradiography (NAA). This technique exposes paintings or other works to a flux of thermal neutrons produced in a nuclear reactor. These neutrons activate some of the elements in pigments to form radioactive isotopes. The radioactive atoms thus formed decay to a stable atom by emitting beta particles, X-rays, or gamma rays. The rate of decay depends on the half-life of the excited isotope. Since the radioactive isotopes of the elements have different half-lives, films left in contact with the painting at various times following exposure to neutrons reveal the location of different elements.

Raman Spectroscopy. Currently, Raman spectroscopy is best suited for identification of pigments, especially inorganic and nonfluorescing pigments. The instrumentation has been most successfully applied to analysis of pigments in water-based media owing to the fluorescence of oil binders.

X-Ray Fluorescence Spectroscopy. Air-path energy dispersive X-ray fluorescence spectrometry (XRF) is widely used in the examination of fine art. The instrumentation is designed so that works of art can be analyzed without taking samples and without any surface preparation at all. There are, however, several limitations to the method. XRF is a surface-analysis technique. The depth of analysis ranges from ~20 μ for metals to >100 μ for objects of low density, eg, paintings, watercolors, and photographs.

Analysis Requiring Samples

Obtaining samples from works of art is undertaken with the utmost consideration of several concerns including maintaining the integrity, both visual and structural, of the work. Samples are removed from an object when it is agreed that the information obtained from them will outweigh the small damage that results from their acquisition. Sample sites are chosen in consultation with conservators and curators to answer specific issues relating to understanding the condition of a work, conservation treatment and analysis for understanding its appearance and preservation. They are usually removed from the edges of old losses or wide cracks. Surgical scalpels and tungsten needles are the two most common tools for obtaining paint and fiber samples and microspatulas have been used for this.

A useful and common approach to examining works of art is the study of minute cross-sections. Cross-sections from corroded metals, deteriorating stone sculptures, murals, and easel paintings all can reveal extremely useful information on the condition of a work. Cross-sections are also used to study artists' working methods.

Light Microscopy. In addition to the use of stereomicroscopes for surface examination, plain and polarized light microscopy has been used for the examination of paint cross-sections, pigments, and metallurgical analysis, as well as for fiber and wood identification. Reflected and transmitted polarized light microscopy remain the most versatile tools for a conservation scientist, although the techniques require much experience. The methods developed for mineralogy and petrology are used for identifying mineral pigments.

Chromatographic Methods. Thin-layer chromatography (tlc) has been used for identification of binding media and dyes, but the comparatively large samples required for these analyses means that tlc has been replaced by higher-performance liquid chromatography (hplc) and other chromatographic methods. Over the last 20 years, advances in chromatographic methods have allowed them to be used to analyze the extremely small samples that are available from works of art.

Mass Spectrometry. The high sensitivity of mass spectrometry makes it useful for analyzing artists' materials.

Laser desorption–ionization mass spectrometry used for analysis of natural and synthetic organic artists colorants, is particularly interesting to conservation scientists because information regarding the distribution of pigments among paint layers may be acquired.

X-Ray Diffraction (XRD). This technique is used for characterization of crystalline materials. Most synthetic inorganic pigments and many organic pigments can be well-characterized using XRD. The extremely small size of samples has meant that until recently the most useful diffractometers were equipped with Gandolfi cameras and the most sensitive detector was photographic film. With very long exposure times (up to 18 h) samples of corrosion products and pigments barely visible to the naked eye could be identified.

SEM-EDS. This technique finds application to the examination of samples from all types of fine art. Imaging samples using back-scatter electron detectors allows pigments to be differentiated according to their average atomic number and energy dispersive spectrometry can be used to identify phases in the sample. Element mapping is extremely useful for this purpose.

PRESERVATION STUDIES

Fine art and other objects in museums, in particular archaeological finds, may have been subjected to harsh environments or treatments and are in poor states of preservation. In order to be preserved and safely exhibited, objects must be in the most stable condition that conservators and conservation scientists can obtain.

Environment

Stable temperature and relative humidity are very important for the preservation of art. Work in 1950s suggested that an indoor environment of 70°F and 50% RH could be achieved and maintained in diverse geographical locations. These values became the standard for the museum environment that remains in effect today.

It is important to keep the humidity constant for several reasons including maintenance of dimensional stability of materials that contain water, eg, wood and paper, control of bacteria and fungi, and slowing hydrolysis reactions.

Light

Most artists' materials are sensitive to the effects of light. Inorganic pigments can change color, while organic pigments, in particular natural and early synthetic pigments and dyes, are susceptible to fading. Ultraviolet light is particularly harmful in the degradation of varnishes and fibers.

Exhibition and Display Cases

Scientific investigations have been undertaken to determine how best to pack and ship valuable works. Guidelines for packing to insulate works from shock as well as temperature and humidity changes have been developed. The design of exhibit cases is very important in the preservation of works of art, as materials used to fabricate display cases can cause damage to art.

Biodeterioration

Biodeterioration of cultural heritage, including fine art, causes huge losses. Bacteria and fungi are microscopic agents that thwart the conservation scientists' goal of preservation. The best way to obviate the deleterious effects of bacteria and fungi is to maintain artworks in stable environmental conditions with well-controlled temperature and humidity and clean air.

Pest control and cleanliness are important steps in preventing insect damage to works of art. Ridding a work of pests once it is infected is undertaken with the aim, of course, of killing the infestation, but the choice of method for accomplishing this is made with a view to the detrimental aesthetic effects that an insecticide might have on organic materials. Freezing has been shown to be effective for some infestations and storing works in anoxic environments for prolonged periods is one of treatments preferred for dealing with works of art that are infested by insects.

MATERIALS–METHODS FOR TREATING AND CLEANING WORKS

Conservators are constrained in the methods they use for treating and cleaning works of art by the tenets of the profession. Specifically, they aim never to remove or alter any part of the artist's work. To preserve the quality of the artist's surface, they avoid harsh physical cleaning methods and chemicals that react with the surface. When the term cleaning is applied to fine art it also implies the removal of old varnishes, protective coatings, and old restorations. When the term is applied to works on paper, it can mean washing out acidic degradation products from the support or old mounting materials (such as acidic paper mats or wooden backings) that disfigure the work, and removing stains. Cleaning metal sculptures involves removing old and often ineffective protective coatings and corrosion products.

Conservators work with the "principle of reversibility," which is that any material used to treat an object can be removed in the future without causing any damage. So, glues and consolidants that can be removed in the future using solvents that do not damage the original material are preferred. New materials should not only be easily removed from a work in the future if wished, but they should remain stable and effective for their purpose over long time frames. Typically, conservators seek materials that have longer lifetimes than the informal industry standard of "permanent" equating to only 15 years.

DEGRADATION STUDIES

All materials deteriorate and the goal of conservation, of preserving works indefinitely, depends on understanding the degradation processes of materials so that they can be mitigated.

Degradation of Paintings, Paint, and Varnish

Paintings are subject to physical and chemical degradation. The common supports for paintings in oil, fabric, and wood, respond more rapidly than paint to changes in relative humidity. The unequal response of the paint film and the wood support can cause delamination of the paint.

The sensitivity of oil paint to solvent cleaning has been examined, and it has been found that older paints have higher proportions of azelaic acid, which forms by chain scission and oxidation. Varnishes discolor and become brittle over time. Thermal and photochemical reactions lead to different products.

Paper

Paper is a polycellulose. It was and is produced from a variety of materials. Slowing the degradation of paper is important for the preservation of archival documents as well as for the conservation of fine art.

Glass

All glass is erroneously considered to be a stable material. While some modern formulations of glass do provide an exceptionally stable product, some old glass is inherently unstable; it may become cloudy or crizzled, suffering from so-called glass disease. It is important to identify these glasses, which may include glazes in ceramics and enamels on metal, and store and exhibit them appropriately. Treatment for hygroscopic reactions is rapid washing to remove the highly alkaline material followed by drying with alcohols. After treatment it is important to maintain the glass in a dry environment. It has been shown that in some instances, efforts to protect the glass with synthetic coatings were misguided, however, this procedure does provide protection to certain glasses in some instances.

Works of Art Made from Metals and Alloys

Corrosion is a major problem for the preservation of works of art made from metal alloys. Both the stability of a work and its appearence are compromised. The corrosion processes involved are varied and complicated and require a deft analyst to diagnose them and stabilize and restore the work of art. Objects buried or submerged in salt water are most rapidly corroded if oxygen is present. On occasion, buried or submerged works of art remain relatively well preserved owing to dryness or to an anaerobic environment. On removing objects from these environments, degradation is mitigated by removal of salts and compacted earths using a variety of techniques. The method for cleaning these objects is chosen to preserve the natural patina on the surface of the objects.

AUTHENTICITY AND PROVENIENCE STUDIES

Dating

Dendrochronology. This dating technique relies on measuring the width of tree rings. While the exact amount of growth is determined by factors that cannot be estimated, over the long term the ratio of growth from season to season is determined by climatic conditions. It has been possible to establish "master chronologies" for various geographical regions. Dendrochronology provides an earliest possible date for the felling of the wood used in the creation of a work of art; this date is useful for discriminating between originals and copies and for detecting forgeries. The caveats of the dating method are uncertainties owing to the indeterminant number of sapwood rings that might have been present on the tree when it was felled, and that while the age of the wood may be known, the date that it was used for making an object is unknown.

Carbon-14 Dating. The long half-life of C-14 makes this dating technique more suitable for archaeological objects rather than those considered fine art.

Thermoluminescence Dating (TL Dating). This dating can be used to establish the date of manufacture of ceramics and core materials that contain certain minerals. Over time, radiation absorbed by these materials allows electrons to move through the crystal lattice until "pinned" at a lattice defect. When the material is heated the electrons are released from these sites and light is emitted. The amount of light emitted depends on the length of time and the radiation dose received by the material since it was last heated. The range of age that can be estimated using TL dating is 300–10,000 years. For a single sample, the error is +/− 15% so the technique must be carefully applied.

Element/Isotope Ratios

The ratios of isotopes present in a sample of lead white can be measured and a ratio index of the isotopes then calculated. Comparison of the ratio index of lead white in a painting can indicate if the layers were all painted at the same time. Marble can be provenanced by the calcium/strontium ratios and glass can be provenanced by the ratios of lanthanide and rare earth elements.

Detection of Forgeries

All the analytical and investigative techniques available are brought to bear on establishing or refuting the authenticity of a work of art. Despite the sophistication of modern analysis, the connoisseur's eye is the most important tool. Fakes, forgeries, and imitations (which were never intended to dupe collectors) are distinguished from originals by their materials and their facture. A thorough knowledge of the history of materials, technology, and artists' practices is vital for authentification. X-radiography and ir reflectography can reveal atypical working methods.

Artists' Pigments: A Handbook of their History and Characteristics, Vols. 1–3, National Gallery of Art, Washington, D.C.

National Gallery Technical Bulletin, Vols. 1–27, National Gallery, London.

W. S. Taft, *The Science of Paintings*, Springer-Verlag, New York, 2000.

P. M. Whitmore, *Contributions to Conservation Science: A Collection of Robert Feller's Published Studies on Artists' Paints, Paper, and Varnishes*, Carnegie Mellon University Press, Pittsburgh, 2002.

BARBARA H. BERRIE
National Galley of Art,
Washington, D.C

FINE CHEMICALS

The requirement for more and more sophisticated organic chemicals and biopharmaceuticals has contributed to the emergence of the fine chemicals industry as a distinct entity. This is backward integrated, production oriented, and supplies advanced intermediates and active substances to the specialty chemicals industries. Custom manufacturing, whereby the customer provides the manufacturing process and is served on an exclusive basis, is an important part of the fine chemicals business. The fine chemicals industry has its own characteristics with regard to R&D, production marketing, and finance.

In the chemical business, products may be described as commodities, fine chemicals, or specialties. Various commodities are also known as petrochemicals, basic chemicals, organic and inorganic chemicals (large volume), monomers, commodity fibers, and plastics. Advanced intermediates, building blocks, bulk drugs, bulk vitamins, and bulk pesticides and active pharmaceutical ingredients (APIs) are typical fine chemicals. "Ready-for-use" adhesives, biocides, catalysts, dyestuffs and pigments, electronic chemicals, imaging/photo chemicals, food and feed additives, flavors and fragrances, ingredients for household and personal care products, pesticides, pharmaceuticals, specialty polymers, veterinary drugs, and water treatment chemicals are all specialties. The added value is highest for specialties.

It is common to both commodities and fine chemicals that they are identified according to specifications, according to what they are. Both are sold within the chemical industry, and customers know better how to use them than suppliers. Specialties are identified according to performance, according to what they can do. In terms of volume the border line comes at ~1,000 t/year, in terms of unit sales prices the line is set at ~$ 10/kg.

RESEARCH AND DEVELOPMENT

Product innovation absorbs considerable resources in the fine chemicals industry, mostly because of the shorter life cycles of fine chemicals compared to commodities. Consequently, research and development (R&D) plays an important role. The main tasks of R&D in fine chemicals are to design and develop the synthesis, to transfer the processes from the laboratory via pilot plant successfully to the industrial scale, and finally to optimize existing processes. At all times during this course of action, it must be ensured that the three critical boundary conditions economy, safety, and ecology are met. R&D has to manage the following functions in order to deliver the requested services: literature and patent research, process research, process development, analytical development, thermal safety, and bench-scale laboratory and pilot plant.

PRODUCTION

General Comments

Typically, fine chemicals are manufactured in batch multipurpose plants. There are, however, a few examples of fine chemicals produced in dedicated or continuous plants. This can be advantageous if the raw materials or products are gases or liquids rather than solids, if the reaction is strongly exothermic or endothermic or otherwise hazardous, and if the high volume requirement for the product warrants a continued capacity utilization.

Given the wide variety of fine chemicals, the requirements to manufacture, handle, and store these compounds varies greatly. However, what they all have in common is the fact that their efficient manufacturing is driven by technology and high quality considerations.

Depending on the specific properties of the both non-cGMP and cGMP substances, severe restrictions have to be applied in the way these substances are manufactured. Highly toxic, nonpharmaceutical materials, such as pesticides, should not be manufactured in buildings and equipment being used for cGMP production. Highly sensitizing substances, materials of an infectious nature, and molecules of high pharmacological activity or high toxicity should be manufactured only in dedicated and completely segregated production areas.

Plant Design

Structure of the Plant. A fine chemicals plant is typically divided into a reaction part, also referred to as "wet section", and a product finishing part, also referred to as "dry section". The logical building block of the wet section is the train. Usually, it consists of three reactors, head tanks, receivers, and a filtration unit (centrifuge or nutsche). By definition, a train is a "chemical manufacturing tool" able to handle one chemical step in a fine chemical's multistep synthesis.

The number of products offered by a fine chemicals manufacturer typically exceeds the number of production trains. Yet, for reasons of economy of scale, the production capacity considerably exceeds the yearly requirement for each product. Furthermore, the product portfolio is regenerated at a fast pace. This set of circumstances leads to the multipurpose plant, as opposed to a dedicated plant. A multipurpose plant has to be capable of handling several types of chemical reactions and performing a series of unit operations.

Piping Concept. The choice of the proper piping concept is key for any competitive multipurpose plant design. The basic requirements for a piping system are, beside corrosion resistance for a wide array of substances, ease of cleanability (due to quality and costs) and of course a high degree of flexibility in order to ensure the needed multipurpose character of the plant. Typically, the following approaches are available: a preinstalled piping system with an adequate number of manifolds and coupling stations, according to the required flexibility and a process specific piping concept. A process specific piping concept is the system of choice in cases where the products to be manufactured are still unknown during the design phase of the plant. This system is also ideal in cases when the campaign lengths are expected to be short, ie, when frequent product changes are likely.

Automation. The complexity of the plant design, the degree of sophistication, and the quality requirements of the fine chemicals to be produced, the necessity to process hazardous chemicals, the sensitivity of product specifications to changes of reaction parameters, and the availability of a skilled work force all determine the degree of automation that is advisable.

Full process control computerization for a multipurpose plant is much more complex and might therefore also be much more expensive than for a dedicated single-product plant. Whenever possible, all efforts have to be made to choose standard process control systems and to apply standard control software.

The fact that automation systems need to be validated has become a critical aspect of all automation systems that are being applied for cGMP productions. Some guidance on this topic can be found in the U.S. Code of Federal Regulations 3.

Material Handling Principles. The material handling in a multipurpose plant is mainly driven by the following considerations: To optimize direct labor costs versus investment costs by the mechanization of material handling operations, and to comply with all pertinent quality requirements regarding safety, hygiene, and cGMP, if applicable.

Special Equipment. Standard reaction conditions and standard materials of construction available in multipurpose plants are usually:

temperature	$-20°C$ to $< 200°C$
pressure	10 mbar to 3 bar
material of construction	stainless steel and glass lined

In order to make a multipurpose plant really fit today's broad market requirements, an extension of the standard conditions (ie, adding special features to enhance the flexibility of a plant) is an absolute must. Flexibility, however, always has its price. Exotic or highly specialized equipment should only be installed in a multipurpose plant if there is a specific need. Excessive flexibility is counterproductive.

Quality Aspects During the Design Phase. In order to ensure the required quality of a project, the entire design phase needs to be highly structured. The feasibility study represents the first step in a design phase. A task force, consisting of process engineers, sales and marketing representatives, and other specialists, led by a project leader, develops the definition of the project and a first

cost estimate. Typically, the project leader will be responsible for implementing the project.

Plant Operation

Safety and Ecology Standards. In today's global economy, it is vital for fine chemicals manufacturers to adhere to international standards for safety and ecology. For that purpose, there are several highly developed systems available like the International Organization for Standardization's *ISO Management System, ISO, Geneva,* the *Responsible Care Trademark of the American Chemistry Council* program, which is of U.S. origin or the European Union *Eco-Management and Audit Scheme* (EMAS), European Commission, Environment DG.

Quality. Because fine chemicals are sold according to stringent specifications, adherence to constant and strict specifications, at risk because of the batchwise production and the use of the same equipment for different products in multipurpose plants, is a necessity for fine chemicals companies. During the course of the past years, the quality and documentation aspects in general have become more and more the success determining factor in the fine chemicals business. This is even more true for cGMP productions.

The *ISO management system standards,* specifically, *ISO 9001* deals primarily with management and focuses on the customer's requirements, regulatory requirements, the customer's satisfaction and continuous improvement on all pertinent processes.

Standards for food-grade chemicals in the United States are published in the *Food Chemicals Codex* (FCC), for laboratory reagents in *Reagent Chemicals—ACS Specifications,* and for electronic grade chemicals in the *Book of SEMI Standards* (BOSS) by Semiconductor Equipment and Materials International (SEMI).

Fine chemicals for the use in pharmaceuticals are to be manufactured according to the guidance for industry *ICH Q7A,* ie, good manufacturing practice for active pharmaceutical ingredients. In addition, the U. S. Code of Federal Regulations represents a specific guidance for the United States.

General standards for drugs are typically published in the so-called national pharmacopoeia. The names of the different national pharmacopoeia are formed by pharmacop(o)eia combined with the name of the country, eg, *United States Pharmacopeia and National Formulary* (USP–NF). Finally, the WHO publishes a *Pharmacopoeia Internationalis.*

A comprehensive training program for all employees is another essential building block to secure adequate quality and safety standards. The program has to incorporate the entire work force involved into any aspect of the manufacturing process and needs to be documented.

All quality aspects within a company are to be controlled by an independent organizational unit. Beside the quality control unit, the quality assurance activities are also part of this operation. Hereby the main aspects to be considered are releasing or rejecting products; reviewing and approving qualification reports; reviewing and approving validation reports (the validation process is a program, what provides a high degree of assurance that a process will consistently produce a result meeting predetermined acceptance criteria); approving all specifications and master production instructions; making sure that critical deviations are investigated and resolved; establishing a system to release or reject raw materials, labeling materials; approving changes that potentially affect intermediate or API quality; making sure that internal quality audits are performed; and making sure that effective systems are used for maintaining and calibrating critical equipment. These criteria are mandatory for cGMP products, however, it is recommended to utilize, whenever possible, the same criteria for non-cGMP products.

Production Planning. Production planning for a fine chemicals company operating one or more multipurpose plants is an extremely demanding task. The goal must be to achieve optimum capacity utilization, which is important for the profitability of the company. However, conflicting interests of marketing, manufacturing, and controlling have to be aligned carefully. Particularly critical is an excellent communication to the marketing and sales group, which determines what quantity of which products can be sold, and manufacturing, which determines how a most advantageous use of the existing equipment can be made and what type of plant is needed in the future. There are both short- and long-term aspects to production planning. A useful tool for the short-term planning is a rolling 18 month sales forecast, which is committing for the first 2–6 months and somewhat more flexible for the rest of the period.

Operating Costs. The main elements determining production costs are identical for fine chemicals and commodities. For a breakdown of typical production costs in its major elements, see Table 1.

In addition, the operating schedule has a significant impact on the production costs. Whereas continuous plants typically run 24 h/day, there is more freedom in establishing operating schedules for multipurpose plants. Depending on the work load and the flexibility of the work force, schedules can be adjusted as needed. Some schedules still include only a one or two shift operation (eg, 8 or 16 h/day for 5 days a week). Frequently, in this case some minimum activity is maintained during the night, such as supervision of reflux reactions, solvent distillations, or dryers. A full 7 days per week operation, consisting of four or five shift crews, each working 8 h/day is becoming the standard. In terms of production costs, this is the most advantageous scheme. Higher salaries for night work is more than offset by lower fixed costs. Also, only part of the work force has to adhere to this scheme.

Examples of State-of-the-Art Multipurpose Plants

Two examples of state-of-the-art multipurpose plants are (1) a large fine chemical plant with an innovative lay-out, and (2), a typical pharmaceutical fine chemicals plant of a midsize custom manufacturer.

Biofine Chemicals Plants

The production of biofine chemicals, by using biotechnological methods, fundamentally follows the same pattern as the one for synthetic fine chemicals: Preparation and charging of the raw material, reaction, liquid/solid (crude product) separation, product purification, and packaging. Depending on the specific bioprocess used, there are, however, more or less substantial differences in the design and operation of the plant. Simple fermentations used for specific steps in low molecular weight fine chemicals (eg, conversion of a carbonyl to an amido group, or of a carbonyl to a chiral hydroxy group) can be carried out in conventional multipurpose plants. This is particularly the case if immobilized enzymes are used as catalysts. The production of modern high molecular weight biopharmaceuticals by recombinant processes requires specifically designed plants, where utmost attention is paid to the safeguard of sterility.

Code of Federal Regulations; Part 210; Current Good Manufacturing Practice in Manufacturing, Processing, Packing, or Holding of Drugs, U.S. Department of Health and Human Services, Food and Drug Administration (21CFR part 11).

Food Chemicals Codex (FCC), 5th ed., Institute of Medicine, Washington D.C., 2003.

A. Kleemann and J. Engel, *Pharmaceutical Substances*, 4th ed., Thieme-Verlag, Stuttgart, New York, 2001, pp. 146–150.

A. Liese, K. Seelbach, and C. Wandrey; *Industrial Biotransformations*; Wiley-VCH, Weinheim, 2000.

PETER POLLAK
Fine Chemicals Business
Consultant, Reinach,
Switzerland
ERICH HABEGGER
Schweizerhall Chemie AG,
Basel, Switzerland

FLAME RETARDANTS

TERMINOLOGY

Some pertinent definitions include *fire retardant (flame retardant)*, used to describe polymers in which flammability has been reduced by some modification as measured by one of the accepted test methods; *fire-retardant chemical*, used to denote a compound or mixture of compounds that when added to or incorporated chemically into a polymer serves

Table 1. Major Cost Elements

Type of costs	Percentage, %
material costs	35
labor costs	15
energy costs	5
ecology costs	5
quality control, quality assurance	5
repair and maintenance	5
research and development	5
general overheads	10
depreciation	15

to slow or hinder the ignition or growth of fire, the foregoing effect occurring primarily in the vapor phase; *materials*, single substances of which things are constructed; and *products*, consumer items made of one or more materials that may be composed of single or blended polymers, may be layered or fiber-reinforced, and generally contain a variety of additives.

MEASURING THE FIRE PERFORMANCE OF PRODUCTS

Laws have been promulgated to improve the fire performance of individual combustibles. Traditionally, these require meeting a particular level of performance on a prescribed fire test. Most of the fire test methods in regulations have been developed by consensus standards organizations in response to a particular fire hazard. The two leading U.S. based entities are ASTM International and NFPA International. International Organization for Standardization (ISO) is the predominant non-U.S. based fire standards organization. In the United States, the methods are then referenced in the model building codes; currently, the two major building codes are the ICC Building Code and NFPA 5000. The methods are also cited in fire codes, such as the National Electrical Code and the Life Safety Code. These code structures are in turn adopted, often with modification, by governmental jurisdictions. In addition, there are a number of voluntary practices. For example, Underwriters Laboratories (UL) allows the use of its endorsement on products that meet their test criteria, and the upholstered furniture industry has adopted voluntary cigarette ignition-resistance standards.

Fire test methods attempt to provide correct information on the fire contribution of a product by exposing the whole product or a small sample of it to conditions intended to replicate the fire scenario(s) of concern. The tests most often measure the resistance of the specimen to ignition, whether the specimen continues to burn beyond an initial ignition, the rate at which the specimen burns, and/or the composition and quantity of the combustion products.

METHODS FOR IMPROVED PERFORMANCE

The materials of attention in promoting fire safety are generally organic polymers, both natural [eg, woods (qv), papers and wools (qv)] and synthetic [eg, nylons (see POLYAMIDES), polyurethanes, and rubbers (qv)]. Less fire-prone products generally are inherently more stable polymeric structures or contain fire-retardant additives. The former are usually higher priced engineering plastics (qv) that achieve increased stability at elevated temperatures by incorporating stronger (often aromatic) chemical bonds in the backbone of the polymer. Examples are the polyimides, polybenzimidazoles, and polyetherketones. There are also some advanced polymers, such as the polyphosphazenes and the polysiloxanes, that have strong inorganic backbones. Thermally stable pendent groups are also necessary. Strongly bonded polymers may, however, be brittle or difficult to process.

Fire-retardant additives are most often used to improve fire performance of low-to-moderate cost commodity polymers. These additives may be physically blended with or chemically bonded to the host polymer. They generally effect either lower ignition susceptibility or, once ignited, lower flammability. Ignition resistance can be improved solely from the thermal behavior of the additive in the condensed phase. Retardants such as inorganic hydroxides [aluminium trihydroxide $(Al(OH)_3$, ATH)] and magnesium hydroxide [$Mg(OH)_2$] add to the heat capacity of the product, thus increasing the enthalpy needed to bring the polymer to a temperature at which fracture of the chemical bonds occurs. The endothermic elimination of water can be a significant component of the effectiveness of this family of retardants. Other additives, such as the organophosphates, change polymer decomposition chemistry. These materials can induce the formation of a cross-linked, more stable solid and can also lead to the formation of a surface char layer. This layer both insulates the product from further thermal degradation and impedes the flow of potentially flammable decomposition products from the interior of the product to the gas phase where combustion would occur.

J. W. Gilman and T. Kashiwagi, *Polymer-Layered Silicate Nanocomposites with Conventional Flame Retardants*, in T. J. Pinnavaia and G. Beall, eds., *Polymer-Clay Nanocomposites*, Wiley & Sons, Ltd., West Sussex, U.K., 2000, pp. 193–206.

J. Green, *Phosphorus-Containing Flame Retardants*, in A. F. Grand and C. A. Wilkie, eds., *Fire Retardancy of Polymeric Materials*, Marcel Dekker, Inc., New York, 2000, pp. 147–170.

J. R. Hall, Jr., *The Total Cost of Fire in the United States*, NFPA International, Quincy, Mass., 2000.

W. E. Horn, Jr., *Inorganic Hydroxides and Hydroxycarbonates: Their Function and Use as Flame-Retardant Additives*, in A. F. Grand and C. A. Wilkie, eds., *Fire Retardancy of Polymeric Materials*, Marcel Dekker, Inc., New York, 2000, pp. 285–352.

RICHARD G. GANN
JEFFREY W. GILMAN
Fire Research Division,
National Institute of
Standards and Technology

FLAME RETARDANTS, HALOGENATED

Halogenated flame-retardants fall into three general classes, additive, reactive and polymeric. Additives are blended into the polymer using common polymer processing equipment at the same time other ingredients such as synergists, stabilizers, pigments, and processing aids are added. This is typically true of polymeric flame-retardants as well but these materials also have the advantage of low blooming and less environmental concerns due to their polymeric nature. Reactive flame-retardants become part of the polymer by either reacting into the polymer backbone or grafting onto it.

Highly brominated aromatic compounds have been developed to address the concerns with diphenyl oxide based flame retardants and ethane-1,2-bis(pentabromo-phenyl) is one such example. Other types of brominated flame retardants have also come into wider use including tetrabromobisphenol-A, which is the largest volume, brominated flame retardant in use today. It can function as either an additive or a reactive flame retardant and is used as a raw material in the manufacture of other flame-retardants. Other newer flame retardants, which exemplify the move to polymeric flame retardants, include brominated polystyrenes, polybromostyrenes and brominated epoxy oligomers (BEOs).

The use of flame retardants came about because of concern over the flammability of synthetic polymers (plastics). Concern over flammability should arise via a proper risk assessment, which takes into account not only the flammability of the material, but also the environment in which it is used.

FUNDAMENTALS OF FLAMMABILITY

In order for a solid to burn it must be volatilized, because combustion is almost exclusively a gas-phase phenomenon. In the case of a polymer, this means that decomposition must occur. Decomposition begins in the solid phase and may continue in the liquid (melt) and gas phases. It produces low molecular weight chemical compounds that eventually enter the gas phase and act as fuel. Heat from combustion of this fuel causes further decomposition and volatilization and, therefore, further combustion. Thus the burning of a solid is like a cyclic chain reaction. For a compound to function as a flame retardant it must interrupt this cycle in some way. There are several mechanisms by which flame retardants modify flammability by braking this cycle and some flame retardants have more than one mode of action. Mechanisms are inert gas dilution, thermal quenching, protective coating, physical dilution and flame poisoning. Halogens and some phosphorus flame retardants act by flame poisoning. The flame retardant dissociates into radical species that compete with chain propagating and branching steps in the combustion process.

Table 1. Flammability Tests

Designation[a,b]	Description or application[c]	Characteristic measured
ASTM E162-02a	radiant panel	flame spread
ASTM E119-00a	building materials	fire endurance
MVSS 302	materials for automotive interiors	burning rate
ASTM D2863-00	limiting oxygen index (LOI)	ease of extinction
ASTM E662-03	NBS smoke chamber	smoke
ASTM E84-03B, UL 723, UL 910	Steiner tunnel	flame spread and smoke
UL 94	vertical burn	ignition resistance
UL 790, ASTM E108-00	roof burn	flame spread
UL 1715	room burn	flashover potential and smoke
CAL 133	furniture	flashover potential and smoke
CAL 117 ASTM E1353-02	cigarette ignition for furniture	ignitability
ASTM E1354-03	cone calorimeter	heat release and smoke
ASTM E906-99	OSU heat release rate calorimeter	heat release and smoke
FMRC 4910	contraction	heat, smoke, toxic and corrosive products release, ignition and flame spread

[a]Designations listed together are not meant to imply equivalency.
[b]CAL = California; MVSS = Motor Vehicle Safety Standard; UL = Underwriter's Laboratory.
[c]NBS = National Bureau of Standards; OSU = Ohio State University; FMRC = Factory Mutual Research Corp.

FLAMMABILITY TESTING

One problem associated with discussing flame retardants is the lack of a clear, uniform definition of flammability. Hence, no clear, uniform definition of decreased flammability exists. The latest American Society for Testing and Materials (ASTM) compilation of fire tests lists over one hundred methods for assessing the flammability of

Table 2. Chlorinated Compounds Used as Additive Flame Retardants

Common name	Molecular formula	Halogen%	Mp, °C
dodecachloropenta-cyclooctadeca-7,15 diene		65.1	350
pentabromochloro-cyclohexane	$C_6H_5ClBr_5$	78	170
chlorinated paraffin		39–70	
bromo/chloro alpha olefin and paraffins		24–35 19–35	liquid

Table 3. Brominated Reactive Flame Retardants and Intermediates

Compound	Structure	Bromine,%	Mp,°C
diester/ether diol of tetrabromophthalic-anhydride		46	liquid
tetrabromobisphenol-A-bis(allyl ether)		51.2	119
tetrabromobisphenol-A		58.4	180
dibromostyrene		59	liquid
dibromoneopentylglycol		61	109
tribromophenyl allyl ether		64.2	74–76
tetrabromophthalic anhydride		68	270
pentabromobenzylacrylate		71	117
2,4,6-tribromophenol		72.5	95.5
tribromoneopentyl alcohol		73.6	62–67

materials. These range in severity from small-scale measures of the ignitability of a material to actual testing in a full-scale fire. Several of the most common tests used on plastics are summarized in Table 1.

FLAME RETARDANTS

Bromine and chlorine compounds have the most commercial significance as flame-retardant chemicals with bromine predominating due to its higher efficiency. Halogenated flame retardants can be broken down into three classes: brominated aliphatic, brominated aromatic and chlorinated aliphatic. As a general rule, the thermal stability decreases as brominated aromatic > chlorinated aliphatic > brominated aliphatic. The thermal stability of the aliphatic compounds is such that with few exceptions, thermal stabilizers (eg, organo tins compounds or hydrotacites) must be used to allow the flame retardant to undergo processing. Brominated aromatic compounds are much more stable and may be processed in thermoplastics at fairly high temperatures without the use of stabilizers and at very high temperatures with stabilizers.

Antimony–Halogen Synergism

Antimony oxide is commonly employed as a fire-retardant supplement for halogen-containing polymer systems as a means of reducing the halogen levels required to obtain a given degree of flame retardancy. This reduction is desirable because the required halogen content may be so high that it affects the physical properties of the final polymer. In many cases, the antimony oxide is used simply to give a more cost-effective system.

Brominated and Chlorinated Additive Flame Retardants

Additive flame retardants are those that do not react in the application designated. There are a few compounds that can be used as an additive in one application and as a reactive in another. Tetrabromobisphenol-A (TBBPA) is the most notable example. Some available bromine are brominated diphenyloxides, decabromodiphenyl oxide, tetrabromobisphenol-A, ethane-1,2-bis(pentabromophenyl), tetradecabromodiphenoxybenzene, octabromo-1-phenyl-1,3,3,-trimethylindan, tris(tribromophenyl) triazine, tetrabromobisphenol-A, bis-(2,3-dibromopropylether), tris-dibromopropyl isocyanurate, hexabromocyclododecane, 1,2-bis(2,4,6-tribromophenoxy) ethane, ethylenebis(tetrabromophthalimide), bis(2-ethylhexyl)tetrabrom phthalate. Bis(hexachlorocyclopentadieno)-cyclooctane, chlorinated paraffins and chlorine containing additive flame retardants are listed in Table 2.

Polymeric/Oligomeric Flame Retardants

There are several halogenated oligomeric/polymeric flame retardants and their high molecular weight provide numerous advantages such as, low volatility, low toxicity, easy handling (no dust) and resistance to bloom and plate-out. In some cases they are used at levels high enough that the resulting flame-retarded resin should properly be viewed as a polymer blend or alloy. The main oligomeric/polymeric flame retardants are brominated carbonate oligomers, brominated epoxy oligomers, homopolymers of a dibromo/tribromo styrene monomer, brominated polystyrene, poly(2,6-dibromophenylene oxide), and poly(2,3,4,5,6-pentabromobenzyl acrylate).

Reactive Flame Retardants

Tables 3 and 4 list the commercially available reactive flame retardants and intermediates.

ECONOMIC ASPECTS

There are a relatively small number of producers of halogenated flame retardants, especially for brominated flame retardants, where three producers account for greater than 80% of world production.

HEALTH AND SAFETY FACTORS

Halogenated flame retardants, which are primarily brominated, come in variety of structural classes. Their potential effects are related to their non-halogenated substructures, and thus should be considered individually with respect to their health and environmental profiles.

Table 4. Chlorinated Reactive Flame Retardants and Intermediates

Compound	Structure	Chlorine, %	Mp, °C
hexachlorocyclo-pentadiene		78.0	11[a]
chlorendic anhydride		57.7	240
chlorendic acid		55.0	b
tetrachlorophthalic anhydride		49.6	255–257

[a] 239°C bp.
[b] Decomposes to the anhydride.

In general, the acute toxicity of the halogenated flame retardants currently in use is quite low. They typically have high oral and dermal LD_{50} values and are not skin or eye irritants. Most are not mutagens. Most brominated flame retardants examined to date have relatively high no adverse effect levels on repeated doses, and do not affect fetal development. Many can be considered persistent in the environment, although there are notable exceptions. Because of persistency, it is generally recommended that they not be released to the environment through waste or water discharges. Brominated flame retardants generally are not toxic to aquatic organisms, again with a few notable exceptions. The few that do show some evidence of bioaccumulation typically do not induce chronic effects.

Specific information on each individual flame retardant can be acquired from the manufacturer. The latest MSDS should always be consulted prior to use, the appropriate personal protective gear worn during use, and proper disposal utilized.

P. W. Dufton, *Flame Retardants for Plastics, Chem Tec*, 2003.

Handbook of Fire Retardant Coatings and Fire Testing Services, Technomic, Lancaster, Pa., 1990.

D. Price, B. Iddon, and B. J. Wakefield, eds., *Bromine Compounds Chemistry and Applications*, Elsevier, Amsterdam, the Netherlands, 1988.

J. Troitzsch, *International Plastics Flammability Handbook*, Hanser Publishers, Munich, Germany, 1990.

ARTHUR G. MACK
Albemarle Technical Cente

FLAME RETARDANTS, PHOSPHORUS

One of the principal classes of flame retardants used in plastics and textiles is that comprising phosphorus, phosphorus–nitrogen, and phosphorus–halogen compounds. Several reviews of phosphorus flame retardants have been published (see also PHOSPHORUS COMPOUNDS).

MECHANISMS OF ACTION

Condensed-Phase Mechanisms

The mode of action of phosphorus-based flame retardants is probably best understood in cellulosic systems. Cellulose (qv) decomposes by a noncatalyzed route to tarry depolymerization products, notably levoglucosan, which then decompose to volatile combustible fragments such as alcohols, aldehydes (qv), ketones (qv), and hydrocarbons (qv). However, when catalyzed by acids, the decomposition of cellulose proceeds primarily as an endothermic dehydration of the carbohydrate to water vapor and char. Phosphoric acid is particularly efficacious in this catalytic role because of its low volatility (see PHOSPHORIC ACID AND PHOSPHATES). Also, when strongly heated, phosphoric acid yields polyphosphoric acid, which is even more effective in catalyzing the cellulose dehydration reaction. The flame-retardant action is believed to proceed by way of initial phosphorylation of the cellulose.

Vapor-Phase Mechanisms

Phosphorus flame retardants can also exert vapor-phase flame-retardant action. Both physical and chemical vapor-phase mechanisms have been proposed for the flame-retardant action of certain phosphorus compounds, such as triphenyl phosphate physical (endothermic) modes of action have been shown to be of dominant importance in the flame-retardant action of a wide range of non-phosphorus-containing volatile compounds.

Interaction with Other Flame Retardants

Some claims have been made for a phosphorus–halogen synergism, ie, activity greater than that predicted by some additivity model. Unlike the firmly established antimony–halogen synergism, however, phosphorus–halogen interactions are often merely additive and in some cases slightly less than additive.

Antagonism between antimony oxide and phosphorus flame retardants has been reported in several polymer systems, and has been explained on the basis of phosphorus interfering with the formation or volatilization of antimony halides, perhaps by forming antimony phosphate.

COMMERCIAL PHOSPHORUS-BASED FLAME RETARDANTS

A large number of phosphorus compounds have been described as having flame-retardant utility. The compounds demonstrating commercial utility are much more limited in number.

Inorganic Phosphorus Compounds

Red Phosphorus. This allotropic form of phosphorus is relatively nontoxic and, unlike white phosphorus, is not spontaneously flammable. Red phosphorus is, however, easily ignited. In finely divided form, it has been found to be a powerful flame-retardant additive.

Ammonium Phosphates. Monoammonium phosphate, $NH_4H_2PO_4$, and diammonium phosphate, $(NH_4)_2HPO_4$, or mixtures of the two, which are more water soluble and nearly neutral, are used in large amounts for nondurable flame retarding of paper (qv), textiles (qv), disposable nonwoven cellulosic fabrics, and wood (qv) products. The advantage is high efficiency and low cost. Ammonium phosphate finishes are resistant to dry-cleaning solvents but not to laundering or even to leaching by water. One general advantage of ammonium phosphates and phosphorus compounds as flame retardants, especially in comparison to borax, which is also used for nondurable cellulosic flame retardancy, is effectiveness in preventing afterglow.

Insoluble Ammonium Polyphosphate. When ammonium phosphates are heated by themselves under ammonia pressure or preferably in the presence of urea, relatively water-insoluble ammonium polyphosphate is produced. There are several crystal forms, depending on heating conditions. Grades are available that are coated with melamineformaldehyde or other thermoset resins to impede hydrolysis.

Phosphoric Acid-Based Systems for Cellulosics. Semidurable flame-retardant treatments for cotton (qv) or wood (qv) can be attained by phosphorylation of cellulose, preferably in the presence of a nitrogenous compound.

Additive Organic Phosphorus Flame Retardants

Melamine and Other Amine Phosphates. Four melamine phosphate are commercial products, these are melamine orthophosphate, $C_3H_6N_6 \cdot H_3O_4P$; dimelamine orthophosphate, $2\,C_3H_6N_6 \cdot H_3O_4P$; melamine pyrophosphate $2\,C_3H_6N_6 \cdot H_4O_7P_2$; and a newly introduced melamine polyphosphate. All four are available as finely divided solids. All are used commercially in flame-retardant coatings (qv) and from patents also appear to have utility in a variety of thermoplastics and thermosets.

Trialkyl Phosphates. Triethyl phosphate is used as an additive for polyester laminates, rigid polyurethane foams, and in cellulosics. It is added as a precure to unsaturated polyester resins and also to the polyol precursor of rigid polyurethane foams, where it functions as a viscosity depressant and remains in the foam as a flame retardant. Higher trialkyl phosphates are relatively flame retardant liquids; tributyl phosphate is used in fire-resistant aircraft hydraulic fluids and trioctyl phosphate is used as a low-flammability low-temperature vinyl plasticizer.

Oligomeric Ethyl Ethylene Phosphate

It has been introduced as a high efficiency flame retardant for flexible or rigid polyurethane and polyisocyanurate foams. Because of its oligomeric character, it has low vapor emission (thus freedom from window fogging) in applications such as automobile seating.

Dimethyl Methylphosphonate. Dimethyl methylphosphonate (DMMP), $C_3H_9O_3P$, a water-soluble liquid, bp 185°C, is made by the Arbuzov rearrangement of trimethyl phosphite. DMMP is a viscosity depressant and flame retardant in alumina trihydrate (ATH) filled unsaturated polyester resins such as these used for bathtubs, shower stalls, and in halogenated resins such as used in ductwork.

Diethyl Ethylphosphonate. A liquid compound introduced for applications similar to those of DMMP is diethyl ethylphosphonate, $C_6H_{15}O_3P$.

Halogenated Alkyl Phosphates and Phosphonates

In this important class of additives, the halogen contributes somewhat to flame retardancy although this contribution is offset by the lower phosphorus content.

2-Chloroethanol Phosphate (3:1). Tris(2-chloroethyl) phosphate, $C_6H_{12}Cl_3O_4P$ [2-chloroethanol phosphate (3:1)], is a low viscosity liquid product. This phosphate is used in rigid polyurethane and polyisocyanurate foams, carpet backing, flame-laminated and rebonded flexible foam, flame-retardant coatings, most classes of thermosets, adhesives (qv), cast acrylic sheet, and wood.

1-Chloro-2-Propanol Phosphate (3:1). Tris(1-chloro-2-propyl) phosphate, $C_9H_{18}Cl_3O_4P$, is much lower in reactivity to water and bases than the 2-chloroethyl homologue. It is used in isocyanurate foam to reduce friability and brittleness, and is also used in flexible urethane foams in combination with melamine.

1,3-Dichloro-2-Propanol Phosphate (3:1). Tris(1,3-dichloro-2-propyl) phosphate, $C_9H_{15}Cl_6O_4P$, is a leading additive for flexible urethane foams, but is also used in rigid foams, and can be added to the isocyanate or the polyol–catalyst mixture. This halogenated phosphate is also useful as a flame retardant in styrene–butadiene and acrylic latices for textile backcoating and binding of nonwovens.

Bis(2-Chloroethyl) 2-Chloroethylphosphonate. Although bis(2-chloroethyl)-2-chloroethyl phosphonate, $C_6H_{12}Cl_3O_3P$, is not as stable as the corresponding phosphate, it is useful as a flame-retardant additive in rigid urethane foams, rebonded foams, adhesives, and coatings.

Diphosphates. 2-Chloroethyl diphosphates have low volatility and good-to-fair thermal stability, and are thus useful in those open cell (flexible) foams that have requirements for improved resistance to dry and humid aging.

Oligomeric 2-Chloroethyl Phosphate. This product low in volatility and useful in resin-impregnated air filters, in flexible urethane foam, rebonded foam, and structural foam.

Tris(tribromoneopentyl) Phosphate. This high melting solid is a promising flame retardant additive for polyolefins, such as polypropylene. Tris(tribromoneopentyl) phosphate is one of the few retardants that appears to be useful in polypropylene fibers.

Oligomeric Cyclic Phosphonates. Antiblaze N and 1045 are used as flame-retardant finishes for polyester fabric. These high-percent phosphorus-content products are also useful flame retardants in polyester resins, polyurethanes, polycarbonates, nylon-6, and textile backcoating.

Pentaerythritol Phosphates. A bicyclic pentaerythritol phosphate, CN-1197, is used in thermosets, preferably in combination with melamine or ammonium polyphosphate.

Cyclic Neopentyl Thiophosphoric Anhydride. This solid additive is remarkably stable, surviving addition to the highly alkaline viscose, the acidic coagulating bath, and also resisting multiple laundering of the rayon fabric.

Phosphinates. Alkaline earth or aluminum salts of phosphinic acids, $RR'PO_2H$, where R and R' are small organic radicals, appear to be effective as flame retardants at unusually low loadings in thermoplastics such as acrylonitrile–butadiene–styrene (ABS), polybutylene terephthalate (PBT), or polyamide.

Aryl Phosphates. Aryl phosphates were introduced for flammable plastics such as cellulose nitrate and later for cellulose acetate. Cellulosics are a significant area of use but are exceeded now by plasticized vinyls. Principal applications are in wire and cable insulation, connectors, automotive interiors, vinyl moisture barriers, plastic greenhouses (Japan), furniture upholstery, conveyer belts (especially in mining), and vinyl foams.

Phosphine Oxides. Triphenylphosphine oxide, $C_{18}H_{15}OP$, is disclosed in many patents as a flame retardant, and may find some limited usage, as such, in the role of a vapor-phase flame inhibitor.

Reactive Organic Phosphorus Compounds

Organophosphorus Monomers. Many vinyl monomers containing phosphorus have been described in the literature, but few have gone beyond the laboratory. Bis(2-chloroethyl) vinylphosphonate, $C_6H_{11}Cl_2O_3P$, has applications as a comonomer imparting flame retardancy for textiles and specialty wood and paper applications.

Phosphorus-Containing Diols and Polyols. The commercial development of several phosphorus-containing diols occurred in response to the need to flame retard rigid urethane foam insulation used in transportation and construction.

Nonreactive additive flame retardants dominate the flexible urethane foam field. Auto seating applications exist, particularly in Europe, for a reactive polyol or nonvolatile additive for flexible foams to avoid windshield fogging, which can be caused by vapors from the more volatile additive flame retardants.

Reactive Phosphate Diol Oligomer. A recently introduced reactive flame retardant for flexible foam, Clariant's OP 550, reacts into the foam structure and provides flame retardancy, eg, in automobile seat cushions, without the disadvantage of windshield fogging caused by additives.

Oligomeric Phosphate–Phosphonate. A commercially used reactive oligomeric alcohol, Akzo Nobel's Fyrol 51, is a water-soluble liquid containing ~21% phosphorus. Fyrol 51, or 58 if diluted with a small amount of isopropyl alcohol, is used along with amino resins to produce a flame-retardant resin finish on paper used for automotive air filters, or for backcoating of upholstery fabric. It can also be incorporated into polyurethanes. Combinations with Fyrol 6 permit the OH number to be adjusted to typical values used in flexible foam, urethane coating, or reaction injection molding (RIM) applications.

Reactive Organophosphorus Compounds in Textile Finishing

Although synthetic fibers can be flame retarded using additives or comonomers, the flame retarding of cotton (qv) requires the application of a textile finish.

Tetrakis(hydroxymethyl)phosphonium Salts. Tetrakis-(hydroxymethyl)phosphonium chloride, is a water-soluble crystalline compound. Finishes based on the methylol groups are durable to numerous launderings.

Phosphonate Finish

A competitive cotton finish, Ciba's Pyrovatex CP, was introduced in the 1970s especially for children's cotton sleepwear, workwear, and other uses.

Pyrovatex CP coreacts on cellulose with an amino resin in the presence of a latent acid catalyst to produce finishes durable to laundering.

Phosphorus-Containing Polymers

A large number of addition and condensation polymers having phosphorus built in have been described, but few have been commercialized. No general statement seems warranted regarding the efficacy of built-in vs. additive phosphorus. However, in textile fibers, there is greater assurance of permanency.

Polyester Fibers Containing Phosphorus. Numerous patents describe poly(ethylene terephthalate) (PET) flame retarded with phosphorus-containing difunctional reactants. At least two of these appear to be commercial.

Hoechst-Celanese's (now Ticona's FR Trevira CS in Europe or KoSa's Avora CS in America) is useful for children's sleepwear, work clothing, and home furnishings. Toyobo's HEIM II (former GH) is copolycondensed with the other reactants in PET manufacture to produce a flame-retardant polyester. The fabric is used mainly for furnishings in public buildings in Japan.

HEALTH, SAFETY, AND ENVIRONMENTAL FACTORS

Toxicology

Two factors should be considered when discussing the toxicity of flame-retardant materials: the toxicity of the compounds themselves and the effect of the flame retardants on combustion product toxicity.

Product Toxicology. A review of the structure–toxicity and structure–biodegradation relationships of organophosphorus compounds shows clearly that toxicity and biodegradation properties differs greatly from compound to compound, and therefore no general statement is valid. The phosphorus-based flame retardants as a class exhibit moderate-to-low toxicity.

A particular mode of neurotoxicity was discovered for tricresyl phosphate that correlated with the presence of the o-cresyl isomer (or certain other specific alkylphenyl isomers) in the triaryl phosphates.

Mutagenic and later carcinogenic properties were found for tris(2,3-dibromopropyl) phosphate, a flame retardant briefly used on polyester fabric in the 1970s. This product was withdrawn from the market when the problem was discovered. The chemically somewhat-related tris(dichloroisopropyl) phosphate (used in foams) has been intensively studied and found not to display significant mutagenic activity. Tris(2-chloroethyl) phosphate appears to be a weak tumor inducer in a susceptible rodent strain while the chloropropyl homologue is not.

The newer tetraphenyl resorcinol diphosphate has been the subject of very thorough toxicological studies and has been found to have a very low degree of inhalation toxicity, rapid metabolism and clearance when fed to test animals, low or no reproductive toxicity, no significant teratogenic effect, and no demonstrable immunotoxicity, so it appears quite safe for its intended flame retardant use in styrenic polymer blends.

Effects on Combustion Toxicology. There appears to be no documented case of any type of fire retardant contributing to human fire casualties. Carbon monoxide appears to be the dominant lethal gas.

Effects on Visible Smoke

Smoke is a main impediment to egress from a burning building. Although some examples are known where specific phosphorus flame retardants increased smoke in small-scale tests, other instances are reported where the presence of the retardant reduced smoke. The effect appears to be a complex function of burning conditions and of other ingredients in the formulation.

Environmental Considerations

The phosphate flame retardants, plasticizers, and functional fluids have come under intense environmental scrutiny. Results published to date on acute toxicity to aquatic algae, invertebrates, and fish indicate substantial differences between the various aryl phosphates. The Environmental Protection Agency (EPA) has summarized this data as well as the apparent need for additional testing.

Tests in pure water, river water, and activated sludge showed that commercial triaryl phosphates and alkyl diphenyl phosphates undergo reasonably facile conversion to inorganic phosphate by hydrolysis and biodegradation. The phosphonates can undergo biodegradation of the carbon-to-phosphorus bond by certain microorganisms.

ECONOMIC ASPECTS

The largest volume use may be in plasticized vinyl. Other large use areas for phosphorus flame retardants are flexible and rigid urethane foams, polyester resins, epoxy and phenolic resins, adhesives, cotton and polyester textiles, polycarbonate–ABS blends, and poly(phenylene oxide) high impact polystyrene (PPO–HIPS) blends. Development efforts are very active on finding phosphorus flame retardants to replace halogen and halogen–antimony systems, especially in Europe and the Far East. To some extent, this has already happened where phosphate-flame-retarded polyphenylene ether–HIPS, or polycarbonate–ABS blends have replaced bromine–antimony–retarded HIPS or ABS. A more difficult challenge is to replace tetrabromobisphenol-A, the largest single brominated flame retardant, in electrical and electronic

circuit boards. To date, only limited penetration of phosphorus flame retardants into this market has occurred.

J. Green, in A. F. Grand and C. A. Wilkie, eds., *Fire Retardancy of Polymeric Materials*, Marcel Dekker, Inc., New York, 2000, pp. 148–170.

E. D. Weil, paper presented at *8th Annual BCC Conference on Recent Advances in Flame Retardancy of Polymeric Materials*, Stamford, Conn., June 2–4, 1997.

E. D. Weil, in A. F. Grand and C. A. Wilkie, eds., *Fire Retardancy of Polymeric Materials*, Marcel Dekker, Inc., New York, 2000, pp. 116–145.

E. D. Weil and M. Lewin, in A. R. Horrocks and D. Price, eds., *Fire Retardant Materials*, Woodhead Publishing Ltd., Cambridge, UK, pp. 31–68, 2001.

EDWARD D. WEIL
Polymer Research Institute,
Polytechnic University of
New York

FLAVOR CHARACTERIZATION

Flavor is the collection of sensations from the taste, olfactory, and trigeminal sensory systems. Taste perceptions include the currently recognized basic tastes (sweet, sour, salty, butter, and umami). Olfactory perceptions provide nearly limitless specific characterization of objects that smell, and trigeminal sensations provide the qualities such as coolness, pungency, and hot pepper burn. All three perceptual systems contribute to the flavor of a food. Within each perceptual system individual sensations interact sometimes enhancing each other, sometimes suppressing each other, and sometimes fusing to produce a character that cannot be mentally separated into its more basic components.

Both sensory and chemical analyses are necessary for thorough flavor characterization. While chemical analyses can tell one the flavor compounds present in a food and their concentrations, the only way one can tell if a collection of chemical components has a strawberry flavor is to have it evaluated by people that can recognize strawberry flavor. Only people can tell how sweet a food tastes. Analytical chemistry can tell how much of which sweeteners are present in a product, which is certainly related to sweetness, but accompanying bitter or sour tastes can suppress the perceived sweetness.

The "Holy Grail" in terms of chemically characterizing flavor would be to have characterized the flavor stimuli in a food such that chemical data could predict human judgments. While one can point to a few examples where chemical data can predict a given sensory note (most commonly an off-note resulting from a single aroma component), this goal is far from being attained.

With the above considerations in mind, there is a compelling need to use chemical data to provide an understanding of the stimuli provided to an individual when eating. The end goal may be to provide the chemical basis of an off flavor in a food, to understand how desirable flavors are formed biologically or through processing, or to understand how flavor changes during storage. While only humans can tell if a food tastes like strawberry or chocolate, or "good" or "bad," they are quite inept at describing the chemical basis for this judgment. Herein lies the value of chemically characterizing the flavor of a food, and thus explaining the fundamental bases of human flavor perception.

Liking, acceptability, or preferences for a flavor do not characterize the flavor. The identical flavor may be liked by some and disliked by others. Liking of most flavors is a matter of opinion and opinions are heavily influenced by socialization and previous experiences with a specific or similar flavor.

SENSORY METHODS FOR FLAVOR CHARACTERIZATION

Quantitative sensory methodology can be grouped into two categories, difference tests and scaling tests. Difference tests are used when the goal is to determine whether people can discriminate between two samples. Scaling tests are used when the goal is to measure the intensity of a sensation or an opinion. Some examples of these tests are ranking, category scaling, magnitude estimation, labeled magnitude scales, scaling bases, and descriptive analysis.

Perceptual Abilities

People make sensory measurements, and people differ markedly in their sensory physiology, their knowledge of sensory attributes and their social and cultural experiences.

Perceptions Differ. People can be divided into three groups: nontasters, tasters, and supertasters, based on their ability to taste the bitter compound 6-*n*-propylithiouracil (PROP). These three "taster groups" differ in their perceived intensity of other taste compounds as well.

Although relatively few people are anosmic (have no ability to smell) many have specific anosmias and are thus unable to smell one or more specific compound(s), but otherwise have normal smelling abilities.

Together these differences in PROP taster status and these specific anosmias suggest that different people will perceive flavors differently. Thus expecting members of a group to agree on the character and intensities of the many attributes present in a flavor is unreasonable.

Knowledge Differs. Expert tasters have apprenticed themselves to other initially more knowledgeable people for long periods of time during which they have been exposed to many flavor variations, described or labeled those flavor variations, and compared-confirmed their perceptions with other experts. Expert tasters are those that can claim to be able to tell the country of origin and the botanical basis of a wine, a coffee, a tea, etc.

Expert tasters and people with descriptive analysis training are assumed to no longer represent consumers.

Mixtures. All flavors are mixtures of chemicals. Thus, understanding the perceptions of chemicals in mixtures is essential. The vast amount of research on mixtures of tastes and odorants has focused on binary mixtures. However, flavors are almost always much more complex than a mixture of two chemicals. When three or more odorants are mixed together, they often form a new coherent quality, and the individual qualities of the individual odorants may no longer be apparent. This synthetic aspect of smell is similar to the synthetic nature of color perception.

CHEMICAL METHODS FOR FLAVOR CHARACTERIZATION

The task of chemically characterizing the flavor of a food involves developing methods to isolate, identify, and quantify the food components that contribute to food flavor, notably taste, trigeminal, and aroma stimuli. This process includes the analysis of volatile (aroma contributors) and nonvolatile (taste and trigeminal stimuli) food components. It must be recognized that not all volatile and nonvolatile components of a food contribute to flavor and thus, this task must include methodologies to distinguish between those food components that contribute to our definition of flavor and those that do not.

Chemical Characterization of Aroma

Hurdles. Isolating and identifying aroma compounds in a food matrix is one of the most formidable tasks faced by an analytical chemist. A primary obstacle is that laboratory instrumentation is not as sensitive to many odors as is the human olfactory system.

The fact that trace quantities of aroma components are distributed throughout a food matrix further complicates the aroma isolation–concentration process. Aroma isolation methods based on volatility are complicated by the fact that water is the most abundant volatile in a food. Thus, any procedure that draws a vacuum or involves distillation will also extract–isolate the water from the sample. Aroma isolation and analysis are also made difficult by the fact that aromas comprise a large number of chemical classes. The absolute number of aroma compounds in a food further complicates aroma analysis. It is a rather simple, natural aroma that has <200 identified constituents. A final problem complicating the instrumental study of aroma is instability. The food product being examined is a dynamic system, readily undergoing aroma changes while being stored awaiting analysis to begin. The aroma isolation process may initiate chemical reactions (eg, thermally induced degradation or oxidations) that alter the aroma profile and introduce artifacts.

Unfortunately, once each of the points above have been considered and one has obtained some instrumental profile of the aroma compounds in a food, one is left with the huge question of attempting to determine the importance of each volatile to the perceived aroma.

Aroma Isolation and Analysis. Most of the techniques used in aroma isolation take advantage of solubility or volatility of the aroma compounds. Inherently, aroma compounds must be volatile to be sensed and thus, it is logical that volatility is a common basis for separation from a food matrix. Likewise, aroma compounds tend to be more soluble in an organic solvent than in an aqueous solution (eg, a food matrix) and thus, aroma isolates may be prepared by solvent extraction processes.

Solvent extraction is often used for aroma isolation when the food does not contain any lipid. When lipid is present, then solvent extraction will extract the lipid as well, providing an aroma isolate in a fat matrix. This extract cannot be concentrated or analyzed without further processing to separate the lipid:aroma fractions.

Analysis of Taste Substances

Taste has generally been thought of as a relatively simple sense being composed of salt, sweet, sour, bitter, and umami sensations. This simplification is not justified since it is clear that each basic taste sensation has many nuances. Sour can be used to illustrate such nuances. There is no single sour perception any more than there is a single sweet, bitter or salt sensation. Each taste compound yields a different and unique taste character that complicates this basic sensory component.

Furthermore, it is worthwhile to note that taste influences aroma perception. Thus, while each acidulant gives a unique sensory character (taste), it also influences our overall flavor perception (interaction with aroma to give an overall flavor perception).

From a chemical analysis perspective, taste can readily be accounted for through well-established analytical techniques.

Nonvolatile components (as a whole—taste and nontaste) in foods play a greater role in food flavor than just defining the taste sensation. Nonvolatiles in foods are known to interact with some aroma compounds (chemically "bind") thereby exerting an additional influence on flavor. Thus, we may be interested in the analysis of nonvolatiles in foods beyond those that contribute to taste perception. Some of these are sweeteners, salt, acidulants, umami, bitter substances, and a selection of taste substances.

The Chemical Analysis of Trigeminal Stimuli

The chemical characterization of substances in food that give a trigeminal response is less problematic than that of aroma or taste, primarily because of the limited number or compounds known to elicit a trigeminal response. Also, these substances in any given food are typically closely related compounds and they have been well researched. Compounds that elicit a trigeminal response can do so in the mouth or olfactory region. While most of these compounds elicit a pain response, menthol also provides a cooling effect to subsequent air or liquid stimuli.

A. Chaintreau, in R. Marsili, ed., "Quantitative Use of Gas Chromatography-olfactometry: The GC-/"SNIF/" Method," in *Flavor, Fragrance, and Odor Analysis*, Marcel Dekker, New York, 2002, p. 333.

S. S. Nielsen, ed., *Food Analysis*, 2nd ed., Aspen Publishing, Inc., Gaithersburg, 1998, p. 151.

H. Stone and J. L. Sidel, *Sensory Evaluation Practices*, Academic Press, San Diego, 1993, p. 338.

A. J. Taylor, R. S. T. Linforth, I. Baek, J. Davidson, M. Brauss, and D. A. Gray, "Flavor Release and Flavor Perception," in *Flavor Chemistry*, Am. Chem. Soc., Washington, D.C., 2000, p. 151.

GANY REINECCIUS
ZATA M. VICKERS
University of Minnesota

FLAVOR DELIVERY SYSTEMS

For several decades, many flavors have been encapsulated, generally in solid matrices and often by spray drying, although other delivery systems and encapsulation techniques are also commercially being used. The primary and classic purposes of encapsulation are to protect volatile flavor compounds from evaporation and those susceptible to oxidation from degradation by atmospheric oxygen during storage in low moisture states. Besides providing a degree of protection, which can be very high, especially for dense, glassy systems, encapsulation converts the liquid flavor into a solid, ideally free-flowing powder that is easier to handle.

During the last one and a half decades, the motivations to encapsulate flavors have become more varied as attention has shifted from protecting the flavor in dry powder form to optimizing the performance of the flavor in complex food matrices. These food matrices are often moist or liquid and the performance of the flavor in such matrices is often unsatisfactory because of a limited chemical stability, or because of a premature, unbalanced, or incomplete release from the food matrix. To counter these negative effects, and to optimize the rate and extent of release of the flavor during food manufacturing, storage, preparation immediately prior to consumption and during consumption of the food product, extensive ranges of so-called controlled release or controlled delivery systems were developed.

Flavor compounds, which comprise both the volatiles that are sensed by the olfactory epithelium in nasal cavities and the nonvolatiles that impact the taste buds in the mouth, differ widely in chemical and physical nature. The volatile compounds, which are known as aroma compounds, are usually of low molecular weight ($M_w < 250$ Da), are liquid at STP (STP = standard pressure and temperature) and most of them are fairly-to-highly hydrophobic. A limited number of important impact compounds are hydrophilic. Encapsulation of flavors is primarily directed toward the volatile compounds, as the major stability and handling issues are experienced with these compounds. In fact, although often developed for the volatile flavor compounds, many encapsulation technologies can also be applied to nonvolatile taste compounds.

Because of the comparatively well-defined release aimed for, the selection and method of application of a

Table 1. Characteristics of the Most Important Glass Encapulation Systems

Characteristic	Capsule type				
	Spray-dried	Spray dried and agglomerated	Extruded	Vacuum-dried	Fluidized-bed dried[a]
size	20–150 μm	100–250 μm	0.4–2 mm	50–400 μm	0.3–2 mm
density	low	medium	high	variable	medium
flavor load	<25%	<25%	10–15%	<10%	<10%
flowability	poor	good	good	variable	good
surface oil	medium	medium	high[b]	high	low
processing	simple	standard	standard	simple	complex[c]
shelf life[d]	1 year	1–2 years	~4 years	1–2 years	1–2 years

[a] Limited controlled-release properties when coated.
[b] Low when washed.
[c] Depending on the number of coatings.
[d] Oxidation sensitive flavor.

controlled release system for flavors is critical and must be tailored for a specific application.

ENCAPSULATION TECHNOLOGIES: GLASS ENCAPSULATION

General Principles of Glass Encapsulation and Material Science of Amorphous Carbohydrates

Although the size, shape, and structure of glass encapsulation systems varies widely (Table 1), the working principle of all these systems is the same and is based on the material science of amorphous carbohydrate glasses. Almost invariably, water-soluble carbohydrates in the amorphous, glassy state are used as the encapsulation matrix, since, under controlled conditions, they combine high physical and chemical stability with very high barrier properties with respect to oxygen and organic molecules. In addition, they are chemically inert with respect to most classes of small organic compounds, which is important because of the wide chemical variety of flavor compounds. They are also fairly cheap and available in highly pure form.

As for all amorphous materials, amorphous carbohydrates exhibit at least two important phases, a rubbery, viscoelastic state, and a glassy, brittle state. Identical from a structural point of view, the distinction between these two states is in the rate and extent of molecular motion. In the rubbery state, translational and rotational motion of the matrix molecules is still possible, but in the glassy state large-scale molecular motion is effectively inhibited.

The glass-rubber transition is an important concept in glass encapsulation. It demarcates the regime in which the encapsulation matrix is physically stable, ie, not undergoing any significant structural changes or molecular rearrangements on the time frame of the experiment (eg, the shelf life of the encapsulated flavor), from the state in which the matrix is soft, sticky, and moldable. Therefore, to ensure that the capsules retain their structure and properties during shelf life, one requires that the encapsulation matrix is in the glassy state during storage. Generally, a compromise is made between the physical stability of the encapsulation matrix, which is

favored by increasing the molecular weight of the encapsulation matrix, and the oxygen permeability, which is usually minimized by reducing the molecular weight of the encapsulation matrix. In most commercial products, as prepared by, eg, melt extrusion and spray drying, the encapsulation matrix is composed of a mixture of intermediate or high molecular weight carbohydrates (eg, starches and maltodextrins of low DE) and low molecular weight carbohydrates (usually disaccharides, eg, sucrose).

The glass-transition temperature (T_g) is not only important during storage or use of the capsules, but also during production, and should in this case be considered in relation to the matrix viscosity. During production, the capsules obtain their final shape and size in the rubbery state. Then, by drying, cooling, or both at the same time, the encapsulation matrix is quenched into the glassy state.

Most of the important flavors are hydrophobic, including all citrus oils and many of the savory top notes, are dissolved in a solvent, which is often an oil like MCT (MCT = medium-chain triglyceride), and their dosage in the capsules is high, often >10% (see Table 1). Therefore, the flavor will not completely dissolve in the essentially hydrophilic carbohydrate matrix, but will remain phase separated. In effect, during encapsulation, a solid matrix is usually formed around liquid droplets of flavor and flavor oil (Fig. 1). To ensure a proper embedding of the flavor in the capsules (ie, to minimize surface oil), and to minimize the leakage of flavor upon rupture of the capsules, the flavor droplets should be as small as possible, preferably <1 μm in diameter. The exception to this rule in glass encapsulation is encapsulation by melt extrusion, where larger droplets (up to ~20 μm) can be tolerated as the capsules are large (typically ~1 mm) and the matrix is very dense (ie, no pores). In addition, in case of extruded products, the surface oil is often washed off during production.

In order to achieve such small oil droplets, emulsifiers are needed. These emulsifiers can either be the encapsulation matrix (eg, gum acacia, which is an arabinogalactan with surface active properties) or can be added to the encapsulation matrix in the relatively small amounts needed to cover only the surface of the oil droplets [eg, starch esterified

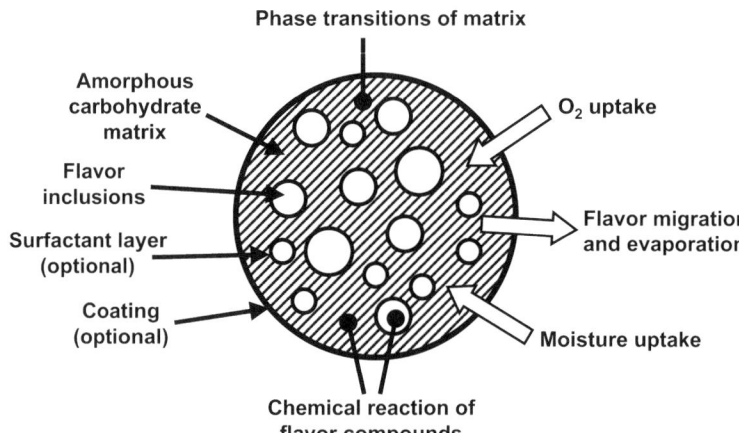

Figure 1. General characteristics of glass encapsulated flavors. Indicated are the main structural features and the principal physical and chemical effects influencing the stability of the encapsulated flavor.

by 1-octenyl succinic anhydride (OSA starches), sucrose esters, and the like]. Important criteria for successful flavor emulsifiers are that they are food-grade, form stable oil-in-water emulsions with a wide variety of hydrophobic compounds, have a relatively simple phase behavior (eg, do not show multiple dispersed phases in mutual equilibrium) and do not interact with the flavor compounds.

Extrusion Encapsulation

The encapsulation of flavors by extrusion was initiated in the 1950s after initial trials on the encapsulation of citrus oils in hard-candy matrices demonstrated the suitability of carbohydrate matrices for the encapsulation of oxygen-sensitive essential oils.

For encapsulation by extrusion, originally a vertical extrusion setup was used which differs considerably from modern extrusion technology. As extrusion screws are absent, the product matrix is forced through the die plate by overpressure in a feeder vessel and the viscosity of the feed consequently needs to be quite low.

The twin-screw process is more recent, but is rapidly becoming the standard melt-extrusion process in the flavor industry. In addition, the extrusion field has been very active in developing matrices for flavor encapsulation. An early improvement was the use of noncrystallizing carbohydrates, which improves the facility of processing as no attention needs to be paid to avoid undesired matrix crystallization.

Encapsulation by Spray Drying

Spray drying is used as an encapsulation technique for food and pharmaceutical ingredients because of its relatively simple, continuous operating conditions, and easily accessible machinery. In spray drying of an oil-based flavor, the flavor is emulsified into an aqueous solution, or dispersion, of an edible carrier material, usually a carbohydrate, and the emulsified material is pumped through a spraying nozzle or atomizer into a high temperature chamber. The ideal carrier should have good emulsifying properties, be a good film former, have low viscosity at high solids levels, exhibit low hygroscopicity, and release the encapsulated ingredients when reconstituted in a finished food product. Furthermore, the infeed solid content is important.

The retention of flavor compounds during spray drying is best understood according to the selective diffusion mechanism. Flavor compounds are preferentially retained in the drying droplet because the diffusion coefficient of organic compounds decreases much faster with decreasing water content than the diffusion coefficient of water. An additional effect slowing down the diffusion of organic compounds is microentrapment of hydrophobic compounds in small inclusions, leading to an effective slowing down of the rates of release of organic volatiles because of the partitioning of the aroma compounds into the hydrophobic inclusions.

Two types of atomizers are widely used in the industry: single-fluid, high pressure spray nozzles and centrifugal wheels.

Encapsulation by Freeze Drying and Vacuum Drying

Two additional technologies are occasionally used to encapsulate flavors in a glassy matrix.

Freeze drying is a multiple-step operation in which the dispersion containing the matrix material (usually largely consisting of carbohydrates) and the flavor is first frozen, then dried by direct sublimation of the frozen solvent, and by desorption of the sorbed or bound solvent under reduced pressure.

Freeze drying is practical only for encapsulating heat-sensitive flavors. Major disadvantages of freeze drying are the high energy cost and the long drying time. In addition, the powders are fragile and protection of oxidation-sensitive flavors is non-optimal because of the extensive open and porous structure.

Vacuum drying is widely used to preserve the aromatic qualities of herbs and spices, but can also be applied to entrap flavors in glassy carbohydrate-based matrices, eg, maltodextrin and starches.

Fluidized-Bed Encapsulation Technologies

Fluidized-bed technologies are used to coat solid powders with a polymer film and usually produce a core-shell morphology or particles with an onion-like structure. The process involves spraying a polymer solution or dispersion onto a fluidized powder and evaporating the solvent to leave a polymer film covering the powder particles.

In the case of flavors, the first motivation to use fluidized-bed encapsulation techniques is to create relatively large particles with regular (usually spherical) shapes and a narrow particle size distribution. A second motivation is to create capsules with controlled release properties by applying one or more coatings with specific functionalities to the core particles. In flavor encapsulation by fluidized-bed techniques, the process generally comprises at least one step in which a flavor emulsion is sprayed onto a suitable core.

In order to ensure a proper embedding of the flavor, the flavor emulsion generally contains a significant fraction of water-soluble carbohydrates like maltodextrins and other starch hydrolysis products and/or low-molecular weight carbohydrates.

This section would not be complete without a brief mention of the essentially three spraying techniques that are commonly employed in fluidized-bed encapsulation. The first spraying technique is top spraying (1), in which the atomizer is placed at the top of the fluidized-bed chamber and is directed downward. With bottom spraying or Wurster coating (2), the atomizer is placed close to the air inlet at the bottom of the fluidized-bed chamber, pointing upward. The third method of spraying is side or tangential spraying.

Comparison of Glass Encapsulation Systems

Although the working principle of all glass encapsulation systems discussed here is the same (ie, protecting sensitive flavor and essential oils in amorphous, glassy carbohydrates during storage in low moisture states), the wide variety in structure, size, and shape, and the varying conditions during capsule production often lead to clear advantages of one type of capsule over another.

- If the shelf life of an oxidation-sensitive flavor is of prime importance, an extrusion-encapsulated product would often be preferred because of its excellent barrier properties with respect to oxygen.
- If the price is of key concern, and if the flavor to be used is not too sensitive toward environmental factors, spray drying is the method of choice.
- Fluidized-bed encapsulation is an emerging technology for flavor encapsulation, albeit for niche applications. The principal reason to apply fluidized-bed encapsulation is that the particles are very regular in shape and usually also rather large. Disadvantages of fluidized-bed encapsulation are their relatively low flavor load, the complexity of the process, and concomitantly the high cost of the final product.
- Freeze drying has found only limited application as a flavor encapsulation technology due to its higher operating cost and longer process time.

ENCAPSULATION TECHNOLOGIES: CONTROLLED RELEASE SYSTEMS

In contrast to glass encapsulation systems, controlled release systems have a variety of objectives. The principal aim is to control the release of the flavor in the food application, but this is a general statement, which in reality covers a large number of situations. Retarded flavor release can be desirable during food processing or food preparation, usually to minimize flavor losses due to excessive volatilization or temperature- or moisture-induced chemical instability. Conversely, in situations where the flavor would strongly, sometimes irreversibly bind to food matrix constituents, and so never be released from the food matrix, a controlled release system can be helpful to speed up the rate of flavor release. In still another situation, a controlled release system might be used to modulate the release of individual flavor compounds from the food product during food consumption, thereby providing a more interesting flavor profile to the consumer.

In this section, a selection is made of the many controlled release systems for flavors that are around. Priority has been given to those systems that have proven merits in actual food applications (coacervate capsules, spray-chilled capsules, and inclusion complexes). In addition, a number of promising or otherwise interesting technologies are discussed, eg, extrusion encapsulation, gel encapsulation, and complex fluids.

Coacervation Encapsulation and Gel Encapsulation

For encapsulation purposes, two phenomena involving polymer phase behavior are of significant importance: gelation and coacervation.

Coacervation is the formation of phase-separated, fluid polymer phases in an aqueous medium. Two types of coacervation are generally distinguished: simple and complex coacervation. Whereas in simple coacervation, a polymer phase, usually but not necessarily composed of a single polymer, phase separates by hydrophobic interactions, in complex coacervation the separated phase is constituted of two polymers which are oppositely charged. The two oppositely charged polymers form an electrostatic complex which is overall close to electrically neutral and separates out of solution in the form of small globules leaving a depleted aqueous phase.

Complex coacervation is used for the encapsulation of hydrophobic substances (active ingredients, excipients or solvents) in either the solid or the liquid state. Encapsulation by complex coacervation has found widespread use for a wide range of active ingredients including flavors and essential oils.

Gel encapsulation is a process that is considerably simpler than encapsulation by complex coacervation. In gelation, the gelating polymer generates a permanent three-dimensional network in solution by the formation of physical bonds (hydrogen bonds, ionic bonds) between polymer segments, often but not always involving a conformational change of the polymer in solution. Gelation is distinct from simple coacervation in that the gel often does not phase separate from the solution.

An essential distinction between encapsulation by gelation and by coacervation is that in the latter case, individual capsules each containing one oil droplet are formed because of the phase separation of the polymers and their affinity for the hydrophobic oil surface.

Extrusion Encapsulation

One of the most suitable and flexible ways to prepare polymeric matrices for controlled release of active ingredients, including flavors, is extrusion. For controlled release applications in the food domain, the polymers that can be used are almost all largely hydrophilic.

Extrusion encapsulation to prepare controlled release systems is in many respects very similar to glass encapsulation by twin-screw extrusion. However, because of the higher molecular weight of the encapsulation matrix, either temperature or plasticizer content or both are generally higher than with extrusion of low molecular weight carbohydrates. Because of these factors, significant degradation of the encapsulated flavor is often induced, either during extrusion (if the processing temperatures are too high) or during postextrusion treatments. Plasticizers other than water, eg, glycerol, are also commonly used. Since they are much less volatile than water, they are retained in the matrix.

One of the most popular food polymers used for preparation of controlled systems by extrusion is starch. Starches are available with highly diverse molecular weights, compositions and structures. Also other hydrocolloids like agar and proteins like whey can be used to obtain matrices exhibiting controlled release properties.

In order to modify the release behavior, either low molecular weight compounds can be added that results in enhanced release rates, or water absorbing compounds, which results in a slower release. Also, the release rates are reduced by adding hydrophobic materials to the extrusion matrix or by coating the final product.

Spray Chilling

In spray chilling, the carrier material is heated to above the point of fusion, mixed with the active ingredient, and atomized into a cooled chamber. The layout of the process is very similar to spray drying, but in spray chilling no solvent needs to be evaporated. Instead, a solid matrix is obtained by rapidly cooling the matrix with the dispersed active ingredient to below its melting point. The technique is variously known as spray chilling if the melting point of the encapsulation matrix is low ($\sim 30–45$ °C) and as spray cooling if a high melting point matrix is selected (> 45°C). Suitable encapsulation matrices are fats and waxes. An important factor to control during spray chilling is the crystallization of the fat.

Encapsulation in Complex Fluids

Under the heading "complex fluids," all materials and structures that combine hydrophobic and hydrophilic properties in one structure and that form, at ambient temperatures, fluid, fluctuating systems are noted. This finding implies that the intermolecular interactions holding the structure together are of the same order of magnitude as thermal energy. The classic example of a complex fluid in food technology is an water-in-oil (W/O) or an oil-in-water (O/W) emulsion. In the field of flavor encapsulation O/W emulsions are traditionally used to aid in the dispersion of water-insoluble flavor compounds in

aqueous systems like beverages. In encapsulation, emulsions are widely used to disperse flavors and flavor solvents in the encapsulation matrix. Emulsions, however, also influence the release of flavors from a food matrix, and thereby in part determine the flavor perception. Although these effects are relatively limited for most flavor compounds, the retardation can be substantial, in particular for very hydrophobic compounds. Emulsions may thus be seen as examples of controlled release systems for flavors.

Inclusion Complexation

Inclusion complexation, also known as molecular encapsulation, primarily uses cyclodextrins to complex and entrap molecules. Other materials that form (inclusion) complexes with flavors and that are of potential interest for flavor delivery are starch and various proteins like albumin.

Encapsulation of flavor compounds in cyclodextrins is carried out by dissolving the cyclodextrin in water and adding the flavor under vigorous stirring. Often, temperatures of $\sim 60–80$°C are used to improve the aquous solubility of cyclodextrin and to accelerate complex formation.

The principal functionalities of cyclodextrins in flavor encapsulation are to protect the flavor from volatilization and chemical reaction, and to modify the release of the flavor in the food product.

APPLICATION OF FLAVOR DELIVERY SYSTEMS AND OUTLOOK

Flavors are used in highly diverse fields, but the principal applications are in beverages, confectionery, dairy, and culinary products.

Given that major applications for flavors, (instant) beverages and culinary products, are generally dry, powdered products in the form in which they are sold to the consumers, or during a certain stage of production or storage, it is clear that glass encapsulation systems still play a dominant role in the whole of the flavor encapsulation field.

Controlled release systems for flavors still occupy a rather small part of the total field for flavor encapsulates, which is partially because few controlled release systems are actually commercially available at an acceptable price and in sufficiently large quantities to have a serious impact on the market. Currently, the main controlled release technologies on the market are coacervate capsules and spray chilling.

As emphasized earlier, controlled release systems need to be carefully tuned for both the flavor they are supposed to deliver and for the food matrix in which they are applied. This means that food development efforts need to be strongly focused on innovative ways to introduce flavor compounds in a wide variety of food matrices, to retain them in these products, and to release them at the required moment.

K. B. De Roos, *Trends Food Sci. Technol.* **17**, 236–243 (2006).

D. Kilburn, J. Claude, T. Schweizer, A. Alam, and J. Ubbink. *Biomacromolecules* **6**, 864–879 (2005).

S. J. Risch and G. A. Reineccius, eds., *Encapsulation and Controlled Release of Food Ingredients.* ACS Symp. Ser. 590. American Chemical Society, Washington, D.C., 1995.

D. D. Roberts and A. J. Taylor, *Flavor Release.* ACS Symp. Ser. 763. American Chemical Society, Washington, D.C., 2000.

W. H. Rulkens and H. A. C. Thijssen, *J. Food Technol.* **7**, 95–105 (1972).

JOB UBBINK
ANNEMARIE SCHOONMAN
Nestlé Research Center

FLAVORS

Flavor is viewed as a division between physical (appearance, texture, and consistency), and chemical sense (smell, taste, and feeling). The Society of Flavor Chemists, Inc. defines flavor as "the sum total of those characteristics of any material taken in the mouth, perceived principally by the senses of taste and smell and also the general senses of pain and tactile receptors in the mouth, as perceived by the brain."

The acceptability of food is determined by its flavor, and a large number of flavoring substances are used. Industrial flavorings are needed for the commercial preparation of foods. However, most of the daily food intake, even in industrialized countries, contains flavor naturally or flavor formed during cooking and preparation for human consumption. Only a minor part of the daily food intake is covered by foods containing added flavorings.

Flavors do several things in food systems. Foremost among these functions is their ability to render food more acceptable and enjoyable. Flavors are often used to create the impression of flavor where little or none exists; to alter the flavor of a product, eg, the flavor of dairy products; to modify, supplement, or enhance an existing flavor, eg, the butter flavor in margarine and the meat or chicken flavor in bouillon. The addition of flavoring is often necessary to compensate for the loss of flavor during the processing of foods, eg, pasteurized foods, concentrated citrus fruit juices, alcoholic beverages, or during the freezing, filtration, pasteurization, and long-term storage of foods.

Food Acceptance

Four features of food are recognized to determine acceptance, ie, flavor, nutritive value, appearance, and mouth-feel. When all four aspects are in proper quantitative proportions, a food finds general acceptance. When all four are interdependent, appearance takes precedence over the others. However, a report by the Food Marketing Institute has shown that consumers placed nutrition second to flavor in importance. A food must have the expected or proper appearance and color before it will be readily consumed.

Taste

Certain basic principles are involved in the physiology of flavor perception. Researchers studying taste generally agree that there are at least five tastes, ie, salty, sour, bitter, sweet, and umami. Umami can be defined to the Japanese as the taste of three broths, ie, Kombu, Shiitake, and Katsuobushi. In English, the narrower definitions of the taste of monosodium L-glutamate (MSG), or the broad but vague concept of savory, meaty, or brothy are used (see FLAVOR CHARACTERIZATION). Tastes are perceived by certain sensory cells or taste buds, contained in the approximately 10,000 papillae located on the tongue.

Taste-active chemicals react with receptors on the surface of sensory cells (taste buds) in the papillae, causing electrical depolarization (drop in the voltage across the sensory cell membrane). The collection of biochemical events that are involved in this process is called transduction. There are several aspects that affect the extent and character of taste and smell; people differ considerably in their sensitivity to and appreciation of smell and taste, and there is a lack of a common language to describe smell and taste experiences.

Odor

The physiology of odor, which is the determining characteristic of flavor, is more complex and less understood than that of taste. It has been claimed that odor is 80% of flavor. A large number of odors are distinguishable, but it is not known how this is accomplished. Olfactory response is only observed when the substance contacts the olfactory membrane, called the olfactory mucosa or olfactory epithelium, which occupies an area of about $2.5\,cm^2$ in each nostril. Above the nasal passages, the two olfactory clefts are separated by the nasal septum. For a substance to have an odor, it must be capable of reaching the olfactory epithelium high up in the nose, and must come in contact with the olfactory cilia membrane.

The odor of a substance is most logically attributed to its molecular structure. As in taste, its perception is preceded by the process called transduction, in which a chemical reaction with a receptor cell excites a nerve center, giving a sensation. Brain functions such as emotion, attention, cognition, etc mediate these sensations into perceptions. Although odor quality appears to be associated with chemical structures, it has not been possible to predict odor type accurately on this basis.

Whatever the physiology of odor perception may be, the sense of smell is keener than that of taste. If flavors are classed into odors and tastes as is common practice in science, it can be calculated that there are probably more than 10^4 possible sensations of odor and only a few, perhaps five, sensations of taste.

Flavor Materials

Materials for flavoring may be divided into several groups. The most common groupings are either natural or artificial flavorings. Natural materials include spices and herbs; essential oils and their extractives, concentrates, and isolates; fruit, fruit juices, and fruit essence; animal and vegetable materials and their extracts; and aromatic chemicals isolated by physical means from natural products, eg, citral from lemongrass and linalool from bois de rose.

Table 1. Pineapple Flavor, Natural and Artificial

Ingredient	Wt%
pineapple juice conc, 60 brix	60.0
pineapple fortifier artificial	1.0
ethyl alcohol, 95%	15.0
water	24.0

Artificial materials include aliphatic, aromatic, and terpene compounds that are made synthetically, as opposed to those isolated from natural sources. Natural and artificial flavors are defined as a combination of natural flavors and artificial flavors. It is assumed that whichever portion is in greater amount becomes the first portion of the name (Table 1).

Artificial flavors are defined in the *Code of Federal Regulations* (CFR) as any substance or substances the function of which is to impart flavor, and which are not derived from natural sources. These items include the list of substances found in CFR 21 parts 172.515, 182.60, and the Flavor and Extracts Manufacturing Association's Generally Recognized as Safe (FEMA GRAS) lists.

In 1992 a large number of flavor materials were allowable on the FEMA and the FDA lists (Tables 2 and 3).

In commerce, several classifications of flavoring and compounded flavorings are listed according to composition to allow the user to conform to state and federal food regulations and labeling requirements, as well as to show their proper application. Both supplier and purchaser are subject to the control of the FDA, USDA, and the Bureau of Alcohol, Tobacco, and Firearms (BATF). The latter regulates the alcoholic content of flavors and the tax drawbacks on alcohol, ie, return of a portion of the tax paid on ethyl alcohol used in flavoring.

One class of flavorings, known as true fruit, is composed of fruit juices, their concentrates, and their essences (Table 4). A second group, fruit flavors with other natural flavors (WONF), contains fruit concentrates or extracts that may be fortified with natural essential oils or extractives (isolates), or other naturally occurring plants;

Table 2. Chemical Classes Approved for Use in Flavors

	Compounds		
Chemical class	1992	1965	Example
sulfur	152	13	thioester, thiol, mercaptan
nitrogen	99	21	amino acids, pyrazine, ester
acids	67	42	
esters	546	372	methyl, ethyl, allyl, terpene
acetals	28	21	
aldehydes	122	21	terpene
ketones	144	64	terpene, ionone, pyrone
alcohols	143	80	terpene, phenols
ethers	52	51	dioxane, furan, oxide
hydrocarbon	8		terpene
miscellaneous	11		
Total	*1415*	*730*	

Table 3. Natural and Food Ingredients Used in Flavors[a]

Compound	Number of items
Natural flavors[b]	
absolutes	19
botanical extracts	114
botanicals	249
concretes	3
essential oils	144
oleoresins	20
miscellaneous	9
Total	*558*
Food ingredients[c]	
emulsifiers	51
preservatives	49
anticaking agents	12
multipurpose	180
flavor	18
Total	*310*

[a]Information courtesy of Flavor Knowledge Systems, Glenview, Illinois.
[b]FEMA and FDA listings.
[c]FDA listing.

Table 4. True Fruit Apple Flavor

Ingredient	Wt%
apple juice cone, 72 brix[a]	80.0
apple essence[b]	5.0
ethyl alcohol, 95%	15.0

[a]Brix = g of sugar per 100 g liquid.
[b]150 = fold, ie, one gallon (3.785 L) of concentrated distillate is obtained from 150 gallons of single-fold juice.

(eg, Apple WONF contains apple juice conc, 72 brix (40.0 wt %), apple essence, 150-fold (20.0 wt%), apple fortifier natural (1.0 wt%), ethyl alcohol, 95% (15.0 wt%), and water (24.0 wt%). A third class, artificial fruit flavors, includes fruit concentrates fortified with synthetic materials. It is important to note that other than fruit flavors can also be fortified with synthetic materials, eg, in the making of an artificial maple or meat flavor (Table 5).

Table 5. Pineapple Flavor, Artificial

Ingredient	Wt%
allyl cyclohexane propionate	1.4
allyl caproate	13.0
methyl-b-methylthiolpropionate	0.2
geranyl propionate	0.5
ethyl isovalerate	1.0
ethyl butyrate	1.0
γ-nonalactone	0.1
maltol	1.0
vanillin	0.5
2,5-dimethyl-2 (OH)-4-(2H)-furanone	0.2
orange oil	1.0
ethyl alcohol, 95%	46.0
propylene glycol	34.1

A flavor is composed of two parts, a flavor portion and a diluent portion. The flavor portion is composed of three parts, a character item, a contributory item, and a differential item. A character item is a material whose smell or taste is reminiscent of the named flavor, a chemical additive, or a blend of chemicals that provides the major part of a flavor's sensory identity (more or less characteristic). A contributory item is an additive which, when smelled and/or tasted, helps to create, enhance, or potentiate the named flavor (not characteristic but essential by virtue of its acting together with the latter to produce a definite character). A differential item is an additive or a combination or additives which when smelled or tasted have little if any character reminiscent of the named flavor (neither characteristic or essential).

It is in this area where the greatest number of examples of creativity occur.

The function of the flavor portion is to give the flavor a name and to add character fixation to the flavor. Character fixation is the use of relatively high boiling point solids, at concentrations that are above their threshold values (at the use level), because once they exceed their threshold concentrations, the perception of the flavor does not change.

The diluent portion is the largest portion of a flavor. Its function is to keep the flavor homogeneous and to determine the form of the flavor. The form of a flavor is essentially its physical appearance, ie, liquid, powder, or paste, ie, that which makes the flavor applicable.

One class of flavorings, known as true fruit, is composed of fruit juices, their concentrates, and their essences. A second group, fruit flavor (WONF), contains fruit concentrates or extracts that may be fortified with natural essential oils or extractives (isolates), or other naturally occurring plants.

The apple fortifier is composed of a blend of botanical extractives and natural chemicals (isolates and those derived via natural processes). Pineapple fortifier artificial is composed in some part, if not altogether, of artificial chemicals, botanical extractives, essential oils, etc.

Compounding

In the compounding technique, constituents are selected or rejected because of their odor, taste, and physical chemical properties, eg, boiling point, solubility, and chemical reactivity, as well as the results of flavor tests in water, milk, or an appropriate medium. A compound considered to be characteristic is then combined with other ingredients into a flavor and tasted as a finished flavor in the final product by an applications laboratory.

A flavor is tried at several different levels and in different mediums until the most characteristic one is selected. This is important because the character of the material is known to change quality with concentration and environment.

Specifications

Specifications for many of the essential oils and artificial flavorings are available. Physical specifications encourage standardization and uniformity in basic flavor and perfume materials. Although compliance with specifications does not guarantee the flavor quality standards will be accepted, the specifications fill a need and provide valuable reference for the flavor industry.

The *Food Chemical Codex* defines food-grade quality for the identity and purity of chemicals used in food products. In the United States, the FDA adopts many of the *Food Chemicals Codex* specifications as the legal basis for food-grade quality of flavor and food chemicals.

Flavor Precursors

The characteristic flavors of foods, such as in fruits and vegetables, are considered to result from enzymatic action upon certain more complex components during the normal developmental or ripening process. It has been found that the flavor of fruit can be increased by a process called precursor atmosphere (PA). When apples were stored in a controlled atmosphere containing butyl alcohol, the butyl alcohol levels increased by a factor of two, and the polar products, butyl esters, and some sesquiterpene products increased significantly. The process offers the possibility of compensating for loss of flavor in fruit handling and processing owing to improper transportation conditions or excessive heat.

Another process employed to increase the formation of volatile compounds in fruit is that of bioregulators. When a bioregulator is applied to lemon trees, an increase in both the aldehyde and alcohol fractions of the lemon oil extracted from the fruit of the treated lemon trees was observed.

Enzymes not only produce characteristic and desirable flavor but also cause flavor deterioration (see ENZYMES, INDUSTRIAL). The latter enzyme types must be inactivated in order to stabilize and preserve a food. Freezing depresses enzymatic action. A more complete elimination of enzymatic action is accomplished by pasteurization.

The creation of flavor by enzymes is used in the fermentation process to prepare products such as alcoholic beverages, cheese, pickles, vinegar, bread, and sauerkraut. In some vegetables, such as Cruciferae (mustard) and Alliacae (onion and garlic), the flavor components are released enzymatically when the tissue is crushed or broken. Several essential oils are also created enzymatically. A more complex flavor development occurs in the production of chocolate. The chocolate beans are first fermented to develop fewer complex flavor precursors; upon roasting, these give the chocolate aroma. The flavor development process with vanilla beans also allows for the formation of flavor precursors. The green vanilla beans, which have little aroma or flavor, are scalded, removed, and allowed to perspire, which lowers the moisture content and retards the enzymatic activity. This process results in the formation of the vanilla aroma and flavor, and the dark-colored beans that after drying are the product of commerce.

The use of dry heat, as in roasting, baking, and frying, develops flavor characteristics not found in the unheated product. The distinctive flavor development in breakfast foods, ie, the crust of baked bread, the aroma of roasted coffee, etc, can be directly attributed to the chemical combinations brought about during the heat-treatment operation. These types of flavor are generally characterized by the presence of pyrazines in the product.

Flavor Regulations

The Pure Food and Drug Act of 1906 introduced federal regulations to combat food and drug adulteration and fraud. The law was superseded by the Food Drug, and Cosmetic Act of 1938 and the Amendments of 1954, 1958, 1960, and 1962 (see FOOD ADDITIVES). Flavor regulations differ from country to country, but progress is being made to harmonize regulations in the interest of international trade. There are at least five different ways countries can control and administer flavor regulations: countries may have a positive list of flavor materials, a negative list of flavor materials, a mixed-system having both a positive and negative list, no flavor regulations, or countries may demand prior government approval before a material can be used.

Sensory Evaluation

The type of food and its processing affect flavoring efficiency; therefore, flavor materials must be taste-tested in the food itself. Because there has been a lack of standardization of testing techniques, a committee on sensory evaluation of the Institute of Food Technologists has offered a guide which is designed to help in developing standard procedures. For each type of problem, appropriate taste tests are suggested together with the type of panel, number of samples per test, and analysis of data.

T. E. Furia and N. Bellanca, *Fenaroli's Handbook of Flavor Ingredients*, 2nd ed., Vol. 2, CRC Press, Boca Raton, Fla., 1975.

H. Heath, *Source Book of Flavors*, Avi Publications Co., Inc., Westport, Conn., 1981.

R. W. Moncrieff, *The Chemical Senses*, CRC Press, Cleveland, Ohio, 1967.

H. W. Schultz, H. Day, and L. M. Libby, eds., *The Chemistry and Physiology of Flavor*, Avi Publications Co., Westport, Conn., 1967.

FRANK FISCHETTI, JR.
Craftmaster Flavor Technology,
Inc.

FLAX FIBER

Flax (*Linum usitatissimum* L; "linen most useful") is a versatile crop that is grown throughout the world and in a variety of climates. Linen, which is used for apparel and interior textiles, comes from the long, strong bast fibers that form in the outer portions of the flax stem. Flax fibers also are used in industrial applications, eg, composites, geo-textiles, insulation, and specialty papers. Flax seeds are the source of linseed oil, which has been widely used in paints, varnishes, cosmetics, and linoleum. More recently, flax seeds are being recognized as a health food, with nutritional benefits from lignans and omega-3 fatty acids. Even the woody core tissue (shive), which is

removed during cleaning of fiber, is used for particleboards and animal bedding. Linen, which is valued for comfort and its distinctive appearance, remains a favorite in the textile industry.

STRUCTURE AND CHEMISTRY OF FLAX

Anatomy

Bast fibers are produced in the outer regions of the stem between the outermost cuticle–epidermis layer and the innermost, woody tissues. The structure of the stem is important in retting, which is the process of separating fiber and nonfiber fractions, and for the quality of the fibers derived for industrial applications. Tissues in the stem cross section (Fig. 1) are identified as follows: outermost cuticle layer covering the epidermis, thin-walled parenchyma cells inside the epidermis and surrounding fiber bundles, bast fibers formed in bundles, cambium, and woody core cells. The fibers exist in bundles of ultimate (ie, individual) fibers in a ring encircling the core tissues. About 20–50 bundles form in cross-sections of flax stems, with 10–40 spindle-shaped ultimate fibers of 2–3 cm long and 15–20 µm in diameter per bundle. Oval-shaped bundles indicate high quality fiber, while irregularly shaped bundles indicate poor quality. A polygonal cross-sectional shape (3–7 sides) and thick cell walls provide better quality fibers. The occurrence of kink bands, which appear similar to nodes, arises from processing methods and has been implicated in failures of compression tests. A thin cambium layer separates fibers and core tissues. These core tissues are comprised of lignified woody cells, which constitute the "shive" fraction produced during fiber cleaning.

Chemistry

The stem cuticle of flax contains waxes, cutin, and aromatics. This structure serves as a barrier to protect plants from invading organisms and water loss. The cuticle closely covers the epidermis, and this relationship constitutes a rigid and formidable structure that influences the ease of retting. The cuticularized epidermis in flax must be breached for microbes and enzymes to reach the inner tissues and loosen fibers from nonfiber tissues. During retting, microorganisms enter the stems through cracks and disruptions in the cuticle, partially degrade tissues, and thereby separate the cuticle/epidermal barrier from the fibers. Incomplete degradation, ie, poor retting, leaves this protective barrier and fibers still attached and contributes to reduced fiber and yarn quality. Amounts of cuticular fragments are inversely related to quality of yarn and fiber.

Thin-walled parenchyma cells, which occur between the epidermis and fibers and between the fiber bundles, are rich in pectin and other matrix polysaccharides. The cambium, a specialized tissue for secondary cell growth, exists between fibers and woody cells and is also rich in pectin. The separation of fibers from the woody core at the cambium can easily occur, especially when stems have been stored in dry climates for an extended time.

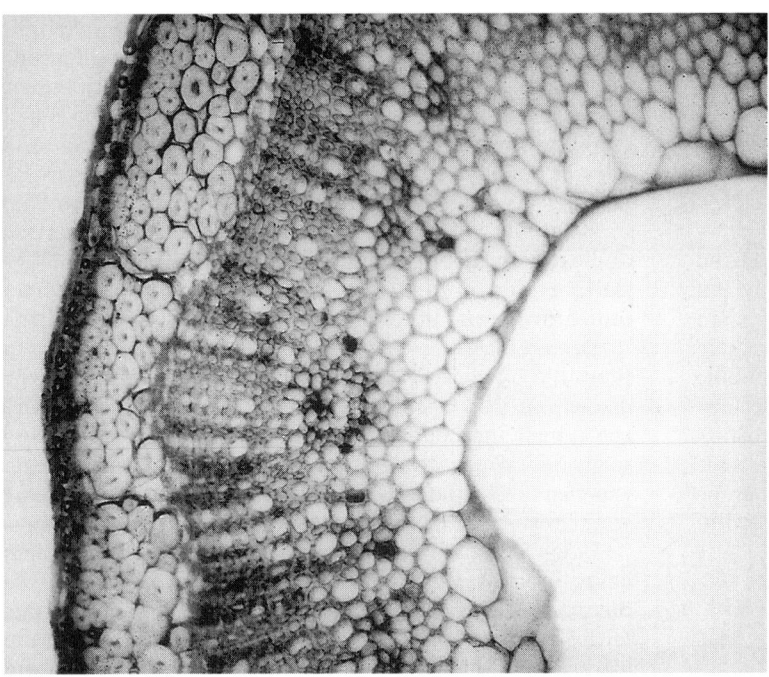

Figure 1. Light microscopy of cross section of flax stem showing the organization of tissues: epidermis with protective cuticle to the outside; parenchyma underneath the epidermis and between fiber bundles; ultimate fibers in bundles; cambium, and woody core cells.

Flax fibers are primarily comprised of cellulose, but pectins, hemicellulose, and phenolic compounds also are present. Compared with cotton fibers, which typically contain ~95% cellulose, flax has a lower percentage of cellulose and more pectin and hemicellulose. The increase in mannose and galactose along with glucose after retting suggests an intimate involvement of noncellulosic sugars in the secondary fiber walls of flax. Further, hemicellulosic constituents such as galacto–gluco–mannans and xylose are often reported as substantial components in flax fibers. The presence of these noncellulosic carbohydrates is thought to impart distinguishing characteristics, such as high moisture regain, to flax. Further complicating the structure of flax fiber is the association of proteins and proteoglycans with secondary walls, which possibly provides a structural dimension to flax fibers. In contrast to cotton, flax fibers stained with Oil Red, which indicates the presence of wax gave no positive reaction, indicating little or no waxes present. Therefore, while flax fiber is considered primarily a cellulosic fiber, its characteristics differ from cotton and many other natural fibers and allow application in a range of industries.

Woody core tissues are the most highly lignified cells in flax stems based on wet chemical, histochemical and spectroscopic comparisons with fiber and other tissues in the stem. Lignin imparts rigidity and strength to plants tissues generally. Guaiacyl and syringyl lignins both occurred in core cells, with the guaiacyl type more prevalent in cultivars examined.

Pectin, which is strategically located in plants, is particularly important to maintain the structure of flax stems; its degradation is of fundamental importance for retting and, therefore, the quality of flax fibers. Parenchyma, cambium, and the middle lamella binding fibers in the bundles are rich in this component. Reports indicate in some cases that pectin in primary walls generally may have a high proportion of oligosaccharide chains on the backbone and longer chains than the pectin in the middle lamellae. A rhamno-galacturonan structure of type I pectin, which is a prominent form in plants, likely forms the backbone of the high molecular weight polysaccharides in flax fiber as shown by nuclear magnetic resonance (nmr).

Lignin, consisting of recalcitrant compounds with a complex polyphenylpropanoid structure, is a major limitation generally to microbial degradation of plant carbohydrates. Often lignin is mentioned as a detriment to the quality of flax fiber. Studies to localize sites of lignin and aromatic compounds within cells, using histochemical stains and ultraviolet (uv) absorption microspectrophotometry, showed that these compounds occurred nonuniformly in middle lamellae between fibers, with the greatest levels in cell corners. Lignin, however, does not appear to impede fiber separation from the core cells, particularly with subsequent processing to clean fiber. Heavily localized areas of aromatics that remain on retted fiber, however, could influence properties or reduce processing efficiency. Chemically extracted aromatics from bast tissues of flax inhibited the activities of pectinase and other enzymes, suggesting a possible role for such compounds in limiting retting. Most of the lignin occurs in the woody core cells and is removed as this shive fraction is separated from the fiber during cleaning. Rapid analysis of lignin and aromatics could provide a useful method for assessing shive content as a contaminant of fiber for grading fiber quality.

Production

Flax can be grown for fiber or linseed. For fiber, seeds from high fiber varieties are densely sown to give a final plant density of ~2000 plants/m^2. Planting in this way gives thin-stemmed, straight, and tall plants for high fiber yield and quality. For optimal quality of fiber, flax plants are harvested before full seed maturity. Fiber yields as well as quality vary due to variety, environment, and agronomic practices, but total fiber yields of 25–30% of straw dry weight are possible. Linseed varieties, in contrast, are sown in low densities to maximize branching for greater seed production. Plants grown in this manner to full seed maturity have thick stems and produce fiber of low quality.

In traditional production of linen such as that practiced in Europe, flax plants are pulled from the soil. Plants can be harvested by mowing when short flax fiber, rather than long line for linen, is the objective. While costs are likely to be less with mowing, fiber remaining in the stubble reduces yield. With linseed, the fiber for paper or low-value composites may be in tangled straw that results from the combine used to harvest seeds. Other methods of harvest, eg, a stripper header to collect seeds, may be used that leave the linseed straw residue in a more suitable state for collection and processing of fiber.

PROCESSING

Retting

Retting, which is the separation or loosening of fiber bundles from nonfibrous tissues, is a major problem in processing flax. In retting, fiber bundles are separated from the cuticularized epidermis and the woody core cells and subdivided to smaller bundles and ultimate fibers. The quality of retting determines both yield and quality of the fiber, and plant development and seasonal variation in turn influence retting. Under-retted flax results in coarser fibers heavily contaminated with shive and cuticular fragments, while overretting can reduce fiber strength due to excessive thinning of bundles or microbial attack on fiber cellulose. Two primary methods for retting, namely water- and dew-retting, have been used traditionally over millennia to separate fibers for textile and other commercial applications.

Water-retting depends on fermentation by anaerobic bacteria, such as *Clostridium felsinium*, to degrade pectins and other matrix substances. In early times, bundled flax stalks were submerged in natural bodies of running or still water (eg, lakes, rivers, dams) for 5–7 days and then dried in the field for 1–2 weeks. Particular reference is often made to the river Lys (for Courtrai flax) in Belgium, where the suitability for cold water-retting and excellent quality fiber led to an active linen industry. Retting pits or tanks were constructed (Fig. 2) where temperature and other conditions could be controlled.

Despite the fact that the highest flax fiber quality is produced by water-retting, this practice has been largely discontinued in western Europe due to the high costs and the stench and pollution arising from fermentation of the plant material. Dew-retting is now the most common practice for separating flax fibers, even though some water-retted fiber is still marketed.

Dew-retting is reportedly the oldest method of retting, being used by Egyptians for millenia. Even though the flax fiber is of lower quality than that from water-retting, lower labor costs and higher fiber yields make dew-retting attractive. Stalks are pulled or mowed, spread in uniform and thin, nonoverlapping swaths (Fig. 3), and left in the field where the moisture and temperature encourage colonization and partial degradation of flax stems by consortia of fungi, yeasts, and aerobic bacteria. Flax plants are turned over on a regular basis to produce more uniform retting. Primarily, indigenous fungi effect dew-retting, and successions of various species and groups occur during the process.

Dew-retting suffers from several disadvantages: the declining quality of flax fiber in the years since dew-retting replaced water-retting. In addition to poor and inconsistent fiber quality, dew-retting requires occupation of agricultural fields for several weeks and restriction to geographical regions that have the appropriate moisture and temperature for effective microbial growth. In western Europe, which reportedly produces the highest quality linen due to a favorable climate for dew-retting,

Figure 2. Pits, or tanks, for water-retting of flax. Flax stems are bundled and submerged. Conditions can be altered in the pits, such as using warm water for more uniform retting, aeration to change the microbial metabolism, inoculation of specific retting microorganisms, and addition of supplements to promote retting.

Figure 3. Swaths of flax pulled and laid in uniform rows by specialized pulling equipment for dew-retting. With proper moisture and temperature, indigenous microorganisms from the soil and plant colonize and partially degrade the flax stems. When fibers have been loosened from the nonfiber cells and before fibers begin to be degraded, flax is harvested in round or square bales for further processing.

crop losses of $\sim$33% often occur. The prolific fungal colonization and contact with the soil during dew-retting result in a heavily contaminated product, which creates another problem for United States textile mills that import flax fiber for blending with cotton and spinning on high efficiency, short staple systems. Despite these problems, dew-retting still remains the method of choice for extracting flax and linen fibers for most of the world.

Considerable research has been undertaken to find a replacement for dew-retting. Chemical-retting has been evaluated using a variety of methods, including EDTA or other chemical chelators at high pH, detergents, strong alkali, and steam explosion. The Reutlingen Steam Explosion Treatment uses impregnation and steam to remove pectins and hemicelluloses from decorticated flax; a fast decompression separates bundles to smaller bundles and ultimate fibers. Steam explosion provides fiber of a constant quality that can be designed for different applications. Successful laboratory results have been reported with chemical-retting methods, but at times fiber properties are less satisfactory than those from other methods. To date, no chemical-retting methods are used commercially.

Enzymes have been considered for some time as a potential replacement for dew-retting flax. Early work with water- and dew-retting microorganisms showed conclusively that pectinases were required for effective retting. Successful enzyme-retting could provide considerable advantages.

Mechanical Cleaning

In traditional linen production, mechanical cleaning follows retting to remove shive and cuticularized epidermis from the fiber. The quality of retting determines the quantity and quality of the fiber remaining after scutching.

From the primitive manual tools such as hammers and boards used to scutch flax, modern equipment, although automated, scutches flax more or less by the same methods. As the long line flax is beaten, a short fiber fraction, called tow, is removed along with contaminants and cleaned separately.

Scutched flax is then cleaned using a combing action called hackling, which removes smaller contaminants, disentangles and aligns the long fibers, and separates the bundles without destroying length. A short fiber fraction, called hackling tow, is produced as a by-product of the long-line fiber. Fibers are then processed into sliver (a continuous strand of loosely assembled fibers) and then roving (sliver with reduced diameter and a slight twist to hold fibers together). From this material, yarns are made using a wet, ring spinning system that is relatively slow and expensive in comparison to cotton spinning. The tow fibers are cleaned and refined for cottonized flax, blended with cotton or other fibers, and spun on efficient dry ring or rotor spinning systems.

"Total fiber" scutching can be carried out to process only one type of fiber from the flax stems rather than long-line and tow for traditional linen. This process is simpler than that for traditional linen in that alignment of stems is not as critical for processing, and nontraditional sources of fiber (eg, linseed straw) may be used.

FLAX FIBER PROPERTIES AND GRADING

Measurement of Fiber Properties

For traditional long-line flax used in textiles, a number of factors are subjectively judged by experienced graders and include weight in hand, strength, fineness, handle (softness, smoothness, pliability), luster, cleanliness, parallelism of fiber bundles, freedom from knots and entanglements, length and shape. High fiber strength reduces breaks during spinning, which improves efficiency, and has been considered the "best single measure of yarn quality". Color also is important. With the advent of dew-retting over water-retting, the color of fibers tended to be darker. These colors also have been used to indicate the degree of retting within certain constraints. Fiber lots having different colors may be blended to provide a final product with a desired color. For blending of tow or total fiber with cotton, fiber length and length distribution affect spinning efficiency and should be uniform in various lots.

As mentioned previously, the nature of fibers and fiber bundles in flax and their propensity to fracture to smaller sizes throughout processing presents difficulties in assessment. Nevertheless, methods to derive objective values for various parameters, such as fiber strength, length, and fineness, are available.

Flax fiber fineness can be calculated gravimetrically as the weight per unit length, as is done with textiles generally. Resistance to airflow for a known fiber mass in a known volume has been used as a quick method to indicate fiber fineness.

Attempts have been made to employ instruments that rapidly and objectively analyze cotton to assess flax fiber as well. Hardware and software modifications allowed some success over time, but optimum performance of the equipment for flax analyses has not been achieved as of yet.

Recent attempts have been made to use rapid, spectroscopic methods to assess flax fiber quality in place of time-consuming older methods. Models using near infrared (nir) reflectance spectroscopy were used to monitor the degree of enzyme-retting. Components in the flax stem (eg, carbohydrates, aromatics, waxes) as well as changes in fiber cellulose have been detected using Fourier transform (ft) Raman spectroscopy. The use of ir imaging for flax has potential to identify the site of specific components that relate to retting efficiency, utilization, and quality.

Standards

Even though flax is considered the oldest textile fiber known, objective standards recognized for the industry, such as for cotton, at this time are limited for flax. Instead, flax is traditionally bought and sold by the subjective judgment of experienced graders who appraise by look and feel, called organoleptic tests. Various classification schemes that include the source (eg, Belgium, France, Russia, or China), processing history (eg, water- or dew-retted), or application (eg, warp or weft yarn) have been used within an industry segment. Grading systems for traditional linen attempt to assess fineness, length and shape of fibers, strength, density, luster, color, handle, parallelism, cleanliness, and freedom from neps and knots.

The development of standards for judging flax fiber quality has been held back by difficulty in assessing flax due to its complex physical structure, inconsistent measuring practices, lack of industry support, and a rather small, confined market for traditional long-line flax and tow. However, international interest in increasing for standards for assessing fiber quality. ISO 2370 for fineness has existed since 1980. More recently and since 2003, subcommittee "Flax and Linen" of ASTM International has developed a terminology standard and three test methods for judging fiber flax quality for: color, fineness, trash content. Work continues for refinement and new standards to assess other properties.

S. K. Batra, in M. Lewin and E. M. Pearce, eds., *Handbook of Fiber Chemistry*, Marcel Dekker, Inc., New York, 1998, pp. 505–575.

J. Janick and A. Whipkey, eds., *Trends in New Crops and New Uses*, ASHS Press, Alexandria, Va., 2002, p. 599.

R. Kozlowski, ed., *Euroflax Newsletter*, No. 16, Institute of Natural Fibres, Poznan, Poland, 2001.

H. S. S. Sharma, J. Lefevre, and J. Boucaud, in H. S. S. Sharma and C. F. Van Sumere, eds., *The Biology and Processing of Flax*, M. Publications, Belfast, Northern Ireland, 1992, pp. 199–212.

DANNY E. AKIN
Russell Research Center,
ARS-USDA

FLOCCULATING AGENTS

Flocculation is defined as the process by which fine particles, suspended in a liquid medium, form stable aggregates called flocs. The degree of flocculation can be defined mathematically as the number of particles in a system before flocculation divided by the number of particles (flocs) after flocculation. Flocculation makes the suspension nonhomogeneous on a macroscopic scale. A complete or partial separation of the solid from the liquid phase can then be made by using a number of different mechanical devices. Flocculating agents are chemical additives, which, at relatively low levels compared to the weight of the solid phase, increase the degree of flocculation of a suspension. They act on a molecular level on the surfaces of the particles to reduce repulsive forces and increase attractive forces.

APPLICATIONS

The principal use of flocculating agents is to aid in making solid–liquid separations, which, for the most part, can be divided into two types: settling and filtration. In the first case, floc formation increases the settling rate of the suspended solids, by increasing the size of the suspended particles. In the case of filtration, flocculation has several effects. First of all, it prevents fine particles from either passing through the filter or clogging the filter. Permeability and cohesiveness of the flocculated material are desirable when a centrifuge is used to make a solid–liquid separation.

The principal uses of flocculating agents are

1. Removing small amounts of suspended inorganic or organic particles from surface water prior to its use as drinking water or industrial process water. This is often called raw water clarification and can involve both inorganic and organic flocculating agents.

2. Concentrating the organic solids in municipal, agricultural, and industrial wastewater to produce a sludge with a minimum volume and water content for incineration or other means of disposal, and a clarified (very low suspended solids) water that can be discharged or recycled. This operation is often called dewatering (qv).

3. Removing suspended inorganic material from waste streams generated in the beneficiation of ores or nonmetallic minerals, to form a concentrated slurry that can be used for reclamation of mined out areas or other uses and a clarified water that can be discharged or recycled.

4. Separating the solid–liquid phases in leaching operations, where a valuable material is contained in the liquid phase, so its recovery is to be maximized.

5. Other industrial solid–liquid separations include juice clarification in the sugar industry and treatment of water used in oil and gas drilling and biotechnology.

6. Polymeric flocculants are used to bind fine cellulose fibers and solid inorganic additives to long cellulose fibers as the paper pulp is being formed into sheets on a paper machine.

7. Polymeric flocculants are used as an additive to irrigation water to control the loss of soil in the run off. Flocculation also stabilizes the soil structure and makes it more permeable to irrigation water. Flocculants can be used to reduce erosion of exposed soil by rainwater at construction sites.

CHEMICAL COMPOSITION

Flocculants can be classified as inorganic or organic.

Inorganic Flocculating Agents

The inorganic flocculating agents are water-soluble salts of divalent or trivalent metals. For all practical purposes these metals are aluminum, iron, and calcium. The principal materials currently in use are: aluminum sulfate (hydrate), aluminum chloride hydroxide, polyaluminum—silicate—sulfate or PASS, ferric chloride ferric sulfate, ferrous sulfate, and calcium hydroxide.

Organic Flocculants

The organic flocculants are all water-soluble natural or synthetic polymers. Since the 1950s the use of natural products as flocculating agents has steadily declined as more effective synthetics have taken their place. The only natural polymers used to a significant degree as flocculants are starch and guar gum. Examples of synthetic polymers include acrylamide–acrylic polymers and their derivatives, polyamines and their derivatives, poly(ethylene oxide), and allylamine polymers.

Mechanism of Flocculation

In order to form flocs, the individual particles in a suspension must collide. Flocculation can be classified as either orthokinetic or perikinetic. In orthokinetic flocculation, particle motion results from turbulence in the suspension, and in the latter from Brownian motion. For all practical purposes, this is the main type of flocculation. At very close distances, polar materials are attracted by dipole-induced dipole interactions commonly called van der Waals forces. In most aqueous suspensions, ionization of surface groups gives the particle an overall negative charge. The charged particles in suspension are surrounded by a group of positive ions referred to as the double layer. When particles approach each other, the resulting electrostatic repulsion of the double layers prevents the particle from joining to form a floc. Increasing the ionic strength of the liquid medium reduces the repulsion until the particles start to aggregate. This ionic strength is called the critical flocculation concentration. The thickness of the double layer can be reduced by deliberately adding higher charged ions to the system. This allows the particles to become closer and be attracted by the van der Waals forces. This mechanism is called double-layer compression or charge neutralization and is often cited as the mechanism for the inorganic flocculating agents. It is the explanation for the empirically derived Schulze-Hardy rule that the critical flocculation concentration of positive ions for a particular system decreases proportionally with the sixth power of the charge. However, application of this rule is somewhat of an oversimplification since individual Fe^{+3} and Al^{+3} ions do not exist in aqueous solutions under most conditions of pH and concentration. When aluminum and ferric salts are added to water under certain conditions of pH and temperature they can hydrolyze to form insoluble precipitates that coat and entrain suspended particles as well as adsorb dissolved organic material. This mechanism is called sweep flocculation.

The mechanisms of organic polymeric flocculating agents are in most cases different from the inorganics. The first of these mechanisms is referred to as the charge patch or electrostatic mechanism. These polymers adsorbed on a negative particle surface in a flat conformation. This promotes flocculation by first reducing the overall negative charge on the particle thus reducing interparticle repulsion. This effect is called charge neutralization and is associated with reduced zeta potential.

The mechanism of high molecular weight polymeric flocculating agents is called bridging. Some individual segments of a very high molecular weight polymer, usually a high molecular weight anionic polyacrylamide, adsorb on a surface. Large segments of the polymer extend into the liquid phase where other segments are adsorbed on other particles, effectively linking the particles together with polymer bridges. In contrast to the first two mechanisms, bridging is strongly affected by molecular weight and the ionic content of the solution.

Flocculant Performance and Selection

There is no comprehensive quantitative theory for predicting flocculation behavior that can be used for flocculant selection. This must ultimately be determined experimentally. There are three variables that affect the results obtained in any particular process that uses flocculation. These are the type of flocculant, type of substrate, and type of mechanical treatment of the flocculated substrate. The size and physical properties of the flocs that form, rather than the degree of flocculation, are the key elements in determining the practical effectiveness of a flocculant in any specific application. The effect of mechanical treatment can be viewed in terms of the type of force applied to the flocs. In thickeners and settling basins, the flocs are acted on by gravity and by the weight of material added on top of them. In vacuum filters, the flocs are subjected to atmospheric pressure. In belt presses and plate-and-frame filters, the flocs are subjected to mechanical pressure and in centrifuges they are subject to centrifugal forces. In a flowing system, such as a continuous paper machine, they are subjected to shear and elongational forces on the same scale as the particle size. In addition to the type of force that is applied to the flocs, the kinetics of floc formation also plays an important role in the results obtained in their application.

There are some general principles that can serve as guidelines for initial screening in terms of both flocculant chemistry and molecular weight. In general, the large flocs formed by high molecular weight polymers tend to settle faster than smaller ones.

In the case of thickeners, the process of compaction of the flocculated material is important. The flocs settle to the bottom and gradually coalesce under the weight of the material on top of them. As the bed of flocculated material compacts, water is released. Usually, the bed is slowly stirred with a rotating rake to release trapped water. The concentrated slurry, called the underflow, is pumped out the bottom. Compaction can often be promoted by mixing coarse material with the substrate because it creates channels for the upward flow of water as it falls through the bed of flocculated material. The amount of compaction is critical in terms of calculating the size of the thickener needed for a particular operation. The process of compaction has been extensively reviewed in the literature.

For most substrates, the operating dosage of flocculant necessary to give the settling rate necessary to operate a thickener is well below the maximum amount that can be adsorbed on the substrate. As more and more polymer is added above this operating dosage, the flocs can become larger and somewhat sticky. The bed of flocculated material then becomes very viscous. The rake mechanism may become overloaded and the flocculated material may not flow into the underflow pump. The dosage response and the sensitivity to overdosing may affect the selection of flocculating agent.

For filter belt presses and centrifuges, resistance to shear and mechanical pressure is the most important parameter. In general, flocs produced by charge patch neutralization are stronger than those produced by inorganic salts alone.

For vacuum filters, both the rate of filtration and the dryness of the cake may be important. The filter cake can be modeled as a porous solid, and the best flocculants are the ones that can keep the pores open.

Retention aid polymers are used in a very high shear environment, so floc strength and the ability for flocs to re-form after being sheared is important. The optimum floc size is a compromise. Larger flocs give better free drainage, but tend to produce an uneven sheet due to air breakthrough in the suction portions of the paper machine. In some cases, the type of floc needed for retention can be seen as similar to that needed for vacuum filtration. Floc size can be controlled by both the type of flocculant and the addition point.

General guidelines concerning the initial selection of flocculant chemistry are (1) suspensions of organic materials, such as municipal waste, are usually treated with a cationic flocculant, either inorganic or organic; and (2) suspensions of inorganic materials such as clay are usually treated with an anionic polymer or a combination of an anionic polymer with a cationic flocculating agent.

Acidic suspensions such as those produced by acid leaching often respond to natural products such as guar as well as nonionic polyacrylamides and anionic polyacrylamides containing sulfonic acid groups.

Laboratory Flocculant Testing. The objective of laboratory testing of flocculants is to determine which chemical composition and molecular weight will give the best cost performance. The usual method is to simulate on a laboratory scale the formation of flocs and then subject them to the same or similar types of forces as would be encountered in a full-scale dewatering device.

OPERATING PARAMETERS AND CONTROL

Flocculating agents differ from other materials used in the chemical process industries in that their effect not only depends on the amount added, but also on the concentration of the solution and the point at which it is added. The process streams to which flocculants are added often vary in composition over relatively short time periods. This presents special problems in process control.

Dilution

In many applications, dilution of the flocculant solution before it is mixed with the substrate stream can improve performance. The mechanism probably involves getting a more uniform distribution of the polymer molecules.

Addition Point

The flocculant addition point in a continuous system can also have a significant effect on flocculant performance.

Automatic Control

In some industries, the waste streams can vary in composition over a relatively short time period. When the solids level of a slurry changes, the entire dosage response may change. Automatic systems are available for thickeners that adjust the dosage according to the incoming solids level, overflow turbidity, and streaming current potential.

ANALYTICAL METHODS

Inorganic flocculants are analyzed by the usual methods for compounds of this type. Residual metal ions in the effluent are measured by spectroscopic techniques such as atomic absorption.

The detection of organic polymers in solution represents a more difficult problem, especially in industrial water and wastewater. In theory, charged polymers react with polymers of the opposite charge in solution and such reactions can be used to titrate the concentration of polymer present. There are a number of techniques using this method.

The molecular weights and molecular weight distributions of lower molecular weight polymeric flocculants are determined by viscosity measurements, such as the intrinsic viscosity, and by size exclusion chromatography. High molecular weight acrylamide-based polymers are characterized by light scattering techniques.

HEALTH AND SAFETY FACTORS

Based on animal studies and mutagenicity studies, trace amounts of organic polymers do not appear to present a toxicity problem in drinking water. The reaction products with both chlorine and ozone also appear to have low toxicity. The principal concern is the presence of unreacted monomer and other toxic and potentially carcinogenic nonpolymeric organic compounds in commercial polymeric flocculants. The principal compounds are acrylamide in acrylamide-based polymers, dimethyldiallyammonium chloride in allylic polymers, and epichlorohydrin and chlorinated propanols in polyamines.

Water soluble polymers have the potential for aquatic toxicity. Anionic polymers show some algicidal effects, probably due to chelation of trace nutrient metals. On the other hand, laboratory studies have shown that cationic polymers are toxic to fish because of the interaction of these polymers with gill membranes. This is especially important in the area of municipal and industrial wastewater treatment, where large amounts of cationic polymer are used and there is a possibility of residual flocculating agents in the effluent.

H. P. Panzer and F. Halverson, in B. M. Moudgil and B. J. Scheiner, eds., *Flocculation and Dewatering*, Engineering Foundation, New York, 1989, pp. 239–249.

R. S. Farinato, S. Y. Huang, and P. Hawkins, in R. S. Farinato and P. Dubin, eds., *Colloid Polymer Interactions*, John Wiley & Sons, Inc., New York, 1999, pp. 3–50.

J. Gregory, in B. M. Moudgil and P. Somasundaran, eds., *Flocculation, Sedimentation and Consolidation*, American Institute of Chemical Engineers, New York, 1985, pp. 125–138.

R. A. Williams, X. Jia, in K. Nishinari, ed., *Hydrocolloids-Part 2*, Elsevier Science, New York, 2000.

Howard I. Heitner
Cytec Industries

FLOW MEASUREMENT

Flow measurement is a broad field covering a spectrum ranging from the minuscule flow rates associated with the pharmaceutical industry to the immense volume flows involved in water transfer and treatment. This measurement is an essential part of the production, distribution, consumption, and disposal of all liquids and gases including fuels, chemicals, foods, and wastes.

FLOWMETER SELECTION

A number of considerations should be evaluated before a flow measurement method can be selected for any application. These considerations can be divided into four general classifications: fluid properties; ambient environment; measurement requirements; and economics.

One important fluid consideration in meter selection is whether the flow is laminar or turbulent in nature. This can be determined by calculating the pipe Reynolds number, Re, a dimensionless number which represents the ratio of inertial to viscous forces within the flow.

$$q = \frac{\pi D^2}{4} V, \qquad Re = \frac{4 \rho q}{\pi \mu D}$$

where ρ is the fluid density; μ is the fluid absolute viscosity; V is the average fluid velocity; D is the pipe or meter inlet diameter; and q is the volumetric flow rate.

A low Reynolds number indicates laminar flow and a parabolic velocity profile. In this case, the velocity of flow in the center of the conduit is much greater than that near the wall. If the operating Reynolds number is increased, a transition point is reached (somewhere over $Re = 2000$) where the flow becomes turbulent and the velocity profile more evenly distributed over the interior of the conduit. This tendency to a uniform fluid velocity profile continues as the pipe Reynolds number is increased further into the turbulent region.

FLOW CALIBRATION STANDARDS

Flow measuring equipment must generally be wet calibrated to attain maximum accuracy, and principal flowmeter manufacturers maintain extensive facilities for this purpose. A number of governments, universities, and large flowmeter users also maintain flow laboratories. Calibrations are generally performed with water or air using one or more of four basic standards: weigh tanks, volumetric tanks, pipe provers, or master flowmeters.

FLOWMETER CLASSIFICATIONS

Flowmeters have traditionally been classified as either electrical or mechanical depending on the nature of the output signal, the power requirements, or both. Improvements in electrical transducer technology have blurred the distinction between these categories. Many flowmeters previously classified as mechanical are now used with electrical transducers.

The flowmeters discussed herein are divided into two groups based on the method by which the basic flow signal is generated. The first group consists of meters in which the signal is generated from the energy of the flowing fluid. The second group of flowmeters comprises those meters that derive their basic signal from the interaction of the flow and an external stimulus. Meters can be further divided into three subgroups depending on whether fluid velocity, the volumetric flow rate, or the mass flow rate are measured.

FLUID ENERGY ACTIVATED FLOWMETERS

Positive-Displacement Flowmeters

Positive-displacement flowmeters separate the incoming fluid into chambers of known volume which, using the energy of the fluid, advance through the meter and discharge into the downstream pipe. The total volume of fluid passing through the meter is the product of the internal-meter swept volume and the number of fillings. Meter sizing is based on the relationship between flowmeter capacity, pressure drop across the meter, and fluid viscosity.

All positive-displacement meters depend on very close clearance dimensions between rotating and moving parts and thus are not suitable for fluids containing abrasive particles. These meters have broad application in the distribution of natural gas for two reasons: the completely mechanical nature and the ability to maintain good accuracy over long periods of time. Wear in positive-displacement meters tends to increase leakage so that errors are in the direction of underregistration, the most acceptable mode of error for commercial billing meters within the gas industry. Meters are normally periodically recalibrated and adjusted to read within 1% of the actual volumetric flow.

Positive-displacement meters also find broad application in the measurement of viscous liquids because high viscosities provide lubrication and minimize seal leakage. They normally do not require specific upstream or downstream piping.

Positive-displacement meters can be constructed for high or low temperature use by adjusting the design clearance to allow for differences in the coefficient of thermal expansion of the parts.

Owing to small operating clearances, filters are commonly installed before these meters to trap entrained particles, minimizing seal wear and resulting loss of accuracy.

There are at least five types of positive-displacement meters commercially available: reciprocating piston meters, bellows or diaphragm meters, nutating disk meters, rotary impeller vane, and gear meters.

Differential-Pressure Flowmeters

Differential-pressure or variable head flowmeters are the oldest, most common group of flow measurement devices. This general category includes orifice plates, venturi tubes, flow nozzles, elbow meters, wedge meters, pitot tubes, and laminar flow elements. All are based on the Bernoulli principle: in a flowing stream, the total energy, ie, the sum of the pressure head, velocity head, and eleva-

tion, remains constant. Differential-pressure devices all create some restriction in the fluid conduit causing a temporary increase in fluid velocity and a corresponding decrease in local head or pressure. For these conditions, the Bernoulli principle can be applied to give a general equation for head meters:

$$q = kA(2gh)^{1/2}$$

where q is the volumetric rate of flow; k is the dimensionless experimentally determined flow coefficient; A is the inside cross-sectional area of the pipe; h is the differential produced by the restriction measured in height of the flowing fluid; and g is the gravitational constant.

An outstanding advantage of common differential pressure meters is the existence of extensive tables of discharge coefficients in terms of beta ratio and Reynolds numbers. These tables, based on historic data, are generally regarded as accurate to within 1–5%, depending on the meter type, the beta ratio, the Reynolds number, and the care taken in manufacture. Improved accuracy can be obtained by running an actual flow calibration on the device.

Examples of differential-pressure flowmeters are oriface plates, venturi tubes, flow nozzles, critical nozzles, elbow meters, wedge meters, pitot tubes, laminar flow meters, and target flowmeters.

Variable-Area Flowmeters

In variable-area meters, the pressure differential is maintained constant and the restriction area allowed to change in proportion to the flow rate. A variable-area meter is thus essentially a form of variable orifice. In its most common form, a variable-area meter consists of a tapered tube mounted vertically and containing a float that is free to move in the tube. When flow is introduced into the small diameter bottom end, the float rises to a point of dynamic equilibrium at which the pressure differential across the float balances the weight of the float less its buoyancy. The shape and weight of the float, the relative diameters of tube and float, and the variation of the tube diameter with elevation all determine the performance characteristics of the meter for a specific set of fluid conditions. A ball float in a conical constant-taper glass tube is the most common design; it is widely used in the measurement of low flow rates at essentially constant viscosity. The flow rate is normally determined visually by float position relative to an etched scale on the side of the tube. Such a meter, with care in manufacture and calibration, can provide readings accurate to within several percent of full-scale flow for either liquid or gas.

A variety of other float shapes are available, some of which are designed to be insensitive to fluid viscosity changes. Tubes having various tapers are made to give linear or logarithmic scales and long slow taper tubes are available for higher resolution. Tubes may contain flutes, triangular flats, or guide rods to center the float and prevent chatter. Metal tubes are available for high pressure service. Other somewhat less common forms of the variable-area meter use a tapered plug riding vertically within the bore of an orifice.

Because of the design, variable-area meters are relatively insensitive to the effects of upstream piping and have a pressure loss which is essentially constant over the whole flow range. These meters have greatest application where direct visual indication of relatively low flow rates of clean liquids or gases are required.

Head-Area Meters

The Bernoulli principle, the basis of closed-pipe differential-pressure flow measurement, can also be applied to open-channel liquid flows. When an obstruction is placed in an open channel, the flowing liquid backs up and, by means of the Bernoulli equation, the flow rate can be shown to be proportional to the head, the exact relationship being a function of the obstruction shape.

Weirs. Weirs are dams or obstructions built across open channels that have, along their top edge, an opening of fixed dimensions and shape through which the stream can flow. This opening is called the weir notch and its bottom edge is designated the crest. Weirs are commonly used in irrigation, water works, wastewater discharge lines, electrical generating facilities, and pollution monitoring.

Flumes. Flumes, open channels that have gradual rather than sharp restrictions, are closely analogous to venturi meters for closed pipes. Weirs are analogous to orifice plates. The flume restriction may be produced by a contraction of the sidewalls, by a raised portion of the channel bed (a low broad-crested weir), or by both.

Cup and Vane Anemometers

A number of flowmeter designs use a rotating element kept in motion by the kinetic energy of the flowing stream such that the speed is a measure of fluid velocity. In general, these meters, if used to measure wind velocity, are called anemometers; if used for open-channel liquids, current meters; and if used for closed pipes, turbine flowmeters.

Current Meters

Various vane designs have been adapted for open-channel flow measurement. The rotating element is partially immersed and rotates rather like a water wheel. Operation is similar to that of vane anemometers.

Turbine Meters

The turbine meter represents a refinement of the anemometer or current-meter design for use in a closed conduit. A typical turbine cross section is shown in Figure 1.

Oscillatory Flowmeters

Three different oscillatory fluid phenomena are used in flow measurement: fluid oscilation, vortex precession, and vortex shedding.

EXTERNAL STIMULUS FLOWMETERS

External stimulus flowmeters are generally electrical in nature. They derive their signal from the interaction of the fluid motion with some external stimulus such as a

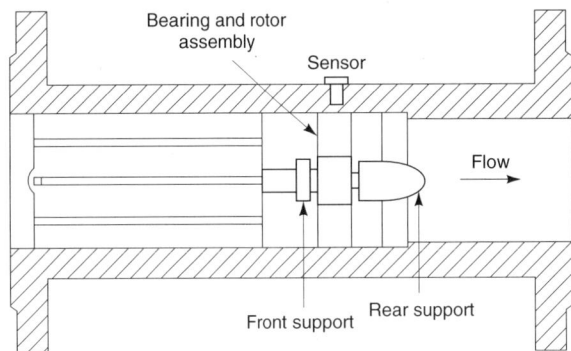

Figure 1. Turbine meter cross-section.

magnetic field, laser energy, an ultrasonic beam, or a radioactive tracer.

ELECTROMAGNETIC FLOWMETERS

Faraday's law of electromagnetic induction is used to measure the flow of conducting liquids using meter designs similar to that of Figure 2.

Electromagnetic flowmeters are available using either alternating current (ac) or pulsed direct current (dc) coil drives. The ac-actuated meters require zero adjustment at full pipe and no flow conditions. These meters provide a high accuracy and wide turndown and are the preferred meter in certain applications. The main limitation on performance is a tendency to zero shift in coating applications. Electromagnetic flowmeters that use a pulsed dc voltage coil excitation, eliminate the zero shift problem. The on–off period is synchronized at a multiple of the line frequency so any ac power noise averages out. Meters of this design function accurately under conditions where sinusoidal excitation meters do not provide acceptable results. Because of their design, pulsed dc meters have a slower speed of response than ac excitation types.

Electromagnetic flowmeters are available with various liner and electrode materials. Liner and electrode selection is governed by the corrosion characteristics of the liquid. For corrosive chemicals, fluoropolymer or ceramic

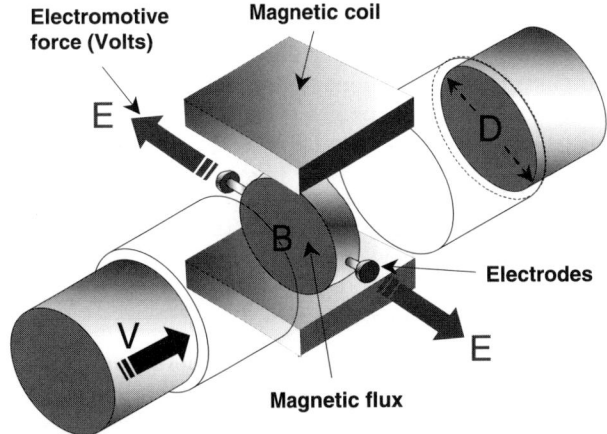

Figure 2. Electromagnetic flow meter.

liners and noble metal electrodes are commonly used. Polyurethane or rubber and stainless steel electrodes are often used for abrasive slurries.

Another approach to the problem of electrode coating is the electrodeless magnetic flowmeter. In this design there are no electrodes in contact with the process. Large plates placed on the outside of a ceramic spool perform the same function and are capacitively coupled to the transmitter through a high impedance amplifier. This meter has been found to provide satisfactory service at very low fluid conductivities and under coating conditions where other magnetic flowmeters required cleaning after short periods of operation.

Momentum Flowmeters

Momentum flowmeters operate by superimposing on a normal fluid motion a perpendicular velocity vector of known magnitude thus changing the fluid momentum.

Coriolis-Type Flowmeters

In Coriolis-type flowmeters the fluid passes through a flow tube being electromechanically vibrated at its natural frequency. The fluid is first accelerated as it moves toward the point of peak vibration amplitude and is then decelerated as it moves from the point of peak amplitude. This creates a force on the inlet side of the tube in resistance to the acceleration and an opposite force on the outlet side resisting the deceleration. The result of these forces is an angular deflection or twisting of the flow tube that is directly proportional to the mass flow rate through the tube.This inherent mass capability is a prime advantage of this technology over most other flowmeters.

Ultrasonic Flowmeters

Ultrasonic flowmeters can be divided into three broad groups: passive or turbulent noise flowmeters, Doppler or frequency-shift flowmeters, and transit time flowmeters.

Passive Detectors. Passive or turbulent noise detectors are ultrasonic microphones clamped on to the flow conduit. Passive detectors can be used for liquids, gases, or slurries to activate flow switches or to provide a low cost general indication of relative flow.

Doppler Flowmeters. Doppler flowmeters sense the shift in apparent frequency of an ultrasonic beam as it is reflected from air bubbles or other acoustically reflective particles that are moving in a liquid flow.

Transit Time Flowmeters. This type of ultrasonic meter depends on measuring the transit time of an ultrasonic beam through the flow.

Laser Doppler Velocimeters

Laser Doppler flowmeters have been developed to measure liquid or gas velocities in both open and closed conduits. Velocity is measured by detecting the frequency shift in the light scattered by natural or added contaminant particles in the flow. The technique has greatest application in open-flow studies such as the determination of engine exhaust velocities and ship wake characteristics.

Correlation Flowmeters

Tracer Type. A discrete quantity of a foreign substance is injected momentarily into the flow stream and the time interval for this substance to reach a detection point, or pass between detection points, is measured. Among the tracers that have historically been used are salt, anhydrous ammonia, nitrous oxide, dyes, and radioactive isotopes. The most common application area for tracer methods is in gas pipelines.

Cross Correlation. Considerable research has been devoted to correlation techniques where a tracer is not used. In these methods, some characteristic pattern in the flow, either natural or induced, is computer-identified at some point or plane in the flow. The correlation signal can be electrical, optical, or acoustical. This technique is used commercially to measure paper pulp flow and pneumatically conveyed solids.

Thermal Flowmeters

Hot-Wire and Hot-Film Anemometers. Hot-wire devices depend on the removal of heat from a heated wire or film sensor exposed to the fluid velocity. Hot-wire anemometers are normally operated in a constant temperature mode. In this mode, the current to the sensor becomes the flow-dependent variable. Hot-wire signals are dependent on the heat transfer from the sensor and thus on both the fluid velocity and density, ie, the mass-flow rate. These signals are also dependent on the thermal conductivity and specific heat of the fluid and are susceptible to any contamination that changes the heat transfer. For these reasons hot-wire and hot-film anemometers are primarily used in clean liquids and gases where they can be calibrated for the exact condition of use. Applications are in the measurement of low air velocities both in the atmosphere and in building ventilation studies.

Differential-Temperature Thermal Flowmeters. Meters of this type inject heat into the fluid and measure the resulting temperature rise or, alternatively, the amount of power required to maintain a constant temperature differential.

T. H. Burgess and co-workers, *Flow 2*, Flow Sensors, Instrument Society of America, Research Triangle Park, N.C., (CDROM), 2003.

Measurement of Fluid Flow by Means of Orifice Plates, Nozzles and Venturi Tubes Inserted in Circular Cross Section Conduits Running Full, ISO 5167-2:2003, International Organization for Standardization, Geneva, Switzerland, 2003.

D. W. Spitzer, ed., *Flow Measurement*, 2nd ed., Instrument Society of America, Research Triangle Park, N.C., 2001.

D. W. Spitzer, *Industrial Flow Measurement*, 3rd ed., Instrument Society of America, Research Triangle Park, N.C., 2004.

THOMAS H. BURGESS
Consultant

FLUID CATALYTIC CRACKING (FCC), CATALYSTS AND ADDITIVES

Fluid catalytic cracking (FCC) is a central technology in modern refining and is used to upgrade a wide variety of gas, oil, and residual feedstocks to gasoline, diesel fuel, and light gases. The first FCC unit, PCLA-1, came on stream in 1942 and though considered a mature technology, both hardware and catalyst developments continue to be made.

Some of the new developments of the past 50 years are a direct response to environmental regulation. Another important driving force is the goal to process heavier crude sources with higher levels of contaminants. With the increasing integration of petroleum refining and chemicals production, the value of specific molecules in the product spectrum is driving FCC catalysts toward specific selectivities. Each of these trends has long-term implications for the development of future catalysts and hardware for the FCC unit.

FCC CATALYST DESIGN FUNDAMENTALS

FCC catalysts are generally comprised of zeolite, clay, active-matrix, and binder. The primary source of cracking activity comes from the zeolite component. Zeolite Y (faujasite) has been used in FCC catalysts for over 40 years; however, a tremendous amount of R&D effort over the years has led to the continuous introduction of zeolite Y derivatives with a wide variety of performance-enhancing properties.

The principal modifications involve zeolite ultrastabilization and rare-earth ion exchange. In the ultrastabilization process, steam calcination is used to increase the framework Si/Al ratio of the zeolite Y.

Rare-earth ion exchange increases the activity of the zeolite and retards crystal destruction and dealumination in the hydrothermal environment of the regenerator. The higher the degree of rare-earth exchange of the zeolite, the higher the unit-cell size of the equilibrated zeolite in the equilibrium catalyst. Increasing the degree of rare earth exchange increases the gasoline selectivity of the FCC catalyst.

Most fluid cracking catalysts, especially those used in resid applications, are manufactured with an *active-matrix* component, usually based on specialty aluminas. Active-matrix contributes a catalytically active surface that derives its activity (and selectivity) predominantly from porous silica-alumina that may have undergone chemical or physical modification to enhance activity as well as other aspects of its performance. Of importance is the maximization of pores in the range 100–600 Å, since these pores are known to be important for coke-selective bottoms upgrading.

Clay is an important component that fulfills the remaining catalyst particle performance requirements. Clay serves as a heat sink and transfer medium. It provides little or no activity to the catalyst, but it does provide mechanical strength and density to the particle for optimum fluidization properties.

The binder is the "glue" that holds all catalyst components together and provides particle physical integrity.

RESIDUAL OIL PROCESSING

Resid feedstocks have a higher boiling range compared with vacuum gas oil's and are characterized by high concentrations of polynuclear naphthenes, aromatics, and asphaltenes as well as high levels of contaminant metals, notably nickel and vanadium and high Conradson carbon residue. Not surprisingly, proper design of the FCC catalyst is crucial for units operating with such feedstocks.

For units processing resid feeds where performance is impacted by high levels of contaminant metals and high feed Conradson carbon residue, catalyst coke selectivity is the most important performance characteristic. Catalyst stability and metals tolerance are also very important features of a catalyst designed for resid applications. To facilitate the cracking of the large hydrocarbon molecules found in resid feeds, the catalyst must be designed to maximize diffusion of these bulky species to the active acid sites. Zeolitic and matrix activity must be optimized for each application and is dependent on feedstock properties along with the dominant bottoms cracking mechanisms characteristic of the feed. Resistance to less common catalyst poisons such as iron is critical for an increasing number of refiners and can be addressed by the correct choice of matrix technology. Finally, catalyst physical properties must be optimized to ensure maximum unit retention.

Contaminant Metals (Ni, V, and Fe)

In the 1990s, a wide range of technologies were developed to improve the tolerance of FCC catalysts to contaminant metals in the feed, primarily vanadium, and nickel. In order to increase the tolerance of FCC catalysts to Ni contamination, special matrix alumina technologies were developed. These react with contaminant Ni, removing it from the surface and forming stable Ni–aluminate phases in the bulk of the alumina. Oxygen TPD measurements provide evidence for the solid-state diffusion of Ni into the alumina away from the surface and therefore the reactants.

Vanadium is particularly harmful to catalyst performance because it deactivates the zeolite component and causes yields to deteriorate. Catalyst manufacturers have developed more stable zeolites (eg, Z-17, Z-28, Z-30, CSSN, and CSX) and a series of V traps to increase the ability of the zeolite to handle vanadium. These traps are based on Ba, Ti, rare earth, and other elements. Some are more effective than others, but the basic idea is the same, ie, to keep the V away from the zeolite by binding to the surface of an inactive particle.

Understanding of the effect of Fe contaminants on the FCC catalyst is relatively new and comes from direct observations in the field. To minimize catalyst deactivation by contaminant Fe, refiners are recommended to focus on the following where possible:

Feedstocks: Use low Fe, Na, and Ca feeds and reduce its acid content to lower equipment corrosion.

FCC unit: Minimize the regeneration temperature.

FCC catalyst: Use Al_2O_3 bound catalysts instead of SiO_2 bound ones.

SHORT CONTACT TIME (SCT) CRACKING

In the FCC version of the "short contact time" mode, catalyst and feed are contacted at the bottom of the riser with cracking reactions proceeding along the length of the riser. Contact time over the length of the riser is ~5–8 s. At the end of the riser, the catalyst and products are separated in the disengager. It is here that product vapor can undergo nonselective (essentially thermal) cracking as hydrocarbon vapors come into contact with hot, coked catalyst. Today advanced riser termination technologies have one objective in common—reduction of nonselective post-riser cracking. New processes are emerging where the catalyst/feed contact time is reduced to <2 s. The so-called millisecond catalytic cracking (MSCC) process has also been introduced.

CLEAN FUELS PRODUCTION

Gasoline Sulfur Reduction

Gasoline vapors and tailpipe emissions contain NO_x and VOCs, which can react in the atmosphere to produce ozone, a major component of smog. Toxic compounds (eg, benzene, a known carcinogen) are also emitted by vehicles. Sulfur recently has come under even further scrutiny. Sulfur in gasoline degrades the performance of catalytic converters by poisoning the active sites.

In order to achieve low levels, refiners have a number of choices including gas oil hydrotreating, FCC feed hydrotreating, gasoline sulfur adsorption processes, and gasoline hydrofinishing. Each of these techniques has debits associated with it. Some, like gasoline hydrofinishing, lead to a reduction in gasoline olefins and thereby reduce gasoline octane. Some, like feed hydrotreating, have a high demand for molecular hydrogen, which is often in short supply in the refinery. Lowering gasoline end point can significantly lower gasoline volume. Furthermore, it may be difficult to find a home for the high sulfur heavy gasoline, without further processing. Thus, there is great incentive for refiners and catalyst manufacturers to find ways to reduce gasoline sulfur directly in the FCC unit.

Gasoline Sulfur Speciation. FCC gasoline contributes to >90% of the total gasoline pool sulfur with the remaining 10% originating from straight run naphtha.

Reaction Mechanism. The reactivity of feed sulfur compounds and the mechanism by which they end up in gasoline is a subject of ongoing study. Mercaptans and sulfides are converted to H_2S and do not significantly increase the amount of sulfur in gasoline. Hydrogen sulfide production from FCC has been found to correlate well with the amount of non-thiophenic sulfur in the feedstock. In addition, sulfur has been correlated in FCC gasoline with the sulfur species in FCC feed. Alkylthiophenes and aromatic sulfides are believed to be the key contributors to gasoline range sulfur.

Gasoline Sulfur Reduction Catalyst/Additive Technologies. Approaches for developing gasoline sulfur reduction (GSR) catalyst/additive technologies may include: (1) direct desulfurization of thiophenes and thiophenols from the product gasoline, (2) adsorption of gasoline sulfur precursors; or (3) alkylation of gasoline sulfur precursors into a higher boiling range fraction. The best alternatives are obviously (1) and (2), since they do not transfer the sulfur to the other liquid products. In addition, it is desirable that the additive converts the sulfur compounds into H_2S and hydrocarbon products rather than to coke.

Sulfur reduction additives such as GSR-1 and its successor, D-PriSM, have been shown to be most effective at reducing sulfur species in the light FCC naphtha and have been used in >20 refineries worldwide.

As an alternative to sulfur reduction additives, catalyst technologies such as SuRCA and SATURN have been successfully commercialized. The SuRCA catalysts are designed to completely replace the conventional FCC catalyst in the circulating inventory, while maintaining or even enhancing existing yields and selectivities. SATURN catalyst technology incorporates a gasoline sulfur reduction functionality that is effective over the entire gasoline boiling range.

Gasoline Olefins Reduction

It is well known that the hydrogen-transfer activity of the FCC catalyst can significantly affect gasoline olefins levels. Typically, increasing the rare-earth content of the catalyst increases hydrogen-transfer activity, thereby decreasing gasoline olefins. However, when using conventional FCC catalysts, the decrease in gasoline olefins is often accompanied by large reductions in LPG olefins and gasoline research octane number (RON), and poorer coke selectivity. The loss of valuable propylene, butylenes, and octane barrels is often not economical, even for the gasoline olefin- constrained refiner.

RFG catalysts have been introduced as a solution to this problem. RFG catalysts are specifically designed with maximum hydrogen-transfer activity to reduce FCC gasoline olefins. Gasoline olefins can be significantly reduced while keeping propylene, butylenes, and gasoline RON constant. Coke selectivity is also maintained, even at very high metals levels.

CONCLUSIONS

The FCC unit remains the primary hydrocarbon conversion unit in the modern petroleum refinery. To further improve profitability, refiners are increasing the amount of lower cost, residual feedstocks processed. These feeds are not only heavier, but also contain high levels of contaminant metals compared with conventional vacuum gas oils. Zeolite and matrix technologies in the FCC catalyst have evolved to maximize conversion of these difficult feeds to valuable transportation fuels and light olefins.

Increasingly stringent environmental regulations have also changed the objective function of the modern refinery. Special gasoline sulfur reduction additive and catalyst technologies are now being widely applied in FCC units to help refiners comply with tighter gasoline specifications. Developments in additives technologies have been made in response to environmental regulations on emissions (eg, NO_x, SO_x). In addition, breakthroughs in the stabilization of ZSM-5-based additives and technologies capable of double-digit propylene yields open up opportunities for refiners to increase margins by participating in the petrochemicals market.

M. Lesemann, J. R. D. Nee, and S. J. Zinger, "The Evolving Role of Fluid Catalytic Cracking in the Petrochemicals Market", NPRA Annual Meeting, Paper AM-04-58, San Antonio, Tex., March 2004.

S. K. Purnell, "IMPACT: A Breakthrough Technology for Resid Processing", Catalagram No. 93, Davison Catalysts Publication, 2003.

J. W. Wilson, *Fluid Catalytic Cracking Technology and Operations,* Penwell Publishing, Oklahoma, 1977.

J. R. D. NEE
R. H. HARDING
G. YALURIS
W. C. CHENG
X. ZHAO
T. J. DOUGAN
J. R. RILEY
W. R. Grace & Co.

FLUID CATALYTIC CRACKING (FCC) UNITS, REGENERATION

Fluid catalytic cracking (FCC) is the central process in a modern, gasoline-oriented refinery. In U.S. refineries, the amount of feed processed by fluid catalytic cracking units (FCCU) is equivalent to 34% of the total crude oil processed in the United States.

The popularity of the catalytic cracker stems from its ability to produce large quantities of high octane gasoline and other valuable light products and to use a wide range of refinery process streams as feed materials. The FCCU feeds include high molecular-weight vacuum gas oils (VGO), which are traditional feeds; atmospheric and vacuum residues; coker gas oils; gas oils from other thermal and hydrocracking operations; lube oil extracts; and deasphalted oils.

The FCC process is highly complex but self-contained. A typical UOP FCC unit of 2002 design is shown in Figure 1a. Figure 1a was originally prepared by UOP for inclusion in the 3rd edition of Meyer's *Handbook of Petroleum Refining Processes.* The 2002 design is contrasted to a mid-1970s FCC unit in Figure 1b.

FCCU HEAT BALANCE

Because of the thermal coupling of reactor and regenerator, any change on the reactor side creates a rapid change on the regenerator side, which, in turn, influences the reactor side, and vice versa. This dynamic interaction rapidly comes to equilibrium, and the catalytic cracker adjusts to a new steady-state, heat-balanced condition. The first law of FCCU catalytic cracking is that the FCCU will always adjust itself to stay in heat balance. If an FCCU is not in heat balance, it is out of control. The operating characteristics of an FCCU regenerator are thus dictated by the constraints of the heat balance (Fig. 2).

The amount of heat that must be produced by burning coke in the regenerator is set by the heat balance requirements and not directly set by the coke-making tendencies of the catalyst used in the catalytic cracker or by the coking tendencies of the feed. Indirectly, these tendencies may cause the cracker operator to change some of the heat-balance elements, such as the amount of heat removed by a catalyst cooler or the amount put into the system with the feed, which would then change the amount of heat needed from coke burning.

If FCCU operations are not changed to accommodate changes in feed or catalyst quality, then the amount of heat required to satisfy the heat balance essentially does not change. Thus the amount of coke burned in the regenerator expressed as a percent of feed does not change. The consistency of the coke yield, arising from its dependence on the FCCU heat balance, has been classified as the second law of catalytic cracking.

COKE FORMATION

The coke burned in an FCCU regenerator is a poorly defined, hydrocarbonaceous material that has either not been desorbed from the catalyst surface or has not been purged from between catalyst particles during the passage of the catalyst through the steam stripper. This coke originates from four different sources: (1) catalytic coke is produced directly from the acid catalyzed cracking reaction; (2) contaminant coke arises from the dehydrogenation activity of contaminating metals (principally nickel and vanadium) in the feed to the FCCU; (3) additive coke is directly related to the high boiling, refractory components in the feed to the FCCU. This coke correlates with the basic nitrogen and molecular weight of the feed and with the carbon residue content of the feed; and (4) cat-to-oil coke is related to the catalyst circulation rate and derives from the hydrocarbons leaving the stripper. These hydrocarbons are potentially strippable but because of incomplete stripping are entrained by the flowing catalyst into the regenerator.

The amount of catalytic coke that is formed depends on the type of catalyst used in the FCCU, the coking tendency of the feed, the degree of conversion of the feed, and the length of time the catalyst is exposed to the feed.

Coke deposition is essentially independent of space velocity. These observations, which were developed from the study of amorphous catalysts during the early days of catalytic cracking, still characterize the coking of modern day zeolite FCC catalysts over a wide range of hydrogen-transfer (H-transfer) capabilities.

Coke on the catalyst is often referred to as delta coke (ΔC), the coke content of the spent catalyst minus the coke content of the regenerated catalyst. Delta coke

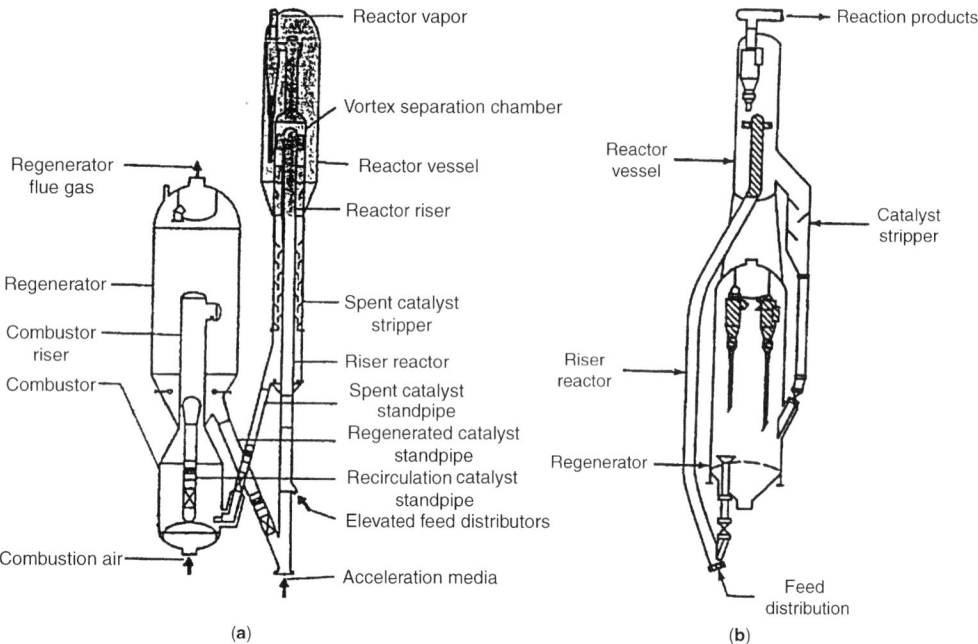

Figure 1. (**a**) Modern UOP FCC unit with: Short contact time riser VSS disengager combustor-style regenerator. (**b**) Mid-1970s UOP FCC unit with: Riser cracking Stacked reactor bubbling-bed regenerator.

directly influences the regenerator temperature and controls the catalyst circulation rate in the FCCU, thereby controlling the ratio of catalyst: hydrocarbon feed (cat/oil ratio, or C/O).

Catalytic coke is also a function of the type of feed being processed. The more aromatic the feed, the higher the coke (on feed) yield at a given conversion. The type of cracking catalyst also influences the coke yield. Increasing the H-transfer characteristics of a Y-type zeolite, by

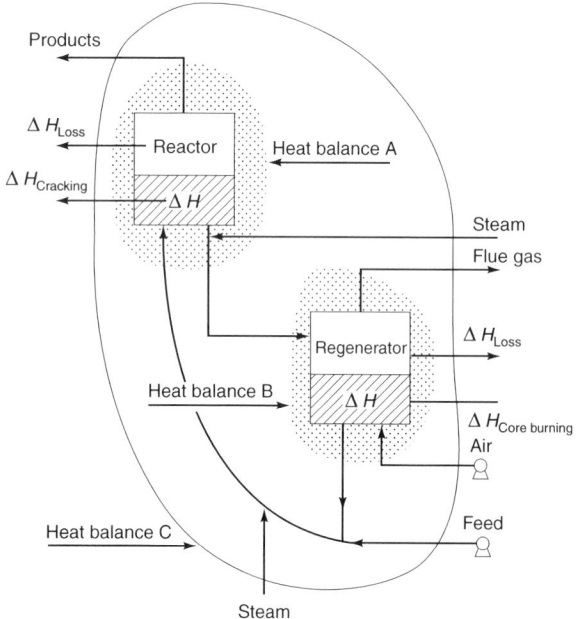

Figure 2. Heat balances of the FCCU. Heat balance around the reactor A, heat balance around the regenerator B, and the overall heat balance around the entire catalytic cracker C must be considered.

increasing the amount of rare earth exchanged on the zeolite, results in an increase in the catalytic coke yield. Modern catalysts for FCC applications frequently contain two distinctly different, active components; an active zeolite component for producing high yields of gasoline while yielding minimum catalytic coke and an active, nonzeolitic component, typically a catalytically active alumina located in the so-called matrix portion of the catalyst surrounding the zeolite. The active alumina matrix provides improved conversion of the heavy portion of the feed but typically increases catalytic coke yield at constant conversion.

Contaminant coke can vary substantially because of large variations in the metals content of the feed being processed and from variations in the manner that cracking catalysts respond to these metals. Nickel is considered the most active dehydrogenation agent, and hence the most troublesome of the contaminant metals. Copper is equally active as a dehydrogenation catalyst, but the amount of copper present is usually too small to have an effect. Vanadium is a less active dehydrogenation agent than nickel, about one-fourth as effective as nickel. The metals content of the equilibrium catalyst in the FCCU is frequently referred to as the equivalent nickel content.

The coking characteristics of a cracking catalyst increase as the equivalent nickel level on the catalyst increases. At high equivalent nickel levels, contaminant coke can be the major coke component on the catalyst. The effect of the contaminant nickel on the catalyst coke content can be mitigated through the use of a metal passivator added to the feed or through the incorporation of an effective metals traps into the cracking catalyst. Coking problems due to contaminant vanadium still remain a problem.

COKE BURNING

The burning of coke in the regenerator provides the heat to satisfy the FCCU heat balance requirements. The heat released from the burning of coke comes from the reaction of carbon and hydrogen to form carbon monoxide, carbon dioxide, and water. The heat generated from burning coke thus depends on the hydrogen content of the coke and the relative amounts of carbon that burn to CO and CO_2, respectively.

Decreasing the specific heat of combustion increases the amount of the coke that must be burned; the coke component that increases is essentially the cat-to-oil coke, which is increased by increasing the catalyst circulation rate. Thus decreasing the specific heat of combustion results in an increase in catalyst circulation rate. Because of this relationship to coke yield, the increase in the catalyst circulation rate results in a decrease in regenerator temperature.

The hydrogen content of the coke can vary considerably, depending on the efficiency of the stripping operation to remove the lighter hydrocarbons from the catalyst flowing into the regenerator.

The coke-burning rate in the regenerator is a function of the oxygen partial pressure, the carbon content on the catalyst leaving the reactor, the temperature in the regenerator, and the residence time in the regenerator. The last three variables are all coupled together through the FCCU heat-balance requirements.

The hydrocarbon feed rate to the reactor also affects the burning kinetics in the regenerator. Increasing the reactor feed rate increases the coke production rate, which in turn requires that the air rate to the regenerator increase. Because the regenerator bed level is generally held constant, the air residence time in the dense phase decreases. This decrease increases O_2 content and consequently CO combustion in the dilute phase, frequently referred to as after burn.

ENVIRONMENTAL ASPECTS

The FCCU regenerator is one of the principal sources of air pollutants from a refinery. Recently, governmental regulations, particularly in the United States, have significantly reduced the allowable FCCU emissions for particulates and sulfur oxides. A new 50 MBPD unit or an existing 50 MBPD unit being significantly revamped in 2002 would be required to meet the following contaminant levels in the regenerator flue gas: catalyst particulates <0.5 t/day, sulfur oxides <0.7 t/day. Nitrogen oxide limits are not yet defined but are expected to be similarly constrained. Carbon monoxide emissions would be limited to <3 t/day. Much research has been done in the past and is ongoing to develop technologies to meet these emissions levels and reduce them even further.

REGENERATOR OPERATING PARAMETERS

To maximize the performance of an FCCU, most units run at one or more unit constraints. Frequently, one of these constraints is the regenerator temperature, which is set by metallurgical limits for safe operations. Process variables on both the reactor and the regenerator side are thus manipulated to keep the regenerator temperature as close as possible to this regenerator temperature limit.

Changes in feed quality, which will affect the coking tendency of the feed and the coke-making characteristics of the circulating catalyst, influence the regenerator temperature. The refiner can rapidly respond to these changes by manipulating certain independent process variables, which include reactor riser temperature, temperature of the hydrocarbon feed entering the riser, fresh catalyst addition rate, and cooling duty if a catalyst cooler exists. If the change in feed quality is of prolonged duration, the refiner can also change catalyst type.

The heavy material in the reactor feed, which remains as a residue after evaporation and pyrolysis of the oil, is designated as Conradson carbon (ASTM test method D189-76). This residue is frequently used as an indicator of the coking tendency of the feed. Incorporating heavy feedstocks such as vacuum resids into the FCCU feed greatly increases the feed's Conradson carbon content. A 2% increase in Conradson carbon can cause a regenerator temperature increase of $\sim28°C$.

The presence of contaminant metals on the equilibrium catalyst can significantly increase the catalyst coking tendency, which in turn results in an increase in regenerator temperature if all other factors remain unchanged.

To decrease the regenerator temperature, heat must be taken out of the system either by

1. Decreasing conversion, which decreases the amount of catalytic coke produced. Decreasing conversion can be achieved by either decreasing reactor temperature or by decreasing equilibrium catalyst activity. For a given decrease in conversion, the latter is more effective in reducing regenerator temperature. Typically, decreasing reactor riser outlet temperature by 8°C decreases conversion 2–2.5% and decreases regenerator temperature by 6–8°C. Reducing equilibrium catalyst activity by four numbers, either by decreasing the fresh catalyst addition rate or by using a lower activity fresh catalyst, also reduces conversion by $\sim2\%$ and decreases regenerator temperature by >42°C, depending on the initial temperature.

2. Through the physical removal of heat. Heat can be removed by using a catalyst-cooling heat exchanger or by cooling the feed entering the reactor. Typically, decreasing the feed temperature by 28°C decreases the regenerator temperature by 6°C.

3. Through a decrease in the relative amount of CO converted to CO_2 in the regenerator. Although burning coke to CO rather than to CO_2 releases 70% less heat, this method is less frequently used. It requires that more feed be converted to coke to satisfy the heat balance, thus decreasing yield profitability, and it also requires that the unconverted CO be burned to CO_2 in an external CO boiler, to prevent any CO emission to the atmosphere.

FCCU REGENERATOR CONFIGURATION

A modern FCCU typically operates continuously for 3–5 years between turnarounds, during which time quantities on the order of 10^{10} kg of feedstock are processed and quantities of catalyst in the range of 7×10^{10} kg are circulated through the regenerator.

In the modern unit design, the main vessel elevations and catalyst transfer lines are typically set to achieve optimum pressure differentials because the process favors high regenerator pressure, to enhance coke-burning kinetics and power recovery from the flue gas, and low reactor pressure to enhance product yields and selectivities.

An alternative two-stage regenerator design was independently developed incorporating two distinct bed-regeneration zones within a single regenerator vessel. Coked catalyst from the catalyst stripper enters the upper regeneration zone. This first-stage is operated in partial combustion, with CO in the exiting flue gas, to produce a lower heat of combustion. Catalyst passes by gravity flow from the first stage to the second zone, located below the first stage. This second stage operates in a complete combustion mode with excess oxygen, to complete the coke burn and provides regenerated catalyst with low carbon-on-catalyst. The excess oxygen-containing flue gas from this second zone passes upward through the first stage. Additional air is also provided to the first stage through a separate air distributor. Togeteher, these two air streams provide the oxygen needed to achieve partial combustion in the first stage. Since the flue gas stream contains CO, it is sent to a CO boiler.

An external dense-phase catalyst cooler was developed specifically to remove heat from catalyst that passes from this upper regeneration zone to the lower regeneration zone. The amount of heat removed can be adjusted to provide better control of the process conditions—a key feature of the catalyst cooler.

The Effect of Fluidization on Regeneration

Several fluidization regimes have been used in FCCU regenerators, depending on the technology evolution and the objectives of the particular design.

Bubbling Bed Regime. Early FCCU regenerators operated in the bubbling–bed regime. The bubbling-bed regenerator design offered limited solids carryover and transport through the freeboard region.

The bubbling-bed design had several major problems, chief among these being poor catalyst–air mixing.

Turbulent-Bed Regime. The commercialization of high alumina catalysts in 1955 was one of the chief reasons for a shift in regenerator design to the turbulent-bed regime. The catalyst was much more active, but highly sensitive to coke deactivation, so the poor regeneration of the bubbling-bed design was no longer acceptable. The need to improve mixing of the combustion air and the catalyst, led designs to change to the turbulent fluidization regime. At the higher gas velocities of the turbulent-bed regime (0.3–1.0 m/s), the distinct bubble phase disappears.

The surface of the upper bed is considerably more diffuse and has reduced pressure fluctuations and substantially higher entrainment of solids into the freeboard region. Because of the higher coke-burning capacity requirements and improved contacting efficiency, the vast majority of commercial regenerators converted to the turbulent-bed regime. In this configuration, the ultimate regeneration capacity is set by the sharp increase in solid entrainment that occurs as the linear velocity approaches 1.0 m/s and the corresponding effect this air velocity has on the cyclone separation efficiency and dipleg hydraulics.

Some improvement efforts include: optimizing fluidized-bed length and diameter ratios, optimizing air distribution through the proper grid pressure drop and grid plugging pattern, and optimizing catalyst distribution via cyclone dipleg discharge orientation, and proper location of the spent catalyst addition and withdrawal points into the regenerator. These addition and withdrawal devices include tangential inlets, swirl distributors, deflector plates, and baffles as well as the entry elevation into the fluidized bed.

Fast Fluidized Regime. The high efficiency combustor style regenerator operates in the fast fluidized regime. The fast fluidized regime (1.0–3.0 m/s superficial velocity) extends into the transport phase, where a sharp increase in the rate of solids carryover occurs as the transport velocity is approached. In the absence of any solid recycle, the bed would rapidly disappear. Beyond this velocity, catalyst fed to the base of the regenerator travels upward in a fully entrained transport flow. The voidage or density of the resulting suspension is dependent not only on the velocity of the gas but also on the solid flow rate. If the solid rate is low, a dilute-phase flow will result. If solids are fed to the regenerator at a sufficiently high rate, eg, by recirculating the solid carried over back to the regenerator, then maintaining a relatively large solid concentration is possible. This condition is referred to as the fast-fluidized bed.

The high velocities in the combustor promote catalyst to air contact. As a result, the catalyst residence time required for catalyst regeneration is greatly reduced.

The overall benefits of this high efficiency combustor over a conventional bubbling- or turbulent-bed regenerator are enhanced and controlled carbon-burn kinetics (carbon on regenerated catalyst at <0.05 wt%); ease of start-up and routine operability; uniform radial carbon and temperature profiles; limited afterburn in the upper regenerator section and uniform cyclone temperatures; and reduced catalyst inventory and air-blower horsepower.

MECHANICAL HARDWARE

From a mechanical point of view, the FCCU regenerator can be divided into two main sections: the regenerator vessel and its internals, which include the air distributor, the cyclones, the plenum chamber, and catalyst coolers; and the flue gas handling section, which includes power, heat, and emission control systems (SO_x, NO_x, and particulates).

D. Selberg and J. R. Riley, *Super DeSO_x Provides Industry Leading Effectiveness*, Grace Davison Catalagram, No. 92, 2003.

M. T. Smith and T. F. Petti, *High Unit Retention Critical to Efficient Operation of Power Recovery Trains*, Grace Davison Catalagram, No. 88, 2001.

J. W. Wilson, *Fluid Catalytic Cracking; Technology and Operation*, PennWell Publishing Co. 1997.

L. L. UPSON
F. S. ROSSER
C. L. HEMLER
P. PALMAS
L. E. BELL
W. J. REAGAN
B. W. HEDRICK
UOP LLC

FLUID MECHANICS

Fluid mechanics is both a descriptive science of the phenomena that occur when fluids flow and a quantitative science showing how these phenomena may be described in mathematical terms and predicted when appropriate conditions are prescribed. To a practicing chemical technologist, fluid mechanics is an entire body of knowledge, theoretical and empirical, qualitative and quantitative, allowing analysis of the performance of complex plant equipment handling moving fluids. Before being in a position to calculate the details of a flow, one needs to understand the phenomena well enough to model the process properly. At times the technologist's needs are best satisfied by an empirical correlation; at other times the necessary skills consist largely of knowing how to apply the idealized mathematical solutions to a practical situation.

MATHEMATICAL DESCRIPTION OF FLUID MOTION

Fundamental Concepts

Fluid materials differ from solid materials in that fluids are capable of unlimited deformation and will keep deforming, as long as they are subjected to shear stresses, while solids will normally deform only by a finite amount. In some instances, as during extrusion of very viscous or plastic materials, one may encounter both fluid-like and solid-like behavior. Although all materials consist of discrete particles, such as molecules and atoms, most fluid mechanical phenomena can be adequately described under the continuum hypothesis, which considers that the material properties of the fluid are not affected by the process of subdivision, thus allowing the definition of a fluid element as a material entity in the fluid having an infinitesimally small volume. This idealization allows one to define derivatives at a mathematical point. The study of flow phenomena under the continuum hypothesis is only appropriate when the changes of interest occur over length scales much larger than molecular sizes or mean free paths between molecular collisions, so that material properties are actually averages over a large number of molecules. There are cases, however, in which the continuum assumption breaks down, and then one must resort to more general analytical models.

Fluid mechanics is an exact science, governed by three universal principles, which have been confirmed by experimentation and are accepted as valid. These are the law of the conservation of mass, including the law of conservation of individual species, electric charge, etc; the energy principle, or first Law of Thermodynamics; and the momentum principle, ie, Newton's second law.

Integral Equations of Motion

Some problems in fluid mechanics can be solved to a sufficient degree of accuracy by examining only the overall balances of mass, momentum, and energy. In applying these balances, an appropriate control volume is established first and the rates of accumulation of the quantities within the volume are balanced against their rates of generation within the volume and their rates of transport through the control surface, which encloses the volume. In most cases, the volume of interest is fixed in space, as for example the space within a pipe or tank. In other cases, as in wave propagation, a moving control volume with an attached coordinate system is more appropriate.

Conservation of Mass. The general equation for the conservation of mass of a chemical species with concentration (mass per unit volume) c_i is the scalar equation:

$$\underbrace{\frac{\partial}{\partial t}\int_B c_i dB}_{\text{rate of change in } B} + \underbrace{\int_S c_i V \cos\theta\, dS}_{\substack{\text{convective flux} \\ \text{through } S}} = \underbrace{-\int_S J_i dS}_{\substack{\text{diffusive flux} \\ \text{through } S}} + \underbrace{\int_B R_i dB}_{\substack{\text{production by} \\ \text{chemical reaction}}} \quad (1)$$

where t is time, B is the control volume, S is the control surface, V is the magnitude of the velocity, θ is the angle between the velocity vector and the outward normal vector on a control surface element, J_i is the boundary flux through difusion and R_i is the rate of production of species i by chemical reaction.

Momentum Equation. The general equation for the balance of momentum is the vector equation:

$$\underbrace{\frac{\partial}{\partial t}\int_B \rho\vec{V}dB}_{\text{rate of change in } B} + \underbrace{\int_S \rho\vec{V}V\cos\theta\, dS}_{\substack{\text{convective flux} \\ \text{through } S}} = \underbrace{\int_B \vec{F}_B dS}_{\text{net body force}} + \underbrace{\int_S \vec{F}_S dB}_{\text{net surface force}} \quad (2)$$

where ρ is the fluid density. The two terms on the right-hand side represent the external forces applied on the control volume. The body forces arise from gravitational, electrostatic, and magnetic fields. The surface forces are the shear and normal forces acting on the fluid at its boundary; diffusion of momentum, as manifested in viscosity, is included in these terms. In practice, the vector equation is usually resolved into its Cartesian components and the normal stresses are set equal to the pressures over those surfaces through which fluid is flowing.

Energy Equation. The specific (ie, per unit mass) energy of a flowing fluid is the sum of its internal, kinetic and potential specific energies. The equation describing the conservation of energy, also expressing the first law of Thermodynamics, is the scalar equation:

$$\underbrace{\frac{\partial}{\partial t_B}\rho\, e\, dB}_{\text{rate of change in } B} + \underbrace{a_S\rho\left(e+\frac{P}{\rho}\right)V\cos\theta\, dS}_{\text{convective flux through } S} = \underbrace{Q}_{\substack{\text{heat transfer rate}\\\text{from surroundings}}}$$
$$-\underbrace{W_{sh}}_{\substack{\text{mechanical power by}\\\text{shear stresses on } S}} - \underbrace{W_{other}}_{\substack{\text{electrochemical/radiation}\\\text{power to surroundings}}} \quad (3)$$

Note that the term containing the pressure P actually represents mechanical power by normal stresses on S, but has been grouped with the specific energy flux for convenience.

Differential Equations of Motion

The differential equations of motion can be derived from the corresponding integral ones, by letting the control volume vanish. These equations may be expressed in a coordinate system moving with the body (the Lagrangian viewpoint) or in a fixed coordinate system (the Eulerian viewpoint).

Conservation of Mass. Neglecting diffusion and chemical reactions, the differential mass conservation equation, also commonly referred to as the continuity equation, is written as

$$\frac{\partial\rho}{\partial t} + \sum_{i=1}^{3}\frac{\partial(\rho U_i)}{\partial x_i} = 0 \quad (4)$$

Momentum Equation. This equation is derived by applying Newton's second law on an infinitesimal control volume, containing a fluid element. In an Eulerian viewpoint formulation, the (total) acceleration of a fluid element is expressed as

$$\frac{DU_i}{Dt} = \frac{\partial U_i}{\delta t} + \sum_{j=1}^{3}U_j\frac{\partial U_i}{\partial x_j}, \quad i=1,2,3 \quad (5)$$

in which the first term on the right-hand side is the local acceleration and the sum of the three other terms is the convective acceleration.

Next, it is necessary to understand the nature of the forces that might be exerted on this volume. There may be body forces, which, in Newtonian mechanics, are pictured as acting at a distance and include gravitational, electrostatic, and magnetic forces; if a noninertial frame of reference is used, one must also consider equivalent forces resulting from accelerations of the frame, eg, the centrifugal force. In addition, there are surface forces, exerted on the faces or edges of the element. Surface stresses arise from intermolecular forces and motions of molecules. At boundaries between phases, additional forces, which are generalizations of surface tension,

must be included. Surface forces per unit surface area are called stresses. When considering all stresses applied on an element with faces normal to the Cartesian axes and with all surface forces decomposed to their Cartesian components, one would define stresses as the components of a symmetrical, second-order tensor.

In thermodynamics, pressure is defined as work by the fluid per unit volume. In fluid mechanics, pressure is defined as the average normal stress. In most cases, but not always, the fluid mechanical pressure may be taken as equal to the thermodynamic pressure.

For flow problems in which the viscosity is vanishingly small, the normal stress component is negligible, but for fluid of high viscosity, eg, polymer melts, it can be significant and even dominant. Assuming that the viscosity is uniform in space, one gets the simplified and most commonly used Navier-stroke (N-S) equation form

$$\rho\frac{DU_i}{Dt} = f_i - \frac{\partial P}{\partial x_i} + \mu\sum_{n=1}^{3}\frac{\partial^2 U_i}{\partial x_n^2}, \quad i=1,2,3 \quad (6)$$

For some materials, the linear constitutive relation of Newtonian fluids is not accurate. Either stress depends on strain in a more complex way, or variables other than the instantaneous rate of strain must be taken into account. Such fluids are known collectively as non-Newtonian.

Dimensional Analysis, Similarity and Modeling

The majority of technological flow problems are solved by carrying out laboratory experiments whose results are correlated, so as to yield useful information about systems that may differ greatly in size and in fluid properties. For the behavior of the experimental model to duplicate that of a system of interest, two criteria must, in principle, be met: first, the experimental apparatus must be geometrically similar to the system of interest; and second, certain dimensionless groupings of variables must be matched on the two systems. There are two basic methods available for determining the dimensionless groups appropriate to a given situation: dimensional analysis, which can be applied when the equations governing the process are not known; and similarity analysis, which proceeds from the governing equations and offers physical insights into the meanings of the groups.

Dimensional analysis is a mathematical technique that proceeds from the general principle that physical laws must be independent of the units of measurement used to express them. If one quantity is related to a group of other quantities, the quantities comprising the group must be related in such a manner that the net units or dimensions of the group are the same as those of the dependent quantity. A dimensionless group can then be formed immediately by division. A useful tool, the Buckingham Pi theorem, asserts that the number of dimensionless groups needed to describe a situation is equal to the total number of variables less the number of fundamental dimensions needed to express them. Fundamental dimensions are generally taken to be length, time, mass, temperature, and heat content. The Pi theorem is a powerful tool because it limits the amount of experimental work needed to establish a general relationship.

The strength of dimensional analysis lies in its ability to limit the number of studies that need to be made and to handle situations in which the governing equations are not known. It can even handle lack of geometrical symmetry by using ratios of important dimensions as additional dimensionless groups. Its weakness lies in the need to know which variables must be included and which can be ignored, and in its awkwardness when handling several variables that have the same dimensions, eg, densities in multiphase mixtures.

Similarity analysis starts from the equation describing a system and proceeds by expressing all of the dimensional variables and boundary conditions in the equation in reduced or normalized form. Velocities, eg, are expressed in terms of some reference velocity in the system, eg, the average velocity. When the equation is rewritten in this manner certain dimensionless groupings of the reference variables appear as coefficients, and the dimensional variables are replaced by their normalized versions. If another physical system could be described by the same equation with the same numerical values of the coefficients, then the solutions to the two equations (normalized variables) would be identical and either system would be an accurate model of the other.

In addition to the above important conclusion, similarity analysis also provides a means of simplifying the solution of a problem by order-of-magnitude analysis. Without claiming mathematical rigor, one may hypothesize that dimensionless properties are all of order one, reflecting the expectation that the scales have been chosen appropriately. Thus, the Reynolds number may be viewed as representing the ratio of inertia force to viscous force. Similarly, the Froude number represents the square root of the ratio of inertia force and gravitational force.

In addition to the Reynolds, Froude, and Euler numbers, a large number of other dimensionless groups have been in use in fluid mechanics, heat transfer and related disciplines. A partial list is given as follows: Mach number; Prandtl number; Schmidt number; Weber number (for liquids); capillary number (for two-phase flows); cavitation number (for liquids); Nusselt number; Grashof number; Péclet number; Rayleigh number (for free thermal convection); Marangoni number (for convection induced by surface tension gradients); Richardson number (for density-stratified flows); Taylor number (for rotating flows); Rossby number (for rotating flows); Strouhal number (for periodic vortex shedding from bluff objects); and Knudsen number for gases.

Turbulent Flows

Although the space-time dependence of velocity and pressure in turbulent flows is accurately described by the same general continuity and momentum equations as in laminar flows, turbulent properties change far too rapidly and randomly to be of interest in their entirety. In most engineering applications, interest lies instead in their statistical averages. The most general type of averaging is ensemble averaging, in which all properties are averages over a large number of repeated realizations of the same experiment.

Any attempt to produce additional equations for the Reynolds stresses, based on the basic equations of motion and statistical procedures, would introduce many more unknowns, such as averages of triple products of fluctuations. Thus, the Reynolds averaging process produces an open hierarchy of equations, which can only be closed by the introduction of turbulence models, namely empirical relationships among the various statistical properties.

SPECIFIC FLOWS

Internal Flows

Flows that are confined by solid walls, except at inlet and outlet ports, are called internal. They include flows in pipes, ducts, channels and various types of machinery. On the boundary, the fluid exercises normal and shear stresses on the wall, which result in a force on the wall and pressure losses in the fluid.

Two distinct states of fully developed flow are observed in a pipe. At low Reynolds numbers, the flow is laminar and the fluid flows in concentric cylindrical sheaths or laminae that do not mix with each other.

When the Reynolds number increases to nominally about Re ≈ 2, 100 for commercial pipes, an abrupt change occurs. Random instabilities, which decay at lower velocities, grow, destroying the laminar flow pattern. At Re ≈ 10,000, the transformation is nearly complete. This process is called transition to turbulence. The velocity over most of the pipe, called the core, becomes fairly uniform, about equal to the average velocity. Only in a thin region near the wall is the radial variation of the velocity substantial. Frictional losses for turbulent pipe flows cannot be predicted theoretically; instead, they are based on empirical correlations.

Pressure losses associated with fully developed pipe flow are termed "major losses." All other losses, associated with the entrance region, flow area changes, direction changes, flow through valves and other devices and loss of kinetic energy at the exit, are collectively termed as "minor losses," even when they represent the majority of total losses. Additional losses occur in the entrance region of the pipe, where the flow velocity profile develops from a nominally uniform one towards its fully developed one. Throughout the entrance region, the velocity gradient at the wall is larger than for the fully developed profile, producing a higher pressure drop and greater heat transfer per unit length than for fully developed flow. Although entrance phenomena are often ignored, these can be of significance, especially in capillary viscometers and compact (short tube) heat exchangers. The entrance losses depend on the shape of the entrance and can be significantly reduced by using a bell-mouth entry. When a flow expands abruptly, separation occurs producing a jet of fluid flanked by recirculating eddies. When separation occurs, the ability of the fluid to recover pressure upon deceleration is seriously impaired because the kinetic energy is lost to friction. When the ratio of duct areas is large, all of the kinetic energy is lost and no pressure is recovered.

Flows Near Solid Walls

Except at very low Reynolds numbers, the flow field past immersed solid boundaries can be divided into two reasonably well-defined regions: a thin region close to the surface, called the boundary layer, in which the gradient of tangential velocity is large and shear stresses are important; and a region away form the boundary, called the free stream, in which velocity gradients are small, resulting in small or negligible viscous stresses. The edge of the boundary layer is taken to be the point at which the velocity differs from that in the free stream by some small amount, say 1%. Although boundary layer analysis is widely applicable, it can fail badly in regions of strong deceleration where the boundary layer flow can separate under the influence of the unfavorable pressure gradient imposed by the main flow. When the flow is bounded on only one side, the boundary layer can, in theory, grow indefinitely, albeit slowly, into the surrounding fluid.

Flows Past Solid Bodies

A fluid moving past a solid body exerts a drag force on the solid. Both shear stresses and normal stresses can contribute to the drag, commonly referred to as skin friction drag and form or pressure drag, respectively. Their relative importance depends on the shape of the body and the Reynolds number. Body shapes (eg, wings) for which skin friction drag predominates are called streamlined, while those shapes (eg, buildings) for which form drag dominates are called bluff. At extremely low Reynolds numbers (eg, < 1), skin friction drag always dominates.

Periodic vortex shedding produces lateral forces of the same period on the cylinder. Should the cylinder be weakly supported and have a natural frequency close to the shedding frequency, it would oscillate strongly in concert with the vortex street, a phenomenon known as vortex induced vibration. Flow induced vibration is much more likely to happen for elongated (eg, cylinders) rather than compact (eg, spheres) bodies.

Besides the drag force, the character of the flow also affects the heat and mass transfer to the surface. At low Reynolds numbers, the thinner boundary layer on the upstream face of the cylinder exhibits a higher heat-transfer coefficient than does the thickened layer on the downstream face.

When the contour of the immersed object is rounded, as in the case of circular cylinders, spheres and the like, separation points are free to move along the surface, depending on the Reynolds number, the surface roughness and the flow disturbances. When, however, the object has sharp corners or edges, as in the case of a rectangular cylinder, separation always occurs at these edges and no critical regime is observed. In most technological processes, flow separation is undesirable and effort is made to eliminate or reduce it by redesigning the object shape (eg, streamlining), applying suction or tangential fluid injection, or using various control devices.

Free Shear Flows

Free shear flows are flows which are confined not by solid boundaries, but by fluid at a different speed. Simple shear flows include jets, in which a stream is injected into slower surroundings, wakes, in which the stream has been retarded by the presence of an upstream object, mixing layers, which are the regions between two parallel streams of different speeds, and plumes, which are fluid columns rising due to buoyancy.

Mass and heat transport within the shear flow follow the same general pattern as does momentum transport. For gases, it is found experimentally that the thermal or concentration jet spreads somewhat faster and decays more rapidly than does the momentum jet.

Compressible Flows

The flow of easily compressible fluids, ie, gases, exhibits features not evident in the flow of substantially incompressible fluid, ie, liquids. These differences arise because of the ease with which gas velocities can be brought to or beyond the speed of sound and the substantial reversible exchange possible between kinetic energy and internal energy. The Mach number, the ratio of the gas velocity to the local speed of sound, plays a central role in describing such flows.

Flow Instability and Secondary Flows

In many flow situations it is found that a mathematically valid solution to the Navier-Stokes equations is closely verified by experiment over some ranges of the variables, but when the variables are changed new flow patterns that are not in keeping with that solution are observed. The change is often rather sudden, and may involve a change from an unstable laminar solution to another, stable, laminar one or may involve transition to turbulence. Such behavior occurs because solutions to the Navier-Stokes equations are generally not unique. When more than one solution exists, it is possible to observe one flow pattern under one set of circumstances and a different pattern under another. At the present stage of development of fluid mechanics, an experiment must be performed to determine whether a given solution applies. In some cases, however, it is possible to determine the criteria by which one flow pattern becomes unstable in favor of another by analyzing the equations of motion. This analysis is the topic of a field called hydrodynamic stability.

Fluids in Motion. Many of the instabilities associated with fluids in motion are of the shear-flow type, in which the velocity varies principally in a direction perpendicular to the flow direction. At low Reynolds number, such flows have laminar solutions, which can be found by solving the Navier Stokes equations. As the Reynolds number increases above a certain value, which is different for each flow configuration, these solutions no longer describe the flow, which becomes unstable and changes possibly to another laminar state but eventually from laminar to turbulent. In boundary layer flows over smooth plane walls with extremely low external disturbances, the instability, usually referred to as Tollmien-Schlichting instability, occurs in essentially four steps: first, small two-dimensional waves form and are linearly amplified; second, the two-dimensional waves develop into finite three-dimensional waves and are amplified by nonlinear interactions; third, a turbulent spot forms at some localized point in the flow; and, finally, the turbulent spot propagates until the spot fills the entire flow field with turbulence.

In free shear flows, such as jets and wakes, there occurs another type of instability, called Kelvin-Helmholtz instability. This instability may be illustrated as follows. Suppose that a small disturbance causes a slight waviness of the boundary between the two flows. The fluid on the convex sides of each flow moves slightly faster and that on the concave sides moves slightly slower. According to the Bernoulli equation, this disturbance decreases the pressure on the convex sides of each flow, and thus the initial disturbance is amplified. Kelvin-Helmholtz instability is also observed in horizontal concurrent flows of stratified immiscible fluids.

An instability involving transition from one laminar flow pattern to another, called Taylor instability, occurs in the flow between coaxial cylinders in relative rotation with respect to each other. When the inner cylinder is rotated at an angular velocity below a critical value, motion is purely circumferential, called circular Couette flow. Above this value, however, centrifugal force destabilizes the flow and a series of laminar, cellular vortices known as Taylor cells are superimposed on the main flow.

Fluids at Rest. Fluids at rest may be set into motion by impressing upon them gradients in body or surface forces. Benard instability refers to the formation of convection cells within a fluid as a result of the action of a gravitational field on density differences induced by a temperature gradient in the fluid. Such behavior is observed, eg, when fluid is confined between two horizontal plates of which the lower one is heated.

Secondary Flows. In many cases, a cursory examination of the flow pattern might indicate a rather simple type of flow with a high degree of symmetry, whereas in fact the flow realized is more complex. In the flow around a bend, eg, one might imagine that the individual streamlines simply follow the general course of the curvature of the pipe; in fact they do not. Instead, a pattern of secondary flow develops that is superimposed on the main flow so that the streamlines are actually helical. Adopting the convention that the main flow is parallel to the tube axis, then at the axis the secondary flow is directed outward toward the section of pipe having the weakest curvature, returning inward along the pipe wall. This type of secondary flow is a consequence of the inertia of the fluid.

Secondary flows also occur in channels that have polygonal cross-sectional shapes, but are otherwise straight. In these cases the secondary flows are directed outward into the corners and return along the walls. Such secondary flows are associated with gradients in turbulent stresses and only occur in turbulent flows.

Flow in Porous Media

Flow of fluids through fixed beds of solids occurs in situations as diverse as oil-field reservoirs, catalyst beds and filters, and absorption towers. The complex interconnected pore structure of such systems makes it necessary to use simplified models to make practical quantitative predictions. An important parameter is the porosity or void fraction ε, defined as the average fraction of the cross-section that is not occupied by the solid. One of the more successful treatments of single-phase pressure drop through such systems employs the results for flow through tubes, using average velocities and tube diameters. The average velocity through the pores V_l is called the interstitial velocity V_l, while the apparent velocity based on the entire cross-section V_S is called the superficial velocity.

For nonspherical particles, equivalent spherical diameters are employed and additional corrections for shape are introduced.

Non-Newtonian Fluids

For many fluids the Newtonian constitutive relation involving only a single, constant viscosity is inapplicable. Either stress depends in a more complex way on strain, or variables other than the instantaneous rate of strain must be taken into account. Such fluids are known collectively as non-Newtonian and are usually subdivided further on the basis of behavior in simple shear flow.

Another common class of non-Newtonian fluids is the Bingham fluids, which have a linear constitutive relationship, similar to Newtonian fluids, but have a nonzero intercept termed the yield stress.

In configurations more complex than pipes, eg, for flow around bodies or through nozzles, additional shearing stresses and velocity gradients must be accounted for.

Two-Phase Flows

Gas-Liquid Flows. When two or more fluids flow together, a much greater range of phenomena occurs as compared to flow of a single phase. In a conduit many of the technically significant phenomena have to do with the positions assumed by the phase boundaries, and these are governed by the flow conditions rather than by the walls of the conduit. In addition to the densities and viscosities, surface properties can be important.

Atomization. A gas or liquid may be dispersed into another liquid by the action of shearing or turbulent impact forces that are present in the flow field. The steady-state drop size represents a balance between the fluid forces tending to disrupt the drop and the forces of interfacial tension tending to oppose distortion and breakup. When the flow field is laminar, the ability to disperse is strongly affected by the ratio of viscosities of the two phases. Dispersion, in the sense of droplet formation, does not occur when the viscosity of the dispersed phase significantly exceeds that of the dispersing medium. More commonly, atomization occurs under turbulent conditions. The mechanism of atomization and its quantitative description are still incompletely understood. It is possible that breakup occurs because of the impact forces exerted by the small, nearly isotropic turbulent eddies.

Flow Past Deformable Bodies. The flow of fluids past deformable surfaces is often important, eg, in flows of liquids containing gas bubbles or drops of another liquid. Proper description of the flow must allow for both the deformation of these bodies from their shapes in the absence of flow and for the internal circulations that may be set up within the drops or bubbles in response to the external flow. Deformability is related to the interfacial tension and density difference between the phases; internal circulation is related to the drop viscosity.

Where surface-active agents are present, the notion of surface tension and the description of the phenomena become more complex. As fluid flows past a circulating drop (bubble), fresh surface is created continuously at the nose of the drop. This fresh surface can have a different concentration of agent, hence a different surface tension, from the surface further downstream that was created earlier. Neither of these values need equal the surface tension developed in a static, equilibrium situation. A proper description of the flow under these circumstances involves additional dimensionless groups related to the concentrations and diffusivities of the surface-active agents.

COMPUTATIONAL FLUID DYNAMICS

Analytical solutions of the equations of motion are known only for a small number of laminar and inviscid flows in fairly simple geometrical configurations subjected to well defined boundary and/or initial conditions. The fact that the majority of problems of practical interest are not amenable to such "exact" solutions has motivated the development of powerful numerical algorithms for their computation. The various methods of numerical solutions of fluid mechanical and related equations are collectively known as computational fluid dynamics (CFD).

The start of all numerical procedures is the identification of appropriate equations, usually partial differential equations, but also algebraic, ordinary differential or integral, which describe the physical problem at hand. The next step is to convert the equations into forms that can be solved digitally.

Computation of Turbulent Flows. The time-dependent, three-dimensional Navier-Stokes equations are generally considered adequate to represent turbulent flows and so their solutions, which would be functions of location and time, should be realistic. CFD analyses that produce such solutions are called Direct Numerical Simulations (DNS) and require the use of a mesh and time step that are fine enough to resolve the smallest motions of dynamic significance. At the same time, the computational domain has to be maintained large enough to encompass the large-scale features of the flow and the integration time large enough for the solution to become insensitive to the applied starting conditions. As the flow Reynolds number increases, however, the fine structure of the turbulence becomes increasingly finer, thus requiring a finer mesh for its simulation. As a result, DNS have only been performed successfully mainly on relatively low Reynolds number flows with relatively

simple geometrical boundaries, including, among others, homogeneous and isotropic turbulence, boundary layers, channel flows, and isothermal and chemically reacting mixing layers.

An alternative approach, far less computationally intensive than DNS, but also based on the unsteady Navier-Stokes equations, is offered by Large Eddy Simulations (LES). The mesh in these simulations is fine enough to describe the large scale motions and part of the activity in the intermediate, "inertial" subrange, but too coarse to precisely resolve the finest motions of turbulence, where most of the kinetic energy dissipation to heat takes place and where mixing my molecular motions occurs. LES model the fine structure through a subgrid scale model and incorporate this model into a set of spatially filtered dynamic equations, which are then solved numerically.

EXPERIMENTAL FLUID MECHANICS

Despite significant developments in CFD, experimental fluid mechanics remain an indispensable source of practical information for designing engineering systems and solving engineering problems as well as a prerequisite in developing and verifying theories, models and algorithms. Quite commonly, however, experimental fluid mechanics utilize laboratory-based apparatus, which in some way simulates the actual or idealized performance of real systems. Sometimes the objective of measurement is to document bulk properties, namely values averaged in space and/or time. Other times, however, the interest is in measuring temporal variations of a property at a specific location, and even the simultaneous temporal variation of local properties throughout the flow domain.

Measurement of Flow Rate. This area encompasses the various flow meters which measure either the mass or the volumetric flow rate through the cross-section of a pipe, duct, etc. They include: direct methods, restriction flow meters, averaging Pitot-tubes, laminar flow elements, rotameters or variable area flow meters, vortex shedding flow meters, drag or target flow meters, turbine flow meters, ultrasonic flow meters, electromagnetic flow meters, Coriolis flow meters, and thermal mass flow meters.

Pressure Measurement. Among the most common pressure-measuring instruments one could list the following: liquid-in-glass manometers, mechanical pressure gages, and electrical transducers.

Temperature Measurement. The most common types of thermometers are: liquid-in-glass thermometers, bimaterial thermometers, thermocouples, resistance temperature detectors (RTD), and thermistors.

Velocity Measurement. Unlike flow rate measurement methods, local flow velocity measurement entails sufficient spatial resolution to essentially identify the velocity at a particular location and often to map the velocity variation over a line, surface or volume of interest. Some velocity measurement methods have relatively slow response and provide time-averaged values, while others have sufficient temporal resolution to measure the instantaneous velocity, as required for the documentation of transient and turbulent flows. The following are the most widely used velocity measurement techniques: pressure tubes, hot wire and hot film anemometry (HWA), laser Doppler velocimetry (LDV or LDA), particle tracking, particle image velocimetry (PIV), and wind velocity.

Composition Measurement. The identification of chemical composition of passive mixtures as well as of reacting flows is a very important task in chemical processes and combustion. At the laboratory level, concentration can be measured with reasonable spatial and temporal resolution by the following techniques: thermal probes, electric conductivity probes, light scattering methods, and laser induced fluorescence methods.

The measurement of suspended particulate (solid particles or droplets) can be obtained by gravimetric analysis of samples removed from the suspension and by photographic, visual, and optical methods.

Void fraction measurement is particularly important in thermonuclear power generation plants.

Flow Visualization. Although of qualitative nature, flow visualization is a very useful component in any fluid mechanics experiment, as it can detect the presence of apparatus malfunctions and unwanted influences, identify the state of the flow (eg, whether laminar or turbulent) and the boundaries of different flow regions, and provide valuable insight into physical and chemical mechanisms that may affect the flow. Because motions in clean air, water and many other fluids cannot be discriminated by the unaided human eye, the visualization of flow patterns requires either the introduction of foreign substances, which act as flow markers when properly illuminated, or the use of optical techniques, which are sensitive to refractive index variations in the fluid.

R. B. Bird, W. E. Stewart, and E. N. Lightfoot, *Transport Phenomena*, John Wiley & Sons, Inc., New York, 2002.

N. De Nevers, *Fluid Mechanics for Chemical Engineers*, 3rd ed., McGraw-Hill, Boston, 2005.

R. W. Fox, A. T. McDonald, and P. J. Pritchard, *Introduction to Fluid Mechanics*, John Wiley & Sons, Hoboken, New Jersey, 2004.

J. O. Wilkes, *Fluid Mechanics for Chemical Engineers*, Prentice Hall, Upper Saddle River, N. J., 1999.

STAVROS TAVOULARIS
University of Ottawa

FLUIDIZATION

Gas–solids fluidization is the levitation of a bed of particles by a gas. Intense solids mixing and good gas–solids contact create an isothermal system having good mass

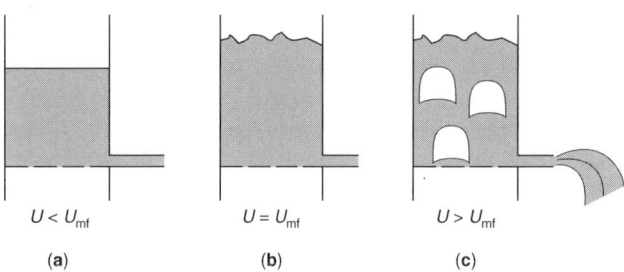

Figure 1. Fluidized-bed behavior where U is the superficial gas velocity and U_{mf} is the minimum fluidization velocity: (**a**) packed bed, no flow; (**b**) fluid bed, uniform expansion; and (**c**) bubbling fluid bed, flow.

transfer. The gas-fluidized bed is ideal for many chemical reactions, drying, mixing, and heat-transfer applications. Solids can also be fluidized by a liquid or by gas and liquid combined. Liquid and gas–liquid fluidization applications are growing in number, but gas–solids fluidization applications dominate the fluidization field. This article discusses gas–solids fluidization. The basic concepts of a gas-fluidized bed are illustrated in Figure 1.

Fluidized-bed applications at present may be separated into catalytic reactions, noncatalytic reactions, and physical processes. Examples of fluidized-bed applications include the following:

Chemical Catalytic Processes
  Catalytic cracking of heavy petroleum fractions (FCC)
  Phthalic anhydride
  Acrylonitrile
  Aniline (hydrogenation of nitrobenzene)
  Synthesis of polyethylene and polypropylene
  Fischer-Tropsch synthesis
  Oxidation of SO_2 to SO_3
  Chlorination or bromination of methane, ethylene, etc
  Maleic anhydride (from butane)
  Pyridine

Chemical Noncatalytic Processes
  Roasting of sulfide and sulfate ores (ZnS, pyrites, Cu_2S, $CuCoS_4$, nickel sulfides)
  Calcination (limestones, phosphates, aluminum hydroxide)
  Incineration of waste liquids and solids refuse
  Coking (thermal cracking)
  Combustion of coal and other fuels
  Gasification of coal, peat, wood wastes
  Carbonization of coal (decomposition without oxygen)
  Fluoridation of UO_2 pellets
  Catalyst regeneration
  Hydrogen reduction of ores
  Titanium dioxide

Physical Processes
  Drying [eg, phosphates, coal, poly(vinyl chloride) (PVC), polypropylene, foods]

Granulation (eg, pharmaceuticals, fertilizers)

Classification

Blending

Coating (eg, polymer coat on metal object)

High temperature baths

Airslide conveying

Absorption (eg, CS_2)

Filtering of aerosols

Medical beds

Quenching, annealing, tempering

PARTICLE PROPERTIES

Fluidized-bed design procedures require an understanding of particle properties. The most important properties for fluidization are particle size distribution, particle density, and sphericity.

Particle Size

The solids in a fluidized bed are never identical in size and follow a particle size distribution. An average particle diameter, d_p, is generally used for design. It is necessary to give relatively more emphasis to the low end of the particle size distribution (fines), which is done by using the surface volume diameter, d_{sv}, to calculate an average particle size:

$$d_{sv} = 1/\Sigma(x_i/d_{pi}) \qquad (1)$$

Particle size distribution is usually plotted on a log-probability scale, which allows for quick evaluation of statistical parameters.

Solid Density

Solids can be characterized by three densities: bulk, skeletal, and particle. Bulk density is a measure of the weight of an assemblage of particles divided by the volume the bed occupies. This measurement includes the voids between the particles and the voids within porous particles. The skeletal, or true solid density, is the density of the solid material if it had zero porosity. Fluid-bed calculations generally use the particle density, ρ_p, which is the weight of a single particle divided by its volume, including the pores. If no value for particle density is available, an approximation of the particle density can be obtained by multiplying the bulk density by 2.

Sphericity

Sphericity, ψ, is a shape factor defined as the ratio of the surface area of a sphere the volume of which is equal to that of the particle, divided by the actual surface area of the particle.

$$\psi = d_{sv}/d_v \qquad (2)$$

Angles of Repose and Internal Friction

The angle of repose is the angle that a pile of solids forms with the horizontal plane. The angle of internal friction is the angle with the horizontal that the flow, no-flow boundary forms when solids are flowing over themselves. This angle is a slight function of the solids flow rate. However, a typical angle of internal friction for a nonsticky material without sharp corners generally exceeds 65°. When designing fluidized-bed internal baffles, the baffles generally are angled at greater than 65° to the horizontal to prevent a zone of stagnant solids forming on top.

Terminal Velocity

A knowledge of terminal velocity is important in fluidized beds because it relates to how long particles are retained in the system. If the operating superficial gas velocity in the fluidized bed far exceeds the terminal velocity of the bed particles, the particles are quickly removed. Large particles will be removed almost immediately when gas velocity exceeds the single particle terminal velocity, whereas small particles will take a long time to entrain because of interactions between neighbouring particles.

Minimum Fluidization Velocity

There is a minimum superficial gas velocity required to just fluidize a bed of solids. The minimum fluidization velocity can be estimated from the Zenz plot, assuming the voidage at minimum fluidization is 0.5. Alternatively, it can be estimated via a correlation that gives a result equivalent to the plot using a voidage of 0.4. Using both methods defines the range within which a measured value for minimum fluidization velocity falls.

Particle Regimes

Particles were classified with respect to how they fluidize in air at ambient conditions into Geldart groups (Fig. 2).

Interparticle Forces

Interparticle forces are often neglected in the fluidization literature, although in many cases these forces are

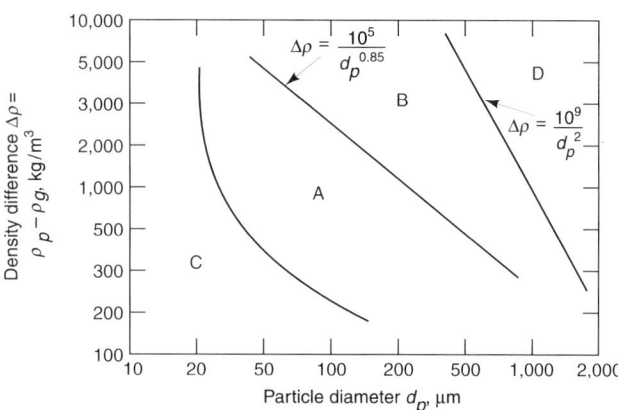

Figure 2. Geldart group particle classification diagram for air at ambient conditions. Group A consists of fine particles; B, coarse particles; C, cohesive, very fine particles; and D, moving and spouted beds.

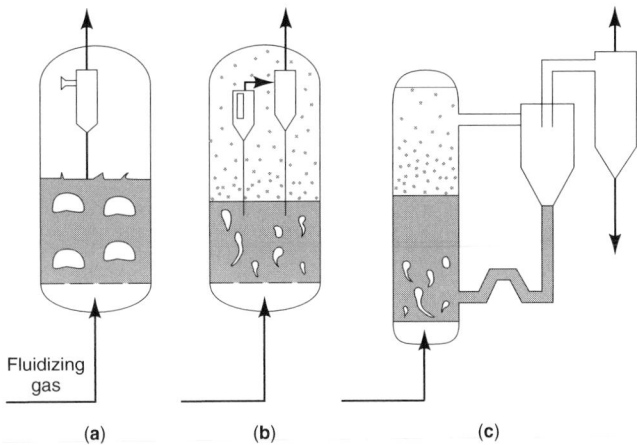

Figure 3. Schematics of commercially used beds, where the shaded area represents the solids: (**a**) vigorously bubbling, (**b**) turbulent, and (**c**) fast fluidized.

stronger than the hydrodynamic ones used in most correlations. The most common interparticle forces encountered in gas fluidized beds are van der Waals, electrostatic, and capillary. Interparticle forces predominate for Group C powders, are important for Group A, and can on rare occasions influence Group B and D powders.

FLUIDIZATION REGIMES

The different fluidized-bed regimes are a function of gas velocity. At a low gas velocity, the solids are in a packed- or fixed-bed state. As the gas velocity is increased, the drag and buoyancy forces eventually overcome the weight of the particles and interparticle forces, and the particles are completely supported by the gas.

Figure 3 shows how three reactors might appear when operating in the three most common commercial fluidization regimes.

Pressure Drop

The pressure drop across a two-phase suspension is composed of various terms, such as static head, acceleration, and friction losses for both gas and solids. For most dense fluid-bed applications, outside of entrance or exit regimes where the acceleration pressure drop is appreciable, the pressure drop simply results from the static head of solids. Therefore, the weight of solids in the bed divided by the height of solids gives the apparent density of the fluidized bed. The measurement of pressure drop across the bed is the most common and useful diagnostic technique employed for control of fluidized beds.

Effects of Temperature and Pressure on Minimum Fluidization Velocity

Many basic fluid-bed properties are affected by temperature and pressure. Pressure has little effect on the minimum fluidization velocity, U_{mf}, of fine Group A particles because frictional resistance is mostly a function of viscos-

ity and interparticle forces. However, the larger Group B and D particles show a decrease in minimum fluidization velocity with increasing pressure. Increasing temperature increases viscosity and, therefore, reduces U_{mf} for Group A and most Group B particles.

Bubbles and Fluidized Beds

Bubbles, or gas voids, exist in most fluidized beds and their role can be important because of the impact on the rate of exchange of mass or energy between the gas and solids in the bed. Bubbles are formed in fluidized beds from the inherent instability of two-phase systems. Bubbles, which are inherently undesirable, can grow to a large size in Group B powders and cause contact inefficiencies brought on by significant gas bypassing. Bubble size control is achieved by controlling particle size distribution or by increasing gas velocity.

Two-Phase Theory. According to the bubble two-phase theory of fluidization, all gas above that required for minimum fluidization passes through the bed in the form of bubbles. This is only an approximation for Group A particles fluidized at low velocities. Because fluid beds containing Group A solids are usually operated at high multiples of U_{mf}, this two-phase theory has little application in the instance of Group A particles.

Bed Expansion and Bed Density

Bed density can readily be determined for an operating unit by measuring the pressure differential between two elevations within the bed. This is a highly useful measurement for control and monitoring purposes.

Solids and Gas Mixing

Solids in an unrestricted fluidized bed can be almost completely backmixed, giving the bed uniform solids properties and a constant temperature throughout. The engine driving the solids mixing and circulation is the drag exerted by the gas on the particles.

Mass Transfer

Mass transfer in a fluidized bed can occur in several ways. Bed-to-surface mass transfer is important in plating applications. Transfer from the solid surface to the gas phase is important in drying, sublimation, and desorption processes. Mass transfer can be the limiting step in a chemical reaction system. In most instances, gas from bubbles, gas voids, or the conveying gas reacts with a solid reactant or catalyst.

Heat Transfer

One of the reasons fluidized beds have wide application is the excellent heat-transfer characteristics. Particles entering a fluidized bed rapidly reach the bed temperature, and particles within the bed are isothermal in almost all commercial situations. Gas entering the bed reaches the bed temperature quickly. In addition, heat transfer to surfaces for heating and cooling is excellent.

DISTRIBUTOR DESIGN

Good gas distribution is necessary for the bed to operate properly, and this requires that the pressure drop over the distributor be sufficient to prevent maldistribution arising from pressure fluctuations in the bed. Because gas issues from the distributor at a high velocity, care must also be taken to minimize particle attrition. Many distributor designs are used in fluidized beds. The most common ones are perforated plates, plates with caps, and pipe distributors.

PIPE DISTRIBUTORS

Jet Penetration

At the high gas velocities used in commercial practice, there are jets of gas issuing from distributor holes. It is essential that jets not impinge on any internals, otherwise the internals may be quickly eroded.

Particle Attrition

Distributor jets are a potential source of particle attrition. Particles are swept into the jet, accelerated to a high velocity, and smash into other particles as they leave. To reduce attrition at distributors, a shroud or larger diameter pipe is often added concentric to the jet hole. This shroud length allows the jet issuing from the orifice to expand and fill the shroud. The gas velocity leaving the shroud should not exceed 70 m/s, to minimize attrition.

Bed Internals

Various types of internals that may be found in commercial fluidized beds include solids and gas distributors; cyclones and cyclone diplegs; solids return and withdrawal lines; heat-transfer tubes; supports, hangers, and guides for heat-transfer tubes; baffles; secondary gas-injection nozzles; and pressure, temperature, and sample probes.

ENTRAINMENT

Entrainment, or elutriation, is the carryover of particles from a fluidized bed with the exiting gas. When the gas velocity exceeds the terminal velocity of a Group B particle, the particle is usually removed from the bed. For Group A and C powders, the gas drag needs to overcome interparticle forces as well, and gas velocities of many times the single-particle velocity are needed to entrain particles from the bed at high rates. Knowledge of the entrainment rate is important in order to estimate cyclone inlet loading, solids loss rates, and to predict bed particle size changes resulting from the selective loss of fines.

Transport Disengaging Height

The height above the bed at which entrainment becomes essentially constant with height is termed the transport disengaging height (TDH). It is desirable to locate

cyclones and vessel outlets above TDH so as to minimize solids loading to the cyclones.

Cyclones

Cyclones are an integral part of most fluidized-bed systems. A cyclone is an inexpensive device having no moving parts that separates solids and gases using centrifugal force. Cyclones are used commercially to remove solids with high efficiency down to ~15 μm.

CIRCULATING FLUIDIZED BEDS

Circulating fluidized beds (CFBs) are high velocity fluidized beds operating well above the terminal velocity of all the particles or clusters of particles. A very large cyclone and seal leg return system are needed to recycle solids in order to maintain a bed inventory. There is a gradual transition from turbulent fluidization to a truly circulating, or fast-fluidized bed, as the gas velocity is increased, and the exact transition point is rather arbitrary. The solids are returned to the bed through a conduit called a standpipe. The return of the solids can be controlled by either a mechanical or a nonmechanical valve.

The bed level is not well defined in a circulating fluidized bed, and bed density usually declines with height. Axial density profiles for different CFB operating regimes show that the vessel does not necessarily contain clearly defined bed and freeboard regimes. The solids may occupy only between 5 and 20% of the total bed volume.

Pressure Balance and Standpipes

The pressure balance around the loop of a circulating fluidized bed is illustrated in Figure 4.

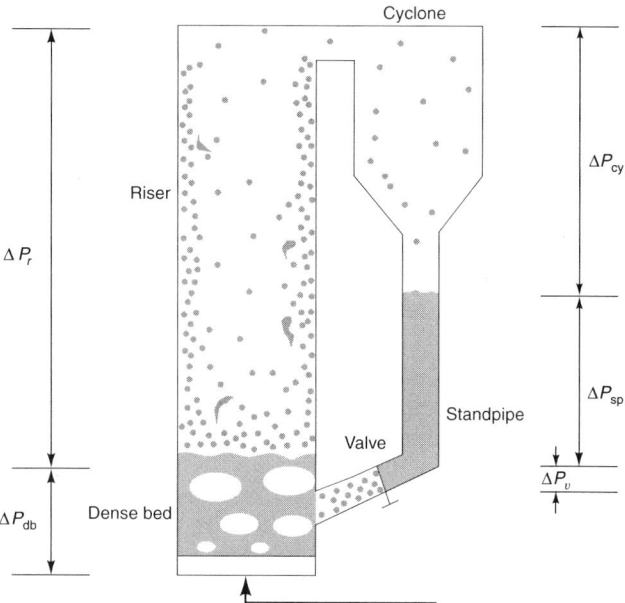

Figure 4. CFB pressure balance, where $\Delta P_{sp} = \Delta P_v + \Delta P_{db} + \Delta P_r + \Delta P_{cy}$. A high gas velocity in a fast bed results in a high solids entrainment rate. Head buildup in the standpipe overcomes head buildup in the dense bed, riser, cyclone, and sealing device (valve), and allows solids to recirculate.

Nonmechanical Valves

Nonmechanical valves, which have no moving parts in the solids flow path, are often used to control the flow of Group B solids. Examples of nonmechanical valves are an L-valve, J-valve, loop seal, and reverse seal.

SCALE UP OF FLUIDIZED BEDS

The greatest problems in scaling up fluidized beds have been encountered using Group B solids where bubbles can grow to very large sizes and bubble size control is more difficult than with Group A solids.

Several attempts have been made to determine proper scaling relationships for Group B fluidized beds. The basic scaling factors are Reynolds number

$$Re_p = d_p U_{\rho_g}/\mu \qquad (3)$$

and the Froude number

$$\text{Froude number} = \frac{U^2}{gd_p} \qquad (4)$$

where density ratio, ρ_p/ρ_g; length ratios, L/d_p and D/d_p; and other factors such as ψ, sphericity, particle size distribution, and bed geometry.

There are some data to suggest that hydrodynamic similarity improves scale-up for two-phase systems such as fluidized beds, even though it is not as convincing as single-phase evidence.

Group A particles cause fewer scale-up problems because fluidized beds of Group A particles generally are operated in the vigorously bubbling or turbulent fluidization regimes. Also, it is not unusual for a maximum stable bubble—gas void to be on the order of 25 mm or less for these particles. Thus a pilot-plant facility can generally be operated using the same gas void size that a commercial unit would experience. An example of scale-up concerns and ways to avoid them for Group A powders is shown in Figure 5. In this case, the efficiency was maintained at 80% for a turbulent fluidized bed, ie, there was

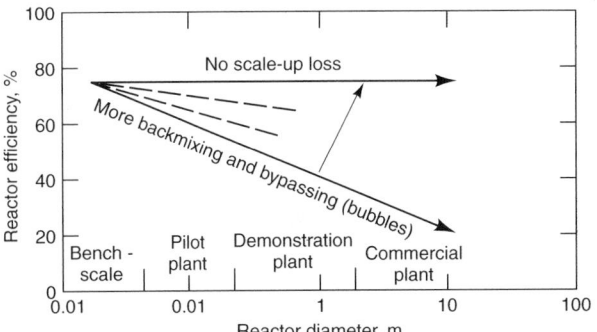

Figure 5. Turbulent and bubbling beds scale-up comparison where increasing gas velocity, fines content, and H/D staging can help maintain reactor efficiency as the reactor diameter increases. A 100% efficiency is equivalent to plug flow.

no scale-up loss, but efficiency decreased with scale up for a bubbling bed. Adding fines and operating at a higher gas velocity in a bubbling bed, ie, moving it toward turbulence, can offset scale-up loss.

D. Geldart, ed., *Gas Fluidization Technology*, John Wiley & Sons, Inc., Chichester, U.K., 1986.

A. C. Hoffman and L. E. Stein, *Gas Cyclones and Swirl Tubes*, Springer-Verlag, Berlin, 2002.

T. M. Knowlton, in Wen-Cheng Yang, ed., *Handbook of Fluidization and Fluid- Particle Systems*, Marcel Dekker, Inc., New York, 2003, Chapt. 21, p. 580.

F. A. Zenz and D. F. Othmer, *Fluidization and Fluid Particle Systems*, Reinhold Publishing Corp., New York, 1960.

AMOS A. AVIDAN
Bechtel Corporation
DESMOND F. KING
Chevron Texaco
TED M. KNOWLTON
Institute of Gas Technology/
  Particulate Solids Research
  Inc.
MEL PELL
E. I. du Pont de Nemours & Co.,
  Inc.

FLUORINE

Fluorine, F_2, is a diatomic molecule existing as a pale yellow gas at ordinary temperatures. Its name is derived from the Latin word fleure, meaning to flow, alluding to the well-known fluxing power of the mineral fluorite, CaF_2, which is the most abundant naturally occurring compound of the element. Although radioactive isotopes between atomic weight 17 and 22 have been artificially prepared and have half-lives between 4 s for ^{22}F and 110 min for ^{18}F, fluorine has a single naturally occurring isotope, ^{19}F, and has an atomic weight of 18.9984. Fluorine, the most electronegative element and the most reactive nonmetal, is located in the upper right corner of the Periodic Table. Its electron configuration is $1s^2 2s^2 2p^5$.

The only commercially feasible method of preparing elemental fluorine is by the electrolysis of molten fluoride-containing salts. The only chemical route, which does not rely on compounds derived from F_2, has more recently been discovered. About 50% of the fluorine produced is used to make UF_6 for the nuclear power industry. Other large uses include the production of nitrogen trifluoride, sulfur hexafluoride, tungsten hexafluoride, carbon tetrafluoride, and hexafluoroethane.

Fluorine, which does not occur freely in nature except for trace amounts in radioactive materials, is widely found in combination with other elements, accounting for ~0.065 wt% of the earth's crust. The most important natural source of fluorine for industrial purposes is the mineral fluorspar, CaF_2, which contains ~49% fluorine.

Table 1. Physical Properties of Fluorine

Characteristic	Value
melting point, °C	-219.61^a
boiling point, °C	-188.13^a
heat of vaporization, ΔH_{vap}, at -188.44°C and 98.4 kPa, J/mol^b	6544
heat capacities, J/(mol · K)^c solid at -223°C	49.338
at -238°C	31.074
density of liquid at bp, kg/m^3	1516^a
density of solid, kg/m^3	1900^c
viscosity, mPa·s($=$ cP) liquid at -187.96°C	0.257
at -203.96°C gas at 0°C and 101.3 kPa^d	0.414
	0.0218
thermal conductivity, gas at 0°C and 101.3 kPa, W/(m · K)	0.02477

^{a}Generally accepted value.
^{b}To convert J to cal, divide by 4.184.
^{c}Mean estimate value.
^{d}To convert kPa to mm Hg, multiply by 7.5.

PHYSICAL PROPERTIES

Fluorine is a pale yellow gas that condenses to a yellowish orange liquid at -188°C, solidifies to a yellow solid at -220°C, and turns white in a phase transition at -228°C. Fluorine has a strong odor that is easily detectable at concentrations as low as 20 ppb. The odor resembles that of the other halogens and is comparable to strong ozone. Physical properties are given in Table 1.

CHEMICAL PROPERTIES

Fluorine is the most reactive element, combining readily with most organic and inorganic materials at or below room temperature. Many organic and hydrogen-containing compounds, in particular, can burn or explode when exposed to pure fluorine. With all elements except helium, neon, and argon, fluorine forms compounds in which it shows a valence of -1. Fluorine reacts directly with the heavier helium-group gases xenon, radon, and krypton to form fluorides (see HELIUM GROUP GASES, HELIUM GROUP COMPOUNDS). Fluorine is the most electronegative element and thus can oxidize many other elements to their highest oxidation state.

The reactivity of fluorine compounds varies from extremely stable, eg, sulfur hexafluoride, and the perfluorocarbons (see FLUORINE COMPOUNDS, ORGANIC); to extremely reactive, eg, chlorine trifluoride and bromine trifluoride. Another unique property of nonionic metal fluorides is great volatility.

Fluorine is the first member of the halogen family. However, many of its properties are not typical of the other halogens. Fluorine has only one valence state, -1, whereas the other halogens also form compounds in which their valences are $+1$, $+3$, $+5$, or $+7$. Fluorine also has the lowest enthalpy of dissociation relative to the other halogens, which is in part responsible for its greater reactivity. Furthermore, the strength of the bond fluorine forms with other atoms is greater than those formed by the other halogens.

Reactions

Metals. At ordinary temperatures, fluorine reacts vigorously with most metals to form fluorides.

Nonmetals. Sulfur reacts with fluorine to yield the remarkably stable sulfur hexafluoride, SF_6.

Silicon and boron burn in fluorine forming silicon tetrafluoride, SiF_4, and boron trifluoride, BF_3, respectively. Selenium and tellurium form hexafluorides, whereas phosphorus forms tri- or pentafluorides. Fluorine reacts with the other halogens to form eight interhalogen compounds (see FLUORINE COMPOUNDS, INORGANIC, HALOGENS).

Water. Fluorine reacts with water to form hydrofluoric acid, HF, and oxygen difluoride, OF_2.

Oxygen. Oxygen does not react directly with fluorine under ordinary conditions, although in addition to oxygen difluoride, three other oxygen fluorides are known. Dioxygen difluoride, O_2F_2, trioxygen difluoride, O_3F_2, and tetraoxygen difluoride, O_4F_2, are produced in an electric discharge at cryogenic temperatures by controlling the ratio of fluorine to oxygen.

Nitrogen. Nitrogen usually does not react with fluorine under ordinary conditions and is often used as a diluent to moderate fluorinations. However, nitrogen can be made to produce nitrogen trifluoride, NF_3, by radiochemistry, glow discharge, or plasma, synthesis.

Noble Gases. Fluorine has the unique ability to react with the heavier noble gases to form binary fluorides.

Hydrogen. The reaction between fluorine and hydrogen is self-igniting and extremely energetic. It occurs spontaneously at ambient temperatures as evidenced by minor explosions that sometimes occur in fluorine-generating cells from the mixing of the H_2 and F_2 streams.

Ammonia. Ammonia (qv) reacts with excess fluorine in the vapor phase to produce N_2, NF_3, N_2F_2, HF, and NH_4F.

Organic Compounds. The reaction of pure or undiluted fluorine and organic compounds is usually accompanied by either ignition or a violent explosion of the mixture because of the very high heat of reaction. However, useful commercial-scale syntheses using fluorine are undertaken. Volatile compounds may be fluorinated in the gas phase by moderating the reaction using an inert gas such as nitrogen, by reducing reaction temperatures (≤ -78°C), and/or by the presence of finely divided packing materials. Solutions or dispersions of higher boiling materials may be fluorinated in inert solvents such as 1,1, 2-trichloro-1,2,2-trifluoroethane or some perfluorocarbon fluids, eg, Fluorinert FC-27 or FC-75 (3M). Efficient removal of the very high reaction heat, which leads to molecular fragmentation and runaway reactions, is the underlying principle in any of the aforementioned approaches.

Polymers. The dilution of fluorine using an inert gas significantly reduces the reactivity, thus allowing controlled reactions to take place with hydrocarbon polymers, even at elevated temperatures. Fluorine may also be used in conjunction with other reactive gases, eg, oxygen and water vapor, to activate polymer surfaces in order to improve chemical bonding and adhesion.

Carbon and Graphite. Fluorine reacts with amorphous forms of carbon, such as wood charcoal, to form carbon tetrafluoride, CF_4, and small amounts of other perfluorocarbons.

Fluorine reacts with high purity carbon or graphite at elevated temperatures under controlled conditions to produce fluorinated carbon, $(CF_x)_n$.

MANUFACTURE

Fluorine is produced by the electrolysis of anhydrous potassium bifluoride, KHF_2 or $KF \cdot HF$, which contains various concentrations of free HF. The fluoride ion is oxidized at the anode to liberate fluorine gas, and the hydrogen ion is reduced at the cathode to liberate hydrogen.

Commercial Cells

All commercial fluorine installations employ medium temperature cells having operating currents of ≥ 5000 A. The C and E type of the Atomic Energy Commission (AEC) (now the Department of Energy) cell designs predominate in the United States and Canada. The other cell type used in the United States is a proprietary design developed by Allied Chemical, Corp. (now Honeywell International, Inc.). This latter cell has a capacity of 5000 A and is used by Honeywell at its Metropolis, Illinois, plant. Table 2 gives the operating characteristics of a typical commercial size cell (AEC E-type).

Anodes. Fluorine cell anodes are the most important cell component, and their design and materials of construction are key factors in determining productivity and cell life. Today, anodes are made from petroleum coke and a pitch binder that is calcined at temperatures below that needed to convert the material to graphite. The anode carbon has low electrical resistance, high physical strength, and is resistant to reaction with fluorine.

Other Cell Components. American fluorine manufacturers use Monel or steel cathodes. Polytetrafluoroethylene (PTFE) provides the most satisfactory electrical insulation.

Table 2. AEC E-Type Cell Operating Characteristics

Characteristic	Value
current, A	6000
operating voltage, V	9–12
cell operating temperature, °C	90–105
hydrogen fluoride in electrolyte, %	40–42
effective anode area, m^2	3.9
anode current density, A/m^2	1500
anodes	32
anode life, $A \cdot h$	$40 - 80 \times 10^6$

Cells must be fitted with mild steel jackets and/or coils to remove heat during cell operation and to provide heat to maintain the electrolyte molten during shutdown. All commercial cells are totally jacketed.

Heat Transfer

A large portion of cell operating voltage is consumed in ohmic processes that generate heat and are a result of the large separation between anode and cathode and the resistivity of the electrolyte. Approximately 34.8 MJ (33,000 Btu) must be removed per kilogram of fluorine produced from any fluorine cell. This is accomplished by jacketing the cell and/or by using cooling tubes.

Raw Materials

The principal raw material for fluorine production is high purity anhydrous hydrofluoric acid. Each kilogram of fluorine generated requires ~1.1 kg HF.

Processes

The generation of fluorine on an industrial scale is a complex operation. The basic raw material, anhydrous hydrogen fluoride, is stored in bulk and charged to a holding tank from which it is continuously fed to the cells. Electrolyte for the cells is prepared by mixing $KF \cdot HF$ with HF to form $KF \cdot 2HF$. The newly charged cells are started up at a low current, which is gradually increased at a conditioning station separate from the cell operating position until full current is obtained at normal voltages. After conditioning, cells are connected in series using ca 12 V provided for each cell by a low voltage, 6000 A dc rectifier. Hydrogen fluoride content is maintained between 40 and 42% by continuous additions. The electrolyte level must be set and controlled at a certain level below the cell head in order to maintain a seal between the fluorine and hydrogen compartments. The cells are operated at 95–105°C and cooled with water at 75°C.

Equipment

Fluorine can be handled using a variety of materials. System cleanliness and passivation are critical to success. Materials such as nickel, Monel, aluminum, magnesium, copper, brass, stainless steel, and carbon steel are commonly used.

All equipment, lines, and fittings intended for fluorine service must be leak-tight, dry, and thoroughly cleansed of all foreign matter before use. After cleaning, the system should be filled with dry nitrogen.

The corrosion resistance of all materials used with fluorine depends on the passivation of the system. This is a pickling operation intended to remove the last traces of foreign matter, and to form a passive fluoride film on the metal surface.

Carbon steel or bronze-body gate valves are commonly used in gaseous fluorine service at low pressure. Plug valves, having Monel bodies and plugs, are recommended for moderate pressure service <500 kPa (<5 atm). For valve-stem packing PTFE polymer is recommended and it must be maintained leaktight. Valves lubricated or packed with grease or other organics should never be used. Bellows-type valves having Monel or stainless steel bellows are recommended for high pressure service, but not ball valves.

Compressors and blowers for gaseous fluorine service vary in design from multistage centrifugal compressors to diaphragm and piston types. Standard commercial instrumentation and control devices are used in fluorine systems. Pressure is measured using Bourdon-type gauges or pressure transducers. Stainless steel or Monel construction is recommended for parts in contact with fluorine. Standard thermocouples are used for all fluorine temperature-measuring equipment, such as the stainless-steel shielded type, inserted through a threaded compression fitting welded into the line. For high temperature service, nickel-shielded thermocouples should be used.

Dilute mixtures (eg, 10 or 20% F_2 in N_2) are generally less hazardous than pure fluorine, but the same precautions and procedures should be employed.

ECONOMIC ASPECTS

Fluorine gas is packaged and shipped in steel cylinders conforming to Department of Transportation (DOT) specifications 3A1000 and 3AA1000 under a pressure of 2.86 MPa (415 psi). Mixtures of 10 and 20% fluorine in nitrogen or other inert gases are commercially available in cylinders and tube trailers from Airgas, Inc.

HEALTH AND SAFETY FACTORS

Fluorine, the most reactive element known, is a dangerous material but may be handled safely using proper precautions. In any situation where an operator may come into contact with low pressure fluorine, safety glasses, a neoprene coat, boots, and clean neoprene gloves should be worn to afford overall body protection. This protection is effective against both fluorine and the hydrofluoric acid which may form from reaction of moisture in the air.

In addition, face shields made of conventional materials or, preferably, transparent, highly fluorinated polymers, should be worn whenever operators approach equipment containing fluorine under pressure. A mask having a self-contained air supply or an air helmet with fresh air supply should always be available. Leaks in high pressure systems usually result in a flame from the reaction of fluorine with the metal. Shields should be provided for valves, pressure-reducing stations, and gauges.

Toxicity

Fluorine is extremely corrosive and irritating to the skin. Inhalation at even low concentrations irritates the respiratory tract; at high concentrations fluorine inhalation may result in severe lung congestion. The American Conference of Governmental Industrial Hygienists (ACGIH) has established the 8-hour time-weighted average TLV as 1 ppm or 1.6 mg/m^3, and the short-term exposure limit TLV as 2 ppm or 3.1 mg/m^3.

Burns

Skin burns resulting from contact with pure fluorine gas are comparable to thermal burns and differ considerably from those produced by hydrogen fluoride. Fluorine burns heal much more rapidly than hydrofluoric acid burns.

Disposal

Fluorine can be disposed of by conversion to gaseous perfluorocarbons or fluoride salts. Because of the long atmospheric lifetimes of gaseous perfluorocarbons (see ATMOSPHERIC MODELING), disposal by conversion to fluoride salts is preferred.

USES

Elemental fluorine is used captively by most manufacturers for the production of various inorganic fluoride. The market for gaseous fluorine in cylinders is small, due to the difficulties involved in handling the gas and the small amount of product that can be packaged into each cylinder. One large use of fluorine is in the manufacture of uranium hexafluoride, UF_6.

Another large use for elemental fluorine is in production of sulfur hexafluoride, SF_6, a gaseous dielectric for electrical and electronic equipment. An important use of fluorine is in the preparation of a polymer surface for adhesives (qv) or coatings (qv).

F. A. Cotton, ed., *Progress in Inorganic Chemistry*, Vol. 2, Interscience, New York, 1960.

Handbook of Compressed Gases, 3rd ed., Van Nostrand Reinhold, New York, 1990.

M. Hudlicky, *Chemistry of Organic Fluorine Compounds*, Ellis Harwood Limited, Sussex, U.K., 1976.

M. Stacey, J. C. Tatlow, and A. G. Sharpe, eds., *Advances in Fluorine Chemistry*, Vol. 2, Butterworths Inc., Washington, D.C., 1961.

GEORGE SHIA
Honeywell Specialty Chemicals

FLUORINE COMPOUNDS, INORGANIC

Fluorine (qv), the most electronegative element, is much more reactive than the other elements. Indeed, fluorine reacts with virtually every other element, including the helium-group elements. Because of unique properties, fluorine has been called a superhalogen and several of its compounds called superacids. The term *superacid* is used for systems having higher acidities than anhydrous sulfuric or fluorosulfuric acid.

The basic fluorine-containing minerals are fluorite, commonly called fluorspar, CaF_2; and fluorapatite, commonly called phosphate rock. The reaction of calcium fluoride and sulfuric acid produces hydrogen fluoride. Fluorosilicic acid is produced from fluorapatite as a by-product in the production of phosphoric acid. The boiling point of hydrogen fluoride, 19.54°C, is much higher than that of HCl, −84.9°C, owing to extensive molecular association via hydrogen bonding in the former. Hydrogen fluoride is the most common reagent for production of fluorine compounds. Elemental fluorine, a pale greenish yellow gas, is produced by electrolysis of anhydrous potassium fluoride–hydrogen fluoride melts.

The fluoride ion is the least polarizable anion. It is small, having a diameter of 0.136 nm, 0.045 nm smaller than the chloride ion. The small size of F^- allows for high coordination numbers and leads to different crystal forms and solubilities, and higher bond energies than are evidenced by the other halides.

A number of elements exhibit the highest oxidation state only because fluorides and oxidation states of +6 and +7 are not uncommon. Fluorine forms very reactive halogen fluorides. Fluorine's special properties lead to many applications. Its complexing properties account for its use as a flux in steelmaking and as an intermediate in aluminum manufacture. The reaction of fluorides with hydroxyapatite, $Ca_5(PO_4)_3OH$, which is found in tooth enamel, to form less soluble and/or more acid-resistant compounds, led to the incorporation of fluorides in drinking water and dentifrices (qv) to reduce dental caries. Many fluorides are volatile and in many cases are the most volatile compounds of an element. This property led to the use of UF_6 for uranium isotope enrichment, critical to the nuclear industry, and the use of metal fluorides in chemical vapor deposition (WF_6, MoF_6, ReF_6), in ion implantation (qv) for semiconductors (qv) BF_3, PF_3, AsF_5, etc), and as unreactive dielectrics (SF_6). Fluorinated steroids, other fluorinated drugs, and anesthetics have medical applications. The stability, lack of reactivity and, therefore, lack of toxicity of some fluorine compounds are also demonstrated by studies reporting survival of animals in an atmosphere of 80% SF_6 and 20% oxygen, and use of perfluorochemicals as short-term blood substitutes because of the ability to efficiently transport oxygen and carbon dioxide. Fluorides including HF, BF_3, SbF_5, PF_5, and several complexes, eg, BF_4^-, PF_6^-, SbF_6^-, and AsF_6^-, are used in many applications in catalysis (qv).

SOURCES AND APPLICATIONS

The earth's crust consists of 0.09% fluorine. Among the elements fluorine ranks about thirteenth in terrestrial abundance. The ores of most importance are fluorspar, CaF_2; fluorapatite, $Ca_5(PO_4)_3F$; and cryolite, Na_3AlF_6. Fluorspar is the primary commercial source of fluorine. Twenty-six percent of the world's high quality deposits of fluorspar are in North America. Most of that is in Mexico. The majority of the fluorine in the earth's crust is in phosphate rock in the form of fluorapatite which has an average fluorine concentration of 3.5%. Recovery of these fluorine values as by-product fluorosilicic acid from phosphate production has grown steadily, partially because of

environmental requirements (see PHOSPHORIC ACID AND PHOSPHATES).

Production of hydrogen fluoride from reaction of CaF_2 with sulfuric acid is the largest user of fluorspar and accounts for approximately 60–65% of total U.S. consumption.

SYNTHESIS

Most inorganic fluorides are prepared by the reaction of hydrofluoric acid with oxides, carbonates, hydroxides, chlorides, or metals. Routes starting with carbonate, hydroxide, or oxide are the most common and the choice is determined by the most economical starting material.

SAFETY, TOXICITY, AND HANDLING

Hazards associated with fluorides are severe. Anhydrous or aqueous hydrogen fluoride is extremely corrosive to skin, eyes, mucous membranes, and lungs; it can cause permanent damage and even death. Detailed information about safety, toxicity, and handling can be obtained from the producers of hydrogen fluoride, eg, Elf Atochem North America, Inc., DuPont, and AlliedSignal. Fluorides susceptible to hydrolysis can generate aqueous hydrogen fluoride. Ingestion of excess fluorides may cause poisoning or damage to bones and/or teeth. Fluorine-containing oxidizers can react with the body in addition to causing burns.

Because hydrogen fluoride is extremely reactive, special materials are necessary for its handling and storage. Anhydrous hydrogen fluoride is produced and stored in mild steel equipment. Teflon or polyethylene are frequently used for aqueous solutions.

The OSHA permissible exposure limit and the American Conference of Governmental Industrial Hygienists (ACGIH) established threshold limit value (TLV) for fluorides is 2.5 mg of fluoride per cubic meter of air. This is the TLV–TWA concentration for a normal 8-h work day and a 40-h work week.

R. E. Banks, *J. Fluorine Chem.* **33**, 3–26 (1986).

R. J. Gillespie, *Acc. Chem. Res.* **1**, 202 (1968).

H. Moissan, *Compt. Rend.* **12**, 1543 (1886).

Threshold Limit Values for Chemical Substances and Physical Agents, 1992–1993, The American Conference of Governmental Industrial Hygienists, Cincinnati, Ohio.

CHARLES B. LINDAHL
TARIQ MAHMOOD
Elf Atochem
North America, Inc.

FLUORINE COMPOUNDS, ORGANIC

Fluorine is the most abundant halogen occurrence in the earth's crust, yet fewer than a dozen fluorinated organic compounds are known in Nature and usually they are found at low concentrations in exotic locations. Exclusive of these, all organic fluorine compounds are manmade.

PHYSICAL PROPERTIES

Substitution of fluorine for hydrogen in an organic compound has a profound influence on the compound's chemical and physical properties. Several factors that are characteristic of fluorine and that underlie the observed effects are the large electronegativity of fluorine, its small size, the low degree of polarizability on the carbon–fluorine bonds, and the weak intermolecular forces.

The replacement of chlorine by fluorine results in a nearly constant boiling point (bp) drop of ~50°C for every chlorine atom that is replaced. A similar boiling point effect with hydrocarbons is apparent, even though the molecular weight of the fluorocarbon is much higher than the corresponding hydrocarbon analogue. An analogous drop in the corresponding fluorocarbon freezing point results in a widened liquid range for applications like lubricating fluids and greases. One other significant property difference, attributed to weak intermolecular forces, can be found in the very low surface tensions of fluorocarbons as compared to hydrocarbons and water.

The low surface tension of highly fluorinated organic compounds is commercially important for their application in surfactants, antisoiling textile treatments, lubricants, and specialty wetting agents.

In contrast, the viscosities of fluorocarbons are higher than those of the corresponding hydrocarbons. This can be explained by the greater stiffness of the fluorocarbon chain arising from the large repulsive forces between molecules, and from the greater density imparted by the more massive fluorine atoms (vs, hydrogen). The fluorocarbon viscosity drops rapidly with increasing temperature and is accompanied by a simultaneous large decrease in density.

The refractive indexes and dielectric constants for the fluorocarbons are both lower than that for the corresponding hydrocarbon analogue.

Preparation

Various methods are briefly described for the synthesis of selected fluorinated compounds. There are many known ways to introduce fluorine into organic compounds, but hydrogen fluoride, HF, is considered to be the most economical source of fluorine for many commercial applications.

Halogen Exchange. The exchange of a non-fluorine halogen atom in an organic compound for a fluorine atom is the most widely used method of fluorination.

Replacement of Hydrogen. Three methods of substitution of a hydrogen atom by fluorine are (1) reaction of a C–H bond with elemental fluorine (direct fluorination), (2) reaction of a C–H bond with a high valence state metal fluoride like AgF_2, CoF_3, or CuF_2, and (3) electrochemical fluorination in which the reaction occurs at the anode of a cell containing a source of fluoride, usually HF.

Electrochemical Fluorination. The electrochemical fluorination (ECF) of highly fluorinated organic compounds involves the electrolysis of an organic reactant in liquid anhydrous HF at a voltage below that for liberation of fluorine. The reaction is limited by temperature (usually done at 0° C) and by the solubility of the reactant in HF. Electrical conductivity is required for current to flow and the reaction to proceed. Fluorination takes place at the nickel anode by a stepwise, free radical process not involving the intermediate formation of elemental fluorine. Hydrogen is liberated at the cathode. Recent studies seem to confirm that reversible formation of a hypervalent nickel fluoride film at the anode surface is the actual fluorinating agent attacking chemisorbed compounds with C–H bonds. The process is used to fluorinate acyl halides, sulfonyl halides, ethers, carboxylic acids, and amines. The product is a fluorocarbon having no residual hydrogen. The ECF yields decrease with increasing number of carbon atoms in the feed structure.

ECF is successfully used on a commercial scale to produce certain perfluoroacyl fluorides, perfluoroalkylsulfonyl fluorides, perfluoroalkyl ethers, and perfluoroalkylamines.

Telomer Formation. Fluorinated compounds with an active C–Br or C–I bond can add to a fluoroolefin to form addition products in high yield. The olefin most often used is tetrafluoroethylene (TFE). Telomerization involves reacting a telogen, or addition agent like $CBrF_3$, CF_3I, or C_2F_5I, with an olefin to form a longer chain addition product called telomer.

Aromatic Ring Fluorination. The selective introduction of fluorine at specific unactivated aryl ring positions is desirable for preparation of certain fluorinated compounds requiring one or more fluorine. The formation of an aryl diazonium fluoride salt, followed by decomposition is a classical reaction (the Schiemann Reaction) for aryl fluoride preparation.

CHEMICAL PROPERTIES AND APPLICATIONS

Substitution of fluorine into an organic molecule results in enhanced chemical stability. The resulting chemical reactivity of adjacent functional groups is drastically altered due to the large inductive effect of fluorine. These effects become more pronounced as the degree of fluorine substitution is increased, especially on the same carbon atom. This effect demonstrates a maximum in the fluorocarbons and their derivatives.

Fluorinated Alkanes

As the fluorine content increases, the chemical reactivity decreases until complete fluorination is achieved, after which the alkane is inert to most chemical attack, including the highly reactive element fluorine. Their lack of reactivity leads to their use in certain commercial applications where stability is valued when in contact with highly reactive chemicals.

Fluorinated Olefins. Certain chlorofluorocarbons (CFCs) are used as raw materials to manufacture key fluorinated olefins to support polymer applications.

Fluorinated Aromatic Hydrocarbons. Many aromatic fluorocarbon derivatives, eg, hexafluorobenzene, octafluorotoluene, and perfluoronaphthalene are examples of compounds that readily undergo nucleophilic ring substitution reactions with loss of one or more fluorine substituents.

Fluorinated Heterocyclic Compounds. The direct action of fluorine on uracil yields the cancer chemotherapy agent, 5-fluorouracil, as one special example of a selective fluorination on a commercial scale.

Fluorinated Acids. Generally, reactions of fluorinated acids are similar to organic acids and they find applications, particularly trifluoroacetic acid and its anhydride, as promoters in the preparation of esters and ketones and in nitration reactions.

Fluorinated Biologically Active Compounds. Many biologically active compounds are prepared from fluorobenzene, difluorobenzene, benzotrifluoride, and fluorinated steroids. The strong interest in these substances is based on the following considerations: (*1*) fluorine most closely resembles bioactive hydrogen analogues with respect to steric requirements at the receptor sites; (*2*) fluorine alters electronic effects, owing to its high electronegativity; (*3*) fluorine imparts improved oxidative and thermal stability to the parent molecule; (*4*) fluorine imparts lipid solubility, thereby increasing the *in vivo* absorption and transport rates across membranes.

Many fluorinated, biologically active agents have been developed and successfully used in the treatment of human diseases. The biological property of fluorinated organics has been further extended to applications in the agrochemical, veterinary and pest management fields. Examples include analgesics, antibiotics, antidepressants, antifungal agents, antiviral agents, appetite depressants, diruetics, inhalation anesthetics, tranquilizers, fungicides, herbicides, insecticides, and rodenticides.

ECONOMIC ASPECTS

Domestic production of CFCs is now nonexistent due to environmental pressures. CFC production outside of the United States is also mostly curtailed or will be so in the third world countries during the next few years. Replacement of CFCs with HCFC and HFC production is widespread. Development of acceptable, alternative products is still ongoing due to the small HCFC ozone depletion effect and general global warming concerns. As the largest global supplier, Du Pont alone has spent well over $500 million cumulatively through 2000 to develop alternatives.

The Minnesota Mining and Manufacturing Company, or 3M, manufactures specialty perfluorochemicals using mainly ECF methods at their St.Paul, Minn. and Decatur, Ala. sites. Fluoroolefin telomerization technology is practiced by Asahi Glass, Du Pont, and Hoechst AG at a variety of their sites to manufacture a line of perfluorinated specialty chemicals for stain-resistant treatment, surfactant applications as well as specialty fluids. The fluoroolefins are also polymerized and copolymerized to manufacture a range of either plastic or elastomeric products.

HEALTH AND SAFETY FACTORS

The safety of fluorine compounds is possibly as varied as the numbers of compounds known that bear fluorine substituents. Aerosol or vapor inhalation is the most likely route of exposure where adverse health effects may occur. All new fluorine compounds should be handled with caution as one would do with any potentially hazardous substance until full toxicological properties are known. Existing fluorine compounds cover the range from biologically inert materials, like fluorocarbon fluids suitable for potential blood substitutes through to biologically active materials like the very highly toxic octafluoroisobutylene. The major commercial fluorinated compounds, like the CFCs, exhibit a very low order of toxicity. The potential cardiotoxicity from inhalation of bronchiodilator aerosols using CFCs as propellants is well documented in the medical literature.

P. V. Ramachandran, ed., *Asymmetric Fluoroorganic Chemistry: Synthesis, Applications and Future Directions*, ACS Symposium Series No. 746, American Chemical Society Publisher, Washington, D.C., 2000.

J. Scheirs, ed., *Modern Fluoropolymers: High Performance Polymers for Diverse Applications*, John Wiley & Sons, Inc., New York, 1997.

J. C. Tatlow and co-workers, eds., *Advances in Fluorine Chemistry*, Vols. 1–7, W. A. Benjamin, Inc., New York, 1960–1973.

J. T. Welch and S. Eswarakrishnan, *Fluorine in Bioorganic Chemistry*, John Wiley & Sons, Inc., New York, 1991.

WILLIAM X. BAJZER
Dow Corning Corporation

FLUOROETHERS AND FLUOROAMINES

Fluoroaliphatic ethers and perfluorotertiary amines together with the perfluoroalkanes and cycloalkanes comprise a class of unreactive materials known in the industry as inert fluids. These fluids are colorless, nearly odorless, essentially nontoxic, nonflammable, dense, and extremely nonpolar. In the electronics industry, the lower molecular weight compounds find application in such areas as thermal management, testing, cleaning solvents, and solvents for coating applications. Higher molecular weight polymers and oligomers are used in a variety of applications, including hazardous-duty vacuum pump fluids, specialty greases, and various specialty cosmetics and lubricants.

Many perfluoroaliphatic ethers and tertiary amines have been prepared by electrochemical fluorination, direct fluorination using elemental fluorine, or, in a few cases, by fluorination using cobalt trifluoride.

More recently, the new class of hydrofluoroethers has been commercialized. These ethers comprise a perfluorinated portion separated by oxygen from a hydrocarbon portion and are prepared by alkylation of perfluorinated acyl fluorides. Their structure gives these ethers extremely useful properties including enhanced solvency and the ability to form many useful nonflammable azeotropes with a variety of hydrocarbon cosolvents. In addition to these hydrofluoroethers, a recent class of hydrofluoropolyethers has been commercialized. These materials contain hydrogen atoms at the alpha and omega positions of a perfluoropolyether chain and have properties intermediate between the perfluorinated fluids and the hydrofluoroethers.

PHYSICAL PROPERTIES

Perfluorinated compounds boil at much lower temperatures and have lower heats of vaporization than the corresponding hydrocarbon analogues even though they have considerably higher molecular weights (Table 1). Many of the unusual properties of the perfluorinated inert fluids are the result of the extremely low intermolecular interactions.

Thermal Stabilities

The perfluoroethers have thermal stabilities comparable to those of the perfluoroalkanes.

Electrical Properties

The low polarizability of perfluorinated liquids makes them excellent insulators.

CHEMICAL PROPERTIES

The inert character of the perfluoroethers and tertiary amines is demonstrated by their lack of basicity or reactivity as compared with their hydrocarbon analogues. Both classes of compounds are nonflammable.

Solvent Properties

In comparison to the more familiar hydrocarbon systems, the solvent properties of the perfluorinated inert liquids are also unusual due to their nonpolar nature and low intermolecular forces. They are generally very poor solvents for most organic compounds.

ECONOMIC ASPECTS

Information on the production levels of the perfluoroethers and perfluorotertiary amines is not disclosed, but the products are available commercially and are marketed, eg, as part of the Fluorinert Electronic Liquids family by 3M Co.

HEALTH AND SAFETY FACTORS

Over the years animal studies have repeatedly shown that perfluorinated inert fluids are non-irritating to the eyes and skin and practically nontoxic by ingestion, inhalation, or intraperitoneal injection. The commercially available hydrofluoroethers and hydrofluoropolyethers have also been shown to be low in toxicity. As with all fluorinated compounds, thermal degradation of the fluorinated ethers and amines can produce toxic decomposition products including hydrogen fluoride and perfluoroisobutene. All of the commercially available fluorinated ethers and amines are nonflammable, exhibiting no open or closed cup flash point.

USES

The unique combination of properties of the perfluorinated fluids makes them useful in a variety of applications in the electronics industry. The lower molecular weight materials are used in two principal areas in this industry: heat transfer, both direct and indirect, and testing applications. Perfluorinated liquids have been used as the total immersion coolant for supercomputers. Testing applications include liquid burn-in testing, gross

Table 1. Physical Properties of Some Fluorinated Liquids

Name	Molecular formula	Bp, °C	d^{25}	Viscosity, cps/25°C	Pour point, °C
perfluoro-4-methylmorpholine	$CF_3N(CF_2)_2O(CF_2)_2$	50	1.70	0.68	−80
perfluorohexane	C_6F_{14}	58	1.68	0.64	−74
Galden HT 170[a]		170	1.77	3.2	−97
Fluorinert FC-75	$C_8F_{16}O^b$ (cyclic)	103	1.76	1.3	−93
perfluorooctane	C_8F_{18}	103	1.77	0.71	−30
perfluorotripropylamine	$(C_3F_7)_3N$	128	1.82	1.40	−50
perfluorotributylamine	$(C_4F_9)_3N$	178	1.86	4.7	−50
perfluoro(diethylamino) ethyl ether	$[(C_2F_5)_2NC_2F_4]_2O$	178			−80
perfluorohexyl ether	$(C_6F_{13})_2O$	181	1.81		−90
K7 Fluid	$C_3F_7O[C_3F_6O]_5C_2F_5$	250	1.82		−80
methoxyheptafluoropropane	$C_3F_7OCH_3$	34	1.40	0.45	−122

[a]Mixture of perfluorinated polyethers marketed by Solvay Solexis.
[b]$C_8F_{16}O$ represents a mixture of isomers of cyclic perfluoroaliphatic ethers, primarily perfluoro-2-butyltetrahydrofuran.

and fine leak testing, and electrical environmental testing.

The higher molecular weight perfluoropolyethers are useful as specialty lubricants. They provide good lubrication under boundary conditions in systems in which the mechanical parts are exposed to high temperatures or aggressive chemical environments. They are typically used as the working fluid in hazardous duty vacuum pumps used in plasma etching. Specialty greases, used in high temperature environments in which a hydrocarbon-based grease fails, have also been formulated by blending perfluoropolyethers with fluorinated polymers. Additionally, perfluoropolyethers are used as lubricants for magnetic media, lubricant and sealing agent for oxygen service, inert hydraulic fluids, etc. They have also found application in cosmetics and as a protective coating for outdoor stone art and masonry. Perfluorinated compounds are also potentially useful as inert reaction media, particularly when one of the reactants is gaseous. The high solubility of oxygen and carbon dioxide in perfluorinated liquids has allowed their use as blood substitutes and as oxygenation media for biotechnology.

The hydrofluoropolyethers have begun to find use as heat-transfer fluids. The hydrofluoroethers are used in a number of applications as replacements for ozone-depleting chlorofluorocarbons (CFCs). These applications include electronics and precision cleaning as well as solvents for the deposition of lubricants and coatings on computer hard disks, electronic components and medical devices. Hydrofluoroethers are also employed as heat transfer fluids in the manufacture of semiconductors and electronic components. In addition, they are used in a number of specialty applications such as an alternative to flammable solvents in fingerprint development and a component in some cosmetic formulations.

G. Caporiccio, in R. E. Banks, D. W. A. Sharp, and J. C. Tatlow, eds., *Fluorine: The First Hundred Years (1886–1986)*, Elsevier Sequoia, New York, 1986, pp. 314–320.

IPCC (Intergovernmental Panel on Climate Change), Climate Change 2001 : The Scientific Basis, J. Houghton, ed., Cambridge University Press, Cambridge, U.K., 2001.

J. Owens, in B. Kanegsberg and E. Kanegsberg, eds., *Handbook for Critical Cleaning*, CRC Press LLC, Boca Raton, Fla., 2000.

Michael G. Costello
Richard M. Flynn
John G. Owens
3M Company

FOAMS

Foam is a nonequilibrium dispersion of gas bubbles in a relatively smaller volume of liquid. An essential ingredient in a liquid-based foam is surface-active molecules. These reside at the interfaces and are responsible for both the tendency of a liquid to foam and the stability of the resulting dispersion of bubbles. Important uses for custom-designed foams vary widely from familiar examples of detergents, cosmetics, and foods, to fire extinguishing, oil recovery, and a host of physical and chemical separation techniques. Unwanted generation of foam, on the other hand, is a common problem affecting the efficiency and speed of a vast number of industrial processes involving the mixing or agitation of multicomponent liquids. In all cases, control of foam rheology and stability is desired. These physical properties, in turn, are determined by both the physical chemistry of their liquid–vapor interfaces and by the structure formed from the collection of gas bubbles.

PHYSICAL CHEMISTRY OF INTERFACES

The chemical composition, physical structure, and key physical properties of a foam, namely, its stability and rheology, are all closely interrelated. Since there is a large interfacial area of contact between liquid and vapor inside a foam, the physical chemistry of liquid–vapor interfaces and their modification by surface-active molecules plays a primary role underlying these interrelationships.

For aqueous solutions, the chemical constituents most commonly responsible for foaming are surfactants, ie, surface-active agents. Such molecules find wide use in other settings, and are distinguished by having both hydrophilic and hydrophobic regions.

Reduced Surface Tension

Just as surfactants self-organize in the bulk solution as a result of their hydrophilic and hydrophobic segments, they also preferentially adsorb and organize at the solution–vapor interface. In the case of aqueous surfactant solutions, the hydrophobic tails protrude into the vapor and leave only the hydrophilic head groups in contact with the solution. The favorable energetics of the arrangement can be seen by the reduction in the interfacial free energy per unit area, or surface tension, σ. In most custom foams, the surfactant concentration in the base liquid is slightly above the CMC. However, the reduced surface tension is not in itself responsible for the foaming; the primary benefit is that less mechanical energy need be supplied to create the large interfacial area in a foam. The prevention of bubble coalescence needed for significant foaming is accomplished through other physical chemical mechanisms involving surfactants, creating repulsive forces between the bubble surfaces and via the viscoelastic properties of these surfaces.

Gibbs Elasticity and Marangoni Flows

The reduction of surface tension with increasing surfactant adsorption gives rise to a nonequilibrium effect that can, in some cases, promote foaming. A sudden increase in the interfacial area by mechanical perturbation or thermal fluctuation results in a locally higher surface tension because the number of surfactant molecules per unit area simultaneously decreases. The Gibbs elasticity, E, is often used to quantify the instantaneous change in surface tension σ with area A, ie, $E = d\sigma/d \ln A$. If the film of liquid separating two neighboring bubbles in a foam develops a thickness variations or a surfactant density fluctuation at their interfaces, the resulting local surface tension gradient will induce a Marangoni flow of liquid toward the direction of higher σ. This flow of liquid toward the fluctuations area helps heal the fluctuation and thus keeps the neighboring bubbles from coalescing.

Interfacial Forces

Neighboring bubble surfaces in a foam interact through a variety of forces that depend on the composition and thickness of liquid between them, and on the physical chemistry of their liquid–vapor interfaces. For a foam to be relatively stable, the net interaction must be sufficiently repulsive at short distances to maintain a significant layer of liquid in between neighboring bubbles. Otherwise two bubbles could approach so closely as to expel all the liquid and fuse into one larger bubble. Interfacial forces include the van der Waals interaction, the electrostatic double layer interaction, and disjoining pressure.

PHYSICAL PROPERTIES OF FOAM

Structure

Foam structure is characterized by the "wetness" of the system. Foams with arbitrarily large liquid to gas ratios can be generated by excessive agitation or by intentionally bubbling gas through a fluid. If the liquid content is sufficiently great, the foam consists of well-separated spherical bubbles that rapidly rise upwards displacing the heavier liquid. Such a system is usually called a froth, or bubbly liquid, rather than a foam.

If there are sufficiently strong repulsive interactions, such as from the electric double-layer force, then the gas bubbles at the top of a froth collect together without bursting. Furthermore, their interfaces approach as closely as these repulsive forces allow; typically on the order of 100 nm. Thus bubbles on top of a froth can pack together very closely and still allow most of the liquid to escape downward under the influence of gravity while maintaining their spherical shape. Given sufficient liquid, such a foam can resemble the random close-packed structure formed by hard spheres.

A dry foam, by contrast, is one with so little liquid that the bubbles are severely distorted into approximately polyhedral shapes. Typically this occurs for foams with <1% liquid by volume.

A complete characterization of the structure of a foam requires a characterization of the structure of the bubbles that comprise the foam. The total liquid content can be readily found from the mass densities of the foam and the liquid from which it was made. However, a more detailed determination of the bubble structure, including their average size, their shape, their structure and their size distribution is much more difficult, and is typically impeded by the problems in visualizing the interior of a foam. Even in the absence of any intrinsic optical absorption of the liquid, the strong mismatch in the indexes of refraction between the gas and the fluid results in a large scattering of light, usually precluding direct visualization of the interior structure of a foam. As a result, other, less direct, methods have been developed, and must be used, except in exceptional cases where the foam structure has been optimized for visualization.

One optical imaging technique that circumvents the problem of multiple light scattering and thus can be used more generally is to estimate the bubble size distribution from the area individual foam bubbles occupy at a glass surface. Such experiments, and the systematic differences between bulk and surface bubble distributions, have been reviewed. Another technique that also directly measures the bubble size distribution is the use of a Coulter counter, where individual bubbles are drawn through a small tube and counted. This yields a direct measure of the bubble size distribution, but it is invasive and cannot probe the structure of the foam.

One technique that does probe the foam structure directly is cryomicroscopy. The foam is rapidly frozen, and the solid structure is cut open and imaged with an optical or electron microscope. Such methods are widely applicable and provide a direct image of the foam structure; however, they destroy the sample and may also perturb the foam structure in an uncontrolled manner during the freezing.

Stability

Control of foam stability is important in all applications, whether degradation of a custom foam is to be minimized or whether excessive foaming is to be prevented. In all cases, the time evolution of the foam structure provides a natural means of quantifying foam stability. There are three basic mechanisms whereby the structure may change: by the gravitational segregation of liquid and bubbles, by the coalescence of neighboring bubbles via film rupture, and by the diffusion of gas across the liquid between neighboring bubbles.

Drainage

All foams and froths consist of liquid and vapor components that have very different mass densities, making them susceptible to gravitationally induced segregation. In very wet froths the vapor bubbles rapidly move upward while the liquid falls. In longer lived foams, the gas fraction is higher and the bubbles are tightly packed. Nevertheless, the heavier fluid may still drain downward through the foam liquid network between the bubbles. As time proceeds, some liquid drains out and accumulates at the bottom of the foam, while the overall liquid fraction of the foam decreases. Provided there is no rupture of the films, this free drainage proceeds until there develops a vertical, hydrostatic pressure gradient to offset gravity. This results in a nonuniform gas:liquid volume fraction with the foam being more wet near the bottom of the container.

Experimentally, as an alternative to the free drainage situation, the forced drainage experimental procedure is often used for determining the typical drainage speeds and rates within a foam. Drainage rates and speed also depend on the gas used. This is due to a coupling between the coarsening process and the drainage flows. The coupling is efficient when the typical times scales of both processes are close, and in that case, this results in an acceleration of the drainage rates. Finally, drainage depends also on the size and shape of the container as well.

On the ground, getting rid of drainage is not an easy task, and no efficient solutions has ever been found.

Film Rupture

Another general mechanism by which foams evolve is the coalescence of neighboring bubbles via film rupture. This occurs if the nature of the surface-active components is such that the repulsive interactions and Marangoni flows are not sufficient to keep neighboring bubbles apart. Bubble coalescence can become more frequent as the foam drains and there is less liquid to separate neighbors. Long-lived foams can be easily formulated in which film rupture is essentially negligible, by ensuring that the surface-active agents provide a sufficiently large barrier that prevents the two films from approaching each other. Then film rupture is probably a thermally activated process in which a large, rare fluctuation away from equilibrium thickness and over an energy barrier is needed. Film rupture can also be enhanced by mechanical shock.

Gas Diffusion

For very long-lived foams, film rupture is negligible and drainage slows to a stop as hydrostatic equilibrium is attained. Nevertheless, the foam is still not in thermodynamic equilibrium and continues to evolve with time. This occurs through an entirely different, though very general, means: gas diffusion. Smaller bubbles have a greater interfacial curvature and hence, by Laplace's law, have a higher internal pressure than larger bubbles. This results in a diffusive flux of gas from smaller to larger bubbles. Thus with time small bubbles shrink while large bubbles grow. This process is known as coarsening, or ripening, and results in the net increase in the average bubble size over time. It is ultimately driven by surface tension and serves to decrease the total interfacial surface area with time.

Any means of characterizing foam structure can be used to study foam evolution provided that the measurement can be made noninvasively and sufficiently rapidly. One technique that has been applied successfully is the measurement of the change in the pressure head over an evolving foam. Multiple light scattering techniques (in both static and dynamic modes) have been used to follow the time evolution of a foam.

A common engineering technique for determining foam stability entails measuring the amount of foam produced. For defoaming applications, this is often a more important measure of stability than the foam structure.

Rheology

The rheology of foam is striking; it simultaneously shares the hallmark rheological properties of solids, liquids, and gases. Like an ordinary solid, foams have a finite shear modulus and respond elastically to a small shear stress. However, if the applied stress is increased beyond the yield stress, the foam flows like a viscous liquid. In addition, because they contain a large volume fraction of gas, foams are quite compressible, like gases. Thus foams defy classification as solid, liquid, or vapor, and their mechanical response to external forces can be very complex.

One simple rheological model that is often used to describe the behavior of foams is that of a Bingham plastic. This applies for flows over length scales sufficiently large that the foam can be reasonably considered as a continuous medium. The Bingham plastic model combines the properties of a yield stress like that of a solid with the viscous flow of a liquid.

While the Bingham plastic model is an adequate approximate description of foam rheology, it is by no means exact. More detailed models attempt to relate the rheological properties of foams to the structure and behavior of the bubbles.

To determine rheological parameters such as the yield stress and effective viscosity of a foam, commercial rheometers are available; rotational and continuous-flow-tube viscometry are most commonly employed (see RHEOLOGICAL MEASUREMENTS). However, obtaining reproducible results independent of the sample geometry is nevertheless a difficult goal which arguably has not been achieved in most of the experiments reported in the scientific literature.

PRODUCTION

Several techniques are available for the generation of special-purpose foam with the desired properties. The simplest method is to disperse compressed gas directly into an aqueous surfactant solution by means of a glass frit. A variation of this method that allows for control of liquid content is to simultaneously pump gas and surfactant solution through a bead pack or steel wool, for example, at fixed rates. In the same spirit, another solution, which can be used for laboratory purposes, consists of pushing a pressurized surfactant solution through a single pinhole, and to add any desired gas flow rate just after that pinhole. Large amount of foams, at any liquid fractions, are then produced by the subsequent turbulent mixing of these fluids inside a final tube of confinement. Less reproducible mechanical means of foam generation include brute force shaking and blending. For highly reproducible foams composed of small bubbles, such as shaving creams, the aerosol technique is especially suitable.

APPLICATIONS

Foams have a wide variety of applications that exploit their different physical properties. The low density, or high volume fraction of gas, enable foams to float on top of other fluids and to fill large volumes with relatively little fluid material. These features are of particular importance in their use for fire fighting. The very high internal surface area of foams makes them useful in many separation processes. The unique rheology of foams also results in a wide variety of uses, as a foam can behave as a solid, while still being able to flow once its yield stress is exceeded. Foams are used in food, oil recovery, detergentry, textiles, cosmeties, and firefighting.

SAFETY, HEALTH, AND ENVIRONMENT

Foams play important roles in environmental issues, both beneficial and detrimental.

Natural Waters

Many water systems have a natural tendency to produce foam upon agitation. The presence of pollutants exacerbates this problem. This was particularly severe when detergents contained surfactants that were resistant to biodegradation. Then, water near industrial sites or sewage disposal plants could be covered with a blanket of stable, standing foam. However, surfactant use has switched to biodegradable molecules, which has greatly reduced the incidence of these problems.

Wastewater Treatment

The treatment of wastewater, either from sewage or from industrial processes, typically entails a preliminary filtration to remove the large volumes of solids, and then a slower settling to remove the sand and gravel (see WATER). The water is then treated by an activated sludge process to remove the remaining dissolved solids and organic colloidal particles. Activated sludge is a biomass that assists in the degradation of the organic waste in the water. The process entails a mixing and aeration of the wastewater with the activated sludge, which can lead to problems of foaming. The foams produced can be quite stable, resulting in additional problems for waste disposal. The foams produced in this process differ from those normally encountered in that the foam producing and stabilizing agents are microbial, primarily including Nocardia, Microthrix parvicella, and Rhodococcus. These foams are more difficult to treat with defoaming agents. Moreover, it is very difficult to predict the degree of foamability of the waste being treated. In other, more specialized wastewater treatments, these problems do not arise, and defoaming agents can be used effectively.

Chlorofluorocarbon Alternatives

There still is no completely satisfactory propellant for use in the aerosol method of foam production. Chlorofluorocarbons, still widely used, are harmful to atmospheric ozone and low molecular weight hydrocarbons, now popular, eg, in producing shaving cream, are explosive and promote the greenhouse effect. The difficulty is in creating a safe, stable liquid that can be readily emulsified and whose vapor pressure at room temperature is roughly 200–300 kPa (2–3 atm).

C. Isenberg, *The Science of Soap Films and Soap Bubbles*, Dover Publications, New York, 1992.

D. Weaire and S. Hutzler, *The Physics of Foams*, Oxford University Press, 1999.

A. J. Wilson, ed., *Foams: Physics, Chemistry, and Structure*, Springer-Verlag, New York, 1989.

ARNAUD SAINT-JALMES
Universite Paris-Sud
DOUGLAS J. DURIAN
University of Pennsylvania
DAVID A. WEITZ
Harvard University

FOOD ADDITIVES

A food additive is any substance that becomes part of a food product either directly or indirectly during some phase of processing, storage, or packaging.

Direct food additives are those that have been intentionally added to food for functional purpose, in controlled amounts, usually at low levels (from parts per million, ppm, to 1–2% by weight). Basic foodstuffs are excluded from the definition, although ingredients that are added to foods (eg, high fructose corn syrup, starches, and protein concentrates) are often included among food additives.

Included in the direct food additive category are

- Inorganic chemicals—phosphates, sulfates, nitrates, etc.
- Synthetic chemicals—dyes, silicones, benzoates, vitamin A, etc.
- Extraction products from natural sources—essential oils, gums, vitamin E, etc.
- Fermentation derived products—enzymes, lactic acid, citric acid, etc.

Indirect or *nonintentional additives* on the other hand, are those entering into food products in small quantities as a result of growing, processing, or packaging. Examples of these are lubricating oils from processing equipment or components of a package that migrate into the food before consumption.

The difference between food additives and ingredients is mainly in the quantity used in any given formulation. Food ingredients can be consumed alone as food (eg, butter, sucrose), while food additives are mainly used in small quantities relative to the total food consumption but which nonetheless play a large part in the production of desirable and safe food products.

FUNCTION OF FOOD ADDITIVES

Direct food additives serve several major functions in foods. Many additives, in fact, are multifunctional. The basic functions are

- Extend shelf-life (eg, retard the onset of rancidity).

- Ensure microbial safety (eg, against botulism, Listeria).
- Enhance appetability and palatability (flavor, color and texture).
- Improve nutritional value (eg, vitamin and trace mineral fortification).
- Facilitate food processing (eg, emulsifiers, anticaking agents).

Substances that come under the general definition of direct food additives number in the thousands and include both inorganic and organic chemicals, natural products, and modified natural and synthetic or artificial materials.

The U.S. Code of Federal Regulations (CFR) provides classification for food additives. In the CFR *direct food additives* are divided into the following eight categories:

1. Food preservatives (eg, sodium nitrate, sorbates).
2. Coatings, films, and related substances (eg, polyacrylamide).
3. Special dietary and nutritional additives (eg, vitamins).
4. Anticaking agents (eg, sodium stearate, silicon dioxide).
5. Flavoring agents and related substances (eg, vanillin).
6. Gums, chewing gum bases, and related substances (eg, xanthan gum).
7. Other specific usage additives (eg, calcium lignosulfonate).
8. Multipurpose additives (eg, glycine).

Secondary direct food additives permitted in food for human consumption are divided into four different types in the CFR, as follows:

1. Polymer substances for food treatment (eg, acrylate, acrylamide resins).
2. Enzyme preparations and microorganisms (eg, rennet, amylase).
3. Solvents, lubricants, release agents, and related substances (eg, hexane).
4. Specific usage additives (eg, boiler water additives, defoaming agents).

Indirect food additives included in the CFR are divided into eight categories, as follows:

1. Components of adhesives (eg, calcium ethyl acetoacetate 1,4-butanediol modified with adipic acid).
2. Components of coatings (eg, acrylate ester copolymer coatings and poly(vinyl fluoride) resins).
3. As components of paper and paperboard (eg, slimicides, sodium nitrate/urea complex, and alkyl ketene dimers).
4. As basic components of single- and repeated use food contact surfaces (eg, cellophane, ethylene–acrylic acid copolymers, isobutylene copolymers and nylon resins).

5. As components of articles intended for repeated use (eg, ultrafiltration membranes and textiles and textile fibers).
6. Controlling growth of microorganisms (eg, sanitizing solutions).
7. Antioxidants and stabilizers (eg, octyltin stabilizers in vinyl chloride plastics).
8. Certain adjuvants and production aids (eg, animal glue, hydrogenated castor oil, synthetic fatty alcohols, and petrolatum).

GOVERNMENT REGULATIONS

The application of food additives is highly regulated worldwide although regulatory philosophy, the approval of specific product, and the level of enforcement differ from country to country. The United States, western Europe, and Japan are the largest consumers of food additives.

According to the legal definition, food additives that are subject to the 1958 Food Additives Amendment to the Food, Drug & Cosmetic (FD&C) Act of 1938 include "any substance the intended use of which results or may reasonably be expected to result directly or indirectly in its becoming a component or otherwise affecting the characteristics of any food". This definition includes any substance used in the production, processing, treatment, packaging, transportation or storage of food. If a substance is added to a food for a specific purpose it is referred to as a *direct additive*.

Indirect food additives are those that become part of the food in trace amounts due to its packaging, storage or other handling.

For regulatory purposes, all food additives fall into one of three categories:

1. Generally Recognized As Safe (GRAS) substances.
2. Prior sanctioned substances.
3. Regulated—direct–indirect additives.

In deciding whether an additive should be approved, the Food and Drug Administration (FDA) considers the composition and properties of the substance, the amount likely to be consumed, its probable long-term effects, and various other safety factors.

All *color additives* are subject to the Color Additive Amendment of 1960. Colors permitted for use in foods are classified either as certified or exempt from certification. Certified colors are manmade, with each batch being tested by the manufacturer and the FDA (certified) to ensure that they meet strict specifications for purity. Color additives that are exempt from certification include pigments derived from natural sources. However, color exempt from certification also must meet certain legal criteria for specifications and purity.

Flavor substances are regulated somewhat differently, and the rules are less restrictive. However, the use of aroma chemicals as flavor ingredients is regulated also under laws that may differ from country to country.

Under FDA, USDA, and BATF regulations, the ingredients of a food or beverage must be stated on the product label in decreasing order of predominance.

The FDAs Food Additives Amendment also contains what is known as the *Delaney Clause*, which mandates the FDA to ban any food additive found to cause cancer in humans or animals, regardless of dose level or intended use.

SWEETENERS POLYOLS AND BULKING AGENTS

Classification of Sweeteners

Sweeteners are used in formulated foods for many functional reasons as well as to impart sweetness. They render certain foods palatable and mask bitterness; add flavor, body, bulk, and texture; change the freezing point and control crystallization; control viscosity, which contributes to body and texture; and prevent spoilage. Certain sweeteners bind the moisture in food that is required by detrimental microorganisms. Alternatively, some sweeteners can serve as food for fermenting organisms that produce acids that preserve the food, thus extending shelf life by retaining moisture. Sweeteners may be classified in a variety of ways:

- Nutritive or Nonnutritive. Materials either are metabolized and provide calories or are not metabolized and thus are noncaloric.
- Natural or Synthetic. Commercial products that are modifications of a natural product (eg, honey or crystalline fructose, is considered natural; saccharin is a synthetic compound).
- Regular or Low Calorie/Dietetic/High Intensity. Although two sweeteners may have the same number of calories per gram, one may be considered low calorie or high intensity if less material is used for equivalent sweetness.
- As Foods. For example, fruit juice concentrates can impart substantial sweetness.

Sucrose

Sucrose, $C_{12}H_{22}O_{11}$, commonly known as table sugar (or refined sugar), is the standard against which all sweeteners are measured in terms of quality of taste and taste profile. It is consumed in the greatest volume of all sweeteners. However, sucrose, high fructose corn syrup, and other natural sweeteners, such as molasses, honey, maple syrup, and lactose sweeteners, are food ingredients rather than additives and are not covered in this section.

Polyols

Polyols or sugar alcohols are a group of low digestible carbohydrates derived from the hydrogenation of their sugar or syrup source (eg, lactitol from lactose). These unique sweeteners taste like sugar but have special advantages. They are a group of sweeteners that provide the bulk of sugars, without as many calories as sugar. Polyols are important sugar substitutes and are utilized where their different sensory, special dietary, and functional properties make them feasible. Polyols are obtained from their parent sugars by catalytic hydrogenation. Food products sweetened with polyols and containing no sucrose can be labeled as "sugarless", "sugar free", or "no sugar".

High Intensity Sweeteners

High intensity sweeteners once used mainly for dietetic purposes, are now used as food additives in a wide variety of products. Termed *"high intensity"* because they are many-folds sweeter than sucrose and closely mimic its sweetness profile. Because of the very low use levels, however, high intensity sweeteners cannot perform other key auxiliary functions in food and often must be used in conjunction with other additives such as low-calorie bulking agents.

Bulking Agents

Bulking agents are substances that add bulk to food products while contributing fewer calories than the ingredients they replace. In applications where sugar is replaced by a high intensity sweetener, bulking agents make up for the lost volume, and ideally provide some or all of the functional properties of the sugar. The most important properties of a bulking agent are reduced calorie content through limited digestibility, solubility, and minimal side effects.

FLAVOR, TASTE, APPEARANCE, AND TEXTURE ENHANCERS

Acidulants

Acidulants complement fruit and other flavors in carbonated beverages preserves, fruit drinks, and desserts. Their ability to lower pH makes them useful as preservatives, and they are used in chemical leavening agents, as gelling agents, defoaming agents, emulsifiers, and in the production of cultured dairy products. In the choice of an appropriate acid, the effect of the acid on the overall flavor system, the rate and degree of solubility of the acid, its hygroscopicity, and its strength must all be considered. Some examples are: acetic acid, adipic acid, citric acid, fumaric acid, glucono-delta-lactone, lactic acid, malic acid, phosphoric acid, tartaric acid and vinegar.

Flavors and Flavor Enhancers

Flavorings are concentrated preparations used to impart a specific aroma to food or beverage. Flavoring ingredients are the most numerous single group of intentional additives utilized by the food industries. Flavors should not be viewed as a single homogeneous class of food additives, but as a composite of closely interrelated and somewhat overlapping sectors with differentiated characteristics, as follows: essential oils and natural extracts, aroma chemicals, flavor compositions (Table 1).

Table 1. Commercial Flavor Compositions

Type of flavor	Classification	Manufacturing process	Raw materials	Product form
compounded flavors	natural or synthetic	blending, mixing	essential oils, natural extracts, fruit juice concentrates, aroma chemicals	liquid, spray-dried, encapsulated
natural extracts	natural	extraction, enzymatic	food substrates (eg, plants, fish, meat, etc)	liquid, paste
reaction flavors (thermally processed)	natural	heating/pressure cooking	amino acids and sugars, hydrolyzed proteins	paste, powder
enzymatically modified flavors	natural	enzymatic/microbial reaction	food substrates (eg, cheese)	paste, powder

Colors

Colors are used in foods to improve appearance and thereby influence the perception of texture and taste. When synthetic colors were introduced in the late nineteenth century, they were immediately adopted by the food industry.

Food colors may be added to food to:

- Give attractive appearance to foods that would otherwise look unattractive or unappetizing (eg, colorless gelatin-based jelly) and thus enhance enjoyment.
- Restore the original appearance of the food where the natural colors have been destroyed by heat processing and with subsequent storage.
- Intensify colors naturally occurring in foods where the color is weaker than that which the consumer associates with a food of that type of flavor (eg, fruit yogurts, sauces, soft drinks).
- Ensure uniformity of color due to natural variations in color intensity, eg, fruits obtained at different times during the season, thereby assuring uniformity in appearance and acceptability.
- Help protect flavor and light-sensitive vitamins during shelf storage by a sunscreen effect.
- Help preserve the identity or character by which foods are recognized, ie, product identification.
- Serve as visual indication of quality—thus, in addition to enhancing the acceptability of foods, colors aid in food manufacture, storage, and quality control.

Certified food colors can be divided into dyes and lakes. Color regulators specify a minimum of 85% pure dye for primary colors, but most dye lots contain from 90 to 93% pure dye. FD&C dyes are also used in the production of lakes, which are pigments prepared by combining a certified dye with an insoluble alumina hydrate substratum. Lakes are both water- and oil-insoluble and impart color through dispersion in food.

Noncertified colors (sometimes called "natural" colors) can be from either natural origins (primary source), or produced synthetically.

Thickeners and Stabilizers

Thickeners and stabilizers (also called hydrocolloids, gums, or water-soluble polymers) provide a number of useful functions to food products. The technical base for these effects results from the ability of these materials to modify the physical properties of water. Some of these are: natural hydrocolloids (unmodified starch), modified starch, agar, alginates, locust bean gum, guar gum, gum arabic, carrageenan, pectin, casein, gelatin, semisynthetic hydrocolloids (cellulose derivatives), xanthan gum, and gellan gum.

Emulsifiers

Emulsifiers, or surfactants, are additives that allow normally immiscible liquids, such as oil and water, to form a stable mixture. Emulsifiers possess both hydrophilic and lipophilic groups within the same molecule; the ratio of hydrophilic to lipophilic groups, known as the HLB value, is a characteristic indicator for emulsifiers, which allows assessment of their action and performance. Emulsifiers are widely used in foods in order to perform one or more of the following functions:

- Increase stability and prevent phase separation in food emulsions (eg, mayonnaise, salad dressings).
- Improve the shelf life of flavors and retard the onset of rancidity in fats and oils containing food emulsions.
- Improve texture, reduce crumb firmness, and complex with starches (baked goods).

Bleaching, Maturing, and Dough-Conditioning Agents

Because some chemicals serve as both bleaching and maturing agents, and other are referred to as dough-conditioning agents or bread improvers, it is perhaps desirable to consider all of them under one heading. Bleaching agents are used in the production of certain cheeses, processed fruits, crude fats and oils and meat products to neutralize color that may be present naturally. Bleaching plays special importance in the flour milling and baking industries.

Firming Agents

Fruits and vegetables contain pectin components that are relatively insoluble and form a firm gel around the fibrous tissues of the fruit and prevent its collapse. Addition of calcium salts causes the formation of calcium pectate gel, which supports the tissues and affords protection against softening during processing.

Glazing and Polishing Agents

These agents are used on coated confections to give luster to the otherwise dull coating.

PRESERVATIVES

Antimicrobials

The choice of a preservative takes into consideration the product to be preserved, the type of spoilage organism endemic to it, the pH of the product, period of shelf life, and ease of application. No one preservative can be used in every product to control all organisms, and therefore combinations are often used.

Preservatives may be divided into two main groups: (a) antioxidants and (b) antimicrobials.

- *Antimicrobial agents* are capable of retarding or preventing growth of microorganisms such as yeast, bacteria, molds, or fungi, and subsequent spoilage of foods. The principal mechanisms are reduced water availability and increased acidity. Sometimes these additives also preserve other important food characteristics such as flavor, color, texture, and nutritional value.
- *Antioxidants* are food additives that retard atmospheric oxidation and its degrading effect, thus extending the shelf life of foods.

The primary food additives used for this function are as follows: Benzoic acid, sorbic acid, propionic acid, parabens, organic acids, sulfur dioxide, nitrates and nitrites, and natural alternatives.

Antioxidants

Antioxidants are food additives that retard atmospheric oxidation and its degrading effects, thus extending the shelf life of food. They are also used to scavenge oxygen and prevent the discoloration of cut or bruised fruits and vegetables. Changes in flavor accompany changes in color. One of the important methods of controlling browning of fruits is by using water-soluble antioxidants. They are also added to the packaging materials of some cereals.

pH Adjusting Agents

A large group of chemical additives that are widely used in foods might be considered under the broad heading of pH adjusting agents. Other terms that describe these chemicals include acids, alkalis, buffers, and neutralizers. These chemicals are used in most segments of the food processing industries.

Quite often, the pH may be difficult to adjust or to maintain after adjustment. Stability of pH can be accomplished by the addition of buffering agents. Within limits buffers can effectively maintain the desired pH even when additional acid or alkali may be added.

Fumigants for Insect and Pest Control

These compounds are volatile substances used for controlling insects or pests.

Gases

Gases provide three basic functions as food ingredients: preservation, carbonation, and aeration.

Sequestering Agents

These agents, also called chelates, combine with polyvalent metal ions to form a soluble metal complex to improve the quality and stability of products as free metallic ions promote oxidation of food. They are used in various aspects of food production and processing chiefly to obviate undesirable properties of metal ions without the necessity of precipitating or removing these ions from solutions.

PROCESSING AIDES

Anticaking Agents

Dry food products that contain hygroscopic substances require the addition of an anticaking agent. These additives must be insoluble in water, have the capacity of absorbing excess moisture, or by coating particles making them water repellent.

Antifoaming (Defoaming) Agents

These agents are substances used to reduce foaming caused by proteins or gases that may be interfere with processing.

Enzymes

Enzymes are biological catalysts that make possible or greatly speed up chemical reaction by combining with the reacting chemicals and bringing them into the proper configuration for the reaction to take place. They are not affected by the reaction. All enzymes are proteins and become inactive at temperatures above ~40°C or in unfavorable conditions of acidity or alkalinity. Some of the specific functions food enzymes perform are to:

- Speed up reactions.
- Reduce viscosity.
- Improve extractions.
- Carry out bioconversions.
- Enhance separations.
- Develop functionality.
- Create/intensify flavor.
- Synthesize chemicals.

Enzymes are produced from animal tissues (eg, pancreatin, tripsin, lipase), plant tissues (eg, ficin, bromelin), and most frequently by microorganisms (eg, pectic, or starch enzymes). Microbial production from a variety of species of molds, yeast, and bacteria is increasingly becoming the predominant source of enzymes.

Humectants

In certain foods, it is necessary to control the amount of water that enters or exits the product. It is for this purpose that humectants are employed.

Leavening Agents

Many bakery products rely on chemical leavening agents to produce the gas that gives them volume. Bicarbonates produce carbon dioxide in the presence of heat and moisture. When used alone, sodium bicarbonate reacts to give products a bitter, soapy flavor. Thus it is always combined with a leavening acid.

Leavening acids are classified according to the rate at which they release carbon dioxide from sodium bicarbonate. Some acids begin producing carbon dioxide as soon as they come into contact with water; others do not begin to react unless heat is present as well. The type of leavening acid needed depends on the product.

Lubricants and Release Agents

These agents are substances added to food processing equipment to prevent food ingredients and finished products from sticking to them. They are also used to prevent pieces or confection from adhering to each other.

Manufacturing Aids

These aids including catalysts, filter aids, clarifying and clouding agents are used to improve the appearance or performance of food products.

Solvents

Solvents are generally used to either extract particular compounds, or to carry additives into a food system.

Water-Correcting Agents

Water used in the beverage industries is often corrected to a uniform mineral salt content that corresponds to water known to give the most satisfactory final product. Some of the chemicals in addition to standardizing the salt contents also control the acidity, thus providing uniform conditions for yeast fermentation in the breweries.

NUTRIENTS

Dietary Fibers

Dietary fibers is a broad term that encompasses the indigestible carbohydrate and carbohydrate-like components of foods that are found predominantly in plant cell walls. Those fibers that have colligative properties, such as gums, are referred to as soluble fibers. They are often used to provide viscosity and texture in processed foods, and have been linked to lowered serum cholesterol levels. Insoluble fibers, such as cereal brans and specialty flour ingredients, tend to cause a laxative effect when con-

sumed in large quantities. Dietary fiber has become an important food additive owing to the link between high fiber intake and the lowering of serum cholesterol, the prevention of cancer, and the avoidance of digestive tract disease.

Vitamins

Vitamins are nutritive substances required for normal growth and maintenance of life. They play an essential role in regulating metabolism, converting fat and carbohydrates into energy, and forming tissues and bones. Vitamins are typically divided into two groups: fat soluble and water-soluble vitamins. The fat-soluble group usually measured in International Units (IU). The water-soluble group usually measured in units of weight.

Thirteen vitamins are recognized as essential for human health, and deficiency diseases occur if any one is lacking. Because the human body cannot synthesize most vitamins, they must be added to the diet. Most vitamins are currently consumed as pharmaceutical preparations, or over-the-counter vitamin supplements.

In addition, vitamins may be used as functional ingredients in foods. Vitamin E (tocopherol) and vitamin C (ascorbic acid) protect foods by serving as antioxidants to inhibit the destructive effects of oxygen.

A. T. Brannen, *Food Additives*, Marcel Dekker, New York, 1980.

Code of Federal Regulations, *Title 21—Food and Drugs. Subchapter B—Food For Human Consumption.* Parts 100–199. U.S. Government Printing Office, Washington, D.C., 2003.

R. J. Lewis, *Food Additives Handbook*, Van Nostrand Reinhold, New York, 1989.

G. Reineccious, *Source Book of Flavors*, 2nd ed., Chapman & Hall, New York, 1994.

Laszlo P. Somogyi
Food Industry Consultant

FOOD PROCESSING

Food processing operations can be grouped into three categories: preparation, assembly, and preservation of foods. Preparation processes are used to convert raw plant or animal tissue into edible ingredients. Assembly processes are used to combine and form ingredients into consumer products. Preservation processes are used to prevent the spoilage of foods. Five sources of food spoilage must be addressed in order to deliver fresh, safe foods and ingredients: microbial contamination, including viruses; enzyme activity from enzymes in the food itself and from external enzymes such as from microbial activity; chemical deterioration such as oxidation and non-enzymatic browning; contamination from animals, insects, and parasites; and losses owing to mechanical damage such as bruising. The processing of foods is regulated by federal food laws that cover good manufacturing practices, nutritional content of foods, and food and ingredient standards.

Plants and animals are the primary sources of food. Genetic engineering (qv), as well as conventional breeding methods, are being used to improve the yield, color, flavor, texture, nutrient content, and resistance to diseases, insect loss, and climatic stress. However, product quality can vary owing to weather, soil, growing practices, harvest methods, and post-harvest handling. Thus food processing unit operations must be designed to accept raw materials having a wide range of qualities. In addition, provision often must be made for profitable use of by-products and waste streams.

REGULATIONS

Food processing operations are usually regulated and mandated by national and international laws, regulations, and standards that define nutritional requirements and the use of certain ingredients, process conditions, and even the composition of some products. Food safety and toxicology regulations include standards for toxic and carcinogenic substances in foods, pathogenic microbes, and physical hazards.

THEORETICAL BASIS

Food preservation theory has yielded mathematical models for predicting the conditions needed to produce and store foods which are safe to consume and have a maximum quality. While safety is largely a function of microbial growth and quality loss a function of biochemical, chemical, and mechanical changes, factors affecting both safety and quality are interrelated and can rarely be considered in isolation.

Spore-forming bacteria are among the most pathogenic and heat resistant organisms known. Research since the early 1920s has been directed toward the development of mathematical models to predict the rate of heat inactivation of *Clostridium botulinum* spores as a function of heating time and temperature, and the composition of the suspending media. Thermal inactivation of spores appears to follow first-order kinetics. Thus, if the inactivation rate of a spore population is known at several temperatures, and the rate of heating of the slowest point in a package can be determined or calculated from heat-transfer theory, then the time needed to sterilize the package can be calculated for any external heating condition. Spore germination can be inhibited by antibiotic substances produced by several types of lactic acid-producing bacteria. These substances, called bacteriosins, are finding increased use in preventing the growth of Gram-positive bacteria and thus must be taken into consideration when developing theoretical models for microbial inactivation.

Enzyme inactivation by heat has also been subjected to mathematical modeling in a manner similar to microbial inactivation. Changes in foods resulting from heating or storage, such as the loss of pigments, flavors, and vitamins, can be approximated by first-order kinetics. Deterioration mechanisms have been studied to allow the prediction of shelf life, particularly the shelf life of foods

susceptible to non-enzymatic browning and lipid oxidation. Thus, the coupling of reaction kinetics, which mathematically describe quality loss via chemical, biochemical or microbial degradation, with process or storage conditions, such as temperature and moisture content, plays an integral role in the successful production of high quality, safe to eat foods.

PRESERVATION OF FOODS

Preservation operations to reduce or eliminate food spoilage can be grouped into five categories: heat treatments; storage near or below the freezing point of water; dehydration and control of water activity; chemical preservation; and use of mechanical operations such as washing, peeling, filtration, centrifugation, grinding, ultrahigh hydrostatic pressure, and most importantly, packaging. Most food preservation technologies use two or more preservation operations because virtually all processed foods are packaged.

Short-Term Storage

Short-term storage operations include packaging followed by refrigeration, mild heat treatment, or a combination of each. Chemical preservatives may be used to further increase storage time.

Long-Term Storage

Inactivation of microbes and enzymes in foods and food ingredients is necessary to ensure a long useful packaged shelf-life. This can be achieved by using one or more preservation operations such as applying heat; using storage temperatures below $-18°C$; drying to water activities below 0.65, and by adding chemical preservatives such as organic acids (acetic or lactic) or table salt.

Thermal Preservation Technology

The heat preservation of foods can be accomplished by various combinations of heating times and temperatures depending on the number and type of heat-resistant microorganisms and/or spores present, composition of the food, physical characteristics of the food and package, and desired storage time. Foods that are free of pathogenic and spoilage microbes are termed commercially sterile and are produced using temperatures above $100°C$, the most common being $121°C$. Mild heat treatments used to inactivate viruses, vegetative pathogenic bacteria, and certain yeasts (qv) and molds, are referred to as pasteurization operations and are typically carried out below $100°C$.

Rapid heating and cooling of liquid foods, such as milk, can be performed in a heat exchanger and is known as high temperature short time (HTST) processing. HTST processing can yield heat-preserved foods of superior quality because heat-induced flavor, color, and nutrient losses are minimized.

Water Activity

The rates of chemical reactions as well as microbial and enzyme activities related to food deterioration have been linked to the activity of water (qv) in food. Water activity, at any selected temperature and moisture content can be measured by determining the equilibrium relative humidity surrounding the food. Thus water activity is different from, but related to the moisture content of the food as measured by standard moisture tests. Microbial growth and many deterioration reactions may be prevented by reducing water activity.

Freezing Preservation

The rate of loss of color, flavor, texture, and nutrients, the growth of microbes, and the activity of enzymes and other life forms are all functions of temperature. Thus lower storage temperatures prolong the useful life of foods.

Equipment for food freezing is designed to maximize the rate at which foods are cooled to $-18°C$ to ensure as brief a time as possible in the temperature zone of maximum ice crystal formation. This rapid cooling favors the formation of small ice crystals, which minimize the disruption of cells and may reduce the effects of solute concentration damage. Rapid freezing requires equipment that can deliver large temperature differences and/or high heat-transfer rates.

Many formulated foods and certain animal products tolerate freezing and thawing well because their structures can accommodate ice crystallization, movement of water, and related changes in solute concentrations. Starches can be modified for freeze–thaw stability against gel breakdown through several freeze–thaw cycles. By contrast, most fruits and vegetables lose significant structural quality on freezing and during storage because their rigid cell structures fail to accommodate to ice crystal formation. Frozen food storage equipment must be designed to minimize temperature fluctuations. Most frozen foods have a useful storage life of 1 year at $-18°C$. However, foods high in fat such as sausage products may become rancid after 2 weeks in frozen storage if not protected from oxygen by special packaging and antioxidants.

Dehydration Processing

Dehydration is one of the oldest means of preserving food. Microbes generally do not grow below a minimum water activity of 0.65.

Foods dried to water activities in the range of 0.65–0.85 are often referred to as intermediate moisture foods. These partially dried foods tend to be soft and rehydrate easily. The remaining water acts as a plasticizer. Because molds and yeast may be able to grow in these partially dried products, they must be preserved by heat, vacuum, or modified atmosphere packaging, refrigeration, or chemical means.

Foods high in sucrose, protein, or starch (qv) tend to bind water less firmly and must be dried to a low moisture content to obtain microbial stability. Fresh plant and animal tissue when dried to a water activity much below 0.97 show irreversible disruption of metabolic processes. Products susceptible to oxidation and oxidative rancidity such as potato chips, can be treated with antioxidants and inert gas packed to minimize exposure to oxygen.

Low temperature storage can further reduce the rate of chemical deterioration.

Chemical Preservation

Food additives (qv) can enhance the effectiveness of food preservation by heat, refrigeration, and drying methods. The addition of a food-grade acid to a low acid food to shift the pH to a value below 4.5 allows heat preservation at or below $100°C$ instead of in the range of $121°C$. Antioxidants such as butylated hydroxyanisole (BHA) can be added to potato chips to reduce the need for expensive oxygen-impermeable flexible packaging. Sulfur dioxide is used in wine (qv) and in dry fruit and vegetable products to preserve colors and flavors and prevent nonenzymatic browning.

Food can be preserved by fermentation (qv) using selected strains of yeast, lactic acid producing bacteria, or molds. The production of ethanol (qv), lactic and other organic acids, and antimicrobial agents in the food, along with the removal of fermentable sugars, can yield a product having an extended shelf-life.

Lactic acid-producing bacteria associated with fermented dairy products have been found to produce antibiotic-like compounds called bacteriocins. Concentrations of these natural antibiotics can be added to refrigerated foods in the form of an extract of the fermentation process to help prevent microbial spoilage. Other natural antibiotics are produced by *Penicillium roqueforti*, the mold associated with Roquefort and blue cheese, and by *Propionibacterium* sp., which produce propionic acid and are associated with Swiss-type cheeses.

Ionizing radiation is considered to be a chemical preservation method and applications must be cleared by the Food and Drug Administration for use, not only on a product-by-product basis, but also on a dose basis.

Other Technologies

Several technologies for the preservation of foods using a minimum of heat are being explored. The application of ultrahigh pressure, in the range of 200–1000 MPa, to the preservation of foods has been commercialized. Capacitance discharge has also been investigated as a means to pasteurize or commercially sterilize foods, which can pass between plates sufficiently close together to allow an electric field of ~25,000 V/cm. Very high intensity pulses of visible light can be used to pasteurize fruit juices using a minimum of heating in a manner that appears to be similar to capacitance discharge.

PROCESS OPTIMIZATION

Food processing operations can be optimized according to the principles used for other chemical processes if the composition, thermophysical properties, and structure of the food is known. However, the complex chemical composition and physical structures of most foods can make process optimization difficult.

A common approach to maximizing overall nutrient quality of a food is to use a chemical marker with well documented kinetic properties and methods for detection. Ascorbic acid, $C_6H_8O_6$, or thiamine can often be used as indicators of process conditions.

COMPUTER INTEGRATED MANUFACTURING, INSTRUMENTATION, AND CONTROLS

The use of computer integrated manufacturing (CIM) by food processing firms has increased significantly in the past several years. Thermal processing controls have been developed to the point where time and temperature process deviations can be corrected on line. Freezer, dryer, and vacuum evaporator operating conditions can be controlled and optimized using systems available to the non-food process industries. This increase in CIM has lead to improvements in food quality and safety, increased processing speed and efficiencies, and reduced downtime and waste.

P. M. Davidson and A. L. Branen, eds., *Antimicrobials in Foods*, 2nd ed., Marcel Dekker, Inc., New York, 1993.

F. A. Paine, ed., *Packaging User's Handbook*, Van Nostrand Reinhold Co., New York, 1991.

M. D. Pierson and D. A. Corlett, Jr., eds., *HACCP-Principles and Applications*, Van Nostrand Reinhold Co., New York, 1992.

A. Teixeira, in D. R. Heldman and D. B. Lund, eds., *Handbook of Food Engineering*, Marcel Dekker, Inc., New York, 1992, Chapt. 11.

BRIAN E. FARKAS
North Carolina State University

DANIEL F. FARKAS
Oregon State University

FORENSIC CHEMISTRY

Forensic chemistry can be defined as the application of chemistry to the law. In American jurisprudence, courts, and judges are established to make factual determinations of matters brought before them. The fact-finding of the courts must often grapple with complex scientific issues and the legal system has a particular way to deal with these technical and scientific matters. To testify as an expert witness in a particular field or area of endeavor, the individual must qualify as an expert in a specific area, ie, have special knowledge, skill, training experience, or education.

Forensic science is an applied science having a focus on practical scientific issues that come up during criminal investigations or at trial. Some components are unique to the field because it is conducted within the legal arena.

PHYSICAL EVIDENCE

Forensic scientists work with evidence, ie, "data presented to a court or jury in proof of the facts in issue and which may include the testimony of witnesses, records, documents or objects". Physical evidence is real or tangible and can literally include almost anything, eg, the transient scent of perfume on the clothing of an assault victim; the metabolite of a drug detected in the urine of an individual in a driving-under-the-influence-of-drugs case; the scene of an explosion; or bullets removed from a murder victim's body.

Examination of physical evidence provides two subtle and different types of conclusion. All members of a class or group have identical characteristics. Types of physical evidence that exhibit class characteristics are paint (qv), glass (qv), fibers (qv), fabric, building material, etc. This type of physical evidence is said to be identified. The best that chemical and physical examinations can ever do is to place items into groups of similarly manufactured items. It is not possible to differentiate one item of evidence as being uniquely distinguishable from another.

Some types of physical evidence, because of the manner in which the material is made, are unique; such evidence can be individualized. Examination can show an item of individualized evidence is unique and comes from one, and only one source. The classic example is fingerprints. Other categories of evidence exhibiting individualization are handwriting, markings on bullets fired from the same gun, and broken pieces of glass or plastic that can be physically fit together again, and forensic deoxyribonuclic acid (DNA) evidence.

Physical evidence serves three purposes. In some cases, it is used to prove a component or element of a crime.

Physical evidence is also used is to develop associative evidence in a case. Physical evidence may help to prove a victim or suspect was at a specific location, or that the two came in contact with one another.

Most of the forensic science or crime laboratories located in North America are associated with law enforcement agencies, medical examiner–coroner departments, or prosecutors' offices. There are a small number of private laboratories that provide forensic testing in specialized areas such a DNA profiling, forensic toxicology and fire debris analysis. There are a large number of independent consultants, also. Laboratories exist at the municipal, county, state, and federal levels of government. There are ~300 government-operated forensic science laboratories in the United States.

Forensic science laboratories are generally divided into separate specialty areas. These typically include forensic toxicology, solid-dose drug testing, forensic biology, trace evidence analysis, firearms and tool mark examination, questioned documents examination, and latent fingerprint examination. Laboratories principally employ chemists, biochemists, and biologists at various degree levels.

The bulk of the scientific testing in crime laboratories involves the analysis and characterization of either synthetic or biochemical organic substances or both. Additionally, there are a number of evidence categories classified as inorganic.

FORENSIC TESTING

Toxicology

Psychoactive substances, illicit and ethical (licit) drugs and alcohol (ethanol), are the greatest source of physical evidence analyzed in state and local crime laboratories. Drug testing falls into two categories: solid dose samples and toxicology (qv) related cases, eg, blood, urine, or tissue specimens in postmortem cases or cases involving driving under the influence of alcohol or drugs, as well as workplace or employee drug testing.

Blood and urine are most often analyzed for alcohol by headspace gas chromatography (gc) (qv) using an internal standard, eg, 1-propanol.

Breath alcohol testing is accomplished by a number of techniques. The oldest reliable procedure involves bubbling a measured volume of deep-lung air containing alcohol through an acidic solution of potassium dichromate, $K_2Cr_2O_7$. Newer instruments rely on infrared spectroscopy to measure the blood alcohol concentration in breath.

Driving under the influence of alcohol cases are complicated because people sometimes consume alcohol with other substances. The most common illicit substances taken with alcohol are marijuana and cocaine. Forensic toxicology laboratories having large caseloads rely on immunoassay (qv) techniques to screen specimens. Immunoassay technology involves the manufacture of antibodies that are specific to particular drugs or to a class of drugs.

There are several immunological techniques in use. In these tests, antibodies combine with the drug or drug metabolites present in blood or urine, in competition with a labeled drug or metabolite that is in the reaction mixture. Radioimmunoassay (RIA) uses reagents tagged with radioactive isotopes such as ^{125}I; enzyme multiplied immunological technique (EMIT) employs an enzyme label and fluorescence polarization immunoassay (FPIA) uses drugs or drug metabolites labeled with fluorescein.

Thin-layer chromatography (tlc) is frequently used. One drawback to tlc, however, is that the technique has a high detection limit and low levels of drugs may be missed. Another drawback is that a single tlc separation does not suffice for an identification of an unknown drug or drug metabolite.

Gas chromatography and gas chromatography–mass spectroscopy (gc/ms) are the most common analytical procedures used in modern forensic toxicology laboratories (see ANALYTICAL METHODS, HYPHENATED INSTRUMENTS). Drugs are separated from their biological matrices, ie, blood, urine, and liver, by liquid–liquid or solid-phase extraction (qv) using the distribution of the suspect drug between an acid or alkaline aqueous solution and an immiscible organic phase.

Solid-Dose Narcotics and Dangerous Drugs

Solid-dose drug testing differs from forensic toxicology in that the solid form of the drug is tested, rather than a biological specimen containing the drug and its metabolite. The typical drugs of abuse in North America are heroin; cocaine, ie, free-base, crack, and the HCl salt; marijuana; hashish, a concentrated form of marijuana; amphetamine; methamphetamine; phencyclidine; and LSD.

Trace Evidence

Trace evidence refers to minute, sometimes microscopic material found during the examination of a crime scene or a victim's or suspect's clothing. Trace evidence often helps police investigators develop connections between suspect and victim and the crime scene. The challenge to the forensic scientist is to locate, collect, preserve, and characterize the trace evidence.

Trace evidence in criminal investigations may consist of hairs; both natural and synthetic fibers (qv); fabrics; glass (qv); plastics; soil; plant material; building material such as cement (qv), paint (qv), stucco, and wood (qv), flammable fluid residues, eg, in arson investigations; and explosive residues, eg, from bombings. Perhaps the simplest examination done is the physical match. Other examinations result only in demonstrating class characteristics.

Microscopy (qv) plays a key role in examining trace evidence owing to the small size of the evidence and a desire to use nondestructive testing (qv) techniques whenever possible. Polarizing light microscopy is a method of choice for man-made textile fibers and crystalline materials such as minerals. Other microscopic procedures involving ir, vis, and ultraviolet (uv) spectroscopy (qv) also are used to examine many types of trace evidence.

More traditional analytical techniques also are used. Capillary column gc is the method of choice for characterizing flammable fluid residues in arson cases. Scanning electron microscopy (sem) and energy dispersive X-ray analysis (edx) are used frequently in gunshot residue examination and to characterize evidence of an inorganic origin. Pattern recognition examinations are important in footwear and tire impression cases. Lasers (qv) and other high intensity or alternative light sources are useful in crime laboratories to visualize latent fingerprints, seminal fluid stains, obliterated writings, and erasures, and to aid in specialized photographic work. Infrared and uv light sources are also used to view items of evidence.

Forensic Biology

Stains of blood and other body fluids can be powerful physical evidence in crimes against the person (eg, homicide and rape). Blood is mixture of cells (red blood cells and white blood cells), cell fragments (platelets), proteins (including a number of enzymes), inorganic salts, and water. The process for the forensic testing of suspected bloodstains proceeds as follows: (1) the suspected stain is identified as blood; (2) the species of animal from which the blood came is determined; and (3) DNA is extracted from the stain and profiled.

Chemical tests are relied upon for the identification of bloodstains. Presumptive tests are first applied to the suspected stain. Presumptive tests have low detection limits but lack complete specificity; they are also simple, quick, and economical so that large numbers of suspect stains can be rapidly screened. The presumptive tests for blood make use of the peroxidase activity of hemoglobin. The results of the presumptive tests are confirmed with a Takayama crystal test for the heme moiety in hemoglobin. The Takayama reagent contains reducing agents that remove oxygen from the reagent and reduce methemoglobin (the oxidized form of hemoglobin) to hemoglobin, hydroxide ions to hydrolyze the globlin proteins and pyridine, which complexes with heme to form pink, birefringent, leaf-like crystals.

Immunological tests are used to determine the species of animal the blood came from. Forensic biology laboratories typically use antisera against whole blood or against hemoglobin. The reactions of the antisera with blood proteins from the evidentiary blood stains can be carried out in a number of ways. Blood proteins are first extracted from the evidentiary stains with isotonic saline or a buffer. In the Ring Test, the blood stain extract is carefully layered on top of a layer of antiserum. In the Ochterlony double diffusion test, the blood proteins and the antibodies diffuse toward one another through a agarose gel; a band of precipitation forms within the gel. Recently, immunochromatography kits for the identification of human hemoglobin have become available. These kits make are of the same technology as home pregnancy test kits. The benefits of immunochromatography are numerous: sample preparation is minimal; no reagent preparation is required; and the shelf-life of the kits is several years.

Blood collected as evidence in criminal acts is usually dried and deposited on a variety of substrates. Sample size is usually on the order of a 2- or 3-mm diameter stain. Traditional typing involves ABO blood grouping, and characterizing stable polymorphic proteins or enzymes present in blood by means of electrophoresis (see BLOOD FRACTIONATION; ELECTROSEPARATIONS, ELECTROPHORESIS).

More recently, the forensic application of DNA testing has dramatically enhanced the ability to determine the source of a blood sample. Two procedures are in forensic use: restriction fragment length polymorphism (RFLP) and polymerase chain reaction (PCR).

R. H. Cravey and R. C. Baselt, *Introduction to Forensic Toxicology*, Biomedical Publications, Davis, Calif., 1981.

K. Inman and N. Rudin, *An Introduction to Forensic DNA Analysis*, CRC Press, Boca Raton, Fla., 1997.

S. H. James, ed., *Forensic Science: An Introduction to Scientific and Investigative Techniques*, CRC Press, Boca Raton, Fla., 2003.

R. Saferstein, *Criminalistics: An Introduction to Forensic Science*, 8th ed., Pearson Education, Inc., Upper Saddle River, N.J., 2004.

WALTER F. ROWE
The George Washington
University

FORMALDEHYDE

Formaldehyde, $H_2C{=}O$, is the first of the series of aliphatic aldehydes. Because of its relatively low cost, high purity, and variety of chemical reactions, formaldehyde has become one of the world's most important industrial and research chemicals.

Table 1. Properties of Monomeric Formaldehyde

Property	Value
density, g/cm^3	
boiling point at 101.3 kPa,^{a} °C	−19
melting point, °C	−118
heat of vaporization,^{b} ΔH_v at 19°C, kJ/mol^c	23.3
heat of formation, ΔH_f° at 25°C, kJ/mol^c	−115.9
heat of combustion, kJ/mol^c	563.5
heat of solution at 23°C kJ/mol^c	
in water	62
in methanol	62.8
flammability in air	

^{a}To convert kPa to mm Hg, multiply by 7.5.
^{b}At 164 to 251 K, $\Delta H_v = (27{,}384 + 14.56T − 0.1207T^2)$ J/mol^d (3).
^{c}To convert J to cal, divide by 4.184.

PHYSICAL PROPERTIES

At ordinary temperatures, pure formaldehyde is a colorless gas with a pungent, suffocating odor. Physical properties are summarized in Table 1.

Formaldehyde is produced and sold as water solutions containing variable amounts of methanol. These solutions are complex equilibrium mixtures of methylene glycol, $CH_2(OH)_2$, poly(oxymethylene glycols), and hemiformals of these glycols. Ultraviolet spectroscopic studies indicate that even in highly concentrated solutions the content of unhydrated HCHO is < 0.04 wt%.

Density and refractive index are nearly linear functions of formaldehyde and methanol concentration. The refractive index may be expressed by a simple approximation for solutions containing 30–50 wt% HCHO and 0–15 wt% CH_3OH: $n_D^{18} = 1.3295 + 0.00125 F + 0.0000113 M$.

Viscosities have been measured for representative commercial formaldehyde solutions. Over the ranges of 30–50 wt% HCHO, 0–12 wt% CH_3OH, and 25–40°C, viscosity in mPa·s(=cP) may be approximated by viscosity = $1.28 + 0.039 F + 0.05 M − 0.024 t$.

In methanol–formaldehyde–water solutions, increasing the concentration of either methanol or formaldehyde reduces the volatility of the other. The flash point varies with composition, decreasing from 83 to 60°C as the formaldehyde and methanol concentrations increase.

Formaldehyde solutions exist as a mixture of oligomers, $HO(CH_2O)_nH$. Their distribution has been determined for 6–50 wt% HCHO solutions with low methanol using nmr and gas chromatographic techniques. The equilibrium constants appear to be nearly independent of temperature over the range of 30–65°C. Hence, methanolic solutions can be stored at relatively low temperatures without precipitation of polymer.

Commercial formaldehyde–alcohol solutions are clear and remain stable above 16–21°C. They are readily obtained by dissolving highly concentrated formaldehyde in the desired alcohol.

CHEMICAL PROPERTIES

Formaldehyde is noted for its reactivity and its versatility as a chemical intermediate. It is used in the form of anhydrous monomer solutions, polymers, and derivatives.

Formaldehyde condenses with itself in an aldol-type reaction to yield lower hydroxy aldehydes, hydroxy ketones, and other hydroxy compounds; the reaction is autocatalytic and is favored by alkaline conditions.

An important synthetic process for forming a new carbon–carbon bond is the acid-catalyzed condensation of formaldehyde with olefins (Prins reaction).

A commercial process based on the Prins reaction is the synthesis of isoprene from isobutylene and formaldehyde through the intermediacy of 4,4-dimethyl-1,3-dioxane.

With acidic catalysts in the liquid phase, formaldehyde and alcohols give formals, eg, dimethoxymethane from methanol.

Monosubstituted acetylenes add formaldehyde in the presence of copper, silver, and mercury acetylide catalysts to give acetylenic alcohols (Reppe reaction).

Primary and secondary amines readily give alkylaminomethanols; the latter condense upon heating or under alkaline conditions to give substituted methyleneamines. With ammonia, the important industrial chemical, hexamine, is produced.

Mono- and dimethylol derivatives are made by reaction of formaldehyde with unsubstituted amides.

Formaldehyde reacts with syn gas (CO, H_2) to produce added value products, eg, glycolaldehyde and glycolic acid.

MANUFACTURE

Most of the world's commercial formaldehyde is manufactured from methanol and air either by a process using a silver catalyst or one using a metal oxide catalyst. Reactor feed to the former is on the methanol-rich side of a flammable mixture and virtually complete reaction of oxygen is obtained; conversely, feed to the metal oxide catalyst is lean in methanol and almost complete conversion of methanol is achieved.

Silver Catalyst Process

The silver-catalyzed reactions occur at essentially atmospheric pressure and 600 to 650°C and can be represented by two simultaneous reactions:

$$CH_3OH + 0.5 O_2 \longrightarrow HCHO + H_2O$$
$$\Delta H = -156 \, kJ (-37.28 \, kcal) \qquad (1)$$

$$CH_3OH \longrightarrow HCHO + H_2$$
$$\Delta H = +85 \, kJ (20.31 \, kcal) \qquad (2)$$

Between 50 and 60% of the formaldehyde is formed by the exothermic reaction and the remainder by endothermic reaction, with the net result of a reaction exotherm. Carbon monoxide and dioxide, methyl formate, and formic acid are by-products. In addition, there are also physical losses, liquid-phase reactions, and small quantities of methanol in the product, resulting in an overall plant yield of 86–90% (based on methanol).

The reaction occurs at essentially adiabatic conditions with a large temperature rise at the inlet surface of the catalyst. The predominant temperature control is thermal ballast in the form of excess methanol or steam, or both, which is in the feed.

Aqueous formaldehyde is corrosive to carbon steel, but formaldehyde in the vapor phase is not. All parts of the manufacturing equipment exposed to hot formaldehyde solutions must be a corrosion-resistant alloy such as type-316 stainless steel. Theoretically, the reactor and upstream equipment can be carbon steel, but in practice alloys are required in this part of the plant to protect the sensitive silver catalyst from metal contamination.

Metal Oxide Catalyst Process

In contrast to the silver process, all of the formaldehyde is made by the exothermic reaction at essentially atmospheric pressure and at 300–400°C. By proper temperature control, a methanol conversion greater than 99% can be maintained. By-products are carbon monoxide and dimethyl ether, in addition to small amounts of carbon dioxide and formic acid. Overall plant yields are 88–92%.

A typical metal oxide catalyst has an effective life of 12 to 18 months. Compared to silver, it is much more tolerant to trace contamination. It requires less frequent change-outs, but a longer down time for each replacement. In contrast to a silver-catalyst plant, there is little economic justification to incinerate the metal oxide plant tail gas for the purpose of generating steam.

The requirements for the material of construction are the same as for the silver catalyst process except the use of alloys to protect the catalyst is not as important.

Development of New Processes

There has been significant research activity to develop new processes for producing formaldehyde. One possible route is to make formaldehyde directly from methane by partial oxidation. Another possible route for producing formaldehyde is by the dehydrogenation of methanol, which would produce anhydrous or highly concentrated formaldehyde solutions. A third possible route is to produce formaldehyde from methylal that is produced from methanol and formaldehyde. However, no commercial units are known to exist.

ECONOMIC ASPECTS

Essentially all formaldehyde is produced as aqueous solutions containing 25–56 wt% HCHO and 0.5–15 wt% CH_3OH. All information on capacity, demand, and prices is reported on a 37 wt% formaldehyde basis. Commercial production is from methanol either using a silver or metal oxide catalyst. The major portion of this production (70–80%) is used captively.

Formaldehyde demand is in part related to the housing and construction sector. Decrease in housing impacts the demand for particleboard and plywood made of urea–formaldehyde and phenol–formaldehyde resins. These two types of construction materials consume more than one-third of the formaldehyde demand.

Formaldehyde capacity has always exceeded demand and this situation is expected to continue.

STORAGE AND HANDLING

As opposed to gaseous, pure formaldehyde, solutions of formaldehyde are unstable. Both formic acid (acidity) and paraformaldehyde (solids) concentrations increase with time and depend on temperature.

Paraformaldehyde solids can be minimized by storing formaldehyde solutions above a minimum temperature for less than a given time period. The addition of methanol as an inhibitor or of another chemical as a stabilizer allows storage at lower temperatures and/or for longer times. Stabilizers for formaldehyde solutions include hydroxypropylmethylcellulose, methyl- and ethylcelluloses, poly(vinyl alcohol)s, or isophthalobisguanamine at concentrations ranging from 10 to 1000 ppm.

Materials of construction preferred for storage vessels are 304-, 316-, and 347-type stainless steels or lined carbon steel.

HEALTH AND SAFETY FACTORS

Sources of human exposure to formaldehyde are engine exhaust, tobacco smoke, natural gas, fossil fuels, waste incineration, and oil refineries. It is found as a natural component in fruits, vegetables, meats, and fish and is a normal body metabolite. Facilities that manufacture or consume formaldehyde must control workers' exposure in accordance with the following workplace exposure limits in ppm: action level, 0.5; TWA, 0.75; STEL, 2 ACGIH TLV TWA, 1 ppm. In other environments such as residences, offices, and schools, levels may reach 0.1 ppm HCHO due to use of particle board and urea–formaldehyde foam insulation in construction.

Formaldehyde causes eye, upper respiratory tract, and skin irritation and is a skin sensitizer. Although sensory irritation, eg, eye irritation, has been reported at concentrations as low as 0.1 ppm in uncontrolled studies, significant eye/nose/throat irritation does not generally occur until concentrations of 1 ppm, based on controlled human chamber studies. Odor detection has commonly been reported to occur in the range of 0.06–0.5 ppm.

Formaldehyde is classified as a probable human carcinogen by the International Agency for Research on Cancer (IARC) and as a suspected human carcinogen by the American Conference of Governmental Industrial Hygienists (ACGIH).

Formaldehyde is not considered a teratogen and has not been reported to cause adverse reproductive effects. In vitro mutagenicity assays with formaldehyde have yielded positive responses, while in vivo assays have been largely negative.

USES

Formaldehyde is a basic chemical building block for the production of a wide range of chemicals finding a wide variety of end uses such as wood products, plastics, and coatings, eg, amino and phenolic resins, 1,4-butanediol, polyols, acetal resins, hexamethylenetetramine, slow-release fertilizers, methylenebis(4-phenyl isocyanate),

chelating agents, formaldehyde–alcohol solutions, paraformaldehyde, and trioxane and tetraoxane.

"Formaldehyde, Chemical Profile," *Chemical Market Reporter* (May 14, 2001).

R. J. Lewis, Sr., *Sax's Dangerous Properties of Industrial Materials*, 10th ed., John Wiley & Sons, Inc., New York, 2000.

G. Reuss and co-workers, in W. Gerhartz, ed., *Ullmann's Encyclopedia of Industrial Chemistry*, 5th ed., VCH Verlagsgesellschaft mbH, Weinheim, Germany, 1988, pp. 619–651.

H. ROBERT GERBERICH
GEORGE C. SEAMAN
Hoechst-Celanese Corporation

FRACTIONATION, PLASMA

Human blood plasma contains >700 different proteins (qv). Some of these are used in the treatment of illness and injury and form a set of pharmaceutical products that have become essential to modern medicine (Table 1). Preparation of these products is commonly referred to as blood plasma fractionation, an activity often regarded as a branch of medical technology, but which is actually a process industry engaged in the manufacture of specialist biopharmaceutical products derived from a natural biological feedstock (see PHARMACEUTICALS).

In 2004, there were ~80 organizations undertaking plasma fractionation worldwide, having plant capacities ranging from 5 to 2900 m^3 plasma/year. Virtually all of these plants use methods based on those originally devised, but with additional purification using modern bioprocess technologies.

MANUFACTURING AND PROCESSING

Plasma fractionation is unusual in pharmaceutical manufacturing because it involves the processing of proteins and the preparation of multiple products from a single feedstock. A wide range of unit operations are utilized to accomplish these tasks. They are listed in Table 2; some are common to a number of products and all must be closely integrated.

PRINCIPAL UNIT OPERATIONS

Protein Precipitation

The separation of proteins according to differences in solubility plays a significant role in plasma fractionation; a number of precipitation steps are used in the processes for albumin, immunoglobulin (immune globulin), and Factor VIII manufacture. Solubility behavior, a unique property of a protein, is determined by size, composition, and conformation, as well as by the environment in which

Table 1. Pharmaceutical Plasma Derivatives[a]

Product	Clinical application	Molecular weight × 10³	Normal plasma concentration, g/L
	Albumin[b]		
human serum albumin	protein and volume replacement	68	31–33
plasma protein fraction	volume replacement	68	36–40
	Coagulation proteins[c]		
Factor VIII	hemophilia A treatment	300	3×10^{-4}
Factor IX complex	treatment of hemophilia B and other coagulation disorders	57	5×10^{-3}
antiinhibitor coagulant complex[d]	hemophilia A treatment where Factor VIII antibodies are present		
	Inhibitors[c]		
α-1-proteinase inhibitor	emphysema treatment	52	1.5
antithrombin III	antithrombin III deficiencies treatment	58	0.1
	Immunoglobulins[e]		
immune globulin intravenous (normal)	immunoglobulin (IgG) replacement; treatment of immune disorders	150	12.5
immune globulin intravenous	treatment of cytomegalo-virus (CMV) infection in immune-suppressed individuals	150	
immune serum globulin (normal)	prevention of hepatitis A and rubella infections	150	12.5
hepatitis B immune globulin	prevention of hepatitis B infection	150	
pertussis immune globulin	prevention of whooping cough infection	150	
rabies immune globulin	prevention of rabies infection	150	
rho(D) immune globulin	prevention of hemolytic disease of the newborn	150	
tetanus immune globulin	treatment or prevention of tetanus infection	150	
vaccinia immune globulin	prevention of small-pox infection	150	
varicella immune globulin	prevention of chicken-pox infection	150	

[a] U.S. Licensed.
[b] Active component is albumin.
[c] Active component is indicated product.
[d] Active component is not known.
[e] Active component is IgG.

the molecules are located. The protein surface can be regarded as mostly hydrophilic and the protein interior as largely hydrophobic. These properties are determined by the nature and distribution of the amino acid residues that make up the protein (see AMINO ACIDS). The ionizable and polar amino acids are involved in charge repulsion, which plays an important role in preventing the aggregation of protein molecules. It is for this reason that proteins normally display a solubility minimum at their iso-ionic pH.

The solubility of a protein also is determined by the physical and chemical nature of its environment. Properties of the solution that influence protein solvation and protein–protein interactions are particularly important. The presence of substances that compete preferentially for water molecules reduces protein solubility; a number of substances of this type, such as neutral salts, organic solvents, and nonionic polymers, have been used to precipitate proteins. Other available precipitation reagents function by interacting directly with the protein, ie, either changing the surface charge or linking protein molecules together to form aggregates that exceed the solubility limit. A large number of parameters are potentially available for the manipulation of protein solubility and many of these have been applied to the separation of plasma proteins.

Solid–Liquid Separation

The separation of proteins by precipitation technology is accomplished when the solid and liquid phases have been separated from one another. Centrifugation, using either tubular bowl or multichamber centrifuges, is used for this purpose. Protein protein precipitates consist of large numbers of small particles, typically 0.1–1.0 µm in diameter, which aggregate together to from a large partical or floc. Both of these types of centrifuge function by retaining the solids within the rotating bowl, while the feed suspension and resultant supernatant flow continuously.

Table 2. Plasma Fractionation Unit Operations[a]

Unit operation	Method/technology	FVIII	FIX	IgG (im)	IgG (iv)	PPF	Albumin
			Protein separation[b]				
fractional precipitation[c]	cold-ethanol precipitation		++	+	+	+	+
fractional extraction[d]	from cold-ethanol ppt			+	+	+	+
solid–liquid separation	centrifugation	+	++	++	++	++	++
	depth filtration			++	++	++	++
selective adsorption	depth filtration			+	+	+	+
selective adsorption/ desorption[e]	ion-exchange chromatography	+	+		++		++
	immuno-affinity chromatography	++	++				
			Virus inactivation, in-process				
heat treatment	carbohydrate stabilized	++	++		++		
	fatty acid stabilized					++	++
chemical treatment[f]	solvent–detergent treated	++	++	++	++		
			Formulation and finishing[g]				
selective adsorption	depth filtration			+	+	+	+
membrane filtration	cross-flow filtration	++	++	++	++	+	+
	dead-end filtration	+	+	+	+	+	+
stabilization	chemical additives	+	++		+	+	+
dispensing	aseptic-dispensing	+	+	+	+	+	+
drying	freeze drying	+	+	++	++		
			Virus inactivation, terminal				
heat treatment	pasteurization					+	+
	dry heating	++	++				

[a] +, method in common use; ++, optional method, depending on procedures used by different manufacturers; im, intramuscular; iv, intravenous.
[b] Size exclusion by gel filtration is an optional method for FVIII.
[c] Charge reduction (pH, temperature) and other precipitation are common methods of FVIII fractionation.
[d] Extraction from other precipitates is an optional method for FVIII fractionation.
[e] Affinity chromatography is an optional method for FIX.
[f] Optional chemical treatments include potassium thiocyanate [333-20-0] for FIX and acid/enzyme treatment for IgG (iv).
[g] Selective proteolysis by acid/enzyme treatment is an optional method for IgG iv.

Protein Adsorption

The selective adsorption (qv) of a protein or group of proteins to a solid-phase reagent, followed in most cases by some form of selective desorption, also constitutes a principal form of protein separations technology in plasma fractionation. Solid-phase reagents can be categorized according to the forces responsible for adsorption.

The forces involved in the separation of proteins by ion-exchange adsorption are believed to be similar to those associated with protein precipitation. However, separation is usually achieved by manipulation of pH and ionic strength alone. Ion-exchange chromatography (qv) has been used in the preparation of Factor IX concentrates since the early 1970s and is also used in the preparation of Factor VIII concentrates.

Affinity chromatography is used in the preparation of more highly purified Factor IX concentrates as well as in the preparation of products such as antithrombin III. Immunoaffinity chromatography utilizes the high specificity of antigen–antibody interactions to achieve a separation.

Membrane Separations

The availability and development of microporous synthetic membranes and associated process technology has had a significant impact on the manufacture of plasma derivatives. Two very different areas of application exist: (1) the concentration of protein solutions and the removal of low molecular weight solutes use membrane systems that retain macromolecular substances, ie, ultrafiltration; (2) the removal of bacteria utilizes membranes sized to retain particles and larger microorganisms while allowing the macromolecular proteins to pass through, ie, sterile filtration.

Freeze-Drying

Plasma derivatives must have a defined shelf-life, usually for a period of 2 years. Products that are not stable in solution for this length of time are normally freeze-dried and reconstituted using sterile water for injection at the time of use. Freeze-drying involves the separation of water from nonvolatile constituents by sublimation from a frozen state. To carry out the process effectively it is necessary to specify the operating conditions carefully at a number of stages.

Inactivation and Removal of Viruses

In developing methods of plasma fractionation, the possibility of transmitting infection from human viruses present in the starting plasma pool has been recognized. Consequently, studies of product stability encompass investigation of heat treatment of products in both solution and dried states to establish virucidal procedures that could be applied to the final product. Salts of fatty acid anions, such as sodium caprylate and the acetyl derivative of the amino acid tryptophan, sodium acetyl-tryptophanate are capable of stabilizing albumin solutions to 60°C for 10 h; this procedure prevents the transmission of viral hepatitis. The degree of protein stabilization obtained and the safety of the product in clinical practice have been confirmed. The procedure has also been shown to inactivate the human immunodeficiency virus.

PROCESS RATIONALE

The products of plasma fractionation must be both safe and efficaceous, having an active component, protein composition, formulation, stability, and dose form appropriate to the intended clinical application. Processing must address a number of specific issues for each product. Different manufacturers may choose a different set or combination of unit operations for this purpose.

Human plasma is collected from donors either as a plasma donation, from which the red cells and other cellular components have been removed and returned to the donor by a process known as plasmapheresis, or in the form of a whole blood donation. These are referred to as source plasma and recovered plasma, respectively. In both instances, the donation is collected into a solution of anticoagulant to prevent the donation from clotting and to maintain the stability of the various constituents. Regulations in place to safeguard the donor specify both the frequency of donation and the volume that can be taken on each occasion.

Following donation, the separated plasma is frozen and transported to a fractionation plant, where it is held in frozen storage before being released for processing. On entering processing, plasma is vulnerable to bacterial contamination and proteolytic degradation. The more labile

constituents are particularly at risk. The early process steps aim for a degree of purification, the creation of a stable environment free from bacterial growth, and, where possible, a significant reduction in process volume. These objectives can be met by precipitation processes. Ideally, a range of intermediate products are produced at this stage that are held in storage pending release for further purification. The subdivision of processes in this manner carries a number of advantages including flexibility in scheduling and batch sizing, as well as in maximizing the utilization of limiting or expensive resources.

Factor VIII, immunoglobulin, and albumin are all held as protein precipitates, the first as cryoprecipitate and the others as the Cohn fractions FI + II + III (or FII + III) and $FIV_4 + V$ (or FV), respectively. Similarly, fractions $FIV_1 + FIV_4$ can provide an intermediate product for the preparation of antithrombin III and α-1-proteinase inhibitor. This ability to reduce plasma to a number of compact, stable, intermediate products, together with the bactericidal properties of cold ethanol, are the principal reasons these methods are still used industrially.

The Factor VIII molecule consists of multiple polypeptides having molecular weights of $\sim 80-210 \times 10^3$. The purified form consists of a light (mol wt = 80×10^3) and a heavy (mol wt = $90-210 \times 10^3$) chain, associated via a calcium linkage. The molecule is also bound noncovalenty to von Willebrand factor (mol wt = 220×10^3) forming complexes in the molecular weight range $1-10 \times 10^6$. Factor VIII is a particularly labile molecule vulnerable to degradation both by proteolysis and by depletion of calcium ions. Factor VIII is contained in cryoprecipitate, the precipitate that forms when frozen plasma is thawed. This has enabled the molecule to be removed from the nonideal environment of the plasma stream at the very beginning of the manufacturing process. Factor VIII is also vulnerable during cryoprecipitation because it is resolubilized as the temperature of the thawed suspension rises. Consequently, processing must be carried out both rapidly and with a high degree of temperature control if loss of Factor VIII is to be minimized.

Fibrinogen and fibronectin are the other principal proteins in cryoprecipitate. Both are poorly soluble, adherent proteins that can limit the capacity of subsequent chromatographic and filtration operations. As for all protein precipitates, some supernatant remains trapped within the amorphous cryoprecipitate particles and is carried over with the mass of solids; other coagulation factors, which in their activated form can degrade Factor VIII, are of particular concern in this regard. The concentration of fibrinogen and fibronectin is normally reduced using further precipitation steps; residual coagulation factors of the prothrombin complex can be removed by adsorption to aluminium hydroxide. It is necessary to maintain a sufficient concentration of ionized calcium throughout the process to prevent the dissociation of Factor VIII.

At this point in the process, a virus inactivation step is normally included, eg, either incubation in the presence of a solvent–detergent mixture or heat treatment. Further purification is subsequently required to remove chemicals, eg, tri-n-butyl phosphate and polysorbate-80 used in the solvent–detergent treatment or stabilizers used during pasteurization. This is achieved chromatographically using either ion-exchange adsorption, immunoadsorption, or size exclusion chromatography.

Direct ion-exchange adsorption is used to recover Factor IX from the supernatant that remains following cryoprecipitation. Alternatively, Factor IX can be recovered from Cohn Fraction III.

A number of other plasma products are entering into clinical use; growth is expected in at least some of these areas. Fibrinogen, previously withdrawn because of the hepatitis risk, can now be supplied in a virally inactivated form suitable either for infusion or as part of a fibrin sealant kit used for wound healing. Fibrinogen can be recovered from cryoprecipitate, Cohn Fraction I, or from side fractions of Factor VIII processing.

Another by-product of Factor VIII processing having clinical value is von Willebrand factor. It has been recovered from side fractions using ion-exchange and affinity chromatography.

Alpha-1-proteinase inhibitor and antithrombin III are used to treat people with hereditary deficiencies of these proteins. Both can be recovered from Cohn Fraction IV using ion-exchange chromatography and affinity chromatography, respectively. Some manufacturers recover antithrombin III directly from the plasma stream by affinity adsorption.

ECONOMIC ASPECTS

Estimates for a number of economic aspects of plasma fractionation can be made. The world capacity for plasma fractionation exceeded 30,000 t of plasma in 1999 and has increased by ~150% since 1980.

Worldwide the clinical use of albumin is almost 400 t/year and IgG iv about 46 t/year, however, the level of use varies widely between regions with commercial products often being imported to meet demand. In 2000, the total world sales of plasma derivatives exceeded $5,000,000,000 with 35% of sales being in North America, 29% in Europe and 26% in Asia/Pacific.

Many countries aspire to supply their requirement for plasma products from their own plasma resource, a position supported by the World Health Organization (WHO). Where multiple products are prepared from a common feedstock, the product in shortest supply dictates the scale of the manufacturing operation. The extent to which the demand for plasma products can be met from national supplies of plasma is determined by the volume of plasma collected as well as by-product yields. The relatively large volume of plasma taken from each paid U.S. donor has led to ~50% of the total world supply of plasma being provided from the United States. In 2001, 6.5 million L of plasma were exported from the United States for fractionation elsewhere.

REGULATION AND CONTROL

The preparation of clinical products from human plasma is regulated as a pharmaceutical manufacturing operation by national authorities who are responsible for giving authorization to distribute a product in their country. This is done by the Food and Drug Administration in the United States and by the Medicines & Health Related Products Agency (MHRA) in the United Kingdom.

HEALTH, SAFETY, AND ENVIRONMENTAL FACTORS

The possibility that infectious donations of plasma may enter the fractionation process places staff at risk; the transmission of hepatitis B to fractionation workers had been reported before the screening of plasma for hepatitis B infection was introduced. The extensive testing of donations that takes place (~1993) reduces this risk substantially. Nevertheless, it is assumed that plasma for fractionation may be contaminated with viruses such as hepatitis B, hepatitis C, and HIV, and appropriate precautions should be taken.

Ethanol (qv), the principal bulk reagent in plasma fractionation, is categorized as a highly flammable material with vapor concentration of 3–19% ethanol being explosive at temperatures above the flash point of 13°C. These properties must be considered in the design and specification of equipment and facilities involved in ethanol fractionation. Once the fractionation process has been completed, there are waste solutions containing up to 40% ethanol that require disposal. Some manufacturers recycle this material using distillation (qv), a procedure that must be regulated and controlled to the satisfaction of local environmental authorities.

J. A. Hooper, M. Alpern, and S. Mankarious, in H. W. Krijnen, P. F. W. Strengers, and W. G. van Aken, eds., *Immunoglobulins*, Netherlands Red Cross, Amsterdam, 1988, pp. 361–380.

Medicines Control Agency, *Rules and Guidance for Pharmaceutical Manufacturers and Distributors*, 6th ed., The Stationery Office, London, 2002.

J. R. Sharp, *Quality in the Manufacture of Medicines and Other Healthcare Products*, Pharmaceutical Press, London, 2000.

D. M. Surgenor, *Edwin J. Cohn and the Development of Protein Chemistry*, Harvard University Press, Boston, 2002.

PETER R. FOSTER
SNBTS Protein Fractionation Centre

FRIEDEL-CRAFTS REACTIONS

In 1877, at the Sorbonne in Paris, Charles Friedel and his American associate, James Mason Crafts, showed that anhydrous aluminum chloride could be used as a condensing agent in a general synthetic method for furnishing an infinite number of hydrocarbons. In work stretching over 14 years, they extended their discoveries of the catalytic effect of aluminum chloride in a variety of organic reactions: (1) reactions of alkyl and acyl halides and unsaturated compounds with aromatic and aliphatic hydrocarbons; (2) reactions of acid anhydrides with

Table 1. Friedel-Crafts Catalyst Activities

Group	Characteristic	Examples
A	very active, high yields but extensive intra- and intermolecular isomerization	$AlCl_3$, $AlBr_3$, AlI_3, $GaCl_3$, $GaCl_2$, $GaBr_3$, GaI_3, $ZrCl_4$, $HfCl_4$, $HfBr_4$, HfI_4, SbF_5, NbF_5, $NbCl_5$, TaF_5, $TaCl_5$, $TaBr_5$, MoF_6, and $MoCl_5$
B	moderately active, high yields without significant side reactions	$InCl_3$, $InBr_3$, $SbCl_5$, WCl_6, $ReCl_5$, $FeCl_3$, $AlCl_3$–RNO_2, $AlBr_3$–RNO_2, $GaCl_3$–RNO_2, SbF_5–RNO_2, and $ZnCl_2$
C	weak, low yields without side reactions	BCl_3, BBr_3, BI_3, $SnCl_4$, $TiCl_4$, $TiBr_4$, $ReCl_3$, $FeCl_2$, and $PtCl_4$
D	very weak or inactive	many metal, alkaline-earth, and rare-earth element halides

aromatic hydrocarbons; (3) reactions of oxygen, sulfur, sulfur dioxide, carbon dioxide, and phosgene with aromatic hydrocarbons; (4) cracking of aliphatic and aromatic hydrocarbons; and (5) polymerization of unsaturated hydrocarbons. The diversity of reactions is astounding.

Many important industrial processes such as the production of high octane gasoline, ethylbenzene (eventually leading to polystyrene), synthetic rubber, plastics, and detergent alkylates are based on Friedel-Crafts chemistry. The scope of the reactions is extremely wide as they form a large part of the more general field of electrophilic reactions, the class of reactions involving electron deficient carbocationic intermediates.

To define Friedel-Crafts reactions, it was necessary to come to a clear understanding that not one but a number of electrophilic reactions are classified as Friedel-Crafts type Friedel-Crafts-type reactions are generally considered to be any substitution, isomerization, elimination, cracking, polymerization, or addition reaction that takes place under the catalytic effect of Lewis acid-type acidic halides (with or without cocatalysts) or Brønsted acids. Friedel-Crafts reactions are not limited to the formation of carbon–carbon bonds but also lead to formation or cleavage of carbon–oxygen, carbon–nitrogen, carbon–sulfur, carbon–halogen, carbon–metals, and many other types of bonds. Friedel-Crafts reactions can be divided into two general categories: alkylations and acylations. Within these two broad areas there is considerable diversity.

Catalysts

Friedel-Crafts catalysts are electron acceptors, ie, Lewis acids. The alkylating ability of benzyl chloride was selected to evaluate the relative catalytic activity of a large number of Lewis acid halides. The results of this study suggest four categories of catalyst activity (Table 1). Catalysts include acid halides (Lewis acids), metal alkyls and alkoxides, protic acids (Brønsted Acid), acidic oxides and sulfides (acids chalcogenides), acid cation-exchange resins, and superacids (Brønsted supera-

cids, super Lewis acids, Brønsted-Lewis superacids, solid superacids, and superacids zeolites).

G. A. Olah, ed., *Friedel-Crafts and Related Reactions*, Vols. 1–4, Wiley-Interscience, New York, 1963–1965.

G. A. Olah, *Friedel-Crafts Chemistry*, John Wiley & Sons, Inc., New York, 1973.

G. A. Olah, G. K. Prakash, and J. Sommer, *Superacids*, John Wiley & Sons, Inc., New York, 1985, pp. 24–27 and references therein.

R. M. Roberts and A. A. Khalaf, *Friedel-Crafts Alkylation Chemistry*, Mercel Dekker, Inc., New York, 1984.

GEORGE A. OLAH
V. PRAKASH REDDY
G. K. SURYA PRAKASH
University of Southern
California

FUEL CELLS

Fuel cells are electrochemical devices that convert the chemical energy of a fuel directly into electrical and thermal energy. In a typical fuel cell, hydrogen is fed to the anode (negative electrode), and oxygen (or air) is fed to the cathode (positive electrode). The electrons removed from the oxidized fuel pass round an external circuit to the cathode. The fuel cell has the theoretical capability of producing electrical energy for as long as the reactants are provided to the electrodes. In reality, degradation or malfunction of the components limits the practical operating life of fuel cells, though systems have run continuously for several years.

Besides the direct production of electricity, heat is also produced in fuel cells. This heat can be used in combined heat and power (CHP) systems, or, in some high temperature fuel cells, for the production of further electrical energy using turbines.

Although the majority of fuel cells use hydrogen as the fuel, other fuels can be used, methanol being a notable example. Also, methane can be used, though this is usually re-formed to hydrogen and carbon monoxide either within the fuel cell itself, or in a fuel reformer adjacent to the fuel cell system.

BASIC PRINCIPLES AND PROBLEMS

For the operation of a fuel cell to be as effective as possible, the electrodes should have a large area, and the electrolyte between the cells should be as thin as possible. A common electrolyte for fuel cells is an acid, with mobile H^+ ions. The reactions at the anode and cathode of such a cell are shown in Figure 1.

Practical fuel cells cannot be this simple. One major problem is that the voltage of each cell is only 1 V or less. This

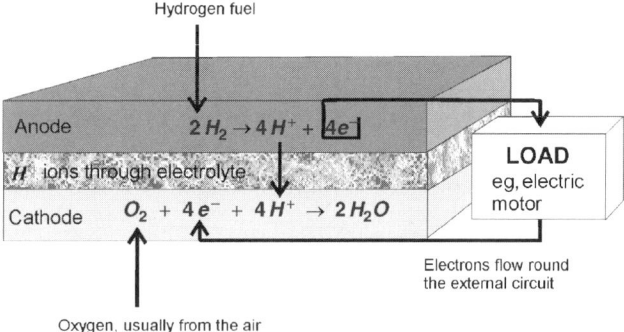

Figure 1. Basis of fuel cell operation with acid electrolyte and hydrogen fuel.

means that for useful levels of power many such cells must be connected in series. If this is to be done efficiently then each cell must be connected to the next in a way that avoids the current being taken off the edge of the electrode (as it is in Fig. 1), but over the whole surface on the electrode. The way this is usually done is to use a bipolar plate as the cell interconnect. The plate must be a good conductor of electricity, and have channels over its surface to allow the fuel to flow over the anode and the air or oxygen over the cathode. At the same time, it must make a good electrical contact with the electrodes, be as thin and light as possible, and also cheap to manufacture. A further requirement of the cell interconnects is that leakage of reactants must be prevented, which is often a very difficult feature of fuel cell stack design.

Fuel cells where the fuel and oxidant gases are fed in at the sides of the stack, as in Figure 2, are said to be "externally manifolded." Many designs do not use this method, and have a solid edge to the cell interconnect, to help in preventing leaks. The reactants are fed through tubes running down the stack, with precisely placed holes that release the gases into the channels of the bipolar plates. This method, known as "internal manifolding", makes sealing the edges of the electrodes easier, but is prone to internal leaks. Some cells use a mixture, with internal manifolding for the fuel, and external manifolding for the air fed to the cathodes.

The efforts of fuel cell developers over recent years has been directed at these key problems:

- To improve the current produced per square centimeter (cm^2) of electrode.
- To reduce manufacturing costs.
- To simplify the problem of fuel supply.
- To extend the life of the cell.
- To maintain the inherent simplicity of the fuel cell operation in the whole system.
- To enable the best use of the heat produced by the fuel cell.

DIFFERENT FUEL CELL TYPES

The key feature of each type of fuel cell is, in most cases, the electrolyte that is used. One important exception to

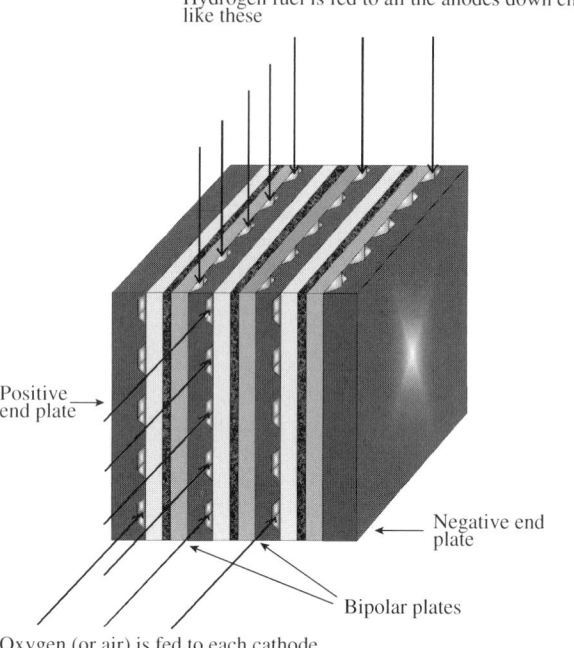

Figure 2. Externally manifolded fuel cells.

this is the "direct methanol fuel cell", which is defined by the fuel used (see Table 1). As well as facing up to different problems, the various fuel types also try to play to the different strengths of fuel cells. The proton exchange membrane (PEM) fuel cell (PEMFC) capitalizes on the essential simplicity of the fuel cell. The electrolyte is a solid polymer, in which protons are mobile. With a solid and immobile electrolyte, this type of cell is inherently very simple. These cells run at quite low temperatures, so the problem of slow reaction rates is addressed by using sophisticated catalysts and electrodes. A weakness of this type of cell is that the only suitable gaseous fuel is quite pure hydrogen.

An important variant of the PEM fuel cell is the direct methanol fuel cell (DMFC). As a fuel, methanol has many advantages over hydrogen—being a liquid it can be readily stored and transported. The main problems with the DMFC is that there is a large overvoltage at the anode, which makes the cell less efficient, and the fact that the methanol tends to diffuse through the proton conducting membrane.

Although PEM fuel cells were used on the first manned spacecraft, the alkaline fuel cell (AFC) was used on the Apollo and Shuttle Orbiter craft. The problem of slow reaction rate is overcome by using highly porous electrodes, with a platinum catalyst, and by operating at quite high pressures. These cells have been designed to operate over a fairly wide temperature range. The major disadvantage is the alkaline electrolyte, which is usually circulated around the fuel cell, making the system more complex.

Table 1. Data for Different Types of Fuel Cells

Fuel cell type	Mobile ion	Operating temperature °C	Applications and notes
alkaline fuel cell (AFC)	OH^-	50–200	used in space vehicles, eg, Apollo, shuttle
proton exchange membrane (PEM)	H^+	50–100	especially suitable for vehicles and mobile applications, but also for lower power CHP systems
direct methanol fuel cell (DMFC)	H^+	20–100	likely early applications in portable electronic equipment, but perhaps also for cars
phosphoric acid fuel cell (PAFC)	H^+	∼ 220	large numbers of 200-kW CHP systems in use
molten carbonate fuel cell (MCFC)	CO_3^{2-}	∼650	suitable for medium to large scale CHP systems, up to megawatt (MW) capacity
solid oxide fuel cell (SOFC)	O^{2-}	∼600–1000	suitable for all sizes of CHP systems, 0.5 kW to multi-MW

The phosphoric acid fuel cell (PAFC) was the first to be produced in commercial quantity and enjoy widespread terrestrial use. Porous electrodes, platinum catalysts, and a fairly high temperature (∼220°C) are used to boost the reaction rate to a reasonable level. The hydrogen fuel problem is solved by "re-forming" methane to hydrogen and carbon dioxide, but the equipment needed to do this adds considerably to the costs, complexity, and size of the fuel cell.

As is the way of things, each fuel cell type solves some problems, but brings new difficulties of its own. The solid oxide fuel cell (SOFC) operates in the region of 600–1000°C, which means that high reaction rates can be achieved without expensive catalysts, and that gases such as natural gas can be used directly, or "internally reformed" within the fuel cell, without the need for a separate unit. However, the ceramic materials that these cells are made from are difficult to handle, so they are expensive to manufacture, and there is still quite a large amount of extra equipment needed to make a full fuel cell system.

The molten carbonate fuel cell (MCFC), has the interesting feature that it needs the carbon dioxide in the oxidant supply for the cathode to work. The high temperature means that a good reaction rate is achieved using a comparatively inexpensive catalyst—nickel. The nickel also forms the electrical basis of the electrode. Like the SOFC, the MCFC can use gases such as methane and coal gas directly, without an external re-former. However, this simplicity is somewhat offset by the nature of the electrolyte, a hot and corrosive mixture of lithium, potassium, and sodium carbonates.

An important feature of fuel cells is that the different types of cell have a very wide range of application. They can be used in small portable electronics equipment (PEMFCs or DMFCs) right through to large multi-MW power stations (SOFCs and MCFCs), and everything in between. As generators of electrical power, fuel cells have a range of possible application that far exceeds any

other type, which explains the very great interest being shown in these devices of late.

FUEL CELL THERMODYNAMICS

Open Circuit Voltage

The energy driving electron transfer in electrochemical cells is the Gibbs free energy. If the overall fuel cell reaction produces a change in molar Gibbs free energy of Δg, and involves the transfer of n electrons, then the cell emf, E, in the absence of any losses is given by the formula:

$$E = \frac{-\Delta g}{nF} \qquad (1)$$

where F is the Faraday constant. In the case of the hydrogen fuel cell, for which the overall reaction is $H_2 + \frac{1}{2} O_2 \rightarrow H_2O$, $n = 2$.

However, the value of the Gibb's free energy is not constant. It varies with temperature and pressure, as well as the purity of the reactants, in accordance with standard chemical thermodynamics. Some example values for the hydrogen fuel cell are given in Table 2. They show that the Gibb's free energy change for the reaction, and hence the open circuit voltage, declines in magnitude with increasing temperature, which has important implications for the efficiency of high temperature fuel cells.

Table 2. The Parameter Δg and Maximum Electromotive Force (emf), for Hydrogen Fuel Cells Using Pure Reactants at Standard Pressure

Form of water product	Temperatures °C	Δg, kJ/mol	Maximum emf, V
liquid	25	−237.2	1.23
gas	100	−225.3	1.17
gas	200	−220.4	1.14
gas	400	−210.3	1.09
gas	600	−199.6	1.04
gas	800	−188.6	0.98
gas	1000	−177.4	0.92

Operating Fuel Cell Voltages

The actual operating voltage of a cell is always less than the no-loss open circuit voltage of a fuel cell, due to irreversibilities in the system. These occur most strongly when a current is drawn from the cell, but even the open circuit fuel cell voltage is usually reduced by various effects. This finding is especially noticeable in low temperature cells, where very small quantities of fuel passing through the electrolyte of the cell to the cathode cause a mixed reaction, which depresses the voltage. However, it is when a current is drawn that a more important drop in voltage occurs.

The difference between the open circuit voltage and the theoretical no loss emf considered in the previous section is normally called the "overvoltage," though the terms "voltage loss," "over potential," "irreversibility" are also used, reflecting the interdisciplinary nature of fuel cells. There are three main causes of voltage loss that increase with current: activation losses, ohmic losses, and mass transport (or concentration) overvoltage.

Fuel Cell Efficiency and Efficiency Limit

One of the most important advantages of fuel cells is that they have potentially very high efficiencies. The efficiency of a fuel cell is usually given in terms of the ratio of the electrical energy output to the specific enthalpy of the fuel used—in other words the ratio of the electrical output to the heat energy that could be obtained by burning the fuel.

Since the theoretical open circuit emf E_0 can be easily calculated, and the cell voltage easily measured or modeled, this makes the efficiency straightforward to find. However, there are also system inefficiencies to consider, of which the most important relates to fuel utilization (eq. 1).

Practical fuel cell efficiencies of ∼80% have been achieved—but this was using pure hydrogen and oxygen reactants, at high pressure, in the Shuttle Orbiter fuel cells. More typical figures are ∼40–45% obtained from cells running on air, though these are higher than the great majority of heat engines.

It is often stated that fuel cells are not subject to the Carnot limit that applies to heat engines, and so have an inherently higher efficiency. This statement is highly misleading, as fuel cells also have a theoretical limit to their efficiency. The Gibb's free energy harnessed by electrochemical cells such as fuel cells is the maximum possible value for the electrical work obtained from a fuel cell.

In practice the actual operating efficiencies of fuel cells tend to rise with temperature, because the losses mentioned previously, especially the activation overvoltage, decrease with temperature and more than compensate for the changes in the Gibb's free energy. Furthermore, the potential advantage of the high temperature fuel cell becomes even clearer if we consider the possibility of combined cycle systems. The energy that is not converted into electrical energy is in the form of heat, and is carried by the hot air, carbon dioxide, and steam that leaves the fuel cell. This hot gas stream can be used to drive a heat engine. This combination of a high temperature fuel cell with a turbine makes a system that is particularly efficient, and their development is being actively pursued by turbine makers such as Rolls Royce.

J. Larminie and A. Dicks, *Fuel Cell Systems Explained*, John Wiley & Sons, Inc., New York, 2003.

Vielstich, Lamm, and Gasteiger, eds., *Handbook of Fuel Cells: Fundamentals Technology, Applications*, Vol. 4, John Wiley & Sons, Inc., Hoboken, N.J., March 2003.

JAMES LARMINIE
Oxford Brookes University

FULLERENES

Fullerenes are closed cage carbon molecules C_{2n} ($n \geq 10$) comprised of a combination of n carbon atoms of sp^2 (trigonal) hybridization arranged into 12 pentagons and $(n - 20)/2$ hexagons, and are sometimes referred to as the "third form of carbon" (after diamond and graphite). A notable difference from these other forms is that fullerenes are soluble in many solvents and undergo many chemical reactions. They are formed by various procedures, especially vaporization of graphite either by use of lasers or electric arc discharge in an inert atmosphere, but do not occur naturally because of their instability; they oxidise quite rapidly in air. For a given value of n, very many isomers (having a wide range of stabilities) are possible. The number increases geometrically with increasing size of n, but only a few are isolable. The most abundant fullerene is the spherical I_h isomer of C_{60} (there are 1819 other C_{60} isomers, all unstable), which has a diameter of ∼10.0 Å. It is axiomatic that stable fullerenes cannot have adjacent pentagons because this introduces too much strain into the cages, and this is known as the "Isolated Pentagon Rule."

The next most abundant fullerene is C_{70}, which is shaped like a kiwi fruit; fullerenes containing >70 atoms are known as "higher fullerenes." Those of the latter that have been isolated in quantities sufficient for physical and chemical studies rather than mere characterization include C_{76}, C_{78} (three isomers) C_{82} and C_{84} (mainly two isomers); C_{74}, C_{80}, C_{86}–C_{96} and other isomers of C_{84} (seven or so are known) have been isolated in single milligram quantities. The geodesic structure of the molecules led to C_{60} being named as Buckminsterfullerene after Richard Buckminsterfullerene who popularized the use of geodesic domes in building construction.

DISCOVERY AND SYNTHESIS

Evidence for fullerenes was first published in 1984 by a team from the Exxon research laboratories in New Jersey, who used mass spectroscopic analysis of the product from laser-vaporized graphite. In 1985, a collaborative team from Rice (U.S.)/Sussex (U.K.) universities were carrying out similar experiments, with the intention of forming

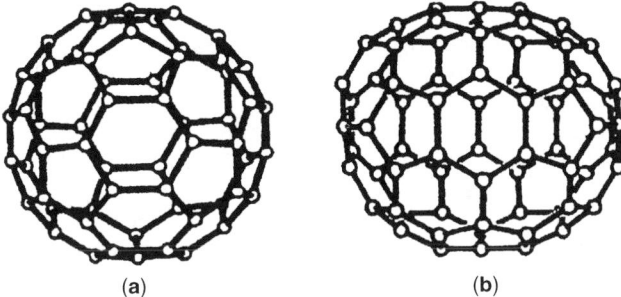

Figure 1. Structures of (**a**) C_{60} and (**b**) C_{70}.

long-chain carbon molecules thought to be produced in vast amounts by carbon stars. They duplicated the previous results, but found that by manipulation of the reaction conditions, C_{60} appeared as the dominant species, with C_{70} formed also in substantial amounts. They deduced that the structure of C_{60} was that of a truncated icosahedral closed cage molecule, consisting of 60 sp^2 (trigonal) hybridized carbon atoms (Fig. 1a), whereas C_{70} (Fig. 1b) contained an additional equatorial band of 10 carbons.

The major breakthrough in fullerene research came in 1990, when macroscopic quantities of fullerenes became available. These were produced by arc discharge of graphite rods in an inert atmosphere of helium at 100 Torr; argon may also be used but the yield is lower. This process produces a fullerene-containing soot (fullerene content ~5%) from which the fullerenes may be extracted with solvents such as toluene, chloroform, etc. The structures of the two main fullerenes were proved by ^{13}C nmr spectroscopy, C_{60} giving just one line (all the carbon atoms are equivalent), whereas C_{70} gives five lines in a 1:2:1:2:1 ratio, due to the presence in this ratio of the five distinct carbons.

It is important to exclude all traces of oxygen from the arc-discharge process (which may use either ac or dc current) otherwise significant amounts of fullerene oxides are produced. Other methods of fullerene formation include burning of benzene in oxygen-deficient flames and pyrolysis of aromatic hydrocarbons such as naphthalene, but to date the arc-discharge process is the one used most widely. When dc is used, large quantities of nanotubes are deposited on the cathode.

Endohedral or Incar fullerenes

Fullerenes with elements or molecules trapped inside them are known as either endohedral or *incar*fullerenes (this latter being the IUPAC term). They are generally prepared by filling hollow graphite rods with, eg, either a metal, metal carbide, or metal oxide, which on arc-discharge produces fullerenes with the element trapped inside, but the yields are very low (~0.1%); typical elements are La, Y, Sc, Ce, Eu.

Multiwall or Nested Fullerenes

Nested fullerenes were first observed in graphitized carbon particles before fullerenes were known. Nested fullerenes can be produced by electron-beam irradiation of carbon nanoparticles, by laser melting of carbon under high pressure, by shock wave treatment of soot, by the high temperature annealing of nanodiamonds, and by hydrocarbon combustion.

Heterofullerenes

In these fullerenes, one or more carbon atoms are replaced by other elements (thus far either boron, nitrogen, or phosphorus). Borafullerenes are obtained by arc discharging between carbon rods packed with boron nitride, and can only be detected spectroscopically. Azafullerenes are derived from the corresponding fullerene by a synthetic procedure.

Phosphafullerenes $C_{59}P^{\bullet}$ and $C_{69}P^{\bullet}$ have been made by coevaporating carbon and phosphorus in a radiofrequency furnace.

Opened Fullerenes

Holes have been made in fullerenes by various procedures, the objective being to incorporate molecules within the cages by the use of high pressure. While this is achievable in principle, useful incorporation of molecules will only be achieved by chemical procedures, yet to be devised.

Elongated Fullerene or Nanotubes

These tube-like carbon structures are effectively greatly elongated fullerenes, and may either be single-wall nanotubes (SWNT) (paralleling empty fullerenes) or multi wall nanotubes (MWNT) (paralleling nested fullerenes). Nanotubes can also be formed during combustion and pyrolysis of hydrocarbons, and when finely divided metals are used as catalysts under the latter conditions, the tubes are frequently spiraled.

Quasifullerenes

Although fullerenes are usually considered as being comprised only of 12 pentagons and any number of hexagons, it is feasible to have fullerenes containing a seven-membered ring, in which case an additional pentagon is required to close the cage. Although there are as yet no characterized examples of parent *quasi*fullerenes, derivatives are known having seven-membered rings and in these cases the presence of adjacent pentagons (forbidden with normal fullerenes) becomes possible.

Dimeric and Fused Fullerenes

Fullerenes may also be joined together by pairs of bonds to give dimeric fullerenes or they can be fused together.

PHYSICAL PROPERTIES

Bond Lengths and Structure

C_{60} has I_h symmetry while that of C_{70} is D_{5h}. C_{60} exists at room temperature as a face-centred cubic structure (fcc, lattice constant 14.17 Å) with a mass density of 1.72 g cm^{-3}, and a phase transfer to a simple cubic

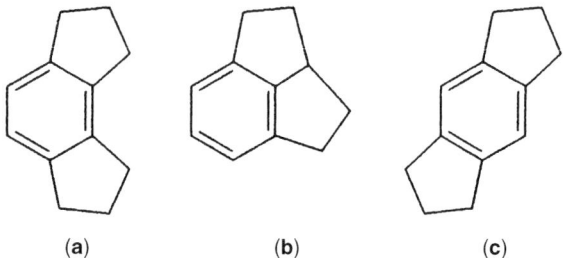

Figure 2. Favorable (**a**) and unfavorable (**b, c**) bond locations in pentagonhexagon combinations.

structure occurs <260 K (36). At room temperature, the molecules in the lattice rotate at a rate of $\sim 1 \times 10^{10}$ s^{-1}. This limits single-crystal X-ray determination of structure either to fullerenes possessing heavy addends, or to those that have solvent interactions that slow the rotation.

The bonds in fullerenes fall into groups each of differing lengths. The lower the symmetry of the cage the higher the number of groups. The differences arise because it is unfavorable to have double bonds in the pentagonal rings (Fig. 2) as this increases strain. In C_{60}, the most symmetrical fullerene, the bond lengths are 1.45 Å (within a pentagon, known as a 6:5 bond because it is common to a six- and a five-membered ring) and 1.40 Å (connecting two pentagons, known as a 6:6 bond because it is common to two six-membered rings). The double-bond character of the latter results in addition reactions taking place readily across this bond. This is the basis of most of the chemistry of C_{60}. For other fullerenes, addition also takes place across the shortest bonds, ie, those with the highest π-electron density. It follows from the above that delocalization of electrons in fullerenes is relatively weak, therefore they are not very "aromatic" molecules. This finding has an important bearing on their reactivities and addition patterns.

The number of isomers that can exist theoretically for a fullerene C_n increases at least geometrically, the higher the value of n. In practise, only a few isomers are stable enough to be isolated, these numbers being for n values. The structures of the isomers are determined by ^{13}C nmr (nuclear magnetic resonence) spectroscopy, but for larger fullerenes the numbers and intensities of the spectral lines usually reduce the number of possibilities to a handful of the same symmetry, but further definitive characterization is not possible.

Solubility

The solubility of fullerenes decreases with increasing size, and, for example the solubility of C_{70} in the above solvents is very roughly 50% of the values for C_{60}. The water solubility is greatly increased by the addition of hydroxyl groups either to the cage (giving fullerenols) or having them present in addends such as tris malonate adducts, $C_{60}[C(CO_2H)_2]_3$.

Stability

The heats of formation of C_{60} and C_{70} are $\sim$10.0 and 9.3 kcal mol^{-1} per atom, respectively, so that they are each less stable than either diamond or graphite for which the respective values are 0.4 and 0 kcal mol^{-1}. C_{70} is the more stable, and a general rule of thumb is that the larger the fullerene, the less reactive it is. The reactivity is due partly to the curvature of the surface that weakens the bonds since orbital overlap is poorer. The heats of sublimation also increase with increasing fullerene size, but the differences in values are too small to permit effective separation of fullerene via sublimation.

Fullerenes are unstable in air, especially when in the form of thin films that increases greatly the surface to volume ratio; C_{60} can absorb up to 4% by weight of oxygen. Fullerenes decompose at elevated temperatures, eg, >750°C for C_{60}.

Electron Affinities and Electrochemistry; Oxidation and Reduction

Fullerenes have substantial electron affinities, and the (gas-phase) values increase fairly regularly from 2.65 and 2.73 eV for C_{60} and C_{70}, respectively, to 3.39 eV for C_{106}. However, there is a sharp discontinuity for C_{72} and C_{74} (fullerenes that are difficult to isolate) that have respective values of 3.09 and 3.28 eV, for reasons that are currently unclear. The electron affinities are increased by electron-withdrawing addends, eg, the value for $C_{60}F_{48}$ is 4.06 eV.

The electron affinities are also manifest in fullerenes having energetically low-lying lowest unoccupied molecular orbitals (LUMO). They are thus readily reduced and can easily accept electrons (up to six in the case of C_{60}) reversibly under electrochemical conditions. It becomes increasingly difficult to add each successive electron, so that the electrochemical potentials for each addition are successively more negative. These values are dependant on temperature, solvent, and supporting electrolyte. The first reduction potentials of C_{60}, C_{70}, C_{76}, and one isomer each of C_{78} and C_{84} correlate linearly with the electron affinities. Just as addition of an electron to the cage makes the reduction potentials more negative for the subsequent additions, so the presence of electron-supplying groups on the cage produce similar effects. The ease of reduction of fullerenes means that they readily form hydrides and such is the oxidizing ability of C_{60} that it will even oxidize H_2S to sulfur.

Electron removal (oxidation) is by contrast difficult because (for C_{60}) the highest occupied molecular orbital (HOMO) is calculated to be energetically low lying, with the ionization potential estimated as 7.8 eV.

A major difficulty with regard to derivatizing fullerenes is that the strong electron withdrawal by the cages makes reaction of them with positive species unfavorable. However, this can be circumvented by converting the cages to anions with varying degree of charge (achieved by control of the electrochemical potential), so that reaction with the desired number of positive species occurs rapidly.

Spectroscopic Properties

The higher the symmetry level of a fullerene, the fewer lines that are observed in the infrared (ir), Raman, and

^{13}C nmr spectra.^{13}C nmr is used generally to identify the structure of a particular fullerene and derivatives.

Superconductivity in Fullerenes

Some fullerenes become superconducting when doped with alkali metals, with transition temperatures up to 33 and 40 K being observed for $RbCs_2C_{60}$ and Cs_3C_{60}, respectively. Numerous other combinations of alkali metal and fullerenes have been examined, but these give lower transition temperatures. It is unclear whether this development will have any long-term potential given that the doped materials are pyrophoric.

CHEMICAL REACTIVITY

Before describing the chemistry of fullerenes, the method of displaying a fullerene as a two-dimensional diagram (Schlegel diagram) is shown. The need for these diagrams arises when there are many groups (addends) attached to the cages, hence it is not possible to show clearly where these addends are on the far side of the cage. Figure 3 shows the Schlegel diagrams for (a) C_{60} and (b) C_{70}, respectively, together with the numbering schemes that are used to denote positions on the cages.

It is also necessary when describing their derivatives, to adopt a nomenclature for the fullerenes that follow the existing rules of organic chemistry. Thus C_{60} is [60]fullerene and C_{70} is [70]fullerene.

Chemistry has been carried out on those isolated fullerenes described earlier, but mostly on C_{60}, much less on C_{70}, and very little on the higher fullerenes. This reflects not only the respective availabilities of the fullerenes but also their solubilities and reactivities, which decrease with fullerene size. A further complication concerns their symmetries. Thus C_{60} gives only one mono-addition product, C_{70} can give up to five, while some higher fullerenes can in principle give very many more. Note that in any addition, two bonds must be formed to the cage, so that two groups must add (either X_2 or XH); X_2 can be

the components of a cyclic group. In the case of radical addition, which involves one group attaching to the cage, the resultant fullerenyl radical becomes stabilized either by abstraction of hydrogen from the surroundings, or if steric conditions permit, through dimerization.

A substantial problem when studying the chemistry of fullerenes is that polyaddition can occur to give, in principle, many thousands of isomers. In practice, electronic and steric constraints usually limit these numbers to single figures, nevertheless separating these isomers from each other, and from isomers of derivatives of different addition levels is a formidable task. The technique that is most widely used in high pressure liquid chromatography (HPLC), but the columns that are particularly useful for separating fullerene derivatives are very expensive. Moreover, the solubilities of the derivatives, though usually greater than the parent fullerenes, are not high. Thus only small quantities of compounds can be produced in this way. A further problem is to decide where the addends have become attached to the cage, and for this nmr (proton, carbon, and fluorine) is widely used. Interpretation of these data in terms of possible structures is daunting with considerable intuition being required in many cases. When large enough crystals can be grown (this is not often the case), single-crystal X-ray analysis can provide definitive structures. Unfortunately, many of the crystals of suitable size do not diffract sufficiently for the acquisition of meaningful data. The information that can be obtained by these structural studies often reveals considerable information about the electronic and steric effects that operate in the cages.

Fullerenes are electron deficient, this deficiency arising from the presence of the electronegative sp^2-hybridized carbon atoms. This electronegativity is also increased by strain. Reaction occurs primarily with electron-rich species (nucleophiles). Strain due to the cage curvature also causes poor overlap of adjacent C–C bonds, thereby facilitating reaction. It follows therefore that the larger fullerenes are less reactive since the cage curvature is less.

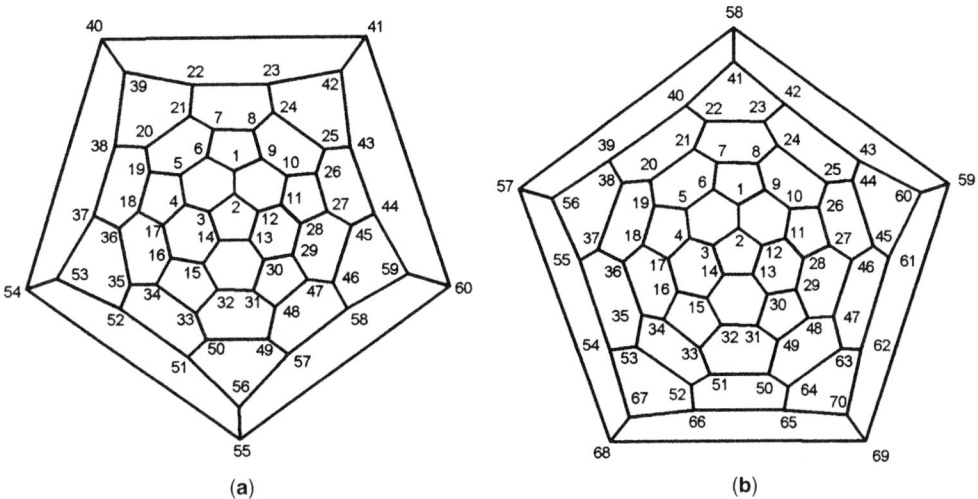

Figure 3. Schlegel diagram for (a) C_{60} and (b) C_{70} respectively.

There is considerable bond fixation in fullerenes, which may be regarded as "superalkenes" rather than aromatic compounds. However, there is some delocalization of electrons. This is most clearly demonstrated by patterns that occur upon addition, The sites of reactivity are therefore the localized double bonds, and for both C_{60} and C_{70}, the 1,2 bonds (see Fig. 3 **a** and **b**) are the most reactive; for C_{70} the next most reactive bond is the 5,6 bond. These are the sites connecting two pentagons. There are only 12 pentagons present whatever the size of the fullerene or nanotube, and for the latter, these are located at the ends of the tube, an area that constitutes less or much less than one-thousandth of the overall molecule. Consequently, nanotube chemistry is effectively precluded by the virtual impossibility of obtaining meaningful spectroscopic evidence for reaction products.

A majority of current chemical studies are devoted toward forming derivatives in which there is potential for the transfer, under light stimulation, of electrons from an electron-rich addend to the electron-deficient cage, (usually [60]fullerene since this is the most electron-deficient fullerene). This electroactive process, sometimes referred to as light-harvesting, is seen as a potential electricity-generating application of fullerenes types of reactions include additions, cycloadditions, reactions with electrophiles and nucleophiles, reaction of anions with electrophiles, formation of radicals, organometallic derivatives, and of polymers.

K. M. Kadish and R. S. Ruoff, eds., *Fullerenes, Chemistry, Physics and Technology*, John Wiley & Sons, Inc., Chichester, U.S., 2000.

R. Taylor, *Lecture Notes on Fullerene Chemistry; A Handbook for Chemists*, Imperial College Press, London, 1999.

R. Taylor and co-workers, *Recent Adv. Chem. Phys. Fullerenes Related Mater.* **7**, 462 (1999); **10** (2002).

ROGER TAYLOR
University of Sussex

FURAN DERIVATIVES

Furan (**1**) is a 5-membered heterocyclic, oxygen-containing, unsaturated ring compound. From a chemical perspective it is the basic ring structure found in a whole class of industrially significant products. The furan nucleus is also found in a large number of biologically active materials. Compounds containing the furan ring (as well as the tetrahydrofuran ring) are usually referred to as furans. From a manufacturing standpoint, however, furfural (**2**) is the feedstock from which all of the commercial furan derivatives are derived.

Furan is produced from furfural commercially by decarbonylation; loss of carbon monoxide from furfural gives furan directly. Tetrahydrofuran (**3**) is the saturated analogue containing no double bonds.

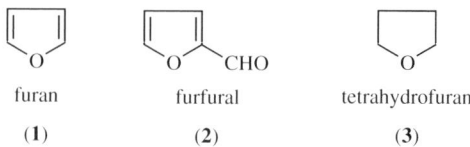

furan	furfural	tetrahydrofuran
(**1**)	(**2**)	(**3**)

Furfural is derived from biomass by a process in which the hemicellulose fraction is broken down into monomeric 5-carbon sugar units which then are dehydrated to form furfural.

This article primarily discussed simple furans in which the nucleus occurs as a free monocycle. Sometimes polychlorinated isobenzofurans are referred to simply as *furans*. These isobenzofurans, comprised of chlorine-containing fused rings, are not simple furans. This presentation is generally limited to compounds (including resins or polymers) derived from furfural.

Tetrahydrofuran (**3**) is produced commercially from furfural by decarbonylation followed by hydrogenation.

The furan nucleus is a cyclic, dienic ether with some aromaticity. It is the least aromatic of the common 5-membered heterocycles.

The balance between aromatic and aliphatic reactivity is affected by the type of substituents on the ring. Furan functions as a diene in the Diels-Alder reaction. With maleic anhydride, furan readily forms 7-oxabicyclo[2.2.1]hept-5-ene-2,3-dicarboxylic anhydride (**4**).

(**4**)

Radicals derived from furan are named similarly to analogous radicals in the benzene series. Typical radicals are 2 (or α)-furyl, 2-furfuryl, 2-furoyl, and 2-furfurylidene.

FURFURAL

Furfural can be classified as a reactive solvent. It resinifies in the presence of strong acid. Furfural is an excellent solvent for many organic materials, especially resins and polymers. On catalyzation and curing of such a solution, a hard rigid matrix results, which does not soften on heating and is not affected by most solvents and corrosive chemicals.

Furfural is formed by a series of reactions when biomass materials containing hemicellulose are treated with acid at an elevated temperature. When treated with aqueous acid, the hemicellulose is depolymerized to give primarily xylose, which under the reaction conditions loses three molecules of water and cyclizes to give furfural.

Physical Properties

When freshly distilled, is a colorless liquid with a pungent, aromatic odor reminiscent of almonds. Furfural is

Table 1. Physical Properties of Furan Derivatives

	Furfural	Furfuryl alcohol	Furan	Tetrahydrofurfuryl alcohol
General Properties				
molecular weight	96.09	98.10	68.08	102.13
boiling point at 101.3 kPa (1 atm), °C	161.7	170	31.36	178
freezing point, °C	−36.5		−85.6	< − 80
refractive index, n_D				
20°C	1.5261	1.4868	1.4214	1.4250
density, d_4, at 20°C	1.1598	1.1285	0.9378	1.0511
vapor density (air=1)	3.3	3.4	2.36	3.5
critical pressure, P_c, MPa	5.502		5.32	
critical temperature, T_c, °C	397		214	
solubility, wt %, in water				
25°C	20°C	8.3	∞	∞
			1	
alcohol; ether	∞	∞	∞	∞
Fluid properties				
viscosity, mPa·s (= cP)				
20°C			0.38	6.24
25°C	1.49	4.62		
surface tension, mN/m (= dyn/cm)				
25°C		ca 38		37
29.9°C	40.7			
Electrical properties				
dielectric constant				
20°C	41.9			
23°C				13.6

^{a}To convert MPa to atm, divide by 0.101.

miscible with most of the common organic solvents, but only slightly miscible with saturated aliphatic hydrocarbons. Inorganic compounds, generally, are quite insoluble in furfural.

Important physical properties of furfural, as well as similar properties for furfuryl alcohol, tetrahydrofurfuryl alcohol and furan are given in Table 1.

Chemical Properties

The chemical properties of furfural are generally characteristic of aromatic aldehydes but with some differences attributable to the furan ring. Furfural resinifies in the presence of acid and heat. Open chain compounds are formed from furfural under strong oxidizing conditions.

Furfural is very thermally stable in the absence of oxygen. At temperatures as high as 230°C, exposure for many hours is required to produce detectable changes in the physical properties of furfural, with the exception of color. However, a temperature above 250°C, in a closed system, furfural will spontaneously and exothermically decompose to furan and carbon monoxide with a substantial increase in pressure.

Furfural can be, reduced to 2-furan-methanol, referred to herein as furfuryl alcohol, or converted to furan by decarbonylation over selected catalysts.

Acetals are readily formed with alcohols and cyclic acetals with 1,2 and 1,3-diols.

Just as most other aldehydes do, furfural condenses with compounds possessing active methylene groups.

Furfural is a resin former under the influence of strong acid. It will self-resinify as well as form copolymer resins with furfuryl alcohol, phenolic compounds, or convertible resins of these.

Manufacture

Furfural is produced from annually renewable agricultural sources such as nonfood residues of food crops and wood wastes. The pentosan polysaccharides, xylan and arabinan, commonly known as hemicellulose, are the principal precursors of furfural and are always found together with lignin and cellulose in plant materials. Theoretically, all pentosan-containing substances are potentially usable for the production of furfural. Only a relatively few, however, are commercially significant.

Furfural is commercially produced in batch or continuous digesters where the pentosans are first hydrolyzed to pentoses (primarily xylose, which are then subsequently cyclodehydrated to furfural.

In all processes, raw material is charged to the digester and heated with high pressure steam. Enough excess steam is used to drive the furfural out of the reaction zone as vapor. The condensed reactor vapors are fed to a stripping column from which an enriched furfural–water distillate mixture is taken overhead and condensed. The liquid passes into a decanter where it separates into two layers. The furfural-rich lower layer containing about 6% water is processed further to obtain the furfural of commerce, and the water-rich layer containing about 8% furfural is recycled back to the stripper column as reflux.

Uses

Furfural is primarily a chemical feedstock for a number of monomeric compounds and resins.

Hydrogenation to furfuryl alcohol is the largest use. Some of the furfuryl alcohol is further hydrogenated to produce tetrahydrofurfuryl alcohol. The next major product is furan, produced by decarbonylation. Furan is a chemical intermediate, most of it is hydrogenated to tetrahydrofuran, which in turn is polymerized to produce polytetramethylene ether glycol (PTMEG). The principal direct application of furfural is as a selective solvent, and it also used as a reactive solvent in certain application.

Furfural has been used as a component in many resin applications, most of them thermosetting.

FURFURYL ALCOHOL

Physical Properties

Furfuryl alcohol (2-furanmethanol) is a liquid, colorless, primary alcohol with a mild odor. On exposure to air, it gradually darkens in color. Furfuryl alcohol is completely miscible with water, alcohol, ether, acetone, and ethyl acetate, and most other organic solvents with the exception of paraffinic hydrocarbons. It is an excellent, highly polar solvent, and dissolves many resins. The physical constants of furfuryl alcohol are listed in Table 1.

Chemical Properties

Furfuryl alcohol undergoes the typical reactions of a primary alcohol such as oxidation, esterification, and etherification. Although stable to strong alkali, furfuryl alcohol is very sensitive to acid, thus imposing limitations on the conditions used in many of the typical alcohol reactions. It is a more reactive solvent than furfural and easily resinifies under the influence of acid and heat.

Under acidic conditions, furfuryl alcohol polymerizes to black polymers, which eventually become crosslinked and insoluble in the reaction medium. Copolymer resins are formed with phenolic compounds, formaldehyde andor other aldehydes. Ethoxylation and propoxylation of furfuryl alcohol provide useful ether alcohols.

The chemistry of furfuryl alcohol polymerization involves reaction that give linear chains or oligomers containing essentially two repeating units (**5, 6**) with predominating.

(5)   (6)

Development of color has been shown to be due to conjugated sequences along linear oligomeric chains (**7**)

(7)

Manufacture

Furfuryl alcohol has been manufactured on an industrial scale by employing both liquid-phase and vapor-phase hydrogenation of furfural. Copper-based catalysts are preferred because they are selective and do not promote hydrogenation of the ring.

Uses

Furfuryl alcohol is widely used as a monomer in manufacturing furfuryl alcohol resins, and as a reactive solvent in a variety of synthetic resins and applications. The final cross-linked products display outstanding chemical, thermal, and mechanical properties. Many commercial resins of various compositions and properties have been prepared by polymerization of furfuryl alcohol and other co-reactants such as furfural, formaldehyde, glyoxal, resorcinol, phenolic compounds and urea.

A number of applications of furfuryl alcohol are based on its reactive solvent properties. When a resin solution is cured, a hard, rigid, thermoset matrix results, often with outstanding properties. For example, furfuryl alcohol is reactive solvent for phenolic resins in the manufacture of refractcries for landles holding molton steel.

The industrial value of furfuryl alcohol is a consequence of its low viscosity, high reactivity, and the outstanding chemical, mechanical, and thermal properties of its polymers, corrosion resistance, nonburning, low smoke emission, and excellent char formation.

FURAN

Physical Properties

Furan, a colorless liquid with a strong ethereal odor, is low-boiling and highly flammable. It is miscible with most common organic solvents but only very slightly soluble in water. The physical properties of furan are listed in Table 1.

Chemical Properties

Furan is a heat-stable compound, although at 670°C in the absence of catalyst, or at 360°C in the presence of nickel, it decomposes to form a mixture consisting mainly of carbon monoxide, hydrogen, and hydrocarbons. Substitution and addition reactions can be effected under controlled conditions, with reaction occurring first in the 2- and 5-positions.

Strong dienophiles also add to furan. Although both endo and exo isomers are formed initially, the former rapidly isomerize to the latter in solution, even at room temperature.

Catalytic hydrogenation of furan to tetrahydrofuran is accomplished in either liquid or vapor phase.

Manufacture

Furan is produced commercially by decarbonylation of furfural in the presence of a noble metal catalyst.

Uses

Furan is utilized as a chemical building block in the production of other industrial chemicals for use as pharmaceuticals, herbicides, stabilizers, and fine chemicals. Furan is readily hydrogenated, hence it is a source of commercial tetrahydrofuran (THF).

Washing and cleaning agents containing salts of maleic acid–furan copolymers form complexes with alkaline-earth ions. These cleaning compositions do not contain phosphorus or nitrogen and find use in metal, foodstuff, and machine dishwashing products.

TETRAHYDROFURFURYL ALCOHOL

Physical Properties

Tetrahydrofurfuryl alcohol (2-tetrahydrofuranmethanol) (8) is a colorless, high-boiling liquid with a mild, pleasant odor. It is completely miscible with water and common organic solvents. Tetrahydrofurfuryl alcohol is an excellent solvent, moderately hydrogen-bonded, essentially nontoxic, biodegradable, and has a low photochemical oxidation potential. Most applications make use of its high solvency. The more important physical properties of tetrahydrofurfuryl alcohol are listed in Table 1.

(8)

Chemical Properties

Without inhibitors, tetrahydro- furfuryl alcohol is susceptible to autoxidation. In the absence of air, however, no observable changes occur even after several years storage. In the presence of air, if a stabilizer such as Naugard is added, tetrahydrofurfuryl alcohol remains colorless after protracted periods of storage.

The reactions of tetrahydrofurfuryl alcohol are characteristic of its structure, involving primary alcohol and cyclic ether functional groups. As a primary alcohol, it undergoes normal displacement or condensation reactions affording new functional groups, (eg, halides, esters, alkoxylates, ethers, glycidyl ethers, cyanoethyl ethers, amines, etc). As a cyclic ether, it is typically unreactive, but the ring can be forced to open by hydrolysis or hydrogenolysis to give a variety of open-chain compounds.

All the common monobasic and dibasic esters of tetrahydrofurfuryl alcohol have been prepared by conventional techniques. Tetrahydrofurfuryl acrylate and methacrylate, specialty monomers, have been produced by carbonylation (nickel carbonyl and acetylene) of the alcohol as well as by direct esterification and ester interchange.

Manufacture

Tetrahydrofurfuryl alcohol is produced commercially by the vapor-phase catalytic hydrogenation of furfuryl alcohol. Liquid phase reduction is also possible.

Uses

Tetrahydrofurfuryl alcohol is of interest in chemical and related industries where low toxicity and minimal environmental impact are important. For many years tetrahydrofurfuryl alcohol has been used as a specialty organic solvent. The fastest growing applications are in formulations for cleaners and paint strippers, often as a replacement for chlorinated solvents. Other major applications include formulations for crop sprays, water-based paints, and the dyeing and finishing of textiles and leathers. Tetrahydrofurfuryl alcohol also finds application as an intermediate in pharmaceutical applications.

OTHER FURAN DERIVATIVES

Other furan compounds, best derived from furfural, are of interest although commercial volumes are considerably less than those of furfural, furfuryl alcohol, furan, or tetrahydrofurfuryl alcohol. Applications include solvents, resin intermediates, synthetic rubber modifiers, therapeutic uses, as well as general chemical intermediates.

Compounds containing the furan ring are generally excellent solvents. Some are miscible with both water and with hexane. Presence of the ether oxygen adds polarity as well as the potential for hydrogen bonding. Ring-substituted derivatives of furfural and furfuryl alcohol are reactive solvents, similar to the parent compounds.

Compounds containing the furan or tetrahydrofuran ring are biologically active and are present in a number of pharmaceutical products.

Derivatives of furan and tetrahydrofurfuryl alcohol are used in the polymerization of synthetic rubber to control stereoregularity and other properties.

HEALTH AND SAFETY FACTORS

As with all chemical compounds, the Material Safety Data Sheet (MSDS) for each of the specific furan derivatives should be reviewed before starting to work with these materials. Additional information on toxic effects of most of these compounds can be found in RTECS (*Registry of Toxic Effects of Chemicals*), HSDB (*Hazardous Substances Data Bank* from the National Library of Medicine), and standard works on toxicology. Toxicology studies are taking place on a continuing basis with many chemicals, including furan derivatives. New data may change the perspective on toxicity of these chemicals.

Precautions should be taken when working with these compounds became furfural, furfuryl alcohol, and furan are moderately toxic, tetrahydrofurfuryl alcohol is less so. Since regulations change from time to time, up-to-date exposure limit recommendations from OSHA (Occupational Safety and Health Agency) or ACGIH (American

Council of Governmental Industrial Hygienists) need to be consulted and followed.

F. M. Dean, *Advances in Heterocyclic Chemistry*, **30**, 168; **31**, 237 (1982).

A. P. Dunlop and F. N. Peters, *The Furans*, ACS Monograph 119, Reinhold Publishing Corp., New York, 1953.

A. Gandini "Furan Polymers" in J. I. Kroschwitz, ed., *Encyclopedia of Polymer Science and Technology*, 2nd ed., Vol. 7, John Wiley & Sons, Inc., New York, 1987.

M. V. Sargent and T. M. Cresp "Furans" in D. Barton and W. D. Ollis, eds., *Comprehensive Organic Chemistry; The Synthesis and Reaction of Organic Compounds*, Vol. 4, Pergamon Press Ltd., Oxford, U.K., 1979.

R. H. KOTTKE
Great Lakes Chemical
Corporation

FURNACES, ELECTRIC

The term electric furnace applies to all furnaces that use electrical energy as their sole source of heat. Electric furnaces are used mainly for heating solid materials to desired temperatures below their melting points for subsequent processing, or melting materials for subsequent casting into desired shapes, ie, electric heating furnaces or electric melting furnaces. The latter includes so-called holding furnaces which store a molten charge received from separate melting furnaces.

Classification is by the manner in which the electrical energy is converted into heat. Thus three distinct types of widely used industrial furnaces can be distinguished: electric resistance furnaces, electric arc furnaces, and electric induction furnaces. The conversion of electrical energy into heat in each type of furnace is schematically illustrated in Figure 1.

RESISTANCE FURNACES

The most widely used and best known resistance furnaces are indirect-heat resistance furnaces or electric resistor furnaces. They are categorized by a combination of four factors: batch or continuous; protective atmosphere or air atmosphere; method of heat transfer; and operating temperature. The primary method of heat transfer in an electric furnace is usually a function of the operating temperature range. The three methods of heat transfer are radiation, convection, and conduction. Radiation and convection apply to all of the furnaces described. Conductive heat transfer is limited to special types of furnaces.

Operating temperature ranges are classified as low, medium, and high; there is no standard or precise definition of these ranges. Generally, a low temperature furnace operates below 760°C, medium temperature ranges from 760–1150°C, and furnaces operating above 1150°C are high temperature furnaces. There is often indiscriminate use of the words furnace and oven. The term oven should be used when temperatures are below 760°C, and the word furnace applied for higher temperatures. The term furnace is used here regardless of operating temperature.

Batch Furnaces

In batch furnaces the desired time–temperature cycle for the product to be processed is accomplished by subjecting the entire furnace and its contents or charge of work to the particular cycle. Batch furnaces are most often used for very large and/or heavy charges, low production rates, infrequent operation, variable time–temperature cycle, and processing material that must be in batches because of previous or subsequent operations. Larger batch furnaces are often of the elevator or car-bottom type.

Medium-sized loads are often processed in a bell furnace. The operation of this furnace is opposite to that of an elevator furnace: the work load is placed on a stationary hearth and the furnace is lowered over the hearth.

Small loads are commonly processed in a box furnace. Box furnaces may be single-ended or double-ended. A single-ended box furnace is usually used in an air atmosphere application where the product can be removed hot from the furnace for cooling. A double-ended box furnace is usually used in a controlled atmosphere application. In this case a water cooler is attached to one end.

Other versions include the pit furnace, which is a box furnace with the door on top and which is often installed in a pit with the top of the furnace near floor level.

Continuous Furnaces

These furnaces are applicable for uniform charges of work that arrive at the furnace continuously, moderate to high production rates, constant time–temperature cycle, and continuous operation over at least one and preferably two or three shifts per day. Continuous furnaces are usually named for the method used to convey the material through the furnace. The roller-hearth furnace the mesh-belt furnace and the pusher furnaces. Roller-hearth furnace, the mesh-belt furnace and the pusher furnace.

Furnace Atmospheres

Electric furnaces can operate either with air in the interior of the furnace or with a protective atmosphere; the choice is dictated by the process requirements of the work. The furnace must be designed for the atmosphere to be used, because the combination of temperature and atmosphere are significant factors in selecting internal materials used in the furnace construction; this applies particularly to the selection of heating element (resistor) material. It is feasible and common to design an electric furnace that can operate in both air and protective atmospheres although shortened element life generally results from frequent alternating between reducing atmospheres and oxidizing atmospheres.

Low Temperature Convection Furnaces

Low temperature convection furnaces are designed to transfer the heat from the heating elements by forced

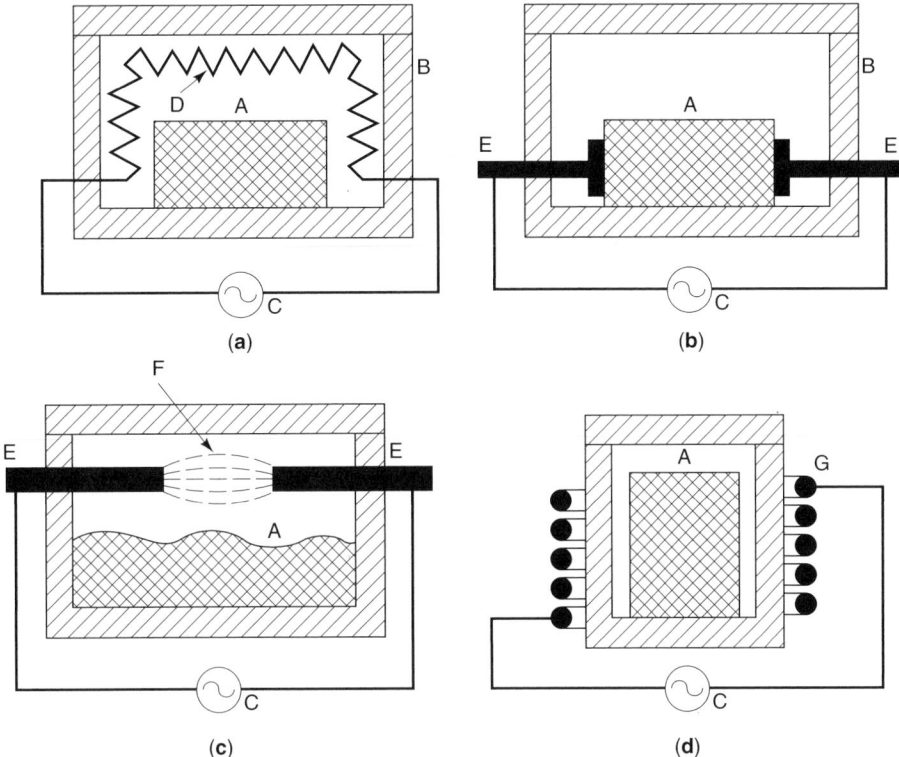

Figure 1. Main types of electric furnaces: (**a**) resistance furnace, indirect heat (resistor furnace); (**b**) resistance furnace, direct heat; (**c**) arc furnace; (**d**) induction furnace. A, charge to be heated or melted; B, refractory furnace lining; C, electric power supply; D, resistors; E, electrodes; F, electric arc; G, induction coil.

convection. Convection is normally used in furnaces operating below 760°C because it is the most effective means of heat transfer that can maintain good uniformity of temperature on various workload configurations. Convection furnaces also are used (in this range of temperatures) where it is important that no part of the work load exceed the controlled temperature.

There are four types of radiation furnaces: low temperature radiation furnaces, medium temperature radiation furnaces, high temperature radiation furnaces, and vacuum radiation furnaces.

Conduction Furnaces

Conduction furnaces utilize a liquid at the operating temperature to transfer the heat from the heating elements to the work being processed. Conduction furnaces are of three general types. One has a pot or crucible with suitable exterior insulation. The salt-bath furnace is another type of conduction furnace. The third type of conduction furnace is a fluidized bed.

Direct-Heat Electric-Resistance Furnaces

Direct-heat electric furnaces use the material to be heated as the resistor, and the furnace consists of an insulated enclosure to retain the heat, a power source of suitable voltage, and means of attaching the power leads to the work. This type of furnace has several limitations that have prevented widespread use. Since the work is the resistor, it must have a uniform cross section between power connection points, and the material must be homo-

geneous. Varying sections or nonuniformities in the material can produce hot or cold spots in proportion to the change in electrical resistance. Also, a given furnace must be designed for work in which each piece to be heated has about the same resistance and power requirements.

ARC FURNACES

Arc furnaces used in electric melting, smelting, and electrochemical operations are of two basic designs: the indirect and the direct arc. The arc of the indirect-arc furnace is maintained between two electrodes and radiates heat to the charge. The arcs of the direct-arc furnace are maintained between the charge and the electrodes, making the charge a part of the electrical power circuit. Not only is heat radiated to the charge, but the charge is heated directly by the arc and the current passing through the charge.

Indirect-Arc Furnaces

The typical indirect-arc furnace is a single-phase furnace utilizing two horizontally mounted graphite electrodes, each of which project into an end of a refractory-lined horizontally mounted cylindrical steel shell.

Although rocking of the furnace to intermittently cover and hence protect up to 90% of the refractory, as well as improved refractories, has done much to make the indirect-arc furnace more viable, these furnaces are becoming less common, primarily due to high operating costs as a result of erosion of the refractory by the intense arc radiation.

Direct-Arc Furnaces

Open-Arc Furnaces. Most of the open-arc furnaces are used in melting and refining operations for steel and iron. Although most furnaces have three electrodes and operate utilizing three-phase a-c power to be compatible with power transmission systems, d-c furnaces are becoming more common. Open-arc furnaces are also used in melting operations for nonferrous metals (particularly copper), slag, refractories, and other less volatile materials.

Refractory Linings. The refractory linings for the hearth and lower walls of furnaces designed for melting ferrous materials may be acidic, basic, or neutral. Silica has been widely used in the past, and is still being used in a number of iron and steel foundries. Alumina, a neutral refractory, is normally used for furnace roofs and in the walls for iron foundries, but basic brick can also be used in roofs.

Electrodes. Almost all the electrodes used in open-arc furnaces are prefabricated and are made of regular or dense graphite.

Voltage. The voltage chosen for open-arc furnaces must be high enough to compensate for the voltage drops caused by the resistance and inductance of the primary and secondary electrical circuits and still have the required power input available to sustain the arcs. In the smaller furnaces, the voltage must be high enough to penetrate any thin oxide coatings on the scrap. Also, it must provide a sufficient area of meltdown; otherwise, the electrodes bore a small hole through the scrap, melting insufficient metal to cover the hearth resulting in high consumption of the bottom refractories. The highest phase-to-phase no-load voltage for a 200 kVA production furnace usually is 200 V, and 1000 V for a 120,000 kVA furnace is not uncommon. Lower voltages are also available for the operator to use during a furnace refining cycle; the lowest voltage is approximately one-third of the highest voltage. However, high productivity operations generally do not make use of the lower voltage taps.

Vacuum-Arc Furnace. Another type of open-arc furnace is the vacuum-arc furnace which is used for melting metals that have high temperature melting points.

Plasma-Arc Furnace. The plasma-arc furnace, sometimes used in the production of castings of high alloy steels and special alloys or for the smelting of fine materials, usually has a furnace shell similar to that of the three-phase conventional open arc-furnace used in the production of iron, steel, or ferroalloys. However, water-cooled nonconsumable electrodes are used to conduct direct current and argon (to serve as the plasma base) to the sealed furnace interior. The plasma torch can be of

either the transferred or nontransferred type. The two types are distinguished by the electric current conduction path.

D-C Arc Furnace. With the advent of more economical thyristor-controlled d-c power supplies, as well as limitations imposed by power companies on arc furnace-generated flicker, d-c furnaces have become more common. These furnaces are nearly identical to their counterparts, except they typically have a single electrode passing through the roof and a means to collect the current through a furnace bottom electrode.

Submerged-Arc Furnace. Furnaces used for smelting and for certain electrochemical operations are similar in general design to the open-arc furnace in that they are usually three-phase, have three vertical electrode columns and a shell to contain the charge, but direct current may also be utilized.

Arc-Resistance Furnace. The arc-resistance furnace is similar to the submerged-arc furnace except the electrodes of the former are most often in direct contact with material, usually slag or a nonmetallic material, but they may also arc to the slag layer. Even when the electrode is in contact with the melt there are still minute arcs between the bottom and sides of the electrode, because it is not wetted by the slag, and the majority of the heat is developed in the melt in the immediate vicinity of the electrode tip. The primary difference between the arc-resistance and submerged-arc furnace is that the former exhibits ohmic conductance.

Electrodes. Graphite electrodes are used primarily in smaller furnaces or in sealed furnaces. Prebaked carbon electrodes, made in diameters of <152 cm or 76 by 61 cm rectangular, are used primarily in smelting furnaces where the process requires them. However, self-baking electrodes are preferred because of their lower cost.

INDUCTION FURNACES

Induction furnaces utilize the phenomena of electromagnetic induction to produce an electric current in the load or workpiece. This current is a result of a varying magnetic field created by an alternating current in a coil that typically surrounds the workpiece. Power to heat the load results from the passage of the electric current through the resistance of the load. Physical contact between the electric system and the material to be heated is not essential and is usually avoided. Nonconducting materials cannot be heated directly by induction fields.

Induction Heating

Design. The coil of an induction heater typically encircles the load. The current intensity within the load is greatest at the surface and diminishes to zero at the center. This crowding of the current close to the surface is known as skin effect.

Power Supplies and Controls. Induction heating furnace loads rarely can be connected directly to the user's electric power distribution system. If the load is to operate at the supply frequency, a transformer is used to provide the proper load voltage as well as isolation from the supply system. Adjustment of the load voltage can be achieved by means of a tapped transformer or by use of a solid-state switch. The low power factor of an induction load can be corrected by installing a capacitor bank in the primary or secondary circuit.

Induction Melting

Induction melting applications almost always contain the liquid metal charge within a hearth formed by a suitable refractory material. It is possible to design the hearth to satisfy a wide variety of application requirements ranging from a few kilograms to hundreds of tons of metal and for operation in normal or hostile environments. The environmental impact of an induction furnace is generally less than that of an equivalent fuel-fired furnace.

Coreless Induction Furnaces

Coreless furnaces derive their name from the fact that the coil encircles the metal charge but, in contrast to the channel inductor described later, the coil does not encircle a magnetic core.

Frequency Selection. When establishing the specifications for a coreless induction furnace, the material to be melted, the quantity of metal to be poured for each batch, and the quantity to be produced per hour must be considered simultaneously.

Operation. Small and medium sized coreless induction furnaces powered from high frequency power supplies can be started with a charge of metal pieces at room temperature, usually scrap material of appropriate alloy. As the metal charge begins to melt, a molten pool is established and the charge compacts, allowing additional charge to be added. Alloy additions and temperature adjustments complete the melting cycle. Large coreless furnaces operating at line frequency are often started with a molten initial charge, although it is possible to start with a charge of solid material.

Hearth. The induction melting coil is almost always round and in the form of a right cylinder. It is highly desirable that the refractory lining within the coil be uniform in thickness, so most hearths are cylindrical whether they hold a few kg or 59 t.

Channel Induction Furnaces

The term channel induction furnace is applied to those in which the energy for the process is produced in a channel of molten metal that forms the secondary circuit of an iron core transformer. The primary circuit consists of a copper coil which also encircles the core. This arrangement is quite similar to that used in a utility transformer.

Inductor. The channel inductor assembly consists of a steel box or case that contains the inductor refractory and the inductor core and coil assembly. The channel is formed within the refractory. Inductor power ratings range from 25 kilowatts for low temperature metals to 5000 kilowatts for molten iron.

Hearth. The hearth of a channel induction furnace must be designed to satisfy restraints that are imposed by the operating inductor, ie, the inductor channels must be full of metal when power is required, and it is also necessary to provide a sufficient level of metal above the channels to overcome the inward electromagnetic pressure on the metal in the channel when power is applied.

Operation. Channel furnaces can be used for melting or holding metal. In either case, the inductor and the hearth refractory are preheated to avoid thermal shock as the liquid metal is introduced at start up. Once the inductor channel has been flooded, it is rarely emptied until the inductor is taken out of service.

HEALTH AND SAFETY FACTORS

Because intense heat is generated in arc furnaces, plus the carbon monoxide generated by both melting and smelting furnaces, all new furnace installations require pollution control equipment. This normally consists of off-gas afterburning (sometimes with energy recovery), and dust collection equipment, typically a baghouse.

For arc furnace worker safety, high power electrical systems require proper design and precautions, and handling of molten materials requires a minimum of fire-retardant clothing and often dust masks. Water must be prevented from coming in contact with the melt. Furthermore, since open-arc furnace noise levels commonly exceed 100 dBA, hearing protection is a necessity. Noise is normally not a problem with smelting furnaces.

APPLICATIONS

Electric furnaces are used for annealing, brazing, carburizing, galvanizing, forging, hardening, melting, sintering, enameling, and tempering metals, most notably aluminum, copper, iron and steel, and magnesium alloys.

Arc Furnaces

Nearly all open-arc furnaces used in foundries and steel mills are three-phase and contain individually controlled jib-type electrode arms, each supporting a vertical column of graphite electrodes. Generally, smaller furnace transformers (<7500 kW) and some larger transformers also contain a multitap reactor to provide sufficient inductive reactance to offset the negative characteristic of the arc so as to provide the desired arc stability.

Induction Furnaces

A unique capability of induction heating is apparent in its ability to heat the surface of a part to a high temperature while the interior remains at room temperature. The ability to precisely control the power and length of the induction heating cycle allows it to be integrated into complex work handling equipment.

Induction heating is used to heat steel reactor vessels in the chemical process industry.

High process temperatures generally not achievable by other means are possible when induction heating of a graphite susceptor is combined with the use of low conductivity high temperature insulation such as flake carbon interposed between the coil and the susceptor. Processes include purification, graphitization, chemical vapor deposition, or carbon vapor deposition to produce components for the aircraft and defense industry.

Small-and-medium sized foundries producing castings for automotive and other similar applications often utilize iron melting channel melting furnaces. A combination of a channel induction holding furnace with an induction heating coil used in a process is called continuous galvanizing. In this installation the alloying of iron and zinc to produce a "galvannealed" strip for automobile bodies with improved fabrication and corrosion resistance characteristics. The control provided by the use of induction furnaces results in a superior product compared to fuel-fired alternatives.

American Foundrymen's Society, Inc., *Refractories Manual*, 2nd ed., Des Plaines, Ill., 1989.

American Society for Metals, *Metals Handbook, Heat Treating*, Vol. 4, 9th ed., Metals Park, Ohio, 1991.

S. L. Semiatin and D. E. Stutz, *Induction Heat Treatment of Steel*, American Society for Metals, Metals Park, Ohio, 1986.

W. Trinks, M. H. Mawhinney, R. A. Shannon, R. J. Reed, and J. R. Garvey, *Industrial Furnaces*, 6th ed., John Wiley & Sons, Inc., Hoboken, N.J., 2004.

Richard A. Sommer
Robert R. Walton
Wellman Furnaces, Inc.
J. Kevin Cotchen
MAN GHH Corporation

FURNACES, FUEL-FIRED

A furnace is a device (enclosure) for generating controlled heat with the objective of performing work. In fossil-fuel furnaces, the work application may be direct (eg, rotary kilns) or indirect (eg, plants for electric power generation). The furnace chamber is either cooled (waterwall enclosure) or not cooled (refractory lining). In this article, furnaces related to metallurgy such as blast furnaces are excluded because they are covered under associated topics.

Waterwall furnaces were employed by the ancient Greeks and Romans for household services. Furnaces, in general, and waterwall furnaces, in particular, were neglected for about the next 1600 years. In part, this may be ascribed to the fact that steam as a working fluid had no application until the invention of the first commercially successful steam engine at the end of the seventeenth century, followed by Newcomen's engine in 1705 with self-regulating steam valves.

In fire-tube furnaces developed in the nineteenth century, such as typified by the Scotch-Marine boiler, thin currents of water contact a multiplicity of tubes; thus, the hot gases transmit heat simultaneously to all regions of the bulk of the water. Therefore, this boiler–furnace combination steams readily and responds promptly to load changes, and is, for a given amount of heating surface, the least expensive of all furnace–boiler installations. The development of seamless, thick-wall tubing for stationary power plants (ie, water-tube furnaces) and other engines for motive power, such as diesel–electric, has in many cases eclipsed the fire-tube boiler. For applications calling for moderate amounts of lower pressure steam, however, the modern fire-tube boiler continues to be the indicated choice.

CLASSIFICATION

Furnaces are either cooled (water- or air-cooled chamber) or not cooled (refractory-lined chamber). In general, the basic structure roughly resembles either a rectangular box or a cylinder with variations for generally good reasons. Hence, the open-hearth furnace is used for steel melting and refining, the reverbatory furnace for copper, and glass tanks for various materials. If the material tends to be lumpy, as in smelting, cupola melting, or lime burning, the furnace is constructed vertically to make gravity feeding possible.

The roof of the typical furnace collapses unless it is built as a sprung or self-supporting arch, or is flat but suspended, along with the rest of the furnace, from an enclosing structural framework. All modern power-plant waterwall furnaces of any significant megawatt rating are of suspended construction.

If a refractory-wall furnace is very large, the walls must be relatively thick to provide the necessary structural strength required to withstand the weight of the roof and thermal stresses as well as to retain heat. Large furnaces are built with two or even three different types of brick, with the outer brick generally the more highly insulating and the inside brick able to withstand the highest temperatures. The middle bricks are, therefore, often of lower quality and carefully selected since their softening temperature must not be exceeded.

A furnace may be direct-fired or indirect-fired. The indirect-fired is known as a muffle furnace, and in such furnaces the combustion gases are separated from the stock being heated to prevent contamination.

FUELS

Fuel-fired furnaces primarily utilize carbonaceous or hydrocarbon fuels. Since the purpose of a furnace is to generate heat for some useful application, flame temperature and heat transfer are important aspects of furnace design. Heat transfer is impacted by the flame emissivity. A high emissivity means strong radiation to the walls.

The carbon/hydrogen ratio of fuels is a variable used widely in fuel technology to estimate emissivity. However, its effect on flame temperature is usually misrepresented.

Gas, fuel oil, coal (anthracite, bituminous, subbituminous, lignite), and wood are generally representative of fuels to be encountered over the range of industrial furnaces and, depending on the type (cooled or refractory wall), exhibit operating temperatures considerably different from adiabatic values. The choice of fuel is dependent on a number of factors including cost, availability, cleanliness, emissions, reliability, and operations. Small furnaces tend to burn cleaner, easier to use fuels. Large furnaces can more effectively use coal, or other solid fuels.

The transition from a choice of multiple fossil fuels to various ranks of coal, with the subbituminous varieties a common choice, does in effect entail a fuel-dependent size aspect in furnace design. A controlling factor of furnace design is the ash content and composition of the coal. If wall deposition thereof (slagging) is not properly allowed for or controlled, the furnace may not perform as predicted. Furnace size varies with the ash content and composition of the coals used.

POWER-PLANT FURNACES

In 2000, 71% of the electric power consumed in the United States was generated by fossil fuel combustion. The bulk of electric power generation will probably be supplied by coal-fired power-plant furnaces supplying steam to turbogenerators. In terms of megawatts supplied, the coal-fired power plant is, therefore, the foremost component of importance in the energy supply system. Power-plant furnaces are of waterwall type and are generally designed for steam pressures in the range of 12.4–24.1 MPa (1800–3500 psi); the latter value is referred to as supercritical, ie, >22.1 MPa abs (3208 psia).

Although furnace sizes (dimensions for a given MW production) do not vary too widely between principal manufacturers, the type of firing employed by each is generally quite distinctive. This finding indicates that the furnace size is not strongly controlled by the type of firing system, particularly for pulverized-coal firing. The furnace needs to be sufficiently large to permit the oxygen enough time to penetrate (diffuse through) the blanketing CO_2 layer evolving from the burning coal particle. The residual ash particles are, of course, considerably smaller than the parent coal particle, on the order of a mean size of ~10 μm before postcombustion agglomeration. Although flame temperatures should be high for combustion efficiency in order to minimize CO formation and combustible carbon loss, it is further required that the combustion products (gases) are sufficiently cooled to enter the convection banks below the temperature at which slagging occurs. These contradictory conditions (aside from pollution con-

trol requirements) influence the furnace size and have led to solutions such as tangential firing.

INDUSTRIAL FURNACES

Generally speaking, industrial furnaces are an order of magnitude smaller than power-plant furnaces since the applications are usually on an individual basis (hospital complex, chemical plant, paper mill, etc) rather than feeding power to a regional electric grid. Like the power-plant furnace, the function of the industrial furnace usually is to generate steam, generally for a chemical process, mechanical power, or heating application, rather than electric power generation. There are also many fired heaters that utilize the hot exhaust gases directly for heating, drying, roasting, calcining, etc. Industrial boilers include package boilers, papermill (recovery) furnaces, large industrial furnaces, (waterwall), and refractory-wall furnaces.

These boilers are also known as shop-assembled boilers. Low capacity units are shipped complete with fuel-burning equipment, safety and combustion controls, and boiler trim. Large capacity units, because of tunnel and overpass shipping clearances and railroad flatcar limitations, are designed as multiple integrated-component packages such as the complete furnace–burner assembly, forced-draft fan assembly, and the like.

FLUIDIZED-BED COMBUSTION

New furnace concepts in development stages include fluidized-bed furnaces, coal gasification furnaces, and MHD furnaces. Of these technologies, fluidized-bed combustion has reached commercial-scale operations, coal gasification has reached the demonstration phase.

For decades, fluidized-bed reactors have been used in noncombustion reactions in which the thorough mixing and intimate contact of the reactants in a fluidized bed result in high product yield with improved economy of time and energy.

ANALYSIS

The rising demand for fuel efficiency and performance guarantees are imposing increasing requirements for analytical complexity. Analytical emphasis has shifted toward a heat-transfer oriented view of predicting furnace performance: (1) wall absorption rate and gas temperature profiles, and (2) pollutant formation. Among the various models, the Hottel zoning method is still popular. Another analytical model that lends itself to engineering analysis of furnaces is generally referred to as PSR theory (perfectly stirred reactor theory) and is, in fact, applied on an industrial basis, where it is important to obtain a quantitative evaluation of the emissivity of the combustion products. Commercial computer codes that calculate complex flow fields with reacting flows are available and being validated. The advent of very high speed computers has brought furnace calculations into managable proportions. The most firmly established mechanism at this time is the prediction of thermal NO_x formation.

J. Barthelemy, J. Pisano, and J. C. Semedard, *Flexible Fuel Choice and Clean Efficient Combustion: Circulating Fluid Bed Technology for the Americas*, 2003 Alstom Power, Velizy, France.

W. Bartok and A. Sarofim, *Fossil Fuel Combustion–A Source Book*, John Wiley & Sons Inc., New York, 1991.

P. Basu, C. Kefa, and L. Jestin, *Boilers and Burners*, Springer, New York, 1999.

H. Rudiger and G. Scheffknecht, *Advanced Steam Power Plant Technology*, PowerGen Europe 2001, May 29–31, 2001, Brussels, Belgium.

CARL R. BOZZUTO
Alstom Power Environment
Sector

G

GALLIUM AND GALLIUM COMPOUNDS

Gallium is a scarce but not a rare element. The concentration of gallium in the earth's crust is estimated at between 15–19 ppm compared with 16 ppm for lead and arsenic. Gallium is more abundant than antimony, silver, bismuth, molybdenum, and tungsten, however, it is present in very low concentrations, with little tendency to concentrate in a particular geologic environment. To date, ores containing gallium have not been economically mined solely for the contained gallium. Gallium is usually present in the 5–200 ppm range in most minerals. Table 1. Gallium is generally concentrated in aluminum minerals (eg, bauxite, corundum).

PROPERTIES

Physical Properties

Gallium, at. wt 69.717, has two stable isotopes, ^{69}Ga, 60.4%, and ^{71}Ga, 39.6%, and twelve unstable isotopes, from mass 63 through 76. The radius of the atom is 0.138 nm, and of the ions Ga^{3+} and Ga^{+}, is 0.062 nm and 0.133 nm, respectively. Solid gallium has a metallic, slightly bluish appearance.

The physical properties of gallium, especially its thermal properties, are exceptional. It has a low mp and vaporizes above 2200°C, ie, it has the longest liquid interval of all the elements. Also, it is easily supercooled. However, it expands during solidification by 3.2%, a property shared by only two other elements, germanium and bismuth. Its crystal structure is unusual for a metal. Gallium crystallizes in the orthorhombic system, and it is very anisotropic. This latter property is attributed to the existence of Ga–Ga covalent bonds along its [001] axis. Principal physical properties of the normal or alpha form are listed in Table 2.

Gallium wets almost all surfaces, especially in the presence of oxygen which promotes formation of a gallium suboxide film. Gallium diffuses quickly into the crystal lattice of certain metals, particularly aluminum. If a line is traced with a piece of solid gallium on a sheet of aluminum, the aluminum quickly becomes brittle on that line as a result of the diffusion of gallium to the grain boundaries of the aluminum. Because of this property of embrittling aluminum, the DOT Office of Hazardous Materials has classified gallium in Hazard Class HM-181 and has placed restrictions on air shipment. IATA (international air transport regulation) classifies gallium as corrosive.

Chemical Properties

In accordance with its normal potential, gallium is chemically similar to zinc and is somewhat less reactive than aluminum. Just as aluminum, it is protected from air-oxidation at ambient temperature by a fine film of oxide. Normally, gallium is trivalent; however, it also may be monovalent. Gallium forms some compounds of mixed valence, eg, gallium (I) tetrachlorogallate(−1), $Ga(GaCl_4)$. Amphoteric character is one of gallium's principal properties.

EXTRACTION

A minor amount of gallium is extracted as a by-product from the zinc industry. The gallium content of sphalerites generally is concentrated in the residues of zinc distillation and in the iron mud resulting from purification of zinc sulfate solutions. Gallium is extracted from these streams by acidic solutions and gallium salts are recovered by liquid-liquid extraction using dialkylphosphates, trialkyl phosphates, hydroxyquinolines, or isopropyl ether, among others.

Recovery from Bayer Liquor

The significant amount of primary gallium is recovered from the alumina industry. The main source is the sodium aluminate liquor from Bayer-process plants that produce large quantities of alumina. Several methods have been developed to recover gallium from Bayer liquor, eg. carbonation, electrolysis, chemical reduction, liquid–liquid extraction, and ion-exchange resins.

Purification

Extraction from aluminum or zinc ores produces crude gallium metal or concentrates. These concentrates are transformed to sodium gallate, gallium chloride, or gallium sulfate solutions which are purified, then electrolyzed. Gallium is deposited as a liquid.

The purification of the gallium salt solutions is carried out by solvent extraction and/or by ion exchange. The most effective extractants are dialkyl-phosphates in sulfate medium and ethers, ketones, alcohols, and trialkyl-phosphates in chloride medium. Electrorefining, ie, anodic dissolution and simultaneous cathodic deposition, is also used to purify metallic gallium.

Ultrahigh (>99.99999% = 7.N) purity metallic gallium is achieved by a combination of several operations such as filtration, electrochemical refining, heating under vacuum, and/or fractional crystallization.

Recycling. A large part of the wastes from the gallium arsenide, GaAs, industry is recovered for both economical and environmental reasons. Several processes are effective and being used to recover both the gallium either as a metal, a salt, or a hydroxide for recycling, and the arsenic in some form for recycling or disposal. Thermal decomposition of gallium arsenide waste is one method which competes with the use of hydrometallurgical routes in caustic soda media.

Table 1. Gallium Abundances in Various Minerals

Mineral	Formula	Ga ppm
Abite	$NaAlSi_3O_8$	40
Almandine	$Fe_3Al_2(SiO_4)_3$	20
Bauxite	$Al_2O_3 \cdot 2H_2O$	90
Biotite	$K(Mg,Fe)_2AlSi_3O_{10}(OH)_2$	40
Calcite	$CaCO_3$	0.1
Chromite	$FeCr_2O_4$	18
Corundum	Al_2O_3	100
Cryolite	Na_2AlF_6	3
Diopside	$CaMg(SiO_3)_2$	5
Franklinite	$Zn(Fe,Mn)_2O_4$	10
Hematite	Fe_2O_3	1
Hornblende	$(Fe^{3+}, Fe^{2+}, Al, Mg, Mn)_5$ $(Ca, Na, K)_2Si_8O_{22}(OH)_2$	10
Jarosite	$KFe_3(SO4)_2(OH)_6$	8–23
Lepidolite	$KLiAl_2Si_3O_{10}(OH,F)_2$	100
Limonite	$2Fe_2O_3 \cdot 3H_2O$	3
Magnetite	$FeO \cdot Fe_2O_3$	30
Microcline	$KAlSi_2O_8$	40
Muscovite	$KAl_3Si_3O_{10}(OH)_2$	200
Nepheline	$(Na,K)AlSiO_4$	20
Olivine	$(Mg,Fe)_2SiO_4$	2
Oligoclase	$(NaAlSi_3O_8)_{7-9}$ $(CaAl_2Si_2O_8)_{3-1}$	10
Orthoclase	$KAlSi_3O_8$	10
Phlogopite	$KMg_3AlSi_3O_{10}(OH)_2$	50
Quartz	SiO_2	<1
Sodalite	$3NaAlSiO_4 \cdot NaCl$	100
Sphalerite	ZnS	1–1000
Spodumene	$LiAl(SiO_3)_2$	60

PRODUCTION

Estimated crude gallium production was 70 metric tons (t) in 2003.

The gallium either comes from mining sources or is recycled from scrap. Scrap-recycling capacity is taking a larger place each year. In the United States, the main producers of gallium are Gallium Compounds LLC, formerly Eagle-Picher Technologies, and Recapture Metals; in Japan, Rasa Industries, Mitsubishi Metals, and Sumitomo Metal Mining; in France, Rhône-Poulenc; and in Germany, Preussag (Metaleurop) and Ingal.

ECONOMIC ASPECTS

Despite very strong growth in the market for many gallium-containing devices used in electronics, achieving

Table 2. Physical Properties of Normal Gallium

Property	Value
melting point, °C	29.7714
boiling point, °C	ca 2200
density at mp, g/cm^3	
solid	5.904
liquid	6.095
vapor pressure, at 1198 K Pa^a	0.14

^{a}To convert Pa to mm Hg, divide by 133.3.

a balance in the gallium industry has been difficult economically. Light-emitting diodes (LED) must face increasing competition from liquid crystal displays (LCD) and other alternatives to cathode ray tubes (CRT) for display devices. However, developments in LED technology have created a new market in large area outdoor and semi-outdoor displays.

Gallium-based laser diodes have also lost some of the market in telecommunications to indium-based devices. Optical fiber systems moved to the use of longer wavelengths for long distance networks.

Nevertheless, the market in compact disc equipment is increasing, and computer data storage has expanded rapidly, as well as telecommunication by satellite and the use of high frequency devices.

In spite of the limited growth of gallium demand, the forecast for use in integrated circuits shows an increasing market. The high speed of GaAs is a great advantage in the military, space and supercomputing, and general computer markets.

Another attractive application for gallium is the use in high performance photovoltaic cells in satellites, but owing to the weight of the substrates, these have been replaced by germanium. The solar cells are manufactured using multiple layers of gallium compounds deposited by metalloorganic chemical vapor deposition (MOCVD) on the Ge substrates.

HEALTH AND SAFETY FACTORS

Elemental Ga is insoluble in water and is therefore poorly absorbed in mammals. Severe intravenous overexposures are lethal because of kidney failure. Renal damage was similar to that seen for mercury.

Very little toxicological data are available on Ga effects in humans. Experiments in general, on the Ga-containing drugs have shown that anorexia, nausea, vomiting, skin rashes, and depression of red and white blood cell counts can occur. Other clinical experiments on Ga-containing drugs have resulted in some bone marrow depression, dermatitis and severe itching and gastrointestinal disturbances.

The toxicity of metallic gallium or gallium salts is very low. The corrosive, poisonous, or irritating nature of some gallium compounds is attributable to the anions or radicals with which it is associated. Gallium metal-organics, such as $Ga(CH_3)_3$, react vigorously with air, and can be explosive. The gallium halides, except the fluoride, hydrolyze in water to form corresponding halogen acids. Gallium phosphide, arsenide, selenide, and telluride react slowly with water, and more vigorously with acids and bases, to liberate toxic compounds.

Processing of gallium arsenide (GaAs) wafers produces fine particulate crystals, which may present a hazard to industrial workers. Previous toxicological studies in animals and culture caused both the Environmental Protection Agency and World Health Organization to classify GaAs as an immunotoxicant and potential carcinogen.

Although health risks to industrial workers exposed to GaAs remains unclear, animal studies have established that chemical exposure causes detrimental biological effects. Intratracheal instillation and inhalation of GaAs results in pulmonary inflammation and edema in rodents. In contrast to systemic immune defects, GaAs has the opposite effect at the exposure site. Hence, the ability of GaAs to up-regulate proinflammatory cytokines may extend to the lungs, which could contribute to pulmonary damage.

USES

More than 95% of the gallium consumed in the United States was in the form of GaAs. GaAs was manufactured into optoelectronic devices (LEDs, laser diodes, photodetectors, and solar cells) and integrated circuits (ICs).

Medical Uses of Gallium

Gallium can be used to detect such diseases as Hodgkin's disease, lymphomas, and interstitial lung disease. Gallium nitrate is also used as an anticancer drug for lymphomas and bones. It can reverse bone degeneration and/or hypercalcemia cancer, osteoporosis, and Paget's syndrome. In the case of therapeutic action gallium halts bone resorption, normalizes serum calcium levels, adds bone mass, and kills cancer cells. Some gallium drugs become acute renal toxins at 10 to 100 times the therapeutical dose. Gallium does not accumulate in tissues, nor does it cause mutations or show other signs of toxicity. In dental applications gallium alloys are nonstaining and used in the fabrication and repair of dental protheses.

Catalysis Application

Gallium is used in catalysts for aromatization in the petroleum (qv) industry.

ALLOYS AND INTERMETALLIC COMPOUNDS

Alloys

Gallium has complete miscibility in the liquid state with aluminum, indium, tin, and zinc. No compounds are formed. However, these binary systems form simple eutectics having the following properties:

Metal	Ga, %	Melting point, °C
Al	96	26.4
In	76	15.7
Sn	91.5	20.6
Zn	96.3	25.0

The systems obtained when gallium is in the presence of bismuth, cadmium, germanium, mercury, lead, silicon, or thallium present miscibility gaps. No intermetallic compounds are formed.

Intermetallic Compounds

Numerous intermetallic gallium-transition element compounds have been reported. Gallium also forms numerous compounds with lanthanides and yttrium as well as actinides.

COMPOUNDS OTHER THAN INTERMETALLIC

Compounds include gallium hydrides; gallium halides (gallium(I), gallium(I)gallium(III), gallium(III); gallium oxyhalides; gallium halogenates; sulfohalides and sulfohalogenates compounds with ammonia gallium oxides (gallium(I), gallium(III) gallates(III) of numerous) metalic elements gallium chalcogenides (mixed chalcogenides, sulfates and selenates gallium nitrogen compounds gallium compounds with phosphorus, arsenic, and antimony and carbon compounds of gallium.

L. F. Borisenko, "Promising Types of Gallium-bearing Deposits," *Lithology and Mineral Resources*, **28**(1), 25–37 (1993).

A. J. Downs, ed., *Chemistry of Aluminium, Gallium, Indium, and Thalium*, Routledge, Chapman and Hall, London, June 1993.

N. N. Greenwood and A. Earnshaw, *Chemistry of the Elements*, 2nd ed., Butterworth and Heinemann, Oxford, U.K., 1998.

M. J. Howes and D. V. Morgan, *Gallium Arsenide Materials, Devices, and Details*, John Wiley & Sons, Ltd., Chichester, U.K., 1985.

FLOYD GRAY
DEBORAH A. KRAMER
JAMES D. BLISS
U. S. Geological Survey

GAS, NATURAL

Natural gas is a mixture of naturally occurring hydrocarbon and nonhydrocarbon gases found in porous geologic formations beneath the earth's surface. Methane is a principal constituent and the mixture may contain higher hydrocarbons such as ethane, propane, butane, and pentane. Gases such as carbon dioxide, nitrogen, hydrogen sulfide, various mercaptans, and water vapor along with trace amounts of other inorganic and organic compounds can also be present. Natural gas is found in a variety of geological formations including sandstones, shales, and coals.

Discussions of natural gas can involve the definitions listed in Table 1.

PROPERTIES

The composition of natural gas at the wellhead depends on the characteristics of the reservoir and is highly variable with respect to both the constituents present and the concentrations of these constituents.

The physical properties of the principal constituents of natural gas are listed in Table 2. These gases are odorless, but for safety reasons, natural gas is odorized before distribution to provide a distinct odor to warn users of possible gas leaks in equipment. Sulfur-containing compounds such as organic mercaptans, aliphatic sulfides, and cyclic sulfur compounds are effective odorants at low concentrations and are added to natural gas at levels ranging from 4 to 24 mg/m^3.

Table 1. Definitions Associated with Natural Gas

Term	Definition
Associated gas	free natural gas in immediate contact, but not in solution, with crude oil in the reservoir.
Dissolved gas	natural gas in solution in crude oil in the reservoir.
Dry gas	gas where the water content has been reduced by a dehydration process or gas containing little or no hydrocarbons commercially recoverable as liquid product.
Liquefied natural gas (LNG)	natural gas that has been liquefied by reducing its temperature to 111 K at atmospheric pressure. It remains a liquid at 191 K and 4.64 MPa (673 psig).
Natural gas liquids (NGL)	a liquid hydrocarbon mixture which is gaseous at reservoir temperatures and pressures, but recover able by condensation or absorption (qv).
Nonassociated gas	free natural gas not in contact with, nor dissolved in, crude oil in the reservoir.
Sour gas	gas found in its natural state containing compounds of sulfur at concentrations exceeding levels for practical use because of corrosivity and toxicity.
Sweet gas	gas found in its natural state containing such small amounts of sulfur compounds that it can be used without purification with no deleterious effect on piping or equipment, and without the potential for health hazards.
Wet gas	unprocessed or partially processed natural gas produced from strata containing condensible hydrocarbons.

The pressure–volume–temperature (PVT) behavior of many natural gas mixtures can be represented over wide ranges of temperatures and pressures by the relationship, $PV = ZnRT$, where P is the absolute pressure; V, the volume; Z, the compressibility factor for the mixture; n, the number of moles of gas; R, the universal gas constant; and T, the temperature in Kelvin.

PRODUCTION

Natural gas is produced from reservoirs containing both oil and gas (associated gas) and from nonassociated reservoirs holding only gas. These reservoirs may be relatively shallow and require wells drilled to depths of a few hundred meters. However, production is also being realized from reservoirs located at substantial depths requiring wells drilled to depths in excess of 6100 m. Production takes place both at onshore installations and on offshore platforms which service wells drilled to provide access to reservoirs located below the floor of the ocean.

PROCESSING

Because of the wide variation in the composition of natural gas as it is recovered at the wellhead and because

Table 2. Physical Constants of Natural Gas Constituents

Compound	Formula	Mol wt	Boiling point, K[a]	Critical pressure, kPa[b]	Critical temperature, K
methane	CH_4	16.043	111.64	4595	190.56
ethane	C_2H_6	30.070	184.55	4871	305.34
propane	C_3H_8	44.097	231.08	4247	369.86
2-butane	C_4H_{10}	58.123	261.37	3640	407.86
n-butane	C_4H_{10}	58.123	272.65	3796	425.17
2-pentane	C_5H_{12}	72.150	301.00	3381	460.44
n-pentane	C_5H_{12}	72.150	309.23	3369	469.71
n-hexane	C_6H_{14}	86.177	341.89	3012	507.38
n-heptane	C_7H_{16}	100.204	371.58	2736	540.21
n-octane	C_8H_{18}	114.231	398.83	2487	568.83
n-decane	$C_{10}H_{22}$	142.285	447.32	2104	617.60
nitrogen	N_2	28.013	77.35	3400	126.21
oxygen	O_2	31.999	90.20	5043	154.59
carbon dioxide	CO_2	44.010	194.68[c]	7384	304.22
hydrogen sulfide	H_2S	34.076	212.88	8963	373.41
water	H_2O	18.015	373.16	22055	647.14
air		28.963	78.83	3771	132.43

[a]At atmospheric pressure, 101.3 kPa (1 atm).
[b]To convert kPa to psi, multiply by 0.145.
[c]Denotes sublimation temperature.

natural gas can be used over a wide range of hydrocarbon contents, any specification for natural gas is usually broadly defined. However, the natural gas obtained at the wellhead usually undergoes some type of treatment or processing prior to its use for safety, economic, or system and material compatibility reasons.

Dehydration

Produced gas is usually saturated with water vapor at the wellhead temperature and pressure. Generally, these water-vapor levels are reduced to concentrations no greater than 112 mg/m^3 (7 lbs/10^6 ft^3) gas to prevent condensation during transmission in high pressure pipelines and to reduce the possibility of corrosion. Usually the process selected for dehydration involves either liquid or solid desiccants. Dehydration may also be accomplished by expansion refrigeration which utilizes the Joule-Thompson effect.

Natural Gas Liquids

Natural gases containing high concentrations of the higher hydrocarbons are processed both to reduce the potential for condensation of these higher molecular-weight compounds during transmission and subsequent use, and to recover the natural gas liquid (NGL) products which can be marketed in both the fuel and petrochemical feedstock market.

Natural gas liquids are recovered from natural gas using condensation processes, absorption processes employing hydrocarbon liquids similar to gasoline or kerosene as the absorber oil, or solid-bed adsorption processes using adsorbants such as silica, molecular sieves, or activated charcoal.

Acid Gas Constituents

There are more than 30 processes available for removing the acid gas constituents such as hydrogen sulfide, carbon dioxide, and other organic sulfur compounds, ie, carbonyl sulfide, organic mercaptans, and disulfides. Because of the toxicity of hydrogen sulfide, requirements for removal are severe.

Both batch processes and continuous processes are used. Batch processes are used when the daily production of sulfur is small and of the order of 10 kg. When the daily sulfur production is higher, of the order of 45 kg, continuous processes are usually more economical. Using batch processes, regeneration of the absorbant or adsorbant is carried out in the primary reactor. Using continuous processes, absorption of the acid gases occurs in one vessel and acid gas recovery and solvent regeneration occur in a separate reactor.

Iron sponge is the oldest and most widely used batch process for removing sulfur compounds from natural gas.

There are numerous chemical and physical solvents available for use in continuous acid gas removal processes. The chemical absorbants include aqueous solutions of organic amines such as monoethanolamine, diethanolamine, triethanolamine, diglycolamine, or methyldiethanolamine. Adsorption systems employing

molecular sieves are available for feed gases having low acid gas concentrations. Another option is based on the use of polymeric, semipermeable membranes which rely on the higher solubilities and diffusion rates of carbon dioxide and hydrogen sulfide in the polymeric material relative to methane for membrane selectivity and separation of the various constituents.

Nitrogen

The separation of nitrogen from natural gas relies on the differences between the boiling points of nitrogen (77.4 K) and methane (91.7 K) and involves the cryogenic distillation of a feed stream that has been preconditioned to very low levels of carbon dioxide, water vapor, and other constituents that would form solids at the low processing temperatures.

SPECIFICATIONS

Whereas there is no universally accepted specification for marketed natural gas, standards addressed in the United States are listed in Table 3.

TRANSMISSION AND STORAGE

As exploration and production activities have expanded both the natural gas resource base and the worldwide proven reserves, long-distance gas transmission pipelines have been constructed to link these resources to the industrialized areas and population centers of the world. The availability of high tensile-strength steel pipe and the development of techniques to construct, weld, and lay large diameter high pressure pipelines make it possible to economically transport natural gas to the marketplace. These transmission systems, coupled with localized, lower pressure distribution networks, bring gas to large segments of the world.

Natural gas can be stored in many ways. The most common is in underground geologic formations. Two types of underground storage facilities are aquifer reservoirs and salt caverns.

Table 3. Natural Gas Pipeline Specifications[a]

Characteristic	Specification	Test method[b]
water content, mg/m^3	64–112	ASTM (1986) D1142
hydrogen sulfide, mg/m^3	5.7	GPA (1968) Std. 2265
gross heating value,[c] MJ/m^3	35.4	GPA (1986) Std. 2172
hydrocarbon dew point at 5.5 MPa,[d] K	264.9	ASTM (1986) D1142
mercaptan content, mg/m^3	4.6	GPA (1968) Std. 2265
total sulfur, mg/m^3	23–114	ASTM (1980) D1072
carbon dioxide, mol%	1–3	GPA (1990) Std. 2261
oxygen, mol%	0–0.4	GPA (1990) Std. 2261

[a]Gas must be commercially free of sand, dust, gums, and free liquid. Delivery temperature, 322.16 K; delivery pressure, 4.83 MPa.
[b]ASTM = American Society for Testing Materials; GPA = Gas Processors' Association.
[c]To convert MJ/m^3, multiply by 26.86.
[d]To convert MPa to psi, multiply by 145.

Liquefied natural gas (LNG) also plays a large role in both the transportation and storage of natural gas. At a pressure of 101.3 kPa (1 atm), methane can be liquefied by reducing the temperature to about $-161°C$. When in the liquid form, methane occupies approximately 1/600 of the space occupied by gaseous methane at normal temperature and pressure.

USES

Fuel

Natural gas is used as a primary fuel and source of heat energy throughout the industrialized countries for a broad range of residential, commercial, and industrial applications.

Chemical Use

Both natural gas and natural gas liquids are used as feedstocks in the chemical industry. The largest chemical use of methane is through its reactions with steam to produce mixtures of carbon monoxide and hydrogen.

Ethylene is produced by steam-cracking the ethane and propane fractions obtained from natural gas, and the butane fraction can be catalytically dehydrogenated to yield 1,3-butadiene, a compound used in the preparation of many polymers. The n-butane fraction can also be used as a feedstock in the manufacture of (MTBE) methyl *tertiary* butyl ether.

Energy Information Administration, www.eia.gov., March 2004.

M. Malveda, "Natural Gas Liquids," *Chemical Economics Handbook*, SRI International, Menlo Park, Calif., Nov. 2001.

T. J. Woods, *The Long-Term Trends in U.S. Gas Supply and Prices: 1992 Edition of the GRI Baseline Projection of U.S. Energy Supply and Demand to 2010*, Gas Research Insights, Gas Research Institute, Chicago, Ill., Dec. 1991.

KERMIT E. WOODCOCK
Consultant
MYRON GOTTLIEB
Gas Research Institute

GASOLINE AND OTHER MOTOR FUELS

Gasoline and other motor fuels comprise the largest single use of energy in the United States, and in 2002 accounted for 22% of all energy usage. The cost of this energy has been and is expected to continue to be a primary factor in the national economy. Moreover, the fraction of resources from which these fuels come that is provided by foreign sources is a matter of political concern. The fraction of total crude oil produced domestically shrunk from 75% in 1970 to 39% in 2002 and is predicted to continue to drop. In the 1970s, two Organization of Petroleum Exporting Countries (OPEC) embargoes resulted in rapid increases in the price of crude oil and therefore motor fuels. These increases triggered programs designed to develop alternative sources of fuels such as coal, oil shale, and natural gas. In 2001, as a result of lower price volatility and improved energy efficiencies, the inflation adjusted cost of driving an average passenger car is about 40% lower than it was in 1960. Today, alternative transportation fuels are being considered as replacements for traditional petroleum based fuels based on their perceived potential to improve the security of energy supply and to reduce emissions of greenhouse gases.

General Aspects of Manufacture of Motor Fuels

All motor fuel in the United States is manufactured by private companies. Many of these are vertically integrated. That is, the same company finds the crude oil or buys it from a producing government, refines it into finished products, and then sells these products to independent retailers who specialize in that company's blended products or sells them at company operated service stations. There are also a significant number of companies that participate in only some aspects of the business cycle such as refining or marketing.

Four groups are involved in the production or use of motor fuels in the United States: (*1*) manufacturers of the vehicles; (*2*) manufacturers and/or marketers of the fuels; (*3*) purchasers and users of fuels and vehicles; and (*4*) federal, state, and local regulatory agencies.

Appropriate specifications are set when the vehicle and the fuel are viewed as a system. The American Society of Testing and Materials (ASTM), now known as ASTM International, was founded in 1902 to promote just such a process. ASTM Committee D-2 provides a forum for regulators, vehicle manufacturers, fuel producers, and consumers to develop and recommend nonbinding standards for petroleum products.

GASOLINE

Gasoline demand is largely determined by the growth in the number of cars, the kilometers of paved roads available for driving, population, and economic growth. A primary factor in moderating the demand for gasoline has been the dramatic improvement in automotive fuel economy from the mid-1970s through the mid-1980s. Future projections include growth in travel and modest improvements in fuel economy. The recent trend of using technology to improve performance rather than efficiency will continue; average horsepower is projected to grow by 24% while fuel economy will improve by only 6%.

Requirements of Good Gasoline

To satisfy high performance automotive engines, gasoline must meet exacting specifications, some of which are varied according to location and based on temperatures or altitudes. The fuel must evaporate easily and burn completely when the spark plug fires in each cylinder. Early detonation of the fuel in the cylinder can cause destructive engine knock. The fuel must be chemically stable. It should not form gums or other polymeric deposit precursors. There should be no particulate contaminants or entrained water. Contaminants must be prevented from the point of manufacture in the refinery all the way through the distribution system until the fuel is metered from the vehicle tank into the engine.

Octane. Octane is probably the single most recognized measure of gasoline quality. Broadly speaking, octane is a measure of the combustion characteristics of gasoline. Low octane gasoline has a tendency to preignite, causing rapid energy release and pressure fluctuations in the cylinder which result in a loud metallic noise commonly called knock. In addition to producing an objectionable sound, knock can reduce the amount of useful work that is extracted from the engine. Under extreme conditions of prolonged knock, overheating and even engine damage can occur.

Chemical Factors. Because knock is caused by chemical reactions in the engine, it is reasonable to assume that chemical structure plays an important role in determining the resistance of a particular compound to knock. The chemical factors affecting knock are shown in Table 1.

Vehicle Factors. Because knock is a chemical reaction, it is sensitive to temperature, change density and reaction time. Engine operating and design factors which affect the tendency to produce knocking are

Increased Knock	Decreased Knock
Higher compression ratio	Increased turbulence
Advanced spark schedule	Exhaust gas recycle
Higher coolant temperature	Cooled air charge
Turbocharging or supercharging	High altitudes
Combustion chamber deposits	High humidity
Engine loading	

Measuring Octane. Two different values need to be considered when discussing octane measurements. One is the knocking tendency of the fuel, called the fuel octane number. The other is the knocking tendency of the vehicle, called octane number requirement.

The octane value of an unknown fuel sample is determined by comparing its knocking tendency to various primary reference fuels (PFRs). Its measured octane is equal to the octane of the PRF which has the same knocking intensity. Knock intensity is controlled to an average value by varying the compression ratio of the cooperative fuel research (CFR) engine. In practice, the exact value of

Table 1. Chemical Factors Affecting Knock

Change in chemical structure	Change in knocking tendency	
	Increase	Decrease
longer paraffin chains	x	
isomerizing normal paraffins		x
aromatizing normal paraffins		x
alkylating aromatics		x
saturating aromatic rings	x	

a fuel's octane number is determined to the nearest 0.1 octane number by interpolation from two PRFs that are no more than two octane numbers apart. The CFR engine is operated at two conditions to simulate typical on-road driving conditions. The less severe condition measures research octane number (RON); the more severe one measures motor octane number (MON). The octane number requirement (ONR) of a car is the octane number which causes barely audible, or trace knock when driven by a trained rater.

Volatility. The properties of a gasoline which control its ability to evaporate are critical to good operation of a vehicle. In an Otto cycle engine, the fuel must be in the vapor state for combustion to take place. The volatility or vaporization characteristics of a gasoline are defined by three ASTM tests: Reid vapor pressure (RVP), the distillation curve, and the vapor/liquid ratio (V/L) at a given temperature, ASTM D 4953, D 86, and D 2533, respectively.

Startability. In order to achieve combustion in an Otto cycle engine, the air/fuel ratio in the combustion chamber must be near the stoichiometric ratio. Unfortunately, when the engine is first started, the walls of the combustion chamber and the intake manifold are not hot enough to vaporize much fuel. The ability of a fuel to achieve good starting can be correlated with RVP and a measure of the front end of the distillation curve, either E70 or T_{10}. Obviously, fuel specifications change with ambient temperatures. At higher temperatures, lower RVP and front-end volatilities are adequate to provide good starting characteristics.

Vapor Lock. At the other end of the spectrum from starting is vapor lock, which occurs when too much of the fuel evaporates and either starves the engine for fuel or provides too much fuel to the engine. It occurs on days that are warmer than usual and when the car has reached full operating temperatures.

Vehicle manufacturers minimize vapor lock by keeping the fuel system cool and under positive pressure. Fuel manufacturers minimize the problems by seasonal volatility blending.

Warm-Up. Warm-up refers to that period of operation beginning immediately after the car has started and continuing until the engine has reached normal operating temperatures, usually after 10 minutes or so of operation. During this period, the vehicle designer wants to get the vehicle equivalence ratio to stoichiometric as soon as possible to minimize emissions. On the other hand, if the mixture is leaned out too soon, the car experiences poor driveability during the warm-up period. Fuel system design is critical during this period. From the fuel's perspective, the middle of the distillation curve plays the largest role in achieving good warm-up performance.

Icing. At temperatures within 5°C of freezing and under conditions of high humidity, ice can form in the intake system of vehicles with carburetors or throttle body fuel injectors. In the extreme, ice can clog the carburetor jets and stall the car completely. After the car is fully warmed-up there is generally enough heat in the intake system to prevent any ice buildup. The universal adoption of port fuel injection has eliminated this concern from today's cars.

Back-End Volatility. The portion of the gasoline that boils above 150°C is referred to as the back end. Generally, as the engine heats up the fuel vaporizes completely. However, if there are too many back ends in the gasoline, then not all may boil off and the performance of the lubricant may be degraded. Very heavy molecules, such as those having more than 12 carbon atoms, may contribute to combustion chamber deposits. Condensed ring aromatics are particularly strong contributors to these deposits.

Cleanliness. Good gasoline must be both chemically and physically clean. Chemical cleanliness means that it does not contain nor react under conditions of storage and use to form unwanted by-products such as gums, sludge, and deposits. Chemical cleanliness is assured by controlling the hydrocarbon composition and by appropriate additives. Physical cleanliness means that there are no undissolved solids or large amounts of free water in the gasoline. Stability Existent Gum is determined by (ASTM D 381). Oxidative Stability (ASTM D25), and Potential Gum (D 873).

Manufacture

Distillation. Petroleum refining begins with the distillation of crude oil into a number of different fractions. In many cases, two distillations are carried out in units called pipestills. One at atmospheric pressure and one under vaccum. Table 2 shows typical boiling ranges for the various crude oil fractions and typical yields from Arab Light, a common crude oil.

Catalytic Cracking. Over 50% of the gasoline in the United States is obtained by catalytic cracking which uses a fluidized bed of powdered or small diameter catalysts that are continuously regenerated in an adjacent vessel called a regenerator. The principal class of reactions in the FCC process converts high boiling, low octane normal paraffins to lower boiling, higher octane olefins, naphthenes (cycloparaffins), and aromatics. FCC naphtha is almost always fractionated into two or three streams.

Thermal Cracking. Certain cracking conversion processes are carried out without catalysts. Heavy residuum

streams in the refinery can be cracked thermally to produce coke and a mixture of lighter products. If naphtha is produced, it may require extensive treating before it can be used directly in gasoline or used as feed to other processes. Heavy distillate may also be thermally cracked using high (~800°C) temperature steam. This process, steam cracking, is used to generate olefins for use in chemicals plants, but also generates material in the naphtha range which, if the quality is appropriate, may be used in gasoline.

Reforming. Catalytic reforming is a process to increase the octane of gasoline components. The feed to a reforming process is naphtha (usually virgin naphtha) boiling in the 80–210°C range. The catalysts are platinum on alumina, normally with small amounts of other metals such as rhenium.

Depending on the catalysts and operating conditions, the following types of reactions occur to a greater or lesser extent: (*1*) heavy paraffins lose hydrogen and form aromatic rings; (*2*) cycloparaffins lose hydrogen to form corresponding aromatics; (*3*) straight-chain paraffins rearrange to form isomers; and (*4*) heavy paraffins are hydrocracked to form lighter paraffins.

Reformers generate highly aromatic, high octane product streams, and a great deal of hydrogen.

Some of the negative aspects of reformate are production of benzene, polynuclear or multiring aromatics (PNAs), and light gas (C_1–C_4).

Alkylation. Alkylation is the chemical combination of two light hydrocarbon molecules to form a heavier one and involves the reaction of butenes and butanes in the presence of a strong acid catalyst such as sulfuric or hydrofluoric acid. The product is a heavier multibranched isoparaffin. Propene and the various pentenes may also be used, to produce C_7 or C_9 isoparaffins, respectively.

Isomerization. Isomerization is a catalytic process which converts normal paraffins to isoparaffins. The feed is usually light virgin naphtha and the catalyst platinum on an alumina or zeolite base. Octanes may be increased by over 30 numbers when normal pentane and normal hexane are isomerized. Another beneficial reaction that occurs is that any benzene in the feed is converted to cyclohexane.

Hydrogen Processing. Hydrogen is probably the most valuable refinery chemical in terms of its ability to improve the quality of refinery streams. It can be used to remove unwanted species such as sulfur and nitrogen from gasoline and diesel.

Blending Agents. Blending agents are components of gasoline that are used at levels up to 20% and which are not natural components of crude oil. As of this writing, all blending agents are oxygenated compounds such as ethers and alcohols and are used for one or more of a variety of reasons, including to increase the octane of the fuel, to reduce vehicle emissions, to reduce dependency on imported crude oil, and/or to reduce emissions of greenhouse gases by using renewable resources.

Table 2. Typical Properties of Crude Oil Fractions

Fraction	Boiling range, °C	Yield from crude, %
gas	<0	<1
light virgin naphtha	0–100	
heavy virgin naphtha	100–200	18 (L + H)
gas oil/kerosene	200–400	33
residue	>400	48

Additives. Gasoline additives are used to improve the performance of the fuel either because the hydrocarbon components themselves contain some deficiency or because it is more effective to add a small amount of additive than to change the composition of the gasoline. Additives are added in parts per million levels to distinguish them from blending agents which are added in the percents. They include dyes, antioxidants, metal deactivators, corrosion inhibitors, antiicing additives, detergent additives. intake valve detergents, and demulsifiers.

Blending and Distribution

When blending gasoline from its components, refinery operators must balance a number of factors in the most economical way. First, the components must be used at the same rate at which these are produced or else the refinery either runs out of material or drowns in excess components. Secondly, each gasoline fuel grade must be produced to the specifications set by marketers and regulators. The specifications should not be exceeded in a way which increases manufacturing cost. Third, blend targets must take into account the fact that the gasoline might not be sold immediately and may travel in pipelines, barges, or tankers and then sit in a distribution terminal. The time between production in the refinery and sale to the customer can be as long as one month.

Many gasoline properties, especially octane and RVP, do not blend linearly. Proper prediction of the octane of refinery blends is important because octane has traditionally been one of the most expensive gasoline properties and raising pool octane often entails significant investment and increased operating costs. Also, it is possible to meet targets for the different grades by properly choosing blend stocks to take advantage of the nonlinear blending characteristics.

Gasoline blends are shipped from the refinery to a storage terminal. At the terminal, additives are metered into the gasoline as it is loaded onto tank trucks for shipping to individual service stations. The trucks, which have 40,000 L capacity, have 4–5 compartments so that they can deliver different grades at the same time. Service station tanks are buried underground and have a capacity of 12,000–15,000 L. High volume service stations may have underground tanks as large as 45,000 L. Submerged turbine pumps transfer the gasoline from the tanks to dispensers at the dispensing islands. Tanks have been made of a number of materials, although the most popular today is reinforced fiberglass. As a result of environmental concerns about gasoline leakage from underground tanks, many new installations have double-wall construction with leak detectors between the two walls.

Fuel Economy

Fuel economy, typically expressed as distance driven per volume of fuel consumed, ie, in km/L (mi/gal), is measured over two driving cycles specified by the Federal Test Procedure. Fuel economy is measured using a carbon balance method calculation. The carbon content of the exhaust is calculated by adding up the carbon monoxide, carbon dioxide, and unburned hydrocarbons concentrations. Then using the percent carbon in the fuel, a volumetric fuel economy is calculated. If the heating value of the fuel is known, an energy specific fuel economy in units such as km/MJ can be calculated as well.

Environmental Issues

The Clean Air Act directs EPA to set National Ambient Air Quality Standards (NAAQS) for a number of pollutnats such as ozone, CO and NOx, and sets or directs EPA to set vehicle emissions standards and fuel composition standards to help meet these NAAQSs. The first of these standards set by EPA went into effect in 1968 and mandated that the vapors from the vehicle crankcase be routed back through the engine and burned. Since then, the standards have continued to grow stricter.

The emissions standards have forced changes in vehicle hardware. Many of these changes in hardware have resulted in changes in fuel as well.

In addition to setting standards for exhaust emissions, the government set standards for evaporative emissions. These refer to hydrocarbons that escape from the vehicle when fuel evaporates; while the car is operating (running losses); while it is sitting and not being operated (diurnal emissions), immediately after operation (hot soak); or while it being refueled. In order to control evaporative emissions, auto manufacturers have installed canisters of activated charcoal in their vehicles since 1972.

Although the charcoal canisters are about 95% effective, fuel volatility still impacts the mass of vapors that break through the canister. Therefore, EPA mandated that starting in the summer of 1992, RVP levels be reduced below the levels specified in ASTM D 4814. Class C regions, generally the northern part of the country, are limited to a maximum RVP of 62 kPa (9.0 psi) vs an ASTM limit of 79 kPa (11 psi), and the southern Class B regions are limited to a maximum RVP of 54 kPa (7.8 psi) vs 69 kPa (9.0 psi) for ASTM.

The Clean Air Act Amendments of 1990 introduced a new concept in emissions reduction: reducing exhaust emissions by controlling the composition of the fuel. Whereas previous regulations had lowered exhaust emissions by setting standards for new vehicles, this law mandated gasoline marketers to change gasoline composition so that emissions from existing vehicles would be reduced. Reduction targets of 15% and at least 20% were set for 1995 and 2000, respectively. Reductions are to be measured against 1990 vehicles and industry average gasoline. The reductions are for hydrocarbons (summer ozone season only), and for air toxics (year round). Air toxics are defined as the sum of the emissions of benzene, 1,3-butadiene, formaldehyde, acetaldehyde, and polycyclic organic material. Reformulated gasoline (RFG) may not result in any increase in NO_x emissions. Additionally, RFG must contain no more than 1% benzene and at least 2% oxygen.

In the winter, the Clean Air Act mandates that gasoline in all areas which exceed the NAAQS for CO must contain at least 2.7% oxygen. This is based on the assumption that adding oxygen to the fuel reduces CO emissions.

Table 3. California Phase 3 Gasoline Compostion

| Property | Limit per liter | Values for averagers | |
		Average	Cap
RVP, kPa[a]	48	NA	50
sulfur, ppm	20	15	30
aromatics, vol%	25	22	35
olefins, vol%	6	4	10
T_{90}, °C	152	146	166
T_{50}, °C	101	95	104
benzene, vol%	0.8	0.7	1.1

[a] To convert kPa to psi, multiply by 0.145.

Compliance with the RFG regulations is measured by using a formula to calculate the emissions from a given fuel. This formula, the Complex Model, predicts average exhaust and evaporative emissions as a function of gasoline chemical and physical parameters. Sulfur levels must average 30 ppm and all blends must be 80 ppm or less.

The state of California has taken a different conceptual approach to reducing emissions through control of gasoline composition. Instead of defining a performance target, ie, 25% reduction, California has defined composition targets which are aimed at achieving emissions reductions. Alternative formulations are allowed if they achieve equal emissions perform as defined by the "Production Model." The third round of regulations, known as Phase 3, are shown in Table 3 and took effect on January 1, 2004.

Gasoline composition regulations are being considered in many other parts of the world as well. The options range from the basic, such as eliminating lead, to full reformulations such as those adopted in the United States and Europe. The choice of appropriate regulations should be based on many factors, including vehicle technology, the status of air quality, and cost effectiveness relative to other emission control steps.

DIESEL FUEL

As a fuel for internal combustion engines, diesel fuel ranks second only to gasoline. Diesel cars, thought at one time to be very promising, have encountered significant customer resistance in the United States. In many countries, diesels engines do not have the negative image that they do in the United States.

Combustion in Diesel Engines

Unlike the spark-ignited gasoline engine, the diesel engine, first used by Rudolf Diesel in the 1890s to burn finely powdered coal dust, employs compression ignition. Liquid fuel was employed soon after.

There are two categories of diesel engines: direct injection (DI) and indirect injection (IDI). In DI engines, the fuel is injected directly into the combustion chamber. In IDI engines, there is a small prechamber into which the fuel is injected. The fuel starts to ignite in the prechamber and the hot burning gases are forced out into the main combustion chamber through a small passage. IDI engines may operate at higher speeds and use lower pressure injec-

tor systems which tend to be less expensive. Most new engines, both light and heavy duty, use DI designs to take advantage of higher power and better fuel economy.

Requirements for Good Diesel Fuel

Diesel fuel is used in a wide variety of vehicular engines ranging from small passenger cars to large trucks and construction equipment. There are actually three grades of diesel fuel defined in ASTM D 975, the specification for diesel fuels. The first is Grade 1-D, suitable for high speed engines which operate under widely varying conditions of speed and load. Grade 1-D also has excellent low temperature properties. Grade 2-D is a general-purpose diesel suitable for use either in automotive or nonautomotive applications. It can be used in high speed engines involving relatively high loads and uniform speeds. Grade 4-D is much more viscous and is used in low and medium speed engines having sustained loads at substantially constant speed. Most cars and trucks use 2-D, a general-purpose grade. 1-D, a more volatile, lower density, lower aromatic fuel, used in cold weather and in municipal buses.

Ignition Quality. The ability of diesel fuel to burn with the proper characteristics is described by its cetane number, a measure of ignition delay. Excessively long ignition delays (low cetane number) cause rough engine operation, misfiring, incomplete combustion, and poor startability.

The procedure for measuring the cetane number of diesel fuel (ASTM D 613) is similar to that used for measuring gasoline octane number. Cetane (n-hexadecane), $C_{16}H_{34}$, is defined as having a cetane number of 100; α-methylnaphthalene, $C_{11}H_{10}$, is defined as having a cetane number of 0. 2,2,4,-4,6,8,8-Heptamethylnonane (HMN), $C_{16}H_{34}$, which can be produced in high purity, is used as the low reference fuel and has a cetane number of 15. Blends of cetane and HMN represent intermediate ignition qualities according to the formula: cetane number = % cetane + 0.15 (% HMN).

The cetane number of a fuel depends on its hydrocarbon composition. In general, normal paraffins have high cetane numbers, isoparaffins and aromatics have low cetane numbers, and olefins and cycloparaffins fall somewhere in between. Most diesel fuels marketed in the United States have cetane numbers ranging between 40 and 50. Most manufacturers specify a minimum cetane number of 40–45.

Cetane number is difficult to measure experimentally. Therefore, various correlation equations have been developed to predict cetane number from fuel properties.

Cold Temperature Properties. Diesel fuel must be able to be pumped and to flow through all filters and injectors at the lowest temperature that may be encountered in use. When the temperature is lowered, wax molecules in the fuel start to crystallize. This temperature is known as the wax appearance point or cloud point. These temperatures, which are generally the same, are measured by ASTM D 2500 and D 3117, respectively. If the temperature is lowered still further, the fuel gels and does not flow. This is the pour point and is measured by ASTM D 97. These tests measure the ability of a fuel to operate in a diesel engine. Generally, the cloud point of a fuel is 4–6°C above the pour point, although fuels having differences of 11°C are not uncommon. The true operability temperature is somewhere in between the two; cloud point is too high and pour point is too low. Many engine manufacturers recommend fuels having pour points of 6°C below the lowest temperature at which the engine is expected to operate. Additives may be used to extend downward the operating range of diesel fuel.

Volatility. Volatile light fractions in diesel fuel help to provide easy engine starting but are generally low in cetane number and energy content. Heavy fractions, which have good cetane and energy content, can contribute to deposit formation and hard starting if present in too high concentrations. Desirable quality characteristics are obtained by careful blending of refinery streams.

Viscosity. For optimum performance of diesel engine injector pumps, the fuel should have the proper viscosity. Too low viscosity results in excessive injector wear and leakage. Viscosity that is too high may cause poor atomization of the fuel upon injection into the cylinders.

Density. The greater the density of diesel fuel, the greater its heat content per unit volume and therefore the greater its power or fuel economy. Because diesel fuel is purchased on a volume basis, density is often stipulated in purchase specifications and measured on delivery.

Flash Point. Specifications for flash point vary with grade; the lowest value is 38°C for grade 1-D. Controlling flash point is important in order to prevent the vapor space in storage and vehicle tanks from being in the explosive range. Setting the flash point at 38°C protects most storage vessels from exploding.

Carbon Residue. The tendency of a diesel fuel to form carbon deposits in an engine can be roughly predicted by the Ramsbottom Coking Method (ASTM D 524), which determines the amount of carbon residue left after evaporation and chemical decomposition of the fuel at elevated temperatures for a specified length of time. For use in high speed diesel engines operating over a range of loads and speeds, ASTM specifications call for no more than 0.15% Ramsbottom carbon residue.

Sulfur. Sulfur in diesel fuel should be kept below set limits for both environmental and operational reasons. Operationally, high levels of sulfur can lead to high levels of corrosion and engine wear owing to emissions of SO_3 that can react with condensed water during start-up to form sulfuric acid. Diesel fuel can contain sulfur concentrations as high as 5000 ppm. As particulate emission standards for diesels became more stringent, engine manufacturers made the case that they could not meet the standards unless sulfur levels were reduced. Starting in 2006, diesel fuels with 15 ppm sulfur must be sold in the U.S.

Ash Content. The fuel injectors of diesel engines are designed to very close tolerances and are sensitive to any abrasive material in the fuel. Therefore, the maximum permissible ash content of the fuel is specified.

Aromatics Content. Aromatic compounds have very poor ignition quality and, although they are not specifically limited in ASTM D 975, there are practical limitations to using high aromatic levels in highway diesel fuel. The federal government began effectively limiting aromatic content to below 40% starting in October, 1993 by specifying a minimum cetane index of 40. California limits aromatic levels below 10% beginning in the same time period, also because of emissions concerns.

Stability. Diesel fuel can undergo unwanted oxidation reactions leading to insoluble gums and also to highly colored by-products. Stability is measured using ASTM D 2274.

Lubricity. Diesel fuel has traditionally had adequate lubricity to protect fuel pumps and injectors from excessive wear. Changes in equipment and fuel processing have caused concern that lubricity may no longer be adequate. Lower viscosity fuels have poorer lubricity, and processing steps that lower sulfur may also remove trace levels of surface active compounds that provide natural lubricity.

Lubricity is measured in two different tests: ASTM D 6078, the scuffing load ball-on-cylinder evaluator (SLBO-CLE); and ASTM D 6079, the high frequency reciprocating rig (HFRR). In response to concerns voiced by engine and equipment manufacturers, ASTM will likely adopt a maximum specification of 520 mm average wear scar in the HFRR test.

Diesel Fuel Manufacture

The biggest factors in determining how diesel fuel is blended in a given refinery are the availability of high cetane stocks. In order of decreasing ignition quality, the hydrocarbon types rank in the following order: normal paraffins, olefins, cycloparaffins, branched paraffins, and aromatics. Because straight-run distillates contain the greatest amount of normal paraffins and cycloparaffins and the least amounts of branched paraffins and aromatics, these are the preferred stocks for diesel blending. Cracked stocks, which are relatively rich in aromatics, are less desirable from the standpoint of ignition quality. However, these have high energy density and good cold temperature properties. In the United States, where a high level of cracking is necessary to meet gasoline demand, the large supply of cracked fractions and the relatively small supply of straight-run distillates make substantial use of cracked stocks economically necessary. This has been made possible through the use of cetane improvers to improve cetane and through the use of hydrogenation to improve stability.

Other additives include ignition improvers, stability improvers, corrosion inhibitors, detergent additives, cold flow improvers, lubricity improvers, dyes, antifoamants, and oxygenates.

Diesel Environmental Regulations

Emission standards have been set for heavy-duty vehicles in much the same manner as they have been set for gasoline engines. Because heavy-duty vehicles are primarily diesels, the focus is on diesel engine emissions. Standards have been written in units of grams per brake-horsepower-hour (g/bhph = g/kW · h × 1.34), which normalize the emissions according to the total energy output of an engine over the specified driving cycle.

It is also generally agreed that in terms of engine design parameters, there is an inverse relationship between NO_x and particulates. Rapid, complete combustion reduces particulate emissions but also promotes the formation of NO_x. Exhaust gas recycle (EGR), which lowers NO_X emissions, generally causes particulate emissions to increase.

California has taken a slightly different approach to diesel fuel composition standards from that of the federal government. In October 1993, California limited diesel fuel to no more than 10% aromatics and 500 ppm sulfur. Alternative formulations are possible if these are shown to have equivalent NO_x emissions to a base reference fuel. In addition to the specifications that apply to the commercial fuel, other aspects of the reference fuel composition are tightly controlled. The fuel must also have a minimum cetane number of 48 without cetane improvers.

ALTERNATIVE FUELS

Alternative fuels fall into two general categories. The first class consists of fuels that are made from sources other than crude oil but that have properties the same as or similar to conventional motor fuels. In this category are fuels made from coal, shale and natural gas through Fischer-Tropsch synthesis. Biofuels such as ethanol and fatty acid methyl esters (biodiesel) also belong in this category, too. In the second category are fuels that are different from gasoline and diesel fuel and which use redesigned or modified engines. These include compressed natural gas (CNG), liquefied petroleum gas (LPG) and hydrogen. The most probable use of hydrogen would be in fuel cells vehicles, both of which will be discussed elsewhere.

Alcohols and Ethers, A Technical Assessment of Their Application as Fuels and Fuel Components, API Publication 4261, 3rd ed., American Petroleum Institute, Washington, D.C., 2001.

Automotive Handbook, 5th Edition, Robert Bosch GmbH, Stuttgart, Germany, 2000.

J. B. Heywood, *Internal Combustion Engine Fundamentals*, McGraw-Hill Book Co., Inc., New York, 1988.

K. Owen and T. Coley, *Automotive Fuels Handbook Second Edition*, Society of Automotive Engineers, Warrendale, Pa., 1995.

ALBERT M. HOCHHAUSER
ExxonMobil Research and
Engineering Co.

GELATIN

Gelatin is a protein obtained by partial hydrolysis of collagen, the chief protein component in skin, bones, hides, and white connective tissues of the animal body. Type A gelatin is produced by acid processing of collagenous raw material; type B is produced by alkaline or lime processing. Because it is obtained from collagen by a controlled partial hydrolysis and does not exist in nature, gelatin is classified as a derived protein.

Uses of gelatin are based on its combination of properties: reversible gel-to-sol transition of aqueous solution; viscosity of warm aqueous solutions; ability to act as a protective colloid; water permeability; and insolubility in cold water, but complete solubility in hot water. It is also nutritious. These properties are utilized in the food, pharmaceutical, and photographic industries. In addition, gelatin forms strong, uniform, clear, moderately flexible coatings which readily swell and absorb water and are ideal for the manufacture of photographic films and pharmaceutical capsules.

CHEMICAL COMPOSITION AND STRUCTURE

Gelatin is not a single chemical substance. The main constituents of gelatin are large and complex polypeptide molecules of the same amino acid composition as the parent collagen, covering a broad molecular weight distribution range. In the parent collagen, the 18 different amino acids are arranged in ordered, long chains, each having ~95,000 mol wt. These chains are arranged in a rodlike, triple-helix structure consisting of two identical chains, called α_1, and one slightly different chain called α_2. These chains are partially separated and broken, ie, hydrolyzed, in the gelatin manufacturing process. Different grades of gelatin have average molecular weight ranging from ~20,000 to 250,000.

Analysis shows the presence of amino acids from 0.2% tyrosine to 30.5% glycine. The five most common amino acids are glycine, 26.4–30.5%; proline, 14.8(18%; hydroxyproline, 13.3–14.5%; glutamic acid, 11.1–11.7%; and alanine, 8.6–11.3%. The remaining amino acids in decreasing order are arginine, aspartic acid, lysine, serine, leucine, valine, phenylalanine, threonine, isoleucine, hydroxylysine, histidine, methionine, and tyrosine.

Stability

Dry gelatin stored in airtight containers at room temperature has a shelf life of many years. However, it decomposes above 100°C. Aqueous solutions or gels of gelatin are highly susceptible to microbial growth and breakdown by proteolytic enzymes. Stability is a function of pH and electrolytes and decreases with increasing temperature because of hydrolysis.

PHYSICAL AND CHEMICAL PROPERTIES

Commercial gelatin is produced in mesh sizes ranging from coarse granules to fine powder. In Europe, gelatin is also produced in thin sheets for use in cooking. It is a vitreous, brittle solid, faintly yellow in color. Dry commercial gelatin contains about 9–13% moisture and is essentially tasteless and odorless with specific gravity between 1.3 and 1.4. Most physical and chemical properties of gelatin are measured on aqueous solutions and are functions of the source of collagen, method of manufacture, conditions during extraction and concentration, thermal history, pH, and chemical nature of impurities or additives.

MANUFACTURE AND PROCESSING

Although new methods for processing gelatin, including ion exchange and crossflow membrane filtration, have been introduced since 1960, the basic technology for modern gelatin manufacture was developed in the early 1920s. Acid and lime processes have separate facilities and are not interchangeable. Most type A gelatin is made from pork skins, yielding grease as a marketable by-product. Type B gelatin is made mostly from bones, but also from bovine hides and pork skins.

Bovine hides and skins are substantial sources of raw material for type B gelatin and are supplied in the form of splits, trimmings of dehaired hide, rawhide pieces, or salted hide pieces. Like pork skins, the hides are cut to smaller pieces before being processed. Sometimes the term calfskin gelatin is used to describe hide gelatin. The liming of hides usually takes a little longer than the liming of ossein from bone.

Most manufacturing equipment should be made of stainless steel. The liming tanks, however, can be either concrete or wood. Properly lined iron tanks are often used for the washing and acidification, ie, souring, operations. Most gelatin plants achieve efficient processes by operating around the clock. The product is tested in batches and again as blends to confirm conformance to customer specifications.

USES

Food Products

Gelatin formulations in the food industry use almost exclusively water or aqueous polyhydric alcohols as solvents for candy, marshmallow, or dessert preparations. In dairy products and frozen foods, gelatin's protective colloid property prevents crystallization of ice and sugar.

Pharmaceutical Products

Gelatin is used in the pharmaceutical industry for the manufacture of soft and hard capsules.

For arresting hemorrhage during surgery, a special sterile gelatin sponge known as absorbable gelatin sponge or Gelfoam is used.

Gelatin can be a source of essential amino acids when used as a diet supplement and therapeutic agent. As such, it has been widely used in muscular disorders, peptic ulcers, and infant feeding, and to spur nail growth.

Photographic Products

Gelatin has been used for over 100 years as a binder in light-sensitive products. The useful functions of gelatin in photographic film manufacture are a result of its protective colloidal properties during the precipitation and chemical ripening of silver halide crystals, setting and film-forming properties during coating, and swelling properties during processing of exposed film or paper. Quality requirements of photographic gelatin may be very elaborate and can include over 40 chemical and physical tests, in addition to photographic evaluation. Most chemical impurities are limited to less than 10 ppm.

Gelatin is also used in so-called subbing formulations to prepare film bases such as polyester, cellulose acetate, cellulose butyrate, and polyethylene-coated paper base for coating by aqueous formulations.

Derivatized Gelatin

Chemically active groups in gelatin molecules are either the chain terminal groups or side-chain groups. In the process of modifying gelatin properties, some groups can be removed. Commercially successful derivatized gelatins are made mostly for the photographic gelatin and micro-encapsulation markets. In both instances, the amino groups are acylated.

A. P. Giraud, in M. DeClercq, ed., "Photographic Gelatin," in *Proceedings of the Seventh IAG Conference*, Louvain-la-Neuve, 1999, The International Working Group for Photographic Gelatin, 1999, pp. 18–23.

C. R. Maxey and M. R. Palmer, in R. J. Cox, ed., *Photographic Gelatin II*, Academic Press, Inc., New York, 1976, pp. 27–36.

O. M. Simion, in M. DeClercq, ed., Photographic Gelatin, in *Proceedings of the Seventh IAG Conference*, Louvain-la-Neuve, 1999, The International Working Group for Photographic Gelatin, Fribourg, Switzerland, 1999, pp. 163–172.

Thomas R. Keenan
Knox Gelatine, Inc.

GENETIC ENGINEERING, ANIMALS

Most cells within an animal contain a complete copy of genetic information. The genetic information within cells is encoded by deoxyribonucleic acid (DNA). The sequence of nucleotides within DNA is paired with its complementary base within the double-stranded DNA molecule. A complete sequence of DNA is known as a genome and the sequence of DNA dictates an animal's genotype. For humans and most mammals, a single cell contains about six billion base pairs of DNA. Surprisingly, most of the genome consists of DNA that does not encode specific genes. The function of the DNA that lies outside known genes is poorly understood. Genes consist of regulatory regions (DNA that controls gene transcription) as well as transcribed regions [DNA that is transcribed into ribonucleic acid (RNA)]. Transcribed regions of DNA consist of introns and exons. Exons contain the DNA that encodes the specific protein for the gene. Introns are regions of DNA between exons whose function is poorly understood. Transcription is the synthesis of RNA from DNA. The RNA is chemically modified after synthesis and introns are spliced out of the RNA molecule. The processed RNA (termed messenger RNA or mRNA) encodes the sequence of amino acids for cellular proteins. The amount of specific proteins within a cell determines its function within the organism. All proteins have some function within a cell and certain proteins are directly involved in growth and disease. Genetic engineering is the process of modifying genes within an animal so that the amount or type of cellular protein is changed. Changing cellular proteins through genetic engineering can have a variety of applications in animal agriculture and human medicine.

CLASSICAL METHODS OF GENETIC ENGINEERING TO IMPROVE ANIMAL GENETICS

Humans have been practicing genetic engineering for a long time. For example, farm animals have been selected for superior traits for several hundred years. The variation in productivity of farm animals is partially dependent on the activity and function of genes within their DNA. Selection for superiority in a certain trait is equivalent to selecting for superior DNA sequence within the animal. A technology that accelerates genetic progress is artificial insemination. Although superior male genetics can be exploited through artificial insemination, it is more difficult to exploit superior female genetics.

Embryo Splitting

Perhaps the oldest method for creating multiple copies of a single individual is embryo splitting. The procedure for embryo splitting is not complicated but requires delicate instrumentation. By using microdissection tools (either a knife or needle), embryos are separated into two to four pieces (Fig. 1). Although the technology showed initial promise, embryo splitting is generally not practiced commercially because it requires additional time and technical expertise. Conception rates after split embryo transfer may also be slightly lower than conception rates after whole embryo transfer.

Embryo Cloning by Nuclear Transfer from Blastomeres

A principal limitation to embryo splitting is that an embryo can be split only a few times before the pieces are too small to continue development. In contrast, embryo cloning can be used theoretically to make an unlimited number of copies of the same embryo. Cloned embryos are not limited by size because cells housing a complete copy of the embryonic genome are transplanted back into an oocyte that has had its nucleus removed. Therefore, the transplanted cell becomes part of a new one-cell embryo.

Embryo Cloning by Nuclear Transfer from Somatic Cells

The entire field of mammalian embryology changed when the cloned sheep named "Dolly" was born. Dolly demon-

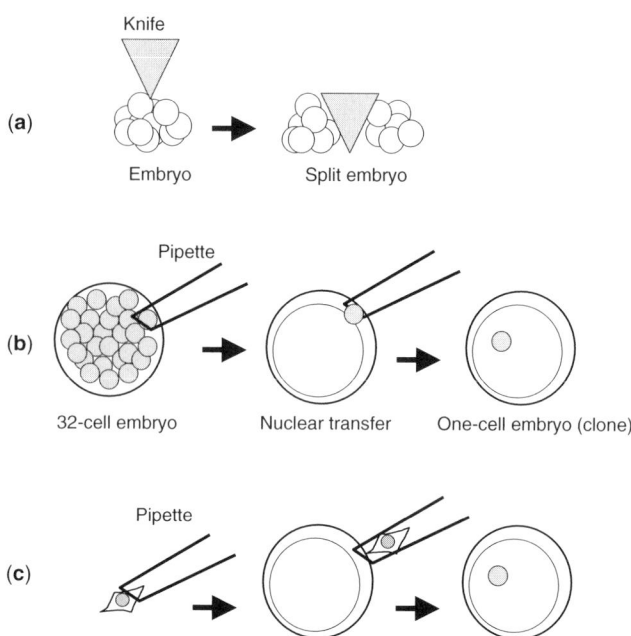

Figure 1. Methods for the production of identical animals. (**a**) Embryo splitting. A microdissection knife is used to cut a morula stage embryo into two identical halves. (**b**) Embryo cloning by nuclear transfer from blastomeres. A cell from a morula stage embryo is removed and fused with an oocyte that has had its nucleus removed. The newly produced one-cell embryo is identical to the original morula. (**c**) Embryo cloning by nuclear transfer from somatic cells. A somatic cell is inserted next to an enucleated oocyte by using a specialized pipette. The somatic cell and the oocyte membranes are then fused. The newly produced one-cell embryo is identical to the original somatic cell.

strated that mammals could be cloned from adult somatic cells. Dolly was created by fusing an adult mammary cell with an enucleated oocyte. Somatic cell nuclear transfer is similar to blastomere nuclear transfer because a karyoplast is fused with an enucleated oocyte. At the time of this writing a variety of somatic cell types have been used for cloning in a number of species (Table 1). Cloning by somatic cell nuclear transfer enables the production of

Table 1. Species and Cell Types Used for Somatic Cell Nuclear Transfer (Cloning) in Animals

Species	Cell types
sheep	embryonic cells
cattle	fetal fibroblasts
goat	mammary gland cells
pig	cumulus cells (ovary)
mouse	oviductal cells
cat	leukocytes
rabbit	granulosa cells (ovary)
	skin fibroblast (newborn and adult)
	newborn liver
	tail tip cells
	fetal germ cells
	fetal ovary
	fetal testicle

an unlimited number of cloned animals from a single adult animal.

GENETIC MODIFICATION OF ANIMALS BY TRANSGENESIS

Genetic selection is the process of manipulating the genome by selecting superior individuals with the best complement of genes in the genome. Animals are selected based on their phenotype (outward expression of the trait of interest). In most cases, the genes that we are selecting for are completely unknown. Transgenesis is the process through which the genome of an animal is modified. A transgenic animal may have a stable modification, a stable deletion, or a stable insertion of a foreign gene into its genome. The foreign gene is called a transgene. Transgenic animals are produced that either over- or underexpress specific proteins within certain cells. The changes in protein expression lead to animals with unique characteristics. Transgenic animals can be classified as traditional transgenics where a gene is inserted at a random location in the genome or gene-targeted transgenics (knock-outs, knock-ins, and conditional knock-outs) where a specific gene within the genome is modified.

Utility of Transgenic Animals

Genes may be randomly inserted into laboratory animal genomes for the purpose of studying basic biological questions or for the purpose of developing models for human diseases. Genes may be randomly inserted into the farm animal genome for the purpose of improving milk production, growth rate, or disease resistance (desirable traits for farmed animals) or for producing recombinant proteins in milk. Transgenes may be a modified gene from the same species (eg, a porcine growth hormone gene engineered to increase blood growth hormone concentrations within a transgenic pig) or may be a gene from an entirely different species (eg, the spider silk protein expressed mammalian cells). Transgenic animals are produced by using a combination of molecular biology techniques (used to synthesize the transgene) and embryo manipulation/embryo culture (used to insert the transgene into the embryo). The most common methods for the production of transgenic animals with a random insertion include pronuclear microinjection of one cell embryos and retroviral infection of oocytes, zygotes, or early cleavage-stage embryos. Transgenes can also be introduced into animals by fusing transgenic embryonic stem cells with early cleavage stage embryos or by nuclear transfer from transgenic embryonic stem cells or transgenic somatic cells into enucleated oocytes.

Creating Transgenic Animals with a Random Gene Insertion

Transfer of foreign genes into animals is done at an early stage of embryonic development (one cell to blastocyst stage) prior to implantation or placentation. Embryos at this stage of development can be grown outside the uterus of the mother (*in vitro* embryo culture) in specialized medium containing nutrients that support their growth. For best results, micromanipulation and gene transfer are performed on one-cell embryos because integration of the transgene into the embryonic DNA theoretically assures that all of the adult cells carry the foreign gene. The goal for modern transgenesis is to create transgenic animals that have controllable transgene expression. The transgene expression may be tissue-specific, developmental-specific, or responsive to specific internal or external signals.

CONSTRUCTION OF A TRANSGENE FOR RANDOM INSERTION

Transgenes are assembled by splicing together DNA from selected genes. Transgene assemblies contain two important parts: a promoter or regulatory region and a protein-coding region. The promoter is a DNA sequence that dictates the activity as well as the tissue specificity for the expression of the protein-coding region of the transgene. A vast array of promoters is available for use in the construction of transgenes. Selection of the appropriate promoter depends on the goals of the project. Promoters may be homologous (similar to the endogenous promoter for the expressed gene) or heterologous (different from the endogenous promoter for the expressed gene). A DNA sequence that encodes mRNA for a protein is spliced next to the promoter. The DNA within this region generally includes exons and introns as well as a polyadenylation signals for the mature mRNA. The DNA is transcribed when the promoter region is activated within certain cells. Activation of the transgene results in the synthesis of heterogeneous nuclear RNA (hnRNA), which is processed into mature mRNA for the production of a specific protein within the transgenic animal. In some cases, the protein from the transgene may have the same biological activity and structure as the naturally occurring protein. However, the protein from the transgene may be expressed in a greater amount or synthesized within a different tissue. The adjoining promoter within the transgene determines the amount and pattern of protein expression.

Methods for random insertion of a transgene into an embryonic genome, include microinjection of a transgene, sperm-mediated gene transfer, retroviral infection of oocytes and embryos, and gene transfer with embryonic stem cells and somatic cells.

Other gene targeting methods include genetic modification of animals by gene targeting, production of targeted deletions, production of conditional targeted deletions (Cre/*loxP* system), and knock-in mice.

APPLICATIONS OF TRANSGENIC AND GENE-TARGETED ANIMALS

The production and use of transgenic and gene targeted animals represents an evolving technology of engineering animal species for specific roles in science and agriculture. Transgenic and gene targeted animals have become mainstays for scientific research. Although the bulk of the genetically modified animals used for scientific research are mice, farm animals are being created when mouse models do not recapitulate the physiology of human disease. Transgenic farm animals are quickly becoming essential for the low-cost production of recombinant proteins that cannot be produced by bacteria. The productivity of farmers may be improved through the use of transgenic farm animals in their herds. Pigs with deleted immune recognition genes may soon provide organs for transplantation into humans.

GENE THERAPY IN HUMANS

Gene therapy is the genetic modification of humans for the purpose of correcting genetic diseases. The procedures for gene therapy in humans are related to those described here. However, gene therapy in humans is typically practiced on somatic cells of children and adults and is intended to relieve the disease state of the treated individual. The genetic modifications are not passed to offspring because germ cells are not targeted. The techniques used for genetic modification of laboratory and farm animals could be applied to human embryos for the purpose of correcting genetic defects. If successful, the genetic defect would be corrected in both somatic and germ cells of the individual arising from the manipulated embryo. However, genetic modification of animals is still an imperfect technology with limitations that are poorly understood. There is a high rate of embryonic loss associated with embryo manipulation and embryo culture. In farm animals, the offspring created by cloning and gene targeting typically have poor health. Second generation animals, however, appear normal. The techniques for genetic modification in animals will need to be optimized before they can be considered safe for the purpose of correcting genetic defects in both somatic and germ cells of humans.

P. Hasty and co-workers, in A. L. Joyner, ed., *Gene Targeting A Practical Approach*, Vol. 1, Oxford University Press, Oxford, U.K., 2000.

S. G. Nonchev and M. K. Maconochie, in I. J. Jackson and C. M. Abbott, eds., *Mouse Genetics and Transgenics A Practical Approach*, Oxford University Press, Oxford, U.K., 2000, pp. 61–86.

V. Papaioannou and R. Johnson, in A. L. Joyner, ed., *Gene Targeting A Practical Approach*, Oxford University Press, Oxford, U.K., 2000, pp. 133–175.

C. A. Pinkert and J. D. Murray, in J. D. Murray, G. B. Anderson, A. M. Oberbauer, and M. M. McGloughlin, eds., *Transgenic Animals in Agriculture*, CAB International, Wallingford, U.K., 1999, pp. 1–18.

MATTHEW C. LUCY
University of Missouri

GENETIC ENGINEERING, MICROBES

Genetic engineering of microbes has undergone a major transformation during the two last decades as many new tools have become available. Genomics, proteomics, fluxomics, and bioinformatics have changed the landscape of molecular biology by providing direct access to entire microbial genomes and insight concerning how global expression of these genomes changes in response to various environmental conditions.

GENE TECHNOLOGIES

The techniques for isolating and manipulating genes from microbes have become routine. Essentially, genes are cloned into plasmid- or phage-based vectors and introduced into the microbe of choice. A well-planned selection scheme or screening method is then necessary to identify host microbes that contain the cloned gene of interest. Most of the reagents that are needed for gene cloning are commercially available and are now frequently supplied as kits for routine tasks such as extraction of DNA, purification of DNA framents for cloning, and isolation of plasmids. In addition, construction and screening of gene libraries are performed by small biotechnology companies.

Gene-Transfer Methods

When bacteria are exposed to an electric field, a number of physical and biochemical changes occur. The bacterial membrane becomes polarized at low electric field strength. The nature of the membrane disturbance is not clearly understood, but bacteria, yeast and fungi are capable of DNA uptake following exposure to approriate electric field conditions. This method, called electroporation, has been used to transform a variety of bacteria and yeast strains that are recalcitrant to other methods.

Gene Amplification

The technique to amplify DNA *in vitro*, known as the polymerase chain reaction (PCR), has turned out to be extremely powerful and is widely used for gene manipulations. The PCR is highly efficient, and low amounts of target DNA are sufficient for amplification of target sequences. As a result, PCR is used in diagnostics and forensic science. PCR can also be used for site-directed mutagenesis, for domain swapping in proteins and for isolating homologous genes.

Gene Alteration

A variety of procedures have been described for introducing specific mutations (base pair changes, deletions, or insertions) into a gene or to subject a specific gene to random mutagenesis. Among these techniques, PCR has proven to be particularly useful for site-directed mutagenesis.

There has been an increase in the number of methods that generate variants of a protein by recombining similar genes with sequences that have diverged because of natural variation or random mutagenesis. One such technique is DNA shuffling, which mimics homologous recombination by using PCR to reassemble random fragments of genes. Another method involves a combinatorial approach to create a library of all possible single base pair deletions in a given piece of DNA by utilizing exonuclease III to incrementally truncate genes, gene fragments or gene libraries.

Genomics

The shotgun approach to sequencing microbial genomes has transformed the landscape of microbial genomics. Comparison of genome sequences is extremely useful in the manipulation of microbes for production of small molecules or proteins. Just as the sequence of the human genome will aid the pharmaceutical industry in discovery of new therapeutics, the sequences of microbial genomes and comparative genomics will accelerate production of industrial chemicals and proteins in a wide range of microbes that so far have not been amenable to engineering by classic methods. The genome sequences are also a rich source for a large number of gene sequences that can be engineered as biocatalysts.

Metabolic Engineering

The development of a variety of new techniques has made it possible to alter entire biochemical pathways that are encoded by multiple genes to overproduce small molecules such as amino acids, vitamins or chemicals. Development of new biocatalysts is being enhanced by coupling the traditional approaches of metabolic engineering with new molecular techniques.

DNA Microarray

DNA microarrays have two major applications, analysis of gene expression and detection of specific nucleotide sequences. The overwhelming advantage of DNA microarrays over other DNA hybridization technologies is that DNA microarrays can be used to analyze thousands of individual genes simultaneously. In addition, DNA microarrays can be used in high throughput approaches for gene discovery, measuring abundance of mRNAs and comparative genomics. Another elegant method to rapidly analyze levels of specific mRNAs in a population of transcripts has been developed. This method involves sequencing cDNA molecules that are tethered to microbeads to generate a unique sequence signature for each type of transcript.

Bioinformatics

Various computational tools have become essential for genetic engineering of microbes. The ability to genetically engineer microbes is greatly enhanced by the tremendous amount of sequence information that is now available. The number of nucleotide and protein sequences in public and private databases is rapidly increasing. There are a variety of software packages available that can be used to access, manipulate and analyze the sequences in these databases.

Sources of Microbial Genes

Rapid sequencing methods and other molecular biology tools have greatly facilitated isolation and characterization of genes and microbes with relevant catalytic activities. With the development of PCR, any gene for which the sequence and species are available can be cloned. Even a gene of unknown sequence can be cloned by designing PCR primers that are based on conserved portions of a consensus sequence derived from related genes of known sequence from other species.

Expression Vectors

A vector that allows optimal expression of cloned genes is required for engineering a microbe to produce a protein or small molecule. Basic expression vectors typically contain an efficient, inducible promoter and convenient restriction sites that allow a coding sequence to be cloned in a proper translational reading frame with a start codon and Shine-Dalgarno sequence. Some expression vectors have additional specialized features.

Display of Peptides and Proteins on Phages

Phage display of peptides and proteins is an extremely versatile method that allows large random libraries to be screened in a relatively short time for a specific function, eg, binding of a hormone to a receptor. Bacteriophages are viruses that specifically infect bacteria.

Phage display is a powerful technique because it is possible to isolate a peptide that has a specific binding affinity from a pool of 10^{12} random peptides in a relatively short period of time.

Expression of Functional Antibodies

Monoclonal and polyclonal antibodies are used in both diagnostics and research. Isolation of a specific monoclonal antibody having high affinity to antigen is expensive and laborious. A combination of gene amplification technology along with phage display provides an excellent method to express a repertoire of antibodies in a short time.

HOST SYSTEMS FOR GENE EXPRESSION

One of the potentials of genetic engineering of microbes is production of large amounts of recombinant proteins and small molecules. A major challenge related to expression of functional proteins in heterologous hosts is that each protein is unique and the stability of the protein depends on the host. Thus, it is not feasible to have a single omnipotent microbial host for the production of all recombinant proteins. Rather, several microbial hosts have to be studied: *Escherichia coli* (a well-developed genetic system and a large base of biochemical information initially made *E. coli* the host organism of choice for gene manipulations but the genetic engineering techniques for *E. coli* have undergone continuous refinement) Expression vectors have to be tailored to the microbe of choice which is dictated by the process economics and feed stock issues.

Erwinia (Gram-negative bacteria that are significant as plant pathogens), Pseudomonads (a diverse group of Gram-negative bacteria that inhabit water and soil and colonize plants), *Bacillus* (a diverse group of Gram-positive, aerobic, endospore-forming bacteria that are primarily found in soil) *Clostridium* (a diverse group of Gram-positive, anaerobic, endospore-forming bacteria that are primarily found in soil), streptomyces (Gram-positive, filamentous soil bacteria are the source of most clinically useful natural antibiotics), *Rhodococcus* (bacteria that are Gram-positive, aerobic, nonmotile actinomycetes), lactic acid bacteria (ubiquitous in Nature, and are found in envrionments that range from plant surfaces to the gastrointestinal tracts of many animals), Yeasts (the primary eukaryotic microbes used for the production of heterologous proteins), filamentous fungi (highly versatile and generate a wide range of commercial products including organic acids such as citric acid secondary metabolites such as antibiotics, and a variety of industrial enzymes).

PRODUCTS FROM GENETICALLY ENGINEERED MICROBES

Genetically engineered microbes are used to produce many commercial products. These products include therapeutic proteins, bulk enzymes, antibiotics, microbe-product chemicals and polymers.

FUTURE PROSPECTS

Biotechnology offers a number of advantages to industry. However, a variety of novel biocatalysts must be developed for biotechnology to deliver on its potential for providing both new and improved processes. The basic tools are in place to take advantage of the rapidity with which recombinant DNA experiments can be performed. In addition, a large amount of DNA sequence is already available from a variety of organisms, with more sequence being deposited every day. Hence, one can expect the number of industrial applications based on genetically engineered microbes to rapidly increase.

R. H. Baltz, G. D. Hegeman, and P. L. Skatrud, *Industrial Microorganisms: Basic and Applied Molecular Genetics*, ASM, Washington, D.C., 1993.

R. H. Doi and M. McGloughlin, *Biology of Bacilli Applications to Industry*, Butterworth-Heinemann, Stoneham, Mass., 1992.

M. A. Innis and co-workers, *PCR Protocols: Guide to Methods and Applications*, Academic Press, Inc., New York, 1990.

J. Sambrook, E. F. Fritsch, and T. Maniatis, *Molecular Cloning*, Vols. 1, 2, 3, Cold Spring Harbor Laboratory Press, Cold Spring Harbor, New York, 1989.

VASANTHA NAGARAJAN
MICHAEL BRAMUCCI
Du Pont

GENETIC ENGINEERING, PLANTS

Several discoveries in the 1980s and 1990s permitted the transition of plant molecular biology from a fledgling science to commercial reality. These discoveries ranged from the identification of biologically important genes to the development of methods to introduce new genes into plants and regulate gene expression. The former process is commonly referred to as transformation. Nearly five dozen plant species have been transformed and the list of plant species subject to transformation include principal field crops such as corn, cotton, rape, rice, soybean, and wheat. In addition, several horticultural species such as tomato, potato, petunia, chrysanthemum, apple, walnut, melons, etc, have been subject to transformation. More than 500 field tests have been conducted and transgenic plants such as transgenic tomato, soybean, corn, rape, potato, petunia, melons, and cucumbers are in the advanced stages of commercial development and regulatory process.

Four methods have been extensively investigated for the introduction of transferred deoxyribonucleic acid (T-DNA) into plants. These include agrobacterium mediated T-DNA transfer, direct uptake of DNA by protoplasts, particle acceleration techniques such as electrostatic discharge or biolistics gun technology, and DNA uptake into partially digested immature embryos. By far the most commonly used method for gene introduction into dicotyledonous plants is the agrobacterium technology.

Expression of genes that have been introduced into plants is regulated by promoters, although the extent of regulation of gene activity by the promoter is influenced at least to some extent by the insertion site of the gene within the chromosome. The choice of promoters is dictated by the tissue and developmental specificity required for gene expression. By far the most commonly used promoter for constitutive gene expression in both mono- and dicotyledonous plants is the Cauliflower mosaic virus (CaMV) 35S promoter.

In order to determine which plant cells have been transformed, selectable marker genes are introduced during transformation. These marker genes permit selective growth of transgenic cells on the medium used for tissue propagation whereas the nontransgenic cells are killed.

Two specific applications of plant biotechnology include engineering tolerance to the widely used herbicide glyphostate, $C_3H_8NO_5P$ and increasing starch biosynthesis in plants. The first application deals with a trait which directly impacts the farmer during the production phase of agriculture; the second application deals with a trait that impacts the consumer of agricultural products. These traits may be referred to as agronomic and quality traits, respectively.

A number of other agronomic and quality traits are being investigated. These include insect, virus, disease, and nematode resistance, fertilizer-use efficiency, ripening control, fruit firmness, etc. Of these traits the most advanced agronomic trait for bioengineering is insect resistance. Insect resistant cotton and corn have been obtained by introduction and expression of a *Bacillus thurigiensis* kurastaki gene (BtK gene). Insect-resistant potato has been obtained by expression of a BtT gene which encodes a protein, selectively toxic to the Colorado potato beetle, a principal pest of potato virus resistance, conferred by expression of the viral coat protein (CP) gene in transgenic plants, has also received considerable attention. Products such as potato, squash, melons, etc, based on this technology are in advanced stages of development and commercialization.

Both tomato fruit ripening and fruit firmness are among the advanced quality traits that are being investigated. A variety of approaches, based on inhibition of ethylene production are being pursued for enhancement of shelf life of tomato. For enhancing fruit firmness, cell wall hydrolytic enzymes such as polygalacturonidase and pectin methylesterase are being investigated.

R. B. Horsch and co-workers, *Science* **223**, 496 (1984).

S. R. Padgette and co-workers, *J. Biol. Chem.* **266**, 22364 (1991).

F. J. Perlak and D. A. Fischhoff, *Advances in Engineered Pesticides*, Marcel Dekker Inc., New York, 1993.

D. M. Stark and co-workers, *Science* **258**, 287 (1992).

JANICE W. EDWARDS
GANESH M. KISHORE
DAVID M. STARK
Monsanto Company

GENETIC ENGINEERING, PROCEDURES

The contemporary meaning of genetic engineering implies a use of the techniques of molecular biology, especially recombinant DNA techniques, rather than breeding in the formation of new genotypes. Recombinant DNA molecules are composed of two parts. First, a *vector* whose function is to provide the biochemical functions necessary for replication of the recombinant DNA molecule. Second, the *passenger* DNA that is joined to the vector and is replicated passively under control of the vector. Recombinant DNA technology allows the construction *in vitro* of DNA molecules that are not found in nature and their subsequent introduction into organisms, resulting in new genotypes and phenotypes of the recipient. Keep in mind that, particularly in plant and animal science, the two techniques are often complementary.

ANALYSIS OF DNA INFORMATION

Molecular biology is an information-based science. In this context, one can define information as the negative logarithm of the probability of a system occupying a particular state, given the total number of states available to it. In a DNA sequence there are four possible states at each position, corresponding to the four nucleic acid bases, adenine, guanine, thymine, and cytosine. A DNA of chain length n therefore has 4^n possible arrangements available to it. Since there are so many potential sequences of a DNA molecule, and since DNA molecules of the same base composition can have similar biochemical properties, but very different sequences, standard biochemical techniques cannot address the most biologically important property of DNA, its information content. Genetic engineering techniques allow the analysis and manipulation of genetic information based on its nucleotide sequence.

Sequence-Dependent Cleavage of DNA by Restriction Enzymes

Bacteria in nature are constantly exposed to exogenous DNA, primarily from bacteriophage (viruses) in the environment. Probably as a defense system, many bacteria contain a two-part DNA restriction and modification system. Restriction enzymes are of several types; the most useful for cloning, the Type II restriction enzymes, recognize specific sequences, usually 4–8 base pairs (bp) in length, and cut DNA molecules within these sequences. In Nature, restriction enzymes serve as a sort of immune mechanism: invading viruses are inactivated by restriction enzyme digestion of their DNA.

Location of Specific Sequences to DNA Restriction Fragments

A second technique that is universally applied to DNAs large and small, is that of Southern blotting. In these experiments, DNA fragments separated by gel electrophoresis are denatured *in situ* and transferred by capillary action to a nitrocellulose or nylon membrane, thereby making a "contact print" of the DNA in the gel. Solution conditions of hybridization can be set up to distinguish exact from inexact matches. The hybridized probe is detected by directly exposing the filter to X-ray film (if the probe is radioactive) or by enzymatic staining (if the probe is labeled by chemical modification). In either case the "contact print" of the gel shows the mobilities of the DNA fragments complementary to the probe. It is important to note that the specificity of Watson-Crick base pairing in DNA allows single fragments to be detected in the midst of a large excess of noncomplementary DNA. This specificity is the basis for, among other techniques, genetic fingerprinting of individual human DNAs and the use of species-specific gene probes for detection of bacterial species.

Restriction Sites as Genetic Markers

In the early 1980s workers recognized that the presence of a restriction enzyme site is a chromosomal marker that can be assayed by Southern blotting of genomic DNA. Thus, if the pattern of a Southern blot is different for the DNA of different individuals in the population, and if one or more patterns cosegregate with a mutant gene causing a disease, the restriction pattern is a surrogate diagnostic marker for the disease. This phenomenon, termed restriction fragment linked polymorphism (RFLP), can be used to predict an inheritance pattern in the absence of any other information about the disease other than its pattern of heredity.

GENE ISOLATION BY RECOMBINANT DNA TECHNIQUES

Workers in the early 1970s recognized that restriction enzymes provided tools not only for DNA mapping but also for construction of new DNA species not found in nature. A collection of recombinant DNA species consisting of many passenger sequences joined to identical vector molecules is called a **library**. Individual recombinant DNAs are isolated from single clones of the library for detailed analysis and manipulation.

Plasmid DNAs

Plasmids are nucleic acid molecules capable of intracellular extrachromosomal replication. Ultimately, a plasmid is defined by its mode of DNA replication. DNA replication is initiated at a single, characteristic sequence, termed the origin. The origin sequence determines the copy number of the plasmid relative to the host chromosome and the host enzymes that are involved in plasmid replication.

Plasmids can be introduced into cells by several methods. The most common method is transformation, where the recipient cells are made competent to receive DNA by washing with a solution of Ca^{2+} or other inorganic ions. Then the naked DNA is added directly; a fraction of the cells take up the DNA and replicate it. Some but not all plasmids also transfer by conjugation, a sexual process where the DNA is donated from one cell to another after physical contact.

Most plasmids are topologically closed circles of DNA. They can be separated from the bulk of the chromosomal DNA by virtue of their resistance to alkaline solution.

Plasmid Vectors for Facile Introduction of Passenger DNA and Selection of Recombinants

Three parts of the vector are key to its utility. The origin sequence, *ori*, allows the replication of plasmid DNA in high copy number relative to the chromosome. A gene, *amp*, encoding the enzyme beta-lactamase, which hydrolyzes penicillin compounds, allows growth of plasmid-containing cells in media containing ampicillin. The third region of the plasmid allows the introduction of passenger DNA.

Construction of a Recombinant Plasmid by Joining Vector and Passenger DNA

The unique restriction sites in the plasmid vector DNA provide sites in the molecule for insertion of restriction-digested DNA fragments.

Identification of the Desired Passenger Sequence in Plasmid Cloning Experiments

The objective of recombinant DNA construction is to obtain a clone of a single DNA sequence. If more than a single restriction fragment is ligated into a vector, the result is a library of clones, all of which have the same vector sequence, but with different passengers. Libraries are often described by the source of the passenger DNA. Genomic DNA libraries contain the total chromosomal DNA inserted into a vector. Copy DNA (cDNA) libraries contain passenger DNA derived by copying messenger RNA into DNA using the enzyme RNA-dependent DNA polymerase (reverse transcriptase).

The number of independent sequences in a DNA population is defined as its complexity. In a cloning experiment, the complexity of a library reflects the complexity of the passenger DNA population.

Given a library of sufficient complexity, it is then necessary to find the clone of interest against the background of recombinant clones containing other passenger sequences. Three strategies are generally employed. The simplest method is to use genetic complementation of a mutant in the host. Alternatively, genes encoding antibiotic resistance have been identified by direct phenotypic selection. The most common screening method uses a radioactively labeled probe to hybridize to DNA from the recombinant bacteria. The agar plate containing colonies of recombinant bacteria is blotted with a sheet of nitrocellulose or nylon filter paper, thereby transferring some of the bacteria in the colony to the filter. These transferred colonies are lysed with alkali *in situ*, thereby also making the DNA single stranded.

Vectors for Cloning Larger Fragments of DNA

Plasmid DNAs used in molecular cloning have a practical limit in the amount of DNA that can be inserted into them. When complex libraries are needed, eg, to isolate a mammalian gene, other cloning strategies are needed. These strategies are based on the replication of the bacteriophage lambda.

Isolation of DNA for Phage Cloning

Because lambda-derived cloning vectors accept only a narrow size range of DNA inserts, a library constructed from completely restriction-digested DNA is unlikely to be representative of the total passenger DNA population. In order to construct representative libraries the passenger DNA population is partially digested with a frequently cutting (4-bp recognition sequence) restriction enzyme under conditions where the average size of the products is close to 20 kb. The passenger DNA fragments are then separated by agarose gel electrophoresis or by sucrose density gradient centrifugation to eliminate those smaller and larger fragments in the digestion products.

Screening of Recombinant Phage by DNA Hybridization or Antibody Recognition

In screening a recombinant phage library, the phage are plated at high density so that plaques are nearly contiguous. Then the plate is blotted with nitrocellulose or nylon filter paper. Dipping the filter into alkaline solution lyses the phage particles and denatures the DNA, making it ready for hybridization with a DNA probe. The phage from the hybridizing region of the plate are diluted to a lower density, plated, and rescreened. After two or three screenings, a clonal population of recombinant phage are present. The passenger DNA can be analyzed by Southern blotting, restriction mapping and sequencing.

Cosmid Vectors

While the amount of information required for growth of a phage is large, that required for replication and selection of a plasmid is much less. Addition of cos sequences to a plasmid (hence the name cosmid) allows recombinant molecules of the appropriate size to be packaged into infectious particles.

Artificial Chromosomes for Insertion of Passenger DNA Molecules $\geq 10^5$ bp

The sizes of the passenger DNA molecules inserted into cosmid and phage vectors are limited by the requirement for packaging of the recombinant DNA into the phage head. In addition, recombinant plasmids in *E. coli* are often unstable if they are >50–100 kb. A system for constructing libraries from complex DNAs that overcomes many of these limitations had been developed. Reasoning that individual eukaryotic chromosomes are extremely large DNA molecules (>10^6–10^7bp in most organisms), they constructed a vector that provides the sequences necessary for faithful replication and segregation of an individual yeast chromosome to the recombinant. Passenger DNA inserted into these vectors is replicated by yeast as an extra, linear chromosome in the nucleus. In addition to a cloning site and sequences allowing the preparation of the vector from *E. coli*, these YAC vectors contain sequences required for chromosomal propagation.

An artificial chromosome library can represent all the sequences of the human genome (3×10^9 bp) with a 99% probability of finding an individual sequence in as few as 50,000 clones. Mapping and sequencing of complex genomes, as in the Human Genome Project, has used artificial chromosome libraries as the first component of the program.

Clones Linked Together to Form Larger Maps

Many eukaryotic genes are larger than the carrying capacity of a single phage or cosmid vector. Genomic DNA from eukaryotes contains noncoding sequences termed introns within the coding sequence; therefore, the DNA required to encode a protein can be much longer than that predicted from the size of the corresponding mRNA.

As a result of these considerations it is usually necessary to identify clones in the library that are linked to the first clone obtained by screening. The library is rescreened using a restriction fragment from one end of the passenger DNA segment in the first clone as a probe. The process, termed chromosome walking, may be reiterated several times until the contiguous inserts, termed contigs, cover the entire chromosomal region of interest.

ANALYSIS OF DNA SEQUENCES

After a desired clone is obtained and mapped with restriction enzymes, further analysis usually depends on the determination of its nucleotide sequence. The nucleotide sequence of a new gene often provides clues to its function and the structure of the gene product. Additionally, the DNA sequence of a gene provides a guidepost for further manipulation of the sequence, eg, leading to the production of a recombinant protein in bacteria.

The sequence of a gene predicts the sequence of the protein it encodes. The relationship between nucleotide sequence of a DNA or its mRNA and the amino acid sequence of the protein it encodes is given by the genetic code. Several strategies are available for identifying protein-coding regions. The frequency at which synonomous codons are used varies in different organisms. Sequences are searched using a database of codon frequency in the organism of interest to identify the most likely coding regions. In addition, a DNA sequence can be translated in all three reading frames. A database of known protein coding sequences is then searched using the predicted amino acid sequences as a query. Statistically significant homologies can provide a clue to the structure and function of the protein(s) encoded by the cloned DNA.

Determination of DNA Sequence Information

Almost all DNA sequence is determined by enzymatic methods that exploit the properties of the enzyme DNA polymerase.

COMPUTER ANALYSIS OF DNA SEQUENCE INFORMATION

The amount of information from a single DNA sequencing project can be staggering. Therefore, it is almost always necessary to analyze the data by computer methods. A number of commercial and purpose-built systems are available for analysis of DNA sequence information, operating on a variety of platforms, ranging from personal computers through workstations and supercomputers, depending on the intensity of the task. The field is not fully developed and research is ongoing in algorithm development, database manipulation and network applications, among others.

Assembly and Analysis of the Results from a Sequencing Project

The size of a gene almost always exceeds the data available from a single DNA sequencing experiment. It is necessary to identify contiguous regions of sequence information (contigs) and assemble these into completed projects. This can be done by relatively simple string matching, followed by highlighting points where two sets of overlapping information disagree. Resolution of discrepancies is then a matter of the investigator's judgment or doing more sequencing experiments across this stretch of DNA.

USES OF SEQUENCE INFORMATION

DNA sequence information is the starting point for other applications, including the expression of a gene product, the search for related sequences in biological samples, *in vitro* mutagenesis of the sequence, and structure–function studies of gene expression.

Specific Amplification of Related Sequences by the Polymerase Chain Reaction

If the sequence of a gene is known, primers that are unique to the gene can be synthesized. An oligonucleotide longer than 15–18 residues is likely to be unique even in a complex genome. Such an oligonucleotide, if hybridized to a single-stranded DNA, can be used to prime DNA polymerase so that the DNA is replicated. If two primers are made, each complementary to one strand of a gene, then the strand of each DNA located between them can be specifically replicated by DNA polymerase. The newly replicated DNA strands can be separated by heating and will then serve as template for another round of primed synthesis, leading to another doubling in the concentration of the original amplified sequence. Since the concentration of the DNA of interest doubles with each cycle, at least a 50,000-fold increase in its concentration is achievable within a few hours. This polymerase chain reaction (PCR) is used in a variety of experimental manipulations and diagnostic procedures. Sequences less than a few hundred base pairs long are most efficiently amplified by PCR but even this relatively limited information can be fruitful.

EXPRESSION OF GENES IN A HETEROLOGOUS HOST

In many cases, it is possible to synthesize the product of a gene in a different organism, eg, bacteria, yeast, or higher eukaryote. The details of these methods are covered in other articles. Recombinant DNAs directing the synthesis of the gene product must contain information specifying a number of biochemical processes: replication of the recombinant DNA, selection of recombinants, transcription of the foreign gene, translation of the foreign gene, and stability and purification of the recombinant protein.

MUTAGENESIS OF CLONED DNA

Genetics begins with mutants; indeed, the primary definition of a gene is a unit of mutation. Mutational analysis of a cloned gene is often essential for identifying structure–function relationships in its expression or in the protein encoded by the cloned gene. Alternatively, expression of a recombinant protein is often dependent on the codon usage optimal for the host. A number of techniques are available for mutagenesis. Randomized treatment of DNA with a chemical mutagen continues to be useful. In addition, a short synthetic DNA can be made with specific or random mutations introduced during synthesis. This altered information can then be incorporated into the cloned gene. A few of the more general approaches are described here. Other methods include mutagenesis by synthetic DNA, linker-scanning mutagenesis, and mutagenic PCR.

REGULATORY AND SAFETY ISSUES

Safety regulations have been modified to recognize that no new hazards will be created by recombinant DNA research, eg, DNA introduction will not make a pathogen out of a nonpathogen.

F. M. Ausubel and co-eds., *Current Protocols in Molecular Biology*, Wiley-Interscience, New York.

B. D. Davis, ed., *The Genetic Revolution*, Johns Hopkins University Press, Baltimore, Md., 1991.

M. A. Innis, D. H. Gelfand, J. J. Sninsky, and T. J. White, *PCR Protocols: A Guide to Methods and Applications*, Academic Press, New York, 1990.

J. Sambrook, E. F. Fritsch, and T. Maniatis, *Molecular Cloning: A Laboratory Manual*, Cold Spring Harbor Laboratory Press, Cold Spring Harbor N.Y., 1989.

Francis J. Schmidt
University of Missouri-
Columbia

GEOTHERMAL ENERGY

Heat emanating from within the earth is the major source of geothermal energy. This vast repository of energy is generated from the decay of natural radioisotopes and heat from the molten core of the earth.

Natural sources of geothermal fluids for heating and bathing have been utilized since prehistoric times. In the 1800s and 1900s applications of hydrothermal resources expanded widely to include space and district heating, agriculture, aquaculture, industrial processing, and electric power generation (qv). Historically, this energy was utilized by diverting surface hot water or steam sources. As technology progressed, wells were drilled to tap geothermal fluids more efficiently; and improvements in drilling technology have enabled access to and recovery of deeper and hotter fluids. In addition, development of techniques to extract geothermal heat from rock in which no natural mobile fluids exist is under way in several countries.

The useful applications of hydrothermal resources depend on the temperature of the extracted fluid. Figure 1 shows the distribution of thermal energy in the United States as a function of temperature. It is clear that relatively low temperature fluids can be effectively applied for purposes such as greenhouse heating, fish farming, and especially space heating. Waters at higher temperatures can be used for a variety of industrial processes and the production of electrical energy. Direct uses of geothermal energy generally require that the point of application be near the source of the hot water. Transportation of hot fluid over more than a few kilometers is often economically impractical. The higher temperature resources can, however, be converted to electricity. It is then possible to apply the power generated by hydrothermal energy in a variety of ways and at distant locations. The efficiency of electrical generation is directly related to the thermal quality of the resource. Using the most advanced power generating equipment, it may be economical to generate electricity from geothermal waters at temperatures as low as 150°C.

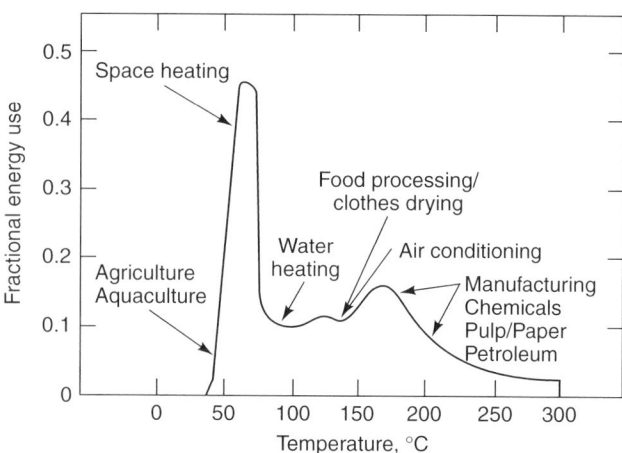

Figure 1. Thermal energy use versus temperature. Electricity generation is practical from thermal energy sources hotter than 150°C.

To successfully compete with the multitude of energy sources available, geothermal energy must be available and retrievable in both a convenient and an economical manner. In most geothermal power developments, these conditions have been met using high grade geothermal resources in the form of hot water and steam but these particular hydrothermal resources are limited. Most of the world's potential geothermal energy is found in rock that is hot but either lacks permeability or fluid. Although research and development have demonstrated success in extracting thermal energy on a limited basis, the vast hot dry rock resource has not yet been shown to be an economically feasible source of energy on a scale large enough for practical use.

GEOTHERMAL ENERGY RESOURCES

Type

Figure 2 is a generalized view of a cross section of the earth indicating the subsurface conditions for geothermal resources.

Magnitude

Whereas the total amount of energy stored within the earth is extremely large, only a very small fraction of that energy is accessible, in part because drilling and energy extraction costs escalate rapidly with depth. Commercial drilling can reach depths of $\sim$9000 m, so all of the thermal energy in the earth to that depth can be considered part of the geothermal resource base. This resource base has been estimated to be on the order of 1×10^7 EJ $(1 \times 10^7$ quads where 1 quad $= 10^{15}$ BTU) in the United States alone. The total world consumption of energy in all forms is only $\sim$300 EJ (300 quads); thus the earth's heat has the potential to supply all energy needs for the foreseeable future. Economic consid-

erations, however, must be factored into a development scenario.

Resource type	Accessible energy, EJ
hydrothermal	130,000
geopressured	540,000
hot dry rock	10,000,000
magma	500,000
Total	*11,170,000*

HYDROTHERMAL RESOURCES

Hydrothermal resources are characterized by the presence of heat relatively close to the earth's surface coincident with adequate permiability and fluid content and provide a mechanism for its transfer to the surface. Hydrothermal resources occur throughout the world in regions where continental plates meet and the upwelling of magma and the earth's crust lead to rock temperatures that are higher than the worldwide average. Within these areas, the places where water is located in geologic formations are scattered and difficult to predict. Most of these areas have been identified by surface exposure of the fluid resource.

Temperatures of hydrothermal reservoirs vary widely, from aquifers that are only slightly warmer than the ambient surface temperature to those that are 300°C and hotter. The lower temperature resources are much more common. The value of a resource for thermal applications increases directly with its temperature. In regions having hotter water more extensive use of geothermal resources has been implemented. Resources in remote areas often go unused unless hot enough to be employed in generating electricity.

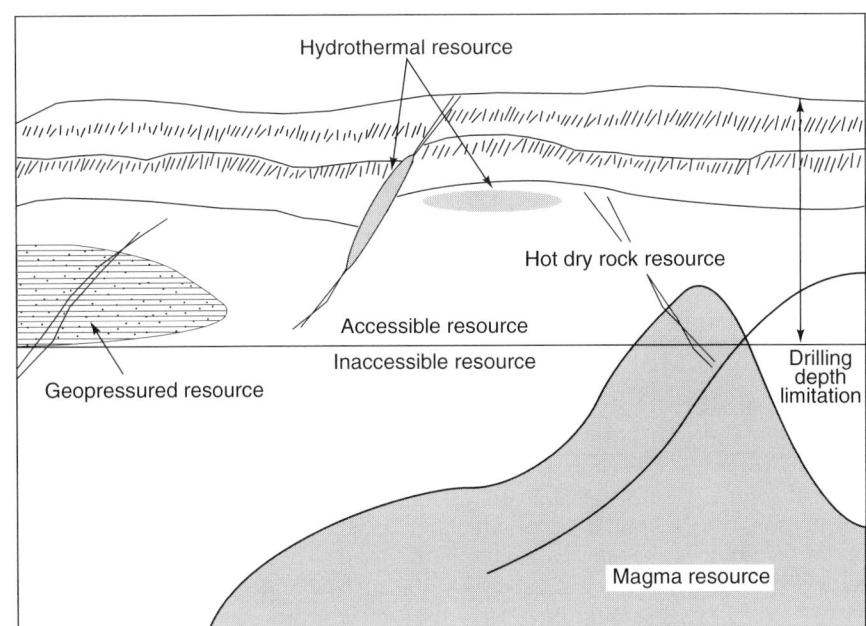

Figure 2. Geothermal Reservoir.

Drilling Field Development

The techniques for drilling hydrothermal wells have been adapted from those in use in the oil, gas, and water well drilling industry (see NATURAL GAS; NOMENCLATURE IN THE PETROLEUM INDUSTRY). Rotary drilling rigs are normally employed along with conventional drilling equipment such as steel casing, drilling lubricants, and casing cements.

Direct Uses of Hydrothermal Energy

Use of low temperature hydrothermal energy for direct thermal applications is widespread. The largest volume use of hydrothermal fluid is also one of the simplest. In regions such as some parts of the state of Wyoming, where hydrothermal fluids are found in close proximity to partially depleted oil fields, the hot hydrothermal fluid is pumped down oil wells at the perimeter of the field to heat the remaining oil. Hydrothermal energy is also employed to supply process heat for agriculture, primarily to heat greenhouses, and in aquaculture applications which involve warming the water in commercial ponds to enhance the growth rate of fish.

The application of geothermal energy in heating and cooling using heat pump technology results in 30–60% more efficiency than conventional electric heating and cooling systems.

Electric Power Generation

Hydrothermal steam and hot water resources having temperatures in excess of ∼150°C are generally suitable for the production of electricity. There has however, been recent interest in using geothermal resources in the production of hydrogen.

Worldwide Hydrothermal Development

Electric generation capacity from hydrothermal energy outside the United States was more than 8000 MW in 21 countries by 2000. Hydrothermal resources have been especially well developed in countries on the Pacific rim.

Environmental Issues

Hydrothermal energy, recognized as one of the clean power sources for the twenty-first century, is not entirely free of environmental problems; such as *atmospheric emissions, aquatic pollution, and terrestrial problems.*

GEOPRESSURED RESOURCES

The Resource

Geopressured resources consist of highly overpressured mixtures of hydrocarbons, predominantly methane, and water, in sedimentary formations. The potentially useful energy in geopressured resources exists as three components: fossil chemical from the methane, heat from the water, and mechanical from the high pressure of the fluid. Geopressured resources are generally found very deep in the earth at levels of 3600–6000 m or more.

The Technology

The wellbore can occur when the structure of the formation sand is disturbed by the turbulent flow, greatly reducing the energy production capacity of the well. The high salinity of geopressured water can lead to a spent fluid disposal problem. The most common solution is to pump the saline water down a nearby well into a formation at a shallower depth than the geopressured resource. The formation of calcium carbonate scale creates significant operational difficulties in utilizing highly saline geopressured fluids. Scale inhibitors and the requirement for frequent removal of accumulated scale from piping and equipment can both add substantially to the maintenance cost of geopressured facilities. In one proposed power plant design, the mechanical power is first utilized in a pressure-reduction turbine, then the hydrocarbon and aqueous fluids are separated, and the water is fed to the heat exchanger of a binary power plant. The gas is used to produce electricity through conventional technology or sold directly to off-site users. No commercial geopressure power plants are in operation at the time of this update.

Economic

The cost of energy from the Pleasant Bayou plant, at 12–18¢/kWh, was not competitive. The high costs can be related to a number of factors: the multiple energy forms, each effectively requiring its own generating plant; the salinity of the fluid and the depth at which the resource exists.

Direct Uses of Geopressured Fluids

Many of the uses typical of hydrothermal energy, such as greenhouse, fishfarm, and space heating, have been proposed for geopressured resources, but none has been commercially developed.

HOT DRY ROCK (HEAT MINING)

The Resource

The largest quantity of accessible geothermal energy exists in the form of hot rock which contains insufficient natural fluids to allow the transport of its energy to the surface or to a well bore. Because hot dry rock (HDR) is widely distributed, it also has the greatest potential for widespread application and is the only technology capable of making geothermal energy available on a national and worldwide basis. Whereas the HDR resource is found almost everywhere, it is not equally easy or economic to reach at every location.

The amount of thermal energy stored within hot dry rock at accessible depths is enormous. Even a minute fraction of this resource could supply all the world's energy needs for decades or even centuries. However, to utilize HDR resources, a practical means of accessing the hot rock and transporting its energy to the surface must be developed. In effect, an underground heat exchanger must be created to transfer the thermal energy of the rock to a mobile fluid. Because of the low thermal conductivity of hard rock, the surface area of the heat exchanger would have to be extremely large.

The Technology

The basic technique for extracting energy from HDR was conceived and patented in the early 1970s. It is based on drilling and hydraulic fracturing technologies developed in the petroleum and geothermal industries. The first step is to drill a well into sufficiently hot and impervious rock, with the exact depth of the well to be determined by local heat-flow and thermal conditions. Wells drilled for HDR applications are similar in many aspects to hydrothermal wells except that these wells generally penetrate into a much greater depth of hard, crystalline rock. After the well has been completed, a segment of the bottom portion of the well is blocked off using a packer which provides pressure isolation. Water under increased pressure is pumped through the packer and forced into joints in the surrounding rock body to form a reservoir consisting of interconnected fractures. The shape and orientation of the reservoir are functions of the natural stress features of the host rock.

To complete the system, a second well is drilled into the reservoir at some distance from the first. In operation, water pumped down one well heats as it flows through the fractures in the reservoir rock and returns to the surface through the second well, where its thermal energy is extracted using binary technology. The water can then be recycled back into the formation. The use of chemical tracers and geophysical surveys can provide valuable supporting data with regard to the multiplicity of flow paths and the transit time of fluid within the reservoir.

Issues related to operation of a HDR geothermal energy system are the efficiency of energy extraction from the rock in the reservoir region; the impedance to flow as the water traverses the reservoir body; and the water losses resulting from leakage from the fracture network.

Economics

The costs of developing HDR resources are closely tied to the depth at which sufficiently hot rock is found. In the eastern United States, it is generally necessary to drill much deeper to reach hot rock. Because drilling is the most expensive single factor in HDR development, the first HDR electric plants are expected to be built in areas of high heat flow.

A number of studies have been conducted to assess the economics of producing electricity from HDR. No commercial HDR facility has been built as of this writing.

Environmental Considerations

When operated as a closed loop, no significant amounts of air, water, or terrestrial pollutants are produced. Because the active reservoir is located thousands of feet below the water table, there is minimal danger of ground or surface water contamination. Finally, when a plant is decommissioned at the end of its useful life, the underground system could be permanently shut in by techniques already well known and proven in the oil, gas, and geothermal industries.

HDR Outside the United States

The extraction of geothermal energy from HDR has been evaluated at a number of locations around the world. All work is based on the same general technical approach which was employed in the United States. In Japan, reservoirs were created in rock at about 200°C at two locations. Two HDR projects have been initiated in other countries. A pilot study was initiated in France in 1999 with the drilling of a 5000 m well. The second (injection) well was drilled in 2002 and permeability stimulation tests initiated. The project envisions circulating water at the 5000 m depth through induced fractures resulting in 200°C water being produced to operate a 6 MWe turbine.

MAGMA

The Resource

The core of the earth is generally believed to consist of molten rock known as magma. The energy content of the core is essentially boundless, but it is unreachable from a practical standpoint because it lies many kilometers below the surface. The technology to drill to those depths does not presently exist. In volcanically active regions, however, magma intrusions can be found relatively close to the surface in some localities. In these areas magma is a potentially useful geothermal resource, but relatively little is known about intrusive magma chambers.

The U.S. Geological Survey estimates that the total amount of magma energy existing within 10 km of the surface is on the order of 50,000–500,000 EJ in molten or partially molten magma. It has been estimated that only 2 km^3 of magma could provide enough energy to operate a 1000-MW electric power plant for 30 years.

Work on the extraction of useful energy from magma has been limited primarily to paper and laboratory studies aimed at understanding the formation, extent, cooling, and other facets of magmatic bodies. Field drilling has been limited.

J. Baumgartner and A. Gerard, "Electricity Production From Underground Heat", Information circular by Geothermie Soultz, F-67250Kutzenhausen, France, 2002.

R. DiPippo, *Geothermal Energy as a Source of Electricity*, U.S. Department of Energy, Washington, D.C., 1980.

C. Stone, ed., *Monograph on The Geyser's Geothermal Field*, Geothermal Resources Council, Davis, Calif., 1992.

J. W. Tester and H. J. Herzog, *Economic Predictions for Heat Mining: A Review and Analysis of Hot Dry Rock (HDR) Geothermal Energy Technology*, MIT Energy Lab. Report #MIT-EL 90-0001, Massachusetts Institute of Technology, Cambridge, Mass., 1990, p. 59.

ROY MINK
United States Department of Energy

GERMANIUM AND GERMANIUM COMPOUNDS

Germanium (Ge) has an atomic number of 32, falling in Group IV of the periodic table of the elements just below silicon (Si).

The semiconductor properties of Ge offered a new field of research opportunities, and led to the development of the first transistor by Bell Laboratories. The usage of Ge in power transistors and telephonic applications fueled Ge's strong growth to usage exceeding 100 tons annually during the 1950s and 1960s. During the latter part of this period, the abundance and processing ease of silicon have allowed Si to fuel the semiconductor, telecommunications, and computer age through the 1990s. However, germanium is still important today in these and many more industries.

SOURCES AND SUPPLIES

Germanium is sourced primarily as a by-product. Since the price has fallen to half the value of 1996, recovery efforts from these sources has been discontinued. Their availability and the continued engineering efforts to lower costs of recovery and processing may indeed establish a long-term price ceiling should a demand surge be sustained for several years.

PROPERTIES

The physical, thermal, and electronic properties of germanium metal are shown in Table 1.

CHEMICAL PROPERTIES

Germanium Metal

Germanium is quite stable in air up to 400°C where slow oxidation begins. The metal resists concentrated hydrochloric acid, concentrated hydrofluoric acid, and concentrated sodium hydroxide solutions, even at their boiling points. Nitric acid attacks germanium at all temperatures more readily than does sulfuric acid. germanium reacts readily with mixtures of nitric and hydrofluoric acids and with molten alkalies and more slowly with aqua regia. In compounds, germanium can have a valence of either 2 or 4.

Germanium halides include germanium tetrachloride, $GeCl_4$, germanium tetrabromide, $GeBr_4$, and germanium tetraiodide, GeI_4, and germanium tetrafluoride, GeF_4.

Germanium oxides include germanium dioxide, GeO_2, and germanium monoxide, GeO. Germanites include sodium heptagermanate, $Na_3HGe_7O_{16}.4H_2O$. Germanides include magnesium germanide, Mg_2Ge. Germanes include germane, GeH_4. Organogermanium compounds include spirogermanium (2-aza-8-germaspiro-[4,5]-decane-2-propanamine-8,8-diethyl-N,N-dimethyl dihydrochloride), $C_{17}H_{36}GeN_2 \cdot 2HCl$, and carboxyethyl germanium sesquioxide (3,3′-germanoic anhydride dipropanoic acid), $C_6H_{10}Ge_2O_7$.

Alloys

Many Ge alloys have been prepared and studied.

Table 1. Properties of Germanium

Parameter	Value
Physical properties	
atomic weight	72.59
density at 25°C, g/cm³	5.323
Mohs' hardness	6.3
Thermal properties	
melting point, °C	937.4
boiling point, °C	2830
heat capacity at 25°C, J/(kg·K)[a]	322
heat of combustion, J/g[a]	7380
vapor pressure at 2080°C, kPa[b]	1.33
coefficient of linear expansion, 10⁻⁶/K	
at 100 K	2.3
at 300 K	6.0
thermal conductivity, W/(m·K)	
at 100 K	232
Electronic properties	
intrinsic resistivity at 25°C, Ω·cm	53
intrinsic conductivity type	N (negative)
band gap, direct at 25°C, minimum eV	0.67

[a]To convert J to cal, divide by 4.184.
[b]To convert kPa to mm Hg, multiply by 7.5.

MANUFACTURING AND PROCESSING

Ore Processing

No mineral is mined solely for its germanium content. Almost all of the Ge recovered worldwide is a by-product of other metals, primarily zinc, copper, and lead. In the United States, zinc concentrates have been roasted and then sintered for zinc recovery. The sinter fume is chemically leached, and the germanium is selectively precipitated from the leach solution by fractional neutralization and sent to the germanium refinery.

Purification

Regardless of the source of Ge, all Ge concentrates are purified by similar techniques. The ease with which concentrated germanium oxides and germanates react with concentrated hydrochloric acid and the convenient boiling point (83.1°C) of the resulting $GeCl_4$ make chlorination a standard refining step.

ECONOMIC ASPECTS

Because germanium is usually recovered as a by-product, its price and availability over the long term are subject to supply and demand considerations for its host products, such as zinc, in addition to the Ge value. This is not the case over the short term because producers, end users, and commodity brokers often have varying stocks of product and Ge-containing scrap. Therefore, short-term pricing is controlled mainly by demand.

HEALTH AND SAFETY FACTORS

The toxicity of germanium compounds is generally of low order. Usually, the toxicity or corrosive nature of the compound is by action of the other part of the compound than from the Ge content. Germane is considered a toxic gas. Users must employ calibrated detectors and adequate safety breathing apparatus and ventilation. Germanium tetrachloride is a corrosive liquid. Germanium tetrafluoride produces hydrogen fluoride in aqueous acidic solution. Users must consider it as similar in nature to silicon tetrachloride and hydrochloric acid, and provide protective clothing, face shields, gloves, and a neutralization chemical. Despite the "popularity" of ^{132}Ge as a health food or Oriental remedy, no producer advocates human or animal consumption of Ge products.

RECYCLING AND SCRAP RECOVERY

Except for the germanium metal and optical industries, which use a pure product that can be most easily recovered and recycled (eg, broken ir windows and lenses), other manufacturers' processes [poly(ethylene terephthalate) and optical fiber production] and recycling effort were minimal until the 1995–1996 price spike. Now, recycling has become common practice, as several of the producers can recover Ge to below 1% content (depending on the matrix). Given the finite nature of germanium, it is important that all processes be examined for all opportunities to recycle this valuable element.

USES

Many of the uses of germanium and Ge compounds have been reviewed. Uses, by market, are listed in Table 2.

New Product Introductions

The most publicized new product is the silicon germanium chip from IBM. Currently SiGe is touted as a low-price, high-speed device designed to substitute for GaAs chips in certain applications, such as cellular telephones. The hetrojunction bipolar transistor technology for SiGe has become practical only since 1989, and now the uses are multiplying into direct broadcast satellite, automobile radar systems, and personal digital assistants.

Two automobile manufacturers have incorporated germanium-based night vision systems in limited editions in their product lines. At least one manufacturer has announced a similar system for buses and trucks. These night vision systems are also finding use in firefighting sytems for thermal imaging.

Table 2. Germanium Usage (%) by Market, 2000

optical fiber production	20
polymerization catalysts	35
infrared optics	25
electronic and solar applications	12
miscellaneous	8
Total	*100%*

Other new applications recently announced include germanium emitters for cleanroom ionization, Ge-containing superconductors, Ge substrates for ultra-bright-light-emitting diodes, and other semiconductor uses, and a revival of the 1960s Ge-transistorized "wah-wah" pedal for rock bands.

F. Glocking, *The Chemistry of Germanium*, Academic Press, Inc., London, 1969.

J. D. Jorgenson, "Germanium," in *Minerals Yearbook*, U.S. Geological Survey, Reston, Va., 2002.

Dennis W. Thomas
Eagle-Picher Technologies, LLC

Tariq Mahmood
Charles B. Lindahl
Elf Atochem, North
America, Inc.

GLASS

The American Society for Testing and Materials ASTM defines glass as "an inorganic product of fusion that has been cooled to a rigid condition without crystallizing." However, this definitions does not explicitly address the character of a noncrystalline structure and the glass-transformation behavior, two characteristics that separate glasses from other solids. In addition, glasses may be made by processes that do not necessarily produce liquids. In principle, any melt forms a glass if cooled so rapidly that insufficient time is provided to allow reorganization of the structure into crystalline (periodic) arrangements.

FUNDAMENTALS

Structural Descriptions of Glass-Forming Systems

Inorganic glasses are readily formed from a wide variety of materials, principally oxides, chalcogenides, halides, salts, and combinations of each.

Silicate Glasses. The structure of silica glass consists of well-defined SiO_4 tetrahedra connected to another neighboring tetrahedron through each corner. The structure of alkali silicate glasses also consists of a network of SiO_4 tetrahedra, but some of the corners are now occupied by nonbridging oxygens that are linked to the modifying polyhedra.

One view of the glass structure is a "modified random network" in which the alkali ions and nonbridging oxygen (NBO) cluster to form alkali-rich regions surrounded by presilicate network. X-Ray and neutron diffraction studies, extended X-ray absorption fine structure (EXAFS) data, and molecular dynamics (MD) simulations give a picture of the glass structure consistent with this view.

Glasses containing <10 mol% alkali oxides are considerably more difficult to melt due to higher viscosities.

Borate Glasses. The structure of vitreous B_2O_3 consists of planar triangular BO_3 units that link to form larger units known as boroxol rings. These well-defined units are connected by oxygens so that the B−O−B angle is variable and twisting out of the plane of the boroxol group can occur, producing the loss of long range order associated with glass. For vitreous boron trioxide (v-B_2O_3) the results by MD and quantum mechanical simulations, nuclear magnetic resonance (nmr), nuclear quantum resonance (nqr), ir and Raman spectroscopic studies, and inelastic neutron scattering all indicate that a large fraction of B atoms ($\sim$80−85%) are in the planar boroxol rings.

Phosphate Glasses. The basic building blocks of crystalline and amorphous phosphates are PO_4-tetrahedra. These tetrahedra link through covalent bridging oxygens to form various phosphate anions. The networks of phosphate glasses can be classified by the oxygen/phosphorus ratio, which sets the number of linkages though bridging oxygens to neighboring P-tetrahedra. Thus, metaphosphate (O/P = 3) and polyphosphate (O/P > 3) glasses have structures that are based on chainlike phosphate anions that are themselves interconnected though terminal oxygens by ionic bonds with modifying cations.

The extent of the network polymerization in silicate and phosphate glasses changes monotonically as a function of composition; however, the compositional dependence of a variety of ultraphosphate glass properties, are anomalous when compared with the silicate analogues.

Germanate Glasses. The structure of GeO_2 glass is very similar to that of SiO_2 glass. Since the Ge^{4+} ion is larger in diameter than the Si^{4+} ion, the Ge−O distance is also larger than the Si−O distance (in silicate glasses), with a bond length of $\sim$0.173 nm and Ge−O−Ge bond angle smaller than the Si−O− Si bond angle. Gas diffusion studies suggest that the open volume of germanate glass is slightly less than that of silicate glass. Recent reports, including neutron diffraction, high energy photon diffraction (using synchrotron radiation), magic angle spinning nuclear resonance (mas nmr) and Raman spectroscopy, suggest that the structure of vitreous germania resembles that of quartz-like GeO_2, with [GeO_4] tetrahedra providing the basic structural units, giving a continuous random network.

Chalcogenide Glasses. Chalcogenide glasses are produced by melting group 16 (VI A) elements (S, Se, and Te) with other elements, generally of group 15 (V A) (eg, Sb, As) and group 14 (IV A) (eg, Ge, Si) to form covalently bonded solids. When melted in an atmosphere particularly deficient in oxygen and water, the glasses have unique optical and semiconducting properties. Structural models for these glasses are based on the high degree of covalent bonding between chalcogenide atoms.

Halide Glasses. Structural models for fluoride glasses based on BeF_2 are directly analogous to those for alkali silicate glasses, with the replacement of nonbridging oxygens by nonbridging fluorines (NBF). Fluoride glasses have been studied for the past 30 years and have found various applications in optics, sensors, is instrumentation, medicine and telecommunications. Of particular importance are the heavy metal fluoride glasses (HMFG) based on ZrF_4 in numerous multicomponent systems in which some fluorides act as glass formers in association with alkali and divalent fluorides. Extensive development work has also been carried out on fluorophosphate glass (5−20% P_2O_5), initially for use as optical glasses but more recently for use in high power lasers.

Organic Glasses. Organic glasses consist of carbon−carbon chains, which are so entangled, that rapid cooling of the melt prevents reorientation into crystalline regions. Like low crystallinity glass−ceramics, the organic glasses presented small regions of oriented chains. Low molecular weight organic glasses are increasingly investigated because they potentially combine several interesting properties such as easy purification, good processability and high gas solubility. Numerous applications are envisaged, eg, in light emitting devices, in nonlinear optics, in optical data storage, and in photovoltaic and photochromic materials.

Metallic Glasses. Structural models for metallic glasses include variations of the random network theory, crystalline theory, and a dense random packing of spheres. Structural methods such as X-ray diffraction, electron microscopy, Mössbauer resonance, nmr, and thermal analysis, have been used to study the structures of glassy metals. Heat capacity data demonstrated that the metals were indeed vitreous and not amorphous with microcrystallization.

Computer Modeling of the Glass Structure

Recent software and hardware developments have produced a new characterization technique for glass structure: atomistic simulations based on MD calculations of silicates, borate glasses, and phosphate glasses. Static lattice simulations cannot be applied in a straightforward way to glasses as in the study of physics and chemistry of crystalline solids. The MD studies of alkali silicates (Na−, K−, Na−K, and Li) provide "snapshot" pictures of the atomic configuration. This allows the identification of key features and correlation of the atomic scale structure with the macroscopic experimental properties.

The distribution of alkali modifiers throughout the glass network is one aspect of technological importance. Studies of alkali silicate glasses reveal that the alkali ions are not randomly distributed within the silica network but rather aggregate in alkali-rich regions on a nanoscale, consistent with the "modified random network" structural model introduced in the section on Silicate Glasses. Lithium-silicates exhibit the greatest degree of aggregation, possibly because of the size and mobility of the ion. The disilicate composition marks the onset of the thermodynamically predicted homogeneous glass-forming region. Such results help relate phase separation and immiscibility tendencies for the alkali silicates to structural and thermodynamic considerations.

Surfaces can be modeled using MD in two ways: by removing the periodicity in one dimension or by increasing the dimension of one of the box edges, without scaling the atomic coordinates. The second method creates a series of 2D slabs with top and bottom surfaces.

Glass–Ceramics

Glass–ceramics are normally obtained by a controlled crystallization process of suitable glass-forming melts. Internal or external nucleation is promoted to develop microheterogeneities from which crystallization can subsequently begin. As a result, the amorphous matrix transforms into a microcrystalline ceramic aggregate. The composition of the crystalline phases and the crystalline sizes define the properties of the final material. Therefore, the major components and the composition of the glass are selected to ensure precipitation of crystals that provide desired properties of the glass–ceramic. By definition, glass–ceramics are >50% crystalline after heat treatment; frequently, the final product is >90% crystalline. Table 1 shows examples of commercial glass–ceramic systems.

Devitrification and Phase Separation

Devitrification is the uncontrolled formation of crystals in glass during melting, forming, or secondary processing in contrast to the controlled crystallization associated with glass–ceramic processing. Devitrification can affect glass properties including optical transparency, mechanical strength, and sometimes the chemical durability. Glass-formation ability (GFA) depends on the avoidance of devitrification.

Glasses that derive their color, optical transparency, or chemical durability from a small amount of a finely dispersed, amorphous, second phase are termed phase-separated glass, distinguished from glass ceramics because they remain predominantly amorphous. Phase separation can occur by processes: (1) nucleation and growth and (2) spinodal decomposition. The morphologies of the phase-separation microstructures obtained by these two different processes are different. Spinodal decomposition produces a composite material with two highly interconnected amorphous phases, whereas phase separation that occurs by the classical nucleation process produces a microstructure in which discrete, spherical droplets are embedded in an amorphous matrix. The most important parameter affecting the morphology of phase separation is the composition of the liquid. Discrete particle morphologies will be observed for compositions near the edges of liquid–liquid miscibility gaps. Morphologies with larger volume fractions of both phases, often with a greater degree of connectivity, will be found for composition near the center of miscibility gaps.

Surfaces of Glasses

The surface of a glass plays a major role in its ability to function in a given application. Four characteristics of the surface make a glass suitable for particular applications: (1) ability to be ground and polished, (2) chemical durability, (3) ability to bond specific molecules, and (4) resistance to mechanical damage (strength is limited by presence of Griffith flaws).

PROPERTIES

The properties of glasses depend on their chemical composition and their structure. Most properties can be discussed from a starting point represented by the material of the crystalline form by considering what modifications, structural disorder, absence of translation periodicity, spatial variations in atomic concentrations or local structure will have on the chosen property. A major advantage of glasses is that their properties can be tailored by adjusting their composition.

Optical Properties

Optical glasses are usually described in terms of their refractive index at the sodium D line (589.3 nm), n_D, and

Table 1. Commercial Glass–Ceramics

Commercial designation	Major crystalline phases	Properties	Application
Corning 9632	β-quartz solid	low expansion, high strength, thermal stability	electrical range tops
Corning 9608	β-spodumene solid solution, $Li_2O \cdot Al_2O_3 \cdot (SiO_2)_4$	low expansion, high chemical durability	cooking utensils
Neoceram (Japan)	β-spodumene	low expansion	cooking ware
Corning 0303	nepheline [12251-37-3], $Na_2O \cdot Al_2O_3 \cdot 2SiO_2$	high strength, bright white	tableware
Corning 9625	α-quartz solid solution (SiO_2); spinel ($MgO \cdot Al_2O_3$); enstatite $MgO \cdot SiO_2$	very high strength	classified
High K (Corning)	$3 Al_2O_3 \cdot 2 SiO_2$; $(Ba, Sr, Pb) Nb_2O_6$	high dielectric constant	capacitors
Corning 9455	β-spodumene solid solution; mullite, $3 Al_2O_3 \cdot 2 SiO_2$	low expansion, high thermal and mechanical stability	heat exchangers

their Abbé number, ν, which is a measure of the dispersion or the variation of index with wavelength. Glasses with n_D < 1.60 and ν < 55 are defined as crown glasses and those with n_D > 1.60 and < 50 are defined as flint glasses. A low dispersion is desirable in optical glasses used for lenses because dispersion causes chromatic aberration. Fluorophosphates, having absorption edges located well into the ultraviolet (uv), are examples of glasses with high Abbé numbers and low refractive indexes.

The addition of alkali or alkaline-earth oxides to a glass-forming oxide shifts the uv absorption edge to lower energies (longer wavelengths). Conversely, the range of uv transmission is enhanced when the cations in the glass have a high charge/radius ratio, indicating a stronger cation–oxygen bond. High purity fused SiO_2 glass has been developed that is highly resistant to optical damage by uv (190–300-nm) radiation. The addition of nitride ions to oxide glasses shifts the uv edge to longer wavelengths, probably because of the greater polarizability of the trivalent nitrogen. Nitride glasses, in contrast to conventional optical glasses, or fluoride optical glasses, posses a remarkable combination of desirable properties, including, high hardness, high refractive index, and high softening temperature.

In the visible region, absorption by additives such as transition metal or lanthanide ions is usually more important than contributions from the glass formers themselves. The coloration of glass by uv radiation from sunlight (solarization) results from the oxidation of transition metal ions in the glass. Optically pumped laser action has been observed for most lanthanide ions in a variety of glass systems. The best glass laser systems have the following qualities: the absorption spectrum of the lasing ion matches the spectrum of the pump radiation; the absorbed radiation efficiently produces excited-state ions; the excited state has a long lifetime; the probability of radiative decay is high; and the line width of the emitted radiation (fluorescence) is narrow. The line width of the fluorescence band of the lanthanide ion is affected by the glass matrix. In general, the smaller the field strength of the anions, the less the perturbation of the coordination shells of the fluorescing ion and the narrower the line width, ie, fluoride and chloride glasses promote narrower line widths than those seen in oxide glasses.

The visible transmission of photochromic glasses decreases with increasing frequency of light, and the effect is reversible. Chalcogenide glasses such as As_2S_3 are colored or even opaque, because of the small difference in energy between the conduction and valence bands. On the other hand, color in reduced amber glasses is the result of a Fe^{3+}–S^{2-} chromophore, not involving Fe^{2+}.

Chemical Durability

The chemical durability of glass is critical for many applications, including the performance of glass containers for food and beverages, pharmaceuticals, and corrosive chemicals; the retention of high transparency for optical components, including, windows, exposed to ambient conditions; the use of glass as a long-term host for radio-

active and hazardous materials; and the performance of bioactive glasses implanted in the body.

Electrical Properties

Ion Conducting Glasses. In alkali containing glasses, charge is carried by alkali ions moving from modifier site to modifier site, and so properties like conductivity are sensitive to composition (ie, the number of charge carrying ions) and structure (the nature of the modifier site).

Mixed-Alkali Effect. In single alkali glass systems, different processes contribute to the electrical conduction at different temperature. The substitution of a second alkali ion, at constant alkali content, in many phosphate, borate, and silicate glasses causes a decrease in the electrical conductivity up to five orders of magnitude. This is called the mixed-alkali effect (MAE), observed in ionic conductive glasses.

Semiconductimg Glasses. Chalcogenide glasses can be switched between low and high conductivity states using an applied voltage. There are two types of switching: threshold and memory. In the case of threshold switches, a small current is required to maintain the ON (high conductivity) state. In contrast, memory switches remain on indefinitely in the absence of a current and require a short, high current pulse to return to the state. Semiconductivity in oxide glasses involves polarons (conducting electrons in an ionic solid together with the induced *polarization* of the surrounding lattice).

Thermal Properties

When a typical liquid is cooled, its volume decreases slowly until it reaches the melting point, where the volume decreases abruptly as the liquid is transformed into a crystalline solid. If a glass-forming liquid is cooled below the melting point without the occurrence of crystallization, it is considered to be a supercooled liquid until the glass-transition temperature is reached. At temperatures below the glass-transition temperature, the material is a solid.

High silica glasses such as Pyrex have low CTE (coefficient of thermal expansion) and are used in applications requiring good resistance to thermal shock. Ultralow expansion SiO_2–TiO_2 glasses have CTEs of practically zero, as do certain lithium-aluminosilicate glass ceramics, like Zerodur. Some applications, such as glass-to-metal seals, require glasses to have higher CTEs to match metals and other materials. Highly modified silicate glasses and glass–ceramics and phosphate glasses have been developed for high CTE (>10×10^{-6}/°C) sealing applications.

The thermal conductivity of glass is dependent on lack of long-range structural order. The mean free path of a phonon in a glass is on the order of a few interatomic spacings, so phonons are damped out over very short distances, making glasses good thermal insulators, at least up to temperatures where radiative processes become dominant. Thermal conductivity increases when glasses are crystallized to form a glass–ceramic. On the other hand, the thermal conductivity of an aerogel is exceptionally low. A combination of this property of silica aerogels with polymer cross-linking has resulted in very high strength and very light materials, for potential applications in aerospace.

Mechanical Properties

High strength glass fibers combine high temperature durability, stability, transparency, and resilience at low cost weight–performance. Various glass compositions have been developed to provide combinations of fiber properties for specific end-use applications.

Strength and Fatigue. The "inert intrinsic strength" of silica fibers is ∼14 GPa. This term has been operationally defined as the strength of flaw-free glass measured under conditions where no delayed failure is allowed.

While the measurement of MOE (modulus of elasticity) of silicate glasses is straightforward, the calculation of strength is not similarly possible as strength is a "weakest link" property. It depends not on the average properties of the sample (ie, properties of the network), but on the weakest portion of the sample. Cracks concentrate the stress so that it may be orders of magnitude greater at the crack tip than the applied stress. If the applied stress is not the critical stress, then failure will not occur instantaneously. If there is moisture present in the environment, subcritical slow crack growth (fatigue) will occur, which is of major consequence in silica lightguide fibers.

Flaw Generation and Strengthening. Glass surfaces can be damaged by either mechanical means or by chemical means, ie, a chemical interaction that leads to mechanical degradation. In this case, a solid, liquid, or gas phase may react with the glass surface forming a new product or developing residual stresses due to bonding materials with different thermal expansion coefficients.

The most common techniques for improving the strength of glass surfaces are based on the fact that failure in glasses occurs in tension that in turn is the result of stress concentrations due to surface flaws. Thus the reduction of tensile stresses at the surface by superposition of a surface compression is usually very effective. The use of coatings is another way of preventing the formation of flaws or flaw growth.

MANUFACTURE

Glasses can be prepared by methods other than cooling from a liquid state, including from the solid–crystalline state (ie, lunar glasses) and vapor phases and by ultrafast quenching procedures: (1) melt spinning, in which molten metal is ejected onto a rapidly spinning cylinder to form thin ribbons; (2) splat quenching, in which the melt is smashed onto an anvil by a compressed-air-driven hammer; (3) twin-roller quenching, in which the melt is forced between two cylinders rotating in opposite directions at the same speed; (4) laser glazing, in which a short, intense laser pulse is focused onto a very small

Table 2. Technical Innovations of the Twentieth Century

basic glass processing	float glass process
	ribbon machine for glass bulbs
	owens suction machine (containers)
	Danner process for making glass tubing
	continuous melting of optical glass
fiberglass	continuous glass fibers
	steam blown glass wool
	rotary fiberizing
specialty glass items	
	glass ceramics
	radiant glass–ceramic cooktops
	glass microspheres
	laminated glass
	borosilicate laboratory and consumer glassware
	large, flat-glass TV tubes
	automotive solar control electrically heated windshield
	automotive tempered window 2.5 mm thick
	photochromic and photosensitive glass
	ceramic and glass foodware safety
glass lasers and fiber optics	glass lasers
	low loss optical fibers
	erbium-doped optical fiber amplifiers
	ultraviolet-induced refractive index changes in glass
	fiber optic sensors
other	bioactive glasses, ceramics and glass–ceramics
	nuclear waste glasses
	chemical tempering of glass products (ion exchange)

volume of a sample; and (5) laser spin melting in which a rapidly rotating rod of the starting material is introduced into a high power laser beam, eg, a CO_2 laser, causing molten droplets to spin off and form into small glass spheres. Table 2 summarizes the technical innovations of the twentieth century concerning glass processing and new glass developments.

Glass manufacture requires four major processing stages: batch preparation, melting and refining, forming, and postforming (Fig. 1). Silica is the basis of most commercial glasses; however, it has a high melt viscosity, even at temperatures close to 2000°C, making melting and working extremely difficult. Container and flat glass compositions are based on the $Na_2O–CaO–SiO_2$ system with addition of other minor components to improve glass formation, lower liquidus temperature, and improved durability.

Glass-Manufacturing Processes

Batch Preparation. This step refers to mixing and blending of raw materials to achieve a desired glass composition.

Melting and Refining. Commercial melting refers to forming a homogeneous molten glass from the raw materials at temperatures between 1430 and 1700°C (2600–3100°F).

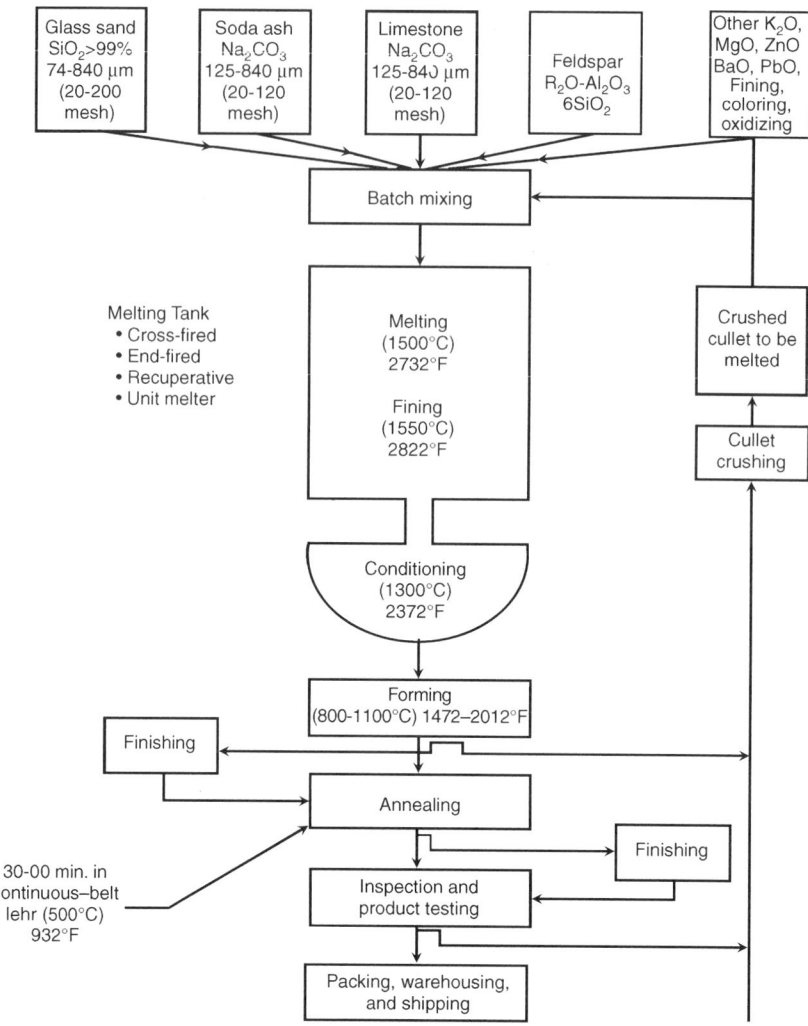

Figure 1. Overview of glass manufacturing.

Glass Forming. Molten glass can be molded, drawn, rolled, cast, blown, pressed, or spun into fibers. Glass containers are formed by transferring the molten glass into molds by a method called gob feeding.

Postforming and Finishing Operations. After taking its final shape, the glass product may be subjected to curing–drying (fiber glass products), tempering, annealing, laminating, and coating, polishing, decorating, cutting, or drilling. Annealing is the process of slow cooling to release stresses by the time the glass product reaches room temperature. Strain and stresses are dependent on how fast the glass is cooled through the glass-transition temperature.

Annealing is done for all types of glasses except fibers. Tempering is used to impart strength to glass sheets.

Glass Melting Tanks

Furnaces. In general, furnaces are classified as discontinuous or continuous. Discontinuous furnaces are used in small glass-melting operations (small blown and pressed tableware and specialty glasses) and are operated for short periods of time. In continuous furnaces, the glass level remains constant, with new batch materials being constantly added as molten glass is removed. Continuous furnaces are classified into four categories: direct-fired, recuperative, regenerative, and electric melting: continuous furnaces can be fired by natural gas, electricity, or a combination of both. In natural gas furnaces, the gas is burned in the combustion space above the molten glass and the transfer of energy occurs through radiation and convection. Electricity is introduced using electrodes that are placed directly into the molten glass.

Refractories for the Glass Industry. Today, continuous furnaces are expected to last up to 10 years in operation. Glass-contact refractories have to be carefully selected to improve furnace life minimizing side reactions that would lead to glass defects. At the flux line (glass level), surface tension and density driven flows at the combustion-atmosphere–refractory–molten-glass interface increases the refractory corrosion.

Sensors and Controls for Glassmakers. On-line sensors provide a direct measure of some molten glass property: glass flow, melting rate, viscosity, strength, color, refractory corrosion, emissions, etc, which need to be controlled to optimize the glass-melting process. For best applications, a sensor should not change the environment or affect the property being measured; and the sensor should not be degraded by the environment. The advent of nontraditional methods of melting glasses will also require nontraditional on-line sensors under very demanding conditions.

Computer Modeling of Glass Melting. Several physical and mathematical modeling techniques are being implemented to investigate glass-making processes, optimize these processes, and evaluate new or different operating conditions, including the following:

- Design new manufacturing installations to reduce costs of plant construction and plant operation while increasing furnace life.
- Investigate and solve day-to-day operation problems.
- Improve process efficiency, fuel efficiency, throughput rates, production yields, and product quality (less defects).
- Develop new products and processes in less time.
- Ensure environmental quality and meet current or projected regulations.

Advanced Melting Techniques

The following are being reevaluated for advanced melters: submerged combustion melting (a 6 ton/day unit is production in Ukraine), the BOC convective glass melting system (CGM), which directs oxyfuel flames vertically down onto the batch surface at the charging end of the furnace, microwave of silicate glasses, induction melting, and plasma melting, a technique that has been used to melt iron silicates in metal recuperation units and is being used in Japan as waste incinerators.

Finally, the U.S. DOE is promoting a program called Next Generation Melter to develop new melting technologies that will significantly increase the efficiency and lower the cost of glass production.

USES

Glass is commonly used in different applications such as architecture, beverage containers, insulation (eg, noise, thermal), and some lesser known applications such as nuclear waste encapsulation. The newer applications of glasses include components in solid-state batteries, electronic switches and memories, electrophotography, solar cells, microspheres for optical strengthening and medical uses, novel glass–ceramics (machinable and bioactive materials), solder and sol–gel glasses, gradient index optics, communication fibers, sensors, and nonlinear, active, and digital optics.

Container, Architecture, Insulation

Silicate glasses are commonly used in beverage containers, window panes, and automobile windshields. However, coatings are used to obtain properties not inherent in

glasses. The most widely used is silver coatings in mirrors. Today, most demand on coatings is on sodalime silica (SLS) glass surfaces and include architectural coatings (to reflect ir wavelengths reducing solar gain, control of light), container coatings (to prevent surface contact damage), and automotive coatings (window defrosters, color enhancement, support for electrical–electronic connections). Recently, coatings have been used in the optical fiber industry, that can strengthen the glass as well as provide lubricity and abrasion resistance. Other developments are thin-film-based products, such as liquid-crystal and electrochromic glazing that provide occupant-adjustable optical properties in automotive applications. Most of these applications demand that the coatings have high abrasion and chemical resistance and adhere strongly to the substrate. Coating research continues to improve cost reduction, coating application, and understanding the role of different treatments. Two general methods are known for preparing SLS substrates to obtain high quality coatings and both depend on an alkali diffusion barrier. In the first method, the glass surface is coated with pure SiO_2 by a sol–gel process. The SiO_2 coated substrate is heated then at 500°C where some densification occurs. A second method is gas-phase dealkalization procedure based on the reactions with acidic gases, HCl, SO_2, SO_3, and DFE (1,1-difluoroethane). The sol–gel coating provides higher quality SnO_2 coatings as compared to dealkalized SLS substrates.

Container Coatings. This surface treatment lubricates glass containers so they can be handled safely. The main process used is a cold-end coating of polyethylene combined with a hot-end coating of SnO_2 or TiO_2. Lubricity is regulated by varying the amount of polyethylene. Hot-end coatings are applied before annealing, using $SnCl_4$ or $TiCl_4$ or organic tin compounds by a CVD process to produce coatings of either SnO_2 or TiO_2. Cold-end coatings are organic materials applied after the annealing lehr. Materials for cold-end coatings include polyethylene, oleic acid, stearates, and silicones.

Automotive Coatings. Thin-film glass coatings are used for various purposes: to reduce interior heat build-up and air conditioner load by reflecting solar ir radiation; to provide heat to melt ice and frost from the windshield; to increase reflection and reduce visible transmission for rear occupant privacy; to reduce glare and enhance driver visibility; to serve as radio and telephone antennae; to provide an enhanced reflective region on the windshield for instrument display; to reduce emissivity to prevent frost build-up; to act as moisture sensors to trigger defrost and wiper operation; and to provide matching colors to enhance styling. Some coatings are being used to perform more than one function. An electrically heated windshield that also reduces solar heat load and electrically conductive coatings for heatings, or for solar reduction, may also be used as radio or mobile phone antennae.

Current coating technologies include on-line (a continuous process, integral with the float glass-making process) vs off-line procedures, pyrolytic (vs ambient temperature, vacuum vs ambient pressure, chemical vs physical deposition, before vs after glass bending, and monolithic vs laminated).

Off-line processes offer greater flexibility in a batch mode; film chemistry; and the control of important parameters such as temperature, pressure, and glass speed. However, the process must be preceded by thorough washing and drying. Off-line processes are typified by vacuum sputtering. Thick-film coatings are applied off-line by silk screen prior to bending or tempering, high temperature steps that serve to fire the coatings.

Architectural Coatings. SPD (suspended particle devices) film allows the production of a "smart" window that provides controllable degrees of light transmission. Used in conjunction with low E glass, which reflects heat and other commercially available materials, SPD smart windows can also block uv light and promote energy efficiency.

Medical Applications

Over the last decade, considerable attention has been directed toward the use of bioactive fixation of implants. Bioactive fixation has been defined as "interfacial bonding of an implant to tissue by the formation of a biologically active HAp layer on the implant surface." Studies of various compositions of bioactive glasses, ceramics and glass–ceramics have established that there are different levels of bioactivity, as measured by rates of bonding to bulk implants or, alternatively the rate of osteoblastic proliferation in the presence of bioactive particulates.

A limited number of bioactive glass compositions containing SiO_2–Na_2O–CaO–P_2O_5 with <55% SiO_2 exhibit a high bioactivity index that bond to both bone and soft connective tissues and have been identified as bioglasses. These materials have been classified as Class A, and are osteoproductive (enhance osteoblastic activity) as well as osteoconductive (bone growth and bond along the material surface). Materials classified as Class B only exhibit osteoconductivity.

New glass-based materials are being developed to repair bone by mixing crushed glass particles with a polymer. The mixture is to be injected into the area of a crushed vertebrae or other damaged bone that then fills the cracks, gluing the broken pieces back together. Once this mixture hardens, it turns into a bonelike substance, bonding itself to the original bone. Another method is being devised to use biodegradable glass spheres that will be used to irradiate arthritis joints. Similar procedures can be used to treat other ailments. Instead of using a solid glass sphere, a hollow sphere or shell filled with a drug and injected into the body, or spread as a cream onto the skin and gradually released into the body's system. This type of treatment releases the drug in a more uniform manner and targets the infection or diseased area.

Communication and Electronics

There are several advantages in using light pulses through silica glass fibers for telecommunications in comparison to copper wires that require repeaters or sig-

nal boosters at intervals of ~2 km; eg, the repeaters in commercial fiber-optic systems are 30 km apart. Also, the glass fibers are small (typically ~100 μm) and more of them fit into a cable of a given size. The glass fibers are not susceptible to electromagnetic interference, so the signal is clearer. Finally, the information carried on optical fibers can be modulated at very high frequencies with more simultaneous transmissions being possible.

Photonic Applications. Optoelectronic applications such as optical switches and modulators require materials having NLO properties; eg, the refractive indexes are nonlinear dependent on the intensity of the applied electric field and are noticeable only high energy sources such as lasers are used. It has been found that glasses containing small amounts of semiconducting microcrystals exhibit large optical nonlinearities. Halides and chalcogenide glasses present potential applications in infrared optics and optoelectronics.

Many organic and inorganic solids have been considered for photonic applications because of their nonlinear optical properties. Chalcogenide glasses with nonlinear refractive index have been theoretically identified to be some such candidate materials. Another new family of glasses with high nonlinear optical properties, so-called quantum dot solids, is formed by nanocomposites made up with microcrystallites of cadmium sulfide and cadmium selenide in a silicate glass matrix.

Optical Fiber Sensors. Advances in optical-fiber temperature and pressure sensors include industrial applications of fiber-optic temperature sensors. Temperature sensing is limited by the maximum service temperature of the fibers. Advanced temperature and pressure sensors are based on Bragg gratings. Optical-fiber sensors based on fiber Bragg gratings (FBGs) provide accurate, nonintrusive, and reliable remote measurements of temperature, strain, and pressure, and they are immune to electromagnetic interference. FBGs are extensively used in telecommunications, and as sensors, FBGs find many industrial applications in composite structures used in the civil engineering, aeronautics, train transportation, space, and naval sectors. Tiny FBG sensors embedded in a composite material can provide *in situ* information about polymer curing (strain, temperature, refractive index) in a nonintrusive way. Additionally, FBGs may be used in instrumentation as composite extensometers primarily in civil engineering applications.

NIF Laser Glass

The National Ignition Facility (NIF) has both the largest laser and the largest optical instrument ever built. The main objective of the NIF optics is to steer 192 laser beams through a 700-ft long building onto a dime-size laser-fusion target, compressing and heating BB-sized capsules of fusion fuel to thermonuclear ignition. The experiments will help scientists sustain confidence in the nuclear weapon stockpile without nuclear tests. It will also produce additional benefits in basic science and fusion energy.

Glasses for Nuclear Waste Disposal

Vitrification is being used to immobilize high level nuclear waste (HLW) in a stable, chemically durable borosilicate glass. Borosilicate glasses have a good chemical durability, but may not be suitable for all HLW compositions, such as, wastes containing phosphates, halides and heavy metals (Bi, U, Pu).

Many phosphate glasses have a chemical durability that is usually inferior to that of most silicate and borosilicate glasses, but iron phosphate glasses are an exception. In addition to their generally excellent chemical durability, iron phosphate glasses have low melting temperature, typically between 950 and 1150°C.

ASTM C162-99, Standard Terminology of Glass and Glass Products, 1999.

G. N. Greaves, in C. R. Catlow, ed., *Defects and Disorder in Crystalline and Amorphous Solids*, Kluwer Academic, Dordrecht, 1994, p. 87.

W. Höland and G. H. Beall, *Glass Ceramic Technology*, American Ceramic Society, Westerville, Ohio, 2002.

J. E. Shelby, *Introduction to Glass Science and Technology*, RSC Paperbacks, The Royal Society of Chemistry, 1997.

J. B. Wachtman, ed., in *Ceramic Innovations in the 20th Century*, The American Ceramic Society, Westerville, Ohio, 1999.

DAVID C. BOYD
PAUL S. DANIELSON
DAVID A. THOMPSON
Corning Incorporated

MARIANO VELEZ
SIGNO T. REIS
RICHARD K. BROW
University of Missouri-Rolla

GLASS–CERAMICS

Glass–ceramics are polycrystalline materials formed by the controlled crystallization of glass. Most commercial glass–ceramic products are formed by highly automated glass-forming processes and converted to a crystalline product by the proper heat treatment. Glass–ceramics can also be prepared via powder processing methods in which glass frits are sintered and crystallized. The range of potential glass–ceramic compositions is therefore extremely broad, requiring only the ability to form a glass and control its crystallization.

Glass–ceramics can provide significant advantages over conventional glass or ceramic materials, by combining the ease and flexibility of forming and inspection of glass with improved and often unique physical properties in the glass–ceramic. They possess highly uniform microstructures, with crystal sizes on the order of 10 μm or less; this homogeneity ensures that their physical properties are highly reproducible.

Unlike conventional ceramic materials, glass-ceramics are fully densified with zero porosity. They generally are at least 50% crystalline by volume and often are >90% crystalline.

More than $500 million in glass–ceramic products are sold yearly worldwide. These range from transparent, zero-expansion materials with excellent optical properties and thermal shock resistance to jadelike highly crystalline materials with excellent strength and toughness. The highest volume is in stovetops and stove windows, cookware and tableware, and architectural cladding. Glass–ceramics are also referred to as Pyrocerams, vitrocerams, devitrocerams, sitalls, slagceramics, melt-formed ceramics, and devitrifying frits.

PROCESSING

Glass-ceramic articles can be fabricated by means of either bulk or powder processing methods. Both methods begin with melting a glass of the desired composition.

Bulk Glass–Ceramic Processing

In this most common method of glass-ceramic manufacture, articles are melted and fabricated to shape in the glass state. Most forming methods can be employed, including rolling, pressing, spinning, casting, and blowing. The article is then crystallized using a heat treatment designed for that material. This process, known as ceramming, typically consists of a low temperature hold to induce nucleation, followed by one or more higher temperature holds to promote crystallization and growth of the primary phase. Because crystallization occurs at high viscosity, article shapes are typically preserved with little or no shrinkage (1–3%) or deformation during the ceramming. A typical heat treatment cycle, as illustrated in Figure 1, comprises both nucleation and crystallization temperature holds, but some glass-ceramics are designed to nucleate and/or crystallize during the ramp itself, eliminating the need for multiple holds.

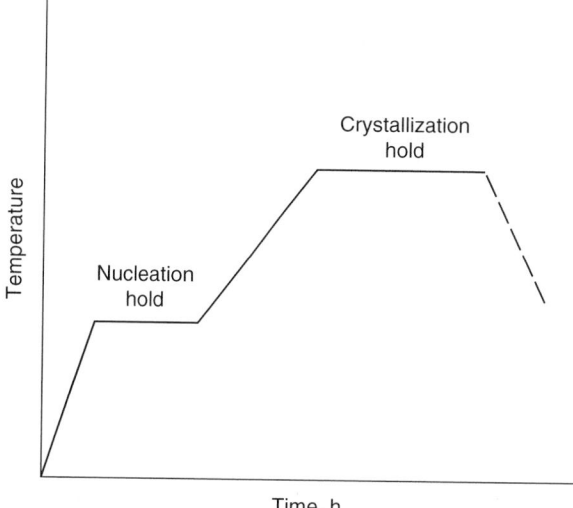

Figure 1. Heat treatment cycle for a glass-ceramic material.

Powder Glass–Ceramic Processing

The manufacture of glass–ceramics from powdered glass, using conventional ceramic processes such as spraying, slip-casting, or extrusion, extends the range of possible glass–ceramic compositions by taking advantage of surface crystallization. In these materials, the surfaces of the glass grains serve as the nucleating sites for the crystal phases. The glass composition and processing conditions are chosen such that the glass softens prior to crystallization and undergoes viscous sintering to full density just before the crystallization process is completed. Given these conditions, the final crystalline microstructure is essentially the same as that produced from the bulk process.

The precursor glass powders can be produced by various methods, the simplest being the milling of quenched glass to an average particle size of 3–15 μm. Sol-gel processes, in which highly uniform, ultrafine amorphous particles are grown in a chemical solution, may be preferable for certain applications.

Such so-called devitrifying frits are employed extensively as sealing frits for bonding glasses, ceramics, and metals. Other applications include cofired multilayer substrates for electronic packaging, matrices for fiber-reinforced composite materials, refractory cements and corrosion-resistant coatings, honeycomb structures in heat exchangers, and bone and dental implants and prostheses.

Secondary Processing (Strengthening)

Because their polycrystalline microstructures provide resistance to crack propagation, glass–ceramics possess mechanical strength inherently superior to that of glass. This strength may be augmented by a number of techniques that impart a thin surface compressive stress to the body.

DESIGN

There are three key variables in the design of a glass-ceramic: the glass composition, the glass-ceramic phase assemblage, and the nature of the crystalline microstructure. The glass–ceramic phase assemblage, ie, the types of crystals and the proportion of crystals to glass, is responsible for many of the physical and chemical properties, such as thermal and electrical characteristics, chemical durability, elastic modulus, and hardness. In many cases these properties are additive; eg, a phase assemblage comprising high and low expansion crystals has a bulk thermal expansion proportional to the amounts of each of these crystals.

Finally, the nature of the crystalline microstructure, ie, crystal size and morphology and the textural relationship among the crystals and glass, is the key to many mechanical and optical properties, including transparency/opacity, strength and fracture toughness, and machinability. These microstructures can be quite complex and often are distinct from conventional ceramic microstructures.

PROPERTIES

Thermal Properties

Many commercial glass-ceramics have capitalized on their superior thermal properties, particularly low or zero thermal expansion coupled with high thermal stability and thermal shock resistance: Properties that are not readily achievable in glasses or ceramics.

Mechanical Properties

Like glass and ceramics, glass-ceramics are brittle materials that exhibit elastic behavior up to the strain that yields breakage. Because of the nature of the crystalline microstructure, however, strength, elasticity, toughness (resistance to fracture propagation), and abrasion resistance are higher in glass-ceramics than in glass. Knoop hardness values of up to 900 can be obtained in glass-ceramics containing particularly hard crystals such as sapphirine.

Optical Properties

Glass-ceramics may be either opaque or transparent. The degree of transparency is a function of crystal size and birefringence, and of the difference in refractive index between the crystals and the residual glass. When the crystals are much smaller than the wavelength of light, as in some mullite and spinel glass-ceramics, or when the crystals have low birefringence and the index of refraction is closely matched, as in some Mg-stuffed β-quartz glass-ceramics, excellent transparency can be achieved.

Certain glass-ceramic materials also exhibit potentially useful electrooptic effects: glasses with microcrystallites of Cd-sulfoselenides, which show a strong nonlinear response to an electric field, as well as glass-ceramics based on ferroelectric crystals such as niobates, titanates, zirconates, and germanates. Such crystals permit electric control of scattering and other optical properties. Ferroelectric glass-ceramics have potential use as optical modulators, switches, and filters.

Chemical Properties

The chemical durability is a function of the durability of the crystals and the residual glass. Generally, highly siliceous glass-ceramics with low alkali residual glasses, such as glass-ceramics based on β-quartz and β-spodumene, have excellent chemical durability and corrosion resistance similar to that obtained in borosilicate glasses.

Electrical Properties

The dielectric properties of glass-ceramics strongly depend on the nature of the crystal phase and on the amount and composition of the residual glass. In general, glass-ceramics have such high resistivities that they are used as insulators.

GLASS–CERAMIC FAMILIES

All commercial as well as most experimental glass–ceramics are based on silicate bulk glass compositions.

Glass–ceramics can be further classified by the composition of their primary crystalline phases, which may consist of silicates, oxides, phosphates, or borates. The principal commercial glass–ceramics are based on silicate crystals.

NEW AND POTENTIAL APPLICATIONS

Although glass–ceramics have been employed for many years as cookware and dinnerware, as cladding and other architectural materials, and in a wide range of industrial components, their future roles are likely to involve entirely new technologies. Glass–ceramics will find numerous applications in the rapidly growing field of optoelectronics.

Glass–ceramics will also play a role in the burgeoning field of information storage and display. Several glass–ceramics have been developed for use as disk substrates in hard disk drives; these are thinner, flatter, and more rugged than traditional aluminum alloy substrates, and their precise, uniform surfaces permit higher data storage capacity and improved reliability. Glass–ceramics have also been used for the read/write heads used in disk drives. Other glass–ceramics may find potential use in liquid crystal and electroluminescent displays. Finally, glass–ceramics will play a key role in the growing arsenal of advanced materials, both alone and in combinations with other materials. Low dielectric constant glass-ceramics are projected as the best candidates for high performance multilayer packaging materials. Other potential products include superconducting glass–ceramics, new types of bioceramics for bone implants and prostheses, durable glass–ceramics for waste disposal, and refractory, corrosion-resistant glass–ceramic coatings for superalloys.

M. H. Lewis, ed., *Glasses and Glass-Ceramics*, Chapman and Hall, London, 1989.

R. Morrell, *Handbook of Properties of Technical and Engineering Ceramics, Part 1: An Introduction for the Engineer and Designer*, HMSO, London, 1985.

Z. Strnad, *Glass-Ceramic Materials, Glass Science and Technology*, Vol. 8, Elsevier, Amsterdam, 1986.

W. Vogel, *Chemistry of Glass*, The American Ceramic Society, Columbus, Ohio, 1985.

LINDA R. PINCKNEY
Corning Inc.

GLYCOLS

ETHYLENE GLYCOLS

Glycols are diols, compounds containing two hydroxyl groups attached to separate carbon atoms in an aliphatic chain. Ethylene glycol, the adduct of water and ethylene oxide, is the simplest glycol and is the principal topic of this section. Diethylene, triethylene, and tetraethylene glycols are oligomers of ethylene glycol. Polyglycols are higher molecular weight adducts of ethylene oxide and are distinguished by intervening ether linkages in the hydrocarbon chain. These polyglycols are commercially important; their properties are significantly affected by molecular weight. They are water soluble, hygroscopic, and undergo reactions common to the lower weight glycols.

Ethylene glycol, EG, is a colorless, practically odorless, low viscosity, hygroscopic liquid of low volatility. It is completely miscible with water and many organic liquids. The uses for ethylene glycol are numerous. Some of the applications are polyester resins for fiber, PET containers, and film applications; all-weather automotive antifreeze and coolants, defrosting and deicing aircraft; heat-transfer solutions for coolants for gas compressors, heating, ventilating, and air-conditioning systems; water-based formulations such as adhesives, latex paints, and asphalt emulsions; manufacture of capacitors; and unsaturated polyester resins. The oligomers also have excellent water solubility but are less hygroscopic and have somewhat different solvent properties. The number of repeating ether linkages controls the influence of the hydroxyl groups on the physical properties of a particular glycol.

Physical Properties

Ethylene glycol and its lower polyglycols are colorless, odorless, high boiling, hygroscopic liquids completely miscible with water and many organic liquids. Physical properties of ethylene glycols are listed in Table 1. Ethylene glycols markedly reduce the freezing point of water.

Chemical Properties

The hydroxyl groups on glycols undergo the usual alcohol chemistry giving a wide variety of possible derivatives. Hydroxyls can be converted to aldehydes, alkyl halides, amides, amines, azides, carboxylic acids, ethers, mercaptans, nitrate esters, nitriles, nitrite esters, organic esters, peroxides, phosphate esters, and sulfate esters. The largest commercial use of ethylene glycol is its reaction with dicarboxylic acids to form linear polyesters.

Manufacture

In 1937 the first commercial application of the Lefort direct ethylene oxidation to ethylene oxide followed by hydrolysis of ethylene oxide became, and remains, the main commercial source of ethylene glycol production. Ethylene oxide hydrolysis proceeds with either acid or base catalysis or uncatalyzed in neutral medium. Acid-catalyzed hydrolysis activates the ethylene oxide by protonation for the reaction with water. Base-catalyzed hydrolysis results in considerably lower selectivity to ethylene glycol. The yield of higher glycol products is substantially increased since anions of the first reaction products effectively compete with hydroxide ion for ethylene oxide. Neutral hydrolysis (pH 6–10), conducted in the presence of a large excess of water at high temperatures and pressures, increases the selectivity of ethylene glycol to 89–91%. In all these ethylene oxide hydrolysis processes the

Table 1. Properties of Glycols

Property	Ethylene glycol	Diethylene glycol	Triethylene glycol	Tetraethylene glycol
formula	$HOCH_2CH_2OH$	$HO(CH_2CH_2O)_2H$	$HO(CH_2CH_2O)_3H$	$HO(CH_2CH_2O)_4H$
mol wt	62.07	106.12	150.17	194.23
sp gr, 20/20°C	1.1155	1.1185	1.1255	1.1247
bp at 101.3 kPa,[a] °C	197.6	245.8	288	dec
mp, °C	−13.0	−6.5	−4.3	−4.1
viscosity at 20°C, mPa·s(= cP)	20.9	36	49	61.9

[a] To convert kPa to mm Hg, multiply by 7.5.

principal by-product is diethylene glycol. The higher glycols, ie, triethylene and tetraethylene glycols, account for the remainder.

Health, Safety, and Environmental Factors

Biodegradability of a product may be evaluated by extended-term biochemical oxygen demand (BOD) tests. This procedure permits comparison of the amount of oxygen consumed by microorganisms in the oxidation of the test material to the theoretical oxygen required to completely oxidize the chemical to carbon dioxide and water. Laboratory BOD tests using unacclimated biomass show that ethylene glycol is readily biodegraded in a system which attempts to simulate the dilute biological conditions of a river or lake. Ethylene glycol can be treated effectively in conventional wastewater treatment plants and does not persist in the environment under expected conditions.

None of the glycols are highly irritating to the mucous membranes or skin. A splash of these neat materials in the eye may produce marked irritation, but permanent damage should not be expected. In general, each glycol is believed to have low acute dermal toxicity and vapor and inhalation toxicity when exposure is in an industrial situation at normal room temperature, since the amount absorbed through the skin is believed to be minimal and the vapor pressures are relatively low.

Ethylene glycol is not recommended for use as an ingredient in food or beverages, or where there is a significant contact with food or potable water. For adult humans, death has occurred after as little as 30 mL ethylene glycol has been ingested; yet survival has been reported following one liter and more. The clinical sequence of toxic response has been described as follows: Phase I: central nervous system (CNS) and metabolic abnormalities; Phase II: cardio and pulmonary abnormalities; Phase III: renal insufficiency.

There is no evidence of genetic toxicity for ethylene, diethylene, and triethylene glycols from a battery of *in vitro* tests. Tetraethylene glycol is not believed to be mutagenic.

Uses

In 2000, uses for ethylene glycol in the U.S. were as followed: PET bottle-grade resins, 34%; antifreeze, 26%; polyesters fibers, 24%; polyester film, 4%; polyester engineering resins, 3%; aircraft deicing fluids, 2%; miscellaneous, 7%. Poly(ethylene terephthalate) provides unique properties for each end use. Wrinkle resistance, strength, durability, and stain resistance are enhanced when polyester is combined with natural fibers to produce apparel, home furnishing fabrics, carpeting, and fiberfill.

Derivatives

In addition to oligomers ethylene glycol derivative classes include monoethers, diethers, esters, acetals, and ketals as well as numerous other organic and organometallic molecules. These derivatives can be of ethylene glycol, diethylene glycol, or higher glycols and are commonly made with either the parent glycol or with sequential addition of ethylene oxide to a glycol alcohol, or carboxylic acid forming the required number of ethylene glycol subunits.

Ethylene glycol monoethers are commercially manufactured by reaction of an alcohol with ethylene oxide.

Glycol monoethers are widely used in cleaning formulations to facilitate the removal of grease and greasy soils, to aid in solubilizing other components in the cleaner formulation, and to improve the storage stability of the formulations. Glycol monoethers are used as jet fuel additives to inhibit icing in fuel systems. They are used as solvents and cosolvents for conventional solvent-based lacquer, enamel, and wood stain industrial coating systems as well as cosolvents for waterborne industrial coating systems. Other applications include dye solvents in the textile, leather, and printing industries; solvents for insecticides and herbicides for agricultural applications; couplers and mutual solvents for soluble oils, hard-surface cleaners, and other soap–hydrocarbon systems; semiconductor manufacture; printed circuit board laminating formulations; freeze–thaw agents in latex emulsions; diluents for hydraulic brake fluids; chemical reaction solvents; and chemical intermediates.

Ethylene glycol diethers are made by derivatizing both glycol hydroxyl groups. Strong solvating and stability properties allow numerous applications for glycol diethers including adhesives and coatings, ink formulations, cleaning compounds, batteries, electronics, polymer solvents, polymer plasticizers, gold refining, and gas purification.

Cyclic polyethers, or crown ethers, are cyclic structures containing ethylene glycol units as $-(CH_2CH_2O)_n-$ with $n > 2$ and generally between 4 and 8. Crown ethers greatly improve the solubility of salts in organic solvents and have a wide range of applications in organic reactions, especially as phase-transfer reagents. Glycols can be used in the manufacture of poly(ethylene glycol) (PEG). PEGs find applications in agriculture products, ceramics, chemical intermediates, coatings and adhesives, cosmetics and toiletries, electronics, foods and feeds, household products, lubricants, metal processing, mining, paper, petroleum, pharmaceuticals, photography, plastics and resins, printing, rubber chemicals, textiles and leather, and wood products. Ethylene glycols can also react with other alkylene oxides as well as other alkylene glycols reacted with ethylene oxide to form mixed oxide polyglycols which can be used in many applications similar to those already mentioned.

Ethylene glycol (as well as the higher glycols) can be esterified with traditional reagents such as acids, acid chlorides, acid anhydrides, and via transesterification with other esters. Low molecular weight glycol esters are good solvents for cellulose esters and printing inks, and are employed in industrial extraction processes and the protective coating industry. Fatty acid esters, together with other surfactants, are good-emulsifying, stabilizing, dispersing, wetting, foaming, and suspending agents.

Ethylene glycol in the presence of an acid catalyst readily reacts with aldehydes and ketones to form cyclic acetals and ketals. Cyclic acetals and ketals are used as protecting groups for reaction-sensitive aldehydes and ketones in natural product synthesis and pharmaceuticals.

Diethylene Glycol. Physical properties of diethylene glycol are listed in Table 1. Diethylene glycol is similar in many respects to ethylene glycol, but contains an ether group. It can be made directly by the reaction of ethylene glycol with ethylene oxide, but this route is rarely used because more than an adequate supply is available from the hydrolysis reaction.

Triethylene Glycol. Physical properties of triethylene glycol are listed in Table 1. Triethylene glycol is a colorless, water-soluble liquid with chemical properties essentially identical to those of diethylene glycol. It is a coproduct of ethylene glycol produced via ethylene oxide hydrolysis. Significant commercial quantities are also produced directly by the reaction of ethylene oxide with the lower glycols.

Triethylene glycol is an efficient hygroscopicity agent with low volatility, and about 53% is used as a liquid drying agent for natural gas. As a solvent triethylene glycol is used in resin impregnants and other additives, steam-set printing inks, aromatic and paraffinic hydrocarbon separations, cleaning compounds, and cleaning poly(ethylene terephthalate) production equipment. Approximately 12% triethylene glycol is used in some form as a vinyl plasticizer. Triethylene glycol esters are important plasticizers for poly(vinyl butyral) resins, nitrocellulose lacquers, vinyl and poly(vinyl chloride) resins, poly(vinyl acetate), and synthetic rubber compounds and cellulose esters. The fatty acid derivatives of triethylene glycol are used as emulsifiers, demulsifiers, and lubricants.

Tetraethylene Glycol. Physical properties of tetraethylene glycol are listed in Table 1. Tetraethylene glycol has properties similar to diethylene and triethylene glycols and may be used preferentially in applications requiring a higher boiling point, higher molecular weight, or

lower hygroscopicity. Tetraethylene glycol is miscible with water and many organic solvents. It is a humectant that, although less hygroscopic than the lower members of the glycol series, may find limited application in the dehydration of natural gases. Other possibilities are in moisturizing and plasticizing cork, adhesives, and other substances.

PROPYLENE GLYCOLS

The propylene glycol family of chemical compounds consists of monopropylene glycol (PG), dipropylene glycol (DPG), and tripropylene glycol (TPG). These chemicals are manufactured as coproducts and are used commercially in a large variety of applications. They are available as highly purified products which meet well-defined manufacturing and sales specifications. All commercial production is via the hydrolysis of propylene oxide.

The propylene glycols are clear, viscous, colorless liquids that have very little odor, a slightly bittersweet taste, and low vapor pressures. The most important member of the family is monopropylene glycol, also known as 1,2-propylene glycol, 1,2-dihydroxypropane, 1,2-propanediol, methylene glycol, and methyl glycol. All of the glycols are totally miscible with water.

Propylene glycol, when produced according to the U.S. Food and Drug Administration good manufacturing practice guidelines at a registered facility, meets the requirements of the U.S. Food, Drug, and Cosmetic Act as amended under Food Additive Regulation CFR Title 21, Parts 170–199. It is listed in the regulation as a direct additive for specified foods and is classified as generally recognized as safe (GRAS). In addition, it meets the requirements of the *Food Chemicals Codex* and the specifications of the *U.S. Pharmacopeia XXII*. Because of its low human toxicity and desirable formulation properties it has been an important ingredient for years in food, cosmetic, and pharmaceutical products.

Physical and Chemical Properties

Table 2 lists various physical and chemical properties and constants for the propylene glycols.

Chemical Properties

Monopropylene glycol (1,2-propanediol) is a difunctional alcohol with both a primary and a secondary hydroxyl.

1,2-Propylene glycol undergoes most of the typical alcohol reactions, such as reaction with a free acid, acyl halide, or acid anhydride to form an ester; reaction with alkali metal hydroxide to form metal salts; and reaction with aldehydes or ketones to form acetals and ketals. The most important commercial application of propylene glycol is in the manufacture of polyesters by reaction with a dibasic or polybasic acid. Polyethers are also products of commercial importance.

Environmental Considerations

The propylene glycols vary in biodegradability, but it is expected that all of the propylene glycols will exhibit

Table 2. Properties of Glycols

Physical properties	Propylene glycol	Dipropylene glycol	Tripropylene glycol
formula	$C_3H_8O_2$	$C_6H_{14}O_3$	$C_9H_{20}O_4$
molecular weight	76.1	134.2	192.3
boiling point at 101.3 kPa[a], °C	187.4	232.2[b]	265.1[b]
vapor pressure, kPa[a], 25°C	0.017	0.0021	0.0003
density at 25°C, g/mL	1.032	1.022	1.019
viscosity at 25°C, mPa·s(=cP)	48.6	75.0	57.2

[a] To convert kPa to mm Hg, multiply by 7.5.
[b] Varies with isomer distribution.

moderate to high biodegradability in a natural environment.

All of the propylene glycols are considered to be practically nontoxic to fish on an acute basis ($LC_{50} < 100$ mg/L) and practically nontoxic to aquatic invertebrates, also on an acute basis.

Uses

Consumption of propylene glycol follows an erratic pattern in the United States. Principal uses of propylene glycol in the United states are in unsaturated polyester resins; functional fluids (antifreeze, deicing, heat transfer); food, drug, and cosmetics; liquid detergents; paints and coatings; humectants; and other.

Dipropylene Glycol. Dipropylene glycol is similar to the other glycols in general properties, and its fields of use are comparable. The greater solvency of dipropylene glycol for castor oil indicates its usefulness as a component of hydraulic brake fluid formulations; its affinity for certain other oils has likewise led to its use in cutting oils, textile lubricants, and industrial soaps. It is also used as a reactive intermediate in manufacturing polyester resins, plasticizers, and urethanes. Fragrance or low odor grades of dipropylene glycol are established standard base formulating solvents in the fragrance industry and for some personal care products such as deodorants.

Tripropylene Glycol. Tripropylene glycol is an excellent solvent in many applications where other glycols fail to give satisfactory results. Its ability to solubilize printing ink resins is especially marked, so much so that it finds its way into creams designed to remove ink stains from the hands. Tripropylene glycol is also used as a reactant to produce acrylate resins which are useful in radiation-cured coatings, adhesives, and inks.

OTHER GLYCOLS

Properties

Glycols such as neopentyl glycol, 2,2,4-trimethyl-1,3-pentanediol, 1,4-cyclohexanedimethanol, and hydroxypivalyl hydroxypivalate are used in the synthesis of polyesters, unsaturated and urethane foams. Their physical properties are shown in Table 3.

Neopentyl Glycol

Neopentyl glycol, or 2,2-dimethyl-1,3-propanediol, is a white crystalline solid at room temperature, soluble in water, alcohols, ethers, ketones, and toluene but relatively insoluble in alkanes.

Chemical Properties. Neopentyl glycol can undergo typical glycol reactions such as esterification, etherification, condensation, and oxidation.

Table 3. Physical Properties of Several Glycols

Properties	Neopentyl glycol	2,2,4-Trimethyl-1,3-pentanediol	1,4-Cyclo hexane-dimethanol[a]	Hydroxypivalyl hydroxypivalate
molecular formula	$C_5H_{12}O_2$	$C_8H_{18}O_2$	$C_8H_{16}O_2$	$C_{10}H_{20}O_4$
mol wt	104.2	146.2	144.2	204.3
melting range, °C	124–130	46–55	45–50[b]	46–50
boiling point, °C, at 101.3 kPa[c]	212	236	286	290
boiling range, °C		215–235		
density, g/cm³				
at 20°C	1.06		1.02	1.02
at 15°C		0.937		
heat of combustion, kJ/mol[d]	−3100	−5050	−4849[e]	

[a] Mixture of isomers, cis/trans ratio (wt%) = ∼32/68.
[b] Mp of cis isomer [3236-47-3] = 41°C; mp of trans isomer [3236-48-4] = 70°C.
[c] To convert kPa to mm Hg, multiply by 7.5.
[d] To convert kJ to kcal, divide by 4.184.
[e] Paar bomb.

Manufacture. Commercial preparation of neopentyl glycol can be via an alkali-catalyzed condensation of isobutyraldehyde with 2 moles of formaldehyde (crossed Cannizzaro reaction).

Uses. Neopentyl glycol is used extensively as a chemical intermediate in the manufacture of polyester resins, polyurethane polyols, synthetic lubricants, polymeric plasticizers, and other polymers. It imparts a combination of desirable properties to properly formulated esterification products, including low color, good weathering and chemical resistance, and improved thermal and hydrolytic stability.

A comparison of coatings formulations based on various glycols to determine the effects of the various glycol structures on the performance properties of the coatings has been made. Properties compared included degree of cure, flexibility, hardness, hydrolytic stability, processibility, chemical and stain resistance, and viscosity.

2,2,4-Trimethyl-1,3-Pentanediol

2,2,4-Trimethyl-1,3-pentanediol is a white, crystalline solid. Trimethylpentanediol is soluble in most alcohols, other glycols, aromatic hydrocarbons, and ketones, but it has only negligible solubility in water and aliphatic hydrocarbons.

Chemical Properties. Trimethylpentanediol, with a primary and a secondary hydroxyl group, enters into reactions characteristic of other glycols.

Manufacture and Processing. 2,2,4-Trimethyl-1,3-pentanediol can be produced by hydrogenation of the aldehyde trimer resulting from the aldol condensation of isobutyraldehyde.

Uses. The versatility of trimethylpentanediol as an intermediate is reflected by the diversity of its commercial applications.

1,4-Cyclohexanedimethanol

1,4-Cyclohexanedimethanol, 1,4-dimethylolcyclohexane, or 1,4-bis(hydroxymethyl) cyclohexane, is a white, waxy solid. The commercial product consists of a mixture of cis and trans isomers.

1,4-Cyclohexanedimethanol is miscible with water and low molecular weight alcohols and appreciably soluble in acetone. It has only negligible solubility in hydrocarbons and diethyl ether.

Chemical Properties. The chemistry of 1,4-cyclohexanedimethanol is characteristic of general glycol reactions; however, its two primary hydroxyl groups give very rapid reaction rates, especially in polyester synthesis.

Manufacture. The manufacture of 1,4-cyclohexanedimethanol can be accomplished by the catalytic reduction under pressure of dimethyl terephthalate in a methanol solution.

Uses. The most important application for 1,4-cyclohexanedimethanol is in the manufacture of linear polyesters for use as fibers.

Hydroxypivalyl Hydroxypivalate

Hydroxypivalyl hydroxypivalate or 3-hydroxy-2,2-dimethylpropyl 3-hydroxy-2,2-dimethylpropionate is a white crystalline solid at room temperature.

Hydroxypivalyl hydroxypivalate is soluble in most alcohols, ester solvents, ketones, and aromatic hydrocarbons. It is partially soluble in water.

Chemical Properties. Both hydroxy groups on hydroxypivalyl hydroxypivalate are primary, which results in rapid reactions with acids during esterification. The absence of hydrogens on the carbon atom beta to the hydroxyls is a feature this glycol shares with neopentyl glycol, resulting in excellent weatherability.

Manufacture. Hydroxypivalyl hydroxypivalate may be produced by the esterification of hydroxypivalic acid with neopentyl glycol or by the intermolecular oxidation–reduction (Tishchenko reaction) of hydroxypivaldehyde using an aluminum alkoxide catalyst.

Uses. Saturated polyesters made from hydroxypivalyl hydroxypivalate are most often used for formulating coatings which have very low initial color and which retain the low color exposure to weathering.

J. Lacson, "Mono-, Di-, and Triethylene Glycols," *Chemical Economics Handbook*, SRI International, Menlo Park, Calif., Nov. 2003.

J. March, *Advanced Organic Chemistry*, 4th ed., John Wiley & Sons, Inc., New York, 1992.

M. W. FORKNER
J. H. ROBSON
W. M. SNELLINGS
Union Carbide Corporation

A. E. MARTIN
F. H. MURPHY
The Dow Chemical Company

T. E. PARSONS
Eastman Chemical Company

GOLD AND GOLD COMPOUNDS

The chemical symbol for gold (the Anglo Saxon and German term) is Au (for *aurum*, Latin for gold) with an atomic number of 79, it is listed as a group Ib metal in the periodic table, having one electron on its outer shell. There are six artificial isotopes having masses between 196 and 200, four of which provide significant radiation; radioactive [198]Au- is actively used for medical analysis.

Although gold has lost its status as the world's numeraire, it will never lose its intrinsic value.

The rigorous performance demands of electronics, telecommunications, and aerospace systems created a new role for gold beyond that of a readily fungible asset.

OCCURRENCE

Gold is widely distributed throughout the earth in very small concentrations. In the crust it averages ~3 parts per billion (ppb), and in seawater it varies from 0.001 to 44 ppb. Gold occurs primarily as a native metal, although often alloyed with other elements, usually silver. The proportion of silver reaches ~20%, the alloy is termed *electrum*. The most abundant natural minerals of gold are the tellurides ($AuTe_2$ and $AuAgTe_4$).

Gold weathered out of rock naturally collects in stream beds as particles or tumbled into nuggets. This gold is readily separated mechanically in water by weight or by amalgamation with mercury. Known since ancient times as placer mining, it continues to this day.

Traditionally, gold deposits have been classified in terms of their host rocks (igneous, metamorphic, and sedimentary) or where they occur (veins, skarns, disseminations, or coproducts with other metals). Discoveries of highly disseminated gold deposits have now led to a new classification based on the oxidation/reduction state of the transporting fluid and its sources. As a consequence, gold geochemistry is the new prospector's tool.

Properties

Gold, Au, atomic number 79, is a third row transition metal in Group 11 (IB) of the Periodic Table. It occurs naturally as a single stable isotope of mass 197, ^{197}Au, which is also formed via the decay of ^{197}Pt (half-life 20 min) formed in the irradiation of platinum with slow neutrons. Selected properties are shown in Table 1. Gold is characterized by high density, high electrical and thermal conductivities, and high ductility. At least 26 unstable gold isotopes have been made; the most frequently used is ^{198}Au which has a half-life of 2.7 d.

Gold is the most noble of the noble metals. Other than in the atomic state, the metal does not react with oxygen, sulfur, or selenium at any temperature. It does, however, react with tellurium at elevated (ca 475°C) temperatures to produce gold ditelluride, $AuTe_2$, which is also found in the naturally occurring mineral, calaverite, AuTe. Gold reacts with the halogens, particularly in the presence of moisture.

Gold reacts with various oxidizing agents at ambient temperatures provided a good ligand is present to lower the redox potential below that of water. Thus, gold is not attacked by most acids under ordinary conditions and is stable in basic media. Gold does, however, dissolve readily in 3:1 hydrochloric—nitric acid (aqua regia) to form $HAuCl_4$ and in alkaline cyanide solutions in the presence of air or hydrogen peroxide to form $(Au(CN)_2)^-$. These reactions are important to the extraction and refining of the metal.

At high temperatures, attack by concentrated sulfuric and nitric acids is slow and is negligible for phosphoric acid. Gold is very resistant to fused alkalies and to most fused salts except peroxides. Gold readily amalgamates with mercury. Gold is very corrosion and tarnish resistant and imparts corrosion resistance to most of the commonly used gold alloys.

Gold alloys also are subject to stress-corrosion cracking, especially alloys below 14 carat (58% gold). Jewelry items may have areas of high local stress that can induce cracking and corrosion in normally harmless environments such as solutions of hydrochloric or nitric acid. Stress-corrosion cracking occurs primarily in single-phase alloys. A remedy against it is stress relief by annealing; however, in the case of multiphase, low carat alloys, care must be exercised so that annealing does not lead to the formation of homogeneous solutions.

EXTRACTION AND REFINING

Primary gold ores are usually classified as refractory (difficult to treat) or nonrefractory. Gold is extracted from nonrefractory ores, which contain free gold in a relatively inert matrix, by grinding or crushing and gravity concentration, then direct cyanidation (in agitated leaching in vats or spray leaching on open dumps or heaps) or by selective flotation (adsorption of gold-bearing ore particles to specific chemicals that float to the surface in an aqueous mix).

Conventional Extraction

The conventional recovery of the gold by the cyanidation process wherein finely ground ore is agitated in vats of 0.1% sodium cyanide solution. An alternative method becoming more common is to conduct the pregnant solution through activated-carbon columns that adsorb the gold. The gold is stripped off the carbon by hot caustic soda.

Refractory ores contain interfering elements or minerals such as an activated type of carbon that tends to adsorb the gold cyanide complex, resulting in a loss in

Table 1. Gold Properties

Property	Value
atomic weight	196.9665
melting point, K	1337.59
boiling point, K	3081
density[a] at 273 K, g/cm^3	19.32
Brinell hardness (10/500/90), annealed at 1013 K, kgf/mm^2	25
tensile strength, annealed at 573 K, MPa[b]	123.6–137.3
elongation, annealed at 573 K, %	39–45
compressibility at 300 K, Pa^{-1}[c]	6.01×10^{-12}
heat of fusion, J/mol[d]	1.268×10^4
vapor pressure, Pa[c] at 1000 K	5.5×10^{-8}
specific heat at 298 K, J/(g·K)[d]	1.288×10^{-1}
thermal conductivity at 273 K, W/(m·K)	311.4
thermal expansion at 273–373 K, K^{-1}	1.416×10^{-7}
electrical resistivity at 273 K, cm	2.05×10^{-6}

[a]The commercially accepted value has been given. Measured values and density calculations from   x-ray data show some variations.
[b]To convert MPa to psi, multiply by 145.
[c]To convert Pa to mm Hg, divide by 1333.3.
[d]To convert Pa to mm Hg, divide by 133.3.

recovery of the metal. For such ores, direct cyanidation cannot be used. Recovery techniques are designed to meet the specific extraction requirements for the ore.

Flotation is an important alternative to the use of cyanidation. Here, finely ground ore particles are treated with chemicals that selectively adsorb the precious-metal values and float them to the top of the vat. The floating matter is skimmed off, the chemicals removed, and the raw concentrate melted into a high precious-metal content bar for shipment to a precious-metal refinery.

Heap Leaching

The discovery of vast quantities of ore with highly disseminated gold particles, and the existence of large quantities of mine tailings from old less-efficient operations in the western United States whose values were too low for recovery by standard methods (under ca. 0.1 oz/ton of ore), led to the development in 1969 of heap/dump leaching by very dilute sodium cyanide solutions.

Oxidation and Other Alternatives to Cyanidation

Oxidation and the use of bacteria to oxidize ores is gaining favor over conventional roasting or pressure oxidation because, although slower, it provides improved gold recovery, is easily controlled by unskilled labor, is environmentally friendly, and has greatly reduced capital cost. Among the most considered alternatives to sodium cyanide is thiourea.

Refining

The gold-bearing material from the various ore concentration methods is smelted with fluxes that capture and remove impurities in molten slag to produce a gold-rich mixture of metals. Borax is added to the molten mixture and chlorine gas bubbled through it to convert the base metals into their chlorides, which are collected and removed in the slag. The gold remaining is about 99.5% pure.

MANUFACTURING AND PROCESSING

It is mainly for electronics applications (such as for microcircuits and their connections) that gold is commercially supplied at extremely high purity (99.999%). For all other uses the pure metal received from the refinery is too soft, possessing unacceptable wear characteristics for brazes, decorative applications, electrical contacts, and jewelry. In order to provide the advantageous properties of gold to these applications, alloying with other metals is necessary.

GRADES, SPECIFICATIONS, AND QUALITY CONTROL

To guard against fradulent gold bars entering international trading markets, the London Bullion Market Association (LBMA) in concert with international bullion market associations worldwide has established requirements for "good delivery bars," which is defined by the LBMA as being the product of a refiner/assayer approved by the LBMA bearing the assayer's stamp (ingot mark) and its serial number, Good delivery bars are at least 995 fine (99.5% pure gold), between 350 and 430 troy ounces, with the precise weight and fineness stamped on each bar, and stamped with the ingot mark and bar number of an "approved refiner" which must have (1) been in the refining business for at least years, (2) an annual production of gold not less than 10 metric tons in the form of 400-oz gold bars, (3) an international reputation as a gold supplier producing at least 10 metric tons per year, and (4) have a net worth of at least 10,000,000 pounds Sterling.

Specifications for gold bullion, brazing alloys, electric contact alloys, and other parameters are published by American Society for Testing and Materials (ASTM), the American Welding Society, Japanese Industrial Standards, Society of Automotive Engineers (SAE) (Aerospace Materials Specifications), and U.S. Dept. Defense (DOD).

The most reliable quantitative method for determining the purity of gold is the fire assay. However, as this is an expensive and time-consuming technique, instrumental methods for routine quality control are often used. For these, the fire assay serves as the referee method.

The high-purity assay required for COMEX certification is usually done using emission spectroscopy.

Dentistry

Most casting alloys meet the composition and properties criteria of specification no. 5 of the American Dental Association which prescribes four types of alloy systems constituted of gold–silver–copper with addition of platinum, palladium, and zinc. Composition ranges are specified, as are mechanical properties and minimum fusion temperatures. Wrought alloys for plates also may include the same constituents. Similarly, specification no. 7 prescribes nickel and two types of alloys for dental wires with the same alloy constituents.

ANALYTICAL METHODS

A total quantitative determination of the gold values in a sample can only be made using the "fire assay". The fire assay has been the standard reference method to determine the proportion of gold in a sample since ancient Egyptian times. Other methods are capable of determining only surface components of a sample, such as surface composition, but not its totality.

When the gold sample contains other noble metals such as Pt or Pd, a reducing agent is used to remove the gold, then the Pt or Pd is dissolved for quantitative evaluation. Pt and Pd are checked by instrumentation to determine whether gold is being carried with them. If so another separation is made to determine how much gold is being carried through. Then the Pt or Pd is checked again to be sure that no gold is present.

Quality Control

For bullion and jewelry, the fire assay is sufficient for quality control of 999 fine. For 9999 or 99999 fine, requirements set by the electronics industry, tests for impurities are made by inductively coupled plasma. When impurities in the parts per billion range must be identified, for such applications as gold sputtering targets for electronics circuitry, glow discharge mass spectroscopy is used.

Mass spectroscopy identifies elements by their reaction to a magnetic field. When used in combination with emission spectroscopic techniques, a quantitative assay of trace elements can be made. If identification of trace elements in the ppb range is required, the glow discharge mass spectroscope is used.

Fingerprinting Gold

The Anglo American Research Laboratories, Ltd., Crown Mines, South Africa, was established in 1993 to develop a reliable means to determine the provenance of gold. The procedure has also found use to identify the geological source of ancient gold, the sources of other precious metals and of nonmetals, such as ivory, rhinoceros horn, and crude oil.

HEALTH AND SAFETY FACTORS

Therapeutic Gold Complexes

Of the remittive and antiinflammatory agents available for the treatment of arthritis, gold salts have provided a most effective means in the manangement of this severe progressive disease. Gold–sulfur complexes have also been found to be beneficial in the treatment of psoriatic arthritis, discoid lupus, pemphigus vulgaris, and bronchial asthma. Gold dicyanide, in combination with other drugs, retards HIV production *in vitro*. Metallic gold has for centuries been used for tooth restorations. Gold is also ideal for coating surgical implants where biological inactivity relative to body fluids is imperative.

The advent of nanogold technology (ca 1–3 millionths of a meter in diameter) has spawned new applications for gold. Nanosized particles of gold on which biologically active material are adsorbed are accelerated to supersonic speed by a helium gas jet in a hand held device. The gold and its adsorbed material is propelled through the human skin or plant walls directly into the target tissue layers. Gold has the required adsorptiveness to firmly hold small pieces of DNA, it is biologically inert, never associated with any toxicity so it is very safe, and it has sufficient density to carry DNA transdermally into the target cell without harming it. Once inside, the cell's natural functions work with the DNA and the gold is abandoned.

USES

The fact that gold possesses a unique combination of advantageous properties [resistance to corrosion (best of all metals), electrical conductivity (third highest of all metals), infrared reflectivity (best of all metals or materials), thermal conductivity (third best of all metals), plus high malleability and ease of plating by several means], explains its application despite its cost.

Jewelry

Its oxidation-free, highly reflective, metallic sunlike luster makes gold the most desirable metal for jewelry and long-lasting personal decoration.

Electronics

The use of gold coatings for corrosion-free surfaces for soldering, and static-free contacts and connections has been standard since the beginning of electronics. Now the use of submicron-sized particles of gold is being applied to the fabrication of subminiature electronic components which, despite their decreasing size, demand equally flawless transmission of digitized data.

Gold's superior electrical conductivity, lubricity, and freedom from oxidation are exploited for sliding contacts in the dual-spin satellites where one portion of the satellite must spin for stability and the other must direct its solar panels to the sun.

Brazing

Gold alloys provide brazing filler metals wetting a wide range of metals and ceramics at temperatures considerably below the melting point of the metals and thermal degradation of the ceramics being brazed. The joints formed are advantageously tough, shock-resistant, and ductile (no cracks to repair), with high electrical and heat conductivity.

Decoration

The use of gold leaf to protect artistic and architectural masterpieces has been practiced since ancient times. Properly applied gold leaf on domes, commercial buildings, and statues will last for over 25 years outdoors before touchup is required; interior decoration will last indefinitely.

Corrosion Control. Gold's inactivity has made it a prime candidate for linings of reactor vessels in which severely caustic chemical reactions take place.

Hydrogen Barrier

The inactivity of gold combined with its large atomic size blocks the flow of hydrogen atoms through it. This characteristic has found a use in space technology to keep the hydrogen from contacting the impeller alloys, which would result in embrittlement.

Solar Heat Control and Space Satellites

In the space vehicle gold is used (*1*) in the conductors in the computer microcircuits, (*2*) for connectors and contacts for circuit boards and components, (*3*) for thermal control to reflect away the heat from the sun that would otherwise overheat and degrade the function of a satellite's instruments and position control devices, (*4*) to conduct away (by electrically grounding) external radiowave or solar electrical bursts from interfering with the satellite's electronics, and (*5*) to provide sliding surfaces, because gold, which does not oxidize, will not cold-weld to oxidized metals.

Medical Uses

Gold has found wide use in dentistry for tooth restorations both as foil and powder.

In medical and biological research, gold particles have revolutionized their procedures. Particles with an average diameter of 1.6 μm will adsorb proteins or other organics allowing researchers to track exactly where drugs react with body tissues to determine how best to provide healthful service.

Infrared Reflective Surfaces

Gold is the most infrared (heat) reflective metal. Gold is widely applied in the space satellites to reject solar heat from components, electronics, and instruments. It is also used on commercial aircraft windshields and fighter plane canopies not only to protect the pilots from intense solar radiation, but to provide a conductive surface to heat the transparent surfaces and clear them of frost. It has been used for architectural glazing to reflect away the heat of the sun and reduce summer heat intake and winter heat loss.

Lubricative Surfaces

The gears in space satellites used to point experiments in specific directions and the slip rings of large control mode gyroscopes have surfaces plated with gold to act as lubricants as conventional lubricants evaporate away in the high vacuum conditions of outer space.

Coins and Medallions

Gold bullion coins minted by national governments have become a convenient, safe way for the public to invest in gold. Commemorative gold coins that celebrate events, such as the Olympics, are also used as investments.

Gold as Catalyst

Supported gold catalysts are very active for ambient temperature CO oxidation. Gold catalysts have been shown to be much more stable than mercuric chloride catalysts for the synthesis of vinyl chloride via ethylene hydrochlorination. Studies show gold nanoparticles can be effective catalysts for the oxidation of alcohols, including diols, and oxidation of glycerol to glucerate. The redox chemistry of gold as a catalyst is in its infancy.

DERIVATIVES

Gold Compounds

The chemistry of nonmetallic gold is predominantly that of Au(I) and Au(III) compounds and complexes. In the former, coordination number two and linear stereochemistry are most common. The majority of known Au(III) compounds are four coordinate and have square planar configurations. In both of these common oxidation states, gold preferably bonds to large polarizable ligands and, therefore, is termed a class b metal or soft acid. Considerable research into organogold chemistry has revealed organogold complexes and gold compounds in general exhibit unusual properties including closer than normal gold to gold distances shorter than the normal van der Waals distances.

Cluster Compounds

More recently, an increasing amount of interest has developed in gold-containing bimetallic cluster compounds which permit investigation at the molecular level of the metal—metal interactions thought to occur in bimetallic catalysts or alloys. Most often, these compounds are prepared from organophosphine stabilized Au(I) compounds such as halides or alkyls and generally contain one to three gold phosphine fragments bonded to one or more transition metal atoms, most often as carbonyl species.

Metals and Alloys in the Unified Numbering System, 6th ed., SAE/ASTM, Warrendale, Pa., 1993.

S. Patai and Z. Rappaport, *The Chemistry Derivatives of Gold and Silver*, Wiley, New York, 1999.

Gold Bulletin, a Quarterly Survey of Research on Gold and its Applications in Industry, World Gold Council, London. Website: www.gold.org.

Gold, Progress in Chemistry, Biochemistry and Technology, H. Schmidbaur, ed., John Wiley & Sons, Ltd., 1999.

The U.S. Gold Industry (1998), a biennial publication of the Nevada Bureau of Mines and Geology, ordered from The Gold Institute, Washington Dc, [Website: www.goldinstitute.org].

J. G. COHN
Engelhard Corporation

ERIC W. STERN
Engelhard Corporation

SAMUEL F. ETRIS
The Gold Institute

GRAPHITE, ARTIFICIAL

In its many varying manufactured forms, carbon and graphite can exhibit a wide range of electrical, thermal, and chemical properties that are controlled by the selection of raw materials and thermal processing during manufacture.

PHYSICAL PROPERTIES

The graphite crystal, the fundamental building block for manufactured graphite, is one of the most anisotropic bodies known. Properties of graphite crystals illustrating this anisotropy are given in Table 1. Anisotropy is the direct result of the layered structure with extremely strong carbon—carbon bonds in the basal plane and weak bonds between planes. The anisotropy of the single crystal is carried over in the properties of commercial graphite, though not nearly to the same degree. By the selection of raw materials and processing conditions, graphites can be manufactured with a very wide range of properties and degree of anisotropy. Manufactured graphite is a composite of coke aggregate (filler particles), binder carbon, and pores.

Electrical Properties

Manufactured graphite is semimetallic in character with the valence and conduction bands overlapping slightly.

Table 1. Properties of Graphite Crystals at Room Temperature

Property	Value in basal plane	Value across basal plane
resistivity, $\mu\Omega \cdot m$	0.40	ca 60
elastic modulus, GPa[b]	1020	36.5
tensile strength (est), GPa[b]	96	34
thermal conductivity, W/(m·K)	ca 2000	10
thermal expansion, $°C^{-1}$	-0.5×10^{-6}	27×10^{-6}

[a]To convert GPa to psi, multiply by 145,000.

Conduction is by means of an approximately equal number of electrons and holes that move along the basal planes. The resistivity of single crystals as measured in the basal plane is approximately 0.40 $\mu\Omega \cdot m$; this is several orders of magnitude lower than the resistivity across the layer planes. Thus the electrical conductivity of formed graphite is dominated by the conductivity in the basal plane of the crystallites and is dependent on size, degree of perfection, orientation of crystallites, and on the effective carbon–carbon linkages between crystallites. Manufactured graphite is strongly diamagnetic and exhibits a Hall effect, a Seebeck coefficient, and magnetoresistance. Graphites made from petroleum coke usually have a room temperature resistivity range of 5 – 15 $\mu\Omega \cdot m$ and a negative temperature coefficient of resistance to about 500°C, above which it is positive. Graphites made from a carbon black base have a resistivity several times higher than those made from petroleum coke, and the temperature coefficient of resistance for the former remains negative to at least 1600°C.

Thermal Conductivity

Compared with other refractories, graphite has an unusually high thermal conductivity near room temperature above room temperature, the conductivity decreases exponentially to approximately 1500°C and more slowly to 3000°C. With the grain, the thermal conductivity of manufactured graphite is comparable with that of aluminum; against the grain, it is comparable to that of brass. However, graphite is similar to a dielectric solid in that the principal mechanism for heat transfer is lattice vibrations. The electronic component of thermal conductivity is less than 1%.

Mechanical Properties

The hexagonal symmetry of a graphite crystal causes the elastic properties to be transversely isotropic in the layer plane; only five independent constants are necessary to define the complete set. Low values of shear and cleavage strengths between the layer planes compared with very high C–C bond strength in the layer planes suggest that graphite always fails through a shear or cleavage mechanism. However, the strength of manufactured graphite depends on the effective network of C–C bonds across any stressed plane in the graphite body. Until these very strong bonds are broken, failure by shear or cleavage cannot take place. Porosity affects the strength of gra-

phite by reducing the internal area over which stress is distributed and by creating local regions of high stress.

Chemical Properties

The impurity (ash) content of all manufactured graphite is low, since most of the impurities originally present in raw materials are volatilized and diffuse from the graphite during graphitization. Iron, vanadium, calcium, silicon, and sulfur are principal impurities in graphite; traces of other elements are also present. Through selection of raw materials and processing conditions, the producer can control the impurity content of graphites to be used in critical applications. Because of its porosity and relatively large internal surface area, graphite contains chemically and physically adsorbed gases. Desorption takes place over a wide temperature range, but most of the gas can be removed by heating in a vacuum at approximately 2000°C.

Graphite reacts with oxygen to form CO_2 and CO, with metals to form carbides, with oxides to form metals and CO, and with many substances to form laminar compounds. Of these reactions, oxidation is the most important to the general use of graphite at high temperatures. Oxidation of graphite depends on the nature of the carbon, the degree of graphitization, particle size, porosity, and impurities present. These conditions may vary widely among graphite grades. Graphite is less reactive at low temperatures than many metals; however, since the oxide is volatile, no protective oxide film is formed. The rate of oxidation is low enough to permit the effective use of graphite in oxidizing atmospheres at very high temperatures when a modest consumption can be tolerated. A formed graphite body alone will not support combustion. The differences in oxidation behavior of various types of graphite are greatest at the lowest temperatures, tending to disappear as the temperature increases.

RAW MATERIALS

The raw materials used in the production of manufactured carbon and graphite largely control the ultimate properties and practical applications of the final product. This dependence is related to the chemical and physical nature of the carbonization and graphitization processes.

Essentially any organic material can be thermally transformed to carbon. In most carbon and graphite processes, the initial polymerization reactions occur in the liquid state. The subsequent stages of crystal growth, heteroatom elimination, and molecular ordering occur in the solid phase. The result is the development of a three-dimensional graphite structure.

Most of the raw materials in the production of bulk carbon and graphite products are derived from petroleum and coal. These precursors are generally residual by-products and exhibit a highly aromatic composition. The two main raw materials for carbon and graphite are pitches and cokes. Pitches are derived from distillation and thermal heat treatment of tars or oils. Pitches are solid at room temperature but soften to a liquid at elevated temperatures.

Cokes are infusible solids with average molecular weights estimated to be on the order of several thousand. Coke can be described as a thermoset aromatic hydrocarbon polymer. The molecular order and structure of coke is determined by the chemical and physical processes that occur in the liquid mesophase state. Subsequent high temperature heat treatment of coke, as in calcination, induces solid-state polymerization and ordering while removing substituent atoms in the form of gaseous by-products.

Filler Materials

Petroleum Coke. Petroleum coke is the largest single precursor material in terms of quantity for manufactured carbon and graphite products. Commercial coke is produced by the delayed coking of heavy petroleum by-products. Delayed coking technology is currently practiced throughout the world. Delayed coke can fall into several categories: fuel-grade, aluminum-grade, and coke for carbon and graphite. The coke employed for carbon and graphite includes regular coke and needle coke. Needle or premium coke is used in the production of graphite electrodes for electric arc furnaces, whereas regular coke is used for other carbon and graphite products.

The key difference in the production of regular and needle petroleum coke is the feedstock precursor. Needle coke is generally produced from highly aromatic starting materials such as decant oil derived from catalytic cracking of petroleum distillate. Regular and aluminum-grade cokes are generally prepared from the residues (resids) of crude oil distillation. Pyrolysis tars from naphtha or gas-oil cracking can also be employed to produce intermediate-grade cokes, which exhibit more or less needlelike character depending on the tar composition and coke preparation parameters.

Although the feedstock exerts the greatest influence, coke properties are also determined by process variables including coking time and temperature, pressure, and recycle ratio. The most important properties in delayed coke are sulfur level, volatile matter content, ash, coefficient of thermal expansion (CTE), and hardness. Premium cokes for graphite electrodes are anisotropic and have very low CTEs in the extrusion direction. Regular cokes are isotropic and give graphite artifacts with high CTE values.

Coal-Tar Pitch Coke. The key to producing needle coke from coal tar or coal-tar pitch is the removal of the high concentrations of infusible solids, or material insoluble in quinoline (QI), which are present in the original tar. The QI inhibits the growth of mesophase and results in an isotropic, high CTE coke from coal-tar pitch. After removal of the QI, very anisotropic and low CTE cokes are obtained from coal-tar-based materials.

Because of very high aromaticity, high coking value, and excellent coking characteristics, coal tar and coal-tar pitches are very attractive premium coke precursors.

Natural Graphite. Natural graphite is a crystalline mineral form of graphite occurring in many parts of the world (see GRAPHITE, NATURAL). It is occasionally used as a component in carbon and graphite production. Interca-lated natural graphite is used to form a flexible graphite product.

Carbon Blacks. Carbon blacks are occasionally used as components in mixes to make various types of carbon products. Carbon blacks are generally prepared by deposition from the vapor phase using petroleum distillate or gaseous hydrocarbon feedstocks.

Anthracite. Anthracite is preferred to other forms of coal in the manufacture of carbon products because of its high carbon-to-hydrogen ratio, its low volatile content, and its more ordered structure. It is commonly added to carbon mixes used for fabricating metallurgical carbon products to improve specific properties and reduce cost.

Synthetic Resins. Various polymers and resins are utilized to produce some specialty carbon products such as glassy carbon or carbon foam and as treatments for carbon products. Typical resins include phenolics, furan-based polymers, and polyurethanes. Because they form little or no mesophase, the ultimate carbon end product is nongraphitizing.

Binders

Pitches. Carbon articles are made by mixing a controlled size distribution of coke filler particles with a binder such as coal-tar or petroleum pitch. The mix is then formed by molding or extruding and is heated in a packed container to control the shape and set the binder. Thus the second most important raw material for making a carbon article is the pitch binder.

A binder used in the manufacture of electrodes and other carbon and graphite products must: (1) have high carbon yield, usually 40–60 wt % of the pitch; (2) show good wetting and adhesion properties to bind the coke filler together; (3) exhibit acceptable softening behavior at forming and mixing temperature, usually in the range of 90–180°C; (4) be low in cost and widely available; (5) contain only a minor amount of ash and extraneous matter that could reduce strength and other important physical properties; and (6) produce binder coke that can be graphitized to improve the electrical and thermal properties.

The principal binder material, coal-tar pitch, is produced by the distillation of coal tar. Petroleum pitch is used to a much lesser extent as a binder in carbon and graphite manufacture.

Additives

In addition to the primary ingredients, the fillers and binders, minor amounts of other materials are added at various steps in the carbon and graphite manufacturing process. Although the amounts of these additives are usually small, they can play an important role in determining the quality of the final product.

Calcining

Nearly all raw coke utilized in carbon manufacture is calcined. Calcination consists of heating raw coke to remove volatiles and to shrink the coke to produce a strong, dense particle. Raw petroleum coke, eg, has 5–15% volatile matter. When the coke is calcined to 1400°C, it shrinks approximately 10–14%. Less than 0.5% of volatile matter in the form of hydrocarbons remains in raw coke after it is calcined to 1200–1400°C. During calcination, the evolving volatiles are primarily methane and hydrogen, which burn during the calcining process to provide much of the heat required. The calcining step is particularly important for those materials used in the manufacture of graphite products, such as electrodes, since the high shrinkages occurring in raw coke during the baking cycle of large electrodes would cause the electrode to crack.

PROCESSING

Crushing and Sizing

Calcined petroleum coke arrives at the graphite manufacturer's plant in particle sizes ranging typically from dust to 50–80 mm diameter. In the first step of artificial graphite production, the run-of-kiln coke is crushed, sized, and milled to prepare it for the subsequent processing steps. The degree to which the coke is broken down depends on the grade of graphite to be made. If the product is to be a fine-grained variety for use in aerospace, metallurgical, or nuclear applications, the milling and pulverizing operations are used to produce sizes as small as a few micrometers in diameter. If, on the other hand, the product is to be coarse in character for products like graphite electrodes used in the manufacture of steel, a high yield of particles up to 25 mm diameter is necessary.

Proportioning

The size of the largest particle is generally set by application requirements. For example, if a smoothly machined surface with a minimum of pits is required, as in the case of graphites used in molds, a fine-grained mix containing particles no larger than 1.6 mm with a high flour content is ordinarily used. If high resistance to thermal shock is necessary, eg, in graphite electrodes used in melting and reducing operations in steel plants, particles up to 25 mm are used to act as stress absorbers in preventing catastrophic failures in the electrode.

Generally, the guiding principle in designing carbon mixes is the selection of the particle sizes, the flour content, and their relative proportions in such a way that the intergranular void space is minimized. If this condition is met, the volume remaining for binder pitch and the volatile matter generated in baking are also minimized. Volatile evolution is often responsible for structural and property deterioration in the graphite product. In practice, most carbon mixes are developed empirically with the aim of minimizing binder demand and making use of all the coke passed through the first step of the system. Typically, a coarse-grained mix may contain a large particle, eg, 13 m diameter, two to three intermediate particle sizes, and flour. In this formulation approximately 25 kg of binder pitch would be used for each 100 kg of coke. Although binder levels increase as particle size is reduced, and they are greatest in all-flour mixes

where surface area is very high, the principle of minimum binder level still applies.

Mixing

Once the raw materials have been crushed, sized, and stored in charging bins and the desired proportions established, the manufacturing process begins with the mixing operation. The purpose of mixing is to blend the coke filler materials and distribute the pitch binder over the surfaces of the filler grains as it melts or is added as a liquid. The intergranular bond ultimately determines the properties and structural integrity of the graphite. Thus the more uniform the binder distribution is throughout the filler components, the greater the likelihood for a structurally sound product.

The degree to which mixing uniformity is accomplished depends on factors such as time, temperature, and batch size. However, a primary consideration in achieving mix uniformity is mix design.

Forming

One purpose of the forming operation is to compress the mix into a dense mass so that pitch-coated filler particles and flour are in intimate contact. For most applications, a primary goal in the production of graphite is to maximize density; this goal begins by minimizing void volume in the formed, green, product. Another purpose of the forming step is to produce a shape and size as near that of the finished product as possible. This reduces raw material usage and cost of processing graphite that cannot be sold to the customer and must be removed by machining prior to shipment. Two important methods of forming are extrusion and molding.

Extrusion. This process is used to form most carbon and graphite products. In essence, extrusion presses comprise a removable die attached by means of an adapter to a hollow cylinder called a mud chamber. The cylinder is charged with mix that is extruded in a number of ways depending on the press design. For one type of press, the cooled mix is introduced into the mud chamber in the form of plugs that are molded in a separate operation. A second type of extruder called a tilting press makes use of a moveable mud chamber-die assembly to eliminate the need for precompacting the cooled mix. A third type of extrusion press makes use of an auger to force mix through the die. This press is used principally with fine-grained mixes because of its tendency to break down large particles.

Molding. Molding is the older of the two forming methods and is used to form products ranging in size from brushes for motors and generators to billets as large as 1.75 m diameter by 1.9 m in length for use in specialty applications.

Several press types are used in molding carbon products. The presses may be single-acting or double-acting, depending on whether one or both platens move to apply pressure to the mix through punched holes in either end of the mold. The use of single-acting presses is reserved for products whose thicknesses are small compared with their cross-sections. As thickness increases, the acting pressure on the mix diminishes with distance from the punch because of frictional losses along the mold wall. Acceptable thicknesses of molded products can be increased by using double-acting presses that apply pressures equally at the top and bottom of the product. Jar molding is another method used to increase the length of the molded piece and keep nonuniformity within acceptable limits.

As with extruded products, molded pieces have a preferred grain orientation. The coke particles are aligned with their long dimensions normal to the molding direction. Thus the molded product has two with-grain directions and one cross-grain direction which coincides with the molding direction. Strength, modulus, and conductivity of molded graphites are higher in both with-grain directions and expansion coefficient is higher in the cross-grain direction. Isostatic molding is a forming technique used to orient the coke filler particles randomly, thereby imparting isotropic properties to the finished graphite. Cold isostatic molding is also used.

Baking

In the next stage, the baking operation, the product is fired to 800–1000°C. One function of this step is to convert the thermoplastic pitch binder to solid coke. Another function of baking is to remove most of the shrinkage in the product associated with pyrolysis of the pitch binder at a slow heating rate. This procedure avoids cracking during subsequent graphitization where very fast firing rates are used. The conversion of pitch to coke is accompanied by marked physical and chemical changes in the binder phase, which if conducted too rapidly, can lead to serious quality deficiencies in the finished product. For this reason, baking is generally regarded as the most critical operation in the production of carbon and graphite.

Impregnation

In some applications, the baked product is taken directly to the graphitizing facility for heat treatment to 3000°C. However, for many high performance applications of graphite, the properties of stock processed in this way are inadequate. The method used to improve those properties is impregnation with coal-tar or petroleum pitches. The function of the impregnation step is to deposit additional pitch coke in the open pores of the baked stock, thereby improving properties of the graphite product.

Further property improvements result from additional impregnation steps separated by rebaking operations. However, the gains realized diminish quickly, for the quantity of pitch picked up in each succeeding impregnation is approximately half of that in the preceding treatment. Many nuclear and aerospace graphites are multiple pitch-treated to achieve the greatest possible assurance of high performance.

Graphitization

Graphitization is an electrical heat treatment of the product to ca 3000°C. The purpose of this step is to cause the carbon atoms in the petroleum coke filler and pitch coke binder to orient into the graphite lattice configuration. This ordering process produces graphite with intermetallic properties that make it useful in many applications.

Puffing

In the temperature range of 1500–2000°C, most petroleum cokes and coal-tar pitch cokes undergo an irreversible volume increase known as puffing. This effect in petroleum cokes has been associated with thermal removal of sulfur and increases with increasing sulfur content. Some mechanisms other than sulfur removal may be more dominant in coal-tar pitch cokes. Because of the recent emphasis on the use of low sulfur fuels, many of the sweet crudes that had been used to produce coker feedstocks are now being processed as fuels. Desulfurization of sour crudes or coker feeds is possible but expensive. The result is an upward trend in the sulfur content of many petroleum cokes, leading to greater criticality in heating rate in the puffing temperature range during graphitization.

USES

Aerospace and Nuclear Reactor Applications

Graphite with its exceptional strength and thermal stability at high temperatures is a prime candidate material for many aerospace and nuclear applications. Its properties, through process modifications, are tailorable to meet an array of design criteria for survival under extremely harsh environmental operations.

Aerospace and nuclear reactor applications of graphite demand high reliability and reproducibility of properties, physical integrity of product, and product uniformity. The manufacturing processes require significant additional quality assurance steps that result in high cost.

Aerospace. Graphite applications in the aerospace industry include rocket nozzle components, nose cones, motor cases, leading edges, control vanes, blast tubes, exit cones, thermal insulation, and any other applications where a rapid temperature rise and unusually high operating temperatures are encountered. Graphite is one of the few materials that can reasonably meet the demands encountered under these conditions. Of particular importance in this type of application are the excellent thermal properties of graphite, eg, high thermal shock resistance, high thermal stress resistance, and a strength increase with temperature increase. In addition, its excellent machinability makes it possible to maintain the required close tolerances for the machining of precision components for aerospace vehicles.

The erosion of graphite in nozzle applications is a result of both chemical and mechanical factors. Changes in temperature, pressure, or fuel-oxidizing ratio markedly affect erosion rates. Graphite properties affecting its resistance to erosion include density, porosity, and pore size distribution.

Nose cones and leading edge components fabricated of graphite are used on both ballistic and glider-type reentry vehicles. Design technology and the ability to control properties of manufactured graphite favorably has increased its use in aerospace applications.

Nuclear Applications. The strength of graphite at high temperatures and its behavior with respect to products of nuclear fission/fusion make it a suitable material for nuclear moderators and reflectors, materials of construction, and thermal columns in various reactors. Since its use in the first nuclear reactor, CP-1, constructed in 1942 at Stagg Field, University of Chicago, many thousands of metric tons of graphite have been used for this purpose. The advanced gas-cooled reactors (AGR), the high temperature gas-cooled reactors (HTGR), the molten salt breeder reactors (MSBR), and liquid metal fueled reactors (LMFR) all use graphite moderators. The thermal stability, resistance to corrosion, and high thermal conductivity of graphite make it a most suitable moderator material for consideration in advanced design, high temperature, atomic energy efficient nuclear reactors.

Graphite is chosen for use in nuclear reactors because it is the most readily available material with good moderating properties and a low neutron capture cross section. Other features that make its use widespread are its low cost, stability at elevated temperatures in atmospheres free of oxygen and water vapor, good heat transfer characteristics, good mechanical and structural properties, and excellent machinability.

Chemical Applications. Carbon and graphite exhibit excellent resistance to the corrosive actions of acids, alkalies, and organic and inorganic compounds, an attribute that has fostered the use of graphite in process equipment where corrosion is a problem. Other than in the chemical process industries, graphite is used extensively in the steel, food, petroleum, pharma-ceutical, and metal finishing industries. The high thermal conductivity and thermal stability of graphite have made it a useful material in heat exchangers and high temperature gas-spray coolers.

Manufactured carbon and graphite exhibit varying degrees of porosity depending on its method of preparation. Equipment fabricated from these materials must be operated essentially at atmospheric pressure; otherwise, some degree of leakage must be tolerated. Carbon used as a liner for tanks and vessels for the handling of highly corrosive inorganic acids such as hydrofluoric, nitric, phosphoric, sulfuric, and hydrochloric is backed up by an impervious membrane of lead or plastic to prevent seepage through the lining. The carbon lining protects the impermeable membrane material from adverse temperature and abrasion effects. Carbon linings have provided indefinite life with a minimum of maintenance.

Self-Supporting Structures. Self-supporting structures of carbon and graphite are used in a variety of ways. Water-cooled graphite towers serve as chambers for the burning of phosphorus in air. The high thermal conductivity of graphite allows rapid heat transfer to a water film on the outside of the tower, thereby maintaining inside wall temperature below 500°C, the oxidation temperature threshold of graphite. Phosphorus combustion chambers six meters in diameter by eleven meters in height have been built using cemented graphite block construction.

Impervious Graphite. For applications where fluids under pressure must be retained, impregnated materials are available. Imperviousness is attained by blocking the pores of the graphite or carbon material with thermosetting resins such as phenolics, furans, and epoxies. Because the resin pickup is relatively small (usually 12–15 wt%), the physical properties exhibited by the original graphite or carbon material are retained. However, the flexural and compressive strengths are usually doubled. Graphite is also made impervious in a vacuum impregnation process.

Low Permeability Graphite. Low permeability graphite materials have been developed by graphite manufacturers in response to the need by the chemical industry to control a equipment corrosion problem in high-temperature processes (370°C and above).

Porous Graphite. Several grades of low density, porous carbon and graphite are commercially available. A controlled combination of high permeability and porosity characterizes these materials. Porous carbon and graphite are used in filtration of hydrogen fluoride streams, caustic solutions, and molten sodium cyanide; in diffusion of chlorine into molten aluminum to produce aluminum chloride; and in aeration of waste sulfite liquors from pulp and paper manufacture and sewage streams.

Mechanical Applications

Carbon–graphite possesses lubricity, strength, dimensional stability, thermal stability, and ease of machining, a combination of properties that has led to its use in a wide variety of mechanical applications for supporting rotating or sliding loads in contact. Its principal applications are in bearings, seals, and vanes, which are in sliding contact with a partner material. Mechanical applications of carbon–graphite include face, ring, and circumferential seals for gases and fluids both corrosive and noncorrosive; carbon cages for roller and ball bearings, carbon sleeve bearings and bushings, carbon thrust bearings or washers, and combination sleeve/thrust bearings; packing rings for steam and water valve shafts and packing rings for compressor tail rods; and nonlubricated compressor parts such as piston rings, wear rings, segments, scuffer shoes, shaft tail-rod packing rings, pistons, and piston skirts. Miscellaneous applications include flat-plate slider parts for supporting machinery and facilitating sliding movement under load; and rotor vanes and metering device parts.

Carbon–graphite materials do not gall or weld even when rubbed under excessive load and speed. Early carbon materials contained metal fillers to provide strength and high thermal conductivity, but these desirable properties can now be obtained in true carbon–graphite materials that completely eliminate the galling tendency and other disadvantages of metals.

Electrode Applications

With the exception of carbon use in the manufacture of aluminum, the largest use of carbon and graphite is as electrodes in electric-arc furnaces. In general, the use of graphite electrodes is restricted to open-arc furnaces of the type used in steel production; whereas, carbon electrodes are employed in submerged-arc furnaces used in phosphorus, ferroalloy, and calcium carbide production.

Metallurgical Applications

Because of their unique combination of physical and chemical properties, manufactured carbons and graphites are widely used in several forms in high temperature processing of metals, ceramics, glass, and fused quartz. A variety of commercial grades is available with properties tailored to best meet the needs of particular applications. Industrial carbons and graphites are available in a broad range of shapes and sizes.

Structural Graphite Shapes. In many metallurgical and other high temperature applications, manufactured graphite is used because it neither melts nor fuses to many common metals or ceramics, exhibits increasing strength with temperature, has high thermal shock resistance, is nonwarping, has low expansion, and possesses high thermal conductivity. However, because of its tendency to oxidize at temperatures above 750 K, prolonged exposure at higher temperatures frequently necessitates use of a nonoxidizing atmosphere. In addition, prolonged contact both with liquid steel and with liquid metals that rapidly form carbides should be avoided.

Electric Heating Elements. Machined graphite shapes are widely used as susceptors and resistor elements to produce temperatures up to 3300 K in applications utilizing nonoxidizing atmospheres. The advantages of graphite in this type of application include its very low vapor pressure (lower than molybdenum), high black body emissivity, high thermal shock resistance, and increasing strength at elevated temperatures with no increase in brittleness. Graphites covering a broad range of electrical resistivity are available and can be easily machined into complex shapes at lower cost than refractory metal elements. Flexible graphite cloth is also used widely as a heating element since its low thermal mass permits rapid heating and cooling cycles.

Carbon and Graphite Powder and Particles. Manufactured graphite powders and particles are used extensively in metallurgical, chemical, and electrochemical applications where the uniformity of physical and chemical characteristics, high purity, and rapid solubility in certain molten metals are important factors. The many grades of carbon and graphite powders and particles are classified on the basis of fineness and purity.

Refractory Applications

Various forms of carbon, semigraphite, and graphite materials have found wide application in the metals industry, particularly in connection with the production of iron, aluminum, and ferroalloys.

Carbon as a Blast Furnace Refractory. The first commercial use of carbon as a refractory for a blast furnace lining took place in France in 1872, followed in 1892 by a carbon block hearth in a blast furnace of the Maryland Steel Co. at Sparrows Point, Maryland. Although initially used only for the hearth bottom of blast furnaces, carbon, semigraphite, semigraphitized carbon, and graphite refractories have been successfully applied to hearth walls, tuyere zones, boshes, and even the lower to midstack of modern, intensely cooled, high performance blast furnaces around the world. Additionally, carbon has also been used extensively for iron trough, iron runner, and slag runner safety linings, especially when external cooling is employed to extend the life of the ceramic working linings.

Carbonaceous and graphitic materials possess important characteristics that make them ideal blast furnace refractories: (1) they do not soften or lose strength at high operating temperatures of approximately 1150–1200°C; (2) they resist attack by molten slag and iron; (3) their relatively high thermal conductivity, when combined with adequate cooling and proper design concepts, promotes the formation of solidified coatings of slag and iron on their hot face. These coatings prevent erosion from the molten materials and process gases, promoting long life; (4) they possess excellent resistance to thermal shock, preventing spalling and cracking which interrupts heat transfer to the cooling system and exposes more refractory surface area to chemical attack; (5) a positive, low coefficient of thermal expansion provides dimensional stability and tightening of joints in the multipiece linings. However, because of their relatively low threshold temperature for oxidation from steam, carbon dioxide, or air, care must be taken to limit their exposure to these elements and maintain proper cooling at all times, to minimize damage from these temperature-dependent reactions.

The prime requirement of any carbonaceous material used in the blast furnace hearth wall or bottom is to contain liquid iron and slag safely within the crucible, throughout extended periods of continuous operation, often up to 15 years.

This requirement is most readily achieved if the lining design concepts employed and the carbonaceous or graphitic materials utilized with these concepts, combine to provide a refractory mass free from cracking caused by mechanical and thermal stress. Additionally, the refractory materials must exhibit thermal conductivities that are high enough to permit the formation of solidifying layers of iron and slag on their hot faces and permeabilities that are low enough to prevent the impregnation of the refractories by alkalies and other process contaminants. It is also helpful if the refractory materials themselves are resistant to attack from alkalies by virtue of the inclusion of various additives during their manufacture. Proper cooling of the materials also contributes to their longevity.

Refractories for Cupolas. Carbon brick and block are used to line the cupola well or crucible. Their resistance to molten iron and both acid and basic slags provides not only insurance against breakouts but also operational flexibility to produce different iron grades without the necessity of changing refractories. Carbon is also widely used for the tap hole blocks, breast blocks, slagging troughs, and dams.

Refractories for Electric Reduction Furnaces. Carbon hearth linings are used in submerged-arc, electric-reduction furnaces producing phosphorus, calcium carbide, all grades of ferrosilicon, high carbon ferrochromium, ferrovanadium, and ferromolybdenum. Carbon is also used in the production of beryllium oxide and beryllium copper where temperatures up to 2273 K are required.

Refractories in the Aluminum Industry. Carbon materials are used in the Hall-Heroult primary aluminum cell as anodes, cathodes, and sidewalls because of the need to withstand the corrosive action of the molten fluorides used in the process. Aluminum smelters generally have an on-site carbon plant for anode production. Anode technology is focused on raw materials (petroleum coke and coal-tar pitch), processing techniques, and rodding practices.

Prebaked cathode blocks used today are electrically calcined anthracite coal, semigraphite, semigraphitized, or graphite composition. Desired cathode operating characteristics include resistance to sodium attack, high operating strength, low porosity, high thermal shock resistance, and low electrical resistance.

A. J. Dzermejko, *Blast Furnace Hearth Design Theory, Materials and Practice*, paper presented at the meeting of the Association of Iron and Steel Engineers, Toronto, Ontario, Canada, 1990.

D. J. Page, *Industrial Graphite Engineering Handbook*, Union Carbide Corp., Carbon Products Division, New York, 1991.

G. A. Saunders, in L. C. F. Blackman, ed., *Modern Aspects of Graphite Technology*, Academic Press, New York, 1970.

T. R. Hupp
I. C. Lewis
J. M. Criscione
R. L. Reddy
C. F. Fulgenzi
D. J. Page
F. F. Fisher
A. J. Dzermejko
J. B. Hedge
UCAR Carbon Company, Inc.

GRAPHITE, NATURAL

Natural graphite, one of three forms of crystalline carbon—the other two being diamond and fullerenes—occurs worldwide. Graphitization of naturally occurring organic or inorganic carbon may take place at temperatures as low as 300–500°C or as high as 800–1,200°C, such as when an igneous intrusion contacts a carbonaceous body. The word graphite is derived from the Greek word "graphein", to write.

The three principal forms of natural graphite—crystalline flake, lump, and amorphous—are distinguished by physical characteristics that are the result of major differences in geologic origin and occurrence. Crystalline flake graphite consists of flat, plate-like particles with angular, rounded, or irregular edges, depending on the abrasion it has undergone, and occurs in such metamorphic rocks as schist, marble, and gneiss. Lump graphite occurs in veins and is believed to be hydrothermal in origin. It is typically massive, ranging in particle size from extremely fine to coarse. Amorphous graphite is formed by metamorphism of coal. Its low degree of crystallinity and very fine particle size make it appear amorphous.

CRYSTALLOGRAPHIC PROPERTIES

Parallel layers of condensed planar C_6 rings constitute the crystal structure of graphite. Each carbon atom joins to three neighboring carbon atoms at 120° angles in the plane of the layer. Weak van der Waals forces pin the carbons in adjoining layers, thus accounting in part for the marked anisotropic properties of the graphite crystal. Figure 1 shows a Laue X-ray diffraction pattern of a single natural graphite crystal.

In the completely graphitized crystallite, the planar C_6 layers stack in ordered parallel spacing 0.33538 nm apart at room temperature. The hexagonal form of graphite contains the most common stacking order: ABABAB. A small percentage of graphites exhibit ABCABCABC stacking order, resulting in the rhombohedral form. Extensive grinding promotes the rhombohedral structure, probably through pressure. Heating >2000°C transforms the rhombohedral structure to hexagonal, suggesting that the latter is more stable. Impact from explosion can convert rhombohedral graphite to cubic-structured carbon, ie, diamond.

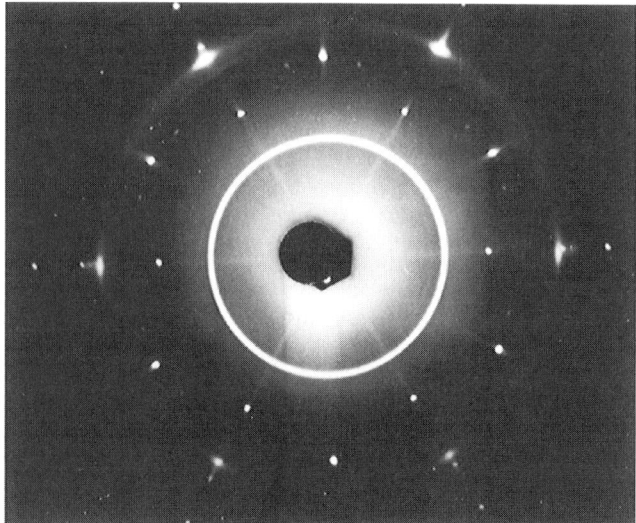

Figure 1. Laue X-ray differaction pattern of a single natural graphite crystal.

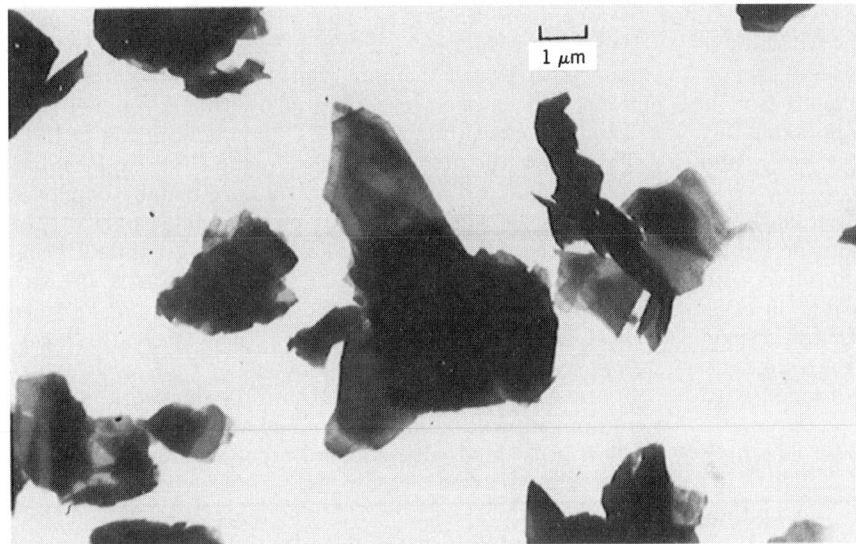

Figure 2. Electron micrograph of Sri Lanka (Ceylon) graphite.

Grinding graphite to particle sizes smaller than ~0.1 µm reduces the crystallite size to <20 nm, at which size two-dimensional 2D ordering replaces the three-dimensional (3D) ordering of graphite. The weakened pinning forces permit the planar layers to move further apart and assume progressively random, though parallel, positions with respect to each other. This turbostratic structure is the characteristic structure of the so-called amorphous carbon that is found in chars.

The physical properties of finely ground but highly ordered natural graphites, such as 5-µm Sri Lanka (Ceylon) graphite, differ from those of such turbostratic carbons as chars, carbon blacks, or carbons formed from heavily ground graphites. Figure 2 is an electron micrograph of 5-µm Sri Lanka graphite; the straight edges and angles of the particles contrast sharply to the rounded shapes of carbon black particles.

PHYSICAL PROPERTIES

Solid articles made of natural graphite always require a binder; ie., they are almost always composites. The binder, the processing, and the type of graphite used, combined with graphite's anisotropy, influence the properties of the composites. Table 1 lists some of the physical properties of natural graphite.

Graphite's strength increases with temperature. Relief of internal stresses up to ~2500°C accounts for this unusual property. Plastic deformation occurs >2500°C.

CHEMICAL PROPERTIES

Graphite burns slowly in air hotter than 450°C, the rate of combustion increasing with temperature. The particle size and morphology govern the ignition temperature. Flake graphites generally resist oxidation better than granular graphites.

Above 800°C, graphite reacts with water vapor, carbon monoxide, and carbon dioxide. Chlorine has a negligible effect on graphite, and nitrogen, none. Many metals and

metal oxides react with graphite to form carbides >1500°C. These reactions occur with the carbon atom and destroy the graphitic structure. A series of compounds in which the graphite structure is retained, known as graphite compounds, consist of two general kinds—crystal and covalent compounds.

Graphite can be regenerated from the crystal compounds because the graphitic structure has not been too greatly altered. The dark crystal compounds are called intercalation compounds, interstitial compounds, or lamellar compounds because they are formed by cations that fit in between the planar carbon networks.

Metal and ammonia compounds of graphite are electron donors; the electrical resistance decreases from that of the original graphite, and the Hall coefficient remains negative. Graphite compounds with the halogens (except fluorine), metal halides, and sulfuric acid (graphite sulfate) are electron acceptors. The electrical resistance decreases, but the Hall coefficient changes from negative to positive.

Table 1. Physical Properties of Natural Graphite

density[a] g/mL	
calculated	2.265
experimental, pure Sri Lanka	~2.25
compressibility, N/m^2 [b], Sri Lanka	
at low pressures	4.5×10^{-11}
at high pressures	$<2 \times 10^{-11}$
average	3.1×10^{-11}
shear modulus, N/m^2 [b]	2.3×10^9
Young's modulus, N/m^2 [b]	1.13×10^{14}
heat of vaporization[d], kJ/mol[c]	711
sublimation point, K	4000–4015
triple point, K	
graphite-liquid-gas, 101.3 kPa[d]	3900 ± 50
graphite–diamond–liquid, 12–13 GPa[d]	4100–4200
surface energy, J/cm^2 [c]	$\sim 1.2 \times 10^{-5}$

[a] The difference between the calculated and experimental values of density is caused by dislocations and imperfections.
[b] To convert N/M^2 to dyn/cm^2, multiply by 10.
[c] To convert J to cal. divide by 4.184.
[d] To convert kPa to atm, divide by 101.3.

Graphite sulfate, long known and early investigated, forms when graphite is warmed in concentrated sulfuric acid containing a small quantity of an oxidizing agent, such as concentrated nitric acid. The graphite swells and becomes blue. The compound, approximately $C_{24}^+(HSO_4)^- \cdot 2\,H_2SO_4$, hydrolyzes at once in water, and the graphite is recovered.

The covalent compounds of graphite differ markedly from the crystal compounds. They are white or lightly colored electrical insulators, have variable formulas, and occur in but one form, unlike the series typical of the crystal compounds. In the covalent compounds, the carbon network is deformed, and the carbon atoms rearrange tetrahedrally as in diamond. Often they are formed with explosive reactions.

Graphite fluoride continues to be of interest as a high temperature lubricant.

The fullerenes are related to graphite even though they are not graphite compounds, but rather are a separate class of carbonaceous materials. Fullerenes have complex configurations of ~60 or more carbon atoms, the best known of which are shaped like a soccerball (C_{60}). A preparation method for fullerenes involving heating graphite rods in a vacuum chamber and collecting the fullerene-rich soot deposited on the chamber surfaces was a significant breakthrough because of the large yields of the materials.

MANUFACTURE AND PROCESSING

Crystalline Flake Graphite

Crystalline flake is usually mined by excavation and open pit methods. Since it is at or near the surface, it is usually highly weathered (particularly in the tropics, where it is lateritized), while underground crystalline flake deposits usually have unweathered ore. After mining, crystalline flake is processed by a number of methods, which can be broadly classified under the flotation and the chemical leaching methods, to remove silicate or carbonate gangues.

Flotation methods yield rather impure graphite concentrate, unless multiple flotation cycles are used. The soft graphite "marking" the gangue particles with a thin layer of graphite, which also floats, causes this lack of a clean separation. The concentrate can further be enriched by chemical leaching or by repeated flotation cycles.

The process itself usually begins with carefully monitored crushing and grinding, to remove oversize pieces and desliming to remove the clay before flotation. This crushing and grinding usually involves a series of steps in which the crude ore passes through a primary crusher and a series of roll crushers and classifiers to remove oversize particles and the gangue. Ball mills commonly do the regrinding between flotation cycles to liberate more gangue minerals. The ore is kept as coarse as possible to prevent breaking of the flakes. The soft large flakes of graphite are cut up during grinding by sharp quartz and other hard mineral fragments. Removing the sharp fragments allows further grinding of the gangue without reducing the flake size. The larger the flake, the more valuable is the product. While high recoveries are common with proper combination of regrindings and flotations, it does not necessarily follow that a commercially usable product (now 90–94% C) is made in this fashion. This shows the attractiveness of using chemical leaching.

The chemical leaching methods are usually employed on graphite concentrates to produce high purity (97–99% or more C) graphite products. If the graphite concentrate has a carbonate gangue, it is leached in a vat with hydrochloric acid. If the graphite concentrate has a silica or silicate gangue (more common), it is leached in a covered hydrofluoric acid resistant vat or tank with hydrofluoric acid (HF). The vat or tank, the ductwork, and the equipment contacting HF is made of mild steel or Monel alloy or is lined with rubber or polytetrafluoroethylene, depending on temperature, HF strength, and method of use. The fluoride acid waste, which is poisonous and reactive, is treated to make it as innocuous as possible and then disposed of according to local environmental regulation. The fluoride acid waste can be neutralized with lime and filtered to yield a disposable waste product mostly composed of calcium fluosilicate. The graphite concentrates can also be leached by strong alkalis to remove silica and silicates.

Amorphous and Lump (High Crystalline) Graphite

Amorphous and lump (also called high crystalline) graphite is mostly mined underground. Amorphous graphite beds are quite thick, since they originated as coal beds, and are drilled, blasted, and loaded by hand into cars. Other conventional methods are also used. Lump graphite, mined only in Sri Lanka, is mined from narrow, steeply sloping veins by overhand stoping and filling, using temporary stulls as required for wall support. The stoping is usually done by hand drilling in order to avoid unwanted fines and product contamination. Preliminary sorting is done at the mine entrance. The processing of both graphites is labor intensive.

The processing of amorphous graphite usually involves grinding and screening to remove coarse impurities. Sometimes air separation is used to separate coarse impurities from amorphous graphite, and sometimes drying is needed.

The processing of lump graphite, a more expensive commodity, depends heavily on human labor. Men do heavier hand cobbing and sorting operations, usually to remove quartz impurities. Women remove the fines by wiping lumps on wet burlap and do light hand sorting.

SPECIFICATIONS AND STANDARDS

The American Society for Testing and Material (ASTM) publishes specifications, recommended practices, and definitions for graphite, as has the U.S. Government. Domestically used flake is classified according to purity with high grade containing 95–96% C, and low grade, 90–94% C. High purity crystalline flake contains 99% or more C. Sri Lankan graphite is classified according to lump, chip, and dust with subclassifications. Amorphous graphite is classified according to locality and carbon content, seldom >85% C. The variety of specifications exists because graphite is found worldwide, is mined by many small establishments, and is subject to keen competition among suppliers.

USES

The uses of graphite derive from its physical and chemical properties. Graphite exhibits the properties of a metal and a nonmetal, which make it suitable for many industrial applications. The metallic properties include thermal and electrical conductivity. The nonmetallic properties include chemical inertness, high thermal resistance, and lubricity. The combination of conductivity and high thermal stability allows graphite to be used in many applications, such as refractories, batteries, and fuel cells. Lubricity and thermal conductivity make it an excellent candidate as a lubricant at friction interfaces while furnishing a thermally conductive matrix to remove heat from the same interfaces. Lubricity and electrical conductivity allow its use as brushes for electric motors. A graphite brush in an electric motor effectively transfers electric current to a rotating armature, and the natural lubricity of the brush minimizes frictional wear.

Today's high technology products, such as friction materials and battery and fuel cells, demand higher purity graphite. Graphite has a high melting point, excellent thermal shock resistance, high thermal and electrical conductivity, low coefficient of friction, and is chemically inert. As such, high temperature applications dominate graphite end uses.

Minor uses of graphite include coating smokeless powder and gunpowder grains to control burning rate and prevent static sparking from friction between grains, roofing granules, packings, gaskets, stove polish, static eliminator, polish for tea leaves and coffee beans, pipe-joint compounds, boiler compounds, wire drawing, welding rod coatings, catalysts, oil-well drilling muds, lock lubrication, coatings for tape cartridges, mechanical mounts in cassettes, mercury and silver dry cells, exfoliated flake for gaskets and packing, aircraft disk brakes, catalyst pellet production, O-rings and oil seal, and interior and exterior coatings for cathode ray tubes.

H. A. Taylor, Jr., "Graphite", *Industrial Minerals and Rocks*, 6th ed., Society for Mining Engineers, New York, 1991.

H. A. Taylor, Jr., "Graphite", *Financial Times Executive Commodity Reports*, Mining Journal Books, London, 2000.

A. R. Ubbelohde and F. A. Lewis, *Graphite and its Crystal Compounds*, Oxford University Press, London, 1960.

P. L. Walker, ed., *Chemistry and Physics of Carbon*, Marcel Dekker, New York, Vols. 1–13, 1965–1977.

Rustu S. Kalyoncu
U.S. Geological Survey,
U.S. Department of the Interior

Harold A. Taylor, Jr.
Basics/Mines

GREEN CHEMISTRY

Green chemistry and green engineering describe the efforts of chemists and engineers to develop processes and products that prevent pollution and are inherently safe for humans and the environment. Implementing green chemistry and engineering will improve the quality of the environment for present and future generations. Manufacturing a chemical product requires a synthetic pathway as well as a safe and efficient chemical manufacturing strategy. Thus, both chemistry and engineering are required to achieve true source reduction.

Green chemistry is a term first proposed by Paul Anastas, formerly of the U.S. Environmental Protection Agency (EPA). Definitions of green chemistry typically include aspects of basic chemistry and applied engineering practice.

Green chemistry, by definition, addresses source reduction. It is the science and technology of preventing wastes before they are generated, and includes every consideration from feedstock origins to end of product life. The term "sustainable chemistry" is sometimes used with essentially the same meaning.

In the past several years, attention has focused on the more enlightened realization that pollution should be prevented before it occurs. Green chemistry and engineering are concepts driven by the emphasis on pollution prevention (P2). The driving force for P2 in the United States is the Pollution Prevention Act of 1990. This act established pollution prevention as the preferred environmental policy of the United States, rather than end of pipe treatment, remediation, or disposal of wastes. The full spectrum of pollution prevention options ranges from true source reduction to treatment and disposal. Following is a hierarchy for pollution prevention and waste management, in decreasing order of preference: (1) Source reduction, (2) in process recycling, (3) on-site recycling, (4) Off-site recycling, (5) waste treatment to render the waste less hazardous, (6) secure disposal, (7) direct release to the environment.

A similar pollution prevention hierarchy, based on practice at the Du Pont Company is given in Fig. 1.

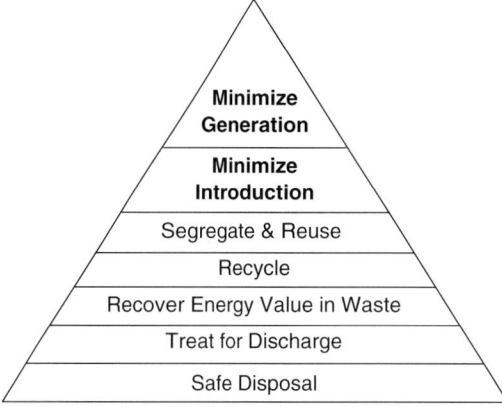

Figure 1. Pollution prevention hierarchy used at Du Pont. Reproduced with permission of the American Institute of Chemical Engineers. Copyright © 1999 AIChE. All right reserved.

THE TWELVE PRINCIPLES OF GREEN CHEMISTRY

Design and implementation of completely green products and processes is an enormous challenge. Because the number of chemical synthesis pathways is enormous, there is not a systematic and failsafe method for ensuring that the chemistry being implemented is green. Indeed, it is more nearly correct to inquire if a proposed chemical manufacturing process is "greener" than other alternatives. Thus, green chemistry recognizes the importance of incremental improvements. P. T. Anastas and J.C. Warner have formulated a set of twelve principles that guide and define the scope of green chemistry. These principles are presented below: prevention (Principle 1) , atom economy (Principle 2) , less hazardous chemical syntheses (Principle 3), design safer chemicals (Principle 4), safer solvents and auxiliaries (Principle 5), design for energy efficiency (Principle 6), use renewable feedstocks (Principle 7), reduce derivatives (Principle 8), catalysis (Principle 9),design for degradation (Principle 10), real-time analysis for pollution prevention (Principle 11), and inherently safer chemistry for accident prevention (Principle 12).

CONCEPTS RELATED TO GREEN CHEMISTRY

A number of concepts and industrial initiatives are related to green chemistry. These related and sometimes competing terms may cause confusion, and related concepts are defined and described below.

Pollution Prevention

As defined by the Pollution Prevention Act of 1990, Pollution Prevention encompasses those activities listed in the Introduction. Activities such as waste treatment and secure disposal, fall outside the concept of green chemistry.

Waste Minimization

Waste minimization, and the related term waste reduction, generally refer to reduction in the amount of solid or liquid waste produced by a process (air pollution being excepted).

Design for the Environment

Design for the Environment (DfE) refers to the design and manufacture of products and processes in a way that has minimal impact upon the environment.

Industrial Ecology

Industrial ecology describes the science of use and reuse of natural resources in manufacturing, rather than the traditional practice of extracting and using resources, then discarding and disposing. The concept is that the manufacture of goods should be a closed-loop system, analogous to the cycles that are observed in natural ecosystems.

Sustainability

Sustainability, or sustainable development, has been defined as meeting the needs of humans today, while not compromising the ability of future generations to meet their needs. Some regard the second half of this definition as a negative statement, and offer an equivalent definition of sustainable development: the ability to meet the needs of society today, while ensuring that future generations will also be able to meet their needs. Interpretation of sustainability from a corporate viewpoint has led to the concept of the so-called "triple bottom line" for industry: (1) economic prosperity and continuity for the business and its shareholders: (2) social well-being and equity for both employees and affected communities; and (3) environmental protection and resource conservation, both local and global.

Responsible Care

Responsible Care encompasses a diversity of elements including process safety, transportation, and interaction with suppliers, distributors, and other stakeholders. With respect to the environment, the Responsible Care program describes broad obligations for members of the American Chemistry Council, including providing chemicals that can be manufactured safely and in a manner that protects the environment and the health of people.

INDUSTRIAL EXAMPLES

The number of publications related to green chemistry and engineering has grown rapidly in the past several years. These cover the spectrum from fundamental chemistry to chemical engineering unit operations. The industrial examples selected below illustrate one or more of the twelve principles of green chemistry: use of catalysts to minimize toxic reagents; inherently safer chemistry: methyl lsocyanate and agricultural chemicals; alternative solvents: supercritical carbon dioxide as a media for chemical reactions atom-efficiency: synthesis of ibuprofen; design safer chemicals: environmentally safe isothiazolone marine antifoulant; biodegradable products: thermal polyaspartate polymers; and renewable resources: use of renewable fats and oils.

GREEN CHEMISTRY EDUCATION

The scope of green chemistry and engineering encompasses most if not all of the traditional educational and research disciplines (eg, organic and inorganic chemistry, catalysis, chemical separations, thermodynamics and energy utilization). The scope also includes financial analysis, safety, health, hazard and risk assessment, as well as the intersection of these with public policy. This makes green chemistry education a considerable challenge. Nevertheless, progress is being made. The U.S. EPA, in cooperation with the ACS, has an active program of workshops and textbook development for green chemistry and engineering. Other educational programs, case studies, and green chemistry and engineering materials are appearing on-line.

SUMMARY AND PROSPECTS

The greening of a chemical product or process may be accomplished through a number of means, including improved synthetic procedures, better catalysts, novel process design, use of renewable raw materials, discovery of less toxic products, use of environmentally benign solvents, or design of recyclable materials. Thus, the concept of green chemistry can be invoked in any aspect of fundamental or applied chemistry and chemical engineering, as well as toxicology and risk assessment.

In reality, it is difficult to declare a product or process as being completely "green"; in practice, a comparison must be made to determine if a choice of product–process is "greener" than other alternatives. Therefore, financial and social metrics must also be brought to bear alongside fundamental scientific data in all efforts to make greener chemical products and processes. Many of these metrics deal with long time frames and may not appear consistent with short economic and financial time horizons.

Achieving the goals of green chemistry requires cooperation among numerous stakeholders. For technology development, green chemistry requires an interdisciplinary approach including chemists, biochemists, biologists, engineers, statisticians, and health care professionals. Implementation of projects requires a cooperation of private industry and policy makers from local, state, and federal government.

D. T. Allen and D. R. Shonnard, *Green Engineering. Environmentally Conscious Design of Chemical Processes*, Prentice Hall PTR, Upper Saddle River, N.J., 2002.

P. T. Anastas and J. C. Warner, *Green Chemistry: Theory and Practice*, Oxford University Press, New York, 1998.

J. Clark and D. Macquarrie, *Handbook of Green Chemistry and Technology*, Blackwell Science Ltd., Oxford, U.K., 2002.

M. M. El-Halwagi, *Pollution Prevention through Process Integration*, Academic Press, San Diego, Calif., 1997.

MICHAEL A. MATTHEWS
University of South Carolina

GRIGNARD REACTIONS

The term Grignard reaction refers to both the preparation of a class of organomagnesium halide compounds and their subsequent reaction with a wide variety of organic and inorganic substrates. Of all the "name reactions" in chemistry, there is arguably none better known than the "Grignard reaction".

The general sequence of the reactions is now embodied in the following generic forms, where RX = an organic halide (most typically a chloride or bromide, although fluorides can be induced to react); S = a coordinating solvent (such as an ether or an amine); and AZ = a substrate with an electronegative group, Z:

$$RX + Mg + nS \longrightarrow RMgX \cdot S_n$$
$$RMgX \cdot S_n + AZ \longrightarrow RAZMgX \cdot S_n$$
$$RAZMgX \cdot S_n \longrightarrow RA + ZMgX \cdot S_n$$

The heterolysis of AZ is dependent on the substrate and does not always occur. The final isolation of the product usually involves a hydrolysis step.

The development of improved industrial procedures, including the substitution of tetrahydrofuran (THF) for diethyl ether and the demonstration that the less reactive, but significantly less expensive, vinyl and aryl chlorides could be successfully used, has greatly expanded the commercial possibilities of this reaction. In the flavor, fragrance, pharmaceutical, and fine chemical industries, its use can generally be regarded as routine. Tens of thousands of metric tons of Grignard reagents are produced annually for captive use or merchant sale.

The great value of the Grignard reaction to the synthetic chemist is its general applicability as a building block for an impressive range of structures and functional groups. The Grignard reagent can act both as a prototypical carbon nucleophile that can undergo addition and substitution reactions and as a strong base that can deprotonate acidic substrates, resulting in the conjugate base or in some cases elimination reactions. Grignard reagents react with most functional groups containing polar multiple bonds (eg, ketones, nitriles, sulfones, and imines), highly strained rings (epoxides), acidic hydrogens (eg, alkynes), and certain highly polar single bonds (eg, carbon–halogen and metal–halogen).

PREPARATION OF GRIGNARD REAGENTS

A Grignard reagent is prepared by first adding magnesium and a partial charge of solvent to the reactor, followed by the gradual addition of RX, in the remaining solvent, to the reaction. The quality of the solvent, magnesium, and RX can have a marked and often deleterious effect on the preparation of the Grignard reagent. Some of the problems are homocoupled product, formation of $RMgO_2X$, and noninitiated reaction of RX with Mg. Therefore, proper preparation and handling of each component must be carried out.

Solvent Preparation

The most critical aspect of the solvent is that it must be dry (<0.02 wt% of H_2O) and free of O_2.

Other considerations for the solvent are the solubility of the Grignard reagent, the temperatures required for initiation and adventitious reactions of the Grignard with the solvent. Based on these three considerations, the best general solvent for the preparation of a Grignard reagent is THF. However, other solvents that are commonly used are diethyl ether, methyl *tert*-butyl ether, di-*n*-butyl ether, glycol diethers, tertiary amines, toluene, dioxane (for the preparation of R_2Mg), and hexane.

Magnesium Preparation

A surface coating resulting from the oxidation or hydration of the metal surface is the principal problem encountered from the magnesium component. Fortunately, there are numerous methods to remove the inert coating, thus activating the magnesium. For industrial use, the best method is using freshly chipped Mg turnings with a small quantity of the desired Grignard added to the reactor before addition of RX.

The Organohalogen Component

Just as for Mg and the solvent, the organic halide must be dry (<0.02 wt% of H_2O) and free of O_2. The relative reactivity of the halogens is reflected in the rate of disappearance of Mg, which follows the general order $I > Br > Cl \gg F$. Unfotunately, the rate of disappearance of Mg does not always correlate with the formation of active Grignard. Typically, the more reactive the RX, the higher the probability of forming a homocoupled product. Therefore, when choosing X, the rate of reactivity, product selectivity, and cost must be taken into account.

Other Methods

There are several alternative methods for making Grignard reagents. Metal-exchange reactions are straightforward and MgR_2 can easily be prepared by this route.

Hydromagnesation reactions allow for the preparation of a Grignard from an olefin.

INDUSTRIAL MANUFACTURING PROCESS

In spite of many years of industrial use, the commercial-scale production of Grignard reagents has not been extensively described. The only practically important method is the batch method described by Grignard in 1900, namely, formation of the Grignard reagent, reaction with a substrate, followed by hydrolysis of the reaction mixture.

The equipment can be constructed of carbon steel except for the hydrolysis vessel, which is usually glass-lined to avoid corrosion by aqueous acids. It is desirable to use stainless steel or, preferably, glass-lined vessels throughout. All vessels must be supplied with an inert gas (nitrogen or argon) for purging and blanketing and are vented to release off-gases. It is imperative that the reaction vessel be protected with a rupture disk.

ANALYSIS OF GRIGNARD REAGENTS

There are three potential problems that may occur during Grignard reagent preparation: (1) oxidation by O_2, (2) hydrolysis by H_2O, or (3) homocoupling during the addition of alkyl or aryl halide. All three of these reactions decrease the active Grignard reagent while maintaining the same equivalents of base. Consequently, the concentration of a Grignard reagent should not be assumed, based on the reactants. The disadvantages of not analyzing the Grignard, reagent are improper stoichiometry, potentially deleterious side reactions, highly exothermic quenching processes, phase splits, waste disposal, and cost problems. The analytical technique must be able to differentiate between active Grignard and total basicity. Many methods are available to measure the active Grignard ranging from titration to electrophilic quenching followed by gas chromatography (gc) analysis. Potentiometric titration

using 2-butanol as the titrant is recommended as the general method for determining the activity of Grignard reagents. The advantages of this method are that it does not titrate Mg–OH or Mg–OR, results are reproducible (3 standard deviations = 0.6% of value), analysis of the solution is direct, and analysis takes <10 min.

Electrophilic quench followed by gc analysis can give the accuracy and precision of a potentiometric titration. However, each Grignard reagent and its product must be calibrated vs a gc standard, which is time, consuming. Also, these methods typically take a minimum of 1 h from the start of the quench procedure to obtaining the final chromatogram. The advantage of the gc method is that a direct measurement of the homocoupled product, oxidized Grignard, hydrolyzed Grignard, and unreacted alkyl or aryl halide can all be made.

Total basicity is measured by standard acid–base titration techniques.

Spectroscopic techniques such as nuclear magnetic resonance (nmr) and crystallography have been employed in structure elucidation. The nmr data are extremely dependent on the solvent, steric bulk of the organic group, temperature, and concentration.

Crystal structures of Grignard reagents do not necessarily correspond to their structure in solution. In general, the crystal structures indicate the reagents are ligated with THF or diethyl ether and are frequently observed to be dimers.

ECONOMIC ASPECTS

The Grignard reaction has been commercially important for >50 years, and for certain industrial processes it remains the favored (or only) practical route to construct various element–carbon bonds.

There are five components to the cost of using a Grignard reagent: (1) magnesium metal, (2) the halide, (3) the solvent, (4) the substrate, and (5) disposal of the by-products. Prices for tetrahydrofuran and diethyl ether, the two most commonly used solvents, have also increased. The cost of the halide depends on its structure, but as a general rule the order of cost is chloride < bromide < iodide.

HEALTH AND SAFETY FACTORS

Fire Hazards

The hazards associated with the manufacture, transport, and use of Grignard reagents are related to the flammability of the solvents employed and the exothermic reactions involved in their preparation and use.

Toxicology

Because of their high reactivity, there is little meaningful information on the health hazards of Grignard reagents *per se*. Rather, consideration needs to be given to the reagents employed, including the solvents and the products (or by-products) of the reaction. Some starting materials, such as organic halides (notably methyl bromide and vinyl chloride), are particularly toxic.

Regulatory Considerations

Commercial use of a Grignard reagent in the United States requires that it appear on the Environmental Protection Agency (EPA) list of Chemical Substances in Commerce. A corresponding registration exists for the European Community, Japan and a number of other countries.

REACTIONS AND APPLICATIONS OF GRIGNARD REAGENTS—RECENT DEVELOPMENTS

Reactions and applications of Grignard reagents include asymmetric syntheses, reactions with inorganic chlorides, reagents as bases, metal-assisted modified reactions, intramolecular reactions, as methacrylate polymerization catalysts, and reagents as supports for the Ziegler-Natta process.

There are several reviews and books that discuss the general chemistry of Grignard reagents. The focus here is the specific commercial growth areas in Grignard chemistry.

ALKYL CROSS-COUPLING REACTIONS

Grignard reagents can be made to undergo several types of transition metal catalyzed coupling reactions. Until recently, catalyzed alkyl coupling was generally unavailable owing to the tendancy for β-elimination or low reactivity. However, several recent reports in the literature describe Grignard-related systems that not only perform alkyl couplings, but do so with inexpensive iron and nickel catalysts or with stoichiometric amounts of cuprates.

Aryl chlorides, alkyl bromides, and various aryl and alkyl iodides have all demonstrated excellent selectivity, often in the presence of sensitive functional groups such as esters and nitriles. For the commercial synthetic chemist there is significant value in reaction systems that avoid the protection/deprotection "gymnastics" frequently required in complex multistep syntheses.

M. Fieser and L. Fieser, *Reagents for Organic Synthesis*, John Wiley & Sons, Inc., New York, 1967, p. 415.

V. Grignard, *Compt. Rend.* **130**, 1322 (1900).

G. Silverman and P. Rakita, *Handbook of Grignard Reagents*, Marcel Dekker, Inc., New York, 1996.

PHILIP E. RAKITA
Armour Associates, Ltd.

GROUNDWATER MONITORING

Groundwater monitoring is used to analyze the impact of a variety of surface and subsurface activities, including seawater intrusion, application of agricultural products such as herbicides, pesticides, and fertilizers, residential septic systems, and industrial waste ponds. Another focus of groundwater monitoring has been contamination associated with waste landfills and ruptured underground petroleum storage tanks.

The design of a groundwater monitoring strategy requires a basic understanding of groundwater flow systems. The majority of groundwater flow occurs in formations known as aquifers. At least two types of data can be retrieved using groundwater wells, ie, groundwater pressure and groundwater quality. A monitoring well allows measurement of these properties at a specific point in an aquifer. *Monitoring wells* come in a variety of sizes and materials, but each is basically a pipe extending from the ground surface to a point in the aquifer at which the pressure or contaminant is to be assessed. Monitoring wells are functional only in the saturated zone of the subsurface. Within the unsaturated soil zone, tensiometers, soil moisture blocks, and psychrometers have been used to assess fluid pressures. Fluid samples are retrieved using suction cup lysimeters for subsequent quality analysis.

AQUIFERS

The *term aquifer* is used to denote an extensive region of saturated material. There are many types of aquifers. The primary distinction between types involves the boundaries that define the aquifer. An unconfined aquifer, also known as a phreatic or water table aquifer, is assumed to have an upper boundary of saturated soil at a pressure of zero gauge, or atmospheric pressure. A confined aquifer has a low permeability upper boundary that maintains the interstitial water within the aquifer at pressures greater than atmospheric. For both types of aquifers, the lower boundary is frequently a low permeability soil or rock formation.

GROUNDWATER PRESSURE AND ENERGY

The energy state of soil water can be defined with respect to the Bernoulli equation, neglecting thermal and osmotic energy as

$$E = z + P/\gamma + v^2/2g \qquad (1)$$

where E is the energy per unit weight (L), P the pressure (F/L^2), γ the specific weight (F/L^3), z the elevation (L), and v the average velocity (L/T). The three energy terms represented by the right-hand side of the equation are pressure energy, potential energy, and kinetic energy, respectively. In most groundwater applications, the kinetic energy term is much less significant than the other two and is neglected. Thermal gradients cause moisture to migrate toward colder regions. However, thermal energy has been neglected in the present formulation and the equation cannot be used to simulate problems where there is a significant temperature gradient present.

When the energy terms are expressed as energy per unit weight, the term head is often used. Therefore, the total head, $h(L)$, is equal to the elevation head, z, plus the pressure head, P/γ:

$$h = z + \frac{P}{\gamma} \qquad (2)$$

CALCULATION OF GROUNDWATER FLOW

The framework for the solution of porous media flow problems was established by the experiments of Henri Darcy in the 1800s. The relationship between fluid volumetric flow rate, Q, hydraulic gradient, and cross-sectional area, A, of flow is given by the Darcy formula:

$$Q = KA\frac{h_1 - h_2}{\Delta l} \qquad (3)$$

Here h_z represents the hydraulic head at location z, whereas Δl is the hydraulic length between points 1 and 2. A is an area perpendicular to the discharge vector. The constant $K(L/T)$, which maintains the equality, has been termed the hydraulic conductivity, permeability, or simply conductivity.

This form of Darcy's law is applicable only to saturated flow. As discussed earlier, there are distinctions between the state of soil water in the saturated and unsaturated regions. These distinctions lead to an alternative form of Darcy's law for the case of unsaturated flow.

The vertical component of flow can be determined if a well is screened at two different elevations. Frequently, nested wells are used instead of a single well and multiple screenings to determine the vertical component of flow. Nested wells must be situated close enough to one another so horizontal gradients do not become a factor. Nested wells can also be used to analyze multilayer aquifer flow. There are many situations involving interaquifer transport owing to leaky boundaries between the aquifers. The primary case of interest involves the vertical transport of fluid across a horizontal semipermeable boundary between two or more aquifers.

MONITORING WELL DESIGN FOR CONTAMINANT TRANSPORT STUDIES

Monitoring wells are installed by first completing a soil boring to the approximate depth of groundwater measurements. Drilling methods for the borehole include auger, mud rotary, cable tool, jetted wells, and driven wells. During the drilling, a boring log is prepared that records details of the subsurface materials encountered as the depth progresses. A well casing is installed in the borehole with a well screen at or near the bottom of the borehole. In the vicinity of the well screen, a filter pack of natural, ie, typically sand or pea gravel, or synthetic materials is used to preclude clogging of the well screen. Often, a secondary filter pack consisting of finer materials is placed above the primary filter. Above this is the virtually impermeable bentonite seal. A neat cement grout above this layer extends to the ground surface.

A variety of techniques can be used to retrieve the groundwater sample once the well is in place. Pumps, bailers, and syringes are among the devices used to draw the sample to the surface.

Design of a groundwater monitoring program minimally includes consideration of materials, location, indicator parameters, and timing.

Data analysis is aided by a variety of statistical techniques to assess significance, highlight trends, and form mathematical models of any correlations developed.

P. Domenico and F. Schwartz, *Physical and Chemical Hydrogeology*, John Wiley & Sons, Inc., New York, 1990.
C. W. Fetter, *Applied Hydrogeology*, 4th ed., Macmillan Publishers, New York, 2001.
D. M. Nielsen, ed., *Practical Handbook of Ground-Water Monitoring*, Lewis Publishers, Inc., Chelsea, Mich., 1991.

CAROL J. MILLER
Wayne State University

GROWTH REGULATORS, ANIMAL

The growth of animals can be defined as an increase in mass of whole body, tissue(s), organ(s), or cell(s) with time.

Improved understanding of the control of metabolic aspects of growth has provided the opportunity to regulate animal growth. Improvement of rate and efficiency of growth benefits the producer. Improvement in composition of meat animals benefits the producer through more efficient gain and greater value, and benefits the processor through less labor requirement for trimming and removal of fat. The consumer benefits by receiving a quality, desirable food at a cost reflective of efficient production.

Four general classes (ca 1993) of growth regulators are approved by the Food and Drug Administration (FDA) for use in food-producing animals in the United States. These include naturally occurring and synthetic estrogens and androgens; ie, anabolic steroids, ionophores, antibiotics, and bovine somatotropin. Compounds in the first class, anabolic steroids, act as metabolism modifiers to alter nutrient partitioning toward greater rates of protein synthesis and deposition, thereby increasing the weight at which 25 to 30% lipid content in the body or carcass is achieved. Ionophores have highly selective antibiotic activity and appear to enhance feed conversion efficiency through effects on ruminal microbes. Antibiotics, administered at subtherapeutic doses, enhance growth through improving feed conversion efficiency and/or growth rate, with no consistent effect on body or carcass composition.

Two other classes of growth regulators—somatotropin or somatotropin secretogogues—and select synthetic phenethanolamines have been investigated for their ability to alter growth. In 1993, the FDA approved administration of recombinant bovine somatotropin for increasing milk production in dairy cows. One phenethanolamine, ractopamine (Paylean), was approved by the FDA in December 1999 for use in finishing pigs. Administration of native or recombinant somatotropin (ST) to growing pigs, cattle, and lambs dramatically enhances rate, efficiency, and composition of gain. Likewise, experimental dietary administration of select synthetic phenethanolamines, most of which are β-adrenergic agonists, also has produced striking changes in rates of skeletal muscle and adipose tissue growth and accretion in growing cattle, lambs, pigs, and poultry.

Somatotropin, the β-adrenergic agonists, and the anabolic steroids are considered metabolism modifiers because these compounds alter protein, lipid, carbohydrate, mineral metabolism, or combinations of these and they partition nutrient use toward greater rates of protein deposition; ie, muscle growth and lesser rates of lipid accretion.

ANABOLIC STEROIDS

Several anabolic steroid implants have been approved for use in beef cattle in the United States, but only one, zeranol, is approved for use in lambs. Anabolic steroids are not used for growth regulation in swine or poultry.

Commercial products approved by the FDA include the naturally occurring hormone estradiol (Compudose); the natural hormone progesterone, used in combination with estradiol or estradiol benzoate ie, Steer-oid, Synovex-S; and Synovex-C for calves; the fungal metabolite zeranol (Ralgro), which has estrogenic properties; the synthetic progestin melengestrol acetate (MGA), testosterone propionate in combination with estradiol benzoate; ie, Synovex-H or Heifer-oid; and a synthetic testosterone analogue, trenbolone acetate (TBA), which is used alone; ie, Finaplix, or in combination with estradiol, ie, Revalor. Ractopamine hydrochloride was also approved by the FDA for use in finishing beef cattle in June, 2003. The commercial product name is Optaflexx.

Economics

Estimates of anabolic steroid use in growing cattle indicate that savings associated with reduced feed costs are approximately $50 per animal. Increased value of the carcass resulting from the increased amount of salable lean meat produced is estimated to range from $15 to $30 per animal.

Withdrawal from anabolic steroid treatment is not required before slaughter because residue levels in edible tissues are negligible, significantly lower than other sources of estradiol such as the normal endogenous production in humans and the phytoestrogens consumed in plant food sources.

IONOPHORES

An ionophore may be defined as an organic substance that binds a polar compound and acts as an ion-transfer agent to facilitate movement of monovalent (eg, sodium and potassium) and divalent (eg, calcium) ions through cell membranes. The change in electrical charge in membranes influences the transport of nutrients and metabolites across the cell membrane, but the exact mechanism by which ionophores improve growth performance in growing ruminants is not known.

Monensin and other ionophores are being fed to over 90% of feedlot cattle grown for beef to enhance efficiency of gain; improvements of 5–10% are common. Ionophores also are used as anticoccidial drugs in poultry production and have similar, but lesser, effects in ruminants.

Doses range from 6 to 33 ppm in the diet, but very little if any ionophore can be measured in the circulation after feeding. Tissue and blood concentrations are very low.

ANTIBIOTICS

Antibiotics approved for use as growth enhancers in livestock and poultry include bacitracins, bambermycins, lincomycin, penicillin, streptomycin, tetracyclines, tiamulin, tylosin, and virginiamycin.

Chemically synthesized antimicrobials used in animal and poultry feeds include arsenicals (eg, arsanilic acid, sodium arsanilate, and roxarsone), sulfa drugs eg, sulfadimethoxine, sulfamethazine, and sulfathiazole carbadox and nitrofurans (eg, furazolidone and nitrofurazone).

GROWTH HORMONE-RELEASING FACTOR

Exogenous administration of the naturally occurring growth hormone-releasing factor (GRF(1-44NH$_2$)) stimulates ST secretion and increases circulating concentrations of ST in growing pigs, cattle, and sheep.

Because administration of GRF is presumed to act through the same mechanisms involved in ST mediation of metabolism and tissue growth, similar interactions with gender, genotype, and nutritional status are expected.

HEALTH AND SAFETY FACTORS

The U.S. Food and Drug Administration's Center for Veterinary Medicine thoroughly evaluates the proposed use of any compound, natural or synthetic, used in food-producing animals for human food safety, safety to the animal of intended use, and safety to the environment. When a compound receives approval by the FDA, the efficacy and safety have been extensively investigated and the necessary labeling, handling, use, and withdrawal time requirements, if any, are determined. The Food Safety and Inspection Service (FSIS) of the USDA is responsible for ensuring that USDA-inspected meat and poultry products are safe, wholesome, and free of adulterating residues.

D. H. Beermann, *Animal Growth Regulation*, Plenum Press, New York, 1989.

J. C. Bouffault and J. P. Willemart, *Anabolics in Animal Production*, Office International des Epizooties, Paris, 1983.

D. L. Hancock, J. F. Wagner, and D. B. Anderson, *Growth Regulation in Farm Animals, Advances in Meat Research*, Vol. 7, Elsevier Science Publishers Ltd., Essex, U.K., 1991.

USDA, *Domestic Residue Data Book*, Food Safety and Inspection Service, National Residue Program, Washington, D.C., 1992; published annually.

DONALD H. BEERMANN
University of Nebraska

GROWTH REGULATORS, PLANT

The availability of new plant growth regulators has not grown appreciably in the past 20 years. In part, this has been due to the high costs associated with the discovery and application of new agrochemicals which, from inception to market, are presently estimated to be in the order of US \$25–35 million and ~8 years for a synthetic, and US \$2–4 million, and ~2–4 years, for a natural product. Into that equation is thrown the cost of registration and the environmental impact, the latter requiring tests across many plant and zoological species. During the past decade, two other factors have entered the picture. The first is the matter of intellectual property and its protection, which translates to the fact that many researchers now withhold disclosure of work to their colleagues in either verbal or written form. The second, closely related to the first, are those times when the structure and activity of the new, potentially blockbuster plant growth regulator is presented by an enthusiastic researcher without benefit of patent protection. Subsequently, no chemical company will underwrite the development of a bioactive compound without patent coverage since both national and international patents, plus maintenance fees, have to be paid. Simply stated, unprotected disclosure only impedes the flow of utilitarian materials to the market place.

Plant growth regulators are, by definition, bioactive compounds. This implies that their homologues and analogues may possess pharmaceutical properties. If 50 g of a bioactive compound, used to treat 1 hm^2 at an on-farm cost of US \$100, and then the same compound where used at 10 mg/dose as a medicinal at, say, US \$5/tablet. It would yield income of US \$25,000. Little wonder that the agrochemical–plant growth regulator market has remained torpid with, perhaps, the exception of natural products.

NATURAL PRODUCTS

The most acceptable plant growth regulators are those that perform well and, when their task has been accomplished, are degraded to nonharmful environmental products. A small number of novel natural product and their derivatives have recently made their way to the marketplace. Because of their relative ease of registration it is anticipated that this will become an expanding enterprise. Additionally, a number of microorganisms that produce profitable plant growth regulators have also been introduced, especially by the Japanese. The "pure" active materials are included in Table 1.

Table 1. Natural Plant Growth Regulators

Product	Structure	Trade name	LD$_{50}$, g/kg
	Available Natural Products		
AVG	 (1)	AVG, Retain, Prestiage	no toxicity
brassinolide	 (2)		mouse, 1.31[a]
24-epibrassinolide	 (3)		rat[a], 2 mouse, 1

Table 1. (*Continued*)

Product	Structure	Trade name	LD$_{50}$, g/kg
carvone	(4)	Carvone, Talent	rat, 1.64
cytokinins		Trigger, Burst, Yield Booster, Jump	rat, 5
kinetin^{a}	(5)		
6-benzyladenine	(6)		
n-decanol	$H_3C-(CH_2)_8-CH_2OH$ (7)	Off-Shhot-T, Antak, Sucker Pluker Contac	rat, 12.8
dikegulac	(8)	Dikegulac, Atrimmec, Atrimol	mouse, 19.5: rat, 3t
ethylene	$H_2C=CH_2$ (9)		mouse, 950/L in air
gamma aminobutyric acid (GABA) and	(10)	Auxigro, GABA	rat, 5/kg
glutamic acid	(11)		
gibberellins			

Table 1. (*Continued*)

Product	Structure	Trade name	LD$_{50}$, g/kg
GA3	(**12**)	G.A., Pro-Gibb, Berelex	mouse, 15
	(**13**)		
GA4 and GA7	(**14**)	Provide, Regulex, Novagib	no toxicity
indole-3 butyric acid	(**15**)	IBA, Hormodin, Rhizopin, Jiffy Grow	mouse, 100
L-lactic acid	(**16**)	L-lactic acid, Propel	
triacontanol	(**17**)	Triacontanol	no toxicity

Available Natural Product Derivatives

atonik	(**18**)	Atonik	rat, 10

Table 1. (*Continued*)

Product	Structure	Trade name	LD$_{50}$, g/kg
N–(phenylmethyl)-1, H-purine-6-amine	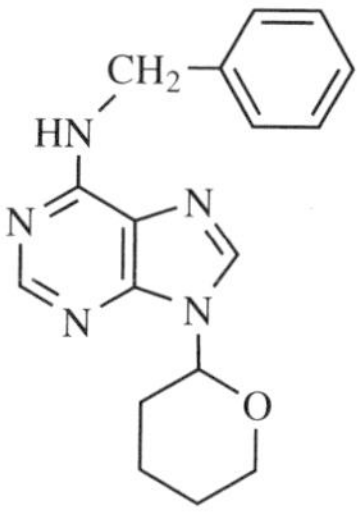 (19)	Accel	rat, 5
ethephon; 2-chloro-ethylphosphonic acid	$Cl-CH_2-CH_2-\overset{\displaystyle O}{\underset{\displaystyle OH}{P}}-OH$ (20)	Ethephon, Ethrel, Prep.	mouse, 3.03
fatty acid methyl ester	$C_n-H_y-\overset{\displaystyle O}{C}-OCH_3$ (21) where $n = 8–12$	Off-Shoot-O	rat, 20.5
N-(phosphonomethyl)- glycine	$HO-\overset{\displaystyle O}{\underset{\displaystyle OH}{P}}-CH_2-NH-CH_2COOH$ (22)	Round up, Glyphosate, Polado	rat, 3.9

Miscellaneous Products

3-hydroxyuridine	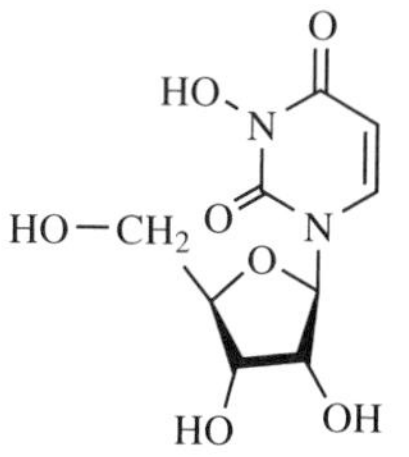 (23)	none	

^{a}Oral.

^{b}Kinetin; 6-(4-hydroxy-1,3-dimethylbut-*trans*-2-enylamino)-9-β-D-ribofuranosylpurine. Sodium chloride: LD$_{50}$ rat, oral, 3.75 g/kg.

SYNTHETIC COMPOUNDS

There are more synthetic plant growth regulators than naturally occurring agents. With the ability of the organic chemist to synthesize novel agents, it became common practice to generate a number of synthetics and test them in various biological systems. With a better understanding of organic synthetic pathways, if a particular structure demonstrated interesting biologi-cal effects, it was used as a "lead" compound in the synthetic exploitation of more active agents. This classic structure–activity relationship led to the development of some of the most interesting agents used in agrochemistry. Much of the information concerning the development of synthetic plant growth regulators remains proprietary except in very successful cases. Table 2 offers information on commercially available synthetic plant growth regulators.

Table 2. Authorized Synthetic Plant Growth Regulators

Product	Structure	Trade name	LD_{50}, g/kg
Alar[a]	(24)	Alar, B-Nine, Daminozide Dazide	rat, 8.4
Amidochlor	(25)	Amidochlor, Limit	rat, 3.1
Ancymidol	(26)	Ancymidol, A-Rest, Reducymol, Slectone	mouse, 5 rat, 4.5
Butralin	(27)	Butralin, Tamex	rat, 1.26
Chlormequat chloride	(28)	CCC, Cycogan, Hormocel, Chlomequat chloride, Cycocel, Arotex 5C	rat (male), 31 rat (female), 18
4-chlorophenyl-4,4-dimethyl-triazol pentenol	(29)	Uniconazole, Sumagic	rat, 1.79
Chlorpropham	(30)	Chlorpropham, CIPC, Sprout Nip, Bud-Nip, ChloroIPC, Taterpix	rat, 1.2
Cloxyfonac[a]	(31)	Cloxyfonac, Tomatlane, CHPA	rat, 5.0
copper hydroxide	$Cu(OH)_2$ (32)	Copper hydroxide, Spinout	

Table 2. (*Continued*)

Product	Structure	Trade name	LD$_{50}$, g/kg
3-CPA	(**33**)	Fruitone CPA, 3-CP, Cloprop	rat, 10
4-CPA	(**34**)	PCPA, Tomato Fix, Sure-Set	rat, 0.85
Cyclanilide	(**35**)	Cyclanilide, Finish	rat, 0.21
Cyclohexane-carboxamide	(**36**)	Cyclohexanecarboxamide, AC 94377	rat, 5.0
Dichloroprop^a	(**37**)	Dichloroprop, 2,4-DP, Dormone	rat, 0.5
2,4-dichlorophenxoy-acetic acid	(**38**)	2,4-D, Citrus Fix, Hivol-44	mouse, 0.368 rat, 0.375
Dimethipin	(**39**)	Dimethipin "N252," Harvade, Dimethipin "UBI-N252"	rat, 1.18
Dormex	$H_2N-C\equiv N$ (**40**)	Dormex, Alzodel	rat, 1.25
Etacelasil^a	(**41**)	Etacelasil, Alsol, CGA 13586	rat, 2.06

Table 2. (*Continued*)

Product	Structure	Trade name	LD$_{50}$, g/kg
ethoxyquin	 (**42**)	Stop Scald, Nix-Scald, Santoquin	rat, 1.92 mouse, 1.73
ethylchloroindazolyl-acetate[a]	 (**43**)	Ethylchlo-zate, Figaron	rat, 4.8
flumetralin	 (**44**)	Primet, CGA 41065	rat, 3.1
flurprimidol	 (**45**)	Cutless, Cutless-TP	rat, 0.914 (male)
Folcysteine[a]	 (**46**)	Ergostim	rat, 0.71 (female) rat, 4.5
Folex	 (**47**)	Folex, Tribufos, Merphos	rat, 0.234
Forchlorfenuron[a]	 (**48**)	Forchlorfenuron, CPPU, Sitofex	rat, 4.92
Inabenfide[a]	 (**49**)	Seritard, CGR 811	rat, 15

Table 2. (*Continued*)

Product	Structure	Trade name	LD$_{50}$, g/kg
maleic hydrazide	(50)	Maleic hydrazide, Sucker-Stuff MH-30, Retard, Fair-plus, Royal MH-30	rat, 4
mefluidide	(51)	Embark 2-S	mouse, 1.92 rat, 4.0
mepiquat chloride[a,b]	(52)	PIX, Bas 08300, Ponnax, Terpal (with Ethephon)[b]	rat, 1.42 (Terpal) rat, 1.5
Morphactin[a]			
Chlorfluren-Me	(53)		
Chlorfurenol- Me	(54)	Morphactin, Maintain-CF125, Curbiset	rat, 4.0
Dichloroflurenol-Me	(55)		
flurenolbutyl	(56)		
naphthaleneacetic acid	(57)	NAA-800, NAA	rat, 1.0

Table 2. (*Continued*)

Product	Structure	Trade name	LD$_{50}$, g/kg
naphthaleneacetamide	(**58**)	Fruitone-N	rat, 6.4
paclobutrazol	(**59**)	Bonzi, Clipper, PP333, Proturf Bounty, Parlay	rat, 1.356
N-phenylthalamic acid[a]	(**60**)	Nevirol	rat, 9.0
prohexadione-calcium	(**61**)	Apogee, Viviful, Medex, Baseline	rat, 2.0
Sevin	(**62**)	Sevin, Carbaryl	rat, 0.56
tetrachloronitrobenzene	(**63**)	Technazene, Fusarex, Chipman 3	rat, 2.047
Tomaset[a]	(**64**)	Tomaset, N-M-T	rat, 5.0
trinexapac-ethyl[c]	(**65**)	Omega, Vision, Primo, Nomow	rat, 4.46

[a]Not available in the United States.
[b]Terpal product available only outside the United States.
[c]May be used only on noncrop plants in the United States.
[d]Used on turf, but only outside the United States.

H. G. Cutler, *Natural Products and Their Potential in Agriculture*, ACS Symposium Series 380, Washington, D.C., 1988.

S. Hayat and A. Ahmad, eds., *Brassinsteroids*, Kluwer Academic Publishers, 2003.

J. W. Mitchell and G. A. Livingston, *Methods of Studying Plant Hormones and Growth-Regulating Substances*, Agriculture Handbook No. 336, USDA, Washington, D.C., 1968.

W. T. Thompson, *Agricultural Chemicals, Book III—Miscellaneous Agricultural Chemicals, 1999–2000 Revision*, Thompson Publications, Fresno, Calif., 2001.

HORACE G. CUTLER
STEPHEN J. CUTLER
Mercer University

GUMS

The term "gums" does not designate a scientific class of substances that can be defined precisely. The term is applied to a variety of substances that produce sticky or slimy, viscous solutions or molecular dispersions or gels in an appropriate solvent or swelling agent. Some are specific chemical substances, some materials that contain several components. As employed in industry, the term most often refers to hydrophilic, natural polymers or modifications of natural polymers that thicken or are gel aqueous systems at low concentrations. More specifically, the term most often refers to water-soluble polysaccharides and derivatized polysaccharides. When used as ingredients in food products, polysaccharidic gums, along with certain proteins (especially gelatin) are sometimes referred to as hydrocolloids.

In the food, petroleum production, ore refining, and many other industries the term is usually defined by the general feature that they are at least partially soluble in room-temperature water, forming viscous solutions, usually at low concentrations, although there are exceptions to the latter. Another commonality is that they have current or potential commercial importance.

Commercial polysaccharidic gums are obtained from the seeds, roots, and tubers of land plants, from seaweeds, from microorganisms (fermentation), and by derivatization of cellulose (from farms and forests). Since polysaccharides are abundant, come from renewable sources, are safe (nontoxic), are amenable to both chemical and biochemical modification, and are biodegradable, they find widespread, extensive use. Industrial gums are most often used because of their ability to thicken or gel aqueous systems at low concentration, but these general behaviors are manifested in a variety of way. In addition, they are often also employed to impart properties other than thickening or gelation to a system. Starches and modified starch products are not considered gums, even though they, like the gums, are composed of water-soluble polysaccharides that can provide viscous solutions or gels at relatively low concentrations and are extensively used in practical applications. Table 1 lists properties of selected gums.

SPECIFIC GUMS

Guar and Locust Bean Gums (Galactomannans)

Commercial guar gum is the ground endosperm of guar seeds. Guar endosperm preparations can be modified via reactions of its hydroxyl groups (as can any other polysaccharide). Derivatives are made to control its rate of hydration, peak viscosity, ash content, insoluble material, heat stability, and compatibility with other materials. Guar gum forms very high viscosity, pseudoplastic (instantaneously shear thinning) solutions at low concentrations. Guar gum is used in food products. Guar gum or modified guar gum products are used in textile printing pastes, to thicken and gel blasting agents and explosive slurries, in water and water–methanol-based fracturing fluids for oil and gas wells, as processing aids in the separation of certain minerals from their ores, and in fabric softeners. For nonfood applications, guar gum and modified guar gum products are often sold with additives that control the rate of hydration, resistance to enzymes, dispersibility, or flow properties of the dry powder.

Like guar gum, commercial locust bean gum (LBG) is the ground endosperm of the seeds of the locust bean (carob) tree. Locust bean (carob) gum has low cold-water solubility and is used when delayed viscosity development is desired. The general properties of LBG are similar to those of guar gum. Locust bean gum is used primarily in food products.

Algins/Alginates

Algins are linear, anionic polysaccharides with block copolymer structures. They are extracted from brown algae. The specific properties exhibited by a solution of an algin preparation depend on the ratio of its monomeric units, the concentration and type of cations in solution, the temperature, and the degree of polymerization. Alginate molecules in aqueous solutions are highly hydrated, linear polyelectrolytes in extended conformations.

An important and useful property of alginates is their ability to form gels by reaction with calcium ions. Alginates with a higher percentage of polyguluronate segments form the more rigid, more brittle gels. Alginates with the higher percentage of polymannuronate segments form the more elastic, more deformable gels that have a reduced tendency to undergo syneresis. Sodium alginate is used extensively in textile printing pastes and in gelled food products.

Agars, Carrageenans, and Furcellarans

Carrageenan is a generic term applied to polysaccharides extracted from a number of closely related species of red algae. Agar and furcellaran are also red seaweed extracts, members of the same larger family. Commercial carrageenans are composed primarily of three types of polymers: κ-,ι and λ-carrageenan. Most carrageenan is used to make gels. Blends of the three general types of carrageenan are often employed, with each preparation usually standardized to a gel characteristic. Most carrageen preparations are used in food products; they are also used in toothpaste and in room freshener gels.

Table 1. Selected Gums and Their Unique Properties

Gum	Unique property
agar	gels require high temperatures for remelting; gels are compatible with high solute concentrations
algins/alginates	gelation with Ca^{2+}
propylene glycol alginates	surface activity
carboxymethylcelluloses	form clear, stable, either pseudoplastic or thixotropic solutions
κ-type carrageenans	gelation with K^+; form complexes with proteins, especially milk proteins, forming soft, thixotropic gels; form gels via synergistic interaction with locust bean gum
ι-type carrageenans	gelation with Ca^{2+}
curdlan	irreversible thermogelation
gellans	form gels with a range of textures with any cation; good suspension stabilizers
guar gum	gelation with borate and titanium ions
gum arabic	low solution viscosity at high concentrations. Newtonian flow of solutions of up to 50% concentration; both an emulsifier and an emulsion stabilizer; emulsions can be spray-dried without conditioning air
hydroxyethylcelluloses	form clear, water-soluble, oil- and grease-resistant films and coatings
hydroxypropylcelluloses	form non-tacky, heat-sealable packaging films
locust bean (carob) gum	gelation via synergistic interactions with κ-carrageenan and xanthan
methylcelluloses, including hydroxypropylmethyl-celluloses	reversible thermogelation; soluble in cold water; insoluble in hot water
high methoxyl pectins	form spreadable gels with ~65% sugar and a solution pH of ~3
low methoxyl pectins	gelation with Ca^{2+}, ie, without sugar and acid
amidated pectins	gelation with very low concentrations of Ca^{2+}
welan	very thermal stable; very good suspension stabilization; solutions will tolerate high concentrations of salts
xanthan	solutions are highly pseudoplastic; temperature has no effect on solution viscosity from 0 to 95°C; stable in highly acidic systems; pH has no effect on solution viscosity from pH 1 to 12. Synergistic thickening with guar gum; synergistic gelation with locust bean gum; very good emulsion and suspension stabilization; imparts freeze–thaw stability to products

Carrageenans are extracted primarily from the *Chondrus* and *Gigartina* species of red algae. Furcellaran is obtained from the *Furcellaria* species. Agars are obtained primarily from *Gelidium* and *Gracilaria* species. A useful property of the red seaweed extracts is their ability to form gels with water and milk. Carrageenans are blended and standardized to provide products that will form a wide variety of gels that do not melt at room temperature, do not require refrigeration, and are freeze–thaw stable.

Agars are the least soluble members of this class of polysaccharides. They can be dissolved only at temperatures >100°C. When hot agar solutions are cooled, strong, brittle, turbid gels form. By far the greatest use of agar in the United States is in the preparation of microbiological culture media. Agar is also used in bakery icings, because of its nonmelting characteristics and its compatibility with high sugar concentrations. Agarose, the linear component of agar, is used in making gels for electrophoresis, to make media for size-exclusion chromatography, and in several biotechnological applications.

Pectins

Pectins are mixtures of polysaccharides that originate from plants. They are water soluble, and their solutions gel under suitable conditions. The commercial importance of pectin is predominantly the result of its unique ability to form spreadable gels (jams, jellies, preserves, etc) in the presence of a solute that competes for water of hydration (almost always sugar) at pH ~3, or in the presence of calcium ions.

Gellan

Gellan is a gelling, bacterial polysaccharide produced by specific species of *Sphingomonas*. By varying the degree of acylation, products that provide a range of gel textures are made available. Gellan has good thermal stability and dilute solutions of gellan have very good suspending power. A primary use is in plant tissue culture media. Commercially, gellan is called gellan gum.

Welan

Welan is a nongelling bacterial polysaccharide produced by a species of *Alcaligenes*. For a polysaccharide, welan is exceptionally heat stable. Temperature has little effect on its low-concentration, high-viscosity solutions, even up to 150°C or higher. Welan solutions are also relatively unaffected by pH. They are pseudoplastic, stable in the presence of high concentrations of salt, and are good suspension stabilizers. These properties are what is needed for drilling, workover, and completion fluids, particularly in deep oil and gas wells where bottom hole temperatures exceed 120°C.

Xanthan

Xanthan (known commercially as xanthan gum) has a main chain identical to the structure of cellulose molecules. However, in xanthan, every other β-D-glucopyranosyl unit in the main chain is substituted with a trisaccharide unit. The molecular weight is probably on the order of 2×10^6. The unusual properties of xanthan undoubtedly result from its structural rigidity.

Xanthan solutions are extremely pseudoplastic and have high yield values, making them almost ideal for the stabilization of aqueous dispersions, suspensions, and emulsions. Whereas other polysaccharide solutions decrease in viscosity when they are heated (with the exception of welan solutions), xanthan solutions containing a low concentration (0.1%) of salt change little in viscosity over the temperature range 0–95°C. Xanthan is used in various aspects of petroleum production, including oil well drilling, hydraulic fracturing, and work over, completion, pipeline cleaning, and enhanced-oil-recovery fluids. It also finds application in a large number and variety of foods and in consumer and agricultural chemical products.

Gum Arabic

Of the gums of ancient commerce, only gum arabic is still in significant use. Gum arabic preparations are mixtures of highly branched, branch-on-branch, acidic polysaccharides, the composition of which varies with the species, season, and climate. Gum arabic is unique among gums because of its high solubility and the low viscosity and Newtonian flow of its solutions. Its main uses in the United States are in the preparation of flavor oil emulsions and dry powders made by spray drying such emulsions, in coating certain confections, and in making encapsulating coacervates for pressure-sensitive transfer record sheets (carbonless copy paper).

Cellulose Derivatives

Cellulose is derivatized to make both water-soluble gums and hydrophobic, thermoplastic polymers, the latter not being gums. All water-soluble cellulose derivatives vary in the extent of their derivatization, the ratio of substituent groups if more than one, and their average degree of polymerization (average molecular weight). In other words, each type of water-soluble cellulose derivative is itself a family of products, each member of which is tailor-made to have specific properties.

General Applications of Water-Soluble Gums

Adhesives: Billboard corrugating, remoistenable, wallpaper
Agriculture: Encapsulation of pesticide formulations
Biotechnology: Microbiological and cell culture media
Ceramics: Binders, glazes, slip agents
Construction: Gypsum spray plaster formulations, cement formulations, tape joint compounds

Cosmetics: Stabilizers, film formers, and emollients in creams and lotions
Explosives: Water-resistant gel formation
Processed foods: Bakery products, beverages, breakfast cereal products, confectionary products, dairy products, dietetic foods, dry mixes, noncook pie fillings, dry flavor powders, frozen foods, icings, jams, jellies, processed meat products, pet foods, preserves, noncook puddings, pourable salad dressings, sauces, spreads, syrups, toppings, whipped products
Metal working: Refractory coating formulations
Mining and minerals: Flocculation of particles, depression of slimes
Oil and gas production: Cementing, drilling fluids, fracturing fluids, enhanced oil recovery, pipeline cleaning, workover fluids, completion fluids, packer fluids
Paint: Latex paint thickener
Personal care products: Creams, denture adhesives, lotions, ointments, shampoos, toothpastes
Pharmaceutical and related applications: Dental impression material, granulating agents, plasma volume expander, sustained release agents, tablet binders, tablet coatings, tablet excipients, treatment for constipation, treatment for diarrhea, wound-healing fibers
Polymer production: Suspension and emulsion polymerization adjuncts
Printing: Ink thickeners
Textile: Printing pastes
Tobacco: Reconstituted sheet
Wildfire control: Rheology modifier, ignition retardation, foam stabilization

J. N. BeMiller, in S. S. Cho and M. L. Dreher, eds., *Handbook of Dietary Fiber*, Marcel Dekker, New York, 2001.

R. Lapasin and S. Pricl, *Rheology of Industrial Polysaccharides*, Chapman & Hall, New York, 1995.

R. L. Whistler and J. N. BeMiller, eds., *Industrial Gums*, 3rd ed., Academic Press, San Diego, 1993.

P. A. Williams and G. O. Phillips, in A. M. Stephen, ed., *Food Polysaccharides and Their Applications*, Marcel Dekker, New York, 1995.

JAMES N. BEMILLER
Purdue University

HAFNIUM AND HAFNIUM COMPOUNDS

Hafnium, Hf, is in Group 4 (IVB) of the Periodic Table as are the lighter elements zirconium and titanium. Hafnium is a heavy gray-white metallic element never found free in nature. It is always found associated with the more plentiful zirconium. The two elements are almost identical in chemical behavior.

Hafnium is obtained as a by-product of the production of hafnium-free nuclear-grade zirconium. Hafnium's primary use is as a minor strengthening agent in high-temperature nickel-base superalloys. Additionally, hafnium is used as a neutron- absorber material, primarily in the form of control rods in nuclear reactors.

PHYSICAL PROPERTIES

Hafnium is a hard, heavy, somewhat ductile metal having an appearance slightly darker than that of stainless steel. The physical properties of hafnium are summarized in Table 1. These data are for commercially pure hafnium, which may contain from 0.2 to 3% zirconium. Although a number of radioactive isotopes have been artificially produced, naturally occurring hafnium consists of six stable isotopes.

CHEMICAL PROPERTIES

Hafnium's aqueous chemistry is characterized by a high degree of hydrolysis, the formation of polymeric species, a very slow approach to true equilibrium, and the multitude of complex ions that can be formed. Partially reduced di- and trihalides have been produced by reducing anhydrous hafnium tetrahalides with hafnium metal.

Hafnium is a highly reactive metal. The reaction with air at room temperature is self-limited by the adherent, highly impervious oxide film which is formed. This film provides oxidation stability at room temperature and resistance to corrosion by aqueous solutions of mineral acids, salts, or caustics.

OCCURRENCE AND MINING

The primary commercial source is zircon (zirconium orthosilicate). Zircon sand is found in heavy mineral sand layers of ancient ocean beaches. Principal zircon-sand-producing countries are Australia, South Africa, the United States, and Ukraine. Zircon is always a coproduct from the mining of rutile and ilmenite mineral sands to supply the titanium oxide pigment industry. Baddeleyite a naturally occurring zirconium oxide, is available from South Africa and Russia.

Most of the heavy mineral sands operations in the world are similar. Typically the quartz sand overburden is bulldozed away to reach the heavy mineral sand layer, which usually has 2 to 8% heavy minerals. The excavation is flooded and the heavy mineral sands layer is mined by a floating dredge with a cutter-head suction. The sand slurry is pumped to a wet-mill concentrator mounted on a barge behind the dredge. Wet concentration using screens, cones, spirals, and sluices removes roots, coarse sand, slimes, quartz, and other light minerals. The tailings are returned to the back end of the excavation. Rehabilitation of worked-out areas is about a 10-year project, which includes replacing the overburden and topsoil to pre-existing levels and contours, and reestablishing the natural vegetation, usually from company-owned nurseries.

MANUFACTURE

Decomposition of Zircon

Zircon sand is inert and refractory. Therefore, the first extractive step is to convert the zirconium and hafnium portions into active forms amenable to the subsequent processing scheme. For the production of hafnium, this is done in the United States by carbochlorination. In Ukraine, fluorosilicate fusion is used. Caustic fusion is the usual starting procedure for the production of aqueous zirconium chemicals, which usually does not involve hafnium separation.

Table 1. Physical Properties of Hafnium

Property	Value
atomic number	72
atomic weight	178.49
density, at 298 K, kg/m^3	13.31×10^3
melting point, K	2504
boiling point, K	4903
specific heat, at 298 K, J/(kg·K)^a	144
latent heat of fusion, J/kg^a	1.53×10^5
electrical resistivity, at 298 K, $\Omega \cdot$m	3.37×10^{-7}
Hall coefficient, at 298 K, V·m/(A·T)	-1.62×10^{-12}
work function, J^a	6.25×10^{-19}
thermal conductivity, W/(m·K)	
at 273 K	23.3
at 1273 K	20.9
Young's modulus, at 293 K, GPa^b	141
shear modulus, at 293 K, GPa^b	56
Poisson's ratio, at 293 K	0.26
thermal expansion coefficient, linear,^{c} from 293 to 1273 K, 10^{-6}/K	6.1

^{a}To convert J to cal, divide by 4.184.
^{b}To convert GPa to psi, multiply by 145,000.
^{c}For random polycrystalline orientation.

Table 2. Physical Properties of Some Hafnium Compounds

Property	HfB$_2$	HfC	HfO$_2$	HfN	HfP	HfS$_2$	HfSe$_2$	HfSi$_2$	HfF$_4$	HfCl$_4$	HfBr$_4$	HfI$_4$
melting point, °C	3370	3830	2810	3330				1750	>968	432[a]	424[a]	449[a]
sublimation point, °C									968	317	322	393
specific gravity, g/cm^3												
theoretical	11.2	12.7		13.84	9.78			8.03				
measured	10.5	12.2	9.68			6.03	7.46	7.2			5.09	
resistivity at RT, μΩ·cm	8.8	37	>10^8	33			20	60			4.90	
color	gray	gray	white	gold		purple-brown	dark brown		white	white	white	yellow-orange
coefficient of thermal expansion, 10^{-6}	5.7	6.59	6.1	6.9								
hardness[b] kgf/mm^2	2900[c]	2300[d]	1050[e]	1640[f]				930[f]				
structure	hexagonal	face-centered cubic	monoclinic[f]	cubic	hexagonal	hexagonal	hexagonal	rhombic[g]	monoclinic	monoclinic	cubic	cubic
lattice parameters, nm												
a	0.3141	0.4640	0.51156	0.452	0.365	0.364	0.375	0.3677	0.957	0.631	1.095	1.176
b			0.51722		1.237	0.584	0.616	1.455	0.993	0.7407		
c	0.3740		0.52948					0.3649	0.773	0.6256		

[a] At 3.34 MPa (33 atm).
[b] 1 kgf/mm^2 = 9.8 MPa.
[c] Vicker's hardness.
[d] Knoop hardness.
[e] Diamond pyramid hardness (DPH), 2 kg.
[f] Microhardness, 50 gf/mm^2 = 460 kPa.
[g] Tetragonal above 1,600°C.

1219

Separation of Hafnium

Many methods have been proposed for the separation of hafnium and zirconium; three different industrial methods are in use: liquid–liquid extraction, molten salt distillation, and fluorozirconate crystallization.

Reduction

Hafnium oxide can be reduced using calcium metal to yield a fine, pyrophoric metal powder.

Refining

Kroll-process hafnium sponge and electrowon hafnium do not meet the performance requirements for the two principal uses of hafnium metal. Further purification is accomplished by the van Arkel-de Boer (ie, iodide bar) process and by electron beam melting.

HEALTH AND SAFETY FACTORS

High surface-area forms of hafnium metal such as foil, fine powder, and sponge are very easily ignited, and fine machining chips can be pyrophoric. Most hafnium compounds require no special safety precautions, because hafnium is nontoxic under normal exposure. Acidic compounds such as hafnium tetrachloride hydrolyze easily to form strongly acidic solutions and to release hydrogen chloride fumes, and these compounds must be handled properly. Whereas laboratory tests in which soluble hafnium compounds were injected into animals did show toxicity, feeding test results indicated essentially no toxicity when hafnium compounds were taken orally.

HAFNIUM COMPOUNDS

Most hafnium compounds have been of slight commercial interest, aside from intermediates in the production of hafnium metal. However, hafnium oxide, hafnium carbide, and hafnium nitride are quite refractory and have received considerable study as the most refractory compounds of the Group 4 (IVB) elements. Physical properties of some of the hafnium compounds are shown in Table 2.

D. J. Cardin, M. F. Lappert, and C. L. Ralston, *Chemistry of Organo-Zirconium and Hafnium Compounds*, Haisted Press, Division of John Wiley & Sons, Inc., New York, 1986.

J. B. Hedrick, "Zirconium and Hafnium," *Mineral Commodity Summaries*, U.S. Geological Survey, Reston, Va., Jan. 2004.

P. C. Wailes, R. S. P. Coutts, and H. Weigold, *Organometallic Chemistry of Titanium, Zirconium, and Hafnium*, Academic Press, Inc., New York, 1974. Excellent for organometallic chemistry of zirconium and hafnium.

RALPH H. NIELSEN
Teledyne Wah Chang
Corporation

N-HALAMINES

N-Halamines are inorganic and organic compounds in which oxidative halogen is attached to nitrogen. They have both research and industrial importance. Numerous *N*-fluoramines have been prepared and are used as selective fluorinating agents. Iodamines are the least stable and least studied. Only the chloro and bromo derivatives are of commercial importance, particularly for disinfection and free-halogen stabilization applications. Generally, the *N*-halamines can be considered as halogen release agents, and many find use in bleaching, disinfecting, and sanitizing applications. Others, such as the halogen derivatives of ammonia, are important because they are industrial process intermediates (monochloramine) or of significance in water treatment (mono-, di-, and trichloramine). Very recent developments concern the covalent attachment of *N*-halamine moieties to insoluble polymers, creating materials with considerable commercial potential for water disinfection and biocidal coatings.

PROPERTIES

The available halogen in an *N*-halamine is the percent of N–X halogen expressed in terms of equivalent molecular halogen; ie, it is a measure of the oxidizing capacity in terms of elemental halogen. For example, the available chlorine (av Cl_2) in trichloroisocyanuric acid (TCCA) is 91.5%. In water treatment, a distinction is made between free available chlorine (FAC) and combined available Cl_2 (CAC). Historically, FAC has referred to $HOCl + ClO^-$, and CAC to ammonia chloramines and other slightly hydrolyzed N–Cl compounds. FAC reacts with *N,N*-dimethyl-*p*-phenylenediamine (DPD), whereas CAC requires the presence of acidic KI. Although TCCA exists predominatly in the form of chloroisocyanurates in aqueous media, it analyzes essentially as FAC, because the hydrolysis reactions are so rapid. Thus, TCCA is a reservoir of HOCl, representing potential FAC.

The halogen in *N*-halamines is formally positive; ie, the oxidation state is + 1. *N*-Halamines hydrolyze, yielding hypohalous acid, which ionizes to a hypohalite ion, depending on pH. The extent of hydrolysis is a function of the polarity of the N-X bond. Bromo compounds hydrolyze to a greater extent than do chloro compounds.

Commercial products such as bleaches, dishwasher detergents, and hard-surface cleaners are formulated with alkaline ingredients such as polyphosphates, silicates, etc., so that chloramines initially hydrolyze to hypochlorite during use. Consumption of hypochlorite forms acidic compounds by reaction with soil, causing the pH to drop, resulting in formation of some HOCl, which increases the bleaching rate. In laundry bleaching this can result in a lowering of the tensile strength of the fabric. In bleaching applications, four variables are important including pH, temperature, contact time, and concentration. In commercial laundries, optimum pH, and temperature are in the 10.2–11.0 and 66–71°C ranges, respectively.

The disinfection efficacy of *N*-halamines is generally related to the extent of hydrolysis to hypohalous acid.

For example, NH_2Cl ($K_h \sim 10^{-12}$) is a poor bactericide compared to HOCl. By contrast, monochloroisocyanurate ($K_h \sim 10^{-6}$) exhibits good bactericidal properties. For optimum disinfection in swimming pools, the pH is maintained in the 7.2–7.6 range, where HOCl represents 69–47% of the FAC. By contrast, the HOBr fraction varies from 97 to 93%. Hypochlorous acid is a superior virucide to HOBr, but HOBr is more effective against certain algae.

In studies with organic *N*-chloramines, the following factors were shown to significantly influence antimicrobial activity: (*1*) the aliphatic chain length; (*2*) the degree of chlorination of the N atom; and (*3*) the nature of a positive charge.

Some chloramines and bromamines exhibit the lack of stability expected of compounds with bonds between two strongly electronegative elements. Decomposition kinetics can thus be rapid, energetic, and explosive in many cases; eg, NCl_3 and NBr_3. However, these compounds can be handled safely in dilute organic or aqueous solution. In general, bromamines are less stable than chloramines. A significant factor responsible for degradation of *N*-haloorganics is dehydrohalogenation across the carbon–nitrogen bond. Commercial organic *N*-bromamines and *N*-chloramines have good stability, which is a function of temperature, moisture, and impurities.

When chlorine is employed for outdoor swimming pool sanitation, it is relatively rapidly decomposed by sunlight. Isocyanuric acid stabilizes chlorine by formation of photostable chloroisocyanurates. By contrast, bromine is not effectively stabilized by isocyanuric acid.

INORGANIC CHLORAMINES AND BROMAMINES

Monochloramine

The most important of the ammonia halamines, monochloramine, is prepared by reaction of equimolar solutions of NH_3 and ClO^-. Monochloramine is the least odorous of the chloramines; the odor and taste threshold is 5 ppm. Chloramination; ie, *in situ* NH_2Cl formation, is increasingly employed to disinfect public water supplies to reduce trihalomethane (THM) formation.

Dichloramine

The least stable inorganic chloramine, dichloramine, has not been prepared in pure form. However, it has sufficient stability in dilute organic or aqueous solutions for determination of some physical and chemical properties. It has a pungent odor and can impart an odor or off taste to water at concentrations >0.8 ppm. Dichloramine is useful for preparation of diazirine. The formation of nitrosodimethylamine from it during chlorination has also been recently studied.

Trichloramine

Nitrogen trichloride, trichloramine, the only stable pure ammonia halamine, is a shock-sensitive dense yellow liquid (bp 71°C) with a volatility similar to chloroform.

Trichloramine has a pungent odor and is a lachrymator. It is the most irritating of the chloramines and can impart an odor or off taste to water at concentrations above only 0.02 ppm. Dilute solutions are relatively stable when protected from light and volatilization. Trichloramine has been used as a bleaching agent for flour, in the manufacture of paper, and as a fungicide for treatment of fruit.

Monobromamine

In organic solvents monobromamine is dark violet. The solutions are relatively unstable. In aqueous media, the decomposition increases with pH and Br/N mol ratio. Monobromamine disproportionates to dibromamine.

Dibromamine

Dibromamine can be prepared in ether by reaction of Br_2 with a slight excess of NH_3. The solution has a strawberry-yellow color and a sharp, irritating odor. Although stable at −70°C, it decomposes rapidly at ≥0°C. Dibromamine is less stable than NH_2Br.

Tribromamine

Pure solid nitrogen tribromide is deep red and explodes even at −100°C. Formation of NBr_3 in aqueous media is favored by lower pH and an excess of Br to N. Tribromamine is more stable than dibromamine.

Sulfamates and Imidosulfonates

Sodium, potassium, and bromine analogues of *N*-chlorosulfamic acid, $ClHNSO_3H$, and *N,N*-dichlorosulfamic acid, Cl_2NSO_3H, have been prepared, and the kinetics of chlorination of sulfamic acid have been studied. Dichlorosulfamate is relatively stable in dilute buffered aqueous solution, but is decomposed by excess av Cl_2, forming NCl_3 and H_2SO_4. Sulfamic acid was once used as a stabilizer for av Cl_2 in swimming pools but, its use was discontinued because of the poor bactericidal properties of mono- and dichlorosulfamate. *N*-Chlorosulfamates are useful in dishwashing compositions, textile bleaching, and vat or sulfur dyeing. *N,N*-Dichlorosulfamate can be used as a bleach and a disinfectant. *N*-Chloroimidodisulfonates can be used for fabric bleaching and stain removal.

Other Inorganic Compounds

N-Halosulfinylamines, O=S=NX, are thermally stable liquids at room temperature but react explosively with water. *N,N'*-Dibromosulfurdiimide is shock sensitive. *N*-Chloroimidodisulfuryl fluoride, $(FSO_2)_2NCl$, rearranges photochemically to a tetrasubstituted hydrazine and adds to unsaturated molecules such as olefins, CO, and cyanogen halides. *N*-Chloroimidosulfuryl fluoride, $F_2S(O)=NCl$, adds photochemically to olefins. Pentafluorosulfanyl-*N,N*-dichloramine, SF_5NCl_2, bp 64°C, is shock sensitive, unstable at 80°C, and hydrolyzes slowly. In the presence of SF_5Cl it photolyzes to the novel hydrazine $(SF_5)_2NN(SF_5)_2$.

ORGANIC CHLORAMINES AND BROMAMINES

Organic chloramines and bromamines can be broadly classified as aliphatic, aromatic, and heterocyclic. N-Halamines are versatile reagents that react with a variety of substrates; add to olefins and acetylenes, providing routes to cyclic compounds and can also cleave certain C–C, C–N, and C–O bonds, and act as aminating, halogenating, dehydrohalogenating, and oxidizing agents. They are useful in the preparation of many types of compounds.

Aliphatic Compounds

The aliphatic compounds of organic chloromines and bromamines can be classified as amines, amides, ureas, cyanamides and their derivatives, amino acids, carbamates, sulfonamides and other aliphatic compounds.

Aromatic Compounds

The aromatic compounds can be divided into sulfonamidates and sulfonamides and other aromatic compounds.

Heterocyclic Compounds.

The heterocyclic compounds consist of chlorinated glycolurils developed in the 1950s and 60s for protection against chemical agents and as bleaches, disinfectants, and foliage protectants. Chlorinated hydantoins are no longer used in home laundering because of changing needs of synthetic fabrics but used to a small extent in commercial laundries where temperatures of $\sim 70^{\circ}$C occur.

Imidazolidinones have the potential for water disinfection and in hard surface cleaners Isocyanurates. Dichloroisocyanuric acid (DCCA, HCl$_2$Cy, where Cy = isocyanurate anion) forms various simple salts and dihydrates. A number of double salts have also been prepared.

Among the melamines trichloromelamine is widely used for sanitation in the food and beverage industry and by the U.S. military kitchen services.

The oxazolidinone, 3-chloro-4,4-dimethyl-2-oxazolidinone, has been extensively evaluated as a disinfectant. Its disinfection effectiveness is significantly lower than that of hypochlorite or chloroisocyanurate, but it may find use in applications for which kill time is not of primary importance; eg, cooling towers. A number of bromine analogues have also been prepared, eg, 3-bromo-4,4-dimethyl-2-oxazolidinone a better disinfectant.

The succinimide, N-chlorosuccinimide, 1-chloropyrrolidine-2,5-dione, is a white solid with a slight chlorine odor, a mp of 150–151°C, and a solubility in water of 1.4% at 25°C used in organic synthesis for highly selective oxidation of primary and secondary alcohols to carbonyl compounds, providing improved synthesis of prostaglandins, and in conversion of allylic and benzylic alcohols to halides. N-Bromosuccinimide is especially useful in organic synthesis for allylic bromination, aromatic ring and side-chain bromination, for oxidation, for dehydrohalogenation, and in commercial production of cortisone and vitamin D$_3$.

Other heterocyclic compounds are useful as laundry bleach, reagents for organic synthesis, as disinfectants bactericide, impregnating clothing for protection against chemical agents such as mustard gas. Used also in organic synthesis for oxidation of alcohols to carbonyl compounds and hydrazo compounds to azo compounds. 1,3-Dihalouracils have utility as pesticides, fungicides, bleaching, and sanitizing agents. N-Chloro- and N-bromopolymaleimides are useful as halogenating agents.

N-HALAMINE POLYMERS

Probably the topic concerning N-halamine chemistry that has received the most attention during the past decade, and the one with the greatest commercial potential in the areas of water disinfection and biocidal coatings, is the derivatization of polymeric materials with N-halamine functional groups.

The polymer-bound N-chloro-N-sodiobenzenesulfonamidates and its derivatives are useful in water disinfection and in removal of cyanide from water. Recently, poly(styrene-co-divinylbenzene) has been functionalized with a hydantoin moiety that can be mono- or dichlorinated before use in an in situ water disinfection application. Possibly the most useful feature of the new polymer is that upon loss of the bound oxidative halogen, regeneration is possible numerous times, simply by exposing the polymer in its cartridge filter to aqueous free chlorine (dilute bleach).

For the textile industry, cellulose has been functionalized with an N-chlorohydantoin moiety to render it biocidal. Grafting techniques are also being employed for cellulose and other fibers to functionalize with a biocidal N-halamine moiety. The biocidal textiles produced are effective against both Gram positive and Gram negative bacteria in contact times ranging from 2 to 30 min.

In other work, an N-halamine-functionalized elastomer has been produced from a poly(styrene-butylene) copolymer for the purpose of creating biocidal rubber gloves and tubing. Several N-halamine copolymer coating materials have been prepared for use in grafting to surfaces and for polyurethane paints. A very recent development is the synthesis of 3-triethoxysilylpropyl-5,5-dimethylhydantoin, which is soluble in aqueous alcohol solutions and can be bonded to a variety of surfaces, such as cellulose, glass, ceramics, and paint. Upon chlorination in situ, the surfaces become biocidal, inactivating pathogens in minutes.

ECONOMIC ASPECTS

The overall growth in chloroisocyanurates the period 1999–2004 was 3.5%/year. The U.S. consumption of bromochloro- and dibromo-dimethylhydantoins, tetrachloroglycoluril, and other specialty N-halamines, is small. The consumption of polymeric N-halamine materials is expected to become significant during the current decade.

HEALTH AND SAFETY FACTORS

Chloramines and bromamines react with moisture, releasing potentially corrosive, toxic, and explosive gases, and should thus be stored, under dry conditions at moderate temperatures, segregated from incompatible materials. Because they are highly reactive, they should not be mixed with other materials such as acids, bases, reducing agents, oxidizing agents, organic compounds, ammonium compounds, etc, since vigorous reactions can occur, accompanied by fire and even explosions, liberating large amounts of heat and potentially toxic gases.

N-Halamines are irritating to the skin, eyes, and mucous membranes. However, they are nonirritating under use conditions in dilute aqueous solution. Even though there have been suggestions that cyanuric acid and the chloroisocyanurates could be carcinogens, recent laboratory studies may refute these suggestions. Monochloramine has been shown to be a weak mutagen; its use in drinking water is under review by the EPA. It has been suggested that free chlorine and N-chloramines are naturally produced DNA repair inhibitors, and the carcinogenic potential of chlorinated water containing free chlorine and chloramines has been discussed. Finally, there is little published information available concerning the degradation of N-halamines in the environment, although an interesting study of the enzymatic degradation of substituted hydantoins to amino acids has been reported.

USES

Monochloramine is used in water treatment and as an intermediate in the manufacture of hydrazine. Chloroisocyanurates are employed primarily for sanitation in swimming pools and spas. They are also used in hard surface cleaners, laundry products, as toilet bowl cleaners, and as shrink-proofing agents in wool finishing. TCCA is the principal product in pool and spa use, SDCC in nonpool/spa applications. Dichlorodimethylhydantoin is employed as a bleaching agent in industrial and institutional cleaning products. Bromochlorodimethylhydantoin is used primarily as a sanitizer in spas and to a smaller extent in swimming pools. Industrial water treatment applications include cooling towers, air washers, pasteurizers, and paper mills. Small amounts of bromochlorodimethyl hydantoin and tetrachloroglycoluril are used as components of Ca(OCl)$_2$ tablets called Sanuril that are employed in wastewater treatment. N-Chlorodimethyloxazolidinone, formed in situ from av Cl$_2$ and dimethyloxazolidinone, is employed as an algistat in cooling towers and swimming pools. Trichloromelamine is employed as a disinfectant; eg, in restaurants and the U.S. military kitchen services. N-Halamines such as N-bromo- and N-chlorosuccinimide, N-bromocaprolactam, N,N-dibromo- and N,N-dichlorodimethylhydantoin, chloramine-T, N,N-dichlorourethane, dibromo- and trichloroisocyanuric acids, etc., are employed as selective reagents for halogenation, oxidation, and other transformations in research, in analysis, and in small-scale production of specialty chemicals, pharmaceuticals, flavors, fragrances, etc. The polymeric N-halamines are being used for odor removal in cutting oils and will soon find use in potable water disinfection and in biocidal coatings for a variety of applications.

"Chlorinated Isocyanurates," in Chemical Economics Handbook, Stanford Research Institute International, Stanford, Calif., Feb. 2001.

R. W. Lowry, Pool Chlorination Facts, 2003, Lowry Consulting Group, Jasper, Ga., p. 118.

J. C. Morris and R. A. Isaac, Water Chlorination; Environmental Impact and Health Effects, Vol. 4, Ann Arbor Science, Ann Arbor, Mich., 1983.

J. L. S. Saguinsin and J. C. Morris, in J. D. Johnson, ed., Disinfection; Water and Wastewater, Ann Arbor Science, Ann Arbor, Mich., 1975, Chapt. 14.

S. DAVIS WORLEY
Auburn University

JOHN A. WOJTOWICZ
Olin Corporation, Retired

HALOGEN FLUORIDES

The halogen fluorides are binary compounds of bromine, chlorine, and iodine with fluorine. Of the eight known compounds, only bromine trifluoride, chlorine trifluoride, and iodine pentafluoride have been of commercial importance. The halogen fluorides are powerful oxidizing agents; chlorine trifluoride approaches the reactivity of fluorine.

The halogen fluorides offer an advantage over fluorine in that they can be stored as liquids in steel containers and, unlike fluorine, high pressure is not required. Bromine trifluoride is used as an oxidizing agent in cutting tools used in deep oil-well drilling, whereas chlorine trifluoride is used to convert uranium to UF$_6$ in nuclear fuel processing. Except for iodine pentafluoride, the halogen fluorides have no commercial importance as fluorinating agents.

PHYSICAL PROPERTIES

The physical properties of the halogen fluorides are given in Table 1.

CHEMICAL PROPERTIES

Reactions with Metals

All metals react to some extent with the halogen fluorides, although several react only superficially to form an adherent fluoride film of low permeability that serves as protection against further reaction. This protective capacity is lost at elevated temperatures. Hence, each metal has a temperature above which it continues to react. Metals that form no protective fluoride film react readily with the halogen fluorides. The rapid reaction of ClF$_3$ and BrF$_3$ with metals is the basis of the commercial use in cutting pipe in deep oil wells. In this application, the pipe is cut by the high-temperature reaction of the halogen fluoride and the metal.

Table 1. Physical Properties of the Halogen Fluorides

Property	BrF	BrF$_3$	BrF$_5$	ClF	ClF$_3$	ClF$_5$	IF$_5$	IF$_7$
boiling point, °C	20	125.7	40.9	−100.1	11.75	−13.1	102	5.5
melting point, °C	−33	8.8	−60.6	−155.6	−76.3	−103	8.5	4.5
liquid density at 25°C, g/mL		2.803	2.463^a	1.620^a	1.825^a	1.790^b	3.252	2.669
critical temperature, °C					154.5	142.6		
$-\Delta H_f$ (g) at 25°C, kJ/mol^c	58.5	255.4	443.9	56.4	164.5	254.6	839.3	961.0
$-\Delta G_f$ (g) at 25°C, kJ/mol^c	73.6	229.1	351.5	57.7	124.4	163.0	771.6	841.4
heat of vaporization, kJ/mol^c		42.8	30.6	20.1	27.50	22.21	35.92	24.7
E_{diss}, kJ/mol^c	254 260			253	105 160			122
heat of fusion, kJ/mol^c		12.01	5.66		7.60		11.21	
specific heat, gas, J/(mol · K)^c	32.9	66.5^d		32.0	65.2		99.1	136.3
specific conductivity, liquid, at 25°C, W · cm		8.0 × 10^{-3}	9.1 × 10^{-8}	1.9 × 10^{-7f}	4.9 × 10^{-9}	1.25 × 10^{-9f}	5.4 × 10^{-6}	10^{-9}

^{a} At boiling point.
^{b} At 20°C.
^{c} To convert J to cal, divide by 4.184.
^{d} The specific heat of the liquid is 124.5 J/(mol · K) (29.8 cal/(mol · K)).
^{e} At 145 K.
^{f} At 256 K.

Reactions with Nonmetals

Few elements withstand the action of interhalogen compounds at elevated temperatures, and many react violently at or below ambient temperatures. In general, reactions of halogens and halogen fluorides yield mixtures. Halogen fluorides react with sulfur, selenium, tellurium, phosphorus, silicon, and boron at room temperature to form the corresponding fluorides.

Reactions with Inorganic Compounds

In an investigation of the reactions of BrF$_3$ with oxides, little or no reaction was found with the oxides of Be, Mg, Ce, Ca, Fe, Zn, Zr, Cd, Sn, Hg, Th, and the rare earths, whereas the oxides of Mo and Re formed stable oxyfluorides.

Water reacts violently with all halogen fluorides. In addition to HF, the products may include oxygen, free halogens (except for fluorine), and oxyhalogen acids.

Salts of halides other than fluorides react with halogen fluorides to produce the corresponding metal fluoride and release the free higher halogen. Filter paper moistened with KI solution darkens readily in the presence of ClF$_3$ and the bromine fluorides. This serves as a sensitive detector for leaks in equipment containing these halogen fluorides.

Reactions with Organic Compounds

Most organic compounds react vigorously, exhibiting incandescence or even explosively with ClF$_3$ and BrF$_3$.

For this reason, only the less reactive iodine pentafluoride is used as a fluorinating agent to any extent.

SHIPPING, SPECIFICATIONS, AND ANALYTICAL METHODS

Bromine trifluoride is commercially available at a minimum purity of 98%. Free Br$_2$ is maintained at less than 2%. Other minor impurities are HF and BrF$_5$. Free Br$_2$ content estimates are based on color, with material containing less than 0.5% Br$_2$ having a straw color, and ca 2% Br$_2$ an amber-red color.

Shipping

Bromine trifluoride is classified as an oxidizer and poison by the DOT. It is shipped as a liquid in steel cylinders in quantities of 91 kg or less that are fitted with either a valve or plug to facilitate insertion of a dip tube.

Chlorine trifluoride is commercially available at 99% minimum purity and is shipped as a liquid under its own vapor pressure in steel cylinders in quantities of 82 kg per cylinder or less. It is also classified as an oxidizer and poison by the DOT.

Iodine pentafluoride is commercially available at a minimum purity of 98%. Iodine heptafluoride is the principal impurity and maintained at less than 2%. Free I$_2$ and HF are minor impurities. Iodine pentafluoride is shipped as a liquid in steel cylinders in various quantities up to the 1,350 kg cylinders. It is classified as an oxidizer and poison by DOT.

Handling

The halogen fluorides are highly reactive compounds that must be handled with extreme caution. The more reactive compounds, such as bromine trifluoride and chlorine trifluoride, are hypergolic oxidizers and react violently and sometimes explosively with many organic and inorganic materials at room temperature. At elevated temperatures, these cause immediate ignition of most organic substances and many metals.

Disposal

Moderate amounts of chlorine trifluoride or other halogen fluorides may be destroyed by burning with a fuel such as natural gas, hydrogen, or propane. The resulting fumes may be vented to water or caustic scrubbers. Alternatively, they can be diluted with an inert gas and scrubbed in a caustic solution.

HEALTH AND SAFETY FACTORS

The time-weighted average (TWA) concentrations for 8-h exposure to bromine trifluoride, bromine pentafluoride, chlorine trifluoride, chlorine pentafluoride, and iodine pentafluoride have been established by the ACGIH and NIOSH on a fluoride basis. No toxicity data have been reported on the other halogen fluorides, but all should be regarded as highly toxic and extremely irritating to all living tissue.

USES

Chlorine trifluoride is utilized in the processing of nuclear fuels to convert uranium to gaseous uranium hexafluoride. Chlorine trifluoride has also been used as a low-temperature etchant for single-crystalline silicon. Bromine trifluoride has been used in gas phase silicon etching. Bromine trifluoride and chlorine trifluoride are used in oil-well tubing cutters. Chemical cutter tools are commercially available for use in wells at any depth.

Iodine pentafluoride is an easily storable liquid source of fluorine having little of the hazards associated with other fluorine sources. It is used as a selective fluorinating agent for organic compounds.

H. C. Fielding, in R. E. Banks, ed., *Organofluorine Chemicals and their Industrial Applications*, Ellis Horwood Publishers, Chichester, UK, 1979.

R. J. Lewis, ed., *Sax's Dangerous Properties of Industrial Materials*, 10th ed., John Wiley & Sons, Inc., New York, 2000.

L. Stein, in V. Gutmann, ed., *Halogen Chemistry*, Vol. 1, Academic Press, Inc., New York, 1967, p. 133.

WEBB I. BAILEY
ANDREW J. WOYTEK
Air Products and Chemicals,
Inc.

HALOGENATED HYDROCARBONS, TOXICITY AND ENVIRONMENTAL IMPACT

Chlorinated biphenyls, chlorinated naphthalenes, benzene hexachloride, and chlorinated derivatives of cyclopentadiene are no longer in commercial use because of their toxicity. However, they still affect the chemical industry because of residual environmental problems. This article discusses the toxicity and environmental impact of these materials.

POLYCHLORINATED BIPHENYLS

Polychlorinated biphenyls (PCBs) typify halogenated aromatic hydrocarbons (HAHs), industrial compounds, or by-products that have been widely identified in the environment and chemical waste dump sites. PCBs were used in industry as heat-transfer fluids, organic diluents, lubricant inks, plasticizers, fire retardants, paint additives, sealing liquids, immersion oils, adhesives, dedusting agents, waxes, and as dielectric fluids for capacitors and transformers.

Chemistry and Environmental Impact

PCBs are synthesized by the chlorination of biphenyl and the resulting products are designated according to their percent (by weight) chlorine content. Over 600 million kg of commercial PCBs were produced in the United States, and the estimated worldwide production is approximately double this quantity.

The identification of PCB residues in fish, wildlife, and human tissues has been reported since the 1970s. The results of these analytical studies led to the ultimate ban on further use and production of these compounds.

Commercial PCBs: Toxic and Biochemical Effects

PCBs and related halogenated aromatic hydrocarbons elicit a diverse spectrum of toxic and biochemical responses in laboratory animals, depending on a number of factors including age, sex, species, and strain of the test animal and the dosing regimen (single or multiple). The toxic responses elicited by most PCB preparations are also observed for other classes of HAHs. They include a progressive weight loss not simply related to decreased food consumption and accompanied by weakness, debilitation, and ultimately death; ie, a wasting syndrome; lymphoid involution, thymic, and splenic atrophy with associated humoral and/or cell-mediated immunosuppression and/or associated bone marrow and hematologic dyscrasia; a skin disorder called chloracne accompanied by acneform eruptions, alopecia, edema, hyperkeratosis, and blepharitis resulting from hypertrophy of the Meibomian glands; hyperplasia of the epithelial lining of the extrahepatic bile duct, the gall bladder, and urinary tract; hepatomegaly and liver damage accompanied by necrosis, hemorrhage, and intrahepatic bile duct hyperplasia; hepatotoxicity also manifested by the development of porphyria and altered metabolism of porphyrins; teratogenesis, developmental and reproductive toxicity observed in several animal species; carcinogenesis as caused by PCBs in laboratory animals and primarily associated with their effects as promoters; and endocrine and reproductive dysfunction; ie, altered plasma levels of steroid and thyroid hormones with menstrual irregularities, reduced conception rate, early abortion, excessive menstrual and postconceptional hemorrhage, and anovulation in females; and testicular atrophy and decreased spermatogenesis in males.

The biochemical responses elicited by PCBs, which are also numerous, include the induction of CYP1A1 and CYP1A2 gene expression and the associated monooxygenase enzyme activities; ie, aryl hydrocarbon hydroxylase (AHH) and ethoxyresorufin *O*-deethylase (EROD), and several other cytochrome P-450-dependent monooxygenases; the induction of steroid-metabolizing enzymes, DT diaphorase, UDP glucuronosyl transferase, epoxide hydrolase, glutathione (*S*)-transferase, and δ-aminolevulinic acid synthetase; increased Ah receptor binding activity; decreased uroporphinogen decarboxylase activity; and decreased vitamin A levels.

Structure–Function Relationships

Since PCBs and related HAHs are found in the environment as complex mixtures of isomers and congeners, any meaningful risk and hazard assessment of these mixtures must consider the qualitative and quantitative structure–function relationships. Several studies have investigated the structure–activity relationships for PCBs that exhibit 2,3,7,8-tetrachlorodibenzo-*p*-dioxin (**1**) (TCDD)-like activity.

(**1**)

The data show that 3,3′,4,4′,5-pentachlorobiphenyl is the most toxic coplanar PCB congener and the 2,3,7,8-TCDD 3,3′,4,4′,5-pentachlorobiphenyl potency ratios are 66:1 (body weight loss, rat); 8.1:1 (thymic atrophy, rat); 10:1 (fetal thymic lymphoid development, mouse); 125:1 (AHH induction, rat); 3.3/1 (AHH induction, hepatoma H-4-II E cells, rat); and 100/1 (embryo hepatocytes, chick). Both the 3,3′,4,4′-tetra- and 3,3′,4,4′,5,5′-hexachlorobiphenyl congeners are considerably less toxic than 3,3′,4,4′,5-pentachlorobiphenyl and their relative potencies are highly variable.

Human Health Effects

Any assessment of adverse human health effects from PCBs should consider the route(s) of and duration of exposure; the composition of the commercial PCB products; ie, degree of chlorination; and the levels of potentially toxic PCDF contaminants. As a result of these variables, it would not be surprising to observe significant differences in the effects of PCBs on different groups of occupationally exposed workers.

Chloracne and related skin problems have been observed in several groups of workers and it was suggested that the air concentrations of commercial PCBs >0.2 mg/m³ were associated with this effect.

The effects of occupational exposure to PCBs on the concentrations of several serum clinical, chemical, and hematological parameters have been reported. Mildly elevated SGOT and γ-glutamyl transpeptidase (GGTP) suggest some liver damage and induction of hepatic monooxygenase enzymes; these results are similar to those observed in animal studies. A relatively high incidence of pulmonary dysfunction in capacitor-manufacturing workers has been reported, with symptoms including coughing, 13.8%; wheezing, 3.4%; tightness in the chest, 10.1%; and upper respiratory or eye irritation, 48.2%. The pulmonary toxicity of PCBs in laboratory animals has not been widely reported.

It is apparent from most reports that workplace exposure to relatively high levels of PCBs results in limited and moderate toxicity in humans. These toxic symptoms appear to be reversible after exposure to PCBs is terminated, and this is accompanied by a decline in serum levels of PCBs.

Environmental exposures to PCBs are significantly lower than those reported in the workplace and are therefore unlikely to cause adverse human health effects in adults. However, it is apparent from the results of several recent studies on children that there was a correlation between *in utero* exposure to PCBs, eg, cord blood levels, and developmental deficits, including reduced birth weight, neonatal behavior anomalies, and poorer recognition memories.

POLYBROMINATED DIPHENYL ETHERS

Polybrominated diphenyl ethers (PBDEs) are synthesized by the bromination of diphenyl ether to give PBDE mixtures that are highly stable and resistant to chemical, thermal, and biological breakdown. Several commercial PBDE formulations have been used as flame retardants in electronic equipment, electrical appliances, and building materials. Recent studies have identified levels of PBDEs throughout the global ecosystem, including air, water, sediments, fish, wildlife, and humans. In contrast to many other halogenated aromatic pollutants, there is evidence that PBDE levels are increasing. The toxicology of PBDEs has not been extensively investigated and their impact on human health is unknown.

HYDROXYLATED PCBs

The widespread environmental contamination with DDT and PCBs triggered a concern for the potential adverse impacts of these persistent environmental compounds. In 1994 it was reported that hydroxylated PCBs were present in human serum and wildlife samples, and subsequent studies have identified hydroxy-PCB congeners in wildlife and humans. Most of the hydroxy-PCBs are metabolites of persistent higher chlorinated biphenyls. The estrogenic activity of most hydroxy PCBs is weak, and some of the congeners identified in humans exhibit both weak estrogen receptor agonist and antagonist activities. In contrast, many of the same hydroxy-PCBs are potent inhibitors of sulfotransferase activities; this response may indirectly increase estrogen levels by preventing sulfation and excretion.

POLYCHLORINATED NAPHTHALENES

Polychlorinated naphthalenes (PCNs) are halogenated aromatic hydrocarbons that are no longer produced. They can be synthesized by the chlorination of naphthalene. The commercial products were graded and sold according to their chlorine content (wt%) and used as waxes and impregnants (for protective coatings), water repellents, and wood preservatives.

Animal and Human Toxicology

The mammalian toxicology of PCNs has not been studied in detail; however, it is believed that these compounds elicit mixture- and structure-dependent biochemical and toxic responses resembling those reported for PCBs and other toxic HAHs.

There have been several reported accidental exposures to commercial PCNs. One of the earliest incidents, the poisoning of cattle, was first reported in 1941 in New York State and became known as X-disease because of its unknown etiology. Eventually it was traced to the use of PCNs as high-pressure lubricants in feed-pelleting machines, which resulted in contamination of the feed and ingestion of the PCNs by the animals. The symptoms exhibited by the cattle included a thickening of the skin referred to as hyperkeratosis, excess lacrimation and salivation, anorexia, depression, and a decrease in plasma vitamin A.

Human incidents have been reported in workers involved in the production or uses of PCNs. In humans the inhalation of hot vapors was the most important route of exposure, resulting in symptoms including rashes or chloracne, jaundice, weight loss, yellow atrophy of the liver, and in extreme cases, death.

LINDANE AND HEXACHLOROCYCLOPENTADIENE

Both lindane (**2**) and hexachlorocyclopentadiene (**3**) are halogenated hydrocarbons; unlike the PCBs and PCNs, they do not contain an aromatic ring.

(2)   (3)

Chemistry and Environmental Impact

Lindane is produced by the photocatalyzed addition of chlorine to benzene to give a mixture of isomers. Lindane

has been produced worldwide for its use as an insecticide and for other minor uses in veterinary, agricultural, and medical products.

The relatively high stability and lipophilicity of lindane and its global use pattern has resulted in significant environmental contamination by this hydrocarbon.

The highly reactive hexachlorocyclopentadiene is rapidly degraded in the environment and is not routinely detected as an environmental pollutant.

Animal and Human Toxicity

The acute toxicity of lindane depends on the age, sex, and animal species and on the route of administration. Some of the toxic responses caused by lindane in laboratory animals include hepato- and nephotoxicity, reproductive and embryotoxicity, mutagenicity in some short-term *in vitro* bioassays, and carcinogenicity The mechanism of the lindane-induced response is not known. Only minimal data are available on the mammalian toxicities of hexachlorocyclopentadiene.

The effects of occupational exposure to lindane have been investigated extensively. These studies indicated that occupational exposure to lindane resulted in increased body burdens of this chemical; however, toxic effects associated with these exposures were minimal and no central nervous system disorders were observed.

K. Ballschmiter, C. Rappe, and H. R. Buser, in R. D. Kimbrough and A. A. Jensen, eds., *Halogenated Biphenyls, Terphenyls, Naphthalenes, Dibenzodioxins and Related Products*, 2nd ed., Elsevier/North-Holland, Amsterdam, The Netherlands, 1989.

L. Hansen, in S Safe and O. Hutzinger, eds., *Polychlorinated Biphenyls (PCBs); Mammalian and Environmental Toxicology*, Vol. 1, Springer-Verlag Publishing Co., Heidelberg, Germany, 1987, p. 15.

O. Hutzinger, S. Safe, and V. Zitko, *The Chemistry of PCBs*, CRC Press, Boca Raton, Fla., 1974.

STEPHEN H. SAFE
Texas A & M University

HAZARD ANALYSIS AND RISK ASSESSMENT

The purpose of hazard analysis and risk assessment in the chemical process industry is to (*1*) characterize the hazards associated with a chemical facility; (*2*) determine how these hazards can result in an accident, and (*3*) determine the risk; ie, the probability and consequences of these hazards. The complete procedure is shown in Figure 1.

HAZARD IDENTIFICATION PROCEDURES

Methods for performing hazard analysis and risk assessment include safety review, checklists, Dow Fire and Explosion Index, Dow Chemical Exposure Index, what-if analysis, hazard and operability analysis (HAZOP), failure modes and effects analysis (FMEA), fault tree analysis, and event tree analysis.

SCENARIO IDENTIFICATION

An important part of hazard analysis and risk assessment is the identification of the scenario, or design basis by which hazards result in accidents. Hazards are constantly present in any chemical facility. It is the scenario, or sequence of initiating and propagating events, which makes the hazard result in an accident. Many accidents have been the result of an improper identification of the scenario.

Most hazard identification procedures have the capability of providing information related to the scenario. This includes the safety review, what-if analysis, HAZOP, failure modes, and FMEA, and fault tree analysis. Using these procedures is the best approach to identifying these scenarios.

SOURCE MODELING AND CONSEQUENCE MODELING

Once the scenario has been identified, a source model is used to determine the quantitative effect of an accident. This includes either the release rate of material, if it is a continuous release, or the total amount of material released, if it is an instantaneous release.

Once the source modeling is complete, the quantitative result is used in a consequence analysis to determine the impact of the release. This typically includes dispersion modeling to describe the movement of materials through the air, or a fire and explosion model to describe the consequences of a fire or explosion. Other consequence models are available to describe the spread of material through rivers and lakes, groundwater, and other media.

PROBABILITY

In order to complete an assessment of risk, a probability must be determined. The easiest method for representing failure probability of a device is an exponential distribution:

$$R(t) = e^{\mu t} \tag{1}$$

where $R(t)$ is the reliability, μ is the failure rate in faults per time, and t is the time.

Once the reliability is defined, the failure probability, $P(t)$, follows:

$$P(t) = 1 - R(t) = 1 - e^{-\mu t} \tag{2}$$

A considerable assumption in the exponential distribution is the assumption of a constant failure rate. Many real devices demonstrate a failure rate curve more like that shown in Figure 2. For a new device, the failure rate is initially high, owing to manufacturing defects, material defects, etc. This period is called infant

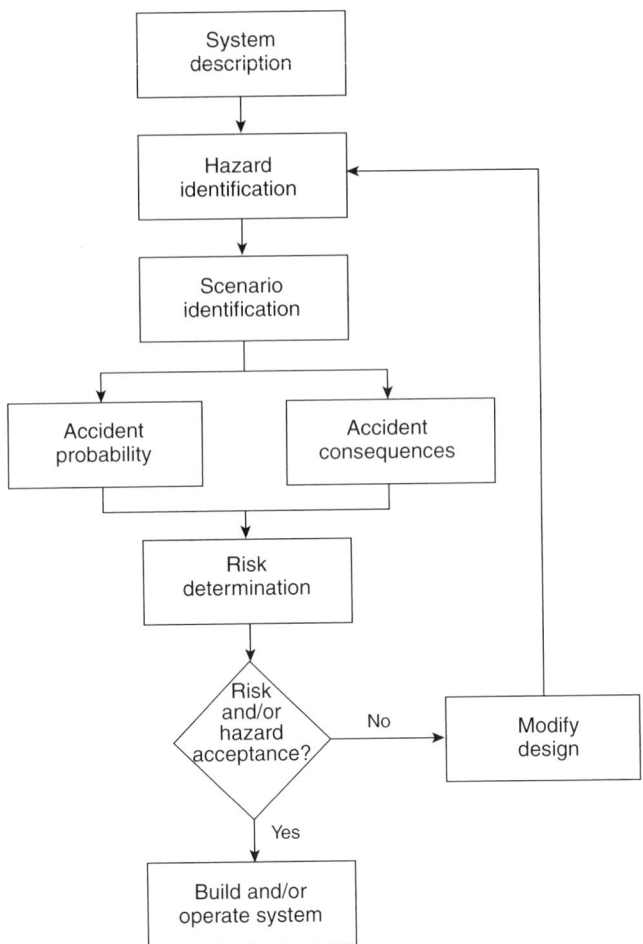

Figure 1. Flow chart representing the complete hazard identification and risk assessment procedure.

mortality. Following this is a period of a relatively constant failure rate. This is the period during which the exponential distribution is most applicable. Finally, as the device ages, the failure rate eventually increases.

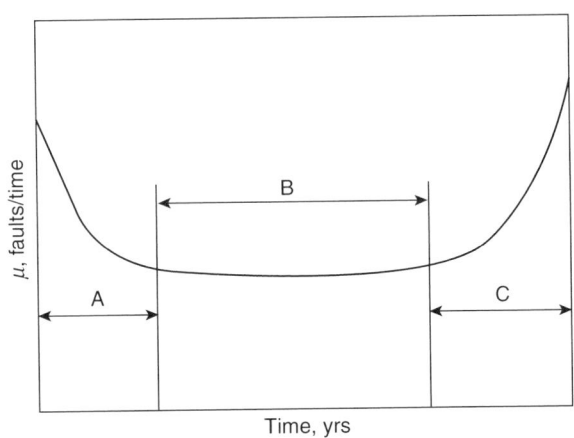

Figure 2. Failure rate curve for real components. A, infant mortality; B, period of approximately constant μ; and C, old age.

The next step is to develop a method to determine the overall reliability and failure probability for systems constructed of a variety of individual components. This requires an understanding of how components are linked. Components are linked either in series or in parallel. For series linkages, overall failure results from the failure of any of the components. For parallel linkages, all of the components must fail.

The computational technique for the two linkages is the followings: for series linkages the reliabilities of the individual components are multiplied together, for parallel linkages the failure probabilities are multiplied together. This method for combining the distributions assumes that the failures of the individual devices are independent of each other, and that the failure of one device does not strain an adjacent device causing it, too, to fail. It also assumes that devices fail hard; that is, the device is obviously failed and not in a partially failed state.

HAZARD ACCEPTANCE AND INHERENT SAFETY

The remaining step in the hazard identification and risk assessment procedure shown in Figure 1 is to decide on risk acceptance. For this step, few resources are available and analysts are left basically on their own.

A relatively recent concept that could have significant impact on future designs is that of inherently safer design. This basic principle states that what is not there cannot be blown up or leak into the environment. Thus, the idea is to avoid the hazard in the first place.

Inherently safer design is performed by three techniques. First, there is substitution. This means substituting a less hazardous material for the material in use and asking whether that flammable solvent is really necessary. Or is that toxic chemical the only possible reaction pathway? The second method for inherently safer design is attenuation; ie, operating the process at lower temperatures and pressures. The last, inherently safer design technique, is intensification. This means using much smaller inventories of hazardous raw and intermediate materials, and reducing process hold-up and inventories. These inventories are readily reducible if the management practices associated with the resources are improved.

D. A. Crowl and J. F. Louvar, *Chemical Process Safety: Fundamentals with Applications*, 2nd ed., Prentice-Hall, Englewood Cliffs, N.J., 2002.

Guidelines for Chemical Process Quantitative Risk Analysis, 2nd ed., American Institute of Chemical Engineers, Center for Chemical Process Safety, New York, 2000.

Guidelines for Consequence Analysis of Chemical Releases, American Institute of Chemical Engineers, Center for Chemical Process Safety, New York, 1999.

Guidelines for Technical Management of Chemical Process, Safety, American Institute of Chemical Engineers, Center for Chemical Process Safety, New York, 1989.

S. R. Hanna and P. J. Drivas, *Guidelines for Use of Vapor Cloud Dispersion Models*, 2nd ed., American Institute of Chemical Engineers, Center for Chemical Process Safety, New York, 1996.

DANIEL A. CROWL
Michigan Technological
University

HAZARDOUS WASTE INCINERATION

The U.S. EPA defines hazardous wastes as by-products of society that possess at least one of four characteristics: ignitability, corrosivity, reactivity, or toxicity. Properly designed and operated incinerators can safely destroy most hazardous organic liquid and solid wastes. The thermal decomposition and oxidation that occur in incinerators are usually ensured by exposing the waste to temperatures of 800°C or higher. Adequate mixing of the waste with air ensures the oxidation of organic compounds.

About 15% of total U.S. wastes generated annually are classified as hazardous waste and 9% of the hazardous waste is incinerated. About 15% of the municipal solid waste generated is disposed of in incineration systems.

U.S. regulations governing the design and operation of incinerators include the Resource Conservation and Recovery Act (RCRA), the Toxic Substances Control Act (TSCA), and the Clean Air Act (CAA). Many states are authorized to regulate hazardous waste and incinerator programs, and state regulations may be more stringent than federal.

OVERVIEW OF THE INCINERATION PROCESS

A general schematic of a hazardous waste incinerator and supporting equipment is shown in Figure 1. The key elements in the process include (1) preparation of the waste for feeding, (2) combustion, (3) air pollution control, and (4) disposal of liquid and solid residues.

TYPES AND OPERATION OF HAZARDOUS WASTE INCINERATORS

Most solid-waste incinerators are rotary kilns, and liquid-waste incinerators are commonly used in on-site facilities. The discussion below focuses on incinerators for the destruction of solid and liquid wastes.

Rotary Kiln Incinerators

Rotary kiln incinerators, predominate for government commercial, and on-site destruction of solid wastes. The rotary kiln is versatile because it can accept a wide variety of liquid and solid hazardous wastes. Any liquid capable of being atomized by steam or air can be incinerated, as well as heavy tars, sludge, pallets, drums, and filter cakes.

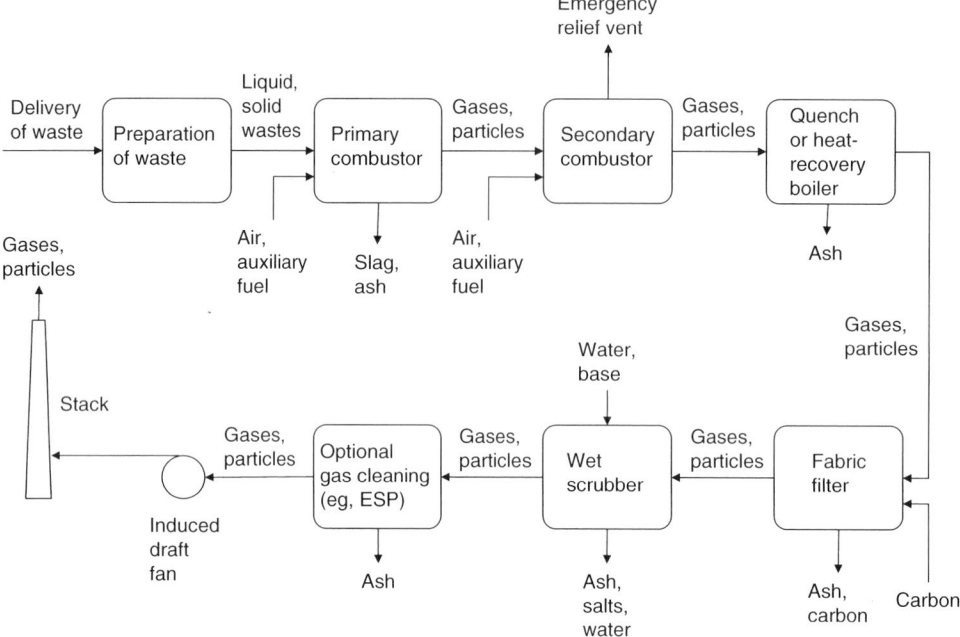

Figure 1. Schematic of a generic hazardous waste incineration process. Not all of the elements shown here are part of every incineration process.

This ability to accept diverse feeds is the outstanding feature of the rotary kiln. Fluxing agents, such as waste glass, are added to the feed of a slagging kiln to help reduce the melting point of the ash. The major components in such a system include the rotary kiln, a secondary combustor, a kiln burner, a fuel and liquid waste injection lances, a solids feed chute, a wet deslagger, and secondary combustor burners.

Fluidized-Bed Incinerator

Fluidized-bed incinerators are employed in the paper and petroleum industries, in the processing of nuclear wastes, and in the disposal of sewage sludge. These units are versatile and can be used for disposal of shredded solids, liquids, and gaseous combustible wastes.

The basic fluid-bed unit consists of a refractory-lined vessel, a perforated plate that supports a bed of granular material and distributes air, a section above the fluid bed referred to as the freeboard area, an air blower to move air through the unit, a "cyclone" to remove all but the smallest particulates and return them to the fluid bed, an air preheater for thermal economy, an auxiliary heater for start-up, and a system to move and distribute the feed in the bed. Auxiliary fuel may be added to the bed. Over a proper range of airflow velocities, usually 0.8–3.0 m/s, the solids become suspended in the air and move freely through the bed.

The fluidized bed has many desirable characteristics. Because of the movement of the particles, the bed operates isothermally and minimizes hot or cold regions. Large fluctuations in fuel quality are damped out because of the high thermal capacity of the bed. Solid particles are

reduced in size in the bed until they become small and light enough to be carried out of the bed. A bed material such as limestone may be chosen to react with acidic gases, such as SO_2, to remove them from the combustion gases. Because of the excellent air-to-solid contact, the fluid bed may be operated at low excess air rates. High heat-transfer rates allow large quantities of heat to be removed by a small heat-transfer area in the bed. Fluid beds are not effective in handling materials that contain components with a low ash-melting or softening temperature.

Fixed Hearth Incinerator

Fixed hearth incinerators, also known as controlled air or starved air incinerators, employ two combustion chambers. Auxilliary fuel and liquid wastes may be fed to both chambers. Waste is pumped or ram fed into the primary chamber and partially burned with 50–80% of the air required for complete combustion. Air is supplied at low velocities to the primary chamber through the grate on which the waste rests. Velocities and burning rates are keep low to prevent the entrainment of particles. The resulting vapors and gases—tars, carbon monoxide, and lighter hydrocarbons—are burned with added air in the secondary chamber. The air added to the secondary chamber is 100–200% of that required for complete combustion, and additional fuel or liquid waste may be added to complete the destruction process.

Fixed hearth units are generally lower in capacity and cost than rotary kilns. The ram feeding of the waste limits their capacity. Their small size and relatively low particulate control requirements make them attractive for smaller, on-site operations.

Moving-Hearth Incinerators

In this type of incinerator, sometimes called a stoker, the waste moves through the furnace on a moving grate that provides support for the waste, admits the underfire air through openings, transports the waste from the feed chute to the ash quench, and agitates the bed to bring a fresh charge to the surface and expose the waste to oxygen and flame. In roller hearths, the waste moves through the furnace in containers on rollers. Auxiliary fuel and liquid wastes may be fed above the grate.

As the waste enters the furnace, radiation from the hot combustion gases and the refractory furnace lining heats and dries it. As it is heated further, it pyrolyzes and ignites. Combustion takes place both in the solid, to burn out the residue, and in the gas space to burn out the pyrolysis products. Overfire air jets greatly assist mixing and combustion in the overfire air space.

Rotary Hearth Furnace

Rotary hearth furnaces are most often used for incineration of municipal and industrial sludge. Their main components are a refractory-lined shell, a central rotating shaft, a series of solid flat hearths, a series of rabble arms having teeth for each hearth, an afterburner (possibly above the top hearth), an exhaust blower, fuel burners, an ash removal system, and a feed system. The primary advantage of this system is the long residence time in the furnace, controlled by the speed of the central shaft and the pitch of the teeth on it that spiral the solids down to the next hearth.

Liquid Incinerators

Liquid incinerators typically provide 0.3–2 seconds gas-phase residence time at 800–1,200°C and 25–250% excess air. The furnace may be vertical or horizontal. Vertical furnaces are normally used for wastes that are high in salts. Nonaqueous wastes are fed through the burner, atomized, and burned in suspension. Aqueous wastes are also atomized and injected near the burner. The factors that govern the efficiency of waste destruction include atomization (mean drop size and size distribution), temperature, residence time, O_2 concentration and mixing.

Cement Kilns

Cement kilns are used to destroy 29% of the hazardous waste treated in the United States. The method of feeding hazardous waste depends on the temperature and residence time required for destruction and on the physical form of the waste. Liquids can be injected at the kiln burner by either blending them with a conventional liquid fuel or by injection through a separate lance. In wet-process kilns, solid wastes may be fed in the middle of the kiln through a hatch in the rotating kiln wall. In the newer, more efficient preheater-precalciner kilns, solid wastes are typically dropped in the feed end (cold end) of the kiln. Solids may also be propelled into the hot end of wet and preheater-precalciner kilns.

Lightweight Aggregate Kilns

Lightweight aggregate is a coarse aggregate used in the manufacture of lightweight concrete materials like concrete block, structural concrete, and pavement. The kilns are fired with coal, coke, natural gas, fuel oil, or liquid hazardous waste. Kilns that burn liquid waste are typcially fired without auxiliary fuels.

Halogen Acid Furnace

A halogen acid furnace is an industrial furnace designed to produced halogen acids from halogenated organic wastes. The wastes are burned to convert the chlorine that they contain to hydrogen chloride. The HCl is absorbed in water and recovered for reuse or neutralization.

AIR POLLUTION CONTROL AND EMISSIONS

The key air pollution control devices, which are downstream of the primary and secondary combustors, may vary from one plant to another, but the first piece of equipment is always a quenching operation or heat recovery boiler that lowers the temperature of the combustion products so that they can enter the particulate control device.

One such may be a baghouse or fabric filter, a series of bags suspended in a vertical orientation through which particulate-laden gases pass. Particulates are filtered out at the bag surface, and the bags are periodically shaken or rapped to allow the dust collected to fall into a hopper below. Particulate removal efficiencies are in the 95–99% range.

A wet scrubber, if used, follows the fabric filter. There are hundreds of different types of scrubbers, which can serve two purposes: to remove particles and to remove acid gases like HCl and SO_2. If the primary purpose of the wet scrubber is to control gaseous pollutants, a packed or plate column will serve.

Venturi scrubbers consist of a convergent section, a throat, and a divergent section. Particulate laden gases enter the convergent section, accelerate to approximately 130 m/s (425 ft/s), and are mixed with water via a spray system at the throat. Removal efficiencies are in the 50–99% range, depending on particulate size distribution and the throat pressure drop.

Both wet and dry electrostatic precipitators are generally used for polishing or small particulate removal. In dry electrostatic precipitators, a positive electrical charge is given to particles in the flowing combustion gases. The charged particles are then collected on oppositely charged plates. Collection efficiencies are a function of particle size, gas velocity, the uniformity of gas flows, both with time and across the field, the electrostatic field strength, and particulate matter electrical resistivity. Wet precipitators work on the same principles but the collection plates are wet.

Heavy metals entering an incinerator can leave the system with the bottom ash, the captured fly ash, or the exhaust gases. The fly ash is typically enriched with heavy metals relative to the entering solid waste. The fraction leaving with the exhaust gases can include metal vapors such as mercury and submicron particles that escape capture in the air pollution control devices.

Metals entering the incinerator as liquid streams usually leave the furnace in the fly ash.

The emissions of polychlorinated dibenzo-*p*-dioxins (PCDDs) and PCDF from incinerators are of interest to the public, scientists, and engineers. The proposed mechanism by which chlorinated dioxins and furans form has shifted from one of incomplete destruction of the waste to one of low temperature, downstream, catalytic formation on fly ash particles.

CHALLENGES IN OPERATION AND MONITORING

Some of the challenges that may be encountered in operating a hazardous waste incinerator include "puffing" associated with the batch feeding of highly volatile, high heating value waste; loss of flame; particle entrainment; encapsulation of unburned solid waste; steam explosions; slag buildup; brick and refractory failures; temperature measurement; stack sampling for mercury and other trace species; nonhomogeneous waste; and poor carbon burnout.

U.S. REGULATIONS AFFECTING DESIGN AND OPERATION OF INCINERATORS

U.S. regulations governing the design and operation of incinerators include the Resource Conservation and Recovery Act (RCRA), the Toxic Substances Control Act (TSCA), and the Clean air Act 1999 (CAA). Many states are authorized to regulate hazardous waste and incinerator programs, and state regulations may be more stringent than the federal.

The Resource Conservation and Recovery Act (RCRA, Subtitle C) regulates hazardous waste disposal. It identifies wastes as being hazardous if it falls into one or more of the following categories: ignitable; eg, having a flash point <60°C; corrosive; eg, having a pH <2 or >12.5; reactive; eg, reacting violently when mixed with water; toxic, as determined by the toxicity characteristic leaching procedure; or listed in Subtitle C as a hazardous waste from nonspecific sources, industry-specific sources, or as an acute hazardous or toxic waste.

W. T. Davis, ed., *Air Pollution Engineering Manual*, 2nd ed., John Wiley & Sons, Inc., New York, 2000.

R. C. Flagen and J. H. Seinfeld, *Fundamentals of Air Pollution Engineering*, Prentice Hall, Englewood Cliffs, N.J., 1988.

J. J. Santoleri, in J. J. Reynolds and L. Theodore, *Introduction to Hazardous Waste Incineration*, 2nd ed., Wiley-Interscience, New York, 2002.

U.S. Environmental Protection Agency, *Terms of the Environment*, http://www.epa.gov/OCEPAterms/hterms.html, last accessed May 2004.

Geoffrey D. Silcox
Joann S. Lighty
University of Utah

Melvin E. Keener
Coalition for Responsible Waste
Incineration

HEAT-EXCHANGER NETWORKS

The temperatures at which heat is being rejected by the hot process streams in a heat-exchanger network and demanded by the cold process streams control the opportunities for heat recovery. These temperatures are in these controlled by the design of the core process. Therefore, changes to the core process can result in increased heat recovery and reduced operating cost; ie the benefits are not restricted to operating cost, changes to the core of the process can also result in reductions in the capital cost of the network.

ECONOMICS OF HEAT RECOVERY

Consider a simple system involving just one hot and one cold stream. Both streams need to be taken from a set supply temperature to a set target temperature. These process objectives can be achieved by using hot and cold utilities or by using heat recovery supplemented by the use of hot and cold utilities. How does an engineer identify the optimum design? The logical way would be to set up the heat recovery system shown in Figure 1 and undertake cost optimization.

When there is no recovery (X equals zero), only the heater and cooler are present. The cold utility consumption equates with the heat rejected by the hot stream. The hot utility consumption equates with the heat demanded by the cold stream. Hence, the annual operating cost can be determined.

The sizes of the two utility exchangers can be estimated using the standard heat exchanger design equation. Assuming that the cost of each unit can be determined from a cost equation, the capital cost of the system can be determined. This cost number is multiplied by an annualization factor in order to set it on the same basis as the operating cost. Then the addition of the operating cost and the annualized capital cost yields the total annual cost of the system that provides the objective function to be minimized.

If heat recovery is now introduced (X is now finite), both heater and cooler loads are reduced by X and the total quantity of heat transferred within the system is reduced by X. This can mean that heat recovery leads to

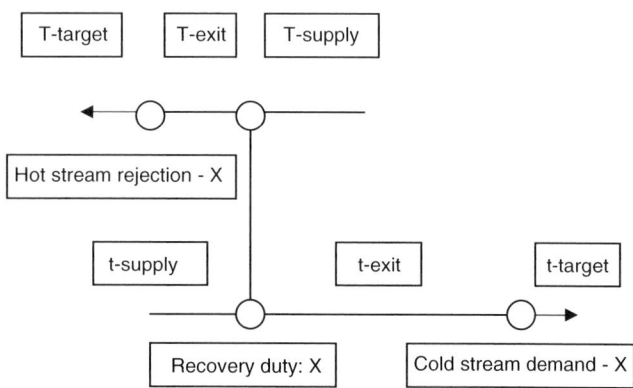

Figure 1. Simple heat recovery system.

a saving in capital. However, at low heat recovery levels the capital cost increases, because an additional heat exchanger is introduced. The reduced heater and cooler loads translate into reduced operating cost.

In the optimization the load on the heat recovery unit is named and the total annual cost of the system is determined.The minimum value is the economic heat recovery level.

Energy Demand

The process streams in the simple system described above are represented on a plot of temperature against heat load (Fig. 2). If the heat capacity flow-rate of a stream does not vary with temperature it appears as a straight line. The slope of the line is the reciprocal of the heat capacity flow-rate of the stream. If the heat capacity flow-rate varies with temperature, it appears as a curved line. When the lines representing the hot stream are superimposed such that the overlap equates with a heat recovery level X, the distance the hot stream overshoots the cold equates with the heat rejected to cold utility, and the distance the cold stream overshoots the hot equates with the heat demanded from the hot utility. The two lines make a close approach on the temperature axis. This is called the minimum temperature approach.

When the heat capacity flow-rate of the cold stream exceeds that of the hot stream (which therefore has a line of a larger slope) the close approach occurs at the cold end of the unit. When the hot stream has the larger heat capacity flow rate, the close approach occurs at the hot end of the unit.

It is common to find that the point of minimum temperature approach occurs partway along the temperature span covered by the process streams. This point has been termed the pinch point.

Process changes that increase the overlap of the two composites give rise to increased heat recovery. Changes that increase the vertical separation between the two composites result in increased temperature, driving forces for the heat exchangers, and thus result in capital cost savings.

Approximate Capital Cost

The capital cost of a heat exchanger network is controlled by a number of factors: the number of heat recovery matches made; the number of individual exchangers used in these matches; the size of the individual exchangers; he cost of installation of the exchangers; the cost of piping.

Algorithms successfully developed which allow the engineer to determine the minimum number of heat recovery matches and the number of exchangers required in these matches do not depend on the design of the exchanger network. They are applied ahead of the design process and are used to guide the design process.

The engineer should pay close attention to the number of heat recovery matches used in the system. The fixed cost released by the removal of a unit pays for a substantial increase in area of another unit.

Although the size of individual units cannot be determined ahead of design, an approximate determination of the total amount of surface the network will require can be made. A network must be structured inorder to achieve a design that is close to this value. Total network area does not fully reflect the exchanger costs, but is a guide to those costs.

Piping costs were separated from local installation cost because they can exceed exchanger costs. They are a function of network design and plant layout.

Network Area. The condition required for the minimization of the amount of heat transfer surface within an individual heat exchanger is pure counter-flow. Any disruption of this arrangement results in a waste of temperature driving force and the need for additional area in order to fulfill the required duty. The condition of pure counterflow extends to thermal systems. In the context of heat exchanger networks it requires the individual streams to be matched such that hot and cold contact temperatures align vertically between the composite curves.

The system can be divided up into a series of heat intervals. Within each interval both the hot and the cold composites are single straight lines (indicating constant heat capacity flow-rate). The hot and cold temperatures at entry to and exit from each interval are fixed by the positioning of the composites and can be estimated.

THE PINCH DESIGN METHOD

The costs associated with a heat-exchanger network have two components: operating costs (dominated by utility consumption) and capital cost. These costs can be combined to yield a total annual cost. In developing a design it is necessary to control costs as a network structure is progressively built up. The problem can be approached by controlling utility cost, controlling capital cost, or controlling total annual cost. The pinch design method uses a staged approach. First, is determined an economic heat recovery level. This sets the optimum minimum temperature approach. Then an initial network structure is developed, using a systematic procedure that both ensures that

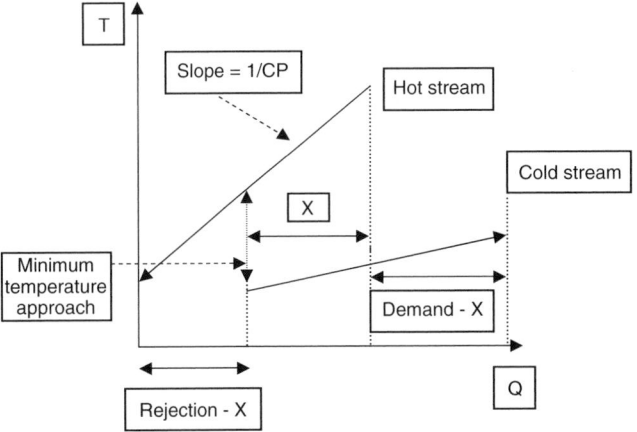

Figure 2. T-Q Plot of simple process system.

the minimum temperature approach constraint will be met and the network will use only the identified amount of hot and cold utility. Finally, this network structure is subjected to optimization aimed at minimizing its total annual cost.

The pinch design method is systematic and elegant but does have pitfalls. The procedure dictates which stream matches should be made but, sometimes these matches are impractical. No guidance is given on how alternatives should be identified. Also, the method is centered on the use of the grid diagram. This diagram does not contain information on where the streams originate and where they are going. The result can be a network design that incurs a great deal of expense in terms of piping and layout-related costs. The method is easy and straightforward to apply in problems of (typically) up to ten streams. However, some processes have in excess of forty hot and cold streams. Finally, the results of the optimization applied at the end of the design process are dependent upon the starting structure. Different structures yield different final designs. Most importantly, the presence of small and unnecessary design elements can prevent the identification of significant structural changes.

Subdividing a system at the pinch provides a means of controlling the minimum temperature approach.

To implement the pinch design method, proceed to divide the problem at the pinch, then next, develop the design for streams positioned below the pinch, making sure that no hot utility is used. Next, develop the design for streams positioned above the pinch, making sure that no cold utility is used. Finally, merge the two subnetworks, then subject the resultant structure to optimization.

SHORTCOMINGS OF THE PINCH DESIGN METHOD

The pinch design method, despite its simplicity and elegavce, contains a number of significant dangers for the unwary. The methodology has four major shortcomings: 1.) it ignores the practical aspects of exchanger design and application; 2.) it make any consideration of plant layout different. 3.) its application becomes more complex as the number of streams involved increases and 4.) the optimization result is dependent upon both the starting structure and the order in which heat recovery matches are eliminated.

In the individual synthesis steps, the division of the synthesis problem into two parts, can result in a need for more than one heat exchanger handling a given stream, which is not always practical or desirable. Next, in the development of network structure at the pinch, it is not always possible to split process streams. Then the designer is required to maximize the heat loads on a .Finally, having dealt with the matches in the pinch region, the designer is asked to finish the problem. Now some duties have a maximum as well as minimum temperature difference. After the pinch matches have been set and loads maximized, the temperature differences for the remaining streams can be so high that this constraint is infringed.

The solution to the above problems is systematic problem decomposition. Here the overall problem is broken down into a number of self-contained problems set by a consideration of both practical and layout issues. Since this process leads to a number of small problems, it avoids the difficulties inherent in applying the methodology to a large number of streams. And since it results in simple structures containing elements that have been subjected to optimization, the approach removes the need for complex final optimization.

PRACTICAL DESIGN CONSIDERATIONS

One problem associated with the use of shell-and-tube exchangers is that at close temperature approaches the designer needs to use a number of shells in series. In some instances the need for a series of shells can be removed through the use of a unit with an internal longitudinal baffle named an F shell. However, a better approach might be to use a plate-and-frame exchanger rather than a shell-and-tube unit.

Plate-and-frame heat exchangers have a major cost and weight advantage over shell-and-tube units, and in the context of network design they have a further distinct advantage. In Figure 3 one sees the plant arrangement necessary if shell-and-tube exchangers are used on a simple heat recovery/trim cooling system. The recovery unit requires four individual exchangers. The trim cooling requires two exchangers. Plate-and-frame exchangers can handle more than one hot and one cold stream in a given unit, however. This results in simpler plants and significant savings in piping costs. In Figure 4 a plate-and-frame exchanger (set up as a multistream exchanger) that can be used to satisfy the duty performed by the system shown in Figure 3 is. The benefit of a shown pure counterflow removes the need for exchangers in series in order to maintain a good temperature-driving force. The ability to multistream eliminates the need for two separate units.

Compact heat exchangers offer a large reduction in size and weight, a pure counterflow arrangement; and

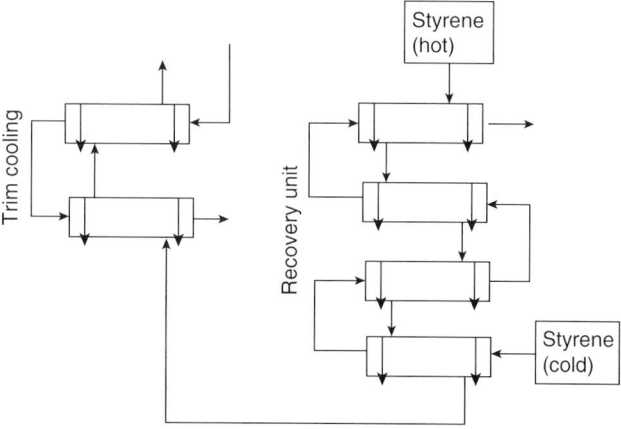

Figure 3. Shell-and-tube design for heat recovery/trim cooling system.

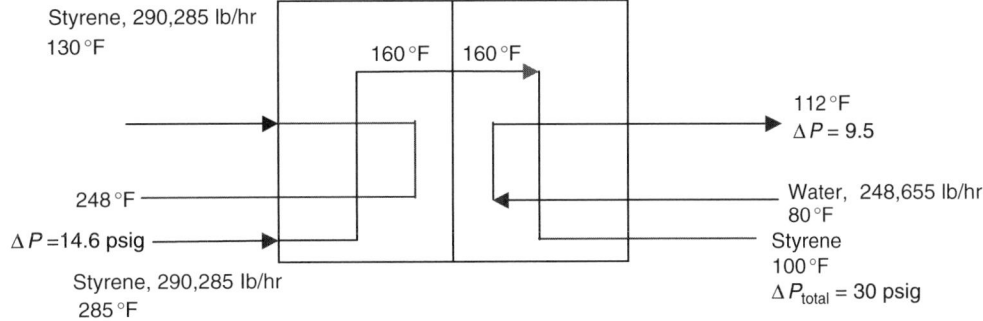

Figure 4. Multistream plate-and-frame design for heat recovery/trim cooling system.

an ability to handle more than one hot and one cold stream in a single unit.

The installed cost of a heat exchanger is very dependent upon exchanger weight and space requirements. Since the installed cost is up to four times the original purchase price, making decisions on the basis of purchase price alone is false economy. Many of the benefits relating to weight and space savings can be realized only if the decision to use a compact design has been made before the plant layout is fixed.

Current design procedures are based on the use of shell-and-tube technology. Applying the cost relationships for this technology results in much lower heat recovery than using costs for plate-and-frame exchangers. Since the structure required for a heat recovery network can change as the heat recovery level changes, the choice of exchanger technology affects network design.

Factors in the Design of a Heat-Exchanger Network

In the eagerness to maximize heat recovery, it is possible to overdo it. It must always be borne in mind that heat recovery is justifiable only to the extent it is economical. The cost of the energy recovered must always be balanced against the equipment cost to recover energy.

The heat exchanger network must be workable from a process viewpoint, especially for different operations. This means that various temperature constraints or requirements such as for pump-around streams must be fulfilled at all operating scenarios, as with different feedstocks. If a hot stream goes from a heat exchanger network to a reboiler, it has to be ensured that it is hot enough for proper reboiler operation under all operating conditions. In order to achieve these goals, partial or total bypassing of one or more heat exchangers has to be anticipated in advance.

Practical equipment operability and maintenance should be ensured. Stream velocities must be maintained sufficiently high so as to minimize fouling under all plant operation. In order to achieve this goal, heat exchangers have often to have multiple units in parallel so that at times of lower through put one or more of the multiple shells can be bypassed.

C. Haslego and G. T. Polley, *Chemical Engineering Progress*, 32–37 (Sept. 2002).

G. F. Hewitt and co-workers, *User Guide on Process Integration for the Efficient Use of Energy*, IChemE, Rugby UK, (1982).

Graham Polley
Mukherjee Rajiv
Consultants

HEAT PIPES

Heat pipes are used to perform several important heat transfer roles in the chemical and closely related industries. Examples include heat recovery, the isothermalizing of processes, and spot cooling in the molding of plastics. Heat pipes are highly efficient, two-phase heat-transfer devices with effective thermal conductivities of up to ~1000 times that of solid copper, depending on the application. As a result, the heat pipe can produce nearly isothermal conditions, making it an almost ideal heat transfer element. In another form, the heat pipe can provide positive, rapid, and precise control of temperature under conditions that vary with respect to time.

The heat pipe is self-contained, has no mechanical moving parts, and requires, in its basic form, no external power other than the heat that flows through it.

PRINCIPLES OF OPERATION

The heat pipe achieves its high performance through the process of two-phase heat transfer. It consists of three sections: evaporator, condenser, and adiabatic region. The evaporator section is mounted to the heat-generating components or submerged in a hot fluid, while the condenser is thermally coupled to a heat sink, radiator, or cold fluid. The adiabatic section allows heat to be transferred from the evaporator to the condenser with very small heat losses and temperature drops. Figure 1 presents a comparison of the operating principles of a basic heat pipe and a thermosyphon.

The unique aspect of the heat pipe lies in the means of returning the condensed working fluid from the condenser section to the evaporator. Condensate return is accomplished by means of a specially designed wick. The surface tension of the liquid is the active force that produces wick

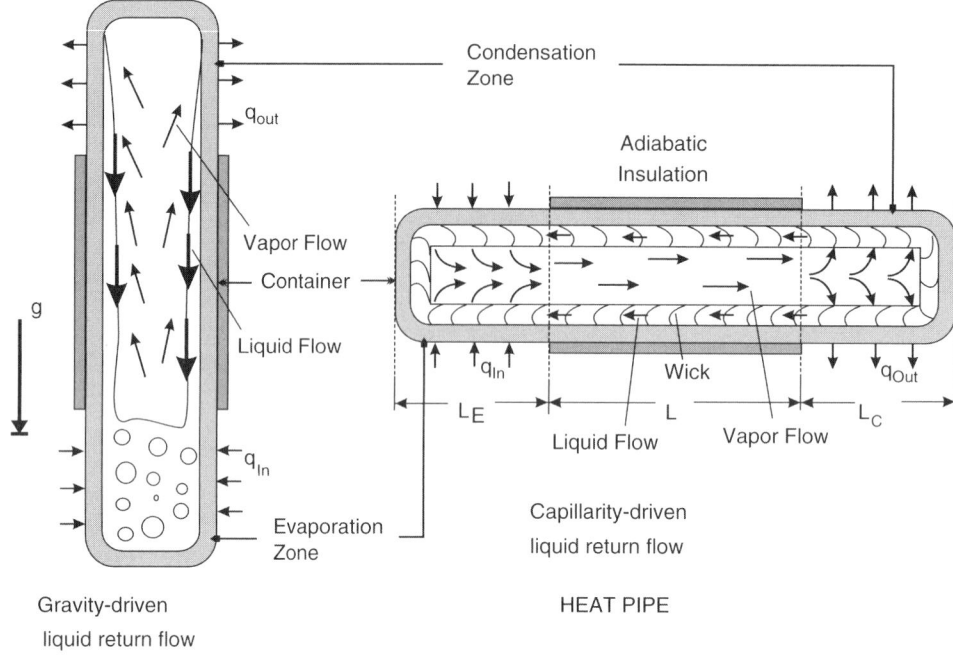

Figure 1. Comparison of the operating principles of a heat pipe and a thermosyphon.

pumping, which is a familiar process in lamp wicks and sponges. Using proper design, a substantial liquid mass flow rate can be sustained against the pressure head of the counterflowing vapor or even against a slight gravitational head. In applications where the heat source is below the heat sink, the condensate returns by gravity; ie, a heat pipe without a wick represents a conventional thermosyphon.

The heat pipe consists, then, of the following components: a closed, evacuated vessel (evacuation is required to establish a contaminant-free system and to prevent air or other gases from interfering with the desired vapor flow), a wick structure of appropriate design, and a working fluid at the desired operating temperature. Heat pipes may also include gas reservoirs (variable conductance/diode heat pipes) and liquid or gas traps (diodes).

In order for a heat pipe to operate, the maximum capillary pumping head, $(\Delta P_c)_{max}$, must be able to overcome the total pressure drop in the heat pipe, which consists of the pressure drop required to move the vapor from the evaporator to the condenser, ΔP_v, the potential head due to difference in elevation between the evaporator and the condenser, ΔP_g, and the pressure drop required to return the liquid from the condenser to the evaporator, ΔP_l. The basic condition for proper operation of a heat pipe can thus be expressed in the form:

$$(\Delta P_c)_{max} \geq \Delta P_v + \Delta P_g + \Delta P_l$$

Under this condition, there is liquid flow toward the evaporator, and heat can be transferred. If this condition is not met, the wick will dry out in the evaporator region and the heat pipe will cease to operate. The pressure difference in the vapor is a direct function of the mass flow rate and an inverse function of the cross-sectional area of the vapor space. The mass flow rate is related directly to the transferred power and inversely to the latent heat of vaporization. The gravitational head can be either positive or negative, depending on whether it aids or opposes the desired flow in the wick.

DESIGN FEATURES AND OPERATIONAL LIMITS

Heat pipes can be designed to operate over a very broad range of temperatures from cryogenic (less than$-243°C$) applications utilizing titanium alloy/nitrogen heat pipes, to high-temperature applications ($>2,000°C$) using tungsten/silver heat pipes. There are many factors to consider when designing a heat pipe: compatibility of materials, operating temperature range, power limitations, thermal resistances, and operating orientation.

Heat pipes can operate in the fixed conductance, variable conductance, or diode mode. The fixed conductance heat pipe can transfer heat in either direction and operates over broad temperature ranges but has no inherent temperature control capability. Constant conduction heat pipes allow isothermalization of shelves, radiators and structures; spread heat from high heat dissipating components; and conduct heat away from heat-generating devices.

The heat pipe has properties of interest to equipment designers. One is the tendency to assume a nearly isothermal condition while transporting useful quantities of thermal power. A typical heat pipe may require as little as one thousandth the temperature differential needed by a copper rod to transfer a given amount of power between two points.

A second property is the ability of the heat pipe to effect heat-flux transformation. As long as the total heat flow is in equilibrium, the fluid streams connecting the evaporating and condensing regions essentially are unaffected by the local power densities in these two regions. Thus the heat pipe can accommodate a high evaporative power density, coupled with a low condensing power density, or vice versa. The heat pipe can be used to accomplish the desired matching of power densities by simply adjusting the heat input and output areas in accordance with the requirements. Heat flux transformation ratios exceeding 12:1 have been demonstrated in both directions.

The third characteristic of interest grows directly from the first, ie, the high thermal conductance of the heat pipe can make possible the physical separation of the heat source and the heat consumer (heat sink). Heat pipes >100 m in length have been constructed and shown to behave predictably. Separation of source and sink is especially important in applications in which chemical incompatibilities exist.

The fourth characteristic, temperature flattening, makes use of all three of the preceding properties. The evaporation region of a heat pipe can be regarded as consisting of many subelements, each receiving heat and an influx of liquid working fluid and each evaporating this fluid at a rate proportional to its power input. Within limitations, each incremental unit of evaporation area operates independently of the others, except that all are fed to a common vapor stream at a nearly common temperature and pressure. The temperature of the elements is, therefore, nearly uniform. It can be seen that the power input to a given incremental area can differ widely from that received by other such areas. Under other circumstances, a nonuniform power profile would produce a nonuniform temperature profile. In the case of the heat pipe, however, uniformity of temperature is preserved; only the local evaporation rate changes. In this fashion, the heat pipe can flatten the very nonuniform power input profile from a flame, delivering heat to the sink with the same degree of uniformity as if the heat source were uniform. Another example is the use of a heat pipe to cool simultaneously, and to nearly the same temperature, a number of electronic components operating at different power levels.

The most important heat pipe design consideration is the amount of power the heat pipe is capable of transferring. Heat pipes can be designed to carry from a few watts up to several kilowatts.

The total thermal power transfer capability of a heat pipe is the product of the latent heat of vaporization and maximum mass flow rate of the fluid that can be sustained by the wick. Operation at greater power results in evaporation rates exceeding the rate of liquid return. The resulting dryness can lead to an uncontrolled rise in temperature in the uncooled section of the evaporator, and ultimate failure. The effect of gravity is similar. If the desired operation requires that the liquid flow be upward against gravity or another accelerating force, operation is affected adversely to the degree that it is a function of the lift height, liquid density, and the mass flow rate.

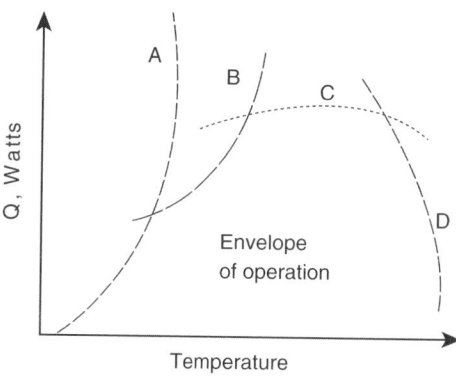

Figure 2. The axial heat flow, Q, versus operating temperature showing the envelope of heat pipe operating limits.

The maximum heat transport capability of the heat pipe is governed by several limiting factors. There are five primary limitations of heat transport inside heat pipe: viscous, sonic, flooding, capillary, and boiling limits. These heat transport limits are a function of the heat-operating temperature. The envelope of these limits is shown generically in Figure 2. Curve A represents limits associated with vapor flow, ie, either insufficient working fluid vapor pressure is available to transport vapor along the length of the heat pipe (viscous limit) or the vapor flow has reached the sonic velocity (sonic limit). Curve B represents the entrainment limit that occurs when friction with the outgoing vapor prevents the returning liquid from reaching the evaporator. Curve C, the wicking limit, occurs when the capillary pressure developed in the wick can no longer support the total pressure drop in the fluid flow path (includes liquid, vapor, and gravitational effects). Curve D is the boiling limit that occurs when vapor is generated within the capillary structure in an uncontrolled manner, much the same as film boiling.

SUITABLE WORKING FLUIDS AND SELECTION OF MATERIALS

In heat pipe design, a high value of surface tension is desirable in order to enable the heat pipe to operate against gravity and to generate a high capillary driving force. A high latent heat of vaporization is also desirable, to transfer large amounts of heat with minimum fluid flow and hence maintain low pressure drops within the heat pipe. The resistance to fluid flow could be minimized by choosing fluids with low values of vapor and liquid viscosities.

The operating lifetime of a given heat pipe is usually determined by corrosion mechanisms. A number of pairs of materials have long (thousands of hours) undegraded life when properly processed.

Several wick structures are in common use. First is a fine-pore [0.14–0.25 mm (100–60 mesh) wire spacing] woven screen that is rolled into an annular structure consisting of one or more wraps inserted into the heat pipe bore. Where high heat transfer in a given diameter is of paramount importance, a fine-pore screen is placed over longitudinal slots in the vessel wall. Where complex geometries are desired, the wick can be formed by powder metallurgy techniques.

The vessel in which a heat pipe is enclosed must be impermeable, to assure against loss of the working fluid or leakage into the heat pipe of air combustion gases or other undesired materials from the external environment. The vessel, as well as the wick, must be compatible with the working fluid. Where possible, the wick and vessel are made of the same material to avoid the formation of galvanic corrosion cells in which the working fluid can serve as the electrolyte.

T. P. Cotter, *Theory of Heat Pipes, LA-3246-MS*, Los Alamos Scientific Laboratory, University of California, Los Alamos, N.M., 1965.

T. P. Cotter, *Heat Pipe Startup Dynamics, LA-DC-9026*, Los Alamos Scientific Laboratory, University of California, Los Alamos, N.M., 1969.

A. Faghri, *Heat Pipe Science and Technology*, Taylor & Francis, Washington, 1995.

M. Ivanovskii, V. Sorokin, and I. Yagodkin, *The Physical Principles of Heat Pipes*, Translated by R. Berman and G. Rice, Clarendon Press, Oxford, 1982.

BORIS KOSOY
Odessa State Academy of
Refrigeration

HEAT TRANSFER

In order to select a proper heat exchanger for a given application, various factors such as pressure, temperature, size, fouling factor, and the use of toxic or corrosive fluids must be considered. These pressure and temperature requirements mainly dictate the type of heat exchanger selected. In general, for high pressures and temperatures, tubular heat exchangers that conform to safety regulations and manufacturing codes are used. For moderate pressures, small but very efficient plate heat exchangers can be employed.

There are three heat-transfer modes, ie, conduction, convection, and radiation, each of which may play a role in the selection of a heat exchanger for a particular application. The basic design principles of heat exchangers are also important, as are the analysis methods employed to determine the right size heat exchanger.

HEAT-TRANSFER THEORY

A heat exchanger is designed and built based on heat-transfer principles; thus, some understanding of these basic principles is essential to design or select a heat exchanger. Efficiency and economics may depend directly on how effectively fundamental heat-transfer principles are applied in the design of the heat exchanger.

Conduction Heat Transfer. When there is a temperature difference in a body, there is an energy transfer from the high temperature region to the low temperature region, a phenomenon called an energy transfer by conduction. Although conduction occurs in liquids and gases, the contribution to heat transfer is relatively small as compared to convection or radiation for these cases. In a solid such as a metal tube wall or a flat wall made of multicomponent materials, however, conduction is the dominant heat-transfer mode. In most conventional heat exchangers, heat transfer occurs between two fluids separated by solid walls, which are either a tube wall in tubular heat exchangers or a plane wall in plate heat exchangers.

Fourier's Law of Heat Conduction. The heat-transfer rate, Q, per unit area, A, in units of $W/m^2(Btu/(ft^2 \cdot h))$ transferred by conduction is directly proportional to the normal temperature gradient:

$$\frac{Q}{A} \sim \frac{dT}{dx} \qquad (1)$$

or in equation form the heat-transfer rate Q becomes,

$$Q = -kA \frac{dT}{dx} \qquad (2)$$

where the proportionality constant, k, is called the thermal conductivity of the material. The minus sign, required in equation 2 to ensure that the direction of the heat transfer is positive when the temperature gradient is negative, is necessary because thermal energy flows in the direction of decreasing temperature.

CONVECTION HEAT TRANSFER

Convective heat transfer occurs when heat is transferred from a solid surface to a moving fluid owing to the temperature difference between the solid and fluid. Convective heat transfer depends on several factors, such as temperature difference between solid and fluid, fluid velocity, fluid thermal conductivity, turbulence level of the moving fluid, surface roughness of the solid surface, etc. Owing to the complex nature of convective heat transfer, experimental tests are often needed to determine the convective heat-transfer performance of a given system. Such experimental data are often presented in the form of dimensionless correlations.

Convective heat transfer is classified as forced convection and natural (or free) convection. The former results from the forced flow of fluid caused by an external means such as a pump, fan, blower, agitator, mixer, etc. In the natural convection, flow is caused by density difference resulting from a temperature gradient within the fluid. An example of the principle of natural convection is illustrated by a heated vertical plate in quiescent air.

Newton's Cooling Law of Heat Convection

The heat-transfer rate per unit area by convection is directly proportional to the temperature difference between the solid and the fluid which, using a proportionality constant called the heat-transfer coefficient, h, becomes

$$Q = hA(T_{\text{fluid}} - T_{\text{solid}}) \qquad (3)$$

Basic Thermal Design Methods for Heat Exchangers

The basic heat-transfer principles of sizing and rating heat exchangers are important to design. Sizing refers to determining the amount of heat-transfer surface area required to transfer a specified quantity of thermal energy from one fluid to another for given fluid conditions and thus usually applies to the design of a new heat exchanger. Rating refers to determining the rate of heat transfer for given fluid-inlet conditions and given heat-exchanger geometry and thus applies to the performance of an existing heat exchanger. However, the sizing and rating calculation methods can be used interchangeably to obtain either piece of heat exchanger information.

Heat-Exchanger Effectiveness Method

The method of heat-exchanger effectiveness is useful in determining or rating the performance of a given heat exchanger. This method can also be used in sizing a new heat exchanger. The heat-exchanger effectiveness, ϵ, is defined as the ratio of the actual rate of heat transfer in a given heat exchanger to the maximum, ie, thermodynamically possible, heat-transfer rate.

Design Margins

The heat-transfer surface area determined using the sizing or rating methods is considered the minimum required area. There are additional surface-area requirements for design margins in the final sizing of a heat exchanger. Considerations should include the effect of uncertainties in thermal design parameters, bypass flow effects, entrance and exit span areas, baffle-tube support plate area, and plugged tube allowance.

Pressure Drop Calculations. There are two principal costs to consider in sizing a heat exchanger: manufacturing costs and operating costs. From a manufacturing standpoint, in general, the less the heat-transfer surface area, the lower the manufacturing cost. The operating cost of a heat exchanger results primarily from the cost of the power to run fluid-moving devices such as pumps and fans, and this power consumption is directly proportional to fluid stream pressure drop. Therefore, an optimum design of a heat exchanger requires a proper balance between thermal sizing and pressure drop.

TYPES OF HEAT EXCHANGERS

The shell-and-tube exchanger is the workhorse of power, chemical, refining, and other industries (Fig. 1). One fluid flows on the inside of the tubes whereas the other fluid is flowing through the shell and over the outside of the tubes. Baffles are used to ensure that the shellside fluid flows across the tubes, thus inducing high heat transfer.

Plate heat exchangers, which are used as an alternative to shell-and-tube heat exchangers in relatively low temperature and pressure applications involving liquids and two-phase flows, have some important advantages over shell-and-tube exchangers. Plate heat exchangers include plate–frame heat exchangers, plate–fin heat

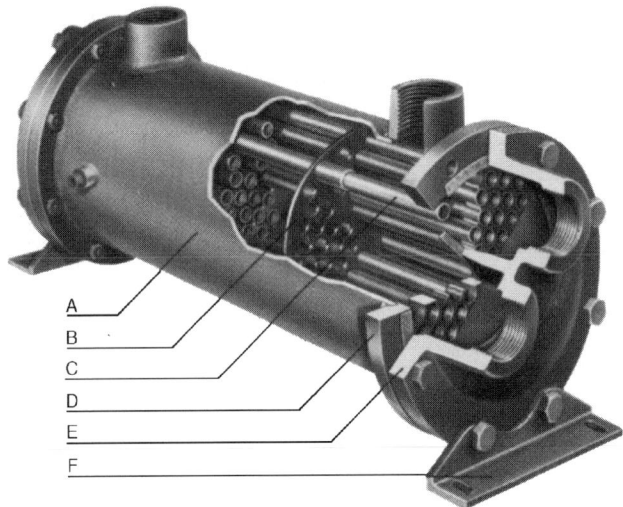

Figure 1. Shell-and-tube heat exchanger: A, shell of high strength; B, tube sheet; C, tubes (normally small diameter tubes are seamless, but large diameter tubes (>1 in.) are welded tubes); D, bonnets; E, baffles to assure more efficient circulation by providing minimum clearance between tubes and tube holes as well as baffles and shells; and F, mounting brackets.Courtesy of Basco.

exchangers, plate–coil heat exchangers. The advantages of using plate heat exchangers are less surface for heat transfer is required, resulting in weight, volume, and cost advantage over shell-and-tube and other noncompact heat exchangers; thermal rating of plate heat exchangers can readily be increased or decreased by varying the number of plates which is important if substantial changes in load occur; and the increased effectiveness of plate heat exchangers reduces the required cooling flow rate, resulting in savings relative to piping, pumps, valves, and operating cost. In spite of these advantages, however, the plate heat exchangers are rarely used, even in relatively low temperature and pressure applications. This may be because of the widespread familiarity with shell-and-tube exchangers and the large number of manufacturers of shell-and-tube exchangers. Plate heat exchangers are not normally used in nuclear applications for safety considerations. Note that ASME Boiler and Pressure Vessel codes do not recognize the plate heat exchangers.

The principal disadvantage of plate–frame heat exchangers is the large number of surfaces that must be sealed by gaskets.

Cold-plate heat exchangers for electronics cooling applications operate at heat-flux levels typically on the order of 2–10 W/cm^2. The electronics industry has targeted heat-flux capacities of up to 25 W/cm^2 for the next generation cold-plate for advanced applications. Achieving this level of cooling requires development of methods of providing high surface-density cooling within coolant passages in the cold plate. Potential scenarios that might provide high heat flux cooling may include high density ribbed or finned surfaces or impinging jet cooling of the cold-plate primary surface.

USE OF HEAT EXCHANGERS

Heat exchangers are used whenever energy has to be transferred, and the proper design and use of heat exchangers are vitally important for efficient operation of an industrial system, for energy conservation, and ultimately for the protection of the environment. Despite decades of continuous research and development, there are numerous design and operating problems originating from a lack of understanding of basic flows and heat-transfer phenomena such as flow distribution in manifolds, flow-induced vibration in two-phase flows, heat-transfer enhancement, fouling, etc.

In order to help companies and organizations overcome problems associated with heat exchangers, Heat Transfer Research Inc. (HTRI) was established in the United States in 1962, and the Heat Transfer and Fluid Flow Service (HTFS) was established in the United Kingdom in 1967. These organizations provide results of heat-transfer and fluid-flow research, design methods, supporting computer programs, and proprietary equipment testing. In addition, the American Society of Mechanical Engineers (ASME), American Society for Testing and Materials (ASTM), and Tubular Exchanger Manufacturers Associations Inc. (TEMA) provide various safety and design codes and technical services. More recent reference books come with software disks, and analyses or design calculations of various energy systems can be conducted.

Fundamental issues involved in the use of various heat exchangers have been summarized in a thermal science workshop sponsored by the National Science Foundation. There are a number of areas that require different types of heat exchangers. Some of the emerging technologies where heat exchangers are expected to play a critical role are electronic cooling, micro and macro gravity applications, ozone depletion, global warming and other environmental issues, biotechnology, high temperature superconductors, and ultrahigh temperature waste-heat recovery.

HEADER DESIGN

Headers, ie, manifolds and tanks, are the chambers or transition ducts at each end of the heat-exchanger core on each fluid side for distributing fluid to the core at the inlet and collecting fluid at the exit. These may be classified broadly as normal, turning, and oblique flow headers. Poor design of headers reduces heat-transfer performance significantly and may also increase pressure drop substantially owing to flow maldistribution, flow separation, and jet effects. Thus header design is an important problem for all heat exchangers where fluid from the inlet pipe is distributed to the exchanger core via manifolds and tanks. If novel heat-exchanger applications are contemplated, the header volume must be a very small fraction of the total exchanger volume, particularly for highly compact heat-exchanger applications.

No design theory and modeling is available to obtain uniform flow for normal headers, ie, diffusers having downstream flow resistance and turning headers, with or without vanes. Only very limited design information is available for oblique flow headers. Manifolds in a heat exchanger can be further classified into four types: dividing, combining, parallel, and reverse flow manifolds. Parallel and reverse flow manifolds are those which combine dividing and combining flow manifolds. In a parallel flow manifold, the flow directions in dividing and combining flow headers are the same; in a reverse flow manifold, the flow directions are opposite. The objective of the manifold design is to obtain a uniform flow distribution in the heat-exchanger core, with the manifold occupying the smallest fraction of volume of the total heat exchanger.

Several investigators have conducted analytical and numerical studies on dividing and combining flow manifolds. Friction was shown always to increase the flow imbalance in a combining flow manifold and friction might either increase or decrease the flow imbalance in a dividing flow manifold depending on the area ratio. The larger the cross-sectional area of dividing and combining flow manifolds, the better the flow distribution has been reported to be.

The flow distribution is a direct consequence of the static pressure difference between dividing and combining flow headers. There are two factors controlling the pressure variations in manifold headers: friction and momentum. In a combing flow header, these two factors lower the pressure along the header in the flow direction. However, in a dividing flow header, these two factors work in opposite directions. The friction effect lowers the pressure along the header, whereas the momentum effect increases the pressure. The flow velocity decreases in the flow direction owing to fluid loss into channels, creating momentum deficiency along the dividing flow header and thus increasing pressure. Furthermore, the pressure increases near the end of the dividing flow header due to the conversion of kinetic energy to stagnation pressure.

PERFORMANCE ENHANCEMENT IN HEAT EXCHANGERS

Static Mixer

To enhance the performance of conventional shell-and-tube heat exchangers, one can use static mixer elements inside tubes as shown in Figure 2. Process fluid is continuously mixed, thus producing performance enhancement.

Advanced Heat-Transfer Fluid

A conventional heat-exchanger system requires a high volumetric flow rate, resulting in the consumption of a large

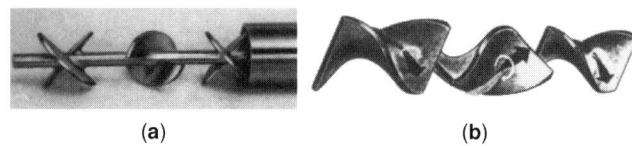

(a)   (b)

Figure 2. Static mixers which provide a continuous mixing and processing unit with a nonmoving part. These static mixers can be easily installed in new and existing pipelines. (**a**) Courtesy of Ross; (**b**) courtesy of Chemineer.

amount of pumping power. The use of an advanced heat-transfer fluid has been proposed to increase the convective heat-transfer coefficient by increasing the effective thermal capacity of working fluids, a technique that would permit the use of a smaller volumetric flow rate and smaller heat exchangers.

E. Choi, Y. I. Cho, and H. G. Lorsch, *Int. J. Heat Mass Trans.* **37**, 207–215 (1994).

B. K. Hodge, *Analysis and Design of Energy Systems*, 2nd ed., Prentice-Hall, Inc., Englewood Cliffs, N.J., 1990.

H. Martin, *Heat Exchangers*, Hemisphere Publishing Corp., New York, 1992.

E. A. D. Saunders, *Heat Exchangers: Selection, Design, and Construction*, John Wiley & Sons, Inc., New York, 1988.

YOUNG I. CHO
Drexel University

S. M. CHO
Foster Wheeler Energy
Corporation

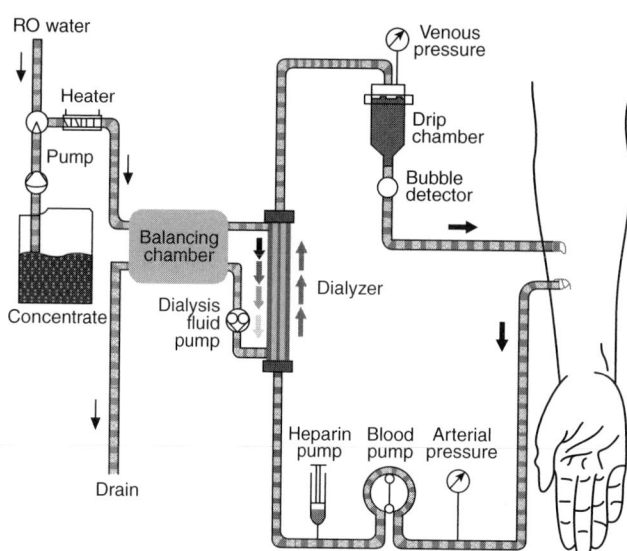

Figure 1. Hemodialysis process. Copyright Fresenius Medical Care (2006). Reproduced with permission.

HEMODIALYSIS

Hemodialysis, a mechanical system to cleanse the kidneys, sustained the lives of approximately one million patients worldwide in 2006. In the last 35 years, hemodialysis membranes have become increasingly efficient, and hemodialysis machines have become increasing sophisticated. Kidneys' primary function is to generate urine for excretion of water and metabolic waste products. The kidneys not only remove accumulated nitrogen products (urea, creatinine, uric acid and others), but also maintain homeostasis of water and electrolytes (sodium, potassium, chloride, calcium, phosphate, magnesium) and regulate acid–base balance. In addition, human kidneys perform a few endocrine and metabolic functions and conversion of vitamin D to its active form.

While historically, dialyzers have been called "artificial kidneys," hemodialysis does not replace the kidneys' endocrine or metabolic functions. Here the focus is on the excretory functions carried out by hemodialysis.

HEMODIALYSIS SYSTEM AND PROCESS

A hemodialysis system consists of three main components: the machine hardware, the disposable blood circuit (including the dialyzer), and the dialysate solution. The primary function of the hemodialysis machine is to move fluids: (*1*) to mix dialysate from its component solutions; (*2*) to pump blood and dialysate to the dialyzer; (*3*) to control fluid removal from the patient; and (*4*) to deliver heparin.

A hemodialysis machine performs a number of monitoring functions, including conductivity monitoring (dialysate composition); pressure monitoring; air bubble detection; detection of hemolysis; detection of kinked lines; detection of disconnected lines; on-line clearance measurement (optional); access flow measurement (optional); blood volume monitoring (optional); and blood temperature monitoring (optional).

Figure 1 shows the essentials of the hemodialysis process. Blood is drawn from the patient using a roller pump, mixed with anticoagulant (depending on the patient's prescription), and delivered to the bottom header of the dialyzer, which distributes the blood flow among the thousands of hollow fibers.

Dialysate is generated from concentrate and water, heated, and delivered to one end of the shell space surrounding the hollow fibers. The dialysate flows among the fibers, collecting toxins, and exits to the drain. A proportioning chamber ensures that the desired amount of excess fluid is removed from the patient, ie, that the dialysate flow rate to the drain is slightly greater than the inlet dialysate flow rate. The dialyzer provides a membrane barrier that permits the passage of metabolic waste products like urea, creatinine, uric acid, and inorganic phosphate to move from the bloodstream of the patient to the dialysate, while at the same time preventing the elimination of important blood proteins like albumin and immunoglobulin.

Hemodialysis patients generally receive treatment three times per week. Over the past 30 years, with the development of more permeable membranes, average treatment times have dropped from 4 or 5 h to around 3.5 h. Typical blood flow rates in the U.S. have risen from 300 mL/min to over 400 mL/min.

DIALYZER MASS TRANSPORT REQUIREMENTS

Dialyzers are generally characterized in terms of an ultrafiltration coefficient (Kuf), solute clearances, and the product of the mass transfer coefficient times the surface area (KoA). The ultrafiltration coefficient of a dialyzer is reported as the volumetric filtration rate (mL/h) per mm Hg transmembrane pressure (TMP) across the membrane when filtering blood. The FDA classifies dialyzers as high flux if Kuf > 8 mL/h/mm Hg. In the U.S. market, over 92% of patients are treated with high flux dialyzers. The clearance rate of a solute, or solute clearance, is defined as the mass removal rate divided by the concentration of the solute in the blood, expressed in units of mL/min. The clearance represents the equivalent volume of blood fully cleared of the solute each min and cannot exceed the blood flow rate to the dialyzer. Another term used to characterize the transport properties of dialysis membranes is the so-called "mass transfer area coefficient" (MTAC), the product of the mass transfer coefficient (Ko) times the membrane surface area (A), or "KoA." Usually the terms MTAC and KoA reported are those for urea. Ko should equal the maximum clearance obtained at high blood and dialysate flow rates. Since mass transport rates depend on solute size and other characteristics, evaluation of dialyzer performance requires identification of the solutes to be removed. While urea (MW = 60 Da) has served as a marker solute for about 40 years, the full spectrum of solutes has yet to be identified.

OTHER DIALYSIS MEMBRANE REQUIREMENTS

While its transport properties play an important role in the selection of a dialyzer membrane, an equally important consideration in the evolution of the dialyzer technology has been biocompatibility, or the compatibility of the dialyzer with blood. Three aspects of biocompatibility important in dialysis are clotting, activation of the complement cascade, and cytokine generation.

Clotting is generally managed through the use of systemic heparinization, although a small percentage (<5%) of dialysis treatments are performed heparin-free due to allergic reactions to heparin. Heparin-coated dialyzers have only recently been introduced.

Complement proteins are so named because they complement antibody activity to eliminate pathogens. The "alternate pathway" of the complement cascade is normally activated by bacterial surface molecules. Complement activation during dialysis was first identified by the rapid drop in white blood cell counts (neutropenia) during the first 30 minutes of dialysis. Regenerated cellulose membranes activate complement through the alternate pathway. Modified cellulose membranes approach the biocompatibility profile of synthetic materials in terms of neutropenia and complement activation.

Whereas membrane materials used in downstream processing of biological products may be selected for their resistance to protein fouling, early dialysis membranes were found to be more blood compatible after adsorption of blood proteins. Endotoxins are bacterial products released from gram-negative bacteria upon death. Because endotoxins cause fever, they are also called pyrogens. Because endotoxin fragments fall in the middle molecule range, they may be inadvertently transported from dialysate to blood during high-flux dialysis, leading to cytokine generation. Adsorption (membrane binding) is one mechanism by which hydrophobic compounds like endotoxins, cytokines, peptides, growth factors, and proteins may be removed during hemodialysis. While adsorption of endotoxins and cytokines is clearly desirable, the adsorptive properties of a dialyzer are not important for removal of solutes such as beta-2 microglobulin, because it does not matter to the patient whether a toxin goes down the drain or is adsorbed within the membrane wall.

Sterilizability

Since dialyzed blood is returned to patients, dialyzers and associated tubing must be sterilized. The most common form of sterilization has been ethylene oxide. Because some patients develop allergic reactions to ethylene oxide-altered human serum albumin (ETO-HSA), other sterilization methods such as gamma irradiation, steam sterilization, and e-beam sterilization have been developed.

By the 1990s, the practice of dialyzer reuse had become commonplace in the United States. With the use of more biocompatible polysulfone membranes, a 2004 study found a 5–10% mortality benefit associated with single-use dialyzers.

MEMBRANE MATERIALS, SPINNING TECHNOLOGY, AND STRUCTURE

Dialyzer membrane performance depends on the biomaterial used, its thickness, and the hydraulic permeability, pore size and density, biocompatibility and the hydrophilic/hydrophobic properties.

Membrane Materials

Current dialyzer membranes can be classified based on their chemical compositions as cellulosic, modified cellulosic, and synthetic. Cellulosic membranes are very effective in removing low molecular weight toxins, but their very low means pore size results in poor middle molecule removal. The membrane has a high density of hydroxyl groups on their glucosan rings, which activates the complement cascade via the alternate pathway. Activation of the complement cascade makes these membranes bioincompatible. Despite being considered the least biocompatible dialyzer material, these membranes are still used in some parts of the world, primarily due to their lower cost.

Modified cellulosic membranes are made more biocompatible by the substitution of hydroxyl groups with other moieties or by coating the membrane with a biocompatible coating. Cellulose diacetate and cellulose triacetate differ in the degree of substitution of the hydroxyl groups with acetate groups. Other groups such as diethylaminoethyl (DEAE) and benzyl groups were added to make the membranes more biocompatible. These substituted membranes are more hydrophobic and also have a larger mean pore size, resulting in higher water permeability and middle molecule clearances compared to unmodified cellulosics. This chemical modification influences membrane properties such as protein absorption, wettability, biocompatibility and clearance of both small and middle molecules.

In the dialysis field, the term "synthetic membrane" is used to denote all polymeric membranes that are not cellulose-based. Synthetic membranes with high water permeability were developed in the 1960s primarily for hemofiltration. These membranes are now manufactured with a range of permeabilities. They have thicker walls ($> 20\,\mu m$) and are either symmetric or asymmetric. The skin layer is in contact with the blood and controls the solute removal. The pore structure of the support layer is much more open and varies among the various synthetic membranes. This layer dictates the thermal and mechanical properties of the membranes.

Hollow-Fiber Spinning Technology

Selected polymers are dissolved with solvents and spun through tube-in-orifice nozzles to form hollow-fiber structures, in either a dry-wet or a dry spinning mode. Because the inner surface of the hollow fiber plays an important role for separation, the hollow fiber is usually spun with an inner liquid to control the pore structure of the lumen surface. Membrane porosities up to around 75% may be employed, while the pore diameters are controlled in the nanometer range. The hollow fibers are wound on a spool or a reel.

Pore Size, Distribution, and Density

The nature of the pore size distribution may significantly influence a membrane's sieving properties. The desirable features for a high-flux membrane include a large number of relatively large pores having a narrow distribution of sizes. This type of distribution leads ideally to a solute-sieving coefficient vs. molecular weight profile with a sharp cut-off at a molecular weight just below that of albumin, similar to that of the native kidney. In actual practice, all highly permeable membranes have measurable albumin sieving coefficient values such that the design of this type of membrane involves striking a balance between optimized large molecular weight toxin removal and minimal albumin losses.

Hollow Fiber Geometry

A study in 2000 suggested that small changes of the inner diameter of the fiber could result in dramatic changes in removal of both urea and middle molecules. High shear rates were also obtained by lowering the inner diameter of the fiber at a given blood flow rate. This leads to a reduction of the protein boundary layer and improves the membrane permeability. Decreasing hollow fiber inner diameter improves diffusive mass transfer by

shortening path length and attenuating boundary layer effects through higher shear rates. Wall thicknesses of cellulosic fibers range from $6\text{--}15\,\mu m$. Noncellulosic membrane thicknesses are greater than $20\,\mu m$. Since wavy hollow fibers provide better dialysate flow distribution, hollow fibers are often axially undulated in a post-spinning process. Other techniques for improving performance include radial variations in wall thickness, spacer yarns, and knitting of hollow fibers, designed to achieve uniform dialysate flow around all hollow fibers.

DIALYZER DESIGN AND PERFORMANCE

Dialyzer performance depends not only on the membrane properties but also on the device properties. Such properties as fiber length, membrane surface area, number of fibers, hollow fiber packing density, and header design all affect solute clearances.

The effectiveness of dialysis as a replacement for kidney function depends on the mass transfer characteristics of the membranes as well as the device parameters.

FUTURE DIRECTIONS

The standard mode of dialysis in most countries today is high-flux dialysis, in which high ultrafiltration rates are counterbalanced by back-filtration. To prevent excessive fluid loss, net ultrafiltration is controlled volumetrically. In this mode, clearances are improved over conventional hemodialysis, where low-flux membranes lead to low ultrafiltration rates and minimal convection.

Several convective therapies have been considered as alternatives to the high-flux dialysis commonly employed today. Hemofiltration is a purely convective filtration process in which large volumes of water are removed from the patient and discarded, and sterile, pyrogen-free replacement fluid is administered. Hemofiltration is often used for acute renal failure, where large quantities of fluid (up to 20 L) must be removed in a short time (1–2 days). One form of hemofiltration, continuous arteriovenous hemofiltration (CAVH) is particularly suited for use in intensive care units due to the simplicity of the process and slow, "gentle" nature of the therapy. In CAVH, blood flow is driven by the arterio-venous pressure difference, rather than by a pump, the filtrate waste stream is generated by a simple gravity drain, and no dialysate is employed. Continuous veno-venous hemofiltration (CVVH) and continuous veno-venous hemodiafiltration (CVVHD) employ a blood-side pump, with and without a dialysate stream, respectively.

Conventional hemodiafiltration (HDF) utilizes large convective transport with ultrafiltration rates above 70 mL/min. Since such ultrafiltration rates result in total ultrafiltration volumes that exceed the desired weight losses in patients, sterile replacement fluid must be administered. An improvement in survival (35%) has recently been reported using high efficiency ($> 15\text{--}25$ L/session) HDF. The major drawbacks of this treatment are the complexity of the system and the increased cost of the therapy over conventional hemodialysis.

One approach for increasing middle molecule removal has been to modify the dialyzer design to improve internal filtration, thereby increasing convective solute removal. Immunoadsorption is another way to remove middle molecules, either specifically or nonspecifically. Adsorptive processes can be carried out either by chemically modifying a hemodialysis membrane to create adsorption sites or by the use of an add-on device (eg, affinity column) during hemodialysis. It should be mentioned that the 1–2 square meter membrane surface area on the hollow-fiber lumen is much smaller than the surface area within the porous membrane structure, and may be insufficient to provide significant toxin removal. However, one could argue that adsorptive sites within the membrane wall offer little benefit unless significant backfiltration of a toxin is taking place, because it makes no difference to the patient whether a toxin is adsorbed within the membrane walls or flushed away with the spent dialysate.

Middle molecule removal can be increased by extending the treatment time with nocturnal dialysis, especially if performed daily. Short daily dialysis may provide some additional middle molecule removal. Removal of relatively small solutes with substantial compartmental effects, such as inorganic phosphate, may benefit the most from short daily dialysis. In the United States, nocturnal dialysis and short daily dialysis are under investigation in studies funded by the National Institutes of Health, but are not generally available because Medicare pays for only standard treatment.

C. K. Colton and E. G. Lowrie, in B. M. B. A. F. C. Rector, ed., *The Kidney*, Saunders Publishing Co., Philadelphia, 1981.

T. Jirka, S. Cesare, A. Di Benedetto, M. Chang, P. Ponce, and N. Richards, *Impact of On-line Haemodiafiltration (HDF) on Patient Survival: Results From a Large Network Database*, 2005.

N. J. Ofsthun, S. Karoor, and M. Suzuki, "Hemodialysis Membranes," in N. Li, T. Fane, T. Matsuura, and W. Ho, eds., *Membrane Science and Technology*, John Wiley & Sons, Inc., Hoboken, N.J., 2006.

R. A. Ward, J. K. Leypoldt, W. R. Clark, C. Ronco, G. J. Mishkin, and E. P. Paganini, *Seminars in Dialysis* **14**, 160–174 (2001).

NORMA J. OFSTHUN
Corporate Research, Fresenius
Medical Care

HERBICIDES

The broadest definition of "herbicides" includes all agents that destroy or inhibit plant growth. Thus, an herbicidal agent may be animal; ie, a home gardener with a hoe, or

a grazing herbivore; vegetable, ie, a parasitic weed or one plant species competing successfully with another; or mineral, ie, chemicals with herbicidal activity. The definition of a weed as "a plant growing where it is not wanted" is convenient, although perhaps not scientific. It focuses on one of the basic problems of weed control; ie, selectively killing weeds without crop damage.

Cultural practices like crop rotation, mowing, and tilling the soil are important, but the use of chemicals for weed control has been adopted globally. The term "pesticide" includes all classes of chemicals used against insects, weeds, plant pathogens, rodents, algae, snails, and other pests. Legally, it also includes growth regulators. The term "herbicide" refers specifically to weedkillers.

Modern agriculture demands that herbicides and other crop-protection chemicals be integrated into a production system that includes the development of pest-resistant and high yielding crop varieties, crop management, plant nutrition, and mechanization of farming methods and pest-control techniques. In this system, chemical control is an important component. Pesticides have been stated to increase production of crops, livestock, and forest products by 25% and thus contribute to the stability of food prices.

The widespread introduction of chemicals for weed control (herbicides) brought about major changes in agriculture affecting not only the economics of farming, but also the communities that were founded and based on crop production. The chemical inputs were, however, expensive, and adverse environmental effects and other problems gradually offset some of the benefits. Reductions in the use of herbicides have been driven to some extent by regulation, but the more significant changes in the patterns of herbicide use are due to progress in the applications of biotechnology to agriculture. An improved understanding of metabolic processes in plants, modes of action of herbicides, and plant genetics, coupled with the ability to manipulate genes and facilitate their expression in plants, are major factors in these changes.

The use of selective herbicides that could kill weeds without damage to crops growing in the same cultivated area is a twentieth-century development that has brought about major changes in agriculture and agricultural communities. Sulfuric acid, sodium chlorate, arsenic compounds, copper sulfate, and other inorganic compounds have been used as weed killers since the early twentieth century. Sodium chlorate was used to control deep-rooted perennial weeds in noncrop areas. Borates also found use for control of weeds in specific locations. The introduction of synthetic organic herbicides that acted selectively against broad-leaved weeds changed the situation irreversibly. The first organic chemical herbicide to be introduced in 1932, was 4,6-dinitro-o-cresol (DNOC) used initially as an insecticide. By 1982, almost 95% of the corn, cotton, and soybean acreage was being treated with pesticides by U.S. farmers.

DEVELOPMENT OF HERBICIDES

As knowledge of biochemical targets has increased through studies of metabolism and mode of action of pesticides, screening techniques have been improved, making it possible to identify candidate compounds that are effective at specific receptor sites. The introduction of newer synthetic techniques, such as combinatorial chemistry, which can generate large numbers of new compounds, made it possible to increase the throughput of compounds. One application of biotechnology is to increase herbicide tolerance in existing crops by genetic modification. Weeds can then be eliminated by conventional herbicides without damage to the growing crop. This favors the use of currently registered chemicals that have been shown to be environmentally acceptable. Other approaches involve the genetic manipulation of crops to introduce genes responsible for generating insecticidal *Bacillus thuringiensis* toxins, resistance to diseases or plant-parasitic nematodes.

MODES OF HERBICIDE ACTION

Photosynthesis is the light-driven, membrane-localized electron/proton transport system by which plants, algae, and some bacteria convert the energy of a quanta of light into the phosphoryl group transfer potential of adenosine triphosphate (ATP) and the redox potential of nicotinamide adenine dinucleotide phosphate (NADPH) while oxidizing water to produce oxygen. In higher plants, the photosynthetic light energy conversion processes are localized in the thylakoid membranes of the grana of chloroplasts and the carbon-fixing processes occur in the stroma. Chloroplasts are chlorophyll-bearing, double-membrane-bound organelles within photosynthetic plant cells; grana consist of stacks of thylakoids, vesiclelike structures that have internal spaces defined by a membrane and that are connected by unstacked stromal thylakoids. The thylakoid membranes contain the light-harvesting pigments and the electron- and proton- translocating components of both Photosystem I (PSI) and Photosystem II (PSII) of photosynthesis.

Both PSI and PSII are necessary for photosynthesis. The terms PSI and PSII have come to represent two distinct but interacting reaction centers in photosynthetic membranes. The two centers are considered in combination with the proteins and electron-transfer processes specific to the separate centers.

Photosystem I Inhibitors

Photosystem I is the reaction center or site in photosynthetic membranes of oxygen-evolving organisms at which light-activated electron transfers lead to a reduction of the iron–sulfur, FeS^-, centers of ferredoxin. PSI cycling is ensured by electrons transferred from PSII or the cytochrome b_6/f complex via the copper-containing plastocyanin (PC) which is the primary electron donor to P700, the specialized chlorophyll a molecule associated with PSI. From P700, electrons are transferred singly to FeS^- and thence to soluble ferredoxin. Ferredoxin nicotinamide adenine dinucleotide phosphate reductase transfers electrons to $NADP^+$ from soluble ferredoxin.

PSI transport processes include both this directional electron transfer to produce NADPH and a cyclic electron transfer that pumps protons into the stroma, resulting in the synthesis of ATP. Production of ATP by photophosphorylation can be inhibited by either uncouplers, eg, phenylhydrazones, carbanilates, diphenylamines, and ethane diamines or by inhibitors at ATP synthase. Since ATP synthesis is not specific to oxygen-producing organisms, PSI inhibitors of the latter type are toxic to both plants and animals. Other herbicides have been reported to act by causing general destruction of the chloroplasts through membrane component peroxidation. Studies indicate that the diphenyl ethers, acifluorfen and oxyfluorfen, inhibit protoporphyrinogen oxidase, the penultimate enzyme in heme synthesis.

The interaction between PSI and PSII reaction centers depends on the thermodynamically favored transfer of electrons from low redox potential carriers to carriers of higher redox potential. This process serves to communicate reducing equivalents between the two photosystem complexes. Photosynthetic and respiratory membranes of both eukaryotes and prokaryotes contain structures that serve to oxidize low potential quinols while reducing high potential metalloproteins.

Photosystem II Inhibitors

The PSII complex is that structural entity capable of light absorption, water oxidation, plastoquinone reduction, and generation of transmembrane charge asymmetry and the chemical potential of hydrogen ions. The so-called core of PSII is a set of five hydrophobic polypeptides, two of which form the reaction center that performs the primary photochemical charge separation of PSII; ie, the 47 and 43 kDa polypeptides.

The PSII complex contains two distinct plastoquinones that act in series, Q_A and Q_B, the latter reversibly associated with a 30–34 kDa polypeptide in the PSII core. This secondary quinone acceptor polypeptide is the most rapidly turned over protein in thylakoid membranes. It serves as a two-electron gate and connects the single-electron transfer events of the reaction center with the pool of free plastoquinone in the membrane. Q_B is probably the most studied protein in thylakoid membranes, since it is the binding site of many, if not most, PSII-inhibiting herbicides.

Many commercial herbicides inhibit electron flow on the reducing side of PSII. Compounds as chemically different as atrazine, metribuzin, diuron, bromacil, ioxynil, and dinoseb all block electron transfer from Q_A to Q_B. Herbicidal PSII inhibitors, such as these and other triazines, ureas, pyrimidines, nitriles, and phenols, all appear to have the same site and mode of action; differences in activity are determined by the lipophilicity of the various side chains. These herbicides prevent electron transfer from Q_A to Q_B by displacing Q_B from its binding site on the Q_B polypeptide. The extensive research effort that has increased understanding of PSII and the binding of herbicides to the various polypeptides in photosynthetic membranes has initiated molecular modeling studies aimed at new inhibitors and herbicides. Quantitative

Structure-Activity Relationship (QSAR) techniques have been applied, usually *a posteriori,* in investigations of various chemical classes.

From the beginning, herbicide developmental research has been focused on PSII and, more recently, the different modes of inhibitor binding that confer specificity and efficacy. But a decreasing tolerance for herbicides which persist in the soil is a contributing factor in lessening use and acceptability. Other modes of action, particularly those relevant to membrane function or enzymatic activity, now show greater potential for producing new herbicides with desirably low application rates.

Bleaching Herbicides

Membrane-based modes of herbicidal action relevant to photosynthesis include bleaching herbicides, which can act at multiple sites in lipid metabolism and are reported to affect chloroplast pigments by interfering with phytoene desaturation.

There are three distinct groups of phytoene desaturase (dehydrogenase) inhibitors. The norflurazon class includes fluridone, flurochloridone, flurtamone, S3442, diflufenican, and difunon. These compounds directly inhibit the conversion of phytoene to the colorless α- and β-carotenes which are the substrates for phytoene desaturase (PD). Norflurazon and fluridone act as reversible noncompetitive PD inhibitors in cell-free systems.

A second class of herbicides primarily affects ζ-carotene desaturase. These herbicides are apparent feedback inhibitors of PD as well. The third class consists of the benzoylcyclohexane-diones, atypical bleaching herbicides that induce phytoene accumulation when applied either pre- or postemergence.

QSAR studies of the typical PD inhibitors from the norflurazon group have been used to elucidate the influence of various substituents of herbicidal activity. The first steps have also been made toward employing QSAR in the construction of a model describing the general features of PD-inhibitors and in the characterization of PD through determination of PD enzymology, amino-acid composition, and genetic markers.

Chlorophyll Biosynthesis Inhibitors

Chemically, the chlorophylls are magnesium-porphyrin complexes in which the four central nitrogen atoms of the pyrrole rings are coordinated with a Mg^{2+} ion to form an extremely stable planar complex. Several herbicides are reported to inhibit chlorophyll biosynthesis, eg, oxadiazon, DTP, and MK-616.

Lipid and Wax Synthesis Inhibitors

Lipids are present in all plant organs and on leaf surfaces. In plant roots and shoots, acyl lipids such as phospho- or glycolipids are structural components of the essential biological membranes of cell compartmentation, enzymology, and bioenergetics. Acyl lipids are constituents of a large variety of different structures with different functions

and are, therefore, promising potential target sites for herbicide action.

The oxyphenoxy propionic acids and the analogous diclofop–methyl; clofop–isobutyl; haloxyfop–methyl; and fenthiaprop–ethyl all inhibit fatty acid synthesis *de novo.* Sethoxydim and alloxydim, also inhibitors of fatty acid synthesis, are cyclohexanedione derivatives. The herbicidal activity of the cyclohexanediones is similar to the oxyphenoxy propionic acids, but alloxydim and sethoxydim also cause necrosis in meristematic regions, and leaf chlorosis. These herbicides are selective against grasses.

Incorporation inhibition of ^{14}C-acetate into lipids is the most rapid and pronounced effect of the α-chloracetamides, alachlor, metazachlor, and metolachlor, suggesting that the acetate-incorporating steps of lipid synthesis are the site of action for these herbicides.

Pyridazinone herbicidal activity depends on inhibition of multiple target sites in plants, as well as changes in fatty acid composition. The oxyphenoxypropionic acids and the cyclohexadiones however, are phytotoxic because they inhibit synthesis *de novo* of fatty acids. Inhibition of lipid synthesis could also produce the other physiological effects attributed to these herbicides.

Radical Damage to Antioxidative Systems and Cellular Components Inducers

The herbicidal activities of many of the inhibitors of PSII; eg, metflurazon, norflurazon, and fluridone, are enhanced by light, possibly through the mechanism of photooxidative destruction. Light and photosynthetic electron transport convert DPEs *p*-nitro- or *p*-chlorodiphenyl ethers into free radicals of undetermined structure. The radicals produced in the presence of the bipyridinium and DPE herbicides decrease leaf chlorophyll and carotenoid content and initiate a general destruction of chloroplasts, with concomitant formation of short-chain hydrocarbons from polyunsaturated fatty acids.

The chemical mechanism of other herbicides also involves peroxidative destruction of polyunsaturated fatty acids by starter radicals. Fenton reactions produce alkoxy radicals which can split into alkyl radicals leading to hydrocarbon gases or can initiate further radical destruction of other chloroplast components. Lipid hydroperoxides also decompose to form cytotoxic malondialdehye (MDA), a compound often used as an index of lipid peroxidation and that can cause intra- and intermolecular cross-linking of sulfhydryl-containing proteins. Proteins can also be fragmented or modified by hydrogen peroxide in the presence of transition metals. The resulting hydroxyl radicals and the alkoxy radical intermediates from lipid peroxidation also attack proteins and individual amino acids.

Herbicidal Inhibition of Enzymes

The list of known enzyme inhibitors contains five principal categories: group-specific reagents; substrate or ground-state analogues, ie, rapidly reversible inhibitors; affinity and photo-affinity labels; suicide substrate, or k_{cat}, inhibitors; and transition-state, or reaction- intermediate, analogues, ie, slowly reversible inhibitors.

The radical-generating herbicides, described above, that attack specific amino acid residues are examples of group-specific enzyme inhibitors. Substrate analogue enzyme inhibitors include the organoarsenicals, MAA and MSMA. Arsenical pesticides have been widely used as herbicides since 1951. The arsonic acid herbicides—MAA, MSMA, DSMA, and AMA—are not technically plant growth regulators, since they act through enzyme systems to inhibit growth and thus kill plants relatively slowly. Cacodylic acid and its sodium salt are used extensively as selective postemergence herbicides in cotton and noncrop areas and orchards. Cacodylate is reportedly a nonspecific competitive inhibitor of adenine nucleotide deaminase and may inhibit other enzymes as well.

Suicide substrate and reaction intermediate inhibitors promise the highest degree of specificity and have drawn increased attention. Heteroatom or radical replacement in reaction–intermediate analogues is a simple pesticide development strategy that offers potential for achieving extremely potent inhibition without high chemical reactivity. This simple design strategy may produce effective intermediate inhibitors for families of mechanistically related enzymes.

A carbonyl reagent, aminooxyacetate is the basis for the herbicides benzadox, benzamidooxyacetic acid and other lipophilic analogues. Benzadox decreases photosynthesis and inhibits both alanine and aspartate aminotransferase. Isonicotinic hydrazide affects glycine–serine aminotransferase, and aminoacetonitrile inhibits glycine decarboxylation in a manner similar to that of gulfosinate.

When ATP-synthase, specifically the plastidic coupling factor CF_0–CF_1, is inhibited, ATP formation by photophosphorylation is blocked. Inhibitors of this energy-transfer process prevent conversion into ATP of the electrochemical potential formed by electron transport during photosynthesis. One of the most studied inhibitors of the plastidic coupling factor is tentoxin, a cyclic tetrapeptide produced by *Alternaria alternata f. tenuis*. Tentoxin is plant-specific, able to pass through membranes, and highly active, binding to the catalytic site of the CF_1. Tentoxin is a naturally occurring compound with potential as a model for synthetic analogues with herbicidal activity.

Amino Acid and Nucleotide Biosynthesis Inhibitors

The metabolism of amino acids is affected by both chemical herbicides and biogenic inhibitors.

Herbicides also inhibit 5-*enol*-pyruvylshikimate synthase, a susceptible enzyme in the pathway to the aromatic amino acids, phenylalanine, tyrosine, and tryptophan, and to the phenylpropanes. Acetolactate synthase, or acetohydroxy acid synthase, a key enzyme in the synthesis of the branched-chain amino acids isoleucine and valine, is also sensitive to some herbicides. Glyphosate, the sulfonylureas, and the imidazoles all inhibit specific enzymes in amino acid synthesis pathways.

In plants and microorganisms, synthesis of aromatic amino acids and ubiquinone proceeds by the shikimate pathway. In plants, this pathway also provides the precursors for indoleacetic acid (IAA), a plant growth regulator, alkaloids, lignin, the flavonoids, and a wide variety of secondary metabolites. Some 20 enzyme-catalyzed reactions are involved in the production of the aromatic amino acids. However, herbicide mode of action studies have focused on the three enzymatic steps that convert shikimic acid to chorismic acid, the branch intermediate for a variety of metabolites.

Two relatively new classes of herbicides, the sulfonylureas and the imidazolinones, have totally different chemical structures but remarkably similar modes of action in plants. Both types of herbicides inhibit the same key enzyme acetolactate synthase, in the biosynthetic pathway to branched-chain amino acids. The branched-chain amino acids, ie, valine, leucine, and isoleucine, are essential amino acids produced by microorganisms and plants only. Their synthesis proceeds from threonine and pyruvate through a common series of reactions.

The imidazolinone herbicides selectively block branched-chain amino acid synthesis through inhibition of ALS. Genetic evidence strengthens the hypothesis that ALS is the site of action of both the imidazolinones and the sulfonylureas.

Amitrole (3-amino-*s*-triazole) blocks histidine synthesis in bacteria by inhibiting imidazole glycerol phosphate dehydrase. In light-grown plants, amitrole increases free amino acids and decreases protein. The effect on protein synthesis may be indirectly caused by interferences with purine metabolism and/or glycine–serine interconversion. Amitrole may also interfere with purine synthesis at the step in which formylglycineamidine ribotide is cyclized to form 4-aminoimidazole. Amitrole also inhibits catalase.

Cell Division Inhibitors

The most common mode of action of soil-applied herbicides is growth inhibition, primarily through direct or indirect interference with cell division. The influences of herbicides on cell division fall into two classes, ie, disruption of the mitotic sequence and inhibition of mitotic entry from interphase (G_1, S, G_2). If cell-cycle analyses indicate increases in abnormal mitotic figures, combined with decreases in one or more of the normal mitotic stages, the effect is upon mitosis. Mitotic effects usually involve the microtubules of the spindle apparatus in the form of spindle depolymerization, blocked tubulin synthesis, or inhibited microtubule polymerization. Herbicides that block mitotic entry decrease or prevent the formation of mitotic figures in meristems. Amino acid, protein, RNA, DNA, and ATP synthesis and/or utilization can all arrest cell growth. Although not registered as herbicides, cycloheximide inhibits mitotic entry by inhibiting protein synthesis.

The best understood of the herbicides inhibiting cell division are the dinitroanilines. Micromolar levels of these herbicides act by disrupting mitosis and producing aberrant mitotic figures in which there is no chromosome movement, due to the absence or dysfunction of the spindle apparatus. Dinitroanilines are reported to bind to higher plant tubulin and to prevent, in a concentration-dependent manner, the polymerization of higher plant tubulin in microtubles. Characteristically, dinitroaniline treatment induces swelling of the cell-elongation zone of the root-tip area, suggesting that the microtubule orientation skeleton necessary for cell wall formation is absent or functioning improperly. Phosphorothioamidates, although not developed commercially as herbicides, also disrupt tubulin function and the mitotic sequence.

Various compounds in the N-phenylcarbamate class of herbicides, act through disruption of mitosis, inhibition of PSII electron transport, or uncoupling of oxidative phosphorylation and photophosphorylation. These compounds affect the microtubule organizing center, causing the formation of multipolar spindle configurations. The N-phenylcarbamates also cause root-tip swelling, a branching of the cell plate during cell-wall formation during telophase, and chromosome abnormalities like bridging between daughter chromosomes.

The growth inhibitory mechanism of the thiocarbamate herbicides is not well defined.

Chloroacetamide herbicides, such as alachlor, allidochlor metolachlor, and propachlor, are general growth inhibitors that inhibit both cell enlargement and mitotic entry. Chloroacetamide herbicides are alkylating agents that could inhibit the cell cycle at interphase through alkylation of essential enzymes.

Whatever the mode of action, herbicides that inhibit amino acid synthesis also cause a rapid inhibition of cell growth, usually through inhibition of mitotic entry.

DCPA (dimethyl 2, 3, 5, 6-tetrachloro-1, 4-benzenedi carboxylate) inhibits the growth of grass species by disrupting the mitotic sequence, probably at entry. DCPA influences spindle formation and function and causes root-tip swelling and brittle shoot tissue. DCPA, like colchicine and vinblastine, arrests mitosis at prometaphase and is associated with formation of polymorphic nuclei after mitotic arrest. Pronamide also inhibits root growth by disrupting the mitotic sequence in a manner similar to the effect of colchicine and the dinitroanilines. Cinmethylin and bensulide prevent mitotic entry.

Before the germination process has begun, dormant seeds are not sensitive to chemical herbicides, and many noxious weeds owe their persistence to seed dormancy survival strategies. Weed seeds lie dormant in the soil until specific conditions of environment and time since dispersal are met, and a portion of the seeds in the soil seed bank germinate. These germinated seedlings then become susceptible to herbicides. Other dormant seeds, with different germination requirements, remain in the soil and eventually germinate after earlier herbicide applications are no longer effective. Programs to control weed species surviving through seed dormancy and parasitic weeds, that germinate only in the presence of host roots frequently include development of germination stimulants that break seed dormancy. The nonparasitic weed seedlings are then susceptible to pre-emergence herbicides such as those that inhibit cell division, an essential process in seed germination. Further, suicide germination in the absence of an obligate host kills the parasitic weed seedlings by starvation.

Plant Growth Regulator Synthesis and Function Inhibitors

In a broad sense, herbicides are exogenous plant growth regulators, ie, plant growth inhibitors. Endogenous plant growth regulators, which can both stimulate and retard growth, are now preferably called plant growth regulators (PGRs). These compounds are common to all higher plants and also for the synthetic compounds that are analogues, competitors, or antagonists of the natural PGR compounds. These PGRs are classified as auxins, gibberellins, cytokinins, abscisic acid, and ethylene.

ENVIRONMENTAL FATE OF HERBICIDES

Herbicide Fates in Plants

Beyond modes of action and structure—activity relationships, developers of new herbicides must also consider uptake by plants, translocation within the plant, and possible deactivation of herbicides by contact with soil. Some of these problematic factors can be addressed as part of the QSAR studies and during the screening process. Considerable attention is also being paid to the use of safeners, which protect the crop from herbicides that specifically target the weeds usually associated with that crop. Environmental protection and pesticide regulation concerns are the driving forces in the current efforts toward minimizing application rates, optimizing delivery through improved formulations and application equipment, and increasing target specificity. These research and development efforts include other important and related areas of interest to chemists; eg, the fate and detection of herbicides in the soil and ground and surface water.

Factors Affecting Environmental Fate

The fate of herbicides in the environment is influenced by many chemical, biological, and physical factors. The principal transport and dissipation pathways include sorption to organic and mineral soil and sediment constituents; transport to groundwater in the solution phase by mass flow and/or diffusion; transport to surface water in either the solution or sorbed phases; loss to the atmosphere through volatilization, with redeposition at a later time and location; transformation or mineralization by biological, chemical, or photochemical processes; and uptake by plant or animal species. These processes do not operate as isolated systems but occur simultaneously and involve significant interaction and feedback. Although the environmental fates of most herbicides are controlled primarily by one or two of the outlined processes, all of these factors influence the fate to some extent.

Measurement of Environmental Fate

Continued concern is expressed over the potential contamination of surface and groundwaters by agricultural chemicals. Herbicides have received much of this attention, due to their widespread use and the large total volume applied. However, this perceived threat to groundwater resources appears to be largely unfounded. A survey of private wells and public water well supplies in the United States has revealed that <1% contain herbicides at levels that would affect human or animal health. In addition, those sources that are contaminated can usually be attributed to point rather than nonpoint sources. A point source of contamination is readily located and thus more easily controlled and remediated, and is generally associated with industrial sources or municipal wastewater plants, although agricultural sources such as herbicide equipment rinsing stations also could be point sources. A nonpoint contamination source is one in which the exact source is unknown. They are typically diffuse, often of a large area extent, and are generally of agricultural origin. Nonpoint sources are generally treated by modifications in agricultural management practices. Typical modifications would include the use of alternative herbicide formulations, the splitting of the herbicide application in time, or the installation of vegetative buffer strips to trap runoff.

A re-evaluation of the water quality problem has revealed that surface water resources, rather than groundwater resources, are at higher risk of contamination from agricultural chemicals.

The public health implications of drinking water contamination by herbicides are unclear. The levels that have been detected in groundwater are generally in the part per billion (ppb) or part per trillion (ppt) range and are below estimated acute toxicity levels. However, the long-term health effects of this exposure are generally unknown. Several studies have demonstrated that the mortality from some types of cancer is significantly higher in rural residents of many corn belt states. The U.S. Environmental Protection Agency (EPA) has developed a classification scheme in an attempt to further evaluate the carcinogenic potential of herbicides and pesticides. In this system, chemicals are placed in one of five groups, A–E, according to their carcinogenic potential, ranging from definite (A) human carcinogens to no evidence of carcinogenicity for humans (E). The principal difference between these groups is the amount of accumulated evidence demonstrating carcinogenic potential.

This classification scheme is used in part in the determination and calculation of health advisory (HA) drinking water levels or carcinogenic risk estimates. The majority of herbicides in use in the United States for which HAs have been issued fall into Group D, with a smaller percentage falling into Group C. This would indicate that there are insufficient data to classify the carcinogenic potential of many herbicides. This does not imply that chemical companies are not adequately testing herbicides. To the contrary, exhaustive toxicological testing of a potential herbicide is required by the U.S. EPA before registration. The lack of data does indicate, however, that further testing will be required before the carcinogenic potential of many herbicides is known. Based on available HAs and the U.S. EPA classification scheme, acifluorfen, alachlor, amitrole, haloxyfop–methyl, lactofen, and oxadiazon have been listed as B2 carcinogens. Further information on carcinogenic risk assessment is available.

Since 1984, dramatic technical advances have been made in the analysis of trace organic chemicals in the environment. Indeed, these advances have been largely responsible for the increased public and governmental awareness of the wide distribution of herbicides in the environment. The ability to detect herbicides at ppb and

ppt levels has resulted in the discovery of trace herbicide residues in many unexpected and unwanted areas. The realization that herbicides are being transported throughout the environment, albeit at extremely low levels, has caused much public and governmental concern. However, the public health implications remain unclear.

Traditionally, herbicides have been analyzed by gas chromatography (gc) or spectrophotometric methods. The method of choice when accuracy and sensitivity are of the utmost importance is gc, especially when combined with mass spectrometry. However, several other methods are used for routine monitoring or screening purposes. High-pressure liquid chromatography (hplc) provides detection limits that nearly rival gc and require significantly less sample preparation and cleanup. Advances in the 1980s have made thin-layer chromatography (tlc) a valuable tool in herbicide analysis. Another analytical tool that has received much attention and shows great promise for routine analysis is enzyme immunoassay (eia). This technique offers the advantages of low cost analysis, few interferences, high specificity and sensitivity, and a minimal amount of sample preparation.

A mobility ranking based on soil thin-layer chromatography (stlc) is used to classify the herbicide leaching potential of various herbicides. The rankings range from I (immobile) to V (very mobile) with intermediate categories of II (low mobility), III (intermediate), and IV (mobile). This method is widely used and has been accepted for submission of leaching data for herbicide registration purposes by the EPA.

HERBICIDE GROUPS

Herbicides can be grouped according to common structural features. Sometimes the assignment is arbitrary when there are a multitude of functional groups; eg, acifluorfen, which is a diphenyl ether (phenoxy compound) as well as a trifluoromethyl compound.

Phenoxyalkanoics

The phenoxyalkanoic herbicide grouping is composed of two subgroups, the phenoxyacetic acids and the phenoxypropionic acids. They are widely used for foliar control of broadleaf weeds. The more heavily functionalized phenoxypropionic acid herbicides are relatively new herbicides compared to the phenoxyacetic acids and are used primarily for selective control of grassy weeds in broadleaf crops.

Considerable concern has been raised over the carcinogenic potential of 2,4-D. However, the World Health Organization (WHO) has evaluated the environmental health aspects of this chemical and concluded that 2,4-D poses an insignificant threat to the environment. They did indicate, however, that only limited data on toxicology in humans are available. An HA has been issued for MCPA. It was found in 4 of 18 SW samples analyzed and in none of GW samples, and has been placed in Group D for carcinogenic potential. The EPA has published two gc methods for the analysis of the phenoxyalkanoic herbicides.

Bipyridiniums

The bipyridinium herbicides (Table 1), paraquat and diquat, are nonselective contact herbicides and crop desiccants. Diquat is also used as a general aquatic herbicide. Paraquat and diquat are much more toxic than most herbicides and ingestion of sufficient quantities can result in death if prompt medical treatment is not obtained.

Benzonitrile, Acetic Acid, and Phthalic Compounds

Benzonitrile herbicides (Table 1) are generally used for pre-emergence and post-emergence control of broadleaf weeds. Dichlobenil also controls grass weeds. Dichlobenil, endothall and fenac are used as aquatic herbicides. Most benzonitriles are selective in their control. Benzonitrile herbicides are acidic in nature; thus their environmental fate is influenced by changes in soil pH. Sorption of these herbicides is expected to increase with decreasing pH.

Dinitroanilines and Derivatives

Dinitroaniline herbicides are used principally for the selective, pre-emergence control of annual grasses and broadleaved weeds.

Acid Amides

The principal use of acid amide herbicides is the selective control of seedling grass and certain broadleaved weeds. The majority of acid amide herbicides are applied pre-emergence or preplant incorporated, except for propanil, which is applied postemergence.

Phenylcarbamates

Phenylcarbamate herbicides represent one of two subgroups of carbamate herbicides, the phenylcarbamates and the thiocarbamates. The carbamate herbicides are used, in general, for the selective pre-emergence control of grass and broadleaved weeds. Exceptions would include barban, desmedipham, and phenmedipham, which are applied postemergence.

Thiocarbamates

Thiocarbamate herbicides are nonionic. Diallate and triallate were strongly sorbed to both cation- and anion-exchange resins, but minimally to kaolinite or montmorillonite. This behavior suggests a physical rather than an ionic mechanism of attraction.

Triazines

Triazine herbicides are one of several herbicide groups that are heterocyclic nitrogen derivatives. Triazine herbicides include the chloro-, methylthio-, and methoxy-triazines. They are used for the selective pre-emergence control and early postemergence control of seedling grass and broadleaved weeds in cropland. In addition, some of the triazines, particularly atrazine, prometon and simazine, are used for the nonselective control of vegetation in noncropland. Simazine may be used for selective control of aquatic weeds.

Pyridines and Pyridazinones

Pyridine herbicides are auxin-type herbicides generally used for selective control of broadleaved weeds in cropland, rangelands, and noncroplands. The pyridazinones are used primarily for the selective pre- and postemergence control of seedling grass and broadleaved weeds in cotton and sugarbeets.

Sulfonylureas

Sulfonylurea herbicides are a relatively new class of herbicides generally used for selective pre- and postemergence control of broadleaved weeds in croplands. Sulfometuron–methyl is used for broad-spectrum selective or nonselective weed control in noncroplands.

Imidazoles

Imidazole herbicides are generally used for selective pre- and postemergence control of grass and broadleaved weeds in croplands. Buthidazole and imazapyr are used for broad-spectrum, nonselective weed control in noncroplands.

Other Heterocyclic Nitrogen Derivative Herbicides

The herbicides in this group are heterocyclic nitrogen derivatives that do not readily fall into one of the previously discussed groups. They have a wide range of uses and properties. Most of these herbicides are used for selective, pre- and/or postemergence weed control. Amitrole is used for post-emergence, nonselective weed control in noncroplands and also as an aquatic herbicide.

Ureas and Uracils

Urea herbicides are generally used for selective pre-emergence and early postemergence control of seedling grass and broadleaved weeds. Uracil herbicides are generally used for selective control of annual and perennial weed control in certain crops and for general weed control in noncrop areas. Bromacil, linuron and tebuthiuron are used for the nonselective control of weeds in noncropland.

Aliphatic–Carboxylics

These are used primarily for the selective control of annual and perennial grass weeds in cropland and noncropland. Dalapon is also used as a selective aquatic herbicide.

Metal Organics and Inorganics

The metal organic herbicides are arsenicals used for the selective, postemergence control of grass and broadleaved weeds in cropland and noncroplands.

Miscellaneous Trifluoromethyl Compounds

The herbicides in this group are used for a wide variety of weed-control purposes. Acifluorfen, lactofen, and oxyfluorfen are used for selective, pre-, and postemergence weed control in croplands. Fluorochloridone is used for selective, pre-emergence weed control in cropland, and fluridone, fomesafen, and mefluidide are used

Table 1. Environmental Health Advisories for Herbicides

Herbicide	Health advisories[a]		Mobility[d]	Carcinogenic potential group[b]	Analytical methods[c]
	SW	GW			
Bipyridinium compounds					
diquatop			immobile		hplc
paraquat		0/843	immobile	E	hplc
Benzonitrile, acetic acid, and phthalic compounds					
chloramben	13/34	1/566	very mobile	D	gc[e]
DCPA	386/1995	12/982		D	gc[e]
dicamba	262/806	2/230	very mobile	D	gc[e]
dichlobenil			low		
endothall	0/3	0/604		D	gc
naptalam					uv
Dinitroaniline and derivatives					
benefin					gc
dinitramine					gc
dinoseb	1/89	0/1270		D	
fluchloralin					gc
oryzalin					uv
pendimethalin					gc
trifluralin	172/2047	1/507	immobile	C	ir
Acid amides					
alachlor					gc
bensulide			immobile		hplc
diphenamide	0/3	0/678	intermediate	D	gc
metolachlor	2091/4161	13/596		C	gc
napropamide					
pronamide	20/391			C	gc
propachlor	34/1690	2/99	intermediate	D	gc
propanil			low		
Phenyl carbamates					
chloropropham			low		
karbutilate					hplc
propham	1/392	0/583	intermediate	D	hplc
Thiocarbamates					
asulam					uv
butylate	91/836	2/152		D	gc, glc
EPTC					hplc
thiobencarb			relatively immobile		gc, glc
triallate					gc, glc
vernolate					hplc
Triazines					
ametryn	2/1190	24/560	intermediate	D	general[f]
atrazine	4123/10,942	343/3208	intermediate	C	gc
cyanazine	1708/5297	21/1821	intermediate	D	ir
hexazinone			relatively immobile	D	gc
metribuzin	938/4651	0/416		D	general[f]
prometon	386/1419	36/746	intermediate	D	gc
prometryn			low		
propazine	33/1097	15/906	intermediate	C	general[f]
simazine	922/5873	202/2654	intermediate	C	gc
terbutryn					general[f]

Table 1. (*Continued*)

Herbicide	Health advisories[a]		Mobility[d]	Carcinogenic potential group[b]	Analytical methods[c]
	SW	GW			
			Pyridines		
clopyralid			minimal[g]		
fluroxypyr			varied		
picloram	420/744	3/64	mobile[h]	D	general[h]
triclopyr			intermediate		hplc
			Pyridazinones		
norflurazon			low		
pyrazon					uv
			Sulfonylureas		
chlorimuron, ethyl			mobile		
chlorsulfuron			intermediate to very mobile		hplc, gc
metsulfuron, methyl					gc
sulfometuron			mobile to very mobile		
			Imidazole compounds		
buthidazole					eia
imazamethabenz					eia
imazapyr					eia
imazaquin			mobile to very mobile[i]		eia
imazethapyr			immobile to mobile[i]		eia
			Other heterocyclic nitrogen derivatives		
amitrole			mobile		vis
bentazon			very mobile	D	hplc
isoxaben			immobile		
			Ureas and uracils		
bromacil	0/3	0/841	mobile	C	glc
chloroxuron			immobile		glc
diuron	0/25	0/1337	low	D	ir
fluometuron	0/14	0/156	intermediate	D	uv
linuron					uv
tebuthiuron			intermediate to very mobile	D	uv
terbacil				E	uv
			Aliphatic-carboxylic		
dalapon	0/14	0/14	very mobile	D	ir
TCA			very mobile		
			Inorganics and metal organics		
AMS				D	titration
			Miscellaneous trifluoromethyl compounds		
acifluorfen				B$_2$	hplc
fluridone					gc
lactofen					hplc
			Amino acid analogues		
glufosinate,			intermediate		
glyphosate	0/6	0/98	immobile to low mobility	D	hplc

Table 1. (*Continued*)

Herbicide	Health advisories[a]		Mobility[d]	Carcinogenic potential group[b]	Analytical methods[c]
	SW	GW			
			Other miscellaneous compounds		
cinmethylin					gc
ethofumesate					
tridiphane					gc

[a] SW = surface water; GW = ground water. Positive results/number of tests.

[b] Group A, human carcinogen; Group B, probable human carcinogen; Group C, possible human carcinogen; Group D, not classifiable; Group E, no evidence of carcinogenicity for humans.

[c] gc = gas chromatography; hplc = high pressure liquid chromatography; ir = infrared spectroscopy; uv = ultraviolet spectroscopy; glc = gas-liquid chromatography; eia = enzyme immunoassay; vis = visible spectroscopy.

[d] Mobility ranking based on soil thin-layer chromatography (stlc).

[e] Gc for chlorinated pesticides can be used.

[f] General draft method for nitrogen- and phosphorus-containing pesticides.

[g] Mobility has been reported to be mobile and minimal in different studies.

[h] General draft method for determination of chlorinated acids in water.

[i] Mobility is a function of soil pH 306.

for postemergence control. Fluridone is also used as an aquatic herbicide.

Amino Acid Analogues

Amino acid analogue herbicides also control a large variety of weeds. Glyphosate and glufosinate are used for the broad-spectrum, nonselective control of grass and broadleaved weeds. Diethatyl is used for selective, pre-emergence control of grass and broadleaved weeds. Flamprop is used to control the growth of wild oats in wheat.

Miscellaneous Other Herbicides

The herbicides in this group are not readily included in any of the preceding groups. Acrolein (2-propenal) is used as a contact, aquatic herbicide. Sethoxydim, clethodim, and tridiphane are used for selective, postemergence weed control. Cinmethylin and clomazone are used for selective pre-emergence control and etholumesate for selective pre- and postemergence weed control.

ECONOMIC ASPECTS

Consumption of herbicides in recent years has risen slightly because of increased planting. However, herbicide use is expected to decline through 2006 because of the introduction of newer herbicides with more highly active ingredients.

Herbicide production has decreased steadily in western Europe since 1989. About 50 major herbicides are used in this area of the world, being produced by European-based companies. Japan's consumption of herbicides has been declining at the rate of 2.6%/yr and the market is not expected to grow. Less rice is being planted and rice farmers are using herbicides that offer more residual weed control.

OTHER WEED MANAGEMENT AGENTS AND TECHNIQUES

Adoption by the agricultural community requires that an innovative weed management agent must be an effective

control of the target species, be cost effective, and be practical to employ. It must not interfere with crop production practices such as crop rotation or the use of other pesticides. Additionally, new weed-control agents cannot pose a significant threat to human health or the environment. Considerable costs are incurred in the development, registration, production, and marketing of weed-control agents. These costs require that an herbicide have sufficient long-term market viability and market niche potential to justify these costs in time and money. The need for safe and effective methods of crop production in an environment that contains competitive weeds is becoming increasingly critical.

Weed Management Strategies

Managers of agroecosystems are being encouraged to manage weed populations at levels that are below their economic optimum thresholds, rather than attempting to eliminate or control all noncrop plants regardless of their actual impact. Decisions concerning management of weed populations should be governed by both agroecological principles and site-specific considerations in the context of an overall integrated pest management program. However, the practical implementation of integrated pest management programs can be difficult.

Nonchemical or traditional practices, such as weed seed removal, optimal crop seeding rates, crop selection, enhanced crop competitiveness, crop rotation, and mechanical weed control, are all important components of an effective weed management program. In the context of modern intensive chemical herbicide application, nonchemical practices may even represent an innovative approach to weed management and should receive careful consideration.

Natural Products and Allelopathic Compounds as Herbicides

There is growing concern that compounds that do not occur in nature may produce unanticipated health and environmental problems. However, plants, fungi, marine organisms, and certain bacteria produce a vast array of organic compounds, and many of these natural products exhibit biological activity. In nature, these compounds are produced in minute quantities and present interesting chemical problems in detection, identification, quantification, and production of active and stable analogues of these natural products. Although these compounds appear to be ecologically safe in naturally occurring amounts, the large quantities required for agricultural applications may cause environmental problems similar to those associated with chemical herbicides.

Investigations of natural product chemistries have aided in the development of bialaphos, cinmethylin, picloram, glufosinate, and other important herbicides. Additional compounds may be found through investigations of natural products that cause plants and other organisms to undergo rapid physiological change, such as plant hormones and phytotoxins. Many plant hormones and phytotoxins are also produced by microorganisms. Additionally, microorganisms have been reported to contain novel natural products that could provide basic structural templates for the development of new herbicides.

Plant Pathogens and Insects as Control Agents

Concerns about accumulations of chemical control agents in the environmental and food resources have also increased interest in microbial weed control agents. Controlling weeds with carefully screened plant pathogens offers several benefits, including a high degree of specificity for a given target weed, low potential for negative human health and environmental impact, inability to accumulate in the food chain, and other advantages. The high degree of host specificity may limit the market size for some biological control agents, but these biocontrol agents can be combined with chemical herbicides and other pathogens to increase the spectrum of weeds controlled. The marketing of biological control agents may also be constrained by slow expression of phytotoxicity, pathogen dependence on optimum environmental conditions, potential resistance of the weed toward the pathogen, and lack of formulation stability under field conditions and during preuse storage. These constraints can be addressed by genetic manipulation of selected pathogenic strains to produce more effective control agents and by the investigation of the mechanisms of disease resistance in plants.

There are two principal approaches to the biological control of weeds. The first is referred to as classical or inoculative biological weed control. The intent of classical biological weed control approaches is to manage introduced weed populations by introducing host-specific pathogens from the weed's native range, thus moderating the growth of weed populations by the reestablishment of an old association between host and pathogen populations in the expanded range.

An additional approach to biological weed control is referred to as the inundative or augment approach to biological weed management. This approach utilizes pathogenic propagules formulated as a weed control agent, eg, mycoherbicides. The mass-inoculation of pathogenic propagules in an effective formulation can enhance the dissemination and survival of the pathogens, overwhelm target weed resistance, and produce results similar to those achieved with chemical herbicides. Mycoherbicides often contain native pathogens that are active against native weeds and are thus highly selective against the target weed species.

Control of Weed Seeds

If agents that control weed seed germination could be applied prior to planting, interference from weeds would be prevented until reintroduction of weed propagules. Additionally, if a very large portion of the weed seed bank could be stimulated to germinate prior to planting, weeds could be controlled by a single cultivation or application of nonselective herbicide.

Biotechnology

Genetic modification may provide plants resistant to disease, nematodes, or insects. Plants resistant to herbicides are being marketed, but their acceptance in some areas is a controversial issue. Internationally, there is no agreement on safety protocols. Introduction of viable organisms produced by genetic modification has generated a number of unanswered questions, and there still remains a need for readily applicable techniques to assess the environmental impact of the new technology. The rapid expansion of this field of science opens many questions of application, ownership, and exploitation of its novel discoveries.

W. H. Ahrens, ed., *Herbicide Handbook*, 7th ed., Weed Science Society of America, Lawrence Kans., 1994.

A. E. Greenberg, L. S. Clesceri, and A. D. Eaton, *Standard Methods for the Examination of Water and Wastewater*, 18th ed., American Public Health Association, Washington, D.C., 1992.

J. R. Plimmer, D. W. Gammon, and N. N. Ragsdale, eds., *Encyclopedia of Agrochemicals*, Vol. 2, Wiley-Interscience, Hoboken, NJ, 2003.

C. D. S. Tomlin, ed., *The Pesticide Manual*, 11th ed., Brit. Crop Prot. Council, Farnham, UK, 1997.

Jack R. Plimmer
Tampa, Florida
Judith M. Bradow
Christopher P. Dionigi
Richard M. Johnson
Suhad Wojkowski
United States Department
   of Agriculture

HERBICIDES, BIOTECHNOLOGY

BIOCONTROL OF WEEDS WITH PLANT PATHOGENS

Because of concerns of health, safety, and sustainability, there is a growing interest in reducing chemical weed control measures in both agricultural and natural systems. This has led to an increased interest in the use of biological agents to control weeds. Insects, pathogens, grazing animals, and allelopathic crops can all be used for to biological control of weeds.

There are several advantages of biological control of weeds over chemical or cultural methods. Biological control methods for weeds usually cause less contamination of soil, water, and food with unwanted synthetic compounds, and they do not contribute to soil erosion, as tillage, the main nonchemical method of weed management, does. Furthermore, they are generally more targeted to specific weeds than are synthetic herbicides. Biocontrol measures are ideal for weeds that escape chemical control, for organic farming, and for weeds that are in areas in which herbicides cannot be used because of environmental sensitivity. Another major concern is evolved herbicide resistance, which now has developed in more than 200 weed biotypes. These factors, coupled with the banning of many herbicides, more stringent registration and regulation, and the need for nonchemical alternatives in environmentally sensitive areas, have promoted the use of plant pathogen as biological weed control agents.

Under the inundative or augmentative approach, a native biocontrol organism is provided in sufficient quantity to overwhelm the defenses of a target population of weeds. Indigenous biocontrol species are generally used with this approach. The use of indigenous plant pathogens with limited host specificity has been the primary emphasis of research and development of microbial herbicides or "bioherbicides." Several microbes have been patented and commercialized as biocontrol agents for various weeds. Table 1 provides a sample of these. Although many have been patented, few have been commercialized, and few of those commercialized have remained on the market.

Commercial herbicides target many weed species, but most bioherbicides are host specific, targeting only one species or a few closely related species. Unless the target weed species is a major problem in rice, the biocontrol agent is likely to be too expensive for use with just one weed. The use of broad-spectrum bioherbicides such as

Myrothecium verrucaria has only recently been considered and shows promise for invasive weeds such as kudzu (*Pueraria lobata*), a weed that covers millions of hectares in the southeastern United States.

Most of the other problems associated with microbial biocontrol agents stem from the argument that it is unpredictable or too poor to be economical. Most of these organisms require a very narrow environmental window compared to most commercial herbicides. Most commonly, they need an extended period of dew or very high humidity in order to infect the host.

Microclimate-related efficacy problems have been solved or reduced with innovative formulations. For example, formulation of mycoherbicide spores in invert emulsions provides the proper microclimate, trapping water in the formulation and increasing the time that the spore has to infect target species.

Wounding the target plant, either mechanically or with a contact herbicide, increases the virulence of most plant pathogens. For example, paraquat applied before the mycoherbicide *Puccinia canaliculata* to *Cyperus esculentus*, resulted in almost complete control of the weed, compared to 10 and 60% control for paraquat and the mycoherbicide, respectively. The problem in most crop situations is that the wounding must be confined to the weed. Thus, a selective herbicide or selective method of wounding the weeds must be used. The wounding approach is ideal for weeds in mowed areas, such as in turf. *X. campestris* pv. *poannua* enters the host through mowing wounds, causing lethal, systemic wilt of the target species.

There is less need for dew with plant pathogens that cause soil-borne diseases. Application can be to moist soil or can be made in granules that are activated after rain or irrigation. Deleterious rhizobacteria do not directly kill the weeds but reduce their growth and competitive ability.

The application technology for foliar application of microbial biocontrol agents can be challenging. Ordinary spray equipment will not work with viscous invert emulsions. Spray systems are available for such formulations,

Table 1. A Sample of Some Commercial Microbial Biocontrol Agents Used for Weed Management

Microbe	Target weed	Trade name
Alternaria cassiae	*Cassia obtusifolia*	Casst
Alternaria sp.	*Cuscuta* spp.	Smolder
Chondrostereum pupureum	various angiosperm trees	Biochon ECO-clear
Colletotrichum gloeosporioides f. sp. *aeschynomene*	*Aeschynomene virginica*	Collego
Colletotrichum gloeosporioides f. sp. cuscutae	*Cuscuta* spp.	Lubao 2
Colletotrichum gloeosporioides f. sp. malvae	*Malva pusilla*	BioMal
Cylindrobasidium laeve	*Acacia* spp.	Stumpout
Fusarium spp.	*Abutlion throphrasi*	Velgo
Phytophthora palmivora	*Morrenia oderata*	DeVine
Puccinia canaculata	*Cyperus esculentus*	Dr. BioSedge
Xanthomonas campestris pv. *Poannua*	*Poa annua*	XPo, Comperico

but the added cost of the specialized application system reduces the probability of adoption by a farmer.

Finally, the limited shelf life and special storage conditions of a living organism are other complications and expenses that limit adoption of many of these products.

Agricultural ecosystems are a web of interacting factors, each influencing the other, often in subtle ways, but sometimes having unexpected, profound effects. In most cases, particularly in production agriculture, biological control strategies must coexist with other weed management technologies. Thus, one should strive to predict what these interactions will be and how they can be used to improve the efficacy of biocontrol. There are indications that some interactions will be antagonistic to biocontrol approaches. Few biocontrol options exist, despite significant research efforts and increasing public pressure to reduce or eliminate dependence on synthetic herbicides.

Biotechnology to Improve Biocontrol Agents

The interest in using biological control agents to control weeds greatly exceeds the availability of efficacious biological control agents, even though numerous weed pests and pathogen have been identified. With weed pathogens, formulation, mass production, and storage life of the pathogens are frequently cited as limitations in their development into commercial products. However, the majority of the described weed pathogens simply lack sufficient virulence or host range to provide economical and efficacious weed control. Without a major effort to genetically modify weed pathogens to modify host range and/or increase virulence, the rate of commercialization of microbial weed biological control agents will probably remain low.

CONTROL OF WEEDS WITH ALLELOPATHY

Plants can interfere with each other through competition for resources or through allelopathy. Allelopathy can be narrowly defined as chemical warfare between different plant species. Both crops and weeds produce phytotoxins that could be allelochemicals that provide an advantage in plant–plant competition. Proving the role of these compounds as allelochemicals has been problematic.

There are several ways that allelopathy could be used in weed management: allelopathic cover or smother crops; allelopathic companion crops; allelopathic mulch or incorporation of phytotoxic crop residues; production of allelopathic crop cultivars with weed-suppressing potential, and use of allelochemicals as sources of natural herbicides.

Biotechnology to Improve Allelopathy

Little research has concentrated on the development of allelopathy as an important trait in major agricultural crops, even though it clearly exists in the germplasm of cucumbers, barley, rice, wheat, rye, and sorghum. At this time, the level of allelopathic activity in these crops is inadequate to provide satisfactory weed management in the field. Standard breeding programs could possibly

be used to enhance allelopathy of these crops. However, as long as allelopathy is considered a value added trait of little economic value and yield remains the major selection criteria of most breeding programs, allelopathy will never be developed in cultivars produced by traditional methods. In most cases allelopathy will act as a quantitative trait that is difficult to select for in breeding programs. Furthermore, traditional breeding methods would probably be insufficient for creating lines that provide adequate weed control without the intervention of some application of commercial herbicides.

The fate of allelopathy, as a practical tool for more environmental friendly agriculture, appears to lie in the hands of biotechnology. This task is likely to be more complicated than creating a herbicide-resistant crop or producing a crop with resistance to insects or pathogens, as the applications presently in use are the result of manipulating one gene. The transfer of an allelopathic trait to a non-allelopathic crop may require the manipulation of several genes.

The ultimate goal of biotechnology research on allelopathy is to either enhance this trait in a crop where this naturally occurs or to transfer this trait to another species. Both of these goals require the production of transgenic plants. A major consideration when producing transgenic plants is to use constructs that are tissue specific in comparison to those that are constitutively expressed. Constitutively expressed promoters can result in autotoxicity or result in unnecessary metabolic costs. In the case of allelopathy, it would be desirable to express the genes solely in the roots or root hairs. To enhance production of allelochemicals in a crop that already has the trait, it would be best to increase the expression of a regulatory gene that is controlling several genes of the biosynthetic pathway.

Because of the selective nature of allelopathy, it should not be expected that allelopathy alone could control all the weeds in a typical agricultural setting. It could, however, function as a component of an overall weed management strategy. Incorporation of allelopathic traits together with other potential plant interference traits (eg, early vigor, leaf size, plant height, and tillering) into commercial cultivars could be a major step toward further development of sustainable crop production systems with less reliance on herbicides.

Allelochemicals as Herbicide Leads

Allelochemicals can interfere with cell elongation and cell division. Abnormal mitotic stages were observed when artemisinin, quassinoids, lignans, and 1,8-cineole were tested in onion seedling bioassay, suggesting interference with mitosis. Some compounds can destabilize membranes. Inhibition of mineral ion uptake, respiration and protein synthesis, amino acid synthesis, and photosynthesis are other important mechanisms of the action of allelochemicals.

In general, the more phytotoxic an allelochemical is, the more probable its autotoxicity may be. Therefore, one might expect that any level of autotoxicity by allelochemicals could reduce the yield of allelopathic crops.

The effectiveness of allelopathic crops and many plant pathogens in killing or suppressing weeds is dependent on natural phytotoxins. At least part of the efficacy of some microbial herbicide preparations has been speculated to be due to the presence of high levels of phytotoxins in the preparation. These compounds could be considered biologically based weed killers.

Biocontrol organisms are a potential source of new phytotoxins for consideration in herbicide discovery. Several commercial herbicides, such as the triketones, have been based on natural product structures, while others are themselves the natural products. One of the most attractive aspects of natural compounds as herbicides is that they often have entirely new molecular target sites.

Although natural product-based herbicides have the potential for being used directly as herbicides or as templates for new herbicides that might have better toxicological or environmental profiles than synthetic compounds, there is no guarantee of this.

Several commercial herbicides have been or will soon be removed from the agrochemical market because of their impact on the environment or the cost of reregistration. Evolution of weed resistance to many commercial herbicides is becoming increasingly problematic because resistance to one herbicide may preclude the use of other classes of chemicals targeting the same site of action. As a result, the number of chemical tools available to manage weeds is becoming limited.

Allelochemicals may also be useful in providing leads for synthetic herbicides, as the diversity of molecular structures from living sources provides novel structures that are unlikely to be produced by traditional pesticide synthesis programs. Most biologically active natural compounds are water soluble, nonhalogenated molecules, whereas most synthetic pesticides are lipophilic, halogenated products. Thus, plant-derived secondary compounds may provide a source of environmentally safer herbicides with novel molecular sites of action. By modifying these natural products, the end product could be made more active, selective, or persistent. The precursor for the end product may be obtained from a natural source, if the economy of this approach is superior to that of chemical synthesis. The discovery of triketones, a new important class of herbicides, represents a successful example of the chemical ecology approach used as strategy to select sources of natural products for the discovery of potential herbicides. This new class of herbicides led to the discovery of a new molecular target site, p-hydroxyphenylpyruvatedioxygenase (HPPD), an enzyme involved in plastoquinone synthesis. Other natural products, such as sorgoleone and usnic acids, are also good inhibitors of HPPD. Once a natural product has been found to have good phytochemical activity, it is necessary to consider how this information can be applied. In exceptional cases a compound may perform sufficiently well to be a product. Generally, the chemical complexity of many secondary products, which often includes multiple chiral centers, prohibits economical production of the compounds; thus, the source of the compound becomes a key issue. Few natural products have all the necessary characteristics to compete with the best synthetic

agrochemicals. It is much likely, therefore, that a plant natural product will be used as a lead for synthesis rather than as a product. In many cases, improvements in both the potency and physical properties are necessary to generate a commercially viable product. However, if the mode of action of the compound is novel, it may provide a source of inspiration to biochemists and result in the development of a new bioassay capable of detecting other, structurally simpler, compounds with the same mode of action.

CROP RESISTANCE TO HERBICIDES VIA BIOTECHNOLOGY

In developed countries, herbicides are the dominant class of pesticides. Genetic engineering has provided alternatives to pesticide use in managing microbial and insect pests in crops, but the first transgenic crops designed for better weed management have been those which resist herbicides. Here its a brief review of the area of herbicide-resistant crops (HRCs) produced by biotechnological methods.

Glyphosate Resistance

Glyphosate is a highly effective, but environmentally and toxicologically safe, herbicide that inhibits a critical enzyme of the shikimate pathway, 5-enolpyruvylshikimate-3-phosphate synthase (EPSPS). It is very effective in controlling perennial weeds in which subterranean tissues must be killed in order to prevent regrowth.

The rapid adoption of glyphosate-resistant crops is due to several factors. First, this technology greatly simplifies weed management. In many cases, it allows farmers to use only one herbicide, and apply only treatments after the weed problem develops. In cases in which glyphosate is the only herbicide used, the farmer is less dependent on consultants for specialized recommendations for several herbicides that are sometimes applied at different times. Weed management with glyphosate-resistant crops generally requires less equipment, time, and energy than with selective herbicides. The efficacy of glyphosate in combination with glyphosate-resistant crops is generally very good. In many cases, it fills weed management gaps that existed with available selective herbicide. Furthermore, the economics of this approach, even with the "technology fee" added to the cost of the seed, are generally good. Most published economic analyses predict an economic advantage for glyphosate-resistant crops over conventional weed management. The declining cost of glyphosate due to the expiration of its patent favors a continued economic advantage for glyphosate-resistant crop-based weed management.

Glufosinate Resistance

Glufosinate is a nonselective herbicide. There are no published cases of evolved resistance to glufosinate or other glutomine synthetase (GS) inhibitors. However, a variety of oats with resistance to tabtoxinine-β-lactam, a natural glutamine analogue that inhibits GS, has been reported.

Resistance to GS inhibitors through the overexpression of GS has been obtained in cell lines of alfalfa (*Medicago* sativa) and rape (Brassica napus) and in regenerated tobacco plants (Nicotiana tabacum). Developing crop lines resistant to glufosinate by developing plants that overexpress GS may result in undesirable traits. A complicating factor to developing glufosinate-resistant plants is that it would probably require the overexpression of all isoforms of GS in the plant.

To date, only glufosinate-resistant maize, canola, rice, and cotton are currently commercially available in the United States. No weeds have evolved resistance to glufosinate as of this date, but, in canola the transgene should readily move to weedy relative, since gene transfer occurs even without the selection pressure of a herbicide.

Bromoxynil Resistance

Bromoxynil (3,5-dibromo-4-hydroxybenzonitrile), an inhibitor of photosynthesis II of photosynthesis, is not a widely used herbicide. The first introduced commercial HRC was bromoxynil-resistant cotton. This product has been valuable for specific, but not widespread, weed problems.

Sulfonylurea and Imidazolinone Resistance

The sulfonylurea and imidazolinone herbicides are very potent inhibitors of the acetolactate synthase (ALS), a key enzyme of branched-chain amino acid synthesis. They represent a large segment of the herbicide market. Differential metabolic degradation is the mechanism of selectivity in crops in all cases, and specific sulfonylurea and imidazolinone herbicides have been designed for particular crops. However, certain weed species rapidly evolved resistance at the target site level to these herbicides. These weeds with a resistant form of ALS appear to pay little or no metabolic penalty for resistance. Thus, crops could be transformed with a resistant form of ALS to broaden the array of compatible ALS inhibitor herbicides and to reduce the potential for phytotoxicity on the crop.

Resistance to Other Herbicides

Resistance to a large number of other selective herbicides has been achieved with transgenes, but most of these will never be commercially available for economic, environmental, toxicological, or other reasons.

The Future of Transgenic, Herbicide-Resistant Crops

To date, there are commercial, transgenic HRCs for three herbicides: bromoxynil, glyphosate, and glufosinate. Only three transgenes are used with these products. Biotechnology-derived HRCs through mutant selection are also available.

Companies will not market a product unless there is a clear economic reward. With an HRC, the ideal situation is production of transgenic crops that are resistant only to an excellent, reasonably inexpensive, nonselective herbicide to which there is an economic link. To some extent, this has been the case with glyphosate- and glufosinate-resistant crops.

The future for HRCs that are resistant to selective herbicides is less certain. Selective herbicides already exist for all major crops. Thus, a crop that is genetically engineered to be resistant to yet another selective herbicide must fulfill a weed management need that is unmet, such as those use niches filled by bromoxynil-resistant crops. Most selective herbicides belong to herbicide classes represented by several commercial analogues, and thus most resistance transgenes are likely to provide resistance to all members of the herbicide class. The economics of profiting from a HRC tied to selective herbicides hinges on several factors, including the cost of producing and developing the transgenic crop; whether or not there are economic links to manufacturers of the members of the herbicide class; and the degree of need for the product.

F. E. Dayan and S. O. Duke, in R. M. Roe, J. D. Burton, and R. J. Kuhr, eds., Herbicide Activity: Toxicology, Biochemistry and Molecular Biology, IOS Press, Amsterdam, The Netherlands, 1997, pp. 11–35.

S. O. Duke, ed., Herbicide-Resistant Crops, CRC Press, Boca Raton, Fla., 1993.

J. R. Plimmer, D. W. Gammon, and N. N. Ragsdale, eds., Encyclopedia of Agrochemicals, Vol. 2, John Wiley & Sons, Inc., New York, 2003.

A. K. Watson, Biological Control of Weeds Handbook, Weed Science Society American, Champaign, Ill., 1993.

STEPHEN O. DUKE
USDA, ARS, Natural Products
    Utilization Res. Unit

BRIAN E. SCHEFFLER
USDA, ARS, Catfish Genetics
    Research

C. DOUGLAS BOYETTE
USDA, ARS, Southern Weed
    Science Research

JOHN LYDON
USDA, ARS, Sustainable
    Agricultural Systems Res.
    Unit

ANNA OLIVA
State University of New York

HIGH PERFORMANCE FIBERS

High performance fibers are generally characterized by remarkably high unit tensile strength and modulus as well as resistance to heat, flame, and chemical agents that normally degrade conventional fibers. Applications include uses in the aerospace, biomedical, civil engineering, construction, protective apparel, geotextiles, and electronic areas.

PREPARATION AND PROPERTIES

The principal classes of high performance fibers are derived from rigid-rod polymers, gel spun fibers, modified carbon fibers, carbon-nanotube composite fibers, ceramic fibers, and synthetic vitreous fibers.

Rigid-Rod Polymers

Rigid-rod polymers are often liquid crystalline polymers classified as lyotropic, such as the aramid, Kevlar, or thermotropic liquid crystalline polymers, such as Vectran.

Liquid Crystallinity. The liquid crystalline state is characterized by orientationally ordered molecules. The molecules are characteristically rod or lathe shaped and can exist in three principal structural arrangements: nematic, cholesteric, smectic, and discotic.

Industrial Lyotropic Liquid Crystalline Polymers (Aramid Fibers). The first polyaramid fiber (MPD-1), based on poly(m-phenylene isophthalamide), was not liquid crystalline but was the first aramid fiber to be commercialized by DuPont, under the trade name Nomex nylon in 1963 and Nomex aramid in 1972. The principal market niche for Nomex was as a heat-resistant material. Teijin also introduced a fiber (trademark Conex) based on MPD-1 in the early 1970s. In 1970, Du Pont introduced an aramid fiber, Fiber B, for use in tires, which was probably based on polybenzamide (PBA) spun from an organic solvent. Fiber B had high strength and an exceptionally high modulus. Another version of Fiber B with an even higher modulus fiber based on PPT was introduced under the name PRD-49 for use in rigid composites. The undrawn and drawn fibers were later announced as Kevlar-29 and Kevlar-49, respectively. In 1975, Akzo of the Netherlands reported the commercialization of an aramid fiber, Twaron, based on PPT.

Nomex. This fiber was commercialized for applications requiring unusually high thermal and flame resistance. It retains useful properties at temperatures as high as 370°C. Nomex has low flammability and has been found to be self-extinguishing when removed from the flame. An outstanding characteristic is low smoke generation on burning.

MPD-1 fibers may be obtained by the polymerization of isophthaloyl chloride and m-phenylenediamine in dimethylacetamide with 5% lithium chloride. Fibers may be either dry spun or wet spun directly from solution.

Kevlar. In the 1970s, researchers at Du Pont reported that the processing of extended-chain all-para aromatic polyamides from liquid crystalline solutions produced ultrahigh-strength, ultrahigh-modulus fibers. The greatly increased order and the long relaxation times in the liquid crystalline state compared to conventional systems led to fibers with highly oriented domains of polymer molecules. The most common lyotropic aramid fiber is PPT, marketed as Kevlar by Du Pont. Aramid fiber is available from Akzo under the trade name Twaron. These fibers are used in body armor, cables, and composites for sports and space applications.

Technora. In 1985, Teijin Ltd. introduced Technora fiber, previously known as HM-50, into the high-performance fiber market. Technora is based on the 1:1 copolyterephthalamide of 3,4'-diaminodiphenyl ether and p-phenylenediamine. Technora is a wholly aromatic copolyamide of PPT, modified with a crankshaft-shaped comonomer, which results in the formation of isotropic solutions that then become anisotropic during shear alignment during spinning. The polymer is synthesized by the low-temperature polymerization of p-phenylenediamine, 3,4'-diaminophenyl ether, and terephthaloyl chloride in an amide solvent containing a small amount of an alkali salt.

Heterocyclic Rigid-Rod Polymers PBO, PBZ, and PIPD. PBZ, a family of p-phenylene-heterocyclic rigid-rod and extended-chain polymers includes poly(p-phenylene-2,6-benzobisthiazole) (trans-PBZT), poly(p-phenylene-2,6-benzobisoxazole) (cis-PBO), and poly[2,6-diimadazo[4,5-b:4',5'-e]pyridinylene-1,4(2,5-dihydroxy)phenylene] (PIPD).

Toyobo (Zylon) has marketed the PBO fiber and Magellan Systems International has brought the M5 fiber to the marketplace. The PBO fibers have the highest reported tensile modulus of any known polymeric fiber, 280–360 GPa($41–52 \times 10^6$ psi). Both PBO and PBZT are among the most radiation-resistant polymers. Although the compressive strengths of PBO and PBZT are approximately an order of magnitude less than the tensile strengths, alloys of these fibers with high compressive strength fibers can be produced. The polymers are now being evaluated for other applications such as nonlinear optics. Possible PBO applications include reinforcing fibers in composites, multilayer circuit boards, athletic equipment, marine applications, woven fabrics, and fire-resistant fibers. Magellan reports a tenacity of 5.3 GPa, a modulus of 350 GPa, and a compressive strength of 1.6 GPa for M5 fiber. Possible M5 applications include advanced lightweight composites, hard and soft ballistic armour, high strength cables, advanced fabrics and textiles, and high performance fire retardant materials.

Polybenzimidazole (PBI) Fibers. Poly[(2,2'-m-phenylene)-5,5'-bisbenzimidazole] is a textile fiber originally marketed by the Celanese Corporation that does not form liquid crystalline solutions due to its bent meta backbone monomeric component. PBI has excellent resistance to high temperature and chemicals. PBI is being marketed as a replacement for asbestos and as a high-tempera ture filtration fabric with excellent textile apparel properties.

Typical properties of stabilized PBI are a tenacity of 0.27 N/tex (3.1 gf/den), a fiber breaking elongation of 30%, an initial modulus of 3.9 N/tex (45 gf/den), a density of 1.43 gf/cm^3, and a moisture regain of 15% (at 21°C and 65% relative humidity).

Solution dyeing of PBI is necessary because the T_g of PBI is >400°C, and as a result dye molecules diffuse only slowly into the PBI fiber structure. Since the pigments are added to the spinning dope, the pigments must be capable of withstanding the high temperatures used in the various fiber-forming processes.

Industrial Thermotropic LCPs. Vectran, poly(6-hydroxy-2-naphthoic acid-co-4-hydroxybenzoic acid), was the first thermotropic fiber to become commercially available. Vectran is synthesized by the melt acidolysis of *p*-acetoxybenzoic acid and 6-acetoxy-2-naphthoic acid. Vectran HS fibers are reported to have typical tensile strength and modulus values of 2 N/tex (23 gf/den) and 46 N/tex (550 gf/den) respectively. The melting point and density are reported to be 330°C and 1.4 g/cm^3. The fibers have excellent chemical resistance except for their resistance to alkali.

Gel Spun Fibers

In the mid-1970s it was discovered at the Dutch States Mines Co. (DSM) that through an ingenious new method of gel spinning ultrahigh molecular weight polyethylene it was possible to produce fibers having twice the tenacity of Kevlar, which was then considered the strongest known fiber. These high-performance polyethylene fibers (HPPE) produced by the DSM subsidiary company Stamicarbon were called Dyneema. Those produced by Allied-Signal in the United States were called Spectra 1000. The commercial products have somewhat lower strengths than the laboratory fibers but are still in the high 2.6 N/tex (30 gf/den) range.

Properties. Fiber property comparisons for the different products are given in Table 1. The attributes of HPPE fibers include high strength; high abrasion resistance; high uv stability as compared to other synthetics; high resistance to acids, alkali, organic chemicals, and solvents; and low density. Disadvantages are a low melting point of ~150°C, which means performance is limited to no >120°C; difficult processing; and poor surface adhesion properties.

Modified Carbon Fibers (Elongatable Carbonaceous Fiber)

It is difficult to weave or knit regular carbon fiber. To overcome this drawback, an exciting new modification of carbon fiber technology was developed; by using less stringent carbonizing conditions and only partially carbonizing the precursor fibers, improved textile fiber properties have been achieved.

Properties. Unlike regular carbon fibers, these new products do not conduct electricity, but do exhibit good textile processing properties and possess exceptional ignition-resistant, flame-retardant, and even fire-blocking properties.

Carbon-Nanotube Fibers

In 2000, results were reported for a carbon-nanotube spinning method that gave fibers several tens of centimeters long with a Young's modulus of the fibers between 9 and 15 GPa. Then in 2003, this technique was improved

Table 1. Properties of Commercial HPPE Fibers

Fiber	Tenacity, N/tex	Initial modulus, N/tex	Elongation at break, %
Dyneema	1.01–3.57	57–128	3–7
Spectra 1000	3.4–3.57	162–171	3–7

to give composite fibers tougher than any natural or synthetic fibers described to date. The composite fibers were ~50 μm in diameter and contained ~60% single wall nanotubes by weight. They reported a tensile strength of 1.8 GPa, (comparable to that of spider silk) and an energy-to-break of 570 J/g that is higher than that of spider dragline silk, and graphite fiber (12 J/g). Suggested potential applications for these carbon-nanotube fibers include distributed sensors, electronic interconnects, electromagnetic shields, and attennas and batteries.

Silicon Carbide Ceramic Fibers

The commercially produced continuous and multifilament Nicalon fiber is produced from polydimethylsilane; however, other organosilicon polymers have been used for the production of silicon carbide fiber.

Silicon carbide has high thermooxidative stability and good thermal and electrical insulation properties. In composite applications, this fiber can be used to reinforce polymer, metal, and ceramic matrices.

Vitreous Fibers

Man-made vitreous fibers (MMVF) comprise a number of glass and specialty glass fibers and also refractory ceramic fibers. The vitreous state in glass is somewhat analogous to the amorphous state in polymers. However, unlike organic polymers, it is not desirable to achieve the crystalline state in glass. Glasses are produced from glass-forming compounds such as SiO_2, P_2O_5, etc., which are mixed with other intermediate oxides such as Al_2O_3, TiO_2, or ZnO and modifiers or fluxes like MgO, Li_2O, BaO, CaO, Na_2O, and K_2O.

A wide range of glass compositions is available to suit many textile fiber needs; the three most common glass compositions are referred to as E, S, and AR glasses. AR glass is a special glass with higher contents of Zr_2O designed to resist the calcium hydroxide in the cementitious products where it is used. S glass is a magnesium–aluminum–silicate cross-linked glass used where high mechanical strength or higher application temperatures are desired. E glass is a member of the calcium–aluminum–silicate family containing <2% alkali and is the predominant glass used to make textile and continuous filament fibers.

Manufacture. Vitreous fibers are produced by several processes, including continuous drawing, the rotary process, and flame attenuation

Properties. Glass fibers made from various compositions have softening points in the range 650–970°C. Fiber length and diameter distributions are significant factors in determining thermal and acoustical insulation properties. Slag wool and rock wool fibers are prepared from the slag from pig iron blast furnaces. Slag and rock wool fibers are used to prevent fires from spreading. In the United States 70% of the slag wool is used for ceiling tiles.

Refractory Ceramic Fibers (RCF). These MMVF materials constitute only ~1% of the vitreous fiber market but

Table 2. Classification of High-Performance Fibers and High-Technology Fibers by Properties

Property	Fiber types	Applications
high tenacity and modulus	aramids, gel spun polyethylene, PBO, PIPD, polyarylate	tires, antiballistics, ropes, optical cables
resistant to heat and flame resistance to chemical agents	aramids, PEEK, PBI, polyimides, EDF fluorocarbons, polyolefins	protective clothing for various applications filters, geotextiles, marine applications
microtex and hollow fibers	most synthetics and regenerated fibers	filtration, leisure, insulation, biomedical fashion, fragrances
intricate shapes and porosities	most synthetics and regenerated fibers	antimicrobial, fiber optics, specialty wipes

have exceptional high-temperature performance characteristics. They are produced by using high percentages of Al_2O_3 ~50:50 with SiO_2 as is or modified with other oxides like ZrO_2 or by using Kaolin clay that has similar high amounts of Al_2O_3. Specially prepared ceramic fibers are used to protect space vehicles on reentry and can withstand temperatures >1,250°C.

APPLICATIONS

Commercial high-performance fibers and high-technology textile products have become an increasingly important segment of fiber and textile consumption worldwide. In some instances, various technologies and concepts are combined or refined to produce a textile product for the

desired application(s). Thus, sophistication and enhancement of properties may be introduced at the fiber, yarn, and/or fabric levels.

Structure/Property Classification

The relationship between structure and properties of textile or fibrous substrates and their applications is one method of classifying nontraditional or high technology textiles (Table 2). At the fiber level, the distinguishing high performance characteristics are high tenacity/ strength fibers, hollow fibers, very fine or microtex (microdenier) fibers (hollow or nonhollow), fibers with unique porosities, bicomponent and biconstituent fibers, and fibers with superior resistance to extreme heat, flame, and/or chemical agents. At the fabric or product level, the classes may

Table 3. Classification of High-Performance and High Technology Textiles by Types of Applications

Application class	Subcategories	Examples of specific products
transportation manufacturing	civilian and military aircraft, traffic electronics, information, and communication	components of aircraft, air bags, seats in planes and cars optical fibers for telecommunications and computers, printed circuit boards, industrial filters and belts
agriculture and forestry	horticulture, erosion control, barriers	greenhouse covers, control of drainage, land nets cultivation of plants
civil engineering and construction	geotextiles and geomembranes architectural	road reinforcement, pond liners, and dams, fabric roofs, soil stabilization
fishery and marine	pollution containment, aquatic life, industrial and leisure marine equipment and vehicles	breeding of corrosion resistant composites, conveyers floating backwaters, screens for fish breeding, speedboat components
protective clothing	chemical/environmental/ biological security, heat and flame	firefighters' uniforms, bulletproof vests, protection from toxins and diseases
sports and leisure	sporting goods and vehicles, spas and pools	golf clubs, tennis rackets, snowmobiles, bicycle frames, spa and pool parts, ski wear and sportswear, luxury apparel
biomedical and health care	devices, artificial organs, sutures, wound care, prostheses	kidneys and artificial limbs, bioimplants, dressings for wounds, hydrogel composites
defense and aerospace	chemical/biological protection, camouflage, components for weapons	chemical/biological warfare protection composites for armour and other weapons, space suits and materials for space travel
energy use and conversion	insulation, containment of hazardous waste, shields from high energy sources	all types of insulation, waste containers, electromagnetic shielding

be described as coated and laminated fabrics, composites and fiber-reinforced materials, three-dimensional fabric structures, and fabrics containing polymers or structural features that impart multifunctional properties or allow the fibrous substrate to act as an intelligent material.

Classification by Types of Application

Another way to classify high performance fibers and high technology textile materials or products is by types of applications. A scheme of 10 main categories has been adopted. They include transportation, manufacturing, agriculture and forestry, civil engineering and construction, fishery and marine uses, protective clothing, sports and leisure, biomedical and health care, defense and aerospace, and energy use and conservation (Table 3).

M. Lewin and J. Preston, eds., *High Technology Fibers, Parts AC*, Marcel Dekker, Inc., New York, 1985-1993.

T. L. Vigo and B. J. Kinzig, eds., *Composite Applications: The Role of Matrix, Fiber, and Interface*, VCH Publishers, New York, 1992.

T. L. Vigo and A. F. Turbak, eds., *High-Tech Fibrous Materials: Composites, Biomedical Materials, Protective Clothing, and Geotextiles*, American Chemical Society, Washington, D.C., 1991.

MALCOLM B. POLK
Georgia Institute of Technology
TYRONE L. VIGO (Deceased)
U.S. Department of Agriculture
ALBIN F. TURBAK
President, Falcon Consultants, Inc.

HIGH-PRESSURE CHEMISTRY

GENERAL

High-pressure (HP) technology for liquid systems can range from a relatively straightforward form of autoclaving to elaborate, highly specialized techniques involving rapid experimental observation following mixing of two liquids under pressure. Spectroscopic methods in which detection is also *in situ*—ie, the sample is monitored while it is under pressure—are also used. It is convenient to separate high pressure investigations into different types.

A Pressure Cycle Involving System or Product Distribution Change

The first category of study is when the objective or expected outcome of pressurizing a sample or a chemical reaction is that after decompression the product is different from the sample prior to compression. It can also be different from the product obtained by allowing a reaction to proceed at ambient pressure. The product can also be different from that obtained by either catalysis or nonconventional methods, such as microwaves or ultrasound.

Obviously, the pressure variable is of value only if the product difference is materially or synthetically useful. Other favorable aspects of pressure application can be faster reaction times, increased yields, favorable isomer distribution, reduction of unwanted pathways that produce wasteful by-products, inactivation of biological spoilage organisms, or an economic processing benefit. The primary focus here is on characterization of materials or products following decompression and exploiting their value.

Organic Chemistry Reactions: Mechanism

The second category of high-pressure chemistry studies involves primarily organic chemistry reactions. Reactions that have been thoroughly investigated at ambient pressure are subsequently studied at various pressures. A principal objective is to determine as many features of the reaction mechanism as possible, thus enabling a tuning of the reaction conditions to yield a more favorable product outcome. Since organic chemistry reactions can be grouped into series of similar reactions, often similar mechanisms prevail for related reactions, so valuable information can be extrapolated throughout a given series.

Inorganic Chemistry Reactions: Mechanism

A third category involves many inorganic reactions whose mechanisms are sought by using the pressure variable in addition to other variables; eg, solvent, concentration, ionic medium (strength and different electrolytes), and temperature variations. Often the scale of the system is small, using materials not available in large quantities, and the interest is more in the realm of academic research. The rapidity of many reactions of transition metal complexes has stimulated the development of measurement technology, and technically sophisticated instruments have been constructed. In the reactions considered here, the significant feature is that upon decompression and conclusion of the reaction the system will be identical to that obtained if the reaction had proceeded to an end point or a position of equilibrium without the application of pressure.

Pressure: Units and Magnitude

The choice of magnitude of pressure or range of pressures in a particular application is often dictated by the specific properties of the system and by the specifications, ie, the pressure rating and range of the appropriate available apparatus or instrument. Of course, the experimental design will also be governed by expectations of the outcome or hypothesis regarding the results that will ensue from application of pressure. For most purposes, the measurement of pressure need not be highly accurate or precise, because a pressure measurement that deviates 1 or 2% percent from the actual value has only a minimal effect on the mechanistic analysis or synthetic product.

BASIC PRINCIPLES AND PRACTICAL ASPECTS

Compressing liquids causes an increase in temperature. Typically, organic solvents, having lower heat capacities, experience a more significant temperature increase than water, under comparable pressures. Undesirable heating effects can be minimized by increasing pressure incrementally in steps with adequate cooling by thermostated circulating fluid around the sample container.

However, stepwise compression can in some cases lead to a change in the overall outcome of pressure application as compared with proceeding to the required value of elevated pressure in one step. From a practical standpoint, it usually takes longer to reach a required operating pressure than it does to permit "instant" decompression. There may be consequences of rapid cooling of the sample from the thermally equilibrated temperature of operation at elevated pressure, upon rapid decompression.

Compression of liquids gives rise to a much smaller volume reduction than compression of gases. Therefore, the potential level of hazard is much reduced in high-pressure practice if the total system is liquid based. Failure of a component in the system that is supposed to maintain hydrostatic pressure will quickly lead to liquid leakage, and rapid overall decompression. Therefore, the safety regime is much less rigorous than it would be in practice with compressed gases. However, if elevated temperatures or toxic liquids are employed, extra precautions must be taken to prevent liquid leakage reaching operators, and appropriate training must be provided.

The means to achieve pressure; ie, the pressurizing device, and the materials chosen to house the sample, depend in part on the magnitude of pressure required, the type of sample, and the hypothesis about the likely events occurring in the system upon pressure application. In addition, in which category the system or reaction falls will determine the selection of apparatus or instrument.

THEORETICAL ASPECTS

For materials processing and some organic synthetic–preparative studies, there is no underlying theory regarding the effect of pressure. Rather, a successful result mostly arises from a combination of precedent from the literature, exploratory, and trial-and-error approaches with the chemistry system at issue, and intuition. In mechanistic studies there is a theoretical basis for design and conduct of experiments. The theory of the effect of pressure on kinetic parameters is developed from transition state theory, commonly described in standard physical chemistry or reaction kinetics texts. The key parameter of interest from this theoretical treatment, that can be obtained experimentally, and is valuable in mechanistic diagnosis, is the volume of activation.

APPARATUS, TECHNIQUES, AND METHODS IN HIGH-PRESSURE CHEMISTRY AND RELATED FIELDS

The choice of apparatus or instrument depends on the system being studied and whether the pressure activity is related to bringing about change in one or more substances, synthesis, or mechanistic determination according to general classifications.

Equipment for High-Pressure Processing

For industrial scale high-pressure processing, Avure Technologies QUINTUS press technology vessels from 2- to 320-dm³ capacity for batch or continuous processing are prestressed using spring steel wire around the cylinder. The vertically orientated cylinder has axially sliding end closures and a patented safety liner. Connections for the pressure media and contents and for the sensors for temperature and pressure are located in the end closures. A closure manipulator permits opening and closing of the pressure vessel. Process temperature (4–35°C) is measured by a thermocouple and the pressure is measured by a pressure transducer. For a 215-dm³ pressure vessel, the cycle time is ~7 min and the upper operating temperature is 25°C. Water is the pressurized fluid. This development has found considerable use in the beverage and food industry, where products can be pressure treated in suitable packaging. Although the medium of the samples is water and the pH range is normally close to neutral, there is no need for contact of the sample with the metal of the high-pressure vessel, as inert plastic containers are used to hold the sample.

The advantages of high pressure treatment are (1) reduction of spoilage organisms without using chemical preservatives; (2) retention of characteristics of fresh products; and (3) elimination in some cases of heating for (bacterial) safety purposes. The technology can also be applied to emulsions or colloids whose organization or microstructure can be changed as a result of pressure-induced physical or physicochemical kinetics, or gel formation for matrices used in a host of personal products and foods.

Organic Chemistry

Organic chemistry reactions that have been pressure treated are divided into two sections. The first has one or more of the following objectives: Improvement is sought in synthetic procedure, in a better yield, in production of a better ratio of desired to less desired or unwanted products, in avoiding the use of costly catalysts, in a lower temperature regime, or in simply accelerating the reaction. The second section covers practical issues associated with mechanistic studies of organic reactions, although there can be considerable overlap with the former section.

Synthesis. A typical apparatus is of the piston-cylinder type, in which the high-pressure chamber is a vertical steel cylinder, sometimes surrounded by external jackets that can be closed and sealed at the top and the piston from which pressure is applied is sealed at the lower end of the cylinder. Clearly, proper sealing of the piston is a key necessity of any version of this general type of apparatus or indeed of all high pressure equipment. For many applications, pressures up to 1,500 MPa are commonly used.

Mechanistic Studies. The basic principle here is again a piston-cylinder, but with the important addition of a valve for release of aliquot samples, for analysis while the sample is under pressure, essential if reaction kinetics are to be studied.

Several conditions must be met both the apparatus and the properties of the homogeneous system being studied. These are that the time taken to withdraw the sample must be very rapid, compared with the rate of progress of the reaction, the analytical method for measuring the concentration of reactant or product must be very rapid, compared with the rate of progress of the reaction; the relevant property of the sample is being measured at ambient pressure, rather than at the high pressure of the reaction the error of the measurement is negligible, provided the first two conditions prevail, and the pressure in the cylinder is reduced as soon as the sampling valve is opened. Therefore, the pressure application system must have the capability of restoring the pressure to that of the presample withdrawal level as soon as the sampling valve is closed. This must be carried out very rapidly, compared with the rate of reaction progress. One advantage of this type of apparatus, unlike several reported below, is that it does-not compel dedication of an instrument (eg, a spectrophotometer) to high-pressure mode.

Inorganic Reactions

Among the inorganic reactions that have benefited from being studied over a range of pressures are some of those of the coordination compounds of the transition metals. In the last few years, solvent exchange reactions of some main group metal ions and increasingly of the lanthanide or first row of the inner transition metals (the latter elements possessing up to 14 f electrons) have also been the subject of high-pressure studies. Whereas many organic reactions are not particularly rapid, often requiring elevated temperatures, the inorganic reactions considered here can span reaction times from 10^{10} to 10^{-10} s at ambient temperature.

Slow Reactions. It is customary in kinetics studies to divide reactions into somewhat arbitrary categories of conventional time range reactions (half-life of greater than a few minutes) and rapid reactions. But for high-pressure kinetics the division has to be altered slightly, because once pressure is applied following a reaction initiation there is a delay period during which the reaction restores to the desired temperature of measurement. A reaction in the "awkward" time window can be studied by the simple expedient of slowing it down by lowering the measurement temperature. However, if the time window cannot be suitably positioned by temperature change, a mixing system has been developed. In many cases, for rapid reactions the experimental method succeeds by applying pressure on the reactants before they are mixed and the reaction is initiated.

Rapid Reaction

Mixing Methods. The critical factor is designing and constructing a method of mixing two reactant solutions rapidly so that the mixed solution has homogeneous composition. Rapid monitoring of the mixed, reacting solution is subsequently required.

The challenge for adapting this technology to high pressures required the reactant solutions to be compressed and thermally equilibrated prior to mixing. A further aspect was to be able to ensure that the reactant syringe-piston and stopping syringe-piston could move synchronously and efficiently under pressure. Effectively sealing components in a high pressure SF(stopped flow) device might not be any more difficult than in an ambient pressure device, as the pressure would be the same in all directions. Sometimes component sealing could be a problem at ambient pressure at temperatures >20°C away from the ambient.

Relaxation Methods. Other methods may need to be introduced when a reaction is so rapid so that it cannot be monitored by the Sf method. It was recognized ~50 years ago that when a reaction is faster than it is physically possible to mix liquid-phase reactants, the solution resides in premixing the reactants to form an equilibrium state, then perturbing that state rapidly by a physical probe. The latter is one upon which the position of equilibrium depends eg, on temperature, pressure, an electrical impulse a photochemical impulse, pulse radiolysis, or ultrasound wave. Following rapid perturbation, the system then "relaxes" to an equilibrium state, and the relaxation process can be monitored and kinetic parameters thereby derived. The temperature-jump method requires administering a very rapid, on the order of a few microseconds (μs), temperature increase to a solution at equilibrium.

A second relaxation technique is the pressure jump method. The mostly dormant recent period of applying high-pressure, pressure jump technology reflects the difficulty of experimental practice and lack of versatility of the method in so far as not many chemical systems are as amenable to study by this technique relative to some other techniques.

Radiation Methods. Pulse radiolysis can be regarded as a pulse of high-energy radiation in the form of a beam of electrons or X-rays. Provided the energy source is suitable and the monitoring system (usually uv–vis spectroscopy) is capable of following very rapid reactions, useful data can be obtained.

Electrochemical Methods. Interest in various oxidation–reduction reactions and their mechanisms has stimulated the development of high-pressure adaptation of ambient pressure apparatus or instruments. For example, self-exchange electron-transfer reactions can be studied in appropriate cases by high-pressure nmr spectroscopy. Other self-exchange reactions and other redox reactions can be studied by high-pressure electrochemistry.

Other High-Pressure Methods. In addition to these techniques, several highly specialized high-pressure methods have been established. This technology has not been used widely, nor indeed applied very much to chemistry itself, but rather more to examine properties of substances and to investigations in biochemistry. Several projects involving neutron radiation on materials at high pressure are ongoing. Both liquids and gases are being studied under the umbrella of the ISIS neutron radiation programme.

Against a background of public health issues, understanding the unfolding and refolding of proteins, particularly *in vivo*, has become a priority. Molecular chaperones help to ensure that a protein has a correct final three-dimensional assembly. High-pressure (by means of the diamond anvil cell) Fourier transform infrared (ftir) and fluorescence spectroscopies have been employed in protein-folding investigations. In addition, the volumes of activation for the protein folding processes in the presence of denaturants have been determined using pressure-jump and HPSF methods. Other high-pressure methods for protein folding studies include a high-pressure jump method in conjunction with a small angle synchrotron X-ray scattering technique and high-pressure Raman spectroscopy. Elastic and inelastic X-ray diffraction experiments upon liquids have been carried out in a special high-pressure cell capable of being operated over a wide temperature range.

High-Pressure Nmr Spectroscopy

Ambient pressure nmr spectroscopy ranges from the routine measurement for structural characterization of small, usually organic, molecules to complement other methods such as ir spectroscopy, elemental analysis, and chromatographic procedures to extremely sophisticated nmr spectroscopy to gain insight into structures of macromolecules such as proteins. Nmr spectroscopy would not be the method of choice unless the properties of the system were such that other techniques could not be exploited, since modern high-resolution nmr spectrometers are very expensive to purchase and operate.

SELECTED APPLICATIONS OF HIGH-PRESSURE CHEMISTRY TECHNIQUES

Commercial Products

Only in the past decade has high-pressure technology become established as one leading to commercial products. A whole range of food products has now been treated by pressure cycles, with the principal objective of removing or severely reducing spoilage microorganisms by pressure inactivation without heat treatment or addition of chemical preservatives. Typical pressures are 500–600 MPa for up to 3 min. The advantages of high pressure treatments are extended safe product life and retention of favorable characteristics that may be reduced by heat treatment. Data have been acquired as a function of pressure variation, pressure duration, pH, subsequent storage conditions, and other variables that indicate the magnitude of pressure inactivation of microorganisms in each of several products.

More-fundamental studies on the effect of pressure on physicochemical characteristics of emulsions, colloids, and on the resulting microstructure have been conducted. Thus, high-pressure technology can be exploited to modify the properties of materials to generate matrices that can form the basis of a gelatinous product, as in cosmetics, personal products and various types of foods. In the area of dairy product research, high pressure has been added recently to other traditional methods such as mild acidification or enzyme treatment for producing a gel or other desirable microstructure. The enzyme chymosin is the active component of the renneting process, the name given to the process inducing the microstructures that generate cheeses. An aqueous dispersion of skimmed milk powder (SMP) forms a colloid. Upon pressure treatment at ambient temperature, provided the pressure is >270 MPa, a suitable cosolute is present and, following pressure treatment, decompression is rapid, the colloid is converted to a gelled state. The final form of the gelled product is not obtained immediately following decompression, thus enabling the physicochemical kinetics of ultimate product formation to be monitored.

High Pressure in Organic Synthesis

The advantage of high pressure over thermally initiated synthesis can be, in appropriate cases, a modification of the direction of the synthesis resulting in different proportions of the reaction products. A classic example of the value of high pressure in organic synthesis is the reaction of isophorone dienamine with acrylonitrile. At ambient pressure the yield of the desired product is 3%, and two other compounds in 10 and 20% yields, respectively, are prepared, But applying 1,500 MPa at room temperature, the same reaction proceeds smoothly to generate the desired product with a 90% yield, a dramatic difference.

A more recent example of a successful use of high pressure in synthetic chemistry is the preparation of "belt" compounds from a diene and a dienophile. At ambient pressure, the yield was 2–3% but at 1,000 MPa it was 30–35%.

Mechanistic Organic Chemistry

One area of mechanistic organic chemistry where high pressure can be exploited is regioselectivity. When two products of a reaction can be formed potentially, pressure can effect the direction of the reaction exclusively or predominantly to one product. An example is the intramolecular hetero-Diels-Alder reaction of the benzylidenebarbituric acid derivative.

Inorganic Chemistry Reactions

Solvent Exchange. This type of reaction is invariably studied by nmr spectroscopy using a magnetically active atom in the solvent, although in principle in some cases it would be possible to study solvent exchange using radiochemical techniques. Solvent exchange is the most simple reaction studied herein, in that there is no net reaction and the reaction volume is zero, and nmr spectroscopy both at ambient and high pressure is the most widely employed method of study.

Reactions of Protein. High-pressure studies involving binding of small molecules to proteins, invariably metalloproteins, have a lengthy history. Considerable additional insight into the binding mechanism and nature of the binding site has been a consequence of these studies. Where less progress has been possible is in providing additional mechanistic delineation in metalloenzyme catalyzed reactions. When a reaction is multistep, as is the case in enzyme catalyzed reactions, separation of observed effects of pressure on steady state or even transient state kinetics parameters into the contributions of the actual kinetic and equilibrium parameters of the enzyme catalyzed pathway is likely to be less than unequivocal.

Other Applications

It is worthwhile citing also the use of high pressure in reactions where gases such as CO and syngas mixtures in solvents are used in hydroformylation reactions catalyzed by rhodium carbonyl clusters. Gas pressures up to 100 MPa have been used along with both ^{13}C and hpnmr spectroscopy, plus high-pressure ir spectroscopy have been applied to identify the form of the cluster that is the actual catalytic agent. The specific technical aspects with respect to the hpnmr in this context have been presented. Another reaction involving gases under pressure in solvents is the copolymerization of styrene with carbon monoxide catalyzed by a palladium(II) complex, studied by hpnmr spectroscopy. These reactions are examples from organometallic chemistry.

THEORETICAL STUDIES

Efforts to calculate changes in reacting systems, including calculating the volume of activation, have been devoted mostly to water exchange reactions and electron self-exchange reactions. Various degrees of success have been achieved and satisfactory agreement with pressure-derived experimental parameters has been recorded in some cases; eg, as discussed above for water exchange on first-row transition metal ions in oxidation state two. More recent calculations have been on water exchange for the hexaaqua ions of rhodium(II) and iridium(III).

J. Hyde, W. Leitner, and M. Poliakoff, in R. van Eldik and F.-G. Klärner, eds., *High Pressure Chemistry, Synthetic, Mechanistic and Supercritical Applications*, Wiley-VCH, New York, 2002, Chapt. 12.

T. J. Mason and J. L. Luche, in R. van Eldik and C. D. Hubbard, eds., *Chemistry under Extreme or Non-Classical Conditions*, Wiley / Spektrum, New York / Heidelberg, 1997, Chapt. 8.

L. F. Tietze and P. L. Steck, in R. van Eldik and F.-G. Klärner, eds., *High Pressure Chemistry, Synthetic, Mechanistic and Supercritical Applications*, Wiley-VCH, 2002, Chapt. 8.

M. L. Tobe and J. Burgess, *Inorganic Reaction Mechanisms*, Addison-Wesley Longman, Harlow, 1999.

Colin D. Hubbard
Rudi van Eldik
University of Erlangen-
Nürnberg

HIGH PURITY GASES

High purity industrial gases are routinely delivered in large quantities having purities exceeding 99.999% (>5 nines pure). There are many applications for gases where purity even higher than 99.999% is required.

There is no universally accepted definition of what purity levels correspond to high purity. However, gases having total impurities specified <1 ppm on a molar or volume basis must be manufactured and handled differently from regular gases if that specification is to be maintained. A good working definition of high purity is gases having certain individual impurities held to levels <0.1 ppm.

Depending on a volume, high-purity gases can be delivered using either bulk systems, where a plantwide distribution system is integrated with central gas storage facilities, or cylinders, where a short local distribution system is supplied from a single high-pressure cylinder.

Gases used in the manufacture of semiconductor materials fall into three principal areas: the inert gases, used to shield the manufacturing processes and prevent impurities from entering; the source gases, used to supply the molecules and atoms that stay behind and contribute to the final product; and the reactive gases, used to modify the electronic materials without actually contributing atoms or molecules.

PRODUCTION AND PURIFICATION

The separations processes used for manufacturing high purity gases are generally the same as those used for making lower purity products. Purification by distillation and adsorption are often used. Chemical conversion processes, where the impurities are converted into more easily separable forms through a selective chemical reaction, are often employed in point-of-use purifiers.

Bulk Gases

The bulk gases are usually characterized by high volume flow requirements in the manufacturing process. Historically, these have consisted of nitrogen, oxygen, argon, hydrogen, and, to a lesser extent, helium.

Nitrogen. Because of numerous applications in semiconductor manufacturing, high-purity nitrogen is produced both at high volumes and at some of the lowest impurity levels seen for any of the high-purity gases.

Distillation. All high-purity nitrogen is manufactured from air, using multistage cryogenic distillation.

Chemical Conversion. In both on-site and merchant air-separation plants, special provisions must be made to remove certain impurities. The main impurity of this type is carbon monoxide, CO. The most common approach for CO removal is chemical conversion to CO_2 using an oxidation catalyst in the feed air to the air separation unit. The CO_2 is removed by a prepurification unit in the air separation unit.

At throughputs below 500 nm^3/h, a wide variety of inert gas purification processes based on chemical conversion can be used to produce high-purity nitrogen. Typically, the impure nitrogen is passed through a bed of reagents, where the conversion reactions occur, causing the impurities to remain behind in the bed of reagents. These processes require that the bulk of the oxygen in the feed be removed by some other method.

Combined Distillation and Chemical Conversion. On-site generators using distillation are almost always combined with chemical conversion purifiers in large bulk high-purity nitrogen supply systems.

Oxygen. High-purity oxygen for use in semiconductor device manufacture is produced in relatively small quantities compared to nitrogen. There are two different purification processes in general use for manufacturing the gas: distillation and chemical conversion plus adsorption.

Distillation. As for nitrogen, all high-purity oxygen is derived from air through the air separation process using cryogenic distillation. Generally, air separation units that manufacture commercial purity oxygen also remove nitrogen and other light impurities to levels low enough for high-purity applications.

Chemical Conversion and Adsorption. Where additional distillation is not practical, hydrocarbons and heavy noble gases can also be removed by combining chemical conversion with adsorption. Commercial purity oxygen is passed through a high-temperature bed of oxidation catalyst, where impurities are oxidized to CO_2 and H_2O. The CO_2 and H_2O products from catalytic oxidation of hydrocarbon impurities are removed using a temperature swing adsorption (TSA) process. The adsorbent is typically one of the molecular sieves.

Argon. High-purity argon has many applications as an inert gas during the manufacture of semiconductor devices. In all these applications, nitrogen is a reactive impurity which, in addition to O_2, H_2O, CO_2, CO, and all hydrocarbons, must be removed from the argon to low levels.

Distillation. Conventional purity argon is separated from air using a combination of distillation and chemical conversion. High-purity argon is made the same way.

Chemical Conversion. Except for control of nitrogen impurity levels, the same chemical conversion methods used for nitrogen purification at low flow rates can also be used for argon purification.

Hydrogen and Helium. Whereas hydrogen and helium are very different chemically, these gases have low boiling points and are normally liquefied during manufacture. Because the boiling points are so low, even very small amounts of trace impurities tend to freeze and form solid deposits. To prevent formation of these deposits,

trace impurities must be removed prior to liquefaction. Similar methods are used to purify hydrogen and helium prior to liquefaction.

Hydrogen. High-purity hydrogen is usually delivered and stored as a cryogenic liquid and vaporized when needed. This vaporized liquid seldom needs any further processing to meet high-purity specifications. High pressure hydrogen gas can be delivered and further purified on-site to meet high purity specifications. This is accomplished using combinations of chemical conversion, cryogenic adsorption, and palladium membrane processes.

Helium. High-purity helium is usually not required in large quantities and is therefore not commonly delivered as a cryogenic liquid. Instead, high-pressure cylinders are filled from a liquid helium source by the gas supplier and then transported to the customer. When high-purity helium is required, the high-pressure gaseous helium is processed through an on-site purifier.

Because helium is not chemically reactive, the same chemical conversion processes used for purification of nitrogen and argon are also applied to helium purification.

Specialty Gases

The specialty gases are generally more reactive than the bulk gases.

Purification of specialty gases can be divided into two areas: purification done by the gas supplier on a bulk scale prior to filling the cylinder or other delivery container, and purification carried out by the consumer on a point-of-use scale generally just prior to use.

Bulk Purification. Many specialty gases originate as byproducts or low-purity intermediate chemicals produced during the course of manufacturing something else. The purification processes tend to utilize standard methods.

Distillation. Processes that utilize either simple liquid vapor flash processes or multistage distillation often used are for purification of bulk specialty gases.

Adsorption and Chemical Conversion. In some cases, removal of moisture or oxygen added by small amounts of air contamination is all that is necessary to make a gas suitable for high-purity applications. With a limited objective, it is usually most effective to use an adsorption process. That may be designed either with or without the capability for repeated regeneration.

Chemical conversion processes can also be used for moisture and oxygen removal.

Point-of-Use Purification. For the user of cylinder quantities of reactive specialty gases, there are only a limited number of ways to remove impurities and obtain high purity. Specialized point-of-use purifiers have thus been developed that purify small streams of many important reactive gases. They are usually effective for removing the contamination added by the users' gas distribution system, mostly air and moisture.

DELIVERY AND CONTROL

Once a gas has been purified, it must be brought to its intended point of use without being degraded by the addition of excessive contamination.

Delivery methods for high-purity gases can be divided both according to chemical reactivity with respect to the containment system and, to a less significant degree, according to volume through put requirements. Many highly flammable gases such as H_2 and SiH_4 are still inert with respect to the containment systems. Even though special provisions must be made because of flammability, the technology used to deliver flammable gases is similar to that employed for inert gases.

Bulk Gases

Attaining high-purity gases where they are used requires a suitable gas distribution system. To achieve a high-purity distribution system, there must be an absence of dead zones, external leakage, outgasing, and particulate contamination.

Specialty Gases

The purity of specialty gases depends on the systems and procedures adopted by the distributors for bulk gas supply and cylinder preparation, filling, and delivery. Most of the precautions taken into consideration in the bulk gases delivery system are also applied for specialty gases, to eliminate recontamination.

ANALYSIS AND CERTIFICATION

Ensuring that the purification and delivery processes are working properly is essential to successful applications of high-purity gases.

Particulates

Separation of particulate impurities is an important part of the process for manufacturing high purity gases. The most common approach is through filtration technology.

Description of Analytical Methods

Procedures for analyzing both gas and particulate impurities in high-purity bulk gas products such as nitrogen, argon, oxygen, helium, and hydrogen are available. Typical gaseous impurities include oxygen, moisture, carbon monoxide-carbon dioxide, total hydrocarbons (THC), argon, and nitrogen analyzed to the low PPB level. Particle impurities are analyzed to the 0.1-µm level.

Gaseous Impurities. Instrument calibration (excluding the apims) typically consists of two steps: zeroing and panning. Zeroing is accomplished by allowing the analyzer to sample a gas the contaminant level of which is below the lower detection limit of the analyzer. This is called a zero gas. A typical method for generating zero gas is to take the actual sample gas and run it through a gas purifier.

Spanning, accomplished using a sample gas containing a known volume concentration of impurity, is performed at levels that are the same order of magnitude as the required detection. The actual span concentration is selected so that the majority of expected measurements fall at or within its value.

The calibration procedures for an atmospheric pressure ionization mass spectrometer (apims) involve the generation of separate calibration curves for each of the monitored impurities.

Particulate Impurities. Particle counters require factory calibration every two years. In addition, the background signal associated with both the instrument and its sampling system must be quantified so that it may be subtracted from sample measurements. To accomplish this, an absolute filter (<0.01 µm rating) is employed. It The removes all particles entering the sampling system, so any particle registered by the counter can be directly attributed to the sampling system or instrument noise.

Sampling and Analysis Guidelines. As a general safety consideration, all gases should be vented to an external area and, whenever possible, inert gases should be used as the test gas for piping systems.

APPLICATIONS OF HIGH-PURITY-GASES

The applications of high-purity gases are primary in the semiconductor industries. In addition to the microelectronics industry, other applications for high-purity oxygen include fiber optics manufacturing, production of pharmaceuticals, and usage as calibration media in research and development laboratories and in the pollution control field. Applications for high-purity hydrogen include oxidation processing and epitaxial growth for both silicon and gallium arsenide.

High-purity argon is likely to be used in the high technology fields of electronics, fiber optics, research and development, powder metal spraying, and hot isostatic pressing.

Other applications of high-purity specialty gases include hydrogen bromide for etching single-crystal silicon, polysilicon, and aluminum; nitrogen trifluoride as a fluorine source for *in situ* cleaning processes for chemical deposition equipment, semiconductor etching and deposition, and high-energy chemical lasers; and sulfur hexafluoride as a key etching material in certain semiconductor manufacturing processes.

The primary driving force for high purity gases has been the increasing purity demands from high technology industries such as ceramics, optical fibers, and silicon wafer fabrication.

R. DiNapoli and A. M. Sass, in J. J. McKetta, ed., *Encyclopedia of Chemical Processing and Design*, Vol. 31, Marcel Dekker, Inc., New York, 1990, p. 236.

W. H. Whitlock, "The Ultra-High Purity Challenge", in *Separation of Gases, Proceeding of the Fifth BOC Priestley Conference*,

Birmingham, U.K., Sept. 19–21, 1989, Royal Society of Chemistry, 1990.

WALTER H. WHITLOCK
EDWARD F. EZELL
SHUEN-CHENG HWANG
The BOC Group, Inc.

HIGH-TEMPERATURE ALLOYS

In general, the strength of most metals decreases with increasing temperature. At elevated temperatures, the increased rates of diffusion controlled and thermally activated processes result in the increased mobility of atoms and increased vacancy concentrations, and can activate additional slip systems or change slip systems. Furthermore, although grain boundaries can strengthen a metal at low temperatures, deformation at and along grain boundaries can occur more easily at high temperatures. Due to the more rapid diffusion rates at high temperature, particles or precipitates can coarsen, eventually resulting in an overaged condition with reduced strength. Lastly, at elevated test temperatures, significant interactions between the metal and the environment can occur. The rapid oxidation of the surface or grain boundary penetration that can occur at elevated temperatures will result in reduced strengths and perhaps reduced ductilities.

Materials that are exposed to stresses at elevated temperatures exhibit creep, an undesirable time-dependent plastic deformation. Generally, creep pertains to deformation at strain rates less than 0.01/minute. The turbine blades on the spinning rotor of an operating gas turbine, which may be at temperatures up to 1200°C, slowly grow in length due to creep. Typical elevated-temperature structural applications include power generation plants, aircraft and rocket propulsion, airframes and automotive engines.

The materials selected for high-temperature applications, typically referred to as high-temperature alloys, are known for their ability to resist, and exhibit useful strengths at, elevated temperatures. In general, the physical properties of the elements that could be utilized as the alloy base for high-temperature materials include melting temperatures, elastic modulus, and density. Some of these properties are more important than others, depending on the application. If the alloy is being considered for a rotating component or for an airframe for an aerospace application, the density becomes very important, because the stresses from centrifugal forces or payload capabilities. In addition, although a high melting temperature would be expected to be an essential requirement, an element with a high melting point does not necessarily make the element a candidate for elevated temperature service, particularly in air.

Additional considerations for the selection of a high-temperature alloy system include its thermal conductivity and thermal expansion coefficient. In most cases it is desirable for an elevated-temperature material to exhibit high thermal conductivity and a relatively low coefficient of thermal expansion. Within a component operating at high temperature, thermal gradients can develop during steady-state operation and during changes in operating conditions. In either case, the thermal gradient can result in significant thermal strains and thermal fatigue. Increased levels of thermal conductivity can lessen these thermal gradients, resulting in reduced thermal strains. In addition, as a component is heated to operating temperature it will expand. The fit-up of components and the gaps designed into elevated temperature systems must incorporate the coefficient of thermal expansion for all the components in the system. Utilizing alloys with reduced coefficients of thermal expansion may simplify the design.

Although it does not display a particularly high melting point, Ni has been, and will continue to be, the preferred alloy base for high-temperature alloys.

CREEP

At an elevated temperature, the strength of a material is strongly dependent on the strain rate, time of exposure, and test temperature. In some respects, metals at elevated temperature exhibit a viscoelastic type of behavior similar to polymeric materials at low temperature. Materials exposed to stresses at elevated temperatures exhibit creep, an undesirable time-dependent plastic deformation. Generally, creep pertains to deformation at very low strain rates (eg, less than 0.01/min).

Although creep can occur at almost any temperature, the full effects of creep deformation are observed at homologous temperatures in excess of about 0.4 $(= T/T_m)$. The creep behavior of a material is generally defined as the time to failure and the fracture strain as measured during experiments. In general, an increase in the temperature results in increased creep deformation rates, since the material strength generally decreases with increasing temperature. Increased stresses will result in increased creep deformation rates. Hence the time to rupture will decrease with increasing test temperature and/or stress. For design purposes, it is necessary to perform numerous creep tests at several stresses, at several temperatures. Frequently, materials selection for an elevated temperature application is based on the creep properties rather than the tensile properties.

The Basic Creep Test

A creep test consists of loading a specimen at a specified load and temperature and measuring the displacement as a function of time and the time to failure, sometimes called rupture. The test can be run with a constant load, usually applied by a dead weight load or by mechanical means. Creep tests can also be performed under constant stress conditions, but the applied load must than be continuously adjusted, decreased so that the stress remains constant as the sample area decreases, due to the plastic deformation when the test is being performed in a tensile mode. It is also possible to perform the test with compressive stresses with modified fixturing. Compression creep testing is frequently done when testing brittle materials or when only very limited amounts of test material are available. During constant stress, a creep test requires the load to increase with the increasing cross-sectional area of the sample. In general, though, most creep testing is performed in tension.

The Basic Stress Rupture Test

The stress rupture test is very similar to the creep test, except that creep strain is not measured during the test. The stress rupture test consists of loading a specimen at a specified load and temperature, but only the time to failure is measured. The rupture ductility is measured by comparing the separation of gage marks, sample length, or some other feature before and after testing. Similar to creep testing, test durations can vary between approximately 10 hours up to 100,000 hours or more, but most tests last between 10 hours and 10,000 hours.

The cost of performing a stress rupture test is significantly less than that of the creep test, since the time to set-up and the equipment utilized during the test are reduced. However, the amount of data produced by the stress rupture test is also significantly less than from the creep test. In most cases, stress rupture testing is used for screening large numbers of samples, materials, and/or test conditions. Creep testing is most commonly done for scientific reasons, to understand deformation mechanisms, and for design purposes.

The Basic Stress Relaxation Test

Creep testing to determine the effects of stress and temperature on the creep rate, time to specified creep strains and time to rupture of a given material in a given condition can be very time consuming and expensive. Numerous samples must be tested under a variety of test conditions. Stress relaxation testing offers the potential to reduce the number of tests and the time to complete the testing by using a single sample to determine a wide range of strain rate versus stress data. Unlike creep testing, where a fixed load (or stress) is applied and the strain monitored, stress relaxation testing applies a fixed strain and the load is monitored. The initial elastic loading of the sample to a fixed length is converted to plastic strain, which results in a reduction of the load. The initial displacement is usually based on the deformation corresponding to a given load (or stress). The load is then monitored as a function of time, which is used to determine the stress versus strain-rate data. A significant amount of data can be accumulated from only a few samples in a very short time. However, creep testing data and the stress relaxation data are frequently not in agreement. Very precise temperature control and load measuring techniques are required. In addition, the loading rate of the sample has a significant effect on the properties and may result in significant errors. Lastly, the stress relaxation testing can determine very low creep rates in a very short period of time. Similar data from creep testing would require significant numbers of samples and long-term testing. However, the short duration of the stress relaxation test may not be a representative test condition for long-term elevated temperature service. Long-term creep tests

may be a more accurate method to evaluate the service condition, since microstructural evolution can occur and any microstructural instabilities may not be observed in the much shorter-term stress relaxation test. In general, a combination of both creep testing and stress relaxation testing would be a more productive way of generating a large amount of data on the creep properties of a material.

FATIGUE

Most materials subjected to repetitive, alternating, vibrational, or fluctuating stresses and temperatures will fail at stresses much lower than that required to cause failure in a single application of load, such as a tensile test. These dynamic loads are considered fatigue loads, since they result in fatigue failures that occur after a considerable period of time. The occurrence of fatigue failures has been more and more common as the number and complexity of engineering systems, such as aircraft, vehicles, pumps, etc., have increased. Approximately 90% of service failures due to mechanical causes can be attributed to fatigue.

Frequently, fatigue failures are particularly difficult to predict and can occur without any warning or indication and at stress levels that are often quite low. Analysis of the fracture surface can also be quite confusing, since the fracture often appears to be brittle, with little no gross plasticity. The fatigue failure can sometimes be recognized by a flat fracture surface with several distinct regions. This is surface can have regions that are essentially smooth, even at high magnification, due to the rubbing action of the fracture surfaces as the crack propagates. In some cases, a series of rings or beach marks can be observed on the fracture surfaces, indicating the progress of the crack as the material was cycled in service or testing. The beach marks can often be followed back to the initiation site, which usually is a point of stress concentration. Typical fatigue crack initiation sites include both microstructural stress concentrators such as inclusions, coarse carbides, grain boundary triple points, and porosity, and macroscopic stress concentrators such as machining marks, sharp corners, and notches.

In order for fatigue to occur the stress imposed on the material must have tensile and cyclic components large enough to cause fatigue damage. In addition, it is necessary to accumulate enough cycles to initiate and propagate a fatigue crack. Other factors that can affect the fatigue behavior of a material include temperature, environmental interactions, stress concentrators, microstructure, overload, residual stresses, and combined stresses. It is not possible to perform tests in a laboratory that can completely replicate the service environment. Therefore, several of these factors are often not included and only a few test variables are evaluated at a time.

Fatigue can be broken down into fatigue crack initiation, fatigue crack growth, and overloading. For design purposes it is necessary to understand how various service environments, the loading spectrum, component design, alloy content, and metallurgical details affect the fatigue properties of the candidate alloys.

Cyclic Stresses and Strains

Fatigue stress has units of force per unit area and is primarily based on the engineering stress, the load divided by the original cross-sectional area, Which can be a tensile stress or a compressive stress. In addition, the stress can be normal stress, acting perpendicular to the plane of interest, or in the plane, a shear stress.

The strain measured in fatigue testing is almost exclusively engineering strain, utilizing the original cross-sectional area and/or the original gauge length. Both compressive and tensile strains are utilized in fatigue testing. The engineering strains are calculated as the change in gauge length divided by the original gauge length or the change in the cross-sectional area divided by the original cross-sectional area. Most commonly the strains are normal strains, but on some occasions shear strains are utilized.

Because It is not possible to recreate the service environment and loading, simplified loading cycles, which can be recreated in a laboratory, are used.

Factors Affecting Fatigue Behavior

The fatigue behavior of a material is effected by (1) the test material and its microstructur; (2) the test temperature (at elevated temperature there can also be an interaction of the temperature, loading wave-pattern, and the environment;) (3) sample size and surface finish, and (4) the stress or strain range of the test.

Practical Aspects of Fatigue Property Characterization

There are three primary types of fatigue testing; high cycle fatigue (HCF), low cycle fatigue (LCF), and fatigue crack growth rate (FCGR) testing. HCF tests are usually performed in load control in which the maximum and minimum loads are selected based on the stress amplitude. The desired load waveform is then applied to the sample and measured by a load cell in the test frame. In most cases, no attempt is made to measure the strain or the crack growth rate. The sample is cycled until failure occurs or until the test is stopped and the sample is considered a run-out. In general, HCF tests last between 10,000 and 10,000,000 cycles. Most testing is performed at frequencies in the range of 20–50 Hz. Therefore, HCF tests that run up to 10^7 cycles, which are considered run-outs, take about one week to complete when run at 20 Hz. Run-out samples are not tested further or, in some cases, the stress amplitudes are increased and the sample is put back into test.

In general, most HCF testing is performed on cylindrical test samples with either button-head loading or threaded ends. However, plate samples or other geometries can be utilized.

Once the sample has failed, fracture surfaces are carefully examined optically and then by using scanning electron microscopy (SEM). The fatigue crack initiation site is determined and the crack growth behavior observed. In some cases, fatigue striations can be observed that represent the step-by-step growth of the fatigue crack. Lastly, the overload when the sample ultimately fails is also characterized. Transmission electron microscopy (TEM) is also frequently performed to evaluate deformation mechanisms.

HCF testing can also be done using ultrasonic fatigue testing, where very high cyclic frequencies are used. Frequencies greater then 1 KHz are used that allow for testing to very high cycle to failure lifetimes within much shorter test times. However, there are significant limitations when performing ultrasonic fatigue testing.

Low cycle fatigue (LCF) testing is performed in strain or displacement control. The strain is measured by a extensometer applied to the samples directly or by an extensometer frame if testing at high temperature. Although the test is controlled by monitoring the strain and/or displacement, the load, and, therefore the stress, are also measured. The cyclic stress and cyclic strain are plotted on a graph for each cycle during the test. During a typical test, the load can either continuously increase if the sample cyclically hardens or decrease if cyclic softening occurs. When a crack forms, the tensile-load-bearing capability of the material is decreased, since the cross-sectional area of the sample decreases as the crack forms and grows. When cyclic hardening or softening occurs, both the maximum and minimum stresses will change. When a crack forms, though, only the tensile side of the cyclic stress–strain curve will change.

In general, most LCF testing utilizes cylindrical test samples with either button-head loading or threaded ends. In some cases, plate samples can also be utilized, but since most LCF testing is fully reversed, the cross-sectional area of the samples must be sufficient to support the compressive load and not buckle.

As in HCF testing, once an LCF sample has failed the fracture surface is completely characterized by optical and SEM techniques. Transmission electron microscopy is also frequently performed to evaluate deformation mechanisms.

Similar to HCF testing, fatigue crack growth rate testing is performed under load control, to control the stress intensity at the crack tip. In FCGR testing a crack is grown in the material and the rate at which the crack advances under cyclic loading is measured. The crack initiation event is not important. A fatigue crack is preferred over a machined notch, since the stress concentration factor is much greater for a sharp fatigue crack and there are no complications from residual stresses from machining. Once the initial crack is grown into the test samples, the crack length is measured and the test started. Under stress, a controlled waveform is applied to the sample and the crack grows in length, slowly at first but then faster as the crack grows. In general, the crack growth rate increase at higher stress amplitude (ie, stress intensity). In addition, increased test temperatures will also result in more rapid crack growth rates.

FCGR tests are usually also run at 20 Hz but require sample precracking and sample preparation and typically take upto two days to run. Most fatigue crack growth tests are easily completed within a day.

As with HCF and LCF testing, once the sample has failed the fracture surface is completely characterized by optical and SEM techniques.

Fatigue Data Analysis and Potential Problems

Regardless of the material, fatigue failures will occur when the components are exposed to cyclic stress, strains, temperatures, or any combination thereof. Since a very large proportion of engineering failures are related at least in part to fatigue, it is important to consider the cyclic behavior of materials being considered for an application. However, there is no single method to evaluate the fatigue properties of a material. The ultimate use, service environment, intended lifetime, cost, and cost of failure must be considered when developing a test plan. There is a substantial database on the fatigue properties on a wide variety of engineering materials.

When comparing fatigue data it is necessary to ensure that the materials have been processed in a similar manner to obtain the appropriate microstructures, prepared in similar manners, and tested under very similar conditions. Comparison of fatigue data can be done with any accuracy only if the test conditions are similar. Representation and comparison of fatigue data generated either at high temperature or in aggressive environments can result in additional difficulties.

HIGH-TEMPERATURE ALLOY DESIGN: STRENGTHENING MECHANISMS

The elements that can be considered for elevated temperature applications must be alloyed with the appropriate element to increase their strength, creep and fatigue resistance, environmental resistance, and/or processability the most common characterestics sought in a material its are in tensile (monotonic), creep (or time dependent), and fatigue properties.

Tensile Strengthening

Both high-and low-temperature applications can benefit from increased tensile strength.

Cold Work. One of the simplest ways to increase the strength of a material is to cold work the sample. Deformation processing at temperatures below the dynamic recover/recrystallization temperature results in a significant increase in dislocation density the point at which a metal deforms under shear stress. This increased dislocation density results in significantly higher flow stresses due to the extensive amount of dislocation interaction, which impedes dislocation motion. The increased flow stress is generally accompanied by a decrease in ductility. As temperature is increased, a cold-worked material will start to go through recovery and recrystallization. The resulting material will have a finer grain size than the starting material, with a lower strength but higher ductility than the cold-worked material. If the sample is held for a longer time, the grain size of the sample will increase, resulting in a further reduction in strength and ductility.

Grain Size Effects. Another simple way to increase the strength of a material is to decrease its grain size, usually by deformation processing. Decreasing the grain size by repeated cold working and recrystallization can result in a significant increase in strength and ductility. The finer grain size have shorter slip distances, which means small dislocation pile-ups at obstacles such as grain boundaries. Since these dislocation pile-ups are shorter and therefore have lower stresses acting at the obstacle, it takes additional stresses to propagate slip beyond the obstacle. Reduction of grain size is one of the few methods to increase strength without decreasing ductility. This technique would be expected to increase fatigue resistance, but may not be useful for increasing creep resistance. The higher tensile strength of the finer-grained material will still be observed at high temperature but will also result in dramatically reduced creep strength. At higher temperatures, the grain boundaries are weakened and grain boundary sliding will occur. This type of deformation will not be obvious at high strain rates but will be obvious during creep rates . Frequently, the high ductility of the fine-grained material can be used to produce superplasticity. However, for high-temperature strength and creep resistance a fine grain size is not desirable.

Alloying, Substitutional Solid Solutions. In substitutional solid solutions, the solute atoms occupies the solvent atom lattice sites. In general, the more similar the solute and solvent atoms, the greater their solubility. The effectiveness of strengthening from adding the solute atom can be evaluated by considering the solubility of the solute in the solvent and the rate of hardening for the solute. In general, the substitutional atom produces a relative strengthening of about 1/10 of the solute atom.The source of the strength is generally attributed to a lattice misfit, modulus misfit, order, stacking fault, interactions and electrical interactions. The strength of the alloy has been observed to increase with increasing size difference of the solute and solvent atoms and/or the resulting lattice parameter.

Alloying, Interstitial Solid Solutions. In interstitial solid solutions, the solute atoms are much smaller and occupy the interstitial sites. Interstitial solute atoms produce a nonspherical lattice distortion and produce a strengthening equivalent to three times the shear modulus of the solute atom. Since the interstitial solute atoms exhibit both dilatational and shear components, they can interact with both screw and edge dislocations.

Modulus Misfit Strengthening. A modulus interaction occurs if the presence of a solute atom locally alters the modulus of the lattice. Softening is observed locally if the solute atom has a smaller modulus that results in a reduction in the strain field of a dislocation. Therefore, the dislocation will be attracted by the solute atom, which will reduce the mobility of the dislocation. A repulsion of the dislocation and the solute atom is observed if the solute atom has a greater shear modulus, which results in local hardening of the lattice and an increase in the strain field. Although the modulus misfit strengthening is similar to the size misfit, both screw and edge dislocations are affected by modulus misfit strengthening.

Stacking Fault Interactions. Stacking fault interactions, sometimes referred to as chemical strengthening, are due to the preferential segregation of solute atoms to the stacking faults in extended dislocations. For this mechanism to occur the solute atom must have a preferential or increased solubility in the HCP structure of the FCC stacking fault. Stacking fault interactions would be considered a viable mechanism at both low and high temperatures, but the contribution at any temperature would likely be relatively small.

Short Range Order (SRO) Hardening. Solute atoms that tend to arrange themselves so that they can maximize the number of dissimilar nearest neighbors result in the formation of short range order or SRO. Strengthening occurs since the motion of a dislocation through a region of SRO will result in a reduction in the degree of SRO. This local disordering will cause an increase in the energy of the alloy, and therefore it is not energetically favorable for the dislocation to move through the region of SRO.

Long Range Order (LRO) Hardening. Long range order, or LRO, hardening occurs in alloys that exhibit a superlattice. In a superlattice there is a long-range periodic arrangement of dissimilar atoms. The movement of dislocations through the superlattice creates a region of disorder called an antiphase boundary (APB), because the atoms across the slip plane become out of phase with respect to the energetically preferred superlattice. This local disorder the APB, and in order to minimize the energy to create the APB, a second dislocation follows the first dislocation and re-creates order in its wake. The combined dislocations with the APB are called a super lattice dislocation or a superdislocation. As deformation occurs, more APB's are formed, reducing the spacing, and causing greater dislocation–dislocation interaction, resulting in increased hardening. Typically, ordered alloys exhibit much higher work-hardening rates from cold work than do disordered alloys, but the yield strength of the ordered alloys is often lower than that of disordered alloys. However, the fatigue strength of ordered alloys is typically very high, due to the planar slip and slowed diffusion resulting from the ordered lattice. The ductility of most ordered alloys is reduced due to the presence of the ordered lattice reducing the number of active slip systems. Therefore, ordered alloys frequently exhibit limited ductility and are brittle at low temperatures, but exhibit increased ductility at higher test temperatures.

Precipitation Hardening

Since only a few alloy systems permit extensive solid solubility between two or more elements and these systems exhibit only a moderate level of solid solution hardening, most commercial alloys contain a heterogeneous microstructure consisting of two or more phases. The vast majority of the se microstructures fall into two classes. The two phases can be present in discrete grains, sometimes referred to as the aggregated type of two-phase microstructure, in which the second phase is on the order of a grain size of the matrix. The other general category of two-phase microstructures is when the second phase is much finer and is dispersed throughout the matrix.

The strengthening from precipitation of a second phase is usually additive to solid solution strengthening. Two-phase alloys produced by heat treatment ensure a maximum solid strengthening component, since the precipitates form from supersaturated solid solution. In addition, the presence of second-phase particles in matrix results in internal stresses that modify the deformation of the matrix.

Not all second-phase particles produce strengthening, in order for particle strengthening to occur there must be a strong particle--matrix bond. Deformation occurs first in the weaker phase and, if very little of the stronger phases is present, mostly continues to occur in the softer phase. As deformation continues, the flow of the softer matrix will occur around the particles of the harder phase. Two-phase alloys with hard particles exhibit different deformation characteristics than the ductile phases. However, the mechanical properties are dependent on the distribution of the second phase in the microstructure. If the second phase is present as a continuous phase along grain boundaries, the alloy will exhibit low ductility. If, however, the second phase is not continuous, the brittleness of the alloy is reduced considerably.

Fine second phase particles distributed throughout the ductile matrix are commonly used to strengthen an alloy. There are two basic types of strengthening observed from precipitates. The first, due to precipitation hardening, is usually produced by appropriate heat treatments. The second, referred, to as dispersion hardening, is attributed to the hardening of a matrix with hard, brittle, incoherent second-phase particles.

Composite Strengthening

In order to further increase the strength and temperature capability of high temperature materials there has been a great deal of interest in the development of composite materials. In general, the matrixes of these composites are relatively weak and the reinforcing phases are high strength, high modulus, but relatively brittle. By incorporating a second phase in a high-temperature alloy matrix a composite with increased capabilities may be developed. Similar techniques have been utilized to produce polymer–matrix and glass–matrix composites. However, the use of composites for high-temperature applications has not yet been realized.

Summary of Strengthening Mechanisms

In most structural high-temperature alloys, several strengthening mechanisms are combined to produce a useful alloy (Table 1). Strengthening from only one or two of the se mechanisms would not produce a useful balance of properties. How these strengthening mechanisms interact to produce the strength observed in the alloys is not obvious. In many cases, the strengths are simplistically assumed to be additive. In a qualitative sense this assumption is probably correct and many attempts have been made to develop techniques to predict the strength of an alloy based on its composition, phase stability, and microstructure predictions. However, most of these techniques are not successful in accurately predicting the strength of alloys outside very narrow ranges.

Table 1. Strengthening Mechanisms for Both High-Temperature and Low-Temperature Applications

Strengthening mechanism	Useful at low temperatures	Useful at high temperatures
cold work	yes	no
grain refinement	yes	no
solid solution alloying	yes	yes
precipitation hardening	yes	yes
long range ordering	yes	yes
dispersion strengthening	yes	yes
composite	yes	yes, if stable

SURFACE STABILITY AND ENVIRONMENTAL RESISTANCE

Structural materials exposed to high temperatures must exhibit surface stability and resistance to the environment. When used in air, the material must resist oxidation, in combustion environments, the attack of a variety of combustion by-products that can be very aggressive species. In general, surface stability is described as resistance to oxidation and hot corrosion. In addition, the environmental resistance that is designed in to high-temperature structural materials can be achieved by alloying to the base metal and/or using coatings.

Oxidation

When a metal is in contact with the environment at elevated temperatures it will react to form surface scale. If the environment contains oxygen, the scale is usually an oxide and the process of formation of the oxide is called oxidation. In some cases the oxide can be protective, slow growing and prevent further damage from environmental interaction. However, in other cases the scale or oxide that forms is not protective, or fast growing and can even be in a gaseous form.

Oxidation damage to a material can take place in relatively inert environments such as air or in very aggressive environments containing sulfur, halogens, and water vapor. Rapid attacks called hot corrosion can be observed when condensed molten salts and oxidizing environments are present simultaneously.

Structural materials without sufficient environmental resistance can fail after a short period of time as the result of rapid oxidation and/or hot corrosion combined with rapid oxide growth and spalling. The adherence of the oxide can be dramatically altered by alloying and by control of impurities (eg, sulfur). Therefore, in addition to having an impact on strength, alloying of high-temperature alloys must also include the consideration of environmental resistance.

Hot Corrosion

The presence of molten salts on the surface of a component at elevated temperature, in addition to an oxidizing environment, results in a more rapid attack of the materials, called hot corrosion or sulfidation. The molten salt interacts with the oxygen in the environment, the surface oxide, and the base metal to reduce the protective nature of the surface oxide scale. In severe cases, the surface oxide is essentially dissolved by the molten salt. The most common type of salt observed to cause hot corrosion is Na_2SO_4. The presence of this salt on superalloys at high temperatures, such as those experienced in the first few stages of a turbine, results in hot corrosion. In general, the formation of the salt is attributed to the presence of impurities in the air the fuel and the alloy. These types of impurities are frequently present in marine or near-coastal applications and when lower quality or coal-derived fuels are used. In addition, common alloying elements in alloy, such as V, Mo and W, can contribute to hot corrosion.

The depth and degree of hot corrosion damage is extremely variable and can be a function of temperature, environment, and alloy content. In most cases, the corrosion rate reaches a maximum at 850–900°C, then decreases at higher and lower temperatures. At lower temperatures, the salt is solid and no hot corrosion is observed. At higher temperatures, oxidation is the primary concern.

Coatings

Since it is necessary for structural materials to exhibit environmental resistance, and the alloying to produce the needed environmental resistance often results in degraded mechanical properties, it has become quite common to apply coatings designed specifically for environmental resistance. The need for coatings increases only as the temperature increases and/or as the environmental resistance of the alloy decreases. Superalloy coatings typically provide a reservoir for aluminum and/or chromium for the growth of protective Al_2O_3 or Cr_2O_3 surface scales to reduce oxidation and corrosion damage.

HIGH-TEMPERATURE ALLOY SYSTEMS

Low-Alloy Steels

The low-alloy steels that can be used at elevated temperatures typically contain up to about 10 wt% Cr and up to about 1.5 wt% Mo, as well as small amounts of other elements. The relatively low alloy content of these alloys results in a reduced cost and, often, reduced processing costs.

These alloys typically contain increased levels of Cr for oxidation and corrosion resistance. The low-alloy steels also contain increased levels of Mo for increased strength and greater pitting resistance. Mo is also one of the most effective elements for increased creep resistance and temper embrittlement. When improved oxidation resistance is needed, Si is often added. Higher-temperature strengths can be obtained by increasing the stability of the carbides with the addition of vanadium or titanium. The addition of Ni is used when increased toughness is needed.

Wrought Stainless Steels

Generally, oxidation/corrosion resistance increases with increasing Cr content.

All of the stainless steels rely on the Cr_2O_3 surface scale for oxidation and corrosion resistance. As long as the Cr content of the alloy is above about 13 wt% Cr, the Cr_2O_3 is protective up to about 1,000°C. However, if the intended application includes temperature cycling, the, oxidation resistance of stainless steels may not be sufficient.

Although most of the stainless steels utilize wrought processing techniques, there are also cast versions that are compositionally very similar to the wrought varieties. These alloys are generally used in metallurgical heat treatment furnaces, power-generating-plant equipment, gas turbines and glass-manufacturing equipment. Alloys with high Ni content generally exhibit increased strength, ductility, and environmental resistance.

Modified Stainless Steels

Several modified versions of 12 wt% Cr, ferritic AISI 403 alloy have been developed that typically contain increased levels of Mo and/or tungsten to solid solution strengthen the matrix. These alloy modifications are intended to produce significant improvements in creep strength without any reduction in environmental resistance. The modified versions of AISI 403 are typically used in the quench and temper condition and often for high-temperature fasteners, turbocharger blades, boilers, superheaters, and re-heater tubes.

Ni-Base Alloys

Ni-base alloys are generally used for corrosion and heat resistant but can also be used for including coinage, battery electrodes, filters, and catalysts. Nickel and Ni-base alloys are vitally important to modern industry, due to their ability to withstand a wide variety of severe operating conditions involving the combinations of corrosive environments, high temperatures, and high stresses in one component or location.

Ni and Ni-base alloys are used in a wide variety of products for consumer, industrial, military, transportation, aerospace, marine and architectural applications (see Table 2).

The majority of the corrosion-resistant Ni-base alloys have compositions and processing to produce microstructures optimized for corrosion/oxidation resistance. In most cases, the environmental resistance is for aqueous corrosion resistance. However, more aggressive environments and high-temperature oxidation and sulfidation may also be encountered that require Ni-base alloys. In addition, many of the alloys developed for heat resistance, or elevated temperature service, may also exhibit excellent environmental resistance. Therefore, some overlap between environmental and heat resistant alloys may occur.

COMMERCIALLY NICKEL AND NI-BASE ALLOY

Commercially Pure Ni Grades

Commercial purity Ni grades such as Nickel 201 typically have minimum Ni contents of 99% and are highly resistant to many corrosive and oxidizing environments. Tight control on impurities and intentional additions is used to maintain their high purity levels. In general, commercial purity alloys exhibit moderate strengths, good ductility, high toughness, and are typically used for chemical processing and the electronics industry.

Low-Alloy Nickel Alloys

Alloys with a minimum Ni content of 94% are considered low-alloy Ni alloys. Additions of up to 5 wt% Mn (Nickel 211) or up to 5 wt% Al and 0.6 wt% Ti (Duranickel) are used to increase the resistance of the alloy to sulfur embrittlement and corrosion/oxidation, respectively. Precipitation hardened alloys, including both Duranickel and alloy 360, exhibit significantly higher strengths and lower ductilities than the commercially pure Ni alloys and the low-alloy Ni alloy, Nickel 211. Joining of the low-alloy Ni alloys can be more difficult for the precipitation-hardened alloys due to the potential for strain age cracking.

Nickel–Copper Alloys

Most Ni-Cu alloys contain approximately 25–40 wt% Cu for solid solution strengthening and were developed for corrosion-resistant applications. Most Ni-Cu alloys

Table 2. Selected Application Employing Nickel and Ni-base Alloys

Aircraft gas turbines: disks, combustors, bolts, casing, shafts, exhaust systems, blades, vanes, afterburners, thrust reversers, etc.
Industrial/power generating gas turbines: disks, combustors, bolts, casing, shafts, exhaust systems, blades, vanes (buckets), etc.
Chemical processing industry: piping, bolts, fans, valves, flanges, reaction vessels, pumps.
Paper mills: tubing, blades, bleaching equipment, scrubbers.
Steam turbines: bolts, blades, tubing.
Metal processing: dies, tolling, fixtures.
Rocket engine: disks, blades, cases, combustors, bolts, shafts.
Heat treating equipment: furnace components, baskets, trays, fans, muffles.
Nuclear power systems: bolts, springs, valve stems, control rod drive mechanisms.
Reciprocating/automotive engines: turbocharger impellers, exhaust valves, glow plugs, catalytic converters.
Medical applications: prosthetic/implant devices, stints.

exhibit moderate strength, high ductility, and high toughness. However, the precipitation-hardened alloys can exhibit much higher strengths but reduced ductility and toughness. In general, the Ni–Cu alloys exhibit good resistance to corrosion and stress corrosion cracking in a variety of environments.

Nickel–Molybdenum Alloys

Approximately 25–30 wt% of Mo is added to Ni-Mo alloys to confer resistance to acid and other reducing environments. The Ni-Mo alloy Hastelloy B exhibits excellent resistance to boiling hydrochloric acid and is widely used in the chemical processing industry. Other alloys have been developed with reduced levels of impurities that exhibit even greater resistance to severe environments, but they exhibit very poor elevated temperature oxidation resistance due to the lack of elements such as Cr and Al.

Nickel–Chromium Alloys

The addition of 15–30 wt% Cr to Ni-Cr-Fe alloys results in the formation of a very protective Cr_2O_3 surface oxide that provieds resistance to both corrosion and oxidation to high temperatures. The Ni-Cr-Fe alloys Alloys 600 and 690 exhibit moderate strength to elevated temperature, excellent corrosion resistance in a variety of environments, immunity to stress corrosion cracking, and resistance to hot corrosion. The Ni-Cr-Fe alloys are widely used for elevated temperature service from gas turbines, steam generators, mechanical property testing equipment, chemical processing, and nuclear reactors.

Iron–Nickel–Chromium Alloys

The Fe-Ni-Cr alloys are similar to the Ni-Cr-Fe alloys but have increased Fe contents for reduced cost. These alloys are also similar to austenitic stainless steels but have significantly higher Ni contents. The Fe-Ni-Cr alloys exhibit good oxidation/corrosion resistance, stress corrosion resistance, and moderate strength to moderate temperatures. These alloys are very fabricable and weldable and are most commonly fabricated by wrought (I/M) processing techniques. The Fe-Ni-Cr alloys are used extensively for chemical processing and power generation.

Nickel–Chromium–Molybdenum(Tungsten) Alloys

The large number of Ni-Cr-Mo(W) alloys that have been developed are widely used in gas turbines, nuclear applications, chemical processing, pollution control, and waste treatment. In most applications, these alloys are used in relatively low-stress applications that require oxidation/corrosion resistance, weldability, and fabricability.

Precipitation-Hardened Ni-base and Ni-Fe-base Alloys

The need for high strength, corrosion-resistant alloys for fastener and drilling equipment applications in sour gas wells led to the development of precipitation-hardened Ni-base and Ni-Fe-base alloys. The maximum strength of these alloys relies on some of the residual work from wrought processing. Welding of these materials is possible, but the strength, and possibly the corrosion resistance, of the weld region will not be equivalent to that of the base metal.

Controlled Expansion Alloys

The coefficient of thermal expansion of most structural alloys ranges from 5 to 25×10^{-6} m/m·K. However, for some applications, a reduced coefficient of thermal expansion would be beneficial to reduce stresses due to heating and cooling, to control tolerances between components, and match the thermal expansion of dissimilar materials (eg, joining metals to ceramics). Typical applications of controlled expansion alloys include pendulums and balance wheels for clocks and watches, glass-to-metal joints, vessels and piping for cryogenic liquids, superconducting systems in power transmission, integrated circuit lead frames, components for radios and other electronic devices and structural components in precision measurement systems. Most controlled-expansion alloys are made-up of nickel, cobalt and iron with small additions of C, Si, Cu, Cr and Mn.

Precipitation-Hardened Ni-base Superalloys

The superalloys are generally high-strength, corrosion resistant Ni-base alloys used at temperatures in excess of about 540°C. There are two major types of precipitation-hardened Ni-base superalloys. The first type achieves high strengths due to the precipitation of γ'' and can be used up to temperatures approaching 750°C. Significantly higher temperature capabilities are observed in the γ' strengthened superalloys, can be used up to temperatures approaching 90% of the melting temperature. The primary applications for the superalloys include industrial, power generating, and aerospace propulsion gas turbines, as well as nuclear reactors, aircraft structures, spacecraft structures, petrochemical processing, and orthopedic and dental prosthesis.

The alloy compositions of the Ni-base superalloys are tailored not just for processing but also for applications. For example, alloys traditionally intended for industrial gas turbine applications must be resistant to hot corrosion and therefore have high Cr content and increased Ti/Al-ratios. Aircraft engine alloys require greater levels of oxidation resistance and will have higher Al content and lower Cr content.

OTHER COMMERCIAL ALLOYS

Co-Base Superalloys

Co-base superalloys are principally used in highly corrosive environments at temperatures ranging from 650–1,000°C and usually at low stress levels. These alloys are frequently used in industrial gas turbine blades/vanes and as combustor liners and afterburner components. Since the chemistry of the Co-base superalloys is simpler than the more highly alloyed Ni-base superalloys, the thermal conductivity of Co-base alloys is significantly greater than that of Ni-base alloys. Therefore, Co-base alloys are more resistant to thermal fatigue and thermal shock than Ni-base alloys. In addition, Co-base alloys typically have significantly higher Cr content and will exhibit greater hot corrosion resistance than Ni-base alloys.

Refractory Metals and Their Alloys

The refractory metal elements are of obvious interest for high temperature applications due to their high melting points.

In many high-temperature applications in the electrical and electronics industry and for many space applications, refractory metals are the best materials. In these applications, the metals are protected by a vacuum or inert gas from oxidation. However, for most other high-temperature applications, poor oxidation resistance has limited the use of refractory metals. The oxides of the refractory metals do not exist as thin adherent and protective scales, in Ni- and Co-base superalloys, but are instead present as rapidly growing, porous oxides at intermediate temperatures and often become volatile at elevated temperatures.

In order to use the refractory metal alloys in oxygen-containing environments at high temperatures the alloys must be coated.

Another problem is joining. Brazing of refractory metals can prevent the occurrence of recrystallization, but braze a joints do not exhibit the same temperature capabilities as the base metal.

Ordered Intermetallic Alloys

In theory, intermetallic compounds offer many attractive properties for high-temperature applications. However, the ductility and toughness of these alloys are still limited and may preclude their use for structural components.

J. R. Davis, "Mechanical Properties at Elevated Temperatures", in J. R. Davis, ed., *ASM Specialty Handbook–Heat Resistant Materials*, ASM, Materials Park, OH, 1997.

G. E. Dieter, "Creep and Stress Rupture", in *Mechanical Metallurgy*, 3rd ed., McGraw-Hill, New York, 1986.

"Fatigue and Fracture", *Metals Handbook*, 10th ed., Vol. 19, ASM International, Materials Park, Ohio, 1996.

"Mechanical Testing", *Metals Handbook*, 10th ed., Vol. 8, Fatigue Testing, ASM International, Materials Park, Ohio, 2000, p. 679.

G. E. FUCHS
Materials Science & Engineering

HYBRID NANOCOMPOSITE MATERIALS

As the particle size of the components in a composite material is reduced, their interface increases, with its area increasing proportionally to the reduction in particle size for a given amount of material. An increased interface/bulk ratio leads to a growing relevance of the interface in determining the properties of the composite, but with each of the components still keeping its own identity as separate phases. Yet, as the degree of dispersion increases further and the components reach nanometric dimensions, one gets into a twilight region between physical mixtures and chemical compounds, the realm of nanocomposite materials. At that point, composite materials break into a new, chemical dimension.

One can distinguish two major approaches for the development of nanometric structures and nanocomposite materials. A downsizing approach strives to keep reducing the particle sizes of bulk solid materials until their properties reach quantum effects. An upsizing approach starts with statistical bundles of quantized atoms or molecules and aims at assembling ever-growing atomic or molecular ordered clusters with cooperative properties.

CROSS-BREEDING OF MATERIALS

It is in this chemical realm of nanocomposite compounds that hybrid materials thrive. In this case, "hybrid" refers to the mixed nature of the composite, normally made of dissimilar organic and inorganic components, each with characteristic chemical properties but interacting so closely that they frequently form a single phase. In combining organic and inorganic components, materials scientists can gain the best of two worlds. But they can gain more than that, because in certain cases these combinations present properties that are superior to the simple combination of properties from the components; ie, they present synergic behavior.

Indeed, being composed of an organic and an inorganic part has turned out to be the most common case among hybrid compounds, and the term "hybrid" has become almost a synonym for organic–inorganic combinations. Taken in a wider sense, certain combinations of dissimilar inorganic–inorganic as well as organic–organic compounds can also be properly considered within the category of hybrid materials.

Nowadays, research and development of hybrid materials spans combinations that harness chemical and biochemical activity, electrochemical or photochemical properties, as well as magnetic, optical or transport properties in materials formed by organic or inorganic matrices and molecular, polymeric, or crystalline species.

Among the many possible criteria to classify a large family of compounds are two major ways to classify hybrid materials; a chemical classification, according to composition, and a technical classification, by properties and applications.

TYPES OF HYBRIDS ACCORDING TO CHEMICAL CRITERIA

From a chemical point of view, hybrid materials can be broadly classified according to the nature of the matrix phase or predominant compound. We will use an arbitrary short-hand notation to facilitate the analysis of related groups of materials. Depending on the dominant structural matrix, we will distinguish and analyze organic–inorganic (OI) and inorganic–organic (IO) hybrids. Thus, we will refer to *OI* materials or compounds to denote hybrids where the *organic* phase predominates and constitutes the matrix where an inorganic species is inserted or dispersed, whereas *IO* materials will be those with *inorganic* matrices and integrated organic "guests." The term "organic–inorganic" (without initials) will be used when referring generically to hybrid materials.

Another classification commonly accepted that will also be used, centers on the type of chemical interaction at the interface between the components of a hybrid material. This bonding criterion divides hybrid materials into two distinct classes. These are Class I hybrids, in which the organic and inorganic compounds are bound together by ionic or weak chemical interactions, and Class II hybrids, in which the organic and inorganic components are linked by strong chemical bonds.

Hybrids Based on Siloxane Bonds

Hybrid materials based on silicon networks (both natural and synthetic) are probably the most widely studied. In this broad category we can make an initial distinction between materials based on silicates and those based on sol–gel materials such as polysiloxanes, polysilsesquioxanes, or even silica.

The intercalation of organic compounds in mineral silicates can be considered the first antecedent of hybrid inorganic–organic materials. Eventually this field led to the preparation of stable intercalation adducts such as those formed by intercalation of crown ethers into phyllosilicates and later to the direct chemical modification of the interlamellar space in layered silicates by a process that came to be known as grafting.

The development of hybrid materials based on the sol–gel growth of Si–O networks opened the field by going beyond the limits imposed by the crystal chemistry of natural silicates, broadening substantially the kind of compounds that could be prepared. Among these sol–gel materials are the single-chain polysiloxanes and the oxygen cross-linked polysilsesquioxanes, both of which could be considered condensation products between molecular organosilanes (ie, SiR_4) and silica (SiO_2). Polysilsesquioxanes, with an extra oxygen bridge between silicon atoms, represent the polymers closer to three-dimensional silica.

Due to their single flexible linear chain, polysiloxanes present low glass-transition temperatures, which can be controlled by the use of adequate substituents. On the other hand, the cross-linked nature of polysilsesquioxanes confers them higher glass-transition temperatures in addition to higher thermal stability.

A further step in the control of molecular structure with these polymers comes from the introduction of bridged polysilsesquioxanes. By using adequate monomer units formed by an organic group with two or more—$Si(OR)_3$ moieties, the connectivity between silicon atoms increases and the organic bridge can be used as a spacer whose size and nature allow to control the design as well as function of the nanostructure.

Hybrids Based on Transition-Metal Oxides

In addition to silicate minerals, transition-metal oxides constitute the other large category of inorganic solids extensively studied as intercalation hosts and later used for the development of hybrid inorganic–organic materials.

Layered oxides of the early transition elements, such as V_2O_5 or MoO_3, were among the earliest and most widely studied in relation to their ability to insert small organic molecules first and organic polymeric materials later between inorganic slabs.

The use of oxides and conducting polymers for the design of hybrid materials allowed for a change in emphasis from structural to functional materials. Whereas the main strength of silicon networks was the control of structure and microstructure, the variety of chemical, magnetic, or electronic properties in transition-metal oxides offered wider opportunities to prepare functional hybrids. Furthermore, not only the inorganic oxides but also the organic ionically or electronically conducting polymers inserted into them could contribute with their own activity and allow for the design of multifunctional materials.

In this respect is the first example of intercalation of the well-known ion-conducting polymer PEO into a solid, the layered silicate montmorillonite.

Concerning hybrids based on intrinsically conducting polymers, some of the systems studied in detail comprise such early transition-metal oxides as V_2O_5 and MoO_3 and conducting polymers (normally polyaniline, polypyrrole, or polythiophene derivatives). These systems have been the subject of basic chemical and structural studies as well as efforts to determine their possible application as lithium-inserting cathodes in rechargeable lithium batteries.

Hybrids Based on Other Mineral Solids

When it comes to layered inorganic phases, oxides are just part of a wider family of inorganic mineral compounds with interlayer spacing suitable for the insertion of organic compounds and polymers. Thus, the group of I–O hybrids includes materials based on a wide variety of inorganic hosts. A true host–guest association would imply that the organic molecule could leave the inorganic structure reversibly, which is normally not the case for the hybrid materials described here. Indeed, the key to the preparation of many of these I–O hybrid materials relies upon *in situ* polymerization of the corresponding monomers once they have been intercalated into the inorganic host or alternatively on to the simultaneous formation of both organic and inorganic polymeric structures. Otherwise, the long chains and large molecular weights of the frequently insoluble polymers inserted would prevent their effective diffusion into the host structure. The irreversible character of the insertion process leading to the permanent integration of the organic and inorganic components is normally an important condition in the design of hybrid materials.

Hybrids Based on Organic Polymers

The category of organic–inorganic hybrids is materials with organic polymers forming the predominant or host matrix where a variety of inorganic species can be embedded. Within this category of O–I hybrid materials are two further types, according to the polymers used, one where conventional (insulating) polymers provide the structural base and another with conducting polymers (electronic or ionic) contributing their peculiar combination of semiconducting and electrochemical behavior to the development of functional materials.

Concerning materials based on conventional, insulating organic polymers, the field has evolved from the simple use of inorganic additives and fillers, in order to get mixtures with certain improved properties, to the design of true hybrid materials with organic and inorganic components interacting at the molecular level. An interesting example is the development of methods to produce exfoliated nanocomposites from *in situ* polymerization within layered silicates. Those same layered silicates intercalated with organic polymers constitute a classical example of inorganic–organic (I–O) hybrids. This exfoliated systems represent the ultimate level of dispersion between extended solids and polymers and constitute organic–inorganic (O–I) systems with single inorganic layers dispersed within a polymer matrix.

Hybrid materials based on conventional polymers are also making it to the market. Polyethylene, used for insulation of electrical wires, presents excellent mechanical and processing properties. To meet the ever-growing demands in performance, the material has been engineered to stand higher temperatures, by modification with pendant alkoxysilanes that cross-link by a hydrolytic mechanism to form silsesquioxane networks. The resulting material can also be used as heat-shrinkable tubing and compression-resistant foams.

Another practical example of a hybrid organic–inorganic composite material is the use of poly(vinyl alcohol) and/or polypropylene fibers in high-performance concrete (HPC). The properties of the concrete can be modify by reducing the size and amount of the crack source at different scales. The hybrid fibers combined with an expansive agent provide better enhancement for shrinkage resistance and impermeability of HPC than the typical method of incorporating hybrid fibers or an expansive agent.

Many of these organic–inorganic hybrid materials based on insulating polymers present remarkable properties such as flexibility, mechanical strength, optical density, and thermal and mechanical properties that makes them appropriate as structural, biomedical, or electrical and thermal insulating materials. For other applications where good electronic or ionic transport properties are required these hybrids are no longer useful. In those cases, organic–inorganic hybrid materials based on conducting polymers come to add the conductivity and functionality of those remarkable polymers toward the design of functional materials.

Within materials based on conducting polymers we can distinguish those based on ion-conducting polymers (with poly(ethylene oxide) or PEO as a typical example) and those based on electronic conducting polymers (either electron or hole conductors, such as polyaniline, polypyrrole, polythiofene, etc). The latter can be used for their intrinsic conducting properties but also for their reversible electroactivity and mixed ion–electron conductivity.

The area of lithium batteries to develop solid electrolytes with improved conductivity has received much attention. Recent research efforts have shown improvement in ionic conductivity by application of novel hybrid organic–inorganic polyelectrolytes.

The design of organic–inorganic hybrids based on conducting organic polymers provide another interesting way to introduce molecular inorganic species into conducting polymer networks, thus integrating the activity of those molecular species into useful extended materials. There are two major ways to accomplish this integration, the covalent linkage of molecules to form Class II hybrids or their incorporation as counterions to form Class I materials.

The simplest way to obtain organic–inorganic hybrids with molecular species integrated in conducting polymers is to take advantage of counterion with incorporation during the doping process of the polymers. Doping of these polymers involves the incorporation of charge-balancing species into the structure, with anions for p-doped and cations for n-doped materials, hence the possibility of using active ionic species as dopants for the design of functional hybrids, either by ionic exchange or by *in situ* incorporation of the active species during polymerization and doping. This doping process leads to less strongly bound inorganic species (ie, class I hybrids) but has the advantage of greater simplicity and versatility and has allowed the systematic preparation and study of hybrids containing a wide variety of anionic transition-metal complexes.

PROPERTIES AND APPLICATIONS

The fields of applications of hybrid materials are as varied as the materials themselves, but in general we could distinguish two main avenues concerning the type of application for which they are targeted. A first one puts the main emphasis on the mechanical properties of the hybrids for their application as structural materials, whereas the second aims at the development of functional materials, where mechanical properties are secondary and chemical activity, physical properties, or a combination are paramount.

Structural Hybrid Materials

The mechanical, structural, or thermal properties of materials can be improved by the use of hybrid organic–inorganic materials, in wire and cable insulation. Hybrid materials are already in the market in the field of structural applications.

A remarkable example is that of POSS nanometric inorganic units polyhedral oligomeric silselsquioxones that are embedded into polymer matrices to form nanostructured hybrid materials. Incorporation of POSS into traditional plastics results in polymeric nanocomposites that show many advantages relative to traditional plastics, such as increased heat resistance and hardness as well as decreased flammability and heat evolution. POSS technology has numerous applications in biological systems, pharmaceuticals, electronics, medical plastics, consumer products, and in construction and transportation markets.

When it comes to structural applications, hybrid materials normally rely upon silicon-based inorganic compounds (silsesquioxanes, silica, or silicates) and organic polymers, as in silsesquioxane hybrids for low-dielectric matrices to interconnect material in integrated circuits. As structural materials silicon-based hybrids have been used their transparency in optical applications or in gradient-refractive index lenses. Applications are extensive in the area of thin films and coatings, as in coatings to avoid abrasion, or as decoration in the case of colored glasses, for corrosion prevention, coatings for electrical components, scratch-resistant coatings for the automobile industry, as adhesives, in cosmetics, as impact modifiers, and in water repellent or cable insulating materials.

The field of hybrid materials promises great new breakthroughs. A remarkable example is the design of "intelligent" hybrid materials that can adapt to a certain environment or modify *in situ* their properties according to the requirements of a particular application. The self-passivating/self-healing behavior of a hybrid nanocomposite made of Nylon 6 and a layered silicate (PLSN) after 10 min of oxygen plasma exposure forms a tough ceramic passivation layer on the polymer surface. This behavior makes the hybrid much more durable than Nylon 6 itself and has led to considering the application of this material in harsh environments such as solid rocket motor exhausts.

Other applications of hybrid organic–inorganic materials related to property modulations is their application in electrorheology (ER). Typically, ER fluids are suspensions composed of micrometer-sized particles and dielectric liquids. An external electric field is applied to change such hybrid properties as viscosity, yield stress, and shear modulus, which affect flow properties. Those changes are caused by the formation of an internal chain-like structure due to the arrangement of the dispersed particles within the fluid. The chains are held together by interparticle forces, which have sufficient strength to inhibit the fluid flow. When the electric field is removed, the particles return to their original random distribution. These properties fulfill the requirements that make ER fluids one of the most important actuators used in smart materials and structures. They are very promising materials for application as active engine mounts, shock absorbers, adaptive structures, fluid drive control, automobile vibration control, or robotic systems. Conventional ER fluids can be composed of carbonaceous particles or zeolites.

Functional Hybrid Materials

Functional materials are those with chemical, electrochemical, or photochemical activity, although a wider definition would also include magnetic, electronic, and optical materials.

Functional Materials Based on Chemical Activity

Both the organic and the inorganic components can be active in chemically or electrochemically active functional hybrids. Electroactive inorganic compounds can be used for the development of hybrid electrodes in combination with conducting organic polymers, but the latter in turn are also electroactive. This electroactivity is behind the development of hybrid materials for electrocatalysis or energy storage applications where the electrochemical activity is paramount.

Other applications include the harnessing of electrochromic, and photoelectrochromic properties, application in display devices, photovoltaics, novel energy-conversion systems, proton-pump electrodes, sensors, or chemiresistive detectors that work as artificial "noses" and outperform human olfaction in detecting biogenic amines.

Sensors are one whole field with many interesting developments related to hybrid materials. Sol-gel techniques are getting a growing relevance in the manufacture of sensor structures. Incorporation, encapsulation, or entrapment of organic or biological species is possible with materials synthesized by this method, thanks in part to their formation at room temperature. These strategies provide sturdy materials carrying active molecules capable of responding to chemical changes in the environment. The simplicity of this sol–gel approach has made of it the method of choice for routine fabrication of optical sensors. Among hybrid organic–inorganic materials, we can find many other applications as sensors, including some dealing with new breakthroughs in miniaturization.

Functional Materials Based on Physical Properties

Electronics and optoelectronics are two of the areas that could benefit more from the properties of hybrid organic inorganic materials, as in hybrids based on the self-assembly of siliceous hybrids and general mesoporous and mesostructured materials. We can also find electronic and optoelectronic applications of hybrids based on conducting organic polymers or other nonsiliceous materials like perovskites with organic molecules.

Hybrid Biomaterials

The fast-growing category of biomaterials includes both structural and functional members. Yet, their very specific focus on biomedical applications have made us consider a separate section for these special materials. The remarkable mechanical characteristics of structural hybrid biomaterials like teeth or bone, such as their high specific strength and modulus and high toughness, make them suitable models for biomimetic synthesis. This type of synthesis tries to imitate the building mechanics of biological systems to obtain composites that are ordered from the nanometer to the microscopic length scale.

A synthesis of biomaterials can be made by applying different techniques such as the sol–gel process, building blocks, self-assembly, etc, all as part of a biomimetic approach. These techniques have been widely used by many different groups with great success for the synthesis of hybrid materials not always related to biological systems.

Some biohybrids have been formed with titania, zirconia, and alumina with aramides (aromatic polyamides), poly(tetramethylene oxide), epoxides, etc. Bone or seashell-layered architectures have inspired the formation of materials with similar fracture resistance, as for example montmorillonite clays and alkyl amonioum ion or aramide-clay composites.

Present applications include bioencapsulation of active biological materials such as enzymes, antibodies, living microbial, plant and animal cells, DNA, etc, within ceramic matrices. In these cases, the combination of delicate functional biostructures with inorganic ceramics is possible, thanks to the fact that the inorganic framework is formed at room temperature by sol–gel techniques. In most cases the final applications of bioencapsulates are the syntheses of nanocomposites for sensors, catalysis, or diagnostics.

An emerging area dealing with biomolecular electronics (BME) will probably experience important growth in the near future. Biomolecules, proposed both by their activity and robustness as compared with other simpler organic molecules, could lead to cheap and easy-to-fabricate devices with self-assembling and self-healing/repairing characteristics. Examples making use of biomolecules as state of the art materials are the use of the metalloprotein azurin for the assembly of single-molecule devices and the fabrication of interconnects with DNA molecules.

Hybrid materials are no longer newcomers to the materials science arena. Examples abound of how industry can improve product quality and reduce costs by applying the hybrid approach to the development of new and improved materials and novel applications. Among all the kinds of hybrid materials mentioned, it seems that siloxane-based materials are closer to marketable applications as structural materials and are also being developed for a variety of functional applications.

C. M. Chan, G. Z. Cao, H. Fong, M. Sarikaya, T. Robinson, and L. Nelson, *Organic-Inorganic Hybrid Materials II.*, San Francisco, Calif., 1999.

L. F. Nazar, T. Kerr, and B. Koene, in J. Pinnavaia and M. F. Thorpe, eds., *Proceeding of Access Nanoporous Material* Plenum, New York, 1995.

C. L. Soles, E. K. Lin, W. L. Wu, C. Zhang, and R. M. Laine, *Organic/Inorganic Hybrid Materials*, MRS Symp. Serv., 2000.

G. Torres-Gomez, M. Lira-Cantu, and P. Gomez-Romero, *J. New Mater. Electrochem. Syst.* **2**, 145 1999.

PEDRO GÓMEZ-ROMERO
MONICA LIRA-CANTÚ
Materials Science Institute of
Barcelona (CSIC)

HYDRAZINE AND ITS DERIVATIVES

Hydrazine (diamide), N_2H_4, a colorless liquid having an ammoniacal odor, is the simplest diamine and unique in its class because of the N–N bond.

Hydrazine and its simple methyl and dimethyl derivatives have endothermic heats of formation and high heats of combustion. Hence, these compounds are used as rocket fuels. Other derivatives are used as gas generators and explosives. Hydrazine, a base slightly weaker than ammonia, forms a series of useful salts. As a strong reducing agent, hydrazine is used for corrosion control in boilers and hot-water heating systems; also for metal plating, reduction of noble-metal catalysts, and hydrogenation of unsaturated bonds in organic compounds. Hydrazine is also an oxidizing agent under suitable conditions. Having two active nucleophilic nitrogens and four replaceable hydrogens, hydrazine is the starting material for many derivatives, among them foaming agents for plastics, antioxidants, polymers, polymer cross-linkers, and chain-extenders, as well as fungicides, herbicides, plant-growth regulators, and pharmaceuticals. Hydrazine is also a good ligand; numerous complexes have been studied. Many heterocyclics are based on hydrazine, where the rings contain from one to four nitrogen atoms as well as other heteroatoms.

The many advantageous properties of hydrazine ensure its continued commercial utility. Hydrazine is produced commercially primarily as as aqueous solutions, typically 35, 51.2, 54.4, and 64 wt% N_2H_4 (54.7, 80, 85, and 100% hydrazine hydrate). Anhydrous hydrazine is produced for use as a rocket propellant and in limited commercial applications.

PHYSICAL PROPERTIES

Anhydrous hydrazine is a colorless, hygroscopic liquid having a musty ammoniacal odor. It fumes in air, owing to the absorption of water and perhaps also of carbon dioxide, forming carbazic acid, $CH_4N_2O_2$. Hydrazine is miscible with water, alcohol, amines, and liquid ammonia but has only limited solubility in other nonpolar solvents. Its physical properties are more like the isoelectronic hydroxylamine or hydrogen peroxide rather than ethane, owing to hydrogen bonding, which is exemplified in its relatively high melting (2°C) and boiling (113.5°C) points as well as an abnormally high [39.079 kJ/mol (9.340 kcal/mol)] heat of vaporization as compared to 14.64 kJ/mol (3.50 kcal/mol) for the isoelectronic ethane.

CHEMICAL PROPERTIES

Thermal Decomposition

Hydrazine is a high energy compound having a high positive heat of formation; however, elevated (>200°C) temperatures are needed before appreciable decomposition occurs. The decomposition temperature is lowered significantly by many catalysts, particularly copper, cobalt, molybdenum, ruthenium, iridium, and their oxides. Iron oxides (rust) also catalyze decomposition. Hydrazine, especially in high concentrations, should be handled with care, using scrupulously clean systems.

Acid–Base Reactions

Anhydrous hydrazine undergoes self-ionization to a slight extent, forming the hydrazinium, $N_2H_5^+$, and the hydrazide, $N^2H_3^-$, ions:

$$2\,N_2H_4 \rightleftharpoons N_2H_5^+ + N_2H_3^- \qquad K_i = 10^{-25}$$

Hydrazinium salts, $N_2H_5^+X^-$, are acids in anhydrous hydrazine, metallic hydrazides, $M^+N_2H_3^-$, are bases.

Hydrazine as Nucleophile

Reaction of hydrazine and carbon dioxide or carbon disulfide gives, respectively, hydrazinecarboxylic acid, $NH_2NHCOOH$, and hydrazinecarbodithioic acid, $NH_2NHCSSH$, in the form of the hydrazinium salts. These compounds are useful starting materials for further synthesis.

Reductions

Hydrazine is a very strong reducing agent. In the presence of oxygen and peroxides, it yields primarily nitrogen and water with more or less ammonia and hydrazoic acid. It is used in metal reductions, hydrogenations, carbonyl reductions, catalytic hydrogenations, diazene reductions, aldehyde syntheses, and olefin syntheses.

Alkylhydrazines

Mono- and higher substituted alkyl hydrazines can be made by alkylation of hydrazine using alkyl halides.

Substituted alkyl hydrazines are prepared from suitable alkylating agents. Epoxides yield hydroxyalkyl-hydrazines; aziridines give β-aminoalkylhydrazines; sultones yield ω-sulfoalkylhydrazines; and acrylonitrile, β-cyano-ethylhydrazine.

A general synthesis for arylhydrazines is via diazotization of aromatic amines, followed by reduction of the resulting diazonium salt.

Hydrazides and Related Compounds

Substitution of the hydroxyl group in carboxylic acids with a hydrazino moiety gives carboxylic acid hydrazides. In this formal sense, a number of related compounds fall within this product class, although they are not necessarily prepared this way. Some of the more common of these compounds include thiohydrazides, sulfonylhydrazides, semicarbazide, thiosemicarbazide, carbohydrazide, thiocarbohydrazide, amidrazones, hydrazidines, aminoguanidine, diaminoguanidine, and triaminoguanidine.

Carboxylic acid hydrazides are prepared from aqueous hydrazine and the carboxylic acid, ester, amide, anhydride, or halide. The reaction usually goes poorly with the free acid.

Hydrazones and Azines. Depending on reaction conditions, hydrazines react with aldehydes and ketones to give hydrazones, azines, and diaziridines. Hydrazones are formed from mono- and N,N-disubstituted hydrazines. Many of these compounds are highly colored and have found use as dyes and photographic chemicals. Several pharmaceuticals and pesticides are members of this class.

Heterocyclics

One of the most characteristic and useful properties of hydrazine and its derivatives is the ability to form heterocyclic compounds. Numerous pharmaceuticals, pesticides, explosives, and dyes are based on these rings.

MANUFACTURE

The commercially feasible processes involve partial oxidation of ammonia or urea using hypochlorite or hydrogen peroxide. Most hydrazine is produced by some variation of the Raschig process, which is based on the oxidation of ammonia using alkaline hypochlorite. Ketazine processes are modifications in which the oxidation is carried out in the presence of a ketone such as acetone or butanone. A process developed by Produits Chimiques Ugine Kuhlmann (PCUK) and practiced by Atofina (France) and Mitsubishi Gas (Japan) involves the oxidation of ammonia by hydrogen peroxide in the presence of butanone and another component that apparently functions as an oxygen-transfer agent. The oxidation of benzophenoneimine has received much attention but does not have commercial applications.

ECONOMIC ASPECTS

World demand for hydrazine solutions is nearly equally divided between captive use and merchant business. Synthesis of chemical blowing agents (CBAs) for the foaming of plastics accounts for ∼30% of demand; another 25% goes into agricultural chemicals (pesticides). Water treatment represents nearly 25% of the market. Pharmaceutical applications are significant but difficult to quantify; the remaining applications are fragmented among many industries.

Hydrazine is a mature product likely to grow worldwide in step with gross national product. The use of hydrazine in chemical blowing agents in southeast Asia has dropped significantly due to the adoption of nonhydrazine routes to azodicarbonamide, the major compound in this application. In water treatment, the growth might be somewhat less, perhaps 2–3%.

SHIPMENT AND SPECIFICATIONS

Shipment of hydrazine solutions is regulated in the United States by the Department of Transportation (DOT), which classifies all aqueous solutions between 64.4 and 37% N_2H_4 as "Corrosive" materials with a subsidiary risk of "Poison." Hydrazine has been identified by both the Environmental Protection Agency and the DOT as a hazardous material and has been assigned a reportable quantity (RQ) of 0.450 kg (1 lb) (on a N_2H_4 basis) if spilled.

HANDLING AND STORAGE

Hydrazine is a base, a reducing agent, and a high-energy compound; it is also volatile and toxic. These properties determine its proper handling, storage, use, and disposal. Inadvertent contact with acids or oxidizing agents must be avoided, because extremely exothermic reactions and evolution of gases may result. Flash points, however, of aqueous and anhydrous hydrazine are relatively high compared to common combustibles such as alcohol or gasoline. It is much easier and safer to handle aqueous solutions than the anhydrous product; aqueous solutions containing <40% hydrazine have no flash point; ie, they cannot be ignited. Liquid hydrazine is not sensitive to shock or friction. In the absence of decomposition catalysts, liquid anhydrous hydrazine has been heated to >200°C without appreciable decomposition. Hydrazine fires are effectively combated using water, because hydrazine is miscible with water in all proportions.

The broad explosive range of hydrazine vapor is a concern. An inert gas blanket helps avoid formation of explosive mixtures. Hydrazine may ignite wood, rags, paper, or other common organic materials. Thus, these should not be used near hydrazine. Use protective clothing to avoid body contact and provide adequate ventilation to reduce inhalation danger.

Materials of Construction

Materials generally considered satisfactory for all N_2H_4 concentrations, including anhydrous, are 304 L and 347 stainless steels having <1.0 wt% molybdenum, a catalyst for the decomposition of hydrazine. For concentrations <10%, cold-rolled steel is satisfactory. Among the nonmetallic materials, poly(tetrafluoroethylene), polyethylene, and polypropylene are suitable; PVC is not recommended. Ethylene–propylene–diene monomer (EPDM) rubber, and polyketones and polyphenylene sulfides are reportedly suitable for use with anhydrous hydrazine.

Disposal

Spills and wastewater containing hydrazines must be contained and treated. Proper disposal methods for the hydrazines make use of their reductive properties. Fuel-grade hydrazines may be burned, but aqueous solutions <50% may require supplementary fuel. Chemical destruction of dilute (preferably 5% or less) solutions can be achieved with various oxidants such as NaOCl, $Ca(OCl)_2$, H_2O_2, and acidified permanganate; however, MMH and UDMH may form mutagenic nitrosamines. A method is described for treating contaminated wastewater in which N_2H_4, MMH, UDMH, and N-nitrosodimethylamine are effectively decomposed at a controlled pH of ~5 by an (uv)-induced chlorination. Ozonation of these three hydrazines yields a variety of products, including methanol, formaldehyde monomethylhydazone, and formaldehyde dimethylhydrazone; and tetramethyl tetrazene from the oxidative coupling of two molecules of UDMH. Methyldiazene was also found as an oxidation product of MMH. The rates of decomposition from aqueous solutions are greatest at pH 9.1 in the presence of uv light. Blowdown from boilers treated with hydrazine for corrosion control is effectively treated by neutralization with lime and chlorination. A vapor suppressant foam system (ASE95 polyacrylic/MSAR combination) has been evaluated for covering hydrazine fuel spills to minimize the release of toxic fumes.

HEALTH AND SAFETY FACTORS

Hydrazine is toxic and readily absorbed by oral, dermal, or inhalation routes of exposure. Contact with hydrazine irritates the skin, eyes, and respiratory tract. Liquid splashed into the eyes may cause permanent damage to the cornea. At high doses it can cause convulsions, but even low doses may result in central nervous system depression. Death from acute exposure results from convulsions, respiratory arrest, and cardiovascular collapse. Repeated exposure may affect the lungs, liver, and kidneys. Evidence is limited as to the effect of hydrazine on reproduction and/or development; however, animal studies demonstrate that only doses that produce toxicity in pregnant rats result in embryotoxicity.

The TLV is set at 0.01 ppm (hydrazine) TWA; 0.2 ppm (MMH); and 0.5 ppm (UDMH). The International Agency for Research on Cancer (IARC) classifies hydrazine as a 2B or possible human carcinogen. The American Conference of Governmental Industrial Hygienists (ACGIH) classifies hydrazine as an A3 or confirmed animal carcinogen with unknown relevance to humans.

USES

The principal applications of hydrazine solutions include chemical blowing agents, 28%; agricultural pesticides, 25%; water treatment, 31%; and pharmaceutical, 10%. The remaining 6% finds use in a variety of fields, including explosives, polymers and polymer additives, antioxidants, metal reductants, hydrogenation of organic groups, photography, xerography, and dyes.

E. Schmidt, *Hydrazine and Its Derivatives*, John Wiley & Sons, Inc., New York, 2001. Contains an exhaustive bibliography.

P. A. S. Smith, *The Chemistry of Open-Chain Nitrogen Compounds*, Vols. 1 and 2, W. A. Benjamin, Inc., Menlo Park, Calif., 1965–1966.

P. A. S. Smith, *Derivatives of Hydrazine and Other Hydronitrogens Having N–N Bonds*, The Benjamin/Cummings Publishing Co., Inc., Reading, Mass., 1983.

The World Market for Organic Derivatives of Hydrazine and Hydroxylamine: A 2004 Gobal Trade Perspective, ICON Group International, San Diego, Calif., 2004, p. 83.

EUGENE F. ROTHGERY
North Branford, Connecticut

HYDRIDES

Hydrogen is a versatile element that forms compounds with most of the other elements in the Periodic Table. Strictly speaking, a hydride is a compound in which hydrogen is present as a negative ion, H-. Only the most electropositive metals, the alkalis and alkaline earths, form the simple hydrides that conform to this strict ionic definition. The complex hydrides containing the borohydride and aluminohydride ions are also generally considered to be hydrides. These two classes of hydrides are used primarily as reducing agents in the metals processing, pharmaceutical, and other fine chemical industries.

The term "hydride" is also frequently applied to the hydrogen compounds with transition and rare-earth elements. These compounds, known as metallic hydrides, are of great interest because of their unique capability to store large quantities of hydrogen safely. Although they are not yet fully commercial, their wide-spread use and application in the transportation and utility industries are envisioned.

SIMPLE HYDRIDES

Ionic Hydrides

The ionic or saline hydrides contain metal cations and negatively charged hydrogen ions. These crystallize in the cubic lattice similar to that of the corresponding metal halide and, when pure, are white solids. When dissolved in molten salts or hydroxides and electrolyzed, hydrogen gas is liberated at the anode. The densities are greater than those of the parent metal, and formation is exothermic. All are strong bases.

Physical properties of the alkali metal hydrides are given in Table 1. Sodium hydride finds commercial use in organic synthesis in condensation and alkylation reactions.

Table 2 gives thermochemical data of alkaline-earth metal hydrides. Calcium hydride, CaH_2, because of its low reactivity with most organic liquids and its high reactivity with water, has found wide use as a superdrying agent of esters, ketones, halides, electrical insulator oils, silicones, solvents for Ziegler-Natta polymerization systems, monomers, and air and other gases. It is also used as an analytical reagent to determine water in organic liquids.

Covalent Hydrides

In all hydrides, hydrogen is bound to an atom of lower electronegativity ($X_H = 2.1$) than itself. In covalent hydrides, the hydrogen–metal bond is effected through a common electron pair. Beryllium and magnesium hydrides included in this group, are polymeric materials, as is aluminum hydride. The simple hydrides of silicon, germanium, tin, and arsenic are gaseous or easily volatile compounds. Table 3 gives some properties of these compounds.

COMPLEX HYDRIDES

The complex hydrides are a large group of compounds in which hydrogen is combined in fixed proportions with two other constituents, generally metallic elements. These compounds have the general formula $M(M'H_4)_n$, where n is the valence of M, and M' is a trivalent Group 3 (IIIA) element such as boron, aluminum, or gallium. The complex hydrides have achieved significant and broad use as reducing agents in many different areas of chemistry. Sodium borohydride in particular and lithium aluminum hydride are the most important commercially. The most important complex hydrides are listed in Table 4.

Borohydrides

The alkali metal borohydrides are the most important complex hydrides. They are ionic, white, crystalline, high-melting solids that are sensitive to moisture but not to oxygen. They include lithium borohydride, $LiBH_4$, and sodium borohydride, $NaBH_4$.

Complete hydrolysis of $NaBH_4$ produces 2.37 L hydrogen (STP) per gram of borohydride; similarly, addition of acid to a cold aqueous solution liberates the theoretical amount of hydrogen. The inorganic reductions with $NaBH_4$ are numerous and varied. Sodium borohydride reacts with boron halides to form diborane B_2H_6, which is more conveniently handled as the monomer BH_3 complexed with an ether, sulfide, or amine. Sodium borohydride is used extensively for the reduction of organic compounds. It is classified as a flammable solid. It is available as a powder, in caplets, and granules and as a 12% solution in caustic soda. The principal uses of $NaBH_4$ are in synthesis of pharmaceuticals and fine organic chemicals; removal of trace impurities from bulk organic chemicals; wood-pulp bleaching, clay leaching, and vat-dye reductions; and removal and recovery of trace metals from plant effluents.

Sodium borohydride is used for the removal of trace impurities, such as carbonyls or peroxides, from bulk organic chemicals (alcohols or glycols). This technique, called process-stream purification, greatly improves stability in products sensitive to deterioration and development of undesirable colors and odors in, for example, plasticizer alcohols. Sodium borohydride is used widely to generate sodium hydrosulfite for reductive bleaching of mechanical pulps for manufacture of several grades of paper, especially newsprint. This technology is also being applied in clay leaching and in vat dye reductions. Sodium

Table 1. Physical Properties of Alkali Metal Hydrides

Hydride	Mp, °C	$\Delta H_{(298)}$, kJ/mol[a]	$\Delta F_{(298)}$, kJ/mol[a]	S, J/(mol·K)[a]	Lattice energy, kJ/mol[a]	Density, g/cm³
LiH	688	−90.7	−70	25	916	0.77
NaH	420 dec	−56.5	−37.7	48	791	1.36
KH	dec	−57.9	−37.3	61	720	1.43
RbH	300 dec					2.60
CsH	dec					3.4

[a]To convert J to cal, divide by 4.184.

Table 2. Physical Properties of Alkaline-Earth Metal Hydrides

Hydride	$\Delta H_{(298)}$, kJ/mol[a]	$\Delta F_{(298)}$, kJ/mol[a]	S, J/(mol·K)[a]	Density, g/cm³
CaH₂	−186.3	−147.4	42	1.90
SrH₂	−180.5	−138.6	54	3.27
BaH₂	−171.2	−132.3	67	4.16

[a]To convert J to cal, divide by 4.184.

borohydride provides a simple, efficient means of removing dissolved trace metals from manufacturing-plant effluents. There is also extensive use of NaBH₄ in the analytical determination of metals that form volatile hydrides (As, Sb, Bi, Sn, Pb, etc).

ALUMINOHYDRIDES

In general, the aluminohydrides are more active and powerful reducing agents than the corresponding borohydrides. They decompose vigorously with water. Reaction also occurs with alcohols, although more moderately, providing a route to substituted derivatives.

Freshly prepared lithium aluminum hydride is a white crystalline solid that tends to become gray during storage, although very little loss in purity occurs. Although lithium aluminum hydride is best known as a nucleophilic reagent for organic reductions, it converts many metal halides to the corresponding hydride; eg, Ge, As, Sn, Sb, and Si. Commercial manufacture of LiAlH₄ uses the original synthetic method, ie, addition of a diethyl ether solution of aluminum chloride to a slurry of lithium hydride.

Sodium aluminum hydride can be prepared from NaH, but direct synthesis from the elements is more economical.

ALUMINOHYDRIDE DERIVATIVES

The few known derivatives of the aluminohydrides are principally alkoxy substitutions, including the trimethoxy,

LiAlH(OCH₃)₃, triethoxy, LiAlH(OC₂H₅)₃, and tri-t-butoxy aluminohydrides, LiAlH(O-t-C₄H₉)₃.

METALLIC HYDRIDES

A number of metal alloys are very useful for safely storing large volumes of hydrogen, because these easily dissolve hydrogen at relatively low temperatures and pressures, forming interstitial hydrides. The hydrogen is subsequently released by applying heat and lowering the pressure. Many metals and binary and ternary alloys have been thoroughly studied for this application.

Hydrogen-storage alloys are commercially available from several companies in the United States, Japan, and Europe. A commercial use has been developed in rechargeable nickel–metal hydride batteries, which are superior to nickel–cadmium batteries by virtue of their improved capacity and elimination of the toxic metal cadmium. Other uses are expected to develop in nonpolluting internal combustion engines and fuel cells, heat pumps and refrigerators, and electric utility peak-load shaving.

HEALTH AND SAFETY FACTORS

In general, hydrides react exothermically with water, resulting in the generation of hydrogen. This hydrolysis reaction is accelerated by acids or heat and, in some instances, by catalysts. Because the flammable gas hydrogen is formed, a potential fire hazard may result unless adequate ventilation is provided. Ingestion of hydrides must be avoided because hydrolysis to form hydrogen could result in gas embolism.

Another aspect of the hydrolysis of hydrides is the alkalinity that results, especially from alkali metal and alkaline-earth hydrides. This alkalinity can cause chemical burns in skin and other tissues. Affected skin areas should be flooded with copious amounts of water.

Although there is little toxicity information published on hydrides, a threshold limit value (TLV) for lithium hydride in air of 25 μg/m³ has been established. More extensive data are available for sodium borohydride in the

Table 3. Properties of Covalent Hydrides

Hydride	Formula	Mp, °C	Bp, °C	Density[a] g/cm³	g/L
beryllium hydride[b]	BeH₂	125 dec	220[c]		
magnesium hydride	MgH₂	280 dec		1.45	
aluminum hydride	AlH₃				
silane[d]	SiH₄	−185[e]	−119.9	0.68[f] (−185)	1.44[g] (20)
germane	GeH₄	−165	−90	1.523[f] (−142)	3.43[g] (0)
stannane	SnH₄	−150	52		
arsine	AsH₃	−116.9[e]	−62	1.604[f] (64)	2.695[g]

[a]Temperature in °C is given in parentheses.
[b]$\Delta H_{298} = 19.3$ kJ/mol (4.6 kcal/mol).
[c]Begins to dissociate.
[d]$\Delta H_{298} = 30.55$ kJ/mol (7.30 kcal/mol).
[e]Freezing point.
[f]Liquid.
[g]Gas at atmospheric pressure.

Table 4. Complex Hydrides

Formula	Density, g/cm^3	Mp, °C
LiBH$_4$	0.66	278
NaBH$_4$	1.074	505
KBH$_4$	1.177	585
Be(BH$_4$)$_2$	0.702	123 dec
Mg(BH$_4$)$_2$		320 dec
Ca(BH$_4$)$_2$		260 dec
Zn(BH$_4$)$_2$		>50 dec
Al(BH$_4$)$_3$	0.549	−64.5^a
Zr(BH$_4$)$_4$	1.13	28.7
Th(BH$_4$)$_4$	2.59	204 dec
U(BH$_4$)$_4$	2.67	100 dec
(CH$_3$)$_4$NBH$_4$	0.84	>310
(C$_2$H$_5$)$_4$NBH$_4$	0.926	225 dec
(C$_4$H$_9$)$_4$NBH$_4$		>300
(C$_8$H$_{17}$)$_3$CH$_3$NBH$_4$	0.9	ca 30
C$_{16}$H$_{33}$(CH$_3$)$_3$NBH$_4$	0.9	ca 160
NaBH$_3$CN	1.20	240 dec
NaBH(OCH$_3$)$_3$	1.24	230 dec
LiAlH$_4$	0.917	190 dec
NaAlH$_4$	1.28	178
Mg(AlH$_4$)$_2$		140 dec
Ca(AlH$_4$)$_2$ >230 dec		
LiAlH(OCH$_3$)$_3$		
LiAlH(OC$_2$H$_5$)$_3$		
LiAlH(OC$_4$H$_9$)$_3$	1.03	>400
NaAlH$_2$(OC$_2$H$_4$OCH$_3$)$_2$	1.122	205 dec
NaAlH$_2$(C$_2$H$_5$)$_2$		85

^{a} Bp, 44.5°C.

powder and solution forms. The acute oral LD$_{50}$ of NaBH$_4$ is 50–100 mg/kg for NaBH$_4$ and 50–1,000mg/kg for the solution. The acute dermal LD$_{50}$ (on dry skin) is 4–8 g/kg for NaBH$_4$ and 100–500 mg/kg for the solution. The reaction or decomposition by-product sodium metaborate is slightly toxic orally (LD$_{50}$ is 2,000–4,000 mg/kg) and nontoxic dermally.

R. M. Adams and A. R. Siedle, *Boron, Metallo-Boron Compounds and Boranes*, Wiley-Interscience, New York, 1964.

R. Bau, "Transition Metal Hydrides," *Advances in Chemistry Series*, Vol. 167, American Chemical Society, Washington, D.C., 1978.

W. N. Lipscomb, *Boron Hydrides*, W. A. Benjamin, Inc., New York, 1963.

A. Pelter, K. Smith, and H. C. Brown, *Borane Reagents*, Academic Press, Inc., London, 1988.

TIM EGGEMAN
Neoterics International

HYDROBORATION

Hydroboration is the addition of a boron–hydrogen bond across a double or triple carbon–carbon bond to give an organoborane:

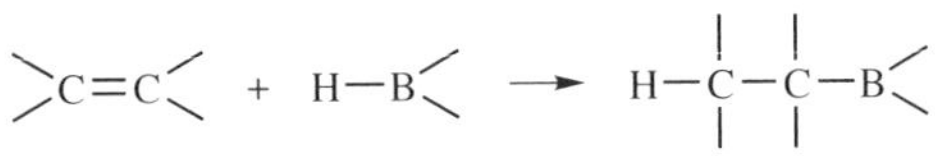

The boron atom in organoboranes can be replaced with other elements, usually with high stereoselectivity; many functional groups are tolerated. Consequently, organoboranes are among the most versatile synthetic intermediates, and their role in organic synthesis is constantly increasing.

One of the newer and more fruitful developments in this area is asymmetric hydroboration giving chiral organoboranes, which can be transformed into chiral carbon compounds of high optical purity. Other new directions focus on catalytic hydroboration, asymmetric allylboration, cross-coupling reactions, enolboration–aldolization, and applications in biomedical research.

THE HYDROBORATION REACTION

Diborane, the first hydroborating agent studied, reacts sluggishly with olefins in the gas phase. In the presence of weak Lewis bases; eg, ethers and sulfides, it undergoes rapid reaction at room temperature or even <0°C. The catalytic effect of these compounds on the hydroboration reaction is attributed to the formation of monomeric borane complexes from the borane dimer.

Mono-, di-, and trialkylboranes may be obtained from olefins and the trifunctional borane molecule.

Mechanism

The characteristic features of hydroboration were originally accounted for in terms of a simple four-center transition-state model serving as a useful working hypothesis. Models based on orbital symmetry considerations were also advanced.

Hydroborating Agents

Mono- and dialkylboranes obtained by controlled hydroboration of hindered olefins and by other methods can serve as valuable hydroborating agents for more reactive olefins. Heterosubstituted boranes are also available and used for this purpose. These borane derivatives show differences in reactivity and selectivity.

Borane Complexes. Borane solutions in tetrahydrofuron (THF) are commercially available or can be prepared by absorbing gaseous diborane in THF. Diborane can be conveniently generated from the reaction of sodium borohydride with boron trifluoride etherate, and recently, improved procedures were reported. Although BH$_3$/THF is a useful reagent, it must be stabilized with small amounts of sodium borohydride for longer storage at 0°C. Borane–dimethyl sulfide complex is free of these inconveniences and is widely used in laboratory practice. Its disadvantage is the unpleasant smell of dimethyl sulfide, which is volatile and water insoluble. Borane–triethylamine complex is used when slow liberation of borane at elevated temperatures is advantageous.

However, some of the organoboranes are labile and should be protected from such temperatures. Fortunately, recent studies revealed a range of tertiary amines and silylamines of intermediate steric requirements strong enough to form a 1:1 adduct with borane but weak enough to supply borane to an olefin at room temperature. The possibility of recovering the carrier amine from the hydroboration or reduction products makes these new borane–amine adducts highly promising and environmentally benign reagents.

Monosubstituted Boranes. Only a few monoalkylboranes are directly available by hydroboration. 2,3-Dimethyl-2-butylborane (thexylborane, ThxBH$_2$), easily prepared from 2,3-dimethyl-2-butene, is the best studied.

Monoisopinocampheylborane, IpcBH$_2$, is an important asymmetric hydroborating agent. It is prepared from α-pinene either directly or, better, by indirect methods.

A number of less hindered monoalkylboranes are available by indirect methods.

Monohalogenoboranes are conveniently prepared from BMS and boron trihalides (BX$_3$ where X = Cl, Br, I) by the redistribution reaction. The products are liquids, soluble in various solvents and stable over prolonged periods.

Disubstituted Boranes. Even slight differences in steric or electronic effects of substituents may have an effect on the hydroboration reaction course. These effects are well demonstrated in disubstituted boranes; consequently, a range of synthetically useful reagents has been developed. Primary dialkylboranes react readily with most alkenes at ambient temperatures and dihydroborate terminal acetylenes.

In contrast to simple unhindered dialkylboranes, borinanes and borepanes do not redistribute readily. These boraheterocyclic reagents can be prepared by hydroboration of the corresponding dienes with borane, 9-borabicyclo[3.3.1]nonane, or monochloroborane, followed by thermal isomerization or reduction, respectively. Dicyclohexylborane, Chx$_2$BH, is prepared in quantitative yield by the same method.

9-Borabicyclo[3.3.1]nonane, 9-BBN, is the most versatile hydroborating agent among dialkylboranes. It is commercially available or can be conveniently prepared by the hydroboration of 1,5-cyclooctadiene with borane, followed by thermal isomerization of the mixture of isomeric bicyclic boranes initially formed. The most hindered of all presently known hydroborating agents is possibly dimesitylborane, Mes$_2$BH, an air-stable white solid, slightly soluble in THF, the best etheral solvent, which is commercially available.

Dihalogenoboranes are conveniently prepared by the redistribution of BDM with boron trihalide–dimethyl sulfide complexes. Dichloroborane complexes with diethyl ether and dimethyl sulfide are so strong that direct hydroboration does not proceed. The addition of a decomplexing agent; eg, boron trichloride, is necessary for hydroboration.

Alkylchloro- and alkylbromoboranes are valuable reagents for the synthesis of di- and trialkylboranes having different alkyl groups. Thexylchloroborane (ThxBHCl), is a very useful reagent. It can be prepared by the reaction of monochloroborane with 2,3-dimethyl-2-butene or from thexylborane and hydrogen chloride.

Among chiral dialkylboranes, diisopinocampheylborane is the most important and best-studied asymmetric hydroborating agent. The most convenient synthesis, providing product of essentially 100% ee, involves the hydroboration of α-pinene with BMS in THF.

Selectivity

Chemoselectivity. Double and triple carbon–carbon bonds are more reactive toward borane and substituted boranes than most other functionalities. Consequently, many functional groups are tolerated in the hydroboration reaction. Using a suitably chosen hydroborating agent; eg, dicyclohexylborane, even aldehydes and ketones may be unprotected. Hydroboration of allylic and vinylic derivatives leads to α-, β-, or γ-substituted organoboranes prone to eliminations or rearrangements. In some cases, such transformations may be synthetically useful.

Conjugated ketones are either reduced to allylic alcohols or undergo 1,4-addition to give enolboranes.

Regioselectivity. Hydroboration of olefins and acetylenes involves predominant placement of the boron atom at the less hindered site of the multiple bond. The direction of addition is governed by polarization of the boron–hydrogen bond and by combination of steric and electronic effects of substituents at the multiple bond.

Functional groups influence the regioselectivity of hydroboration by inductive, mesomeric, and steric effects, their magnitude depending on the proximity of the double bond and the functional group.

Stereoselectivity. The addition of a boron–hydrogen bond across the double bond proceeds cleanly in a cis fashion, leading to simple diastereoselection for suitably substituted double bonds. Double bonds are approached by the hydroborating agent from the less sterically hindered face. The thermodynamically less stable addition products may result, as has been demonstrated for β-pinene and camphene. Borane discriminates well between faces differing significantly in steric hindrance. When the difference is small, low selectivity results. Bulky, sterically demanding hydroborating agents show higher stereoselectivity.

Catalytic Hydroboration

Hydroboration of alkenes with relatively stable dialkoxyboranes; eg, catecholborane (CBH) at room temperature is sluggish. The reaction rate is dramatically increased in the presence of a catalytic amount of Wilkinson's catalyst.

This result prompted intensive studies on the catalytic hydroboration of alkenes and alkynes. Catecholborane and pinacolborane are the reagents most often used.

REACTIONS OF ORGANOBORANES

Organoboranes available by hydroboration and also by other methods are versatile synthetic intermediates. The organic groups attached to boron can be transferred, usually with high stereoselectivity, to hydrogen, oxygen, nitrogen, halogens, sulfur, selenium, metal atoms, and carbon. Consequently, carbon–heteroatom and carbon–carbon bonds can be constructed. The combination of hydroboration and functionalization corresponds overall to anti-Markovnikov addition of the elements of HX to a double bond:

$$RCH{=}CH_2 \xrightarrow{HB{<}} RCH_2CH_2B{<} \longrightarrow RCH_2CH_2X$$

Replacement of Boron by Hydrogen or a Heteroatom

These reactions include protonolysis, oxidation, halogenolysis, replacement of boron by nitrogen, replacement of boron by sulfur and selenium, and replacement of boron by a metal.

Carbon–Carbon Bond Formation

These reactions include coupling of organic groups atteched to boron, cross-coupling of organoboranes with organic halides and triflates, organoborate rearrangements, single–carbon insertion reactions (carbonylation, cyanidation, and related reactions) reactions with acyl carbanion equivalents, α-alkylation of carbonyl compounds and derivatives, α-bromination transfer, addition to carbonyl compounds, allylboration, allenylboration, and propargylboration, and reactions of boron-stabilized carbanions.

Thermal Isomerization of Organoboranes

Trialkylboranes undergo isomerization under the action of heat, generally at temperatures >100°C, the boron atom moving to the least hindered site of the alkyl group.

Concerted Reactions of Organoboranes

Allylic organoboranes react via cyclic transition states not only with aldehydes and ketones but also with alkynes, allenes, and electron-rich or strained alkenes. Bicyclic structures, which can be further transformed into boraadamantanes, are obtained from triallyl- or tricrotylborane and alkynes. The addition proceeds in three discrete steps and the intermediates can be isolated. Unsaturated organoboranes react the Diels-Alder reaction.

Polymerization

Hydroboration of α,ω-dienes with monoalkylboranes gives reactive organoboron polymers that can be transformed into polymeric alcohols or polyketones by carbonylation, cyanidation, or the DCME reaction followed by oxidation.

Synthesis of Isotopically Labeled Compounds

Organoborane reactions have been applied for the synthesis of isotopically labeled compounds important in chemical and biological research and in modern medical imaging techniques, such as positron emission tomography (PET) and magnetic resonance imaging (MRI). Organoboranes tolerate a wide range of physiologically active functionalities and hence are wellsuited intermediates in radiopharmaceutical pathways.

ASYMMETRIC SYNTHESIS VIA CHIRAL ORGANOBORANES

Asymmetric induction in the hydroboration reaction may result from the chirality present in the olefin (asymmetric substrate), the reagent (asymmetric hydroboration), or the catalyst (catalytic asymmetric hydroboration).

Synthesis of Chiral Alcohols, Ketones, Halides, Deuterated Hydrocarbons, and Amines

The chiral organoboranes produced by asymmetric hydroboration can be transformed into the heterosubstituted chiral products by applying methodologies developed for achiral organoboron compounds. Thus, the organoboranes obtained from alkenes are oxidized to the corresponding chiral alcohols. Among other examples are the alcohols derived from heterocyclic olefins, dienes, functionalized olefins, and deuterium- or tritium-labeled chiral alcohols.

Synthesis of Chiral Alkanes, Alkenes, and Alkynes

An efficient general synthesis of α-chiral (Z)- and (E)-alkenes in high enantiomeric purity is based on the hydroboration of alkynes and 1-bromoalkynes, respectively, with enantiomerically pure IpcR*BH readily available by the hydroboration of prochiral alkenes with monoisopinocampheylborane, followed by crystallization.

It is also possible to prepare α-chiral acetylenes and alkanes by this method. In a shorter synthesis of α-chiral alkynes, a prochiral disubstituted (Z)-alkene is hydroborated with diisopinocampheylborane and the trialkylborane produced is treated with alkynyllithium, followed by iodine.

Synthesis of α-Chiral and Homologated Aldehydes, Acids, and β-Chiral Alcohols

A general approach to these compounds is based on the reaction of dichloromethyllithium with boronic esters. Rearrangement of the complex followed by reduction with potassium triisopropoxyborohydride provides the homologated boronic ester, which can be oxidized to the corresponding alcohol or transformed into the homologated aldehyde by reaction with methoxy-(phenylthio) methyllithium. β-Chiral alcohols not available in high optical purity by asymmetric hydroboration of terminal alkenes are readily prepared by this method.

Synthesis of α- and β-Chiral Ketones, Esters, and Nitriles

Chiral boronic esters are convenient precursors of α-chiral ketones (R*COR′), which can be prepared via the dialkylborinic ester or dialkylthexyl route. The conversion of chiral boronic esters into optically pure B-alkyl-9-BBN derivatives, followed by reaction with α-bromoketones, α-bromoesters, or α-bromonitriles, leads to the homologated β-chiral ketones, esters, and nitriles, respectively.

Asymmetric Allylboration

Optically active allylic boronates and allylic dialkylboranes transfer the allylic group to aldehydes enantioselectively. The asymmetric allyl- and crotylboration of aldehydes has emerged as an effective alternative to the aldol methodology in reactions involving acyclic stereoselection.

Homologation of Boronic Esters

A convenient general method of enantioselective carbon–carbon bond formation, not involving hydroboration, is based on the homologation reaction of boronic esters derived from optically active 1,2-diols, eg, 2,3-pinanediol.

Enolboration

The aldol reaction is one of the most powerful methodologies for the formation of carbon–carbon bonds in a stereodefined manner. Boron enolates are important intermediates for this transformation, since transition states of boron-mediated aldol reactions appear tightly organized, transmitting well the spatial arrangement to the aldol product.

H. C. Brown and M. Zaidlewicz, "Recent Developments", *Organic Syntheses Via Boranes*, Vol. 2, Aldrich Chemical Co., Milwaukee, Wisc., 2001.

Yu. N. Bubnov and co-workers, in M. G. Davidson, A. K. Hughes, T. B. Marder, and K. Wade, eds., *Contemporary Boron Chemistry*, Royal Society of Chemistry, Cambridge, 2000, p. 446.

T. Hayashi, in E. N. Jacobsen, A. Pfaltz, and H. Yamamoto, eds., *Comprehensive Asymmetric Catalysis*, Vol. 1, Springer, Berlin-Heidelberg, 1999.

A. Pelter, K. Smith, and H. C. Brown, *Borane Reagents*, Academic Press, London, 1988.

MAREK ZAIDLEWICZ
Nicolaus Copernicus University

HYDROCARBONS

Hydrocarbons, compounds of carbon and hydrogen, are structurally classified as aromatic and aliphatic; the latter includes alkanes (paraffins), alkenes (olefins), alkynes (acetylenes), and cycloparaffins. Crude petroleum oils, which span a range of molecular weights of these compounds, excluding the very reactive olefins, have been classified according to their content as paraffinic, cycloparaffinic (naphthenic), or aromatic.

Hydrocarbons are important sources for energy and chemicals and are directly related to the gross national product. The United States has led the world in developing refining and petrochemical processes for hydrocarbons from crude oil and natural gas. Hydrocarbons from crude oil have become the energy sources of the industrial world, largely replacing wood and even displacing coal. However, in the United States, crude oil production peaked at 1.3×10^6 t/d (9.6×10^6 bbl/d) (conversion factors vary, depending on the oil source) in 1970, causing increased reliance on foreign oil sources. Since the crude oil embargo in 1973, a number of alternative energy sources have been investigated to reduce the U.S. international trade deficit. The fossil-fuel era may turn out to have been a brief interlude between the wood-burning era of the nineteenth century and the renewable energy sources era of the twenty-first century.

Hydrocarbon resources can be classified as organic materials, which are either mobile such as crude oil or natural gas, or immobile materials, including coal, lignite, oil shales, and tar sands. Most hydrocarbon resources occur as immobile organic materials that have a low hydrogen-to-carbon ratio. However, most hydrocarbon products in demand have an H:C higher than 1.0.

Products	Molar H:C ratio
natural gas	4.0
LPG	2.5
gasoline	2.1
fuel oil	
light	1.8
heavy	1.3
coal	0.8

Immobile hydrocarbon sources require refining processes involving hydrogenation. Additional hydrogen is also required to eliminate sources of sulfur and nitrogen oxides that would be emitted to the environment. Resources can be classified as mostly consumed, proven but still in the ground, and yet to be discovered. Since 1950 the dominance of reserves has been in the Eastern Hemisphere and in offshore fields. Proved world gas reserves are nearly 4×10^{12} trillion metric feet, with 31% in the Middle East.

HYDROCARBON USE

Energy Sources

Hydrocarbons from petroleum are still the principal energy source for the United States. About 60% of the world's energy is supplied by gas and oil and about 27% from coal.

A significant obstacle to increased gas use is the lack of sufficient transportation and distribution systems. Environmental concerns have encouraged reliance on natural gas as a cleaner burning fuel. Combustion of natural gas emits about half the CO_2 that coal generates at equivalent heat output.

Natural gas imports have grown more slowly than other sources because imports from overseas require governmental licenses and cryogenic liquefaction plants are very expensive. Natural gas imports are chiefly by pipeline from Canada.

Gas and oil are the principal energy sources, even though the United States has large reserves of coal. Although the use of coal and lignite is being encouraged as an energy source, economic and environmental considerations have kept petroleum consumption high. The use of compressed natural gas (CNG) is expected to grow in response to the Clean Air Act of 1990. Reliance on foreign imports has remained high.

Raw Materials

Petroleum and its lighter congener, natural gas, are the predominant sources of hydrocarbon raw materials, accounting for over 95% of all such materials.

Synthesis Gas Chemicals. Hydrocarbons are used to generate synthesis gas, a mixture of carbon monoxide and hydrogen, for conversion to other chemicals. The primary chemical made from synthesis gas is methanol, though acetic acid and acetic anhydride are also made by this route. Carbon monoxide is produced by partial oxidation of hydrocarbons or by the catalytic steam reforming of natural gas. About 96% of synthesis gas is made by steam reforming, followed by the water gas shift reaction to give the desired H_2/CO ratio.

Aliphatic Chemicals. The primary aliphatic hydrocarbons used in chemical manufacture are ethylene, propylene, butadiene, acetylene, and the n-paraffins.

Cyclic Hydrocarbons. The cyclic hydrocarbon intermediates are derived principally from petroleum and natural gas, though small amounts are derived from coal. Most cyclic intermediates are used in the manufacture of more advanced synthetic organic chemicals and finished products such as dyes, medicinal chemicals, elastomers, pesticides, and plastics and resins.

End Use Chemicals

Lubricants. Petroleum lubricants continue to be the mainstay for automotive, industrial, and process lubricants. Synthetic oils are used extensively in industry and for jet engines; they, of course, are made from hydrocarbons.

Lubricating oils are also used in industrial and process applications such as hydraulic and turbine oils, machine oil and grease, marine and railroad diesel, and metalworking oils. Process oils are used in the manufacture of rubber, textiles, leather, and electrical goods.

Synthetic lubricants are tailored molecules with a higher viscosity index and a lower volatility for a given viscosity than lube oils from petroleum. Synthetic oils have the following advantages: energy conservation, extended drain periods, fuel economy, oil economy, high-temperature performance, easier cold starting, cleaner engines, cleaner intake valves, and reduced wear.

Synthetic oils have been classified by ASTM into synthetic hydrocarbons, organic esters, others, and blends. Synthetic oils may contain the following compounds: dialkylbenzenes, poly(α-olefins); polyisobutylene, cycloaliphatics, dibasic acid esters, polyol esters, phosphate esters, silicate esters, polyglycols, polyphenyl ethers, silicones, chlorofluorocarbon polymers, and perfluoroalkyl polyethers.

Agriculture/Food. Large quantities of hydrocarbons are used in agriculture, particularly as energy sources.

Although solar energy is a cheap alternative fuel, the convenience and reproducibility of drying with LPG has made hydrocarbon-derived energy the drying method of choice for such diverse applications as curing tobacco and drying peanuts, corn, and soybeans. In addition to these uses, hydrocarbons are used as the feedstock for a large variety of pesticides.

Hydrocarbons are also used extensively in packaging, particularly in plastic films and to coat boxes with plastic and (to a much lesser extent) wax. Polymeric resins derived from hydrocarbons are also used to make trays and cases for delivery of packaged foodstuffs.

Highly pure n-hexane is used to extract oils from oilseeds such as soybeans, peanuts, sunflower seed, cottonseed, and rapeseed. There has been some use of hydrocarbons and hydrocarbon-derived solvents such as methylene chloride to extract caffeine from coffee beans, though this use was rapidly supplanted by supercritical water and/or carbon dioxide, which are natural and therefore more acceptable to the public.

Feedstocks for Protein. Certain microorganisms, such as some bacteria, fungi, molds, and yeasts, can metabolize hydrocarbons and hydrocarbon-derived materials. Because single-cell proteins (SCPs) are about 50% protein by weight, it was believed early in the development phase that the economics of SCP production would be favorable. That belief has proven essentially correct, but acceptance of SCPs as a primary source of food protein has been slow. Except for limited uses as flavor enhancers and similar additives, SCPs have not made a significant impact on the markets for proteins. The future for hydrocarbons as a feedstock for SCPs is not bright, as the original n-paraffin feeds have been largely supplanted by alcohols derivable from nonpetroleum sources.

Surfactants. Surfactants are natural or synthetic chemicals that reduce the surface tension of water or other solvents and are used chiefly as soaps, detergents, dispersing agents, emulsifiers, foaming agents, and wetting agents. Surfactants may be produced from natural fats and oils, from silvichemicals such as lignin, rosin, and tall oil, and from chemical intermediates derived from coal and petroleum.

The greatest amount of surfactant consumption is in packaged soaps and detergents for household and industrial use. The remainder is used in processing textiles and leather, in ore flotation and oil-drilling operations, and in the manufacture of agricultural sprays, cosmetics, elastomers, food, lubricants, paint, pharmaceuticals, and a host of other products.

Coatings. Protective and decorative coatings for homes, vehicles, and a variety of industrial uses provide a large market for hydrocarbons. At one time, most paints, varnishes, and other coatings utilized organic chemical solvents. However, due to environmental concerns and solvent cost, approximately 40% of all coatings are now waterborne or even dispense with solvent altogether (powder coating).

Vinyl, alkyd, and styrene–butadiene latexes are used as film formers in most architectural coatings. Because alkyd resins require organic solvents, their use has decreased substantially for architectural coatings but is still holding up in industrial applications, where their greater durability justifies the added expense.

Polymers. Hydrocarbons from petroleum and natural gas serve as the raw material for virtually all polymeric materials commonly found in commerce, with the notable exception of rayon, which is derived from cellulose extracted from wood pulp. Even with rayon, however, the cellulose is treated with acetic acid, much of which is manufactured from ethylene.

Synthetic Fibers. Virtually all synthetic fibers are produced from hydrocarbons, as follows:

Fiber	Hydrocarbon precursor
nylon	cyclohexane
cellulose acetate	ethylene, methane
acrylics	propylene
polyesters	p-xylene, ethylene
polyolefins	propylene, ethylene
carbon fibers	pitch

Elastomers. Elastomers are polymers or copolymers of hydrocarbons. Natural rubber is essentially polyisoprene, whereas the most common synthetic rubber is a styrene–butadiene copolymer. Moreover, nearly all synthetic rubber is reinforced with carbon black, itself produced by partial oxidation of heavy hydrocarbons. The two most important elastomers, styrene–butadiene rubber and polybutadiene rubber, are used primarily in automobile tires.

Plastics and Resins. Plastics and resin materials are high molecular weight polymers that at some stage in their manufacture can be shaped or otherwise processed by application of heat and pressure. Some 40–50 basic types of plastics and resins are available commercially, but literally thousands of different mixtures (compounds) are made by the addition of plasticizers, fillers, extenders, stabilizers, coloring agents, etc.

The two primary types of plastics, thermosets and thermoplastics, are made almost exclusively from hydrocarbon feedstocks. Thermosetting materials are those that harden during processing (usually during heating, as the name implies) such that in their final state they are substantially infusible and insoluble. Thermoplastics may be softened repeatedly by heat, than hardened again by cooling.

METHANE, ETHANE, AND PROPANE

Physical Properties

Methane, ethane, and propane are the first three members of the alkane hydrocarbon series having the composition C_nH_{2n+2}. Selected properties of these alkanes are summarized in Table 1.

Production and Shipment

The large-scale use of natural gas requires a sophisticated, extensive pipeline system. In many underdeveloped areas, large quantities of natural gas are being flared because they must be produced with crude oil. However, the opportunity for utilizing the streams or for bringing the gas to industrial markets is being developed. Several large-scale ammonia plants have been built in developing countries. A third possibility is liquefaction of the methane and shipment in specially designed refrigerated tanker ships. Liquefied natural gas (LNG) plays a large role in both the transportation and storage of natural gas. At a pressure of 101.3 kPa (1 atm), methane can be liquefied by reducing the temperature to about $-161°C$. In the liquid form, methane occupies approximately 1/600th of the space occupied by gaseous methane at normal temperature and pressure. In spite of the very low temperature of the liquid, LNG offers advantages for both shipping and storing natural gas.

Economic Aspects

Ethane is primarily extracted from natural gas, then consumed as a feedstock in the production of industrial fuel. In 2000, 14.8 million metric tons were consumed for ethylene manufacture in the United States. The annual growth rate for ethylene was expected to be 5% through 2005. Natural gas is expected to be cheaper in the future; thus, the U.S. ethylene-based petrochemical industry should remain competitive.

Health and Safety Factors

Methane, ethane, and propane are all on the Environmental Protection Agency Toxic Substances Control Act Chemical Inventory and the Test Submission Data Base. Industrially, ethane is handled similarly to methane.

Uses

The largest use of methane is for synthesis gas, a mixture of hydrogen and carbon monoxide. Synthesis gas, in turn, is the primary feed for the production of ammonia and methanol. Synthesis gas is produced by steam reforming of methane over a nickel catalyst.

Methane is also used for the production of several halogenated products, principally the chloromethanes. Due to environmental pressures, this outlet for methane is decreasing rapidly.

Much interest has been shown in the direct conversion of methane to higher hydrocarbons, notably ethylene. Development of such a process would allow utilization of the natural gas from remote wells that is currently flared from such wells because the pipeline gathering systems needed tend to be prohibitively expensive. If the gas could be converted on-site to a condensable gas or pumpable liquid, bringing those hydrocarbons to market would be facilitated.

Table 1. Selected Properties of Methane, Ethane, and Propane

Property	Methane	Ethane	Propane
molecular formula	CH_4	C_2H_6	C_3H_8
molecular weight	16.04	30.07	44.09
mp, K	90.7	90.4	85.5
bp, K	111	185	231
explosivity limits, vol%	5.3–14.0	3.0–12.5	2.3–9.5
autoignition temperature, K	811	788	741
flash point, K	85	138	169
heat of combustion, kJ/mol[a]	882.0	1541.4	2202.0
heat of formation, kJ/mol[a]	84.9	106.7	127.2
heat of vaporization, kJ/mol[a]	8.22	14.68	18.83
vapor pressure at 273 K, MPa[b]		2.379	0.475
specific heat, J/(mol·K)[a]			
at 293 K	37.53	54.13	73.63
at 373 K	40.26	62.85	84.65
density, kg/m³[c]			
at 293 K	0.722	1.353	1.984
at 373 K	0.513	0.992	1.455
critical point			
pressure, MPa[b]	4.60	4.87	4.24
temperature, K	190.6	305.3	369.8
density, kg/m³[c]	160.4	204.5	220.5
triple point			
pressure, MPa[b]	0.012	1.1×10^{-6}	3.0×10^{-10}
temperature, K	90.7	90.3	85.5
liquid density, kg/m³[c]	450.7	652.5	731.9
vapor density, kg/m³[c]	0.257	4.51×10^{-5}	1.85×10^{-8}
dipole moment	0	0	0
hazards	fire, explosion, asphyxiation[d]	fire, explosion, asphyxiation[d]	fire, explosion, asphyxiation[d]

[a]To convert J to cal, divide by 4.184.
[b]To convert MPa to atm, divide by 0.101.
[c]To convert kg/m³ to lb/ft³, divide by 16.0.
[d]No significant toxic effects.

The most important commercial use of ethane and propane is in the production of ethylene by way of high-temperature (ca 1,000 K) thermal cracking. Ethane has been investigated as a feedstock for production of vinyl chloride, at scales up to a large pilot plant, but nearly all vinyl chloride is still produced from ethylene.

Propane's largest use outside of steam cracking is as a fuel, since it is the chief constituent of NGL (natural gas liquids). Historically, NGLS have been used for homes and businesses located away from natural gas systems. Recently, environmental concerns coupled with the clean-burning nature of NGL have stimulated research on and field trials of propane as a fuel source for internal combustion engines in cars, buses, and so on. Propane's main competition in the replacement fuel market is compressed natural gas (CNG). Compared to CNG, NGLs have better driveability, longer range, and more simple conversion from gasoline.

BUTANES

Butanes are naturally occurring alkane hydrocarbons that are produced primarily in association with natural gas processing and certain refinery operations such as catalytic cracking and catalytic reforming. Butanes include the two structural isomers n-butane, $CH_3CH_2CH_2CH_3$ and the isobutane, $(CH_3)_2CHCH_3$ (2-methylpropane).

Properties

The alkanes have low reactivities as compared to other hydrocarbons. Much alkane chemistry involves free-radical chain reactions that occur under vigorous conditions; eg, combustion and pyrolysis. Isobutane exhibits a different chemical behavior than n-butane, owing in part to the presence of a tertiary carbon atom and to the stability of the associated free radical.

Shipment

Butanes are shipped by pipeline, rail car, sea tanker, barge, tank truck, and metal bottle throughout the world. All U.S. container shipments must meet Department of Transportation regulations. Domestic water shipments are regulated by the U.S. Coast Guard.

Economic Aspects

The world butane business is operated by the oil and gas industries. Usually, butane data are included as part of the liquefied petroleum gas stream and separate data are not always available. In 2001, North America consumed 55 thousand metric tons of butanes.

Specifications, Standards, Quality Control, and Storage

Large quantities of butane are shipped under contract standards rather than national or worldwide specifications. Most of the petrochemical feedstock materials are sold at purity specifications of 95–99.5 mol%. Butane and butane–petroleum mixtures intended for fuel use are sold worldwide under specifications defined by the Gas Processors Association. Butanes may be readily detected by gas chromatography and commonly are stored in caverns or refrigerated tanks.

Health and Safety Factors

n-Butane and isobutane are colorless, flammable, nontoxic gases. They are simple asphyxiants, irritants, and anaesthetics at high concentrations. Isobutane causes drowsiness in a short time in concentrations of 1 vol%; however, there are no apparent injuries from either hydrocarbon after 2 h exposures at concentrations of up to 5%. Occupational exposure limits for butane for NIOSH is 800 ppm and the ACGIH TLV is 800 ppm also (40). The extreme flammability of these hydrocarbons necessitates handling and storage precautions. Storage in well-ventilated areas away from heat and ignition sources is recommended. Because they are heavier than air, they should not be used near sparking motors or other nonexplosion-proof equipment. Contact of the liquid form of the hydrocarbons with the skin can cause frostbite. Both butane and isobutane form solid hydrates with water at low temperatures. Hydrate formation in liquefied light petroleum product pipelines and certain processing equipment can lead to plugging and associated safety problems.

Uses

Butanes are used as gasoline-blending components, liquefied gas fuel, and in the manufacture of chemicals. n-Butane and small amounts of isobutane are blended directly into motor fuel to control its volatility. Larger amounts of butanes are used in the winter, particularly in cold climates, to make engine starting easier.

In addition to its use as a motor fuel alkylate, isobutane is a reactant in the production of propylene oxide by peroxidation of propylene.

Liquid petroleum gas (LPG) is a mixture of butane and propane, typically in a ratio of 60:40 butane–propane; however, the butane content can vary from 100 to 50% and less. LPG is consumed as fuel in engines and in home, commercial, and industrial applications. Increasing amounts of LPG and butanes are used as feedstocks for substitute natural gas (SNG) plants. n-Butane, propane, and isobutane are used alone or in mixture as hydrocarbon propellents in aerosols.

Production of maleic anhydride by oxidation of n-butane represents one of butane's largest markets. Butane and LPG are also used as feedstocks for ethylene production by thermal cracking. A relatively new use for butane of growing importance is isomerization to isobutane, followed by dehydrogenation to isobutylene for use in MTBE synthesis. Smaller chemical uses include production of acetic acid and by-products. Methyl ethyl ketone (MEK) is the principal by-product. n-Butane is also used as a solvent in liquid–liquid extraction of heavy oils in a deasphalting process.

PENTANES

There are three isomeric pentanes, commonly called n-pentane, isopentane (2-methylbutane), and neopentane (2,2-dimethylpropane).

Properties

Each isomer has its own individual set of physical and chemical properties; however, these properties are similar. The fundamental chemical reactions for pentanes are sulfonation to form sulfonic acids, chlorination to form chlorides, nitration to form nitropentanes, oxidation to form various compounds, and cracking to form free radicals. Many of these reactions are used to produce intermediates for the manufacture of industrial chemicals.

Occurrence and Recovery

Pentanes occur chiefly in straight-run gasoline, natural gasoline, and in certain refinery streams. Straight-run gasoline is the gasoline boiling range material recovered from crude oil by distillation. Natural gasoline is the C_5+ fraction of the liquids recovered from natural gas. Appreciable quantities of pentanes are produced in catalytic cracking, while smaller amounts come from hydrocracking and catalytic reforming.

Health and Safety Factors

Pentanes are only slightly toxic. Because of their high volatilities and, consequently, their low flash points, they are highly flammable. Pentanes are classified as nonreactive; ie, they do not react with fire-fighting agents.

The threshold limit value for the time-weighted average (8-h) exposure to pentanes is 600 ppm or 1770 mg/m^3 (51 mg/SCF); the short-term exposure limit (15 min) is 750 ppm or 2250 mg/m^3 (64 mg/SCF). Pentanes are classified as simple asphyxiants and a anaesthetics. The ICC classifies all three pentanes as flammable liquids and requires that they be affixed with a red label for shipping.

Uses

The main use for pentanes has been in motor fuel, though regulations limiting fuel vapor pressure are decreasing the amount of pentanes, particularly isopentane, present in gasoline during warm parts of the year. At one time, significant quantities of pentane were used as feedstock for ethylene units. However, most U.S. ethylene capacity is now based on ethane–propane feedstock.

Some outlet has to be found for the increasing amount of pentane displaced from gasoline by vapor pressure regulation, and it is likely that much of that pentane will find its way into alkylation streams.

HEXANES

Properties

The flash point of n-hexane is −21.7°C and its autoignition temperature is 225°C. The explosive limits of hexane vapor in air are 1.1–7.5%. Above 2°C the equilibrium mixture of hexane and air above the liquid is too rich to fall within these limits.

Health and Safety Factors

Hexane is classified as a flammable liquid by the ICC, and normal handling precautions for this type of material should be observed. The ACGIH TLV the maximum concentration of hexane vapor in air to which a worker may be exposed without danger of adverse health effects is 150 ppm; the NIOSH PEL is 50 ppm; the OSHA PEL is 500 ppm.

n-Hexane can be grouped with the general anaesthetics in the class of central nervous system depressants. Hexane vapors are mildly irritating to mucous membranes. Exposure to concentrations in excess of 1% hexane may cause dizziness, unconsciousness, prostration, and death. Prolonged skin contact with hexane results in irritation and dermatitis. Direct contact with lung tissue can result in chemical pneumonitis, pulmonary edema, and hemorrhage.

Uses

Other than fuel, the largest-volume application for hexane is in extraction of oil from seeds; eg, soybeans, cottonseed, safflower seed, peanuts, rapeseed, etc. Hexane has been found ideal for these applications because of its high solvency for oil, its low boiling point, and its low cost. Its narrow boiling range minimizes losses, and its low benzene content minimizes toxicity. These same properties also make hexane a desirable solvent and reaction medium in the manufacture of polyolefins, synthetic rubbers, and some pharmaceuticals. The solvent serves as a catalyst carrier and, in some systems, assists in molecular weight regulation by precipitation of the polymer as it reaches a certain molecular size.

CYCLOHEXANE

Cyclohexane, C_6H_{12}, is a clear, essentially water-insoluble, noncorrosive liquid that has a pungent odor. It is easily vaporized, readily flammable, and less toxic than benzene. Structurally, it is a cycloparaffin.

Properties

The predominant stereochemistry of cyclohexane has no influence in its use as a raw material for nylon manufacture or as a solvent. The most important commercial reaction of cyclohexane is its oxidation (in liquid phase) with air in the presence of a soluble cobalt catalyst or boric acid to produce cyclohexanol and cyclohexanone. Cyclohexanol is dehydrogenated with zinc or copper catalysts to cyclohexanone, which is used to manufacture caprolactam.

Occurrence

Cyclohexane is present in all crude oils in concentrations of 0.1–1.0%. The cycloparaffinic crude oils, such as those from Nigeria and Venezuela, have high cyclohexane concentrations, and the highly paraffinic crude oils, such as those from Indonesia, Saudi Arabia, and Pennsylvania, have low concentrations. Concentrations of cycloparaffins in crude oils from Texas, Oklahoma, and Louisiana tend to fall in between.

Economic Aspects

The principal costs in cyclohexane manufacture are maintenance expenses, interest and return charges on the plant and working capital, and the cost of benzene and high purity hydrogen.

The cyclohexane market is considered mature, but demand for nylon fiber is expected to grow, especially in China and Taiwan. The annual growth rate through 2006 was expected to be at 3%.

Specifications

For nylon manufacture, a typical purity specification of cyclohexane is 99.8%. A sulfur level of less than 1 ppm is specified for high purity cyclohexane, as measured by ASTM method D3120.

Health and Safety Factors

The threshold limit value (TLV) for cyclohexane is 300 ppm (1050 mg/m^3). With prolonged exposure at 300 ppm and greater, cyclohexane may cause irritation to eyes, mucous membranes, and skin. At high concentrations it is an anaesthetic, and narcosis may occur. Because of its relatively low chemical reactivity, toxicological research has not been concentrated on cyclohexane.

Uses

Almost all of the cyclohexane that is produced in concentrated form is used as a raw material in the first step of nylon-6 and nylon-6,6 manufacture. Cyclohexane also is an excellent solvent for cellulose ethers, resins, waxes, fats, oils, bitumen, and rubber. When used as a solvent, it usually is in an admixture with other hydrocarbons.

The cyclohexane in crude oil has three primary dispositions. Some of it is included in a light fraction (35–75°C) that is distilled from crude oil and is blended with other materials into motor gasoline. Alternatively, this fraction is used as a feed to ethylene manufacture. The third, and most important, disposition is as a feed to a catalytic reformer where the naturally occurring cyclohexane is converted to benzene by dehydrogenation.

M. W. Ball, D. Ball, and O. S. Turner, *The Fascinating Oil Business*, Bobbs Merrill Co., Inc., Indianapolis, Ind., 1965.

R. C. Reid, J. M. Prausnitz, and T. K. Sherwood, *The Properties of Gases and Liquids*, 3rd ed., McGraw-Hill, New York, 1977.

DAVID E. MEARS
Unocal

ALAN D. EASTMAN
Phillips Petroleum

HYDROFLUOROCARBONS

Substitution of fluorine for hydrogen in an organic compound to produce hydrofluorocarbons (HFCs) has a profound influence on the compound's chemical and physical properties. A stronger C–F bond is the main contribution to the pronounced chemical and thermal stability of HFCs. The substitution of fluorine for hydrogen also reduces the flammability and increases the heat capacity of the materials. In general, compounds that have more than one-half of the hydrogen replaced by fluorine are nonflammable. Because of these properties, fluorochemicals and chlorofluorochemicals (CFCs) have been used in applications where heat capacity, chemical and thermal stability, and/or nonflammability are important properties. These traditionally include refrigeration fluids, foam blowing agents, solvents, and firefighting agents.

In 1974 chlorofluorochemicals were found to deplete the stratospheric ozone layer. Because of this, a phase-out program for production and applications of these chemicals was developed under the United Nations Montreal Protocol in 1987. Since that time, the industry has been seeking alternative materials that provide the benefits afforded by CFCs.

The ideal CFC substitute must not harm the ozone layer or contribute to other detrimental atmospheric phenomena such as global warming. It also must be nontoxic, nonflammable, and thermally and chemically stable under normal use conditions and be reasonably priced. The fluorocarbon producers have found substitutes that match many, but not all, of these criteria.

HYDROFLUOROCARBONS AS BLOWING AGENTS

Physical Properties

Physical properties of the new hydrofluorocarbons are listed in Table 1. For insulation foams, the most important property is thermal conductivity. One of the advantages of fluorochemical blowing agents is that most of the material remains entrapped in the foam cells; thus, the low thermal conductivity of these blowing agents contributes to the insulating properties of the final product over long periods of time. Other important properties of the foam are total density; dimensional stability; and a high percentage of closed foam cells, which helps retain cell gases. The final thermal conductivity of the foam will be a result of a combination of the thermal conductivity of the solid polymer, the percent of closed foam cells and the thermal conductivity of the cell gas, including that of the foam-blowing agent.

Closed foam cell structures should have good aging properties and the foam should have a lower density. Because of the higher costs of the newly developed HFCs, the foam industry has become willing to consider blowing agents of a totally different nature, such as simple hydrocarbons like cyclopentane or isopentane.

Use as Foam-Blowing Agents

Chlorofluorocarbons (CFCs) such as CFC 11 ($CFCl_3$) and CFC 12 (CF_2Cl_2) were used extensively as foam-blowing agents prior to 1996. Hydrochlorofluorocarbons such as 141b (CH_3CFCl_2) and 123 (CF_3CHCl_2) were developed as blowing agents for polyurethane foams as an interim solution under the Montreal Protocol. While HCFC 123 was later found to be toxic, and was therefore never used commercially to any great extent, other fluorochemicals such as HCFC-22 (CHF_2Cl) and mixtures of HCFC-22 and HCFC-142b (CH_3CF_2Cl) have been used in urethane foams. Because of its high solubility, HCFC-142b has been used extensively in polystyrene foam (XPS) since the phase-out of CFC 12. However, the impending 2010 phase-out of HCFC 142b in the United States is causing XPS manufacturers to search for a new blowing agent.

HYDROFLUOROCARBONS AS REFRIGERANTS

Physical Properties

A summary of the physical properties of these new refrigerants, together with their environmental properties, can be found in Table 2.

Use as Refrigerants

Between differences in operating conditions of the interior and exterior heat exchangers, refrigeration applications cover a wide variety of conditions. These differences often dictate the use of different fluids. While a variety of HFC substitutes have been developed for these applications, in many cases it has not been possible to find any single material that provides all the properties of the CFCs or HCFCs that are being replaced. Thus, for many applications the industry is switching to blends of two to as many as four components.

In addition to efficient performance, the other important properties of refrigerants include flammability, operating pressures, and compatibility with a variety of materials used in manufacturing the refrigeration

Table 1. Physical Properties of HFC Blowing Agents

Properties	134a	245fa	365mfc
chemical formula	CF_3CH_2F	$CF_3CH_2CHF_2$	$CF_3CH_2CF_2CH_3$
molecular weight	102.03	134	148
boiling point, °C	−25.9	15.3	40
freezing point, °C	−96.6	<−160	−35
critical temperature, °C	101.06		
critical pressure PSI	588.7		
vapor density, g/cm³	.00526		
liquid density, g/cm³	1.207	1.32	1.27
solubility of water, ppm	1100	1600	840
flammability limit in air, vol.%	none	none	3.5–13
flash point, °C	none	none	−27
ozone depletion potential ODP	0	0	0
global warming potential GWP	1300	950	890
atmospheric lifetime, years	13.8	7.2	9.9

Table 2. Physical Properties of the New HFC Refrigerants

Properties	HFC-134a	HFC-32	HFC-152a	HFC-125
chemical formula	CF_3CH_2F	CH_2F_2	CH_3CHF_2	CF_3CHF_2
molecular weight	102	52.02	66	120
boiling point, °C	−26.5	−51.6	−24.7	−48.5
critical T, °C	100.6	78.4	113.5	66.3
critical P, atm	40.03	57.8	44.4	34.7
liquid density at 20°C, g/cm^3	1.203	0.977	0.911	1.23
flammability range by vol / air	none	14–31	3.9–16.9	none
ODP^a	zero	zero	zero	zero
GWP^b	1300	650	140	2800

^{a}By definition, all HFCs have zero ODP.
^{b}EPA Scientific Assessment of Ozone Depletion 1998.

equipment, such as hoses, seals, etc. In addition, the refrigerant must be chemically and thermally stable. In many of today's applications, refrigerants have been in constant use for decades. Another important issue in refrigeration is the fact that the compressor oil must be capable of being carried out of the compressor with the refrigerant fluid and returned to the compressor.

While a large variety of blends have been developed, a 50:50 (by weight) mixture of HFC-32 and 125, which has been designated R-410A by ASHRAE, is becoming the most widely accepted of these materials for new equipment installations.

The use of HFCs in new refrigeration equipment is not required in the United States until 2010. Until then, HCFCs such as HCFC-22 (CHClF$_2$) will continue to dominate the market. The HCFCs can be used for maintenance of installed equipment until January 1, 2030. Although HFC-152a, 1,1-difluoroethane has a lower Global Warming Potential GWP than HFC-134a and was found to have a more efficient cooling cycle, the fact that it is a flammable gas has limited its use as a refrigerant.

HYDROFLUOROCARBONS AS SOLVENTS

Physical Properties

The physical properties of the industrial products offered in this industry are usually complex, proprietary mixtures.

Use as Solvents

Perhaps the most important application of fluorinated solvents is to remove oil, grease, and related contaminants from plastics and metal; hence its use in the electrical and electronic industries where metal–plastic combinations pose numerous cleaning problems. Another important application of HFC solvents is in the area of precision engineering; eg, in aerospace and military uses, where the solvents need to have high thermal and chemical stability, be noncorrosive, and remove readily from equipment without leaving a residue. While advances in the design of fluxes and polishing compounds often allow the use of aqueous cleaning agents, even in these cases, final water removal and drying can be a problem. For articles that may be damaged at temperatures

>100°C, drying can consume much time and still fail to remove water from blind holes or crevices.

While the hydrophobicity introduced by the fluorine facilitates the evaporation of the solvents at low temperatures, it also hinders its ability to absorb and azeotrope the water. This problem can be overcome by using mixtures of hydrophilic solvents.

HYDROFLUOROCARBONS AS FIRE-FIGHTING AGENTS

Use as Fire-Fighting Agents

The three halons, 1211 (CF$_2$ClBr), 1301 (CF$_3$Br), and 2402 (CF$_2$BrCF$_2$Br), have been used as fire-fighting agents to replace the highly toxic substances methyl bromide and carbon tetrachloride under the Montreal protocol. The current selected replacements for these substances are HFC-227ea, 1,1,1,2,3,3,3-heptafluoropropane, and HFC-ketone nonafluoro-4-trifluoromethylpent-3-one. These two products extinguish fires by physically cooling the flame and removing heat from the flame to the extent that the combustion reaction cannot sustain itself. Both are suited for extinguishing fires in areas containing high-value equipment, such as telecommunications facilities, computer rooms, electronic and data processing equipment, record storage facilities, art galleries, museums, and electrical control rooms.

K. Lorenz, *Proceeding of PU Latin America 2001*, Rigid Foam Session, Paper 3, p. 1.

W. A. Sheppard and C. M. Sharts, *Organo Fluorine Chemistry*, W. A. Benjamin, New York, 1969, p. 62.

MAHER Y. ELSHEIKH
CHRISTOPHER A. BERTELO
ARKEMA King of Prussia

WILLIAM R. DOLBIER JR.
University of Florida

HYDROGELS

Hydrogels are hydrophilic polymer networks that are able to swell and retain large amounts of water and maintain three-dimensional (3D) swollen structures. These hydrogels do not dissolve in water. Upon swelling, hydrogels increase in volume but keep their shape. In general, the amount of water absorbed by the hydrogel is at least 20% of its total weight; for superabsorbent hydrogel it is usually >95% of the total weight. Due to the high water content, the mechanical strength of swollen hydrogels is usually poor. The ability of a hydrogel to imbibe water is determined by the hydrophilic groups in the hydrogels network chains and the degree of cross-linking. In general, hydrogels can be cross-linked chemically or physically. Chemically cross-linked hydrogels are cross-linked by covalent bonds and do not dissolve in water in any condition, while physically cross-linked hydrogels can be reversible in shape because they are cross-linked through noncovalent bonds such as van der Waals interactions, ionic interactions, hydrogen bonding, or hydrophobic interactions. These physical hydrogels can show sol–gel reversibility. Because of the 3D network structures, the molecular weight of hydrogels is considered to be infinite. Responsive hydrogels can reversibly change volume in response to slight changes in the properties of the medium, including pH, temperature, electric field, ionic strength, salt type, solvent, external stress, or light. In the past four decades, hydrogels have been a topic of extensive research because of their unique bulk and surface properties. Various hydrogels have been synthesized by using a variety of monomers and methods.

THE COMMON MONOMERS USED FOR SYNTHESIZING HYDROGELS

Table 1 shows common monomers for preparing hydrogels. Each monomer contains a carbon double bond through which polymerization propagates to produce polymer chains. A cross-linker that contains two double bonds is added to the monomers to obtain a 3D cross-linked structure. The polymerization usually is carried out using heat (with thermal-initiators), light (with photoinitiators), γ radiation, or an electron beam.

Radiation polymerization does not require initiators and can be used with almost any monomer. Initiators provide free radicals that initiate a chain reaction among the monomer and cross-linker molecules. Thermal initiators include peroxides and azo compounds that undergo cleavage at a rate that is markedly temperature dependent, and redox systems comprised of reducing agents such as ferrous salts, sodium metabisulfite, or tetramethylenediamine (TEMED), plus the oxidizing agents ammonium persulfate or hydrogen peroxide. Radiation sources include Co-60, Ce-137, or electron beams.

GENERAL METHODS OF PREPARATION

Polymerization

The techniques used for the production of hydrogels include bulk, solution, suspension, emulsion, gaseous, and plasma polymerizations (Table 2). Emulsion and suspension polymerizations provide good control over the shape and particle size distribution of the hydrogel.

Hydrogels Prepared Using Radiation

Ionizing radiation has long been recognized as a useful tool for the synthesis of hydrogels. A major advantage is its ability to carry out a hydrogel synthesis and sterilization all in one step without the use of an initiator or cross-linker. A hydrogel can be obtained by irradiation of a solid polymer, monomer (bulk or solution), or aqueous solution of polymer. Upon irradiation of the polymer solution, reactive macromolecular intermediates are formed, due to the direct action of radiation on the polymer chains or from the reaction of the intermediates generated in water with the polymer molecules. Irradiation of hydrophilic polymers in a dry form requires special sample preparation such as pressing or melting. It is difficult to obtain homogeneous hydrogels. The monomer irradiation method in which polymerization is followed by cross-linking is the most frequently used. Because many monomers are harmful or even toxic, special care should be taken when

Table 1. Common Monomers Used for Hydrogel Production

Monomer	Abbreviation	Chemical structure
hydroxyethyl methacrylate	HEMA	$CH_2{=}C(CH_3)COOCH_2CH_2OH$
hydroxyethoxyethyl methacrylate	HEEMA	$CH_2{=}C(CH_3)COOCH_2CH_2OCH_2CH_2OH$
ethylene glycol dimethacrylate	EGDMA	$CH_2{=}C(CH_3)COOCH_2CH_2OCOC(CH_3){=}CH_2$
acrylic acid	AA	$CH_2{=}CHCOOH$
methacrylic acid	MAA	$CH_3CH{=}CHCOOH$
N-vinyl-2-pyrrolidone	NVP	$CH_2{=}CHNCOCH_2CH_2CH_2$
vinyl acetate	VAC	$CH_2{=}CHCOOCH_3$
N-substituted acrylamide	n-AM	$CHR_3{=}CHCONR_1R_2$
hydroxydiethoxyethyl methacrylate	HDEEMA	$CH_2{=}(CH_3)COOCH_2CH_2OCH_2CH_2OCH_2CH_2OH$
2-acrylamido-2-methylpropanesulfonic acid	AMPS	$CH_2{=}CHCONH(CH_3)CH_2SO_3H$
acrylonitrile	AN	$CH_2{=}CH{-}CN$

Table 2. The Common Polymerization Methods for Preparing Hydrogels

Polymerization method	Important features	Problems related to polymer preparation and purity
bulk (mass)	initiator and monomer needed, cross-linking agent can be added	high viscosity, difficult agitation lead to non-uniformity of products; residual monomers
solution	initiator, solvent, and monomer needed; easy agitation: controlled heat transfer; polymer soluble or insoluble in solvent	chain transfer frequently gives broad molecular weight distribution products; difficulty in removing solvent
suspension	initiator, solvent, monomer, and suspending agent needed; cross-linking agent can be added; polymer production in spherical or irregular particles depending on monomer–suspending agent interfacial tension	
emulsion	initiator, solvent, monomer, suspending agent, and emulsifier needed	residual emulsifier, etc.
gaseous	reaction in gaseous phase; high pressure; unknown kinetics	pure polymers; technique not applied to many systems
plasma	glow discharge; unknown kinetics	new technique; ultrapure polymers, high cost of manufacture

using this method for the formation of hydrogels for biomedical uses. During irradiation of the monomer, multiple consecutive and parallel reactions occur. Typical hydrogels prepared by irradiation include poly(vinyl alcohol) (PVA), poly(N-vinyl-2-pyrrolidone) (PVP), poly(ethylene oxide) (PEO), polyacrylamide (PAM), poly(acrylic acid) (PAA), and poly(methyl vinyl ether) (PMVE) and poly(acrylamide/maleic acid) [P(AM–MA)]. Cross-linking agents and monomers can be added to increase the extent of cross-linking, modify the gel structure, or initiate grafting reactions.

The presence of oxygen plays an important role in the irradiation process. The initially generated carbon-centered macroradicals react with oxygen to form corresponding peroxyl radicals that can cause chain scissions. In the initial period, the degradation is dominant. Cross-linking or gel formation occurs only after the oxygen is used up. Many polymers that can be expected to form cross-linked structures degrade if irradiated in air under slow dose rates, due to the formation of weak peroxidic bonds in the main polymeric chain. These bonds decompose and cause oxidative degradation of the main chain.

PROPERTIES

Hydrogels Based on Natural Polymers

The natural polymers that can be cross-linked include the macromolecules extracted from animal collagen, plants, and seaweed. Polysaccharides are classified on the basis of their main monosaccharide components and the sequences and linkages between them. Polysaccharides copolymerized with acrylic acid or acrylamide form superabsorbent hydrogels, with water absorption as high as 420 times its weight. Alginate can spontaneously form a hydrogel by cross-linking through the gluronic acid

residues. The ionically cross-linked alginate hydrogels have been used in controlled release of several proteins.

Gelatin is a natural physical hydrogel that is prepared by partial hydrolysis of water-insoluble collagen fibers from connective tissues. Ionic polysaccharides are also capable of being gelled in situ.

Poly(2-hydroxy ethyl methacrylate) Hydrogels (PHEMA)

PHEMA hydrogels advantages include hydrolysis stability, nonantigenicity, and nonirritability. Its drawbacks include inertness to cell adhesion, low swelling in water, poor mechanical strength, and nondegradablity.

When a HEMA is copolymerized with hydrophobic compositions, hydrogels with superior mechanical and tensile strength can be obtained. For example, hydroxypropyl methacrylates (HPMA) can be incorporated into HEMA to increase its mechanical strength.

Poly(acrylic/methacrylic acid) Hydrogels

Acrylic/methacrylic acid hydrogels exhibit extremely high swelling capacities; eg, poly(acrylic acid) hydrogels can have water content as high as 99%, and hence are used as a superabsorbing material. The commercially important superabsorbent polymers are cross-linked polymers of partially neutralized acrylic acid or terpolymers of acrylic acid, sodium acrylate, and a cross-linker, or graft terpolymers with starch or poly(vinyl alcohol) (PVA). The swelling and elasticity of these polymers depend on the precise structure of the polymer network and the cross-link density.

Hydrogels with a high swelling ability can be prepared with a monomer conversion as high as 100% when acrylamide–acrylic acid mixtures with cross-linkers are irradiated in a ^{60}Co-γ source. This hydrogel has been tested for the removal of some textile dyes from aqueous solutions.

PVA Hydrogels

PVA is usually cross-linked for several applications, especially for medical and pharmaceutical uses. PVA can be cross-linked by a difunctional agent that condenses with the organic hydroxyl groups, including formaldehyde, glutaradehyde, acetadehyde, or maleic acid, in which acetal bridges form between the pendant hydroxyl groups of the PVA chains.

The PVA hydrogels have been used for a number of biomedical and pharmaceutical applications, due to their advantages such as being nontoxic, noncarcinogenic, and having bioadhesive characteristics with ease of processing. In addition to blood contact, artificial kidney, and drug delivery applications, PVA show as potential applications for soft tissue replacement, articular cartilage, artificial organs, and membranes.

PEG Hydrogels

The chemistry and biological application of PEG have been the subject of intense study in both academia and industry. PEG is a nontoxic, water soluble polymer that resists recognition by the immune system. Traditionally, PEG has been used in biological research as a precipitating agent for proteins. It has been approved for a wide range of biomedical applications, as it is biocompatible, nontoxic, and nonimmunogenic.

Hydrogels based on PEG derivatives have been widely used in covalent attachment to proteins to reduce immunogenicity, proteolysis, and kidney clearance, attached to low molecular weight drugs for enhanced solubility, reduces toxicity, and to alter biodistribution.

Enzymes have long been used in the biomedical field as diagnostic tools or disease markers. Unfortunately, they are quite unstable biopolymers and denature quickly. To overcome its shortcomings, PEG hydrogels have been used to improve the life span of enzymes in the organism by immobilizing the enzyme in the matrix of PEG-modified hydrogels. PEG-BSA hydrogels have interesting characteristics for enzyme immobilization useful in biomedical applications.

Poly(acrylamide)-based Hydrogels

A large number of polymers based on N-alkyl acrylamide and its copolymers with acidic and basic comonomers have been synthesized that have a lower critical solution temperature (LCST). Poly(N-isopropylacrylamide) is the most widely studied thermosensitive acrylamide hydrogel. Water behaves as a good solvent through hydrogen bonding with the amide groups at room temperature. This hydrogen bonding with water is increasingly disrupted on heating, rendering water to act as a poor solvent, leading to gradual chain collapse. Inter- and intrapolymer hydrogen bonding and polymer–polymer hydrophobic interactions become dominant above the LCST. The introduction of ionic comonomers makes it possible to obtain hydrogels sensitive to changes in both temperature and pH.

Poly(N-vinyl 2-pyrrolidone) Hydrogels

Hydrogels based on NVP can be synthesized by free-radical polymerization. NVP is probably the most effective comonomer used to increase the water uptake ability of an HEMA hydrogel, as in contact lenses. The graft copolymer of NVP onto silicone rubber exhibits hydrophilicity, high oxygen permeability, and improved wettability. NVP polymers are also known to have excellent biocompatibility that is designed for controlled nonburst degradation in the vitreous body. Cross-linked PNVP has the potential to be used as a vitreous substitute.

Polyurethane Hydrogels

Polyurethane hydrogels are widely used in soft contact lenses, controlled release devices, semipermeable membranes and hydrophilic coatings. Because of their excellent mechanical and physical properties, polyurethanes are widely used in medical applications, such as coatings for medical devices to prevent protein adsorption.

Biodegradable Hydrogels

Biodegradable hydrogels are of interest in pharmaceutical, veterinary, agricultural, and environmental applications. The use of biodegradable hydrogels is desirable because the dosage is degraded and eliminated from the body after drug delivery. These are specially advantageous as delivery systems for large molecular weight drugs, such as peptides and proteins, which are not easily delivered using general polymers.

The biodegradable hydrogels are usually classified into three categories: (1) hydrogels with a degradable polymer backbone; (2) hydrogels with degradable cross-linking agents; and (3) hydrogels with degradable pendant groups. These can also be classified into either natural degradable and synthetic degradable hydrogels or cross-linked and noncross-linked hydrogels.

Smart Hydrogels

Polymers that respond to external stimuli are termed smart, intelligent, stimuli-responsive, or environmentally sensitive. Among water-soluble polymers and hydrogels the term "smart" may be applied to systems that respond reversibly to slight changes in the properties of the medium. The properties can be pH, temperature, ionic strength, illumination, or electric field. The response is readily observed optically because of new-phase formation in a hitherto homogeneous solution, by sudden swelling or contraction of the hydrogel.

Microgels

A microgel is a cross-linked latex microparticle that is swollen by a good solvent and has a globular structure. Since the beginning of the 1970s, the research on microgels has increased steadily, due to its growing industrial and commercial importance. Due to their compact structure, the intrinsic viscosities of the microgels are much lower than those of corresponding linear or branched polymers.

The deformable nature of microgel particles has important implications for its rheological properties. Compared to hard-sphere particles, swollen microgel particles have a much higher dispersion viscosity, due to the large effective hydrodynamic diameter of the swollen particles.

Microgel particle dispersions are shear thinning and provide rheological control for automotive surface coatings. The particles also have good film-forming properties and favor the alignment of added metallic flakes parallel to the substrate surface. Microgel particles also show promise in the printing and pharmaceutical industries. Their high surface area and good surface-coating characteristics have allowed functionalized microgel particles to be coated on offset plates with impressive results. Alternatively, microgel particles have potential applications as drug delivery systems designed to swell in the vicinity of the target sites inside the body and release.

APPLICATIONS

Molecular Separation

Temperature sensitive hydrogels have been used to separate large from small molecules. In the separation, a collapsed hydrogel is added to the aqueous solution that is to be concentrated. Its hydrogel swells by absorbing water and small molecular solutes, leaving the larger molecules behind. After equilibrium swelling is reached, the solid hydrogel is physically removed from the solution. The hydrogel is then reactivated for another deswelling use, by raising the temperature a little over the LCST. This technology has been studied for the separation of soy proteins, enzymes, cellulases, nonionic surfactants, lignin, and bacteria dispersions. Since most thermosensitive hydrogels have an LCST $<50^{\circ}$C, they require low energy to operate.

Although smart hydrogels have several advantages in the separation of biomolecules over conventional methods, a number of properties like swelling rate, mechanical strength, and surface adsorption of hydrogels need to be considered or improved before large scale use.

Protein Isolation by Conjugating

The ability of responsive polymers in aqueous solutions to form a separate phase after a slight change in conditions has been used to isolate and purify proteins. A responsive polymer can be conjugated with a ligand that has an affinity to the target protein. When the polymer is added to the mixture containing the protein and the conditions are changed, the polymeric conjugate, together with the target protein, segregates as a separate phase, while the impurities remain in the solution. The polymer-rich phase is filtered from the supernatant. Then the target protein is either eluted from the polymeric phase or is redissolved, altering the conditions in such a way that the complex dissociates. The polymeric conjugate is reprecipitated—but this time without the target protein, which remains in the solution. Instead of precipitation, reversible flocculation may also be used for the purification of biomolecules.

Hydrogel-based Drug Delivery

Over the past few decades, there has been increasing attention devoted to the development of controlled-release systems using polymeric carriers. The polymer used for drug delivery must be biocompatible and degradable, and the degradation products of the polymer must be nontoxic and not create any inflammatory response also, the degradation of the polymer should occur within a reasonable period of time.

In general, the hydrogels are suitable for the controlled release of most low molecular weight, water-soluble drugs. In addition, the biodegradable hydrogel systems are useful for the delivery of macromolecular drugs, such as peptides and proteins, which are entrapped in the hydrogel network until the hydrogel is degraded.

Hydrogels as Artificial Organs

The physical properties of hydrogels resemble living tissue more than any other kind of synthetic biomaterials. In particular, the high water content and the soft, rubbery consistency give them a strong, superficial resemblance to living soft tissue. PHEMA and its copolymers have been used as hemodialysis membranes that act as an artificial kidney. The artificial kidney works like a hemodialysis machine, which cleanses the blood of people with kidney failure outside the body. Some artificial kidneys incorporate living kidney cells into its design, so that it can produce important hormones, process metabolites, and provide immune functions that dialysis cannot.

Recently, hydrogel membranes have been used for cell microencapsulation in biohybrid artificial organs. Hydrogels can also be used as an alternative to ultrafiltration membranes for macroencapsulation.

Porous Hydrogels for Tissue Engineering

The use of biocompatible hydrogels for tissue engineering is an area of intense research activity. The candidate materials include both natural polymers (such as fibrin, collagen, and gelatin) and synthetic polymers (PLA and PLGA).

The freeze-drying technique has been used to prepare porous gelatin hydrogels for tissue engineering. This method provides a promising way to prepare porous scaffolds for cell growth, extracellular matrix production, and for regeneration of damaged or lost tissues. Animal experiments have proven that the biological performance of growth factors is enhanced by their sustained release, in marked contrast to the growth factors administered in the solution form.

Wound Dressings

Throughout the healing process, wounds produce a variety of fluids, generally known as wound exudate. Wound dressings can help control the wound. There are two types of dressings: dry and wet. Healing with a wet environment is faster than with a dry environment, due to the fact that renewed skin, forms during healing in a wet environment.

Hydrogel dressings, originally invented as wound burn dressings, have many interesting properties: immediate pain control, easy replacement, transparency to allow healing follow-up, absorption and loss prevention of body fluids, a barrier against bacteria, good adhesion, good handling, oxygen permeability, and control of drug release. Also, hydrogels produced by radiation polymerization are fully sterile. The hydrogel allows the permeation of drugs and oxygen to the wound and sticks to healthy skin surrounding the wound, but not to the newly forming dermis. Hydrogel wound dressings are made in sheets with an impermeable polymeric backing sheet. The backing sheet prevents the partially hydrated hydrogel from dehydrating and drying onto the wound bed.

Collagen hydrogels have a high tensile strength, low extensibility, controllable cross-linking, and a low antigenicity. These hydrogels can be produced in a variety of forms such as sheets, tubes, sponges, and powder. *In vivo* studies showed that the use of a collagen-based hydrogel sponge allowed cell migration, inhibited wound contraction, and accelerated wound repair. PVA/PVP hydrogel is promising for use as a burn wound covering.

Fire Protection

Due to its high water content, hydrogels can be used as burn-free materials for fire protection. One commercial fire protection product pure virgin wool woven with a unique interlinking cell construction that allows the blanket to hold up to 14 times its weight of the hydrogel. Burn-free blankets reduce physiological and psychological trauma by providing immediate cooling, soothing, and moistening of the burn, and reduce the pain and trauma of wounds from fires, flames, scalds, and chemical and electrical injuries. Hydrogels are also used as a dry fire extinguishing agent or mixed with water as an aqueous extinguishing agent.

Hydrogel as Superabsorbent Materials

In the past 20 years, superabsorbent polymers (SAPs) have achieved a worldwide market. In contact with water, SAP's hydrophilic backbone interacts through hydration and hydrogen bonding with the solvent, accompanied by an energy decrease and entropy increase.

Diapers

The first commercial production of a superabsorbent polymer was as feminine napkins, using cross-linked starch-grafted polyacryalte, then in baby diapers. Commercial diaper products are now packed with poly(acrylic acid) resins.

Water-sealing Construction Materials

SAPs can be used as a water-blocking construction filler with cement and an asphalt emulsion. The main benefit of the fill material is its high ductility compared to conventional construction backfills such as gravel or sand. Similar principles are employed in making and using sealing compounds for electrical and optical cables.

Agricultural Applications

Superabsorbent polymers have found limited utility as additives to soil to improve its water-holding ability. The hydrogel absorbs the water and controls its release back into the soil as conditions become drier. Poly(ammonium acrylate) and poly(acrylamide) hydrogels have been studied for spray application to the soil and plants. The hydrogel reduces water consumption and increases both water and fertilizer efficiency; hence, the product can be used for conserving irrigation water and increasing the agricultural potentialities of sandy soils.

Biodegradable hydrogels based on polysaccharide or polyester–polyurea–polyurethane are used as an artificial soil for plant material capable of growth. Besides a direct addition to growth media, polymer hydrogels can also be used as coatings for seeds and bare roots.

HEALTH AND SAFETY FACTORS

Although most monomers and organic solvents involved in hydrogel synthesis are harmful or toxic to direct touch, polymerization and cross-linking suppresses the toxicity. Hydrogels are highly biocompatible and have extremely low toxicity to the human body. Most hydrogel products such as contact lenses, medicines, surgical dressings, and foods can be used directly. Biodegradable hydrogels can degrade into harmless small molecules within the human body. Some hydrogel breast implants have been removed from the market, due to the lack of long-term toxicity data or clinical follow-up. The long-term risks of implants are not always obvious during the first few years of use. Acrylate, acrylamide monomers are eye and skin irritants; substituted acrylamides are less toxic. Acrylonitrile monomer is a neurotoxin, inhalation and ingestion should especially be avoided.

N. A. Peppas, ed., *Hydrogels in Medicine Pharmacy*, CRC press, Boca Raton, Fla., Vol I, 1987.

W. Rhee, J. Rosenblatt, M. Castro, J. Schroeder, P. R. Rao, C. F. H. Harner, and R. A. Berg, in J. M. Harris and S. Zalipsky, eds., *Poly(ethylene glycol) Chemistry and Biological Applications*, ACS, Washington, D.C., 1997.

C. Valenta, A. Walzer, A. E. Clausen, and A. Bernkop-Schnurch, *27th Proceedings of the International Symposium on Controlled Release of Bioactive Materials (2000)*, 2000, pp. 930–931.

WENSHENG CAI
RAM B. GUPTA
Auburn University

HYDROGEN

Hydrogen, the lightest element, has three isotopes: hydrogen, H, at wt 1.0078; deuterium, D, at wt 2.0141; and tritium, T, at wt 3.0161. Hydrogen is very abundant, being one of the atoms composing water; deuterium and tritium occur naturally on earth, but at very low levels. Tritium, a radioactive low-energy beta emitter with a half-life of 12.26 yr, is useful as a tracer in hydrogen reactions.

Whereas hydrogen atoms exist under certain conditions, the normal state of pure hydrogen is the hydrogen molecule, H_2, the lightest of all gases. Molecular hydrogen is a product of many reactions but is present at only low levels (0.1 ppm) in the earth's atmosphere. The hydrogen molecule exists in two forms, designated *ortho*-hydrogen and *para*-hydrogen, depending on the nuclear spins of the atoms. Many physical and thermodynamic properties of H_2 depend on the nuclear spin orientation, but the chemical properties of the two forms are the same.

Hydrogen is a very stable molecule having a bond strength of 436 kJ/mol(104 kcal/mol) and is not particularly reactive under normal conditions. However, at elevated temperatures and with the aid of catalysts, H_2 undergoes many reactions. Hydrogen forms compounds with almost every other element, often by direct reaction of the elements. The explanation for its ability to form compounds with such chemically dissimilar elements as alkali metals, halogens, transition metals, and carbon lies in the intermediate electronegativity of the hydrogen atom.

Hydrogen can be liquefied and stored as a cryogenic liquid (LHy). Hydrogen is viewed as an important player in the future energy equation for ultraclean transportation fuels. Fuel cells are receiving significant R&D funding by the world's automobile manufacturers and some of the major oil companies. Upon combustion, hydrogen returns to water, accompanied by virtually no pollution and no greenhouse gas production, in contrast to hydrocarbon-based fuels.

PHYSICAL AND THERMODYNAMIC PROPERTIES

Tables 1, 2, and 3 outline many of the physical and thermodynamic properties of *para*- and normal hydrogen in the solid, liquid, and gaseous states, respectively.

Table 1. Physical and Thermodynamic Properties of Solid Hydrogen

Property	Hydrogen	
	para-	Normal
mp, K (triple point)	13.803	13.947
vapor pressure at mp, kPa[a]	7.04	7.20
vapor pressure at 10 K, kPa[a]	0.257	0.231
density at mp, (mol/cm³) × 10³	42.91	43.01
heat of fusion at mp, J/mol[b]	117.5	117.1
heat of sublimation at mp, J/mol[b]	1023.0	1028.4
C_p at 10 K, J/(mol·K)[b]	20.79	20.79
enthalpy at mp, J/mol[b,c]	−740.2	321.6
internal energy at mp, J/mol[b,c]	−740.4	317.9
entropy at mp, J/(mol·K)[b,c]	1.49	20.3
thermal conductivity at mp, mW/(cm·K)	9.0	9.0
dielectric constant at mp	1.286	1.287
heat of dissociation at 0 K, kJ/mol[b]	431.952	430.889

[a] To convert kPa to mm Hg, multiply by 7.5.
[b] To convert J to cal, divide by 4.184.
[c] Base point (zero values) for enthalpy, internal energy, and entropy are 0 K for the ideal gas at 101.3 kPa (1 atm) pressure.

Table 2. Physical and Thermodynamic Properties of Solid Hydrogen

Property	Hydrogen	
	para-	Normal
mp, K (triple point)	13.803	13.947
normal bp, K	20.268	20.380
critical temperature, K	32.976	33.18
critical pressure, kPa[a]	1292.8	1315
critical volume, cm³/mol	64.144	66.934
density at bp, mol/cm³	0.03511	0.03520
density at mp, mol/cm³	0.038207	0.03830
compressibility factor, $Z = PV/RT$		
at mp	0.001606	0.001621
bp	0.01712	0.01698
critical point	0.3025	0.3191
adiabatic compressibility, $(-\partial V/V\partial P)_s$, MPa^{-1}[b]		
at triple point	0.00813	0.00813
bp	0.0119	0.0119
coefficient of volume expansion, $(-\partial V/V\partial T)_p$, K^{-1}		
at triple point	0.0102	0.0102
bp	0.0164	0.0164
heat of vaporization, J/mol[c]		
at triple point	905.5	911.3
bp	898.3	899.1
C_p, J/(mol·K)[c]		
at triple point	13.13	13.23
bp	19.53	19.70
C_v, J/(mol·K)[c]		
at triple point	9.50	9.53
bp	11.57	11.60
enthalpy, J/mol		
at triple point	−622.7	438.7
bp	−516.6	548.3
internal energy, J/mol[c,d]		
at triple point	−622.9	435.0
bp	−519.5	545.7
entropy, J/(mol·K)[c,d]		
at triple point	10.00	28.7
bp	16.08	34.92
velocity of sound, m/s		
at triple point	1273	1282
bp	1093	1101
viscosity, mPa·s(= cp)		
at triple point	0.026	0.0256
bp	0.0133	0.0133
thermal conductivity, mW/(cm·K)		
at triple point	0.73	0.73
bp	0.99	0.99
dielectric constant		
at triple point	1.252	1.253
bp	1.230	1.231
surface tension, mN/m (= dyn/cm)		
at triple point	2.99	3.00
bp	1.93	1.94
isothermal compressibility, $1/V(\partial V/V\partial P)_T$, MPa^{-1}[b]		
at triple point	−0.0110	−0.0110
bp	−0.0199	−0.0199

[a] To convert kPa to mm Hg, multiply by 7.5.
[b] To convert MPa to atm, divide by 0.101.
[c] To convert J to cal, divide by 4.184.
[d] Base point (zero values) for enthalpy, internal energy, and entropy are 0 K for the ideal gas at 101.3 kPa (1 atm) pressure.

Table 3. Physical and Thermodynamic Properties of Gaseous Hydrogen[a]

Property	Hydrogen	
	para-	Normal
density at 0°C, (mol/cm^3) × 10^3	0.05459	0.04460
compressibility factor, $Z = PV/RT$, at 0°C	1.0005	1.00042
adiabatic compressibility, $(-\partial V/V\partial P)_s$, at 300 K, MPa^{-1b}	7.12	7.03
coefficient of volume expansion, $(\partial V/V\partial P)_p$, at 300 K, K^{-1}	0.00333	0.00333
C_p at 0°C, J/(mol·K)[c]	30.35	28.59
C_v at 0°C, J/(mol·K)[c]	21.87	20.30
enthalpy at 0°C, J/mol[c,d]	7656.6	7749.2
internal energy at 0°C, J/mol[c,d]	5384.5	5477.1
entropy at 0°C, J/(mol·K)[c,d]	127.77	139.59
velocity of sound at 0°C, m/s	1246	1246
viscosity at 0°C, mPa·s(= cP)	0.00839	0.00839
thermal conductivity at 0°C, mW/(cm·K)	1.841	1.740
dielectric constant at 0°C	1.00027	1.000271
isothermal compressibility $1/V(\partial V/\partial P)_T$, at 300 K, MPa^{-1b}	−9.86	−9.86
self-diffusion coefficient at 0°C, cm^2/s		1.285
gas diffusivity in water at 25°C, cm^2/s		4.8×10^{-5}
Lennard-Jones parameters		
collision diameter, σ, m × 10^{10}		2.928
interaction parameter, ϵ/k, K		37.00
heat of dissociation at 298.16 K, kJ/mol[c]	435.935	435.881

[a] All values at 101.3 kPa (1 atm).
[b] To convert MPa to atm, divide by 0.101.
[c] To convert J to cal, divide by 4.184.
[d] Base point (zero values) for enthalpy, internal energy, and entropy are 0 K for the ideal gas at 101.3 kPa (1 atm) pressure.

Bonding of Hydrogen to Other Atoms

The hydrogen atom can either lose the 1s valence electron when bonding to other atoms, to form the H$^+$ ion, or conversely can gain an electron in the valence shell to form the hydride ion, H$^-$.

Most hydrogen compounds are formed through covalent bonding of hydrogen to the other atoms.

Reactions of Synthesis Gas

The main hydrogen manufacturing processes produce synthesis gas, a mixture of H$_2$ and CO. Synthesis gas can have a variety of H$_2$-to-CO ratios. The water gas shift reaction is used to reduce the CO level and produce additional hydrogen, or to adjust the H$_2$-to-CO ratio to one more beneficial to subsequent processing. Synthesis gas is used mainly to produce ammonia and methanol.

Other Reactions of Hydrogen

Sulfur, nitrogen, and oxygen are heteroatoms, which are abundant in many fuel sources such as petroleum, coal, and oil shale. These elements are considered pollutants and detriments to the refining process. Hydrogen is used to reduce the levels of these contaminants. Hydrogen reacts with a number of metal oxides at elevated temperatures to produce the metal and water.

Reactions of Hydrogen and Other Elements

Hydrogen forms compounds with almost every other element. Direct reaction of the elements is possible in many cases.

Hydrogen reacts directly with a number of metallic elements to form hydrides. The ionic or saline hydrides are formed from the reaction of hydrogen with the alkali metals and with some of the alkaline-earth metals.

Other metals also form compounds with hydrogen, either through direct heating of the elements or during electrolysis with the metal as an electrode. These metallic hydrogen compounds, also called hydrides, are in fact covalently bonded and do not contain H$^-$. Many metallic hydrides are nonstoichiometric in nature and appear to be metal alloys, having properties typical of metals, such as high electrical conductivity. Some compounds, such as MgH$_2$, are intermediate in properties between the saline hydrides and the metallic hydrides.

Reactions of Atomic Hydrogen

Atomic hydrogen is a very strong reducing agent and a highly reactive radical that can be produced by various means.

Absorption of Hydrogen in Metals

Many metals and alloys absorb hydrogen in large amounts. The absorption is largely reversible for palladium and for some other metals and alloys. Hydrogen diffuses and absorbs in many metals, with detrimental effects. Hydrogen exposure, under certain conditions, can seriously weaken and embrittle steel and other metals. Hydrogen at elevated temperatures can also attack the carbon in steel, forming methane bubbles that can link to form cracks. Alloying materials such as molybdenum and chromium combine with the carbon in steel to prevent decarburization by hydrogen.

SOURCES AND SUPPLIES

There is a broad range of hydrogen feedstock sources, from light hydrocarbons to solids and end products, which use pure hydrogen through mixtures of HyCO syngas to pure carbon monoxide. A principal source of hydrogen in refineries is the by-product production of hydrogen from catalytic reforming to produce high-octane gasoline.

Hydrogen is also a significant by-product of several chemical processes. A principal commercial source utilized extensively today is the steam cracking of ethane heavier feedstocks (propane, butane, naphtha). Generally, the hydrogen by-product from an ethylene plant can be used directly by a refiner within its refinery hydrogen loop.

Hydrogen is also a significant by-product of the chlor-alkali processes, since chlorine manufacture is significant around the world. There are three types of chlor-alkali processes: (*1*) mercury cells; (*2*) diaphragm; and (*3*) the newer membrane cells. The hydrogen purity ranges from 95 to 98+%, with impurities of air, water, carbon oxides, methane, and HCl. In addition, the mercury cells contain amounts of mercury that must be removed prior to use.

Several other chemical processes yield a by-product of hydrogen stream, which is commercially recovered today. These include styrene manufacture from ethylbenzene and the gasoline additive methyl *tert*-butyl ether (MTBE).

The Kvaerner process, a relatively new manufacturing process to produce carbon black, yields a hydrogen stream of approximately $0.14 \times 10^6 m^3/d$ (5×10^6 SCF/d). This process involves the thermal reduction of hydrocarbons to carbon black and hydrogen.

MANUFACTURE AND PRODUCTION

The main commercial processes for the on-purpose production of hydrogen are steam reforming, partial oxidation (coal, coke, resid), or electrolysis of water.

These processes all produce hydrogen from hydrocarbons and water:

steam reforming	$CH_4 + 2 H_2O \longrightarrow CO_2 + 4 H_2$
naphtha reforming	$C_n H_{2n} + 2 + n H_2O \longrightarrow n CO + (2 n + 1) H_2$
resid partial oxidation	$CH_{1.8} + 0.98 H_2O + 0.51 O_2 \longrightarrow CO_2 + 1.88 H_2$
coal gasification	$CH_{0.8} + 0.6 H_2O + 0.7 O_2 \longrightarrow CO_2 + H_2$
water electrolysis	$2 H_2O \longrightarrow 2 H_2 + O_2$

New Production Methods

Recent advances in small-scale hydrogen production via steam methane reforming, partial oxidation, and autothermal reforming of hydrocarbons such as natural gas, propane, methanol, etc, offer the promise of economic on-site hydrogen production at capacities below $0.028 \times 10^6/d$ $1 \times 10^6/d$ SCF pd. These units have been developed as low-cost fuel processors to provide a hydrogen-rich feed gas for fuel cells and are now being integrated with advanced small-scale hydrogen purification units to enable the production of pure hydrogen.

Electrolysis, a well-proven process to convert water to hydrogen, is used industrially on a limited scale. Electrolysis is a clean, reliable process and the hydrogen produced is very pure. The main drawback is that the electricity used to drive the process is usually 3–5 times more expensive than the fossil-fuel energy used to generate it. However, electrolysis combined with electric energy produced from renewable sources such as solar photovoltaic cells, wind, or hydroelectric power can become important in the future to produce hydrogen in a sustainable way some further new techniques to produce hydrogen are aqueous alkaline electrolysis, solid polymer electrolytes, high-temperature steam electrolysis, photolysis, and biomass conversion.

SHIPMENT AND STORAGE

A full range of hydrogen supply options is generally available to industry in the U.S. and Europe and increasingly in Latin America and the Asia-Pacific region.

Hydrogen Storage

The methods for storage of hydrogen include compressed gas or a cryogenic liquid, adsorbed on solid materials such as carbon and as chemically bound hydrogen in metal hydrides.

A novel approach to hydrogen containment is the use of high-strength glass microspheres 25–500 μm in diameter and with a 1 μm wall thickness. The glass microspheres are permeable to hydrogen only at high temperatures (300–400°C) when hydrogen is charged at high pressure up to 250–600 atm. On cooling, the hydrogen inside the microspheres is trapped at high pressure. The microspheres store about 10–15 wt% hydrogen (on glass weight alone) and can be safely carried in nonpressurized containers. To release the hydrogen from the microspheres, they must be heated once again. These materials are in the early stages of the laboratory development but offer the promise of the very efficient storage and transport of hydrogen.

A recently discovered option for hydrogen storage is adsorption on carbon nanotubes. These tiny tubelike configurations of carbon have a very large surface area per unit weight. These materials have the potential to store large amounts of hydrogen.

ECONOMIC ASPECTS

The estimated total world production/ consumption of on-purpose hydrogen and syngas is approximately 420×109 m^3/yr [15,000 billion standard cubic feet per year (BSCFY)]. Ammonia and methanol producers, petroleum refiners, and third-party hydrogen suppliers are the largest producers of hydrogen on a world-wide scale.

ENVIRONMENTAL CONCERNS

There are two environmental considerations associated with hydrogen: (*1*) nitrogen oxide (NO_x) formation when high hydrogen content gases are burned in a fuel system and (*2*) emissions (NO_x and SO_x) associated with the production of hydrogen.

The first concern can be mitigated by the recovery of hydrogen and its reuse within the refining or chemical facility, thereby reducing the volume of hydrogen burned in the fuel system. Stricter environmental regulations are requiring facilities to use low NO_x burners in their furnaces or heaters.

The second concern can be addressed by using the latest environmental controls for NO_x control and sulfur removal in the design of a steam methane reforming (SMR) or particle oxidation facility. The SMR process is the more environmentally acceptable of the two processes to produce hydrogen.

RECYCLING AND DISPOSAL

Many purge, off-gas or by-product streams in refineries and related petrochemical and chemical operations contain hydrogen in significant volume and concentrations to warrant recovery and recycling of the hydrogen to other processes versus disposal as a fuel gas stream to recover the Btu value only. A good example of hydrogen management within a refinery is the cascading of hydrogen purges from high-severity, high-pressure hydrocrackers or hydrorefining units to lower-pressure, less-severity hydrotreaters. Hydrogen recovery technologies include membranes, adsorption systems, and cryogenics.

HEALTH AND SAFETY FACTORS

Hydrogen gas is not considered toxic, but it can cause suffocation by the exclusion of air. The main hazard associated with the use of hydrogen lies in its extremely wide flammability limits in oxygen or air.

Contact with liquid hydrogen can cause tissue freezing and severe cryogenic burns. When spilled, liquid hydrogen will vaporize rapidly, forming a potentially explosive and oxygen vapor cloud. Visibility may be obscured in this vapor cloud.

Mandatory regulations governing the distribution of liquid or gaseous hydrogen are listed in the *Code of Federal Regulations*. The National Fire Protection Association has published specific requirements for liquid and gaseous hydrogen storage and delivery systems at consumer sites.

Hydrogen does not cause any adverse ecological effects.

USES

The demand for and supply of hydrogen is currently concentrated in three main markets worldwide: refining, petrochemicals, and chemicals. Hydrogen is one of the most important industrial commodities used to desulfurize or hydrogenate various petroleum products derived from crude oil into transportation fuels.

The metals industry, which historically has used small quantities of hydrogen for annealing or reducing atmospheres, is now moving forward with implementing several new direct reduction of iron ore (DRI) processes, which require large quantities of hydrogen or HyCO syngas.

Hydrogen is used in small volumes in the electronics industry in semiconductor manufacture, quartz melting, polysilicone manufacture, and fiber optic manufacture. The very high purity that is required and is generally provided by gas supplied via vaporization of liquid hydrogen.

Hydrogen is also used in the reducing atmosphere in float glass production plants. It is used to cool generators in electric power plants and to prevent corrosion in nuclear power plants. The only current commercial use of hydrogen as a fuel today is for rocket propulsion.

THE ECONOMY FOR HYDROGEN

Driven by concerns about oil dependence, deteriorating urban air quality, and recognizing the benefits of a hydrogen-based energy system, active R&D programs in hydrogen energy applications are under way in more than thirty countries. As part of current efforts, efficient water electrolyzers, large-scale hydrogen liquefaction plants, liquid hydrogen carrier ships, and hydrogen gas turbines are being designed.

Catalytic combustion of hydrogen is adequate for cooking, water heating, and space heating but may be a non-uniform temperature distribution at the catalyst surface, rapid changes in the operational state, and relatively small flux densities.

Use of hydrogen as a fuel for road and air transportation is being actively developed, driven by stringent new clean air standards around the world that encourage and in some cases require the use of clean alternatives to petroleum-based fuels. A transition from gasoline and diesel first to natural gas, followed by the gradual introduction of hydrogen, is envisioned.

The hydrogen hybrid electric vehicle takes advantage of high-efficiency electric drivetrains and the low emissions of hydrogen IC ingines. The concept involves the use of relatively small IC engines operating on hydrogen to power electric generators that charge batteries.

The use of liquid hydrogen as a rocket fuel for use in space flights is well established. It is now being seriously considered for use as an aircraft fuel.

The most attractive energy conversion technology that uses hydrogen is fuel cells. A fuel cell is an energy conversion device that combines hydrogen and oxygen in an electrochemical process to produce electric power, some low-temperature heat, and water vapor as the only emissions. The electricity produced can be used for any useful purpose.

Applications for fuel cells that are being actively developed include portable power systems (25 W to 1 KW), distributed stationary power generation (1 KW up to MW scale), and as a propulsion system for road vehicles, ships, and submarines. Although fuel cell cars are still under development, hydrogen fuel cell buses are becoming a reality and small fleets have begun operation.

H.-W Pohl and D. Wildner, "Hydrogen Demonstrator Aircraft," *Hydrogen Energy Progress XI, Proceedings of the 11th World Hydrogen Energy Conference*, Stuttgart, Germany, Vol. 2, 1996, pp. 1779–1786.

V. Raman and co-workers, "A Rapid Fill Hydrogen Fuel Station for Fuel Cell Buses," *Hydrogen Energy Progress XII, Proceedings of the 12th World Hydrogen Energy Conference*, Buenos Aires, Argentina, Vol. 2, 1998 pp. 1629–1642.

M. Ross and C. Shishkevish, *Molecular and Metallic Hydrogen*, R-2056-ARPA, Rand Corp., Santa Monica, Calif., 1977.

WILLIAM F. BAADE
UDAY N. PAREKH
VENKAT S. RAMAN
Air Products and Chemicals, Inc.

HYDROGEN CHLORIDE

Hydrogen chloride, HCl, exists in solid, liquid, and gaseous states and is very soluble in water.

Hydrochloric acid is found naturally in the gases evolved from volcanoes, particularly those in Mexico and South America. Its formation is attributed to high-temperature reaction of water with the salts found in seawater. The original atmosphere of the earth is considered to have contained water, carbon dioxide and hydrogen chloride in the ratio of 20:3:1. Hydrogen chloride was also detected in the atmosphere of the planet Venus. The dissociation of HCl is considered the source of chlorine detected in the spectra of distant stars.

Hydrochloric acid is also present in the digestive system of most mammals. The gastric mucosa lining the human stomach produces about 1.5 L/d of gastric juices, containing an acid concentration in the range of 0.05 to 0.1 N. A deficiency of hydrochloric acid impairs the digestive process, particularly of carbohydrates and proteins, and excess acid causes gastric ulcers.

PHYSICAL AND THERMODYNAMIC PROPERTIES

Anhydrous Hydrogen Chloride

Anhydrous hydrogen chloride is a colorless gas that condenses to a colorless liquid and freezes to a white crystalline solid. The physical and thermodynamic properties of HCl are summarized in Table 1 for selected temperatures and pressures. The high thermal stability of hydrogen chloride is a consequence of the large enthalpy of its formation.

Hydrogen Chloride–Water System

Hydrogen chloride is highly soluble in water, and this aqueous solution does not obey Henry's law at all concentrations.

Hydrogen chloride and water form four hydrates. The dihydrate is formed when a saturated solution is cooled at atmospheric pressure. The monohydrate has a melting of $-15.35°C$; the trihydrate has a melting point of $-24.9°C$; the hexahydrate is very unstable and has a melting point of $-70°C$. Addition of hydrogen chloride to pure water lowers the freezing point until a eutectic temperature of about $-85°C$ is reached at 25% HCl. Hydrogen chloride and water form constant boiling mixtures.

Hydrogen chloride is completely ionized in aqueous solutions at all but the highest concentrations.

The viscosity of hydrochloric acid solutions, η, increases slightly with increasing concentration.

The specific heat of aqueous solutions of hydrogen chloride decreases with acid concentration. The electrical conductivity of aqueous hydrogen chloride increases with temperature.

Hydrogen Chloride–Water–Inorganic Compound Systems

Salting out metal chlorides from aqueous solutions by the common ion effect upon addition of HCl is utilized in many practical applications. The properties of the $FeCl_2 \cdot HCl \cdot H_2O$

Table 1. Physical and Thermodynamic Properties of Anhydrous Hydrogen Chloride

Property	Value
melting point, °C	-114.22
boiling point, °C	-85.05
heat of fusion at $-114.22°C$, kJ/mol[a]	1.9924
heat of vaporization at $-85.05°C$, kJ/mol[a]	16.1421
entropy of vaporization, J/(mol·K)[a]	85.85
triple point, °C	-114.25
critical temperature, T_c,°C	51.54
critical pressure, P_c, MPa[b]	8.316
critical volume, V_c L/mol	0.069
critical density, g/L	424
critical compressibility factor, Z_c	0.117
$\Delta H°_f$ at 198 K, kJ/mol[a]	-92.312^c
	-100.4^d
$\Delta G°_f$ at 298 K, kJ/mol[a]	-95.303
$S°$ at 298 K, J/(mol·K)[a]	186.786
dissociation energy at 298 K, kJ[a]	431.62^c
	427.19^d
compressibility coefficient	0.00787
internuclear separation, nm	0.12510
dipole moment, C·m[e]	3.716^c
	3.74^d
ionization potential, J[a]	20.51^c
	20.45^d
heat capacity, C_p, J/(mol·K)[a]	
vapor (constant pressure)	
at 273.16 K	29.162
at 973.2 K	30.554
liquid at 163.16 K	60.378
solid at 147.16 K	48.98
surface tension at 118.16 K, mN/cm(=dyn/cm)	23
viscosity, mPa·s(=cP)	
liquid at 118.16 K	0.405
vapor at 273.06 K	0.0131
vapor at 523.2 K	0.0253
thermal conductivity, mW/(m·K)	
liquid at 118.16 K	335
vapor at 273.16 K	13.4
density, g/cm^3	
liquid	
at 118.16 K	1.045
at 319.15 K	0.630
solid	
rhombic at 81 K	1.507
cubic at 98.36 K	1.48
cubic at 107 K	1.469
refractive index	
liquid at 283.16 K	1.254
gas at 273.16 K	1.0004456
dielectric constant	
liquid at 158.94 K	14.2
gas at 298.16 K	1.0046
electrical conductivity, $(\Omega \cdot m)^{-1}$	
at 158.94 K	1.7×10^{-7}
at 185.56 K	3.5×10^{-7}

[a] To convert J to cal, divide by 4.184.
[b] To convert MPa to atm, divide by 0.101.
[c] Measured value.
[d] Calculated value.
[e] To convert C·m to debye, divide by 3.34×10^{-30}.

system are important to the steel-pickling industry. Other metal chlorides that are salted out by the addition of hydrogen chloride to aqueous solutions include those of magnesium, strontium, and barium.

Metal chlorides that are not readily salted out by hydrochloric acid can require high concentrations of HCl for precipitation. This property is used to recover hydrogen chloride from azeotropic mixtures. The solubility of chlorine in hydrochloric acid is an important factor in the purification of by-product hydrochloric acid.

Hydrogen Chloride-Organic Compound Systems

The solubility of hydrogen chloride in many solvents follows Henry's law. Notable exceptions are HCl in polyhydroxy compounds such as ethylene glycol, which have characteristics similar to those of water.

CHEMICAL PROPERTIES

Reactions of Anhydrous Hydrogen Chloride

Hydrogen chloride reacts with inorganic compounds by either heterolytic or homolytic fission of the H—Cl bond. However, anhydrous HCl has high kinetic barriers to either type of fission, and hence this material is relatively inert.

Anhydrous HCl protonates the Group 15 (V) hydrides. The heavier transition-metal oxides require a higher reaction temperature, and the primary reaction product is usually the corresponding oxychlorides. Thermodynamic considerations for the reaction $M + nHCl \longrightarrow MCl_n + n/2H_2$ indicate that most metals should react with HCl. However, this reaction is kinetically slow at all but elevated temperatures. Hydrogen chloride and oxygen react in the gaseous state to liberate chlorine. Anhydrous HCl forms addition compounds at lower temperatures with halogen acids such as HBr and HI, and also with HCN. Hydrogen chloride reacts with sulfur trioxide, yielding liquid chlorosulfuric acid.

Reaction with Organic Compounds. Hydrogen chloride adds to carbon–carbon double and triple bonds in a variety of organic compounds. Acetylene and hydrogen chloride historically were used to make chloroprene. The olefin reaction is used to make ethyl chloride from ethylene and to make 1,1-dichloroethane from vinyl chloride. Lower alcohols such as methanol can be converted to the corresponding alkyl chlorides by carrying out the reaction $ROH + HCl \longrightarrow RCl + H_2O$, using either a liquid or a solid catalyst.

The introduction of the chloromethyl group to both aliphatic and aromatic compounds is carried out by reaction of paraformaldehyde and hydrogen chloride. This method is used for synthesizing methyl chloromethyl ether benzyl chloride and chloromethyl acetate.

Hydrochloric Acid

Most metals and alloys react with aqueous hydrochloric acid via

$$M + nH_3O^+ \longrightarrow M^{n+} + nH_2O + n/2H_2.$$

This is essentially a corrosion reaction involving anodic metal dissolution where the conjugate reaction is the hydrogen evolution process.

Oxides and hydroxides react with HCl to form a salt and water, as in a simple acid–base reaction. However, reactions with low solubility or insoluble oxides and hydroxides are complex and the rate is dependent on many factors similar to those for reactions with metals.

HCl can be electrolyzed to produce H_2 and chlorine. Many organic reactions are catalyzed by acids such as HCl. Typical examples of the use of HCl in these processes include conversion of lignocellulose to hexose and pentose, sucrose to inverted sugar, esterification of aromatic acids, transformation of acetaminochlorobenzene to chloroanilides, and inversion of methone.

MANUFACTURING AND PROCESSING

Hydrogen chloride is produced by the direct reaction of hydrogen and chlorine, by reaction of metal chlorides and acids, and as a by-product from many chemical manufacturing processes such as chlorinated hydrocarbons.

Hydrogen Chloride Produced from Incineration of Waste Organics

Environmental regulations regarding the disposal of chlorine-containing organic wastes has motivated the development of technologies for burning or pyrolyzing the waste organics and recovering the chlorine values as hydrogen chloride. Several processes, catalytic and noncatalytic, have been developed to treat these wastes to produce hydrogen chloride.

Hydrogen Chloride from Hydrochloric Acid Solutions

Gaseous hydrogen chloride is obtained by partially stripping concentrated hydrochloric acid using an absorber–desorber system.

Purification

Gaseous HCl from all the manufacturing processes described invariably contains moisture, and sometimes organic species. H_2SO_4 drying can be used to remove small amounts of water, reducing the residual water content to less than 0.02%. If the water content is high, it can be removed as concentrated hydrochloric acid by cooling the gas mixture before drying, with sulfuric acid. Addition of chlorosulfuric acid to this stream reduces the water content to less than 10 ppm. This mixture also removes unsaturated organics such as ethylene and vinyl chloride and certain organic compounds such as monochloroacetic acid.

Chlorine can be removed by either activated carbon adsorption or by reaction with olefins such as ethylene over-activated carbon at temperatures of 30–200°C. Addition of liquid high-boiling paraffins can reduce the chlorine content in the HCl gas to less than 0.01% Solid absorbents generally remove the organics from HCl.

Crude HCl recovered from production of chlorofluorocarbons by hydrofluorination of chlorocarbons contains unique impurities which can be easily removed.

Use of air or purified HCl gas as stripper is practiced to remove volatile dissolved organics and chlorine from aqueous HCl.

Materials of Construction

Cast iron, mild steel, and steel alloys are resistant to attack by dry, pure HCl at ambient conditions and can be used at temperatures up to the dissociation temperature of HCl.

Aqueous Hydrochloric Acid. Tantalum and zirconium exhibit the highest corrosion resistance to HCl. However, the corrosion resistance of zironium is severely impaired by the presence of ferric or cupric chlorides. Tantalum–molybdenum alloys containing more than 50% tantalum are reported to have excellent corrosion resistance. Common plastics and elastomers show excellent resistance to hydrochloric acid within the temperature limits of the materials. Carbon and graphite rendered impervious with 10–15% phenolic, epoxy, or furan resin are among the most important materials for hydrochloric acid service up to 170°C. The most important applications of these materials for hydrochloric acid service are heat exchangers and centrifugal pumps.

Glass and ceramic-coated equipment are widely used for handling hydrochloric acid.

ECONOMIC ASPECTS

The market for hydrogen chloride in the United States, western Europe, and Japan amounted to 16.6×10^6 tons in 2000. Hydrochloric acid is used primarily to produce chlorinated organics such as ethylene dichloride and chlorinated inorganics such as calcium chloride. Generation of HCl from 1995 to 2000 increased by 8.7%, due in large part to the fluorocarbons and isocyanate sectors. Projected demand for HCl for 2005 was 2,295 tons.

STORAGE AND HANDLING

All Department of Transportation (DOT), Environmental Protection Agency (EPA), and Occupational Safety and Health Act (OSHA) rules and regulations should be reviewed prior to handling hydrochloric acid, and all regulations must be followed. All employees handling HCl must be trained to ensure that they are familiar with the appropriate materials safety data sheets and applicable regulations.

The DOT classifies HCl as a corrosive material and requires that it be transported in DOT-approved delivery vessels.

HEALTH AND SAFETY FACTORS

Hydrogen chloride in air is an irritant, severely affecting the eye and the respiratory tract. The vapor in the air, normally absorbed by the upper respiratory mucous membranes, is lethal at concentrations of over 0.1% in air, when exposed for a few minutes. The maximum allowable concentration under normal working conditions has been set at 5 ppm per OSHA and ACGIH.

Hydrogen chloride in the lungs can cause pulmonary edema, a life-threatening condition.

Hydrogen chloride in air can also be a phytotoxicant. Tomatoes, sugar beets, and fruit trees of the Prunus family are sensitive to HCl in air. Exposure of concentrated hydrochloric acid to the skin can cause chemical burns or dermatitis.

USES

Hydrogen chloride and the aqueous solution, muriatic acid, find application in many industries. In general, anhydrous HCl is consumed for its chlorine value, whereas aqueous hydrochloric acid is often utilized as a nonoxidizing acid. The latter is used in metal cleaning operations, chemical manufacturing, petroleum well activation, and in the production of food and synthetic rubber.

Most of the HCl produced is consumed captively; ie, at the site of production, either in integrated operations such as ethylene dichloride–vinyl chloride monomer (EDC/VCM) plants and chlorinated methane plants or in separate HCl consuming operations at the same location.

Anhydrous Hydrogen Chloride

In the United States, all ethylene dichloride (EDC) is produced from ethylene, either by chlorination or oxychlorination (oxyhydrochlorination).

Most of the HCl consumed in the manufacture of methyl chloride from methanol is a recycled product.

Several methods are available for generating chlorine from HCl. These include electrolysis of metallic chloride solutions, electrolysis of hydrochloric acid, oxidation of hydrogen chloride to chlorine with nitric acid, and oxidation of hydrogen chloride to chlorine using oxygen in the presence of catalysts (The Deacon process and modified Deacon process).

Perchloroethylene (PCE) and trichloroethylene (TCE) can be produced either separately or as a mixture in varying proportions by reaction of C_2-chlorinated hydrocarbons.

Most ethyl chloride is produced by the hydrochlorination of ethylene using anhydrous HCl.

Other uses for anhydrous HCl include use in cottonseed delinting and disinfecting, as a catalyst promotor for petroleum isomerization, in the production of agricultural chemicals, and in the preparation of hydrochloride salts in the pharmaceutical industry.

Aqueous Hydrochloric Acid

The largest captive use of aqueous HCl is for brine acidification prior to electrolysis in chlorine/caustic cells and the largest merchant markets for HCl are steel pickling and oil-well acidizing.

S. Austin and A. Glowacki, *Ullmann's Encyclopedia of Industrial Chemistry*, 5th ed., Vol. A13, VCH, Weinheim, Germany, 1989, p. 283.

J. E. Buice, R. L. Bowlin, K. W. Mall, and J. A. Wilkinson, *Encyclopedia of Chemical Processing and Design*, Vol. 26, Marcel Dekker, Inc., New York, 1987, p. 396.

Chlorine and Hydrogen from HCl by Electrolysis, Uhde, Dortmund, Germany, 1987.

"Hydrochloric Acid," *Chemical Economics Handbook*, SRI International, Menlo Park, Calif., Nov. 2001.

MOHAMED W. M. HISHAM
TILAK V. BOMMARAJU
Occidental Chemical
Corporation

HYDROGEN ENERGY

Fossil fuels, including coal, natural gas, and petroleum, provide much of the energy used today. However, their supply is inherently and geographically limited, and there are significant environmental impacts to the continued and increased conventional use of fossil fuels. To address social equity, global climate change, urban air pollution, energy security through diversity, and economic growth issues, new energy solutions are needed. One of the most promising solutions is a system using hydrogen and electricity as primary energy carriers.

The simplest element, hydrogen, H_2, has the potential to provide all of our energy services with little or no impact on the environment, locally or globally. It can be made from domestic resources, offering opportunities for energy independence to all countries and regions. Hydrogen can be used in all applications that currently use the more familiar hydrocarbon fuels.

Like electricity, hydrogen can be produced from many different resources, including fossil fuels, renewable resources, and nuclear energy. Unlike electricity, hydrogen can be produced at one point in time and used at a later date, that is, it can be stored for long periods of time. Hydrogen is an effective storage medium for large amounts of energy, particularly for more than a few days.

Hydrogen offers benefits as an on-board fuel for "zero emission" transportation needs that can never be met with hydrocarbon fuels. However, significant advances in hydrogen storage capacities are needed to give hydrogen vehicles driving ranges comparable to those of conventional and hybrid gasoline-fueled vehicles.

FUNDAMENTALS OF HYDROGEN ENERGY SYSTEMS

Molecular hydrogen (H_2) is not found in substantial quantities in nature: hydrogen atoms (H) are almost always associated with other elements, principally oxygen (ie, water, (H_2O)) and carbon (ie, methane, CH_4; coal and other fossil fuels; and organic matter). Therefore, energy must be used to produce hydrogen. This energy, which can be thermal, photonic, or electrical, can be provided by renewable, fossil, or nuclear resources.

Once produced, hydrogen needs to be stored. Although hydrogen has the highest energy density of common fuels on a weight basis, it has the lowest energy density on a volumetric basis, at standard conditions.

In order to store the same amount of energy in a given volume, significantly higher pressures are required for compressed hydrogen gas systems than for other gases. In order for hydrogen to be considered a viable fuel for transportation needs, advanced storage systems must be developed.

Using hydrogen to provide energy services requires an electrochemical device (eg, a fuel cell) or a thermochemical device (eg, a combustion engine). In these devices, hydrogen combines with oxygen to form water as the main product and to release useful energy.

Both electrical and thermal energy can be released via these reactions. High overall energy efficiencies are possible, depending on the design of the system.

The water cycle is an important aspect of hydrogen energy systems: that is, water is a feedstock for the production of hydrogen (even when using fossil fuels) and is also the principal product of the use of hydrogen. This characteristic is important in understanding the role of hydrogen in sustainable energy systems.

HYDROGEN PRODUCTION

Today, hydrogen is produced primarily from fossil fuels (natural gas, petroleum, coal) using well-known commercial processes. Worldwide, the predominant feed is natural gas (48%), followed by oil–petroleum (30%), coal (18%), and electricity (4%). "Captive" hydrogen is produced at the consumer site where it is used in ammonia–fertilizer and methanol production and in petroleum refining. These represent the vast majority of hydrogen production and consumption (98.8%).

Large-scale catalytic steam reformation of natural gas is a well-known, high-energy efficiency, commercial process for the production of hydrogen. The other sources of hydrogen are from petroleum refining, coal gasification, electrolysis, and the chloralkali process.

Research is being conducted to develop improved and innovative hydrogen production technologies. For the long term, there are technical, economic, and energy efficiency improvements required to make hydrogen a pervasive commodity fuel. Many current research efforts are focused on small-scale, on-site production of hydrogen. In addition to process innovations, nonconventional resources present opportunities and challenges for hydrogen production. Sources of methane-rich gas, such as landfill gas, volcanic offgas, and coalbed methane; and organic waste streams, such as municipal solid waste (MSW), paper, plastics, animal waste, and sewage sludge, represent important resource streams for hydrogen production.

HYDROGEN STORAGE

Commercial Hydrogen Storage Technologies

Practically, hydrogen can be stored as a liquid or compressed gas to increase the volumetric energy density.

Liquefaction of hydrogen is an energy-intensive process, with a third or more of the energy content of the stored hydrogen needed to condense the hydrogen. Proper storage tank design and vehicle integration may provide the driving range that the consumer desires without compromising vehicle. Hydrogen can also be stored in a solid-state form by reaction with certain metals and metal alloys.

Future Hydrogen Storage Technologies

On-board hydrogen storage remains a critical technical barrier for hydrogen-fueled vehicles. This is a particular challenge for PEM fuel-cell vehicle applications, where hydrogen must be liberated at temperatures compatible with the waste heat of the fuel cell ($<80\,^\circ$C).

Chemical hydrides are also being considered as hydrogen carriers and hydrogen storage media. The ultimate usefulness of this approach for widespread hydrogen use will depend on the ability and the cost to regenerate the spent material, preferably at a centralized processing plant.

Sodium borohydride has been proposed as a chemical carrier for hydrogen. However, the economics of this system must also include the bidirectional delivery system: $NaBH_4$ must be delivered to the end user; and $NaBO_2$ returned to a centralized processing facility for regeneration. This bidirectional transport of the energy carrier adds to the cost and complexity of the energy storage system.

Adsorption of hydrogen molecules on activated carbon, C_{60} structures such as fullerenes, and carbon foams has been studied as a medium for hydrogen storage. Although the amount of hydrogen stored is comparable to a system at high pressure, these carbon systems require low temperatures.

Recent work has also identified nanostructured carbon as an interesting and promising material for hydrogen storage, although the mechanisms are not yet fully understood. In addition, research is focused on modifying carbon nanotubes in an attempt to enhance the hydrogen storage capabilities using metal dopants and other additives.

HYDROGEN DELIVERY AND INFRASTRUCTURE

Currently, hydrogen is transported from central production facilities via pipeline or in tube trailers as a compressed gas, or in tanker trucks as a cryogenic liquid. As the role of hydrogen changes from that of an important industrial chemical, produced on site, to that of an energy carrier on a par with electricity, we will need a highly efficient infrastructure optimized for delivering hydrogen. The character of that infrastructure is unknown: will it be a centralized production with vast distribution networks similar to our current infrastructure for electricity, natural gas, and gasoline; will it rely on hydrogen produced on site; or will it be some combination? At present, delivery depends on pipelines, tanker trucks and tube trailers, and other options, such as eliminating delivery entirely by using on-site electrolysis to produce hydrogen where needed.

HYDROGEN USE

Conventional reciprocating internal combustion engines (ICEs) and gas turbines can be designed to run on hydrogen or hydrogen-blended fuels. These devices, with >100 years of development, are currently powered by hydrocarbon fuels and used in commercial applications throughout the current energy economy. The ICEs provide opportunities for near- and mid-term use of hydrogen as an energy carrier by utilizing the unique combustion characteristics of hydrogen; ie, its low flammability limit and high octane rating. These optimized conventional conversion devices can achieve extraordinarily low emissions without the need for postclean-up technologies. Hydrogen as a fuel has many favorable combustion properties when compared to hydrocarbons. These properties allow combustion strategies for hydrogen that are not possible with hydrocarbon flames.

For many reasons, a correctly designed hydrogen-fueled reciprocating ICE is more efficient than its hydrocarbon counterpart. Because of hydrogen's wide flammability range, hydrogen-fueled reciprocating ICEs are frequently designed without a throttle. To modify the power output from these engines, one tailors the mass of injected fuel to match the power demands on the engine. The lower limit on the mass injected is limited by the lean flammability limit or idle requirements, and the limit on the upper end for the amount of mass injected is limited by the limiting equivalence ratio. (A significant efficiency loss term in a conventional hydrocarbon-fueled spark-ignited reciprocating internal combustion engine is the air induction throttle). Such combustion strategies require no unique hardware to implement.

Highly efficient hydrogen ICEs with zero emissions, having power densities commensurate with their spark-ignited hydrocarbon counterparts, are a practical end-use of hydrogen. While this technology has been demonstrated, work remains. Research is underway to investigate the fundamental issues associated with direct in-cylinder injection of hydrogen. If this can be achieved without increased emissions, further improvements in efficiency, power density, and engine control can be achieved. Research into injection systems, materials compatibility, and other hydrogen-specific issues is also being performed to accelerate the introduction of this technology into the market place. The same combustion strategies can be applied to gas turbines for power production using hydrogen as the fuel.

Electrochemical Processes

Fuel cells are electrochemical devices that convert chemical energy into electrical energy. In principle similar to a battery, a fuel cell will, however, continue to produce electricity (and heat) as long as fuel and an oxidant are supplied. The only "recharging" required for a fuel cell is to refill the fuel supply. For a fuel cell that is fueled directly with hydrogen, the only products are electricity, water (or water vapor) and heat. Electrochemical devices can theoretically provide higher energy efficiencies that those that rely on combustion reactions, such as a heat engine. At low temperatures, the

Table 1. Fuel Cell Types and Their Characteristics

Type	Mobile ion	Temperature of operation, °C	Uses	Benefits	Issues
alkaline (AFC)	OH^-	150–200	manned space flights (starting w/ Apollo missions) some interest in vehicle use	produces drinking water for astronauts and heat for the spacecraft highly reliable	extremely sensitive to CO_2
phosphoric acid (PAFC)	H^+	150–200	stationary power combined heat and power (CHP)	electricity-only efficiency of ~40% improved efficiency with CHP reliable operation commercial history	cost reductions have stagnated ($3500–4500/kW)
molten carbonate (MCFC)	CO_3^{2-}	~650	stationary power CHP	electricity-only efficiency of ~60% high efficiency (80–85%) with CHP fuel flexibility noble metal catalyst not required	long startup time durability cost
solid oxide (SOFC)	O^{2-}	650–1000	stationary power CHP auxiliary power units (APUs)	electricity-only efficiency of 50–60% 80–85% efficiency for CHP fuel flexibility noble metal catalyst not required solid electrolyte	long startup time durability cost
proton exchange membrane (PEMFC)	H^+	50–80	vehicles portable appliances stationary power	fast startup high power density	cost durability sensitivity to CO and H_2S

theoretical maximum efficiency of an electrochemical device (fuel cell) operating at low current density can approach 80%, although practical efficiencies of a fuel cell power system will be much less (on the order of 40–45%). The theoretical maximum efficiency of heat engines is typically significantly less than that of an electrochemical device, especially at low and moderate temperatures.

Table 1 of summarizes the main characteristics of the various types of fuel cells, characterized by electrolyte.

ECONOMIC AND ENVIRONMENTAL ASPECTS

In attempting to gain acceptance of hydrogen systems, we must go beyond the technical aspects and point to the social, economic, political, and environmental benefits.

Prior to September 11, 2001, the primary driver behind hydrogen technologies was local air quality and global climate change: a hydrogen fuel cell vehicle would emit only water, and no CO_2. The stark reality of 9/11 once again brought energy security (and global vulnerability to oil supply disruption) to the forefront. Given that hydrogen can be produced from so many different resources, a shortage in one or even two resources can be compensated for by using other resources to produce the hydrogen we need for energy services. Hydrogen allows us to remove energy from the security as well as environmental equation. This key difference from other energy sources is an essential component in understanding why a transition to hydrogen energy is important.

Annual Energy Outlook 2003, U.S. Department of Energy, Energy Information Administration, DOE/EIA-0383, 2003.

EPRI Journal Online, "Outlook Promising for Advanced Batteries for Electric Vehicles", http://www.epri.com/journal, 2003.

P. Hoffmann, *Tomorrow's Energy*, MIT Press, Cambridge, Mass., 2001.

J. Larminie and A. Dicks, *Fuel Cell Systems Explained*, John Wiley & Sons, Ltd., U.K., 1999.

Union of Concerned Scientists, "Energy Security—Solutions to Protect Americas Power Supply and Reduce Oil Dependence", http://www.ucsusa.org, 2002.

CATHERINE E. GRÉGOIRE PADRÓ
Los Alamos National Laboratory

JAY O. KELLER
Sandia National Laboratories

HYDROGEN FLUORIDE

Hydrogen fluoride, HF, is the most important manufactured fluorine compound. The largest in terms of volume, it serves as the raw material for most other fluorine-containing chemicals. It is available either in anhydrous form or as an aqueous solution (usually 70%). Anhydrous hydrogen fluoride is a colorless liquid or gas having a boiling point of 19.5°C. It is a corrosive, hazardous material, fuming strongly, which causes severe burns upon contact. Rigorous safety precautions are the standard throughout the industry, but in practice hydrogen fluoride can be handled quite safely.

Table 1. Properties of Anhydrous Hydrogen Fluoride[a]

Property	Value
formula weight	20.006
composition, wt%	
H	5.038
F	94.96
boiling point at 101.3 kPa, °C	19.54
critical pressure, MPa	6.48
critical temperature, °C	188.0
critical density, g/mL	0.29
critical compressibility factor	0.117
melting point, °C	−83.55
density, liquid, 25°C, g/mL	0.958
heat of vaporization, 101.3 kPa, kJ/mol	7.493
heat of fusion, −83.6°C, kJ/mol	3.931
heat capacity, constant pressure, liquid at 16°C, J/(mol·K)	50.6
heat of formation, ideal gas, 25°C, kJ/mol	−272.5
free energy of formation, ideal gas, 25°C, kJ/mol	−274.6
entropy, ideal gas, 25°C, J/(mol·K)	173.7
vapor pressure, 25°C, MPa	122.9
viscosity, liquid, 0°C, mPa·s (=cP)	0.256
surface tension, mN/m (=dyne/cm), 0°C	10.2
refractive index, liquid, 25°C, 589.3 nm	1.1574
molar refractivity, cm^3	2.13
dielectric constant, at 0°C	83.6
dipole moment, C·m	6.104×10^{-30}
thermal conductivity, at 25°C, J(s·cm·°C)	
liquid	4.1×10^{-3}
vapor	2.1×10^{-4}
cryoscopic constant, K_f, mol/(kg·°C)	1.52
ebullioscopic constant, K_b, mol/(kg·°C)	1.9

[a] To convert kPa to psi, multiply by 0.145; to convert J to cal, divide by 4.184; to convert C·m to debye, divide by 3.336×10^{-30}.

PROPERTIES

Physical Properties

The physical properties of anhydrous hydrogen fluoride are summarized in Table 1. The specific gravity of the liquid decreases almost linearly from 1.1 at −40°C to 0.84 at 80°C.

Hydrogen fluoride is unique among the hydrogen halides in that it strongly associates to form polymers in both the liquid and gaseous states. At high temperatures or low partial pressures, HF gas exists as a monomer. At lower temperatures and higher partial pressures hydrogen bonding leads to the formation of chains of increasing length, and molecular weights of 80 and higher are observed. Electron diffraction study of the gas has shown the hydrogen to fluorine distances to be about 0.10 nm and 0.155 nm, the F–H···F distance to be 0.255 nm, and the polymer to have a linear zigzag configuration. The angle H–F–H is reported to be about 120°. Monomeric HF has an H·F distance of 0.0917 nm. Cyclical polymers (possibly H_6F_6) also probably occur. In general, polymers of differing molecular weights are present in equilibrium at a given temperature and pressure, and an average molecular weight encompasses many different actual molecules.

This high degree of association results in highly nonideal physical properties. For example, the heat effects resulting from vapor association may be significantly larger than the latent heat of vaporization.

Chemical Properties

Hydrogen fluoride, characterized by its stability, has a dissociation energy of 560 kJ (134 kcal), which places it among the most stable diatomic molecules. HF is, however, highly reactive and has a special affinity for oxygen compounds, reacting with boric acid to form boron trifluoride and with sulfur trioxide and sulfuric acid to form fluorosulfonic acid. The latter reaction demonstrates the dehydrating power of anhydrous hydrogen fluoride. HF belongs to the only class of compounds that readily react with silica and silicates With organic compounds, HF acts as a dehydrating agent, a fluorinating agent, a polymerizing agent, a catalyst for condensation reactions, and a hydrolysis catalyst. Hydrogen fluoride reacts with alcohols and unsaturated compounds to form fluorides and with alkylene oxides to give alkylene fluorohydrins.

The strong catalytic activity of anhydrous hydrogen fluoride results from its ability to donate a proton, as in its dimerization of isobutylene. Anhydrous hydrogen fluoride is an excellent solvent for ionic fluorides. The soluble fluorides act as simple bases, becoming fully ionized and increasing the concentration of HF_2^-. Because of the small size of the fluoride ion, F^- participates in coordination structures of a high rank. Tantalum and niobium form stable hexafluorotantalate and hexafluoroniobate ions and hydrogen fluoride attacks these usually acid-resistant metals. Hydrogen fluoride in water is a weak acid.

Whereas hydrogen fluoride is a fairly weak acid as a solute, it is strongly acidic as a solvent. As the concentration of hydrogen fluoride increases in aqueous mixtures, the system becomes more acidic, with water acting as a strong base. In dilute aqueous solution, an isolated hydrogen fluoride molecule donates a proton to an aggregate of water molecules and forms an aquated fluoride ion. When small amounts of water are present in the system, a proton is transferred to an isolated water molecule from polymeric hydrogen fluoride. The fluoride ion thus formed is part of a stable polymeric anionic complex. This difference in the solvation of the fluoride ion at the extremes of composition in the H_2O–HF system is probably the principal factor affecting the ease of proton transfer.

For anhydrous hydrogen fluoride, the Hammett acidity function H_0 approaches −11. The high negative value of H_0 shows anhydrous hydrogen fluoride to be in the class of superacids. Addition of antimony pentafluoride to make a 3 M solution in anhydrous hydrogen fluoride raises the Hammett function to −15.2, nearly the strongest of all acids.

MANUFACTURE

Essentially all hydrogen fluoride manufactured worldwide is made from fluorspar and sulfuric acid, according to the reaction $CaF_2(s) + H_2SO_4 \rightarrow CaSO_4(s) + 2 HF(g)$.

Generally, yields on both fluorspar and sulfuric acid are greater than 90% in commercial plants.

The key piece of equipment in a hydrogen fluoride manufacturing plant is the reaction furnace. The reaction between calcium fluoride and sulfuric acid, which is endothermic (1400 kJ/kg of HF) (334.6 kcal/kg), must for good yields be carried out at a temperature in the range of 200°C. Most industrial furnaces are horizontal rotating kilns, externally heated by, for example, circulating the combustion gas in a jacket. Other heat sources are possible; eg, supplying the sulfuric acid value as SO_3 and steam, which then react and condense, forming sulfuric acid and releasing heat.

In all HF processes, the HF leaves the furnace as a gas, contaminated with small amounts of impurities such as water, sulfuric acid, SO_2, or SiF_4. Various manufacturers utilize different gas-handling operations, which generally include scrubbing and cooling. Crude HF is condensed with refrigerant and is further purified by distillation. Plant vent gases are scrubbed with the incoming sulfuric acid stream to remove the bulk of the HF. The sulfuric acid is then fed to the furnace. Water or alkali scrubbers remove the remainder of the HF from the plant vent stream. Some manufacturers recover by-products from the process.

Because of the large quantity of phosphate rock reserves available worldwide, recovery of the fluoride values from this raw material source has frequently been studied. Strategies involve recovering the fluoride from wet-process phosphoric acid plants as fluosilicic acid, H_2SiF_6, then processing this acid to form hydrogen fluoride. Numerous processes have been proposed, but none has been commercialized on a large scale. Other technologies proceeding via intermediates such as NH_4F or KHF_2 are also possible. The future for these technologies is uncertain. As long as fluorspar supplies remain abundant, there is little justification to proceed with such processes.

ECONOMIC ASPECTS

Production costs of hydrogen fluoride are heavily dependent on raw materials, particularly fluorspar, and significant changes have occurred in this area. Fluorspar is marketed in three grades: acid, ceramic, and metallurgical. Metallurgical grade is commonly sold as lump or gravel, ceramic grade as a dried flotation filter cake or as briquettes or pellets. Acid grade, used for HF manufacture, is the purest form, having a minimum CaF_2 content of 97%. Most of the acid-grade spar used for HF production in the United States is imported, from China and Africa. Purity is expected to become a significant concern as reserves are depleted.

SPECIFICATIONS, SHIPPING, AND ANALYSIS

Hydrogen fluoride is shipped in bulk in tank cars and tank trucks. A small volume of overseas business is shipped in ISO tanks. Bulk shipments are made of anhy-

drous HF as well as 70% aqueous solutions. A small amount of aqueous solution may be shipped as 50%. Cars and trucks used for anhydrous HF transport are of carbon steel construction. It is possible to ship 70% aqueous HF in steel from a corrosion standpoint; however, rubber lining is commonly used to eliminate iron pickup, which is detrimental to product quality in a number of applications. Hydrogen fluoride of less than 60% strength must always be shipped in lined containers. An improved method for safe storage, transporation, and handling of hydrogen fluoride. It involves lowering the vapor pressure of HF using a sulfone diluent. Anhydrous hydrogen fluoride is also available in cylinders, and aqueous hydrogen fluoride, either 50% or 70%, is also shipped in polyethylene bottles and carboys.

ENVIRONMENTAL CONCERNS

Although it is widely recognized as a hazardous substance, large volumes of HF are safely manufactured, shipped, and used, and have been for many years. The hydrogen fluoride industry has undertaken a significant effort to investigate the behavior of HF releases so as better to define the risks associated with an accidental spill and to design effective mitigation systems. Water spray curtains or water monitors were found to remove between 25 and 95% of HF released in field tests.

HEALTH AND SAFETY FACTORS

Mild exposure to HF via inhalation can irritate the nose, throat, and respiratory system. The onset of symptoms may be delayed for several hours. Severe exposure via inhalation can cause nose and throat burns, lung inflammation, and pulmonary edema, and can also result in other systemic effects, including hypocalcemia (depletion of body calcium levels), which if not promptly treated can be fatal. Permissible air concentrations are OSHA PEL, 3 ppm (2.0 mg/m^3) as F; OSHA STEL, 6 ppm (5.2 mg/m^3) as F; and ACGIH TLV, 3 ppm (2.6 mg/m^3) as F. Ingestion can cause severe mouth, throat, and stomach burns, and may be fatal. Hypocalcemia is possible even if exposure consists of only small amounts or dilute solutions of HF.

Both liquid HF and its vapor can cause severe skin burns that may not be immediately painful or visible. HF can penetrate the skin and attack underlying tissues, and large (over 160 cm^2) burns may cause hypocalcemia and other systemic effects that may be fatal. Even very dilute solutions may cause burns. Both liquid and vapor can cause irritation to the eyes, corneal burns, and conjunctivitis.

Unlike other acid burns, HF burns always require specialized medical care. The fluoride ion is extremely mobile and easily penetrates deeply into the skin. Immediate first aid consists of flushing the affected area with copious quantities of water for at least 20 minutes. Subsequently, the area needs to be immersed in iced 0.13% benzalkonium chloride solutions or massaged with 2.5% calcium gluconate gel. For larger burns, subcutaneous injection of 5% calcium gluconate solution beneath the affected area may be required. Eye exposure requires flushing with water for at least 15 minutes, and subsequent treatment by an eye specialist.

Exposure to HF vapor should be treated by moving the victim to fresh air, followed by artificial respiration if required, and administration of oxygen if the victim is having difficulty breathing. A qualified physician must be called, and the victim should be held under observation for at least 24 hours. First aid for swallowed HF consists of drinking large quantities of water; milk or several ounces of milk of magnesia may be given. Vomiting should not be induced.

Hydrogen fluoride is not a carcinogen. However, HF is highly reactive, and heat or toxic fumes may be evolved. Reaction with certain metals may generate flammable and potentially explosive hydrogen gas.

Many opportunities for dermal and inhalation exposure exist in the high technology industries. Numerous reports of injuries due to HF have occurred.

USES

In the North American HF market, approximately 57% of it goes into the production of fluorocarbons, 3% to the nuclear industry, 3% to alkylation processes to produce a very high octane gasoline, 3% to steel pickling, and 20% to other markets requiring fluorinated chemicals that are relatively complet and of high value. This does not include the HF going to aluminum fluoride, the majority of which is produced captively for this purpose. Large amounts of HF are consumed in the production of aluminum fluoride, AlF_3, and cryolite (sodium aluminum fluoride), used by the aluminum industry. Both of these compounds are used in the fused alumina bath from which aluminum is produced by the electrolytic method.

Hydrofluoric Acid, Anhydrous—Technical, Properties, Uses, Storage, and Handling, E. I. du Pont de Nemours & Co., Inc., Wilmington, Del., 1984.

M. Kilpatrick and J. G. Jones, in J. J. Lagowski, ed., *The Chemistry of Nonaqueous Solvents*,Vol. 2, Academic Press, Inc., New York, 1967, pp. 43–49.

M. Kirschner, "Chemical Profiles," *Chem. Market Reporter* (Oct. 7, 2002).

M. M. Miller, "Fluorspar" *Mineral Commodity Summaries*, U. S. Geological Survey, Reston, Va., Jan. 2002.

ROBERT A. SMITH
Allied Signal Inc.

HYDROGEN-ION ACTIVITY

Hydrogen ions are involved in a wide variety of natural and industrial reactions, and the equilibrium positions as well as the rates of these reactions are therefore dependent on hydrogen-ion concentration. The hydrogen ion is more correctly termed hydronium ion. The unhydrated proton does not exist in aqueous solution but rather is bound to several molecules of water. This ion, sometimes represented as $H(H_2O)_n^+$, is usually written simply as H^+. More important is the distinction between the hydrogen ion concentration and its activity. The hydrogen ion concentration, or total acidity, is obtained by titration and corresponds to the total concentration of hydrogen ions available in a solution, ie, free, unbound hydrogen ions as well as hydrogen ions associated with weak acids. The hydrogen ion activity refers to the effective concentration of unbound hydrogen ions, ie, the form which affects physicochemical reaction rates and equilibria. The effective concentration of hydrogen ion in solution is expressed in terms of pH, which is the negative logarithm of the hydrogen-ion activity, a_{H^+}

$$pH = -\log_{10} a_{H^+} \tag{1}$$

The relationship between activity, a, and concentration, c, is

$$a = \gamma c \tag{2}$$

where the activity coefficient γ is a function of the ionic strength of the solution and approaches unity as the ionic strength decreases; ie, the difference between the activity and the concentration of hydrogen ion diminishes as the solution becomes more dilute. The pH of a solution may have little relationship to the titratable acidity of a solution that contains weak acids or buffering substances; the pH of a solution indicates only the free hydrogen-ion activity. If total acid concentration is to be determined, an acid–base titration must be performed.

PH DETERMINATION

Two methods are used to measure pH: electrometric and chemical indicator. The most common is electrometric and uses the commercial pH meter with a glass electrode. This procedure is based on the measurement of the difference between the pH of an unknown or test solution and that of a standard solution.

More recently, two different types of nonglass pH electrodes have been described which have shown excellent pH-response behavior. In the neutral-carrier, ion-selective electrode type of potentiometric sensor, synthetic organic ionophores, selective for hydrogen ions, are immobilized in polymeric membranes. Another type of pH sensor is based on an integrated ion-selective electrode and insulated-gate field-effect transistor. These sensors, usually termed ion-selective field-effect transistors (ISFETS), are based on the modulation of the transistor source-drain current by a potential (or charge) applied to the transistor gate region.

The second method for measuring pH, the indicator method, has more limited applications. The success of this procedure depends on matching the color that is produced by the addition of a suitable indicator dye to a portion of the unknown solution with the color produced by adding the same quantity of the same dye to a series of standard solutions of known pH. The indicator dyes can also be immobilized onto paper strips (eg, litmus paper) or, more recently, have been placed onto the distal end of fiber-optic probes which, when combined with spectrophotometric readout, provide more quantitative indicator-dye pH determinations.

Accuracy and Interpretation of Measured pH Values

To define the pH scale and permit the cailbration of pH measurement system, a series of reference buffer solution have been certified by U.S. National Institute of Standards and technology (NIST). The acidity finction which is experimental basis for the assignment of pH, is reproducible with about 0.003 pH unit from 10 to 40°C. However, error in the standard potential of the cell, in the compossion of the buffer materials, and in the preparation of the solutions may raise the uncertaninty to 0.005 pH unit. The accuracy of the practical scale may be further reduced to 0.008–0.01 pH unit as a result of variations in the liquid-junction potential.

Sources of Error

Several common causes of measurement problems are electrode interferences and/or fouling of the pH sensor, sample matrix effects, reference electrode instability, and improper calibration of the measurement system.

PH MEASUREMENT SYSTEMS

Glass Electrodes

The glass electrode is the hydrogen-ion sensor in most pH-measurement systems. The pH-responsive surface of the glass electrode consists of a thin membrane formed from a special glass that, after suitable conditioning, develops a surface potential that is an accurate index of the acidity of the solution in which the electrode is immersed. To permit changes in the potential of the active surface of the glass membrane to be measured, an inner reference electrode of constant potential is placed in the internal compartment of the glass membrane. The inner call commonly consists of a silver–silver chloride electrode or calomel electrode in a buffered chloride solution. Immersion electrodes are the most common common glass electrodes. Miniature and microelectrodes are alaso used widely, particularly in physiological studies. Capillary electrodes permit the use of small samples and provide protection from exposure to air during the measurements. The composition of the glass has a profound effect on the electrical resistance, the chemical durability of the pH-senstive surface, and the assuracy of the pH response in alkaline solutions.

Reference Electrodes and Liquid Junctions

The electrical circuit of the pH cell is completed through a salt bridge that usually consists of a concentrated solution of potassium chloride. The solution makes contact at one end with the test solution and at the other with a reference electrode of constant potential. The liquid junction is formed at the area of contact between the salt bridge and the test solution.

The commercially used reference electrode–salt bridge combination usually is of the immersion type. Some provision is made to allow a slow leakage of the bridge solution out of the tip of the electrode to establish the liquid junction with the test solution.

Combination electrodes have increased in use and are a consolidation of the glass and reference electrodes in a single probe, usually in a concentric arrangement, with the reference electrode compartment surrounding the pH sensor. The advantages of combination electrodes include the convenience of using a single probe and the ability to measure small volumes of sample solution or in restricted-access containers.

Theoretical considerations favor liquid junctions by which cylindrical symmetry and a steady state of ionic diffusion are achieved.

Samples that contain suspended matter are among the most difficult types from which to obtain accurate pH readings because of the so-called suspension effect, ie, the suspended particles produce abnormal liquid-junction potentials at the reference electrode. Internal consistency is achieved by pH measurement using carefully prescribed measurement protocols, as has been used in the determination of soil pH.

Another effect that may result in spurious pH readings is caused by streaming potentials. Presumably, these are attributable to changes in the reference electrode liquid junction that are caused by variations in the flow rate of the sample solution. This problem may be avoided by maintaining constant flow and geometry characteristics and calibrating the system under operating conditions that are identical to those of the sample measurement.

pH Instrumentation

The pH meter is an electronic voltmeter that provides a direct conversion of voltage differences to differences of pH at the measurement temperature.

Because of the very large resistance of the glass membrane in a conventional pH electrode, an input amplifier of high impedance (usually 10^{12}–$10^{14}\Omega$) is required to avoid errors in the pH (or mV) readings.

In addition, most devices provide operator control of settings for temperature and/or response slope, isopotential point, zero or standardization, and function (pH, mV, or monovalent–bivalent cation–anion). Microprocessors are incorporated in advanced-design meters to facilitate calibration, calculation of measurement parameters, and automatic temperature compensation.

Temperature Effects

The emf, E, of a pH cell may be written

$$E = E_g^{o'} - k\text{pH} \tag{3}$$

where k is the Nernst factor (2.303 RT)/F, and $E^{o'}{}_g$ includes the liquid-junction potential and the half-cell emf on the reference side of the glass membrane. Changes of temperature alter the scale slope because k is proportional to T. The scale position also is changed because the standard potential is temperature dependent: $E^{o'}_g$ is usually a quadratic function of the temperature.

The objective of temperature compensation in a pH meter is to nullify changes in emf from any source except changes in the true pH of the test solution. Nearly all pH meters provide automatic or manual adjustment for the change of k with T.

NONAQUEOUS SOLVENTS

The activity of the hydrogen ion is affected by the properties of the solvent in which it is measured. Scales of pH only apply to the medium, ie, the solvent or mixed solvents, eg, water–alcohol, for which the scales are developed. The comparison of the pH values of a buffer in aqueous solution to one in a nonaqueous solvent has neither direct quantitative nor thermodynamic significance. Consequently, operational pH scales must be developed for the individual solvent systems.

Other difficulties of measuring pH in nonaqueous solvents are the complications that result from dehydration of the glass pH membrane, increased sample resistance, and large liquid-junction potentials. These effects are complex and highly dependent on the type of solvent or mixture used.

INDICATOR PH MEASUREMENTS

The indicator method is especially convenient when the pH of a well-buffered colorless solution must be measured at room temperature with an accuracy no greater than 0.5 pH unit. Under optimum conditions an accuracy of 0.2 pH unit is obtainable.

Because they are weak acids or bases, the indicators may affect the pH of the sample, especially in the case of a poorly buffered solution. Variations in the ionic strength or solvent composition, or both, also can produce large uncertainties in pH measurements, presumably caused by changes in the equilibria of the indicator species. Specific chemical reactions also may occur between solutes in the sample and the indicator species to produce appreciable pH errors.

INDUSTRIAL PROCESS CONTROL

The pH meter and electrodes for process control do not differ matrially from those used for measurements in the laboratory, but the emphasis in industrial applications is on rugged construction with stand mechanical stresses and extremess in ambient conditions.

The pH meter usually is coupled to a data recording device and often to a pneumatic or electric controller. The controller governs the addition of reagent so that the pH of the process stream is maintained at the desired level.

R. G. Bates, *Determination of pH, Theory and Practice*, 2nd ed., Wiley-Interscience, New York, 1973.

G. Eisenman, ed., *Glass Electrodes for Hydrogen and Other Cations*, Marcel Dekker, New York, 1967.

H. Galster, *pH Measurement: Fundamentals, Methods, Applications, Instrumentation*, VCH, New York, 1991.

Y. C. Wu, W. F. Koch, and R. A. Durst, *Standardization of pH Measurements*, National Bureau of Standards Special Publication 260-53, U.S. Government Printing Office, Washington, D.C., 1988.

RICHARD A. DURST
Cornell University

ROGER G. BATES
University of Florida

HYDROGEN PEROXIDE

Hydrogen peroxide, H_2O_2, mol wt 34.016, is a strong oxidizing agent commercially available in aqueous solution over a wide range of concentrations. It is a weakly acidic, nearly colorless clear liquid that is miscible with water in all proportions. The atoms are covalently bound in a nonpolar $H-O-O-H$ structure having association (hydrogen bonding) somewhat less than that found in water. Today it is manufactured primarily in large, strategically located anthraquinone autoxidation processes. Its many uses include bleaching wood pulp and textiles, preparing other peroxygen compounds, and serving as a nonpolluting oxidizing agent.

PHYSICAL PROPERTIES

The most important physical properties of pure hydrogen peroxide and aqueous H_2O_2 solutions are summarized in Tables 1 and 2.

In aqueous solution the hydrogen bonds (association) between water and H_2O_2 molecules are appreciably more stable than those between molecules of the individual species. This increase in attraction forces is evidenced from many properties such as heat of mixing, vapor pressure, viscosity, and dielectric constant. Physical constants of H_2O_2 and H_2O are compared in Table 3.

Table 1. Properties of Pure Hydrogen Peroxide

Property	Value
mp, °C[a]	−0.41
bp, °C	150.2
density at 25°C, g/mL	1.4425
viscosity at 20°C, mPa·s (=cP)	1.245
surface tension at 20°C, mN/m (=dyn/cm)	80.4
specific conductance at 26°C, $(\Omega \cdot cm)^{-1}$	4×10^{-7}
heat of fusion, J/g[b]	367.52
specific heat at 25°C, J/(g·K)[b]	2.628
heat of vaporization at 25°C, kJ/g[b]	1.517
dissociation constant[c] at 20°C	1.78×10^{-12}
heat of dissociation, kJ/mol[b]	34.3
refractive index, mp[20]	1.4084

[a] Tendency to supercooling.
[b] To convert J to cal, divide by 4.184.

Table 2. Physical Properties of Commercial Hydrogen Peroxide Solutions

Parameter	10%	35%	50%	60%	70%	90%
density at 20°C, kg/m³	1.034	1.113	1.195	1.2364	1.288	1.387
freezing point, °C	−6.0	−33.0	−52.2	−55.5	−40.3	−11.9
boiling point, °C	102	108	114	119	125	141
viscosity at 20°C, mPa·s	1.01	1.11	1.17		1.24	1.26
surface tension at 20°C, N/m	0.0731	0.0746	0.0757		0.0773	0.0792
$\Delta H_{vap.}$ at 25°C, kJ/mol	2.357		2.017	1.928	1.832	1.627
total vapor pressure at 30°C, Pa		3.200	2.400	1.867	1.467	0.667
partial vapor pressure at 30°C, Pa		48	99	120	175	333
Henry's law constant at 20°C, Pa·m³/mol			1×10^{-3}			
refractive index n_D^{20}		1.3563	1.3672	1.3745	1.3827	1.3995

CHEMICAL PROPERTIES

Hydrogen peroxide is a weak acid having a pK_a of 11.75 at 20°C:

$$H_2O_2 + H_2O \rightarrow H_3O^+ + HO_2^-$$

Decomposition

The decomposition of hydrogen peroxide may be homogeneous or heterogeneous and can occur in the vapor or the condensed phase. Although there is considerable evidence that the decomposition occurs as a chain reaction involving free radicals, the products of the decomposition are water and oxygen gas. The mechanism and rate of hydrogen peroxide decomposition depend on many factors, including temperature, pH, and the presence or absence of a catalyst such as metal ions, oxides, and hydroxides.

Table 3. Physical Properties of Hydrogen Peroxide and Water

Property	H_2O_2	H_2O
mp, °C	−0.43	0
bp (101.3 kPa), °C	150,2	100
heat of melting, J/g	368	334
heat of vaporization, J/g		
at 25°C	1519	2443
at bp	1387	2258
specific heat, J g⁻¹ K⁻¹		
liquid (25°C)	2.629	4.182
gas (25°C)	1.352	1.865
relative density, g/m³		
at 0°C	1.4700	0.9998
at 20°C	1.4500	0.9980
at 25°C	1.4425	0.9971
viscosity, mPa·s		
at 0°C	1.819	1.792
at 20°C	1.249	1.002
critical temperature, °C	457	374.2
critical pressure, MPa	20.99	21.44
refractive index n_D^{20}	1.4084	1.3330

Free-Radical Formation

Hydrogen peroxide can react directly or after it has first ionized or dissociated into free radicals. Hydrogen peroxide can form free radicals by homolytic cleavage of either an O–H or the O–O bond.

Stabilization

Pure hydrogen peroxide solutions are relatively stable and can be stored for extended periods in clean passive containers. Commercial solutions, however, invariably contain or may be exposed to varying amounts of catalytic impurities and must therefore contain reagents that deactivate these impurities, either by adsorption or through formation of complexes.

Molecular Addition

Oxyacid salts, metal peroxides, nitrogen compounds, and others form crystalline peroxohydrates in the presence of hydrogen peroxide.

Substitution

A variety of peroxygen compounds can be formed through substitution reactions of hydrogen peroxide with organic reagents. Inorganic peroxygen compounds can be prepared through similar reactions with inorganic reagents.

Oxidation

Hydrogen peroxide is a strong oxidant. Hydrogen peroxide oxidizes a wide variety of inorganic compounds, ranging from iodide ions to the various color bodies of unknown structure in cellulosic fibers.

Reduction

Hydrogen peroxide reduces stronger oxidizing agents such as sodium hypochlorite, potassium permanganate, and ceric sulfate.

Antimicrobial Properties

Hydrogen peroxide exhibits good bacteriostatic characteristics at 20–40 ppm but only moderate bactericidal properties. High concentrations (0.1–3%) and/or high temperatures (up to 80°C) are required to get good kill rates for bacteria, yeasts, and viruses.

MANUFACTURE

Hydrogen peroxide, composed of equimolar amounts of hydrogen and oxygen, can be formed directly by catalytically combining these gaseous elements. It can also be formed from compounds that contain the peroxy group; from water and oxygen by thermal, photochemical, electrochemical, or similar processes; and by the uncatalyzed reaction of molecular oxygen with appropriate hydrogen-containing species. It has been manufactured commercially by processes based on the reaction of barium peroxide or sodium peroxide with an acid; the electrolysis of sulfuric acid and related compounds; the autoxidation of 2-alkylanthrachinones, isopropyl alcohol, and hydrazobenzene; and more recently by the Huron–Dow process through the cathodic reduction of oxygen in an electrolytic cell using dilute sodium hydroxide as the electrolyte. By far the majority of hydrogen peroxide produced since 1957 has been based on the autoxidation of 2-alkylanthrahydroquinones (see Fig. 1).

Purification and Concentration

The crude product from any hydrogen peroxide process can be used as such, but commercial grades are further purified, concentrated, and stabilized.

Procedures include solvent extraction followed by optional air stripping to remove residual solvent and treatment with synthetic resins, polyethylene, waxes, carbon, and aluminum and magnesium hydroxides and alumina. Active ion-exchange resins have been used to remove both metallic and acidic impurities. More recent patented methods for purifying crude hydrogen peroxide

include further contact with an aromatic gasoline in static mixers, followed by a series of coalescing steps to effect phase separation, passing the solution through a hydrocyclone packed with halogen-containing porous styrene–divinylbenzene copolymer resin and passing the solution through an anion-exchange resin pretreated with various chelating agents to remove metal ions and organic impurities then passing the solution through a zeolite.

Concentration of hydrogen peroxide prepared by the autoxidation processes can be carried out safely and conveniently by distillation at reduced pressure.

STORAGE AND TRANSPORTATION

Materials of construction suitable for transport and storage of concentrated H_2O_2 are specific stainless steels, aluminum (99.5% min.), and certain aluminum–magnesium alloys. Contamination of the H_2O_2 with even small quantities of chloride ion might cause severe pitting of aluminum. Stainless steel must be passivated before its use. Up to a 60% H_2O_2 concentration can be stored in polyethylene (PE) tanks. Other compatible materials are poly(vinylidene fluoride) (PVDF), Teflon, poly(vinyl chloride) (PVC), and glass.

Commercial H_2O_2 solution are supplied in bulk by railroads cars, trucks, and ISO containers. PE containers of various capacities, plus plastic drums and IBCs are available.

Up to a concentration of 8%, H_2O_2 solutions are not regulated by the DOT. Solutions of H_2O_2 >40% concentration are prohibited from air transport. In the U.S., shipment of H_2O_2 in polyethylene drums is permitted up to a concentration of 52% only. H_2O_2 solutions with <20% concentration are not considered dangerous. The Bureau of Explosives regulations in the United States classify all H_2O_2 solutions >20% as oxidizing and corrosive.

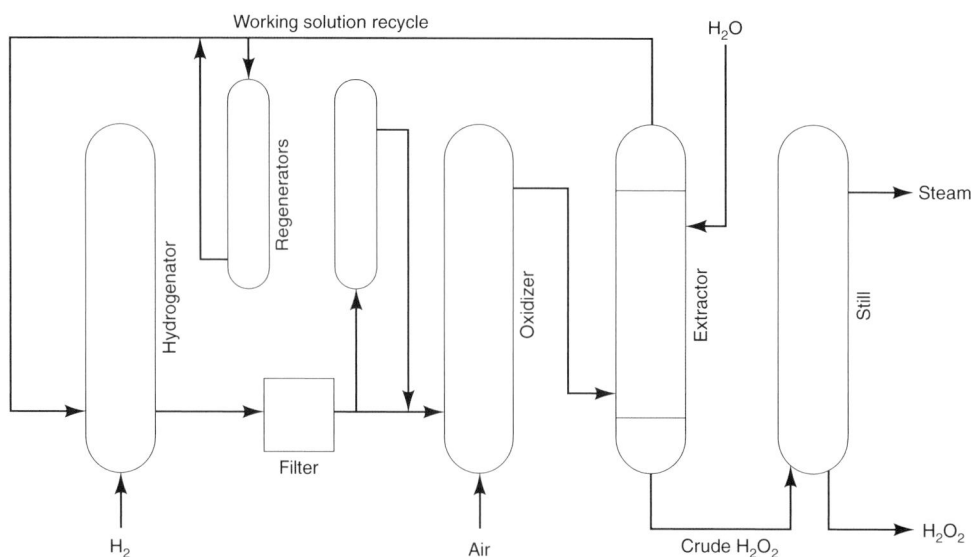

Figure 1. Anthrahydroquinone autoxidation, Riedel–Pfleiderer process.

ECONOMIC ASPECTS

The driving force behind the accelerated growth in hydrogen peroxide use since the mid-1970s has been the pulp and paper industry, with its demand for an environmentally sound replacement for chlorine and chlorine dioxide bleaching of chemical pulps, the growth of recycled fiber processing, and the establishment of bleached chemo-thermo mechanical pulp (BCTMP), which can be bleached only with hydrogen peroxide. Other growth segments include the textile industry (cotton bleaching), chemical industry (epoxidized soybean oils, organic peroxides, amine oxides, specialty chemicals and intermediates, plus downstream peroxide products), environmental applications, and specialty markets such as food processing, the cosmetic industry, liquid nonchlorine detergent bleaches, and the electronics industry.

HEALTH AND SAFETY FACTORS

Hydrogen peroxide, especially in high concentrations, is a high-energy material and a strong oxidant. The Comprehensive Environmental Response, Compensation, and Liability Act (CERCLA) reportable spill quantity for greater than 52% hydrogen peroxide is 1 lb (0.45 kg). It is considered an acute, reactive, and pressure hazard under Superfund and Reauthorization Amendments (SARA) Title III. However, it can be handled safely if proper personal protective equipment is worn and the proper precautions are observed. Some generally applicable control measures and precautions for handling hydrogen peroxide include the use of adequate ventilation to keep airborne concentrations below occupational exposure limits, 8 h TWA, 1 ppm (1.4 mg/m^3), use of coverall chemical splash goggles in combination with a full-length face shield if spraying is a potential occurrence, use of a NIOSH/MSHA-approved respirator if airborne concentration can exceed exposure limits, and use of neoprene or other impervious and compatible gloves. Short-term exposure should not exceed 3 mg/m^3 for 15 min. Other clothing items such as impervious aprons, pants, jackets, hoods, boots, and totally encapsulating chemical suits with a breathable air supply should be available for use as necessary.

Health and Physiological Effects

Hydrogen peroxide is irritating to the skin, eyes, and mucous membranes. However, low concentrations (3–6%) are used in medicinal and cosmetic applications.

Decomposition and Explosive Hazards

The principal hazards associated with hydrogen peroxide are (1) decomposition of H_2O_2, with unrelieved pressure buildup; (2) spontaneous combustion of mixtures of H_2O_2 and readily oxidizable material; (3) inadvertent admission of incompatible materials into a tank containing H_2O_2 or vice versa; (4) decomposition of H_2O_2 to form an oxygen-rich vapor phase; (5) deflagration or detonation of a condensed-phase mixture of H_2O_2 and organic compounds initiated by shock or thermal effects; and (6) explosive reaction of H_2O_2 vapor.

USES

Bleaching

The largest single use for hydrogen peroxide worldwide is wood-pulp bleaching. Environmental concerns have led the pulp-and-paper industry in Europe and North America to turn to alkaline solutions of hydrogen peroxide as a replacement for chlorine and hypochlorite and chlorine dioxide in bleaching applications.

Environmental Applications

Hydrogen peroxide is an ecologically desirable pollution-control agent because it does not form toxic by-products and yields only water or oxygen on decomposition. It has been used in increasing amounts to convert domestic and industrial effluents to an environmentally compatible state.

Chemical Uses

Hydrogen peroxide or a peroxycarboxylic acid made from H_2O_2 is used in the manufacture of a number of organic and inorganic chemicals. The electrophilic epoxidation of soybean oil, linseed oils, and related unsaturated esters with peroxyformic or peroxyacetic acid formed *in situ* from aqueous hydrogen peroxide has been used to prepare plasticizers and stabilizers and a range of oxides.

H_2O_2 is also used to produce other compounds, including magnesium silicate sols for fiber coating and forming, hydroxyimidazoles for use as antimycotics and herbicides, plus 5-hydroxy-hydantoin for use in penicillin and cephalosporin synthesis.

Other chemical uses for hydrogen peroxide include the preparation of cyanogen, cyanogen chloride, bromine, and iodic acid. H_2O_2 finds increasing use in bromination reactions as an alternative to chlorine for the regeneration of Br_2 from HBr, such as in the manufacture of the flame-retardant tetrabromobisphenol A and to avoid methylbromide by-product formation. H_2O_2 is also employed in the manufacture and regeneration of N-methylmorpholine N-oxide, which is used as a solvent for cellulose in the manufacture of cellulose fibers.

Derivative Formation

Hydrogen peroxide is an important reagent in the manufacture of organic peroxides, including *tert*-butylhydroperoxide, benzoylperoxide, di-*tert*-butylperoxy cyclohexanes, peroxyacetic acid, esters such as *tert*-butyl peroxyacetate, and ketone derivatives such as methyl ethyl ketone peroxide. These are used as polymerization catalysts, crosslinking agents, and oxidants.

Mining and Hydrometallurgy

Hydrogen peroxide, in combination with various carbonates or bicarbonates, is used as an oxidant for the in-place solution mining of low-grade uranium ores.

Propellants

The catalytic decomposition of hydrogen peroxide 70% or greater proceeds rapidly and with sufficient heat release that the products are oxygen and steam. The thrust developed from this reaction can be used to propel torpedoes and other small missiles.

Other Uses

There are numerous small specialty uses for hydrogen peroxide. such as oxidizing metal ions to a higher valence state to facilitate subsequent removal, chemical polishing metal surfaces, and other metal surface treatments. By catalytically decomposing hydrogen peroxides, the oxygen can be used as an *in situ* blowing agent for preparing certain foam rubbers and plastics. Minor amounts are also used to prepare aseptic packaging for foods, tripe bleaching, and as a direct human food ingredient.

Hydrogen peroxide is used as a topical disinfectant and an antimycotic to sterilize contact lenses. Small amounts are used in cosmetic preparations such as hair colorations and bleaching formulations. H_2O_2 is the active ingredient in color-safe household bleaches and carpet and hard-surface cleaners. H_2O_2 and other inorganic peroxo compounds are employed as a peroxide source in dentifrice products. USP topical solutions of 3% H_2O_2 are sold in North America as an over-the-counter drug for disinfection.

GENERAL REFERENCES

W. Büchner, R. Schliebs, G. Winter, and K.-H. Büchel, *Industrial Inorganic Chemistry*, VCH, Weinheim, 1989.

G. Goor, W. Kunkel, and O. Weiberg, eds., in *Ullmann's Encyclopedia of Technical Chemistry*, 5th ed., Vol. A13, Verlag Chemie GmbH, Weinheim, Germany, 1989.

K. Weissermel and H.-J. Arpe, *Industrial Organic Chemistry*, 5th ed., VCH, Weinheim, Germany, 1998.

W. S. Wood, *Hydrogen Peroxide*, Monograph No. 2., Royal Institute of Chemistry, London, 1954.

W. Eul
Degussa AG Frankfurt Germany

A. Moeller
N. Steiner
Degussa AG Hanan Germany

HYDROTHERMAL PROCESSING

Hydrothermal processing can be defined as any heterogeneous reaction in the presence of aqueous solvents or mineralizers under high pressure and temperature conditions to dissolve and recrystallize (recover) materials that are relatively insoluble under ordinary conditions.

The hydrothermal processing of advanced materials has many advantages, such as high product purity and homogeneity, crystal symmetry, metastable compounds with unique properties, narrow particle-size distribution, a lower sintering temperature, a wide range of chemical compositions, a single-step process, dense sintered powders, submicron particles with a narrow size distribution, simple equipment, lower energy requirements, fast reaction times, growth of crystals with polymorphic modifications, growth of crystals with very low solubility, and a host of others. Here the term "hydrothermal" is used to describe all the heterogeneous chemical reactions taking place in a closed system in the presence of a solvent.

In nature, hydrothermal circulation has always been assisted by bacterial activity. Hydrothermal processing as such is thus a part of solution processing and can be described as superheated aqueous solution processing. The solution processing is located in the pressure–temperature range characteristic of conditions of life on earth; the hydrothermal processing method becomes a part of this solution processing. All other processing routes are connected with increasing temperature and/or increasing (or decreasing) pressure. Therefore, they are environmentally stressed. Thus, hydrothermal processing can be considered as environmentally benign.

In the last decade, the hydrothermal technique has offered several advantages, eg, homogeneous precipitation using metal chelates under hydrothermal conditions, decomposition of hazardous and/or refractory chemical substances; monomerization of high polymers; eg, poly (ethylene terephthalate); and a host of other environmental engineering and chemical engineering issues dealing with the recycling of rubbers and plastics instead of burning. The solvation properties of supercritical solvents are being extensively used for detoxifying organic and pharmaceutical wastes and also to replace toxic solvents commonly used for chemical synthesis. Similarly, the technique is being used to remove caffeine and other food-related compounds selectively. In fact, the food and nutrition experts are now using the new term "hydrothermal cooking."

During the twenty-first century, hydrothermal technology will on the whole not be limited to crystal growth or the leaching of metals but will take a very broad shape covering several interdisciplinary branches of science. Further, the growing interest to enhance hydrothermal reaction kinetics using microwave, ultrasonic, mechanical, and electrochemical reactions will be distinct. With the ever-increasing demand for composite nanostructures, the hydrothermal technique offers a unique method for coating various compounds on metals polymers and ceramics, as well as fabrication of powders or bulk ceramic bodies.

NATURAL HYDROTHERMAL SYSTEMS

Hydrothermal research on the whole began with the study of natural systems during mid 19th century. Recently some researchers are discussing about a new concept for carrying out hydrothermal reactions under natural hydrothermal conditions, viz. geothermal reactor. The principles of geothermal reactor include the direct use of geothermal energy as a heat source or driving force for chemical reactions. It helps to produce a hydrothermal synthesis of minerals and a host of inorganic materials, to extract useful chemical elements contained in crustal materials, such as basalt, and use them as raw materials for hydrothermal synthesis. Thus, the concept of a geothermal reactor leads to the construction of a high-temperature, high-pressure autoclave underground. This has several advantages over conventional autoclave technology. Its main disadvantage is that the flow characteristics of a high-temperature slurry accompanied by a chemical reaction must be well understood to control the reaction.

The spectacular nature of the submarine hydrothermal ecosystem that is independent of sunlight as a source of reducing power has focused much interest on hydrothermal processes to explain an array of geochemical processes and phenomena. The presence of supercritical fluids such as H_2O, CO_2, or CH_4, is the main constituent of any hydrothermal system. These fluids serve as excellent solvents of organic compounds and would probably be of great potential for several of the chemical reactions, eventually leading to a discovery of the origin of life. Further, the pressure and temperature gradients existing in natural hydrothermal systems have a dramatic effect on the properties of the hydrothermal fluids.

PHYSICAL CHEMISTRY OF HYDROTHERMAL PROCESSING OF ADVANCED MATERIALS

Physical chemistry of hydrothermal processing of materials is perhaps the least known aspect in the literature. The behaviour of the solvent under hydrothermal conditions dealing with the aspects like structure at critical, supercritical and sub-critical conditions, dielectric constant, pH variation, viscosity, coefficient of expansion, density, etc are to be understood with respect to the pressure and temperature. In the recent years thermochemical modeling is greatly assisting to intelligently engineer the materials processing under hydrothermal conditions. The modeling can be successfully applied to very complex

aqueous electrolyte and non-aqueous systems over wide ranges of temperature, concentration and is widely used in both industry and academia. Such a rational approach has been used quite successfully to predict optimal synthesis conditions for controlling phase purity, particle size, morphology and size distribution of PZT, hydroxyapatite and other related systems.

APPARATUS USED IN HYDROTHERMAL PROCESSING OF MATERIALS

Materials processing under hydrothermal conditions requires a pressure vessel capable of containing a highly corrosive solvent at high temperature and pressure. Designing a suitable hydrothermal apparatus—popularly known as an autoclave, reactor, pressure vessel, or high-pressure bomb—is a difficult task. An ideal hydrothermal autoclave should have the following characteristics: (1) inertness to acids, bases, and oxidizing agents; (2) easy to assemble and disassemble; (3) of sufficient length to obtain a desired temperature gradient; (4) leakproof with unlimited capabilities to the required temperature and pressure.; and (5) rugged enough to bear high pressure and temperature experiments for a long time, so that no machining or treatment is needed after each experimental run.

Figure 1 shows the most popular autoclaves designs, such as the Morey, modified Bridgman, and Tuttle-Roy autoclaves. In most of these autoclaves, pressure can be either directly measured or be calculated or PVT relations for water. Table 1 lists commonly used autoclaves.

Safety and maintenance of the autoclaves is the prime factor in carrying out experiments under hydrothermal conditions. It is estimated that for a 100-cm^3 vessel at 20,000 psi, the stored energy is $\sim$15,000 foot-1b. The hydrothermal solutions—either acidic or alkaline—at high temperatures are hazardous to humans if the autoclave explodes. Therefore, the vessels should have rupture disks calibrated to burst above a given pressure. The most important arrangement is that provision should be made for venting live volatiles out in the event of rupture.

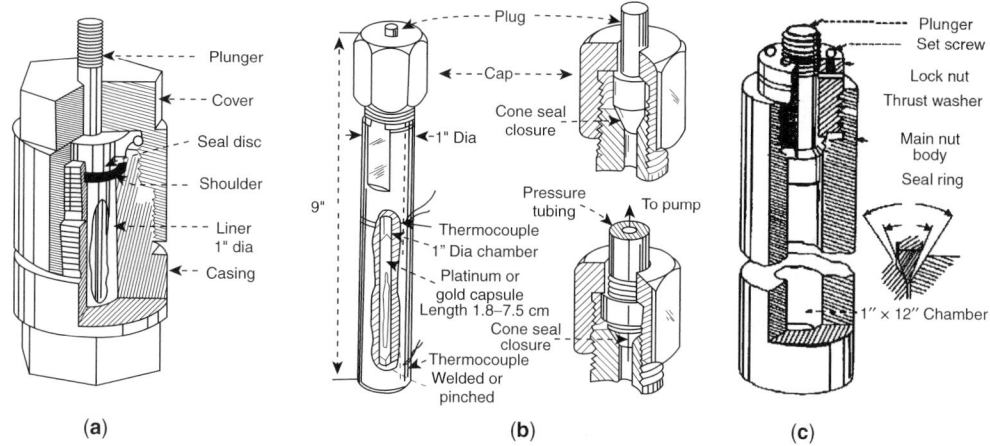

(a)   (b)   (c)

Figure 1. Commonly used autoclaves in the hydrothermal processing of materials. (**a**) Morey autoclave. (**b**) Tuttle–Roy autoclave. (**c**) Modified Bridgman autoclave.

Table 1. Commonly Used Autoclaves

Type	Characteristic data
Pyrex tube 5-mm i.d., 2-mm wall thickness	6 bar at 250°C
quartz tube 5-mm i.d., 2-mm wall thickness	6 bar at 300°C
flat plate seal, Morey type	400 bar at 400°C
welded Walker-Buehler closure 2600 bar at 350°C	2000 bar at 480°C
delta ring, unsupported area	2300 bar at 400°C
modified Bridgman, unsupported area	3700 bar at 500°C
full Bridgman, unsupported area	3700 bar at 750°C
cold-cone seal, Tuttle-Roy type	5000 bar at 750°C
piston cylinder	40 kbar, 1000°C
belt apparatus	100 kbar, > 1500°C
opposed anvil	200 kbar, > 1500°C
opposed diamond anvil	up to 500 kbar, > 2000°C

Proper shielding of the autoclave should be in place to divert corrosive volatiles away from personnel. In the case of a large autoclave, the vessels must be placed in a pit with proper shielding.

HYDROTHERMAL CRYSTAL GROWTH OF QUARTZ AND RELATED MATERIALS

The hydrothermal technique is the only one of producing the most important piezoelectric materials, such as α-quartz. Today the electronic industries are largely inclined to use synthetic quartz, because natural quartz crystals are generally irregular in shape, automatic cutting is cumbersome, and the yield is low. Over 3,000 tons of quartz are produced annually, for a variety of applications.

The type of crystal to be grown depends on the application. The more precise the need, the more stringent the requirements. For most applications, a truly high quality material is not needed. For high-precision uses, such as in navigational devices and satellites, a very high quality material must be used. Most of the recent research on quartz growth is for improved resonator performance, which requires the growth of high quality, low dislocation quartz.

In comparison with α-quartz, there are two more materials like berlinite and gallium phosphate, which are also produced hydrothermally. On the whole, the quality of the crystals grown and their growth rate depend on the solvent, solvent concentration, growth temperature, type of nutrient, seed orientation, and temperature gradient.

In recent years there has been a growing interest in the hydrothermal growth of gallium nitride crystals. The energy difference between the valence and conducting energy bands in GaN make it an attractive material for photonic and electronic devices. Gallium nitride laser diodes offer over a sixfold increase in storage capacity for the next generation of DVDs, while gallium nitride LED-based illumination sources are replacing less energy efficient lighting. Further, transistors made from gallium nitride exhibit 10 to 100 times the power capacity of transistors made of silicon or gallium arsenide.

By using the hydrothermal technique, a wide variety of technologically important crystals, such as potassium titanyl phosphate (KTP), potassium titanyl arsenate, calcite, metal oxides, phosphates, tungstates, vanadates, and several gemstones like ruby, corundum, emerald, and colored quartz have been grown both as bulk and small crystals. This method facilitates the growth of more perfect crystals with a lesser degree of defects.

HYDROTHERMAL SYNTHESIS OF ADVANCED INORGANIC MATERIALS

The hydrothermal technique has proved its efficiency in the synthesis of a great variety of inorganic materials like zeolites, complex coordinated compounds with mixed-framework structures covering a wide range of silicates, phosphates, vanadates, arsenates, molybdates, tungstates, fluorides, sulfates, selenides, borates, etc. The process of their formation under hydrothermal conditions leads to phase purity, tight control over their morphology, size, and incorporation of special physical and chemical properties, which is not possible in other conventional techniques. For example, the synthesis of zeolites based molecular sieves and catalysts are possible only under hydrothermal conditions. Generally three processes are widely used to produce zeolites and all the three processes utilize the hydrothermal conditions.

On the whole, the process of zeolite synthesis involves considerable trial and error in the design of a chemical process for a zeolite having specific properties. The synthesis is carried out under subcritical water temperatures and pressures using batch reactors. The experimental duration lasts from several hours to several days, depending on the process and product composition. However, the recent microwave hydrothermal process has significantly reduced the reaction time for zeolite formation.

HYDROTHERMAL PROCESSING OF ADVANCED CERAMICS

The term "advanced ceramics processing" usually refers to the hydrothermal technique. Most of the ceramic oxide powders are made by a ball-milling and calcination processes, resulting in the formation of less dense and uncontrolled particle distribution.

The hydrothermal technique is ideal for processing very fine powders having high purity, controlled stoichiometry, high quality, narrow particle size distribution, controlled morphology, uniformity, less defects, dense particles, high crystallinity, excellent reproducibility, controlling of microstructure, high reactivity–sinterability, etc. Further, the technique facilitates such issues as energy saving, use of a large volume of equipment, better nucleation control, is pollution free, and provides higher dispersion, high rate of reaction, better shape control, and low temperature operation in the presence of the solvent. In nanotechnology, the hydrothermal technique has an edge over the other materials processing technologies,

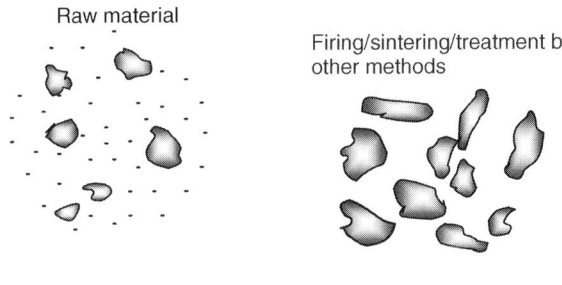

Raw material

Firing/sintering/treatment by other methods

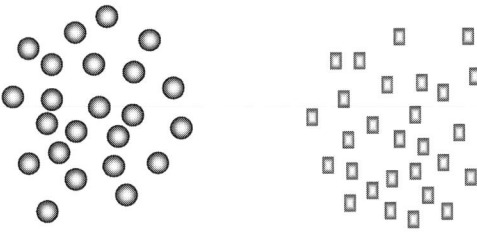

Hydrothermal processing

Figure 2. The differences in particle processing by hydrothermal and conventional ball-milling or sintering and firing techniques.

since it is ideal for processing designer particulates, particles with high purity, high crystallinity, high quality, monodispersion, with the desired physical and chemical characteristics. Today such particles are in great demand in industry. Figure 2 shows the major differences in the products obtained by ball milling or sintering or firing and hydrothermal methods.

The hydrothermal technique is also popularly used in the synthesis of the PZT family of materials and of TiO_2 and related metal oxide powders, such as ZnO, Fe_3O_4, SnO_2, α-MnO_2, etc. Among these, TiO_2 is the most extensively studied material, owing to its unique heterogeneous photocatalytic properties. It is nontoxic and chemically inert. Several organic pollutants present in industrial waste waters, phenol, nitrophenol isomers, and their derivatives, aromatic hydrocarbons, cyanide, sulfites, etc, have been completely removed by using this TiO_2-based heterogeneous photocatalysis.

Hydrothermal Processing of Advanced Bioceramics

Bioceramics represent a broad spectrum of ceramic materials designed for chemical compatibility and optimal mechanical strength with the physiological environment. Applications include replacements for hips, knees, teeth, tendons, and ligaments as well as the repair of periodontal disease, maxillofacial reconstruction, augment and stabilization of the jawbone, spinal fusion, and bone fillers after tumor surgery.

The most significant bioceramics today are of a complex material called HAp, $Ca_{10}(PO_4)_6(OH)_2$, which is the main mineral constituent of teeth and bones, representing 43% by weight. It has the physicochemical advantages of stability, inertness, and biocompatibility. HAp ceramics do not exhibit any cytoxic effects, and HAp can bond directly to bone. Unfortunately, due to its low reliability,

especially in wet environments, HAp ceramics cannot be used for heavy load-bearing applications, as in artificial teeth or bones.

Hydrothermal Processing of Composites

In recent years there has been a growing tendency to impregnate the active metal oxides onto the surface layers of activated carbon to prepare a highly effective photocatalytic support-based composite, for environmental issues. Preparation of such a composite increases the active sites for the reaction by bringing the reactant molecules close to the catalyst surface. The activated carbon has a high surface area and a well-developed porosity, which are essential for achieving large metal dispersions, giving high catalytic activity. The surface area and porosity available in the activated carbon is usually much larger than those in alumina or silica, and a large proportion of the surface area is contained within micropores. Hydrothermal technology has been efficiently used to impregnate TiO_2 or ZnO onto surface layers and the pores of activated carbon. Other interesting composite materials are the coatings of ZnO or Nd_2O_3 onto the TiO_2 particulates. When Nd_2O_3 or ZnO is coated onto the TiO_2, the efficiency of the prepared composites (Nd_2O_3:TiO_2) increases.

Related Methods of Hydrothermal Processing of Materials

The modern methods of hydrothermal processing of materials have special applications in the processing of some selected technological materials, such as hydrothermal transformation, alteration, recycling, densification, solidification, strengthening, sintering, etc. The most commonly used processing techniques are HHP, HIP, hydrothermal sintering, microwave hydrothermal, hydrothermal leaching, hydrothermal decomposition of toxic organic materials, etc.

Hydrothermal Treatment–Recycling–Alteration

This is probably one of the most important areas of research in the field of hydrothermal technology, wherein the supercritical water (SCW) properties are exploited for effective detoxification and disposal of problematic industrial, nuclear, military, and municipal wastes. In SCW at 450–700°C, many organic compounds are rapidly (0.1–100s) and efficiently (99.9–99.99 + %) oxidized to supercritical water oxidation (SCWO), with their carbon, hydrogen, and nitrogen almost completely converted to CO_2, H_2O (mineralization), and N_2. These attributes make SCW an attractive medium for chemical reactions and physical separations, ie, for hydrothermal processing. Environmental applications include rapid, efficient destruction of hazardous organic substances. Thus, supercritical water oxidation is an emerging technology for the treatment of aqueous waste streams, so that they can be recycled as process streams and for the ultimate destruction of organic wastes. Recycling in this manner of waste plastics, such as polyethylene, polystyrene, polypropylene and poly(ethylene terephthalate), radioactive waste, concrete wastes, etc., has received special attention.

A new trend is being set in hydrothermal technology. This technology is going to be highly cost effective.

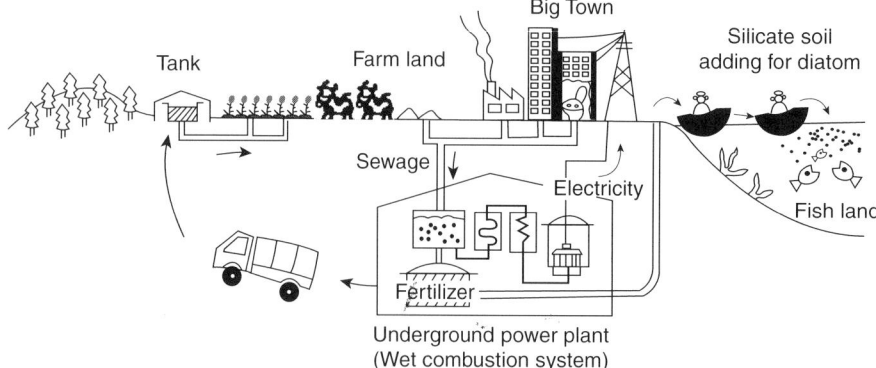

Figure 3. An imaginary new generation complex industry constructed undergrounds, which exhausts no toxic or hazardous waste.

Already, groups all over the world are searching for new avenues in materials synthesis and processing under a new concept called green materials and green processing, which deals with industrial ecology, environmentally friendly methods, recycling, etc.

Recovery, recycling, decomposition, and treatment processes under hydrothermal conditions are going to play a major role in the twenty-first century. One may now imagine a new generation complex industry constructed underground, which exhausts no toxic or hazardous waste (Fig. 3). It is an ideal closed system that deals with the recovery process of human waste to energy resource using hydrothermal technology. It decomposes the night soil without any insecticides and deodorizers, using hydrothermal technology.

K. Byrappa and M. Yoshimura, *Handbook of Hydrothermal Technology*, Noyes Publications, N. J., 2001.

K. Byrappa, *Growth of Quartz Crystals, Bulk Crystal Growth*, John Wiley & Sons, Inc., 2005, pp. 387–406.

K. Byrappa and T. Ohachi, eds., *Handbook of Crystal Growth Technology*, Springer, Germany and William Andrew, U.S., 2003.

K. Byrappa, "Hydrothermal Growth of Crystals, in D. T. J. Hurle, ed., *Handbook of Crystal Growth*, Vol. 2, Elsevier Science Publishers, UK, 2004, Chapt. 9, pp. 441–539.

K. BYRAPPA
University of Mysore

HYDROXYCARBOXYLIC ACIDS

LACTIC ACID

Lactic acid (2-hydroxypropanoic acid), $CH_3CHOHCOOH$, is the most widely occurring hydroxycarboxylic acid. Lactic acid is a naturally occurring organic acid that can be produced by fermentation or chemical synthesis. It is present in many foods both naturally or as a product of *in situ* microbial fermentation, as in sauerkraut, yogurt, buttermilk, sourdough breads, and many other fermented foods. Lactic acid is also a principal metabolic intermediate in most living organisms, from anaerobic prokaryotes to humans.

Although lactic acid is ubiquitous in nature and has been produced as a fermentation by-product in many industries (eg, corn steep liquor, a principal by-product of the multi-million-ton per year corn wet-milling industry, contains ~25 wt% lactic acid), it has not been a large-volume chemical. Lactic acid has been considered a relatively mature fine chemical in that only its use in new applications; e.g., as a monomer in plastics or as an intermediate in the synthesis of high volume oxygenated chemicals, would cause a significant increase in its anticipated demand. In the last decade lactic acid production has grown considerably, mainly due to the development of new uses. The production technology is now primarily based on carbohydrate fermentation.

Physical Properties

Pure, anhydrous lactic acid is a white, crystalline solid with a low melting point. However, it is difficult to prepare the pure anhydrous form of lactic acid; generally, it is available as a dilute or concentrated aqueous solution. A few important physical and thermodynamic properties of lactic acid are summarized in Table 1.

Lactic acid is also the simplest hydroxy acid that is optically active. L (+)-Lactic acid (**1**) occurs naturally in blood and in many fermentation products. The chemically produced lactic acid is a racemic mixture. Some fermentations also produce the racemic mixture or an enantiomeric excess of D (−)-lactic acid (**2**).

$$\begin{array}{cc} \text{COOH} & \text{COOH} \\ | & | \\ \text{HO}-\text{C}-\text{H} & \text{H}-\text{C}-\text{OH} \\ | & | \\ \text{CH}_3 & \text{CH}_3 \\ (\mathbf{1}) & (\mathbf{2}) \end{array}$$

Many of the physical properties are not affected by the optical composition, with the important exception of the melting point of the crystalline acid, which is estimated to be 52.7–52.8°C for either optically pure isomer, whereas the reported melting point of the racemic mixture ranges from 17 to 33°C.

Table 1. Physical and Thermodynamic Properties of Lactic Acid

Property	Value
density, g/mL at 20°C	1.2243
viscosity,[a] mPa(=cP)	36.9
dissociation constant, pK_a at 25°C	3.862
heat capacity, J/(gK)[b]	
crystalline	1.41
liquid, 25°C	2.34
heat of dissociation at 25°C, J/mol[b]	−263
free energy of dissociation, kJ/mol[b]	20.9
heat of solution, L(+) at 25°C, kJ/mol[b]	7.79
heat of fusion, kJ/mol[b]	
racemic	11.33
L(+)	16.86
heat of combustion, MJ/mol[b]	
racemic	−1.355
L(+)	−1.343
heat of formation, MJ/mol[b]	
crystalline L(+)	−0.693
dilute solution	−0.686
free energy of formation, MJ/mol[b]	
crystalline L(+)	−0.522
liquid racemic	−0.529

[a] 88.6 wt% solution at 25°C.
[b] To convert J to cal, divide by 4.184.

Chemical Properties

Its two functional groups permit a wide variety of chemical reactions for lactic acid. The primary classes of these reactions are oxidation, reduction, condensation, and substitution at the alcohol group.

Manufacturing and Processing

Lactic acid can be manufactured either by chemical synthesis or by carbohydrate fermentation; both are used for commercial production. In the last decade, lactic acid production has grown considerably, mainly due to the development of new uses, and the production technology is now primarily based on carbohydrate fermentation (see Fig. 1).

Economic Aspects

By 2003, Sterling had a exited the business and two new manufacturers using carbohydate fermentation technology, Archer Daniels Midland (ADM) and Cargill Dow, had entered it. ADMs focus has been on lactic acid and its derivatives for conventional and other uses, whereas Cargill Dow has been the primary leader in the lactic-based polymer business. In the future, the incentive for economical production of lactic acid will continue to come from the development of new, large-volume uses of lactic acid, particularly as feedstocks for biodegradable polymers, "green" solvents, and oxygenated chemicals.

Specifications, Quality Control, and Analytical Methods

Lactic acid is primarily sold as a *heat-stable fermentation product*: a highly refined, heat-stable product from esterification of fermentation-derived lactic acid, followed by hydrolysis of the recovered ester to produce the acid. Other categories include food grade fermentation, handled with food-grade standards; and synthetic, a highly purified product from a chemical synthesis process. The latter is water white, has excellent heat stability, and can be used in both food and industrial applications. The technical is a crude product from either a synthetic or fermentation process, used in industrial applications where high purity is not required.

Lactic acid is generally recognized as safe (GRAS) for multipurpose food use. Lactate salts such as calcium and sodium lactates and esters such as ethyl lactate used in pharmaceutical preparations are also considered safe and nontoxic.

Uses

Traditionally, the principal use of lactic acid has been in food and food-related applications, which in the United States accounted for ~85% of the demand. Currently, with the development and commercialization of the biopolymers, lactic acid use has increased considerably, with 20–30% of the 2003 production, estimated has been in these new applications.

HYDROXYACETIC ACID

Hydroxyacetic acid (glycolic acid), $HOCH_2COOH$, is the first and simplest member of the family of hydroxycarboxylic acids. It occurs naturally as the chief acidic constituent of sugar-cane juice and also occurs in sugar beets and unripe grape juice. It is widely used as a cleaning agent for a variety of industrial applications, and also as a specialty chemical and biodegradable copolymer feedstock.

Glycolic acid is a colorless, translucent solid; mp = 10°C; bp = 112°C; d at 25°C = 1.26 g/mL; K_a at 25°C = 1.5×10^{-4}; pH at 25°C = 0.5; heat of combustion = 697.1 kJ/mol (166.6 kcal/mol); heat of solution = −11.55 kJ/mol; and flash point >300° C.

Glycolic acid is soluble in water, methanol, ethanol, acetone, acetic acid, and ethyl acetate. It is slightly soluble in ethyl ether and sparingly soluble in hydrocarbon solvents.

Because it contains both a carboxyl and a primary hydroxyl group, glycolic acid can react as an acid or an alcohol or both. Thus, some of the important reactions it can undergo are esterification, amidation, salt formation, and complexation with metal ions, which lead to many of its uses. As a fairly strong acid it can liberate gases (often toxic) when it reacts with the corresponding salts.

Hydroxyacetic acid is produced commercially in the United States as an intermediate by the reaction of formaldehyde with carbon monoxide and water.

OTHER HYDROXY ACIDS

Apart from lactic and hydroxyacetic acids, other α- and β-hydroxy acids have been small-volume specialty products produced in a variety of methods for specialized uses.

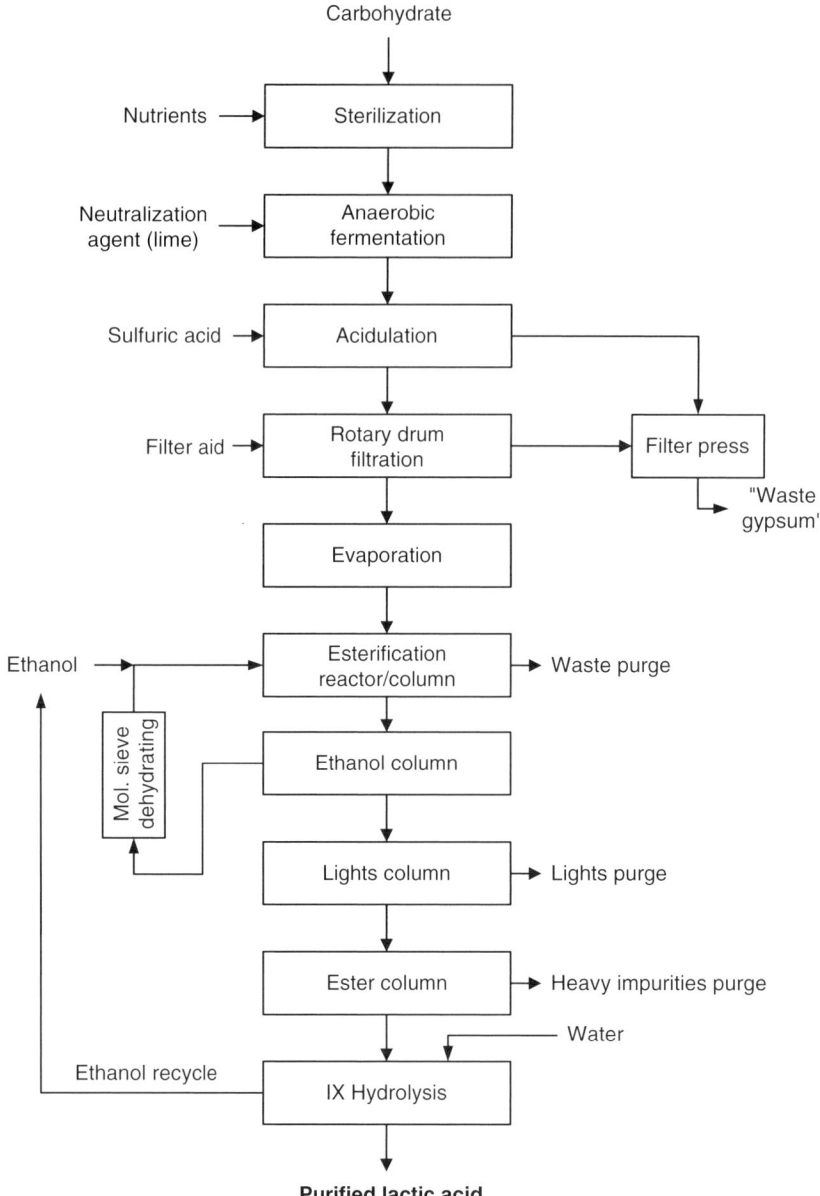

Figure 1. Conventional process for lactic acid production from carbohydrate fermentation.

The general preparation of α-hydroxy acids is by the hydrolysis of an α-halo acid or by the acid hydrolysis of the cyanohydrins of an aldehyde or a ketone. β-hydroxy acids may be made by catalytic reduction of β-keto esters followed by hydrolysis. β-Hydroxy acids can also be prepared by the Reformatsky reaction. γ-Hydroxy acids are seldom obtained in the free state because of the ease with which they form monomeric inner esters, which form stable five-membered rings. Thus, the lactones of these acids are the common chemical forms. Among these lactones, γ-butyrolactone is one of the larger-volume specialty chemicals derived from dehydrogenation of 1,4-butanediol.

The common reactions that α-hydroxy acids undergo, such as self- or bimolecular esterification to oligomers or cyclic esters, hydrogenation, oxidation, etc, are the same as those in connection with lactic and hydroxyacetic acid. A reaction that is of value for the synthesis of higher aldehydes is decarbonylation under boiling sulfuric acid with loss of water.

β-Hydroxy acids lose water, especially in the presence of an acid catalyst, to give α,β-unsaturated acids, and frequently β,γ-unsaturated acids. γ-hydroxybutyric acid and its derivatives, particularly its sodium salt, have been studied and used as anaesthetics, tranquilizers, sedatives, and hypnotics in surgery and general obstetrics.

Certain bacterial species produce polymers of γ-hydroxybutyric acid and other hydroxyalkanoic acids as storage polymers. These are biodegradable polymers with some desirable properties for the manufacture of biodegradable packaging materials.

γ-Butyrolactone undergoes amination reactions with methylamine or ammonia to produce *N*-methyl-2-pyrrolidinone (NMP) or 2-pyrrolidinone (PDO), respectively, both of which are commercially important derivatives.

Other multifunctional hydroxycarboxylic acids are the mevalonic and aldonic acids, which can be prepared for specialized uses as aldol reaction products (mevalonic acid) and mild oxidation of aldoses (aldonic acids).

C. H. Holten, A. Muller, and D. Rehbinder, *Lactic Acid*, International Research Association, Verlag Chemie, Copenhagen, Denmark, 1971.

Lactic Acid and Lactates, product bulletin, Purac Inc., Arlington Heights, Ill., 1989.

E. S. Lipinsky and R. G. Sinclair, *Chem. Eng. Prog.* **82**(8), 26 (1986).

S. C. Prescott and C. G. Dunn, *Industrial Microbiology*, 3rd ed., McGraw-Hill Book Co., Inc., New York, 1959.

RATHIN DATTA
Consultant

I

IMMUNOASSAY

Immunoassay is a method that identifies and quantifies unknown analytes using antibody–antigen reactions. Techniques are based in immunochemistry, analytical chemistry, and biochemistry, with a history of development paralleling advances in microbiology and immunology.

The success of immunoassay in clinical diagnostics, and the generic nature of immunoassay technology, has resulted in the application of the method in other areas. By the late 1980s commercial immunoassay products and systems were available for detection and diagnosis in environmental, food, and chemical processing applications. In the 1990s immunoassays were adapted for the high throughput screening used in new drug discovery. Whereas the application of immunoassays in these areas is small in relation to clinical diagnostic immunoassays, nonclinical applications of immunoassays have become commonplace.

THE ANTIBODY–ANTIGEN REACTION

Immunoassays are based on the binding and complexing of an antigen to an antibody, and the use of some physical or chemical means to measure and quantify the antigen–antibody complex. The antibody–antigen reaction is a typical reversible bimolecular reaction having rate constants for the forward and backward reactions that are dependent on the concentration of the antigen (Ag) and antibody (Ab), affinity for the antigen as defined by the association constant of the antibody for its antigen, temperature, pH, and other environmental conditions. This reaction is represented by an equation common to reversible receptor–ligand assays:

$$Ab + Ag \rightleftarrows AbAg \qquad (1)$$

The equilibrium constant for the reaction is determined by the mass action equation:

$$K = \frac{[AbAg]}{[Ab][Ag]} \qquad (2)$$

where [Ab] is the concentration of antibody sites for antigen and [Ag] is the concentration of free antigen. The association, or affinity, constant for the antibody–antigen reaction is then further defined as the equilibrium constant at half-saturation of the antibody with antigen. Because at half-saturation AbAg and Ab are equal, these cancel in the above equation and the association constant, K_a, is equal to the reciprocal of the free antigen concentration:

$$K_a = \frac{1}{[Ag]} \qquad (3)$$

Thus, if the antibody has a high affinity for the antigen, it has a high association constant. Typical association constants range from 10^6 to 10^{10} L/mol and as high as 10^{13} L/mol for some monoclonal antibodies.

The definition of an association constant for an antibody–antigen reaction can become more complex if the antibody–antigen reaction involves a multivalent antigen, as is the case when a polyclonal antiserum is used for detection of an antigen. This type of multivalent binding, termed "avidity", is defined by the equation

$$x\,Ab + y\,Ag \longrightarrow Ab_xAg_y \qquad (4)$$

A definition of the association (or avidity) constant for such multivalent antibody–antigen reactions must consider not only the heterogeneity of the antibodies and the antigen determinant site(s) but also an apparent additive effect of binding two antigen molecules to a single antibody. Such effects lead to a multiplying of the individual association constants and an apparent large increase in the total association–avidity constant. This multiplication of avidity through multivalent binding has been exploited to increase the sensitivity of many immunoassays.

BASIC TECHNOLOGY

The principal approach to immunoassay is illustrated in Figure 1, which shows a basic sandwich immunoassay. In this type of assay, an antibody to the analyte to be measured is immobilized onto a solid surface, such as a bead or a plastic (microtiter) plate. The test sample suspected of containing the analyte is mixed with the antibody beads or placed in the plastic plate, resulting in the formation of the antibody–analyte complex. A second antibody, which carries an indicator reagent, is then added to the mixture. This indicator may be a radioisotope for RIA; an enzyme, for EIA; or a fluorophore, for fluorescence immunoassay (FIA). The antibody indicator binds to the first antibody–analyte complex, the free second antibody indicator is washed away, and the two-antibody–analyte complex is quantified using a method compatible with the indicator reagent, such as quantifying radioactivity or enzyme-mediated color formation.

In fact, most RIAs (and many nonisotopic immunoassays) use a competitive binding format. In this approach, the analyte in the sample to be measured competes with a known amount of added analyte that has been labeled with an indicator that binds to the immobilized antibody. After reaction, the free analyte–analyte-indicator solution is washed away from the solid phase. The analyte-indicator on the solid phase or remaining in the wash solution is then used to quantify the amount of analyte present in the sample, as measured against a control assay, using only an analyte indicator. There are many

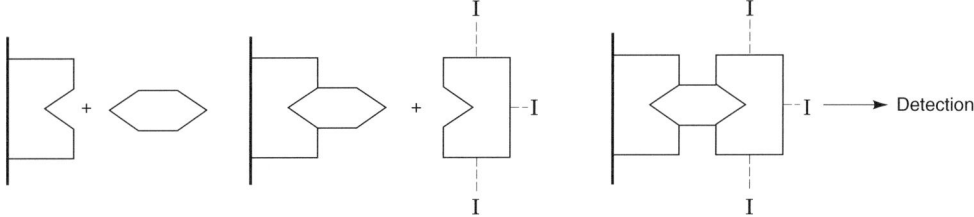

Figure 1. A principal approach to immunoassay, the sandwich immunoassay, where the thick line represents the solid matrix, (⋈) the antibody, (◇) the antigen, and I an indicator molecule such as an enzyme, fluorophore, or radioisotope.

variations on these two basic approaches for immunoassays.

Immunoassay Design

The basic reagent and design requirements of an immunoassay are an antibody, antigen, conjugates of either or both, and a means for separating bound and unbound reagents.

There are many possible means for quantification of the antigen–antibody reaction. Immunoassays may be classified according to the technology used for detection and quantification of the analyte being detected, eg, turbidimetric agglutination immunoassays, radioimmunoassay, enzyme immunoassay, fluorescence immunoassay, and chemiluminescent immunoassay.

Comparison of Methodologies

A heterogenous immunoassay is a multistep assay requiring the sequential addition of reagents with washing steps between reagent additions. Most immunoassay kits and many commercial immunoassay analyzers are based on heterogenous EIA or FIA.

During the 1980s, a number of homogenous immunoassays were developed and commercialized. Homogeneous immunoassays occur in one vessel, requiring no separation of components prior to quantification. The advantages of homogeneous immunoassays are simple formats and rapid data output, producing user-friendly, cost-effective products. The technical challenges to consider are the necessity to remove or minimize background interference from the reagents and nonspecific binding reactions.

Monoclonal vs Polyclonal Antibodies. A continuing question facing the developer of an immunoassay is whether to use monoclonal (MAb) or polyclonal (PAb) antibodies in the assay. Polyclonal antibodies are the natural mixture of antibodies resulting from the immune response to an antigen. A family of antibodies results, each binding specifically to a different antigenic determinant (or part of a determinant) on the same antigen.

Whereas such diversity in the immune response may have evolved as a protective means to the host animal, PAbs present problems to the immunoassay developer looking for high antibody specificity and low total protein.

In 1975, the first successful production of MAbs was reported. By fusing normal antibody-producing cells with a B-cell tumor (myeloma), hybridoma cell lines resulted that produced antibodies having a specificity to only one determinant on an antigen; ie, all the antibodies produced from the cell line were identical.

The singularity of MAbs and the ease of mass production appeared to be the answer to rapid development of highly specific immunoassays. Whereas MAbs appear to be the choice for use in immunoassays, a majority of immunoassay developers and suppliers use polyclonal antibodies. The primary reason for this choice lies in the investment of time and costs required to fuse, clone, and screen thousands of hybridomas to discover those producing MAbs having the high avidity required for an assay. In addition, MAb-producing hybridoma cells can be extremely unstable, losing antibody production capabilities or simply dying out in a few passages (generations).

The question of whether to use MAbs or PAbs in an assay is a matter both of assay requirements (specificity and sensitivity) and economics and cannot be answered on technical merit alone.

IMMUNOASSAY–DNA PROBE HYBRID ASSAYS

Nucleic acid [deoxyribonucleic acid (DNA) and ribonucleic acid (RNA)] probes utilize labeled; ie, radioactive, enzymatic, or fluorescent, fragments of DNA or RNA (the probe) to detect complimentary DNA or RNA sequences in a sample. Because the probe is tailored for one specific nucleic acid, these assays are highly specific and very sensitive.

As the result of high specificity and sensitivity, nucleic acid probes are in direct competition with immunoassay for the analytes of some types of clinical analytes, such as infectious disease testing. Assays are being developed, however, that combine both probe and immunoassay technology. In such hybrid probe–immunoassays, the immunoassay portion detects and amplifies the specific binding of the probe to a nucleic acid. Either the probe per se or probe labeled with a specific compound is detected by the antibody, which in turn is labeled with an enzyme or fluorophore that serves as the basis for detection.

Assays using the technology you have now been developed for a number of bacteria, viruses, and human chromosome segments.

IMMUNO(BIO)SENSORS

Immunoassay technology is also being applied in the development of antibody-based biosensors, or immunosensors. A biosensor is an electronic detection device containing a biological molecule such as an antibody, enzyme, or receptor as its basic detection element. The ideal biosensor employs a homogeneous (one-step, no-prep) format, real-time detection (results in <1 min) in a cost-effective, portable, user-friendly design. Immunosensors have been designed that use both direct and indirect immunoassay technology to detect specific analytes within a minute or less in a variety of matrices. Indirect immunosensors may employ EIA, FIA, or CLIA principles whereby an enzyme-, fluorophore- or chemiluminescent-labeled analyte competes with the target (nonlabeled) analyte for binding sites on the immobilized antibody.

Measurements may be based on the perturbation of an electrical field by the antibody–antigen binding event; changes in light scattering fluorescence or chemiluminescence on an optical fiber; or changes in the weight of an antibody–antigen complex as compared to the weight of the antibody alone on a piezoelectric crystal.

Immunosensors have become principal players in chemical, diagnostic, and environmental analyses. They are being used as rapid-screening devices in noncentralized clinical laboratories, in intensive care facilities, as bedside monitors, in physicians' offices, and in environmental and industrial settings. Industrial applications for immunosensors include then use as the basis for automated online or flow-injection analysis systems to analyze and control pharmaceutical, food, and chemical processing lines and as drug-screening assays in new drug discovery. Immunosensors, are not expected to replace laboratory-based immunoassays but to open up new applications for immunoassay-based technology.

G. E. Abraham, ed., *Handbook of Radioimmunoassay*, Marcel Dekker, Inc., New York, 1977.

J. Clausen, *Immunochemical Techniques for the Identification and Estimation of Macromolecules*, 3rd ed., Elsevier, Amsterdam, The Netherlands, 1988.

R. F. Taylor, *Worldwide Immunoassay Markets, Technologies and Applications, 2000–2005*, TC Associates Inc., Boxford Mass., 2000.

R. F. Taylor and J. S. Schultz, eds., *Handbook of Chemical and Biological Sensors*, IOP Publishing Ltd, Bristol, 1996.

R. F. TAYLOR
Tc Associates Inc.

INCLUSION COMPOUNDS

Notwithstanding the immense number and great variety of inclusion compounds, all of them may be classified into three main categories being either a complex, a cavitate, or a clathrate according to the criteria given in Figure 1. Typical examples for each class of inclusion compounds are the crown complexes, the calix–cavitates, and the hydroquinone clathrates but in many of the recently known inclusion situations there are borderline cases treated as complex=clathrate hybrids (coordinatoclathrates or clathratocomplexes depending on the dominant inclusion

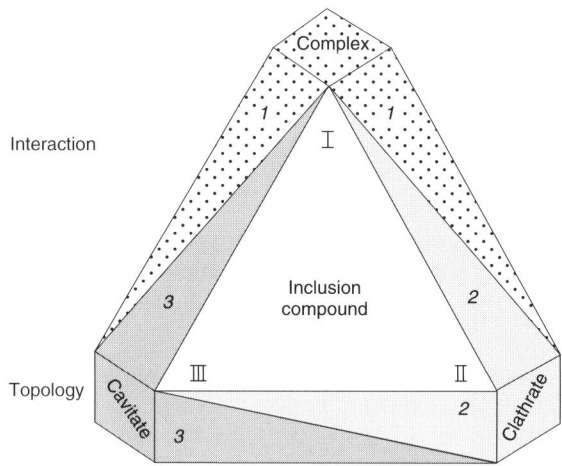

Figure 1. Classification/nomenclature of host–guest type inclusion compounds, definitions and relations: (*1*) coordinative interaction, (*2*) lattice barrier interaction, (*3*) monomolecular shielding interaction; (I) coordination-type inclusion compound (inclusion complex), (II) lattice-type inclusion compound (multimolecular/extramolecular inclusion compound, clathrate), (III) cavitate-type inclusion compound (monomolecular/intramolecular inclusion compound).

character). By way of contrast, the description addition compound (adduct) may be used to the best advantage if a cavity does not exist either at the host molecule or in the lattice build-up. Inclusion compound, therefore, is the generic term of choice which refers to the presence of any not precisely defined cavity. In a more detailed topological characterization, there are two-dimensional open intercalates (layer- or sandwich-type inclusions), one-dimensional open channel inclusions (tubulates), and totally enclosed cage inclusions (cryptates).

INTRAMOLECULAR CAVITY INCLUSIONS: CAVITATES

Cavitates include crown macroning inclusion compounds (coronates),cryptates, podates, cyclophane host inclusion compounds, calixarene inclusion compounds, cyclodextrin and amylose inclusion compounds cucurbituril inclusion compounds, moleculer cleft inclusion compounds, and anionic quest inclusion compounds.

Extramolecular Cavity Inclusions: Lattice-Type Inclusion Compounds, Clathrates

These compounds include Holfmann- and Werner-type inclusion compounds, inclusion compounds of urea, thiourea and selenourea, inclusion compounds of gossypol, inclusion compounds of phenolic hosts inclusion compounds of deoxycholic acid (choleic acids), inclusion compounds of macrocyclic and oligocyclic lattice hosts and recently disigned organic hast lattice.

PREPARATION AND CHARACTERIZATION OF INCLUSION COMPOUNDS

There are several ways to prepare inclusion compounds. In solution, they may simply be formed by dissolving together host and guest in a common solvent. Inclusion formation in solution applies only for intramolecular cavity inclusions and complexes. Crystalline inclusion compounds may be prepared by crystallization from the guest solvent or by cocrystallization of host and guest from an inert solvent. Solid inclusion compounds are also formed by direct exposure of the host to the vapor or liquid guest or, sometimes, by grinding solid host and guest together. Moreover, replacement of an included guest has been demonstrated in particular cases.

Appropriate guest molecules are those that have a suitable size and shape to accommodate the host cavity and that complement the host cavity chemically.

Stabilities of inclusion compounds span a wide range. Some are very stable at ambient conditions and require heating to considerable temperatures or treatment under high vacuum to cause decomposition. Others are only stable when in contact with mother liquor or excess guest solvent from which the inclusion compound was grown. A simple yet informative way for estimation of inclusion stabilities is to relate the decomposition point of the inclusion compound to the usual boiling point of the respective guest liquid.

USES

Inclusion compounds open up a wide area of applications. An important aspect in this connection is the specific microenvironment created by the host enclosure of the guest which exerts an influence on the physical, spectroscopic, chemical, and other properties of the guest.

Retardation and Control

This influence may manifest itself in a reduced volatility and therefore lower possible storage and handling problems of a compound when included; toxic and hazardous substances become safer.

Shielding and Stabilization

Inclusion compounds may be used as sources and reservoirs of unstable species. The inner phases of inclusion compounds uniquely constrain guest movements, provide a medium for reactions, and shelter molecules that self-destruct in the bulk phase or transform and react under atmospheric conditions.

Solubilization and Activation

Compounds included in a host take solubility properties of the host shell and thus become more soluble when trapped in polar or apolar media, depending on the nature of the host. This leads to important uses in chemical synthesis known as the phase-transfer principle.

Organized Media Effects

Another general reason for using host–guest inclusion chemistry in synthesis is controlled selectivity and artificial enzyme mimicry.

Sensing

Crown compounds modified by responsible chromogenic groups (chromoionophores) proved valuable tools for measuring metal ions and even enantiomeric guest concentrations in solution. Ion selective electrodes based on crown compounds and podands as the sensitive component have broad analytical applications from industrial wastewater control to clinical bedside monitoring of blood.

J. L. Atwood, J. E. D. Davies, and D. D. MacNicol, eds., *Inclusion Compounds*, Vols. 1–3, Academic Press, Inc., London, 1984; Vols. 4–5, Oxford University Press, Oxford, U.K., 1991.

J.-M. Lehn, *Supramolecular Inclusion Chemistry—Concepts and Perspectives*, Weinheim, Germany, 1995.

F. Vögtle, *Supramolecular Chemistry—An Introduction*, John Wiley & Sons, Ltd., Chichester, U.K., 1991.

E. Weber, ed., *Molecular Inclusion and Molecular Recognition — Clathrates I and II*, Springer, Berlin-Heidelberg, 1987 and 1988.

EDWIN WEBER
Technische Universität
Bergakademic Freiberg

INDIUM AND INDIUM COMPOUNDS

Indium is a very soft, silvery metal. The element is found in the periodic table of the elements under Group IIIA. The atomic number is 49, the atomic weight 114.82. Indium highly malleable and ductile, having a face-centered tetragonal crystalline structure.

The abundance of indium in the earth's crust is about 0.1 ppm, similar to that of silver, except that indium is very widely distributed and never in concentrations high enough to justify mining it.

The principal commercial source of indium is sphalerite (ZnS) but is found in lesser quantities in association with other sulfides such as galena (PbS), stannite, (Cu_2FeSnS), and cassiterite (SnO_2). Indium follows zinc through flotation concentration. Commercial recovery of the metal is achieved by treating residues, flue dusts, slags, and metallic intermediates in zinc smelting and associated lead and copper smelting.

PROPERTIES

Table 1 lists many of the atomic, electrical, nuclear, physical, and thermal properties of indium. The highly plastic nature of indium, probably its most noted feature, results from mechanical twinning. Indium retains this plasticity at cryogenic temperatures. Indium does not work harden, can endure considerable deformation through compression, cold welds easily, and has a distinctive cry on bending, as do tin and cadmium.

Indium metal is not oxidized by air at ordinary temperatures but at red heat burns to form the trioxide In_2O_3. On

Table 1. Physical Properties of Indium

Property	Value
atomic weight	114.82
atomic number	49
melting point, °C	156.61
boiling point, °C	2080
latent heat of fusion, kJ/kg[a]	28.47
latent heat of vaporization at bp, kJ/kg[a]	1959.42
specific heat at 25°C, kJ/(kg · K)[a]	0.233
coefficient of linear expansion, 0–100°C, $\times 10^6$°C^{-1}	24.8
electrial resistivity, $\Omega \cdot$ m	
at 3.38 K	superconducting
20°C	84
154°C	291
181°C	301
222°C	319
280°C	348
electrode oxidation—reduction potential,[b] V	0.38
density, kg/m^3	
at 20°C	7.300
164°C	7.026
volume increase on melting, %	2.5
thermal conductivity at 0°C, W/(m · K)	83.7
surface tension at temperature, T in K between mp and bp, mN/m (= dyn/cm)	$602 - 0.1\,T$
relative isotopic abundance, wt %	
113 In	4.23
115 In	95.77
vapor pressure, p, at temperature, T in K between mp and bp, kPa[c]	$\log_{10} p = 9.835 - 12{,}860/T$ $-0.7 \log_{10} T$
thermal neutron cross section at 2200 m/s, $\times 10^{-28}$ m^2/atom[d]	
absorption	190 ± 10
scattering	22 ± 0.5
Brinell hardness number, HB	0.9
tensile strength, MPa[c]	
at 295 K	1.6
76 K	15.0
4 K	31.9
elongation, %	22
modulus of elasticity, GPa[c]	12.74
Poisson's ratio at 20°C	0.4498

[a] To convert J to cal, divide by 4.184.
[b] Versus the hydrogen electrode at 0.0 V.
[c] To convert kPa to atm, divide by 101.3; to convert MPa to atm, divide by 0.101; to convert GPa to atm, divide by 1.01×10^{-4}.
[d] To convert m^2 to barn, multiply by 10^{28}.

heating, indium also reacts directly with the metalloids—arsenic, antimony, selenium, and tellurium—and with halogens, sulfur, and phosphorus. Indium dissolves in mineral acids and amalgamates with mercury but is minimally affected by alkalis, boiling water, and most organic acids. The metal surface passivates easily. Indium usually exhibits a valence of +3 but may also have valences of +2 and +1. In general, the chemistry of indium in its trivalent compounds stems from its nonionic or covalent bonding characteristics. Indium is electroplated easily from a variety of baths, including cyanide, sulfate, fluoborate, and sulfamate.

SOURCES AND SUPPLIES

Indium, a minor metal, is a by-product of other mining and refining operations, typically zinc or lead. Indium is recovered from fumes, dusts, slags, residues, and alloys from zinc or lead–zinc smelting.

The most important mines producing indium are found in South America, Canada, the former Soviet Union, and China. Canada has the potential to eventually become the largest source of indium in the world.

USES

In the 130+ years since its discovery, indium has found numerous uses in a wide variety of applications, specifically, metallurgical, chemical, and as a fusible alloy.

The characteristics and properties of indium make it useful for a wide variety of metallurgical applications. Pure indium metal has a number of uses by itself. When used as a metallurgical additive, indium provides certain benefits to the other metals with which it is alloyed.

Indium's wetting ability makes it useful as a sealing material for glass-to-glass, glass-to-metal, ceramic-to-metal, and other seals. Indium foil, strip, and wire as well as indium–tin alloys are usually used for such sealing applications.

Indium's sealing ability, along with its low vapor pressure, make it ideal for use in vacuum applications. Because it retains its ductility at very low temperatures, indium is ideal for use in cryogenic applications for making seals. It can be conformed to any space to be sealed. As the temperature drops, the tensile strength increases, improving the performance of the seal.

Indium's wetting ability makes it useful for bonding in the manufacture of sputtering targets. The fact that it is soft and melts at a low temperature makes this application quite attractive, as the targets can be removed for replacement easily.

The low melting point and low coefficient of friction of indium make it a suitable candidate for applications requiring a metal lubricant. A thin layer of indium applied to a surface greatly reduces the coefficient of friction and works as a lubricant.

Indium is sputtered in a very thin transparent coating onto glass for use in windscreens. The sputtered indium is, in fact, an electrically conductive coating of indium oxide. By applying electric current to this conductive coating, it is possible to heat the glass to prevent icing or fogging.

A similar type of coating of indium oxide on structural glass can be used to limit the transfer of radiant energy through the windows of a building while letting sunlight in. The difference with this coating is that it is not electrically conductive; instead, it is infrared reflecting. This application has the great potential of saving energy on heating and cooling costs.

Added to various alloys, indium brings strength and corrosion resistance. Added to aluminum and zinc alloys used in sacrificial anodes, it increases the efficiency and potential of the anodes, which is particularly useful for offshore oil platforms.

Indium is used in coating bearings for use in high-performance engines. The addition of indium improves corrosion and abrasion resistance and helps retain an oil film on the bearing surface.

Because of its high cross-section for capturing thermal neutrons, indium is added into the nuclear control rods that are introduced into the nuclear reactor core to moderate the progress of the nuclear fission reaction.

Indium Alloys

Indium containing low-melting-point alloys has become particularly useful in the optical industry for securing lenses for grinding processes. For processing plastic lenses, the indium content of the alloy serves as a form of temperature control.

Today's electronics are manufactured with delicate circuit boards that often cannot stand high temperatures that typically are produced by soldering with conventional tin/lead solders. The addition of indium to these solders lowers the melting range of the material.

The addition of small amounts of indium enhances the tensile strength and ductility of dental alloys and improves tarnish resistance.

Indium Compounds

Indium oxide is a bright yellow powder used in, for example, the communications industry, as in the bases of cordless phase.

Indium tin oxide, considered the best transparent electrode, is widely used in flat-panel displays. Indium trichloride, $InCl_3$, is widely used to coat the inside of glass for use in street-lamp production.

Indium hydroxide is often used in the production of alkaline batteries.

MANUFACTURE, PROCESSING, AND SHIPMENT

The normal purity of commercial-grade indium is 99.99%, although lower grades are sometimes available. Indium ingots are generally used for metallurgical additives, sacrificial anodes, manufacturing of alloys, and bonding applications.

ECONOMIC ASPECTS

World demand, production, and stock values for indium are difficult to gauge because there is no definitive source for these values.

The supply of indium is not easily changed, because it is dependent on the supply of the associated metals with which it is mined.

ENVIRONMENTAL CONCERNS

Indium metal in and of itself poses little or no environmental risk. However, the form that indium takes or the other metals associated with it may pose some dangers to the environment. For example, indium alloys that contain lead or cadmium could be considered hazardous to the environment because of the lead or cadmium.

RECYCLING

Since indium metal is a minor metal, the supply cannot be considered unlimited. Any scrap that is generated in the processing or use or indium metal or chemicals should be safely collected, stored, and returned to the supplier if possible. In particular, the alloys that can be made containing indium are readily reclaimed because of their low melting point.

HEALTH AND SAFETY FACTORS

Physiologically, indium is a nonessential element. In commercial use, it can be considered nonhazardous. There have been no reported cases of systemic effects in human exposure to indium. The threshold limit value (TLV) of the American Conference of Governmental Industrial Hygienists (ACGIH) is 0.1 mg/m^3. The primary toxic effects of ionic indium are on the kidneys. Indium has no demonstrated irritating effects on skin, but there may be effects on the respiratory system. Absorption through digestion is about 0.5% of intake, through respiration, about 5%. There have been no reports of accidental or industrial poisoning. Handling indium with bare hands will leave a gray-to-black residue on the skin.

R. P. Elliot, *Constitution of Binary Alloys, First Supplement*, McGraw-Hill Book Co. Inc., New York, 1965.

H. A. Taylor, Jr., Basic Mines, *Indium Advocate News*, www.basicsmines.com/indium/

R. C. Weast, ed., *CRC Handbook of Chemistry and Physics*, 68th ed., CRC Press, Boca Raton, Fla., 1988.

Mark J. Chagnon
Atlantic Metals, Alloys

INDUSTRIAL HYGIENE

Industrial hygiene is devoted to the anticipation, recognition, evaluation, and control of environmental factors or stresses arising in or from the workplace that may cause sickness, impaired health and well-being, or significant discomfort and inefficiency among workers or among the citizens of the community. It is a profession practiced by >11,000 industrial hygienists in the United States and many more worldwide. U.S. industrial hygienists are typically members of the American Industrial Hygiene Association (AIHA), the largest industrial hygiene organization; the American Conference of Governmental Industrial Hygienists (ACGIH); and the American Academy of Industrial Hygiene (AAIH). Many are certified industrial hygienists (CIHs) as a result of meeting the requirements of the American Board of Industrial

Hygiene (ABIH). Outside the United States, industrial (also called occupational) hygienists are members of such professional associations as the British Occupational Hygiene Society (BOHS) and the International Occupational Hygiene Association (IOHA).

Industrial hygienists work closely with members of several other professions concerned with workplace health and safety; eg, occupational medicine, occupational health nursing, and safety engineering. All of these groups are involved in the implementation of the laws that regulate workplace health and safety. In the United States the principal law is the Occupational Safety and Health Act (OSHA) enforced by the U.S. Department of Labor (U.S. DOL). Similar laws are in place in almost every other country in the world and are proposed by such international organizations as the World Health Organization (WHO) and the International Labor Organization (ILO).

A partial list of the hazards or conditions arising from the workplace and with which industrial hygienists are concerned includes the following.

Chemical	microwave radiation
carcinogens	extremely low frequency
acute poisons	
reproductive hazards	vibration
corrosives	magnetic fields
irritants	ulraviolet radiation
pneumoconiosis producing	infrared radiation
	laser radiation
neurotoxins	
nephro (kidney) toxins	*Ergonomic*
	repetitive strain injury (RSI)
Physical	carpal tunnel syndrome
noise	back injury
heat	lifting hazards
cold	visual display units
ionizing radiation	human/machine interaction

Industrial hygienists must be able to anticipate what workplace materials or events may give rise to any of these hazards, to recognize the hazards that occur, to evaluate a hazard to determine the degree of risk it presents, and to control hazards so as to reduce risk. The industrial hygienist's job begins when a new chemical or process is conceived. Based on data from animal experiments and/or human epidemiology relating to a substance or an analogous chemical, it is possible to estimate the toxicity of the substance. Whenever possible, it is best to avoid using potentially dangerous chemicals. Similarly, potentially hazardous processes that produce excessive noise, heat, or other stress-related situations should be anticipated and avoided. However, the industrial hygienist can usually devise ways to use potentially dangerous chemicals safely.

RECOGNITION OF POTENTIAL HAZARDS

The process of recognition of potential hazards is based on extensive knowledge of what kinds of hazards may occur in any industry, process, or job activity. Chemical hazard sources includs the following.

Fugitive Emissions

Fugitive emissions or leaks occur wherever there are breaks in a barrier that maintains containment.

Process Operations

In a modern chemical plant there are a few actions the operators may need to take which can involve contact with process materials. Sampling of process streams is one such task.

Material Handling

Some material-handling steps are difficult to accomplish with total containment. Solids handling is often done by open means, both because the hazard is perceived to be less and because it is more difficult to design totally closed solids handling systems.

Maintenance

Open system maintenance can add to exposure by disturbing and dispersing deposits of materials in equipment. Most maintenance is done while a plant is in operation. Thus, the maintenance workers are in close proximity to operating equipment for long periods of time. Maintenance that exposes workers to health risks includes welding, painting, sandblasting, insulation, chemical cleaning, and catalyst handling.

Waste Handling

Housekeeping procedures in general can have a significant impact on employee exposure, and certain waste handling procedures can result in very serious exposure if proper precautions are not taken.

Air-cleaning systems are often used to remove dust or vapors from plant or process exhaust streams. It is necessary to enter the air cleaner periodically for inspection or repair. Dust deposits inside the equipment are likely to be stirred up and inhaled by unprotected workers.

Wastewater treatment facilities may receive chemical process wastes and spills. These wastes may volatize on emerging from a closed sewer system into open waste-treatment tanks, particularly if hot streams have heated the tank. These releases can occur without warning and result in unexpected employee exposure.

HAZARD EVALUATION

The evaluation phase of industrial hygiene is the process of making measurements on some set of samples that permits a conclusion about the risk of harm resulting from exposure to a hazardous substance. Before conducting an evaluation, it is necessary to make a number of choices: what and where to sample, when to sample, how long to sample, how many samples to take, what sampling and analytical methods to use, what exposure criteria to use in the analysis of the data, and how to report the results.

These choices as a whole constitute the evaluation plan. The object is to find if one or more workers have an unacceptable probability of being exposed in excess of some established limit.

Decision Process

In many cases, the decision regarding the need for exposure reduction measures is obvious and no formal statistical procedure is necessary. However, as exposure criteria are lowered and control becomes more difficult, close calls become more common, and a logical decision-making process is needed. A typical process is shown in Figure 1.

GENERIC EXPOSURE ASSESSMENT

The United Kingdom and, to an increasing degree, European governments are implementing less quantitative means of assessing and controlling workplace hazards, particularly for small and medium size establishments (SMEs). In the United Kingdom the Control of Substances Hazardous to Health Regulations (COSHH) requires employers to assess the risks to health from chemicals and decide what controls are needed; use those controls and make sure workers use them; make sure the controls are working properly; inform workers about the risks to their health; and train workers.

OTHER AGENTS

Evaluations of occupational exposure to physical agents such as noise, radiation or heat, biological agents, and multiple chemical agents are similar to the process for single chemical substances but have some key differences.

Control

The evaluation phase should be planned to yield the data needed to draw accurate conclusions about control needs. The need for certainty depends on how difficult it is to achieve control.

Although the evaluation phase comes chronologically between the recognition and control phases, the control options play a considerable role in the extent or intensity of the evaluation phase.

Options

Traditional control options for overexposure are materials substitution, process change, containment, enclosure, isolation, source reduction, ventilation, providing personal protection, changing work practices, and improving housekeeping. A simple way of looking at the selection of control options is to find the cheapest option that results in the desired amount of exposure reduction. It is not actually that simple, however, because the various options differ in ways other than cost and degree of control. Some of the other factors to consider in selection of control options are operability, reliability, and acceptability.

American Industrial Hygiene Association, *Mathematical Models for Estimating Occupational Exposure to Chemicals*, AIHA Press, Fairfax, Va., 2000.

R. R. Fullwood, *Probabilistic Safety Assessment in the Chemical and Nuclear Industries*, Butterworth-Heinemann, Boston, Mass., 2000.

D. J. Paustenbach, *Human and Ecological Risk Assessment*, Wiley-Interscience, New York, 2002.

J. M. Samet and J. D. Spengler, eds., *Indoor Air Pollution*, Johns Hopkins University Press, Baltimore, Md., 1991.

JEREMIAH LYNCH
Exxon Chemical Co. (retired)

INFRARED SPECTROSCOPY

Infrared (ir) spectroscopy is a technique based on the vibrations of atoms of a molecule. An ir spectrum is commonly obtained by passing ir radiation through a sample and determining what fraction of the incident radiation is absorbed at a particular energy. The energy at which any peak in an absorption spectrum appears corresponds to the frequency of a vibration of a part of a sample molecule.

For a molecule to show ir absorptions, an electric dipole moment of the molecule must change during the movement. The interactions of ir radiation with matter may be understood in terms of changes in the molecular dipoles associated with vibrations and rotations. The atoms in molecules can move relative to one another; ie, bond lengths can vary or one atom can move out of its present plane. These stretching and bending movements are collectively referred to as vibrations.

There will be many different vibrations for even fairly simple molecules. The complexity of an ir spectrum arises

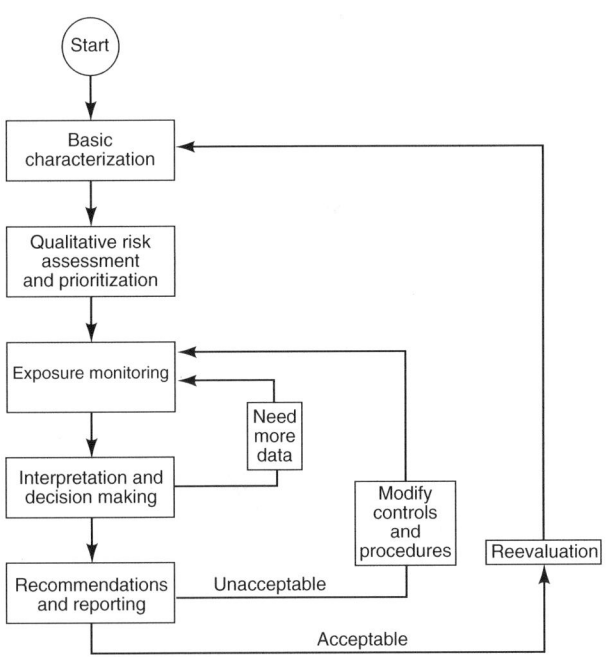

Figure 1. The decision-making process.

from the coupling of vibrations over a large part of or over the complete molecule, called skeletal vibrations. Bands associated with skeletal vibrations are likely to conform to a pattern or fingerprint of the molecule as a whole rather than a specific group within the molecule.

INSTRUMENTATION

The most significant advances in ir spectroscopy, have come about as a result of the introduction of Fourier transform (ft) spectrometers. This type of instrument employs an interferometer and exploits the well-established mathematical process of Fourier transformation. Fourier transform infrared (ftir) spectroscopy has dramatically improved the quality of ir spectra and minimized the time required to obtain data.

Fourier transform spectroscopy is based on the idea of the interference of radiation between two beams to yield an interferogram, a signal produced as a function of the change of pathlength between the two beams. The two domains of distance and frequency are interconvertible by the mathematical method of Fourier transformation.

In an ftir spectrometer, the radiation emerging from a source is passed through an interferometer to the sample before reaching a detector. Upon amplification of the signal, in which high-frequency contributions have been eliminated by a filter, the data are converted to digital form by an analog-to-digital converter and transferred to the computer for Fourier transformation. The most common interferometer used in ftir spectrometry is a Michelson interferometer, which consists of two perpendicularly plane mirrors, one of which can travel in a direction perpendicular to the plane (Fig. 1). A semireflecting film, the beam splitter, bisects the planes of these two mirrors. The beam splitter's material has to be chosen according to the region to be examined. Materials such as germanium or iron oxide are coated onto an ir transparent substrate

such as potassium bromide or cesium iodide to produce beam splitters for the mid- or near-ir regions. Thin organic films, such as poly(ethylene terephthalate), are used in the far-ir region.

If a collimated beam of monochromatic radiation of wavelength λ (cm) is passed into an ideal beam splitter, 50% of the incident radiation will be reflected to one of the mirrors and 50% transmitted to the other mirror. The two beams are then reflected from these mirrors, returning to the beam splitter, where they recombine and interfere. Fifty percent of the beam reflected from the fixed mirror is transmitted through the beam splitter and 50% is reflected back in the direction of the source. The beam that emerges from the interferometer at $90°$ to the input beam is called the transmitted beam. This is the beam detected in ftir spectrometry. The moving mirror produces an optical path difference between the two arms of the interferometer. For path differences of $(n + \frac{1}{2})\lambda$, the two beams interfere destructively in the case of the transmitted beam and constructively in the case of the reflected beam.

Ftir spectrometers use a Globar or Nernst source for the mid-infrared region. If the far-infrared region is to be examined, then a high-pressure mercury lamp can be used. For the near-infrared, tungsten–halogen lamps are used as sources. There are two commonly used detectors used in the mid-infrared region. The normal detector for routine use is a pyroelectric device incorporating deuterium tryglycine sulfate (DTGS) in a temperature-resistant alkali halide window. For more sensitive work, mercury cadmium telluride (MCT) can be used, but has to be cooled to liquid nitrogen temperatures. In the far-infrared germanium or indium-antimony detectors are employed, operating at liquid helium temperatures. For the near-infrared detectors are generally lead sulfide photoconductors.

The essential equations for a Fourier transformation relating the intensity falling on the detector, $I(\delta)$, to the

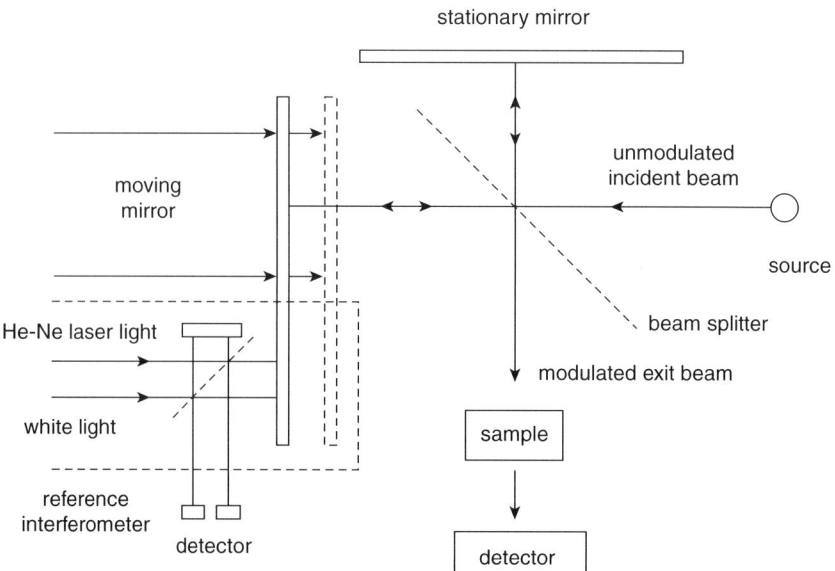

Figure 1. Michelson interferometer.

spectral power density at a particular wavenumber, $\bar{v}$, given by $B(\bar{v})$ are as follows:

$$I\delta = \int_0^{+\infty} B(\bar{v}) \cos 2\pi \bar{v} \, \delta \, d\bar{v}$$

which is one-half of a cosine FT pair, the other being

$$B(\bar{v}) = \int_{-\infty}^{+\infty} I(\delta) \cos 2\pi \bar{v} \, \delta \, d\delta$$

These two interconvertible equations are known as an ft pair. The first shows the variation in power density as a function of the difference in path length, which is an interference pattern. The second shows the variation in intensity as a function of wave number. Each can be converted into the other by the mathematical method of Fourier transformation.

The main advantages of rapid scanning instruments is their ability to increase the signal-to-noise ratio (SNR) by signal averaging, leading to an increase in SNR that is proportional to the square root of the time. There are diminishing returns for signal averaging in that it takes an increasingly longer time to achieve greater and greater improvement. The accumulation of a large number of repeat scans makes greater demands on an instrument if it is to exactly reproduce the conditions. To offset this difficulty it is normal to incorporate a laser monochromatic source in the beam of the continuous source, which produces standard fringes that can accurately line- up successive scans and determine and control the displacement of the moving mirror.

SAMPLING METHODS

Transmission Methods

Transmission spectroscopy is the oldest, most straightforward ir method. It is based upon the absorption of ir radiation at specific wavelengths as it passes through a sample. Using this approach it is possible to analyze samples in liquid, solid, or gaseous form.

There are several different types of transmission liquid cells available. Fixed pathlength sealed cells are useful for volatile liquids, but cannot be taken apart for cleaning.

Semi-permanent cells are demountable so that the windows can be cleaned. The spacer is usually made of polytetrafluoroethylene (PTFE, Teflon) and is available in a variety of thicknesses, allowing one cell to be used for various pathlengths. All these cell types are filled using a syringe and the syringe ports are sealed with PTFE plugs before sampling.

There are three general methods for examining solid samples in transmission infrared spectroscopy: alkali halide discs, mulls and films. The use of alkali halide discs involves mixing a solid sample with a dry alkali halide powder. The mixture is usually ground with an agate mortar and pestle and subjected to pressure in an evacuated die. This sinters the mixture and produces a clear transparent disc and the most commonly used alkali halide is KBr. The mull method for solid samples involves grinding the sample then suspending (about 50 mg) in 1–2 drops of a mulling agent. This is followed by further grinding until a smooth paste is obtained. The most commonly used mulling agent is Nujol (liquid paraffin). Films can be produced by either solvent casting or by melt casting. In solvent casting the sample is dissolved in an appropriate solvent (the concentration depends on the required film thickness). A solvent needs to be chosen which not only dissolves the sample, but will also produce a uniform film.

Gases have densities which are several orders of magnitude less than liquids, hence pathlengths must be correspondingly greater, usually 10 cm or longer. The walls are of glass or brass with the usual choice of windows. The cells can be filled by flushing or from a gas line.

Reflectance Methods

Reflectance techniques may be used for samples that are difficult to analyze by conventional transmittance methods. Attenuated total reflectance spectroscopy (ATR) utilizes the phenomenon of total internal reflection (Fig. 2). A beam of radiation entering a crystal will undergo total internal reflection when the angle of incidence at the interface between the sample and crystal is greater than the critical angle. The critical angle is a function of the refractive indexes of the two surfaces. The beam penetrates a fraction of a wavelength beyond the reflecting surface. When a material that selectively absorbs radiation is in close contact with the reflecting surface, the beam loses energy at the wavelength where the material absorbs.

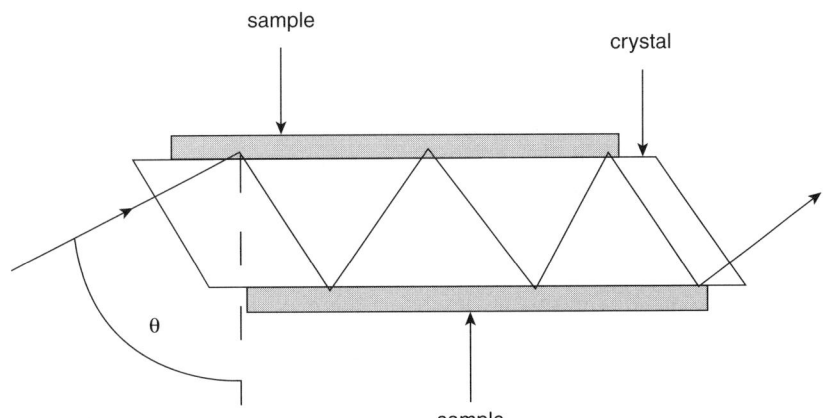

Figure 2. Attenuated total reflectance cell.

The resultant attenuated radiation can then be measured and plotted as a function of wavelength by a spectrometer, giving rise to the absorption spectral characteristics of the sample. The depth of penetration in atr is a function of the wavelength, λ, the refractive index of the crystal, n_2, and the angle of incident radiation, θ. The depth of penetration, d_p, for a nonabsorbing medium is given by the following formula:

$$d_p = (\lambda/n_1)/ \, (2\pi[\sin\theta - (n_1/n_2)^2]^{1/2})$$

where n_1 is the refractive index of the sample. The crystals used in atr cells are made from materials that have low solubility in water and are of a very high refractive index. Such materials include zinc selenide (ZnSe), germanium (Ge), and thallium/iodide (KRS-5). Different designs of atr cells allow both liquid and solid samples to be examined. It is also possible to set up a flow-through an atr cell that allows for a continuous flow of solutions through the cell and permits spectral changes to be monitored over time.

Specular reflectance occurs when the reflected angle of radiation equals the angle of incidence. The amount of light reflected depends on the angle of incidence, the refractive index, surface roughness, and absorption properties of the sample. For most materials the reflected energy is only 5–10%, but in regions of strong absorptions, the reflected intensity is greater. The resultant data appears different from normal transmission spectra, as derivative-like bands result from the superposition of the normal extinction coefficient spectrum with the refractive index dispersion (based upon Fresnel's relationships). However, the reflectance spectrum can be corrected using a Kramers-Kronig transformation (or K-K transformation). The corrected spectrum appears like the familiar transmission spectrum. Increased pathlengths through thin coatings can be achieved using grazing angles of incidence (up to 85°). Grazing angle sampling accessories allow measurements to be made on samples over a wide range of angles of incidence. Solid samples, particularly coatings on reflective surfaces, are simply placed on a flat surface. The technique is also commonly used for liquid samples that can be poured into a Teflon trough. Oriented films on the liquid surface can be investigated using this method.

In external reflectance, the energy that penetrates one or more particles is reflected in all directions and this component is called diffuse reflectance. In the diffuse reflectance technique, commonly called DRIFT, a powdered sample is usually mixed with KBr powder. The DRIFT cell reflects radiation to the powder and collects the energy reflected back over a large angle. Diffusely scattered light can be collected directly from material in the sampling cup or, alternatively, from material collected using an abrasive sampling pad. DRIFT is particularly useful for sampling powders or fibers.

Microsampling Methods

It is possible to combine an ir spectrometer with a microscope facility in order to study very small samples. In recent years there have been considerable advances in ftir microscopy with samples of the order of microns being characterized.

If a microscope facility is not available, there are other special sampling accessories available to allow examination of microgram or microlitre amounts. This is accomplished using a beam condenser so that as much as possible of the beam passes through the sample. Microcells are available with volumes of around 4 µls and pathlengths up to 1 mm. A diamond anvil cell (DAC) uses two diamonds to compress a sample to a thickness suitable for measurement and increase the surface area.

Infrared imaging using ftir microspectroscopic techniques has emerged as an effective approach to studying complex or heterogeneous specimens. The technique can be used to produce a two- or three-dimensional picture of the properties of a sample. Each pixel (or spatial location) is represented by an ir spectrum.

This is possible because, instead of reading the signal of only one detector as in conventional ftir spectroscopy, a large number of detector elements are read during the acquisition of spectra. This is possible due to the development of focal plane array (FPA) detectors.

SPECTRAL ANALYSIS

Group Frequencies

The ir spectrum can be divided into three main regions: the far-ir (<400 cm^{-1}), the mid-ir (4000–400 cm^{-1}), and the near-ir (13,000–4,000 cm^{-1}). Many ir applications employ the mid-ir region, but the near- and far-ir also provide important information about certain materials. The mid-ir spectrum (4,000–400 cm^{-1}) can be approximately divided into four regions. The nature of a group frequency may generally be determined by the region in which it is located. The regions are generalized as the X–H stretching region (4,000–2,500 cm^{-1}), the triple-bond region (2,500–2,000 cm^{-1}), the double-bond region (2,000–1,500 cm^{-1}), and the fingerprint region (1,500–600 cm^{-1}).

So far it has been assumed that each band in an ir spectrum can be assigned to a particular deformation of the molecule, the movement of a group of atoms, or the bending or stretching of a particular bond. This is possible for many bands, particularly stretching vibrations of multiple bonds that are well behaved. However, many vibrations are not so well behaved and may vary by hundreds of wave numbers even for similar molecules.

APPLICATIONS

One of the most common applications of ir spectroscopy is in the identification of organic compounds. Generally, the ir bands for inorganic materials are broader, fewer in number, and appear at lower wave numbers than those for organic materials. The bands in the spectra of ionic or coordination compounds will depend on the structure and orientation of the ion or complex. The ir spectra of metal complexes and minerals have also been widely reported.

Infrared spectroscopy is a popular method for characterizing polymers. This spectroscopy may be used to identify the composition of polymers, to monitor polymerization processes, characterize polymer structure, examine polymer surfaces, and investigate polymer degradation processes.

Infrared spectroscopy has also proven to be a powerful tool for the study of biological molecules. The application of this technique to biological problems is continually expanding, particularly with the advent of increasingly sophisticated sampling techniques such as ir imaging. Biological systems, including lipids, proteins, peptides, biomembranes, nucleic acids, animal tissue, microbial cells, plants, and clinical samples, have all been successfully studied using ir spectroscopy. Infrared spectroscopy has been employed for a number of decades to characterize isolated biological molecules, particularly proteins and lipids. Microscopic techniques combined with sophisticated analytical methods now allow for complex samples of micron size to be investigated.

This technique also finds use in a wide range of industrial applications. Infrared spectroscopy has been extensively used in both qualitative and quantitative pharmaceutical analysis. The technique is important for the evaluation of raw materials used in production, active ingredients, and excipients. Both the mid- and near-ir techniques may be used to obtain qualitative and quantitative information about food samples. Commercial grains are commonly analyzed using nir spectroscopy. Infrared spectroscopy also plays an important role in quality control in the pulp and paper industry and the paint industry for quality control, product improvement, and failure analysis, as well as for forensic identification purposes. In addition, the technique has been applied to a broad range of environmental sampling problems, including air, water, and soil analysis. The development of remote-sensing ir techniques has been particularly useful in this field. Simple, practical ir techniques have been developed for measuring trace gases and atmospheric composition.

J. M. Chalmers and P. R. Griffiths, eds., *Handbook of Vibrational Spectroscopy*, John Wiley & Sons Inc., New York, 2002.

H. Günzler and H. U. Gremlick, *IR Spectroscopy: An Introduction*, Wiley-VCH, Weinheim, 2002.

R. G. Messerschmidt and M. A. Harthcock, eds., *Infrared Microspectroscopy: Theory and Applications*, Marcel-Dekker, New York, 1998.

B. Stuart, *Infrared Spectroscopy: Fundamentals and Applications*, John Wiley & Sons, Inc., Chichester, 2004.

BARBARA STUART
University of Technology
Sydney

INITIATORS, ANIONIC

In anionic polymerization, the reactive propagating intermediate generated by the initiation reaction is an anion; ie, a species that carries a formal negative charge, with a corresponding positively charged counterion. In living anionic polymerization, the kinetic steps of chain termina-

tion and chain transfer are absent. This unique aspect of many anionic polymerizations provides a methodology for preparing polymers with control of the significant variables affecting polymer properties, including molecular weight, molecular weight distribution, block copolymer composition, copolymer sequence distribution, and microstructure, as well as molecular architecture (linear, branched, and cyclic macromolecules). An important consideration for preparation of polymers with well-defined structures and low degrees of compositional heterogeneity is the choice of a suitable initiator.

In general, an appropriate initiator is a species that has approximately the same structure and reactivity as the propagating anionic species; ie, the pK_a of the conjugate acid of the propagating anion should correspond closely to the pK_a of the conjugate acid of the initiating species. If the initiator is too reactive, side reactions between the initiator and monomer can occur; if the initiator is not reactive enough, the initiation reaction may be slow or inefficient.

The general relationship between monomer structural type, pK_a, and appropriate initiating species is shown in Table 1. The monomers that form the least stable propagating anions; ie, which have the largest values of pK_a for the corresponding conjugate acids, are the least reactive monomers in anionic polymerization and require the use of the most reactive initiators, as shown in Table 1.

ALKALI METALS

The use of alkali metals for anionic polymerization of diene monomers is primarily of historical interest. The mechanism of the anionic polymerization of styrenes and 1,3-dienes initiated by alkali metals has been described in detail, the first step being an electron-transfer reaction from the metal to the lowest unoccupied molecular orbital of an adsorbed monomer molecule to form radical anion intermediates that rapidly dimerize to form dianions.

Aromatic Radical Anions

Many aromatic hydrocarbons react with alkali metals in polar aprotic solvents to form stable solutions of the corresponding radical anions. These solutions can be analyzed by uv–visible spectroscopy and stored for further use.

Sodium naphthalene and other aromatic radical anions react with monomers such as styrene by reversible electron transfer to form the corresponding monomer radical anions, which rapidly dimerize.

Monomers that can be polymerized with aromatic radical anions include styrenes, dienes, epoxides, and cyclosiloxanes. Aromatic radical anions that are too stable do not efficiently initiate polymerization of less reactive monomers; thus, the anthracene radical anion cannot initiate styrene polymerization.

ALKYLLITHIUM COMPOUNDS

Anionic polymerization of vinyl monomers can be effected with a variety of organometallic compounds; alkyllithium

Table 1. Relationships Between Monomer Reactivity, Carbanion Stability, and Suitable Initiators

| Monomer type | $pK_a{}^a$ | | Initiators[b] |
	In DMSO	In H_2O	
ethylene	56		RLi
dienes and	44		NH_2^-, RLi, RMt
styrenes	43		aromatic radical anions,[c] cumyl K, Mt, RMgX
acrylonitrile	32		
alkyl methacrylates, alkyl acrylates	30–31	27–28	fluorenyl$^-$, $Ar_2C^\bullet$, ketyl radical anions[d]
vinyl ketones	26	19	
oxiranes	29–32	16–18	RO^-
thiiranes	17	12–13	
nitroalkenes	17	10–14	
siloxanes		10–14	RO^-, OH^-
β-lactones	12	4–5	$RCOO^-$
alkyl cyanoacrylates	12.8		HCO_3^-, H_2O
vinylidene cyanide	11	11	

[a] The pK_a of the conjugate acid of the anionic propagating intermediate.
[b] Mt refers generally to alkali metals (Li, Na, K, Rb, Cs).

[c] For example, naphthalene radical anion with counterion (Li^+, Na^+, K^+).
[d] Ar2CO•

compounds are the most useful class. A variety of simple alkyllithium compounds are available commercially. Most simple alkyllithium compounds are soluble in hydrocarbon solvents such as hexane and cyclohexane and can be prepared by reaction of the corresponding alkyl chlorides with lithium metal.

Simple alkyllithium compounds are aggregated in solution, in the solid state, and even in the gas phase. The important differences between the various alkyllithium compounds are their degrees of aggregation in solution and their relative reactivity as initiators for anionic polymerization of styrene and diene monomers.

The kinetics of initiation reactions of alkyllithium compounds often exhibit fractional kinetic order dependence on the total concentration of initiator, consistent with the reaction of the unassociated form of the alkyllithium.

The use of aliphatic solvents causes profound changes in the observed kinetic behavior for the alkyllithium initiation reactions with styrene, butadiene, and isoprene; ie, the inverse correspondence between the reaction-order dependence for alkyllithium and degree of organolithium aggregation is generally not observed. Also, initial rates of initiation in aliphatic solvents are several orders of magnitude less than those observed, under equivalent conditions, in aromatic solvents. Furthermore, pronounced induction periods are observed in aliphatic hydrocarbon solvents.

The relative reactivities of alkyllithiums as polymerization initiators are intimately linked to their degree of association. In the following, the average degree of association in hydrocarbon solution, where known, is indicated in brackets after the alkyllithium. For styrene polymerization, the relative reactivity of alkyllithium initiators is menthyllithium $>sec$-C_4H_9Li $>i$-C_3H_7Li $>i$-C_4H_9Li $>n$-C_4H_9Li $>t$-C_4H_9Li. For diene polymerization, menthyllithium $>$sec-C_4H_9Li $>i$-C_3H_7Li $>t$-C_4H_9Li $>i$-$C_4H_9Li>n$-C_4H_9Li.

Alkyllithium compounds are primarily used as initiators for polymerizations of styrenes and dienes in hydrocarbon solutions.

Quantitative Analysis of Alkyllithium Initiator Solutions

The amount of carbon-bound lithium is calculated from the difference between the total amount of base determined by acid titration and the amount of base remaining after the solution reacts with either benzyl chloride, allyl chloride, or ethylene dibromide.

Copolymerization Initiators and the Effects of Alkali Metal Alkoxides

The copolymerization of styrenes and dienes in hydrocarbon solution with alkyllithium initiators produces a tapered block copolymer structure, because of the large differences in monomer reactivity ratios for styrene ($r_s < 0.1$) and dienes ($r_d > 10$). In order to obtain random copolymers of styrenes and dienes, it is necessary to either add small amounts of a Lewis base such as THF or an alkali metal alkoxide (MtOR, where Mt = Na, K, Rb, or Cs).

Difunctional and Trifunctional Initiators

These initiators are of considerable interest for the preparation of triblock copolymers, telechelic polymers, and macrocyclic polymers. Aromatic radical anions, such as lithium naphthalene or sodium naphthalene, are efficient difunctional initiators. However, the necessity of using polar solvents for their formation and use limits their utility for diene polymerization.

The methodology for preparation of hydrocarbon-soluble, dilithium initiators is generally based on the reaction of an aromatic divinyl precursor with 2 mol of butyllithium.

Although a plethora of divinyl aromatic compounds have been investigated as precursors for hydrocarbon-soluble dilithium initiators, the best system demonstrated to produce a hydrocarbon-soluble dilithium initiator is based on 1,3-bis(1-phenylethenyl)benzene.

Functionalized Initiators

The use of alkyllithium initiators that contain functional groups provides a versatile method for the preparation of end functionalized polymers and macromonomers. For a living anionic polymerization, each functionalized initiator molecule produces one macromolecule with the functional group from the initiator residue at one chain end and the active carbanionic propagating species at the other chain end.

OTHER INITIATORS

Other initiators include organosodium and organopotassium compounds, organomagnesium compounds and Mixed organometallic compounds, organobarium compounds, 1,1-diphenylmethylcarbanions, fluorenyl carbanions, enolate initiators, metal-free initiators, and alkoxides and related initiators.

HEALTH AND SAFETY FACTORS

The hazards associated with organolithium compounds and other reactive organometallic compounds such as cumyl potassium are corrosivity, flammability, and pyrophoricity. The corrosive nature of organolithium compounds can cause both chemical and thermal burns. Organolithium compounds are flammable and are typically supplied in flammable hydrocarbon solutions. n-Butyllithium, *sec*-butyllithium, and *tert*-butyllithium are pyrophoric; ie, they spontaneously ignite on exposure to air, oxygen, or moisture. Users of these reactive organometallic compounds are advised to strictly follow the handling instructions recommended by their suppliers. Because hydrocarbon solutions of alkyllithium compounds are air and moisture sensitive, they should be either handled in an inert atmosphere or with syringes, using recommended procedures for handling air-sensitive compounds. Alkyllithium reagents react with acidic compounds that contain reactive hydrogens such as water, alcohols, phenols, acids, and even primary and secondary amines. The reaction of butyllithium with water produces butane and lithium hydroxide, which can lead to spontaneous ignition in the presence of oxygen. Carbon dioxide fire extinguishers must not be used, because carbon dioxide reacts exothermically with alkyllithium compounds. It is prudent to have an all-purpose fire extinguisher available when working with organometallic compounds. Suitable fire-extinguishing chemicals include powdered limestone and powders containing sodium chloride and sodium bicarbonate.

H. Hsieh and R. P. Quirk, *Anionic Polymerization: Principles and Practical Applications*, Marcel Dekker, New York, 1996.

M. Szwarc, *Carbanions, Living Polymers and Electron Transfer Processes*, Wiley-Interscience, New York, 1968.

RODERIC P. QUIRK
The University of Akron

VICTOR M. MONROY
Monroy Technology Concepts, LLC

INITIATORS, CATIONIC

Cationic polymerization may be induced by a variety of physical (high-energy radiation, direct or indirect ultraviolet radiation, and electroinitiation;) and chemical (protic acids, Friedel-Crafts acids, and stable cation salts methods). The most important initiating system is the cation donor (initiator)/Friedel-Crafts acid (coinitiator) system. Butyl rubber, a copolymer of isobutylene with 0.5–2.5% isoprene to make vulcanization possible, is the most important commercial polymer made by cationic polymerization. Another important commercial application of cationic polymerization is the manufacture of polybutenes, low molecular weight copolymers of isobutylene, and a smaller amount of other butenes used in adhesives, sealants, lubricants, viscosity improvers, etc.

Unless one is working with superdried systems or in the presence of proton traps, adventitious water is always present as a proton source. Polymerization rates, monomer conversions, and to some extent polymer molecular weights are dependent on the amount of protic impurities; therefore, well-established drying methods should be followed to obtain reproducible results.

In place of a proton source; ie, a Brønsted acid, a cation source such as an alkyl halide, ester, or ether can be used in conjunction with a Friedel-Crafts acid. Initiation with the ether-based initiating systems in most cases involves the halide derivative that arises upon fast halidation by the Friedel-Crafts acid MX_n. The efficiency of the initiator–coinitiator system depends greatly on the monomer in question. As a general rule the stability (reactivity) of the initiating cation should be close to that of the propagating chain end. Since initiation involves two subsequent events; ie, ion generation and cationation, species on the two extremes are less active or may be completely inactive, because they form ionic species very slowly and/or in extremely low concentration, or form ions in high concentration that are, however, too stable to cationate the monomer.

The activity of an initiating system is also affected by the nature of the Friedel-Crafts (Lewis) acid. The following acidity scale can be established: $BF_3 < AlCl_3 < TiCl_4 < BCl_3 < SbF_5 < SbCl_5 < BBr_3$. The activity of the boron halide-based system is greatly solvent dependent; ie, sufficient activity occurs only in a polar solvent.

Solvent polarity and temperature also influence the results. The dielectric constant and polarizability are of little predictive value for the selection of solvents relative to polymerization rates and behavior. In cationic polymerization of vinyl monomers, chain transfer is the most significant chain-breaking process. The activation energy of chain transfer is relatively high; consequently, the molecular weight of the polymer increases with decreasing temperature.

Initiation by a carbocation source provides control of the head group (controlled initiation) when used in conjunction with a Friedel-Crafts acid; eg, $(C_2H_5)_3Al$, $(CH_3)_3Al$, $(C_2H_5)_2AlCl$, $TiCl_4$, BCl_3 for isobutylene, or I_2 and zinc halides for vinyl ethers, where a chain transfer to monomer is absent or negligible, or in the presence of a proton trap to abort a chain transfer to monomer. That is, initiation from tertiary, allylic, and benzylic halides gives rise to macromolecules carrying tertiary, allylic, and benzylic head groups. Initiation by halogens results in head groups carrying the halogen. Controlled initiation, however, is achieved only when polymer formation from adventitious protic impurities is also absent or negligible.

A special case of controlled initiation is the inifer method. The word inifer (from *ini*tiator trans *fer* agents) describes compounds that function simultaneously as initiators and as chain-transfer agents. The inifer technique provided the first carbocationic route toward the synthesis of telechelic (α,γ-functional) polyisobutylenes and more recently telechelic poly(p-chlorostyrenes).

Although direct initiation by metal halides has been postulated with Friedel-Crafts acids, it was proven only for aluminum halides and more recently for BCl_3. Kinetic investigation of polymerizations by BCl_3 suggests that initiation is by haloboration according to the Sigwalt-Olah theory.

Many initiating systems used in the cationic polymerization of vinyl monomers can also be used to initiate the ring-opening polymerization of cyclic monomers such as cyclic ethers, acetals, lactams, lactones, and siloxanes. Polymerization of cyclic monomers may involve different types of ionic as well as covalent growing species. Under certain conditions termination processes may be absent. The polymerization of cyclic monomers, however, is almost always complicated by a inter- and intramolecular chain transfer to polymer, resulting in cyclic oligomer formation. The extent of cyclic oligomer formation can be minimized in the polymerization of epoxides by the recently discovered activated monomer mechanism. Cyclic ether and acetal polymerizations are also important commercially. Polymerization of tetrahydrofuran (THF) is used to produce polyether diol, and polyoxymethylene, an excellent engineering plastic, is obtained by the ring-opening polymerization of trioxane with a small amount of cyclic ether or acetal comonomer to prevent depolymerization.

A variety of initiating systems have been described that allow not only controlled initiation but also controlled propagation in the polymerization of vinyl monomers. In these living polymerization systems, chain breaking (chain transfer and irreversible termination) is absent. The key to living carbocationic polymerizations is a rapid, dynamic equilibrium between a very small amount of active and a large pool of dormant species. For a specific monomer, the rate of exchange as well as the position of the equilibrium depend on the nature of the counteranion in addition to temperature and solvent polarity. Therefore, initiator–coinitiator systems that bring about living polymerization under a certain set of experimental conditions are largely determined by monomer reactivity.

Since the discovery of living cationic systems, cationic polymerization has progressed to a new stage where the synthesis of designed materials is now possible.

G. Bradley, ed., *Photoinitiators for Free Radical, Cationic and Anionic Photopolymerization*, Wiley, New York, 1998.

K. Matyjaszewski, ed., *Cationic Polymerization. Mechanism, Synthesis, and Application*, Dekker, New York, 1996. Kennedy and E. Marechal, *Carbocationic Polymerization*, Wiley-Interscience, New York, 1982.

P. Sigwalt and G. A. Olah, *Makromol. Chem.* **175**, 1077 and 1039 (1974).

RUDOLF FAUST
University of Massachusetts,
Lowell

INITIATORS, FREE RADICAL

Free-radical initiators are chemical substances that, under certain conditions, initiate chemical reactions by producing free radicals.

Initiators contain one or more labile bonds that cleave homolytically when sufficient energy is supplied to the molecule. The energy must be greater than the bond dissociation energy (BDE) of the labile bond. Radicals are reactive chemical species possessing a free (unbonded or unpaired) electron. Radicals may also be positively or negatively charged species carrying a free electron (ion radicals). Initiator-derived radicals are very reactive chemical intermediates and generally have short lifetimes; ie, half-life times less than 10^{-3} s.

The principal commercial initiators used to generate radicals are peroxides and azo compounds. Lesser amounts of carbon–carbon initiators and photoinitiators, and high-energy ionizing radiation are also employed commercially to generate radicals.

FREE-RADICAL FORMATION AND USE

There are three general processes for supplying the energy necessary to generate radicals from initiators: thermal processes, microwave or ultraviolet (uv) radiation processes, and electron transfer (redox) processes. Radicals can also be produced in high-energy radiation processes. Once formed, radicals undergo two basic types of reactions: propagation reactions and termination reactions.

Radicals are employed widely in the polymer industry, where their chain-propagating behavior transforms vinyl monomers into polymers and copolymers. The mechanism of addition polymerization involves all three types of

Table 1. Bond Dissociation Energies

Precursor	BDE, kJ/mol[a]
$(R)_3C-H$	381
$(R)_2CH-H$	406
RCH_2-H	418
CH_3-H	439
$RO-H$	439
$RCOO-H$	444
C_6H_5-H	469
$HO-H$	498

[a] To convert kJ/mol to kcal/mol, divide by 4.184.

reactions: initiation, propagation by addition to carbon–carbon double bonds, and termination.

Structure–Reactivity Relationships

Much has been written about the structure reactivity of radicals. No single unifying concept has satisfactorily explained all the radical reactions reported in the literature. A longstanding correlation of structure and reactivity involves comparisons of the energies required to homolytically break covalent bonds to hydrogen. It is assumed that this energy, the hydrogen bond dissociation energy, reflects the stability and the reactivity of the radical coproduced with the hydrogen atom. However, this assumption, which should really be limited to radical reactivity and selectivity in hydrogen atom abstraction reactions, can be particularly misleading for reactions with polar transition states, in which radicals can behave either as nucleophiles or electrophiles. Nevertheless, the correlation of radical reactivity with BDE is quite useful. Table 1 shows some general BDE values for the formation of various carbon and oxygen radicals from various precursors. According to the theory, the higher the BDE, the higher the reactivity and the lower the stability of the radical formed by removal of a hydrogen atom. The choice of an initiator for a given radical process depends on the reaction conditions and the reactivity of the initiator. These two factors must be balanced so that the reaction is successful.

Activation Parameters

Thermal processes are commonly used to break labile initiator bonds in order to form radicals. The amount of thermal energy necessary varies with the environment, but the absolute temperature, T, is usually the dominant factor. The energy barrier, the minimum amount of energy that must be supplied, is called the activation energy, E_a. A third important factor, known as the frequency factor, A, is a measure of bond motion freedom (translational, rotational, and vibrational) in the activated complex or transition state. The relationships of A, E_a, and T to the initiator decomposition rate (k_d) are expressed by the equation below, where R is the gas constant and A and E_a are known as the activation parameters.

$$k_d = Ae^{(-E_a/RT)} \text{ or } \ln k_d = \ln A - E_a/RT$$

Half-Life

Once these activation parameters have been determined for an initiator, half-life times at a given temperature; ie, the time required for 50% decomposition at a selected temperature, and half-life temperatures for a given period; ie, the temperature required for 50% decomposition of an initiator over a given time, can be calculated. Half-life data are useful for comparing the activity of one initiator with another when the half-life data are determined in the same solvent and at the same concentration and, preferably, when the initiators are of the same class.

Commercial initiators are primarily organic and inorganic peroxides, aliphatic azo compounds, certain organic compounds with labile carbon–carbon bonds, and photoinitiators.

ORGANIC PEROXIDES

Organic peroxides are compounds possessing one or more oxygen–oxygen bonds. They have the general structure ROOR′ or ROOH and decompose thermally by the initial cleavage of the oxygen–oxygen bond to produce two radicals:

$$ROOR' \longrightarrow RO\cdot + \cdot OR'$$

Following radical generation, the radicals produced (RO· and R′O·) can initiate the desired reaction. However, when the radicals are generated in commercial applications, they are surrounded by a solvent, monomer, or polymer "cage." When the cage is solvent, the radical must diffuse out of this cage to react with the desired substrate. When the cage is monomer, the radical can react with the cage wall or diffuse out of the cage. When the cage is polymer, reaction with the polymer can occur in the cage. Unfortunately, other reactions can occur within the cage and can adversely affect the efficiency of radical generation and radical reactivity. If the solvent reacts with the initiator radical, solvent radicals may participate in the desired reaction.

Two secondary propagating reactions often accompany the initial peroxide decomposition: radical-induced decompositions and β-scission reactions. Both reactions affect the reactivity and efficiency of the initiation process.

Approximately 100 different organic peroxide initiators, in well over 300 formulations (liquid, solid, paste, powder, solution, dispersion), are commercially produced throughout the world, primarily for the polymer and resin industries.

The eight classes of organic peroxides that are produced commercially for use as initiators are listed in Table 2. Included are the 10-h half-life temperature ranges (nonpromoted) for the members of each peroxide class.

INORGANIC PEROXIDES

Inorganic peroxide–redox systems have been employed for initiating emulsion homo- and copolymerizations of vinyl

Table 2. Commercial Organic Peroxide Classes[a]

Organic peroxide class	Structure	10-h $t_{1/2}$[b,c], °C
diacyl peroxides	$\underset{\substack{\| \\ R-C-OO-C-R}}{O \qquad\quad O}$	21–75
dialkyl peroxydicarbonates	$\underset{\substack{\| \\ RO-C-OO-C-OR}}{O \qquad\quad O}$	49–51[d]
tert-alkyl peroxyesters	$\underset{\substack{\| \\ R-C-OO-t\text{-}R}}{O}$	38–107
O-tert-alkyl O-alkyl monoperoxycarbonates	$\underset{\substack{\| \\ RO-C-OO-t\text{-}R}}{O}$	99–100
di(tert-alkylperoxy)ketals	$\underset{R}{\overset{R'}{>}}C\underset{OO-t\text{-}R}{\overset{OO-t\text{-}R}{<}}$	92–110
di-tert-alkyl peroxides tert-alkyl hydroperoxides	$t\text{-}R\text{—}OO\text{—}t\text{-}R$ $t\text{-}R\text{—}OO\text{—}H$	115–128 —[e]
ketone peroxides[a,d]	$HOO-\underset{R}{\overset{R'}{C}}\left(OO-\underset{R}{\overset{R'}{C}}\right)_x OOH$	—[e]

+ other structures

[a] $x = 0$ or 1.
[b] Temperature at which $t_{1/2} = 10$ h.
[c] In benzene, unless otherwise noted.
[d] In trichloroethylene (TCE).
[e] Not applicable.

monomers. These systems include hydrogen peroxide–ferrous sulfate, hydrogen peroxide–dodecyl mercaptan, potassium peroxydisulfate–sodium bisulfite, and potassium peroxydisulfate–dodecyl mercaptan. Potassium peroxydisulfate, $K_2S_2O_8$, (or the corresponding sodium or ammonium salt), is an inorganic peroxide that is used widely in emulsion polymerization (eg, latexes, rubbers, etc), usually in combination with a reducing agent.

When handling and using peroxide initiators, care should be exercised, since they are thermally sensitive and decompose (sometimes violently) when exposed to excessive temperatures, especially when they are in their pure or highly concentrated states. However, they are useful as initiators because of their thermal instability. What may be a safe temperature for one peroxide can be an unsafe temperature for another, since peroxide initiators encompass a wide activity range. Because some peroxides are shock or friction sensitive in the pure state, they are generally desensitized by formulating them into solutions, pastes, or powders with inert diluents and dispersions or emulsions with an aqueous diluent. All manufacturers' literature should be carefully scrutinized and the peroxide safety literature reviewed before handling and using specific peroxide initiator compositions.

AZO COMPOUNDS

Generally, the commercially available azo initiators are of the symmetrical azonitrile type:

$$R-\underset{CN}{\overset{R'}{C}}-N=N-\underset{CN}{\overset{R'}{C}}-R$$

The symmetrical azonitriles are solids with limited solubilities in common solvents. Some commercial aliphatic azo compounds and their 10-h halflife temperatures are listed in Table 3.

Care should be exercised in handling and using azo initiators in their pure and highly concentrated states, because they are thermally sensitive and can decompose rapidly when overheated. Although azonitriles are generally less sensitive to contaminants, the same cautions that apply to peroxides should also be applied to handling and using azo initiators. The manufacturers' safety literature should be read carefully. The potential toxicity hazards of decomposition products must be considered when using azonitriles.

CARBON–CARBON INITIATORS

Carbon–carbon initiators are hexasubstituted ethanes that undergo carbon–carbon bond scission when heated

Table 3. Commercial Azo Initiators

Name	Structure
2,2′-azobis[4-methoxy-2, 4-dimethylpentanenitrile]	$CH_3-C(OCH_3)(CH_3)-CH_2-C(CH_3)(C\equiv N)-N=N-C(CH_3)(C\equiv N)-CH_2-C(OCH_3)(CH_3)-CH_3$
2,2′-azobis[2, 4-dimethylpentanenitrile]	$HC(CH_3)_2-CH_2-C(CH_3)(C\equiv N)-N=N-C(CH_3)(C\equiv N)-CH_2-CH(CH_3)_2$
2,2′-azobis[isobutyronitrile]	$CH_3-C(CH_3)(C\equiv N)-N=N-C(CH_3)(C\equiv N)-CH_3$
2,2′-azobis[2- methylbutyronitrile]	$CH_3CH_2-C(CH_3)(C\equiv N)-N=N-C(CH_3)(C\equiv N)-CH_2CH_3$
1,1′-azobis[cyclohexanecarbonitrile]	cyclohexane-C(C≡N)-N=N-C(C≡N)-cyclohexane
4,4′-azobis[4-cyanovaleric acid]	$CH_2CH_2(COOH)-C(CH_3)(C\equiv N)-N=N-C(CH_3)(C\equiv N)-CH_2CH_2(COOH)$
dimethyl 2,2′-azobis[2-methylpropionate]	$CH_3OOC-C(CH_3)_2-N=N-C(CH_3)_2-COOCH_3$
azobis[2-acetoxy-2-propane]	$CH_3CO(O)-C(CH_3)_2-N=N-C(CH_3)_2-O(O)CCH_3$
2,2′-azobis[2-amidinopropane] dihydrochloride	$H_2N-C(=NH)-C(CH_3)_2-N=N-C(CH_3)_2-C(=NH)-NH_2 \cdot 2\,HCl$

^{a}Temperature at which $t_{1/2} = 10$ h.

to produce radicals. The thermal stabilities of the hexasubstituted ethanes decrease rapidly with increasing size of the alkyl groups. The 10-h half-life temperature range of this class of initiators is very broad, extending from about 100°C to well above 600°C. An extensive compilation of half-life data on carbon–carbon initiators has been published. The commercially available carbon–carbon initiators are tetrasubstituted 1,2-diphenylethanes that undergo homolyses to generate low-energy, *tert*-aralkyl radical pairs. Three carbon–carbon initiators are currently available commercially: 2,3-dimethyl-2,3-diphenylbutane (**1**), 3,4-dimethyl-3,4-diphenylhexane (**2**), and poly(1,4-diisopropylbenzene) (**3**).

$$C_6H_5-C(CH_3)_2-C(CH_3)_2-C_6H_5$$
(**1**)

$$C_6H_5-C(CH_3)(C_2H_5)-C(C_2H_5)(CH_3)-C_6H_5$$
(**2**)

$$[-C_6H_4-C(CH_3)_2-C(CH_3)_2-]_x$$
(**3**)

OTHER RADICAL-GENERATING SYSTEMS

There are many chemical methods for generating radicals reported in the literature that involve unconventional initiators. Most of these radical-generating systems cannot broadly compete with the use of conventional initiators in industrial polymer applications, owing to cost or efficiency considerations. However, some systems may be well-suited for initiating specific radical reactions or polymerizations; eg, grafting of monomers to cellulose using a cericion.

INITIATION THROUGH RADIATION AND PHOTOINITIATORS

High-energy ionizing radiation sources (eg, X-rays, γ-rays, α-particles, β-particles, fast neutrons, and accelerator-generated electrons) can generate radical sites on organic substrates. If the substrate is a vinyl monomer, radical polymerization can occur. If the substrate consists of a polymer and a vinyl monomer, then polymer cross-linking, degradation, grafting of the monomer to the polymer, and polymerization of the monomer can all occur. Radical polymerizations of vinyl monomers with ionized plasma gases have been reviewed.

Initiation of radical reactions with uv radiation is widely used in industrial processes. In contrast to high-energy radiation processes where the energy of the radiation alone is sufficient to initiate reactions, initiation by uv irradiation usually requires the presence of a photoinitiator; ie, a chemical compound or compounds that generate initiating radicals when subjected to uv radiation. There are two types of photoinitiator systems: those that produce initiator radicals by intermolecular hydrogen abstraction and those that produce initiator radicals by photocleavage.

ECONOMIC ASPECTS

The principal worldwide producers of organic azo initiators are DuPont, Atofina Chemical, and Wako Pure Chemical Industries. The worldwide market for organic azo initiators is small, being only about 10% of the market for organic peroxide initiators. Because most of the consumption of organic peroxides and azo initiators is in the developed countries, market growth is expected to continue to be modest, 2–3% annually.

C. H. Bamford, in H. F. Mark, N. M. Bikales, C. G. Overberger, G. Menges, J. I. Kroschwitz, eds., *Encyclopedia of Polymer Science and Engineering*, 2nd ed., Vol. 13, Wiley-Interscience, New York, 1988, pp. 708–867.

J. Fossey, D. Lefort, and J. Sorba, *Free Radicals in Organic Chemistry*, John Wiley & Sons, Inc., New York, 1995.

G. Moad and D. H. Solomon, *The Chemistry of Free Radical Polymerization*, Pergamon, London, 1995, pp. 7–41.

TERRY N. MYERS
Elf Atochem North America, Inc.

INKS

Writing inks differ from printing inks in that the latter are generally applied to a substrate by means of a printing press. Printing inks as supplied to the graphic arts industry are used in much greater volume by far as compared to writing inks. This article is divided into a discussion of printing inks, followed by some miscellaneous categories of ink, including ink for ball-point pens, with which the greatest amount of ink writing is done. The number of printing-ink manufacturing establishments in the United States is approximately 270, including some 30 captive ink manufacturers.

PRINTING INKS

Printing ink is a mixture of coloring matter dispersed or dissolved in a vehicle or carrier, which forms a fluid or paste that can be printed on a substrate and dried. The colorants used are generally pigments, toners, dyes, or combinations of these materials, which are selected to provide the desired color contrast with the background on which the ink is printed. The vehicle used acts as a carrier for the colorant during the printing operation and in most cases also serves to bind the colorant to the substrate. Printing inks are applied in thin films on many substrates such as paper, paper board, metal sheets and metallic foil, plastic films and molded plastic articles, textiles, and glass. Printing inks can be designed to have decorative, protective, or communicative functions. In some cases, combinations of these functions are achieved.

There are many classes of printing ink, which vary considerably in physical appearance, composition, method of application, and drying mechanism. These also fall into three general types of consistency or viscosity, paste, liquid, and solid. The classes are letterpress and lithographic (litho) inks, which are called paste inks, and flexographic (flexo), rotogravure (gravure) inks and jet inks, which are called liquid inks.

The four key properties of inks are drying, rheology, color, and end use properties. Use properties are considerations that determine how printed substrates function throughout all processing and usage, from the time of printing through the useful life of the printed product.

Drying

Drying may be defined as any process that results in the transformation of a fluid-printing ink into a very high viscosity or solid film. An ink is considered dry when a print does not stick or transfer to another surface pressed into contact with it. Drying is accomplished by one or more of the following physical or chemical mechanisms: absorption, solvent evaporation, precipitation, oxidation, polymerization, cold setting, gelation, and radiation curing.

Rheology

The rheology or flow being of inks as they print or set is their primary physical property. In a Newtonian liquid, any stress produces a flow, the rate of flow being with proportional to the stress. But most inks are generally non-Newtonian and have a nonlinear flow curve. The common terms used to describe ink rheology are viscosity (the resistance to flow); yield value (stress at which a liquid starts to flow); shear thinning (decreasing viscosity with increasing agitation); and shear thickening (the opposite of shear thinning). Mayonnaise is an example of a material having a high yield value but a low viscosity. Ketchup is an example of a shear thinning fluid.

Color and Coloring Materials

Another key property of all types of ink is color, which may very well be the most important one to the consumer, because it has such a great psychological impact. Color has three different attributes: hue or shade, saturation or chroma, and lightness or value. Pigments are used for color but also affect such physical properties as bulk, opacity, specific gravity, viscosity, yield value, and printing qualities. Different pigments of the same hue can have varying fade, heat, chemical, and bleed resistance. Judicious selection of pigment is also required, according to the use of the ink, in considering subsequent operations such as varnishing, waxing, lacquering, or laminating. The end use of the printed product makes demands on the resistance properties of the colored pigments chosen. Inks also make use of inorganic pigments and of dyes, most commonly in the form of organic pigments, dyes rendered insoluble in one of a number of ways. Five dye families that are of interest to ink manufacturers are azo dyes, triphenylmethane dyes, anthraquinone dyes, vat dyes, and phthalocyanine compounds.

Toners are full-strength undiluted pigments used to strengthen tinctorially weak batches of pigments. Occasionally, dyes are utilized as toners. The most common pigments used in ink manufacture are black, white, red, yellow, blue, purple, and green pigments, plus organic and inorganic pigments, daylight fluorescent and metallic powders. In addition, there are driers, waxes, antioxidants, and miscellaneous other additives.

LETTERPRESS AND LITHO NEWSPRINT INKS

The U.S. news ink industry, with a sales potential of over $490 million annually, represents a dynamic, ever-changing segment of the graphic arts market. Of the three printing modes currently available, the dominant web offset lithographic process continues to grow, at the expense of letterpress printing. The change in the market share is governed by a growing demand for color and improved print quality, which the letterpress process cannot deliver. Water-based flexo has made some penetration of the newspaper market, but its growth has essentially plateaued.

The printing of newspapers is conducted at very high speeds, often reaching 3,000 feet per minute. Inks dry by absorption of liquid into the porosity of the substrate. Some evaporation of water in a flexo publication ink can accelerate the drying process.

Web Offset

Web offset is, by far, the dominant type of publication and commercial printing. Web offset lithographic printing uses planographic, aluminum-based printing plates, a fountain solution, and an ink formulated to accept and emulsify water.

Web offset inks are somewhat highly pigmented, to yield the desired print density at a thin printing film ($\sim 1\mu m$). The viscosity of offset inks is relatively high but varies with the press configuration. A variety of additives are used to control the properties of wetting and dispersion of pigments, flow, lithography, and rub-off of inks. These additives belong to classes of materials such as surfactants, bentonite clays, alkyds, functional resins, polymers, etc.

Letterpress

This process, also called typography, is the oldest printing process still in use. Inks in the printing process are transferred directly from a raised area to a substrate. The printing plates contain a thick layer of photopolymer deposited over a plastic or aluminum base.

Basic raw materials for letterpress inks, such as vegetable oils, resins, and pigments, are essentially the same as those used in web offset inks. Inks are tinctorially weaker than offset, relatively fluid, and their low and high shear viscosities are lower than for offset.

Flexography

Printing is conducted with a typographic plate similar to letterpress. However, the chemistry of the photopolymer makes the plate water-insensitive. A high quality of print has been demonstrated by newspapers utilizing this printing process. The printed matter is virtually smudge-free.

Typical inks are water-based, with acrylic emulsion resins as the main binder. Inks of this type occasionally use natural products such as starches, lignins, and lignin derivatives. Hence, ecologically this process is more desirable. Press-ready inks are very fluid and of low viscosity. Inks contain a variety of additives for the elimination of foaming, dispersion of pigments, rheological modifiers, slip agents, etc.

Web Offset Heat-Set Publication and Commercial Inks

Almost all heat-set inks are now printed on web offset presses and are based on vehicles containing synthetic resins and/or some natural resins. These are dissolved in hydrocarbon solvent fractions that are specially fractionated for use in the ink industry. They dry in less than one second by means of solvent evaporation in a heatset oven.

Sheet-Fed Offset Inks

Inks for these presses are based on vehicles containing phenolic-modified, maleic-modified, or unmodified rosin-ester resins dissolved in vegetable drying oils and diluted with hydrocarbon solvents. Some inks also contain alkyds, which may be modified with other polymers, such as urethanes, styrene, and the like. On coated stocks, sheet-fed inks can be formulated to quick-set to a tack-free state by precipitation and solvent separation and then dry fully by oxidation. The most commonly used oils are linseed, soya, and tall oils. Special acrylic resins have been developed for use in quickset inks, and offer nonskinning properties and excellent press stability.

Duplicator and Business Form Inks

Duplicator sheet-fed machines require very press-stable yet quick-setting inks. They must also possess good lithographic properties of wide tolerance for fountain solution and provide good printing properties on a wide variety of uncoated papers. The inks can contain drying oil alkyds along with hydrocarbon resins and high boiling (200–370°C) hydrocarbon solvents. Business form inks closely resemble the lithographic heatset or quick-set inks. Business forms have also been printed by the inkjet method. These inks are usually based on water, glycols, and dyes.

Folding-Carton Inks

The majority of folding-carton inks are based on various quick-set vehicles. However, when maximum gloss, good rub, and product resistance are required, they contain mainly oleoresinous vehicles. They dry by oxidation to form tough, glossy films. Hybrid uv curing inks that are capable of running on presses with conventional roller materials are finding growth in the sheetfed market. This allows a printer to run a high-gloss uv coating over the hybrid uv drying ink inline without glossing back as the uv coating would do if applied over a wet conventional ink. This process leads to significant reduction in process time and better economics.

Metal Container Inks

Ink vehicles for metal containers that are printed on special flat sheet-fed litho presses are based mainly on blends of oleoresinous varnishes containing alkyds, polyesters, and melamine resins. These inks dry during a 5–15 min cure at 150–250°C in long gas-fired ovens. Polymerization, oxidation, and cross-linking reactions accounts for their drying and hardening. Ultraviolet inks also have been used in metal decorating but are not as popular in this type of printing.

Plastics

Vehicles in offset inks for plastics (polyethylene, polystyrene, vinyl) are based on hard drying oleoresinous varnishes that are sometimes are diluted with hydrocarbon solvents. Uv inks are widely used for decoration of preformed plastic containers.

Manufacture

Paste inks are produced in two ways: (1) by mixing predispersed (preground) or flushed pigment concentrates with vehicles, solvents, oils, and compounds, and filtering; or (2) by mixing dry pigments or resin-coated pigments with vehicles and compounds and then dispersing them with various types of ink mills. The more fluid inks (news, flexo, or gravure) usually are delivered in tank trucks directly to the printer. Ink vehicles are usually produced in separate resin/varnish plants.

FLEXOGRAPHIC, ROTOGRAVURE, AND JET INKS

Flexo and gravure inks are both known as liquid inks, because of their low viscosity. The inks for both systems have basic components in common with inks for other printing processes. Vehicles disperse and carry the pigment, and also contribute most to the end use properties. Colorants provide color. Solvents dissolve resins in the vehicle and determine the drying rate. Additives modify ink properties to overcome deficiencies.

The vehicle is composed of resins, solvent, and additives. Solvents are required for two reasons. The first requirement a solvent must satisfy is to dissolve the resin; this results in a low-viscosity ink suitable for printing. Secondly, the solvent must evaporate quickly and completely from the printed film.

Both flexo and gravure inks are delivered in the form of a virgin ink concentrate, which retards the speed of pigment settling and reduces shipping costs. Solvent is used press-side to reduce the ink to a correct printing viscosity.

Additives are used to provide a specific property. For example, a wax provides rub resistance in the printed film or a special surfactant reduces foam generation in the fountain.

Manufacture

Manufacturing processes consist of two general operations, vehicle preparation and pigment dispersion. Vehicle preparation can be as basic as polymerization of resins or as simple as cold dissolving of vehicle solids in appropriate solvents. Pigment dispersion can be done in a ball mill, which lends itself to volatile fluid formulations, or in a vertical or horizontal media mill. Much of the pigment dispersion done currently is in a horizontal or vertical media or shot mill.

Rotogravure Inks

Since there are no rubber or plastic components in contact with the solvents contained in gravure ink formulations, it is permissible to use more aggressive solvents such as ketones and aromatic hydrocarbons, which cannot be tolerated in flexo inks. This provides the gravure ink formulator much greater latitude in regard to binder selection. In other respects the compositions generally are similar.

There are 10 gravure ink types categorized by the binders or solvents used: A, aliphatic hydrocarbon; B, aromatic hydrocarbon; C, nitrocellulose; D, polyamide resins; E, SS nitrocellulose; M, polystyrene; T, chlorinated rubber; V, vinyls; W, water-based; and X, miscellaneous.

Ketones and esters are required for C-type inks. The usual solvent as for D-type inks are mixtures of an alcohol, such as ethyl alcohol or isopropyl alcohol, with either aliphatic or aromatic hydrocarbons. The alcohols, proprietary denatured ethyl alcohol and isopropyl alcohol, are commonly used for E-type inks. Aromatic hydrocarbon solvents are used for M-type inks. T-type inks are also reduced with aromatic hydrocarbons. Acrylic resins are used to achieve specific properties for V-type inks. W-type inks use water, or mixtures of water and alcohol, as the solvent. Inks which are not of a recognized type are classified as X-type.

The main advantages of water-based inks include excellent press stability, printing quality, heat resistance, an absence of fire hazard, and the convenience and economy of the for reduction and wash-up. The main disadvantage of these inks is the increased energy required for drying, due to the high latent heat of water.

The majority of Type A and Type B inks are used for gift wraps, newspaper supplements, catalogs, advertising inserts, and similar publication work. Inks in the Type C group are used for printing on foil, paper, cellophane, paperboard, coated and uncoated paper, glassine, acetate, metallized paper, and some specialized fabrics. Type C inks are the dominant group used in packaging gravure. Type D inks have excellent adhesion to many plastic films. They are used in foil, paper, and paperboard as well as on a variety of films. Type E inks are often used on paper and paperboard, some grades of cellophane shellac, or nitrocellulose-primed foil-pouch stock glassine, and many specialty coated papers and boards.

Water inks are primarily used in packaging gravure on board and paper. Publication gravure printers are actively testing Type W inks for various publication applications. Type V inks are used for printing vinyl films and Saran.

InkJet Printing

This is a noncontact form of printing that has the advantage of including all the information on a printed page in digitized form in computer memory, thereby eliminating the need for a plate. One principle of operation involves the issuance of liquid ink from an orifice at very high speed to form a jet that is then broken up by ultrasonic energy to produce uniform droplets that can be charged electrically. These droplets can be deflected electrostatically into a catcher, while the uncharged droplets continue in flight to form dots on the printing surface to construct images. This is called continuous inkjet. Another principle of operation is the drop-on-demand, DOD, jet which emits drop-lets of ink only when energized by the computer. Inkjet technology provides a means of fast, dependable, quality, single or personalized copy printing. Its nonimpact nature permits printing on uneven surfaces and delicate materials. Computer operation permits the encoding of repetitive and nonrepetitive information. Since the information sent to the inkjet is computer driven, the information can be changed from image to image, unlike other technologies, allowing for individual images and information to be printed for each print.

Lamination Inks

This class of ink is a specialized group. In addition to conforming to the constraints described for flexo and gravure inks, these inks must not interfere with the bond formed when two or more films—eg, polypropylene and polyethylene—are joined with the use of an adhesive in order to obtain a structure that provides resistance properties not found in a single film. In addition to polyamide, lamination inks ordinarily contain modifiers such as polyketone resin, plasticizer, and wax to impart specific properties such as block resistance and increased bond strength. Because laminating inks are usually reverse-side printed and end-up sandwiched between films, gloss is not a primary requirement.

Screen-Process Inks

These inks, often known in the past as silk-screen inks, are printed on the substrate by being forced through a screen stencil by means of a squeegee. For many years this was a hand-printing operation but has now become largely mechanized. Screen-process inks are dispersions of pigments in vehicles that are, for the most part, solutions of resins in solvents of the boiling range of VM&P naphtha. Drying of solvent-based inks is usually by evaporation but in some cases is a combination of oxidation and evaporation. Various types of binders are used, such as rosin esters, phenolics, cellulose derivatives, vinyls, and oleoresinous varnishes, depending on the film properties desired. uv inks are also widely used for screen-process printing. After premixing, the ink is ground on a three-roll or media mill. The resulting ink should be short and soft, so as not to drag on the squeegee and to release the substrate cleanly after the print is made, without excessive stringing.

MISCELLANEOUS INKS

Stamp-Pad Inks

These inks are impregnated into a cloth or foam-rubber pad and transferred by pressure to rubber type, which is then stamped or impressed against the substrate. The inks must be completely nondrying in the pad and yet dry by rapid penetration into the paper. Since it is desirable that the total ink soak into the stock, dyes rather than pigments are used. The vehicles used are usually glycols.

Ball-Point Inks

These inks are medium-viscosity semi-Newtonian fluids of high tinctorial strength, which must be slow drying and free of particles so that they continue to feed to the paper without clogging. Drying on the paper is accomplished by rapid penetration and some evaporation. These properties are obtained by strong dye solutions and pigment dispersions in vehicles containing oleic acid and castor oil or a sulfonamide plasticizer. The rheology of these inks exhibits modest thixotropy, which prevents their leakage through the openings around the ball.

Water-Based Writing Inks

These consist of very fine pigment dispersions in an aqueous media containing small amounts of glycol or glycerol and a dispersing aid. They dry mainly by evaporation and quick wetting of cellulosic fibers in paper substrates.

Engraving Inks

Owing to the thick film that can be deposited (ca 10–40 μm), high strength formulations are not required, but the body of the ink is quite short so as to wipe cleanly from the plate. Drying is by a combination of oxidation or polymerization and by evaporation of the solvent. The pigment, including a large percentage of colorless extender pigment, is dispersed on a three-roll mill in a vehicle composed of a of heat-bodied drying of oil or an oleoresinous vehicle, sometimes in combination with a resin–solvent type vehicle. Web engraving presses using heat or electron beam curing have been developed. They use appropriate polymerizing vehicles.

Electrographic Inks

Electrographic printing is accomplished by causing charged colored particles to move in an electrostatic field to a substrate in the form of an image. The image may be formed by a screen stencil or by a gravure cylinder. One of the primary advantages of this process is its ability to print across gaps and thus without pressure. The ink, also called an electrographic toner, is a powder composed of pigment dispersed in a solid resin. After the image is deposited on the substrate, it is heat- or solvent-fused to a continuous film.

Decal Inks

In decalcomania, a transfer method of printing, the inks must dry completely on the surface by oxidation or solvent evaporation, because the treated paper has no ink absorbency. After the initial printing, the design is transferred to the permanent substrate by direct contact and soaking with water. The formulation of decalcomania inks is governed by the particular printing process employed in printing the transfer paper. Decalcomanias for ceramics require pigments that may be heated to high temperatures. Further, most decalcomanias should use pigments that are fast to light, because many are subsequently transferred to outdoor signs or to store windows. Vehicles consist of oleoresinous varnishes containing metallic driers or are resin–solvent types.

Heat-Transfer Inks

Heat-transfer printing is similar to the decalcomania process in that the printing is first done on a temporary substrate, but heat is used as the transfer mechanism rather than water solubility.

Heat-Sublimation Inks

Another type of heat-transfer ink is used for transferring sublimable dyes from preprinted paper substrates to various types of synthetic fabrics using high temperatures (typically 190–210°C). The inks can be either lithographic, flexographic, screen, jet, or rotogravure. The inks are formulated with a sufficient content of a special dye, which turns to a vapor when heated sufficiently. The dye vapors transfer to the fabric and bonds to it. The inks are preprinted on a coated paper of 65–90 pounds and dried. The paper is placed in close contact with the fabric and heated to transfer the image to the substrate.

QUALITY CONTROL AND TESTING

Control of inks is done by examining their color strength, hue, tack, rheology, drying rate, stability, and product resistance. Elaborate control equipment and laboratory testing procedures are employed to test the finished inks. Proofing presses and sometimes pilot presses are utilized by ink manufacturers to control production and test new formulations. New regulations, availability and cost of raw materials, new or modified substrates, and faster presses require constant updating of the formulations. Ultraviolet light, electron beam, and water-based inks for various applications are some of the areas being actively pursued. Trends to a more universal use of renewable resources have led to increased amounts of natural products such as vegetable oils and rosin in litho inks of all types, particularly in sheet-fed litho inks.

ENVIRONMENTAL CONSIDERATIONS AND REGULATORY COMPLIANCE

In general, environmental concerns affect the ink-manufacturing companies, the converting industry that uses the products to produce packaging and printed matter of all types, and even extend to issues related to the ultimate end user companies whose printed or packaged products are often distributed worldwide. Common environmental aims such as reducing air pollution, use of renewable resources rather than crude-oil-based chemistries, biodegradable inks and coatings, the pressure to recycle waste materials back into the raw material supply, etc. all impact printing ink technology. Over the past twenty years, new ink products have been developed in response to these environmental concerns and will continue to emerge as new issues supplant existing ones.

United States regulations encompass federal, state, and local guidelines. In addition, there are numerous voluntary industry guidelines affecting ink making.

H. Kipphan, ed., *Handbook of Print Media*, Springer-Verlag, Heidelberg, Germany ISBN#: 3-540-67326-1.

Printing Ink Handbook, National Association of Printing Ink Manufacturers, Hasbrouck Heights, N.J., 1988.

W. E. Rusterholz, *Focus on Inks, High Volume Printing, Environment* 2000, Apr. 1990.

RadTech Primer published by RadTech International North America (www.radtech.org)

L. Wilson, *What the Printer Should Know about Paper*, GATF Press, Sewickley, Pa., 1997.

R. W. Bassemir
A. Bean
Sun Chemical Corporation

INSECTICIDES

There are nearly 1 million described species of insects, constituting approximately 72% of all animal species. About 1% of them are considered significant pests. This article summarizes the chemistry, properties, uses, and advantages and disadvantages of many chemicals used for insect control.

Approximately 70% of all insecticide use is in agriculture. Applications are generally made directly to raw agricultural commodities to protect plants and animals from insect attacks. With the exception of microbial insecticides, nearly all of the uses of insecticides result in residues of the various chemicals and their degradation products.

Insects attack humans or domestic animals; transmit human, animal, and plant diseases; destroy structures, and compete for the available supplies of food and fiber. In the United States there are more than 10,000 species of insects, mites, and ticks that cause losses to agriculture, but only about 600–700 species require annual applied control measures.

ROLE OF CHEMICALS IN INSECT CONTROL

The role of chemicals in the control of insect pests constantly undergoes searching reevaluation in terms of benefits versus adverse effects. Benefits and the long-term and short-term implications for ecological systems, human health, and the environment are constantly reassessed as each new insecticide discovery is factored into the balance.

Plant-derived insecticides—eg, nicotine, rotenone, veratrine, and pyrethrum—have been used to kill insect pests since antiquity. The first practical synthetic organic insecticide was developed in Germany in 1892, but it was the discovery of the insecticidal properties of DDT in 1939 that began an era of chemical pest control resulting in the synthesis and evaluation of hundreds of thousands of synthetic organic chemicals as insecticides. Dichlorodiphenyltrichloroethane (DDT), with its efficient contact insecticidal action together with long residual persistence and relative safety to humans and domestic animals, largely replaced the arsenicals. By 1976, more than 200 chemical compounds were marketed as insecticides and the total application to primary crops was 58,000 t.

The value of insecticides in controlling human and animal diseases spread by insects has been dramatic. Between 1942 and 1952, the use of DDT in public health measures to control the mosquito vectors of malaria and the human body louse vector of typhus saved 5 million lives and prevented 100 million illnesses. Insecticides have provided the means to control such important human diseases as filariasis, transmitted by *Culex* mosquitoes, and onchocerciasis, transmitted by *Simulium* blackflies.

Although the economic benefits of insecticide use became rapidly apparent to the farmer, it was observed that after several years of repeated applications of an insecticide the onset of resistance in a pest species could rapidly nullify its beneficial effects. Pest management systems in which chemicals could be used more effectively were thus developed, as were alternative practices to control insects. Amounts applied per acre were reduced by the introduction of chemicals that were effective at low rates. Table 1 shows the amounts of insecticides used in 1995. The top ten have been categorized in order of weight of active ingredient used, but if the number of acres treated by a particular pesticide were to be used as a basis it is likely that several pyrethroids and perhaps avermectins would now be included in the top ten.

Table 1. Insecticides Used by American Farmers in 1995

Name	Active ingredients	Class
Oil	23,181	petroleum oils
Chlorpyrifos	6,697	organophosphate
Terbufos	3,942	organophosphate
Methyl parathion	2,704	organophosphate
Carbofuran	2,314	carbamate
Carbaryl	2,073	carbamate
Phorate	2,020	organophosphate
Cryolite	1,839	inorganic
Aldicarb	1,825	carbamate
Propargite	1,646	propynyl sulfite (acaricide)
Total	67,594	

INTEGRATED PEST MANAGEMENT

Although employment of chemicals for insect pest control is essential to modern society, the extensive and injudicious use of chemical insecticides since 1946 has resulted in many problems, including (1) widespread insect resistance, (2) emergence of resurgent and secondary pests whose regulating natural enemies have been adversely affected, (3) hazards to human health, (4) environmental pollution, and (5) exponentially increasing costs of new insecticides. Many of these unintended consequences of chemical pest control arose from the pervasive eradication philosophy that resulted from euphoria about the effectiveness of successive generations of organochlorine, organophosphorus, carbamate, and pyrethroid insecticides.

The primary goals of IPM are (1) to determine how the life system of the pest needs to be modified to reduce its numbers to tolerable levels; ie, below the economic threshold; (2) to apply biological knowledge and current technology to achieve the desired modification; ie, applied ecology; and (3) to devise procedures for pest control compatible with economic and environmental control aspects; ie, economic and social acceptance.

IPM practices rely heavily on the protection and conservation of natural enemies, parasites, predators, and diseases that regulate or balance populations of insect pests. IPM programs are based on two important parameters: the economic injury level, defined as the population density of a pest that causes enough injury to justify the cost of remedial treatment; and the economic threshold, defined as the pest density at which control measures should be applied to prevent an increasing insect population from reaching the economic injury level. Whenever applied, IPM practices have consistently resulted in decreases in insecticide applications of 50% to 90% over conventional spray programs.

Insecticide management is concerned with the safe, efficient, and economical handling of insecticides during manufacture, utilization, and disposal. The essential components are selection of the proper insecticide for the IPM program; selection of the mode, timing, and dosage of application; consideration of the problems of resistance and resurgence, the possible effects of insecticide residues on food crops and in the environment; and the impact of these on humans, domestic animals, and wildlife.

I. Ishaaya, ed., *Biochemical Sites Important in Insecticide Action and Resistance*, Springer, Berlin, 2001.

R. L. Metcalf and R. A. Metcalf, *Destructive and Useful Insects*, 5th ed., McGraw-Hill, New York, 1993.

A. S. Perry, I. Yamamoto, I. Ishaaya, and R. Y. Perry, *Insecticides in Agriculture and Environment*, Springer, Berlin, 1998.

J. R. Plimmer, ed., *The Encyclopedia of Agrochemicals*, Wiley, New York, 2003.

JACK R. PLIMMER
Tampa

DEREK W. GAMMON
Sacramento

IODINE AND IODINE COMPOUNDS

Iodine (I), atomic number 53, atomic weight 126.9044, is a nonmetallic element belonging to the halogen family in Group 17 VII of the periodic table. The only known stable species has a mass number of 127. There are 22 other iodine isotopes having masses between 117 and 139; 14 of these yield significant radiation.

Iodine is a bluish black, crystalline solid having a metallic luster. It is obtained in blocks or lumps like other elements, but in shiny flakes or prills that can be easily crushed to powder. Iodine crystallizes in rhomboidal plates belonging to the triclinic system.

OCCURRENCE IN NATURE

Iodine is, indeed, one of the scarcest of nonmetallic elements in the total composition of the earth. Although not abundant in quantity, iodine is distributed almost everywhere, in rocks, soils, water bodies, plants, animal tissues, and food-stuffs. Except for the possible occurrence of elemental iodine vapor in the air near certain iodine-rich springs, iodine never occurs free in nature, always found in combination with other elements.

Wherever iodine occurs, its quantities are generally exceedingly small, and very sophisticated chemical methods are required to detect it. Only a few substances characteristically contain iodine in relatively large quantities. These are seaweeds, sponges, and corals; the underground waters from certain deep oil-well boring and mineral springs; and, most impressive of all, the vast natural deposits of sodium nitrate ("caliche" ore) found in the northern part of Chile. Even in these, however, the proportion of iodine is small, rarely exceeding 1 part in 500.

PHYSICAL PROPERTIES

Iodine crystallizes in the orthorhombic system and has a unit cell of eight atoms arranged as a symmetrical bipyramid. The cell constants at $18°C$ are given in Table 1, along with other physical properties.

Table 1. Physical Properties of Iodine

Properties	Solid	Liquid	Gaseous
color	bluish black	bluish black	violet
melting point	113.6		
boiling point, $°C$		185	
critical temperature, $°C$		553	
critical pressure, kPa^c		11753.7	
density, g/mL			
$20°C$	4.93		
$60°C$	4.89		
$120°C$		3.960	
$180°C$		3.736	
vapor density at 101.3 $kPa,^d$ g/L			6.75
cubic coefficient of expansion $0-113.6°C$, $°C^{-1}$	2.81×10^{-4}		
crystal structure, 4 mol I_2 per unit cell	orthorhombic		
unit cell dimensions at $18°C$, pm			
a	477.61		
b	725.01		
c	977.11		
entropy at $25°C$, $J/(mol \cdot K)^e$	116.81		62.25
specific heat, $J/(g \cdot K)^e$			
$25-113.6°C$ $0.1582 + 1.9628 \times 10^{-4} T^f$			
$113.6-184°C$		0.3165	
$25-1200°C$			0.1465
heat of fusion at $113.6°C$, J/g^e	62.17		
heat of sublimation at $113.6°C$, J/g^e	238.40		
heat of vaporization at bp, J/g^e		164.45	
viscosity, mPa(=cP)			
at $116°C$		2.268	
at $185°C$		1.414	
vapor pressure, kPa^d			
at $25°C$	0.04133^a		
at $113.6°C$	12.0655	b	
thermal conductivity at $24.4°C$, $W/(m \cdot K)$	0.4581		
electrical resistivity, $\Omega \cdot cm$			
at $25°C$	5.85×10^6		
at $110°C$	8.33×10^5		
at $140°C$		1.1×10^5	
dielectric constant			
at $23°C$	10.3		
at $118°C$		11.08	
refractive index, n_D	3.34		

^{a}For the solid, between 0 and $113.6°C$, $\log p_{kPa} = -(3410.71/T) - 0.3523 \log T - 1.301 \times 10^{-3} T + 14.3140$ where T is in Kelvin.
^{b}For the liquid, between 113.6 and $186°C$, $\log p_{kPa} = -(2300.24/T) + 10.025$, where T is in Kelvin.
^{c}To convert kPa to atm, divide by 101.
^{d}To convert kPa to mm Hg, multiply by 7.50.
^{e}To convert J to cal, divide by 4.184.
^{f}T is in Kelvin.

Iodine dissolves in many organic solvents, and the color of the resulting solutions varies with the nature of the solvent.

Iodine vapor is characterized by the familiar violet color and by its unusually high specific gravity, approximately nine times that of air.

CHEMICAL PROPERTIES

The electron configuration of the iodine atom is $[Kr]4d^{10}5s^25p^5$ and its ground state is $2p^0_{3/2}$. Principal oxidation states are -1, $+1$, $+3$, $+5$, and $+7$, but the oxide IO_2, where iodine has an oxidation state of $+4$, is also known. Iodine forms thermodynamically stable compounds in all these oxidation states, except the $+4$. Iodine is the heaviest of all the common halogens and the least electronegative. It is usually less violent in its reactions than the other members of the halogen family. Iodine presents mild oxidizing properties in acidic solutions.

Iodine forms compounds with all the elements except sulfur, selenium, and the noble gases. It reacts only indirectly with carbon, nitrogen, oxygen, and some noble metals such as platinum.

The chemistry of aqueous iodine has been extensively studied because of the role of iodine as a disinfectant. The system is vey complex, owing to the number of oxidation states available to iodine under ambient conditions.

MANUFACTURE AND PROCESSING

The production processes are basically related to the raw materials containing iodine: seaweeds, mineral deposits, and oil-well or natural-gas brines.

The earliest successful manufacture of iodine started in 1817 using certain varieties of seaweeds. The seaweed was dried, burned, and the ash was lixiviated to obtain iodine and potassium and sodium salts.

The only iodine obtained from minerals has been a by-product of the processing of nitrate ores in Chile. There are two ways for producing iodine from caliche iodates: (1) from solutions containing more equivalent iodine than its solubility as elemental iodine in the same solution ($\sim0.4\,g/L$ at $25°C$) and (2) from more diluted equivalent iodine solutions.

About 50% of the iodine currently consumed in the world comes from brines processed in Japan, the United States, and the former Soviet Union. The predominant production process for iodine from brines is the blow-out method, which was first used in Japan. Iodine is present in subsurface brines as sodium and/or potassium iodide, and aconcentration that varies from about 10 to 150 ppm. The recovery process can be divided into brine cleanup, iodide, oxidation to iodine, followed by air blowing out and recovery, and iodine finishing.

The newest process uses ion-exchange resins on brines already oxidized to liberate iodine. The liberated iodine, in the form of polyiodide, is adsorbed on Amberlite IRA-400, an anion-exchange resin. When the ion-exchange resin is saturated, it is discharged from the bottom of the column and then transferred to the elution column, if iodine is eluted (desorbed) using caustic solution followed by

sodium chloride. The regenerated resin is then returned to the adsorption column. The elutriant, rich in iodide and iodate ions, is acidified and oxidized to precipitate iodine. The crude iodine is then separated in a centrifuge and purified with hot sulfuric acid or refined by sublimation.

ECONOMIC ASPECTS

Iodine has traditionally seen boom and bust cycles. Tight supply has sent prices up, followed by price increases that encouraged fast expansion of the capabilities of the iodine-producing companies, where upon prices fell, putting margin pressure on some higher-cost brine-based producers, possibly forcing them to curtail production.

Plants in the United States are basically iodine producers that need to extract the solutions from deep wells and then reinject the processed solutions because of environmental reasons and maintain the underground pressure of the exploitation area. In Japan, iodine is mainly a by-product of the natural-gas production, and the wells are less deep. Depleted solutions are either discarded into the ocean or reinjected. For Chilean iodine, which is associated mainly with nitrate production, plant location is usually adjacent to nitrate plants.

GRADE SPECIFICATIONS, STANDARDS, AND SHIPPING

Commercial iodine has a minimum purity of 99.8%. In the past the requirements were attained basically only by sublimation, whereas now a days the specifications can be met by direct-production iodine.

Iodine is packed in fiber drums lined with double-ply polyethylene containing 10, 25, and 50 kg. There are no specific transportation, shipping, or safety requirements. As low temperatures reduce vapor pressure and therefore decrease the rate of loss, a cool, well-ventilated storage area is recommended. There is no specific freight classification; iodine is shipped as "chemicals." Special labels are subject to specific country regulations.

HEALTH AND SAFETY FACTORS

Iodine is much safer to handle at ordinary temperatures than the other halogens, because it is a solid and its vapor pressure is only 1 kPa (7.5 mm Hg) at $25°C$, compared to 28.7 kPa (215 mm Hg) for bromine and 700 kPa (6.91 atm) for chlorine. When handling properly packed containers, the usual work clothes are sufficient. In the handling of solid, unpacked iodine, rubber gloves, a rubber apron, and safety goggles are recommended. Respirators or masks are also recommended.

The U.S. Occupational Safety and Health Administration (OSHA) has set for iodine a ceiling level of 0.1 ppm in air. The American Conference of Governmental Industrial Hygienists (ACGIH) has established 0.1 ppm as the threshold limit value TLV [Time-weighted Average (TWA)] for iodine. The maximum allowable concentration (MAK) value is also 0.1 ppm.

Empty containers may be destroyed in an incinerator or decontaminated by washing with a dilute thiosulfate or sulfite solution. Bulk wastes should be treated by controlled iodine recovery processes.

Emergency treatment includes irrigating the eyes with water or washing the contaminated body area with a 5% thiosulfate solution followed by saline catharsis. If swallowed, gastric lavage with 5% solution of thiosulfate, followed by saline catharis, should be accomplished, and medical attention should be provided. If pulmonary signs are severe, oxygen should be supplied with an intermittent positive-pressure breathing apparatus.

Chronic absorption of iodine causes *iodism* characterized by insomnia, inflammation of the eyes and nose, bronchitis, tremor, diarrhea, and weight loss.

Iodine is not combustible itself but can react very vigorously with reducing materials.

ENVIRONMENTAL CONCERNS

Pure iodine sublimates when heated and toxic fumes are released: thus, iodine must be stored in sealed containers, away from heat and sunlight. In case of a large spillage of material, use of an aqueous sodium thiosulfate solution or dry sodium carbonate is recommended to neutralize the spilled iodine.

Most common iodine derivatives are not harmful to the environment and are quickly degraded into the basic elements. Inorganic iodides and iodates of calcium and potassium, which represent the main iodine use worldwide, are ingested by human beings and animals, as additives in edible salt and components in pharmaceuticals, for the prevention and treatment of different diseases.

Some processes for the production of chemical intermediates containing iodine produce waste streams that are recycled for both environmental and economic purposes. Recycling normally burns the impurities that should not be discharged into the environment and recovers the iodine in its elemental form or as a derivative.

USES

Iodine has a wide range of uses in chemical and related industries. A high percentage of the initial use of iodine lies in the production of intermediates, which are frequently marketed as such before reaching their ultimate end use. However, the best information available from the market suggests a breakdown of world iodine consumption, as follows:

Iodine consumption	%
Catalysts	8
Stabilizers	6
X-rays contrast media	20
Sanitizers and disinfectants	17
Pharmaceuticals	8
Animal and fowl feeds	10
Photography	6
Herbicides	6
Ink, colorant and dyes	4
Other uses	15

The main specific uses of iodine are as catalysts for many reactions, stabilizes, (primarily in nylon and rubber), photographic and X-ray films, X-ray contact media, dyes, inks, colorants, animal feeds to counter goiter, and extensively in health and sanitary applications, including sanitizers, cleaning products, water treatment chemicals, disinfectants, pharmaceuticals, radiation protection, and medical research.

INORGANIC COMPOUNDS

Iodides and iodates are the principal iodine inorganic derivatives. Some of the more important inorganic iodine intermediates and derivatives are potassium iodide (KI), MW 166.02, mp 686°C, I 76.45%, sodium iodide, (NaI), MW 149.92, mp 662°C, I 84.66%) and hydrogen iodide, (HI), MW 127.93, mp −50.9°C, bp −35.1°C, I 99.21%. The iodates, which are stable at room temperatures but lose oxygen on heating, are potassium iodate (KIO$_3$, MW 214.02, I 59.30%) sodium iodate, (NaIo$_3$), MW 197.90, I 64.13%.

ORGANIC COMPOUNDS

The organic iodine compounds have lower heats of formation and greater reactivities than do their chlorine and bromine analogues. As in the case of the inorganic iodides, their indiexs of refraction and specific gravities are higher than the corresponding chlorine and bromine derivatives.

The aliphatic iodine derivatives are usually prepared by reaction of an alcohol with hydroiodic acid or phosphorous triiodide; by reaction of iodine, an alcohol, and red phosphorous; addition of iodine monochloride, monobromide, or iodine to an olefin; or replacement reaction by heating the chlorine or bromine compound with an alkali iodide. The aromatic iodine derivatives are prepared by reacting iodine and the aromatic system with oxidizing agents such as nitric acid, fuming sulfuric acid or mercuric oxide. They include methyl iodide (CH$_3$I, MW 141.95, I 89.41%), methylene iodide, (CH$_2$I$_2$, MW 267.87, I 94.76%, mp 6.0°C and bp 181°C), thymol iodide (C$_2$OH$_{24}$I$_2$O$_2$, MW 550.23, I 46.13%), iodoform (CHI$_3$, MW 393.78, I 96.69%, mp ∼ 120°C and 4.008 density at 20°C), and ethyl iodide (C$_2$H$_5$I), a colorless liquid with a density of 1.933 at 20°C and a boiling point of 72.2°C.

American Chemical Society Specifications, Reagent Chemicals, 8th ed., American Chemical Society, Washington, D.C., 1993.

Roskill Information Services Ltd., *The Economics of Iodine*, 6th ed., London, Sept. 1994.

ARMIN LAUTERBACH
GUSTAVO OBER
SQM Chemicals

ION EXCHANGE

Ion exchange is a process in which cations or anions in a liquid are exchanged with cations or anions on a solid sorbent. Cations are interchanged with other cations, anions are exchanged with other anions, and electroneutrality is maintained in both the liquid and solid phases. The process is reversible, which allows extended use of the sorbent resin before replacement is necessary.

Many naturally occurring inorganic and organic materials have ion-exchange properties. This article places emphasis on the styrenic and acrylic resins that are made as small beads.

The primary application for ion exchange is the softening and deionization of water. The remaining applications include waste treatment, catalysis, purification of chemicals, plating, hydrometallurgy, food processing, and pharmaceutical uses. Because ion-exchange resins are insoluble polymeric acids and bases, these resins are also useful in removing acids and bases from gaseous streams via the neutralization of functional groups.

Weak and strong acid-type resins are for removal of cations and are called cation exchangers. Weak and strong base resins remove anions and are called anion exchangers. In addition to these four resin types, there are specialty resins used in applications where higher specificity for certain ions under challenging conditions is a critical factor.

Continuous columnar operation of ion-exchange systems is preferred over batch operation. Each column must be taken off-stream periodically to remove the adsorbed ions and restore the resin to the ionic form required for the adsorption step. In this sense, a columnar ion-exchange operation is not continuous. Continuous operation has been approached in a number of designs by moving resin, or vessels containing resin, in a direction opposite to the flow of liquid. Some of these approaches have been abandoned; others are increasing in popularity.

MANUFACTURE OF RESINS

The production of ion-exchange resins is a multiple step process. It begins with the polymerization of monomers to form solid intermediate copolymers that are insoluble in both water and solvents. The copolymers are functionalized during additional steps in different reactors from those used for copolymer production. Conversion to another ionic form may be required after functionalization is completed. Packaging in fiber drums is common. Alternative containers include metal drums, bulk boxes and bags, and smaller plastic, paper, and burlap bags. Polyethylene liners are used as a barrier between containers and water-containing resin.

Manufacture of ion-exchange resins has traditionally been a batch process. Significant progress was made more recently in the development of a continuous process for the manufacture of copolymer beads.

Copolymerization

The chemistry of the resin matrix, the type and degree of porosity, the particle size, and the particle size distribution are established in the copolymerization step. Formulations and operating procedures must be strictly followed. Reaction vessels must be well designed. Mistakes made during copolymerization are rarely corrected during functionalization.

Functionalization

Copolymers do not have the ability to exchange ions. Such properties are imparted by chemically bonding acidic or basic functional groups to the aromatic rings of styrenic copolymers, or by modifying the carboxyl groups of the acrylic copolymers. There does not appear to be a continuous functionalization process on a commercial scale.

PHYSICAL AND CHEMICAL PROPERTIES

Ion-exchange resins are used repeatedly in a cyclic manner over many years, and deterioration of both physical and chemical properties can be anticipated. Comparison of the properties of used resin with those of new resin is helpful to learning more about the nature and cause of deterioration. Corrective action frequently extends the life of the resin. Comparison of properties must always be made with the resin in the same ionic form.

Particle Shape and Size

With few exceptions, resins are supplied as small, round beads having a diameter between 0.3 and 1.2 mm. Some resins are reduced to a smaller size by grinding to satisfy specific requirements in applications for electric power generation and pharmaceuticals.

Density and Specific Gravity

Density generally pertains to the bulk, or pack-out, weight of wet resin per unit volume. The density is characteristic of the resin and is dependent on the copolymer structure, the degree of cross-linking, the nature of the functional groups, and the ionic form of those groups. A change in density after extended use is a signal that chemical degradation has occurred. The density of most cation exchangers is in the 800-900 g/L range, whereas most anion exchangers are in the range 640-740 g/L.

The specific gravity generally refers to the value determined for wet resin when using a pycnometer. Values range from about 1.04 to about 1.25. Cation exchangers have a greater specific gravity than anion exchangers.

Porosity

The structure of ion-exchange resins is either microporous or macroporous. Microporous resins are more commonly referred to as gel or gelular-type resins.

Capacity

Capacity is a measure of the quantity of ions, acid, or base removed (adsorbed) by an ion-exchange material. The quantity removed is directly correlated with the number of functional groups. Operating capacity, also called the working capacity or column capacity, is a measure of the quantity of ions, acids, or bases adsorbed, or exchanged, under the conditions existing during batch or columnar operation.

Selectivity

A significant exchange of ions does not occur unless the functional group of the resin has a greater selectivity for ions in solution than for ions occupying the functional group, or unless there is a mass action effect, as in regeneration. Selectivity coefficients have been reported in numerous publications for both cations and anions.

The need to know selectivity coefficients precisely is rarely necessary in industrial applications. However, knowledge of relative differences is important when deciding if the reaction is favorable or not.

Selectivity differences increase as the degree of crosslinking of a resin increases, but these differences are relatively minor. Structural composition of the functional groups has a much greater effect on the magnitude of selectivity differences.

Kinetics

The degree to which an ion-exchange reaction is completed depends on a number of factors which include contact time, ionic concentration, degree of cross-linking, and temperature.

Moisture and Water Content

Each resin has a characteristic water content dependent on the resin matrix, the structure of the functional groups, and the ionic form of those groups. Resins are packaged by weight and sold by volume. The dewatering operation prior to packaging is a critical step since removal of too little is costly to the buyer. Analyzing for water content is important to both the seller and user. The quantity of water contained by the resin is recorded on a percentage basis and determined by two methodologies. In each procedure, a small (ca 15 g) sample is removed from a larger composite sample collected during pack-out. In one procedure, the sample is accurately weighed before and after placing in a $105°C$ oven for at least 8 h. This procedure yields the moisture content typical of resin contained in the shipping containers. In the other procedure, a similar sample is soaked in water, then filtered under vacuum in a Buchner funnel prior to weighing before and after oven drying. The moisture content reported is a pseudoequilibrium value typical of the specific resin and its ionic form.

Swelling and Shrinking

Ion-exchange resins shrink or swell reversibly as they are converted from one ionic form to another. The degree of change is dependent on the resin matrix, the functional group, and the ions adsorbed by the functional groups.

The degree of swelling and shrinking is important for design of ion exchange columns, especially for the location of the distributors used to disperse incoming fluids, and collect outgoing ones, evenly over the cross-sectional area of the resin bed.

Hydraulic Properties

Both the resistance to liquid flow through a resin bed and the degree to which a resin bed expands during a backwashing step are important design factors for ion-exchange systems. These characteristics are also critical to those using the resins because movements of resins not only signal the existence of a problem but give indications as to the nature of the problem. Pressure drop and hydraulic information for new resins are available from the resin manufacturer and the supplier of equipment.

Factors which have the greatest impact on pressure drop are the depth of the bed, flow rate, viscosity, temperature, and particle size.

Backwashing is the upward flow of water through a bed of resin at a flow rate sufficient to fluidize the resin, but not so great that resin is carried out of the column with the exiting water. Resins are backwashed to remove dirt and resin fragments, to classify resin particles by size, and to relieve any packing that may have occurred with previous use.

Oxidants, such as dissolved chlorine in water supplies, react with synthetic ion exchangers to cause a loss of capacity, physical weakening of the resin, and partial solubilization of the resin. Anion-exchange resins are most prone to loss of functionality as the oxidant attacks and severs the linkage between nitrogen and carbon on the polymeric structure. In addition to this form of degradation, some the functional groups of strong base anion exchangers are convert to weak base groups. The overall effect is a loss of both strong base and total capacity with increase in weak base capacity. Loss of functional groups with cation-exchange resins by oxidative attack is uncommon. The rate of oxidative attack is enhanced by the presence of metals such as iron and copper which serve as catalysts, by higher temperatures, and by higher concentrations of oxidants. Aside from low concentrations of oxidants found in most water supplies, the processing of chemical streams with much higher levels of oxidizing chemicals is practiced occasionally on an industrial basis. The potential dangers are generally recognized.

Ion-exchange resins should not be used at temperatures above those recommended by the manufacturer. Exceptions are made when frequent replacement of resin is an economic advantage over the operating and capital cost of cooling and reheating the process stream. Functional groups are lost from both cation- and anion-exchange resins when the temperature limit is exceeded.

Excessive pressure drop across the resin bed causes fragmentation of the beads. Resins shrink and swell as they are alternately put through adsorption and regeneration cycles. The larger the volume change and the shorter the time involved, the greater the potential for physical damage to the resin particles. The appearance of cracks is the first sign of physical deterioration. Fragmentation into smaller irregularly shaped particles is a sign of further deterioration.

Resins should always be protected from freezing, although that may not always be possible. Generally, a few freeze–thaw cycles do not result in visual damage (cracking or fragmentation). Nevertheless, some weakening of the physical structure occurs because fragmentation is apparent if cycling continues.

Cation and anion exchangers lose weight and capacity, cross-linking is removed, and water-soluble components are released if the radiation tolerance limits have been exceeded. The effects of gamma radiation have been studied more than other types of radiation.

EQUIPMENT

Ion-exchange systems in process applications may be batch, semicontinuous, or continuous. Batch operations are not common but, where used, involve a kettle with mechanical agitation. Injecting with air or an inert gas is an alternative. A screened siphon or drain valve is required to prevent resin from leaving with the product stream.

Semicontinuous and continuous systems are, with few exceptions, practiced in columns. Most columnar systems are semicontinuous since flow of the stream being processed must be interrupted for regeneration. Columnar installations almost always involve the process stream flowing down through a resin bed. Those that are upflow use a flow rate that either partially fluidizes the bed, or forms a packed bed against an upper porous barrier or distributor for process streams.

SYSTEMS

Ion-exchange systems vary from simple one-column units, as used in water softening, to numerous arrays of cation and anion exchangers which are dependent upon the application, quality of effluent required, and design parameters.

CYCLIC OPERATION

Resins are seldom used once and discarded. Whether the system is run batchwise or in columns, the resin must be periodically removed from service and regenerated. An exception is the use of a resin as a catalyst in organic reactions. Each cycle consists of two principal steps, rinse and backwash. Failure to use good practices results in poor cyclic performance.

SHIPPING

Shipping resins in a water wet condition is standard practice. Removal of water by evaporative methods is expensive and not necessary for the majority of applications since they take place in aqueous systems. Dry resins (almost always strong acid cation exchangers) are required in several catalytic applications in the chemical industry.

HEALTH AND SAFETY FACTORS

Ion-exchange resins are not considered hazardous. However, cation exchangers when in the hydrogen form, and anion exchangers when in the hydroxide form, yield acidic and basic solutions, respectively, when in contact with neutral salt solutions. The corrosive potential should not be overlooked, and skin sensitivity has been reported occasionally, especially when gloves are not used when handling resin. Resins which have been used to remove toxic substances may slowly release these materials if the toxic substances are still attached to the resin.

C. Calmon, *React. Polym.* **4**(2), 131–146 (1986).

K. Dorfner, *Ion Exchangers*, Ann Arbor Science Publishers, Ann Arbor, Mich., 1973.

R. Kunin, *Ion Exchange Resins*, Robert E. Krieger Publishing Co., Huntington, N.J., 1972.

F. C. Nachod and J. Schubert, eds., *Ion Exchange Technology*, Academic Press, Inc., New York, 1956.

CHARLES DICKERT
Consultant

ION IMPLANTATION

Modern technology depends on materials with precisely controlled properties. Ion beams are a favored method (and in integrated circuit technology the prime method) to achieve controlled modification of surfaces and near-surface regions. In every integrated circuit production line there are ion implantation systems. In addition to integrated circuit technology, ion beams are used to modify the mechanical, tribological, and chemical properties of metals, intermetallics, and ceramics without altering their bulk properties. The term "tribological" describes the science of interacting surfaces in relative motion.

Ion implantation of materials results from the introduction of atoms into the surface layer of a solid substrate by bombardment of the solid with ions in the electronvolt (eV) to megaelectronvolt (MeV) energy range. Several ballistic-like atomic processes occur during ion implantation. The ballistic interactions of an energetic ion with a solid are shown schematically in Figure 1. The figure shows sputtering events at the surface, single-ion/single-atom recoil events, the development of a collision cascade involving a large number of displaced atoms, and the final position of the incident ion. The solid-state aspects of ion-implanted materials are particularly broad because of the range of physical properties that are sensitive to the presence of trace amounts of foreign atoms. Electrical, mechanical, optical, magnetic, and superconducting properties are all affected and indeed may even be dominated by the presence of such foreign atoms. The use of energetic ions affords the possibility of introducing a wide range of atomic species, independent of thermodynamic factors, thus making it possible to obtain impurity concentrations and distributions of particular interest; in many cases, these distributions would not be otherwise attainable.

The implantation system shown in Figure 2 illustrates a conventional ion implantation system in widespread use within the semiconductor industry. This directed-beam

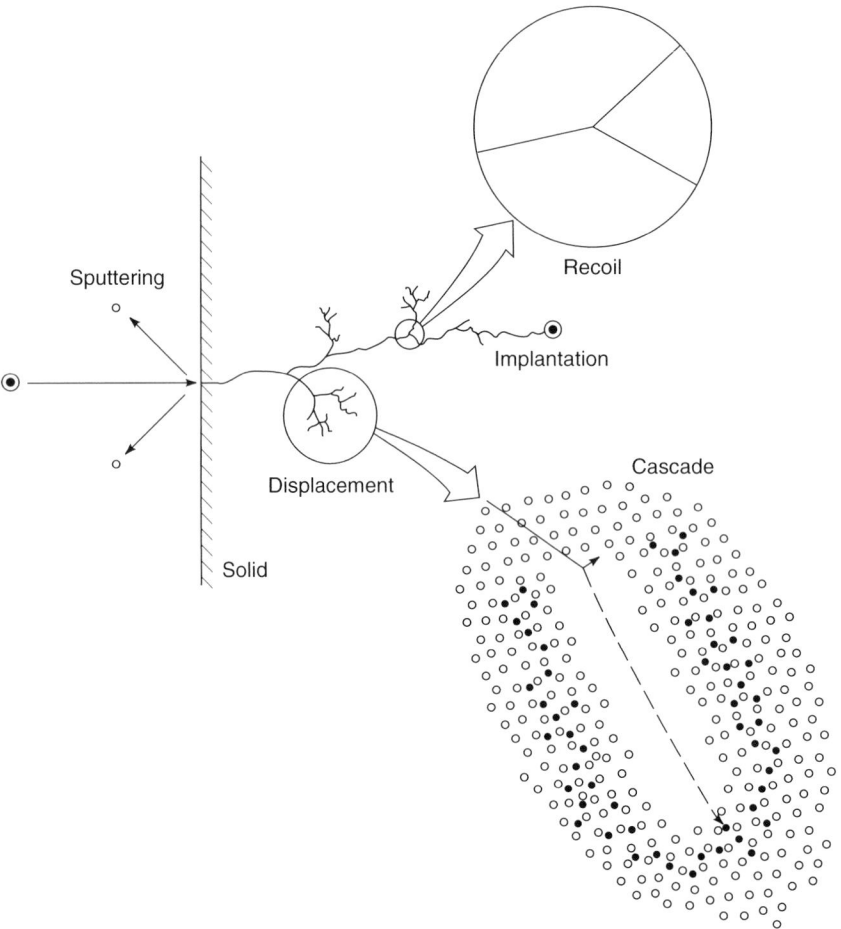

Figure 1. The ballistic interactions of an energetic ion with a solid. Depicted are sputtering events at the surface, single-ion/single-atom recoil events, the development of a collision cascade involving a large number of displaced atoms, and the final position of the incident ion.

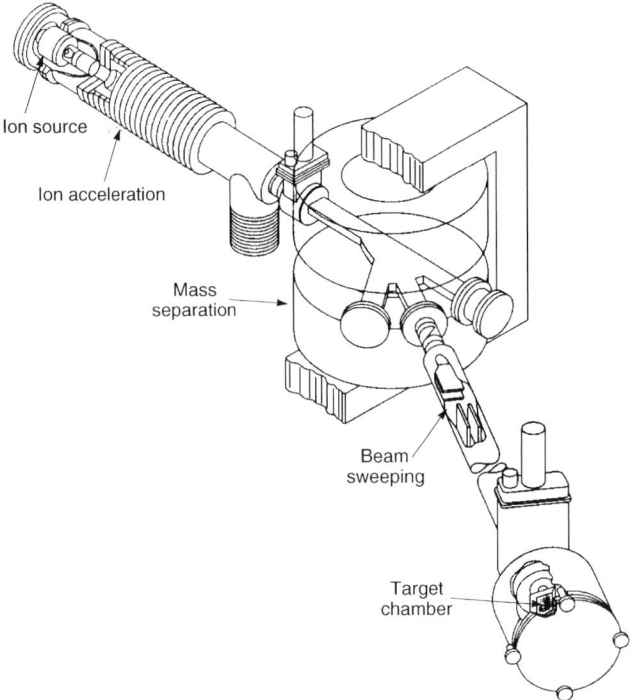

Figure 2. Schematic drawing of a conventional, directed-beam ion implantation system.

system has a mass-separating magnet (for mass analysis) that is almost mandatory for semiconductor processing to eliminate unwanted species that often contaminate the extracted beam.

One ion implantation system that does not use mass analysis is the plasma immersion ion implantation (PIII) system. This system does not use the extraction and acceleration scheme found in traditional mass-analyzing implanters, but rather the sample to be implanted is placed inside a plasma. The sample is repetitively pulsed at negative voltages (typically between a few kV and 10's of kVs) to envelop the surface with a flux of energetic plasma ions. Because the plasma surrounds the sample, and because the ions are accelerated normal to the sample surface, plasma-source implantation occurs over the entire surface, thereby eliminating the need to manipulate nonplanar samples in front of the ion beam. Ion implantation systems that implant all surfaces simultaneously are referred to as omnidirectional systems.

Ion implantation outside the traditional semiconductor applications is used for the controlled modification of surface-sensitive properties, which has had two principal thrusts: (1) as a metallurgical tool for studying basic mechanisms in areas such as aqueous corrosion, high-temperature oxidation, and metallurgical phenomena (eg, impurity trapping) and (2) as a means of beneficially modifying the mechanical or chemical properties of materials. Ion implantation can modify the mechanical, chemical, and/or optical–electronic properties of a surface. Optical–electrical properties, the traditional industrial application of ion implantation, such as the refractive index, reflectance, conductivity, and magnetic properties, can be modified. Ion implantation can also modify chemical properties relevant to the fields of electrochemistry (corrosion), catalysis, and oxidation resistance. The fastest-growing research application of ion implantation modifies the mechanical and tribological properties; eg, hardness, modulus, friction, wear resistance and fatigue resistance, of a material surface.

Some of the advantages of ion implantation in comparison to other surface treatments, such as coatings, are (1) surface properties can be optimized independently of the bulk properties; (2) the process is not limited by thermodynamic constraints, so solid solubility limits can be exceeded by several orders of magnitude, alloy compositions are not limited by diffusion, and metastable compounds can be produced; (3) the process modifies existing surfaces, so there are no interfaces to degrade mechanical properties and original dimensions are retained; (4) low process temperatures avoid thermally related degradation in surface finish and bulk mechanical properties; and (5) the process is highly controllable and reproducible.

Ion implantation processes also have limitations. An intrinsic basic limitation of directed-beam ion implantation is that it is a line-of-sight process; it is not feasible to apply it to samples having complicated geometries. Second, the range of ions in solids is generally low, which leads to shallow penetration and a thin modified layer. Finally, ion implantation as a surface modification tool is generally unfamiliar to most users of other surface modification processes.

These limitations can be addressed in a number of ways. First, plasma immersion ion implantation techniques have the ability to treat complicated geometries. The shallow penetration of ion implantation would in itself make it appear useless as a technique for engineering applications; however, there are several situations involving both physical and chemical properties in which the effect of the implanted ion persists to depths far greater than the initial implantation range. The thickness of the modified zone can also be extended by combining ion implantation with a deposition technique or if deposition occurs spontaneously during the ion implantation process. In addition, ion implantation at elevated temperatures (but below temperatures at which degradation of mechanical properties could occur) has been shown to increase the penetration depths substantially.

ION–SOLID INTERACTIONS

Ion Stopping

Ion–solid interactions are the foundation that underlies the broad application of ion implantation to the modification of materials. The major features governing the successful exploitation of ion implantation are the range distribution of the energetic ions, the amount and nature of the lattice disorder that is created, and the location of the energetic ions in the crystal lattice. At high dose levels (used to incorporate >5–10 atomic % of implanted species to modify the composition of the target) other phenomena become important: sputtering, ion-induced phase formation, and the transformation of one phase to another (ie, the transformation of a crystalline material into an amorphous material).

Nuclear collisions can involve large discrete energy losses and significant angular deflection of the trajectory of the ion. In nuclear stopping, the average energy loss results from elastic collisions with target atoms. This process is responsible for the production of lattice disorder by the displacement of atoms from their lattice position. Electronic collisions occur continuously and involve much smaller energy losses per collision, negligible deflection of the ion trajectory, and negligible lattice disorder. Electronic stopping is an inelastic process that results from energy transferred from the ion to the target electrons. Typical units for the energy loss rate are electronvolts per nanometer (eV/nm) or kiloelectronvolts per micrometer (keV/μm).

A proper understanding of the mechanisms of energy loss is important not only in controlling the depth profile of implanted dopant atoms but also in determining the nature of the lattice disorder produced during ion implantation or ion irradiation of the solid.

Range

For the energy regime normally used in heavy ion implantation (ie, tens to hundreds of keV) the nuclear contribution to the stopping process normally dominates, which this will be reflected in the particular ion trajectories as the ion comes to rest within the solid.

In range theory, the range distribution is regarded as a transport problem describing the slowing down of energetic ions in matter. Two general methods for obtaining range quantities, one using simulations and the other employing analytical methods, have been developed. At intermediate energies, $0.05 < \epsilon < 10$, a rather useful rule of thumb for predicting heavy-ion ranges, usually with an accuracy of $30-40\%$, is the equation

$$R(\text{nm}) = \frac{6E(\text{keV})}{\rho(\text{g/cm}^3)} \frac{M_2}{Z_2} \frac{M_1 + M_2}{M_1} \frac{(Z_1^{2/3} + Z_2^{2/3})^{1/2}}{Z_1}$$

where ρ is the mass density of the target.

Implanted Species Concentration

The peak atomic density N_p in the ion implantation distribution is estimated using

$$N_p = \frac{0.4\phi_i}{\Delta R_p}$$

where N_p is in units of atoms/cubic centimeter, φ_i is the ion dose in units of atoms/square centimeters, and ΔR_p is in units of centimeters.

To obtain the peak atomic concentration C_p resulting from this peak number of implanted ions requires knowing N, the atomic density of the substrate. The general relation for the concentration of the implanted species at the peak of the distribution is given by

$$C_p = \frac{N_p}{N_p + N}$$

Channeling

The crystal orientation influence on ion penetration is called channeling or the channeling effect. When an ion trajectory is aligned along atomic rows, the positive atomic potentials of the line of atoms steer the positively charged ion within the open space, or channels, between the atomic rows. These channeled ions do not make close-impact collisions with the lattice atoms and have a much lower rate of energy loss, and hence a greater range than those of nonchanneled ions. The depth distribution of channeled ions is difficult to characterize under routine implantation conditions. The channeling distribution depends on surface preparation, substrate temperature, beam alignment, and disorder introduced during the implantation process itself. The channeling effect requires that the incident ions be aligned within a critical angle of the crystal axes or planes. The critical angle depends on the ion energy, ion species, and substrate, but is typically $<5°$.

Radiation Damage

It has been known for many years that bombardment of a crystal with energetic (keV to MeV) heavy ions produces regions of lattice disorder. An implanted ion entering a solid with an initial kinetic energy of 100 keV will come to rest in the timescale of $\sim 10^{-13}$ s, due to both electronic and nuclear collisions. As an ion slows down and comes to rest in a crystal, it makes a number of collisions with the lattice atoms. In these collisions, sufficient energy may be transferred from the ion to displace an atom from its lattice site. Lattice atoms that are displaced by an incident ion are called primary knock-on atoms (PKA). A PKA can in turn displace other atoms, secondary knock-ons, etc. This process creates a cascade of atomic collisions, collectively referred to as the collision cascade. The disorder can be directly observed by techniques sensitive to lattice structure, such as electron-transmission microscopy, MeV-particle channeling, and electron diffraction.

Radiation-Enhanced Diffusion

Ion irradiation is quite efficient in forming vacancy-interstitial pairs. The atomic displacements resulting from energetic recoiling atoms can be highly concentrated into small localized regions containing a large concentration of defects well in excess of the equilibrium value. If the defects are produced at temperatures where they are mobile and can in part anneal out, the balance between the rate of formation versus the rate of annihilation leads to a steady state of excess concentration of defects. Since the atomic diffusivity is proportional to the defect concentration, an excess concentration of defects leads to an enhancement in the diffusional process.

Sputtering

In the process called sputtering, surface atoms are removed by collisions between the incoming particles and the atoms in the near surface layers of a solid (see Fig. 1). Sputtering sets the limit of the maximum concentration of atoms that can be implanted and retained in a target material.

Simulations

The previous sections have used analytical approaches to describe ion–solid interactions. Two different types of computer simulations are also used: Monte Carlo (MC) and molecular dynamics (MD). The Monte Carlo method relies on a binary collision model. Molecular dynamics solves the many-body problem of Newtonian mechanics for many interacting particles.

The MC methods, applied to ion–solid interactions, have a number of distinct advantages over analytical calculations based on transport theory. The MC approach allows for a more rigorous treatment of elastic scattering and of the determination of angular and energy distributions. As the name suggests, the results require averaging over many simulated particle trajectories.

To examine a solid as it approaches equilibrium; ie, atom energies of 0.025 eV, requires molecular dynamic (MD) simulations. Such simulations follow the spatial and temporal evolution of atoms in a cascade as the atoms regain thermal equilibrium in ~ 10 ps. By use of MD, one can follow the physical and chemical effects that influence the final cascade state. Molecular dynamics have been used to solve a variety of cascade phenomena:

defect evolution, dynamics of recombination, liquid-like core effects, and final defect states. The MD programs have also been used to model sputtering processes.

Other Processes Utilizing Ion Beams

Materials under ion irradiation undergo significant atomic rearrangement. The most obvious example of this phenomenon is the atomic intermixing and alloying that can occur at the interface separating two different materials during ion irradiation. This process is known as ion beam mixing (IBM). A related process uses ions to bombard material as it is being deposited onto a substrate. This process is called ion beam assisted deposition (IBAD) or ion assisted deposition (IAD).

ION IMPLANTATION SYSTEMS

Ion implanters are quite complicated machines that place high demands on process control and productivity. Their use of high voltages and toxic gases has also made safety a prominent consideration in equipment development. The wide spectrum of implantation doses and energies required in production have meant that no single machine strategy has been considered profitable in terms of cost of ownership and overall equipment effectivensss. Ion implantation systems are large: Typical dimensions are $5 \times 3 \times 3$ m^3, with weights ranging from 900 to 1600 kg.

The most reproducible and effective way to achieve a uniform and controlled introduction of dopants in integrated circuit technology is by ion implantation. In the ion implantation of semiconductors a beam of dopant ions of fixed energy, typically between 30 and 100 keV, is rastered across the surface of the semiconductor. Proper rastering assures a uniform flux of dopants over the surface of the semiconductor and the ions have sufficiently high velocity, $\sim 10^7$ cm/s, to penetrate through the surface and any thin surface contaminates and come to rest in the semiconductor some 10–100 nm below the surface. By choosing the appropriate ion energy, the location of the dopants below the surface can be precisely controlled. Another advantage of ion implantation is that selected areas can be implanted by the use of masks that leave well-defined areas of the semiconductor exposed to the beam and other areas masked (protected) from the beam.

Ion implantation became the dominant form of introducing dopants into silicon in the early 1980s with the transition from bipolar to CMOS transistors for the majority of electronic devices. Ion implantation processing is now used for both deep and shallow junctions in CMOS transistors.

During ion implantation, each ion produces a region of disorder around the ion track (Fig. 1). As the implantation proceeds, the amount of disorder builds up until all the atoms have been displaced. In some materials an amorphous layer may be produced over a depth R_p. The materials most susceptible to this ion beam amorphization are compound materials and elemental materials where there is significant covalent bonding.

It is also possible to induce solid-phase epitaxy of an amorphous layer by ion irradiation, a process commonly referred to as ion beam induced epitaxy crystallization (IBIEC). Studies on IBIEC are performed by heating a pre-existing amorphous layer onto a single-crystal substrate at a fixed temperature and irradiating it with ion beams having low current densities, to avoid further heating. Beam energies are chosen such that the projected range of the irradiating ions is well beyond the original crystal–amorphous interface. This allows one to discriminate between the damage clustering, which is typically produced at the end of the range, and the effects of a passing beam on a preexisting amorphous layer. The main result is the large enhancement of the crystallization kinetics induced by the ion beam irradiation. In addition to IBIEC being a low-temperature process, the regrowth activation energy is significantly lower than that measured by thermal annealing alone.

A new use of ion implantation is in the ion–cut process, which provides a way to cleave thin layers of semiconductor materials that can then be transferred onto a host of other substrates. This process uses ion implanted gas atoms such as H to promote cleavage of thin surface layers that are transferred from bulk substrates onto a host of other substrates. In the future, the ion-cut process may allow for greater integration of dissimilar materials, making possible the development of new microelectronics and 3D electronics.

Manostructured materials such as nanocomposites, nanocrystals, nanoclusters (NC), and quantum dots often possess unique size-dependent electronic, optical, magnetic, catalytic, and high strength properties. While there are many proven methods for producing NCs, ion implantation provides a unique way of producing encapsulated NCs, where the encapsulation may take place in substrates that provide additional functionality. For example, one can produce light-emitting Si NCs in a SiO$_2$ film that is integrated with an electronic device fabricated on Si. The synthesis of NCs by ion implantation takes advantage of the fact that virtually any element can be implanted into any substrate independent of thermodynamic factors, thus making it possible to obtain impurity concentrations and distributions that would not be attainable by traditional alloying methods. Thus, by ion implantation it is possible to introduce immiscible elements into a substrate, such as Au into SiO$_2$ or produce concentrations well above the solubility limit, for example by implanting Si into SiO$_2$, thereby producing metastable alloy solutions.

In general, the wear properties of materials are more important than their friction properties. Ion implantation has thus been mostly studied for improving wear resistance. The application of ion implantation to these areas has not only yielded surfaces with improved properties but has been important for the study of basic mechanisms. The problem areas impacted by implantation can be grouped into three main categories: (1) cutting and slitting operations; (2) corrosive applications and adhesive wear; and (3) extrusion operations and applications where large surface forces occur.

Ion implantation of metals is becoming more routine on a commercial scale, mainly with nitrogen implantation as an antiwear treatment of high value critical components. The primary use to date is for the antiwear treatment of surgical prostheses such as hips and knees. In use, these implanted components are in articulating contact against a mating ultrahigh molecular weight polyethylene cup, and wear of either component is a prime concern for the longevity of the prosthesis. A large number of knee, hip, and other joint prostheses are being treated each year in the United States. Ion implantation for such medical devices is attractive, since there is no concern regarding delamination, as for sharp interfaces, and nitrogen is considered benign in the human body.

Ion implantation appears to be an attractive technique for treating industrial components by (1) the stabilization of microstructure, preventing a change in wear mode; (2) the stimulation of transformations to a wear-resistant mode; or (3) the creation of chemical passivity to prevent a corrosive wear mode. Components benefiting from nitrogen ion implantation for improving wear resistance of tool steel alloys include plastic injection-molding tools, metal rolls, piercing tools, forming tools, and other components used in mild-wear applications. Successful utilization of nitrogen implantation requires relatively low tool surface operating temperatures, since the nitrogen/defect structures attributed to improvement of wear resistance are not stable at high temperatures.

The gap between laboratory wear testing and industrial application trials is extremely difficult to bridge, since there is often little or no control over testing in the industrial environment. Despite these limitations, several examples of industrial successes involving ion-implanted tools have been reported, and blind tests of nitrogen-implanted machine tools have been performed, including tool taps, dies, punches, and TiN-coated WC cutting inserts. The implanted tools showed lifetime improvements ranging from $1.5\times$ to $4\times$. No unimplanted tool demonstrated better performance than an implanted tool. Improvements were also observed for implanted tool dies and punches.

Fatigue represents a singularly dangerous mode of material failure in that no obvious prior warning is given of impending fracture. Generally, such failure occurs upon the cyclic loading at some stress below the static fracture stress. High loading amplitudes give rise to short lifetimes (low-cycle fatigue), whereas relatively low loads yield longer lifetimes (high cycle fatigue). Ion implantation has been employed to improve high cycle fatigue in copper, steel, nickel, and titanium alloys. The improvements for low cycle fatigue are smaller. In both cases; ion implantation changes near surface slip, promoting reversibility and increasing the homogenization or suppressing surface slip. Strengthening mechanisms involved include solid solution hardening, precipitation hardening, and compressive stress. The failure mode of implanted surfaces is seen to shift from slip band cracking to grain boundary cracking.

High temperature fatigue and fretting fatigue behavior has also been improved by implantation. This has been achieved by using species that inhibit oxidation or harden the surface. Fretting behavior is closely connected to oxidation resistance, perhaps due to third-party effects of

oxidation products. Oxidation resistance alone has also been improved by ion implantation.

Ion implantation may be used to modify either the local or generalized aqueous corrosion behavior of metals and alloys. Metallic systems have been doped with suitable elements in order to systematically modify the nature and rate of the anodic and/or cathodic half-cell reactions that control the rate of corrosion.

The following mechanisms in corrosion behavior have been affected by implantation, (1) expansion of the passive range of potential; (2) enhancement of resistance to localized breakdown of passive film; (3) formation of amorphous surface alloy to eliminate grain boundaries and stabilize an amorphous passive film; (4) shift open circuit (corrosion) potential into passive range of potential; (5) reduce/eliminate attack at second-phase particles; and (6) inhibit cathodic kinetics.

The nature of the microstructure of a surface alloy can have a significant influence on its corrosion behavior. It is well known that multiphase alloys tend to be susceptible to localized galvanic corrosion between phases of different chemical reactivity. Thus it is always desirable to produce single-phase alloys to avoid such effects. Chemical homogeneity in single-phase alloys is also desirable. Ion implantation may be used to form single-phase solid solutions often far in excess of the equilibrium composition. This is one major advantage of the use of ion implantation as a surface alloying technique.

In addition to the conventional approach to designing corrosion resistant alloys, ion implantation offers some scope for the formation of amorphous surface alloys. One advantage of such alloys is that the absence of grain boundaries allows for the formation of a continuous passive film that is not disrupted at the grain boundary region. The most researched application for corrosion resistance by ion implantation has been for high-cost, high-precision aerospace bearings.

Ion implantation and sputtering in general are useful methods for preparing catalysts on metal and insulator substrates. Ion implantation should be considered in cases where one needs good adhesion of the active metal to the substrate or wants to produce novel materials with catalytic properties different from either the substrate or the pure active metal.

Ion implantation has also been used for the creation of novel catalytically active materials. Ruthenium oxide is used as an electrode for chlorine production, because of its superior corrosion resistance. There are, however, continuing difficulties for catalytic applications of ion implantation. One is possible corrosion of the substrate of the implanted or sputtered active layer; this is the main factor in the long-term stability of the catalyst. Ion-implanted metals may be buried below the surface layer of the substrate and hence show no activity. Preparation of catalysts with high surface areas presents problems for ion beam techniques. While it is apparent that ion implantation is not suitable for the production of catalysts in a porous form, the results to date indicate its strong potential for the production and study of catalytic surfaces that cannot be fabricated by more conventional methods.

The ultrafast "quenching" times associated with the decay of ion collision cascades have also been utilized for producing various classes of metastable compounds, including diamondlike carbon coatings. These DLC films are highly adherent and insulating and demonstrate high chemical resistance to acids, bases, and solvents.

B. Brown, T. E. Alford, M. Nastasi, and M. C. Vella, eds., *14th International Conference on Ion Implantation Technology*, IEEE, Piscataway, N.J., 2003.

M. Nastasi, J. W. Mayer, and J. K. Hirvonen, *Ion Solid Interactions: Fundamentals and Applications*, Cambridge University Press, Cambridge, 1996.

M. Nastasi and J. W. Mayer, *Ion Implantation and Synthesis of Materials*, Springer-Verlag, Heidelberg, 2005.

J. F. Ziegler, ed., *Ion Implantation: Science and Technology*, Ion Implantation Technology Co., Maryland, 2000.

MICHAEL NASTASI
Los Alamos National Laboratory

JAMES W. MAYER
Arizona State University

KEVIN C. WALTER
The Essex Technology
Group, LLC

IONIC LIQUIDS

This article deals with the field of ionic liquids (ILs), with a focus on nomenclature, preparation, properties, and handling. These low-melting salts are sometimes also referred to as liquid organic salts, fused salts, molten salts, ionic melts, nonaqueous ionic liquids (NAILs), room-temperature molten salts, room-temperature ionic liquids (RTILs), organic ionic liquids (OILs) and ionic fluids, but the simpler description of ionic liquids is both adequate and accurate.

The term 'ionic liquid' should be literally understood as a liquid that consists entirely of ions, as opposed to an ionic *solution*, which is a solution of a salt in a molecular solvent. Similarly, eutectic mixtures of ionic liquids with molecular organic species are not ionic liquids, either. In recent years, many ionic liquids have emerged as environmentally benign alternatives to volatile organic compounds (VOCs, such as trichloromethane, ethanenitrile and dimethylmethanamide), which cause emissions and effectively damage the ecological balance. The merit of ionic liquids, in this respect, is their negligibly small vapor pressure; therefore they are easily retained in a process. This concept has been realized, especially in preparative chemistry, as shown in many examples in this review: when used as solvents in catalytic processes, separation of the solvent (and the catalyst) from the product is facilitated, and recycling is easily possible. However, their "green" aspects have been grossly distorted in the literature; ionic liquids are not intrinsically green

(many are, for example, highly toxic), but they can be designed to be green.

In order to distinguish between room-temperature molten salts (ionic liquids) and high-temperature molten salts (melts), a melting point of $100°C$ appears to be generally accepted temperature as arbitrary cut-off temperature for a salt to qualify as ionic liquid. From a practical point of view, ionic liquids that fall within this working definition can be divided into three categories: first-generation, second-generation, and third-generation ionic liquids.

First-generation ionic liquids consist of bulky cations such as 1,3-dialkylimidazolium (or N,N'-dialkylimidazolium) or 1-alkylpyridinium (or N-alkylpyridinium), and anions based mostly on haloaluminate(III); these have been studied extensively. The merit of these ionic liquids is their tuneable Lewis acidity: the addition of aluminium(III) chloride to 1-ethyl-3-methylimidazolium chloride at a mole ratio of 1:1, for example, leads to the formation of Lewis neutral $[AlCl_4]^-$. Larger amounts of added aluminium(III) chloride bring about the formation of anions such as $[Al_2Cl_7]^-$ and $[Al_3Cl_{10}]^-$, which are Lewis acids. Historically, these ionic liquids were developed for battery applications, but their potential as solvents and promoters in typical Lewis acid catalysis was investigated. The drawback of these systems is their great sensitivity towards water, which forms hydroxoaluminate(III) species with the aluminium(III) chloride and therefore decomposes the ionic liquid.

The ionic liquids of the second category are the cynosure of this article: they are usually air-stable and water-stable, and can be used on the bench-top. Second-generation ionic liquids are based on large organic cations derived from the alkylation of organic nitrogen-, phosphorus-, or sulfur-bases. Although there are almost limitless possibilities of devising such cations, the most frequently used to date are 1-alkyl-3-methylimidazolium salts, $[C_n\text{mim}]X$. Less research has been carried out on tetraalkylphosphonium-, tetraalkylammonium- or trialkylsulfonium-based ionic liquids. By combining these cations with smaller anions such as trifluoromethylsulfonate (or triflate), bis{(trifluoromethyl)sulfonyl}amide (or bistriflamide), ethanoate, nitrate, or halide, to name a few, low melting salts (see Fig. 1) are obtained.

It is generally accepted that this category of ionic liquids was established in 1992, although some rare examples of compounds qualifying nowadays as second-generation ionic liquids can be found in the literature from the beginning of last century.

Recently, a third generation of "task-specific" ionic liquids has emerged: these novel ionic liquids feature chemical functionalities which have been designed for specific applications. Very little is known to date about their physical properties, preparative methods, etc, and the future will show if they can bring about an ecological or economic benefit – but that future does look bright.

Although the above brief summary is far from comprehensive, it highlights the versatility and tunability of ionic liquids: depending on the application, the ionic liquid can be designed for performance (*designer solvents*). In fact, it has been predicted that at least one million simple ionic liquids and at least one trillion (10^{18}) ternary ionic

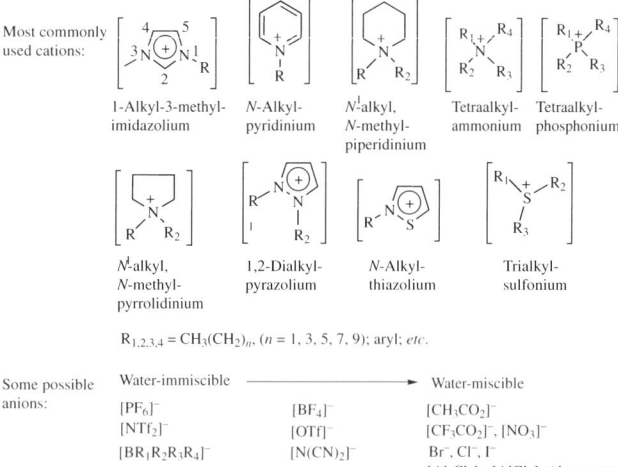

Most commonly used cations:

1-Alkyl-3-methyl-imidazolium   N-Alkyl-pyridinium   N¹-alkyl, N-methyl-piperidinium   Tetraalkyl-ammonium   Tetraalkyl-phosphonium

N¹-alkyl, N-methyl-pyrrolidinium   1,2-Dialkyl-pyrazolium   N-Alkyl-thiazolium   Trialkyl-sulfonium

$R_{1,2,3,4} = CH_3(CH_2)_n$, $(n = 1, 3, 5, 7, 9)$; aryl; etc.

Some possible anions:	Water-immiscible ⟶ Water-miscible		
$[PF_6]^-$		$[BF_4]^-$	$[CH_3CO_2]^-$
$[NTf_2]^-$		$[OTf]^-$	$[CF_3CO_2]^-$, $[NO_3]^-$
$[BR_1R_2R_3R_4]^-$		$[N(CN)_2]^-$	Br^-, Cl^-, I^-
			$[Al_2Cl_7]^-$, $[AlCl_4]^-$ (decomp.)

Figure 1. Some commonly used ionic liquid systems.

liquids can be prepared from the combination of anions, cations, and substituents, and over 30,000 entries for 1,3-functionalized imidazolium derivatives are already recorded in the CAS database. Both advantages and pitfalls lie in this enormous choice: although there is certainly an optimal ionic liquid for any given application, it requires a lot of experience to find the needle in the haystack. The choice of the anion and cation not only influences the physical properties, but also, if used as solvent, the thermodynamics and the kinetics of a reaction. Thus, by optimizing the solvent, the reaction outcome can be improved and controlled. At a zeroth level approximation, the anion controls the chemistry, and the cation fine-tunes the physical properties such as solubility, density, viscosity, etc. As an example, the nature of the anion may determine if the ionic liquid will be water-miscible (eg, I^-) or not (eg, $[NTf_2]^-$). Thus tuning the properties of the solvent to suit a particular application is feasible, and in many instances, ionic liquids have emerged as environmentally benign alternatives to volatile organic compounds by reducing emissions and damage to the ecological balance. The merit of ionic liquids, in this respect, is their negligibly small vapor pressure; they are easily retained in a process, and recycling is easily accomplished.

PREPARATION AND PURITY

In the laboratory, the common first step in the synthesis of both first- and second-generation ionic liquids is identical: the organic base, eg, 1-methylimidazole, is alkylated in a Menschutkin-type reaction to yield the corresponding halide intermediate, eg, 1-alkyl-3-methylimidazolium chloride. This intermediate salt is then either used in an ion exchange, or treated with acid, and the halide thus exchanged with the respective anion. Depending on the hydrophobicity of the anion, more or less extensive work-up strategies follow to remove halide ions, which may include multiple extraction or precipitation steps. Hydrophilic ionic liquids can not be purified by these

methods, so that a silver(I) salt has to be used instead of a Group 1 metal salt, and residual silver ions are removed quantitatively by electrolysis. However, this method is expensive and involves yet another clean-up stage. Due to the fact that water-immiscible ionic liquids are easier to prepare than water-miscible ones, these are many more investigated in the literature at present.

Some other alkylating agents are available which permit the direct preparation of halide-free ionic liquids. Examples include diethyl sulfate, alkyl trifluoroethanoate, alkyl trifluoromethylsulfonate and alkyl bis{(trifluoromethyl)sulfonyl}amide.

Less often employed methods of preparation are also feasible, and include the simple neutralization of organic bases with acids. Depending on the nucleophilicity of the resulting anion and basicity of the base, surprisingly stable and low melting salts with sufficient stability for electrochemical applications can be obtained. For imidazolium-based ionic liquids, the addition of a base leads to carbene formation, which, after distillation, can be reacted with acids to yield virtually halide-free ionic liquids.

Very little attention has been paid to analytical issues regarding the purity of ionic liquids. This is mainly due to the fact that their negligibly low vapor pressure excludes analysis by gas-phase based methods, eg, gas chromatography. Another convenient indicator for purity, the melting point of a compound, is also not viable, because ionic liquids are often liquids at room-temperature, and often form glasses on freezing, rather than crystals.

In general, impurities stem from the preparation the ionic liquid (water, halide, and amine) or from the absorption of water from the atmosphere. Critically, they have a strong influence on both physico-chemical and chemical properties. Thus, they may deactivate transition metal catalysts, and lead to a reduced rate of reaction or selectivity.

For the quantitative determination of halide impurities, methods such as the Volhard titration, ion selective electrode, electrophoresis, ion chromatography or electrochemical techniques have been established, thus allowing for the detection of impurities on a ppb level. The water content can be determined very conveniently by automated Karl-Fischer titration, and residual amine content can be determined either by uv-vis spectroscopic or chromatographic methods.

Although theoretically colorless, ionic liquids often have a yellow to brown tint (or, in extreme cases, black!), the origin of which is still unknown. The nature of this impurity (or, more likely, these impurities) can not be determined spectroscopically, but has no apparent influence on reaction outcomes or physico-chemical properties. However, some methods were developed for its removal using decolorizing carbon and neutral alumina. The impurities are believed to be intensely coloured and only present at parts per billion levels.

THERMAL AND CHEMICAL STABILITY

Apart from their physical properties, the thermal and chemical stability of an ionic liquid will determine its applicability in a reaction. At present, little is known about the instability of ionic liquids, although from an industrial application point of view, this aspect might be one of the most important factors.

Many researchers have observed that ionic liquids often alter in their colour, depending on the method of preparation, especially if heated above 80°C at reduced pressure to remove traces of water. As mentioned above, the nature of this tint is still unknown; however, it has been reported that under high thermal stress, 1-butyl-3-methylimidazolium chloride decomposes to, among others, chlorobutane, chloromethane, and imidazole decomposition products such as buta-1,3-diene, butadi-1,3-yne, ethanenitrile/isocyanomethane, 2-methylpropane, 2-propenenitrile and pent-3-en-1-yne. The reversed Menschutkin reaction and the Hofmann degradation, especially for ionic liquids containing strong nucleophilic anions, may be the preferred mechanisms.

In fact, many reports overestimate the thermal stability (placing them at 400°C and higher) of many 1,3-dialkylimidazolium ionic liquids by thermogravimetric analysis. Since the heat transfer in ionic liquids is slow, the sample temperature lags behind the measured temperature by between 75 and 150°C, at the high heating rates (between 10 and 20°C min^{-1}) that are frequently used. Under these conditions, the decomposition reaction does not reach equilibrium before the end of the experiment. The most accurate data require isothermal studies. Under isothermal conditions, however, [C$_4$mim][OTf] shows long-term stability at 200°C, and for [C$_2$mim][NTf$_2$] a weight-loss of 1% was found over 10 h at 250°C.

As for the chemical stability, two important degradation mechanisms exist for second generation ionic liquids, which are often not taken into account during experimental design. Firstly, 1-alkyl-3-methylimidazolium ionic liquids possess acidic ring-protons at the C-2 position which may be removed by bases to yield very stable carbenes, which in turn may be either responsible for side-reactions or the deactivation of a catalyst. Thus, for base-catalysed reactions, 1,2,3-trialkylimidazolium ionic liquids can be used, in which the C-2 position is "blocked". Alternatively, inert cations are chosen, such as pyridinium.

Likewise, it should be noted that ionic liquids based on the hexafluorophosphate or tetrafluoroborate anions are an exception from the general rule that second generation ionic liquids are water stable: these anions hydrolyse quantitatively to hydrofluoric acid, and phosphate and borate, respectively. There are many erroneous reports in the literature of "catalytic effects" in these ionic liquids which are simply due to the presence of hydrogen fluoride! It is recommended that, in most circumstances, these ionic liquids are unsuitable media for reactive chemistry.

PHYSICAL AND CHEMICAL PROPERTIES

Because of the many variations possible for both the anion and the cation, and due to their differences in shape and size, an immense array of different ionic liquids is ima-

ginable. In this way, ionic liquids constitute the largest group of solvents available. The properties of the solvent can be specifically designed by the deliberate combination of anion and cation. Thus, the physical properties such as melting point, viscosity and density can be tuned, and secondary properties, like solubility and reactivity of a specific substrate, are influenced as well. Ionic liquids possess therefore the potential to substitute for any molecular solvent, especially the banned halocarbons.

Phase Behavior and Liquid Range

The phase behavior of ionic liquids is governed by the bulkiness and low symmetry of their cations and/or anions, and the rate of attainment of the thermodynamic equilibrium between the liquid and solid state is kinetically restrained, eg, by the freezing or thawing of Brownian motion of longer chain segments. If the viscosity of such compounds increases too quickly upon cooling, the ions cannot achieve the optimal position necessary for crystallization, and super-cool into a glass. This phenomenon is also known as frustration of crystallisation, and leads to polymorphism in many ionic liquids. For example, 1-alkylpyridinium derivatives, [C$_n$py]X, melt at considerably higher temperatures than the imidazolium derivatives, due to the fact that the 1-alkylpyridinium cation possesses two mirror planes and a C$_2$ rotation axis that are not present in the 1-alkyl-3-methylimidazolium cation. Thus, the melting point of [C$_4$py][BF$_4$] is 15.3°C, whereas the analogous [C$_4$mim][BF$_4$] melts at −71°C (glass transition).

It is therefore not surprising that many ionic liquids super-cool. In fact, all salts belonging to the [C$_n$mim][BF$_4$] series ($n = 2$–18; see Figure 2) have a tendency to super-cool before crystallizing ($n = 10$–18) or forming glasses ($n = 2$–9). Small variations in the length of the alkyl chain attached to the imidazolium moiety lead to huge differences in the melting/freezing/glass transition points. The trend of rising transition temperatures as n increases beyond 8 is independent of the anion, and can be attributed to increased van der Waals forces between the alkyl chains, which override the effect of lower symmetry.

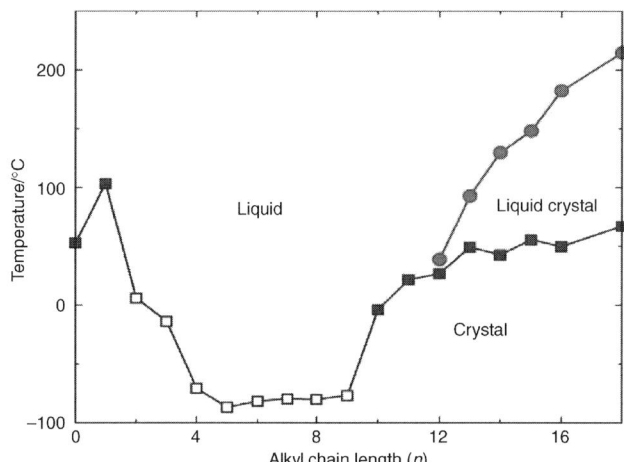

Figure 2. Phase diagram of [C$_n$mim][BF$_4$]. Melting points (filled squares), glass (open squares) and clearing points (circles).

Upon heating, these higher analogues exhibit liquid crystalline phases over a specific temperature range, which disrupt upon further heating to form true isotropic liquid phases. The thermal range of the liquid crystalline phase markedly increases with increasing chain length.

Various ionic liquids were investigated with respect to finding a direct relationship between the extent of hydrogen-bonding and their melting point. However, ionic liquids which exhibit a higher degree of hydrogen-bonding (eg, ethanoates) show similar (low) glass-transition temperatures to those which are less prone to hydrogen-bonding (eg, bis{(trifluoromethyl)sulfonyl}amides). Also, methylation in the C-2 position of the imidazolium cation leads to an increase in melting point and formation of crystalline solids, rather than suppressing hydrogen-bonding and thus reducing the melting point. This phenomenon was explained by an improvement of packing efficiency for the methylated derivatives, which masks the asymmetry of the 1,3-dialkylimidazolium cation, and highlights the complexity for predicting melting points, and designing new ionic liquids.

In summary, due to their complex phase behavior and high decomposition temperatures, ionic liquids often feature an exceptionally large liquid range. For example, the melting point of $[C_4mim][OTf]$ is $16\,^\circ C$, with long-term stability at $200\,^\circ C$, and $[C_2mim][NTf_2]$ melts at $-3\,^\circ C$ and slowly decomposes at temperatures $>250\,^\circ C$. This large range appears to be a generic feature of ionic liquids. (There are few accurate decomposition temperatures recorded in the literature.) This contrasts to the standard pressure liquid ranges of water ($100\,^\circ C$), liquid ammonia ($45\,^\circ C$), or ethanoic acid ($100\,^\circ C$).

Vapor Pressure

As highlighted above, the extremely low vapor pressure of ionic liquids is one of their main assets, which is due to their ionic nature, and fact that their chemical and thermal stability can be designed to be high. Owing to this property, ionic liquids cannot evaporate from the reaction vessel and are retained quantitatively, unlike other industrially used solvents, such as dichloromethane or dimethyl sulfoxide. Therefore, the impact on the environment and the process operation personnel is minimal. It means that volatile products, and even products with conventionally low vapor pressures, can be isolated from ionic liquids by distillation. However, a negligibly small vapor pressure is not the same as zero vapor pressure, and there has been recent evidence that some ionic liquids can actually be distilled, albeit at high temperatures and low pressures (eg, $300\,^\circ C$ and 0.1 mbar), the vapor consisting of tightly bound ion pairs.

Conductivity

Due to their ionic nature, ionic liquids are highly conducting materials. In fact, it is this property which historically led to the development of ionic liquids, making them ideal candidates for electrochemical applications. Their conductivity, just like other properties, can be fine-tuned by deliberate choice of the anion and cation: both moieties contribute to the conductivity. However, high viscosities lead to low diffusion rates, which can be a disadvantage in electrosynthetic applications.

Electrochemical Potential Windows

The electrochemical potential window is the potential range specific to each ionic liquid in which it is not substantially destroyed by oxidation of the anion or reduction of the cation. Typically, the window for ionic liquids is between 3.0 and 6.4 V. All imidazolium-based ionic liquids have a cathodic limit of about -2 V vs. saturated calomel electrode, and the anodic limit is determined by the nature of the anion. The large electrochemical windows thus obtained, especially for the fluorine-containing anions, show that ionic liquids are ideal candidates not only as solvents and electrolytes for electrochemical redox reactions, but also for battery and fuel cell applications.

It should be noted here that the presence of traces of water or halides in an ionic liquid decreases both the anodic and cathodic limits, and also influence the diffusion coefficient of dissolved species. Furthermore, the choice of electrode plays a role: 1,3-dialkylimidazolium ionic liquids have been reported to undergo carbene formation at the surface of platinum-electrodes, while glassy carbon electrodes are inert.

The trend of cation stability follows the order of: pyridinium $<$ pyrazolium $\leq$ imidazolium $\leq$ sulfonium $\leq$ ammonium, while a series of anions can be arranged in the following order: Cl^-, F^-, $Br^- <$ chloroaluminates $\leq$ $[BF_4]^-$, $[PF_6]^-$, $[AsF_6]^- \leq [OTf]^-$, $[NTf_2]^-$, $[(C_2F_5SO_2)_2N]^-$, $[CTf_3]^-$.

Polarity

Solvents are known to influence the chemical reactivity of a substrate both thermodynamically and kinetically, due to differences in the solvation of substrates and products and the formation of activated complexes. The 'solvent polarity' is defined as the overall solvation capability of solvents derived from all possible, non-specific and specific intermolecular interactions between the solute and solvent molecules. Interactions thus include dispersive, di- or multipolar, and electrostatic, as well as hydrogen bonding. However, since an overall value of very diverse parameters (such as refractive index, electric permittivity, Hildebrand's solubility parameter, dipole moment, etc) is impossible to establish, ranking of different solvents is generally achieved by comparing the effect of one of their properties to a reference solute. Given their structure and diversity of functionality, complex solvents such as ionic liquids, which are capable of multiple types of interactions, are even more difficult to probe than molecular solvents.

For example, systematic studies conducted using various organic probe molecules in $[C_4mim][PF_6]$ indicated with Reichardt's dye an $E_T(30)$ value similar to that of ethanol, whereas pyrene and 1-pyrenecarbaldehyde give a higher polarity (similar to ethanenitrile and dimethyl sulfoxide). The Stokes' shift value obtained from the dansylamide probe for $[C_4mim][PF_6]$ is close to the one obtained for ethanenitrile. The positive solvatochromic dye Nile Red, on the other hand, points to a higher polarity of the ionic liquid, which is located between the polarities of 90 wt% glycerol in water and pure water.

It was found that the anion dependent polarity (nucleophilicity) decreases with increasing anion size, probably due to the higher charge delocalization, in the order of $[NO_2]^- > [NO_3]^- > [BF_4]^- > [NTf_2]^- > [PF_6]^-$, and $[OTf]^- > [NTf_2]^- > [PF_6]^-$.

In addition to the lack of a method which measures the overall solvating capacity, variations found between publications using the same probe molecule may be due to differences in the recording technique (emission *vs.* absorbance), solvent purity, recording temperature or water content of the ionic liquid sample.

Solvent-solute interactions were studied with the solvation parameter model developed by Abraham, which uses a large number of probe molecules which exhibit various solute-solvent interactions due to acidic, basic, electron-donating, electron-withdrawing, and aromatic functional groups. For this purpose, capillary tubing was coated with various ionic liquids, and the retention of the probes on these gas chromatographic columns was analyzed by multiple linear regression analysis. It was found that the most dominant types of interaction were dipolarity, hydrogen-bond basicity, and dispersion forces. The dispersion forces were very similar for all ionic liquids, but the hydrogen-bond basicity depends mostly on the anion of the ionic liquid. The substitution pattern on the imidazolium cation played a crucial role for the interaction with probes containing nonbonding or π-electron systems. If the substituents are capable of inducing a higher electron density at the ring, stronger interactions with such probes were achieved.

The keto-enol tautomerism of pentane-2,4-dione is known to be influenced by the nature of the solvent, and the ratios can be determined using ^1H nmr spectroscopy. The equilibrium is dependent on both the concentration and the temperature. The results obtained from various ionic liquids based on the 1-butyl-3-methylimidazolium cation were compared to thirty molecular solvents, and strong correlations with parameters such as reorganisation energy, Snyder polarity index and dielectric constant were observed. Data obtained for the dielectric constant suggested that the ionic liquids investigated possess a higher polarity than both methanol and ethanenitrile.

In conclusion, these studies suggest that ionic liquids are **not** as highly polar as they are often assumed to be, but as the results have proven to be technique dependent, the question remains open.

Viscosity

For imidazolium-based cations, the viscosity of ionic liquids is dependent on the alkyl chain length of the cation, as well as the nature of the anion. It increases with increasing alkyl chain length, due to the increased possibility of van der Waals-type interactions between the cations. Ionic liquids with short chains ($n < 12$) exhibit Newtonian behaviour, whereas those derivatives that exhibit liquid crystalline phases (see, for example, Figure 2) show non-Newtonian behavior. These compounds give curves with abrupt discontinuities; at lower temperatures, their viscosity is high, due to their order, which, with increased temperature, is disrupted. This leads to the formation of a liquid crystal mesophase, which turns into an isotropic melt at even higher temperatures.

Besides the length of the alkyl chain, the symmetry of the anion, its molar mass and the ability to form hydrogen bonds all influence the viscosity. The reported viscosities of second-generation ionic liquids range from between 20 to 2000 cP, which appears to be quite high from an engineering point of view. As well as a dramatic drop with increasing temperature, addition of organic solvents with varying properties has a similar effect, even if present in quantities <10 mol%. Nevertheless, one of the future challenges of ionic liquid research is to reduce the viscosity of ionic liquids by at least one order of magnitude.

Density

The density of $[C_n mim]^+$ and $[C_n py]^+$ ionic liquids containing inorganic anions such as $[BF_4]^-$ or $[PF_6]^-$, is higher than that of water, ie, if one of these ionic liquids is immiscible with water, it will form the heavier, lower phase. Within a homologous series, eg, $[C_2\text{-}C_{10}mim][BF_4]$, the ionic liquids with shorter alkyl chain possess higher densities than the ones with longer chains, due to the ability of the former to achieve better packing of the ions. Recently, it has been observed that phosphonium-based ionic liquids possess very low densities: at $30°C$, the density of trihexyltetradecylphosphonium chloride is 0.88 g cm^{-3}, while that of the corresponding $[BF_4]^-$ was found to be 0.93 g cm^{-3}. In contrast, the density $[C_2 mim]Br\text{-}AlBr_3$ ($X = 0.5$) is >2.4 g cm^{-3} at 22°C. Thus, a large range of densities is covered by ionic liquids, and this could be easily extended by design.

Refractive Index

The refractive index of an ionic liquid increases linearly with increasing alkyl chain for a given cation type. Of more importance, however, is the ability to design ionic liquids with high refractive indices, significantly higher than can be achieved with organic molecular solvents. Guided by the parachor, novel ionic liquids were designed to permit the refractive indices of gems and minerals to be measured safely, without exposing the researchers to noxious fumes.

Water Solubility

The solubility of water in ionic liquids is strongly influenced by their composition and in particular by the nature of the anion. Hydrophobic ionic liquids are usually based on fluorinated anions, most commonly $[PF_6]^-$, $[PF_3(CF_3)_3]^-$, $[OTf]^-$, $[NTf_2]^-$ and $[BF_4]^-$. The properties of the cation play a role as well, albeit somewhat subordinate: the hydrophobicity of a homologous series, for example $[C_n mim][BF_4]$, increases with increasing alkyl chain length, so that $[C_n mim][BF_4]$ ($n = 2\text{-}5$) are fully miscible with water at room-temperature, but the higher derivatives are not. Similarly, all reported second-generation ionic liquids based upon $[P_{6\ 6\ 6\ 14}]^+$ are hydrophobic too. $[C_nmim][OTf]$ with $n > 4$ or $[NTf_2]^-$ also give water-immiscible ionic liquids.

Solubility is, as would be expected, also dependent on the temperature. Thus, $[C_4mim][BF_4]$ is miscible with water at 25°C; however, phase separation can be achieved at temperatures <5°C. This phenomenon has been used to advantage: in the rhodium-catalyzed hydrogenation of polar alkynes, $[C_8mim][BF_4]$, water, the starting material and the product mix to give a homogeneous phase at 80°C and 60 bar, which phase-separates upon cooling.

It should be noted that although ionic liquids that form biphasic systems with water are often referred to as "hydrophobic ionic liquids", but they **DO** dissolve water to some extent and are often hygroscopic. The driving force for hygroscopicity must be related to a change of internal order, leading to a more favorable structure of lower energy. The physico-chemical properties of ionic liquids, such as their solubility, polarity, conductivity and viscosity are altered by the amount of water absorbed. This will change the rates of chemical reactions and efficiencies when ionic liquids are used as solvents in reaction. In catalytic reactions, for example, the coordination ability of a solvent with a catalyst is known to influence the efficiency of the catalyst. Also, traces of water present will influence the solubility of other substances in the ionic liquid medium. As an example, the solubility of carbon dioxide was first underestimated, as no precautions were taken to exclude traces of water in the ionic liquid.

In conclusion, the ability to absorb water from the atmosphere is a critical issue: although these second-generation ionic liquids may be used on the bench top, extreme care must be taken to avoid absorption and accumulation of water. Even when the ionic liquid investigated is water immiscible, it needs to be thoroughly dried, and handled in a dry-box, or under a dry dinitrogen or argon atmosphere.

Solubility

Ionic liquids can be designed to dissolve most materials (such as polymers, organic and inorganic compounds, and transition metal catalysts) to a higher degree than conventional solvents. This may lead to process intensification, since a smaller reactor volume implies less energy for heating to reaction temperature or cooling. In one approach, the solubility of a reactant and product can be fine-tuned by alteration of the lipophilic character of the ionic liquid, eg, by increasing the alkyl chain length. In this way, biphasic systems can be obtained, in which the catalyst is dissolved in the ionic liquid, while the product forms a second layer. Thus, the advantages of homogeneous catalysis are sustained, and the catalyst and solvent is easily recycled by simple phase separation.

Only few data for alkene solubility are in the literature. Generally speaking, it can be concluded that lengthening the alkyl chain of alkenes leads to a decrease in solubility in a given ionic liquid, which can be attributed to the increase of the non-polar character of these substrates. Likewise, if the alkyl chain length on the cation of the ionic liquid is increased, the solubility of a given non-polar organic substrate, such as an alkene, increases. The solubility of a polar molecule such as water, on the other hand, decreases in such a case.

It is often difficult to correlate the influence of the anion of the ionic liquid with the solubility of a given substrate. However, the solubility of hex-1-ene increases for the same cation, eg, $[C_4mim]^+$, as follows: $[BF_4]^- < [PF_6]^- < [OTf]^- < [CF_3CO_2]^- < [NTf_2]^-$. Similarly, the solubility of alcohols increases in the following order: $[PF_6]^- < [BF_4]^- < [NTf_2]^- < [OTf]^- < [N(CN)_2]^-$. These findings correspond approximately to the trends established for the polarity (*vide supra*).

The solubility of different solid and liquid organic substrates was measured in $[C_4mim][PF_6]$ at ambient temperature and pressure. In general, solids were less soluble than liquid organics. The solubility of substances with the potential for strong intermolecular interactions (eg, those exhibiting a large dipole moment) is higher than that of non-polar molecules. Also, aromatic substances are soluble to a higher degree than nonaromatic compounds of equivalent molecular weight and polarity.

Several ternary phase diagrams have been determined for systems where two molecular solutes are present, especially for ionic liquid-water-alcohol mixtures and ionic liquid-ethanol-(1,1-dimethylpropyl ethyl ether). The presence of alcohol increases the solubility of water in the ionic liquid, up to a molar ratio of 1:1 (alcohol:water). Thus, this ratio can be used as a tool to fine-tune the solubility, so that either total or partial solubility or essentially complete phase separation can be achieved.

As for organics, few data for gas solubilities are available at present. In general, the solubility of gases increases with increasing pressure, and decreases with increasing temperature. The solubility of CO_2 (20 mole% at 10 bar) in $[C_4mim][PF_6]$ is considerably greater than that of the other gases (less than 5 mole%). The Henry's law constants obtained are: 50 bar for CO_2, 190 bar for CH_4, 303 bar for CO, 281 bar for N_2 and 353 bar for H_2. Other authors report Henry coefficients of 0.003 mol l^{-1} bar^{-1} H_2 for $[C_4mim][BF_4]$ and 0.00088 mol l^{-1} bar^{-1} for $[C_4mim][PF_6]$. These values indicate that gas solubilities are dependent on the type of anion of the ionic liquid, which has been shown to have implications in the outcome of reactions: in asymmetric hydrogenations, for example, a higher solubility of dihydrogen leads to higher enantiomeric excesses as well as higher conversions.

Detailed investigations in $[C_4mim]^+$ ionic liquids have shown that at 25°C, the solubility of CO_2 increases in the following order: $[NO_3]^- < [N(CN)_2]^- < [BF_4]^- \sim [PF_6]^- < [OTf]^- < [NTf_2]^- < [CTf_3]^-$. For a given anion, the increase of the alkyl chain substituent on the 1-alkyl-3-methylimidazolium cation also increases the CO_2 solubility slightly, while the introduction of a methyl substituent in the C-2 position decreases its solubility. The solubility of carbon monoxide, as determined by high-pressure NMR spectroscopy, It increases in the order: $[BF_4]^- < [PF_6]^- < [SbF_6]^- < [OTf]^- < [NTf_2]^-$. Similar to the trend observed for carbon dioxide, and a longer alkyl substituent on the cation also improves the solubility.

Moisture present in the ionic liquid decreases the CO_2-solubility. Since the ionic liquid does not dissolve in the

CO_2-phase, as opposed to conventional organic liquids, cross-contamination and leaching are not encountered, so that CO_2 can be used for the extraction of products from the ionic liquid phase.

HANDLING, SAFETY AND TOXICOLOGY

When working with *first-generation* ionic liquids, close attention needs to be paid to the materials of containment, since low-density as well as high-density polyethenes are decomposed at temperatures >100°C. To the best of our knowledge, most *second-generation* ionic liquids based on anions other than $[PF_6]^-$ and $[BF_4]^-$ are stable in the presence of water and dioxygen. $[PF_6]^-$-based ionic liquids have been found to etch glass in the presence of traces of water at elevated temperatures, due to the formation of toxic HF. The use of such ionic liquids is thus being phased out. Additionally, residual halide impurities, from the preparation of the ionic liquid may corrode metal vessels (eg, copper and brass) so that high-grade stainless steel, Hastelloy, Teflon, or glass equipment should be used.

Very little is presently known about the environmental behavior, safety and health issues (eg, flammability/compatibility) and disposal strategies for ionic liquids, although an environmental risk assessment strategy for toxicity determination has been outlined. The acute oral toxicity of the ionic liquid 3-hexyloxymethyl-1-methylimidazolium tetrafluoroborate was assessed by the Gadumm method, and found that the $LD_{50} = 1400$ mg kg^{-1} for female and the $LD_{50} = 1370$ mg kg^{-1} for male Wistar rats. $[C_2mim]Cl$ was found to possess a low aquatic toxicity (*daphnia magna*: $EC_{50} > 100$ mg l^{-1}), and its potential for bioaccumulation is low (log K_{OW}: −0.31), resulting however also in a poor biodegradability. Generally, alkoxy-substituted ionic liquids display antimicrobial activity against cocci, rods and fungi. The alkyl chain length greatly affected these activities: shorter substituents render the ionic liquid less active, while substituents with 10–14 carbon atoms in the alkoxy-group gave rise to very high anti-microbial activities, while relatively little influence of the anion was found. Using the 'Closed Bottle Test' (OECD), it was found that some ester-functionalised ionic liquids biodegrade easier than 1,3-dialkylimidazolium derivatives.

The marine bacterium *Vibrio fischeri*, as well as two rat cell types, were exposed to dialkylimidazolium-based ionic liquids. While the bacterium luminescence was not influenced by the choice of anion, a stronger inhibition of luminescence was found for derivatives with longer alkyl substituents, which also decreased the cell viability of mammalian cells, and therefore the EC_{50}. However, the effect concentrations of the test systems were lower than the toxicity of solvents such as propanone, ethanenitrile, methanol or methyl *t*-butyl ether.

Although some biocatalytic processes using enzymes have been reported to proceed well in the presence of ionic liquids, acetylcholinesterase was found to be increasingly more inhibited with lengthening of the alkyl-substituent on the 1,3-dialkylimidazolium cation. As for the type of cation, phosphonium and pyridinium derivatives were less inhibitory than imidazolium ionic liquids.

Concluding from these studies, it appears that the higher toxicity of longer chain alkyl-substituted derivatives is related to an increased hydrophobicity which allows for the permeation of the ionic liquid into the cell membrane.

No data have been published on the procedures for the safe disposal of ionic liquids. However, recently a study of the adsorption of $[C_4mim]Cl$ onto bacterial and mineral surfaces, such as the Gram-positive soil bacterial species *bacillus subtilis*, gibbsite, quartz and Na-montmorillonite, was conducted to elucidate the fate and transport in environmental systems. The results suggest that the geologic retardation of this ionic liquid will occur if interlayer clays are present in the subsurface, but little interaction with other common surfaces occurs. Therefore, the mobility of this ionic liquid, once reaching the water table, is expected to be high and viable disposal strategies should be devised.

STRUCTURAL STUDIES

Pure Ionic Liquids

Although at present, there is very little detail known about the structure of ionic liquids in their pure state, and even less about ordering of their mixtures, more and more researchers have recently become interested in this aspect. As a result of this tendency, this literature discussion focuses strongly on the review of such interactions.

Owing to the fact that ionic liquids are non-crystalline over a wide temperature range, it is difficult to obtain crystal structures of 1-alkyl-3-methylimidazolium salts of medium alkyl chain length. The few crystal structures of short chain homologues available show that hydrogen bonding between the anions and cations can occur to varying degrees, if the anion is nucleophilic (eg, halides); the cationic ring C-H protons are significant hydrogen-bond donors. However, since the phase behaviour of long-chain 1-alkyl-3-methylimidazolium salts is much more complex (eg, liquid crystal polymorphism, amphiphilicity, etc), it is questionable if predictions of the structure in the liquid state can be made from data derived in the solid state, even if their structures are formally very similar.

The $[C_4mim]^+$ cation was investigated by various techniques including Raman spectroscopy, X-ray crystallography and X-ray powder diffraction, and two polymorphs of the chloride salt were identified both in solid and liquid state, one being a metastable form of the other. They differ in the angle in which the butyl moiety protrudes from the imidazolium plane, which brings about changes in the hydrogen bond network. Both structures represent minima of the potential surface energy, and give distinctive crystals. The formation of a "eutectic" liquid region below the melting point of both polymorphs can be anticipated.

Nmr spectroscopy was used to study structural arrangements in the liquid or solid state. For $[C_2mim]$ $[BF_4]$, a strong dependence of the hydrogen bonding between the ions on the temperature was observed. Below 24°C, the ions behave in a "quasi-molecular" manner as discrete ion pairs, while at higher temperatures, individual ions are present, which may participate in an extended hydrogen-bonded network.

Neutron diffraction data, crystallography and molecular dynamics calculations showed that adjacent cations are hydrogen bonded to one another *via* $CH_3...\pi$ interactions in a dimeric fashion, and are influenced by variations in anion size and hydrogen-bonding ability, leading to both different packing of the dimers and an expansion of the liquid network. Due to the high hydrogen-bonding ability of chloride anions, the H-2 of the cation was the most likely position of interaction. On the other hand, the hexafluorophosphate anions are most likely positioned above or below the imidazolium ring (Coulombic forces), and little interaction occurs axial to the H-2 proton. Although such an anion-cation stacking had been proposed, other experimental data (from ir and nmr spectroscopy) indicates that the hexafluorophosphate anion preferably interacts with the H-2 hydrogen, similarly to the chloride anion, albeit less pronounced. The cation-cation distribution in the chloride salt showed an even distribution for the chloride salt. On the other hand, essentially no cation-cation interaction was found to exist in the $[PF_6]^-$ homologue, due to the $[PF_6]^-$ anion obstructing the space facing the imidazolium ring at high probability.

Ionic Liquid–Cosolvent Mixtures

Ionic Liquid–Organic Cosolvent Mixtures. Ionic liquid/CO_2 mixtures were investigated at high pressures and elevated temperatures using attenuated total reflectance (atr) infrared spectroscopy. It was found that the dissolved CO_2 interacts with the anions of the ionic liquids ($[C_4mim][BF_4]$ or $[C_4mim][PF_6]$), with the axis of the $O=C=O$ molecule perpendicularly orientated towards the P–F or B–F bonds. The interaction with the tetrafluoroborate was stronger than with the hexafluorophosphate, indicating that the former anion acts as a stronger Lewis base towards CO_2.

Another interesting type of interaction between some ionic liquids and aromatic organic substrates (eg, benzene, toluene or xylene) is the formation of clathrates (liquid inclusion compounds). Many aromatics have a remarkably high solubility in ionic liquids of the 1-alkyl-3-methylimidazolium type, but in most instances phase separation occurs at high concentrations of the substrate. In such clathrates, the cation-anion interactions of the pure ionic liquids is disturbed by the presence of the aromatic compound, and a new internal structure, based on associative interactions between the aromatic compound and the salt ions, is formed. An energetic equilibrium between the forces driving the crystallization of the pure salt (eg, cation-anion association) and the solvation interactions of the ions with the aromatic compound must exist to obtain a stable clathrate. Liquid clathrates exhibit low viscosities (as compared to the pure ionic liquid), and constant, but non-stoicheiometric, compositions. In one instance ($[C_1mim][PF_6]$), it was possible to crystallize

and analyze such an inclusion compound with a ratio of ionic liquid:benzene of 1:2.

The existence of inclusion compounds highlights yet again the importance of understanding the interactions possible in ionic liquids. For example, when reactions involving aromatic compounds are carried out in an ionic liquid, this phenomenon will have a strong impact on the separation of the product from the solvent, and possibly on the reactivity of the aromatic compound as well. Furthermore, it can be predicted that ionic liquids may be used to favourably influence the separation of aromatics from aliphatic compounds.

Ionic Liquid–Water Mixtures. In voltammetric studies, water had a strong influence on the diffusion rate of ionic and neutral species, leading to the suggestion that a nanostructure consisting of polar and nonpolar regions is present in the ionic liquids. Such an inhomogeneity allows the neutral molecules to reside in less polar regions, whereas ionic species undergo fast diffusion in the more polar ("wet") regions.

In a dynamic nmr spectroscopy study of $[C_4mim][BF_4]$-water mixtures, it was found that a continuous increase in the water/ionic liquid ratio leads to a replacement of the hydrogen bonds between the ring hydrogen atoms and the fluorine atoms of the tetrafluoroborate with hydrogen bonds between the ring hydrogen atoms and the water molecules. This replacement does, however, not occur in an on/off switch fashion, but rather continuously. The structural change leads to a breaking of the imidazolium-imidazolium cationic associations, thus increasing the distance between neighboring cations. Increasing amounts of water also gave rise to augmented interactions of the methyl moiety with the butyl moiety, indicating that an orientational change occurred, which allows the methyl moiety to be accommodated in closer vicinity to the butyl moiety.

Attenuated total reflectance and transmission infrared spectroscopy showed that at any water concentration up to the saturation point of $[C_4mim]$-based ionic liquids, water has distinct interactions *via* hydrogen bonding with the anions under study ($[PF_6]^-$, $[SbF_6]^-$, $[NTf_2]^-$, $[BF_4]^-$, $[ClO_4]^-$, and $[OTf]^-$), and areas of clusters or pools of water are not present. Instead, 2:1 complexes, in which both protons of water are bound to two discrete anions, $A^-...H-O-H...A^-$ exist. An additional type of state of water was found with more nucleophilic anions, and their ATR-IR spectra indicate the presence of clusters of water molecules in addition to the 2:1 complexes. The enthalpies of interaction of water with different anions were determined from the spectral shifts. They increase in the following order: $[PF_6]^- < [SbF_6]^- < [BF_4]^- < [NTf_2]^- < [ClO_4]^- < [OTf]^- < [NO_3]^-$. From this study, no estimate as to the state of water at concentrations higher than the saturation point can be derived, and clusters or aggregates may be present in such a case. In fact, excess molar volume studies of $[C_4mim][BF_4]$–water mixtures have shown a sharp increase once water is present in excess over the ionic liquid mole fraction.

In conclusion, although there are, as yet, very little data available on the structure of pure ionic liquids or

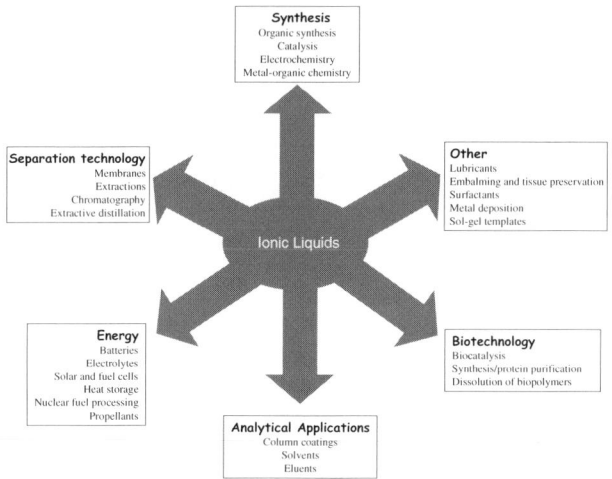

Figure 3. Some applications of ionic liquids.

ionic liquid–cosolvent mixtures, it is obvious that it is influenced by several factors, which determine the physico-chemical properties. These factors are the size/nucleophilicity of the anion, the type of cation, the substituents (length and functionalities) on the cation, temperature, and the nature and concentration of additives, such as water or organic solutes.

LABORATORY APPLICATIONS FOR IONIC LIQUIDS

In recent years, ionic liquids have been used for a wide variety of applications (Figure 3), such as the recovery of biofuels, and the deep desulfurisation of diesel oil by extraction of butanol or sulfur-containing compounds, respectively, into the ionic liquid. Ionic liquids also have potential as lubricants, in solar cells, for heat storage, in nuclear fuel processing, in membrane technology, as sol-gel templates, and in the dissolution of cellulose.

Ionic liquids have also shown promise in some startling applications, such as embalming and tissue preservation at one extreme, and in thrusters for NASA space projects at the other.

Ionic Liquids as Solvents for Separations

Ionic liquids can be combined with other fluids, such as water or supercritical fluids, to give potentially benign systems for separation technology, eg, extraction or chromatography. The combination of their advantageous properties may offer new opportunities for the clean-up of existing chemical extraction processes, where they substitute for toxic, flammable and volatile organic solvents. In the following, interesting, and often unexpected, solvent-solute interactions are reviewed.

Extraction from ionic liquids with a supercritical fluid is the ultimate method for 'green' partitioning of phases, since commonly used volatile organic solvents as extracting media may cause harmful emissions. The first example of the combination of ionic liquids with supercritical fluids was the extraction of naphthalene from an ionic liquid. This study showed that extractions with ionic

liquid-CO_2 systems are particularly suitable for substances that are hydrophilic or water-sensitive, and therefore do not permit aqueous extraction from ionic liquid media, or for poorly volatile or thermally labile products, for which distillation as a method of separation is not applicable. Further studies showed that supercritical CO_2 can be used to quantitatively extract an array of organic solutes from [C_4mim][PF_6]. Substrates with high volatility and low polarity possess a higher solubility in supercritical CO_2, whereas such substrates with high polarity and aromaticity favor the ionic liquid phase. A correlation between the dipole moment and the amount of CO_2 necessary for the recovery was established.

Recently, various reports of catalytic reactions carried out in ionic liquids, from which the products are removed after the reaction using supercritical fluids, have been published. These publications illustrate that catalyst and ionic liquid recycling are feasible using this method.

Distribution ratios of uncharged and charged compounds in [C_4mim][PF_6]/water to octan-1-ol/water systems have been determine, using radiochemical tracers, and found a close relationship between the two systems, the former being approximately one order of magnitude lower than the latter. Charged, polar compounds have a lower partition coefficient than uncharged or nonpolar substances, ie, they are better soluble in the ionic liquid (or octan-1-ol) than in water.

For acids (eg, benzoic acid) as well as for bases (eg, aniline), the partition coefficient can be controlled *via* alteration of the pH of the aqueous phase, as it is classically accomplished in separation science. Much like molecular solvents, ionic liquids tend to preferably dissolve non-dissociated species. The same applies to more complicated molecules, such as the antibiotic erythromycin-A.

An additional advantage of ionic liquids in separation technology is the possibility of fine-tuning the solubility of a substrate by alteration of the alkyl chain length, eg, from butyl ([C_4mim][PF_6]) to octyl ([C_8mim][PF_6]). For example, thymol blue, which is distributed between the ionic liquid [C_4mim][PF_6] and the aqueous phase, enriches in the aqueous layer at pH 12 as mentioned above. However, if [C_8mim][PF_6] is used under the same conditions, the dye possesses a higher solubility in the ionic liquid phase than in water.

Although phase separation is generally fast, in some instances it can be further facilitated by choosing an ionic liquid with a melting point above room-temperature; in such a case, the extraction is carried out at elevated temperatures in a liquid-liquid system. Upon cooling, the ionic liquid solidifies together with the extracted substrate, which is easily isolated after decanting the aqueous phase.

Extraction of Inorganic Compounds

Water-immiscible imidazolium-based ionic liquids have been used to extract alkali and alkaline earth metals from aqueous media with the help of crown ethers. In summary, higher distribution coefficients were found for ionic liquids than for conventional organic solvents, due to the limited solubilities of ionic species in non-ionic organic solvents. In competitive metal extraction experiments, the anion of the metal salts does not exhibit an influence on their distribution. Such behavior is commonly observed in octan-1-ol extraction, whereas with chloroform, an anion effect of the metal salt is apparent. This finding was attributed to anion stabilization by hydrogen bonding. However, both the nature of the cation and anion of the ionic liquid play an important role. Thus, the distribution coefficient of the strontium(II) nitrate extraction is higher if the imidazolium cation is 2-methylated rather than possessing a proton in C-2 position, indicating that hydrogen bonding is an important factor. By alteration of the alkyl chain length of the cation, the solubility of the crown ether complex can be indirectly tuned, as it is dependent on the amount of water which co-dissolves in the respective ionic liquid.

Imidazolium-based hexafluorophosphate ionic liquids with thiourea, thioether or urea side chains have been designed for the extraction of mercury(II) and cadmium(II) from aqueous solutions, allowing for enrichment of the metals in the ionic liquid independent of the pH of the aqueous phase. Additionally, doping the much cheaper $[C_4mim][PF_6]$ with the thioether-derived ionic liquid often leads to comparable results as when using the thioether-based extractant in its pure form.

Gas/Liquid Extractions

Ionic liquids have been designed for the extraction of carbon dioxide from gas mixtures, eg, sour gas. By integrating a secondary amine into the alkyl chain on the imidazolium cation, it is possible to sequester carbon dioxide by formation of a carbamate. This reaction is reversible; the CO_2 may be driven off by heating under reduced pressure. The remaining ionic liquid can then be reused without loss of activity. The development of such application-specific materials demonstrates well that ionic liquids can be regarded as designer solvents.

Ionic Liquids in Chromatography

Due to their high thermal stability, ionic liquids with melting points close to $100\,^{\circ}C$ have been investigated for application as stationary phases in gas chromatography. The ionic liquid combinations studied include tetraalkylammonium, 1-alkyl-3-methylimidazolium or 1-alkylpyridinium cations with anions such as benzoate, halides, hexafluorophosphate or tetrafluoroborate, immobilized on Chromosorb or silica.

Lower melting ionic liquids, in particular tetraalkylammonium nitrate and thiocyanate, can also be used as mobile phases. In cases where the viscosity or the melting point are too high, co-solvents, such as dichloromethane, water, ethanenitrile, methanol or tetrahydrofuran can be added to achieve suitable flow-properties.

An interesting usage of ionic liquids in chromatography was presented by Novartis, who patented the use of ionic liquids as solvents in headspace gas chromatography. Unlike molecular solvents, chromatograms obtained using ionic liquids do not feature broad solvent peaks, limited temperature range and carry-overs from consecutive injections. Due to their negligibly low vapor pressure, interfering solvent peaks are not generated, lower detection limits achieved, and the application range of headspace gas chromatography extended.

Electrochemical and Battery Applications

Ionic liquids have been widely investigated for application as electrolytes. In fact, the development of electrolytes with low melting point such as 1-ethylpyridinium bromide/aluminium(III) chloride, and subsequently, the air- and moisture stable second-generation salts, triggered the current intense interest in ionic liquids. Their high conductivities, large electrochemical windows, and wide liquid ranges lend ideally to their use as electrolytes in batteries. For example, the application of pyridinium- and imidazolium-derived ionic liquids in conjunction with inexpensive graphite electrodes has advantages over the conventional 'rocking-chair' battery, which is complex in its synthesis, purification and assemblage.

Although many electrolytes developed for battery use are based on haloaluminate molten salts, some newer developments have shown that the second-generation ionic liquids, such as $[C_2mim][NTf_2]$, have excellent properties for this purpose as well.

Conducting polymers such as polypyrrole and polythiophene or poly(paraphenylene) can be made by electrochemical oxidative dehydrogenation. These polymers, which are conductive for both, ions and electrons, find application in display devices, mechanical actuators or in corrosion prevention.

Electrodeposition is an important tool in the purification of metals, manufacture of nanocrystalline metals and nanoscale semiconductors, and is the historic origin of ionic liquids. A wide electrochemical window is a prerequisite for the effective deposition of highly electropositive elements such as aluminium, silicon or lithium, transition metals, and alloys. Ionic liquids show advantages over aqueous solutions: deposits with superior mechanical properties are obtained, and the risk of dihydrogen evolution, as it is frequently observed in aqueous solution, is avoided.

Dye-sensitised solar cells are used to directly generate electricity from light. A dye absorbs light and releases electrons which are collected at a counter electrode. The dye is regenerated by electrons from an electrolyte containing a redox couple. Due to their negligibly low vapor pressure, ionic liquids featuring thermal and electrochemical stability, high conductivity, and low viscosity have been used as both redox couples and electrolytes. Ionic liquids based on the trifluoroethanoate or bis{(trifluoromethyl)sulfonyl}amide anions can serve as electrolytes, while ionic liquids based on iodide, such as the $[C_3mim]I$, are employed to generate the $[I_3]^-/I^-$ redox couple while simultaneously acting as electrolytes.

Reactions in Ionic Liquids

Many organic syntheses incorporating ionic liquids have been reported in the literature. They can be categorised into reactions that occur without the interaction of an added catalyst, and those in which a transition metal facilitates the reaction. Neither will be considered in this highly condensed account, but extensive reviews can be found elsewhere.

INDUSTRIAL APPLICATIONS FOR IONIC LIQUIDS

The huge industrial interest in ionic liquids is reflected in the patent literature, and, to-date, there are more than a dozen processes that have reached commercialisation. A large number of ionic liquids are now available from gram to multi-ton scale, and so there is no perceived barrier to further commercialization.

The first ionic liquid process which was developed to be implemented in existing technology is the Difasol® process of the Institut Français du Pétrole. The nickel-catalysed dimerization of butenes is carried out in Lewis acidic organochloroaluminate ionic liquids, in a continuously run biphasic mode. In this reaction, the product phase is insoluble in the catalyst-containing ionic liquid phase. The separation of the product is thus facilitated and the catalyst can be reused. Furthermore, the selectivity to dimers is improved when compared to the conventional Dimersol® process, and the catalyst is used more efficiently.

In the BASIL™ (Biphasic Acid Scavenging Utilising Ionic Liquids) process by BASF AG, 1-methylimidazole is used as a base to neutralise acid liberated during the reaction of phenylchlorophosphines with alcohols to yield alkyoxyphenylphosphines, used as photoinitiators in the manufacture of printing inks, glass fiber and wood coatings. As opposed to the old process, which employed a tertiary amine, a protonated ionic liquid (1-H-3-methylimidazolium chloride) is formed, which can be easily phase-separated from the product, making time-consuming filtration steps superfluous. Additionally, the space-time yield is increased from 8 kg m^{-3} h^{-1} to 690,000 kg m^{-3} h^{-1} using the BASIL™ process, and the yield increased from 50% to 98%. 1-Methylimidazole is recycled by thermal decomposition of 1-H-3-methylimidazolium chloride.

Eastman Chemical Company ran a plant for the conversion of 3,4-epoxy-1-butene to 2,5-dihydrofuran from 1996–2004 using phosphonium-based ionic liquids, producing 1400 metric tons per year of product. Newer developments include a process for the synthesis of organo-modified siloxane polymers *via* a hydrosilylation route, developed by Degussa AG. The products are used in a number of industrial applications, such as stabilisers for polyurethane foams, antifoaming agents, and emulsifiers. The ionic liquid serves as a solvent for the platinum catalyst, which can be easily recycled. Another sector of interests is the use of ionic liquids as secondary stabiliser (TEGO 662C) in universal pigment pastes for all kinds of varnishes and coatings, reducing the consumption of volatile organic solvents. Finally, the use of ionic liquids as electrolytes in lithium-ion batteries has been investigated. Conventionally, organic solvents such as ethene carbonate and dimethyl carbonate, together with lithium hexafluorophosphate, are used to produce batteries with high charge density. However, under extreme temperature conditions, the performance of the electrolyte

decreases, and improper handling of the charging process can lead to safety issues. The use of ionic liquids addresses these problems.

Air Products have incorporated ionic liquids in to gas storage cylinders (their GASGUARD Sub-Atmospheric Systems technology) to provide a safer storage and delivery system. Iolitec has developed ionic liquid systems for cleaning spray nozzles, and Central Glass Co., Ltd. (Japan) have commercialized the Sonogashira reaction for the synthesis of pharmaceutical intermediates.

Eli Lilly and Co. showed that molten pyridinium chloride can be used on multikilogram scale to demethylate 4-methoxyphenylbutanoic acid to 4-hydroxyphenylbutanoic acid, a precursor in the manufacture of a preclinical candidate.

OVERALL SUMMARY AND FUTURE CHALLENGES

Although ionic liquids have received enormous attention in the literature, and hold a great potential for improving industrial processes ecologically or economically, there are three main issues which have to be addressed soon to quicken their (already impressively fast) implementation in industry.

Firstly, there is still a lack of reliable physical data, due to the fact that the purity of ionic liquids has only recently been recognised to be of utmost importance for any measurements. However, IUPAC have created a freely accessible database of physical and thermochemical properties, which will enable rapid testing and retrieval of extant data.

The second issue, which hampers a widespread use as industrial solvents, arises from the obvious current deficiency of toxicological and ecotoxicological data, and only one preliminary study has dealt with the life-cycle assessment of ionic liquids to date. Obtaining toxicological data is costly, and the sheer number of ionic liquids complicates pre-marketing notification processes, since most ionic liquids are not included in the European Inventory of Existing Chemical Substances (EINECS). Also, the question of how ionic liquids are to be disposed off has not been fully addressed.

The final issue is the growing mythology about ionic liquids – ingrained assumptions about them which people believe, but have no reason to believe.

Nevertheless, a huge step towards the industrial use of ionic liquids has been taken in the last five years, when the first ionic liquids became commercially available from companies such as Merck, Cytec, SACHEM, DuPont, BASF, Fluka, Scionix, Solvent Innovation, Acros, Chemada, Degussa, etc, and a number are already available on the multi-ton scale.

The history of ionic liquid research shows a development at an interface between industrial and academic collaboration, in the fields of chemistry, engineering and catalysis. These collaborations are facilitated by academic research centers with industrial sponsorship, such as QUILL at the Queen's University of Belfast and the Centre for Green Manufacturing at the University of Alabama. QUILL, indeed, has published two reports aimed at providing general information on both the industrialization of ionic liquids, and sourcing them. Typically, processes take 10–15 years to move from initial experiments to industrial plant; with ionic liquids, this transition is being made in 3–4 years. The next decade of ionic liquid research should be even more fascinating and fruitful than the last one.

M. Deetlefs and K. R. Seddon, *Chimica Oggi-Chemistry Today*, **24**, 16–23 (2006).

H. Ohno, ed., *Electrochemical Aspects of Ionic Liquids*, Wiley-Interscience, Hoboken, 2005.

R. D. Rogers and K. R. Seddon, eds., *Ionic Liquids IIIB: Fundamentals, Progress, Challenges, and Opportunities - Transformations and Processes*, ACS Symp. Ser., Vol. 902, American Chemical Society, Washington D.C., 2005; *Ionic Liquids IIIA: Fundamentals, Progress, Challenges, and Opportunities - Properties and Structure*, ACS Symp. Ser., Vol. 901, American Chemical Society, Washington D.C., 2005.

F. van Rantwijk, R. M. Lau, and R. A. Sheldon, *Trends in Biotechnology*, **21**, 131–138 (2003).

P. Wasserscheid and T. Welton, eds., *Ionic Liquids in Synthesis*, Wiley VCH, Weinheim, Germany, 2003.

ANNEGRET STARK
Friedrich-Schiller-University of
Jena, Institute for Technical
Chemistry and
Environmental Chemistry

KENNETH R. SEDDON
The QUILL Centre, The Queen's
University of Belfast

IONOMERS

The word "ionomer" was first coined in 1965, to describe a class of ionic thermoplastics consisting of ethylene and partly neutralized methacrylic acid units. Later it was recognized that sometimes such ionomers as the polyacryamide ionomers in solution showed polyelectrolyte behavior. Thus, another definition was proposed by. If the bulk properties of polymers were governed by ionic interactions in ionic aggregates, the polymers would be classified as ionomers. When polymers are dissolved in solvents of high dielectric constants and the solution properties are governed by electrostatic interactions over a distance longer than that is typical of molecular dimensions, the polymers are polyelectrolytes.

In the 1950s it was found that the properties of elastomers could be changed significantly by the introduction of carboxylate ionic groups to the non ionic elastomers. Since then the introduction of ionic groups to non-ionic polymers has been used as a powerful tool for the modification of polymer properties. The typical examples of the effects of ionic interactions on the properties of ionomers can be found in studies on glass-transition temperatures, mechanical properties, transport properties, and melt viscosities.For example, the glass-transition temperatures

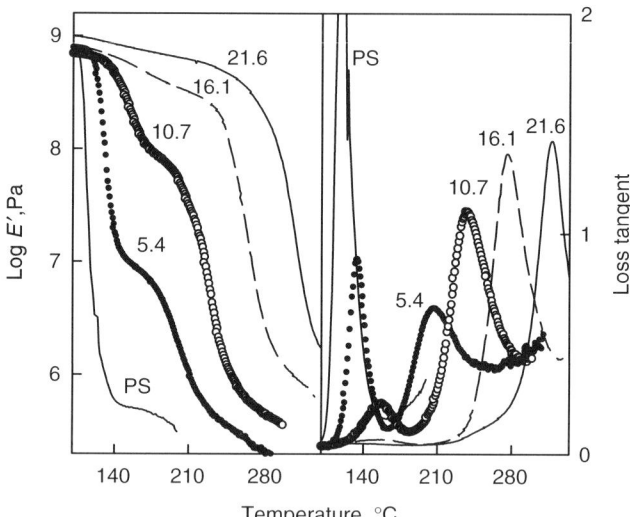

Figure 1. Storage moduli (E') and loss tangent measured at 1 Hz as a function of temperature for PSMANa ionomers with ion contents marked near each plot.

(T_g) of poly(phosphoric acid), $(HPO_3)_x$, increase from -10 to $280°C$, and to $520°C$ upon the neutralization of the acid groups with Na^+ and Ca^{2+} respectively. This indicates that the ionization as well as ionic interactions between anions and cations are surely responsible for the increase in the T_g.

To envisage the mechanical properties of ionomers, the storage modulus (E') and loss tangent (tan δ) of poly (styrene-*co*-Na methacrylate) (PSMANa) ionomers are shown in Fig. 1 as a function of temperature. It is seen that the modulus curve changes significantly with ion contents due to an increase in the glass transition (T_g) of the ionomer in part. For the ionomers of intermediate ion content, the modulus changes from a glassy modulus, through a T_g, through a "ionic" plateau, through another glass transition, through a rubbery plateau, to a modulus for flow as the temperature increases.

In regard to the transport properties of ionomers, above a certain ion content the ionic groups of the perfluorosulfonate and perfluorocarboxylate ionomer films in water become hydrated, resulting in the formation of water channel, in which ionic domains contain water. The formation of a percolative water channel allows the ionomers high conductance. Similar trends can also be found in methyl methacrylate- and styrene-based ionomer films immersed in water.

The increasing melt viscosity of ionomers is another example of the striking effects of ionic interactions on polymer properties. Melt viscosity also increases with increasing ion content. Polymer properties can be changed significantly through ionic interactions and ion aggregation.

GLASS TRANSITIONS

At certain ion contents, amorphous random ionomers, such as polystyrene-based ones, show two glass transitions (T_gs).

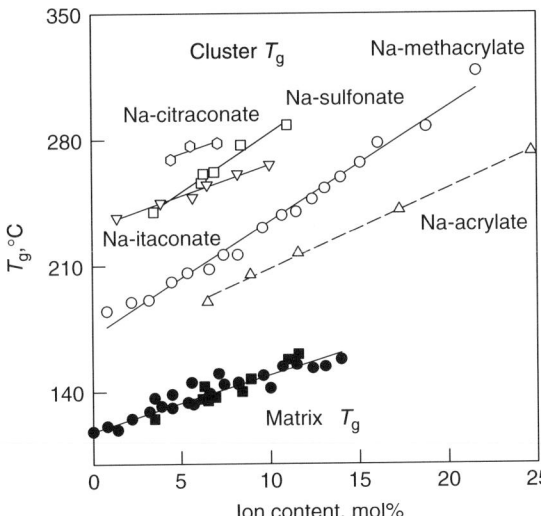

Figure 2. Glass-transition temperatures of matrix and cluster regions versus ion content for sodium neutralized poly(styrene-*co*-methacrylate), poly(styrene-*co*-acrylate), poly(styrene-*co*-styrenesulfonate), poly(styrene-*co*-itaconate), and poly(styrene-*co*-citraconate) ionomers.

As shown in Fig. 1, the non ionic polystyrene (PS) homopolymer and the PSMANa ionomer containing 21.6 mol% of ions exhibit only one modulus drop as well as one loss tangent peak at ~120 and at ~320°C, which are the matrix and cluster loss tangent peaks for the PS and the PSMANa ionomer, respectively. Interestingly enough, the activation energies for the glass-transitions calculated using an Arrhenius equation with the data of frequencies and temperatures of the peak maximum are found to be comparable each other The ionomers of intermediate ion contents, show two modulus drops and two loss tangent peaks. Two glass transitions in some of the PS-based ionomers have also been observed by using a differential scanning calorimeter and by measuring the volume of ionomers as a function of temperature.

In the case of PSMANa ionomers of low ion contents, the cluster T_g is ~60°C higher than the matrix T_g and increases almost in parallel with the matrix T_g (Fig. 2). By increasing ion content further, however, the cluster T_g shifts to higher temperatures more rapidly; thus, the cluster T_g of the 15 mol% ionomer becomes ~110°C higher than the matrix T_g. At this point, it should be mentioned that whereas the mechanism of the matrix T_g of the ionomers is the same as that of non-ionic polymers, the mechanism of the cluster T_g is still rather ambiguous.

Since the strength of ionic interaction is directly related to the temperature for ion hopping, the types of ionic groups (ie, including the strength, size, oxidation number of ions as well as number of ionic groups per ionic repeat unit) affects the cluster T_g significantly. For example, the ionic interaction between Na-sulfonate ionic groups is stronger than that between Na-carboxylate ionic groups owing to the difference in the 3D arrangement of the cation and the anion. At this point, the activation energies for ion hopping of the Na sulfonates polystyrene (NaSPS) system are generally lower than those of the PSMANa system, despite the opposite trend in the strength of ionic interaction. This indicates that the strength of the ionic interaction is directly related the cluster T_g, not with the activation energies for ion hopping.

The plasticization of polymers is effective for the decreasing T_g of the polymer. Thus, one can use the plasticization method to decrease the matrix T_g of ionomers. If the matrix T_g were to decrease sufficiently, more multiplets would form, leading to noticeable clustering in the ionomer system. Therefore, sometimes unclustered ionomers can be converted into clustered materials through polar, nonpolar, or amphiphilic plasticization.

When ionic repeat unit has two ionic groups, only a very weak cluster loss tangent peak (indicating weak clustering) is present at a much higher temperature than the cluster T_g of a well-clustered PSMANa ionomer; the weak cluster peak shifts to higher temperatures with increasing ion content. This weak clustering is probably due to the fact that the presence of two ion pairs in one ionic repeat unit is more effective in stiffening polymer chains, compared to the methacrylate ionomer. In addition, the higher cluster T_g is owing to the presence of two ion pairs in an ionic repeat unit, which makes ion hopping difficult and thus results in a higher cluster T_g. Morphological study of this ionomer reveals that, as ion content increases, more ionic groups participate in the formation of multiplets.

PROPERTIES

For the stress relaxation results of ionomers it was found that when the individual curves of modulus versus time were shifted to make one master curve, the time–temperature superposition of the curves was applicable for ionomers containing less than ~ 6 mol% of ions. However, above that ion content the time–temperature superposition of the individual curves was found to fail. As the ion content of ionomers increases, the nature of the ionomer changes from a matrix-dominant material to a cluster-dominant material progressively. Thus, at a certain ion content the percolation threshold for the connectivity of clustered regions occurs. Below that ion content, the continuous matrix regions are responsible for ionomer behavior, but above the percolation threshold both the clustered and unclustered matrix regions are attributed to the ionomer properties. Thus, the time–temperature superposition is not applicable above the percolation threshold for the connectivity of the clustered regions. At very high ion content the cluster regions, being a continuous phase, determine the ionomer properties.

Other Ionomers

Various ionomers based on polymers other than polystyrene have also been investigated extensively. Among them the most well-known ionomers are based on polyethylene (PE) and PTFE, which are partly crystalline. The crystallinity of these two ionomers makes interpretation of experimental results difficult, compared to the amorphous ionomers such as styrene ionomers.

Because the description of the properties and morphologies of all the ionomer families is not possible, only a few of the more extensively studied will be named here. These include block ionomers, telechelic ionomers, star ionomers, polyurethane ionomers, ionenes, ethyl (or butyl) acrylate zwitterionomers , ethyl acrylate ionomers, methyl methacrylate ionomers, and liquid crystalline ionomers.

Ionomer Blends

In view of the significant improvement in material properties that can be achieved in polymer blends over their individual components, it is not surprising that extensive efforts have been made to study miscibility in polymers.

The entropy of the mixing of polymers is very small because of the very high molecular weights. Thus, even a very small positive enthalpy change caused by unfavorable mixing makes most polymer pairs immiscible. In the case of miscible polymer pairs, sometimes one finds specific interactions between the polymer pairs, which reduce the enthalpy change even to negative values. Therefore, the addition of specific additives to polymer pairs or the physical and chemical modification of polymer pairs are required for miscibility enhancement. The first two methods include the addition of interfacial agents such as emulsifiers, reactive processing, high stress shearing, cocross-linking, the formation of interpenetrating networks, and others. The last method, chemical modification, is the introduction of interacting groups to polymer pairs. The interactions include hydrogen bonding, acid–base interactions, ionic interactions, dipole–dipole interactions, and the formations of donor–acceptor complexes or charge-transfer complexes. Since the early 1980s, the introduction of ionic interactions to non-ionic polymers have received considerable attention as a tool for the enhancement of polymer miscibility.

Various types of ionic interactions can be used to enhance polymer miscibility. A simple ion–ion interaction, which arises between groups such as N-methylpyridinium and benzenesulfonate, is the first candidate. Ionomer blends having this type of interaction can be obtained by mixing a first copolymer containing sulfonic acid neutralized with a large cation, such as an alkyl ammonium ion, and a second copolymer containing pendent pyridinium ions with a large counteranion, such as iodide. In this case the micro-ion, alkyl ammonium iodide, still remains in a solvent, and the polymer blend precipitates as a gel because of strong interactions between ionic groups.

An ion pair–ion pair interaction can also be used as one of the tools for miscibility enhancement. When a sodium carboxylate ion pair attached to one polymer chain interacts with a quaternary ammonium halide ion pair on the other polymer, miscibility of blends can be achieved. In this blend, the microcounter-ions remain in the blend and, in some cases, would crystallize in the form of microcrystals, which act as a filler.

The third and fourth types of ionic interactions used in ionomer blends are ion–dipole interactions involving the alkali and alkaline earth metal cations, and ion coordina-tion with the presence of transition metal ions. For example, the lithium cation from the PSMALi ionomer can interact with a polar polymer, such as poly(ethylene oxide), PEO, through ion–dipole interactions.

The strength of ionic interactions and the ion content of ionomers affect the enhancement of the miscibility of ionomer blends strongly. For example, when the interactions between ionic groups generated in an ionomer blend upon the mixing of two non-ionic copolymers are relatively strong, the ionomer blend becomes more homogeneous with increasing ion content. However, at high ion contents the melt viscosities of the ionomer blends become very high, making thir processing difficult. In addition, the increasing ion content naturally means that the chemical structure and physical nature of ionomers become very different from those of the non ionic polymers (ie, copolymerization effects).In addition, when the strengths of the ionic interactions of two ionomers are different due to the presence of two different ion pairs, sometimes each ionomer would like to form its own phase-separated regions, thus leading to immiscible blends. Therefore, the type of ionic interactions and the amount of functional groups should also be taken into account for specific applications of ionomer blends.

PLASTICIZATION OF IONOMERS

External plasticizers used for ionomers can be classified into polar, nonpolar, and amphiphilic plasticizers. These three different types of additives can be used to plasticize either polar or nonpolar regions selectively or both regions, depending on the polarities of the additives and the ionomers. In the case of the ionomers, they usually have both polar and nonpolar characteristics. Thus, it is possible to plasticize either the ion-rich regions by using a polar plasticizer or the regions of low polarity by using a nonpolar plasticizer. An early study on dual plasticization found that the ion-rich regions in the SPS ionomers could be plasticized by using a polar plasticizer such as glycerol, which weakened the ionic interactions. As a result, fast rate of ion hopping increased, which decreased the cluster T_g and thus the melt viscosity of the ionomer decreased significantly. It was also found that a nonpolar plasticizer such as dioctylphthalate, interacting with the nonpolar hydrocarbon regions of ionomers, would be distributed more or less evenly throughout both the matrix and cluster regions. Thus, it lowered both the matrix and cluster T_g values. These results imply that appropriate plasticization is a useful menu for the modification of the thermal behavior of ionomers and processing methods.

When polar plasticizers are added to ionomers, the plasticizers go into ionic aggregates and cover ionic groups, which reduces the strength of ionic interactions to a point where the ionic groups become relatively mobile. Then the energy for ion hopping decreases, and thus at relatively low temperatures the ions can move from one multiplet to another at a fast rate. As a result, the cluster T_g shifts to lower temperatures. If the amount of plasticizer were sufficiently large, the multiplets might be shattered completely. At this point, the polar plasticizers in ionomers reside not in the nonpolar matrix but in the multiplets. Thus, the polar plasticization has only little effect on the matrix T_g. It was also found that the intensity of a small-angle X-ray scattering peak decreased, and the peak shifted to lower angles upon the addition of methanol to SPS ionomers. As expected, the small-angle X-ray scattering peak disappeared completely at high plasticizer contents. The addition of polar plasticizer to the ionic aggregates leads to the increase in the size of the aggregates which in turn, decreases the electron density of ionic aggregates. However, when water was added to the Zn-SPS ionomer, it was found that the small-angle X-ray scattering peak shifted to a lower angle with increasing peak height. Thus, it was suggested that the ionic aggregates were swollen by the water, which induced a rearrangement of ionic groups, leading to the changes in the number of the ionic aggregates and in the fraction of ionic groups that formed ionic aggregates.

Plasticizers of low polarity such as diethylbenzene or dioctylphthalate can be mixed with polymers of low polarity. In the case of styrene ionomers, a nonpolar plasticizer could exist relatively evenly in the cluster and matrix regions because both the two regions consist primarily of nonpolar polystyrene matrix. For example, when the amount of diethylbenzene in PSMANa ionomers increases, both the matrix and cluster loss tangent peaks of the ionomers shift to lower temperatures. As expected, the small-angle X-ray scattering peak a found not to change much with the amounts of plasticizer. While a polar plasticizer would lower the melt viscosity strongly, it would have little effect on the matrix T_g. The nonpolar plasticizer would lower the melt viscosity only slightly, but lower the matrix T_g significantly.

From the above it shoud be clear that plasticization effects on the properties and morphologies of ionomers can be changed drastically by the types of ionomers and plasticizers used. In other words, a plasticizer can reside only in multiplets or in the matrix in a well distributed form or as a phase-separated filler, or be present in both matrix regions and multiplets, depending on the polarity, crystallinity, and types of functional groups of plasticizer and ionomer. Thus, the nature of the ionomer and plasticizer can change plasticizations effects significantly.

APPLICATIONS

Ionomers have been used in a wide range of applications. In the 1950s, B F. Goodrich and DuPont introduced poly (butadiene-co-acrylonitrile-co-acrylate) elastomers and sulfonated chlorinated polyethylene (Hypalon), respectively, to markets. In the mid-1960s, partially neutralized poly(ethylene-co-methacrylate) ionomers (Surlyn) where made available by Du Pont. Since then, substantial studies have focused on the synthesis of new classes of ionomers, new synthetic routes, and new applications.

Membranes

One of the most useful applications of ionomers can be found in the ionomer membrane of superperselectivity such as poly(tetrafluoroethylene) (PTFE)-based ionomers,

eg, Nafion, which are copolymers of tetrafluoroethylene and a perfluorinated monomer containing a sulfonate group at the end of a long side chain. Since a (PTFE) matrix is known for its high strength, thermal stability, and chemical stability, the PTFE ionomers find extensive applications in the chlor-alkali industry, in which PTFE ionomer membranes are used to separate the cathode and anode compartments for the electrolysis process of NaOH from brine. The use of these membranes saves the cost of electricity considerably and produces pure NaOH and H_2 on the cathode side and Cl_2 and spent brine on the anode side.

In fuel cells, the conversion of energy generated from chemical reactions into electrical energy takes place; thus, the fuel cells can be used as electric power sources in space crafts and submarines and for the extra electricity needed in new electrical or hybrid automobile applications, such as power batteries, TV, and air conditioning. The fuel cell consists of anode and cathode compartments, are separated by an electrolyte, at which chemical reactions between H_2 and O_2 (a fuel and an oxidant) occur to produce water and electricity. The electrolyte for fuel cells should be stationary and have high energy efficiency per unit mass and long-term stability. Ionomers as proton exchange membranes can be used as membrane electrolytes.

Molded Materials

Ionomers have found extensive application as coatings and molded parts, which include golf ball and bowling pin covers, bumper guards and body side molding strips of automobiles, shoe parts, ski boot shells, and bottle stoppers. Perhaps the most well known use for polyethylene ionomers is golf ball covers. Examples of applications of ethylene-based ionomers can be found in packaging, impact modifiers, rheology modifiers, modifiers for glass-reinforced thermoplastics, ionomer foamed objects, and elastomeric materials.

Fertilizer Coating

Ionomers are also used as coating for agricultural fertilizers, which improves release properties. Zn sulfo-EPDM ionomers are used as a coating material for the slow release of fertilizers. The volatile polar cosolvent is added to the ionomer solution to plasticize multiplets, which leads to a significant decrease in solution viscosity. This in turn permits easy spraying of ionomer solutions on the spherical urea fertilizers. As the cosolvent and solvent evaporate very rapidly, the aggregation of the ionic groups and the formation of a strong thin film take place. Using this technique, one could obtain various patterns of fertilizer release by changing the size of fertilizer particles and film thickness. When ionomer-coated urea fertilizers are exposed to water, the water diffuses into the core, consisting of urea, through the outer ionomer layer, and dissolves the urea particles, resulting in an increase in the osmotic pressure of water inside the coating layer. As the amount of water in the fertilizer cores increases, the pressure on the coating layer increases as well, to a point where cracking of the coating layer occurs. Then the urea solution is released through the crack. Because of variation in the thickness of the coating layers, the release of the urea solution takes place over long times. Thus, the ionomer-deposition technique on the fertilizer particles allows the appropriate amount of fertilizers be released at the height of the growing season, when fertilizers are most needed.

A. Eisenberg and J.-S. Kim, *Introduction to Ionomers*, John Wiley & Sons, Inc., New York, 1998.

S. Schlick, ed., *Ionomers: Characterization, Theory, and Applications*, CRC Press, Inc., Boca Raton, Fla., 1996.

M. R. Tant, K. A. Mauritz, and G. L. Wilkes, eds., *Ionomers: Synthesis, Structure, Properties and Applications*, Blackie Academic Professional, New York, 1996.

JOON-SEOP KIM
Department of Polymer Science
& Engineering, Chosun
University

IRON

Iron, Fe, from the Latin *ferrum*, atomic number 26, is the fourth most abundant element in the earth's crust, outranked only by aluminum, silicon, and oxygen. It is the world's least expensive and most useful metal. Although gold, silver, copper, brass, and bronze were in common use before iron, it was not until humans discovered how to extract iron from its ores that civilization developed rapidly.

Pure iron is a silvery white, relatively soft metal and is rarely used commercially. Typical properties are listed in Table 1.

Iron is alloyed with other elements for commercial applications. The most important alloying element is carbon. Small amounts of carbon alloyed with iron lower the melting point. The distinction between steels and other irons is based on properties and defined by the iron–carbon phase diagram. Steel is generally classified as iron–carbon alloys (0–2% C) that have a high melting point and can be hot rolled. Iron with carbon up to about 2% can be heated to a temperature at which only one phase (gamma iron) exists. Gamma iron is face-centered cubic (fcc) in structure and therefore is plastic, or malleable, which allows hot rolling. Cast irons contain sufficient quantities of the eutectic (about 2–5% C) to make the metal too brittle to hot roll; thus the requirement that it be cast. Pig iron from a blast furnace is liquid iron saturated with carbon (>4.3% C), depending on the temperature corresponding to the liquidus line.

Iron is indispensable in the human body. The average adult body contains 3 grams of iron. About 65% is found in hemoglobin, which carries oxygen from the lungs to the various parts of the body. Iron is also needed for the proper functioning of cells, muscles, and other tissues.

Table 1. Properties of Iron

Property	Value
atomic mass	55.847
isotopic abundance	
mass	54 56 57 58
abundance, %	6.04 91.57 2.11 0.28
melting point, °C	1537
boiling point, °C	3000
crystal structure[a]	bcc
density,[b] g/cm^3	7.87
thermal conductivity at 0°C, W/(m·K)	79
electrical resistivity at 20°C, μΩ·cm	9.71
tensile strength, MPa[c]	240–280
yield strength, MPa[c]	70–140
Young's modulus of elasticity, GPa[c]	195
Poisson's ratio	0.3
elongation in 5 cm at 20°C, %	40–60
reduction of area, %	65–78
Brinell hardness	82–100
impact strength (izod notched bar)	
longitudinal, J/m[d]	4859
transverse, J/m[d]	2990
thermal expansion, K^{-1}	
from 0–300°C	12.6×10^{-6}
0–600°C	14.6×10^{-6}
specific heat, J/(g·K)[d]	
at 100°C	0.50
500°C	0.67
800°C	1.26
transition from magnetic to paramagnetic, °C	ca 770

[a] Room temperature.
[b] Hot rolled.
[c] To convert MPa to psi, multiply by 145.
[d] To convert J to cal, divide by 4.184.

IRON ORES

Minerals

Iron-bearing minerals are numerous and are present in most soils and rocks. However, only a few minerals are important sources of iron and are thus called ores. Hematite is the most plentiful iron mineral mined, followed by magnetite, goethite, siderite, ilmenite, and pyrite.

Sources

Iron ore deposits were formed by many different processes; eg, weathering, sedimentation, hydrothermal, and chemical. Iron ores occur in igneous, metamorphic, and sedimentary deposits. Normally, as-mined iron ore contains 25 to 68% iron. The main iron ore deposits in the United States lie near Lake Superior in Minnesota (Mesabi range) and Michigan (Marquette range).

Canada's chief deposits occur along the borders between Quebec and Newfoundland in an area called the Labrador Trough and in an area north of Lake Superior. Most of the deposits are similar to those found in Minnesota and Michigan.

Other countries that have large iron ore deposits include Brazil (Carajas and Quadrilatero Ferrifero deposits), Australia (Pilbara deposits), Ukraine (Krivi Rog deposit), Russia (Kursk deposit), Venezuela (Cerro Bolivar deposit), India (Bihar-Orissa, Hospet, Kudremukh, and Goa deposits), South Africa (Sishen and Thabazimbi deposits), and Sweden (Kiruna, Svappavaara, and Malmberget deposits).

Beneficiation

Iron ore coming from the mine must be properly sized. A gyratory crusher is normally used for primary crushing down to approximately 300 mm. Secondary crushing down to 25 mm can be done in a cone crusher. Fine grinding can be done by rod mills, followed by either ball or pebble mills. In some cases, autogenous grinding can be used to replace the cone crusher and rod mills.

Iron ores of different characteristics and compositions can be blended to a more uniform composition. This can be accomplished during handling operations involved in transporting ore to its point of use, or through special blending facilities, such as stacking and reclaiming. Sand and clay can be removed from iron ore by washing in a log washer or classifier, followed by screening. Low-intensity magnetic separators are used to upgrade iron ores containing magnetite. High-intensity magnetic separators are used to upgrade iron ores containing hematite or ilmenite.

Agglomeration

Iron ore concentrates are often too fine to be used directly in the iron-making processes; therefore, they must be agglomerated. The agglomerating methods typically used in the iron ore industry are pelletizing, sintering and, to a limited extent, briquetting and nodulizing.

IRONMAKING

Blast Furnace

The blast furnace is the predominant method for making iron. In essence, the blast furnace is a large, counter-current, chemical reactor in the form of a vertical shaft which is circular in cross-section (see Figure 1). Iron ore, coke, and fluxes constitute the burden, which is charged continually into the throat at the top. Pressures in the shaft are controlled to 100–300 kPa (1–3 atms) gauge. Preheated air (hot blast) is blown in through water-cooled nozzles (tuyeres) around the circumference of the furnace near the bottom. The oxygen in the air reacts with the coke to form hot reducing gases (mostly carbon monoxide) which ascend through the burden and (1) provide heat for melting; (2) react with the iron ore to reduce it to iron; and (3) heat the ore, coke, and fluxes to reaction temperatures. Nitrogen in the hot blast is heated by the coke combustion and aids in heat transfer to the burden. The gases leaving the top of the furnace (top gas) are cleaned, cooled, and used as fuel to preheat the air for the hot blast.

The molten iron (hot metal or pig iron) and slag (molten oxides) that are produced accumulate in the bottom of the furnace. The hot metal and slag are drained semicontinuously through a taphole (tapping, or casting) into a

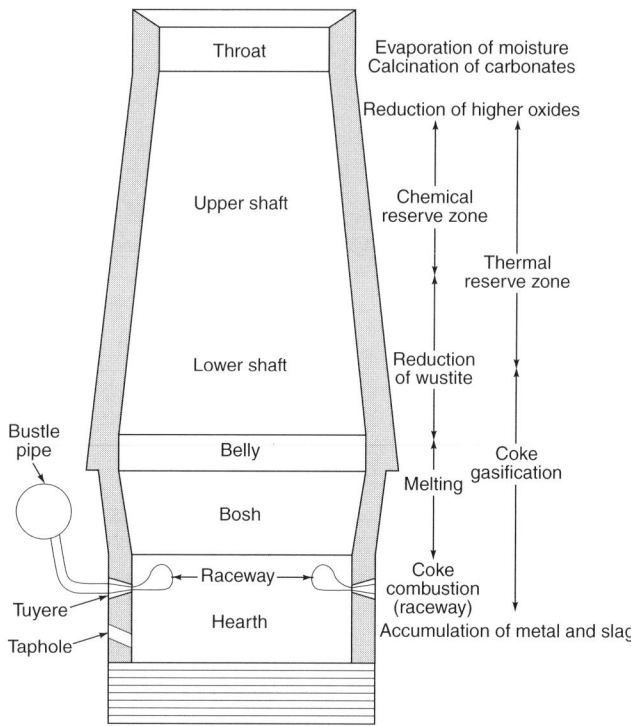

Figure 1. Schematic of a blast furnace.

trough. The hot metal is separated from the slag by a weir/dam arrangement at the end of the trough, then flows through runners to a refractory-lined rail car. The hot metal is then transported to a nearby site for further processing. Most of all the pig iron produced in the United States is used for steelmaking. The remainder is cast into pigs for remelting or used directly for iron castings.

Figure 2 shows the material flow diagram for a blast furnace plant. The ore and fluxes are stockpiled in a large open yard, from which these are reclaimed by crane (ore bridge) and transferred by conveyor to the stockhouse. Coke is delivered by rail or conveyor from the coke plant. The stockhouse consists of a row of bins from which the raw materials are weighed out in the desired order and amount, and conveyed by either skip car or conveyor to the top of the furnace. Special additives to the charge, such as upgraded yard scrap, manganese ore, or calcium chloride, for flushing out alkalies, are added at this location.

Direct Reduction

Direct reduction processes are distinguished from other ironmaking processes in that iron oxide is converted to metallic iron without melting.

Direct Smelting

Direct smelting processes use coal directly instead of coke. Several processes are under development which effectively divide the functions of the blast furnace into two separate but connected unit operations. First, the iron ore is prereduced in a shaft furnace or a fluidized bed, depending on the process and the type of ore used. Second,

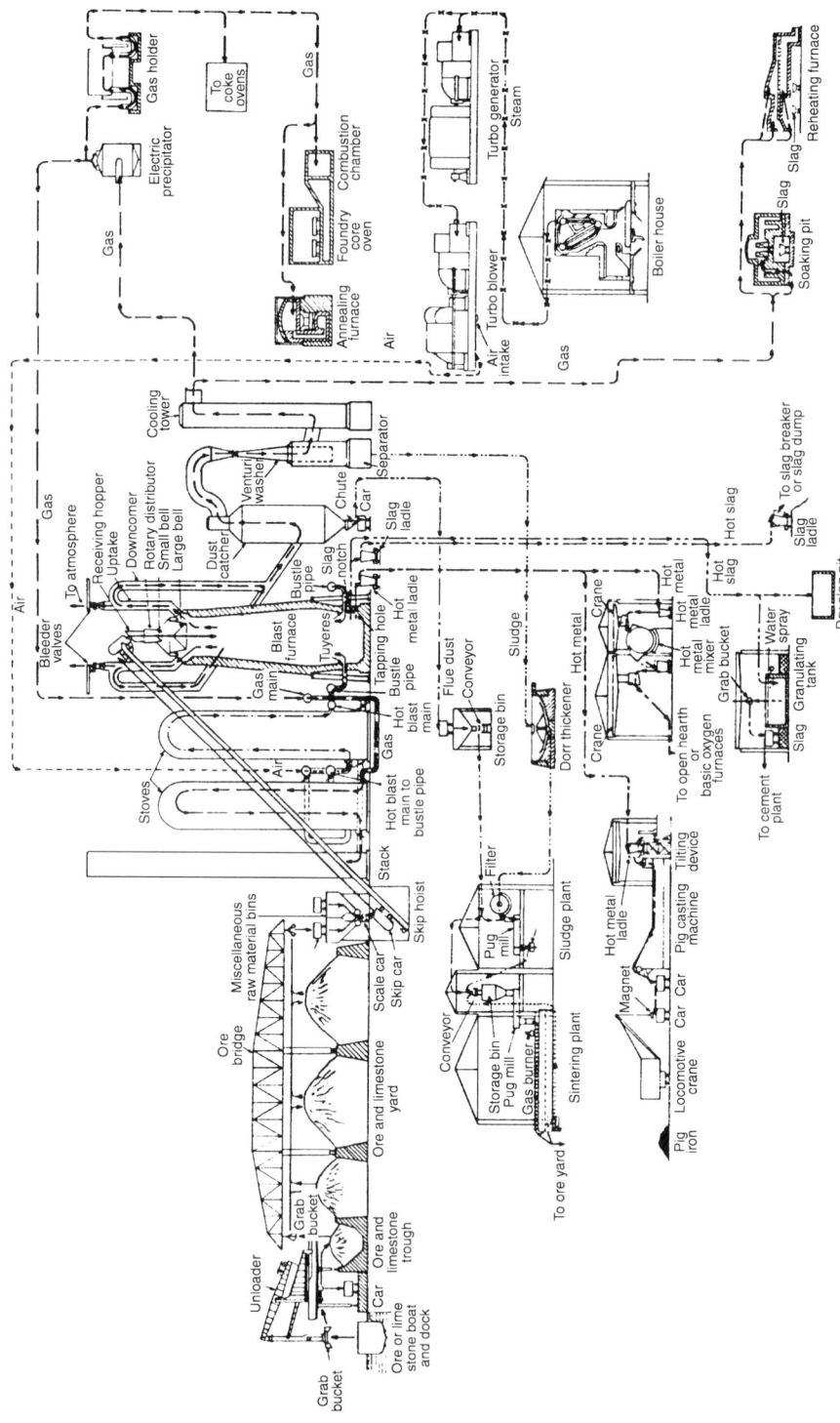

Figure 2. Flow diagram depicting the principal units and auxiliaries in a modern blast furnace plant, showing the steps in the manufacture of pig iron from receipt of raw materials to disposal of pig iron and slag as well as the methods for utilizing the furnace gases. (——), Miscellaneous raw material; (– – –), cold-blast air; (■), hot-blast air; (———), blast furnace gas; (—×—), steam; (— — —), hot metal; (— — — —), hot slag; (– – –) flue dust; (–·–·–), sludge; and (·), sinter.

the prereduced ore is charged into a molten bath into which coal and oxygen or air are also introduced. The gases leaving the smelter are used to perform the reduction in the prereduction vessel. The Corex process is the only one of the newer iron-making processes operating on a commercial scale.

CAST-IRON PRODUCTION

Cast irons are normally produced by melting iron or steel scrap along with pig iron. The carbon and silicon levels are adjusted to obtain the desired properties. Melting is done in cupolas, electric furnaces, or air furnaces.

HANDLING, SHIPPING, AND STORING

In handling, shipping, and storing direct reduced iron (DRI), care should be taken to avoid oxidation. Millions of tons of DRI in pellet and lump form have been shipped by barge, ocean vessel, truck, and rail. The key to avoiding oxidation is simply to keep the material cool and dry. In general, oxidation of DRI takes place in two forms: reoxidation and corrosion. Reoxidation occurs when the metallic iron in hot DRI reacts with oxygen in the air to form either Fe_3O_4 or Fe_2O_3. The reaction continues as long as the DRI remains hot and sufficient oxygen is available. Corrosion occurs when the metallic iron in DRI is wetted with fresh or salt water and reacts with oxygen from air to form rust, $Fe(OH)_3$. The corrosion reactions continue as long as water is present.

ECONOMIC ASPECTS

In 2003 the United States produced 5% of the world's iron ore output and consumed about 7%. China has become the dominant source of growth in demand for iron ore.

About 98% of iron ore is used to produce pig iron, this being the best indicator of iron ore consumption worldwide. In 2002, China produced 170 million tons of pig iron, Japan 81 million.

ENVIRONMENTAL CONCERNS

The most difficult environmental problem has been pollution from recovery coke oven batteries, which emit particulates and hydrocarbon compounds from doors during coking and when the coke is pushed from the oven on the way to the quench tower. Although most of the batteries now in operation are operating under agreements with the various pollution control agencies, there have been no new batteries built for a long time and it would be difficult to get permission to build a new one, although rebuilding existing ones is possible.

The water systems of all steel plants continuously lose water, due to evaporation in cooling towers, since a major function of the water is equipment cooling. However, since much of the water comes in contact with the rolling and casting equipment, which use lubricants, and is also treated with corrosion inhibiting chemicals, the build-up of

these oils and chemicals requires that an amount of water be continuously removed from the system. This "blow down" has to be treated before being discarded. This is an expensive process. Also, since the hardness of the water has to be kept low in order to avoid scaling the inside of cooling channels (thus decreasing in their effectiveness) the incoming make-up water to replace the evaporation and blow-down losses has to be softened, leading to a waste stream of concentrated salt, which has to be disposed of. In general, water system management is a complex and expensive subsystem operation that has a direct effect on the cost and quality of the steel made.

The largest future environmental problem facing the world's steel industry is that of greenhouse gas emissions, specifically carbon dioxide. The preparation of sinter fines or pellets uses a large amount of electricity, in addition to hydrocarbon fuel. The production of electricity is primarily based on coal and oil, which end up as CO_2 and water. The reduction of iron ore is largely based on the use of carbon, which ultimately ends up as CO_2. The oxygen used in steelmaking is produced from air using electricity, and the limestone used ultimately dissociates into CaO and CO_2. Thus, the industry produces a tremendous amount of carbon dioxide. The result is that the industry is one of the largest contributors to greenhouse gas emissions. Unfortunately, there is no economic substitute for the reductant and the energy requirements of the industry at this point in history, so the only choice to reduce these emissions is to improve incrementally the energy efficiency of the existing plants and processes.

HEALTH AND SAFETY FACTORS

Iron presents minimal health risks. Skin contact should not result in any adverse health effect. Excessive inhalation of dust may be irritating to the respiratory tract. Dust may also cause mechanical irritation on eye contact. Extremely large oral doses would be required to cause gastrointestinal disturbance. Iron dust does present a moderate fire and explosion hazard when exposed to heat and flame.

USES

Over 95% of the world's DRI production is consumed in electric arc furnace steelmaking. The remaining 5% is split among blast furnaces, oxygen steelmaking, foundries, and ladle metallurgy facilities.

The primary use of DRI is as a clean supplement or replacement for the ferrous scrap charge in high-quality-oriented electric arc furnace (EAF) steelmaking. By controlling the level of residual elements in the charge, steelmakers can upgrade their product mix and reduce off-grade heats. Also, a low level of residual elements in carbon steel changes its physical properties for the better.

Hot briquetted iron (HBI) is used as a trim coolant or scrap replacement in oxygen steelmaking. In the oxygen steelmaking process, the molten steel often is overheated. Trim coolant is then, fed to the furnace to cool the molten steel to the desired temperature. HBI is preferred for this application because its high density ensures effective slag penetration and complete melting in the molten steel bath. Steel yield is increased when HBI is used as a trim coolant instead of iron ore. Also, the violent reactions that can occur when using iron ore are eliminated.

HBI is an effective trim coolant for molten steel in ladle metallurgy facilities, ladle refiners, ladle furnaces, and vacuum degasers. It provides cold iron units in an ideal size and density for penetrating the ladle slag and cooling the metal. HBI can be a valuable charge material for ductile and malleable irons as well as steel. It is of particular value in making ductile iron castings because of its very low residual element content.

G. S. Brady and H. R. Clauser, *Materials Handbook*, 11th ed., McGraw-Hill Book Co., Inc., New York, 1977.

R. M. Brick and A. Phillips, *Structure and Properties of Alloys*, McGraw-Hill Book Co., Inc., New York, 1949.

"Iron Ore," *Mineral Commodity Summaries*, U.S Geological Survey, Washington, D.C., Jan. 2004.

W. T. Lankford, Jr. and co-workers, *The Making, Shaping, and Treating of Steel*, 10 th ed., Association of Iron and Steel Engineers, Pittsburgh, Pa., 1985.

J. A. LEPINSKI
PT Perkasa Indobaja

JEFFREY C. MYERS
Midrex Direct Reduction
    Corporation

GORDON H. GEIGER
Consultant

IRON COMPOUNDS

Iron is thought to be the major component of the earth's core. After oxygen, silicon, and aluminum, iron is the fourth most abundant element in the earth's crust, more abundant than any other transition metal element.

Iron is the lightest element of Group 8 (VIIIB) of the periodic table and is the first metallic element in the table that fails to attain an oxidation state equal to the number of electrons in the valence shell; ie, no Fe(VIII) is known.

The standard aqueous reduction potentials of the Fe(II)/Fe(0) couple is $-0.44\,V$ and that of the Fe(I II)/Fe(II) couple is $+0.77\,V$. Consequently, iron metal reacts readily with most nonmetals and dissolves in dilute acids to afford the iron(II) cation. Dissolution does not occur in chromic acid, concentrated nitric acids, or hydrogen peroxide, H_2O_2, because the metal is protected by formation of a passivating oxide film, which can be removed mechanically or by acids of coordinating anions such as HCl. The facile interconversion of iron(II) and iron(III) and the ability of the coordination environment to fine tune the redox potential of the couple are reflected in the large variety of functions that iron performs in biological systems.

Salts of iron(II) are known for almost all of the common anions. The exceptions, including NO_2^-, result from redox incompatibilities. Many of the salts are hydrates and are subject to either efflorescence or hydration.

SALTS AND SIMPLE COORDINATION COMPOUNDS

Acetates

Anhydrous iron(II) acetate, $Fe(C_2H_3O_2)_2$, is a colorless compound that can be recrystallized from water to afford hydrated species. Iron(II) acetate is used in the preparation of dark shades of inks and dyes and is used as a mordant in dyeing. An iron acetate salt that is a mixture of indefinite proportions of iron(II) and iron(III) it is used as a catalyst of acetylation and carbonylation reactions.

Iron(III) acetate, $Fe(C_2H_3O_2)_3$, is prepared industrially by treatment of scrap iron with acetic acid followed by air oxidation. Iron(III) acetate is used as a catalyst in organic oxidation reactions, as a mordant, and as a starting material for the preparation of other iron-containing compounds.

Basic iron(III) acetate is a brown-red material that is soluble in alcohols and acids but insoluble in water. Basic iron acetate is used as a mordant in dyeing and printing and for the weighting of silk and felt. It is reported to affect the oxidation of saturated hydrocarbons in the presence of oxygen, acetic acid, aqueous pyridine, and zinc powder.

Alkoxides

Iron(III) alkoxides like ferric ethoxide $Fe(C_2H_5O)_3$, and ferric are commercially available. Iron(III) alkoxides have been used in the sol-gel synthesis of metal oxides and as catalysts for the polymerization of lactide.

Carbonates

Iron(II) carbonate, $FeCO_3$, precipitates as a white solid when air-free solutions of alkali metal carbonates and iron(II) salts are mixed. Ferrous carbonate occurs naturally as siderite or spathic iron ore. The compound is used as a flame retardant and as an iron supplement in animal feed. Ferrous carbonate redissolves in water in the presence of carbon dioxide to yield iron(II) hydrogen carbonate, $Fe(HCO_3)_2$, which also can be formed from iron and carbon dioxide saturated water in the absence of oxygen. It undergoes air oxidation that evolves carbon dioxide and a precipitate of hydrated iron(III) oxide. This reaction accounts for the precipitation of iron from the water of many springs on exposure to air.

Citrates

Iron citrate is a compound that contains citric acid and iron(II) and iron(III) in indefinite ratios. All of the iron citrate compounds are used as supplements to soils and animal diets.

Iron(III) ammonium citrate is of indefinite stoichiometry. Iron ammonium citrates are water soluble but are insoluble in alcohol. The compounds are used to fortify bread, milk, and other foods.

Cyanides

As a monodentate ligand, the cyanide ion coordinates to metal ions almost exclusively through the carbon atom. In this mode, it is very high in the spectrochemical series, thus most cyanide complexes are low spin. As a bidentate ligand the cyanide ion can bridge two metal ions by coordinating to metal ions through either the carbon or nitrogen atoms. Bonding to iron by cyanide involves synergistic σ-donation and π-acceptance by the ligand. Owing to the negative charge, the cyanide anion is a somewhat stronger donor and a weaker acceptor than isoelectronic, neutral carbon monoxide. The cyanide anion yields a variety of penta-1 cyano-complexes, pentacyano cayno complexes, and the finely divided, intensely blue precipitate Prussian Blue (tetrairon(III) tris(hexakiscyanoferrate)), $Fe_4[Fe(CN)_6]_3$, also known as Berliner Blau, is identical to Turnbull's Blue. Prussian Blue compounds are used as pigments in inks and paints and its formation on sensitized paper is utilized in the production of blueprints.

The intense blue color of Prussian Blue is attributed to electron transfer between the $[Fe(CN)_6]^{4-}$ and Fe(III) ions. A related pigment called Berlin Green is obtained by oxidation of Prussian Blue.

Formates

Iron(II) formate dihydrate $Fe(HCO_2)_2 \cdot 2H_2O$, is a green salt which can be prepared from iron(II) sulfate and sodium formate in an inert atmosphere. The compound is slightly soluble in water and fairly resistant to air oxidation. An anhydrous salt is known.

Iron(III) formate [555-76-0], $Fe(HCO_2)_3$, can be obtained from iron(III) nitrate and formic acid in alcohol solution. The red compound is soluble in water but only slightly soluble in alcohol. Up to two waters of hydration may be included, in which event the color of the compound is more yellow. Aqueous solutions hydrolyze to afford basic iron(III) formates (analogous to basic acetates) and eventually a precipitate of iron hydroxide and free formate.

Fumarates

Iron(II) fumarate, $Fe(C_4H_2O_4)$, is prepared by mixing hot aqueous solutions of sodium fumarate and iron(II) sulfate, followed by filtration of the resulting slurry. It has limited solubility in water but is more soluble in acid solution. The compound is red-orange to red-brown and finds uses as a hematinic. A nonstoichiometric compound and iron(III) fumarate, $Fe_2(C_4H_2O_4)_3$, are also available.

Halides

All of the anhydrous and hydrated binary halides of iron(II) and iron(III) are known, with the exception of the hydrated iodide of iron(III). A large number of complex iron halides have been prepared and characterized.

Gluconates

Iron(II) gluconate dihydrate $Fe[HO-CH_2(CHOH)_4-CO_2]_2 \cdot 2H_2O$, is prepared from barium or calcium gluconate and iron(II) sulfate. It is a yellow-green powder and has a slight odor of caramel. The compound is quite soluble in water but nearly insoluble in alcohol. It is used as a hematinic, in the treatment of anemia, and to color, fortify, and flavor foods. Isotonic solutions are available. An anhydrous salt is also known. Iron(III) gluconate, $Fe[HOCH_2(CHOH)_4CO_2]_3$, has been examined as a nutritional supplement.

Nitrates

Iron(II) nitrate hexahydrate, $Fe(NO_3)_2 \cdot 6H_2O$, is a green crystalline material that is very soluble in water. Crystallization at temperatures below $-12°C$ affords an nonahydrate. Iron(II) nitrate is a useful reagent for the synthesis of other iron-containing compounds and is used as a catalyst for reduction reactions.

Iron(III) nitrate nonahydrate, $Fe(NO_3)_3 \cdot 9H_2O$), is prepared by dissolving iron in nitric acid that has a specific gravity of at least $1.115 \, g/cm^3$. Acid of too high concentration passivates the iron, however. The hygroscopic, monoclinic, colorless-to-pale violet crystals are very soluble in water and soluble in alcohol and acetone. Iron(III) nitrate hexahydrate, $Fe(NO_3)_3 \cdot 6H_2O$, forms colorless, cubic crystals. It is also very soluble in water. Iron(III) nitrate is used as a mordant in dyeing, weighting silks, leather tanning, as a catalyst for oxidation reactions, and as a reagent for the synthesis of other iron-containing compounds.

Oxides and Hydroxides

Iron(II) oxide, FeO, is a black solid that affords a fine, pyrophoric powder that can decompose water. Strong heating of the freshly prepared powder decreases its state of division and its reactivity. FeO occurs naturally as the mineral wüstite. Crystalline FeO has a cubic, rock salt structure, but is always deficient in iron because of the presence of some iron(III). The solid is easily oxidized in air, is a strong base, and absorbs carbon dioxide. It is insoluble in water, alcohol, or alkali but readily soluble in acids. Iron(II) oxide is used in the manufacture of green, heat-absorbing glass, in ceramic mixtures, and in a variety of catalyst preparations, notably those used in ammonia synthesis and methanation.

Iron(III) oxide, Fe_2O_3, exists in two different crystalline forms. The α-form occurs naturally as the mineral hematite, the principal ore of iron. The γ-form occurs naturally as the mineral maghemite and can be prepared synthetically by careful oxidation of Fe_3O_4. γ-Fe_2O_3 has a spinel structure in which the oxide ions form cubic closepacked layers and the iron(III) ions are randomly distributed over the tetrahedral and octahedral holes. Unlike the α-form, the γ-form is ferrimagnetic and is used as a magnetic material in the production of magnetic recording media. Iron(III) oxide is insoluble in water but dissolves in hydrochloric or sulfuric acids. The color and appearance of iron(III) oxide depend on the size and shape of the particles, and the identity and amount of impurities and water present. Yellow, orange, or red pigments are known. It is used in large quantities as a red pigment for paint, rubber, ceramics, and paper, and in coatings for steel and other metals. Iron(III) oxide also finds use in the preparation of rare-earth/iron garnets and other ferrites; as a polishing agent for glass, diamonds, and precious metals (jeweler's rouge); and as a catalyst for oxidation reactions.

Triiron tetroxide (iron(II, III) oxide), Fe_3O_4, is a mixed Fe^{II}/Fe^{III} oxide which occurs naturally as the mineral magnetite (lodestone). It is an important ore of iron. The compound is strongly ferrimagnetic and has a Curie point of 860 K, at which temperature the effective magnetic moment is 3.9×10^{-23} J/T ($4.2 \mu_B$). Iron(II,III) oxide is insoluble in water, alcohol, ether, and dilute acids but dissolves in concentrated acids. It is a fairly good conductor of electricity, owing to electron transfer between iron(II) and iron(III). Blue steel has a surface coating of iron(II,III) oxide as a corrosion-resistant film. The compound is used as a pigment for glass, ceramics, and paint; in magnetic recording media; as a polishing compound; and in many catalytic preparations. Ferrofluids, which the fluid properties of a liquid and the magnetic properties of a solid, are used in rotating shaft seals in computer hard disk drives, in voice coil gaps of loudspeakers to damp undesired vibrations and for cooling, and to separate metals from ores.

Iron(II) hydroxide, $Fe(OH)_2$, is prepared by precipitation of an iron(II) salt solution by a strong base in the absence of air. It occurs as pale green, hexagonal crystals or a white amorphous powder. It is practically insoluble in water, fairly soluble in ammonium salt solutions, and soluble in acids and in concentrated NaOH solution. It is slowly oxidized by air. Conversion to $Fe_2O_3 \cdot xH_2O$ is eventually complete.

Iron(III) hydroxide, $FeHO_2$, is a red-brown amorphous material that forms when a strong base is added to a solution of an iron(III) salt. It is also known as hydrated iron(III) oxide. The fully hydrated $Fe(OH)_3$ has not been isolated. It is insoluble in water and alcohol but redissolves in acid. Iron(III) hydroxide loses water to form Fe_2O_3. Iron(III) hydroxide is used as an absorbent in chemical processes, as a pigment, and in abrasives. Salt-free iron(III) hydroxide is the highly colored precipitate known as yellow boy that deposits in streams affected by acid mine drainage.

Ferrites, Garnets, and Ferrates

Iron in oxidation states +3 and higher forms numerous oxide compounds that formally appear to contain oxo ligands (O^{2-}). Many of these have interesting magnetic and chemical properties. The ferrites (FeO_2^-) and garnets are really mixed metal oxides. Ferrite compounds are spinels and have the general formula $M^{2+}Fe^{III}_2O_2$. Some adopt normal spinel structures in which the M(II) ions occupy tetrahedral sites in the cubic oxide lattice and the Fe(III) ions occupy octahedral sites. Others have the inverse spinel structure in which one half of the Fe(III) ions occupy tetrahedral sites and the other half occupies octahedra sites. Inverse spinels are ferrimagnetic. One

such material is iron(II,III) oxide. The inverse spinel ferrites are used in magnetic recording media, as cores in high-frequency transformers, and in computer memory systems. Hexagonal ferrites, such as $BaFe_{12}O_{19}$, are used to construct permanent magnets. Garnets have the general formula $M_3Fe_5O_{12}$, where M is trivalent, and are useful in microwave applications.

Mixed oxides of Fe(IV) can be prepared by heating iron(III) oxide with a metal oxide or hydroxide in oxygen at elevated temperatures and are readily decomposed by mineral acids to iron(III) and oxygen.

Compounds of iron(V) are extremely rare.

Perchlorates

Iron(II) perchlorate hexahydrate $Fe(ClO_4)_2 \cdot 6H_2O$, crystallizes in hygroscopic, light green hexagonal prisms which are stable in dry air and extremely soluble (0.978 g/mL H_2O at 0°C) in water and alcohol. It is susceptible to air oxidation in aqueous solution and decomposes above 100°C. Yellow iron(III) perchlorate hexahydrate, $Fe(ClO_4)_3 \cdot 6H_2O$, is also extremely soluble in water (1.198g/mL H_2O at 0°C).

Sulfates

Iron(II) sulfate heptahydrate $FeSO_4 \cdot 7H_2O$, forms blue-green monoclinic crystals that are very soluble in water and somewhat soluble in alcohols. It is known by many other names, including cupperas, green vitriol, and iron vitriol. The compound is efflorescent in dry air. In moist air, the compound oxidizes to yellow-brown basic iron(III) sulfate. Aqueous solutions tend to oxidize. The rate of oxidation increases with an increase in pH, temperature, and light. The compound loses three waters of hydration to form iron(II) sulfate tetrahydrate, $FeSO_4 \cdot 4H_2O$, at 56°C. Further warming to 65°C forms white iron sulfate monohydrate, $FeSO_4 \cdot H_2O$, which is stable to 300°C. Strong heating results in decomposition with loss of sulfur dioxide. Solutions of iron(II) sulfate reduce nitrate and nitrite to nitric oxide, whereupon the highly colored $[Fe(H_2O)_5(NO)]^{2+}$ ion is formed. This reaction is the basis of the brown ring text for the qualitative determination of nitrate or nitrite.

Iron(II) sulfate forms double salts of formula $M_2SO_4 \cdot FeSO_4 \cdot 6H_2O$ with alkali sulfates. Iron(II) ammonium sulfate (Mohr's salt), $FeSO_4 \cdot (NH_4)_2SO_4 \cdot 6H_2O$, is used as a primary standard for iron. It is soluble in water but insoluble in alcohols. Both the solid and its solution are more stable to oxidation than iron(II) sulfate. Most iron(II) sulfate is a by-product of the steel (qv) industry.

Iron(III) sulfate $Fe_2(SO_4)_3$, is a gray-white material that forms hygroscopic rhombic or rhombohedral crystals. It is slightly soluble and dissolves slowly in cold water and decomposes in hot water. Several hydrates are known including the monohydrate, the hexahydrate, the heptahydrate, and the nonahydrate. These can be difficult to obtain pure. The commercially available hydrate contains about 20% water by weight and is yellowish in color. Alum compounds of the general formula $MFe(SO_4)_2 \cdot 12H_2O$, M monovalent, are known. Iron(III) ammonium alum dode-cahydrate, $(NH_4)Fe(SO_4)_2 \cdot 12H_2O$, and iron(III) potassium alum dodecahydrate, $KFe(SO_4)_2 \cdot 12H_2O$, are important as mordants in dyeing. Iron(III) sulfate is used in pigments, as a coagulant in water and sewage treatment, and as a mordant.

Sulfides

Three sulfides of iron are known. Iron(II) sulfide FeS, is a gray nonstoichiometric material obtained by direct reaction of iron and sulfur. The actual stoichiometry is typically $Fe_{0.9}S$. It is found in nature as the mineral pyrrhotite which usually contains nickel as well. FeS has a NiAs structure. It is almost insoluble in water, oxidizes readily in air, and dissolves in aqueous acids with the evolution of H_2S. The above reactions represent a reasonable route for the synthesis of H_2S gas. Iron disulfide, FeS_2, can be prepared by heating Fe_2O_3 in H_2S. FeS_2 is found in nature as the minerals pyrite (fool's gold) and marcasite, both of which have a brassy yellow color and a metallic luster. Pyrite is frequently found in large, well-formed crystals, is composed of iron(II) and S_2^{2-} ions in a distorted rock salt structure. Heating solid pyrite affords Fe_2O_3 and SO_2 in air or FeS and sulfur in a vacuum. Roasting of pyrites has been used in the past to produce SO_2 for sulfuric acid production and iron(III) oxide for use as an iron ore. Marcasite is less stable than pyrite and therefore is more reactive. Fe_2S_3 is an unstable black precipitate produced when aqueous iron(III) solutions are treated with S^{2-}. It is rapidly decomposed in moist air to Fe_2O_3 and sulfur. A purer material can be obtained by reaction of anhydrous $FeCl_3$ with bistrimethylsilylsulfide. Iron(III) sulfide occurs in nature in the form of the double sulfide minerals chalcopyrite, $CuFeS_2$, and bornite, Cu_3FeS_3, which can be represented as $Cu_2S \cdot Fe_2S_3$ and $3Cu_2S \cdot Fe_2S_3$, respectively. Iron(III) sulfide finds applications in cathodes in secondary Li batteries, coal liquefaction, and desulfurization.

Chelate Compounds

A chelate is a multidentate ligand which binds to a metal atom at more than one coordination site resulting in a complex having a closed-ring structure. Chelate complexes are more stable, ie, have greater formation constants, than analogous complexes of unidentate ligands, wherein no rings are formed. The enhanced stability, called the chelate effect, is thought to result predominantly from favorable entropic effects. In general, chelates that contain five-membered rings are more stable than this containing six-membered rings.

ORGANOMETALLIC COMPOUNDS

The organometallic compounds include such carbonyls as iron pentacarbonyl, $Fe(CO)_5$, a toxic, yellow-orange, oily liquid which does not react with air at room temperature, an antiknock agent in gasoline is suitable for use in pigments and polishing compound; diiron nonacarbonyl $Fe_2(CO)_9$, reactions of which give a variety of products;

triiron dodecacarbonyl, $Fe_3(CO)_{12}$, a useful compound for synthesis of iron carbonyl derivatives because it is more reactive than $Fe(CO)_5$ and more soluble and stable than $Fe_2(CO)$, bis(cyclopentadienyl)iron or ferrocene use a as an internal standard in electrochemistry; and bis(cyclopentadienyldicarbonyliron), $[Fe(CO)_2(C_5H_5)]_2$, a purple-red, air-sensitive solid, mononuclear compound that has had have considerable utility in organic syntheses.

COMPOUNDS OF BIOCHEMICAL RELEVANCE

Iron is perhaps the most important of the transition elements that play a role in biochemistry. It is an essential element for all organisms. The functions of iron-containing metalloproteins include electron transfer, dioxygen transport and storage, activation of dioxygen and hydrogen peroxide with concurrent oxidation of substrates, dismutation of superoxide and peroxide, activation and production of dihydrogen, reduction and rearrangement of substrates, and phosphate hydrolysis, among others. Because of the near total insolubility of iron under physiological conditions, iron metalloproteins and chelate compounds function in the solubilization, uptake, transport, and storage of iron.

Iron Sulfur Compounds

A large number of compounds are known in which iron is tetrahedrally coordinated by a combination of thiolate and sulfide donors. Figure 1 shows four particularly important classes of iron sulfur compounds that are known to occur in proteins.

(μ-Oxo)bis(μ-carboxylato) Diiron Complexes

Several nonheme iron proteins of widely varying function contain a binuclear iron site as a common structural feature. The proteins include hemerythrin, the O_2-transport protein of marine invertebrates; ribonucleotide reductase, an enzyme which catalyzes the deoxygenation of ribonucleoside diphosphates to deoxyribonucleosides; methane monooxygenase, an enzyme which catalyzes the oxidation

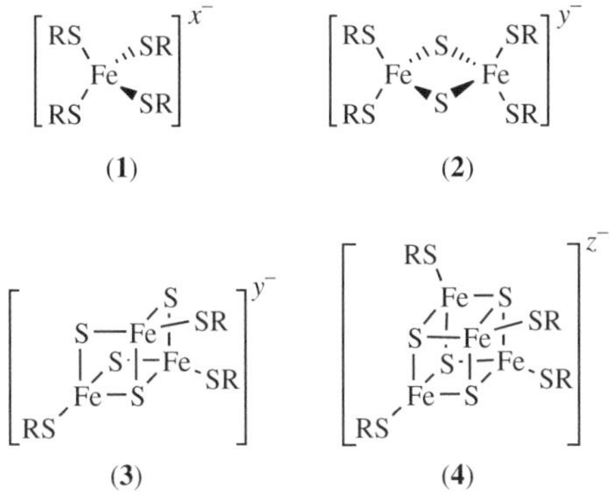

Figure 1. Four important classes of iron compounds where x, y, and z represent 1 or 2, 2 or 3, and 1, 2, or 3, respectively. Structure (12) is a 2Fe–2S center; (13), a 3Fe–4S; and (14) a 4Fe–4S.

of methane to methanol; and purple acid phosphatases, which catalyze the dephosphorylation of phosphoproteins and nucleotides. The site contains two antiferromagnetically coupled iron atoms that are coupled by a bridging oxo or hydroxo group and two bridging carboxylate groups and is recognizable as a portion of the basic ferric acetate structure. The enzymes differ in the nature of the terminal ligands to each iron.

Siderophores

Iron is not readily available at physiological pH because it is present as the insoluble hydrated iron(III) oxide, which has $K_{sp} \sim 10^{-39}$. Bacteria synthesize chelating agents to facilitate the solubilization of iron from the environment, transport it into the organism, and release of iron. Most contain negatively charged oxygen-donor groups which preferentially complex iron(III) and afford octahedral, high-spin complexes called siderophores. The two principal classes of donor groups employed are catecholates (**5**) and hydroxamates (**6**).

(5)   (6)

ECONOMIC ASPECTS

Transparent iron oxide pigments are used for automotive applications. Transparent pigments in coatings protect against uv light exposure.

Ferric chloride's biggest use is in water treatment for municipal wastewater treatment (60%); municipal potable water treatment (20%), and industrial water treatment (8%). Use in treatment of industrial water has advanced at a modest rate of 0.5%. Ferric chloride is acidic, less corrosive materials are preferred in this market.

Dry ferrous sulfate's primary uses are in animal feeds (30%), water treatment (30%), and fertilizers (25%). Moist and liquid ferrous sulfate is used in the production of iron oxide (60%) and water treatment (30%). The uses in animal feeds and as fertilizers represent stable markets.

HEALTH AND SAFETY FACTORS

Most iron salts and compounds may be safely handled following common safe laboratory practices. Some compounds are irritants. A more serious threat is ingestion of massive quantities of iron salts, which results in diarrhea, hemorrhage, liver damage, heart damage, and shock. A lethal dose is 200–250 mg/ kg of body weight. The majority of the victims of iron poisoning are children under five years of age.

Two compounds associated with particular industrial risks are iron(III) oxide, Fe_2O_3, and iron pentacarbonyl, $Fe(CO)_5$. Chronic inhalation of iron(III) oxide leads to siderosis. Adequate ventilation and mechanical filter respirators should be provided to those exposed to the oxide. Iron pentacarbonyl is volatile and highly toxic.

F. A. Cotton and G. Wilkinson, *Advanced Inorganic Chemistry*, 5th ed., John Wiley & Sons, Inc., New York, 1988.

N. N. Greenwood and A. Earnshaw, *Chemistry of the Elements*, Pergamon Press, Oxford, U.K., 1984.

G. Wilkinson, R. D. Gillard, and J. A. McCleverty, eds., *Comprehensive Coordination Chemistry*, Pergamon Press, Oxford, U.K., 1987, Vols. 1,2,4, and 6.

ALAN M. STOLZENBERG
West Virginia University